ESCHERICHIA COLI
AND SALMONELLA TYPHIMURIUM

CELLULAR AND MOLECULAR BIOLOGY

VOLUME 1

ESCHERICHIA COLI AND SALMONELLA TYPHIMURIUM

CELLULAR AND MOLECULAR BIOLOGY

VOLUME 1

Editor in Chief

Frederick C. Neidhardt
Department of Microbiology and Immunology
University of Michigan, Ann Arbor, Michigan

Editors

John L. Ingraham
Department of Bacteriology
University of California
Davis, California

K. Brooks Low
Radiobiology Laboratories
Yale University
New Haven, Connecticut

Boris Magasanik
Biology Department
Massachusetts Institute of Technology
Cambridge, Massachusetts

Moselio Schaechter
Department of Molecular Biology and Microbiology
Tufts University School of Medicine
Boston, Massachusetts

H. Edwin Umbarger
Department of Biological Sciences
Purdue University
West Lafayette, Indiana

AMERICAN SOCIETY FOR MICROBIOLOGY
Washington, D.C.

Library of Congress Cataloging-in-Publication Data

Escherichia coli and Salmonella typhimurium.
    Includes index.
    1. Escherichia coli.   2. Salmonella typhimurium.
I. Neidhardt, Frederick C.
QR82.E6E83 1987   589.9′5   87-1065

ISBN 0-914826-89-1
ISBN 0-914826-85-9 (soft)

CONTRIBUTORS

Sankar Adhya, Laboratory of Molecular Biology, National Cancer Institute, Bethesda, Maryland 20892

Barbara J. Bachmann, Department of Human Genetics, Yale University School of Medicine, New Haven, Connecticut 06510

Jon Beckwith, Department of Microbiology and Molecular Genetics, Harvard Medical School, Boston, Massachusetts 02115

Ronald Bentley, Department of Biological Sciences, University of Pittsburgh, Pittsburgh, Pennsylvania 15260

Claire M. Berg, Department of Molecular and Cell Biology, The University of Connecticut, Storrs, Connecticut 06268

Douglas E. Berg, Department of Microbiology and Immunology and Department of Genetics, Washington University School of Medicine, St. Louis, Missouri 63110

Thomas A. Bickle, Department of Microbiology, Biozentrum, Basel University, CH-4056 Basel, Switzerland

Glenn R. Björk, Department of Microbiology, University of Umeå, S-901 87 Umeå, Sweden

Philip L. Bloch, Department of Microbiology and Immunology, The University of Michigan Medical School, Ann Arbor, Michigan 48109-0620

Hans Bremer, Department of Biology, University of Texas at Dallas, Richardson, Texas 75080

Gene M. Brown, Massachusetts Institute of Technology, Cambridge, Massachusetts 02139

Michael Cashel, Section on Molecular Regulation, Laboratory of Molecular Genetics, National Institute of Child Health and Human Development, Bethesda, Maryland 20892

Dominique A. Caugant, Department of Biology, University of Rochester, Rochester, New York 14627

Georges N. Cohen, Unité de Biochimie Cellulaire, Département de Biochimie et Génétique Moléculaire, Institut Pasteur, 75724 Paris Cedex 15, France

Nancy L. Craig, Department of Microbiology and Immunology and The G. W. Hooper Foundation, University of California, San Francisco, California 94143

Irving P. Crawford, Department of Microbiology, The University of Iowa, Iowa City, Iowa 52242

John E. Cronan, Jr., Department of Microbiology, University of Illinois, Urbana, Illinois 61801

Walter B. Dempsey, Department of Medical and Microbial Genetics, Veterans Administration Medical Center, and Department of Biochemistry, University of Texas Health Science Center, Dallas, Texas 75216

Patrick P. Dennis, Department of Biochemistry, Faculty of Medicine, The University of British Columbia, Vancouver, British Columbia, Canada V6T 1W5

Richard C. Deonier, Molecular Biology, Department of Biological Sciences, University of Southern California, University Park, Los Angeles, California 90089-1481

William D. Donachie, Department of Molecular Biology, University of Edinburgh, Edinburgh EH9 3JR, Scotland

Karl Drlica, The Public Health Research Institute of the City of New York, New York, New York 10016

Max Eisenberg, Department of Biochemistry and Molecular Biophysics, College of Physicians and Surgeons, Columbia University, New York, New York 10032

Eric Eisenstadt, Department of Cancer Biology and Laboratory of Toxicology, Harvard School of Public Health, Boston, Massachusetts 02115

Barry I. Eisenstein, Departments of Microbiology and Medicine, University of Texas Health Science Center, San Antonio, Texas 78284

J. C. Escalante-Semerena, Department of Biology, University of Utah, Salt Lake City, Utah 84112

Dan G. Fraenkel, Department of Microbiology and Molecular Genetics, Harvard Medical School, Boston, Massachusetts 02115

Clement E. Furlong, Departments of Genetics and Medicine, Division of Medical Genetics, Center for Inherited Diseases, University of Washington, Seattle, Washington 98195

O. Gabriel, Department of Biochemistry, Schools of Medicine and Dentistry, Georgetown University, Washington, D.C. 20007

Robert B. Gennis, Departments of Chemistry and Biochemistry, University of Illinois, Urbana, Illinois 61801

Costa Georgopoulos, Department of Cellular, Viral and Molecular Biology, University of Utah Medical Center, Salt Lake City, Utah 84132

Nicolas Glansdorff, Research Institute of the Centre d'Enseignement et de Recherches des Industries Alimentaires et Chimiques, and Microbiology, Vrije Universiteit Brussel, B-1070 Brussels, Belgium

Larry Gold, Department of Molecular, Cellular, and Developmental Biology, University of Colorado, Boulder, Colorado 80309

Susan Gottesman, Laboratory of Molecular Biology, National Cancer Institute, Bethesda, Maryland 20892

Marianne Grunberg-Manago, Institut de Biologie Physico-Chimique, 75005 Paris, France

Douglas Hanahan, Cold Spring Harbor Laboratory, Cold Spring Harbor, New York 11724

Flemming G. Hansen, Department of Microbiology, The Technical University of Denmark, 2800 Lyngby-Copenhagen, Denmark

Charles E. Helmstetter, Department of Experimental Biology, Roswell Park Memorial Institute, Buffalo, New York 14263

John W. B. Hershey, Department of Biological Chemistry, School of Medicine, University of California, Davis, California 95616

Bruce Holloway, Department of Genetics, Monash University, Clayton, Victoria 3168, Australia

Barbara C. Hoopes, Department of Biological Sciences, Carnegie Mellon University, Pittsburgh, Pennsylvania 15213

Jane S. Hurley, Department of Biology, University of Calgary, Calgary, Alberta, Canada T2N 1N4

W. John Ingledew, Department of Biochemistry and Microbiology, University of St. Andrews, St. Andrews KY16 9AL, Scotland

John Ingraham, Department of Bacteriology, University of California, Davis, California 95616

R. Jeter, Department of Biology, University of Utah, Salt Lake City, Utah 84112

Sue Jinks-Robertson, Department of Molecular Genetics and Cell Biology, The University of Chicago, Chicago, Illinois 60637

Eugene P. Kennedy, Department of Biological Chemistry, Harvard Medical School, Boston, Massachusetts 02115

Thomas C. King, Washington University School of Medicine, St. Louis, Missouri 63110

Nancy Kleckner, Department of Biochemistry and Molecular Biology, Harvard University, Cambridge, Massachusetts 02138

Joachim Knappe, Institut für Biologische Chemie, Universität Heidelberg, D-6900 Heidelberg, Federal Republic of Germany

Arthur L. Koch, Department of Biology, Indiana University, Bloomington, Indiana 47405

Steven Krawiec, Department of Biology, Lehigh University, Bethlehem, Pennsylvania 18015

Nicholas M. Kredich, Department of Medicine, Division of Rheumatic and Genetic Diseases, Duke University Medical Center, Durham, North Carolina 27710

D. R. Kuritzkes, Infectious Disease Unit, Massachusetts General Hospital, Boston, Massachusetts 02114

Sidney R. Kushner, Department of Genetics, University of Georgia, Athens, Georgia 30602

Robert Landick, Department of Biological Sciences, Stanford University, Stanford, California 94305

Thomas Leisinger, Mikrobiologisches Institut Eidgenössische Technische Hochschule Zurich, ETH-Zentrum, CH-8092 Zurich, Switzerland

E. C. C. Lin, Department of Microbiology and Molecular Genetics, Harvard Medical School, Boston, Massachusetts 02115

K. Brooks Low, Radiobiology Laboratories, Yale University School of Medicine, New Haven, Connecticut 06511

S. E. Luria, Department of Biology, Massachusetts Institute of Technology, Cambridge, Massachusetts 02139

P. Ronald MacLachlan, Salmonella Genetic Stock Centre, Department of Biology, University of Calgary, Calgary, Alberta, Canada T2N 1N4

Robert M. Macnab, Department of Molecular Biophysics and Biochemistry, Yale University, New Haven, Connecticut 06511

Boris Magasanik, Department of Biology, Massachusetts Institute of Technology, Cambridge, Massachusetts 02139

Peter C. Maloney, Department of Physiology, The Johns Hopkins University School of Medicine, Baltimore, Maryland 21205

Stanley R. Maloy, Department of Microbiology, University of Illinois, Urbana, Illinois 61801

Paul Margolin, Department of Biology, City College of New York, New York, New York 10031

M. G. Marinus, Department of Pharmacology, University of Massachusetts Medical School, Worcester, Massachusetts 01605

William R. McClure, Department of Biologic Sciences, Carnegie Mellon University, Pittsburgh, Pennsylvania 15213

Elizabeth McFall, Department of Microbiology, New York University School of Medicine, New York, New York 10016

Roger McMacken, Department of Biochemistry, The Johns Hopkins University, Baltimore, Maryland 21205

R. Meganathan, Department of Biological Sciences, Northern Illinois University, DeKalb, Illinois 60115

Charles G. Miller, Department of Molecular Biology and Microbiology, Case Western Reserve University, Cleveland, Ohio 44106

Frederick C. Neidhart, Department of Microbiology and Immunology, The University of Michigan Medical School, Ann Arbor, Michigan 48109-0620

O. M. Neijssel, Laboratorium voor Microbiologie, Universiteit Amsterdam, 1018 WS Amsterdam, The Netherlands

Jan Neuhard, Enzyme Division, University Institute of Biological Chemistry B, DK-1307 Copenhagen, Denmark

Hiroshi Nikaido, Department of Microbiology and Immunology, University of California, Berkeley, California 94720

Hugh G. Nimmo, Department of Biochemistry, University of Glasgow, Glasgow G12 8QQ, Scotland

Harry F. Noller, Thimann Laboratories, University of California, Santa Cruz, California 95064

Masayasu Nomura, Department of Biological Chemistry, University of California, Irvine, California 92717

William D. Nunn, Department of Molecular Biology and Biochemistry, University of California, Irvine, California 92717

Per Nygaard, Enzyme Division, University Institute of Biological Chemistry B, DK-1307 Copenhagen, Denmark

Howard Ochman, Department of Biochemistry, University of California, Berkeley, California 94720

Donald B. Oliver, Department of Microbiology, State University of New York at Stony Brook, Stony Brook, New York 11794

B. Olivera, Department of Biology, University of Utah, Salt Lake City, Utah 84112

James T. Park, Department of Molecular Biology and Microbiology, Tufts University School of Medicine, Boston, Massachusetts 02111

Teresa A. Phillips, Department of Microbiology and Immunology, The University of Michigan Medical School, Ann Arbor, Michigan 48109-0620

A. J. Pittard, Department of Microbiology, University of Melbourne, Parkville, Victoria 3052, Australia

Robert K. Poole, Department of Microbiology, Kings College London, Kensington Campus, London W8 7AH, England

P. W. Postma, Laboratory of Biochemistry, University of Amsterdam, 1018 TV Amsterdam, The Netherlands

Christian R. H. Raetz, Department of Biochemistry, College of Agricultural and Life Sciences, University of Wisconsin, Madison, Wisconsin 53706

Lawrence J. Reitzer, Department of Biology, Massachusetts Institute of Technology, Cambridge, Massachusetts 02139

Paul D. Rick, Department of Microbiology, Uniformed Services University of the Health Sciences, Bethesda, Maryland 20814-4799

Monica Riley, Department of Biochemistry, State University of New York at Stony Brook, Stony Brook, New York 11790

Arthur C. Robinson, Department of Molecular Biology, University of Edinburgh, Edinburgh EH9 3JR, Scotland

Charles O. Rock, Department of Biochemistry, St. Jude Children's Research Hospital, Memphis, Tennessee 38101

D. Roof, Department of Biology, University of Utah, Salt Lake City, Utah 84112

Barry P. Rosen, Department of Biological Chemistry, University of Maryland School of Medicine, Baltimore, Maryland 21201

J. Roth, Department of Biology, University of Utah, Salt Lake City, Utah 84112

Kenneth E. Rudd, Section on Molecular Regulation, Laboratory of Molecular Genetics, National Institute of Child Health and Human Development, Bethesda, Maryland 20892

Isabelle Saint-Girons, Unité de Biochimie Cellulaire, Département de Biochimie et Génétique Moléculaire, Institut Pasteur, 75724 Paris Cedex 15, France

Kenneth E. Sanderson, Salmonella Genetic Stock Centre, Department of Biology, University of Calgary, Calgary, Alberta, Canada T2N 1N4

Robert Schleif, Biochemistry Department, Brandeis University, Waltham, Massachusetts 02254

David Schlessinger, Washington University School of Medicine, St. Louis, Missouri 63110

Maxime Schwartz, Unité de Génétique Moléculaire, Institut Pasteur, 75724 Paris Cedex 15, France

Robert K. Selander, Department of Biology, University of Rochester, Rochester, New York 14627

Lynn Silver, Department of Basic Microbiology, Merck and Co., Rahway, New Jersey 07065

Ron Skurray, Department of Microbiology, Monash University, Clayton, Victoria 3168, Australia

George V. Stauffer, Department of Microbiology, University of Iowa, Iowa City, Iowa 52242

B. A. D. Stocker, Department of Medical Microbiology, Stanford University, School of Medicine, Stanford, California 94305

Gary Stormo, Department of Molecular, Cellular, and Developmental Biology, University of Colorado, Boulder, Colorado 80309

Joan L. Suit, Department of Biology, Massachusetts Institute of Technology, Cambridge, Massachusetts 02139

D. W. Tempest, Laboratorium voor Microbiologie, Universiteit Amsterdam, 1018 WS Amsterdam, The Netherlands

Gerald J. Tritz, Department of Microbiology and Immunology, Kirksville College of Osteopathic Medicine, Kirksville, Missouri 63501

H. E. Umbarger, Department of Biological Sciences, Purdue University, West Lafayette, Indiana 47907

Martti Vaara, National Public Health Institute, SF-00280 Helsinki, Finland

Ruth A. VanBogelen, Department of Microbiology and Immunology, The University of Michigan Medical School, Ann Arbor, Michigan 48109-0620

Vicki Vaughn, Department of Microbiology and Immunology, The University of Michigan Medical School, Ann Arbor, Michigan 48109-0620

R. T. Vinopal, Department of Molecular and Cell Biology, The University of Connecticut, Storrs, Connecticut 06268

Peter H. von Hippel, Institute of Molecular Biology and Department of Chemistry, University of Oregon, Eugene, Oregon 97403

Kaspar von Meyenburg, Department of Microbiology, The Technical University of Denmark, 2800 Lyngby-Copenhagen, Denmark

Graham C. Walker, Biology Department, Massachusetts Institute of Technology, Cambridge, Massachusetts 02139

Barry L. Wanner, Department of Biological Sciences, Purdue University, West Lafayette, Indiana 47907

George M. Weinstock, Department of Biochemistry and Molecular Biology, University of Texas Medical School, Houston, Texas 77225

Robert A. Weisberg, Section on Microbial Genetics, Laboratory of Molecular Genetics, National Institute of Child Health and Human Development, Bethesda, Maryland 20892

Thomas S. Whittam, Department of Biology, Pennsylvania State University, University Park, Pennsylvania 16802

Neil Willetts, Biotechnology Australia, Roseville, New South Wales 2070, and Department of Biological Sciences, University of Sydney, Sydney, New South Wales 2006, Australia

Joanne M. Williamson, Merck Sharp and Dohme Research Laboratories, Rahway, New Jersey 07065

Allan C. Wilson, Department of Biochemistry, University of California, Berkeley, California 94720

Malcolm E. Winkler, Department of Molecular Biology, Northwestern University Medical School, Chicago, Illinois 60611

Thomas D. Yager, Institute of Molecular Biology and Department of Chemistry, University of Oregon, Eugene, Oregon 97403

Charles Yanofsky, Department of Biological Sciences, Stanford University, Stanford, California 94305

PREFACE

The absence of a single, comprehensive treatment of the molecular and cellular biology of *Escherichia coli* (and *Salmonella typhimurium*) has been more than just an inconvenience to the research community. It has been difficult for investigators, not to mention students, to envision the existing body of knowledge. As a consequence, opportunities to make the most of new information by integrating it with what is already known have been missed; this book addresses that problem. It is the outcome of an invitation extended in 1980 by David Schlessinger as Chair of the Books Committee of the American Society for Microbiology. The enthusiasm of Schlessinger and his committee for what they called "Project *E. coli*" led to its approval in the absence of a written outline, and with the understanding that there would be a delay before work could begin.

A general outline was developed in 1983, and the Editorial Board was selected in 1984. The six Editors developed notions of the intended function, general style, and scope of the projected book. The important decision was made to include *S. typhimurium*. A Table of Contents (closely resembling the final version) was drawn up, and potential authors were identified. The feasibility of the project was then put to the test: would investigators be willing to contribute the necessary time and effort? The results were warmly gratifying. Of the nearly 100 individuals originally invited, only a handful were unable to commit themselves to the task, and they supported the search for appropriate alternates. It deserves mention that authors and Editors alike worked without compensation (or even reimbursement); each receives a free copy of the completed work, together with the undying gratitude of the ASM (which, as the saying goes, together with a 110 priority score might get them a funded grant from the U.S. Public Health Service). In some instances considerably revised drafts of their manuscripts were required to achieve coordination with chapters on closely related subjects.

As for the intended nature of the book, the hope was that it could provide both information and perspective. This view was emphasized in instructions to the authors. They were urged to provide all the important maps, pathways, structures, numbers, and processes that exist scattered throughout dozens of primary research journals, review series, and treatises. Topics and items not explicitly included (technical methods, recipes, protein and nucleic acid sequences, and restriction maps fell into this category) were expected to be referenced. Although there was no intention to provide a comprehensive bibliography containing every paper written on *E. coli* and *S. typhimurium*, key papers were to be cited individually, and reference was to be made to published topical reviews to guide the reader into the primary literature on a given subject. In short, the book was intended to serve as the logical first resource for anyone seeking information about these organisms.

At the same time, the work was never seen solely as a "handbook" of data. We wanted an integrated collection of thoughtful, narrative reviews that would provide coherence and integration of information and would point out areas in which new information and understanding are needed. Authors were reminded that one of the primary functions of the book would be to highlight our ignorance. Although space limitations would prevent most authors from tracing the evolution of ideas in their topic, it was hoped that some historical perspective would be provided, perhaps while indicating the directions for future work.

The six Editors worked jointly to develop the general approach and intended style of the work, construct the outline, and select topics and authors. Review and scientific editing of manuscripts were done primarily but not exclusively on a section basis by individual Editors (part I, M.S.; part II, H.E.U.; part III, K.B.L.; part IV, B.M.; part V, J.L.I.; and part VI, F.C.N.), and some manuscripts were reviewed by more than one editor.

At several points decisions had to be made by the Editors to expedite completion of the volume. A number of subjects, either overlooked until too late in the project or scheduled for chapters not completed on time, had to be omitted. We anticipate that these subjects will appear in a second edition of the work. The Editors welcome suggestions of topics and approaches for future editions.

Besides gratitude to our approximately 125 fellow authors, the Editors would like to express thanks to the staff of the ASM Publications Department, and especially to Sara Joslyn, Managing Editor for Books, for her skill, dedication, and patience throughout this project, and to her predecessor, Susan Birch, for helping guide the early phases.

F. C. NEIDHARDT
J. L. INGRAHAM
K. B. LOW
B. MAGASANIK
M. SCHAECHTER
H. E. UMBARGER

Volume 1

CONTENTS

PART I. MOLECULAR ARCHITECTURE AND ASSEMBLY OF CELL PARTS

PART II. METABOLISM AND GENERAL PHYSIOLOGY

Section A. Class I Reactions: Generation of Precursor Metabolites and Energy

A1. Central Metabolic Routes: Metabolism of Glucose

A2. Energy Production

A3. Entry of Carbon from Sources Other than Glucose into Central Metabolic Routes

Section B. Class II Reactions: Conversion of Precursor Metabolites to Small-Molecule Building Blocks

B1. Biosynthesis of Amino Acids

B2. Biosynthesis and Conversions of Nucleotides

Volume 2

CONTENTS

1. Introduction

MOSELIO SCHAECHTER AND FREDERICK C. NEIDHARDT

*Department of Molecular Biology and Microbiology, Tufts University School of Medicine, Boston,
Massachusetts 02111, and Department of Microbiology and Immunology, University of Michigan,
Ann Arbor, Michigan 48109*

Escherichia coli and *Salmonella typhimurium* are gram-negative rods of the family *Enterobacteriaceae*. They resemble each other in most ways but differ in some essential details. Like many of the eubacteria, neither species is well delineated. *E. coli* represents a wide cluster of biotypes, whereas *S. typhimurium* is more circumscribed. Both species have been known since the early days of bacteriology, *E. coli* as a common member of the intestinal flora and *S. typhimurium* as a frequent agent of gastroenteritis.

These organisms have been preeminent in research laboratories for nearly a century. At first glance this may seem surprising because neither is a particularly good model for studying some of the more exciting phenomena in cell biology. They do not differentiate (like *Caulobacter*), sporulate (like *Bacillus*), fix nitrogen (like the closely related *Klebsiella*), photosynthesize (like *Rhodospirillum*), chemosynthesize (like *Nitrosomonas*), excrete large amounts of protein (like *Bacillus*), or grow in exotic environments (like *Thermus* or *Thiobacillus*).

The reason for the popularity of these two species in microbiological research is not quite lost in antiquity, but is somewhat complex. Since the early days, bacterial physiologists chose organisms that were easily accessible, were not highly virulent, and grew readily on defined media. *E. coli* emerged only gradually as the undisputed winner. By the late 1930s, it was the organism of choice, thanks to the work done on its bacteriophages by investigators like the Wollmans and Bronfenbrenner. This was followed by the crucial research of Monod on growth physiology and enzymatic adaptation and that of Delbruck and Luria on phage genetics. In the late 1940s its fate was sealed by the discoveries of conjugation by Lederberg and of transduction by Zinder and Lederberg. *S. typhimurium* became the popular organism for genetic and biochemical experiments that relied on transduction. Genetic analysis of cellular function became possible and was used to take full advantage of other convenient properties of these species. These include relative ease of preparing enzymatically active cell extracts and the flexible ways with which both small and large molecules can be tagged with radioactive isotopes. Labeling with different precursors was carried out extensively by a group of investigators at the Department of Terrestrial Magnetism of the Carnegie Institution of Washington. They published the earliest "*E. coli* bible" (1).

With these major advantages it is no wonder that *E. coli*, *S. typhimurium*, and their bacteriophages led to major triumphs of biology such as the elucidation of most biosynthetic pathways, the refinement of the concept of the gene, the solution of the genetic code, the discovery of molecular mechanisms of gene regulation, and the molecular portrayal of viral morphogenesis.

Today, *E. coli* and *S. typhimurium* are special to us because more is known about them than about any other cellular form of life. To give an idea of the current extent of this knowledge, about 1/3 of the gene products of *E. coli* have been studied in some biochemical detail, and their genes have been identified; of the order of 10% of the *E. coli* genome has been sequenced; and map positions have been determined for approximately half its genes. The rest are not far behind. We know perhaps 80% of the organism's metabolic pathways. We have a remarkably detailed picture of how its macromolecules are made, and we can use purified cell components to construct reasonable facsimiles of these molecules in the test tube. Less complete is our knowledge of how gene expression is regulated, for this turns out to be quite a rococo business. Nonetheless the main features of regulation can be described for many operons, and the list of control mechanisms that has been described is impressive, if somewhat daunting. Considerably less is known about how the component parts "talk" to each other to result in regulated growth or in rapid responses to changes in the environment.

Much of the work done with *E. coli* and *S. typhimurium* has been under defined laboratory conditions. Increasingly, we have become aware that this gives only a partial picture and that to understand many of the properties of these organisms we must take into account how they function in their natural environments. Alas, these environments are very complex, and it is difficult to assess the properties that are relevant from the eye view of the organism. In broad strokes, we know that strains of *E. coli* are found habitually in the large intestine of vertebrates, usually as minority members of the normal flora. They can cause diseases in other body sites, and they spread between individual hosts, often after a relatively short extracorporeal residence. Almost never do they find themselves in a constant, unchanging environment. Rather, they feed only intermittently in the large intestine and undergo partial desiccation when voided from their host. If deposited in a body of water they face the problems of nutrition at low concentrations of foodstuff. Between hosts they undergo shifts in temperature, some subtle, some drastic. These circumstances, plus some new ones, are sometimes faced by these organisms when they are grown in laboratory

1

media and intermittently saved in Bijou bottles, dunked in glycerol, and placed in a freezer, or lyophilized.

Given such a varied existence, it can be expected that *E. coli* and *S. typhimurium* are highly adaptive. This is, in fact, the case. These species and many of their relatives can synthesize the preponderance of their constituents, often from a single organic compound and a few minerals. Their repertoire of nutritional abilities is quite impressive. If anything, it is surprising that they have not yet learned to live on Tris buffer. They respond to changes in temperature and available nutrients by making rapid adjustments in the synthesis of regulatory molecules. As a consequence, they can vary considerably in size and chemical composition. Unamuno's dictum, "Yo soy yo y mi circunstancia" ("I am I and my circumstance"), applies to them very well.

As expected from organisms that grow in competition with others, *E. coli* and *S. typhimurium* are also highly efficient in the way they husband their energy resources. When presented with different kinds of nutrients, they grow at different rates. The richer the array of nutrients provided, the faster they grow, sparing themselves the task of synthesizing many of the compounds provided exogenously. Under conditions of active growth they synthesize the quantities of small and large molecules they require and very little extra. Only under conditions of starvation or other stresses do they accumulate constituents that they cannot use immediately. Under such conditions, a certain amount of energy is expended "inefficiently" for the sake of adaptation. For instance, nongrowing cells contain a small but significant number of ribosomes that are not engaged in protein synthesis. Although the making of these ribosomes is at a cost, their presence at the time of refeeding allows the cells to resume growth faster.

The grand strategy that allows these organisms to be both highly adaptive and very efficient is difficult to perceive. Unlike cells of metazoans, bacteria and other normally unicellular organisms have evolved an extraordinarily sensitive and complex set of controlling mechanisms that sense both the external and the internal milieux. A common characteristic of these sensing and controlling mechanisms is that they operate very fast. Thus, *E. coli* or *S. typhimurium* can go from a nongrowth state to one of active growth within seconds of the addition of nutrients to their culture. This is just what is expected from organisms that must compete with others for their food.

These organisms must often cope with strong environmental stresses, some life-threatening. They have evolved a series of "alarm reactions" which permit them to repair damaged DNA, shut off the synthesis of unwanted RNA, or protect their membranes. Most of these responses lead to the selective shut-off or turn-on of specific macromolecules. The number of regulatory devices used is very large and possibly beyond our current comprehension. Shut-off of gene expression at transcription or translation take on a bewildering number of versions. Even after proteins are made, their activity or stability can be modified. We know that our knowledge in this area is incomplete and that we have much to learn.

The impression gained is that the demands from selective pressure for efficiency and adaptability have resulted in a complex network of regulatory interactions. The constraints of space and speed do not permit the physical separation of different metabolic activities that is seen in higher cells. "Spaceship *E. coli*" is not a large space station with many compartments and with the luxury of many redundant parts and back-up systems. It must make do with fewer parts and make multiple use of many of them. If this is so, individual macromolecules can be expected to function in several ways. We have a few examples that suggest this: many proteins have not only an enzymatic or controlling function but they also regulate their own synthesis (autorepression); some polypeptides act as subunits of different enzymes or structural proteins. If this turns out to be true on a large scale, it will make it difficult to ascribe the regulatory properties of a given protein or nucleic acid to their proper place in the overall scheme. Pleiotropy reigns supreme! For this reason alone the task ahead is not trivial, although in our opinion it can be faced with optimism.

A central goal in biology is the achievement of a cellular paradigm. Until we understand thoroughly the growth of one cell, we shall be handicapped in imagining the nature of our understanding of any cell. There is little question that *E. coli/S. typhimurium* offer the best opportunity of achieving that paradigm. We propose two reasons: these organisms have a seemingly insuperable head start, and they can be studied by unmatched genetic and molecular techniques. Thus, they occupy a central role in current biology. Not everyone is mindful of it, but all cell biologists have two cells of interest: the one they are studying, and *Escherichia coli*!

LITERATURE CITED

1. **Roberts, R. B., D. B. Cowie, P. H. Abelson, E. T. Bolton, and R. V. Britten.** 1955. Studies of biosynthesis in *Escherichia coli*. Publ. 607. Carnegie Institute of Washington, Washington, D.C.

2. Chemical Composition of *Escherichia coli*

FREDERICK C. NEIDHARDT

Department of Microbiology and Immunology, The University of Michigan Medical School, Ann Arbor, Michigan 48109-0620

INTRODUCTION

The need for determining the chemical composition of bacterial cells, especially that of *Escherichia coli*, has long been recognized. In the modern era, notable measurements of the total composition of this organism were made in 1946 by Taylor (15) and in the early 1950s by a group of biophysicists at the Department of Terrestrial Magnetism of the Carnegie Institution of Washington (12). The latter group, making full use of then newly available radioisotopes, mounted a monumental study of the composition of *E. coli*, its metabolism, and its control of biosynthetic pathways. Their report (12) has been used extensively for the past three decades. It has been such a valuable sourcebook and guide that it is often called the *E. coli* bible.

The possibility of an accurate inventory of the molecules of the cell is greater today than it was in the past. The chemical structures of the more complex components of the cell are closer to being known (parts I and IV). Advances have been made in both the isolation of cellular structures (part IV) and the resolution of complex mixtures of cellular molecules by gas chromatography, high-pressure liquid chromatography, and two-dimensional gel electrophoresis.

The value of an inventory is greater now as well. Most of the fueling and biosynthetic pathways of the cell have been described (part I, sections A and B), and the general mechanisms for the synthesis of the major macromolecules are reasonably well understood (part I, section C). Therefore, it is possible to calculate flow through pathways and to estimate growth requirements for energy, reducing power, and metabolites under defined conditions (see references 7 and 16). Such calculations help uncover areas in which knowledge is skimpy (energy costs for assembly, processing, and proofreading, for example) and can turn up discrepancies that otherwise would lay hidden. The single most important value of a chemical inventory of the cell, however, is that it frequently provides the basis for in vivo tests of theories of cell growth and regulation (see references 3 and 7).

PROBLEMS IN SPECIFYING CELL SIZE AND COMPOSITION

Aside from technical difficulties inherent in various analytical procedures, and there are many, there are some problems uniquely related to the nature of bacteria. Specifically, these include their size and compositional variation and the heterogeneity of growing cell populations.

It is obviously meaningless to talk about the size or composition of a bacterial cell without specifying the strain, the growth conditions, and the phase of growth. The stationary phase is not a unique state that can be reproducibly achieved in the laboratory, nor are the culture phases that are transitional between different stages in growth and between different states of balanced growth. Therefore, measurements intended to be compared with those made by others are ideally made on cultures in steady-state, exponential growth. At the risk of pedantry, it should be pointed out that phrases such as early-log-phase culture or late-log-phase culture are simply confessions that the investigator has not followed with care the prescriptions necessary to achieve steady-state growth and therefore to prepare an experimentally reproducible culture (e.g., see p. 267–270 in reference 7).

Since the size and the composition of the *E. coli* cell are such sensitive functions of growth rate (see chapter 6 in reference 7), it is recommended that the growth rate of a culture always be specified. The following items are suggested specifications of a particular culture used to measure a parameter of interest: (i) the organism identified by strain and source (e.g., *E. coli* B/r, obtained from S. Cooper), (ii) the medium (e.g., glucose-morpholinepropanesulfonic acid [MOPS] minimal medium as described by [cite reference]), (iii) the aeration (e.g., grown aerobically, with shaking [200 rpm, a 50-ml culture in a 250-ml Erlenmeyer flask with a Morton closure]), (iv) the temperature (e.g., 37°C), (v) the growth phase (e.g., balanced exponential growth, achieved by serial subcultivation through five mass doublings), and (vi) the growth rate (e.g., generation time = 42 min).

TABLE 1. Composition of an average *E. coli* B/r cell[a]

Component(s)	% Total dry wt[b]	Amt (g, 10^{15})/cell[c]	Mol wt	Molecules/cell	No. of different kinds of molecules[d]
Protein	55.0	156	4.0×10^4	2,350,000	1,850
RNA	20.5	58			
23 S rRNA		31.0	1.0×10^6	18,700	1
16 S rRNA		15.5	5.0×10^5	18,700	1
5 S rRNA		1.2	3.9×10^4	18,700	1
Transfer		8.2	2.5×10^4	198,000	60
Messenger		2.3	1.0×10^6	1,380	600
DNA	3.1	8.8	2.5×10^9	2.1	1
Lipid	9.1	25.9	705	22,000,000	
Lipopolysaccharide	3.4	9.7	4,070	1,430,000	1
Peptidoglycan	2.5	7.1	(904)n	1	1
Glycogen	2.5	7.1	1.0×10^6	4,300	1
Polyamines	0.4	1.1			
Putrescine		0.83	88	5,600,000	1
Spermidine		0.27	145	1,100,000	1
Metabolites, cofactors, ions	3.5	9.9			800+

[a] Calculated for an average cell in a population of *E. coli* B/r in balanced growth at 37°C in aerobic glucose minimal medium with a mass doubling time of 40 min. The cell is defined by dividing the total biomass, or the amount of any of its measured components, by the total number of cells in the population. This average cell, therefore, is approximately 44% through its division cycle (see reference 10 for the function describing the distribution of cell ages in a population), and, if increase in cell mass is exponential, is approximately 33% larger than when it was born. This table is modified from data in reference 7, Table 1.

[b] Relative amounts of the major components based on information in references 3, 12, and 16 and on unpublished experiments of F. C. Neidhardt (see the text). In some cases, data from strains other than B/r, from growth conditions other than the reference one, or from both had to be used (see references concerning glycogen [4], polyamine [9], and lipid [15]).

[c] Based on measurements of the total dry mass and the number of cells measured in portions of a reference culture (unpublished observations). The wet weight is calculated from the assumption that 70% of *E. coli* protoplasm is water. The total dry weight per cell is 2.8×10^{-13} g; the water content (assuming that 70% of the cell is water) is 6.7×10^{-13} g; the total weight of one cell is 9.5×10^{-13} g.

[d] Based on the following components: protein, examination of two-dimensional O'Farrell gels (T. A. Phillips and F. C. Neidhardt, unpublished observations); stable RNA, chapter 85; mRNA, assuming three genes per average transcriptional unit; lipid, an indeterminant number of species because of the variety of fatty acids associated with the following four major types of phospholipids exclusive of lipopolysaccharide: 76% phosphatidylethanolamine, 20% phosphatidylglycerol, and small amounts of cardiolipin and unidentified species (1, 11); and metabolites, cofactors, and ions, roughly estimated as described in reference 7, Table 3.

COMPOSITION OF AN AVERAGE *E. COLI* B/r CELL

The B and B/r strains of *E. coli* have been the subject of extensive biochemical and metabolic studies, more so than even the K-12 strains popular with geneticists. (Rapid growth in minimal medium, serving as host to T-even and other phages, and tight variance of cell division are some of the reasons that B and B/r strains have been favored by physiologists.) The information in Table 1 has been compiled from several sources listed in the footnotes to the table. The compilation was guided by that of Umbarger (16), and began with the overall percent composition data of Roberts et al. (12) for macromolecules. These data were then adjusted to match other information on glycogen (4), lipid (15), polyamines (9), and stable RNA/protein/DNA ratios (3). The dry weight per cell was determined in my laboratory by weighing the dried cells removed by filtration from samples of a culture which had also been assayed for total cell count with the aid of an electronic particle counter. The calculated weight per average cell was found to be consistent with the size of the DNA genome and the number of copies of DNA predicted for a cell of average age. The water content was assumed from the value commonly cited in textbooks.

RESIDUE COMPOSITION OF E. COLI B/r PROTOPLASM

The information in Table 2 is derived heavily from the primary data of Roberts et al. (12) (cited in reference 16), though the amino acid analysis has been replaced by similar results from measurements in my laboratory.

The footnotes to the table contain the sources of the other analytical data or assumptions used in the compilation or both.

FUTURE MOLECULAR INVENTORY

The tables in this chapter contain current information useful for computing various parameters for *E. coli*. It is hoped, however, that the main function served by them will be to highlight the softness and incompleteness of the information and the need for a more detailed, as well as more accurate, inventory. Many powerful tools are now available to construct an accurate inventory. In the context of ongoing efforts to identify the individual genes and proteins of this organism (see chapters 53 and 55), an explicit program to refine the overall chemical analysis of the cell makes great sense. Current interest in the assembly reactions of the cell, in the process of cell division, and

TABLE 2. Residue composition of *E. coli* B/r protoplasm[a]

Residues	Amt (μmol/g of dried cells)	Residues	Amt (μmol/g of dried cells)
Protein amino acids[b]		Lipid components[e]	
Alanine	488	Glycerol	161
Arginine	281	Ethanolamine	97
Asparagine	229	$C_{16:0}$ fatty acid (43%)	
Aspartate	229	$C_{16:1}$ fatty acid (33%)	
Cysteine	87	$C_{18:1}$ fatty acid (24%)	
Glutamate	250	Average fatty acid	258
Glutamine	250		
Glycine	582	LPS components[f]	
Histidine	90	Glucose	16.8
Isoleucine	276	Glucosamine	16.8
Leucine	428	Ethanolamine	25.2
Lysine	326	Rhamnose	8.4
Methionine	146	Heptose	25.2
Phenylalanine	176	KDO	25.2
Proline	210	Hydroxymyristic acid	33.6
Serine	205	Fatty acid ($C_{14:0}$)	16.8
Threonine	241		
Tryptophan	54		
Tyrosine	131	Peptidoglycan components[g]	
Valine	402	*N*-Acetylglucosamine	27.6
		N-Acetylmuramic acid	27.6
RNA nucleotides[c]		Alanine	55.2
AMP	165	Diaminopimelate	27.6
GMP	203	Glutamate	27.6
CMP	126		
UMP	136		
		Glycogen components (glucose)[h]	154
DNA nucleotides[d]			
dAMP	24.6	Polyamines[i]	
dGMP	25.4	Putrescine	34.1
dCMP	25.4	Spermidine	7.0
dTMP	24.6		

[a] Compiled and calculated for *E. coli* B/r in balanced growth at 37°C in aerobic glucose minimal medium, mass doubling time 40 min. This table is modified from data in reference 6, Table 10.

[b] There is 550 mg of total protein per g of dried cells (Table 1). With an average residue molecular weight of 108, there is a total of 5,081 μmol of amino acid residues. The amino acid composition is based on an analysis of *E. coli* B/r protein by T. A. Phillips, except that cysteine and tryptophan values are from reference 12, and no distinction is made between glutamate and glutamine and between aspartate and asparagine. The data have been corrected to exclude peptidoglycan (murein) amino acids. An alternative analysis is given in reference 12 and cited in reference 16.

[c] There is 205 mg of total RNA per g of dried cells (Table 1). This consists of 197 mg of stable RNA (167 mg of rRNA and 30 mg of tRNA) plus 8.3 mg of mRNA. With an average nucleotide residue molecular weight of 325, there is a total of 630 μmol of nucleotide residues. The A/G/C/U ratios are based on the analysis in reference 12, corrected for DNA.

[d] There is 31 mg of DNA per g of dried cells (Table 1). With an average residue molecular weight of 309, there is a total of 100 μmol of nucleotide residues. The ratios of individual residues are based on (A + T)/(G + C) = 0.97 (5, 12).

[e] There is 91 mg of total phospholipid exclusive of lipid A per g of dried cells (Table 1). With an average molecular weight of 705 (calculated as if all were phosphatidylethanolamine), there is 129 μmol of total phospholipid. The simplifying assumption is made that, of the total, 97 μmol is phosphatidylethanolamine and 32 μmol is phosphatidylglycerol, ignoring the small amounts of cardiolipin and minor, unidentified lipids (1, 11). The fatty acid composition is based on the analysis in reference 13, corrected for myristic and hydroxymyristic acid from lipid A.

[f] There is 34 mg of lipopolysaccharide per g of dried cells (Table 1). With an average molcular weight of 4,070, there is 8.4 μmol of lipopolysaccharide (LPS), assuming the following structure for the rough LPS of strain B/r: lipid A (4 hydroxymyristic acid residues, 2 saturated fatty acids assumed to be $C_{14:0}$, 2 glucosamines, 1 phosphoryl group, and 1 ethanolamine in pyrophosphate linkage); inner core (3 2-keto-3-deoxyoctulosonic acid [KDO], 1 rhamnose, 1 phosphoethanolamine, 3 heptose, 1 phosphoryl, and 1 ethanolamine in pyrophosphate linkage); and an outer core (2 glucose residues). This composition is compiled from information in several sources, including reference 2 (for lipid A) and reference 8 (for the polysaccharide portion). Different strains of *E. coli* differ slightly in LPS structure, particularly in the outer core. Many K-12 strains will have in the outer core 1 galactose, 1 glucose, and 1 acetylglucosamine residue in addition to the residues for B/r shown here (8).

[g] There is 25 mg of peptidoglycan per g of dried cells (Table 1). With an average molecular weight of 904 for a disaccharide subunit, there is 27 μmol of subunits consisting of 1 *N*-acetylglucosamine, 1 *N*-acetylmuramic acid, 1 D-glutamate, 1*m*-diaminopimelic acid, 1 D-alanine, and 1 L-alanine residue (for a review, see reference 6).

[h] There is 25 mg of glycogen per g of dried cells (Table 1). With a glucosyl molecular weight of 162, there is 154 μmol of glucosyl residues.

[i] There are 3 mg of putrescine (molecular weight, 88) and 1 mg of spermidine (molecular weight, 145) per g of dried cells (reference 9, Table 1).

in the operation of regulatory networks can benefit greatly from such an undertaking.

LITERATURE CITED

1. **Ames, G. F.** 1968. Lipids of *Salmonella typhimurium* and *Escherichia coli*: structure and metabolism. J. Bacteriol. **95**:833–843.
2. **Bulawa, C. E., and C. Raetz.** 1984. The biosynthesis of gram-negative endotoxin. Identification and function of UDP-2,3-diacylglucosamine in *Escherichia coli*. J. Biol. Chem. **258**:4846–4851.
3. **Dennis, P. P., and H. Bremer.** 1974. Macromolecular composition during steady-state growth of *Escherichia coli* B/r. J. Bacteriol. **119**:270–281.
4. **Dietzler, D. N., M. P. Leckie, and C. J. Lais.** 1973. Rates of glycogen synthesis and the cellular levels of ATP and FDP during exponential growth and the nitrogen-limited stationary phase of *Escherichia coli* W4597 (K). Arch. Biochem. Biophys. **156**:684–693.
5. **Dunn, D. B., and J. D. Smith.** 1958. The occurrence of 6-methyladenine in deoxyribonucleic acid. Biochem. J. **68**:627–636.
6. **Ghuysen, J.-M.** 1968. Use of bacteriolytic enzymes in determination of wall structure and their role in cell metabolism. Bacteriol. Rev. **32**:425–464.
7. **Ingraham, J. L., O. Maaløe, and F. C. Neidhardt.** 1983. Growth of the bacterial cell. Sinauer Associates, Sunderland, Mass.
8. **Jansson, P.-E., A. A. Lindberg, B. Lindberg, and R. Wollin.** 1981. Structural studies on the hexose region of the core in lipopolysaccharides from Enterobacteriaceae. Eur. J. Biochim. **115**:571–577.
9. **Morris, D. R., and C. M. Jorstad.** 1970. Isolation of conditionally putrescine-deficient mutants of *Escherichia coli*. J. Bacteriol. **101**:731–737.
10. **Powell, E. O.** 1956. Growth rate and generation time of bacteria, with special reference to continuous culture. J. Gen. Microbiol. **15**:492–511.
11. **Randle, C. L., P. W. Albro, and J. C. Dittmer.** 1969. The phosphoglyceride composition of gram-negative bacteria and the changes in composition during growth. Biochim. Biophys. Acta **187**:214–220.
12. **Roberts, R. B., R. H. Abelson, D. B. Cowie, E. T. Bolton, and R. J. Britten.** 1955. Studies of biosynthesis in *Escherichia coli*. Carnegie Inst. Wash. Publ. **607**:1–521.
13. **Shaw, M. K., and J. L. Ingraham.** 1965. Fatty acid composition of *Escherichia coli* as a possible controlling factor of the minimal growth temperature. J. Bacteriol. **90**:141–146.
14. **Smith, J. D., and G. R. Wyatt.** 1951. The composition of some microbial deoxypentose nucleic acids. Biochem. J. **49**:144–148.
15. **Taylor, A. R.** 1946. Chemical analysis of the T2 bacteriophage and its host, *Escherichia coli* strain B. J. Biol. Chem. **165**:271–284.
16. **Umbarger, H. E.** 1977. A one-semester project for the immersion of graduate students in metabolic pathways. Biochem. Education **5**:67–71.

3. Outer Membrane

HIROSHI NIKAIDO[1] AND MARTTI VAARA[2]

Department of Microbiology and Immunology, University of California, Berkeley, California 94720,[1] and National Public Health Institute, SF-00280 Helsinki, Finland[2]

INTRODUCTION

The outer membrane of gram-negative bacteria has been studied intensively during the last decade, and this chapter can give only a brief outline of what has been discovered. More details can be found in the following reviews. For general treatment of the subject, see Inouye (67), Lugtenberg and van Alphen (86), Nikaido and Nakae (118), and Nikaido and Vaara (124); concerning the proteins, see DiRienzo et al. (34) and Osborn and Wu (130); on transport functions, see Benz (17), Konisky (73), and Nikaido (114; Methods Enzymol., in press); on genetics, see Mäkelä and Stocker (91); and on export of proteins into the outer membrane, see Benson et al. (16), Michaelis and Beckwith (98), and Silhavy et al. (157).

ISOLATION

Early studies of the composition of the cell wall fractions from gram-positive and gram-negative bacteria revealed the former contained large amounts of proteins and lipids that were absent from the latter (148). Electron microscopic examination of thin sections of gram-negative bacteria showed that these cells are covered by an extra membrane layer, outer membrane, which is located outside the peptidoglycan layer (46). The idea that proteins and lipids were present in this membranous structure was borne out by the isolation of the *Escherichia coli* and *Salmonella typhimurium* outer membranes by equilibrium density centrifugation in sucrose (99, 129, 154), which took advantage of the fact that the outer membrane has a

TABLE 1. Major components of the outer membrane[a]

Component	Molecules/cell (10^5)	Fatty acyl chains/cell (10^5)	Surface area (μm^2)	
LPS	34.6	242	4.9	6.7 = Outer surface
Porins + OmpA	2		1.8	
Lipoprotein	7	21	0.5	6.4 = Inner surface
Phospholipids	87	174	4.1	

[a] The data on LPS and phospholipids are from Smit et al. (158). However, the number of molecules of LPS was multiplied by 3, because LPS, which was then believed to exist as a trimer, is now known to exist as a monomer. Each LPS, lipoprotein, and phospholipid molecule contains 7, 3, and 2 fatty acyl residues, respectively. For the area per fatty acyl chain, the values of 0.202 and 0.234 nm^2, calculated from the coordinates of Rietschel et al. (144), were used for LPS and phospholipids (and also for murein lipoprotein). For porins and OmpA, the surface area was calculated by assuming cylindrical shapes for the proteins, with a length of 5 nm (corresponding to the assumed thickness of the bilayer) and with the partial specific volume taken as 0.72. Note that the calculated values are in an excellent agreement with the surface area per cell, approximately 6 μm^2 (158).

higher buoyant density (close to 1.22) than the cytoplasmic or inner membrane (1.15).

Equilibrium density centrifugation still remains the most reliable method for the isolation of outer membrane. The disruption of the cells can be carried out either by EDTA-lysozyme lysis (129) or with a French press (154, 158). The latter method has several advantages: it can be used on any strain; it avoids the use of EDTA that may remove some of the components, especially lipopolysaccharides (LPS); and it can easily handle large amounts of cells. On the other hand, recovery of inner membrane vesicles may be poor because of their small size. A modification that avoids the use of a preliminary rate sedimentation step has been proposed (69). Sonication can be used if done judiciously (H. Nikaido, unpublished data), but extensive treatment is dangerous because it may result in the formation of hybrid vesicles. If one wishes to collect only the outer membrane fraction, this can be done rapidly by rate centrifugation by taking advantage of the much larger size of outer membrane vesicles after French press rupture (H. Nikaido, unpublished data) or by acid aggregation of outer membranes obtained with lysozyme-EDTA treatment (180).

When one is interested only in protein composition, extraction of cell envelope with mild detergents such as sodium dodecyl sarcosinate (Sarkosyl) can be used because these detergents extract preferentially the cytoplasmic membrane proteins (39). Conclusions based on this method, however, must be interpreted with care, because some outer and inner membrane proteins may show atypical behavior.

COMPONENTS

The outer membrane contains at least two types of lipids, LPS and phospholipids, as well as a set of characteristic proteins (Table 1). In addition, the outer membrane from members of the family *Enterobacteriaceae* contains a unique polysaccharide, enterobacterial common antigen.

Phospholipids

The phospholipid composition of the outer membrane is similar to that of the cytoplasmic membrane, with a slight enrichment in phosphatidylethanolamine (129).

LPS

LPS is a unique constituent of the bacterial outer membrane and is not found anywhere else in the cell except as biosynthetic intermediates. It will be described in detail elsewhere in this volume, but the following points should be emphasized in relation to its structural and functional roles in the outer membrane. (i) LPS is composed of three parts: the proximal, hydrophobic lipid A region; the distal, hydrophilic O antigen polysaccharide region that protrudes into the medium; and the core polysaccharide region that connects the two (Fig. 1). (ii) The lipid A region is unusual in that a single backbone structure, corresponding to glucosaminyl-β(1→6)glucosamine, is substituted with six or seven fatty acid residues, all of which are saturated. (iii) The more proximal part of the core region is extremely rich in charged groups, especially negatively charged groups. (iv) The properties of mutants producing incomplete LPS molecules suggest the nature of biological functions performed by these various parts of the LPS molecule. Loss of O antigen results in loss of virulence, suggesting that this portion is important in host-parasite interactions. (It should be noted here that the K-12 strain of *E. coli* completely lacks the O antigen and that the B strain lacks in addition the more distal part of the core [68]). Loss of the more proximal part of the core, as in the "deep rough" mutants (i.e., in Rd1, Rd2, and Re mutants [Fig. 1]), makes the strains exceptionally sensitive to a wide range of hydrophobic compounds, including dyes, antibiotics, bile salts, other detergents, and mutagens (for references, see Nikaido [113]); thus this area, containing large numbers of charged groups, must be important in maintaining the barrier property of the outer membrane (see also below). Mutants in the assembly of lipid A cannot be isolated except as conditional lethal mutants, and thus this region may be essential for the cell, presumably for the successful assembly of the outer membrane.

Enterobacterial Common Antigen and Capsular Polysaccharides

The enterobacterial common antigen is an acidic polysaccharide containing *N*-acetyl-D-glucosamine, *N*-acetyl-D-mannosaminuronic acid, and 4-acetamido-4,6-dideoxy-D-galactose (84). It is reported to repre-

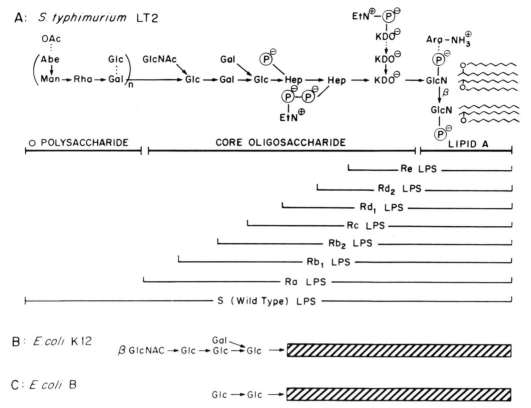

FIG. 1. Structure of LPS. (A) LPS from *S. typhimurium* LT2 and its mutants. Ra through Re refer to the chemotypes of the mutant LPS produced. (B) LPS from *E. coli* K-12. (C) LPS from *E. coli* B. Dotted lines represent partial substitution. The arrangement of the 2-keto-3-deoxyoctonic acid residues is after the results of Brade and Rietschel (21). In B and C, the hatched areas are assumed to have a structure very similar to the corresponding area in the *S. typhimurium* LPS (68). However, minor differences apparently exist. In *E. coli* K-12, the core portion is different by having a rhamnosyl residue on one of the 2-keto-3-deoxyoctonic acid residues. Furthermore, the lipid A portion shows several differences: the reducing glucosamine residue carries only two, instead of three, fatty acid chains; the lipid A often carries a pyrophosphate group instead of a phosphate group at C-1; and the 4-aminoarabinose substituent seems to be largely absent (144). Abbreviations: Abe, abequose; Man, D-mannose; Rha, L-rhamnose; Gal, D-galactose; OAc, O-acetyl; GlcNAc, N-acetyl-D-glucosamine; Glc, D-glucose; Hep, L-glycero-D-mannoheptose; KDO, 2-keto-3-deoxyoctonic acid (3-deoxy-D-mannooctulosonic acid); EtN, ethanolamine; P, phosphate; GlcN, D-glucosamine; AraNH$^{\oplus}_3$, 4-aminoarabinose. All the hexose and heptose residues in the core region are α-linked, unless otherwise indicated. The wavy lines represent fatty acyl residues.

sent up to 0.2% of the dry weight of *E. coli* (95) and apparently is anchored to the outer membrane through a covalently linked phospholipid moiety (75) in most strains, i.e., in those strains containing enterobacterial common antigen in the "nonimmunogenic" form.

Two types of capsular polysaccharides are found in enteric bacteria. One, the M antigen or colanic acid, is distributed widely among these organisms (48) and is composed of glucose, galactose, fucose, and glucuronic acid (Fig. 2). This polysaccharide is made only under certain physiological conditions, such as high osmolarity of the medium, low temperature, and low humidity, the last factor favoring its production on solid but not in liquid media (2). These environmental conditions suggest that colanic acid is made by the enteric bacteria when they are out of the animal intestinal tract, probably as a means of protection against desiccation. The other type of polysaccharide includes the serotype-specific K antigen, i.e., the classical, specific, capsular polysaccharides that are elaborated by most freshly isolated strains of *E. coli*, but are absent in *S. typhimurium*. Polysaccharides of this

type are also absent from the commonly used laboratory strains of *E. coli*, such as K-12 and B. These capsular polysaccharides are discussed elsewhere in

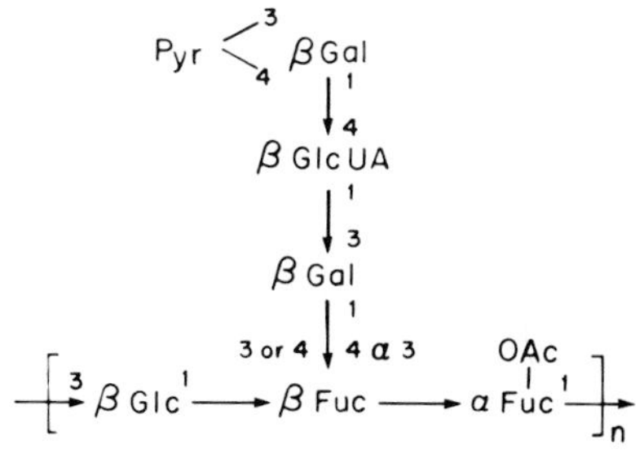

FIG. 2 Structure of M (mucoid) antigen. Abbreviations: GlcUA, D-glucuronic acid; Pyr, pyruvate; Fuc, L-fucose; Gal, D-galactose; Glc, D-glucose; OAc, O-acetyl. Note that a pyruvate residue is attached through a ketal (3,4-ethylidene) linkage to the nonreducing terminal galactose residue of the branch. In one strain the pyruvate is replaced by an acetone residue. This structure is based on the results of Garegg et al. (45).

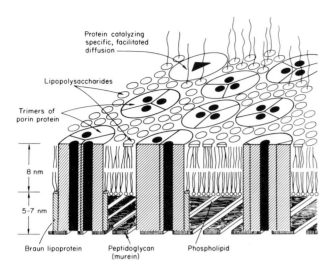

Protein catalyzing
specific, facilitated
diffusion

Lipopolysaccharides

Trimers of
porin protein

8 nm

5-7 nm

Braun lipoprotein   Peptidoglycan   Phospholipid
                     (murein)

FIG. 3. Schematic representation of the structure of the *E. coli* and *S. typhimurium* outer membrane-peptidoglycan complex. The specific channel is drawn as a monomer for simplicity. From Nikaido and Nakae (118), with permission.

this volume, but one should note that at least some of them are known to be anchored to the outer membrane through lipid tails (47, 153).

Proteins

The outer membrane has a characteristic protein composition, which is entirely different from that of the cytoplasmic membrane. In earlier studies it was thought that the outer membrane contained only a few proteins, but more recent studies have revealed the presence of a number of minor proteins. Nevertheless, the protein pattern is dominated by a few major proteins, which are described briefly below.

Murein lipoprotein. Murein lipoprotein is a small (7,200-dalton) protein that exists in a large number of copies, 7×10^5 per cell. About one-third of the population of this protein occurs in a form bound covalently to the peptidoglycan layer through the ε-amino

group of its C-terminal lysine, whereas the rest occurs as free proteins. The N-terminal residue, cysteine, is modified in an unusual manner: its sulfhydryl group is substituted with a diglyceride, and its amino group is substituted with a fatty acid residue through an amide linkage (23). The amino acid sequence of this protein is known (23), and the polypeptide chain appears to exist mostly in α-helical form (24). It has been difficult to determine its aggregation state because of nonspecific association caused by the hydrophobic lipid moiety. However, a recent chemical cross-linking study (182) of a hybrid lipoprotein lacking the lipid moiety showed that it exists as trimers. There is no convincing evidence that the protein portion of the lipoprotein either is exposed on the cell surface or penetrates significantly into the outer membrane (Fig. 3).

Mutants with deletions through the structural gene for the lipoprotein (*lpp*) are viable in laboratory media (59). Their outer membrane also shows unaltered diffusion rates for hydrophilic solutes (117). However, the cell envelope of these mutants apparently is unstable, resulting in the release of outer membrane vesicles and periplasmic enzymes into the growth medium (59). Thus, it is thought that the main function of the lipoprotein is structural in that it stabilizes the outer membrane-peptidoglycan complex.

Recently, a series of minor lipoproteins which presumably share a similar lipid structure in their N termini has been identified (64). Their functions are unknown.

Porins. The proteins coded for by the *ompF*, *ompC*, and *phoE* genes in *E. coli* and those coded for by the *ompF*, *ompC*, and *ompD* genes in *S. typhimurium* are called porins because they produce relatively nonspecific pores or channels that allow the passage of small hydrophilic molecules across the outer membrane (Table 2). (Recently, a protein that appears to be the PhoE equivalent in *S. typhimurium* also has been described [5]). Among the porins, the PhoE protein is unique in that it is produced only under conditions of phosphate starvation (4, 168). Thus, in the usual culture media, only the OmpF and OmpC (and in *S. typhimurium* also OmpD) porins are produced, but the relative abundance of these porins is under an efficient

TABLE 2. Porins of *E. coli* K-12 and B and *S. typhimurium* LT2[a]

Name	Mol wt[b]	Alternative names K-12	Alternative names LT2	Gene	Map position	Regulation of synthesis	Presence in: B	Presence in: K-12	Presence in: LT2	Estimated diam (nm)	Receptors for phage/colicin in: E. coli	Receptors for phage/colicin in: S. typhimurium
OmpF	38,306	Ia, 1a, O-9, b	35K	*ompF*	21	Repressed by high osmolarity	+	+	+	1.16	TuIa, T2, TP1, TP2, TP5, ColA	
OmpC	37,083	Ib, 1b, O-8, c	36K	*ompC*	47	Derepressed by high osmolarity	−	+	+	1.08	TuIb, Mel, PA2, 434, SS1, TP2, TP5, TP6	PH42, PH105, PH221
OmpD	(38,000)		34K	*ompD*	28	Cyclic AMP dependent	−	−	+			PH31, PH42, PH51
PhoE	36,782	Ic, e, E		*phoE*	6	Derepressed by P_i starvation	?	+	+	1.16	TC23, TC45	

[a] For references, see Lugtenberg and van Alphen (86), Nikaido and Vaara (124), and Mizuno et al. (100).
[b] Calculated molecular weights from the amino acid sequences of K-12 proteins. The number for OmpD protein is an estimate from mobility in SDS-PAGE.

TABLE 3. Outer membrane proteins involved in specific transport[a]

Name	Gene	Map position	Mol wt[b]	Receptors for phages and colicins	TonB requirement	Solutes transported	Approx K_d
Phage λ receptor	lamB	91	47,393	λ, K10, TP1, TP5, SS1	−	Maltose, maltodextrin	1 mM[c]
Phage T6 receptor	tsx	9	(26,000)	T6, colicin K	−	Nucleosides	
BtuB	btuB	89	66,400	BF23, E colicins	+	Vitamin B_{12}	3 nM[d]
Cir	cir	44	(74,000)	Colicins I and V	?	Ferric iron?	
Iut	iut	—[e]	(74,500)	Cloacin DF13	+	Fe^{3+}-aerobactin	
FhuE	fhuE	16	(76,000)		+	Fe^{3+}-coprogen	
FhuA (TonA)	fhuA	3	(78,000)	T1, T5, φ80, colicin M	+	Ferrichrome	<10 μM[f]
FecA	fecA	7	(80,500)		+	Ferric citrate	
FepA	fepA	13	(81,000)	Colicins B and D	+	Fe^{3+}-enterobactin	0.3 μM[g]
Fiu	fiu	18	(83,000)		?	Ferric iron?	

[a] For most proteins, references can be found in Lugtenberg and van Alphen (86) and Nikaido and Vaara (124). For FhuE and Fiu, see Hantke (56); for Iut, see Grewal et al. (49).
[b] The molecular weights of BtuB and phage λ receptor are based on the gene sequence; others are estimates from SDS-PAGE.
[c] See Luckey and Nikaido (82).
[d] See Holroyd and Bradbeer (61).
[e] The gene is on the ColV plasmid.
[f] See Luckey et al. (83) and Wayne and Neilands (176).
[g] See Hollifield and Neilands (60).

regulation by the osmotic activity of the media, as well as by temperature (see below). The total amount of these porins present is relatively constant and is very large (Table 2), making porins one of the most abundant proteins in E. coli and S. typhimurium in terms of mass; they can represent up to 2% of the total protein of the cell.

The OmpF porin of E. coli B (27) and the ompF, ompC, and phoE genes of E. coli K-12 (66, 100, 132) have been sequenced. There is a very strong homology among the sequences of the three proteins (100). The sequences do not show long stretches of hydrophobic amino acid residues. However, the secondary-structure prediction shows many 11- to 15-residue stretches which are predicted to exist as β-sheets and contain only a few charged residues. Since β-sheet is a much more extended conformation than α-helix and can cross the thickness of the membrane in 11 or 12 residues, it is tempting to imagine that the protein crosses the membrane many times by using these stretches. Indeed, infrared spectroscopy has shown that many of the β-sheet structures, which are extremely abundant in porins (109, 146), are oriented perpendicular to the surface of the membrane (44).

When the outer membrane-peptidoglycan complex is extracted with sodium dodecyl sulfate (SDS) at temperatures below 60°C, much of the porin is left behind in the insoluble fraction, frequently as two-dimensional crystals (146). Although these crystals were extremely useful in the structural studies of porin (see below), they are most probably artifacts of extraction (124, 147). Porins can be brought into solution by including 1 M NaCl in the SDS solution (57), but even under these conditions they exist as tightly associated trimers (108). When the trimers are heated in SDS at 100°C, they finally become dissociated into monomers, which then migrate in SDS-polyacrylamide gel electrophoresis (SDS-PAGE) at rates close to what are expected from their molecular weights. Electron microscopic studies with image enhancement suggested that each of the monomeric units contained a channel at the outer surface, but the three channels became merged toward the inner surface of the

membrane (36). However, a more recent study with PhoE protein indicates that the three channels remain separate without merging with each other (B. Jap, personal communication) (Fig. 3). Recently, mutants apparently producing larger pores have been isolated (15).

Porins act as receptors for various phages (Table 2). They can also be modified, in intact cells, by a macromolecular reagent that cannot penetrate the outer membrane (71). Although their association with peptidoglycan is likely to be artifactual (86), their transport function, which is described below, clearly necessitates their transmembrane orientation.

Some strains of E. coli are known to produce alternative or additional porins. Thus, Schnaitman (156) found that several clinical isolates of E. coli produce an additional porin, called protein 2 or Lc, which does not exist in strain K-12 and appears to be coded by a prophage (138). Encapsulated strains of E. coli are frequently found to produce a characteristic porin, often called protein K (164, 178).

Proteins involved in specific diffusion processes. The proteins involved in specific diffusion processes are summarized in Table 3. The LamB protein of E. coli is involved in allowing the passage of maltose and maltodextrins through the outer membrane. Its structural gene, lamB, is a part of the maltose regulon. When fully induced, the protein is made in very large copy numbers, comparable to those of porins. Although S. typhimurium is not sensitive to phage lambda, an outer membrane protein inducible with maltose and with a similar molecular weight has been reported (134). The lamB gene has been completely sequenced (29). This protein resembles porin in many ways: it forms tightly associated trimers stable even in SDS (134), is rich in β-sheet structure (44), and can become associated with the peptidoglycan sheets when other proteins are extracted with detergents (42). The analysis of phage-resistant missense mutations by DNA sequencing, as well as protease digestion of the protein coupled with the use of monoclonal antibodies, produced results so far compatible with a model in which the polypeptide chain weaves through the lipid bilayer many times by using the β-sheet regions to cross the membrane (24a, 149). The transport functions of this protein are described further below.

The phage T6 receptor protein of *E. coli*, coded for by gene *tsx*, is an outer membrane protein. Two laboratories found that *tsx* mutants are defective in the active transport of nucleosides when they exist at low external concentrations (55, 96). This is a characteristic phenotype for strains with low outer membrane permeability for the particular substrate, and it led to the hypothesis that the Tsx protein forms a specific pore for nucleosides. This hypothesis is reasonable because nucleosides are not expected to diffuse rapidly enough through the porin channels to saturate the exceptionally high V_{max} of the nucleoside transport system of *E. coli* (72). However, attempts to reconstitute the Tsx protein into liposomes or planar lipid membranes have not been successful so far.

E. coli outer membrane contains several other proteins that are involved in specific diffusion processes. At least some of these proteins bind their substrates quite tightly, in contrast to the very loose or nondemonstrable substrate binding of LamB and Tsx proteins (Table 3). Furthermore, again in contrast to the LamB or Tsx protein, the proper function of these other proteins requires functional TonB product (137). Although TonB is often assumed to be involved in "energy coupling," its mode of action is not clear. The gene *tonB* has been sequenced, and the amino acid sequence deduced is quite unusual (137). It has been reported that the collaboration between TonB and the outer membrane protein BtuB results in the accumulation of very large numbers of vitamin B_{12} molecules in the periplasmic space against the concentration gradient (142), but the mechanism proposed in a recent paper (61) appears inadequate to explain these results. The gene *btuB* has been sequenced (58).

There are a number of outer membrane proteins that are involved in the transport of various chelates of ferric ion (Table 3); this is also expected because iron chelates (usually more than 700 daltons) are too large to go through the porin channels easily. (The transport of iron chelates in bacteria has been reviewed recently by Neilands [110].) All of the proteins involved in the passage of iron chelates across the outer membrane appear to be induced by iron starvation and sometimes become major proteins of the outer membrane under these conditions (110).

OmpA protein. The OmpA proteins have monomer molecular weights (35,159 in *E. coli* K-12), similar to porins, but they behave very differently upon solubilization in SDS at low temperature. Porins remain oligomeric under these conditions, and their mobility in SDS-PAGE increases when the samples are heated at 100°C in SDS so as to convert them into monomers. In contrast, the OmpA protein extracted at room temperature apparently is incompletely denatured and thus runs much faster in SDS-PAGE than expected from its true molecular weight. Complete denaturation through heating in SDS actually reduces its mobility (155).

The OmpA protein is almost as abundant as porins, and thus its copy number can approach 10^5 per cell. Although OmpA protein can be cross-linked chemically to other proteins, especially murein lipoprotein (133), there is little evidence that it exists as homogeneous oligomers or as stoichiometric complexes with other proteins.

The OmpA protein is rich in β-sheet structure (109). It appears to span the thickness of the membrane, as it can be labeled by a nonpenetrating reagent in intact cells (71) and serves as receptor for several phages (86). At the same time it can be cross-linked chemically to the underlying peptidoglycan layer (37, 139). An *ompA* mutant of *E. coli* showed reduced overall transport rates for amino acids (94), and two independent *ompA* mutants were recently found to be defective in peptide uptake (J. W. Payne, personal communication). On the other hand, an *ompA* mutant of *S. typhimurium* showed an unaltered permeability to cephaloridine (122). It is not possible to reconcile these data, but it seems unlikely that the OmpA protein forms pores by itself, because outer membranes of mutants with intact OmpA protein but without porin show only traces of permeability toward hydrophilic solutes (see below). Mutants lacking OmpA protein are extremely poor recipients in conjugation (93), and those lacking both OmpA and lipoprotein produce unstable outer membrane (160).

The OmpA protein (28) and the *ompA* gene (102) have been sequenced. A striking finding is the Ala-Pro-Val-Val-Ala-Pro-Ala-Pro-Ala-Pro-Ala-Pro sequence at residues 176 through 187, a sequence resembling the "hinge" region of immunoglobulins. Indeed, this exposed, protease-sensitive sequence appears to separate the protein into two large domains, the N-terminal domain inserted into the outer membrane and the C-terminal domain presumably located in the periplasmic space (28).

Other proteins. A few enzymes have been located in the outer membrane. They include phospholipase A1 (126), the structural gene for which has been sequenced (33), and proteases (40, 89).

STRUCTURE AND FUNCTIONS

Cell Surface Functions

Because the outer surface of the outer membrane constitutes the outermost area of the cell, this region is in immediate contact with the environment and thus has developed various properties that are characteristic of the cell surface of unicellular organisms. These properties include the presence of hydrophilic, usually negatively charged carbohydrate chains to avoid surface phagocytosis. Clearly the polysaccharide chains of LPS and capsular polysaccharides fulfill this requirement, and one notes that the major outer membrane proteins are all acidic proteins. Furthermore, as is common in microorganisms with extensive contact with higher animals, the surface structure has undergone extensive structural diversification, as seen in the diversity of O antigen and capsular polysaccharide structure; this presumably helps in avoiding attack by preexisting antibodies and digestive enzymes (90).

In addition, the cell surface of these symbiotic and parasitic bacteria has developed machinery for attachment to the components of the host cell surface. For this topic, there are extensive reviews available (3, 11).

The Lipid Bilayer and Its Barrier Function

Asymmetric bilayer. Clearly, the lipid bilayer forms the basic continuum of the outer membrane just as in most other biological membranes, as indicated by the typical trilaminar morphology of the thin sections of outer membrane (46). The X-ray diffraction data show the characteristic 0.42-nm spacing between the hydrocarbon chains in directions parallel to the surface of the membrane (169), a large portion of the hydrocarbon chains in the outer membrane shows cooperative thermal melting (131), and the membrane cleaves in the middle by freeze-fracturing procedures (38, 174).

However, the distribution of lipids in the bilayer is highly asymmetric. Most of the LPS molecules are clearly located in the outer leaflet of the bilayer, as shown by electron microscopy after antibody labeling (103, 105) and by enzymatic modification of LPS in intact cells (41). Given this fact, and given the observation that the number of hydrocarbon chains in phospholipids is about equal to that in LPS in the outer membrane, it follows that most of the inner leaflet must be occupied by phospholipids and that there should be few, if any, phospholipid molecules in the outer leaflet. This hypothesis (158) was tested by Kamio and Nikaido (70) by treating intact *S. typhimurium* cells with an impermeable reagent, CNBr-activated dextran. There was no labeling of phosphatidylethanolamine, which should have occurred if some of these molecules were located in the outer leaflet. However, in deep rough mutants there were significant amounts of phospholipid molecules in the outer leaflet (see Fig. 4B and C), and these molecules were efficiently labeled. The asymmetric distribution of LPS and phospholipids in wild-type cells has been supported also by experiments using spin-labeled probes (123) and is consistent with other lines of evidence (124).

Lateral interaction between LPS molecules appears to be very strong, at least in the presence of divalent cations that neutralize the electrostatic repulsion between these polyanionic molecules. Thus, if one produces mixed bilayers initially containing domains of both pure LPS and pure phospholipids, the LPS molecules do not diffuse out readily into the neighboring phospholipid domains (165). Such a strong interaction will contribute to the lower permeability of the LPS monolayer (see below) and may also help to stabilize the asymmetric bilayer structure.

Asymmetric bilayer as permeation barrier. Phospholipid bilayers are very permeable to hydrophobic molecules, and the permeability is higher the more hydrophobic the solutes (30, 161). In contrast, the outer membranes of *E. coli* and *S. typhimurium* do not show a high degree of permeability to hydrophobic molecules, as suggested by the strong resistance of wild-type strains to hydrophobic antibiotics (e.g., macrolides, novobiocin, rifamycins, actinomycin D), detergents (e.g., SDS, bile salts, Triton X-100), and hydrophobic dyes (e.g., eosin, methylene blue, brilliant green) (79, 118). In fact, the resistance to dyes and bile salts has been exploited in the fabrication of a number of selective media favoring the growth of these organisms but suppressing that of gram-positive bacteria. In addition, the low permeability of wild-type outer membrane to crystal violet (50) and to a

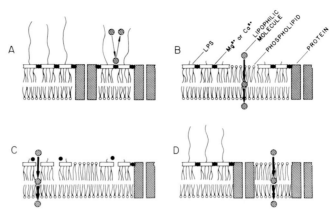

FIG. 4. Hypothetical structure of outer membrane in wild-type, deep rough mutant, and EDTA-treated wild-type cells of *E. coli* and *S. typhimurium*. (A) Untreated wild-type cells in which the outer leaflet is almost entirely composed of LPS and proteins. Hydrophobic (lipophilic) molecules (⊘) cannot enter the bilayer, presumably because of the strong lateral interaction between LPS molecules. (B) Deep rough mutants, in which the decrease in protein content creates space which is filled in with phospholipids. Hydrophobic molecules are hypothesized to penetrate through the phospholipid bilayer domains created in this manner. (C) Deep rough mutants. The structure is the same as in panel B, but the lipophilic solutes here are assumed to penetrate predominantly through the LPS monolayer, taking advantage of the weakened lateral interaction between defective LPS molecules. (D) EDTA-treated wild-type cells. Phospholipid molecules are assumed to fill in the void left by the selective removal of LPS by EDTA. From Nikaido and Vaara (124).

hydrophobic penicillin, nafcillin (113), has been demonstrated by direct measurement of penetration kinetics. The simplest hypothesis to explain these data is that the low permeability is caused by the asymmetric structure of the bilayer, in other words, by the low permeability of the LPS monolayer that comprises the outer half of the bilayer. Since LPS contains only saturated fatty acid chains, the hydrocarbon interior of the monolayer would be much less fluid than that of the usual phospholipid membrane. A recent X-ray diffraction study (76) has shown the very rigid, nearly crystalline nature of the hydrocarbon portion of the LPS. It would be much more difficult for hydrophobic solutes to penetrate into such an ordered, rigid matrix than into the fluid interior of the more usual biological membrane. Furthermore, if the penetration of solutes is induced by the temporary dislocation of lipids to produce "cavities," as is generally believed, then both the strong lateral interaction between LPS molecules and the large size of LPS molecules will make the production of such cavities very difficult. The important role LPS plays in the barrier functions of the outer membrane is confirmed by the study of situations in which the barrier properties are largely lost because of changes in the LPS monolayer. These situations are described below under the headings Deep rough mutants, Effects of EDTA, Effects of divalent cations, and Effects of polycations.

We emphasize, however, that the asymmetric bilayer is not totally impermeable to hydrophobic compounds. In fact, it is expected to show a significant permeability toward very hydrophobic compounds

(97), although not much experimental evidence for this exists (124).

Deep rough mutants. The permeability of the outer membrane does not become altered when the structure of LPS becomes defective by the loss of the entire O polysaccharide and even of the distal portion of the core. However, when the sugars in the more proximal portion of the core are lost (chemotypes Rd1, Rd2, and Re [Fig. 1]), the *E. coli* and *S. typhimurium* mutants (called deep rough mutants) become dramatically more sensitive to hydrophobic dyes, detergents, and antibiotics. They also become hypersensitive to fatty acids, phenol, and polycyclic hydrocarbons (for references, see Nikaido and Vaara [124]). Although these results clearly indicate the importance of LPS in preventing the diffusion of hydrophobic solutes in the wild-type outer membrane, the mechanism of permeability increase in the deep rough mutants is not entirely clear. This is because the deep rough mutations are pleiotropic and apparently hamper the insertion of several major outer membrane proteins (1, 74). This is followed by a compensatory increase in the phospholipid content of the outer membrane (158) and by the appearance of some phospholipid molecules in the outer leaflet of the membrane (70). Thus, there are two possible explanations for higher permeability of deep rough outer membrane: (i) the presence of phospholipid in the outer leaflet allows the penetration of the solutes into the hydrophobic interior (Fig. 4B); (ii) strong lateral interaction between LPS molecules, necessary for preventing penetration by solutes, becomes much weaker with the incomplete, deep rough LPS (Fig. 4C). These possibilities are discussed in detail by Nikaido and Vaara (124).

Some other types of mutations alter the structure of LPS and also increase the permeability of the outer membrane, presumably that of the bilayer region; these mutants are discussed in a recent review (124). In addition, there are several genetically defined mutations that produce hypersensitivity to hydrophobic agents and detergents but do not appear to alter LPS or outer membrane proteins, at least with current analytical methods (163).

Effects of EDTA. Leive (78) found that treatment of *E. coli* cells with EDTA in Tris buffer made them much more sensitive to actinomycin D, a hydrophobic agent, and that the procedure liberated about one-half of the LPS but little else from the cells. Furthermore, she found that such EDTA-treated cells became hypersensitive to a number of other hydrophobic agents, including dyes and detergents (79). Clearly, removal of divalent cations that neutralized the electrostatic repulsion between LPS molecules resulted in the destabilization of the LPS monolayer portion of the membrane. Again, in principle the increase in permeability could have been caused either by the filling in of the space, formerly occupied by LPS, by phospholipids, or by the weakening of the lateral LPS-to-LPS interaction. We favor the former possibility, because it fits better with the observation that the cells must continue biosynthesis of macromolecules (presumably LPS) in order to reestablish the barrier property of the outer membrane (79).

In another use of EDTA in bacterial physiology, one rapidly dilutes *E. coli* or *S. typhimurium* cells, preincubated in 20% sucrose containing EDTA and Tris buffer, in order to get the release of periplasmic proteins (111). In this "osmotic shock" procedure, the periplasmic space first becomes filled with 20% sucrose, which produces plasmolysis. Upon dilution, the system will attempt to correct the osmotic imbalance through the influx of water into the periplasm as well as through the efflux of sucrose from the periplasm into the medium. Because the latter process is limited by the poor permeability of porin channels toward sucrose (see below), the outer membrane, which has become weakened by the EDTA-Tris treatment, becomes ruptured momentarily by the massive influx of water into the periplasm.

In both of the above procedures, Tris, a bulky primary amine, contributes to the weakening of the lateral interaction between LPS molecules by partially replacing other cations tightly bound to LPS. Amines like Tris presumably inhibit the tight association between LPS molecules through steric hindrance; it is known that triethylamine is an excellent agent for dissociating the intermolecular aggregation of purified LPS (43). We note that Tris alone (0.1 M, pH 7.2) is sufficient, without EDTA, to cause the release of about 20% of the LPS from *S. typhimurium* (R. Hukari and M. Vaara, unpublished data). Further, Tris alone is able to sensitize smooth bacteria to complement. (141).

Effects of divalent cations. As stated above, the LPS monolayer portion of the outer membrane must be stabilized by divalent cations, and LPS was shown to have a high-affinity binding site(s) for Ca^{2+} (152). In mutants producing unstable outer membranes, such as mutants lacking murein lipoprotein or deep rough mutants, the addition of millimolar concentrations of Mg^{2+} was found to stabilize the outer membrane structure, as judged by the decreased release of periplasmic enzymes into the growth medium (25, 117).

In contrast, at low temperature, very high concentrations of divalent cations, especially Ca^{2+}, appear to disrupt the outer membrane. Thus, DNA is not expected to penetrate through the normal outer membrane, but treatment of cells with 20 mM or higher concentrations of Ca^{2+} at 0°C can make *E. coli* and *S. typhimurium* become recipients in transformation and transfection (92). The maltose-binding protein in the periplasmic space also becomes accessible for antibody added from the outside (22). It is known that Ca^{2+} "freezes" acidic phospholipids by raising their thermal transition temperature (135); most probably Ca^{2+} influences the LPS monolayer in the same manner and produces "cracks" by the freezing of the LPS monolayer.

Effects of polycations. Polymyxin is a decapeptide antibiotic with a fatty acid tail, with five positively charged groups and no negative charge. It first binds to the outer membrane of *E. coli* or *S. typhimurium* (166), presumably by binding to LPS, and then goes through the outer membrane by disrupting this membrane barrier. This conclusion is supported by the study of a polymyxin-resistant (*pmrA*) mutant of *S. typhimurium*. This mutant contains four- to sixfold higher amounts of a positively charged sugar, 4-aminoarabinose, in the lipid A portion of its LPS (173). Consequently, the mutant LPS is less acidic and binds polymyxin much less readily than the wild-type LPS. Among members of the *Enterobacteriaceae*, there are also naturally occurring "mutants" that are intrinsically resistant to polymyxin. *Proteus mirabilis*, one of these species, is known to produce LPS that contains much higher amounts of 4-aminoarabinose substituent than does that of *E. coli* or *S. typhimurium* (144). These results show clearly that the binding to LPS, followed by the disruption of the barrier property of the outer membrane bilayer, is an essential step in the action of polymyxin.

Polymyxin is thus a prototype of polycationic agents which disorganize and disrupt the outer membrane. However, its action is complicated by the presence of the hydrophobic fatty acid tail. Recently Vaara and Vaara (170–172) studied the action of polymyxin nonapeptide, a papain-cleaved derivative of polymyxin that has lost the N-terminal diaminobutyric acid residue and the fatty acyl residue attached to it. This polycation, as well as a polylysine, Lys$_{20}$, were remarkably active in sensitizing the wild-type strains of *E. coli* and *S. typhimurium* to a number of hydrophobic agents, such as novobiocin, fusidic acid, erythromycin, clindamycin, rifampin, actinomycin D, cloxacillin, and nafcillin. Lys$_{20}$ released about 30% of the LPS from the cell, and thus its sensitizing action could have the same basis as that of EDTA (see above). In contrast, polymyxin nonapeptide did not cause any release of LPS. Polymyxin nonapeptide produced long, fingerlike projections involving only the outer leaflet of the outer membrane. This expansion of the outer leaflet undoubtedly is the cause of the increase in permeability, although the precise molecular mechanism of the process is not yet clear.

Hancock (53) has proposed that aminoglycosides cross the *Pseudomonas aeruginosa* outer membrane mainly by a similar mechanism, i.e., by binding to LPS followed by the disorganization of the membrane. This is a plausible mechanism for *P. aeruginosa*, which produces an outer membrane with extremely low hydrophilic permeability. In enteric bacteria, however, the porin channels are expected to have a reasonably high permeability, at least for most aminoglycosides, especially in view of the net positive charge of these molecules. Thus, the relative significance of the membrane disorganization pathway in these bacteria is unclear at present.

DIFFUSION OF HYDROPHILIC COMPOUNDS

Diffusion through Porin Channels

The outer membranes of *E. coli* and *S. typhimurium* act as a molecular sieve that allows the passage of small hydrophilic molecules. Since most hydrophilic solutes available for experimental use are flexible molecules, it is difficult to determine the cutoff size with precision. However, with sugars and peptides the cutoff seems to be around 600 daltons (32, 136). That this diffusion occurs mainly through the channels made by the porin proteins was first suggested by reconstituting purified porins with phospholipids and LPS (106, 107) and was confirmed by the properties of mutants that produce greatly reduced amounts of porin (7, 10, 87, 122).

In order to understand the role of outer membrane permeability in bacterial physiology, it is essential to have a quantitative estimate of permeability or to know the permeability coefficient of the outer membrane toward the solute of interest. Qualitative knowledge that a certain compound is permeable or not tells us very little (114, 124). The permeability of the porin channel can be measured in reconstituted systems (for example, in liposomes) (81, 115) or with planar lipid bilayer (150, 151) or black lipid films (17). In intact cells, the best approach is to combine the influx of solutes through the outer membrane with their enzymatic hydrolysis in the periplasmic space (184); this and other approaches have been reviewed (H. Nikaido, Methods Enzymol, in press).

Extensive quantitative studies on the permeability of *E. coli* porin channels have shown that the rate of diffusion of solutes through these channels is affected greatly by the gross physicochemical properties of the solutes but that there is no sign of true specificity. The three major factors that influence the permeability are the size, electrical charge, and hydrophobicity of the solute molecule, as described in detail by Nikaido and Vaara (124).

Size of the solute. As predicted for diffusion through narrow channels, the chance that the solute molecule successfully enters the channel by the process of random collision (and therefore the magnitude of the macroscopic permeability coefficient toward that particular kind of solute) is strongly influenced by the size of the molecule. Thus, the rate of diffusion of disaccharides through the *E. coli* OmpF and OmpC channels is nearly two orders of magnitude lower than the rate of diffusion of a pentose, although the molecular weight of disaccharides, 342, places them well within the "exclusion limit" of 600 daltons (119, 120). This example emphasizes the importance of quantitative considerations on permeability. In fact, the dependence of the permeability on solute size is somewhat different for the *E. coli* K-12 OmpF and the OmpC channels. From the slopes of the permeability-versus-solute size curves one can calculate the equivalent diameters of the OmpF and OmpC pores as, respectively, 1.16 and 1.08 nm (120), using an equation proposed by Renkin (140). Similar studies have not been carried out with *S. typhimurium* porins, but their pore sizes appear to be similar to those in *E. coli* (17, 106), and there is evidence suggesting that OmpF porin again makes a channel slightly larger than that of OmpC porin (see Porin channels in bacterial physiology, below). *E. coli* PhoE porin seems to produce a channel with a diameter in a similar range (120). These conclusions are in agreement with the results of black lipid film studies, which suggested a slightly higher single-channel conductance, and therefore a larger diameter, for the OmpF than for the OmpC porin (19).

The small pore size, which produces huge differences in rates of penetration even among relatively small solutes, apparently is a characteristic of the enteric bacteria. Many other bacteria, such as *P. aeruginosa* and *Rhodopseudomonas sphaeroides*, produce porins with significantly larger channels (19, 54, 177). As expected with these larger channels, the permeability of solutes is not as strongly influenced by their size.

Hydrophobicity of the solute. Studies with monoanionic cephalosporins showed that the permeability through the OmpF channel of *E. coli* was affected negatively by the hydrophobicity of the solute, with a 10-fold increase in the 1-octanol/water partition coefficient of the uncharged form of the solute producing a four- to fivefold reduction in the penetration rate (121). Although such a negative effect is likely to be seen with all kinds of solutes, the magnitude of the effect is expected to be related to the size of the molecule in relation to the diameter of the pore: hydrophobicity will affect the penetration of larger molecules much more than that of smaller molecules. This point, however, awaits experimental verification.

Electrical charge. OmpF and OmpC porins showed a preference for cations in black lipid film studies (20). Furthermore, liposome studies showed that the diffusion of uncharged sugars through these channels was several times faster than that of corresponding negatively charged sugar acids and that the addition of another negative charge further slowed penetration (120). The retardation effect of the negative charges is further enhanced in intact cells: thus, in the liposome system cephacetrile and cefsulodin, cephalosporins with one and two negative charges, respectively, had penetration rates corresponding to 60 and 23% of that of the zwitterionic cephaloridine in the liposome system (181), whereas through the outer membrane of intact cells their penetration rates were only 23 and 3%, respectively, of that of cephaloridine (121). Possibly the Donnan potential across the outer membrane, which is negative inside (162), is responsible for this more severe retardation of anions in intact cells.

In contrast to the behavior of OmpF and OmpC (and possibly OmpD) channels, the presence of negative charges does not hinder and sometimes accelerates the diffusion of solutes through the PhoE porin (120). The PhoE channel also shows clear anion selectivity in black lipid film studies (18). Although some workers have hypothesized that this channel is specific for phosphate esters, it seems more likely that the channel simply favors any negatively charged compound.

Porin channels in bacterial physiology. Enteric bacteria have to survive in an environment that is full of powerful detergents (the bile salts), and the properties of the porin channels are ideal for excluding these large, negatively charged, and hydrophobic compounds. Furthermore, the synthesis of OmpF porin is known to be repressed by high-osmolarity media (57) as well as by high temperature (85). At least in some strains, this means that practically no OmpF porin is synthesized at 37°C in the presence of about 1% NaCl (159), i.e., under conditions that mimic the interior of the bodies of animals. By switching from the larger OmpF channel to the narrower OmpC channel, the bacteria lose only 50% or so of the permeability toward small nutrients with molecular weights of 100 to 200, while permeability toward larger, more hydrophobic, or negatively charged compounds (or toward compounds with a combination of these properties) is much more drastically reduced because of the more restrictive properties of the OmpC channel (124). Thus, the switching to the narrower channel has obvious ecological advantages in an environment full of inhibitory substances, such as the intestinal tract or other parts of higher animals. As a proof that this is indeed happening, A. A. Medeiros (personal communication) has observed that cephalosporin therapy of a patient infected by a strain of *S. typhimurium* original-ly possessing only OmpF and OmpC porins resulted in the selection of a more resistant mutant lacking the OmpC porin. In media of low osmotic activity the mutant is as sensitive as the parent strain because of the production of the OmpF porin, but it becomes more resistant in media containing about 1% NaCl owing to the virtual absence of functional porin under these conditions. These results suggest that in the body of the patient the original strain must have been producing only the OmpC porin when chemotherapy was instituted.

On the other hand, the OmpF porin is probably beneficial when the bacteria are out of the animal body. In ponds and streams, for example, the low osmolarity and the low temperature will result in the strong induction of OmpF porin, whose wider channel will allow the more efficient uptake of nutrients from a very dilute environment. The PhoE porin, which is induced by phosphate starvation (4, 168), may be also useful in a similar environment, by allowing a more rapid diffusion of phosphorylated compounds.

The molecular mechanism of osmotic regulation has been studied in several laboratories. Hall and Silhavy (52) identified a regulatory gene, *ompR*, which apparently produces an osmotic sensor. The reciprocal rates of production of OmpF and OmpC proteins may be related to the simultaneous synthesis of the *ompC* mRNA and a small RNA (called *mic*RNA) with a sequence complementary to the beginning of the *ompF* gene; such as "antimessage" is expected to bind to the *ompF* mRNA and decrease its translation (101).

One aspect of porin physiology that puzzles many people is why such a large number of porin molecules, up to 10^5 per cell, are necessary for the cell. This can be understood readily when one remembers that porin channels (i) are nonspecific and (ii) mediate simple diffusion processes. Because the channels are narrow but nonspecific, diffusion rates of even moderately large molecules, such as disaccharides, become very slow in comparison with those of very small molecules (see above). Furthermore, because the rate of simple diffusion is proportional to the concentration difference between the external medium and the periplasm, the rate, which could be quite high in, for example, laboratory media containing a 0.01 M carbon source, drops more than three orders of magnitude at micromolar concentrations of nutrients, the kind of environment for which most of the active transport systems of *E. coli* seem to have been designed (72). Thus, we can calculate that 10^5 molecules of porin per cell is absolutely necessary for the transport of moderately large molecules at external concentrations in the micromolar range, although this would be a large excess for smaller molecules or for solutes that exist in higher concentrations (124). In fact, diffusion through the porin channels becomes the rate-limiting step in the high-V_{max} transport systems for large molecules such as disaccharides or nucleosides. This is why specific diffusion systems are necessary for these compounds in spite of their ability to diffuse through the porin channels.

Experiments with planar bilayer systems have suggested that the porin channels are voltage regulated (150, 151). However, no voltage dependence was observed by workers using black lipid films (17), and the physiological significance of this phenomenon is unclear, especially in view of the failure to observe the closing of the porin channels in intact cells even in the presence of an elevated Donnan potential across the outer membrane (J. Hellman and H. Nikaido, unpublished data).

Diffusion through Specific Diffusion Channels

Only the LamB channel has been studied in detail both in intact cells and in reconstituted systems. The isolated LamB protein produces a channel which allows the diffusion of small, unrelated molecules, but apparently becomes more specific with the maltose series of oligosaccharides as the solute molecules become larger (81, 82). A more recent study with black lipid films suggests that the nonspecific permeability toward smaller molecules may have been produced either by contaminating proteins or by denatured forms of the LamB protein (R. Benz, personal communication). In intact cells, however, the efficient transport of maltose as well as maltodextrins across the outer membrane seems to require the cooperation of maltose-binding protein, the product of the *malE* gene (175). The isolated MalE protein was indeed shown to interact specifically with the LamB protein (6). The MalE protein was also reported to produce striking changes in the alkali cation permeability to the LamB channel in planar lipid films (112). However, the primary physiological function of the LamB-MalE complex cannot be to allow the diffusion of alkali cations, and the physiological significance of this observation is not clear.

ASSEMBLY

Junctions between Inner and Outer Membranes

Bayer (8) discovered sites at which inner membranes are apparently fused to the outer membrane in thin sections of *E. coli* cells plasmolyzed in 20% sucrose. He has emphasized the role of these junctions in various physiological functions of the cell, including the process of injection of phage nucleic acids. As described below, these fusion sites apparently play important roles in the assembly of the outer membrane. More recently, Bayer et al. (9) showed that fractions of intermediate density between the outer and inner membranes of *Salmonella anatum* contained structures in which outer membrane vesicles were joined to inner membrane vesicles. This work thus represents the first isolation of a membrane fraction presumably enriched in the fusion sites; however, no specific component of this fraction has been found so far, presumably because such sites constitute only a very small part of the preparation, which is dominated by outer and inner membrane vesicles. Although phospholipase A has been claimed to be a marker enzyme for these sites, its specific activity in the intermediate-density preparation is only slightly higher than in the outer membrane (9).

Fusion sites of a potentially different type (and function) have been discovered by MacAlister et al. (88), who showed by electron microscopy of serial sections of *S. typhimurium* cells that zones of adhesion which continuously encircle the cell exist on both sides of the cell division site. These circular "perisep-

tal annuli" are likely to be involved in the cell division process, and they appear to be different from the pointlike "Bayer fusion sites." Interestingly, penicillin-binding protein 3, an enzyme implicated in the biosynthesis of septal peptidoglycan, was strongly enriched in the membrane preparations of intermediate density when precautions were taken to limit the extent of peptidoglycan hydrolysis during the membrane preparation (145). This suggests that the intermediate-density fractions may be enriched not only for Bayer fusion sites but also for fragments of the periseptal annuli, which could contain also the nascent septal peptidoglycan together with its biosynthetic enzyme.

Phospholipids

By introducing labeled phospholipid molecules into the outer membrane of intact cells, Jones and Osborn (69) showed that there is a rapid exchange between the phospholipids of the outer and the inner membranes. It is not clear how this exchange occurs. However, attempts to detect the phospholipid exchange (or carrier) proteins in the periplasmic space have been unsuccessful (69), and it seems most likely at present that the exchange occurs through the Bayer fusion sites. The kinetics of translocation of newly synthesized phospholipids from the inner to the outer membrane was studied by Donohue-Rolfe and Schaechter (35). Phosphatidylglycerol and cardiolipin are translocated much more rapidly than phosphatidylethanolamine. Interestingly, the translocation of the latter compound is inhibited under conditions that decrease the proton motive force.

LPS

As described elsewhere in this book, LPS is synthesized on the cytoplasmic membrane. Core oligosaccharide is built by successive extension by transfer from nucleotide-sugars to incomplete LPS molecules presumably anchored onto the inner surface of the inner membrane. The repeating unit of the O antigen is made independently in a form anchored to the membrane via a carrier lipid, undecaprenol pyrophosphate. There are several possibilities as to how polymerization of the O-antigen polysaccharide and its transfer to the core LPS may be linked with the export of the LPS molecule to the outer membrane (128). Munford and Osborn (105) showed by immunoelectron microscopy that in mutants defective in core biosynthesis, the newly made O chain, presumably still linked to the undecaprenol pyrophosphate carrier, is located on the outer surface of the inner membrane. This supports the idea that the polymerization of the O-antigen repeating unit is coupled to the translocation of the saccharide moiety from the inner to the outer side of the cytoplasmic membrane and that the transfer of O antigen to the core LPS is likely to occur on the outer surface of the cytoplasmic membrane.

In strains capable of synthesizing the core, the LPS molecules containing the newly made O chains eventually appear on the outer surface of the outer membrane as patches at places under which Bayer fusion sites appear to be located (104). The molecular mechanism of this transport process, however, is not yet clear. In contrast to phospholipids, the translocation of LPS is apparently unidirectional. LPS added to the outer membrane from the external medium was not found to move to the inner membrane (69).

Proteins

All outer membrane proteins so far studied are known to be synthesized initially with signal peptides, and evidence suggests that their biosynthesis shares at least the initial part of the pathway with that of periplasmic proteins exported into the periplasm (98, 157). Since protein export is dealt with elsewhere in this book, only the peculiarities of the assembly of outer membrane proteins will be discussed here.

How does the cell distinguish proteins destined for outer membrane from those destined for periplasm? In principle, outer membrane proteins could first be exported into the periplasm and then move into the outer membrane. Such a mechanism may indeed be utilized by some proteins. However, among many mutants with impaired export processes for outer membrane proteins, there seems to be no well-established case in which the mutant protein accumulates in the periplasmic space. On the other hand, for porins at least, the export appears to take place at spots corresponding to the Bayer fusions sites (159), a result favoring the possibility that some, if not most, outer membrane proteins are exported through these sites without going through the periplasmic space. Thus, the recognition of these proteins is likely to take place fairly early in the biosynthetic process. The simplest hypothesis is that a linear, primary sequence is recognized. However, such a sequence cannot be within the signal sequence, as signal sequences of the outer membrane proteins by themselves are not enough to push the fused proteins into the outer membrane (157). With the *lamB* gene, studies of LamB-LacZ hybrid proteins containing different lengths of the N-terminal portion of LamB protein showed that amino acids 1 through 49 of the mature protein sequence are sufficient to direct the export of the hybrid protein to the outer membrane, whereas amino acids 1 through 43 are not enough (14). This is consistent with the result of a computer search that identified a short region of homology between LamB and the outer membrane proteins OmpA and OmpF in the area corresponding to the residues 36 through 45 of the mature LamB sequence (125). On the other hand, OmpA protein containing deletions in this homologous region apparently is exported to the outer membrane (U. Henning, personal communication).

Most outer membrane proteins show a strong affinity toward LPS (124). Therefore, another attractive possibility is that association with LPS leads to the export to the outer membrane. This idea is indeed consistent with the finding that deep rough mutants, with severely defective LPS, do not incorporate normal amounts of outer membrane proteins (see above). Furthermore, Beher et al. (12) showed that OmpA proteins are hardly exported in the presence of LPS from unrelated strains. This point is controversial, however, since Cole et al. (31) found that OmpA proteins are efficiently exported even in different species, which presumably contain LPS with signifi-

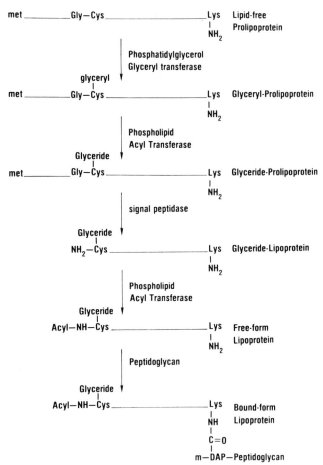

FIG. 5. Pathway of synthesis of murein lipoprotein. m-DAP, *meso*-Diaminopimelic acid. From Tokunaga et al. (167), with permission.

cantly different structures. Finally, when the synthesis of LPS is inhibited at an even earlier step, by using a temperature-sensitive mutant in 2-keto-3-deoxyoctonicacid synthesis, the synthesis of OmpA protein is transiently increased rather than decreased, as expected from the hypothesis discussed above (143).

An outer membrane protein, LamB, has been reported to be synthesized by the cell only during a short period in the cell cycle (127). Since porins are made continuously during the cell cycle, this observation seems to suggest the existence of two different modes of synthesis or insertion (or both) of outer membrane proteins. However, the original observation was obtained by synchronizing a temperature-sensitive mutant population by heat treatment, and recently a similar treatment was shown to produce an inactivation of catabolite activator protein in another strain due to an unnoticed temperature-sensitive mutation in the gene *crp* (13). Thus, the cell-cycle-dependent synthesis of LamB could also have been due to a similar artifact; indeed, we have not been able to observe cell-cycle- or cell-size-dependent synthesis of LamB in unsynchronized, unheated populations of the strain used by Ohki (H. Nikaido, unpublished data).

Lipoprotein has to undergo extensive covalent modifications before its insertion into the outer membrane. The relationship between these modifications and translocation has been studied in several laboratories. The protein is synthesized as a precursor containing a 20-residue-long signal peptide, prolipoprotein (51). A breakthrough was made with the discovery that an antibiotic, globomycin, induces the accumulation of a precursor in which the sulfhydryl group of the cysteine residue is already substituted by a diglyceride group but the cleavage of signal peptide has not occurred (62, 63). This and· studies with isolated envelope fractions as well as intact cells led to the establishment of the pathway shown in Fig. 5 (26, 63, 77, 167). Cleavage is catalyzed by an enzyme different from the signal peptidase involved in the processing of other exported proteins (179), and the gene coding for this enzyme (lspA) has been sequenced (65, 183). Despite this wealth of knowledge, the mechanism of export of lipoprotein is still not clear. The globomycin-induced precursor (glyceride-prolipoprotein, Fig. 5) is associated entirely with the inner membrane. However, a mutant in which the Gly-14 residue of the signal sequence was changed to Asp was found to export a substantial amount of prolipoprotein into the outer membrane, without cleavage of the signal sequence or attachment of lipid substituents (80). Possibly the balance between the hydrophobicity of the lipid substituents and the signal sequence and the hydrophilicity of the rest of the molecule plays a decisive role in the process.

LITERATURE CITED

1. **Ames, G. F., E. N. Spudich, and H. Nikaido.** 1974. Protein composition of the outer membrane of *Salmonella typhimurium*: effect of lipopolysaccharide mutations. J. Bacteriol. **117:**406–416.

2. **Anderson, E. S., and A. H. Rogers.** 1963. Slime polysaccharides of the Enterobacteriaceae. Nature (London) **198:**714–715.

3. **Arbuthnott, J. P., and C. J. Smith.** 1979. Bacterial adherence in host/pathogen interaction in animals, p. 165–198. *In* D. C. Ellwood, J. Melling, and P. Rutter (ed.), Adhesion of microorganisms to surfaces. Academic Press, Inc., London.

4. **Argast, M., and W. Boos.** 1980. Co-regulation in *Escherichia coli* of a novel transport system for *sn*-glycerol-3-phosphate and outer membrane protein Ic (e,E) with alkaline phosphatase and phosphate-binding protein. J. Bacteriol. **143:**142–150.

5. **Bauer, K., R. Benz, J. Brass, and W. Boos.** 1985. *Salmonella typhimurium* contains an anion-selective outer membrane porin induced by phosphate starvation. J. Bacteriol. **161:**813–816.

6. **Bavoil, P., and H. Nikaido.** 1981. Physical interaction between the phage receptor protein and the carrier-immobilized maltose-binding protein of *Escherichia coli*. J. Biol. Chem. **256:**11385–11388.

7. **Bavoil, P., H. Nikaido, and K. von Meyenburg.** 1977. Pleiotropic transport mutants of *Escherichia coli* lack porin, a major outer membrane protein. Mol. Gen. Genet. **158:**23–33.

8. **Bayer, M. E.** 1979. The fusion sites between outer membrane and cytoplasmic membrane of bacteria: their role in membrane assembly and virus infection, p. 167–202. *In* M. Inouye (ed.), Bacterial outer membranes. John Wiley & Sons, Inc., New York.

9. **Bayer, M. H., G. P. Costello, and M. E. Bayer.** 1982. Isolation and partial characterization of membrane vesicles carrying markers of the membrane adhesion sites. J. Bacteriol. **149:**758–767.

10. **Beacham, I. R., D. Haas, and E. Yagil.** 1977. Mutants of *Escherichia coli* "cryptic" for certain periplasmic enzymes: evidence for an alteration of the outer membrane. J. Bacteriol. **129:**1034–1044.

11. **Beachey, E. H. (ed.).** 1980. Bacterial adherence. Chapman & Hall, Ltd., London.

12. **Beher, M., A. Pugsley, and C. Schnaitman.** 1980. Correlation between the expression of an *Escherichia coli* cell surface protein and the ability of the protein to bind to lipopolysaccharide. J. Bacteriol. **143:**403–410.

13. **Benner, D., N. Müller, and W. Boos.** 1985. Temperature-sensitive catabolite activator protein in *Escherichia coli* BUG6. J. Bacteriol. **161:**347–352.

14. **Benson, S. A., E. Bremer, and T. J. Silhavy.** 1984. Intragenic regions required for LamB export. Proc. Natl. Acad. Sci. USA **81:**3830–3834.

15. **Benson, S. A., and A. Decloux.** 1985. Isolation and characterization of outer membrane permeability mutants in *Escherichia coli* K-12. J. Bacteriol. **161:**361–367.

16. **Benson, S. A., M. N. Hall, and T. J. Silhavy.** 1985. Genetic analysis of protein export in *Escherichia coli* K12. Annu. Rev. Biochem. **54:**101–134.

17. **Benz, R.** 1985. Porin from bacterial and mitochondrial outer membranes. Crit. Rev. Biochem **19:**145–190.

18. **Benz, R., R. P. Darveau, and R. E. W. Hancock.** 1984. Outer-membrane protein PhoE from *Escherichia coli* forms anion-selective pores in lipid-bilayer membranes. Eur. J. Biochem. **140:**319–324.

19. **Benz, R., and R. E. W. Hancock.** 1981. Properties of the large ion-permeable pores formed from protein F of *Pseudomonas aeruginosa* in lipid bilayer membranes. Biochim. Biophys. Acta **646:**298–308.

20. **Benz, R., K. Janko, and P. Lauger.** 1979. Ionic selectivity of pores formed by the matrix protein (porin) of *Escherichia coli*. Biochim. Biophys. Acta **551:**238–247.

21. **Brade, H., and E.-T. Rietschel.** 1984. α-2→4-Interlinked 3-deoxy-D-manno-octulosonic acid disaccharide. A common constituent of enterobacterial lipopolysaccharides. Eur. J. Biochem. **145:**231–236.

22. **Brass, J. M., W. Boos, and R. Hengge.** 1981. Reconstitution of maltose transport in *malB* mutants of *Escherichia coli* through calcium-induced disruption of the outer membrane. J. Bacteriol. **146:**10–17.

23. **Braun, V.** 1975. Covalent lipoprotein from the outer membrane of *Escherichia coli*. Biochim. Biophys. Acta **415:**335–377.

24. **Braun, V., H. Rotering, J.-P. Ohms, and H. Hagenmeier.** 1976. Conformational studies on murein lipoprotein from outer membrane of *Escherichia coli*. Eur. J. Biochem. **70:**610–610.

24a.**Charbit, A., J.-M. Clément, and M. Hofnung.** 1984. Further sequence analysis of the phage lambda receptor site. Possible implications for the organization of the LamB protein in *Escherichia coli* K12. J. Mol. Biol. **175:**395–401.

25. **Chatterjee, A. K., H. Ross, and K. E. Sanderson.** 1976. Leakage of periplasmic enzymes from lipopolysaccharide-defective mutants of *Salmonella typhimurium*. Can. J. Microbiol. **22:**1549–1560.

26. **Chattopadhyay, P. K., and H. C. Wu.** 1977. Biosynthesis of the covalently linked diglyceride in murein lipoprotein of *Escherichia coli*. Proc. Natl. Acad. Sci. USA **74:**5318–5322.

27. **Chen, R., C. Kramer, W. Schmidmayr, U. Chen-Schmeisser, and U. Henning.** 1982. Primary structure of major outer-membrane protein I (ompF protein, porin) of *Escherichia coli* B/r. Biochem. J. **203:**33–43.

28. **Chen, R., W. Schmidmayr, C. Kramer, U. Chen-Schmeisser, and U. Henning.** 1980. Primary structure of major outer membrane protein II (ompA protein) of *Escherichia coli* K-12. Proc. Natl. Acad. Sci. USA **77:**4592–4596.

29. **Clément, J. M., and M. Hofnung.** 1981. Gene sequence of the λ receptor, an outer membrane protein of *Escherichia coli* K12. Cell **27:**507–514.

30. **Cohen, B. G., and A. D. Bangham.** 1972. Diffusion of small nonelectrolytes across liposome membranes. Nature (London) **236:**173–174.

31. **Cole, S. T., I. Sonntag, and U. Henning.** 1982. Cloning and expression in *Escherichia coli* K-12 of the genes for major outer membrane protein OmpA from *Shigella dysenteriae*, *Enterobacter aerogenes*, and *Serratia marscenscens*. J. Bacteriol. **149:**145–150.

32. **Decad, G. M., and H. Nikaido.** 1976. Outer membrane of gram-negative bacteria. XII. Molecular-sieving function of cell wall. J. Bacteriol. **128:**325–336.

33. **de Geus, P., H. M. Verheij, N. H. Riegman, W. P. M. Hoekstra, and G. H. de Haas.** 1984. The pro- and mature forms of the *E. coli* K-12 outer membrane phospholipase A are identical. EMBO J. **3:**1799–1802.

34. **DiRienzo, J. M., K. Nakamura, and M. Inouye.** 1978. The outer membrane proteins of gram-negative bacteria: biosynthesis, assembly, and functions. Annu. Rev. Biochem. **47:**481–532.

35. **Donohue-Rolfe, A. M., and M. Schaechter.** 1980. Translocation of phospholipids from the inner to the outer membrane of *Escherichia coli*. Proc. Natl. Acad. Sci. USA **77:**1867–1871.

36. **Dorset, D. L., A. Engel, A. Massalski, and J. P. Rosenbusch.** 1984. Three dimensional structure of a membrane pore. Electron microscopical analysis of *Escherichia coli* outer membrane matrix protein. Biophys. J. **45:**128–129.

37. **Endermann, R., C. Kramer, and U. Henning.** 1978. Major outer

membrane proteins of *Escherichia coli* K-12. Evidence for protein II* being a trans-membrane protein. FEBS Lett. **86**:21–24.

38. **Fiil, A., and D. Branton.** 1969. Changes in the plasma membrane of *Escherichia coli* during magnesium starvation. J. Bacteriol. **98**:1320–1327.

39. **Filip, C., G. Fletcher, J. L. Wulff, and C. F. Earhart.** 1977. Solubilization of the cytoplasmic membrane of *Escherichia coli* by the ionic detergent sodium-lauryl sarcosinate. J. Bacteriol. **115**:717–722.

40. **Fiss, E. M., W. C. Hollifield, Jr., and J. B. Neilands.** 1979. Absence of ferric enterobactin receptor modification activity in mutants of *Escherichia coli* K-12 lacking protein a. Biochem. Biophys. Res. Commun. **91**:29–34.

41. **Funahara, Y., and H. Nikaido.** 1980. Asymmetric localization of lipopolysaccharides on the outer membrane of *Salmonella typhimurium*. J. Bacteriol. **141**:1463–1465.

42. **Gabay, J., and K. Yasunaka.** 1980. Interaction of the LamB protein with the peptidoglycan layer in *Escherichia coli* K-12. Eur. J. Biochem. **104**:13–18.

43. **Galanos, C., and O. Lüderitz.** 1975. Electrodialysis of lipopolysaccharides and their conversion to uniform salt forms. Eur. J. Biochem. **54**:603–610.

44. **Garavito, R. M., J. A. Jenkins, J. M. Neuhaus, A. P. Pugsley, and J. P. Rosenbusch.** 1982. Structural investigations of outer membrane proteins from *Escherichia coli*. Ann. Microbiol. (Paris) **133A**:37–41.

45. **Garegg, P. J., B. Lindberg, T. Onn, and T. Holme.** 1971. Structural studies on the M-antigen from two mucoid mutants of *Salmonella typhimurium*. Acta Chem. Scand. **25**:1185–1194.

46. **Glauert, A. M., and M. J. Thornley.** 1969. The topography of the bacterial cell wall. Annu. Rev. Microbiol. **23**:159–198.

47. **Gotschlich, E. C., B. A. Fraser, O. Nishimura, J. Robbins, and T.-Y. Liu.** 1981. Lipid on capsular polysaccharides of gram-negative bacteria. J. Biol. Chem. **256**:8915–8921.

48. **Grant, W. D., I. W. Sutherland, and J. F. Wilkinson.** 1969. Exopolysaccharide colanic acid and its occurrence in the *Enterobacteriaceae*. J. Bacteriol. **100**:1187–1193.

49. **Grewal, K. K., P. J. Warner, and P. H. Williams.** 1982. An inducible outer membrane protein involved in aerobactin mediated iron transport by colV strains of *Escherichia coli*. FEBS Lett. **140**:27–30.

50. **Gustafsson, P., K. Nordström, and S. Normark.** 1973. Outer penetration barrier of *Escherichia coli* K-12: kinetics of the uptake of gentian violet by wild type and envelope mutants. J. Bacteriol. **116**:893–900.

51. **Halegoua, S., J. Sekizawa, and M. Inouye.** 1977. A new form of structural lipoprotein of outer membrane of *Escherichia coli*. J. Biol. Chem. **252**:2324–2330.

52. **Hall, M. N., and T. J. Silhavy.** 1981. Genetic analysis of the *ompB* locus in *Escherichia coli*. J. Mol. Biol. **151**:1–15.

53. **Hancock, R. E. W.** 1981. Aminoglycoside uptake and mode of action—with special reference to streptomycin and gentamycin. II. Effects of aminoglycosides on cells. J. Antimicrob. Chemother. **8**:429–445.

54. **Hancock, R. E. W., G. M. Decad, and H. Nikaido.** 1979. Identification of the protein producing transmembrane diffusion pores in the outer membrane of *Pseudomonas aeruginosa* PAO1. Biochim. Biophys. Acta **554**:323–331.

55. **Hantke, K.** 1976. Phage T6-colicin K receptor and nucleoside transport in *Escherichia coli*. FEBS Lett. **70**:109–112.

56. **Hantke, K.** 1983. Identification of an iron uptake system specific for coprogen and rhodotorulic acid in *Escherichia coli* K12. Mol. Gen. Genet. **191**:301–306.

57. **Hasegawa, Y., H. Yamada, and S. Mizushima.** 1976. Interactions of outer membrane proteins O-8 and O-9 with peptidoglycan sacculus of *Escherichia coli* K-12. J. Biochem. (Tokyo) **80**:1401–1409.

58. **Heller, K., and R. J. Kadner.** 1985. Nucleotide sequence of the gene for the vitamin B_{12} receptor protein in the outer membrane of *Escherichia coli*. J. Bacteriol. **161**:904–908.

59. **Hirota, Y., H. Suzuki, Y. Nishimura, and S. Yasuda.** 1977. On the process of cellular division in *Escherichia coli*: a mutant of *E. coli* lacking a murein-lipoprotein. Proc. Natl. Acad. Sci. USA **74**:1417–1420.

60. **Hollifield, W. G., Jr., and J. B. Neilands.** 1978. Ferric enterobactin transport system in *Escherichia coli* K-12. Extraction, assay, and specificity of the outer membrane receptor. Biochemistry **17**:1922–1928.

61. **Holroyd, C. D., and C. Bradbeer.** 1984. Cobalamin transport in *Escherichia coli*, p. 21–23. *In* L. Leive and D. Schlessinger (ed.), Microbiology—1984. American Society for Microbiology, Washington, D.C.

62. **Hussain, M., S. Ichihara, and S. Mizushima.** 1980. Accumulation of glyceride-containing precursor of the outer membrane lipoprotein in the cytoplasmic membrane of *Escherichia coli* treated with globomycin. J. Biol. Chem. **255**:3707–3712.

63. **Hussain, M., S. Ichihara, and S. Mizushima.** 1982. Mechanism of signal peptide cleavage in the biosynthesis of the major lipoprotein of the *Escherichia coli* outer membrane. J. Biol. Chem. **257**:5177–5182.

64. **Ichihara, S., M. Hussain, and S. Mizushima.** 1981. Characterization of new membrane lipoproteins and their precursors of *Escherichia coli*. J. Biol. Chem. **256**:3125–3129.

65. **Innis, M. A., M. Tokunaga, M. E. Williams, J. M. Loranger, S.-Y. Chang, S. Chang, and H. C. Wu.** 1984. Nucleotide sequence of the *Escherichia coli* prolipoprotein signal peptidase (*lsp*) gene. Proc. Natl. Acad. Sci. USA **81**:3708–3712.

66. **Inokuchi, K., N. Mutoh, S. Matsuyama, and S. Mizushima.** 1982. Primary structure of the *ompF* gene that codes for a major outer membrane protein of *Escherichia coli* K-12. Nucleic Acids Res. **10**:6957–6968.

67. **Inouye, M. (ed.).** 1979. Bacterial outer membranes. John Wiley & Sons, Inc., New York.

68. **Jansson, P., A. A. Lindberg, B. Lindberg, and R. Wollin.** 1981. Structural studies on the hexose region of the core in lipopolysaccharide from Enterobacteriaceae. Eur. J. Biochem. **115**:571–577.

69. **Jones, N. C., and M. J. Osborn.** 1977. Translocation of phospholipids between the outer and inner membrane of *Salmonella typhimurium*. J. Biol. Chem. **252**:7405–7412.

70. **Kamio, Y., and H. Nikaido.** 1976. Outer membrane of *Salmonella typhimurium*: accessibility of phospholipid head groups to phospholipase C and cyanogen bromide activated dextran in the external medium. Biochemistry **15**:2561–2570.

71. **Kamio, Y., and H. Nikaido.** 1977. Outer membrane of *Salmonella typhimurium*. Identification of proteins exposed on cell surface. Biochim. Biophys. Acta **464**:589–601.

72. **Koch, A. L.** 1971. The adaptive response of *Escherichia coli* to a feast and famine existence. Adv. Microb. Physiol. **6**:147–217.

73. **Konisky, J.** 1979. Specific transport systems and receptors for colicins and phages, p. 319–359. *In* M. Inouye (ed.), Bacterial outer membranes. John Wiley & Sons, Inc., New York.

74. **Koplow, J., and H. Goldfine.** 1974. Alterations in the outer membrane of the cell envelope of heptose-deficient mutants of *Escherichia coli*. J. Bacteriol. **117**:527–543.

75. **Kuhn, H. M., E. Neter, and H. Mayer.** 1983. Modification of the lipid moiety of the enterobacterial common antigen by the "*Pseudomonas* factor." Infect. Immun. **40**:696–700.

76. **Labischinski, H., G. Barnickel, H. Bradaczek, D. Naumann, E. T. Rietschel, and P. Giesbrecht.** 1985. High state of order of isolated bacterial lipopolysaccharide and its possible contribution to the permeation barrier property of the outer membrane. J. Bacteriol. **162**:9–20.

77. **Lai, J.-S., W. M. Philbrick, and H. C. Wu.** 1980. Acyl moieties in phospholipids are the precursors for the fatty acids in murein lipoprotein of *Escherichia coli*. J. Biol. Chem. **255**:5384–5387.

78. **Leive, L.** 1965. Release of lipopolysaccharide by EDTA treatment of *E. coli*. Biochem. Biophys. Res. Commun. **21**:290–296.

79. **Leive, L.** 1974. The barrier function of the gram-negative envelope. Ann. N.Y. Acad. Sci. **235**:109–127.

80. **Lin, J. J. C., H. Kanazawa, J. Ozols, and H. C. Wu.** 1978. An *E. coli* mutant with an amino acid alteration within the signal sequence of outer membrane prolipoprotein. Proc. Natl. Acad. Sci. USA **75**:4891–4895.

81. **Luckey, M., and H. Nikaido.** 1980. Specificity of diffusion channels produced by λ phage receptor protein of *Escherichia coli*. Proc. Natl. Acad. Sci. USA **77**:167–171.

82. **Luckey, M., and H. Nikaido.** 1980. Diffusion of solutes through channels produced by phage lambda receptor protein of *Escherichia coli*: inhibition by higher oligosaccharides of maltose series. Biochem. Biophys. Res. Commun. **93**:166–171.

83. **Luckey, M., R. Wayne, and J. Neilands.** 1975. In vitro competition between ferrichrome and phage for the outer membrane T5 receptor complex of *Escherichia coli*. Biochem. Biophys. Res. Commun. **64**:687–693.

84. **Lugowski, C., E. Romanowska, L. Kenne, and B. Lindberg.** 1983. Identification of a trisaccharide repeating unit in the enterobacterial common antigen. Carbohydr. Res. **118**:173–181.

85. **Lugtenberg, B., R. Peters, H. Bernheimer, and W. Berendsen.** 1976. Influence of cultural conditions and mutations on the composition of the outer membrane proteins of *Escherichia coli*. Mol. Gen. Genet. **147**:251–262.

86. **Lugtenberg, B., and L. van Alphen.** 1983. Molecular architecture and functioning of the outer membrane of *Escherichia coli* and

other gram-negative bacteria. Biochim. Biophys. Acta **737:** 51–115.

87. **Lutkenhaus, J. F.** 1977. Role of a major outer membrane protein in *Escherichia coli*. J. Bacteriol. **131:**631–637.

88. **MacAlister, T. J., B. MacDonald, and L. I. Rothfield.** 1983. The periseptal annulus: an organelle associated with cell division in gram-negative bacteria. Proc. Natl. Acad. Sci. USA **80:** 1372–1376.

89. **MacGregor, C. H., C. W. Bishop, and J. E. Blech.** 1979. Localization of proteolytic activity in the outer membrane of *Escherichia coli*. J. Bacteriol. **137:**574–583.

90. **Mäkelä, P. H., D. J. Bradley, H. Brandis, M. M. Frank, H. Hahn, W. Henkel, K. Jann, S. A. Marse, J. B. Robbins, L. Rosenstreich, H. Smith, K. Timmis, A. Tomasz, M. J. Turner, and D. C. Wiley.** 1980. Evasion of host defenses group report, p. 174–197. *In* H. Smith, J. J. Skehel, and M. J. Turner (ed.), The molecular basis of microbial pathogenicity. Dahlem Konferenzen 1980. Verlag Chemie GmbH, Weinheim, Federal Republic of Germany.

91. **Mäkelä, P. H., and B. A. D. Stocker.** 1981. Genetics of the bacterial cell surface. Symp. Soc. Gen. Microbiol. **31:**219–264.

92. **Mandel, M., and A. Higa.** 1970. Calcium-dependent bacteriophage DNA infection. J. Mol. Biol. **53:**159–162.

93. **Manning, P. A., and M. Achtman.** 1979. Cell-to-cell interaction in conjugating *Escherichia coli*: the involvement of the cell envelope, p. 409–447. *In* M. Inouye (ed.), Bacterial outer membranes. John Wiley & Sons, Inc., New York.

94. **Manning, P. A., A. P. Pugsley, and P. Reeves.** 1977. Defective growth functions in mutants of *Escherichia coli* K12 lacking a major outer membrane protein. J. Mol. Biol. **116:**285–300.

95. **Mayer, H., and G. Schmidt.** 1979. Chemistry and biology of the enterobacterial common antigen (ECA). Curr. Top. Microbiol. **85:**99–153.

96. **McKeown, M., M. Kahn, and P. Hanawalt.** 1976. Thymidine uptake and utilization in *Escherichia coli*: a new gene controlling nucleoside transport. J. Bacteriol. **126:**814–822.

97. **McMurry, L. M., J. C. Cullinane, and S. B. Levy.** 1982. Transport of the lipophilic analog minocycline differs from that of tetracycline in susceptible and resistant *Escherichia coli* strains. Antimicrob. Agents Chemother. **22:**791–799.

98. **Michaelis, S., and J. Beckwith.** 1982. Mechanism of incorporation of cell envelope proteins in *Escherichia coli*. Annu. Rev. Microbiol. **36:**435–465.

99. **Miura, T., and S. Mizushima.** 1968. Separation by density gradient centrifugation of two types of membranes from spheroplast membrane of *Escherichia coli* K12. Biochim. Biophys. Acta **150:**159–161.

100. **Mizuno, T., M.-Y. Chou, and M. Inouye.** 1983. A comparative study on the genes for three porins of the *Escherichia coli* outer membrane: DNA sequence of the osmoregulated *ompC* gene. J. Biol. Chem. **258:**6932–6940.

101. **Mizuno, T., M.-Y. Chou, and M. Inouye.** 1984. A unique mechanism regulating gene expression: translational inhibition by a complementary RNA transcript (*mic*RNA). Proc. Natl. Acad. Sci. USA **81:**1966–1970.

102. **Movva, N. R., K. Nakamura, and M. Inouye.** 1980. Gene structure of the OmpA protein, a major surface protein of *Escherichia coli* required for cell-cell interaction. J. Mol. Biol. **147:**317–328.

103. **Mühlradt, P. F., and J. R. Golecki.** 1975. Asymmetrical distribution and artifactual reorientation of lipopolysaccharide in the outer membrane bilayer of *Salmonella typhimurium*. Eur. J. Biochem. **51:**343–352.

104. **Mühlradt, P. F., J. Menzel, J. R. Golecki, and V. Speth.** 1973. Outer membrane of Salmonella. Site of export of newly synthesized lipopolysaccharide on the surface. Eur. J. Biochem. **35:**471–481.

105. **Munford, C. A., and M. J. Osborn.** 1983. An intermediate step in translocation of lipopolysaccharide to the outer membrane of *Salmonella typhimurium*. Proc. Natl. Acad. Sci. USA **80:**1159–1163.

106. **Nakae, T.** 1976. Outer membrane of *Salmonella*. Isolation of protein complex that produces transmembrane channels. J. Biol. Chem. **251:**2176–2178.

107. **Nakae, T.** 1976. Identification of the outer membrane protein of *E. coli* that produces transmembrane channels in reconstituted vesicle membranes. Biochim. Biophys. Res. Commun. **71:**877–884.

108. **Nakae, T., J. Ishii, and M. Tokunaga.** 1979. Subunit structure of functional porin oligomers that form permeability channels in the outer membrane of *Escherichia coli*. J. Biol. Chem. **254:**1457–1461.

109. **Nakamura, K., and S. Mizushima.** 1976. Effects of heating in dodecyl sulfate solution on the conformation and electrophoretic mobility of isolated major outer membrane proteins from *Escherichia coli* K-12. J. Biochem. (Tokyo) **80:**1411–1422.

110. **Neilands, J. B.** 1982. Microbial envelope proteins related to iron. Annu. Rev. Microbiol. **36:**285–309.

111. **Neu, H. C., and L. A. Heppel.** 1965. The release of enzymes from *Escherichia coli* by osmotic shock and during the formation of spheroplasts. J. Biol. Chem. **240:**3685–3692.

112. **Neuhaus, J.-M., H. Schindler, and J. Rosenbusch.** 1983. The periplasmic maltose-binding protein modifies the channel-forming characteristics of maltoporin. EMBO J. **2:**1987–1991.

113. **Nikaido, H.** 1976. Outer membrane of *Salmonella typhimurium*: transmembrane diffusion of some hydrophobic substances. Biochim. Biophys. Acta **433:**118–132.

114. **Nikaido, H.** 1979. Nonspecific transport through the outer membrane, p. 361–407. *In* M. Inouye (ed.), Bacterial outer membranes. John Wiley & Sons, Inc., New York.

115. **Nikaido, H.** 1983. Proteins forming large channels from bacterial and mitochondrial outer membranes: porins and phage lambda receptor protein. Methods Enzymol. **97:**85–100.

116. **Nikaido, H.** 1985. Role of permeability barriers in resistance of β-lactam antibiotics. Pharmacol. Therapeut. **27:**197–231.

117. **Nikaido, H., P. Bavoil, and Y. Hirota.** 1977. Outer membranes of gram-negative bacteria. XV. Transmembrane diffusion rates in lipoprotein-deficient mutants of *Escherichia coli*. J. Bacteriol. **132:**1045–1047.

118. **Nikaido, H., and T. Nakae.** 1979. The outer membrane of gram-negative bacteria. Adv. Microb. Physiol. **20:**163–250.

119. **Nikaido, H., and E. Y. Rosenberg.** 1981. Effect of solute size on diffusion rates through the transmembrane pores of the outer membrane of *Escherichia coli*. J. Gen. Physiol. **77:**121–135.

120. **Nikaido, H., and E. Y. Rosenberg.** 1983. Porin channels in *Escherichia coli*: studies with liposomes reconstituted from purified proteins. J. Bacteriol. **153:**241–252.

121. **Nikaido, H., E. Y. Rosenberg, and J. Foulds.** 1983. Porin channels in *Escherichia coli*: studies with β-lactams in intact cells. J. Bacteriol. **153:**232–240.

122. **Nikaido, H., S. A. Song, L. Shaltiel, and M. Nurminen.** 1977. Outer membrane of *Salmonella*. XIV. Reduced transmembrane diffusion rates in porin-deficient mutants. Biochem. Biophys. Res. Commun. **76:**324–330.

123. **Nikaido, H., Y. Takeuchi, S. Ohnishi, and T. Nakae.** 1977. Outer membrane of *Salmonella typhimurium*. Electron spin resonance studies. Biochim. Biophys. Acta **465:**152–164.

124. **Nikaido, H., and M. Vaara.** 1985. Molecular basis of bacterial outer membrane permeability. Microbiol. Rev. **49:**1–32.

125. **Nikaido, H., and H. C. P. Wu.** 1984. Amino acid sequence homology among the major outer membrane proteins of *Escherichia coli*. Proc. Natl. Acad. Sci. USA **81:**1048–1052.

126. **Nishijima, M., S. Nakaike, Y. Tamori, and S. Nojima.** 1977. Detergent-resistant phospholipase A of *Escherichia coli* K-12. Eur. J. Biochem. **73:**115–124.

127. **Ohki, M.** 1979. The cell cycle-dependent synthesis of envelope proteins in *Escherichia coli*, p. 293–315. *In* M. Inouye (ed.), Bacterial outer membranes. John Wiley & Sons, Inc., New York.

128. **Osborn, M. J.** 1979. Biosynthesis and assembly of the lipopolysaccharide of the outer membrane, p. 15–34. *In* M. Inouye (ed.), Bacterial outer membranes. John Wiley & Sons, Inc., New York.

129. **Osborn, M. J., J. E. Gander, E. Parisi, and J. Carson.** 1972. Mechanism of assembly of the outer membrane of *Salmonella typhimurium*. Isolation and characterization of cytoplasmic and outer membrane. J. Biol. Chem. **247:**3962–3972.

130. **Osborn, M. J., and H. C. P. Wu.** 1980. Proteins of the outer membrane of Gram-negative bacteria. Annu. Rev. Microbiol. **34:**369–422.

131. **Overath, P., M. Brenner, T. Gulik-Krzywicki, E. Shechter, and L. Letellier.** 1975. Lipid phase transitions in cytoplasmic and outer membranes of *Escherichia coli*. Biochim. Biophys. Acta **389:**358–369.

132. **Overbeeke, N., H. Bergmans, F. van Mansfeld, and B. Lugtenberg.** 1983. Complete nucleotide sequence of *phoE*, the structural gene for the phosphate limitation inducible outer membrane pore protein of *Escherichia coli* K12. J. Mol. Biol. **163:**513–532.

133. **Palva, E. T.** 1979. Protein interactions in the outer membrane of *Escherichia coli*. Eur. J. Biochem. **93:**495–503.

134. **Palva, E. T., and P. Westermann.** 1979. Arrangement of the maltose-inducible major outer membrane proteins, the bacteriophage λ receptor in *Escherichia coli* and the 44K protein in *Salmonella typhimurium*. FEBS Lett. **99:**77–80.

135. **Papahadjopolos, D., and A. Portis.** 1978. Calcium-induced lipid phase transitions and membrane fusion. Ann. N.Y. Acad. Sci. **308:**50–65.

136. **Payne, J. W., and C. Gilvarg.** 1968. Size restriction on peptide utilization in *Escherichia coli*. J. Biol. Chem. **243:**6291–6299.

137. **Postle, K., and R. F. Good.** 1983. DNA sequence of *Escherichia coli tonB* gene. Proc. Natl. Acad. Sci. USA **80:**5235–5239.

138. **Pugsley, A. P., and C. A. Schnaitman.** 1978. Outer membrane proteins of *Escherichia coli*. VII. Evidence that bacteriophage-directed protein 2 functions as a pore. J. Bacteriol. **133:**1181–1189.

139. **Reithmeier, R. A. F., and P. D. Bragg.** 1977. Cross-linking of the proteins in the outer membrane of *Escherichia coli*. Biochim. Biophys. Acta **466:**245–256.

140. **Renkin, E. M.** 1954. Filtration, diffusion, and molecular sieving through porous cellulose membranes. J. Gen. Physiol. **38:**225–232.

141. **Reynolds, B. L., and H. Pruul.** 1971. Sensitization of complement-resistant smooth gram-negative bacterial strains. Infect. Immun. **3:**365–372.

142. **Reynolds, P. R., G. P. Mottur, and C. Bradbeer.** 1980. Transport of vitamin B12 in *Escherichia coli*. Some observations on the roles of the gene products of *btuC* and *tonB*. J. Biol. Chem. **255:**4313–4319.

143. **Rick, P. D., B. A. Neumeyer, and D. A. Young.** 1983. Effect of altered lipid A synthesis on the synthesis of the OmpA protein in *Salmonella typhimurium*. J. Biol. Chem. **258:**629–635.

144. **Rietschel, E. T., H.-W. Wollenweber, H. Brade, U. Zähringer, B. Lindner, U. Seydel, H. Bradaczek, G. Barnickel, H. Labischinski, and P. Giesbrecht.** 1984. Structure and conformation of the lipid A component of lipopolysaccharides, p. 187–220. *In* E. T. Rietschel (ed.), Chemistry of endotoxin. Elsevier/North-Holland Publishing Co., Amsterdam.

145. **Rodríguez-Tébar, A., J. A. Barbas, and D. Vázquez.** 1985. Location of some proteins involved in peptidoglycan synthesis and cell division in the inner and outer membranes of *Escherichia coli*. J. Bacteriol. **161:**243–248.

146. **Rosenbusch, J. P.** 1974. Characterization of the major envelope protein from *Escherichia coli*. Regular arrangement on the peptidoglycan and unusual dodecylsulfate binding. J. Biol. Chem. **249:**8019–8029.

147. **Rosenbusch, J. P., A. Steven, M. Alkan, and M. Regenass.** 1979. Matrix protein: a periodically arranged porin protein in the outer membrane of *Escherichia coli*, p. 1–10. *In* W. Baumeister and W. Vogell (ed.), Electron microscopy of molecular dimensions, Springer-Verlag KG, Berlin.

148. **Salton, M. R. J.** 1964. Bacterial cell wall. Elsevier/North-Holland Publishing Co., Amsterdam.

149. **Schenkman, S., A. Tsugita, M. Schwartz, and J. P. Rosenbusch.** 1984. Topology of phage λ receptor protein. Mapping targets of proteolytic cleavage in relation to binding sites for phage or monoclonal antibodies. J. Biol. Chem. **259:**7570–7576.

150. **Schindler, H., and J. P. Rosenbusch.** 1978. Matrix protein from *Escherichia coli* outer membranes forms voltage-controlled channels in lipid bilayers. Proc. Natl. Acad. Sci. USA **75:**3751–3755.

151. **Schindler, H., and J. P. Rosenbusch.** 1981. Matrix protein in planar membranes: clusters of channels in a native environment and their functional reassembly. Proc. Natl. Acad. Sci. USA **78:**2302–2306.

152. **Schindler, M., and M. J. Osborn.** 1979. Interaction of divalent cations and polymyxin B with lipopolysaccharide. Biochemistry **18:**4425–4430.

153. **Schmidt, M. A., and K. Jann.** 1982. Phospholipid substitution of capsular (K) polysaccharide antigens from *Escherichia coli* causing extraintestinal infections. FEMS Microbiol. Lett. **14:**69–74.

154. **Schnaitman, C. A.** 1970. Protein composition of the cell wall and cytoplasmic membrane of *Escherichia coli*. J. Bacteriol. **104:**890–901.

155. **Schnaitman, C. A.** 1973. Outer membrane proteins of *Escherichia coli*. I. Effect of preparation conditions on the migration of protein in polyacrylamide gels. Arch. Biochem. Biophys. **157:**541–552.

156. **Schnaitman, C. A.** 1974. Outer membrane proteins of *Escherichia coli*. IV. Differences in outer membrane proteins due to strain and cultural differences. J. Bacteriol. **118:**454–464.

157. **Silhavy, T. J., S. A. Benson, and S. D. Emr.** 1983. Mechanisms of protein localization. Microbiol. Rev. **47:**313–344.

158. **Smit, J., Y. Kamio, and H. Nikaido.** 1975. Outer membrane of *Salmonella typhimurium*: chemical analysis and freeze-fracture studies with lipopolysaccharide mutants. J. Bacteriol. **124:**942–958.

159. **Smit, J., and H. Nikaido.** 1978. Outer membrane of gram-negative bacteria. XVIII. Electron microscopic studies on porin insertion sites and growth of cell surface of *Salmonella*

typhimurium. J. Bacteriol. **135:**687–702.

160. **Sonntag, I., H. Schwarz, Y. Hirota, and U. Henning.** 1978. Cell envelope and shape of *Escherichia coli*: multiple mutants missing the outer membrane lipoprotein and other major outer membrane proteins. J. Bacteriol. **136:**280–285.

161. **Stein, W. D.** 1967. The movement of molecules across cell membranes, p. 65–125. Academic Press, Inc., New York.

162. **Stock, J. B., B. Rauch, and S. Roseman.** 1977. Periplasmic space in *Salmonella typhimurium* and *Escherichia coli*. J. Biol. Chem. **252:**7850–7861.

163. **Sukupolvi, S., M. Vaara, I. M. Helander, P. Viljanen, and P. H. Mäkelä.** 1984. New *Salmonella typhimurium* mutants with altered outer membrane permeability. J. Bacteriol. **159:**704–712.

164. **Sutcliffe, J., R. Blumenthal, A. Walter, and J. Foulds.** 1983. *Escherichia coli* outer membrane protein K is a porin. J. Bacteriol. **156:**867–872.

165. **Takeuchi, Y., and H. Nikaido.** 1981. Persistence of segregated phospholipid domains in phospholipid-lipopolysaccharide mixed bilayers: studies with spin-labeled phospholipids. Biochemistry **20:**523–529.

166. **Teuber, M., and J. Bader.** 1976. Action of polymyxin on bacterial membranes. Binding capacities for polymyxin B of inner and outer membranes isolated from *Salmonella typhimurium* G30. Arch. Microbiol. **109:**51–58.

167. **Tokunaga, M., H. Tokunaga, and H. C. Wu.** 1982. Post-translational modification and processing of *Escherichia coli* prolipoprotein in vitro. Proc. Natl. Acad. Sci. USA **79:** 2255–2259.

168. **Tommassen, J., and B. Lugtenberg.** 1980. Outer membrane protein e of *Escherichia coli* K-12 is co-regulated with alkaline phosphatase. J. Bacteriol. **143:**151–157.

169. **Ueki, T., T. Mitsui, and H. Nikaido.** 1979. X-ray diffraction studies of outer membrane of *Salmonella typhimurium*. J. Biochem. (Tokyo) **85:**173–182.

170. **Vaara, M., and T. Vaara.** 1983. Sensitization of Gram-negative bacteria to antibiotics and complement by a nontoxic oligopeptide. Nature (London) **303:**526–528.

171. **Vaara, M., and T. Vaara.** 1983. Polycations sensitize enteric bacteria to antibiotics. Antimicrob. Agents Chemother. **24:**107–113.

172. **Vaara, M., and T. Vaara.** 1983. Polycations as outer membrane-disorganizing agents. Antimicrob. Agents Chemother. **24:**114–122.

173. **Vaara, M., T. Vaara, M. Jensen, I. Helander, M. Nurminen, E. T. Rietschel, and P. M. Mäkelä.** 1981. Characterization of the lipopolysaccharide from the polymyxin-resistant *pmrA* mutants of *Salmonella typhimurium*. FEBS Lett. **129:**145–149.

174. **van Gool, A. P., and N. Nanninga.** 1971. Fracture faces in the cell envelope of *Escherichia coli*. J. Bacteriol. **108:**474–481.

175. **Wandersman, C., M. Schwartz, and T. Ferenci.** 1979. *Escherichia coli* mutants impaired in maltodextrin transport. J. Bacteriol. **140:**1–13.

176. **Wayne, R., and J. B. Neilands.** 1975. Evidence for common binding sites for ferrichrome compounds and bacteriophage φ80 in the cell envelope of *Escherichia coli*. J. Bacteriol. **121:**497–503.

177. **Weckesser, J., L. S. Zalman, and H. Nikaido.** 1984. Porin from *Rhodopseudomonas sphaeroides*. J. Bacteriol. **159:**199–205.

178. **Whitfield, C., R. E. W. Hancock, and J. W. Costerton.** 1983. Outer membrane protein K of *Escherichia coli*: purification and pore-forming properties in lipid bilayer membranes. J. Bacteriol. **156:**873–879.

179. **Wolfe, P. B., W. Wickner, and J. M. Goodman.** 1983. Sequence of the leader peptidase gene of *Escherichia coli* and the orientation of leader peptidase in bacterial envelope. J. Biol. Chem. **258:**12073–12080.

180. **Wolf-Watz, H., S. Normark, and G. D. Bloom.** 1973. Rapid method for isolation of large quantities of outer membrane from *Escherichia coli* K-12 and its application to the study of envelope mutants. J. Bacteriol. **115:**1191–1197.

181. **Yoshimura, F., and H. Nikaido.** 1985. Diffusion of β-lactam antibiotics through the porin channels of *Escherichia coli* K-12. Antimicrob. Agents Chemother. **27:**84–92.

182. **Yu, F., H. Furukawa, K. Nakamura, and S. Mizushima.** 1984. Mechanism of localization of major outer membrane lipoprotein in *Escherichia coli*. Studies with the OmpF-lipoprotein hybrid protein. J. Biol. Chem. **259:**6013–6018.

183. **Yu, F., H. Yamada, K. Daishima, and S. Mizushima.** 1984. Nucleotide sequence of the *lspA* gene, the structural gene for lipoprotein signal peptidase of *Escherichia coli*. FEBS Lett. **173:**264–268.

184. **Zimmermann, W., and A. Rosselet.** 1977. Function of the outer membrane of *Escherichia coli* as a permeability barrier to beta-lactam antibiotics. Antimicrob. Agents Chemother. **12:**368–372.

4. The Murein Sacculus

JAMES T. PARK

Department of Molecular Biology and Microbiology, Tufts University School of Medicine, Boston, Massachusetts 02111

INTRODUCTION

The murein sacculus is a truly unique example of molecular architecture in nature because it changes shape during the cell division cycle while it is held together by covalent bonds. It was recognized over 20 years ago that morphogenesis of a bacterial cell during growth and division must involve hydrolysis of covalent bonds in the murein sacculus and insertion of new polymer (41). Only then could both elongation and septation occur. During the past 20 years a great deal has been learned about the structure of murein and the enzymes of *Escherichia coli* involved in murein synthesis. However, relatively little progress has been made on understanding the biochemistry of morphogenesis, i.e., the specific tasks performed by individual enzymes during morphogenesis and how the processes are controlled to maintain an intact sacculus during growth and division.

This chapter is concerned with the structure and assembly of the murein sacculus. For this reason only the final stages in the synthesis of murein and what is currently known about the biochemistry of morphogenesis of *E. coli* will be considered. A more detailed discussion of the enzymology of murein synthesis is presented in chapter 42.

TERMINOLOGY

The rigid, shape-determining structure in bacterial cell walls is a complex polymer consisting of two amino sugars and at least four amino acids. The term murein, derived from *murus*, the Latin word for wall, was introduced by Weidel and Pelzer (41) as a trivial name for these newly recognized cell wall polymers. Since the polymer contains roughly equal amounts of polysaccharide and peptide, it belongs to the general class of polymers called peptidoglycans. Thus, mureins represent a large group of closely related peptidoglycans found exclusively in bacterial cell walls. Strictly defined, the term murein bears the same relation to the term peptidoglycan as collagen does to protein. However, because practically all known peptidoglycans are mureins, the two terms have come to be used interchangeably.

From the point of view of enzymologists, mureins are complex heteropolysaccharides. The repeating unit of this heteropolysaccharide (and slight modifications thereof) has been termed muropeptide (41). Weidel and Pelzer (41) also introduced the term murein sacculus to indicate that the murein of a cell was actually present as one giant molecule existing in the form of "a rigid bag of the volume and shape of the cell."

CHEMICAL STRUCTURE OF THE MUREIN

The murein sacculus of *E. coli* is composed of *N*-acetylglucosamine, *N*-acetylmuramic acid (*N*-acetylglucosamine with D-lactic acid ether substituted at C-3), L-alanine, D-glutamic acid, *meso*-diaminopimelic acid (DAP), and D-alanine in roughly equimolar amounts. These components form the basic muropeptide repeating unit of the murein (Fig. 1). The sugars are linked together by 1→4 glycosidic bonds. Attached to the carboxyl group of each muramic acid by an amide linkage is a short peptide, L-alanyl-D-isoglutamyl-L-*meso*-diaminopimelyl-D-

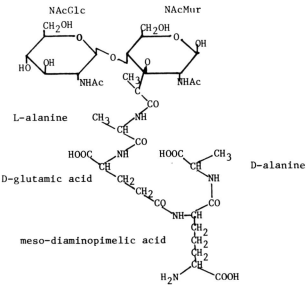

FIG. 1. Chemical structure of the principal muropeptide monomer present in the murein of *E. coli* and other gram-negative bacteria (C6 of Weidel and Pelzer [41]), showing the favored ringlike structure of the peptide proposed by Barnickel et al. (3). NAcGlc, *N*-Acetylglucosamine; NAcMur, *N*-acetylmuramic acid; Ac, acetyl.

alanine. In the murein sacculus of *E. coli* a small percentage of the peptides lack D-alanine, and an even smaller percentage terminate with an additional D-alanine. This composition is shared by nearly all gram-negative bacteria as well as by a few gram-positive rods (32). In some species the carboxyl groups of glutamic acid or DAP (or both) may be amidated, but in *E. coli* neither is amidated. A unique feature of the peptide backbone of all mureins, as exemplified in the muropeptide of *E. coli*, is the alternating sequence of optical isomers D-L-D-L-D, beginning with the D-lactyl group of muramic acid and including the L configuration of DAP.

The arrangement of muropeptides in the murein of *E. coli* is shown in Fig. 2. The sugars form linear chains of alternating units of *N*-acetylglucosamine

and *N*-acetylmuramic acid linked β-1→4. The *N*-acetylmuramic acid at the end of each strand is present as a nonreducing 1,6-anhydro sugar (21). The average glycan strand, according to recent estimates of the 1,6-anhydromuramic acid content of sacculi, is about 30 muropeptides in length (10, 17, 19). Recent evidence indicates that some strands are much longer initially and become shorter with time (U. Schwarz, personal communication). A molecule of lipoprotein is attached about every 10th muropeptide (8). The covalent link is from the L-carboxyl group of DAP to the epsilon amino group of the carboxy-terminal lysine of the lipoprotein. Adjacent strands are cross-linked to each other through the peptide side chains. Most of these cross-links are between the carboxyl group of the D-alanine in position 4 of one peptide and the free amino group of DAP of a muropeptide from the adjacent strand. However, it has recently been shown that about 20% of the cross-links do not involve D-alanine but, rather, directly link the DAP residues of neighboring chains (17). About half of the muropeptides of the sacculus are involved in cross-links, and another 5% are involved in links that hold three muropeptides together as originally observed by Gmeiner (18). These estimates are based on the proportions of muropeptide monomers, dimers, and trimers present in the digests of murein sacculi treated with egg white lysozyme or *Chalaropsis* β-*N*-acetylmuramidase. These enzymes hydrolyze the β-1→4 bonds between C-1 of *N*-acetylmuramic acid and C-4 of *N*-acetylglucosamine, thereby reducing the glycan chains to disaccharide units. After enzymatic hydrolysis, muropeptides that are cross-linked are present as muropeptide dimers or trimers.

In early studies in which the components of lysozyme digests were characterized, eight components were isolated by paper chromatography. They were designated C1 through C8. The two principal components were C6, the muropeptide monomer consisting of a disaccharide tetrapeptide, and C3, a dimer composed of two C6 muropeptides cross-linked to each other. Also present were lesser amounts of C5, a disaccharide tripeptide, and C4, a dimer identical in composition to C3 but differing in that the two disaccharides of the dimer are glycosidically linked. The cyclic dimer is now known to be an artifact of digestion with egg white lysozyme, since the glycan chains of sacculi are completely hydrolyzed to disaccharides by *Chalaropsis* β-*N*-acetylmuramidase. Glauner and Schwarz (17) have fractionated *Chalaropsis* digests of *E. coli* sacculi into 60 components by high-pressure liquid chromatography. They have characterized 33 of them, or nearly all of the components that represent at least 0.1% each of the total muropeptides of the digest. This great number of components is largely accounted for by the fact that monomers, dimers, and trimers are not only present in unmodified form but are also present at the ends of chains. Hence they contain a disaccharide with 1,6-anhydromuramic acid. Lipoprotein is also found attached to monomers, dimers, and trimers, and following treatment with pronase, these fragments retain arginine and lysine from the lipoprotein. Adding to the complexity of the *Chalaropsis* digests is the fact that the peptides of any given component may contain two, one, or no D-

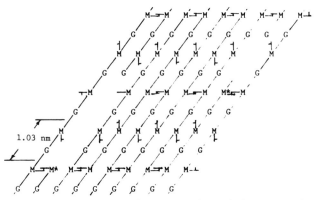

FIG. 2. Arrangement of muropeptides and glycan strands in the murein sacculus. M, *N*-acetylmuramic acid; G, *N*-acetylglucosamine; M*, 1,6-anhydro-*N*-acetylmuramic acid; T, peptide.

alanine residues. It thus appears that chain termination and lipoprotein attachment can occur more or less randomly, irrespective of the location of cross-links. On the other hand, there is a significant enrichment of DAP-DAP linked muropeptide dimers containing lipoprotein and a complete absence of normal cross-linked dimers containing lipoprotein (17). This suggests that the DAP-DAP dimer is the normal acceptor for lipoprotein. These dimers may subsequently break down, as most of the lipoprotein was shown to be linked to monomers with a tripeptide (4.7% of the total muropeptides versus 2.1% attached to a muropeptide in dimers and trimers). Five of the characterized components contained glycine instead of D-alanine in either position 4 or position 5 of the peptide, indicating a lack of absolute specificity in the synthesis of the precursor pentapeptide.

MOLECULAR ARCHITECTURE OF THE MUREIN

Based on the DAP content and size of the murein sacculus of *E. coli*, the sacculus is believed to consist of a single monomolecular sheet of murein (7). The orientation of the sugars and peptide side chains in strands of peptidoglycan remains a matter of conjecture, however. This is due primarily to the fact that the material does not have a crystalline structure, so X-ray diffraction patterns lack detail. Nonetheless, Debye-Scherrer rings indicating periodicities of about 1 nm and 0.44 nm have been reported (16). These patterns are believed to reflect the length of the disaccharide unit in the strand, 1.03 nm, and the distance between strands. However, the polymer is remarkably flexible. Even a multilayered, highly cross-linked murein such as that found in staphylococci can undergo considerable swelling and shrinking in response to the osmotic strength of the medium in which the bacteria are suspended (24). The scatter corresponding to 0.44-nm repeat distances almost corresponds to the interchain distance found in chitin, where the chains are tightly hydrogen bonded to each other. However, in murein this is probably the minimum distance between chains when sacculi are in the dry state. In fact, a more extended structure for the muropeptide cross-links seems likely in an aqueous environment. If the estimates of the actual muropeptide content of sacculi by Braun et al. (7) are accurate, the average distance between chains would be 1.25 nm. Until recently most models for the structure of murein have assumed a chitinlike conformation in which the sugar rings present a flat surface with all the peptide side chains pointing in one direction and the glycan chains are closely stacked and stabilized by hydrogen bonds (for a review, see reference 30). It has been recognized for some time that certain properties of murein are not readily compatible with this type of model, namely, the elasticity of the polymer and its susceptibility to hydrolysis by egg white lysozyme and other β-muramidases.

A new model recently proposed by Barnickel et al. (4) seems to overcome these objections and, in addition, provides a plausible explanation for the facts that in *E. coli* only half of the muropeptides participate in cross-links and that in gram-positive bacteria the sacculi consist of multilayered sheets of murein cross-linked to each other.

MOLECULAR ARCHITECTURE OF A GLYCAN STRAND OF THE MUREIN

In the proposed model of Barnickel et al. (4) the sugars do not lie in a flat plane. Instead, they are tilted with respect to each other to accommodate the lactic acid moiety of muramic acid. Simply put, as a consequence of this, muropeptides are twisted in such a way that each strand of murein is actually a spiral with each peptide side chain being exposed approximately 90° around the spiral relative to its neighbors. With this configuration only every fourth muropeptide would be positioned to form a cross-link with a strand on its left. Similarly, a cross-link would occur with a strand on the right only with every fourth muropeptide whose peptide is pointing to the right (Fig. 2). Actually, three of every four muropeptides are on opposite sides of a given strand relative to their immediate neighbors and hence cannot be linked. Considering the strands on each side, this permits 50% of the muropeptides to form cross-links, which is roughly the number observed in *E. coli*.

ARRANGEMENT OF GLYCAN STRANDS IN THE MUREIN SACCULUS

It is implicit from Fig. 2 that glycan strands run parallel to each other. Although this arrangement is consistent with the formation of an orderly cross-linked monolayer, there is no direct evidence for it. However, there is indirect evidence indicating not only that the glycan strands are parallel to each other, but also that they are perpendicular to the axis of the cell and hence follow the circumference of the cell's cylinder. This was revealed by partial digestion of sacculi with a penicillin-insensitive endopeptidase which hydrolyzes the DD cross-links (40) and by demonstration that sonication ruptures sacculi perpendicular to their axis (39). Since individual strands average only 30 nm in length, about 80 strands aligned end to end are needed to span the circumference once. As adjacent strands are cross-linked to each other, neighboring strands must overlap in order to form a continuous rigid layer. It is not known whether the short strands are exactly perpendicular to the axis of the cell and hence appear to form rings around the circumference or whether slight pitch exists so that, in effect, they form a spiral.

The arrangement of murein strands at the poles of the cell is unknown. It seems probable that as the septa grow inwards to form the new poles of daughter cells, strands are anchored to the existing sacculus by transpeptidation. This would form a template for continued ingrowth by laying down a series of ever-smaller concentric rings of murein. Since septa consist of two poles which are formed simultaneously, the question arises as to whether the two poles are covalently linked to each other at any time during their formation. The few published electron micrographs of septa do not provide a definite answer to this question. At some point in the division process, the new cell poles are clearly separated by a mysterious space of uniform width observed in cells of the *envA* mutant, a chain-forming mutant (9). Burdett and Murray (9) also demonstrated that the outer membrane does not necessarily invaginate simultaneously with the inner

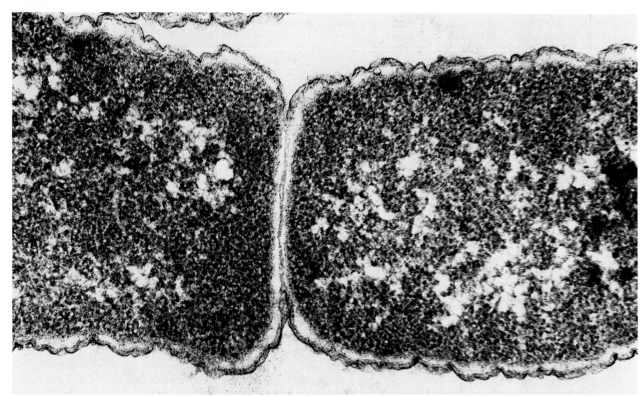

FIG. 3. *E. coli* PM61 (*envC*), showing apparent fusion of the murein of the two cell poles in the septum.

membrane and the sacculus. The space separating the new poles, if occupied by a few molecules of a murein-associated enzyme, might account for the remarkably fixed distance they observed between the two poles formed in the septum.

Electron micrographs of another chain-forming, division-defective mutant of *E. coli* provide an entirely different picture. The septa of the *envC* mutant, *E. coli* PM61, appear fused in some regions and irregularly separated in other parts of the same septum (Fig. 3) (28). This suggests a sequence for septum formation in which the two poles are initially fused and are then separated by a specific enzyme (amidase or endopeptidase) to allow insertion of new outer membrane. In normal *E. coli* cells these processes would occur in rapid succession, so suggestions for such a sequence are only seen in these rare chain-forming strains.

RELATIONSHIP OF THE MUREIN SACCULUS TO THE OUTER MEMBRANE AND THE CYTOPLASMIC MEMBRANE

The murein sacculus is located in the periplasm, the space between the outer membrane and the cytoplasmic membrane. When examined by electron microscopy of thin sections fixed and stained by conventional methods, the sacculus of *E. coli* is usually seen as a dense layer 2 to 3 nm thick, associated with the outer membrane and separated from the cytoplasmic membrane. However, recent micrographs of specimens prepared by less disruptive methods indicate that the murein layer may be from 7 nm to as much as

15 nm thick and that the space between the two membranes is of constant thickness (1, 15, 20, 25) (Fig. 4). This could indicate a double layer of murein or, alternatively, may simply be a reflection of the native state of hydration of the polymer. The outer membrane is presumed to be firmly anchored to the sacculus by the covalently linked lipoprotein molecules embedded in the outer membrane. Additional tight association of the outer membrane and the sacculus may be provided by the outer membrane porins, especially the *ompC* and *ompF* gene products, which are known to remain associated with the murein sacculus even after envelopes are heated at 60°C with sodium dodecyl sulfate (31). Altogether, the lipoproteins and porins provide over 400,000 contacts between the murein sacculus and the outer membrane.

For many years it has been recognized that the cytoplasmic membrane of growing cells appears to be attached to the outer membrane at a few hundred sites termed adhesion or fusion sites (5). Actual fusion of the two membranes at these sites has never been seen by electron microscopy and may not occur. It therefore seems reasonable to suppose that many of these "fusion sites" represent places where new strands of murein are being covalently attached to the murein sacculus while their growing ends are still attached to bactoprenolpyrophosphate, a component of the cytoplasmic membrane (27). As is shown below, about 100 such connections between the sacculus and the cytoplasmic membrane should exist in growing cells. Similarly, other "fusion sites" may represent places of synthesis of lipopolysaccharides or outer membrane proteins.

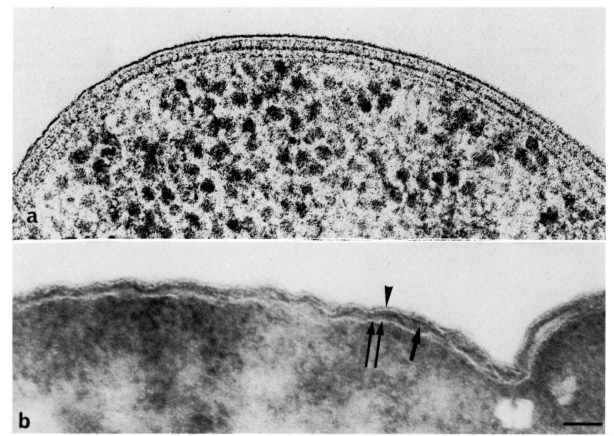

FIG. 4. (a) Thin section of *E. coli*, showing the uniform distance between the outer and inner membranes and the thick, space-filling layer containing the murein in the periplasmic space. Reprinted by permission from reference 20. (b) Thin section of *E. coli*, again showing, by a different method, a relatively thick murein layer. Reprinted by permission from reference 25.

ASSEMBLY OF THE MUREIN SACCULUS

During the cell division cycle, an *E. coli* cell doubles in length and forms a septum which becomes the two new polar caps of the daughter cells. The two processes, elongation of the sacculus and formation of polar caps, appear to be quite separate events, each involving synthesis of murein. One reason for assuming that elongation and septation involve separate pathways for incorporation of murein is that septation can be blocked by a variety of means without affecting elongation (34).

Elongation

A working model

Heteropolysaccharide synthesis generally involves assembly on the inner side of the cytoplasmic membrane of a lipid-linked repeating unit of the polymer, followed by polymerization and finally release of the polysaccharide on the other side of the membrane. The end result is random extrusion of material to form a slime or capsule, or, with some degree of organization or self-assembly, to produce a rigid, structural unit such as cellulose or chitin. Murein, in contrast, though initially synthesized by the rules of heteropolysaccharide synthesis, must be assembled according to

a specific plan which allows a covalently closed sacculus to enlarge (i.e., elongate) and then divide into two separate sacculi without loss of integrity. Recent data have revealed some aspects of the plan by which *E. coli* achieves elongation (11, 12).

Growing cells were pulse-labeled with [^{14}C]DAP, dimers were isolated from the sacculi, and the distribution of [^{14}C]DAP between the two halves of the dimers was determined. The two muropeptides of the dimer are termed donor and acceptor. The donor represents the newly formed strand of murein, since pentapeptide is needed to drive the cross-linking reaction and the half-life of pentapeptides in murein is less than 1 min (13). The data indicated that during an initial period of 6 to 10 min, 80% of the [^{14}C]DAP was in the donor position. The proportion of [^{14}C]DAP in the acceptor position of dimers then gradually increased. Let us assume that the percentage of [^{14}C]DAP present in the donor position remained constant for 8 min. During the first 8 min about 400,000 muropeptides are inserted into the sacculus, so the relatively constant ratio of donor to acceptor was surprising. Lars Burman and I have interpreted this finding in the light of a tentative model (Fig. 5). The essential features of the model are that (i) each of about 100 separate enzyme complexes independently synthesizes and inserts new strands of murein; (ii) each enzyme complex inserts two strands (Fig. 5A); and (iii)

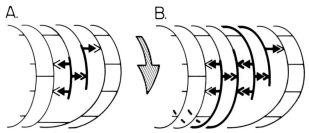

FIG. 5. Structure of the murein in a growth site for elongation of the murein sacculus as it may exist during a pulse-labeling experiment. The thick lines represent the new pair of radioactive strands of murein inserted between pre-existing strands (thin vertical lines) during the first few minutes of the pulse (A) and 8 min later when radioactive strands begin to contact other radioactive strands as a result of continued insertion along a spiral path (B). The arrows between strands indicate the direction of cross-links in dimers from donor to acceptor strands.➔, Radioactive donor; >, radioactive acceptor; >, nonradioactive acceptor. The regions in panels A and B with arrows indicating the direction of the cross-links represent about 1/300 of the circumference of the sacculus. Dimers shown in panel A contain one radioactive acceptor and four radioactive donors (acceptor/donor radioactivity ratio = 0.25). After insertion of two complete circumference lengths of pairs of radioactive strands adjacent to each other, the situation shown by the arrow-linked dimers in panel B should exist. Dimers shown in panel B contain four radioactive acceptors for every seven radioactive donors (acceptor/donor radioactivity ratio = 0.57). Reprinted by permission from reference 12.

the enzymes involved, which are integral membrane proteins, move unidirectionally around the circumference of the cell about once every 8 min, inserting new strands. After 8 min, new strands begin to be cross-linked to the radioactive strands inserted 8 min earlier, and hence relatively more radioactivity appears in acceptors (Fig. 5B).

Enzymes that may be involved in elongation

Although we do not know precisely which enzymes are responsible for elongation, a great deal has been learned about the variety of enzymes which may participate. These are the so-called penicillin-sensitive or penicillin-binding proteins (PBPs). The enzymatic activities of the penicillin-binding proteins are described in chapter 42. To summarize briefly, PBPs 1a, 1b, and 3 are all two-headed enzymes capable of polymerizing the glycan chains and forming the cross-links by transpeptidation. PBPs 4, 5, and 6 all have DD carboxypeptidase activity; that is, they can hydrolyze the D-alanyl-D-alanine bond present in the pentapeptide of precursors and of newly formed strands of murein. PBP 4 also acts as an endopeptidase which can hydrolyze the cross-links between the D-alanine and the amino group of the D-isomeric center of DAP. In addition, E. coli contains a penicillin-insensitive endopeptidase and a penicillin-insensitive polymerase. Thus, multiple enzymes with similar activities are present, although it seems likely that their specificities may well differ under natural conditions.

My co-workers and I have suggested that PBPs 1a and 1b make and cross-link the two strands as shown in Fig. 5. This is in agreement with the early report of Spratt (35) that PBP 1 was involved in elongation. We also have suggested that PBP 4 is the endopeptidase normally involved in breaking cross-links to allow insertion of the new strands. Why are two strands inserted together? There are two possible reasons to account for this. As shown in the model (Fig. 5), it is visualized that, e.g., PBP 1a synthesizes one strand and cross-links it to the existing strand on its left. Since PBP 1a is anchored to the cytoplasmic membrane, it is unlikely that it could cross-link in the other direction since this would require the transpeptidase to rotate 180° at a time when the polymerase site of the same protein presumably is extending the glycan strand. Thus, a second enzyme, e.g., PBP 1b, would make the other strand and form the cross-links in the other direction. This rationale does not predict how the two new strands become linked to each other. PBP 2 and PBP 1a are possible candidates for this reaction. PBP 2 has been shown to have transpeptidase activity (23), and PBP 1a can form trimers (38).

The second possible reason for believing that two strands are inserted at a time stems from the model of Barnickel et al. (4) for the three-dimensional structure of a strand of peptidoglycan, as discussed above. In their model each strand forms a fourfold helix, with every fourth peptide pointing in a given direction. This model is very attractive because it explains why only about 50% of the muropeptides are involved in cross-links in gram-negative bacteria. If we apply the concept of a helical strand to the model, the cell could not insert one strand at a time, since once a strand is aligned and cross-linked on one side the peptides cannot be aligned properly on the other side. They are, in fact 2 nm farther apart than normal (the length of two disaccharide units), and this may prevent transpeptidation from taking place as illustrated in Fig. 6A. If, on the other hand, two strands are inserted (Fig.

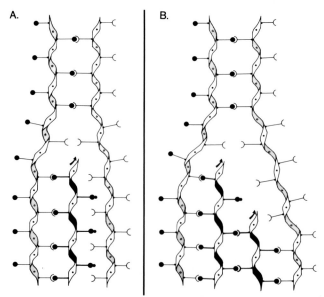

FIG. 6. Alignment of glycan strands to permit cross-linking during elongation. If one strand is inserted at a time (A), the peptides cannot be properly aligned; insertion of two strands at a time (B) leads to proper alignment.

6B), the peptides are properly aligned on both sides of all strands.

In summary, elongation of the murein sacculus is a multisite process, pairs of strands being inserted at about 100 separate locations simultaneously. The enzymes responsible move continuously in one direction around the circumference. Because the enzymes are membrane bound, their specificity may be more restricted than would otherwise be the case. Thus, if PBP 1a formed left-handed cross-links, PBP 1b would form right-handed cross-links. Also, because the specificity of PBP 4 may be similarly restricted, no more than half of the cross-links can be cleaved to allow insertion of new strands of murein.

Unanswered questions

Viability of PBP mutants. The model appears inconsistent with the observation that mutants lacking PBP 1a, 1b, or 4 are viable. Mutation in these cases is defined as the inability to bind radioactive penicillin. It is possible that mutants are viable if they retain some enzymatic activity. Deletion of the genes in question will be necessary to prove whether or not a particular PBP is essential. A double mutant of PBP 1b and PBP 1a^{ts} was found to be lethal at the restrictive temperature, suggesting that PBP 1a can substitute for PBP 1b (37). However, PBP 1a and PBP 1b clearly have different functions, as seen by the differences in morphology when the mutants are grown in the presence of furazlocillin to inhibit cell division (33). PBP 1a mutants increased in length two- to threefold and then ceased growing. PBP 1b mutants grew to form much longer filaments which lysed abruptly after about 150 min.

Location of PBPs. A recent paper by Rodriguez-Tebar et al. (29) questions the assumption that PBPs are located in the cytoplasmic membrane. In fact, their results show that roughly equal amounts of most PBPs are associated with the inner and outer membrane fractions as isolated by isopycnic centrifugation in sucrose gradients. The reason why the same PBPs are associated with both the inner and outer membranes is not clear. Certainly there is no documented example of an exported protein being randomly distributed in this manner. The most likely explanation seems to be that PBPs are true inner membrane proteins, but because of their association with murein (an outer membrane component), small fragments of inner membrane enriched for PBPs remain associated with outer membrane vesicles and hence sediment with the outer membrane fraction.

Role of PBP 2. It was recognized over 10 years ago that PBP 2 is required for continued elongation of the cylindrical wall (36). Loss of PBP 2 function by mutation or by specific inhibition with mecillinam causes *E. coli* to grow as spherical cells which may eventually lyse (35). Because most cells complete one cell division cycle in the presence of mecillinam before changing shape, it has been suggested that PBP 2 is required for initiation of cell wall growth at new sites. However, it is not understood how a specific transpeptidase, i.e., PBP 2, can initiate elongation or preserve the cylindrical shape of the sacculus. Thus, PBP 2 remains an enigma and may well hold the key to eventual understanding of the biochemistry of the elongation process.

Septation: Formation of Polar Caps

Septation is a very complex process involving both temporal and topological controls. Two entirely new polar caps of murein must be formed, and the murein sacculi of the two daughter cells must be cleaved from each other by one or more murein hydrolases without loss of integrity to either. As summarized in an excellent review by Donachie et al. (14), about 15 genes are required for septum formation, two for separation, and two more for initiation of septation (see also chapter 99). Only one of these genes, *pbpB*, codes for a PBP (35). *pbpB*, also known as *sep* or *ftsI* (2), is the structural gene for PBP 3. The amino acid sequence derived from the DNA sequence of this gene suggests that it is an ectoprotein anchored in the cytoplasmic membrane near its carboxy-terminal end and possessing a signal sequence at its amino terminus (26). Ishino and Matsuhashi (22) have shown that it has both transpeptidase and transglycosylase activity. It is the only PBP that is clearly essential for survival, in the sense that all *pbpB* mutants that do not bind penicillin to PBP 3 at the restrictive temperature are unable to form colonies under restrictive conditions. The mutants are unable to divide under these conditions and hence form long filaments, indicating that elongation continues but septation is blocked. Certain β-lactam antibiotics such as piperacillin and furazlocillin, which selectively bind to PBP 3 at the MIC, appear to specifically inhibit synthesis of that murein laid down during septum formation (6).

It is not appropriate to discuss here the evidence that 18 other genes are required for normal cell division (see chapter 99). Determination of the biochemical function of these gene products in cell division is one of the more difficult but fascinating chapters that remain to be written about *E. coli*. At present, it is a little like knowing the members of the orchestra, but not what instrument each plays or how they cooperate to play music. Let us hope that some understanding of the symphony called cell division will be forthcoming before the turn of the century.

LITERATURE CITED

1. **Amako, K., K. Murata, and A. Umeda.** 1983. Structure of the envelope of *Escherichia coli* observed by the rapid-freezing and substitution fixation method. Microbiol. Immunol. **27:**95–99.
2. **Bachmann, B. J.** 1983. Linkage map of *Escherichia coli* K-12, edition 7. Microbiol. Rev. **47:**180–230.
3. **Barnickel, G., H. Labischinski, H. Bradaczek, and P. Giesbrecht.** 1979. Conformational energy calculation of the peptide part of murein. Eur. J. Biochem. **95:**157–165.
4. **Barnickel, G., D. Naumann, H. Bradaczek, H. Labischinski, and P. Giesbrecht.** 1983. Computer aided molecular modeling of the three-dimensional structure of bacterial peptidoglycan, p. 61–66. *In* R. Hakenbeck, J.-V. Holtje, and H. Labischinski (ed.), The target of penicillin. Walter de Gruyter & Co., Berlin, Federal Republic of Germany.
5. **Bayer, M. E.** 1979. The fusion sites between outer membrane and cytoplasmic membrane of bacteria: their role in membrane assembly and virus infection, p. 167–202. *In* M. Inouye (ed.), Bacterial outer membranes. John Wiley & Sons, Inc., New York.
6. **Botta, G., and J. T. Park.** 1981. Evidence for involvement of penicillin-binding protein 3 in murein synthesis during septation but not during cell elongation. J. Bacteriol. **145:**333–340.
7. **Braun, V., H. Gnirke, U. Henning, and K. Rehn.** 1973. Model for the structure of the shape-maintaining layer of the *Escherichia coli* cell envelope. J. Bacteriol. **114:**1264–1270.
8. **Braun, V., and K. Rehn.** 1969. Chemical characterization, spatial distribution and function of a lipoprotein (murein-lipoprotein) of the *E. coli* cell wall. The specific effect of trypsin on the membrane structure. Eur. J. Biochem. **10:**426–438.

9. **Burdett, I. D. J., and R. G. E. Murray.** 1974. Septum formation in *Escherichia coli*: characterization of septal structure and the effects of antibiotics on cell division. J. Bacteriol. **119**:303–324.

10. **Burman, L. G., and J. T. Park.** 1983. Changes in the composition of *Escherichia coli* murein as it ages during exponential growth. J. Bacteriol. **155**:447–453.

11. **Burman, L. G., and J. T. Park.** 1983. A new model for growth of the murein sacculus, p. 119–127. *In* R. Hakenbeck, J.-V. Holtje, and H. Labischinski (ed.), The target of penicillin. Walter de Gruyter & Co., Berlin, Federal Republic of Germany.

12. **Burman, L. G., and J. T. Park.** 1984. Molecular model for elongation of the murein sacculus of *Escherichia coli*. Proc. Natl. Acad. Sci. USA **81**:1844–1848.

13. **de Pedro, M. A., and U. Schwarz.** 1981. Heterogeneity of newly inserted and pre-existing murein in the sacculus of *Escherichia coli*. Proc. Natl. Acad. Sci. USA **78**:5856–5860.

14. **Donachie, W. D., K. J. Begg, and N. F. Sullivan.** 1984. Morphogenes of *Escherichia coli*, p. 27–62. *In* R. Losick and L. Shapiro (ed.), Microbial development. Cold Spring Harbor Laboratory, Cold Spring Harbor, N.Y.

15. **Dubochet, J., A. W. McDowell, B. Mange, E. N. Schmidt, and K. G. Lickfeld.** 1983. Electron microscopy of frozen-hydrated bacteria. J. Bacteriol. **155**:381–390.

16. **Formanek, H.** 1982. Possible models of mureins and their Fourier transforms. Z. Naturforsch. **37C**:226–235.

17. **Glauner, B., and U. Schwarz.** 1983. The analysis of murein composition with high-pressure-liquid chromatography, p. 29–34. In R. Hakenbeck, J.-V. Holtje, and H. Labischinski (ed.), The target of penicillin. Walter de Gruyter & Co., Berlin, Federal Republic of Germany.

18. **Gmeiner, J.** 1980. Identification of peptide-cross-linked trisaccharide peptide trimers in murein of *Escherichia coli*. J. Bacteriol. **143**:510–512.

19. **Gmeiner, J., P. Essig, and H. H. Martin.** 1982. Characterization of minor fragments after digestion of *Escherichia coli* murein with endo-N,O-diacetylmuramidase from *Chalaropsis*, and determination of glycan chain length. FEBS Lett. **138**:109–112.

20. **Hobot, J. A., E. Carlemalm, W. Villiger, and E. Kellenberger.** 1984. Periplasmic gel: new concept resulting from reinvestigation of bacterial cell envelope ultrastructure by new methods. J. Bacteriol. **160**:143–152.

21. **Holtje, J.-V., D. Mirelman, N. Sharon, and U. Schwarz.** 1975. Novel type of murein transglycosylase in *Escherichia coli*. J. Bacteriol. **124**:1067–1076.

22. **Ishino, F., and M. Matsuhashi.** 1981. Peptidoglycan synthetic enzyme activities of highly purified penicillin-binding protein 3 in *Escherichia coli*: a septum-forming reaction sequence. Biochem. Biophys. Res. Commun. **101**:905–911.

23. **Ishino, F., S. Tanaki, B. G. Spratt, and M. Matsuhashi.** 1982. A mecillinam sensitive peptidoglycan cross-linking reaction in *Escherichia coli*. Biochem. Biophys. Res. Commun. **109**:689–696.

24. **Labischinski, H., G. Barnickel, and D. Naumann.** 1983. The state of order of bacterial peptidoglycan, p. 49–54. *In* R. Hakenbeck, J.-V. Holtje, and H. Labischinski (ed.), The target of penicillin.

Walter de Gruyter & Co., Berlin, Federal Republic of Germany.

25. **Leduc, M., C. Frehel, and J. van Heijenoort.** 1985. Correlation between degradation and ultrastructure of peptidoglycan during autolysis of *Escherichia coli*. J. Bacteriol. **161**:627–635.

26. **Nakamura, M., I. N. Maruyama, M. Soma, J.-I. Kato, H. Suzuki, and Y. Hirota.** 1983. On the process of cellular division in *Escherichia coli*: nucleotide sequence of the gene for penicillin-binding protein 3. Mol. Gen. Genet. **191**:1–9.

27. **Park, J. T., and L. G. Burman.** 1985. Elongation of the murein sacculus of *Escherichia coli*. Ann. Inst. Pasteur Microbiol. **136A**:51–58.

28. **Rodolakis, A., P. Thomas, and J. Starka.** 1973. Morphological mutants of *Escherichia coli*. Isolation and ultrastructure of a chain-forming *envC* mutant. J. Gen. Microbiol. **75**:409–416.

29. **Rodriguez-Tebar, A., J. A. Barbas, and D. Vazquez.** 1985. Location of some proteins involved in peptidoglycan synthesis and cell division in the inner and outer membranes of *Escherichia coli*. J. Bacteriol. **161**:243–248.

30. **Rogers, H. J., H. R. Perkins, and J. B. Ward.** 1980. Microbial cell walls and membranes. Chapman & Hall, Ltd., London.

31. **Rosenbush, J. P.** 1974. Characterization of the major envelope protein from *Escherichia coli*. Regular arrangement on the peptidoglycan and unusual dodecyl sulfate binding. J. Biol. Chem. **249**:8019–8029.

32. **Schleifer, K. H., and O. Kandler.** 1972. Peptidoglycan types of bacterial cell walls and their taxonomic implications. Bacteriol. Rev. **36**:407–477.

33. **Schmidt, L. S., G. Botta, and J. T. Park.** 1981. Effects of furazlocillin, a β-lactam antibiotic which selectively binds to penicillin-binding protein 3, on *Escherichia coli* mutants deficient in other penicillin-binding proteins. J. Bacteriol. **145**:632–637.

34. **Slater, M., and M. Schaechter.** 1974. Control of cell division in bacteria. Bacteriol. Rev. **38**:199–221.

35. **Spratt, B. G.** 1975. Distinct penicillin binding proteins involved in the division, elongation, and shape of *Escherichia coli*. Proc. Natl. Acad. Sci. USA **72**:2999–3003.

36. **Spratt, B. G., and A. B. Pardee.** 1975. Penicillin-binding proteins and the cell shape in *E. coli*. Nature (London) **254**:516–517.

37. **Suzuki, H., Y. Nishimura, and Y. Hirota.** 1978. On the process of cell division in *Escherichia coli*: a series of mutants altered in the penicillin binding proteins. Proc. Natl. Acad. Sci. USA **75**:664–668.

38. **Tomioka, S., F. Ishino, S. Tamaki, and M. Matsuhashi.** 1982. Formation of hyper-crosslinked peptidoglycan with multiple crosslinkages by a penicillin-binding protein, 1A, of *Escherichia coli*. Biochem. Biophys. Res. Commun. **10**:1175–1182.

39. **Verwer, R. W. H., E. H. Beachey, W. Keck, A. M. Stoub, and J. E. Poldermans.** 1980. Oriented fragmentation of *Escherichia coli* sacculi by sonication. J. Bacteriol. **141**:327–332.

40. **Verwer, R. W. H., N. Nanninga, W. Keck, and U. Schwarz.** 1978. Arrangement of glycan chains in the sacculus of *Escherichia coli*. J. Bacteriol. **136**:723–729.

41. **Weidel, W., and H. Pelzer.** 1964. Bagshaped macromolecules—a new outlook on bacterial cell walls. Adv. Enzymol. **26**:193–232.

5. Cytoplasmic Membrane

JOHN E. CRONAN, JR.,[1] ROBERT B. GENNIS,[2] AND STANLEY R. MALOY[1]

Departments of Microbiology,[1] Chemistry,[2] and Biochemistry,[2] University of Illinois, Urbana, Illinois 61801

INTRODUCTION

Although electron microscopic studies had long indicated that gram-negative bacteria have two external membranes (38), it was not until the outer and cytoplasmic membranes were separated by Miura and Mizushima (176) that the study of the cytoplasmic and outer membranes of *Escherichia coli* and *Salmonella typhimurium* become possible. The methods to separate the two membranes were developed and refined by Schnaitman (225) and Osborn et al. (194). The outer membrane has received much study, due in part to its relatively simple protein composition, but it has become clear that the outer membrane is a specialized and rather atypical membrane (193), the properties of which may not be fully applicable to other systems.

The predominant membrane structure is a phospholipid bilayer, which provides the hydrophobic barrier needed to allow the differential concentration of small molecules and ions relative to the growth medium. The bilayer is permeable only to water (166) and small hydrophobic molecules such as medium-chain fatty acids (153). The essential nature of this barrier is demonstrated by the bactericidal action of molecules which introduce a pore into the cytoplasmic membrane (e.g., colicins Ia, Ib, E1, and K [124]). Different protein species can be resolved from cytoplasmic membranes by two-dimensional electrophoretic techniques. Many of these proteins are involved in bioenergetic and biosynthetic reactions, whereas other proteins are involved in transport of specific solutes. Although the bioenergetic components of eucaryotic cells are generally found in intracellular organelles, the equivalent components are located in the procaryotic cytoplasmic membrane. Some proteins are tightly bound to the membrane (integral membrane proteins), some are loosely bound (peripheral membrane proteins), and others are thought to interact only transiently with the membrane.

The outer membrane of *E. coli* and *S. typhimurium* serves to protect the cytoplasmic membrane from surface-active agents (e.g., bile salts) present in the natural environment and from host defense mechanisms (e.g., complement) (149). In the intact cell, stabilization against the turgor pressure of cytosol is provided by the peptidoglycan layer located between the two membranes (174). The peptidoglycan also determines the contour of the cell and, hence, that of the cytoplasmic membrane.

In this review we will first concentrate on the structural organization of the components of the cytoplasmic membrane and related functional aspects. We will then discuss what is known concerning the dynamics of the membrane components and possible mechanisms of cytoplasmic membrane genesis.

BULK COMPOSITION OF THE CYTOPLASMIC MEMBRANE

The cytoplasmic membranes of *E. coli* and *S. typhimurium* contain 75% of the cellular phospholipid (the remainder being found in the outer membrane) and about 6 to 9% of the total cellular protein (about 60% of the cell envelope protein) (6, 125, 194, 244). Three species of phospholipids are present (chapter 30), and at least 100 major protein species are present in aerobically grown cells. The latter value is at best a lower limit estimated by the two-dimensional gel electrophoric patterns of cell envelopes reported by Ames and Nikaido (6). A definitive analysis of the number of proteins and their relative abundance has not yet been reported (F. Neidhardt, personal communication) and is overdue. However, since a number of proteins are only loosely bound to the membrane, the values obtained will depend on the method of membrane purification and the conditions of growth.

LIPID COMPONENTS

The Lipids of Cytoplasmic Membrane Are Arranged in a Bilayer

Direct analyses of purified inner membranes by a variety of physical techniques demonstrate that the intra- and intermolecular properties of the cytoplasmic membrane lipids are essentially identical to those seen in synthetic lipid bilayers made from purified phospholipids. The strongest evidence is that from X-ray diffraction, the only direct and unambiguous method to assay the packing arrangement of phospholipid hydrocarbon chains (see references 17 and 208). Several groups have reported the high-angle diffraction patterns of *E. coli* inner membranes to demonstrate that the lipids are packed into the lamellar arrangement characteristically found in lipid bilayers (54, 80, 137, 142, 145, 184, 195, 209). Low-angle X-ray diffraction of partially oriented membrane samples also gives values for the thickness of the cytoplasmic membrane as the distance between polar head groups of the phospholipids (17). Two values, 5.4 to 5.5 nm and 6.4 to 6.5 nm, have been reported (80, 145). The larger value (145) is that seen in membranes below the onset of the lipid phase transition (see below), whereas the smaller value is seen above the transition temperature. Differential scanning calorimetry, the method first used to detect phase transitions in natural membranes (246), is less direct than X-ray diffraction but gives the same conclusion. The lipids of *E. coli* cytoplasmic membranes undergo the cooperative melt characteristic of phospholipids packed in a lamellar array (see below) (100, 112, 167, 245).

Early results suggested that only 80% of the inner membrane phospholipids undergo this cooperative melt (167, 195, 245), whereas later estimates often approach 100% (100, 137, 142, 145, 224). This discrepancy may be due to the inherent insensitivity of both the calorimetric and the X-ray methods; neither method is able to detect the clusters of lipid molecules smaller than a few hundred molecules which may exist in membranes of certain fatty acid compositions (see below). A very sensitive technique, deuterium nuclear magnetic resonance (NMR), indicates that >97% of the cytoplasmic membrane lipids go through the phase transition (187). Other less direct techniques (fluorescence and electron spin resonance [ESR] probes) have also been used to study *E. coli* membranes. Some of the data obtained with such probes are consistent with those obtained by direct methods, whereas others give artifactual or uninterpretable results (see references 41 and 196).

Two other approaches to the arrangement of the cytoplasmic membrane lipids have been reported. The first, a straightforward calculation from (i) the number of lipid molecules, (ii) the packing cross-section of lipids in a lipid monolayer, and (iii) the surface area of *S. typhimurium*, shows that the lipid content is that expected for a lipid bilayer (244). The results of freeze-fracture electron microscopy also demonstrate that the lipid bilayer is the dominant structural component of the cytoplasmic membrane for both *E. coli* (13, 26) and *S. typhimurium* (244).

It should be noted that two groups have reported the presence of a nonbilayer phase in the lipids of the cytoplasmic membrane of *E. coli* (26, 44). These NMR results were interpreted as the presence of inverted lipid micelles (e.g., an H type II phase). However, a later report (187) from one of these laboratories stated that their earlier work (44) was in error due to the degradation of the sample during analysis. Moreover, Gally et al. (64), using deuterium NMR, reported the complete absence of nonlamellar phases in *E. coli* cytoplasmic membrane vesicles, and thus the possibility of the presence of nonlamellar phases seems remote.

In conclusion, there is overwhelming evidence that the lipid component of cytoplasmic membranes is structured into a lipid bilayer. It should be noted that not only is the interchain packing of the phospholipid molecules qualitatively and quantitatively very similar to that of known lipid bilayers, but the intrachain packing (that is, the conformations of the lipid acyl chains and head groups) also agrees closely with that seen in model membranes (232). Deuterium NMR experiments have shown that the conformations of the acyl chains of the lipids of the cytoplasmic membrane of *E. coli* are essentially identical to those of pure bilayer lipids (44, 63, 64, 108, 187). This is also true of the head group of phosphatidylglycerol (18).

Physical Properties of the Lipid Bilayer

The major function of the lipid bilayer of the cytoplasmic membrane is to act as a hydrophobic barrier allowing the intracellular concentration of small molecules and proteins. However, the membrane must also provide a suitable milieu for function of the transport proteins required for accumulation of essential molecules and for proper function of the proteins and other components involved in bioenergetic and biosynthetic reactions.

Phase transitions

The barrier function of the cytoplasmic membrane is known to depend critically on the physical state of the lipid bilayer. Phospholipid bilayers undergo reversible changes of state dependent on temperature and lipid composition. At low temperatures in membranes of a given lipid composition, the acyl chains are in a closely packed ordered array that is quite solid, whereas at higher temperatures the lipids "melt" to form a phase in which the acyl chains are in a disordered state approximating that of a liquid hydrocarbon (see references 41, 166, and 206). The phospholipids remain arranged in a lipid bilayer in both states due to the driving force of hydrophobic interactions, but the bilayer thins (ca. 15%) upon melting due to greater movement of the acyl chains (i.e., the chains are less often in an extended conformation) (143). In addition, the average area occupied by a phospholipid molecule increases from about 4.7 nm^2 to 6.5 to 7.0 nm^2. Thus, the phase transition has a large effect on the structure of the lipid bilayer, the dominant structural component of the cytoplasmic membrane.

The temperature range over which the order-disorder transition occurs can be very abrupt (1 to 2°C) in model bilayers formed from a single species of a synthetic phospholipid, but in biological membranes such as *E. coli* cytoplasmic membranes, the transitions occur over a wide temperature range (10 to 20°C) (119). This is due to the diversity of lipid classes found in natural membranes and is attributed to phase separations (see below). The temperature range over which the transition occurs depends on the fatty acid composition and the polar head group composition of the phospholipids. As discussed in detail earlier (39, 41), the presence of unsaturated fatty acids lowers the phase transition whereas saturated fatty acyl chains have the opposite effect. The chain length of the acyl chains also affects the transition, short chains giving lower transitions. (However, chains of >12 carbons are needed to form a stable bilayer.) The effect of the double bond is purely steric (41). A useful (although overly simplistic) visualization of the effect of a double bond on the transition is that the kink in the acyl chain (due to the double bond) lessens close packing such that the bilayer has some fluid character even below the onset of the phase transition.

The barrier properties of synthetic phospholipid bilayers reflect the physical state of the bilayer. Fluid (disordered) bilayers are much more permeable to small molecules than are ordered bilayers (257). In the case of water movement, this permeability has been attributed to the formation of transient pores in fluid bilayers (166). In *E. coli* the production of an abnormally fluid bilayer is known to result in loss of the barrier properties of the cytoplasmic membrane (ii; for review see references 39 and 41). A similar result is seen when *E. coli* cells are manipulated to have an abnormally ordered bilayer, probably due to the onset of lipid phase separations (39).

Phase separations

Phase separations were first seen in model membranes composed of two different lipids. Such bilayers undergo two distinct phase transitions rather than a single broad transition, and this is called a phase separation. Each lipid associates largely with its own species, and a membrane consisting of alternating patches of the two lipids is formed. However, similar lipid species are miscible and therefore give a single broad transition. The rules determining miscibility of lipid species are still not completely clear in model systems, but the formation of patches of ordered lipid in a fluid bilayer (or vice versa) has been observed in *E. coli* cytoplasmic membranes (for review see reference 39). Under extreme conditions, phase separation can disrupt membrane function. For example, growth of an *E. coli* fatty acid auxotroph in the absence of a supply of unsaturated fatty acids results in the synthesis of phospholipids containing two saturated chains (239). From model system studies, these lipids are expected to be immiscible with the lipids normally synthesized (which contain one saturated chain and one unsaturated chain). The expected phase separation could account for the loss of inner membrane function in these membranes (41). Phase separations could be physiologically harmful for several reasons. For example, phase separation could lead to a decrease in permeability of the bilayer to water due to the lower degree of permeability of phospholipids containing two saturated fatty acids (166). Other changes could result from the partition of certain membrane proteins into either the ordered or disordered patches (see reference 39). In should be noted that many laboratories have reported lipid phase separations in *E. coli* based on results with ESR or fluorescent probes. However, most of this evidence is correlational (at best) since the location of the probes is only assumed and not demonstrated.

PROTEIN COMPONENTS

Components Involved in Electron Transport and Oxidative Phosphorylation

The cytoplasmic membrane contains all the apparatus required for energy transduction and oxidative phosphorylation (for reviews see references 21 and 99). This includes the aerobic respiratory chain and the ATP synthase as well as the electron transport proteins induced when the organisms are grown anaerobically. The respiratory system couples the oxidation of organic substrates (or hydrogen) to the generation of a proton motive force which is then used to drive ATP synthesis, flagellar motion, or the uptake of nutrients such as lactose. A major feature is the apparent "modular" nature of the system. The modules are of several types: (i) those enzymes which oxidize substrates such as NADH, D-lactate, or succinate and reduce quinone; (ii) those enzymes which oxidize quinone and reduce a water-soluble terminal electron acceptor, such as oxygen, fumarate, or nitrate; (iii) the ATP synthase and other proteins designed to utilize the proton motive force.

Generally, the enzymatic activities associated with enzymes in the first class are measured using water-soluble oxidants such as dichlorophenolindophenol or ferricyanide. These are called dehydrogenase activities and are not dependent on the presence of the other components of the respiratory chain. Alternatively, assays can be performed which require the intact respiratory chain, such as oxidase assays, where, for example, an oxygen electrode can be used to measure oxygen utilization rates dependent on the oxidation of a particular substrate.

The dehydrogenases are mostly flavoproteins or metalloflavoproteins, whereas the enzymes in the second group are metalloproteins with a variety of prosthetic groups, including hemes. Taken together as a single complex system, there are about 20 enzymes involved in electron transport, some constitutive, but many inducible under specific growth conditions. However, these proteins do not appear to be physically associated with each other. Rather, they are functionally linked through small diffusible molecules. In particular, ubiquinone-8 and menaquinone-8 act as hydrogen carriers within the membrane to couple the flavoproteins to the terminal enzymes in the electron transport chains, and protons couple the electron transport system to the ATP synthase (H^+-translocating ATPase). What little evidence there is suggests that each enzyme in this system is probably freely diffusing within the bilayer and randomly distributed in the plane of the membrane (65; but see reference 111).

Components Present during Aerobic Growth

Dehydrogenases

Table 1 lists electron transport flavoprotein dehydrogenases which feed electrons into the respiratory chain. In all cases where data are available, the enzymes catalyze the two-electron oxidation of the organic substrate on the inside surface of the cytoplasmic membrane. In many cases, it is clear that the proteins are not transmembranous, but are bound through hydrophobic interactions to the membrane surface. In some cases the binding appears to be nonspecific to the phospholipid bilayer, though in other instances, specific proteins may be present to "anchor" the catalytic subunits to the membrane, as is apparently the case for succinate dehydrogenase (270). Only pyruvate oxidase (121) and D-lactate dehydrogenase (K. Matsushita and H. Kaback, personal communication) have been shown to directly reduce ubiquinone-8, the major species of quinone present in the membrane. Many of the purified dehydrogenases can be added back to E. coli cytoplasmic membrane vesicles to reconstitute oxidase activity and create a proton motive force. One salient feature is that the dehydrogenases can be added to either inverted or right-side-out vesicles with similar results (e.g., reference 78). This is consistent with nonspecific binding and reduction of the mobile quinone. The simultaneous reconstitution of D-amino acid oxidase and D-lactate dehydrogenase suggests there are no specific binding sites on the E. coli membrane for these flavoproteins (78). However, a previous study was interpreted to show that there are a limited number of binding sites for some dehydrogenases in the cytoplasmic membrane (132). A brief discussion of the better characterized systems follows.

Succinate dehydrogenase. Succinate dehydrogenase catalyzes the oxidation of succinate to fumarate and is a critical component of the tricarboxylic acid cycle (for a review, see reference 83). Unfortunately, it has not yet been purified to homogeneity from E. coli. However, the sdh genes encoding this enzyme have been sequenced, revealing an operon encoding four polypeptides with molecular weights of 64,300, 26,000, 14,200, and 12,800 (43, 270). The two larger subunits can be identified as the flavoprotein and iron-sulfur-containing components of the enzyme, and the smaller subunits are very hydrophobic. Analysis of immunoprecipitin arcs exhibiting succinate dehydrogenase activity by sodium dodecyl sulfate-polyacrylamide gel electrophoresis (SDS-PAGE) has clearly shown the two high-molecular-weight subunits, but the published data are equivocal concerning the presence of the two hydrophobic peptides in the Triton X-100–solubilized form of the enzyme (105, 198). Apparently, the association between the large, hydrophilic subunits, which possess the succinate dehydrogenase active site, and the hydrophobic peptides is labile, though under certain circumstances the four-subunit form of this enzyme can be solubilized (37a). Succinate dehydrogenase has also been isolated from other organisms in forms containing subunits in addition to the flavoprotein and iron-sulfur proteins (see reference 83). In some cases, a b-type cytochrome is part of this complex, and this also appears to be true in E. coli (180). Sidedness studies showing the extent of exposure of the subunits on either side of the membrane have not been reported, though it is assumed that the two hydrophobic subunits serve to anchor the catalytic subunits on the inner surface of the membrane. Possibly, these subunits are important for facilitating the reduction of ubiquinone-8 in the bilayer.

Succinate dehydrogenase is induced during aerobic growth on nonfermentable substrates and is reportedly not present during anaerobic growth in the pres-

TABLE 1. Components of the cytoplasmic membrane involved in energy transduction

Components	Subunits (mol wt)	Prosthetic groups	Topology	Comments	References
Class I: Dehydrogenases					
NADH dehydrogenase	1. 47,300	FAD	Cytoplasmic	Cloned, sequenced	102, 103
Succinate dehydrogenase	1. 64,300 2. 26,000 3. 14,200 4. 12,800	FAD (covalently linked to subunit 1) heme *b*; nonheme iron	Cytoplasmic (subunits 1, 2)	Cloned, sequenced	43, 270
D-Lactate dehydrogenase	1. 64,600	FAD	Cytoplasmic	Cloned, sequenced	28, 217
L-Lactate dehydrogenase	1. 43,000	FMN	Presumably cytoplasmic		61, 113
Pyruvate oxidase[a]	1. 60,000 (tetramer)	FAD	Cytoplasmic	Cloned, sequenced; lipid-requiring enzyme, peripheral membrane protein	31, 121
sn-Glycerol-3-phosphate dehydrogenase (aerobic)	1. 58,000 (dimer)	FAD	Cytoplasmic	Peripheral membrane protein, cloned	212, 231, 266
sn-Glycerol-3-phosphate dehydrogenase (anaerobic)	1. 62,000 2. 43,000	FAD Nonheme iron	Cytoplasmic	Cloned	229, 230
D-Amino acid dehydrogenase	1. 55,000 2. 45,000	FAD Nonheme iron	Cytoplasmic	D-Alanine is best substrate	192
Proline oxidase[a]	1. 130,000 (dimer?)	FAD	Cytoplasmic	Peripheral membrane enzyme, cloned	71, 169, 170
Malate oxidase[a]	Not purified	FAD dependent	Probably cytoplasmic	Peripheral membrane enzyme	185
Hydrogenase	1. 56,000 (dimer)	Nonheme iron	Transmembrane	May be several different hydrogenases	2
Formate dehydrogenase	1. 110,000 2. 32,000 3. 20,000	Heme *b* (iron-protoporphyrin IX) Nonheme iron; molybdenum-pterin cofactor; selenium	Transmembrane	May be more than one formate dehydrogenase	19, 70
Dihydroorotate dehydrogenase	1. 67,000	?	Cytoplasmic		8, 109
Class II: Terminal reductases					
Cytochrome *o* terminal oxidase complex	1. 66,000 2. 35,000 3. 22,000 (?) 4. 17,000 (?)	Heme *b* (iron-protoporphyrin IX) Copper (?)	Probably transmembrane	Cloned	9, 116, 157, 158
Cytochrome *d* terminal oxidase complex	1. 57,000 2. 43,000	Heme *b* (iron-protoporphyrin IX) Heme *d* (chlorin)	Probably transmembrane	Cloned, sequenced	73, 117, 172
Trimethylamine *N*-oxide reductases	Not purified	?	?	2 to 4 enzymes; not purified	134
Nitrate reductase	1. 155,000 (α) 2. 60,000 (β) 3. 20,000 (γ)	Nonheme iron Heme *b* (iron-protoporphyrin IX) Molybdenum-pterin cofactor	Transmembrane		19, 32, 177
Fumarate reductase	1. 64,000 2. 27,100 3. 15,000 4. 14,000	FAD (covalently linked to subunit 1) Nonheme iron	Cytoplasmic (subunits 1, 2)	Requires menaquinol as reductant; cloned, sequenced	34, 37, 138, 241, 267
Class III: Enzymes utilizing the proton motive force					
ATP synthase	F_1 1. 55,300 (α) 2. 50,100 (β) 3. 31,400 (γ)	None	F_1, cytoplasmic F_0, transmembrane	Cloned, sequenced	22, 60, 120

TABLE 1—*Continued*

Components	Subunits (mol wt)	Prosthetic groups	Topology	Comments	References
ATP synthase	4. 19,300 (δ) 5. 14,900 (ε) F₀ 1. 30,300 (a or χ) 2. 17,200 (b or ψ) 3. 8,200 (c or Ω)				
Flagellar "motor"	Not purified			Very complex apparatus extending through both membranes	152
Numerous permeases	See Table 2				
Transhydrogenase	Not purified	?	Presumably transmembrane	Cloned	33a; see reference 21 for a review
Miscellaneous					
Cytochrome b_{556}	14,200	Heme *b* (iron-protoporphyrin IX)	Transmembrane	One of the small protein "anchors" of succinate dehydrogenase; sequenced	118, 180
Cytochrome b_{561}	18,000	Heme *b* (iron-protoporphyrin IX)	?	Unknown function; not a major component in wild-type cells; cloned	179, 181

[a] These enzymes are actually dehydrogenases and not true oxidases since they do not directly interact with oxygen.

ence of fumarate (258). The subunit structure of this enzyme is very similar to that of fumarate reductase, with primary sequence homology between the two large catalytic subunits (43, 270).

D-Lactate dehydrogenase. D-Lactate dehydrogenase is a single-polypeptide flavoprotein which appears to be constitutive and catalyzes the oxidation of D-lactate to pyruvate. D-Lactate dehydrogenase has been purified and extensively studied by several laboratories (59, 123, 204). The *ldh* gene that encodes D-lactate dehydrogenase has been cloned (273) and sequenced (28, 217). The protein does not have a low average polarity and does not contain any hydrophobic stretches which could be identified as likely to be involved in binding to the membrane (28). The purified enzyme, however, does bind to *E. coli* membrane preparations (78, 205, 237) and to phospholipids (e.g., reference 126), which greatly stimulate the catalytic velocity. It appears that this enzyme probably binds to the membrane through nonspecific hydrophobic interactions with the phospholipid, rather than to an "anchor" provided by another protein component. Membrane localization experiments clearly show that D-lactate dehydrogenase is present on the inside surface of the cytoplasmic membrane (237). Reconstitution experiments have shown that D-lactate dehydrogenase can reduce ubiquinone-8 in the lipid bilayer in the absence of other proteins (K. Matsushita and H. Kaback, personal communication).

There is also an L-lactate dehydrogenase which is induced when cells are grown aerobically or anaerobically on D- or L-lactate (e.g., reference 190). This enzyme has been purified, but is less well characterized than the D-lactate dehydrogenase (113).

NADH dehydrogenase. NADH dehydrogenase has also been isolated as a single-subunit flavoprotein (102, 103). Purification was greatly facilitated by using a strain which contained the cloned *ndh* gene on a multicopy plasmid (272). The gene has been sequenced; like that of D-lactate dehydrogenase, the deduced protein sequence does not contain obvious transmembrane stretches (274), though there is a hydrophobic region near the N terminus. Crossed immunoelectrophoresis of solubilized *E. coli* membranes indicates two antigens with "diaphorase" activity (i.e., NADH-dye oxidoreductase activity). Both were localized to the inside of the membrane, and one was identified as the respiratory NADH dehydrogenase (199). However, this arc contains nonheme iron, whereas the purified NADH dehydrogenase does not. The relationships between these enzymes and other preparations (254) of the respiratory NADH dehydrogenase are not known. Possibly, the enzyme is a single-subunit NADH-ubiquinone oxidoreductase, but the existence of additional subunits, so far unidentified, cannot be ruled out. If electron flow from NADH to quinone is accompanied by proton translocation (e.g., coupling site), as has often been claimed (see reference 21), then

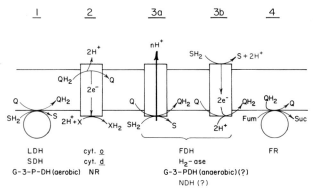

FIG. 1. Schematic diagram showing the strategies used by cytoplasmic membrane components for carrying out electron transfer reactions coupled to proton transloca- tion. (Type 1) Simple dehydrogenases which reduce quinone (Q represents either ubiquinone or menaquinone) on the inside surface of the bilayer. The lactate dehydrogenases (D-LDH, L-LDH), succinate dehydrogenase (SDH), and probably the aerobic sn-glycerol 3-phosphate dehydrogenase (G-3-PDH) conform to this model. These dehydrogenases are not coupling sites. (Type 2) "Cou- pling" provided by the net transfer of charge and protons across the bilayer during quinol oxidation due to the separation of the active sites for the oxidative and reduc- tive half-reactions on different sides of the membrane. This is probably how both terminal oxidases (cyt. o, cyt. d) and the nitrate reductase (NR) function. (Type 3) In some cases, proton translocation appears to accompany the dehydrogenase reaction, as is the case for formate dehy- drogenase (FDH) and hydrogenase (H$_2$-ase). It is speculat- ed that both the NADH dehydrogenase (NDH) and anaer- obic sn-glycerol 3-phosphate dehydrogenase (G-3-PDH) function this way, though biochemical studies have yet to confirm this. Scheme 3a is most probable for all these enzymes, though for formate dehydrogenase and hydro- genase the possibility that the active sites are periplasmic has not been ruled out, in which case model 3b would apply. (Type 4) Fumarate reductase (FR) appears only to function on the inner surface of the membrane. Hence, any dehydrogenase coupled to fumarate reductase which drives oxidative phosphorylation must be type 3 and not type 1.

additional components would be required beyond the isolated flavoprotein, as speculated in Fig. 1, or another enzyme may be involved.

Pyruvate oxidase. Although apparently unimpor- tant physiologically, pyruvate oxidase has been exten- sively studied. It is loosely bound to the membrane and is purified in a water-soluble form (191). The enzyme catalyzes the oxidative decarboxylation of pyruvate to acetate and reduces ubiquinone-8 in the membrane. The affinity of this enzyme for the mem- brane, and for lipids generally, is enhanced when the flavin is reduced, i.e., upon addition of the substrate (121, 227, 228). The membrane binding appears to be regulated allosterically in this manner. Binding of lipids alters the active site (210) and accelerates the catalytic velocity by over 20-fold as determined using artificial electron acceptors (16). The poxB gene which encodes pyruvate oxidase has been cloned (68) and sequenced (C. Grabau and J. E. Cronan, Jr., unpub- lished data). Like D-lactate dehydrogenase and NADH dehydrogenase, there are no apparent transmem-

brane hydrophobic stretches. Pyruvate oxidase has, similarly, been shown to be located on the inner surface of the membrane (236). Recently isolated mutants deficient in lipid activation may help eluci- date how this enzyme and possibly others bind to the membrane (31).

Quinones

An essential part of the E. coli respiratory chain are the quinones, located in the cytoplasmic membrane. Ubiquinone-8 is the predominant species found when cells are grown aerobically, and menaquinone-8 is a major component in anaerobically grown cells (e.g., see reference 263). (The number "8" in this notation designates the number of isoprene side chain units.) Other minor species are also present. The manner in which these long-chain isoprenoid compounds re- side in the bilayer is not clear, but they clearly can rapidly diffuse laterally (57, 76) and catalyze reactions across the bilayer (62). Thus, quinones can couple the various components of the respiratory chain. Mena- quinone-8 is generally a poor substitute for ubiqui- none-8 as a carrier between the components present in aerobic cells (263), but menaquinone is absolutely required for the anaerobic trimethylamine N-oxide reductases (134, 164) and the fumarate reductase system (see reference 130). The quinones are present in at least 10-fold molar excess over the cytochromes (263).

Cytochromes

Electrochemical and spectroscopic methods have been used for many years to analyze the heme pro- teins in the E. coli membrane (e.g., see references 23, 77, and 147). Biochemical studies have revealed that the major cytochromes of aerobically grown cells are organized into three separable species: the cyto- chrome o complex, the cytochrome d complex, and cytochrome b_{556}. The two complexes function as ter- minal oxidases, and cytochrome b_{556} is part of the succinate dehydrogenase complex (180). Under anaer- obic conditions, cytochromes are induced which are parts of the formate dehydrogenase (b^{fdh}) and nitrate reductase (b^{nr}) enzymes (77). Cytochromes are also specifically induced by growth in the presence of trimethylamine N-oxide (23). The cytochromes in E. coli are distinct from those found in the mitochond- rion. For example, there is no cytochrome c involved in aerobic respiration in E. coli.

Cytochrome o complex (b_{562}-b_{555} or b_{562}-o). Cells grown with high aeration (e.g., harvested in early log phase, low cell density) contain only b-type cyto- chrome, i.e., absorption peaks in the reduced-minus- oxidized spectrum near 560 nm. Biochemically, these resolve into the cytochrome o complex and cyto- chrome b_{556} (e.g., references 115 and 127). The former derives its name from the fact that it binds CO and functions as an oxidase. The enzyme has been purified to homogeneity and contains two (116) or four (157, 158) subunits by SDS-PAGE analysis. The two sub- units with low molecular weight do not stain well with Coomassie blue, but are clearly seen in immuno- precipitation experiments (128). The enzyme contains iron-protoporphyrin IX (heme b) and possibly copper

(116) as well. Spectroscopic studies indicate two peaks at 555 and 562 nm which could represent two separate cytochrome components or could represent a single type of redox center with a split α-band in the reduced-minus-oxidized spectrum. Reconstitution experiments show that this enzyme functions as a ubiquinol-8 oxidase and must be transmembranous (30). Quinol oxidase activity results in generating a voltage difference across the bilayer. No information is published on the detailed structure of this enzyme or its subunit stoichiometry.

Cytochrome d complex (b_{558}-a_1-d). When cells are grown under oxygen-limited conditions, it appears that the biosynthesis of the cytochrome o complex is reduced and a second oxidase is induced, the cytochrome d complex (128, 211). This oxidase is more resistant to oxidase inhibitors such as cyanide (72) and has a higher affinity for oxygen than does the cytochrome o complex (211). Other growth conditions also result in the appearance of the cytochrome d complex (see reference 99). Mutants lacking either of these two terminal oxidases have been isolated (10, 72), and under normal laboratory aerobic growth conditions, either enzyme can support growth. Presumably, under conditions of stress (e.g., low aeration) the cytochrome d complex confers an advantage to the cells, but this has not been demonstrated.

The cytochrome d complex has been purified and shown to contain two subunits by SDS-PAGE analysis (117, 172). Potentiometric studies show three types of redox active centers (122). The larger of the two subunits has been identified as cytochrome b_{558} (74) and appears to contain the site for quinol oxidation (129). The smaller subunit contains the a_1 and d components.

Cytochrome a_1 appears to contain iron-protoporphyrin IX and not heme a (122). Cytochrome d is involved directly in oxygen binding (122, 203).

It is likely that both terminal oxidases act as "coupling sites" by physically separating the active site for quinol oxidation, on the outer surface of the membrane, from the active site of oxygen reduction on the inner surface of the membrane (see Fig. 1). The result is the generation of a proton electrochemical gradient across the cytoplasmic membrane accompanying electron flow through the system. It is the orientation of these oxidases in the membrane which determines the direction of the electrochemical gradient, whereas the simple dehydrogenases can, in principle, operate to reduce quinone from either side.

Cytochrome b_{556}. Cytochrome b_{556} appears to be present during aerobic growth at both high and low oxygen levels (147). In situ the cytochrome can be reduced by substrates such as NADH or succinate and is oxidized by air (115, 117). Cytochrome b_{556} has been purified and shown to be a single subunit by SDS-PAGE analysis (118). Partial sequence data have been used to identify this cytochrome as the product of the $sdhC$ gene; therefore, it is clear that this cytochrome is part of succinate dehydrogenase (180).

Cytochrome b_{561}. Cytochrome b_{561} appears to be a minor cytochrome component in the membrane. It was discovered by accident during attempts to clone cytochrome b_{556} (179, 182). Cytochrome b_{561} is a single-subunit, diheme protein that has been purified to homogeneity from an overproducing strain (181). The role of this cytochrome is unknown.

ATP synthase (H$^+$-translocating ATPase)

ATP synthase is responsible for utilizing the proton electrochemical gradient across the cytoplasmic membrane for ATP synthesis in cells grown aerobically or anaerobically. The enzyme has been extensively studied and is structurally similar to the related ATP synthase present in the eucaryotic mitochondrion and chloroplast (coupling factor) (e.g., reference 262). The bacterial enzyme consists of eight subunits (for reviews, see references 22 and 60). Three of them, a, b, and c (in 1:2:10–12 stoichiometry), make up the F_0 portion of the enzyme which extends through the cytoplasmic membrane and contains the "proton channel." The other five subunits (α, β, γ, δ, and ε in 3:3:1:1:1 stoichiometry) make up the F_1 portion of the enzyme. This portion contains the catalytic site for ATP synthesis (or hydrolysis when the reverse reaction is operative) and is located on the inner surface of the cytoplasmic membrane attached to the F_0 portion. All of the subunits are encoded by the atp operon (formerly the unc operon), which has been completely sequenced (see reference 60). An additional (ninth) open reading frame exists in the atp operon, which codes for a protein but whose function is not known (66). The polypeptide sequences have been used to predict structures for several of the subunits (234). One of the subunits in F_0 (subunit c) reacts with N,N'-dicyclohexylcarbodiimide, resulting in blocking of the proton channel, and many mutations have been isolated which also are defective for proton conductance (see reference 60). Results of recent in vivo (120) and reconstitution (226) experiments indicate that all three F_0 subunits are required for formation of the proton channel. However, the presence of F_1 subunits α and β, required in vivo (120), is not required for reconstitution of the channel in vitro (226).

Strains containing the atp operon cloned onto a multicopy plasmid overproduce the ATP synthase, but may also produce extra cytoplasmic membrane which is elaborated within the cells (261).

Components Present during Anaerobic Growth

$E.$ $coli$ can also carry out oxidative phosphorylation in the absence of oxygen by using either nitrate or fumarate as terminal oxidant (see reference 99). Trimethylamine N-oxide, the compound responsible for the smell of decaying fish, can also be used as a terminal electron acceptor (see references 99 and 134). Many of the enzymes in $E.$ $coli$ which are involved in anaerobic electron transfer are repressed during aerobic growth, and a number, including the nitrate and fumarate reductases, require the fnr (also called $nirR$ and $nirA$) gene product for expression (235). A similar regulatory locus has been found in $S.$ $typhimurium$ (247). As is the case for aerobic growth, the enzymes present in the cytoplasmic membrane during anaerobic growth depend on the carbon source (e.g., glucose, glycerol) as well as on the terminal oxidant. In addition to the induction of enzymes required for the reduction of various terminal oxidants, there are several dehydrogenases which are usually studied in anaerobically grown cells.

Dehydrogenases

Formate dehydrogenase. Formate dehydrogenase oxidizes formate to CO_2 and has usually been studied as a component of the membranes of cells grown anaerobically on nitrate. However, formate oxidation can also be coupled to oxygen or fumarate. When formate oxidation is coupled to the production of H_2 (proton reduction), the cooperation of formate dehydrogenase and hydrogenase is required, and the enzymes carrying out this formate-hydrogen lyase reaction may be distinct. That is, there may be more than one formate dehydrogenase and more than one hydrogenase (see reference 99).

The formate dehydrogenase found in nitrate-grown cells has been purified to homogeneity and shown to contain three subunits, a molybdenum cofactor, selenium, and a cytochrome (b^{fdh}) (19, 56, 70). The two largest subunits span the membrane (70), and studies of the H^+/e^- ratios suggest that the enzyme can pump protons across the membrane concomitant with electron flow through the system (see Fig. 1) (104). The active site for formate oxidation appears to be cytoplasmic, though definitive proof is lacking (104). The equivalent enzyme in *Vibrio succinogenes* is reportedly periplasmic (131).

Glycerol-3-phosphate dehydrogenase. There are two enzymes which oxidize *sn*-glycerol 3-phosphate to dihydroxyacetone phosphate. One is present in aerobically grown cells (266), and the other is induced when cells are grown anaerobically in glycerol plus fumarate (114). The anaerobic enzyme is required for electron transfer to fumarate. Both dehydrogenases have been purified (229, 266) and shown to function on the inner surface of the cytoplasmic membrane (258, 265). The anaerobic enzyme is isolated as a water-soluble protein, and it has been postulated to complex with another protein on the cytoplasmic membrane (133), though this has not been demonstrated. The gene (called *glpA*) encoding the anaerobic enzyme has been cloned (230). The electron transfer pathway and the extent of oxidative phosphorylation and proton translocation which accompanies glycerol-fumarate oxidoreductase activity are not clear (see reference 99). It can be speculated (see Fig. 1) that the anaerobic enzyme may be part of a transmembrane complex which functions as a proton pump, which would be necessary to drive oxidative phosphorylation using fumarate as an oxidant.

Nitrate reductase

Nitrate reductase is one of the better-characterized enzymes in the *E. coli* cytoplasmic membrane. It has been purified in numerous laboratories and contains three subunits (e.g., reference 56). The enzyme has been isolated in numerous forms but appears to have a basic subunit stoichiometry of $\alpha_2\beta_2\gamma_4$ (32). The three subunits of this enzyme are coded by a single *nar* operon, which has been cloned (163, 214). The smallest subunit (γ) is exposed on the periplasmic side of the membrane, and the other two subunits (α and β) are exposed on the inner surface of the membrane (see reference 19). The γ subunit is a *b* cytochrome, b^{nr} (32).

The enzyme also contains a molybdenum cofactor, identical to that in formate dehydrogenase, and iron-sulfur centers. It is likely that the enzyme functions by oxidizing quinone on the periplasmic side of the membrane via the *b* cytochrome and reducing nitrate to nitrite on the inside of the cell (177). This separation of the active sites across the bilayer is similar to the models proposed for both of the *E. coli* terminal oxidases (121, 158). The result is that the oxidation of quinol by the enzyme results in generation of a proton motive force. It has been reported that in heme-deficient strains, the $\alpha\beta$ subunits do not bind to the membrane, suggesting a role for the *b* cytochrome as an anchor in the membrane (151).

Fumarate reductase

Fumarate reductase is repressed in the presence of oxygen or nitrate and carries out the reduction of fumarate to form succinate. Not surprisingly, the enzyme is structurally very similar to succinate dehydrogenase. As is the case with various succinate dehydrogenases, fumarate reductase has been isolated in a two-subunit form capable of reducing fumarate by using artificial electron donors (48). However, characterization of the *frd* operon (see reference 37) and subsequent biochemical studies (138) have revealed the presence of two smaller hydrophobic subunits, similar to those putatively associated with *E. coli* succinate dehydrogenase. The sequence of the *frd* operon has been determined (35, 75). There are no sequence homologies between the small hydrophobic peptides of succinate dehydrogenase and fumarate reductase. Regions of the catalytic subunits do appear similar, and it is likely that these two enzymes have resulted from divergent evolution, presumably following a gene duplication event. The K_m values for fumarate and succinate are quite different for the two enzymes, consistent with their functions.

Overproduction of fumarate reductase by cloning the operon on a multicopy plasmid results in excessive synthesis of cytoplasmic membrane, which builds up inside the cells (267), similar to the observations made with the ATP synthase (see below). In addition, the two catalytic subunits can be observed in a water-soluble cytoplasmic form, perhaps due to an absence of a sufficient amount of the anchor proteins, the two hydrophobic subunits (36).

The catalytic subunits are localized on the inner surface of the cytoplasmic membrane (e.g., reference 258). The sequences of the hydrophobic subunits suggest they may span the membrane, but this has not been demonstrated. There is no indication that any of the subunits of *E. coli* fumarate reductase is a *b* cytochrome. It is likely that all the chemistry catalyzed by fumarate reductase occurs on the inner surface of the membrane, in contrast to the two oxidases and nitrate reductase, where quinol is probably oxidized at the periplasmic face of the membrane. If this is correct, then, unless fumarate reductase functions as a proton pump (e.g., the hydrophobic subunits serve a role analogous to the F_0 portion of the ATP synthase), the enzyme does not function as a coupling site per se. Additional studies are required to clarify the number of coupling sites in fumarate-grown cells.

Conclusions

Unlike the mitochondrial system, the electron transport proteins in *E. coli* are present neither in fixed stoichiometry nor necessarily at very high concentrations in the cytoplasmic membrane. In general, the electron transport proteins appear to be coupled via a pool of mobile quinone in the bilayer, rather than by direct interactions through a stable complex of associated proteins. Many dehydrogenases, some constitutive and many inducible, can feed electrons into the quinone pool. Generally, the dehydrogenase active sites are located on the inside of the cell, and in most cases these enzymes are not transmembranous. Some can be isolated as water-soluble, cytoplasmic species (e.g., proline oxidase [1, 169], pyruvate oxidase [191], and anaerobic glycerol-3-phosphate dehydrogenase [266]). Only formate dehydrogenase (69) and hydrogenase (70) have been shown to span the membrane. Formate dehydrogenase, NADH dehydrogenase, the anaerobic glycerol-3-phosphate dehydrogenase, and hydrogenase may catalyze proton translocation reactions and function as coupling sites (see Fig. 1 and reference 99). However, this has yet to be unambiguously demonstrated for the NADH and glycerol-3-phosphate dehydrogenases.

The quinone pool can apparently carry reducing equivalents to any of the enzymes interacting with the terminal oxidants. The two terminal oxidases and nitrate reductase appear to function by oxidizing quinone at the periplasmic surface and transferring electrons to a separate active site on the cytoplasmic side where the oxidant is reduced. This creates a proton motive force and drives oxidative phosphorylation via the ATP synthase. Fumarate reductase apparently operates differently, and the location of the coupling sites in the fumarate system is uncertain.

By regulating which dehydrogenases and which terminal reductases are present in the cytoplasmic membrane, oxidative phosphorylation can be supported by a very diverse group of substrates under many growth conditions.

Proteins Involved in Solute Transport

The cytoplasmic membrane forms a hydrophobic permeability barrier to most hydrophilic molecules. Aside from a few molecules that seem to diffuse freely across the cytoplasmic membrane (e.g., water [166] and medium-chain fatty acids [153]), entry of most molecules into the cell requires specific inner membrane proteins that mediate their uptake.

Certain proteins simply mediate diffusion of a substrate across the membrane. Although such facilitated diffusion is a common transport mechanism in eucaryotes, it is rare in procaryotes (96). In *E. coli* and *S. typhimurium*, glycerol uptake is the only known example of facilitated diffusion across the cytoplasmic membrane. The glycerol facilitator protein (the *glpF* gene product) forms a specific pore in the cytoplasmic membrane which allows diffusion of glycerol and related polyols across the membrane (84). Entry of glycerol into the cell is very rapid (about 6×10^5 molecules per s per pore) and temperature independent and does not require metabolic energy (84). The *glpF* pore allows both influx and efflux of glyerol. Hence, the internal and external concentrations of glycerol are rapidly equilibrated, but there is no net concentration of glycerol inside the cell. However, in wild-type cells the uptake of glycerol is rapidly coupled to its conversion to *sn*-glycerol 3-phosphate by glycerol kinase. The *sn*-glycerol 3-phosphate cannot pass through the *glpF* pore and thus is trapped inside the cell in a form that can be used as a carbon and energy source (144). The *glpF* pore is relatively specific for glycerol and related polyols, but the basis for this specificity is not yet known.

Most molecules enter the cell by active transport systems. There are three active transport mechanisms in *E. coli* and *S. typhimurium*: (i) active transport driven by the chemiosmotic gradient, (ii) binding protein-dependent transport, and (iii) group translocation. The three mechanisms differ with respect to the form of energy required, the number and location of proteins required, and the form of substrate transported (Fig. 2). Active transport systems driven by the chemiosmotic gradient require a single integral membrane transport protein. The substrate is transported by ion symport or antiport. The energy required for concentration of the substrate inside the cell is obtained either directly (e.g., proton symport) or indirectly (e.g., sodium symport) from the chemiosmotic gradient. About 40% of the substrates transported by *E. coli* enter the cell by ion-driven transport system (268). Binding protein-dependent transport systems require a high-affinity, periplasmic, substrate-binding protein which transfers the substrate to integral membrane proteins, which then transport the substrate across the membrane into the cytoplasm (4). The energy for binding protein-dependent transport seems to be obtained from substrate-level phosphorylation, although the precise mechanism remains unclear (91). Another 40% of the substrates transported by *E. coli* enter the cell by binding protein-dependent transport systems (268). Neither ion-driven transport nor binding protein-dependent transport systems modify the substrate during transport. Group translocation, on the other hand, couples transport of the substrate to its chemical modification (e.g., by attaching a phosphate or coenzyme A group to the substrate) (81). This traps the substrate in the cell in a different form from the exogenous substrate so that the concentration gradient of unmodified substrate never equilibrates. Thus, the energy for group translocation is obtained from the chemical modification of the substrate. Group translocation systems share an integral membrane protein, a membrane-associated enzymatic activity, and often several loosely associated membrane proteins (49). The properties of some of the best-characterized transport systems for *E. coli* and *S. typhimurium* are shown in Table 2.

Most substrates are transported by more than one system (e.g., galactose is transported by five different systems [268] and proline is transported by at least three different systems [27, 42, 168]. The separate transport systems for a given substrate usually have different affinities or are induced under different growth conditions, possibly providing a physiological advantage to bacteria required to grow under highly variable conditions. For example, a high-affinity (K_m = 0.3 μM) proline transport system (the *putP* gene product) is required for *E. coli* or *S. typhimurium* to use proline as a sole carbon or nitrogen source, but

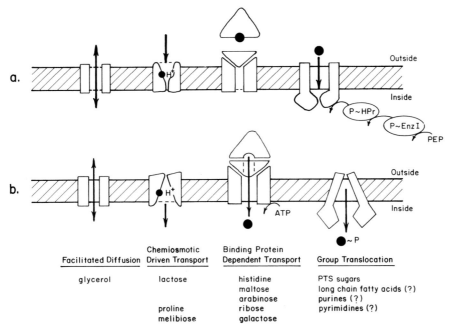

FIG. 2. Possible models for cytoplasmic membrane transport systems. Each of the systems probably has one or more membrane-spanning proteins that forms a specific channel in the cytoplasmic membrane. Facilitated diffusion and ion-driven transport systems require only one gene product, whereas binding protein-dependent transport and group translocation systems require several gene products. Facilitated diffusion allows entry and exit of substrate through a specific pore. Binding of substrate to specific sites accessible to the periplasmic side of the active transport systems coupled with an appropriate energy source (a) allows a conformational change in the carrier proteins and releases the substrate inside the cell (b). These cartoons are highly speculative; the precise mechanism of transport is not yet understood for any of these systems.

this system is only induced when high concentrations of exogenous proline are present (154); a constitutive, low-affinity (K_m = 300 µM) proline transport system (the *proP* gene product) can supply adequate proline for cellular protein synthesis (168); and a third proline transport system (the *proU* gene product) is induced by osmotic stress and allows accumulation of high intracellular proline pools required for cells to grow under these conditions (42).

Organization and Topology of Transport Proteins

Most early models for membrane transport systems envisioned mobile or rotating carriers that bind the substrate and then flip-flop across the membrane. However, this type of mechanism is thermodynamically unfeasible because it requires exposed hydrophilic domains of the transport protein to pass through the hydrophobic lipid bilayer (243). It is much more likely that transport proteins are arranged in the membrane to form a specific channel which the substrate transverses (243). Although the molecular structure and function have not yet been fully elucidated for any of the active transport systems, there have been recent advances in understanding each of the three transport mechanisms. The advances in a few key systems are discussed below.

Ion-driven active transport systems

Lactose transport by *E. coli* is the best understood ion-driven active transport system. Lactose and related galactosides are transported by the *lacY* gene

product, lactose permease (see review on the *lac* operon in this volume). Lactose permease is a highly hydrophobic, integral membrane protein (25, 106). Its apparent molecular weight determined by SDS-PAGE is 33,000, but the molecular weight calculated from the DNA sequence of the *lacY* gene is 46,500 (25). Such anomalous electrophoretic behavior on SDS-polyacrylamide gels has also been observed for several other integral membrane proteins and seems to be due to binding of excess SDS to the highly hydrophobic regions of these proteins. Circular dichroism analysis of lactose permease in synthetic membrane vesicles indicates that 70 to 85% of the polypeptide has an α-helix secondary structure (58, 260). Furthermore, the secondary structure predicted from the *lacY* DNA sequence suggests that 8 (197) or 12 (58) hydrophobic, α-helical segments of the protein span the membrane. Each of the α-helical domains has about 20 amino acids. These α-helices may form a channel in the membrane with hydrophobic regions that interact with surrounding membrane lipids and a hydrophilic interior (197).

Several types of experiments demonstrate that lactose permease spans the membrane: (i) lactose permease is susceptible to protease in both right-side-out and inside-out membrane vesicles (106); (ii) certain sulfhydryl reagents (e.g., *N*-ethylmaleimide and *p*-chloromercuribenzoate), which bind to lactose permease and block the active site, inactivate transport when added to either side of the membrane (106); (iii) active transport of lactose can occur in both directions, depending upon the direction of the chemiosmotic gradient (107, 135); and (iv) monoclonal antibodies

TABLE 2. Some defined transport systems in *E. coli* and *S. typhimurium*

Substrate	Energy source	No. of required proteins	Comments	References
Facilitated diffusion				
Glycerol	None required	1	*glpF* gene product	84
Ion-driven transport systems				
Lactose	Proton motive force	1	*lacY* gene product	106, 197
Proline	Sodium motive force	1	*putP* system	27
Melibiose	Sodium motive force	1	*melB* gene product	49, 255
Arabinose	Proton motive force	1	*araE* system	218
Glycerol phosphate	Proton motive force	1	*glpT* system	148
Galactose	Proton motive force	1	*galP* system	183
Phosphate	Proton motive force	1	*pit* system	216
Binding protein-dependent transport systems				
Histidine	ATP	4	*hisJ, P, M, Q*	3
Lysine, ornithine, arginine	ATP	4	*argT, hisP, M, Q*	3
Maltose	Complex (?)	4	*malE, F, G, K*	24, 86
Arabinose		2	*araF, G*	92
Ribose	ATP (?)	2–3	*rbsB, T*	98, 146
Leucine, isoleucine, valine		5	Two binding proteins, three shared membrane components	186
Peptides	ATP	4	*oppA, B, C, D*	7, 89
Succinate, malate, aspartate	Complex (?)	3	*dctA, B, cbt*	110
Phosphate	Complex (?)	3–4	*pts* system	248
Sulfate		?		240
Vitamin B$_{12}$	Complex (?)	5	Requires outer membrane receptor and *tonB* protein in addition to a binding protein and inner membrane transport components	93
Group translocation systems				
PTS sugars (mannitol, fructose, mannase, *N*-acetylglucosamine, sorbitol, galactitol)	PEP	3–4	Common enzyme I and HPr; unique enzymes II and III for each substrate	49
Long-chain fatty acids	Coenzyme A + ATP	2	*fadL, D*	15, 153
Short-chain fatty acids	Acetyl coenzyme A	?	*ato* system	110

have been isolated which bind to specific domains of lactose permease: some only bind to the cytoplasmic domain and some only bind to the periplasmic domain (29, 87). Whereas the "active site" of lactose permease is symmetrical, the protein itself is asymmetrical in the membrane: both the C-terminal and N-terminal ends of the protein are on the cytoplasmic side of the membrane (197). Thus, the symmetry of the substrate binding site can be most easily explained by a membrane channel model. The substrate binding site seems to lie deep within a hydrophilic channel (probably at a different position than the binding site for the sulfhydryl inhibitors), since the rate of interaction of bound substrate with specific fluorescent probes is slower than that predicted by diffusion to a site on the surface of the protein (175).

Recently, lactose permease has been purified and reconstituted into proteoliposomes in a fully active form (259, 271). The reconstituted lactose permease catalyzes all known functions of the permease in vivo with kinetics similar to those seen in membrane vesicles, although possibly at a somewhat lower rate than in whole cells (197). Many questions about lactose permease function have been clarified using the reconstituted proteoliposomes. Although several early

studies indicated that lactose permease must interact with other membrane proteins to obtain energy from the chemiosmotic gradient (94, 202), reconstitution of lactose permease activity from a single purified polypeptide indicates that no other protein is required for lactose transport (259). Furthermore, proteoliposomes reconstituted simultaneously with purified lactose permease and the cytochrome *o* oxidase complex directly couple lactose transport to the oxidation of electron donors (158). This argues strongly against a requirement for a second protein which couples proton and lactose symport in vivo.

Lactose permease transports galactosides by proton symport. Under most conditions, one proton is transported per lactose molecule (275). Binding of protons to the permease has not been demonstrated directly, but is inferred from the coupled transport of protons with galactosides (197). Each molecule of lactose permease has a single galactoside binding site (251). Kinetic models predict that lactose permease has an ordered reaction mechanism. The permease first binds a proton which activates it to bind a lactose molecule; the complex then undergoes a conformational change, releasing the galactose and then the proton inside the cell (106, 197). Release of the proton may be the rate-limiting step for active transport since active transport is considerably slower than lactose counterflow (106). Lactose counterflow is the simple exchange of an internal lactose molecule for an external lactose molecule; it does not require energy and does not result in the net concentration of lactose across the membrane. Counterflow seems to involve the rapid binding and release of lactose from the carrier without dissociation of the bound proton; however, release of the proton is required to concentrate lactose inside the cell (106). Thus, lactose permease directly transforms the chemiosmotic gradient ($[H^+]_{outside} > [H^+]_{inside}$) into a lactose gradient ($[lactose]_{inside} > [lactose]_{outside}$).

Although not yet fully exploited, genetic dissection of the lactose permease may provide direct evidence for the functional domains involved in transport (90). Those mutations that reveal the most about the structure and mechanism of lactose permease are unique mutations that alter a specific transport function without destroying the permease per se. For example, mutants have been isolated that change the sugar specificity of lactose permease (171). Most mutations that decrease the affinity of the permease for lactose map in the C-terminal end of the gene and are clustered into the more hydrophilic regions of the polypeptide (171). Sequence analysis of these mutations may indicate which amino acids actually form the lactose-binding site in the channel. Previously, a single mutant was isolated that could not actively transport lactose but still allowed facilitated diffusion of lactose (269), possibly due to an inability of lactose permease to release protons. Isolation and fine-structure mapping of more mutations of this type may define the proton-binding domain of lactose permease.

Some lacY mutations show negative dominant complementation ($lacY^{-d}$): that is, diploid strains carrying one $lacY^{-d}$ gene and one wild-type lacY gene are phenotypically Lac$^-$ (171). Most lacY mutations are recessive to the lacY$^+$ gene. The rare $lacY^{-d}$ mutations map at several positions in the lacY gene. The most common explanation for such intercistronic negative dominance is that the mutant protein interferes with aggregation of the normal gene product. Thus, $lacY^{-d}$ mutations were interpreted as an indication that the lacY gene product must dimerize in the membrane to function properly (171). Electron inactivation analysis also suggested that dimers of lactose permease may be required for high-affinity active transport but not for facilitated diffusion (67). However, rotational diffusion measurements suggest that lactose permease acts as a monomer under both conditions (50). Furthermore, the $lacY^{-d}$ mutations only cause small changes in the kinetics of lactose transport (J. K. Wright, personal communication), indicating that this phenotype may not be due to a defect in protein dimerization. None of the current models for lactose permease explain the requirement for dimerization.

It is evident that by combining in vitro biochemical analysis and in vivo genetic analysis, a good molecular model for the structure and function of lactose permease should be made available soon. No other ion-driven transport system is nearly as well characterized as that of lactose permease. The kinetics of transport have been determined in vivo for a number of other systems, and recently the genes for several

other ion-driven transport proteins have been cloned. Comparison of the properties of these other transport proteins with lactose permease may yield a general model for proton-driven transport systems. It will be particularly interesting to compare those transport proteins that are driven by sodium symport (e.g., the melibiose transport system [49] and the major proline permease [27]) with the proton symport systems discussed above. Sodium symport depends upon a sodium gradient generated by sodium-proton antiport (240). The reasons why some transport systems use sodium symport rather than proton symport are not understood.

Binding-protein-dependent transport system

The high-affinity histidine permease is a binding-protein-dependent transport system that has been thoroughly characterized by biochemical, genetic, and molecular techniques. Four genes are required for the high-affinity histidine transport system: the hisJ gene, which encodes the histidine-binding protein; the hisQ and hisM genes, which encode integral membrane proteins; and the hisP gene, which encodes another membrane-associated protein (4). The hisJ gene product is a soluble periplasmic protein. It has a N-terminal signal sequence for secretion across the inner membrane into the periplasm and can be released from the periplasm by cold osmotic shock (4). Genetic analysis of the hisJ gene has indicated that the histidine-binding protein has two discrete domains, a domain that binds histidine and a domain that interacts with the hisP protein (5). The membrane proteins are present in much lower amounts than the binding protein (3). The hisQ and hisM genes have very similar DNA sequences, suggesting that they arose by duplication of an ancestral gene (3). The hisQ and hisM proteins are very hydrophobic. Based upon the secondary structure predicted from the DNA sequence of these genes, both have three α-helical membrane spanning regions of about 20 amino acids each (88). The hisP protein is not overly hydrophobic (88). The genetic evidence that the hisP protein interacts with the periplasmic binding protein suggests that the hisP protein may be a peripheral membrane protein. Neither the hisQ, hisM, nor hisP gene has an N-terminal signal sequence found on exported proteins (88).

Binding-protein-dependent transport seems to require energy from substrate-level phosphorylation (possibly ATP [14] or acetyl-phosphate [95, 97]). Neither the substrate itself nor any of the transport proteins seem to be posttranslationally modified (e.g., phosphorylated) during the transport process (3). However, the hisM and hisP proteins have been recently shown to react with 8-azido-ATP, a photo-affinity probe resembling ATP (91), suggesting that these proteins bind ATP at least transiently. Inhibition of the chemiosmotic gradient decreases but does not eliminate binding-protein-dependent transport (86). It is possible that this is an indirect effect. However, the energy requirements may be more complex than previously thought. It is not apparent how energy is coupled to transport in any binding-protein-dependent transport system. Recently several binding-protein-dependent transport systems have been reconstituted in membrane vesicles in vitro (24). It is hoped that studies in reconstituted systems where the energy source can be readily manipulated will clarify the mode of energy coupling to binding-protein-dependent transport systems.

Physical analysis of the binding protein-substrate complex suggests that the substrate binds to a cleft in the binding protein and the cleft then closes around the substrate (i.e., the substrate is no longer able to interact with the solvent) (155). This is the form of the binding protein believed to interact with the membrane proteins. In most systems, mutants lacking the binding protein are completely defective for transport; hence, the binding protein does not simply increase the local concentration of substrate, but plays a direct role in transport of the substrate (86). It is possible to obtain mutants of the maltose transport system in which one of the integral membrane proteins is altered, yielding a low-affinity, binding-protein-independent transport system (238). Thus, the membrane proteins probably directly mediate the transport of the substrate into the cell.

It is not known how the substrate is ultimately transported. Several models have been proposed, mainly focusing on conformational changes of the membrane proteins upon interaction with the binding protein-substrate complex (see reference 91). It is not known whether the membrane proteins form a channel in the membrane or whether the substrate is passed through the membrane in a "bucket-brigade" manner. However, since transport by binding-protein-dependent systems is unidirectional, a simple channel model may not be adequate.

Many other binding-protein-dependent transport systems have also been thoroughly characterized (see Table 2). All of the different systems have many common features: the binding protein generally has a high affinity for its specific substrate (K_d = 0.1 to 1 µM), the substrate is not modified during transport, transport is unidirectional, energy seems to be obtained from substrate-level phosphorylation, and the numbers and locations of proteins required are similar. However, a number of questions remain. What is the direct energy source and how is energy utilization coupled to transport? How do the different proteins interact with each other and what is the precise function of each of the different proteins? Why is transport strictly unidirectional?

Group translocation

The best characterized group translocation systems are the phosphoenol pyruvate (PEP)-dependent sugar phosphotransferase systems (PTS). There are at least 10 sugar-specific transport systems in E. coli (221). The mannitol-specific transport system is the best-characterized PTS in E. coli and S. typhimurium. Energy for this transport system is derived from PEP (49). Two proteins, enzyme I (ptsI) and HPr (ptsH), initiate a sequential phosphoryl group transfer from PEP to the membrane-bound transport protein (49). Enzyme I and HPr are proteins required for uptake of all PTS sugars. These proteins seem to be peripheral membrane proteins, forming a functional association with membrane-bound PTS components (49). The high-energy phosphoryl group from PEP is transferred sequentially to enzyme I, then to HPr. The phospho-

HPr binds to the cytoplasmic domain of the membrane-bound, mannitol-specific enzyme II (enzyme II^{mtl}, the mtlA gene product). Thus, when mannitol binds to the periplasmic side of enzyme II^{mtl}, it is simultaneously transported and phosphorylated, with the release of mannitol 1-phosphate in the cytoplasm.

Enzyme II^{mtl} is an integral membrane protein which completely spans the membrane (222). Vectorial, PEP-dependent sugar transport requires enzyme I and HPr in addition to enzyme II^{mtl}, but by itself enzyme II^{mtl} can catalyze exchange group phosphorylation, as shown (223): ^{14}C-sugar$_{outside}$ + sugar$_{inside}$ = ^{14}C-sugar$_{inside}$ + sugar$_{outside}$. Thus, enzyme II is probably the actual sugar permease.

Enzyme II^{mtl} has been solubilized in nonionic detergents and purified to homogeneity (101). It has a single polypeptide chain with a molecular weight of 60,000. The purified enzyme II^{mtl} catalyzes both PEP-dependent mannitol phosphorylation (the vectorial transport reaction) and PEP-independent transphosphorylation (the phosphoryl exchange reaction) in vitro (101, 141). The PEP-dependent phosphorylation reaction is highly dependent upon the protein concentration, suggesting that it is catalyzed by dimers (222).

The mtlA gene has recently been cloned and sequenced; the secondary structure of the protein has been predicted from the DNA sequence (136). The N-terminal half of enzyme II^{mtl} has seven highly hydrophobic regions, each about 20 amino acids long, that may span the membrane, forming a channel. However, the C-terminal half of the protein is relatively hydrophilic and may form a bulky cytoplasmic domain involved in binding phospho-HPr.

Mutants have been isolated that specifically alter only one function of enzyme II^{mtl} (140). Some mutants prevent the PEP-dependent phosphorylation reaction but retain the transphosphorylation reaction (140, 222). These mutants may have normal sugar-binding sites but are unable to bind phospho-HPr. Other mutations prevent both reactions and may affect the sugar-binding sites. Fine-structure mapping and sequence analysis of such mutations may demonstrate which amino acids are involved in binding phospho-HPr, sugar binding, and the protein-protein interactions required for transphosphorylation.

The other sugar transport PTS differ from the mannitol-specific PTS in some details. For example, many sugar transport PTS also require a peripheral membrane protein, enzyme III, which accepts the phosphoryl group from HPr and interacts with enzyme II. However, the actual transport process seems very similar for each of the PTS sugars (49).

Several other substrates (e.g., long-chain fatty acids [153], uracil [81], purines [81], and nucleosides [81]) are believed to be transported by a vectorial group translocation mechanism. However, none of these systems has been extensively characterized yet. A general feature of each of the proposed group translocation systems is that the modified substrate is the form readily metabolized by the cell, and often translocation is the rate-limiting step in metabolism of the substrate.

Chemotaxis

Bacterial chemotaxis is another membrane function closely tied to transport. By specifically sensing the concentration gradient of attractants or repellents, *E. coli* and *S. typhimurium* can modify the flagellar response and thus move toward or away from a substance. Both the proteins responsible for sensing the chemical gradient (sensory transducers) and the proteins that drive the flagella (the flagellar motor) are integral membrane proteins (201). In some cases the binding proteins associated with binding protein-dependent transport systems are the chemoreceptors (e.g., galactose, maltose, and ribose) (201). In addition to specifically interacting with the integral membrane transport proteins, these binding proteins also interact with chemotaxis-specific sensory transducer proteins (the *tar*, *tsr*, and *trg* gene products). The interaction of the binding proteins with the sensory transducer proteins probably involves a specific domain on the binding protein: unique mutations in the maltose-binding protein can prevent chemotaxis to maltose without affecting maltose transport, and other mutations prevent transport without affecting chemotaxis (82). Interaction of the binding protein-substrate complex with the sensory transducer seems to cause a conformational change in the sensory transducer, making it susceptible to methylation or demethylation by cytoplasmic enzymes. The methylation state of the sensory transducer, in turn, elicits a signal that directs the membrane-bound proteins that drive the flagella motor (201).

None of the other periplasmic binding proteins seem to be used as chemoreceptors (201). Exogenous serine and aspartate bind directly to the chemotaxis sensory transducer proteins. Chemotaxis toward these substrates does not require transport or metabolism of the substrate.

In contrast to the above examples, chemotaxis to certain substances does seem to require that the substrate enter the cell (250). For example, chemotaxis to PTS sugars requires the PEP-dependent phosphate-transferring enzymes in addition to the specific enzyme II, indicating that sugar binding to the external surface of enzyme II does not elicit chemotaxis (222). In addition, aerotaxis and phototaxis seem to be directly mediated by the chemiosmotic gradient (250).

It is not clear what the actual cellular signal is that elicits changes in the flagellar motor, but mutations in over 30 genes can cause motility defects (201). The structure of the flagellar motor has been recently reviewed (152), and chemotaxis is more thoroughly described elsewhere in this volume.

Regulation

The transport proteins present in the membrane at any given time are regulated by a number of factors. Most transport proteins are specifically induced in cells grown in the presence of their substrate, and many are subject to catabolite repression. Thus, the synthesis of most transport proteins is regulated in such a way that the proteins are made only when the substrate is available and needed by the cell. Since the substrate or a metabolic derivative of it is usually the inducer, a certain basal level of the transport protein must be present in the membrane even under non-induced conditions. Induction can be elicited either internally or externally: the substrate can be trans-ported and interact with a cytoplasmic repressor or activator protein (e.g., lactose transport or arabinose transport, respectively [see chapters on *lac* and *ara* regulation in this volume]), or binding of the substrate to the external surface of a membrane protein can cause induction (e.g., the *uhp* transport system [49, 56a]). Thus, the protein composition of the membrane varies dramatically with the growth conditions.

The activity of transport proteins is also controlled by a number of factors. The activity of many transport proteins is affected by the chemiosmotic gradient (49) or the phosphorylation state of the glucose-specific PTS enzyme II (222a). These conditions may cause conformational changes in the membrane transport proteins. In addition, the membrane fluidity may affect the function of some membrane proteins (39, 159). All three types of transport systems have been reported to be sensitive to the membrane fatty acid and phospholipid composition: in most cases, altering the membrane composition decreased transport three- to fourfold (139). Altered membrane composition seems to decrease the activity of the membrane-bound permeases rather than simply preventing insertion of newly formed protein into the membrane. However, in many cases the decrease in permease activity could be due to a secondary consequence of altered membrane structure on the energy coupling to transport (see reference 41 and below). In addition, in many cases the effects of membrane composition on permease function were studied before it was realized that multiple permeases can transport a single substrate (e.g., "proline permease"). Thus, it would be useful to reexamine this problem in proteoliposomes containing single permeases and with manipulatable energy supplies. The lactose transporter discussed above seems an ideal candidate for such an investigation due to the high quality of the proteoliposome preparations and the very abundant data on the effects of membrane fatty acid composition on lactose transport in vivo.

INTERACTIONS OF PROTEINS AND LIPIDS WITHIN THE CYTOPLASMIC MEMBRANE

A number of years ago, ESR labeling results suggested that a substantial portion of membrane lipids were immobilized through interaction with membrane proteins. These lipids (called boundary lipids) were thought to exchange only slowly with the lipids of the membrane bilayer. More recent results using deuterium NMR indicate the absence of boundary lipid in *E. coli* cytoplasmic membranes. Indeed, Nichol et al. (187) have reported that <3% of the lipids of the *E. coli* cytoplasmic membrane could be bound to membrane proteins. Rationalization of the conflicting results of the ESR and NMR experiments is generally believed to lie in the differing time scales of the techniques (see references 40 and 232). In the ESR experiments, lipids undergoing lipid-protein interactions lasting $<10^7$ s would be considered as immobilized lipids, whereas the deuterium NMR time scale is 10^3- to 10^5-fold greater. The deuterium NMR time scale may be closer to physiological time scales in that it is similar to time required for conformational changes in enzymes (79). In this regard, Keniry et al. (112) have reported that the motions of the amino acid side chains of *E. coli* cytoplasmic membrane proteins are similar to those seen in protein crystals and are unaffected by the physical state of the lipid bilayer.

It should be noted that the interrelationships among the lipid phase transition, the conformation of phospholipid molecules, and the viscosity of the lipid bilayer are complex, and thus intuitive deductions are often incorrect (232). For example, the introduction of a protein into a lipid bilayer will alter the conformation of those phospholipid molecules which contact the protein. Those lipid molecules will be less often able to assume the preferred conformation, which results in a decrease in the order parameter of most segments of the molecule. With regard to the acyl chains, this decreased probability of achieving the preferred conformation is equivalent to an increase in temperature. However, the presence of the protein results in an increased rather than a decreased microviscosity of the lipid phase, since the lateral and rotational movement of lipid molecules is impeded by interaction with the irregular (as compared to lipids) surface of the protein.

DYNAMICS OF MEMBRANE COMPONENTS

Phospholipid Mobility

As discussed earlier for the outer membrane (40), ESR labeling studies of both *E. coli* (219) and *S. typhimurium* (189) cytoplasmic membranes give a lateral diffusion rate of 2×10^{-8} to 3×10^{-8} cm^2/s, a value similar to that found in eucaryotic cell plasma membranes (33) (both at temperatures above the phase transition). However, recent results of Borle and Seelig (18) suggest that these results are overestimates due not only to the usual problems of probe perturbations but also to the fact that the ESR probes would fail to sense the viscous environment of the phospholipid head groups (see below). Unfortunately, the small size of bacteria precludes use of the technique of fluorescence recovery from photobleaching so successfully used in larger cells, and thus no firm values can be given for the rate of lateral motion of the cytoplasmic membrane lipids. However, it does seem clear that the lipids undergo rapid lateral diffusion. If diffusion were slow, the physical properties of the lipids in the intact membrane would not agree so closely with those of the model systems (where diffusion is known to be rapid). This is particularly true of the NMR results, where rapid motion and equilibration of lipid phases are required for informative data to be obtained (232). Also, recent X-ray diffraction results using a very high-intensity synchrotron X-ray source demonstrate that the order-disorder transition of cytoplasmic membrane lipids is complete within 0.3 s (209). Since lateral lipid diffusion may be a limiting factor in the rate of the transition (208, 209), these data argue for rapid lateral diffusion of lipids. Rapid diffusion was also required to explain the results of biological experiments concerned with mixing of lipid phases (253). It should be noted that no direct values for the rate of rotation (in the plane of the bilayer) of phospholipid molecules in the cytoplasmic membranes are available, but again the NMR results, particularly ^{31}P NMR, indicate that rapid rotational motion must occur.

The question of transbilayer movement ("flip-flop") of phospholipid molecules has not been addressed in these organisms. It seems essential that newly synthesized phospholipids must somehow be able to flip from the cytoplasmic face of the inner membrane (the presumed site of synthesis) to the periplasmic face to allow membrane growth (see reference 40). However, no measurements have been reported, and it remains unclear whether or not flip-flop could occur rapidly and spontaneously in a functioning biological membrane in contrast to the very slow flip-flop seen in synthetic lipid bilayers (see references 40 and 162).

Protein Mobility

No direct measurements of the lateral or rotational mobility of protein molecules in the cytoplasmic membranes of *E. coli* or *S. typhimurium* have been reported, due to the small size of these bacteria. The only numerical value for diffusion of bacterial proteins (3×10^{-10} cm^2/s for a 10^5-dalton protein) in the literature is dependent on certain models based on a large extrapolation from ESR labeling results on the lipids (219). This value is considerably slower than the diffusion rate measured for proteins in eucaryotic plasma membranes (1×10^{-9} to 5×10^{-9} cm^2/s). Kell (111) has argued that diffusion in bacterial membranes must be 10 to 100 times slower than that seen in eucaryotic membranes to account for the efficiency of certain energy coupling processes, but no plausible mechanism to slow diffusion in bacterial membranes has been proposed. However, deuterium NMR work of Borle and Seelig (18) indicated that the viscosity of the membrane-water interface of *E. coli* cytoplasmic membranes (measured by the relaxation time of the nonacylated glycerol moiety of phosphatidylglycerol) is much greater than that generally assumed. Borle and Seelig found that the viscosity is 2.9 P rather than the 0.01 P used in most calculations (33, 220, 264). This 300-fold difference in viscosity results in at least a 20-fold difference in diffusion rate, using the generally accepted equations to calculate lateral diffusion rates. The interface viscosity is significantly greater in *E. coli* cytoplasmic membranes than in synthetic membranes made from the purified membrane lipids, suggesting that proteins are involved in the hydrogen-bonding network thought to cause interface viscosity (18). The findings of Borle and Seelig indicate that diffusion of these membrane proteins which protrude from the bilayer may be impeded at least as much by the viscosity of the interface as by the viscosity of the hydrophobic portion of the membrane (generally thought to be a few poise) (18, 242). It should be noted that the phosphate groups of the *E. coli* cytoplasmic membrane phosphatidylglycerol molecules are not accessible to divalent cations added to the suspending medium (in contrast to the behavior of synthetic membrane bilayers) (63), which suggests that these groups are somehow shielded from the solution.

A corollary of the Borle and Seelig findings (18) is that membrane proteins of similar size and conformation may have different lateral mobilities depending on the degree of interaction with the hydrogen bond network. Indeed, various rates of lateral diffusion have been reported for proteins of eucaryotic plasma membranes, although some of the differences seem due to interaction with cytoskeleton components (33).

BIOGENESIS OF THE CYTOPLASMIC MEMBRANE

The major structure of the cytoplasmic membrane is the lipid bilayer, and thus, to a first approximation, the synthesis of lipids is synonymous with membrane biogenesis. However, the insertion of various membrane proteins into the bilayer is required for a functional membrane, and thus the interrelationships between phospholipid synthesis and the synthesis and insertion of membrane proteins have been the subject of many investigations. Early work argued that insertion of cytoplasmic membrane proteins (e.g., the lactose permease) required the concomitant synthesis of certain phospholipids. However, subsequent experiments showed the early results to be due to invalid transport assays and nonspecific cellular damage (for reviews, see references 39, 41, and 196). The work of Bell and co-workers (160, 161) showed that the lack of coupling observed with the lactose permease was also true for bulk protein synthesis. Using mutants specifically blocked in the first step of phospholipid synthesis, these workers found that both the inner and outer membranes became enriched with protein (160). The protein-to-phospholipid ratio of both membranes increased approximately 60%. This work demonstrated that the membrane is not normally saturated with protein and that the synthesis of membrane phospholipid is not required for the synthesis and insertion of most membrane proteins. The protein-enriched membranes are not lethal to the cell, and the enrichment with protein was quickly normalized upon the restoration of phospholipid synthesis (161). Likewise, phospholipid synthesis does not depend on bulk membrane protein synthesis (for a review, see reference 213). Although phospholipid synthesis in *E. coli* is inhibited when protein synthesis is blocked by starvation for a required amino acid due to the accumulation of the nucleotide ppGpp, this mechanism cannot specifically couple the synthesis of membrane proteins and phospholipids, since (i) ppGpp is not accumulated in normally growing cells and (ii) the synthesis of all protein (not just membrane proteins) is inhibited. Moreover, mutants (*relA*) are known which fail to accumulate ppGpp and synthesize lipids normally (see chapter 30).

From these experiments, the question of coupling between the synthesis of membrane proteins and phospholipids seemed a dead issue. However, two recent reports indicate that synthesis of discrete cytoplasmic membrane proteins can markedly increase the amount of phospholipid synthesized by *E. coli*. In the first paper, Weiner et al. (267) reported that *E. coli* cells harboring a recombinant plasmid carrying the fumarate reductase (*frd*) genes overproduced the reductase 30-fold. Upon overproduction, the enzyme comprised over 50% of the total inner membrane protein. Concurrently, the amount of phospholipid increased so that the cellular lipid/protein ratio remained constant. The excess membrane was localized in novel tubular structures which seemed to branch from the cytoplasmic membrane into the cell cyto-

plasm. These tubules were oriented in the long axis of the cell and appeared to be composed of an aggregate of fumarate reductase and phospholipid. Despite this overproduction of membrane, the composition of the various lipid components in these cells was quite normal, as were the levels of membrane-bound enzymes other than the reductase. The synthesis of extra membranes did not begin immediately upon shift to anaerobic conditions (thus inducing reductase synthesis), but only after a lag which was interpreted as the time needed to saturate the pre-existing membrane (267). In the second paper, the overproduced protein was the F_1/F_0 proton-translocating ATPase of the inner membrane (261). Cloning the segment of DNA which carries the eight *atp* genes encoding the subunits of this enzyme into plasmid pBR322 resulted in a 10- to 12-fold overproduction of enzyme activity in cells harboring the recombinant plasmid. These cells grew significantly more poorly than cells carrying only pBR322. Morphologically, the cells were strikingly similar to those overproducing fumarate reductase, although the tubular membranes were less abundant. Cells that overproduced the enzyme to a lesser extent (fivefold) were found to contain significantly fewer tubular structures. Taken together, the Weiner et al. (267) and von Meyenberg et al. (261) reports suggest a new level of regulation of membrane phospholipid synthesis.

Would massive overproduction of any *E. coli* inner membrane protein trigger the increased phospholipid synthesis necessary for the formation of the intracytoplasmic membrane, or do fumarate reductase and ATP synthase have special attributes? This is an open question because most workers do not assay for membrane formation. The only other cells which overproduce inner membrane proteins in which phospholipid synthesis has been monitored are those *E. coli* strains which carry cloned phospholipid biosynthetic genes (see reference 207). However, the increase in total cytoplasmic membrane protein produced by these clones was small (<1% of the total inner membrane protein) since these are minor membrane proteins. Thus, the overproduced protein could have been accommodated without extra lipid synthesis. It seems possible that a greater overproduction would trigger increased membrane production. Indeed, R. M. Bell and co-workers (personal communication) have found that cells having a massive overproduction of *sn*-glycerol 3-phosphate acyltransferase tend to accumulate intracytoplasmic membrane vesicles. On the other hand, Rule et al. (217) have reported that strains which produce D-lactate dehydrogenase to levels of 35% of the total cellular protein do not contain intracellular membranes.

In regard to other well-studied membrane proteins, the *lacY* gene product (the proton-lactose transporter) has been produced to a level of 10 to 15% of the membrane protein (251), all of which is membrane bound and active. No morphological or lipid compositional work has been reported on this strain, although permease overproduction is toxic to cell growth (200). The *Halobacterium* bacteriorhodopsin gene has been cloned and expressed in *E. coli* (see reference 53), but the level of expression is poor (0.02 to 0.05% of the total membrane protein). No characterization of the cells (other than an implication that uncontrolled overproduction was detrimental [53]) has been reported. However, since this is the only protein for which we have detailed information on how it transverses the lipid bilayer, further efforts seem justified.

In conclusion, it seems likely that the extra membrane synthesis observed upon overproduction of fumarate reductase, the F_0/F_1 ATPase, and perhaps sn-glycerol 3-phosphate acyltransferase is due to the sheer mass of membrane protein produced. The first two enzymes are major cytoplasmic membrane proteins in normal cells, and upon overproduction, the protein capacity of the membrane is exceeded, somehow triggering the formation of additional membrane. However, it should be noted that there are discrepancies in our knowledge of this phenomenon. First, a paper by Cole and Guest (36) seemed to conflict with the report of Weiner et al. (267). Cole and Guest overproduced fumarate reductase 30-fold through isolation of chromosomal duplications of the frd gene. In agreement with Weiner and co-workers (267), Cole and Guest (36) found that the overproduced reductase exceeded the capacity of the membrane, but reported the excess enzyme to be present in a soluble rather than in a membrane-bound state. However, the result might be due to fragmentation of the intracellular tubules (267) since the "soluble" enzyme sedimented significantly lower in the gradient than lactamase. Thus, although direct experiments are needed, it seems probable that the work of these two groups does not conflict.

A shortcoming of the work on both the reductase and the ATPase is that neither laboratory has excluded the possibility that the increased membrane may not be the result of overproduction, but rather a result of the selection of mutant strains from the transformed population that possessed extra membrane and thus were able to tolerate the enzyme overproduction resulting from introduction of the plasmid. However, plasmid clones carrying either gene cluster have been independently isolated in several laboratories, and thus the transformation frequency seems unlikely to have been extraordinarily low. On the other hand, R. Simoni (personal communication) and co-workers have not observed intracellular membranes in their strains which harbor the cloned atp operon, thus suggesting that bacterial strain differences may exist.

Models of Membrane Biogenesis

We can conceive of two distinct models for formation of the new membranes discussed above. The first model proposes that biogenesis of the new membranes begins in the cell cytosol as the result of physical association of the overproduced membrane proteins with monomeric lipid molecules present in the cytosol. The notion rests mainly on the experimental evidence from model synthesis that phospholipids, either as single molecules or small (when compared with the average vesicle size) micelles, can move spontaneously between lipid bilayers (51, 52, 156, 188, 215, 233, 252). The rate of monomeric phospholipid movement is not rapid (half-life of 0.5 to 1 day) but is of the same order of magnitude as the amount of lipid needed for the synthesis of intracytoplasmic

membranes upon induction of fumarate reductase (half-life of about 8 h, assuming the lag is due to nucleation).

The net increase in lipid could be due to binding of the lipids by the proteins (as demonstrated for the ATPase), but since phospholipid-phospholipid contacts must be involved for membrane formation, a second factor must be involved. This factor could be the small radius of curvature of the new membranes (diameter of 16 nm) compared with that of the inner membrane (ca. 300 nm at the cell poles). Due to the larger surface energy of small vesicles (215), the rate of loss of phospholipid from these vesicles would be much slower than that from large vesicles (51, 188, 215, 252). Thus, this model proposes that growth of the extra membranes begins in the cytosol, and growth of the membranes proceeds toward the inner membrane.

The second model proposes that some geometrical attribute(s) of fumarate reductase and the ATPase allows branching of the cytoplasmic membrane when the protein molecules become crowded. This freedom to branch would stimulate phospholipid synthesis and result in the formation of the tubules as extensions of the cytoplasmic membrane. Both models assume that the rate of phospholipid synthesis is somehow regulated by the density of packing of molecules in the membrane, high packing density inhibiting phospholipid biosynthetic enzymes. This is a reasonable hypothesis which is discussed elsewhere in this volume (chapter 30).

It should be noted that the proposal of the first model, the formation de novo of a membrane in the cell cytosol, is a situation similar to that inferred for the phages PR4 and PRD1, two lipid-containing phages which grow on E. coli and S. typhimurium. The hosts of these closely related phages (12) are various gram-negative bacteria, including E. coli and S. typhimurium, which harbor R-factors required to provide a phage adsorption site. The phages are icosahedral particles with a diameter of 65 nm (20) in which the external icosahedral protein coat encloses a lipid bilayer membrane (46, 150) composed of phospholipid and protein. This membrane in turn surrounds the genome, a double-stranded linear DNA of 14.5 kilobase pairs (47, 150). Although the phospholipids of the phage are derived from those present in the host membrane (178), electron microscopic observations give no evidence of assembly of the phage membrane at the host membrane (150). Indeed, the nascent phage particles (which contain the lipid bilayer) are first seen in the nuclear region of the cell. It should also be noted that the formation of the phage lipid bilayer seems tightly coupled to formation of the icosahedral shell since neither in wild type (150) nor in cells infected with phage nonsense mutants (45, 173) are membranes seen that lack the external protein shell. This finding suggests that assembly of the shell is required for appropriation of host lipids and thus has similarities with the overproduced cytoplasmic membrane proteins discussed above. Indeed, evidence for specific interactions of several proteins with the phage phospholipids has been reported (46). It seems possible that assembly of the phage bilayer involves diffusion transport of phospholipids from the host membrane to the cytosolically located assembly site, as proposed above for the assembly of intracellular membranes in *E. coli*. Indeed, membranous structures reminiscent of the *E. coli* intracellular membranes are seen in cells infected with certain nonsense mutants of phage PRD1 (173).

OUTLOOK

The combined approaches of biochemical and genetic analysis have greatly increased our knowledge of the function of the cytoplasmic membrane. However, our knowledge of the structure of the membrane has lagged behind. We have detailed knowledge only of the least intricate structure of the membrane, the phospholipid bilayer. Although the amino acid sequences of a number of cytoplasmic membrane proteins are known from gene sequencing, we have only the most rudimentary ideas of how these proteins associate with the membrane, whether or not they interact with other membrane proteins, and their dynamics within the membrane. No cytoplasmic membrane protein has yet been crystallized, and thus our only knowledge of the structure of these proteins is indirect and of extremely low resolution.

The prospects of answering most of these questions seem good. The isolation of mutant enzymes specifically deficient in binding to the membrane, such as the pyruvate oxidase mutants isolated by Chang and Cronan (31), should give information on the mechanisms of association of various proteins with the membrane. Likewise, the isolation of second-site revertants, mutant proteins which suppress the defects of other mutant proteins, has given information on protein-protein interactions (201) and should provide further information, particularly in combination with in vitro mutagenesis. The purification of functional complexes from the membrane is another useful approach, as is reconstitution from purified components. Studies of protein motion and dynamics within the membrane seem the least approachable problem at present. Not only do we lack techniques of the needed resolution, but the presence of the outer membrane limits such investigations to isolated or reconstituted membranes. The study of the structure of specific proteins and complexes seems very likely to be experimentally accessible. Several peripheral membrane proteins are water-soluble molecules, and thus crystallization by standard techniques seems likely. Moreover, several integral membrane proteins have recently been crystallized from detergent solutions (55). In this regard, the intracellular membranes observed upon overproduction of ATP synthase and fumarate reductase may be sufficiently well ordered to allow the use of image reconstruction techniques such as those used for bacteriorhodopsin (85, 256). Antibodies raised against synthetic peptides synthesized from information gained by DNA sequencing may become a powerful method to locate structural domains of various proteins now that the relationship between immunogenicity and the type of protein domain is becoming clear (249). In conclusion, it seems that the cytoplasmic membrane of *E. coli* and *S. typhimurium* will continue to be the best understood plasma membrane in terms of structure, function, and role in overall cell physiology.

LITERATURE CITED

1. **Abrahamson, J. L. A., L. G. Baker, J. T. Stephenson, and J. M. Wood.** 1983. Proline dehydrogenase from *Escherichia coli* K-12: properties of the membrane-associated enzyme. Eur. J. Biochem. **134:**77–82.
2. **Adams, M. W. W., and D. O. Hall.** 1978. Purification of the membrane-bound hydrogenases of *Escherichia coli*. Biochem J. **183:**11–22.
3. **Ames, G.** 1984. Histidine transport system of *Salmonella typhimurium*, p. 13–16. *In* L. Leive and D. Schlessinger (ed.), Microbiology—1984. American Society for Microbiology, Washington, D.C.
4. **Ames, G., and C. Higgins.** 1983. The organization, mechanism of action, and evolution of periplasmic transport systems. Trends Biochem. Sci. **8:**97–100.
5. **Ames, G., and E. Spudich.** 1976. Protein-protein interaction in transport: periplasmic histidine-binding protein J interacts with P protein. Proc. Natl. Acad. Sci. USA **73:**1877–1881.
6. **Ames, G. F., and K. Nikaido.** 1976. Two-dimensional gel electrophoresis of membrane proteins. Biochemistry **15:**616–623.
7. **Andrews, J. C., and S. A. Short.** 1985. Genetic analysis of *Escherichia coli* oligopeptide transport mutants. J. Bacteriol. **161:**484–492.
8. **Andrews, S., G. B. Cox, and F. Gibson.** 1977. The anaerobic oxidation of dihydro-orotate by *Escherichia coli* K-12. Biochim. Biophys. Acta **462:**153–160.
9. **Au, D. C-.T., G. N. Green, and R. B. Gennis.** 1984. The role of quinones in the branch of the *Escherichia coli* respiratory chain which terminates in cytochrome *o*. J. Bacteriol. **157:**122–125.
10. **Au, D. C-.T., R. M. Lorence, and R. B. Gennis.** 1985. Isolation and characterization of an *Escherichia coli* mutant lacking the cytochrome *o* terminal oxidase. J. Bacteriol. **161:**123–127.
11. **Baldassare, J. J., K. B. Rhinehart, and D. F. Silbert.** 1976. Modification of membrane lipid: physical properties in relation to fatty acid structure. Biochemistry **15:**2986–2994.
12. **Bamford, D. H., L. Rouhiainen, K. Takkinen, and H. Soderlund.** 1984. Comparison of lipid-containing bacteriophages PRD1, PR3, PR4, PR5 and L17. J. Gen. Virol. **57:**365–373.
13. **Bayer, M. E., J. Kaplow, and H. Goldfine.** 1975. Alterations in envelope structure of heptose-deficient mutants of *Escherichia coli* as revealed by freeze-etching. Proc. Natl. Acad. Sci. USA **72:**5145–5149.
14. **Berger, E., and L. Heppel.** 1974. Different mechanisms of energy coupling for the shock-sensitive and shock resistant amino acid permeases of *Escherichia coli*. J. Biol. Chem. **249:**7747–7755.
15. **Black, P. N., S. Kianian, C. D. Russo, and W. D. Nunn.** 1985. Long chain fatty acid transport in *Escherichia coli* cloning, mapping and expression of the *fadL* gene. J. Biol. Chem. **260:**1780–1789.
16. **Blake, R., II, L. P. Hager, and R. B. Gennis.** 1978. Activation of pyruvate oxidase by monomeric and micellar amphiphiles. J. Biol. Chem. **253:**1963–1971.
17. **Blaurock, A. E.** 1982. Evidence of bilayer structure and of membrane interrelation from X-ray diffraction analysis. Biochem. Biophys. Acta **650:**167–207.
18. **Borle, F., and J. Seelig.** 1983. Structure of *Escherichia coli* membranes. Deuterium magnetic resonance studies of the phosphoglycerol head group in intact cells and model membranes. Biochemistry **22:**5536–5544.
19. **Boxer, D. H., A. Malcolm, and A. Graham.** 1982. *Escherichia coli* formate to nitrate respiratory pathway: structural analysis. Biochem. Soc. Trans. **10:**480–481.
20. **Bradley, D. E., and E. L. Rutherford.** 1975. Basic characterization of a lipid-containing bacteriophage specific for plasmids of the P, N, and W compatibility groups. Can. J. Microbiol. **21:**152–163.
21. **Bragg, P.** 1980. The respiratory system of *Escherichia coli*, p. 115–136. *In* C. J. Knowles (ed.), Diversity of bacterial respiratory systems. CRC Press, Boca Raton, Fla.
22. **Bragg, P. D.** 1985. The ATPase complex of *Escherichia coli*. Can. J. Biochem. Cell Biol. **62:**1190–1197.
23. **Bragg, P. D., and N. R. Hackett.** 1983. Cytochromes of the trimethylamene N-oxide anaerobic pathway of *Escherichia coli*. Biochim. Biophys. Acta **725:**168–177.
24. **Brass, J., U. Ehmann, and B. Bukau.** 1983. Reconstitution of maltose transport in *Escherichia coli*: conditions affecting import of maltose-binding protein into the periplasm of calcium-treated cells. J. Bacteriol. **155:**97–106.
25. **Buchel, D., B. Gronenborn, and B. Mueller-Hill.** 1980. Sequence of the lactose permease gene. Nature (London) **283:**541–545.
26. **Burnell, E. L. van Alphen, A. Verkleig, and B. de Kruijff.** 1980.

[31]P nuclear magnetic resonance and freeze-fracture electron microscopy studies of *Escherichia coli*. 1. Cytoplasmic membrane and total phospholipids. Biochim. Biophys. Acta **597**:492–501.

27. **Cairney, J., C. F. Higgins, and I. R. Booth.** 1984. Proline uptake through the major transport system of *Salmonella typhimurium* is coupled to sodium ions. J. Bacteriol. **160**:22–27.

28. **Campbell, H. D., B. L. Rogers, and I. G. Young.** 1984. Nucleotide sequence of the respiratory D-lactate dehydrogenase gene of *Escherichia coli*. Eur. J. Biochem. **144**:367–373.

29. **Carrasco, N., D. Herzlinger, R. Mitchell, S. DeChiara, W. Danho, T. Gabriel, and H. Kaback.** 1984. Intramolecular dislocation of the COOH terminus of the *lac* carrier in reconstituted proteoliposomes. Proc. Natl. Acad. Sci. USA **81**:4672–4676.

30. **Carter, K., and R. Gennis.** 1985. Reconstitution of the ubiquinone-dependent pyruvate oxidase system of *Escherichia coli* with the cytochrome *o* terminal oxidase complex. J. Biol. Chem. **260**:10986–10990.

31. **Chang, Y.-Y., and J. E. Cronan, Jr.** 1984. An *Escherichia coli* mutant deficient in pyruvate oxidase activity due to altered phospholipid activation of the enzyme. Proc. Natl. Acad. Sci. USA **81**:4348–4352.

32. **Chaudhry, G. R., and C. H. MacGregor.** 1983. Cytochrome *b* from *Escherichia coli* nitrate reductase: its properties and association with the enzyme complex. J. Biol. Chem. **258**:5819–5827.

33. **Cherry, R. J.** 1979. Rotational and lateral diffusion of membrane proteins. Biochim. Biophys. Acta **559**:289–327.

33a.**Clarke, D. A., and P. D. Bragg.** 1985. Cloning and expression of the transhydrogenase gene of *Escherichia coli*. J. Bacteriol. **162**:367–373.

34. **Cole, S. T.** 1982. Nucleotide sequence coding for the flavoprotein subunit of the fumarate reductase of *Escherichia coli*. Eur. J. Biochem. **122**:479–484.

35. **Cole, S. T., T. Grundstrom, B. Jaurin, J. J. Robinson, and J. H. Weiner.** 1982. Location and nucleotide sequence of *frdB*, the gene coding for the iron-sulphur protein subunit of the fumarate reductase of *Escherichia coli*. Eur. J. Biochem. **126**:211–216.

36. **Cole, S. T., and J. R. Guest.** 1979. Production of a soluble form of fumarate reductase by multiple gene duplication in *Escherichia coli* K-12. Eur. J. Biochem. **102**:65–71.

37. **Cole, S. T., and J. R. Guest.** 1982. Molecular genetic aspects of the succinate:fumarate oxidoreductases of *Escherichia coli*. Biochem. Soc. Trans. **10**:473–475.

37a.**Condon, C., R. Cammack, D. S. Patel, and P. Owen.** 1985. The succinate dehydrogenase complex of *Escherichia coli*: immunochemical resolution and biophysical characterization of a 4-subunit enzyme complex. J. Biol. Chem. **260**:9427–9434.

38. **Costerton, J. W., J. M. Ingram, and K.-J. Cheng.** 1974. Structure and function of the cell envelope of gram-negative bacteria. Bacteriol. Rev. **38**:87–110.

39. **Cronan, J. E., Jr.** 1978. Molecular biology of bacterial membrane lipids. Annu. Rev. Biochem. **47**:163–189.

40. **Cronan, J. E., Jr.** 1979. Phospholipid synthesis and assembly, p. 35–65. *In* M. Inouye (ed.), Bacterial outer membranes. John Wiley & Sons, Inc., New York.

41. **Cronan, J. E., Jr., and E. P. Gelmann.** 1975. An estimate of the minimum amount of unsaturated fatty acid required for growth of *Escherichia coli*. J. Biol. Chem. **248**:1188–1195.

42. **Csonka, L. N.** 1982. Third L-proline permease in *Salmonella typhimurium* which functions in media of elevated osmotic strength. J. Bacteriol. **151**:1433–1443.

43. **Darlison, M. G., and J. R. Guest.** 1984. Nucleotide sequence encoding the iron-sulphur protein subunit of the succinate dehydrogenase of *Escherichia coli*. Biochem. J. **223**:507–517.

44. **Davis, J. H., C. P. Nichol, G. Weeks, and M. Bloom.** 1979. Study of the cytoplasmic and outer membranes of *Escherichia coli* by deuterium magnetic resonance. Biochemistry **18**:2103–2112.

45. **Davis, T. N., and J. E. Cronan, Jr.** 1983. Nonsense mutants of the lipid containing bacteriophage PR4. Virology **126**:600–613.

46. **Davis, T. N., and J. E. Cronan, Jr.** 1985. An alkyl imidate labeling study of the organization of phospholipids and proteins in the lipid-containing bacteriophage PR4. J. Biol. Chem. **260**:663–671.

47. **Davis, T. N., E. D. Muller, and J. E. Cronan, Jr.** 1982. The virion of the lipid-containing bacteriophage PR4. Virology **120**:287–306.

48. **Dickie, P., and J. H. Weiner.** 1979. Purification and characterization of membrane-bound fumarate reductase from anaerobically grown *E. coli*. Can. J. Biochem. **57**:813–821.

49. **Dills, S., A. Apperson, M. Schmidt, and M. Saier.** 1980. Carbohydrate transport in bacteria. Microbiol. Rev. **44**:385–418.

50. **Dornmair, K., A. Corin, J. Wright, and F. Jähnig.** 1985. The size

of lactose permease derived from rotational diffusion measurements. EMBO J. **4**:3633–3638.

51. **Duckwitz-Peterlein, G., G. Eilenberger, and P. Overath.** 1977. Phospholipid exchange between bilayer membranes. Biochim. Biophys. Acta **469**:311–325.

52. **Duckwitz-Peterlein, G., and H. Moraal.** 1978. Transport of lipids through water as exchange mechanism between two liposome populations. Biophys. Struct. Mech. **4**:315–326.

53. **Dunn, R. J., N. R. Hackett, K.-S. Huong, S. Jones, H. G. Khorana, D.-S. Lee, M. J. Liao, K.-M. Lo, J. McCoy, S. Noguchi, R. Radhakrishan, and U. L. RajBhandary.** 1978. Studies on the light-transducing pigment bacteriorhodopsin. Cold Spring Harbor Symp. Quant. Biol. **48**:853–862.

54. **Dupont, G. A. Gabriel, M. Chabre, T. Guli-Krzwicki, and E. Schechter.** 1972. Use of a new detector for X-ray diffraction and kinetics of the ordering of the lipids of *Escherichia coli*: membranes and model systems. Nature (London) **238**:331–333.

55. **Eisenberg, D.** 1984. Three-dimensional structure of membrane and surface proteins. Annu. Rev. Biochem. **53**:595–623.

56. **Enoch, H. G., and R. L. Lester.** 1975. The purification and properties of formate dehydrogenase and nitrate reductase in *Escherichia coli*. J. Biol. Chem. **250**:6693–6705.

56a.**Epstein, W.** 1983. Membrane mediated regulation of gene expression in bacteria, p. 281–292. *In* J. Beckwith, J. Davies, and J. Gallant (ed.), Gene function in prokaryotes. Cold Spring Harbor Laboratory, Cold Spring Harbor, N.Y.

57. **Fato, R., M. Battino, G. P. Castelli, and G. Lenaz.** 1985. Measurement of the lateral diffusion coefficients of ubiquinones in lipid vesicles by fluorescence quenching of 12-(9-anthroyl) stearate. FEBS Lett. **179**:238–242.

58. **Foster, T., M. Boublik, and H. Kaback.** 1983. Structure of the *lac* carrier protein of *Escherichia coli*. J. Biol. Chem. **258**:31–34.

59. **Futai, M.** 1973. Membrane D-lactate dehydrogenase from *Escherichia coli*. Purification and properties. Biochemistry **12**:2468–2474.

60. **Futai, M., and H. Kanazawa.** 1983. Structure and function of proton-translocating adenosine triphosphatase (F_0F_1): biochemical and molecular biological approaches. Microbiol. Rev. **47**:285–312.

61. **Futai, M., and H. Kimura.** 1977. Inducible membrane-bound L-lactate dehydrogenase from *Escherichia coli*: purification and properties. J. Biol. Chem. **252**:5820–5827.

62. **Futarni, A., E. Hunt, and G. G. Hanska.** 1979. Vectorial redox reactions of physiological quinones; requirement of a minimum length of the isoprenoid side chain. Biochim. Biophys. Acta **547**:583–596.

63. **Gally, H. U., G. Pluschke, P. Overath, and J. Seelig.** 1980. Structure of *Escherichia coli* membranes and cells as studied by deuterium magnetic resonance. Biochemistry **18**:5605–5610.

64. **Gally, H. U., G. Pluschke, P. Overath, and J. Seelig.** 1980. Structure of *Escherichia coli* membranes. Fatty acyl chain order parameters in inner and outer membranes and derived liposomes. Biochemistry **19**:1638–1643.

65. **Garland, P. B., K. Johnson, and G. R. Reid.** 1982. The diffusional mobility of proteins in the cytoplasmic membrane of *Escherichia coli*. Biochem. Soc. Trans. **10**:484–485.

66. **Gay, N. J.** 1984. Construction and characterization of an *Escherichia coli* strain with a *uncI* mutation. J. Bacteriol. **158**:820–825.

67. **Goldkorn, T., G. Rimon, E. Kemper, and H. Kaback.** 1984. Functional molecular weight of the *lac* carrier protein from *Escherichia coli* as studied by radiation inactivation analysis. Proc. Natl. Acad. Sci. USA **81**:1021–1025.

68. **Grabau, C., and J. E. Cronan, Jr.** 1984. Molecular cloning of the gene (*poxB*) encoding the pyruvate oxidase of *Escherichia coli*, a lipid-activated enzyme. J. Bacteriol. **160**:1088–1092.

69. **Graham, A.** 1981. The organization of hydrogenase in the cytoplasmic membrane of *Escherichia coli*. Biochem. J. **197**:283–291.

70. **Graham, A., and D. H. Boxer.** 1981. The organization of formate dehydrogenase in the cytoplasmic membrane of *Escherichia coli*. Biochem. J. **195**:627–637.

71. **Graham, S. B., J. T. Stephenson, and J. M. Wood.** 1984. Proline dehydrogenase from *Escherichia coli*: reconstitution of a functional membrane association. J. Biol. Chem. **259**:2656–2661.

72. **Green, G. N., and R. B. Gennis.** 1983. Isolation and characterization of an *Escherichia coli* mutant lacking cytochrome *d* terminal oxidase. J. Bacteriol. **154**:1269–1275.

73. **Green, G. N., R. E. Kranz, and R. B. Gennis.** 1984. Cloning the *cyd* locus coding for the cytochrome *d* complex of *Escherichia coli*. Gene **32**:99–106.

74. **Green, G. N., R. G. Kranz, R. M. Lorence, and R. B. Gennis.** 1984.

Identification of subunit I as the cytochrome b_{558} component of the cytochrome d terminal oxidase complex of *Escherichia coli*. J. Biol. Chem. **259**:7994–7997.

75. **Grundstrom, T., and B. Jaurin.** 1982. Overlap between *ampC* and *frd* operons on the *Escherichia coli* chromosome. Proc. Natl. Acad. Sci. USA **79**:1111–1115.

76. **Hackenbrock, C. R.** 1981. Lateral diffusion and electron transfer in the mitochondrial inner membrane. Trends Biochem. Sci. **6**:151–154.

77. **Hackett, N. R., and R. D. Bragg.** 1983. Membrane cytochromes of *Escherichia coli* grown aerobically and anaerobically with nitrate. J. Bacteriol. **154**:708–718.

78. **Haldar, K. P., Y. Oslsiewski, C. Walsh, G. J. Kaczarowski, A. Bhaduri, and H. R. Kaback.** 1982. Simultaneous reconstitution of *Escherichia coli* membrane vesicles with D-lactic and D-amino acid dehydrogenase. Biochemistry **21**:4590–4606.

79. **Hammes, G. G., and P. R. Schimmel.** 1970. Rapid reactions and transient states, p. 67–141. *In* P. D. Boyer (ed.), The enzymes, vol. 2. Academic Press, Inc., New York.

80. **Harder, M. E., and L. J. Banaszak.** 1979. Small angle X-ray scattering from the inner and outer membranes from *Escherichia coli*. Biochim. Biophys. Acta **552**:89–102.

81. **Hays, J.** 1978. Group translocation transport systems, p. 43–102. *In* B. Rosen (ed.), Bacterial transport. Marcel Dekker, New York.

82. **Hazelbauer, G.** 1975. Maltose chemoreceptor of *Escherichia coli*. J. Bacteriol. **122**:206–214.

83. **Hederstedt, L., and L. Rutberg.** 1981. Succinate dehydrogenase—a comparative review. Microbiol. Rev. **45**:542–555.

84. **Heller, K., E. Lin, and T. Wilson.** 1980. Substrate specificity and transport properties of the glycerol facitator of *Escherichia coli*. J. Bacteriol. **144**:274–278.

85. **Henderson, R.** 1977. The purple membrane from *Halobacterium halobium*. Annu. Rev. Biophys. Bioeng. **6**:87–109.

86. **Hengge, R., and W. Boos.** 1983. Maltose and lactose transport in *Escherichia coli*. Examples of two different types of concentrative transport systems. Biochim. Biophys. Acta **737**:443–478.

87. **Herzlinger, D., P. Viitanen, N. Carrasco, and H. Kaback.** 1984. Monoclonal antibodies against the *lac* carrier protein from *Escherichia coli*. 2. Binding studies with membrane vesicles and proteoliposomes reconstituted with purified *lac* carrier protein. Biochemistry **23**:3688–3693.

88. **Higgins, C., P. Haag, K. Nikaido, F. Ardeshir, G. Garcia, and G. Ames.** 1982. Complete nucleotide sequence and identification of membrane components of the histidine transport operon of *Salmonella typhimurium*. Nature (London) **298**:723–727.

89. **Higgins, C. F., M. Hardie, D. Jamieson, and L. M. Powell.** 1983. Genetic map of the *opp* (oligopeptide permease) locus of *Salmonella typhimurium*. J. Bacteriol. **153**:830–836.

90. **Hobson, A. C., D. Gho, and B. Müller-Hill.** 1977. Isolation, genetic analysis, and characterization of *Escherichia coli* mutants with defects in the *lacY* gene. J. Bacteriol. **131**:830–838.

91. **Hobson, A., R. Weatherwax, and G. Ames.** 1985. ATP-binding sites in the membrane components of histidine permease, a periplasmic transport system. Proc. Natl. Acad. Sci. USA **81**:7333–7337.

92. **Hogg, R.** 1984. High affinity L-arabinose transport: the *araF,G* operon in *Escherichia coli*, p. 38–41. *In* L. Leive and D. Schlessinger (ed.), Microbiology—1984. American Society for Microbiology, Washington, D.C.

93. **Holroyd, C. D., and C. Bradbeer.** 1984. Cobalamin transport in *Escherichia coli*, p. 21–23. *In* L. Leive and D. Schlessinger (ed.), Microbiology—1984. American Society for Microbiology, Washington, D.C.

94. **Hong, J.** 1977. An *ecf* mutation in *Escherichia coli* pleiotropically affecting energy coupling in active transport but not generation or maintenance of energy potential. J. Biol. Chem. **252**:8582–8588.

95. **Hong, J., A. Hunt, P. Masters, and M. Lieberman.** 1979. Requirement for acetyl-phosphate for the binding protein-dependent transport systems in *Escherichia coli*. Proc. Natl. Acad. Sci. USA **76**:1213–1217.

96. **Houslay, M., and K. Stanley.** 1982. Dynamics of biological membranes. John Wiley & Sons, Inc., New York.

97. **Hunt, A., and J. Hong.** 1983. Properties and characterization of binding protein dependent active transport of glutamine in isolated membrane vesicles of *Escherichia coli*. Biochemistry **22**:844–850.

98. **Iida, A., S. Harayama, T. Iino, and G. Hazelbauer.** 1984. Molecular cloning and characterization of genes required for ribose transport and utilization in *Escherichia coli* K-12. J. Bacteriol. **158**:674–682.

99. **Ingledew, W. S., and R. K. Poole.** 1984. The respiratory chains of *Escherichia coli*. Microbiol. Rev. **48**:222–271.

100. **Jackson, M. B., and J. M. Sturtevant.** 1977. Studies of the lipid phase transitions of *Escherichia coli* by high sensitivity differential scanning calorimetry. J. Biol. Chem. **252**:4749–4751.

101. **Jacobson, G., C. Lee, J. Leonard, and M. Saier.** 1983. Mannitol-specific enzyme II of the *Escherichia coli* phosphotransferase system. J. Biol. Chem. **258**:10748–10756.

102. **Jaworowski, A., H. D. Campbell, M. I. Poulis, and J. G. Young.** 1981. Genetic identification of the respiratory NADH dehydrogenase of *Escherichia coli*. Biochemistry **20**:2041–2047.

103. **Jaworowski, A., G. Mayo, D. C. Shaw, H. D. Campbell, and I. G. Young.** 1981. Characterization of the respiratory NADH dehydrogenase of *Escherichia coli* and reconstitution of NADH oxidase in *ndh* in mutant membrane vesicles. Biochemistry **20**:3621–3628.

104. **Jones, R. W.** 1980. Proton translocation by the membrane-bound formate dehydrogenase of *Escherichia coli*. FEMS Microbiol. Lett. **8**:167–171.

105. **Jones, R. W., J. H. Weiner, and R. B. Gennis.** 1984. Immunological analysis of membrane-bound antigens from *Escherichia coli* which have succinate dehydrogenase activity. Biochem. Soc. Trans. **12**:800–801.

106. **Kaback, H.** 1983. The *lac* carrier protein in *Escherichia coli*. J. Membr. Sci. **76**:95–112.

107. **Kaczorowski, G., and H. Kaback.** 1979. Mechanism of lactose translocation in membrane vesicles from *Escherichia coli*. 1. Effect of pH on efflux, exchange, and counterflow. Biochemistry **18**:3691–3697.

108. **Kang, S. Y., H. S. Gutowsky, and E. Oldfield.** 1979. Spectroscopic studies of specifically deuterium-labeled membrane systems. Nuclear magnetic resonance investigation of protein-lipid interactions in *Escherichia coli* membranes. Biochemistry **18**:3268–3272.

109. **Karlibian, D., and P. Conchould.** 1974. Dihydro-oratate oxidase of *Escherichia coli* K-12: purification, properties, and relation to the cytoplasmic membrane. Biochim. Biophys. Acta **364**:218–232.

110. **Kay, W. W.** 1978. Transport of carboxylic acids, p. 385–411. *In* B. P. Rosen (ed.), Bacterial transport. Marcel Dekker, New York.

111. **Kell, D. B.** 1984. Diffusion of protein complex in prokaryotic membrane: fast, free, random or directed? Trends Biochem. Sci. **9**:86–87.

112. **Keniry, M. A., A. Kintanar, R. L. Smith, H. S. Gutowisky, and E. Oldfield.** 1984. Nuclear magnetic resource studies of amino-acids and proteins. Deuterium nuclear magnetic resonance relaxation of deuteromethyl-labeled amino-acids in crystals and in *Halobacterium halobium* and *Escherichia coli* membranes. Biochemistry **23**:288–298.

113. **Kimura, H., and M. Futai.** 1978. Effects of phospholipids on L-lactate dehydrogenase from membranes of *Escherichia coli*: activation and stabilization of the enzyme with phospholipids. J. Biol. Chem. **253**:1095–1100.

114. **Kistler, W. S., and E. C. C. Lin.** 1972. Purification and properties of the flavine-stimulated anaerobic L-α-glycerophosphate dehydrogenase of *Escherichia coli*. J. Bacteriol. **112**:539–547.

115. **Kita, K., and Y. Anraku.** 1981. Composition and sequencing of the *b* cytochrome in the respiratory chain of aerobically grown *Escherichia coli* K-12 in the early exponential phase. Biochem. Int. **2**:105–112.

116. **Kita, K., K. Konishi, and Y. Anraku.** 1984. Terminal oxidase of *Escherichia coli* aerobic respiratory chain. 1. Purification and properties of cytochrome b_{556}–o. Complex from cells in the early exponential phase of aerobic growth. J. Biol. Chem. **259**:3368–3374.

117. **Kita, K., K. Konishi, and Y. Anraku.** 1984. Terminal oxidase of *Escherichia coli* aerobic respiratory chain. 2. Purification and properties of cytochrome b_{558}-d complex from cells grown with limited oxygen and evidence of branched electron carrying system. J. Biol. Chem. **259**:3375–3381.

118. **Kita, K., I. Yamoto, and Y. Anraku.** 1978. Purification and properties of cytochrome b_{556} in the respiratory chain of aerobically grown *Escherichia coli* K-12. J. Biol. Chem. **253**:8910–8915.

119. **Kleeman, W., and H. M. McConnell.** 1974. Lateral phase separations in *Escherichia coli* membranes. Biochim. Biophys. Acta **345**:220–230.

120. **Klionsky, D. J., W. S. Brusilow, and R. D. Simoni.** 1983. Assembly of a functional Fo of the proton-translocating ATPase of *Escherichia coli*. J. Biol. Chem. **258**:10136–10143.

121. **Koland, J. G., M. J. Miller, and R. B. Gennis.** 1984. Reconstitution of the membrane-bound, ubiquinone-dependent pyruvate oxidase respiratory chain of *Escherichia coli* with the cytochrome *d* terminal oxidase. Biochemistry **23**:445–453.

122. **Koland, J. G., M. J. Miller, and R. B. Gennis.** 1984. Potentiometric analysis of the purified cytochrome *d* terminal oxidase complex from *Escherichia coli*. Biochemistry **23:**1051– 1056.

123. **Kohn, L. D., and H. R. Kaback.** 1973. Mechanisms of active transport in isolated bacterial membrane vesicles. Purification and properties of the membrane-bound D-lactate dehydrogenase from *Escherichia coli*. J. Biol. Chem. **248:**7012–7017.

124. **Konisky, J.** 1979. Specific transport system and receptors for colicins and phages, p. 319–359. *In* M. Inouye (ed.), Bacterial outer membranes. John Wiley & Sons, Inc., New York.

125. **Koplow, J., and H. Goldfine.** 1974. Alterations in the outer membrane of the cell envelope of heptose-deficient mutants of *Escherichia coli*. J. Bacteriol. **117:**527–543.

126. **Kovatchev, S. W., L. C. Vaz, and H. Eibl.** 1981. Lipid dependence of the membrane bound D-lactate dehydrogenase of *Escherichia coli*. J. Biol. Chem. **256:**10369–10374.

127. **Kranz, R. G., C. A. Barassi, M. J. Miller, G. N. Green, and R. B. Gennis.** 1983. Immunological characterization of an *Escherichia coli* strain which is lacking cytochrome *d*. J. Bacteriol. **156:**115–121.

128. **Kranz, R. G., and R. B. Gennis.** 1983. Immunological characterization of the cytochrome *o* terminal oxidase from *Escherichia coli*. J. Biol. Chem. **258:**10614–10621.

129. **Kranz, R. G., and R. B. Gennis.** 1984. Characterization of the cytochrome *d* terminal oxidase complex of *Escherichia coli* using polyclonal and monoclonal antibodies. J. Biol. Chem. **259:**7998–8003.

130. **Kroger, A.** 1978. Fumarate as terminal acceptor of phosphorylative electron transport. Biochim. Biophys. Acta **505:**129–145.

131. **Kroger, A., E. Dorres, and E. Winkler.** 1980. The orientation of the substrate sites of formate dehydrogenase and fumarate reductase in the membrane of *Vibrio succinogenes*. Biochim. Biophys. Acta **589:**118–136.

132. **Kung, H., and U. Henning.** 1972. Limited availability of binding sites for dehydrogenases on the cell. Proc. Natl. Acad. Sci. USA **69:**925–929.

133. **Kuritzkes, D. R., X.-Y. Zhang, and E. C. C. Lin.** 1984. Use of ϕ(*glp-lac*) in studies of respiratory regulation of the *Escherichia coli* anaerobic *sn*-glycerol-3-phosphate dehydrogenase genes (*glpAB*). J. Bacteriol. **157:**591–598.

134. **Kwan, H. S., and E. L. Barrett.** 1983. Roles for menaquinone and the two trimethylamine oxide (TMAO) reductases in TMAO respiration in *Salmonella typhimurium*: Mu *d*(Apʳ *lac*) insertion mutations in *men* and *tor*. J. Bacteriol. **155:**1147–1155.

135. **Lancaster, J., and P. Hinkle.** 1977. Studies of β-galactoside transporter in inverted membrane vesicles of *Escherichia coli*. 1. Symmetrical facilitated diffusion and proton gradient coupled transport. J. Biol. Chem. **252:**7657–7661.

136. **Lee, C., and M. Saier.** 1983. Mannitol-specific enzyme II of the bacterial phosphotransferase system. 3. The nucleotide sequence of the permease gene. J. Biol. Chem. **258:**10761–10767.

137. **Legendre, S., L. Letellier, and E. Schechter.** 1980. Influence of lipids with branched-chain fatty acids on the physical, morphological and functional properties of *Escherichia coli* cytoplasmic membrane. Biochim. Biophys. Acta **602:**491–505.

138. **Lemire, B. D., J. J. Robinson, and J. H. Weiner.** 1982. Identification of membrane anchor polypeptides of *Escherichia coli* fumarate reductase. J. Bacteriol. **152:**1126–1131.

139. **Leonard, J., C. Lee, A. Apperson, S. Dills, and M. Saier.** 1981. The role of membranes in the transport of small molecules, p. 1–52. *In* B. Ghosh (ed.), Organization of prokaryotic cell membranes, vol. I. CRC Press, Boca Raton, Fla.

140. **Leonard, J., and M. Saier.** 1981. Genetic dissection of catalytic activities of the *Salmonella typhimurium* mannitol enzyme II. J. Bacteriol. **145:**1106–1109.

141. **Leonard, J., and M. Saier.** 1983. Mannitol-specific enzyme II of the bacterial phosphotransferase system. 2. Reconstitution of vectorial transphosphorylation in phospholipid vesicles. J. Biol. Chem. **258:**10757–10760.

142. **Letellier, L., and E. Schechter.** 1977. Lipid and protein segregation in *Escherichia coli* membrane: different cytoplasmic membrane fractions. Proc. Natl. Acad. Sci. USA **74:**452–456.

143. **Lewis, B. A., and D. M. Engleman.** 1983. Lipid bilayer thickness varies linearly with acyl chain length in fluid phosphatidylcholine vesicles. J. Mol. Biol. **166:**211–217.

144. **Lin, E. C. C.** 1976. Glycerol dissimilation and its regulation in bacteria. Annu. Rev. Microbiol. **30:**535–578.

145. **Linden, C. D., J. K. Blaisie, and C. F. Fox.** 1977. A confirmation of the phase behavior of *Escherichia coli* cytoplasmic membrane lipids by X-ray diffraction. Biochemistry **16:**1621–1625.

146. **Lopilato, J., J. Garwin, S. Emr, T. Silhavy, and J. Beckwith.** 1984. D-Ribose metabolism in *Escherichia coli* K-12: genetics,

147. regulation, and transport. J. Bacteriol. **158:**665–673.

147. **Lorence, R. L., G. N. Green, and R. B. Gennis.** 1984. Potentiometric analysis of the cytochromes of *Escherichia coli* utilizing a mutant strain lacking the cytochrome *d* terminal oxidase complex. J. Bacteriol. **157:**115–121.

148. **Ludtke, D., T. Larson, C. Beck, and W. Boos.** 1982. Only one gene is required for the *glpT*-dependent transport of *sn*-glycerol-3-phosphate in *Escherichia coli*. Mol. Gen. Genet. **186:**540–547.

149. **Lugtemberg, B., and L. Von Alphen.** 1983. Molecular architecture and functioning of the outer membrane of *Escherichia coli* and other gram-negative bacteria. Biochim. Biophys. Acta **737:**51–115.

150. **Lundstrom, K. H., H. D. Bamford, E. T. Palva, and K. Lounatmaa.** 1979. Lipid-containing bacteriophage PR4: structure and life cycle. J. Gen. Virol. **43:**583–592.

151. **MacGregor, C. H.** 1976. Biosynthesis of a membrane-bound nitrate reductase in *Escherichia coli*: evidence for a soluble precursor. J. Bacteriol. **126:**122–131.

152. **Macnab, R. M., and S.-I. Aizawa.** 1984. Bacterial motility and the bacterial flagellar motor. Annu. Rev. Biophys. Bioeng. **13:**51–83.

153. **Maloy, S., C. Ginsburgh, R. Simons, and W. Nunn.** 1981. Transport of long and medium chain fatty acids by *Escherichia coli* K-12. J. Biol. Chem. **356:**3735–3742.

154. **Maloy, S. R., and J. R. Roth.** 1983. Regulation of proline utilization in *Salmonella typhimurium*: characterization of *put*::Mu *d*(Ap, *lac*) operon fusions. J. Bacteriol. **154:**561–568.

155. **Mao, B., M. Pear, J. McCammon, and F. Quiocho.** 1982. Hinge-bending in L-arabinose binding protein. The Venus fly trap model. J. Biol. Chem. **257:**1131–1133.

156. **Martin, F. J., and R. C. MacDonald.** 1976. Phospholipid exchange between bilayer membrane vesicles. Biochemistry **15:**321–327.

157. **Matsushita, K., L. Patel, R. B. Gennis, and H. R. Kaback.** 1983. Reconstitution of active transport in proteoliposomes containing cytochrome *o* oxidase and *lac* carrier protein purified from *Escherichia coli*. Proc. Natl. Acad. Sci. USA **80:**4889–4893.

158. **Matsushita, K., L. Patel, and H. R. Kaback.** 1984. Cytochrome *o* type oxidase from *Escherichia coli*. Characterization of the enzyme and mechanism of electrochemical proton gradient generation. Biochemistry **23:**4703–4714.

159. **McElhaney, R. N.** 1982. Effects of membrane lipids on transport and enzymic activities. Curr. Top. Membr. Transp. **17:**317–380.

160. **McIntyre, T. M., and R. M. Bell.** 1975. Mutants of *Escherichia coli* defective in membrane phospholipid synthesis. Effect of cessation of net phospholipid synthesis on cytoplasmic and outer membranes. J. Biol. Chem. **250:**9053–9059.

161. **McIntyre, T. M., B. K. Chamberlain, R. E. Webster, and R. M. Bell.** 1977. Mutants of *Escherichia coli* defective in membrane phospholipid synthesis. Effects of cessation and reinitiation of phospholipid synthesis on macromolecular synthesis and phospholipid turnover. J. Biol. Chem. **252:**4487–4493.

162. **McLean, L. R., and M. C. Phillips.** 1984. Kinetics of phosphatidylcholine and lysophosphatidylcholine exchange between unilamellar vesicles. Biochemistry **23:**4624–4630.

163. **McPherson, M. J., A. J. Barron, D. J. C. Pappin, and J. C. Wootton.** 1984. Respiratory nitrate reductase of *Escherichia coli*: sequence identification of the large subunit gene. FEBS Lett. **177:**260–264.

164. **Meganathan, R.** 1984. Inability of *men* mutants of *Escherichia coli* to use trimethylamine-N-oxide as an electron acceptor. FEMS Microbiol. Lett. **24:**57–62.

165. **Melchoir, D. L.** 1982. Lipid phase transitions and regulation of membrane fluidity in prokaryotes. Curr. Top. Membr. Transp. **17:**263–316.

166. **Melchoir, D. L., and A. Carruthers.** 1983. Studies of the relationship between bilayer water permeability and bilayer physical state. Biochemistry **22:**5797–5807.

167. **Melchoir, D. L., and J. M. Steim.** 1976. Thermotropic transitions in biomembranes. Annu. Rev. Biophys. Bioeng. **5:**205–238.

168. **Menzel, R., and J. Roth.** 1980. Identification and mapping of a second proline permease in *Salmonella typhimurium*. J. Bacteriol. **141:**1064–1070.

169. **Menzel, R., and J. Roth.** 1981. Purification of the *putA* gene product. J. Biol. Chem. **256:**9755–9761.

170. **Menzel, R., and J. Roth.** 1981. Enzymatic properties of the purified *putA* protein from *Salmonella typhimurium*. J. Biol. Chem. **256:**9762–9766.

171. **Mieschendahl, M., D. Buchel, H. Bocklage, and B. Müller-Hill.** 1981. Mutations in the *lacY* gene of *Escherichia coli* define functional organization of lactose permease. Proc. Natl. Acad. Sci. USA **78:**7652–7656.

172. **Miller, M. J., and R. B. Gennis.** 1983. Purification and charac-

terization of the cytochrome *d* terminal oxidase complex from *Escherichia coli*. J. Biol. Chem. **248**:9159–9165.

173. **Mindich, L., D. Bamford, T. McGraw, and G. Mackenzie.** 1982. Assembly of bacteriophage PRD1: particle formation with wild-type and mutant viruses. J. Virol. **44**:1021–1030.

174. **Mirelman, D.** 1979. Biosynthesis and assembly of cell wall peptidoglycan, p. 115–166. *In* M. Inouye (ed.), Bacterial outer membranes. John Wiley & Sons, Inc., New York.

175. **Mitaku, S., J. Wright, L. Best, and F. Vahnig.** 1984. Localization of the galactose binding site in the lactose carrier of *Escherichia coli*. Biochim. Biophys. Acta **776**:247–258.

176. **Miura, T., and S. Mizushima.** 1968. Separation by density gradient centrifugation of two types of membranes from spheroplast membrane of *Escherichia coli* K-12. Biochim. Biophys. Acta **150**:159–164.

177. **Morpeth, F. F., and D. H. Boxer.** 1985. Kinetic analysis of respiratory nitrate reductase from *Escherichia coli*. Biochemistry **24**:40–46.

178. **Muller, E. D., and J. E. Cronan, Jr.** 1983. Lipid-containing bacteriophage PR4: effects of altered lipid composition of the virion. J. Mol. Biol. **165**:109–124.

179. **Murakami, H., K. Kita, and Y. Anraku.** 1984. Cloning of *cybB*, the gene for cytochrome b_{556} of *Escherichia coli* K-12. Mol. Gen. Genet. **198**:1–6.

180. **Murakami, H., K. Kita, and Y. Anraku.** 1985. The *Escherichia coli* b_{556} gene, *cybA*, is assignable as *sdhC* in the succinate dehydrogenase gene cluster. FEMS Microbiol. Lett. **30**:307–312.

181. **Murakami, H., K. Kita, and Y. Anraku.** 1986. Purification and properties of a diheme cytochrome b_{561} of the *Escherichia coli* respiratory chain. J. Biol. Chem. **261**:540–551.

182. **Murakami, H., K. Kita, H. Oya, and Y. Anraku.** 1984. Chromosomal location of the *Escherichia coli* cytochrome b_{556} gene, *cybA*. Mol. Gen. Genet. **196**:1–5.

183. **Nagelkerke, F., and P. W. Postma.** 1978. 2-Deoxygalactose, a specific substrate of the *Salmonella typhimurium* galactose permease: its use for the isolation of *galP* mutants. J. Bacteriol. **133**:607–613.

184. **Nakayama, H., T. Mitsui, M. Nishihara, and M. Kito.** 1980. Relation between growth temperature of *Escherichia coli* and phase transition temperatures of its cytoplasmic and outer membranes. Biochim. Biophys. Acta **601**:1–10.

185. **Narindrasorasak, S., A. H. Goldie, and B. D. Sanwal.** 1979. Characteristics and regulation of a phospholipid-activated malate oxidase from *Escherichia coli*. J. Biol. Chem. **254**:1540–1545.

186. **Nazos, P., T. Su, R. Landick, and D. Oxender.** 1984. Branched chain amino acid transport in *Escherichia coli*, p. 24–28. *In* L. Leive and D. Schlessinger (ed.), Microbiology—1984. American Society for Microbiology, Washington, D.C.

187. **Nichol, C. P., J. H. Davis, and M. Bloom.** 1980. Quantitative study of the fluidity of *Escherichia coli* membranes using deuterium magnetic resonance. Biochemistry **19**:451–457.

188. **Nichols, J. W., and R. E. Pagano.** 1981. Kinetics of soluble lipid monomer diffusion between vesicles. Biochemistry **20**:2783–2784.

189. **Nikaido, H., Y. Takeuchi, S.-I. Ohnrski, and T. Nalsea.** 1977. Outer membrane of *Salmonella typhimurium*. Electron spin resonance studies. Biochim. Biophys. Acta **465**:152–157.

190. **Nishimura, Y., I. K. P. Tan, Y. Ohgami, K. Kohgami, and T. Kamihara.** 1983. Induction of membrane-bound L-lactate dehydrogenase in *Escherichia coli* under condition of nitrate respiration, fumarate reduction and trimethylamine-N-oxide reduction. FEMS Microbiol. Lett. **17**:283–286.

191. **O'Brien, T. A., H. L. Schrock, P. Russell, R. Blake II, and R. B. Gennis.** 1976. Preparation of *Escherichia coli* pyruvate oxidase utilizing a thiamine pyrophosphate affinity column. Biochim. Biophys. Acta **452**:13–29.

192. **Olsiewski, P. J., G. J. Kaczorowski, and C. Walsh.** 1980. Purification and properties of D-amino acid dehydrogenase, an inducible membrane-bound non-sulfur flavoenzyme from *Escherichia coli*-B. J. Biol. Chem. **255**:4487–4494.

193. **Osborn, M. J.** 1979. Biosynthesis and assembly of the lipopolysaccharide of the outer membrane, p. 15–34. *In* M. Inouye (ed.), Bacterial outer membranes. John Wiley & Sons, Inc., New York.

194. **Osborn, M. J., J. E. Gander, E. Parisi, and J. Carson.** 1972. Mechanism of assembly of the outer membrane of *Salmonella typhimurium*. Isolation and characterization of cytoplasmic and outer membranes. J. Biol. Chem. **247**:3962–3972.

195. **Overath, P., M. Brenner, T. Gulik-Krzywicki, E. Schechter, and L. Letellier.** 1975. Lipid phase transitions in cytoplasmic and outer membranes of *Escherichia coli*. Biochim. Biophys. Acta **389**:358–369.

196. **Overath, P., and L. Thilo.** 1978. Structural and functional aspects of biological membranes revealed by lipid phase transitions. Int. Rev. Biochem. **19**:1–44.

197. **Overath, P., and J. Wright.** 1983. Lactose permease: a carrier on the move. Trends Biochem. Sci. **8**:404–408.

198. **Owen, P., and C. Condon.** 1982. The succinate dehydrogenase of *Escherichia coli*: subunit composition of the Triton X-100 solubilized antigen. FEMS Microbiol. Lett. **14**:223–227.

199. **Owen, P., G. J. Kaczorowski, and H. R. Kaback.** 1980. Resolution and identification of iron-containing antigens in membrane vesicles from *Escherichia coli*. Biochemistry **19**:596–600.

200. **Padan, E., T. Arbel, A. Rimon, A. B. Shira, and A. Cohen.** 1983. Biosynthesis of the lactose permease in *Escherichia coli* minicells and effect of carrier amplification on cell physiology. J. Biol. Chem. **258**:5666–5673.

201. **Parkinson, J. S., and G. Hazelbauer.** 1983. Bacterial chemotaxis: molecular genetics of sensory transduction and chemotactic gene expression, p. 293–318. *In* J. Beckwith, J. Davies, and J. Gallant (ed.), Gene function in prokaryotes. Cold Spring Harbor Laboratory, Cold Spring Harbor, N.Y.

202. **Plate, C., and J. Suit.** 1981. The *eup* locus of *Escherichia coli* and its role in H$^+$/solute symport. J. Biol. Chem. **256**:12974–12980.

203. **Poole, R. K., C. Kumar, I. Salmon, and B. Chance.** 1983. The 650nm chromophore in *Escherichia coli* is an oxygenated compound, not the oxidized form of cytochrome oxidase *d*: a hypothesis. J. Gen. Microbiol. **129**:1335–1344.

204. **Pratt, E. A., L. W.-M. Fung, J. A. Flowers, and C. Ho.** 1979. Membrane-bound D-lactate dehydrogenase from *Escherichia coli*: purification and properties. Biochemistry **18**:312–316.

205. **Pratt, E. A., J. A. Jones, P. F. Cottam, S. R. Dowd, and C. Ho.** 1983. A biochemical study of the reconstitution of D-lactate dehydrogenase-deficient membrane vesicles using fluorine-labeled components. Biochim. Biophys. Acta **729**:167–175.

206. **Quinn, P. J.** 1981. The fluidity of cell membranes and its regulation. Prog. Biophys. Mol. Biol. **38**:1–104.

207. **Raetz, C. R. H.** 1982. Genetic control of phospholipid bilayer assembly, p. 435–477. *In* J. N. Hawthorne and G. B. Burell (ed.), Phospholipids. Elsevier/North-Holland Publishing Co., Amsterdam.

208. **Ranck, J. L.** 1983. X-ray diffraction studies of the phase transitions of hydrocarbon chains in bilayer systems: status and dynamics. Chem. Phys. Lipids **32**:251–270.

209. **Ranck, J. L., L. Letellier, E. Schechter, B. Kron, P. Pernot, and A. Terdieu.** 1984. X-ray analysis of the kinetics of *Escherichia coli* lipid and membrane structural transitions. Biochemistry **23**:4955–4961.

210. **Recny, M., and L. P. Hager.** 1983. Isolation and characterization of the protein-activated form of pyruvate oxidase: evidence for a conformational change in the environment of the flavin prosthetic group. J. Biol. Chem. **258**:5189–5195.

211. **Rice, C. W., and W. P. Hempfling.** 1978. Oxygen-limited continuing culture and respiratory energy conservation in *Escherichia coli*. J. Bacteriol. **134**:115–124.

212. **Robinson, J. J., and J. H. Weiner.** 1980. The effect of amphipaths on the flavin-linked aerobic glycerol-3-phosphate dehydrogenase from *Escherichia coli*. Can. J. Biochem. **58**:1172–1178.

213. **Rock, C. O., and J. E. Cronan, Jr.** 1982. Regulation of bacterial membrane lipid synthesis. Curr. Top. Membr. Transp. **17**:209–233.

214. **Rondeau, S. S., P.-Y. Hsu, and J. A. DeMoss.** 1984. Construction in vitro of a cloned *nar* operon from *Escherichia coli*. J. Bacteriol. **159**:159–166.

215. **Roseman, M. A., and T. E. Thompson.** 1980. Mechanism of the spontaneous transfer of phospholipids between bilayers. Biochemistry **19**:439–444.

216. **Rosenberg, H., L. Russell, P. Jacomb, and K. Chegwiden.** 1982. Phosphate exchange in the *pit* transport system in *Escherichia coli*. J. Bacteriol. **149**:123–130.

217. **Rule, G. S., E. A. Pratt, C. Chin, F. Wold, and C. Ho.** 1985. Overproduction and nucleotide sequence of the respiratory D-lactate dehydrogenase of *Escherichia coli*. J. Bacteriol. **161**:1059–1068.

218. **Russo, R., J. Lee, P. Clarke, and G. Wilcox.** 1984. Identification of the *araE* gene product of *Salmonella typhimurium* LT2, p. 42–46. *In* L. Leive and D. Schlessinger (ed.), Microbiology—1984. American Society for Microbiology, Washington, D.C.

219. **Sackmann, E., H. Trauble, H.-J. Galla, and P. Overath.** 1973. Lateral diffusion, protein mobility, and phase transitions in *Escherichia coli* membranes. A spin-label study. Biochemistry **12**:5360–5369.

220. **Saffman, P. G., and M. Delbruck.** 1975. Brownian motion in biological membranes. Proc. Natl. Acad. Sci. USA **72**:3111–3113.

221. **Saier, M.** 1977. Bacterial phosphoenolpyruvate:sugar phospho-

transferase systems: structural, functional, and evolutionary relationships. Microbiol. Rev. **41:**856–871.

222. **Saier, M., and J. Leonard.** 1983. The mannitol enzyme II of the bacterial phosphotransferase system: a functionally chimaeric protein with receptor, transport, kinase, and regulatory activities, p. 11–30. *In* J. Kane (ed.), Multifunctional proteins: catalytic/structural and regulatory. CRC Press, Boca Raton, Fla.

222a.**Saier, M., M. Novatny, D. Comey-Fuhrman, T. Osumi, and J. Desai.** 1983. Cooperative binding of the sugar substrates and allosteric regulatory protein (enzyme III^{Glc} of the phosphotransferase system) to the lactose and melibiose permeases in *Escherichia coli* and *Salmonella typhimurium.* J. Bacteriol. **155:**1351–1357.

223. **Saier, M., and M. Schmidt.** 1981. Vectorial and nonvectorial transphosphorylation catalyzed by enzymes II of the bacterial phosphotransferase system. J. Bacteriol. **145:**391–397.

224. **Schechter, E., L. Letellier, and T. Gulik-Krzywicki.** 1974. Relations between structure and function in cytoplasmic membrane vesicles isolated from an *Escherichia coli* fatty-acid auxotroph. Eur. J. Biochem. **49:**61–76.

225. **Schnaitman, C. A.** 1970. Protein composition of the cell wall and cytoplasmic membrane of *Escherichia coli.* J. Bacteriol. **104:** 890–901.

226. **Schneider, E., and K. Altendorf.** 1985. All three subunits are required for the reconstitution of an active proton channel (Fo) of *Escherichia coli* ATP synthase (F1Fo). EMBO J. **4:**515–518.

227. **Schrock, H. L., and R. B. Gennis.** 1977. High affinity lipid binding sites on the peripheral membrane enzyme pyruvate oxidase: specific ligand effects on detergent binding. J. Biol. Chem. **252:**5990–5995.

228. **Schrock, H. L., and R. B. Gennis.** 1980. Specific ligand enhancement of the affinity of *Escherichia coli* pyruvate oxidase for dipalmitoyl lecithin. Biochim. Biophys. Acta **614:**215–220.

229. **Schryvers, A., and J. H. Weiner.** 1981. The anaerobic *sn*-glycerol-3-phosphate dehydrogenase of *Escherichia coli.* Purification and characterization. J. Biol. Chem. **256:**9959–9966.

230. **Schryvers, A., and J. Weiner.** 1982. The anaerobic *sn*-glycerol-3-phosphate dehydrogenase: cloning and expression of the *glp*A gene of *Escherichia coli* and identification of the *glp*A products. Can. J. Biochem. **60:**224–231.

231. **Schweizer, H., W. Boos, and T. J. Larson.** 1985. Repressor for the *sn*-glycerol-3-phosphate regulation of *Escherichia coli* K-12: cloning of the *glp*R gene and identification of its product. J. Bacteriol. **161:**563–566.

232. **Seelig, J., and A. Seelig.** 1980. Lipid conformation in model membranes and biological membranes. Q. Rev. Biophys. **13:**19–61.

233. **Sengupta, P., E. Sackmann, W. Kuhnle, and H. P. Scholz.** 1976. An optical study of the exchange kinetics of membrane bound molecules. Biochim. Biophys. Acta **436:**869–878.

234. **Senior, A. E.** 1983. Secondary and tertiary structure of membrane proteins involved in proton translocation. Biochim. Biophys. Acta **728:**81–95.

235. **Shaw, D. J., and J. R. Guest.** 1982. Nucleotide sequence of the *fnr* gene and primary structure of the Fnr protein of *Escherichia coli.* Nucleic Acids Res. **10:**6119–6130.

236. **Shaw-Goldstein, L. A., R. B. Gennis, and C. Walsh.** 1978. Identification, localization and function of the TPP- and FAD-dependent pyruvate oxidase in isolated membrane vesicles in *Escherichia coli.* Biochemistry **17:**5606–5613.

237. **Short, S. A., H. R. Kaback, and L. D. Kohn.** 1975. Localization of D-lactate dehydrogenase in natural and reconstituted *Escherichia coli* membrane vesicles. J. Biol. Chem. **250:**4291–4296.

238. **Shuman, H.** 1982. Active transport of maltose in *Escherichia coli* K-12. J. Biol. Chem. **257:**5455–5461.

239. **Silbert, D. F.** 1970. Arrangement of fatty acyl groups in phosphatidylethanolamine from a fatty acid auxotroph of *Escherichia coli.* Biochemistry **9:**3631–3640.

240. **Silver, S.** 1978. Transport of cations and anions, p. 221–323. *In* B. P. Rosen (ed.), Bacterial transport. Marcel Dekker, New York.

241. **Simpkin, D., and W. J. Ingledew.** 1984. Location of the catalytic site of the respiratory fumarate reductase of *Escherichia coli.* J. Gen. Microbiol. **130:**2851–2855.

242. **Sinensky, M.** 1974. Homoeoviscous adaptation—a homeostatic process that regulates the viscosity of membrane lipids in *Escherichia coli.* Proc. Natl. Acad. Sci. USA **71:**522–525.

243. **Singer, S.** 1977. Thermodynamics, the structure of integral membrane proteins and transport. J. Supramolec. Struct. **6:**313–323.

244. **Smit, J., Y. Kamio, and H. Nikaido.** 1975. Outer membrane of *Salmonella typhimurium:* chemical analysis and freeze-fracture studies with lipopolysaccharide mutants. J. Bacteriol. **124:**942–958.

245. **Steim, J. M.** 1974. Differential scanning calorimetry. Methods Enzymol. **32B:**262–272.

246. **Steim, J. M., M. E. Tourtellotte, J. C. Reinert, R. N. McElhaney, and R. L. Rader.** 1969. Calorimetric evidence for the liquid-crystal-line state of lipids in a biomembrane. Proc. Natl. Acad. Sci. USA **63:**104–109.

247. **Strauch, K. L., J. B. Lenk, B. L. Gamble, and C. G. Miller.** 1985. Oxygen regulation in *Salmonella typhimurium.* J. Bacteriol. **161:**673–680.

248. **Surin, B. P., H. Rosenberg, and G. B. Cox.** 1985. Phosphate specific transport system of *Escherichia coli:* nucleotide sequence and gene-polypeptide relationships. J. Bacteriol. **161:**189–198.

249. **Tainer, J. A., E. D. Getzoff, H. Alexander, R. A. Houghten, A. J. Olson, R. A. Lerner, and W. A. Hendrickson.** 1985. The reactivity of anti-peptide antibodies is a function of the atomic mobility of sites in a protein. Nature (London) **312:**127–134.

250. **Taylor, B.** 1983. Role of proton motive force in sensory transduction in bacteria. Annu. Rev. Microbiol. **37:**551–573.

251. **Teather, R., J. Bramhall, I. Riede, J. Wright, M. Furst, G. Aichele, U. Wilhelm, and P. Overath.** 1980. Lactose carrier protein of *Escherichia coli.* Structure and expression of plasmids carrying the Y gene of the *lac* operon. Eur. J. Biochem. **108:**223–231.

252. **Thilo, L.** 1977. Kinetics of phospholipid exchange between bilayer membranes. Biochim. Biophys. Acta **469:**326–334.

253. **Thilo, L., and P. Overath.** 1976. Randomization of membrane lipids in relation to transport system assembly in *Escherichia coli.* Biochemistry **15:**328–334.

254. **Thompson, J. W., and B. M. Shapiro.** 1981. The respiratory chain NADH dehydrogenase of *Escherichia coli:* isolation of an NADH:quinone oxidoreductase from membranes and comparison with the membrane-bound NADH:dichlorophenolindophenol oxidoreductase. J. Biol. Chem. **256:**3077–3084.

255. **Tsuchiya, T., K. Ottina, Y. Moriyama, M. Newman, and T. Wilson.** 1982. Solubilization and reconstitution of the melibiose carrier from a plasmid-carrying strain of *Escherichia coli.* J. Biol. Chem. **257:**5125–5128.

256. **Unwin, N., and R. Henderson.** 1984. The structure of proteins in biological membranes. Sci. Am. **250:**78–94.

257. **van Deenan, L. L. M.** 1965. Phospholipids and biomembranes, p. 1–115. *In* R. T. Holman (ed.), Progress in the chemistry of fats and other lipids, vol. 8, part I. Pergamon Press, Inc., Elmsford, N.Y.

258. **van der Plas, J., K. J. Hellingwerf, H. G. Seijen, J. R. Guest, J. H. Weiner, and W. N. Konings.** 1983. Identification and localization of enzymes of the fumarate reductase and nitrate respiration systems of *Escherichia coli* by crossed immunoelectrophoresis. J. Bacteriol. **153:**1027–1037.

259. **Viitanen, P., M. Garcia, and H. Kaback.** 1984. Purified reconstituted *lac* carrier protein from *Escherichia coli* is fully functional. Proc. Natl. Acad. Sci. USA **81:**1629–1633.

260. **Vogel, H., J. K. Wright, and F. Jähnig.** 1985. The structure of lactose permease derived from Raman spectroscopy and prediction methods. EMBO J. **4:**3625–3631.

261. **von Meyenburg, K., B. B. Jorgensen, and B. Deurs.** 1984. Physiological and morphological effects of overproduction of membrane-bound ATP synthase in *Escherichia coli* K-12. EMBO J. **3:**1791–1797.

262. **Walker, J. E., G. Falk, N. J. Gay, and V. L. J. Tybulewicz.** 1984. Genes for bacterial and mitochondrial ATP synthase. Biochem. Soc. Trans. **12:**234–235.

263. **Wallace, B. J., and I. G. Young.** 1977. Role of quinone in electron transport to oxygen and nitrate in *Escherichia coli:* studies with *ubiA⁻* and *menA⁻* double quinone mutant. Biochim. Biophys. Acta **461:**84–100.

264. **Weaver, D. L.** 1982. Note on the interpretation of lacteral diffusion coefficients. Biophys. J. **38:**311–313.

265. **Weiner, J. H.** 1974. The localization of glycerol-3-phosphate dehydrogenase in *Escherichia coli.* J. Membr. Biol. **15:**1–14.

266. **Weiner, J. H., and L. A. Heppel.** 1972. Purification of the membrane-bound and pyridine nucleotide independent L-glycerol-3-phosphate dehydrogenase from *Escherichia coli.* Biochem. Biophys. Res. Commun. **47:**1360–1365.

267. **Weiner, J. H., B. D. Lemire, M. L. Elmes, R. D. Bradley, and D. G. Scraba.** 1984. Overproduction of fumarate reductase in *Escherichia coli* induces a novel intracellular lipid-protein organelle. J. Bacteriol. **158:**590–596.

268. **Wilson, D.** 1978. Cellular transport mechanisms. Annu. Rev. Biochem. **47:**933–965.

269. **Wilson, T., and M. Kusch.** 1972. A mutant of *Escherichia coli* K-12 energy-uncoupled for lactose transport. Biochim. Biophys. Acta **255:**786–797.

270. **Wood, D., M. G. Darlison, R. J. Wilde, and J. R. Guest.** 1984.

Nucleotide sequence encoding the flavoprotein and hydrophobic subunits of the succinate dehydrogenase of *Escherichia coli*. Biochem. J. **222**:519–534.

271. **Wright, J. K., and P. Overath.** 1984. Purification of the lactose:H$^+$ carrier of *Escherichia coli* and characterization of galactoside binding and transport. Eur. J. Biochem. **138**:497–508.

272. **Young, I. G., A. Jaworowski, and M. I. Poulis.** 1978. Amplification of the respiratory NADH dehydrogenase of *Escherichia coli* by gene cloning. Gene **4**:25–36.

273. **Young, I. G., A. Jaworowski, and M. Poulis.** 1982. Cloning of the gene for the respiratory D-lactate dehydrogenase of *Escherichia coli*. Biochemistry **21**:2092–2095.

274. **Young, I. G., B. L. Rogers, H. D. Campbell, A. Jaworowski, and D. C. Shaw.** 1981. Nucleotide sequence coding for the respiratory NADH dehydrogenase of *Escherichia coli*: UUG initiation codon. Eur. J. Biochem. **116**:165–170.

275. **Ziberstein, D., S. Schuldiner, and E. Padan.** 1979. Proton electrochemical gradient in *Escherichia coli* cells and its relation to active transport of lactose. Biochemistry **18**:669–673.

6. Periplasm and Protein Secretion

DONALD B. OLIVER

Department of Microbiology, State University of New York at Stony Brook, Stony Brook, New York 11794

INTRODUCTION

The periplasmic space lies between the inner and outer membranes of gram-negative bacteria. Because of this location, this space should not be thought of as a single homogenous compartment but rather as consisting of several distinct microenvironments created by the two boundary membranes and the peptidoglycan layer. Periplasmic proteins localized within these regions fulfill important functions in the processing of essential nutrients and their transport into the cell and in the biogenesis of the cell envelope. Periplasmic polysaccharides and other small molecules serve to buffer the cell from changing osmotic and ionic environments and thus help to preserve the more constant internal environment needed for cell growth and viability. Clearly, the periplasmic space is a dynamic structure at the crossroads between the anabolic and catabolic activities of the cell.

The purpose of this chapter is twofold: to review what is currently known about the structure and contents of the periplasmic space and to summarize our current understanding of the biogenesis of periplasmic proteins. Since the biogenesis of periplasmic proteins shares similarities with that of outer and trans-inner membrane proteins, studies of *Escherichia coli* protein secretion will be discussed collectively. For more comprehensive reading and references, see reviews on the periplasm (12, 69) and on protein secretion (16, 35, 83, 96).

STRUCTURE OF THE PERIPLASMIC SPACE

The structure of the periplasmic space of *E. coli* and *Salmonella typhimurium* has been largely inferred from electron microscopic studies. Unfortunately, the routine procedures used in specimen preparation often cause a release of the periplasmic contents and an increased separation of the inner and the outer membranes, making the periplasmic space appear enlarged and relatively empty. However, cryoelectron microscopic techniques result in an improved picture of this structure, in agreement with the better electron micrographs obtained by more conventional methods (2, 33, 34, 47, 70, 78). Thin sections reveal the cell envelope to be a multilayered structure, each layer being of a uniform thickness. The inner and outer membranes are approximately 7.5 nm thick, with each membrane having a typical double-track appearance due to the deposition of the heavy metal stain on each side. There is a 7.5-nm-thick layer next to the inside face of the outer membrane; this layer is lysozyme sensitive. In better micrographs it can be seen that the inner portion of this layer is composed of an electron-dense peptidoglycan layer approximately 2.0 nm thick. The peptidoglycan in this layer is highly cross-linked, forming the meshwork of the cell wall and thus dividing the periplasmic space into inner and outer periplasmic regions. It has been proposed that relatively un-cross-linked peptidoglycan polymers are also found in the inner periplasmic space, forming a periplasmic gel (47). The peptidoglycan layer is, in fact, attached to the outer membrane by covalent bonds with the Braun lipoprotein (20) and by strong ionic interactions with the matrix protein which forms hexagonal arrays on the peptidoglycan surface (103). These interactions explain why the peptidoglycan layer remains attached to and at a fixed distance from the outer membrane when cells are plasmolyzed. In contrast, the inner membrane shrinks away from the rest of the envelope during plasmoly-

56

sis, leaving a relatively empty inner periplasmic space between these two structures. However, under normal physiological conditions the inner periplasmic space is approximately 4 nm thick in cross section and has a content similar in density to that of the cytoplasm (34). The volume and hence the thickness of the inner periplasmic compartment, in fact, change somewhat depending on the osmolarity of the surrounding medium (see below).

This simple layered structure of the cell envelope is complicated by the existence of adhesion zones between the inner and outer membranes which appear to correspond to export sites for components needed for outer membrane growth. Such adhesion zones or "Bayer patches" appear as attachment sites between the inner and outer membranes when cells are plasmolyzed and examined in the electron microscope (10, 11). The extent of fusion between the two membranes is not known. There are approximately 200 to 400 such sites distributed over the membrane of actively growing cells, covering an estimated 5% of the membrane surface. Adhesion sites are not seen in cells grown to stationary phase, indicating that their appearance is growth phase dependent. Using various types of pulse-labeling techniques, it has been possible to show that adhesion sites correspond to export sites for newly synthesized polysaccharides, lipopolysaccharides, and outer membrane proteins (10, 87). Whether the export of periplasmic components also occurs at or adjacent to such sites is unknown. It appears that only a fraction (10%) of the adhesion sites may be active in the export of outer membrane components during a given period. Regulation of the formation, activity, and specificity of adhesion sites could help to explain the complex topological organization of the cell envelope during growth and division.

Physiological and electron microscopic measurements show that the periplasmic space of E. coli and S. typhimurium is approximately 20 to 40% of the total cell volume in normal growth media (110). Physiological measurements were done on cell suspensions by measuring the distribution of radioactive substances capable of penetrating both the inner and outer membranes (water), the outer membrane only (sucrose), or neither membrane (inulin). From such measurements cytoplasmic, periplasmic, and total cellular volumes were determined. The volume of the periplasmic space determined by this method is in agreement with a similar measurement made from electron micrographs of unstained, fixed cells.

Physiological studies have also established that the periplasmic volume and osmolarity respond to changes in the osmotic strength of the external medium. For example, high concentrations of solutes which can penetrate the outer but not the inner membrane increase the osmolarity and volume of the periplasm, with a concomitant shrinkage of the cytoplasmic compartment. Since the cytoplasmic membrane is in fact flexible and unable to support an osmotic gradient, the periplasm and cytoplasm are iso-osmolar. The osmotic strength of these two compartments is approximately 170 mosM for cells in water and increases to 300 mosM in M63 minimal medium (110). This range reflects primarily changes in cell volume and not the loss of intracellular solutes. Since M63 minimal medium is 145 mosM, there is normally an

osmotic gradient between the periplasm and the external environment, and this osmotic pressure (3.5 atm [ca. 350 kPa]) is exerted against the peptidoglycan layer.

Biochemical studies indicate that the osmolarity of the periplasm may be regulated at least in part by the biosynthesis and export of so-called membrane-derived polysaccharides to the periplasmic space. These periplasmic polysaccharides contain 8 to 10 glucose units in a branched structure linked by $\beta(1\rightarrow2)$ and $\beta(1\rightarrow6)$ bonds and are multiply substituted with 1-phosphoglycerol residues derived from membrane phosphatidylglycerol, as well as with O-succinyl ester residues. This results in an average molecular weight of 2,200 to 2,600 and a net charge of -5. In one study it has been reported that cells can correct for external osmolar fluctuations by regulating the synthesis of these polymers and that they can account for up to 7% of the dry weight of the cell in media of low osmolarity (63). In another study, synthesis of membrane-derived polysaccharides also was found to be regulated by the osmolarity of the medium, but in the sense opposite to that found in the first study (25). This contradiction appears to be due to strain differences (25). Regulation of the synthesis of the membrane-derived polysaccharides requires de novo protein synthesis, but the details of this regulation remain to be determined. The location and identity of the osmosensor and its tie-in with the biosynthesis of these polymers need to be elucidated. A putative regulatory locus has recently been identified (25).

The presence of the cationic, membrane-derived polysaccharides in the periplasm creates a significant Donnan equilibrium across the outer membrane, resulting in an outer membrane potential. This potential can be measured by the unequal distributions of radioactive Na^+ and Cl^- ions across the outer membrane and amounts to approximately 30 mV in cells growing in M63 minimal medium (110). The physiological relevance of this outer membrane potential remains uncertain. However, it is clear that the Donnan equilibrium would create a particular ionic composition in the periplasm which could help to regulate the activities of this compartment as well as those of the two surrounding membranes.

PROTEIN CONTENT OF THE PERIPLASM

Periplasmic proteins are probably localized differentially in the periplasmic space, since they may peripherally associate with the inner membrane, the outer membrane, or the peptidoglycan layer or may freely diffuse within this compartment. Periplasmic proteins have been largely defined by methods that selectively release the contents of the periplasmic compartment from the rest of the cell. Obviously, such methods represent a compromise between conditions that are stringent enough to disrupt the peripheral associations of periplasmic proteins with cell envelope structures and those that are mild enough to prevent the release of integral membrane or cytoplasmic proteins. Thus, certain methods release only a subset of periplasmic proteins, while others release both periplasmic and certain nonperiplasmic proteins. Clearly, additional criteria are needed to sub-

TABLE 1. Proteins released by spheroplasting or osmotic shock

Protein	Reference(s)[a]
Binding proteins	
Arabinose[b]	90
Arginine specific[b]	101
Cystine and diaminopimelic acid[b]	18
Galactose-glucose[b,c]	77
Glutamate-aspartate[b]	121
Glutamine[b,c]	119, 122
Histidine[c]	46
Leucine specific[b]	88
Leucine-isoleucine-valine[b,c]	91, 122
Lysine-arginine-ornithine[b]	100
Maltose[b]	62
Phosphate[b]	72
Ribose[b,c]	1, 120
Sulfate[c]	89
Thiamine[b]	43
Vitamin B_{12}[b]	112
Xylose[b]	26
Scavenging enzymes	
Acid phosphatase (pH 2.5)	30
Acid phosphatase (pH 4.5)	
ADP-glucose hydrolase	
Alkaline phosphatase	
L-Asparaginase	
Carboxypeptidase II	
Cyclic phosphodiesterase (3′-nucleotidase)	
Endonuclease I	
Nicotinate phosphoribosyl transferase	
Polygalacturonic acid transeliminase[d]	
Polyphosphatase	
Sugar phosphate phosphohydrolase[d]	
UDP-glucose hydrolase (5′-nucleotidase)	
Detoxifying enzymes	
Alkylsulfohydrolase	
Aminoglycoside 3′-phosphotransferase II[d]	
β-Lactamase	60
Streptomycin adenylating enzyme[d]	
Other proteins	
Cytochrome c	
Hydrogenase	
Nitrite reductase	
Phosphoglucomutase	
Phosphoglucose isomerase	
Nonperiplasmic proteins (released by osmotic shock only)	
Aminopeptidase N[e]	
Cytidine deaminase	
Deoxyriboaldolase	
Deoxyribomutase	
Elongation factor Tu	
Purine deoxynucleoside phosphorylase	
Purine phosphoribosyl transferase	
Thymidine phosphorylase	
Uridine phosphorylase	

[a] See Table 1 of reference 12 if no reference is given.
[b] *E. coli* protein.
[c] *S. typhimurium* protein.
[d] Protein that is probably periplasmic, but release by spheroplasting has not been reported.
[e] Released from only one *E. coli* strain.

histochemical studies using electron microscopy, (ii) lack of crypticity (for enzymes) in intact cells, (iii) inhibition of activity (for enzymes) or selective labelling by reagents that do not permeate the inner membrane, and (iv) selective release by certain envelope mutants. Most of these criteria only substantiate an envelope location and are not useful in confirming a periplasmic location.

Two methods have been commonly employed to release periplasmic proteins: spheroplast formation (19, 73) and osmotic shock (79). Spheroplasts are usually prepared in a concentrated sucrose solution by treatment of cells with lysozyme-EDTA. Presumably, removal of the peptidoglycan layer allows periplasmic proteins to leach out through the outer membrane which has been breached by EDTA. Osmotic shock involves pretreatment of cells with a concentrated sucrose solution and EDTA followed by a rapid dilution into medium of low osmotic strength. Periplasmic proteins presumably are expelled during this procedure by the sudden expansion of the inner membrane against the cell wall. Centrifugation is then used to remove the cells from the resulting periplasmic fraction. The profiles of periplasmic proteins released by these two methods differ somewhat, since osmotic shock releases a number of proteins that are not released during spheroplast formation (Table 1). Furthermore, certain of these proteins are not periplasmic but probably are derived from the cytoplasmic membrane. Therefore, release during spheroplast formation appears to provide the better operational definition for periplasmic proteins. Recently, a third method for releasing periplasmic proteins has been described which involves treatment of a cell suspension with chloroform and subsequent dilution and removal of cells (3). In addition to these major methods, some of the periplasmic proteins can be released simply by washing cells with Tris-KCl, Tris-EDTA, or sucrose-Tris-EDTA (for references see Beacham [12]).

A list of the known proteins that are released during spheroplast formation and osmotic shock in *E. coli* is given in Table 1. The known periplasmic proteins can be divided into several groups, which include (i) binding proteins that function in the transport of small molecules and in chemotaxis, (ii) "scavenging" enzymes which break down complex molecules into simpler precursors or which function in cell wall biogenesis, (iii) detoxifying enzymes that inactivate toxic molecules, and (iv) other proteins that do not fall into these simple groups.

Periplasmic binding proteins represent a relatively homogenous group that have been more extensively studied than other periplasmic proteins. Binding proteins for sugars, amino acids, vitamins, and ions have been characterized. They are generally abundant and have high affinities for their respective substrates ($K_m = 10^{-7}$ to 10^{-6} M). Binding proteins interact with their respective inner membrane permeases, thereby allowing for the translocation of small molecules across the inner membrane. Certain carbohydrate-binding proteins have a dual function since they also interact with inner membrane proteins that are signal transducers for chemotaxis (e.g., maltose-binding protein binds to the *tar* gene product transducer, and ribose-binding protein and galactose-binding protein

stantiate the periplasmic location of a given protein. These criteria have included (i) localization based on bind to the *tsr* gene product transducer [see chapter 49]). Several binding proteins have been crystallized, and their structures have been determined (i.e., arabinose [80, 93], galactose [95], maltose [94], and sulfate [92]). They appear to have similar structures, since the molecules are ellipsoidal with two distinct globular domains. At least for the arabinose- and sulfate-binding proteins, the substrate-binding site is located deep within a cleft formed by the close packing of these two domains. Further work is needed to determine whether the other binding proteins share these similarities.

ROLE OF THE SIGNAL PEPTIDE IN DIRECTING EXPORT

Periplasmic proteins, like other envelope proteins, are made as larger precursors containing an amino-terminal signal peptide of 20 to 40 amino acid residues which is cleaved during export. The amino acid sequence for signal peptides of 13 periplasmic proteins for *E. coli* is known. These proteins include alkaline phosphatase (50), eight binding proteins (arabinose and galactose [105], histidine and lysine-arginine-orithine [45], isoleucine-valine [69], leucine [88], maltose [13], and phosphate [72]), two β-lactamases (AmpC [60] and TEM [111]), and the two subunits of the heat-labile enterotoxin (27, 109). These signal peptides, like other procaryotic signal peptides, contain at least three conserved features: (i) the amino terminus has one or two positively charged amino acid residues; (ii) the amino terminus is followed by a stretch of 14 to 20 neutral, primarily hydrophobic amino acids known as the hydrophobic core; and (iii) a stretch of approximately six amino acid residues after the hydrophobic core is predicted to form a reverse turn. This segment ends in a consensus processing site denoted AXB, where B is the last amino acid residue of the signal peptide and is alanine, glycine, or serine and A includes these amino acids as well as leucine, valine, and isoleucine. It is not known if this sequence around the processing site is required for recognition by the signal peptidase or if the preference for small amino acid residues is merely for steric considerations. In addition, signal peptides are predicted to assume an α-helical or β-sheet secondary structure (4) according to the rules developed by Chou and Fasman (24). Such secondary structures have been corroborated using model synthetic peptides (102).

Since signal peptides from periplasmic proteins are structurally similar to their inner and outer membrane protein counterparts, it seems likely that the signal peptide plays no role in determining the ultimate location of a given protein. Rather, analysis of mutants in this sequence indicates that it serves to promote early steps in protein secretion. Signal sequence mutations that affect translation, secretion, and processing have been isolated and are caused by amino acid alterations in the basic amino terminus, the hydrophobic core, and the processing site, respectively. These alterations are shown in Fig. 1. Since similar results have been obtained with all envelope proteins, these data will be discussed collectively. For purposes of discussion, mutations in the signal sequence will be numbered according to the amino acid

residue that is altered, starting with the amino terminus.

Basic Amino Terminus

The charge of the amino terminus of the signal peptide may be important in the coupling of translation to secretion. Mutants in which the amino terminus of the signal peptide of the major outer membrane lipoprotein (Lpp) is uncharged show nearly wild-type levels of Lpp synthesis and secretion (53, 116). However, mutants in which the amino terminus of the Lpp signal peptide has a net negative charge show a two-to fivefold reduction in Lpp synthesis and slower rates of Lpp secretion and processing (Fig. 1). A mutant in which the amino terminus of the lambda receptor (LamB) signal peptide is missing one of two basic amino acid residues is similarly reduced for LamB expression (Fig. 1) (44). It was possible to rule out an effect of mRNA secondary structure on LamB translation. This implies that the amino-terminal region of this signal peptide may also be involved in coupling of translation to secretion.

Hydrophobic Core Region

The hydrophobic core region of the signal peptide appears to be important in the association of the exported protein with the inner membrane and its translocation through the bilayer. Mutations in the hydrophobic core of the signal peptide can result in the accumulation of unsecreted protein precursors within the cytoplasm and cytoplasmic membrane of the cell. Three explanations have been offered for such defects, depending on the type of alteration present.

(i) Charged amino acids and deletions disrupt the required hydrophobicity of the core region, either in a site-specific way or by reducing the length below a certain minimal value needed for function (14). Most mutations that strongly block export of the TEM β-lactamase (Bla), LamB, maltose-binding protein (MalE), and alkaline phosphatase (PhoA) are of this type (Fig. 1). Suppressor analysis has been used to show that the length of the hydrophobic core region is critical in MalE signal peptide function. Mal^+ intragenic suppressor mutations have been isolated from strains that have charged amino acid substitutions or a deletion within the hydrophobic core region of this signal peptide. Most mutations that restore secretion of MalE lengthen the disrupted hydrophobic core region by insertion of additional hydrophobic amino acid residues or by extension of the amino-terminal boundary of the hydrophobic core by at least two amino acid residues (Fig. 2) (8; P. Bassford, personal communication).

(ii) Certain amino acid alterations destabilize a secondary structure necessary for signal peptide function. The presence of the helix-destabilizing amino acids proline and glycine in mutant signal peptides has been interpreted in these terms. The best documented example of a secondary structure defect is a small deletion in the LamB signal peptide that removes amino acid residues 10 through 13, which would have a net effect of destabilizing an α-helix by bringing the helix-destabilizing amino acids Pro-9

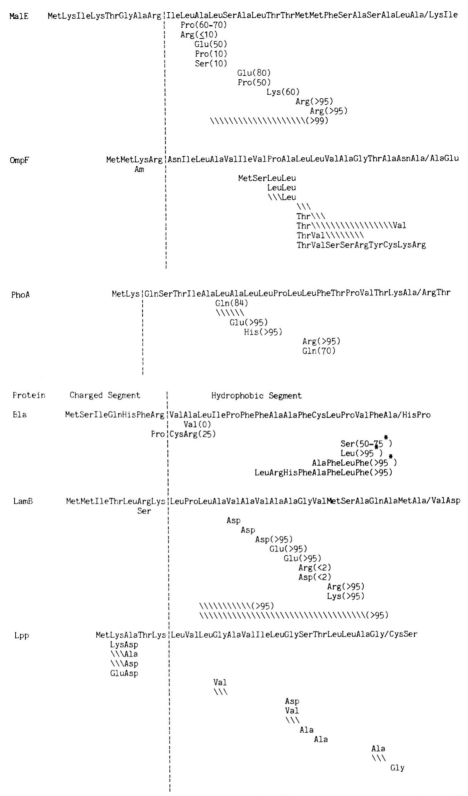

FIG. 1. *E. coli* signal sequence mutations. The amino acid(s) altered by a given mutation is indicated below the wild-type signal sequence. Deletions are indicated by hatching; Am indicates an amber mutation. Numbers in parentheses indicate the percentage of export or processing that is blocked by a given mutation; asterisks indicate a processing defect only. Slashes within sequences indicate the signal peptide processing site. References are as follows: Bla (66), LamB (38, 39, 44), Lpp (51–54, 71, 115, 116), MalE (8, 9, 13), OmpF (E. Sodergren, personal communication), PhoA (74; S. Michaelis, J. Hunt, and J. Beckwith, personal communication).

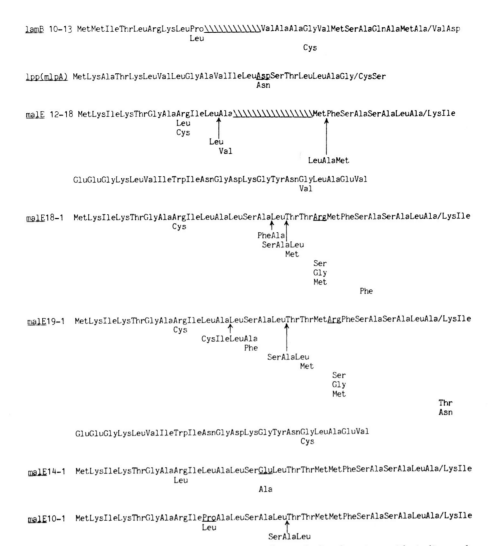

FIG. 2. Intragenic suppressors of signal sequence mutations. The underlined amino acids indicate the position of the primary mutation. Deletions are indicated by hatching. The amino acids altered by the second site suppressor mutations are given below the mutant signal sequence. Upward arrows indicate the insertion of additional amino acids into the signal sequence. Slashes indicate the signal sequence processing site. References are as follows: *lamB*10-13 (40), *lpp* (114), *malE* 12-18 (8), *malE*18-1, *malE*19-1, *malE*14-1, and *malE*10-1 (P. Bassford, personal communication).

and Gly-17 close together (Fig. 1). In support of this hypothesis, revertants of this deletion mutant that restore LamB export are found to have changed either of these two residues to an alternate amino acid (Fig. 2) (40).

(iii) Certain amino acid alterations have a positional effect because they disrupt a recognition site within the signal peptide that interacts with export machinery components. Since a biochemical system for studying this interaction in *E. coli* has not yet been perfected, this explanation remains speculative. However, there are numerous examples of positional effects of signal sequence mutations. For example, certain structurally similar and adjacent mutations within the MalE signal peptide (Arg-10 and Glu-11, Pro-10 and Pro-11 [Fig. 1]) block MalE export to very different extents. Clearly, further genetic and biochemical characterization of these mutants is required in order to determine the cause of such positional effects.

Processing Site

The region of the signal peptide around the processing site appears to play a role in processing only. Mutations in this region, either within the signal sequence or early in the mature sequence, can slow or eliminate processing, but they generally do not affect export of the mutant protein. The known processing mutations span a region around the cleavage site that includes the last four amino acid residues of the signal sequence and up to the first two amino acid residues of the mature sequence (66, 104). However, a systematic mutant analysis of the limits of this region has not been carried out. At least in certain cases such mutations directly affect the cleavage reaction of the precursor protein with the signal peptidase, as this can be demonstrated directly in vitro (104). Although export to the correct cellular compartment does not require processing, the maintenance of the uncleaved signal peptide can have dramatic effects on the solubility and topology of the secreted protein precursor (66).

ROLE OF MATURE PROTEIN SEQUENCES IN DIRECTING EXPORT

Since proteins destined for export to different locations have signal sequences of similar structure and function, it is likely that the more distal steps in export depend on sequences located in the mature portion of the secreted protein. Sequences that are important in the localization of proteins have been termed topogenic sequences. A gene fusion approach has been developed to locate the topogenic sequences present on the polypeptide chain of secreted proteins. By fusing the *lacZ* gene that encodes the normally cytoplasmic enzyme β-galactosidase to increasingly larger amounts of a gene coding for an exported protein, a nested set of hybrid proteins can be produced. These contain increased amounts of the amino-terminal portion of the secreted protein fused to a constant amount of enzymatically active β-galactosidase at the carboxy terminus (9). The intracellular distribution of the different-sized hybrid proteins can indicate where the localization information resides on the polypeptide chain of the secreted protein. This approach assumes that β-galactosidase is a passive carrier which does not positively or negatively affect the export information to which it is fused. As discussed below, this is not always the case.

The most extensive gene fusion analysis of a periplasmic protein has been done with MalE, which is expressed at high levels only in the presence of maltose. A *malE-lacZ* fusion encoding 14 amino acid residues of the MalE signal sequence fused to β-galactosidase does not initiate export and is found in the cytoplasm (13). In contrast, larger *malE-lacZ* fusions which encode an intact signal sequence and from 23 to 300 amino acid residues of mature MalE are exported out of the cytoplasm and are found in the inner membrane (98). Such secretion is abortive since the β-galactosidase portion of the hybrid protein apparently cannot pass through the membrane into the periplasm, but instead gets stuck in the secretion sites in the inner membrane (9). Consistent with this notion, synthesis of large amounts of such fusion proteins by the addition of maltose interferes with export and results in the cytoplasmic accumulation of precursors to many normally secreted proteins. In fact, in such maltose-sensitive strains the export of all of the major outer membrane proteins and most of the periplasmic proteins is blocked during induction, implying that envelope proteins probably share a common early step(s) in their secretion (56; L. Liss and D. Oliver, unpublished data). Since here β-galactosidase interferes with secretion, the gene fusion approach cannot be used to locate the export information contained on MalE or other periplasmic proteins.

An alternative approach to determine whether mature sequences are required for the secretion of periplasmic proteins is to study the secretion of truncated proteins produced by chain-terminating mutants. In one study, two *malE* amber mutants were used which made proteins that were 30 and 90% of the size of wild-type malE protein. Although both proteins were processed, only the larger one was found in the periplasm (57). However, protease protection experiments were used to show that at least a portion of the smaller amber fragment was exterior to the cytoplasmic membrane. Whether the additional sequences present on the larger protein are needed for completion of traversal through or release from the inner membrane or whether they allow the larger protein to assume a water-soluble conformation in the periplasm is unclear. Similar studies have been done with chain-terminating mutants in the *bla* gene (65) and in the gene encoding arginine-binding protein (22). Taken together, these studies indicate that certain carboxy-terminal sequences are not required for export to the periplasmic space.

Three lines of evidence support the idea that export-specific information is located within the mature portion of MalE protein. However, this evidence is indirect and needs more direct confirmation.

(i) MalE signal sequence mutations in the hydrophobic core can be weakly suppressed by amino acid alterations in the mature region (Fig. 2) (8). However, the suppressor mutations are sufficiently close to the signal peptide that they may restore correct folding without normally facilitating the export of wild-type MalE.

(ii) The synthesis of secreted proteins with signal sequence defects can interfere with the export of normal envelope proteins, causing a delay in their export and processing (6). Interference requires not only a mutation in the signal sequence of the interfering protein, but also the presence of certain carboxy-terminal sequences (V. Bankaitis and P. Bassford, personal communication).

(iii) Mature sequences within MalE determine the rate at which MalE-LacZ hybrid proteins are abortively exported and processed (98). For example, the MalE portion of a shorter hybrid protein (signal sequence plus 23 amino acid residues of mature MalE) is secreted into the membrane and processed posttranslationally with slow kinetics. On the other hand, very rapid kinetics, similar to those in wild-type MalE export, are seen with a longer hybrid protein (signal sequence plus 189 amino acid residues of mature MalE). However, the proximity of β-galactosidase sequences to the signal sequence could also explain these results.

Since the presence of export-specific information within the mature portion of a periplasmic protein has not yet been proven, it is possible that this region is largely devoid of such information. In contrast, topogenic sequences have been identified in the mature portion of inner and outer membrane proteins. Certain proteins that span the inner membrane possess a stop-transfer or membrane-anchoring sequence made up of 19 to 23 uncharged, primarily hydrophobic amino acid residues flanked by charged amino acid residues (32). This sequence apparently functions to stop export of the protein through the membrane as well as to serve as a membrane anchor, since deletion of this sequence results in a periplasmic location of the truncated protein. Topogenic sequences have also been identified in the mature portion of LamB; the sequences are required to target this protein to the outer membrane. There is a sequence within the first 49 mature amino acid residues of LamB which allows this protein to achieve an outer membrane location

(15). A second sequence further in LamB appears to improve the efficiency of export (17).

One problem with defining a given topogenic sequence is the difficulty in clearly discriminating whether the sequence is required for export to a given compartment or for the thermodynamic stability of the protein once it has been properly exported. Thus, even if a mutant protein is correctly secreted, it may be unable to associate stably with the correct cellular compartment. However, the studies cited above tend to indicate that periplasmic proteins may be devoid of additional topogenic signals within their mature sequence. This would allow them to completely traverse the inner membrane in an uninterrupted fashion and be released into the periplasmic space. Clearly this interpretation is speculative, and further work is needed to resolve this point.

PHYSIOLOGY OF PROTEIN EXPORT

The timing of protein secretion with respect to translation has been the subject of a number of studies. These studies were aimed at understanding whether protein synthesis plays a role in protein secretion and determining whether the translocation of the polypeptide chain through the inner membrane occurs in a conformationally extended or folded form. It has been demonstrated that a number of periplasmic and outer membrane proteins are preferentially synthesized on membrane-bound polysomes (96). Direct proof of cotranslational secretion of PhoA has been obtained by labeling spheroplasts with a membrane-impermeable reagent and subsequently completing the synthesis of these externally labeled, nascent chains in vitro (108). In contrast, the synthesis of Bla protein is complete before translocation and processing are detected (65). Therefore, it appears that secreted proteins fall into at least two categories, depending on whether their synthesis is complete before their export and processing commences. It has been possible to study such kinetics in detail by using a gel system that determines the proportion of nascent chains that have crossed the inner membrane enough either to be sensitive to externally added protease or to have had their signal peptide processed. As expected, certain proteins are processed completely cotranslationally (AmpC), while others are processed completely posttranslationally (Bla), but most show a mixture of both modes of processing (arabinose-binding protein, MalE, PhoA, LamB, and OmpA) (61). Of interest is that all proteins examined needed to reach 80% of their full length before any traversal or processing was detected (97). Although these studies do not rule out the possibility that polypeptide chain traversal occurs in an extended rather than in a folded form, they do indicate that traversal of even a nascent protein chain occurs in a nonlinear fashion, seemingly independent of chain elongation.

Studies on the energetics of protein secretion indicate than an energized membrane or proton motive force is required for proteins to cross the cytoplasmic membrane. Either the membrane potential or the pH gradient appears to be sufficient for this requirement (5). Reagents such as the proton ionophore carbonyl cyanide m-chlorophenylhydrazone or the potassium ionophore valinomycin at concentrations that reduce the membrane proton motive force also block secretion of a large number of envelope proteins in a reversible manner (28, 41, 124). This export block does not appear to be due to a reduction in the intracellular ATP levels. It remains unknown whether the proton motive force plays a direct role in protein export (for example, as an energy source) or an indirect role (for example, in maintaining membrane structure).

EXPORT MACHINERY

The existence of a complex export machinery that is required for the secretion of many envelope proteins is clearly suggested by both biochemical and genetic studies. Progress with the in vitro secretion system for E. coli has been slow compared with its mammalian counterpart due to difficulty in preparing protein-synthesizing extracts and membrane vesicles that are active in protein translocation. On the other hand, progress in genetically defining the components required for protein secretion has been rapid, yet we are unsure as to the precise function that these components serve in this process. Clearly, future progress in this area will require an interplay between these two different approaches.

Biochemical Studies

In vitro protein secretion systems have been developed which combine in vitro protein synthesis with secretion into inverted vesicles prepared from membranes. Authentic secretion and processing of Bla, OmpA, and PhoA have been demonstrated by several criteria (23, 75, 99). These systems mimic the in vivo situation in their sensitivity to agents that reduce the membrane proton motive force or which perturb membrane structure (ethanol and phenethyl alcohol). Posttranslational secretion takes place at high efficiency in these systems, although the rate of translocation is extremely slow (approximately 15 min) compared with normal in vivo rates, making it uncertain whether the normal and complete export pathway is being utilized. Export has been shown to require a soluble factor sedimenting at about 12S (76), as well as a protein factor(s) localized on the membranes (23). Further characterization of these and other factors is needed.

Certain secreted proteins apparently do not require export machinery components during their biogenesis. A prime example is the trans-inner membrane coat protein of bacteriophage M13. It has been shown that radiochemically pure coat protein precursor integrates into liposomes containing only E. coli phospholipids and the highly purified processing enzyme, that the precursor is correctly processed, and that up to 70% of the mature protein correctly spans the membrane (81). These in vitro experiments coincide with in vivo findings that coat protein is secreted posttranslationally (59) and is not blocked in mutants showing general secretion defects (see below) (123). Thus, certain proteins apparently completely catalyze their own export, although this is probably not the general case for most envelope proteins in E. coli.

Two distinct signal-peptide-processing enzymes have been identified and purified from E. coli inner membranes. Their genes have been mapped, cloned, and sequenced. Signal peptidase I, or leader pepti-

TABLE 2. Genetic loci implicated in protein secretion

Gene	Min	Wild-type or mutant phenotype(s)[a]	Reference(s)
expA	22	Ts on minimal media; decreases amount of certain envelope proteins	29
lepB	55.5	Gene for signal peptidase I; essential gene	124, 125
lspA	0.5	Gene for signal peptidase II; essential gene; Ts mutant available	49, 127, 128
perA (envZ)	75	Decreases amount of certain envelope proteins; transcriptional regulation of OmpC and OmpF	117
prlA (secY)	72	Ts allele shows general export defect; other alleles suppress signal sequence mutations or secA(Ts) mutants	21, 37, 107
prlB (rbsB)	84	Deletion in ribose-binding protein; suppresses lamB signal sequence mutations without processing	16, 36, 37
prlC	68	Suppressor of lamB and malE signal sequence mutations	16, 36, 37
prlD	2.5	Suppressor of certain malE and lamB signal sequence mutations; Synergistic export defects with certain prlA alleles	7
prlE	8.5	Slight Cs phenotype; reduces export of MalE, PhoA, and LamB; probably equivalent to secD and ssaD	See text
prlF	70	Relieves maltose sensitivity of lamB-lacZ and malE-lacZ fusions; slight Cs phenotype	64
secA	2.5	Ts alleles block export of most envelope proteins; Am allele blocks translation of MalE	84, 85
secB	81	Tn5 insertions (nonessential gene); reduces export of MalE, LamB, and OmpF but not PhoA and RbsB	67
secC (rpsO)	69	Gene for ribosome protein S15; Cs alleles suppress secA(Ts) mutants; translational block of MalE, RbsB, LamB, and OmpF	42
ssaD	10	Cs allele suppresses secA(Ts) mutant; decreased expression of MalE; probably equivalent to secD and prlE	82
ssaE	50	Cs allele suppresses secA(Ts) mutant; decreased expression of MalE	82
ssaF	83	Cs allele suppresses secA(Ts) mutant; decreased expression of MalE; probably rpmH coding for ribosomal protein L34	82
ssaG	41	Cs allele suppresses secA(Ts) mutant; decreased expression of MalE	82
ssaH	94.5	Cs allele suppresses secA(Ts) mutant; decreased expression of MalE	82
ssyA	54	Cs allele suppresses secY(Ts) mutant; slower protein synthesis	106
ssyB	10	Cs allele suppresses secY(Ts) mutant	See text
ssyC	69	Cs allele suppresses secY(Ts) mutant	See text
ssyD	3	Cs allele suppresses secY(Ts) mutant	See text
ssyE	72	Cs allele suppresses secY(Ts) mutant	See text
ssyF	20	Cs allele suppresses secY(Ts) mutant	See text

[a] Am, Amber; Cs, cold sensitive; Ts, temperature sensitive.

dase, appears to be the general enzyme for processing precursors to periplasmic as well as inner and outer membrane proteins (leucine- and isoleucine-valine-binding proteins, MalE, LamB, OmpA, OmpF, and M13 coat protein) (118). The purified enzyme consists of a single polypeptide chain of 36,000 daltons and is found in both the inner and outer membranes (129). The enzyme is anchored in the inner membrane by a short amino-terminal tail and contains a large periplasmic region (125). Thus, it is itself a secreted protein, although it does not possess a cleaved signal peptide. However, its secretion is blocked by agents which reduce the membrane potential as well as in mutants showing general secretion defects (see below) (123, 124). No mutations in this enzyme have yet been reported, although the gene appears to be essential for cell survival (31).

Signal peptidase II, or lipoprotein-specific signal peptidase, is the processing enzyme for glyceride-modified lipoprotein precursors. Using an inhibitor of this enzyme, the antibiotic globomycin, or a mutant in the structural gene for the enzyme, it has been possible to show that signal peptidase II is responsible for the processing of the Braun lipoprotein as well as seven additional minor lipoproteins found in the cell envelope (126, 127). The purified enzyme consists of a single polypeptide chain of 18,000 daltons and is a highly hydrophobic protein found exclusively in the inner membrane (49, 113, 128). In keeping with its specificity, it only processes the glyceride-modified form of lipoprotein precursors.

In addition to signal peptidases, the cell must possess an enzyme(s) that degrades the processed signal peptide. The inner membrane enzyme, protease IV, has been shown to be responsible for the degradation of the Lpp signal peptide, but only after its cleavage by signal peptidase II (48). Whether this is the only enzyme that hydrolyzes signal peptides in E. coli remains to be determined.

Genetic Studies

Two general types of genetic selections have been successfully applied to the isolation of mutations in the cellular export machinery. The first selection relies on the existence of a mutation within the signal sequence of a secreted protein that blocks its export and therefore its function. By selecting for restoration of function, it has been possible to isolate extragenic suppressor mutants which compensate for the original signal sequence defect apparently by altering export machinery components. The prlA, prlB, prlC, and prlD mutations have been isolated in this fashion (Table 2) (7, 37). The second selection relies on the fact that hybrid proteins containing a carboxy-terminal β-galactosidase moiety have a Lac⁻ phenotype when they are exported to the inner or outer membrane. By selecting for Lac$^+$ derivatives, it has been possible to isolate extragenic suppressor mutants which apparently partially inactivate the export machinery so that the hybrid protein is not exported from the cytoplasm and remains enzymatically active. The *secA, secB, secD,* and *prlE* mutations have been isolated in this fashion (67, 84; C. Cardel, J. Hunt, S. Michaelis, and J. Beckwith, personal communication; S. Benson, D. Kiino, and C. Cardel, personal communication).

It has also been possible to use some of these secretion mutants as a starting point for defining additional genes involved in the export process by isolating extragenic suppressor mutations of conditional lethal secretion mutants. Suppressors of temperature-sensitive mutants in *secA* and *secY* have been isolated using this approach (Table 2) (82, 106).

By using these genetic methods, a large number of loci have been defined. They are listed in Table 2. The data indicating that these loci are directly involved in protein secretion are strong in certain cases and weak in others. Certain loci, in fact, do not appear to be directly involved in protein secretion (e.g., *prlB* [Table 2]). In general, physiological criteria have been used to infer that a given mutation directly affects protein secretion. Secretion mutants have been inferred when the following defects have been found: (i) accumulation of protein precursors, (ii) restored secretion of an envelope protein containing a defective signal peptide, (iii) specific reduction in the synthesis of several exported proteins under different regulatory controls, and (iv) reduction in the synthesis of an exported protein, but normal synthesis of that protein when it contains a signal sequence mutation. Clearly, these physiological defects could also be due to indirect effects, and biochemical characterization of the function of these gene products is sorely needed. Nonetheless, there has been extensive characterization of certain secretion mutants. This has allowed important inferences to be drawn about the general requirement for the export machinery in envelope protein biogenesis and its potential tie-in with the translation machinery. These data are discussed below.

***prlA* and *prlD* mutants.** The *prlA* and *prlD* mutants were isolated as extragenic suppressors of signal sequence mutations in particular envelope proteins. This suppression is more general, since both mutants have been shown to suppress signal sequence mutations in the hydrophobic core of a variety of envelope proteins (LamB, MalE, PhoA, and OmpF for *prlA* [36]; LamB and MalE for *prlD* [7]), resulting in the secretion and processing of the normally export-defective protein to the correct cellular location. The *prlA* mutants show a range in suppression efficiencies, depending on the signal sequence allele, but in general are strong suppressors of signal sequence defects. In contrast, the single *prlD* mutant isolated is a weak suppressor of certain signal sequence defects only and can, in fact, exacerbate such defects. Such allele specificity has been interpreted as implying that these two proteins interact directly with the signal peptide. Although neither mutant alone shows export defects with normal envelope proteins, certain *prlA prlD* double mutants show severe secretion defects which result in the accumulation of precursors of periplasmic

and outer membrane proteins (MalE, RbsB, LamB, and OmpA [7]).

A conditional lethal temperature-sensitive mutant in the *prlA* gene has been isolated using localized mutagenesis (107). When this mutant is shifted to the nonpermissive temperature, it accumulates precursors to a number of exported proteins in the cytoplasm. These proteins include MalE, OmpA, OmpF, and Lpp. In a different study it has been shown that reducing the level of wild-type PrlA also results in a similar export-defective phenotype (58). Since both export-defective and signal sequence suppressor mutations have been obtained in *prlA*, the evidence that this gene is directly involved in protein export is compelling. Biochemical characterization of the function of the recently identified PrlA protein (55) should help to clarify the role of the protein in secretion and its potential interaction with PrlD and the signal peptide.

***secA* and *ssa* mutants.** Temperature-sensitive mutants in the *secA* gene were isolated using a genetic selection that demands a partial defect in export at the permissive temperature (84). Growth at the nonpermissive temperature results in a complete block of secretion, with the cytoplasmic accumulation of a number of envelope proteins. A more thorough analysis of a *secA* mutant under these conditions showed that export is blocked for all of the major outer membrane proteins and for most, but not all, of the periplasmic proteins (L. Liss and D. Oliver, unpublished data). In addition, under these conditions export of the transmembrane protein, signal peptidase I, is also blocked, while M13 coat protein biogenesis is not (123). Thus, SecA appears to be an essential component of the major secretion system in *E. coli* which functions in an early step in periplasmic, inner, and outer membrane protein biogenesis. This does not preclude other, minor export pathways that do not utilize SecA.

In order to identify additional components of the export machinery which interact with SecA, extragenic suppressors of a *secA*(Ts) mutant have been isolated. This analysis has been confined to essential genes by isolating suppressors which not only allow growth at the high temperature, but also render the cell cold sensitive for growth at a low temperature. These suppressors fall into at least seven genes (Table 2). Some of these suppressors may be directly involved in protein export since they (i) restore secretion in the *secA*(Ts) mutant at the high temperature, (ii) still require normal SecA levels, and (iii) often show export-related defects as cold-sensitive single mutants.

One of the suppressors falls into the *prlA* gene, which was previously implicated in protein export (21). This suggests that PrlA and SecA probably are components of a common export machinery. This conclusion is further supported by the fact that *prlA*-mediated suppression of signal sequence mutations depends on the presence of normal SecA levels in the cell (86). Two other suppressors are in ribosomal genes, indicating that ribosomes may interact with the export machinery during secretion. In this category are *secC* (*rpsO* which codes for S15) (42) and *ssaF* (*rpmH* which codes for L34) (82). Not only do these mutants suppress the *secA*(Ts) export defect at the high temperature, but also, as single cold-sensitive mutants, they show synthetic defects for exported proteins but not for cytoplasmic proteins at the low temperature. One interpretation of these results is that certain portions of the ribosome interact with the export machinery to couple protein synthesis with secretion. Whether this or another explanation is correct will require additional genetic and biochemical studies.

CONCLUSION

A preliminary understanding of the structure and contents of the periplasmic space has been obtained. Areas for further research include (i) identification of additional periplasmic proteins which are not detected by simple enzymatic and binding assays (some of these proteins may play important roles in cell envelope biogenesis and cell division; (ii) clarification of the extent to which various periplasmic proteins are confined to particular microenvironments within this region and whether this serves to regulate their activities; (iii) clarification of the physiology of membrane-derived polysaccharide synthesis, its mechanism of regulation, and its role in conditioning the periplasmic environment; and (iv) elucidation of the different forms of communication between the periplasm and the cytoplasm which help to regulate the different activities in these two compartments.

Considerable progress has been made in understanding the biogenesis of periplasmic proteins and secreted proteins in general. Areas of progress include (i) elucidation of the essential nature of the signal sequence and a preliminary understanding of some of the structural features necessary for signal peptide function; (ii) definition of certain topogenic sequences present in the mature portion of inner and outer membrane proteins that are required in more distal steps in protein secretion; (iii) identification of two major signal peptidases that process different envelope proteins; and (iv) identification via mutant analysis of a number of genes affecting protein secretion, at least some of which should define components of a general protein export pathway in *E. coli*. Further research will be needed in order to dissect biochemically the various steps in export that are mediated by the signal sequence and other topogenic sequences and to elucidate the role of the various components of the export machinery in executing these steps. An interplay between the existing genetic and biochemical approaches should allow rapid progress in this field.

ACKNOWLEDGMENTS

I thank my many colleagues who sent manuscripts in advance of publication, and Phyllis Leder and Sandra Burns for preparing the manuscript.

This work was supported by Public Health Service grant GM32958 from the National Institute of General Medical Sciences.

LITERATURE CITED

1. **Aksamit, R., and B. Koshland, Jr.** 1972. A ribose binding protein of *Salmonella typhimurium*. Biochem. Biophys. Res. Commun. **48:**1348–1353.
2. **Amaeo, K., K. Murata, and A. Umeda.** 1983. Structure of the envelope of *Escherichia coli* observed by the rapid-freezing and substitution fixation method. Microbiol. Immunol. **27:**95–99.
3. **Ames, G. F.-L., C. Prody, and S. Kustu.** 1984. Simple, rapid, and quantitative release of periplasmic proteins by chloroform. J. Bacteriol. **160:**1181–1183.
4. **Austen, B. M.** 1979. Predicted secondary structures of amino-terminal extension sequences of secreted proteins. FEBS Lett. **103:**308–318.
5. **Bakker, E. P., and L. L. Randall.** 1984. The requirement for

6. **Bankaitis, V. A., and P. J. Bassford, Jr.** 1984. The synthesis of export-defective proteins can interfere with normal protein export in *Escherichia coli*. J. Biol. Chem. **259:**12193–12200.
7. **Bankaitis, V. A., and P. J. Bassford, Jr.** 1985. Proper interaction between at least two components is required for efficient export of proteins to the *Escherichia coli* cell envelope. J. Bacteriol. **161:**169–178.
8. **Bankaitis, V. A., B. A. Rasmussen, and P. J. Bassford, Jr.** 1984. Intragenic suppressor mutations that restore export of maltose binding protein with a truncated signal peptide. Cell **37:**243–252.
9. **Bankaitis, V. A., J. P. Ryan, B. A. Rasmussen, and P. J. Bassford, Jr.** 1985. The use of genetic techniques to analyze protein export in *Escherichia coli*. Curr. Top. Membr. Transp. **24:**105–150.
10. **Bayer, M. E.** 1979. The fusion sites between outer membrane and cytoplasmic membrane of bacteria: their role in membrane assembly and virus infection, p. 167–202. *In* M. Inouye (ed.), Bacterial outer membranes. John Wiley & Sons, Inc., New York.
11. **Bayer, M. H., G. P. Costello, and M. E. Bayer.** 1982. Isolation and partial characterization of membrane vesicles carrying markers of the membrane adhesion sites. J. Bacteriol. **149:**758–767.
12. **Beacham, I. R.** 1979. Periplasmic enzymes in gram-negative bacteria. Int. J. Biochem. **10:**877–881.
13. **Bedouelle, H., P. J. Bassford, Jr., A. V. Fowler, I. Zabin, J. Beckwith, and M. Hofnung.** 1980. Mutations which alter the function of the signal sequence of the maltose binding protein of *Escherichia coli*. Nature (London) **285:**78–81.
14. **Bedouelle, H., and M. Hofnung.** 1981. On the role of the signal peptide in the initiation of protein exportation, p. 361–372. *In* B. Pullman (ed.), Intermolecular forces. D. Reidel Publishing Co., Dordrecht, The Netherlands.
15. **Benson, S. A., E. Brenner, and T. J. Silhavy.** 1984. Intragenic regions required for LamB export. Proc. Natl. Acad. Sci. USA **81:**3830–3834.
16. **Benson, S. A., M. N. Hall, and T. J. Silhavy.** 1985. Genetic analysis of protein export in *Escherichia coli* K12. Annu. Rev. Biochem. **54:**101–134.
17. **Benson, S. A., and T. J. Silhavy.** 1983. Information within the mature LamB protein necessary for localization to the outer membrane of *E. coli* K12. Cell **32:**1325–1335.
18. **Berger, E. A., and L. A. Heppel.** 1972. A binding protein involved in the transport of cystine and diaminopimelic acid in *Escherichia coli*. J. Biol. Chem. **247:**7684–7694.
19. **Birdsell, D. C., and E. H. Cota-Robles.** 1967. Production and ultrastructure of lysozyme and ethylenediaminetetraacetate-lysozyme spheroplasts of *Escherichia coli*. J. Bacteriol. **93:**427–437.
20. **Braun, V.** 1975. Covalent lipoprotein from the outer membrane of *Escherichia coli*. Biochim. Biophys. Acta **415:**335–337.
21. **Brickman, E. R., D. B. Oliver, J. L. Garwin, C. Kumamoto, and J. Beckwith.** 1984. The use of extragenic suppressors to define genes involved in protein export in *Escherichia coli*. Mol. Gen. Genet. **196:**24–27.
22. **Celis, R. T. F.** 1981. Chain-terminating mutants affecting a periplasmic binding protein involved in the active transport of arginine and ornithine in *Escherichia coli*. J. Biol. Chem. **256:**773–779.
23. **Chen, L., D. Rhoads, and P. C. Tai.** 1985. Alkaline phosphatase and OmpA protein can be translocated posttranslationally into membrane vesicles of *Escherichia coli*. J. Bacteriol. **161:**973–980.
24. **Chou, P. Y., and G. D. Fasman.** 1978. Prediction of the secondary structure of proteins from their amino acid sequence. Adv. Enzymol. **47:**45–148.
25. **Clark, D. P.** 1985. Mutant of *Escherichia coli* deficient in osmoregulation of periplasmic oligosaccharide synthesis. J. Bacteriol. **161:**1049–1053.
26. **Copeland, B. R., R. J. Richter, and C. E. Furlong.** 1982. Renaturation and identification of periplasmic proteins in two-dimensional gels of *Escherichia coli*. J. Biol. Chem. **257:**15065–15071.
27. **Dallas, W. S., and S. Falkow.** 1980. Amino acid sequence homology between cholera toxin and *Escherichia coli* heat-labile toxin. Nature (London) **228:**499–501.
28. **Daniels, C. J., D. G. Bole, S. C. Quay, and D. L. Oxender.** 1981. Role for membrane potential in the secretion of protein into the periplasm of *Escherichia coli*. Proc. Natl. Acad. Sci. USA **78:**5396–5400.

energy during export of β-lactamase in *Escherichia coli* is fulfilled by the total protonmotive force. EMBO J. **3:**895–900.

29. **Dassa, E., and P.-L. Boquet.** 1981. *expA*: a conditional mutation affecting the expression of a group of exported proteins in *Escherichia coli* K-12. Mol. Gen. Genet. **181**:192–200.

30. **Dassa, E., C. Tetu, and P. L. Boquet.** 1980. Identification of the acid phosphatase (optimum pH 2.5) of *Escherichia coli*. FEBS Lett. **116**:275–278.

31. **Date, T.** 1983. Demonstration by a novel genetic technique that leader peptidase is an essential enzyme of *Escherichia coli*. J. Bacteriol. **154**:76–83.

32. **Davis, N. G., J. D. Boeke, and P. Model.** 1985. Fine structure of a membrane anchor domain. J. Mol. Biol. **181**:111–121.

33. **DePetris, S.** 1967. Ultrastructure of the cell wall of *Escherichia coli* and chemical composition of its constituent layers. J. Ultrastruct. Res. **19**:45–83.

34. **Dubochet, J., A. W. McDowall, B. Menge, E. N. Schmid, and K. G. Lickfeld.** 1983. Electron microscopy of frozen-hydrated bacteria. J. Bacteriol. **155**:381–390.

35. **Duffaud, G. D., S. K. Lehnhardt, P. E. March, and M. Inouye.** 1985. Structure and functions of the signal peptide. Curr. Top. Membr. Transp. **24**:65–104.

36. **Emr, S. D., and P. J. Bassford, Jr.** 1982. Localization and processing of outer membrane and periplasmic proteins in *Escherichia coli* strains harboring export-specific suppressor mutations. J. Biol. Chem. **257**:5852–5860.

37. **Emr, S. D., S. Hanley-Way, and T. J. Silhavy.** 1981. Suppressor mutations that restore export of a protein with a defective signal sequence. Cell **23**:79–88.

38. **Emr, S. D., J. Hedgpeth, J.-M. Clement, T. J. Silhavy, and M. Hofnung.** 1980. Sequence analysis of mutations that prevent export of λ receptor, and *Escherichia coli* outer membrane protein. Nature (London) **285**:82–85.

39. **Emr, S. D., and T. J. Silhavy.** 1982. Molecular components of the signal sequence that function in the initiation of protein export. J. Cell Biol. **95**:689–695.

40. **Emr, S. D., and T. J. Silhavy.** 1983. Importance of secondary structure in the signal sequence for protein secretion. Proc. Natl. Acad. Sci. USA **80**:4599–4603.

41. **Enequist, H. G., T. R. Hirst, S. Harayama, S. J. S. Hardy, and L. L. Randall.** 1981. Energy is required for maturation of exported proteins in *Escherichia coli*. Eur. J. Biochem. **116**:227–233.

42. **Ferro-Novick, S., M. Honma, and J. Beckwith.** 1984. The product of gene *secC* is involved in the synthesis of exported proteins in *E. coli*. Cell **38**:211–217.

43. **Griffith, T. W., and F. R. Leach.** 1973. The effect of osmotic shock on vitamin transport in *Escherichia coli*. Arch. Biochem. Biophys. **159**:658–663.

44. **Hall, M. N., J. Gabay, and M. Schwartz.** 1983. Evidence for a coupling of synthesis and export of an outer membrane protein in *Escherichia coli*. EMBO J. **2**:15–19.

45. **Higgins, C. F., and G. F.-L. Ames.** 1981. Two periplasmic transport proteins which interact with a common membrane receptor show extensive homology: complete nucleotide sequences. Proc. Natl. Acad. Sci. USA **78**:6038–6042.

46. **Higgins, C. F., P. D. Haag, K. Nikaido, F. Ardeshir, G. Garcia, and G. F.-L. Ames.** 1982. Complete nucleotide sequence and identification of membrane components of the histidine transport operon of *S. typhimurium*. Nature (London) **298**:723–727.

47. **Hobot, J. A., E. Carlemalm, W. Villiger, and E. Kellenberger.** 1984. Periplasmic gel: new concept resulting from the reinvestigation of bacterial cell envelope ultrastructure by new methods. J. Bacteriol. **160**:143–152.

48. **Ichihara, S., N. Beppu, and S. Mizushima.** 1984. Protease IV, a cytoplasmic membrane protein of *Escherichia coli*, has signal peptide peptidase activity. J. Biol. Chem. **259**:9853–9857.

49. **Innis, M. A., M. Tokunaga, M. E. Williams, J. M. Loranger, S. Y. Chang, S. Chang, and H. C. Wu.** 1984. Nucleotide sequence of the *Escherichia coli* prolipoprotein signal peptidase (lsp) gene. Proc. Natl. Acad. Sci. USA **81**:3708–3712.

50. **Inouye, H., W. Barnes, and J. Beckwith.** 1982. Signal sequence of alkaline phosphatase of *Escherichia coli*. J. Bacteriol. **149**:434–439.

51. **Inouye, S., T. Franceschini, M. Sato, K. Itakura, and M. Inouye.** 1983. Prolipoprotein signal peptidase of *Escherichia coli* requires a cysteine residue at the cleavage site. EMBO J. **2**:87–91.

52. **Inouye, S., C.-P. S. Hsu, K. Itakura, and M. Inouye.** 1983. Requirement for signal peptide cleavage of *Escherichia coli* prolipoprotein. Science **221**:59–61.

53. **Inouye, S., X. Soberon, T. Franceschini, K. Nakamura, K. Itakura, and M. Inouye.** 1982. Role of positive charge on the amino-terminal region of the signal peptide in protein secretion across the membrane. Proc. Natl. Acad. Sci. USA **79**:3438–3441.

54. **Inouye, S., G. P. Vlasuk, H. Hsiung, and M. Inouye.** 1984. Effects of mutations at glycine residues in the hydrophobic region of the *Escherichia coli* prolipoprotein signal peptide on the secretion across the membrane. J. Biol. Chem. **259**:3729–3733.

55. **Ito, K.** 1984. Identification of the *secY* (*prlA*) gene product involved in protein export in *Escherichia coli*. Mol. Gen. Genet. **197**:204–208.

56. **Ito, K., P. J. Bassford, Jr., and J. Beckwith.** 1981. Protein localization in *E. coli*: is there a common step in the secretion of periplasmic and outer-membrane proteins? Cell **24**:707–717.

57. **Ito, K., and J. R. Beckwith.** 1981. Role of the mature protein sequence of maltose-binding protein in its secretion across the *E. coli* cytoplasmic membrane. Cell **25**:143–150.

58. **Ito, K., D. P. Cerretti, H. Nashimoto, and M. Nomura.** 1984. Characterization of an amber mutation in the structural gene for ribosomal protein L15, which impairs the expression of the protein export gene, *secY*, in *Escherichia coli*. EMBO J. **3**:2319–2324.

59. **Ito, K., G. Mandel, and W. Wickner.** 1979. Soluble precursor of an integral membrane protein: synthesis of procoat protein in *Escherichia coli* infected with bacteriophage M13. Proc. Natl. Acad. Sci. USA **76**:1199–1203.

60. **Jaurin, B., and T. Grindstrom.** 1981. *ampC* cephalosporinase of *Escherichia coli* K-12 has a different evolutionary origin from that of β-lactamases of the penicillinase type. Proc. Natl. Acad. Sci. USA **78**:4897–4901.

61. **Josefsson, L.-G., and L. L. Randall.** 1981. Different exported proteins in *E. coli* show differences in the temporal mode of processing *in vivo*. Cell **25**:151–157.

62. **Kellermann, O., and S. Szmelcman.** 1974. Active transport of maltose in *Escherichia coli*: involvement of a periplasmic maltose-binding protein. Eur. J. Biochem. **47**:139–149.

63. **Kennedy, E. P.** 1982. Osmotic regulation and the biosynthesis of membrane-derived oligosaccharides in *Escherichia coli*. Proc. Natl. Acad. Sci. USA **79**:1092–1095.

64. **Kiino, D. R., and T. J. Silhavy.** 1984. Mutation *prlF1* relieves the lethality associated with export of β-galactosidase hybrid proteins in *Escherichia coli*. J. Bacteriol. **158**:878–883.

65. **Koshland, D., and D. Botstein.** 1982. Evidence for posttranslational translocation of β-lactamase across the bacterial inner membrane. Cell **30**:893–902.

66. **Koshland, D., R. T. Sauer, and D. Botstein.** 1982. Diverse effects of mutations in the signal sequence on the secretion of β-lactamase in *Salmonella typhimurium*. Cell **30**:903–914.

67. **Kumamoto, C. A., and J. Beckwith.** 1983. Mutations in a new gene, *secB*, cause defective protein localization in *Escherichia coli*. J. Bacteriol. **154**:253–260.

68. **Kumamoto, C. A., D. B. Oliver, and J. Beckwith.** 1984. Signal sequence mutations disrupt feedback between secretion of an exported protein and its synthesis in *E. coli*. Nature (London) **308**:863–864.

69. **Landick, R., and D. L. Oxender.** 1982. Bacterial periplasmic binding proteins, p. 81–88. *In* A. Martonosi (ed.), Membranes and transport, vol. 2. Plenum Publishing Corp., New York.

70. **Leduc, M., C. Frehel, and J. van Heijenoort.** 1985. Correlation between degradation and ultrastructure of peptidoglycan during autolysis of *Escherichia coli*. J. Bacteriol. **161**:627–635.

71. **Lin, J. J. C., H. Kanazawa, J. Ozols, and H. C. Wu.** 1978. An *Escherichia coli* mutant with an amino acid alteration within the signal sequence of outer membrane prolipoprotein. Proc. Natl. Acad. Sci. USA **75**:4891–4895.

72. **Magota, K., N. Otsuji, T. Miki, T. Horiuchi, S. Tsunasawa, J. Kondo, F. Sakiyama, M. Amemura, T. Morita, H. Shinagawa, and A. Nakata.** 1984. Nucleotide sequence of the *phoS* gene, the structural gene for the phosphate-binding protein of *Escherichia coli*. J. Bacteriol. **157**:909–917.

73. **Malamy, M. H., and B. L. Horecker.** 1964. Release of alkaline phosphatase from cells of *Escherichia coli* upon lysozyme spheroplast formation. Biochemistry **3**:1889–1893.

74. **Michaelis, S., H. Inouye, D. Oliver, and J. Beckwith.** 1983. Mutations that alter the signal sequence of alkaline phosphatase in *Escherichia coli*. J. Bacteriol. **154**:366–374.

75. **Müller, M., and G. Blobel.** 1984. *In vitro* translocation of bacterial proteins across the plasma membrane of *Escherichia coli*. Proc. Natl. Acad. Sci. USA **81**:7421–7425.

76. **Müller, M., and G. Blobel.** 1984. Protein export in *Escherichia coli* requires a soluble activity. Proc. Natl. Acad. Sci. USA **81**:7737–7741.

77. **Muller, N., H.-G. Heine, and W. Boos.** 1982. Cloning of *mglB*, the structural gene for the galactose-binding protein of *Salmonella typhimurium* and *Escherichia coli*. Mol. Gen. Genet. **185**:473–480.

78. **Murray, R. G. E., P. Steed, and H. E. Elson.** 1965. The location

of the mucopeptide in sections of the cell wall of *Escherichia coli* and other gram-negative bacteria. Can. J. Microbiol. **11**:547–560.

79. **Neu, H. C., and L. A. Heppel.** 1965. The release of enzymes from *Escherichia coli* by osmotic shock and during the formation of spheroplasts. J. Biol. Chem. **240**:3685–3692.

80. **Newcomer, M. E., D. M. Miller III, and F. A. Quiocho.** 1979. Location of the sugar-binding site of L-arabinose-binding protein. J. Biol. Chem. **254**:7529–7533.

81. **Ohno-Iwashita, Y., and W. Wickner.** 1983. Reconstitution of rapid and asymmetric assembly of M13 procoat protein into liposomes which have bacterial leader peptidase. J. Biol. Chem. **258**:1895–1900.

82. **Oliver, D. B.** 1985. Identification of five new essential genes involved in the synthesis of a secreted protein in *Escherichia coli*. J. Bacteriol. **161**:285–291.

83. **Oliver, D.** 1985. Protein secretion in *Escherichia coli*. Annu. Rev. Microbiol. **39**:615–648.

84. **Oliver, D. B., and J. Beckwith.** 1981. *E. coli* mutant pleiotropically defective in the export of secreted proteins. Cell **25**:765–772.

85. **Oliver, D. B., and J. Beckwith.** 1982. Regulation of a membrane component required for protein secretion in *Escherichia coli*. Cell **30**:311–319.

86. **Oliver, D. B., and L. R. Liss.** 1985. *prlA*-mediated suppression of signal sequence mutations is modulated by the *secA* gene product of *Escherichia coli* K-12. J. Bacteriol. **161**:817–819.

87. **Osborn, M. J.** 1979. Biosynthesis and assembly of lipopolysaccharide of the outer membrane, p. 15–34. *In* M. Inouye (ed.), Bacterial outer membranes. John Wiley & Sons, Inc., New York.

88. **Oxender, D. L., J. J. Anderson, C. J. Daniels, R. Landick, R. P. Gunsalus, G. Zurawski, and C. Yanofsky.** 1980. Amino-terminal sequence and processing of the precursor of the leucine-specific binding protein, and evidence for conformational differences between the precursor and the mature form. Proc. Natl. Acad. Sci. USA **77**:2005–2009.

89. **Pardee, A. B.** 1966. Purification and properties of a sulfate-binding protein from *Salmonella typhimurium*. J. Biol. Chem. **241**:5886–5892.

90. **Parsons, R. G., and R. W. Hogg.** 1974. Crystallization and characterization of the L-arabinose-binding protein of *Escherichia coli* B/r. J. Biol. Chem. **249**:3602–3607.

91. **Penrose, W. R., G. E. Nicholalds, J. R. Piperno, and D. L. Oxender.** 1968. Purification and properties of the leucine-binding protein from *Escherichia coli*. J. Biol. Chem. **243**:5921–5928.

92. **Pflugrath, J. W., and F. A. Quiocho.** 1985. Sulfate sequestered in the sulfate-binding protein of *Salmonella typhimurium* is bound solely by hydrogen bonds. Nature (London) **314**:257–260.

93. **Quiocho, F. A., G. L. Gilliland, and G. N. Phillips, Jr.** 1977. The 2.8Å resolution structure of the L-arabinose-binding protein from *Escherichia coli*. J. Biol. Chem. **252**:5142–5149.

94. **Quiocho, F. A., W. E. Meador, and J. W. Pflugrath.** 1979. Preliminary crystallographic data of receptors for transport and chemotaxis in *Escherichia coli*: D-galactose and maltose-binding proteins. J. Mol. Biol. **133**:181–184.

95. **Quiocho, F. A., and J. W. Pflugrath.** 1980. The structure of D-galactose-binding protein at 4.1Å resolution looks like L-arabinose-binding protein. J. Biol. Chem. **255**:6559–6561.

96. **Randall, L., and S. J. Hardy.** 1984. Export of protein in bacteria: dogma and data, p. 1–20. *In* B. H. Sitir (ed.), Modern cell biology 3. Alan R. Liss, New York.

97. **Randall, L. L.** 1983. Translocation of domains of nascent periplasmic proteins across the cytoplasmic membrane is independent of elongation. Cell **33**:231–240.

98. **Rasmussen, B. A., V. A. Bankaitis, and P. J. Bassford, Jr.** 1984. Export and processing of MalE-LacZ hybrid proteins in *Escherichia coli*. J. Bacteriol. **160**:612–617.

99. **Rhoads, D. B., P. C. Tai, and B. D. Davis.** 1984. Energy-requiring translocation of the OmpA protein and alkaline phosphatase of *Escherichia coli* into inner membrane vesicles. J. Bacteriol. **159**:63–70.

100. **Rosen, B. P.** 1971. Basic amino acid transport in *Escherichia coli*. J. Biol. Chem. **246**:3653–3662.

101. **Rosen, B. P.** 1973. Basic amino acid transport in *Escherichia coli*. II. Purification and properties of an arginine-specific binding protein. J. Biol. Chem. **248**:1211–1218.

102. **Rosenblatt, M., N. V. Beaudette, and G. D. Fasman.** 1980. Conformational studies of the synthetic precursor-specific region of preproparathyroid hormone. Proc. Natl. Acad. Sci. USA **77**:3983–3987.

103. **Rosenbusch, J. P.** 1974. Characterization of the major envelope protein from *Escherichia coli*. J. Biol. Chem. **249**:8019–8029.

104. **Russel, M., and P. Model.** 1981. A mutation downstream from the signal peptidase cleavage site affects cleavage but not membrane insertion of phage coat protein. Proc. Natl. Acad. Sci. USA **78**:1717–1721.

105. **Scripture, J. B., and R. W. Hogg.** 1983. The nucleotide sequences defining the signal peptides of the galactose-binding protein and the arabinose-binding protein. J. Biol. Chem. **258**:10853–10855.

106. **Shiba, K., K. Ito, and T. Yuru.** 1984. Mutation that suppresses the protein export defect of the *secY* mutation and causes cold-sensitive growth of *Escherichia coli*. J. Bacteriol. **160**:696–701.

107. **Shiba, K., K. Ito, T. Yura, and D. Cerretti.** 1984. A defined mutation in the protein export gene within the *spc* ribosomal protein operon of *Escherichia coli*: isolation and characterization of a new temperature-sensitive *secY* mutant. EMBO J. **3**:631–635.

108. **Smith, W. P., P.-C. Tai, R. C. Thompson, and B. D. Davis.** 1977. Extracellular labeling of nascent polypeptides traversing the membrane of *Escherichia coli*. Proc. Natl. Acad. Sci. USA **74**:2830–2834.

109. **Spicer, E. K., and J. A. Noble.** 1982. *Escherichia coli* heat-labile enterotoxin. J. Biol. Chem. **257**:5716–5721.

110. **Stock, J. B., B. Rauch, and S. Roseman.** 1977. Periplasmic space in *Salmonella typhimurium* and *Escherichia coli*. J. Biol. Chem. **252**:7850–7861.

111. **Sutcliffe, J. G.** 1978. Nucleotide sequence of the ampicillin resistance gene of *Escherichia coli* plasmid pBR322. Proc. Natl. Acad. Sci. USA **75**:3737–3741.

112. **Taylor, R. T., S. A. Norrell, and M. L. Hanna.** 1972. Uptake of cyanocobalamin by *Escherichia coli* B: some characteristics and evidence for a binding protein. Arch. Biochem. Biophys. **148**:366–381.

113. **Tokunaga, M., J. M. Loranger, and H. C. Wu.** 1983. Isolation and characterization of an *Escherichia coli* clone overproducing prolipoprotein signal peptidase. J. Biol. Chem. **258**:12102–12105.

114. **Tokunaga, H., and H. C. Wu.** 1984. Studies on the modification and processing of prolipoprotein in *Escherichia coli*. J. Biol. Chem. **259**:6098–6104.

115. **Vlasuk, G. P., S. Inouye, and M. Inouye.** 1984. Effects of replacing serine and threonine residues within the signal peptide on the secretion of the major outer membrane lipoprotein of *Escherichia coli*. J. Biol. Chem. **259**:6195–6200.

116. **Vlasuk, G. P., S. Inouye, H. Ito, K. Itakura, and M. Inouye.** 1983. Effects of the complete removal of basic amino acid residues from the signal peptide on secretion of lipoprotein in *Escherichia coli*. J. Biol. Chem. **258**:7141–7148.

117. **Wanner, B. L., A. Sarthy, and J. Beckwith.** 1979. *Escherichia coli* pleiotropic mutant that reduces amounts of several periplasmic and outer membrane proteins. J. Bacteriol. **140**:229–239.

118. **Watts, C., W. Wickner, and R. Zimmermann.** 1983. M13 procoat and a pre-immunoglobulin share processing specificity but use different membrane receptor mechanisms. Proc. Natl. Acad. Sci. USA **80**:2809–2813.

119. **Weiner, J. H., and L. A. Heppel.** 1971. A binding protein for glutamine and its relation to active transport in *Escherichia coli*. J. Biol. Chem. **246**:6933–6941.

120. **Willis, R. C., and C. E. Furlong.** 1974. Purification and properties of a ribose-binding protein from *Escherichia coli*. J. Biol. Chem. **249**:6926–6929.

121. **Willis, R. C., and C. E. Furlong.** 1975. Purification and properties of a periplasmic glutamate-aspartate binding protein from *Escherichia coli* K12 strain W3092. J. Biol. Chem. **250**:2574–2580.

122. **Willis, R. C., R. G. Morris, C. Cirakoglu, G. D. Schellenberg, N. H. Gerber, and C. E. Furlong.** 1974. Preparation of the periplasmic binding proteins from *Salmonella typhimurium* and *Escherichia coli*. Arch. Biochem. Biophys. **161**:64–75.

123. **Wolfe, P. B., M. Rice, and W. Wickner.** 1985. Effects of two *sec* genes on protein assembly into the plasma membrane of *Escherichia coli*. J. Biol. Chem. **260**:1836–1841.

124. **Wolfe, P. B., and W. Wickner.** 1984. Bacterial leader peptidase, a membrane protein without a leader peptide, uses the same export pathway as pre-secretory proteins. Cell **36**:1067–1072.

125. **Wolfe, P. B., W. Wickner, and J. M. Goodman.** 1983. Sequence of the leader peptidase gene of *Escherichia coli* and the orientation of leader peptidase in the bacterial envelope. J. Biol. Chem. **258**:12073–12080.

126. **Yamada, H., H. Yamagata, and S. Mizushima.** 1984. The major outer membrane lipoprotein and new lipoproteins share a common signal peptidase that exists in the cytoplasmic membrane of *Escherichia coli*. FEBS Lett. **166**:179–182.

127. **Yamagata, H., C. Ippolito, M. Inukai, and M. Inouye.** 1982. Temperature-sensitive processing of outer membrane lipoprotein in an *Escherichia coli* mutant. J. Bacteriol. **152:**1163–1168.

128. **Yu, F., H. Yamada, K. Daishima, and S. Mizushima.** 1984. Nucleotide sequence of the *lspA* gene, the structural gene for lipoprotein signal peptidase of *Escherichia coli*. FEBS Lett. **173:**264–268.

129. **Zwizinski, C., T. Date, and W. Wickner.** 1981. Leader peptidase is found in both the inner and outer membranes of *Escherichia coli*. J. Biol. Chem. **256:**3593–3597.

7. Flagella

ROBERT M. MACNAB

Department of Molecular Biophysics and Biochemistry, Yale University, New Haven, Connecticut 06511

INTRODUCTION

A salient feature of many true wild-type strains of *Escherichia coli* or *Salmonella typhimurium*, when observed by electron microscopy (Fig. 1) or—with suitable optics—light microscopy (66) (Fig. 2), is a set of extremely thin helical appendages that are utilized in motility. These have traditionally been called flagella, but, strictly speaking, they are only the external portions of the flagella and are more correctly called flagellar filaments. The structure and assembly of the entire flagellum, including this external filament, is the subject of this chapter; only passing reference will be made to the manner in which flagella function. The latter aspect, including the modulation of function by environmental information, is dealt with elsewhere in this volume (chapter 49).

Although the term flagellum is used for the propulsive organelle of eucaryotes also (80), the bacterial flagellum is completely distinct from that organelle in several regards. (i) The external filament of the bacterial flagellum does no chemomechanical work, but is

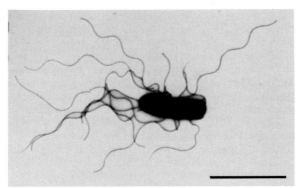

FIG. 1. Electron micrograph of a cell of *S. typhimurium*, showing peritrichous arrangement of flagella. Bar, 5 μm (Courtesy of S. Aizawa.)

passively driven by a motor at its base. (ii) The mechanism is one of rotation (7, 63, 85), not bending (and hence the term "flagellum," meaning whip, is inaccurate). (iii) The motor is switchable, operating in both the counterclockwise and clockwise rotational senses (7, 63, 85). (iv) The energy source is the transmembrane proton potential (proton motive force), not ATP (51, 62, 70, 74). (v) The external filament contains only one protein, whereas the eucaryotic flagellum contains many and is physically much larger (procaryotic filament, 20 nm in diameter; eucaryotic flagellum, ca. 200 nm in diameter).

It should be recognized that flagellar motility is not the only form of motility displayed by bacteria, and also that the pattern of flagellation varies considerably among species (e.g., reference 67). In the case of *E. coli* and *S. typhimurium*, the points of attachment of the flagellar filaments to the cell body show no defined pattern; such cells are described as peritrichously flagellated.

This chapter begins with the morphology and biochemistry of the flagellum, continues with the genetics underlying its assembly, structure, and function, and then considers the process of assembly itself.

Throughout, where a statement is made without further qualification, it is either known or presumed to apply equally to both *E. coli* and *S. typhimurium*. Where a gene symbol refers to one of the species only,

it will be subscripted by an E for *E. coli* (for example, $flaU_E$) or by an S for *S. typhimurium* ($flaFI_S$).

For further information and discussion regarding bacterial flagella, see references 8, 36–38, 68, and 89.

MORPHOLOGICAL AND BIOCHEMICAL DESCRIPTION

The bacterial flagellum, as it is currently recognized morphologically (1, 2, 13, 17, 19), consists of the external filament, a short curved segment called the hook, and a complex structure of rings and rods called the basal body (Fig. 3). Such a structure is obtained by dissolution of the peptidoglycan layer and inner and outer membranes (2, 16) and has been called the "intact flagellum." For reasons that will be presented briefly in this chapter and expanded upon in chapter 49, this term is likely to be a misnomer, since there are several critically important components that are not present in this structure (2, 15, 27).

Filament

The filament is not a part of the chemomechanical energy-transducing apparatus, but it is the propulsive component of the flagellum that performs hydrodynamic work on the medium. It represents by far the major component in terms of mass (and therefore biosynthetic cost), possesses remarkable structural properties, is significant clinically in terms of its potency and variability as an antigen, and has been extensively studied.

Geometry. The flagellar filament is of ill-defined length, typically in the range from 5 to 10 μm. It has a constant diameter of ca. 20 nm throughout its length (75, 82). The helical parameters vary slightly between the species and even among different strains; typical values are 2 to 2.5 μm for the helical wavelength (pitch) and 0.4 to 0.6 μm for the helical diameter (4). For any given filament, these values are constant throughout its length; in other words, the structure has a strictly conserved geometry and may be regarded as a unidimensional crystal.

Composition. The filament consists of an indefinite quaternary assembly of several thousand copies of a single protein, flagellin, whose molecular weight is

FIG. 2. Dark-field light micrograph of a flagellated cell of *S. typhimurium*. During swimming the flagella, shown dispersed here, coalesce into a bundle. Bar, 5 μm. (From reference 69, with permission.)

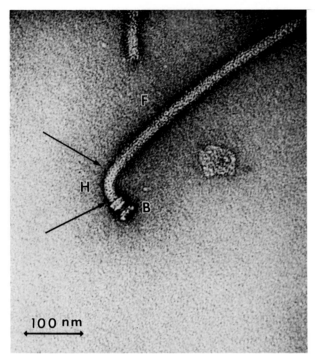

FIG. 3. "Intact flagellum" isolated from *S. typhimurium*. The propulsive helical filament (F) is connected by the hook (H) to the basal body (B). As is discussed in the text, this structure is not likely to comprise the entire flagellar apparatus. (Courtesy of S. Aizawa and H. Kagawa. From reference 68, with permission.)

typically around 55,000 (Table 1) (59) but can vary greatly from strain to strain (64). Flagellin has the unusual characteristic (although one shared by several other mechano-enzyme proteins) of containing methyl-modified residues, in this case, *N*-methyl-lysines (3). No functional significance appears to derive from this modification, since mutants lacking the modification function normally.

Flagellin is a quite elongated protein; three-dimensional image reconstruction suggests a trilobar structure with an overall axial ratio of ca. 3 or 4 (82). It also possesses a high dipole moment (ca. 860 Debye as the monomer) (25). Melting and circular dichroism properties of monomeric flagellin versus filament suggest that the full ordering of the subunit requires its incorporation into the filament (23). It has also been suggested that the outermost domain in the intact filament may not be required for quaternary interactions and may therefore be free to diverge antigenically (see below).

Mechanical properties. A striking characteristic of the flagellar filament is its rigidity; its coefficient of flexural rigidity is ca. 10^{-15} dyne cm^2 (24), making flagellar filament about two orders of magnitude stiffer than F-actin, for example. Detached filaments in buffer show essentially no flexing Brownian motion, and even during active motility lateral forces induce only rather gentle bending (R. Macnab, unpublished data). Torsional rigidity is also estimated to be high (ca. 10^{11} dyne cm^{-2}; 32), in agreement with the absence of appreciable distortion of the waveform during flagellar rotation (69).

The capacity of flagellar filaments to undergo inelastic conformational changes will be discussed below.

Self-assembly characteristics. Isolated flagellar filaments can be readily depolymerized by heat, acid pH, or other treatments (4). The reverse process of polymerization into filament occurs spontaneously under suitable conditions, yielding a structure that is indistinguishable from the original (4, 49); normally, this process requires initiation by short "seed" fragments of filament. In vitro, the rate of elongation is constant (33), with monomer being added at the end that would be distal to the attachment point of the filament to the cell (4); the assembly process in vivo will be discussed in a later section.

Basic subunit organization. The flagellin subunits are arranged on a cylindrical lattice, with approximately hexagonal close-packing (60, 75). One of the directions of the lattice is almost, but not quite, parallel to the filament axis, and there are 11 columns ("fibrils") of subunits in this direction. Each subunit is tilted at an angle of ca. 45° to the filament axis (82), giving rise to a pronounced notch and arrowhead appearance for the distal and proximal ends, respectively, of filaments grown in vitro.

Quasi-equivalence and polymorphism. As mentioned above, the filament is made from subunits of a single protein. If all the subunits in the cylindrical lattice were in truly identical environments, the filament would be straight, whereas it is observed to be helical; indeed, it must be helical to convert rotational force (torque) into translational force (thrust). There must therefore be a quasi-equivalence (53), rather than a true equivalence, of subunit environments. All of the available evidence is consistent with the idea that subunits in a given fibril are in identical environments, but that the fibrils differ among themselves, with a shorter subunit-subunit distance for fibrils on one side of the filament than for those on the other side. Since the fibrils are not quite parallel to the filament axis, the variation of fibril length around the circumference introduces macroscopic helicity to the filament (4).

The situation is in fact even more complex than this. For a given molecular species of flagellin, a variety of helical forms can occur depending on physical conditions such as pH or ionic strength (Fig. 4); in other words, flagellar filaments display polymorphism (49). Theory predicts 11 + 1 = 12 forms, of which about 9 have been observed experimentally; they include 2 extreme forms that are straight and intermediate right-handed and left-handed helical forms such as "curly," "coiled," etc. The effect can also be obtained by alteration of the flagellin primary sequence, even by a single amino acid substitution (40, 41, 71); it can also be obtained by incorporation of an amino acid analog such as *p*-fluorophenylalanine (50, 92). The phenomenon can be explained in terms of the numbers of the 11 fibrils that are in the extended versus the compressed states (10). For a more detailed consideration of the molecular events surrounding flagellar polymorphism, see references 11 and 49.

Interconversion among polymorphic forms can also be caused by mechanical force and appears to be an important feature of tumbling in the motility pattern of *E. coli* and *S. typhimurium* (69; chapter 49).

TABLE 1. Genes and gene products involved in flagellation and motility of *E. coli* and *S. typhimurium*[a]

Gene symbol	Gene product		
	Mol wt[b]	Location	Function and comments
Region I genes			
flaU$_E$	?	?	Necessary for addition of P ring; repressor of late operons
flaFI$_S$	?		
flbA$_E$	?	?	Necessary for any detectable flagellar structure; genetic evidence suggests a motor component
flaFII$_S$	?		
flaW$_E$	?	?	Necessary for any detectable flagellar structure
flaFIII$_S$	14×10^3	Basal body	
flaV$_E$	11×10^3	Basal body?	Completes rod structure; necessary for addition of L ring and outer cylinder
flaFIV$_S$	?		
flaK$_E$	42×10^3	Flagellar	Structural protein for hook, which connects flagellar filament to basal body complex and may be a "universal joint"
flaFV$_S$	42×10^3	hook	
flaX$_E$	30×10^3	Basal body	Necessary for any detectable flagellar structure; rod structural protein?; absent in ring preparations
flaFVI$_S$	32×10^3		
flaL$_E$	27×10^3	Basal body	Necessary for any detectable flagellar structure; rod structural protein?; absent in ring preparations
flaFVII$_S$	30×10^3		
flaY$_E$	?	Basal body	Necessary for L ring and outer cylinder; present in ring preparations; probably L ring structural protein
flaFVIII$_S$	27×10^3		
flaM$_E$	38×10^3	Basal body	Necessary for P ring assembly; present in ring preparations; probably P ring structural protein; very basic (pI ca. 9).
flaFIX$_S$	38×10^3		
flaZ$_E$	?	?	Necessary for any detectable flagellar structure; positive regulation of flagellar filament synthesis
flaFX$_S$	?		
flaS$_E$	60×10^3	Hook/filament	Caps hook; necessary for filament addition; hook accessory protein (HAP1)
flaW$_S$	60×10^3	junction	
flaT$_E$	35×10^3	Hook/filament	Caps hook, following HAP1; necessary for filament addition; hook accessory protein (HAP3)
flaU$_S$	35×10^3	junction	
Region II genes			
flaH$_E$	?	?	Necessary for any detectable flagellar structure; interaction with outer membrane
flaC$_S$	?		
flaG$_E$	?	?	Necessary for any detectable flagellar structure
flaM$_S$	?		
motB$_E$	34,147	Cell membrane	Necessary for motor rotation; can activate pre-existing flagella; site-limited when overproduced
motB$_S$	?		
motA$_E$	31,974	Cell membrane	Necessary for motor rotation; extremely hydrophobic; contains regions of high positive and negative charge density; not site-limited
motA$_S$	?		
flaI$_E$	21,519	Cytoplasm	Necessary for any detectable flagellar structure; necessary for expression of all other motility/taxis operons; high density of symmetry elements in gene
flaE$_S$	?		
flbB$_E$	13,570	Cytoplasm	Necessary for any detectable flagellar structure; necessary for expression of all other motility/taxis operons; operon is under cyclic AMP/CAP control
flaK$_S$	?		
?$_E$			Necessary for any detectable flagellar structure; position with respect to flaK$_S$ not known precisely
flaY$_S$	?	?	
Region III genes			
flaD$_E$	?	?	Necessary for any detectable flagellar structure; necessary for filament addition; activator of late operon expression
flaL$_S$	?		
?$_E$			
nml$_S$	?	?	*N*-Methylation of certain lysine residues of flagellin

TABLE 1—*Continued*

Gene symbol	Gene product		
	Mol wt	Location	Function and comments
hag_E $H1_S$	ca. 55×10^3 ca. 55×10^3	Flagellar filament	Filament structural protein (flagellin); molecular weight dependent on serotype of strain; $H1_S$ and $H2_S$ alternately expressed
$flbC_E$ $flaV_S$	? 53×10^3	Flagellar filament	Distal end of filament?; necessary for filament addition; prevents loss of flagellin to medium; hook accessory protein (HAP2)
$flaN_E$ $flaAI_S$	? ?	?	Necessary for any detectable flagellar structure
$flaBI_E$ $flaAII.1_S$	60,521 65×10^3	Basal body	Necessary for any detectable flagellar structure; probably M ring
$flaBII_E$ $flaAII.2_S$	36,788 ?	Peripheral to cell membrane?	Necessary for any detectable flagellar structure; switch component of motor; interacts with CheY, CheZ, FlaN$_S$, and FlaQ$_S$ proteins; gene also known as $motC_S$, $cheV$
$flaBIII_E$ $flaAII.3_S$	26,108 ?	?	Necessary for any detectable flagellar structure
$flaC_E$ $flaAIII_S$	56×10^3 ?	?	Necessary for any detectable flagellar structure
$flaO_E$ $flaS_S$	17×10^3 ?	?	Necessary for any detectable flagellar structure
$flaE_E$ $flaR_S$	54×10^3 ?	Cytoplasm?	Hook length control (polyhook mutant phenotype); necessary for filament addition (*phf* alleles excepted)
$flaAI_E$ $?_S$	17,201	Cell membrane?	Necessary for any detectable flagellar structure; very hydrophobic
$flaAII_E$ $flaQ_S$	37,806 ?	Peripheral to cell membrane	Necessary for any detectable flagellar structure; switch component of motor; interacts with CheY, CheZ, FlaAII.2$_S$, and FlaN$_S$ proteins; gene also known as $motE$, $cheC$ ($cheC_S$ formerly called $cheU_S$)
$motD_E$ $flaN_S$	14,388 ?	Peripheral to cell membrane	Necessary for any detectable flagellar structure; switch component of motor; FlaN$_S$ protein interacts with CheY$_S$, FlaAII.2$_S$, and FlaQ$_S$ protein; $flaN_S$ gene also known as $motD_S$
$flbD_E$ $flaP_S$	? ?	?	Necessary for any detectable flagellar structure; correspondence between $flbD_E$ and $flaP_S$ not established
$flaR_E$ $flaB_S$	? ?	?	Necessary for any detectable flagellar structure
$flaQ_E$ $flaD_S$	? ?	?	Necessary for any detectable flagellar structure
$flaP_E$ $flaX_S$	? ?	?	Necessary for any detectable flagellar structure
Salmonella H2 region $rh1_S$	?	Cytoplasm?	Repressor of $H1_S$ gene; $rh1$ only expressed in one polarity of hin_S region
$H2_S$	ca. 55×10^3	Flagellar filament	Filament structural protein (flagellin); molecular weight dependent on serotype of strain; $H1_S$ and $H2_S$ alternately expressed; $H2_S$ only expressed in one polarity of hin_S region
hin_S	21,332	Cytoplasm?	Mediates inversion of the region containing the hin_S operon and the $H2_S$ promoter, activating or deactivating $H2_S$ and $rh1_S$ gene expression

[a] Genes in region I map at 24 min (*E. coli*) or 23 min (*S. typhimurium*); genes in region II map at 41 min (*E. coli*) or 40 min (*S. typhimurium*); genes in region III map at 42 to 43 min (*E. coli*) or 40 min (*S. typhimurium*); and genes in the *S. typhimurium* H2 region map at 56 min. For further detail regarding the genomic organization, see Table 3 of chapter 49. Data were gathered from a number of sources, including references 2, 6, 29, 31, 37, 42, 56, 57, 61, 72, 73, 83, 90a, 94, 95, 99a, and 100.

[b] Where a molecular weight is given an exact value, it has been obtained by translation of the corresponding gene sequence; such values are available for several flagellins and are strain and (in *S. typhimurium*) phase dependent. Otherwise, values given refer to the apparent molecular weight obtained by sodium dodecyl sulfate-polyacrylamide gel electrophoresis.

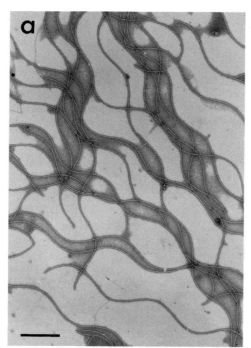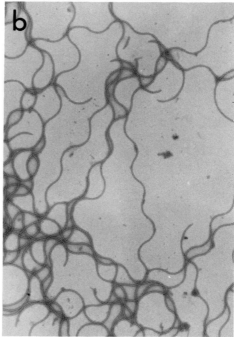

FIG. 4. Two of the helical polymorphs available to flagellar filaments of *S. typhimurium*. The samples were prepared identically, except that the pH for the sample in panel a was 7.0 and that for the sample in panel b was 4.5. Different polymorphs are generated by the cell during tumbling (see chapter 49). Bar, 1 μm. (Courtesy of R. Kamiya and S. Asakura. From reference 48, with permission.)

Antigenicity. Flagellin and flagellar filaments are potent antigens, and flagellar serotype has provided a major basis for distinguishing strains, particularly of *S. typhimurium*, from a clinical and epidemiological perspective (20). The antigenic determinants appear to lie at outer radii of the filament (64), since it was found in a study of different flagellar serotypes of *E. coli* that they corresponded to a wide range of flagellin molecular weights and a variety of different superficial appearances. Based partly on genetic fine mapping and partly on peptide sequence information, it has been concluded that (i) the N and C termini of flagellin are highly conserved, (ii) the central region is much more free to diverge without interfering with the assembly process, and (iii) such divergence is likely to be responsible for antigenic variation (reviewed in reference 36). This conclusion has now been further supported by comparative sequencing of different flagellin genes of *S. typhimurium* (99); the N and C termini of the predicted translation products are highly conserved, while the middle third of the protein sequence shows considerable variation.

Hook

The filament is connected to the cell by the hook, which is structurally similar to the filament but contains a quite distinct subunit, the hook protein, with a molecular weight of ca. 42,000 (Table 1) and an elongated shape like that of flagellin (98). The same 11-fibril structure exists (98), as does the quasi-equivalence that produces macroscopic helicity and polymorphism (44). The helical wavelength of the hook is quite short (ca. 130 nm), but even so the hook is only

of sufficient length (ca. 80 nm) for a partial turn of the helix. Largely by hypothesis, but with some supporting structural evidence (98), the hook is believed to act as a flexible coupling, or universal joint, between filament and cell. The flexibility may not be elastic but could involve cycling of the quasi-equivalent phase of the hook, an event which is presumed to be forbidden in the case of the filament (see discussion in reference 68).

In contrast to the filament, the hook has a rather well-defined length (ca. 80 nm). Between hook and filament are two junction or hook accessory proteins (HAP1 and HAP3; Table 1; 29–31). Given the similarities between filament and hook in terms of lattice, and subunit shape and disposition (98), it is unclear why such proteins should represent a structural requirement. It seems more likely that they are required for a proper control of filament assembly (see below).

Antigenically, the hook proteins of *E. coli* and *S. typhimurium* are rather similar to each other (45), in contrast to the rather wide range of antigenicity exhibited among flagellins even within one species; this may reflect the fact that the flagellar filament constitutes a much greater antigenic challenge to the host organism and that the bacterium has therefore developed the potential for antigenic variability in response.

Basal Body

Geometry and composition. The hook is connected to a structural complex known as the basal body, which is embedded in the cell surface (1, 2, 13, 17, 19). A particle called the hook-basal body complex (Fig. 5)

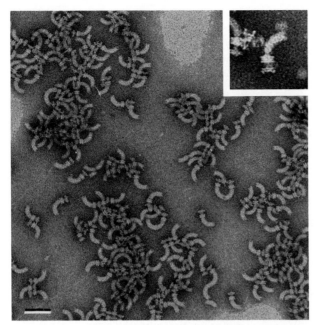

FIG. 5. A preparation of flagellar hook-basal body complexes used for biochemical analysis. Inset shows the structure at higher magnification and may be compared with the cartoon of the flagellar apparatus shown in Fig. 6. Bar, 100 nm. (Modified from reference 2, with permission.)

is obtained by depolymerizing the filament from the "intact flagellum."

The basal body is seen to be of cylindrical symmetry, consisting of a rod joined to the hook plus a set of four rings. The outer two rings (called L and P) are spaced apart but appear to be linked by a cylindrical wall. The inner two rings (S and M) appear to abut directly. Computerized image analysis is under way in an attempt to further refine our knowledge of the basal body (B. Stallmeyer and D. DeRosier, personal communication). Both geometrical and biochemical considerations (18) suggest that, in the intact cell, the L ring is in the plane of the outer (lipopolysaccharide) membrane, the P ring is in the plane of the peptidoglycan layer, and the S and M rings (standing for supramembrane and membrane, respectively) are in the vicinity of the cell membrane, probably towards its outer face. A cartoon that illustrates the organization of the filament, hook, and basal body, as well as other predicted features that will be discussed later, is given in Fig. 6.

There are at least seven proteins associated with the basal body (excluding the hook and hook accessory proteins) (2, 57; C. Jones, unpublished data); in *S. typhimurium*, these are the 65,000-molecular-weight (65K) FlaAII.1 protein, the 38K FlaFIX protein, the 32K FlaFVI protein, the 30K FlaFVII protein, the 27K FlaFVIII protein, and the 14K FlaFIII protein (Table 1), plus the 16K protein for which the corresponding gene is not known. There may be as many as 15 proteins in the basal body, on the basis of preliminary differential radiolabeling experiments of basal bodies from conditional mutants (C. Jones, unpublished data).

Postulated function. There is no firm information regarding the function of the various morphological features of the basal body. It may be largely a passive structure (see next section). The L, P ring pair together with its cylindrical wall is likely to provide a bushing through which the rod, as "transmission shaft," passes. Which of the remaining features are rotating and which are stationary is not known; specifically, it is not known whether the rod passes through the S ring and is engaged to the M ring, although this is commonly assumed to be the case. The M ring may act as a mounting plate for components essential for motor rotation and switching (see below).

Other Undetected Motor Structures

Although the basal body is a major component of the bacterial flagellum, it does not contain several proteins that are known to be important for motor function (2, 12, 27). These are the MotA and MotB proteins and several Fla proteins involved in switching (Table 1). Their properties are discussed briefly here and more extensively in chapter 49.

MotA and MotB proteins. The MotA and MotB proteins are integral to the cell membrane and are necessary for motor rotation, but not for motor switching or for flagellar assembly (15, 22, 27, 79, 90a). In fact, pre-existing basal bodies can be made to rotate by virtue of newly synthesized MotB protein (9). These observations suggest that the MotA and MotB proteins

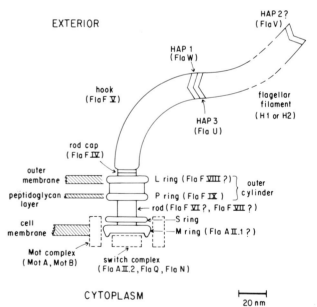

FIG. 6. Current knowledge of the structure of the flagellum of *E. coli* or *S. typhimurium*, as deduced from electron microscopy and genetic and biochemical analysis. Principal features are the helical flagellar filament (truncated in this representation), the hook, and the basal body (see Fig. 3 and 5). The basal body is embedded in the layers of the cell surface as shown. Where the identity of a gene product is known or suspected, it is given in parentheses (using *S. typhimurium* nomenclature). The switch and Mot complexes have not been detected by electron microscopy and are therefore indicated in dashed outline; there are a number of lines of evidence to suggest they are in the locations shown. (Based on data from references 2, 17, 18, 29, 31, 37, 57, 94, and 95.)

may be arranged circumferentially around the S, M ring pair (Fig. 6).

Switch proteins. At least three non-basal body proteins (FlaBII$_E$/FlaAII.2$_S$, FlaAII$_E$/FlaQ$_S$, and MotD$_E$/FlaN$_S$) that are needed for flagellar assembly are also involved in the rotation of the motor and its switching (14, 22, 52, 76, 96, 97, 101). When overproduced, they appear in the cytoplasmic fraction (12), yet, at least in the case of FlaAII.2$_S$ and FlaQ$_S$, they are retained in cytoplasm-free cell envelopes (78). They may be peripheral to the cell membrane, perhaps at the cytoplasmic face of the M ring (Fig. 6), and may form a switch complex that provides the interface between the energy-transducing mechanism and the sensory information from the chemotaxis system. The existence of such a switch complex is supported by intergenic suppression analysis (S. Aizawa, S. Yamaguchi, M. Kihara, C. Jones, and R. Macnab, unpublished data).

GENETICS

Mutants

Mutants have provided much useful information about the flagellar system. We refer here only to those mutants defective in flagellar structure and function; chapter 49 contains similar information for mutants defective in the sensory transduction apparatus.

Nonflagellate or partially flagellate (Fla⁻) mutants. The non- or partially flagellate mutant phenotype is by far the most common with respect to motility. As an illustration of this fact, we may note that many "wild-type" laboratory strains that are under study for reasons unrelated to motility are nonflagellate. Indeed, spontaneous nonflagellate mutants rapidly (ca. 10 days) overtake a motile population in stirred culture because of the metabolic cost of synthesizing the flagellar apparatus. (The growth disadvantage of this synthesis is estimated to be about 2% [R. Macnab, unpublished data].)

Originally defined in terms of loss of the highly visible external filament, the Fla⁻ phenotype is now defined as any substantial diminution of the flagellar apparatus. Of the many genes needed for complete assembly of the apparatus, a large fraction are necessary for any detectable flagellar structure (94, 95). Defective genes for later components of the basal body, for the hook, and for the filament can, however, permit production of partial flagellar structures.

Polyhook mutants. A special example of partial flagellation is exhibited by flaE$_E$/flaR$_S$ mutants, which display the polyhook phenotype, whereby the normal process of hook length control fails and excessively long hooks are produced (77, 93; see below).

Paralyzed (Mot⁻) mutants. In Mot⁻ mutants, no morphological defects are manifest in the flagellar apparatus, yet it does not rotate. In some cases (motA and motB mutants), the gene product can be totally lacking; in other cases (certain missense mutations in flaBII$_E$/flaAII.2$_S$, flaAII$_E$/flaQ$_S$, and motD$_E$/flaN$_S$), the paralyzed phenotype occurs because of defective function of a component whose presence is nonetheless necessary for flagellar structure and assembly.

Switch-defective (Che⁻) mutants. Switch-defective (Che⁻) mutants possess rotating flagella but the switching properties are abnormal, with either a high counterclockwise or a high clockwise rotational bias compared to wild type. Although many such mutants are actually defective in components of the sensory apparatus (see chapter 49), the three fla genes described above in the context of paralyzed mutants can generate a switch-defective phenotype also. Interestingly, a detailed mapping of fla, mot, and che alleles of these genes indicates a considerable segregation of function within the primary sequence of the proteins (99a).

Genes and Genomic Organization

Conventions for gene symbols. The following conventions are used for naming flagellar and motility genes. (i) Genes whose products are essential for assembly of the morphologically recognizable flagellar apparatus are given the symbol fla (or flb). (ii) A special exception to this convention is made for the flagellin structural gene (or genes, in the case of S. typhimurium; see below), which for historical reasons is given the symbol hag$_E$ or H1$_S$ (or H2$_S$), for "Hauch antigen." (Hauch, meaning "breath" in German, described the swarm morphology of motile strains on soft plates.) (iii) Genes whose products are not necessary for assembly of the flagellum (as currently recognized), but are necessary for motor rotation, are given the symbol mot. (iv) More loosely, the symbols mot and che (for chemotaxis) have been applied to certain fla genes whose products, if present but defective, can prevent normal motor rotation or switching, respectively; a preferable terminology is fla(Mot) and fla(Che) (99a).

Genes involved in motility. The genes currently known to be involved in flagellation and motility are summarized in Table 1. A complete compilation of all genes involved in flagellation, motility, and chemotaxis, together with their organization into operons, is given in Table 3 of chapter 49. For further description of the original genetic analyses, see references 36 and 86; for the establishment of homologies between the two species, see reference 61.

There are about 33 fla (flb) genes, either one (E. coli) or two (S. typhimurium) flagellin genes, and two mot genes. S. typhimurium has two additional genes that control mutually exclusive expression of its two flagellin genes (see below). In view of the intensive studies of flagellar genetics, there are unlikely to be many undiscovered genes, although a few of those listed in Table 1 have only been detected recently as a result of DNA sequencing (e.g., the subdivision of the flaA$_E$ and flaB$_E$ loci; 6) or large-scale mutant analysis (e.g., flaAII.3$_S$; 99a). Motility-linked genes are highly clustered on the chromosome. A large cluster, termed region I, contains 12 genes and occurs at about 24 min (E. coli) or 23 min (S. typhimurium); most of the known basal body genes are located in this cluster. Two other clusters (regions II and III) occur at around 41 to 43 min (E. coli) or 40 min (S. typhimurium); region II also contains sensory-linked genes. (The justification for considering II and III to be separate regions is the existence of non-motility-related genes such as uvrC in between.) In the case of S. typhimurium, the operon for expression of one of the two flagellin genes, H2 (see section on phase variation below), is in a remote location at 56 min.

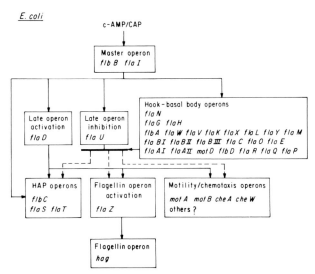

FIG. 7. Control of gene expression within the flagellar/motility regulons of *E. coli*. (i) Cyclic AMP binds to CAP (or CRP) protein, and the complex binds in the promoter region of the master operon, acting as a positive regulator. (ii) Expression of the *flbB*$_E$ and *flaI*$_E$ genes activates expression of all other flagellar/motility genes, directly or indirectly. Direct activation occurs with most of the hook-basal body operons and with two one-gene operons (*flaD*$_E$ and *flaU*$_E$). (iii) *flaD*$_E$ expression activates operons for later components of the flagellar apparatus (hook accessory proteins, flagellin [with *flaZ*$_E$ expression as an intermediate], and Mot and Che proteins). (iv) *flaU*$_E$ acts on the same operons as *flaD*$_E$, but as a negative regulator. (v) Where all of the basal body operons are being expressed, it is postulated that FlaU protein is incorporated in the basal body, preventing it from acting as an inhibitor of late gene expression. (Scheme based on results from reference 54 and from Y. Komeda [unpublished data]; a related scheme applies in *S. typhimurium* [37].)

Gene Product Functions

For a number of the genes, products have been identified with flagellar structure and function (Table 1). These include the genes for flagellin, hook protein, hook-filament junction proteins, and several basal body proteins, as well as the *mot* genes and those *fla* genes whose products appear to constitute the flagellar switch.

A few regulatory genes have also been identified. The most noteworthy of these are the two genes of a "master operon" for flagellar assembly. There is also a gene involved in regulating hook length and genes involved in the variation of flagellin phase in *S. typhimurium*. These topics are all covered below.

For the remainder of the genes (approximately 17), no function has yet been clearly identified. Some may code for presently unknown structural components, but many may prove to be involved in the complex logistical process of assembly of the organelle into the cell surface, perhaps coding for scaffolding proteins or enzymes.

Control of Expression

For this discussion of control of expression of flagellar and motility genes, refer to Fig. 7.

Overall control of the flagellar/motility regulon. All genes known to relate exclusively to motility exist within operons dedicated exclusively to that function. The entire system constitutes a regulon, since expression of all of the operons is under the positive control of a master regulatory operon, i.e., the *flbB/flaI* operon in *E. coli* or the *flaK/flaE* operon in *S. typhimurium* (54, 58, 84). This operon is itself positively regulated by a cyclic AMP/CAP protein complex; the CAP (also known as CRP) binding site, in the promoter region of the operon, is referred to in the literature as the *cfs* locus (for constitutive for flagellar synthesis). The actual role of the two gene products in activating expression of motility/chemotaxis operons is not known, but in several cases the presumed control regions of the target operons show sequence homology that may indicate a common activator binding site (P. Matsumura, personal communication). The sequence of this operon in *E. coli* is now available (P. Matsumura, personal communication); the *flaI*$_E$ gene shows extensive symmetries (R. Macnab, unpublished data) that may reflect self-regulation.

Secondary control. There is also evidence for secondary control, especially for what may be regarded as functionally late operons, notably those for flagellin synthesis (a biosynthetically expensive item for the cell) and for chemotaxis (54; Y. Komeda, unpublished data). Negative regulation may involve use of a structural protein (the *flaU*$_E$ gene product) as repressor in the event that it is not incorporated into the flagellar apparatus; thus normal expression of the basal body genes relieves the system from *flaU*$_E$ inhibition (Y. Komeda, unpublished data). Positive control of the *mocha*$_E$, *flbC*$_E$, *flaS*$_E$, and *flaZ*$_E$ operons is accomplished by the *flaD*$_E$ gene, while the *flaZ*$_E$ gene in turn activates *hag*$_E$ expression. The *flaL*$_S$ gene, homologous to *flaD*$_E$, appears to play a similar role in *S. typhimurium* (37, 91).

It is evident that there is a fairly complex cascade of control, as might be expected for a system that is architecturally complex and biosynthetically costly.

Flagellin phase variation. For reasons that are not really understood but which may relate to survival against host defense systems, *Salmonella* spp. (but not *E. coli*) have two genes, at different locations on the chromosome, that code for the filament protein flagellin (65). The genes are highly homologous but not identical, and so their products are antigenically distinct. In wild-type cells, either product is perfectly functional in assembling into a helical flagellar filament that can propel the cell.

At any given time, only one of the two genes is being expressed and therefore a cell possesses homogeneous filaments. In a stochastic fashion, on a time scale on the order of 10^3 to 10^5 generations, the genes change roles so that the nonexpressed one becomes expressed and vice versa.

The mechanism of this unusual phenomenon has been shown to be as follows (Fig. 8) (39, 90). The *H1* gene is subject to repression by the product of the *rh1* gene, which is contained within the *H2* operon. Expression of the *H2* operon therefore precludes expression of the *H1* gene. Alternating expression and nonexpression of the *H2* operon is caused by the inversion of a 970-base-pair region of DNA upstream of the *H2* gene that contains the *H2* promoter; in one

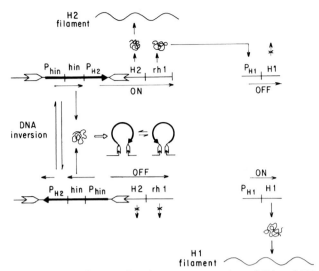

FIG. 8. Mechanism for alternate expression of *H1* and *H2* structural genes for flagellin in *S. typhimurium*. The two genes are in separate locations on the genome. The promoter for *H2* is part of a region of the genome that is bounded by an inverted repeat which can be inverted by homologous recombination mediated by the Hin (or Vh2) protein, a product of a gene within the invertible region. (i) In the orientation shown at the top of the figure, the promoter for the *H2* operon (P$_{H2}$) is correctly placed and H2 flagellin is synthesized together with the repressor of the *H1* operon, the *rh1* gene product. (ii) In the other orientation, shown at the bottom of the figure, P$_{H2}$ is separated from the *H2-rh1* operon it promotes and so neither H2 flagellin nor H1 repressor is synthesized. The *H1* operon is therefore expressed, and H1 flagellin is synthesized. (Based on references 39 and 87 and studies quoted therein.)

orientation the promoter is in the correct position and orientation for transcription initiation, whereas in the other it is not. The inversion region is bounded by an inverted repeat that permits homologous recombination. It also contains the promoter and coding sequence of a gene, *hin* (also known as *vh2*), whose product is necessary for the inversion event; *hin* is expressed in either orientation, possibly under negative autoregulation to ensure that inversion is a relatively infrequent event.

Interestingly, the Hin protein shows extensive sequence homology with TnpR protein, which performs a related recombinational function, resolution of the cointegrate structure of Tn*3* and the host chromosome. It is suggested that the flagellin phase variation mechanism may have evolved from a host-associated transposon (90).

ASSEMBLY

The flagellum is itself a reasonably complex organelle, and furthermore, it must be assembled in a strict relationship to the surface structures of the cell. The situation presumably requires quite sophisticated assembly mechanisms, a presumption that is supported by the large number of flagellar genes for which no structural role has been identified. Although there is some information regarding the order in which the organelle is assembled, knowledge of the mechanism

involved is virtually nonexistent. One of the very few biochemical clues is that ubiquinone (5) (or, under anaerobic conditions, menaquinone; 26) is a requirement for flagellar assembly, suggesting that a membrane-like reduction reaction may be involved.

Relationship to Cell Surfaces and Growth

Flagellation in *E. coli* and *S. typhimurium* is random with respect to both number of flagella (a typical value is around 8) and sites of origin. Nothing is known of the initiation of flagellar assembly, the earliest event which determines that at some site on the cell surface the first flagellar component will be inserted. It is a reasonable hypothesis that flagellar synthesis occurs at regions of cell surface growth, but there is no clear evidence in this regard. Early ideas of a specialized ribosome dedicated to flagellar synthesis have not received any experimental support.

That the processes of flagellar and cell surface assembly are interdependent is indicated by the fact that *galU*$_E$ mutants, defective in outer membrane assembly, are nonflagellate. A second-site mutation, mapping to the *flaH*$_E$ gene, was capable of suppressing the nonflagellate phenotype (55).

As far as is known, flagellar synthesis is a continuous process in growing cells; in contrast to the situation with certain other species (notably, *Caulobacter crescentus*; 81), there is no obvious temporal relationship to the cell growth and division cycle. This is consistent with the spatially random origin of the flagella.

Order of Flagellar Assembly: Partial Structures

As a result of electron microscopic examination of nonflagellate mutants, it has become evident that a number of these are not totally nonflagellate but possess partial structures (94, 95). From such studies, it has been possible to describe a series of structures of increasing complexity that presumably represent intermediates in the flagellar assembly process in wild-type cells. It has also been possible to specify which genes are needed to reach any given stage in the process.

A summary of these data is given in Fig. 9. The overall process is seen to proceed from proximal structure to distal, with the flagellar filament being the last component to be assembled. The earliest structures have almost certainly escaped detection thus far and are hypothesized in the figure. The first recognizable structure (which already requires at least 16 genes; 37) is a complex containing the M ring, the S ring, and the rod; presumably, in the intact cell, the switch complex proteins are also present at this stage. Then the rod is capped or in some way modified and the outer rings are added, followed by the hook, the hook accessory proteins, and finally the filament. It seems likely on the basis of mutant phenotype and other evidence (9) that the Mot proteins do not have to be added in a strict temporal relationship to this pathway. They may be capable of associating with the motor at any time beyond its early stages, regardless of the presence or absence of distal structure.

This largely linear scheme may be an oversimplification, since alternative structures are detected in

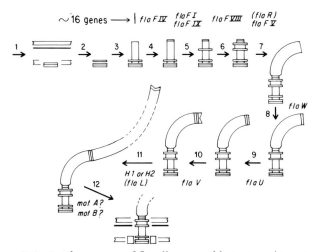

FIG. 9. The process of flagellar assembly in *S. typhimurium* (refer also to Fig. 6). A similar scheme exists for *E. coli*. Stages 1, 2, and 12 are hypothetical but are supported by circumstantial evidence; the partial structures depicted following stages 3 through 11 have been observed by electron microscopy. Where a given gene is required to achieve an incremental stage of assembly, it is indicated; the simplest partial structure that has actually been observed (the "rivet" following stage 3) requires at least 16 genes (*flaAI*, *flaAII.1*, *flaAII.2*, *flaAIII*, *flaB*, *flaC*, *flaD*, *flaFII*, *flaFIII*, *flaFVI*, *flaFVII*, *flaFX*, *flaM*, *flaN*, *flaP*, *flaQ*). The three layers of the cell surface, namely, the cell membrane, the peptidoglycan layer, and the outer membrane, are depicted at the beginning and end of the pathway, but are omitted in the intermediate stages, partly for reasons of clarity, but also because it is not known how insertion into the cell surface is accomplished. Stage 1: The M ring (FlaAII.1 protein?) inserts into the cell membrane, with the switch complex proteins (FlaAII.2, FlaQ, FlaN) believed to be peripherally attached. Stage 2: The S ring (gene unknown) is added. Stage 3: The rod (FlaFVI? and FlaFVII?) is added. Stage 4: The distal end of the rod is capped (FlaFIV). Stage 5: The P ring (FlaFIX) is added. Stage 6: The L ring (FlaFVIII?) and the cylindrical surface of the outer cylinder are added, yielding the complete basal body. Stage 7: Hook protein (FlaFV) is added to the capped rod and grows distally, presumably by extrusion through a central channel. At a fairly well-defined length, hook growth stops; this process is controlled by the *flaR* gene. The structure at this stage is called the hook-basal body. Stage 8: HAP1 (FlaW) is added to the distal end of the hook. Stage 9: HAP3 (FlaU) is added to the distal end of HAP1. Stage 10: HAP2 (FlaV) is added to the distal end of HAP3, giving the hook a notched appearance and enabling filament growth to commence. Stage 11: Under *flaL* control, flagellin (H1 or H2) is synthesized, extrudes through the filament core, and assembles into filament at the growing distal end; HAP2 may be at the growing end and prevent excretion into the medium. The structure depicted at this stage has been called the intact flagellum. Stage 12: The motility-enabling proteins (MotA and MotB) are inserted into the membrane and are believed to circumferentially associate with the basal body to yield the true complete flagellum; Mot protein addition may be possible at any point after stage 3. (Based on data from references 2, 29, 31, 37, 57, 94, 95. Figure modified and extended from reference 37.)

certain mutants (for example, an "L-ring-less" flagellum in *flaFVIII*_S mutants; 94), albeit with a lower probability than the structure indicated on the direct pathway. Preliminary experiments involving differential radiolabeling of basal bodies from conditional mutants also suggest that a simple linear scheme may be inadequate (C. Jones, unpublished data).

The Logistical Problem

With the exception of the most cytoplasmic-proximal components, such as the switch proteins, the M ring, and the Mot proteins, flagellar components must be assembled at sites that are external to the cytoplasm. How, for example, is the L-ring/P-ring outer cylinder complex assembled, and how does it incorporate itself into the outer membrane and cell wall? Virtually no information is available on such issues. One suspects that the basal body may be capable of self-assembly; it is an extremely durable structure when isolated, but there presumably are no covalent linkages since it can be dissociated into its protein subunits by sufficiently extreme treatments (boiling in sodium dodecyl sulfate; 2, 57). There may, on the other hand, be covalent linkages between the basal body and the cell surface, especially the peptidoglycan layer.

Polarity of Filament Growth

The rod, hook, and filament may be viewed in a sense as one continuous, though heterogeneous, filamentous structure. It is known that the hook and filament resemble each other structurally (98), and it is likely that the rod may be similar also. The assembly of this structure proceeds in the sequence rod, hook, and then filament, a proximal-to-distal process.

Much is known of the self-assembly characteristics of the filament and, to a lesser degree, the hook. The striking fact is that the proximal-to-distal character is preserved at the level of individual subunit addition: each flagellin monomer traverses the entire length of the currently existing filament structure and is incorporated at the distal end (21, 34). In contrast to the constant rate of filament elongation in vitro (43), the rate of filament elongation decreases exponentially with filament length (35), indicating either a diffusional or perhaps even a frictional resistance to export of the subunits. It is believed that they travel down a central channel or core within the filament and presumably within the rod and hook also. A similar mechanism is inferred for the hook and may also apply to the rod.

Control of Hook Length

Control of the size of structures in the flagellum appears to fall into three categories. In the first, exemplified by the rings, the oligomeric structure is closed upon itself and therefore size is presumably self-determined; there is structural (17) and physiological (9) evidence to suggest that there may be approximately 16 subunits in such closed structures. In the second category, for which the filament is the only example in the flagellum, there is no control of length other than the relative rates of flagellin synthesis, new organelle initiation, and filament breakage; filament length under these circumstances becomes a statistical matter. In the third category, exemplified by the hook and perhaps the rod, elongation could perhaps proceed indefinitely as far as structure is concerned, but appears to be under assembly control so that the length is in fact quite well defined.

Hook length control involves the product of the $flaE_E/flaR_S$ gene (77, 88, 93). The protein is found in the cytoplasm, at least in the amplified genetic constructions where it can be detected (6). In the absence of this protein, the hook length is indeterminate, often as much as 10 times the normal value of about 80 nm. Mutants of this type are called "superhook" or "polyhook" (a rather misleading term that suggests multiple end-to-end hooks, which is not the case). Improper termination of the hook in $flaE_E/flaR_S$ mutants is also apparent from the absence of the hook accessory proteins that form the junction to the filament in wild-type cells (31) and from the fact that the polyhook structure in general cannot assemble a filament at its distal end (93).

Role of the Hook Accessory Proteins

Mutants lacking any of the three hook accessory proteins (HAP1, HAP2, and HAP3, coded for by the $flaS_E/flaW_S$, $flbC_E/flaV_S$, and $flaT_E/flaU_S$ genes, respectively; 30, 61) do synthesize flagellin, but fail to assemble it into filament. Instead, it is simply excreted into the medium (28). However, mutants defective only in HAP2 are capable of assembling exogenously supplied flagellin onto the hook (28, 46, 47). Conversely, if HAP2 is supplied exogenously, only endogenous flagellin can be assembled (37). From these and other data, Iino and his colleagues have proposed (see Fig. 9) that HAP1 and HAP3 complete the hook structure, and then, after HAP2 is added, growth of filament from endogenous flagellin is enabled in an ever-growing zone between HAP1/HAP3 at the proximal end and HAP2 at the distal end, with HAP2 in some way controlling the addition of flagellin monomer and preventing its loss into the medium (37). The latter aspects of this model are supported by the finding that the distal end of a growing filament in vivo, or the intact distal end of a filament broken from the cell, is incapable of accepting exogenous flagellin and has a morphologically distinct capping structure rather than the notched appearance seen at the distal end of a reconstituted filament or at the broken distal end of a native filament (42).

CONCLUSIONS

The flagellum is a fairly complex part of the architecture of *E. coli* and *S. typhimurium*, although one possessing a high degree of spatial order and symmetry. The external helical filament and the hook have been studied extensively with respect to structure; low- to medium-resolution information on subunit organization is available. Both structures take remarkable advantage of the phenomenon of quasi-equivalence. The basal body has also been studied structurally, but there is no information yet regarding subunit organization. The basal body may be a passive structure to which the actual energy-transducing and switching components (identified genetically, biochemically, but not yet morphologically) are attached.

Flagellar genetics is complex, but has been studied extensively. The entire system constitutes a regulon, or coordinately expressed unit.

Some information is available regarding the sequence of flagellar assembly, but almost nothing is understood of the mechanism of assembly or incorporation into the cell surface.

ACKNOWLEDGMENTS

I would like to acknowledge the many colleagues who have communicated manuscripts and results prior to publication. Thanks are also due to May Kihara for assistance in assembling the data for this chapter.

Work in my laboratory has been supported by Public Health Service grant AI12202 from the National Institutes of Health.

LITERATURE CITED

1. **Abram, D., H. Koffler, and A. E. Vatter.** 1965. Basal structure and attachment of flagella in cells of *Proteus vulgaris*. J. Bacteriol. **90:**1337–1354.
2. **Aizawa, S.-I., G. E. Dean, C. J. Jones, R. M. Macnab, and S. Yamaguchi.** 1985. Purification and characterization of the flagellar hook-basal body complex of *Salmonella typhimurium*. J. Bacteriol. **161:**836–849.
3. **Ambler, R. P., and M. W. Rees.** 1959. ε-N-Methyl-lysine in bacterial flagellar protein. Nature (London) **184:**56–57.
4. **Asakura, S.** 1970. Polymerization of flagellin and polymorphism of flagella. Adv. Biophys. **1:**99–155.
5. **Bar Tana, J., B. J. Howlett, and D. E. Koshland, Jr.** 1977. Flagellar formation in *Escherichia coli* electron transport mutants. J. Bacteriol. **130:**787–792.
6. **Bartlett, D. H., and P. Matsumura.** 1984. Identification of *Escherichia coli* region III flagellar gene products and description of two new flagellar genes. J. Bacteriol. **160:**577–585.
7. **Berg, H. C.** 1974. Dynamic properties of bacterial flagellar motors. Nature (London) **249:**77–79.
8. **Berg, H. C., M. D. Manson, and M. P. Conley.** 1982. Dynamics and energetics of flagellar rotation in bacteria, p. 1–31. *In* W. B. Amos and J. G. Duckett (ed.), Prokaryotic and eukaryotic flagella. Cambridge University Press, Cambridge, U.K.
9. **Block, S. M., and H. C. Berg.** 1984. Successive incorporation of force-generating units in the bacterial rotary motor. Nature (London) **309:**470–472.
10. **Calladine, C. R.** 1978. Change of waveform in bacterial flagella: the role of mechanics at the molecular level. J. Mol. Biol. **118:**457–479.
11. **Calladine, C. R.** 1982. Construction of bacterial flagellar filaments, and aspects of their conversion to different helical forms, p. 33–51. *In* W. B. Amos and J. G. Duckett (ed.), Prokaryotic and eukaryotic flagella. Cambridge University Press, Cambridge, U.K.
12. **Clegg, D. O., and D. E. Koshland, Jr.** 1985. Identification of a bacterial sensing protein and effects of its elevated expression. J. Bacteriol. **162:**398–405.
13. **Cohen-Bazire, G., and J. London.** 1967. Basal organelles of bacterial flagella. J. Bacteriol. **94:**458–465.
14. **Dean, G. E., S.-I. Aizawa, and R. M. Macnab.** 1983. *flaAII (motC, cheV)* of *Salmonella typhimurium* is a structural gene involved in energization and switching of the flagellar motor. J. Bacteriol. **154:**84–91.
15. **Dean, G. E., R. M. Macnab, J. Stader, P. Matsumura, and C. Burks.** 1984. Gene sequence and predicted amino acid sequence of the *motA* protein, a membrane-associated protein required for flagellar rotation in *Escherichia coli*. J. Bacteriol. **159:**991–999.
16. **DePamphilis, M. L., and J. Adler.** 1971. Purification of intact flagella from *Escherichia coli* and *Bacillus subtilis*. J. Bacteriol. **105:**376–383.
17. **DePamphilis, M. L., and J. Adler.** 1971. Fine structure and isolation of the hook-basal body complex of flagella from *Escherichia coli* and *Bacillus subtilis*. J. Bacteriol. **105:**384–395.
18. **DePamphilis, M. L., and J. Adler.** 1971. Attachment of flagellar basal bodies to the cell envelope: specific attachment to the outer, lipopolysaccharide membrane and the cytoplasmic membrane. J. Bacteriol. **105:**396–407.
19. **Dimmitt, K., and M. I. Simon.** 1971. Purification and partial characterization of *Bacillus subtilis* flagellar hooks. J. Bacteriol. **108:**282–286.
20. **Edwards, P. R., and W. H. Ewing.** 1972. Identification of Enterobacteriaceae, 3rd ed., p. 146–258. Burgess Publishing Co., Minneapolis.
21. **Emerson, S. U., K. Tokuyasu, and M. I. Simon.** 1970. Bacterial flagella: polarity of elongation. Science **169:**190–192.
22. **Enomoto, M.** 1966. Genetic studies of paralyzed mutants in *Salmonella*. II. Mapping of three *mot* loci by linkage analysis. Genetics **54:**1069–1076.

23. **Fedorov, O. V., N. N. Khechinashvili, R. Kamiya, and S. Asakura.** 1984. Multidomain of flagellin. J. Mol. Biol. **175**:83–87.

24. **Fujime, S., M. Maruyama, and S. Asakura.** 1972. Flexural rigidity of bacterial flagella studied by quasielastic scattering of laser light. J. Mol. Biol. **68**:347–359.

25. **Gerber, B. R., L. M. Routledge, and S. Takashima.** 1972. Self-assembly of bacterial flagellar protein: dielectric behavior of monomers and polymers. J. Mol. Biol. **71**:317–337.

26. **Hertz, R., and J. Bar-Tana.** 1977. Anaerobic electron transport in anaerobic flagellum formation in *Escherichia coli.* J. Bacteriol. **132**:1034–1035.

27. **Hilmen, M., and M. Simon.** 1976. Motility and the structure of bacterial flagella, p. 35–45. *In* R. Goldman, T. Pollard, and J. Rosenbaum (ed.), Cell motility. Cold Spring Harbor Laboratory, Cold Spring Harbor, N.Y.

28. **Homma, M., H. Fujita, S. Yamaguchi, and T. Iino.** 1984. Excretion of unassembled flagellin by *Salmonella typhimurium* mutants deficient in the hook-associated proteins. J. Bacteriol. **159**:1056–1059.

29. **Homma, M., and T. Iino.** 1985. Locations of hook-associated proteins in flagellar structures of *Salmonella typhimurium.* J. Bacteriol. **162**:183–189.

30. **Homma, M., K. Kutsukake, and T. Iino.** 1985. Structural genes for flagellar hook-associated proteins in *Salmonella typhimurium.* J. Bacteriol. **163**:464–471.

31. **Homma, M., K. Kutsukake, T. Iino, and S. Yamaguchi.** 1984. Hook-associated proteins essential for flagellar filament formation in *Salmonella typhimurium.* J. Bacteriol. **157**:100–108.

32. **Hoshikawa, H., and R. Kamiya.** 1985. Elastic properties of bacterial flagellar filaments. II. Determination of the modulus of rigidity. Biophys. Chem. **22**:159–166.

33. **Hotani, H., and S. Asakura.** 1974. Growth-saturation in vitro of *Salmonella* flagella. J. Mol. Biol. **86**:285–300.

34. **Iino, T.** 1969. Polarity of flagellar growth in *Salmonella.* J. Gen. Microbiol. **56**:227–239.

35. **Iino, T.** 1974. Assembly of *Salmonella* flagellin in vitro and in vivo. J. Supramolec. Struct. **2**:372–384.

36. **Iino, T.** 1977. Genetics of structure and function of bacterial flagella. Annu. Rev. Genet. **11**:161–182.

37. **Iino, T.** 1985. Structure and assembly of flagella, p. 9–37. *In* N. Nanninga (ed.), Molecular cytology of *Escherichia coli.* Academic Press, London.

38. **Iino, T.** 1985. Genetic control of flagellar morphogenesis in *Salmonella*, p. 83–92. *In* M. Eisenbach and M. Balaban (ed.), Sensing and response in microorganisms. International Science Services, Rehovot, Israel.

39. **Iino, T., and K. Kutsukake.** 1983. Flagellar phase variation in *Salmonella*: a model system regulated by flip-flop DNA inversions, p. 395–406. *In* K. Mizobuchi, I. Watanabe, and J. D. Watson (ed.), Nucleic acid research. Future developments. Academic Press, Inc., New York.

40. **Iino, T., and M. Mitani.** 1966. Flagella-shape mutants in *Salmonella.* J. Gen. Microbiol. **44**:27–40.

41. **Iino, T., and M. Mitani.** 1967. A mutant of *Salmonella* possessing straight flagella. J. Gen. Microbiol. **49**:81–88.

42. **Ikeda, T., S. Asakura, and R. Kamiya.** 1985. "Cap" on the tip of *Salmonella* flagella. J. Mol. Biol. **184**:735–737.

43. **Ishihara, A., and H. Hotani.** 1980. Micro-video study of discontinuous growth of bacterial flagellar filaments in vitro. J. Mol. Biol. **139**:265–276.

44. **Kagawa, H., S.-I. Aizawa, and S. Asakura.** 1979. Transformations in isolated polyhooks. J. Mol. Biol. **129**:333–336.

45. **Kagawa, H., S. Asakura, and T. Iino.** 1973. Serological study of bacterial flagellar hooks. J. Bacteriol. **113**:1474–1481.

46. **Kagawa, H., H. Morishita, and M. Enomoto.** 1981. Reconstitution in vitro of flagellar filaments onto hook structures attached to bacterial cells. J. Mol. Biol. **153**:465–470.

47. **Kagawa, H., T. Nishiyama, and S. Yamaguchi.** 1983. Motility development of *Salmonella typhimurium* cells with *flaV* mutations after addition of exogenous flagellin. J. Bacteriol. **155**:435–437.

48. **Kamiya, R., and S. Asakura.** 1976. Helical transformations of *Salmonella* flagella *in vitro.* J. Mol. Biol. **106**:167–186.

49. **Kamiya, R., H. Hotani, and S. Asakura.** 1982. Polymorphic transition in bacterial flagella, p. 53–76. *In* W. B. Amos and J. G. Duckett (ed.), Prokaryotic and eukaryotic flagella. Cambridge University Press, Cambridge, U.K.

50. **Kerridge, D.** 1959. The effect of amino acid analogues on the synthesis of bacterial flagella. Biochim. Biophys. Acta **31**:579–581.

51. **Khan, S., and R. M. Macnab.** 1980. Proton chemical potential, proton electrical potential and bacterial motility. J. Mol. Biol. **138**:599–614.

52. **Khan, S., R. M. Macnab, A. L. DeFranco, and D. E. Koshland, Jr.** 1978. Inversion of a behavioral response in bacterial chemotaxis: explanation at the molecular level. Proc. Natl. Acad. Sci. USA **75**:4150–4154.

53. **Klug, A.** 1967. The design of self-assembling systems of equal units. Symp. Int. Soc. Cell Biol. **6**:1–18.

54. **Komeda, Y.** 1982. Fusions of flagellar operons to lactose genes on a Mu *lac* bacteriophage. J. Bacteriol. **150**:16–26.

55. **Komeda, Y., T. Icho, and T. Iino.** 1977. Effects of *galU* mutation on flagellar formation in *Escherichia coli.* J. Bacteriol. **129**:908–915.

56. **Komeda, Y., K. Kutsukake, and T. Iino.** 1980. Definition of additional flagellar genes in *Escherichia coli* K12. Genetics **94**:277–290.

57. **Komeda, Y., M. Silverman, P. Matsumura, and M. Simon.** 1978. Genes for the hook-basal body proteins of the flagellar apparatus in *Escherichia coli.* J. Bacteriol. **134**:655–667.

58. **Komeda, Y., H. Suzuki, J. Ishidsu, and T. Iino.** 1975. The role of cAMP in flagellation of *Salmonella typhimurium.* Mol. Gen. Genet. **142**:289–298.

59. **Kondoh, H., and H. Hotani.** 1974. Flagellin from *Escherichia coli* K12: polymerization and molecular weight in comparison with *Salmonella* flagellins. Biochim. Biophys. Acta **336**:117–139.

60. **Kondoh, H., and M. Yanagida.** 1975. Structure of straight flagellar filaments from a mutant of *Escherichia coli.* J. Mol. Biol. **96**:641–652.

61. **Kutsukake, K., T. Iino, Y. Komeda, and S. Yamaguchi.** 1980. Functional homology of *fla* genes between *Salmonella typhimurium* and *Escherichia coli.* Mol. Gen. Genet. **178**:59–67.

62. **Larsen, S. H., J. Adler, J. J. Gargus, and R. W. Hogg.** 1974. Chemomechanical coupling without ATP: the source of energy for motility and chemotaxis in bacteria. Proc. Natl. Acad. Sci. USA **71**:1239–1243.

63. **Larsen, S. H., R. W. Reader, E. N. Kort, W.-W. Tso, and J. Adler.** 1974. Change in direction of flagellar rotation is the basis of the chemotactic response in *Escherichia coli.* Nature (London) **249**:74–77.

64. **Lawn, A. M.** 1977. Comparison of the flagellins from different flagellar morphotypes of *Escherichia coli.* J. Gen. Microbiol. **101**:112–130.

65. **Lederberg, J., and T. Iino.** 1956. Phase variation in *Salmonella.* Genetics **41**:744–757.

66. **Macnab, R. M.** 1976. Examination of bacterial flagellation by dark-field microscopy. J. Clin. Microbiol. **4**:258–265.

67. **Macnab, R. M.** 1979. Bacterial flagella, p. 207–223. *In* W. Haupt and M. E. Feinleib (ed.), Encyclopedia of plant physiology, new series, vol. 7. Springer-Verlag, Berlin.

68. **Macnab, R. M., and S.-I. Aizawa.** 1984. Bacterial motility and the bacterial flagellar motor. Annu. Rev. Biophys. Bioeng. **13**:51–83.

69. **Macnab, R. M., and M. K. Ornston.** 1977. Normal-to-curly flagellar transitions and their role in bacterial tumbling. Stabilization of an alternative quaternary structure by mechanical force. J. Mol. Biol. **112**:1–30.

70. **Manson, M. D., P. Tedesco, H. C. Berg, F. M. Harold, and C. van der Drift.** 1977. A protonmotive force drives bacterial flagella. Proc. Natl. Acad. Sci. USA **74**:3060–3064.

71. **Martinez, R. J., A. T. Ichiki, N. P. Lundh, and S. R. Tronick.** 1968. A single amino acid substitution responsible for altered flagellar morphology. J. Mol. Biol. **34**:559–564.

72. **Matsumura, P., M. Silverman, and M. Simon.** 1977. Cloning and expression of the flagellar hook gene on hybrid plasmids in minicells. Nature (London) **265**:758–760.

73. **Matsumura, P., M. Silverman, and M. Simon.** 1977. Synthesis of *mot* and *che* gene products of *Escherichia coli* programmed by hybrid ColE1 plasmids in minicells. J. Bacteriol. **132**:996–1002.

74. **Matsuura, S., J.-I. Shioi, Y. Imae, and S. Iida.** 1979. Characterization of the *Bacillus subtilis* motile system driven by an artificially created proton motive force. J. Bacteriol. **140**:28–36.

75. **O'Brien, E. J., and P. M. Bennett.** 1972. Structure of straight flagella from a mutant *Salmonella.* J. Mol. Biol. **70**:133–152.

76. **Parkinson, J. S., S. R. Parker, P. B. Talbert, and S. E. Houts.** 1983. Interactions between chemotaxis genes and flagellar genes in *Escherichia coli.* J. Bacteriol. **155**:265–274.

77. **Patterson-Delafield, J., R. J. Martinez, B. A. D. Stocker, and S. Yamaguchi.** 1973. A new *fla* gene in *Salmonella typhimurium*—*flaR*—and its mutant phenotype—superhooks. Arch. Mikrobiol. **90**:107–120.

78. **Ravid, S., and M. Eisenbach.** 1984. Direction of flagellar rotation in bacterial cell envelopes. J. Bacteriol. **158**:222–230.

79. **Ridgway, H. F., M. Silverman, and M. I. Simon.** 1977. Localization of proteins controlling motility and chemotaxis in *Esche-*

richia coli. J. Bacteriol. **132:**657–665.

80. **Satir, P., and G. K. Ojakian.** 1979. Plant cilia, p. 224–249. *In* W. Haupt and M. E. Feinleib (ed.), Encyclopedia of plant physiology, new series, vol. 7. Springer-Verlag, Berlin.

81. **Shapiro, L.** 1976. Differentiation in the *Caulobacter* cell cycle. Annu. Rev. Microbiol. **30:**377–407.

82. **Shirakihara, Y., and T. Wakabayashi.** 1979. Three-dimensional image reconstruction of straight flagella from a mutant *Salmonella typhimurium.* J. Mol. Biol. **131:**485–507.

83. **Silverman, M., P. Matsumura, R. Draper, S. Edwards, and M. I. Simon.** 1976. Expression of flagellar genes carried by bacteriophage lambda. Nature (London) **261:**248–250.

84. **Silverman, M., and M. Simon.** 1974. Characterization of *Escherichia coli* flagellar mutants that are insensitive to catabolite repression. J. Bacteriol. **120:**1196–1203.

85. **Silverman, M., and M. Simon.** 1974. Flagellar rotation and the mechanism of bacterial motility. Nature (London) **249:**73–74.

86. **Silverman, M., and M. I. Simon.** 1977. Bacterial flagella. Annu. Rev. Microbiol. **31:**397–419.

87. **Silverman, M., J. Zieg, M. Hilmen, and M. Simon.** 1979. Phase variation in *Salmonella:* genetic analysis of a recombinational switch. Proc. Natl. Acad. Sci. USA **76:**391–395.

88. **Silverman, M. R., and M. I. Simon.** 1972. Flagellar assembly mutants in *Escherichia coli.* J. Bacteriol. **112:**986–993.

89. **Simon, M., M. Silverman, P. Matsumura, H. Ridgway, Y. Komeda, and M. Hilmen.** 1978. Structure and function of bacterial flagella. Symp. Soc. Gen. Microbiol. **28:**271–286.

90. **Simon, M., J. Zieg, M. Silverman, G. Mandel, and R. Doolittle.** 1980. Phase variation: evolution of a controlling element. Science **209:**1370–1374.

90a.**Stader, J., P. Matsumura, D. Vacante, G. E. Dean, and R. M. Macnab.** 1986. Nucleotide sequence of the *Escherichia coli motB* gene and site-limited incorporation of its product into the cytoplasmic membrane. J. Bacteriol. **166:**244–252.

91. **Suzuki, H., and T. Iino.** 1975. Absence of messenger ribonucleic acid specific for flagellin in non-flagellate mutants of *Salmonella.* J. Mol. Biol. **95:**549–556.

92. **Suzuki, T., and T. Iino.** 1977. Appearance of straight flagellar filaments in the presence of *p*-fluorophenylalanine in *Pseudomonas aeruginosa.* J. Bacteriol. **129:**527–529.

93. **Suzuki, T., and T. Iino.** 1981. Role of the *flaR* gene in flagellar hook formation in *Salmonella* spp. J. Bacteriol. **148:**973–979.

94. **Suzuki, T., T. Iino, T. Horiguchi, and S. Yamaguchi.** 1978. Incomplete flagellar structures in nonflagellate mutants of *Salmonella typhimurium.* J. Bacteriol. **133:**904–915.

95. **Suzuki, T., and Y. Komeda.** 1981. Incomplete flagellar structures in *Escherichia coli* mutants. J. Bacteriol. **145:**1036–1041.

96. **Tsui-Collins, A. L., and B. A. D. Stocker.** 1976. *Salmonella typhimurium* mutants generally defective in chemotaxis. J. Bacteriol. **128:**754–765.

97. **Vary, P. S., and B. A. D. Stocker.** 1973. Nonsense motility mutants in *Salmonella typhimurium.* Genetics **73:**229–245.

98. **Wagenknecht, T., D. J. DeRosier, S.-I. Aizawa, and R. M. Macnab.** 1982. Flagellar hook structures of *Caulobacter* and *Salmonella* and their relationship to filament structure. J. Mol. Biol. **162:**69–87.

99. **Wei, L. N., and T. M. Joys.** 1985. Covalent structure of three phase-1 flagellar filament proteins of *Salmonella.* J. Mol. Biol. **186:**791–803.

99a.**Yamaguchi, S., H. Fujita, A. Ishihara, S.-I. Aizawa, and R. M. Macnab.** 1986. Subdivision of flagellar genes of *Salmonella typhimurium* into regions responsible for assembly, rotation, and switching. J. Bacteriol. **166:**187–193.

100. **Yamaguchi, S., H. Fujita, T. Taira, K. Kutsukake, M. Homma, and T. Iino.** 1984. Genetic analysis of three additional *fla* genes in *Salmonella typhimurium.* J. Gen. Microbiol. **130:**3339–3342.

101. **Yamaguchi, S., T. Iino, T. Horiguchi, and K. Ohta.** 1972. Genetic analysis of *fla* and *mot* cistrons closely linked to *H1* in *Salmonella abortusequi* and its derivatives. J. Gen. Microbiol. **70:**59–75.

8. Fimbriae

BARRY I. EISENSTEIN

Departments of Microbiology and Medicine, University of Texas Health Science Center, San Antonio, Texas 78284

INTRODUCTION

Terminology

Three types of proteinaceous appendages project from the surface of *Escherichia coli* cells: (i) flagella, which are the thickest and longest appendages and are the effectors of bacterial locomotion; (ii) organelles involved in conjugal transfer of genetic material; and (iii) organelles that act as adherence factors in the colonization of eucaryotic surfaces. The terminology for the latter two types is confused because upon their discovery of these organelles in 1955, Duguid et al. referred to both types as fimbriae (10), whereas Brinton, at the time of his extensive biochemical and biophysical characterizations, referred to them as pili (4, 5). Presently both terms are in active use, although, following the practice of Ottow (35), many who employ the term fimbriae limit that use to the adhesive organelles (adhesins) and reserve the term pili for the sexual appendages. Since this usage gives credit to both pioneer workers and has the advantage of giving distinctive labels to entities that are distinguishable both structurally and functionally, this review will utilize the Ottow terminology.

Ecology

Fimbriae of various types are made by virtually every species of gram-negative bacterium that has been examined (9). They are usually classified functionally by their receptor avidity, as determined by the ability of either fimbriate bacteria or their purified fimbriae to agglutinate erythrocytes from different species. Among these hemagglutinins, a simple dichotomy exists: fimbriae whose agglutination is prevented by coincubation with D-mannose are designated mannose sensitive, and all the rest are termed mannose resistant. Although the majority of clinically isolated *E. coli* strains possess mannose-sensitive (type 1) fimbriae, many also (or instead) possess fimbriae of the mannose-resistant type (34). In recent years many clinical and animal model studies have demonstrated the importance of fimbriae in the pathogenesis of *E. coli* (and other gram-negative) infections, particularly in the urinary and gastrointestinal tracts (3). In contrast to many of the clinical isolates, the nonpathogenic K-12 strain line of *E. coli* possesses only type 1 fimbriae. Since K-12 strains are so well characterized genetically and biochemically, the limited repertoire of their fimbrial expression is actually an advantage in research devoted to the molecular biology of these organelles. The type 1 fimbria binds in a lectinlike fashion to a reasonably well-defined receptor moiety and is thus an excellent model for studies involving structure-function, immunochemistry, genetic regulation, protein secretion, and organelle morphogenesis.

MOLECULAR ARCHITECTURE

Structure

A typical fimbriate bacterium, e.g., most K-12 strains of *E. coli*, contains approximately 100 to 300 of these organelles, each with a width of ~7 nm and a length of 0.2 to 2 μm, arranged peritrichously around the cell (Fig. 1). A single fimbria consists of ~1,000 repeating subunits of a single polypeptide of ~17 kilodaltons (kDa). X-ray crystallographic studies have determined that the subunits are arranged in a simple helix such that each turn consists of ~3.14 subunits; the axial rise per unit is 0.809 nm (26). Brinton calculated the inner diameter to be 2.0 to 2.5 nm (too small to permit this axial hole to act as a conduit for

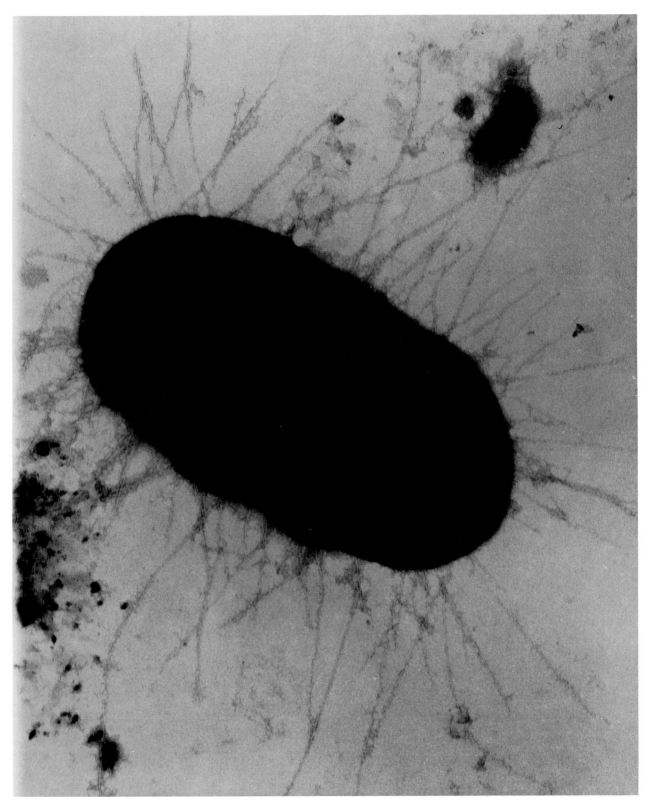

FIG. 1. *E. coli* K-12 strain CSH50 with type 1 fimbriae. Courtesy of Stanley Holt.

subunit traversal, as occurs in flagella) and the helical pitch distance to be 2.32 nm (5).

The primary structure of the fimbrial subunit has been determined by amino acid analysis (5, 38), ami-no-terminal sequencing (20a, 24, 40), and DNA sequencing of the gene encoding the subunit (23, 33). On the basis of these data, it has been found that the molecule contains ~160 amino acid residues, a high proportion of which contain nonpolar side chains and few of which contain basic groups. Thus, the molecule is, on the whole, acidic (pI, ~3.9) and extremely hydrophobic. The latter trait may explain the difficulty of solubilizing fimbriae in an aqueous milieu and may be partly responsible for the organelle's ability to bind so avidly to eucaryotic membranes. A single subunit also contains two residues of cysteine, which most likely form an intrachain disulfide bond.

Structure-Function and Immunochemical Studies

Type 1 fimbriae are both receptor-specific ligands and antigens (7). Whether their unique binding specificity for mannose-containing receptors depends on the quaternary structure of the organelle is not known, although molecular details of the receptor have been pursued successfully with hapten inhibition studies (18, 30, 38, 39). Recent attempts to analyze the nature of the lectinlike activity of the fimbriae themselves have resorted to dissociation-renaturation studies. One rationale of such an approach is to remove any minor, undetected component of the native organelle that might be responsible for this important biological property. Brinton found that depolymerization could be achieved with the use of agents that break hydrogen or hydrophobic bonds, including hot acid or guanidine hydrochloride. Upon removal of these agents, subunits repolymerize into structures that are indistinguishable from native fimbriae, as determined by electron microscopy (5). It is not known if renatured fimbriae reacquire mannose-binding activity. The dissociation-renaturation procedure has also been used to examine the need for quaternary determinants for binding activity. Eshdat and co-workers dissociated fimbriae with saturated guanidine and then, after dialysis, determined the ability of the stable (dimeric) subunits to reacquire mannose-binding activity. When dimers were applied to a mannan-Sepharose column, only 25% of the protein bound to the column (17), implying either that the guanidine had denatured irreversibly the majority of subunits or that only a minor population of native subunits possessed the lectinlike activity seen with the intact organelle. Moreover, monomers were not stable enough in solution to test directly, leaving unanswered the requirement of quaternary structure for lectinlike activity.

Clearly, though, the quaternary structure of the native fimbriae endows them with properties different from those of their subunits. Upon dissociation, the usual UV absorption peak seen at 287 nm disappears, suggesting that tyrosine residues have had their immediate environment altered, indicating in turn a change in secondary structure (5, 25). Dissociation also disrupts the immunochemistry of the intact organelle; a subset of antifimbrial monoclonal antibodies detects an epitope(s) present in the organelle but not in the subunit (2, 13). These studies, taken together, do not yet allow discrimination among at least three different possibilities concerning the binding properties of type 1 fimbriae. (i) Subunits themselves are uniformly capable of binding. Relatively harsh methods of disaggregation prevent this property from being readily demonstrated. (ii) Subunits are

heterogeneous, a finding supported by the appearance of multiple bands on both sodium dodecyl sulfate gels (25) and isoelectric focusing gels (38). Perhaps only a portion of subunits bind (see below, Genetics). (iii) The quaternary structure of intact fimbriae is needed to provide the necessary lectinlike sites.

GENETICS

The genetic organization of *E. coli* fimbrial expression is similar to that of *Salmonella* flagella in at least two respects: (i) the expression is under phase variation control so that one cell of a thousand will spontaneously switch to the opposite phenotype (i.e., Fim$^+$→Fim$^-$→Fim$^+$, etc.) (5), and (ii) in addition to the gene for the fimbrial subunit, several genes are required for organelle expression (45, 46). Recent genetic research has been of great help in our understanding of organelle morphogenesis and structure-function relationships.

Phase Variation

The genetic and molecular details of phase variation recently have been worked out. Operon fusion technology has permitted the in vivo observation that the oscillation of fimbrial expression is under transcriptional control (11) and that, at least in *E. coli* K-12 strains, it is not influenced by environmental conditions (14). Nevertheless, the switching frequency is so rapid that when there is selective growth pressure for one phase or the other (e.g., Fim$^+$ bacteria in static broth), the state of fimbriation appears to respond genetically to environmental signals. This capability is itself probably a virulence factor (12); Fim$^+$-phase bacteria are more capable of adhering to mucosal surfaces and thereby initiating infection, whereas Fim$^-$-phase bacteria, because they are not recognized by circulating phagocytes, are more capable of surviving in the bloodstream (42, 43). Thus, bacteria do better with fimbriae during the colonization stage of infection and better without fimbriae during the tissue invasion stage.

The phase switch property maps at 98 min, adjacent to the known *fim* (or *pil*) genes, and regulates a promoter that directs transcription clockwise (20). Merodiploid analysis performed with strains containing both *fimD-lac* and *fimD*$^+$ operons showed that phase switch is actually in the DNA and adjacent to the operon fusion. Moreover, a specialized λ phage that contains the switch requires a chromosomally determined *trans*-active factor to express its switching phenotype (19). The *trans*-active gene maps away from the DNA switch and is not cross-complemented by *hin* (C. S. Freitag, personal communication), distinguishing it from the other class of procaryotic DNA switches (36, 44, 47) (see chapter 7). Recently, my co-workers and I have subcloned the switch, sequenced the DNA, and determined the molecular basis for its activity (1). The switch is an invertible element different from that controlling *Salmonella* flagella; it is small, consisting of 314 base pairs bounded by 9-base-pair inverted repeats, and it is driven by a different recombinase. It is just upstream of the fimbrial structural gene (which we now refer to as *fimA* rather than *fimD*) and contains a consensus *E. coli* promoter when in the "on" orien-

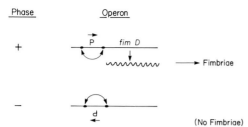

FIG. 2. The invertible element controlling phase variation of type 1 fimbriae. Plus-phase cells (Fim⁺) and minus-phase cells (Fim⁻) both contain the structural gene for the fimbrial subunit, but the gene is only expressed in the plus-phase state. The control is transcriptional; a 314-base-pair invertible element contains a promoter that is permitting the transcription of *fimD* (now called *fimA*) in one orientation only. The switch recombines at 9-base-pair inverted repeats that bound the element. (Data from reference 1.)

tation (Fig. 2). The DNA in the "on" orientation might be more difficult to clone, since the two published sequences of the *E. coli* fimbrial gene show the adjacent invertible element in the "off" orientation (23, 33).

Operon Organization

The multiple operons and genes required for fimbrial (organelle) expression are reminiscent of the operon organization of the K88 adhesin (22), at least as determined by molecular cloning of DNA from a clinical isolate of *E. coli* (31, 32). Orndorff and Falkow have found that four genes, encoding 86-, 30-, 23-, and 17-kDa polypeptides, are required for expression of type 1 organelles. Since the 17-kDa polypeptide is the fimbrial subunit, it was suggested that the other polypeptides are needed for assembly of the organelle. Deletional or transposon-mediated mutations in either of the genes encoding the 86- or 30-kDa protein result in continued intracellular production of fimbrial antigen (i.e., the 17-kDa subunit) without organelle morphogenesis. Recently, mutagenesis studies have led to the identification of an additional gene, *pilE*, which is required to convert antigenically fimbriate cells into those that can also hemagglutinate (D. Schauer, L. Mauer, and P. E. Orndorff, Abstr. Annu. Meet. Am. Soc. Microbiol. 1985, B88, p. 32). Thus, the lectinlike binding activity of type 1 fimbriae may be dissociable from the organelle structure, as has been recently found in the Pap-pilus (P-fimbria) organelle (29). These observations may also explain why the growth of *E. coli* in subinhibitory concentrations of streptomycin results in the synthesis of lectin-deficient organelles (15). Possibly the gene encoding the lectinlike phenotype is differentially susceptible to the translation arrest induced by the antibiotic.

FIMBRIAL SUBUNIT SECRETION AND PROCESSING

In *E. coli* virtually all proteins so far examined that are destined to be localized in the periplasmic space or outer membrane are initially synthesized as precursor polypeptides and are then processed at their amino terminus to give rise to the mature proteins (for

reviews, see references 37 and 41). Typically this amino-terminal "signal sequence" consists of 15 to 30 predominantly hydrophobic amino acids; it is thought to interact specifically with membrane structures that guide the mature protein toward its destination.

Organelles that project from the bacterial cell (i.e., flagella, fimbriae, and pili) are special cases of exported proteins in that they are assembled after the secretion step. Flagellar subunits are possibly unique in that they are not synthesized as precursors containing signal sequences (47). Unlike fimbrial and pilin subunits, they are added to the growing flagellar tip after having first traversed from the cytoplasm directly through the cell envelope through an internal channel in the flagellum, which is anchored in the cytoplasmic membrane (16). Probably because they are first secreted into a cytoplasmic membrane pool (28), F-pilin subunits are synthesized as 14-kDa precursors that are converted by the *traQ* product to mature 7-kDa pilin subunits (21). Ippen-Ihler and co-workers (21) feel that it is likely that the *tra* operon products form a specialized transmembrane complex needed for secretion and assembly. The precise precursor-product relationship between the 14- and 7-kDa polypeptides is not known.

Type 1 fimbrial subunits, in contrast to those of the other, more complex, organelles, behave more like typical outer membrane proteins in their initial secretion-processing pathway. Based on radioimmunoprecipitation studies using monospecific antifimbrial sera in a strain containing a *malE-lacZ* protein fusion (Fig. 3), my co-workers and I have found that the 17-kDa polypeptide is first synthesized as a 19-kDa precursor that is cotranslationally processed (6). Under conditions of maltose induction, the hybrid protein, with the signal-sequence-containing amino terminus of the MalE protein, competes for processing sites at the cytoplasmic membrane. In all signal-sequence-containing proteins studied to date, such competition results in the accumulation of the precursor molecule of the protein studied. Our results have been confirmed by DNA sequence data, which show the presence of a typical signal peptide of 23 amino acid residues (23, 33). We also found that fimbrial synthesis and processing depends on normal protein export in two other respects: both the SecA protein

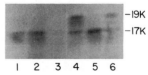

FIG. 3. Radioimmunoprecipitation of mature fimbrial subunits and their signal-sequence-containing precursors by rabbit antifimbrial serum. Lanes 1 and 2, Fim⁺ *E. coli* K-12 controls at 30°C, showing only the mature subunit. Lane 3, Fim⁺ control at 42°C, showing neither precursor nor mature subunit, probably due to rapid assembly. Lane 4, Isogenic derivative with a temperature-sensitive *secA* mutation grown at 42°C. Note the 19-kDa precursor. Lanes 5 and 6, Isogenic derivative with *malE-lacZ* gene fusion grown in the absence and presence, respectively, of maltose, the inducer of the hybrid protein. Note the 19-kDa precursor when the MalE-LacZ protein is induced. (From reference 6.)

A B

FIG. 4. Specificity of a monoclonal antibody for fimbrial organelles and of antifimbrial serum for both organelles and fimbrial subunits, as shown by radioimmunoprecipitation. A culture of strain CSH50 was pulse-labeled with [³H]leucine, chased for 30 s with unlabeled leucine, and precipitated with trichloroacetic acid to prevent further metabolic processes. Lane A, Immunoprecipitation with the monoclonal antibody. After immunoprecipitation, the immunoprecipitates were analyzed by sodium dodecyl sulfate-polyacrylamide gel electrophoresis (13% separating gel) and fluorography. Only sodium-dodecyl-sulfate-stable material (intact organelles) was immunoprecipitated. Lane B, Immunoprecipitation with rabbit antifimbrial serum after the removal of intact fimbriae by the monoclonal antibody. A small amount of material was caught at the stacking gel-separating gel interface (arrow), but most of the immunoprecipitated material migrated as subunits. (From reference 8.)

(Fig. 4) and an active proton motive force (6) are required.

ORGANELLE ASSEMBLY

Kinetics

Once the nascent precursors have been processed into mature protein subunits, what happens to them? Do they form a large pool, like sex pilin, in the inner membrane? By making use of a quaternary-structure-specific monoclonal antibody for radioimmunoprecipitation analyses (Fig. 4) and of pulse-chase labeling, we have been able to characterize the kinetics of assembly and determine the size of the subunit pool (8). We found that the assembly of subunits into structures that contain the quaternary-specific epitope requires 1 to 3 min in log-phase cultures grown at 37°C (Fig. 5). Assembly of already synthesized subunits continues at the same rate even after blockage of protein synthesis (Fig. 5), and it does so until the pool is depleted. In contrast to the large subunit pool seen with F-pilin (28), the fimbrial subunit pool is quite small. Chloramphenicol, which stops translation but not assembly, prevents organelle elongation. This has been determined by radioimmunoprecipitation (Fig. 6) and by studies employing whole cells at various times after defimbriation to inhibit a fimbria-specific enzyme-linked immunosorbent assay. Assembly into mature fimbriae requires 1 to 3 min during log phase but significantly longer as the bacterial growth rate slows. Unlike what is seen with F-pili and non-type-1

fimbriae, lowered temperature (even to 18°C) has a minimal effect on type 1 fimbrial assembly relative to total protein synthesis (8).

A Model of Morphogenesis

These results, along with those related to subunit processing, permit a model of organelle morphogenesis that is consistent with that proposed by Mooi and co-workers for K88 fimbriae (27). Their model is based on a genetic and biochemical evaluation of the genes required for K88 synthesis and assembly, which include accessory (or assembly) genes encoding 81-, 27-, and 17-kDa polypeptides, in addition to the 26-kDa fimbrial subunit gene (27). Basically, the mature subunit is exported across the cytoplasmic membrane, where it encounters and is stabilized by the 17- and 27-kDa proteins. The subunit–carrier-protein aggregate is transported through the periplasmic space to the 81-kDa protein, which is embedded in the outer membrane.

Given the similarity of the genetic organization of the type 1 and K88 fimbriae, Orndorff and Falkow (31) have proposed similar assembly functions for their 86- and 30-kDa proteins, as well as for a 14-kDa polypeptide encoded by a gene mapping in the same region. Taken together with our data, a morphogenesis model similar to that proposed for the K88 fimbria emerges (Fig. 7). First, the 19-kDa precursors are synthesized,

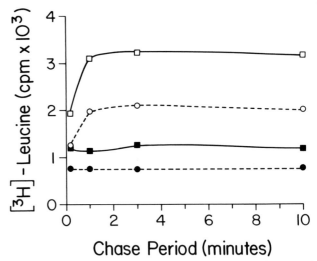

FIG. 5. Effect of chloramphenicol on the synthesis and assembly of fimbriae and on the synthesis of the maltose-binding protein. Immunoprecipitations were performed after different chase periods with either anti-maltose-binding protein serum or quaternary-specific antifimbrial monoclonal antibody. The Fim⁺ K-12 strain was pulse-labeled with [³H]leucine for 30 s and then chased with unlabeled leucine for the indicated times in either the absence or presence of 100 μg of chloramphenicol. Shown are total counts immunoprecipitated by the monoclonal antibody in the absence (□) or presence (○) of chloramphenicol or by anti-maltose-bindingprotein serum in the absence (■) or presence (●) of chloramphenicol. Note that although the total amount of fimbriae synthesized is reduced by the antibiotic, the reduction is no greater than that seen with the rapidly synthesized maltose-binding protein, and that the kinetics of fimbrial assembly (as measured by acquisition of the quaternary-specific epitope) is unchanged. (From reference 8.)

A B C D

FIG. 6. Effect of chloramphenicol on the elongation of fimbriae. Fimbriae were separated by size on a 5% acrylamide gel after immunoprecipitation of solubilized *E. coli* cells that had been pulse-labeled for 20 s and chased for either 10 s or 20 min in the absence or presence of chloramphenicol. Lane A, 10-s chase, with no chloramphenicol added. Lane B, 20-min chase, with no chloramphenicol added. Note the disappearance of the label in the smaller-sized fimbriae. Lane C, 20-min chase, with chloramphenicol added after 10 s of chase. Note the persistence of label, as in lane A, in the smaller-sized fimbriae, which indicates that the subunit pool must be small enough to prevent detectable further growth of the organelles. Lane D, 20-min chase, with chloramphenicol, at the same concentration as in lane C, added 1 min before pulse-labeling. Note inhibition of label uptake, demonstrating efficacy of chloramphenicol at this concentration. (From reference 8.)

processed cotranslationally, and translocated across the cytoplasmic membrane, where they form a small pool; this step is typical of other membrane-bound proteins. Next, the mature subunits are transported by specific assembly proteins to the outer membrane in a process that does not require new protein synthesis.

Questions that need exploring at the present include the following. Where in the envelope is the organelle anchored? Do fimbriae grow from the tip out (like flagella) or from the base out (like F-pili)? We also wish to explore more of the molecular details of phase variation, the ultimate determinant of fimbrial expression.

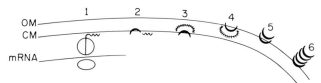

FIG. 7. Schematic representation of the multistep process of fimbrial synthesis, secretion, and assembly. Step 1, Precursors are synthesized on membrane-bound ribosomes. Step 2, Mature subunit, after cotranslational processing of the signal sequence, is embedded in the cytoplasmic membrane (CM). Step 3, Assembly proteins (hatched crescent) make contact with fimbrial subunit. Step 4, Transport of complex across the periplasmic space to the outer membrane (OM). Step 5, Early stage of assembly; quaternary epitopes first appear. Step 6, Mature organelle.

ACKNOWLEDGMENTS

I thank my co-workers John Abraham, Janice Clements, Doug Dodd, and Cynthia Freitag, as well as numerous other colleagues, for their help with the work summarized here.

This work is supported in part by Public Health Service research grant AI-19670, an Institutional Training Grant, and a Research Career Development Award from the National Institutes of Health.

LITERATURE CITED

1. **Abraham, J. M., C. S. Freitag, J. R. Clements, and B. I. Eisenstein.** 1985. An invertible element of DNA controls phase variation of type 1 fimbriae of *Escherichia coli*. Proc. Natl. Acad. Sci. USA **82:**5724–5727.
2. **Abraham, S. N., D. L. Hasty, W. A. Simpson, and E. H. Beachey.** 1983. Antiadhesive properties of a quaternary structure-specific hybridoma antibody against type 1 fimbriae of *Escherichia coli*. J. Exp. Med. **158:**1114–1128.
3. **Beachey, E. H., B. I. Eisenstein, and I. Ofek.** 1982. Current concepts: bacterial adherence in infectious diseases. Scope monograph. The Upjohn Co., Kalamazoo, Mich.
4. **Brinton, C. C., Jr.** 1959. Non-flagellar appendages of bacteria. Nature (London) **183:**782–786.
5. **Brinton, C. C., Jr.** 1965. The structure, function, synthesis and genetic control of bacterial pili and a molecular model for DNA and RNA transport in gram-negative bacteria. Trans. N.Y. Acad. Sci. **27:**1003–1054.
6. **Dodd, D. C., P. J. Bassford, Jr., and B. I. Eisenstein.** 1984. Dependence of secretion and assembly of type 1 fimbrial subunits of *Escherichia coli* on normal protein export. J. Bacteriol. **159:**1077–1079.
7. **Dodd, D. C., and B. I. Eisenstein.** 1982. Antigenic quantitation of type 1 fimbriae on the surface of *Escherichia coli* cells by an enzyme-linked immunosorbent inhibition assay. Infect. Immun. **38:**764–773.
8. **Dodd, D. C., and B. I. Eisenstein.** 1984. Kinetic analysis of the synthesis and assembly of type 1 fimbriae of *Escherichia coli*. J. Bacteriol. **160:**227–232.
9. **Duguid, J. P., and D. C. Old.** 1980. Adhesive properties of Enterobacteriaceae, p. 185–217. *In* E. H. Beachey (ed.), Bacterial adherence. Chapman & Hall, Ltd., London.
10. **Duguid, J. P., I. W. Smith, G. Demster, and P. N. Edwards.** 1955. Non-flagellar filamentous appendages ('fimbriae') and haemagglutinating activity in *Bacterium coli*. J. Pathol. Bacteriol. **70:**335–348.
11. **Eisenstein, B. I.** 1981. Phase variation of type 1 fimbriae in *Escherichia coli* is under transcriptional control. Science **214:**337–339.
12. **Eisenstein, B. I.** 1982. Operon fusion of the phase variation switch, a virulence factor in *Escherichia coli*. Infection **10:**112–115.
13. **Eisenstein, B. I., J. R. Clements, and D. C. Dodd.** 1983. Isolation and characterization of a monoclonal antibody directed against type 1 fimbriae organelles from *Escherichia coli*. Infect. Immun. **42:**333–340.
14. **Eisenstein, B. I., and D. C. Dodd.** 1982. Pseudocatabolite repression of type 1 fimbriae in *Escherichia coli*. J. Bacteriol. **151:**1560–1567.
15. **Eisenstein, B. I., I. Ofek, and E. H. Beachey.** 1981. Loss of lectin-like activity in aberrant type 1 fimbriae of *Escherichia coli*. Infect. Immun. **31:**792–797.
16. **Emerson, S. U., K. Tokuyasu, and M. I. Simon.** 1970. Bacterial flagella: polarity of elongation. Science **169:**190–192.
17. **Eshdat, Y., F. J. Silverblatt, and N. Sharon.** 1981. Dissociation and reassembly of *Escherichia coli* type 1 pili. J. Bacteriol. **148:**308–314.
18. **Firon, N., I. Ofek, and N. Sharon.** 1982. Interaction of mannose-containing oligosaccharides with the fimbrial lectin of *Escherichia coli*. Biochem. Biophys. Res. Commun. **105:**1426–1432.
19. **Freitag, C. S., J. M. Abraham, J. R. Clements, and B. I. Eisenstein.** 1985. Genetic analysis of the phase variation control of expression of type 1 fimbriae in *Escherichia coli*. J. Bacteriol. **162:**668–675.
20. **Freitag, C. S., and B. I. Eisenstein.** 1983. Genetic mapping and transcriptional orientation of the *fimD* gene. J. Bacteriol. **156:**1052–1058.
20a.**Hermodson, M. A., K. C. S. Chen, and T. M. Buchanan.** 1978. *Neisseria* pili proteins: amino-terminal amino acid sequences and identification of an unusual amino acid. Biochemistry **17:**442–445.
21. **Ippen-Ihler, K., D. Moore, S. Laine, D. A. Johnson, and N. S. Willetts.** 1984. Synthesis of F-pilin polypeptide in the absence of F *traJ* product. Plasmid **11:**116–129.

22. **Kehoe, M., R. Sellwood, P. Shipley, and G. Dougan.** 1981. Genetic analysis of K88-mediated adhesion of enterotoxigenic *Escherichia coli.* Nature (London) **291:**122–126.
23. **Klemm, P.** 1984. The *fimA* gene encoding the type 1 fimbrial subunit of *Escherichia coli:* nucleotide sequence and primary structure of the protein. Eur. J. Biochem. **143:**395–399.
24. **Klemm, P., I. Ørskov, and F. Ørskov.** 1982. F7 and type 1-like fimbriae from three *Escherichia coli* strains isolated from urinary tract infections: protein chemical and immunological aspects. Infect. Immun. **36:**462–468.
25. **McMichael, J. C., and J. T. Ou.** 1979. Structure of common pili from *Escherichia coli.* J. Bacteriol. **138:**969–975.
26. **Mitsui, Y., F. P. Dyer, and R. Langridge.** 1973. X-ray diffraction studies of bacterial pili. J. Mol. Biol. **79:**57–64.
27. **Mooi, F. R., A. Wijfjes, and F. K. de Graaf.** 1983. Identification and characterization of precursors in the biosynthesis of the K88ab fimbria of *Escherichia coli.* J. Bacteriol. **154:**41–49.
28. **Moore, D., B. A. Sowa, and K. Ippen-Ihler.** 1981. Location of an F-pilin pool in the inner membrane. J. Bacteriol. **146:**251–259.
29. **Norgren, M., S. Normark, D. Lark, P. O'Hanley, G. Schoolnik, S. Falkow, C. Svanborg-Edén, M. Båga, and B. E. Uhlin.** 1984. Mutations in *E. coli* cistrons affecting adhesion to human cells do not abolish Pap pili fiber formation. EMBO J. **3:**1159–1165.
30. **Ofek, I., D. Mirelman, and N. Sharon.** 1977. Adherence of *Escherichia coli* to human mucosal cells mediated by mannose receptors. Nature (London) **265:**623–625.
31. **Orndorff, P. E., and S. Falkow.** 1984. Organization and expression of genes responsible for type 1 piliation in *Escherichia coli.* J. Bacteriol. **159:**736–744.
32. **Orndorff, P. E., and S. Falkow.** 1984. Identification and characterization of a gene product that regulates type 1 piliation in *Escherichia coli.* J. Bacteriol. **160:**61–66.
33. **Orndorff, P. E., and S. Falkow.** 1985. Nucleotide sequence of *pilA,* the gene encoding the structural component of type 1 pili in *Escherichia coli.* J. Bacteriol. **162:**454–457.
34. **Ørskov, I., and F. Ørskov.** 1983. Serology of *Escherichia coli* fimbriae. Prog. Allergy **33:**80–105.
35. **Ottow, J. C. G.** 1975. Ecology, physiology, and genetics of fimbriae and pili. Annu. Rev. Biochem. **29:**79–108.
36. **Plasterk, R. H. A., and P. van de Putte.** 1984. Genetic switches by DNA inversions in prokaryotes. Biochim. Biophys. Acta **782:**111–119.
37. **Randall, L. L., and S. J. S. Hardy.** 1984. Export of protein in bacteria. Microbiol. Rev. **48:**290–298.
38. **Salit, I., and E. C. Gotschlich.** 1977. Hemagglutination by purified type 1 *Escherichia coli* pili. J. Exp. Med. **146:**1169–1181.
39. **Salit, I. E., and E. C. Gotschlich.** 1977. Type I *Escherichia coli* pili: characterization of binding sites on monkey kidney cells. J. Exp. Med. **146:**1182–1194.
40. **Salit, I. E., J. Vavougios, and T. Hofmann.** 1983. Isolation and characterization of *Escherichia coli* pili from diverse clinical sources. Infect. Immun. **42:**755–762.
41. **Silhavy, T. J., S. A. Benson, and S. D. Emr.** 1983. Mechanisms of protein localization. Microbiol. Rev. **47:**313–344.
42. **Silverblatt, F. J., and I. Ofek.** 1978. Influence of pili on the virulence of *Proteus mirabilis* in experimental hematogenous pyelonephritis. J. Infect. Dis. **138:**664–667.
43. **Silverblatt, F. J., and I. Ofek.** 1979. Effects of pili on susceptibility of *Proteus mirabilis* to phagocytosis and on adherence to bladder cells, p. 49–59. *In* E. H. Kass and W. Brumpitt (ed.), Infections of the urinary tract. Proceedings of the Third International Symposium on Pyelonephritis. The University of Chicago Press, Chicago.
44. **Silverman, M., and M. Simon.** 1980. Phase variation: genetic analysis of switching mutants. Cell **19:**845–854.
45. **Swaney, L. M., Y.-P. Liu, K. Ippen-Ihler, and C. C. Brinton, Jr.** 1977. Genetic complementation analysis of *Escherichia coli* type 1 somatic pilus mutants. J. Bacteriol. **130:**506–511.
46. **Swaney, L. M., Y.-P. Liu, C.-M. To, C.-C. To, K. Ippen-Ihler, and C. C. Brinton, Jr.** 1977. Isolation and characterization of *Escherichia coli* phase variants and mutants deficient in type 1 pilus production. J. Bacteriol. **130:**495–505.
47. **Zieg, J., and M. Simon.** 1980. Analysis of the nucleotide sequence of an invertible controlling element. Proc. Natl. Acad. Sci. USA **77:**4196–4200.

9. The Nucleoid

The Public Health Research Institute of the City of New York, New York, New York 10016

INTRODUCTION

In *Escherichia coli* the genomic DNA is a large circular molecule which is compacted more than 1,000-fold into a microscopically distinct structure called a nucleoid. The nucleoid is functionally analogous to the eucaryotic nucleus, but there are intriguing structural differences. For example, the nucleoid is not surrounded by a membrane as is the eucaryotic nucleus (97). Moreover, assays that detect superhelical tension in bacterial DNA fail to do so in the DNA of higher cells (100), and chromatin structure appears to be more labile in bacteria (33). Since the nucleoid, not naked DNA, is the substrate for the many DNA-binding proteins that govern chromosomal processes, it is important that we understand the details of nucleoid structure. This is becoming increasingly obvious as we learn more about DNA supercoiling and the possibility that some of the DNA may have fewer base pairs per turn than naked DNA (3a).

Bacterial nucleoids have been examined both by microscopy of whole or sectioned cells and by cell fractionation strategies. In the first section below several microscopic studies are outlined, and in subsequent sections the focus shifts to information gained from the examination of isolated nucleoids. In the later sections a brief description of nucleoid isolation is followed by a discussion of ideas that have evolved from the use of agents that perturb nucleoid structure. Some of these agents appear to alter nucleoid structure in living cells; thus it has been possible to find correlations between changes in nucleoid function and presumed changes in nucleoid structure. Most of these correlations involve DNA supercoiling and topoisomerase activities. For earlier reviews readers are referred to Pettijohn (78, 79), Kleppe et al. (50), Pettijohn and Carlson (80), and Pettijohn and Sinden (86).

NUCLEOIDS INSIDE CELLS

Nucleoids can be seen in growing cells by light microscopy if the refractive index of the medium is properly adjusted (66, 91, 99). Although these low-resolution studies provide little detail about how the DNA is organized, they do establish that in living cells the bacterial chromosome is a discrete, compact structure. Sharper images are now being obtained by confocal scanning light microscopy, and it appears that the nucleoid has a lobular shape with clefts (113), a conclusion that has also been reached by electron microscopy (see below).

Nucleoids are readily visualized after fixation. A number of stains have been used for light microscopy, including Feulgen-type stains and fluorescent dyes (for a review see Bjornsti et al. [2]); an example using ethidium bromide to stain the DNA is shown in Fig. 1. In general, the appearance of fixed nucleoids, examined either by light or by low-resolution electron microscopy (for examples see reference 2), is similar to that of nucleoids in living cells. Since a variety of agents, such as antibiotics, phage infection, and ionizing radiation, alter the shape of the nucleoid (7, 49, 60, 61, 96), it is likely that the integrity and metabolic activity of the DNA are important in determining the gross structure of the nucleoid.

High-resolution electron microscopy of thin sections reveals additional details, but it has been difficult to rule out artifacts of fixation (for discussion see reference 40). Kellenberger and his associates have

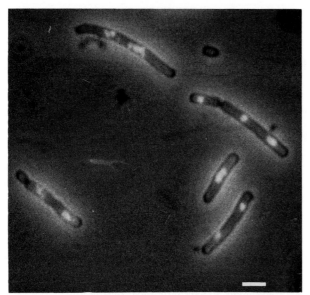

FIG. 1. Nucleoids in *E. coli* cells. Cultures of a temperature-sensitive *dnaA* mutant of *E. coli* were grown at 30°C and then shifted to 43°C for 60 min. A sample of cells was heat fixed onto a glass slide and stained with ethidium bromide as described by Hecht et al. (37). The photograph was taken by phase-contrast microscopy with UV illumination. Bar, 1 μm. Photo courtesy of T. Steck.

bypassed some of the fixation problems by using a rapid-chilling, cryosubstitution procedure (Fig. 2). The region occupied by the DNA has been identified by immunolabeling (2), and it corresponds to the ribosome-free region shown in Fig. 2. The lobular nature of the nucleoid is readily apparent.

The metabolically active regions of the nucleoid appear to be at the edges of the structure, for it is there that single-stranded nucleic acids (2), nascent RNA (98), and ribosomes (Fig. 2; 40) are located. We do not know how nucleotide sequences in the interior of the nucleoid gain access to the replication and gene expression machinery. Perhaps there is continuous movement of the DNA, as suggested by Ryter and Chang (98), or perhaps the clefts among the lobes of the nucleoid are so deep that no region of the genome is effectively buried in the interior.

NUCLEOID ISOLATION

The addition of nonionic detergents and molar NaCl to lysozyme-treated *E. coli* cells results in nonviscous cell lysates (81, 108). These cell lysates can be pipetted or gently mixed without shearing the chromosomal DNA, and nucleoids can be isolated from the lysate by centrifugation into a sucrose density gradient. The cellular DNA sediments as a distinct band (Fig. 3), and when removed from the gradient, it remains compact: nucleoids resediment at the same rate, and the folded DNA resists breakage by gentle shear forces. Measurement of DNA mass indicates that isolated nucleoids contain the entire bacterial genome (37), and electron microscopic analyses reveal few if any DNA breaks (9, 10). The dimensions of the nucleoids are close to the dimensions of the bacterial nucleoid seen in vivo, as calculated from their sedimentation properties (108,

119) and as observed by light (37) and electron (84) microscopy. Thus it appears likely that some of the constraints that fold and compact the DNA in vivo are also present in nucleoids isolated in this manner.

A high counterion concentration is important for maintaining the folded conformation of bacterial nucleoids, particularly at the time of cell lysis (108). This is in contrast to eucaryotic systems, in which DNA in chromatin remains packaged at moderate ionic strength. It has been suggested that some of the factors stabilizing intracellular DNA packaging are easily dissociated from the nucleoid during cell lysis, even at low salt concentrations (3a, 84); thus, the high counterion concentration may substitute for these stabilizing components by neutralizing DNA-phosphate charge repulsions. The need for a high counterion concentration makes it important to stress that some of the macromolecular interactions present in isolated nucleoids undoubtedly differ from those present in intracellular nucleoids. It is also important to point out that during cell lysis macromolecules may fortuitously bind to nucleoids or macromolecular interactions may become altered (68). Thus the in-depth study of complex structures such as nucleoids requires a variety of approaches to bypass potential artifacts.

Several cell lysis procedures have been devised for obtaining nucleoids, and their distinguishing characteristics have been summarized (86). Here it is necessary only to point out that folded chromosomal DNA can be isolated either relatively free of cell envelopes or associated with them. When cell lysis occurs in molar NaCl between 20 and 25°C, isolated nucleoids contain little material derived from the cell envelopes (84, 121). Cell lysis at less than 4°C results in nucleoids that contain large envelope fragments (84, 95, 120). Envelope-associated nucleoids can also be obtained if during cell lysis spermidine is substituted for the high concentration of salt (53) or if the detergent conditions are altered (18, 19).

NUCLEOID STRUCTURE

The major component of the bacterial chromosome is a large circular DNA molecule (5) of about 4,400 kilobase pairs (kbp) (5, 51). Since replication (for a review, see Davern [8]) initiates at a single origin and proceeds bidirectionally to a termination site 180° away (Fig. 4a), chromosomes isolated from exponentially growing cells are theta structures (Fig. 4b). The amount of DNA in a given chromosome depends on the extent it has replicated. In rapidly growing cells a round of replication initiates before the previous one finishes; consequently, chromosomes can become complex structures (Fig. 4c). In one case (114) the DNA content of nucleoids isolated from rapidly growing cells was found to be equivalent to 2.8 genomes and corresponded to the DNA content of an *E. coli* cell at this growth condition. Thus nucleoids have a diameter of only 0.5 μm (37), but their DNA is several thousand micrometers long. Understanding how this DNA is folded and packaged has been the goal of most studies of nucleoid structure. The majority of these studies have focused on envelope-free nucleoids because these nucleoids are easily and reproducibly isolated from a variety of *E. coli* strains and because

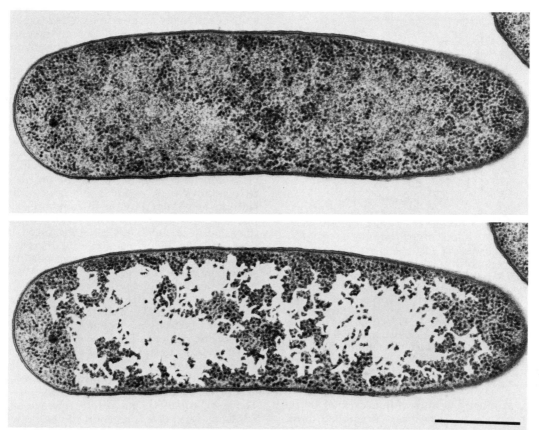

FIG. 2. Cryosubstituted *E. coli* B showing the organization of the bacterial nucleoid. Exponentially growing bacterial cells were harvested by filtration and immediately frozen without any pretreatment on a copper block cooled with liquid helium. They were then substituted at $-90°C$ in a 2.5% OsO_4–100% acetone mixture for 64 h, after which the temperature was raised slowly to room temperature and the cells were embedded in Epon. Thin sections were stained with uranyl acetate and lead citrate, and a typical section is shown in the upper portion of the figure. Bar, 0.5 μm. The bacterial nucleoid is seen as a ribosome-free area localized within the cytoplasm. The nucleoid has many clefts, and it contains rather grainy structures intermingled with fibrous elements. The same cell is shown in the lower portion of the figure, but the ribosome-free area has been whitened for clarity. Osmium-amines staining (a Feulgen-type DNA stain) demonstrated that this area contains DNA (40). This was further corroborated by immunolabeling studies using the protein A-gold technique with antibodies specific for double-stranded DNA and antibodies directed against single-stranded DNA (2). The double-stranded DNA is present throughout the ribosome-free area, while single-stranded DNA is present only along the ribosome-nucleoid border. The histonelike protein HU and topoisomerase I are also concentrated along the nucleoid border. Harvesting cells by aerated filtration followed by rapid freezing and cryosubstitution is thought to be important in maintaining the in vivo conformation of the nucleoid. Earlier studies had shown that harvesting by centrifugation and fixing with OsO_4 or glutaraldehyde cause ion leakage from the cells; as a result, the ionic concentration of the growth medium affects nucleoid morphology. Such is not the case with the filtration-rapid freezing procedure (for additional discussion see reference 40). Photo courtesy of J. Hobot.

they represent the purest form of folded DNA obtained so far.

Measurement of Conformational Changes in Nucleoid Structure

Hydrodynamic properties of isolated nucleoids are very dependent on changes in shape, and much of our knowledge of nucleoid structure has been obtained by determining how a variety of perturbations unfold the DNA and change the sedimentation and viscometric properties of the nucleoid. One of the special features of large DNA is that the rate of sedimentation during centrifugation depends on the speed of the centrifuge rotor: as speed increases, sedimentation rate decreases. Zimm (123) explained this phenomenon for linear DNA in the following way. During sedimentation, the ends of a DNA molecule experience a greater frictional force

than the middle, producing an extended DNA conformation; in a sense the molecule has a V-like shape. At higher rotor speeds the rate of DNA transport increases, so the degree of extension and the frictional coefficient increase. This causes the sedimentation rate to decrease. Although isolated nucleoids lack free ends (10, 47, 48), they do exhibit a reversible dependence of sedimentation rate on rotor speed, analogous to that observed with purified DNA (36, 83). Apparently the large DNA loops of the nucleoid (see below) become extended and behave like linear DNA (83). Consequently, accurate measurement of the sedimentation coefficient requires extrapolation to zero rotor speed. Rotor speed dependence also increases as the DNA unfolds (36); thus speed dependence can serve as an additional way to measure unfolding.

A second peculiarity of large DNA is observed during viscometric measurements: the DNA molecules

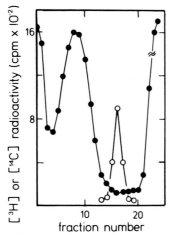

FIG. 3. Sedimentation of nucleoids in a sucrose density gradient. *E. coli* cells grown in the presence of [³H]thymidine were lysed in 1 M NaCl at 20°C by the method of Stonington and Pettijohn (108). Small samples of the cell lysate and ¹⁴C-labeled bacteriophage T4 were sedimented into a 10 to 30% (wt/vol) sucrose density gradient. After centrifugation, fractions were collected from the bottom of the centrifuge tube and the radioactivity in each fraction was determined. Sedimentation is from right to left. Symbols: ●, ³H-labeled chromosomal DNA; ○, ¹⁴C-labeled bacteriophage T4 (S = 1,000).

tend to orient along the lines of flow created by the viscometer. Soon after shear stress is applied to unfolded nucleoids, the solution exhibits an increase in viscosity. This increase probably arises from the DNA being stretched. A few minutes later the apparent viscosity decreases dramatically, possibly reflecting the alignment of the DNA molecules. The magnitude of this transient change in apparent viscosity corresponds empirically with shape changes observed by sedimentation analyses (17). Since this method is very sensitive to nucleoid aggregation, it has found limited application.

DNA Supercoiling

DNA in isolated nucleoids is negatively supercoiled (82, 119). Supercoiling is generally described in terms of relaxed DNA, a closed circular DNA created by ligation of a nicked circle. Supercoiling represents a

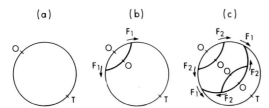

FIG. 4. Replication of bacterial chromosomes. (a) Replication begins at a site called the origin (O) and stops at a site called the terminus (T). (b) A theta structure is created by two replication forks (F) moving in opposite directions from the origin toward the terminus. (c) A complex structure is created when a second round of replication initiates before the previous one finishes. F1, Replication forks from the first round of replication; F2, replication forks from the second round of replication.

state of relatively greater free energy (ΔG) in a closed circular DNA and arises when an excess or a deficiency of duplex turns arises. Negatively supercoiled DNA has a deficiency of duplex turns relative to relaxed DNA. The striking effect of an intercalating dye (ethidium bromide) on sedimentation of supercoiled nucleoids is illustrated in Fig. 5 (comparable hydrodynamic changes can also be observed by using viscometric techniques [17]). At low dye concentrations the sedimentation coefficient decreases as the dye titrates (relaxes) the negative supercoils, reaching a minimum at the dye concentration which removes all of the supercoils. At this point the chromosomal DNA is completely relaxed; this same sedimentation rate can be obtained by the introduction of nicks into the DNA with DNase I (119). At higher dye concentrations, positive supercoils are introduced into intact DNA and the sedimentation rate increases. At low DNA concentrations, the dye concentration at the sedimentation minimum or viscosity maximum is proportional to the superhelical density of the DNA. Consequently, dye titration curves similar to the one shown in Fig. 5 can be used to measure changes in the level of supercoiling in chromosomal DNA. For example, a shift in the sedimentation minimum to lower dye concentrations would indicate a decrease in negative supercoiling. In the remainder of this chapter the term "titratable supercoiling" is used to specify experimental measurements made on extracted DNA. The term "superhelical tension" refers to intracellular DNA topology. It is important to distinguish the two parameters because a direct relationship between the two has not been extensively documented.

The discovery of enzymes that either introduce (gyrase; 30) or remove (topoisomerase I; 117) titratable supercoils in vitro suggested that chromosomal DNA may be under superhelical tension in vivo. This concept gained support from the observation that nucleoids have a lower level of titratable supercoiling

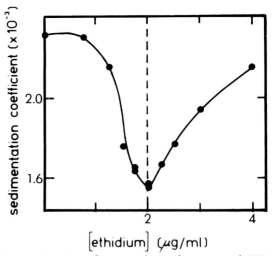

FIG. 5. Titration of supercoils in chromosomal DNA. *E. coli* cells were lysed and centrifuged into sucrose density gradients as described in the legend to Fig. 3. Gradients contained various concentrations of ethidium bromide as indicated on the abscissa. The means of the radioactive distribution in the nucleoid and bacteriophage peaks were used to determine relative sedimentation rates.

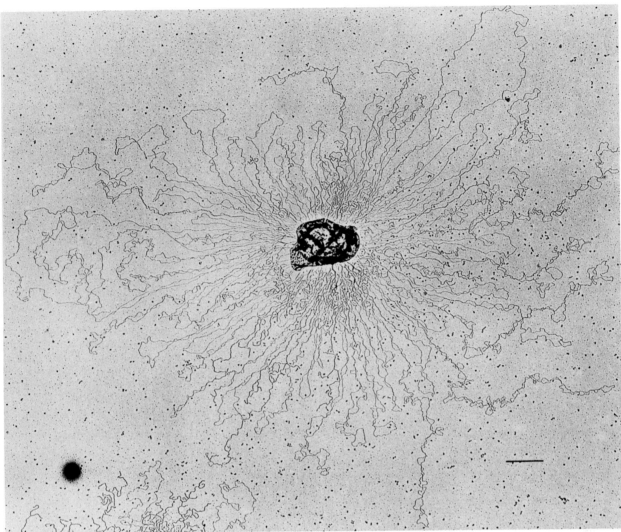

FIG. 6. Membrane-associated nucleoid showing DNA in loops. Nucleoids were isolated from *E. coli* cells and were spread for electron microscopy in the presence of spermidine. Bar, 1 μm. The figure is reprinted from reference 48 by permission of Springer-Verlag, Berlin; O. Ryder; R. Kavenoff; and Designer Genes, Box 100, Del Mar, Calif.

when isolated from cells treated with a specific inhibitor of gyrase (16). A direct demonstration that the intracellular DNA is under superhelical tension came from differences in psoralen binding between nicked and intact DNA (100). DNA in cells treated with radioactive psoralen binds less psoralen if the DNA is nicked by gamma irradiation or if cells are treated with an inhibitor of gyrase. Subsequent analyses of catenanes formed during site-specific recombination indicate that intracellular supercoiling is in the form of plectonemic supercoils and that the level of superhelical tension is about 40% of that measured with purified DNA (J. Bliska and N. Cozzarelli, submitted for publication).

Long-Range DNA Folding

Both sedimentation and electron microscopic studies indicate that the supercoils in isolated nucleoids cannot be relaxed by a single nick. This indicates that the chromosome is segregated into topologically independent domains (9, 82, 119). These domains may correspond to the large DNA loops that have been observed by electron microscopic examination of isolated nucleoids (see Fig. 6; 47, 48). The number of domains has been estimated from the number of nicks required to relax all of the supercoils. In living cells the numer of domains is about 50 per genome equivalent of DNA (101). Isolated nucleoids appear to contain more domains (about 100 per genome equivalent) than nucleoids in living cells (63); thus it is possible that some of the constraints observed with isolated nucleoids are due to in vitro artifacts. At present we can say little else about the molecular interactions responsible for establishing the topologically independent domains.

One function of the domains may be to localize the relaxing effects that strand scissions are expected to have on superhelical tension in chromosomal DNA. One could imagine that segregation into domains would allow the majority of the DNA to remain under superhelical tension even though DNA repair and replication create nicks and gaps.

RNA-Mediated Constraints

Treatment of isolated nucleoids with RNase or treatment of growing cells with rifampin, an inhibitor of RNA polymerase, produces striking changes in the hydrodynamic properties of isolated nucleoids. Sedimentation rates decrease and viscosity increases (17, 36, 82, 108). The hydrodynamic changes cannot be explained by the loss of RNA mass; thus RNA probably contributes to constraining DNA, at least in isolated nucleoids. Since partial digestion with RNase produces nucleoids with intermediate sedimentation rates (17, 84), there are probably a number of RNA-mediated constraints.

The RNA associated with the nucleoid is nascent RNA (84). Efforts to identify a new, stable RNA species or any unique DNA-associated RNA have met with no success (35). Even the small fraction of RNA (18 to 25 molecules per genome equivalent of DNA) that remains DNA associated after phenol and sodium dodecyl sulfate treatment appears to be composed largely of nascent species (35). These observations, coupled with the finding that only a few minutes of rifampin treatment is required to produce unfolded DNA (20, 45, 82), suggest that nascent RNA contributes to DNA folding in isolated nucleoids. Whether RNA-mediated constraint of DNA operates in vivo is unknown; in vitro these constraints could arise from RNA entanglements. It is clear, however, that RNA-mediated constraints are technically important for the isolation of compact nucleoids.

In the previous section it was pointed out that chromosomal DNA inside living cells is organized into topologically independent domains. Nascent RNA appears to play little role in establishing these domains in vivo, for the number of domains is not changed by treatment of cells with rifampin (101). In vitro, however, RNA may participate in creating domains since nucleoids appear to have more domains in vitro than in vivo. This hypothesis would explain why RNase treatment reduces the number of nicks required to relax supercoiling in isolated nucleoids (82). So far little progress has been made in studying topological domains in vitro after treatment of the nucleoids with RNase. The unfolding caused by RNase is so extensive that the nucleoids cannot be handled easily, and the sedimentation techniques lack the necessary precision to detect folding that might remain after removal of RNA (36, 63, 114).

Protein-Mediated Constraints

By analogy with eucaryotic chromatin we predict that proteins must play important roles in bacterial chromosomes as well. However, our understanding of this aspect of nucleoid structure lacks detail. Protein denaturants, such as ionic detergents, cause extensive changes in the hydrodynamic properties of isolated nucleoids (17, 108), consistent with DNA folding being one type of DNA constraint imposed by proteins. Detergent-induced unfolding is blocked by protein cross-linking agents such as formaldehyde (84); however, formaldehyde treatment does not block nucleoid unfolding caused by RNase (84). Thus, in vitro the putative protein constraints on DNA folding are insufficient to maintain the DNA in a folded conformation.

From these observations it has been suggested that a major portion of the DNA in isolated nucleoids is constrained by protein-RNA cross-links (84). Since the RNA appears to be nascent RNA, at least one of the protein moieties may be RNA polymerase. So far proteases have not been used successfully to study DNA folding because the hydrodynamic properties of intact nucleoids are affected little by these enzymes (108, 119). If proteins are involved in DNA folding, they must not be susceptible to protease attack when nucleoids are intact. Little else is known about proteins that might constrain the DNA folds. For this type of study it will probably be more profitable to examine Bacillus licheniformis: with this organism nucleoids can be isolated which exhibit a large sedimentation decrease and a loss of titratable supercoils upon protease treatment (102).

Another level of DNA constraint may be mediated by histonelike proteins wrapping DNA into nucleosomelike structures. Such constraints could account for the observation that nicking followed by ligation fails to relax all of the supercoiling in F plasmids (85). Moreover, bacterial chromatin having a particulate structure has been detected by electron microscopy (33). By combining estimates of particle diameter with measurements of contour length, Griffith estimated that bacterial chromatin averages one nucleosomelike particle per 220 to 290 base pairs (bp) of DNA. Unfortunately, these particulate structures are very labile, dissociating within minutes after cell lysis, and so they have not been characterized biochemically.

Bacterial cells do contain histonelike proteins. The best studied is one called HU (93). This small (9,700-dalton), basic, heat-stable, DNA-binding protein has physical properties and an amino acid composition similar to that of eucaryotic histones (93). Like eucaryotic histones, HU appears to be highly conserved in a number of species: many amino acid sequence homologies are found among HU proteins isolated from eubacteria (for tabulation see reference 109), and immunologically cross-reacting proteins are found in E. coli, Salmonella typhimurium, Bacillus subtilis (94), blue-green algae (cyanobacteria; 34), and plant chloroplasts (3). In addition, HU is able to wrap DNA into nucleosomelike particles in vitro (3a, 94), and HU appears to be associated with the chromosome since it copurifies with membrane-associated nucleoids isolated at low salt concentration (92, 115).

Efforts to isolate chromatin have resulted in an enrichment of HU. One approach has been to partially digest DNA in membrane-associated nucleoids (115) or in cell lysates (59, 62) to release protein-bound DNA fragments. Three statements have been made about the fragments from isolated nucleoids (115). First, two proteins, HU and protein 1 (a 16,000-dalton protein), constitute the majority of the protein associated with the DNA. Second, on average there is one each of the two proteins per 150 to 200 bp of DNA. This corresponds to about 24,000 copies of DNA-bound HU per genome equivalent of DNA, a number that is in reasonable agreement with minimum estimates of 17,000 copies per genome equivalent derived from HU purification studies (92). Third, micrococcal nuclease digestion of the "chromatin" fragments produces two small DNA-protein particles detectable by gel electrophoresis. The larger contains a 160-bp DNA molecule and the smaller contains a 120-bp one. Both particles seem to contain equal amounts of HU and protein 1, but the number of copies of each protein per particle is unknown. Many more proteins are found associated with "chromatin" isolated directly from lysates, and no claim is made for HU being an integral component (59, 62).

Another approach has been to examine plasmid-protein complexes extracted from cells. HU appears to be associated with the plasmids (4, 122). These plasmid-protein complexes contain several other proteins including RNA polymerase and a small (15,500-dalton), abundant, neutral protein called H1 (104). It has been argued that H1 and protein 1 (described above) are identical (56); whether HU and H1 (protein 1) interact is not known.

Although the studies described above are intriguing, a number of questions remain. For example, which of the protein-DNA interactions found in vitro also occur in vivo? Is the isolated chromatin representative of the majority of the chromosomal DNA or only a small, specialized subset of it? How are the DNA-binding proteins distributed along the DNA? They do not seem to form a regular repeating array since partial digestion with micrococcal nuclease does not generate the type of oligomeric series found with eucaryotic chromatin (115; J. Rouviere-Yaniv, unpublished data). Why does HU appear to be concentrated around the periphery of the nucleoid rather than spread throughout it when immunolabeling of nucleoids is examined by electron microscopy (2)? Why is protein H, another abundant, histonelike protein (41), not a major component of fragmented chromatin (115)?

Still another level of DNA constraint may involve DNA interactions with proteins of the cell envelope. Twenty years ago, Jacob et al. (43, 44) postulated that bacteria may contain envelope-chromosome attachments which allow the envelope to serve as a mitotic apparatus: growth of the envelope between daughter chromosomes would pull the chromosomes apart. Although subsequent examination of membrane and envelope growth has not revealed a pattern that would segregate chromosomes in a simple manner, several cell fractionation procedures have led to the isolation of envelope-associated DNA (for review see Ogden and Schaechter [75]). Envelope-DNA complexes have been obtained which are enriched for nucleotide sequences located at the origin of replication, and when purified DNA fragments are mixed with envelope fractions, those containing the origin preferentially bind to envelope components (38, 52, 72). By cutting *oriC*-containing DNA with restriction nucleases it has been possible to localize several sites which bind strongly to the outer membrane of *E. coli* (55, 118). Thus the cell envelope may be intimately involved in the initiation phase of DNA replication. Whether these interactions are associated with chromosome segregation or other aspects of nucleoid structure is not known.

Cell fractionation procedures also produce envelope-DNA complexes that contain nucleotide sequences other than those of the origin of replication. There appear to be about 20 envelope-DNA attachments per genome (1, 14, 20). Several intriguing numerical similarities have emerged: in addition to about 20 envelope-DNA attachments per genome there are about 50 topologically independent domains of DNA (101), about 50 DNA cleavage events generated by an inhibitor of DNA gyrase (103), between 20 and 50 S1 nuclease-sensitive regions (1; S. H. Manes and K. Drlica, unpublished data), and about 100 microscopically visible DNA loops that radiate from envelope remnants associated with isolated nucleoids (see Fig. 6; 10, 48). Whether any of these parameters are functionally related is unknown.

Several additional points should be noted when considering hypotheses about DNA-envelope interactions. First, the nucleotide sequences remaining with the envelope after restriction nuclease digestion of isolated nucleoids are largely random (14). Second, treatment of cells with rifampin before cell lysis eliminates most of the attachments between the cell envelope and DNA (20). Third, envelope fragments cannot be released from isolated nucleoids by the same methods used to prepare envelope-free nucleoids from cells (68). The next step is to demonstrate clearly that the associations between DNA and envelope are present in living cells; without that we are left with the nagging possibility that the associations may be artifacts occurring during cell lysis.

Multiple Genomes

Although few details are available concerning the segregation of daughter chromosomes after a round of replication, several general ideas are emerging. The first is that segregation appears to be slow, and isolated nucleoids are often composed of two or more daughter chromosomes. This phenomenon is particularly apparent in nucleoids isolated from rapidly growing cells in which multiple rounds of replication are occurring simultaneously; in some strains more than half of the nucleoids appear as dumbbell-shaped doublets when examined by light microscopy (37) or by scanning electron microscopy (114). Second, proteins may be involved in holding the daughter chromosomes together. Addition of sodium dodecyl sulfate to nucleoids isolated from a wild-type strain doubles the number of DNA-containing particles (114). Third, gyrase is probably involved in decatenating interlinked daughter chromosomes. We have observed that an abnormally high percentage of the nucleoids isolated from a temperature-sensitive gyrase mutant contain multiple genomes if the cells are incubated at restrictive temperature (105). This phenomenon is discussed in more detail below.

FUNCTIONAL ASPECTS OF CHROMOSOMAL DNA SUPERCOILING

Topoisomerases and DNA Supercoiling

DNA supercoiling is the structural property of nucleoids that has been most accessible to experimental manipulation in vivo. Many details are now known about the biochemistry of the enzymes (topoisomerases) that appear to regulate supercoiling (reviewed in references 6 and 27), and a number of ways have been discovered for altering topoisomerase activities (reviewed in reference 13). These activities change DNA linking numbers (for discussion of DNA topology, see reference 27). Since linking numbers are unaffected by DNA isolation procedures that avoid DNA breakage, changes in titratable supercoiling, an in vitro measurement, reflect changes in intracellular topoisomerase activities. Thus we expect changes in titratable supercoiling also to reflect changes in intracellular superhelical tension, a supposition that is supported by studies involving an inhibitor of gyrase (16, 100). It should not be assumed, however, that the absolute level of intracellular superhelical tension is directly related to the level of titratable supercoiling, because proteins may relieve or constrain some of the superhelical tension in vivo. Indeed, it appears that intracellular superhelical tension may contribute to less than half of the titratable supercoiling measured in vitro (85; Bliska and Cozzarelli, submitted).

The best studied topoisomerases are gyrase and topoisomerase I. Gyrase is composed of two types of subunit, the gyrA and the gyrB gene products, each of which can be selectively inhibited by treating cells with specific antibiotics or by means of mutations which render gyrase temperature sensitive. The gyrA protein is affected by nalidixic acid, oxolinic acid, and a number of related compounds. Novobiocin, coumermycin A₁, and chlorobiocin inhibit the gyrB protein. No specific inhibitors have been found for topoisomerase I, but this activity is eliminated by mutations in the topA gene.

Treatment of cells with inhibitors of gyrase has two effects on titratable supercoiling measured in extracted DNA: (i) specific inactivation of either subunit of gyrase blocks the introduction of titratable supercoils into relaxed, circular bacteriophage lambda DNA during superinfection of a lysogen (29, 31); and (ii) inhibition of either subunit leads to a loss of titratable supercoils from the bacterial chromosome (16, 64) and from plasmid DNA molecules (46, 57, 85). Gyrase is thought to be the target of the drugs because mutants with drug-resistant gyrase genes show no drug-induced change in titratable supercoiling. Inhibition of DNA gyrase by conditional-lethal mutations gives results similar to those obtained using antibiotic inhibitors (24, 28, 42, 57, 70, 106, 116). Thus, gyrase both introduces and maintains DNA supercoiling in bacterial cells.

E. coli extracts contain both DNA-relaxing activities and supercoiling activities, suggesting that in vivo the two types of activities compete to produce the proper level of superhelical tension. Indeed, closed circular DNA molecules treated with gyrase in vitro can become more highly supercoiled than DNA molecules extracted from wild-type cells (31). When mutations became available in topA, the gene encoding topoisomerase I (71, 107, 110, 111), it became possible to test this hypothesis. Higher-than-normal levels of titratable supercoiling are observed in chromosomal and plasmid pBR322 DNA from E. coli cells containing a point mutation in topA, and titratable supercoiling in the chromosome returns to normal when a wild-type topA gene is introduced into the mutant strain (88). Similarly, certain plasmids isolated from topA mutants of S. typhimurium have higher-than-normal levels of titratable supercoiling (87a, 90). Thus, it appears that topoisomerase I is involved in modulating the level of bacterial DNA superhelical tension by opposing gyrase.

Not all topA mutants fit the pattern described above. A topA deletion mutant exhibits a 25% reduc-

tion in titratable supercoiling relative to wild-type cells, a result opposite to that described above (88). Introduction of a wild-type topA gene into this strain failed to change titratable supercoiling back to normal levels, suggesting that a suppressor mutation was present. The presence of suppressor mutations, some of which map in gyrA and some in gyrB (11, 88, 89), reduces the level of titratable supercoiling (88). These suppressor mutations appear to mitigate the detrimental effects of the loss of topoisomerase activity, as judged by cell growth (11). For example, deletions in topA are easily transduced into E. coli strains containing a suppressor mutation by selecting for flanking markers, and the ΔtopA transductants have normal growth rates. In contrast, similar transductions into wild-type E. coli strains produce no ΔtopA transductants within the normal incubation period. Very slowly growing colonies which contain the topA deletion appear only after incubation for two to three times longer than normal. Upon subculturing, these ΔtopA transductants grow more rapidly, presumably because members of the clones have acquired a suppressor mutation (11).

The same type of study was subsequently performed with S. typhimurium. Measurement of plasmid supercoiling indicates that the interplay of topoisomerase mutations and supercoiling is similar in E. coli and in S. typhimurium (90). However, the two bacterial species may differ in one aspect: deletion mutations in topA may not require the acquisition of second-site compensatory mutations for growth of S. typhimurium (90).

To summarize, the simplest interpretation of the data is that gyrase increases DNA superhelical tension and topoisomerase I carries out the opposite reaction. The two activities are balanced to produce an optimal level of superhelical tension. A mutation in topA that partially inactivates topoisomerase I activity allows gyrase to introduce higher-than-normal levels of superhelical tension into chromosomal DNA. If topoisomerase I activity is completely eliminated, however, the level of superhelical tension may become unacceptably high (in E. coli), halting cell growth. Compensatory mutations then arise in gyrA and gyrB that lower the level of superhelical tension and allow cell growth. This is not a complete explanation, however, since compensatory mutations have also been mapped in a locus (toc) that does not appear to affect gyrase activity (89).

The inhibitors of gyrase and the several types of mutations in gyrA, gyrB, and topA have been used to alter DNA supercoiling in vivo, and a number of physiological processes have been found to be affected by changes in supercoiling and topoisomerase action (for reviews, see references 6, 13, and 27). The following sections briefly consider two of these processes, DNA replication and gene expression.

Role of Topoisomerases and Supercoiling in Chromosome Replication

DNA under superhelical tension has a greater free energy than relaxed DNA: the tension is lost spontaneously when a nick occurs in the DNA. In a sense, supercoiled DNA is under strain and is energetically activated. Processes which unwind DNA, such as DNA replication and transcription, are expected to be energetically favored in DNA molecules under negative superhelical tension because these processes relieve the strain. Supporting this idea are several physiological studies in which inactivation of gyrase inhibits chromosome replication.

The concept that DNA supercoiling is involved in the initiation of replication developed primarily from in vitro studies of bacteriophage φX174 replication in which the gene A protein requires a supercoiled substrate (65). Subsequently, an in vitro oriC, dnaA-dependent plasmid replication system was developed (26). Replication of the plasmid requires gyrase activity (26), and the specificity of the system for oriC and dependence on dnaA protein requires topoisomerase I (25). Moreover, this system is stimulated by the presence of the histonelike protein HU (12). Thus it is likely that nucleoid structure in the vicinity of the origin is important for initiation of replication.

A requirement for gyrase activity during initiation of replication is also supported by physiological studies. Some temperature-sensitive mutants exhibit the gradual decline in DNA synthesis at restrictive temperature characteristic of temperature-sensitive initiation mutants (22, 76). In another type of study the germination of spores of B. subtilis was examined. In a thymine-requiring strain, germinating spores develop the potential to initiate replication, but replication does not begin until thymine is supplied. Treatment with coumermycin A$_1$, an inhibitor of gyrase, before addition of thymine leads to a loss of initiation potential (74). Since shifting gyrase mutants to restrictive temperatures or treating cells with coumermycin leads to a loss of titratable supercoiling in chromosomal DNA (16, 106, 116), gyrase may act indirectly on replication through its effect on superhelical tension.

A role for supercoiling in the elongation phase of DNA replication has not been clearly demonstrated in vitro. Nevertheless, the results of several physiological studies are most easily interpreted if this is the case. For example, two temperature-sensitive gyrase mutations in E. coli (gyrA43 [54] and gyrB402 [23]) rapidly inhibit DNA synthesis when cells are shifted to restrictive temperature. It is not clear, however, how the gyrA43 and gyrB402 mutations differ functionally from other temperature-sensitive gyrase mutations which affect only initiation of replication, since all temperature-sensitive mutants examined have reduced levels of titratable supercoiling after incubation at restrictive temperatures (106, 116). Perhaps cells having gyrase mutations other than gyrA43 and gyrB402 retain enough residual activity to allow replication fork movement (elongation) but not enough to allow initiation.

Antibiotic inhibitors of gyrase also appear to block the elongation phase of replication. The two classes of inhibitors (represented by coumermycin A$_1$/novobiocin and nalidixic/oxolinic acid) both lower DNA synthesis rates rapidly. However, each class has unique features. Coumermycin A$_1$, for example, causes a loss of titratable supercoils from the bacterial chromosome and a decrease in the rate of DNA synthesis (16). The gradual loss of superhelical tension associated with coumermycin A$_1$ treatment may eliminate part of the driving force needed for fork movement. In contrast, low concentrations of oxolinic acid decrease DNA synthesis rates with little or no loss of titratable

supercoils (103). With this drug even partial inhibition of DNA synthesis is very rapid, as if inactivation of gyrase by oxolinic acid blocks replication fork/DNA movement (15, 21, 103). Studies of the mechanism of action of oxolinic and nalidixic acids provide a plausible candidate for such a physical block to replication. In vitro these two drugs trap a gyrase-DNA reaction intermediate in which gyrase has cleaved its DNA substrate and the gyrA protein has become covalently bound to the DNA (for a review see reference 77). Since this reaction also appears to occur in vivo (39, 87, 103), formation of drug-gyrase-DNA complexes is probably responsible for inhibition of DNA replication by oxolinic and nalidixic acids.

Studies using gyrase mutants provide evidence for gyrase participation in daughter chromosome resolution as well as in other aspects of the replication cycle (105). The sedimentation coefficients of nucleoids isolated from a temperature-sensitive gyrase mutant increase after cells are shifted to nonpermissive temperature (105, 106). Microscopic examination of nucleoids isolated from the mutant revealed that 90% of them become dumbbell-shaped doublets (in the comparable wild-type strain fewer than 20% of the nucleoids are doublets). These doublet nucleoids probably correspond to daughter chromosomes that have failed to segregate. Incubation of isolated doublet nucleoids with purified gyrase causes many of the doublets to separate into singlets; gyrase activity appears to be sufficient to separate the daughter chromosomes in vitro under conditions in which no DNA synthesis should occur. Since gyrase is the major decatenating activity in E. coli (Bliska and Cozzarelli, submitted), the simplest interpretation of these results is that the decatenating activity of gyrase is required for proper chromosome segregation.

Role of Topoisomerases and Supercoiling in Gene Expression

The expression of a number of genes appears to be influenced by supercoiling/topoisomerase activity (for list, see reference 13). Since this topic has been reviewed recently (13), only a few general comments will be offered here. Two points emerge from a number of in vitro and in vivo studies. First, negative superhelical tension is probably a way to energetically activate the chromosome for processes, such as the initiation of transcription, that involve DNA strand separation. Second, the response of individual genes to changes in titratable supercoiling varies considerably. The thought is now emerging that superhelical tension affects not only strand separation but also RNA polymerase-promoter binding (for a review, see reference 66a).

The availability of topoisomerase mutations that slightly alter steady-state levels of titratable supercoiling makes it possible to determine whether most or only a few genes are sensitive to changes in supercoiling. We have initiated a survey of the relative abundance of large numbers of proteins in E. coli strains in which titratable chromosomal supercoiling is either 15% higher or 25% lower than that found in DNA from wild-type cells (T. R. Steck, R. Franco, and K. Drlica, unpublished data). So far we have measured the relative abundance for about 50 proteins, and we find that 40% of them exhibit a change greater than 25%. For about 10% the change is more than twofold. For those proteins whose abundance appears to be affected by supercoiling, the majority increase in relative abundance when titratable supercoiling is either above or below that in DNA from wild-type cells. Although many indirect effects may influence this type of measurement, the net effect of supercoiling on gene expression appears to be substantial. If this proves to be generally true, we might expect that for a given growth condition, intracellular superhelical tension will be actively maintained at a set level. Experimentally, correlations between titratable supercoiling and expression of the genes encoding gyrase and topoisomerase I indicate that these genes may participate in the homeostatic control of superhelical tension (67, 112). The system may be finely tuned since even the small changes in supercoiling expected to arise from temperature changes appear to be compensated for by topoisomerase activity (32).

Several general points should also be noted concerning topoisomerase mutations and inhibitors as tools for exploring physiological relationships between superhelical tension and gene expression. First, we are unable to perturb superhelical tension without also perturbing topoisomerase activity. Thus in vitro measurements are required to determine whether expression of a gene is affected by supercoiling or by local effects of topoisomerases. We suspect that local effects may be important because the relative abundance of some proteins differs in two strains that have the same level of average titratable chromosomal supercoiling but carry different alleles of topA (R. Franco, T. R. Steck, and K. Drlica, unpublished data). A second point involves the measurement of titratable supercoiling. In chromosomal measurements, only averages are determined; we do not know how the supercoiling in a given topological domain relates to that average. Obviously the best correlations between titratable supercoiling and gene expression are made with plasmids because plasmids consist of a single domain. The level of superhelical tension in vivo appears to be about 40% of that measured in vitro (Bliska and Cozzarelli, submitted), so the appropriate plasmids can be adjusted to that level in vitro to examine transcription parameters. A third point concerns topA mutants. We know that mutations in the genes encoding gyrase, as well as in other genes, accumulate in these strains (89). Thus caution must be exercised when interpreting genetic experiments. A fourth point involves gyrase inhibitors. The action of nalidixic acid and related compounds on gene expression has been interpreted in terms of decreasing supercoiling. We now know that this class of drug forms drug-gyrase-DNA complexes and that it can have significant effects on RNA synthesis even when it increases titratable supercoiling (64). Thus coumermycin A_1 and temperature-sensitive gyrase mutations are the best tools for rapidly decreasing supercoiling.

CONCLUDING REMARKS

The bacterial nucleoid is a compact, lobular structure containing clefts (40, 113). Transcription and translation appear to take place near the edges of the nucleoid, and presumably the clefts and the absence of a surrounding membrane allow most, if not all, of the chromosomal DNA access to processes occurring in the cytoplasm. Although it is not known whether the nucleoid has a dynamic or a static structure, striking morphological changes occur when protein synthesis is inhibited (2, 7, 49). One speculation is that active genes are normally pulled out of the bulk of the nucleoid; if translation is blocked, those genes return, and the nucleoid assumes a condensed configuration.

DNA topology is an important aspect of nucleoid structure, and physiological studies with DNA topoisomerases have provided the following general view. In E. coli, chromosomal DNA is circular (5), so topoisomerases are able to energetically activate the DNA by maintaining it under negative superhelical tension. Gyrase introduces tension by lowering DNA linking numbers (30), and topoisomerase I modulates the level by increasing them (11, 88, 117). The level of tension in wild-type cells is probably optimal for growth (90) and may be tightly controlled: small alterations in superhelical tension in plasmid DNA expected from changes in temperature seem to be corrected by topoisomerase action (32). This tight control of superhelical tension appears to be important for gene expression since transcription of certain genes is sensitive to changes in titratable supercoiling, topoisomerase activity, or both.

The chromosomal DNA is divided into about 50 topologically independent domains; as a consequence we expect relaxation associated with DNA repair to be confined to small (100-kbp) regions (101, 119). Gyrase is distributed along the DNA at roughly 100-kbp intervals (as determined by oxolinic acid-induced DNA cleavage [103]), which corresponds to about one gyrase per domain. We do not know how this distribution is determined; the low frequency of cleavage is especially puzzling since DNA molecules contain large numbers of cleavage sites (plasmid pBR322 contains at least 74 cleavage sites [58, 73]). The distribution of gyrase on the chromosome does not change when DNA, RNA, or protein synthesis is inhibited; thus the enzyme appears to be able to maintain superhelical tension throughout the chromosome under a variety of conditions. Since the domains restrict relaxation associated with discontinuous DNA synthesis to only 2% of the chromosome at any given time, initiation of a new round of replication, which probably requires that the origin of replication be under negative superhelical tension, can begin before the previous round finishes.

The ability of gyrase to decatenate DNA is probably responsible for unlinking circular daughter chromosomes after a round of replication. The same activity could also remove entanglements that might arise as replication forks pass through the many chromosomal loops, thus reducing the need for special sorting mechanisms.

Although we are beginning to understand some of the interrelationships among supercoiling, topoisomerases, and chromosome function, a number of questions are still unanswered. For example, we do not know how topological domains are established, whether specific nucleotide sequences are involved, or whether envelope-DNA interactions participate. Nor do we know whether superhelical tension is the same in each. Also unanswered is whether cells respond to changes in environmental conditions by changing superhelical tension so that certain genes can be more easily induced or repressed. Our knowledge of other aspects of chromosome structure is also incomplete. We do not know what role, if any, nascent RNA and the cell envelope play in nucleoid structure, although it is becoming increasingly likely that the envelope has a role in initiation of replication. Nor do we know what role histonelike proteins, polyamines, and nucleosomelike particles play in compacting bacterial DNA. Understanding nucleoid structure and function is still a formidable task.

ACKNOWLEDGMENTS

I thank the following for stimulating discussions, for providing experimental data before publication, and for critical comments on the manuscript: M. Bjornsti, R. Franco, J. Hobot, E. Kellenberger, I. Lossius, D. Pettijohn, G. Pruss, J. Rouviere-Yaniv, M. Schaechter, and T. Steck.

My work has been supported by a Research Career Development Award from the National Cancer Institute, Public Health Service research grants GM 32005 and GM 24320 from the National Institutes of Health, and an EMBO Fellowship.

LITERATURE CITED

1. **Abe, M., C. Brown, W. G. Hendrickson, D. H. Boyd, P. Clifford, R. H. Cole, and M. Schaechter.** 1977. Release of *Escherichia coli* DNA from membrane complexes by single-strand endonucleases. Proc. Natl. Acad. Sci. USA **74**:2756–2760.
2. **Bjornsti, M. A., J. Hobot, A. Kelus, W. Villiger, and E. Kellenberger.** 1985. New electron microscopy data on the structure of the nucleoid and their functional consequences. *In* C. Gualerzi (ed.), Bacterial chromatin. Springer-Verlag, New York.
3. **Briat, J.-F., S. Letoffe, R. Mache, and J. Rouviere-Yaniv.** 1984. Similarity between the bacterial histone-like protein HU and a protein from spinach chloroplasts. FEBS Lett. **172**:75–79.
3a. **Broyles, S. S., and D. E. Pettijohn.** 1986. Interaction of the *Escherichia coli* HU protein with DNA. Evidence for formation of nucleosome-like structures with altered DNA helical pitch. J. Mol. Biol. **187**:47–60.
4. **Busby, S., A. Kolb, and H. Buc.** 1979. Isolation of plasmid-protein complexes from *Escherichia coli*. Eur. J. Biochem. **99**:105–111.
5. **Cairns, J.** 1963. The chromosome of *Escherichia coli*. Cold Spring Harbor Symp. Quant. Biol. **28**:43–46.
6. **Cozzarelli, N. R.** 1980. DNA gyrase and the supercoiling of DNA. Science **207**:953–960.
7. **Daneo-Moore, L., and M. L. Higgins.** 1972. Morphokinetic reaction of *Streptococcus faecalis* (ATCC 9790) cells to the specific inhibition of macromolecular synthesis: nucleoid condensation of the inhibition of protein synthesis. J. Bacteriol. **109**:1210–1220.
8. **Davern, C. I.** 1979. Replication of the prokaryotic chromosome with emphasis on bacterial chromosome replication in relation to the cell cycle, p. 131–169. *In* D. M. Prescott and L. Goldstein (ed.), Cell biology: a comprehensive treatise, vol. 2. Academic Press, Inc., New York.
9. **Delius, H., and A. Worcel.** 1973. Electron microscopic studies on the folded chromosome of *Escherichia coli*. Cold Spring Harbor Symp. Quant. Biol. **38**:53–58.
10. **Delius, H., and A. Worcel.** 1974. Electron microscopic visualization of the folded chromosome of *Escherichia coli*. J. Mol. Biol. **82**:107–109.
11. **DiNardo, S., K. A. Voelkel, R. Sternglanz, A. E. Reynolds, and A. Wright.** 1982. *Escherichia coli* DNA topoisomerase I mutants have compensatory mutations in DNA gyrase genes. Cell **31**:43–51.
12. **Dixon, N., and A. Kornberg.** 1984. Protein HU in the enzymatic replication of the chromosomal origin of *Escherichia coli*. Proc. Natl. Acad. Sci. USA **81**:424–428.
13. **Drlica, K.** 1984. Biology of bacterial deoxyribonucleic acid topoisomerases. Microbiol. Rev. **48**:273–289.
14. **Drlica, K., E. Burgi, and A. Worcel.** 1978. Association of the folded chromosome with the cell envelope of *Escherichia coli*: nature of the membrane-associated DNA. J. Bacteriol. **134**:1108–1116.
15. **Drlica, K., E. C. Engle, and S. H. Manes.** 1980. DNA gyrase on the bacterial chromosome: possibility of two levels of action. Proc. Natl. Acad. Sci. USA **77**:6879–6883.
16. **Drlica, K., and M. Snyder.** 1978. Superhelical *Escherichia coli*

DNA: relaxation by coumermycin. J. Mol. Biol. **120**:145–154.
17. **Drlica, K., and A. Worcel.** 1975. Conformational transitions in the *Escherichia coli* chromosome: analysis by viscometry and sedimentation. J. Mol. Biol. **98**:393–411.
18. **Dworsky, P.** 1975. A mild method for the isolation of folded chromosomes from *Escherichia coli*. Z. Allg. Mikrobiol. **15**:231–241.
19. **Dworsky, P.** 1976. Comparative studies on membrane-associated, folded chromosomes from *Escherichia coli*. J. Bacteriol. **126**:64–71.
20. **Dworsky, P., and M. Schaechter.** 1973. Effect of rifampin on the structure and membrane attachment of the nucleoid of *Escherichia coli*. J. Bacteriol. **116**:1364–1374.
21. **Engle, E. C., S. H. Manes, and K. Drlica.** 1982. Differential effect of antibiotics inhibiting gyrase. J. Bacteriol. **149**:92–98.
22. **Filutowicz, M., and P. Jonczyk.** 1981. Essential role of *gyrB* gene product in the transcriptional event coupled to *dnaA*-dependent initiation of *Escherichia coli* chromosome replication. Mol. Gen. Genet. **183**:134–138.
23. **Filutowicz, M., and P. Jonczyk.** 1983. The *gyrB* gene product functions in both initiation and chain polymerization of *Escherichia coli* chromosome replication: depression of the initiation deficiency in *gyrB*-ts mutants by a class of the *rpoB* mutations. Mol. Gen. Genet. **191**:282–287.
24. **Friedman, D. I., L. C. Pantefaber, E. J. Olson, D. Carver, M. H. O'Dea, and M. Gellert.** 1984. Mutations in the DNA *gyrB* gene that are temperature sensitive for lambda site-specific recombination, Mu growth, and plasmid maintenance. J. Bacteriol. **157**:490–497.
25. **Fuller, R., L. Bertsch, N. Dixon, J. Flynn, J. Kaguni, R. Low, T. Ogawa, and A. Kornberg.** 1983. Enzymes in the initiation of replication at the *E. coli* chromosomal origin, p. 275–288. *In* N. Cozzarelli (ed.), Mechanisms of DNA replication and recombination. Alan R. Liss, Inc., New York.
26. **Fuller, R., J. Kaguni, and A. Kornberg.** 1981. Enzymatic replication of the origin of the *Escherichia coli* chromosome. Proc. Natl. Acad. Sci. USA **78**:7370–7374.
27. **Gellert, M.** 1981. DNA topoisomerases. Annu. Rev. Biochem. **50**:879–910.
28. **Gellert, M., R. Menzel, K. Mizuuchi, M. H. O'Dea, and D. Friedman.** 1983. Regulation of DNA supercoiling in *E. coli*. Cold Spring Harbor Symp. Quant. Biol. **47**:763–767.
29. **Gellert, M., K. Mizuuchi, M. H. O'Dea, T. Itoh, and J.-I. Tomizawa.** 1977. Nalidixic acid resistance: a second genetic character involved in DNA gyrase activity. Proc. Natl. Acad. Sci. USA **74**:4772–4776.
30. **Gellert, M., K. Mizuuchi, M. H. O'Dea, and H. Nash.** 1976. DNA gyrase: an enzyme that introduces superhelical turns into DNA. Proc. Natl. Acad. Sci. USA **73**:3872–3876.
31. **Gellert, M., M. H. O'Dea, T. Itoh, and J.-I. Tomizawa.** 1976. Novobiocin and coumermycin inhibit DNA supercoiling catalyzed by DNA gyrase. Proc. Natl. Acad. Sci. USA **73**:4474–4478.
32. **Goldstein, E., and K. Drlica.** 1984. Regulation of bacterial DNA supercoiling: plasmid linking numbers vary with growth temperature. Proc. Natl. Acad. Sci. USA **81**:4046–4050.
33. **Griffith, J. D.** 1976. Visualization of prokaryotic DNA in a regularly condensed chromatin-like fiber. Proc. Natl. Acad. Sci. USA **73**:563–567.
34. **Haselkorn, R., and J. Rouviere-Yaniv.** 1976. Cyanobacterial DNA-binding protein related to *Escherichia coli* HU. Proc. Natl. Acad. Sci. USA **73**:1917–1920.
35. **Hecht, R. M., and D. E. Pettijohn.** 1976. Studies of DNA bound RNA molecules isolated from nucleoids of *Escherichia coli*. Nucleic Acids Res. **3**:767–788.
36. **Hecht, R. M., D. Stimpson, and D. Pettijohn.** 1977. Sedimentation properties of the bacterial chromosome as an isolated nucleoid and as an unfolded DNA fiber. J. Mol. Biol. **111**:257–277.
37. **Hecht, R. M., R. T. Taggart, and D. E. Pettijohn.** 1975. Size and DNA content of purified *E. coli* nucleoids observed by fluorescence microscopy. Nature (London) **253**:60–62.
38. **Hendrickson, W. G., T. Kusano, H. Yamaki, R. Balakrishnan, M. King, J. Murchie, and M. Schaechter.** 1982. Binding of the origin of replication of *Escherichia coli* to the outer membrane. Cell **30**:915–923.
39. **Hill, W. E., and W. L. Fangman.** 1973. Single-strand breaks in deoxyribonucleic acid and viability loss during deoxyribonucleic acid synthesis inhibition in *Escherichia coli*. J. Bacteriol. **116**:1329–1335.
40. **Hobot, J. A., W. Villiger, J. Escaig, M. Maeder, A. Ryter, and E. Kellenberger.** 1985. Shape and fine structure of nucleoids ob-

served on sections of ultrarapidly frozen and cryosubstituted bacteria. J. Bacteriol. **162:**960–971.

41. **Hubscher, U., H. Lutz, and A. Kornberg.** 1980. Novel histone H2A-like protein of *Escherichia coli.* Proc. Natl. Acad. Sci. USA **77:**5097–5101.

42. **Isberg, R. R., and M. Syvanen.** 1982. DNA gyrase is a host factor required for transposition of Tn5. Cell **30:**9–18.

43. **Jacob, F., S. Brenner, and F. Cuzin.** 1963. On the regulation of DNA replication in bacteria. Cold Spring Harbor Symp. Quant. Biol. **28:**329–348.

44. **Jacob, F., A. Ryter, and F. Cuzin.** 1966. On the association between DNA and membrane in bacteria. Proc. R. Soc. London **169:**267–278.

45. **Jones, N. C., and W. D. Donachie.** 1974. Protein synthesis and the release of the replicated chromosome from the cell membrane. Nature (London) **257:**252–254.

46. **Kano, Y., T. Miyashita, H. Nakamura, K. Kuroki, A. Nagata, and F. Imamoto.** 1981. *In vivo* correlation between DNA supercoiling and transcription. Gene **13:**173–184.

47. **Kavenoff, R., and B. Bowen.** 1976. Electron microscopy of membrane-free folded chromosomes from *Escherichia coli.* Chromosoma (Berl.) **59:**89–101.

48. **Kavenoff, R., and O. Ryder.** 1976. Electron microscopy of membrane-associated folded chromosomes of *Escherichia coli.* Chromosoma (Berl.) **55:**13–25.

49. **Kellenberger, E., A. Ryter, and J. Sechaud.** 1958. Electron microscopy study of DNA-containing plasms. II. Vegetative and mature phage DNA as compared with normal bacterial nucleoids in different physiological states. J. Biophys. Biochem. Cytol. **4:**671–678.

50. **Kleppe, K., S. Ovrebo, and I. Lossius.** 1979. The bacterial nucleoid. J. Gen. Microbiol. **112:**1–13.

51. **Klotz, L. C., and B. Zimm.** 1972. Size of DNA determined by viscoelastic measurements: results on bacteriophages, *Bacillus subtilis,* and *Escherichia coli.* J. Mol. Biol. **72:**779–800.

52. **Korn, R., S. Winston, T. Tanaka, and N. Sueoka.** 1983. Specific *in vitro* binding of a plasmid to a membrane fraction of *Bacillus subtilis.* Proc. Natl. Acad. Sci. USA **80:**574–578.

53. **Kornberg, T., A. Lockwood, and A. Worcel.** 1974. Replication of the *Escherichia coli* chromosome with a soluble enzyme system. Proc. Natl. Acad. Sci. USA **71:**3189–3193.

54. **Kreuzer, K. N., and N. R. Cozzarelli.** 1979. *Escherichia coli* mutants thermosensitive for deoxyribonucleic acid gyrase subunit A: effects on deoxyribonucleic acid replication, transcription, and bacteriophage growth. J. Bacteriol. **140:**424–435.

55. **Kusano, T., D. Steinmetz, W. G. Hendrickson, J. Murchie, M. King, A. Benson, and M. Schaechter.** 1984. Direct evidence for specific binding of the replicative origin of the *Escherichia coli* chromosome to the membrane. J. Bacteriol. **158:**313–316.

56. **Laine, B., P. Sautiere, A. Spassky, and S. Rimsky.** 1984. A DNA-binding protein from *E. coli.* Isolation, characterization, and its relationship with proteins H1 and B1. Biochem. Biophys. Res. Commun. **119:**1147–1153.

57. **Lockshon, D., and D. R. Morris.** 1983. Positively supercoiled plasmid DNA is produced by treatment of *Escherichia coli* with DNA gyrase inhibitors. Nucleic Acids Res. **11:**2999–3017.

58. **Lockshon, D., and D. Morris.** 1985. Sites of reaction of *Escherichia coli* DNA gyrase on pBR322 *in vivo* as revealed by oxolinic acid-induced plasmid linearization. J. Mol. Biol. **181:**63–74.

59. **Lossius, I., A. Holck, R. Aasland, L. Haarr, and K. Kleppe.** 1985. Proteins associated with chromatin from *Escherichia coli. In* C. Gualerzi (ed.), Bacterial chromatin. Springer-Verlag, New York.

60. **Lossius, I., and K. Kleppe.** 1981. Influence of alkylating agents on the structure of the bacterial nucleoid, p. 41–47. *In* E. Seeberg and K. Kleppe (ed.), Chromosome damage and repair. Plenum Publishing Corp., New York.

61. **Lossius, I., P. Kruger, R. Male, and K. Kleppe.** 1983. Mitomycin-C-induced changes in the nucleoid of *Escherichia coli* K12. Mutat. Res. **109:**13–20.

62. **Lossius, I., K. Sjastad, L. Haarr, and K. Kleppe.** 1984. Two-dimensional gel electrophoretic separations of the proteins present in chromatin of *Escherichia coli.* J. Gen. Microbiol. **130:**3153–3157.

63. **Lydersen, B. K., and D. E. Pettijohn.** 1977. Interactions stabilizing DNA tertiary structure in the *Escherichia coli* chromosome investigated with ionizing radiation. Chromosoma (Berl.) **62:**199–215.

64. **Manes, S. H., G. J. Pruss, and K. Drlica.** 1983. Inhibition of RNA synthesis by oxolinic acid is unrelated to average DNA supercoiling. J. Bacteriol. **155:**420–423.

65. **Marians, K. J., J.-E. Ikeda, S. Schlagman, and J. Hurwitz.** 1977. Role of DNA gyrase in φX replicative-form replication *in vitro.* Proc. Natl. Acad. Sci. USA **74:**1965–1968.

66. **Mason, D., and D. M. Powelson.** 1956. Nuclear division as observed in live bacteria by a new technique. J. Bacteriol. **71:**474–479.

66a.**McClure, W.** 1985. Mechanism and control of transcription initiation in prokaryotes. Annu. Rev. Biochem. **54:**171–204.

67. **Menzel, R., and M. Gellert.** 1983. Regulation of the genes for *E. coli* DNA gyrase: homeostatic control of DNA supercoiling. Cell **34:**105–113.

68. **Meyer, M., M. DeJong, C. Woldringh, and N. Nanninga.** 1976. Factors affecting the release of folded chromosomes from *Escherichia coli.* Eur. J. Biochem. **63:**469–475.

69. **Miller, O. C., B. A. Hamkalo, and C. A. Thomas.** 1970. Visualization of bacterial genes in action. Science **169:**392–395.

70. **Mirkin, S. M., and Z. Schmerling.** 1982. DNA replication and transcription in a temperature-sensitive mutant of *E. coli* with a defective DNA gyrase subunit. Mol. Gen. Genet. **188:**89–95.

71. **Mukai, F. H., and P. Margolin.** 1963. Analysis of unlinked suppressors of an 0° mutation in *Salmonella.* Proc. Natl. Acad. Sci. USA **50:**140–148.

72. **Nagai, K., W. Hendrickson, R. Balakrishnan, H. Yamaki, D. Boyd, and M. Schaechter.** 1980. Isolation of a replication origin complex from *Escherichia coli.* Proc. Natl. Acad. Sci. USA **77:**262–266.

73. **O'Connor, M., and M. Malamy.** 1985. Mapping of DNA gyrase cleavage sites *in vivo:* oxolinic acid induced cleavages in plasmid pBR322. J. Mol. Biol. **181:**545–550.

74. **Ogasawara, N., M. Seiki, and H. Yoshikawa.** 1981. Initiation of DNA replication in *Bacillus subtilus.* V. Role of DNA gyrase and superhelical structure in initiation. Mol. Gen. Genet. **181:**332–337.

75. **Ogden, G. B., and M. Schaechter.** 1985. Chromosomes, plasmids, and the bacterial cell envelopes, p. 282–286. *In* L. Leive (ed.), Microbiology—1985. American Society for Microbiology, Washington, D.C.

76. **Orr, E., N. F. Fairweather, I. B. Holland, and R. H. Pritchard.** 1979. Isolation and characterization of a strain carrying a conditional lethal mutation in the *cou* gene of *Escherichia coli* K12. Mol. Gen. Genet. **177:**103–112.

77. **Peebles, C. L., N. P. Higgins, K. N. Kreuzer, A. Morrison, P. O. Brown, A. Sugino, and N. R. Cozzarelli.** 1979. Structure and activities of *Escherichia coli* DNA gyrase. Cold Spring Harbor Symp. Quant. Biol. **43:**41–52.

78. **Pettijohn, D. E.** 1976. Prokaryotic DNA in nucleoid structure. Crit. Rev. Biochem. **4:**175–202.

79. **Pettijohn, D. E.** 1982. Structure and properties of the bacterial nucleoid. Cell **30:**667–669.

80. **Pettijohn, D. E., and J. D. Carlson.** 1979. Chemical, physical, and genetic structure of prokaryotic chromosomes, p. 1–57. *In* D. M. Prescott and L. Goldstein (ed.), Cell biology: a comprehensive treatise, vol. 2. Academic Press, Inc., New York.

81. **Pettijohn, D. E., K. Clarkson, C. R. Kossman, and O. G. Stonington.** 1970. Synthesis of ribosomal RNA on a protein-DNA complex isolated from bacteria: a comparison of ribosomal RNA synthesis *in vitro* and *in vivo.* J. Mol. Biol. **52:**281–300.

82. **Pettijohn, D. E., and R. M. Hecht.** 1973. RNA molecules bound to the folded bacterial genome stabilize DNA folds and segregate domains of supercoiling. Cold Spring Harbor Symp. Quant. Biol. **38:**31–41.

83. **Pettijohn, D., R. M. Hecht, D. Stimpson, and S. van Scoyk.** 1978. An explanation for rotor speed effects observed during sedimentation of large folded DNA molecules. J. Mol. Biol. **119:**353–359.

84. **Pettijohn, D. E., R. M. Hecht, O. G. Stonington, and T. D. Stomato.** 1973. Factors stabilizing DNA folding in bacterial chromosomes, p. 145–162. *In* R. D. Wells and R. B. Inman (ed.), DNA synthesis *in vitro.* University Park Press, Baltimore.

85. **Pettijohn, D. E., and O. Pfenninger.** 1980. Supercoils in prokaryotic DNA restrained *in vivo.* Proc. Natl. Acad. Sci. USA **77:**1331–1335.

86. **Pettijohn, D. E., and R. R. Sinden.** 1985. Structure of the nucleoid. *In* N. Nanninga (ed.), Molecular cytology of *Escherichia coli.* Academic Press, Inc., New York.

87. **Pisetsky, D., I. Berkower, P. Wickner, and J. Hurwitz.** 1972. Role of ATP in DNA synthesis in *Escherichia coli.* J. Mol. Biol. **71:**557–571.

87a.**Pruss, G. J.** 1985. DNA topoisomerase mutants. Increased heterogeneity in linking number and other replicon-dependent changes in DNA supercoiling. J. Mol. Biol. **185:**51–63.

88. **Pruss, G. J., S. H. Manes, and K. Drlica.** 1982. *Escherichia coli* DNA topoisomerase I mutants: increased supercoiling is corrected by mutations near gyrase genes. Cell **31:**35–42.

89. **Raji, A., D. J. Zabel, C. S. Laufer, and R. E. DePew.** 1985. Genetic analysis of mutations that compensate for loss of *Esch-*

erichia coli DNA topoisomerase I. J. Bacteriol. **162**:1173–1179.

90. **Richardson, S. M. H., C. F. Higgins, and D. M. J. Lilley.** 1984. The genetic control of DNA supercoiling in *Salmonella typhimurium*. EMBO J. **3**:1745–1752.

91. **Robinow, C. F.** 1956. The chromatin bodies of bacteria. Bacteriol. Rev. **20**:207–242.

92. **Rouviere-Yaniv, J.** 1978. Localization of the HU protein on the *Escherichia coli* nucleoid. Cold Spring Harbor Symp. Quant. Biol. **42**:439–447.

93. **Rouviere-Yaniv, J., and F. Gros.** 1975. Characterization of a novel, low-molecular-weight DNA-binding protein from *Escherichia coli*. Proc. Natl. Acad. Sci. USA **72**:3428–3432.

94. **Rouviere-Yaniv, J., M. Yaniv, and J.-E. Germond.** 1979. *E. coli* DNA binding protein HU forms nucleosome-like structure with circular double-stranded DNA. Cell **17**:265–274.

95. **Ryder, O., and D. Smith.** 1974. Isolation of membrane-associated folded chromosomes from *Escherichia coli*: effect of protein synthesis inhibition. J. Bacteriol. **120**:1356–1363.

96. **Ryter, A.** 1960. Etude au microscope electronique des transformations nucléaires de *E. coli* K12S et K12S(λ26) après irradiation aux rayons ultraviolets et aux rayons X. J. Biophys. Biochem. Cytol. **8**:399–412.

97. **Ryter, A.** 1968. Association of the nucleus and the membrane of bacteria: a morphological study. Bacteriol. Rev. **32**:39–54.

98. **Ryter, A., and A. Chang.** 1975. Localization of transcribing genes in the bacterial cell by means of high resolution autoradiography. J. Mol. Biol. **98**:797–810.

99. **Schaechter, M., J. Williamson, J. R. Hood, and A. L. Koch.** 1962. Growth, cell, and nuclear divisions in some bacteria. J. Gen. Microbiol. **29**:421–434.

100. **Sinden, R. R., J. Carlson, and D. E. Pettijohn.** 1980. Torsional tension in the DNA double helix measured with trimethylpsoralen in living *E. coli* cells. Cell **21**:773–783.

101. **Sinden, R. R., and D. E. Pettijohn.** 1981. Chromosomes in living *Escherichia coli* cells are segregated into domains of supercoiling. Proc. Natl. Acad. Sci. USA **78**:224–228.

102. **Sloof, P., A. Maagdelijn, and E. Boswinkel.** 1983. Folding of prokaryotic DNA. Isolation and characterization of nucleoids from *Bacillus licheniformis*. J. Mol. Biol. **163**:277–297.

103. **Snyder, M., and K. Drlica.** 1979. DNA gyrase on the bacterial chromosome: DNA cleavage induced by oxolinic acid. J. Mol. Biol. **131**:287–302.

104. **Spassky, A., S. Rimsky, H. Garreau, and H. Buc.** 1984. H1a, an *E. coli* DNA binding protein which accumulates in stationary phase, strongly compacts DNA *in vitro*. Nucleic Acids Res. **12**:5321–5340.

105. **Steck, T. R., and K. Drlica.** 1984. Bacterial chromosome segregation: evidence for DNA gyrase involvement in decatenation. Cell **36**:1081–1088.

106. **Steck, T. R., G. J. Pruss, S. H. Manes, L. Burg, and K. Drlica.** 1984. DNA supercoiling in gyrase mutants. J. Bacteriol. **158**:397–403.

107. **Sternglanz, R., S. DiNardo, K. A. Voelkel, Y. Nishimura, Y.** Hiroto, K. Becherer, L. Zumstein, and J. C. Wang. 1981. Mutations in the gene coding for *Escherichia coli* DNA topoisomerase I affecting transcription. Proc. Natl. Acad. Sci. USA **78**:2747–2751.

108. **Stonington, O. G., and D. E. Pettijohn.** 1971. The folded genome of *Escherichia coli*. J. Mol. Biol. **68**:6–9.

109. **Tanaka, I., K. Appelt, J. Dijk, S. W. White, and K. S. Wilson,** 1984. 3-A resolution structure of a protein with histone-like properties in prokaryotes. Nature (London) **310**:376–381.

110. **Trucksis, M., and R. DePew.** 1981. Identification and localization of a gene that specifies production of *Escherichia coli* DNA topoisomerase I. Proc. Natl. Acad. Sci. USA **78**:2164–2168.

111. **Trucksis, M., E. Golub, D. Zabel, and R. DePew.** 1981. *Escherichia coli* and *Salmonella typhimurium supX* genes specify deoxyribonucleic acid topoisomerase I. J. Bacteriol. **147**:679–681.

112. **Tse-Dinh, Y.** 1985. Regulation of the *Escherichia coli* DNA topoisomerase I gene by DNA supercoiling. Nucleic Acids Res. **13**:4751–4763.

113. **Valkenburg, J. A. C., C. L. Woldringh, G. J. Brakenhoff, H. T. M. van der Voort, and N. Nanninga.** 1985. Confocal scanning light microscopy of the *Escherichia coli* nucleoid: comparison with phase-contrast and electron microscope images. J. Bacteriol. **161**:478–483.

114. **Van Ness, J., and D. E. Pettijohn.** 1979. A simple autoradiographic method for investigating long range chromosome substructures: size and number of DNA molecules in isolated nucleoids of *Escherichia coli*. J. Mol. Biol. **129**:501–508.

115. **Varshavsky, A., S. Nedospasov, V. Bakayev, T. Bakayev, and G. Georgiev.** 1977. Histone-like proteins in the purified *Escherichia coli* deoxyribonucleoprotein. Nucleic Acids Res. **8**:2725–2745.

116. **Von Wright, A., and B. A. Bridges.** 1981. Effect of *gyrB*-mediated changes in chromosome structure on killing of *Escherichia coli* by ultraviolet light: experiments with strains differing in deoxyribonucleic acid repair capacity. J. Bacteriol. **146**:18–23.

117. **Wang, J. C.** 1971. Interaction between DNA and an *Escherichia coli* protein. J. Mol. Biol. **55**:523–533.

118. **Wolf-Watz, H.** 1984. Affinity of two different regions of the chromosome to the outer membrane of *Escherichia coli*. J. Bacteriol. **157**:968–970.

119. **Worcel, A., and E. Burgi.** 1972. On the structure of the folded chromosome of *Escherichia coli*. J. Mol. Biol. **71**:127–147.

120. **Worcel, A., and E. Burgi.** 1974. Properties of a membrane-attached form of the folded chromosome of *Escherichia coli*. J. Mol. Biol. **82**:91–105.

121. **Worcel, A., E. Burgi, J. Robinton, and C. L. Carlson.** 1973. Studies on the folded chromosome of *Escherichia coli*. Cold Spring Harbor Symp. Quant. Biol. **38**:43–51.

122. **Wu, F., A. Kolb, and H. Buc.** 1982. A transcriptionally active plasmid-protein complex isolated from *Escherichia coli*. Biochim. Biophys. Acta **696**:231–238.

123. **Zimm, B.** 1974. Anomalies in sedimentation. IV. Decrease in sedimentation coefficients of chains at high fields. Biophys. Chem. **1**:279–291.

10. Ribosomes

HARRY F. NOLLER[1] AND MASAYASU NOMURA[2]

Thimann Laboratories, University of California, Santa Cruz, California 95064,[1] and Department of Biological Chemistry, University of California, Irvine, California 92717[2]

INTRODUCTION

Ribosomes are large ribonucleoprotein particles that are responsible for translation of the genetic code. Their molecular architecture is vast and complex, as is the system of pathways leading to their assembly. During the past 25 years a great deal has been learned about these particles, and we shall attempt to summarize in this chapter some highlights of the present state of our knowledge in these areas. As is often the case in molecular biology, *Escherichia coli* has been, and continues to be, the reference organism for ribosome research. Any discussion of *E. coli* ribosomes will thus be representative of the state of ribosome research in general. By contrast, comparatively few studies have utilized *Salmonella* ribosomes, which, however, are likely to resemble closely those of *E. coli*.

In the first section, we review the properties of the individual macromolecular components of ribosomes: the ribosomal proteins (r-proteins) and rRNA. Ribosomal structure is discussed insofar as the folding of these individual macromolecules is understood. In the second section, we discuss ribosome assembly, as it has been studied by in vitro approaches. RNA-protein interactions in the ribosome and the process of in vitro assembly of functional ribosomes are summarized. In vivo assembly is discussed only in connection with information obtained from the in vitro assembly reaction. The present state of our understanding of ribosomal architecture is summarized in the third section. This includes the overall morphology of *E. coli* ribosomes and their subunits, as well as the location of specific protein and RNA features in the structure. The fourth section brings together ribosome morphology, rRNA secondary structure, and rRNA–r-protein assembly to give a sense of how information obtained from a number of quite different approaches is beginning to converge on a structure which must eventually account for the functional complexity of these particles.

For a more extensive description of these topics, the reader is referred to two classic volumes on ribosome structure and function (13, 100) as well as review articles on r-proteins and RNA (6, 92, 96, 151, 153, 158), ribosome assembly (86, 87, 98), and structure (74, 81, 95, 152).

THE MACROMOLECULAR COMPONENTS OF *E. COLI* RIBOSOMES

r-Proteins

Identification, isolation, and stoichiometry. The structural complexity of ribosomes first became evident in early studies on r-proteins. Waller (149) found dozens of distinct protein bands by using starch gel electrophoresis on *E. coli* 30S and 50S subunit proteins. The notion that these represented individual, distinct protein species was met with some skepticism due to the unprecedented structural complexity implied. Biochemical characterization of purified r-proteins finally established that the observed bands, for the most part, indeed represented individual protein species with distinct sizes, N-termini, and amino acid sequences (144). It is now generally accepted that the *E. coli* ribosome contains 52 different proteins; 21 are found in the 30S subunit, and 31 are in the 50S subunit (reviewed by Wittmann [151]).

Isolation of r-proteins has usually been accomplished by cation-exchange chromatography in 6 M urea followed by gel filtration (reviewed by Wittmann [150]). By consensus, the presently used nomenclature system, based on a widely used two-dimensional gel electrophoresis method (61), has been adopted (156). In this system, small (30S) subunit proteins are called S1, S2, S3, ... S21, and large (50S) subunit proteins are called L1, L2, ... L34. Subsequently, it was discovered that L8 is in fact an aggregate of proteins L7/L12 and L10 (111) and that L26 is identical to S20 (157). Furthermore, proteins L7 and L12 are identical, except for the N-terminal acetylation of L7 (137).

Once the proteins were isolated and characterized, another important question arose. Do all ribosomes in a population have the same protein composition, or are there different classes of ribosomes, with different structures and possibly different functional roles? Protein stoichiometry is difficult to measure because of the ease with which certain r-proteins become dissociated from the particles during isolation, particularly during high-salt washing steps. At present, it is generally believed that all of the proteins and RNA molecules are present in 1:1 stoichiometry with each other, except for L7/L12, which is probably represented at the level of four copies per ribosome (48, 132). This implies that all functional ribosomes have essentially equivalent structures and that "specialized ribosomes" probably do not exist in *E. coli* cells. This view is supported by the results of in vitro reconstitution studies (52; see also below). Proteins which bind independently to rRNA reach saturation stoichiometries of 1:1 with the RNA (164). Single-component omission experiments show that, in many instances, omission of an r-protein appears to affect the assembly or activity of all particles in a population, rather than a specific fraction of them (52, 99).

Properties of r-proteins. All of the *E. coli* r-proteins

TABLE 1. *E. coli* r-proteins[a]

Protein	Residues	Mol wt	Protein	Residues	Mol wt
S1	557	61,159	L7	120	12,220
S2	240	26,613	L9	147	15,431
S3	232	25,852	L10	165	17,737
S4	203	23,137	L11	141	14,874
S5	166	17,151	L12	120	12,178
S6	135	15,704	L13	142	16,019
S7	177	19,732	L14	123	13,541
S8	129	13,996	L15	144	14,981
S9	128	14,569	L16	136	15,296
S10	103	11,736	L17	127	14,364
S11	128	13,728	L18	117	12,770
S12	123	13,606	L19	114	13,002
S13	117	12,968	L20	117	13,366
S14	97	11,063	L21	103	11,565
S15	87	10,001	L22	110	12,227
S16	78	9,191	L23	99	11,013
S17	83	9,573	L24	103	11,185
S18	74	8,896	L25	94	10,694
S19	91	10,299	L26=S20	86	9,553
S20	86	9,553	L27	84	8,993
S21	70	8,369	L28	77	8,875
L1	233	24,599	L29	63	7,274
L2	269	29,416	L29	63	7,274
L3	209	22,258	L31	62	6,971
L4	201	22,087	L32	56	6,315
L5	178	20,171	L33	54	6,255
L6	176	18,832	L34	46	5,381

[a] From reference 151.

have been sequenced (151) either by direct protein sequencing, by DNA sequencing of their genes, or both. Table 1 lists their calculated molecular weights, which average 14,800. Besides their generally small size, r-proteins tend to be quite basic, typically containing about 20% Lys + Arg, as might be expected for proteins that exist in a stable complex with RNA. Notable exceptions include the rather acidic proteins S1 (molecular weight, 61,000) and L7/L12 (molecular weight, 12,000), which are both strongly implicated in functional roles in the translation process (see chapters 40 and 77).

With the sequences of all 52 r-proteins in hand, the structural complexity of the ribosome is even more apparent. Computer-assisted searches have turned up surprisingly little sequence homology between different r-proteins (151, 153). Although it is difficult to rule out possible cryptic sequence homologies between proteins, this result suggests that each r-protein originated independently, or at least diverged from the others at a very early evolutionary stage. Such a view is consistent with proposals that the ribosome originated as a structure that was composed mainly, or even completely, of RNA (96). It is then easy to imagine the stepwise acquisition of individual proteins during evolution, gradually refining the processes of assembly and translation. (It is also possible that some of the r-proteins are involved in additional functions unrelated to translation per se.) If the ribosome evolved in this way, we would not necessarily expect a priori to find any sequence homology between different r-proteins.

There is presently some controversy concerning the overall shapes of the r-proteins. Several studies have reported evidence supporting highly elongated struc-

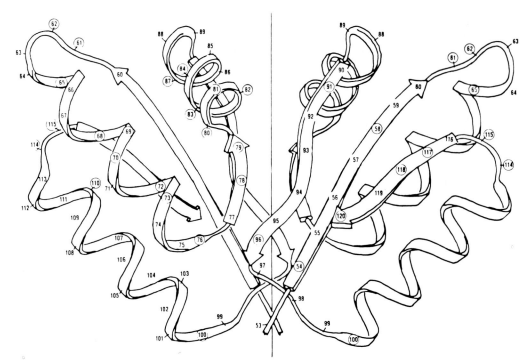

FIG. 1. Structure of the dimer of the C-terminal fragment of r-protein L7/L12. Positions of the residues are indicated, with the invariant ones circled. The vertical line is a crystallographic twofold axis. Residues 77 to 79 form a fourth β-strand, continuing the β-sheet into the symmetry-related molecule. From reference 74, with permission.

tures, while other evidence seems to indicate that they are compact and globular (reviewed by Liljas [74]). This is a difficult question to resolve, since the relevant structure is that of an r-protein in its functional state, i.e., in its complexed form within the ribosome. Short of a high-resolution crystal structure of the ribosome, neutron scattering studies provide one of the few informative approaches to this problem. Radii of gyration have been measured in situ by Ramakrishnan et al. (114) for several 30S r-proteins. Although the uncertainty of the measurements significantly limits interpretation of these data, most radii of gyration are consistent with globular shapes except for proteins S1, S4, and S8, which appear to be significantly elongated.

Three-dimensional structures of isolated r-proteins. As will be discussed below, progress toward an X-ray crystallographic structure for the ribosome has been slow, due to the difficulty of preparing suitable crystals. It has also proven difficult to obtain good crystals for the individual proteins from *E. coli* ribosomes, although those from *Bacillus stearothermophilus* appear to be more amenable to crystallization (reviewed by Liljas [74]). As for the solution studies, an important consideration is whether an isolated r-protein maintains the same conformation as in its assembled functional state in the ribosome.

The most advanced crystal structure determination for a ribosomal component is that of protein L7/L12 (reviewed by Liljas [74]). This protein is of particular interest because it is implicated in functions related to GTP-dependent involvement of the elongation factors EF-G and EF-T in protein synthesis (reviewed by Möller [80]). Studies on the isolated protein indicate a highly elongated structure. A current model describes

its maximum dimension as 130 Å (13.0 nm), with the N-terminal domain (residues 1 through 40) and plum-shaped globular C-terminal domain (residues 50 through 120) separated by a flexible connecting region (residues 40 through 50). The molecule is particularly susceptible to proteolytic cleavage within the connecting region, giving rise to N-terminal and C-terminal fragments which have been the subject of detailed crystallographic analysis (72, 74, 75).

The structure of the L7/L12 C-terminal fragment has been solved to 0.17-nm resolution (74) and represents the most detailed structural information presently available for any ribosomal component. This fragment crystallizes as a dimer, the monomers of which are related by a crystallographic dyad axis (Fig. 1). There is good evidence to support the notion that the functional form of this protein has a similar dimeric structure. The globular structure of the C-terminal fragment monomer is composed of three α-helical segments and three antiparallel β-strands, in a βααβαβ arrangement (Fig. 1).

The N-terminal fragment structure has been solved at 0.35-nm resolution. Dimerization of the intact L7/L12 molecule appears to be mainly dependent on interactions involving its N-terminal domain. Further quaternary organization is believed to consist of interaction of the N-terminal domains of two L7/L12 dimers with a single molecule of L10, which in turn interacts with 23S rRNA (74).

rRNA

E. coli ribosomes are composed of 38% protein and 62% RNA. It is an intriguing problem to understand why these particles, unlike other cellular polymerases, contain RNA as a major structural component. The possibility that rRNA is a functional, rather than a mere structural, component seems less implausible in light of the recent discovery of RNA enzymes. The ultimate resolution of this problem depends on a detailed understanding of the nature of rRNA. Progress in this area has been discussed in several recent reviews (6, 46, 92, 96, 158).

Three species of rRNA are found as integral structural components of the ribosomes of *E. coli* and other bacteria. The 30S subunit contains 16S rRNA, whereas 5S and 23S rRNA are found in the 50S subunit. At present, primary structures are known for all three molecules and reliable secondary structure models have been deduced. Clues to their tertiary folding are beginning to emerge, and several regions of interaction with r-proteins and functional ligands have been identified. These findings are summarized below.

Primary structures. There are seven independent rRNA transcriptional units in *E. coli* (see chapter 85). Although small differences in primary structure are known to exist between the genes in different transcriptional units, the sequences are virtually (> 99%) identical. The complete DNA sequence has been determined for the *rrnB* transcriptional unit, which serves as a model for the gene organization and primary structure for *E. coli* rRNAs (9). The arrangement of genes, spacers, promoters, and terminators for *rrnB* is shown in Fig. 2.

The smaller rRNA, 5S RNA, was the first to be sequenced by isotopic methods (10). It is composed of 120 nucleotides and contains no posttranscriptionally modified nucleotides (Fig. 3).

Over a decade elapsed before a complete sequence was obtained for 16S rRNA by DNA sequencing of the cloned gene (7) and by direct RNA sequencing (12). The 16S rRNA chain contains 1,542 nucleotides, 10 of which (positions 527, 966, 967, 1207, 1402, 1407, 1498, 1516, 1517, and 1518) are methylated (35). Analysis of phylogenetic conservation of sequence in 16S rRNA has provided insight into the relative biological significance of different regions of this molecule (46, 158). The occurrence of posttranscriptional modifications in regions of high sequence conservation suggests that they are functionally important. The fact that they show characteristic differences between the three major lines of descent would imply that they may be involved in "fine tuning" of crucial structural features

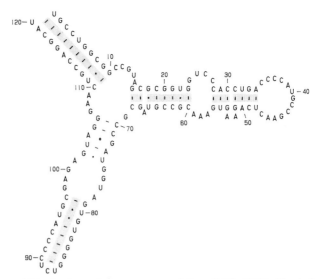

FIG. 3. Secondary structure of *E. coli* 5S rRNA. Shaded helices are considered proven by comparative sequence analysis. From reference 92.

(96). Another possibility is that this may reflect some differences in the regulation of translation and other roles that ribosomes may play.

The 23S rRNA consists of 2,904 nucleotides in *E. coli* (8) and, like 16S rRNA, contains numerous posttranscriptionally modified nucleotides, including pseudouridines and methylated bases (35). Many, but not all, of the modified positions have been placed in the sequence (8). As for the r-proteins, little or no significant sequence homology has been detected between the three rRNAs, nor has any been found between different regions of the individual rRNAs, apart from what would be anticipated to occur by chance. The absence of significant structural symmetry in the ribosome is again apparent. As we shall see below, this appears to extend to secondary and higher order levels of structural organization.

Secondary structures. In the decade after the publication of the 5S rRNA primary structure (10), dozens of different secondary structure models were proposed for this molecule in spite of considerable relevant biochemical and physical data and of free energy prediction rules for estimation of RNA helix stabilities (reviewed by Erdmann [33]). Resolution of this dilem-

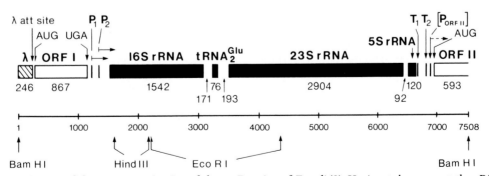

FIG. 2. Schematic map of the gene organization of the *rrnB* region of *E. coli* (9). Horizontal arrows at the rRNA promoters P1 and P2 and putative promoter P_{ORFII} indicate the direction of transcription. The hatched bar corresponds to λ DNA, and the open bars indicate open reading frames. From reference 9.

ma came from the realization that, as in the case of tRNA, 5S rRNAs from different species are likely to have very similar structures, despite significant sequence differences (36, 90). Figure 3 shows a model that incorporates most of the currently accepted refinements to the original Fox-Woese proposal for *E. coli* 5S rRNA (92). Its five helices exemplify the kinds of local features seen in the large rRNAs. There are examples of single- and double-base bulges and compound helices connected by internal loops. The small size of 5S rRNAs makes it especially amenable to computer analysis. One such method (130), which, significantly, incorporates a comparative algorithm, has produced a credible secondary structure for this molecule. Numerous groups have recently discussed the secondary structure of 5S rRNA in detail (4, 23, 25, 38, 77, 85).

Secondary structure models for *E. coli* 16S and 23S rRNAs have been proposed by three different groups of investigators (reviewed by Noller [92], Brimacombe et al. [6], and Ebel et al. [31]). We will discuss the secondary structure of *E. coli* 16S and 23S rRNAs according to the models presented in Fig. 4 and 5 (46, 92, 94, 158). Models for 16S rRNA proposed by two other groups are in general agreement with that presented in Fig. 4; detailed discussions of specific differences between the various models may be found in references 6 and 158. The model for the secondary structure of 23S rRNA shown in Fig. 5 (92, 94) differs in several significant aspects from those proposed by the other two groups.

Evidence for the existence of each helix comes from several approaches, but most crucially from comparative sequence analysis. Significantly, every one of the helices that was considered proven by comparative analysis in the first published version of the structure (91) has survived to the present version, having been tested by some 40 additional complete 16S-like rRNA sequences and an abundance of experimental information. Helices considered proven by comparative criteria (two or more phylogenetically independent base pair changes occurring within a given helix) are indicated in Fig. 4 by shading. Few possibilities for additional base pairing remain, and virtually all of the helices depicted in Fig. 4 are considered proven.

The 16S rRNA molecule is subdivided into three major structural domains and one minor domain by three sets of long-range base paired interactions (Fig. 4). The 5' domain (residues ca. 26 through 557) is defined by the 27-37/547-556 helix, the central domain (residues ca. 564 through 912) is defined by the 564-570/880-886 helix, and the 3' major (residues ca. 926 through 1391) and 3' minor (residues ca. 1392 through 1542) domains are defined by the 926-933/1384-1392 helix. The extent to which these domains, defined here only in terms of secondary structure, correspond to true structural domains is relevant to our understanding of ribosomal architecture. Fragments of 16S rRNA corresponding closely to the three major domains have actually been isolated in conjunction with studies on r-protein binding sites, as discussed below. The ability of certain r-proteins to remain bound to these fragments, or even to rebind to RNA fragments from which proteins have been removed (32, 82, 162, 165), supports the suggestion that these domains are true structural entities. This is not to imply that there may not be extensive interactions between domains.

Within each domain, the structure is organized into series of simple and compound helices, separated by various interior loops and bulges. Many of these represent structural types not found in tRNA. Singly bulged nucleotides are found at positions 31, 55, 94, 397, 746, 1042, 1049, 1227, and 1441. The bulged adenosine at position 1227 is an extremely conserved feature found in all 16S-like rRNAs thus far sequenced. The bulged nucleotides at positions 31, 397, and 746 are nearly always present, although their precise positions may shift, and in some cases they are replaced by mismatches or other irregularities. Multiple G·U pairs are almost always found in the 829-840/846-857 helix, although their positions are variable.

Somewhat unexpected was the absence of long, regular helices in rRNA. Instead, the structure is formed by the joining of many short helices, the junctions of which tend to create a variety of structural irregularities. The reason for this kind of architecture has been rationalized in two ways (96). First, multiple small helices allow a much more complex three-dimensional structure, which could approach that of a globular protein. Another reason may be that long, stable helices could create kinetic "traps," interfering with correction of nonproductive folding errors during assembly. Significantly, no "knots" (11) are found in the secondary structure of the rRNAs (apart from the 9-13/21-25 versus 17-20/915-918 helices), i.e., where the loop contained by a helix is involved in pairing with a sequence outside the helix. This could also be part of a strategy to avoid dead ends in assembly.

Figure 5 shows the most recent refinement of a secondary structure model for *E. coli* 23S rRNA (92, 94). The 5'- and 3'-terminal sequences are base paired, giving the whole the form of a closed loop. As in the 16S rRNA, further long-range base-paired interactions partition the chain into readily identifiable structural domains. There are six major domains in 23S rRNA, defined by their respective long-range interactions: domain I (16-25/515-524), domain II (579-585/1255-1261), domain III (1295-1298/1642-1645), domain IV (1656-1664/1997-2004), domain V (2043-2054/2615-2625), and domain VI (2630-2637/2781-2788). These domains project from the central loop created by pairing of the ends of the molecule.

Experimental evidence for higher-order structure. Comparative analysis provides convincing evidence for the existence of a given helical element in vivo. rRNAs may be dynamic structures with structural "switches" which permit the interconversion of different conformers during ribosome assembly or in the course of protein biosynthesis. Thus, experimental approaches that provide information bearing on rRNA conformation not only help to test proposed structural models, but may also indicate whether or not all of the helical elements defined by the comparative approach actually exist at the same time. It is worth mentioning that there are as yet no convincing examples of mutually exclusive helices that are supported by comparative analysis.

A wide variety of experimental methods has been used to study rRNA conformation. The most successful of these approaches include: (i) the use of single- and double-strand-specific chemical and enzymatic

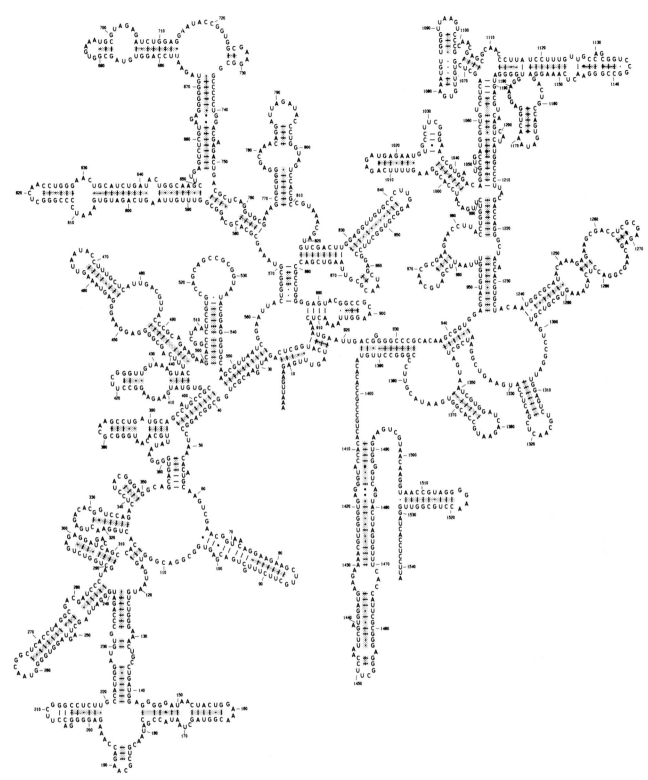

FIG. 4. Secondary structure of *E. coli* 16S rRNA. Shaded helices are considered proven by comparative sequence analysis. From reference 46, with permission.

probes, including cross-linking agents (e.g., references 79a, 147, 148, and 160); (ii) two-dimensional gel analysis, in which RNA fragments from partial nuclease digests, associated by base pairing in the first dimen-sion, are resolved under denaturing conditions in the second dimension (e.g., reference 117); (iii) oligonucle-otide probes, which presumably bind to free, unpaired regions of the rRNA (e.g., reference 78); (iv) direct

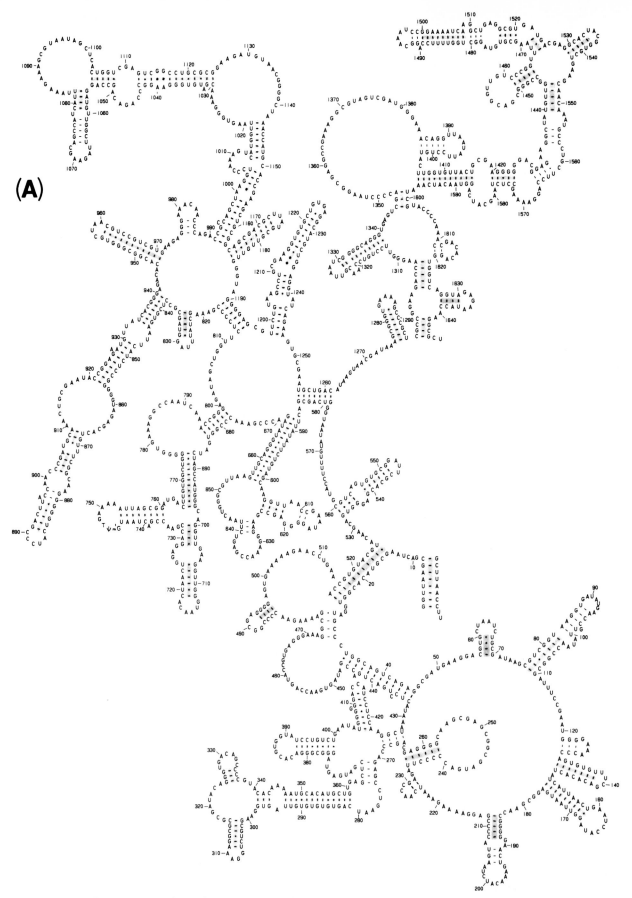

(A)

FIG. 5. Secondary structure of *E. coli* 23S rRNA. Shaded helices are considered proven by comparative sequence analysis. From reference 92.

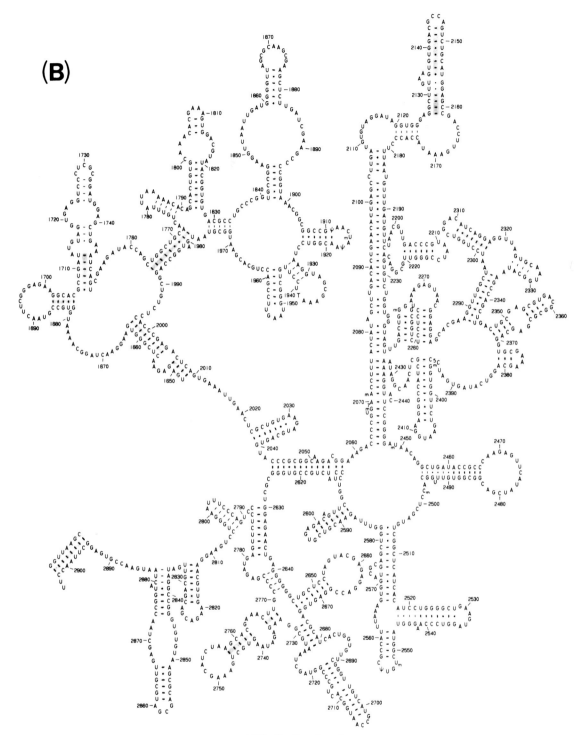

FIG. 5—*Continued*

observation of structural features in partially unfolded rRNA by electron microscopy (e.g., reference 64); and (v) nuclear magnetic resonance techniques (e.g., reference 63). For a more detailed discussion of each of these approaches, see Brimacombe et al. (6) and Noller (92).

These approaches have been used most extensively with *E. coli* 16S rRNA. They have yielded evidence for the existence of 34 helices from the two-dimensional

gel approach (42, 66, 117), 11 helices by psoralen cross-linking (138, 139, 146, 159, 160), and 6 helices by UV cross-linking (167). Analogous evidence has been obtained in support of the secondary structure of 23S rRNA (reviewed by Noller [92]).

Recently developed methodology has permitted systematic chemical probing of bases at virtually every position of *E. coli* 16S rRNA. One approach (147) ex tends the method of Peattie and Gilbert (110) to RNA molecules of any size. Modified RNA is hybridized to restriction fragments of the 16S rRNA gene, excess RNA is trimmed away with RNase T_1, and the RNA in the hybrids is 3' end labeled. Individual RNA fragments are then recovered from the hybrids and subjected to aniline-induced strand scission at the positions of modification, which are then identified by gel electrophoresis. A second approach uses primer extension with reverse transcriptase, which pauses or stops at modified bases (79a). The results of probing naked 16S rRNA are in close accord with the structure depicted in Fig. 4 and give further support for the power and accuracy of the comparative approach. In certain cases, discrepancies between the probe data and the proposed structure are resolved when 30S ribosomal subunits are probed instead of 16S rRNA. Thus, certain aspects of the biologically relevant RNA conformation (which is presumably what is identified by the comparative method) appear to depend on interactions between the 16S rRNA and the 30S r-proteins.

Finally, certain experimental results, not easily explained by consideration of secondary interactions alone, suggest the existence of tertiary or quaternary interactions. The possible significance of these results will be discussed below in connection with higher-order structure and assembly.

ASSEMBLY OF RIBOSOMES

In Vitro Reconstitution of Ribosomes

Even though the structure of ribosomes is very complex, functionally active ribosomes can be reconstituted from purified molecular components. The first successful reconstitution of ribosomes was achieved with *E. coli* 30S subunits by incubating 16S rRNA and a mixture of 30S r-proteins (141). Some important factors for the success were the use of moderately high ionic strength (the optimum being 0.37), presumably to prevent non-specific ionic interactions between RNA and basic r-proteins and yet to allow specific RNA-protein interactions; the use of high concentrations of Mg^{2+} ion; and the use of moderately high incubation temperatures (the optimum being about 40°C) (142). The reconstitution of 50S subunits was achieved first with 50S subunits from a thermophilic bacterium, *B. stearothermophilus*, using essentially the same conditions as those used for *E. coli* subunits except for higher temperatures, usually 60°C (97). It was subsequently carried out with *E. coli* 50S subunits by means of a two-step incubation method (27, 89; see also references 1 and 104). The successful reconstitution of ribosomes demonstrates that all of the information needed for correct assembly is contained in the structure of the ribosome molecular components. Reconstitution of 30S subunits has been demonstrated even with denatured 16S rRNA molecules (2, 133). The reconstitution systems have been used extensively to study the structure of ribosomes, the functional roles of individual molecular components, and the mechanism of assembly of ribosomes in vitro. Most of the studies on ribosome reconstruction have been reviewed previously (for the 30S subunit reconstitution, see reference 98; for the 50S subunit, see reference 86).

Assembly Mapping of r-Proteins

In the case of 30S subunits, the in vitro assembly reaction takes place with 16S rRNA and a mixture of 21 purified r-proteins under a fixed optimum condition. It was first shown that the assembly reaction is cooperative and sequential (100, 142; see below). These characteristics were subsequently studied using purified proteins. Since the reaction proceeds to completion to produce the final product, the 30S subunit with all the r-proteins stably bound to RNA, a simple logic was used to construct an "assembly map": if protein B will not bind well to RNA (or an RNA-protein complex) until protein A does, then protein B must follow protein A in

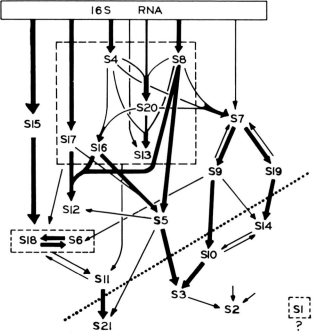

FIG. 6. Assembly map of 30S r-proteins. Arrows between proteins indicate facilitating effect of one protein on the binding of another; a thick arrow indicates a major facilitating effect. The thick arrows from 16S RNA to S4, S8, S15, S17, and S20 indicate that each of these proteins binds directly to 16S RNA, but its binding is weaker and is stimulated by other proteins as indicated in the map. Direct binding of S13 to 16S RNA was observed by some workers, but not by others. The arrow from the large box to a small box containing S18 and S6 was originally placed (and is retained in this figure) because some unidentified proteins in the large box stimulated the binding of S18 and S6. (Recent studies by Gregory et al. [45] show that S8 [together with S15] stimulates binding of S18 and that S8 [together with S15 and S18] also stimulates binding of S6.) The binding of S2 takes place at a later stage in the assembly sequence and is stimulated by the presence of S3 and probably by several other proteins. Proteins above the dotted line are those either required for the formation of RI* particles or found in the isolated 21S RI particles. (The actual isolated 21S RI particles contained only traces of S5, S19, and S12; it was suggested that these proteins are important for the temperature-dependent rate-limiting reaction but bind very weakly to RI particles at low temperatures [53]). The map is taken from reference 50. Although the position of S1 was not determined in the original studies, it is known that some r-proteins among the group, S2, S9, S10 and S14, are required for the binding of S1 (71).

the assembly reaction. It was found that, under reconstruction conditions, only seven of the 30S r-proteins bind individually to 16S rRNA to make a stable complex (see the legend to Fig. 6). Certain other proteins bind only after some of the first six proteins are bound. Binding of the remaining proteins requires the presence of several proteins in addition to some or all of the initial ones. In this way, the "sequence" of addition of proteins to 16S rRNA has been analyzed and a map was constructed (50, 79; Fig. 6). It should be noted that the arrows connecting proteins in the assembly map represent only the interactions (direct or indirect) that have been detected in the experiments used to construct the map. Since not all the combinations were tested, other interactions that stabilize the ribosome structure may have been missed. In addition, some (hypothetical) weak interactions which are important for the early assembly reaction but do not lead to an immediate incorporation of the participating proteins into stable particles must have been missed. As emphasized originally (79), it is also likely that many, perhaps all, of the proteins in addition to the initial binding proteins also interact with 16S rRNA in the finished ribosome structure (see, e.g., reference 55).

The assembly map shows the interdependence of r-proteins in the assembly reaction but does not necessarily show the temporal sequence of the assembly. Studies on intermediate particles in the in vitro assembly reaction gave results which are consistent with the assembly map (53). It appears that the "sequence" shown in the assembly map corresponds roughly to the actual temporal sequence of assembly (for further discussion, see below).

Interdependence of binding of two proteins in the assembly map does not necessarily indicate their physical proximity. Protein A may help the binding of protein B by directly interacting with B, or may help it indirectly by creating a correct binding site for B through some conformational alteration. Yet, as described in a later section, a good correlation has been observed between assembly interdependence and physical proximity (as determined directly by physical measurement).

An assembly map for *E. coli* 50S r-proteins was constructed by Nierhaus and his co-workers using the same approach (115, 116, 118; Fig. 7). It is evident that construction of a 50S assembly map is inherently more complex and difficult. First, many more components are involved in the assembly. Second, the standard procedure for reconstitution of functionally active 50S particles involves two steps with different reaction conditions, and hence, the assumption that the interactions observed in the first incubation conditions are all pertinent to "genuine" assembly reaction may have to be taken somewhat more cautiously. In connection with this second problem, it should be noted that Zimmermann and his co-workers (125, 126) used a different incubation condition, namely, that used for the reconstruction of 30S subunits, and constructed a partial assembly map which showed some discrepancies from that of Röhl and Nierhaus (115). These discrepancies may be at least in part a result of the different conditions used by the two groups. It is possible that the interactions observed by Zimmermann's group, but not shown in Fig. 6, are still "genuine" ones and may be taking place during the second step of the standard reconstitution system (and in the finished ribosome structure).

Intermediate Particles in Assembly

The assembly of 30S ribosomal subunits involves 22 molecular reactants, yet the overall reaction follows

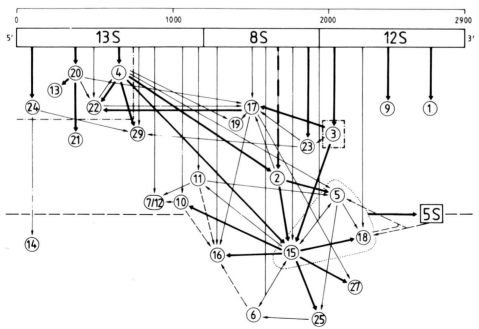

FIG. 7. Assembly map of 50S r-proteins. The three main fragments of 23S RNA (13S, 8S, and 12S RNAs) are indicated, and proteins are arranged according to their binding regions on 23S RNA. Proteins above the dashed line are those important for the conformational change RI50 [I] → RI*50 [I]. L5, L18, and L15, circled by the dotted line, are the proteins important for mediating the binding of 5S RNA to 23S RNA. The map is taken from Röhl and Nierhaus (115).

simple first-order reaction kinetics. This indicates that the rate-limiting step is a unimolecular reaction (142). In fact, by carrying out the reaction at suboptimal temperatures, one can observe the appearance of intermediate particles sedimenting at about 21S (called RI particles) which are slowly converted to the final products, the reconstituted 30S particles (53, 101). RI particles were isolated from reaction mixtures kept at low temperatures (10°C or below) as functionally inactive 21S particles which contain some but not all of the 30S r-proteins. When these particles are heated (40°C) in the absence of free r-proteins, new particles (RI* particles) are formed which sediment at about 26S and are capable of binding the remaining r-proteins to form active 30S subunits at lower temperatures. Thus, the following reaction sequence can be recognized:

$$16S \text{ rRNA} \xrightarrow{+ \text{ protein}} RI \longrightarrow RI^* \xrightarrow{+ \text{ protein}} 30 \text{ subunits}$$

The step from RI to RI* is the major rate-determining step that requires high activation energy and represents a unimolecular rearrangement of the intermediate particle (53, 142). The initial binding of the proteins on the 16S rRNA at low temperatures does not markedly affect the folding of the RNA molecule, but the step from RI to RI* at higher temperatures involves significant folding of RNA and results in a structure considerably more compact than that of the 16S rRNA or the RI particles (136).

The protein composition of the RI particles was inferred from the isolated 21S particles and from determination of the identity of the proteins required for the essential conformational change in the RI → RI* reaction. These proteins are shown above the dotted line in the assembly map in Fig. 6 (see the comments in the legend to the figure). Some of the interactions shown in the assembly map are really essential, e.g., omission of protein S7 in the reconstitution mixtures leads to formation of particles which are missing proteins S19, S14, and S10 completely and probably S9, S3, and S2 (79). Other interactions might be dispensable thanks to the extensive cooperative interactions among ribosomal components (see a later section). For example, even though S16 plays an important role in the assembly reaction according to the map, 30S-like particles with almost full activity are slowly formed even in its absence (54). Another example concerns S6, which has an important role in the binding of S18 (79; see Fig. 6) but can be omitted from the reconstitution mixture without significant effects on assembly kinetics or the function of the assembled particles (reference 99 and other results cited in reference 98). It appears that many other cooperative interactions enable binding of S18 and other "downstream" proteins, S11 and S21.

The sequence of addition of proteins in vivo is not well known. Some "intermediate precursor particles" exist in small amounts in growing cells and can be detected by pulse-labeling techniques (for review, see Schlessinger [119]). In addition, there are several E. coli mutants in ribosome assembly which accumulate defective ribosomal particles which resemble the in vitro RI particles (e.g., reference 83). However, mutations or some other conditions (such as addition of low concentrations of chloramphenicol) lead to accumula-

tion of incomplete particles and so may affect some undefined steps in the assembly. Thus, it is difficult to evaluate the exact relationship between these particles and "true" intermediates in the assembly in vivo. "Precursor particles" present in small amounts in exponentially growing cells have been isolated, and their protein composition has been analyzed (56, 88). The 30S precursor particles were found to have proteins similar to those found in the RI particles isolated as 21S particles in vitro, the main difference being that S7, S9, and S19 were absent from the in vivo intermediates. Thus, the rate-limiting step in the in vivo 30S assembly is apparently different from that observed in vitro. It is known that ribosome assembly in vivo starts with precursor rRNA and that rRNA processing and maturation is completed during ribosome assembly. However, this does not explain why the binding of S7 (and S9 and S19) to the precursor RNA does not take place in vivo, unless the synthesis of S7 is rate-limiting in vivo (119). It has been speculated that in vivo assembly may involve some unique mechanisms which would help decrease the kinetic barrier for the assembly reaction. For example, assembly may take place in association with membrane (49), or it may occur simultaneously with transcription of rRNA. In either case, the sequence of protein addition may be different from that observed in vitro. Regarding this possibility, de Narvaez and Schaup (24) analyzed the amounts of r-proteins present in isolated nucleoid preparations and suggested a sequence of protein addition which is different from that observed in vitro. They suggest that precursor particles isolated from growing cells may be artifacts. It has also been suggested that some (hypothetical) non-ribosomal factors participate in the assembly reaction in vivo (134). Clearly, further studies are needed for understanding the detailed sequences of events in the in vivo assembly.

The sequence of events during in vitro E. coli 50S reconstitution was also studied in a way similar to that used for 30S reconstitution, and the following reaction scheme was proposed (27, 123):

$$23S \xrightarrow{5S + \text{ proteins}} RI_{50}(I) \longrightarrow RI_{50}^*(I) \xrightarrow{+ \text{ proteins}} RI_{50}(II) \longrightarrow 50S \text{ subunits}$$

(Essentially the same reaction scheme was proposed earlier for the reconstitution of B. stearothermophilus 50S ribosomal subunits [34].) Proteins that participate in the formation of $RI_{50}(I)$ as well as those that participated in the later reaction were classified by Nierhaus and his co-workers (Fig. 7). Like S16 in the case of the 30S assembly reaction proteins, L24 (127) and L20 (102) appear to be essential for the assembly reaction but not for protein synthesis by the assembled 50S subunits.

E. coli Mutants That Lack r-Proteins in Their Ribosomes

Although reconstitution experiments indicated that most of the r-proteins are required for assembly or activity of ribosomes or both, some r-proteins appear to be dispensable. As mentioned above, S6 appears not to be needed in the reconstitution of active 30S subunits, as examined in various known functional assays. Of course, this does not mean that S6 does not have any significant role in the structure and function of ribosomes in vivo. It is now known that ribosomes and proteins have functions other than in protein synthesis. For example, some r-proteins function as translational repressors as free r-proteins (see chapter 85). At least one r-protein, S10, has been shown to participate in antitermination during lytic growth of phage lambda, either as a free protein or as a part of ribosomes (37). When not engaged in protein synthesis, ribosomes also apparently function as a negative feedback regulator in the regulation of rRNA synthesis (60; see chapter 85). Thus, r-proteins whose function has not been demonstrated by in vitro reconstitution may have some unknown important function in vivo. We would then expect that all of the r-proteins are likely to be indispensable. It was therefore surprising to find mutants of E. coli that have some r-proteins completely missing from their ribosomes.

Mutants missing S20 were isolated as a temperature-sensitive mutant (58), as suppressors of a temperature-sensitive valyl-tRNA synthetase mutant (155), and as suppressors of an erythromycin-dependent mutant (20). Mutants missing L11, L15, L28, L29, or L30 were also isolated as suppressors of the erythromycin-dependent mutant (20). Those missing L1 or S9 were isolated as suppressors of a kasugamycin-dependent mutant (19, 21). As expected, most of these grow at slow rates. In addition, the nature of the mutations has not been characterized, and many of the mutants have one or more additional mutational alterations in their ribosomes (e.g., see references 17–21). Thus, interpretation of the assembly and the functional capability of these mutant ribosomes is difficult. There is at least one case where the nature of a mutation leading to a loss of an r-protein has been well characterized. Starting with a strain carrying a temperature-sensitive amber suppressor [supF(TS)], a mutant was isolated with an amber mutation in the structural gene for L15, as confirmed by DNA sequencing. This mutant produced ribosomes that lack L15 at higher temperatures (59). With a proper genetic complementation to suppress a polar effect of the mutation on distal genes, it was shown that the mutant cells could grow and synthesize proteins at normal rates at the higher temperatures. Thus, these results, together with those obtained with a different mutant (76), strongly suggest that L15 is dispensable for protein synthesis and cell growth even though in vitro reconstitution studies showed that it is important in assembly (115, 116; see Fig. 7) and in peptidyl transferase activity (120). It is probable that extensive cooperativity among ribosomal components permits assembly as well as the activity of the assembled particle in the absence of L15, and that L15 is still important in these processes. Its importance may be manifested only in the absence of some other protein components.

Cooperativity of the Assembly Reaction

It is evident from the assembly maps shown in Fig. 6 and 7 that cooperative binding of proteins to rRNA is an essential feature of the assembling reaction; binding of certain r-proteins depends on prior binding or a simultaneous binding of other r-proteins. In an early study using unfractionated 30S r-proteins, cooperativity of the assembly reaction was examined by systematically changing the ratio of the 30S r-protein mixture to 16S rRNA (101). It was found that, when the ratio of r-proteins to rRNA becomes smaller (0.2 to 0.4), cooperativity is not complete, but the number of sites or groups of sites per 16S rRNA molecule that can bind r-proteins independently of each other (noncooperatively) is still at most three. Subsequent studies showed that at least seven 30S r-proteins bind to 16S rRNA individually at their respective specific binding sites in the absence of other proteins (see Fig. 6 and a later section). Although not generally emphasized, it is evident that even these seven protein binding sites are not independent of each other; binding of proteins tends to take place on the same RNA molecule in the presence of excess RNA. It is not known how and what kind of dead-end particles are produced at these independent "nucleation sites" when 16S rRNA is present in large excess. Available information on the assembly reaction as well as on the protein binding sites on 16S rRNA suggests that one such nucleation site is perhaps at (and near) the S7 binding site in the 3' domain of 16S rRNA, where S9 and S19 also cooperatively bind. Another site is perhaps at the S8-S15 binding site in the central domain of the 16S rRNA, where S8, S15, S18, and S6 bind cooperatively. Thus, each of the nucleation sites may correspond to a specific region in each of the domains of 16S rRNA.

It should also be noted that once RI* particles are formed, the binding of the remaining functionally important proteins (S21, S14, S10, S3, S2, and perhaps S1) appears to be completely cooperative; that is, these proteins seem to participate together in the assembly reaction even in the presence of a large excess of RI* particle (101). Since the early study was performed with a protein mixture, this striking conclusion should be confirmed using purified proteins.

Experiments similar to those done with the 30S ribosomal reconstitution system were also done with the 50S reconstitution system by Nowotny and Nierhaus (103). The results indicate the presence of two independent nucleation sites, one of them involving L24 and another involving L3.

Strong cooperativity in ribosome assembly may be important in vivo to achieve efficient production of ribosomes under conditions where excess rRNA is synthesized, such as during slow-growth conditions (see chapter 85). Without a strong cooperativity, even a small excess production of rRNA over r-proteins would lead to a drastic decrease in the production of the complete functional ribosomes, resulting in the accumulation of various kinds of defective incomplete particles. Available data suggest strong cooperativity in ribosome assembly in vivo, but do not establish whether there are multiple independent nucleation sites for each subunit assembly as observed in vitro, or whether in vivo assembly initiates from a single nucleation site cooperatively as a result of some additional conditions.

Interactions between the 30S and 50S Ribosomal Assembly Reactions

In vitro reconstitution experiments have demonstrated that the assembly of 30S subunits can take place in the absence of 50S ribosomal components. Conversely, assembly of 50S subunits can take place in the absence of 30S ribosomal components. Yet, some observations indicate that in vivo assembly of one subunit is apparently coupled with that of the other. For example, in a cold-sensitive mutant, a single mutation in the gene for S5 affects not only the assembly of 30S subunits but also that of 50S subunits (84). Conversely, there is an instance in which a single mutation in the gene for a 50S r-protein, e.g., L22, affects not only the assembly of 50S subunits, but also that of 30S subunits (108). There are two ways to explain these observations. One is to assume that ribosome assembly in vivo takes place only on a single large precursor rRNA (30S rRNA) and that both 30S and 50S subunit assembly reactions interact with each other (108; see also reference 30). The second explanation is that failure of the assembly of one ribosomal subunit causes accumulation of free r-proteins for that subunit, some of which may be repressors that inhibit the synthesis not only of proteins for that subunit but also of some for the other subunit via translational feedback (see chapter 85). Thus the apparent interaction of the two assembly reactions may be an indirect result of regulation of the synthesis of 30S and 50S r-proteins at the level of translation. The available data suggest that the second explanation is probably correct. There is no good evidence for ribosome assembly in vivo on the large 30S precursor rRNA. In fact, there are cases in which the assembly of 30S subunits takes place in the absence of simultaneous assembly of 50S subunits (for example, see reference 135). In this case, 30S subunits are assembled from newly synthesized 16S rRNA and 30S r-proteins in the pool produced by a prior degradation of 30S subunits by high-temperature treatment; the 23S rRNA portion synthesized simultaneously with 16S rRNA is degraded without forming 50S subunits.

Reconstitution with Altered rRNA

The in vitro reconstitution system has been used extensively to study the role of individual r-proteins in the assembly and function of ribosomes. Studies on rRNA were also done by substituting, for intact 16S rRNA, 16S rRNA from other bacterial species, chemically altered 16S rRNA, and mutationally or physiologically altered 16S rRNA (reviewed by Nomura and Held [98]). In fact, one of the first indications of the importance of 16S rRNA in recognizing translation initiation sites on mRNA was obtained by analyzing the ability of hybrid 30S ribosomal subunits, consisting of *B. stearothermophilus* 16S rRNA and *E. coli* r-proteins, to translate R17 phage RNA relative to synthetic mRNAs (51). However, manipulation of large rRNAs was difficult in the past, and studies of rRNAs by using in vitro reconstitution systems has been limited. The recent development of genetic engineering techniques, combined with isolation of rRNA operons and elucidation of their nucleotide sequences, has now made it possible to make various mutations in rRNA genes (43, 128). Since very efficient expression of rRNA genes can now be achieved from a repressible promoter (44), even isolation of mutations that would otherwise be lethal has become possible.

Thus, the ribosome assembly reaction can be studied both in vivo and in vitro, using a variety of specifically designed mutant rRNA molecules expressed from engineered mutant rRNA genes.

RNA-PROTEIN INTERACTIONS

The mechanism of RNA-protein recognition is only poorly understood at present. Although considerable progress has recently been made in the area of DNA-protein interaction (107), there is reason to suspect that recognition of RNA may differ in several fundamental respects. DNA, with rare exceptions, has a simple, continuous helical structure; recognition of specific sites on a DNA molecule by a protein thus depends on sequence recognition and possibly on subtle, sequence-dependent variations in helical geometry (26). In contrast, RNA structure is complex, containing not only helical elements but also single-stranded hairpin loops, bulge loops, and interior loops, and has the ability to achieve compact, globular structures by means of rather complex tertiary interactions, as in tRNA (62). Furthermore, RNA helices are rigidly confined to the A or A' geometry, in which the major groove is deep and narrow, excluding any possibility of penetration by a protein α-helix, the means by which DNA sequences appear to be "read" by their cognate proteins (107). (An exception is the recent finding that RNA can attain a left-handed "Z" helical conformation under certain rather extreme solvent conditions [47].)

Ribosomes provide a wealth of model systems for the study of specific RNA-protein interactions. Several protein binding sites have been localized on rRNA, and it is worthwhile to review briefly what has been learned from these studies. For a more extensive discussion of this topic, the reader is referred to a review by Zimmermann (164).

Discovery of the in vitro reconstitution method (141) provided the means to study the mechanism of binding of r-proteins to rRNA. The fact that fully functional particles were obtained argued that any intermolecular interactions observed in this system are likely to be authentic. Criteria for the specificity of binding are: (i) a given protein should bind only to one species of RNA; (ii) in protein excess, the molar ratio of protein to RNA should plateau at a stoichiometric ratio (e.g., 1:1); (iii) interaction of one protein with the RNA should not hinder the binding of another; and (iv) a complexed protein should be incorporated into the ribosomal structure upon addition of the remaining proteins without prior dissociation from the RNA (163).

Detailed localization of the RNA binding sites for r-proteins has been studied by several general approaches. (i) The classical "bind-and-chew" method utilizes gentle nuclease digestion of protein-RNA complexes, followed by isolation of ribonucleoproteins containing subfragments of the RNA (reviewed by Zimmermann [164]). In several cases, the ability of the protein to rebind to the purified RNA fragment confirms that at least part of its RNA binding site is contained in the fragment. This approach has proven to be generally quite applicable and useful. However, a number of potential difficulties have been recognized. In some cases, the protected RNA fragment has been obtained even in the absence of protein, suggesting that it is the structure of the RNA itself, rather than interaction with protein, that stabilizes the protected fragment (40). Conversely, features of RNA binding sites that have lower inherent stability may not survive nuclease treatment and so may go undetected. (ii) Protection of RNA from chemical or enzymatic probes by bound protein ("footprinting") has been successfully exploited, particularly in studies of 5S rRNA, where analysis of the RNA is relatively easy (28, 39, 109). (iii) "Damage/selection" experiments identify RNA sites whose chemical modification interferes with protein binding (109). (iv) Chemical or UV cross-linking of bound proteins to RNA indicates sites that are proximal to the protein; in some cases the cross-linked site may not be part of the binding site proper (161). (v) Protein-RNA complexes may be examined by electron microscopy (16). (vi) Site-directed mutagenesis of cloned rRNA genes in sequences corresponding to protein binding regions in the RNA has the potential to define structural and sequence requirements for protein recognition (128).

Protein binding site regions in 16S rRNA obtained by the first approach above (bind-and-chew method) are shown in Fig. 8. Particularly striking is the way in which the RNA fragments obtained correspond closely to domains or subdomains of the RNA structure. This observation argues that these domains, inferred mainly from comparative sequence analysis, are true physical entities; their ability to rebind to their cognate proteins suggests that they possess some degree of structural autonomy in the ribosome.

What structural features of rRNA are actually recognized or bound by the proteins? It is rather doubtful that all of the S4 binding fragment depicted in Fig. 8 is important for binding, since the mass of the RNA moiety outweighs S4 itself by more than a factor of

six. Yet, attempts to isolate smaller fragments that retain their binding ability have not been successful. It is more informative to consider proteins whose binding sites have been defined within smaller regions of rRNA, which may actually represent something approaching their true binding regions. Most useful for this purpose are the binding sites for S8 and S15 on 16S rRNA and L18 and L25 on 5S rRNA, all of which have been localized to relatively small regions (40 to 100 nucleotides).

Proteins S8 and S15 bind to the central domain of 16S rRNA, to positions 583-610/632-653 and 654-672/733-756, respectively (82, 165, 166). L18 and L25 bind to 5S rRNA within positions 15-69 and 70-109, respectively (29, 39, 109). One may begin to draw generalizations about the nature of r-protein binding sites from the structural features common to these four regions. Each binding region contains a central "core" of about 40 nucleotides, typically arranged into a pair of 7- to 8-base-pair helices joined by an internal loop. Nucleotides in the base-paired regions are not generally phylogenetically conserved (46, 109), but certain nucleotides in the internal loop and flanking regions tend to be conserved. Single- and double-base bulge loops are present in the S8, S15, and L18 helices. The L25 site, in contrast, appears to be highly irregular in its pairing, containing several G·U and possibly A·G pairs. The structure of the L25 binding site region behaves nevertheless very much like a typical helical structure, from X-ray scattering data (73) and chemical and enzymatic probe studies (28, 93, 109). Garrett et al. (41) have termed these two kinds of structures type I (binding sites with single- or double-base bulges) and type II (binding sites with extensive noncanonical pairing). It should be cautioned that the present collection of well-defined binding regions is small, and any generaliza-

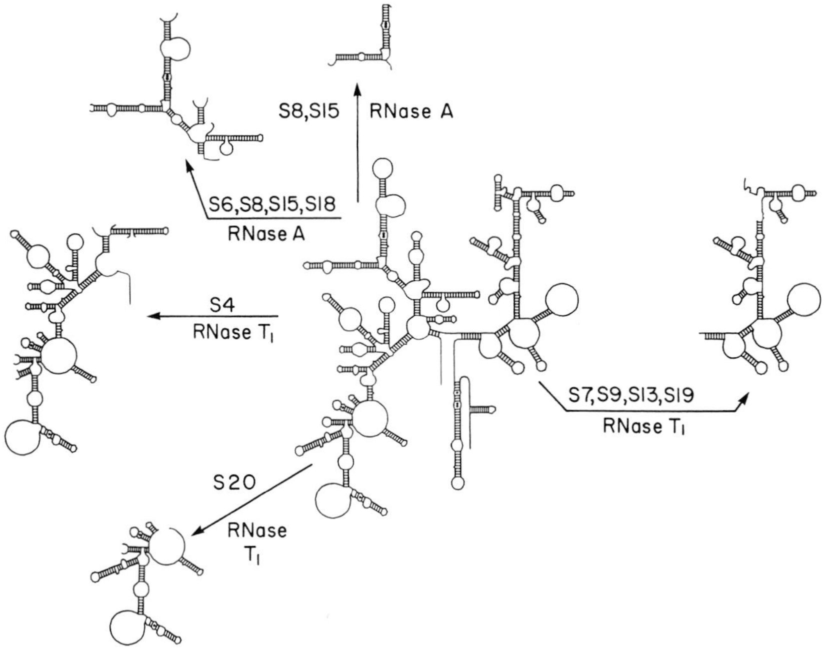

FIG. 8. RNA fragments found in ribonucleoprotein complexes after nuclease digestion of r-protein–rRNA complexes (32, 82, 165) or gently unfolded 30S ribosomal subunits (162). Note how the fragments correspond closely to predicted structural domains or subdomains. From reference 158.

tions about what typifies an r-protein binding site should be considered as working models.

There are additional scattered clues to the nature of the structural features recognized by the r-proteins. Certain unpaired nucleotides in the internal loop and flanking regions are strongly protected from chemical and enzymatic attack in studies on S15 (82) and L18 (28, 39). Cleavage at base-paired positions by cobra venom RNase is prevented by binding of L18 and L25 (28), suggesting that the proteins also recognize the helical structure. Nuclear magnetic resonance and other spectroscopic studies with L18 and L25 have shown that their binding site helices are not disrupted by protein binding (3, 63, 124). The role of bulged nucleotides is less clear; damage/selection experiments using diethylpyrocarbonate have shown that carbethoxylation of the singly bulged A66 (Fig. 3) causes a significant decrease in the binding affinity of L18 for 5S rRNA (109). The prevalence of bulged nucleotides in protein binding site helices is very intriguing; their high phylogenetic conservation in many cases suggests an important role in assembly, function, or both.

ARCHITECTURE OF RIBOSOMES AND THEIR SUBUNITS

General Morphology of Ribosomal Subunits

From the preceding discussion, it is clear that structural information concerning the individual molecular components, particularly the rRNAs, is beginning to provide clues about the overall structural organization of the *E. coli* ribosome. At the same time, studies on the overall shape of ribosomal subunits and the location of specific structural features are beginning to converge on the information obtained by the first approach. Our current perception of the shape of ribosomes derives mainly from electron microscopy and, to some extent, from X-ray and neutron scattering and diffraction studies. More extensive discussions of these results can be found in reviews by Liljas (74), Wittmann (152), and Ramakrishnan et al. (114).

Several groups have studied the shapes of ribosomal subunits by electron microscopy using either negatively stained (5, 68, 129) or freeze-dried and shadowed preparations (122) or dark-field electron microscopy (67). Discrepancies between the models obtained from the different groups, due mainly to differences in interpretation, have for the most part gradually given way to general agreement on models for the 30S and 50S subunits typified by that of Lake (68) (Fig. 9 and 10). A critical discussion of the various models and resolution of the problems encountered in interpretation of the electron microscopy data can be found in reference 113.

The 30S subunit (Fig. 9) is divided into a "body," containing ca. two-thirds of the mass of the particle, and a "head," containing the other one-third of the mass. A "platform"-like projection extends from the body of the particle near the head-body junction, and a "cleft" is seen (or not seen) between the head and platform (depending on the particular method of specimen preparation and interpretation). Detailed descriptions of the models currently proposed by each group are presented by Wittmann (152). The largest dimension of the particle is about 24 to 25 nm.

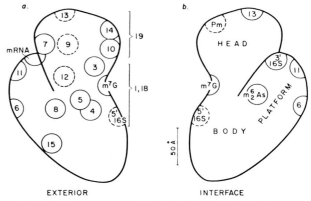

FIG. 9. Consensus model of the *E. coli* 30S ribosomal subunit. (a) Exterior surface; (b) subunit interface. Circles designate positions of protein and RNA sites on two-dimensional projections of each subunit surface. Closed or broken circles are site locations with greater or lesser certainty, respectively. rRNA, Entry/exit site of mRNA; Pm, puromycin. From reference 113, with permission.

The 50S subunit has a more spherical or globular shape (Fig. 10) and contains three recognizable projections: the "L7/L12 stalk" on the right; the "central protuberance" in the middle; and the "L1 ridge" on the left (68, 95). There is generally better agreement between the different groups concerning the shape of the 50S particle. The largest dimension of the 50S subunit is again about 24 to 25 nm.

The relative orientation of the 30S and 50S subunits in the 70S particle has also been a matter of controversy (see discussion in reference 113). Consensus appears to have gravitated to the arrangement proposed by Lake (69) (Fig. 11). In this model, the platform side of the 30S subunit faces the concave side of the 50S subunit, with the head of the 30S particle positioned between the L1 ridge and the central protuberance. This arrangement is supported by single- and double-antibody labeling of 70S ribosomes (69).

Both X-ray and neutron scattering methods have been employed to study the shape of ribosomes and their subunits in solution (reviewed by Moore [81] and Liljas [74]). Although the results of solution scattering experiments are much more subtle to interpret than

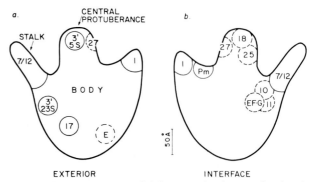

FIG. 10. Consensus model for *E. coli* ribosomal subunit. (a) Exterior view; (b) interface view. E, Exit site for nascent polypeptide. See legend to Fig. 9. From reference 113, with permission.

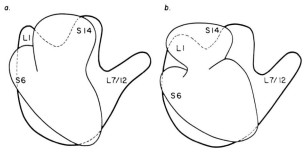

FIG. 11. Relative orientation of subunits in the 70S ribosome. Site locations of four proteins are shown for comparison. (a) After Lake (69); (b) after Stöffler and co-workers (discussed in reference 113). From reference 113, with permission.

those from electron microscopy, they are nevertheless of great importance because they measure the properties of fully hydrated particles under conditions that are approximately physiological. It is therefore reassuring that a model for the 50S subunit derived from neutron scattering data (131) bears striking resemblance to that obtained by electron microscopy.

Besides shape, information concerning the relative spatial distribution of RNA and protein can be obtained from measurements of their respective aggregate radii of gyration in contrast matching experiments. There is presently general agreement that the protein and RNA distributions in the 50S subunit have roughly coincident centers of mass, but that the protein component has a much larger radius of gyration (9 to 10 nm) than the RNA component (5.5 to 6 nm) (18). In the 30S subunit, the distributions of the protein and RNA components are again roughly concentric, with a tendency for the surface to be protein rich; the radii of gyration are ca. 7.8 nm for the protein and 6.1 nm for the RNA (reviewed by Damaschun et al. [22]).

Structural details of macromolecules are traditionally solved by X-ray crystallography. Unfortunately, ribosomes, particularly those from *E. coli*, have proven difficult to crystallize. Some progress has been made in recent years, however, in obtaining crystalline sheets of *E. coli* 30S and 50S subunits (14, 15) and three-dimensional crystals of *E. coli* 70S ribosomes (154). However, the latter are not suitable for diffraction studies. The crystalline sheets have been studied using three-dimensional reconstruction of electron microscopic images.

Internal Architecture

Possible clues to internal structural features have been suggested by results from electron microscopy and from low-angle X-ray scattering. Three-dimensional reconstructions of electron micrographs of tilt series of individual 30S and 50S subunits have been aligned by computer correlation methods (65, 106). Analysis of the averaged structures shows stained channel-like interior structural features which may correspond to positively stained segments of the rRNA.

It has long been known that in low-angle X-ray scattering studies on ribosomes, weak maxima are observed at about 4.5, 3, and 2.5 nm (70). Using protein-depleted particles and contrast variation methods, Serdyuk et al. (121) have shown that these maxima are due to structural features of the rRNA. The data are consistent with parallel packing of RNA helices within the ribosome, as originally suggested by Langridge (70).

Location of Proteins and RNA in the Ribosomes

During the past decade, much has been learned about the spatial arrangement of r-proteins in the ribosome. Most crucial has been information obtained from immunoelectron microscopy (IEM) and neutron diffraction studies (reviewed by Lake [68], Stöffler et al. [129], Moore [81], and Ramakrishnan et al. [114]). Important contributions have also come from singlet-singlet energy transfer (57), protein-protein cross-linking (143), and in vitro assembly interactions (17, 98). Positioning of the RNA in the ribosome represents a different sort of problem. Apart from positioning the relatively small 5S rRNA, locating the rRNA component might at first be considered trivial, because its mass probably coincides roughly with that of the whole particle. Actually, the information of interest is the location of specific sites or regions of the rRNA in the ribosome. In effect, this amounts to determination of aspects of its tertiary structure. Among the relevant experimental approaches to this problem are IEM (e.g., reference 112), singlet-singlet energy transfer (105), RNA-protein or RNA-RNA cross-linking (6), and protein binding site studies.

IEM. The relative position of a protein in the ribosome structure can be inferred from the position of its antibody attachment site in relation to recognizable structural features on the ribosome. Although this method is straightforward in principle, a great deal of controversy over its various results has ensued over the past several years. Happily, this situation has largely been resolved (see discussion by Prince et al. [113]). Many of the r-proteins have been located by this method on their respective subunits, and the placements are in quite good agreement with the results obtained from other approaches (Fig. 9 and 10).

Features of the RNA have also been localized in the electron microscopy model using the IEM method. Antibodies to methylated nucleotides have allowed positioning of the dimethyladenosines 1518 and 1519 of 16S rRNA on the platform (112) and of $^{7}_{m}G$ 527 on the body (145) (Fig. 9). Derivatization of the 5′ and 3′ termini of the rRNAs with haptens has made possible the location of the 5′ and 3′ ends of 16S rRNA and the 3′ ends of 5S and 23S rRNA (reviewed by Noller [92]).

Neutron diffraction. A three-dimensional map of the positions of 15 of the 21 30S subunit proteins has been determined by neutron diffraction (114). Selective deuterium labeling of pairs of proteins that are reconstituted with the other (nondeuterated) components into a 30S subunit allows measurement of the distance between the centers of mass of the two deuterated proteins. Measurement of numerous pairwise distances eventually permits triangulation of their spatial positions relative to one another. This three-dimensional map (Fig. 12) correlates well with the results of the IEM approach and permits correla-

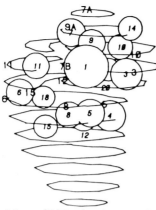

FIG. 12. Positions of 30S subunit r-proteins from neutron diffraction. From reference 114, with permission.

tion of the neutron data with the structure for the 30S particle obtained by electron microscopy (Fig. 9).

Singlet-singlet energy transfer. Distance measurements can also be obtained from radiationless singlet-singlet energy transfer between two fluorescent probes. Two different fluorescent groups, an acceptor and a donor, are attached covalently to two different proteins, which are then reconstituted into a 30S subunit. Particles are irradiated at a wavelength which the donor but not the acceptor absorbs. The excited donor molecule decays, transferring its fluorescent energy to the acceptor, which absorbs in the region of the emission spectrum of the donor. The acceptor then fluoresces, and the emitted energy is measured. Distances can be inferred because the quantum yield of the transfer is proportional to the inverse sixth power of the distance between the probes. These measurements have been performed for many of the possible pairwise combinations of 30S subunit proteins (57). The results are in quite good agreement with neutron diffraction measurements that have been performed on the same pairs (Table 2), with the possible exception of the S4-S15 and S4-S20 distances. Because S4 is known to have an elongated structure in the ribosome (114), the discrepancies could be accounted for by an asymmetric distribution of the fluorescent probes on S4.

Singlet-singlet energy transfer has also been used to measure inter- and intramolecular distances between the ends of the rRNAs (105). Distances between probes attached to the 3' end of 16S rRNA and 5S or 23S rRNA were estimated to be about 5.5 and 7.1 nm,

respectively. The corresponding distance between 5S rRNA and 23S rRNA was too large to be measured accurately but was estimated to be greater than 6.5 nm.

Cross-linking. Neighboring molecules in the ribosome have been identified by covalent cross-linking, giving protein-protein, protein-RNA, or RNA-RNA proximity information. Both chemical and photochemical strategies have been employed, and results have come from many different groups of investigators. Protein-protein cross-linking has been reviewed by Traut et al. (143), and RNA-RNA and RNA-protein studies have been reviewed by Brimacombe et al. (6).

Among the many bifunctional reagents that have been employed, 2-iminothiolane (143) is one of the more useful and elegant ones. This cyclic reagent reacts with amino groups and, in so doing, opens to provide an available sulfhydryl moiety. Oxidation then creates disulfide bridges between neighboring sulfhydryls, forming covalent cross-links. Cross-linked proteins are then identified by two-dimensional diagonal gel electrophoresis, in which they migrate as pairs in off-diagonal positions. Some results of protein-protein cross-linking are summarized in Table 2, in comparison with results from other protein proximity studies. There is a strong correlation with the neutron diffraction and fluorescence studies; proximal proteins are cross-linked, and distal ones (> 4 nm apart) are not.

Wower and Brimacombe (161) have further exploited this reagent for RNA-protein cross-linking. Ribosomal subunits are treated with 2-iminothiolane as before, but after the initial treatment they are photolysed with UV light. This causes photoreaction to occur between free sulfhydryl groups and adjacent uridine residues in 16S rRNA. After nuclease digestion, protein-oligonucleotide adducts are analyzed to identify the cross-linked protein and RNA moieties.

RNA-RNA cross-linking has been performed on rRNA by several strategies, one of the best known being photoinduced cross-linking in the presence of psoralen derivatives (reviewed by Thompson and Hearst [140]). Psoralens intercalate into nucleic acids and photoreact with adjacent pyrimidines on opposite helical strands, effectively cross-linking the two strands of a helix. Sites of cross-linking in rRNA have been identified by electron microscopy, more recently by sequencing the cross-linked oligonucleotides (reviewed by Thompson and Hearst [140]).

Assembly interactions. Protein-protein contacts are undoubtedly made during ribosome assembly, and it

TABLE 2. Some protein-protein proximities in the 30S ribosomal subunit[a]

Protein pair	Distance (nm) by method:		Cross-link (143)	Assembly map (17, 98)	Protein-protein protection (45)
	Neutron diffraction (114)	Fluorescence (57)			
S4-S15	9.0	6.6	−	−	+
S4-S18	8.7	>8.1	−	−	
S4-S20	6.3	4.6	−	±	
S6-S18	3.3	3.9	+	+	
S7-S9	3.4	3.1	+	+	+
S8-S15	3.5	3.2	+	−	+
S15-S18	7.1	6.9	−	+	

[a] Parentheses in heads indicate source (reference) of data.

is likely that at least some of the assembly map dependencies reflect this (Fig. 6 and 7; 98). Studies by Craven et al. (17) show that strong protection of one protein by another often occurs during assembly, and this is likely to reflect protein-protein contacts in the ribosome. It is possible that either type of phenomenon can be caused by an indirect mechanism due to allosteric perturbations involving, for example, rRNA, and so any conclusions are necessarily tentative. Some results from these studies are summarized in Table 2. In general, the correlation is good; proteins showing strong assembly interactions were almost always found to be close by direct measurement, although assembly mapping failed to detect interdependence of some closely located protein pairs (see discussion in previous section).

CONCLUSION

Early clues to the structural complexity of the ribosome have by now been amply confirmed. Primary structures for the 52 r-proteins and the three rRNAs, and accurate secondary structure models for the latter, show little evidence for structural symmetry. Solution of the three-dimensional structure of the ribosome will therefore likely prove to be an awesome task. In vivo assembly of ribosomes, based on what is known from extensive in vitro studies, will probably reflect their structural complexity.

Our current knowledge of ribosome structural organization is presently most advanced for the *E. coli* 30S subunit. The extent to which the various approaches described above are coalescing to give a coherent view of this particle is summarized in Fig. 13. The secondary structure model for 16S rRNA and the electron microscopy model for the 30S subunit are shown drawn to approximately the same scale. Numbers on the subunit model show the positions of 30S subunit proteins, determined by IEM and neutron diffraction

studies, and certain features of 16S rRNA, determined by IEM. The positions of these features of 16S rRNA, and approximate positions of the RNA binding sites for several proteins, are indicated in the secondary structure model. This information represents the beginnings of a low-resolution structural map of the ribosome. Thus, the 3'-major domain of 16S rRNA must for the most part be located in the head of the electron microscopic model; certain parts of the central and 3'-minor domains are found on the platform. Taking into account recent cross-linking results and stereochemical information, one can begin to formulate more detailed models for the packing of 16S rRNA in the subunit and the way in which the individual proteins are accommodated. A preliminary exercise of this kind has been presented recently (95).

Corresponding data for the 50S subunit are less advanced, but efforts on its structure will very likely continue to benefit from approaches used in studying the small subunit. Even if high-resolution crystallographic data were forthcoming, they would undoubtedly benefit greatly from the kinds of results that are now being obtained with the indirect approaches described here. In any case, the next decade should provide many exciting insights into the nature of this ancient and ubiquitous biological structure.

LITERATURE CITED

1. **Amils, R., E. A. Matthews, and C. R. Cantor.** 1978. An efficient *in vitro* total reconstitution of the *Escherichia coli* 50S ribosomal subunit. Nucleic Acids Res. **5**:2455–2470.
2. **Barritault, D., M. F. Guerrin, and D. H. Hayes.** 1979. Reconstitution of active 30S ribosomal subunits *in vitro* using heat-denatured 16S rRNA. Eur. J. Biochem. **98**:567–571.
3. **Bear, D. G., T. Schleich, H. F. Noller, and R. A. Garrett.** 1977. Alteration of 5S RNA conformation by ribosomal proteins L18 and L25. Nucleic Acids Res. **4**:2511–2526.
4. **Böhm, S., H. Fabian, and W. Welfle.** 1982. Universal structure features of prokaryotic and eukaryotic ribosomal 5S RNA derived from comparative analysis of their sequences. Acta Biol. Med. Ger. **41**:1–16.
5. **Boublik, M., N. Robakis, and W. Hellmann.** 1982. Electron microscopic study of free and ribosome bound rRNAs from *E. coli*. Eur. J. Cell. Biol. **27**:177–184.
6. **Brimacombe, R., P. Maly, and C. Zwieb.** 1983. The structure of ribosomal RNA and its organization relative to ribosomal protein. Prog. Nucleic Acid Res. Mol. Biol. **28**:1–48.
7. **Brosius, J., M. L. Palmer, P. J. Kennedy, and H. F. Noller.** 1978. Complete nucleotide sequence of a 16S ribosomal RNA gene from *Escherichia coli*. Proc. Natl. Acad. Sci. USA **75**:4801–4805.
8. **Brosius, J., T. J. Dull, and H. F. Noller.** 1980. Complete nucleotide sequences of a 23S ribosomal RNA gene from *Escherichia coli*. Proc. Natl. Acad. Sci. USA **77**:201–204.
9. **Brosius, J., T. J. Dull, D. D. Sleeter, and H. F. Noller.** 1981. Gene organization and primary structure of a ribosomal RNA operon from *Escherichia coli*. J. Mol. Biol. **148**:107–124.
10. **Brownlee, G. G., F. Sanger, and B. G. Barrell.** 1967. Nucleotide sequence of 5S ribosomal RNA from *Escherichia coli*. Nature (London) **215**:735–736.
11. **Cantor, C. R.** 1980. Physical and chemical techniques for the study of RNA structure, p. 23–50. *In* G. Chambliss, G. R. Craven, J. Davies, K. Davis, L. Kahan, and M. Nomura (ed.), Ribosomes: structure, function and genetics. University Park Press, Baltimore.
12. **Carbon, P., C. Ehresmann, B. Ehresmann, and J. P. Ebel.** 1979. The complete nucleotide sequence of the ribosomal 16S RNA from *Escherichia coli*. Experimental details and cistron heterogeneities. Eur. J. Biochem. **100**:399–410.
13. **Chambliss, G., G. R. Craven, J. Davies, K. Davis, L. Kahan, and M. Nomura (ed.).** 1980. Ribosomes: structure, function and genetics. University Park Press, Baltimore.
14. **Clark, M. W., M. Hammons, J. A. Langer, and J. A. Lake.** 1979. Helical arrays of *Escherichia coli* small ribosomal subunits produced *in vitro*. J. Mol. Biol. **135**:507–512.
15. **Clark, M. W., K. Leonard, and J. A. Lake.** 1982. Ribosomal

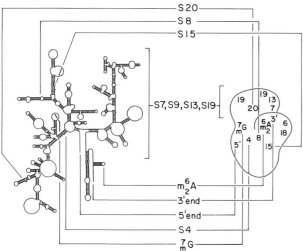

FIG. 13. Location of some structural features of 16S RNA in the electron microscope model for the 30S ribosomal subunit (see Fig. 9). Positions of r-proteins and RNA sites are based on immunoelectron microscopy (see Fig. 9) and neutron diffraction (see Fig. 12). From reference 158, with permission.

crystalline arrays of large subunits from *Escherichia coli*. Science **216**:999–1001.

16. **Cole, M. D., T. Koller, W. A. Strycharz, and M. Nomura.** 1978. Electron microscopic determination of the binding sites of ribosomal proteins S4 and S8 on 16S RNA. Proc. Natl. Acad. Sci. USA **75**:270–274.

17. **Craven, G. R., B. Rigby, and L. M. Changchien.** 1974. Chemical approaches to the analysis of ribosome architecture, p. 559–572. *In* M. Nomura, A. Tissieres, and P. Lengyel (ed.), Ribosomes. Cold Spring Harbor Laboratory, Cold Spring Harbor, N.Y.

18. **Crichton, R. R., D. M. Engelman, J. Haas, M. H. J. Koch, P. B. Moore, R. Parfait, and H. B. Stuhrmann.** 1977. Contrast variation study of specifically deuterated *Escherichia coli* ribosomal subunits. Proc. Natl. Acad. Sci. USA **74**:5547–5550.

19. **Dabbs, E. R.** 1978. Kasugamycin-dependent mutants of *Escherichia coli*. J. Bacteriol. **136**:994–1001.

20. **Dabbs, E. R.** 1979. Selections for *Escherichia coli* mutants with proteins missing from the ribosome. J. Bacteriol. **140**:734–737.

21. **Dabbs, E. R.** 1980. The ribosomal components responsible for kasugamycin-dependence, and its suppression, in a mutant of *Escherichia coli*. Mol. Gen. Genet. **177**:271–276.

22. **Damaschun, G., J. J. Müller, and H. Bielka.** 1979. Scattering studies of ribosomes and ribosomal components. Methods Enzymol. **59**:706–750.

23. **Delihas, N., and J. Anderson.** 1982. Generalized structures of the 5S ribosomal RNAs. Nucleic Acids Res. **10**:7323–7344.

24. **de Narvaez, C. C., and H. W. Schaup.** 1979. *In vivo* transcriptionally coupled assembly of *Escherichia coli* ribosomal subunits. J. Mol. Biol. **134**:1–22.

25. **de Wachter, R., M. W. Chen, and A. Vandenberghe.** 1982. Conservation of secondary structure in 5S ribosomal RNA: a uniform model for eukaryotic, eubacterial, archaebacterial and organelle sequences is energetically favorable. Biochemie **64**:311–329.

26. **Dickerson, R. E., H. R. Drew, B. N. Conner, R. M. Wing, A. V. Fratini, and M. L. Kopka.** 1982. The anatomy of A-, B- and Z-DNA. Science **216**:475–485.

27. **Dohme, F., and K. H. Nierhaus.** 1976. Total reconstitution and assembly of 50S subunits from *Escherichia coli* ribosomes *in vitro*. J. Mol. Biol. **107**:585–599.

28. **Douthwaite, S., A. Christensen, and R. A. Garrett.** 1982. Binding sites of ribosomal proteins on prokaryotic 5S ribonucleic acids: a study with ribonucleases. Biochemistry **21**:2313–2320.

29. **Douthwaite, S., R. A. Garrett, R. Wagner, and J. Feunteun.** 1979. A ribonuclease-resistant region of 5S RNA and its relation to the RNA binding sites of proteins L18 and L25. Nucleic Acids Res. **6**:2543–2470.

30. **Duncan, M. J., and L. Gorini.** 1975. A ribonucleoprotein precursor of both 30S and 50S ribosomal subunits of *Escherichia coli*. Proc. Natl. Acad. Sci. USA **72**:1533–1537.

31. **Ebel, J. P., C. Branlant, P. Carbon, B. Ehresmann, C. Ehresmann, A. Krol, and P. Stiegler.** 1983. Structure of ribosomal RNA, p. 177–183. *In* C. Helene (ed.), Structure, dynamics, interactions and evolution of biological macromolecules. D. Reidel, Boston.

32. **Ehresmann, C., P. Stiegler, P. Carbon, E. Ungewickell, and R. A. Garrett.** 1981. The topography of the 5′ end of 16S RNA in the presence and absence of ribosomal proteins S4 and S20. Eur. J. Biochem. **103**:439–446.

33. **Erdmann, V.** 1976. Structure and function of 5S and 5.8S RNA. Prog. Nucleic Acids Res. Mol. Biol. **18**:45–90.

34. **Fahnestock, S. V., W. Held, and W. Nomura.** 1972. The assembly of bacterial ribosomes, p. 179–217. *In* R. Markham, J. B. Bancroft, D. R. Davies, D. A. Hopwood, and R. W. Horne (ed.), First John Innes Symposium on Regulation of Subcellular Structures. Elsevier/North-Holland Publishing Co., Amsterdam.

35. **Fellner, P., and F. Sanger.** 1968. Sequence analysis of specific areas of the 16S and 23S ribosomal RNAs. Nature (London) **219**:236–238.

36. **Fox, G. E., and C. R. Woese.** 1975. 5S RNA secondary structure. Nature (London) **256**:505–507.

37. **Friedman, D. I., A. T. Schauer, M. R. Baumann, L. S. Baron, and S. L. Adhya.** 1981. Evidence that ribosomal protein S10 participates in control of transcription termination. Proc. Natl. Acad. Sci. USA **78**:1115–1119.

38. **Garrett, R. A., S. Douthwaite, and H. F. Noller.** 1981. Structure and role of 5S RNA-protein complexes in protein biosynthesis. Trends Biochem. Sci. **6**:137–139.

39. **Garrett, R. A., and H. F. Noller.** 1979. Structure of complexes of 5S RNA with ribosomal proteins L5, L18 and L25 from *Escherichia coli*: identification of kethoxal-reactive sites on the 5S RNA. J. Mol. Biol. **132**:637–648.

40. **Garrett, R. A., E. Ungewickell, V. Newberry, J. Hunter, and R.**

Wagner. 1977. An RNA core in the 30S ribosomal subunit of *Escherichia coli* and its structural and functional significance. Cell Biol. Int. Rep. **1**:487–502.

41. **Garrett, R. A., B. Vester, H. Leffers, P. M. Sørensen, J. Kjems, S. O. Oleson, A. Christensen, J. Christiansen, and S. Douthwaite.** 1984. Mechanisms of protein-RNA recognition and assembly in ribosomes, p. 331–352. *In* B. F. C. Clark and H. U. Petersen (ed.), Gene expression. Munksgaard, Copenhagen.

42. **Glotz, C., and R. Brimacombe.** 1980. An experimentally derived model for the secondary structure of the 16S ribosomal RNA from *Escherichia coli*. Nucleic Acids Res. **8**:2377–2395.

43. **Gourse, R. L., M. J. R. Stark, and A. Dahlberg.** 1982. Site directed mutagenesis of ribosomal RNA. Construction and characterization of deletion mutants. J. Mol. Biol. **159**:397–416.

44. **Gourse, R. L., Y. Takebe, R. Sharrock, and M. Nomura.** 1985. Feedback regulation of rRNA and tRNA synthesis and accumulation of free ribosomes after conditional expression of rRNA genes. Proc. Natl. Acad. Sci. USA **82**:1069–1073.

45. **Gregory, R. J., M. L. Zeller, D. L. Thurlow, R. L. Gourse, M. J. R. Stark, A. E. Dahlberg, and R. A. Zimmermann.** 1984. Interaction of ribosomal proteins S6, S8, S15, and S18 with the central domain of 16S ribosomal RNA from *Escherichia coli*. J. Mol. Biol. **178**:287–302.

46. **Gutell, R. R., B. Weiser, C. R. Woese, and H. F. Noller.** 1985. Comparative anatomy of 16S-like ribosomal RNA. Prog. Nucleic Acid Res. Mol. Biol. **32**:155–216.

47. **Hall, K., P. Cruz, I. Tinoco, T. M. Jovin, and J. H. van de Sande.** 1984. Z-RNA—a left-handed RNA double helix. Nature (London) **311**:584–586.

48. **Hardy, S. J. S.** 1975. The stoichiometry of the ribosomal proteins of *Escherichia coli*. Mol. Gen. Genet. **140**:253–274.

49. **Hayward, A. M.** 1971. Cellular site of *Escherichia coli* ribosomal RNA synthesis. Proc. Natl. Acad. Sci. USA **68**:435–439.

50. **Held, W. A., B. Ballou, S. Mizushima, and M. Nomura.** 1974. Assembly mapping of 30S ribosomal proteins from *Escherichia coli*. J. Biol. Chem. **249**:3103–3111.

51. **Held, W. A., W. R. Gette, and M. Nomura.** 1974. Role of 16S ribosomal ribonucleic acid and the 30S ribosomal protein S12 in the initiation of natural messenger ribonucleic acid translation. Biochemistry **13**:2115–2122.

52. **Held, W. A., S. Mizushima, and M. Nomura.** 1973. Reconstitution of *Escherichia coli* 30S ribosomal subunits from purified molecular components. J. Biol. Chem. **248**:5720–5730.

53. **Held, W. A., and M. Nomura.** 1973. Rate-determining step in the reconstitution of *Escherichia coli* 30S ribosomal subunits. Biochemistry **12**:3273–3281.

54. **Held, W. A., and M. Nomura.** 1975. *Escherichia coli* 30S ribosomal proteins uniquely required for assembly. J. Biol. Chem. **250**:3179–3184.

55. **Hochkeppel, H. K., E. Spicer, and G. R. Craven.** 1976. A method of preparing *Escherichia coli* 16S RNA possessing previously unobserved 30S ribosomal protein binding sites. J. Mol. Biol. **101**:155–170.

56. **Homann, H. H., and K. H. Nierhaus.** 1971. Protein composition of biosynthetic precursors and artificial subparticles from ribosomal subunits in *Escherichia coli* K12. Eur. J. Biochem. **20**:249–257.

57. **Huang, K. H., R. H. Fairclough, and C. R. Cantor.** 1975. Singlet energy transfer studies of the arrangement of proteins in the 30S *Escherichia coli* ribosome. J. Mol. Biol. **97**:443–470.

58. **Isono, K., A. G. Cumberlidge, S. Isono, and Y. Hirota.** 1978. Further temperature-sensitive mutants of *Escherichia coli* with altered ribosomal proteins. Mol. Gen. Genet. **152**:239–243.

59. **Ito, K., D. P. Ceretti, H. Nashimoto, and M. Nomura.** 1984. Characterization of an amber mutation in the structural gene for ribosomal protein L15, which impairs the expression of the protein transport gene, *secY* in *Escherichia coli*. EMBO J. **3**:2319–2324.

60. **Jinks-Robertson, S., R. L. Gourse, and M. Nomura.** 1983. Expression of rRNA and tRNA genes in *Escherichia coli*. Evidence for feedback regulation by products of rRNA operons. Cell **33**:865–876.

61. **Kaltschmidt, E., and H. G. Wittmann.** 1970. Number of proteins in small and large ribosomal subunits of *Escherichia coli* as determined by two-dimensional electrophoresis. Proc. Natl. Acad. Sci. USA **67**:1276–1282.

62. **Kim, S. H., F. L. Suddath, G. J. Quigley, A. McPherson, J. L. Sussman, A. Wang, N. C. Seeman, and A. Rich.** 1974. Three-dimensional tertiary structure of yeast phenylalanine transfer RNA. Science **185**:435–440.

63. **Kime, M. J., and P. B. Moore.** 1983. Nuclear Overhauser experiments at 500 MHz on the downfield proton spectrum of a

ribonuclease-resistant fragment of 5S ribonucleic acid. Biochemistry 22:2615–2622.

64. **Klein, B. K., T. C. King, and D. Schlessinger.** 1983. Structure of partially denatured *Escherichia coli* 23S ribosomal RNA determined by electron microscopy. J. Mol. Biol. **168:**809–830.

65. **Knauer, V., R. Hegerl, and W. Hoppe.** 1983. Three-dimensional construction and averaging of 30S ribosomal subunits of *Escherichia coli* from electron micrographs. J. Mol. Biol. **163:**409–430.

66. **Kop, J., A. M. Kopylov, R. Siegel, R. Gupta, C. R. Woese, and H. F. Noller.** 1984. Probing the structure of 16S ribosomal RNA from *Bacillus brevis.* J. Biol. Chem. **259:**15287–15293.

67. **Korn, A. P., P. Spitnik-Elson, and D. Elson.** 1982. The topology of the 30S ribosomal subunit and a proposal for the surface distribution of its RNA by dark field electron microscopy. J. Biol. Chem. **257:**7155–7160.

68. **Lake, J. A.** 1980. Ribosome structure and functional sites, p. 207–236. *In* G. Chambliss, G. R. Craven, J. Davies, K. Davis, L. Kahan, and M. Nomura (ed.), Ribosomes: structure, function and genetics. University Park Press, Baltimore.

69. **Lake, J. A.** 1982. Ribosomal subunit orientations determined in the monomeric ribosome by single- and by double-labeling immune electron microscopy. J. Mol. Biol. **161:**89–106.

70. **Langridge, R.** 1963. Ribosomes: a common structural feature. Science **140:**1000.

71. **Laughrea, M., and P. B. Moore.** 1978. On the relationship between the binding of ribosomal protein S1 to the 30S subunit of *Escherichia coli* and 3′ terminus of 16S RNA. J. Mol. Biol. **121:**411–430.

72. **Leijonmarck, M., S. Ericksson, and A. Liljas.** 1980. Crystal structure of a ribosomal component at 2.6A resolution. Nature (London) **286:**824–826.

73. **Leontis, N. B., and P. B. Moore.** 1984. A small angle x-ray scattering study of a fragment derived from *E. coli* 5S RNA. Nucleic Acids Res. **12:**2193–2203.

74. **Liljas, A.** 1982. Structural studies of ribosomes. Prog. Biophys. Mol. Biol. **40:**161–228.

75. **Liljas, A., S. Eriksson, D. Donner, and C. G. Kurland.** 1978. Isolation and crystallization of stable domains of the protein L7/L12 from *Escherichia coli* ribosomes. FEBS Lett. **88:**300–304.

76. **Lotti, M., E. R. Dabbs, R. Hasenbank, M. Stöffler-Meilicke, and G. Stöffler.** 1983. Characterization of a mutant from *Escherichia coli* lacking protein L15 and localization of protein L15 by immuno-electron microscopy. Mol. Gen. Genet. **192:**295–300.

77. **Mackay, R. M., D. F. Spencer, M. N. Schnare, W. F. Doolittle, and M. W. Gray.** 1982. Comparative sequence analysis as an approach to evaluating structure, function and evolution of 5S and 5.8S ribosomal RNAs. Can. J. Biochem. **60:**480–489.

78. **Mankin, A. S., E. A. Skripkin, N. V. Chichkova, A. M. Kopylov, and A. A. Bogdanov.** 1981. An enzymatic approach for localization of oligodeoxyribonucleotide binding sites on RNA. Application to studying rRNA topography. FEBS Lett. **131:**253–256.

79. **Mizushima, S., and M. Nomura.** 1970. Assembly mapping of 30S ribosomal proteins from *E. coli*. Nature (London) **226:**1214–1218.

79a.**Moazed, D., S. Stern, and H. Noller.** 1986. Rapid chemical probing of conformation in 16S ribosomal RNA and 30S ribosomal subunits using primer extension. J. Mol. Biol. **187:**399–416.

80. **Möller, W.** 1974. The ribosomal components involved in EF-G and EF-Tu-dependent GTP hydrolysis, p. 711–732. *In* M. Nomura, A. Tissieres, and P. Lengyel (ed.), Ribosomes. Cold Spring Harbor Laboratory, Cold Spring Harbor, N.Y.

81. **Moore, P. B.** 1980. Scattering studies of the three-dimensional organization of the *E. coli* ribosome, p. 111–134. *In* G. Chambliss, G. R. Craven, J. Davies, K. Davis, L. Kahan, and M. Nomura (ed.), Ribosomes: structure, function and genetics. University Park Press, Baltimore.

82. **Müller, R., R. A. Garrett, and H. F. Noller.** 1979. The structure of the RNA binding site of ribosomal proteins S8 and S15. J. Biol. Chem. **254:**3873–3878.

83. **Nashimoto, H., W. Held, E. Kaltschmidt, and M. Nomura.** 1971. Structure and functions of bacterial ribosomes. XII. Accumulation of 21S particles by some cold-sensitive mutants of *Escherichia coli*. J. Mol. Biol. **62:**121–138.

84. **Nashimoto, H., and M. Nomura.** 1970. Structure and function of bacterial ribosomes. XI. Dependence of 50S ribosomal assembly on simultaneous assembly of 30S subunits. Proc. Natl. Acad. Sci. USA **67:**1440–1447.

85. **Nazar, R. N.** 1982. The eukaryotic 5.8 and 5S ribosomal RNAs and related rDNAs. Cell Nucleus **10:**1–28.

86. **Nierhaus, K. H.** 1980. Analysis of the assembly and function of the 50S subunit from *Escherichia coli* ribosomes by reconstitu-

tion, p. 267–294. *In* G. Chambliss, G. R. Craven, J. Davies, K. Davis, L. Kahan, and M. Nomura (ed.), Ribosomes: structure, function and genetics. University Park Press, Baltimore.

87. **Nierhaus, K. H.** 1982. Structure, assembly and function of ribosomes. Curr. Top. Microbiol. Immunol. **97:**81–155.

88. **Nierhaus, K. H., K. Bordasch, and H. Homann.** 1973. Ribosomal proteins. XLIII. *In vivo* assembly of *Escherichia coli* ribosomal proteins. J. Mol. Biol. **74:**587–597.

89. **Nierhaus, K. H., and F. Dohme.** 1974. Total reconstitution of functionally active 50S ribosomal subunits from *Escherichia coli*. Proc. Natl. Acad. Sci. USA **71:**4713–4717.

90. **Nishikawa, K., and S. Takemura.** 1974. Structure and function of 5S ribosomal ribonucleic acid from *Torulopsis utilis*. J. Biochem. **76:**935–947.

91. **Noller, H. F.** 1980. Structure and topography of ribosomal RNA, p. 3–22. *In* G. Chambliss, G. R. Craven, J. Davies, K. Davis, L. Kahan, and M. Nomura (ed.), Ribosomes: structure, function and genetics. University Park Press, Baltimore.

92. **Noller, H. F.** 1984. Structure of ribosomal RNA. Annu. Rev. Biochem. **53:**119–162.

93. **Noller, H. F., and R. A. Garrett.** 1979. Structure of 5S ribosomal RNA from *Escherichia coli*: identification of kethoxal-reactive sites in the A and B conformations. J. Mol. Biol. **132:**621–636.

94. **Noller, H. F., J. Kop, V. Wheaton, J. Brosius, R. R. Gutell, A. M. Kopylov, F. Dohme, W. Herr, D. A. Stahl, R. Gupta, and C. R. Woese.** 1981. Secondary structure model for 23s ribosomal RNA. Nucleic Acids Res. **9:**6167–6189.

95. **Noller, H. F., and J. A. Lake.** 1984. Ribosome structure and function: localization of rRNA, p. 217–297. *In* E. E. Bittar (ed.), Membrane structure and function, vol. 6. John Wiley and Sons, Inc., New York.

96. **Noller, H. F., and C. R. Woese.** 1981. Secondary structure of 16S ribosomal RNA. Science **212:**403–411.

97. **Nomura, M., and V. A. Erdmann.** 1970. Reconstitution of 50S ribosomal subunits from dissociated molecular components. Nature (London) **228:**744–748.

98. **Nomura, M., and W. A. Held.** 1974. Reconstitution of ribosomes: studies of ribosome structure, function and assembly, p. 193–223. *In* M. Nomura, A. Tissieres, and P. Lengyel (ed.), Ribosomes. Cold Spring Harbor Laboratory, Cold Spring Harbor, N.Y.

99. **Nomura, M., S. Mizushima, M. Ozaki, P. Traub, and C. V. Lowry.** 1969. Structure and function of ribosomes and their molecular components. Cold Spring Harbor Symp. Quant. Biol. **34:**49–61.

100. **Nomura, M., A. Tissieres, and P. Lengyel (ed.).** 1974. Ribosomes. Cold Spring Harbor Laboratory, Cold Spring Harbor, N.Y.

101. **Nomura, M., P. Traub, C. Guthrie, and H. Nashimoto.** 1969. The assembly of ribosomes. J. Cell. Physiol. **74:**241–252.

102. **Nowotny, V., and K. H. Nierhaus.** 1980. Protein L20 from the large subunit of *Escherichia coli* ribosomes is an assembly protein. J. Mol. Biol. **137:**391–399.

103. **Nowotny, V., and K. H. Nierhaus.** 1982. Initiation proteins for the assembly of the 50S subunit from *Escherichia coli* ribosomes. Proc. Natl. Acad. Sci. USA **79:**7238–7242.

104. **Nowotny, V., H. J. Rheinberger, K. H. Nierhaus, B. Tesche, and R. Amils.** 1980. Preparation procedures of the proteins and RNA influence the total reconstitution of 50S subunits from *E. coli* ribosomes. Nucleic Acids Res. **8:**989–998.

105. **Odom, O. W., D. J. Robbins, D. Dottavia-Martin, G. Kramer, and B. Hardesty.** 1980. Distances between 3′ ends of ribosomal ribonucleic acids reassembled into *Escherichia coli* ribosomes. Biochemistry **19:**5947–5954.

106. **Oettl, H., R. Hegerl, and W. Hoppe.** 1983. Three-dimensional reconstruction and averaging of 50S ribosomal subunits of *Escherichia coli* from electron micrographs. J. Mol. Biol. **163:**431–450.

107. **Pabo, C. O., and R. T. Sauer.** 1984. Protein-DNA recognition. Annu. Rev. Biochem. **53:**293–322.

108. **Pardo, D., C. Vola, and R. Rosset.** 1979. Assembly of ribosomal subunits affected in a ribosomal mutant of *E. coli* having an altered L22 protein. Mol. Gen. Genet. **174:**53–58.

109. **Peattie, D. A., S. Douthwaite, R. A. Garrett, and H. F. Noller.** 1981. A "bulged" double helix in a RNA-protein contact site. Proc. Natl. Acad. Sci. USA **78:**7331–7335.

110. **Peattie, D. A., and W. Gilbert.** 1980. Chemical probes for higher-order structure in RNA. Proc. Natl. Acad. Sci. USA **77:**4679–4682.

111. **Petersson, I., S. J. S. Hardy, and A. Liljas.** 1976. The ribosomal protein L8 is a complex of L7/L12 and L10. FEBS Lett. **64:**135–138.

112. **Politz, S. M., and D. G. Glitz.** 1977. Ribosome structure: localization of N6,N6-dimethyladenosine by electron microscopy of a ribosome-antibody complex. Proc. Natl. Acad. Sci. USA **74:**1468–1472.

113. **Prince, J. B., R. R. Gutell, and R. A. Garrett.** 1983. A consensus model of the *Escherichia coli* ribosome. Trends Biochem. Sci. **8:**359–363.

114. **Ramakrishnan, V., M. Capel, M. Kjeldgaard, D. M. Engelman, and P. B. Moore.** 1984. Positions of proteins S14, S18 and S20 in the 30S ribosomal subunit of *Escherichia coli.* J. Mol. Biol. **174:**265–284.

115. **Röhl, R., and K. H. Nierhaus.** 1982. Assembly map of the large subunit (50S) of *Escherichia coli* ribosomes. Proc. Natl. Acad. Sci. USA **79:**729–733.

116. **Röhl, R., H. E. Roth, and K. H. Nierhaus.** 1982. Inter-protein dependences during assembly of the 50S subunit from *Escherichia coli* ribosomes. Hoppe-Seyler's Z. Physiol. Chem. **363:**143–157.

117. **Ross, A., and R. Brimacombe.** 1979. Experimental determination of interacting sequences in ribosomal RNA. Nature (London) **281:**271–276.

118. **Roth, H. E., and K. H. Nierhaus.** 1980. Assembly map of the 50S subunit from *Escherichia coli* ribosomes, covering the proteins present in the first reconstitution intermediate particle. Eur. J. Biochem. **103:**95–98.

119. **Schlessinger, D.** 1974. Ribosome formation in *Escherichia coli,* p. 393–416. *In* M. Nomura, A. Tissieres, and P. Lengyel (ed.), Ribosomes. Cold Spring Harbor Laboratory, Cold Spring Harbor, N.Y.

120. **Schulze, H., and K. H. Nierhaus.** 1982. Minimal set of ribosomal components for reconstitution of the peptidyl-transferase activity. EMBO J. **5:**609–613.

121. **Serdyuk, I. N., A. K. Grenader, and V. E. Koteliansky.** 1977. Study of 30-S ribosomal subparticle protein-deficient ribonucleoprotein derivatives by x-ray diffusion-scattering. Eur. J. Biochem. **79:**504–508.

122. **Shatsky, I. N., A. G. Evstafieva, T. F. Bystrova, A. A. Bogdanov, and V. D. Vasiliev.** 1980. Topography of RNA in the ribosome: localization of the 3'-end of the 23S RNA on the surface of the 50S ribosomal subunit by immune electron microscopy. FEBS Lett. **122:**251–255.

123. **Sieber, G., and K. H. Nierhaus.** 1978. Kinetic and thermodynamic parameters of the assembly *in vitro* of the large subunit form *Escherichia coli* ribosomes. Biochemistry **17:**3505–3511.

124. **Spierer, P., A. A. Bogdanov, and R. A. Zimmermann.** 1978. Parameters for the interaction of ribosomal proteins L5, L18 and L25 with 5S RNA from *Escherichia coli.* Biochemistry **17:**5394–5398.

125. **Spierer, P., C. Wang, T. L. Marsh, and R. A. Zimmermann.** 1979. Cooperative interactions among protein and RNA components of the 50S ribosomal subunit of *Escherichia coli.* Nucleic Acids Res. **6:**1669–1682.

126. **Spierer, P., and R. A. Zimmermann.** 1976. RNA-protein interactions in the ribosome. VIII. Cooperative interactions in the 50S subunit of *Escherichia coli.* J. Mol. Biol. **103:**647–653.

127. **Spillman, S., and K. H. Nierhaus.** 1978. The ribosomal protein L24 of *Escherichia coli* is an assembly protein. J. Biol. Chem. **253:**7047–7050.

128. **Stark, M. J. R., R. J. Gregory, R. L. Gourse, D. L. Thurlow, C. Zwieb, R. A. Zimmermann, and A. E. Dahlberg.** 1984. Effects of site-directed mutations in the central domain of 16S ribosomal RNA upon ribosomal protein binding, RNA processing and 30S subunit assembly. J. Mol. Biol. **178:**303–322.

129. **Stöffler, G., R. Bald, B. Kastner, R. Lührmann, M. Stöffler-Meilecke, G. Tischendorf, and B. Tesche.** 1980. Structural organization of the *Escherichia coli* ribosome and localization of functional domains, p. 171–206. *In* G. Chambliss, G. R. Craven, J. Davies, K. Davis, L. Kahan, and M. Nomura (ed.), Ribosomes: structure, function and genetics. University Park Press, Baltimore.

130. **Studnicka, G. M., F. A. Eiserling, and J. A. Lake.** 1981. A unique secondary folding pattern for 5S RNA corresponds to the lowest energy homologous secondary structure in 17 different prokaryotes. Nucleic Acids Res. **9:**1885–1904.

131. **Stuhrmann, H. B., M. H. J. Koch, R. Parfait, J. Haas, K. Ibel, and R. R. Crichton.** 1977. Shape of the 50S subunit of *Escherichia coli* ribosomes. Proc. Natl. Acad. Sci. USA **74:**2316–2320.

132. **Subramanian, A. R.** 1975. Copies of proteins L7 and L12 and heterogeneity of the large subunit of *Escherichia coli* ribosome. J. Mol. Biol. **95:**1–8.

133. **Sypherd, P. S.** 1971. Ribosomal proteins and the conformation of ribonucleic acid. J. Mol. Biol. **56:**311–318.

134. **Sypherd, P., R. Bryant, K. Dimmitt, and T. Fujisawa.** 1974. Genetic control of ribosome assembly. J. Supramolec. Struct. **2:**166–177.

135. **Tal, M., A. Silberstein, and K. Myner.** 1977. *In vivo* reassembly of 30S ribosomal subunits following their specific destruction by thermal shock. Biochim. Biophys. Acta **479:**479–496.

136. **Tam, M. F., and W. E. Hill.** 1981. Physical characteristics of the reconstitution intermediates (RI_{30} and RI_{30*}) from the 30S ribosomal subunit of *Escherichia coli.* Biochemistry **30:**6480–6484.

137. **Terhorst, C. P., W. Möller, R. Laursen, and B. Wittmann-Liebold.** 1973. The primary structure of an acidic protein from 50S ribosomes of *Escherichia coli* which is involved in GTP hydrolysis dependent on elongation factors G and T. Eur. J. Biochem. **34:**138–152.

138. **Thammana, P., C. R. Cantor, P. L. Wollenzien, and J. E. Hearst.** 1979. Crosslinking studies on the organization of the 16S ribosomal RNA within 30S *Escherichia coli* subunit. J. Mol. Biol. **135:**271–283.

139. **Thompson, J. F., and J. E. Hearst.** 1983. Structure and function of *E. coli* 16S RNA elucidated by psoralen crosslinking. Cell **32:**1355–1365.

140. **Thompson, J. F., and J. E. Hearst.** 1983. Structure-function relations in *E. coli* 16S RNA. Cell **33:**19–24.

141. **Traub, P., and M. Nomura.** 1968. Structure and function of *E. coli* ribosomes. V. Reconstitution of functionally active 30S ribosomal particles from RNA and protein. Proc. Natl. Acad. Sci. USA **59:**777–784.

142. **Traub, P., and M. Nomura.** 1969. Structure and function of *Escherichia coli* ribosomes. VI. Mechanism of assembly of 30S ribosomes studied *in vitro.* J. Mol. Biol. **40:**391–413.

143. **Traut, R. R., J. M. Lambert, G. Boileau, and J. W. Kenny.** 1980. Protein topography of *Escherichia coli* ribosomal subunits are inferred from protein crosslinking, p. 89–110. *In* G. Chambliss, G. R. Craven, J. Davies, K. Davis, L. Kahan, and M. Nomura (ed.), Ribosomes: structure, function and genetics. University Park Press, Baltimore.

144. **Traut, R. R., P. B. Moore, H. Delius, H. Noller, and A. Tissieres.** 1967. Ribosomal proteins of *E. coli.* I. Demonstration of different primary structures. Proc. Natl. Acad. Sci. USA **57:**1294–1301.

145. **Trempe, M. R., K. Ohgi, and D. G. Glitz.** 1982. Ribosome structure. Localization of 7-methylguanosine in the small subunits of *Escherichia coli* and chloroplast ribosomes by immunoelectron microscopy. J. Biol. Chem. **257:**9822–9829.

146. **Turner, S., J. F. Thompson, J. E. Hearst, and H. F. Noller.** 1982. Identification of a site of psoralen crosslinking in *E. coli* 16S ribosomal RNA. Nucleic Acids Res. **10:**2839–2849.

147. **Van Stolk, B. J., and H. F. Noller.** 1984. Chemical probing of conformation in large RNA molecules. Analysis of 16S ribosomal RNA using diethylpyrocarbonate. J. Mol. Biol. **180:**151–177.

148. **Vassilenko, S. K., P. Carbon, J. P. Ebel, and C. Ehresmann.** 1981. Topography of 16S RNA in 30S subunits and 70S ribosomes. Accessibility to cobra venom ribonuclease. J. Mol. Biol. **152:**699–721.

149. **Waller, J. P.** 1964. Fractionation of the ribosomal protein from *Escherichia coli.* J. Mol. Biol. **10:**319–336.

150. **Wittmann, H. G.** 1974. Purification and identification of *Escherichia coli* ribosomal proteins, p. 93–114. *In* M. Nomura, A. Tissieres, and P. Lengyel (ed.), Ribosomes. Cold Spring Harbor Laboratory, Cold Spring Harbor, N.Y.

151. **Wittmann, H. G.** 1982. Components of bacterial ribosomes. Annu. Rev. Biochem. **51:**155–183.

152. **Wittmann, H. G.** 1983. Architecture of prokaryotic ribosomes. Annu. Rev. Biochem. **52:**35–65.

153. **Wittmann, H. G., J. A. Littlechild, and B. Wittmann-Liebold.** 1980. Structure of ribosomal proteins, p. 51–88. *In* G. Chambliss, G. R. Craven, J. Davies, K. Davis, L. Kahan, and M. Nomura (ed.), Ribosomes: structure, function and genetics. University Park Press, Baltimore.

154. **Wittmann, H. G., J. Müssig, J. Piefke, H. S. Gewitz, H. J. Rheinberger, and A. Yonath.** 1982. Crystallization of *Escherichia coli* ribosomes. FEBS Lett. **146:**217–220.

155. **Wittmann, H. G., G. Stöffler, D. Geyl, and A. Böck.** 1975. Alteration of ribosomal proteins in revertants of a valyl-tRNA synthetase mutant of *Escherichia coli.* Mol. Gen. Genet. **141:**317–329.

156. **Wittmann, H. G., G. Stöffler, I. Hindennach, C. G. Kurland, L. Randall-Hazelbauer, E. A. Birge, M. Nomura, E. Kaltschmidt, S. Mizushima, R. R. Traut, and T. A. Bickle.** 1971. Correlation of 30s ribosomal proteins of *Escherichia coli* isolated in different laboratories. Mol. Gen. Genet. **111:**327–333.

157. **Wittmann, H. G., and B. Wittmann-Liebold.** 1974. Chemical structure of bacterial ribosomal proteins, p. 115–140. *In* M. Nomura, A. Tissieres, and P. Lengyel (ed.), Ribosomes. Cold Spring Harbor Laboratory, Cold Spring Harbor, N.Y.

158. **Woese, C. R., R. Gutell, R. Gupta, and H. F. Noller.** 1983. Detailed analysis of the higher-order structure of 16S-like ribosomal ribonucleic acids. Microbiol. Rev. **47:**621–669.

159. **Wollenzien, P. L., and C. R. Cantor.** 1982. Marking the polarity of RNA molecules for electron microscopy by covalent attachment of a psoralen-DNA restriction fragment. J. Mol. Biol. **159:**151–166.

160. **Wollenzien, P. L., J. E. Hearst, P. Thammana, and C. R. Cantor.** 1979. Base pairing between distant regions of the *Escherichia coli* 16S rRNA in solution. J. Mol. Biol. **135:**255–269.

161. **Wower, I., and R. Brimacombe.** 1983. The localization of multiple sites on 16S RNA which are cross-linked to proteins S7 and S8 in *Escherichia coli* 30S ribosomal subunits by treatment with 2-iminothiolane. Nucleic Acids Res. **11:**1419–1437.

162. **Yuki, A., and R. Brimacombe.** 1975. Nucleotide sequences of *Escherichia coli* 16S RNA associated with ribosomal proteins S7, S9, S10, S14 and S19. Eur. J. Biochem. **56:**23–34.

163. **Zimmermann, R. A.** 1974. RNA-protein interactions in the ribosome, p. 225–270. *In* M. Nomura, A. Tissieres, and P. Lengyel (ed.), Ribosomes. Cold Spring Harbor Laboratory, Cold Spring Harbor, N.Y.

164. **Zimmermann, R. A.** 1980. Interactions among protein and RNA components of the ribosome, p. 135–170. *In* G. Chambliss, G. R. Craven, J. Davies, K. Davis, L. Kahan, and M. Nomura (ed.), Ribosomes: structure, function and genetics. University Park Press, Baltimore.

165. **Zimmermann, R. A., G. A. Mackie, A. Muto, R. A. Garrett, E. Ungewickell, C. Ehresmann, P. Stiegler, J. P. Ebel, and P. Fellner.** 1975. Location and characteristics of ribosomal protein binding sites in the 16S RNA of *Escherichia coli*. Nucleic Acids Res. **2:**279–302.

166. **Zimmermann, R. A., and K. Singh-Bergmann.** 1979. Binding sites for ribosomal proteins S8 and S15 in the 16S RNA of *Escherichia coli*. Biochim. Biophys. Acta **563:**422–431.

167. **Zwieb, C., and R. Brimacombe.** 1980. Localisation of a series of intra-RNA cross-links in 16S RNA induced by ultraviolet irradiation of *Escherichia coli* 30S ribosomal subunits. Nucleic Acids Res. **8:**2397–2411.

PART II. METABOLISM AND GENERAL PHYSIOLOGY

Section A. Class I Reactions: Generation of Precursor Metabolites and Energy

A1. Central Metabolic Routes: Metabolism of Glucose

11. Phosphotransferase System for Glucose and Other Sugars

P. W. POSTMA

Laboratory of Biochemistry, University of Amsterdam, 1018 TV Amsterdam, The Netherlands

INTRODUCTION

Sugars, amino acids, vitamins, and other nutrients are taken up by bacteria via specific transport systems localized in the cytoplasmic membrane. This chapter deals with a complex transport system, the phospho-enolpyruvate (PEP):carbohydrate phosphotransferase system (PTS), which is involved in the uptake and concomitant phosphorylation of a large number of carbohydrates. Apart from their role in transport, components of the PTS are also involved in chemotaxis and in the regulation of the synthesis and activity of many non-PTS transport systems.

Accumulation of carbohydrates by *Escherichia coli* and *Salmonella typhimurium* follows one of two types of pathways: the substrate either is accumulated in an unchanged form or is covalently modified during transport. A well-known example of the first class of substrates is lactose, which is accumulated via the lactose permease (M protein), making use of an electrochemical proton gradient as the driving force. The second class of substrates consists of a number of carbohydrates that are accumulated as the corresponding phosphate esters. The phosphoryl group is derived from PEP. The system responsible for the uptake and phosphorylation of this class of carbohydrates, the PEP:carbohydrate PTS, is the subject of this chapter.

The PTS, discovered in 1964 by Roseman and coworkers (64), was found in *E. coli* extracts as an activity involved in the phosphorylation of a number of carbohydrates at the expense of PEP. It soon became evident from genetic, transport, and growth studies that this enzymatic activity represented part of a bacterial carbohydrate transport system. Identification and purification of its components have resulted in the scheme shown in Fig. 1. Two phosphoproteins, enzyme I and HPr, common to all PTS carbohydrates, catalyze the sequential transfer of a phosphoryl group from PEP to the carbohydrate-specific, membrane-bound enzymes II and finally to the carbohydrate. In a number of cases another protein, enzyme III, is positioned between phospho-HPr and a particular enzyme II. To distinguish the various enzymes II and III, a three-letter superscript indicating which carbohydrate is the preferred substrate is assigned to each (see also Table 1). In the remainder of this chapter I will discuss the properties of the various components of the PTS, its transport activity, its genetics, and finally its role in the regulation of cellular metabolism. These subjects have been covered in a number of reviews (24, 103, 105, 106, 112, 121).

COMPONENTS OF THE PTS

Soluble Components Enzyme I and HPr

Enzyme I and HPr, the two general proteins of the PTS, are both phosphoproteins. The phosphoryl group is bound to a histidine residue in both cases, via the N-3 position of the imidazole ring in enzyme I and via

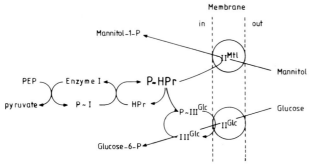

FIG. 1. The PTS. Enzyme I and HPr are the general PTS proteins. Of the many different enzymes II only two are shown. II^{Mtl} is specific for mannitol, and II^{Glc}, together with III^{Glc}, is specific for glucose. P~I, P~HPr, and P~III^{Glc} are the phosphorylated forms of enzyme I, HPr, and III^{Glc}, respectively.

the N-1 position in HPr. Both proteins have been purified from *E. coli* (6, 113, 148) and *S. typhimurium* (11, 151). Enzyme I consists of two identical polypeptides, the molecular weights of which have been estimated to be between 70,000 and 58,000 by using sodium dodecyl sulfate-polyacrylamide gel electrophoresis (SDS-PAGE). Sedimentation equilibrium analysis yields a molecular weight of 58,000 for the monomer (63). It is not yet clear whether one or two phosphoryl groups are incorporated per enzyme I dimer, which is the catalytically active form of the enzyme (81). Whereas Weigel et al. (149) reported that two phosphoryl groups are incorporated in an enzyme I dimer of *S. typhimurium*, Misset and Robillard (82) found only one group in *E. coli* enzyme I. An indication that the subunits may behave differently is the observation that incorporation of one molecule of 3-bromopyruvate per *E. coli* enzyme I dimer does not affect the activity at all, whereas incorporation of a second molecule is accompanied by inactivation (45). Phosphorylated enzyme I has been isolated and can transfer its phosphoryl group to purified HPr.

HPr, the second general protein of the PTS, has also been purified, and the amino acid sequence of the enzyme from *S. typhimurium* has been determined (150). It is a small protein, with a molecular weight of approximately 9,000. Since the *ptsH* gene of *E. coli*, specifying HPr, has been cloned and its nucleotide sequence has been determined (23), a comparison of the deduced amino acid sequence with that of HPr from *S. typhimurium* is possible. There were some significant differences in the carboxy-terminal part (positions 63–71), but a correction to the previously determined amino acid sequence has been published (109) and shows that the HPr's from *E. coli* and *S. typhimurium* are identical. Phosphorylated HPr has been isolated and the binding of the phosphoryl group has been studied in several ways, including by ^{1}H nuclear magnetic resonance studies (54). The results show that the enteric bacteria have very similar HPr molecules which are different than those from gram-positive organisms. However, a sequence of amino acids around the histidine of the active site shows considerable conservation in all HPr molecules studied.

From genetic studies (see below) it has become clear that there is another protein that can replace HPr in

the phosphoryl transfer from P~enzyme I to the various enzymes II and III (129, 130). This protein, called pseudo-HPr or FPr, is specified by a gene belonging to the *fru* operon. Since the *fru* operon is normally repressed and is induced only by growth in the presence of fructose, FPr is not present in cells under most growth conditions. In cells in which the *fru* operon is expressed constitutively (*fruR* strains), FPr is synthesized and can replace HPr (33). The molecular weight of FPr has been estimated to be 8,000 (146). FPr might resemble HPr (for a discussion of the fructose PTS as the ancestor of the other PTSs, see reference 124). In these studies it was also shown that HPr can replace FPr in the PEP-dependent phosphorylation of fructose via II^{Fru}/III^{Fru} (33); thus, HPr and FPr can catalyze the same reactions. A different conclusion was drawn by Kornberg (58). From the behavior of an *E. coli* mutant impaired in FPr activity, it was concluded that the phosphoryl group is transferred mainly via FPr when fructose is phosphorylated via II^{Fru}/III^{Fru}. It has not been excluded, however, that the mutation is in the *fruB* gene, which specifies III^{Fru} and the first gene of the *fru* operon. Such a mutation may affect FPr synthesis also, owing to a polar effect.

PEP and the phosphorylated proteins enzyme I and HPr are in equilibrium as measured during in vitro phosphorylation studies. The pathways can be reversed; i.e., pyruvate can be converted into PEP by phospho-HPr. The equilibrium constant of the reaction

$$PEP + \text{enzyme I} \rightleftharpoons \text{phospho-enzyme I} + \text{pyruvate}$$

is 1.5 ± 1. The equilibrium constant of the overall reaction

$$PEP + \text{enzyme I} \rightleftharpoons \text{phospho-enzyme I} + \text{pyruvate}$$
$$\text{phospho-enzyme I} + HPr \rightleftharpoons \text{phospho-HPr} + \text{enzyme I}$$
$$\overline{PEP + HPr \rightleftharpoons \text{phospho-HPr} + \text{pyruvate}}$$

is reported to be 11 ± 7.7 (149). These results suggest that the apparent standard free energy of both phosphoproteins is high and comparable to that of PEP.

Enzymes III

Phosphorylated HPr can react in two different ways, either directly with an enzyme II or first with an enzyme III positioned between HPr and its associated membrane-bound enzyme II (Fig. 1). As an example of the former class, Fig. 1 shows II^{Mtl}, the enzyme II specific for mannitol. The III^{Glc}/II^{Glc} system, specific for glucose, is an example of the latter class. The various enzymes III that have been identified are listed in Table 1.

The best-studied example of an enzyme III is III^{Glc}, which is involved in glucose and methyl α-glucoside uptake and phosphorylation via enzyme II^{Glc}. Enzyme III^{Glc} has been purified from *S. typhimurium* (138) and characterized in some detail. The apparent molecular weight of the III^{Glc} monomer is 20,000 as determined by SDS-PAGE. It accepts one phosphoryl group, linked to a histidine residue (His-91) at the N-3 position of the imidazole ring (78); ^{1}H nuclear magnetic resonance studies support this conclusion (26). The mono-

TABLE 1. Enzymes II and III of the PTS

Enzyme II/III	Genetic symbol[a]	Map position (min)[a,b]	Substrate(s)[c]	Reference(s)
II^{Nag}/?	nagE	16	*N-Acetylglucosamine*	152
II^{Glc}/III^{Glc}	ptsG/crr	24 (25)/52 (49)	*Glucose*, methyl α-glucoside, thioglucose, glucosamine, sorbose, mannose, 2-deoxyglucose	19, 140, 142
II^{Man}/III^{Man}	ptsM/ptsL	40	*Mannose, glucose*, 2-deoxyglucose, fructose, N-acetylglucosamine, methyl α-glucoside	19, 30, 50, 142
II^{Fru}/III^{Fru}	fruA/fruB	46	*Fructose*, glucose, sorbose, xylitol	31, 33, 140
II^{Gat}/?	gatA	47	*Galactitol*, glucitol	67, 68
II^{Gut}/III^{Gut}	gutA/gutB	58 (58)	*Glucitol*, galactitol, mannitol, fructose	67, 68
II^{Mtl}	mtlA	81 (78)	*Mannitol*, glucitol	67, 68
II^{Bgl}/?	bglC	83	*β-Glucosides*, glucose, methyl α-glucoside	135
II^{Scr}/III^{Glc}	scrA/crr	—	*Sucrose*	69
II^{Lac}/?	—	—	*Lactose*	41
II^{Sor}/?	—	—	*Sorbose*	141
II^{Dha}/?	—	—	*Dihydroxyacetone*	49

[a] The genetic symbols and map positions on the *E. coli* and *S. typhimurium* chromosomes are taken from Bachmann (7) and Sanderson and Roth (133), respectively. —, No genetic symbol available; no map position known yet.

[b] Positions in parentheses are for *S. typhimurium*; other positions are for *E. coli*.

[c] The underlined compounds are recognized by the enzyme II with high affinity. They are most likely the "natural" substrates. The other compounds listed are recognized with a much lower affinity.

mer can aggregate to form dimers, trimers, and hexamers (139), but whether this aggregation is a prerequisite for activity is not known. Certain *crr* mutations result in enzymes III^{Glc} with altered aggregation behavior (139), suggesting that *crr* is the structural gene for III^{Glc}. The N terminus of III^{Glc} is essential for its enzymatic activity. Both derivatization with fluorescein isothiocyanate and the cleavage of the first seven N-terminal amino acids decrease the enzymatic activity very strongly (46, 78); both altered forms can still accept the phosphoryl group from phospho-HPr, but they are unable to transfer it to II^{Glc}.

The *crr* genes from *S. typhimurium* (89) and *E. coli* (79) have been cloned, and the nucleotide sequence of the *crr* gene from *S. typhimurium* has been determined (89). Based on the observation of a remaining III^{Glc}-like activity in membranes of *crr* deletion strains of *S. typhimurium*, which is inhibitable by an antibody against the soluble III^{Glc}, there might be a gene specifying an enzyme that can partly replace III^{Glc} (139). In a later section I will discuss III^{Glc} more extensively since it also plays a central role in the regulation of non-PTS uptake systems.

A number of other enzymes III exist which are involved in mannose (30), fructose (33, 146), and glucitol (36) uptake and phosphorylation. These enzymes III have been identified also by incorporation of ^{32}P from PEP into polypeptides and autoradiography after SDS-PAGE (76, 146). In at least one case, that of mannitol, no enzyme III seems to be required for the phosphorylation, since the purified components enzyme I, HPr, and enzyme II^{Mtl} are sufficient for PEP-dependent mannitol phosphorylation (48).

Membrane-Bound Enzymes II

Table 1 lists the various enzymes II found in *E. coli* and *S. typhimurium*. Identification is based mainly on genetic and biochemical studies using intact cells. Most carbohydrate substrates can be recognized by more than one enzyme II, and some enzymes II can recognize a large variety of carbohydrates (Table 1). Except for II^{Mtl}, II^{Glc}, and II^{Man}, which have been purified and reconstituted with respect to in vitro phosphorylation (see below), knowledge about enzymes II is limited. Two of the enzymes II mentioned in Table 1, those for lactose and sorbose, are in fact only found in the closely related organism *Klebsiella aerogenes*. Enzyme II^{Sor} is expressed in *E. coli* from an R'*sor* plasmid. Enzyme II^{Scr}, specific for sucrose, is also expressed in *E. coli* from a plasmid; its origin is not clear. The last PTS substrate, dihydroxyacetone, is somewhat surprising. The closely related substrate glycerol is transported unchanged into the cell via a facilitator, and glycerol metabolism is in fact regulated via the PTS (see below).

II^{Mtl} is a polypeptide of molecular weight 60,000 (as measured by SDS-PAGE) that spans the cytoplasmic membrane. It probably exists as a dimer (116). Based on inactivation studies with membrane-impermeable reagents, antibodies, and proteases, it has been concluded that the bulk of the protein is oriented toward the cytoplasmic side (47). From the nucleotide sequence of the cloned *mtlA* gene, specifying II^{Mtl}, a similar conclusion has been drawn (65); i.e., a hydrophobic part and a large hydrophilic part have been identified. II^{Mtl} catalyzes the PEP-dependent phosphorylation of mannitol via a ping-pong mechanism, during which II^{Mtl} becomes phosphorylated (115).

II^{Glc} from *S. typhimurium* has also been purified (29). It is a polypeptide of 43,000 molecular weight, and evidence points again to phosphorylation of the protein. Studies employing ^{31}P nuclear magnetic resonance to determine the change in chirality of the phosphoryl group in the phosphoryl donor (PEP) relative to that in the final product (methyl α-glucoside 6-phosphate) (9), studies on the production of glucose 6-phosphate from glucose by membranes phosphorylated by PEP and PTS proteins (80), and direct labeling studies with [^{32}P]PEP (98) suggest that II^{Glc} is covalently phosphorylated, as are enzyme I, HPr, and III^{Glc}. Phosphorylated II^{Glc} from *S. typhimurium* has recently been isolated (28). The phosphoryl group from P~II^{Glc} can be transferred to glucose, and the presence of nonphosphorylated III^{Glc} accelerates this phosphoryl transfer. A complex of II^{Glc} and III^{Glc}, in a ratio of 1:2, can be precipitated with anti-II^{Glc} monoclonal antibodies. II^{Glc} is a dimer, like II^{Mtl} (28).

II^{Man} has been purified from *E. coli*. It is a 27-kilodalton transmembrane protein that requires III^{Man} for phosphorylation of mannose and 2-deoxyglucose by phosphorylated HPr. III^{Man}, a 35-kilodalton protein, is found both as a cytoplasmic protein and associated with the membrane (30). II^{Man} might also be involved in the injection of lambda DNA after absorption of the phage particles to the host cell, since *pel* mutants, defective in lambda DNA injection, have a defective II^{Man} (27). The *pel* gene might be identical to *ptsM*, which specifies II^{Man} (30, 95).

CELLULAR CONCENTRATIONS OF PTS COMPONENTS

Depending on growth conditions, the cellular concentration of the general PTS proteins enzyme I and HPr varies at most two- to threefold (111, 142). PTS carbohydrates generally give rise to the highest activities as measured by in vitro phosphorylation. These results have been confirmed by studies with antibodies specific for enzyme I and HPr (77). In contrast, the level of III^{Glc} is virtually constant under all growth conditions (138). The number of enzyme I and HPr molecules per cell has been estimated to be 10,000 to 30,000 and 100,000 to 300,000, respectively (76). A *Salmonella* cell contains approximately 20,000 molecules of III^{Glc} (138). The levels of the enzymes II are more variable than those of the general proteins of the PTS. Some enzymes II are inducible 10-fold or more, for example, those for the hexitols mannitol, glucitol, and galactitol (71). The activities of II^{Man} and II^{Glc}, on the other hand, vary only two- to threefold in *S. typhimurium*, as do the general PTS proteins, as measured by in vitro phosphorylation. These results are also dependent on the strain studied. For example, II^{Glc} is claimed to be an inducible enzyme in *E. coli*. A mutation in *umgC*, a gene closely linked to the *ptsG* gene which specifies II^{Glc}, results in constitutive expression of II^{Glc} (50). A more definitive answer awaits measurement with antibodies, since transport studies using intact cells to determine the level of the enzymes II suggest in some cases a larger induction than was determined from in vitro phosphorylation studies.

IN VITRO PHOSPHORYLATION

As discussed above, PTS activity can be assayed in vitro by measuring the transfer of the phosphoryl group from PEP to carbohydrates. In general, the formation of carbohydrate phosphate from carbohydrate is determined by separating substrate and product by anion-exchange chromatography (147). Alternatively, one can measure the formation of pyruvate from PEP by quantitatively converting pyruvate to lactate in an NADH-linked reaction catalyzed by lactate dehydrogenase. In this way, the overall phosphorylating activity of the PTS has been measured in toluene-treated cells (32) made permeable to small molecules such as PEP and carbohydrates, and in bacterial extracts. PTS activities measured in extracts are generally lower than in such treated cells, probably because of dilution of the proteins involved since they act as both enzyme and substrate. PTS activities in toluene-treated cells are close to the rates of carbohydrate uptake measured in intact cells (see below).

In bacterial extracts the orientation of the enzymes II with respect to both the soluble PTS components and the carbohydrate constitutes a problem that is recognized but not yet solved. In intact cells, the carbohydrate to be taken up and the soluble PTS enzymes face opposite sides of the membrane and its associated enzyme II, whereas during in vitro assays all components are added to the same side of the membrane, which was originally facing the inside of the cell since the orientation of the membrane inverts during preparation of bacterial extracts. It is not clear whether a carbohydrate has to cross the membrane before reacting with its enzyme II. This problem is related to the question of whether PTS carbohydrates already within the cell can be phosphorylated via the enzymes II; no clear-cut answer is yet possible. It has been suggested that *S. typhimurium* vesicles, oriented right side out (like intact cells), can take up and phosphorylate methyl α-glucoside via II^{Glc} only when provided with intravesicular enzyme I. Phosphorylation of methyl α-glucoside by these vesicles in the presence of extravesicularly added enzyme I, HPr, and III^{Glc} occurs with approximately 20% of the rate of uptake in these vesicles (10). Sonic disruption increases phosphorylation of methyl α-glucoside, which has been interpreted as an indication of inversion of the membrane orientation and increased accessibility of II^{Glc} to its substrate. II^{Man}, on the other hand, is able to take up and phosphorylate 2-deoxyglucose when the soluble PTS enzymes are present intravesicularly and to phosphorylate this carbohydrate when enzyme I and HPr are added at the outside. However, the evidence that II^{Glc} and II^{Man} behave differently is not very convincing since it relies on the observation that phosphorylation by right-side-out vesicles via II^{Glc} is stimulated by sonic disruption, whereas no stimulation, but rather inhibition, is seen in the case of II^{Man}. Since it cannot be excluded that II^{Man} is inactivated during the sonic oscillation procedure, it might well be that a similar stimulation of II^{Man} would in fact have been found. In conclusion, there is no convincing evidence that an enzyme II can phosphorylate its substrate at both sides of the membrane (see also the next section).

Another reaction catalyzed by the enzyme II has been described by Saier and co-workers (123). Either membranes containing an enzyme II or purified enzymes II are able to catalyze transphosphorylation; i.e., a PTS carbohydrate is phosphorylated by the corresponding sugar phosphate, as follows.

$$\text{Sugar(1)} + \text{sugar(2)-phosphate}$$
$$\rightleftharpoons \text{sugar(2)} + \text{sugar(1)-phosphate}$$

The reaction is favored by low sugar and high sugar phosphate concentrations. Its rate is very low relative to the rate of PEP-dependent carbohydrate phosphorylation. This difference might be due to the fact that this reaction is far out of equilibrium when a phosphorylated enzyme II intermediate has to be formed during the exchange of a phosphoryl group from a carbohydrate phosphate to another carbohydrate. In the case of II^{Glc}, kinetic evidence suggests that, indeed, $P{\sim}II^{Glc}$ is an obligatory intermediate during transphosphorylation (80). Whether this reaction occurs in intact cells is discussed in the next section.

TRANSPORT OF PTS CARBOHYDRATES

Central in the functioning of the PTS is the uptake and concomitant phosphorylation of carbohydrates. No free carbohydrate is found in the cell initially, and subsequent metabolism requires in all cases the carbohydrate phosphate.

Uptake of the various PTS carbohydrates via the respective enzymes II has been measured in different ways. Uptake of labeled compounds is relatively easy to measure but suffers from the drawback that subsequent metabolism results in the formation of compounds that can leave the cell (for instance, acetate, lactate, pyruvate, or CO_2) and thus leads to an underestimation of the rate of uptake. The use of nonmetabolizable analogs or of mutants blocked in the first metabolic step circumvents this problem, but the initial rate of uptake is often so high that the steady-state level of accumulation (i.e., the point at which there is no further net uptake) is reached within seconds. For example, the maximal rate of uptake of 2-deoxyglucose via II^{Man} is approximately 400 nmol/min per mg at 25°C (142). Since the intracellular pool does not exceed 20 nmol/mg, the process lasts only a few seconds. Alternatively, the disappearance of a PTS carbohydrate from the medium can be measured, but this method is relatively insensitive. Finally, uptake has been measured in bacterial vesicles obtained by lysis of osmotically fragile cells (53).

By using one or more of these methods, several PTS uptake systems have been characterized. In general, *E. coli* and *S. typhimurium* take up and phosphorylate PTS carbohydrates with a high apparent affinity. The apparent K_m ranges from 10 to 20 μM in the case of mannose and glucose uptake by *S. typhimurium* (142) and from 0.2 to 10 μM for the uptake of hexitols by *E. coli* (68). The following questions have been asked. (i) Is transport possible in the absence of PTS-mediated phosphorylation? (ii) Is transphosphorylation occurring in intact cells to any extent? (iii) Can PTS carbohydrates be phosphorylated intracellularly via the PTS? (iv) Can PTS carbohydrates leave the cell, and if so, how?

Mutants defective in enzyme I, HPr, or both are unable to grow on PTS carbohydrates. Transport studies with tight *ptsI* mutants or *ptsHI* deletion strains, lacking all enzyme I and HPr, have shown that no transport of the hexoses glucose, mannose, and fructose is possible (132). Even equilibration of the free sugar, present at high extracellular concentrations, does not occur, provided that all other non-PTS transport systems that might have some affinity for the PTS carbohydrates under study have been eliminated (108). Glucose, mannose, and fructose, for example, can be taken up quite well via the galactose permease, provided that this transport system is expressed constitutively (108, 122). However, even basal, uninduced levels of galactose permease are sufficient to allow uptake of 2-deoxyglucose (a substrate of

II^{Man}) in *ptsHI* deletion mutants of *S. typhimurium*, at about 1% of the PTS uptake rate via II^{Man}. A mutation in the galactose permease lowers the residual rate of uptake in *pts* mutants by another factor of 50 (108). Although no facilitated diffusion seems possible via the enzymes II, mutants have been isolated in which II^{Glc} can equilibrate glucose in the absence of PTS-dependent phosphorylation (101). Although the apparent affinity of the altered II^{Glc} for glucose is about 1,000-fold lower ($K_m = 10$ mM) than that of the wild-type protein ($K_m = 10$ μM), the maximal rate of glucose uptake is quite similar. It has been proposed that II^{Glc} functions as a pore that is closed in the absence of phosphorylation but opens upon phosphorylation. Some mutations in the gene for II^{Glc} might result in a permanently open pore. Furthermore, the altered II^{Glc} protein is unable to catalyze the phosphorylation of glucose by phosphorylated III^{Glc} in an in vitro assay. Curiously, it has been suggested that uptake of the non-PTS carbohydrate galactose can be catalyzed in both *E. coli* and *S. typhimurium* by an enzyme II in the absence of phosphorylation. In *S. typhimurium*, II^{Man} is thought to be involved in this process (99), and in *E. coli*, II^{Glc} is (59).

Uptake of nonmetabolizable analogs such as methyl α-glucoside reveals an interesting feature of the PTS. Although a steady-state level of uptake is reached rapidly, continuous exchange occurs (32). The rate of uptake and phosphorylation of methyl α-glucoside equals the rate of exit of nonphosphorylated methyl α-glucoside. Although (vectorial) transphosphorylation has been shown to occur in bacterial extracts and phospholipid vesicles reconstituted with II^{Mtl} (72), it is by no means clear that this process is involved in the phenomenon described above, because transphosphorylation rates are low as measured in these in vitro systems. Intracellular dephosphorylation of the carbohydrate phosphate before exit by a phosphatase has been proposed (40). It is quite unclear, however, how the PTS carbohydrate leaves the cell. The experiments described above, indicating the absence of enzyme II-catalyzed facilitated diffusion, suggest that exit occurs via another, still unknown pathway. Alternatively, the enzymes II possibly act as functionally asymmetrical enzymes, allowing facilitated diffusion in one direction but not in the other. One obvious difference between these conditions is the presence of phosphorylated PTS proteins in the case of efflux from wild-type cells in contrast to influx in *ptsHI* deletion strains. Possibly the presence of phosphorylated HPr (or III^{Glc}) affects the catalytic activity of II^{Glc} in facilitated diffusion. There is no information about the phosphorylation state of the enzyme II under these conditions.

Glucose and other PTS carbohydrates are normally present at the outside of the cell. Phosphorylation occurs during translocation. One might ask whether glucose (generated intracellularly from maltose or melibiose, for instance) can be phosphorylated via the PTS inside the cell without first leaving the cell. Again the data for *E. coli* and *S. typhimurium* are not conclusive, although most of the evidence argues against this possibility. A recent paper on *Streptococcus lactis* gives rather good evidence that, at least in the case of glucose in this organism, II^{Man} is able to phosphorylate this carbohydrate at the inside (143).

REGULATION OF PTS ACTIVITY

The PTS activity can be regulated in a number of ways. In a previous section I discussed how the levels of the various PTS components vary with growth conditions. A similar variation of PTS activities is found in *cya* and *crp* mutants, which have lower enzyme I and HPr activities than the corresponding wild-type strains. Of course, this reduction affects the rate of uptake and phosphorylation of PTS carbohydrates.

When more than one PTS carbohydrate is present, competition is observed between the different substrates. One reason for this might be that they compete for a common pool of phosphorylated PTS proteins. Experiments with intact and toluene-treated cells suggest that enzyme I or HPr, or both, is rate limiting (137).

There are other possible mechanisms by which the PTS activity can be regulated. In particular, one has to consider the possibility that the intracellular accumulation of carbohydrate phosphates might inhibit some enzymes II. For instance, evidence has been presented that uptake via II^{Gut}, the enzyme II specific for glucitol, is very sensitive to carbohydrate phosphates such as glucose 1-phosphate, glucose 6-phosphate, and fructose 6-phosphate (70). II^{Mtl}, on the other hand, is not inhibited. Genetic evidence has been presented to show that inhibition of uptake of a PTS carbohydrate by a second one can be exerted at the level of the enzyme II. Amaral and Kornberg (5) studied fructose uptake by *E. coli* and found that glucose inhibited fructose uptake and metabolism, but that a mutation in the gene coding for II^{Fru} abolished this inhibition.

Finally, regulation of PTS activity by the electrochemical potential should be mentioned. It was observed very early that uncouplers of oxidative phosphorylation stimulate the net uptake of PTS carbohydrates, whereas the presence of oxidizable substrates inhibits this uptake (39). Recently, a more detailed examination revealed the importance of the redox state of SH groups in the enzymes II (114). The presence of oxidizable substrates also inhibits phosphorylation of PTS carbohydrates by membrane preparations. It has been proposed that the affinity of, e.g., enzyme II^{Glc} for its substrate, methyl α-glucoside, is affected by changes in the redox potential or activity of the respiratory chain, but that its maximal velocity is not affected. Similar effects were observed in the case of non-PTS transport systems, for example, the lactose carrier (57). In the latter case, however, it was shown under experimentally more favorable conditions (high levels of the lactose carrier) that the affinity was not altered but instead the number of active carriers decreased (92). A similar explanation has been proposed for the enzymes II. Grenier et al. (37) showed that II^{Mtl} is oxidized to an inactive rather than a low-affinity form. In the cases of II^{Glc} and II^{Man} it was shown that the latter enzyme II is less sensitive to inhibition by oxidizing agents (38). This differential sensitivity may explain the change in apparent K_m observed earlier by Robillard and Konings (114), since II^{Glc} and II^{Man} have widely different affinities for methyl α-glucoside (142).

GENETICS OF THE PTS

Genes for the PTS Proteins

The genes for enzyme I (*ptsI*) and HPr (*ptsH*) are linked and are located at approximately 52 min on the *E. coli* chromosome (7) and at 49 min on that of *S. typhimurium* (133). The genes for enzymes II and III are scattered, except for the *crr* gene, specifying III^{Glc}, which is closely linked to the *ptsHI* genes. The known locations of the genes for enzymes II and III are indicated in Table 1. Nomenclature is not uniform since in the past a number of genes specifying enzymes II have been given the *pts* symbol, although none of these form an operon or regulon with *ptsHI*. For this reason, I prefer the nomenclature proposed by Lin (74), i.e., separate symbols for each of the genes specifying enzymes II and III (and their related metabolic enzymes if they constitute an operon). For example, the genes specifying II^{Mtl} and mannitol phosphate dehydrogenase will be called *mtlA* and *mtlD*, respectively. I shall continue to use *ptsG* and *ptsM* for II^{Glc} and II^{Man}, respectively, because of their frequent use.

The *ptsH* and *ptsI* genes in *E. coli* and *S. typhimurium* form an operon together with the promoter *ptsHp*. A large number of *ptsH* and *ptsI* mutations have been identified in both *E. coli* (119, 120) and *S. typhimurium* (17, 18). A second, weak promoter just in front of *ptsI* has been suggested for *E. coli* (15). In *S. typhimurium*, *ptsHp* mutants have lower levels of both HPr and enzyme I, but the level of III^{Glc} is not affected (16, 18).

The *crr* gene, specifying III^{Glc}, is closely linked to the *pts* operon. The gene order in *E. coli* is *purC . . . cysA crr ptsI ptsH ptsHp . . . dsd*. Studies with *S. typhimurium* suggested an inversion in this area: *purC . . . cysA cysK ptsHp ptsH ptsI crr . . . dsd* (18). Recent studies, however, have shown that the correct order in *S. typhimurium* is similar to that in *E. coli*; i.e., the *crr* gene is localized between *cysA* and *pts*: *purC . . . cysA crr ptsI ptsH ptsHp cysK . . . dsd* (89).

What is the role of the *ptsHp* promoter in *crr* expression? Study of *S. typhimurium* strains containing deletions extending from *cysK* into the *ptsHI* genes showed that these strains contain normal levels of III^{Glc} (18). This finding suggests that the *crr* gene has its own promoter, since the *ptsHp* promoter was deleted. From complementation studies with plasmids containing the *crr* gene, Danchin and co-workers (22) concluded that the *ptsI* and *crr* genes in *E. coli* are transcribed from a common promoter and constitute a single transcriptional unit that also includes *ptsH*. They also suggest that a second promoter for *crr* might be located inside *ptsI*. This second promoter is most likely the major one, because *ptsHp* mutations do not affect the levels of III^{Glc}.

The *ptsHI* (14, 66) and *crr* (79, 89) genes have been cloned. The nucleotide sequences of the *E. coli ptsH* gene (23) and the *S. typhimurium crr* gene (89) have been determined.

As mentioned above, the genes specifying the various enzymes II and III are scattered across the chromosome. In a number of systems they form an operon together with the gene coding for the first enzyme involved in subsequent metabolism, e.g., the operons for the hexitols, β-glucosides, and fructose. The *E. coli mtlA* gene, specifying II^{Mtl}, has been cloned, as have the *ptsM* and *ptsL* genes, specifying II^{Man} and III^{Man}, respectively (30). The nucleotide sequence of the *mtlA* gene has been determined (65). In this case, both genetic and biochemical data suggest that enzyme II^{Mtl} consists of a single polypeptide and that no protein other than II^{Mtl}, HPr, and enzyme I is required for PEP-dependent mannitol phosphorylation. In other cases the answer is less clear. Recently some genetic evidence for a gene specifying a soluble III^{Gut}, involved in glucitol phosphorylation, has been obtained (134). Finally, in the case of the fructose PTS, the genes coding for enzymes II^{Fru} and III^{Fru} form an operon that also contains genes coding for fructose-1-phosphate kinase and pseudo-HPr (or FPr) (33).

Phenotype of Mutants Defective in Components of the PTS

Mutants defective in the common proteins of the PTS, enzyme I and HPr, and those defective in an enzyme II or III have different phenotypes. Whereas mutants of the former class are unable to grow on all PTS carbohydrates, those in the latter class exhibit defective growth only on the carbohydrate for which the missing enzyme II or III is specific. Since a number of PTS carbohydrates are substrates of more than one enzyme II (Table 1), it is found that many enzyme II mutants continue to grow, even on the substrate for which that enzyme II is specific. Thus, enzyme II^{Glc}, specified by *ptsG*, and enzyme II^{Man}, the product of the *ptsM* gene, can transport both glucose and mannose (although the affinity of II^{Glc} for mannose is rather low). Consequently, both *ptsG* and *ptsM* mutants of *S. typhimurium* grow on glucose and mannose.

Although the phenotype of *ptsI* mutants is clear-cut in principle, some complications arise when *ptsH* mutants and the role of HPr in overall PTS phosphorylation are studied. Walter and Anderson (145) showed that in *K. aerogenes* HPr is not required for the fructose PTS. *ptsH* strains of *S. typhimurium* are known to grow reasonably well on the PTS carbohydrate fructose (129, 130). Furthermore, pregrowth of a *ptsH* strain on fructose allows growth on other PTS carbohydrates for a generation or so. This suggests that some protein induced by growth on fructose can substitute for HPr, not only during fructose uptake and phosphorylation, but also during uptake and phosphorylation of other PTS carbohydrates. Presumably, the cellular concentration of this protein decreases during subsequent cell divisions. Genetically, it was found that *ptsH* mutations are easily suppressed by a second mutation, outside the *pts* operon. This so-called pseudo-HPr suppressor mutation allows growth of a *ptsH* mutant on all PTS carbohydrates and occurs at high frequency (130). The suppressor mutation, *fruR*, has recently been mapped at 3 min on the *S. typhimurium* chromosome, closely linked to *ilvIH*. The *fruR* gene probably specifies a repressor of the *fru* operon which contains, in addition to the genes for II^{Fru} (*fruA*), III^{Fru} (*fruB*), and fructose-1-phosphate kinase (*fruK*), a gene coding for pseudo-HPr or FPr (*fruF*). FPr can replace HPr and is expressed constitutively in *fruR* strains (33).

The phenotype of *pts* mutants is variable for yet another reason. Just as PTS carbohydrates often can

be transported by more than one enzyme II, some can also be taken up via non-PTS transport systems. In such cases, however, subsequent phosphorylation by a kinase is generally required before the carbohydrates can be metabolized. For example, glucose is an excellent substrate of the galactose permease but is unable to induce this transport system. Constitutive expression of the galactose permease allows a *ptsHI* deletion strain of *S. typhimurium* to grow on glucose (100, 122). Growth on fructose and mannose is also possible if the strain contains in addition a manno(fructo)kinase to convert the intracellular mannose (fructose) to its phosphate ester (108, 132).

K. aerogenes ptsI mutants have an unexpected phenotype. These mutants, which lack enzyme I, are still able to grow on glucose (and to a lesser extent on mannose and fructose), in contrast to *E. coli* and *S. typhimurium ptsI* mutants. Growth on the hexitols mannitol and glucitol is completely abolished, however. The growth on glucose is due to a membrane-bound glucose dehydrogenase that converts glucose to gluconate in the periplasmic space. The gluconate is subsequently taken up by the cell and metabolized. The glucose dehydrogenase contains pyrroloquinoline quinone as a prosthetic group (85). A similar glucose dehydrogenase apoenzyme is also present in *E. coli* and *S. typhimurium*, but these organisms are unable to synthesize pyrroloquinoline quinone. Growth of *E. coli* and *S. typhimurium ptsI* mutants on glucose is restored by addition of pyrroloquinoline quinone (44).

My discussion of the phenotype of mutants defective in components of the PTS has been restricted until now to PTS carbohydrates. Surprisingly, mutants lacking enzyme I, HPr, or both are also unable to grow on a large number of non-PTS carbohydrates. This unexpected phenotype is the subject of the next section.

REGULATION BY THE PTS

As mentioned in the previous section, *ptsHI* mutants are defective in growth on both PTS and a number of non-PTS carbon sources. Table 2 shows the extent of this defect. For reasons that will be made clear below, we can divide the non-PTS carbon sources that do not support growth of the *ptsHI* mutants into two classes (102). Class I includes well-known non-PTS carbohydrates such as lactose, maltose, melibiose, and glycerol. Class II consists of Krebs cycle intermediates (succinate, malate, and citrate), xylose, rhamnose, and galactose (when transported via the methyl-β-galactoside permease). One can ask what these carbon sources have in common that makes them sensitive to *pts* mutations. Whereas this question has been approached in different ways for *E. coli* and *S. typhimurium*, temporarily leading to conflicting views, it is now clear that the same mechanisms operate in both species.

An early connection was made between the defect in *pts* mutants and that in adenylate cyclase (*cya*) or cyclic-AMP (cAMP)-binding protein (*crp*) mutants (for a review of the role of cAMP in bacteria, see reference 144). In particular, the observation that, at least in *E. coli*, the phenotype of a *pts* mutant with respect to growth on non-PTS carbohydrates could be reversed by the addition of cAMP (97) suggested that the PTS

TABLE 2. Phenotype of *ptsH*, *ptsHI*, and *crr* mutants

Relevant genotype	Growth[a] on:			
	PTS sugars[b]	Class I compounds[c]	Class II compounds[d]	Galactose
Wild type	+	+	+	+
ptsI (leaky)	−	+	+	+
ptsI (leaky) + PTS sugar	−	−	−	+
pts(HI)	−	−	−	+
pts(HI) crr	−	+	−	+
pts(HI) crp (Crp*)	−	+	+	+
pts(HI) + cAMP	−	+	+	+
crr	+	+	−	+
crr crp (Crp*)	+	+	+	+
crr + cAMP	+	+	+	+

[a] +, Growth after 48 h at 37°C; −, no growth.
[b] PTS sugars include glucose, mannose, fructose, and hexitols.
[c] Class I compounds include maltose, melibiose, glycerol, and lactose.
[d] Class II compounds include xylose, rhamnose, and Krebs cycle intermediates.

was involved in the regulation of cAMP metabolism. An important discovery was made by Gershanovitch and co-workers, who found that induction of the *lac* operon was impaired in *pts* mutants (34). In studies with *S. typhimurium*, more emphasis was placed on the entry of the inducer into the cell to explain the phenotype of *pts* mutants. Saier and co-workers (127–130) found a suppressor mutation that restores growth of tight *pts* mutants on class I carbon sources (Table 2); this mutation also restores growth of leaky *ptsI* mutants (less than 1% remaining enzyme I activity) on class I compounds in the presence of nonmetabolizable PTS analogs. The role of the suppressor mutation in the regulation of the inducer entry was demonstrated in transport studies. In leaky *ptsI* strains, uptake of class I compounds is strongly inhibited by PTS carbohydrates (so-called inducer exclusion). The suppressor mutation abolishes this inhibition of uptake of class I compounds by PTS carbohydrates. The gene was called *crr*, for carbohydrate repression resistance; the biochemical defect associated with the *crr* mutation is a lowered activity of III^{Glc}, the protein involved in the transport of glucose via II^{Glc} (127). The *crr* mutation restores growth of *pts* mutants on all class I compounds at the same time (Table 2) and prevents inhibition of uptake of non-PTS carbohydrates by several PTS carbohydrates. The *crr* mutation does not restore growth on class II carbon sources, nor is the transport of class II compounds inhibited by PTS carbohydrates (107) (Table 2).

Later studies showed that similar *crr* mutations could be isolated in *E. coli* (60, 61). Conversely, it was found that cAMP restores growth of *S. typhimurium ptsHI* mutants on both class I and class II compounds (136). A similar effect was found when a mutation was introduced that results in an altered cAMP-binding protein that is active in the absence of cAMP.

On the basis of these findings, a hypothesis has been formulated in which the phosphorylation state of one of the PTS proteins plays a central regulatory role (127, 128). This model is schematically depicted in

Fig. 2. Two processes are postulated to be instrumental in regulation by the PTS: (i) the entry of inducers, transported via class I transport systems, is inhibited by the nonphosphorylated form of III^{Glc}, and (ii) adenylate cyclase is activated by phosphorylated III^{Glc}. This proposal is based on the following reasoning. In mutants defective in the general protein enzyme I or HPr, III^{Glc} will be in the nonphosphorylated form. The same may happen when a cell is growing in the presence of a PTS carbohydrate. The steady-state phosphorylation state of III^{Glc} will be determined by the rate of phosphorylation via P~HPr and the rate of dephosphorylation via the enzymes II and their PTS substrates. Inhibition of non-PTS uptake systems in *ptsHI* mutants and in wild-type strains growing in the presence of a PTS carbohydrate will be relieved by *crr* mutations since the inhibitor, nonphosphorylated III^{Glc}, is eliminated. The rate of phosphorylation has become irrelevant. In the case of adenylate cyclase, the situation is different. First, it has been found that in intact cells and in toluene-treated cells, adenylate cyclase activity is inhibited by PTS carbohydrates (43). This inhibition requires the enzymes II (42). A *crr* mutation does not relieve this inhibition of adenylate cyclase by PTS carbohydrates, however. In contrast, mutants lacking only III^{Glc}, but having a normal amount of enzyme I and HPr, also exhibit low adenylate cyclase activity (88). It has been concluded that phosphorylated III^{Glc} is an activator of adenylate cyclase rather than III^{Glc} being an inhibitor.

This elaborate regulatory mechanism has been proposed on the basis of genetic studies and experiments with whole cells. Thus, no evidence for the direct interaction of III^{Glc} with the various target proteins was available until recently, and different models have been proposed in which proteins other than III^{Glc}

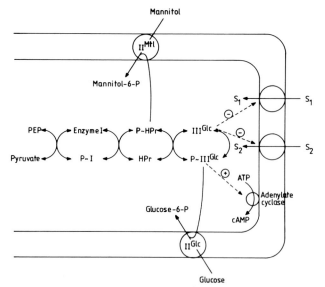

FIG. 2. Model for regulation by the PTS. In addition to the general proteins of the PTS, two enzymes II are shown; these are specific for mannitol (II^{Mtl}) and glucose (II^{Glc}). Activation (+) of adenylate cyclase by phosphorylated III^{Glc} (P~III^{Glc}) and inhibition (−) of two different non-PTS uptake systems by III^{Glc} are indicated. S1 and S2 represent lactose, melibiose, maltose, or glycerol.

were supposed to interact. Before I consider in more detail the consequences of the regulatory model just described, I shall present the recent evidence that III^{Glc} is indeed the regulatory molecule.

Two uptake systems that are regulated by the PTS have been studied in more detail, namely, those for lactose and glycerol. With membrane vesicles that contain elevated levels of the lactose carrier, binding of purified III^{Glc} to the lactose carrier (dependent on the presence of a substrate of the lactose carrier) can be shown to occur (91, 94). The apparent affinity of III^{Glc} for the lactose carrier is approximately 10 μM, and between 1 and 1.5 molecules of III^{Glc} are bound per molecule of the lactose carrier (91). Phosphorylated III^{Glc} is unable to bind, as predicted by the model. III^{Glc} is also able to inhibit uptake and exchange of β-galactosides in membrane vesicles. Finally, III^{Glc} inhibits the active uptake of a β-galactoside in liposomes reconstituted with purified lactose permease (91). From these studies it can be concluded that interaction between III^{Glc} and the lactose carrier molecule is sufficient to explain inducer exclusion of lactose.

In the second system, that for glycerol uptake, matters are slightly different. Facilitated diffusion of glycerol through the glycerol facilitator does not seem to be inhibited by III^{Glc}. Rather, it is the inhibition of glycerol kinase by nonphosphorylated III^{Glc} that impairs glycerol metabolism in pts mutants and wild-type strains in the presence of a PTS carbohydrate (104). Phosphorylated III^{Glc} has no effect on glycerol kinase activity, just as it is unable to bind to the lactose carrier. Binding of III^{Glc} to glycerol kinase requires the substrate glycerol (20), again just as the binding of III^{Glc} to the lactose carrier requires a β-galactoside. Although only the second step in glycerol metabolism is regulated by III^{Glc}, induction of the glp regulon is nevertheless impossible since the true inducer is glycerol 3-phosphate, the product of glycerol kinase, rather than glycerol iself. Interestingly, Berman and Lin (12) have described a mutation in the gene for glycerol kinase that suppresses the effect of a ptsI mutation in E. coli. At that time only the resistance of the mutant glycerol kinase to inhibition by fructose 1,6-diphosphate, a feedback inhibitor of glycerol kinase, was noted. It has been shown recently that this mutant glycerol kinase indeed is also more resistant to III^{Glc} (20). Similar mutations in the glpK gene of S. typhimurium have been described (93). Although other class II systems have not been studied in such detail, it is reasonable to assume that in these systems also III^{Glc} will be the direct effector (83, 126).

The situation with adenylate cyclase is less clear. Rates of cAMP synthesis in cells can differ considerably, depending on growth conditions (see, for example, reference 51). Furthermore, in crp strains lacking the cAMP-binding protein (CRP), the in vivo rate of cAMP synthesis can be several hundredfold higher than in the corresponding wild type (51, 75). Regulation could be on the level of enzyme synthesis or enzyme activity. Most studies that deal with expression of the cya gene have used cya-lac fusions, and conflicting results have been obtained. Bankaitis and Bassford (8) and Roy et al. (118) concluded that crp mutations have no or almost no effect on cya expression. Jovanovich (52) and Aiba and co-workers (1, 55, 84) concluded, however, that transcription of the cya gene is under negative control of the cAMP-CRP complex. It should be emphasized, however, that even in those cases in which regulation of transcription is found, the differences are at most 10-fold in in vitro studies and even smaller under in vivo conditions in intact cells. Consequently, regulation of cAMP synthesis most likely also involves regulation at the level of enzyme activity. A four- to fivefold stimulation of E. coli adenylate cyclase by potassium and phosphate was recently reported (73); the effect is observed in toluene-treated cells but not in broken-cell preparations. It was concluded that this activation is mediated via the PTS proteins, since stimulation is absent in pts mutants.

The biochemical evidence obtained up to now to show that phosphorylated III^{Glc} activates adenylate cyclase is not very convincing. It was recently reported that phosphorylated III^{Glc} in the presence of enzyme I and HPr stimulates adenylate cyclase activity in bacterial extracts; this stimulation is at most twofold, however (110). Other attempts have been negative (21). One reason might be that still another protein is required. From genetic studies it has been concluded that possibly the cAMP-binding protein is involved. One model proposes that the cAMP-binding protein is an inhibitor of adenylate cyclase (25). The role of phosphorylated III^{Glc} might be removal of the inhibitor, the cAMP-binding protein, from adenylate cyclase, but the recent observation that adenylate cyclase activity is still regulated by the PTS in crp strains lacking the cAMP-binding protein makes this proposal less likely (21). Roy et al. (117) have demonstrated that adenylate cyclase contains two domains. Truncated proteins, lacking the carboxy terminus, are still catalytically active but have become insensitive to glucose inhibition. Another complication stems from the observation that in certain ptsH and ptsI mutants of E. coli, repression is observed even though the adenylate cyclase activity does not seem to be affected (35). Nevertheless, the proposal of an activator is attractive, since adenylate cyclase activities in bacterial extracts are quite low compared with those in either toluene-treated or intact cells.

Possibly adenylate cyclase is a more complex enzyme than expected from the properties of the purified enzyme (153). Sequence studies of the cloned cya gene (2) have shown a second gene, cyaX, next to the gene coding for the protein with the catalytic activity. The amino acid sequence deduced from the nucleotide sequence of cyaX points to a very hydrophobic protein with a molecular weight of approximately 15,000 (G. Lenzen and A. Danchin, personal communication). Possibly this CyaX protein acts as an anchor protein for the soluble adenylate cyclase molecule, and it may be required for high catalytic activity, regulated by the PTS. It should be remembered that, depending on the preparation, adenylate cyclase activity has been found associated with the membrane (75, 144).

The various examples of PTS-mediated regulation discussed above show clearly that III^{Glc} plays a central role in this regulatory process. Depending on its phosphorylation state it interacts with many different proteins, including both PTS and non-PTS proteins and both soluble and membrane-bound proteins. Ta-

TABLE 3. Proteins that interact with enzyme III^{Glc}

Class	Proteins
PTS proteins	HPr, pseudo-HPr (FPr), enzyme II^{Glc}, enzyme II^{Suc}
Non-PTS uptake systems	Lactose carrier (lacY), melibiose carrier (melB), maltose transport system (malK), glycerol kinase (glpK), adenylate cyclase

ble 3 summarizes the proteins with which III^{Glc} is thought to interact.

With the background sketched above I shall now discuss in some detail other mutations that suppress the ptsHI phenotype. In addition, I shall explain the consequence of the model for a "normal" cell with an intact PTS.

As pointed out earlier, the addition of cAMP to a ptsHI mutant restores growth on both class I and class II compounds (Table 2). Suppressor mutations that have the same effect have been identified in both E. coli and S. typhimurium (3, 136). Genetic studies showed these mutations to be linked to cysG; they also resulted in an altered cAMP-binding protein, making this protein independent from cAMP. The same mutations also suppressed a cya mutation that results in a defective adenylate cyclase (136). These crp mutations (Crp*) allow growth of crr mutants on class II compounds (Table 2). Thus, there exist at least two classes of general suppressor mutations, crr and crp (Crp*), that restore growth of a ptsHI mutant on non-PTS carbon sources. With the first class, inducer exclusion is abolished, but adenylate cyclase activity is still low. With the second class, the expression of catabolic genes becomes independent of cAMP, but inducer exclusion is still possible. I shall discuss below how these mutants can grow again when only part of the defect is suppressed.

In addition to these general suppressor mutations, specific mutations that restore growth only on single carbon sources have been described. In S. typhimurium mutations that restore growth of ptsI strains on glycerol, maltose, melibiose, or lactose have been described. These mutations were shown to be located in genes closely linked to or belonging to the respective glp, mal, mel, or lac operons (131). Most likely they alter the target protein such that it becomes resistant to III^{Glc}. I have already mentioned the mutation in the glycerol kinase gene of E. coli that allows growth of a ptsI mutant on glycerol and renders glycerol kinase more resistant to inhibition by III^{Glc} (20). Mutations also exist in lacY that abolish binding of III^{Glc} to the lactose carrier (91). Other specific suppressor mutations are not in the structural gene but affect the level of gene expression. Two such suppressor mutations were described by Lin and co-workers (12, 13); one resulted in the constitutive expression of the glp regulon owing to a glpR mutation, and the other achieved the same result but was shown to be a promoter-up mutation. In general, expression of class I operons to a high level suppresses the effect of a pts mutation.

Another type of regulatory mutation should be mentioned. Kornberg and co-workers (60–62) identified mutations in E. coli like the crr mutation in S. typhimurium. These so-called iex (inducer exclusion) mutations also restore growth on class I carbon sources,

but they are different from crr mutations in that transport and phosphorylation of methyl α-glucoside, dependent on III^{Glc}, seems quite normal. The observation that the iex mutation renders cells resistant to inhibition by methyl α-glucoside on class I carbohydrates but not on PTS carbohydrates like fructose also suggested that III^{Glc} is catalytically active. On the basis of these results, it was proposed that in E. coli an Iex protein, rather than III^{Glc}, is responsible for the inhibition of the various class I uptake systems (96). Regulation of adenylate cyclase was still thought to occur via III^{Glc}. Recent studies have resolved this apparent contradiction between S. typhimurium and E. coli (86). An E. coli iex mutant indeed contains wild-type levels of III^{Glc}, with almost the same specific activity in glucose transport. The mutant III^{Glc}, however, is temperature sensitive, suggesting that the strain in fact contains a crr mutation. Thus, iex is an allele of crr. The failure to accomplish inducer exclusion is due to the fact that III^{Glc} from an iex strain is unable to bind to and inhibit the lactose permease (86). This example is thus the reverse of the one discussed above, in which the (altered) target protein becomes resistant to III^{Glc} (91). In an iex strain, the mutated III^{Glc} is unable to bind to its target proteins. The phosphorylated form of this mutant III^{Glc} is still able to catalyze phosphoryl transfer to III^{Glc} and to activate adenylate cyclase.

Table 4 summarizes how the effect of a ptsHI mutation can be suppressed. Some suppressor mutations or conditions have a general effect; i.e., the defect on all or most carbon sources affected is suppressed. In other cases a specific suppressor mutation restores growth on only one carbon source. The last suppressor listed in Table 4, elimination of an enzyme II, does not restore growth of ptsHI mutants; it prevents inhibition by a particular PTS carbohydrate, however, when leaky ptsI mutants are studied.

GROWTH AND REGULATION

The previous sections have described how the uptake and metabolism of certain compounds can be regulated by others. The main findings originate from studies with pts mutants, however. One may ask whether these regulatory mechanisms play any role in "normal" cells, i.e., cells with an intact PTS and growing on one or several carbon sources. Still unanswered is the question of how pts strains that have acquired either a crr or a crp (Crp*) mutation can grow again on class I or class II compounds (or both) when only part of the defect caused by the pts mutation has been relieved.

To solve these problems, I will try to answer the following questions. (i) Does inducer exclusion occur

TABLE 4. Suppression of the effects of a ptsHI mutation

Suppression mechanism	Effect
Elimination of III^{Glc}	General
Constitutive operon expression (lac, glp)	Specific
Uptake systems insensitive to III^{Glc} (lac, mal, glpK)	Specific
Gene expression independent of cAMP (crp [Crp*])	General
Addition of cAMP	General
Elimination of an enzyme II	Specific

in Pts$^+$ cells? (ii) Is inducer exclusion still possible in Crp* strains? (iii) What are the consequences of a stoichiometric interaction between III^{Glc} and its target proteins when limiting amounts of III^{Glc} are available?

In wild-type cells containing an intact PTS and fully induced for a class I compound, inducer exclusion of this class I compound by PTS substrates is in general weak or nonexistent (128). However, when the uptake systems for these class I compounds are only partially induced, uptake of, for example, maltose or glycerol can be inhibited by PTS carbohydrates (88, 102). This observation can be explained as follows. In the cell there is a fixed and limited number of III^{Glc} molecules. III^{Glc} interacts stoichiometrically with several other cell components (e.g., the lactose carrier, glycerol kinase). As long as the number of uptake systems does not exceed the number of III^{Glc} molecules, inhibition of uptake is possible. If the number of induced uptake systems is too large, however, some of the transport systems can escape inhibition, owing to the lack of free III^{Glc}. Decreased sensitivity of the lactose carrier to inducer exclusion has also been observed (83). If this model is correct, the following prediction should hold: if more than one class I uptake system is induced at the same time in a cell, inhibition of the first system by PTS carbohydrates should be relieved by addition of a substrate of the second non-PTS system. This is exactly what has been found (87). In *S. typhimurium* cells induced partially for the maltose system and completely for the glycerol system, inhibition of maltose uptake by 2-deoxyglucose was relieved by the addition of glycerol. Both the substrate, glycerol, and the target protein, glycerol kinase in this case, are required. A consequence of this model, in which III^{Glc} interacts with various target proteins and is available only in limited amounts, is that cells that contain high levels of a III^{Glc}-sensitive transport system are not sensitive to PTS-mediated repression. This explains why promoter-up mutations or mutations resulting in constitutive expression of an operon can suppress *ptsHI* mutations.

The ratio between phosphorylated and nonphosphorylated III^{Glc} is important and should change in cells that encounter PTS carbohydrates. It has been possible to determine in intact cells the amounts of P~III^{Glc} and III^{Glc}, based on the observation that P~III^{Glc} has a slightly lower mobility in SDS-PAGE than III^{Glc}. Wild-type cells contain predominantly P~III^{Glc}, whereas in an *S. typhimurium* mutant containing 1% residual enzyme I activity, 50% of III^{Glc} is phosphorylated. In both cases, addition of a PTS carbohydrate results in dephosphorylation of P~III^{Glc} (90).

The various factors important in PTS-mediated inducer exclusion are summarized in Table 5.

I discussed above the observation that both *crr* and *crp* (Crp*) mutations can suppress *ptsHI* mutations (Table 2). This finding is somewhat surprising since in each case only part of the defect, that originating from the *pts* mutation, is relieved. According to the model shown in Fig. 2, either inducer exclusion or the cAMP defect is abolished. However, experiments with *ptsHI* mutants with a Crp* phenotype or growing in the presence of cAMP show that in these cells inducer exclusion is absent (125, 136). Although it has been

TABLE 5. Factors important for PTS-mediated inducer exclusion

Factor	Consequences
Phosphorylation state of III^{Glc}	III^{Glc} inhibits; P~III^{Glc} does not
No. of III^{Glc} molecules	III^{Glc} < non-PTS uptake systems results in escape from inhibition
No. of molecules of non-PTS uptake systems	III^{Glc} < non-PTS uptake systems results in escape from inhibition
Presence of substrates of the non-PTS uptake systems	No inhibition in the absence of the substrate

proposed that an alternative regulatory mechanism, desensitization (i.e., uncoupling of the class II uptake systems and the regulatory protein) is at work (125), a simpler explanation is available. Owing to the presence of cAMP or the *crp* (Crp*) mutation, expression of the inducible operon becomes maximal. It should be remembered that inducer exclusion is never complete (128); i.e., inducer molecules can always enter, although more slowly than in a Pts$^+$ strain. Conversely, in a *pts crr* double mutant the adenylate cyclase activity is low but is not zero. Furthermore, in these double mutants inducer exclusion is completely absent. From these results one may conclude that in a cell the occurrence of either inducer exclusion or a low adenylate cyclase activity still allows the cell to grow on class I compounds. However, if both defects occur at the same time in the cell, they prevent growth.

I have sketched in this section how certain carbohydrates can regulate the uptake of others both by controlling gene expression and by controlling enzyme activity. At least four classes of compounds can be discriminated in *E. coli* and *S. typhimurium*: PTS carbohydrates, class I and class II compounds, and "others." The last class consists of compounds that are affected neither by *pts* nor by *crr* mutations; examples are lactate and gluconate. Class II compounds are unable to support growth of either *pts* or *crr* strains. The defect is due to the strict dependency of class II operons on cAMP. The remaining adenylate cyclase activity, approximately 20%, in those mutants or in wild-type strains in the presence of a PTS carbohydrate is not sufficient. Class I compounds are different in that they also require cAMP for expression of their operons, but the requirement is less than in class II compounds. A differential requirement of different operons for cAMP has been demonstrated both in vivo (4) and in vitro (56). To allow, nevertheless, regulation by PTS carbohydrates, the uptake systems of class I compounds are sensitive to III^{Glc}. Although the residual adenylate cyclase activity in *pts* mutants allows expression of operons for class I compounds, the level of induction will be less and the remaining activity becomes more sensitive to inhibition by III^{Glc}. One reason why PTS carbohydrates should not inhibit adenylate cyclase completely is that expression of the various operons coding for PTS enzymes is itself dependent on cAMP. *cya* mutants, defective in adenylate cyclase, cannot grow on hexitols, for example, and grow much slower than wild-type strains on fructose and mannose. Finally, PTS carbohydrates can regulate each other via competition for the P~HPr pool, as discussed above. There thus exists a hierarchy of compounds in which each can inhibit the uptake and catabolism of the compounds lower in that hierarchy but is inhibited by compounds higher up.

CONCLUSIONS

In this chapter I have described the PTS, a complex uptake system that is involved in both the transport and phosphorylation of many carbohydrates as well as in the regulation of many metabolic pathways. The sugar-specific, membrane-bound enzymes II act also as receptors that are involved in chemotaxis toward these carbohydrates. A central role in regulation is played by one of the PTS proteins, III^{Glc}, which can exist in two forms, phosphorylated and nonphosphorylated. III^{Glc} interacts with many different proteins, both PTS and non-PTS and membrane-bound as well as soluble. Whereas the role of nonphosphorylated III^{Glc} in the inhibition of various non-PTS uptake systems is well established, the role of phosphorylated III^{Glc} in the regulation of adenylate cyclase and cAMP metabolism is still obscure. Until this system has been reconstituted, the second half of the model originally proposed remains hypothetical.

LITERATURE CITED

1. **Aiba, H.** 1985. Transcription of the *Escherichia coli* adenylate cyclase gene is negatively regulated by cAMP-cAMP receptor protein. J. Biol. Chem. **260:**3063–3070.
2. **Aiba, H., K. Mori, M. Tanaka, T. Ooi, A. Roy, and A. Danchin.** 1984. The complete nucleotide sequence of the adenylate cyclase gene of *Escherichia coli*. Nucleic Acids Res. **12:**9427–9440.
3. **Alexander, J. K., and B. Tyler.** 1975. Genetic analysis of succinate utilization in enzyme I mutants of the phosphoenolpyruvate:sugar phosphotransferase system in *Escherichia coli*. J. Bacteriol. **124:**252–261.
4. **Alper, M. D., and B. N. Ames.** 1978. Transport of antibiotics and metabolite analogs by systems under cyclic AMP control: positive selection of *Salmonella typhimurium cya* and *crp* mutants. J. Bacteriol. **133:**149–157.
5. **Amaral, D., and H. L. Kornberg.** 1975. Regulation of fructose uptake by glucose in *Escherichia coli*. J. Gen. Microbiol. **90:**157–168.
6. **Anderson, B., N. Weigel, W. Kundig, and S. Roseman.** 1971. Sugar transport. III. Purification and properties of a phosphocarrier protein (HPr) of the phosphoenolpyruvate-dependent phosphotransferase system of *Escherichia coli*. J. Biol. Chem. **246:**7023–7033.
7. **Bachmann, B. J.** 1983. Linkage map of *Escherichia coli* K-12, edition 7. Microbiol. Rev. **47:**180–230.
8. **Bankaitis, V. A., and P. J. Bassford, Jr.** 1982. Regulation of adenylate cyclase synthesis in *Escherichia coli* K-12: studies with *cya-lac* operon and protein fusion strains. J. Bacteriol. **151:**1346–1357.
9. **Begley, G. S., D. E. Hansen, G. R. Jacobson, and J. R. Knowles.** 1982. Stereochemical course of the reactions catalyzed by the bacterial phosphoenolpyruvate: glucose phosphotransferase system. Biochemistry **21:**5552–5556.
10. **Beneski, D. A., T. P. Misko, and S. Roseman.** 1982. Sugar transport by the bacterial phosphotransferase system. Preparation and characterization of membrane vesicles from mutant and wild type *Salmonella typhimurium*. J. Biol. Chem. **257:**14565–14575.
11. **Beneski, D. A., A. Nakazawa, N. Weigel, P. E. Hartman, and S. Roseman.** 1982. Sugar transport by the bacterial phosphotransferase system. Isolation and characterization of a phosphocarrier protein HPr from wild type and mutants of *Salmonella typhimurium*. J. Biol. Chem. **257:**14492–14498.
12. **Berman, M., and E. C. C. Lin.** 1971. Glycerol-specific revertants of a phosphoenolpyruvate-phosphotransferase mutant: suppression by the desensitization of glycerol kinase to feedback inhibition. J. Bacteriol. **105:**113–120.
13. **Berman, M., N. Zwaig, and E. C. C. Lin.** 1970. Suppression of a pleiotropic mutant affecting glycerol dissimilation. Biochem. Biophys. Res. Commun. **38:**272–278.
14. **Bitoun, R., H. de Reuse, D. Touati-Schwartz, and A. Danchin.** 1983. The phosphoenolpyruvate dependent carbohydrate phosphotransferase system of *Escherichia coli*. Cloning of the *ptsHI-crr* region and studies with a *pts-lac* operon fusion. FEMS Microbiol. Lett. **16:**163–167.
15. **Britton, P., L. G. Lee, D. Murfitt, A. Boronat, M. C. Jones-Mortimer, and H. L. Kornberg.** 1984. Location and direction of transcription of the *ptsH* and *ptsI* genes on the *Escherichia coli* K12 genomes. J. Gen. Microbiol. **130:**861–868.
16. **Cordaro, J. C., R. P. Anderson, E. W. Grogan, Jr., D. J. Wenzel, M. Engler, and S. Roseman.** 1974. Promoter-like mutation affecting HPr and Enzyme I of the phosphoenolpyruvate:sugar phosphotransferase system in *Salmonella typhimurium*. J. Bacteriol. **120:**245–252.
17. **Cordaro, J. C., T. Melton, J. P. Stratis, M. Atagun, C. Gladding, P. E. Hartman, and S. Roseman.** 1976. Fosfomycin resistance: selection method for internal and extended deletions of the phosphoenolpyruvate:sugar phosphotransferase genes of *Salmonella typhimurium*. J. Bacteriol. **128:**785–793.
18. **Cordaro, J. C., and S. Roseman.** 1972. Deletion mapping of the genes coding for HPr and enzyme I of the phosphoenolpyruvate:sugar phosphotransferase system in *Salmonella typhimurium*. J. Bacteriol. **112:**17–29.
19. **Curtis, S. J., and W. Epstein.** 1975. Phosphorylation of D-glucose in *Escherichia coli* mutants defective in glucosephosphotransferase, mannosephosphotransferase, and glucokinase. J. Bacteriol. **122:**1189–1199.
20. **de Boer, M., C. P. Broekhuizen, and P. W. Postma.** 1986. Regulation of glycerol kinase by enzyme III^{Glc} of the phosphoenolpyruvate:carbohydrate phosphotransferase system. J. Bacteriol. **167:**393–395.
21. **den Blaauwen, J. L., and P. W. Postma.** 1985. Regulation of cyclic AMP synthesis by enzyme III^{Glc} of the phosphoenolpyruvate:sugar phosphotransferase system in *crp* strains of *Salmonella typhimurium*. J. Bacteriol. **164:**477–478.
22. **de Reuse, H., E. Huttner, and A. Danchin.** 1984. Analysis of the *ptsH-ptsI-crr* region in *Escherichia coli* K-12: evidence for the existence of a single transcriptional unit. Gene **32:**31–40.
23. **de Reuse, H., A. Roy, and A. Danchin.** 1985. Analysis of the *ptsH-ptsI-crr* region in *Escherichia coli* K-12: nucleotide sequence of the *ptsH* gene. Gene **35:**199–207.
24. **Dills, S. S., A. Apperson, M. R. Schmidt, and M. H. Saier, Jr.** 1980. Carbohydrate transport in bacteria. Microbiol. Rev. **44:**385–418.
25. **Dobrogosz, W. J., G. W. Hall, D. O. Silva, D. K. Sherba, J. G. Harman, and T. Melton.** 1983. Regulatory interactions among the *cya*, *crp* and *pts* gene products in *Salmonella typhimurium*. Mol. Gen. Genet. **192:**477–486.
26. **Dörschug, M., R. Frank, H. R. Kalbitzer, W. Hengstenberg, and J. Deutscher.** 1984. Phosphoenolpyruvate-dependent phosphorylation site in enzyme III^{Glc} of the *Escherichia coli* phosphotransferase system. Eur. J. Biochem. **144:**113–119.
27. **Elliott, J., and W. Arber.** 1978. *E. coli* K-12 *pel* mutants, which block phage λ DNA injection, coincide with *ptsM*, which determines a component of a sugar transport system. Mol. Gen. Genet. **161:**1–8.
28. **Erni, B.** 1986. Glucose-specific permease of the bacterial phosphotransferase system: phosphorylation and oligomeric structure of the glucose-specific II^{Glc}-III^{Glc} complex of *Salmonella typhimurium*. Biochemistry **25:**305–312.
29. **Erni, B., H. Trachsel, P. W. Postma, and J. P. Rosenbusch.** 1982. Bacterial phosphotransferase system. Solubilization and purification of the glucose-specific Enzyme II from membranes of *Salmonella typhimurium*. J. Biol. Chem. **257:**13726–13730.
30. **Erni, B., and B. Zanolari.** 1985. The mannose-permease of the bacterial phosphotransferase system. Gene cloning and purification of the Enzyme II^{Man}/III^{Man} complex of *Escherichia coli*. J. Biol. Chem. **260:**15495–15503.
31. **Ferenci, T., and H. L. Kornberg.** 1974. The role of phosphotransferase syntheses of fructose 1-phosphate and fructose 6-phosphate in the growth of *Escherichia coli* on fructose. Proc. R. Soc. London B **187:**105–119.
32. **Gachelin, G.** 1970. Studies on the α-methylglucoside permease of *Escherichia coli*. A two-step mechanism for the accumulation of α-methylglucoside 6-phosphate. Eur. J. Biochem. **16:**342–357.
33. **Geerse, R. H., C. R. Ruig, A. R. J. Schuitema, and P. W. Postma.** 1986. Relationship between pseudo-HPr and the PEP:fructose phosphotransferase system in *Salmonella typhimurium* and *Escherichia coli*. Mol. Gen. Genet. **203:**435–444.
34. **Gershanovitch, V. N., G. I. Bourd, N. V. Jurovitzkaya, A. G. Skavronskaya, V. V. Klyutchova, and V. P. Shabolenko.** 1967. β-Galactosidase induction in cells of *Escherichia coli* not utiliz-

ing glucose. Biochim. Biophys. Acta 134:188–190.

35. Glesyna, M. L., T. N. Bolshakova, and V. N. Gershanovitch. 1983. Effect of ptsI and ptsH mutations on initiation of transcription of the Escherichia coli lactose operon. Mol. Gen. Genet. 190:417–420.

36. Grenier, F. C., I. Hayward, M. J. Novotny, J. E. Leonard, and M. H. Saier, Jr. 1985. Identification of the phosphocarrier protein enzyme III^gut: essential component of the glucitol phosphotransferase system in Salmonella typhimurium. J. Bacteriol. 161:1017–1022.

37. Grenier, F. C., E. B. Waygood, and M. H. Saier, Jr. 1985. Bacterial phosphotransferase system: regulation of mannitol enzyme II activity by sulfhydryl oxidation. Biochemistry 24:47–51.

38. Grenier, F. C., E. B. Waygood, and M. H. Saier, Jr. 1985. Bacterial phosphotransferase system: regulation of the glucose and mannose enzymes II by sulfhydryl oxidation. Biochemistry 24:4872–4876.

39. Hagihara, H., T. H. Wilson, and E. C. C. Lin. 1963. Studies on the glucose-transport system in Escherichia coli with α-methylglucoside as substrate. Biochim. Biophys. Acta 78:505–515.

40. Haguenauer, R., and A. Kepes. 1971. The cycle of renewal of intracellular α-methylglucoside accumulation by the glucose permease of E. coli. Biochimie 53:99–107.

41. Hall, B. G., K. Imai, and C. P. Romano. 1982. Genetics of the lac-PTS system of Klebsiella. Genet. Res. 39:287–302.

42. Harwood, J. P., C. Gazdar, C. Prasad, A. Peterkofsky, S. J. Curtis, and W. Epstein. 1976. Involvement of the glucose Enzymes II of the sugar phosphotransferase system in the regulation of adenylate cyclase by glucose in Escherichia coli. J. Biol. Chem. 251:2462–2468.

43. Harwood, J. P., and A. Peterkofsky. 1975. Glucose-sensitive adenylate cyclase in toluene-treated cells of Escherichia coli B. J. Biol. Chem. 250:4656–4662.

44. Hommes, R. W. J., P. W. Postma, O. M. Neijsel, D. W. Tempest, P. Dokter, and J. A. Duine. 1984. Evidence of a quinoprotein glucose dehydrogenase apoenzyme in several strains of Escherichia coli. FEMS Microbiol. Lett. 24:329–333.

45. Hoving, H., R. ten Hoeve-Duurkens, and G. T. Robillard. 1984. Escherichia coli phosphoenolpyruvate-dependent phosphotransferase system. Functional asymmetry in Enzyme I subunits demonstrated by reaction with 3-bromopyruvate. Biochemistry 23:4335–4340.

46. Jablonski, E. G., L. Brand, and S. Roseman. 1983. Sugar transport by the bacterial phosphotransferase system. Preparation of a fluorescein derivative of the glucose-specific phosphocarrier protein III^Glc and its binding to the phosphocarrier protein HPr. J. Biol. Chem. 258:9690–9699.

47. Jacobson, G. R., D. M. Kelly, and D. R. Finlay. 1983. The intramembrane topography of the mannitol-specific Enzyme II of the Escherichia coli phosphotransferase system. J. Biol. Chem. 258:2955–2959.

48. Jacobson, G. R., C. A. Lee, J. E. Leonard, and M. H. Saier, Jr. 1983. Mannitol-specific Enzyme II of the bacterial phosphotransferase system. I. Properties of the purified permease. J. Biol. Chem. 258:10748–10756.

49. Jin, R. Z., and E. C. C. Lin. 1984. An inducible phosphoenolpyruvate:dihydroxyacetone phosphotransferase system in Escherichia coli. J. Gen. Microbiol. 130:83–88.

50. Jones-Mortimer, M. C., and H. L. Kornberg. 1980. Amino-sugar transport systems of Escherichia coli. J. Gen. Microbiol. 117:369–376.

51. Joseph, E., C. Bernsley, N. Guiso, and A. Ullmann. 1982. Multiple regulation of the activity of adenylate cyclase in Escherichia coli. Mol. Gen. Genet. 185:262–268.

52. Jovanovich, S. B. 1985. Regulation of a cya-lac fusion by cyclic AMP in Salmonella typhimurium. J. Bacteriol. 161:641–649.

53. Kaback, H. R. 1968. The role of the phosphoenolpyruvate-phosphotransferase system in the transport of sugars by isolated membrane preparations of Escherichia coli. J. Biol. Chem. 243:3711–3724.

54. Kalbitzer, H. R., W. Hengstenberg, P. Rosch, P. Muss, P. Bernsmann, R. Engelmann, M. Dörschug, and J. Deutscher. 1982. HPr proteins of different microorganisms studied by hydrogen-1-high-resolution nuclear magnetic resonance: similarities of structures and mechanisms. Biochemistry 21:2879–2885.

55. Kawamukai, M., J. Kishimoto, R. Utsumi, M. Himeno, T. Komano, and H. Aiba. 1985. Negative regulation of adenylate cyclase gene (cya) expression by cyclic AMP-cyclic AMP receptor protein in Escherichia coli: studies with cya-lac protein and operon fusion plasmids. J. Bacteriol. 164:872–877.

56. Kolb, A., A. Spassky, C. Chapon, B. Blazy, and H. Buc. 1983. On the different binding activities of CRP at the lac, gal, and malT promoter regions. Nucleic Acids Res. 11:7833–7852.

57. Konings, W. N., and G. T. Robillard. 1982. Physical mechanism for regulation of proton solute symport in Escherichia coli. Proc. Natl. Acad. Sci. USA 79:5480–5484.

58. Kornberg, H. 1986. The roles of HPr and FPr in the utilization of fructose by Escherichia coli. FEBS Lett. 194:12–15.

59. Kornberg, H. L., and C. Riordan. 1976. Uptake of galactose into Escherichia coli by facilitated diffusion. J. Gen. Microbiol. 94:75–89.

60. Kornberg, H. L., and P. D. Watts. 1978. Roles of crr-gene products in regulating carbohydrate uptake by Escherichia coli. FEBS Lett. 89:329–332.

61. Kornberg, H. L., and P. D. Watts. 1979. tgs and crr: genes involved in catabolite inhibition and inducer exclusion in Escherichia coli. FEBS Lett. 104:313–316.

62. Kornberg, H. L., P. D. Watts, and K. Brown. 1980. Mechanisms of "inducer exclusion" by glucose. FEBS Lett. 117(Suppl.):K28–K36.

63. Kukuruzinska, M. A., W. F. Harrington, and S. Roseman. 1982. Sugar transport by the bacterial phosphotransferase system. Studies on the molecular weight and association of Enzyme I. J. Biol. Chem. 257:14470–14476.

64. Kundig, W., S. Ghosh, and S. Roseman. 1964. Phosphate bound to histidine in a protein as an intermediate in a novel phosphotransferase system. Proc. Natl. Acad. Sci. USA 52:1067–1074.

65. Lee, C. A., and M. H. Saier, Jr. 1983. Mannitol-specific Enzyme II of the bacterial phosphotransferase system. III. The nucleotide sequence of the permease gene. J. Biol. Chem. 258:10761–10767.

66. Lee, L. G., P. Britton, F. Parra, A. Boronat, and H. L. Kornberg. 1982. Expression of the ptsH^+ gene of Escherichia coli cloned on plasmid pBR322. A convenient means for obtaining the histidine-containing carrier protein HPr. FEBS Lett. 149:288–292.

67. Lengeler, J. 1975. Mutations affecting transport of the hexitols D-mannitol, D-glucitol, and galactitol in Escherichia coli K-12: isolation and mapping. J. Bacteriol. 124:26–38.

68. Lengeler, J. 1975. Nature and properties of hexitol transport systems in Escherichia coli. J. Bacteriol. 124:39–47.

69. Lengeler, J., R. J. Mayer, and K. Schmid. 1982. The phosphoenolpyruvate-dependent phosphotransferase system Enzyme III and plasmid-encoded sucrose transport in Escherichia coli. J. Bacteriol. 151:468–471.

70. Lengeler, J., and H. Steinberger. 1978. Analysis of the regulatory mechanisms controlling the activity of the hexitol transport systems in Escherichia coli K12. Mol. Gen. Genet. 167:75–82.

71. Lengeler, J., and H. Steinberger. 1978. Analysis of the regulatory mechanisms controlling the synthesis of the hexitol transport systems in Escherichia coli K12. Mol. Gen. Genet. 164:163–169.

72. Leonard, J. E., and M. H. Saier, Jr. 1983. Mannitol-specific Enzyme II of the bacterial phosphotransferase system. II. Reconstitution of vectorial transphosphorylation in phospholipid vesicles. J. Biol. Chem. 258:10757–10760.

73. Liberman, E., P. Reddy, C. Gazdar, and A. Peterkofsky. 1985. The Escherichia coli adenylate cyclase complex. Regulation by potassium and phosphate. J. Biol. Chem. 260:4075–4081.

74. Lin, E. C. C. 1970. The genetics of bacterial transport systems. Annu. Rev. Genet. 4:225–262.

75. Majerfeld, I. H., D. Miller, E. Spitz, and H. V. Rickenberg. 1981. Regulation of the synthesis of adenylate cyclase in Escherichia coli by the cAMP-cAMP receptor protein complex. Mol. Gen. Genet. 181:470–475.

76. Mattoo, R. L., R. L. Khandelwal, and E. B. Waygood. 1984. Isoelectrophoretic separation and the detection of soluble proteins containing acid-labile phosphate: use of the phosphoenolpyruvate:sugar phosphotransferase system as a model system for N^1-P-histidine- and N^3-P-histidine-containing proteins. Anal. Biochem. 138:1–16.

77. Mattoo, R. L., and E. B. Waygood. 1983. Determination of the levels of HPr and Enzyme I of the phosphoenolpyruvate-sugar phosphotransferase system in Escherichia coli and Salmonella typhimurium. Can. J. Biochem. Cell. Biol. 61:29–37.

78. Meadow, N. D., and S. Roseman. 1982. Sugar transport by the bacterial phosphotransferase system. Isolation and characterization of a glucose-specific protein (III^Glc) from Salmonella typhimurium. J. Biol. Chem. 257:14526–14537.

79. Meadow, N. D., D. W. Saffen, R. P. Dottin, and S. Roseman. 1982. Molecular cloning of the crr gene and evidence that it is the structural gene for III^Glc, a phosphocarrier protein of the bacterial phosphotransferase system. Proc. Natl. Acad. Sci. USA 79:2528–2532.

80. Misset, O., M. Blaauw, P. W. Postma, and G. T. Robillard. 1983.

Bacterial phosphoenolpyruvate-dependent phosphotransferase system. Mechanisms of the transmembrane sugar translocation and phosphorylation. Biochemistry 22:6163–6170.

81. **Misset, O., M. Brouwer, and G. T. Robillard.** 1980. *Escherichia coli* phosphoenolpyruvate-dependent phosphotransferase system. Evidence that the dimer is the active form of Enzyme I. Biochemistry 19:883–890.

82. **Misset, O., and G. T. Robillard.** 1982. *Escherichia coli* phosphoenolpyruvate-dependent phosphotransferase system: mechanism of phosphoryl-group transfer from phosphoenolpyruvate to HPr. Biochemistry 21:3136–3142.

83. **Mitchell, W. J., T. P. Misko, and S. Roseman.** 1982. Sugar transport by the bacterial phosphotransferase system. Regulation of other transport systems (lactose and melibiose). J. Biol. Chem. 257:14553–14564.

84. **Mori, K., and H. Aiba.** 1985. Evidence for negative control of *cya* transcription by cAMP and cAMP receptor protein in intact *Escherichia coli* cells. J. Biol. Chem. 260:14838–14843.

85. **Neijssel, O. M., D. W. Tempest, P. W. Postma, J. A. Duine, and J. Frank, Jzn.** 1983. Glucose metabolism by K$^+$-limited *Klebsiella aerogenes*: evidence for the involvement of a quinoprotein glucose dehydrogenase. FEMS Microbiol. Lett. 20:35–39.

86. **Nelson, S. O., J. Lengeler, and P. W. Postma.** 1984. Role of III^{Glc} of the phosphoenolpyruvate-glucose phosphotransferase system in inducer exclusion in *Escherichia coli*. J. Bacteriol. 160:360–364.

87. **Nelson, S. O., and P. W. Postma.** 1984. Interactions in vivo between III^{Glc} of the phosphoenolpyruvate:sugar phosphotransferase system and the glycerol and maltose uptake systems of *Salmonella typhimurium*. Eur. J. Biochem. 139:29–34.

88. **Nelson, S. O., B. J. Scholte, and P. W. Postma.** 1982. Phosphoenolpyruvate:sugar phosphotransferase system-mediated regulation of carbohydrate metabolism in *Salmonella typhimurium*. J. Bacteriol. 150:604–615.

89. **Nelson, S. O., A. R. J. Schuitema, R. Benne, L. H. T. van der Ploeg, J. J. Plijter, F. Aan, and P. W. Postma.** 1984. Molecular cloning, sequencing and expression of the *crr* gene: the structural gene for III^{Glc} of the bacterial PEP:glucose phosphotransferase system. EMBO J. 3:1587–1593.

90. **Nelson, S. O., A. R. J. Schuitema, and P. W. Postma.** 1985. The phosphoenolpyruvate:glucose phosphotransferase system of *Salmonella typhimurium*. The phosphorylated form of III^{Glc}. Eur. J. Biochem. 154:337–341.

91. **Nelson, S. O., J. K. Wright, and P. W. Postma.** 1983. The mechanism of inducer exclusion. Direct interaction between purified III^{Glc} of the phosphoenolpyruvate:sugar phosphotransferase system and the lactose carrier of *Escherichia coli*. EMBO J. 2:715–720.

92. **Neuhaus, J. M., and J. K. Wright.** 1983. Chemical modification of the lactose carrier of *Escherichia coli* by plumbagin, phenyl arsinoxide or diethylpyrocarbonate affects the binding of galactoside. Eur. J. Biochem. 137:615–621.

93. **Novotny, M. J., W. L. Frederickson, E. B. Waygood, and M. H. Saier, Jr.** 1985. Allosteric regulation of glycerol kinase by enzyme III^{Glc} of the phosphotransferase system in *Escherichia coli* and *Salmonella typhimurium*. J. Bacteriol. 162:810–816.

94. **Osumi, T., and M. H. Saier, Jr.** 1982. Regulation of lactose permease activity by the phosphoenolpyruvate:sugar phosphotransferase system: evidence for direct binding of the glucose-specific enzyme III to the lactose permease. Proc. Natl. Acad. Sci. USA 79:1457–1461.

95. **Palva, E. T., P. Saris, and T. J. Silhavy.** 1985. Gene fusions to the *ptsM/pel* locus of *Escherichia coli*. Mol. Gen. Genet. 199:427–433.

96. **Parra, F., M. C. Jones-Mortimer, and H. L. Kornberg.** 1983. Phosphotransferase-mediated regulation of carbohydrate utilization in *Escherichia coli* K12: the nature of the *iex (crr)* and *gsr (tgs)* mutations. J. Gen. Microbiol. 129:337–348.

97. **Pastan, I., and R. L. Perlman.** 1969. Repression of β-galactosidase synthesis by glucose in phosphotransferase mutants of *Escherichia coli*. Repression in the absence of glucose phosphorylation. J. Biol. Chem. 244:5836–5842.

98. **Peri, K. G., H. L. Kornberg, and E. B. Waygood.** 1984. Evidence for the phosphorylation of Enzyme II^{Glucose} of the phosphoenolpyruvate:sugar phosphotransferase system of *Escherichia coli* and *Salmonella typhimurium*. FEBS Lett. 178:55–58.

99. **Postma, P. W.** 1976. Involvement of the phosphotransferase system in galactose transport in *Salmonella typhimurium*. FEBS Lett. 61:49–53.

100. **Postma, P. W.** 1977. Galactose transport in *Salmonella typhimurium*. J. Bacteriol. 129:630–639.

101. **Postma, P. W.** 1981. Defective enzyme II-B^{Glc} of the phosphoenolpyruvate:sugar phosphotransferase system leading to un-coupling of transport and phosphorylation in *Salmonella typhimurium*. J. Bacteriol. 147:382–389.

102. **Postma, P. W.** 1982. Regulation of sugar transport in *Salmonella typhimurium*. Ann. Microbiol. (Paris) 133A:261–267.

103. **Postma, P. W.** 1986. Catabolite repression and related processes. Symp. Soc. Gen. Microbiol. 39:21–49.

104. **Postma, P. W., W. Epstein, A. R. J. Schuitema, and S. O. Nelson.** 1984. Interaction between III^{Glc} of the phosphoenolpyruvate:sugar phosphotransferase system and glycerol kinase of *Salmonella typhimurium*. J. Bacteriol. 158:351–353.

105. **Postma, P. W., and J. W. Lengeler.** 1985. Phosphoenolpyruvate:carbohydrate phosphotransferase system of bacteria. Microbiol. Rev. 49:232–269.

106. **Postma, P. W., and S. Roseman.** 1976. The bacterial phosphoenolpyruvate:sugar phosphotransferase system. Biochim. Biophys. Acta 457:213–257.

107. **Postma, P. W., A. Schuitema, and C. Kwa.** 1981. Regulation of methyl β-galactoside permease activity in *pts* and *crr* mutants of *Salmonella typhimurium*. Mol. Gen. Genet. 181:448–453.

108. **Postma, P. W., and J. B. Stock.** 1980. Enzymes II of the phosphotransferase system do not catalyze sugar transport in the absence of phosphorylation. J. Bacteriol. 141:476–484.

109. **Powers, D. A., and S. Roseman.** 1984. The primary structure of *Salmonella typhimurium* HPr, a phosphocarrier protein of the phosphoenolpyruvate:glycose phosphotransferase system. A correction. J. Biol. Chem. 259:15212–15214.

110. **Reddy, P., N. Meadow, S. Roseman, and A. Peterkofsky.** 1985. Reconstitution of regulatory properties of adenylate cyclase in *Escherichia coli* extracts. Proc. Natl. Acad. Sci. USA 82:8300–8304.

111. **Rephaeli, A. W., and M. H. Saier, Jr.** 1980. Regulation of genes coding for enzyme constituents of the bacterial phosphotransferase system. J. Bacteriol. 141:658–663.

112. **Robillard, G. T.** 1982. The enzymology of the bacterial phosphoenolpyruvate-dependent sugar transport systems. Mol. Cell. Biochem. 46:3–24.

113. **Robillard, G. T., G. Dooijewaard, and J. Lolkema.** 1979. *Escherichia coli* phosphoenolpyruvate dependent phosphotransferase system. Complete purification of Enzyme I by hydrophobic interaction chromatography. Biochemistry 18:2984–2989.

114. **Robillard, G. T., and W. N. Konings.** 1981. Physical mechanism for regulation of phosphoenolpyruvate-dependent glucose transport activity in *Escherichia coli*. Biochemistry 20:5025–5032.

115. **Roossien, F. F., M. Blaauw, and G. T. Robillard.** 1984. Kinetics and subunit interaction of the mannitol-specific Enzyme II of the *Escherichia coli* phosphotransferase system. Biochemistry 23:4934–4939.

116. **Roossien, F. F., and G. T. Robillard.** 1984. Mannitol-specific carrier protein from *Escherichia coli* phosphoenolpyruvate-dependent phosphotransferase system can be extracted as a dimer from the membrane. Biochemistry 23:5682–5685.

117. **Roy, A., A. Danchin, E. Joseph, and A. Ullmann.** 1983. Two functional domains in adenylate cyclase of *Escherichia coli*. J. Mol. Biol. 165:197–202.

118. **Roy, A., C. Haziza, and A. Danchin.** 1983. Regulation of adenylate cyclase synthesis in *Escherichia coli*: nucleotide sequence of the control region. EMBO J. 2:791–797.

119. **Rusina, O. Y., and V. N. Gershanovitch.** 1983. Mapping of mutations within genes coding for Enzyme I and HPr of the phosphoenolpyruvate-dependent phosphotransferase system of *Escherichia coli* K12. II. Mapping of *ptsH* mutations within the gene. Genetika 19:397–405.

120. **Rusina, O. Y., T. S. Ilyina, and V. N. Gershanovitch.** 1981. Mapping of mutations within genes coding for Enzyme I and HPr of the phosphoenolpyruvate-dependent phosphotransferase system of *Escherichia coli* K12. I. Mapping of *ptsI* intragenic mutations. Genetika 17:1771–1783.

121. **Saier, M. H., Jr.** 1977. Bacterial phosphoenolpyruvate:sugar phosphotransferase systems: structural, functional, and evolutionary interrelationships. Bacteriol. Rev. 41:856–871.

122. **Saier, M. H., Jr., F. G. Bromberg, and S. Roseman.** 1973. Characterization of constitutive galactose permease mutants in *Salmonella typhimurium*. J. Bacteriol. 113:512–514.

123. **Saier, M. H., Jr., B. U. Feucht, and W. K. Mora.** 1977. Sugar-phosphate: sugar transphosphorylation and exchange group translocation catalyzed by the Enzyme II complexes of the bacterial phosphoenolpyruvate:sugar phosphotransferase system. J. Biol. Chem. 252:8899–8907.

124. **Saier, M. H., Jr., F. C. Grenier, C. A. Lee, and E. B. Waygood.** 1985. Evidence for the evolutionary relatedness of the proteins of the bacterial phosphoenolpyruvate:sugar phosphotransferase system. J. Cell. Biochem. 27:43–56.

125. **Saier, M. H., Jr., D. K. Keeler, and B. U. Feucht.** 1982. Physiological desensitization of carbohydrate permeases and adenylate cyclase to regulation by the phosphoenolpyruvate:sugar phosphotransferase system in *Escherichia coli* and *Salmonella typhimurium.* J. Biol. Chem. **257:**2509–2517.

126. **Saier, M. H., Jr., M. J. Novotny, D. Comeau-Fuhrman, T. Osumi, and J. D. Desai.** 1983. Cooperative binding of the sugar substrates and allosteric regulatory protein (enzyme III^Glc of the phosphotransferase system) to the lactose and melibiose permeases in *Escherichia coli* and *Salmonella typhimurium.* J. Bacteriol. **155:**1351–1357.

127. **Saier, M. H., Jr., and S. Roseman.** 1976. Sugar transport. The *crr* mutation: its effect on repression of enzyme synthesis. J. Biol. Chem. **251:**6598–6605.

128. **Saier, M. H., Jr., and S. Roseman.** 1976. Sugar transport. Inducer exclusion and regulation of the melibiose, maltose, glycerol, and lactose transport systems by the phosphoenolpyruvate:sugar phosphotransferase system. J. Biol. Chem. **251:**6606–6615.

129. **Saier, M. H., Jr., R. D. Simoni, and S. Roseman.** 1970. The physiological behaviour of Enzyme I and heat-stable protein mutants of a bacterial phosphotransferase system. J. Biol. Chem. **245:**5870–5873.

130. **Saier, M. H., Jr., R. D. Simoni, and S. Roseman.** 1976. Sugar transport. Properties of mutant bacteria defective in proteins of the phosphoenolpyruvate:sugar phosphotransferase system. J. Biol. Chem. **251:**6584–6597.

131. **Saier, M. H., Jr., H. Straud, L. S. Massman, J. J. Judice, M. J. Newman, and B. U. Feucht.** 1978. Permease-specific mutations in *Salmonella typhimurium* and *Escherichia coli* that release the glycerol, maltose, melibiose, and lactose transport systems from regulation by the phosphoenolpyruvate: sugar phosphotransferase system. J. Bacteriol. **133:**1358–1367.

132. **Saier, M. H., Jr., W. S. Young, and S. Roseman.** 1971. Utilization and transport of hexoses by mutant strains of *Salmonella typhimurium* lacking Enzyme I of the phosphoenolpyruvate-dependent phosphotransferase system. J. Biol. Chem. **246:**5838–5840.

133. **Sanderson, K. E., and J. R. Roth.** 1983. Linkage map of *Salmonella typhimurium*, edition VI. Microbiol. Rev. **47:**410–453.

134. **Sarno, N. V., L. G. Tenn, A. Desai, A. M. Chin, F. C. Grenier, and M. H. Saier, Jr.** 1984. Genetic evidence for glucitol-specific enzyme III, an essential phosphocarrier protein of the *Salmonella typhimurium* glucitol phosphotransferase system. J. Bacteriol. **157:**953–955.

135. **Schaefler, S.** 1967. Inducible system for the utilization of β-glucoside in *Escherichia coli.* J. Bacteriol. **93:**254–263.

136. **Scholte, B. J., and P. W. Postma.** 1980. Mutation in the *crp* gene of *Salmonella typhimurium* which interferes with inducer exclusion. J. Bacteriol. **141:**751–757.

137. **Scholte, B. J., and P. W. Postma.** 1981. Competition between two pathways for sugar uptake by the phosphoenolpyruvate-dependent sugar phosphotransferase system in *Salmonella typhimurium.* Eur. J. Biochem. **114:**51–58.

138. **Scholte, B. J., A. R. Schuitema, and P. W. Postma.** 1981. Isolation of III^Glc of the phosphoenolpyruvate-dependent glucose phosphotransferase system of *Salmonella typhimurium.* J. Bacteriol. **148:**257–264.

139. **Scholte, B. J., A. R. Schuitema, and P. W. Postma.** 1982. Char-

acterization of factor III^Glc in catabolite repression-resistant (*crr*) mutants of *Salmonella typhimurium.* J. Bacteriol. **149:**576–586.

140. **Slater, A. C., M. C. Jones-Mortimer, and H. L. Kornberg.** 1981. L-Sorbose phosphorylation in *Escherichia coli* K-12. Biochim. Biophys. Acta **646:**365–367.

141. **Sprenger, G. A., and J. W. Lengeler.** 1984. L-Sorbose metabolism in *Klebsiella pneumoniae* and Sor⁺ derivatives of *Escherichia coli* K-12 and chemotaxis toward sorbose. J. Bacteriol. **157:**39–45.

142. **Stock, J. B., E. B. Waygood, N. D. Meadow, P. W. Postma, and S. Roseman.** 1982. Sugar transport by the bacterial phosphotransferase system. The glucose receptors of the *Salmonella typhimurium* phosphotransferase system. J. Biol. Chem. **257:**14543–14552.

143. **Thompson, J., and B. M. Chassy.** 1985. Intracellular phosphorylation of glucose analogs via the phosphoenolpyruvate:mannose-phosphotransferase system in *Streptococcus lactis.* J. Bacteriol. **162:**224–234.

144. **Ullmann, A., and A. Danchin.** 1983. Role of cyclic AMP in bacteria, p. 2–53. *In* P. Greengard and G. A. Robison (ed.), Advances in cyclic nucleotide research, vol. 15. Raven Press, New York.

145. **Walter, R. W., Jr., and R. L. Anderson.** 1973. Evidence that the inducible phosphoenolpyruvate: D-fructose 1-phosphate phosphotransferase system of *Aerobacter aerogenes* does not require "HPr." Biochem. Biophys. Res. Commun. **52:**93–97.

146. **Waygood, E. B., R. L. Mattoo, and K. G. Peri.** 1984. Phosphoproteins and the phosphoenolpyruvate:sugar phosphotransferase system in *Salmonella typhimurium* and *Escherichia coli*: evidence for III^Mannose, III^Fructose, III^Glucitol and the phosphorylation of Enzyme II^Mannitol and Enzyme II^N-acetylglucosamine. J. Cell. Biochem. **25:**139–159.

147. **Waygood, E. B., N. D. Meadow, and S. Roseman.** 1979. Modified assay procedures for the phosphotransferase system in enteric bacteria. Anal. Biochem. **95:**293–304.

148. **Waygood, E. B., and T. Steeves.** 1980. Enzyme I of the phosphoenolpyruvate: sugar phosphotransferase system (PTS) of *Escherichia coli*—purification to homogeneity and some properties. Can. J. Biochem. **58:**40–48.

149. **Weigel, N., M. A. Kukuruzinska, A. Nakazawa, E. B. Waygood, and S. Roseman.** 1982. Sugar transport by the bacterial phosphotransferase system. Phosphoryl transfer reactions catalyzed by Enzyme I of *Salmonella typhimurium.* J. Biol. Chem. **257:**14477–14491.

150. **Weigel, N., D. A. Powers, and S. Roseman.** 1982. Sugar transport by the bacterial phosphotransferase system. Primary structure and active site of a general phosphocarrier protein (HPr) from *Salmonella typhimurium.* J. Biol. Chem. **257:**14499–14509.

151. **Weigel, N., E. B. Waygood, M. A. Kukuruzinska, A. Nakazawa, and S. Roseman.** 1982. Sugar transport by bacterial phosphotransferase system. Isolation and characterization of Enzyme I from *Salmonella typhimurium.* J. Biol. Chem. **257:**14461–14469.

152. **White, R. J.** 1970. The role of the phosphoenolpyruvate phosphotransferase system in the transport of N-acetyl-D-glucosamine by *Escherichia coli.* Biochem. J. **118:**89–92.

153. **Yang, J. K., and W. Epstein.** 1983. Purification and characterization of adenylate cyclase from *Escherichia coli* K-12. J. Biol. Chem. **258:**3750–3758.

12. Glycolysis, Pentose Phosphate Pathway, and Entner-Doudoroff Pathway

DAN G. FRAENKEL

Department of Microbiology and Molecular Genetics, Harvard Medical School, Boston, Massachusetts 02115

INTRODUCTION

Glycolysis and the pentose phosphate pathway are the two central, and constitutive, routes of intermediary carbohydrate metabolism in enteric bacteria (Fig. 1). A third central but inducible route, the Entner-Doudoroff pathway, is also mentioned. The text mainly considers function of the various pathways and reactions, with a final section on general problems. Table 1 lists, by reaction, some key or recent papers on the genetics, enzymology, and molecular biology of the individual reactions.

THE GLYCOLYTIC PATHWAY

Glucokinase and Phosphoglucomutase

Although the phosphoenolpyruvate phosphotransferase system (chapter 11) is the usual route of glucose uptake and phosphorylation, enteric bacteria also contain a glucokinase (*glk*). The normal role of this enzyme is unclear; it might act on glucose generated internally (e.g., from lactose). The slow growth on glucose of mutants defective in phosphotransferase components depends on *glk*, since its further loss then causes almost complete inability to grow on glucose (28, 51). The slow growth on glucose of phosphofruc-

tokinase (*pfkA*) mutants also depends on glucokinase (115). In a wild-type background, *glk* mutants grow almost normally on glucose (28).

Catabolism of galactose and maltose involves formation of glucose 1-phosphate (glucose-1-P), which is converted to glucose-6-P by phosphoglucomutase (*pgm*). In growth on other carbon sources, the same reaction is used biosynthetically to make glucose-1-P (and hence UDP-glucose). Surprisingly, mutants lacking most phosphoglucomutase activity were not greatly affected in growth on galactose (2).

Phosphoglucose Isomerase and Phosphomannose Isomerase

The interconversion of glucose-6-P and fructose-6-P is catalyzed by phosphoglucose isomerase. Mutants (*pgi*) grow slowly on glucose, and this growth depends on the pentose phosphate pathway (54). Phosphoglucose isomerase is responsible for synthesis of glucose-6-P in growth on substances not yielding it by direct catabolism (e.g., glycerol, gluconate, etc.), but growth is adequate because glucose-6-P and derivatives are apparently not essential, as shown by lipopolysaccharide analysis (55), DNA glucosylation (69), or acid-hydrolyzable glucose content, even in a nonsense mutant (132). Purification has revealed a minor form

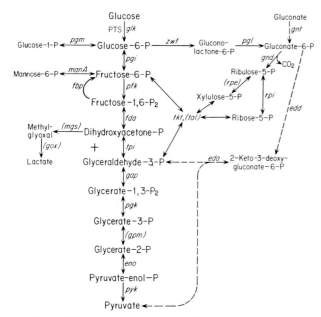

FIG. 1. Glycolysis, the pentose phosphate pathway, and the Entner-Duodoroff pathway. The reactions are schematized, and cofactors and cosubstrates (ADP, P_i, NAD, etc.) are not shown. Gene symbols are those used (or, in parentheses, suggested) for *E. coli*.

of the same enzyme with altered charge, of unknown significance (119), and normal location of some phos-

phoglucose isomerase to the periplasm has also been reported (59).

Phosphomannose isomerase (*manA, pmi*), like phosphoglucomutase, is a side reaction to the glycolytic pathway used both in sugar catabolism (growth on mannose) and polysaccharide synthesis. Mutants lacking the isomerase are blocked in growth on mannose and are defective in mannose-containing polysaccharides (93, 116).

Phosphofructokinase (Fructose-6-P-1-Kinase) and Fructose Bisphosphatase

Phosphofructokinase catalyzes the phosphorylation of fructose-6-P by ATP to yield fructose-1,6-bisphosphate (fructose-1,6-P_2). The main phosphofructokinase of *Escherichia coli*, Pfk-1, is an allosteric enzyme activated by nucleoside diphosphates and inhibited by phosphoenolpyruvate (16). *pfkA* is the structural gene (128), and *pfkA* mutants are impaired in their growth on substances broken down via fructose-6-P (115, 131).

E. coli K-12 also contains a minor phosphofructokinase activity (ca. 10% of the total), Pfk-2, coded by *pfkB* (29). This activity is needed for the residual growth of *pfkA* mutants, but has no clear function in *pfkA*⁺ strains. The *pfkB1* mutation was recognized as a suppressor of the glucose negativity of a *pfkA* mutant (97); it increases expression of Pfk-2 by ca. 25-fold (5) and is a point mutation in the promoter (29). Originally thought to be nonallosteric, Pfk-2 was later shown (unlike Pfk-1) to be inhibited by ATP (85). Strains with

TABLE 1. Some references[a]

Reaction	Gene symbol and map position (min)	References
Glycolytic pathway		
Glucokinase (EC 2.7.1.2)	*glk* (52)	M, 28; E, 61; C, 60
(Phosphoglucomutase [EC 2.7.5.1])	*pgm* (15)	M, 2; E, 81
Phosphoglucose isomerase (EC 5.3.1.9)	*pgi* (91; 91)	M, 47; E, 119; C, 129
(Phosphomannose isomerase [EC 5.3.1.8])	*manA* (36); *pmi* (31)	M, 93, 135; C, S, 95
Phosphofructokinase-1(EC 2.7.1.11)	*pfkA* (88)	M, 3, 106; E, 16, 86; C, 3, 122
Phosphofructokinase-2 (EC 2.7.1.11)	*pfkB* (38)	M, 32; E, 86; C, 29; S, 30
Fructose-1,6-bisphosphatase (EC 3.1.3.11)	*fbp* (96)	M, 47; E, 6; C, 121
Fructose-1,6-bisphosphate aldolase (EC 4.1.2.13)	*fda* (63)	M, 23; E, 12, 13; C, 129
(Methylglyoxal synthase [EC 4.2.99.11])	*mgsA*	E, 74
Triose-phosphate isomerase (EC 5.3.1.1)	*tpiA* (88)	M, 3, 106; C, 122; S, 108
Glyceraldehyde-3-phosphate dehydrogenase (EC 1.2.1.12)	*gap* (39)	M, 75; E, 70; C, 20
Phosphoglycerate kinase (EC 2.7.2.3)	*pgk* (63)	M, 75; E, 34; C, 129
Phosphoglycerate mutase (EC 2.7.5.3)	*gpm*	E, 34
Enolase (EC 4.2.1.11)	*eno* (60)	M, 71; E, 126; C, 129
Pyruvate kinase (EC 2.7.1.40)	*pykA, pykF* (36)	M, 64; E, 62, 92
Pentose phosphate pathway		
Glucose-6-phosphate dehydrogenase (EC 1.1.1.49)	*zwf* (41)	M, 50; E, 14; C, 129
6-Phosphogluconolactonase (EC 3.1.1.31)	*pgl* (17)	M, 88
6-Phosphogluconate dehydrogenase (EC 1.1.1.44)	*gnd* (44; 42)	M, 99, 137; E, 140; C, 15, 101; S, 100
Ribose-5-phosphate isomerase (EC 5.3.1.6)	*rpiA* (63)	M, 124
Transketolase (EC 2.2.1.1)	*tkt*	M, 79
Entner-Duodoroff pathway		
6-Phosphogluconate dehydrase (EC 4.2.1.12)	*edd* (41)	M, 50
2-Keto-3-deoxy-6-phosphogluconate aldolase (EC 4.1.2.14)	*eda* (41)	M, 50; E, 109

[a] This table collects some references to the genetic mapping (M), enzymology (E), cloning (C), or DNA sequencing (S) for most of the reactions in Fig. 1. Many references here are not mentioned in the text, and vice versa. Gene symbols and map positions are given first for *E. coli* (9) and, after a semicolon, for *Salmonella typhimurium* (117). EC numbers are according to *Enzyme Nomenclature 1978* (102). Reactions in parentheses are not directly in the listed pathway.

high levels of a mutant form, Pfk-2*, insensitive to ATP inhibition (66), are impaired in growth on gluconeogenic substances but slightly improved in growth on sugars (31, 33).

Fructose bisphosphatase is used to form fructose-6-P from fructose-1,6-P_2 in growth on substances such as glycerol, succinate, and acetate, and *fbp* mutants fail to grow on these carbon sources (53). *fbp* is the structural gene (121). The enzyme, present constitutively, is sensitive to inhibition by 5'-AMP (6). A mutant strain with AMP-insensitive enzyme in normal amount was not clearly impaired in growth (120).

Fructose-1,6-P_2 Aldolase

The enzyme splitting fructose-1,6-P_2 to dihydroxyacetone phosphate and glyceraldehyde-3-P is, as in other bacteria, a metal-dependent class II aldolase (aldolase 2; 13). Mutants (*fda*) are unable to grow on sugars and sugar derivatives, but still grow on substances such as glycerol or succinate. The most thoroughly studied mutant was temperature sensitive (17, 18), but nonconditional mutants have also been reported (24, 76). Growth of the mutant was inhibited by glucose and other substances giving fructose-1,6-P_2, and secondary blocks preventing the accumulation allowed growth, e.g., on gluconate in the *fda gnd* double mutant (118).

Growth of *fda* mutants on glycerol and other substances requires the formation of fructose-1,6-P_2 from triose phosphate, perhaps employing a minor aldolase activity still found in the mutants (24); this activity may be due to the class I (Schiff base-forming) aldolase that has been characterized from lactate-grown Crookes strain of *E. coli* (12). Differences in aldolase, depending on aeration, have been reported (39).

Triosephosphate Isomerase and the Methylglyoxal Pathway

Triosephosphate isomerase (*tpi*) interconverts dihydroxyacetone phosphate and glyceraldehyde-3-P, and mutants are unable to grow on sugars, glycerol, or succinate alone, but grow on succinate and other substances entering metabolism after the block if appropriately supplemented with ribose (4) or glycerol (76). Such strains are very sensitive to accumulation of dihydroxyacetone phosphate, and perhaps differences in sensitivity, or leakiness, account for the ability (4) or inability (76) of different strains to grow on gluconate. Dihydroxyacetone phosphate toxicity is related to its enzymatic conversion (74) to the bactericidal agent methylglyoxal (26). Although there is a glyoxylase that converts methylglyoxal to D-lactate, and thus a pathway for dihydroxyacetone phosphate dissimilation, it is of low capacity and insufficient to prevent methylglyoxal toxicity in the *tpi* mutant (26).

There are other examples in which excess dihydroxyacetone phosphate causes toxic levels of methylglyoxal (1, 58, 94), and the fact of there being a constitutive enzyme, methylglyoxal synthase, catalyzing its formation seems surprising. Hopper and Cooper (74) have speculated that inhibition by P_i might relate to a normal function for the enzyme, in conditions of limited phosphate, in a sequence bypassing the lower portion (phosphate requiring) of the glycolytic pathway. Mutants with elevated glyoxylase can grow on 1 mM methylglyoxal (58). Methylglyoxal has been reviewed (25).

From Glyceraldehyde-3-P to Phosphoenolpyruvate

For the conversion of glyceraldehyde-3-P to phosphoenolpyruvate, the sequence of reactions is glyceraldehyde-3-P dehydrogenase (*gap*), phosphoglycerate kinase (*pgk*), phosphoglycerate mutase (*gpm*), and enolase (*eno*). A mutational block would be expected to prevent growth on sugars or other materials entering the pathway above (e.g., glucose or glycerol) or below (e.g., succinate or pyruvate) the block, and *gap*, *pgk*, and *eno* mutants were obtained as glycerol negative, succinate negative, but glycerol + succinate positive (71, 75, 76). For these three steps, at least, single enzymes seem to catalyze the reaction in both directions; phosphoglycerate mutase mutants are not known. A thorough study of *gap*, *pgk*, and *eno* mutants (76) reported metabolite accumulations and some nutritional differences between them, and titration of the glycerol requirement showed triose phosphate derivatives to be about 1/20 of total catabolic and anabolic needs. The extreme sensitivity of such mutants to glucose and other sugars was shown to involve both osmotically preventable lysis and particularly severe catabolite repression, but not necessarily mere metabolite accumulation, and not methylgloxal formation.

Pyruvate Kinase

There are two pyruvate kinase isoenzymes, Pyk-I or -F and Pyk-II or -A. They differ kinetically, Pyk-F being activated by fructose-1,6-P_2 (19, 133) and Pyk-A being activated by 5'-AMP and other metabolites (131), and also differ structurally (92), in expression (83, 87), and genetically (64). Double mutants fail to grow more than feebly on sugars except (aerobically) on substrates of the phosphotransferase system, where pyruvate formation during uptake can bypass the block (63). Mutants lacking either one of the isoenzymes grow adequately (63, 64).

THE PENTOSE PHOSPHATE PATHWAY

Oxidative Branch

Glucose-6-P dehydrogenase, 6-phosphogluconolactonase, and 6-phosphogluconate dehydrogenase. Mutants lacking phosphoglucose isomerase depend on the pentose phosphate pathway (oxidative formation of pentose phosphate from glucose-6-P, followed by the reactions of the nonoxidative branch yielding fructose-6-P and glyceraldehyde-3-P). Thus, mutants lacking glucose-6-P dehydrogenase (*zwf*) were obtained as glucose-negative derivatives of *pgi* mutants (48); such double mutants are completely blocked in productive glucose metabolism (49). There are deletions of *zwf* (50) and closely linked *cis*-acting mutations increasing its expression (56). The second reaction of the oxidative branch, 6-phosphogluconolactonase, has been barely studied. *pgi pgl* double mutants were recognized as leaky glucose-negative derivatives of *pgi* mutants (88). The leakiness probably reflects the fact that the lactonase reaction also can proceed nonenzymatically, as confirmed in part by metabolic studies (89). Double mutants with lesions in *pgi* and *gnd* (6-phosphogluconate dehydrogenase) are also unable to grow on glucose (89), but *gnd* mutants were first obtained as gluconate-negative derivatives of mutants (*edd*) lacking the first enzyme of the Entner-Doudoroff pathway (48).

Mutants blocked in any one of the three reactions of the oxidative branch grow at near-normal rates on glucose in batch culture (57). However, in mixed culture with the wild type in glucose-limited chemostats, a *gnd* mutant was rapidly counterselected (68), whereas a *zwf* mutant was almost stable (40).

The *gnd* gene has been subject to more detailed study than any other one discussed in this chapter. Specialized transducing bacteriophages were described (136, 138), cell-free enzyme synthesis was reported (77), and high-level enzyme strains have been employed for enzyme purification (104, 140). Subcloning, fine-structure mapping, and sequence analysis of the gene have also been accomplished (100, 101).

The expression of *gnd* (enzyme activity or protein) is proportional to growth rate over about a fivefold range (45, 139). In *gnd-lac* operon fusions, *lac* expression was not proportional to growth rate (10), but in *gnd-lac* protein fusion strains, proportionality was observed in fusions containing at least the amino-terminal part of the *gnd* gene (11). These and other results seem to show that growth rate dependence is posttranscriptional and requires a sequence early in the structural gene.

Nonoxidative Branch

The interconversion of the three pentose phosphates depends on ribosephosphate epimerase (*rpe*) and ribosephosphate isomerase (*rpi*). Any one of the pentose phosphates may be the initial metabolite (e.g., ribulose-5-P from the *gnd* reaction, ribose-5-P from ribose, and xylulose-5-P from xylose and L-arabinose). Both the epimerase and the isomerase are needed to make other substrate for the transketolase reaction. Two ribosephosphate isomerase activities, A and B, have been found in *E. coli* (35, 43, 123). *rpiA* mutants lack activity A and are ribose auxotrophs, but contain the activity B in growth on ribose, adequate for ribose catabolism (123); suppression of the ribose requirement of *rpiA* strains was caused by mutation to high-level expression of activity B (43).

As usually written, transformation of pentose phosphates into fructose-6-P and glyceraldehyde-3-P involves the sequential action of transketolase, transaldolase, and transketolase again, 3 pentose phosphates giving 2½ hexose phosphates. At least two of the intermediates are used in biosynthesis, erythrose-4-P for the aromatic pathway and sedoheptulose-7-P for heptose derivatives. Transketolase (*tkt*) mutants are aromatic auxotrophs also unable to grow on pentoses (79, 80) and deficient in heptose (41).

Functions of the Pentose Phosphate Pathway

Pentose synthesis may occur both by the oxidative branch and by the nonoxidative branch, and hence, mutants defective in either branch (*rpiA* excepted) are not pentose auxotrophs, whereas a double *gnd tkt* mutant is (80). A variety of labeling experiments support this view of pentose synthesis (21, 78, 80) and also indicate that the oxidative pathway is used much less anaerobically than aerobically.

A second possible function of the oxidative branch is to provide NADPH for biosynthesis, since both glucose-6-P dehydrogenase and 6-phosphogluconate dehydrogenase use NADP. Results of labeling experiments accord with such a function, but the oxidative branch is probably not the major source of NADPH (27).

The third suggested function for the pentose pathway is to operate cyclically for complete oxidation of hexose monophosphate to CO_2 and triose phosphate (i.e., the "hexose monophosphate shunt"). In *E. coli* it is unlikely to function readily in this way, for mutants blocked in the phosphofructokinase reaction are almost completely unable to grow on glucose or arabinose (32). Noncyclic use of the pathway as a route from glucose-6-P to fructose-6-P, however, can occur, as in *pgi* mutants growing on glucose, and it is likely that there is some cyclic use of the shunt in a wild-type strain growing on glucose (82).

The nonoxidative branch of the pentose pathway clearly does have a major catabolic role in growth on pentoses, as well as being used for the portion of gluconate metabolism not employing the Entner-Doudoroff pathway (104).

THE ENTNER-DOUDOROFF PATHWAY

The Entner-Doudoroff pathway has two reactions, a dehydrase (*edd*) forming 2-keto-3-deoxygluconate-6-P from gluconate-6-P and an aldolase (*eda*) giving pyruvate and glyceraldehyde-3-P. The pathway is inducible and employed for metabolism of gluconate, which is converted to gluconate-6-P by gluconokinase (42, 52, 54). The aldolase is also involved in catabolism of glucuronate and galacturonate (chapter 18). *edd* mutants (141) grow at a reduced rate on gluconate, using the pentose phosphate pathway. *eda* mutants (44, 46, 109) grow on neither gluconate nor uronic acids; the gluconate negativity is related to toxicity, since *eda edd* double mutants do grow.

edd and *eda* are closely linked (50). Their expressions are differently controlled, but with an interesting symmetry. The aldolase (*eda*), which has a high basal level, is induced in growth on the glucuronic acids as governed, at least in part, by *kdgR*, and *kdgR* mutants are constitutive for aldolase but not dehydrase (110). The dehydrase (*edd*), which has a low basal level, is induced during growth on gluconate (see below), which also induces the aldolase but not the other functions governed by *kdgR* (110). Induction of the dehydrase as well as the gluconokinase activity is governed by *gntR*; mutants are constitutive (8, 142). Conditions for induction of dehydrase but not kinase were also observed (84).

The role of the Entner-Doudoroff pathway for glucose metabolism in *E. coli* is uncertain. Its employment by a wild-type strain was minimal (54), while for phosphoglucose isomerase (*pgi*) mutants data are equivocal: although the dehydrase was uninduced, labeling and metabolite measurements suggested some use of the pathway (52, 54, 104). Double mutants, *pgi gnd*, were variously found to grow (84, 118) or not (88, 104) on glucose or glucose-6-P. However, the recent finding of glucose oxidase activity in *E. coli* dependent on added pyrrolo-quinoline quinone (73) suggests that in nature Entner-Doudoroff metabolism of glucose by *E. coli* might be significant.

PROBLEMS

Gene Expression

Many of the enzymes are apparently constitutively expressed, their levels not being sharply dependent on obvious metabolic need, as illustrated by 6-phosphogluconate dehydrogenase. The difference in enzyme level between a situation in which it is almost certainly not used (anaerobic growth) and one where it is used (aerobic growth on gluconate) is not more than twofold (45, 127). Furthermore, in growth of an *edd* mutant on gluconate, in which higher expression of *gnd* would increase growth rate (96), and gluconate-6-P levels are high (104), the dehydrogenase level was the same as in the wild type (45). Likewise, in a *zwf* mutant, which had no detectable gluconate-6-P in growth on glucose (104), the dehydrogenase levels were the same in growth on glucose and gluconate (48). Although the *edd* and *zwf* mutants are not normal cases, they do show that the expression of *gnd* is probably unrelated to the level of gluconate-6-P, and, in general, similar arguments against substrate-dependent expression can be made for many of the other genes.

A second consideration with respect to gene expression is that in contrast with highly inducible enzymes, specific regulatory genes, in which mutations lead to higher or lower expression, are not known. The reported genes are almost all defined by deficiency mutations and, where this information is known (through nucleotide sequence, temperature sensitivity, etc.), are structural. (Cases, such as *zwfL1* [56], of *cis*-acting mutations increasing constitutive expression may be promoter alterations, as known for *pfkB1* [29]). The idea that such genes do not require special positive factors for their expression also fits with the relative ease of obtaining efficient cell-free expression (as for *gnd* [77] and *pfkA* [128]). It also accords with there being no operons, as the various genes are unlinked or, if closely linked (*pfkA-tpi, pgk-fda*; 129), with there being no evidence for coexpression. (*pfkA* and *tpi* are now known to be separated by at least one unrelated gene [3].) Furthermore, the fact that cloning on multicopy plasmids gives high enzyme levels argues against titratable elements necessary for expression (see reference 45) as well as against autogenous control.

It should be emphasized, however, that for most of the genes systematic studies of these points have not been made and very few mutants have been obtained. Furthermore, in spite of the above comments, there is considerable evidence for some regulation of gene expression, such as growth rate dependence, common in *E. coli* proteins and studied in detail for *gnd* (see above); data on growth rate dependence in chemostat culture are also available (72). (Growth rate might also be expected to affect relative gene dose according to chromosomal position, but studies have not been reported for those enzymes.)

There have also been many observations of levels of certain glycolytic enzymes being higher in anaerobic than in aerobic culture, e.g., in glucose-limited chemostats (112, 127) and in batch culture (87, 125). In general, the differences are less than threefold in the steady state, but the relative rates of synthesis after shifts between the two conditions show considerably larger differences (38, 125). There is barely any knowledge of the mechanisms governing these changes or of their physiological function. One problem has been to resolve whether different enzyme levels reflect the same enzyme. Both pyruvate kinase isoenzymes are present in substantial amount but are clearly expressed somewhat differently, whereas for phosphofructokinase the predominant Pfk-1 isoenzyme found aerobically is the same one found at higher levels anaerobically (7, 87). Covalent modifications are also conceivable, but are not yet known for these enzymes.

Finally, studies of these genes with respect to other global control mechanisms (e.g., heat shock, SOS, starvations) have not been reported. On the basis of similar or higher enzyme levels in growth on glucose compared with glycerol, most of the genes are probably not subject to carbon catabolite repression. It is not known whether cyclic nucleotides or other alarmones are involved either. Cyclic AMP has been implicated with *fnr* as being needed for expression of a number of anaerobic functions (130), but *fnr* mutants do grow anaerobically on glucose (90).

Metabolism

Metabolic studies are surprisingly scarce. There are two types of questions to address: assessment of metabolic fluxes, and relating fluxes to properties of the enzymes. There is one major source of data on general metabolic flow in wild-type *E. coli* (113) and several on metabolite levels in a variety of situations (e.g., references 91 and 98). However, with some exceptions (e.g., chemostat studies on glucose metabolism; 72), the question which has received most attention is the more specialized one of estimating the relative use of one pathway versus another, e.g., the pentose phosphate pathway versus glycolysis. Various methods, usually involving comparison of the fate of the C-1 and C-6 positions of glucose (see Pentose Phosphate Pathway, above, and reference 103), have shown its fractional use to be 25% or less. However, estimation of net and gross rates, particularly with respect to fluxes in the nonoxidative pathway and the likely equilibration of glucose-6-P and fructose-6-P, requires more information (82).

The natural question of how glucose metabolism is governed, on the other hand, has been difficult even to pose clearly. Metabolic rate certainly varies with exogenous factors such as aerobiosis, growth rate, and limiting nutrient, and in a range of situations it has been closely fitted to a single equation with the only variables being glucose-6-P as inhibitor and fructose-1,6-P_2 as activator (36, 37). Dietzler et al. suggested that levels of the two metabolites were primarily governed by ATP changes through an inhibition of phosphofructokinase and that they in turn act at uptake.

The idea that it is the irreversible steps of the glycolytic pathway that are controlled seems reasonable in view of the fact that two of those steps depend on enzymes, phosphofructokinase and pyruvate kinase, that are sensitive to allosteric effectors (control of the phosphotransferase system is considered in chapter 11), and studies of metabolite levels after metabolic perturbations (67, 91) offer some support. Another way of studying these reactions is to assess them in vitro or in cells made permeable to substrates, using the in vivo metabolite and effector concentrations, as done with phosphofructokinase (111). Attempts to relate in vivo fluxes in the pentose pathway with metabolite levels and in vitro characteristics of the two dehydrogenases of the pathway have also been reported (104, 105).

Another approach to the problem of metabolic control of flux in these pathways is to obtain mutants altered in allosteric characteristics. Again, examples are few. There are cases of clearly deleterious effects, such as Pfk-2* impairing gluconeogenesis (31) and a mutant fructose bisphosphatase in high amount seeming to impair glycolysis (120). Perhaps of equal interest, however, are the several cases showing the relative tolerance of cells to such enzymic differences (22, 33, 114, 120).

Finally, another complication is offered by the possibility that the pathways might be organized as complexes with channeling of metabolites from enzyme to enzyme, i.e., compartmentation. There has been some consideration of E. coli glycolysis in this regard (e.g., reference 65).

The general conclusion from these comments about metabolism is that there is much to be learned.

Isoenzymes

Four examples of isoenzymes were cited: Pfk-1/Pfk-2, Fda-1/Fda-2, Pyk-A/Pyk-F, and Rpi-A/Rpi-B. Each one likely involves separate genes for the separate enzymes, but the quality of evidence varies. Also, as discussed, the functional roles of the individual isoenzymes are usually not clear. Pfk-2 in E. coli K-12, for example, seems entirely dispensable and is normally present at low levels which, to date, have only been increased by mutation. Fda-1 might serve a gluconeogenic function in a mutant lacking Fda-2, but assessing its role in a strain having Fda-2 is another matter, and there are no Fda-1 mutants. For Rpi-B, it is not even clear that the enzyme is expressed at all when a strain has Rpi-A, and Rpi-B⁻ mutants are not known. Pyk-A and Pyk-F present a more symmetrical situation in that either seems to be expressed in adequate level in the absence of the other, a result that also does not clarify individual functions.

Inhibitions

As mentioned, it is characteristic of many of the mutants that their growth on permissive substances is inhibited by substances that cannot be metabolized. Inhibition is sometimes accompanied by substantial accumulation of metabolites before the block, but accumulation of a metabolite is not always toxic (e.g., glucose-6-P in a pgi zwf mutant; 49). There is no general explanation for toxicities. Aside from the possible specific direct effect of high levels of the accumulated metabolite (e.g., on phosphotransferase

system function; 115), methylglyoxal formation (see above), catabolite repression, and osmotically remedied effects (76) have all been implicated. The phenomenon is a useful probe of bacterial physiology, and it sometimes allows positive mutant selections, such as gnd in an fda strain (118). In that case, a substance (gluconate) inhibits growth because of a blocked secondary pathway even though the primary pathway (Entner-Doudoroff) is available; in the complementary case, blockage of the main pathway (eda in the Entner-Doudoroff pathway) prevents use of the pentose pathway, but the inhibition is reversed by an earlier block (edd) preventing accumulation of the intermediate (50).

Such phenomena complicate studies of metabolism with mutants, for the inability to grow on one substrate may reflect a toxicity rather than the primacy of the blocked pathway. The cases of glyoxylase and 6-phosphogluconolactonase, mentioned earlier, relate to the importance of enzymes in reducing substrate concentrations.

ACKNOWLEDGMENTS

Work from this laboratory has been supported by the National Science Foundation and the National Institutes of Health.

LITERATURE CITED

1. **Ackerman, R. S., N. R. Cozzarelli, and W. Epstein.** 1974. Accumulation of toxic concentrations of methylglyoxal by wild-type Escherichia coli K-12. J. Bacteriol. **119:**357–362.
2. **Adhya, S., and M. Schwartz.** 1971. Phosphoglucomutase mutants of Escherichia coli. J. Bacteriol. **108:**621–626.
3. **Albin, R., and P. M. Silverman.** 1984. Physical and genetic structure of the glpK-cpxA interval of the Escherichia coli chromosome. Mol. Gen. Genet. **197:**261–271.
4. **Anderson, A., and R. A. Cooper.** 1969. Gluconeogenesis in Escherichia coli. The role of triose phosphate isomerase. FEBS Lett. **4:**19–20.
5. **Babul, J.** 1978. Phosphofructokinase from Escherichia coli. Purification and characterization of the non-allosteric isozyme. J. Biol. Chem. **253:**4350–4355.
6. **Babul, J., and V. Guixé.** 1983. Fructose bisphosphatase from Escherichia coli. Purification and characterization. Arch. Biochem. Biophys. **225:**944–949.
7. **Babul, J., J. P. Robinson, and D. G. Fraenkel.** 1977. Are the aerobic and anaerobic phosphofructokinases in Escherichia coli different? Eur. J. Biochem. **74:**533–537.
8. **Bächi, B., and H. L. Kornberg.** 1975. Genes involved in the uptake and catabolism of gluconate by Escherichia coli. J. Gen. Microbiol. **90:**321–335.
9. **Bachmann, B. J.** 1983. Linkage map of Escherichia coli K-12, edition 7. Microbiol. Rev. **47:**180–230.
10. **Baker, H. V., II, and R. E. Wolf, Jr.** 1983. Growth rate-dependent regulation of 6-phosphogluconate dehydrogenase level in Escherichia coli K-12: β-galactosidase expression in gnd-lac operon fusion strains. J. Bacteriol. **153:**771–781.
11. **Baker, H. V., II, and R. E. Wolf, Jr.** 1984. An essential site for growth rate-dependent regulation within the Escherichia coli gnd structural gene. Proc. Natl. Acad. Sci. USA **81:**7669–7673.
12. **Baldwin, S. A., and R. N. Perham.** 1978. Novel kinetic and structural properties of the class-I D-fructose 1,6-bisphosphate aldolase from Escherichia coli (Crookes' strain). Biochem. J. **169:**643–652.
13. **Baldwin, S. A., R. N. Perham, and D. Stribling.** 1978. Purification and characterization of the class-II D-fructose-1,6-bisphosphate aldolase from Escherichia coli (Crookes' strain). Biochem. J. **169:**633–641.
14. **Banerjee, S., and D. G. Fraenkel.** 1972. Glucose-6-phosphate dehydrogenase from Escherichia coli and from a "high-level" mutant. J. Bacteriol. **110:**155–160.
15. **Bhaduri, S., T. Kasai, D. Schlessinger, and H. J. Raskas.** 1980. pMB-9 plasmids bearing the Salmonella typhimurium his operon and gnd gene. Gene **8:**239–250.
16. **Blangy, D., H. Buc, and J. Monod.** 1968. Kinetics of the allosteric interactions of phosphofructokinase from Escherichia coli. J. Mol. Biol. **31:**13–35.

17. **Böck, A., and F. C. Neidhardt.** 1966. Isolation of a mutant of *Escherichia coli* with a temperature-sensitive fructose-1,6-diphosphate aldolase activity. J. Bacteriol. **92:**464–469.

18. **Böck, A., and F. C. Neidhardt.** 1966. Properties of a mutant of *Escherichia coli* with a temperature-sensitive fructose-1,6-diphosphate aldolase. J. Bacteriol. **92:**470–476.

19. **Boiteux, A., M. Markus, T. Pleser, and B. Hess.** 1983. Analysis of progress curves. Interaction of pyruvate kinase from *Escherichia coli* with fructose 1,6-bisphosphate and calcium ions. Biochem. J. **211:**631–640.

20. **Branlant, G., G. Flesch, and C. Branlant.** 1983. Molecular cloning of the glyceraldehyde-3-phosphate dehydrogenase genes of *Bacillus stereothermophilus* and *Escherichia coli*, and their expression in *Escherichia coli*. Gene **25:**1–7.

21. **Caprioli, G., and D. Rittenberg.** 1969. Pentose synthesis in *Escherichia coli*. Biochemistry **8:**3375–3384.

22. **Chambost, J.-P., and D. G. Fraenkel.** 1980. The use of 6-labelled glucose to assess futile cycling in *Escherichia coli*. J. Biol. Chem. **255:**2867–2869.

23. **Clark, D. P., and J. E. Cronan, Jr.** 1980. Acetaldehyde coenzyme A dehydrogenase of *Escherichia coli*. J. Bacteriol. **144:**179–184.

24. **Cooper, R. A.** 1969. The conversion of lactate to carbohydrate in *Escherichia coli*. FEBS Symp. **19:**99–106.

25. **Cooper, R. A.** 1984. Metabolism of methylglyoxal in microorganisms. Annu. Rev. Microbiol. **38:**49–68.

26. **Cooper, R. A., and A. Anderson.** 1970. The formation and metabolism of methylglyoxal during glycolysis in *Escherichia coli*. FEBS Lett. **11:**273–276.

27. **Csonka, L., and D. G. Fraenkel.** 1977. Pathways of NADPH formation in *Escherichia coli*. J. Biol. Chem. **152:**3382–3391.

28. **Curtis, S. J., and W. Epstein.** 1975. Phosphorylation of D-glucose in *Escherichia coli* mutants defective in glucosephosphotransferase, mannosephosphotransferase, and glucokinase. J. Bacteriol. **122:**1189–1199.

29. **Daldal, F.** 1983. Molecular cloning of the gene for phosphofructokinase-2 of *Escherichia coli* and the nature of a mutation, *pfkB1*, causing a high level of the enzyme. J. Mol. Biol. **168:**285–305.

30. **Daldal, F.** 1984. Nucleotide sequence of gene *pfkB* encoding the minor phosphofructokinase of *Escherichia coli* K12. Gene **28:**337–342.

31. **Daldal, F., J. Babul, V. Guixé, and D. G. Fraenkel.** 1982. An alteration in phosphofructokinase 2 of *Escherichia coli* which impairs gluconeogenic growth and improves growth on sugars. Eur. J. Biochem. **126:**373–379.

32. **Daldal, F., and D. G. Fraenkel.** 1981. Tn*10* insertions in the *pfkB* region of *Escherichia coli*. J. Bacteriol. **147:**935–943.

33. **Daldal, F., and D. G. Fraenkel.** 1983. Assessment of a futile cycle involving reconversion of fructose 6-phosphate to fructose 1,6-bisphosphate during gluconeogenic growth of *Escherichia coli*. J. Bacteriol. **153:**390–394.

34. **D'Alessio, G., and J. Josse.** 1971. Glyceraldehyde phosphate dehydrogenase, phosphoglycerate kinase, and phosphoglyceromutase of *Escherichia coli*. Simultaneous purification and physical properties. J. Biol. Chem. **246:**4319–4325.

35. **David, J., and H. Wiesmeyer.** 1970. Regulation of ribose metabolism in *E. coli*. II. Evidence for two ribose 5-phosphate isomerase activities. Biochim. Biophys. Acta **208:**56–67.

36. **Dietzler, D. N., M. Leckie, P. E. Bergstein, and M. J. Sughrue.** 1975. Evidence for the coordinate control of glycogen synthesis, glucose utilization, and glycolysis in *Escherichia coli*. J. Biol. Chem. **250:**7188–7193.

37. **Dietzler, D. N., M. P. Leckie, J. W. Lewis, S. E. Porter, T. L. Taxman, and C. J. Lais.** 1979. Evidence for new factors in the coordinate regulation of energy metabolism in *Escherichia coli*. Effects of hypoxia, chloramphenicol succinate, and 2,4-dinitrophenol on glucose utilization, glycogen synthesis, adenylate energy charge, and hexose phosphates during the first two periods of nitrogen starvation. J. Biol. Chem. **254:**8295–8307.

38. **Doelle, H. W., and N. W. Hollywood.** 1978. Transitional steady-state investigations during aerobic-anaerobic transition of glucose utilization by *Escherichia coli* K-12. Eur. J. Biochem. **83:**479–484.

39. **Doelle, H. W., N. W. Hollywood, and A. W. Westwood.** 1974. Effects of glucose concentration on a number of enzymes involved in the aerobic and anaerobic utilization of glucose in turbidostat-cultures of *Escherichia coli*. Microbios **9:**221–232.

40. **Dykhuizen, D. E., and D. L. Hartl.** 1983. Functional effects of *pgi* allozymes in *Escherichia coli*. Genetics **105:**1–18.

41. **Eidels, L., and M. J. Osborn.** 1971. Lipopolysaccharide and aldoheptose biosynthesis in transketolase mutants of *Salmonella typhimurium*. Proc. Natl. Acad. Sci. USA **68:**1673–1677.

42. **Eisenberg, R. C., and W. J. Dobrogosz.** 1967. Gluconate metabolism in *Escherichia coli*. J. Bacteriol. **93:**941–949.

43. **Essenberg, M. K., and R. A. Cooper.** 1975. Two ribose-5-phosphate isomerases from *Escherichia coli* K12: partial characterization of the enzymes and consideration of their possible physiological roles. Eur. J. Biochem. **55:**323–332.

44. **Faik, P., H. L. Kornberg, and E. McEvoy-Bowe.** 1971. Isolation and properties of *Escherichia coli* mutants defective in 2-keto 3-deoxy 6-phosphogluconate aldolase activity. FEBS Lett. **19:**225–228.

45. **Farrish, E. E., H. V. Baker II, and R. E. Wolf, Jr.** 1982. Different control circuits for growth rate-dependent regulation of 6-phosphogluconate dehydrogenase and protein components of the translational machinery in *Escherichia coli*. J. Bacteriol. **152:**584–594.

46. **Fradkin, J. E., and D. G. Fraenkel.** 1971. 2-Keto-3-deoxygluconate 6-phosphate aldolase mutants of *Escherichia coli*. J. Bacteriol. **108:**1277–1283.

47. **Fraenkel, D. G.** 1967. Genetic mapping of mutations affecting phosphoglucose isomerase and fructose diphosphatase in *Escherichia coli*. J. Bacteriol. **93:**1582–1587.

48. **Fraenkel, D. G.** 1968. Selection of *Escherichia coli* mutants lacking glucose-6-phosphate dehydrogenase or gluconate-6-phosphate dehydrogenase. J. Bacteriol. **95:**1267–1271.

49. **Fraenkel, D. G.** 1968. The accumulation of glucose 6-phosphate and its effect in an *Escherichia coli* mutant lacking phosphoglucose isomerase and glucose 6-phosphate dehydrogenase. J. Biol. Chem. **243:**6451–6457.

50. **Fraenkel, D. G., and S. Banerjee.** 1972. Deletion mapping of *zwf*, the gene for a constitutive enzyme, glucose 6-phosphate dehydrogenase, in *Escherichia coli*. Genetics **71:**481–489.

51. **Fraenkel, D. G., F. Falcoz-Kelly, and B. L. Horecker.** 1964. The utilization of glucose-6-phosphate by glucokinaseless and wild type strains of *Escherichia coli*. Proc. Natl. Acad. Sci. USA **52:**1207–1213.

52. **Fraenkel, D. G., and B. L. Horecker.** 1964. Metabolism of glucose in *Salmonella typhimurium*. Study of a mutant deficient in phosphohexose isomerase. J. Biol. Chem. **239:**2765–2771.

53. **Fraenkel, D. G., and B. L. Horecker.** 1965. Fructose-1,6-diphosphatase and acid hexose phosphatase of *Escherichia coli*. J. Bacteriol. **90:**837–842.

54. **Fraenkel, D. G., and S. R. Levisohn.** 1967. Glucose and gluconate metabolism in an *Escherichia coli* mutant lacking phosphoglucose isomerase. J. Bacteriol. **93:**1571–1578.

55. **Fraenkel, D. G., M. J. Osborn, B. L. Horecker, and S. M. Smith.** 1963. Metabolism and cell wall structure of a mutant of *Salmonella typhimurium* deficient in phosphoglucose isomerase. Biochem. Biophys. Res. Commun. **11:**423–428.

56. **Fraenkel, D. G., and A. Parola.** 1972. "Up-promoter" mutations of glucose-6-phosphate dehydrogenase in *Escherichia coli*. J. Mol. Biol. **71:**107–111.

57. **Fraenkel, D. G., and R. T. Vinopal.** 1973. Carbohydrate metabolism in bacteria. Annu. Rev. Microbiol. **27:**69–100.

58. **Freedberg, M. B., W. S. Kistler, and E. C. C. Lin.** 1971. Lethal synthesis of methylglyoxal by *Escherichia coli* during unregulated glycerol catabolism. J. Bacteriol. **108:**137–144.

59. **Friedberg, I.** 1972. Localization of phosphoglucose isomerase in *Escherichia coli* and its relation to induction of the hexose phosphate transport system. J. Bacteriol. **112:**1201–1205.

60. **Fukuda, Y., Y. Shotaro, M. Shimosaka, M. Kuosaku, and A. Kimura.** 1983. Cloning of the glucokinase gene in *Escherichia coli*. J. Bacteriol. **156:**922–925.

61. **Fukuda, Y., S. Yamaguchi, M. Shimosaka, K. Murata, and A. Kimura.** 1984. Purification and characterization of glucokinase in *Escherichia coli* B. Agric. Biol. Chem. **48:**2541–2548.

62. **Garcia-Olalla, C., J. P. Barrco, and A. Garrido-Pertierra.** 1982. Isolation and kinetic properties of pyruvate kinase activated by fructose-1,6-bisphosphate from *Salmonella typhimurium* LT-2.1. Rev. Esp. Fisiol. **38:**409–416.

63. **Garrido-Pertierra, A., and R. A. Cooper.** 1977. Pyruvate formation during the catabolism of simple hexose sugars by *Escherichia coli*: studies with pyruvate kinase-negative mutants. J. Bacteriol. **12:**1208–1214.

64. **Garrido-Pertierra, A., and R. A. Cooper.** 1983. Evidence for two distinct pyruvate kinase genes in *Escherichia coli* K-12. FEBS Lett. **162:**420–422.

65. **Gorringe, D. M., and V. Moses.** 1980. Organization of the glycolytic enzymes in *Escherichia coli*. Int. J. Biol. Macromol. **2:**161–173.

66. **Guixé, V., and J. Babul.** 1985. Effect of ATP on phosphofructokinase-2 from *Escherichia coli*. A mutant enzyme altered in the

allosteric site for MgATP^{2-}. J. Biol. Chem. **260**:1101–1105.

67. **Harrison, D. E. F., and P. K. Maitra.** 1969. Control of respiration and metabolism in growing *Klebsiella aerogenes*. The role of adenine nucleotides. Biochem. J. **112**:647–656.

68. **Hartl, D. L., and D. E. Dykhuizen.** 1981. Potential of selection among nearly neutral allozymes of 6-phosphogluconate dehydrogenase in *Escherichia coli*. Proc. Natl. Acad. Sci. USA **78**:6344–6348.

69. **Hattman, S.** 1964. The functioning of T-even phages with unglucosylated DNA in restricting *Escherichia coli* host cells. Virology **24**:333–348.

70. **Hillman, J. D.** 1979. Mutant analysis of glyceraldehyde-3-phosphate dehydrogenase in *Escherichia coli*. Biochem. J. **179**:99–107.

71. **Hillman, J. D., and D. G. Fraenkel.** 1975. Glyceraldehyde 3-phosphate dehydrogenase mutants of *Escherichia coli*. J. Bacteriol. **122**:1175–1179.

72. **Hollywood, N., and H. W. Doelle.** 1976. Effect of specific growth rate and glucose concentration on growth and glucose metabolism of *Escherichia coli* K-12. Microbios **17**:23–33.

73. **Hommes, R. W. J., P. W. Postma, O. M. Neijssel, D. W. Tempest, P. Dokter, and J. A. Duine.** 1984. Evidence of a quinoprotein glucose dehydrogenase apoenzyme in several strains of *Escherichia coli*. FEMS Microbiol. Lett. **24**:329–333.

74. **Hopper, D. J., and R. A. Cooper.** 1972. The purification and properties of *Escherichia coli* methylglyoxal synthase. Biochem. J. **128**:321–329.

75. **Irani, M. H., and P. K. Maitra.** 1976. Glyceraldehyde-3-P dehydrogenase, glycerate 3-P kinase, and enolase mutants of *Escherichia coli*: genetic studies. Mol. Gen. Genet. **145**:65–71.

76. **Irani, M. H., and P. K. Maitra.** 1977. Properties of *Escherichia coli* mutants deficient in enzymes of glycolysis. J. Bacteriol. **132**:398–410.

77. **Isturiz, T., and R. E. Wolf, Jr.** 1975. *In vitro* synthesis of a constitutive enzyme of *Escherichia coli*: 6-phosphogluconate dehydrogenase. Proc. Natl. Acad. Sci. USA **72**:4381–4384.

78. **Johnson, R., A. I. Krasna, and D. Rittenberg.** 1973. ^{18}O studies on the oxidative and non-oxidative pentose phosphate pathways in wild type and mutant *Escherichia coli* cells. Biochemistry **12**:1969–1977.

79. **Josephson, B. J., and D. G. Fraenkel.** 1969. Transketolase mutants of *Escherichia coli*. J. Bacteriol. **100**:1289–1295.

80. **Josephson, B. J., and D. G. Fraenkel.** 1974. Sugar metabolism in transketolase mutants of *Escherichia coli*. J. Bacteriol. **118**:1082–1089.

81. **Joshi, J. G., and P. Handler.** 1964. Phosphoglucomutase. I. Purification and properties of phosphoglucomutase from *Escherichia coli*. J. Biol. Chem. **239**:2741–2751.

82. **Katz, J., and R. Rognstad.** 1967. The labeling of pentose phosphates from glucose-^{14}C and estimation of the rates of transaldolase, transketolase, the contribution of the pentose cycle, and ribose-phosphate synthesis. Biochemistry **6**:2227–2247.

83. **Kornberg, H. L., and M. Malcovati.** 1973. Control *in situ* of the pyruvate kinase activity of *Escherichia coli*. FEBS Letters **32**:257–259.

84. **Kornberg, H. L., and A. K. Soutar.** 1973. Utilization of gluconate by *Escherichia coli*. Induction of gluconate kinase and 6-phosphogluconate dehydratase activities. Biochem. J. **134**:489–498.

85. **Kotlarz, D., and H. Buc.** 1981. Regulatory properties of phosphofructokinase 2 from *Escherichia coli*. Eur. J. Biochem. **117**:569–574.

86. **Kotlarz, D., and H. Buc.** 1982. Phosphofructokinases from *Escherichia coli*. Methods Enzymol. **90**:60–70.

87. **Kotlarz, D., H. Garreau, and H. Buc.** 1975. Regulation of the amount of phosphofructokinases and pyruvate kinases in *Escherichia coli*. Biochim. Biophys. Acta **381**:257–268.

88. **Kupor, S. R., and D. G. Fraenkel.** 1969. 6-Phosphogluconolactonase mutants of *Escherichia coli* and a maltose blue gene. J. Bacteriol. **100**:1296–1301.

89. **Kupor, S. R., and D. G. Fraenkel.** 1972. Glucose metabolism in 6-phosphogluconolactonase mutants of *Escherichia coli*. J. Biol. Chem. **247**:1904–1910.

90. **Lambden, P. R., and J. R. Guest.** 1976. Mutants of *Escherichia coli* unable to use fumarate as an anaerobic electron acceptor. J. Gen. Microbiol. **97**:145–160.

91. **Lowry, O. H., J. Carter, J. B. Ward, and L. Glaser.** 1971. The effect of carbon and nitrogen sources on the level of metabolic intermediates in *Escherichia coli*. J. Biol. Chem. **246**:6511–6521.

92. **Malcovati, M., and G. Valentini.** 1982. AMP- and fructose-1,6-bisphosphate-activated pyruvate kinases from *Escherichia coli*. Methods Enzymol. **90**:170–179.

93. **Markovitz, A., R. J., Sydiskis, and M. M. Lieberman.** 1967. Genetic and biochemical studies on mannose-negative mutants that are deficient in phosphomannose isomerase in *Escherichia coli* K-12. J. Bacteriol. **94**:1492–1496.

94. **Melton, T., L. L. Snow, C. S. Freitag, and W. J. Dobrogosz.** 1981. Isolation and characterization of cAMP suppressor mutants in *Escherichia coli* K12. Mol. Gen. Genet. **182**:480–489.

95. **Miles, J. S., and J. R. Guest.** 1984. Nucleotide sequence and transcriptional start point of the phosphomannose isomerase gene (*manA*) of *Escherichia coli*. Gene **32**:41–48.

96. **Miller, R. D., D. E. Dykhuizen, L. Green, and D. L. Hartl.** 1984. Specific deletion occurring in the directed evolution of 6-phosphogluconate dehydrogenase in *Escherichia coli*. Genetics **108**:765–772.

97. **Morrissey, A. T. E., and D. G. Fraenkel.** 1972. Suppressor of phosphofructokinase mutations of *Escherichia coli*. J. Bacteriol. **112**:183–187.

98. **Moses, V., and P. B. Sharp.** 1972. Intermediary metabolite levels in *Escherichia coli*. J. Gen. Microbiol. **71**:181–190.

99. **Murray, M. L., and T. Klopotowski.** 1968. Genetic map position of the gluconate-6-phosphate dehydrogenase gene in *Salmonella typhimurium*. J. Bacteriol. **95**:1279–1282.

100. **Nasoff, M. S., H. V. Baker II, and R. E. Wolf, Jr.** 1984. DNA sequence of the *Escherichia coli* gene *gnd* for 6-phosphogluconate dehydrogenase. Gene **27**:253–264.

101. **Nasoff, M. S., and R. E. Wolf, Jr.** 1980. Molecular cloning, correlation of genetic and restriction maps, and determination of the direction of transcription of *gnd* of *Escherichia coli*. J. Bacteriol. **143**:731–741.

102. **Nomenclature Committee of the International Union of Biochemistry.** 1979. Enzyme nomenclature 1978. Academic Press, Inc., New York.

103. **Ogino, T., Y. Arata, and S. Fujiwara.** 1980. Proton correlation nuclear magnetic resonance study of metabolic regulations and pyruvate transport in anaerobic *Escherichia coli* cells. Biochemistry **19**:3684–3691.

104. **Orozco de Silva, A., and D. G. Fraenkel.** 1979. The 6-phosphogluconate dehydrogenase reaction in *Escherichia coli*. J. Biol. Chem. **254**:10237–10242.

105. **Orthner, C. L., and L. I. Pizer.** 1974. An evaluation of regulation of the hexose monophosphate shunt in *Escherichia coli*. J. Biol. Chem. **249**:3750–3755.

106. **Pahel, G., F. R. Bloom, and B. Tyler.** 1979. Deletion mapping of the *polA-metB* region of the *Escherichia coli* chromosome. J. Bacteriol. **138**:653–656.

107. **Pedersen, S., P. L. Bloch, S. Reeh, and F. C. Neidhardt.** 1978. Patterns of protein synthesis in *E. coli*: a catalog of the amount of 140 individual proteins at different growth rates. Cell **14**:179–190.

108. **Pichersky, E., L. D. Gottlieb, and J. F. Hess.** 1984. Nucleotide sequence of the triose phosphate isomerase gene of *Escherichia coli*. Mol. Gen. Genet. **195**:314–320.

109. **Pouysségur, J. M., and F. R. Stoeber.** 1971. Étude du rameau dégradatif commun des hexuronates chez *Escherichia coli* K12. Purification, propriétés et individualité de la 2-céto-3-désoxy-6-phospho-D gluconate aldolase. Eur. J. Biochem. **21**:363–373.

110. **Pouysségur, J. M., and F. Stoeber.** 1972. Mécanisme d'induction des enzymes assurant le métabolisme du 2-céto-3-déoxy-D-gluconate chez *Escherichia coli* K12. Eur. J. Biochem. **30**:479–494.

111. **Reeves, R. E., and A. Sols.** 1973. Regulation of *Escherichia coli* phosphofructokinase *in situ*. Biochem. Biophys. Res. Commun. **50**:459–465.

112. **Reichelt, J. L., and H. W. Doelle.** 1971. The influence of dissolved oxygen concentration on PFK and the glucose metabolism of *Escherichia coli* K-12. Antonie van Leeuwenhoek J. Microbiol. Serol. **37**:497–506.

113. **Roberts, R. B., D. B. Cowie, P. H. Abelson, E. T. Bolton, and R. J. Britten.** 1955. Studies of biosynthesis in *Escherichia coli*. Publication no. 607. Carnegie Institution of Washington, Washington, D.C.

114. **Robinson, J. P., and D. G. Fraenkel.** 1978. Allosteric and non-allosteric *E. coli* phosphofructokinases: effects on growth. Biochem. Biophys. Res. Commun. **81**:858–863.

115. **Roehl, R. A., and R. T. Vinopal.** 1976. Lack of glucose phosphotransferase function in phosphofructokinase mutants of *Escherichia coli*. J. Bacteriol. **126**:852–860.

116. **Rosen, S. M., L. D. Zeleznick, D. G. Fraenkel, I. M. Weiner, M. J. Osborn, and B. L. Horecker.** 1965. Characterization of the cell wall lipopolysaccharide of a mutant of *Salmonella typhimurium* lacking phosphomannose isomerase. Biochem. Z. **342**:375–386.

117. **Sanderson, K. E., and J. R. Roth.** 1983. Linkage map of *Salmonella typhimurium*, edition VI. Microbiol. Rev. **47**:410–453.

118. **Schreyer, R., and A. Böck.** 1973. Phenotypic suppression of a fructose-1,6-diphosphate aldolase mutation in *Escherichia coli.* J. Bacteriol. **115:**268–276.

119. **Schreyer, R., and A. Böck.** 1980. Phosphoglucose isomerase from *Escherichia coli* K10: purification, properties and formation under aerobic and anaerobic conditions. Arch. Microbiol. **127:**289–298.

120. **Sedivy, J. M., J. Babul, and D. G. Fraenkel.** 1986. AMP-insensitive fructose bisphosphatase in *Escherichia coli* and its consequences. Proc. Natl. Acad. Sci. USA **83:**1656–1659.

121. **Sedivy, J. M., F. Daldal, and D. G. Fraenkel.** 1984. Fructose bisphosphatase of *Escherichia coli:* cloning of the structural gene *(fbp)* and preparation of a chromosomal deletion. J. Bacteriol. **158:**1048–1053.

122. **Shimosaka, M., Y. Fukuda, K. Murata, and A. Kimura.** 1982. Application of hybrid plasmids carrying glycolysis genes to ATP production by *Escherichia coli.* J. Bacteriol. **152:**98–103.

123. **Skinner, A. J., and R. A. Cooper.** 1971. The regulation of ribose-5-phosphate isomerization in *Escherichia coli* K12. FEBS Lett. **12:**293–296.

124. **Skinner, A. J., and R. A. Cooper.** 1974. Genetic studies on ribose 5-phosphate isomerase mutants of *Escherichia coli* K-12. J. Bacteriol. **118:**1183–1185.

125. **Smith, M. W., and F. C. Neidhardt.** 1983. Proteins induced by anaerobiosis in *Escherichia coli.* J. Bacteriol. **154:**336–343.

126. **Spring, T. G., and F. Wold.** 1971. The purification and characterization of *Escherichia coli* enolase. J. Biol. Chem. **246:** 6797–6802.

127. **Thomas, A. D., H. W. Doelle, A. W. Westwood, and G. L. Gordon.** 1972. Effect of oxygen on several enzymes involved in the aerobic and anaerobic utilization of glucose in *Escherichia coli.* J. Bacteriol. **112:**1099–1105.

128. **Thomson, J.** 1977. *E. coli* phosphofructokinase synthesized *in vitro* from a Col E1 hybrid plasmid. Gene **1:**347–356.

129. **Thomson, J., P. D. Gerstenberger, D. E. Goldberg, E. Gociar, A. Orozco de Silva, and D. G. Fraenkel.** 1979. ColE1 hybrid plasmids for *Escherichia coli* genes of glycolysis and the hexose monophosphate shunt. J. Bacteriol. **137:**502–506.

130. **Unden, G., and J. R. Guest.** 1984. Cyclic AMP and anaerobic gene expression in *E. coli.* FEBS Lett. **170:**321–325.

131. **Vinopal, R. T., and D. G. Fraenkel.** 1974. Phenotypic suppression of phosphofructokinase mutations in *Escherichia coli* by constitutive expression of the glyoxylate shunt. J. Bacteriol. **118:**1090–1100.

132. **Vinopal, R. T., J. D. Hillman, H. Schulman, W. S. Reznikoff, and D. G. Fraenkel.** 1975. New phosphoglucose isomerase mutants of *Escherichia coli.* J. Bacteriol. **122:**1172–1174.

133. **Waygood, E. B., J. S. Mort, and B. D. Sanwal.** 1976. The control of pyruvate kinase of *Escherichia coli.* Binding of substrate and allosteric effectors to the enzyme activated by fructose-1,6-bisphosphate. Biochemistry **15:**277–282.

134. **Waygood, E. B., M. K. Rayman, and B. D. Sanwal.** 1975. The control of pyruvate kinases of *Escherichia coli.* II. Effectors and regulatory properties of the enzyme activated by ribose-5-phosphate. Can. J. Biochem. **53:**444–454.

135. **Wilkinson, R. G., P. Genski, and B. A. D. Stocker.** 1972. Nonsmooth mutants of *Salmonella typhimurium:* differentiation by phage sensitivity and genetic mapping. J. Gen. Microbiol. **70:**527–554.

136. **Wolf, R. E., Jr.** 1980. Integration of specialized transducing bacteriophage λ *cI857 St68 h80 dgnd his* by an unusual pathway promotes formation of deletions and generates a new translocatable element. J. Bacteriol. **142:**588–602.

137. **Wolf, R. E., Jr., and J. A. Cool.** 1980. Mapping of insertion mutations in *gnd* of *Escherichia coli* with deletions defining the ends of the gene. *J. Bacteriol.* **141:**1222–1229.

138. **Wolf, R. E., Jr., and D. G. Fraenkel.** 1974. Isolation of specialized transducing bacteriophages for gluconate 6-phosphate dehydrogenase *(gnd)* of *Escherichia coli.* J. Bacteriol. **117:**468–476.

139. **Wolf, R. E., Jr., D. M. Prather, and F. M. Shea.** 1979. Growth-rate-dependent alteration of 6-phosphogluconate dehydrogenase and glucose 6-phosphate dehydrogenase levels in *Escherichia coli* K-12. J. Bacteriol. **139:**1093–1096.

140. **Wolf, R. E., Jr., and F. M. Shea.** 1979. Combined use of strain construction and affinity chromatography in the rapid, high-yield purification of 6-phosphogluconate dehydrogenase from *Escherichia coli.* J. Bacteriol. **138:**171–175.

141. **Zablotny, R., and D. G. Fraenkel.** 1967. Glucose and gluconate metabolism in a mutant of *Escherichia coli* lacking gluconate-6-phosphate dehydrase. J. Bacteriol. **93:**1579–1581.

142. **Zwaig, N., R. Nagel de Zwaig, T. Isturiz, and M. Wecksler.** 1973. Regulatory mutations affecting the gluconate system in *Escherichia coli.* J. Bacteriol. **114:**469–473.

13. Anaerobic Dissimilation of Pyruvate

JOACHIM KNAPPE

Institut für Biologische Chemie, Universität Heidelberg, D-6900 Heidelberg, Federal Republic of Germany

INTRODUCTION

A considerable number of fermentation end products are observed when *Escherichia coli* cells grow anaerobically with glucose as carbon and energy source. The list includes succinate, lactate, acetate, ethanol, formate, carbon dioxide, and dihydrogen. Although the route to succinate is initiated at the stage of phosphoenolpyruvate, all the other compounds are derived from pyruvate (Fig. 1). Product patterns vary considerably, depending upon the specific conditions. With dioxygen completely excluded and the pH maintained at about pH 7, batch-cultured *E. coli* K-12 (in glucose minimal medium) does not form dihydrogen and produces only traces of lactate (2). Under these conditions, the products formed (in millimoles per 100 mmol of glucose) are about as follows: succinate, 12; acetate, 75; ethanol, 87; and formate, 113; 3 g (dry weight) of cell mass (corresponding to 115 meq of carbon) is also formed.

The emphasis in this chapter will be on the enzymology of the steps of pyruvate conversion to acetate and formate, as follows (CoA is coenzyme A).

$$\text{Pyruvate} + \text{CoA} \rightleftharpoons \text{acetyl-CoA} + \text{formate} \quad (1)$$

$$\text{Acetyl-CoA} + P_i \rightleftharpoons \text{acetyl phosphate} + \text{CoA} \quad (2)$$

$$\text{Acetyl phosphate} + \text{ADP} \rightleftharpoons \text{acetate} + \text{ATP} \quad (3)$$

This reaction sequence represents the backbone of the cellular machinery for the anaerobic life of *E. coli*. The enzyme catalyzing the reaction in equation 1, pyruvate formate-lyase (EC 2.3.1.54), is tightly controlled on the transcriptional, as well as the posttranslational, level. Its high catalytic activity in the anaerobic cell makes this enzyme a key element in governing the fermentation route (and anaerobic metabolism in general).

The succession of the three reactions to be discussed, with individual $\Delta G^{0\prime}$ values of -3.9, $+2.2$, and -3.1 kcal/mol (1 cal = 4.184 J), respectively (data from reference 29), also represents the most lucid example of substrate level phosphorylation since there is no redox chemistry involved. In fact, the *E. coli* cell is sustained solely by the single ATP gained in this manner when growing anaerobically with pyruvate as sole carbon and energy source (19). In bioenergetic terms, pyruvate has to be ranked as a high-energy compound. The (acetyl) group transfer "potential" $\Delta G^{0\prime}$ (of pyruvate hydrolysis to acetate and formate; calculated from the values given above and -7.6 kcal for ATP hydrolysis) is -12.4 kcal/mol. Pyruvate is thus energetically equivalent to its usual metabolic precursor, phosphoenolpyruvate.

The properties and cellular levels of the proteins involved in conversion of pyruvate to acetate are summarized in Table 1.

PYRUVATE FORMATE-LYASE

Interconvertible Forms

Pyruvate formate-lyase, a homodimeric protein, is subject to posttranslational interconversion and occurs in an inactive form, E_i, and an active form, E_a. The latter form carries a free radical that gives rise to a characteristic electron paramagnetic resonance signal (Fig. 2).

The radical center of E_a is extremely sensitive to destruction by dioxygen. Radical content and catalytic activity are both lost when hypophosphite, a formate analog, acts on E_a (and becomes covalently linked to the polypeptide chain) (14). The question of the nature of the radical-bearing amino acid residue in E_a is not settled. Electron paramagnetic resonance and electron nuclear double resonance spectra show couplings of the unpaired electron with three sets of hydrogen nuclei; the one giving rise to the major doublet splitting of the electron paramagnetic resonance signal is an acidic proton that exchanges in D_2O (V. Unkrig, N. Helle, H. Kurreck, and J. Knappe, unpublished data).

Conversion of E_i to E_a, i.e., introduction of the radical, comprises the mechanistically sophisticated abstraction of a hydrogen atom from the protein in an

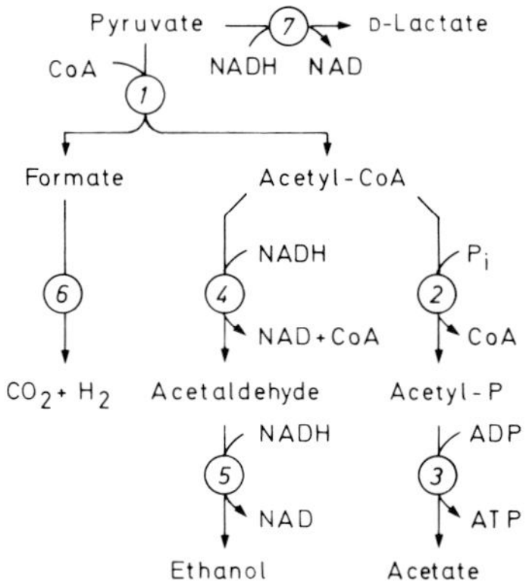

FIG 1. Terminal reactions of glycolysis in *E. coli* cells. 1, Pyruvate formate-lyase (CoA-acylating); 2, phosphotransacetylase; 3, acetate kinase; 4, acetaldehyde dehydrogenase (CoA-acylating); 5, alcohol dehydrogenase; 6, formate hydrogen-lyase complex; 7, lactate dehydrogenase.

anaerobic environment (14). *S*-Adenosylmethionine serves as the hydrogen-accepting substrate to give, in conjunction with an electron from (dihydro)flavodoxin, 5'-deoxyadenosine and methionine. The process is catalyzed by an Fe^{2+}-dependent converter enzyme (pyruvate formate-lyase-activating enzyme) and is de-

pendent on the presence of pyruvate, which acts as an allosteric effector (Fig. 3). It is suggested that an Fe-adenosyl intermediate serves as the direct hydrogen atom abstractor.

Pyruvate formate-lyase can be classified, like ribonucleotide reductase (22), as a radical enzyme, i.e., an enzyme that carries an unpaired electron in its resting state. This property has only more recently been recognized (14). This finding terminated a long search for the enzyme chemical basis of the unique pyruvate cleavage conversion reaction that had been discovered in cell extracts of *E. coli* 40 years earlier (11). At that time, work from the laboratories of C. H. Werkman and F. Lipmann (6, 27, 30) delineated the reaction sequence, equations 1 to 3. The ready exchange of [^{14}C]formate with the pyruvate carboxyl group, also discovered early, was the basis of a convenient assay during later work on the pyruvate formate-lyase system. Experiments in several laboratories through the 1950s and 1960s indicated the involvement of various proteins and cofactors, among them *S*-adenosylmethionine (13), and led eventually to the identification of the catalyst as a single dioxygen-sensitive protein that required previous processing (7, 15, 18). The precursor form (E_i) and the converter enzyme, both being dioxygen insensitive, were purified, as was the flavodoxin system (31), another essen-

TABLE 1. Proteins involved in pyruvate dissimilation

Enzyme/gene	Composition[a]		Functional properties			Cellular level	
			Catalytic parameters				
	Subunit structure/mol wt, kDa (reference[s])	Cofactor(s)	Sp act (µmol of substrate converted/ min per mg)	Substrate K_m(s) or E_m	Allosteric effector(s)	% of cytoplasmic protein constituting the anaerobic cell	Anaerobic induction (fold)[b] (reference[s])
Pyruvate formate-lyase/*pfl* (20 min)	Dimer/2 × 85 (8)	E_i, none; E_a, radical	200	Pyr, 2 mM; CoA, 7 µM		2.7	10 [G70/74][c] (8, 20, 25)
Pyruvate formate-lyase-activating enzyme	Monomer/ 29.5 (8)	Covalently linked organic factor, Fe^{2+}	0.05	AdoMet, 8 µM	Pyr, positive	0.05	Constitutive[d]
Flavodoxin	Monomer/ 20.5 (3, 31)	FMN		E_m, −455 mV		0.1	Constitutive
Ferredoxin[e]	Monomer/12 (17)	[2Fe-2S]		E_m, −380 mV		0.1	Constitutive
Pyruvate:flavodoxin/ ferredoxin oxidoreductase	Unknown/ ~200 (3)	Thiamine diphosphate, Fe-S (?)		Pyr, 1.6 mM; Fld, 12 µM; CoA, 16 µM		~0.1	Constitutive
NADPH:flavodoxin/ ferredoxin oxidoreductase	Monomer/27 (3)	FAD	0.3	NADPH, 4 µM		0.1	Constitutive
Phosphotransacetylase/*pta* (50 min)	Dimer/2 × ~100 (24, 28)	None	~1,000	AcCoA, 130 µM[f]; P_i, 10 mM[f]	NADH, negative; Pyr, positive	0.4	2.6 [E79] (25)
Acetate kinase/*ackA* (50 min)	Monomer/46 (1, 23)	None	1,500	ADP, 1.5 mM; AcP, 5 mM		~1	3.1 [G41.3] (25)

[a] References indicate sources for structural and functional data.

[b] Items in brackets refer to the alphanumeric designation of polypeptides in reference 25. Abbreviations: Pyr, pyruvate; AdoMet, *S*-adenosylmethionine; Fld, flavodoxin; AcCoA, acetyl-CoA; AcP, acetyl phosphate; FMN, flavin mononucleotide; FAD, flavin adenine dinucleotide.

[c] Polypeptide G70 is a proteolytic fragment.

[d] Constitutively expressed according to activity measurements in cell extracts.

[e] Ferredoxin is not actually involved in pyruvate dissimilation (see text).

[f] H. P. Blaschkowski, unpublished data.

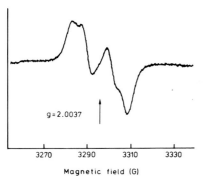

FIG. 2. Electron paramagnetic resonance spectrum of pyruvate formate-lyase (active form, E_a).

tial component required as electron donor. (The latter was replaceable by artificial reductants, most conveniently by [photo]reduced 5-deazaflavin.) This then allowed detection of the unusual reaction of S-adenosylmethionine and formulation of the conversion process operationally as E_i + S-adenosylmethionine + $Fl_{red} \rightarrow E_a$ + 5'-deoxyadenosine + methionine + Fl_{ox} (16) (Fl is flavodoxin). Finally, electron paramagnetic resonance spectroscopy was required to detect the novel mode of covalent enzyme modification occurring in this system, i.e., H atom abstraction to yield a free radical.

The reverse process, converting E_a to E_i (8), is only partially characterized. A pyridine nucleotide-dependent protein is involved, and the reaction is inhibited by pyruvate and stimulated by CoA (D. Kessler and J. Knappe, unpublished data).

Catalytic Mechanism

The radical enzyme catalyzes the reaction in equation 1 in both directions very efficiently; turnover numbers for the forward and reverse processes (at 30°C) are 1.1×10^3 and 4×10^2/s, respectively. That the chemical equilibrium lies far to the right is kinet-

ically determined mainly by the poor affinity of the enzyme for formate (K_m = 25 mM) (12).

The overall process (K_{eq} = 750) is divided into two discrete half-reactions (K_1 = 50; K_2 = 15) mutually linked by an acetyl enzyme intermediate (12), which also contains the radical (14):

$$H_3C\text{-}C\text{-}C\text{-}O^- + HS\text{-}\dot{E} \rightleftharpoons H_3C\text{-}C\text{-}S\text{-}\dot{E} + H\text{-}C\text{-}O^-$$

$$H_3C\text{-}C\text{-}S\text{-}\dot{E} + HS\text{-}CoA \rightleftharpoons H_3C\text{-}C\text{-}S\text{-}CoA + HS\text{-}\dot{E}$$

The acetyl-transferring SH group of the enzyme is identified as Cys-419. It is the second of two adjacent cysteinyl residues in the following active site fragment: Arg-Pro-Asp-Phe-Asn-Asn-Asp-Asp-Tyr-Ala-Ile-Ala-Cys-Cys-Val-Ser-Pro-Met (W. Plaga, W. Rödel, R. Frank, and J. Knappe, unpublished data).

The first half-reaction of pyruvate formate-lyase has the appearance of a straightforward nucleophilic displacement process in which the enzyme is acting as an α-thiolase. However, such a (heterolytic) mechanism is very unlikely. It would require stabilization of the leaving acyl anion, e.g., by a coenzyme (such as that exhibited by thiamine in the common pyruvate reactions in which the bond cleavage occurs with the opposite polarity to yield CO_2). Pyruvate formate-lyase is devoid of a coenzyme. Instead, it has the radical, and this, with no other conceivable function, could participate directly in catalysis. The reversible carbon-carbon bond cleavage may well turn out to be homolytic.

Flavodoxin and Ferredoxin

E. coli cells harbor both types of the widely distributed small proteins that function as carriers for one-electron transfer from a highly negative potential and contain either flavin mononucleotide (flavodoxin) or Fe-S (ferredoxin) in their redox centers. The flavodoxin of *E. coli* was discovered, as outlined above, via its function in pyruvate formate-lyase activation, whereas the [2Fe-2S]ferredoxin (17) was detected merely by its color as a by-product in large-scale procedures for isolating flavodoxin.

Two reductases are known that achieve reduction of these electron carriers (3) (Fig. 4). One, a flavoprotein,

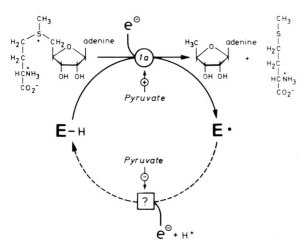

FIG. 3. Posttranslational interconversion of pyruvate formate-lyase. E-H and E·, inactive (E_i) and active (E_a) form, respectively; 1a, activating enzyme which receives an electron (e^-) from dihydroflavodoxin; ?, (uncharacterized) deactivating enzyme.

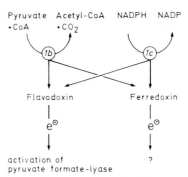

FIG. 4. Electron donor system. 1b, Pyruvate:flavodoxin/ferredoxin oxidoreductase (EC 1.2.7.1); 1c, NADPH:flavodoxin/ferredoxin oxidoreductase (EC 1.18.1.2).

uses NADPH or (less efficiently) NADH as electron donor substrate. The other one, not yet purified to homogeneity, has been identified from functional assays as thiamine diphosphate-dependent (CoA-acetylating) pyruvate:"redoxin" oxidoreductase. It closely resembles the well-characterized enzymes from clostridia or halobacteria. Surprisingly, *E. coli* cells contain this additional enzyme for pyruvate-to-acetyl-CoA conversion in addition to pyruvate formate-lyase and the pyruvate dehydrogenase complex.

In addition to providing an electron for the generation of the radical form of pyruvate formate-lyase, flavodoxin, under the designation component F, also plays a role in the activation of methionine synthase (9). No electron donor function of the ferredoxin is known (3). Constitutive expression, albeit at a low level, of all four proteins of the reducing system (Fig. 4) strongly suggests additional, as yet unknown, functions.

PHOSPHOTRANSACETYLASE

The reversible acetyltransfer reactions catalyzed by phosphotransacetylase (EC 2.3.1.8) (equation 2) occur, according to results with the clostridial enzyme (10), through direct displacement mechanisms of the substrates. Arsenate can replace phosphate, and any acetyl arsenate formed is spontaneously hydrolyzed. This property is exploited in a classical assay of CoA and phosphotransacetylase (CoA-dependent "arsenolysis" of acetyl phosphate) (27).

The dimeric *E. coli* enzyme (24) is cold labile, possibly through its dissociation into subunits. The enzyme is allosterically regulated in a K-type manner: pyruvate increases the affinity for acetyl-CoA, whereas NADH decreases it (28).

ACETATE KINASE

Acetate kinase (EC 2.7.2.1), reversibly catalyzing the reaction in equation 3 (23), appears to be unorthodox among phosphotransferases (and transferases in general) with respect to the mechanism. Net steric inversion determined for the γ phosphoryl group of ATP during transit to acetate (4) is in accordance with the stereochemistry of most phosphotransfer reactions and is normally interpreted as a one-step group transfer. However, a phosphorylated enzyme intermediate has been detected along with ATP-ADP and acetate-acetyl phosphate isotope-exchange reactions (1). Thus, the overall process could in fact follow a triple displacement mechanism (26). The broad substrate specificity gives acetate kinase the additional property (of unknown relevance) of a general purine nucleoside diphosphate kinase.

The high K_m value for acetate (100 mM) makes acetate kinase most suitable for affording the release of acetate as a fermentation end product. It should be noted that acetate uptake in *E. coli*, for aerobic growth on that substrate, is mainly accomplished by an inducible acetyl-CoA synthetase (5).

PHYSIOLOGICAL ASPECT

Except for the formate hydrogen-lyase complex (21), all enzymes of the terminal steps of glycolysis depict-ed in Fig. 1 are already expressed in *E. coli* cells during aerobic growth in glucose medium (19). However, pyruvate formate-lyase is totally arrested in the inactive form (E_i) in such cells (8). Shift to (strict) anaerobiosis changes dramatically the status of this enzyme. Triggered by either a changed redox potential or a possibly elevated cytoplasmic pyruvate concentration, the posttranslational processing of the (aerobically presynthesized) E_i form to E_a will take place. Furthermore, the rate of *pfl* gene transcription is highly derepressed (by an unrecognized mechanism) (20, 25). The exponentially growing anaerobic K-12 (wild type) cell hence contains a 10-fold-elevated level of this enzyme (about 16,000 molecules), 90% or more of which is in the E_a form (8).

The possible glucose fermentation route to lactate is apparently completely bypassed by the high activity of pyruvate formate-lyase during strict anaerobiosis. Recycling of NADH to NAD, then, has to be exclusively achieved by the aldehyde and alcohol dehydrogenase reactions. Mutants with defects in these latter enzymes require, for anaerobic growth, an exogenous electron acceptor (nitrite or nitrate) (19). Cells with a defective pyruvate formate-lyase have an absolute requirement for acetate; they use lactate fermentation for the generation of energy (19).

SUMMARY AND CONCLUSION

The *E. coli* cell harbors an elegant, nonoxidative means to achieve, during anaerobiosis, the central metabolic step of pyruvate-to-acetyl-CoA conversion. The pertinent (radical) enzyme, pyruvate formate-lyase, is clearly distinct from the other known enzymes that use the oxidative way. The principal questions and directions for necessary further research work, not only for a better understanding of *E. coli* metabolism per se, are quite obvious and may be stated in short as evolution of the system and principles of biological radical chemistry.

LITERATURE CITED

1. **Anthony, R. S., and L. B. Spector.** 1970. A phosphoenzyme intermediary in acetate kinase action. J. Biol. Chem. **245:**6739–6741.
2. **Belaich, A., and J. P. Belaich.** 1976. Microcalorimetric study of the anaerobic growth of *Escherichia coli*: growth thermograms in a synthetic medium. J. Bacteriol. **125:**14–18.
3. **Blaschkowski, H. P., G. Neuer, M. Ludwig-Festl, and J. Knappe.** 1982. Routes of flavodoxin and ferredoxin reduction in *Escherichia coli*. CoA-acetylating pyruvate: flavodoxin and NADPH:flavodoxin oxidoreductases participating in the activation of pyruvate formate-lyase. Eur. J. Biochem. **123:**563–569.
4. **Blättler, W. A., and J. R. Knowles.** 1979. Stereochemical course of phosphokinases. The use of adenosine [γ-(S)-^{16}O, ^{17}O, ^{18}O]triphosphate and the mechanistic consequences for the reactions catalyzed by glycerol kinase, hexokinase, pyruvate kinase, and acetate kinase. Biochemistry **18:**3927–3933.
5. **Brown, T. D. K., M. C. Jones-Mortimer, and H. L. Kornberg.** 1977. The enzymic interconversion of acetate and acetyl-coenzyme A in *Escherichia coli*. J. Gen. Microbiol. **102:**327–336.
6. **Chantrenne, H., and F. Lipmann.** 1950. Coenzyme A dependence and acetyl donor function of the pyruvate-formate exchange system. J. Biol. Chem. **187:**757–767.
7. **Chase, T., Jr., and J. C. Rabinowitz.** 1968. Role of pyruvate and S-adenosylmethionine in activating the pyruvate formate-lyase of *Escherichia coli*. J. Bacteriol. **96:**1065–1078.
8. **Conradt, H., M. Hohmann-Berger, H.-P. Hohmann, H. P. Blaschkowski, and J. Knappe.** 1984. Pyruvate formate-lyase (inactive form) and pyruvate formate-lyase activating enzyme of *Escherichia coli*: isolation and stuctural properties. Arch. Biochem. Biophys. **228:**133–142.

9. **Fujii, K., and F. M. Huennekens.** 1974. Activation of methionine synthetase by a reduced triphosphopyridine nucleotide-dependent flavoprotein system. J. Biol. Chem. **249:**6745–6753.
10. **Henkin, J., and R. H. Abeles.** 1976. Evidence against an acyl-enzyme intermediate in the reaction catalyzed by clostridial phosphotransacetylase. Biochemistry **15:**3472–3479.
11. **Kalnitsky, G., and C. H. Werkman.** 1943. The anaerobic dissimilation of pyruvate by a cell-free extract of *Escherichia coli*. Arch. Biochem. Biophys. **2:**113–124.
12. **Knappe, J., H. P. Blaschkowski, P. Gröbner, and T. Schmitt.** 1974. Pyruvate formate-lyase of *Escherichia coli*: the acetyl-enzyme intermediate. Eur. J. Biochem. **50:**253–263.
13. **Knappe, J., E. Bohnert, and W. Brümmer.** 1965. S-Adenosyl-L-methionine, a component of the clastic dissimilation of pyruvate in *Escherichia coli*. Biochim. Biophys. Acta **107:**603–605.
14. **Knappe, J., F. A. Neugebauer, H. P. Blaschkowski, and M. Gänzler.** 1984. Post-translational activation introduces a free radical into pyruvate formate-lyase. Proc. Natl. Acad. Sci. USA **81:**1332–1335.
15. **Knappe, J., J. Schacht, W. Möckel, T. Höpner, H. Vetter, Jr., and R. Edenharder.** 1969. Pyruvate formate-lyase reaction in *Escherichia coli*. The enzymatic system converting an inactive form of the lyase into the catalytically active enzyme. Eur. J. Biochem. **11:**316–327.
16. **Knappe, J., and T. Schmitt.** 1976. A novel reaction of S-adenosyl-L-methionine correlated with the activation of pyruvate formate-lyase. Biochim. Biophys. Res. Commun. **71:**1110–1117.
17. **Knoell, H.-E., and J. Knappe.** 1974. *Escherichia coli* ferredoxin, an iron-sulfur protein of the adrenodoxin type. Eur. J. Biochem. **50:**245–252.
18. **Nakayama, H., G. G. Midwinter, and L. O. Krampitz.** 1971. Properties of the pyruvate formate-lyase reaction. Arch. Biochem. Biophys. **143:**526–534.
19. **Pascal, M. C., M. Chippaux, A. Abou-Jaoudé, H. P. Blaschkowski, and J. Knappe.** 1981. Mutants of *Escherichia coli* K12 with defects in anaerobic pyruvate metabolism. J. Gen Microbiol. **124:**35–42.
20. **Pecher, A., H. P. Blaschkowski, J. Knappe, and A. Böck.** 1982. Expression of pyruvate formate-lyase of *Escherichia coli* from the cloned structural gene. Arch. Microbiol. **132:**365–371.
21. **Pecher, A., F. Zinoni, C. Jatisatienr, R. Wirth, H. Hennecke, and A. Böck.** 1983. On the redox control of synthesis of anaerobically induced enzymes in enterobacteriaceae. Arch. Microbiol. **136:**131–136.
22. **Reichard, P., and A. Ehrenberg.** 1983. Ribonucleotide reductase—a radical enzyme. Science **221:**514–519.
23. **Rose, I. A., M. Grunberg-Manago, S. R. Korey, and S. Ochoa.** 1954. Enzymatic phosphorylation of acetate. J. Biol. Chem. **211:**737–756.
24. **Shimizu, M., T. Suzuki, K.-Y. Kameda, and Y. Abiko.** 1969. Phosphotransacetylase of *Escherichia coli* B, purification and properties. Biochim. Biophys. Acta **191:**550–558.
25. **Smith, M. W., and F. C. Neidhardt.** 1983. Proteins induced by anaerobiosis in *Escherichia coli*. J. Bacteriol. **154:**336–343.
26. **Spector, L. B.** 1982. Covalent catalysis by enzymes, p. 92–96. Springer-Verlag, New York.
27. **Stadtman, E. R., G. D. Novelli, and F. Lipmann.** 1951. Coenzyme A function in and acetyl transfer by the phosphotransacetylase system. J. Biol. Chem. **191:**365–376.
28. **Suzuki, T.** 1969. Phosphotransacetylase of *Escherichia coli* B, activation by pyruvate and inhibition by NADH and certain nucleotides. Biochim. Biophys. Acta **191:**559–569.
29. **Thauer, R. K., K. Jungermann, and K. Decker.** 1977. Energy conservation in chemotrophic anaerobic bacteria. Bacteriol. Rev. **41:**100–180.
30. **Utter, M. F., C. H. Werkman, and F. Lipmann.** 1944. Reversibility of the phosphoroclastic split of pyruvate. J. Biol. Chem. **154:**723–724.
31. **Vetter, H., Jr., and J. Knappe.** 1971. Flavodoxin and ferrodoxin of *Escherichia coli*. Hoppe-Seyler's Z. Physiol. Chem. **352:**433–446.

14. The Tricarboxylic Acid Cycle and Anaplerotic Reactions

HUGH G. NIMMO

Department of Biochemistry, University of Glasgow, Glasgow G12 8QQ, Scotland

INTRODUCTION

The central metabolic pathways are those into which flow all carbon compounds used by the cell and from which all new cell material and waste products are derived. The discovery of the tricarboxylic acid (TCA) cycle in 1937 (57) and its subsequent acceptance as the terminal pathway for the oxidation of foodstuffs in all respiring animal tissues was one of the most significant events in the development of modern biochemistry. Attempts to demonstrate the operation of the TCA cycle in microorganisms were initially unsuccessful owing to various technical difficulties. These problems, and the evidence that finally established the cycle as the major pathway of terminal respiration in microorganisms, have been covered in earlier reviews (53, 58).

The reactions of the TCA cycle and its operation under aerobic conditions are summarized in Fig. 1. The cycle affords the oxidation of acetyl units to CO_2 and the generation of reduced nucleotides that are used for reductive biosynthesis or for trapping energy in the form of ATP. The cycle also provides many of the precursors required for biosynthesis (Fig. 1). This function, of course, dictates the requirement for the so-called anaplerotic reactions, the role of which is to replenish the intermediates of the cycle (see below). The nature of the carbon source dictates which anaplerotic reaction or pathway is used.

In bacteria under anaerobic conditions, the TCA cycle does not operate as shown in Fig. 1, but rather as two separate limbs emanating from oxaloacetate that generate the 2-oxoglutarate and succinyl-coenzyme A (CoA) required for biosynthesis (1). This branched, noncyclic pathway also seems to operate during aerobic growth on glucose, when the energy requirements of the cell are largely satisfied by glycolysis. Amarasingham and Davis (1) proposed that succinate formation from oxaloacetate involved reversal of flux through the TCA cycle enzymes malate dehydrogenase and fumarase, while the reduction of fumarate was catalyzed by a fumarate reductase (46), specifically induced by anaerobiosis, rather than by succinate dehydrogenase. Courtright and Henning (17) subsequently showed that malate dehydrogenase is not necessary for the anaerobic generation of succinate from oxaloacetate and suggested that fumarate formation from oxaloacetate occurs via aspartate. However, there is no evidence that this is the only pathway that leads to anaerobic generation of fumarate. Both of the possibilities described above are shown in Fig. 2.

The basic outline of the TCA cycle and its metabolic roles has thus been clear for many years. In this review, I concentrate on more recent findings in two main areas: first, the organization and expression of the genes encoding TCA cycle enzymes; second, the short-term control of the cycle and anaplerotic reactions.

MOLECULAR GENETICS OF THE TCA CYCLE

Gene Localization and Cloning

The genes encoding the enzymes of the TCA cycle are summarized in Table 1. This table includes details for fumarate reductase, which is involved in the anaerobic generation of succinate, and pyruvate dehy-

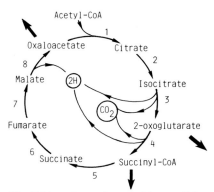

FIG. 1. The TCA cycle under aerobic conditions. The enzymes are as follows. 1, Citrate synthase; 2, aconitase; 3, isocitrate dehydrogenase; 4, 2-oxoglutarate dehydrogenase; 5, succinyl-CoA synthetase; 6, succinate dehydrogenase; 7, fumarase; 8, malate dehydrogenase. Heavy arrows represent fluxes to biosynthesis.

drogenase. Although the latter enzyme is not formally part of the TCA cycle, it shares one subunit with 2-oxoglutarate dehydrogenase, and the control of expression of the two complexes is of considerable interest. All of the genes listed in Table 1 have now been cloned, barring that for aconitase, and the nucleotide sequences of several have been determined. The locations of the genes on the genetic map of *Escherichia coli* are also shown in Table 1.

The most striking feature to emerge from these data is the existence of a cluster of TCA cycle genes at approximately 17 min on the *E. coli* linkage map. Conventional genetic mapping had indicated that the TCA cycle genes were in close proximity to each other but that several other genes were interspersed between them (e.g., reference 3). However, the complete nucleotide sequence of this region has now been established (10, 22, 23, 75, 90, 109), and it is clear that nine TCA cycle genes are in fact contiguous. These genes, *gltA–sdhCDAB–sucABCD*, encode four enzymes or enzyme complexes of the TCA cycle, apparently in three different transcription units. Of the other genes encoding TCA cycle enzymes, that for aconitase has

TABLE 1. Genes encoding enzymes of the TCA cycle

Enzyme (gene)	Map location[a] (min)
Citrate synthase (*gltA*)	17
Aconitase	Not known
Isocitrate dehydrogenase (*icd*)	25
2-Oxoglutarate dehydrogenase complex	
2-Oxoglutarate dehydrogenase (*sucA*)	17
Dihydrolipoamide succinyltransferase (*sucB*)	17
Lipoamide dehydrogenase (*lpd*)	3
Succinyl-CoA synthetase	
α Subunit (*sucD*)	17
β Subunit (*sucC*)	17
Succinate dehydrogenase	
Flavoprotein (*sdhA*)	17
Iron-sulfur protein (*sdhB*)	17
Membrane anchor protein (*sdhC*)	17
Membrane anchor protein (*sdhD*)	17
Fumarase (*fumC*[b])	35.5
Malate dehydrogenase (*mdh*)	70
Pyruvate dehydrogenase complex	
Pyruvate dehydrogenase (*aceE*)	3
Dihydrolipoamide acetyltransferase (*aceF*)	3
Lipoamide dehydrogenase (*lpd*)	3
Fumarate reductase	
Flavoprotein (*frdA*)	94
Iron-sulfur protein (*frdB*)	94
Membrane anchor protein (*frdC*)	94
Membrane anchor protein (*frdD*)	94

[a] Data taken from reference 3, except *fumC* (41).
[b] For the possible roles of *fumA* and *fumB*, see Gene Localization and Cloning.

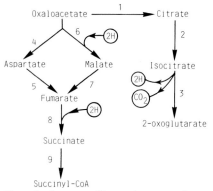

FIG. 2. Branched noncyclic pathway under anaerobic conditions. The enzymes are as follows. 1, Citrate synthase; 2, aconitase; 3, isocitrate dehydrogenase; 4, glutamate-oxaloacetate aminotransferase; 5, aspartase; 6, malate dehydrogenase; 7, fumarase; 8, fumarate reductase; 9, succinyl-CoA synthetase.

not been located, and those for isocitrate dehydrogenase, fumarase, and malate dehydrogenase are well separated from each other and from the nine-gene cluster. The four genes encoding the components of fumarate reductase are found in a single transcription unit at 94 min on the linkage map (15, 16, 35) (Table 1).

The detailed molecular analysis of the TCA cycle gene cluster is largely the work of Guest and his colleagues. In 1981, Guest (38) selected recombinant plasmids from the Clarke-Carbon gene bank (14) that carried the citrate synthase *gltA* gene by complementation of a *gltA* mutant. Subsequently, the *gltA* gene was subcloned into phage vectors by in vitro recombination, and this segment of DNA was extended by prophage integration and aberrant excision. Derivatives carrying the *sdh* genes, *sucA*, and *sucB* were identified by transduction and complementation of appropriate mutants (91). Finally, subcloning of the region beyond *sucB* revealed the presence of two further genes, *sucC* and *sucD*, which encode, respectively, the β and α subunits of succinyl-CoA synthetase (10). The analysis of this cluster is discussed further below.

The gene *mdh*, encoding malate dehydrogenase, has been cloned by Sutherland and McAlister-Henn (97). Three plasmids in the Clarke-Carbon gene bank (14) were known to contain inserts originating close to the *mdh* gene, but only one directed overexpression of malate dehydrogenase activity. Subsequently, the *mdh* gene was identified and then subcloned from this plasmid by screening restriction fragments with a synthetic oligodeoxynucleotide probe synthesized on the basis of the known amino-terminal sequence of the protein. Partial sequencing of the *mdh* gene gave results exactly in accord with the protein sequence (97). The gene *icd*, encoding isocitrate dehydrogenase, has also been selected from the Clarke-Carbon gene bank by complementation of an *icd* glutamate auxotroph (64), but this gene has not yet been characterized further.

The number and location of fumarase genes are not yet fully resolved. Two presumptive fumarase genes, *fumA* and *fumB*, were identified by their ability to complement the deficiency of a mutant, *fumA1*, that lacks fumarase activity (42). The *fumB* gene was only able to complement the *fumA1* mutation in high-copy-number situations. The *fumA* and *fumB* genes are located at 35.5 and 93.5 min, respectively, on the *E. coli* linkage map (41, 42). Hybridization and maxicell analysis showed that these genes encode homologous 61-kilodalton polypeptides. Neither *fumA* nor *fumB* is homologous with *citG*, the structural gene for the fumarase of *Bacillus subtilis*. However, Guest et al. (42) showed that another *E. coli* gene termed *fumC*, immediately adjacent to *fumA*, is very highly homologous to *citG*. This finding suggests that *fumC* may be a structural gene encoding a fumarase in *E. coli*. There seems to be no good evidence for the existence of isoenzymes of fumarase in *E. coli*. The question thus arises concerning the functions of *fumA* and *fumB*. The *fumA* and *fumC* genes are expressed independently (42). It has been suggested that the *fumA* gene product may be a positive regulator for the expression of *fumC*; *fumB* may encode an analogous product that is only capable of activating *fumC* expression when it is present at high levels (42). Other explanations of the results are clearly possible (42), and further work, including isolation and enzymological characterization of the gene products, is urgently required.

Nucleotide Sequence and Protein Structure

The nucleotide sequence of a region of 13,061 base pairs covering the cluster of nine TCA cycle genes has now been established (10, 22, 23, 75, 90, 109), and the fumarate reductase operon has also been sequenced (15, 16, 35). Several interesting features of the proteins have been revealed by this analysis. Succinate dehydrogenase and fumarate reductase are homologous membrane-bound flavoprotein complexes. The succinate dehydrogenase of *E. coli* has proved very difficult to purify, and structural studies at the protein level have not been possible. In contrast, fumarate reductase proved easier to purify, and the isolated enzyme was found to contain equal amounts of two subunits, one a flavoprotein and one an iron-sulfur protein (24). These are encoded by *frdA* and *frdB*, respectively (15, 16, 67). Sequence analysis of the *frd* operon revealed the presence of two additional genes downstream of *frdAB*, termed *frdC* and *frdD* (35) These are now thought to encode two small hydrophobic subunits that are essential for anchoring the flavoprotein and iron-sulfur subunits to the cytoplasmic membrane (66). It is possible that the *frdC* and *frdD* gene products also play other roles, in proton translocation or other aspects of electron transport.

Nucleotide sequence analysis of the *sdh* region and analysis of the proteins produced in maxicell or in vitro transcription/translation experiments showed that this region contained four genes, $\overline{sdhCDAB}$ (22, 109). The flavoprotein subunit encoded by *sdhA* is strikingly homologous to the flavoprotein subunits of *E. coli* fumarate reductase and the succinate dehydrogenases of bovine heart and *Rhodospirillum rubrum*. It has thus been possible to identify the AMP-binding region and the isoalloxazine attachment site of the *E. coli* succinate dehydrogenase. The iron-sulfur protein subunit encoded by *sdhB* is very similar to the corresponding subunit of *E. coli* fumarate reductase, but in both cases the number and location of the iron-sulfur centers are still uncertain. The genes *sdhC* and *sdhD* encode small (M_r, 14,167 and 12,792, respectively), very hydrophobic proteins. These are similar in size and amino acid composition, but not sequences, to the *frdC* and *frdD* gene products. It is assumed that the *sdhC* and *sdhD* gene products play the role of anchor proteins, but further biochemical work will be required to confirm this.

The nucleotide sequences of the genes encoding citrate synthase and the α and β subunits of succinyl-CoA synthetase have also been determined (10, 75). This work has allowed the nucleotide binding site of succinyl-CoA synthetase to be identified and catalytically important residues in citrate synthase to be defined by comparison with the known sequence and structure of pig heart citrate synthase. Weitzman (106) noted that citrate synthase and succinyl-CoA synthetase are related functionally and that they show similarities in molecular size pattern (between "large" and "small" enzymes) across a wide spectrum of organisms. He therefore speculated that they may be related evolutionarily. However, the nucleotide sequences of the *E. coli* genes do not provide any support for this interesting idea.

The pyruvate dehydrogenase and 2-oxoglutarate dehydrogenase multienzyme complexes catalyze homologous reactions (Fig. 3). Each complex comprises three components, a specific dehydrogenase (E1p or E1o), a dihydrolipoamide acyltransferase (E2p or E2o), and a lipoamide dehydrogenase (E3). The E3 component is common to both complexes and is encoded by a single gene, *lpd* (36). In each case the E2 component forms the structural core of the complex, comprising 24 polypeptide chains arranged with oc-

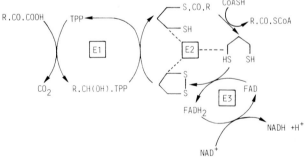

FIG. 3. Reactions catalyzed by the ketoacid dehydrogenase complexes. E1, Dehydrogenase component. E2, Acyltransferase component. E3, Lipoamide dehydrogenase. TPP, Thiamine pyrophosphate. CoASH, Coenzyme A.

tahedral symmetry (86) and providing binding sites for the E1 and E3 components, which do not interact directly with each other. Early work on the stoichiometry of the pyruvate dehydrogenase complex, based on molecular weight estimations, suggested that the E1:E2:E3 ratio was 2:2:1 (25). However, a more rigorous analysis using radioamidination gave a ratio of 1.6:1.0:0.8 (19). For the 2-oxoglutarate dehydrogenase complex, the reported E1:E2:E3 ratio was 1:2:1 (84).

The complete nucleotide sequences of all of the genes involved, *aceE*, *aceF*, *sucA*, *sucB*, and *lpd*, encoding, respectively, E1p, E2p, E1o, E2o, and E3, have now been established (23, 90, 93–95). This work revealed that, perhaps surprisingly, there is very little homology between E1p and E1o, suggesting that the two components are not closely related in evolutionary terms (23). In contrast, E2p and E2o show significant homologies (90). Both proteins comprise C-terminal catalytic and subunit-binding domains and N-terminal lipoyl-binding domains. The homologies between the former domains (34% sequence identity) extend over most of their length and include a span of eight identical residues, including one histidine which may play a role in catalysis. The acetyltransferase component of pyruvate dehydrogenase, E2p (M_r, 65,938), includes three tandemly repeated homologous sequences of some 100 residues, each containing a lipoyl-binding site. In contrast, E2o (M_r, 43,607) contains only one lipoyl-binding region. The three such regions of E2p are very highly conserved, with 70 to 80% identities between them (94). The single region of E2o is only 25 to 29% identical to these three regions. The lipoyl groups of the acyltransferases are thought to form the conformationally flexible (82, 83) "swinging arms" that visit the catalytic centers of the three components of each complex; the lipoyl groups are also thought to facilitate transacylase reactions between neighboring E2 chains as part of an active-site coupling mechanism (e.g., reference 18). The difference in the number of lipoyl-binding domains between E2p and E2o has not yet been rationalized in these terms. However, the availability of the cloned genes and the complete sequence makes the question of the possible role of the number of lipoyl-binding regions an obvious target for protein engineering studies.

Gene Expression

Work in the 1960s showed that expression of the genes encoding the TCA cycle enzymes is neither constitutive nor coordinate (1, 34, 56). The data of Gray et al. (34) suggest that the enzymes can be placed in three main groups: those metabolizing, respectively, the tricarboxylic acids, the 5-carbon dicarboxylic acids, and the four-carbon dicarboxylic acids. In general, the enzymes are at least partially repressed by anaerobiosis, this repression being particularly strong for 2-oxoglutarate dehydrogenase. This enzyme is also strongly repressed during aerobic growth on glucose, apparently owing to catabolite repression (1, 60, 89). There has recently been considerable interest in the control of expression of the TCA cycle genes, especially the *gltA-sdh-suc* cluster, aimed at understanding the regulation of enzyme synthesis at the molecular level. The pyruvate dehydrogenase–2-oxoglutarate dehydro-

genase system is particularly fascinating, for several interrelated reasons. First, the gene *lpd* encodes the E3 component that is common to both complexes. Second, the complexes are regulated independently; for example, pyruvate dehydrogenase is induced 2-fold by growth on pyruvate and repressed 3.5-fold by growth on acetate, relative to growth on 10 mM glucose, and only partly repressed by anaerobiosis (60). In contrast, 2-oxoglutarate dehydrogenase is induced to high levels by growth on acetate and severely repressed by anaerobiosis or high glucose concentrations (1, 60, 89). Third, the subunit stoichiometries of the complexes (see above) would seem to require that unequal amounts of the components be synthesized. There are thus several fascinating questions regarding the control of gene expression, and these have recently been addressed by a variety of techniques.

The genes *aceE*, *aceF*, and *lpd*, which encode the E1, E2, and E3 components of the pyruvate dehydrogenase complex, are adjacent, and the complete nucleotide sequence of this region has been established. Early genetic studies suggested that the genes *aceE* and *aceF* composed an operon with $\overrightarrow{aceEF}$ polarity (44, 60). This was confirmed by postinfection labeling studies with appropriate transducing phage (39). Studies of *ace* deletion and polar mutants suggested that *lpd* could be expressed independently from its own promoter (60); this was later confirmed directly in studies of the cloned gene (39, 43). Guest (37, 40) proposed an autoregulatory model in which transcription of *lpd* is feedback repressed by free E3 subunits.

Smith and Neidhardt (89) used two-dimensional separation of cell proteins to determine the amounts of the various complex subunits in different conditions. They found that on a glucose-salts medium the cellular content of E1o and E2o was exactly in agreement with the 1:2 ratio expected from the subunit stoichiometry of the complex. The data for E1p and E2p were subject to greater experimental error, but the E1p:E2p ratio was some 1.6:1, in good agreement with the stoichiometry reported by Danson et al. (19) (see above). The E3 component, however, was synthesized significantly in excess of the theoretical demand. For each complex, the absolute level of complex showed considerable variation in different conditions, but the E1:E2 ratio was remarkably constant. Smith and Neidhardt (89) also made use of the fact that the E3 component can be largely partitioned towards either pyruvate dehydrogenase or 2-oxoglutarate dehydrogenase by judicious choice of growth conditions. They showed that insertion mutations in *aceE* or *aceF* had little effect on the level of E3 under conditions in which most of the E3 is associated with 2-oxoglutarate dehydrogenase. However, under conditions in which most of the E3 is partitioned towards pyruvate dehydrogenase, these mutations greatly reduced E3 levels. This supports the idea that the *lpd* gene is regulated by autogenous repression by E3.

Spencer and Guest (92) took the analysis of this system a stage further in transcript mapping studies. Using single-stranded M13 probes, they observed independent transcripts corresponding to the *aceEF*, *lpd*, and *sucAB* genes. In addition, *aceEF-lpd* and *sucABCD* readthrough transcripts were observed. The existence of independent *aceEF* transcripts can account for the excess of E1p and E2p over E3 in the pyruvate dehydrogenase complex (25), but not for the excess of E1p over E2p in the complex reported by Danson et al. (19). Quantitative determination of the transcripts on different carbon sources suggested that most of the E3 components supplying the pyruvate dehydrogenase complex were synthesized from *aceEF-lpd* readthrough transcripts. This idea would provide a simple molecular explanation for the reduction of E3 caused by *ace* insertion mutations (see above). Conversely, most of the E3 required for the 2-oxoglutarate dehydrogenase complex appeared to be synthesized from independent *lpd* transcripts. The mechanism controlling this independent expression of *lpd* is not yet understood; direct testing of the idea that free E3 is an autogenous repressor (see above) should soon be possible. The codon usages for all of these genes are typical of strongly expressed *E. coli* genes (90). However, the *sucA* gene differs from the others in its comparatively frequent use of rare codons and its comparatively infrequent use of optimal energy codons (90). This may be significant in view of the fact that the E1o:E2o ratio in the intact 2-oxoglutarate dehydrogenase complex is 1:2.

Sequence analysis of the *sdh-suc* region not only facilitated transcriptional analysis but also revealed interesting features that may play roles in the expression of these genes. Between *sdhB* and *sucA* (23) are three conserved palindromic elements, termed repeated extragenic palindromes (96), that make up an insertion sequence-like region. Four similar repeated extragenic palindromes are found between *sucB* and *sucC* (10, 90). These sequences all show a high degree of homology with the sequences proposed by Higgins et al. (45) as intercistronic transcriptional attenuators. This finding suggested the existence of a combined *sdh-suc* operon (23), an idea that is attractive because the two sets of genes are expressed under the same conditions. However, Spencer and Guest (92) found no evidence of readthrough transcription from *sdh* to *suc*. Thus, the repeated extragenic palindromes between *sdh* and *suc* do not appear to be intercistronic; however, the first palindromic unit may play a role in Rho-dependent termination of the *sdh* transcript. In contrast, *sucABCD* and *sucAB* transcripts have been observed (10, 92), so the repeated extragenic palindromes between *sucB* and *sucC* are intercistronic. It is not yet known whether they act as an attenuator to reduce the expression of *sucCD*.

Another attractive theory is that the palindromes may have played a role in chromosome or operon evolution. Buck et al. (10) speculated that the *sucAB* genes may have been inserted into their present site by transposition and may thereby have disrupted a preexisting *sdh-sucCD* operon. This is an attractive idea, particularly in view of Gest's proposal (32) that evolution of the TCA cycle involved linkage of two pathways, the oxidative synthesis of 2-oxoglutarate and glutamate and the reductive formation of succinate by the introduction of 2-oxoglutarate dehydrogenase and reversal of the reduction of fumarate.

Several TCA cycle genes are subject to catabolite repression. Such genes are activated by the complex formed between cyclic AMP and its receptor protein. This interacts with a conserved sequence close to the appropriate RNA polymerase-binding site (12, 99). Consistent with the fact that citrate synthase and succinate dehydrogenase are subject to catabolite repression, two potential cyclic AMP receptor protein binding sites were observed upstream of the *gltA* and *sdhC* genes (109). However, no similar binding site was observed upstream of *sucA* (23). This is perhaps surprising since 2-oxoglutarate dehydrogenase is also subject to catabolite repression. Spencer and Guest (92) speculated that cyclic AMP may overcome termination after *sdhB* and thus may allow the *suc* genes to be expressed from the *sdh* promoter. This area of uncertainty illustrates the point that the regulation of the TCA cycle genes at the molecular level is still far from understood.

Much less is known about the control of expression of the other TCA cycle genes, but the possible existence of regulatory genes controlling the expression of TCA cycle genes is worthy of mention. The possibility that *fumC* may be the structural gene encoding fumarase and may be under the control of *fumA* has been mentioned above. Courtright and Henning (17) reported that a mutant with a lesion near the histidine operon was deficient in malate dehydrogenase activity. However, the structural gene *mdh* has been shown convincingly to lie at 70.6 min (97). The product of the locus identified by Courtright and Henning (17) may regulate the expression (or perhaps the activity) of malate dehydrogenase.

CONTROL OF THE TCA CYCLE

Studies of the regulation of metabolic pathways in eucaryotic systems have been greatly facilitated by application of the crossover theorem and, more recently, by consideration of control and elasticity coefficients. These approaches have seldom been applied to *E. coli* because of the technical difficulties involved. The first approach is complicated by the facts that the amounts of intracellular metabolites in exponentially growing cultures are extremely small and that these metabolite pools turn over very rapidly. Until recently, the methodology required for the second approach has not been available. One potential advantage of a bacterial system, that it should be possible to examine the behavior of mutants containing enzymes with altered regulatory properties, has not been fully exploited in studies of the TCA cycle. Consequently, attention has been focused on studies of the regulatory properties of enzymes in vitro and on inferences drawn from these.

Studies of citrate synthase, largely by Weitzman and colleagues, have strongly implicated this enzyme in control of the TCA cycle (e.g., reference 106). The enzyme responds sigmoidally to acetyl-CoA (28, 110). It is inhibited by ATP, but this inhibition results simply from competition for the acetyl-CoA binding site and is not thought to be of regulatory significance. However, both NADH and 2-oxoglutarate are allosteric inhibitors, as shown by desensitization experiments (21). NADH inhibition is observed only for the citrate synthases of gram-negative bacteria (106). Among this group, enzymes from strict aerobes can be reactivated by AMP, whereas enzymes from facultative anaerobes (including *E. coli*) cannot. These striking differences in the regulatory properties of citrate synthase between different classes of organisms suggest that the inhibition by NADH may be physiologically significant, perhaps as a feedback mechanism for control of the cycle. Sanwal (88) pointed out that NADH also inhibits several other enzymes of the central metabolic pathways and that the NADH/NAD$^+$ ratio can be an indicator of the rate of flux through glycolysis. Thus, the inhibition of citrate synthase by NADH could be significant in preventing unnecessary flux through citrate synthase (and generation of further NADH) during aerobic growth on glucose.

The inhibition of citrate synthase by 2-oxoglutarate can be rationalized in terms of Fig. 2, which shows that during anaerobic growth a branched noncyclic pathway operates in a strictly biosynthetic mode, producing succinyl-CoA and 2-oxoglutarate. The survey of Weitzman and Dunmore (107) showed that inhibition by 2-oxoglutarate is restricted to the citrate synthases of gram-negative facultative anaerobes. This is consistent with the view that the inhibition is physiologically significant and that its role is one of end product inhibition to control the supply of biosynthetic precursors. This view was further strengthened by Danson et al. (20), who isolated a mutant citrate synthase (produced by a revertant of a citrate synthase-deficient mutant) that was not inhibited by NADH or 2-oxoglutarate. The revertant was apparently able to overproduce a metabolite that was capable of cross-feeding a glutamate auxotroph. The metabolite concerned was presumed to be glutamate (or a close relative), overproduced as a result of the loss of the feedback control.

Direct measurements of the control coefficients of bacterial enzymes can now be made by using recombinant DNA technology. The first application of these methods was by Walsh and Koshland (102) on *E. coli* citrate synthase. They constructed a multicopy plasmid, a derivative of pBR322, containing the *gltA* gene behind the *tac* promoter and also the *lacI*^{q} gene, to produce high levels of the *lac* repressor. This plasmid was used to transform a *recA gltA* host. The expression of citrate synthase was then controlled by varying the concentration of isopropyl-β-D-thiogalactopyranoside in the medium. Walsh and Koshland (102) measured the rate of flux through the TCA cycle (see below) during growth on acetate at different concentrations of isopropyl-β-D-thiogalactopyranoside and thus at different levels of citrate synthase activity. They concluded that, under these conditions, citrate synthase was rate limiting in the TCA cycle (i.e., had a control coefficient approaching unity) provided that its activity was less than 160% of the wild-type activity. However, during growth on glucose plus acetate, the rate of flux through the cycle did not vary as the activity of citrate synthase was changed over the range from 8 to 105% of the activity of the wild type. These results suggest that the relative importance of citrate synthase in regulating flux through the TCA cycle is critically dependent on the growth conditions. However, this analysis is complicated by the fact that the manipulations involved may have affected not only the level of citrate synthase activity but also the phosphorylation state and activity of isocitrate dehydrogenase, another potential control site (see below). It would be very interesting to analyze the control coefficient of citrate synthase under other conditions, such as growth on glycerol, in which phosphorylation of isocitrate dehydrogenase is not involved.

Sanwal (88) has stressed that NADH inhibits not only citrate synthase but also another TCA cycle enzyme, malate dehydrogenase. This fact was taken to support the view that NADH may be one of the central signals for the cycle, acting as an indicator of the availability of energy. However, further work should be carried out to assess the significance of the inhibition. It would be interesting to estimate the control coefficient for malate dehydrogenase under different growth conditions, and the availability of the cloned *mdh* gene (97) makes this a possibility.

In eucaryotic systems, isocitrate dehydrogenase is thought to play a key role in the regulation of the TCA cycle, and the existence of NAD-linked isocitrate dehydrogenases allosterically regulated by AMP or ADP is well known (e.g., reference 85). In contrast, *E. coli* contains only a NADP-linked isocitrate dehydrogenase that is not under allosteric control. The enzyme is subject to concerted inhibition by glyoxylate and oxaloacetate (e.g., reference 106), and for some time this effect was thought to be of physiological significance. However, recent studies show that the inhibition is a complex phenomenon, partly caused by condensation of glyoxylate and oxaloacetate to yield the unstable compound oxalomalate, which is an extremely potent competitive inhibitor of isocitrate dehydrogenase, and partly caused by slow binding of glyoxylate and oxaloacetate to the isocitrate-binding site of the enzyme (79). However, the instability of oxalomalate and the relatively high intracellular concentration of isocitrate during growth on acetate (26, 65, 68, 79) indicate that neither mechanism is physiologically significant. There is no good evidence that isocitrate dehydrogenase is important in control of the TCA cycle in *E. coli* under any circumstances other than growth on acetate (or equivalent carbon sources), which is discussed below.

CONTROL OF ISOCITRATE DEHYDROGENASE AND THE GLYOXYLATE BYPASS DURING GROWTH ON ACETATE

Growth of *E. coli* on acetate requires the two enzymes of the glyoxylate bypass, isocitrate lyase and malate synthase, to generate intermediates of the TCA cycle, as well as phosphoenolpyruvate carboxykinase, to generate phosphoenolpyruvate (Fig. 4). The studies leading to these conclusions have been elegantly reviewed by Kornberg (54, 55), and it is not necessary to repeat the arguments here.

Isocitrate lyase and malate synthase A are induced during growth on acetate (a distinct isoenzyme, malate synthase B, is induced during growth on glycolate), and under these conditions, there is therefore competition for the available isocitrate between isocitrate dehydrogenase and isocitrate lyase. Ashworth and Kornberg (2) reported that isocitrate lyase was allosterically inhibited by phosphoenolpyruvate and suggested that this effect was physiologically significant in the regulation of flux through the glyoxylate bypass. However, detailed kinetic studies with the isocitrate lyase from *Pseudomonas indigofera* showed that phosphoenolpyruvate was merely an analog of succinate and that allosteric effects were not

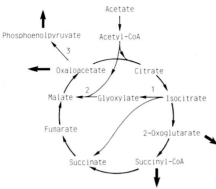

FIG. 4. The TCA cycle and the glyoxylate bypass. The enzymes are as follows. 1, Isocitrate lyase; 2, malate synthase A; 3, phosphoenolpyruvate carboxykinase. Heavy arrows represent fluxes to biosynthesis.

involved (108). Similar results have been observed for the enzyme from *E. coli*, and consideration of the intracellular concentrations of isocitrate and phosphoenolpyruvate (26, 65, 68, 74) suggests that the inhibition of isocitrate lyase by phosphoenolpyruvate is not physiologically significant (C. MacKintosh and H. G. Nimmo, unpublished results). A similar conclusion was reached by Lowry et al. (68). In this section, I review the evidence that phosphorylation of isocitrate dehydrogenase plays a major role in controlling flux at the metabolic branch point at the level of isocitrate.

Discovery of the Phosphorylation of Isocitrate Dehydrogenase

The observations leading to the discovery that isocitrate dehydrogenase can be reversibly inactivated in vivo were made by Bennett and Holms (5, 49) during a study of the levels of TCA cycle enzymes in *E. coli* ML308 during growth on different carbon sources. After cessation of growth on limiting glycerol, the specific activities of malate dehydrogenase, 2-oxoglutarate dehydrogenase, and isocitrate dehydrogenase in crude cell extracts remained constant for several hours into stationary phase. After cessation of growth on limiting glucose, however, the specific activities of malate dehydrogenase and 2-oxoglutarate dehydrogenase remained stable, whereas that of isocitrate dehydrogenase fell to 20% of its original value over 2 h and then rose again to 75% over the next 2 h (49). The explanation of this behavior involves the fact that *E. coli* excretes acetate during growth on glucose but not on glycerol (9, 49). After growth on glucose ceased, the enzymes of the glyoxylate bypass were induced, and the acetate was then oxidized. The specific activity of isocitrate dehydrogenase declined during the period of adaptation to and use of acetate, but rose again after exhaustion of the acetate (49).

Further examination of the system showed that the specific activity of isocitrate dehydrogenase in cells growing exponentially on acetate was only some 30% of the value obtained during growth on glucose or glycerol (5). However, the isocitrate dehydrogenase activity of acetate-grown cells could be increased rapidly in various ways. For example, the addition of

1 mM pyruvate gave a rapid fourfold activation that persisted until the supply of pyruvate was exhausted, whereupon the activity reverted to the value characteristic of growth on acetate. Control experiments showed that this effect was independent of protein synthesis and was not caused by low-molecular-weight effectors of isocitrate dehydrogenase. Other carbon compounds, such as glucose and some dicarboxylic acids, could also cause reactivation of the enzyme (5). Bennett and Holms (5) concluded that a reversible covalent modification of isocitrate dehydrogenase might be involved.

Phosphorylation was first implicated in the control of isocitrate dehydrogenase activity by the work of Garnak and Reeves (30, 31). They added $^{32}P_i$ to a culture of *E. coli* K-12 grown to stationary phase on limiting glycerol in a low-phosphate medium. The addition of acetate to the culture caused a partial inactivation of isocitrate dehydrogenase and incorporation of ^{32}P into a protein that comigrated with pure dehydrogenase on sodium dodecyl sulfate-polyacrylamide gel electrophoresis (30). In further experiments (31), isocitrate dehydrogenase was purified to homogeneity from similar cultures and was shown to contain ^{32}P as phosphoserine.

The ^{32}P-containing isocitrate dehydrogenase was isolated by Garnak and Reeves (31) by using a procedure involving binding of the enzyme to and its elution from a dye-ligand matrix. The purified protein exhibited considerable isocitrate dehydrogenase activity. Subsequent work (6, 29, 63, 76) has made it clear that fully phosphorylated isocitrate dehydrogenase is totally inactive and cannot bind coenzyme or dye-ligand matrices. The protein isolated by Garnak and Reeves (31) therefore probably comprised monophosphorylated isocitrate dehydrogenase dimers rather than fully phosphorylated enzyme. It is instructive to consider that, had the monophosphorylated species not existed, the phosphorylation of isocitrate dehydrogenase might not have been demonstrated so easily.

Wang and Koshland (103–105) demonstrated the existence of at least four different protein kinases in *Salmonella typhimurium* and reported that the substrate for one of these kinases was isocitrate dehydrogenase. They then investigated the conditions required for the phosphorylation of isocitrate dehydrogenase in vivo, using pulse-labeling techniques, but they made no attempts to measure concomitant changes in the activity of the dehydrogenase. This work has not been followed up, and the details of the phosphorylation system have been elucidated by using various strains of *E. coli*, as discussed below.

Components of the Phosphorylation System

LaPorte and Koshland (62) made an enormous contribution when they showed that the activities responsible for catalyzing the phosphorylation and dephosphorylation of isocitrate dehydrogenase, namely, isocitrate dehydrogenase kinase and isocitrate dehydrogenase phosphatase, copurify. They isolated a protein with a subunit M_r of 66,000 that contained both activities; this protein is referred to as the kinase-phosphatase, and the two activities considered in isolation are referred to as the kinase and the phosphatase. Subsequent work elsewhere (76) confirmed the results of LaPorte and Koshland and revealed no evidence to suggest the existence of other, perhaps monofunctional, enzymes that can phosphorylate or dephosphorylate isocitrate dehydrogenase.

This work at the level of the protein did not prove conclusively that a single polypeptide carries both kinase and phosphatase activities. (For example, the kinase-phosphatase might comprise two different subunits with the same M_r value which would not be separated by gel electrophoresis.) However, there is now strong evidence from recombinant DNA work that a single subunit does indeed carry both activities. LaPorte and Chung (61) have cloned the gene aceK (see below) which encodes the kinase-phosphatase. Deletion mapping analysis showed that the aceK-coding region comprises some 1,800 base pairs (61). This region can only code for a single protein of M_r 66,000. Thus, the kinase and phosphatase reactions must both be catalyzed by the same polypeptide chain. It is not yet clear, however, whether this polypeptide contains a single active site, capable of catalyzing both reactions, or two distinct active sites.

Studies with purified components have shown that isocitrate dehydrogenase kinase transfers the γ-phosphoryl group of ATP to a serine residue of isocitrate dehydrogenase with a stoichiometry of one per subunit (63, 76). This phosphorylation results in essentially complete loss of the activity of the dehydrogenase, measured under V_{max} conditions. In agreement with this, an essentially inactive, phosphate-containing form of the dehydrogenase has been isolated from E. coli cells grown on acetate (6). The sequences around the phosphorylation sites in the dehydrogenases from E. coli strains ML308 (8) and K-12 (70) have now been reported; the sequences are almost identical.

Isocitrate dehydrogenase phosphatase requires a divalent metal ion and either ADP or ATP for activity (62, 76). It catalyzes the release of P_i from phosphorylated isocitrate dehydrogenase (76). Its effect on dehydrogenase activity depends on the incubation conditions. In the presence of ADP, or of ATP plus an inhibitor of the kinase, the phosphatase catalyzes full reactivation of the dehydrogenase. On the other hand, when ATP alone is used to activate the phosphatase, both the kinase and the phosphatase are active; as a consequence, $^{32}P_i$ is released from ^{32}P-phosphorylated isocitrate dehydrogenase, but little reactivation of the enzyme occurs (76). The adenine nucleotide requirement of the phosphatase is surprising and has led to the speculation that the active sites of the kinase and phosphatase are distinct and that the phosphatase domain arose through a duplication of a primordial kinase gene followed by a fusion event (62). After the cloning of the aceK gene (61), the complete nucleotide sequence of the gene should soon be available, and this will clearly afford a direct test of the model described above.

The molecular mechanism by which phosphorylation inactivates isocitrate dehydrogenase seems unusual. Many of the eucaryotic enzymes regulated by phosphorylation are allosteric proteins, and phosphorylation typically affects K_m, K_i, or K_a values rather than V_{max}. Phosphorylated isocitrate dehydrogenase is essentially completely inactive, and as judged by fluorescence titration experiments, it is unable to bind NADPH. Its conformation is, however, similar to that of active enzyme containing bound coenzyme (29, 78). It has been suggested that phosphorylation occurs close to or at the coenzyme-binding site and that it elicits a conformational change similar to that induced by the binding of coenzyme (29, 78). Determination of the three-dimensional structure of the various forms of isocitrate dehydrogenase will be required to test this hypothesis fully.

Factors Contributing to Control of the Phosphorylation State of Isocitrate Dehydrogenase

Isocitrate dehydrogenase kinase and phosphatase can be active simultaneously, both in vitro (76) and in intact cells (7). Thus, the phosphorylation state (and hence the activity) of the deydrogenase represents the steady-state balance achieved by the kinase and phosphatase. This emphasizes the importance of effectors of these two activities.

Several metabolites both inhibit the kinase and activate the phosphatase (63, 77), as follows: ADP, AMP, isocitrate, oxaloacetate, 2-oxoglutarate, phosphoenolpyruvate, 3-phosphoglycerate, and pyruvate. Each of these compounds inhibits the kinase and activates the phosphatase. Effective concentrations are given in references 63 and 77. In addition, citrate, fructose 6-phosphate, glyoxylate, and $NADP^+$ inhibit the kinase but have little effect on the phosphatase at the concentrations tested (77).

The inhibition of the kinase by isocitrate is sigmoid (77), but the other effects are hyperbolic. Each kinase inhibitor alone is capable of giving essentially complete inhibition, while the phosphatase activators give two- to threefold activation at saturation. Combinations of effectors have not been tested. There is indirect evidence that the effects of the metabolites listed above are mediated by binding to the kinase-phosphatase rather than to the dehydrogenase itself (77). However, this remains to be tested directly, and the number of regulator binding sites on the kinase-phosphatase is not yet known. In addition to the metabolites listed above, $NADP^+$ and NADPH both inhibit the kinase. This may be caused by binding of the coenzymes to the dehydrogenase. The hypothetical model for the molecular mechanism of the inactivation described above affords an obvious explanation of this inhibition in structural terms. Of the nicotinamide nucleotides, only NADPH is a strong inhibitor of the phosphatase, and this may be mediated by specific binding of NADPH to the kinase-phosphatase.

Since several metabolites each affect the activities of the kinase and phosphatase in opposite directions, the system should respond very sensitively to small changes in the intracellular concentrations of these effectors. Another factor that may play a significant role in increasing the sensitivity of the system has been highlighted by LaPorte and Koshland (63). This involves the concept of "zero-order ultrasensitivity" first discussed by Goldbeter and Koshland (33). If either or both of the converter enzymes in a covalent modification system are close to saturation with their protein substrate (i.e., in the zero-order range), the system can respond (in terms of the degree of modification of the protein substrate) much more sensitively than if both of the converter enzymes are far from saturation (i.e., in the first-order range). LaPorte and Koshland (63) showed that zero-order ultrasensitivity can occur in the isocitrate dehydrogenase system in vitro. The K_m of the phosphatase for the dehydrogenase is ca. 3 µM. The sensitivity of the phosphorylation state of the enzyme to changes in the concentration of 3-phosphoglycerate was much greater at 18 than at 0.7 µM isocitrate dehydrogenase (63). Since the intracellular concentration of isocitrate dehydrogenase is ca. 35 µM (63), zero-order ultrasensitivity could contribute markedly to the behavior of the system in intact cells. This hypothesis might be tested by application of the techniques developed by Walsh and Koshland (102) (see above) to alter the activity of a specified enzyme in intact cells.

Regulation and Role of the Phosphorylation System in Intact Cells

The reversible activation of isocitrate dehydrogenase caused by the addition of pyruvate to E. coli ML308 growing on acetate (see above) has been used to investigate the occurrence and significance of the phosphorylation system in vivo. Using cells prelabeled with $^{32}P_i$, Borthwick et al. (7) showed that the phosphorylation state of the dehydrogenase in vivo was inversely related to its activity and that the reversible activation and deactivation of the enzyme resulted entirely from dephosphorylation and rephosphorylation, respectively. Moreover, the only serine residue phosphorylated in the enzyme in intact cells was identical to the residue labeled by the purified kinase in vitro. This work lays an important foundation in that these changes in isocitrate dehydrogenase activity observed after addition of pyruvate can safely be interpreted in terms of changes in the activity of the kinase-phosphatase, presumably caused by changes in the intracellular concentrations of its effectors. LaPorte et al. (64, 65) and Walsh and Koshland (101) have observed changes in isocitrate dehydrogenase activity after the addition of glucose to acetate-grown cultures. Since in this case the dehydrogenase activity in the acetate cultures could be increased by treatment of extracts with the phosphatase in vitro, it is reasonable to assume that reversible phosphorylation catalyzed by the kinase-phosphatase was responsible for the activity changes observed in vivo.

Holms and Bennett (49) first put forward the idea that the role of the partial inactivation of isocitrate dehydrogenase during growth on acetate was to facilitate flux through the glyoxylate bypass, and this hypothesis has been widely accepted. Reeves and Malloy (87) observed that the dehydrogenase could be phosphorylated even in mutants lacking a functional glyoxylate bypass. However, since the kinase and phosphatase respond to a wide variety of metabolites and not merely to those unique to the bypass (see above), this observation is quite compatible with the suggestion of Holms and Bennett.

Two groups independently proposed very similar mechanisms that explain how phosphorylation of isocitrate dehydrogenase affects the division of flux between the TCA cycle and the glyoxylate bypass (65, 77, 78, 100, 101). The key postulates are that phosphorylation, by reducing the activity of isocitrate dehydrogenase, renders it rate limiting in the TCA cycle (i.e., causes an increase in its control coefficient to a value close to unity), reduces the rate of flux through the cycle, and thus causes an increase in the intracellular concentration of isocitrate. Because the K_m of isocitrate lyase for isocitrate is high relative to the free intracellular concentration of this metabolite, this results in an increased rate of flux through isocitrate lyase and the glyoxylate bypass. These postulates have now largely been verified. The total intracellular content of isocitrate is much higher during growth on acetate than on glucose or glycerol (4, 26, 68) or after the addition of glucose to acetate-grown cells (101). There are some differences between the values reported by the different groups; these could be due either to strain differences or to the different methodologies used. (It should also be remembered that the free intracellular concentration of isocitrate is likely to be lower than the total intracellular content owing to binding of isocitrate to the active sites of the enzymes that metabolize it [65].) Kinetic studies have shown that the K_m of isocitrate lyase for isocitrate is much higher than that of isocitrate dehydrogenase and that it is high relative to the intracellular content of isocitrate (4, 65).

The hypothesis that isocitrate dehydrogenase is rate limiting in the TCA cycle during growth on acetate has been addressed in two ways. Walsh and Koshland (101) showed that the ratio of the V_{max} of isocitrate dehydrogenase (estimated from in vitro assays) to the measured flux through the enzyme rose from 1.1 for cells growing on acetate to 32 upon the addition of glucose. These data are compatible with the view that the dehydrogenase is rate limiting during growth on acetate. A comparison of the maximal activity of the enzyme in E. coli ML308 growing on acetate (6) with the flux through the enzyme calculated by Holms (47, 48) leads to a similar conclusion.

In experiments mentioned above (Control of the TCA Cycle), however, Walsh and Koshland (102) produced evidence to suggest that citrate synthase had a high control coefficient during growth on acetate. This implies that isocitrate dehydrogenase does not play a major role in the regulation of the TCA cycle during growth of this strain on acetate. However, as has been noted above, the manipulations involved may not only have affected the level of expression of citrate synthase but may also have perturbed the control of isocitrate dehydrogenase (for example, by changing the intracellular concentrations of effectors of the kinase-phosphatase). There is in fact some evidence that this may have occurred: Walsh and Koshland (102) noted that addition of acetate to a citrate synthase-overproducing strain growing on glucose completely prevented growth. One explanation for this is that addition of acetate caused essentially complete phosphorylation of isocitrate dehydrogenase and thus prevented growth. Walsh and Koshland (102) did not measure the phosphorylation state and activity of isocitrate dehydrogenase in their experiments, and so their work does not by itself rule out the hypothesis that the dehydrogenase plays an important role in regulating the TCA cycle during growth on acetate. The general approach of studying the role of an enzyme in metabolic control by systematically varying its activity in vivo using recombinant DNA technology is, however, an ingenious and powerful tool.

Koshland's group has carried out a careful analysis of the division of flux between the TCA cycle and the glyoxylate bypass in both theoretical and practical terms. LaPorte et al. (65) analyzed the fluxes through the two limbs of a branch point as a function of the kinetic parameters of the two competing enzymes. They showed that the partitioning of flux at a branch point was ultrasensitive to changes in the activity of one of the enzymes if its K_m for the common substrate was much lower than that of the competing enzyme. This is referred to as the "branch-point effect." These conditions, of course, apply to the competition between isocitrate dehydrogenase and isocitrate lyase. This is yet another factor that may increase the sensitivity with which the division of flux between the TCA cycle and the glyoxylate bypass is controlled.

Walsh and Koshland (100) developed the methodology necessary to determine the net rates of carbon flux through the central metabolic pathways during growth on acetate. This involved measurement of the rates of incorporation of labeled substrates into products and ^{13}C nuclear magnetic resonance studies of the distribution in cellular glutamate of label from [2-^{13}C]acetate. The ratio of the fluxes through isocitrate dehydrogenase and isocitrate lyase was 2.6:1. This value was in good agreement with the kinetic properties of the enzymes and the experimentally determined intracellular content of isocitrate. It is also in agreement with the ratio obtained by Holms from his calculations of the fluxes through each step of the central metabolic pathways of *E. coli* ML308 (48). The same experimental methods were then applied to the transition caused by addition of glucose to cells growing on acetate (65, 101). This perturbation caused dramatic decreases in the flux through isocitrate lyase (by a factor of 150) and the intracellular content of isocitrate. These changes were apparently caused by a 5.5-fold decrease in the flux through citrate synthase and a 4-fold increase in the activity of isocitrate dehydrogenase, resulting from its dephosphorylation.

This work is fully consistent with the view that the role of the phosphorylation of isocitrate dehydrogenase is to divert carbon flux through the glyoxylate bypass. It is interesting that the control system is a very flexible one; for example, it responds to 25-fold changes in the total level of isocitrate dehydrogenase by altering the phosphorylation state of the protein so that the total expressed dehydrogenase activity remains roughly constant (64). However, the work of Koshland's group has thrown little light on the nature of the metabolic "signals" that are responsible for controlling the phosphorylation state of the dehydrogenase. Walsh and Koshland (101) measured the intracellular contents of several TCA cycle metabolites during the transition induced by addition of glucose to cells growing on acetate. The isocitrate content fell slowly over this period; one would expect from the properties of isocitrate dehydrogenase kinase-phosphatase (discussed above) that this would promote increased phosphorylation of the dehydrogenase, but the opposite effect was observed. Other effectors of the kinase-phosphatase, such as pyruvate, phosphoenolpyruvate, and the adenine nucleotides, were not measured (101). In any event, the transition induced by glucose is not an ideal experimental system because of changes, such as the induction of components of the glucose uptake system, that must occur relatively slowly after the addition of glucose.

In contrast, El-Mansi et al. (26, 27) analyzed the metabolic signals responsible for the reversible activation of isocitrate dehydrogenase after the addition of pyruvate to cells growing on acetate. The advantages of this experimental system are, first, that induction of specific protein synthesis does not seem to be relevant, and, second, that the changes are rapid and reversible; once the pyruvate has been exhausted, the values of growth rate and isocitrate dehydrogenase activity and phosphorylation state revert to those characteristic of growth on acetate. We have proposed that one of the major roles of the phosphorylation system is to maintain the intracellular concentration of isocitrate at a value high enough to sustain the necessary flux through isocitrate lyase, and that isocitrate itself is a physiologically relevant effector of the kinase-phosphatase (77, 78). The reversible activation of isocitrate dehydrogenase induced by addition of pyruvate is in fact accompanied by a transient twofold increase in the isocitrate concentration (26). Moreover, the rates of the changes in isocitrate were compatible with the view that isocitrate was partially or fully responsible for the changes in phosphorylation state of the enzyme.

We have recently analyzed the effects of pyruvate on acetate-grown mutants of *E. coli* ML308 lacking pyruvate dehydrogenase or phosphoenolpyruvate synthase or both (27). In each case, the addition of pyruvate caused an activation of isocitrate dehydrogenase that persisted until the supply of pyruvate was exhausted. Since the mutants had very different abilities to generate effectors of the kinase-phosphatase, the conclusion was drawn that pyruvate itself must be a physiologically important signal. The intracellular content of pyruvate could not, of course, be measured because of the pyruvate in the external medium.

These studies have provided evidence that at least two of the effectors of isocitrate dehydrogenase kinase-phosphatase are physiologically relevant. The question of the roles of the other effectors listed above has been carefully considered by Holms (48). Among the metabolites that activate the phosphatase and inhibit the kinase are several biosynthetic precursors. Increases in the concentrations of these compounds should increase the activity of isocitrate dehydrogenase and thus reduce flux through the glyoxylate bypass. These effects can therefore be viewed as a feedback mechanism for controlling the flux through the bypass, which is responsible for generation of the precursors. The effects of AMP and ADP can be seen as a mechanism to increase flux through the TCA cycle (and hence energy generation) when the energy charge of the cell is low. It is probable that the relative importance of the different signals varies under different conditions. However, the point to be stressed is that the phosphorylation of isocitrate dehydrogenase is a system that can integrate a wide variety of metabolic information to optimize the division of flux between the TCA cycle and the glyoxylate bypass.

The *aceK* Gene and Properties of *aceK* Mutants

The genes encoding isocitrate lyase (*aceA*) and malate synthase A (*aceB*) are located at 90 min on the *E. coli* K-12 map and comprise an operon with the polarity *aceBA*. Preliminary work, showing that expression of isocitrate dehydrogenase kinase-phosphatase paralleled that of isocitrate lyase and malate synthase (64, 76), led to the suggestion that the gene encoding the kinase-phosphatase, termed *aceK*, might be part of the *ace* operon. This was supported by the observation that *aceK* is closely linked to *metA*, which is itself adjacent to the *ace* operon (64). LaPorte et al. (64) then isolated and studied several *ace* mutants obtained by phage Mu insertion. Of these mutants, one had lost only the kinase and phosphatase activities. Two others had lost isocitrate lyase activity in addition and also had reduced levels of malate synthase activity. These results suggest that *aceK* is indeed part of the *ace* operon, downstream of *aceA*, but the detailed transcript mapping required to confirm this unambiguously has yet to be carried out. It should be noted that the cellular level of the *aceK* gene product is roughly 1,000-fold less than those for the *aceA* and *aceB* products (64). The mechanism responsible for this great disparity in gene expression has yet to be determined.

The *ace* operon has been cloned from *E. coli* K-12 by Nunn and co-workers (W. D. Nunn, personal communication) by selection for complementation of an *aceA* mutant. LaPorte and Chung (61) then subcloned the *aceK* gene in an expression vector, by selection for complementation of an *aceK* mutant, and located the coding region by deletion mapping. The *aceK* gene has also been cloned from *E. coli* ML308 as part of the *ace* operon by complementation of an *aceA* mutant (E. M. T. El-Mansi, C. MacKintosh, K. Duncan, W. H. Holms, and H. G. Nimmo, unpublished results). The nucleotide sequence of the *aceK* gene should be available shortly and will allow tests of various hypotheses concerning the organization and evolution of the protein (see above, Components of the Phosphorylation System).

We already understand in some detail how flux through the glyoxylate bypass is controlled at the enzyme level (see above), and the mechanisms that regulate expression of the *ace* operon are also of great interest. Expression is controlled by the genes *iclR* and *fadR*, which seem to code for repressor proteins (e.g. see references 72 and 73 and chapter 19). However, the metabolites involved are not yet known. Neither acetate nor acetyl-CoA is a direct inducer (54, 71). Kornberg (54) proposed that synthesis of the enzymes is derepressed by falls in the concentrations of pyruvate and phosphoenolpyruvate during growth on acetate or fatty acids. However, measurement of metabolite levels suggests that this theory is unsatisfactory (59, 68), and the intracellular signals that control expression of the *ace* operon remain to be identified.

LaPorte et al. (64) studied the phenotypic properties conferred by some *aceK* mutations and noted that these depended on the genetic background of the strain carrying the mutation. The effect of the *aceK* mutations in two strains was loss of the ability to grow on acetate, apparently because of a three- to fourfold increase in isocitrate dehydrogenase activity which presumably severely reduced flux through the glyoxylate bypass. In another strain, however, the *aceK* mutation slowed but did not abolish growth on acetate, apparently because the level of isocitrate dehydrogenase activity in this strain was unusually low. Thus, as might be expected from the role of the system, phosphorylation of isocitrate dehydrogenase

is not required for growth on acetate provided that the dehydrogenase activity is already low.

LaPorte et al. (64) also obtained several phenotypic revertants of *aceK* mutants. One of these was extremely interesting. It had very low isocitrate dehydrogenase activity, 100-fold lower than the wild type, yet its ability to grow on acetate was unimpaired. This is puzzling because the isocitrate dehydrogenase activity of the pseudorevertant obviously cannot sustain the flux through the TCA cycle that occurs in the parent strain. Holms (47) has proposed one possible explanation of this, namely that glyoxylate generated by isocitrate lyase is oxidized in a cyclic pathway, termed the glyoxylate-oxidizing cycle, involving malic enzyme and pyruvate dehydrogenase. In this scheme, the only requirement for flux through isocitrate dehydrogenase is for the generation of 2-oxoglutarate for biosynthesis. Holms (47) has also speculated that the glyoxylate-oxidizing cycle may have preceded the TCA cycle in evolutionary terms.

OTHER ANAPLEROTIC SEQUENCES AND THEIR CONTROL

It is well known that growth on carbohydrates requires an anaplerotic reaction in which C_4 acids are synthesized by the fixation of CO_2 to C_3 acids. Evidence for the identity of the anaplerotic enzyme in *E. coli* was obtained by isolation of mutants unable to grow on carbohydrates unless supplemented with intermediates of the TCA cycle. These mutants were devoid of phosphoenolpyruvate carboxylase activity (e.g., reference 55), which converts phosphoenolpyruvate to oxaloacetate. Growth on C_3 acids requires another enzyme in addition to phosphoenolpyruvate carboxylase, namely phosphoenolpyruvate synthase (e.g., reference 55), which catalyzes the reaction pyruvate + ATP $\rightarrow$ phosphoenolpyruvate + AMP + P_i. Since phosphoenolpyruvate is a glycolytic intermediate, this enzyme has a dual function, both gluconeogenic and anaplerotic. Conclusions about the regulation of these steps have been drawn mainly from studies of the allosteric properties of the enzymes in vitro.

Phosphoenolpyruvate carboxylase is activated by acetyl-CoA (11, 69), which reduces its K_m for phosphoenolpyruvate, fructose 1,6-bisphosphate (50), GTP and ppGpp (98), and long-chain fatty acids and their CoA derivatives (52). It is inhibited by aspartate and malate (69, 80, 81). This is clearly a complex system, and it is difficult to assess how significant the various effectors are in vivo. Izui et al. (51) and Morikawa et al. (74) have attempted to do this by measuring the intracellular concentrations of the effectors under different growth conditions and by assessing in vitro the response of the enzyme to each effector at physiological concentrations of the others.

Consideration of the range of effector concentrations encountered in intact cells led to the conclusion that changes in the concentrations of phosphoenolpyruvate (to which the enzyme responds sigmoidally), acetyl-CoA, fructose 1,6-bisphosphate, aspartate, and malate were likely to be important for regulation in vivo. The activation of phosphoenolpyruvate carboxylase by acetyl-CoA and fructose 1,6-bisphosphate is synergistic. The effects can be rationalized in terms of

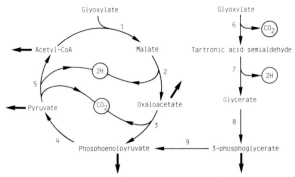

FIG. 5. Catabolism of glyoxylate: the dicarboxylic acid cycle and the glycerate pathway. The enzymes are as follows. 1, Malate synthase B; 2, malate dehydrogenase; 3, phosphoenolpyruvate carboxykinase; 4, pyruvate kinase; 5, pyruvate dehydrogenase; 6, glyoxylate carboligase; 7, tartronic acid semialdehyde reductase; 8, glycerate kinase; 9, enolase. Heavy arrows represent fluxes to biosynthesis.

feed-forward activation of oxaloacetate formation during carbohydrate utilization, and Sanwal (88) has proposed that the observed synergy of activation is a necessary device to stabilize oxaloacetate formation. The effects of malate and aspartate on phosphoenolpyruvate carboxylase can clearly be regarded as an example of feedback inhibition of an anaplerotic reaction by biosynthetic precursors. Much less attention has been paid to the control of phosphoenolpyruvate synthase. There is some evidence that it can be inhibited by biosynthetic precursors, including phosphoenolpyruvate, oxaloacetate and 2-oxoglutarate, and AMP and ADP (13), but the significance of these effects is not yet clear.

Growth of *E. coli* on glyoxylate or glycolate involves catabolism of glyoxylate via a dicarboxylic acid cycle (Fig. 5) (55; see also chapter 19), and in these circumstances the TCA cycle plays a purely anabolic role (55). However, precisely because the pathway of glyoxylate catabolism is cyclic, there must exist an anaplerotic reaction sequence that affords net formation of intermediates of the dicarboxylic acid cycle from glyoxylate. This is achieved by the glycerate pathway (55), which converts glyoxylate to phosphoenolpyruvate, an intermediate of the dicarboxylic acid cycle (Fig. 5). Since 3-phosphoglycerate is an intermediate in the glycolytic pathway, the enzymes glyoxylate carboligase and tartronic acid semialdehyde reductase have dual functions, both anaplerotic and gluconeogenic, whereas enolase plays a purely anaplerotic role. The first enzyme of the glycerate pathway, glyoxylate carboligase, seems to be induced during growth on glyoxylate (55), but little is known of the fine control of this pathway.

Thus, we have a much clearer picture of the regulation of the glyoxylate bypass than about the regulation of other anaplerotic reactions. This reflects first the interest in the phosphorylation of isocitrate dehydrogenase since its discovery (30, 31) in 1979, and second the difficulties involved in establishing the physiological significance of allosteric controls that are most readily studied in vitro. It is clear that *S. typhimurium* (103, 104) (and presumably also *E. coli*) contains several proteins that can be phosphorylated other than isocitrate dehydrogenase. It will be interesting to see whether any of the anaplerotic reactions other than the glyoxylate bypass are also controlled via protein phosphorylation.

ACKNOWLEDGMENTS

Work from this laboratory has been carried out in collaboration with W. H. Holms and many colleagues, to whom I express my thanks. This research has also been supported by the Science and Engineering Research Council, London. I hold a Senior Research Leave Fellowship awarded by the Medical Research Council, London.

LITERATURE CITED

1. **Amarasingham, C. R., and B. D. Davis.** 1965. Regulation of α-ketoglutarate dehydrogenase formation in *Escherichia coli*. J. Biol. Chem. **240:**3664–3668.
2. **Ashworth, J. M., and H. L. Kornberg.** 1963. Fine control of the glyoxylate cycle by allosteric inhibition of isocitrate lyase. Biochim. Biophys. Acta **73:**519–522.
3. **Bachmann, B. J.** 1983. Linkage map of *Escherichia coli* K-12, edition 7. Microbiol. Rev. **47:**180–230.
4. **Bautista, J., J. Satrustegui, and A. Machado.** 1979. Evidence suggesting that the NADPH/NADP⁺ ratio modulates the splitting of the isocitrate flux between the glyoxylic and tricarboxylic acid cycles in *Escherichia coli*. FEBS Lett. **105:**333–336.
5. **Bennett, P. M., and W. H. Holms.** 1975. Reversible inactivation of the isocitrate dehydrogenase of *Escherichia coli* ML308 during growth on acetate. J. Gen. Microbiol. **87:**37–51.
6. **Borthwick, A. C., W. H. Holms, and H. G. Nimmo.** 1984. Isolation of active and inactive forms of isocitrate dehydrogenase from *Escherichia coli* ML308. Eur. J. Biochem. **141:**393–400.
7. **Borthwick, A. C., W. H. Holms, and H. G. Nimmo.** 1984. The phosphorylation of *Escherichia coli* isocitrate dehydrogenase in intact cells. Biochem. J. **222:**797–804.
8. **Borthwick, A. C., W. H. Holms, and H. G. Nimmo.** 1984. Amino acid sequence round the site of phosphorylation in isocitrate dehydrogenase from *Escherichia coli* ML308. FEBS Lett. **174:**112–115.
9. **Britten, R. J.** 1954. Extracellular metabolic products of *Escherichia coli* during rapid growth. Science **118:**578.
10. **Buck, D., M. E. Spencer, and J. R. Guest.** 1985. Primary structure of the succinyl-CoA synthetase of *Escherichia coli*. Biochemistry **24:**6245–6252.
11. **Canovas, J. L., and H. L. Kornberg.** 1965. Fine control of phosphopyruvate carboxylase activity in *Escherichia coli*. Biochim. Biophys. Acta **96:**169–172.
12. **Chapon, C., and A. Kolb.** 1983. Action of CAP on the *malT* promoter in vitro. J. Bacteriol. **156:**1135–1143.
13. **Chulavatnatol, M., and D. E. Atkinson.** 1973. Phosphoenolpyruvate synthase from *Escherichia coli*. J. Biol. Chem. **248:**2712–2715.
14. **Clarke, L., and J. Carbon.** 1976. A colony bank containing synthetic ColE1 hybrid plasmids representative of the entire *E. coli* genome. Cell **9:**91–99.
15. **Cole, S. T.** 1982. Nucleotide sequence coding for the flavoprotein subunit of the fumarate reductase of *Escherichia coli*. Eur. J. Biochem. **122:**479–484.
16. **Cole, S. T., T. Grundstrom, B. Jaurin, J. J. Robinson, and J. H. Weiner.** 1982. Location and nucleotide sequence of *frdB*, the gene coding for the iron-sulphur protein subunit of the fumarate reductase of *Escherichia coli*. Eur. J. Biochem. **126:**211–216.
17. **Courtright, J. B., and U. Henning.** 1970. Malate dehydrogenase mutants in *Escherichia coli* K-12. J. Bacteriol. **102:**722–728.
18. **Danson, M. J., A. R. Fersht, and R. N. Perham.** 1978. Rapid intramolecular coupling of active sites in the pyruvate dehydrogenase complex of *Escherichia coli*: mechanism for rate enhancement in a multimeric structure. Proc. Natl. Acad. Sci. USA **75:**5386–5390.
19. **Danson, M. J., G. Hale, P. Johnson, R. N. Perham, J. Smith, and P. Spragg.** 1979. Molecular weight and symmetry of the pyruvate dehydrogenase multienzyme complex of *Escherichia coli*. J. Mol. Biol. **129:**603–617.
20. **Danson, M. J., S. Harford, and P. D. J. Weitzman.** 1979. Studies on a mutant form of *Escherichia coli* citrate synthase desensitized to allosteric effectors. Eur. J. Biochem. **101:**515–521.
21. **Danson, M. J., and P. D. J. Weitzman.** 1973. Functional groups in the activity and regulation of *Escherichia coli* citrate synthase. Biochem. J. **135:**513–524.
22. **Darlison, M. G., and J. R. Guest.** 1984. Nucleotide sequence encoding the iron-sulphur protein subunit of the succinate dehydrogenase of *Escherichia coli*. Biochem. J. **223:**507–517.

23. **Darlison, M. G., M. E. Spencer, and J. R. Guest.** 1984. Nucleotide sequence of the *sucA* gene encoding the 2-oxoglutarate dehydrogenase of *Escherichia coli* K12. Eur. J. Biochem. **141**:351–359.

24. **Dickie, P., and J. H. Weiner.** 1979. Purification and characterization of membrane-bound fumarate reductase from anaerobically grown *Escherichia coli*. Can. J. Biochem. **57**:813–821.

25. **Eley, M. H., G. Namihara, L. Hamilton, P. Munk, and L. J. Reed.** 1972. α-Keto acid dehydrogenase complexes. Arch. Biochem. Biophys. **152**:655–669.

26. **El-Mansi, E. M. T., H. G. Nimmo, and W. H. Holms.** 1985. The role of isocitrate in control of the phosphorylation of isocitrate dehydrogenase in *Escherichia coli* ML308. FEBS Lett. **183**:251–255.

27. **El-Mansi, E. M. T., H. G. Nimmo, and W. H. Holms.** 1986. Pyruvate metabolism and the phosphorylation state of isocitrate dehydrogenase in *Escherichia coli*. J. Gen. Microbiol. **132**:797–806.

28. **Faloona, G. R., and P. Srere.** 1969. *Escherichia coli* citrate synthase. Purification and the effect of potassium on some properties. Biochemistry **8**:4497–4503.

29. **Garland, D., and H. G. Nimmo.** 1984. A comparison of the phosphorylated and unphosphorylated forms of isocitrate dehydrogenase from *Escherichia coli* ML308. FEBS Lett. **165**:259–264.

30. **Garnak, M., and H. C. Reeves.** 1979. Phosphorylation of isocitrate dehydrogenase of *Escherichia coli*. Science **203**:1111–1112.

31. **Garnak, M., and H. C. Reeves.** 1979. Purification and properties of phosphorylated isocitrate dehydrogenase of *Escherichia coli*. J. Biol. Chem. **254**:7915–7920.

32. **Gest, H.** 1981. Evolution of the citric acid cycle and respiratory energy conversion in prokaryotes. FEMS Microbiol. Lett. **12**:209–215.

33. **Goldbeter, A., and D. E. Koshland, Jr.** 1981. An amplified sensitivity arising from covalent modification in biological systems. Proc. Natl. Acad. Sci. USA **78**:6840–6844.

34. **Gray, C. T., J. W. T. Wimpenny, and M. R. Mossman.** 1966. Regulation of metabolism in facultative bacteria. Biochim. Biophys. Acta **117**:33–41.

35. **Grundstrom, T., and B. Jaurin.** 1982. Overlap between *ampC* and *frd* operons on the *Escherichia coli* chromosome. Proc. Natl. Acad. Sci. USA **79**:1111–1115.

36. **Guest, J. R.** 1974. Gene-protein relationships of the α-ketoacid dehydrogenase complexes of *Escherichia coli*: chromosomal location of the *lpd* gene. J. Gen. Microbiol. **80**:523–531.

37. **Guest, J. R.** 1978. Aspects of the molecular biology of lipoamide dehydrogenase. Adv. Neurol. **21**:219–244.

38. **Guest, J. R.** 1981. Hybrid plasmids containing the citrate synthase gene (*gltA*) of *Escherichia coli* K12. J. Gen. Microbiol. **124**:17–23.

39. **Guest, J. R., S. T. Cole, and K. Jeyaseelan.** 1981. Organisation and expression of pyruvate dehydrogenase complex genes of *Escherichia coli* K12. J. Gen. Microbiol. **127**:65–79.

40. **Guest, J. R., and I. T. Creaghan.** 1973. Gene-protein relationships of the α-ketoacid dehydrogenase complexes of *Echerichia coli* K12: isolation and characterization of lipoamide dehydrogenase mutants. J. Gen. Microbiol. **106**:103–117.

41. **Guest, J. R., J. S. Miles, R. E. Roberts, and S. A. Woods.** 1985. The fumarase genes of *Escherichia coli*: location of the *fumB* gene and discovery of a new gene (*fumC*). J. Gen. Microbiol. **131**:2971–2984.

42. **Guest, J. R., and R. E. Roberts.** 1983. Cloning, mapping, and expression of the fumarase gene of *Escherichia coli* K-12. J. Bacteriol. **153**:588–596.

43. **Guest, J. R., R. E. Roberts, and P. E. Stephens.** 1983. Hybrid plasmids containing the pyruvate dehydrogenase complex genes and the gene-DNA relationships in the 2 to 3 minute region of the *Escherichia coli* chromosome. J. Gen. Microbiol. **129**:671–680.

44. **Henning, U., G. Dennart, R. Hertel, and W. S. Shipp.** 1966. Translation of the structural genes of the *E. coli* pyruvate dehydrogenase complex. Cold Spring Harbor Symp. Quant. Biol. **31**:227–234.

45. **Higgins, C. F., G. F.-L. Ames, W. M. Barnes, J. M. Clement, and M. Hofnung.** 1982. A novel interastronic regulatory element of prokaryotic operons. Nature (London) **298**:760–762.

46. **Hirsch, C. A., M. Rasminsky, B. D. Davis, and E. C. C. Lin.** 1963. A fumarate reductase in *Escherichia coli* distinct from succinate dehydrogenase. J. Biol. Chem. **238**:3770–3774.

47. **Holms, W. H.** 1986. Evolution of the glyoxylate bypass in *Escherichia coli*—an hypothesis which suggests an alternative to the Krebs cycle. FEMS Microbiol. Lett. **34**:123–127.

48. **Holms, W. H.** 1986. The central metabolic pathways of *Escherichia coli*—relationship between flux and control at a branch point, efficiency of conversion to biomass and excretion of acetate. Curr. Top. Cell. Regul. **28**:69–105.

49. **Holms, W. H., and P. M. Bennett.** 1971. Regulation of isocitrate dehydrogenase activity in *Escherichia coli* on adaptation to acetate. J. Gen. Microbiol. **65**:57–68.

50. **Izui, K., T. Nishikido, K. Ishihara, and H. Katsuki.** 1970. Studies on the allosteric effectors and some properties of phosphoenolpyruvate carboxylase from *Escherichia coli*. J. Biochem. (Tokyo) **68**:215–226.

51. **Izui, K., M. Taguchi, M. Morikawa, and H. Katsuki.** 1981. Regulation of *Escherichia coli* phosphoenolpyruvate carboxylase by multiple effectors *in vivo*. J. Biochem. (Tokyo) **90**:1321–1331.

52. **Izui, K., T. Yoshinaga, M. Morikawa, and H. Katsuki.** 1970. Activation of phosphoenolpyruvate carboxylase of *Escherichia coli* by free fatty acids or their coenzyme A derivatives. Biochem. Biophys. Res. Commun. **40**:949–956.

53. **Kornberg, H. L.** 1959. Aspects of terminal respiration in microorganisms. Annu. Rev. Microbiol. **13**:49–78.

54. **Kornberg, H. L.** 1966. The role and control of the glyoxylate cycle in *Escherichia coli*. Biochem. J. **99**:1–11.

55. **Kornberg, H. L.** 1966. Anaplerotic sequences and their role in metabolism. Essays Biochem. **2**:1–31.

56. **Kornberg, H. L.** 1970. The role and maintenance of the tricarboxylic acid cycle of *Escherichia coli*. Biochem. Soc. Symp. **30**:155–171.

57. **Krebs, H. A., and W. A. Johnson.** 1937. The role of citric acid in intermediate metabolism in animal tissues. Enzymologia **4**:148–156.

58. **Krebs, H. A., and J. M. Lowenstein.** 1960. The tricarboxylic acid cycle, p. 129–203. *In* D. M. Greenberg (ed.), Metabolic pathways, 2nd ed., vol. 1. Academic Press, Inc., New York.

59. **Lakshmi, T. M., and R. B. Helling.** 1978. Acetate metabolism in *Escherichia coli*. Can. J. Microbiol. **24**:149–153.

60. **Langley, D., and J. R. Guest.** 1978. Biochemical genetics of the α-keto acid dehydrogenase complexes of *Escherichia coli* K12: genetic characterization and regulatory properties of deletion mutants. J. Gen. Microbiol. **106**:103–117.

61. **LaPorte, D. C., and T. Chung.** 1985. A single gene codes for the kinase and phosphatase which regulate isocitrate dehydrogenase. J. Biol. Chem. **260**:15291–15297.

62. **LaPorte, D. C., and D. E. Koshland, Jr.** 1982. A protein with kinase and phosphatase activity involved in Krebs cycle regulation. Nature (London) **300**:458–460.

63. **LaPorte, D. C., and D. E. Koshland, Jr.** 1983. Phosphorylation of isocitrate dehydrogenase as a demonstration of enhanced sensitivity in covalent regulation. Nature (London) **305**:286–290.

64. **LaPorte, D. C., P. Thorsness, and D. E. Koshland, Jr.** 1985. Compensatory phosphorylation of isocitrate dehydrogenase. A mechanism for adaptation to the intracellular environment. J. Biol. Chem. **260**:10563–10568.

65. **LaPorte, D. C., K. Walsh, and D. E. Koshland, Jr.** 1984. The branch point effect. Ultrasensitivity and subsensitivity to metabolic control. J. Biol. Chem. **259**:14068–14075.

66. **Lemire, B., J. J. Robinson, and J. H. Weiner.** 1982. Identification of membrane anchor polypeptides of *Escherichia coli* fumarate reductase. J. Bacteriol. **152**:1126–1131.

67. **Lohmeier, E., D. S. Hagen, P. Dickie, and J. H. Weiner.** 1985. Cloning and expression of the fumarate reductase gene of *Escherichia coli*. Can. J. Biochem. **59**:158–164.

68. **Lowry, O. H., J. Carter, J. B. Ward, and L. Glaser.** 1971. The effect of carbon and nitrogen sources on the level of metabolic intermediates in *Escherichia coli*. J. Biol. Chem. **246**:6511–6521.

69. **Maeba, P., and B. D. Sanwal.** 1969. Phosphoenolpyruvate carboxylase of *Salmonella*. Some chemical and allosteric properties. J. Biol. Chem. **244**:2549–2557.

70. **Malloy, P. J., H. C. Reeves, and J. Spiess.** 1984. Amino acid sequence of the phosphorylation site of isocitrate dehydrogenase from *Escherichia coli*. Curr. Microbiol. **11**:37–41.

71. **Maloy, S. R., M. Bohlander, and W. D. Nunn.** 1980. Elevated levels of glyoxylate shunt enzymes in *Escherichia coli* strains constitutive for fatty acid degradation. J. Bacteriol. **143**:720–725.

72. **Maloy, S. R., and W. D. Nunn.** 1981. Role of gene *fadR* in *Escherichia coli* acetate metabolism. J. Bacteriol. **148**:83–90.

73. **Maloy, S. R., and W. D. Nunn.** 1982. Genetic regulation of the glyoxylate shunt in *Escherichia coli* K-12. J. Bacteriol. **149**:173–180.

74. **Morikawa, M., K. Izui, M. Taguchi, and H. Katsuki.** 1980. Regulation of *Escherichia coli* phosphoenolpyruvate carboxyl-

ase by multiple effectors *in vivo*. J. Biochem. (Tokyo) **87**:441–449.

75. **Ner, S. S., V. Bhayana, A. W. Bell, I. G. Giles, H. W. Duckworth, and D. P. Bloxham.** 1983. Complete sequence of the *gltA* gene encoding citrate synthase in *Escherichia coli*. Biochemistry **22**:5243–5249.

76. **Nimmo, G. A., A. C. Borthwick, W. H. Holms, and H. G. Nimmo.** 1984. Partial purification and properties of isocitrate dehydrogenase kinase and isocitrate dehydrogenase phosphatase from *Escherichia coli* ML308. Eur. J. Biochem. **141**:401–408.

77. **Nimmo, G. A., and H. G. Nimmo.** 1984. The regulatory properties of isocitrate dehydrogenase kinase and isocitrate dehydrogenase phosphatase from *Escherichia coli* ML308 and the roles of these activities in the control of isocitrate dehydrogenase. Eur. J. Biochem. **141**:409–416.

78. **Nimmo, H. G.** 1984. Control of *Escherichia coli* isocitrate dehydrogenase: an example of protein phosphorylation in a prokaryote. Trends Biochem. Sci. **9**:475–478.

79. **Nimmo, H. G.** 1986. Kinetic mechanism of *Escherichia coli* isocitrate dehydrogenase and its inhibition by glyoxylate and oxaloacetate. Biochem. J. **234**:317–323.

80. **Nishikido, T., K. Izui, A. Iwatani, H. Katsuki, and S. Tanaka.** 1965. Inhibition of the carbon dioxide fixation in *E. coli* by compounds related to the TCA cycle. Biochem. Biophys. Res. Commun. **21**:94–99.

81. **Nishikido, T., K. Izui, A. Iwatani, H. Katsuki, and S. Tanaka.** 1968. Control of carbon dioxide fixation in *E. coli* by compounds related to the TCA cycle. J. Biochem. (Tokyo) **63**:532–541.

82. **Perham, R. N., H. W. Duckworth, and G. C. K. Roberts.** 1981. Mobility of polypeptide chain in the pyruvate dehydrogenase complex revealed by proton NMR. Nature (London) **292**:474–477.

83. **Perham, R. N., and G. C. K. Roberts.** 1981. Limited proteolysis and proton n.m.r. spectroscopy of the 2-oxoglutarate dehydrogenase multienzyme complex of *Escherichia coli*. Biochem. J. **199**:733–740.

84. **Pettit, F. H., L. Hamilton, P. Munk, G. Namihara, M. H. Eley, C. R. Wilms, and L. J. Reed.** 1973. α-Keto acid dehydrogenase complexes. XIX. Subunit structure of the *Escherichia coli* α-ketoglutarate dehydrogenase complex. J. Biol. Chem. **248**:5282–5290.

85. **Plaut, G. W. E.** 1970. DPN-linked isocitrate dehydrogenase of animal tissues. Curr. Top. Cell. Regul. **2**:1–27.

86. **Reed, L. J.** 1974. Multienzyme complexes. Acc. Chem. Res. **7**:40–46.

87. **Reeves, H. C., and P. J. Malloy.** 1983. Phosphorylation of isocitrate dehydrogenase in *Escherichia coli* mutants with a nonfunctional glyoxylate cycle. FEBS Lett. **158**:239–242.

88. **Sanwal, B. D.** 1970. Allosteric controls of amphibolic pathways in bacteria. Bacteriol. Rev. **34**:20–39.

89. **Smith, M. W., and F. C. Neidhardt.** 1983. 2-Oxoacid dehydrogenase complexes of *Escherichia coli*: cellular amounts and patterns of synthesis. J. Bacteriol. **156**:81–88.

90. **Spencer, M. E., M. G. Darlison, P. E. Stephens, I. K. Duckenfield, and J. R. Guest.** 1984. Nucleotide sequence of the *sucB* gene encoding the dihydrolipoamide succinyltransferase of *Escherichia coli* K12 and homology with the corresponding acetyltransferase. Eur. J. Biochem. **141**:361–374.

91. **Spencer, M. E., and J. R. Guest.** 1982. Molecular cloning of four tricarboxylic acid cycle genes of *Escherichia coli*. J. Bacteriol. **151**:542–552.

92. **Spencer, M. E., and J. R. Guest.** 1985. Transcriptional analysis of the *sucAB*, *aceEF* and *lpd* genes of *Escherichia coli*. Mol. Gen. Genet. **200**:145–154.

93. **Stephens, P. E., M. G. Darlison, H. M. Lewis, and J. R. Guest.** 1983. The pyruvate dehydrogenase complex of *Escherichia coli* K12. Nucleotide sequence encoding the pyruvate dehydrogenase component. Eur. J. Biochem. **133**:155–162.

94. **Stephens, P. E., M. G. Darlison, H. M. Lewis, and J. R. Guest.** 1983. The pyruvate dehydrogenase complex of *Escherichia coli* K12. Nucleotide sequence encoding the dihydrolipoyl acetyltransferase component. Eur. J. Biochem. **133**:481–489.

95. **Stephens, P. E., H. M. Lewis, M. G. Darlison, and J. R. Guest.** 1983. The pyruvate dehydrogenase complex of *Escherichia coli* K12. The nucleotide sequence of the lipoamide dehydrogenase component. Eur. J. Biochem. **135**:519–527.

96. **Stern, M. J., G. F.-L. Ames, N. H. Smith, C. E. Robinson, and C. F. Higgins.** 1984. Repetitive extragenic palindromic sequences: a major component of the bacterial genome. Cell **7**:1015–1026.

97. **Sutherland, P., and L. McAlister-Henn.** 1985. Isolation and expression of the *Escherichia coli* gene encoding malate dehydrogenase. J. Bacteriol. **163**:1074–1079.

98. **Taguchi, M., K. Izui, and H. Katsuki.** 1977. Activation of *Escherichia coli* phosphoenolpyruvate carboxylase by guanosine-5′-diphosphate-3′-diphosphate. FEBS Lett. **77**:270–272.

99. **Valentin-Hansen, P.** 1982. Tandem CRP binding sites in the *deo* operon of *Escherichia coli* K12. EMBO J. **1**:1049–1054.

100. **Walsh, K., and D. E. Koshland, Jr.** 1984. Determination of flux through the branch point of two metabolic cycles. The tricarboxylic acid cycle and the glyoxylate shunt. J. Biol. Chem. **259**:9646–9654.

101. **Walsh, K., and D. E. Koshland, Jr.** 1985. Branch point control by the phosphorylation state of isocitrate dehydrogenase. A quantitative examination of fluxes during a regulatory transition. J. Biol. Chem. **260**:8430–8437.

102. **Walsh, K., and D. E. Koshland, Jr.** 1985. Characterization of rate-controlling steps *in vivo* by use of an adjustable expression vector. Proc. Natl. Acad. Sci. USA **82**:3577–3581.

103. **Wang, J. Y. J., and D. E. Koshland, Jr.** 1978. Evidence for protein kinase activities in the prokaryote *Salmonella typhimurium*. J. Biol. Chem. **253**:7605–7608.

104. **Wang, J. Y. J., and D. E. Koshland, Jr.** 1981. The identification of distinct protein kinases and phosphatases in the prokaryote *Salmonella typhimurium*. J. Biol. Chem. **256**:4640–4648.

105. **Wang, J. Y. J., and D. E. Koshland, Jr.** 1982. The reversible phosphorylation of isocitrate dehydrogenase of *Salmonella typhimurium*. Arch. Biochem. Biophys. **218**:59–67.

106. **Weitzman, P. D. J.** 1981. Unity and diversity in some bacterial citric acid-cycle enzymes. Adv. Microb. Physiol. **22**:185–244.

107. **Weitzman, P. D. J., and P. Dunmore.** 1969. Regulation of citrate synthase by α-ketoglutarate. Metabolic and taxonomic significance. FEBS Lett. **3**:265–267.

108. **Williams, J. O., T. E. Roche, and B. A. McFadden.** 1971. Mechanism of action of isocitrate lyase from *Pseudomonas indigofera*. Biochemistry **10**:384–1390.

109. **Wood, D., M. G. Darlison, R. J. Wilde, and J. R. Guest.** 1984. Nucleotide sequence encoding the flavoprotein and hydrophobic subunits of the succinate dehydrogenase of *Escherichia coli*. Biochem. J. **222**:519–534.

110. **Wright, J. A., and B. D. Sanwal.** 1971. Regulatory mechanisms involving nicotinamide adenine nucleotides as allosteric effectors. J. Biol. Chem. **246**:1689–1699.

15. Pathways of Electrons to Oxygen

ROBERT K. POOLE[1] AND W. JOHN INGLEDEW[2]

Department of Microbiology, Kings College London, Kensington Campus, London W8 7AH, England,[1] *and Department of Biochemistry and Microbiology, University of St. Andrews, St. Andrews KY16 9AL, Scotland*[2]

INTRODUCTION

Scope of the Chapter

This review considers the respiratory electron transfer chains that terminate in oxygen as electron acceptor in *Escherichia coli* and *Salmonella typhimurium*. In the natural environments of these facultative bacteria, energy for growth is generated not only by aerobic electron transport but also by anaerobic electron transfer to other acceptors and by substrate level phosphorylation. Although the gut is essentially an anaerobic habitat, the ability to use oxygen may afford advantages close to epithelial cells, where gas passes from blood through the epithelium to the microbial populations attached to it (215).

In common with other bacterial respiratory systems, those in *E. coli* and *S. typhimurium* are bound to or associated with the bacterial cytoplasmic membrane, and they consist of a sequence of components, each of which is capable of undergoing oxidation and reduction. Overall, this sequence is responsible for the conversion of the redox potential energy available between substrates and respiratory oxidants into a biologically useful form, such as the esterification of the terminal phosphate in ATP or direct ion and metabolic transport. In this chapter, we survey the function, structure, and organization of the oxygen-terminated respiratory chains and their component parts.

In recent years, during which the field has been reviewed often (19, 20, 106, 180, and papers in reference 186), there have been major developments in understanding the structure, function, membrane organization, and molecular biology of the components of the *E. coli* respiratory chains and how they relate to the phenomenon of respiration-driven proton translocation. There is a relative paucity of information on *S. typhimurium*, although the available data point to functionally identical arrangements of the respiratory systems in both organisms. In the spirit of this book, we have attempted to make the review as comprehensive and up to date as possible; the literature survey was completed in February 1985 and updated in April 1986.

Overview of the Respiratory Chain Phenotypes and Their Component Parts

Major changes in cellular protein composition accompany the transition from anaerobiosis to aerobiosis and vice versa (225, 226, 240). The number of anaerobically regulated genes may be about 50 (42). With respect to respiratory chain components, the most readily observed change is that of the cytochrome content resulting from changing the respiratory oxidant. The oxidase activities, iron-sulfur centers, and other individual components are also subject to change, but the most useful approach to studying these modifications is perhaps to consider the respiratory chains as modular, comprising a number of dehydrogenase and oxidase complexes (Fig. 1). Dramatic phenotypic changes can be elicited by varying the concentration of the terminal respiratory acceptor, as well as other changes in the medium (85, 127, 180, 188).

Cells grown with high levels of oxygen as respiratory oxidant contain respiratory pigments in relatively low abundance. The cytochromes are predominantly of the b type, which can be resolved into at least three types. These cytochromes have α band absorption maxima at about 560 nm in the reduced form. One of the b-type cytochromes, called cytochrome o, binds CO and acts as the major terminal oxidase. This cytochrome appears to have a split α band with maxima at 555 and 562 nm. An additional b-type cytochrome present in these aerobic respiratory chains has an absorption maximum at 556 nm (values obtained at 77 K). Among the iron-sulfur centers, those attributable to succinate dehydrogenase predominate in most growth substrates.

As the availability of oxygen decreases (e.g., by experimental manipulation of the oxygen transfer rate in culture), absorbances, in addition to those at about 560 nm, are seen at about 590 to 595 nm and at 630 nm in the visible spectrum of reduced cells. The former absorbance value is generally attributed to cytochrome a_1 but is probably due to a heme B-containing protein of uncertain function, whereas the latter absorbance value is due to cytochrome d (formerly cytochrome a_2), an oxidase. Additional b-type

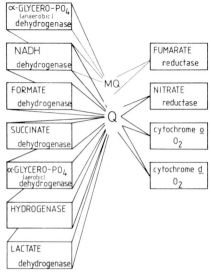

FIG. 1. Overview of the *E. coli* respiratory systems. The major primary dehydrogenases are shown in blocks on the left side, and the major oxidases are shown in blocks on the right. The two are linked, for the purposes of electron transport, by a quinone, in most cases ubiquinone (Q), but in some cases menaquinone (MQ) (see appropriate sections in the text). One potential oxidase, hydrogenase, is also a dehydrogenase, which functions with formate dehydrogenase to give formate:hydrogen-lyase activity. Each block in this diagram may contain a number of individual respiratory chain components (flavins, iron-sulfur centers, cytochromes, and nickel or molybdenum). Not all systems will be present in any one cell type; the presence and concentrations of the different respiratory complexes are regulated by the growth conditions. Reproduced with permission from Ingledew and Poole (106).

cytochromes are also detected (α bands at 555 and 558 nm at 77 K). Under conditions of oxygen limitation, cytochrome *o* persists, but with modified spectral properties.

Anaerobic growth with fumarate as respiratory oxidant results in high levels of the cytochrome *d*-type oxidase and the cytochrome b_{555} and cytochrome b_{558} components. Anaerobic growth with nitrate as terminal oxidant probably leads to expression of a similar set of cytochromes, except that the spectrum of cytochrome *d* is modified, possibly owing to ligand binding, and an additional cytochrome b_{555} (at 77 K), specifically associated with nitrate reductase, is expressed. A cytochrome b_{555} is also associated with the formate dehydrogenase present in nitrate-grown cells. The expression of these anaerobic systems, covered in detail in other chapters, appears to be under the positive control of the *fnr* gene product (47, 218).

Changes in the cytochrome composition are accompanied by changes in the identities and concentrations of quinones. Ubiquinone-8 predominates at high oxygen tension and during nitrate respiration, but menaquinone-8 becomes more evident at low oxygen tensions and during fumarate respiration.

A simplified overview of the *E. coli* respiratory complexes and their organization into branched pathways is shown in Fig. 1. The major primary dehydrogenases (i.e., those directly linked to the respiratory chain and not requiring the mediation of a soluble

cofactor, such as NAD$^+$) are shown on the left, and the oxidase complexes are shown on the right. Quinone pools link these two sets of complexes. Not all of the components shown will be present in the cytoplasmic membrane simultaneously, and a number exist as isoenzymes.

The respiratory complexes are organized in the cytoplasmic membrane in such a way that they catalyze proton translocation. The argument as to whether this is primarily a classical chemiosmotic event or a more localized phenomenon is outside the scope of this review. The mechanism of proton translocation is not known, but there are two alternatives. (i) The sequence, redox chemistry, and vectorial organization of the oxidation-reduction components are such that proton translocation is an inevitable consequence of electron transport (the loop mechanism of Mitchell [160]). (ii) The respiratory complexes pump protons via conformational changes linked to oxidation-reduction reactions. *E. coli*, with its phenotypically and genotypically modifiable respiratory chains, probably offers the greatest potential among the available experimental systems for resolving these important problems.

Consequences of General Microbiological Significance of the Oxygen-Terminated Respiratory Pathways in *E. coli* and *S. typhimurium*

The avidity with which *E. coli* consumes O_2 is well known; the anaerobic conditions that it produces and maintains in the gut allow the growth of strict anaerobes, and in the laboratory this ability has been exploited to remove residual oxygen from media for studies of strict anaerobes (e.g., reference 224).

Specific information on the substrates and oxidants used by *E. coli* in its natural environment (where the mean generation time is about 12 h [250] is scanty, but it is certain that the conditions used in laboratory culture seldom approximate those in the gut, where *E. coli* and other "enteric bacteria are forced by their environment to grow slowly much of the time" (133). Thus, during rapid growth under laboratory conditions on a single substrate, such as succinate (which is unlikely to be present at high concentrations in the gut), it is not surprising that the kinetic limitation on even faster growth is imposed by the capacity of the respiratory chain to transport electrons to excess oxygen (5, 27).

Some of the characteristics acknowledged by taxonomists reflect features of the oxygen metabolism of *E. coli* and *S. typhimurium*. Both organisms are cyanide sensitive, failing to grow in the presence of ≃1 mM KCN (161), and catalase positive, rapidly producing oxygen from added hydrogen peroxide. They are also generally regarded as being oxidase negative (see The Cytochrome *d* Complex, however). This test measures the ability of an organism to catalyze the oxidation of *N,N,N',N'*-dimethyl-*p*-phenylenediamine or *N,N,N',N'*-tetramethyl-*p*-phenylenediamine (TMPD) plus α-naphthal to indophenol blue. A positive reaction within ≃30 s generally reflects the presence of a membrane-bound, high-potential cytochrome *c* linked to an active cytochrome oxidase (121) and is frequently also indicative of the presence of energy-coupling site 3 (117). Jones (116) has suggested that the lack of both cytochrome c and energy-coupling site 3 reflects the relative abundance of energy and reducing power in those environments in which the oxidase-negative species (e.g., the *Enterobacteriaceae*, many pathogens, and some *Bacillus* spp.) are found. *E. coli* and *S. typhimurium* appear to be indistinguishable in these and other respects, although some sources suggest that *Salmonella* spp. vary significantly in their resistance to cyanide.

THE DEHYDROGENASES

The Primary Dehydrogenases Developed during Aerobic Growth

Under aerobic conditions, *E. coli* produces a range of primary dehydrogenases. These are often distinct from the set of enzymes produced during anaerobic growth. The isoenzymes produced include *sn*-glycerol-3-phosphate dehydrogenase, hydrogenase, and probably formate dehydrogenase. Very little work has been reported on *S. typhimurium* dehydrogenases beyond the report of NADH and succinate oxidase activities (55). Membranes isolated from aerobically grown *E. coli* contain, to various extents, the following respiratory activities: succinate oxidase, NADH oxidase, hydrogenase, *sn*-glycerol-3-phosphate oxidase, formate oxidase, lactate oxidase, and D-amino acid oxidase. All of these dehydrogenases contain oxidation-reduction prosthetic groups which may be of the following types: flavins, iron-sulfur centers, hemes, molybdenum, or nickel. The primary dehydrogenases can be further divided into two types: integral membrane proteins and membrane-associated proteins. As a general rule, the former types are likely to contain an intrinsic coupling site, whereas the latter types are not.

Transhydrogenase

Transhydrogenase (EC 1.6.1.1) catalyzes the exchange of reducing equivalents between the pyridine nucleotide pools, (NADH/NAD$^+$) and (NADPH/NADP$^+$). Although it is not correct to think of this enzyme as a functioning primary dehydrogenase, it is placed in this section because it is an integral membrane protein that can catalyze proton translocation; it is site 0. There is probably only one type of transhydrogenase present in *E. coli*, but two types of activity have been reported: non-energy-linked and energy-linked (20). The energy-linked phenomenon stems from the enzyme being an integral membrane protein involved in proton translocation. It is likely that only this enzyme is present in *E. coli* and that this appears to be a non-energy-linked process when activity is measured in uncoupled preparations. Evidence for this conclusion is that a single mutation leads to the loss of both activities (253). The coexistence of both activities would lead to a futile, Δp (proton electrochemical potential)-consuming cycle.

The midpoint potentials of the two pyridine nucleotide couples are similar (NAD$^+$/NADH, -320 mV; NADP$^+$/NADPH, -327 mV). The purpose of the energy-linked transhydrogenase, under normal metabolic conditions, is to generate NADPH for anabolic processes to an extent that considerably lowers the po-

tential of the NADPH/NADP$^+$ couple relative to the NADH/NAD$^+$ couple. Operating in this fashion, the reaction consumes Δp. The reverse reaction, the reduction of NAD$^+$ by NADPH, has been demonstrated in vitro with concomitant generation of Δp (38). The transhydrogenase of *E. coli* is thus a coupled site in respiration and can be designated site 0 by analogy with mitochondrial systems. This role is represented diagrammatically in Fig. 2.

The transhydrogenase has recently been cloned, purified, partially characterized, and reconstituted (41). The detergent-extracted enzyme comprises two subunits of M_r 50,000 and 47,000. When the enzyme was reconstituted in lipid vesicles, it was capable of catalyzing the generation of a transmembrane pH gradient concomitant with the transfer of reducing equivalents from NADPH to the NAD$^+$ analog 3-acetylpyridine adenine dinucleotide. The enzyme activities were sensitive to *N,N'*-dicyclohexylcarbodiimide, which bound preferentially to the 50,000-M_r subunit.

NADH Dehydrogenase

NADH dehydrogenase (EC 1.6.99.3), although important in *E. coli*, has a less important role in bacterial respiration than does its counterpart in mammalian mitochondria. This is because different catabolic pathways predominate in *E. coli*, and many of the dehydrogenases involved in these pathways are themselves primary. The literature on *E. coli* NADH dehydrogenase is indicative of a complex story. The enzyme(s) that concerns us here is that which is linked with the other components of the aerobic respiratory chain to support NADH oxidase activity. In membranes derived from aerobically grown cells, the NADH oxidase activity varies from 226 to 507 ng-atoms of oxygen min^{-1} mg of protein^{-1}, depending on the carbon source (127). Upon detergent fractionation of the respiratory chain as a preliminary step to isolation of the enzyme, its integration with the other components is dislocated, and only dye reductase activity remains. In early work with detergent-treated membranes, several fractions with NADH-dye reductase activities were isolated, but the fractions appeared to be heterogeneous and were not readily comparable (21, 81, 97, 98). The interest in these convolutions is largely historical and has been recently reviewed (106).

Recent indirect approaches to resolving problems associated with the *E. coli* NADH dehydrogenase have proven to be more readily rewarding, although these are not without their own contradictions. Crossed immunoelectrophoresis studies on membrane proteins extracted from aerobically grown cells revealed three dye-staining arcs (NADH zymograms); two of these arcs showed partial coidentity (174–177) and were subsequently shown to be due to dihydrolipoyl dehydrogenase. The remaining arc was number 15 on the numbering system of the investigators and was found to result from a component exclusively located in the cytoplasmic membrane, with its antigenic determinants located on the inner face. The enzyme was shown to contain nonheme iron (by ^{59}Fe autoradiography), and its antibody was shown to cross-react with *S. typhimurium* extracts. Interestingly, extracts

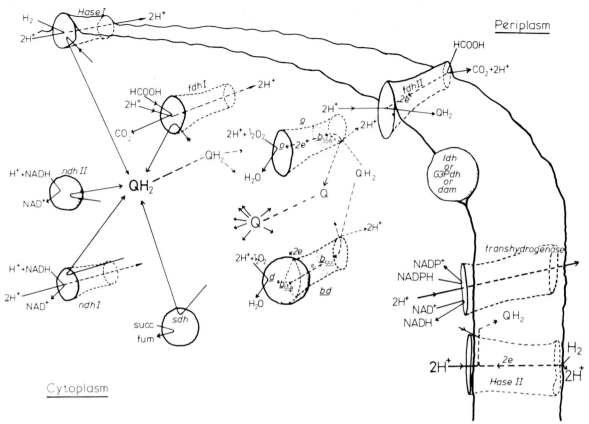

FIG. 2. Schematic representation of the organization and function of the dehydrogenase and oxidase complexes in the aerobic *E. coli* membrane. The diagram indicates the involvement of the complexes in transmembrane electron or proton translocation or both. The cytoplasmic side of the membrane is uppermost (left and bottom). Q and QH_2 denote the oxidized and reduced forms, respectively, of ubiquinone and menaquinone. *ndhI* and *ndhII* represent NADH dehydrogenase in normal and sulfate-limited conditions; *ndhI* is associated with a functional site 1, whereas *ndhII* is not. This does not mean that there are two completely different NADH enzymes (see text). *HaseI* and *HaseII* represent two models for the enzyme hydrogenase (not two different hydrogenases). Model I has the oxidation of hydrogen located on the cytoplasmic side of the membrane, whereas model II has the oxidation site located on the periplasmic side of the membrane. In the former case, transmembrane H^+ translocation is required; in the latter case, only transmembrane electron transfer is required. *fdhI* and *fdhII* represent the aerobic formate dehydrogenase. Two models are shown (as with hydrogenase) because it is not known if the active site is cytoplasmic or periplasmic. *sdh*, succinate dehydrogenase, has no site 1 and is a peripheral membrane protein located on the cytoplasmic side of the membrane, where it reduces quinone. *G3Pdh* (*sn*-glycerol-3-phosphate dehydrogenase), *ldh* (lactate dehydrogenase), and *dam* (D-amino acid dehydrogenase) are similar in location and coupling to succinate dehydrogenase. The two oxidase complexes are shown associated with an additional cytochrome *b* (see Intermediate Components), forming a transmembrane electron transfer complex. The oxidases are shown being reduced by quinol at the periplasmic surface and reducing O_2 at the cytoplasmic surface.

of cells grown on glucose with fumarate or nitrate each gave one major arc of NADH-dye reductase activity, but with apparently different electrophoretic mobilities (237). The NADH dehydrogenase arc position is markedly altered with respect to the ATPase and nitrate reductase arcs (which appear in similar positions in each case). A comparison of these profiles with those of aerobic membranes by Owen and Kaback (174, 175) is difficult, but the relative positions of the ATPase and NADH dehydrogenase arcs in this case bear closer resemblance to that found in the membranes derived from cells grown anaerobically on glycerol with fumarate as oxidant. Comparison of the electrophoretic mobilities of these enzymes with the aerobic enzyme cannot be readily made.

Young and Wallace (252) have isolated a number of phenotypically similar mutants in which the respiratory NADH oxidase activity is negligible. These mu-

tants have lesions (*ndh*) that lie at 22 min on the chromosome and are cotransduced with the *pyrC* gene. The *ndh* gene has been cloned, and, when amplified, leads to overexpression of a dye reductase activity (60-fold) and a sixfold amplification of the NADH oxidase activity (249, 251). From this strain, an enzyme with NADH dye reductase activity has been purified (112). It can be reconstituted into membranes of an *ndh* mutant to restore cyanide-sensitive NADH oxidase activity (113). This enzyme has been synthesized in vitro (196), and the nucleotide sequence, and from this the amino acid sequence, has been determined. The enzyme is a single polypeptide of M_r 47,304 with tightly (but not covalently) bound flavin adenine dinucleotide (FAD). As isolated, this enzyme contains no copper, iron, molybdenum, tungsten, or phosphate but is in association with 70% of its own weight of lipid. A turnover number of 500 s^{-1} was determined, with benzoquinone as acceptor, a value that, if approached in situ with the endogenous quinone, is clearly kinetically competent enough to account for the NADH oxidase activity observed. The amino acid sequence of the protein has a polarity of 43.4%, which is not unusually low overall, but there are hydrophobic regions; these regions may play a part in the assembly of the protein into the membrane (251). The lack of iron in this active isolate appears to be emphatic (31). This finding leads to a comparison of the iron-containing NADH dehydrogenase zymogram arc and the amplified gene product of the *ndh* gene. One possibility is that, in situ, the enzyme is associated with additional iron-containing proteins; this is shown in the zymograms but not in the isolates, especially not those representing strains with amplified expression of the dehydrogenase. However, the *ndh* gene product can reduce quinone without a requirement for additional components. The problem may be related to the phenomenon of the adaptive loss of site 1. *E. coli* can synthesize an NADH dehydrogenase that has site 1 and one that does not (188); these may be the same enzyme, and the loss may be due to the loss of associated iron-sulfur-containing proteins. The evidence for the adaptive loss of proton translocation has come from measurements of H^+/O stoichiometries (the number of protons translocated across the coupling membrane per oxygen atom reduced) in sulfate-limited cells. In these cells, the H^+/O ratio for malate oxidation is 2, whereas in normal cells it is 4. The energy-linked reversed electron transport through the NADH dehydrogenase could not be demonstrated in membrane vesicles prepared from the sulfate-limited cells (188). This association of the NADH dehydrogenase with an optional iron-sulfur protein conferring site 1 activity may explain some of the aforementioned inconsistencies. An alternative possibility is that *E. coli* produces more than one NADH dehydrogenase.

In conclusion, a respiratory NADH dehydrogenase from *E. coli* has been extensively characterized after cloning, amplification, and isolation from an overproducing strain. This enzyme has an M_r of 47,304 and, as isolated, contains flavin but not iron. In this form, the enzyme can rapidly catalyze the reduction of quinones by NADH. This protein may be associated with iron-containing proteins in immunoprecipitated NADH dehydrogenase zymogram arcs and in the membrane when site 1 is operative. The antigenic determinants of this enzyme appear to be located solely on the cytoplasmic aspect of the membrane, where the catalytic site for NADH oxidation must be located. This has implications in the organization of the respiratory chain components for the purpose of generating a Δp and raises the question of whether the NADH dehydrogenase section of the respiratory chain directly translocates protons (i.e., does it have a site 1?). Certainly protons are translocated, and a Δp is generated when NADH is oxidized by oxygen in inverted vesicles, but it is necessary to differentiate between protons translocated by the quinone-oxidase region and protons transported by the dehydrogenase. In aerobically grown cells, an H^+/O ratio of 4 for respiration-driven proton translocation for malate oxidation (NAD$^+$ linked) has been reported. This, taken with the determination of values of 2 for each of succinate, D-lactate, and DL-glycerol-3-phosphate oxidation, strongly suggests a proton-translocating site associated with NADH dehydrogenase (145). In membranes derived from cells grown anaerobically on glycerol and fumarate, the NADH dehydrogenase can also be demonstrated to be proton translocating. NADH oxidation by fumarate in inverted membrane particles from anaerobically grown cells generates a Δp, as measured by quenching of quinacrine (Atebrin) fluorescence (89, 223). In this latter case, the situation is not complicated by possible proton translocation in the terminal section of the respiratory chain. Further evidence for the presence of site 1 in the NADH dehydrogenase in aerobic membranes comes from a study of reversed electron transport through this site. Uncoupler-sensitive, ATP-dependent reduction of NAD$^+$ by succinate, lactate, and ascorbate (plus phenazine methosulfate) has been reported in particles derived from aerobically grown cells (187, 230). This shows that the NADH dehydrogenase has a site 1 but also indicates that lactate dehydrogenase does not. (Succinate would not be expected to have one.) Thus, the NADH dehydrogenase region of this respiratory chain normally has a site 1, but this is not always the case. Under certain growth conditions, e.g., sulfate limitation (188), *E. coli* synthesizes an NADH dehydrogenase that is no longer proton translocating, as discussed in the previous paragraph. This dehydrogenase may be the same enzyme, site 1 activity being an option conferred by additional iron-sulfur proteins. These two options are represented diagrammatically in Fig. 2, in which no attempt is made to distinguish between a loop mechanism and a conformational pump mechanism for proton translocation.

Formate Dehydrogenase

Introduction. Formate oxidase activity is induced in *E. coli* under a range of growth conditions. The activity is present in membranes derived from cells grown aerobically on a range of nonfermentable carbon sources (11 to 219 ng-atoms of oxygen min^{-1} mg of protein^{-1}; reference 127). Membranes derived from cells grown anaerobically with nitrate or fumarate also contain high levels of the activity (58, 105). Although formate dehydrogenase is present in aerobically grown cells, it has not been substantially studied therein. Most of the work has been done on anaerobically grown systems and is covered in detail elsewhere. The picture is further complicated by the suggestion, still largely unresolved, that there are two formate dehydrogenase enzymes. *E. coli* utilizes different oxidants for formate (e.g., oxygen, nitrate, fumarate, and protons), and these activities are dependent on growth conditions. In the last case (formate:hydrogen lyase activity), reducing equivalents are fed through the formate dehydrogenase to a hydrogenase that acts as the terminal enzyme. Formate dehydrogenase is involved in all of these pathways, but is it the same formate dehydrogenase?

Genetics. To date, there are five classes of mutants specifically lacking formate dehydrogenase; i.e., they do not also lack nitrate reductase activity because of lesions in the processing of the molybdenum cofactor common to these two enzymes. These specific lesions are in *fdhA* (80 min), *fdhB* (38 min), *fdhC* (82 min), *fdhD* (87 min), and *fdhE* (87 min) (90). The *fdhA* and *fdhB* lesions produce no immunologically detectable formate dehydrogenase polypeptides (72, 73).

Purification and properties of the enzyme. Formate dehydrogenase was first purified by Enoch and Lester (58) from cells grown on glucose with nitrate as respiratory oxidant. This enzyme has been purified and extensively characterized (58, 71). It consists of three subunits: α, M_r 110,000; β, M_r 32,000; and γ, M_r 20,000, and also contains a cytochrome b_{555}, iron-sulfur centers, a molybdenum cofactor, and covalently bound selenium. The presence of molybdenum and selenium in the active enzyme explains the long-established requirements (178) for these metals in the growth media if formate dehydrogenase is to be developed. Selenium is associated with the 110,000-M_r peptide. Genetic studies indicate that the molybdenum cofactor is the same as that found in nitrate reductase. Cytochrome b_{555}^{fdh} differs from cytochrome b_{555}^{nr} in having a lower midpoint potential, −105 mV at pH 7 (82). By analogy with the nitrate reductase structure, it has been suggested that cytochrome b_{555}^{fdh} is the 20,000-M_r subunit (58). Boxer et al. (18), however, have reported a formate dehydrogenase preparation containing cytochrome b_{555} but lacking the 20,000-M_r subunit.

Organization and energy-coupling considerations. The organization of formate dehydrogenase in the membrane has been studied by labeling exposed tyrosine residues with lactoperoxidase and ^{125}I. The result shows that the α subunit is transmembranous. The β subunit is unlabeled by lactoperoxidase (71). By using [^{35}S]diazobenzenesulfonate and [^{125}I]diiodosulfanilic acid as protein-modification reagents, it was demonstrated that the β subunit is also transmembranous. Although not subunit specific, antibody studies also show the transmembranous nature of the formate dehydrogenase (74, 237). To date, there has been no report on the location of the γ subunit.

There is strong evidence that formate dehydrogenase pumps protons across the cytoplasmic membrane, concomitant with oxidoreduction reactions (69, 118). Stoichiometries (of protons translocated per pair of electrons traversing the respiratory chain to the oxidant) for the overall formate-to-oxygen or formate-to-nitrate reaction have been measured at up to $4H^+/2e^-$ and $3H^+/2e^-$, respectively. The use of menadione and ubiquinone-1 as artificial oxidants for formate oxidation (which will bypass site 2) by E. coli spheroplasts gave stoichiometries of 1.73 and 1.58 $H^+/2e^-$, respectively, suggesting that the formate-to-quinone segment of the respiratory chain translocates $2H^+/2e^-$. This process is inhibited by 2-heptyl-4-hydroxyquinoline N-oxide. The site of formate oxidation with respect to the two surfaces of the cytoplasmic membrane has not been demonstrated. Thus, although formate dehydrogenase has site 1, it may not involve vectorial proton translocation. Chemiosmotically, the same result could be obtained by locating the site of formate oxidation periplasmically and transporting electrons in across the membrane (Fig. 2).

Hydrogenase

Introduction. Hydrogenases (EC 1.12.-.-) catalyze the reversible oxidation of dihydrogen to protons. In E. coli, such enzymes are involved in two major processes. The first process is hydrogen production through the formate hydrogen lyase system, consisting of a formate dehydrogenase and a hydrogenase; this pathway is repressed in the presence of oxygen or nitrate and so does not concern us in this section. The second pertinent role of hydrogenase is in the utilization of H_2 as a respiratory reductant.

Enzymes. E. coli has been reported to be able to produce a multiplicity of hydrogenases (1). Although multiplicity of activity stains on polyacrylamide or similar gels is a common phenomenon for membrane proteins even when no isoenzymes are present, there is strong evidence for two, and possibly three, membrane-bound hydrogenase isoenzymes. Aerobically grown cells have been reported to contain a membrane-bound hydrogenase with an M_r of 56,500; the isolated enzyme (M_r, 113,000) is a dimer (2). When cells are grown on glycerol with fumarate as terminal respiratory oxidant, a hydrogenase with an M_r of 19,000 is developed (17). When cells are grown anaerobically on glucose with nitrate as oxidant, a 64,000-M_r hydrogenase is produced (70). The situation has been partly clarified by the discovery of two distinct hydrogenases in E. coli cells grown fermentatively on glucose. One of these hydrogenases is the 64,000-M_r enzyme (isoenzyme I), and the other hydrogenase appears to be the same as the aerobically produced 56,500-M_r enzyme (isoenzyme II) (9). The former hydrogenase is present in greater abundance, but the activity of the latter hydrogenase, in hydrogen benzyl viologen reductase activity assays, is greater.

Immunological studies (237) failed to detect hydrogenase in membranes from cells grown on glucose with nitrate as oxidant (although a hydrogenase activity is known to be present) but detected a major and minor arc with membranes from cells grown on glycerol with fumarate as oxidant. Crossed immunoelectrophoresis of antiserum from anaerobically grown cells with the aerobic enzyme indicates that the aerobic enzyme is the same as isoenzyme II from the anaerobic cells and distinct from isoenzyme I (communication from D. Boxer cited in reference 93).

The aerobic enzyme (M_r, 56,500) has been extracted with deoxycholate and pancreatin and purified by conventional chromatographic procedures (2). It is an iron-sulfur-containing protein (12 Fe:12 S per molecule). Nickel is also present; although at present the role of nickel is unclear, it does appear to be required for activity and to be a growth requirement if hydrogenase activity is to be developed in certain bacteria. Hallohan and Hall (93) have shown that the oxygen lability of the E. coli aerobic enzyme may be attributable to loss of nickel. It is becoming apparent that most, if not all, hydrogenases contain nickel, the role of which, at least in E. coli, is not well understood. Hydrogenases from other sources have been more thoroughly investigated (e.g., references 4 and 30), and three redox states of the metal have been reported, namely Ni(III), Ni(II), and Ni(I). Only the Ni(III) and Ni(I) states are observable by electron paramagnetic resonance (EPR). The oxidized form of the enzyme from aerobic cells shows a narrow EPR signal at low temperatures, characteristic of a three- or four-iron type of iron-sulfur-cluster species. This center has an E_{m7} of +25 mV, rather high for involvement at the catalytic site (E_{m7} H_2/H^+ = −420 mV) (29). No signals were reported for the iron-sulfur clusters in the reduced form. It was assumed that they were nondetectable by EPR because of spin coupling or a large zero-field splitting. However, it has been pointed out that the spectra may have been so broad that they were overlooked (30). The 64,000-M_r *E. coli* enzyme also contains nickel (9).

The current overview is that there is an assimilatory hydrogenase in cells grown anaerobically and that there is a dissimilatory hydrogenase in aerobically grown cells and in some anaerobically grown cells. The situation with regard to the third reported enzyme, the 19,000-M_r protein isolated from cells grown anaerobically with fumarate, requires some clarification.

Chemiosmotic aspects. Hydrogenase activities have been studied with regard to their contribution to the generation of a Δp. Unfortunately, the systems that best have the potential for understanding this respect do not correspond to the isozymes most studied. It is clear that, during fumarate respiration, hydrogenase activity generates a Δp, as cells can grow anaerobically on hydrogen with fumarate (malate) as terminal oxidant. Thus, site 1 coupling must occur. Unfortunately, the hydrogenase in these cells is not well characterized. The other two hydrogenases are better characterized. The 64,000-M_r (NO_3^- respiration) enzyme is an intrinsic membrane protein and is probably the hydrogenase that has been shown to be transmembranous by antibody and labeling studies (70). The 54,500-M_r (aerobic) enzyme appears to be a membrane-associated species located on the cytoplasmic aspect of the cytoplasmic membrane. A hydrogenase from fumarate-grown cells has antigenic determinants on the cytoplasmic aspect of the cytoplasmic membrane only (237), but it is unclear which enzyme this is. The 19,000-M_r enzyme may not be the sole species present in these membranes.

Measurement of the quenching of quinacrine fluorescence (indicative of ΔpH) in inverted membrane vesicles has shown that H_2-dependent reduction of fumarate apparently translocates protons inwards (118, 119); this behavior would be an apparent outward translocation in whole cells. We say apparent proton translocation because (as shown in Fig. 2) the location of the substrate site on the periplasmic surface of the membrane causes the (scalar) protons from the oxidation to be generated there, contributing to the ΔpH. The location of the substrate site on the cytoplasmic side of the membrane gives rise to a need for transmembrane proton translocation. These two possibilities are represented in Fig. 2. In experiments with spheroplasts, stoichiometries for $H^+/2e^-$ for H_2 oxidation by fumarate and nitrate were 1.85 and 3.3, respectively (116). In this latter case, the higher stoichiometry is attributable to proton translocation by the quinone-nitrate reductase portion of the respiratory chain. Thus, it appears clear that in anaerobic cells H_2 oxidation has a site 1. More work is required on the aerobic system before a similar conclusion can be reached.

Succinate Dehydrogenase

Introduction. *E. coli* has two distinct, membrane-bound flavin- and iron-sulfur-containing enzymes that catalyze the interconversion of succinate and fumarate (100, 228). The oxidizing enzyme, succinate dehydrogenase (EC 1.3.99.1), is a primary dehydrogenase and part of the tricarboxylic acid cycle. The reducing enzyme, fumarate reductase, functions as the terminal enzyme of certain anaerobic respiratory chains. Under nonphysiological conditions, both enzymes can catalyze both reactions (100), and fumarate reductase has been shown to be capable of functionally substituting for succinate dehydrogenase (79). The reverse has not been demonstrated, although there is no reason to suppose it is not possible.

Structurally, succinate dehydrogenase and fumarate reductase are very similar, although antigenically distinct (45), differing only in the regulation of their expression and perhaps in the fine tuning required to meet different kinetic demands. Genes for both enzymes have been cloned, and their nucleotide sequences have been determined. Both enzymes also closely resemble succinate-fumarate oxidoreductases from other sources (96). In aerobically grown *E. coli*, only succinate dehydrogenase is normally expressed; thus, only it is covered in detail in this section. Succinate dehydrogenase has never been satisfactorily isolated and purified from *E. coli*. It has, however, been well characterized by indirect means.

Expression. Succinate dehydrogenase is derepressed aerobically or during growth with nitrate, subject to the catabolite repression system, and is repressed by anaerobiosis (228). Succinate dehydrogenase is induced during nitrate respiration (G. Cumming, unpublished data), a condition in which fumarate reductase is not present (237).

Succinate oxidase activity is present at relatively high levels (122 to 315 ng-atoms of oxygen min^{-1} mg of protein^{-1}) in membranes derived from cells grown aerobically on mannitol, fumarate, glycerol, malate, succinate, acetate, lactate, and hydrolyzed casein (35, 127) but not usually in membranes from cells grown aerobically on glucose (211). Glucose represses all tricarboxylic acid cycle enzymes, including succinate dehydrogenase, and it represses the synthesis of the enzyme even when added to cultures already growing aerobically on succinate (211). This repression can be partially countermanded when the cycle is required for anabolic processes, through growth in minimal media; this explains why, in observing catabolite repression of succinate dehydrogenase, the same attenuation of enzyme levels is not always obtained (35, 43, 75).

Structure. The mammalian succinate dehydrogenase is composed of four polypeptides: a 70,000-M_r flavin-containing subunit, which may also contain iron-sulfur clusters; a 27,000-M_r subunit known as the iron-sulfur protein moiety; and two small (13,500- and 7,000-M_r) polypeptides thought to be involved in membrane attachment (for a review, see reference 15). Attempts to purify the *E. coli* enzyme have met with only limited success (126, 202). However, immunoelectrophoretic and genetic studies have shown that the *E. coli* enzyme has a very similar composition, i.e., a 64,300-M_r flavin-containing subunit, a 26,000-M_r iron-sulfur subunit, and two hydrophobic subunits with M_rs of 14,200 and 12,800 (values taken from sequence analysis). The *E. coli* enzyme, however, raises an antibody that cross-reacts only with the enzyme from other *Enterobacteriaceae*, including *S. typhimurium* (46).

In crossed immunoelectrophoresis studies on *E. coli*, succinate dehydrogenase has been resolved by radiolabeling of the covalently attached flavin mononucleotide (by growth in the presence of radiolabeled riboflavin), zymogram staining, and comparison of mutant and parent strains (45, 46, 106, 120, 172, 173–177). Excision of the material in the arc attributable to succinate dehydrogenase and subsequent analysis on denaturing gel electrophoresis showed that a polypeptide with an M_r of approximately 71,000 contained the covalently bound flavin (172). The full M_r range of polypeptides detected by these analyses was 71,000, 31,000, 29,000, and 13,000 (172) and 73,000 and 26,000 (120). In addition, some association was observed, in parts of the precipitin arc, with an additional 19,000-M_r peptide suggested to be a cytochrome *b* (172).

The prosthetic groups of *E. coli* succinate dehydrogenase have not been analyzed in the enzyme as isolated by conventional procedures. However, iron-sulfur clusters with spectral properties similar to those reported for other succinate-fumarate oxidoreductases have been detected in membranes from such cells (107). Three centers were assigned to succinate dehydrogenase: a ferredoxin center (center 1) with a midpoint potential of −20 mV (pH 7.0), a center detectable in its oxidized form (center 3; midpoint potential, +100 mV), and a center designated center 2 (indicated by its effect on the relaxation rate of center 1; midpoint potential, −220 mV). Limited analysis of the prosthetic groups in the enzyme isolated by immunoprecipitation has been possible, however. First, the presence of covalently bound flavin has been demonstrated by crossed immunoelectrophoretic analysis and subsequent excision and sodium dodecyl sulfate-polyacrylamide gel electrophoretic analysis, combined with radiolabeling of the flavin and zymogram staining, as discussed above. An *E. coli* succinate dehydrogenase preparation, isolated by immunoprecipitation from a detergent-treated membrane extract, has been examined by EPR (44). Spectroscopic measurements indicate the presence of a two-iron ferredoxin center (center 1), a three-iron center (center 3; EPR detectable in the oxidized state), and a center designated center 2 (not visualized directly but via an enhancement of the relaxation rate of center 1) similar to the properties observed in membrane preparations. A *b*-type cytochrome is also associated with the immunoprecipitation and might be b_{556}, specified by *cybA*, now assignable as *sdhC* (165).

Genetics. The genes specifying succinate dehydrogenase lie at 16 min on the current *E. coli* genetic map. This region specifies all four subunits of the enzyme and includes also the genes for three other tricarboxylic acid cycle enzymes, citrate synthase (*gltA*), and the E1 and E2 components of the 2-oxoglutarate dehydrogenase (*sucA* and *sucB*, respectively). These genes have been cloned (229), and the nucleotide sequences have been determined (80). The order of these genes is *gltA-sdh-sucA-sucB*, and the polarities of expression are counterclockwise for *gltA* and clockwise for *sdh*, *sucA*, and *sucB*, relative to the *E. coli* linkage map (229). The nucleotide sequence has revealed an operon of four structural *sdh* genes in the order *sdhC-sdhD-sdhA-sdhB*. These, from the structural data available and by comparison with other succinate-fumarate oxidoreductases, are the genes for the two small hydrophobic polypeptides (M_r 14,200 and 12,800), the large flavin-containing subunit (M_r 64,300), and the so-called iron-sulfur subunit (M_r 26,000), respectively (80).

Succinate dehydrogenase in *S. typhimurium*. Of necessity, the previous discussion has been limited to *E. coli*. The enzyme from *S. typhimurium* is apparently very similar to the *E. coli* enzyme and exhibits greater homology than is evident among most succinate-fumarate oxidoreductases. The material in crossed immunoelectrophoresis arc 22 of *E. coli*, later shown to be due to succinate dehydrogenase, was shown to contain an antigen capable of cross-reacting with a component of *S. typhimurium* membrane extract (174). A highly specific antibody prepared against *E. coli* succinate dehydrogenase by excision and reinjection of the material in the succinate dehydrogenase crossed immunoelectrophoresis arcs cross-reacts strongly with *S. typhimurium* succinate dehydrogenase but does not react with the *E. coli* fumarate reductase (46). The anti-*E. coli* succinate dehydrogenase serum reacted with the succinate dehydrogenases of all *Enterobacteriaceae* tested but with none tested outside of this group.

Coupling of succinate oxidation to the generation of Δp. The H^+/O ratio for succinate oxidation by oxygen in aerobically grown cells is 2.26 (145). This value may be a slight overestimate, as there will be a contribution from the oxidation of endogenous substrate (H^+/O = 3.86). These values suggest that no Q cycle, or alternative mechanism for increasing the stoichiometry, is operating in this region of the bacterial respiratory chain. However, it is not possible to discount an underestimate of the ratio caused by a specific proton-phoretic carrier, such as was encountered in mitochondria (23).

Lactate Dehydrogenase

E. coli produces four lactate dehydrogenases, two of which are NADH-linked, soluble lactate dehydrogenases. In this discussion, we are concerned only with the two respiratory primary dehydrogenases: D(−)- and L(+)-lactate dehydrogenases. The former dehydrogenase is constitutive; the latter one is induced by growth on lactate (D or L).

D-Lactate dehydrogenase, purification and properties. D-Lactate dehydrogenase has been extracted with detergent, purified, and characterized by Kohn and Kaback (134) and by Futai (65) from different *E. coli* strains. Properties of the purified preparations obtained by these groups and by others were compared by Ingledew and Poole (106). Enzymes from different strains are very similar, but details of purification differ. Slight differences in heat inactivation and pH optima suggest that there may be some amino acid substitutions. The enzyme has an M_r of 71,000 to 75,000 (231), and the prosthetic group is noncovalently bound FAD. The K_m for D-lactate varies with the presence or absence of detergent, being lowest with the aggregated enzyme and highest in the membrane (231).

Molecular biology. Amplification of the respiratory D-lactate dehydrogenase has been made possible by the cloning of specific DNA sequences. Two plasmids were isolated from an *E. coli* chromosomal library on the basis of their ability to complement an *ndh* mutant defective in NADH dehydrogenase. The property of one such plasmid (pIY1) could be attributed to its carrying the gene specifying NADH dehydrogenase, whereas the second plasmid (pIY2), derived from *Hin*dIII-digested chromosomal DNA, did not carry this gene but resulted in overproduction of the membrane-bound D-lactate dehydrogenase (250). Complementation by pIY2 of the *ndh* mutant probably arose from the ability of the elevated D-lactate oxidase activities to participate in regeneration of NAD$^+$. Thus, in *ndh* mutants D-lactate probably accumulates as a result of the reduction of pyruvate by the soluble, pyridine nucleotide-linked D-lactate dehydrogenase (232, 233). The amplified respiratory D-lactate oxidase system, reconverting the lactate to pyruvate, thus provides a cyclic system for reoxidation of NADH independent of the NADH dehydrogenase. More recently, a recombinant DNA plasmid has been constructed that, on temperature induction, produces a 300-fold increase in D-lactate dehydrogenase levels. The nucleotide sequence, and thence the primary amino acid sequence of the protein, have been determined (212). The nucleotide sequence of the gene contains features common to bacterial genes: promoter sequence and putative ribosome-binding site. The D-lactate dehydrogenase does not contain a signal sequence, nor is the enzyme processed after translation (except for the removal of formyl-methionine). There are only two short regions in the amino acid sequence that are particularly hydrophobic (regions 46 to 56 and 140 to 156), and there is one highly polar region. Hydrophobic moment analysis indicates several putative amphipathic α helices and β sheets.

Synthesis of lactate dehydrogenase has been studied by using a recombinant plasmid containing the *dld* gene. Expression of the cloned gene, achieved either in vivo with transformed minicells or in vitro with a fractionated transcription and translation system, led to a product identified as the membrane-bound D-lactate dehydrogenase (213). Interestingly, the protein was catalytically active and bound to membrane vesicles during or after synthesis, suggesting that, like NADH dehydrogenase, it is synthesized in the mature form and binds to the membrane without a leading peptide sequence. Addition of FAD (the cofactor) in vitro elicited a twofold increase in synthesis of the enzyme.

Chemiosmotic considerations and membrane localization. Antiserum raised to the D-lactate dehydrogenase extracted and purified from membranes is an effective inhibitor of dehydrogenase activity and lactate-dependent active transport (220) but does not react with the soluble, pyridine nucleotide-dependent enzyme or the partially purified L-lactate dehydrogenase (see below). D-Lactate dehydrogenase is inaccessible to antibody in right-side-out vesicles from strain MS308-225, indicating that the antigenic determinants are on the cytoplasmic aspect of the membrane. Antibody inhibition (220, 221) and immunoadsorption studies have shown that the protein is associated with the cytoplasmic aspect of the inner membrane. In crossed immunoelectrophoretic studies of detergent-extracted inner membranes, lactate de-

hydrogenase has been identified in the immunoprecipitation patterns (70, 174, 175, 227). Despite the predominantly membranous location of D-lactate dehydrogenase activity, mutants constitutive for the synthesis of other dehydrogenases contain an increased proportion of nonsedimentable D-lactate dehydrogenase activity, indicative of competition among dehydrogenases for a common binding site (143). Contradictory results (92), however, showed that purified preparations of D-amino acid dehydrogenase and D-lactate dehydrogenase bind independently to right-side-out and inverted vesicles from *E. coli*, as well as to phosphatidylcholine liposomes, without detectable competition. Each enzyme can feed electrons to a common rate-determining redox component that precedes a site of proton translocation.

The oxidation of D-lactate by lactate dehydrogenase is accompanied by H$^+$ translocation in intact cells (145, 203). Stoichiometries of the 2.26H$^+$/O ratio for lactate oxidation compare with the same value for succinate oxidation and 3.86 for malate oxidation. The value of 2.26 may be a slight overestimate of the true value because endogenous respiration may contribute (H$^+$/O ratio of 3.36). Thus, lactate oxidation can be coupled to active solute transport in membrane vesicles (10, 11), but without the operation of a site 1. Crude or purified (66, 201, 204) preparations of the enzyme are capable of reconstituting lactate-dependent oxygen uptake and active transport in vesicles from *E. coli* mutants (*dld*) that are defective in D-lactate dehydrogenase activity (101, 122, 222).

L-Lactate dehydrogenase. Early studies indicated that L-lactate dehydrogenase was induced in cells grown on DL-lactate. Futai and Kimura (67) have confirmed that enzyme activity is 100-fold higher in such cells than in cells grown aerobically on glycerol or anaerobically on glucose. In contrast, activities of D-lactate dehydrogenase (described above), NADH dehydrogenase, and, to a lesser extent, succinate dehydrogenase were relatively invariant under the growth conditions tested (127, 187). A report by Nishimura et al. (167) indicates that, under fermentative as opposed to oxidative conditions, L-lactate dehydrogenase is not induced, not even in the presence of lactate.

The cholate-extracted L-lactate dehydrogenase has been purified to homogeneity by conventional procedures (67). The properties of the isolated enzyme are quite distinct from those of the D-lactate dehydrogenase, especially with regard to specificity and flavin content. Antibody raised against the enzyme inhibited L-lactate dehydrogenase and oxidase activity in inverted membrane vesicles, indicating that (i) the enzyme is a primary dehydrogenase in the *E. coli* respiratory chain (Fig. 2) and (ii) the enzyme is localized (in the intact cell) on the inner aspect of the cytoplasmic membrane.

D-Amino Acid Dehydrogenase

The role of D-amino acid dehydrogenases in *E. coli* has been considered recently (106). In *E. coli*, D-amino acid dehydrogenase is a primary dehydrogenase, since the enzyme is membrane bound and does not require the intermediation of NAD$^+$. There are a number of differences between the *E. coli* B and the *E. coli* K-12 enzyme in substrate specificity and control of induction (61).

The enzyme from *E. coli* B has been extracted from membranes with Triton X-100 and purified to approximately 65% homogeneity (170, 171). This enzyme consists of two 55,000- and 45,000-M_r subunits and contains noncovalently bound FAD and nonheme iron in the ratio 1:2 to 1:3. Preliminary EPR studies confirmed the presence of an iron-sulfur center. The enzyme in *E. coli* K-12 is probably similar in structure. Mutants (*alnA* or *dad*, a locus lying at 1.5 min) in this strain lack D-alanine dehydrogenase activity and a protein of M_r 55,000 to 60,000, which may be the larger subunit (60). A second locus (*dadA*, at 26 min [61, 244]) may specify the smaller subunit. A third locus, *alnR*, at 99 min, is believed to be involved in positive control of *dad* expression (14).

Chemiosmotic aspects. Reconstitution of D-alanine-dependent active transport has been achieved by incubating the purified dehydrogenase with membrane vesicles (171). The binding itself is nonspecific, since the enzyme is bound to the exposed surface, although its normal site has been shown by immunological studies to be on the inner surface of the cytoplasmic membrane. Coupled to D-alanine oxidase activity, reconstituted vesicles of both types can generate a membrane potential ($\Delta\psi$) of approximately 100 mV. It was suggested that proton translocation occurs distal to the site at which electrons enter the chain from the dehydrogenase and that the D-amino acid dehydrogenase is not part of a proton-translocating loop. This is reasonable, given the similar results obtained from the two orientations of the protein on the membrane. It is most likely that the enzyme releases reducing equivalents directly into the quinone pool.

sn-Glycerol-3-Phosphate Dehydrogenase

E. coli can produce three distinct dehydrogenases of this type: a soluble, pyridine nucleotide-linked enzyme, glycerol-3-phosphate dehydrogenase (EC 1.1.1.8), and two primary respiratory dehydrogenases, a so-called aerobic enzyme and an anaerobic enzyme. In this section, we are interested only in the (primary) aerobic enzyme, which is induced during aerobic growth or during growth with nitrate on glycerol or glycerol-3-phosphate. The enzyme is membrane bound and is specified by the *glpD* gene and, in constitutive strains, can comprise up to 10% of the inner membrane protein. Glycerol-3-phosphate is taken up by a specific transport system; the operons specifying this transport system, the glycerol kinase, and the dehydrogenases constitute a regulon (*glp*) under the control of a single repressor and are induced by glycerol-3-phosphate. A partial purification of the enzyme has been reported by Weiner and Heppel (242), who reported an apparent molecular weight of 80,000, comprising of two subunits of M_r 35,000. The cofactor is noncovalently bound FAD. Activity of the enzyme is dependent on the presence of phospholipids.

Chemiosmotic aspects. The evidence available with regard to energy coupling during glycerol-3-phosphate oxidation indicates that the enzyme is not involved in site 1 phosphorylation. Glycerol-3-phosphate can donate reducing equivalents for the ATP-dependent reversed electron transport to NAD^+. A value of 2.36 has been reported (145) for the H^+/O

ratio for glycerol oxidation in cells that had been grown aerobically on glycerol (the value for endogenous respiration was 3.53). It has been argued that this value is principally due to glycerol-3-phosphate oxidation (88), indicating the absence of site 1. Nor is the anaerobic enzyme involved in site 1, as shown by the failure of glycerol-3-phosphate-mediated reduction of fumarate to support quenching of quinacrine fluorescence in membranes derived from cells grown anaerobically on glycerol with fumarate (89, 223).

INTERMEDIATE COMPONENTS

The *E. coli* respiratory chain can be considered to be modular, comprising a number of dehydrogenase and oxidase complexes. Diffusion of these complexes in the membrane, although sometimes relatively rapid (68), is probably too slow to account for the observed electron transfer rates by a collision mechanism among these complexes. Additional components function as oxidation-reduction mediators among the complexes; these are quinones and *b*-type cytochromes. The role of the quinones is well established; that of the *b*-type cytochrome is less well established.

Quinones

E. coli can synthesize two types of quinone: ubiquinone (mainly Q_8, with eight terpenoid units in the phytol side chain) and menaquinone (mainly MQ_8, also called vitamin K). The quinones are thought to function as mobile carriers of hydrogen between the large and relatively slow-moving dehydrogenase and oxidase complexes. The long phytol side chains (32 carbons long, 8 methyl side groups) impart hydrophobicity and lipid solubility. *S. typhimurium* also contains Q_8 as the predominating species in aerobically grown cells (56).

The ratio of the two quinones in *E. coli* membranes is variable. Under conditions of high aeration and in logarithmic phase, ubiquinone-8 predominates over menaquinone-8 by a ratio of 22:1; in an 18-h aerobic stationary-phase culture, the ratio is 1.5:1, and in an anaerobic stationary-phase culture ubiquinone-8 was undetectable, whereas menaquinone-8 was at relatively high concentrations (Table 1). Other growth conditions for which data are available are nitrate and fumarate respiration; in the former ubiquinone predominates, but in the latter menaquinone predominates.

A variety of mutants that lack one or both quinones have been isolated, and these have proved to be valuable tools in delimiting respective functions and

TABLE 1. Role of quinones in electron transport: quinones under different growth conditions

Growth condition	Amt (nmol/mg [dry wt])		Ratio of ubiquinone to menaquinone in membranes
	Ubiquinone	Menaquinone	
Aerobic (early logarithmic phase)	0.55	0.025	22:1
Stationary phase (18 h)	0.17	0.11	1.5:1
Anaerobic (40 h)	<0.0003	0.25	0.001:1

TABLE 2. Role of quinones in electron transport: restoration of oxidase activities in mutants defective in both ubiquinone and menaquinone

Dehydrogenase/oxidant	Restoration of oxidase activities[a] by addition of:	
	Ubiquinone	Menaquinone
NADH		
O_2	+++	−
Nitrate	+++	−
Glycerol-3-phosphate		
O_2	+++	+++
Nitrate	+++	+++
D-Lactate		
O_2	+++	+
Nitrate	+++	+
Succinate/O_2	+++	−
Formate/nitrate	+++	+++

[a] Activity ranged from none (−) to high (+++).

resolving their biosynthetic pathways. Use of these mutants has shown that the quinones function as carriers of reducing equivalents between the dehydrogenases and the various terminal enzyme complexes. Some activities can be supported by either quinone, but others show marked specificities. Two reasons for the specificities for the quinone species can be proposed; the difference in midpoint potential between ubiquinone (+70 mV at pH 7.0) and menaquinone (−74 mV) may make one more or less suitable as an intermediate between an individual donor and acceptor complex. Alternatively, or in addition, the functional specificity could be for the purpose of compartmentalization of the respiratory chain reactions, with acceptor and donor complexes discriminating between the two quinones. Compartmentalization is feasible in the presence of both quinones, as redox equilibration between the two pools is predicted to be slow (209).

Wallace and Young (241) have studied the role of the two quinones in electron transport to oxygen and to nitrate by using such mutants. NADH, D-lactate, glycerol-3-phosphate, and succinate oxidase activities were measured in membranes derived from aerobically grown cells of the double mutant (*ubi men*). Ubiquinone could restore activity in all four cases, but menaquinone could only fully restore the glycerol-3-phosphate activity and partly support D-lactate oxidase activity. Menaquinone was ineffective for restoring succinate and NADH oxidase activities. In cells grown anaerobically in the presence of nitrate, formate-to-nitrate electron transfer could be supported by either quinone, but glycerol-3-phosphate, D-lactate, and NADH oxidation by nitrate exhibited the same dependencies as in the aerobic system (Table 2). Growth studies support these observations; the mutant *ubi*+ *menA* grows aerobically on glucose as well as does the wild type, but the *ubiA men*+ strain has only 30% of the growth yield, and the double mutant is more severely affected. The anaerobic fumarate respiration system is a menaquinone-utilizing pathway.

Au et al. (6) have studied the steady-state levels of cytochrome *b* reduction in strains lacking the ability to synthesize cytochrome *d* and lacking one or both quinones. These investigators conclude that the qui-

nones operate at only one site in this exclusively cytochrome *o* oxidase-based system, and that site is on the dehydrogenase side of the *b*-cytochromes. (Some dehydrogenases have their own cytochrome *b*, but these are probably present in low abundance in the cells used.) It was also demonstrated that succinate, lactate, and NADH dehydrogenases in these cells strongly preferred ubiquinone over menaquinone as an intermediate, in agreement with the findings of Wallace and Young (241) (Table 2).

Q cycles are relatively well established in mitochondrial and photosynthetic systems and have been suggested to play a role in respiration in *E. coli* (20). At present, however, all of the available data on proton translocation and associated phenomena can be accommodated without invoking a Q cycle.

Cytochromes *b*

A number of spectroscopic investigations of the α bands of cytochromes *b* in *E. coli* indicate a multiplicity of components (206, 217, 219, 238). Multiple *b* cytochromes can also be resolved kinetically (87). This complexity is dealt with in a number of other sections, where the *b* cytochromes are associated with dehydrogenase complexes (The Dehydrogenases) or oxidase complexes (The Cytochrome Oxidase Complexes) or are peripheral to the main respiratory pathways (Other Oxidation-Reduction Components). Here, we are interested in those *b* cytochromes that may function in the main respiratory chain as intermediates between the dehydrogenase and oxidase complexes.

Cytochrome b_{556}. Cytochrome b_{556} is a membrane-bound component developed under aerobic growth conditions. It is distinct from the cytochrome *o* complex and has a lower midpoint potential than *o* (+129 to +46 mV; Fig. 3). Recent reconstitution experiments with pyruvate oxidase and the cytochrome *o* complex cast doubt on the essentiality of cytochrome b_{556} in electron flow from quinone to the oxidase (32).

The nature of the immediate electron donor to cytochrome *o* has been in question for many years. Two studies indicated a role for quinones between cytochrome b_{556} and cytochrome *o* in *E. coli*. The schemes proposed differ in detail but are based on similar experimental approaches (Fig. 4). Downie and Cox (54) placed cytochrome b_{562} on the substrate side of cytochrome b_{556}, whereas Kita and Anraku (128) placed cytochrome b_{562} adjacent to cytochrome *o*. However, the finding that cytochrome *o* itself is probably b_{562} and that with a split α band it may also absorb at 556 nm casts some doubt on the interpretation of these results. Au et al. (6) have studied the steady-state levels of cytochrome *b* reduction in strains lacking the ability to synthesize cytochrome *d* and lacking one or both quinones. These investigators conclude that the quinones operate only at one site and that it is on the dehydrogenase side of the *b* cytochromes.

Stopped-flow kinetic studies by Haddock et al. (87) on aerobically grown cells and membranes derived therefrom indicate the presence of two kinetically distinct pools of cytochrome *b*. The *b* cytochrome accounting for the more rapid of the two phases observed (half-life [$t_{1/2}$] <3.3 ms) accounts for 50% of

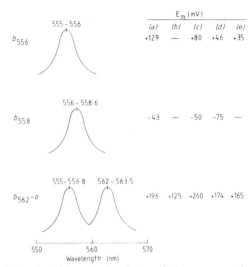

FIG. 3. Possible relationships of the measured redox potentials of b-type cytochromes in early-logarithmic-phase, aerobic *E. coli* to spectrally distinguishable forms. Idealized absorption spectra (reduced minus oxidized) are shown, with the range of reported wavelength maxima at 77 K, for b_{556}, b_{558}, and the b_{562}-o complex. The natural band widths shown (about 3 nm) are arbitrary. The midpoint potentials are those measured at room temperature, near pH 7, and in the absence of CO by (a) Hackett and Bragg (83), (b) Kita et al. (129), (c) Reid and Ingledew (206), (d) Van Wielink et al. (238), and (e) Lorence et al. (152). The b_{558}, the potentials of which are shown here, and which is readily extracted (206), is probably not the same cytochrome that is copurified with cytochrome *d* and which has a more positive potential (e.g., +10 mV; reference 130). The names given (on the left) to the cytochromes are not necessarily those used in the references cited.

the observed absorption change and is thought to correspond to cytochrome *o*. The second phase has a $t_{1/2}$ of approximately 25 ms and accounts for the remaining 50% of the absorbance change in the Soret band. This latter component most probably corresponds to cytochrome b_{556}. The turnover number of this component appears to be adequate to support the observed respiration rates.

Cytochrome b_{556} has been purified (131) from Sarkosyl-solubilized membranes of *E. coli*. The highly hydrophobic cytochrome is an oligomer composed of identical polypeptides each with an $M_r \approx 17,500$ and does not bind CO. It contains equimolar amounts of heme and polypeptide. The α absorption peak is at 556 nm (at 77 K), and its midpoint potential is −45 mV in this solubilized form. A value of approximately +80 mV was obtained for the midpoint potential of the 556-nm component in aerobic membranes (Fig. 3). It is immunologically unrelated to the cytochrome b_{556}^{NR} associated with nitrate reductase (137).

The *cybA* gene for cytochrome b_{556} lies in the region near min 16 (131, 164) on the *E. coli* chromosome. Sequence homology between cytochrome b_{556} and the amino acid residues predicted from the nucleotide sequence of the *sdhC* gene strongly suggests that *cybA* and *sdhC* are synonymous (165). In the strain lacking this cytochrome, another b-type cytochrome, with an M_r of 18,000, was detected. This diheme hemoprotein has now been purified; it has a main α peak at 561 nm and a smaller peak at 555 nm (at 77 K) and has been

named cytochrome b_{561} (164, 165). Like b_{556}, it is reducible by NADH or lactate but not by ascorbate plus phenazine methosulfate, is presumably of low potential, and may act as a functional replacement for b_{556}. A small amount of cytochrome b_{561} has been reported in a wild-type strain (cited in reference 163). The structural gene for cytochrome b_{561} (*cybB*) has been cloned (163).

Cytochrome b_{555}. In *E. coli* grown at low O_2 tension or anaerobically, the cytochrome *bd* oxidase is developed (see The Cytochrome Oxidase Complexes), along with an additional b-type cytochrome with a midpoint potential in the region of +140 mV (206). This component appears to be a functional member of the respiratory chain and perhaps has a relationship to the cytochrome *bd* complex similar to that which the cytochrome b_{556} (see Cytochrome b_{556}) has to the cytochrome *o* complex. There is a lack of information on this cytochrome. In the stopped-flow kinetic analyses of Haddock et al. (87) on whole cells grown anaerobically on glycerol with fumarate as oxidant, a multiphasic *b* cytochrome oxidation was observed. The fastest component observed ($t_{1/2}$, <3.3 ms) probably corresponds to the b_{558} directly associated with the cytochrome oxidase (cytochrome $b_{558}d$). This component was responsible for approximately 40% of the absorbance change at 560 minus 575 nm. A second component had a $t_{1/2}$ of approximately 100 ms, contributed 40% of the total absorption change, and is most probably cytochrome b_{555}. Two phases of cytochrome *b* oxidation also occur during the reaction of cytochrome *d* with O_2 at low temperatures (190). Both of these cytochromes have turnover numbers that are greater than, or within experimental error of, the values required to account for the observed respiration rates. (Only the faster component was analyzed

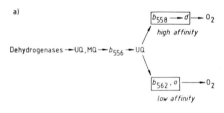

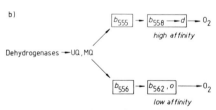

FIG. 4. Arrangement of membrane-bound cytochromes and quinones in the O_2-terminated respiratory chains of *E. coli*. The cytochromes are arranged in order of increasing midpoint potential. (a) The involvement of quinones after cytochrome b_{556} is controversial. Components in boxes have been purified as complexes (except for b_{555}). The b-cytochrome(s) constituting the cytochrome *o*-containing complex (see Fig. 3) has been neither separated nor resolved potentiometrically or genetically. UQ, ubiquinone-8; MQ, menaquinone-8. (b) Model in which quinones are not involved at two sites. For details and references, see text.

in this way in the paper; the kinetic competence of the second component is derived from the available data.) A third, very slow phase of cytochrome b oxidation is observed, with a $t_{1/2}$ of approximately 2 s, corresponding to about 20% of the total change. This last component is too slow to have a role in the main respiratory pathway; in fact, it is probably one of the soluble b cytochromes, as it is not observed in stopped-flow kinetic studies on membrane preparations from these cells (D. S. Wariabharaj and W. J. Ingledew, unpublished data). At present, there are no strong indications that cytochrome b_{555} functions before, after, or between the quinones, although its midpoint potential (+140 mV [206]) would suggest a role on the oxygen side of the predominating quinone (menaquinone; E_{m7}, −74 mV).

THE CYTOCHROME OXIDASE COMPLEXES

Oxidases catalyze the reduction of oxygen by transfer of one, two, or four electrons to the enzyme-bound molecule, giving, respectively, superoxide (O_2^-), peroxide (O_2^{2-}), or water as primary products. More than 90% of biological oxygen consumption can be accounted for by these enzymes.

The model system for studies of cytochrome oxidases is the cytochrome c oxidase, or cytochrome aa_3 (EC 1.9.3.1.), found in the inner mitochondrial membrane of eucaryotic cells (243). Description of its cardinal features allows comparison with the cytochrome oxidases of $E.$ $coli$ and $S.$ $typhimurium$. The mitochondrial cytochrome c oxidase catalyzes the four-electron reduction of O_2 to water and contains four metal centers, viz. two hemes of type A (in cytochromes a and a_3) and, associated with each, a copper atom, Cu_A and Cu_B, respectively. The hemes of cytochrome a_3 and Cu_B constitute the O_2-reactive site, which appears to be situated on the matrix aspect of the molecule. The enzyme comprises 7 to 13 polypeptides, only 2 of which appear to be involved in binding of the four metal centers and in oxygen reduction. The question of whether this oxidase catalyzes vectorial proton translocation is controversial, but there is now strong support for the view that the mitochondrial enzyme does act as a proton pump.

Some bacteria have cytochrome oxidases that appear to be analogous to the mitochondrial oxidase in having similar spectral properties, characteristic of cytochromes a and a_3. $E.$ $coli$ and $S.$ $typhimurium$ are, however, not among these bacteria (179, 180), and their cytochrome oxidases are quite distinct from the aa_3 type. Indeed, it is questionable whether these organisms contain heme A at all (see Cytochrome a_1-Like Pigments). In the subsections that follow, we describe the properties of cytochromes o and d, probably the only cytochrome oxidases in these bacteria. We also cover the hydroperoxidases (peroxidase and catalase), the primary role of which is the disproportionation of a toxic and highly reactive partial-reduction product of O_2, namely peroxide. Catalase is of special interest, since its absorption spectrum may be confused with that of the CO-binding oxidases. We also describe the little that is known of the so-called cytochrome a_1. To avoid confusion in the text, and because the term cytochrome a_1 has long been adopted to describe the heme A-like absorbance near 590

nm of the reduced form, we continue to use the term in places. However, the term hemoprotein b-590 is suggested as an interim name pending better description of its function.

The Cytochrome o Complex

Definition and distribution. Of the four or five major classes of cytochrome oxidase found in bacteria, cytochrome o is the most widespread (180, 183). Cytochrome o was discovered simultaneously in $E.$ $coli$ and in other bacteria with the appearance of bands appropriate to the CO compound of a b-type cytochrome in the photochemical action spectrum for the relief of CO inhibition of respiration (33, 34). The enzyme was called cytochrome o (for oxidase). It is noteworthy that although this terminology has been proven to be appropriate for $E.$ $coli$ (33, 34, 57, 180, 183) and more recently for $S.$ $typhimurium$ (144), such is not so for many bacteria in which cytochrome o has been reported solely on the basis of the presence of a CO-binding protoheme (245). Based on cross-reactivity between monospecific antibodies to the $E.$ $coli$ purified oxidase and membranes from $S.$ $typhimurium$, there seem to be close similarities between the oxidases of these species (141). Similarly, although a CO-binding b-like cytochrome has been reported in $E.$ $coli$ under various laboratory growth conditions, there is no conclusive evidence that this pigment is always cytochrome o or that it is functional when an alternative oxidase (i.e., cytochrome d; see The Cytochrome d Complex) is present. Further photochemical action spectra are required.

Purification. Cytochrome o is tightly membrane bound, and, until recently, its separation from membranes and purification had proved refractory. Two groups have now purified cytochrome o-containing preparations from $E.$ $coli$. Kita et al. (129) used a strain reported to have high levels of the oxidase and extracted the enzyme from inner membranes with 1% Triton X-100 to give a "cytochrome b_{562}-o complex" (Table 3). Matsushita et al. (155) used a cytochrome d-deficient mutant (see The Cytochrome d Complex) and subjected the membrane to sequential extraction with 5 M urea and 6% cholate before dissolving oxidase activity in 1.25% octylglucoside. The most striking difference between the two preparations (Table 3) is the detection (by using a sensitive silver stain) of two smaller subunits in the preparation of Matsushita et al. (155). Support for the presence of four subunits comes from determination of the native M_r as 140,000 (155) and from immunoelectrophoretic studies (140), in which four subunits (M_rs of 51,000, 28,500, 18,000, and 12,000) were found. No clear precedent exists in studies of the other purified bacterial o-type cytochromes; one, two, and four subunits have been described for different bacteria (183).

Spectral properties. In intact cells or membranes of either $E.$ $coli$ or $S.$ $typhimurium$, the presence of cytochrome o is generally determined from spectra of the difference between a reduced sample sparged with CO and a similar sample with no CO (reduced + [CO − reduced]). Peaks corresponding to the absorption of the carbon monoxide-cytochrome o adduct are seen at about 416, 535 to 538, and 567 nm. A major trough in the Soret region in such spectra at 428 to 436 nm is

TABLE 3. Properties of purified cytochrome o-containing preparations from E. coli

Reference	Subunit sizes (M_r)	Native enzyme size (M_r)	nmol of redox center/mg of protein		α Absorbance bands (nm)	Bands in CO + (reduced − reduced-difference spectrum) (nm)	Substrates oxidized	Quinol oxidation inhibitors
			Heme b	Cu				
129	55,000, 33,000	ND	19.5	16.8	558, 562	416 (peak), 430 (trough)	Q_1H_2, TMPD with or without ascorbate	KCN, HQNO, NaN₃, Cd^{2+}, Zn^{2+}, piericidin A, Co^{2+}
155	55,000, 34,000, 22,000, 17,000 (66,000, 35,000, 22,000, 17,000)ᵇ	140,000	17	ND	558, 563	419, 426 (peak), 430 (trough)	Q_1H_2, TMPD, D-amino acid dehydrogenase, phenazine methosulfate, DCI	KCN, HQNO, UHDBT

ᵃ Abbreviations: ND, Not determined; Q_1H_2, ubiquinol-1; HQNO, 2-heptyl-4-hydroxyquinoline N-oxide; DCI, dichlorophenol indophenol; UHDBT, 5-n-undecyl-6-hydroxy-4,7-diaxobenzothiazole.
ᵇ Values in parentheses are corrected values for retardation coefficients in acrylamide.

attributed to loss in absorbance, at this wavelength, of the reduced form on binding CO. (CO binding causes a shift in the wavelengths of maximum absorption leading to a difference spectrum consisting of peaks and troughs and resembling a first-derivative spectrum.) Photodissociation spectra suggest that the α band is small or that it is only slightly affected by CO (185, 191, 194, 195, 216), and attempts, by higher derivative analysis, to determine which of the α bands attributable to b-type cytochromes is (or are) due to cytochrome o were unsuccessful (217).

The purified preparations of cytochrome o (see Intermediate Components) each reveal a split α band in the reduced state, with maxima at 555 to 558 and 562 to 563 nm. The longer wavelength in each case is that described by Matsushita et al. (155), who do not state the temperature of observation, but it is likely that at 77 K the maxima are at about 555 and 562 nm. With each purified preparation, the two peaks are of similar height. It has been assumed that the two maxima arise from two different hemes, but this has not been proven, and it is possible that the two peaks constitute a split α band, common for purified hemoproteins at

77 K (e.g., heart cytochrome c) and reported for the high-potential b cytochrome (o) in membranes from aerobically grown cells (206). Both bands respond to titration with the same midpoint potential in the purified preparation (129) and in the membranes (206).

The CO-difference spectra of the purified enzyme from both groups (Table 4) show an apparently unimodal Soret peak at 416 nm and a trough at 430 nm. This region is the same as, but inverted with respect to, the low-temperature photodissociation spectra of intact cells (e.g., references 183, 194, 195, 216). The α regions of the CO-difference spectra are more difficult to interpret. Both peaks and troughs are broad, and the extinction coefficients are small. In the second-order finite-difference spectrum of the reduced, CO-bound form, the intensity of the trough at lower wavelength (555 nm, measured from reference 129) is more suppressed by CO than is the trough at 561 mn (also measured from reference 129). (Note that in second-order finite difference spectra, the trough position corresponds only approximately to those of the peaks in the undifferentiated spectrum. Better agree-

TABLE 4. Properties of cytochrome d-containing preparation from E. coli

Reference(s)	Subunit sizes (M_r)	Native enzyme size (M_r)	pI	nmol of redox center/mg of protein			α Absorbance peaks at 77 K (reduced minus oxidized) (nm)	CO-difference spectra (room temp)	Substrate(s)	UQ_1H_2 oxidation inhibitors (but at high concn)
				Fe	Heme b	Heme d				
59, 207	43,000, 70,000	150,000	4.8	ND	7.1	ND	558, 595, 629	420, 625, 647 (peak); 433, 444 (trough)	2,3,5,6-TMPD	
158	43,000,ᵇ 57,000	ND	5.3	25.5 −34.1	18.9	None found	558.5, 591, 624.5	420, 642 (peak); 430.5, 444, 622 (trough)	Q_1H_2, 2,3,5,6-TMPD	CN⁻, azide, HQNO
130	26,000, 51,000	77,000 (assumed)	ND	26.6	12.3	9.54	558, 594,ᶜ 624	420, 642 (peak); 430, 442, 560, 622 (trough)	Menadiol, Q_1H_2, ascorbate-TMPD	CN⁻, azide, HQNO, piericidin A, H_2O_2, Zn^{2+}

ᵃ Abbreviations: ND, Not determined; Q_1H_2, ubiquinol-1; HQNO, 2-heptyl-4-hydroxyquionline N-oxide.
ᵇ At 28 K in 12.5% acrylamide.
ᶜ Value at 77 K not given; 594 nm at room temperature.

ment is obtained between the peaks of the fourth-order spectrum and those of the undifferentiated spectrum.) This observation was taken as evidence by Kita et al. (129) that it is the shorter-wavelength component that corresponds to cytochrome o, by definition the CO-binding component. However, this is weak evidence, since the intensities of the bands in higher-order finite-difference spectra are highly dependent on the natural band widths of the components. However, other terminal oxidases that have been studied in detail do contain two different hemes in close association (e.g., aa_3, cd_1, and bd), and it is tempting to endorse the terminology bo for this E. coli enzyme (245). However, the "o" and "b_{562}" components have so far resisted resolution potentiometrically, kinetically, and genetically, and an open mind should be kept concerning the number of types of cytochrome b present.

The near-infrared absorption spectrum has not been examined in detail. However, observations to about 900 nm, after photodissociation of the CO-bound species in cells containing both cytochromes o and d, failed to demonstrate (189, 190) a band analogous to those near 655, 740, or 820 to 840 nm attributed to copper in the mitochondrial enzyme (243). It has been claimed recently (95, 129) that the purified o-b_{562} complex contains copper in amounts almost stoichiometric with heme (Table 3).

Most heme proteins that react with O_2 have a high-spin reduced state, but the spin state of E. coli cytochrome o is unclear. As Wood (245) points out, the CO-difference spectrum, in its shape and the sharpening of the α features at 77 K, suggests a low-spin heme. On the other hand, the rather high (γ peak − γ trough)/(α peak − α trough) ratio in the CO difference spectrum and the spectrum of the oxidized form, with a faint shoulder at 630 nm and continuous absorption from 460 to 530 nm, point to the presence of some high-spin form.

A CO-binding pigment spectrally resembling cytochrome o has been purified from E. coli incubated with excess δ-amino-laevulinic acid (109). This pigment deserves further study.

The recent purification of the b_{562}-o complex from E. coli and its potentiometric titration (129) aids the interpretation of the many previous and conflicting attempts to resolve the spectral and potentiometric properties of the multiple b-type cytochromes in E. coli membranes. The finding that both the 556- and the 562-nm bands (77 K) respond to titration in a pH-dependent fashion with the same E_m (+125 mV, pH 7) in the purified oxidase is consistent with the results of Reid and Ingledew (206). The latter workers reported an E_{m7} of +260 mV for a component with a split α band at 556 and 562 nm in a membrane preparation, and the sensitivity of the measured E_m value to CO. Van Wielink et al. (238, 239) and Lorence et al. (152) also reported comparable E_m values for the bands at about 555 to 557 and 562 to 563 nm, respectively. There is also general agreement that, in membranes, (at least) two potentiometrically distinguishable components have an absorption near 556 nm; it is the one of higher potential that is (or is part of) the b_{562}-o complex. Hackett and Bragg (83) also found two very close bands (555 and 556 nm) with widely differing E_m values, and the midpoint of the higher poten-

tial component (+129 mV) approached that of the 562-nm band.

An attempt to rationalize the relationships between absorption maxima and redox properties of the E. coli membrane-bound cytochromes b present in cells grown under high aeration is shown in Fig. 3. Higher-derivative analyses (28) also support the existence of three major classes of b-type cytochromes; however, interestingly, all such analyses (217, 219, 238) give higher λ_{max} values for the low-potential component. It is surprising that the numerical analyses agree so closely with the data of Fig. 3, especially since spectral results in this field are frequently reported with no description of crucial experimental conditions, such as spectral band width and even temperature. The b_{562}-o complex appears to contain low- and high-spin heme types, as determined from resonance Raman studies (236) and EPR studies (95). The effects of KCN and NaN_3 suggest that the high-spin species is attributable to the ligand-binding cytochrome o.

Catalytic functions. Low apparent K_m values for O_2 have been reported for E. coli containing cytochrome o as a major putative oxidase (48, 208). The most recent estimates of K_m of cell membranes or the purified oxidized complex lie in the range of 1.4 to 2.9 μM. The K_m for O_2 of respiring S. typhimurium cells in which cytochrome o was the only detectable CO-binding cytochrome was 0.74 μM (144).

The electron acceptor of the cytochrome o complex is certainly O_2 (see below), but the nature of the immediate electron donor is still in question (see Intermediate Components). The ultimate product of O_2 reduction on ubiquinol-1 oxidation by the purified o-containing complex from E. coli is water (129). The rate of quinol oxidation exceeded that of O_2 consumption by twofold, suggesting a four-electron transfer to the acceptor. This cannot rule out the reduction of O_2 being carried out in distinct steps. D. O'Hara and R. K. Poole (unpublished data) have failed to detect peroxide as a product or free intermediate in NADH oxidation by membrane particles. Although Matsushita et al. (154, 155) noted that proteoliposomes containing high ratios (5:1) of cytochrome o to the lac carrier exhibited depressed capacity for transport, this effect could be largely prevented by adding superoxide dismutase but not peroxidase or catalase. Superoxide anion, therefore, must be one product of either ubiquinol oxidation or oxygen reduction. In fact, it is known that the artificial reductant used in these experiments (ubiquinone-1) is auto-oxidizable, giving rise to oxygen radicals.

A b-type cytochrome, assumed to be cytochrome o, was demonstrated by Haddock et al. (87), in stopped-flow spectrophotometric studies of E. coli, to be kinetically competent to act as a terminal oxidase ($t_{1/2} < 3.3$ ms). More recently, the kinetics of the reactions of b-type cytochromes in E. coli, including that of cytochrome o with O_2, have been studied at sub-zero temperatures, and they provided the first evidence that an oxygenated compound (reduced cytochrome o-O_2) is an early, probably primary, intermediate in the reaction of a bacterial oxidase with O_2. The triple-trapping technique (36) requires that the sample be prepared in 30% ethylene glycol to allow oxygen to be introduced into the oxidase-CO compound while the sample is still fluid at about −25°C. Flash photolysis of the CO-bound, substrate-reduced cytochrome *o* in intact cells of aerobically grown *E. coli* and comparison with the prephotolysis state yield the photodissociation difference spectrum of the oxidase (194, 195, 216). At temperatures below about −90°C, and in the absence of O_2, recombination of CO is immeasurably slow; at this temperature, however, oxygen reacts to form an early intermediate species (see below).

In the presence of O_2, multiwavelength recordings after flash photolysis of CO-treated cells reveal heterogeneous changes in absorption of *b*-type cytochromes between −79 and −102°C (195); the reaction proceeds via formation of an intermediate which has spectral properties similar to those of the CO compound but which is light insensitive. The reaction affords an interesting comparison with the analogous formation of "compound A" by mitochondrial cytochrome a_3 (179, 243). The formation of the cytochrome *o* intermediate proceeds virtually irreversibly at about −100°C and may be regarded to be an effective, oxygen-trapping strategy (179). This intermediate is believed to be an oxy or oxygenated species (that is, $Fe^{2+} \cdot O_2$ or $Fe^{3+} \cdot O_2^-$ [234]), on the basis of its optical absorbance properties and by analogy with the oxygen compounds of other hemoproteins. Strikingly similar kinetic parameters have been described for the O_2 reaction of the membrane-bound cytochrome *o'* in *Vitreoscilla* spp. (51). No oxygenated form of the purified b_{562}-*o* complex could be found by Kita et al. (129), although the details of their attempts (at room temperature?) are not given. A CO-binding pigment with a γ absorption band at 436 nm, seen in oxygen-limited *E. coli* cells, has been shown (185) to be an *o*-like cytochrome, not cytochrome *d* (see reference 208), on the basis of the insensitivity of its CO complex to laser irradiation at 633 nm. This cytochrome o_{436} reacts with oxygen at sub-zero temperatures approximately 10-fold faster than does cytochrome *o* in cells grown aerobically, suggesting that its synthesis or modification represents an adaptation to lowered oxygen availability.

The subsequent reactions of the oxygenated intermediate of cytochrome *o* are poorly understood owing largely to the spectral similarity of the various *b*-type cytochromes that constitute the cytochrome *o*-terminated electron transport chain. Some success has been obtained by forming a mixed-valence respiratory chain (see reference 37) in which all *b*-type cytochromes, except the CO-bound cytochrome *o*, are oxidized by ferricyanide before photolysis (R. K. Poole, unpublished data). If the scheme of Fig. 4a is correct, existing quinone-deficient mutants could be gainfully used to study electron transfer from cytochrome b_{556} to the oxidase complex.

Genetics. The task of isolating a cytochrome *o*-deficient mutant has been facilitated by the availability of *E. coli* mutants already defective in cytochrome *d* (see The Cytochrome *d* Complex). Thus, a mutant lacking both oxidases was found by Au et al. (7) among a number of mutants which were unable to grow aerobically on the nonfermentable substrates succinate and lactate but which retained the ability to grow anaerobically with nitrate. Spectroscopic and immunological studies show the mutant to be deficient in cytochrome *o*. The reduced minus oxidized spectrum revealed no cytochrome b_{562}, and the CO difference spectrum lacked the α bands characteristic of cytochrome *o*. A redox titration in the presence of CO showed the highest potential component to be missing. The effects of the mutation on the 556-nm band at 77 K have not been reported and will be of interest in view of the inability, to date, to separate this band from the 562-nm band. Cotransduction experiments indicate that the lesion, designated *cyo*, lies between *phoR* and *acrA* at about min 10 on the *E. coli* chromosome.

Chemiosmotic aspects. The number of protons translocated per O atom reduced (H^+/O ratio) for the terminal segment of the *E. coli* aerobic respiratory chain has been reported to be 2.26 (145). This value is the stoichiometry for succinate oxidation, but, as previously explained (see Succinate Dehydrogenase), this may be a slight overestimate owing to a contribution from endogenous respiration. The integer value of 2 has been assumed; this number is critically important because at this value models for respiration-driven proton translocation can be drawn that require neither a Q cycle nor a proton-pumping oxidase (both of which are current concepts in mitochondrial studies). The picture is complicated with respect to the role of cytochrome *o* in this process by the likely presence of both *o* and *d* oxidases in the cells used to determine this ratio. This problem is especially acute in view of the supposed lower K_m for O_2 of cytochrome *d* than that of cytochrome *o*. We cannot, therefore, exclude the possibility that the *d* pathway is operating in the experiments in which this ratio was determined. The experiments should be repeated in a strain lacking the ability to synthesize cytochrome *d*. Accepting, for the time being, a stoichiometry of $2H^+/O$ for succinate oxidation through the cytochrome *o* oxidase leaves two alternative models. In the model that we prefer, the succinate dehydrogenase reduces quinone (this reaction is on the cytoplasmic aspect of the membrane), and the quinol is reoxidized on the periplasmic aspect by an electron carrier, with the two protons being released. The electrons are then transferred to O_2, and the (scalar) oxidase protons are consumed on the cytoplasmic side of the membrane (illustrated in Fig. 2). This leads to proton translocation by a simple loop mechanism, with a stoichiometry of $2H^+/O$. If, on the other hand, the oxidase (O_2^- and H^+-consuming) reaction is located on the periplasmic aspect of the membrane, the respiratory chain must translocate protons by another means.

Matsushita et al. (154, 155) have studied the generation of proton motive force by cytochrome *o* by monitoring internal and external pH changes and the redistribution of lipophilic ions in reconstituted phospholipid vesicles or by monitoring lactose transport in vesicles in which the *lac* carrier protein was also reconstituted. Proteoliposomes were formed with pure oxidase and lipid and, during turnover (ubiquinol-1 as substrate), generated a Δp of −115 to 140 mV (interior alkaline and negative) with a small amount of respiratory control. Measuring pH changes both inside and outside of the vesicles, these investigators showed that, during ubiquinol (hydrogen donor) oxidation, protons are released on the external surface and consumed on the internal surface. An H^+/O ratio of between 0.9 and 1.4 was observed; without separation of the two half-reactions, the scalar proton-consuming and -generating reactions cancel out. These values, which for experimental reasons are likely to be underestimates, suggest an integer value of 2. In oxidant pulse experiments with the electron donor TMPD, little change in the external pH is observed until an uncoupler (protonophore) is added, when alkalinization is observed. Thus, the reconstituted oxidase cannot translocate protons across the membrane but can generate a Δp by the separation (on opposite sides of the membrane) of the scalar proton-generating (quinol oxidation) and -consuming (O_2 reduction) reactions, as indicated in Fig. 2. In the case of quinol oxidation, the H^+ translocation expected from this mechanism is $2H^+/O$.

The Cytochrome d Complex

Definition and distribution. Cytochrome d (previously called cytochrome a_2) is the alternative major oxidase of *E. coli* and appears to be synthesized coordinately with cytochrome a_1 and cytochrome b_{558}. Antigenically identical enzymes are found in *S. typhimurium* and other gram-negative bacteria (141). Cytochrome o persists under some of the conditions that favor synthesis of cytochrome d, including low oxygen tension and attainment of the stationary phase of growth on nonfermentable carbon sources (106). The repression of cytochrome d synthesis during growth on nitrate has recently been shown to be reversed in some *chl* mutants (84). The biosynthesis of cytochrome o is repressed when cytochrome d is induced by lowering the dissolved-O_2 tension during growth (140).

Purification. Only recently has cytochrome d from *E. coli* been purified (Table 4), although its extraction with deoxycholate was reported much earlier and a green deoxycholate-soluble material (202) was active in reconstitution of electron transfer from succinate to oxygen. The first reported purification of a cytochrome bd complex was by Reid and Ingledew (207); the complex contained at least three or four cytochrome types. Subsequently, Finlayson and Ingledew (59) isolated a pure cytochrome bd preparation by two-stage Triton X-100 extraction of membranes. The purified complex contained cytochrome b_{558} and cytochrome d, and its spectrum exhibited a small peak at 595 nm. It comprised two subunits of molecular weight 70,000 and 43,000, and the measured M_r values remained constant to within 5% on 7, 10, and 12.5% (wt/vol) acrylamide. The molecular weight of the native complex in Triton X-100 was approximately 350,000, or approximately 160,000 when corrected for Triton X-100 binding. Miller and Gennis (158) extracted, with a dipolar ionic detergent, a cytochrome d-containing complex from cells grown under O_2-limited conditions. The complex contained cytochromes b_{558}, a_1, and d, yet consisted of only two subunits, with M_r values of about 57,000 and 43,000. These values showed some dependence on acrylamide concentration. Only one type of cytochrome b was detected, but a small shoulder at 547 to 550 nm was observed which, it was supposed, could be an optical transition related to cytochrome a_1. This has not been substantiated in subsequent reports (135). CO bound to cytochrome d, perhaps to cytochrome a_1, and also to part of the cytochrome b. Binding to a b-type

cytochrome could reflect the presence of some cytochrome o or perhaps denaturation of cytochrome b. Interestingly, CO also perturbed the spectrum of an air-treated sample, shifting the α peak of the cytochrome d from 646.5 to 636 nm. This finding is consistent with the designation (189) (see below) of the 646.5- to 652-nm form as an oxy (i.e., $Fe^{2+} \cdot O_2$) species. In the presence of detergents, the complex oxidized ubiquinol, TMPD, and diaminodurol. Protoheme IX and iron were detected, but neither nonheme iron nor copper was detected. The inability to detect copper in the purified preparation, consistent with the previously observed absence of signals near 840 nm during cytochrome d oxidation in vivo (190), appears to preclude a role for copper in the oxygen-reduction mechanism.

The subsequent report of a purified cytochrome d-containing complex by Kita et al. (130) confirmed the main features of the preparation of Miller and Gennis (158) and provided additional information. (i) A somewhat lower M_r for the small subunit was reported. (ii) The amount of heme d was determined and found to be approximately equimolar with that of protoheme, as suggested by Reid and Ingledew (206). (iii) There are a few interesting spectral differences, which are discussed in detail below. Both preparations lack significant amounts of Cu, quinones, phospholipid, and nonheme iron and oxidize ubiquinol-1 and TMPD. Inhibition of quinol oxidation is achieved only with concentrations of cyanide, azide, or HQNO far in excess of those required for inhibition of quinol oxidation by purified cytochrome o.

One of the most puzzling features to remain is the presence of only two subunits but the apparent presence of three spectrally distinguishable cytochrome types, viz. b_{558}, d, and cytochrome a_1 [see Introduction under The Cytochrome Oxidase Complexes and Cytochrome a_1-Like Pigment(s)]. Miller and Gennis (158) proposed that the ratio of types might be 2:1:1, respectively, whereas Kita et al. (130) suggested that the concentration of the a_1-like component is very low, being 3% of the total and present at an a_1/d ratio 10-fold lower than in the membrane. Is the a_1 a contaminant? If not, with what subunit is it associated?

There is now genetic evidence that subunit I (M_r 57,000) bears cytochrome b_{558} (78) [see Cytochrome a_1-Like Pigment(s)]. Briefly, the *cydB* phenotype is the lack of cytochromes d and a_1 and subunit II, but the presence of heme b and subunit I. This may be the subunit that Reid et al. (205) suggested is assembled into the membrane in the absence of heme synthesis and detectable by in vitro reconstitution of membranes from a δ-amino-laevulinic acid-requiring strain with ATP and hematin. Both subunits are required for TMPD oxidase activity (138).

Optical spectral properties. The spectral properties of cytochrome d are distinctive and arise from the presence of the chlorinlike heme (12). Reduced preparations show an absorbance maximum at 628 to 632 nm, which is shifted approximately 5 nm further to the red in the presence of CO. Aeration of cell or membrane suspensions containing cytochrome d results in a further shift of the peak to 648 to 652 nm. It has been widely assumed that this latter narrow band arises from the ferric form (125), but more recent evidence (189, 190) suggests that it should be attributed to a stable oxygenated compound of cytochrome d (182, 199).

Early evidence that the Soret band of cytochrome d is weak and diffuse has been supported by (i) photodissociation spectra at 4 K (193); (ii) selective photolysis by a He-Ne laser of the CO complex of cytochrome d (185); (iii) lack of correlation between the intensities of the α and putative Soret bands in photodissociation spectra in the absence or presence of oxygen (185); and (iv) the assignment of a band at 448 nm to cytochrome a_1, not d (192). A further absorption band at 675 to 680 nm is possibly also attributable to cytochrome d or the d-containing oxidase complex. This signal appears during the course of oxidation of $E.$ $coli$ "cytochrome d_{650}" at sub-zero temperatures (190), and it can also be observed in reduced minus oxidized difference spectra. Its origin is unknown.

In the presence of oxygen, cyanide binds to the reduced form of cytochrome d, eliminating its α band (198). Cyanide also reacts with the form of cytochrome d that absorbs around 650 nm, without the formation of the reduced form. The original interpretation was that the 650-nm (oxidized) form reacted with cyanide to give cyanocytochrome d, which had little or no absorbance in the α band region (198). However, other data led to the proposal that there exists an "invisible" cyanide-binding form of cytochrome d (d^*), intermediate between the oxidized (648 to 650 nm) and reduced (628 to 630 nm) species.

In addition to binding cyanide, ferrous cytochrome d binds CO and other ligands. The reaction with CO is unusual in that the CO liberated from photolysis of the CO complex at sub-zero temperatures immediately recombines with the reduced enzyme unless the photolysis is conducted at temperatures close to that of liquid He (193). This feature (the very rapid recombination of CO), unique among cytochrome oxidases, may be due to the absence in cytochrome d of copper, which in cytochrome aa_3 is believed to be the binding site for the photodissociated CO (3).

Meyer (156) reported that nitrate, when added to reduced membrane particles, gave a difference spectrum that was indistinguishable from that produced by NO in the 610- to 720-nm region. Differences below 610 nm and especially in the Soret band, however, indicate reaction with cytochrome o, a_1, or both. Recently, Hubbard et al. (102, 103) have confirmed and extended these findings. NO_3^-, NO_2^-, $N_2O_3^{2-}$ (trioxodinitrate), and NO all react with reduced membrane particles, causing diminution and shifting of the 630-nm peak to 641 to 645 nm. It is likely that all four compounds result, albeit at different rates, in the formation of the nitrosyl species (106). In contrast, Bragg and Rainnie (22) found that $AgNO_3$, but not $NaNO_3$, reacted with cytochrome d; the discrepancy is unresolved. The finding that NO_3^- and its reduction products react with cytochrome d may explain the position of the α_{max} of cytochrome d (near 634 nm) in cells from nitrate-containing cultures (e.g., reference 87).

Studies with purified preparations (Table 4) have in general been consistent with the conclusions drawn from cells and membranes, discussed above. The b-type cytochrome of the oxidase complex accounts for the α absorbance at 560 to 562 nm in absolute (reduced) and difference (reduced minus oxidized) spectra at room temperature. The β band is at about 531 to 532 nm (room temperature), and the γ band is at 429 to 430 nm. The α and β band positions have been confirmed by potentiometric resolution of the reduced minus oxidized spectrum (135). Low temperature (77 K) shifts all bands 1 to 4 nm to the blue; it most diagnostically shifts the α band of the low-spin cytochrome b to 558 nm.

For the cytochrome d component itself, the α band of the reduced form is very characteristic, lying at 628 to 629 nm in the absolute reduced spectrum at room temperature. Miller and Gennis (158) showed that the spectrum of the air oxidized form of the purified complex had an α band at 646.5 nm but presented no evidence that the heme d is actually in the oxidized, (i.e., ferric) state. For reasons given elsewhere, oxygenated or oxy-form might be more appropriate terms to describe this species with an unusually red-shifted absorbance maximum. Work with the purified enzyme has not yet clarified the contribution of heme d to the Soret region. Reduced (absolute) and reduced minus oxidized (difference) spectra show a distinct shoulder on the red side of the cytochrome b Soret peak (e.g., reference 130), whereas the CO difference spectra in this region are characteristically W shaped, with troughs at about 430 nm and 442 to 444 nm (room temperature). The longer-wavelength features have been assigned to cytochrome d, notwithstanding the substantial evidence that cytochrome d has only a very weak γ band. We suggest [for reasons given in Cytochrome a_1-Like Pigment(s)] that it is the a_1-like component that is most responsible for this band near 440 nm. Interestingly, the preparation of Kita et al. (130), which is claimed to have only 3% contamination by the a_1-like species, has the γ peak (reduced minus oxidized) at 429 nm and only a shoulder at higher wavelengths, whereas the preparation of Miller and Gennis (158) shows the γ peak at 432 nm, possibly by fusion of the 429-nm band with a significantly intense band at a higher wavelength. The 442- to 444-nm trough in CO-difference spectra is also less intense compared with the 429- to 430-nm trough in the preparation of Kita et al. (130), an observation lending support to the idea that the former band emanates largely from the a_1-like component.

Finally, the origin of other bands in the CO-difference spectra of purified preparations is unclear. The published spectra suggest the presence of a CO-binding b-type heme, with features like that of cytochrome o (158). Kita et al. (129) point out that one such feature, the 560-nm trough, is not seen in the spectra of inner membranes or of purified oxidase with excess soybean phospholipids and suggest that the band reflects in some way the environment of the purified complex.

EPR spectral properties. A ferric heme EPR signal has been observed in purified cytochrome bd preparations (59). This signal is of high spin, $g \simeq 6.0$, and no copper (Cu^{2+}) signal is observed. The fully oxidized preparation contained a mixture of both axial and rhombic high-spin signals. Redox titrations performed on the oxidase showed that the signal intensity exhibited a midpoint potential of approximately +180 mV, and marked changes in lineshape responded to titration with a midpoint potential of approximately +300 mV (probably corresponding to cytochrome d). These lineshape changes are complex but are qualitatively similar to those observed in membranes. In addition, the lineshape of the high-spin EPR signal in membranes has been shown to be sensitive to O_2 and pH (W. J. Ingledew, unpublished data). In *Azotobacter* spp., ferric heme resonances at $g \simeq 6.20$, 5.80, and 6.48 (53) have been resolved into axial and rhombic components and attributed to cytochrome *d*. Use of the triple-trapping technique by Kumar et al. (142) has provided a clear kinetic distinction between high-spin axial and rhombic species in the reaction with O_2 at sub-zero temperatures and lent support to the assignment described above of the allocation of the axial component to cytochrome *d* and the rhombic component to cytochrome b_{558} (see also reference 158). In contrast, EPR studies of the purified *bd* complex by Hata et al. (95), although confirming the high redox potential (described above) and the binding of O_2 under air-oxidizing conditions (189), seem to suggest that the ligand-binding cytochrome *d* is of low spin. Further detailed magneto-optical studies are required.

Potentiometric studies. The midpoint potential of cytochrome *d* in membranes, measured at the absorbance maximum of the reduced form, is +260 to +280 mV ($n = 1$) (152, 200, 206). For the purified complexes, the values are in good agreement, being +232 mV (135) and +240 mV (130), but are influenced by the detergent used to extract the complex and by pH (152). Even at potentials as high as +440 mV, the 650-nm form does not appear unless oxygen is present (200) or unless the sample is cycled through a preliminary phase of air oxidation, followed by substrate reduction (99). Under the latter condition, the absorption bands at about 630 and 650 nm exhibit very unusual and complex behavior that could be interpreted only by invoking concerted four-electron transport by cytochrome(s) *d*. An alternative explanation (190) is that the 650-nm species does not arise directly from oxidation of the ferrous state to ferric but that it is attributable to an oxy-ferrous form (see Other Oxidation-Reduction Components).

Catalytic function. Cytochrome *d* in *E. coli* has been proven to be a functional terminal oxidase by photochemical action spectroscopy (34, 144). In addition, stopped-flow spectrophotometry of the reaction of O_2 with cytochrome *d* in membranes and cells has shown the cytochrome to be kinetically competent ($t_{1/2}$, <3.3 ms) to support measured oxidase rates. Unusual multiphasic reaction kinetics, however, were observed when cells of *E. coli*, grown anaerobically with glycerol and nitrate, were mixed with oxygen and monitored at 655 minus 630 nm (87). These studies require reexamination in view of the evidence that the component absorbing at about 655 nm is not a direct oxidation product of the reduced form absorbing at 630 nm, making problematic the use of 655 nm as a measuring wavelength.

Rice and Hempfling (208) measured the apparent K_m of O_2 for *E. coli* containing high levels of cytochrome *d* developed during O_2-limited growth and obtained a value of 0.024 μM. More recent values for cells, membranes, and the purified b_{558}-*d* complex are 0.27, 0.23, and 0.38 μM, respectively (130). Both groups made measurements with a membrane-covered O_2 electrode, and the values are open to the criticism that the unstirred space adjacent to the membrane in such systems causes serious errors. Nevertheless, there is agreement that the cytochrome *d* complex has an O_2 affinity higher than that of cytochrome *o*, thus rationalizing the appearance of cytochrome *d* during growth in O_2-limited situations. Growth competition experiments in chemostats show that organisms containing cytochrome *d* displace potential competitors that have cytochrome oxidase *o* or aa_3 (115). In contradiction to the view that cytochrome *d* is a high-affinity oxidase, Crispin et al. (48) have described kinetic experiments in which cytochrome *d* (reduction monitored at 630 minus 615 nm) exhibited an apparent K_m of 32 μM. Although these results do support the conclusion (87) that cytochrome *d* in nitrate-grown cells exhibits curious reaction kinetics with oxygen, it would be desirable to make independent measurements of the oxygen affinities of the oxidase in such cells with a more conventional approach. Furthermore, the kinetics may have been complicated by the binding of the formate to the oxidase as it does to cytochrome *c* oxidase (243).

The primary events in O_2 binding to the ferrous cytochrome *d* have been described by low-temperature trapping and ligand-exchange procedures (see The Cytochrome *o* Complex). A small proportion of the carbonmonoxy ferrous cytochrome *d* is oxidized during the addition of oxygen at $-25°C$. Initiation of the reaction at $-130°C$ by photolysis of the (remaining) CO adduct has been studied spectrophotometrically and by EPR spectroscopy. At this temperature, the first observed species has a sharp, symmetrical band at 650 to 652 nm (where the reference spectrum is the CO-liganded form). This species (cytochrome d_{650}), although having spectral properties generally attributed to the oxidized cytochrome, is not in the ferric state, as judged by EPR spectroscopy of the trapped intermediate (189). The interpretation of the optical, EPR, and resonance Raman spectra favored by Poole and his collaborators is that cytochrome d_{650} in *E. coli* is a ferrous oxygenated form and an early (first?) intermediate in the oxidation of cytochrome *d*. The oxidized ferric form should be equated with the "invisible species" (d_x or d^*) postulated by others (180) to have little characteristic absorbance in the red region of the spectrum and to be analogous to methemoglobin. This hypothesis (Fig. 5) is the converse of previous proposals, that invoke the existence of a hypothetical, invisible intermediate between the ferrous and ferric forms. The argument is advanced in detail elsewhere (180, 189) and provides an explanation of, for example, the curious potentiometric and kinetic behavior of the 650-nm form. It also rationalizes the ability of CO to react with d_{650} in which, in our proposal, the heme is formally ferrous. The hypothesis is also supported by resonance Raman spectroscopy (182) of aerated cell suspensions or extracted enzyme in which the d_{650} form is quite stable. With laser excitation at 647.1 nm, the spectrum shows resonances at 1,078 to 1,105 cm^{-1}, attributable to the O-O stretching frequency of the oxidase-oxygen adduct and reminiscent of oxyhemoglobin and oxymyoglobin.

The implications of designating cytochrome d_{650} as an oxygenated form are far-reaching. It casts doubt on all measurements of the extent of oxidation or reduction of cytochrome *d* in the presence of O_2 in which redox state has been quantified from peak-minus-

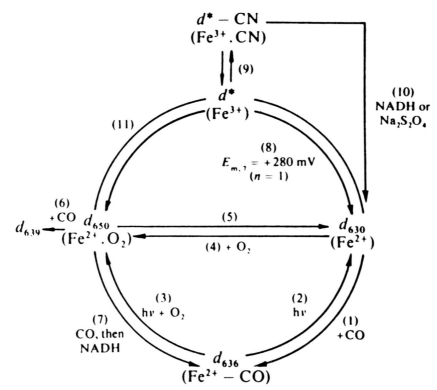

FIG. 5. Proposed interconversions of the forms of cytochrome d and some of their reactions with ligands. Reduced cytochrome d reacts with CO to give a spectrally distinct species (reaction 1). The reaction is reversible with light (reaction 2). In the presence of O_2, photolysis of the CO complex (reaction 3) yields an oxygenated compound that is analogous to oxyhemoglobin. The same compound can be formed (reaction 4) by aerating the reduced oxidase. The reverse reaction (reaction 5) can be observed upon exhaustion of O_2. Addition of CO to d_{650} (reaction 6) forms a species that is not spectrally identical to the form that results from adding CO and then NADH to d_{650} (reaction 7). The E_{m7} for oxidoreduction (reaction 8) of cytochrome d shown (206) is typical of reported values. The oxidized form (previously d^*) reacts with cyanide. Treatment of cyanocytochrome d with reductant (reaction 10) regenerates the reduced form (in *Acinetobacter vinelandii*). The formation of the oxidized d from d_{650} (reaction 11) proceeds via the formation of optically and EPR-detectable intermediates. The reactions of cytochrome d with low-molecular-weight compounds of nitrogen are not shown. Reproduced with permission from Poole et al. (reference 190, which also provides further information).

trough (i.e., approximately 630 minus 650 nm) and permits the reinterpretation of many experiments (189). The extreme stability of the presumptive oxyform is notable by comparison with the elusive oxygen intermediates formed by cytochromes aa_3, and by cytochrome o in *E. coli*, but not unprecedented. A similar stable oxygenated intermediate has been described for the cytochrome o of *Vitreoscilla* spp. (for a review, see reference 180).

The progress of oxidation of the oxygenated form has been monitored spectrophotometrically during the triple-trapping procedure. On photolysis at temperatures between -132 and $-88°C$, the first scan of the sample shows a loss of the absorbance at 650 nm and a slow increase at 675 to 680 nm. The chromophore responsible for the latter change is also seen in the *Azotobacter* cytochrome d but has not been identified. A similar sequence of events is seen when a single sample is scanned repetitively after photolysis at $-91°C$. At higher temperatures, oxidation of at least two b-type cytochromes occurs sequentially (190). Selected photolysis of the cytochrome d-CO complex (λ_{max}, ~636 nm) with a He-Ne laser (line at 633 nm) shows that neither of these is the CO-binding cytochrome o_{436}.

No significant absorbance changes occur between 720 and 860 nm in such experiments over a wide temperature range, in which changes in the oxidation states of the cytochrome d complex are thought to occur. This behavior is consistent with the absence of copper from the purified complex. EPR studies of the reaction are discussed in a previous section.

Genetics. Green and Gennis (76) devised a screening procedure that permitted the isolation of mutants unable to oxidize TMPD. One such mutant studied in detail lacked cytochromes d and b_{558}, the a_1-like component, and both subunits of the oxidase. The gene, designated *cyd*, lies near min 17 on the *E. coli* chromosome. Taking advantage of the increased sensitivity of cytochrome d-deficient mutants to cyanide and azide, they used localized mutagenesis to generate new mutants sensitive to azide (78). Two classes of mutations, *cydA* and *cydB*, were obtained. Both occur in the same 7-kilobase locus as does the original *cyd* mutation. The *cydA* phenotype is the lack of all three chromophores of the complex, as judged by spectrophotometry, and the absence of subunits I and II of the oxidase complex. *cydB* strains lacked cytochrome a_1-like heme and subunit II (at 43 K) but had a normal amount of cytochrome b_{558} and retained subunit I. On this basis, subunit I has been proposed to be the cytochrome b_{558} component of the complex.

Green et al. (77) have isolated from the Clarke and Carbon *E. coli* DNA bank two plasmids carrying the structural genes (*cyd*) for the cytochrome *bd* complex and have subcloned a 5.4-kilobase fragment into pBR322. Colonies containing *cyd*$^+$ plasmids had a yellow-green color arising from the cytochrome *d* chromophore. *cyd*$^+$ subclones specify two polypeptides with mobilities identical to subunits I and II of the purified oxidase complex. Strain PLJ01 has been isolated from a *cydA* mutant, without further mutation, as one that formed cytochrome *d* when grown on glucose but not on succinate or other substrates (114). The gene responsible lies close to *cydA* itself, and one interpretation is that there are two mutations in the original strain (GR19N).

Chemiosmotic considerations. The first question to be asked in considering the role of an oxidase in proton translocation is, what are the observed H$^+$/O ratios in the respiratory chain when this oxidase is functioning? Unfortunately, this question cannot be answered directly because no in vivo studies have been reported in which cytochrome *d* has been characterized as the sole oxidase. The H$^+$/O ratio for the terminal segment of the *E. coli* aerobic respiratory chain has been reported to be an integer of 2 (145) (see also The Cytochrome *o* Complex). This value probably includes contributions from both *o* and *d* oxidases. The closest we have come to a resolution of this problem is a report from Brice et al. (24) in which the H$^+$-to-O stoichiometries were compared in early-logarithmic-phase, O$_2$-limited cells; cytochrome *o* predominated in the former, and cytochrome *d* predominated in the latter. The H$^+$/O ratios were not higher in the cytochrome *d*-containing cells; in fact, they were slightly lower than in cells in which cytochrome *o* predominated. The implication is that the same models pertain to the cytochrome *d* case and the cytochrome *o* case (see The Cytochrome *o* Complex) and that cytochrome *d* neither needs to pump H$^+$ nor requires a Q cycle. The simplest model is one that places the oxidase site on the cytoplasmic aspect of the membrane with the oxidase functioning as the electron-carrying arm of a conventional proton-translocating loop, the quinone acting as the hydrogen-carrying arm (Fig. 2). The value of the H$^+$/O ratios is critical to an understanding of the organization of the respiratory chain and should be determined in a strain lacking the ability to synthesize cytochrome *o*. The sidedness of both *d* and *o* oxidase is also a critical question in considering possible models to explain proton translocation. A cytochrome-*d*-NO compound can be shown to be closer to the cytoplasmic aspect of the membrane than to the periplasmic aspect by EPR studies using the interaction of the paramagnetic center with an extrinsic paramagnetic probe (W. J. Ingledew and N. Bradbury, unpublished data).

Evidence that cytochrome *d* serves as a coupling site comes from studies of the purified complex after reconstitution (136, 159) into proteoliposomes. Electron transfer in such systems from either pyruvate (with pyruvate oxidase and ubiquinone-8) or ubiquinol-8 to oxygen is accompanied by formation of a membrane potential. Such systems also support the concept of an H$^+$/O ratio of 2 for this region and have demonstrated the topology of cytochrome b_{558} in the reconstituted system.

Cytochrome a_1-Like Pigment(s)

Both *E. coli* and *S. typhimurium* exhibit, when reduced, a broad band lying between 585 and 596 nm. Because the position and breadth of this band superficially resemble those of the *Acetobacter* sp. band, to which the name cytochrome a_1 was given (125), the same nomenclature was adopted for the *E. coli* and *S. typhimurium* pigments. However, in marked contrast to the *Acetobacter* enzyme, which was long ago identified as a terminal oxidase (125), no clear function for the a_1-like pigment has emerged in those organisms that form the subject of this treatise. Should the pigment in these organisms continue to be called a_1? To our knowledge, there is no really satisfactory definition of such a hemoprotein, but a minimal requirement for designating a pigment a_1 thus should surely be that it contains heme A. There is no evidence that this is the case in *E. coli* or *S. typhimurium*. Recent attempts to detect heme A in the *E. coli* have been unsuccessful (158, 181). In this section, we review what is known about the a_1-like pigment in these organisms and suggest a new nomenclature.

Spectral and potentiometric properties of cytochrome a_1 in cells, membranes, and the purified *bd* complex. The presence of cytochrome a_1 in *E. coli* and *S. typhimurium* is generally inferred from a broad band at 585 to 596 nm in reduced minus oxidized spectra. Probably because (other?) *a*-type cytochromes have a strong γ band in the reduced state near 440 nm, it is generally assumed that the pigment also contributes with cytochrome *d* to an adsorption shoulder at such wavelengths, poorly separated from cytochromes *b*, although there have been no systematic studies of this region. However, evidence presented in The Cytochrome *d* Complex suggests that the contribution of cytochrome *d* in this region is small, implying that that of cytochrome a_1 is large.

Cytochrome a_1 is generally assumed to bind CO (87, 235). Although CO was reported to be ineffective in influencing the redox titration of this component (206), Poole et al. (192) showed that CO decreased the intensity of the α and γ bands of a cytochrome a_1-like pigment in intact cells from O$_2$-limited cultures. At sub-zero temperatures, a flash-dissociable CO complex with a broad band near 595 nm was attributed to the same component. It is unknown whether these signals emanated from a soluble or membranous component. Cytochrome a_1 in anaerobically or aerobically grown cells apparently titrated as two components that contributed equally to the total spectral change, with E_{m7} values ($n = 1$) of +260 and +160 nm (206). The higher potential may be attributed to interference of the small spectral changes in cytochrome a_1 caused by the neighboring, intense absorption of cytochrome *d* (see The Cytochrome *d* Complex). Much lower values (−9 to −25 mV; $n = 1$) have also been reported (99). Apart from differences in the growth conditions used and slight differences in the choice of redox mediators, there appears to be no apparent cause for these markedly different results.

From a set of spectra of the purified bd complex poised at different ambient redox potentials, Koland et al. (135) have resolved the reduced minus oxidized spectrum of each electrochemically active species. One component, designated a_1, had peaks at 594 and 560 nm, a trough at 645 nm, and an E_m of +113 mV (n = 1), in reasonable agreement with the value obtained by Reid and Ingledew (206).

The soluble hemoprotein b-590. Soluble fractions obtained by high-speed centrifugation of ultrasonically disrupted $E.\ coli$, grown anaerobically with glycerol and fumarate, contain c-type cytochromes (with a typical low-spin absorbance spectrum [245]) plus a hemoprotein that appears to have all of the spectral features of the a_1-like component of membranes and of the bd complex. No cytochrome d is present in such supernatant fluids, however. The partially purified a_1-like component (8) shows peaks in reduced minus oxidized difference spectra at about 444, 568, and 595 nm (room temperature) and reacts with ethyl hydrogen peroxide, NO, cyanide, and CO. The CO reduced minus reduced difference spectrum is particularly striking, with peaks at 425, 540, and 578 nm and troughs at 447, 564, and 598 nm. The preparation has high catalase and cytochrome c peroxidase activities and contains iron but no copper. A purified preparation has confirmed these findings (180a).

Significantly, heme extraction of either the soluble preparation or whole cells yields heme B but not heme A (181). The a_1-like protein is spectrally very similar to certain proteins with a high-spin b-type heme, notably horseradish peroxidase and tryptophan 2,3-dioxygenase. In such proteins, the band near 560 nm is the β band (though lying in the position of the α band of proteins with low-spin heme b), and the α band is the shoulder, often barely detectable, near 590 nm. In view of these similarities, we proposed that soluble $E.\ coli$ cytochrome a_1 contains b-type heme in an environment leading to a distinct α band, as it does in tryptophan 2,3-dioxygenase. On the basis of conventional procedures for nomenclature, and having identified the heme B prosthetic group, the recommended name (181) is hemoprotein b-590 ($Escherichia\ coli$); 590 is the α band wavelength maxima in nanometers (see above), determined at room temperature from absolute spectra of the reduced pure protein. Spectroscopy of a highly purified preparation fully justifies the assignment (180a). The term hemoprotein is preferable to cytochrome until the function is defined (see below). For a cytochrome, the characteristic mode of action involves a single electron and reversible equilibrium between the Fe(II) and Fe(III) states. However, proteins with high-spin hemes seem invariably to have functions that involve ligand binding, in contrast to electron transfer (245).

Possible functions. Because it has long been suspected that cytochrome a_1 or hemoprotein b-590 binds CO, several workers have commented on the possibility that it may act as a third terminal oxidase in $E.\ coli$. It has been stated that "there is no known example of a CO-binding pigment that does not also react with oxygen" (125). However, as Ingledew (104) points out, a pigment can function as a terminal oxidase only if it is kinetically and thermodynamically competent to participate in electron flow to oxygen. Photochemical action spectra (34) did not identify cytochrome a_1 as a functional oxidase in $E.\ coli$, even though an a_1-like pigment does bind CO in a light-reversible fashion in intact cells (192). Evidence for cytochrome a_1 in action spectra was obtained by Edwards et al. (57), but the pigment observed in that study cannot be the soluble hemoprotein b-590 because of differences in the absorption maxima of the CO compounds (180a). Kinetic studies also failed to identify cytochrome a_1 in $E.\ coli$ as an oxidase, although such conclusions are complicated by the possibility of simultaneous oxidation and reduction of the putative oxidase (87).

Despite such a paucity of knowledge on its composition and function, the a_1-like hemoprotein has frequently been depicted in schemes of the respiratory chain (e.g., reference 19) and described as an oxidase (121). Since most a_1-type cytochromes described thus far in $E.\ coli$ and other organisms bind CO, whereas only a few appear to be oxygen reactive, other ligands that could act as terminal electron acceptors require investigation. There is some evidence that nitrite (or a ligand derived therefrom) binds to cytochrome a_1 in $E.\ coli$ membranes (156); nitrate and nitrite appear to bind to cytochrome a_1 in other bacteria (180).

Further work (180a) has revealed more details of the catalytic capabilities and structural properties of the soluble hemoprotein b-590 and has shown it to resemble strongly the hydroperoxidase I from $E.\ coli$ described by Claiborne and Fridovich (39). Koland et al. (135) have also noted the resemblance between cytochrome c peroxidase and the cytochrome a_1 resolved potentiometrically in the purified bd complex. We note that the 645-nm trough in their resolved, reduced minus oxidized spectrum is characteristic of a high-spin heme (245), as proposed for the soluble hemoprotein b-590. Although it is premature to assume that the oxidase-associated and soluble a_1-like components are equivalent, it is interesting that a membrane-bound hydroperoxidase (i.e., catalase/peroxidase) detected immunologically has a subunit molecular weight similar to that of the soluble hydroperoxidase I (137) and to that of hemoprotein b-590 (180a). An a_1b preparation described by J. Barrett and P. Sinclair (Abstr. 7th Int. Congr. Biochem. 1967, H-107) as having peroxidase activity probably consisted largely of hemoprotein b-590 (180a).

In summary, the current weight of evidence is against an oxidase role for the a_1-like protein and suggests to us that the soluble and membrane (oxidase-?)-bound proteins are closely related, each exhibiting catalatic and peroxidatic activities. If this speculation is correct and if the oxidase-associated form is shown not to be a contaminant, an attractive explanation of its role is that it serves to remove peroxide that may be generated by the primary reaction of cytochrome d with O_2 (108). We suggest that the use of the term cytochrome a_1 be discontinued. Hemoprotein b-590 describes this enigmatic component adequately without specifying too restrictively its function. Finally, we are aware of the dangers of suggesting that our comments on the $E.\ coli\ a_1$-like component will apply to other groups of organisms. Although the $S.\ typhimurium$ protein may be shown to resemble that of $E.\ coli$, it is clear that "true" cytochromes a_1 (i.e., with heme A and terminal oxidase function) occur in certain genera (180, 184).

OTHER OXIDATION-REDUCTION COMPONENTS

b-Type Cytochromes

Fourth-order finite-difference analysis of low-temperature spectra showed that the α band previously attributed to cytochrome b_1 contains the bands of five or more different pigments, tentatively assigned to at least two different c-type and three b-type cytochromes (219). There is now agreement from three independent studies using these methods (217, 219, 238) that the α absorption band observed in reduced minus oxidized spectra of aerobically grown cells, or membranes derived therefrom, is composite. Spectral and potentiometric studies of the purified oxidase complexes have, as described in The Cytochrome o Complex and in The Cytochrome d Complex, clarified the resolution of the α band region (Fig. 3). In order of low to high potential, the cytochromes are b_{556}, b_{558}, and the cytochrome o-containing complex with a split α band (in the o oxidase-based respiratory chain). A cytochrome b_{555} is associated with the induction of the cytochrome $b_{558}d$ oxidase under O_2-limited conditions. Multiple b-type cytochromes can also be demonstrated kinetically (87, 190, 195). In addition to those b cytochromes constituting the oxidase complexes, a number of b cytochromes are associated with dehydrogenase complexes (see The Dehydrogenases), and two further b cytochromes are involved in electron transport in the quinone region of the respiratory chain (see Cytochromes b). This section is primarily concerned with additional (unallocated) components; greater detail is provided in our more comprehensive review (106).

Cytochrome b_{562}. A cytochrome b_{562} has been purified from soluble subcellular fractions (91) and extensively characterized. This cytochrome appears to have no relationship to those with bands near 562 nm in membranes, including the b_{562} associated with cytochrome o (137). It is a small molecule ($M_r = 12,000$; 110 amino acids [111]) and contains a single noncovalently bound iron protoporphyrin IX as prosthetic group. Its midpoint potential is approximately +113 mV (110) lower than that of the membrane-bound b_{562}, and it has a pI of 7 to 8. The most recent crystallographic study at 2.5 Å showed the oxidized protein to consist of four nearly parallel α helices, with the heme group inserted between the helices at one end and the heme face partially exposed to solvent (153). In confirmation of earlier work (246), the two ligands to the heme are histidine and methionine. Other spectroscopic methods have yielded information on the spin states of the heme and conformational properties during redox and pH changes (26, 157, 162). Its function is probably unrelated to the aerobic respiratory chain.

Cytochrome b_1. A component referred to as cytochrome b_1 has been extracted from membranes by prolonged sonic treatment to yield a detergent-free, soluble preparation that has been purified and crystallized (50). The molecular weight is 500,000, and the protein contains 8 mol of iron protoporphyrin IX, suggesting a minimum monomeric M_r of approximately 60,000 (see below). The reduced form of cytochrome b_1 has absorption maxima at 425, 527.5, and 557.5 nm. The prosthetic group of cytochrome b_1 can be removed with acidified acetone and then reconstituted (63). Reconstitution is accompanied by spectral changes and acquisition of reactivity with CO. It is possible that cytochrome b_1 corresponds in part to a low-potential cytochrome b which was removed from the membranes by a washing in buffer of low ionic strength (206). This component exhibits an α band centered at 558 nm (at 77 K). A spectrally and potentiometrically similar component was detected by Van Wielink et al. (238). Scott and Poole (217) also described a loosely bound cytochrome b. Yariv (247), however, has pointed out that this hemoprotein is indistinguishable from bacterioferritin in size, redox potential, and optical spectra. The low occupancy of heme binding sites (potentially 24) suggests a subunit M_r of 15,000, not 60,000.

c-Type Cytochromes

The earliest studies (125) suggested that cytochrome c was not present in E. coli, but in subsequent studies workers reported the existence of a c-like cytochrome in E. coli and S. typhimurium (see references in reference 62). The composite α absorption band in reduced minus oxidized difference spectra of aerobically grown E. coli contains one or two peaks attributable to c-type cytochromes. The fourth-order finite difference analysis of Shipp (219) and Scott and Poole (217) suggested such bands at 548 and 551.5 to 552 nm. Van Wielink et al. (238, 239) also resolved the lower-wavelength band and attributed it to a midpoint potential of +67 to +71 mV. The roles, if any, of these c-type cytochromes in the pathways of electrons to oxygen are unknown.

The best-studied cytochrome c in E. coli is cytochrome c_{552} (the subscript referring here to room temperature spectra), found in aerobically grown cells. Barrett & Sinclair (13) also described such a cytochrome in anaerobically grown cells; its concentration was highest under microaerophilic conditions; i.e., when the gas composition was 1 O_2:500,000 He. Strict anaerobiosis, vigorous aeration, or the presence of NO_3^- decreased the cytochrome c_{552} content (see also reference 64). The purified cytochrome had an estimated pI of 4.1 to 4.7, existed in two forms, and bound CO more extensively than did the cytochrome in intact cells. Its midpoint potential (150 mV; reference 13) appears to be distinct from the cytochrome c_{552} of anaerobically grown cells, which have a redox potential of about −200 mV (62). The cytochrome c_{552} of microaerophilic conditions appears not to be a substrate for peroxidase but may be involved in sulfite reduction.

A mutant with an elevated level of cytochrome c_{552} has been briefly described by Kajie et al. (123). Although the mutant grew poorly on succinate aerobically, this defect seems likely to be the result of a mutation in a gene (sox) other than that involved in cytochrome c synthesis.

The 503-nm Pigment

Under certain growth conditions, a pigment absorbing at about 503 nm can be observed in reduced minus oxidized spectra of E. coli cells. Such conditions include growth aerobically in minimal media with glucose, lactate, or succinate as carbon source; growth anaerobically with glucose (169); and growth under iron-limiting conditions (J. A. M. Hubbard, M. N. Hughes, and R. K. Poole, unpublished data). The absence of this pigment in complex media may be due to repression by amino acids (168). The pigment is bleached by cyanide, azide, hydrazine, and dithionite and does not react with CO (169). These investigators suggest that the compound was coprotetrahydroporphyrin and that, although in kinetic equilibrium with flavoprotein, it did not participate in the major electron flux of the respiratory chain. Streptomycin-dependent *E. coli* lack, or contain low levels of, the 503-nm pigment (124, 197) attributed by these investigators to catabolite repression of protoporphyrinogen oxidase. It remains unclear what role the pigment plays in the electron transport chain, but in air-oxidized cells provided with gluconate as substrate, the 503-nm pigment appears before cytochromes become reduced, suggesting that it might act as an electron acceptor from NADH (124).

Mutants Blocked in Heme Biosynthesis

In addition to those mutants defective in specific components of the oxygen-dependent electron transport chain, namely the oxidase complexes, cytochrome b_{556}, dehydrogenases, and quinones, discussed in the relevant sections here and also by Haddock (85, 86), particularly useful mutants are those affected in the *hemA* gene (lying at 26 min) and deficient in the synthesis of Δamino-laevulinic acid, an intermediate in heme synthesis. Such mutants have been isolated in several laboratories and used to probe both the composition and function of the oxygen-dependent respiratory chain and the dependence on this chain of other presumptive energy-linked reactions. Examples are the identification of a proton-translocating, cytochrome-independent segment of the respiratory chain; cytochrome assembly; and the role of cytochromes in iron uptake and O_2 sensing (for a list of examples, see reference 106).

Hydroperoxidases

Catalases and peroxidases, collectively referred to as hydroperoxidases, are responsible for removing the deleterious alkyl and hydrogen peroxides. Their activities are fundamentally similar; peroxidases reduce

peroxide in a single electron transfer from a reduced donor, such as cytochrome *c*, whereas catalases disproportionate peroxides, the oxidative power of one peroxide molecule being used to oxidize another in an electron pair transfer or a dismutation of hydrogen peroxide (for a survey, see reference 214). Although not directly involved in electron transport to dioxygen, it is clear that at least one of the cytochrome a_1-like hemoproteins in *E. coli* is a hydroperoxidase [see Cytochrome a_1-Like Pigment(s)]. Furthermore, it has already been shown that peroxide is a product of O_2 reduction by cytochrome *o* of species of the myxobacterium *Vitreoscilla* (for a survey see reference 180); thus, hydroperoxidase may have a role in completing O_2 reduction in *E. coli* and *S. typhimurium*. Indeed, the synthesis of hydroperoxidase is linked with that of certain components of the electron transport chain and is independent of O_2 or H_2O_2 per se (94).

E. coli possesses three electrophoretically distinct hydroperoxidases, only two of which, HP-I and HP-II, have been well characterized (Table 5). HP-I (of lower anodic mobility) is constitutive and present even in anaerobically grown cells, whereas HP-II is synthesized during growth in which electron transport to a terminal acceptor, such as oxygen or nitrate (94), occurs. The kinetics of induction of dianisidine peroxidase (HP-I; Table 4) and of catalase activities have suggested a possible product-precursor relationship between HP-I and HP-II (94). However, the purified enzymes show no serological cross-reactivity, and the polyacrylamide gel electrophoretic patterns of peptides obtained by treating HP-II with CNBr or chymotrypsin appear to be unrelated to the corresponding patterns obtained with HP-I (40).

The cytochrome a_1-like hemoprotein *b*-590 (8), and probably the a_1b complex of Barrett and Sinclair (Abstr. 7th Int. Congr. Biochem.), share several features in common with HP-I (184). Thus, the cytochrome a_1-like hemoprotein *b*-590 resembles HP-I in its subunit molecular weight (78,000 to 84,000), heme composition, and optical spectrum of the ferric state (180a). In addition, a membrane-associated protein with a *b*-type heme, exhibiting catalase and peroxidase activities and having a subunit molecular weight of 70,000, has a pI of 4.7 (137, 139), which is close to the pI value of the pure hemoprotein *b*-590 (180a). In addition, glucose considerably suppresses the activity of catalase and peroxidase (94, 248) and the amount of the membrane-bound catalase (137). Thus, there is

TABLE 5. Properties of hydroperoxidases from *E. coli*

Enzyme (reference[s])	M_r	No. (size [M_r]) of subunits	K_m for H_2O_2	Sp catalatic act (U/mg)	Peroxidase substrates	Heme	Peak(s) (nm) in spectrum of ferric form	Genetic locus
HPI (39)	337,000	4 (84,000)	3.9	1,486	*o*-Dianisidine, guaiacol, pyrogallol *p*-phenylenediamine	~2 mol of protoheme IX/tetramer	407	*katG*?
HPII (40)	312,000	4 (78,000)	ND[a]	8,925	Negligible peroxidase activity	2 mol of protoheme IX plus unidentified heme?	407, 591	*katG*?
HPIII (148, 210)	ND	ND	30.9	ND	No peroxidase activity	ND	ND	*katE* (37.88 min), *katF* (59 min)

[a] ND, Not determined.

strong circumstantial evidence that all of these proteins are closely related and that they may indeed be identical. No intracellular substrate for the peroxidase activity of HP-I has yet been identified (39), and HP-I and related proteins may have catalase activity as their primary role. A number of loci involved in catalase synthesis in *S. typhimurium* have been reported; one, *katB*, lies near min 100 (146). There have been recent advances in genetic studies of *E. coli* catalases; several catalase-deficient mutants have been isolated and show a 60-fold greater susceptibility to killing by H_2O_2 (147). Mutations in the *xthA* and *recA* genes also result in hypersensitivity to H_2O_2 (e.g., reference 52), but to a lesser extent, indicating that catalase is the primary defense against peroxide. One class of catalase mutants (*katE*) was characterized, and the lesions were mapped at 37.8 min on the *E. coli* chromosome (loci *A* to *D* have been named in *S. typhimurium*).

A colE1 plasmid containing a catalase gene has been identified in the Clarke and Carbon colony bank (150). Strains harboring the plasmid have amplified catalase levels, and in a minicell system the plasmid is responsible for synthesis of a protein with a molecular weight of 84,000, characteristic of catalase. Both HP-I and HP-II appear to be encoded by this plasmid. However, the third catalase, HP-III (Table 5), is not encoded in this way. A further locus, *katF*, lies at min 59 on the *E. coli* chromosome (148) and is involved, together with *katE* (described above), in the synthesis of HP-III. An unlinked locus (*katG*) localized at 89.2 min on the genome affects HP-I and HP-II (149). It should be noted that HP-II is now regarded by some to be an isoenzyme of HP-II, and it has been suggested that the terms HP1A and HP1B be used in the future.

CONCLUSIONS

In this chapter, we have attempted to summarize the published information on the respiratory chains leading to oxygen in *E. coli* and *S. typhimurium*. The body of information from which we had to draw is so great that our product cannot be comprehensive. We have provided some references, however, to serve as a key to further examples in the literature (mainly older references) that we have not cited. The integration of the pathways described here with those to electron acceptors other than oxygen can be understood by further reference to another chapter of this section (Pathways of Electrons to Acceptors Other Than Oxygen) and to our recent, more comprehensive review (106).

In some respects, the cytochrome components of the aerobic respiratory chains of these bacteria appear to be simple. Studies with *E. coli* suggest that pathways to oxygen are terminated by only two complexes. The only other membrane-bound cytochrome that can be placed in such schemes at present is cytochrome b_{556}. However, the roles of other cytochromes found in aerobically grown cells are still unidentified. Although many of these cytochromes appear to be soluble, this description, based on quite arbitrary centrifugation conditions and generally applicable after disintegrative cell disruption, should not be taken as evidence that such proteins do not have roles in membrane-bound electron transport processes. Soluble components with unidentified roles include hemoprotein *b*-590, at least three catalases, cytochrome b_{562}, the *c*-type cytochromes, and the 503-nm pigment. In some cases, the significance of an apparent dual subcellular location of a component, such as catalase and cytochrome b_1 (bacterioferritin), is not understood. New components are still being found (e.g., cytochrome b_{561}), and others may be undiscovered.

The quinone pool is reducible by a large number of primary dehydrogenases. The regulation of electron flow from a given dehydrogenase to one of the two oxidases is unclear.

The methods of molecular biology are making spectacular contributions to the understanding of the structure and biosynthesis of many of the protein components of the respiratory chains. Succinate dehydrogenase is a good example, but similar approaches are under way for the oxidase complexes and the membrane-bound *b* cytochromes. These studies will allow resolution of the protein at the level of amino acid residues; description of their secondary and higher levels of organization; and, from site-directed mutagenesis, identification of domains critical to enzymatic function. In addition, amplification of protein levels, beginning with NADH dehydrogenase, quinone biosynthetic enzymes, and some of the cytochromes, will facilitate extension of more traditional methods of biochemical analysis. Functional studies will be aided by biophysical techniques, the application of which, to the aerobic respiratory chains, is still in its infancy.

We believe that such integrated studies will yield answers of general relevance to students of respiration and oxidative phosphorylation.

ACKNOWLEDGMENTS

Our own work in this area has been financed by the Royal Society, the Nuffield Foundation, the Science and Engineering Research Council, United Kingdom (to W. J. I. and R. K. P.), and by the University of London Central Research Fund (to R. K. P.).

LITERATURE CITED

1. **Ackrell, B. A. C., R. N. Asato, and H. F. Mower.** 1966. Multiple forms of bacterial hydrogenases. J. Bacteriol. **92:**828–838.
2. **Adams, M. W. W., and D. O. Hall.** 1978. Purification of the membrane-bound hydrogenase of *Escherichia coli*. Biochem. J. **183:**11–22.
3. **Alben, J. O., P. P. Moh, F. G. Fiamingo, and R. A. Altschuld.** 1981. Cytochrome oxidase (a_3) heme and copper observed by low-temperature Fourier transform infrared spectroscopy of the CO complex. Proc. Natl. Acad. Sci. USA **78:**234–237.
4. **Albracht, S. P. J.** 1985. The use of electron paramagnetic resonance spectroscopy to establish the properties of nickel and the iron-sulphur cluster in hydrogenase from *Chromatium vinosum*. Biochem. Soc. Trans. **13:**582–585.
5. **Andersen, K. B., and K. Von Meyenburg.** 1980. Are growth rates of *Escherichia coli* in batch cultures limited by respiration? J. Bacteriol. **144:**114–123.
6. **Au, D. C.-T., G. N. Green, and R. B. Gennis.** 1984. The role of quinones in the branch of the *E. coli* respiratory chain which terminates in cytochrome *o*. J. Bacteriol. **157:**122–125.
7. **Au, D. C.-T., R. M. Lorence, and R. B. Gennis.** 1985. Isolation and characterization of an *Escherichia coli* mutant lacking the cytochrome *o* terminal oxidase. J. Bacteriol. **161:**123–127.
8. **Baines, B. S., H. D. Williams, J. A. M. Hubbard, and R. K. Poole.** 1984. Partial purification and characterization of a soluble haemoprotein having spectral similarities to cytochrome a_1 from anaerobically grown *Escherichia coli*. FEBS Lett. **171:**307–314.
9. **Ballantine, S. P., and D. H. Boxer.** 1985. Nickel-containing hy-

drogenase isoenzymes from anaerobically grown *Escherichia coli* K-12. J. Bacteriol. **163**:454–459.

10. **Barnes, E. M., and H. R. Kaback.** 1970. Beta galactoside transport in bacterial membrane preparations. Energy coupling via membrane bound D-lactic dehydrogenase. Proc. Natl. Acad. Sci. USA **66**:1190–1198.

11. **Barnes, E. M., and H. R. Kaback.** 1971. Mechanisms of active transport in isolated membrane vesicles. Part 1. The site of energy coupling between D-lactic dehydrogenase and beta galactoside transport in *Escherichia coli* membrane vesicles. J. Biol. Chem. **246**:5518–5522.

12. **Barrett, J.** 1956. The prosthetic group of cytochrome a$_2$. Biochem. J. **64**:626–639.

13. **Barrett, J., and P. R. Sinclair.** 1967. The cytochrome c(552) of aerobically grown *Escherichia coli* str. McElroy and its function. Biochim. Biophys. Acta **143**:279–281.

14. **Beelen, R. H. J., A. M. Feldmann, and H. J. W. Wijsman.** 1973. A regulatory gene and a structural gene for alaninase in *Escherichia coli*. Mol. Gen. Genet. **121**:369–374.

15. **Beinert, H., and S. P. J. Albracht.** 1982. New insights, ideas and unanswered questions concerning iron-sulphur clusters in mitochondria. Biochim. Biophys. Acta **683**:245–277.

16. **Bennett, R., D. R. Taylor, and A. Hurst.** 1966. D- and L-Lactate dehydrogenases in *Escherichia coli*. Biochim. Biophys. Acta **118**:512–521.

17. **Bernhard, T., and G. Gottschalk.** 1978. The hydrogenase of *Escherichia coli*. Purification, some properties and the function of the enzyme, p. 199–208. *In* H. G. Schlegel and K. Schneider (ed.), Hydrogenases: their catalytic activity, structure and function. Druck Goltze-Druckere und Verlag Erich Goltze GmbH & Co. KG, Göttingen, Federal Republic of Germany.

18. **Boxer, D. H., A. Malcolm, and A. Graham.** 1982. *Escherichia coli* formate to nitrate respiratory pathway: structural analysis. Biochem. Soc. Trans. **10**:480–481.

19. **Bragg, P. D.** 1979. Electron transport and energy-transducing systems of *Escherichia coli*, p. 341–449. *In* R. A. Capaldi (ed.), Membrane proteins in energy transduction. Marcel Dekker AG, Basel.

20. **Bragg, P. D.** 1980. The respiratory system of *Escherichia coli*, p. 115–136. *In* C. J. Knowles (ed.), Diversity of bacterial respiratory systems, vol. 1. CRC Press, Inc., Boca Raton, Fla.

21. **Bragg, P. D., and C. Hou.** 1967. Reduced nicotinamide adenine dinucleotide oxidation in *Escherichia coli* particles. II. NADH dehydrogenases. Arch. Biochem. Biophys. **119**:202–208.

22. **Bragg, P. D., and D. J. Rainnie.** 1974. The effect of silver ions on the respiratory chain of *Escherichia coli*. Can. J. Microbiol. **20**:883–889.

23. **Brand, M. D., B. Reynafarje, and A. L. Lehninger.** 1976. Stoichiometric relationship between energy-dependent proton ejection and electron transport in mitochondria. Proc. Natl. Acad. Sci. USA **73**:437–441.

24. **Brice, J. M., J. F. Law, D. J. Meyer, and C. W. Jones.** 1974. Energy conservation in *Escherichia coli* and *Klebsiella pneumoniae*. Biochem. Soc. Trans. **2**:523–526.

25. **Brock, T. D.** 1971. Microbial growth rates in nature. Bacteriol. Rev. **35**:39–58.

26. **Bullock, P. A., and Y. P. Meyer.** 1978. Circular dichroism and resonance Raman studies of cytochrome b$_{562}$ from *Escherichia coli*. Biochemistry **17**:3084–3091.

27. **Burnstein, C., L. Tiankova, and A. Kepes.** 1979. Respiratory control in *Escherichia coli* K12. Eur. J. Biochem. **94**:387–392.

28. **Butler, W. L.** 1979. 4th Derivative spectra. Methods Enzymol. **56**:501–515.

29. **Cammack, R., B. Crowe, and P. Owen.** 1982. Characterization of succinate dehydrogenase from *Micrococcus lysodeikticus*. Biochem. Soc. Trans. **10**:261–262.

30. **Cammack, R., D. S. Patil, and V. M. Fernandez.** 1985. Electron spin resonance/electron paramagnetic resonance spectroscopy of iron-sulphur enzymes. Biochem. Soc. Trans. **13**:572–578.

31. **Campbell, H. D., and I. G. Young.** 1983. Stereospecificity and requirements for activity of the respiratory NADH dehydrogenase of *Escherichia coli*. Biochemistry **22**:5754–5760.

32. **Carter, K., and R. B. Gennis.** 1985. Reconstitution of the ubiquinone-dependent pyruvate oxidase system of *Escherichia coli* with the cytochrome o terminal oxidase complex. J. Biol. Chem. **260**:10986–10990.

33. **Castor, L. N., and B. Chance.** 1955. Photochemical action spectra of carbon-monoxide-inhibited respiration. J. Biol. Chem. **217**:453–465.

34. **Castor, L. N., and B. Chance.** 1959. Photochemical determinations of the oxidases of bacteria. J. Biol. Chem. **234**:1587–1592.

35. **Cavari, B. Z., Y. Avi-Dor, and N. Grossowicz.** 1968. Induction by

36. **Chance, B.** 1978. Cytochrome kinetics at low temperatures: trapping and ligand exchange. Methods Enzymol. **54**:102–111.

37. **Chance, B., C. Saronio, and J. S. Leigh.** 1979. Compound C$_2$, a product of the reaction of oxygen and the mixed-valance state of cytochrome oxidase. Optical evidence for a type I copper. Biochem. J. **177**:931–941.

38. **Chetkauskaite, A. V., and L. L. Grinyus.** 1979. Transhydrogenase as an additional site of energy accumulation in the *Escherichia coli* respiratory chain. Biokhimiya **44**:869–876.

39. **Claiborne, A., and I. Fridovich.** 1979. Purification of the o-dianisidine peroxidase from *Escherichia coli* B. Physico-chemical characterization and analysis of its dual catalase and peroxidatic activities. J. Biol. Chem. **254**:4225–4252.

40. **Claiborne, A., D. P. Malinowski, and I. Fridovich.** 1979. Purification and characterization of hyperoxidase II of *Escherichia coli* B. J. Biol. Chem. **254**:11664–11668.

41. **Clarke, D. M., and P. D. Bragg.** 1985. Purification and properties of reconstitutively active nicotinamide nucleotide transhydrogenase of *Escherichia coli*. Eur. J. Biochem. **149**:517–523.

42. **Clark, D. P.** 1984. The number of anaerobically regulated genes in *Escherichia coli*. FEMS Microbiol. Lett. **24**:251–254.

43. **Cole, S. T., and J. R. Guest.** 1982. Molecular genetic aspects of the succinate:fumarate oxidoreductases of *Escherichia coli*. Biochem. Soc. Trans. **10**:473–475.

44. **Condon, C., R. Cammack, P. Patil, and P. Owen.** 1985. The succinate dehydrogenase of *Escherichia coli*. Immunochemical resolution and biophysical characterization of a 4-subunit enzyme complex. J. Biol. Chem. **60**:9427–9430.

45. **Condon, C., and P. Owen.** 1982. The succinate dehydrogenase of *Escherichia coli*: resolution as a major membrane-bound immunogen possessing covalently bound flavin. FEMS Microbiol. Lett. **14**:217–221.

46. **Condon, C., and P. Owen.** 1982. Succinate dehydrogenase: a major cross-reacting antigen in the Enterobacteriaceae. FEMS Microbiol. Lett. **15**:109–113.

47. **Creaghan, I., and J. R. Guest.** 1972. Amber mutants of the (alpha)-ketoglutarate dehydrogenase gene of *Escherichia coli* K12. J. Gen. Microbiol. **71**:207–220.

48. **Crispin, J. A. S., M. Dubourdieu, and M. Chippaux.** 1979. Localization and characterization of cytochromes from membrane vesicles of *Escherichia coli* K12 grown in anaerobiosis with nitrate. Biochim. Biophys. Acta **547**:198–210.

49. **Darlison, M. G., and J. R. Guest.** 1984. Nucleotide sequence encoding the iron-sulphur protein subunit of the succinate dehydrogenase of *Escherichia coli*. Biochem. J. **223**:507–517.

50. **Deeb, S. S., and L. P. Hager.** 1964. Crystalline cytochrome b$_1$ from *Escherichia coli*. J. Biol. Chem. **239**:1024–1031.

51. **De Maio, R. A., D. A. Webster, and B. Chance.** 1983. Spectral evidence for the existence of a second cytochrome o in whole cells of *Vitreoscilla*. J. Biol. Chem. **258**:13768–13771.

52. **Demple, B., J. Halbrook, and S. Linn.** 1983. *Escherichia coli* xth mutants are hypersensitive to hydrogen peroxide. J. Bacteriol. **153**:1079–1082.

53. **Dervartanian, D. V., L. K. Iburg, and T. V. Morgan.** 1973. EPR studies on phosphorylating particles from *Azotobacter vinelandii*. Biochim. Biophys. Acta **305**:173–178.

54. **Downie, J. A., and G. B. Cox.** 1978. Sequence of b cytochromes relative to ubiquinone in the electron transport chain of *Escherichia coli*. J. Bacteriol. **133**:477–484.

55. **Drabikowska, A. K.** 1969. Subcellular distribution and function of ubiquinone in *Salmonella typhimurium*. Acta Biochim. Pol. **16**:135–140.

56. **Drabikowska, A. K.** 1970. Electron transport system of *Salmonella typhimurium* cells. Acta Biochim. Pol. **17**:89–98.

57. **Edwards, C., S. Beer, A. Siviram, and B. Chance.** 1981. Photochemical action spectra of bacterial a- and o-type oxidases using a dye laser. FEBS Lett. **128**:205–207.

58. **Enoch, H. G., and R. L. Lester.** 1975. The purification and properties of formate dehydrogenase and nitrate reductase in *Escherichia coli*. J. Biol. Chem. **250**:6693–6705.

59. **Finlayson, S. D., and W. J. Ingledew.** 1985. Cytochrome bd of *Escherichia coli*: its isolation and study by electron paramagnetic resonance. Biochem. Soc. Trans. **13**:632–633.

60. **Franklin, F. C. H., and W. A. Venables.** 1976. Biochemical, genetic and regulatory studies of alanine catabolism in *Escherichia coli* K12. Mol. Gen. Genet. **149**:229–237.

61. **Franklin, F. C. H., W. A. Venables, and H. J. W. Wijsman.** 1981. Genetic studies of D-alanine dehydrogenaseless mutants of *Escherichia coli* K12. Genet. Res. **38**:197–208.

62. **Fujita, T.** 1966. Studies on soluble cytochromes in Enterobac-

teriaceae. I. Detection, purification and properties of cytochrome c_{552} in anaerobically grown cells. J. Biochem. **60**:204–215.

63. **Fujita, T., E. Itagaki, and R. Sato.** 1963. Purification and properties of cytochrome b_1 from *Escherichia coli*. J. Biochem. **53**:282–290.

64. **Fujita, T., and R. Sato.** 1966. Studies on soluble cytochromes in Enterobacteriaceae. IV. Possible involvement of cytochrome c_{552} in anaerobic nitrite metabolism. J. Biochem. **60**:691–700.

65. **Futai, M.** 1973. Membrane D-lactate dehydrogenase from *Escherichia coli*. Purification and properties. Biochemistry **12**:2468–2474.

66. **Futai, M.** 1974. Reconstitution of transport dependent on D-lactate or glycerol-3-phosphate in membrane vesicles of *Escherichia coli* deficient in the corresponding dehydrogenase. Biochemistry **13**:2327–2333.

67. **Futai, M., and H. Kimura.** 1977. Inducible membrane bound D-lactate dehydrogenase from *Escherichia coli*. Purification and properties. J. Biol. Chem. **252**:5820–5827.

68. **Garland, P. B., M. T. Davison, and C. H. Moore.** 1979. Rotational mobility of membrane-bound cytochrome *o* of *Escherichia coli* and cytochrome a_1 *Thiobacillus ferro-oxidans*. Biochem. Soc. Trans. **7**:1112–1114.

69. **Garland, P. B., J. A. Downie, and B. A. Haddock.** 1975. Proton translocation and the respiratory nitrate reductase of *Escherichia coli*. Biochem. J. **152**:547–559.

70. **Graham, A.** 1981. The organization of hydrogenase in the cytoplasmic membrane of *Escherichia coli*. Biochem. J. **197**:283–291.

71. **Graham, A., and D. H. Boxer.** 1981. The organisation of formate dehydrogenase in the cytoplasmic membrane of *Escherichia coli*. Biochem. J. **195**:627–637.

72. **Graham, A., D. H. Boxer, B. A. Haddock, M. A. Mandrand-Berthelot, and R. W. Jones.** 1980. Immunochemical analysis of the membrane-bound hydrogenase of *Escherichia coli*. FEBS Lett. **113**:167–173.

73. **Graham, A., H. E. Jenkins, N. H. Smith, M. A. Mandrand-Berthelot, B. A. Haddock, and D. H. Boxer.** 1980. The synthesis of formate dehydrogenase and nitrate reductase protein in various *fdh* and *chl* mutants of *Escherichia coli*. FEMS Microbiol. Lett. **7**:145–151.

74. **Graham, A., A. D. Tucker, and N. H. Smith.** 1981. The formate-nitrate respiratory chain of *Escherichia coli*: localization of proteins of immunoadsorption studies. FEMS Microbiol. Lett. **11**:141–147.

75. **Gray, C. T., J. W. T. Wimpenny, D. E. Hughes, and M. R. Mossman.** 1966. Regulation of metabolism in facultative bacteria. I. Structural and functional changes in *Escherichia coli* associated with shifts between the aerobic and anaerobic states. Biochim. Biophys. Acta **117**:22–32.

76. **Green, G. N., and R. B. Gennis.** 1983. Isolation and characterization of a mutant in *Escherichia coli* lacking the cytochrome *d* terminal oxidase. J. Bacteriol. **154**:1269–1275.

77. **Green, G. N., J. E. Kranz, and R. B. Gennis.** 1984. Cloning the *cyd* gene locus coding for the cytochrome *d* complex of *Escherichia coli*. Gene **32**:99–106.

78. **Green, G. N., R. G. Kranz, R. M. Lorence, and R. B. Gennis.** 1984. Identification of subunit 1 as the cytochrome b_{556} component of the cytochrome *d* terminal oxidase complex of *E. coli*. J. Biol. Chem. **259**:7994–7997.

79. **Guest, J. R.** 1981. Partial replacement of succinate dehydrogenase function by phage and plasmid-specified fumarate reductase in *Escherichia coli*. J. Gen. Microbiol. **122**:171–179.

80. **Guest, J. R., M. G. Darlison, R. J. Wilde, and D. Wood.** 1985. Structural comparison of the succinate dehydrogenase and fumarate reductase of *Escherichia coli*, p. 225–228. *In* R. C. Bray (ed.), Flavins & flavoproteins. Proceedings of the 8th International Symposium. De Gruyter, Berlin.

81. **Gutman, M., A. Schejter, and Y. Avi-dor.** 1968. The preparation and properties of the membranal DPNH dehydrogenase of *Escherichia coli*. Biochim. Biophys. Acta **162**:506–517.

82. **Hackett, N. R., and P. D. Bragg.** 1982. The association of 2 distinct *b* cytochromes with the respiratory nitrate reductase of *Escherichia coli*. FEMS Microbiol. Lett. **13**:213–217.

83. **Hackett, N. R., and P. D. Bragg.** 1983. Membrane cytochromes of *Escherichia coli* grown aerobically and anaerobically with nitrate. J. Bacteriol. **154**:708–718.

84. **Hackett, N. R., and P. D. Bragg.** 1983. Membrane cytochromes of *Escherichia coli chl* mutants. J. Bacteriol. **154**:719–727.

85. **Haddock, B. A.** 1977. The isolation of phenotypic and genotypic variants for the functional characterization of bacterial oxidation phosphorylation, p. 95–120. *In* B. A. Haddock and W. A. Hamilton (ed.), Microbial energetics. Cambridge University Press, Cambridge.

86. **Haddock, B. A.** 1982. The genetics of electron transport in *Escherichia coli*, p. 459–463. *In* A. N. Martonosi (ed.), Membranes and transport, vol. 1. Plenum Publishing Corp., New York.

87. **Haddock, B. A., J. A. Downie, and P. B. Garland.** 1976. Kinetic characterization of the membrane-bound cytochromes of *Escherichia coli* grown under a variety of conditions by using a stopped-flow dual-wavelength spectrophotometer. Biochem. J. **154**:285–294.

88. **Haddock, B. A., and C. W. Jones.** 1977. Bacterial respiration. Bacteriol. Rev. **41**:47–99.

89. **Haddock, B. A., and M. W. Kendall-Tobias.** 1975. Functional anaerobic electron transport linked to the reduction of nitrate and fumarate in membranes from *Escherichia coli* as demonstrated by quenching of atebrin fluorescence. Biochem. J. **152**:655–659.

90. **Haddock, B. A., and M. A. Mandrand-Berthelot.** 1982. *Escherichia coli* formate-to-nitrate respiratory chain: genetic analysis. Biochem. Soc. Trans. **10**:478–480.

91. **Hager, L. P., and E. Itagaki.** 1967. The preparation and properties of cytochrome b_{562} from *Escherichia coli*. Methods Enzymol. **10**:373–378.

92. **Haldar, K., P. Y. Olsiewski, C. Walsh, G. J. Kazarowski, A. Bhaduri, and H. R. Kaback.** 1982. Simultaneous reconstitution of *Escherichia coli* membrane vesicles with D-lactic and D-amino acid dehydrogenases. Biochemistry **21**:4590–4596.

93. **Hallohan, D. L., and D. O. Hall.** 1985. Association of nickel with the hydrogenase from aerobically grown *Escherichia coli*. Biochem. Soc. Trans. **13**:631–632.

94. **Hassan, H. M., and I. Fridovich.** 1978. Regulation of the synthesis of catalase and peroxidase in *Escherichia coli*. J. Biol. Chem. **253**:6445–6450.

95. **Hata, A., Y. Kirino, K. Matsuura, S. Itoh, T. Hiyama, K. Konishi, K. Kita, and Y. Anraku.** 1985. Assignment of ESR signals of *Escherichia coli* terminal oxidase complexes. Biochim. Biophys. Acta **810**:62–72.

96. **Hederstedt, L., and L. Rutberg.** 1981. Succinate dehydrogenase—a comparative review. Microbiol. Rev. **45**:542–555.

97. **Hendler, R. W., and A. H. Burgess.** 1972. Respiration and protein synthesis in *Escherichia coli* membrane-envelope fragments. VI. Solubilization and characterization of the electron transport chain. J. Cell Biol. **55**:261–281.

98. **Hendler, R. W., and A. H. Burgess.** 1974. Fractionation of the electron transport chain of *Escherichia coli*. Biochim. Biophys. Acta **357**:215–230.

99. **Hendler, R. W., and R. I. Schrager.** 1979. Potentiometric analysis of *Escherichia coli* cytochromes in the optical absorbance range of 500 nm to 700 nm. J. Biol. Chem. **254**:11288–11299.

100. **Hirsch, C. A., M. Rasminsky, B. D. Davis, and E. C. C. Lin.** 1963. A fumarate reductase in *Escherichia coli* distinct from succinate dehydrogenase. J. Biol. Chem. **238**:3770–3774.

101. **Hong, J. S., and H. R. Kaback.** 1972. Mutants of *Salmonella typhimurium* and *Escherichia coli* pleiotropically defective in active transport. Proc. Natl. Acad. Sci. USA **69**:3336–3340.

102. **Hubbard, J. A. M., M. N. Hughes, and R. K. Poole.** 1983. Nitrate but not silver ions induce spectral changes in *Escherichia coli* cytochrome *d*. FEBS Lett. **164**:241–243.

103. **Hubbard, J. A. M., M. N. Hughes, and R. K. Poole.** 1985. Reactions of some nitrogen oxyanions and nitric oxide with cytochrome oxidase *d* from oxygen limited *Escherichia coli* K12, p. 231–236. *In* R. K. Poole and C. S. Dow (ed.), Microbial gas metabolism: mechanistic, metabolic and biotechnological aspects. Academic Press, Inc. (London), Ltd., London.

104. **Ingledew, W. J.** 1978. Cytochrome a_1 as an oxidase?, p. 79–87. *In* H. Degn, D. Lloyd, and G. C. Hill (ed.), Functions of alternative terminal oxidases. Pergamon Press, Oxford.

105. **Ingledew, W. J.** 1983. The electron transport chain of *Escherichia coli* grown anaerobically with fumarate as terminal electron acceptor: an electron paramagnetic resonance study. J. Gen. Microbiol. **129**:1651–1659.

106. **Ingledew, W. J., and R. K. Poole.** 1984. The respiratory chains of *Escherichia coli*. Microbiol. Rev. **48**:222–271.

107. **Ingledew, W. J., G. A. Reid, R. K. Poole, H. Blum, and T. Ohnishi.** 1980. The iron-sulphur centres of aerobically-grown *Escherichia coli* K12. An electron paramagnetic resonance study. FEBS Lett. **111**:223–227.

108. **Ingledew, W. J., and M. Saraste.** 1979. The reaction of cytochrome cd_1 with oxygen and peroxides. Biochem. Soc. Trans. **7**:167–168.

109. **Ishida, A.** 1977. A carbon monoxide-binding hemoprotein formed by heme accumulation in *Escherichia coli*. J. Biochem. **81**:1869–1878.

110. **Itagaki, E., and L. P. Hager.** 1966. Studies on cytochrome b_{562} of *Escherichia coli*. I. Purification and crystallization of cytochrome b_{562}. J. Biol. Chem. **241:**3687–3695.
111. **Itagaki, E., and L. P. Hager.** 1968. The amino acid sequence of cytochrome b_{562} of *Escherichia coli*. Biochem. Biophys. Res. Commun. **32:**1013–1019.
112. **Jaworowski, A., H. D. Campbell, M. I. Poulis, and I. G. Young.** 1981. Genetic identification and purification of the respiratory NADH dehydrogenase of *Escherichia coli*. Biochemistry **20:**2041–2047.
113. **Jaworowski, A., G. Mayo, D. C. Shaw, H. D. Campbell, and I. G. Young.** 1981. Characterization of the respiratory NADH dehydrogenase of *Escherichia coli* and reconstitution of NADH oxidase in *ndh* mutant membrane vesicles. Biochemistry **20:**3621–3628.
114. **Johnson, P. L., and P. D. Bragg.** 1985. Control of formation of the cytochrome *d* complex by growth substrate in a mutant of *Escherichia coli*. FEMS Microbiol. Lett. **26:**185–189.
115. **Jones, C. W.** 1978. Microbial oxidative phosphorylation. Biochem. Soc. Trans. **6:**361–363.
116. **Jones, C. W.** 1980. Cytochrome patterns in classification and identification including their relevance to the oxidase test. Soc. Appl. Bacteriol. Symp. Ser. **8:**127–138.
117. **Jones, C. W., J. M. Brice, and C. Edwards.** 1977. The effect of respiratory chain composition on the growth efficiencies of aerobic bacteria. Arch. Microbiol. **115:**85–93.
118. **Jones, R. W.** 1979. The topography of the membrane-bound hydrogenase of *Escherichia coli* explored by non-physiological electron acceptors. Biochem. Soc. Trans. **7:**724–725.
119. **Jones, R. W.** 1979. Hydrogen-dependent proton translocation by membrane vesicles from *Escherichia coli*. Biochem. Soc. Trans. **7:**1136–1137.
120. **Jones, R. W., R. G. Kranz, and R. B. Gennis.** 1982. Immunochemical analysis of the membrane-bound succinate dehydrogenase of *Escherichia coli*. FEBS Lett. **142:**81–87.
121. **Jurtshuk, P., T. J. Mueller, and W. C. Acord.** 1975. Bacterial terminal oxidases. Crit. Rev. Microbiol. **3:**399–468.
122. **Kaback, H. R.** 1974. Transport studies in bacterial membrane vesicles. Science **186:**882–892.
123. **Kajie, S., K. Miki, E. C. C. Lin, and Y. Anraku.** 1984. Isolation of an *Escherichia coli* mutant defective in cytochrome biosynthesis. FEMS Microbiol. Lett. **24:**25–29.
124. **Kamitakahara, J. R., and W. J. Polglase.** 1970. The 503 nanometer pigment of *Escherichia coli*. Biochem. J. **120:**771–775.
125. **Keilin, D.** 1970. The history of cell respiration and cytochrome, p. 269–288. Cambridge University Press, Cambridge.
126. **Kim, I. C., and P. D. Bragg.** 1971. Some properties of the succinate dehydrogenase of *Escherichia coli*. Can. J. Biochem. **49:**1098–1104.
127. **Kirby, M. C., and W. J. Ingledew.** 1984. The iron-sulphur centres of the *Escherichia coli* electron transport chain: changes in response to aerobic growth on different carbon sources. Biochem. Soc. Trans. **12:**501–502.
128. **Kita, K., and Y. Anraku.** 1981. Composition and sequence of *b* cytochromes in the respiratory chain of aerobically-grown *Escherichia coli* K12 in the early exponential phase. Biochem. Int. **2:**105–112.
129. **Kita, K., K. Konishi, and Y. Anraku.** 1984. Terminal oxidase of *E. coli* aerobic respiratory chain. I. Purification and properties of cytochrome b_{556}-*o* complex from cells in the early exponential phase of aerobic growth. J. Biol. Chem. **259:**3368–3374.
130. **Kita, K., K. Konishi, and Y. Anraku.** 1984. Terminal oxidases of *Escherichia coli* aerobic respiratory chain. II. Purification and properties of cytochrome b_{558}-*d* complex from cells grown with limited oxygen and evidence of branched electron carrying systems. J. Biol. Chem. **259:**3375–3381.
131. **Kita, K., I. Yamato, and Y. Anraku.** 1978. Purification and properties of cytochrome b_{556} in the respiratory chain of aerobically grown *Escherichia coli* K12. J. Biol. Chem. **253:**8910–8915.
132. **Kline, E. S., and H. R. Mahler.** 1965. The lactic dehydrogenases of *Escherichia coli*. Ann. N.Y. Acad. Sci. **119:**905–917.
133. **Koch, A. L.** 1976. How bacteria face depression, recession and derepression. Perspect. Biol. Med. **20:**44–63.
134. **Kohn, L. D., and H. R. Kaback.** 1973. Mechanisms of active transport in isolated bacterial membrane vesicles. Part 15. Purification and properties of the membrane bound D-lactate dehydrogenase from *Escherichia coli*. J. Biol. Chem. **248:**7012–7017.
135. **Koland, J. G., M. J. Miller, and R. B. Gennis.** 1984. Potentiometric analysis of the purified cytochrome *d* terminal oxidase complex from *E. coli*. Biochemistry **23:**1051–1056.
136. **Koland, J. G., M. J. Miller, and R. B. Gennis.** 1984. Reconstitution of the membrane-bound, ubiquinone-dependent pyruvate oxidase respiratory chain of *Escherichia coli* with the cytochrome *d* terminal oxidase. Biochemistry **23:**445–458.
137. **Kranz, R. G., C. A. Barassi, and R. B. Gennis.** 1984. Immunological analysis of the heme proteins present in anaerobically grown *Escherichia coli*. J. Bacteriol. **158:**1191–1194.
138. **Kranz, R. G., C. A. Barassi, M. J. Miller, G. N. Green, and R. B. Gennis.** 1983. Immunological characterization of an *Escherichia coli* strain which is lacking cytochrome *d*. J. Bacteriol. **156:**115–121.
139. **Kranz, R. G., and R. B. Gennis.** 1982. Isoelectric focusing and crossed immunoelectrophoresis of heme proteins in the *Escherichia coli* cytoplasmic membrane. J. Bacteriol. **150:**36–45.
140. **Kranz, R. G., and R. B. Gennis.** 1983. Immunological characterization of the cytochrome *o* terminal oxidase from *Escherichia coli*. J. Biol. Chem. **258:**10614–10621.
141. **Kranz, R. G., and R. B. Gennis.** 1985. Immunological investigation of the distribution of cytochromes related to the two terminal oxidases of *Escherichia coli* in other gram-negative bacteria. J. Bacteriol. **161:**709–713.
142. **Kumar, C., R. K. Poole, I. Salmon, and B. Chance.** 1985. The oxygen reaction of cytochrome *d* terminated respiratory chain of *Escherichia coli* at sub-zero temperatures: kinetic resolution by EPR spectroscopy of two high spin cytochromes. FEBS Lett. **190:**227–232.
143. **Kung, H.-F., and U. Hennings.** 1972. Limiting availability of binding sites for dehydrogenases on the cell membrane of *Escherichia coli*. Proc. Natl. Acad. Sci. USA **69:**925–929.
144. **Laszlo, D. J., B. L. Fandrich, A. Sivaram, B. Chance, and B. L. Taylor.** 1984. Cytochrome *o* as a terminal oxidase and receptor for aerotaxis in *Salmonella typhimurium*. J. Bacteriol. **159:**663–667.
145. **Lawford, H. G., and B. A. Haddock.** 1973. Respiration-driven proton translocation in *Escherichia coli*. Biochem. J. **136:**217–220.
146. **Levine, S. A.** 1977. Isolation and characterization of catalase deficient mutant of *Salmonella typhimurium*. Mol. Gen. Genet. **150:**205–209.
147. **Loewen, P. C.** 1984. Isolation of catalase-deficient *Escherichia coli* mutants and genetic mapping of *katE*, a locus that affects catalase activity. J. Bacteriol. **157:**622–626.
148. **Loewen, P. C., and B. L. Triggs.** 1984. Genetic mapping of *katF*, a locus that with *katE* affects the synthesis of a second catalase species in *Escherichia coli*. J. Bacteriol. **160:**668–675.
149. **Loewen, P. C., B. L. Triggs, C. S. George, and B. E. Hrabarchuk.** 1985. Genetic mapping of *katG*, a locus that affects synthesis of the bifunctional catalase-peroxidase hydroperoxidase I in *Escherichia coli*. J. Bacteriol. **162:**661–667.
150. **Loewen, P. C., B. L. Triggs, G. R. Klassen, and J. H. Weiner.** 1983. Identification and physical characterization of a ColE1 hybrid plasmid containing a catalase gene of *Escherichia coli*. Can. J. Biochem. **61:**1315–1321.
151. **Lorence, R. M., G. N. Green, and R. B. Gennis.** 1984. Potentiometric analysis of the cytochromes of an *Escherichia coli* mutant strain lacking the cytochrome *d* terminal oxidase complex. J. Bacteriol. **157:**115–121.
152. **Lorence, R. M., M. J. Miller, A. Borochov, R. Faiman-Weinberg, and R. B. Gennis.** 1984. Effects of pH and detergent on the kinetic and electrochemical properties of the purified cytochrome *d* terminal oxidase complex of *Escherichia coli*. Biochim. Biophys. Acta **790:**148–153.
153. **Mathews, F. S., P. H. Bethge, and E. W. Czerwinski.** 1979. The structure of cytochrome b_{562} from *Escherichia coli* at 2.5 Å resolution. J. Biol. Chem. **254:**1699–1706.
154. **Matsushita, K., L. Patel, R. B. Gennis, and H. R. Kaback.** 1983. Reconstitution of active transport in proteoliposomes containing cytochrome *o* oxidase and *lac* carrier protein purified from *Escherichia coli*. Proc. Natl. Acad. Sci. USA **80:**4889–4893.
155. **Matsushita, K., L. Patel, and H. R. Kaback.** 1984. Cytochrome *o* type oxidase from *Escherichia coli*. Characterization of the enzyme and mechanism of electrochemical proton gradient generation. Biochemistry **23:**4703–4714.
156. **Meyer, D. J.** 1973. Interaction of cytochrome oxidases aa_3, and *d* with nitrite. Nature (London) New Biol. **254:**276–277.
157. **Myer, Y. P., and P. A. Bullock.** 1978. Cytochrome b_{562} from *Escherichia coli*: conformational and spin-state characterization. Biochemistry **17:**3723–3729.
158. **Miller, M. J., and R. B. Gennis.** 1983. The purification and characterization of the cytochrome *d* terminal oxidase complex of the *Escherichia coli* aerobic respiratory chain. J. Biol. Chem. **258:**9159–9165.

159. **Miller, M. J., and R. B. Gennis.** 1985. The cytochrome d complex is a coupling site in the aerobic respiratory chain of *Escherichia coli*. J. Biol. Chem. **260:**14003–14008.

160. **Mitchell, P.** 1966. Chemiosmotic coupling in oxidative and photosynthetic phosphorylation. Glynn Research, Bodmin, Cornwall, United Kingdom.

161. **Møller, V.** 1954. Diagnostic use of the Braun KCN test within the Enterobacteriaceae. Acta Pathol. Microbiol. Scand. **34:**115–116.

162. **Moore, G. R., R. J. P. Williams, J. Peterson, A. J. Thomson, and F. S. Mathews.** 1985. A spectroscopic investigation of the structure and redox properties of *Escherichia coli* cytochrome b-562. Biochim. Biophys. Acta **829:**83–90.

163. **Murakami, H., K. Kita, and Y. Anraku.** 1984. Cloning of $cybB$, the gene for cytochrome b_{561} of *Escherichia coli* K12. Mol. Gen. Genet. **198:**1–6.

164. **Murakami, H., K. Kita, H. Oya, and Y. Anraku.** 1984. Chromosomal location of the *Escherichia coli* cytochrome b_{556} gene $cybA$. Mol. Gen. Genet. **196:**1–5.

165. **Murakami, H., K. Kita, H. Oya, and Y. Anraku.** 1985. The *Escherichia coli* cytochrome b_{556} gene $cybA$ is assignable as $sdhC$ in the succinate dehydrogenase gene cluster. FEMS Microbiol. Lett. **30:**307–311.

166. **Murakami, H., K. Kita, and Y. Anraku.** 1986. Purification and properties of a diheme cytochrome b_{561} of the *Escherichia coli* respiratory chain. J. Biol. Chem. **261:**548–551.

167. **Nishimura, Y., I. K. P. Tan, Y. Ohgami, K. Kohgami, and T. Kamihara.** 1983. Induction of membrane-bound L-lactate dehydrogenase in *Escherichia coli* under conditions of nitrate respiration, fumarate reduction and trimethylamine-N-oxide reduction. FEMS Microbiol. Lett. **17:**283–286.

168. **Nosoh, Y., and M. Itoh.** 1965. The 503 mu pigment in bacteria and yeast. Plant Cell Physiol. **6:**771–774.

169. **Olden, K., and W. P. Hempfling.** 1973. The 503-nm pigment *Escherichia coli* B: characterization and nutritional conditions affecting its accumulation. J. Bacteriol. **113:**914–921.

170. **Olsiewski, P. J., G. J. Kaczorowski, and C. Walsh.** 1980. Purification and properties of D-amino-acid dehydrogenase, an inducible membrane-bound iron-sulfur flavoenzyme from *Escherichia coli* B. J. Biol. Chem. **255:**4487–4494.

171. **Olsiewski, P. J., G. J. Kaczorowski, C. T. Walsh, and H. R. Kaback.** 1981. Reconstitution of *Escherichia coli* membrane vesicles with D-amino-acid dehydrogenase. Biochemistry **20:**6272–6279.

172. **Owen, P., and C. Condon.** 1982. The succinate dehydrogenase of *Escherichia coli*: subunit composition of the Triton X-100 solubilized antigen. FEMS Microbiol. Lett. **14:**223–227.

173. **Owen, P., and H. R. Kaback.** 1978. Molecular structure of membrane vesicles from *Escherichia coli*. Proc. Natl. Acad. Sci. USA **78:**3148–3152.

174. **Owen, P., and H. R. Kaback.** 1979. Immunochemical analysis of membrane vesicles from *Escherichia coli*. Biochemistry **18:**1413–1422.

175. **Owen, P., and H. R. Kaback.** 1979. Antigenic architecture of membrane vesicles from *Escherichia coli*. Biochemistry **18:**1422–1426.

176. **Owen, P., H. R. Kaback, and K. A. Graeme-Cook.** 1980. Identification of antigen 19/27 as dihydrolipoyl dehydrogenase and its probable involvement in ubiquinone-mediated NADH-dependent transport phenomena in membrane vesicles of *Escherichia coli*. FEMS Microbiol. Lett. **7:**345–349.

177. **Owen, P., G. J. Kaczorowski, and H. R. Kaback.** 1980. Resolution and identification of iron-containing antigens in membrane vesicles from *Escherichia coli*. Biochemistry **19:**596–600.

178. **Pinsett, J.** 1954. The need for selenite and molybdate in the formation of formic dehydrogenase by members of the *Coli-aerogenes* group of bacteria. Biochem. J. **57:**10–16.

179. **Poole, R. K.** 1982. The oxygen reactions of bacterial cytochrome oxidases. Trends Biochem. Sci. **7:**32–34.

180. **Poole, R. K.** 1983. Bacterial cytochrome oxidases. A structurally and functionally diverse group of electron-transfer proteins. Biochim. Biophys. Acta **726:**205–243.

180a. **Poole, R. K., B. S. Baines, and C. A. Appleby.** 1986. Haemoprotein b-590 (*Escherichia coli*), a reducible catalase and peroxidase: evidence for its close relationship to hydroperoxidase I and a 'cytochrome a_1b' preparation. J. Gen. Microbiol. **132:**1525–1539.

181. **Poole, R. K., B. S. Baines, S. J. Curtis, H. O. William, and P. M. Wood.** 1984. Haemoprotein b-590 (*Escherichia coli*); redesignation of a bacterial "cytochrome a_1". J. Gen. Microbiol. **130:**3055–3058.

182. **Poole, R. K., B. S. Baines, J. A. M. Hubbard, N. M. Hughes, and N. J. Campbell.** 1982. Resonance Raman spectroscopy of an oxygenated intermediate species of cytochrome oxidase d from *Escherichia coli*. FEBS Lett. **150:**147–150.

183. **Poole, R. K., B. S. Baines, J. A. M. Hubbard, and H. D. Williams.** 1985. Microbial metabolism of oxygen: the binding and reduction of oxygen by bacterial cytochrome oxidases, p. 31–62. *In* R. K. Poole and C. S. Dow (ed.), Microbial gas metabolism: mechanistic, metabolic and biotechnological aspects. Academic Press, Inc. (London), Ltd., London.

184. **Poole, R. K., B. S. Baines, and H. D. Williams.** 1985. Sensor sensationalism? Alternative views on the nature and role of "cytochrome a_1" in bacteria. Microbiol. Sci. Microbiol. Sci. **2:**21–24.

185. **Poole, R. K., and B. Chance.** 1982. The reaction of cytochrome o in *Escherichia coli* K12 with oxygen. Evidence for a spectrally and kinetically distinct cytochrome o in cells from oxygen-limited culture. J. Gen. Microbiol. **126:**277–287.

186. **Poole, R. K., and C. S. Dow (ed.).** 1985. Microbial gas metabolism: mechanistic, metabolic and biotechnological aspects. Academic Press, Inc. (London), Ltd., London.

187. **Poole, R. K., and B. A. Haddock.** 1974. Energy-linked reduction of nicotinamide adenine dinucleotide in membranes derived from normal and various respiratory-deficient mutant strains of *Escherichia coli* K12. Biochem. J. **144:**77–85.

188. **Poole, R. K., and B. A. Haddock.** 1975. Effects of sulphate-limited growth in continuous culture on the electron-transport chain and energy conservation in *Escherichia coli* K12. Biochem. J. **152:**537–546.

189. **Poole, R. K., C. Kumar, I. Salmon, and B. Chance.** 1983. The 650 nm chromophore in *Escherichia coli* is an oxygenated compound, not the oxidized form of cytochrome oxidase d: an hypothesis. J. Gen. Microbiol. **129:**1335–1344.

190. **Poole, R. K., I. Salmon, and B. Chance.** 1983. The reaction with oxygen of cytochrome oxidase (cytochrome d) in *Escherichia coli* K12: optical studies of intermediate species and cytochrome b oxidation at sub-zero temperatures. J. Gen. Microbiol. **129:**1345–1355.

191. **Poole, R. K., R. I. Scott, and B. Chance.** 1980. Low-temperature spectral and kinetic properties of cytochromes in *Escherichia coli* K12 grown at lowered oxygen tension. Biochim. Biophys. Acta **591:**471–483.

192. **Poole, R. K., R. I. Scott, and B. Chance.** 1981. The light-reversible binding of carbon monoxide to cytochrome a_1 in *Escherichia coli* K12. J. Gen. Microbiol. **125:**431–438.

193. **Poole, R. K., A. Sivaram, I. Salmon, and B. Chance.** 1982. Photolysis at very low temperatures of CO-ligated cytochrome oxidase (cytochrome d) in oxygen-limited *Escherichia coli*. FEBS Lett. **141:**237–241.

194. **Poole, R. K., A. J. Waring, and B. Chance.** 1979. Evidence for a functional oxygen-bound intermediate in the reaction of *Escherichia coli* cytochrome o with oxygen. FEBS Lett. **101:**56–58.

195. **Poole, R. K., A. J. Waring, and B. Chance.** 1979. The reaction of cytochrome o in *Escherichia coli* with oxygen. Low temperature kinetics and spectral studies. Biochem. J. **184:**379–389.

196. **Poulis, M. I., D. C. Shaw, H. D. Campbell, and I. G. Young.** 1981. *In vitro* synthesis of the respiratory NADH dehydrogenase of *Escherichia coli*. Biochemistry **20:**4178–4185.

197. **Poulson, R., K. J. Whitlow, and W. J. Polglase.** 1976. Catabolite repression of protoporphyrin IX biosynthesis in *Escherichia coli* K12. FEMS Lett. **62:**351–353.

198. **Pudek, M. R., and P. D. Bragg.** 1974. Inhibition by cyanide of the respiratory chain oxidases of *Escherichia coli*. Arch. Biochem. Biophys. **164:**682–693.

199. **Pudek, M. R., and P. D. Bragg.** 1976. Trapping of an intermediate in the oxidation-reduction cycle of cytochrome d in *Escherichia coli*. FEBS Lett. **62:**330–333.

200. **Pudek, M. R., and P. D. Bragg.** 1976. Redox potentials of the cytochromes in the respiratory chain of aerobically-grown *Escherichia coli*. Arch. Biochem. Biophys. **174:**546–552.

201. **Raunio, R. P., and W. T. Jenkins.** 1973. D-Alanine oxidase from *Escherichia coli*: localization and induction by L-alanine. J. Bacteriol. **115:**560–566.

202. **Reddy, T. L. P., and R. W. Hendler.** 1978. Reconstitution of *Escherichia coli* succinoxidase from soluble components. J. Biol. Chem. **253:**7972–7979.

203. **Reeves, J. P.** 1971. Transient pH changes during D-lactate oxidation by membrane vesicles. Biochem. Biophys. Res. Commun. **45:**931–936.

204. **Reeves, J. P., J. S. Hong, and H. R. Kaback.** 1973. Reconstitution of D-lactate-dependent transport in membrane vesicles from a D-lactate dehydrogenase mutant of *Escherichia coli*. Proc. Natl. Acad. Sci. USA **70:**1917–1921.

205. **Reid, G. A., B. A. Haddock, and W. J. Ingledew.** 1981. Assembly

of functional b-type cytochromes in membranes from a 5-aminolaevulinic acid-requiring mutant of *Escherichia coli*. FEBS Lett. **131**:346–351.

206. **Reid, G. A., and W. J. Ingledew.** 1979. Characterization and phenotypic control of the cytochrome content of *Escherichia coli*. Biochem. J. **182**:465–472.

207. **Reid, G. A., and W. J. Ingledew.** 1980. The purification of a respiratory oxidase complex from *Escherichia coli*. FEBS Lett. **109**:1–4.

208. **Rice, C. W., and W. P. Hempfling.** 1978. Oxygen-limited continuous culture and respiratory energy conservation in *Escherichia coli*. J. Bacteriol. **134**:115–124.

209. **Rich, P. R.** 1982. The organization of the quinone pool. Biochem. Soc. Trans. **10**:482–484.

210. **Richter, A. E., and P. C. Loewen.** 1981. Induction of catalase in *Escherichia coli* by ascorbic acid involves hydrogen peroxide. Biochem. Biophys. Res. Commun. **100**:1039–1046.

211. **Ruiz-Herrera, J., and L. G. Garcia.** 1972. Regulation of succinate dehydrogenase in *Escherichia coli*. J. Gen. Microbiol. **7**:29–35.

212. **Rule, G. S., E. A. Pratt, C. C. Q. Chin, F. Wold, and C. Ho.** 1985. Overproduction and nucleotide sequence of the respiratory D-lactate dehydrogenase of *Escherichia coli*. J. Bacteriol. **161**:1059–1068.

213. **Santos, E., H.-F. Kung, I. G. Young, and H. R. Kaback.** 1982. *In vitro* synthesis of the membrane-bound D-lactate dehydrogenase of *Escherichia coli*. Biochemistry **21**:2085–2091.

214. **Saunders, B. C., A. G. Holmes, and B. P. Stark.** 1964. Peroxidase. The properties and uses of a versatile enzyme and of some related catalysts. Butterworth & Co. (Publishers) Ltd., London.

215. **Savage, D. C.** 1977. Microbial ecology of the gastrointestinal tract. Annu. Rev. Microbiol. **31**:107–133.

216. **Scott, R., R. K. Poole, and B. Chance.** 1981. Respiratory biogenesis during the cell cycle of aerobically-grown *Escherichia coli* K12. The accumulation and ligand binding of cytochrome *o*. J. Gen. Microbiol. **122**:255–261.

217. **Scott, R. I., and R. K. Poole.** 1982. A re-examination of the cytochromes of *Escherichia coli* using fourth-order finite difference analysis: their characterization under different growth conditions and accumulation during the cell cycle. J. Gen. Microbiol. **128**:1685–1696.

218. **Shaw, D. J., and J. R. Guest.** 1982. Amplification and product identification of the *fnr* gene of *Escherichia coli*. J. Gen. Microbiol. **128**:2221–2228.

219. **Shipp, W. S.** 1972. Cytochromes of *Escherichia coli*. Arch. Biochem. Biophys. **150**:459–472.

220. **Short, S. A., H. R. Kaback, T. Hawkins, and L. D. Kohn.** 1975. Immunochemical properties of the membrane-bound D-lactate dehydrogenase from *Escherichia coli*. J. Biol. Chem. **250**:4285–4290.

221. **Short, S. A., H. R. Kaback, and L. D. Kohn.** 1975. Localization of D-lactate dehydrogenase in native and reconstituted *Escherichia coli* membrane vesicles. J. Biol. Chem. **250**:4291–4296.

222. **Simoni, R. D., and M. K. Shallenberger.** 1972. Coupling of energy to active transport of amino acids in *Escherichia coli*. Proc. Natl. Acad. Sci. USA **69**:2663–2667.

223. **Singh, A. P., and P. D. Bragg.** 1975. Reduced nicotinamide adenine dinucleotide-dependent reduction of fumarate coupled to membrane energization in a cytochrome-deficient mutant of *Escherichia coli* K12. Biochim. Biophys. Acta **396**:229–241.

224. **Smith, P. H., and R. E. Hungate.** 1958. Isolation and characterization of a *Methanobacterium ruminantium* n. sp. J. Bacteriol. **75**:713–718.

225. **Smith, M. W., and F. C. Neidhardt.** 1983. Proteins induced by anaerobiosis in *Escherichia coli*. J. Bacteriol. **154**:336–343.

226. **Smith, M. W., and F. C. Neidhardt.** 1983. Proteins induced by aerobiosis in *Escherichia coli*. J. Bacteriol. **154**:344–350.

227. **Smyth, C. J., J. Siegel, M. R. I. Salton, and P. Owen.** 1978. Immunochemical analysis of inner and outer membranes of *Escherichia coli* by crossed immunoelectrophoresis. J. Bacteriol. **133**:306–319.

228. **Spencer, M. E., and J. R. Guest.** 1973. Isolation and properties of fumarate reductase mutants of *Escherichia coli*. J. Bacteriol. **114**:563–570.

229. **Spencer, M. E., and J. R. Guest.** 1982. Molecular cloning of four tricarboxylic acid cycle genes of *Escherichia coli*. J. Bacteriol. **151**:542–552.

230. **Sweetman, A. J., and D. E. Griffiths.** 1971. Studies on energy-linked reduction of oxidized nicotinamide-adenine dinucleotide by succinate in *Escherichia coli*. Biochem. J. **121**:117–124.

231. **Tanaka, Y., Y. Anraku, and M. Futai.** 1976. *Escherichia coli* membrane D-lactate dehydrogenase, isolation of the enzyme in

aggregated form and its activation by Triton X-100 and phospholipids. J. Biochem. **80**:821–830.

232. **Tarmy, E. M., and N. O. Kaplan.** 1968. Chemical characterization of D-lactate dehydrogenase from *Escherichia coli* B. J. Biol. Chem. **243**:2579–2586.

233. **Tarmy, E. M., and N. O. Kaplan.** 1968. Kinetics of *Escherichia coli* B D-lactate dehydrogenase and evidence for pyruvate-controlled change in conformation. J. Biol. Chem. **243**:2587–2596.

234. **Thomson, A. J.** 1977. Ligand binding properties of the haem group. Nature (London) **256**:15–16.

235. **Trutko, S. M., N. P. Golovchenko, and V. K. Akhimenko.** 1978. Changes in the electron transport chain in *Escherichia coli* as a function of the conditions of culturing and phase of growth. Mikrobiologiya **47**:1–5.

236. **Uno, T., Y. Nishimura, M. Tsuboi, K. Kita, and Y. Anraku.** 1985. Resonance Raman study of cytochrome b_{562}-o complex, a terminal oxidase of *Escherichia coli* in its ferric, ferrous, and CO-ligated states. J. Biol. Chem. **260**:6755–6760.

237. **Van der Plas, J., K. J. Hellingwerf, H. G. Seijen, J. R. Guest, J. H. Weiner, and W. N. Konings.** 1983. Identification and localization of enzymes of the fumarate reductase and nitrate respiration systems of *Escherichia coli* by crossed immuno-electrophoresis. J. Bacteriol. **153**:1027–1037.

238. **Van Wielink, J. E., L. F. Oltmann, F. J. Leeuwerik, J. A. de Hollander, and A. H. Stouthamer.** 1982. A method for *in situ* characterisation of *b*- and *c*-type cytochromes in *Escherichia coli* and in complex III from beef heart mitochondria by combined spectrum deconvolution and potentiometric analysis. Biochem. Biophys. Acta **681**:177–190.

239. **Van Wielink, J. E., W. N. M. Reijnders, L. F. Oltmann, and A. H. Stouthamer.** 1983. The characterization of the membrane-bound *b*- and *c*-type cytochromes of differently grown *Escherichia coli* cells by means of coupled potentiometric analysis and spectrum deconvolution. FEMS Microbiol. Lett. **18**:167–172.

240. **Visser, R. G. F., K. J. Hellingwerf, and W. N. Konings.** 1984. The protein composition of the cytoplasmic membrane of aerobically and anaerobically grown *Escherichia coli*. J. Bioenerg. Biomembr. **16**:295–307.

241. **Wallace, B. J., and I. G. Young.** 1977. Role of quinones in electron transport to oxygen and nitrate in *Escherichia coli*. Studies with a ubiA⁻ menA⁻ double quinone mutant. Biochem. Biophys. Acta **461**:84–100.

242. **Weiner, J. H., and L. A. Heppel.** 1972. Purification of the membrane-bound and pyridine nucleotide independent L glycerol-3-phosphate dehydrogenase from *Escherichia coli*. Biochem. Biophys. Res. Commun. **47**:1360–1365.

243. **Wikström, M., K. Krab, and M. Saraste.** 1981. Cytochrome oxidase. A synthesis. Academic Press, Inc. (London), Ltd., London.

244. **Wild, J., and T. Klopotowski.** 1981. D-Amino acid dehydrogenase of *Escherichia coli* K12: positive selection of mutants defective in enzyme activity and localization of the structural gene. Mol. Gen. Genet. **181**:373–378.

245. **Wood, P. M.** 1984. Bacterial proteins with CO-binding *b* or *c*-type haem: functions and absorption spectroscopy. Biochim. Biophys. Acta **768**:293–317.

246. **Xavier, A. V., E. W. Czerwinski, P. H. Bethge, and F. S. Mathews.** 1978. Identification of the haem ligands of cytochrome b_{562} by X-ray and NMR methods. Nature (London) **295**:245–247.

247. **Yariv, J.** 1983. The identity of bacterioferritin and cytochrome b_1. Biochem. J. **211**:527.

248. **Yoshpe-Purer, Y., Y. Henis, and J. Yashphe.** 1977. Regulation of catalase level in *Escherichia coli* K12. Can. J. Microbiol. **23**:84–91.

249. **Young, I. G., A. Jaworowski, and H. I. Poulis.** 1978. Amplification of the respiratory NADH dehydrogenase of *Escherichia coli* by gene cloning. Gene **4**:25–36.

250. **Young, I. G., A. Jaworowski, and M. Poulis.** 1982. Cloning of the gene for the respiratory D-lactate dehydrogenase of *Escherichia coli*. Biochemistry **21**:2092–2095.

251. **Young, I. G., B. L. Rogers, H. D. Campbell, A. Jaworowski, and D. C. Shaw.** 1981. Nucleotide sequence coding for the respiratory NADH dehydrogenase of *Escherichia coli*. Eur. J. Biochem. **116**:165–170.

252. **Young, I. G., and B. J. Wallace.** 1976. Mutations affecting the reduced nicotinamide adenine dinucleotide dehydrogenase complex of *Escherichia coli*. Biochim. Biophys. Acta **449**:376–385.

253. **Zahl, K. J., C. Rose, and R. L. Hanson.** 1978. Isolation and partial characterization of a mutant of *Escherichia coli* lacking pyridine nucleotide transhydrogenase. Arch. Biochem. Biophys. **190**:598–602.

16. Pathways for Anaerobic Electron Transport

E. C. C. LIN[1] AND D. R. KURITZKES[2]

Department of Microbiology and Molecular Genetics, Harvard Medical School, Boston, Massachusetts 02115,[1] and Infectious Disease Unit, Massachusetts General Hospital, Boston, Massachusetts 02114[2]

INTRODUCTION

Many substrate oxidations are exploited by the cell to generate metabolic energy. A wealth of different mechanisms permit the utilization of diverse sources of potential energy. These mechanisms fall into two broad categories: fermentative reactions that occur in solution and respiratory reactions that are membrane associated and vectorial. The central goal of respiration is to obtain reducing power with which to energize the proton-translocating systems. For this purpose there is no better terminal electron acceptor than O_2, which functions through the aerobic electron transport chain (chapter 15).

As will be made apparent from the different redox systems described, electron transfer reactions are initiated by specific modular units. The electrons are then collected by the quinone pool, which acts as a universal adaptor. From that pool, the electrons are passed on to the appropriate acceptor module. With respect to the nomenclature used in different fields of research, it is useful to note that students of aerobic respiration, especially those who work with mitochondria, customarily see the last enzyme in the electron transport chain as the agent mediating the final step in the oxidative process and hence refer to that enzyme as the terminal oxidase. In contrast, the bacterial physiologists working with anaerobic respiration generally see the last enzyme in the electron transport chain as the agent by which the terminal electron acceptor gets reduced and hence refer to that enzyme as the terminal reductase. Two devices serve to channel the electron flow for maximal energetic benefit to the organism: the choice of carriers with the appropriate redox potential and physiological controls. The physiological controls include regulation at both the level of protein synthesis and the level of catalytic activity. In each module, coenzymes and metal ions play the part of electron transmitters; the lipid microenvironment serves as the insulator. In this way, electrons may pass along a series of cascading redox potentials, with large potential drops occurring only at steps in which coupling to proton translocation is feasible.

A schematic presentation of the anaerobic redox chains that have been studied is shown in Fig. 1. A primary dehydrogenase module typically comprises a proximal electron acceptor (Mo, Se, or flavin adenine dinucleotide [FAD]), a nonheme iron-sulfur protein cluster of relatively negative potential, and a cytochrome b of relatively negative potential. A terminal reductase module typically contains a cytochrome b of relatively positive potential, a nonheme iron-sulfur protein cluster of relatively positive potential, and the cofactor(s) Mo, Se, or FAD. The functional grouping of each of the branches radiating from the quinone pool predicts that the genes coding for the protein elements of each limb evolved into an operon or regulon that is inducible by the primary donor or terminal acceptor. For example, glycerol 3-phosphate induces its chain consisting of the flavoprotein and nonheme iron-sulfur protein subunits of anaerobic glycerol-3-phosphate dehydrogenase (a cytochrome b is postulated for this chain but has not been demonstrated), whereas nitrate induces its

chain consisting of the molybdenum-iron-sulfur protein subunits and cytochrome b_{556}^{NR}.

FORMATE DEHYDROGENATION

Formate derived anaerobically from pyruvate by the action of pyruvate formate-lyase or from an exogenous source can be oxidized to CO_2 in two ways. The fermentative route disposes of the formate by passing electrons from it to protons, with the formation of hydrogen gas. This process is not directly coupled to proton extrusion. The respiratory route disposes of formate by passing electrons from it to the quinone pool. This process energizes the cell by proton translocation across the cytoplasmic membrane.

Escherichia coli

The formate hydrogenlyase system

The fermentative production of CO_2 and H_2 from formate (and indirectly from glucose) was described for *E. coli* in 1901 (192, 326). The process, illustrated in equation 1, depends on the action of a formate hydrogenlyase system (201, 323, 412, 424, 456) which to this day has not been defined in full:

$$HCOO^- + H^+ \underset{FDH_H + \text{hydrogenase}}{\overset{\substack{\text{Formate hydrogenlyase} \\ \text{complex}}}{\rightleftarrows}} CO_2 + H_2 \quad (1)$$

In addition to a formate dehydrogenase associated with hydrogen production, FDH_H, and a hydrogenase, two intermediate components are postulated (145, 171, 336). This would explain why the complete system is more labile than either FDH_H or hydrogenase alone and why the system has not yet been constituted in vitro (176). Although iron (144) and a soluble factor (21) seem to be required, neither a cytochrome nor a quinone has been implicated (364, 442). The reaction catalyzed by the formate hydrogenlyase system is reversible, a feature that accounts for the equilibration of the deuterium in heavy water with the gas (125, 460).

Insofar as the disposal of a formate molecule by its decomposition, instead of by its excretion, consumes an intracellular proton, the formate hydrogenlyase system can be considered the most primitive anaerobic respiratory pathway that creates a proton gradient across the cell membrane as a result of a dehydrogenation. Energization of the membrane, however, is limited by the fact that the sum of the redox potentials for equation 1 at pH 7 under standard physiochemical conditions is zero (10). Nonetheless, under experimental conditions that allowed the escape of H_2 and CO_2 as gases, exogenous formate was shown to help the colonies maintain a higher local pH on agar with phenol red as indicator (36). The formate hydrogenlyase system is induced by formate (in rich broth medium) and repressed anaerobically by nitrate or aerobically (39, 82, 336, 337, 343, 364, 413, 414, 417, 452, 468). In the absence of exogenous formate, endogenous induction depends on the presence of pyruvate formate-lyase, the synthesis of which, in turn, depends on the fnr^+ (also known as $nirA$ or $nirR$) product, a pleiotropic activator protein for genes encoding anaerobic redox systems (111, 383).

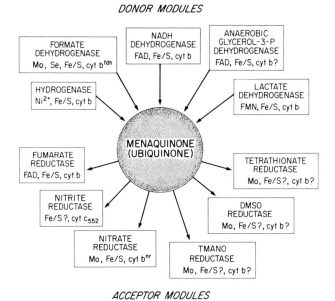

DONOR MODULES

ACCEPTOR MODULES

FIG. 1. Organization of anaerobic respiratory modules in *E. coli* and *S. typhimurium*. Fe/S, Nonheme iron-sulfur protein; cyt b, cytochrome *b*; TMANO, trimethylamine *N*-oxide; DMSO, dimethyl sulfoxide. Tetrathionate reductase has not been reported for *E. coli*, and dimethyl sulfoxide reductase has not been reported for *S. typhimurium*. Scheme adapted from that of Wilson and Lin (450).

Synthesis of the formate hydrogenlyase system is also directly or indirectly dependent on cyclic AMP (cAMP) (332).

FDH_H. The fermentative formate dehydrogenase of the formate hydrogenlyase system reacts with benzyl viologen, which has an $E_0' = -360$ mV, as an artificial electron acceptor. The enzyme does not react with methylene blue, which has an $E_0' = +11$ mV, or with phenazine methosulfate, which has an $E_0' = -145$ mM (336). Selenium and the molybdate cofactor are required for activity (156, 271, 397). The enzyme, which includes a selenopeptide of 80,000 daltons, has a K_m of about 1.5 mM for formate (98, 365). Chlorate resistance mutations affecting the synthesis or attachment of the Mo cofactor abolish FDH_H activity, as well as the activity of respiratory FDH (FDH_N), described below (322, 341, 366).

The *fdhF* gene at min 92.4 encodes the 80,000-dalton subunit (333). FDH_H activity is anaerobically induced by formate and repressed by nitrate (82, 336, 365). A high concentration of formate reverses nitrate repression but not aerobic repression (334).

Hydrogenases. Hydrogenase activity, first demonstrated by Stephenson and Stickland in 1931 (412), catalyzes the reaction shown in equation 2, in which methylene blue or benzyl viologen can serve as an artificial electron acceptor (142, 174, 228, 255, 335):

$$H_2 \leftrightarrow 2H^+ + 2e^- \qquad (2)$$

H_2 is believed to be activated on the cytoplasmic surface of the enzyme (230). Since knowledge of the number of hydrogenase complexes (some of which might share a subunit) and their genetic regulation is still quite fragmentary and uncertain, the summary

given below must be regarded as tentative and partly conjectural.

Although several isoenzymes with hydrogenase activity have been detected (6, 10, 464), >90% of the in vitro activity of cells grown anaerobically on glucose is attributed to a fermentative hydrogenase (isoenzyme 3) associated with formate hydrogenlyase (383, 471). The enzyme appears to require both Ni and Fe for activity (138, 471).

Four genes in the min 58–59 region appear to be concerned with the formate hydrogenlyase system. Products of the *hydA* and *hydB* operons are necessary for functional hydrogenase activity. The *fhl* gene is believed to encode an electron coupling protein of the formate hydrogenlyase system, and the *fdv* gene is believed to be a structural gene for FDH_H. Mutants affected in the formate hydrogenlyase system fail to produce hydrogen gas during anaerobic growth. The anaerobic growth impairment is not manifest when an exogenous electron acceptor, such as nitrate, is available (159, 160, 168, 240, 267, 330, 379). A different gene in this region, *ant* (for anaerobic electron transport), seems also to be part of the formate hydrogenlyase system (466).

The hydrogenase is specifically induced by formate. Enzyme synthesis is not strongly dependent on the *fnr*$^+$ gene product (334, 383). Aerobic growth, anaerobic growth in the presence of nitrate, or growth in an environment of high redox potential has strong repressive effects (39, 82, 173, 337, 451, 452). The anaerobic repression by nitrate is antagonized by formate (334). *cis*-Dominant regulatory mutations can render the expression of a *hyd* structural gene resistant to anaerobic repression by nitrate. In these mutants, but not in the wild-type strain, aerobic repression can be overcome by 30 mM formate. Full derepression still depends on the *fnr*$^+$ product (471).

A respiratory hydrogenase (isoenzyme 2), the synthesis of which is strongly dependent on the *fnr*$^+$ gene product and enhanced by growth on H_2 plus fumarate or on glycerol plus fumarate, is probably responsible for the utilization of H_2 as the electron donor for fumarate reduction (253, 265, 383). This enzyme is a dimer of a polypeptide of about 56,000 daltons, containing Ni^{2+} and iron-sulfur groups (8, 29, 168). Deficiency in this enzyme may account for the phenotype of mutants that fail to grow anaerobically on H_2 plus fumarate. Such mutants retain the ability to grow on glycerol plus fumarate and have a normal level of hydrogenase activity (presumably attributable to isoenzyme 3) when grown anaerobically on glucose (254). Mutations at min 65 impair the H_2-fumarate redox reaction without affecting formate hydrogenlyase activity (267).

Another Ni-containing hydrogenase (isoenzyme 1), with a polypeptide subunit of 64,000 daltons, has been identified. Its synthesis is dependent on the *fnr*$^+$ gene product and inducible by formate, but the function of this protein remains to be defined (29, 162, 383). Anaerobic growth in the presence of nitrate appears to repress the synthesis of isoenzyme 1 but merely to inactivate isoenzyme 2 (383).

FDH_N

Biochemistry and enzymology. The FDH associated with nitrate reduction, FDH_N, can be assayed by reduction of methylene blue or phenazine methosulfate, but not by reduction of benzyl viologen (139, 336, 418). The site of formate oxidation is on the cytoplasmic side with a K_m of about 30 μM (140, 365). FDH_H has been resolved from FDH_H. The two dehydrogenases are also immunochemically distinct (152).

FDH_N complex contains Se, Mo cofactor, heme, and nonheme iron-sulfur protein and consists of three proteins that can be purified as a unit (53, 121, 123, 207, 271, 342, 390, 397). Subunit A is a selenopeptide of 110,000 daltons (98, 123). Subunit B is a protein of 32,000 daltons (123). Subunit C is generally considered to be a cytochrome b (cytochrome b^{fdh}) of 20,000 daltons with an E_0' of about −110 mM (48, 123, 188, 189, 209, 366, 375). Subunits A and B span the cytoplasmic membrane (48, 167, 170). In the absence of Mo, either the synthesis or the stability of the enzyme is impaired (150).

The dehydrogenation of formate to CO_2, coupled to the reduction of nitrate to nitrite, is mediated by a quinone pool. The FDH_N-quinone system provides a proton-translocating site extruding about 1.6 protons per 2 electrons (232).

Genetics and regulation. Mutations in *fdhA* at min 80 result in loss of FDH_N, its associated cytochrome b, and FDH_H (28, 169, 189, 191, 291). Mutations in *fdhB* at min 38, in *fdhC* at min 82, and in *fdhD* at min 87 also result in the absence of activity of both FDHs (191, 291). In contrast, mutations in *fdhE* at min 87 result in the loss of FDH_N with retention of FDH_H (191, 238, 366).

Mutations in *chlA*, *chlB*, and *chlE* prevent the formation of active FDH_N, as well as that of active nitrate reductase. Although *chlA* and *chlB* mutants produce the apoproteins of nitrate reductase, they are deficient in the apoproteins of FDH_N (169, 189). Mixing the soluble fractions extracted from the *chl* mutants partially reconstitutes the FDH_N activity associated with the membrane (431). Although *chlG* mutants lack nitrate reductase, they retain FDH_N activity. This finding suggests that the processing of the Mo cofactor is different for the two enzymes (191). In a mutant blocked in ubiquinone biosynthesis, the aerobic level of FDH_N was virtually abolished (149).

FDH_N is induced by nitrate and repressed by aerobic metabolism (83, 342, 364, 365, 368, 452). The presence of this enzyme in cells reported to be grown aerobically in rich glucose medium was probably the result of inadequate supply of oxygen in dense cultures (273, 461).

Salmonella typhimurium

Genetics and regulation of FDHs and hydrogenases

Mutations in *fhlD* at min 81 affect the FDH_H (68, 378). Mutations in *fhl* at min 93 abolish hydrogenase activity without reducing FDH_H activity (32). Mutations in *hyd* (*fhlB*, *fhlC*) at min 59 abolish hydrogenase activity and reduce the activity of FDH_H (32, 330, 378). Anaerobically, the formate hydrogenlyase system is best induced at pH 6 to pH 7 (419). The *hyd* and *fhl* genes are inducible anaerobically by formate. The basal anaerobic expression is reduced by the presence of nitrate, an effect antagonized by formate (32). Formate hydrogenlyase is subject to aerobic repression.

Mutations in *fdhA* at min 77, in *fdhB* at min 18, or in *fdhC* at min 80 affect both phenazine methosulfate- and benzyl viologen-linked FDHs (31, 66, 68, 329, 378). The activity of both enzymes depends on the $chlD^+$ product (61). Mutations in *fdnA* and *fdnB* at min 85 affect FDH_N but not FDH_H. The mutants retain the FDH_N protein but lack the cytochrome b associated with it (31, 34, 378).

GLYCEROL 3-PHOSPHATE

E. coli

Two glycerol-3-phosphate dehydrogenases

Two distinct membrane-bound flavoproteins of the *glp* regulon can serve as electron donors by converting glycerol 3-phosphate to dihydroxyacetone-phosphate (chapter 18). The enzyme complex specified by the *glpDEG* operon at 75.3 donates electrons aerobically to molecular oxygen or anaerobically to nitrate, whereas the enzyme complex specified by the *glpABC* operon at min 48.6 donates electrons anaerobically to nitrate or fumarate (100, 101, 248, 251, 260, 303; T. J. Larson, M. Ehrmann, and H. Schweizer, personal communication, 1986). The relative contributions of the two dehydrogenases to anaerobic electron transport from glycerol 3-phosphate to nitrate in wild-type cells remain to be determined.

Aerobic glycerol-3-phosphate dehydrogenase has an apparent substrate K_m in the range of 2 mM. A purified protein (a dimer of a 58,000-dalton subunit) has been found to contain noncovalently bound FAD. This catalytically active protein is probably encoded by the *D* gene (193, 359, 386, 444, 447). The *G* gene might encode a nonheme iron-sulfur protein subunit, and the gene *E* might encode an anchor subunit. Anaerobic glycerol-3-phosphate dehydrogenase has an apparent substrate K_m in the range of 0.3 mM (in the presence of flavin mononucleotide as an activator). Gene *A* probably encodes the 62,000-dalton subunit binding FAD. Gene *B* might encode a subunit with membrane-anchoring function. Gene *C* might encode the 43,000-dalton subunit (probably a nonheme iron-sulfur protein) isolated in a 1:1 molar ratio with the gene *A* product. The catalytic component of the anaerobic dehydrogenase is less tightly attached to the plasma membrane than that of the aerobic dehydrogenase (247, 249, 260, 387, 388; Larson et al., personal communication, 1986).

Regulation

The level of aerobic glycerol-3-phosphate dehydrogenase activity in aerobically grown repressor-negative (*glpR*) cells is about three times higher than that in anaerobically grown cells. Nitrate increases the anaerobic level twofold (133).

The level of anaerobic glycerol-3-phosphate dehydrogenase activity in aerobically grown *glpR* cells is about five times lower than that in anaerobically grown cells. Nitrate decreases the anaerobic level more than fourfold. Fumarate, which has no significant effect on aerobic glycerol-3-phosphate dehydrogenase level, increases the anaerobic glycerol-3-phosphate dehydrogenase level about twofold (133; S.

Iuchi, unpublished data). Aerobic induction of anaerobic glycerol-3-phosphate dehydrogenase can be increased three- to fourfold by placing 0.15 mM KCN in the growth medium or by blocking heme synthesis. A mutation in *fnr*, in contrast, lowers the induced level threefold (260).

Although anaerobic glycerol-3-phosphate dehydrogenase is stable aerobically, it cannot function in vivo. Growth of *glpD⁻A⁺* cells in a glycerol-nitrate medium ceases immediately upon a shift from anaerobiosis to aerobiosis (133).

OTHER DONOR SYSTEMS

NADH, D-lactate, and L-lactate flavodehydrogenases also serve as electron donors to most of the anaerobic electron transport chains (chapter 15).

FUMARATE REDUCTASE

E. coli

Fumarate as hydrogen acceptor

In 1924, Quastel and Whetham (353) noted that resting cells of *Bacillus coli* oxidize *leuco*-methylene blue in the presence of fumarate. Soon afterwards, it was realized that fumarate can replace oxygen or nitrate as a terminal hydrogen acceptor for respiration and for anaerobic growth on glycerol or lactate (256, 351, 352, 354, 355). Fumarate has since been shown to serve also as a terminal electron acceptor for growth on formate and molecular hydrogen (38, 265, 289, 464). In addition, fumarate appears to function as a hydrogen acceptor in the anaerobic biosynthesis of uracil (16, 317) and protoporphyrin (213–217).

Enzymology

Both fumarate reductase and succinate dehydrogenase catalyze the interconversion of the two dicarboxylates. In accordance with their physiological roles, however, the reducing enzyme has a lower apparent K_m for fumarate than for succinate, and the reverse is true for the oxidizing enzyme (198). The purified reductase has an apparent K_m of 0.4 mM for fumarate, but the activity is dependent on the nature of the anionic environment (110, 275, 360). Nonetheless, when fumarate reductase is overproduced as a result of gene amplification, it can partially replace succinate dehydrogenase for aerobic growth on succinate (181). During anaerobic growth on glucose, succinate excretion is increased (161).

Fumarate reductase is a membrane-associated enzyme and has its active site facing the cytoplasm (234, 300, 405, 410, 434). Four kinds of subunits contribute to the complex at an equimolar ratio: subunit A of 69,000 daltons, with FAD attached to the polypeptide by an 8-[*N*(3)histidyl] linkage; subunit B of 25,000 daltons, with nonheme iron-sulfur protein clusters of the ferredoxin type; the anchor subunit C of 15,000 daltons; and the anchor subunit D of 13,000 daltons (85, 110, 177, 270, 361, 404, 446, 449). In addition to the [2Fe-2S] cluster in subunit B, a high-potential 3Fe center has been demonstrated but not yet located (204, 227, 306).

A cytochrome *b* may provide additional membrane anchorage to fumarate reductase, although no direct evidence is yet available. Two other cytochrome-associated membrane proteins were shown to accumulate in the cytoplasmic fraction of cells blocked in heme synthesis. In *hemA* mutants of *Bacillus subtilis*, increasing amounts of succinate dehydrogenase antigen accumulate in the cytoplasm when the cells are grown in medium without δ-aminolevulinate. Upon addition of the heme precursor, the antigen shifts to the membrane-bound form (200). Similar results were obtained with nitrate reductase in *hemA* mutants of *E. coli* (279, 281). Given the homologies between fumarate reductase and succinate dehydrogenase, it is likely that cytochromes play a similar role in the membrane binding of Frd proteins.

Genetics and regulation

The structural genes. The *frdABCD* operon encoding this complex is transcribed counterclockwise and located at min 94.4 (89, 90, 409). A number of features of the products can be inferred from the base sequence data. The *A* gene encodes an FAD subunit of 602 amino acids (44% polar) which is synthesized without a membrane signal sequence (84, 89, 90, 275). The *B* gene encodes a nonheme iron-sulfur protein subunit of 244 amino acid residues (43% polar). There is no stretch of hydrophobic residues sufficiently long to pass through a lipid bilayer (86). The *frdC* and *frdD* genes encode proteins rich in hydrophobic amino acids, and the disposition of the hydrophobic stretches allows each polypeptide to traverse the lipid bilayer three times (177, 449).

Coding economy of the *frdABCD* operon and overlap with *ampC*. The *frd* region of the chromosome provides a remarkable illustration of sequence economy in the spacing of individual genes. There is partial overlap between the *frdA* and *frdB* cistrons, as well as overlap of the *frdC* and *frdD* ribosome-binding sites with the *ochre* translation termination codons of the preceding coding sequences (86, 91). Surprisingly, the promoter for *ampC* (the next contiguous gene on the chromosome) overlaps the C-terminal coding sequence of the *frdD* gene. Furthermore, the *ampC* attenuator also serves as terminator for the *frd* operon (37, 85, 177). It will be challenging to learn what evolutionary force brought these two sets of genes with different functions into this interlocking position.

Homology with succinate dehydrogenase. Fumarate reductase shares considerable homology with succinate dehydrogenase of *E. coli* and other organisms (194, 205). The *frdABCD* and *sdhCDAB* operons show striking evolutionary kinship. Within homologous stretches of the *A* genes totaling 563 codons, 246 (44%) are identical, and 112 (20%) are conservative. The 10 conserved positions for cysteine in the *B* genes are arranged in three clusters, two of which resemble the nonheme iron-sulfur protein centers of several bacterial ferredoxins. The *C* genes resemble each other in size and composition but not in sequence. The same is true for the *D* genes (86, 103, 458). Similarity in overall amino acid composition (44%) and polarity (43%) can be discerned even between the FAD subunits of *E. coli* fumarate reductase and beef heart succinate dehydrogenase. A sequence of nine amino acid residues surrounding one of the histidyl residues of the *E. coli* FAD subunit is identical to the sequence surrounding the FAD-bearing histidyl residue of the corresponding subunit of the mitochondrial enzyme (84). Three nonheme iron-sulfur protein centers are present in succinate dehydrogenase of beef heart mitochondria: two Fe_2S_2 and one Fe_4S_4 per mole of flavin (95). The catalytic subunit of the beef heart enzyme copurifies with two other subunits with membrane-anchoring function and a cytochrome *b* (7).

Overproduction of fumarate reductase. The physical connection of *frd* and *ampC* offers a convenient way of isolating mutants amplified in *frd*, since concomitant duplication of the *ampC* gene confers ampicillin resistance (87, 320, 321). Tandem duplications of this sort may provide the cell with up to 80 copies of *frd* per genome. The hypersynthesis of fumarate reductase during anaerobic growth is associated with the appearance of a soluble form of the enzyme (88). These results suggest an upper limit to the number of membrane sites available for Frd anchorage.

Fumarate reductase levels are increased over 30-fold when *frdABCD* is cloned into a pBR322 plasmid. The anaerobic growth rate on glycerol and fumarate, however, is not changed. The inner surface of the plasma membrane becomes covered with knoblike structures (4 nm in diameter), and tubular membranous structures (25 nm in diameter) pack the cell. The membrane also becomes enriched in cardiolipin. The large increase in fumarate reductase does not significantly reduce the membrane attachment of aerobic glycerol-3-phosphate dehydrogenase and D-lactate dehydrogenase (269, 448). Aerobic glycerol-3-phosphate dehydrogenase, D-lactate dehydrogenase, and L-lactate dehydrogenase, on the other hand, seem to compete with each other for membrane attachment but not with succinate dehydrogenase (259). Perhaps the extra subunit encoded by *sdhCDAB* and *frdABCD*, missing in *glpDEG*- and *glpABC*-encoded complexes (and presumably also in the complexes of the two lactate dehydrogenases), is responsible for avid membrane binding. Consequently, the glycerol-3-phosphate and lactate dehydrogenases have to compete for a common anchor protein.

Expression of *frd*. The operon is transcribed as a single unit (229). Derepression occurs in cells grown anaerobically on a variety of carbon sources in the absence of hydrogen acceptors which are more redox positive than fumarate ($E_0' = +30$ mV). Under anaerobic conditions, exogenous fumarate further induces the synthesis of the reductase (210). Growth in the presence of molecular oxygen ($E_0' = +816$ mV) or nitrate ($E_0' = +420$ mV) represses enzyme synthesis (83, 173, 198, 409, 452). In a mutant disrupted in *narL* (encoding the transcriptional activator of the structural genes for nitrate reductase), the presence of nitrate no longer prevents the induction by fumarate. It appears that the *narL* protein, when combined with nitrate and Mo-X (see *chlD* mutation in chlorate resistance, below), acts as a repressor of *frdABCD* (S. Iuchi and E. C. C. Lin, Abstr. Annu. Meet. Am. Soc. Microbiol. 1986, H-80, p. 140; S. Iuchi and E. C. C. Lin, unpublished data). The increased basal enzyme level associated with multicopies of the *frd* region was

interpreted to result from titration of a specific repressor encoded by an unlinked gene (87, 90, 181). However, control by a closely linked activator gene might give the same result. Full expression of *frdABCD* depends on activation by the pleiotropic regulatory protein encoded by *fnr*+ (264). In addition, cAMP, but not its receptor protein, is required (432).

Intermediate components of the electron transport chain

Menaquinone. The products of at least five genes (*menA* at min 88.5 and *men(CEB)D* at min 49) are necessary for menaquinone synthesis (179, 180, 393, 467) (chapter 15). With fumarate as the hydrogen acceptor, strains defective in menaquinone biosynthesis are unable to grow anaerobically on compounds such as glycerol, lactate, and formate. Since such mutants also fail to dehydrogenate dihydroorotate, they require uracil for anaerobic growth on fermentable sugars. Yet growth of fumarate reductase mutants (including those sustaining deletions) on fermentable sugars did not require uracil supplementation (96, 180, 264, 317; S. Iuchi, personal communication). Thus, fumarate is not an obligatory electron acceptor for uracil synthesis. Growth of *men* mutants on glycerol with nitrate or oxygen as terminal electron acceptor remains unimpaired. Ubiquinone, which supports anaerobic electron transport to nitrate (442), cannot function with the fumarate respiratory chain because of its relatively high redox potential ($E_0' = +100$ mV) compared with fumarate ($E_0' = +30$ mV) (199, 258, 318).

Cytochromes. *E. coli* cells grown anaerobically on glycerol-fumarate medium synthesize two *b*-type cytochromes not present in aerobically grown cells (356). These cytochromes have alpha-band absorption maxima at 558 and 555 nm and midpoint potentials at pH 7 of +250 and +40 mV, respectively. Membrane vesicles of cytochrome-deficient strains are unable to form proton gradients or energize anaerobic amino acid transport with fumarate as electron acceptor. Formate-fumarate and NADH-fumarate oxidoreductase activities are also impaired in these vesicles (44, 407).

Electron donors for the fumarate reductase chain

H_2-coupled reduction of fumarate, which promotes anaerobic growth on peptone, can translocate two protons per two electrons (231, 233, 464). NADH can reduce fumarate, but the activity is low relative to the reactions with H_2, formate, or glycerol 3-phosphate as donor (44, 45, 300) and does not seem to require the participation of cytochromes (190, 264, 406).

The reduction of fumarate by formate, glycerol 3-phosphate, D-lactate, or endogenous substrates has been demonstrated to energize the membrane or result in net ATP synthesis (35, 45, 47, 182, 183, 197, 252, 301, 363, 407).

FDH_N can also mediate electron transfer from formate to fumarate reductase (463). The resulting growth increment was reported to be at least as great with fumarate as with nitrate (256). An electrochemical proton gradient of about 100 mV (inside negative) can be generated by the formate-fumarate system (195).

Anaerobic glycerol-3-phosphate dehydrogenase is physically complexed with fumarate reductase in such a way that membrane particles (vesicles) catalyze the coupling of glycerol 3-phosphate dehydrogenation to fumarate reduction without any added cofactor. This coupling activity is much more sensitive to low concentrations of detergents than are the individual activities of the two component enzymes (300). Up to two protons are extruded per fumarate molecule reduced by glycerol 3-phosphate (183, 184, 304). A mutation in the FrdC anchor subunit, which did not detach the FrdA and FrdB catalytic subunits from the membrane, disrupted the cytochrome-dependent electron flow to fumarate and prevented anaerobic growth on glycerol and fumarate. This mutation, however, did not affect the H_2-fumarate pathway for electron transport (445). In contrast, cells incapable of producing functional cytochromes (*hemA*) can grow slowly on glycerol and fumarate through a bypass that does not generate membrane energy. Growth of such cells, however, cannot occur on glycerol and nitrate (406).

NITRATE REDUCTASES

E. coli

Nitrate as hydrogen acceptor and nitrogen source

That oxidation of organic compounds coupled to reduction of nitrate could provide energy for anaerobic growth of microorganisms was perceived by Beijerinck in 1903 (see reference 352). Quastel et al. (352) observed that lactate, succinate, and glycerol afforded anaerobic growth with nitrate as electron acceptor. Nitrate is reduced to nitrite, which becomes growth inhibitory at 0.35%. Resting cells also reduce chlorate to chlorite, which is toxic at extremely low concentrations (352). Nitrate reduction is noncompetitively and reversibly inhibited by O_2, which appears to exert its effect mainly by diverting the electron flow from the substrates (224, 376, 418). Under appropriate anaerobic conditions, nitrate, nitrite, nitrous oxide, or hydroxylamine can be reduced to ammonia and thereby serve as a sole source of nitrogen (293, 294, 469; see also Nitrite Reductases). Aerobically, nitrate cannot serve as a source of nitrogen, although nitrite can do so (250). Cells grown anaerobically with nitrate accumulate elevated concentrations of nitrite but not nitrate (455). The mechanism of nitrate transport has not yet been established (140, 237, 257, 377).

Respiratory nitrate reductase

The enzyme complex catalyzing nitrate reduction is bound to the cytoplasmic membrane (202, 429, 462). The K_m for nitrate is about 0.3 mM (384). The complex consists of three protein subunits with Mo, heme, and nonheme iron-sulfur protein as catalytic groups (42, 51, 129, 141, 203, 226, 271, 390, 430). As a side reaction, the enzyme appears to reduce nitrite to nitrous oxide (408).

The catalytic subunit A (about 145,000 daltons) bears the Mo cofactor and nonheme iron (63, 109, 122, 123, 131, 164, 278, 288).

The physiological role of subunit B (60,000 daltons) is not yet defined (122, 123, 278, 288). Subunit C (20,000 daltons) has been identified as cytochrome b_{556}^{NR} (64, 72, 118, 122, 123, 279, 366, 373, 375, 380, 381, 385). Two heme components, however, have been detected spectrophotometrically: one with a midpoint redox potential of +20 mV and the other with a midpoint redox potential of about +120 mV (187, 188).

Subunits A and B are synthesized in precursor forms A′ and B′, respectively. A′ and B′ appear to be reversibly acylated with a fatty acid by a membrane-associated enzyme. The modified complex AB associates with two subunits C in the plasma membrane to become the holoenzyme $A_2B_2C_4$ (62, 63, 73, 74, 106, 108, 122, 148, 186, 276, 279, 288). Membrane assembly of the holoenzyme is impaired in a *hemA* or *chlE* mutant (281, 284). A highly purified preparation of the enzyme (presumably the AB complex) was reported to contain residues of glucose and glucosamine (132). Subunits A and B are accessible at the cytoplasmic face of the membrane, although subunit B might be partly buried in the membrane. Subunit C, in contrast, spans the membrane (48, 49, 163, 165, 166, 245, 283).

Molybdenum cofactor

Molybdenum is present in subunit A as part of a pteridine-derived Mo cofactor of low molecular mass, the precise structure of which is undetermined. The cofactor itself (<2,000 daltons) is present in cells grown aerobically or anaerobically in the presence or absence of nitrate (13, 14, 225). Unlike other Mo-proteins with negative redox potentials (e.g., at pH 8.2, the $E_0′$ for Mo(VI)-Mo(V) is −440 mV, and the $E_0′$ for Mo(V)-Mo(IV) is −480 mV for xanthine oxidase), the Mo cofactor of nitrate reductase is maintained at a relatively positive potential (at pH 7.1, the $E_0′$ of Mo(VI)-Mo(V) is +220 mV, and the $E_0′$ of Mo(V)-Mo(IV) is +180 mV), making it still a competent electron donor to nitrate, which has an $E_0′$ of +420 mV (438, 439). The catalytic role played by Mo and its atomic surroundings is still somewhat conjectural (9, 102, 109, 141, 235, 236).

Chlorate resistance

A complex machinery exists for the processing of Mo and its insertion into nitrate reductase and other Mo-proteins (e.g., formate hydrogenase, tertiary amine *N*-oxide reductase, tetrathionate reductase, and biotin sulfoxide reductase). Therefore, it is not surprising that mutations in several unlinked loci can cause the production of an apo-nitrate reductase missing only the Mo cofactor. Since nitrate reductase reduces chlorate (also perchlorate) to the highly toxic chlorite, mutants lacking the enzyme activity can be selected on the basis of their resistance (26, 158, 185, 285, 338, 340, 341, 385). The selection works anaerobically, since nitrate reductase is repressed aerobically (339). Besides the nitrate reductase genes, six chlorateresistance loci were reported.

The *chlA* locus at min 17.7 specifies the synthesis of Mo factor (11, 12, 169, 347–349, 437). The locus contains two complementation groups (435). One gene product was identified as protein PA (72,000 daltons). The cellular concentration of this gene product is independent of the presence of O_2 or nitrate during growth (146, 151, 155, 175). Insertion mutants in *chlA* totally lack Mo cofactor (305).

The *chlB* gene at min 86.3 appears to specify a product that is involved in the attachment of Mo cofactor to the enzyme proteins (57, 169, 242, 280, 286, 347). The locus contains three complementation groups (305, 435). One product is believed to be a protein of 35,000 daltons which catalyzes the maturation of the nitrate reductase complex (358). The Mo cofactor concentration is normal in *chlB* insertion mutants, although the concentration in the membrane fraction is slightly elevated (305). Mixing extracts of *chlA* and *chlB* mutants allows the reconstitution of catalytically active nitrate reductase associated with the lipid membrane (19, 22–24, 27, 153, 277, 287, 292, 314, 315, 357).

The *chlD* gene at min 17.2 appears to be involved in forming Mo-X before the incorporation of Mo into the pteridine in Mo-cofactor synthesis (203a, 431a). Defects in *chlD* can be remedied by providing 1 mM molybdate in the growth medium (11, 113, 157, 411, 436). Cells with an insertion in *chlD* synthesized <25% of the Mo factor and the demolybdenum factor found in wild-type cells, unless they were supplemented with molybdate during growth (305).

The *chlE* gene at min 18.2 is probably also involved in Mo cofactor synthesis (114, 169, 178, 280, 416, 436). The locus contains two complementation groups (435). Insertion mutants in *chlE* totally lack Mo cofactor (305).

The *chlF* gene at about min 27 affects FDH_N (158).

The *chlG* gene is located at min 0.5 (114, 158, 169, 416). For reasons which are unclear, the mutation impairs nitrate reductase activity much more than FDH_N (222, 223). Unless supplemented with 1 mM molybdate during growth, the Mo cofactor is practically undetectable, although the level of demolybdenum cofactor is near normal. The *chlG* product may be involved in the insertion of Mo into the cofactor (305).

Wild-type cells grow very slowly on lactate and fumarate anaerobically. Unexpectedly, among clones selected for improved growth rate are *chlA*, *chlB*, *chlD*, or *chlE* mutants (263).

Genetics and regulation

The locus *chlC* at min 27, containing the structural genes for nitrate reductase (178, 347, 350, 368), has been resolved into five genes: *C, H, I, K,* and *L*. The enzyme complex is specified by a single operon: *narCHI* encoding subunits A, B, and C (30, 43, 107, 115, 188, 280, 296, 324, 362, 415, 416). Expression of the operon is dependent on the *fnr+* product (67, 264) and on the specific activator protein encoded by the *narL* product (415). It appears that the *fnr* site is on the 3' side (downstream) of the *narL* site (272). The gene *narK* also appears to encode a regulatory protein affecting anaerobic respiration (416). Mutations in a gene at the *chlC* locus can cause up to 60-fold overproduction of ethanol dehydrogenase. Unlike wild-type cells, the ethanol dehydrogenase is no longer anaerobically repressed by nitrate (71).

The reductase is induced about 20-fold by nitrate (25, 69, 83, 367, 396). Synthesis of the enzyme is subject to aerobic repression. Even the basal level of the enzyme appears to be completely repressed aerobically (127, 128, 130, 429, 440). Evidence from *narC-lac* operon fusions suggests that a Mo compound has a regulatory role in the expression of the nitrate reductase operon (328, 331). Analysis of mutants suggested that the Mo compound is Mo-X and that Mo-X and nitrate act together as coeffectors which combine with the *narL* product to form the transcription activator of the *narCHI* operon (Iuchi and Lin, unpublished data). Mutants in which a hybrid *narI-lacZ* operon becomes partially inducible aerobically or in which the operon becomes constitutive anaerobically still depend on the *fnr+* product for expression (65). Among mutants that synthesize increased aerobic basal levels of nitrate reductase (generally less than 10% of the anaerobically induced level found in wild-type cells), one class is defective in the synthesis of ubiquinone. Although aerobic repression is reduced, nitrate shows an inductive effect only anaerobically. A similar phenotype is exhibited when heme synthesis is blocked (20, 147, 149, 154, 403). Mutations in *chlA*, *chlB*, or *chlE* result in partial or total constitutive expression of *narC-lacZ* and the cytochrome *b* (156, 328, 331). Molecular oxygen appears also to interfere with the formation of active enzyme posttranscriptionally (239).

Electron coupling

Transmission of electrons to nitrate from donor substrates, such as NADH, formate, lactate, succinate, and glycerol 3-phosphate, requires the mediation of a quinone (18, 196, 207, 208, 298, 319, 430, 441). Either ubiquinone or menaquinone can serve the purpose (20, 97, 264, 441). The oxidation of the quinol and the reduction of nitrate occur at two separate sites (307). Formate-dependent nitrate reduction has been shown to be an energy-yielding process (428). The reaction is coupled to proton expulsion before and after the quinone link (46, 47, 234, 237). With H_2, NADH (malate), or endogenous metabolite as donor, four protons are expelled per nitrate molecule reduced to nitrite, but only about two protons are expelled with succinate, lactate, and glycerol 3-phosphate as donors (45, 46, 52, 140, 233, 252, 325). An electrochemical proton gradient of about 160 mV (inside negative) can be generated by the formate-nitrate system (46). Curiously, a higher yield of ATP was observed when glycerol 3-phosphate was dehydrogenated by the *glpD+* enzyme than when it was dehydrogenated by the *glpA+* enzyme (302).

Secondary nitrate reductase

A second nitrate reductase has been described in chlorate-resistant mutants. This enzyme is induced anaerobically by nitrate and repressed by organic nitrogen sources and by ammonia (308). It is probably responsible for the phosphomolybdate reducing activity of the cell (56). A study failed to reveal immunochemical differences between the enzymes in cells grown on nitrate as nitrogen source or as electron acceptor (313).

S. typhimurium

Nitrate reductase

The *chlC* (*narA*) locus at min 34 encodes nitrate reductase (58, 60, 124, 378).

Chlorate resistance

Several chlorate resistance loci are located in the region of 18 to 19 min: in clockwise order, they are *chlD* and *chlA*, affecting nitrate reductase, tetrathionate reductase, and formate hydrogenlyase; *chlE*, affecting nitrate reductase and tetrathionate reductase; and *chlF*, the function of which is not yet defined (12, 378, 420, 421). Mutations in *chlG* at min 54 (378, 422) and mutations in *chlB* at min 84 affect nitrate reductase, tetrathionate reductase, and formate hydrogenlyase activities (61, 378).

Secondary nitrate reductase

A nitrate reductase with electrophoretic mobility different from that of the primary enzyme has been found in a *chlC* mutant which grows on nitrate as a nitrogen source and which has derepressed levels of nitrite reductase and a cytochrome (33). Synthesis of the enzyme, affected by a gene at min 43, is not as strongly repressed aerobically as the primary enzyme. Electron transport from formate to nitrate, mediated by this enzyme, is dependent on FDH_N and menaquinone. The absence of the secondary nitrate reductase is associated with a prolonged lag period for the induction of the primary enzyme (R. E. Unger and E. L. Barrett, Abstr. Annu. Meet. Am. Soc. Microbiol. 1986, K-47, p. 201).

NITRITE REDUCTASES

E. coli

Three enzymes are found in *E. coli* that catalyze the six-electron reduction of nitrite to ammonia under anaerobic conditions.

NADPH-dependent reduction of nitrite

Nitrite is reduced fortuitously by a soluble NADPH-specific sulfite reductase which plays a physiological role in the utilization of sulfite as the source of sulfur and which is repressible by cysteine (119, 120, 244, 266, 290). The enzyme (eight subunits A of 54,000 daltons each and four subunits B of 60,000 daltons each), catalyzing a reduction of sulfite to sulfide and of nitrite to ammonia, contains an unusual array of prosthetic groups. Associated with subunit A are flavin mononucleotide and FAD; associated with subunit B are Fe_4S_4 and siroheme (1, 70, 220, 221, 297, 310–312, 389, 399, 400, 402). The contribution of this enzyme to the total nitrite-reducing activity of the cell, however, is less than 5% (1, 345).

The NADH-nitrite oxidoreductase

Biochemistry and enzymology. The principal nitrite reductase (generally accounting for >50% of total nitrite-reducing activity of the cell) is NADH linked and serves both to convert the substrate to a nitrogen source and to remove excess reducing equivalents during anaerobic growth, as shown in the following reaction (77): $3NADH + 5H^+ + NO_2^- \rightarrow 3NAD^+ + NH_4^+ + 2H_2O$. No membrane potential is generated from this process (344, 427). Nevertheless, anaerobic growth yield is increased because the regeneration of NAD^+ from NADH allows the use of acetyl coenzyme A for substrate level phosphorylation instead of as a hydrogen acceptor to produce ethanol. Indeed, provision of nitrite not only results in the excretion of surplus ammonia but also increases the release of acetate into the fermentation medium (76, 78). In addition, internal consumption of protons should contribute to the proton motive force of the cell, thereby contributing to metabolic energy.

The reductase is a dimer of an 88,000-dalton polypeptide with one noncovalently bound FAD (but no flavin mononucleotide), two [2Fe-2S] clusters, and one siroheme per subunit (55, 92, 93, 212). Since the K_m for nitrite is about 5 μM and since the K_m for hydroxylamine is about 5 mM, the hydroxylamine formed as an intermediate in the reduction of nitrite to ammonia apparently remains enzyme bound. NADH-nitrite reductase is activated by the product NAD at low concentrations and inhibited at high concentrations (94). Under experimental conditions, the NADH-nitrite oxidoreductase activity is independent of cytochrome c_{552} despite the ease with which the two proteins are copurified (93, 274).

Genetics and regulation. Mutations in *nirB* (min 74), *nirC* (min 26.4), *nirD* (min 74; might be *nirB*), *nirE* (min 49.5), and *nirF* (about min 52) result in loss of NADH-nitrite oxidoreductase without concomitant disappearance of cytochrome c_{552} (3, 28, 241). The enzyme activity is missing in *hemA* and *cysG* (*nirB*) mutants. Mutations in the latter gene prevent the synthesis of siroheme, an observation that explains why the mutants lack the oxidoreductase activity but not cytochrome c_{552} (3, 4, 67, 77, 80, 111, 211, 212, 241, 316).

The enzyme is induced by nitrite anaerobically (75, 243, 469). Nitrate at concentrations below 20 to 30 mM induces enzyme synthesis (probably indirectly by forming nitrite) and, at concentrations above 30 mM, counteracts the induction (83). Synthesis of nitrite reductase depends on *fnr⁺*.

The formate-nitrite oxidoreductase

A formate-nitrite oxidoreductase, which is membrane bound, contributes about 20 to 50% (depending on strains) of the overall nitrite-reducing activity of *E. coli* (2, 3, 5, 345). FDH_N appears to be involved in the activation of formate for nitrite reduction, since nitrate competes with nitrite as an electron acceptor and since the acceptor activity of both nitrite and nitrate is blocked by an *fdhA* mutation (1, 5). Cytochrome c_{552}, a protein readily released from the periplasmic side of the membrane and induced anaerobically by nitrite, appears to participate in the electron transport chain (5, 135–137, 172, 374). The cytochrome has a low E_0' of about −200 mV (134). This electron carrier, purified from a mutant producing the protein at 10 times the basal anaerobic level, showed nitrite (K_m of about 110 μM) and hydroxylamine (K_m of about 18 mM) reductase activities with benzyl viologen as an artificial electron donor. Reduced flavin mononucleotide and reduced FAD also served as electron donors for nitrite reduction. Cytochrome c_{552} has a relative molar mass of about 70,000 and contains 6 mol of heme per mol of protein (238, 238a).

In contrast to formate-nitrate oxidoreductase, which depends on either ubiquinone or menaquinone for activity, formate-nitrite reductase is only partially dependent on a quinone (5). Transfer of electrons from formate to nitrite by the membrane system is estimated to produce up to one ATP per formate (309, 344). The H^+-ATPase is implicated in proton extrusion associated with nitrite reduction dependent on endogenous metabolites. It is not clear whether the formate-nitrite system is involved (382).

Although the structural gene for the membrane-associated nitrite reductase has not yet been identified, mutants lacking formate-nitrite activity but retaining normal formate-nitrate oxidoreductase activity have been isolated (4). Cytochrome c_{552} is induced anaerobically by formate. Nitrate has an optimal inducing effect in the range of 20 to 30 mM (probably by giving rise to nitrite), but beyond that concentration range it has a repressing effect (75, 82, 83, 454). Synthesis of the cytochrome is prevented by 20 mM ammonia or by *fnr* (*nirA*) mutations (1, 79, 81). Concentrations of both NADH-nitrite oxidoreductase and cytochrome c_{552} were elevated by a mutation believed to be in *chlA* (437).

S. typhimurium

Nitrite is reduced by the NADPH-specific sulfite reductase which is involved in the biosynthesis of cysteine (112, 398, 401).

TMANO, DIMETHYL SULFOXIDE, AND TETRATHIONATE REDUCTASES

E. coli

TMANO reductase

Trimethylamine *N*-oxide (TMANO) serves as an anaerobic electron acceptor with molecular hydrogen, NADH, formate, or lactate as donor (105, 206, 370, 457, 463, 464): $(CH_3)_3NO + AH_2 \rightarrow (CH_3)_3N + A + H_2O$.

Four electrophoretic forms of TMANO reductase were identified. Enzyme I (200,000 daltons), enzyme II (70,000 daltons), and enzyme III (70,000 daltons) are induced by TMANO under growth conditions which repress the constitutive enzyme IV (100,000 daltons). Enzyme I has a K_m of 1.5 mM for TMANO. It also reduces various *N*-oxides of purine and pyrimidine derivatives, as well as TMANO and hydroxylamine. Enzyme formation is repressed by aerobic growth or anaerobic growth in the presence of nitrate (369, 369a, 395). The major enzyme (probably isoenzyme I) accounts for 93% of the total activity and is encoded by *torA* at min 28. Expression of this gene is reduced about twofold in an *fnr* background (327, 425). Cells grown anaerobically in the presence of TMANO are induced in both *b*- and *c*-type cytochromes which mediate electron transfer from NADH and formate to TMANO (50, 99, 206, 371, 372).

The Mo prosthetic group for nitrate reductase, the formation of which is dependent on the products of *chlA*, *chlB*, and *chlD*, is also required by TMANO reductase (99, 156, 426). Analysis of mutants revealed the obligatory participation of menaquinone in the electron transport chain (99, 299). The redox chain reaction is an energy-conserving process resulting in three to four protons translocated per TMANO reduced (426, 464).

Dimethyl sulfoxide reductase

Of 151 different strains and isolates tested, all were found to reduce dimethyl sulfoxide anaerobically (15). The reduction of dimethyl sulfoxide to dimethyl sulfide has a midpoint redox potential of +160 mV (459): $(CH_3)_2SO + 2H^+ + 2e^- \rightarrow (CH_3)_2S + H_2O$. The membrane-associated reductase acts on the substrate with a K_m of about 0.2 mM. The Mo cofactor processed by the *chl* gene products is required for enzyme activity. H_2, NADH, glycerol, and formate serve as anaerobic electron donor to dimethyl sulfoxide. Reduced flavin mononucleotide and reduced FAD also serve as electron donors; methionine sulfoxide can serve as an acceptor. About three protons are translocated per acceptor molecule reduced. The enzyme is induced anaerobically by dimethyl sulfoxide and is repressed by nitrate. The fnr^+ product appears to be involved in the gene expression (40, 41, 470).

S. typhimurium

TMANO reductase

TMANO serves as an anaerobic electron acceptor for growth on glycerol. In contrast to *E. coli*, anaerobic growth stimulation by TMANO is inferior to that by nitrate (246). One major and two minor TMANO reductases are induced by TMANO. In addition, there is an enzyme that is derepressed during growth in the presence of both TMANO and thiosulfate and that is repressed by nitrate. The electrophoretic activity of this enzyme is distinct from that of the thiosulfate reductase inducible by thiosulfate. The major inducible TMANO reductase is a tetramer or dimer of an 84,000-dalton subunit. Reduced flavin mononucleotide (but not reduced FAD) serves as an electron donor. The K_m for TMANO is about 0.9 mM. The enzyme reduced chlorate with a K_m of about 2 mM and a V_{max} 20% of that observed with TMANO (261).

The major inducible enzyme is encoded by *tor* in the 80 to 83 min region (262). The enzyme activity depends on Mo cofactor, and the function of its electron transport chain requires menaquinone (104, 262).

Reductases of sulfur compounds

Tetrathionate reductase catalyzes the following reaction: $[2H] + S_4O_6^{2-} \rightarrow 2H^+ + 2S_2O_3^{2-}$. The gene *ttr*, encoding tetrathionate reductase, is at min 35 (59, 268, 378).

Thiosulfate reductase activity is attributable to two isoenzymes (D. L. Riggs and E. L. Barrett, Abstr. Annu. Meet. Am. Soc. Microbiol. 1986, K-246, p. 218). Reduction of thiosulfate to hydrogen sulfide depends on Mo cofactor and menaquinone (104, 262).

Among 40 *Salmonella* isolates tested, only one was positive for dimethyl sulfoxide reductase (15).

THE PLEIOTROPIC REGULATOR

In *E. coli*, *fnr* (for fumarate and nitrate reduction, also known as *nirA* or *nirR*) is located at 24.3 min. The gene product serves as a transcription activator protein for a number of genes that encode enzymes participating in anaerobic respiration (65, 67, 81, 264, 316, 391). Mutations in *fnr* not only affect the various terminal reductases but also impair the synthesis of cytochromes with midpoint potential more negative than +100 mV (189).

The Fnr protein has been isolated in pure form from cells harboring the *fnr*+ gene in an expression vector. In vivo, the protein is believed to exist as a dimer of a 28,000-dalton subunit (433). Nucleotide sequence determination indicates that this protein and the cAMP receptor protein have regions of homology corresponding to the nucleotide-binding domain (394). The further observation that the expression of the target *frd* gene depends on cAMP but not on its receptor protein led to the consideration of cAMP also as the effector of the Fnr protein (432).

In *S. typhimurium*, the *fnr* (*oxrA*) gene is located at min 29.5. Mutations in this gene prevent the utilization of nitrate, but surprisingly not of fumarate, as electron acceptor (219). The gene product is also required for the synthesis of a peptidase that has no direct role in anaerobic respiration (423).

DISCUSSION

Just as the choice of carbon source is hierarchically controlled by catabolite repression and inducer exclusion mediated by the phosphoenolpyruvate phosphotransferase system, so too is the choice of respiratory pathway regulated by "global" respiratory control. Two kinds of models for preferential utilization of the terminal electron acceptor with the most positive redox potential may be considered.

According to one model, a common regulatory element acts differentially on the various genes encoding respiratory function. For instance, the intracellular redox potential might influence the relative concentrations of oxidized or reduced forms of pyridine nucleotides, flavin compounds, quinones, hemes, or other cofactors of redox function (295, 451, 452). The concentration of such a component might then act as a signal for the *fnr* system. The regulatory agent need not be confined to this system. Since DNA topoisomerase I appears to be essential for aerobic growth and since DNA gyrase appears to be essential for anaerobic growth (465), it is also conceivable that these two proteins effect the graded regulation of respiratory enzymes by controlling the degree of DNA supercoiling. Control at the level of translation might also be operative in some cases, since modification of the adenosine residue adjacent to the anticodon in some tRNAs varies with the degree of aerobiosis (54). In favor of this model is the evidence that O_2 itself does not serve directly as the effector of aerobic repression. It has been observed that the presence of KCN par-

tially overcomes the aerobic repressive effect on the synthesis of menaquinone (17) and the expression of the *glpA* operon encoding anaerobic glycerol-3-phosphate dehydrogenase (260). The redox potential model, however, would have difficulty in accounting for the almost all-or-none effects exerted by O_2 or nitrate on their subordinate systems without invoking cooperativity. Moreover, there is no evidence for large variations in the steady-state concentrations of redox cofactors after a shift from aerobic to anaerobic growth. In particular, changes in the $NAD^+/NADH$ ratios appear insufficient to provide a reliable regulatory signal (453). Furthermore, O_2 repressed aerobic synthesis of nitrate reductase in cytochrome-deficient *hemA* mutants, despite the fact that the respiratory chain cannot function in these mutants (282).

A chain command model postulates pyramidal control by a set of regulatory proteins. For example, the presence of O_2 leads to the synthesis of a repressing element that prevents the expression of all of the other respiratory systems (337), the presence of nitrate prevents the expression of all of the other anaerobic systems, the presence of TMANO prevents the expression of all of the lower-ranking systems in the redox scale, and so on. In favor of this model is the observation that the *narL* regulatory protein has a pleiotropic function. In the presence of nitrate and Mo-X, it acts as the activator for the nitrate reductase operon but as a repressor for the fumarate reductase operon (Iuchi and Lin, Abstr. Annu. Meet. Am. Soc. Microbiol. 1986; S. Iuchi, unpublished data). This mechanism, however, is unlikely to account for all of the regulatory network of the respiratory system, since the genes at the bottom of the pecking order would have to recognize a large number of higher-order regulators. Much work is needed to clarify the picture.

In addition to discovering all of the basic elements for respiratory control and how they are integrated, another challenge lies in determining the number and kind of protein components in each of the redox modules and the interaction of the protein subunits. A deeper understanding of the respiratory program may also depend upon our ability to reconstruct its natural history. In this regard, a general view is beginning to emerge. One of the first electron acceptor modules to have evolved might have been a soluble enzyme system for "fumarate fermentation," such as the fumarate reductase of *Veillonella alcalescens*, which serves as a hydrogen sink for the oxidation of fermentative intermediates (443). Association of these soluble enzymes with nonheme iron-sulfur proteins and the inner membrane would then evolve membrane-bound respiratory systems capable of electron translocation. The cytochrome-independent fumarate reductase of *Streptococcus faecalis* var. *zymogenes* exemplifies this intermediate stage of development (126, 218, 346). Ultimately, the incorporation of cytochromes would complete the evolution of the respiratory electron transport chain (450). More-complex respiratory systems could then be added to utilize inorganic molecules, such as nitrate, nitrite, and sulfate (116, 117). With the basic respiratory components in place and with O_2 made available by photosynthesis, the stage would be set for the elaboration of the aerobic electron transport chain (for more complete treatment, see references 143 and 450). The evolution of novel electron transport systems need not cease with aerobiosis. For example, the TMANO reductase complex might be a relatively late acquisition, since its substrate is of eucaryotic origin.

ACKNOWLEDGMENTS

We thank Shiro Iuchi for helpful discussion and Alison Ulrich for assistance in reference work.

Research of E.C.C.L. was supported by Public Health Service grant GM11983 from the National Institute of General Medical Sciences and by grant DMB83-14312 from the National Science Foundation.

LITERATURE CITED

1. **Abou-Jaoude, A., M. Chippaux, and M.-C. Pascal.** 1979. Formate-nitrite reduction in *Escherichia coli* K12. 1. Physiological study of the system. Eur. J. Biochem. **95:**309–314.
2. **Abou-Jaoude, A., M. Chippaux, M.-C. Pascal, and F. Casse.** 1977. Formate: a new electron donor for nitrite reduction in *Escherichia coli* K12. Biochem. Biophys. Res. Commun. **78:**579–583.
3. **Abou-Jaoude, A., M. Lepelletier, J., Ratouchniak, M. Chippaux, and M.-C. Pascal.** 1978. Nitrite reduction in *Escherichia coli*: genetic analysis of *nir* mutants. Mol. Gen. Genet. **167:**113–118.
4. **Abou-Jaoude, A., M.-C. Pascal, and M. Chippaux.** 1978. Isolation and phenotypes of mutants from *Escherichia coli* K12 defective in nitrite-reductase activity. FEMS Microbiol. Lett. **3:**235–239.
5. **Abou-Jaoude, A., M.-C. Pascal, and M. Chippaux.** 1979. Formate-nitrite reduction in *Escherichia coli* K12. 2. Identification of components involved in the electron transfer. Eur. J. Biochem. **95:**315–321.
6. **Ackrell, B. A. C., R. N. Asato, and H. F. Mower.** 1966. Multiple forms of bacterial hydrogenase. J. Bacteriol. **92:**828–838.
7. **Ackrell, B. A. C., M. B. Ball, and E. B. Kearney.** 1980. Peptides from complex II active in reconstitution of succinate-ubiquinone reductase. J. Biol. Chem. **255:**2761–2769.
8. **Adams, M. W. W., and D. O. Hall.** 1979. Purification of the membrane-bound hydrogenase of *Escherichia coli*. Biochem. J. **183:**11–22.
9. **Adams, M. W. W., and L. E. Mortenson.** 1982. The effect of cyanide and ferricyanide on the activity of the dissimilatory nitrate reductase of *Escherichia coli*. J. Biol. Chem. **257:**1791–1799.
10. **Adams, M. W. W., L. E. Mortenson, and J. S. Chen.** 1981. Hydrogenase. Biochim. Biophys. Acta **594:**105–176.
11. **Adhya, S., P. Clearly, and A. Campbell.** 1968. A deletion of prophage lambda and adjacent genetic regions. Proc. Natl. Acad. Sci. USA **61:**956–962.
12. **Alper, M. D., and B. N. Ames.** 1975. Positive selection of mutants with deletions of *gal-chl* region of the *Salmonella* chromosome as a screening procedure for mutagens that cause deletions. J. Bacteriol. **121:**259–266.
13. **Amy, N. K.** 1981. Identification of the molybdenum cofactor in chlorate-resistant mutants of *Escherichia coli*. J. Bacteriol. **148:**274–282.
14. **Amy, N. K., and K. V. Rajagopalan.** 1979. Characterization of molybdenum cofactor from *Escherichia coli*. J. Bacteriol. **140:**114–124.
15. **Ando, H., M. Kumagai, T. Karashimada, and H. Iida.** 1957. Diagnostic use of dimethylsulfoxide reduction test within *Enterobacteriaceae*. Jpn. J. Microbiol. **1:**335–338.
16. **Andrews, S., G. B. Cox, and F. Gibson.** 1977. The anaerobic oxidation of dihydroorotate by *Escherichia coli* K12. Biochim. Biophys. Acta **462:**153–160.
17. **Ashcroft, J. R., and B. A. Haddock.** 1975. Synthesis of alternative membrane-bound redox carriers during aerobic growth of *Escherichia coli* in the presence of potassium cyanide. Biochem. J. **148:**349–352.
18. **Asnis, R. E., V. G. Vely, and M. C. Glick.** 1956. Some enzymatic activities of a particulate fraction from sonic lysates of *Escherichia coli*. J. Bacteriol. **72:**314–319.
19. **Azoulay, E., P. Couchoud-Beaumont, and J. M. LeBeault.** 1972. Etude des mutants chlorate-resistants d'*Escherichia coli* K12. IV. Isolement, purification et etude de la nitrate-reductase *in vitro* par complementation. Biochim. Biophys. Acta **256:**670–680.
20. **Azoulay, E., G. Giordano, L. Grillet, R. Rosset, and B. A. Haddock.** 1978. Properties of *Escherichia coli* K-12 mutants that are sensitive to chlorate when grown aerobically. FEMS Microbiol. Lett. **4:**235–240.
21. **Azoulay, E., and B. Marty.** 1970. Etude du systeme multienzymatique hydrogene lyase chez *Escherichia coli* K12 et ses mutants chlorate-resistants. Eur. J. Biochem. **13:**168–173.
22. **Azoulay, E., J. Pommier, and C. Riviere.** 1975. Membrane reconstitution in *chl-r* mutants of *Escherichia coli* K12. IX. Part played by phospholipids in the complementation process. Biochim. Biophys. Acta **389:**236–250.
23. **Azoulay, E., and J. Puig.** 1968. Reconstitution of enzymatically active particles from inactive soluble elements in *Escherichia coli* K12. Biochem. Biophys. Res. Commun. **33:**1019–1024.
24. **Azoulay, E., J. Puig, and P. Couchoud-Beaumont.** 1969. Etude des mutants chlorate-resistants chez *Escherichia coli* K12. I. Reconstitution *in vitro* de l'activite nitrate-reductase particulaire chez *Escherichia coli* K12. Biochim. Biophys. Acta **171:**238–252.
25. **Azoulay, E., J. Puig, and M. L. Martins Rosada de Sousa.** 1969. Regulation de la synthese de la nitrate-reductase chez *Escherichia coli*. Ann. Inst. Pasteur (Paris) **117:**474–485.
26. **Azoulay, E., J. Puig, and F. Pichinoty.** 1967. Alterations of respiratory particles by mutations in *Escherichia coli* K12. Biochem. Biophys. Res. Commun. **27:**270–274.
27. **Azoulay, E., C. Riviere, G. Giordano, J. Pommier, M. Denis, and G. Ducet.** 1977. Participation of cytochrome *b* to the in-vitro reconstitution of the membrane-bound formate-nitrate reductase of *Escherichia coli* K12 and the possible role of sulfhydryl groups and temperature in the reconstitution process. FEBS Lett. **79:**321–326.
28. **Bachmann, B. J.** 1983. Linkage map of *Escherichia coli* K-12, edition 7. Microbiol. Rev. **47:**180–230.
29. **Ballantine, S. P., and D. H. Boxer.** 1985. Nickel-containing hydrogenase isoenzymes from anaerobically grown *Escherichia coli* K-12. J. Bacteriol. **163:**454–459.
30. **Barr, G. C., and S. E. Palm-Nicholls.** 1981. Cloning the *chlC* gene for nitrate reductase of *Escherichia coli*. FEMS Microbiol. Lett. **11:**213–216.
31. **Barrett, E. L., C. E. Jackson, H. T. Fukumoto, and G. W. Chang.** 1979. Formate dehydrogenase mutants of *Salmonella typhimurium*: a new medium for their isolation and new mutant classes. Mol. Gen. Genet. **177:**95–101.
32. **Barrett, E. L., H. S. Kwan, and J. Macy.** 1984. Anaerobiosis, formate, nitrate, and *pyrA* are involved in the regulation of formate hydrogenlyase in *Salmonella typhimurium*. J. Bacteriol. **158:**972–977.
33. **Barrett, E. L., and D. L. Riggs.** 1982. Evidence for a second nitrate reductase activity that is distinct from the respiratory enzyme in *Salmonella typhimurium*. J. Bacteriol. **150:**563–571.
34. **Barrett, E. L., and D. L. Riggs.** 1982. *Salmonella typhimurium* mutants defective in the formate dehydrogenase linked to nitrate reductase. J. Bacteriol. **149:**554–560.
35. **Bar Tana, J., B. J. Howlett, and D. E. Koshland, Jr.** 1977. Flagellar formation in *Escherichia coli* electron transport mutants. J. Bacteriol. **130:**787–792.
36. **Begg, Y. A., J. N. Whyte, and B. A. Haddock.** 1977. The identification of mutants of *Escherichia coli* deficient in formate dehydrogenase and nitrate reductase activities using dye indicator plates. FEMS Microbiol. Lett. **2:**47–50.
37. **Bergström, S., F. P. Lindberg, O. Olsson, and S. Normark.** 1983. Comparison of the overlapping *frd* and *ampC* operons of *Escherichia coli* with the corresponding DNA sequences in other gram-negative bacteria. J. Bacteriol. **155:**1297–1305.
38. **Bernhard, T., and G. Gottschalk.** 1978. Cell yields of *Escherichia coli* during anaerobic growth on fumarate and molecular hydrogen. Arch. Microbiol. **116:**235–238.
39. **Billen, D.** 1951. The inhibition by nitrate of enzyme formation during growth of *Escherichia coli*. J. Bacteriol. **62:**793–797.
40. **Bilous, P. T., and J. H. Weiner.** 1985. Dimethyl sulfoxide reductase activity by anaerobically grown *Escherichia coli* HB101. J. Bacteriol. **162:**1151–1155.
41. **Bilous, P. T., and J. H. Weiner.** 1985. Proton translocation coupled to dimethyl sulfoxide reduction in anaerobically grown *Escherichia coli* HB101. J. Bacteriol. **163:**369–375.
42. **Blum, H., and R. K. Poole.** 1982. The molybdenum and iron-sulphur centres of *Escherichia coli* nitrate reductase are nonrandomly oriented in the membrane. Biochem. Biophys. Res. Commun. **107:**903–909.
43. **Bonnefoy-Orth, V., M. Lepelletier, M.-C. Pascal, and M. Chippaux.** 1981. Nitrate reductase and cytochrome *b* (*nitrate reductase*) structural genes as parts of the nitrate reductase operon. Mol. Gen. Genet. **181:**535–540.
44. **Boonstra, J., J. A. Downie, and W. N. Konings.** 1978. Energy supply for active transport in anaerobically grown *Escherichia coli*. J. Bacteriol. **136:**844–853.
45. **Boonstra, J., M. T. Huttunen, W. N. Konings, and H. R. Kaback.**

1975. Anaerobic transport in *Escherichia coli* membrane vesicles. J. Biol. Chem. **250**:6792–6798.

46. **Boonstra, J., and W. N. Konings.** 1977. Generation of an electrochemical proton gradient by nitrate respiration in membrane vesicles from anaerobically grown *Escherichia coli*. Eur. J. Biochem. **78**:361–368.

47. **Boonstra, J., H. J. Sips, and W. N. Konings.** 1976. Active transport by membrane vesicles from anaerobically grown *Escherichia coli* energized by electron transfer to ferricyanide and chlorate. Eur. J. Biochem. **69**:35–44.

48. **Boxer, D., A. Malcolm, and A. Graham.** 1982. *Escherichia coli* formate to nitrate respiratory pathway: structural analysis. Biochem. Soc. Trans. **10**:480–481.

49. **Boxer, D. H., and R. A. Clegg.** 1975. A transmembrane location for the proton-translocating reduced ubiquinone → nitrate reductase segment of the respiratory chain of *Escherichia coli*. FEBS Lett. **60**:54–57.

50. **Bragg, P. D., and N. R. Hackett.** 1983. Cytochromes of the trimethylamine *N*-oxide anaerobic respiratory pathway of *Escherichia coli*. Biochim. Biophys. Acta **725**:168–177.

51. **Bray, R. C., S. P. Vincent, D. J. Lowe, R. A. Clegg, and P. B. Garland.** 1976. Electron-paramagnetic-resonance studies on the molybdenum of nitrate reductase from *Escherichia coli* K12. Biochem. J. **155**:201–203.

52. **Brice, J. M., J. F. Law, D. J. Meyer, and C. W. Jones.** 1974. Energy conservation in *Escherchia coli* and *Klebsiella pneumoniae*. Biochem. Soc. Trans. **2**:523–526.

53. **Brown, T. A., and A. Schrift.** 1982. Selective assimilation of selenite by *Escherichia coli*. Can. J. Microbiol. **28**:307–310.

54. **Buck, M., and B. N. Ames.** 1984. A modified nucleotide in tRNA as a possible regulator of aerobiosis: synthesis of *cis*-2-methylthioribosylzeatin in the tRNA of *Salmonella*. Cell **36**:523–531.

55. **Cammack, R., R. H. Jackson, A. Cornish-Bowden, and J. A. Cole.** 1982. Electron-spin-resonance studies of the NADH-dependent nitrite reductase from *Escherichia coli* K12. Biochem. J. **207**:333–339.

56. **Campbell, A. M., A. del Campillo-Campbell, and D. B. Villaret.** 1985. Molybdate reduction by *Escherichia coli* K-12 and its *chl* mutants. Proc. Natl. Acad. Sci. USA **82**:227–231.

57. **Casse, F.** 1970. Mapping of the gene *chlB* controlling membrane bound nitrate reductase and formic hydrogen-lyase activities in *Escherichia coli* K12. Biochem. Biophys. Res. Commun. **39**:429–436.

58. **Casse, F., M. Chippaux, and M. C. Pascal.** 1973. Isolation from *Salmonella typhimurium* LT2 of mutants lacking specifically nitrate reductase activity and mapping of the *chl-C* gene. Mol. Gen. Genet. **124**:247–251.

59. **Casse, F., M.-C. Pascal, and M. Chippaux.** 1972. A mutant of *Salmonella typhimurium* deficient in tetrathionate reductase activity. Mol. Gen. Genet. **119**:71–74.

60. **Casse, F., M.-C. Pascal, and M. Chippaux.** 1973. Comparison between the chromosomal maps of *Escherichia coli* and *Salmonella typhimurium*. Length of the inverted segment of the *trp* region. Mol. Gen. Genet. **124**:253–257.

61. **Casse, F., M.-C. Pascal, M. Chippaux, and J. Ratouchniak.** 1972. Mapping of the *chlB* gene in *Salmonella typhimurium* LT2. Mol. Gen. Genet. **119**:67–70.

62. **Chaudhry, G. R., I. M. Chaiken, and C. H. MacGregor.** 1983. An activity from *Escherichia coli* membranes responsible for the modification of nitrate reductase to its precursor form. J. Biol. Chem. **258**:5828–5833.

63. **Chaudhry, G. R., and C. H. MacGregor.** 1983. *Escherichia coli* nitrate reductase subunit A: its role as the catalytic site and evidence for its modification. J. Bacteriol. **154**:387–394.

64. **Chaudhry, G. R., and C. H. MacGregor.** 1983. Cytochrome *b* from *Escherichia coli* nitrate reductase: its properties and association with the enzyme complex. J. Biol. Chem. **258**:5819–5827.

65. **Chippaux, M., V. Bonnefoy-Orth, J. Ratouchniak, and M.-C. Pascal.** 1981. Operon fusions in the nitrate reductase operon and study of the control gene *nirR* in *Escherichia coli*. Mol. Gen. Genet. **182**:477–479.

66. **Chippaux, M., F. Casse, and M.-C. Pascal.** 1972. Isolation and phenotypes of mutants from *Salmonella typhimurium* defective in formate hydrogenlyase activity. J. Bacteriol. **118**:766–768.

67. **Chippaux, M., D. Giudici, A. Abou-Jaoude, F. Casse, and M.-C. Pascal.** 1978. A mutation leading to the total lack of nitrite reductase activity in *Escherichia coli* K12. Mol. Gen. Genet. **160**:225–229.

68. **Chippaux, M., M.-C. Pascal, and F. Casse.** 1977. Formate hydrogenlyase system in *Salmonella typhimurium* LT2. Eur. J. Biochem. **72**:149–155.

69. **Chippaux, M., and F. Pichinoty.** 1970. Les nitrate-reductase bacteriennes. V. Induction de la biosynthese de l'enzyme A par l'azoture. Arch. Microbiol. **71**:361–366.

70. **Christner, J. A., E. Munck, P. A. Janick, and L. M. Siegel.** 1983. Mossbauer evidence for exchange-coupled siroheme and [4Fe-4S] prosthetic groups in *Escherichia coli* sulfite reductase. Studies of the reduced states and of a nitrite turnover complex. J. Biol. Chem. **258**:11147–11156.

71. **Clark, D., and J. E. Cronan, Jr.** 1980. *Escherichia coli* mutants with altered control of alcohol dehydrogenase and nitrate reductase. J. Bacteriol. **141**:177–183.

72. **Clegg, R. A.** 1975. The nitrate respiration complex of *Escherichia coli*: subunit composition. Biochem. Soc. Trans. **3**:689–691.

73. **Clegg, R. A.** 1975. The size of nitrate reductase in *Escherichia coli*. Biochem. Soc. Trans. **3**:691–694.

74. **Clegg, R. A.** 1976. Purification and some properties of nitrate reductase (EC 1.7.99.4) from *Escherichia coli* K12. Biochem. J. **153**:533–541.

75. **Cole, J. A.** 1968. Cytochrome c_{552} and nitrite reductase in *Escherichia coli*. Biochim. Biophys. Acta **162**:356–368.

76. **Cole, J. A.** 1978. The rapid accumulation of large quantities of ammonia during nitrite reduction by *Escherichia coli*. FEMS Microbiol. Lett. **4**:327–329.

77. **Cole, J. A.** 1982. Independent pathways for the anaerobic reduction of nitrite to ammonia in *Escherichia coli*. Biochem. Soc. Trans. **10**:476–478.

78. **Cole, J. A., and C. M. Brown.** 1980. Nitrite reduction to ammonia by fermentative bacteria: a short circuit in the biological nitrogen cycle. FEMS Microbiol. Lett. **7**:65–72.

79. **Cole, J. A., K. J. Coleman, B. E. Compton, B. M. Kavanagh, and C. W. Keevil.** 1974. Nitrite and ammonia assimilation by anaerobic continuous cultures of *Escherichia coli*. J. Gen. Microbiol. **85**:11–22.

80. **Cole, J. A., B. M. Newman, and P. White.** 1980. Biochemical and genetic characterization of *nirB* mutants of *Escherichia coli* K12 pleiotropically defective in nitrite and sulphite reduction. J. Gen. Microbiol. **120**:475–483.

81. **Cole, J. A., and F. B. Ward.** 1973. Nitrite reductase-deficient mutants of *Escherichia coli* K12. J. Gen. Microbiol. **76**:21–29.

82. **Cole, J. A., and J. W. T. Wimpenny.** 1966. The inter-relationships of low redox potential cytochrome c_{552} and hydrogenase in facultative anaerobes. Biochim. Biophys. Acta **128**:419–425.

83. **Cole, J. A., and J. W. T. Wimpenny.** 1968. Metabolic pathways for nitrate reduction in *Escherichia coli*. Biochim. Biophys. Acta **162**:39–48.

84. **Cole, S. T.** 1982. Nucleotide sequence coding for the flavoprotein subunit of the fumarate reductase of *Escherichia coli*. Eur. J. Biochem. **122**:479–484.

85. **Cole, S. T.** 1984. Molecular and genetic aspects of the fumarate reductase of *Escherichia coli*. Biochem. Soc. Trans. **12**:237–238.

86. **Cole, S. T., T. Grundström, B. Jaurin, J. J. Robinson, and J. H. Weiner.** 1982. Location and nucleotide sequence of *frdB*, the gene coding for the iron-sulphur protein subunit of the fumarate reductase of *Escherichia coli*. Eur. J. Biochem. **126**:211–216.

87. **Cole, S. T., and J. R. Guest.** 1979. Amplification and aerobic synthesis of fumarate reductase in ampicillin-resistant mutants of *Escherichia coli* K-12. FEMS Microbiol. Lett. **5**:65–67.

88. **Cole, S. T., and J. R. Guest.** 1979. Production of a soluble form of fumarate reductase by multiple gene duplication in *Escherichia coli* K12. Eur. J. Biochem. **102**:65–71.

89. **Cole, S. T., and J. R. Guest.** 1980. Genetic and physical characterization of lambda transducing phages (lambda*frdA*) containing the fumarate reductase gene of *Escherichia coli* K12. Mol. Gen. Genet. **178**:409–418.

90. **Cole, S. T., and J. R. Guest.** 1980. Amplification of fumarate reductase synthesis with lambda*frdA* transducing phages and orientation of *frdA* gene expression. Mol. Gen. Genet. **179**:377–385.

91. **Cole, S. T., and J. R. Guest.** 1982. Molecular genetic aspects of the succinate:fumarate oxidoreductases of *Escherichia coli*. Biochem. Soc. Trans. **10**:473–475.

92. **Coleman, K. J., A. J. Cornish-Bowden, and J. A. Cole.** 1976. Purification and properties of nitrite reductase from *Escherichia coli* K-12. Proc. Soc. Gen. Microbiol. **3**:84.

93. **Coleman, K. J., A. Cornish-Bowden, and J. A. Cole.** 1978. Purification and properties of nitrite reductase from *Escherichia coli*. Biochem. J. **175**:483–493.

94. **Coleman, K. J., A. Cornish-Bowden, and J. A Cole.** 1978. Activation of nitrite reductase from *Escherichia coli* K12 by oxidized nicotinamide-adenine dinucleotide. Biochem. J. **175**:495–499.

95. **Coles, C. J., R. H. Holm, D. M. Kurtz, Jr., W. H. Orme-Johnson, J. Rawlings, T. P. Singer, and G. B. Wong.** 1979. Characterization of the iron-sulfur centers in succinate dehydrogenase. Proc.

Natl. Acad. Sci. USA **76**:3805–3808.

96. **Cox, G. B., and F. Gibson.** 1974. Studies on electron transport and energy-linked reactions using mutants of *Escherichia coli.* Biochim. Biophys. Acta **346**:1–25.

97. **Cox, G. B., N. A. Newton, F. Gilson, A. M. Snoswell, and J. A. Hamilton.** 1970. The function of ubiquinone in *Escherichia coli.* Biochem. J. **117**:551–562.

98. **Cox, J. C., E. S. Edwards, and J. A. DeMoss.** 1981. Resolution of distinct selenium-containing formate dehydrogenases from *Escherichia coli.* J. Bacteriol. **145**:1317–1324.

99. **Cox, J. C., and R. Knight.** 1981. Trimethylamine *N*-oxide (TMAO) reductase activity in chlorate-resistant or respiration-deficient mutations of *Escherichia coli.* FEMS Microbiol. Lett. **12**:249–252.

100. **Cozzarelli, N. R., W. B. Freedberg, and E. C. C. Lin.** 1968. Genetic control of the L-alpha-glycerophosphate system in *Escherichia coli.* J. Mol. Biol. **31**:371–387.

101. **Cozzarelli, N. R., J. P. Koch, S. Hayashi, and E. C. C. Lin.** 1965. Growth stasis by accumulated L-alpha-glycerophosphate in *Escherichia coli.* J. Bacteriol. **90**:1325–1329.

102. **Cramer, S. P., L. P. Solomonson, M. W. W. Adams, and L. E. Mortenson.** 1984. Molybdenum sites of *Escherichia coli* and *Chlorella vulgaris* nitrate reductase: a comparison by EXAFS. J. Am. Chem. Soc. **106**:1467–1471.

103. **Darlison, M. G., and J. R. Guest.** 1984. Nucleotide sequence encoding the iron-sulphur protein subunit of the succinate dehydrogenase of *Escherichia coli.* Biochem. J. **223**:507–517.

104. **Davidson, A. E., H. E. Fukimoto, C. E. Jackson, E. L. Barrett, and G. W. Chang.** 1979. Mutants of *Salmonella typhimurium* defective in the reduction of trimethylamine oxide. FEMS Microbiol. Lett. **6**:417–420.

105. **Debevere, J. M., and J. P. Voets.** 1974. A rapid selective medium for the determination of trimethylamine oxide-reducing bacteria. Z. Allg. Mikrobiol. **14**:655–658.

106. **DeMoss, J. A.** 1977. Limited proteolysis of nitrate reductase purified from membranes of *Escherichia coli.* J. Biol. Chem. **252**:1696–1701.

107. **DeMoss, J. A.** 1978. Role of the *chlC* gene in formation of the formate-nitrate reductase pathway in *Escherichia coli.* J. Bacteriol. **133**:626–630.

108. **DeMoss, J. A., T. Y. Fan, and R. H. Scott.** 1981. Characterization of subunit structural alterations which occur during purification of nitrate reductase from *Escherichia coli.* Arch. Biochem. Biophys. **206**:54–64.

109. **Der Vartanian, D. V., and P. Forget.** 1975. The bacterial nitrate reductase. EPR studies on the enzyme A of *Escherichia coli* K12. Biochim. Biophys. Acta **379**:74–80.

110. **Dickie, P., and J. H. Weiner.** 1979. Purification and characterization of membrane-bound fumarate reductase from anaerobically grown *Escherichia coli.* Can. J. Biochem. **57**:813–821.

111. **Douglas, M. W., F. B. Ward, and J. A. Cole.** 1974. The formate hydrogenlyase activity of cytochrome c_{552}-deficient mutants of *Escherichia coli.* J. Gen. Microbiol. **80**:557–560.

112. **Dreyfuss, J., and K. J. Monty.** 1963. The biochemical characterization of cysteine-requiring mutants of *Salmonella typhimurium.* J. Biol. Chem. **238**:1019–1024.

113. **Dubourdieu, M., E. Andrade, and J. Puig.** 1976. Molybdenum and chlorate resistant mutants in *Escherichia coli* K12. Biochem. Biophys. Res. Commun. **70**:766–773.

114. **Dykhuizen, D.** 1973. Genetic analysis of the system that reduces biotin-*d*-sulfoxide in *Escherichia coli.* J. Bacteriol. **115**:662–667.

115. **Edwards, E. S., S. S. Rondeau, and J. A. DeMoss.** 1983. *chlC(nar)* operon of *Escherichia coli* includes structural genes for alpha and beta subunits of nitrate reductase. J. Bacteriol. **153**:1513–1520.

116. **Egami, F.** 1973. A comment to the concept on the role of nitrate fermentation and nitrate respiration in an evolutionary pathway of energy metabolism. Z. Allg. Mikrobiol. **13**:177–181.

117. **Egami, F.** 1977. Anaerobic respiration and photoautotrophy in the evolution of prokaryotes. Origins Life **8**:169–171.

118. **Egami, F., Y. Hayase, and S. Taniguchi.** 1960. Tentative d'induction de la nitrate-reductase chez un mutant d'*Escherichia coli* auxotrophe pour l'hemine. Ann. Inst. Pasteur (Paris) **98**:429–438.

119. **Ellis, F. J.** 1964. A rapid assay for sulphite reductase. Biochim. Biophys. Acta **85**:335–338.

120. **Ellis, F. J.** 1966. Sulphur metabolism: the usefulness of *N*-ethylmaleimide. Nature (London) **211**:1266–1268.

121. **Enoch, H. G., and R. L. Lester.** 1972. Effects of molybdate, tungstate, and selenium compounds on formate dehydrogenase and other enzyme systems in *Escherichia coli.* J. Bacteriol. **110**:1032–1040.

122. **Enoch, H. G., and R. L. Lester.** 1974. The role of a novel cytochrome *b*-containing reductase and quinone in the *in vitro*

reconstruction of formate-nitrate reductase. Biochem. Biophys. Res. Commun. **61**:1234–1241.

123. **Enoch, H. G., and R. L. Lester.** 1975. The purification and properties of formate dehydrogenase and nitrate reductase from *Escherichia coli.* J. Biol. Chem. **250**:6693–6705.

124. **Enomoto, M.** 1972. Genetic studies of chlorate-resistant mutants of *Salmonella typhimurium.* Jpn. J. Genet. **47**:227–235.

125. **Farkas, A., L. Farkas, and J. Yudkin.** 1934. The decomposition of sodium formate by *Bacterium coli* in the presence of heavy water. Proc. R. Soc. Lond. B Biol. Sci. **115**:373–379.

126. **Faust, P. J., and P. J. Vandemark.** 1970. Phosphorylation coupled to NADH oxidation with fumarate in *Streptococcus faecalis* 10Cl. Arch. Biochem. Biophys. **137**:392–398.

127. **Fimmel, A. L., and B. A. Haddock.** 1979. Use of *chlC-lac* fusions to determine regulation of gene *chlC* in *Escherichia coli* K-12. J. Bacteriol. **138**:726–730.

128. **Fimmel, A. L., and B. A. Haddock.** 1981. Characterisation of the *Escherichia coli chlC* regulatory region in a cloned *chlC-lac* gene fusion. FEMS Microbiol. Lett. **12**:125–129.

129. **Forget, P.** 1974. The bacterial nitrate reductases. Solubilization, purification and properties of the enzyme A of *Escherichia coli* K12. Eur. J. Biochem. **42**:325–332.

130. **Forget, P.** 1979. Effect of growth conditions on the synthesis of nitrate reductase components in chlorate resistant mutants of *Escherichia coli* K12. Biochem. Biophys. Res. Commun. **89**:659–663.

131. **Forget, P., and M. Dubourdieu.** 1982. Evidence for the presence of a small subunit as the principal component of the nitrate reductase of *Escherichia coli* K12. Biochem. Biophys. Res. Commun. **105**:450–456.

132. **Forget, P. A., and R. Rimassa.** 1977. Evidence for the presence of carbohydrate units in the nitrate reductase A of *Escherichia coli* K12. FEBS Lett. **77**:182–186.

133. **Freedberg, W. B., and E. C. C. Lin.** 1973. Three kinds of controls affecting the expression of the *glp* region in *Escherichia coli.* J. Bacteriol. **115**:816–823.

134. **Fujita, T.** 1966. Studies on soluble cytochromes in *Enterobacteriaceae.* I. Detection, purification, and properties of cytochrome *c*-552 in anaerobically grown cells. J. Biochem. **60**:204–215.

135. **Fujita, T., and R. Sato.** 1966. Studies on soluble cytochromes in *Enterobacteriaeae.* III. Localization of cytochrome *c*-552 in the surface of the cells. J. Biochem. **60**:568–577.

136. **Fujita, T., and R. Sato.** 1966. Studies on soluble cytochromes in *Enterobacteriaceae.* IV. Possible involvement of cytochrome *c*-552 in anaerobic nitrite metabolism. J. Biochem. **60**:691–700.

137. **Fujita, T., and R. Sato.** 1966. Studies on soluble cytochromes in *Enterobacteriaceae.* V. Nitrite-dependent gas evolution in cells containing cytochrome c_{552}. J. Biochem. **62**:230–238.

138. **Fukuyama, T., and E. J. Ordal.** 1965. Induced biosynthesis of formic hydrogenlyase in iron-deficient cells of *Escherichia coli.* J. Bacteriol. **90**: 673–680.

139. **Gale, E. F.** 1939. Formic dehydrogenase of *Bacterium coli*: its inactivation by oxygen and its protection in the bacterial cell. Biochem. J. **33**:1012–1027.

140. **Garland, P. B., J. A. Downie, and B. A. Haddock.** 1975. Proton translocation and the respiratory nitrate reductase of *Escherichia coli.* Biochem. J. **152**:547–559.

141. **George, G. N., R. C. Bray, F. F. Morpeth, and D. H. Boxer.** 1985. Complexes with halide and other anions of the molybdenum centre of nitrate reductase from *Escherichia coli.* Biochem. J. **227**:925–931.

142. **Gest, H.** 1952. Properties of cell-free hydrogenases of *Escherichia coli* and *Rhodospirillum rubrum.* J. Bacteriol. **63**:111–121.

143. **Gest, H.** 1980. The evolution of biological energy-transducing systems. FEMS Microbiol. Lett. **7**:73–77.

144. **Gest, H., and M. Gibbs.** 1952. Preparation and properties of cell-free formic hydrogenlyase from *Escherichia coli.* J. Bacteriol. **63**:661–664.

145. **Gest, H., and H. D. Peck, Jr.** 1955. A study of the hydrogenlyase reaction with systems derived from normal and anaerogenic coli-aerogenes bacteria. J. Bacteriol. **70**:326–334.

146. **Giordano, G., A. L. Fimmel, L. M. Powell, B. A. Haddock, and G. C. Barr.** 1981. Cloning and expression of the *chlA* gene of *Escherichia coli* K12. FEMS Microbiol. Lett. **12**:61–64.

147. **Giordano, G., A. Graham, D. H. Boxer, B. A. Haddock, and E. Azoulay.** 1978. Characterization of the membrane-bound nitrate reductase activity of aerobically grown chlorate-sensitive mutants of *Escherichia coli* K12. FEBS Lett. **95**:290–294.

148. **Giordano, G., L. Grillet, J. Pommier, C. Terriere, B. A. Haddock, and E. Azoulay.** 1980. Precursor forms of the subunits of nitrate reductase *chlA* and *chlB* mutants of *Escherichia coli* K12. Eur. J. Biochem. **105**:297–306.

149. **Giordano, G., L. Grillet, R. Rosset, J. H. Dou, E. Azoulay, and B. A. Haddock.** 1978. Characterization of an *Escherichia coli* K12 mutant that is sensitive to chlorate when grown aerobically. Biochem. J. **176**:553–561.

150. **Giordano, G., B. A. Haddock, and D. H. Boxer.** 1980. Molybdenum-limited growth achieved either phenotypically or genotypically and its effect on the synthesis of formate dehydrogenase and nitrate reductase by *Escherichia coli* K12. FEMS Microbiol. Lett. **8**:229–235.

151. **Giordano, G., C.-L. Medani, D. H. Boxer, and E. Azoulay.** 1982. Effects of tungstate and molybdate on the *in vitro* reconstitution of nitrate reductase in *Escherichia coli* K-12. FEMS Microbiol. Lett. **13**:317–323.

152. **Giordano, G., C.-L. Medani, M.-A. Mandrand-Bertholot, and D. H. Boxer.** 1983. Formate dehydrogenases from *Escherichia coli*. FEMS Microbiol. Lett. **17**:171–177.

153. **Giordano, G., C. Riviere, and E. Azoulay.** 1973. Membrane reconstitution in *chl-r* mutants of *Escherichia coli* K12. V. ATPase incorporation into particles formed by complementation. Biochim. Biophys. Acta **307**:513–524.

154. **Giordano, G., R. Rosset, and E. Azoulay.** 1977. Isolation and study of mutants of *Escherichia coli* K12 that are sensitive to chlorate and derepression for nitrate reductase. FEMS Microbiol. Lett. **2**:21–26.

155. **Giordano, G., L. Saracino, and L. Grillet.** 1985. Identification in various chlorate-resistant mutants of a protein involved in the activation of nitrate reductase in the soluble fraction of a *chlA* mutant of *Escherichia coli* K-12. Biochim. Biophys. Acta **839**:181–190.

156. **Giordano, G., M. Violet, C.-L. Medani, and J. Pommier.** 1984. A common pathway for the activation of several molybdoenzymes in *Escherichia coli* K 12. Biochim. Biophys. Acta **798**:216–225.

157. **Glaser, J. H., and J. A. DeMoss.** 1971. Phenotypic restoration by molybdate of nitrate reductase activity in *chlD* mutants of *Escherichia coli*. J. Bacteriol. **108**:854–860.

158. **Glaser, J. H., and J. A. DeMoss.** 1972. Comparison of nitrate reductase mutants of *Escherichia coli* selected by alternative procedures. Mol. Gen. Genet. **116**:1–10.

159. **Glick, B. R., P. Y. Wang, H. Schneider, and W. G. Martin.** 1980. Identification and partial characterisation of an *Escherichia coli* mutant with altered hydrogenase activity. Can. J. Biochem. **58**:361–367.

160. **Glick, B. R., J. Zeisler, A. M. Banaszuk, J. D. Friesen, and W. G. Martin.** 1981. The identification and partial characterization of a plasmid containing the gene for the membrane-associated hydrogenase from *E. coli*. Gene **15**:201–206.

161. **Goldberg, I., K. Lonberg-Holm, E. A. Bagley, and B. Stieglitz.** 1983. Improved conversion of fumarate to succinate by *Escherichia coli* strains amplified for fumarate reductase Appl. Environ. Microbiol. **45**:1838–1847.

162. **Graham, A.** 1981. The organization of hydrogenase in the cytoplasmic membrane of *Escherichia coli*. Biochem. J. **197**:283–291.

163. **Graham, A., and D. H. Boxer.** 1978. Immunochemical localization of nitrate reductase in *Escherichia coli*. Biochem. Soc. Trans. **6**:1210–1211.

164. **Graham, A., and D. H. Boxer.** 1980. Implication of alpha-subunit of *Escherichia coli* nitrate reductase in catalytic activity. Biochem. Soc. Trans. **8**:329–330.

165. **Graham, A., and D. H. Boxer.** 1980. The membrane localization of the beta-subunit of nitrate reductase from *Escherichia coli*. Biochem. Soc. Trans. **8**:331.

166. **Graham, A., and D. H. Boxer.** 1980. Arrangement of respiratory nitrate reductase in the cytoplasmic membrane of *Escherichia coli*: location of beta-subunit. FEBS Lett. **113**:15–20.

167. **Graham, A., and D. H. Boxer.** 1981. The organization of formate dehydrogenase in the cytoplasmic membrane of *Escherichia coli*. Biochem. J. **195**:627–637.

168. **Graham, A., D. H. Boxer, B. A. Haddock, M.-A. Mandrand-Berthelot, and R. W. Jones.** 1980. Immunochemical analysis of the membrane-bound hydrogenase of *Escherichia coli*. FEBS Lett. **113**:167–173.

169. **Graham, A., H. E. Jenkins, N. H. Smith, M.-A. Mandrand-Berthelot, B. A. Haddock, and D. H. Boxer.** 1980. The synthesis of formate dehydrogenase and nitrate reductase proteins in various *fdh* and *chl* mutants of *Escherichia coli*. FEMS Microbiol. Lett. **7**:145–151.

170. **Graham, A., A. D. Tucker, and N. H. Smith.** 1981. The formate-nitrate respiratory chain of *Escherichia coli*: localisation of proteins by immunoadsorption studies. FEMS Microbiol. Lett. **11**:141–147.

171. **Gray, C. T., and H. Gest.** 1965. Biological formation of molecular hydrogen. Science **148**:186–192.

172. **Gray, C. T., J. W. T. Wimpenny, D. E. Hughes, and M. R. Mossman.** 1966. Regulation of metabolism in facultative bacteria. I. Structural and functional changes in *Escherichia coli* associated with shifts between the aerobic and anaerobic states. Biochim. Biophys. Acta **117**:22–32.

173. **Gray, C. T., J. W. T. Wimpenny, D. E. Hughes, and M. Ranlett.** 1963. A soluble c-type cytochrome from anaerobically grown *Escherichia coli* and various *Enterobacteriaceae*. Biochim. Biophys. Acta **67**:157–160.

174. **Green, D. E., and L. H. Stickland.** 1934. Studies on reversible dehydrogenase systems. I. The reversibility of the hydrogenase system of *Bact. coli*. Biochem. J. **28**:898–900.

175. **Grillet, L., and G. Giordano.** 1983. Identification and purification of a protein involved in the activation of nitrate reductase in the soluble fraction of a *chlA* mutant of *Escherichia coli* K 12. Biochim. Biophys. Acta **749**:115–124.

176. **Grunberg-Manago, M., J. Szulmajster, and A. Prouvost.** 1951. Hydrogenlyase, formicodeshydrogenase et hydrogenase chez *Escherichia coli*. C. R. Acad. Sci. **233**:1690–1692.

177. **Grundström, T., and B. Jaurin.** 1982. Overlap between *ampC* and *frd* operons on the *Escherichia coli* chromosome. Proc. Natl. Acad. Sci. USA **79**:1111–1115.

178. **Guest, J. R.** 1969. Biochemical and genetic studies with nitrate reductase C-gene mutants of *Escherichia coli*. Mol. Gen. Genet. **105**:285–297.

179. **Guest, J. R.** 1977. Menaquinone biosynthesis: mutants of *Escherichia coli* K-12 requiring 2-succinylbenzoate. J. Bacteriol. **130**:1038–1046.

180. **Guest, J. R.** 1979. Anaerobic growth of *Escherichia coli* K12 with fumarate as terminal electron acceptor. Genetic studies with menaquinone and fluoroacetate-resistant mutants. J. Gen. Microbiol. **115**:259–271.

181. **Guest, J. R.** 1981. Partial replacement of succinate dehydrogenase function by phage- and plasmid-specified fumarate reductase in *Escherichia coli*. J. Gen. Microbiol. **122**:171–179.

182. **Gutowski, S. J., and H. Rosenberg.** 1976. Energy coupling to active transport in anaerobically grown mutants of *Escherichia coli* K12. Biochem. J. **154**:731–734.

183. **Gutowski, S. J., and H. Rosenberg.** 1976. Effects of dicyclohexylcarbodiimide on proton translocation coupled to fumarate reduction in anaerobically grown *Escherichia coli* K12. Biochem. J. **160**:813–816.

184. **Gutowski, S. J., and H. Rosenberg.** 1977. Proton translocation coupled to electron flow from endogenous substrates to fumarate in anaerobically grown *Escherichia coli* K12. Biochem. J. **164**:265–267.

185. **Hackenthal, E., W. Mannheim, R. Hackenthal, and R. Becher.** 1964. Die Reduktion von Perchlorat durch Bakterien. I. Untersuchungen an Intakten Zellen. Biochem. Pharmacol. **13**:195–206.

186. **Hackett, C. S., and C. H. MacGregor.** 1981. Synthesis and degradation of nitrate reductase in *Escherichia coli*. J. Bacteriol. **146**:352–359.

187. **Hackett, N. R., and P. D. Bragg.** 1982. The association of two distinct *b* cytochromes with the respiratory nitrate reduction of *Escherichia coli*. FEMS Microbiol. Lett. **13**:213–217.

188. **Hackett, N. R., and P. D. Bragg.** 1983. Membrane cytochromes of *Escherichia coli* grown aerobically and anaerobically with nitrate. J. Bacteriol. **154**:708–718.

189. **Hackett, N. R., and P. D. Bragg.** 1983. Membrane cytochromes of *Escherichia coli chl* mutants. J. Bacteriol. **154**:719–727.

190. **Haddock, B. A., and M. W. Kendall-Tobias.** 1975. Functional anaerobic electron transport linked to the reduction of nitrate and fumarate in membranes from *Escherichia coli* as demonstrated by quenching of atebrin fluorescence. Biochem. J. **152**:655–659.

191. **Haddock, B. A., and M.-A. Mandrand-Berthelot.** 1982. *Escherichia coli* formate-to-nitrate respiratory chain: genetic analysis. Biochem. Soc. Trans. **10**:478–480.

192. **Harden, A.** 1901. The chemical action of *Bacillus coli communis* and similar organisms on carbohydrates and allied compounds. J. Chem. Soc. **79**:610–628.

193. **Hayashi, S., and E. C. C. Lin.** 1965. Capture of glycerol by cells of *Escherichia coli*. Biochim. Biophys. Acta **94**:479–487.

194. **Hederstedt, L., and L. Rutberg.** 1981. Succinate dehydrogenase—a comparative review. Microbiol. Rev. **45**:542–555.

195. **Hellingwerf, K. J., J. G. M. Bolscher, and W. N. Konings.** 1981. The electrochemical proton gradient generated by the fumarate-reductase system in *Escherichia coli* and its bioenergetic implications. Eur. J. Biochem. **113**:369–374.

196. **Heredia, C. F., and A. Medina.** 1960. Nitrate reductase and related enzymes in *Escherichia coli*. Biochem. J. **77**:24–30.

197. **Hertz, R., and J. Bar-Tana.** 1977. Anaerobic electron transport

in anaerobic flagellum formation in *Escherichia coli*. J. Bacteriol. **132**:1034–1035.

198. **Hirsch, C. A., M. Rasminsky, B. D. Davis, and E. C. C. Lin.** 1963. A fumarate reductase in *Escherichia coli* distinct from succinate dehydrogenase J. Biol. Chem. **238**:3770–3774.

199. **Holländer, R.** 1976. Correlation of the function of demethylmenaquinone in bacterial electron transport with its redox potential. FEBS Lett. **72**:98–100.

200. **Holmgren, E., L. Hederstedt, and L. Rutberg.** 1979. Role of heme synthesis and membrane binding of succinate dehydrogenase in *Bacillus subtilis*. J. Bacteriol. **138**:377–382.

201. **Hughes, D. E.** 1951. A press for disrupting bacteria and other micro-organisms. Br. J. Exp. Pathol. **32**:97–109.

202. **Iida, K., and S. Taniguchi.** 1959. Studies on nitrate reductase system of *Escherichia coli*. I. Particulate electron transport to nitrate and its solubilization. J. Biochem. **46**:1041–1055.

203. **Iida, K., and K. Yamasaki.** 1960. Spectrographic determination of molybdenum in nitrate reductase from *Escherichia coli*. Biochim. Biophys. Acta **44**:352–353.

203a. **Imperial, J., R. A. Ugalde, V. K. Shah, and W. J. Brill.** 1985. Mol⁻ mutants of *Klebsiella pneumoniae* requiring high levels of molybdate for nitrogenase activity. J. Bacteriol. **163**:1285–1287.

204. **Ingledew, W. J.** 1983. The electron transport chain of *Escherichia coli* grown anaerobically with fumarate as terminal electron acceptor: an electron paramagnetic resonance study. J. Gen. Microbiol. **129**:1651–1659.

205. **Ingledew, W. J., and R. K. Poole.** 1984. The respiratory chains of *Escherichia coli*. Microbiol. Rev. **48**:222–271.

206. **Ishimoto, M., and O. Shimokawa.** 1978. Reduction of trimethylamine N-oxide by *Escherichia coli* as anaerobic respiration. Z. Allg. Mikrobiol. **18**:173–181.

207. **Itagaki, E., T. Fujita, and R. Sato.** 1961. Solubilization and some properties of formic dehydrogenase from *Escherichia coli*. Biochem. Biophys. Res. Commun. **5**:30–34.

208. **Itagaki, E., T. Fujita, and R. Sato.** 1961. Cytochrome b₁-nitrate reductase interaction in a solubilized system from *Escherichia coli*. Biochim. Biophys. Acta **51**:390–392.

209. **Itagaki, E., T. Fujita, and R. Sato.** 1962. Solubilization and properties of formic dehydrogenase and cytochrome b₁ from *Escherichia coli*. J. Biochem. **52**:131–141.

210. **Iuchi, S., D. R. Kuritzkes, and E. C. C. Lin.** 1985. *Escherichia coli* mutant with altered respiratory control of the *frd* operon. J. Bacteriol. **161**:1023–1028.

211. **Jackson, R. H., J. A. Cole, and A. Cornish-Bowden.** 1981. The steady-state kinetics of the NADH-dependent nitrite reductase from *Escherichia coli* K12. Biochem. J. **199**:171–178.

212. **Jackson, R. H., A. Cornish-Bowden, and J. A. Cole.** 1981. Prosthetic groups of the NADH-dependent nitrite reductase from *Escherichia coli* K12. Biochem. J. **193**:861–867.

213. **Jacobs, J. M., and N. J. Jacobs.** 1977. The late steps of anaerobic heme biosynthesis in *Escherichia coli*: role for quinones in protoporphyrinogen oxidation. Biochem. Biophys. Res. Commun. **78**:429–433.

214. **Jacobs, N. J., and J. M. Jacobs.** 1975. Fumarate as alternate electron acceptor for the late steps of anaerobic heme synthesis in *Escherichia coli*. Biochem. Biophys. Res. Commun. **65**:435–441.

215. **Jacobs, N. J., and J. M. Jacobs.** 1976. Nitrate, fumarate and oxygen as electron acceptors for a late step in microbial heme synthesis. Biochim. Biophys. Acta **449**:1–9.

216. **Jacobs, N. J., and J. M. Jacobs.** 1977. Evidence for involvement of the electron transport system at a late step of anaerobic microbial heme synthesis. Biochim. Biophys. Acta **459**:141–144.

217. **Jacobs, N. J., and J. M. Jacobs.** 1978. Quinones as hydrogen carriers for a late step in anaerobic heme biosynthesis in *Escherichia coli*. Biochim. Biophys. Acta **544**:540–546.

218. **Jacobs, N. J., and P. J. VanDemark.** 1960. Comparison of the mechanism of glycerol oxidation in aerobically and anaerobically grown *Streptococcus faecalis*. J. Bacteriol. **79**:532–538.

219. **Jamieson, D. J., and C. F. Higgins.** 1984. Anaerobic and leucine-dependent expression of a peptide transport gene in *Salmonella typhimurium*. J. Bacteriol. **160**:131–136.

220. **Janick, P. A., D. C. Rueger, R. J. Krueger, M. J. Barber, and L. M. Siegel.** 1983. Characterization of complexes between *Escherichia coli* sulfite reductase hemoprotein subunit and its substrates sulfite and nitrite. Biochemistry **22**:396–408.

221. **Janick, P. A., and L. M. Siegel.** 1983. Electron paramagnetic resonance and optical evidence for interaction between siroheme and Fe₄S₄ prosthetic groups in complexes of *Escherichia coli* sulfite reductase hemoprotein with added ligands. Biochemistry **22**:504–515.

222. **Jenkins, H. E., A. Graham, and B. A. Haddock.** 1979. Characterization of a *chlG* mutant of *Escherichia coli* K12. FEMS Microbiol. Lett. **6**:169–173.

223. **Jenkins, H. E., and B. A. Haddock.** 1980. A specific method for the isolation of *chlG* mutants of *Escherichia coli* K12. FEMS Microbiol. Lett. **9**:293–296.

224. **John, P.** 1977. Aerobic and anaerobic bacterial respiration monitored by electrodes. J. Gen. Microbiol. **98**:231–238.

225. **Johnson, J. L., B. E. Hairline, and K. V. Rajagopalan.** 1980. Characterization of the molybdenum cofactor of sulfite oxidase, xanthine oxidase, and nitrate reductase. Identification of a pteridine as a structural component. J. Biol. Chem. **255**:1783–1786.

226. **Johnson, M. K., D. E. Bennet, J. E. Morningstar, M. W. W. Adams, and L. E. Mortenson.** 1985. The iron-sulfur cluster composition of *Escherichia coli* nitrate reductase. J. Biol. Chem. **260**:5456–5463.

227. **Johnson, M. K., J. E. Morningstar, G. Cecchini, and B. A. C. Ackrell.** 1985. In vivo detection of a three iron cluster in fumarate reductase from *Escherichia coli*. Biochem. Biophys. Res. Commun. **131**:653–658.

228. **Joklik, W. K.** 1950. The hydrogenase of *E. coli* in the cell-free state. I. Concentration, properties and activation. Australian J. Exp. Biol. Med. Sci. **28**:321–329.

229. **Jones, H. M., and R. P. Gunsalus.** 1985. Transcription of the *Escherichia coli* fumarate reductase genes (*frdABCD*) and their coordinate regulation of oxygen, nitrate, and fumarate. J. Bacteriol. **164**:1100–1109.

230. **Jones, R. W.** 1979. The topography of the membrane-bound hydrogenase of *Escherichia coli* explored by non-physiological electron acceptors. Biochem. Soc. Trans. **7**:724–725.

231. **Jones, R. W.** 1979. Hydrogen-dependent proton translocation by membrane vesicles from *Escherichia coli*. Biochem. Soc. Trans. **7**:1136–1137.

232. **Jones, R. W.** 1980. Proton translocation by the membrane-bound formate dehydrogenase of *Escherichia coli*. FEMS Microbiol. Lett. **8**:167–171.

233. **Jones, R. W.** 1980. The role of the membrane-bound hydrogenase in the energy-conserving oxidation of molecular hydrogen by *Escherichia coli*. Biochem. J. **188**:345–350.

234. **Jones, R. W., and P. B. Garland.** 1977. Sites and specificity of the reaction of bipyridilyium compounds with anaerobic respiratory enzymes of *Escherichia coli*. Effects of permeability barriers imposed by the cytoplasmic membrane. Biochem. J. **164**:199–211.

235. **Jones, R. W., and P. B. Garland.** 1978. The proton-consuming site of the respiratory nitrate reductase of *Escherichia coli* is on the cytoplasmic aspect of the cytoplasmic membrane. Biochem. Soc. Trans. **6**:416–418.

236. **Jones, R. W., W. J. Ingledew, A. Graham, and P. B. Garland.** 1978. Topography of nitrate reductase of the cytoplasmic membrane of *Escherichia coli*: the nitrate-reducing site. Biochem. Soc. Trans. **6**:1287–1289.

237. **Jones, R. W., A. Lamont, and P. B. Garland.** 1980. The mechanism of proton translocation driven by the respiratory nitrate reductase complex of *Escherichia coli*. Biochem. J. **190**:79–94.

238. **Kajie, S.-I., K. Miki, E. C. C. Lin, and Y. Anraku.** 1984. Isolation of an *Escherichia coli* mutant defective in cytochrome biosynthesis. FEMS Microbiol. Lett. **24**:25–29.

238a. **Kajie, S.-I., and Y. Anraku.** 1986. Purification of a hexaheme cytochrome c₅₅₂ from *Escherichia coli* K12 and its properties as a nitrite reductase. Eur. J. Biochem. **154**:457–463.

239. **Kapralek, F., E. Jechova, and M. Otavova.** 1982. Two sites of oxygen control in induced synthesis of respiratory nitrate reductase in *Escherichia coli*. J. Bacteriol. **149**:1142–1145.

240. **Karube, I., M. Tomiyama, and A. Kikuchi.** 1984. Molecular cloning and physical mapping of the *hyd* gene of *Escherichia coli* K-12. FEMS Microbiol. Lett. **25**:165–168.

241. **Kavanaugh, B. M., and J. A. Cole.** 1976. Characterization of mutants of *Escherichia coli* K12 defective in nitrite assimilation. Proc. Soc. Gen. Microbiol. **3**:84.

242. **Kelley, W. S., K. Chalmers, and N. E. Murray.** 1977. Isolation and characterization of a lambda *polA* transducing phage. Proc. Natl. Acad. Sci. USA **74**:5632–5636.

243. **Kemp, J. D., and D. E. Atkinson.** 1966. Nitrite reductase of *Escherichia coli* specific for reduced nicotinamide adenine dinucleotide. J. Bacteriol. **92**:628–634.

244. **Kemp, J. D., D. E. Atkinson, A. Ehret, and R. A. Lazzarini.** 1963. Evidence for the identity of the nicotinamide adenine nucleotide phosphate-specific sulfite and nitrite reductases of *Escherichia coli*. J. Biol. Chem. **238**:3466–3471.

245. **Kemp, M. B., B. A. Haddock, and P. B. Garland.** 1975. Synthesis and sidedness of membrane-bound respiratory nitrate reductase

(EC 1.7.99.4) in *Escherichia coli* lacking cytochromes. Biochem. J. **148**:329–333.

246. **Kim, K. E., and G. W. Chang.** 1974. Trimethylamine oxide reduction by *Salmonella*. Can. J. Microbiol. **20**:1745–1748.
247. **Kistler, W. S., C. A. Hirsch, N. R. Cozzarelli, and E. C. C. Lin.** 1969. Second pyridine nucleotide-independent L-alpha-glycerophosphate dehydrogenase in *Escherichia coli* K-12. J. Bacteriol. **100**:1133–1135.
248. **Kistler, W. S., and E. C. C. Lin.** 1971. Anaerobic L-alpha-glycerophosphate dehydrogenase of *Escherichia coli*: its genetic locus and its physiological role. J. Bacteriol. **108**:1224–1234.
249. **Kistler, W. S., and E. C. C. Lin.** 1972. Purification and properties of the flavine-stimulated anaerobic L-alpha-glycerophosphate dehydrogenase of *Escherichia coli*. J. Bacteriol. **112**:539–547.
250. **Kobayashi, M., and M. Ishimoto.** 1973. Aerobic inhibition of nitrate assimilation in *Escherichia coli*. Z. Allg. Mikrobiol. **13**:405–413.
251. **Koch, J. P., S.-I. Hayashi, and E. C. C. Lin.** 1964. The control of dissimilation of glycerol and L-alpha-glycerophosphate in *Escherichia coli*. J. Biol. Chem. **239**:3106–3108.
252. **Konings, W. N., and H. R. Kaback.** 1973. Anaerobic transport in *Escherichia coli* membrane vesicles. Proc. Natl. Acad. Sci. USA **70**:3376–3381.
253. **Krasna, A. I.** 1980. Regulation of hydrogenase activity in enterobacteria. J. Bacteriol. **144**:1094–1097.
254. **Krasna, A. I.** 1984. Mutants of *Escherichia coli* with altered hydrogenase activity. J. Gen. Microbiol. **130**:779–787.
255. **Krasna, A. I., and D. Rittenberg.** 1956. Comparison of the hydrogenase activities of different microogranisms. Proc. Natl. Acad. Sci. USA **42**:180–185.
256. **Krebs, H. A.** 1937. The role of fumarate in the respiration of *Bacterium coli commune*. Biochem. J. **31**:2095–2124.
257. **Kristjansson, J. K., and T. C. Hollocher.** 1979. Substrate-binding site for nitrate reductase of *Escherichia coli* is on the inner aspect of the membrane. J. Bacteriol. **137**:1227–1233.
258. **Kröger, A.** 1978. Fumarate as terminal acceptor of phosphorylative electron transport. Biochim. Biophys. Acta **505**:129–145.
259. **Kung, H.-F., and U. Henning.** 1972. Limiting availability of binding sites for dehydrogenases on the cell membrane of *Escherichia coli*. Proc. Natl. Acad. Sci. USA **69**:925–929.
260. **Kuritzkes, D. R., X.-Y. Zhang, and E. C. C. Lin.** 1984. Use of φ(*glp-lac*) in studies of respiratory regulation of the *Escherichia coli* anaerobic sn-glycerol-3-phosphate dehydrogenase genes (*glpAB*). J. Bacteriol. **157**:591–598.
261. **Kwan, H. S., and E. L. Barrett.** 1983. Roles for menaquinone and the two trimethylamine oxide (TMAO) reductases in TMAO respiration in *Salmonella typhimurium*: Mu d(Ap^r *lac*) insertion mutations in *men* and *tor*. J. Bacteriol. **155**:1147–1155.
262. **Kwan, H. S., and E. L. Barrett.** 1983. Purification and properties of trimethylamine oxide reductase from *Salmonella typhimurium*. J. Bacteriol. **155**:1455–1458.
263. **Lambden, P. R., and J. R. Guest.** 1976. A novel method for isolating chlorate-resistant mutants of *Escherichia coli* K12 by anaerobic selection on a lactate plus fumarate medium. J. Gen. Microbiol. **93**:173–176.
264. **Lambden, P. R., and J. R. Guest.** 1976. Mutants of *Escherichia coli* K12 unable to use fumarate as an anaerobic electron acceptor. J. Gen. Microbiol. **97**:145–160.
265. **Lascelles, J., and J. L. Still.** 1947. The utilization of fumarate and malate by *Escherichia coli* in the presence of molecular hydrogen. Proc. Linn. Soc. N. S. W. **47**:50–57.
266. **Lazzarini, R. A., and D. E. Atkinson.** 1961. A triphosphopyridine nucleotide-specific nitrite reductase from *Escherichia coli*. J. Biol. Chem. **236**:3330–3335.
267. **Lee, J. H., P. Patel, P. Sankar, and K. T. Shanmugam.** 1984. Isolation and characterization of mutant strains of *Escherichia coli* altered in H₂ metabolism. J. Bacteriol. **162**:344–352.
268. **Le Minor, L., M. Chippaux, F. Pichinoty, C. Coynault, and M. Pechaud.** 1970. Methodes simples permettant de rechercher la tetrathionate-reductase en cultures liquides ou sur colonies isolees. Ann. Inst. Pasteur (Paris) **119**:733–737.
269. **Lemire, B. D., J. J. Robinson, R. D. Bradley, D. G. Scraba, and J. H. Weiner.** 1983. Structure of fumarate reductase on the cytoplasmic membrane of *Escherichia coli*. J. Bacteriol. **155**:391–397.
270. **Lemire, B. D., J. J. Robinson, and J. H. Weiner.** 1982. Identification of membrane anchor polypeptides of *Escherichia coli* fumarate reductase. J. Bacteriol. **152**:1126–1131.
271. **Lester, R. L., and J. A. DeMoss.** 1971. Effects of molybdate and selenite on formate and nitrate metabolism in *Escherichia coli*. J. Bacteriol. **105**:1006–1014.
272. **Li, S., T. Rabi, and J. A. DeMoss.** 1985. Delineation of two

distinct regulatory domains in the 5' region of the *nar* operon of *Escherichia coli*. J. Bacteriol. **164**:25–32.
273. **Linnane, A. W., and C. W. Wrigley.** 1963. Fragmentation of the electron transport chain of *Escherichia coli*. Preparation of a soluble formate dehydrogenase-cytochrome b₁ complex. Biochim. Biophys. Acta **77**:408–418.
274. **Liu, M.-C., H. D. Peck, Jr., A. Abou-Jaoude, M. Chippaux, and J. LeGall.** 1981. A reappraisal of the role of the low potential c-type cytochrome (cytochrome c₅₅₂) in NADH-dependent nitrite reduction and its relationship with a co-purified NADH oxidase in *Escherichia coli* K-12. FEMS Microbiol. Lett. **10**:333–337.
275. **Lohmeier, E., D. S. Hagen, P. Dickie, and J. H. Weiner.** 1981. Cloning and expression of the fumarate reductase gene of *Escherichia coli*. Can. J. Biochem. **59**:158–164.
276. **Lund, K., and J. A. DeMoss.** 1976. Association-dissociation behavior and subunit structure of heat-released nitrate reductase from *Escherichia coli*. J. Biol. Chem. **251**:2207–2216.
277. **MacGregor, C., and C. Schnaitman.** 1974. Nitrate reductase in *E. coli*: properties of the enzyme and *in vitro* reconstitution from enzyme-deficient mutants. J. Supramol. Struct. **2**:715–727.
278. **MacGregor, C. H.** 1975. Solubilization of *Escherichia coli* nitrate reductase by a membrane-bound protease. J. Bacteriol. **121**:1102–1110.
279. **MacGregor, C. H.** 1975. Anaerobic cytochrome b₁ in *Escherichia coli*: association with and regulation of nitrate reductase. J. Bacteriol. **121**:1111–1116.
280. **MacGregor, C. H.** 1975. Synthesis of nitrate reductase components in chlorate-resistant mutants of *Escherichia coli*. J. Bacteriol. **121**:1117–1121.
281. **MacGregor, C. H.** 1976. Biosynthesis of membrane-bound nitrate reductase in *Escherichia coli*: evidence for a soluble precursor. J. Bacteriol. **126**:122–131.
282. **MacGregor, C. H., and C. W. Bishop.** 1977. Do cytochromes function as oxygen sensors in the regulation of nitrate reductase biosynthesis? J. Bacteriol. **131**:372–373.
283. **MacGregor, C. H., and A. R. Christopher.** 1978. Asymmetric distribution of nitrate reductase subunits in the cytoplasmic membrane of *Escherichia coli*: evidence derived from surface labelling studies and transglutaminase. Arch. Biochem. Biophys. **185**:204–213.
284. **MacGregor, C. H., and G. E. McElhaney.** 1981. New mechanism for postranslocational processing during assembly of a cytoplasmic membrane protein? J. Bacteriol. **148**:551–558.
285. **MacGregor, C. H., and C. A. Schnaitman.** 1971. Alterations in the cytoplasmic membrane proteins of various chlorate-resistant mutants of *Escherichia coli*. J. Bacteriol. **108**:564–570.
286. **MacGregor, C. H., and C. A. Schnaitman.** 1972. Restoration of NADPH-nitrate reductase activity of a *Neurospora* mutant by extracts of various chlorate-resistant mutants of *Escherichia coli*. J. Bacteriol. **112**:388–391.
287. **MacGregor, C. H., and C. A. Schnaitman.** 1973. Reconstitution of nitrate reductase activity and formation of membrane particles from cytoplasmic extracts of chlorate-resistant mutants of *Escherichia coli*. J. Bacteriol. **114**:1164–1176.
288. **MacGregor, C. H., C. A. Schnaitman, D. E. Normansell, and M. G. Hodgins.** 1974. Purification and properties of nitrate reductase from *Escherichia coli* K12. J. Biol. Chem. **249**:5321–5327.
289. **Macy, J., H. Kulla, and G. Gottschalk.** 1976. H₂-dependent anaerobic growth of *Escherichia coli* on L-malate: succinate formation. J. Bacteriol. **125**:423–428.
290. **Majer, J.** 1960. A TPNH-linked sulfite reductase and its relation to hydroxylamine reductase in enterobacteriaceae. Biochim. Biophys. Acta **41**:553–555.
291. **Mandrand-Berthelot, M.-A., M. Y. K. Wee, and B. A. Haddock.** 1978. An improved method for the identification and characterization of mutants of *Escherichia coli* defective in formate dehydrogenase activity. FEMS Microbiol. Lett. **4**:37–40.
292. **Marcot, J., and E. Azoulay.** 1971. Obtention et etude de doubles mutants chlorate-resistants chez *Escherichia coli* K12. FEBS Lett. **13**:137–139.
293. **McNall, E. G., and D. E. Atkinson.** 1956. Nitrate reduction. I. Growth of *Escherichia coli* with nitrate as sole source of nitrogen. J. Bacteriol. **72**:226–229.
294. **McNall, E. G., and D. E. Atkinson.** 1957. Nitrate reduction. II. Utilization of possible intermediates as nitrogen sources and as electron acceptors. J. Bacteriol. **74**:60–66.
295. **McPhedran, P., B. Sommer, and E. C. C. Lin.** 1961. Control of ethanol dehydrogenase levels in *Aerobacter aerogenes*. J. Bacteriol. **81**:852–857.
296. **McPherson, M. J., A. J. Baron, D. J. Pappin, and J. C. Wootton.** 1984. Respiratory nitrate reductase of *Escherichia coli*. Se-

quence identification of the large subunit gene. FEBS Lett. **177**:260–264.

297. **McRee, D. E., and D. C. Richardson.** 1982. Preliminary X-ray diffraction studies on the hemoprotein subunit of *Escherichia coli* sulphite reductase. J. Mol. Biol. **154**:179–180.

298. **Medina, A., and C. F. de Heredia.** 1958. Vitamin K-dependent nitrate reductase in *Escherichia coli*. Biochim. Biophys. Acta **28**:452–453.

299. **Meganathan, R.** 1984. Inability of *men* mutants of *Escherichia coli* to use trimethylamine-*N*-oxide as an electron acceptor. FEBS Microbiol. Lett. **24**:67–62.

300. **Miki, K., and E. C. C. Lin.** 1973. Enzyme complex which couples glycerol-3-phosphate dehydrogenation to fumarate reduction in *Escherichia coli*. J. Bacteriol. **114**:767–771.

301. **Miki, K., and E. C. C. Lin.** 1975. Anaerobic energy-yielding reaction associated with transhydrogenation from glycerol-3-phosphate to fumarate by an *Escherichia coli* system. J. Bacteriol. **124**:1282–1287.

302. **Miki, K., and E. C. C. Lin.** 1975. Electron transport chain from glycerol-3-phosphate to nitrate in *Escherichia coli*. J. Bacteriol. **124**:1288–1294.

303. **Miki, K., and E. C. C. Lin.** 1980. Use of *Escherichia coli* operon-fusion strains for the study of glycerol-3-phosphate transport activity. J. Bacteriol. **143**:1436–1443.

304. **Miki, K., and T. H. Wilson.** 1978. Proton translocation associated with anaerobic transhydrogenation from glycerol-3-phosphate to fumarate in *Escherichia coli*. Biochem. Biophys. Res. Commun. **83**:1570–1575.

305. **Miller, J. B., and N. K. Amy.** 1983. Molybdenum cofactor in chlorate-resistant and nitrate reductase-deficient insertion mutants of *Escherichia coli*. J. Bacteriol. **155**:793–801.

306. **Morningstar, J. E., M. K. Johnson, G. Cecchini, B. A. C. Ackrell, and E. B. Kearney.** 1985. The high potential iron-sulfur center in *Escherichia coli* fumarate reductase is a three-iron cluster. J. Biol. Chem. **260**:13631–13638.

307. **Morpeth, F. F., and D. H. Boxer.** 1985. Kinetic analysis of respiratory nitrate reductase from *Escherichia coli* K 12. Biochemistry **24**:40–46.

308. **Motohara, K., M. Kobayashi, and M. Ishimoto.** 1976. Assimilatory nitrate reductase in a chlorate-resistant mutant of *Escherichia coli*. Z. Allg. Mikrobiol. **16**:543–550.

309. **Motteram, P. A. S., J. E. G. McCarthy, S. J. Ferguson, J. B. Jackson, and J. A. Cole.** 1981. Energy conservation during the formate-dependent reduction of nitrate by *Escherichia coli*. FEMS Microbiol. Lett. **12**:317–320.

310. **Murphy, M. J., and L. M. Siegel.** 1973. Siroheme and sirohydrochlorin. The basis for a new type of porphyrin-related prosthetic group common to both assimilatory and dissimilatory sulfite reductases. J. Biol. Chem. **248**:6911–6919.

311. **Murphy, M. J., L. M. Siegel, and H. Kamin.** 1973. Reduced nicotinamide adenine dinucleotide phosphate-sulfite reductase of enterobacteria. II. Identification of a new class of heme prosthetic group: an iron-tetrahydroporphyrin (isobacteriochlorin type) with eight carboxylic acid groups. J. Biol. Chem. **248**:2801–2814.

312. **Murphy, M. J., L. M. Siegel, S. R. Tove, and H. Kamin.** 1974. Siroheme: a new prosthetic group participating in six-electron reduction reactions catalyzed by both sulfite and nitrite reductases. Proc. Natl. Acad. Sci. USA **71**:612–616.

313. **Murray, E. D., and B. D. Sanwal.** 1963. An immunological enquiry into the identity of assimilatory and dissimilatory nitrate reductase from *Escherichia coli*. Can J. Microbiol. **9**:781–790.

314. **Mutaftschiev, S., and E. Azoulay.** 1973. Membrane reconstitution in *chl-r* mutants of *Escherichia coli* K12. VI. Morphological study of membrane assembly during complementation between extracts of *chl-r* mutants. Biochim. Biophys. Acta **307**:525–540.

315. **Mutaftschiev, S., J. Olive, and E. Azoulay.** 1976. Ultrastructure des particules reconstituees par complementation d'extracts de mutant *chl-r* d'*Escherichia coli* K12. C. R. Acad. Sci. Ser. D **283**:825–828.

316. **Newman, B. M., and J. A. Cole.** 1978. The chromosomal location and pleiotropic effects of mutations of the *nirA*⁺ gene of *Escherichia coli* K12: the essential role of *nirA*⁺ in nitrite reduction and in other anaerobic redox reactions. J. Gen. Microbiol. **106**:1–12.

317. **Newton, N. A., G. B. Cox, and F. Gibson.** 1971. The function of menaquinone (vitamin K_2) in *Escherichia coli* K-12. Biochim. Biophys. Acta **244**:155–166.

318. **Newton, N. A., G. B. Cox, and F. Gibson.** 1972. Function of ubiquinone in *Escherichia coli*: a mutant strain forming a low level of ubiquinone. J. Bacteriol. **109**:69–73.

319. **Nicholas, D. J. D., and A. Nason.** 1955. Diphosphopyridine nucle-

otide-nitrate reductase from *Escherichia coli*. J. Bacteriol. **69**:580–583.

320. **Normark, S., and L. G. Burman.** 1977. Resistance of *Escherichia coli* to penicillins: fine-structure mapping and dominance of chromosomal beta-lactamase mutations. J. Bacteriol. **132**:1–7.

321. **Normark, S., R. Edlund, T. Grundström, S. Bergström, and H. Wolf-Watz.** 1977. *Escherichia coli* K-12 mutants hyperproducing chromosomal beta-lactamase by gene repetitions. J. Bacteriol. **132**:912–922.

322. **O'Hara, J., C. T. Gray, J. Puig, and F. Pichinoty.** 1967. Defects in formate hydrogenlyase in nitrate-negative mutants of *Escherichia coli*. Biochem. Biophys. Res. Commun. **28**:951–957.

323. **Ordal, E. J., and H. O. Halvorson.** 1939. A comparison of hydrogen production from sugars and formic acid by normal and variant strains of *Escherichia coli*. J. Bacteriol. **38**:199–220.

324. **Orth, V., M. Chippaux, and M.-C. Pascal.** 1980. A mutant defective in electron transfer to nitrate in *Escherichia coli* K12. J. Gen. Microbiol. **117**:257–262.

325. **Ota, A., T. Yamanaka, and K. Okunuki.** 1964. Oxidative phosphorylation coupled with nitrate respiration. II. Phosphorylation coupled with anaerobic nitrate reduction in a cell-free extract of *Escherichia coli*. J. Biochem. **55**:131–135.

326. **Pakes, W. C. C., and W. H. Jollyman.** 1901. The bacterial decomposition of formic acid into carbon dioxide and hydrogen. J. Chem. Soc. **79**:386–391.

327. **Pascal, M.-C., J.-F. Burini, and M. Chippaux.** 1984. Regulation of the trimethylamini *N*-oxide (TMAO) reductase in *Escherichia coli*: analysis of *tor::Mudl* operon fusion. Mol. Gen. Genet. **195**:351–355.

328. **Pascal, M.-C., J.-F. Burini, J. Ratouchniak, and M. Chippaux.** 1982. Regulation of the nitrate reductase operon: effect of mutations in *chlA*, *B*, *D*, and *E* genes. Mol. Gen. Genet. **188**:103–106.

329. **Pascal, M.-C., F. Casse, M. Chippaux, and M. Lepelletier.** 1972. Genetic analysis of mutants of *Salmonella typhimurium* deficient in formate dehydrogenase activity. Mol. Gen. Genet. **120**:337–340.

330. **Pascal, M.-C., F. Casse, M. Chippaux, and M. Lepelletier.** 1975. Genetic analysis of mutants of *Escherichia coli* K12 and *Salmonella typhimurium* LT2 deficient in hydrogenase activity. Mol. Gen. Genet. **141**:173–179.

331. **Pascal, M.-C., and M. Chippaux.** 1982. Involvement of a gene of the *chlE* locus in the regulation of the nitrate reductase operon. Mol. Gen. Genet. **185**:334–338.

332. **Patrick, J. M., and W. J. Dobrogosz.** 1973. The effect of cyclic AMP on anaerobic growth of *Escherichia coli*. Biochem. Biophys. Res. Commun. **54**:555–561.

333. **Pecher, A., F. Zinoni, and A. Bock.** 1985. The seleno-polypeptide of formic dehydrogenase (formate hydrogen-lyase linked) from *Escherichia coli*: genetic analysis. Arch. Microbiol. **141**:359–363.

334. **Pecher, A., F. Zinoni, C. Jatisatienr, R. Wirth, H. Hennecke, and A. Böck.** 1983. On the redox control of synthesis of anaerobically induced enzymes in enterobacteriaceae. Arch. Microbiol. **136**:131–136.

335. **Peck, H. D., Jr., and H. Gest.** 1956. A new procedure for assay of bacterial hydrogenases. J. Bacteriol. **71**:70–80.

336. **Peck, H. D., Jr., and H. Gest.** 1957. Formic dehydrogenase and the hydrogenlyase enzyme complex in coli-aerogenes bacteria. J. Bacteriol. **73**:706–721.

337. **Pichinoty, F.** 1962. Inhibition par l'oxygene de la biosynthese et de l'activite de l'hydrogenase et de l'hydrogenlyase chez les bacteries anaerobies facultatives. Biochim. Biophys. Acta **64**:111–124.

338. **Pichinoty, F., and M. Chippaux.** 1969. Recherches sur des mutants bacteriens ayant perdu les activites catalytiques liees a la nitrate-reductase A. III. Caracteres biochimiques. Ann. Inst. Pasteur (Paris) **117**:145–178.

339. **Pichinoty, F., J. Puig, M. Chippaux, J. Bigliardi-Rouvier, and J. Gendre.** 1969. Recherches sur des mutants bacteriens ayant perdu les activites catalytiques liees a la nitrate-reductase A. II. Comportement envers le chlorate et le chlorite. Ann. Inst. Pasteur (Paris) **116**:409–432.

340. **Piechaud, M., F. Pichinoty, E. Azoulay, P. Couchoud-Beaumont, and J. Gendre.** 1969. Recherches sur des mutant bacteriens ayant perdu les activities catalytiques liees a la nitrate-reductase A. I. Description des methodes d'isolement. Ann. Inst. Pasteur (Paris) **116**:276–287.

341. **Piechaud, M., J. Puig, F. Pichinoty, E. Azoulay, and L. LeMinor.** 1967. Mutations affectant la nitrate-reductase et d'autres enzymes bacteriennes d'oxydoreduction. Etude preliminaire. Ann. Inst. Pasteur (Paris) **112**:24–37.

342. **Pinsent, J.** 1954. The need for selenite and molybdate in the

formation of formic dehydrogenase by members of the *coli-aerogenes* group of bacteria. Biochem. J. **57:**10–16.

343. **Pinsky, M. J., and J. L. Stokes.** 1952. Requirements for formic hydrogenlyase adaptation in nonproliferating suspensions of *Escherichia coli.* J. Bacteriol. **64:**151–161.

344. **Pope, N. R., and J. A. Cole.** 1982. Generation of a membrane potential by one of two independent pathways for nitrite reduction by *Escherichia coli.* J. Gen. Microbiol. **128:**219–222.

345. **Pope, N. R., and J. A. Cole.** 1984. Pyruvate and ethanol as electron donors for nitrite reduction by *Escherichia coli* K 12. J. Gen. Microbiol. **130:**1279–1284.

346. **Pugh, S. Y. R., and C. J. Knowles.** 1982. Growth of *Streptococcus faecalis* var. *zymogenes* on glycerol: the effect of aerobic and anaerobic growth in the presence and absence of haematin on enzyme synthesis. J. Gen. Microbiol. **128:**1009–1017.

347. **Puig, J., and E. Azoulay.** 1967. Etude genetique et biochimique des mutants resistant au ClO₃⁻ (genes *chlA, chlB, chlC*). C. R. Acad. Sci. Ser. D **264:**1916–1918.

348. **Puig, J., E. Azoulay, J. Gendre, and E. Richard.** 1969. Etude genetique des mutants de la region *chlA* chez l'*Escherichia coli* K12. C. R. Acad. Sci. Ser. C **268:**183–184.

349. **Puig, J., E. Azoulay, and F. Pichinoty.** 1967. Etude genetique d'une mutation a effet pleiotrope chez l'*Escherichia coli* K12. C. R. Acad. Sci. Ser. D **264:**1507–1509.

350. **Puig, J., E. Azoulay, F. Pichinoty, and J. Gendre.** 1969. Genetic mapping of the *chlC* gene of the nitrate reductase A system in *Escherichia coli* K12. Biochem. Biophys. Res. Commun. **35:**659–662.

351. **Quastel, J. H., and M. Stephenson.** 1925. Further observations on the anaerobic growth of bacteria. Biochem. J. **19:**660–666.

352. **Quastel, J. H., M. Stephenson, and M. D. Whetham.** 1925. Some reactions of resting bacteria in relation to anaerobic growth. Biochem. J. **19:**304–317.

353. **Quastel, J. H., and M. D. Whetham.** 1924. The equilibria existing between succinic, fumaric, and malic acids in the presence of resting bacteria. Biochem. J. **18:**519–534.

354. **Quastel, J. H., and M. D. Whetham.** 1925. Dehydrogenations produced by resting bacteria. I. Biochem. J. **19:**520–531.

355. **Quastel, J. H., and M. D. Whetham.** 1925. Dehydrogenations produced by resting bacteria. II. Biochem. J. **19:**645–651.

356. **Reid, G. A., and W. J. Ingledew.** 1979. Characterization and phenotypic control of the cytochrome content of *Escherichia coli.* Biochem. J. **182:**465–472.

357. **Riviere, C., and E. Azoulay.** 1971. Reconstitution *in vitro* of membranous particles by complementation between extracts of *chl-r* mutants in *Escherichia coli.* Biochem. Biophys. Res. Commun. **45:**1608–1614.

358. **Riviere, C., G. Giordano, J. Pommier, and E. Azoulay.** 1975. membrane reconstitution in *chl-r* mutants of *Escherichia coli* K12. VIII. Purification and properties of the Fₐ factor, the product of the *chlB* gene. Biochim. Biophys. Acta **389:**219–235.

359. **Robinson, J. J., and J. H. Weiner.** 1980. The effects of amphipaths on the flavin-linked aerobic glycerol-3-phosphate dehydrogenase from *Escherichia coli.* Can. J. Biochem. **58:**1172–1178.

360. **Robinson, J. J., and J. H. Weiner.** 1981. The effects of anions on fumarate reductase isolated from the cytoplasmic membrane of *Escherichia coli.* Biochem. J. **199:**473–477.

361. **Robinson, J. J., and J. H. Weiner.** 1982. Molecular properties of fumarate reductase isolated from the cytoplasmic membrane of *Escherichia coli.* Can. J. Biochem. **60:**811–816.

362. **Rondeau, S. S., P.-Y. Hsu, and J. A. DeMoss.** 1984. Construction *in vitro* of a cloned *nar* operon from *Escherichia coli.* J. Bacteriol. **159:**159–166.

363. **Rosenberg, H., G. B. Cox, J. D. Butlin, and S. J. Gutowski.** 1975. Metabolite transport in mutants of *Escherichia coli* K12 defective in electron transport and coupled phosphorylation. Biochem. J. **146:**417–423.

364. **Ruiz-Herrera, J., and A. Alvarez.** 1972. A physiological study of formate dehydrogenase, formate oxidase and hydrogenlyase from *Escherichia coli* K-12. Antonie van Leeuwenhoek J. Microbiol. Serol. **38:**479–491.

365. **Ruiz-Herrera, J., A. Alvarez, and I. Figueroa.** 1972. Solubilization and properties of formate dehydrogenases from the membrane of *Escherichia coli.* Biochim. Biophys. Acta **289:**254–261.

366. **Ruiz-Herrera, J., and J. A. DeMoss.** 1969. Nitrate reductase complex of *Escherichia coli* K-12: participation of specific formate dehydrogenase and cytochrome *b*₁ components in nitrate reduction. J. Bacteriol. **99:**720–729.

367. **Ruiz-Herrera, J., and I. Salas-Vargas.** 1976. Regulation of nitrate reductase at the transcriptional and translational levels in *Escherichia coli.* Biochim. Biophys. Acta **425:**492–501.

368. **Ruiz-Herrera, J., M. K. Showe, and J. A. DeMoss.** 1969. Nitrate

reductase complex of *Escherichia coli* K-12: isolation and characterization of mutants unable to reduce nitrate. J. Bacteriol. **97:**1291–1297.

369. **Sagai, M., and M. Ishimoto.** 1973. An enzyme reducing adenosine *N*-oxide in *Escherichia coli,* amine *N*-oxide reductase. J. Biochem. **73:**843–859.

369a.**Sakaguchi, M., and A. Kawai.** 1976. Effect of oxygen on formation and activity of trimethylamine *N*-oxide reductase in *Escherichia coli.* Bull. Jpn. Soc. Sci. Fish. **42:**563–569.

370. **Sakaguchi, M., and A. Kawai.** 1977. Electron donors and carriers for the reduction of trimethylamine *N*-oxide in *Escherichia coli.* Bull. Jpn. Soc. Sci. Fish. **43:**437–442.

371. **Sakaguchi, M., and A. Kawai.** 1978. The participation of cytochromes in the reduction of trimethylamine *N*-oxide by *Escherichia coli.* Bull. Jpn. Soc. Sci. Fish. **44:**511–516.

372. **Sakaguchi, M., and A. Kawai.** 1978. Presence of *b*- and *c*- type cytochromes in the membrane of *Escherichia coli* induced by trimethylamine *N*-oxide. Bull. Jpn. Soc. Sci. Fish. **44:**999–1002.

373. **Sanchez-Crispin, J. A., and F. Brito.** 1984. Aislamiento y caracterizacion del segmento de cadena respiratoria especifico del sistema nitrato reductasa de *Escherichia coli* K-12. Acta Cient. Venez. **35:**356–362.

374. **Sanchez-Crispin, J. A., and M. Dubourdieu.** 1984. Reconstitution "*in vitro*" of a formate dependent nitrite reductase activity starting from cytochrome C₅₅₂ and membrane vesicles from *Escherichia coli* K-12. Acta Cient. Venz. **35:**350–355.

375. **Sanchez-Crispin, J. A., M. Dubourdieu, and M. Chippaux.** 1979. Localization and characterization of cytochromes from membrane vesicles of *Escherichia coli* K-12 grown in anaerobiosis with nitrate. Biochim. Biophys. Acta **547:**198–210.

376. **Sanchez-Crispin, J. A., M. Dubourdieu, M. Chippaux, and J. Puig.** 1983. Relationship between the nitrate and oxygen respiratory systems in membrane vesicles of *Escherichia coli* K-12. Effect of 2-*N*-heptyl-4-hydroxyquinoline-*N*-oxide and ultraviolet light. Acta Cient. Venez. **34:**329–335.

377. **Sanchez-Crispin, J. A., M. Dubourdieu, and J. Puig.** 1984. Chlorate metabolism by whole cells and membrane vesicles of *Escherichia coli* K-12. Acta Cient. Venez. **35:**363–368.

378. **Sanderson, K. E., and J. R. Roth.** 1983. Linkage map of *Salmonella typhimurium,* edition VI. Microbiol. Rev. **47:**410–453.

379. **Sankar, P., J. H. Lee, and K. T. Shanmugam.** 1985. Cloning of hydrogenase genes and fine structure analysis of an operon essential for H2 metabolism in *Escherichia coli.* J. Bacteriol. **162:**353–360.

380. **Sato, R., and F. Egami.** 1949. Nitrate reductase III. Bull. Chem. Soc. Japan. **22:**137–143.

381. **Sato, R., and M. Niwa.** 1952. Studies on nitrate reductase. VII. Reinvestigation on the identity of the enzyme with cytochrome b. Bull. Chem. Soc. Jpn. **25:**202–210.

382. **Sawada, M.-T., and M. Ishimoto.** 1985. Proton translocation coupled to nitrite reduction in anaerobically grown *Escherichia coli.* J. Biochem. **97:**205–211.

383. **Sawers, R. G., S. P. Ballantine, and D. H. Boxer.** 1985. Differential expression of hydrogenase isoenzymes in *Escherichia coli* K-12: evidence for a third isoenzyme. J. Bacteriol. **164:**1324–1331.

384. **Schild, J., and J. H. Klemme.** 1985. Enzymatic nitrate assay by a kinetic method employing *Escherichia coli* nitrate reductase. Z. Naturforsch. **40:**134–137.

385. **Schnaitman, C. A.** 1969. Alteration of membrane proteins in a chlorate-resistant mutant of *Escherichia coli* Biochem. Biophys. Res. Commun. **37:**1–5.

386. **Schryvers, A., E. Lohmeier, and J. H. Weiner.** 1978. Chemical and functional properties of the native and reconstituted forms of the membrane-bound, aerobic glycerol-3-phosphate dehydrogenase of *Escherichia coli.* J. Biol. Chem. **253:**783–788.

387. **Schryvers, A., and J. H. Weiner.** 1981. The anaerobic *sn*-glycerol-3-phosphate dehydrogenase of *Escherichia coli.* Purification and characterization. J. Biol. Chem. **256:**9959–9965.

388. **Schryvers, A., and J. H. Weiner.** 1982. The anaerobic *sn*-glycerol-3-phosphate dehydrogenase: cloning and expression of the *glpA* gene of *Escherichia coli* and identification of the *glpA* products. Can. J. Biochem. **60:**224–231.

389. **Scott, A. I., A. J. Irwin, L. M. Siegel, and J. N. Shoolery.** 1978. Sirohydrochlorin. Prosthetic group of sulfite and nitrite reductases and its role in the biosynthesis of vitamin B₁₂. J. Am. Chem. Soc. **100:**7987–7994.

390. **Scott, R. H., and J. A. DeMoss.** 1976. Formation of the formate-nitrate electron transport pathway from inactive components in *Escherichia coli.* J. Bacteriol. **126:**478–486.

391. **Shaw, D., and J. R. Guest.** 1981. Molecular cloning of the *fnr* gene of *Escherichia coli* K12. Mol. Gen. Genet. **181:**95–100.

392. **Shaw, D., and J. R. Guest.** 1982. Amplification and product identification of the *fnr* gene of *Escherichia coli*. J. Gen. Microbiol. **128:**2221–2228.

393. **Shaw, D. J., J. R. Guest, R. Meganathan, and R. Bentley.** 1982. Characterization of *Escherichia coli men* mutants defective in conversion of *o*-succinylbenzoate to 1,4-dihydroxy-2-naphthoate. J. Bacteriol. **152:**1132–1137.

394. **Shaw, D. J., D. W. Rice, and J. R. Guest.** 1983. Homology between CAP and Fnr, a regulator of anaerobic respiration in *Escherichia coli*. J. Mol. Biol. **166:**241–247.

395. **Shimokawa, O., and M. Ishimoto.** 1979. Purification and some properties of inducible tertiary amine *N*-oxide reductase from *Escherichia coli*. J. Biochem. **86:**1709–1717.

396. **Showe, M. K., and J. A. DeMoss.** 1968. Localization and regulation of synthesis of nitrate reductase in *Escherichia coli*. J. Bacteriol. **95:**1305–1313.

397. **Shum, A. C., and J. C. Murphy.** 1972. Effects of selenium compounds on formate metabolism and coincidence of selenium-75 incorporation and formic dehydrogenase activity in cell-free preparations of *Escherichia coli*. J. Bacteriol. **110:**447–449.

398. **Siegel, L. M., E. M. Click, and K. J. Monty.** 1964. Evidence for a two-step electron flow from TPNH to sulfite and hydroxylamine in extracts of *Salmonella typhimurium*. Biochem. Biophys. Res. Commun. **17:**125–129.

399. **Siegel, L. M., and P. S. Davis.** 1974. Reduced nicotinamide adenine dinucleotide phosphate-sulfite reductase of enterobacteria. IV. The *Escherichia coli* hemoflavoprotein: subunit structure and dissociation into hemoprotein and flavoprotein components. J. Biol. Chem. **249:**1587–1598.

400. **Siegel, L. M., P. S. Davis, and H. Kamin.** 1974. Reduced nicotinamide adenine dinucleotide phosphate-sulfite reductase of enterobacteria. III. The *Escherichia coli* hemoflavoprotein catalytic parameters and the sequence of electron flow. J. Biol. Chem. **249:**1572–1586.

401. **Siegel, L. M., and K. J. Monty.** 1964. Kinetic properties of the TPNH-specific sulfite and hydroxylamine reductase of *Salmonella typhimurium*. Biochem. Biophys. Res. Commun. **17:**201–205.

402. **Siegel, L. M., M. J. Murphy, and H. Kamin.** 1973. Reduced nicotinamide adenine dinucleotide phosphate-sulfite reductase of enterobacteria. I. The *Escherichia coli* hemoflavoprotein: molecular parameters and prosthetic groups. J. Biol. Chem. **248:**251–264.

403. **Simoni, R. D., and M. K. Shallenberger.** 1972. Coupling of energy to active transport of amino acids in *Escherichia coli*. Proc. Natl. Acad. Sci. USA **69:**2663–2667.

404. **Simpkin, D., and W. J. Ingledew.** 1984. An electron paramagnetic resonance study of *Escherichia coli* fumarate reductase: its iron-sulphur centres and their organization. Biochem. Soc. Trans. **12:**500–501.

405. **Simpkin, D., and W. J. Ingledew.** 1984. Location and the catalytic site of the respiratory fumarate reductase of *Escherichia coli*. J. Gen. Microbiol. **130:**2851–2855.

406. **Singh, A. P., and P. D. Bragg.** 1975. Reduced nicotinamide adenine dinucleotide dependent reduction of fumarate coupled to membrane energization in a cytochrome deficient mutant of *Escherichia coli* K12. Biochim. Biophys. Acta **396:**229–241.

407. **Singh, A. P., and P. D. Bragg.** 1976. Anaerobic transport of amino acids coupled to the glycerol-3-phosphate-fumarate oxidoreductase system in a cytochrome-deficient mutant of *Escherichia coli*. Biochim. Biophys. Acta **423:**450–461.

408. **Smith, M. S.** 1983. Nitrous oxide production by *Escherichia coli* is correlated with nitrate reductase activity. Appl. Environ. Microbiol. **45:**1545–1547.

409. **Spencer, M. E., and J. R. Guest.** 1973. Isolation and properties of fumarate reductase mutants of *Escherichia coli*. J. Bacteriol. **114:**563–570.

410. **Spencer, M. E., and J. R. Guest.** 1974. Proteins of the inner membrane of *Escherichia coli*: changes in composition associated with anaerobic growth and fumarate reductase amber mutation. J. Bacteriol. **117:**954–959.

411. **Sperl, G. T., and J. A. DeMoss.** 1975. *chlD* gene function in molybdate activation of nitrate reductase. J. Bacteriol. **122:**1230–1238.

412. **Stephenson, M., and L. H. Stickland.** 1931. Hydrogenase: a bacterial enzyme activating molecular hydrogen. I. The properties of the enzyme. Biochem. J. **25:**205–214.

413. **Stephenson, M., and L. H. Stickland.** 1932. Hydrogenlyases. Bacterial enzymes liberating molecular hydrogen. Biochem. J. **26:**712–724.

414. **Stephenson, M., and L. H. Stickland.** 1933. Hydrogenlyases. III. Further experiments on the formation of formic hydrogenlyase by *Bact. coli*. Biochem. J. **27:**1528–1532.

415. **Stewart, V.** 1982. Requirement of *fnr* and *narL* functions for nitrate reductase expression in *Escherichia coli* K-12. J. Bacteriol. **151:**1320–1325.

416. **Stewart, V., and C. H. MacGregor.** 1982. Nitrate reductase in *Escherichia coli* K-12: involvement of *chlC*, *chlE*, and *chlG* loci. J. Bacteriol. **151:**788–799.

417. **Stickland, L. H.** 1929. The bacterial decomposition of formic acid. Biochem. J. **23:**1187–1198.

418. **Stickland, L. H.** 1931. The reduction of nitrates by *Bact. coli*. Biochem. J. **25:**1543–1554.

419. **Stokes, J. L.** 1956. Enzymatic aspects of gas formation by *Salmonella*. J. Bacteriol. **72:**269–275.

420. **Stouthamer, A. H.** 1969. A genetical and biochemical study of chlorate-resistant mutants of *Salmonella typhimurium*. Antonie van Leeuwenhoek J. Microbiol. Serol. **35:**505–521.

421. **Stouthamer, A. H.** 1970. Genetics and biochemistry of reductase formation in *Enterobacteriaceae*. Antonie van Leeuwenhoek J. Microbiol. Serol. **36:**181.

422. **Stouthamer, A. H., and C. W. Bettenhaussen.** 1970. Mapping of a gene causing resistance to chlorate in *Salmonella typhimurium*. Antonie van Leeuwenhoek J. Microbiol. Serol. **36:**555–565.

423. **Strauch, K. L., J. B. Lenk, B. L. Gamble, and C. G. Miller.** 1985. Oxygen regulation in *Salmonella typhimurium*. J. Bacteriol. **161:**673–680.

424. **Swim, H. E., and H. Gest.** 1954. Synergistic effects between soluble and particulate preparations from *Escherichia coli*. J. Bacteriol. **68:**755–756.

425. **Takagi, M., and M. Ishimoto.** 1983. *Escherichia coli* mutants defective in trimethylamine *N*-oxide reductase. FEMS Microbiol. Lett. **17:**247–250.

426. **Takagi, M., T. Tsuchiya, and M. Ishimoto.** 1981. Proton translocation coupled to trimethylamine *N*-oxide reduction in anaerobically grown *Escherichia coli*. J. Bacteriol. **148:**762–768.

427. **Takagi-Sawada, M., and M. Ishimoto.** 1985. Proton translocation coupled to nitrite reduction in anaerobically grown *Escherichia coli*. J. Biochem. **97:**205–211.

428. **Takahashi, H., S. Taniguchi, and F. Egami.** 1957. Nitrate reduction in aerobic bacteria and that in *Escherichia coli* coupled in phosphorylation. J. Biochem. **43:**223–233.

429. **Taniguchi, S., and E. Itagaki.** 1959. Solubilization and purification of particulate nitrate reductase of anaerobically grown *Escherichia coli*. Biochim. Biophys. Acta **31:**294–295.

430. **Taniguchi, S., and E. Itagaki.** 1960. Nitrate reductase of nitrite respiration type from *E. coli*. I. Solubilization and purification from the particulate system with molecular characterization as a metalloprotein. Biochim. Biophys. Acta **44:**263–279.

431. **Terriere, C., G. Giordano, C.-L., Medani, D. H. Boxer, B. A. Haddock, and E. Azoulay.** 1981. Precursor forms of formate dehydrogenase in *chlA* and *chlB* mutants of *Escherichia coli*. FEMS Microbiol. Lett. **11:**287–293.

431a. **Ugalde, R. A., J. Imperial, V. K. Shah, and W. J. Brill.** 1985. Biosynthesis of the iron-molybdenum cofactor and the molybdenum cofactor in *Klebsiella pneumoniae*: effect of sulfur source. J. Bacteriol. **164:**1081–1087.

432. **Unden, G., and J. R. Guest.** 1984. Cyclic AMP and anaerobic gene expression in *E. coli*. FEBS Lett. **170:**321–325.

433. **Unden, G., and J. R. Guest.** 1985. Isolation and characterization of the Fnr protein, the transcriptional regulator of anaerobic electron transport in *Escherichia coli*. Eur. J. Biochem. **146:**193–199.

434. **van der Plas, J., K. J. Hellingwerf, H. G. Seijen, J. R. Guest, J. H. Weiner, and W. N. Konings.** 1983. Identification and localization of the fumarate reductase and nitrate respiration systems of *Escherichia coli* by crossed immunoelectrophoresis. J. Bacteriol. **153:**1027–1037.

435. **Venables, W. A.** 1972. Genetic studies with nitrate reductaseless mutants of *Escherichia coli*. I. Fine structure analysis of the *narA*, *narB*, and *narE* loci. Mol. Gen. Genet. **114:**223–231.

436. **Venables, W. A., and J. R. Guest.** 1968. Transduction of nitrate reductase loci of *Escherichia coli* by phage P1 and lambda. Mol. Gen. Genet. **103:**127–140.

437. **Venables, W. A., J. W. T. Wimpenny, and J. A. Cole.** 1968. Enzymic properties of a mutant of *Escherichia coli* K12 lacking nitrate reductase. Arch. Mikrobiol. **63:**117–121.

438. **Vincent, S. P.** 1979. Oxidation-reduction potentials of molybdenum and iron-sulfur centres in nitrate reductase from *Escherichia coli*. Biochem. J. **177:**757–759.

439. **Vincent, S. P., and R. C. Bray.** 1978. Electron-paramagnetic-resonance studies on nitrate reductase from *Escherichia coli* K12. Biochem. J. **171:**639–647.

440. **Visser, R. G. F., K. J. Hellingwerf, and W. N. Konings.** 1984. The protein composition of the cytoplasmic membrane of aerobically and anaerobically grown *Escherichia coli*. J. Bioenerg. Biomembr. **16:**295–307.

441. **Wainwright, S. D.** 1955. Menadione derivatives and ferrous iron cofactors of the nitrate reductase system of a coliform organism. Biochim. Biophys. Acta **18**:583–585.

442. **Wallace, B. J., and I. G. Young.** 1977. Role of quinones in electron transport to oxygen and nitrate in *Escherichia coli*. Studies with a *ubiA⁻ menA⁻* double quinone mutant. Biochim. Biophys. Acta **461**:84–100.

443. **Warringa, M. G. P. J., and A. Guiditta.** 1958. Studies on succinic dehydrogenase. IX. Characterization of the enzyme from *Micrococcus lactilyticus*. J. Biol. Chem. **230**:111–123.

444. **Weiner, J. H.** 1974. The localization of glycerol-3-phosphate dehydrogenase in *Escherichia coli*. J. Membr. Biol. **15**:1–14.

445. **Weiner, J. H., R. Cammack, S. T. Cole, C. Condon, N. Honore, B. D. Lemire, and G. Shaw.** 1986. A mutant of *Escherichia coli* fumarate reductase decoupled from electron transport. Proc. Natl. Acad. Sci. USA **83**:2056–2060.

446. **Weiner, J. H., and P. Dickie.** 1979. Fumarate reductase of *Escherichia coli*. Elucidation of the covalent-flavin component. J. Biol. Chem. **254**:8590–8593.

447. **Weiner, J. H., and L. A. Heppel.** 1972. Purification of the membrane-bound and pyridine nucleotide-independent L-glycerol 3-phosphate dehydrogenase from *Escherichia coli*. Biochem. Biophys. Res. Commun. **47**:1360–1365.

448. **Weiner, J. H., B. D. Lemire, M. L. Elmes, R. D. Bradley, and D. G. Scraba.** 1984. Overproduction of fumarate reductase in *Escherichia coli* induces a novel intracellular lipid-protein organelle. J. Bacteriol. **158**:590–596.

449. **Weiner, J. H., B. D. Lemire, R. W. Jones, W. F. Anderson, and D. G. Scraba.** 1984. A model for the structure of fumarate reductase in the cytoplasmic membrane of *Escherichia coli*. J. Cell. Biochem. **24**:207–216.

450. **Wilson, T. H., and E. C. C. Lin.** 1980. Evolution of membrane bioenergetics. J. Supramol. Struct. **13**:421–446.

451. **Wimpenny, J. W. T.** 1969. The effect of E_h on regulatory processes in facultative anaerobes. Biotechnol. Bioeng. **11**:623–629.

452. **Wimpenny, J. W. T., and J. A. Cole.** 1967. The regulation of metabolism in facultative bacteria. III. The effect of nitrate. Biochim. Biophys. Acta **148**:233–242.

453. **Wimpenny, J. W. T., and A. Firth.** 1972. Levels of nicotinamide adenine dinucleotide and reduced nicotinamide adenine dinucleotide in facultative bacteria and the effect of oxygen. J. Bacteriol. **111**:24–32.

454. **Wimpenny, J. W. T., M. Ranlett, and C. T. Gray.** 1963. Repression and derepression of cytochrome *c* biosynthesis in *Escherichia coli*. Biochim. Biophys. Acta **73**:170–172.

455. **Wimpenny, J. W. T., and A. M. H. Warmsley.** 1968. The effect of nitrate on Krebs cycle enzymes in various bacteria. Biochim. Biophys. Acta **156**:297–303.

456. **Wolf, J., L. H. Stickland, and J. Gordon.** 1954. Enzymes concerned with gas formation by some coliform bacteria. J. Gen. Microbiol. **11**:17–26.

457. **Wood, A. J., and E. A. Baird.** 1943. Reduction of trimethylamine oxide by bacteria. J. Fish. Res. Board Can. **6**:194–201.

458. **Wood, D., M. G. Darlison, R. J. Wilde, and J. R. Guest.** 1984. Nucleotide sequence encoding the flavoprotein and hydrophobic subunits of the succinate dehydrogenase of *Escherichia coli*. Biochem. J. **222**:519–534.

459. **Wood, P. M.** 1981. The redox potential for dimethyl sulphoxide reduction to dimethyl sulphide: evaluation and biochemical implications. FEBS Lett. **124**:11–14.

460. **Woods, D. D.** 1936. Hydrogenlyases. IV. The synthesis of formic acid by bacteria. Biochem. J. **30**:515–527.

461. **Wrigley, C. W., and A. W. Linnane.** 1961. Formic acid dehydrogenase-cytochrome b_1 complex from *Escherichia coli*. Biochem. Biophys. Res. Commun. **4**:66–70.

462. **Yamagata, S.** 1938. Uber die Nitratreduktase von *Bacterium coli*—Untersuchungen über die biologischen Reduktionen. I. Acta Phytochim. (Jpn.) **10**:283–295.

463. **Yamamoto, I., and M. Ishimoto.** 1977. Anaerobic growth of *Escherichia coli* on formate by reduction of nitrate, fumarate and trimethylamine *N*-oxide. J. Gen. Mikrobiol. **17**:235–242.

464. **Yamamoto, I., and M. Ishimoto.** 1978. Hydrogen-dependent growth of *Escherichia coli* in anaerobic respiration and the presence of hydrogenases with different functions. J. Biochem. **84**:673–679.

465. **Yamamoto, N., and M. L. Droffner.** 1985. Mechanisms determining aerobic or anaerobic growth in the facultative anaerobe *Salmonella typhimurium*. Proc. Natl. Acad. Sci. USA **82**:2077–2081.

466. **Yerkes, J. H., L. P. Casson, A. K. Honkanen, and G. C. Walker.** 1984. Anaerobiosis induces expression of *ant*, a new *Escherichia coli* locus with a role in anaerobic electron transport. J. Bacteriol. **158**:180–186.

467. **Young, I. G.** 1975. Biosynthesis of bacterial menaquinones. Menaquinone mutants of *Escherichia coli*. Biochemistry **14**:399–406.

468. **Yudkin, J.** 1932. Hydrogenlyases. II. Some factors concerned in the production of the enzymes. Biochem. J. **26**:1859–1871.

469. **Zarowny, D. P., and B. D. Sanwal.** 1963. Characterization of a nicotinamide adenine dinucleotide-specific nitrite reductase from *Escherichia coli*, strain K12. Can. J. Microbiol. **9**:531–539.

470. **Zinder, S. H., and T. D. Brock.** 1978. Dimethyl sulphoxide reduction by micro-organisms. J. Gen. Microbiol. **105**:335–342.

471. **Zinoni, F., A. Beier, A. Pecher, R. Wirth, and A. Böck.** 1984. Regulation of the synthesis of hydrogenase (formate hydrogenlyase linked) of *E. coli*. Arch. Microbiol. **139**:299–304.

17. Coupling to an Energized Membrane: Role of Ion-Motive Gradients in the Transduction of Metabolic Energy

Department of Physiology, The Johns Hopkins University School of Medicine, Baltimore, Maryland 21205

PERSPECTIVE

Introduction

The term "energized membrane" no longer suffices to describe membrane transport function in bacteria, for we now recognize that the energized membrane includes operation of a variety of different molecular mechanisms; some of these mechanisms have quite distinct and unusual properties, and many of them are understood in fair detail. In describing this general topic, the objectives have been to outline some features of historical importance, to recount some of the relevant experimental findings, most particularly as tested in *Escherichia coli*, and to summarize the current diversity in a way that indicates the direction of future experiments. In a real sense, the field has

arrived at an important juncture. By and large we know what happens during generation, maintenance, and dissipation of the energized membrane. We must now decide whether to pursue this physiological understanding to its equivalent at the level of chemistry; the new genetics offers extraordinarily positive advantages in such an endeavor. Or should we turn to discovering how these individual units—an electron transport chain, an ion-motive ATPase, or a solute carrier—are interconnected, to each other and to the remainder of metabolic activity, and how these connections might be exploited to support more complex organization? This latter emphasis on the large scale is essentially a concern with intra- and intercellular regulation, and that focus could draw on traditional strengths in the choice of bacteria as an experimental system. Over the past 20 years, as our view of mem-

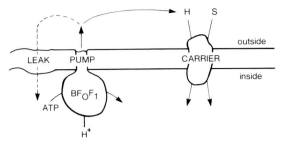

FIG. 1. A chemiosmotic cycle. H^+ extrusion by the bacterial H^+-translocating ATPase (BF_0F_1) establishes an electrochemical proton gradient. H^+ reenters by either a nonspecific (leak) pathway or a specific reaction in which reentry is coupled to transport of solute, S.

brane energization gradually took definition, the field has allowed one to mix chemical and biological intuition. It would be unfortunate if it were not possible to maintain some of this flexibility in the next phases of experimental study.

A Simple Chemiosmotic Cycle

It will be convenient to frame initial discussion around the idea of a chemiosmotic circuit, one that allows some energy-dissipating reaction to drive an energy-consuming step (Fig. 1). There are several forms of such cycles in bacteria, cycles that differ mainly in the way in which ion extrusion occurs. In Fig. 1, the bacterial form of the H^+-translocating ATPase (BF_0F_1) mediates the dissipative reaction, one that is termed primary because ion movement is associated with chemical transformation (ATP hydrolysis). There are several classes of primary transport events in bacteria: ion-motive ATPases, electron transport chains, and light-driven cation or anion pumps (71, 131), but the ATP-linked events are emphasized in this article, since they are found in virtually every cell type and since they operate under both aerobic and anaerobic conditions. The second transport step in Fig. 1 is secondary in nature; no chemical transformation takes place, and only spatial arrangements change, as reactants (products) distribute over the membrane. Just as there are many examples of primary ion pumps in bacteria, one may cite many instances of secondary, ion-coupled carriers. We now realize, as a result of chemiosmotic theory (141, 144, 145, 147), why there need not be direct physical contact between these two very different kinds of reactions for metabolic energy, represented by ATP, to support solute accumulation. Instead, coupling of the two reactions is entirely indirect, mediated only by way of the ion (in this case, H^+) that circulates between its points of exit and reentry.

Defining Criteria for Ionic Coupling

Consideration of the diagram in Fig. 1 reveals that the following three experimental criteria define a chemiosmotic cycle. First, the individual reactions must themselves show an appropriate polarity so that net ion (H^+) translocation occurs in the directions specified. During ATP hydrolysis, BF_0F_1 must move H^+ to the outside; H^+ and solute must move in the

same direction during substrate accumulation. Second, the insulating membrane must be sufficiently impermeable to the coupling ion (and solute, etc.) so that H^+ reentry is more likely through the specific pathway rather than through nonspecific leaks. A leak component is inevitable, but it must be relatively small. Third, the thermodynamic parameters must be appropriately balanced so that one reaction can drive the other in the physiological setting. The "poise" of such reactions is discussed in more detail later, but it should be noted here that this last attribute represents a distinct virtue in coupling by chemiosmotic circuits. They allow the practical coupling of thermodynamically mismatched and biochemically unrelated reactions without requiring molecular contact on the part of participating species.

In this context, what is the evidence relevant to bacterial systems? Does the membrane-bound Ca^{2+}, Mg^{2+}-dependent ATPase (BF_0F_1) actually translocate H^+? Scholes et al. (182) were the first to address this question, in studies of chromatophores prepared from *Rhodospirillum rubrum*, but the experiments in *E. coli* were not available until 1974, when two reports (81, 208) gave results as illustrated in Fig. 2A. In that case, everted membrane particles were prepared in the presence of a well-buffered medium. On resuspension in a poorly buffered medium, H^+ movement into the particles was indicated by an alkalinization of the medium when ATP was hydrolyzed. (The experiment was done at an external pH of 6.2 so that, of itself, ATP hydrolysis did not generate or consume H^+.) In such experiments, it is important to make the membrane permeable to electric charge. This is commonly achieved by use of a high-K^+ medium and addition of valinomycin and also by use of permeant anions (Cl^-, SCN^-, etc.). If this precaution is not taken, the initial ATP-driven H^+ translocation may generate a mem-

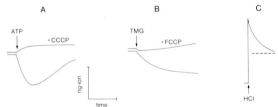

FIG. 2. Changes of external H^+ concentration by operation of chemiosmotic proteins. Appearance and disappearance of H^+ are noted by upward and downward deflection, respectively. (A) Everted vesicles of *E. coli* were suspended anaerobically at pH 6.14 in a KCl medium with valinomycin, with (upper line) or without (lower line) a protonophore, carbonylcyanide *m*-chlorophenylhydrazone (CCCP). ATP was added at the arrow. Calibration: 3 ng-ion H^+; 30 s. Redrawn from reference 81. (B) Intact cells of *E. coli* were at pH 7.0 in a medium with 0.2 M sucrose, 0.03 M choline chloride, 0.25 mM KCl, and valinomycin, with (upper line) and without (lower line) a protonophore, carbonylcyanide-*p*-trifluoromethoxyphenylhydrazone (FCCP). After anaerobic incubation, a lactose permease substrate, thiomethylgalactoside (TMG), was added to 5 mM. Calibration: 77 ng-ion H^+; 78 s. Redrawn from reference 206. (C) Resting cells of *S. lactis* were in 0.3 M KCl–0.06 M KSCN, in the presence of valinomycin. 100 ng-ion H^+ (as HCl) was added at the arrow. The dotted line indicates the final net change of H^+ concentration. Calibration: 206 ng-ion H^+; 190 s. Redrawn from reference 130.

brane potential (electrically positive inside the particle) large enough to limit subsequent net H^+ transfer by a thermodynamic effect. Model calculations (see below) show that this could happen after only a few dozen turnovers by the cell population of BF_0F_1. Also shown in Fig. 2A is that H^+ translocation is inhibited, not potentiated, when the membrane is made H^+ permeable by a protonophore (an uncoupler of oxidative phosphorylation). If the observed H^+ movements were passively responding to some other event, H^+ flux should have been accelerated by the protonophore as the retarding membrane permeability barrier was eliminated. Thus, the judicious use of two ionophores forces the conclusion that H^+ is actively pumped as a result of ATP hydrolysis. Dependence on valinomycin shows that electric charge is carried (but not as K^+); inhibition by uncoupler demonstrates that H^+ itself is translocated. This logical sequence is useful, because (as it happens) membrane-bound ATPases often translocate ions, and in those instances the same arguments can be used to support general conclusions regarding the electrical effect of ion movement and the ions which actually move. Lee et al. (122) have given an especially instructive explication of these principles in their interpretation of H^+ and K^+ movements associated with an H^+/K^+-ATPase exchange pump.

Demonstration of such a coupled reaction (Fig. 2A) is possible in other settings as well, and in a similar fashion it has been possible to examine the reverse reaction, H^+ movement toward the membrane surface at which ATP synthesis occurs. In E. coli, for example, one may drive that reaction with separately imposed electrical (211) or chemical (129) gradients. It is clear, then, that the polarity of ATP-coupled H^+ movements is accurately described in Fig. 1.

The equivalent experiments with H^+/solute cotransport (Fig. 1) would show that H^+ and substrate move in the same direction, a demonstration that was first documented adequately by West and Mitchell (206, 207) in the lactose transport system of E. coli; this lactose carrier continues to be the most carefully examined bacterial transporter. The tracing in Fig. 2B indicates that points made earlier are relevant here, too. In this instance, intact cells were suspended in a lightly buffered, high-K^+ medium under anaerobic conditions to prevent respiratory activity and in the presence of valinomycin to allow charge compensation by countermovement of K^+. The coupling of H^+ with substrate was evident on addition of saturating levels of sugar which promptly induced the inflow of H^+. As before, the permissive effect of valinomycin and the inhibition by a proton conductor showed that H^+ flux moved away from rather than toward its equilibrium position. Although the experiment shown (Fig. 2B) used intact cells, it may be done in any suitably arranged system; most recently, this protocol was reproduced with the same result by using purified lactose transporter reconstituted in proteoliposomes (54). Similar work has been presented for a variety of H^+-coupled transport systems (169), and it is clear that the operative principles extend to other mechanisms as well, provided certain complicating factors are taken into account. For example, when weak acids (bases) are tested, there will necessarily be changes of pH if there is selectivity in

favor of the protonated or deprotonated species (compare data in references 44 and 133). Moreover, coupling at the molecular level may use some other cation, Na^+ for example; and if $Na^+:H^+$ exchange is present, Na^+ entry could be reflected secondarily by pH changes (H^+ entry and Na^+ exit), and measurement of pH alone may give the wrong impression (199, 200).

Experimental verification is also required to support the assertion of a relative membrane impermeability to the coupling ion, H^+. This issue is generally not approached directly, although a few good examples are available from bacterial systems. Work from Mitchell's laboratory first explored this question in microorganisms (180, 181), and Rosen (168) was the first to offer experiments using E. coli. But the most comprehensive tests have not used E. coli; for technical reasons, the anaerobe Streptococcus lactis is more convenient (130), and that is the example shown in Fig. 2C. Once again, the principal advantage of the experiment is a direct focus on the measurement of H^+ fluxes. Thus, cells were suspended at rather high cell density in a poorly buffered medium rich in potassium. (Metabolic inhibitors were not used, but they may be added if required.) The addition of a small quantity of acid caused an immediate titration of external buffering power, but because the cell membrane is relatively impermeable to H^+, subsequent inward diffusion of H^+ to titrate internal buffer groups was delayed, as reflected in the slow relaxation of external pH in the direction of its original value. The kinetics of this process are most easily analyzed if inward-driving force on H^+ is solely a chemical (pH) gradient, and for that reason one includes valinomycin (and K^+) to ensure that the flow of H^+ does not establish an electrically positive interior; inflow of H^+ can then be analyzed to give information about the net H^+ conductance (148). Such measurements indicate that passive H^+ movement across the membrane (leak) will be small compared with the rate at which H^+ moves by primary or secondary reactions; indeed, success in the other experiments of Fig. 2 depended on this relatively low basal H^+ conductance. The passive basal H^+ flux is somewhat, but not greatly, affected by absolute pH, and it appears that the nature of the driving force, whether electrical or chemical, has a similarly small effect. Although such phenomena are not well understood, they do seem to be characteristic of H^+ movement across both biological and synthetic membranes (34, 69, 115, 130).

The values obtained for H^+ conductance in bacteria do not depend strongly on the particular membrane studied, and most experiments yield a figure near 10^{-6} S/cm^2 of surface area (1 S = 1 A/V or 6.2×10^{18} unit charges/s for a 1-V driving force). If cell surface area is about $\pi \times 10^{-8}$ cm^2 (see below), this would correspond to movement of $\pi \times 10^{-2}$ pS per cell. In turn, this would mean that for each 1 mV of driving force about 62π H^+ per s would move passively in response. Steady-state gradients are never more than 200 mV, giving a leak of $< 4 \times 10^4$ H^+ per s, whereas H^+ pumps or porters, operating at ca. 10^2/s, are present at many thousands of copies per cell. The leak component, then, is usually small compared with rates feasible in coupled reactions.

The Central Role of H⁺

Taken together, the examples in Fig. 2 give qualitative and quantitative arguments to support the idea that an indirect coupling between the energy-producing and energy-consuming reactions might use H^+ as an intermediate. This is generally true of bacterial membranes, but it must be realized that chemiosmotic cycles are, in vivo, considerably more complex than outlined here. It is also clear that other ions can serve as coupling intermediates, and one should not eliminate the possibility that larger units (amino acids, sugars, etc.) will eventually be assigned a similar function. Despite these complexities, it is likely that H^+ movements will dominate bacterial membrane transport, if only for the following reasons. (i) Bacterial respiratory chains use many different initial electron donors and terminal acceptors, but the intervening components most often show chemical specificity for hydrogen atoms, electrons, or both (90, 139 [but see reference 198 for a possible exception]). Clearly, then, redox chains will pump protons (or electrons), and oxidizing cells should always initiate a circulation of H^+. (ii) Intracellular pH is well controlled, and when a cell such as *E. coli* grows between pH 5.5 and 8.5, the internal pH is very nearly constant at pH 7.5 to 7.9 (116, 157, 193). This means that maintenance of a pH gradient (usually alkaline inside) is a normal part of the bacterial lifestyle. (iii) Most anaerobes and facultative anaerobes must depend on H^+ circulations, since BF_0F_1 is commonly found on their membranes (135). BF_0F_1 is highly conserved (see below), and it too may have chemical specificity for H^+, at least for part of its operation (146, 147). If charge is transferred (respiration), if a pH gradient is generated (pH regulation), and if H^+ is moved during ATP synthesis and hydrolysis, then elements contributing to a H^+ chemiosmotic circuit are always present.

The Canonical Cell and Considerations of Scale

In concluding these introductory remarks, it seems valuable to set up a model, a canonical cell, to illustrate some further attributes of bacterial energetics based on ion movement. For the sake of these arguments and calculations, it will be assumed that the canonical cell has a radius of 0.5 μm. In that case, cell surface area is $\pi \times 10^{-8}$ cm², and cell volume is $\pi/6 \times 10^{-15}$ liters.

At this point one should pause to note that, if internal pH is controlled at pH 7.5 (described above), there are, on the average, only 10 protons per cell! And since there are thousands (at least) of H^+-coupled pumps or carriers in a single cell, one might legitimately wonder about the role of H^+ per se. Escape from this apparent paradox requires that one recall first that biological materials sample their world with a time scale on the order of milliseconds but that the acid-base dissociation and association reactions of aqueous buffer groups have a picosecond time scale. As a result, enzymatically mediated disappearance or appearance of H^+ may be readily compensated for by environmental chemistry and the large number of buffer molecules present in a cell. Second, one must realize that, in an aqueous medium, the availability of H^+ to a pump or carrier depends not only on free H^+

concentration, but also (more importantly) on the interaction of water with protein-bound H^+ donors and acceptors, i.e., on the pKs of relevant H^+ binding sites. There is no lack of residues with pKs between 6 and 9 in any biological system. And in fact, for some examples, the effective pK for H^+ binding may be so high that the problem is to understand how the system is deprotonated (156).

A second consequence of scale deals with the time in which electrical gradients might be established or dissipated. For example, the synthesis of BF_0F_1 in *E. coli* occurs to an equal degree during growth on either fermentable or oxidizable substrates, and studies of purification (52) indicate that the enzyme comprises about 4 to 6% of total membrane protein, or 0.5 to 1% cell protein (membrane protein makes up 10 to 15% of total protein). At a molecular mass of 528,000 daltons (53), this would correspond to 1,500 to 3,000 molecules per cell, nearly as many as for fully induced β-galactosidase (540,000 daltons; 10,000 tetramers per cell). Purified BF_0F_1 hydrolyzes ATP at 10 to 15 μmol/min per mg of protein (30°C), suggesting a turnover number (at 37°C) of >250 to 400/s relative to ATP and >750 to 1,200/s relative to H^+ transport ($3H^+$ per ATP); comparable estimates are made for other forms of this enzyme (179). Therefore, the canonical cell has an ATP-driven H^+ pump (BF_0F_1) sufficient to hydrolyze ATP at about 10^6/s and in doing so to extrude 3×10^6 H^+/s; respiratory-driven H^+ extrusion, which might also have been chosen for this calculation, should have the same order-of-magnitude capacity. Such ion transport could rapidly polarize the membrane to its maximal physiological value (150 to 200 mV), since bacterial membranes, like other membranes, have an electrical capacitance of 10^{-6} F/cm². Given this capacitance, it is clear that to establish a 200-mV membrane potential over a membrane surface of $\pi \times 10^{-8}$ cm² one needs to transfer only $12\pi \times 10^3$ charges. And how long might this take? At 2,000 BF_0F_1 per cell, each with a turnover number near 1,000 H^+/s, a 200-mV electrical potential might arise as soon as 50 ms after initiation of pumping. (The actual time course is, of course, unpredictable.) Just as important, if the pumps supporting a steady-state potential of 200 mV were to stop suddenly, degradation of this potential by the passive leak component for H^+ (62π/s per mV) would have a much longer time scale; complete dissipation could not occur before 1 s.

Such calculations illustrate several things about coupling that depend on ionic circulations in a small cell. One now appreciates the experimental difficulty of measuring parameters the values of which may change so radically in such a short time. However, the biological implication is perhaps more important. This numerology forces the conclusion that bacterial life is organized in a way that is qualitatively different from that of a larger, eucaryotic cell. For example, channels are commonly (always) found on eucaryotic plasma membranes, but in bacteria they are restricted to the outer membrane of gram-negative cells. The reason is clear, given the numbers just cited. Channels pass ions quickly, 10^6 to 10^9/s (120), and the small size of bacteria places them at risk of an uncoupling if even a single modestly active channel were to open for only part of a second (there are, for example, only about 5 $\times 10^7$ K^+ ions in the canonical cell). In fact, just such an uncoupling is the basis for the toxicity of certain colicins (178). This commentary does not imply that channels (as outer membrane pores) have no role in bacterial physiology (see reference 153); it implies only that discussions of the bacterial plasma membrane are necessarily simpler than those of its eucaryotic counterpart.

The Size and Composition of Ion-Motive Gradients

In the context of coupling, it is of some importance to appreciate that factors that determine the size of an ionic gradient are quite different from those that determine its composition. This difference has physiological significance in that some elements participating in a chemiosmotic cycle are sensitive to an electrochemical gradient, whereas others may respond to only the electrical or chemical component.

The limiting size of the ionic gradient established by the pump of Fig. 1 is usually determined by three factors: (i) the rate at which the ion is pumped, (ii) the stoichiometric ratio between ions pumped and energy input, and (iii) a low but finite membrane permeability to the coupling ion (see above). If, for example, one considers BF_0F_1 during anaerobic growth of E. coli, then the following reaction may be written: $ATP + H_2O + nH^+_{in} \leftrightarrow ADP + P_i + nH^+_{out}$. From standard considerations, and neglecting the participation of water and the appearance of scalar protons (mH^+_{in}) owing to the differing pKs of ATP, ADP, and P_i ($m = 0$ at pH 6.2 if Mg^{2+} is present), one may write an equilibrium constant (K_{eq}) for the reaction as

$$K_{eq} = [ADP][P_i]/[ATP] \cdot \{H^+_{out}\}^n/\{H^+_{in}\}^n$$

where the brackets indicate chemical activity (concentration) and the curly brackets denote electrochemical activity. Note that because H^+ actually moves between inner and outer phases, its tendency to participate in the reaction as a charged particle is properly given by its electrochemical potential (activity), not just by its chemical potential (activity). If one uses natural logarithms and rearranges and multiplies both sides by RT (where R is the gas constant and T is absolute temperature), then

$$nRT \ln\{H^+_{in}\}/\{H^+_{out}\} = RT \ln K_{eq} + RT \ln[ATP]/[ADP][P_i]$$

Terms to the right of the equals sign may be replaced by the expression ΔG_p, which represents Gibbs free energy made available during ATP hydrolysis. In the same way, the term to the left may be replaced by its equivalent, $n\Delta\bar{\mu}_H$, the difference in electrochemical potential for H^+ (65). Accordingly, at equilibrium,

$$n\Delta\bar{\mu}_H = \Delta G_p \tag{1}$$

Clearly, the value of the electrochemical H^+ gradient generated by ATP hydrolysis is set by the stoichiometry of the reaction (nH^+/ATP) and by the ambient concentrations of ATP, ADP, and P_i. (Mg^{2+} and absolute pH play roles in their effect on the value of $RT \ln K_{eq}$ [171].) Similar reasoning applies when H^+ is moved by any other pump, and given that the unavoidable leak component is small (described above), it is apparent that the value of stoichiometry can be a

major determinant to the size of the electrochemical gradient in any real situation. It seems that this equilibrium limit may be closely approached, at least in some experimental situations (132), and for this reason study at the physiological level gives information (nH^+/ATP) relevant to the biochemical reaction.

Certain of these relationships are often expressed in a way that emphasizes the electrical nature of the parameters. Thus, the difference in electrochemical potential for H^+ is also written

$$n\Delta\bar{\mu}_H = zFV_m + RT \ln[H^+]_{in}/[H^+]_{out} \tag{2}$$

where new terms give the membrane potential (V_m), Faraday's constant (F), and valence (z) for the ion in question (here, $z = +1$). Rearrangement yields

$$\Delta\bar{\mu}_H/F = V_m - (RT/F)\Delta pH = \Delta p \tag{3}$$

which stresses the electrical unit (mV). In this case, by analogy to electromotive force, Mitchell (141, 142, 144) coined the term proton-motive force to indicate the parameter Δp.

Three sets of measurements of the electrochemical H^+ gradient (proton-motive force), in cells, spheroplasts, and right-side-out membrane vesicles of E. coli, are summarized in Table 1. Such measurements were obtained under optimal conditions (28, 107, 165; see also references 45, 62, 166, and 223), and the fact that similar findings were made with very different techniques in very different settings illustrates why such values have been accepted with confidence. The data in Table 1 probably represent maximal values one might realistically expect for a cell under oxidizing conditions when respiratory reactions underlie H^+ pumping, the most often used experimental condition. But anaerobic life is just as relevant to E. coli, and for that state Δp is generated by hydrolytic operation of BF_0F_1 (Fig. 1) using ATP derived from substrate-level phosphorylations. This is associated with a somewhat lower maximal value for Δp (62, 106, 108, 132), as

TABLE 1. Electrochemical proton gradient of E. coli[a]

Prepn (reference)	V_m (mV)	$-(RT/F)\Delta pH$ (mV)	Δp (mV)
Cells[b] (107)	94	99	193 (175)[c]
Spheroplasts[d] (28)	132	97	229
Membrane vesicles[e] (165)	73	100	173 (173)

[a] For aerobic conditions at an external pH of 6 to 6.5. Separate assays of membrane potential and pH gradient allowed calculation of Δp by equation 3; the temperature was 25°C ($RT/F = 59$ mV).

[b] Membrane potential from distributions of tetraphenylphosphonium; pH gradient from distributions of benzoic acid.

[c] Values in parentheses give estimates of Δp from lactose accumulation (equation 4), assuming equilibrium distributions at a stoichiometry of $1H^+$/galactoside.

[d] Membrane potential from K^+ distributions in the presence of valinomycin; this method may overestimate V_m if energy-dependent K^+ accumulation also occurs. The pH gradient was calculated from changes of pH after addition of detergent by using known inner and outer buffering powers.

[e] Membrane potential from distributions of triphenylmethylphosphonium; pH gradient from distributions of acetate.

might have been expected. Nevertheless, for both the aerobic and anaerobic condition, Δp rises to near 160 to 220 mV. As discussed below, this places limits on the allowable values for the stoichiometry, $n\text{H}^+/\text{ATP}$.

The preceding discussion makes it clear that factors determining the maximal size of an ionic gradient are reasonably well understood. Note, however (equation 3), that this ion-motive gradient may take on an electrical or chemical character, or some unspecified mixture of the two, according to an entirely different set of rules. In these latter examples, the mechanism is not at all well described, except for the conclusion that the central controlling element reflects a regulation of intracellular pH. The need for control over internal pH is apparent from the earlier comments regarding membrane capacitance. Suppose that a cell resting at pH 7 were to initiate ion pumping by ATP hydrolysis. In a short time (<1 s), V_m would approach its maximum value of 200 mV (Table 1), and net H^+ extrusion would cease as the equilibrium position (equation 1) is approached. Because of low membrane capacitance, this would require (excluding the leak component) about 10^4 hydrolytic events at $3\text{H}^+/\text{ATP}$, or about 10 turnovers of each BF_0F_1 molecule in the cell. ATP would be diminished very little (a 40 μM loss), and because the buffering power of such a resting cell is probably equivalent to ≥ 50 mM P_i (28, 130), the effect of H^+ pumping on the value of internal pH would be even smaller. The same argument would hold for ion pumping that begins at a more acid pH, in which resting buffering power is even stronger, but here a problem presents itself. The low electrical capacitance of the membrane forces the electrical component (V_m, equation 3) to dominate the ion-motive force when ions are initially pumped, and extrinsic mechanisms must come into play to set the value of the chemical gradient so as to rescue internal pH. That such mechanisms operate is unquestioned, for internal pH is always near 7.6 whether outside pH is as high as 8 or as low as 5.5 (see also Table 1), and even beyond these limits there is a reasonable control near pH 7. The best estimates in *E. coli* have been offered by Slonczewski et al. (193), who used ^{31}P nuclear magnetic resonance after incorporation of methylphosphonate to monitor internal pH.

Clearly, the initial ionic movement tending to establish the membrane potential is compensated for by an extrinsic exchange that has the effect of carrying positive charge inward (lowering V_m) and extruding net acid (establishing ΔpH, alkaline inside). The net change is to preserve the free energy represented by $F\Delta p$ (kcal/mol) but to distribute it appropriately into its electrical (FV_m) and chemical ($RT\Delta$pH) equivalents. This trade has been a characteristic finding in the measurement of such gradients in bacterial systems (106–108, 132, 166), with the result that, as external pH is lowered, the membrane potential moves toward zero to allow the increased pH gradient demanded by regulation of internal pH. Such control is one of the most powerful homeostatic mechanisms available to bacteria, and although the relevant molecular mechanism is not known, it presumably involves exchanges with other cations, K^+ and Na^+ (1, 157).

With regard to the model arrangement under discussion (Fig. 1), the significance of pH regulation is twofold. On one hand, given this regulation it would make sense for the physiology of growth if ATP formation by reversal of BF_0F_1 occurred at comparable rates with either the electrical or chemical components of driving force; this would mean that aerobic growth could occur over an extended pH range (as is observed). This, however, makes unusual demands on the biochemistry of ATP synthesis (131, 144), so that when pH is regulated to the benefit of cellular physiology there are accompanying kinetic and thermodynamic consequences that must be accommodated.

It is reasonable to end this discussion on a cautionary note. One should not assume that all membrane events depend on these ionic circulations. In an important experiment, Harold and Brundt (72) showed that growth and cell division (of *Streptococcus faecalis*) continued unabated when cation-motive gradients for H^+, K^+, and Na^+ were collapsed by the ionophore, gramicidin. As expected, growth required a medium of favorable composition (low Ca^{2+}, pH 7.5, rich in amino acids, etc.), but the significant implication is that none of the major events—synthesis and insertion of protein into membranes, DNA replication, cell wall construction, etc.—were obligated by membrane potential or ionic gradients. Although these observations do not entirely rule out a direct link between ion transport and the biosynthesis of macromolecules, it makes such linkage unlikely. In turn, such findings suggest that if collapse of V_m or ΔpH alters some process, one should entertain the idea that the effect is secondarily mediated by other things, perhaps changes of internal pH or changes of internal Ca^{2+}.

ION AND SOLUTE COUPLING

Background

The study of bacterial systems has provided some of the most rigorous experimental tests of the idea that the fluxes of two substrates might be mutually engaged in support of solute accumulation. The initial hypotheses embodying such coupling were formulated mainly by R. Crane (31), whose focus was in mammalian systems, and by P. Mitchell (140, 141), who had interests in both bacteria and eucaryotic organelles. In the work of Crane, for example, an explicit role for sodium was suggested by the striking inhibition of intestinal sugar accumulation when sodium was left out of the incubation medium. In discussing that observation, Crane (31) predicted ". . . a mobile carrier system for actively transported sugars that need be different from that in the red cell membrane [facilitated diffusion] only in that the carrier-sugar complex is not mobile unless Na^+ ion(s) are present and move with it." The analogous idea offered by Mitchell was that proton-conducting uncouplers of oxidative phosphorylation (e.g., dinitrophenol) blocked substrate accumulation by the lactose permease of *E. coli*. The idea that this permease mediated $\text{H}^+/\text{lactose}$ cotransport (142) was therefore a natural extension of the earlier suggestion (141) that H^+-coupled cation exchanges were essential to mitochondrial function. These independent propositions regarding ion-linked solute transport have proven to be correct in their essence. Current research emphasizes the underlying biochemical and biophysical mechanisms.

In the microbial world, progress to a general acceptance of such ideas is particularly well illustrated by studies of lactose transport in *E. coli*. This transport system, and a phosphate self-exchange in *Micrococcus pyogenes (Staphylococcus aureus)*, were the first well-characterized transport reactions in bacteria (140, 167). As it turns out, these two reactions each have a physiological responsibility to carbohydrate transport, yet they involve very different molecular mechanisms; the former is the exemplar of symport; the latter is an intriguing case of antiport (see below).

It has always been clear that provision of metabolic energy to transport is an unusual aspect of lactose (galactoside) movements. For example, even early work indicated that only substrate accumulation, not its entry, required the expenditure of energy (212; for a summary, see reference 110). The demonstration that proton conductors blocked transport without affecting ATP levels (160) made it unlikely that high-energy phosphate played a direct role, and the finding that membrane-active colicins had the same negative effect (46) directly pointed to some aspect of the energized membrane. In this setting, when it was discovered that membrane vesicles would transport lactose in the presence of oxidizable substrates (10), attention abruptly centered on the role of the respiratory chain and on then-controversial ideas deriving from chemiosmotic theory (70). Over the next few years, there was gradual emergence of the principal elements constituting proof of chemiosmotic coupling. (i) It was established that a proton motive force was generated by respiration (or by BF_0F_1, absent respiration), especially during the oxidation reactions that supported lactose transport by membrane vesicles; this task was largely the work of Harold and Kaback and their associates (2, 71, 94). (ii) A second essential step identified the lactose permease as a specifically H^+-coupled system, in which the flux of one substrate is necessarily accompanied by and dependent on movement of the other (Fig. 2B); this was mainly the contribution of West (206, 207, 209). (iii) It was also crucial to establish the quantitative relationship between the H^+ motive driving force and the accumulation of substrate, a point which was made most forcefully by Kashket and Wilson (109). This field naturally drew benefit from the parallel studies of other chemiosmotic proteins, especially BF_0F_1 (134, 168, 177), and the end of the period 1974 to 1976 saw a reasonably wide acceptance of a new overall perspective to solute transport in bacterial systems (71).

Uniport, Symport, and Antiport

Three general categories of secondary carriers are recognized in bacteria (Table 2). When substrate movement is independent of any coupling ion, transport is classed as uniport, or facilitated diffusion. When two (or more) substrates move in the same direction, the reaction is one of symport (cotransport), and when movements are in opposing directions, substrates are engaged in antiport (countertransport, or exchange diffusion). Each of these classes is represented in *E. coli*, but new examples are still being discovered, and in some cases the traditional assignments have been modified in significant ways. The examples summarized in Table 2 represent cases in

TABLE 2. Representative chemiosmotic porters in *E. coli*[a]

Uniport (reference)	Symport (reference[s])	Antiport (reference[s])

Glycerol (175)	H^+/amino acid (8, 93)
	  H^+/glycine
	  H^+/histidine
	  H^+/lysine
	  H^+/phenylalanine
	H^+/organic acid
	  H^+/DL-lactate (73)
	  H^+/pyruvate (138)
	  H^+/succinate (138)
	  H^+/gluconate (117)
	H^+/sugar
	  H^+/arabinose (78)
	  H^+/galactose (78, 79)
	  H^+/lactose (94, 217)
	H^+/inorganic anion
	  H^+/phosphate (170)
	Na^+/amino acid
	  Na^+,H^+/glutamate (56, 57)
	  $Na^+(H^+)$/proline (22, 201)
	Na^+/sugar
	  $Na^+(H^+)$/melibiose (196, 200)
	Mg^{2+}/organic acid
	  Mg^{2+},H^+/citrate[c] (210)
		H^+:cation
		  H^+:Ca^{2+} (6, 16)
		  H^+:$CaHPO_4$ (6)
		  H^+:K^+ (16)
		  H^+:Na^+ (116)
		K^+:cation
		  K^+:$CH_3NH_3^+$ (92)
		$H_2PO_4^-$:organic anion
		  $H_2PO_4^-$:hexose 6-phosphate (3, 133)
		  $H_2PO_4^-$:glycerol 3-phosphate (3, 42)
		  $H_2PO_4^-$:phospho-enolpyruvate[b] (5, 174)

[a] In a few cases, important transporters in other organisms are cited. The stoichiometries of coupling are not specified.
[b] *Salmonella typhimurium.*
[c] *Bacillus subtilis.*

which the reaction type is reasonably well understood.

By and large, the experimental findings are consistent with what one might expect. It is not surprising that bacteria move glycerol by uniport (175), since investment in secondary coupling to this lipid-soluble substrate would be unrealistic. It is perhaps also not surprising that, as a class, uniport is poorly represented. Microorganisms are well adapted to scavenge (or recapture) nutrients from low external concentrations, and uniport, which does not allow accumulation above and beyond electrochemical equilibrium, would not fit this pattern well. On the other hand, symport reactions are well designed for the accumulation of material from the medium, and this class satisfies the physiological demands of many amino acid- and carbohydrate-transport systems. Also noted in Table 2 are several examples of symport reactions that do not depend directly on H^+ but that use Na^+ or Mg^{2+} instead; such diversity is now taken for granted, and one might anticipate even more unusual cases in later studies. Finally, antiport is also well described, particularly as cation exchange. It is often assumed that the study of such exchange (H^+:Ca^{2+}; Na^+:H^+, K^+:H^+, etc.) will add importantly to the understanding of pH regulation, but so far this prediction has not been fulfilled. Instead, we understand cation exchange best in the overall context of coupling. Thus, it is clear that something like Na^+:H^+ exchange is essential to continued operation of Na^+/melibiose symport, or else Na^+ entering in a 1:1 ratio with sugar would rise to toxic levels. In this way, such antiport occupies a central position by serving to connect different ionic circulations (Na^+, H^+, etc.).

The Poise of Ion-Coupled Solute Transport

The listing of Table 2 suggests that if one searches diligently, a chemiosmotic transport reaction may be found to satisfy virtually any physiological need or biochemical curiosity. Interactions among these systems, then, are bound to be complex, and it becomes important to find organizing principles that can indicate the likely behavior. The following expression (35, 169) may be useful in this regard, since it describes the necessary thermodynamic limits. Thus,

$$0 = n(-\Delta\bar{\mu}_{Ion})/F + z_2V_m + (RT/F)(\ln[S]_{in}/[S]_{out}) \quad (4)$$

where it is understood (equation 3) that $\Delta\bar{\mu}_{Ion}/F = z_1V_m + (RT/F)\ln[Ion]_{in}/[Ion]_{out}$. In such cases, z_1 is the valence of the coupling ion, z_2 is the valence of substrate, S, and other terms are as used before. This formalism merely states that, at equilibrium, the distribution of substrate, $[S]_{in}/[S]_{out}$, will reflect a balance between its own tendency to distribute to an equilibrium $[z_2V_m + (RT/F) \ldots]$ and the opposing tendency, as determined by the coupling to some ionic circulation at stoichiometry, n. Thus, when $n = 0$ (uniport), substrate will distribute passively, and if one considers a nonelectrolyte (glycerol; $z_2 = 0$), internal levels of unaltered substrate would never rise above the medium concentration ($[S]_{in}/[S]_{out} = 1$). During antiport, n would take on a negative value, and in this case it is relevant that the term ($-\Delta\bar{\mu}_{Ion}$) is itself a composite reflecting electrical and chemical forces acting to move the coupling ion. Thus, when there is neutral exchange (e.g., Na^+:H^+ antiport), the terms in V_m will cancel ($-z_1V_m + z_2V_m$), and only the respective chemical gradients will determine substrate (and ion) distributions. Finally, when n takes on a positive value during symport, the extent to which substrate distribution is influenced by chemical or electrical gradients depends again on both the stoichiometry of coupling, n, and valences, z_1 and z_2.

This thermodynamic analysis has been important to the rational interpretation of energetics in both procaryotic and eucaryotic cells. The first adequate study dealt with galactoside distributions in bacteria (109), and in that instance the correlations of measurements of membrane potential, pH gradient, and accumulated substrate all documented a coupling ratio of $1H^+$/sugar over a substantial range of the individual parameters. But that case was one of fortunate simplicity: a nonelectrolyte substrate ($z_2 = 0$), a single ionic species used in coupling (H^+) at the simplest of stoichiometries ($n = 1$). A more balanced view is that such coupling becomes quite as involuted as biology demands, and for this reason a more realistic example might be that of H^+,Na^+/glutamate symport as described by Fujimura et al. (56, 57). In these situations, the formalism of equation 4 is only slightly modified. To treat H^+,Na^+/glutamate symport, for example, one would indicate the multiplicity of coupling by substituting $n(-\Delta\bar{\mu}_{Ion\ 1}) + m(-\Delta\bar{\mu}_{Ion\ 2})$ for $n(-\Delta\bar{\mu}_{Ion})$ and incorporate the anionic glutamate

by the valence, z_2. One may also accommodate coupling that involves an anion as the catalytic agent. This circumstance is rather common in animal cells, but it is not yet established in microorganisms, save for a complex series of anion-linked exchanges that serves to drive the accumulation of sugar phosphate (see below).

Aspects of Mechanism Illustrated in Well-Studied Examples

The relationships in equation 3 are thermodynamic in nature, and the kinetic behavior of individual uniporters, antiporters, and symporters is not predictable. For those interested in general, kinetic formalisms, there are several valuable presentations (143, 183), but in any particular case there are usually enough surprises to suggest that generalizations are to be avoided in favor of specific example. In that spirit, then, the following treatment will cite individual cases to describe mechanistic variety. For example, because of its long history, studies of the lactose permease show best how complex the kinetic interpretation becomes if (as is likely) one finds subtle behavioral differences associated with use of one or another substrate. And clearly, the Na^+,H^+/glutamate or Na^+(H^+)/melibiose symporters should be especially favorable to considerations of the role of Na^+ in bacterial transport. These examples will, in the future, also prove useful in answering questions related to ionic selectivity. Finally, P_i:glucose 6-phosphate (G6P) antiport, the newest class of chemiosmotic porter in bacteria, illustrates three issues: the general problem of misclassification, the complexity which arises when stoichiometry is not simple, and the requirement that transport be carefully integrated with the remainder of metabolism.

The lactose transport system of E. coli. The bacterial system best analyzed in terms of kinetic behavior is the H^+/lactose symporter in E. coli (158, 159, 212, 216–218). From the perspective of thermodynamics, this is one of the simpler chemiosmotic reactions, yet both the interpretation and the design of experiments have proven exceedingly difficult when the focus is a purely kinetic study. In part, this is because any two-substrate two-product reaction of arbitrary reaction order gives many possible kinetic solutions. But in addition, since this reaction involves H^+, initial-rate studies are always compromised by the presence of product; assays can be done with zero sugar at one surface, but H^+ is never absent. Accordingly, one must include an accounting for product inhibition, and this requirement has not always been obvious (158, 159). These complicating factors are common to any number of H^+-linked systems, and, for reasons that are less obvious, a third finding is probably of equal general concern. Thus, it appears that the kinetic behavior of H^+/galactoside symport differs considerably depending on which of the commonly used substrates is tested. For the physiological substrate, lactose, it has been clear for some time (93, 94, 110, 217) that tests based on active transport and tests of simple binding give very different expectations for the strength of interaction between lactose and its carrier; binding assays indicate low affinity, whereas tests during sugar accumulation suggest a relatively high affinity, at least on the outer surface of the plasma membrane (i.e., the surface of relatively low proton electrochemical potential). On the other hand, the substrate analog, thiodigalactoside (TDG), shows high-affinity interactions during all assays, whether of binding or transport. This discrepancy (lactose versus TDG) appears to be related to an action by the proton motive force (or one of its components), which considerably strengthens interactions at the influx step if lactose is involved but which has little or no effect on carrier-substrate interactions for TDG (216, 217). This circumstance is entirely unpredicted by the general kinetic schemes (143, 183), although certain arguments based in physiology might have been useful in this regard (see below). Surely the fundamental aspects of the coupling of ion and substrate fluxes should be revealed by studies of either substrate, but both practical and theoretical reasons indicate that it will be as important to understand the molecular basis of such differences.

The description of TDG transport (216, 217) appears to offer a particularly simple biochemical strategy in active transport, since protein-substrate interactions are very nearly the same in the presence and absence of the driving ion gradient. But such simplicity may be physiologically unreliable as a general solution if one argues (as described above) that bacterial porters are generally adapted to reclaim scarce nutrients from low external concentration. This implies that the required cellular concentration of substrate will be considerably higher than that found externally, and this is what poses the problem. A substrate (TDG-like) that binds strongly (high affinity) at both membrane surfaces may nearly saturate the efflux step before internal substrate rises to the value set by thermodynamics. Approach to thermodynamic equilibrium (as in equation 4) will, of course, continue to occur and will reflect a true difference between influx and efflux velocities, but this differential may become so small that for practical purposes it does not take place. This phenomenon is well described in studies of glucose transport by the mammalian erythrocyte (123, 194), but its presence in bacterial systems was only recently appreciated. In this same context, it may now be noted that for conditions that are otherwise the same, a substrate such as lactose can achieve higher internal concentrations in a shorter time, since driving force (Δp) at the same time imposes an asymmetry of interaction constants—the influx reaction saturates at low concentration (μM), whereas efflux does not begin to saturate until internal levels are relatively high (mM); thus, the overall reaction might more rapidly approach its equilibrium position in real time. This has practical implication, for one simple way to estimate Δp in bacteria is to apply equation 4 to a galactoside distribution maintained by the H^+/lactose symporter (62, 107, 108; Table 1). That technique assumes an equilibrium in the symport reaction, of course, and the arguments described above suggest that only lactoselike, not TDG-like, substrates are generally suitable as indirect probes for Δp. At the level of physiology, it may also be noted that the interactions of lactose with its carrier are just those required in a system designed to accumulate at maximal rates in minimal time. In a sense, then, the peculiar kinetic effects of Δp (TDG versus lactose)

present us with the result of experiments in natural molecular biology whereby *E. coli* has matched physiological demand to biochemical constraint.

One object of recent study has been to use the lactose transport system to understand the general structural basis of membrane transport. This was the first bacterial system to offer information at the level of DNA sequence (20), and although studies of secondary and tertiary structure are in their formative period (for the field overall), some useful information has been forthcoming. Interpretations of nucleotide sequence data suggest that the lactose porter is a complex transmembrane structure with about a dozen helical segments which span the lipid bilayer (51, 94; Fig. 3). This pattern seems to be common among chemiosmotic porters, and similar multiple penetrations are evident from the analysis of systems as closely related as the *E. coli* melibiose transporter (219) and as unrelated (?) as the glucose carrier or HCO_3:Cl antiporter of mammalian erythrocytes (114, 151). It also appears to be common for a single polypeptide to specify the behavior of such systems, in both procaryote and eucaryote examples, whether one considers uniport, symport, or antiport.

Other generalizations that draw on studies of H^+/lactose symport may not have found quite such uniform application. There is debate over the question of a monomer or dimer as the functional unit of H^+/lactose symport (63, 159, 217), and it is not clear that this question will resolve easily (94) or that the lactose permease will serve as the appropriate general model. What does seem to be true is that this symport paradigm will be essential to defining the physical and biophysical correlates that link the activity of ion- and substrate-binding sites in a tightly coupled reaction. Several lines of study suggest that these regions are separable by genetic modification (209 [but see reference 218]), and studies of reconstitution confirm that substrate movement can be separated experimentally from a coupling to H^+ flux (23, 192). An appropriate combination of mutagenesis that is site- or selection-directed is now the most promising technique for generating new and informative structures that retain function (17, 94, 137; Fig. 3). Given the prior investment and present excitement associated with the lactose transporter, it seems likely that it will provide molecular answers to questions about flux coupling.

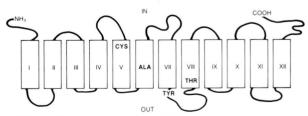

FIG. 3. Organization of the H^+/lactose symport protein. Analysis of the DNA sequence shows that the protein (417 amino acids) has 12 helical segments, each 18 to 22 residues in length, which might span the membrane (51, 94). Bound TDG will protect Cys-148 from alkylation by *N*-ethylmaleimide (217). Also indicated are Ala-177, Tyr-236, and Thr-266; mutation at any one of these sites alters substrate specificity (17, 137), suggesting that galactoside binds in the central third of the molecule.

The role of Na⁺ in bacterial transport. There are, of course, other questions of interest, and other systems may prove better suited to assess these problems. For example, H^+,Na^+/glutamate and $Na^+(H^+)$/melibiose symport (*E. coli*) together should provide answers to questions about the role of Na^+ and the kinetic and biochemical aspects of multiple ionic selectivity. Recent work has highlighted an increasingly important participation of Na^+ in bacterial ionic circulations, and in *E. coli* several systems once classified as H^+ coupled are probably Na^+ coupled (22, 24, 201). It is already established in several cell types, such as *Halobacterium halobium* (119), *Klebsiella aerogenes* (36), *Propionigenium modestum* (83), various alkalophiles (116), and various marine species (*Vibrio alginolyticus* [198]), that Na^+ may be the dominant circulating ion, at least for some conditions. In most of these cases, it is also assumed that primary ion transport is restricted to that of H^+ (see above) and that $Na^+:nH^+$ antiport stands as the intermediary between primary H^+ movements and secondary Na^+-coupled porters (116, 119). It may also be noted that the finding of such Na^+-linked systems correlates with a recent appreciation that Na^+- and H^+-coupled solute movements are easily confused if the affinity for Na^+ is reasonably high, since it is nearly impossible to reduce contaminating Na^+ below 20 μM or so.

As well as giving support to these broad generalizations, studies of Na^+-coupled reactions in bacteria hold specific promise in two instances. In regard to questions of cation selectivity, $Na^+(H^+)$/melibiose symport is now the favored example, since it shows a selectivity that varies according to substrate. Thus, at pH 7, Na^+ is preferred when the substrate is melibiose itself, yet H^+ is preferred if thiomethylgalactoside is used (199, 200). The two cations compete with each other, so that one presumes that a single cation-binding region is involved. This is clearly analogous to operation of the H^+/lactose porter, in which a kinetic effect of Δp is variable depending on the nonelectrolyte substrate (described above).

The situation is considerably more complex if two cations are handled at the same time, as might occur during Na^+,H^+/glutamate coupling. This is at present the most complicated bacterial example to be approached with kinetic tools (56, 57), and regardless of whether the current model (Fig. 4) is correct in detail, it does provide useful ground for discussion. It should be noted that this kinetic model is constructed as a cluster of interfacial binding reactions in series with a pair of reorientation steps (rate or permeability constants k_{+1}, k_{-1}, k_{+2}, k_{-2}). It should also be noted that one of the reorientation steps is associated with charge movement; the unloaded carrier is viewed as neutral, whereas the loaded carrier carries a full unit charge of $+1$. Thus, as specified, the model itself seems to incorporate the conclusion that the kinetic effect of a driving membrane potential is manifested only as a bias to the normal ratio, k_{+1}/k_{-1}. This is the usual way in which kinetic models of bacterial systems are set up. This approach yields satisfactory results, by and large, since the experimental match with theory does suggest that it can be appropriate to assign a net charge movement to one or another form of the carrier (56, 57, 216, 217). In principle, however, this need not be the case (121, 131, 144), for binding reactions might

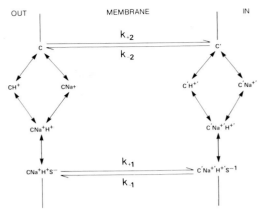

FIG. 4. Kinetic model for Na^+,H^+/glutamate symport by *E. coli* (adapted from reference 57).

follow, rather than precede, the movement of coupling ions down (along) the electric field. This is believed to happen in some ion channels (89, 215). If this were to occur in solute transport, the kinetic effect of a membrane potential might appear as an alteration in the ionic concentration terms when modeled appropriately. At present, there is little convincing evidence that such an "ion well" participates in coupled transport reactions (but see reference 184). Even so, there are reasons to believe that this idea has relevance to BF_0F_1 operation (see below), so the possibility of its use in other settings should not be dismissed out of hand. In general, the voltage sensitivity of these reactions is well described but not well understood (95, 184). In designing an experiment to address such areas, one should recall that membrane voltage might have its driving effect in either (or both) of two fashions. The common assumption is that voltage acts on the protein, either loaded or unloaded with substrate(s) (e.g., Fig. 4), but an acceptable alternative is that voltage might act solely on the ion.

Anion circulations. Discussion to this point has centered on cations and their circulations. Indeed, there is often an unstated assumption that chemiosmotic circuits use pumps or carriers linked to cation movement and that anions do not take part as catalytic agents. This generalization has not been unjustified, since direct evidence supporting an equivalent role for anions has been available only in the past few years (3, 133). In these new examples, the role of anion movement is best understood with regard to neutral exchanges among phosphate and any of several organic phosphates (Table 2); the best-described examples involve one or another sugar phosphate (e.g., P_i:G6P antiport [3, 4]). For example, the operation of a P_i pump (perhaps nH^+/P_i symport; other mechanisms would serve the same end) in series with P_i:G6P exchange constitutes a classical chemiosmotic cycle built on anion movement, an unusual feature in bacteria although commonly found in eucaryote organelles (50, 77, 118). The use of P_i as the circulating element also emphasizes that such exchange must be closely integrated with the remainder of metabolism, since unregulated activity would place at risk those reactions of intermediary metabolism that draw on or contribute to the cellular P_i pool. Such inferences suggest metabolic interrelations yet to be discovered, and in this instance (P$_i$:G6P antiport) one should look closely at the influence of the phosphotransferase system (and vice versa) because in some instances the substrates of anion exchange are also products (sugar 6-phosphates) of the phosphotransferase system (5, 133).

It is also noteworthy that anion exchange can be unbalanced in stoichiometry (e.g., 2P$_i$:G6P); this is certain in the gram-positive systems and seems likely in the *E. coli* examples. Thus, the exchange is best understood as a neutral event involving any suitable combination of monovalent P$_i$, monovalent sugar phosphate, and divalent sugar phosphate (5, 133, 140). Such a complex ionic selectivity is bound to make for difficult kinetic analysis, but for interesting biochemical and physiological studies, and although the picture is most complete with regard to the gram-positive examples, one suspects that such behavior is also characteristic of P$_i$-linked anion exchange in *E. coli* (3, 42). Perhaps most provocative, such variable selectivity predicts a pH-dependent variable stoichiometry, since bulk exchange ratios would reflect an average of neutral molecular events which are either 2:2 or 2:1. Moreover, such anion exchange could be misclassified as H$^+$/anion symport. Consider, for example, what happens for neutral exchange in the presence of a pH gradient when affinity for the organic substrate is much higher than for P$_i$. In that case there will be different ratios for monovalent and divalent forms of the organic substrate on either side of the membrane, with the divalent species dominating in the more alkaline compartment (usually the inside). Thus, mass action dictates that efflux would use a divalent organic anion, whereas influx would use a pair of monovalent species. Accordingly, the overall ratio would approach 1[divalent]:2[monovalent], implying net inflow of as much as 2H$^+$/substrate; the example of 1(G6P^{2-}):2(HG6P^{1-}) antiport is shown in Fig. 5. Until the fundamental role of exchange is appreciated, the net reaction would suggest H$^+$/anion symport as the molecular basis of transport (44).

The distribution of anion exchange among bacterial transport reactions is still not clear, but judging from the known cases it is a likely mechanism for any anionic compound that is also a cellular metabolite. There are obvious reasons for thinking that anion exchange would be the preferred mechanism for trans-port of materials that are also part of some metabolic pool ([5]; e.g., cyclic nucleotides, phosphorylated sugars and other metabolic intermediates, and oxidizable compounds such as succinate, malate, etc.), and, in such cases, to provide for net transport, unusual substrate specificity and transport stoichiometry may also be anticipated.

An Expanded Chemiosmotic Circuit

The foregoing discussions emphasize that chemiosmotic systems in bacteria are ready for detailed characterization at the molecular level, since much of the basic physiology, biochemistry, and kinetics of these proteins are understood, at least in principle. New work will exploit the few examples well suited to exploring specific issues, and it is probable that the systems introduced above will contribute importantly to understanding the structural factors that determine substrate specificity, ionic selectivity, stoichiometry, and the coupling phenomenon itself. On a larger scale, we also now appreciate that chemiosmotic organizations should help in the conceptual integration of transport and metabolism. Thus, in Fig. 5, the simple scheme of Fig. 1 has been expanded to include the subsidiary circulations as represented by the porters discussed above. To different degrees, these other cycles are connected to the dominant H$^+$ flux, and for each case the important role of antiport is evident. For example, the Na$^+$ motive force [$V_m - (RT/F)\Delta$pNa], which drives Na$^+$/solute symport, will parallel the size and composition of the H$^+$ motive force [$V_m - (RT/F)\Delta$pH], provided that there is Na$^+$:H$^+$ antiport to ensure that the respective chemical gradients come into equilibrium. In this way, Na$^+$:H$^+$ antiport ensures adequate driving force to Na$^+$-coupled symport despite the variability of membrane potential during regulation of internal pH. More subtle interactions are seen during exchange of anions. In the example shown, the neutral two-for-one exchange of monovalent and divalent anions illustrates that substrate can play both catalytic and stoichiometric roles in transport. On reflection, this would make physiological sense for compounds that are also central metabolites, and for this reason one may expect anion exchange to be more frequent than is now documented. Clearly, many variants of these patterns are feasible, and because one believes that these same organizations are important in other contexts in other cells, further study of the procaryote examples should have broad significance.

ATP-LINKED ION-MOTIVE PUMPS

E$_1$E$_2$ and F$_0$F$_1$ Pumps

Although not explicitly indicated in the expanded scheme of Fig. 5, the primary active transport of ions in bacteria can involve several different kinds of ATP-linked pumps (there are important exceptions in other cells [36, 83]). These ATP-linked reactions are usually placed in either an E$_1$E$_2$ or an F$_0$F$_1$ category (135), and although the F$_0$F$_1$ class clearly dominates bacterial energetics, the E$_1$E$_2$ examples should not be dismissed; they represent important evolutionary links to ion transport at the eucaryote plasma membrane.

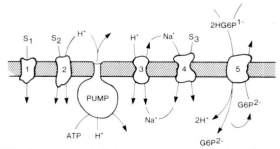

FIG. 5. An expanded chemiosmotic circuit. The scheme of Fig. 1 is enlarged to indicate how subsidiary circulations might be linked indirectly to that of H$^+$. The chemiosmotic porters are (1) uniport, (2) H$^+$/S$_2$ symport, (3) H$^+$:Na$^+$ antiport, (4) Na$^+$/S$_3$ symport, and (5) neutral anion exchange. G6P, Glucose 6-phosphate.

Typically, E_1E_2 pumps use ATP to transport one or more cations (H^+, Na^+, K^+, or Ca^{2+}), with a polypeptide of about 100 kilodaltons as the catalytic subunit. This catalytic subunit is phosphorylated (by the ATP substrate) as a necessary part of the enzymatic reaction, and the term E_1E_2 recognizes that the overall mechanism involves cyclic transformation of the enzyme between phosphorylated and dephosphorylated species. These ATPases are of relatively simple composition, with never more than three distinct polypeptides per functional unit; most often there is only the single catalytic subunit.

E_1E_2 in *E. coli* is best represented by a K^+-translocating ATPase described by Hesse et al. (82); here, a most provocative finding is homology between this ATPase and the Ca^{2+}-ATPase of muscle sarcoplasmic reticulum, and other examples of E_1E_2 (191). In the context of the present discussion, it may be recalled that electrogenic H^+ extrusion mediated by either respiration (aerobic) or BF_0F_1 (anaerobic) is the normal pattern in bacterial life and that maintenance and utilization of a membrane potential depends on the continual operation of such dedicated H^+ pumps. Accordingly, it would seem unlikely that net charge is moved via E_1E_2 for any significant time, for if that were to occur it might endanger central metabolic events (oxidative phosphorylation, pH regulation, etc.). On the whole, then, one expects E_1E_2 ATPases (in bacteria) to be neutral exchangers or to be controlled so as not to directly perturb membrane voltage, except in a transitory way. As yet, the evidence on this point is obscure (76), but it seems true that E_1E_2 pumps do not participate extensively as charge carriers in the overall ionic circuit under the usual conditions. The recent demonstration of K^+:$CH_3NH_3^+$ antiport (92), however, suggests conditions under which a K^+ chemical gradient, established by K^+ incorporation via an E_1E_2 pump, can play a prominent role during nutrient accumulation.

General Features of BF_0F_1, the H^+-Translocating ATPase

The central ion pump of bacterial membranes belongs to the F_0F_1 family of H^+-translocating ATPases, and an understanding of these enzymes has been a central goal in studies of anaerobic and aerobic energy transductions in both procaryote and eucaryote systems. There are, consequently, a number of recent reviews that cover various aspects of the physiology, biochemistry, and genetics of the bacterial form of this enzyme, BF_0F_1 (7, 41, 47, 58, 61, 96, 97, 131, 146, 147, 187). This ion transport system is an extraordinary biological machine. Altogether, the functioning complex has at least eight distinct polypeptides, several in multiple copy, so that the coordinated operation of about two dozen elements is required to couple the transmembrane flux of H^+ to ATP synthesis or hydrolysis.

Physiological position. With only a few exceptions (see reference 135), BF_0F_1 is found on the membranes of all bacteria. In obligate or facultative aerobes the enzyme functions to mediate synthesis of ATP during oxidative phosphorylations, whereas under anaerobic conditions the enzyme itself initiates an H^+ circulation (e.g., Fig. 1 and 5). Acceptance of the idea that

BF_0F_1 plays this dual role in cell physiology has required four kinds of experimental verification. As noted earlier (e.g., Fig. 2 and the text), one important element was demonstration that respiratory activity was associated with the generation of both membrane potential and pH gradient (2, 66). For that case, the companion experiments showed that BF_0F_1 mediated ATP synthesis in response to imposed electrical or chemical gradients of size comparable to those described in vivo (134, 211) and that such synthesis was directly linked to H^+ entry (128, 129). Parallel studies of the anaerobic reaction demonstrated that ATP hydrolysis by BF_0F_1 was coupled to H^+ translocation (81, 182, 208) and that this hydrolysis supported membrane potential and pH gradient of the required sizes (74, 75). Such observations have proven the capacity of BF_0F_1 to participate in an H^+ circulation, and contemporary research at a physiological level addresses certain quantitative issues that are otherwise difficult to approach, particularly the relationship between thermodynamic parameters and the value of stoichiometry (nH^+/ATP), the interpretation of ATP synthetic rates as driven by either electrical or chemical gradients, and the in vivo assembly of the BF_0F_1 complex. Each of these topics will be introduced at an appropriate point in later discussion.

Sector organization. The organization of this sophisticated structure is remarkably similar for the enzymes of bacteria (BF_0F_1), mitochondria (F_0F_1), and chloroplasts (CF_0F_1). The enzyme of *E. coli* and the thermophile, PS3 (EF_0F_1 and TF_0F_1, etc.), are the best characterized of bacterial examples, and although somewhat simpler structures have been found in a few other cell types, especially anaerobes (12, 27), the enzyme in *E. coli* or PS3 is usually taken to be typical of the procaryote example.

The overall architecture of BF_0F_1 is shown in Fig. 6. The enzyme is constructed in two parts, the BF_0 and BF_1 sectors, each of which is a distinct biochemical entity with its characteristic activity. BF_1 is easily released from the membrane by washings at low ionic strength or low pH, and, remarkably, this physical separation is paralleled by a functional division, so

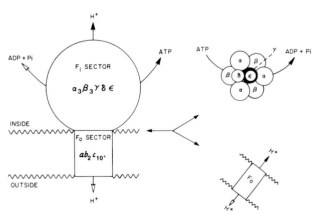

FIG. 6. Sector organization of BF_0F_1. BF_0 and BF_1 sectors are in dynamic equilibrium with BF_0F_1. Each of the separated sectors has its characteristic activity, as indicated at the right. In the diagram at top right, F_1 has been rotated so that the surface interacting with the F_0 sector faces out of the page. Adapted from reference 131.

that the separated sectors retain their individual properties. Thus, free BF_1 hydrolyzes ATP, whereas membranes stripped of F_1 reveal the catalytic role of BF_0 by their newly elevated permeability to H^+. The rebinding of BF_1 to BF_0 restores the coordination of the two half reactions to that found in vivo, where the two activities are functionally coupled. The properties of the isolated or recombined sectors illustrate the central problem; that is, during oxidative phosphorylation, how does BF_0F_1 "couple" the partial reactions so that H^+ flux driven through BF_0 reverses the ATP hydrolysis associated with isolated BF_1. Conversely, when BF_0F_1 operates to extrude H^+ under anaerobic conditions (Fig. 1 and 5), one asks how energy from ATP hydrolysis is made available to move H^+ outward, in an unfavorable direction and against opposing electrical or chemical gradients.

Stoichiometry of coupling with H^+. In probing these two modes of ATPase operation and in studies of chemiosmotic transport proteins at large, it is essential to establish the stoichiometry of ion coupling; an answer to this question immediately restrains possible biochemical mechanisms. This has been an especially significant and often controversial topic in the field of electron transport (see reference 124). Only recently has the same issue been settled with regard to the F_0F_1-type enzymes, and in these cases the work with bacterial systems is as good as any available.

The best estimates of nH^+/ATP stoichiometry for BF_0F_1 derive from experiments based on the thermodynamic inequalities:

$$n\Delta p \geq \Delta G_p/F \tag{5}$$

$$n\Delta p \leq \Delta G_p/F \tag{6}$$

Clearly, equation 5 refers to oxidizing cells, when H^+ movements drive ATP synthesis, whereas equation 6 describes the anaerobic state, when ATP hydrolysis supports H^+ transport. In the case of *E. coli*, Kashket (107, 108) has applied these relationships in experiments using independent measurements of both Δp and ΔG_p. For oxidizing cells, the result was that stoichiometry was near $3H^+/ATP$ ($n \geq 2.7$), in agreement with the results of using anaerobic cells ($n \leq 3.1$). Maloney (132) examined the same parameters in the anaerobe *S. lactis* and concluded that BF_0F_1 operation used $3H^+/ATP$ (the relationship in equation 6) over the entire range of Δp and $\Delta G_p/F$ expected under physiological conditions. This general analysis has also been used to characterize mitochondrial function, and there, too, stoichiometry is best understood as being near $3H^+/ATP$ (11); in the case of CF_0F_1, stoichiometry near $3H^+/ATP$ is also found (164). But among these measurements, the analysis of *E. coli* (107, 108) has a special appeal, because the value of the H^+ motive force could be established by two very different approaches. In one case, Δp was calculated from assays that used the distribution of indirect probes that separately estimated membrane potential and pH gradient (equation 2). The supplemental experiments used lactose accumulation via H^+/lactose symport to infer this driving force as a single quantity (see equation 4 and text). The correspondence between the two approaches is strong evidence that Δp was accurately mea-

sured, a reassurance which has been difficult to obtain in other systems.

It should also be appreciated that fractional stoichiometry is feasible, at least in principle. A complete reaction cycle for $(B)F_0F_1$-type enzymes probably entails synthesis (hydrolysis) of three molecules of ATP (see below), so that molecular stoichiometry would approximate $9H^+/3ATP$. Unfortunately, it is doubtful that current techniques can distinguish among 8, 9, or $10H^+/3ATP$, which means that the most valuable contribution of current studies of stoichiometry has been to rule out a mechanistic class that requires $2H^+/ATP$ ($6H^+/3ATP$) (132, 146).

Genetic Organization—the *unc* Operon

The genetic study of BF_0F_1 began with the work of Butlin et al. (21), who, noting the uncertain state of reconstituted preparations, concluded that "In theory, the study of microbial mutants that lack specific enzymic activities concerned in the coupling reactions could be used to overcome the problem of lack of specificity inherent in other approaches" (21). These investigators then reported on the isolation of *E. coli* mutants defective in the coupling of oxidation to phosphorylation. Such uncoupled (*unc*) mutants did not grow in media requiring ATP from oxidative phosphorylation (e.g., succinate minimal media), but they could survive when presented with a fermentable carbon source (glucose). These isolations initiated a series of elegant genetic experiments (for reviews, see references 61 and 190) which have located the *unc* operon (Fig. 7) near 83 min on the *E. coli* linkage map and which have used F'-mediated complementation to partition various mutations into at least seven of the nine *unc* genes (37, 61). Correspondence between specific genes and specific BF_0F_1 polypeptides was first established for the major F_1 subunits (39, 99, 188), and the current studies of in vitro protein synthesis have now identified each gene with its appropriate gene product (18, 19, 38, 68). And with the recent identification of *uncH* derivatives (88), there are defective mutants available to aid the study of each BF_0F_1 subunit.

The eventual aim of such work has been to apply genetics as an adjunct to the biochemical study of BF_0F_1, and to this end the necessary intermediate goal of defining the nucleotide sequence of the *unc* operon has been achieved by independent work in three laboratories (59, 60, 100–103, 126, 152, 176), with essentially complete agreement. Together with known gene and polypeptide correlations, the nucleotide sequence analysis has highlighted three problems, none as yet satisfactorily resolved. Most puzzling is that the

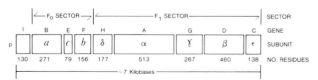

FIG. 7. Genetic organization of the *unc* operon. The *unc* operon lies at 82.5 min on the *E. coli* chromosome (clockwise to the left). Sector, gene, and subunit designations are shown; the number of amino acid residues in each polypeptide is indicated.

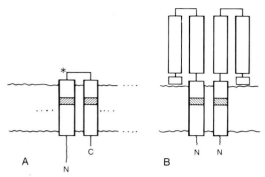

FIG. 8. Proteins of the membrane sector, BF_0. Hypothetical organizations of subunits c and b are shown, using predictions from the amino acid (DNA) sequence (see references 87, 185, and 187); regions of predicted helical content are represented by rectangles. (A) Shaded regions in the subunit c helices contain residues which alter H^+ conduction when covalently modified or when changed by mutation (residue numbers 28, 31, 61, and 64); the asterisk indicates Gln-42 (see text). (B) A subunit b dimer is indicated; the shaded regions would contain Lys-28 (see Fig. 9 and text).

unc operon includes a gene (*uncI*) the product of which is of unknown function. On the basis of its physical behavior (19) and its predicted amino acid content (60), the *uncI* gene product is classified as a membrane protein. But the *uncI* protein has no discernible role in the biochemical properties of BF_0F_1, in its assembly (in vivo or in vitro), or in BF_0F_1 expression. This circumstance is not unprecedented in bacterial systems (e.g., the *lacA* gene product), but the operator-proximal placement of *uncI* (Fig. 8), and its clear expression (19) predicts that a search for *uncI* function will continue. A second issue that can now be sensibly discussed concerns the expression of individual genes. The stoichiometry of BF_0F_1 subunits is highly variable; some polypeptides are present in single copy, and some are present in multiple copy (see below), but the evidence so far indicates that the *unc* operon is transcribed as a single polycistronic message and that the separate genes are translated in simple sequence. For reasons of economy, one naively expects that relative polypeptide synthesis should reflect eventual biochemical use, but if this is true it is certainly not clear how the eventual result is accomplished. Differential expression that exaggerates synthesis of subunit c (more than 10 copies per mole of F_0) is intrinsic to operon structure in the region *uncB-C* (18), and this lends support to a conjecture (18) that mRNA secondary structure might regulate final translation. Clearly, this idea, along with the observation that codon usage correlates roughly with tRNAs of appropriately major or minor abundance (100, 101), should help direct future experiments, but there is presently little direct evidence bearing on the discrepancy between the genetic simplicity and the biochemical variability. An understanding of gene structure has also prompted comparisons among the family of F_0F_1-type enzymes, and those surveys have yielded a particularly striking result, that is, that the β-subunit is nearly the same (ca. 70% homology in amino acid sequence) in every example (58, 173, 187, 203, 204). For some time it has been clear that subunit c is functionally identical in

all cases (87), but the strict conservation of amino acid sequence for subunit β comes as a surprise. Although it is evident that other subunits of BF_0F_1 have their analogs in different systems (Table 3), no other example illustrates the principle of conservation quite so well as subunit β. Overall, such findings reinforce the common assumption that mechanism is equally conserved and that general questions should be addressed with the biological system best suited to the experimental design.

Assembly of BF_0F_1

When BF_0F_1 is properly assembled, the hydrolysis (synthesis) of ATP is immediately coupled to H^+ movements. However, because in vitro the separated BF_0 and BF_1 sectors continue to display their characteristic activities, one wonders whether the partial reactions are regulated in vivo during the interval between assembly of the individual sectors and their eventual binding together. Without such a mechanism, it would appear that the cell risks an elevated H^+ leak (BF_0; encoded by operator-proximal genes) or an unproductive ATP hydrolysis (BF_1; encoded by distal genes). In this context, two specific proposals have emerged. Arguing from the nucleotide sequence, Gay and Walker (60) speculate that *uncI*, the first gene, specifies a "pilot" protein which properly directs in vivo assembly. The present evidence on this point is not convincing, since, despite clear synthesis of the *uncI* protein (19), its presence is not required for coordinated expression of *uncB-C* gene products (202). An alternative offered by Cox et al. (29) provides a solution based on the analysis of *unc* mutants. Briefly, cells carrying a polar *uncD* allele failed to express F_0 capable of binding F_1 (in vitro). This observation suggested that assembly of F_0 might require the presence of the *uncD* protein or a distal gene product or both. The specific proposal invoked both a role for F_1 subunits in coordinating F_0 assembly and a dependence on the assembled (but not operational) F_0 for completion of F_1 (29, 61, 187), so that the proper organization of early gene products (F_0) would require late products (F_1), and vice versa. Despite the appeal of this scenario, recent information from studies of protein synthesis in minicells suggests that other (additional) explanations are needed. Thus, F_1 can be

TABLE 3. Correlation of F_0 and F_1 subunits in H^+-translocating ATPases[a]

E. coli sector/ subunit	Mitochondrial subunit(s)
F_1	
α	α
β	β
γ	γ
δ	OSCP[b]
ϵ	δ
F_0	
a	Subunit 6
b	Unknown
c	Proteolipid, DCCD-binding protein

[a] Homology by sequence, as reported in references 87, 173, and 203 and in references 185 and 187.

[b] OSCP, Oligomycin-sensitivity-conferring protein.

assembled without participation of F_0 (113), in just the fashion predicted by in vitro reconstitution; that is, active ATPase requires at least the $\alpha\beta\gamma$ complex, whereas F_1 competent in energy-linked reactions requires all five of its subunits. Similarly, synthesis of competent BF_0 is demonstrable when BF_1 subunits are absent (9). Of course, these experiments need not reflect truly in vivo behavior, but they do suggest that one might rethink the need for a special mechanism to regulate BF_0F_1 assembly. For example, the data of Friedl et al. (55) suggest that K_{eq} for the reaction BF_0F_1 $\leftrightarrow BF_0 + BF_1$ is about 25 nM (equivalent to seven particles per cell). Given this mutually tight binding and the fact that in vitro binding at high ionic strength nears completion in minutes rather than in hours, it is not clear that assembly need be regulated for any except rapidly growing cells. And if one only seeks to modulate F_0 activity (the H^+ leak), it may be reasonably suggested that protein synthesis should underproduce (modestly) subunit c in relation to its required stoichiometry. This would be an automatic brake to the presence of unoccupied BF_0 owing to a correspondingly modest elevation in the steady-state pool of free BF_1. Clearly, such ideas require more experimental work, and in this case it will be especially interesting to note the kinetics of assembly and sector binding for likely in vivo conditions.

Biochemical Structure of BF_0F_1

The BF_1 sector. As noted earlier, BF_0F_1 contains two dissimilar sectors, BF_0 and BF_1. BF_1 is by far the larger; it contributes about 70% of the total mass and is large enough to appear in electron micrographs as a 95-Å (9.5-nm)-diameter complex of roughly hexagonal shape (98, 197). BF_1 has five different subunits, termed α through ϵ in order of decreasing molecular weight, and it is clear that these subunits are arranged in the stoichiometry $\alpha_3\beta_3\gamma_1\delta_1\epsilon_1$ (15, 52, 53, 125, 197, 222). This is also the best guess as to subunit stoichiometry in other forms of the enzyme, although the evidence is not as complete for F_1 or CF_1 (7). Topological relations among the BF_1 subunits have not yet been completely clarified, but most studies are consistent with a general picture in which the major subunits alternate in quasihexagonal fashion with their centers of mass staggered above and below a central plane (Fig. 6). The alternation of α and β is an important element in several current models, although the supporting evidence is still indirect (7).

Other information regarding α and β is more secure, and much of this relies on the initial efforts in Kagawa's (96, 97) laboratory to reconstitute the various TF_1 subunits of PS3 in all possible combinations. This approach was the first to show that α and β were the subunits most relevant to nucleotide interactions (96, 97); in *E. coli*, $\alpha\beta\gamma$ is the minimal hydrolytic unit (40, 41). The idea that α and β are each important is further reinforced by the finding that BF_1 has six nucleotide-binding sites, one on each of these (six) major subunits (nucleotide binding is summarized in reference 190). The study of *E. coli*, both in vitro and in vivo, also suggests that only α binds nucleotide with high affinity (40) but that this bound nucleotide is not required for energy-linked reactions (127, 162). For this reason, it is believed that nucleotide bound to

α serves to regulate or otherwise modify a catalytic site on β or at the α/β interface.

There is considerable evidence that the β subunit has regions that act during catalysis. Evidence unique to bacterial studies comes from the analysis of *uncD* (β) mutants of *E. coli*. For example, although MgATP is the preferred substrate for hydrolysis by BF_1, some *uncD* mutants show preference for a Ca salt, indicating an altered metal (Ca^{2+}/Mg^{2+}) selectivity (105, 155, 189). Other evidence is deduced from the amino acid sequence and its strong homology to the β subunit of the mitochondria (173). These considerations suggest that the β subunits of *E. coli* and mitochondria each have a stretch of alternating α-helices and β-sheets (residues 240 to 330 in the *E. coli* sequence) that correlate with nucleotide-binding sites in other systems (100, 105, 203, 204). This "Rossmann" fold (172) lies between two residues also believed to be associated with the catalytic region: Tyr-354, which is the site of labeling with a reactive nucleotide analog (43), and Glu-193, which is the location of a Mg^{2+}-protected DCCD (dicyclohexylcarbodiimide) attachment (220).

Because γ is required (*E. coli*) to reconstitute the minimal hydrolytic unit (described above), it is also generally assumed that this smaller polypeptide provides a frame around which α and β organize themselves, and here the evidence gathered in vitro is clearly supported by the study of mutations in *E. coli*. The latter work shows that cells lacking the gene for γ fail to express (assemble) the major F_1 subunits (α + β), whereas the $\alpha\beta\gamma$ triad can be formed by cells lacking genes for either of the two remaining F_1 subunits (58, 105, 113). Kagawa has suggested that, as well as acting to organize BF_1, the γ subunit may play a prominent role in the coupling phenomenon itself. Thus, H^+ movements through the F_0 sector are suppressed when the three minor F_1 subunits are bound (i.e., $\gamma\delta\epsilon$, but not $\delta\epsilon$) (221) supporting speculation that γ acts to monitor H^+ movement between F_0 and F_1 (96, 97). As yet there do not appear to be genetic experiments that bear directly on this point.

Both the studies of reconstitution (see references 41 and 58) and the analysis of mutants shows that δ is essential to proper attachment of the $\alpha\beta\gamma$ core to BF_0. In this respect, then, the δ subunit acts much like the oligomycin-sensitivity-conferring protein of mitochondrial F_0F_1; sequence homology between δ and the oligomycin-sensitivity-conferring protein supports this view (Table 3). Finally, it is quite clear that ATP synthesis (hydrolysis) coupled to H^+ movements requires the additional participation of ϵ, whether one examines in vivo or in vitro behavior. It has also been suggested that ϵ function is analogous to that of the mitochondrial F_1-inhibitor protein, since the (re)binding of ϵ to soluble and ϵ-depleted BF_1 will somewhat inhibit ATPase activity (112, 195). The observation is unquestioned, but one might recall that in mitochondria and chloroplasts the F_0F_1 enzyme is under complex regulation by both intrinsic and extrinsic elements, including the F_1-inhibitor protein (25). As yet, there is no substantial evidence pointing to similar kinetic control over BF_0F_1.

The BF_0 sector. Subunit organization in the BF_0 sector is less well understood than in BF_1. BF_0 has three subunits, usually termed a, b, and c in order of decreasing mass (an alternate nomenclature uses chi, psi, and omega for the same polypeptides [47]). All three subunits are required to form the H^+ permeability pathway. The active BF_0 complex contains a single copy of subunit a, two copies of b, and multiple (>10) copies of c (9, 41, 53, 80).

Based on predictions from the primary sequence (see reference 185), the a subunit is likely to have multiple helices spanning membrane thickness, but the precise number of spans and the sidedness of the N and C termini have yet to be determined biochemically. Analysis of a chain-terminating mutation also suggests some role for a in the binding of BF_1 (48), as might be expected for this largest of BF_0 subunits. The b protein has been the object of several recent experiments, in part because the primary sequence predicts a structure conveniently arranged for study; this polypeptide has a short (30-residue) hydrophobic N-terminal tail, localized in the membrane (85), and a long (130-residue) hydrophilic body which presumably reaches into the aqueous phase (Fig. 8). This hydrophilic extension is clearly important to BF_1 binding; proteases applied at the cytoplasmic surface readily cleave subunit b and abolish BF_1 binding without affecting H^+ conduction via BF_0 (80, 84, 161). At a minimum, then, b has a structural role consistent with the finding that nonsense mutations in $uncF$ (subunit b) yield BF_0 sectors unable to bind BF_1 (for further interpretation, see reference 163). A recent model for the complete enzyme (see below) proposes a more active role for subunit b.

Early speculation on the role of F_0 used the observation that, under appropriate conditions, the inhibitor DCCD blocked H^+ flux through $(B)F_0$ and at the same time inhibited ATP synthesis (or hydrolysis) by $(B)F_0F_1$, all without affecting the properties of isolated $(B)F_1$. This behavior prompted the suggestion that the membrane sector might behave as an H^+ carrier or channel (see reference 144). For this reason, when subunit c (also known as the proteolipid) was identified as the target of DCCD (47), there was the exciting possibility that H^+ conduction might be understood by study of this single polypeptide.

Subunit c has been isolated, and its nucleotide sequence has been determined from many sources, with the result that structural predictions all indicate a hairpinlike organization in which two extremely hydrophobic and membrane-embedded α-helices (136) are separated by a short (10-residue) hydrophilic stretch exposed to cytoplasmic BF_1 (87, 150, 185). In all cases, inhibition of H^+ conduction by DCCD follows a specific attack on an acidic amino acid (Asp-61 in *E. coli*; other examples have Glu) in the C-terminal hydrophobic domain; only a few of the available Asp-61's need be modified to give this inhibition. It is also clear that similar inhibitions of H^+ movement follow when this same residue is replaced by a neutral one (Asp-61 → Gly or Asn) (86, 87; for other effects of these mutants, see reference 149). Further, DCCD-resistant mutations affect residues at the equivalent position in the N-terminal hydrophobic domain, and other mutants that modify H^+ movement without disrupting integration of subunit c map in this same vicinity (data on mutants are summarized in references 87 and 186). All of this evokes a simple picture in which Asp-61 (Glu) plays an obligatory role in BF_0 function, possibly by acting as a proton donor-acceptor in a structure that is readily perturbed by neighboring residues (91) (Fig. 8). This evidence clearly implicates subunit c as one (the only?) central player in the mediation of H^+ flux, but the argument, which has been possible for some time, remains disappointingly circumstantial in nature.

The proton well. The analysis of BF_0 has particular relevance to a theoretical structure known as the proton well, an idea (144) that plays an essential role in certain "direct" coupling mechanisms (131, 146). At issue is how a chemical reaction (ATP synthesis) might draw efficiently on an electrical or chemical driving force or both, a problem continually faced by bacteria during regulation of their internal pH, since both membrane potential and pH gradient must change as outside pH varies. A proton well allows both the electrical and chemical driving forces to be expressed in the same way, as an increased local acidity in the region of the active site within the enzyme (for a review, see reference 131). Relevance to mechanism arises in the following way. The stereochemistry of F_1-mediated ATP hydrolysis (205) points to the synthetic step as an in-line displacement, during which progress from the intermediate state might be represented as

$$\ldots \leftrightarrow [\text{ADPO}^- \cdots \text{P} \cdots \text{O}^-] + 2H^+ \leftrightarrow \text{ADPOPO}_3 + H_2O$$

Briefly, in the simple direct model of Mitchell (146), $2H^+$ provided by $(B)F_0$ would serve as acceptors of the terminal oxygen from phosphate, yielding the products, ATP and water. In this case, the principal role of the proton well of $(B)F_0$ is to make H^+ available, at a reasonable rate, to serve as a reactant even when net driving force is entirely composed of an electrical component (equation 3).

In its simplest version (described above), this mechanism requires a stoichiometry of $2H^+/\text{ATP}$, but the more general formulation gives allowable stoichiometry as $(2 + n)(H^+/\text{ATP})$, where n represents additional ions (of any type, not only protons) that may be needed to drive other parts of the reaction cycle (146). The distinctive feature of such models is that transported H^+ actually participates in the chemical reaction, and this stands in contrast to "indirect" models that use H^+ only to drive changes of protein conformation. Clearly, then, understanding BLF_0 has benefits that extend beyond that of understanding H^+ conduction; the status of the proton (ion) well in particular and of direct coupling mechanisms in general depends importantly on the behavior of this sector. At this point, the kinetic evidence is consistent with the view that BF_0 is a proton well (111, 131), but more positive support is required.

Mutants of Sector and Subunit Interaction

In the absence of new findings concerning the direct participation of subunit c in H^+ conduction, the recent study by Mosher et al. (150) may now shift attention from the intramembranous site of DCCD binding to the short hydrophilic stretch connecting N- and C-terminal membrane domains. Along with Asp-61 (Glu), this short polar strip is conserved in all examples. Mosher et al. (150) describe an *uncE* mutant (*uncE114*) in which a Gln-42 → Glu substitution in this polar segment has disrupted the normal coupling event. BF_0 assembled by the mutant has no defect in H^+ translocation or in its reactivity with DCCD to block H^+ flux. BF_1 binds the mutant BF_0 with normally high affinity, and the bound BF_1 displays the expected ATP hydrolysis. Thus, properties usually associated with the isolated BF_0 (BF_1) sector have been preserved. Nevertheless, all measures of BF_0F_1 interaction indicate profound deficit; modification of BF_0 with DCCD does not suppress ATP hydrolysis by BF_0F_1, as it usually does, nor does the binding of BF_1 block H^+ movements via BF_0, as it should. Clearly, this biochemical phenotype forces one to reconsider the simplistic notion that subunit *c* plays its only (primary) role in H^+ translocation through the membrane (30, 87). Some role in coupling may also be anticipated, and such data predict that this process might be illuminated in a special way by the judicious study of mutants.

Indeed, the analysis of this case suggests an interesting class of mutants, one in which subunit-subunit (sector-sector) interactions have been modified to a greater extent than the catalytic properties of the isolated polypeptide. In several recent examples, the analysis of *unc* mutants has indicated that such interactive events may be uniquely probed in bacterial systems. The case of *uncE114* illustrates the principle with regard to BF_0-BF_1 interaction, but one as easily might have cited the somewhat better-characterized *uncA401*, in which a Ser-373 → Phe substitution alters the α subunit (154). In this latter instance, interactive defects were first indicated by studies of ADP-induced fluorescence of bound aurovertin (214). This phenomenon requires that both aurovertin and ADP bind to the β subunit, but because binding to dissociated subunit(s) does not cause fluorescence changes, it is believed that subunit interaction is instrumental to the nucleotide-induced fluorescence signal. In the *uncA401* enzyme this response is not observed, despite the binding of normal amounts of aurovertin and ADP to BF_1 (162, 213, 214). Perhaps most interesting, it transpires that *uncA401* BF_1 splits ATP at normal rates when substrate is at very low concentration (unisite hydrolysis; see below), but the mutant lacks the usual acceleration of hydrolysis at elevated substrate levels (multisite behavior). This implies that *uncA401* has its dominant effect on the cooperative subunit interactions that contribute to this remarkable behavior (104, 213). Because these sorts of cooperations are a central feature of several models of normal activity (13, 30, 32, 146, 190), further study in the bacterial setting may prove strikingly productive. It is in just this arena that the use of microorganisms may have its greatest impact.

Models and Mobility

Models describing the mechanism of F_0F_1-type enzymes have undergone a qualitative change in the past few years, and one may expect that work with bacteria will soon offer a unique perspective. For the moment, however, studies of the mitochondrial enzyme are still responsible for the more dramatic advances. This is most apparent in a recent kinetic analysis, which has revealed a remarkable cooperativity among (B)F_1 subunits (32, 33, 64, 67). Thus, it has been established that at low substrate concentration ([ATP < [F_1]), ATP is bound very much more tightly and hydrolyzed very much more slowly (by factors of 10^6 to 10^8) than expected on the basis of activity at high substrate concentration ([ATP] ≫ [F_1]). Modeling exercises have indicated that three levels of activity could account for this behavior, and it is presumed that this corresponds to the participation (per mole of F_1) of one, two, and three catalytic sites (uni-, bi-, and tri-site behavior, respectively). More important, assays of ATP binding and hydrolysis and of ADP and P_i release have also suggested that the reaction ATP ↔ ADP + P_i is nearly at equilibrium for the case when ATP, ADP, and P_i are present on the enzyme surface at the same time, giving independent corroboration of the earlier finding (14, 26) that the F_1-ATPase reaction is reversible at low substrate levels. Finally, the kinetic modeling predicts that the basis of rate acceleration (cooperativity) in going from uni- to bi-site behavior involves a dramatic (≥10^6-fold) acceleration of product release from the first site when a second one is occupied (32, 33). Thus, increased turnover may require a kind of give and take in which substrate occupancy at one site promotes product release from its neighbor, and vice versa.

These observations have had two justifiable effects on current thinking. On one hand, direct models for ATP synthesis (146; see above) are less attractive than before, since for suitably arranged conditions F_1 quite obviously mediates ATP synthesis (albeit slowly) with no immediate participation of H^+. Accordingly, the role of H^+ is now most often considered as being directed to the unbinding of preformed ATP, as had been suggested earlier (13, 14). But, of course, such distinctions between direct and indirect formulations are not quite so easily drawn (7, 131, 146), and separating them may require more sophisticated tools than are now used. A second pronounced change in recent attitude concerns how one might understand the cooperativity associated with F_1 operation. Early studies of reconstitution had shown that subunit interactions were important (see above), but the intensity of the cooperativity actually observed is unprecedented and has generated a variety of ideas about how individual subunits, subunit pairs, or subunit triads might coordinate their behavior. Here, the work in bacterial systems may be especially important, for BF_0F_1 is currently the best resource of structural information to support speculation. In fact, in a somewhat more subtle way, bacterial work has already had undeniable influence, since one way to incorporate added complexity and to account for cooperativity has been to invoke mobility in subunit interactions, a true mobility in the sense of the rotation of subunits against one another (30, 146). Clearly, these ideas owe much to the demonstration that bacterial flagella actually do rotate (see chapter 7).

A particularly interesting speculation is made by Cox et al. (30), who propose rotation of a complex involving a subunit *b* dimer and the minor subunits (γδε) of BF_1 (Fig. 9). In this example, rotation would

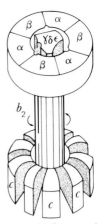

FIG. 9. A rotating model of BF_0F_1 function. The central rotating shaft is a composite of a subunit b dimer and the minor F_1 subunits ($\gamma\delta\epsilon$). The b dimer (Fig. 8) interacts with an encircling array of subunits c. (Subunit a is not shown.) Within F_1, the asymmetric $\gamma\delta\epsilon$ triad comes into sequential contact with each of three surrounding $\alpha\beta$ pairs (patterned after reference 30).

occur within two encircling arrays, one in BF_0 and the other in BF_1. Within the BF_0 sector (Fig. 9), the short hydrophobic arms of the b dimer would be located inside a ring composed of subunit c helices, so that rotation could be driven by the sequential electrostatic attractions between Asp-61 (subunit c) and Lys-23 (subunit b) during proton transfer. At the same time, in the cytoplasmic BF_1, the asymmetric $\gamma\delta\epsilon$ core would be arranged so as to present a different surface to each of three $\alpha\beta$ pairs. Thus (à la Wankel), a full rotation allows each of three otherwise identical catalytic sites to assume a characteristic attribute (substrate binding, bond formation, product release, etc.) (Fig. 9). In the same spirit, but with an emphasis on the (B)F_1 subunits, Mitchell (146) imagines that rotation of such a central core induces sympathetic rotations of the three $\alpha\beta$ pairs, so that efficient operation of any one catalytic site demands occupancy and operation of all others.

These speculations have been cited not because they are strongly supported by experiment but because they indicate new thrust and direction to current thinking. (B)F_0F_1 is sufficiently complex to be considered a "machine," and it is perhaps appropriate that a contemporary perspective emphasizes this by explicit analogy to electrical motors and chemical engines. At the very least, it adds an unexpected excitement to the next generation of study.

ACKNOWLEDGMENTS

I thank the editors of this series for the opportunity to write this essay. I also thank H. R. Kaback and A. Senior for providing material before publication.

Work in this laboratory has been supported by grant GM24195 from the National Institutes of Health.

LITERATURE CITED

1. **Altendorf, K., and F. M. Harold.** 1974. Cation transport in bacteria: K^+, Na^+, and H^+. Curr. Top. Membr. Transp. **5**:1–50.
2. **Altendorf, K., H. Hirata, and F. M. Harold.** 1975. Accumulation of lipid-soluble ions and of rubidium as indicators of the electrical potential in membrane vesicles of *Escherichia coli*. J. Biol. Chem. **250**:1405–1412.
3. **Ambudkar, S. V., T. J. Larson, and P. C. Maloney.** 1986. Reconstitution of sugar phosphate transport systems from *Escherichia coli*. J. Biol. Chem. **261**:9083–9086.
4. **Ambudkar, S. V., and P. C. Maloney.** 1984. Characterization of phosphate:hexose 6-phosphate antiport in membrane vesicles of *Streptococcus lactis*. J. Biol. Chem. **259**:12576–12585.
5. **Ambudkar, S. V., L. A. Sonna, and P. C. Maloney.** 1986. Variable stoichiometry of phosphate-linked anion exchange in *Streptococcus lactis*: implications for the mechanism of sugar phosphate transport by bacteria. Proc. Natl. Acad. Sci. USA **83**:280–284.
6. **Ambudkar, S. V., G. W. Zlotnick, and B. P. Rosen.** 1984. Calcium efflux from *Escherichia coli*. Evidence for two systems. J. Biol. Chem. **259**:6142–6146.
7. **Amzel, L. M., and P. L. Pedersen.** 1983. Proton ATPases: structure and mechanism. Annu. Rev. Biochem. **52**:801–824.
8. **Anraku, Y.** 1978. Active transport of amino acids, p. 171–219. *In* B. P. Rosen (ed.), Bacterial transport. Marcel Dekker, Inc., New York.
9. **Aris, J. P., D. J. Klionsky, and R. D. Simoni.** 1985. The F_0 subunits of *Escherichia coli* F_0F_1-ATP synthase are sufficient to form a functional proton pore. J. Biol. Chem. **260**:11207–11215.
10. **Barnes, E. M., Jr., and H. R. Kaback.** 1970. β-Galactoside transport in bacterial membrane preparations: energy coupling via membrane-bound D-lactic dehydrogenase. Proc. Natl. Acad. Sci. USA **66**:1190–1198.
11. **Berry, E. A., and P. C. Hinkle.** 1983. Measurement of the electrochemical proton gradient in submitochondrial particles. J. Biol. Chem. **258**:1474–1486.
12. **Biketov, S. F., V. N. Kasho, I. A. Kozlov, Y. I. Mileykovskaya, D. N. Ostrovsky, V. P. Skulachev, G. V. Tikhonova, and V. L. Tsuprun.** 1982. F_1-like ATPase from anaerobic bacterium *Lactobacillus casei* contains six similar subunits. Eur. J. Biochem. **129**:241–250.
13. **Boyer, P. D.** 1979. The binding-change mechanism of ATP synthesis, p. 461–479. *In* C. P. Lee, G. Schatz, and L. Ernster (ed.), Membrane bioenergetics. Addison-Wesley Publishing Co., Inc., Reading, Mass.
14. **Boyer, P. D., R. L. Cross, and W. Momsen.** 1973. A new concept for energy coupling in oxidative phosphorylation based on a molecular explanation of the oxygen exchange reactions. Proc. Natl. Acad. Sci. USA **70**:2837–2839.
15. **Bragg, P. D., and C. Hou.** 1980. A cross-linking study of the Ca^{2+},Mg^{2+}-activated adenosine triphosphatase of *Escherichia coli*. Eur. J. Biochem. **106**:495–503.
16. **Brey, R. N., J. C. Beck, and B. P. Rosen.** 1978. Cation/proton antiport systems in *Escherichia coli*. Biochem. Biophys. Res. Commun. **83**:1588–1594.
17. **Brooker, R., and T. H. Wilson.** 1986. Isolation and nucleotide sequencing of lactose carrier mutants that transport maltose. Proc. Natl. Acad. Sci. USA **82**:3959–3963.
18. **Brusilow, W. S. A., D. J. Klionsky, and R. D. Simoni.** 1982. Differential polypeptide synthesis of the proton-translocating ATPase of *Escherichia coli*. J. Bacteriol. **151**:1363–1371.
19. **Brusilow, W. S. A., A. C. G. Porter, and R. D. Simoni.** 1983. Cloning and expression of *uncI*, the first gene in the *unc* operon of *Escherichia coli*. J. Bacteriol. **155**:1265–1270.
20. **Buchel, D. E., B. Gronenborn, and B. Muller-Hill.** 1980. Sequence of the lactose permease gene. Nature (London) **283**:541–545.
21. **Butlin, J. D., G. D. Cox, and F. Gibson.** 1971. Oxidative phosphorylation in *Escherichia coli* K12. Mutations affecting magnesium ion- or calcium ion-stimulated adenosine triphosphatase. Biochem. J. **124**:75–81.
22. **Cairney, J., C. F. Higgins, and I. R. Booth.** 1984. Proline uptake through the major transport system of *Salmonella typhimurium* is coupled to sodium ions. J. Bacteriol. **160**:22–27.
23. **Chen, C.-C., and T. H. Wilson.** 1984. The phospholipid requirement for activity of the lactose carrier of *Escherichia coli*. J. Biol. Chem. **259**:10150–10158.
24. **Chen, C.-C., and T. H. Wilson.** 1986. Solubilization and functional reconstitution of the proline transport system of *Escherichia coli*. J. Biol. Chem. **261**:2599–2604.
25. **Chernyak, B. V., and I. A. Kozlov.** 1986. Regulation of H^+-ATPases in oxidative- and photophosphorylation. Trends Biochem. Sci. **11**:32–35.
26. **Choate, G. L., R. L. Hutton, and P. D. Boyer.** 1979. Occurrence and significance of oxygen exchange reactions catalyzed by mitochondrial adenosine triphosphatase preparations. J. Biol. Chem. **254**:286–290.
27. **Clarke, D. J., F. M. Fuller, and J. G. Morris.** 1979. The proton-translocating adenosine triphosphatase of the obligately anaerobic bacterium *Clostridium pasteurianum*. 1. ATP phosphohydrolase activity. Eur. J. Biochem. **98**:597–612.

28. **Collins, S. H., and W. A. Hamilton.** 1976. Magnitude of the proton motive force in respiring *Staphylococcus aureus* and *Escherichia coli.* J. Bacteriol. **126**:1224–1231.

29. **Cox, G. B., J. A. Downie, L. Langman, A. E. Senior, G. Ash, D. R. H. Fayle, and F. Gibson.** 1981. Assembly of the adenosine triphosphatase complex in *Escherichia coli*: assembly of F_0 is dependent on the formation of specific F_1 subunits. J. Bacteriol. **148**:30–42.

30. **Cox, G. B., D. A. Jans, A. L. Fimmel, F. Gibson, and L. Hatch.** 1984. Hypothesis. The mechanism of ATP synthase. Conformational change by rotation of the b-subunit. Biochim. Biophys. Acta **768**:201–208.

31. **Crane, R.** 1962. Hypothesis for mechanism of intestinal active transport of sugars. Fed. Proc. **21**:891–895.

32. **Cross, R. L.** 1981. The mechanism and regulation of ATP synthesis by F_1-ATPases. Annu. Rev. Biochem. **50**:681–714.

33. **Cross, R. L., C. Grubmeyer, and H. S. Penefsky.** 1982. Mechanism of ATP hydrolysis by beef heart mitochondrial ATPase. Rate enhancements resulting from cooperative interactions between multiple catalytic sites. J. Biol. Chem. **257**:12101–12105.

34. **Deamer, D. W., and J. W. Nichols.** 1983. Proton-hydroxide permeability of liposomes. Proc. Natl. Acad. Sci. USA **80**:165–168.

35. **DeVoe, R. D., and P. C. Maloney.** 1980. Principles of cell homeostasis, p. 3–45. *In* V. B. Mountcastle (ed.), Medical physiology, 14th ed., vol. 1. The C. V. Mosby Co., St. Louis, Mo.

36. **Dimroth, P.** 1982. The role of biotin and sodium in the decarboxylation of oxaloacetate by the membrane-bound oxaloacetate decarboxylase from *Klebsiella aerogenes.* Eur. J. Biochem. **121**:435–441.

37. **Downie, J. A., F. Gibson, and G. B. Cox.** 1979. Membrane adenosine triphosphatases of prokaryotic cells. Annu. Rev. Biochem. **48**:103–131.

38. **Downie, J. A., L. Langman, G. B. Cox, C. Yanofsky, and F. Gibson.** 1980. Subunits of the adenosine triphosphatase complex translated in vitro from the *Escherichia coli unc* operon. J. Bacteriol. **143**:8–17.

39. **Dunn, S. D.** 1978. Identification of the altered subunit in the inactive F_1 ATPase of an *Escherichia coli uncA* mutant. Biochem. Biophys. Res. Commun. **82**:596–602.

40. **Dunn, S. D., and M. Futai.** 1980. Reconstitution of a functional coupling factor from the isolated subunits of *Escherichia coli* F_1 ATPase. J. Biol. Chem. **255**:113–118.

41. **Dunn, S. D., and L. A. Heppel.** 1981. Properties and functions of the subunits of the *Escherichia coli* coupling factor ATPase. Arch. Biochem. Biophys. **210**:421–436.

42. **Elvin, C. M., C. M. Hardy, and H. Rosenberg.** 1985. P_i exchange mediated by the GlpT-dependent *sn*-glycerol-3-phosphate transport system in *Escherichia coli.* J. Bacteriol. **161**:1054–1058.

43. **Esch, F. S., and W. S. Allison.** 1978. Identification of a tyrosine residue at a nucleotide binding site in the β subunit of the mitochondrial ATPase with *p*-fluorosulfonyl ^{14}C-benzoyl-5′-adenosine. J. Biol. Chem. **253**:6100–6106.

44. **Essenberg, R. D., and H. L. Kornberg.** 1975. Energy coupling in the uptake of hexose phosphates by *Escherichia coli.* J. Biol. Chem. **250**:939–945.

45. **Felle, H., J. S. Porter, C. L. Slayman, and H. R. Kaback.** 1980. Quantitative measurements of membrane potential in *Escherichia coli.* Biochemistry **19**:3583–3590.

46. **Fields, K. L., and S. E. Luria.** 1969. Effects of colicins E1 and K on transport systems. J. Bacteriol. **97**:57–63.

47. **Fillingame, R. H.** 1981. Biochemistry and genetics of bacterial H^+-translocating ATPases. Curr. Top. Bioenerg. **11**:35–106.

48. **Fillingame, R. H., M. E. Mosher, L. S. Negrin, and L. K. Peters.** 1983. H^+-ATPase of *Escherichia coli*—*uncB402* mutation leads to loss of χ subunit of F_0 sector. J. Biol. Chem. **258**:604–609.

49. **Fillingame, R. H., L. K. Peters, L. K. White, M. E. Mosher, and C. R. Paule.** 1984. Mutations altering aspartyl-61 of the omega subunit (*uncE* protein) of *Escherichia coli* H^+-ATPase differ in effect on coupled ATP hydrolysis. J. Bacteriol. **158**:1078–1083.

50. **Fliege, R., U. I. Flugge, K. Werdan, and H. W. Heldt.** 1978. Specific transport of inorganic phosphate, 3-phosphoglycerate and triose-phosphates across the inner membrane of the envelope of spinach chloroplasts. Biochim. Biophys. Acta **502**:232–247.

51. **Foster, D. L., M. Boublik, and H. R. Kaback.** 1983. Structure of the *lac* carrier protein of *Escherichia coli.* J. Biol. Chem. **258**:31–34.

52. **Foster, D. L., and R. H. Fillingame.** 1979. Energy-transducing H^+-ATPase of *Escherichia coli*: purification, reconstitution, and subunit composition. J. Biol. Chem. **254**:8230–8236.

53. **Foster, D. L., and R. H. Fillingame.** 1982. Stoichiometry of subunits in the H^+-ATPase complex of *Escherichia coli.* J. Biol. Chem. **257**:2009–2015.

54. **Foster, D. L., M. L. Garcia, M. J. Newman, L. Patel, and H. R. Kaback.** 1982. Lactose-proton symport by purified *lac* carrier protein. Biochemistry **21**:5634–5638.

55. **Friedl, P., J. Hoppe, R. P. Gunsalus, O. Michelsen, K. von Meyenburg, and H. U. Schairer.** 1983. Membrane integration and function of the three F_0 subunits of the ATP synthase of *Escherichia coli* K12. EMBO J. **2**:99–103.

56. **Fujimura, T., I. Yamoto, and Y. Anraku.** 1983. Mechanism of glutamate transport in *Escherichia coli* B. 1. Proton-dependent and sodium ion-dependent binding of glutamate to a glutamate carrier in the cytoplasmic membrane. Biochemistry **22**:1954–1959.

57. **Fujimura, T., I. Yamoto, and Y. Anraku.** 1983. Mechanism of glutamate transport in *Escherichia coli* B. 2. Kinetics of glutamate transport driven by artificially imposed proton and sodium ion gradients across the cytoplasmic membrane. Biochemistry **22**:1959–1965.

58. **Futai, M., and H. Kanazawa.** 1983. Structure and function of proton-translocating adenosine triphosphate (F_0F_1): biochemical and molecular biological approaches. Microbiol. Rev. **47**:285–312.

59. **Gay, N. J., and J. E. Walker.** 1981. The *atp* operon: nucleotide sequence of the region encoding the α-subunit of *Escherichia coli* ATP-synthase. Nucleic Acids Res. **9**:2187–2194.

60. **Gay, N. J., and J. E. Walker.** 1981. The *atp* operon: nucleotide sequence of the promoter and the genes for the membrane proteins, and the δ subunit of *Escherichia coli* ATP-synthase. Nucleic Acids Res. **9**:3919–3926.

61. **Gibson, F.** 1983. Biochemical and genetic studies on the assembly and function of the F_0F_1 adenosine triphosphatase of *Escherichia coli.* Biochem. Soc. Trans. **11**:229–240.

62. **Gober, J. W., and E. R. Kashket.** 1985. Measurement of the proton motive force in *Rhizobium meliloti* with the *Escherichia coli lacY* gene product. J. Bacteriol. **164**:929–931.

63. **Goldkorn, T., G. Rimon, E. S. Kempner, and H. R. Kaback.** 1984. Functional molecular weight of the *lac* carrier protein from *Escherichia coli* as studied by radiation inactivation analysis. Proc. Natl. Acad. Sci. USA **81**:1021–1025.

64. **Gresser, M. J., J. A. Myers, and P. D. Boyer.** 1982. Catalytic site cooperativity of beef heart mitochondrial F_1 adenosine triphosphatase. Correlations of initial velocity, bound intermediate, and oxygen exchange measurements with an alternating three-site model. J. Biol. Chem. **257**:12030–12038.

65. **Greville, G. D.** 1969. A scrutiny of Mitchell's chemiosmotic hypothesis. Curr. Top. Bioenerg. **3**:1–78.

66. **Griniuviene, B., V. Chimieliauskaite, and L. Grinius.** 1974. Energy-linked transport of permeant ions in *Escherichia coli* cells: evidence for membrane potential generation by proton pumps. Biochem. Biophys. Res. Commun. **56**:206–213.

67. **Grubmeyer, C., R. L. Cross, and H. S. Penefsky.** 1982. Mechanism of ATP hydrolysis by beef heart mitochondrial ATPase. Rate constants for elementary steps in catalysis at a single site. J. Biol. Chem. **257**:12092–12100.

68. **Gunsalus, R. P., W. S. A. Brusilow, and R. D. Simoni.** 1982. Gene order and gene-polypeptide relationships of the proton-translocating ATPase operon (*unc*) of *Escherichia coli.* Proc. Natl. Acad. Sci. USA **79**:320–324.

69. **Gutknecht, J.** 1984. Proton/hydroxide conductance through lipid bilayer membranes. J. Membr. Biol. **82**:105–112.

70. **Harold, F. M.** 1972. Conservation and transformation of energy by bacterial membranes. Bacteriol. Rev. **36**:172–230.

71. **Harold, F. M.** 1977. Membranes and energy transduction in bacteria. Curr. Top. Bioenerg. **6**:84–151.

72. **Harold, F. M., and J. Brunt.** 1977. Circulation of H^+ and K^+ across the plasma membrane is not obligatory for bacterial growth. Science **197**:372–373.

73. **Harold, F. M., and E. Levin.** 1974. Lactic acid translocation: terminal step in glycolysis by *Streptococcus faecalis.* J. Bacteriol. **117**:1141–1148.

74. **Harold, F. M., and P. Papineau.** 1972. Cation transport and electrogenesis by *Streptococcus faecalis.* I. The membrane potential. J. Membr. Biol. **8**:27–44.

75. **Harold, F. M., E. Pavlasova, and J. R. Baarda.** 1970. A transmembrane pH gradient in *Streptococcus faecalis*: origin and dissipation by proton conductors and N,N′-dicyclohexylcarbodiimide. Biochim. Biophys. Acta **196**:235–244.

76. **Heefner, D. L.** 1982. Transport of H^+, K^+, Na^+ and CA^{++} in *Streptococcus.* Mol. Cell. Biochem. **44**:81–106.

77. **Heldt, H. W., and U. I. Flugge.** 1986. Transport of metabolites across the chloroplast envelope. Methods Enzymol. **125**:705–716.

78. **Henderson, P. J. F.** 1974. Application of the chemiosmotic theory to the transport of lactose, D-galactose and L-arabinose by

Escherichia coli, p. 409–424. *In* L. Bolis, K. Block, S. E. Luria, and F. Lyen (ed.), Comparative biochemistry and physiology of transport. North-Holland Publishing Co., Amsterdam.

79. **Henderson, P. J. F., Y. Kagawa, and H. Hirata.** 1983. Reconstitution of the *GalP* galactose transport activity of *Escherichia coli* into liposomes made from soybean phospholipids. Biochim. Biophys. Acta **732:**204–209.

80. **Hermolin, J., J. Gallant, and R. H. Fillingame.** 1983. Topology, organization, and function of the psi subunit in the F_0 sector of the H^+-ATPase of *Escherichia coli*. J. Biol. Chem. **258:**14550–14555.

81. **Hertzberg, E. L., and P. C. Hinkle.** 1974. Oxidative phosphorylation and proton translocation in membrane vesicles prepared from *Escherichia coli*. Biochem. Biophys. Res. Commun. **58:**178–184.

82. **Hesse, J. E., L. Wieczorek, K. Altendorf, A. S. Reicin, E. Dorus, and W. Epstein.** 1984. Sequence homology between two membrane transport ATPases, the Kdp-ATPase of *Escherichia coli* and the Ca^{2+}-ATPase of sarcoplasmic reticulum. Proc. Natl. Acad. Sci. USA **81:**4746–4750.

83. **Hilpert, W., B. Schink, and P. Dimroth.** 1984. Life on a new decarboxylation-dependent energy conservation mechanism with Na^+ as coupling ion. EMBO J. **3:**1655–1670.

84. **Hoppe, J., P. Friedl, H. U. Schairer, W. Sebald, K. von Meyenburg, and B. B. Jorgensen.** 1983. The topology of the proton translocating F_0 component of the ATP synthase from *E. coli* K12: studies with proteases. EMBO J. **2:**105–110.

85. **Hoppe, J., C. Montecucco, and P. Friedl.** 1983. Labeling of subunit b of the ATP synthase from *Escherichia coli* with a photoreactive phospholipid analogue. J. Biol. Chem. **258:**2882–2885.

86. **Hoppe, J., H. U. Schairer, and W. Sebald.** 1980. The proteolipid of a mutant ATPase from *Escherichia coli* defective in H^+-conduction contains a glycine instead of the carbodiimide-reactive aspartyl residue. FEBS Lett. **109:**107–111.

87. **Hoppe, J., and W. Sebald.** 1984. The proton conducting F_0-part of bacterial ATP synthases. Biochim. Biophys. Acta **768:**1–27.

88. **Humbert, R., W. S. A Brusilow, R. P. Gunsalus, D. J. Klionsky, and R. D. Simoni.** 1983. *Escherichia coli* mutants defective in the *uncH* gene. J. Bacteriol. **153:**416–422.

89. **Iijima, T., S. Ciani, and S. Hagiwara.** 1986. Effects of the external pH on Ca channels: experimental studies and theoretical considerations using a two-site, two-ion model. Proc. Natl. Acad. Sci. USA **83:**654–658.

90. **Ingledew, W. J., and R. K. Poole.** 1984. The respiratory chain of *Escherichia coli*. Microbiol. Rev. **48:**222–271.

91. **Jans, D. A., L. Hatch, A. L. Fimmel, F. Gibson, and G. B. Cox.** 1984. An acidic or basic amino acid at position 26 of the *b* subunit of *Escherichia coli* F_1F_0-ATPase impairs membrane proton permeability: suppression of the *uncF469* nonsense mutation. J. Bacteriol. **160:**764–770.

92. **Jayakumar, A., W. Epstein, and E. M. Barnes, Jr.** 1985. Characterization of ammonium (methylammonium)/potassium antiport in *Escherichia coli*. J. Biol. Chem. **260:**7528–7532.

93. **Kaback, H. R.** 1972. Transport across isolated bacterial cytoplasmic membranes. Biochim. Biophys. Acta **265:**367–416.

94. **Kaback, H. R.** 1986. Active transport in *Escherichia coli*: passage to permease. Annu. Rev. Biophys. **15:**279–319.

95. **Kaczorowski, G. J., D. E. Robertson, and H. R. Kaback.** 1979. Mechanism of lactose translocation in membrane vesicles from *Escherichia coli*. 2. Effect of imposed $\Delta\psi$, ΔpH, and $\Delta\mu_{H^+}$. Biochemistry **18:**3698–3704.

96. **Kagawa, Y.** 1978. Reconstitution of the energy transformer, gate and channel, subunit reassembly, crystalline ATPase and ATP synthesis. Biochim. Biophys. Acta **505:**45–93.

97. **Kagawa, Y.** 1984. Proton motive ATP synthesis, p. 149–186. *In* L. Ernster (ed.), Bioenergetics. Elsevier Science Publishing, Inc., New York.

98. **Kagawa, Y., N. Sone, M. Yoshida, H. Hirata, and H. Okamoto.** 1976. Proton translocating ATPase of a thermophilic bacterium: morphology, subunits, and chemical composition. J. Biochem. **80:**141–151.

99. **Kanazawa, H., Y. Horiuchi, M. Takagi, Y. Ishino, and M. Futai.** 1980. Coupling factor F_1 ATPase with defective β subunit from a mutant of *Escherichia coli*. J. Biochem. **88:**695–703.

100. **Kanazawa, H., T. Kayano, T. Kayasu, and M. Futai.** 1982. Nucleotide sequence of the genes for β and ε subunits of proton-translocating ATPase from *Escherichia coli*. Biochem. Biophys. Res. Commun. **105:**1257–1264.

101. **Kanazawa, H., T. Kayano, K. Mabuchi, and M. Futai.** 1981. Nucleotide sequence of the genes coding for α, β, and γ subunits of the proton-translocating ATPase of *Escherichia*

102. **Kanazawa, H., K. Mabuchi, T. Kayano, T. Noumi, T. Sekiya, and M. Futai.** 1981. Nucleotide sequence of the genes for F_0 components of the proton-translocating ATPase from *Escherichia coli*: prediction of the primary structure of F_0 subunits. Biochem. Biophys. Res. Commun. **103:**613–620.

coli. Biochem. Biophys. Res. Commun. **103:**604–612.

103. **Kanazawa, H., K. Mabuchi, T. Kayano, F. Tamura, and M. Futai.** 1981. Nucleotide sequence of gene coding for dicyclohexylcarbodiimide-binding protein and the α subunit of proton-translocating ATPase of *Escherichia coli*. Biochem. Biophys. Res. Commun. **100:**219–225.

104. **Kanazawa, H., T. Noumi, I. Matsuoka, T. Hirata, and M. Futai.** 1984. F_1-ATPase of *Escherichia coli*: a mutation (*uncA401*) located in the middle of the α subunit affects the conformation essential for F_1 activity. Arch. Biochem. Biophys. **228:**258–269.

105. **Kanazawa, H., T. Noumi, N. Oka, and M. Futai.** 1983. Intracistronic mapping of the defective site and the biochemical properties of β subunit mutants of *Escherichia coli* H^+-ATPase: correlation of structural domains with functions of the β subunit. Arch. Biochem. Biophys. **227:**596–608.

106. **Kashket, E. R.** 1981. Proton motive force in growing *Streptococcus lactis* and *Staphylococcus aureus* cells under aerobic and anaerobic conditions. J. Bacteriol. **146:**369–376.

107. **Kashket, E. R.** 1982. Stoichiometry of the H^+-ATPase of growing and resting aerobic *Escherichia coli*. Biochemistry **21:**5534–5538.

108. **Kashket, E. R.** 1983. Stoichiometry of the H^+-ATPase of *Escherichia coli* cells during anaerobic growth. FEBS Lett. **154:**343–346.

109. **Kashket, E. R., and T. H. Wilson.** 1973. Proton-coupled accumulation of galactoside in *Streptococcus lactis* 7962. Proc. Natl. Acad. Sci. USA **70:**2866–2869.

110. **Kennedy, E. P.** 1970. The lactose permease system of *Escherichia coli*, p. 49–92. *In* J. R. Beckwith and D. Zipser (ed.), The lactose operon. Cold Spring Harbor Laboratory, Cold Spring Harbor, N.Y.

111. **Kahn, S. and H. C. Berg.** 1983. Isotope and thermal effects in chemiosmotic coupling to the flagellar motor of Streptococcus. Cell **32:**913–919.

112. **Klionsky, D. J., W. S. A. Brusilow, and R. D. Simoni.** 1984. In vivo evidence for the role of the ε subunit as an inhibitor of the proton-translocating ATPase of *Escherichia coli*. J. Bacteriol. **160:**1055–1060.

113. **Klionsky, D. J., and R. D. Simoni.** 1985. Assembly of a functional F_1 of the proton-translocating ATPase of *Escherichia coli*. J. Biol. Chem. **260:**11200–11206.

114. **Kopito, R. R., and H. F. Lodish.** 1985. Primary structure and transmembrane orientation of the murine anion exchange protein. Nature (London) **316:**234–238.

115. **Krishnamoorthy, G., and P. C. Hinkle.** 1984. Non-ohmic proton conductance of mitochondria and liposomes. Biochemistry **23:**1640–1645.

116. **Krulwich, T. A.** 1983. Na^+/H^+ antiporters. Biochim. Biophys. Acta **726:**245–264.

117. **Lagarde, A. E., J. M. Pouyssegur, and F. R. Stoeber.** 1973. A transport system for 2-keto-3-deoxy-D-gluconate uptake in *Escherichia coli* K12. Biochemical and physiological studies in whole cells. Eur. J. Biochem. **36:**328–341.

118. **LaNoue, K. F., and A. C. Schoolwerth.** 1979. Metabolite transport in mitochondria. Annu. Rev. Biochem. **48:**871–922.

119. **Lanyi, J. K.** 1979. The role of Na^+ in transport processes of bacterial membranes. Biochim. Biophys. Acta **559:**377–397.

120. **Latorre, R., and C. Miller.** 1983. Conduction and selectivity in potassium channels. J. Membr. Biol. **71:**11–30.

121. **Lauger, P.** 1979. A channel mechanism for electrogenic ion pumps. Biochim. Biophys. Acta **552:**143–161.

122. **Lee, J., G. Simpson, and P. Scholes.** 1974. An ATPase from dog gastric mucosa: changes of outer pH in suspensions of membrane vesicles accompanying ATP hydrolysis. Biochem. Biophys. Res. Commun. **60:**825–832.

123. **LeFevre, P. G., and G. F. McGinniss.** 1960. Tracer exchange *vs* net uptake of glucose through human red cell surface. J. Gen. Physiol. **44:**87–103.

124. **Lehninger, A. L., A. Reynafarje, A. Alexandre, and A. Villalobo.** 1980. Respiration-coupled H^+ ejection by mitochondria. Ann. N.Y. Acad. Sci. **341:**585–592.

125. **Lunsdorf, H., K. Ehrig, P. Friedl, and H. U. Shairer.** 1984. Use of monoclonal antibodies in immuno-electron microscopy for the determination of subunit stoichiometry in oligomeric enzymes. There are three α-subunits in the F_1-ATPase of *Escherichia coli*. J. Mol. Biol. **173:**131–136.

126. **Mabuchi, K., H. Kanazawa, T. Kayano, and M. Futai.** 1981.

Nucleotide sequence of the gene coding for the δ subunit of proton-translocating ATPase of *Escherichia coli*. Biochem. Biophys. Res. Commun. **102**:172–179.

127. **Maeda, M., H. Kobayashi, M. Futai, and Y. Anraku.** 1977. Studies on the turnovers *in vivo* of adenoside di- and triphosphates in a coupling factor of *Escherichia coli*. J. Biochem. **82:** 311–314.

128. **Maloney, P. C.** 1977. Obligatory coupling between proton entry and the synthesis of adenosine 5'-triphosphate in *Streptococcus lactis*. J. Bacteriol. **132**:564–575.

129. **Maloney, P. C.** 1978. Coupling between H⁺ entry and ATP formation in *Escherichia coli*. Biochem. Biophys. Res. Commun. **83**:1496–1501.

130. **Maloney, P. C.** 1979. Membrane H⁺ conductance of *Streptococcus lactis*. J. Bacteriol. **140**:197–205.

131. **Maloney, P. C.** 1982. Energy coupling to ATP synthesis by the proton-translocating ATPase. J. Membr. Biol. **67**:1–12.

132. **Maloney, P. C.** 1983. Relationship between phosphorylation potential and electrochemical H⁺ gradient during glycolysis in *Streptococcus lactis*. J. Bacteriol. **153**:1461–1470.

133. **Maloney, P. C., S. V. Ambudkar, J. Thomas, and L. Schiller.** 1984. Phosphate/hexose 6-phosphate antiport in *Streptococcus lactis*. J. Bacteriol. **158**:238–245.

134. **Maloney, P. C., E. R. Kashket, and T. H. Wilson.** 1974. A protonmotive force drives ATP synthesis in bacteria. Proc. Natl. Acad. Sci. USA **71**:3896–3900.

135. **Maloney, P. C., and T. H. Wilson.** 1985. The evolution of ion pumps. BioScience **35**:43–48.

136. **Mao, D., E. Wachter, and B. A. Wallace.** 1982. Folding of the mitochondrial proton adenosine triphosphatase proteolipid channel in phospholipid vesicles. Biochemistry **21**:4960–4968.

137. **Markgraf, M., H. Bocklage, and B. Muller-Hill.** 1985. A change of threonine 266 to isoleucine in the lac permease of *Escherichia coli* diminishes the transport of lactose and increases the transport of maltose. Mol. Gen. Genet. **198**:473–475.

138. **Matin, A., and W. N. Konings.** 1973. Transport of lactate and succinate by membrane vesicles of *Escherichia coli*, *Bacillus subtilis* and a *Pseudomonas* species. Eur. J. Biochem. **34**:58–67.

139. **Matsushita, K., L. Patel, and H. R. Kaback.** 1984. Cytochrome *o* type oxidase from *Escherichia coli*. Characterization of the enzyme and mechanism of electrochemical proton gradient generation. Biochemistry **23**:4703–4714.

140. **Mitchell, P.** 1954. Transport of phosphate across the osmotic barrier of *Micrococcus pyogenes*: specificity and kinetics. J. Gen. Microbiol. **11**:73–82.

141. **Mitchell, P.** 1961. Coupling of phosphorylation to electron and hydrogen transfer by a chemi-osmotic type of mechanism. Nature (London) **191**:144–148.

142. **Mitchell, P.** 1963. Molecule, group and electron translocation through natural membranes. Biochem. Soc. Symp. **22**:142–169.

143. **Mitchell, P.** 1967. Translocations through natural membranes. Adv. Enzymol. **29**:33–87.

144. **Mitchell, P.** 1969. Chemiosmotic coupling and energy transduction. Theor. Exp. Biophys. **2**:159–216.

145. **Mitchell, P.** 1979. Compartmentation and communication in living systems. Ligand conduction: a general catalytic principle in chemical, osmotic and chemiosmotic reaction systems. Eur. J. Biochem. **95**:1–20.

146. **Mitchell, P.** 1985. Molecular mechanics of protonmotive F₀F₁ ATPases. Rolling well and turnstile hypothesis. FEBS Lett. **182**:1–7.

147. **Mitchell, P.** 1985. The correlation of chemical and osmotic forces in biochemistry. J. Biochem. **97**:1–18.

148. **Mitchell, P., and J. Moyle.** 1967. Acid-base titration across the membrane system of rat-liver mitochondria: catalysis by uncouplers. Biochem. J. **104**:588–600.

149. **Mosher, M. E., L. K. Peters, and R. H. Fillingame.** 1983. Use of lambda *unc* transducing bacteriophages in genetic and biochemical characterization of H⁺-ATPase mutants of *Escherichia coli*. J. Bacteriol. **156**:1078–1092.

150. **Mosher, M. E., L. K. White, J. Hermolin, and R. H. Fillingame.** 1985. H⁺-ATPase of *Escherichia coli*. An *uncE* mutation impairing coupling between F₁ and F₀ but not F₀-mediated H⁺ translocation. J. Biol. Chem. **260**:4807–4814.

151. **Mueckler, M., C. Caruso, S. A. Baldwin, M. Panico, I. Blench, H. R. Morris, W. J. Allard, G. E. Lienhard, and H. F. Lodish.** 1985. Sequence and structure of a human glucose transporter. Science **229**:941–945.

152. **Nielson, J., F. G. Hansen, J. Hoppe, P. J. Friedl, and K. von Meyenburg.** 1981. The nucleotide sequence of the *atp* genes coding for the F₀ subunits a, b, c and the F₁ subunit δ of the membrane bound ATP synthase of *Escherichia coli*. Mol. Gen. Genet. **184**:33–39.

153. **Nikaido, H., and E. Y. Rosenberg.** 1981. Effect of solute size on diffusion rates through the transmembrane pores of the outer membrane of *Escherichia coli*. J. Gen. Physiol. **77**:121–135.

154. **Noumi, T., M. Futai, and H. Kanazawa.** 1984. Replacement of serine 373 by phenylalanine in the α subunit of *Escherichia coli* F₁-ATPase results in loss of steady-state catalysis by the enzyme. J. Biol. Chem. **259**:10076–10079.

155. **Noumi, T., M. E. Mosher, S. Natori, M. Futai, and H. Kanazawa.** 1984. A phenylalanine for serine substitution in the β subunit of *Escherichia coli* F₁-ATPase affects dependence of its activity on divalent cations. J. Biol. Chem. **259**:10071–10075.

156. **Overath, P., and J. K. Wright.** 1984. Lactose permease: a carrier on the move. Trends Biochem. Sci. **8**:404–408.

157. **Padan, E., and S. Schuldiner.** 1986. Intracellular pH regulation in bacterial cells. Methods Enzymol. **125**:337–365.

158. **Page, M. G. P., and I. C. West.** 1981. The kinetics of β-galactoside-proton symport in *Escherichia coli*. Biochem. J. **196**:721–731.

159. **Page, M. G. P., and I. C. West.** 1982. Alternative-substrate inhibition and the kinetic mechanism of the β-galactoside/proton symport of *Escherichia coli*. Biochem. J. **204**:681–688.

160. **Pavlasova, E., and F. M. Harold.** 1969. Energy coupling in the transport of β-galactosides by *Escherichia coli*: effect of proton conductors. J. Bacteriol. **98**:198–204.

161. **Perlin, D. S., D. N. Cox, and A. E. Senior.** 1983. Integration of F₁ and the membrane sector of the proton-ATPase of *Escherichia coli*. Role of subunit "b" (*uncF* protein). J. Biol. Chem. **258**: 9793–9800.

162. **Perlin, D. S., L. R. Latchney, J. G. Wise, and A. E. Senior.** 1984. Specificity of the proton adenosine triphosphatase of *Escherichia coli* for adenine, guanine, and inosine nucleotides in catalysis and binding. Biochemistry **23**:4998–5003.

163. **Porter, A. C. G., C. Kumamoto, K. Aldape, and R. D. Simoni.** 1985. Role of the *b* subunit of the *Escherichia coli* proton-translocating ATPase. J. Biol. Chem. **260**:8182–8187.

164. **Portis, A. R., Jr., and R. E. McCarty.** 1976. Quantitative relationships between phosphorylation, electron flow, and internal hydrogen ion concentrations in spinach chloroplasts. J. Biol. Chem. **251**:1610–1617.

165. **Ramos, S., and H. R. Kaback.** 1977. The relationship between the electrochemical proton gradient and active transport in *Escherichia coli* membrane vesicles. Biochemistry **16**:854–859.

166. **Ramos, S., S. Schuldiner, and H. R. Kaback.** 1976. The electrochemical gradient for H⁺ and its relationship to active transport in *Escherichia coli* membrane vesicles. Proc. Natl. Acad. Sci. USA **73**:1892–1896.

167. **Rickenberg, H. V., G. N. Cohen, G. Butlin, and J. Monod.** 1956. La galactoside-permease d'*Escherichia coli*. Ann. Inst. Pasteur. (Paris) **91**:829–857.

168. **Rosen, B.** 1973. β-Galactoside transport and proton movements in an adenosine triphosphatase-deficient mutant of *Escherichia coli*. Biochem. Biophys. Res. Commun. **53**:1289–1296.

169. **Rosen, B. P., and E. R. Kashket.** 1978. Energetics of active transport, p. 559–620. *In* B. P. Rosen (ed.), Bacterial transport. Marcel Dekker, Inc., New York.

170. **Rosenberg, H., R. G. Gerdes, and F. M. Harold.** 1979. Energy coupling to the transport of inorganic phosphate in *Escherichia coli*. Biochem. J. **178**:133–137.

171. **Rosing, J., and E. C. Slater.** 1972. The value of ΔG⁰ for the hydrolysis of ATP. Biochim. Biophys. Acta **267**:275–290.

172. **Rossmann, M. G., and P. Argos.** 1981. Protein folding. Annu. Rev. Biochem. **50**:497–532.

173. **Runswick, M. J., and J. E. Walker.** 1983. The amino acid sequence of the β-subunit of ATP synthase from bovine heart mitochondria. J. Biol. Chem. **258**:3081–3089.

174. **Saier, M. H., Jr., D. L. Wentzel, B. U. Feucht, and J. J. Judice.** 1974. A transport system for phosphoenolpyruvate, 2-phosphoglycerate and 3-phosphoglycerate in *Salmonella typhimurium*. J. Biol. Chem. **250**:5089–5096.

175. **Sanno, Y., T. H. Wilson, and E. C. C. Lin.** 1968. Control of permeation to glycerol in cells of *Escherichia coli*. Biochem. Biophys. Res. Commun. **32**:344–349.

176. **Sarastre, M., N. J. Gay, A. Eberle, M. J. Runswick, and J. E. Walker.** 1981. The *atp* operon: nucleotide sequence of the genes for the δ, β, and ε subunits of *Escherichia coli* ATP synthase. Nucleic Acids Res. **9**:5287–5296.

177. **Schairer, H. U., and B. A. Haddock.** 1972. β-galactoside accumulation in a Mg²⁺,Ca²⁺-activated ATPase deficient mutant of *E. coli*. Biochem. Biophys. Res. Commun. **48**:544–551.

178. **Schein, S. J., B. L. Kagan, and A. Finkelstein.** 1978. Colicin K acts by forming voltage-dependent channels in phospholipid bilayer membranes. Nature (London) **276**:159–163.

179. **Schmidt, G., and P. Graber.** 1985. The rate of ATP synthesis by reconstituted CF_0F_1 liposomes. Biochim. Biophys. Acta **808**:46–51.
180. **Scholes, P., and P. Mitchell.** 1970. Acid-base titration across the plasma membrane of *Micrococcus denitrificans*: factors affecting the effective proton conductance and the respiratory rate. J. Bioenerg. **1**:61–72.
181. **Scholes, P., and P. Mitchell.** 1970. Respiration-driven proton translocation in *Micrococcus denitrificans*. J. Bioenerg. **1**:309–323.
182. **Scholes, P., P. Mitchell, and J. Moyle.** 1969. The polarity of proton translocation in some photosynthetic microorganisms. Eur. J. Biochem. **8**:450–454.
183. **Schultz, S. G., and P. F. Curran.** 1970. Coupled transport of sodium and organic solutes. Physiol. Rev. **50**:637–718.
184. **Schwab, G. W., and E. Komor.** 1978. A possible mechanistic role of the membrane potential in proton-sugar cotransport of *Chlorella*. FEBS Lett. **87**:157–160.
185. **Senior, A. E.** 1983. Secondary and tertiary structure of membrane proteins involved in proton translocation. Biochim. Biophys. Acta **726**:81–95.
186. **Senior, A. E.** 1984. Disposition of polar and nonpolar residues on outer surfaces of transmembrane helical segments of proteins involved in proton translocation. Arch. Biochem. Biophys. **234**:138–143.
187. **Senior, A. E.** 1985. The proton-ATPase of *Escherichia coli*. Curr. Top. Membr. Transp. **23**:135–151.
188. **Senior, A. E., D. R. H. Fayle, J. A. Downie, F. Gibson, and G. B. Cox.** 1979. Properties of membranes from mutant strains of *Escherichia coli* in which the β-subunit of the adenosine triphosphatase is abnormal. Biochem. J. **180**:111–118.
189. **Senior, A. E., L. Langman, G. B. Cox, and F. Gibson.** 1983. Oxidative phosphorylation in *Escherichia coli*. Characterization of mutant strains in which F_1-ATPase contains abnormal β-subunits. Biochem. J. **210**:395–403.
190. **Senior, A. E., and J. G. Wise.** 1983. The proton-ATPase of bacteria and mitochondria. J. Membr. Biol. **73**:105–124.
191. **Serrano, R., M. C. Kielland-Brandt, and G. R. Fink.** 1986. Yeast plasma membrane ATPase is essential for growth and has homology with $(Na^+ + K^+)$, K^+- and Ca^{2+}-ATPases. Nature (London) **319**:689–693.
192. **Seto-Young, D., C.-C. Chen, and T. H. Wilson.** 1985. Effect of different phospholipids on the reconstitution of two functions of the lactose carrier of *Escherichia coli*. J. Membr. Biol. **84**:259–267.
193. **Slonczewski, J. L., B. P. Rosen, J. R. Alger, and R. M. Macnab.** 1981. pH homeostasis in *E. coli*: measurement by ^{31}P nuclear magnetic resonance of methylphosphonate and phosphate. Proc. Natl. Acad. Sci. USA **78**:6271–6275.
194. **Stein, W. D.** 1967. The movement of molecules across cell membranes. Academic Press, Inc., New York.
195. **Sternweis, P. C., and J. B. Smith.** 1980. Characterization of the inhibitory (ε) subunit of the proton-translocating adenosine triphosphatase from *Escherichia coli*. Biochemistry **19**:526–531.
196. **Stock, J., and S. Roseman.** 1971. A sodium-dependent sugar cotransport system in bacteria. Biochem. Biophys. Res. Commun. **44**:132–138.
197. **Tiedge, H., G. Schafer, and F. Mayer.** 1983. An electron microscopic approach to the quaternary structure of mitochondria F_1-ATPase. Eur. J. Biochem. **132**:37–45.
198. **Tokuda, H., and T. Unemoto.** 1982. Characterization of the respiratory-dependent Na^+ pump in the marine bacterium *Vibrio alginolyticus*. J. Biol. Chem. **257**:10007–10014.
199. **Tsuchiya, T., J. Lopilato, and T. H. Wilson.** 1978. Effect of lithium ion on melibiose transport in *Escherichia coli*. J. Membr. Biol. **42**:45–59.
200. **Tsuchiya, T., and T. H. Wilson.** 1978. Cation-sugar cotransport in the melibiose transport system of *Escherichia coli*. Membr. Biochem. **2**:63–79.
201. **Tsuchiya, T., Y. Yamane, S. Shiota, and T. Kawasaki.** 1984. Cotransport of proline and Li^+ in *Escherichia coli*. FEBS Lett. **168**:327–330.
202. **von Meyenburg, K., B. B. Jorgensen, J. Nielsen, and F. G. Hansen.** 1982. Promoters of the *atp* operon coding for the membrane-bound ATP synthase of *Escherichia coli* mapped by Tn 10 insertion mutation. Mol. Gen. Genet. **188**:240–248.
203. **Walker, J. E., M. J. Runswick, and M. Saraste.** 1982. Subunit equivalence in *Escherichia coli* and bovine heart mitochondrial F_0F_1 ATPases. FEBS Lett. **146**:393–396.
204. **Walker, J. E., M. Saraste, M. J. Runswick, and N. J. Gay.** 1982. Distantly related sequences of α- and β-subunits of ATP synthase, myosin, kinases, and other ATP-requiring enzymes and common nucleotide binding fold. EMBO J. **1**:945–951.
205. **Webb, M. R., C. Grubmeyer, H. S. Penefsky, and D. R. Trentham.** 1980. The stereochemical course of phosphoric residue transfer catalyzed by beef heart mitochondrial ATPase. J. Biol. Chem. **255**:11637–11639.
206. **West, I. C., and P. Mitchell.** 1972. Proton-coupled β-galactoside translocation in non-metabolizing *Escherichia coli*. J. Bioenerg. **3**:445–462.
207. **West, I. C., and P. Mitchell.** 1973. Stoichiometry of lactose-proton symport across the plasma membrane of *Escherichia coli*. Biochem. J. **132**:587–592.
208. **West, I. C., and P. Mitchell.** 1974. The proton-translocating ATPase of *Escherichia coli*. FEBS Lett. **40**:1–4.
209. **West, I. C., and T. H. Wilson.** 1973. Galactoside transport dissociated from proton movement in mutants of *Escherichia coli*. Biochem. Biophys. Res. Commun. **50**:551–558.
210. **Willecke, K., E.-M. Grier, and P. Oehr.** 1973. Coupled transport of citrate and magnesium in *Bacillus subtilis*. J. Biol. Chem. **248**:807–814.
211. **Wilson, D. M., J. F. Alderete, P. C. Maloney, and T. H. Wilson.** 1976. Proton motive force as the source of energy for adenosine 5'-triphosphate synthesis in *Escherichia coli*. J. Bacteriol. **126**:327–337.
212. **Winkler, H. H., and T. H. Wilson.** 1966. The role of energy coupling in the transport of β-galactosides by *Escherichia coli*. J. Biol. Chem. **241**:2200–2211.
213. **Wise, J. G., L. R. Latchney, A. M. Ferguson, and A. E. Senior.** 1984. Defective proton ATPase of *unc*A mutants of *Escherichia coli* 5'-adenylyl imidodiphosphate binding and ATP hydrolysis. Biochemistry **23**:1426–1432.
214. **Wise, J. G., L. R. Latchney, and A. E. Senior.** 1981. The defective proton-ATPase of *unc*A mutants of *Escherichia coli*. Studies of nucleotide binding sites, bound aurovertin fluorescence, and labeling of essential residues of the purified F_1-ATPase. J. Biol. Chem. **256**:10383–10389.
215. **Woodhull, A.** 1973. Ionic blockade of sodium channels in nerve. J. Gen. Physiol. **61**:687–708.
216. **Wright, J. K.** 1986. Experimental analysis of ion/solute cotransport by substrate binding and facilitated diffusion. Biochim. Biophys. Acta **854**:219–230.
217. **Wright, J. K., K. Dornmair, S. Mitaku, T. Moroy, J. M. Neuhaus, R. Seckler, H. Vogel, U. Weigel, F. Jahnig, and P. Overath.** 1985. Lactose:H^+ carrier of *Escherichia coli*: kinetic mechanism, purification and structure. Ann. N.Y. Acad. Sci. **456**:326–341.
218. **Wright, J. K., and R. Seckler.** 1985. The lactose/H^+ carrier of *Escherichia coli*: lacY^{un} mutation decreases the rate of active transport and mimics an energy-uncoupled phenotype. Biochem. J. **227**:287–297.
219. **Yazyu, H., S. Shiota-Niiya, T. Shimamoto, H. Kanazawa, M. Futai, and T. Tsuchiya.** 1984. Nucleotide sequence of the melB gene and characteristics of deduced amino acid sequence of the melibiose carrier in coli. J. Biol. Chem. **259**:4320–4326.
220. **Yoshida, M., and W. S. Allison.** 1982. The specificity of carboxyl group modification during the inactivation of the coli F_1-ATPase with dicyclohexyl-[^{14}C]carbodiimide. J. Biol. Chem. **257**:10033–10037.
221. **Yoshida, M., H. Okamoto, N. Sone, H. Hirata, and Y. Kagawa.** 1977. Reconstitution of thermostable ATPase capable of energy coupling from its purified subunits. Proc. Natl. Acad. Sci. USA **74**:936–940.
222. **Yoshida, M., N. Sone, H. Hirata, and Y. Kagawa.** 1978. Evidence for three α subunits in one molecule of F_1 ATPase from thermophilic bacterium, PS3. Biochem. Biophys. Res. Commun. **84**:117–122.
223. **Zilberstein, D., S. Schuldiner, and E. Padan.** 1979. Proton electrochemical gradient in *Escherichia coli* cells and its relation to active transport of lactose. Biochemistry **18**:669–673.

A3. Entry of Carbons from Sources Other than Glucose into Central Metabolic Routes

18. Dissimilatory Pathways for Sugars, Polyols, and Carboxylates

E. C. C. LIN

Department of Microbiology and Molecular Genetics, Harvard Medical School, Boston, Massachusetts 02115

INTRODUCTION 245
OLIGOSACCHARIDES AND DISACCHARIDES 245
   E. coli.. 245
     Maltose... 245
     Raffinose 246
     β-Galactoside utilization through the *lac* system................... 246
     β-Galactoside utilization through the *ebg* system.................. 247
     Melibiose and α-galactosides 248
     Trehalose 248
     Sucrose... 249
   S. typhimurium 249
     Maltose... 249
     Lactose 249
     Melibiose.. 249
     Trehalose 249
     Sucrose... 249
β-GLUCOSIDES... 249
   E. coli.. 249
HEXOSES 250
   E. coli.. 250
     D-Glucose 250
     D-Fructose 250
     D-Mannose 250
     L-Sorbose 250
     D-Galactose 250
   S. typhimurium 251
     D-Mannose 251
     D-Galactose 252
THE AMINO SUGARS 252
   E. coli.. 252
     N-Acetyl-D-glucosamine and D-glucosamine........................... 252
METHYLPENTOSES 253
   E. coli.. 253
     L-Fucose 253
     L-Rhamnose 254
   S. typhimurium 254
     L-Fucose and L-rhamnose.. 254
PENTOSES AND TRIOSES.. 254
   E. coli K-12 254
     L-Arabinose... 254
     D-Arabinose... 255
     D-Xylose 255
     D-Xylose system for growth on novel substrates 255
     D-Ribose 255
     Trioses.. 255
   S. typhimurium 255
     L-Arabinose... 255
     D-Arabinose... 256
     D-Xylose 256
     D-Ribose 256
     2-Deoxy-D-ribose 256
     Trioses.. 256

244

INTRODUCTION

It is a remarkable phenomenon that the various carbohydrates discovered to be utilizable by *Escherichia coli* and *Salmonella typhimurium* can all support growth in a mineral medium as the sole source of carbon and energy at a maximal rate seldom longer than 2 to 3 h per doubling at 37°C, since even in highly enriched media the doubling time cannot be shortened significantly below 0.5 h, at which point macromolecular synthesis and assembly become rate limiting. This is impressive because in natural environments a carbon and energy source virtually never appears singly. Hence, one would not expect strong selective pressure for the development of large capacities for individual pathways. Simple starvation would favor mutants that are amplified only in transport functions, not in entire pathways.

The catabolic repertoires for using carbon and energy sources are largely the same in *E. coli* and *S. typhimurium*. Nonetheless, significant differences are found even among the strains and substrains of each species. A few of the variations found among substrains may have resulted from selection under laboratory conditions or mutagenic treatment and cloning of the stocks. More likely, however, the variations reflect the polymorphisms that exist naturally in a strain. Numerous as the catabolites might be, the fundamental biochemical mechanisms for their breakdown are few. These mechanisms include phosphorylation, phosphorolysis, keto-enol isomerizations, removal or addition of hydrogen pairs, and aldol cleavage. It would not be surprising if future nucleotide

sequence data showed that the majority of the genes specifying the special pathways came from duplicated genes for the glycolytic system and the pentose phosphate shunt at a time when life was anaerobic.

OLIGOSACCHARIDES AND DISACCHARIDES

E. coli

Maltose. Over 80% of *E. coli* isolates ferment maltose (212). Certain strains can also grow on amylopectin [a backbone of glucosyl units in $\alpha(1{\rightarrow}4)$ linkage with branches of 24 to 30 glucosyl units in $\alpha(1{\rightarrow}6)$ linkage] through the action of an extracellular α-amylase that releases maltodextrins [linear glucosyl units in $\alpha(1{\rightarrow}4)$ linkage], maltose, and glucose. However, strain K-12 utilizes only maltose and maltodextrins (646), as shown in Fig. 1. The substrates first traverse a selectively permeable channel across the outer membrane provided by the *lamB* protein (maltose porin, a trimer of 47,392 daltons) which is anchored to the peptidoglycan layer. The exploitation of the external surface of the *lamB* product by phage lambda as an attachment site made the protein an object of special interest (110, 151, 218, 242, 243, 356, 378, 480, 537, 650, 694, 782, 784, 787, 840, 854, 880). Mutants lacking this protein are impaired in the transport of maltose at concentrations below 10 μM, whereas growth on maltotriose and larger oligomers is severely hampered (109, 840, 841). Inside the periplasm, the substrates are sequestered by a binding protein, the *malE* gene product (a monomer of 40,661 daltons), which also serves as a receptor for chemotaxis (12, 107, 108, 208,

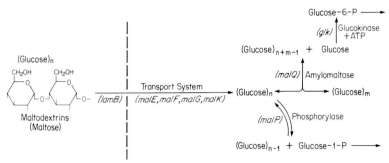

FIG. 1. Catabolic pathway for maltodextrins and maltose.

236, 329–331, 426, 687, 704, 806). The substrates are then delivered into the cytoplasm against a concentration gradient (904) with the participation of the *malF* (56,947 daltons), *malG* (23,000 daltons), and *malK* (40,700 daltons) products (59, 258, 277, 808, 809, 810). Maltose and maltodextrins of up to seven glucopyranose units can be transported at about the same rate. Maltodextrins too large to be transported are nonetheless bound to the outer membrane (233). The entire transport process is believed to work in a bucket-brigade fashion, with the periplasmic protein coming into physical contact with the outer membrane protein and the inner membrane protein complex, giving little chance for the substrate to wander away (60, 89, 95, 234, 338, 889). The energy-requiring step of transport across the plasma membrane is not well understood (235, 705). An acetyl-coenzyme A-dependent enzyme (not encoded by the *mal* regulon) acetylates excess maltose in amylomaltase-defective mutants, and the product is excreted. Such a mechanism might serve as a safety valve during normal utilization of maltose and maltodextrins (90, 256).

After the uptake of maltose or maltodextrin, a series of reshuffles are catalyzed by amylomaltase (a monomer of 71,000 daltons) encoded by *malQ*. Maltose or larger oligomers act as acceptor molecules for the transfer of a 4-α-glucanosyl or glucosyl group from a donor no smaller than maltotriose, liberating free glucose from the reducing end of the donor (323, 574, 575, 645, 780, 791, 859, 902, 903). The glucose is probably converted by glucokinase to glucose 6-phosphate. Alternatively, maltodextrin phosphorylase (a dimer of 81,000 daltons) encoded by *malP* can release glucose 1-phosphate from the sugar chains by phosphorolysis. The enzyme is most active on short linear α(1→4)-linked oligoglucosides and has the interesting property of being associated with two molecules of pyridoxal phosphate (like a number of other phosphorylases), the function of which is unknown in this context (206, 766, 786, 853).

Genes of the *mal* system, under positive control, cluster in two regions. The *malA* group at min 75 contains the *malPQ* operon and the *malT* activator gene, and the *malB* group at min 91 contains the *malEFG* operon and the operon comprising *malK*, *lamB*, and *molA* (function of gene product not yet discovered). Additional genes under the control of the *malT* activator proteins are being discovered. Because the *malB* region is involved in the substrate-uptake process, it would not be surprising if the *molA* gene product has a role in that function (355; see chapter 92).

Raffinose. The majority of *E. coli* isolates, although not strain K-12, grow on raffinose. Raffinose-negative strains with a *lac* system can mutate to grow on this trisaccharide (*O*-α-D-galactopyranosyl-[1→6]-α-D-glucopyranosyl-β-D-fructofuranoside) through constitutive expression of the *lac* operon, because β-galactoside permease acts on the trisaccharide as a fortuitous substrate. The α-galactosidase encoded by *melA* then liberates the galactose from the trisaccharide (40, 167, 212, 447; S. Schaefler, Bacteriol. Proc., p. 54, 1967; W. H. Holmes, Biochem. J. **106**:31, 1968). Sucrose, the other product, is presumably excreted.

The ability of *E. coli* to grow rapidly on raffinose, however, is usually conferred by a special plasmid bearing a cluster of *raf* genes: *rafD* encoding an invertase, *rafB* encoding a permease, and *rafA* encoding an α-galacotsidase (637, 819). Raffinose is first converted to galactose and sucrose by α-galactosidase, and the sucrose is then hydrolyzed by invertase. The plasmid-encoded α-galactosidase (a tetramer of an 82,000-dalton subunit), distinct from the *melA* gene product, hydrolyzes methyl-α-D-galactoside, melibiose, and raffinose (773).

β-Galactoside utilization through the *lac* system. The great majority of *E. coli* strains are lactose positive (212). Ironically, the various mutants of strain ML (merde Lwoff by legend), which contributed so much to the understanding of enzyme induction through studies of its lactose system, were derived from a lactose-negative parent (572). The *lac* operon (see chapter 89) encodes proteins for the utilization of not only lactose but a variety of β-galactosides, including galactosyl-β-1,4-D-fructose and galactosyl-β-1,4-D-arabinose (35). The pathway (Fig. 2) starts with the reaction catalyzed by β-galactoside permease (*lacY* product), which was described in a landmark paper on bacterial membrane transport (707) and remains a classical model for the study of sugar permeation (410, 640, 921; see chapter 17). Also, the permease was the first sugar transport system in bacteria shown to be driven by proton motive force (896, 897). Mutants with an energy-uncoupled permease cannot grow on lactose even though such a protein still catalyzes facilitated diffusion of the substrate. With such a limited function, autocatalytic induction fails probably for two reasons. (i) The substrate cannot be concentrated against a gradient. (ii) The true inducer derived from the substrate cannot be retained. If the *lac* operon is rendered constitutive, cells with the defective permease will grow normally on lactose at a concentration above 5 mM, the K_m of β-galactosidase

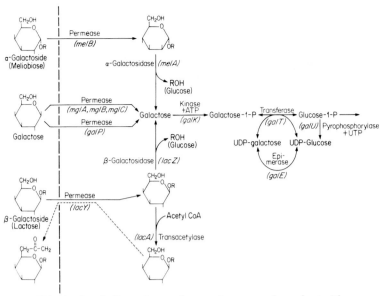

FIG. 2. Catabolic pathways for D-galactose and D-galactosides.

(465, 916, 917, 933). Induction failure also occurs if the affinity of the permease for lactose is lowered (245). On the other hand, excessive lactose-transporting activity can lethally dissipate the proton motive force of a cell (209, 364, 656, 877, 913, 914). The broad specificity of the permease made it possible to detect β-galactosidase in vivo with chromogenic substrates, such as *ortho*-nitrophenyl-β-D-galactoside. This convenience greatly accelerated the genetic studies of the *lac* system (483). The wild-type permease can also transport sucrose against a concentration gradient (335), but mutations are necessary for the permease to transport maltose (807).

Internalized substrates are hydrolyzed by β-galactosidase encoded by *lacZ*. The high molecular weight (a tetramer of subunits of 1,021 amino acid residues) partly explains why this enzyme can account for as much as 5% of the total cytoplasmic proteins in induced cells (278, 887, 952). Although the substrate specificity of the enzyme with respect to the aglycone moiety is remarkably broad, α-lactose (the hemiacetal of glucose in the α-anomeric configuration) is attacked with a lower K_m and a higher V_{max} than β-lactose (885). It is predominantly in the α configuration that lactose is found in milk. With lactose as substrate, the enzyme is more strongly activated by K^+ than by Na^+. The converse is true with *ortho*-nitrophenyl-β-D-galactoside as the substrate. The relative effects of the cations, however, depend on the strain from which the enzyme is isolated (160, 456, 483, 700, 886, 888). The enzyme catalyzes not only the hydrolysis of β-galactosides but also the rearrangement of lactose (probably some of the other substrates as well) to give allolactose (1-6-*O*-β-D-galactopyranosyl-D-glucose), which is a true inducer of the *lac* operon (120, 396, 586). Such transfer reactions can also result in the formation of several different trisaccharides (368).

A third protein (a dimer of about 270 amino acid residues) of the system, which transfers the acetyl group from acetyl-coenzyme A to the 6-OH of galactosides and thiogalactosides, is encoded by *lacA*, the

gene most distal to the promoter. Substrates not readily attacked by β-galactosidase tend to be good substrates for the transacetylase, and vice versa. Indeed, lactose itself is not a substrate for acetylation (29, 115, 116, 246, 280, 598, 948–951). Since acetylated galactosides escape from the cell and are not reaccumulated, the suggestion was made that the transacetylase serves as a detoxifying enzyme. Growth competition between isogenic *lacA*+ and *lacA* cells in lactose media with or without a nonmetabolizable analog supported the hypothesis (36, 343, 915).

β-Galactoside utilization through the *ebg* system. Strain K-12 with a nonpolar deletion within *lacZ* can regain full ability to grow on lactose by a series of mutations recruiting a substitute β-galactosidase. The genetically altered enzyme is specified by the *ebg* (evolving β-galactosidase) locus at min 67.8 (44, 51, 135). Further studies revealed a cluster of three genes arranged and transcribed in clockwise order as the chromosome map is drawn: *ebgR* (encoding a repressor with a predicted molecular weight of 50,000), *ebgA* (encoding an evolved β-galactosidase that is a hexamer of six identical 120,000-dalton subunits), and *ebgB* (encoding a 79,000-dalton protein of unknown function). Although the *ebgA*-encoded enzyme is a very poor lactase, lactose is the most effective inducer (over 200-fold). Even at this level of induction, there is insufficient enzyme activity to permit growth on lactose. Mutations altering both the structure of the enzyme and the regulation of its synthesis are necessary to confer growth ability. Among the regulatory mutations, the majority enhanced the inducibility of the enzyme by lactose, whereas the minority caused constitutive synthesis of the novel enzyme. Since the *ebg* system does not provide a permease for lactose, growth requires the *lacY*-encoded permease, the presence of which is assured by supplying isopropyl 1-thio-β-D-galactoside in the medium (294, 296, 303–305, 307, 320). Mutants that grow on other galactosides were isolated. Selection on phenylgalactoside, a substrate but not an inducer, yielded mutants (*ebgA*+

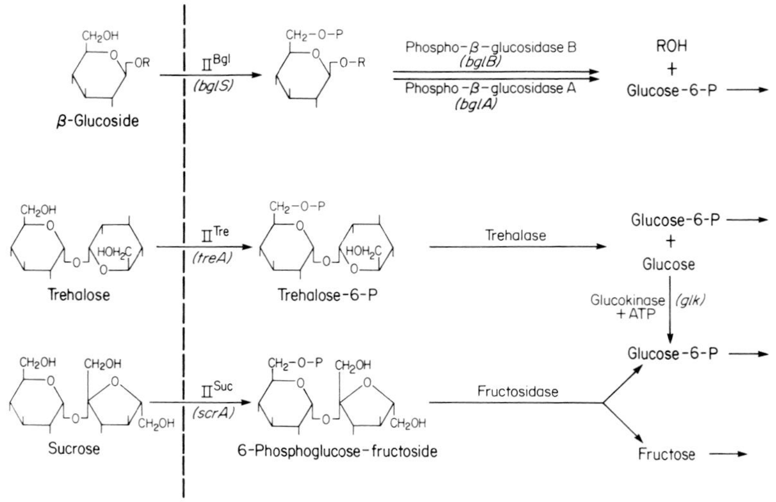

FIG. 3. Catabolic pathways for D-glucosides.

$ebgR^-$) that synthesize the native enzyme constitutively. From such a mutant, which grew extremely slowly on methyl-β-D-galactoside, a derivative [$ebgA$(mg) $ebgR^-$] with moderate growth rate on the compound was selected. A rich variety of other specificity changes was observed with the $ebgA$-encoded enzyme. For instance, mutations in class I sites of $ebgA$ can raise the V_{max} on lactose to a high level which is accompanied by a drop of the V_{max} on lactulose and a slight increase in the V_{max} on galactosyl-β-D-arabinose. Mutations in class II sites (separated from class I sites by 1 kilobase, or about one-third of the gene) can enhance the V_{max} on lactose and lactulose to moderately high levels and on galactosyl-β-D-arabinose to a modest level. When class I mutants were selected for growth on lactulose and when class II mutants were selected for growth on galactosyl-D-arabinose, the resultant class IV mutants grew well on lactose, lactulose, and galactosyl-β-D-arabinose. These mutants were found to harbor both a mutation of the class I site and a mutation of the class II site. Only from a class IV mutant can a derivative that grows on lactobionate (galactosyl-D-gluconate) be isolated in a single step (295, 297, 299, 306). Even more impressive is the evolution of an $ebgA$-encoded enzyme acquiring the ability to catalyze the formation of allolactose from lactose by transgalactosylation. In such a mutant, growth on lactose is no longer dependent on the induction of the $lacY$ permease by isopropyl 1-thio-β-D-galactoside (301, 727). Mutants in which the $ebgR$-encoded repressor increased its sensitivity to lactose as the inducer or responded to lactulose as an inducer were also isolated (298, 300). Nucleotide sequence analysis revealed a high degree of homology between $lacI$ $lacZY$ and $ebgR$ $ebgAB$ (831, 832). The ebg system has thus proven to be a fruitful model for studying the maturation of a regulated operon.

Melibiose and α-galactosides. $E.$ $coli$ K-12 and B, but not ML, grow on melibiose (6-O-α-D-galactopyranosyl-D-glucose), epimelibiose (6-O-α-D-galactopyranosyl-D-mannose), galactinol (1-O-α-D-galactopyranosyl-mesoinositol), melibiitol (6-O-α-D-galactopyranosyl-D-glucitol), galactosylglycerol (2-O-α-D-galactopyrano-sylglycerol), and probably many other α-galactosides with the aid of an inducible permease and an α-galactosidase, as depicted in Fig. 2 (513, 651). Among the α-galactosides, melibiose seems exceptional in that it can also induce and be transported by β-galactoside permease encoded by $lacY$. However, the compound cannot be hydrolyzed by β-galactosidase (512, 573, 682, 799). Melibiose permease (not inducible by lactose) is stable at 37°C in $E.$ $coli$ B, is labile at that temperature in strain K-12, and is apparently absent altogether at that temperature in strain ML (66, 651, 682). Cation-sugar cotransport mechanisms in procaryotes are generally energized by the electrochemical gradient of protons, whereas those of eucaryotes are energized by the gradient of sodium ions. Melibiose permease of $E.$ $coli$ has the unusual property of being able to harness downhill electrochemical gradients of either protons or sodium ions for the uphill transport of the sugar (534, 861, 863–866, 920) and thus appears to be at a branch point of evolution (918, 919). By selection for growth on melibiose in the presence of Li^+, which inhibits H^+-melibiose cotransport, mutant permeases that accept Li^+ or Na^+ but not H^+ as the cation for symport were obtained (609, 801, 802, 846, 862). Enzymic hydrolysis of the α-galactosides is catalyzed by an enzyme (about 200 kilodaltons) having the unusual property of requiring NAD and Mn^{2+} for activity (121, 122, 776). The two-step pathway (Fig. 2) is specified by the mel operon at min 93.4: the promoter-proximal gene A specifying the hydrolase and the promoter-distal gene B specifying the permease with a predicted molecular weight of about 50,000 (316, 775, 776, 947).

Trehalose. Trehalose, 1-(α-D-glucopyranosyl)-α-D-glucopyranoside, is metabolized by an inducible pathway initiated by an inducible phosphoenolpyruvate phosphotransferase system (PTS) enzyme II. The product, trehalose 6-phosphate, is converted to glucose and glucose 6-phosphate by a specific hydrolase (Fig. 3). There is a constitutive hydrolase that acts on trehalose, but its true function is not known (553). The genetic locus specifying trehalose utilization is at min 26 (63). In strains able to utilize trehalose, the sugar also induces the mal system (782). The nature of the physiologically important pathway remains to be established (J. W. Lengeler, personal communication).

Sucrose. Almost one-half of the natural isolates of *E. coli* were reported to be sucrose positive (212). Some strains can give rise to sucrose-positive mutants, the phenotype of which is associated with an inducible α-glucosidase activity (43). Other strains are positive by virtue of harboring a conjugative *scr*+-bearing plasmid (644, 819). The disaccharide is vectorially phosphorylated by enzyme II^{Scr}, and the product is converted to fructose and glucose 6-phosphate by a specific hydrolase (Fig. 3). The system is induced by fructose and is under repressor control (506, 774). Chromosomal integration of the *scr*+ genes for sucrose utilization tend to occur at the *dsd* (min 50.5) locus specifying the structure of D-serine deaminase and the regulatory protein for its synthesis. When insertion occurs, the cell loses the ability to utilize D-serine as a carbon and energy source and becomes sensitive to growth inhibition by the amino acid. The relationship between the sucrose and D-serine traits is intriguing; the diploid F' *sac*+/*dsd*+ cannot synthesize D-serine deaminase (23).

S. typhimurium

Maltose. Except for the fact that the *lamB* protein is not recognized by phage lambda as a receptor, the *mal* regulon appears to be similar to that of *E. coli*. The genes *lamB* (encoding the outer member protein) and *malE* (encoding the periplasmic protein) cluster at min 91 (647, 649, 650). The genes *malQ* (encoding amylomaltase) and *malT* (encoding the activator protein) cluster at min 74 (7, 648).

Lactose. Most strains of *S. typhimurium* are lactose negative. Lactose-positive mutants were observed to arise, and some of these also acquired the ability to ferment cellobiose, salicin, and arbutin (770). However, the lactose-positive trait is generally attributable to plasmid-encoded genes. In a number of cases, the trait is found in pathogenic strains isolated from humans and domestic animals (53, 54, 201, 211, 230, 231, 281, 398, 460, 497, 839). Although the *lac* system from a number of different plasmids was found to be inducible and catabolite repressible and to include a β-galactoside permease and a β-galactosidase similar to those of *E. coli*, no thiogalactoside transacetylase was ever detected (287). This deficiency holds also for the *lac* operon in a plasmid from a clinical isolate of *Yersinia enterocolitica*, despite the homology with the *E. coli* operon (166). It would be interesting to find out whether the transposon Tn951 present in that plasmid is responsible for spreading the lactose-positive character to various enteric organisms not normally fermenting the sugar.

Melibiose. The permease for melibiose of *S. typhimurium* was the first sodium-dependent sugar transport system discovered in bacteria (828, 858). The functionally associated α-galactosidase is also inducible by melibiose and is catabolite repressible (515).

Trehalose. Trehalose is used through the sequential action of a specific enzyme II of the PTS and a trehalose-6-phosphate hydrolase in an inducible pathway (553). The genetic locus specifying trehalose utilization is at min 37 (755).

Sucrose. *Salmonella* strains are typically sucrose negative, but plasmids that bear no known antibiotic resistance genes were observed to confer this trait to both *S. typhimurium* and *E. coli* (498, 927).

β-GLUCOSIDES

E. coli

E. coli shows strain variation in the ability to utilize β-glucosides. It seems that most strains have functional genes for the metabolic pathway (Fig. 3), but in some strains, such as K-12, the structural genes cannot be expressed. Mutants that can grow on arbutin (4-hydroxyphenyl-β-D-glucopyranoside) or salicin (2-[hydroxymethyl]phenyl-β-D-glucopyranoside) arise spontaneously at a frequency of about 10^{-5}. In such cells, *cis*-dominant mutations occurred in the *bglR* site near the promoter, converting the *bglCSB* operon from the cryptic R^+ state to the inducible R state. The operon is located at min 83.4, and the three genes are transcribed counterclockwise from *bglC*, which appears to specify a positive regulator. The *bglS* locus encodes a specific enzyme II-enzyme III complex of the PTS which is believed to function also as a negative regulator of operon expression by interacting with the *bglC* product, thereby blocking its activating function. In cells with the inducible phenotype, the blocking effect can be overcome physiologically by the effector. The blocking effect can also be abolished genetically, by mutations in *bglC* (rare) or *bglS* (frequent), resulting in constitutive expression. The nature of the true inducer is not known. The gene *bglB* encodes phospho-β-glucosidase B. Mutations in this gene abolish the ability to use salicin but not the ability to use arbutin, which can be hydrolyzed by another enzyme, phospho-β-glucosidase A (247, 569, 680, 732, 750, 767–769; A. Wright, personal communication). The spontaneous activations of the cryptic *bgl* operon in strain K-12 often result from the introduction of the insertion elements IS*1* or IS*5* upstream of the existing promoter sequence. Evidently, fruitful insertions can occur between different neighboring base pairs, since Bgl+ mutants vary in their levels of operon expression. By a different mechanism, a specific base sequence upstream of the *bgl* operon makes it expressible under certain degrees of DNA superhelicity. Irrespective of the mechanism of its activation, expression of the functional operon still depends on inducer and cyclic AMP receptor protein-cAMP complex (cAMP-CRP complex). One may ask why the operon should fluctuate between two genetic states in a population. A suggestion was made that metabolically toxic substrates, such as those that can produce cyanides upon hydrolysis, may appear in the environment. When this occurs, it would be advantageous to have the *bgl* operon in the cryptic state (199, 703). The complexity of the system may be even greater if it is true that a product of the *bglY* gene (min 27) directly or indirectly causes the repression of the *bglCSB* operon and that mutation to the Bgl+ phenotype could only occur in strains lacking that repression (186, 187). Some strains are known to grow on esculin (6,7-dihydroxycoumarin-6-glucoside), presumably also through the mediation of the *bgl* system (118).

In strain K-12, phospho-β-glucosidase A, encoded by *blgA*, is synthesized constitutively. This enzyme hydrolyzes phosphorylated arbutin, not phosphorylated salicin. Although in some strains *bglA* is closely situated on the counterclockwise side of the *bgl* operon, the gene is not a part of the *bgl* operon. The expression of *bglA* is regulated by the product of the unlinked *bglT*, which is located at min 84.4. It remains to be discovered whether, in the absence of an active *bgl* operon, phospho-β-glucosidase A has a function of its own (681).

HEXOSES

E. coli

D-Glucose. Vectorial phosphorylation by the PTS is responsible for the uptake of D-glucose (see chapter 11). The rate of D-glucose uptake can increase severalfold by growth on D-glucose or D-fructose, but the magnitude of induction is strain dependent (449, 450). The membrane-associated enzymes II for D-glucose and all of the other PTS substrates also serve as receptors for chemotaxis (11, 508; see chapter 49).

D-Fructose. Growth on D-fructose induces an enzyme II^{Fru}-enzyme III^{Fru} complex of the PTS that phosphorylates the substrate at C-1. A second phosphorylation catalyzed by an ATP-dependent kinase then converts fructose 1-phosphate to fructose 1,6-bisphosphate, as indicated by Fig. 4 (249). The *fruA* (or *ptsF*) locus (see references 617 and 667 for the recommended genetic symbols for the individual pathways initiated by the PTS) at min 45.8 probably consists of an operon of three genes: *fruA*, encoding the protein II^{Fru}; *fruB*, encoding the protein III^{Fru}; and *fruK*, encoding the kinase. D-Fructose 1-phosphate entering through preinduced hexose phosphate permease (see Phosphorylated Sugars and Carboxylates, below) also causes induction. A mutant lacking the enzyme II complex fails to grow at low concentrations of D-fructose (<2 mM), and a mutant lacking the kinase is subject to growth inhibition by D-fructose as a result of fructose 1-phosphate accumulation (237–239, 399, 891).

D-Mannose. An enzyme II^{Man}-enzyme III^{Man} complex of rather broad specificity is mostly responsible for growth on D-mannose (Fig. 4). This complex also catalyzes the phosphorylation of D-glucose and D-fructose. In a mutant defective in the system, some ability to grow on D-mannose is retained because of the side specificity of the enzyme II complex for D-glucose. On the other hand, in a *fru(AB)* mutant some ability to grow on D-fructose (>2 mM) is retained because of the side specificity of the enzyme II complex for D-mannose. In this case, D-fructose is phosphorylated at position 6. The enzyme II complex for mannose is encoded by the *ptsLM* operon at min 40.2: the gene *L* specifies protein III^{Man} and the gene *M* specifies protein II^{Man} (174, 224a, 240, 399, 400, 448, 649a, 892). The transmembrane enzyme II^{Man} serves as the entry channel for the DNA of phage lambda (216, 224a, 649a). Mannose 6-phosphate is converted to fructose 6-phosphate by an amphibolic isomerase that also has the function of synthesizing mannose 6-phosphate for capsular polysaccharide during growth on other carbon sources. Thus, it is not strange that the synthesis of the isomerase is coregulated with those of other enzymes involved in the construction of wall polysaccharides and that it is not inducible by D-mannose or catabolite repressible by glucose. The enzyme is encoded by *manI* (formerly *manA*) at min 35.7 (554, 555). Utilization of D-mannose, however, is catabolite repressible and dependent on the cAMP-CRP complex (253), but the specific regulation of *ptsLM* transcription has not been well characterized.

L-Sorbose. Of 40 *E. coli* strains surveyed, 9 were found to grow on L-sorbose. Strains K-12, B, and C were not among the positives. However, the genetic determinant for the ability to grow on the sugar is transducible and found to be near *metA* at min 91 (934). The *sor* operon introduced into strain K-12 on a plasmid conferred the ability to utilize L-sorbose (Fig. 4), as follows. The sugar is first converted by an enzyme II^{sor} (encoded by *sorA*) to L-sorbose 1-phosphate. The phosphorylated compound is then converted by an NAD(P)H-linked enzyme (encoded by *sorB*) to D-glucitol 6-phosphate, and the reduced compound is converted by an NAD-linked enzyme (encoded by *sorD*) to D-fructose 6-phosphate (822). Although strain K-12 does not grow on L-sorbose, two of its enzymes II (*ptsG* and *fruA* products) can act on the substrate (817).

D-Galactose. Active transport of D-galactose occurs mostly through two inducible systems with different K_ms: one in the range of 5 μM, and the other in the range of 0.5 μM (Fig. 2) (362, 363, 634, 742, 909). The low-affinity transport system, or D-galactose permease, is driven by proton motive force and is not associated with a periplasmic binding protein (179, 336, 430). The permease (a 37,000-dalton integral membrane protein) is encoded by *galP* at min 64 (540, 709). Expression of the gene is under the control of the repressor encoded by *galR* at min 61.5 (128, 908), the same repressor that controls the expression of the *galETK* operon (see below).

The high-affinity transport system is more complex and is referred to as the β-methylgalactoside (an excellent substrate), or *mgl*, system (97, 738, 739). It includes a periplasmic binding protein with high substrate affinity, which also plays an important role in the effective recapture of endogenous D-galactose leaking out of the cytoplasm (37–39, 86, 939). The binding protein attracted even greater interest when it was found to be the signal receptor for chemotaxis (331, 417, 418, 792, 954). This protein, the function of which depends on substrate-induced conformational change, is a polypeptide of 309 amino acids (87, 91, 92,

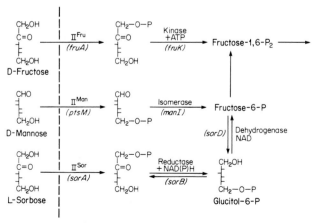

FIG. 4. Catabolic pathways for D-fructose, D-mannose, and L-sorbose.

541, 561). Its three-dimensional structure was established by X-ray diffraction (883). Like other transport mechanisms dependent on a periplasmic protein, the *mgl* system seems to operate directly at the expense of a high-energy phosphate (88, 179, 336, 653, 882; see also chapter 50). It may be noted that β-methylgalactoside can serve as a sole source of carbon and energy, since the compound is able to induce the *lac* system. The induced β-galactoside permease serves as an auxiliary (low affinity) transport system for β-methylgalactoside, and β-galactosidase hydrolyzes the compound to liberate D-galactose, which in turn induces the *mgl* system (715; B. Rotman, personal communication). At high substrate concentrations, the low-affinity *galP* permease and the high-affinity *mgl* permease appear to interact in such a way that most of the flow is directed through the *galP* permease (908).

Proteins of the high-affinity transport system are specified by the *mgl* gene cluster (min 45) with three identified complementation groups: *mglB* encoding the periplasmic binding protein, *mglA* encoding the integral inner membrane protein of 52,000 daltons, and *mglC* encoding a protein of 38,000 daltons apparently loosely associated with the inner membrane. The three genes apparently belong to a single operon, *mglBAC*, with counterclockwise transcription (94, 96, 318, 633, 634, 714, 740, 741). The following three genes have been implicated in regulating *mglBAC* expression: *mglD*, located on the clockwise side of the structural genes, which appears to code for a specific repressor of 19,000 daltons (713, 740; B. Rotman, personal communication); *mglR*, closely linked to *galK* at min 17 (268); and a second *mglR*, situated between min 56 and 74 (504). Mutations in all three genes can lead to constitutive expression of *mglBAC*. It is possible that products of the two unlinked genes influence *mglBAC* expression indirectly; e.g., a deficiency of one of the products may cause internal induction.

Once inside the cytoplasm, D-galactose is metabolized by reactions mediated sequentially by galactose kinase, galactose-1-phosphate uridyl transferase, and UDP-galactose 4-epimerase (415, 461, 462, 800, 910). The net result of the three reactions is the conversion of D-galactose to glucose-1-phosphate. The catalytic amount of UDP-glucose required is supplied by another reaction mediated by UDP-glucose pyrophosphorylase. Most of the mutations affecting D-galactose utilization were noted to be clustered at a locus close to the prophage lambda before the exact nature of the association of the viral genome with the host genome was discovered (482, 578, 579). This cluster included the genes for the kinase, the transferase, and the epimerase (13, 481, 577, 820). These genes belong to the *galETK* operon at min 17 (10, 10a, 124–126, 128, 198, 345, 542, 596, 597, 795a, 851). The operon is unusual in two respects. First, it has two overlapping promoters; the cAMP-CRP complex stimulates transcription from the upstream one and inhibits that of the downstream one. Second, there are two operators, one of which is within the structural gene *galE* (19, 376, 377, 542; see chapter 93). Transcription is under the negative control of a repressor encoded by *galR* at min 61.5, and D-galactose is the true inducer (9, 127, 376, 748, 881). Despite the fact that the structural genes are members of the same operon, synthesis of the enzymes is not always coordinate. There is a relatively high basal level of the epimerase which is necessary for D-galactose biosynthesis when the cells are growing on other carbon sources. It is conceivable that the promoter which is inhibited by cAMP initiates the transcript favoring the synthesis of the epimerase. The rate of basal synthesis of the enzymes programmed by the *gal* operon is determined by the delicate balance between the rate of liberation of free D-galactose from UDP-galactose, the rate of the escape of the sugar from the cytoplasm, the rate of the recapture of the sugar by the high-affinity transport system, and the rate of phosphorylation by the kinase (405, 419, 739a, 938, 940, 941). Early reports that expression of the *gal* operon is also under repressive control of the *lon* (*capR*) gene product were not confirmed (860).

Mutants lacking the kinase fail to grow on D-galactose, but growth on other sources of carbon and energy is not inhibited by the sugar. Such mutants also express the remaining enzymes of the D-galactose pathway (including the β-methylgalactoside permease) constitutively. Mutants lacking the transferase are arrested in their growth by D-galactose, which can cause the accumulation of galactose 1-phosphate to a level of 2 mM. For reasons still unknown, this growth inhibition can be overcome by the addition of yeast extract or casein digest to the culture medium, without reducing the galactose 1-phosphate level in the cells. Glucose is also protective and can reverse the inhibition even 1 h after exposure to D-galactose. (It is now clear that the effect is related to exclusion of D-galactose by glucose; see chapter 11.) Genetic loss of the epimerase, in contrast, has drastic consequences. Exposure of growing cells to D-galactose results in the accumulation of UDP-galactose and galactose 1-phosphate, terminating in bacteriolysis. When grown in the absence of D-galactose, the lipopolysaccharides of the mutant are deficient in this sugar component, since there is no method for its biosynthesis. As a consequence, they become susceptible to attack by phage C-21 (406, 416, 419, 463, 945).

Although the main role of UDP-glucose pyrophosphorylase is the biosynthesis of polysaccharide moieties in the cell wall, deficiency in the activity of this enzyme also prevents growth on D-galactose and renders the cell sensitive to D-galactose inhibition during growth on other carbon and energy sources (836). The enzyme is encoded by *galU* at min 27.2 and is under repressor control (117, 795). When mutants lacking this enzyme are infected by T-even phages, the DNA of their progeny is deficient in the glucosylation of hydroxymethylcytosine. The DNA of such viral crops is thus rendered vulnerable to restriction by certain bacterial hosts in the next round of infection (324, 798). On the other hand, because of the alteration in their cell surface, *galU* mutants (like *galE* mutants) are more resistant to P1 infection (261).

S. typhimurium

D-Mannose. The enzyme II^{Man} of broad specificity also acts on *N*-acetylglucosamine and glucosamine (702). Curiously, growth on D-mannose decreases the activity level of phosphomannose isomerase. Yet mu-

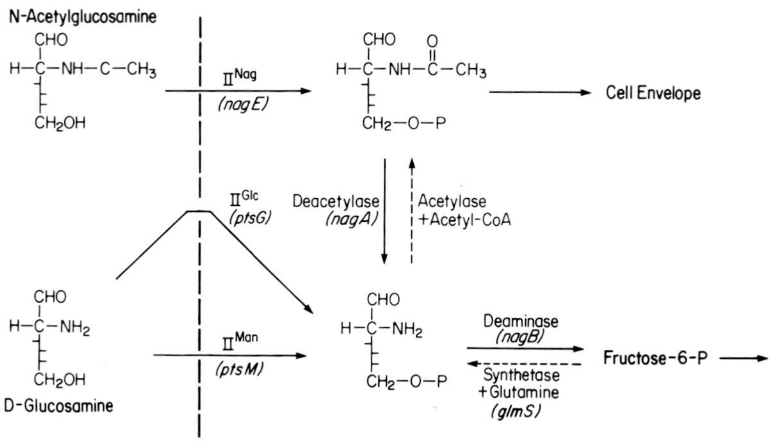

FIG. 5. Metabolic pathways of *N*-acetylglucosamine and D-glucosamine.

tational loss of this enzyme prevents growth on the sugar. However, growth of the mutant on other carbon sources is not inhibited by D-mannose; this is unexpected because accumulation of phosphorylated carbohydrates is generally deleterious to the cell. As anticipated, the isomerase mutants make imperfect wall lipopolysaccharides when grown in the absence of D-mannose. Indeed, they can be selected for their resistance to phage P22 because the altered cell surface makes adsorption difficult (733).

D-Galactose. The low-affinity of the proton-motive-force-driven permease, encoded by *galP* and controlled by the *galR* gene product, serves as the principal transport system for D-galactose, since the sugar fails to induce the *mgl* system in wild-type cells (600, 749, 873). The *mgl* system does serve for growth on β-methylgalactoside and is encoded by the *mglBAEC* operon, as follows. The *B* gene encodes a periplasmic binding protein of 33,000 daltons, the *A* gene encodes a membrane protein of 51,000 daltons, the *E* gene encodes a less well-characterized protein of 21,000 daltons, and the *C* gene encodes a membrane protein of 29,000 daltons (24, 570, 583–585, 954).

D-Galactose-induced cell lysis was first observed in pleiotropic mutants of *Salmonella* species. The sugar induces no lysis in a glucose-containing medium or under conditions not permitting growth. If growth is allowed to take place in a hypertonic medium, protoplasts are formed (262). These mutants were later discovered to lack the epimerase; their exposure to galactose led to the accumulation of UDP-galactose and galactose 1-phosphate. The envelopes of these mutants contain only glucose and lack the usual components of galactose, mannose, rhamnose, and abequose. Among the consequences of the anomalous wall structure is the interference with the adsorption of phages such as P22 (263–265, 610, 638, 811). The positions of the *gal* genes correspond approximately to those of their counterparts in *E. coli*; *galK*, *galT*, and *galE* constitute an operon at min 18, *galU* constitutes one at min 34, and *galR* constitutes one at min 62 (224, 266, 611, 758). Although *E. coli* is known to have only a single form of UDP-glucose pyrophosphorylase, *S. typhimurium* has three forms of this enzyme, each with distinct kinetic properties. It is believed that the protein specified by *galU* is subsequently modified by the product of *galF* (min 42). This process would

explain why a mutation in *galU* can abolish all three activities (260, 601–603).

THE AMINO SUGARS

E. coli

N-Acetyl-D-glucosamine and D-glucosamine. Each amino sugar can enter the cell by at least two pathways initiated by the PTS (Fig. 5). *N*-Acetyl-D-glucosamine is phosphorylated by a specific enzyme II (encoded by *nagE* or *ptsN* at min 15.6) of its own and by the major enzyme II for glucose (*ptsG* product), although *N*-acetylglucosamine cannot induce the latter enzyme. D-Glucosamine is phosphorylated by the enzyme II for D-glucose (encoded by *ptsG*) and the enzyme II for D-mannose (encoded by *ptsM*). Since both of these systems act on D-glucose, it is not surprising that this sugar powerfully inhibits the uptake of D-glucosamine (402, 502, 899, 900). In addition to serving as a general source of carbon and energy, the amino sugars contribute directly to the synthesis of peptidoglycan and lipopolysaccharides (200). The two pathways converge when *N*-acetylglucosamine 6-phosphate is converted by a deacetylase to give glucosamine 6-phosphate (898), which in turn is converted by a deaminase to yield fructose6-phosphate (161, 275, 928–930; M. Soodak, Bacteriol. Proc., p. 131, 1955). In vitro, it was found that the deacetylase was inhibited by glucosamine 6-phosphate and that the deaminase was stimulated by *N*-acetylglucosamine 6-phosphate (162, 901). It would be worthwhile to test whether these kinetic properties are also of physiological significance. The deacetylase is inducible by *N*-acetyl-D-glucosamine but not by D-glucosamine, whereas the deaminase is inducible by both carbon sources. *N*-Acetyl-D-glucosamine is more effective than D-glucosamine in repressing glucosamine-6-phosphatesynthetase(L-glutamine:D-fructose-6-phosphate aminotransferase), the activity of which is necessary for the biosynthesis of macromolecules of the cell envelope during growth on other carbon sources. The *nagA* gene encoding the deacetylase and the *nagB* gene encoding the deaminase are 98% linked by transduction and map at min 15.6 (361). Mutants lacking the deacetylase lyse in the presence of *N*-acetyl-D-glucosamine in the medium, whereas the growth of

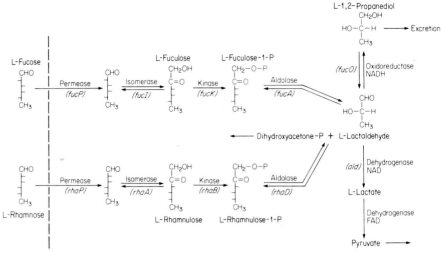

FIG. 6. Catabolic pathways for methylpentoses.

mutants lacking the deaminase is inhibited by either N-acetyl-D-glucosamine or D-glucosamine. The deleterious effect of N-acetylglucosamine can be overcome by addition of uridine (73, 737, 898). It may be noted that mutants lacking glucosamine-6-phosphate synthetase require N-acetyl-D-glucosamine or D-glucosamine for growth; deprivation causes rapid loss of viability which is not preventable by sucrose as an osmotic stabilizer (374, 761, 937).

METHYLPENTOSES

E. coli

L-Fucose. The pathway for the utilization of L-fucose (Fig. 6) is mediated sequentially by L-fucose permease (291), L-fucose isomerase (282), L-fuculose kinase (210, 272, 333), and L-fuculose-1-phosphate aldolase (271, 273). The aldolase cleaves the C_6 substrate to dihydroxyacetone-phosphate and L-lactaldehyde. Anaerobically, L-lactaldehyde is completely reduced to L-1,2-propanediol by an NAD-linked oxidoreductase. For each mole of fucose fermented, 1 mol of L-1,2-propanediol is excreted and irretrievably lost (152, 824). There is suggestive evidence that the product

crosses the plasma membrane by facilitated diffusion (290, 943). The sacrifice of one-half of the carbon skeleton of L-fucose in this way permits more dihydroxyacetone-phosphate to be assimilated as a carbon source. Aerobically, L-lactaldehyde is completely oxidized by an NAD-linked dehydrogenase to L-lactate, which is converted to pyruvate by a dehydrogenase (inducible by L-lactate) of the flavoprotein class (134, 152, 823).

The genes specifying L-fucose utilization are all clustered at min 60.5 in clockwise order: fucO for the propanediol oxidoreductase, fucA for the aldolase, fucP for the permease, fucI for the isomerase, fucK for the kinase, and fucR for the activator protein (48, 138, 519, 815, 816). L-Fuculose 1-phosphate is the inducer (J. M. Bartkus and R. P. Mortlock, Abstr. Annu. Meet. Am. Soc. Microbiol. 1983, K238, p. 216). Although the fucO gene is induced both aerobically and anaerobically, fully catalytically active oxidoreductase is synthesized only anaerobically (100, 141, 142). Because of the lack of activity of the aerobically synthesized fucO product and because of the failure of L-1,2-propanediol to induce the necessary enzymes, wild-type E. coli cannot salvage the excretion product in the presence of molecular oxygen. However, repeated aerobic se-

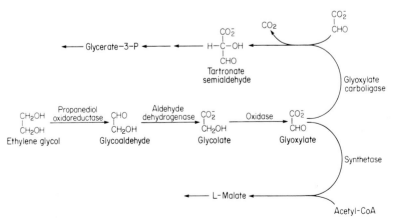

FIG. 7. Catabolic pathway for ethylene glycol.

lection on L-1,2-propanediol eventually gives rise to mutants that constitutively express an altered *fucO*. In such mutants, *fucPIK* becomes noninducible, and *fucA* becomes constitutive. The enzyme patterns of the various mutants thus indicate that the *fuc* system is a regulon containing at least three operons: *fucO*, *fucA*, and *fucPIK* (99, 140, 291, 292, 824). From an L-1,2-propanediol–positive strain, a mutant that uses the constitutive oxidoreductase for growth on xylitol was isolated (942). That same L-1,2-propanediol–positive strain also gave rise to mutants able to grow on ethylene glycol (Fig. 7) by successive involvement of the constitutive oxidoreductase, L-lactaldehyde dehydrogenase, and enzymes of the glycolate pathway (101, 134, 635).

L-Rhamnose. About three-fourths of the *E. coli* strains examined grew on L-rhamnose (212). In parallel with the pathway for L-fucose, the utilization of L-rhamnose (Fig. 6) involves the sequential action of a permease (J. Power, personal communication, 1966), an isomerase (843, 911), a kinase (145, 844, 912), and an aldolase (144, 146, 147, 762, 788). The products are dihydroxyacetone-phosphate and L-lactaldehyde. The *rha* system itself, however, does not include a gene for L-1,2-propanediol oxidoreductase. Growth on L-rhamnose induces the enzyme encoded by *fucO* (98, 143). The report of an L-1,2-propanediol oxidoreductase in a mutant with an extensive deletion covering the *fuc* region (731) has not been confirmed. The genes *rhaD* (encoding the aldolase), *rhaA* (encoding the isomerase), *rhaB* (encoding the kinase), *rhaC* (encoding the activator protein), and *rhaP* (encoding the permease) are clustered at min 87.7 in clockwise order (643, 679; J. Power, personal communication, 1966). There might be a periplasmic binding protein associated with the *rha* system, since L-rhamnose, in contrast to L-fucose, is chemotactic (12).

S. typhimurium

L-Fucose and L-rhamnose. A *fuc* system probably similar to that of *E. coli* is responsible for growth on L-fucose (629). The growth abilities of different strains of L-rhamnose vary. L-Rhamnose-positive mutants were isolated from strains without the growth ability (20, 207, 221, 576). Strain LT2 grows on the sugar; its genes for L-rhamnose utilization are clustered in the order *rhaD-rhaA-rhaB-rhaC-rhaT* (or *rhaP*) at min 86.5 (30, 758). Anaerobic growth on the sugar is associated with equimolar excretion of L-1-2-propanediol (52).

PENTOSES AND TRIOSES

E. coli K-12

L-Arabinose. The pathway for L-arabinose utilization collects the substrate from two inducible systems of transport (Fig. 8). The low-affinity permease (K_m, about 0.1 mM) is encoded by *araE* at min 61.3 (223, 620). The high-affinity system (K_m, 1 to 3 µM) is specified by the *araFG* operon at min 44.8. The *araF* gene encodes a periplasmic binding protein (306 amino acids) with chemotactic receptor function (114, 148, 259, 276, 357–359, 550–552, 607, 608, 654, 655, 686, 688, 772, 792), and the *araG* locus encodes at least

one inner membrane protein (149, 438, 453). Both high- and low-affinity transport are under the control of the *araC* gene product and are thus part of the *ara* regulon (223, 437). The *araE* system is energized by proton motive force, whereas the *araF* system seems to operate at the expense of a high-energy covalent bond (179, 539).

Internally, the sugar is metabolized by a set of enzymes encoded by the *araBAD* operon at min 1.4: an isomerase which reversibly converts the aldose to L-ribulose; a kinase which phosphorylates the ketose to L-ribulose 5-phosphate; and L-ribulose-5-phosphate-4-epimerase, which catalyzes the formation of D-xylose 5-phosphate (104, 219, 222, 491, 492). The genes are transcribed in counterclockwise order: *B* (encoding the kinase), *A* (encoding the isomerase), and *D* (encoding the epimerase). Expression of the operon is under the specific control of the product of *araC* on the clockwise side of *araBAD*. The two operons are divergently transcribed. Whereas *araC* is autogenously repressed by the regulator protein in the presence or absence of the inducer L-arabinose, *araBAD* is activated by the regulator protein-inducer complex and repressed by the regulator protein alone. Expression of both operons is dependent on the cAMP-CRP binding site (493, 625; see chapter 91). Growth of isomerase or kinase mutants on casein hydrolysate is inhibited by the addition of L-arabinose. In the case of the kinase mutant, L-ribulose accumulates in the culture. Growth of epimerase mutants is severely inhibited by L-arabinose, and the accumulated L-ribulose 5-phosphate can attain a concentration of 5 mM (calculation based on 2.7 ml of cell water per g [dry weight] according to Maloney et al. [544]). The growth inhibition is prevented or relieved by the addition of glucose but not by the addition of D-fructose or D-mannose (106, 220, 286). In contrast to *E. coli* K-12, *E.*

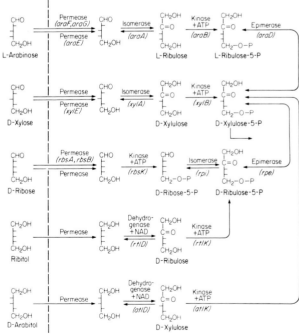

FIG. 8. Catabolic pathways for pentoses and pentitols.

coli B requires a mutation before it can grow on L-arabinose (155).

D-Arabinose. Most strains of *E. coli* are incapable of growing on D-arabinose. Since the enzymes of the L-fucose pathway have side specificities to convert D-arabinose to dihydroxyacetone phosphate and glycoaldehyde (105, 158, 271, 282, 333, 477, 479), all that is required for the additional growth ability is a mutation rendering the *fuc* system inducible by D-arabinose (478, 815, 816). An interesting exception is *E. coli* B/r, which does not grow on L-fucose but does grow on D-arabinose. This organism lacks L-fuculose-1-phosphate aldolase. The D-ribulose arising from D-arabinose is phosphorylated to D-ribulose 5-phosphate, which is isomerized to D-xylulose 5-phosphate (104).

D-Xylose. Most strains of *E. coli* grow on D-xylose, but a mutation is necessary for strain K-12 to grow on the compound (31, 118, 212). Utilization of this pentose is through an inducible and catabolite-repressible pathway (Fig. 8) involving transport across the cytoplasmic membrane (not active on D-ribose or D-arabinose), isomerization to D-xylose, and ATP-dependent phosphorylation of the pentulose to yield D-xylulose 5-phosphate (183). As in the case of D-galactose and L-arabinose utilization, D-xylose can enter through two inducible permeases. The high-affinity (K_m, 0.3 to 3 μM) system depends on a periplasmic binding protein (37,000 daltons) and is probably driven by a high-energy compound. The low-affinity (K_m, about 170 μM) system is energized by proton motive force. This D-xylose-proton-symport system is encoded by *xylE* at min 91.4 (14, 184, 473, 793). The main gene cluster specifying D-xylose utilization, *xylAB(RT)*, is at min 79.7. The nucleotide sequence of the region has been determined. The genes *xylA* (encoding the isomerase, 54,000 daltons) and *xylB* (encoding the kinase, 52,000 daltons) belong to an operon with a second transcriptional start preceding the coding region of *xylB*. It is believed, based partly on studies of *S. typhimurium*, that *xylR* encodes a regulator protein of 10,000 daltons and that *xylT* encodes a transport protein of 16,000 daltons (56, 476, 543, 736, 771, 852, 936). Since the low-affinity permease is specified by the unlinked *xylE*, the *xylT* locus probably codes for the high-affinity transport system and therefore should contain at least two genes (one for a periplasmic protein and one for an integral membrane protein). Further work is required to identify these two genes and the effect of the regulator protein.

D-Xylose system for growth on novel substrates. Constitutive expression of the D-xylose pathway permits growth on xylitol if the cells also constitutively synthesize a catalytically active L-1,2-propanediol oxidoreductase under aerobic conditions (see Methylpentoses, above). Xylitol is transported by a D-xylose permease and is then converted by the L-1,2-propanediol oxidoreductase to D-xylose (942). A mutant which acquired the ability to grow on D-lyxose also uses a D-xylose permease (inducible) for transport. D-Lyxose is then converted to D-xylulose by an isomerase (active also on D-mannose) determined by the locus *mni* in the region of min 86. The gene is apparently cryptic in the wild-type parent (826). D-Xylulose kinase contributes to a novel pathway for D-arabitol in a mutant that

converts the pentitol to D-xylulose by an NAD-linked dehydrogenase (active also on D-galactose) which became constitutive (943).

D-Ribose. At least two transport systems mediate the entry of D-ribose into the cytoplasm (Fig. 8); the one with low affinity is apparently constitutive, and the one with high affinity is inducible. High-affinity transport (K_m of uptake, about 1 μM) is dependent upon a periplasmic binding protein that serves also as a receptor for chemotaxis (267, 270, 331, 906, 907). The D-ribose-binding protein shows substantial amino acid sequence homology with the D-galactose-binding protein, in contrast to the lack of homology between the D-galactose- and L-arabinose-binding proteins (42, 285). The conservation of amino acid sequence between the D-ribose- and the D-galactose-binding proteins is believed to be the consequence of their having to interact with a common signalling component in the plasma membrane (see chapter 49). Transport by the high-affinity system appears to be driven by high-energy bonds (173).

Internal D-ribose is phosphorylated by an inducible kinase that catalyzes the formation of D-ribose 5-phosphate and is feedback inhibitable by the product (180–182, 213, 332, 533). This pathway is specified by the *rbsACBK* operon located at min 84.3 and is transcribed clockwise, as follows: *rbsA* encodes a 50,000-dalton protein required for high-affinity transport, *rbsC* encodes a 27,000-dalton protein presumed to be a transport protein association with the inner membrane, *rbsB* encodes the 29,000-dalton binding protein, and *rbsK* encodes a 34,000-dalton kinase. Expression of the operon is under the control of a repressor encoded by the *rbsR* gene, which is on the clockwise side of the structural genes and which is apparently transcribed independently. Ribose, rather than D-ribose 5-phosphate, which is an intermediate of the pentose phosphate cycle, is the true inducer. The location and true physiological role of the gene(s) specifying the low-affinity D-ribose transport (in the 10 mM range) are not known, except that cAMP is not required for gene expression (31, 32, 372, 535, 878, 879).

E. coli B, which is D-ribose negative (155), has two nonfunctional *rbs* operons (which perhaps arose by duplication and transposition). The one at min 84 has intact structural genes which are not expressible; the one at min 2 is constitutively expressed but has a defective *rbsK*. D-ribose-positive revertants can arise by either a mutation that allows the inducible expression of the min 84 operon or a mutation that restores the kinase gene of the min 2 operon. Expression of the *rbs* operons, in contrast to the K-12 system, seems to be under positive control (1–6).

Trioses. There exists a specific inducible enzyme II for dihydroxyacetone (394). It is not known whether D-glyceraldehyde can be similarly utilized.

S. typhimurium

L-Arabinose. Kinetic studies indicate that there is only a low-affinity permease for L-arabinose uptake. A permease protein of 40,000 daltons is encoded by *araE* at min 62, and transcription is counterclockwise (487, 490, 747a). However, a binding protein was isolated, and its sequence was found to be highly similar to that of the *araF* protein of *E. coli* (360). The metabolic pathway, as well as the regulatory mechanisms for gene expression, seem to be similar to the system in *E. coli* (76–78). The *ara* genes encoding the enzymes in the pathway map at min 2.7 (139, 429). Mutants lacking the epimerase are sensitive to L-arabinose (488, 684, 685, 747). The *araBAD* and *araC* operons of *S. typhimurium* resemble those of *E. coli*. The structural genes are transcribed counterclockwise, and the regulator gene is transcribed clockwise (150, 283, 365, 489).

D-Arabinose. Mutants that acquire the ability to grow on D-arabinose exhibit dual inducibility of the L-fucose pathway (629).

D-Xylose. Although kinetic studies suggested the presence of only one low-affinity transport system (K_m, 0.4 mM) indicative of a proton-motive-force-driven mechanism, a D-xylose-binding protein, expected to be associated with a high-affinity transport system and driven by a high-energy phosphate compound, was detected. Expression of the inducible pathway is subject to catabolite repression, which can be reversed by the addition of cAMP (793). The possibility that the high-affinity transport system in *S. typhimurium* is impaired by a mutation in one of its structural genes deserves examination. The genes for D-xylose utilization cluster at min 78 in clockwise order: *xylT* (encoding a permease)-*xylR* (encoding the activator protein)-*xylB* (encoding the kinase)-*xylA* (encoding the isomerase). The *trans*-activating effect of the *xylR* product was demonstrated by merodiploid analysis (274, 794). Limiting permease activity accounts for the defective abilities of some strains of *S. typhimurium* to utilize D-xylose (581, 630).

D-Ribose. A D-ribose-binding protein, serving as a component for substrate transport and a receptor for chemotaxis, shares a signalling component with the D-galactose-binding protein (21, 22, 582, 833).

2-Deoxy-D-ribose. 2-Deoxy-D-ribose supports growth of strain LT-2 (288). (For a further description of the metabolic and genetic system, see Pentoses in Nucleosides and Deoxynucleosides.)

Trioses. The growth of strain LT-2 on dihydroxyacetone (288) presumably depends on an inducible enzyme II^{Dha}, as in *E. coli* K-12.

PENTOSES IN NUCLEOSIDES AND DEOXYNUCLEOSIDES

E. coli

An early observation that *E. coli* utilized the D-ribose moiety of adenosine faster than it did the free pentose implied that the nucleoside was taken up by the cell directly (825). Later studies revealed that the cell can utilize a number of purine and pyrimidine nucleosides and deoxynucleosides as sole source of carbon and energy. Movement of these compounds (except cytidine and deoxycytidine) across the outer membrane is facilitated by the product of the *tsx* gene (min 9.4). An aqueous pore seems to be provided by the protein (25,000 daltons) which is also exploited by phage T6 and colicin K as receptors (317, 547, 562, 591). Synthesis of the outer membrane protein is under multiple regulation by the cAMP-CRP complex, the *cytR* repressor, and the *deoR* repressor. These regulatory proteins also control the expression of the

genes for the further metabolism of the substrates (455, 457). At least three transport systems are involved in the uptake of the compounds into the cytoplasm (591, 657, 658). A model requiring phosphorolytic cleavage of the nucleosides during translocation was proposed (346, 347, 692) but was not substantiated (123, 642). Instead, transport of nucleosides was shown to precede their enzymatic attack (514, 587). Two major concentrating systems have been well characterized by analysis of mutants. The C system (encoded by *nupC* at min 52), first revealed by its activity on the antibiotic showdomycin [2-(β-D-ribofuranosyl)maleimide] which is a nucleoside analog capable of alkylating sulfhydryl groups, transports cytidine and other ribonucleosides and deoxyribonucleosides (except those of guanine and hypoxanthine). The G system (encoded by *nupG* at min 64) has a broad specificity with preference for the guanine compounds (203, 204, 439, 440, 443, 444, 591, 592, 743–745, 875, 876). Both systems appear to be driven by proton motive force. In the case of the G system, the lowering of the ATP pool increases the rate of nucleoside transport activity. This inverse relationship would indicate the existence of some kind of kinetic feedback mechanism (445, 594, 746). Synthesis of the C system is regulated by the *cytR* repressor and the cAMP-CRP complex, whereas synthesis of the G system is under the double control of the *cytR* and *deoR* products as well as the cAMP-CRP complex (202, 590, 599). A third transport system was found to act on nucleosides of adenine, guanine, and hypoxanthine (441, 442).

The catabolic pathways for the nucleosides might be viewed as a way of extracting from them the pentose or deoxypentose for use as carbon and energy sources. This purpose is achieved by the combined action of eight inducible enzymes that catalyze the following reactions, in which P is phosphate:

$$\text{Cytidine} + H_2O \xrightarrow[\text{(cdd product)}]{\substack{\text{cytidine} \\ \text{deaminase}}} \text{uridine} + NH_3$$
$$\text{Deoxycytidine} + H_2O \longrightarrow \text{deoxyuridine} + NH_3 \quad (1)$$

$$\text{Adenosine} + H_2O \xrightarrow[\text{(add product)}]{\substack{\text{adenosine} \\ \text{deaminase}}} \text{inosine} + NH_3$$
$$\text{Deoxyadenosine} + H_2O \longrightarrow \text{deoxyinosine} + NH_3 \quad (2)$$

$$\text{Uridine} + P_i \xrightleftharpoons[\text{(udp product)}]{\substack{\text{uridine} \\ \text{phosphorylase}}} \text{ribose-1-P} + \text{uracil} \quad (3)$$

$$\text{Xanthosine} + P_i \xrightleftharpoons[\text{(xapA product)}]{\substack{\text{xanthosine} \\ \text{phosphorylase}}} \text{ribose-1-P} + \text{xanthine} \quad (4)$$

$$\text{Thymidine} + P_i \xrightleftharpoons[\text{(deoA product)}]{\substack{\text{thymidine} \\ \text{phosphorylase}}} \text{deoxyribose-1-P} + \text{thymine}$$
$$(\text{Deoxyuridine}) \longleftrightarrow (\text{deoxyribose-1-P} + \text{uracil}) \quad (5)$$

$$\text{Purine (adenine, hypoxanthine, and guanine) ribonucleosides and deoxyribonucleosides} + P_i \underset{\text{(deoD product)}}{\overset{\text{phosphorylase}}{\rightleftharpoons}} \text{ribose-1-P + purine deoxyribose-1-P + purine} \quad (6)$$

$$\text{Deoxyribose-1-P (Ribose-1-P)} \underset{\text{(deoB product)}}{\overset{\text{deoxyribomutase}}{\rightleftharpoons}} \text{deoxyribose-5-P (ribose-5-P)} \quad (7)$$

$$\text{Deoxyribose-5-P} \underset{\text{(deoC product)}}{\overset{\text{deoxyriboaldolase}}{\longrightarrow}} \text{glyceraldehyde-3-P + acetaldehyde} \quad (8)$$

Deamination is not needed for the utilization of all of the nucleosides, and cleavage of the pentose is only needed for the further utilization of 2-deoxy-D-ribose. The utilization of cytidine and deoxycytidine requires their conversion to uridine and deoxyuridine, respectively, by a specific deaminase of about 54,000 daltons encoded by *cdd* (transcribed clockwise at min 45), as shown in equation 1 (153, 154, 157, 159, 366, 408, 409, 420, 593, 890). For adenosine and deoxyadenosine utilization, deamination by an inducible enzyme encoded by *add* (min 36), as shown in equation 2, may occur before phosphorolysis (397, 420, 435, 548, 624, 701).

Phosphorolysis of uridine liberates its uracil moiety with the formation of the ribose 1-phosphate, as shown in equation 3. Reactions of this kind are reversible. Indeed, at neutral pH the synthesis of nucleosides is favored, irrespective of whether the base is a pyrimidine or a purine (103, 652). The specific phosphorylase comprises eight subunits of 22,000 daltons encoded by *udp* at min 85.5 (454, 495, 641, 683). Both *cdd* and *udp* are controlled by cAMP-CRP and by a repressor (37,000 daltons) specified by *cytR* at min 76.5 and responding to cytidine or adenosine as inducers (312, 313, 315, 593, 813, 872).

Another phosphorylase (a dimer of a 100,000-dalton subunit) acts on xanthosine, as shown in equation 4. The enzyme is encoded by *xapA* (close to *nupC* at min 52), which is activated by the closely linked *xapR* product with xanthosine as the probable inducer. A class of *xapR* mutants was isolated for which adenosine, deoxyadenosine, and inosine (but not guanosine or deoxyguanosine) also became inducers (133, 310).

Catabolism of thymidine and purine ribonucleosides and deoxyribonucleosides also begins with the release of their pentose moieties by phosphorolysis (549). The reactions catalyzed by thymidine phosphorylase (353) and purine phosphorylase (373, 434, 639) are shown by equations 5 and 6, respectively. Thymidine phosphorylase (a dimer of an approximately 45,000-dalton subunit) is highly specific for deoxyribose 1-phosphate (111, 689, 697, 698, 783, 785). However, the enzyme can be inhibited by uridine (119). Purine nucleoside phosphorylase (a hexamer of 23,700-dalton subunits), in contrast, has a broad substrate specificity that includes the ability to cleave both the ribonucleosides and the deoxyribonucleosides of adenine, guanine, and hypoxanthine (205, 389, 390, 420).

Transfer of the phosphoryl group of both deoxyribose 1-phosphate and ribose 1-phosphate to C-5 is catalyzed by a single mutase, as indicated in equation 7 (353). The activity of the enzyme (45,000 daltons) is dependent on Mn^{2+} and is stimulated about 10-fold by ribose 1,5-biphosphate or deoxyribose 1,5-biphosphate (311, 494). Ribose 5-phosphate is further metabolized by the nonoxidative pentose phosphate pathway; deoxyribose 5-phosphate, as indicated in equation 8, is converted to glyceraldehyde 3-phosphate and acetaldehyde by a class I deoxyriboaldolase (capable of forming a Schiff's base) active as a monomer or dimer of a polypeptide of 259 amino acids (690, 691, 869). Free 2-deoxy-D-ribose was reported to be utilized by a kinase that acts on C-5 of the sugar (403, 404), but this was not confirmed (352, 587). The enzymes for nucleoside catabolism appear to be intimately associated with the inner surface of the plasma membrane in that they are readily, though only partially, released from the cells by osmotic shock but nonetheless were well retained by spheroplasts (62, 642, 845).

Wild-type cells grow on cytidine, uridine, adenosine, inosine, guanosine, xanthosine, deoxycytidine, deoxyuridine, thymidine, deoxyadenosine, deoxyinosine, and deoxyguanosine. Growth phenotypes of mutants, with the exception of growth phenotypes on adenosine and deoxyadenosine, are predictable on the basis of the eight reactions outlined in the equations above.

Mutants lacking cytidine deaminase fail to grow on cytidine or deoxycytidine (equation 1). Growth of mutants lacking adenosine deaminase is inhibited by adenosine or deoxyadenosine (the activity of purine nucleoside phosphorylase is apparently inadequate) (equation 2). Mutants lacking uridine phosphorylase fail to grow on cytidine and uridine (equation 3). Mutants lacking xanthosine phosphorylase fail to grow only on the particular substrate (equation 4). Mutants lacking thymidine phosphorylase fail to grow on deoxycytidine, deoxyuridine, and thymidine (equation 5). Interestingly, the absence of thymidine phosphorylase could be genetically suppressed by a 30- to 45-fold increase in the level of uridine phosphorylase activity (834). Mutants lacking purine ribonucleoside phosphorylase fail to grow on the ribosides and deoxyribosides of adenine, hypoxanthine, and guanine (equation 6). Mutants lacking the mutase fail to grow on all 12 ribonucleosides and deoxyribonucleosides mentioned (equation 7). Mutants lacking the aldolase (equation 8) fail to grow on only the deoxynucleosides (15, 232, 309, 314). Aldolase mutants are, in addition, sensitive to deoxyribonucleosides as a result of deoxyribose 5-phosphate accumulation. The inhibition of growth by thymidine can be released after the exhaustion of the exogenous supply and the excretion of 2-deoxy-D-ribose and thymine into the medium (28, 215, 532, 588, 589). The absence of the aldolase or mutase activity lowers the concentration of thymine required by thymineless strains, since trapping the free base depends on deoxyribose 1-phosphate, which would be depleted by the activity of these enzymes (112, 319, 626; A. Munch-Petersen, personal communication).

The *deoA* gene (*deo* for deoxynucleoside) encoding thymidine phosphorylase, the *deoD* gene encoding

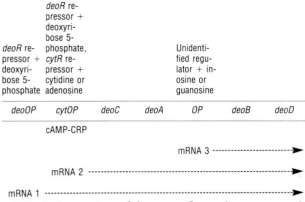

deoR re-pressor + deoxyri-bose 5-phosphate	deoR repressor + deoxyri-bose 5-phosphate, cytR repressor + cytidine or adenosine				Unidenti-fied regu-lator + in-osine or guanosine		
deoOP	cytOP	deoC	deoA	OP	deoB	deoD
	cAMP-CRP					

mRNA 3 --------------------------►

mRNA 2 --►

mRNA 1 --►

FIG. 9. Interactions of three specific regulator gene products with three promoters of the *deoA-deoB-deoC-deoD* gene cluster.

purine nucleoside phosphorylase, the *deoB* gene encoding the mutase, and the *deoC* gene encoding the aldolase are clustered at min 99.5 (15, 16, 177, 953). A regulator gene, *deoR* at min 18.7, codes for a 30,500-dalton repressor which is inactivated by deoxyribose 5-phosphate (55, 113, 532, 593, 813). Effects of polar mutations and discoordinate control of the synthesis of the enzymes suggested that the four enzymes are specified by two operons (17, 18, 58, 83, 129, 587). Further studies, however, revealed that the complex pattern of gene expression reflects the interactions of three specific regulator gene products with three promoters (all transcribed clockwise) of the gene cluster (Fig. 9). The first promoter region, *deoOP*, contains a site for the *deoR* repressor and is independent of the cAMP-CRP complex for expression. Transcription from the nearby second promoter, *cytOP*, is influenced by the *deoR* repressor, the *cytR* repressor, and the cAMP-CRP complex. There are two tandem CRP-binding sites; the one upstream seems to overlap with the binding site for the *cytR* repressor. The lack of either the *deoR* repressor (inactivated by deoxyribose 5-phosphate) or the *cytR* repressor (inactivated by cytidine or adenosine) leads to increased basal expression of the entire *deo* operon; a double regulatory defect leads to constitutive expression of the operon to a level higher than the sum of the effects of single mutations (25, 130, 178, 244, 313, 593, 805, 837, 838, 867, 868, 872). Transcripts from *deoOP* and *cytOP* are tetracistronic. A third promoter, between *deoA* and *deoB*, gives dicistronic transcripts. Although the regulatory gene product controlling this transcription has not yet been identified, induction was observed with inosine or guanosine, neither of which requires expression of *deoC* and *deoA* for utilization (26, 131, 132, 407, 870, 871).

S. typhimurium

Uridine (not an inducer in *E. coli*) or cytidine induces cytidine deaminase (encoded by *cdd* at min 44), uridine phosphorylase (encoded by *udp* at min 84), and the enzymes of the *deo* operon (64, 65, 384, 606, 623). Based on what is known, the organization of the *deo* operon is similar to that of *E. coli*; *deoC* encodes the deoxyriboaldolase (165, 350, 388) *deoA*

encodes thymidine phosphorylase (82, 351), *deoB* encodes phosphodeoxyribomutase (352), and *deoD* encodes purine nucleoside phosphorylase (348, 354, 389, 390, 724), all of which are clustered at min 98.8 and are transcribed clockwise. A second promoter exists between *deoA* and *deoB*. The regulatory gene, *deoR* (min 19 to 20), responds to deoxyribose-5-phosphate as an inducer (61, 81, 269, 352, 386, 387, 725).

Exogenous 2-deoxy-D-ribose supports growth through an inducible pathway involving a permease encoded by *deoP* (min 19 to 20) and an ATP-dependent kinase encoded by *deoK* (min 19 to 20) (349, 758). On the basis of locality, it seems that *deoP* and *deoK* are also under the control of *deoR*. Complex regulatory interactions of the nucleoside and deoxynucleoside pathways with the central glycolytic pathway are suggested by the observation that, among mutants selected for failure to grow on thymidine, 37% were found to have reduced expression of the *deo* operon associated with deficient fructose-1,6-bisphosphatase activity (385). Unlike *E. coli*, *S. typhimurium* lacks xanthosine phosphorylase (310).

HEXURONIDES, HEXURONATES, AND HEXONATES

E. coli

Hexuronides, hexuronates, and 2-keto-3-deoxy-D-gluconate. β-Glucuronides support growth by yielding D-glucuronate through the successive action of an inducible permease (829) and a hexuronidase (Fig. 10). Although the hydrolase is also capable of acting on β-galacturonides, these compounds cannot be utilized, because they fail to induce the system. Mutants selected for growth on a β-galacturonide synthesize the hydrolase constitutively (193, 194).

External D-glucuronate can be utilized via a permease for hexuronates or via a permease for D-2-keto-3-deoxy-D-gluconate (KDG) if it is constitutive. External D-galacturonate can be utilized only via the hexuronate permease (49, 57, 156, 393, 605). A common isomerase reversibly converts D-glucuronate and D-galacturonate to D-fructuronate and D-tagaturonate, respectively (45, 47, 615, 842, 884). These products can also be supplied exogenously by distinct permeases not yet characterized (605). Separate NADH-linked enzymes reduce D-fructuronate and D-tagaturonate to D-mannonate and D-altronate, respectively (45, 344, 661, 664, 665), and separate enzymes catalyze the dehydration of D-mannonate and D-altronate to yield KDG (45, 672, 716, 719, 720, 818). The high substrate specificity of the two dehydratases in vivo was shown by the observation that in a mutant blocked in the cleavage of KDG 6-phosphate the growth stasis caused by D-glucuronate is relieved by a mutation that inactivates D-mannonate dehydratase, whereas the growth stasis caused by D-galacturonate is relieved by a mutation that inactivates D-altronate dehydratase (663, 722).

Exogenous KDG can be delivered by a permease with side specificity on D-glucuronate. The transport is apparently driven by proton motive force (466–471). KDG is phosphorylated by an ATP-dependent kinase (45, 175, 673). The pathway then converges with the Entner-Doudoroff pathway, and KDG 6-phosphate is

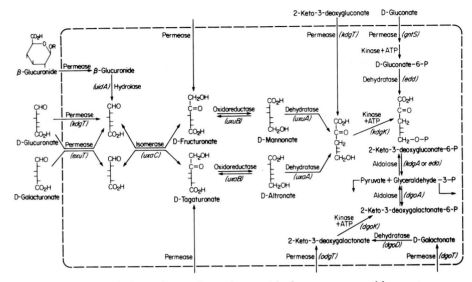

FIG. 10. Catabolic pathways for D-glucuronide, hexuronates, and hexanates.

cleaved by a specific aldolase to glyceraldehyde 3-phosphate and pyruvate (46, 677).

The genes for the utilization of hexuronides and hexuronates (Fig. 11) were shown to involve regulons and operons controlled in a complex manner by four different repressors (830). Utilization of β-glucuronides was found to be affected by mutations in five genes. The gene *uidA* (*gurA*), encoding β-glucuronidase (transcribed counterclockwise), and the gene *uidR*, encoding a specific repressor, are at min 35.8 in clockwise order (57, 80, 613, 616, 618). The genes *gurB* (min 74, possibly *crp*), *gurC* (min 18), and *gurD* (min 68) were also found to affect hexuronide utilization, but the biochemical basis of their effect has yet to be elucidated (614). The *uidA* operon is strongly repressed by the *uidR* product and weakly repressed by the *uxuR* product. Full induction of β-glucuronidase appears to require the cooperative action of both D-glucuronate or a β-glucuronide to lift the repression of the *uidR* product and D-fructuronate to lift the repression of the *uxuR* product (612, 617, 619).

The convergent metabolism of the hexuronates that terminates in the formation of KDG is specified by the *exu* regulon and the *uxu* operon. The *exu* regulon includes four operons: the *uxaB* operon (min 51) encoding D-altronate oxidoreductase; the *uxaCA* operon (min 67.9 and transcribed counterclockwise), with gene *C* encoding the isomerase and gene *A* encoding D-altronate dehydratase; the *exuT* operon (min 68 and transcribed clockwise) encoding hexuronate permease; and the *exuR* operon (min 68.1 and transcribed clockwise) encoding a specific repressor which autogenously represses its own synthesis. The *exuR* repressor has a greater affinity for the *uxaB* than for the *uxaCA* operator. D-Tagaturonate and D-fructuronate appear to be true inducers (79, 176, 369–371, 556, 557, 605, 660, 662, 663, 712, 717).

The metabolic branch converting D-fructuronate to KDG is specified by the *uxu* system consisting of the *uxuR* operon (min 97.7 and transcribed counterclockwise) encoding a specific repressor and the *uxuAB* operon (min 97.8 and transcribed counterclockwise), in which the *A* gene encodes D-mannonate dehydratase and the *B* gene encodes D-mannonate oxidoreductase. The *uxuR* operon is self repressed and catabolite repressed. The *uxuAB* operon is doubly repressed by the *uxuR* and the *exuR* products; both products are inactivated by D-fructuronate (379, 558, 710, 711, 717, 718, 721, 722).

KDG derived from the hexuronates or from an exogenous source is converted to glyceraldehyde 3-phosphate and pyruvate by proteins encoded by the *kdg* regulon under negative control (see chapter 12). Exogenous KDG, however, can serve as the sole source of carbon and energy only if a mutation confers constitutive synthesis of the permease which is strongly repressed and only weakly inducible in strain K-12 (469, 674). The repressor gene, *kdgR*, is at min 40.3, the KDG-phosphate aldolase gene, *kdgA* (*eda*), is at min 40.8, the KDG kinase gene, *kdgK*, is at min 78.3 and the KDG permease gene, *kdgT*, is at min 88.1. The true inducer is KDG (229, 248, 250, 472, 545, 546, 668, 669, 674, 675, 678). The operons are differentially con-

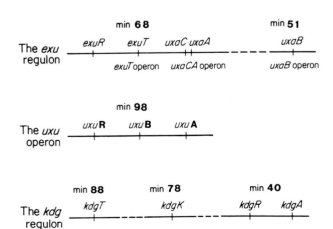

FIG. 11. Genetic systems for the utilization of hexuronates and 2-keto-3-deoxy-D-gluconate.

trolled with respect to both specific and catabolite repression. It appears that the operators of *kdgA*, *kdgK*, and *kdgT* have, in order, increasing affinities for the repressor. Sensitivity to catabolite repression increases in the same order. These features presumably account for both the preferential synthesis of the aldolase when D-gluconate is presented and the dual induction of the aldolase and the kinase when a hexuronate is presented. An unusual feature of catabolite repression of this system is its high sensitivity to pyruvate (469, 671, 675, 676).

When presented with D-glucuronate, D-galacturonate, or D-gluconate, growth of aldolase mutants is arrested, and they excrete KDG 6-phosphate. The stasis can be relieved with glucose (668, 670).

D-Gluconate and D-galactonate. A permease encoded by *gntS* at min 95 provides a route for the entry of D-gluconate (50). An ATP-dependent kinase converts the substrate to gluconate 6-phosphate, which then enters general metabolism via KDG 6-phosphate as well as via ribulose 5-phosphate (see chapter 12). D-Galactonate is transported by a permease encoded by *dgoT* and is converted to 2-deoxy-3-keto-D-galactonate by a dehydrase encoded by *dgoD*. The product is then acted upon by a kinase specified by *dgoK*, and the phosphorylated intermediate is cleaved by an aldolase specified by *dgoA* to yield glyceraldehyde 3-phosphate and pyruvate. All of the genes, including the regulatory gene *dgoR*, cluster at min 82.5. D-Galactonate itself is the true inducer. D-Galactonate is toxic to *dgoA* mutants because of 2-keto-3-deoxy-D-galactonate 6-phosphate accumulation. Strain K-12 can acquire the ability to grow on 2-keto-3-deoxy-D-galactonate as a result of two mutations, one rendering the expression of the *dgo* system constitutive and the other mobilizing a permease encoded by an unlinked gene, *odgT* (164, 185).

PHOSPHORYLATED SUGARS AND CARBOXYLATES

E. coli

Hexose phosphates. Evidence for utilization of hexose phosphates without prior hydrolysis first came as an incidental finding that when cells were grown on hexose phosphates as carbon and energy source in the presence of inorganic phosphate they derived more nucleic acid phosphorus from hexose phosphates than from P_i (723). Studies focused directly on the problem showed that mutants impaired in the utilization of glucose grew normally on glucose 6-phosphate (251, 293). Direct-transport assays revealed the existence of an inducible permease that is active on a number of phosphorylated sugars, including glucose 6-phosphate, glucose 1-phosphate, fructose 6-phosphate, fructose 1-phosphate, mannose 6-phosphate, and glucosamine 6-phosphate (for a review, see reference 197). Analysis of mutants showed that a single permease encoded by the *uhp* gene located at min 82.1 is responsible for the growth abilities on hexose phosphates (451, 659, 923). Transport of the substrate is driven by proton motive force (195, 225, 693, 926). Wide-ranging K_m values (0.02 to 0.5 mM) have been reported for glucose 6-phosphate. Apparently, the values were influenced by the concentration of P_i (likely

to be a competitive inhibitor) present in the assay medium (228). The permease has a side specificity for glycerol 3-phosphate (287a).

Surprisingly, the permease is induced by low concentrations of exogenous glucose 6-phosphate (<0.5 mM) but not by high concentrations of endogenous glucose 6-phosphate (up to 60 mM) generated in a mutant lacking glucose 6-phosphate dehydrogenase and phosphoglucose isomerase (196, 341, 924, 925). Since physiological concentrations of cellular glucose 6-phosphate can vary from 0.5 to 3 mM (367), the external induction mechanism prevents gratuitous production of the permease. Among the substrates tested, only glucose 6-phosphate acts as its true inducer. Although fructose 6-phosphate and mannose 6-phosphate can also support growth, it is believed that induction of the permease depends upon the generation of external glucose 6-phosphate by a periplasmic phosphoglucose isomerase (257, 923). The presence of such an enzyme outside the plasma membrane is a curious phenomenon. It is unknown whether the protein is specified by the same gene that encodes the intracellular enzyme. Full induction of the permease is dependent upon the cAMP-CRP complex (228).

Regulatory genes for this permease map close to the structural gene (226, 241, 411, 413). Fine-structure analysis revealed a cluster of at least three genes (in clockwise order) at the *uhp* locus: *uhpT*, *uhpR*, and *uhpA*. The gene *uhpT*, encoding the permease, is transcribed counterclockwise. It was suggested that *uhpA* encodes a transcriptional activator and that *uhpR* encodes a transmembrane protein that serves as an external receptor for the inducer glucose 6-phosphate. The phenotypes of certain mutants in the *uhpR* region and of their revertants led to the further suggestion that there is a *uhpC* gene encoding a protein which masks the function of the activator protein. When the inducer binds to the transmembrane receptor, it sequesters the C protein, thereby releasing the activator protein and allowing it to function. With a mutant bearing a *uhpT-lac* operon fusion, it was confirmed that induction did not depend on the internal glucose 6-phosphate. External induction occurred with a K_m of about 1 μM. Furthermore, the specificity of induction is more stringent than that of transport; sugar phosphates that are substrates for transport did not inhibit induction, even when present in a 25-fold molar excess (412, 796, 797). In membrane vesicles prepared from uninduced cells, a putative receptor was found that bound glucose 6-phosphate (K_d, 16 μM) but not fructose 6-phosphate (279). It is not known whether the expression of *uhpR* and *uhpA* itself is under regulation. Growth on a hexose phosphate as sole source of carbon and energy creates a problem of phosphate excess.

***sn*-Glycerol 3-phosphate.** For a discussion about *sn*-glycerol 3-phosphate, see Glycerol.

Lactate. On the basis of mutual competition for transport into membrane vesicles, D-lactate and L-lactate were suggested to share a permease distinct from the dicarboxylate transport system (559). D-Lactate and L-lactate are both converted to pyruvate, but by separate enzymes (see chapter 15).

Dicarboxylates. Succinate, fumarate, or L-malate induces a dicarboxylate transport system that acts on

L-aspartate, in addition to the three compounds, but not on oxaloacetate (342, 424, 425, 595). Synthesis of the transport system is subject to catabolite repression. The transport process seems to be driven by proton motive force, with apparent K_m values within the range of 10 to 20 μM (289). A dicarboxylate-binding protein (dissociation constants in the 30 to 50 μM range for fumarate, malate, and succinate) is postulated to mediate substrate translocation across the outer membrane in conjunction with the porin protein (74, 75, 526–529, 696). Three genes determined the transport system: *dctA* (encoding an inner membrane protein) at min 79.6, *dctB* (encoding an inner membrane protein) at about min 16.3, and *cbt* (encoding the binding protein) at about min 16.5. For reasons not clear, *cbt* (carboxylate transport) mutations also impair the transport of D-lactate (524, 525, 530, 531). Some *E. coli* strains were observed to utilize D-tartrate but not L-tartrate (421). The pathway has not yet been described.

Tricarboxylic acids. Although *E. coli* strains are generally citrate negative (452, 473a), a mutant of strain K-12 that grew on the compound was isolated as a consequence of mutations in *citA* and *citB* at about min 17. The ability to grow on isocitrate was acquired concomitantly. Utilization of *cis*- or *trans*-aconitate, however, required preinduction of the novel semiconstitutive transport system (302). The ability of natural isolates to grow on citrate is attributable to plasmid-encoded genes (377a–377f).

S. typhimurium

Phosphorylated substrates. A hexose phosphate permease was also characterized in *S. typhimurium*. The transport is induced by glucose 6-phosphate and has a substrate K_m of 0.13 mM. Cells preinduced or constitutive in this permease also transported D-arabinose 5-phosphate and D-sedoheptulose 7-phosphate (214). Unlike *E. coli*, *S. typhimurium* can utilize phosphoenolpyruvate, 2-phosphoglycerate, and 3-phosphoglycerate through a common permease encoded by *pgt* at min 74 (288, 754).

Carboxylates. A dicarboxylate permease was also reported for *S. typhimurium* (425). In contrast to *E. coli*, *S. typhimurium* can grow on citrate, isocitrate, and tricarballylate (1,2,3-propanetricarboxylate) as sole carbon and energy source (288). *cis*-Aconitate can probably also serve as sole carbon and energy source. Three inducible transport systems can be distinguished. The first is inducible by citrate, isocitrate, or *cis*-aconitase but effectively transports only citrate and isocitrate. The second is inducible by the same three compounds and transports *cis*-aconitate. The third is inducible by tricarballylate and transports tricarballylate, citrate, and *cis*-aconitate (375, 422, 423, 425a).

POLYOLS

E. coli

D-Mannitol. Like the hexoses, D-mannitol and the other utilizable hexitols (Fig. 12) are trapped by vectorial phosphorylation (458). Moreover, the enzymes II for the hexitols also serve as receptors for chemo-

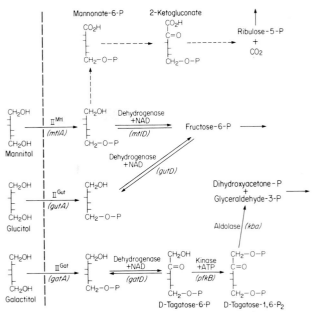

FIG. 12. Metabolic pathways of hexitols.

taxis (503). Enzyme II^{Mtl} is an integral membrane protein consisting of a polypeptide of 637 amino acids. The purified enzyme has a substrate K_m of 11 μM (284, 380–383, 484–486, 510, 728–730). The translocation product, D-mannitol 1-phosphate, is converted by an NAD-linked enzyme dehydrogenase to D-fructose 6-phosphate, as described (932) many years before the discovery of the PTS. This dehydrogenase (apparently a monomer of a 40,000-dalton polypeptide) has a K_m of 0.8 mM for the substrate and a K_m of 0.2 mM for NAD (622). The genes specifying D-mannitol utilization are at min 80.7. The *mtlAD* operon is transcribed in clockwise order, *mtlA* encoding enzyme II^{Mtl} and *mtlD* encoding the dehydrogenase. On the clockwise side of the operon lies *mtlR*, the repressor gene. Induction occurs with D-mannitol but not with D-glucitol or D-galactitol. Internalized D-mannitol (rather than D-mannitol 1-phosphate) seems to be the effector. The transport system acts on D-mannitol, with a K_m of 0.37 μM (and a growth K_m of 0.2 μM), and on D-glucitol, with a K_m of 2.5 mM. Thus, the constitutive expression of the *mtl* operon restores the ability of *gutA* (encoding the enzyme II^{Gut}) mutants to grow on D-glucitol (see below). Mutations in *mtlA* prevent growth on D-mannitol, whereas mutations in *mtlD* subject the cells to growth inhibition by the compound (499, 500, 507, 508, 521, 667, 821). Expression of the *mtl* operon is highly dependent upon the cAMP-CRP complex despite its strong resistance to catabolite repression (571, 946). D-Mannitol 1-phosphate is produced endogenously, even when the cells are growing on carbon sources other than D-mannitol, and contributes to the D-ribose pool through a pathway in which D-mannonate and D-2-ketogluconate are intermediates (734, 735).

D-Glucitol (or D-sorbitol). Utilization of D-glucitol (Fig. 12) occurs through an inducible pathway initiated by a specific enzyme II-enzyme III complex which converts the substrate to D-glucitol 6-phosphate (505, 750). The product is then converted to D-fructose

262   LIN

6-phosphate by an NAD-linked dehydrogenase (a tetramer of a 26,000-dalton subunit) with a substrate K_m of 3.3 mM and an NAD K_m of 0.2 mM (622, 931). The genes specifying D-glucitol utilization are at min 58.3. The *gutAD* (or *srlAD*) operon is transcribed in clockwise order, *gutA* encoding enzyme II(III)^{Gut} and *gutD* encoding the dehydrogenase. The gene for the repressor, *gutR*, lies on the clockwise side of the operon (102, 171, 172, 499, 560, 667, 905). Induction occurs with D-glucitol but not with D-mannitol or D-galactitol. Internal D-glucitol (rather than D-glucitol 6-phosphate) seems to be the effector. The operon is highly sensitive to catabolite repression by glucose. The transport system acts on D-glucitol with a K_m of 12 μM (and a growth K_m of 7 μM) and on D-mannitol with a K_m of 33 mM. Thus, the constitutive expression of the *gut* system restores the ability of *mtlA* mutants to grow on D-mannitol. Constitutivity of the *gut* operon also permits a double *fruA* and *ptsM* mutant to grow on D-fructose. Mutations in *gutA* prevent growth on D-glucitol, whereas mutations in *gutD* subject the cells to growth inhibition by the compound (401, 500, 505, 507).

Galactitol (or dulcitol). About one-half of *E. coli* isolates ferment galactitol (118, 212, 421). Those that do so use a pathway (Fig. 12) involving a specific enzyme II of the PTS which phosphorylates the substrate vectorially to give galactitol 1-phosphate. The product is then converted to D-tagatose 6-phosphate by an NAD-linked dehydrogenase (931). The enzyme II has a transport K_m of 3.3 μM for galactitol and a K_m of 3.7 mM for D-glucitol. The dehydrogenase has a K_m of 0.3 mM for galactitol 1-phosphate. The genes specifying D-galactitol utilization are at min 47.2. The *gatAD* operon is transcribed in clockwise order, *gatA* encoding enzyme II^{Gat} and *gatD* encoding the dehydrogenase. The repressor gene, *gutR*, lies on the clockwise side of the operon. However, two other gene products are required to complete the degradative pathway. The 6-phosphofructokinase II encoded by *pfkB* at min 38 converts D-tagatose 6-phosphate to D-tagatose 1,6-biphosphate, which is then cleaved by a specific aldolase encoded by *kba* at min 69. Mutations in *gatA* prevent growth on galactitol, whereas mutations in *gatD* subject the cells to growth inhibition by the compound. Interestingly, even among substrains of *E. coli* K-12 there are differences in their Gat phenotypes.

Most isolates do not grow on galactitol at any temperature. Galactitol-positive mutants can be isolated from these strains. Some isolates show growth impairment on the compound at temperatures above 30°C. The temperature sensitivity reflects the thermolability of the specific aldolase. Mutants selected for growth on galactitol at 42°C produce either more of the aldolase or a more stable form of it (189, 499–501, 667).

D-Arabitol, ribitol, and xylitol. Less than 10% of *E. coli* isolates grow on ribitol (212). This small group includes strain C (the standard host strain for studies of phage P2) but not strain K-12 or B. Despite the great similarity of their genetic maps, there are morphological, as well as metabolic, differences between strains K-12 and C (516, 922). Strain C can grow on both ribitol and D-arabitol. The metabolism of each pentitol is initiated by an NAD-linked dehydrogenase to give a pentulose which is then phosphorylated by an ATP-dependent kinase (Fig. 8). Enzymes of the two parallel pathways are encoded at min 47 by two divergent operons which probably have overlapping regulatory regions and which are surrounded by inverted repeats believed to be vestiges of a transposable element. It was suggested that strains possessing the dual operons acquired them through a transposable element. The gene encoding D-arabitol dehydrogenase, *atlD* (or *dalD*), is promoter-proximal to the gene encoding D-xylulose kinase, *atlK* (or *dalK*). Similarly, the gene encoding ribitol dehydrogenase, *rtlD* (or *rbtD*), is promoter-proximal to the gene encoding D-ribulose kinase *rtlK* (or *rbtK*). Each operon is under repressor control. Whereas the *atl* operon is induced by the substrate, D-arabitol, the *rtl* operon is induced by the product, D-ribulose. Strain C lacks the *gat* operon, which is present in strain K-12. When the *atl* operon is introduced into strain K-12 by transduction, it is integrated at min 47, replacing the *gat* operon as though these two operons were allelic (522, 523, 699, 764, 935). Mutants lacking D-arabitol dehydrogenase are sensitive to growth inhibition by the pentitol, because it is phosphorylated by the D-xylulose kinase to give a dead-end product. A similar situation holds for mutants lacking ribitol dehydrogenase and producing D-ribulose kinase constitutively (765).

Mutants of strain C can acquire the ability to grow on xylitol by a mutation that derepresses the *rtlDK*

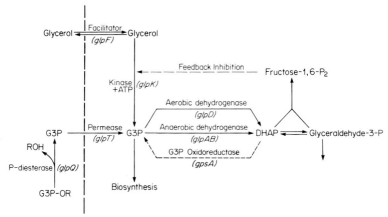

FIG. 13. Metabolic pathways of glycerol and G3P.

operon (763). In similar mutants of *Klebsiella pneumoniae*, the constitutively synthesized ribitol dehydrogenase converts xylitol to D-xylulose. The D-xylulose is partially reduced to D-arabitol by the basal level of D-arabitol oxidoreductase and causes induction of both the *atlDK* operon and the gene encoding D-arabitol permease. The permease has a side specificity for xylitol and accelerates its entry, and the *atlK*-encoded kinase phosphorylates the D-xylulose produced from xylitol by ribitol dehydrogenase (321, 322, 511, 580, 944).

Glycerol. The proteins encoded by the *glp* regulon can be considered to be a system for salvaging the glycerol moiety of phospholipids and triglycerides after their breakdown. The degraded fragments can be utilized in the form of glycerophosphodiesters, *sn*-glycerol 3-phosphate (G3P), or glycerol (Fig. 13). Unlike other carbohydrates, glycerol enters the cytoplasm by facilitated diffusion across the cytoplasmic membrane. The facilitator protein provides a selective channel with an estimated pore size of 0.4 nm (27, 227, 334, 706, 759). Internal glycerol is trapped as G3P by the action of an ATP-dependent kinase that can also phosphorylate dihydroxyacetone (328). As a catabolic enzyme (a tetramer of a 55,000-dalton subunit), the kinase has the unusual feature of being subject to noncompetitive allosteric inhibition by fructose 1,6-bisphosphate and the nonphosphorylated form of enzyme III^{Glc}, a feature which would account for the extreme effectiveness of glucose utilization in preventing glycerol consumption (190–192, 621, 855–857, 956, 957; W. Edgar, I. S. Forrest, W. H. Holmes, and B. Jasani, Biochem. J. **127**:59, 1972) (see chapter 11). Although glycerol kinase is liberated as a soluble enzyme upon rupture of the cell, it is possible that in vivo the protein is physically associated with the facilitator protein. Glycerophosphodiesters entering the periplasmic compartment are hydrolyzed to G3P by a phosphodiesterase, a tetramer of a 40,000-dalton subunit (41, 93, 474, 812). Whether liberated by hydrolysis in the periplasm or derived from the external environment, G3P is actively transported into the cytoplasm by a permease that functions as an oligomer of a subunit of about 33,000 daltons (325, 475, 518, 781, 895). When the cells grow on G3P as the carbon and energy source, the excess phosphate derived is excreted by the same permease. The counterexchange of the organic and inorganic phosphates prevents toxic accumulation of P_i and spares the proton motive force required for G3P transport (217, 789; W. Boos, personal communication, 1984). G3P permease can transport a number of substrate analogs: L-glyceraldehyde 3-phosphate (847); fosfomycin (1,2-epoxypropyl-phosphonate), which is a bactericidal antibiotic (337, 414, 874); 3,4-dihydroxybutyl 1-phosphonate; and glycerol 3-phosphorothioate (308, 496, 636, 803, 804).

Internal G3P can be converted to dihydroxyacetone-phosphate by two membrane enzyme complexes with flavin adenine dinucleotide serving as the coenzyme. The aerobic G3P dehydrogenase (a dimer of a 58,000-dalton subunit) conducts the electrons to the cytochrome oxidases or to the nitrate reductase complex (169, 726, 777, 893, 894). Competition of this enzyme with other dehydrogenases for membrane attachment sites suggests the existence of specialized anchor units

(459). The anaerobic G3P dehydrogenase conducts the electrons to the fumarate reductase complex or to the nitrate reductase complex (431, 433, 563). The extracted enzyme is a complex of a 62,000-dalton subunit and a 43,000-dalton subunit and contains one noncovalently bound flavin adenine dinucleotide and two nonheme irons. For functional attachment to the inner surface of the membrane, a third subunit is required (464, 778, 779). Metabolic energy is generated when G3P dehydrogenation is coupled to fumarate (approximately two protons extruded per fumarate molecule reduced) or nitrate reduction (84, 85, 446, 564, 565, 568, 814).

The *glp* structural genes comprising the regulon are not all linked. (i) The *glpF* gene encoding the facilitator and the *glpK* gene encoding the kinase are at min 88.4. The fact that mutations at this locus can lead to the simultaneous loss or increase of both glycerol kinase and glycerol permeability suggests that the two genes are in the same operon (72, 163, 170). (ii) The counterclockwise-transcribed *glpTQ* operon, with *T* encoding an integral membrane protein and *Q* encoding a glycerophosphodiesterase, is at min 48.6 (168, 475, 538, 567; W. Boos, personal communication, 1984). (iii) The clockwise-transcribed *glpABC operon* is a neighbor of *glpTQ* on the clockwise side. The *A* gene encodes anaerobic G3P dehydrogenase, and the *B* gene apparently encodes a membrane anchor protein (432, 464, 475, 566, 567). In addition, a promoter-distal open reading frame, gene *C*, was found (T. J. Larson, M. Ehrmann, and H. Schweizer, personal communication, 1986). The *glpD* operon encoding aerobic G3P dehydrogenase is at min 75.3 (168). In addition, two promoter-distal open reading frames, genes *E* and *G*, were found (T. J. Larson, M. Ehrmann, and H. Schweizer, personal communication, 1986).

Mutations in *glpR*, which is counterclockwise adjacent to *glpD* and encodes a 33,000-dalton repressor, can lead to constitutivity or noninducibility of the *glp* regulon (168, 790). In wild-type cells, both glycerol and G3P induce the *glp* regulon, but in mutants lacking glycerol kinase only G3P has this effect (436). That the expression of the *glp(KF)* operon itself is also induced by G3P, rather than by glycerol, was demonstrated with a mutant that produces a glycerol kinase protein without catalytic activity. Thus, G3P seems to be the true inducer of the entire *glp* regulon (327). There is a possibility that D-galactose 1-phosphate acts as an anti-inducer (835a). Titration of a thermolabile repressor in a mutant by growth at increasing temperatures indicated that the repressor affinity of the *glpD* operator is about an order of magnitude higher than those of the other *glp* operators. This difference explains the relatively low basal level of aerobic G3P dehydrogenase, which offers the double advantage of curbing wasteful degradation of endogenous G3P and assuring rapid accumulation of G3P during induction. In repressor-constitutive mutants, *glp(KF)* and *glpTQ* are many times more sensitive to glucose repression than is *glpD*. Thus, catabolite repression acts primarily on the operons dealing with substrate input (168, 255, 436). As with other inducible catabolic pathways, mutations blocking the function of the phosphoenolpyruvate phosphotransferase system severely impede the induction of the *glp* system by causing cAMP deficiency and hindering inducer accumulation. Mutants lacking enzyme I of the PTS fail to grow on glycerol. The growth impairment can be physiologically remedied by cAMP supplementation or can be genetically suppressed by (i) abolition of glpR function; (ii) rendering of glycerol kinase insensitive to feedback inhibition by fructose 1,6-bisphosphate, thus allowing the inducer G3P to attain higher levels; or (iii) a mutation increasing the expression of glp(KF) (68, 69, 72). It can be predicted that mutations desensitizing the kinase from inhibition by III^{Glc} would also have a suppressing effect. An apparent glp(KF) promoter mutation allowed growth on glycerol without cAMP (252).

In a glpR mutant, the glpD-encoded dehydrogenase is synthesized at a high level during aerobic growth. The glpA-encoded dehydrogenase is synthesized at a high level during anaerobic growth, and this level is further increased by the addition of fumarate. The activity ratios of the two dehydrogenases can vary over 10-fold. Full anaerobic expression of the glpAB operon requires the pleiotropic Fnr protein. On the other hand, aerobic repression of the glpAB operon is partially relieved by the presence of cyanide or the blocking of heme synthesis. Anaerobic growth with nitrate, which can serve as a terminal electron hydrogen acceptor for either enzyme, gives intermediate levels of both dehydrogenases (255, 464). Probably as a result of inhibition of the nitrate electron-transport chain, the growth of a glpD mutant on glycerol-nitrate is instantly arrested upon the shift from anaerobiosis to aerobiosis (432).

In contrast to wild-type cells, which show a one-half maximal growth rate on glycerol at about 1 μM, mutants lacking glycerol facilitator show reduced growth rate on the substrate at concentrations below 5 mM (326, 706). In the presence of the facilitator, glycerol kinase activity controls the pace of substrate utilization, but it is important for the cell to keep this enzyme activity below certain limits. For instance, if catabolite repression is largely relieved by prior growth on casein hydrolysate, a constitutively produced glycerol kinase which is insensitive to fructose 1,6-bisphosphate inhibition renders the cell vulnerable to glycerol exposure. Under such a condition, the rapid accumulation of dihydroxyacetone-phosphate results in lethal formation of the highly chemically reactive methylglyoxal. The glycerol-susceptible cells can be rendered immune by selection for resistance to methylglyoxal. The resistant mutants produce an elevated level of a glutathione-dependent glyoxalase system and are actually able to grow on 1 mM methylglyoxal as sole carbon and energy source (254, 955). Lethal synthesis of methylglyoxal is not limited to unregulated glycerol catabolism; excessive uptake of other carbohydrates caused by lifting catabolite repression has a similar consequence (8). If a feedback-insensitive kinase is produced inducibly, the cell is not vulnerable to glycerol, even though the compound is utilized at an increased rate. Lethal synthesis of methylglyoxal is prevented by strong self-catabolite repression of glp(KF). In such cells, glycerol exerts a catabolite repressive effect on other inducible systems approaching that of glucose (955).

In addition to the importance of avoiding dihydroxyacetone-phosphate overproduction, it is vital for the cell to maintain its intracellular concentrations of G3P within a certain range under all growth conditions; its overproduction is bacteriostatic, and its overdegradation threatens the biosynthetic pool (517a). In mutants lacking aerobic G3P dehydrogenase, exposure to glycerol results in growth inhibition. The inhibitory effect, associated with depletion of intracellular ATP and other nucleoside triphosphates, can be overcome by glucose (169, 340).

Although glycerol is not chemostatic, at a threshold concentration of 1 mM it can act as a repellant, a process that involves all three methyl-accepting proteins (12, 631, 632). An NAD-linked dehydrogenase of broad specificity and uncertain physiological function, when constitutively expressed, is able to replace glycerol kinase for growth on glycerol (136, 137, 188, 395, 427, 428, 536, 827, 848–850).

S. typhimurium

Hexitols. The mtl operon encoding enzyme II^{Mtl} and mannitol 1-phosphate dehydrogenase maps at min 78. D-Mannitol inhibits the growth of mutants lacking the dehydrogenase, in one case resulting in cell lysis. Mutations in mtlA can have differential effects on transport, phosphorylation, and chemotactic response (67, 391, 509, 751). The specific membrane complex for the utilization of D-glucitol involves two specific proteins: II and III (760). Although the gat operon was found to be inducible (189, 507), detailed analyses of the system have yet to be undertaken.

Inositol. Most strains grow on meso-inositol, although many can do so at 25 but not 37°C (288, 627, 628). Vectorial phosphorylation is not known to be involved in the utilization of the cyclic hexitol. Instead, the first step of its degradation involves an NADP$^+$-dependent dehydrogenation by 2-keto-myo-inositol oxidoreductase. The keto compound is acted on by a dehydratase, and the resultant product is cleaved to 2-deoxy-5-keto-D-gluconate (835). With the exception of the dehydrogenation step requiring NAD instead of NADP, the rest of the pathway should be analogous to that in K. pneumoniae; the ketogluconate is phosphorylated at C-6 by an ATP-dependent kinase, and the product is then cleaved by an aldolase to dihydroxyacetone-phosphate and malonic semialdehyde (33, 34, 70, 71).

Glycerol. All five operons of the glp regulon occupy positions homologous to those of E. coli: glpT and glpA at min 45, glpD and glpR at min 74, and glpK at min 87 (7, 339, 520, 695, 708, 756, 757). Glycerol kinase is feedback inhibited by fructose 1,6-bisphosphate and III^{Glc} (604, 621, 666, 752, 753).

CONCLUDING REMARKS

The carbohydrate moieties of ubiquitous macromolecules, e.g., DNA, RNA, phospholipids, glycogen, and oligosaccharides of cell surfaces, have all been shown to serve E. coli and S. typhimurium as sole carbon and energy sources, and most of the pathways involved have been completely characterized biochemically. Henceforth it will be a challenge to discover additional pathways (cryptic or in operation) in these organisms for carbon and energy sources that are metabolized by enzymes specified by chromosomal genes. Much, however, remains to be uncovered in the area of regulation. For most of the pathways, even the controls at the level of transcription have not been sufficiently characterized for the rules of induction to be understood. The process of uncovering translational controls and covalent modification of proteins that exist in some pathways is just beginning.

Although the catabolic pathways probably all have sufficiently evolved so that during the adapted state the intermediates will not accumulate unduly to burden the cell further with osmotic pressure or to cause interference with other metabolic processes, it is conceivable that temporary excesses are not always avoidable. Such excesses might occur when nutritional supplies change suddenly. There is no evidence that kinetic feedback is as well evolved in catabolic pathways as it is in biosynthetic pathways. Perhaps mechanisms for discharging excess catabolites have not received sufficient attention. The fact that the growth of mutants blocked at various reactions in catabolism is in general not noticeably inhibited by the substrate when the truncated pathway is induced would indicate the existence of discharge mechanisms that act as safety valves. (Even excessive biosynthetic intermediates can be excreted by auxotrophic mutants.) The notable exception is the growth arrest caused by a substrate whenever the further metabolism of a phosphorylated intermediate is mutationally blocked. This general phenomenon implicates the disturbance of a central regulatory mechanism (340) which deserves further study.

A wealth of important information about bacterial population genetics can be expected from comparative studies of the catabolic pathways. The judicious choice of systems studied under the right conditions might also increase our insight into the evolutionary forces that shaped the various kinds of catalytic and regulatory mechanisms.

ACKNOWLEDGMENTS

I thank S. Adhya, D. G. Fraenkel, J. Lengeler, A. Munch-Petersen, P. W. Postma, B. Rotman, H. A. Shuman, F. Stoeber, and A. Wright for helpful comments, and A. Ulrich for assistance in the preparation of this work. Studies in this laboratory were supported by Public Health Service grant 5-RO1-GM11983 from the National Institute of General Medical Sciences and by grant DMB8314312 from the National Science Foundation.

LITERATURE CITED

1. **Abou-Sabe, M.** 1971. Isolation and characterization of *Salmonella typhimurium* D-ribose-positive revertants of *Escherichia coli* B-r. J. Gen. Microbiol. **65:**375–377.
2. **Abou-Sabe, M., J. Pilla, D. Hazuda, and A. Ninfa.** 1982. Evolution of the D-ribose operon on *Escherichia coli* B/r. J. Bacteriol. **150:**762–769.
3. **Abou-Sabe, M., and P. L. Ratner.** 1977. Genetic regulation of the constitutive D-ribose operon in *Escherichia coli* B/r. Biochim. Biophys. Acta **476:**321–332.
4. **Abou-Sabe, M., and J. Richman.** 1973. On the regulation of D-ribose metabolism in *Escherichia coli* B/r. I. Isolation and characterization of D-ribokinaseless and D-ribose permeaseless mutants. Mol. Gen. Genet. **122:**291–301.
5. **Abou-Sabe, M., and J. Richman.** 1973. On the regulation of D-ribose metabolism in *Escherichia coli* B/r. II. Chromosomal location and fine structure analysis of the D-ribose permease and D-ribokinase structural genes by P1 transduction. Mol. Gen. Genet. **122:**303–312.
6. **Abou-Sabe, M. A., and P. L. Ratner.** 1973. Genetic regulation of a constitutive operon. Biochem. Biophys. Res. Commun. **55:**1015–1020.
7. **Aceves-Pina, E., M. V. Ortega, and M. Artis.** 1974. Linkage of the *Salmonella typhimurium* chromosomal loci encoding for the cytochrome-linked L-alpha-glycerophosphate dehydrogenase and amylomaltase activities. Arch. Microbiol. **101:**59–70.
8. **Ackermann, R. S., N. R. Cozzarelli, and W. Epstein.** 1974. Accumulation of toxic concentration of methylglyoxal by wild-type *Escherichia coli* K-12. J. Bacteriol. **119:**357–362.
9. **Adhya, S., and H. Echols.** 1966. Glucose effect and the galactose enzymes of *Escherichia coli*: correlation between glucose inhibition of induction and inducer transport. J. Bacteriol. **92:**601–608.
10. **Adhya, S., and W. Miller.** 1979. Modulation of the two promoters of the galactose operon of *Escherichia coli*. Nature (London) **279:**492–494.
10a. **Adhya, S. L., and J. A. Shapiro.** 1969. The galactose operon of *E. coli* K-12. I. Structural and pleiotropic mutations of the operon. Genetics **62:**231–247.
11. **Adler, J., and W. Epstein.** 1974. Phosphotransferase-system enzymes as chemoreceptors for certain sugars in *Escherichia coli* chemotaxis. Proc. Natl. Acad. Sci. USA **71:**2895–2899.
12. **Adler, J., G. L. Hazelbauer, and M. M. Dahl.** 1973. Chemotaxis toward sugars in *Escherichia coli*. J. Bacteriol. **115:**824–847.
13. **Adler, J., and A. D. Kaiser.** 1963. Mapping of the galactose genes of *Escherichia coli* by transduction with phage P1. Virology **19:**117–126.
14. **Ahlem, C., W. Huisman, G. Neslund, and A. S. Dahms.** 1982. Purification and properties of a periplasmic D-xylose-binding protein from *Escherichia coli* K-12. J. Biol. Chem. **257:**2926–2931.
15. **Ahmad, S. I., P. T. Barth, and R. H. Pritchard.** 1968. Properties of a mutant of *Escherichia coli* lacking purine nucleoside phosphorylase. Biochim. Biophys. Acta **161:**581–583.
16. **Ahmad, S. I., and R. H. Pritchard.** 1969. A map of four genes specifying enzymes involved in catabolism of nucleosides in *Escherichia coli*. Mol. Gen. Genet. **104:**351–359.
17. **Ahmad, S. I., and R. H. Pritchard.** 1971. A regulatory mutant affecting the synthesis of enzymes involved in the catabolism of nucleosides in *Escherichia coli*. Mol. Gen. Genet. **111:**77–83.
18. **Ahmad, S. I., and R. H. Pritchard.** 1973. An operator constitutive mutant affecting the synthesis of two enzymes involved in the catabolism of nucleosides in *Escherichia coli*. Mol. Gen. Genet. **124:**321–329.
19. **Aiba, H., S. Adhya, and B. de Crombrugghe.** 1981. Evidence for two functional *gal* promoter in intact *Escherichia coli* cells. J. Biol. Chem. **256:**11905–11910.
20. **Akhy, M. T., C. M. Brown, and D. C. Old.** 1984. L-Rhamnose utilisation in *Salmonella typhimurium*. J. Appl. Bacteriol. **56:**269–274.
21. **Aksamit, R., and D. E. Koshland, Jr.** 1972. A ribose binding protein of *Salmonella typhimurium*. Biochem. Biophys. Res. Commun. **48:**1348–1353.
22. **Aksamit, R. R., and D. E. Koshland, Jr.** 1974. Identification of the ribose binding protein as the receptor for ribose chemotaxis in *Salmonella typhimurium*. Biochemistry **13:**4473–4478.
23. **Alaeddinoglu, N. G., and H. P. Charles.** 1979. Transfer of a gene for sucrose utilization into *Escherichia coli* K12, and consequent failure of expression of genes of D-serine utilization. J. Gen. Microbiol. **110:**47–59.
24. **Alber, T., M. Fahnestock, S. Mowbray, and G. Petsko.** 1981. Preliminary X-ray data for the galactose binding protein from *Salmonella typhimurium*. J. Mol. Biol. **147:**471–474.
25. **Albrechtsen, H., and S. I. Ahmad.** 1980. Regulation of the synthesis of nucleoside catabolic enzymes in *Escherichia coli*: further analysis of a *deoO*ᶜ mutant strain. Mol. Gen. Genet. **179:**457–460.
26. **Albrechtsen, H., K. Hammer-Jespersen, A. Munch-Petersen, and N. Fiil.** 1976. Multiple regulation of nucleoside catabolizing enzymes: effects of a polar *dra* mutation on the deo enzymes. Mol. Gen. Genet. **146:**139–145.
27. **Alemohammad, M. M., and C. J. Knowles.** 1974. Osmotically induced volume and turbidity changes of *Escherichia coli* due to salts, sucrose and glycerol, with particular reference to the rapid permeation of glycerol into the cell. J. Gen. Microbiol. **82:**125–142.
28. **Alikhanian, S. I., T. S. Iljina, E. S. Kaliaeva, S. V. Kameneva, and V. V. Sukhodolec.** 1966. A genetic study of thymineless mutants of *Escherichia coli* K12. Genet. Res. **8:**83–100.
29. **Alpers, D. H., S. H. Appel, and G. M. Tomkins.** 1965. A spectrophotometric assay for thiogalactoside transacetylase. J. Biol. Chem. **240:**10–13.
30. **Al-Zarban, S., L. Heffernan, J. Nishitani, L. Ransone, and G. Wilcox.** 1984. Positive control of the L-rhamnose genetic system in *Salmonella typhimurium* LT2. J. Bacteriol. **158:**603–608.
31. **Anderson, A., and R. A. Cooper.** 1969. The significance of ribokinase for ribose utilization by *Escherichia coli*. Biochim. Biophys. Acta **177:**163–165.

32. **Anderson, A., and R. A. Cooper.** 1970. Biochemical and genetical studies on ribose catabolism in *Escherichia coli* K12. J. Gen. Microbiol. **62:**335–339.

33. **Anderson, W. A., and B. Magasanik.** 1971. The pathway of *myo*-inositol degradation in *Aerobacter aerogenes*. Identification of the intermediate 2-deoxy-5-keto-D-gluconic acid. J. Biol. Chem. **246:**5653–5661.

34. **Anderson, W. A., and B. Magasanik.** 1971. The pathway of *myo*-inositol degradation in *Aerobacter aerogenes*. Conversion of 2-deoxy-5-keto-D-gluconic acid to glycolytic intermediates. J. Biol. Chem. **246:**5662–5675.

35. **Andrews, K. J., and E C. C. Lin.** 1976. Selective advantages of various bacterial carbohydrate transport mechanisms. Fed. Proc. **35:**2185–2189.

36. **Andrews, K. J., and E. C. C. Lin.** 1976. Thiogalactoside transacetylase of the lactose operon as an enzyme for detoxification. J. Bacteriol. **128:**510–513.

37. **Anraku, Y.** 1968. Transport of sugars and amino acids in bacteria. I. Purification and specificity of the galactose- and leucine-binding proteins. J. Biol. Chem. **243:**3116–3122.

38. **Anraku, Y.** 1968. Transport of sugars and amino acids in bacteria. II. Properties of galactose- and leucine-binding proteins. J. Biol. Chem. **243:**3123–3127.

39. **Anraku, Y.** 1968. Transport of sugars and amino acids in bacteria. III. Studies on the restoration of active transport. J. Biol. Chem. **243:**3128–3135.

40. **Arditti, R. R., J. G. Scaife, and J. R. Beckwith.** 1968. The nature of mutants in the *lac* promoter region. J. Mol. Biol. **38:**421–426.

41. **Argast, M., G. Schumacher, and W. Boos.** 1977. Characterization of a periplasmic protein related to sn-glycerol-3-phosphate transport in *E. coli*. J. Supramol. Struct. **6:**135–153.

42. **Argos, P., W. C. Mahoney, M. A. Hermodson, and M. Hanei.** 1981. Structural predictions of sugar binding proteins functional in chemotaxis and transport. J. Biol. Chem. **256:**4357–4361.

43. **Arr, M., T. Perenyi, and E. K. Novak.** 1970. Sucrose and raffinose breakdown by *Escherichia coli*. Acta Microbiol. Acad. Sci. Hung. **17:**117–126.

44. **Arraj, J. A., and J. H. Campbell.** 1975. Isolation and characterization of the newly evolved *ebg* β-galactosidase of *Escherichia coli* K-12. J. Bacteriol. **124:**849–856.

45. **Ashwell, G.** 1962. Enzymes of glucuronic and galacturonic acid metabolism in bacteria. Methods Enzymol. **5:**190–208.

46. **Ashwell, G., A. J. Wahba, and J. Hickman.** 1958. A new pathway of uronic acid metabolism. Biochim. Biophys. Acta **30:**186–187.

47. **Ashwell, G., A. J. Wahba, and J. Hickman.** 1960. Uronic acid metabolism in bacteria. I. Purification and properties of uronic acid isomerase in *Escherichia coli*. J. Biol. Chem. **235:**1559–1565.

48. **Atherly, A. G.** 1979. *Escherichia coli* mutant containing a large deletion from *relA* to *argA*. J. Bacteriol. **138:**530–534.

49. **Autissier, F., and A. Kepes.** 1972. Segregation de marqueurs membranaires au cours de la croissance et de la division d'*Escherichia coli*. III. Utilisation de marqueurs varies: permeases, phosphotransferases, oxydoreductases membranaires. Biochimie **54:**93–101.

49a. **Babul, J.** 1978. Phosphofructokinases from *Escherichia coli*: purification and characterization of the nonallosteric isozyme. J. Biol. Chem. **253:**4350–4355.

50. **Bächi, B., and H. L. Kornberg.** 1973. Genes involved in the uptake and catabolism of gluconate by *Escherichia coli*. J. Gen. Microbiol. **90:**321–335.

51. **Bachmann, B. J.** 1983. Linkage map of *Escherichia coli* K-12, edition 7. Microbiol. Rev. **47:**180–230.

52. **Badia, J., J. Ros, and J. Aguilar.** 1985. Fermentation mechanism of fucose and rhamnose in *Salmonella typhimurium* and *Klebsiella pneumoniae*. J. Bacteriol. **161:**435–437.

53. **Baron, L. S., W. F. Carey, and W. M. Spilman.** 1959. Characteristics of a high frequency of recombination (Hfr) strain of *Salmonella typhosa* compatible with *Salmonella*, *Shigella*, and *Escherichia* species. Proc. Natl. Acad. Sci. USA **45:**1752–1757.

54. **Baron, L. S., W. F. Carey, and W. M. Spilman.** 1959. Genetic recombination between *Escherichia coli* and *Salmonella typhimurium*. Proc. Natl. Acad. Sci. USA **45:**976–984.

55. **Barth, P. T., I. R. Beacham, S. I. Ahmad, and R. H. Pritchard.** 1968. The inducer of the deoxynucleoside phosphorylases and deoxyriboaldolase in *Escherichia coli*. Biochim. Biophys. Acta **161:**554–557.

56. **Batt, C. A., M. S. Bodis, S. K. Pacataggio, M. C. Claps, S. Jamas, and A. J. Sinskey.** 1985. Analysis of xylose operon regulation by Mud (Ap^r, lac) fusion: trans effect of plasmid coded xylose operon. Can. J. Microbiol. **31:**930–933.

57. **Baudouy-Robert, J., M.-L. Didier-Fichet, J. Jimeno-Abendano, G. Novel, R. Portalier, and F. Stoeber.** 1970. Modalites de l'induction des six premieres enzymes degradant les hexuronides et les hexuronates chez *Escherichia coli* K-12. C.R. Acad. Sci. Ser. D. **271:**255–258.

58. **Baumanis, G. E., Y. V. Smirnov, and V. V. Sukhodoletz.** 1974. Production and study of polar mutants for nucleoside catabolism linked genes in *Escherichia coli*. Genetika **10:**81–88.

59. **Bavoil, P., M. Hofnung, and H. Nikaido.** 1980. Identification of a cytoplasmic membrane-associated component of the maltose transport system of *Escherichia coli*. J. Biol. Chem. **255:** 8366–8369.

60. **Bavoil, P., and H. Nikaido.** 1981. Physical interaction between the phage lambda receptor protein and the carrier-immobilized maltose binding protein of *Escherichia coli*. J. Biol. Chem. **256:**11385–11388.

61. **Beacham, I. R., A. Eisenstark, P. T. Barth, and R. H. Pritchard.** 1968. Deoxynucleoside-sensitive mutants of *Salmonella typhimurium*. Mol. Gen. Genet. **102:**112–127.

62. **Beacham, I. R., E. Yagil, K. Beacham, and R. H. Pritchard.** 1971. On the localization of enzymes of deoxynucleoside catabolism in *Escherichia coli*. FEBS Lett. **16:**77–80.

63. **Becerra de Lares, L., J. Ratouchniak, and F. Casse.** 1977. Chromosomal location of genes governing the trehalose utilization in *Escherichia coli* K-12. Mol. Gen. Genet. **152:**105–108.

64. **Beck, C. F., and J. L. Ingraham.** 1971. Location on the chromosome of *Salmonella typhimurium* of genes governing pyrimidine metabolism. Mol. Gen. Genet. **111:**303–316.

65. **Beck, C. F., J. L. Ingraham, J. Neuhard, and E. Thomassen.** 1972. Metabolism of pyrimidines and pyrimidine nucleosides by *Salmonella typhimurium*. J. Bacteriol. **110:**219–228.

66. **Beckwith, J.** 1963. Restoration of operon activity by suppressors. Biochim. Biophys. Acta **76:**162–164.

67. **Berkowitz, D.** 1971. D-Mannitol utilization in *Salmonella typhimurium*. J. Bacteriol. **105:**232–240.

68. **Berman, M., and E. C. C. Lin.** 1971. Glycerol-specific revertants of a phosphoenolpyruvate phosphotransferase mutant: suppression by the desensitization of glycerol kinase to feedback inhibition. J. Bacteriol. **105:**113–120.

69. **Berman, M., N. Zwaig, and E. C. C. Lin.** 1970. Suppression of a pleiotropic mutant affecting glycerol dissimilation. Biochem. Biophys. Res. Commun. **38:**272–278.

70. **Berman, T., and B. Magasanik.** 1966. The pathway of *myo*-inositol degradation in *Aerobacter aerogenes*. Dehydrogenation and dehydration. J. Biol. Chem. **241:**800–806.

71. **Berman, T., and B. Magasanik.** 1966. The pathway of *myo*-inositol degradation in *Aerobacter aerogenes*. Ring scission. J. Biol. Chem. **241:**807–813.

72. **Berman-Kurtz, M., E. C. C. Lin, and D. P. Richey.** 1971. Promoter-like mutant with increased expression of the glycerol kinase operon of *Escherichia coli*. J. Bacteriol. **106:**724–731.

73. **Bernheim, N. J., and W. J. Dobrogosz.** 1970. Amino sugar sensitivity in *Escherichia coli* mutants unable to grow on *N*-acetylglycosamine. J. Bacteriol. **101:**384–391.

74. **Bewick, M. A., and T. C. Y. Lo.** 1979. Dicarboxylic acid transport in *Escherichia coli* K-12: involvement of a binding protein in the translocation of dicarboxylic acids across the outer membrane of the cell envelope. Can. J. Biochem. **57:**653–661.

75. **Bewick, M. A., and T. C. Y. Lo.** 1980. Localization of the dicarboxylate binding protein in the cell envelope of *Escherichia coli* K12. Can. J. Biochem. **58:**885–897.

76. **Bhattacharya, A. K., and M. Chakravorty.** 1971. Induction and repression of L-arabinose isomerase in *Salmonella typhimurium*. J. Bacteriol. **106:**107–112.

77. **Bhattacharya, A. K., and M. Chakravorty.** 1974. Effect of antibiotics and antimetabolites on the induction of L-arabinose isomerase in *Salmonella typhimurium*. Curr. Sci. **43:**499–503.

78. **Bhattacharya, A. K., and M. Chakravorty.** 1975. Isolation and characterization of an L-arabinose negative mutant of *Salmonella typhimurium*. Indian J. Exp. Biol. **13:**244–246.

79. **Blanco, C., M. Mata-Gilsinger, and P. Ritzenthaler.** 1983. Construction of hybrid plasmids containing the *Escherichia coli uxaB* gene: analysis of its regulation and direction of transcription. J. Bacteriol. **153:**747–755.

80. **Blanco, C., P. Ritzenthaler, and M. Mata-Gilsinger.** 1982. Cloning and endonuclease restriction analysis of *uidA* and *uidR* genes in *Escherichia coli* K-12: determination of transcription direction for the *uidA* gene. J. Bacteriol. **149:**587–594.

81. **Blank, J., and P. Hoffee.** 1972. Regulatory mutants of the *deo* regulon in *Salmonella typhimurium*. Mol. Gen. Genet. **116:**291–298.

82. **Blank, J. G., and P. A. Hoffee.** 1975. Purification and properties

of thymidine phosphorylase from *Salmonella typhimurium*. Arch. Biochem. Biophys. **168**:259–265.

83. **Bonney, R. J., and H. Weinfeld.** 1971. Regulation of thymidine metabolism in *Escherichia coli* K-12: studies on the inducer and the coordinateness of induction of the enzymes. J. Bacteriol. **106**:812–818.

84. **Boonstra, J., M. T. Huttunen, W. N. Konings, and H. R. Kaback.** 1975. Anaerobic transport in *Escherichia coli* membrane vesicles. J. Biol. Chem. **250**:6792–6898.

85. **Boonstra, J., H. J. Sips, and W. N. Konings.** 1976. Active transport by membrane vesicles from anaerobically grown *Escherichia coli* energized by electron transfer to ferricyanide and chlorate. Eur. J. Biochem. **69**:35–44.

86. **Boos, W.** 1969. The galactose binding protein and its relationship to the beta-methylgalactoside permease from *Escherichia coli*. Eur. J. Biochem. **10**:66–73.

87. **Boos, W.** 1972. Structurally defective galactose-binding protein isolated from a mutant negative in the beta-methylgalactoside transport system of *Escherichia coli*. J. Biol. Chem. **247**:5414–5424.

88. **Boos, W.** 1974. Pro and contra carrier proteins: sugar transport via the periplasmic galactose-binding protein. Curr. Top. Membr. Transp. **5**:51–136.

89. **Boos, W.** 1982. Aspects of maltose transport in *Escherichia coli*: established facts and educated guesses. Ann Inst. Pasteur Microbiol. **133A**:145–151.

90. **Boos, W., T. Ferenci, and H. A. Shuman.** 1981. Formation and excretion of acetylmaltose after accumulation of maltose in *Escherichia coli*. J. Bacteriol. **146**:725–732.

91. **Boos, W., and A. S. Gordon.** 1971. Transport properties of the galactose-binding protein of *Escherichia coli*. J. Biol. Chem. **246**:621–628.

92. **Boos, W., A. S. Gordon, R. Z. Hall, and H. D. Price.** 1972. Transport properties of the galactose-binding protein of *Escherichia coli*. J. Biol. Chem. **247**:917–924.

93. **Boos, W., I. Hartig-Beecken, and K. Altendorf.** 1977. Purification and properties of a periplasmic protein related to sn-glycerol-3-phosphate transport in *Escherichia coli*. Eur. J. Biochem. **72**:571–581.

94. **Boos, W., and M. O. Sarvas.** 1970. Close linkage between a galactose-binding protein and the beta-methylgalactoside permease in *Escherichia coli*. Eur. J. Biochem. **13**:526–533.

95. **Boos, W., and L. Staehelin.** 1981. Ultrastructural localization of the maltose-binding protein within the cell envelope of *Escherichia coli*. Arch. Mikrobiol. **129**:240–246.

96. **Boos, W., I. Steinacher, and D. Engelhardt-Altendorf.** 1981. Mapping of *mglB*, the structural gene of the galactose-binding protein of *Escherichia coli*. Mol. Gen. Genet. **184**:508–518.

97. **Boos, W., and K. Wallenfels.** 1968. Untersuchungen zur Induktion der Lac-Enzyme. 2. Die Permeation von Galaktosylglyzerin in *Escherichia coli*. Eur. J. Biochem. **3**:360–363.

98. **Boronat, A., and J. Aguilar.** 1979. Rhamnose-induced propanediol oxidoreductase in *Escherichia coli*: purification, properties, and comparison with the fucose-induced enzyme. J. Bacteriol. **140**:320–326.

99. **Boronat, A., and J. Aguilar.** 1981. Experimental evolution of propanediol oxidoreductase in *Escherichia coli*. Comparative analysis of the wild-type and mutant enzymes. Biochim. Biophys. Acta **672**:98–107.

100. **Boronat, A., and J. Aguilar.** 1981. Metabolism of L-fucose and L-rhamnose in *Escherichia coli*: differences in induction of propanediol oxidoreductase. J. Bacteriol. **147**:181–185.

101. **Boronat, A., E. Caballero, and J. Aguilar.** 1983. Experimental evolution of a metabolic pathway for ethylene glycol utilization by *Escherichia coli*. J. Bacteriol. **153**:134–139.

102. **Boronat, A., M. C. Jones-Mortimer, and H. L. Kornberg.** 1982. A specialized transducing phage, lamba*psrlA*, for the sorbitol phosphotransferase of *Escherichia coli* K12. J. Gen. Microbiol. **128**:605–611.

103. **Bose, R., and E. W. Yamada.** 1974. Uridine phosphorylase, molecular properties and mechanism of catalysis. Biochemistry **13**:2051–2056.

104. **Boulter, J., B. Gielow, M. McFarland, and N. Lee.** 1974. Metabolism of D-arabinose by *Escherichia coli* B/r. J. Bacteriol. **117**:920–923.

105. **Boulter, J. R., and W. O. Gielow.** 1973. Properties of D-arabinose isomerase purified from two strains of *Escherichia coli*. J. Bacteriol. **113**:687–696.

106. **Boyer, H., E. Englesberg, and R. Weinberg.** 1962. Direct selection of L-arabinose negative mutants of *Escherichia coli* strain B/r. Genetics **47**:417–425.

107. **Brass, J. M., K. Bauer, U. Ehmann, and W. Boos.** 1985. Maltose-

108. **Brass, J. M, and M. D. Manson.** 1984. Reconstitution of maltose chemotaxis in *Escherichia coli* by addition of maltose-binding protein to calcium-treated cells of maltose regulon mutants. J. Bacteriol. **157**:881–890.

109. **Braun, V., and H. J. Krieger-Brauer.** 1977. Interrelationship of the phage lambda receptor protein and maltose transport in mutants of *Escherichia coli* K12. Biochim. Biophys. Acta **469**:89–98.

110. **Braun-Breton, C., and M. Hofnung.** 1981. In vivo and in vitro functional alterations of the bacteriophage lambda receptor in *lamB* mutants. J. Bacteriol. **148**:845–852.

111. **Breitman, T. R., and R. M. Bradford.** 1964. The induction of thymidine phosphorylase and excretion of deoxyribose during thymine starvation. Biochem. Biophys. Res. Commun. **17**:786–791.

112. **Breitman, T. R., and R. M. Bradford.** 1967. The absence of deoxyriboaldolase activity in the thymineless mutant of *Escherichia coli* strain 15. A possible explanation for the low thymine requirement of some thymineless strains. Biochim. Biophys. Acta **138**:217–220.

113. **Breitman, T. R., and R. M. Bradford.** 1968. Inability of low thymine-requiring mutants of *Escherichia coli* lacking phosphodeoxyribomutase to be induced for deoxythymidine phosphorylase and deoxyriboaldolase. J. Bacteriol. **95**:2434–2435.

114. **Brown, C. E., and R. W. Hogg.** 1972. A second transport system for L-arabinose in *Escherichia coli* B-r controlled by the *araC* gene. J. Bacteriol. **111**:606–613.

115. **Brown, J. L., D. M. Brown, and I. Zabin.** 1967. Thiogalactoside transacetylase: physical and chemical studies of subunit structure. J. Biol. Chem. **242**:4254–4258.

116. **Brown, J. L., S. Koorajian, and I. Zabin.** 1967. Thiogalactoside transacetylase: amino- and carboxyl-terminal studies. J. Biol. Chem. **242**:4259–4264.

117. **Buchanan, C. E., and A. Markowitz.** 1973. Depression of uridine diphosphate-glucose pyrophosphorylase (*galU*) in *capR* (*lon*), *capS*, and *capT* mutants and studies on the *galU* repressor. J. Bacteriol. **115**:1011–1020.

118. **Buchanan, R. E., and N. E. Gibbons (ed.).** 1974. Bergey's manual of determinative bacteriology, 8th ed. The Williams & Wilkins Co., Baltimore.

119. **Budman, D. R., and A. B. Pardee.** 1967. Thymidine and thymine incorporation into deoxyribonucleic acid: inhibition and repression by uridine of thymidine phosphorylase of *Escherichia coli*. J. Bacteriol. **94**:1546–1550.

120. **Burstein, C., M. Cohn, A. Kepes, and J. Monod.** 1965. Role du lactose et de ses produits metaboliques dans l'induction de l'operon lactose chez *Escherichia coli*. Biochim. Biophys. Acta **95**:634–639.

121. **Burstein, C., and A. Kepes.** 1966. Mise en evidence d' α-galactosidase dans des extraits acellulaires de *Escherichia coli*. C.R. Acad. Sci. Ser. D **262**:227–229.

122. **Burstein, C., and A. Kepes.** 1971. The beta-galactosidase from *Escherichia coli* K12. Biochim. Biophys. Acta **230**:52–63.

123. **Burton, K.** 1977. Transport of adenine, hypoxanthine and uracil into *Escherichia coli*. Biochem. J. **168**:195–204.

124. **Busby, S., H. Aiba, and B. de Crombrugghe.** 1982. Mutations in the *Escherichia coli* operon that define two promoters and the binding site of the cyclic AMP receptor protein. J. Mol. Biol. **154**:211–227.

125. **Busby, S., M. Irani, and B. de Crombrugghe.** 1982. Isolation of mutant promoters in the *Escherichia coli* galactose operon using local mutagenesis on cloned DNA fragments. J. Mol. Biol. **154**:197–209.

126. **Buttin, G.** 1962. Sur la structure de l'operon galactose chez *Escherichia coli* K12. C.R. Acad. Sci. Ser. D **255**:1233–1235.

127. **Buttin, G.** 1963. Mecanismes regulateurs dans la biosynthese des enzymes du metabolisme du galactose chez *Escherichia coli* K-12. I. La biosynthese induite de la galactokinase et l'induction simultanee de la sequence enzymatique. J. Mol. Biol. **7**:164–182.

128. **Buttin, G.** 1963. Mecanismes regulateurs dans la biosynthese des enzymes du metabolisme du galactose chez *Escherichia coli* K-12. II. Le determinisme genetique de la regulation. J. Mol. Biol. **7**:183–205.

129. **Buxton, R. S.** 1975. Genetic analysis of thymidine-resistant and low-thymine-requiring mutants of *Escherichia coli* K-12 induced by bacteriophage Mu-1. J. Bacteriol. **121**:475–484.

130. **Buxton, R. S.** 1979. Fusion of the *lac* genes to the proximal promoters of the *deo* operon of *Escherichia coli* K 12. J. Gen. Microbiol. **112**:241–250.

131. **Buxton, R. S., H. Albrechtsen, and K. Hammer-Jespersen.** 1977. Overlapping transcriptional units in the *deo* operon of *Esch-*

binding protein does not modulate the activity of maltoporin as a general porin in *Escherichia coli*. J. Bacteriol. **161**:720–726.

erichia coli K 12: evidence from phage Mu-1 insertion mutants. J. Mol. Biol. **114**:287–300.

132. **Buxton, R. S., K. Hammer-Jespersen, and T. D. Hansen.** 1978. Insertion of bacteriophage lambda into the *deo* operon of *Escherichia coli* K-12 and isolation of plaque-forming lambda *deo⁺* transducing bacteriophage. J. Bacteriol. **136**:668–681.

133. **Buxton, R. S., K. Hammer-Jespersen, and P. Valentin-Hansen.** 1980. A second purine nucleoside phosphorylase in *Escherichia coli* K-12. I. Xanthosine phosphorylase regulatory mutants isolated as secondary-site revertants of a *deoD* mutant. Mol. Gen. Genet. **179**:331–340.

134. **Caballero, E., L. Baldoma, J. Ros, A. Boronat, and J. Aguilar.** 1983. Identification of lactaldehyde dehydrogenase and glycolaldehyde dehydrogenase as functions of the same protein in *Escherichia coli*. J. Biol. Chem. **258**:7788–7792.

135. **Campbell, J. J., J. Lengyel, and J. Langridge.** 1973. Evolution of a second gene for beta-galactosidase in *Escherichia coli*. Proc. Natl. Acad. Sci. USA **70**:1841–1845.

136. **Campbell, R. L., and E. E. Dekker.** 1973. Formation of D-1-amino-2-propanol from L-threonine by enzymes from *Escherichia coli* K-12. Biochem. Biophys. Res. Commun. **53**:432–438.

137. **Campbell, R. L., R. R. Swain, and E. E. Dekker.** 1978. Purification, separation, and characterization of two molecular forms of D-1-amino-2-propanol:NAD⁺ oxidoreductase activity from extracts of *Escherichia coli* K-12. J. Biol. Chem. **253**:7282–7288.

138. **Chakrabarti, T., Y.-M. Chen, and E. C. C. Lin.** 1984. Clustering of genes for L-fucose dissimilation by *Escherichia coli*. J. Bacteriol. **157**:984–986.

139. **Chelala, C. A., and P. Margolin.** 1984. Effects of deletions on cotransduction linkage in *Salmonella typhimurium*: evidence that bacterial chromosome deletions affect the formation of transducing DNA fragments. Mol. Gen. Genet. **131**:97–112.

140. **Chen, Y.-M., T. Chakrabarti, and E. C. C. Lin.** 1984. Constitutive activation of L-fucose genes by an unlinked mutation in *Escherichia coli*. J. Bacteriol. **159**:725–729.

141. **Chen, Y.-M., E. C. C. Lin, J. Ros, and J. Aguilar.** 1983. Use of operon fusions to examine the regulation of the L-1,2-propanediol oxidoreductase gene of the fucose system in *Escherichia coli*. J. Gen. Microbiol. **129**:3355–3362.

142. **Chen, Y.-M., and E. C. C. Lin.** 1984. Post-transcriptional control of L-1,2-propanediol oxidoreductase in the L-fucose pathway of *Escherichia coli* K-12. J. Bacteriol. **157**:341–344.

143. **Chen, Y.-M., and E. C. C. Lin.** 1984. Dual control of a common L-1,2-propanediol oxidoreductase by L-fucose and L-rhamnose in *Escherichia coli*. J. Bacteriol. **157**:828–832.

144. **Chiu, T.-H., K. L. Evans, and D. S. Feingold.** 1975. L-Rhamnulose-1-phosphate aldolase. Methods Enzymol. **42**:264–269.

145. **Chiu, T. H., and D. S. Feingold.** 1964. The purification and properties of L-rhamnulokinase. Biochim. Biophys. Acta **92**:489–497.

146. **Chiu, T. H., and D. S. Feingold.** 1965. Substrate specificity of L-rhamnulose 1-phosphate aldolase. Biochem. Biophys. Res. Commun. **19**:511–516.

147. **Chiu, T. H., and D. S. Feingold.** 1969. L-rhamnulose 1-phosphate aldolase from *Escherichia coli*. Crystallization and properties. Biochemistry **8**:98–108.

148. **Clark, A. F., T. A. Gerken, and R. W. Hogg.** 1982. Proton nuclear magnetic resonance spectroscopy and ligand binding dynamics of the *Escherichia coli* L-arabinose binding protein. Biochemistry **21**:2227–2233.

149. **Clark, A. F., and R. W. Hogg.** 1981. High-affinity arabinose transport mutants of *Escherichia coli*: isolation and gene location. J. Bacteriol. **147**:920–924.

150. **Clarke, P., H.-C. Lin, and G. Wilcox.** 1982. The nucleotide sequence of the *araC* regulatory gene in *Salmonella typhimurium* LT2. Gene **18**:157–163.

151. **Clement, J. M., and M. Hofnung.** 1981. Gene sequence of the lambda receptor, an outer membrane protein of *E. coli* K12. Cell **27**:507–514.

152. **Cocks, G. T., J. Aguilar, and E. C. C. Lin.** 1974. Evolution of L-1,2-propanediol catabolism in *Escherichia coli* by recruitment of enzymes for L-fucose and L-lactate metabolism. J. Bacteriol. **118**:83–88.

153. **Cohen, R. M., and R. Wolfenden.** 1971. Cytidine deaminase from *Escherichia coli*. J. Biol. Chem. **246**:7561–7565.

154. **Cohen, R. M., and R. Wolfenden.** 1971. The equilibrium of hydrolytic deamination of cytidine and N⁴-methylcytidine. J. Biol. Chem. **246**:7566–7568.

155. **Cohen, S., and R. Raff.** 1951. Adaptive enzymes in the metabolism of gluconate, D-arabinose and D-ribose. J. Biol. Chem. **188**:501–508.

156. **Cohen, S. S.** 1949. Adaptive enzyme formation in the study of

uronic acid utilization by the K-12 strain of *Escherichia coli*. J. Biol. Chem. **177**:607–619.

157. **Cohen, S. S.** 1953. Studies on controlling mechanisms in the metabolism of virus-infected bacteria. Cold Spring Harbor Symp. Quant. Biol. **18**:221–235.

158. **Cohen, S. S.** 1953. Studies on D-ribulose and its enzymatic conversion to D-arabinose. J. Biol. Chem. **201**:72–84.

159. **Cohen, S. S., and H. D. Barner.** 1957. The conversion of 5-methyldeoxycytidine to thymidine *in vitro* and *in vivo*. J. Biol. Chem. **226**:631–642.

160. **Cohn, M., and J. Monod.** 1951. Purification et proprietes de la beta-galactosidase (lactase) d'*Escherichia coli*. Biochim. Biophys. Acta **7**:153–174.

161. **Comb, D. G., and S. Rosemand.** 1956. Glucosamine-6-phosphate deaminase. Biochim. Biophys. Acta **21**:193–194.

162. **Comb, D. G., and S. Rosemand.** 1958. Glucosamine metabolism. IV. Glucosamine-6-phosphate deaminase. J. Biol. Chem. **232**:807–827.

163. **Conrad, C. A., G. W. Stearns III, W. E. Prater, J. A. Rheiner, and J. R. Johnson.** 1984. Characterization of a *glpK* transducing phage. Mol. Gen. Genet. **193**:376–378.

164. **Cooper, R. A.** 1978. The utilisation of D-galactonate and D-2-oxo-3-deoxygalactonate by *Escherichia coli* K-12. Arch. Microbiol. **118**:199–206.

165. **Corina, D. L., and D. C. Wilton.** 1976. An apparent lack of stereospecificity in the reaction catalyzed by deoxyribose 5-phosphate aldolase due to methyl-group rotation and enolization before product release. Biochem. J. **157**:573–576.

166. **Cornelis, G., D. Ghosal, and H. Saedler.** 1978. Tn951: a new transposon carrying a lactose operon. Mol. Gen. Genet. **160**:215–224.

167. **Cornelis, G., R. K. J. Luke, and M. H. Richmond.** 1978. Fermentation of raffinose by lactose-fermenting strains of *Yersinia enterocolitica* and by sucrose-fermenting strains of *Escherichia coli*. J. Clin. Microbiol. **7**:180–183.

168. **Cozzarelli, N. R., W. B. Freedberg, and E. C. C. Lin.** 1968. Genetic control of the L-alpha-glycerophosphate system in *Escherichia coli*. J. Mol. Biol. **31**:371–387.

169. **Cozzarelli, N. R., J. P. Koch, S. Hayashi, and E. C. C. Lin.** 1965. Growth stasis by accumulated L-alpha-glycerophosphate in *Escherichia coli*. J. Bacteriol. **90**:1325–1329.

170. **Cozzarelli, N. R., and E. C. C. Lin.** 1966. Chromosomal location of the structural gene for glycerol kinase in *Escherichia coli*. J. Bacteriol. **91**:1763–1766.

171. **Csonka, L. N., and A. J. Clark.** 1979. Deletions generated by the transposon Tn*10* in the *srl recA* region of the *Escherichia coli* K-12 chromosome. Genetics **93**:321–343.

172. **Csonka, L. N., and A. J. Clark.** 1980. Construction of an Hfr strain useful for transferring *recA* mutations between *Escherichia coli* strains. J. Bacteriol. **143**:529–530.

173. **Curtis, S. J.** 1974. Mechanism of energy coupling for transport of D-ribose in *Escherichia coli*. J. Bacteriol. **120**:295–303.

174. **Curtis, S. J., and W. Epstein.** 1975. Phosphorylation of D-glucose in *Escherichia coli* mutants defective in glucose phosphotransferase, mannose phosphotransferase, and glucokinase. J. Bacteriol. **122**:1189–1199.

175. **Cynkin, M. A., and G. Ashwell.** 1960. Uronic acid metabolism in bacteria. IV. Purification and properties of 2-keto-3-deoxy-D-gluconokinase in *Escherichia coli*. J. Biol. Chem. **235**:1576–1579.

176. **Dabbs, E. R.** 1980. The gene for ribosomal protein S21, *rpsU*, maps close to *dnaG* at 66.5 min on the *Escherichia coli* chromosomal linkage map. J. Bacteriol. **144**:603–607.

177. **Dale, B., and G. R. Greenberg.** 1967. Genetic mapping of a mutation in *Escherichia coli* showing reduced activity of thymidine phosphorylase. J. Bacteriol. **94**:778–779.

178. **Dandanell, G., and K. Hammer.** 1985. Two operator sites separated by 599 base pairs are required for *deoR* repression of the *deo* operon of *Escherichia coli*. EMBO J. **4**:3333–3338.

179. **Daruwalla, K. R., A. T. Paxton, and P. J. F. Henderson.** 1981. Energization of the transport systems for arabinose and comparison with galactose transport in *Escherichia coli*. Biochem. J. **200**:611–627.

180. **David, J., and H. Wiesmeyer.** 1970. Regulation of ribose metabolism in *Escherichia coli*. I. The ribose catabolic pathway. Biochim. Biophys. Acta **208**:45–55.

181. **David, J., and H. Wiesmeyer.** 1970. Regulation of ribose metabolism in *Escherichia coli*. II. Evidence for two ribose-5-phosphate isomerase activities. Biochim. Biophys. Acta **208**:56–67.

182. **David, J., and H. Wiesmeyer.** 1970. Regulation of ribose metabolism in *Escherichia coli*. III. Regulation of ribose utilization *in vivo*. Biochim. Biophys. Acta **208**:68–76.

183. **David, J. D., and H. Wiesmeyer.** 1970. Control of xylose metabolism in *Escherichia coli*. Biochim. Biophys. Acta **201**:497–499.
184. **Davis, E. O., M. C. Jones-Mortimer, and P. J. F. Henderson.** 1984. Location of a structural gene for xylose-H⁺ symport at 91 min on the linkage map of *Escherichia coli* K12. J. Biol. Chem. **259**:1520–1525.
185. **Deacon, J., and R. A. Cooper.** 1977. D-Galactonate utilization by enteric bacteria. The catabolic pathway in *Escherichia coli*. FEBS Lett. **77**:201–205.
186. **Defez, R., and M. De Felice.** 1981. Cryptic operon for beta-glucoside metabolism in *Escherichia coli* K12: genetic evidence for a regulatory protein. Genetics **97**:11–25.
187. **Defez, R., and M. De Felice.** 1982. The metabolism of beta-glucosides in *Escherichia coli* K-12. Ann. Microbiol. (Paris) **133A**:347–350.
188. **Dekker, E. E., and R. R. Swain.** 1968. Formation of D_g-1-amino-2-propanol by a highly purified enzyme from *Escherichia coli*. Biochim. Biophys. Acta **158**:306–307.
189. **Delidakis, C. E., M. C. Jones-Mortimer, and H. L. Kornberg.** 1982. A mutant inducible for galactitol utilization in *Escherichia coli* K12. J. Gen. Microbiol. **128**:601–604.
190. **de Riel, J. K., and H. Paulus.** 1978. Subunit dissociation in the allosteric regulation of glycerol kinase from *Escherichia coli*. 1. Kinetic evidence. Biochemistry **17**:5134–5140.
191. **de Riel, J. K., and H. Paulus.** 1978. Subunit dissociation in the allosteric regulation of glycerol kinase from *Escherichia coli*. 2. Physical evidence. Biochemistry **17**:5141–5145.
192. **de Riel, J. K., and H. Paulus.** 1978. Subunit dissociation in the allosteric regulation of glycerol kinase from *Escherichia coli*. 3. Role in desensitization. Biochemistry **17**:5146–5150.
193. **Didier-Fichet, M.-L., and F. Stoeber.** 1968. Sur les activites glucuronidasique et galacturonidasique d'*Escherichia coli*. C. R. Acad. Sci. Ser. D **266**:1894–1897.
194. **Didier-Fichet, M.-L., and F. Stoeber.** 1968. Sur les proprietes et la biosynthese de la beta-glucuronidase d'*Escherichia coli* K-12. C.R. Acad. Sci. Ser. D **266**:2021–2024.
195. **Dietz, G. W.** 1972. Dehydrogenase activity involved in the uptake of glucose 6-phosphate by a bacterial membrane system. J. Biol. Chem. **247**:4561–4565.
196. **Dietz, G. W., and L. A. Heppel.** 1971. Studies on the uptake of hexose phosphates. II. The induction of the glucose-6-phosphate transport system by exogenous but not by endogenously formed glucose-6-phosphate. J. Biol. Chem. **246**:2885–2890.
197. **Dietz, G. W., Jr.** 1976. The hexose phosphate transport system of *Escherichia coli*. Advances Enzymol. **44**:237–259.
198. **diLauro, R., T. Taniguchi, R. Musso, and B. de Crombrugghe.** 1979. Unusual location and function of the operator in the *Escherichia coli* galactose operon. Nature (London) **279**:494–500.
199. **DiNardo, S., K. A. Voelkel, and R. Sternglanz.** 1982. *Escherichia coli* DNA topoisomerase I mutants have compensatory mutations in DNA gyrase genes. Cell **31**:43–51.
200. **Dobrogosz, W. J.** 1968. N-acetylglucosamine assimilation in *Escherichia coli* and its relation to catabolite repression. J. Bacteriol. **95**:585–591.
201. **Dombrovsky, A. M., and T. I. Serzhantova.** 1978. Identification of lac⁺ Salmonella. Zh. Mikrobiol. Epidemiol. Immunobiol. **5**:48–53.
202. **Doskocil, J.** 1974. Inducible nucleoside permease in *Escherichia coli*. Biochem. Biophys. Res. Commun. **56**:997–1003.
203. **Doskocil, J.** 1976. The role of thiol groups in nucleoside transport. Mol. Cell. Biochem. **10**:137–143.
204. **Doskocil, J., and A. Holy.** 1974. Inhibition of nucleoside-binding sites by nucleoside analogues in *Escherichia coli*. Nucleic Acids Res. **1**:491–502.
205. **Doskocil, J., and A. Holy.** 1977. Specificity of purine nucleoside phosphorylase from *Escherichia coli*. Collect. Czech. Chem. Commun. **42**:370–383.
206. **Doudoroff, M., W. Z. Hassid, E. W. Putnam, A. L. Putnam, A. L. Potter, and J. Lederberg.** 1949. Direct utilization of maltose by *Escherichia coli*. J. Biol. Chem. **179**:921–934.
207. **Duguid, J. P., D. C. Old, and V. B. M. Hume.** 1962. Transduction of fimbriation and rhamnose-fermentation characters in *Salmonella typhimurium*. Heredity **17**:301–302.
208. **Duplay, P., H. Bedouelle, A. Fowler, I. Zabin, W. Saurin, and M. Hofnung.** 1984. Sequences of the malE gene and of its product, the maltose-binding protein of *Escherichia coli* K12. J. Biol. Chem. **259**:10606–10613.
209. **Dykhuizen, D., and D. Hartl.** 1978. Transport by the lactose permease of *Escherichia coli* as the basis of lactose killing. J. Bacteriol. **135**:876–882.
210. **Eagon, R. G.** 1961. Bacterial dissimilation of L-fucose and L-rhamnose. J. Bacteriol. **82**:548–550.
211. **Easterling, S. B., E. M. Johnson, J. A. Wohlhieter, and L. S. Baron.** 1969. Nature of lactose-fermenting Salmonella strains obtained from clinical sources. J. Bacteriol. **100**:35–41.
212. **Edwards, P. R., and W. H. Ewing.** 1972. Identification of Enterobacteriaceae, 3rd ed., p. 68. Burgess Publishing Co., Minneapolis.
213. **Eggleston, L. V., and J. A. Krebs.** 1959. Permeability of *Escherichia coli* to ribose and ribose nucleotides. Biochem. J. **73**:264–270.
214. **Eidels, L., P. D. Rick, N. P. Stimler, and M. J. Osborn.** 1974. Transport of D-arabinose-5-phosphate and sedoheptulose-7-phosphate by the hexose phosphate transport system of *Salmonella typhimurium*. J. Bacteriol. **119**:138–143.
215. **Eisenstark, A., R. Eisenstark, and S. Cunningham.** 1968. Genetic analysis of thymineless (thy) mutants of S. typhimurium. Genetics **58**:493–506.
216. **Elliott, J., and W. Arber.** 1978. E. coli K-12 pel mutants, which block phage DNA injection, coincide with ptsM, which determines a component of a sugar transport system. Mol. Gen. Genet. **161**:1–8.
217. **Elvin, C. M., C. M. Hardy, and H. Rosenberg.** 1985. Pᵢ exchange mediated by the GlpT-dependent sn-glycerol-3-phosphate transport system in *Escherichia coli*. J. Bacteriol. **161**:1054–1058.
218. **Endermann, R., I. Hindennach, and U. Henning.** 1978. Major proteins of the *Escherichia coli* outer cell envelope membrane. Preliminary characterization of the phage lambda receptor protein. FEBS Lett. **88**:71–74.
219. **Englesberg, E.** 1961. Enzymatic characterization of 17 L-arabinose negative mutants of *Escherichia coli*. J. Bacteriol. **81**:996–1006.
220. **Englesberg, E., R. L. Anderson, R. Weinberg, N. Lee, P. Hoffe, G. Huttenbauer, and H. Boyer.** 1962. L-Arabinose-sensitive, L-ribulose 5-phosphate 4-epimerase-deficient mutants of *Escherichia coli*. J. Bacteriol. **37**:146–147.
221. **Englesberg, E., and L. S. Baron.** 1959. Mutation to L-rhamnose resistance and transduction to L-rhamnose utilization in *Salmonella typhosa*. J. Bacteriol. **78**:675–686.
222. **Englesberg, E., J. Irr, J. Power, and N. Lee.** 1965. Positive control of enzyme synthesis by gene C in the L-arabinose system. J. Bacteriol. **90**:946–957.
223. **Englesberg, E., and G. Wilcox.** 1974. Regulation: positive control. Annu. Rev. Genet. **8**:219–242.
224. **Enomoto, M., and B. A. D. Stocker.** 1974. Transduction by phage P1Kc in *Salmonella typhimurium*. Virology **60**:503–514.
224a.**Erni, B., and B. Zanolari.** 1985. The mannose-permease of the bacterial phosphotransferase system: gene cloning and purification of the enzyme IIᴹᵃⁿ/IIIᴹᵃⁿ complex of *Escherichia coli*. J. Biol. Chem. **260**:15495–15503.
225. **Essenberg, R. C., and H. L. Kornberg.** 1975. Energy coupling in the uptake of hexose phosphates by *Escherichia coli*. J. Biol. Chem. **250**:939–945.
226. **Essenberg, R. C., and H. L. Kornberg.** 1977. Location of the gene specifying hexose phosphate transport (uhp) on the chromosome of *Escherichia coli*. J. Gen. Microbiol. **99**:157–169.
227. **Eze, M. O., and R. N. McElhaney.** 1981. The effect of alterations in the fluidity and phase state of the membrane lipids on the passive permeation and facilitated diffusion of glycerol in *Escherichia coli*. J. Gen. Microbiol. **124**:299–307.
228. **Ezzell, J. W., and W. J. Dobrogosz.** 1978. Cyclic AMP regulation of the hexose phosphate transport system in *Escherichia coli*. J. Bacteriol. **133**:1047–1049.
229. **Faik, P., H. L. Kornberg, and E. McEvoy-Bowe.** 1971. Isolation and properties of E. coli mutants defective in 2-keto 3-deoxy-6-phosphogluconate aldolase activity. FEBS Lett. **19**:225–228.
230. **Falcao, D. P., L. R. Trabulsi, F. W. Hickman, and J. J. Farmer.** 1975. Unusual Enterobacteriaceae: lactose-positive *Salmonella typhimurium* which is endemic in Sao Paulo, Brazil. J. Clin. Microbiol. **2**:349–353.
231. **Falkow, S., and L. S. Baron.** 1962. Episomic element in a strain of S. typhosa. J. Bacteriol. **84**:581–589.
232. **Fangman, W. L., and A. Novick.** 1966. Mutation bacteria showing efficient utilization of thymidine. J. Bacteriol. **91**:2390–2391.
233. **Ferenci, T.** 1980. The recognition of maltodextrins by *Escherichia coli*. Eur. J. Biochem. **108**:631–636.
234. **Ferenci, T., and W. Boos.** 1980. The role of the *Escherichia coli* lambda receptor in the transport of maltose and maltodextrins. J. Supramol. Struct. **13**:101–116.
235. **Ferenci, T., W. Boos, M. Schwartz, and S. Szmelcman.** 1977. Energy-coupling of the transport system of *Escherichia coli* dependent on maltose-binding protein. Eur. J. Biochem. **75**:187–193.
236. **Ferenci, T., and U. Klotz.** 1978. Affinity chromatographic isolation of the periplasmic maltose-binding protein of *Escherichia coli*. FEBS Lett. **94**:213–217.

237. **Ferenci, T., and H. L. Kornberg.** 1971. Pathway of fructose utilization by *Escherichia coli*. FEBS Lett. **13**:127–130.

238. **Ferenci, T., and H. L. Kornberg.** 1971. Role of fructose-1,6-diphosphatase in fructose utilization by *Escherichia coli*. FEBS Lett. **14**:360–363.

239. **Ferenci, T., and H. L. Kornberg.** 1973. The utilization of fructose by *Escherichia coli*. Properties of a mutant defective in fructose 1-phosphate kinase activity. Biochem. J. **132**:341–347.

240. **Ferenci, T., and H. L. Kornberg.** 1974. The role of phosphotransferase synthesis of fructose 1-phosphate and fructose 6-phosphate in the growth of *Escherichia coli* on fructose. Proc. R. Soc. Lond. B Biol. Sci. **187**:105–119.

241. **Ferenci, T., H. L. Kornberg, and J. Smith.** 1971. Isolation and properties of a regulatory mutant in the hexose phosphate transport system of *Escherichia coli*. FEBS Lett. **13**:133–136.

242. **Ferenci, T., and K.-S. Lee.** 1982. Directed evolution of the lambda receptor of *Escherichia coli* through affinity chromatographic selection. J. Mol. Biol. **160**:431–444.

243. **Ferenci, T., M. Schwentorat, S. Ullrich, and J. Vilmart.** 1980. Lambda receptor in the outer membrane of *Escherichia coli* as a binding protein for maltodextrins and starch polysaccharides. J. Bacteriol. **142**:521–526.

244. **Fischer, M., and S. A. Short.** 1982. The cloning of the *Escherichia coli* K-12 deoxyribonucleoside operon. Gene. **17**:291–298.

245. **Flagg, J. L., and T. H. Wilson.** 1976. *lacY* Mutant of *Escherichia coli* with altered physiology of lactose induction. J. Bacteriol. **128**:701–707.

246. **Fox, C. F., and E. P. Kennedy.** 1967. A micro radiochemical assay for thiogalactoside transacetylase. Anal. Biochem. **18**:286–294.

247. **Fox, F., and G. Wilson.** 1968. The role of phosphoenolpyruvate-dependent kinase system in beta-glucoside catabolism in *Escherichia coli*. Proc. Natl. Acad. Sci. USA **59**:988–995.

248. **Fradkin, J., and D. G. Fraenkel.** 1971. 2-Keto-3-deoxygluconate 6-phosphate aldolase mutants of *E. coli*. J. Bacteriol. **108**:1277–1283.

249. **Fraenkel, D. G.** 1968. The phosphoenolpyruvate-initiated pathway of fructose metabolism in *Escherichia coli*. J. Biol. Chem. **243**:6458–6463.

250. **Fraenkel, D. G., and S. Banerjee.** 1972. Deletion mapping of *zwf*, the gene for a constitutive enzyme glucose 6-phosphate dehydrogenase in *Escherichia coli*. Genetics **71**:481–489.

251. **Fraenkel, D. G., F. Falcos-Kelly, and B. L. Horecker.** 1964. The utilization of glucose 6-phosphate by glucokinaseless and wild-type strains of *Escherichia coli*. Proc. Natl. Acad. Sci. USA **52**:1207–1213.

252. **Fraser, A. D. E., and H. Yamazaki.** 1980. Characterization of an *Escherichia coli* mutant which utilizes glycerol in the absence of cyclic adenosine monophosphate. Can. J. Microbiol. **26**:393–396.

253. **Fraser, A. D. E., and H. Yamazaki.** 1980. Mannose utilization in *Escherichia coli* requires cyclic AMP but not an exogenous inducer. Can. J. Microbiol. **26**:1508–1511.

254. **Freedberg, W. B., W. S. Kistler, and E. C. C. Lin.** 1971. Lethal synthesis of methylglyoxal by *Escherichia coli* during unregulated glycerol metabolism. J. Bacteriol. **108**:137–144.

255. **Freedberg, W. B., and E. C. C. Lin.** 1973. Three kinds of controls affecting the expression of the *glp* regulon in *Escherichia coli*. J. Bacteriol. **115**:816–823.

256. **Freundlieb, S., and W. Boos.** 1982. Maltose transacetylase of *Escherichia coli*: a preliminary report. Ann. Inst. Pasteur Microbiol. **133A**:181–189.

257. **Friedberg, I.** 1972. Localization of phosphoglucose isomerase in *Escherichia coli* and its relation to the induction of the hexose phosphate transport system. J. Bacteriol. **112**:1201–1205.

258. **Froshauer, S., and J. Beckwith.** 1984. The nucleotide sequence of the gene for *malF* protein, an inner membrane component of the maltose transport system of *Escherichia coli*. Repeated DNA sequences are found in the *malE-malF* intercistronic region. J. Biol. Chem. **259**:10896–10903.

259. **Fukada, H., J. M. Sturtevant, and F. A. Quiocho.** 1983. Thermodynamics of the binding of L-arabinose and of D-galactose to the L-arabinose-binding protein of *Escherichia coli*. J. Biol. Chem. **258**:13193–13198.

260. **Fukasawa, T., K. Jokura, and K. Kurahashi.** 1962. A new enzymatic defect of galactose metabolism in *Escherichia coli* K 12 mutants. Biochem. Biophys. Res. Commun. **7**:121–125.

261. **Fukasawa, T., K. Jokura, and K. Kurahashi.** 1963. Mutations in *Escherichia coli* that affect uridine diphosphate glucose pyrophosphorylase activity and galactose fermentation. Biochim. Biophys. Acta **74**:608–620.

262. **Fukasawa, T., and H. Nikaido.** 1959. Formation of "protoplasts" in mutant strains of *Salmonella* induced by galactose. Nature (London) **183**:1131–1132.

263. **Fukasawa, T., and H. Nikaido.** 1959. Galactose-sensitive mutants of *Salmonella*. Nature. (London) **184**:1168–1169.

264. **Fukasawa, T., and H. Nikaido.** 1960. Formation of phage receptors induced by galactose in a galactose-sensitive mutant of *Salmonella*. Virology **11**:508–510.

265. **Fukasawa, T., and H. Nikaido.** 1961. Galactose-sensitive mutants of *Salmonella*. II. Bacteriolysis induced by galactose. Biochim. Biophys. Acta **48**:470–483.

266. **Fukasawa, T., and H. Nikaido.** 1961. Galactose mutants of *Salmonella typhimurium*. Genetics **46**:1295–1303.

267. **Galloway, D. R., and C. E. Furlong.** 1977. The role of ribose-binding protein in transport and chemotaxis in *Escherichia coli* K12. Arch. Biochem. Biophys. **184**:496–504.

268. **Ganesan, A. K., and B. Rotman.** 1965. Transport systems for galactose and galactosides in *Escherichia coli*. I. Genetic determination and regulation of the methylgalactoside permease. J. Mol. Biol. **16**:42–50.

269. **Garber, B. B., and J. S. Gots.** 1980. Utilization of 2,6-diaminopurine by *Salmonella typhimurium*. J. Bacteriol. **143**:864–871.

270. **Garwin, J. L., and J. Beckwith.** 1982. Secretion and processing of ribose-binding protein in *Escherichia coli*. J. Bacteriol. **149**:789–792.

271. **Ghalambor, M. A., and E. C. Heath.** 1962. The metabolism of L-fucose. II. The enzymatic cleavage of L-fuculose-1-phosphate. J. Biol. Chem. **237**:2427–2433.

272. **Ghalambor, M. A., and E. C. Heath.** 1966. L-Fuculokinase. Methods Enzymol. **9**:461–464.

273. **Ghalambor, M. A., and E. C. Heath.** 1966. L-Fuculose 1-phosphate aldolase. Methods Enzymol. **9**:538–542.

274. **Ghangas, G. S., and D. B. Wilson.** 1984. Isolation and characterization of the *Salmonella typhimurium* LT2 xylose regulon. J. Bacteriol. **157**:158–164.

275. **Ghosh, S., and S. Roseman.** 1962. L-Glutamine-D-fructose 6-phosphate transamidase from *Escherichia coli*. Methods Enzymol. **5**:414–422.

276. **Gilliland, G. L., and F. A. Quiocho.** 1981. Structure of the L-arabinose-binding protein from *Escherichia coli* at 2.4 Å resolution. J. Mol. Biol. **146**:341–362.

277. **Gilson, E., H. Nikaido, and M. Hofnung.** 1982. Sequence of the *malK* gene in *Escherichia coli* K12. Nucleic Acids Res. **10**:7449–7458.

278. **Goldberg, M. E.** 1969. Tertiary structure of *Escherichia coli* beta-D-galactosidase. J. Mol. Biol. **46**:441–446.

279. **Goldenbaum, P. E., and K. S. Farmer.** 1980. *uhp*-Directed, glucose 6-phosphate membrane receptor in *Escherichia coli*. J. Bacteriol. **142**:347–349.

280. **Goldwasser, E.** 1963. Sedimentation constant and molecular weight of thiogalactoside transacetylase. J. Biol. Chem. **238**:3306.

281. **Gonzalez, A. B.** 1959. Lactose-fermenting *Salmonella*. J. Bacteriol. **91**:1661–1662.

282. **Green, M., and S. S. Cohen.** 1956. Enzymatic conversion of L-fucose to L-fuculose. J. Biol. Chem. **219**:557–568.

283. **Greenfield, L., T. Boone, and G. Wilcox.** 1978. DNA sequence of the *araBAD* promoter in *Escherichia coli* B/r. Proc. Natl. Acad. Sci. USA **75**:4724–4728.

284. **Grenier, F. C., E. B. Waygood, and M. H. Saier, Jr.** 1985. Bacterial phospho-transferase system: regulation of mannitol enzyme II activity by sulfhydryl oxidation. Biochemistry **24**:47–51.

285. **Groarke, J. M., W. C. Mahoney, J. N. Hope, C. E. Furlong, F. T. Robb, H. Zalkin, and M. A. Hermodson.** 1983. The amino acid sequence of D-ribose-binding protein from *Escherichia coli* K12. J. Biol. Chem. **258**:12952–12956.

286. **Gross, J., and E. Englesberg.** 1959. Determination of the order of mutational sites governing L-arabinose utilization in *Escherichia coli* B/r by transduction with phage P1bt. Virology **9**:314–331.

287. **Guiso, N., and A. Ullmann.** 1976. Expression and regulation of lactose genes carried by plasmids. J. Bacteriol. **127**:691–697.

287a. **Guth, A., R. Engel, and B. E. Tropp.** 1980. Uptake of glycerol 3 phosphate and some of its analogs by the hexose phosphate transport system of *Escherichia coli*. J. Bacteriol. **143**:538–539.

288. **Gutnick, D., J. M. Calvo, T. Klopotowski, and B. N. Ames.** 1969. Compounds which serve as the sole source of carbon or nitrogen for *Salmonella typhimurium* LT-2. J. Bacteriol. **100**:215–219.

289. **Gutowski, S. J., and H. Rosenberg.** 1975. Succinate uptake and related proton movements in *Escherichia coli* K12. Biochem. J. **152**:647–654.

290. **Hacking, A. J., J. Aguilar, and E. C. C. Lin.** 1978. Evolution of propanediol utilization in *Escherichia coli*: mutants with improved substrate-scavenging power. J. Bacteriol. **136**:522–530.

291. **Hacking, A. J., and E. C. C. Lin.** 1976. Disruption of the fucose pathway as a consequence of genetic adaptation to propanediol

as a carbon source in *Escherichia coli*. J. Bacteriol. **126:**1166–1172.

292. **Hacking, A. J., and E. C. C. Lin.** 1977. Regulatory changes in the fucose system associated with the evolution of a catabolic pathway for propanediol in *Escherichia coli*. J. Bacteriol. **130:**832–838.

293. **Hagihira, H., T. H. Wilson, and E. C. C. Lin.** 1963. Studies on the glucose-transport system in *Escherichia coli* with alpha-methylglucoside as substrate. Biochim. Biophys. Acta **78:**505–515.

294. **Hall, B. G.** 1976. Experimental evolution of a new enzymatic function. Kinetic analysis of the ancestral (*ebg°*) and evolved (*ebg*$^+$0) enzymes. J. Mol. Biol. **107:**71–84.

295. **Hall, B. G.** 1976. Methylgalactosidase activity: an alternative evolutionary destination for the *ebgA°* gene. J. Bacteriol. **126:**536–538.

296. **Hall, B. G.** 1977. The number of mutations required to evolve a new lactase function in *Escherichia coli*. J. Bacteriol. **129:**540–543.

297. **Hall, B. G.** 1978. Experimental evolution of a new enzymatic function. II. Evolution of multiple functions for *EBG* enzyme in *E. coli*. Genetics **89:**453–465.

298. **Hall, B. G.** 1978. Regulation of newly evolved enzymes. IV. Directed evolution of the *EBG* repressor. Genetics **90:**673–681.

299. **Hall, B. G.** 1981. Changes in the substrate specificities of an enzyme during directed evolution of new functions. Biochemistry **20:**4042–4049.

300. **Hall, B. G.** 1982. Evolution of a regulated operon in the laboratory. Genetics **101:**335–344.

301. **Hall, B. G.** 1982. Transgalactosylation activity of EBG beta-galactosidase synthesizes allolactose from lactose. J. Bacteriol. **150:**132–140.

302. **Hall, B. G.** 1982. Chromosomal mutation for citrate utilization by *Escherichia coli* K-12. J. Bacteriol. **151:**269–273.

303. **Hall, B. G., and N. D. Clarke.** 1977. Regulation of newly evolved enzymes. III. Evolution of the *ebg* repressor during selection for enhanced lactase activity. Genetics **85:**193–201.

304. **Hall, B. G., and D. L. Hartl.** 1974. Regulation of newly evolved enzymes. I. Selection of a novel lactase regulated by lactose in *Escherichia coli*. Genetics **76:**391–400.

305. **Hall, B. G., and D. L. Hartl.** 1975. Regulation of newly evolved enzymes. II. The *EBG* repressor. Genetics **81:**427–435.

306. **Hall, B. G., and T. Zuzel.** 1980. Evolution of a new enzymatic function by recombination within a gene. Proc. Natl. Acad. Sci. USA **77:**3529–3533.

307. **Hall, B. G., and T. Zuzel.** 1980. The *ebg* operon consists of at least two genes. J. Bacteriol. **144:**1208–1211.

308. **Hammelburger, J. W., and G. A. Orr.** 1983. Interaction of *sn*-glycerol 3-phosphorothioate with *Escherichia coli*: effect on cell growth and metabolism. J. Bacteriol. **156:**789–799.

309. **Hammer-Jespersen, K.** 1983. Nucleoside catabolism, p. 203–258. *In* A. Munch-Petersen (ed.), Metabolism of nucleotides, nucleosides and nucleobases in microorganisms. Academic Press, Inc. (London), Ltd., London.

310. **Hammer-Jespersen, K., R. S. Buxton, and T. D. Hansen.** 1980. A second purine nucleoside phosphorylase in *Escherichia coli* K-12. II. Properties of xanthosine phosphorylase and its induction by xanthosine. Mol. Gen. Genet. **179:**341–348.

311. **Hammer-Jespersen, K., and A. Munch-Petersen.** 1970. Phosphodeoxyribomutase from *Escherichia coli*. Purification and some properties. Eur. J. Biochem. **17:**397–407.

312. **Hammer-Jespersen, K., and A. Munch-Petersen.** 1973. Mutants of *Escherichia coli* unable to metabolize cytidine: isolation and characterization. Mol. Gen. Genet. **126:**177–186.

313. **Hammer-Jespersen, K., and A. Munch-Petersen.** 1975. Multiple regulation of nucleoside catabolizing enzymes: regulation of the *deo* operon by the *cytR* and *deoR* gene products. Mol. Gen. Genet. **137:**327–335.

314. **Hammer-Jespersen, K., A. Munch-Petersen, P. Nygaard, and M. Schwartz.** 1971. Induction of enzymes involved in the catabolism of deoxyribonucleosides and ribonucleosides in *Escherichia coli* K 12. Eur. J. Biochem. **19:**533–538.

315. **Hammer-Jespersen, K., and P. Nygaard.** 1976. Multiple regulation of nucleoside catabolizing enzymes in *Escherichia coli*: effects of 3:5′ cyclic AMP and CRP protein. Mol. Gen. Genet. **148:**49–55.

316. **Hanatani, M., H. Yazyu, S. S. Niiya, Y. Moriyama, H. Kanzawa, M. Futai, and T. Tsuchiya.** 1984. Physical and genetic characterization of the melibiose operon and identification of the gene products in *Escherichia coli*. J. Biol. Chem. **259:**1807–1812.

317. **Hantke, K.** 1976. Phage T6-colicin K receptor and nucleoside transport in *Escherichia coli*. FEBS Lett. **70:**109–112.

318. **Harayama, S., J. Bollinger, T. Iino, and G. L. Hazelbauer.** 1983.

319. **Harrison, A. P., Jr.** 1965. Thymine incorporation and metabolism by various classes of thymine-less bacteria. J. Gen. Microbiol. **41:**321–333.

320. **Hartl, D., and B. G. Hall.** 1974. A second naturally occurring beta-galactosidase in *Escherichia coli*. Nature (London) **248:**152–153.

321. **Hartley, B. S.** 1984. Experimental evolution of ribitol dehydrogenase, p. 23–54. *In* R. P. Mortlock (ed.), Microorganisms as model systems for studying evolution. Plenum Publishing Corp., New York.

322. **Hartley, B. S.** 1984. The structure and control of the pentitol operons, p. 55–107. *In* R. P. Mortlock (ed.), Microorganisms as model systems for studying evolution. Plenum Publishing Corp., New York.

323. **Häselbarth, V., G. V. Schulz, and H. Schwinn.** 1971. Untersuchungen über amylomaltase. II. Molekulare konstanten und wirkungsweise des enzymes. Biochim. Biophys. Acta **227:**296–312.

324. **Hattman, S., and T. Fukasawa.** 1963. Host-induced modification of T-even phages due to defective glucosylation of their DNA. Proc. Natl. Acad. Sci. USA **50:**297–300.

325. **Hayashi, S.-I., J. P. Koch, and E. C. C. Lin.** 1964. Active transport of L-alpha-glycerophosphate in *Escherichia coli*. J. Biol. Chem. **239:**3098–3105.

326. **Hayashi, S.-I., and E. C. C. Lin.** 1965. Capture of glycerol by cells of *Escherichia coli*. Biochim. Biophys. Acta **94:**479–487.

327. **Hayashi, S.-I., and E. C. C. Lin.** 1965. Product induction of glycerol kinase in *Escherichia coli*. J. Mol. Biol. **14:**515–521.

328. **Hayashi, S.-I., and E. C. C. Lin.** 1967. Purification and properties of glycerol kinase from *Escherichia coli*. J. Biol. Chem. **242:**1030–1035.

329. **Hazelbauer, G. L.** 1975. The maltose chemoreceptor of *Escherichia coli*. J. Bacteriol. **122:**206–214.

330. **Hazelbauer, G. L.** 1975. Role of the receptor for bacteriophage lambda in the functioning of the maltose chemoreceptor of *Escherichia coli*. J. Bacteriol. **124:**119–126.

331. **Hazelbauer, G. L., and J. Adler.** 1971. Role of galactose-binding protein in chemotaxis of *Escherichia coli* toward galactose. Nature (London) **230:**101–104.

332. **Heald, K., and C. Long.** 1955. Studies involving enzymic phosphorylation. 3. The phosphorylation of D-ribose by extracts of *Escherichia coli*. Biochem. J. **59:**316–322.

333. **Heath, E. C., and M. A. Ghalambor.** 1962. The metabolism of L-fucose. I. The purification and properties of L-fuculose kinase. J. Biol. Chem. **237:**2423–2426.

334. **Heller, K. B., E. C. C. Lin, and T. H. Wilson.** 1980. Substrate specificity and transport properties of the glycerol facilitator of *Escherichia coli*. J. Bacteriol. **144:**274–278.

335. **Heller, K. B., and T. H. Wilson.** 1979. Sucrose transport by the *Escherichia coli* lactose carrier. J. Bacteriol. **140:**395–399.

336. **Henderson, P. J. F., R. A. Giddens, and M. C. Jones-Mortimer.** 1977. Transport of galactose, glucose and their molecular analogues by *Escherichia coli* K12. Biochem. J. **162:**309–320.

337. **Hendlin, D., E. O. Stapley, M. Jackson, H. Wallick, A. K. Miller, F. J. Wolf, T. W. Miller, L. Chaiet, F. M. Kahan, E. L. Foltz, H. B. Woodruff, J. M. Mata, S. Hernandez, S. Mochales.** 1969. Phosphonomycin, a new antibiotic produced by strains of *Streptomyces*. Science **166:**122–123.

338. **Hengge, R., and W. Boos.** 1983. Maltose and lactose transport in *Escherichia coli*. Examples of two different types of concentrative transport systems. Biochim. Biophys. Acta **737:**443–478.

339. **Hengge, R., T. J. Larson, and W. Boos.** 1983. *sn*-Glycerol-3-phosphate transport in *Salmonella typhimurium*. J. Bacteriol. **155:**186–195.

340. **Hennen, P. E., H. B. Carter, and W. D. Nunn.** 1978. Changes in macromolecular synthesis and nucleoside triphosphate levels during glycerol-induced growth stasis of *Escherichia coli*. J. Bacteriol. **136:**929–935.

341. **Heppel, L. A.** 1969. The effect of osmotic shock on release of bacterial proteins and an active transport. J. Gen. Physiol. **54:**95–109.

342. **Herbert, A. A., and J. R. Guest.** 1971. Two mutations affecting utilization of C$_4$-dicarboxylic acids by *Escherichia coli*. J. Gen. Microbiol. **63:**151–162.

343. **Herzenberg, L. A.** 1961. Isolation and identification of derivatives formed in the course of intracellular accumulation of thiogalactosides by *Escherichia coli*. Arch. Biochem. Biophys. **93:**314–315.

344. **Hickman, J., and G. Ashwell.** 1960. Uronic acid metabolism in bacteria. II. Purification and properties of D-altronic acid and

Characterization of the *mgl* operon of *Escherichia coli* by transposon mutagenesis and molecular cloning. J. Bacteriol. **153:**408–415.

D-mannonic acid dehydrogenases in *Escherichia coli*. J. Biol. Chem. **235**:1566–1570.

345. **Hill, C. W., and H. Echols.** 1966. Properties of a mutant blocked in inducibility of messenger RNA for the galactose operon. J. Mol. Biol. **19**:38–51.

346. **Hochstadt-Ozer, J.** 1972. The regulation of purine utilization in bacteria. IV. Roles of membrane-localized and pericytoplasmic enzymes in the mechanism of purine nucleoside transport across isolated *Escherichia coli* membranes. J. Biol. Chem. **247**:2419–2426.

347. **Hochstadt-Ozer, J., and E. R. Stadtman.** 1971. The regulation of purine utilization in bacteria. II. Adenine phosphoribosyltransferase in isolated membrane preparations and its role in transport of adenine across the membrane. J. Biol. Chem. **246**:5304–5311.

348. **Hoffee, P., P. Snyder, C. Sushak, and P. Jargiello.** 1974. Deoxyribose-5-P aldolase: subunit structure and composition of active site lysine region. Arch. Biochem. Biophys. **164**:736–742.

349. **Hoffee, P. A.** 1968. 2-Deoxyribose gene-enzyme complex in *Salmonella typhimurium*. I. Isolation and enzymatic characterization of 2-deoxyribose-negative mutants. J. Bacteriol. **95**:449–457.

350. **Hoffee, P. A.** 1968. 2-Deoxyribose-5-phosphate aldolase of *Salmonella typhimurium*. Purification and properties. Arch. Biochem. Biophys. **126**:795–802.

351. **Hoffee, P. A., and J. Blank.** 1978. Thymidine phosphorylase from *Salmonella typhimurium*. Methods Enzymol. **51**:437–442.

352. **Hoffee, P. A., and B. C. Robertson.** 1969. 2-Deoxyribose gene-enzyme complex in *Salmonella typhimurium*: regulation of phosphodeoxyribomutase. J. Bacteriol. **97**:1386–1396.

353. **Hoffman, C. E., and J. P. Lampen.** 1952. Products of deoxyribose degradation by *Escherichia coli*. J. Biol. Chem. **198**:885–893.

354. **Hoffmeyer, J., and J. Neuhard.** 1971. Metabolism of exogenous purine bases and nucleosides by *Salmonella typhimurium*. J. Bacteriol. **106**:14–24.

355. **Hofnung, M. (ed.).** 1982. The maltose system as a tool in molecular biology. Ann. Inst. Pasteur Microbiol. **133A**:5–273.

356. **Hofnung, M., A. Jezierska, and C. Braun-Breton.** 1976. *lamB* mutations in *E. coli* K12: growth of lambda host range mutants and effects of nonsense suppressors. Mol. Gen. Genet. **145**:207–213.

357. **Hogg, R., and M. A. Hermodson.** 1977. Amino acid sequence of the L-arabinose-binding protein from *Escherichia coli* B/r. J. Biol. Chem. **252**:5135–5141.

358. **Hogg, R. W.** 1977. L-arabinose-transport and the L-arabinose binding protein of *Escherichia coli*. J. Supramol. Struct. **6**:411–417.

359. **Hogg, R. W., and E. Englesberg.** 1969. L-arabinose binding protein from *Escherichia coli*. J. Bacteriol. **100**:423–432.

360. **Hogg, R. W., H. Isihara, M. A. Hermodson, D. Koshland, Jr., J. W. Jacobs, and R. A. Bradshaw.** 1977. A comparison of the amino-terminal sequences of several carbohydrate binding proteins from *Escherichia coli* and *Salmonella typhimurium*. FEBS Lett. **80**:377–379.

361. **Holmes, R. P., and R. R. Russell.** 1972. Mutations affecting amino sugar metabolism in *Escherichia coli* K-12. J. Bacteriol. **111**:290–291.

362. **Horecker, B. L., J. Thomas, and J. Monod.** 1960. Galactose transport in *Escherichia coli*. I. General properties as studied with a galactokinaseless mutant. J. Biol. Chem. **235**:1580–1585.

363. **Horecker, B. L., J. Thomas, and J. Monod.** 1960. Galactose transport in *Escherichia coli*. II. Characteristics of the exit process. J. Biol. Chem. **235**:1586–1590.

364. **Horiuchi, T., J. Tomizawa, and A. Novick.** 1962. Isolation and properties of bacteria capable of high rates of beta-galactosidase synthesis. Biochim. Biophys. Acta **55**:152–163.

365. **Horwitz, A. H., L. Heffernan, C. Morandi, J.-H. Lee, J. Timko, and G. Wilcox.** 1981. DNA sequence of the *araBAD-araC* controlling region in *Salmonella typhimurium* LT2. Gene **14**:309–319.

366. **Hosono, H., and S. Kuno.** 1973. The purification and properties of cytidine deaminase from *Escherichia coli*. J. Biochem. **74**:797–803.

367. **Hsie, A. W., H. V. Rickenberg, D. W. Schulz, and W. M. Kirsch.** 1969. Steady-state concentrations of glucose-6-phosphate, 6-phosphogluconate, and reduced nicotinamide adenine dinucleotide phosphate in strains of *Escherichia coli* sensitive and resistant to catabolite repression. J. Bacteriol. **98**:1407–1408.

368. **Huber, R. E., G. Kurz, and K. Wallenfels.** 1976. A quantitation of the factors which affect the hydrolase and transgalactosylase activities of beta-galactosidase (*E. coli*) on lactose. Biochemistry **15**:1994–2001.

369. **Hugouvieux-Cotte-Pattat, N., and J. Robert-Baudouy.** 1981. Isolation of fusions between the *lac* genes and several genes of the *exu* regulon: analysis of their regulation, determination of the transcription direction of the *uxaC-uxaA* operon, in *Escherichia coli* K-12. Mol. Gen. Genet. **182**:279–287.

370. **Hugouvieux-Cotte-Pattat, N., and J. Robert-Baudouy.** 1982. Determination of the transcription direction of the *exuT* gene in *Escherichia coli* K-12: divergent transcription of the *exuT-uxaCA* operons. J. Bacteriol. **151**:480–484.

371. **Hugouvieux-Cotte-Pattat, N., and J. Robert-Baudouy.** 1982. Regulation and transcription direction of *exuR*, a self-regulated repressor in *Escherichia coli* K-12. J. Mol. Biol. **156**:221–228.

372. **Iida, A., S. Harayama, T. Iino, and G. L. Hazelbauer.** 1984. Molecular cloning and characterization of genes required for ribose transport and utilization in *Escherichia coli* K-12. J. Bacteriol. **158**:674–682.

373. **Imada, A., and S. Igarasi.** 1967. Ribosyl and deoxyribosyl transfer by bacterial enzyme systems. J. Bacteriol. **94**:1551–1559.

374. **Imada, A., Y. Nozaki, F. Kawashima, and M. Yonida.** 1977. Regulation of glucosamine utilization in *Staphylococcus aureus* and *Escherichia coli*. J. Gen. Microbiol. **100**:329–337.

375. **Imai, K., T. Iijima, and T. Hasegawa.** 1973. Transport of tricarboxylic acids in *Salmonella typhimurium*. J. Bacteriol. **114**:961–965.

376. **Irani, M., L. Orosz, S. Busby, T. Taniguchi, and S. Adhya.** 1983. Cyclic AMP-dependent constitutive expression of *gal* operon: use of repressor titration to isolate operator mutations. Proc. Natl. Acad. Sci. USA **80**:4775–4779.

377. **Irani, M. H., L. Orosz, and S. Adhya.** 1983. A control element within a structural gene: the *gal* operon of *Escherichia coli*. Cell **32**:783–788.

377a.**Ishiguro, N., K. Hirose, M. Asagi, and G. Sato.** 1981. Incompatability of citrate utilization plasmids isolated from *Escherichia coli*. J. Gen. Microbiol. **123**:193–196.

377b.**Ishiguro, N., C. Oka, Y. Hanazawa, and G. Sato.** 1979. Plasmids in *Escherichia coli* controlling citrate-utilizing ability. Appl. Environ. Microbiol. **38**:956–964.

377c.**Ishiguro, N., C. Oka, Y. Hanazawa, and G. Sato.** 1980. Isolation of citrate utilization plasmid from a bovine *Salmonella typhimurium* strain. Microbiol. Immunol. **24**:757–760.

377d.**Ishiguro, N., C. Oka, and G. Sato.** 1978. Isolation of citrate-positive variants of *Escherichia coli* from domestic pigeons, pigs, cattle, and horses. Appl. Environ. Microbiol. **36**:217–222.

377e.**Ishiguro, N., and G. Sato.** 1979. The distribution of plasmids determining citrate utilization in citrate-positive variants of *Escherichia coli* from humans, domestic animals, feral birds and environments. J. Hyg. **83**:331–344.

377f.**Ishiguro, N., G. Sato, C. Sasakawa, H. Danbara, and M. Yoshikawa.** 1982. Identification of citrate utilization transposon Tn*3411* from a naturally occurring citrate utilization plasmid. J. Bacteriol. **149**:961–968.

378. **Ishii, J. N., Y. Okajima, and T. Nakae.** 1981. Characterization of *lamB* protein from the outer membrane of *Escherichia coli* that forms diffusion pores selective for maltose-maltodextrins. FEBS Lett. **134**:217–220.

379. **Isono, K., and M. Kitakawa.** 1978. Cluster of ribosomal protein genes in *Escherichia coli* containing genes for proteins S6, S18, and L9. Proc. Natl. Acad. Sci. USA **75**:6163–6167.

380. **Jacobson, G. R., D. M. Kelly, and D. R. Finlay.** 1983. The intramembrane topography of the mannitol-specific enzyme II of the *Escherichia coli* phosphotransferase system. J. Biol. Chem. **258**:2955–2959.

381. **Jacobson, G. R., C. A. Lee, J. E. Leonard, and M. H. Saier, Jr.** 1983. Mannitol-specific enzyme II of the bacterial phosphotransferase system. I. Properties of the purified permease. J. Biol. Chem. **258**:10748–10756.

382. **Jacobson, G. R., C. A. Lee, and M. H. Saier, Jr.** 1979. Purification of the mannitol-specific enzyme II of the *Escherichia coli* phosphoenolpyruvate:sugar phosphotransferase system. J. Biol. Chem. **254**:249–252.

383. **Jacobson, G. R., L. E. Tanney, D. M. Kelly, K. B. Palman, and S. B. Corn.** 1983. Substrate and phospholipid specificity of the purified mannitol permease of *Escherichia coli*. J. Cell Biochem. **23**:231–240.

384. **Janion, C.** 1977. On the ability of *Salmonella typhimurium* cells to form deoxycytidine nucleotides. Mol. Gen. Genet. **153**:179–183.

385. **Jargiello, P.** 1976. Simultaneous selection of mutants in gluconeogenesis and nucleoside catabolism in *Salmonella typhimurium*. Biochim. Biophys. Acta **444**:321–325.

386. **Jargiello, P., and P. Hoffee.** 1972. Orientation of the *deo* genes and the *serB* locus in *Salmonella typhimurium*. J. Bacteriol. **111**:296–297.

387. **Jargiello, P., M. D. Stern, and P. Hoffee.** 1974. 2-Deoxyribose 5-phosphate aldolase: genetic analyses of structure. J. Mol. Biol. **88:**671–691.
388. **Jargiello, P., C. Sushak, and P. Hoffee.** 1976. 2-Deoxyribose 5-phosphate aldolase: isolation and characterization of proteins genetically modified in the active site region. Arch. Biochem. Biophys. **177:**630–641.
389. **Jensen, K. F.** 1976. Purine-nucleoside phosphorylase from *Salmonella typhimurium* and *Escherichia coli.* Initial velocity kinetics, ligand binding, and reaction mechanism. Eur. J. Biochem. **61:**377–386.
390. **Jensen, K. F., and P. Nygaard.** 1975. Purine nucleoside phosphorylase from *Escherichia coli* and *Salmonella typhimurium.* Purification and some properties. Eur. J. Biochem. **51:**253–265.
391. **Jensen, P., C. Parkes, and D. Berkowitz.** 1972. Mannitol sensitivity. J. Bacteriol. **111:**351–355.
392. **Jergensen, P., J. Collins, and P. Valentin-Hansen.** 1977. On the structure of the *deo* operon of *Escherichia coli.* Mol. Gen. Genet. **155:**93–102.
393. **Jimeno-Abendano, J., and A. Kepes.** 1973. Sensitization of D-glucuronic acid transport system of *Escherichia coli* to protein group reagents in presence of substrate or absence of energy source. Biochem. Biophys. Res. Commun. **54:**1342–1346.
394. **Jin, R. Z., and E. C. C. Lin.** 1984. An inducible phosphoenolpyruvate: dihydroxyacetone phosphotransferase system in *Escherichia coli.* J. Gen. Microbiol. **130:**83–88.
395. **Jin, R. Z., J. C.-T. Tang, and E. C. C. Lin.** 1983. Experimental evolution of a novel pathway for glycerol dissimilation in *Escherichia coli.* J. Mol. Evol. **19:**429–436.
396. **Jobe, A., and S. Bourgeois.** 1972. *lac* repressor-operator interaction. VI. The natural inducer of the *lac* operon. J. Mol. Biol. **69:**397–408.
397. **Jochimsen, B., P. Nygaard, and T. Vestergaard.** 1975. Location on the chromosome of *Escherichia coli* of genes governing purine metabolism. Adenosine deaminase (*add*), guanosine kinase (*gsk*) and hypoxanthine phosphoribosyltransferase (*hpt*). Mol. Gen. Genet. **143:**85–91.
398. **Johnston, K. G., and R. T. Jones.** 1976. Salmonellosis in calves due to lactose fermenting *Salmonella typhimurium.* Vet. Rec. **98:**276–278
399. **Jones-Mortimer, M. C., and H. L. Kornberg.** 1974. Genetical analysis of fructose utilization by *Escherichia coli.* Proc. R. Soc. Lond. B Biol. Sci. **187:**121–131.
400. **Jones-Mortimer, M. C., and H. L. Kornberg.** 1976. Order of genes adjacent to *ptsX* on the *E. coli* genome. Proc. R. Soc. Lond. B Biol. Sci. **193:**313–315.
401. **Jones-Mortimer, M. C., and H. L. Kornberg.** 1976. Uptake of fructose by the sorbitol phosphotransferase of *Escherichia coli* K12. J. Gen. Microbiol. **96:**383–391.
402. **Jones-Mortimer, M. C., and H. L. Kornberg.** 1980. Amino-sugar transport systems of *Escherichia coli* K12. J. Gen. Microbiol. **117:**369–376.
403. **Jonsen, J., and S. Laland.** 1957. Adaptation of *E. coli* to 2-deoxy-D-ribose. Acta Chem. Scand. **11:**1095–1096.
404. **Jonsen, J., S. Laland, and A. Strand.** 1959. Degradation of deoxyribose by *E. coli.* Studies with cell-free extract and isolation of 2-deoxy-D-ribose 5-phosphate. Biochim. Biophys. Acta **32:**117–123.
405. **Jordan, E., and M. B. Yarmolinsky.** 1963. Control of internal induction of galactose pathway enzymes in an *Escherichia coli* mutant. J. Gen. Microbiol. **30:**357–364.
406. **Jordan, E., M. B. Yarmolinsky, and H. M. Kalckar.** 1962. Control of inducibility of enzymes of the galactose sequence in *Escherichia coli.* Proc. Natl. Acad. Sci. USA **48:**32–40.
407. **Jørgensen, P., J. Collins, and P. Valentin-Hensen.** 1977. On the structure of the *deo* operon of *Escherichia coli.* Mol. Gen. Genet. **155:**93–102.
408. **Josephsen, J., and K. Hammer-Jespersen.** 1981. Fusion of the *lac* genes to the promoter for the cytidine deaminase gene of *Escherichia coli* K-12. Mol. Gen. Genet. **182:**154–158.
409. **Josephson, J., K. Hammer-Jespersen, and T. D. Hansen.** 1983. Mapping of the gene for cytidine deaminase (*cdd*) in *Escherichia coli* K-12. J. Bacteriol. **154:**72–75.
410. **Kaback, H. R.** 1983. The *lac* carrier protein in *Escherichia coli.* J. Membr. Biol. **76:**95–112.
411. **Kadner, R. J.** 1973. Genetic control of the transport of hexose phosphates in *Escherichia coli:* mapping of the *uhp* locus. J. Bacteriol. **116:**764–770.
412. **Kadner, R. J., and D. M. Shattuck-Eidens.** 1983. Genetic control of the hexose phosphate transport system of *Escherichia coli:* mapping of deletion and insertion mutations in the *uhp* region. J. Bacteriol. **155:**1052–1061.
413. **Kadner, R. J., and H. H. Winkler.** 1973. Isolation and characterization of mutations affecting the transport of hexose phosphates in *Escherichia coli.* J. Bacteriol. **113:**895–900.
414. **Kahan, F. M., J. S. Kahan, P. J. Cassidy, and H. Kropp.** 1974. The mechanism of action of fosfomycin (phosphonomycin). Ann. N.Y. Acad. Sci. **235:**364–386.
415. **Kalckar, H. M.** 1958. Uridine diphospho galactose: metabolism, enzymology and biology. Advances Enzymol. **20:**111–134.
416. **Kalckar, H. M.** 1965. Galactose metabolism and cell "sociology." Science **150:**305–313.
417. **Kalckar, H. M.** 1971. The periplasmic galactose binding protein of *Escherichia coli.* Science **174:**557–565.
418. **Kalckar, H. M.** 1976. The periplasmic galactose receptor protein of *Escherichia coli* in relation to galactose chemotaxis. Biochimie **58:**81–85.
419. **Kalckar, H. M., K. Kurahashi, and E. Jordan.** 1959. Hereditary defects in galactose metabolism in *Escherichia coli* mutants. I. Determination of enzyme activities. Proc. Natl. Acad. Sci. USA **45:**1776–1786.
420. **Karlström, O.** 1968. Mutants of *Escherichia coli* defective in ribonucleoside and deoxyribonucleoside catabolism. J. Bacteriol. **95:**1069–1077.
421. **Kauffmann, F.** 1966. The bacteriology of *Enterobacteriaceae;* collected studies of the author and his co-workers, 3rd ed. The Williams & Wilkins Co., Baltimore.
422. **Kay, W. W., and M. J. Cameron.** 1978. Citrate transport in *Salmonella typhimurium.* Arch. Biochem. Biophys. **190:**270–280.
423. **Kay, W. W., and M. J. Cameron.** 1978. Transport of C$_4$-dicarboxylic acids in *Salmonella typhimurium.* Arch. Biochem. Biophys. **190:**281–289.
424. **Kay, W. W., and H. L. Kornberg.** 1969. Genetic control of the uptake of C$_4$-dicarboxylic acids by *Escherichia coli.* FEBS Lett. **3:**93–96.
425. **Kay, W. W., and H. L. Kornberg.** 1971. The uptake of C$_4$-dicarboxylic acids by *Escherichia coli.* Eur. J. Biochem. **18:**274–281.
425a. **Kay, W. W., J. M. Somers, G. D. Sweet, and K. A. Widenhorn.** 1984. Tricarboxylate transport systems: the *tct* operon in *Salmonella typhimurium,* p. 34–37. *In* L. Leive and D. Schlessinger (ed.), Microbiology—1984. American Society for Microbiology, Washington, D.C.
426. **Kellerman, O., and S. Szmelcman.** 1974. Active transport of maltose in *Escherichia coli* K-12: involvement of a "periplasmic" maltose binding protein. Eur. J. Biochem. **47:**139–149.
427. **Kelley, J. J., and E. E. Dekker.** 1984. D-1-Amino-2-propanol:NAD$^+$ oxidoreductase. J. Biol. Chem. **259:**2124–2129.
428. **Kelley, J. J., and E. E. Dekker.** 1985. Identity of *Escherichia coli* D-1-amino-2-propanol:NAD$^+$ oxidoreductase with *E. coli* glycerol dehydrogenase but not with *Neisseria gonorrhoeae* 1,2,-propanediol:NAD$^+$ oxidoreductase. J. Bacteriol. **162:**170–175.
429. **Kemper, J.** 1974. Gene order and co-transduction in the *leu-ara-fol-pyr-A* region of the *Salmonella typhimurium* linkage map. J. Bacteriol. **117:**94–99.
430. **Kerwar, G. K., A. S. Gordon, and H. R. Kaback.** 1972. Mechanisms of active transport in isolated membrane vesicles. IV. Galactose transport by isolated membrane vesicles from *Escherichia coli.* J. Biol. Chem. **247:**291–297.
431. **Kistler, W. S., C. A. Hirsch, N. R. Cozzarelli, and E. C. C. Lin.** 1969. Second pyridine nucleotide-independent L-alpha-glycerophosphate dehdyrogenase in *Escherichia coli* K-12. J. Bacteriol. **100:**1133–1135.
432. **Kistler, W. S., and E. C. C. Lin.** 1971. Anaerobic L-alpha-glycerophosphate dehydrogenase of *Escherichia coli:* its genetic locus and its physiological role. J. Bacteriol. **108:**1224–1234.
433. **Kistler, W. S., and E. C. C. Lin.** 1972. Purification and properties of the flavine-stimulated anaerobic L-alpha-glycerophosphate dehdyrogenase of *Escherichia coli.* J. Bacteriol. **142:**539–547.
434. **Koch, A. L., and W. A. Lamont.** 1956. The metabolism of methylpurines by *Escherichia coli.* J. Biol. Chem. **219:**189–201.
435. **Koch, A. L., and G. Vallee.** 1959. The properties of adenosine deaminase and adenosine nucleoside phosphorylase in extracts of *Escherichia coli.* J. Biol. Chem. **234:**1213–1218.
436. **Koch, J. P., S.-I. Hayashi, and E. C. C. Lin.** 1964. The control of dissimilation of glycerol and L-alpha-glycerophosphate in *Escherichia coli.* J. Biol. Chem. **239:**3106–3108.
437. **Kolodrubetz, D., and R. Schleif.** 1981. Regulation of the L-arabinose transport operons in *Escherichia coli.* J. Mol. Biol. **151:**215–227.
438. **Kolodrubetz, D., and R. Schleif.** 1981. L-arabinose transport systems in *Escherichia coli* K-12. J. Bacteriol. **148:**472–479.
439. **Komatsu, Y.** 1971. Mechanism of action of showdomycin. IV. Interactions between the mechanisms for transport of

showdomycin and various nucleosides in *Escherichia coli*. Agric. Biol. Chem. **35**:1328–1339.

440. **Komatsu, Y.** 1971. Mechanism of action of showdomycin. V. Reduced ability of showdomycin-resistant mutants of *Escherichia coli* K-12 to take up showdomycin and nucleosides. J. Antibiot. **24**:876–883.

441. **Komatsu, Y.** 1973. Adenosine uptake by isolated membrane vesicles from *Escherichia coli* K-12. Biochim. Biophys. Acta. **330**:206–221.

442. **Komatsu, Y.** 1981. A highly showdomycin-resistant mutant of *Escherichia coli* K-12 with altered nucleoside transport characteristics. Agric. Biol. Chem. **45**:609–618.

443. **Komatsu, Y., and K. Tanaka.** 1970. Mechanism of action of showdomycin. II. Effect of showdomycin on the synthesis of deoxyribonucleic acid in *Escherichia coli*. Agric. Biol. Chem. **34**:891–899.

444. **Komatsu, Y., and K. Tanaka.** 1972. A showdomycin-resistant mutant of *Escherichia coli* K-12 with altered nucleoside transport character. Biochim. Biophys. Acta **288**:390–403.

445. **Komatsu, Y., and K. Tanaka.** 1973. Deoxycytidine uptake by isolated membrane vesicles from *Escherichia coli* K 12. Biochim. Biophys. Acta **311**:496–506.

446. **Konings, W. N., and H. R. Kaback.** 1973. Anaerobic transport in *Escherichia coli* membrane vesicles. Proc. Natl. Acad. Sci. USA **70**:3376–3381.

447. **Koppel, J. L., C. J. Porter, and B. F. Crocker.** 1953. The mechanism of the synthesis of enzymes. I. Development of a system for studying this phenomenon. J. Gen. Physiol. **36**:703–722.

448. **Kornberg, H. L., and M. C. Jones-Mortimer.** 1975. *PtsX*: a gene involved in the uptake of glucose and fructose by *Escherichia coli*. FEBS Lett. **51**:1–4.

449. **Kornberg, H. L., and R. E. Reeves.** 1972. Correlation between hexose transport and phosphotransferase activity in *Escherichia coli*. Biochem. J. **126**:1241–1243.

450. **Kornberg, H. L., and R. E. Reeves.** 1972. Inducible phosphoenolpyruvate-dependent hexose phosphotransferase activities in *Escherichia coli*. Biochem. J. **128**:1339–1344.

451. **Kornberg, H. L., and J. Smith.** 1969. Genetic control of hexose phosphate uptake by *Escherichia coli*. Nature (London) **224**:1261–1262.

452. **Koser, S. A.** 1923. Utilization of the salts of organic acids by colon-aerogenes group. J. Bacteriol. **8**:493–520.

453. **Kosiba, B. E., and R. Schleif.** 1982. Arabinose-inducible promoter from *Escherichia coli*. Its cloning from chromosomal DNA, identification as the *araFG* promoter and sequence. J. Mol. Biol. **156**:53–66.

454. **Krenitzky, T. A.** 1976. Uridine phosphorylase from *Escherichia coli*. Kinetic properties and mechanism. Biochim. Biophys. Acta **429**:352–358.

455. **Krieger-Brauer, H. J., and V. Braun.** 1980. Functions related to the receptor protein specified by the *tsx* gene of *Escherichia coli*. Arch. Microbiol. **124**:233–242.

456. **Kuby, S. A., and H. A. Lardy.** 1953. Purification and kinetics of β-galactosidase from *Escherichia coli*, strain K-12. J. Am. Chem. Soc. **75**:890–896.

457. **Kumar, S.** 1976. Properties of adenyl cyclase and receptor cyclic adenosine 3′, 5′-monophosphate protein-deficient mutants of *Escherichia coli*. J. Bacteriol. **125**:545–555.

458. **Kundig, W., S. Ghosh, and S. Roseman.** 1964. Phosphate bound to histidine in a protein as an intermediate in a novel phosphotransferase system. Proc. Natl. Acad. Sci. USA **52**:1067–1074.

459. **Kung, H.-F., and U. Henning.** 1972. Limiting availability of binding sites for dehydrogenases on the cell membrane of *Escherichia coli*. Proc. Natl. Acad. Sci. USA **69**:925–929.

460. **Kunz, L. J., and W. H. Ewing.** 1965. Laboratory infection with a lactose-fermenting strain of *Salmonella typhimurium*. J. Bacteriol. **89**:1629.

461. **Kurahashi, K.** 1957. Enzyme formation in galactose-negative mutants of *Escherichia coli*. Science **125**:114–116.

462. **Kurahashi, K., and A. Sugimura.** 1960. Purification and properties of galactose 1-phosphate uridyl transferase from *Escherichia coli*. J. Biol. Chem. **235**:940–946.

463. **Kurahashi, K., and A. J. Wahba.** 1958. Interference with growth of certain *Escherichia coli* mutants by galactose. Biochim. Biophys. Acta **30**:298–302.

464. **Kuritzkes, D. R., X.-Y. Zhang, and E. C. C. Lin.** 1984. Use of phi(*glp-lac*) in studies of respiratory regulation of the *Escherichia coli* anaerobic *sn*-glycerol-3-phosphate dehydrogenase genes (*glpAB*). J. Bacteriol. **157**:591–598.

465. **Kusch, M., and T. H. Wilson.** 1973. Defective lactose utilization by a mutant of *Escherichia coli* energy-uncoupled for lactose transport. Biochim. Biophys. Acta **311**:109–122.

466. **Lagarde, A. E.** 1977. Evidence for an electrogenic 3-deoxy-2-keto-D-gluconate-proton co-transport driven by the protonmotive force in *Escherichia coli* K-12. Biochem. J. **168**:211–221.

467. **Lagarde, A. E., and B. A. Haddock.** 1977. Proton uptake linked to the 3-deoxy-2-oxo-D-gluconate-transport system of *Escherichia coli*. Biochem. J. **162**:183–187.

468. **Lagarde, A., J. Pouyssegur, and F. Stoeber.** 1972. Accumulation du D-glucuronate par le systeme de transport du 2-ceto-3-desoxy-D-gluconate chez *Escherichia coli* K 12. C.R. Acad. Sci. Ser. D **275**:1831–1834.

469. **Lagarde, A., J. Pouyssegur, and F. Stoeber.** 1973. A transport system for 2-keto-3-deoxy-D-gluconate uptake in *Escherichia coli* K 12. Biochemical and physiological studies in whole cells. Eur. J. Biochem. **36**:328–341.

470. **Lagarde, A., and F. Stoeber.** 1974. Transport of 2-keto-3-deoxy-D-gluconate in isolated membrane vesicles of *Escherichia coli* K 12. Eur. J. Biochem. **43**:197–208.

471. **Lagarde, A. E., and F. R. Stoeber.** 1975. The energy-coupling controlled efflux of 2-keto-3-deoxy-D-gluconate in *Escherichia coli* K-12. Eur. J. Biochem. **55**:343–354.

472. **Lagarde, A. E., and F. R. Stoeber.** 1977. *Escherichia coli* K-12 structural *kdgT* mutants exhibiting thermosensitive 2-keto-3-deoxy-D-gluconate uptake. J. Bacteriol. **129**:606–615.

473. **Lam, V. M. S., K. R. Daruwalla, P. J. F. Henderson, and M. C. Jones-Mortimer.** 1980. Proton-linked D-xylose transport in *Escherichia coli*. J. Bacteriol. **143**:396–402.

473a.**Lara, F. J. S., and J. L. Stokes.** 1952. Oxidation of citrate by *Escherichia coli*. J. Bacteriol. **63**:415–420.

474. **Larson, T. J., M. Ehrmann, and W. Boos.** 1983. Periplasmic glycero-phosphodiester phosphodiesterase of *Escherichia coli*, a new enzyme of the *glp* regulon. J. Biol. Chem. **258**:5428–5432.

475. **Larson, T. J., G. Schumacher, and W. Boos.** 1982. Identification of the *glpT*-encoded *sn*-glycerol-3-phosphate permease of *Escherichia coli*, an oligomeric integral membrane protein. J. Bacteriol. **152**:1008–1021.

476. **Lawlis, V. B., M. S. Dennis, E. Y. Chen, D. H. Smith, and D. J. Henner.** 1984. Cloning and sequencing of the xylose isomerase and xylulose kinase genes of *Escherichia coli*. Appl. Environ. Microbiol. **47**:15–21.

477. **LeBlanc, D. J., and R. P. Mortlock.** 1971. Metabolism of D-arabinose: origin of a D-ribulokinase activity in *Escherichia coli*. J. Bacteriol. **106**:82–89.

478. **LeBlanc, D. J., and R. P. Mortlock.** 1971. Metabolism of D-arabinose: a new pathway in *Escherichia coli*. J. Bacteriol. **106**:90–96.

479. **LeBlanc, D. J., and R. P. Mortlock.** 1972. The metabolism of D-arabinose: alternate kinases for the phosphorylation of D-ribulose in *Escherichia coli* and *Aerobacter aerogenes*. Arch. Biochem. Biophys. **150**:774–781.

480. **Lederberg, E.** 1955. Pleiotropy for maltose fermentation and phage resistance in *Escherichia coli* K-12. Genetics **40**:580–581.

481. **Lederberg, E.** 1960. Genetic and functional aspects of galactose metabolism in *Escherichia coli* K-12. Symp. Soc. Gen. Microbiol. **10**:115–136.

482. **Lederberg, E. M., and J. Lederberg.** 1953. Genetic studies of lysogenicity in *Escherichia coli*. Genetics **38**:51–64.

483. **Lederberg, J.** 1950. The beta-D-galactosides of *Escherichia coli*, strain K-12. J. Bacteriol. **60**:381–392.

484. **Lee, C. A., G. R. Jacobson, and M. H. Saier, Jr.** 1981. Plasmid-directed synthesis of enzymes required for D-mannitol transport and utilization in *Escherichia coli*. Proc. Natl. Acad. Sci. USA **78**:7336–7340.

485. **Lee, C. A., and M. H. Saier, Jr.** 1983. Mannitol-specific enzyme II of the bacterial phosphotransferase system. III. The nucleotide sequence of the permease gene. J. Biol. Chem. **258**:10761–10767.

486. **Lee, C. A., and M. H. Saier, Jr.** 1983. Use of cloned *mtl* genes of *Escherichia coli* to introduce *mtl* deletion mutations into the chromosome. J. Bacteriol. **153**:685–692.

487. **Lee, J.-H., S. Al-Zarban, and G. Wilcox.** 1981. Genetic characterization of the *araE* gene in *Salmonella typhimurium* LT2. J. Bacteriol. **146**:298–304.

488. **Lee, J.-H., L. Heffernan, and G. Wilcox.** 1980. Isolation of *ara-lac* gene fusions in *Salmonella typhimurium* LT-2 by using transducing bacteriophage Mu *d*(Ap^r *lac*). J. Bacteriol. **143**:1325–1331.

489. **Lee, J.-H., J. Nishitani, and G. Wilcox.** 1984. Genetic characterization of *Salmonella typhimurium* LT2 *ara* mutations. J. Bacteriol.**158**:344–346.

490. **Lee, J.-H., R. J. Russo, L. Heffernan, and G. Wilcox.** 1982. Regulation of L-arabinose transport in *Salmonella typhimurium* LT2. Mol. Gen. Genet. **185**:136–141.

491. **Lee, N., and I. Bendet.** 1967. Crystalline L-ribulokinase from

Escherichia coli. J. Biol. Chem. **242**:2043–2050.

492. **Lee, N., and E. Englesberg.** 1962. Dual effects of structural genes in *Escherichia coli.* Proc. Natl. Acad. Sci. USA **48**:335–348.

493. **Lee, N. L., W. O. Gielow, and R. G. Wallace.** 1981. Mechanism of *araC* autoregulation and the domains of two overlapping promoters, P_C and P_BAD, in the L-arabinose regulatory region of *Escherichia coli.* Proc. Natl. Acad. Sci. USA **78**:752–756.

494. **Leer, J. C., and K. Hammer-Jespersen.** 1975. Multiple forms of phosphodeoxyribomutase from *Escherichia coli.* Physical and chemical characterization. Biochemistry **14**:599–607.

495. **Leer, J. C., K. Hammer-Jespersen, and M. Schwartz.** 1977. Uridine phosphorylase from *Escherichia coli.* Physical and chemical characterization. Eur. J. Biochem.**75**:217–224.

496. **Leifer, Z., R. Engel, and B. E. Tropp.** 1977. Transport of 3,4-dihydroxybutyl-1-phosphate, an analogue of sn-glycerol 3-phosphate. J. Bacteriol. **130**:968–971.

497. **Le Minor, L., C. Coynault, and G. Pessoa.** 1974. Determinisme plasmidique du caractere atypique "lactose positif" de souches de *Salmonella typhimurium* et de *Salmonella oranienbourg* isolees du Bresil lors d'epidemies de 1971 a 1973. Ann. Microbiol. (Paris) **125A**:261–285.

498. **Le Minor, L., C. Coynault, R. Rohde, B. Rowe, and S. Aleksic.** 1973. Localisation plasmidique du determinant genetique du caractere atypique "saccharose⁺" des *Salmonella.* Ann. Inst. Pasteur Microbiol. **124B**:295–306.

499. **Lengeler, J.** 1975. Mutations affecting transport of the hexitols D-mannitol, D-glucitol, and galactitol in *Escherichia coli* K-12: isolation and mapping. J. Bacteriol. **124**:26–38.

500. **Lengeler, J.** 1975. Nature and properties of hexitol transport systems in *Escherichia coli.* J. Bacteriol. **124**:39–47.

501. **Lengeler, J.** 1977. Analysis of mutations affecting the dissimilation of galactitol (dulcitol) in *Escherichia coli* K12. Mol. Gen. Genet. **152**:83–91.

502. **Lengeler, J.** 1980. Characterization of mutants of *Escherichia coli* K-12, selected by resistance to streptozotocin. Mol. Gen. Genet. **179**:49–54.

503. **Lengeler, J., A.-M. Auburger, R. Mayer, and A. Pecher.** 1981. The phosphoenolpyruvate-dependent carbohydrate: phosphotransferase system enzymes II as chemoreceptors in chemotaxis of *Escherichia coli* K12. Mol. Gen. Genet. **183**:163–170.

504. **Lengeler, J., K. O. Hermann, H. J. Unsöld, and W. Boos.** 1971. The regulation of the beta-methylgalactoside transport system and of the galactose binding protein of *Escherichia coli* K-12. Eur. J. Biochem. **19**:457–470.

505. **Lengeler, J., and E. C. C. Lin.** 1972. Reversal of the mannitol-sorbitol diauxie in *Escherichia coli.* J. Bacteriol. **112**:840–848.

506. **Lengeler, J., R. J. Mayer, and K. Schmid.** 1982. Phosphoenolpyruvate-dependent phosphotransferase system enzyme III and plasmid-encoded sucrose transport in *Escherichia coli* K-12. J. Bacteriol. **151**:468–471.

507. **Lengeler, J., and H. Steinberger.** 1978. Analysis of regulatory mechanisms controlling the synthesis of the hexitol transport systems in *Escherichia coli* K-12. Mol. Gen. Genet. **164**:163–169.

508. **Lengeler, J., and H. Steinberger.** 1978. Analysis of regulatory mechanisms controlling the activity of the hexitol transport systems in *Escherichia coli* K-12. Mol. Gen. Genet. **167**:75–82.

509. **Leonard, J., and M. H. Saier, Jr.** 1981. Genetic dissection of catalytic activities of the *Salmonella typhimurium* mannitol enzyme II. J. Bacteriol. **145**:1106–1109.

510. **Leonard, J. E., and M. H. Saier, Jr.** 1983. Mannitol-specific enzyme II of the bacterial phosphotransferase system. II. Reconstitution of vectorial transphosphorylation in phospholipid vesicles. J. Biol. Chem. **258**:10757–10760.

511. **Lerner, S. A., T. T. Wu, and E. C. C. Lin.** 1964. Evolution of catabolic pathway in bacteria. Science **146**:1313–1315.

512. **Lester, G.** 1952. The beta-galactosidase of lactose mutants of *Escherichia coli* K-12. Arch. Biochem. Biophys. **40**:390–401.88.

513. **Lester, G., and D. M. Bonner.** 1952. The occurrence of beta-galactosidase in *Escherichia coli.* J. Bacteriol. **63**:759–769.

514. **Leung, K.-K., and D. W. Visser.** 1977. Uridine and cytidine transport in *Escherichia coli* B and transport-deficient mutants. J. Biol. Chem. **252**:2492–2497.

515. **Levinthal, M.** 1971. Biochemical studies of melibiose metabolism in wild-type and *mel* mutant strains of *Salmonella typhimurium.* J. Bacteriol. **105**:1047–1052.

516. **Lieb, M., J. J. Weigle, and E. Kellenberger.** 1955. A study of hybrids between two strains of *Escherichia coli.* J. Bacteriol. **69**:468–471.

517. **Lin, E. C. C.** 1970. The genetics of bacterial transport systems. Annu. Rev. Genet. **4**:225–262.

517a. **Lin, E. C. C.** 1976. Glycerol dissimilation and its regulation in bacteria. Annu. Rev. Microbiol. **30**:535–578.

518. **Lin, E. C. C., J. P. Koch, T. M. Chused, and S. E. Jorgensen.** 1962. Utilization of L-alpha-glycerophosphate by *Escherichia coli* without hydrolysis. Proc. Natl. Acad. Sci. USA **48**:2145–2150.

519. **Lin, E. C. C., and T. T. Wu.** 1984. Functional divergence of the L-fucose system in mutants of *Escherichia coli*, p. 135–164. *In* R. P. Mortlock (ed.), Microorganisms as model systems for studying evolution. Plenum Publishing Corp., New York.

520. **Lin, J. J.-C., and H. C. P. Wu.** 1976. Biosynthesis and assembly of envelope lipoprotein in a glycerol-requiring mutant of *Salmonella typhimurium.* J. Bacteriol. **125**:892–904.

521. **Lin, R.-J., and C. W. Hill.** 1983. Mapping the *xyl, mtl*, and *lct* loci in *Escherichia coli* K-12. J. Bacteriol. **156**:914–916.

522. **Link, C. D., and A. M. Reiner.** 1982. Inverted repeats surround the ribitol-arabitol genes of *E. coli* C. Nature (London) **298**:94–96.

523. **Link, C. D., and A. M. Reiner.** 1983. Genotypic exclusion: a novel relationship between ribitol-arabitol and galactitol genes of *E. coli.* Mol. Gen. Genet. **189**:337–339.

524. **Lo, T. C. Y.** 1977. The molecular mechanism of dicarboxylic acid transport in *Escherichia coli* K12. J. Supramol. Struct. **7**:463–480.

525. **Lo, T. C. Y., and M. A. Bewick.** 1978. The molecular mechanisms of dicarboxylic acid transport in *Escherichia coli* K12. J. Biol. Chem. **253**:7826–7831.

526. **Lo, T. C. Y., and M. A. Bewick.** 1981. Use of a nonpenetrating substrate analogue to study the molecular mechanism of the outer membrane dicarboxylate transport system in *Escherichia coli* K12. J. Biol. Chem. **256**:5511–5517.

527. **Lo, T. C. Y., M. K. Rayman, and B. D. Sanwal.** 1972. Transport of succinate in *Escherichia coli.* I. Biochemical and genetic studies of transport in whole cells. J. Biol. Chem. **247**:6323–6331.

528. **Lo, T. C. Y., M. K. Rayman, and B. D. Sanwal.** 1974. Transport of succinate in *Escherichia coli.* III. Biochemical and genetic studies of the mechanism of transport in membrane vesicles. Can. J. Biochem. **52**:854–866.

529. **Lo, T. C. Y., and B. D. Sanwal.** 1975. Isolation of the soluble substrate recognition component of the dicarboxylate transport system of *Escherichia coli.* J. Biol. Chem. **250**:600–1602.

530. **Lo, T. C. Y., and B. D. Sanwal.** 1975. Genetic analysis of mutants of *Escherichia coli* defective in dicarboxylate transport. Mol. Gen. Genet. **140**:303–307.

531. **Lo, T. C. Y., and B. D. Sanwal.** 1975. Membrane bound substrate recognition compounds of the dicarboxylic transport system in *Escherichia coli.* Biochem. Biophys. Res. Commun. **63**:278–285.

532. **Lomax, M. S., and G. R. Greenberg.** 1968. Characteristics of the *deo* operon: role in thymine utilization and sensitivity to deoxyribonucleotides. J. Bacteriol. **96**:501–514.

533. **Long, C.** 1955. Studies involving enzyme phosphorylation. 4. Conversion of D-ribose into D-ribose 5-phosphate by extract of *Escherichia coli.* Biochem. J. **59**:322–329.

534. **Lopilato, J., T. Tsuchiya, and T. H. Wilson.** 1978. Role of Na⁺ and Li⁺ in thiomethylgalactoside transport by the melibiose transport system of *Escherichia coli.* J. Bacteriol. **134**:147–156.

535. **Lopilato, J. E., J. L. Garwin, S. D. Emr, T. J. Silhavy, and J. R. Beckwith.** 1984. D-ribose metabolism in *Escherichia coli* K-12: genetics, regulation, and transport. J. Bacteriol. **158**:665–673.

536. **Lowe, D. A., and J. M. Turner.** 1970. Microbial metabolism of amino ketones: D-1-aminopropan-2-ol and aminoacetone metabolism in *Escherichia coli.* J. Gen. Microbiol. **63**:49–61.

537. **Luckey, M., and H. Nikaido.** 1980. Specificity of diffusion channels produced by lambda phage receptor protein of *Escherichia coli.* Proc. Natl. Acad. Sci. USA **77**:167–171.

538. **Ludtke, D., T. J. Larson, C. Beck, and W. Boos.** 1982. Only one gene is required for the *glpT*-dependent transport of sn-glycerol-3-phosphate in *Escherichia coli.* Mol. Gen. Genet. **186**:540–547.

539. **Macpherson, A. J. S., M. C. Jones-Mortimer, and P. J. F. Henderson.** 1981. Identification of the *araE* transport protein of *Escherichia coli.* Biochem. J. **196**:269–283.

540. **Macpherson, A. J. S., M. C. Jones-Mortimer, P. Horne, and P. J. F. Henderson.** 1983. Identification of the GalP galactose transport protein of *Escherichia coli.* J. Biol. Chem. **258**:4390–4396.

541. **Mahoney, W. C., R. W. Hogg, and M. A. Hermodson.** 1981. The amino acid sequence of the D-galactose-binding protein from *Escherichia coli* B/r. J. Biol. Chem. **256**:4350–4356.

542. **Majumdar, A., and S. Adhya.** 1984. Demonstration of two operator elements in *gal*: in vitro repressor binding studies. Proc. Natl. Acad. Sci. USA **81**:6100–6104.

543. **Maleszka, R., P. Y. Wang, and H. Schneider.** 1982. A colE1 hybrid plasmid containing *Escherichia coli* genes complement-

ing D-xylose negative mutants of *Escherichia coli* and *Salmonella typhimurium*. Can. J. Biochem. **60**:144–151.

544. **Maloney, P. C., E. R. Kashket, and T. H. Wilson.** 1975. Methods for studying transport in bacteria, p. 1–49. *In* E. D. Korn (ed.), Methods in membrane biology, vol. 5. Plenum Publishing Corp., New York.

545. **Mandrand-Berthelot, M.-A., and A. E. Lagarde.** 1982. Altered transport properties in *Escherichia coli* mutants selected for pH-conditional growth on 3-deoxy-2-oxo-D-gluconate. J. Biol. Chem. **257**:8806–8816.

546. **Mandrand-Berthelot, M.-A., P. Ritzenthaler, and M. Mata-Gilsinger.** 1984. Construction and expression of hybrid plasmids containing the structural gene of the *Escherichia coli* K-12 3-deoxy-2-oxo-D-gluconate transport system. J. Bacteriol. **160**:600–606.

547. **Manning, P. A., and P. Reeves.** 1978. Outer membrane proteins of *Escherichia coli* K-12: isolation of a common receptor protein for bacteriophage T6 and colicin K. Mol. Gen. Genet. **158**:279–286.

548. **Mans, R. J., and A. L. Koch.** 1960. Metabolism of adenosine and deoxyadenosine by growing cultures of *Escherichia coli*. J. Biol. Chem. **235**:450–456.

549. **Manson, L. A., and J. O. Lampon.** 1951. The metabolism of desoxyribose nucleosides in *Escherichia coli*. J. Biol. Chem. **193**:539–547.

550. **Mao, B., and J. A. McCammon.** 1983. Theoretical study of hinge bending in L-arabinose-binding protein. Internal energy and free energy changes. J. Biol. Chem. **258**:12543–12547.

551. **Mao, B., and J. A. McCammon.** 1984. Structural study of hinge bending in L-arabinose-binding protein. J. Biol. Chem. **259**:4964–4970.

552. **Mao, B., M. R. Pear, J. A. McCammon, and F. A. Quiocho.** 1982. Hinge-bending in L-arabinose-binding protein. J. Biol. Chem. **257**:1131–1133.

553. **Marechal, L. R.** 1984. Transport and metabolism of trehalose in *Escherichia coli* and *Salmonella typhimurium*. Arch. Microbiol. **137**:70–73.

554. **Markovitz, A., R. J. Sydiskis, and M. M. Lieberman.** 1967. Genetic and biochemical studies on mannose-negative mutants that are deficient in phosphomannose isomerase in *Escherichia coli* K-12. J. Bacteriol. **94**:1492–1496.

555. **Markovitz, A. M., M. Lieberman, and N. Rosenbaum.** 1967. Derepression of phosphomannose isomerase by regulator gene mutations involved in capsular polysaccharide synthesis in *Escherichia coli* K-12. J. Bacteriol. **94**:1497–1501.

556. **Mata, M., M. Delstanche, and J. Robert-Baudouy.** 1978. Isolation of specialized transducing bacteriophages carrying the structural genes of the hexuronate system in *Escherichia coli* K-12: *exu* region. J. Bacteriol. **133**:549–557.

557. **Mata-Gilsinger, M., and P. Ritzenthaler.** 1983. Physical mapping of the *exuT* and *uxaC* operators by use of *exu* plasmids and generation of deletion mutants in vitro. J. Bacteriol. **155**:973–982.

558. **Mata-Gilsinger, M., P. Ritzenthaler, and J. Robert-Baudouy.** 1978. Identification de plasmides transportant la region *uxu* du systeme des hexuronates chez *Escherichia coli* K-12 a partir de la collection de Clarke et Carbon. C.R. Acad. Sci. Ser. D **286**:237–239.

559. **Matin, A., and W. N. Konings.** 1973. Transport of lactate and succinate by membrane vesicles of *Escherichia coli*, *Bacillus subtilis* and a *Pseudomonas* species. Eur. J. Biochem. **34**:58–67.

560. **McEntee, K.** 1977. Genetic analysis of the *Escherichia coli* K-12 *srl* region. J. Bacteriol. **132**:904–911.

561. **McGowan, E. B., T. J. Silhavy, and W. Boos.** 1974. Involvement of a tryptophan residue in the binding site of *Escherichia coli* galactose-binding protein. Biochemistry **13**:993–999.

562. **McKeown, M., M. Kahn, and P. Hanawalt.** 1976. Thymidine uptake and utilization in *Escherichia coli*: a new gene controlling nucleoside transport. J. Bacteriol. **126**:814–822.

563. **Miki, K., and E. C. C. Lin.** 1973. Enzyme complex which couples glycerol-3-phosphate dehdyrogenation to fumarate reduction in *Escherichia coli*. J. Bacteriol. **114**:767–771.

564. **Miki, K., and E. C. C. Lin.** 1975. Anaerobic energy-yielding reaction associated with transhydrogenation from glycerol 3-phosphate to fumarate by an *Escherichia coli* system. J. Bacteriol. **124**:1282–1287.

565. **Miki, K., and E. C. C. Lin.** 1975. Electron transport chain from glycerol 3-phosphate to nitrate in *Escherichia coli*. J. Bacteriol. **124**:1288–1294.

566. **Miki, K., and E. C. C. Lin.** 1980. Use of *Escherichia coli* operon-fusion strains for the study of glycerol 3-phosphate transport activity. J. Bacteriol. **143**:1436–1443.

567. **Miki, K., T. J. Silhavy, and K. J. Andrews.** 1979. Resolution of *glpA* and *glpT* loci into separate operons in *Escherichia coli* K-12 strains. J. Bacteriol. **138**:268–269.

568. **Miki, K., and T. H. Wilson.** 1978. Proton translocation associated with anaerobic transhydrogenation from glycerol 3-phosphate to fumarate in *Escherichia coli*. Biochem. Biophys. Res. Commun. **83**:1570–1575.

569. **Miki, T., S. Hiraga, T. Nagata, and T. Yura.** 1978. Bacteriophage lambda carrying the *Escherichia coli* chromosomal region of the replication origin. Proc. Natl. Acad. Sci. USA **75**:5099–5103.

570. **Miller, D. M., III, J. S. Olson, and F. A. Quiocho.** 1980. The mechanism of sugar binding to the periplasmic receptor for galactose chemotaxis and transport in *Escherichia coli*. J. Biol. Chem. **255**:2465–2471.

571. **Monod, J.** 1947. The phenomenon of enzymatic adaptation. Growth **11**:223–289.

572. **Monod, J., and A. Audureau.** 1946. Mutation et adaptation enzymatique chez *Escherichia coli*-mutabile. Ann. Inst. Pasteur (Paris) **72**:868–878.

573. **Monod, J., G. Cohen-Bazire, and M. Cohn.** 1951. Sur la biosynthese de la beta-galactosidase (lactase) chez *Escherichia coli*. La specificite de l'induction. Biochim. Biophys. Acta **7**:585–599.

574. **Monod, J., and A.-M. Torriani.** 1948. Synthese d'un polysaccharide du type amidon aux depons du maltose, en presence d'un extrait enzymatique d'origine bacterionne. C.R. Acad. Sci. Ser. D **227**:240–242.

575. **Monod, J., and A. M. Torriani.** 1950. De l'amylomaltase d'*Escherichia coli*. Ann. Inst. Pasteur (Paris) **78**:65–77.

576. **Morgenroth, A., and J. P. Duguid.** 1968. Demonstration of different mutational sites controlling rhamnose fermentation in FIRN and non-FIRN rha-strains of *Salmonella typhimurium*: an essay in bacterial archaeology. Genet. Res. **11**:151–169.

577. **Morse, M. L.** 1962. Preliminary genetic map of seventeen galactose mutations in *Escherichia coli* K-12. Proc. Natl. Acad. Sci. USA **48**:1314–1318.

578. **Morse, M. L., E. M. Lederberg, and J. Lederberg.** 1956. Transduction in *Escherichia coli* K-12. Genetics **41**:142–156.

579. **Morse, M. L., E. M. Lederberg, and J. Lederberg.** 1956. Transductional heterogenotes in *Escherichia coli*. Genetics **41**:758–779.

580. **Mortlock, R. P., D. D. Fossitt, and W. A. Wood.** 1965. A basis for utilization of unnatural pentoses and pentitols by *Aerobacter aerogenes*. Proc. Natl. Acad. Sci. USA **54**:572–579.

581. **Mortlock, R. P., and D. C. Old.** 1979. Utilization of D-xylose by wild-type strains of *Salmonella typhimurium*. J. Bacteriol. **137**:173–178.

582. **Mowbray, S. L., and G. A. Petsko.** 1982. Preliminary X-ray data for the ribose binding protein from *Salmonella typhimurium*. J. Mol. Biol. **160**:545–547.

583. **Mowbray, S. L., and G. A. Petsko.** 1983. The X-ray structure of the periplasmic galactose binding protein from *Salmonella typhimurium* at 3.0-Å resolution. J. Biol. Chem. **258**:7991–7997.

584. **Müller, N., H.-G. Heine, and W. Boos.** 1982. Cloning of *mglB*: the structural gene for the galactose-binding protein of *Salmonella typhimurium* and *Escherichia coli*. Mol. Gen. Genet. **185**:473–480.

585. **Müller, N., H.-G. Heine, and W. Boos.** 1985. Characterization of the *Salmonella typhimurium mgl* operon and its gene products. J. Bacteriol. **163**:37–45.

586. **Müller-Hill, B., H. V. Rickenborg, and K. Wallenfels.** 1964. Specificity of the induction of the enzymes of the *lac* operon in *Escherichia coli*. J. Mol. Biol. **10**:303–318.

587. **Munch-Petersen, A.** 1968. On the catabolism of deoxyribonucleosides in cells and cell extracts of *Escherichia coli*. Eur. J. Biochem. **6**:432–442.

588. **Munch-Petersen, A.** 1968. Thymineless mutants of *E. coli* with deficiencies in deoxyribomutase and deoxyriboaldolase. Biochim. Biophys. Acta **161**:279–282.

589. **Munch-Petersen, A.** 1970. Deoxyribonucleoside catabolism and thymine incorporation in mutants of *Escherichia coli* lacking deoxyriboaldolase. Eur. J. Biochem. **15**:191–202.

590. **Munch-Petersen, A., and B. Mygind.** 1976. Nucleoside transport systems in *Escherichia coli* K-12: specificity and regulation. J. Cell. Physiol. **89**:551–559.

591. **Munch-Petersen, A., and B. Mygind.** 1983. Transport of nucleic acid precursors, p. 259–309. *In* A. Munch-Petersen (ed.), Metabolism of nucleotides, nucleosides and nucleobases in microorganisms. Academic Press, Inc. (London), Ltd., London.

592. **Munch-Petersen, A., B. Mygind, A. Nicolaisen, and N. J. Pihl.** 1979. Nucleoside transport in cells and membrane vesicles from *Escherichia coli* K12. J. Biol. Chem. **254**:3730–3737.

593. **Munch-Petersen, A., P. Nygaard, K. Hammer-Jespersen, and N. Fiil.** 1972. Mutants constitutive for nucleoside catabolizing enzymes in *Escherichia coli* K 12. Isolation, characterization and mapping. Eur. J. Biochem. **27**:208–215.

594. **Munch-Petersen, A., and N. J. Pihl.** 1980. Stimulatory effect of low ATP pools on transport of purine nucleosides in cells of *Escherichia coli.* Proc. Natl. Acad. Sci. USA **77**:2519–2523.

595. **Murakawa, S., K. Izaki, and H. Takahashi.** 1971. Succinate transport in isolated membrane preparations from *Escherichia coli.* Agric. Biol. Chem. **35**:1992–1993.

596. **Musso, R., R. Di Lauro, M. Rosenberg, and B. de Crombrugghe.** 1977. Nucleotide sequence of the operator-promoter region of the galactose operon of *Escherichia coli.* Proc. Natl. Acad. Sci. USA **74**:106–110.

597. **Musso, R. E., R. diLauro, S. Adhya, and B. de Crombrugghe.** 1977. Dual control for transcription of the galactose operon by cyclic AMP and its receptor protein at two interspersed promoters. Cell **12**:847–854.

598. **Musso, R. E., and I. Zabin.** 1973. Substrate specificity and kinetic studies on thiogalactoside transacetylase. Biochemistry **12**:553–557.

599. **Mygind, B., and A. Munch-Petersen.** 1975. Transport of pyrimidine nucleosides in cells of *Escherichia coli* K 12. Eur. J. Biochem. **59**:365–372.

600. **Nagelkerke, F., and P. W. Postma.** 1978. 2-Deoxygalactose, a specific substrate of the *Salmonella typhimurium* galactose permease: its use for the isolation of *galP* mutants. J. Bacteriol. **133**:607–613.

601. **Nakae, T.** 1971. Multiple molecular forms of uridine diphosphate glucose pyrophosphorylase from *Salmonella typhimurium.* III. Interconversion between various forms. J. Biol. Chem. **246**:4404–4411.

602. **Nakae, T., and H. Nikaido.** 1971. Multiple molecular forms of uridine diphosphate glucose pyrophosphorylase from *Salmonella typhimurium.* I. Catalytic properties of various forms. J. Biol. Chem. **246**:4386–4396.

603. **Nakae, T., and H. Nikaido.** 1971. Multiple molecular forms of uridine diphosphate glucose pyrophosphorylase from *Salmonella typhimurium.* II. Genetic determination of multiple forms. J. Biol. Chem. **246**:4397–4403.

604. **Nelson, S. O., and P. W. Postma.** 1984. Interactions *in vivo* between III^Glc of the phosphoenolpyruvate:sugar phosphotransferase system and the glycerol and maltose uptake systems of *Salmonella typhimurium.* Eur. J. Biochem. **139**:29–34.

605. **Nemoz, G., J. Robert-Baudouy, and F. R. Stoeber.** 1976. Physiological and genetic regulation of the aldohexuronate transport system in *Escherichia coli.* J. Bacteriol. **127**:706–718.

606. **Neuhard, J., and J. L. Ingraham.** 1968. Mutants of *Salmonella typhimurium* requiring cytidine for growth. J. Bacteriol. **95**:2431–2433.

607. **Newcomer, M. E., G. L. Gilliland, and F. A. Quiocho.** 1981. L-arabinose-binding protein-sugar complex at 2.4 Å resolution. J. Biol. Chem. **256**:13213–13217.

608. **Newcomer, M. E., D. M. Miller III, and F. A. Quiocho.** 1979. Location of the sugar-binding site of L-arabinose-binding protein. Sugar derivative syntheses, sugar binding specificity, and different Fourier analyses. J. Biol. Chem. **254**:7529–7533.

609. **Niiya, S., K. Yamasaki, T. H. Wilson, and T. Tsuchiya.** 1982. Altered cation coupling to melibiose transport in mutants of *Escherichia coli.* J. Biol. Chem. **257**:8902–8906.

610. **Nikaido, H.** 1961. Galactose-sensitive mutants of *Salmonella.* I. Metabolism of galactose. Biochim. Biophys. Acta **48**:460–469.

611. **Nikaido, H., and T. Fukasawa.** 1961. The effect of mutation in a structural gene on the inducibility of the enzymes controlled by other genes of the same operon. Biochem. Biophys. Res. Commun. **4**:338–342.

612. **Novel, G., M. L. Didier-Fichet, and F. Stoeber.** 1974. Inducibility of beta-glucuronidase in wild-type and hexuronate-negative mutants of *Escherichia coli* K-12. J. Bacteriol. **120**:89–95.

613. **Novel, G., and M. Novel.** 1973. Mutants d'*Escherichia coli* K-12 affectes lour leur croissance sur methyl-beta-D-glucuronide: localisation du gene de structure de la beta-D-glucuronidase (*uidA*). Mol. Gen. Genet. **120**:319–335.

614. **Novel, G., M. Novel, M.-L. Didier-Fichet, and F. Stoeber.** 1970. Etude genetique de mutants du systeme de degradation des hexuronides chez *Escherichia coli* K 12. C.R. Acad. Sci. Ser. D **271**:457–460.

615. **Novel, G., and F. Stoeber.** 1973. Individualite de la D-glucuronate-cetol isomerase d'*Escherichia coli* K12. Biochimie **55**:1057–1070.

616. **Novel, M., and G. Novel.** 1971. Mutations *gur*: localisation precise du locus *gurA* gene de structure de la beta-glucuroni-

617. **Novel, M., and G. Novel.** 1974. Mutants d'*Escherichia coli* K12 capables de croitre sur methyl-beta-D-galacturonide: mutants simples constitutifs pour la synthese de la beta-glucuronidase et mutants doubles dereprimes aussi pour la synthese de deux enzymes d'utilisation du glucuronate. C.R. Acad. Sci. Ser. D **279**:695–698.

618. **Novel, M., and G. Novel.** 1976. Regulation of beta-glucuronidase synthesis in *Escherichia coli* K-12: constitutive mutants specifically derepressed for *uidA* expression. J. Bacteriol. **127**:406–417.

619. **Novel, M., and G. Novel.** 1976. Regulation of beta-glucuronidase synthesis in *Escherichia coli* K-12: pleiotropic constitutive mutations affecting *uxu* and *uidA* expression. J. Bacteriol. **127**:418–432.

620. **Novotny, C. P., and E. Englesberg.** 1966. The L-arabinose permease system in *Escherichia coli* B/r. Biochim. Biophys. Acta **117**:217–230.

621. **Novotny, M. J., W. L. Frederickson, E. B. Waygood, and M. H. Saier, Jr.** 1985. Allosteric deregulation of glycerol kinase by enzyme III Glc of the phosphotransferase system in *Escherichia coli* and *Salmonella typhimurium.* J. Bacteriol. **162**:810–816.

622. **Novotny, M. J., J. Reizer, F. Esch, and M. H. Saier, Jr.** 1984. Purification and properties of D-mannitol-1-phosphate dehydrogenase and D-glucitol-6-phosphate dehydrogenase from *Escherichia coli.* J. Bacteriol. **159**:986–990.

623. **Nygaard, P.** 1973. Nucleoside-catabolizing enzymes in *Salmonella typhimurium.* Induction by ribonucleosides. Eur. J. Biochem. **36**:267–272.

624. **Nygaard, P.** 1978. Adenosine deaminase from *Escherichia coli.* Methods Enzymol. **51**:508–512.

625. **Ogden, S., D. Haggerty, C. M. Stoner, D. Koldrubetz, and R. Schleif.** 1980. The *Escherichia coli* L-arabinose operon: binding sites of the regulatory proteins and a mechanism of positive and negative regulation. Proc. Natl. Acad. Sci. USA **77**:3346–3350.

626. **Okada, T.** 1966. Mutational site of the gene controlling quantitative thymine requirement in *Escherichia coli* K 12. Genetics **54**:1329–1336.

627. **Old, D. C.** 1972. Temperature-dependent utilization of meso-inositol: a useful biotyping marker in the genealogy of *Salmonella typhimurium.* J. Bacteriol. **112**:779–783.

628. **Old, D. C., P. F. H. Dawes, and R. M. Barker.** 1980. Transduction of inositol-fermenting ability demonstrating phylogenic relationships among strains of *Salmonella typhimurium.* Genet. Res. **35**:215–224.

629. **Old, D. C., and R. P. Mortlock.** 1977. The metabolism of D-arabinose by *Salmonella typhimurium.* J. Gen. Microbiol. **101**:341–344.

630. **Old, D. C., and R. P. Mortlock.** 1979. Phylogenetic relationships between different D-xylose biogroups in wild-type *Salmonella typhimurium* strains and a suggested evolutionary pathway. J. Appl. Bacteriol. **47**:167–174.

631. **Oosawa, K., and Y. Imae.** 1983. Glycerol and ethylene glycol: members of a new class of repellents of *Escherichia coli* chemotaxis. J. Bacteriol. **154**:104–112.

632. **Oosawa, K., and Y. Imae.** 1984. Demethylation of methyl-accepting chemotaxis proteins in *Escherichia coli* induced by the repellents glycerol and ethylene glycol. J. Bacteriol. **157**:576–581.

633. **Ordal, G. W., and J. Adler.** 1974. Isolation and complementation of mutants in galactose taxis and transport. J. Bacteriol. **117**:509–516.

634. **Ordal, G. W., and J. Adler.** 1974. Properties of mutants in galactose taxis and transport. J. Bacteriol. **117**:517–526.

635. **Ornston, L. N., and M. K. Ornston.** 1969. Regulation of glyoxylate metabolism in *Escherichia coli* K-12. J. Bacteriol. **98**:1098–1108.

636. **Orr, G. A., J. W. Hammelburger, and G. Heney.** 1983. Interaction of sn-glycerol 3-phosphorothioate with *Escherichia coli.* *In vitro* and *in vivo* incorporation into phospholipids. J. Biol. Chem. **258**:9237–9244.

637. **Ørskov, I., and F. Ørskov.** 1973. Plasmid-determined H_2S character in *Escherichia coli* and its relation to plasmid-carried raffinose fermentation and tetracycline resistance characters. Examination of 32 H_2S-positive strains isolated during the years 1950 to 1971. J. Gen. Microbiol. **77**:487–499.

638. **Osborn, M. J., S. M. Rosen, L. Rothfield, and B. L. Horecker.** 1962. Biosynthesis of bacterial lipopolysaccharide. I. Enzymatic incorporation of galactose in a mutant strain of *Salmonella.* Proc. Natl. Acad. Sci. USA **48**:1831–1838.

639. **Ott, J. L., and C. H. Werkman.** 1957. Coupled nucleoside

phosphorylase reactions in *Escherichia coli*. Arch. Biochem. Biophys. **69:**264–276.

640. **Overath, P., and J. K. Wright.** 1983. Lactose permease: a carrier on the move. Trends Biochem. Sci. **8:**404–408.

641. **Paege, L. M., and F. Schlenk.** 1950. Pyrimidine riboside metabolism. Arch. Biochim. Biophys. **28:**348–358.

642. **Page, M. G. P., and K. Burton.** 1978. The location of purine phosphoribosyl-transferase activities in *Escherichia coli*. Biochem. J. **174:**717–725.

643. **Pahel, G., F. R. Bloom, and B. Tyler.** 1979. Deletion mapping of the *polA-metB* region of the *Escherichia coli* chromosome. J. Bacteriol. **138:**653–656.

644. **Palchaudhuri, S., S. Rahn, D. S. Santos, and W. K. Maas.** 1977. Characterization of plasmids in a sucrose-fermenting strain of *Escherichia coli*. J. Bacteriol. **130:**1402–1403.

645. **Palmer, T. N., B. E. Ryman, and W. J. Whelan.** 1968. The action pattern of amylomaltase. FEBS Lett. **1:**1–3.

646. **Palmer, T. N., G. Wöber, and W. J. Whelan.** 1973. The pathway of exogenous and endogenous carbohydrate utilization in *Escherichia coli*: a dual function for the enzymes of the maltose operon. Eur. J. Biochem. **39:**601–612.

647. **Palva, E. T.** 1978. Major outer membrane protein in *Salmonella typhimurium* induced by maltose. J. Bacteriol. **136:**286–294.

648. **Palva, E. T.** 1979. Relationship between *ompB* genes of *Escherichia coli* and *Salmonella typhimurium*. FEMS Microbiol. Lett. **5:**205–209.

649. **Palva, E. T., P. Liljeström, and S. Harayama.** 1981. Cosmid cloning and transposon mutagenesis in *Salmonella typhimurium* using phage lambda vehicles. Mol. Gen. Genet. **181:**153–157.

649a. **Palva, E. T., P. Saris, and T. J. Silhavy.** 1985. Gene fusions to the *pstM/pel* locus of *Escherichia coli*. Mol. Gen. Genet. **199:**427–433.

650. **Palva, E. T., and P. Westermann.** 1979. Arrangement of the maltose-inducible major outer membrane proteins, the bacteriophage lambda receptor in *Escherichia coli* and the 44K protein in *Salmonella typhimurium*. FEBS Lett. **99:**77–80.

651. **Pardee, A. B.** 1957. An inducible mechanism for accumulation of melibiose in *Escherichia coli*. J. Bacteriol. **73:**376–385.

652. **Parks, R. E., Jr., and R. P. Agarwal.** 1972. Purine nucleoside phosphorylase, p. 483–514. *In* P. D. Boyer (ed.), The enzymes, vol. 7. Academic Press, Inc., New York.

653. **Parnes, J. R., and W. Boos.** 1973. Energy coupling of the beta-methylgalactoside transport system of *Escherichia coli*. J. Biol. Chem. **248:**4429–4435.

654. **Parsons, R. G., and R. W. Hogg.** 1974. Crystallization and characterization of the L-arabinose-binding protein of *Escherichia coli* B/r. J. Biol. Chem. **249:**3602–3607.

655. **Parsons, R. G., and R. W. Hogg.** 1974. A comparison of the L-arabinose and D-galactose-binding proteins of *Escherichia coli* B/r. J. Biol. Chem. **249:**3608–3614.

656. **Peterkofsky, A., and C. Gazdar.** 1979. *Escherichia coli* adenylate cyclase complex: regulation by the proton electrochemical gradient. Proc. Natl. Acad. Sci. USA **76:**1099–1103.

657. **Peterson, R. N., J. Boniface, and A. L. Koch.** 1967. Energy requirements, interactions and distinctions in the mechanisms for transport of various nucleosides in *Escherichia coli*. Biochem. Biophys. Acta **135:**771–783.

658. **Peterson, R. N., and A. L. Koch.** 1966. The relationship of adenosine and inosine transport in *Escherichia coli* Biochim. Biophys. Acta **126:**129–145.

659. **Pogell, B. M., B. R. Maity, S. Frumkin, and S. Shapiro.** 1966. Induction of an active transport system for glucose 6-phosphate in *Escherichia coli*. Arch. Biochem. Biophys. **116:**406–415.

660. **Portalier, R., J. Robert-Baudouy, and F. Stoeber.** 1980. Regulation of *Escherichia coli* K-12 hexuronate system genes: *exu* regulon. J. Bacteriol. **249:**1095–1107.

661. **Portalier, R., and F. Stoeber.** 1972. Dosages colorimetriques des oxydoreductases aldoniques d'*Escherichia coli* K12: applications. Biochim. Biophys. Acta **289:**19–27.

662. **Portalier, R. C., J. M. Robert-Baudouy, and G. M. Nemoz.** 1974. Studies of mutations in the uronic isomerase and altronic oxidoreductase structural genes of *Escherichia coli* K-12. Mol. Gen. Genet. **128:**301–319.

663. **Portalier, R. C., J. M. Robert-Baudouy, and F. R. Stoeber.** 1972. Localisation genetique et caracterisation biochimique de mutations affectant le gene de structure de l'hydrolyase altronique chez *Escherichia coli* K12. Mol. Gen. Genet. **118:**335–350.

664. **Portalier, R. C., and F. R. Stoeber.** 1972. La D-altronate: NAD-oxydoreductase d'*Escherichia coli* K-12: purification, proprietes et individualite. Eur. J. Biochem. **26:**50–61.

665. **Portalier, R. C., and F. R. Stoeber.** 1972. La D-mannonate:NAD-oxydoreductase d'*Escherichia coli* K12: purification, proprietes et individualite. Eur. J. Biochem. **26:**290–300.

666. **Postma, P. W., W. Epstein, A. R. Schuitema, and S. O. Nelson.** 1984. Interaction between III^Glc of the phosphoenolpyruvate: sugar phosphotransferase system and glycerol kinase of *Salmonella typhimurium*. J. Bacteriol. **158:**351–353.

667. **Postma, P. W., and J. W. Lengeler.** 1985. Phosphoenolpyruvate: carbohydrate phosphotransferase system of bacteria. Microbiol. Rev. **49:**232–269.

668. **Pouyssegur, J.** 1971. Localisation genetique de mutations 2-ceto-3-desoxy-6-P-gluconate aldolase negatives chez *E. coli* K12. Mol. Gen. Genet. **113:**31–42.

669. **Pouyssegur, J.** 1972. Mutations affectant le gene de structure de la 2-ceto-3-desoxyphosphogluconate aldolase chez *Escherichia coli* K12. Mol. Gen. Genet. **114:**305–311.

670. **Pouyssegur, J., and F. Stoeber.** 1970. Production de 2-ceto-3-deoxy-6-phosphogluconate par un mutant d'*Escherichia coli* K12. Bull. Soc. Chim. Biol. **52:**1407–1419.

671. **Pouyssegur, J., and F. Stoeber.** 1970. Sur la biosynthese induite des deux dernieres enzymes de la sequence degradative des hexuronates chez *Escherichia coli* K 12. C.R. Acad. Sci. Ser. D **271:**370–373.

672. **Pouyssegur, J., and F. Stoeber.** 1970. Synthese enzymatique du 2-ceto-3-desoxy-D-gluconate. Bull. Soc. Chim. Biol. **52:**1419–1428.

673. **Pouyssegur, J., and F. Stoeber.** 1971. Etude du rameau degradatif commun des hexuronates chez *Escherichia coli* K12. Purification, proprietes et individualite de la 2-ceto-3-desoxy-D-gluconokinase. Biochimie **53:**771–781.

674. **Pouyssegur, J., and F. Stoeber.** 1972. Controle physiologique et genetique du metabolisme du 2-ceto-3-desoxy-gluconate chez *E. coli* K12. C.R. Acad. Sci. Ser. D **274:**2249–2252.

675. **Pouyssegur, J. M., and A. Lagarde.** 1973. Systeme de transport du 2-ceto-3-desoxy-gluconate chez *E. coli* K12: localization d'un gene de structure et de son operateur. Mol. Gen. Genet. **121:**163–180.

676. **Pouyssegur, J. M., and F. Stoeber.** 1974. Genetic control of the 2-keto-3-deoxy-D-gluconate metabolism in *Escherichia coli* K-12: *kdg* regulon. J. Bacteriol. **117:**641–651.

677. **Pouyssegur, J. M., and F. R. Stoeber.** 1971. Etude du rameau degradatif commun des hexuronates chez *Escherichia coli* K12. Purification, proprietes et individualite de la 2-ceto-3-desoxy-6-phospho-D-gluconate aldolase. Eur. J. Biochem. **21:**363–373.

678. **Pouyssegur, J. M., and F. R. Stoeber.** 1972. Rameau degradatif commun des hexuronates chez *Escherichia coli* K12. Mecanisme d'induction des enzymes assurant le metabolisme du 2-ceto-3-desoxy-gluconate chez *Escherichia coli* K12. Eur. J. Biochem. **30:**479–494.

679. **Power, J.** 1967. The L-rhamnose genetic system in *Escherichia coli* K-12. Genetics **55:**557–568.

680. **Prasad, I., and S. Schaefler.** 1974. Regulation of the beta-glucoside system in *Escherichia coli* K-12. J. Bacteriol. **120:**638–650.

681. **Prasad, I., B. Young, and S. Schaefler.** 1973. Genetic determination of the constitutive biosynthesis of phospho-alpha-glucosidase A in *Escherichia coli* K-12. J. Bacteriol. **114:**909–915.

682. **Prestidge, L. S., and A. B. Pardee.** 1965. A second permease for methyl-thio-beta-D-galactoside in *Escherichia coli*. Biochim. Biophys. Acta **100:**591–593.

683. **Pritchard, R. H., and S. I. Ahmad.** 1971. Fluoracil and the isolation of mutants lacking uridine phosphorylase in *Escherichia coli*: location of the gene. Mol. Gen. Genet. **111:**84–88.

684. **Pueyo, C.** 1978. Forward mutations to arabinose resistance in *Salmonella typhimurium* strains. A sensitive assay for mutagenicity testing. Mutat. Res. **54:**311–321.

685. **Pueyo, C., and J. Lopez-Barea.** 1979. The L-arabinose resistance test with *Salmonella typhimurium* strain SV-3 selects forward mutations at several 6*ara* genes. Mutat. Res. **64:**249–258.

686. **Quiocho, F. A., G. L. Gilliland, and G. N. Phillips, Jr.** 1977. The 2.8 Å resolution structure of the L-arabinose-binding protein from *Escherichia coli*. Polypeptide chain folding, domain similarity, and probable location of sugar-binding site. J. Biol. Chem. **252:**5142–5149.

687. **Quiocho, F. A., W. E. Meador, and J. W. Pflugrath.** 1979. Preliminary crystallographic data of receptors for transport and chemotaxis in *Escherichia coli*: D-galactose and maltose binding proteins. J. Mol. Biol. **133:**181–184.

688. **Quiocho, F. A., and N. K. Vyas.** 1984. Novel stereospecificity of the L-arabinose-binding protein. Nature (London) **310:**381–386.

689. **Rachmeler, M., J. Gerhart, and J. Rosner.** 1961. Limited thymidine uptake in *Escherichia coli* due to an inducible thymidine phosphorylase. Biochim. Biophys. Acta **49:**222–225.

690. **Racker, E.** 1951. Enzymatic synthesis of deoxypentose phosphate. Nature (London) **167:**408–409.

691. **Racker, E.** 1952. Enzymatic synthesis and breakdown of desoxyribose phosphate. J. Biol. Chem. **196**:347–365.

692. **Rader, R. L., and J. Hochstadt.** 1976. Regulation of purine utilization in bacteria. VII. Involvement of membrane-associated nucleoside phosphorylases in the uptake and the base-mediated loss of the ribose moiety of nucleosides by *Salmonella typhimurium* membrane vesicles. J. Bacteriol. **128**:290–301.

693. **Ramos, S., and H. R. Kaback.** 1977. pH-dependent changes in proton-substrate stoichiometries during active transport in *Escherichia coli* membrane vesicles. Biochemistry **16**:4271–4275.

694. **Randall-Hazelbauer, L. L., and M. Schwartz.** 1973. Isolation of the bacteriophage lambda receptor from *Escherichia coli* K-12. J. Bacteriol. **116**:1436–1446.

695. **Ravdonikas, L. E.** 1976. Production and characteristics of *Salmonella typhimurium* glycerin mutants. Zh. Mikrobiol. Epidemiol. Immunobiol. **12**:29–32.

696. **Rayman, M. K., T. C. Y. Lo, and B. D. Sanwal.** 1972. Transport of succinate in *Escherichia coli*. II. Characteristics of uptake and energy coupling with transport in membrane preparations. J. Biol. Chem. **247**:6332–6339.

697. **Razzell, W. E., and P. Casshyap.** 1964. Substrate specificity and induction of thymidine phosphorylase in *Escherichia coli*. J. Biol. Chem. **239**:1789–1793.

698. **Razzell, W. E., and G. G. Khorana.** 1958. Purification and properties of a pyrimidine deoxyriboside phosphorylase from *Escherichia coli*. Biochim. Biophys. Acta **28**:562–566.

699. **Reiner, A. M.** 1975. Genes for ribitol and D-arabitol catabolism in *Escherichia coli*: their loci in C strains and absence in K-12 and B strains. J. Bacteriol. **123**:530–536.

700. **Reithel, F. J., and J. C. Kim.** 1960. Studies on the beta-galactosidase from *Escherichia coli* ML 308.1. The effect of some ions on enzymic activity. Arch. Biochem. Biophys. **90**:271–277.

701. **Remy, C. N., and S. H. Love.** 1968. Induction of adenosine deaminase in *Escherichia coli*. J. Bacteriol. **96**:76–85.

702. **Rephaeli, A. W., and M. H. Saier, Jr.** 1980. Substrate specificity and kinetic characterization of sugar uptake and phosphorylation catalyzed by the mannose enzyme II of the phosphotransferase system in *Salmonella typhimurium*. J. Biol. Chem. **255**:8585–8591.

703. **Reynolds, A. E., J. Felton, and A. Wright.** 1981. Insertion of DNA activates the cryptic *bgl* operon in *E. coli* K12. Nature (London) **293**:625–629.

704. **Richarme, G.** 1982. Associative properties of the *Escherichia coli* galactose binding protein and maltose binding protein. Biochem. Biophys. Res. Commun. **105**:476–481.

705. **Richarme, G.** 1985. Possible involvement of lipoic acid in the binding protein-dependent transport systems in *Escherichia coli*. J. Bacteriol. **162**:286–293.

706. **Richey, D. P., and E. C. C. Lin.** 1972. Importance of facilitated diffusion for effective utilization of glycerol by *Escherichia coli*. J. Bacteriol. **112**:784–790.

707. **Rickenberg, H. V., G. N. Cohen, and G. Buttin, and J. Monod.** 1956. La galactoside-permease d'*Escherichia coli*. Ann. Inst. Pasteur (Paris) **91**:829–857.

708. **Riddle, D. L., and J. R. Roth.** 1972. Frameshift suppressors. II. Genetic mapping and dominance studies. J. Mol. Biol. **66**:483–493.

709. **Riordan, C., and H. L. Kornberg.** 1977. Location of *galP*, a gene which specifies galactose permease activity, on the *Escherichia coli* linkage map. Proc. R. Soc. Lond. B Biol. Sci. **198**:401–410.

710. **Ritzenthaler, P., and M. Mata-Gilsinger.** 1982. Use of in vitro gene fusions to study the *uxuR* regulatory gene in *Escherichia coli* K-12: direction of transcription and regulation of its expression. J. Bacteriol. **150**:1040–1047.

711. **Ritzenthaler, P., M. Mata-Gilsinger, and F. Stoeber.** 1980. Construction and expression of hybrid plasmids containing *Escherichia coli* K-12 *uxu* genes. J. Bacteriol. **143**:1116–1126.

712. **Ritzenthaler, P., M. Mata-Gilsinger, and F. Stoeber.** 1981. Molecular cloning of the *Escherichia coli* K-12 hexuronate system genes: the *exu* region. J. Bacteriol. **145**:181–190.

713. **Robbins, A. R.** 1975. Regulation of the *Escherichia coli* methylgalactoside transport system by gene *mglD*. J. Bacteriol. **123**:69–74.

714. **Robbins, A. R., R. Guzman, and B. Rotman.** 1976. Roles of individual *mgl* gene products in the beta-methylgalactoside transport system of *Escherichia coli* K12. J. Biol. Chem. **251**:3112–3116.

715. **Robbins, A. R., and B. Rotman.** 1975. Evidence for binding protein-independent substrate translocation by the methylgalactoside transport system of *Escherichia coli*. Proc. Natl. Acad. Sci. USA **72**:423–427.

716. **Robert-Baudouy, J., J. Jimeno-Abendano, and F. Stoeber.** 1971. Individualite des hydro-lyases mannonique et altronique chez *Escherichia coli* K-12. C.R. Acad. Sci. Ser. D **272**:2740–2743.

717. **Robert-Baudouy, J., R. Portalier, and F. Stoeber.** 1974. Regulation du metabolisme des hexuronates chez *Escherichia coli* K-12: modalites de l'induction des enzymes du systeme hexuronate. Eur. J. Biochem. **43**:1–15.

718. **Robert-Baudouy, J., R. Portalier, and F. Stoeber.** 1981. Regulation of hexuronate system genes in *Escherichia coli* K-12: multiple regulation of the *uxu* operon by *exuR* and *uxuR* gene products. J. Bacteriol. **145**:211–220.

719. **Robert-Baudouy, J., and F. Stoeber.** 1973. Purification et proprietes de la D-mannonate hydrolyase d'*Escherichia coli* K 12. Biochim. Biophys. Acta **309**:473–485.

720. **Robert-Baudouy, J. M., J. M. Jimeno-Abendano, and F. R. Stoeber.** 1975. Individualite des hdyrolyases mannonique et altronique chez *Escherichia coli* K-12. Biochimie **57**:1–8.

721. **Robert-Baudouy, J. M., and R. C. Portalier.** 1974. Studies of mutations in glucuronate catabolism in *Escherichia coli* K-12. Mol. Gen. Genet. **131**:31–46.

722. **Robert-Baudouy, J. M., R. C. Portalier, and F. R. Stoeber.** 1972. Genetic mapping and biochemical characterization of mutations in the mannonic hydrolyase structural gene of *Escherichia coli* K-12. Mol. Gen. Genet. **118**:351–362.

723. **Roberts, R. B., and I. Z. Roberts.** 1950. Potassium metabolism in *Escherichia coli*. III. Interrelationship of potassium and phosphorus metabolism. J. Cell. Comp. Physiol. **36**:15–39.

724. **Robertson, B. C., and P. A. Hoffee.** 1973. Purification and properties of purine nucleoside phosphorylase from *Salmonella typhimurium*. J. Biol. Chem. **248**:2040–2043.

725. **Robertson, B. C., P. Jargiello, J. Blank, and P. A. Hoffee.** 1970. Genetic regulation of ribonucleoside and deoxyribonucleoside catabolism in *Salmonella typhimurium*. J. Bacteriol. **102**:628–635.

726. **Robinson, J. J., and J. H. Weiner.** 1980. The effect of amphipaths on the flavin-linked aerobic glycerol-3-phosphate dehydrogenase from *Escherichia coli*. Can. J. Biochem. **80**:1172–1178.

727. **Rolseth, S., V. Fried, and B. G. Hall.** 1980. A mutant *ebg* enzyme that converts lactose into an inducer of the *lac* operon. J. Bacteriol. **142**:1036–1039.

728. **Roossien, F. F., M. Blaauw, and G. T. Robillard.** 1984. Kinetics and subunit interaction of the mannitol-specific enzyme II of the *Escherichia coli* phosphoenolpyruvate-dependent phosphotransferase system. Biochemistry **23**:4934–4939.

729. **Roossien, F. F., and G. T. Robillard.** 1984. Vicinal dithiol-disulfide distribution in the *Escherichia coli* mannitol specific carrier enzyme II Mtl. Biochemistry **23**:211–215.

730. **Roossien, F. F., and G. T. Robillard.** 1984. Mannitol-specific carrier protein from the *Escherichia coli* phosphoenolpyruvate-dependent phosphotransferase system can be extracted as a dimer from the membrane. Biochemistry **23**:5682–5685.

731. **Ros, J., and J. Aguilar.** 1984. Genetic and structural evidence for the presence of propanediol oxidoreductase isoenzymes in *Escherichia coli*. J. Gen. Microbiol. **130**:687–692.

732. **Rose, S. P., and C. F. Fox.** 1971. The beta-glucoside system of *Escherichia coli*. II. Kinetic evidence for a phosphoryl-enzyme II intermediate. Biochem. Biophys. Res. Commun. **45**:376–380.

733. **Rosen, S. M., L. D. Zeleznick, D. Fraenkel, I. M. Weiner, M. J. Osborn, and B. L. Horecker.** 1965. Characterization of the cell wall lipopolysaccharide of a mutant *Salmonella typhimurium* lacking phosphomannose isomerase. Biochem. Z. **342**:375–386.

734. **Rosenberg, H., and C. M. Hardy.** 1984. Conversion of D-mannitol to D-ribose: a newly discovered pathway in *Escherichia coli*. J. Bacteriol. **158**:69–72.

735. **Rosenberg, H., S. M. Pearce, C. M. Hardy, and P. A. Jacomb.** 1984. Rapid turnover of mannitol-1-phosphate in *Escherichia coli*. J. Bacteriol. **158**:63–68.

736. **Rosenfeld, S. A., P. E. Stevis, and N. W. Y. Ho.** 1984. Cloning and characterization of the *xyl* genes from *Escherichia coli*. Mol. Gen. Genet. **194**:410–415.

737. **Ross, J. P., and C. W. Shuster.** 1972. Amino sugar assimilation by *Escherichia coli*. J. Bacteriol. **112**:894–902.

738. **Rotman, B.** 1959. Separate permeases for the accumulation of methyl-beta-D-galactoside and methyl-beta-D-thiogalactoside in *Escherichia coli*. Biochim. Biophys. Acta **32**:599–601.

739. **Rotman, B., A. K. Ganesan, and R. Guzman.** 1968. Transport systems for galactose and galactosides in *Escherichia coli*. II. Substrate and inducer specificities. J. Mol. Biol. **36**:247–260.

739a. **Rotman, B., and R. Guzman.** 1961. Transport of galactose from the inside to the outside of *Escherichia coli*. Pathol. Biol. **9**:806–810.

740. **Rotman, B., and R. Guzman.** 1982. Identification of the *mglA* gene product in the beta-methylgalactoside transport system of

Escherichia coli using plasmid DNA deletions generated *in vitro*. J. Biol. Chem. **257:**9030–9034.

741. **Rotman, B., and R. Guzman.** 1984. Galactose-binding protein-dependent transport in reconstituted *Escherichia coli* membrane vesicles, p. 57–60. *In* L. Leive and D. Schlessinger (ed.), Microbiology–1984. American Society for Microbiology, Washington, D.C.

742. **Rotman, B., and J. Radojkovic.** 1964. Galactose transport in *Escherichia coli*. The mechanism underlying the retention of intracellular galactose. J. Biol. Chem. **239:**3153–3156.

743. **Roy-Burman, S., and D. W. Visser.** 1972. Transport studies of showdomycin, nucleosides and sugars in *Escherichia coli* B and showdomycin-resistant mutants. Biochim. Biophys. Acta **282:**383–392.

744. **Roy-Burman, S., and D. W. Visser.** 1975. Transport of purines and deoxyadenosine in *Escherichia coli*. J. Biol. Chem. **250:** 9270–9275.

745. **Roy-Burman, S., and D. W. Visser.** 1981. Uridine and uracil transport in *Escherichia coli* and transport-deficient mutants. Biochim. Biophys. Acta **646:**309–319.

746. **Roy-Burman, S., P. J. von Dipper, and D. W. Visser.** 1978. Mechanism of energy coupling for transport of deoxycytidine, uridine, uracil, adenine and hypoxanthine in *Escherichia coli*. Biochim. Biophys. Acta **511:**285–296.

747. **Ruiz-Vazquez, R., C. Pueyo, and E. Cerda-Olmedo.** 1978. A mutagen assay detecting forward mutations in an arabinose-sensitive strain of *Salmonella typhimurium*. Mutat. Res. **54:**121–129.

747a.**Russo, R. J., J.-M. Lee, P. Clarke, and G. Wilcox.** 1984. Identification of the *araE* gene product of *Salmonella typhimurium* LT2, p. 42–46. *In* L. Leive and D. Schlessinger (ed.), Microbiology—1984. American Society for Microbiology, Washington, D.C.

748. **Saedler, H., A. Gullon, L. Fiethen, and P. Starlinger.** 1968. Negative control of the galactose operon in *Escherichia coli*. Mol. Gen. Genet. **102:**79–88.

749. **Saier, M. H., Jr., F. G. Bromberg, and S. Roseman.** 1973. Characterization of constitutive galactose permease mutants in *Salmonella typhimurium*. J. Bacteriol. **113:**512–523.

750. **Saier, M. H., Jr., F. C. Grenier, C. A. Lee, and E. B. Waygood.** 1985. Evidence for the evolutionary relatedness of the proteins of the bacterial phosphoenolpyruvate:sugar phosphotransferase system. J. Cell. Biochem. **27:**43–56.

751. **Saier, M. H., Jr., and M. J. Newman.** 1976. Direct transfer of the phosphoryl moiety of mannitol 1-phosphate to [^{14}C]mannitol catalyzed by the enzyme II complexes of the phosphoenolpyruvate:mannitol phosphotransferase systems in *Spirochaeta aurantia* and *Salmonella typhimurium*. J. Biol. Chem. **251:**3834–3837.

752. **Saier, M. H., Jr., and S. Roseman.** 1976. Sugar transport. Inducer exclusion and regulation of the melibiose, maltose, glycerol, and lactose transport systems by the phosphoenolpyruvate:sugar phosphotransferase system. J. Biol. Chem. **251:**6606–6615.

753. **Saier, M. H., Jr., H. Straud, L. S. Massman, J. J. Judice, M. J. Newman, and B. U. Feucht.** 1978. Permease-specific mutations in *Salmonella typhimurium* and *Escherichia coli* that release the glycerol, maltose, melibiose, and lactose transport systems from regulation by the phosphoenolpyruvate:sugar phosphotransferase system. J. Bacteriol. **133:**1358–1367.

754. **Saier, M. H., Jr., D. L. Wentzel, B. U. Feucht, and J. J. Judice.** 1975. A transport system for phosphoenolpyruvate, 2-phosphoglycerate, and 3-phosphoglycerate in *Salmonella typhimurium*. J. Biol. Chem. **250:**5089–5096.

755. **Saint Pierre, M. L.** 1968. Isolation and mapping *Salmonella typhimurium* mutants defective in the utilization of trehalose. J. Bacteriol. **95:**1185–1186.

756. **Sanderson, K. E.** 1972. Linkage map of *Salmonella typhimurium*. Edition IV. Bacteriol. Rev. **36:**558–586.

757. **Sanderson, K. E., and P. E. Hartman.** 1978. Linkage map of *Salmonella typhimurium*. Edition V. Microbiol. Rev. **42:**471–519.

758. **Sanderson, K. E., and J. R. Roth.** 1983. Linkage map of *Salmonella typhimurium*. Edition VI. Microbiol. Rev. **47:**410–453.

759. **Sanno, Y., T. H. Wilson, and E. C. C. Lin.** 1968. Control of permeation to glycerol in cells of *Escherichia coli*. Biochem. Biophys. Res. Commun. **32:**344–349.

760. **Sarno, M. V., L. G. Tenn, A. Desai, A. M. Chin, F. C. Grenier, and M. H. Saier, Jr.** 1984. Genetic evidence for glucitol-specific enzyme III, an essential phosphocarrier protein of the *Salmonella typhimurium* glucitol phosphotransferase system. J. Bacteriol. **157:**953–955.

761. **Sarvas, M.** 1971. Mutant of *Escherichia coli* K-12 defective in D-glucosamine biosynthesis. J. Bacteriol. **105:**467–471.

762. **Sawada, H., and Y. Takagi.** 1964. The metabolism of L-rham-nose in *Escherichia coli*. III. L-Rhamnulose-phosphate aldolase. Biochim. Biophys. Acta **92:**26–32.

763. **Scangos, G. A., and A. M. Reiner.** 1978. Acquisition of ability to utilize xylitol: disadvantage of a constitutive catabolic pathway in *Escherichia coli*. J. Bacteriol. **134:**501–505.

764. **Scangos, G. A., and A. M. Reiner.** 1978. Ribitol and D-arabitol catabolism in *Escherichia coli*. J. Bacteriol. **134:**492–500.

765. **Scangos, G. A., and A. M. Reiner.** 1979. A unique pattern of toxic synthesis in pentitol catabolism: implications for evolution. J. Mol. Evol. **12:**189–195.

766. **Schächtele, K.-H., E. Schiltz, and D. Palm.** 1978. Amino acid sequence of the pyridoxal phosphate binding site in *Escherichia coli* maltodextrin phosphorylase. Eur. J. Biochem. **92:**427–435.

767. **Schaefler, S.** 1967. Inducible system for the utilization of beta-glucosides in *Escherichia coli*. I. Active transport and utilization of beta-glucosides. J. Bacteriol. **93:**254–263.

768. **Schaefler, S., and W. K. Maas.** 1967. Inducible system for the utilization of beta-glucosides in *Escherichia coli*. II. Description of mutant types and genetic analysis. J. Bacteriol. **93:**264–272.

769. **Schaefler, S., and A. Malamy.** 1969. Taxonomic investigations on expressed and cryptic phospho-beta-glucosidases in *Enterobacteriaceae*. J. Bacteriol. **99:**422–433.

770. **Schäfler, S., and L. Mintzer.** 1959. Acquisition of lactose fermenting properties by *Salmonellae*. I. Interrelationship between the fermentation of cellobiose and lactose. J. Bacteriol. **78:** 159–163.

771. **Schellenberg, G. D., A. Sarthy, A. E. Larson, M. P. Backer, J. W. Crabb, M. Lidstrom, B. D. Hall, and C. E. Furlong.** 1984. Xylose isomerase from *Escherichia coli*. Characterization of the protein and the structural gene. J. Biol. Chem. **259:**6826–6832.

772. **Schleif, R.** 1969. An inducible L-arabinose binding protein and arabinose permeation in *Escherichia coli*. J. Mol. Biol. **46:** 185–196.

773. **Schmid, K., and R. Schmitt.** 1976. Raffinose metabolism in *Escherichia coli* K12: purification and properties of a new alpha-galactosidase specified by a transmissible plasmid. Eur. J. Biochem. **67:**95–104.

774. **Schmid, K., M. Schupfner, and R. Schmitt.** 1982. Plasmid-mediated uptake and metabolism of sucrose by *Escherichia coli* K-12. J. Bacteriol. **151:**68–76.

775. **Schmitt, R.** 1968. Analysis of melibiose mutants deficient in alpha-galactosidase and thiomethylgalactoside permease II in *Escherichia coli* K-12. J. Bacteriol. **96:**462–471.

776. **Schmitt, R., and B. Rotman.** 1966. Alpha-galactosidase activity in cell-free extracts of *Escherichia coli*. Biochem. Biophys. Res. Commun. **22:**473–479.

777. **Schryvers, A., E. Lohmeier, and J. H. Weiner.** 1978. Chemical and functional properties of the native and reconstituted forms of the membrane-bound, aerobic glycerol-3-phosphate dehydrogenase of *Escherichia coli*. J. Biol. Chem. **253:**783–788.

778. **Schryvers, A., and J. H. Weiner.** 1981. The anaerobic *sn*-glycerol-3-phosphate dehydrogenase of *Escherichia coli*. J. Biol. Chem. **256:**9959–9965.

779. **Schryvers, A., and J. H. Weiner.** 1982. The anaerobic *sn*-glycerol-3-phosphate dehydrogenase: cloning and expression of the *glpA* gene of *Escherichia coli* and identification of the *glpA* products. Can. J. Biochem. **60:**224–231.

780. **Schulz, G. V., V. Häselbarth, H. E. Keller, and H. A. Schwinn.** 1966. Das Durch das Enzym Amylomaltase eingestellte Gleichgewicht oligomerer Amylosen. Makromol. Chem. **92:**91–104.

781. **Schumacher, G., and K. Bussmann.** 1978. Cell-free synthesis of proteins related to *sn*-glycerol-3-phosphate transport in *Escherichia coli*. J. Bacteriol. **135:**239–250.

782. **Schwartz, M.** 1967. Sur l'existence chez *Escherichia coli* K12 d'une regulation commune a la biosynthese des recepteurs du bacteriophage lambda et au metabolisme du maltose. Ann. Inst. Pasteur (Paris) **113:**685–704.

783. **Schwartz, M.** 1971. Thymidine phosphorylase from *Escherichia coli*. Eur. J. Biochem. **21:**191–198.

784. **Schwartz, M.** 1975. Reversible interaction between coliphage lambda and its receptor protein. J. Mol. Biol. **99:**185–201.

785. **Schwartz, M.** 1978. Thymidine phosphorylase from *Escherichia coli*. Methods Enzymol. **51:**442–445.

786. **Schwartz, M., and M. Hofnung.** 1962. La maltodextrine phosphorylase d'*Escherichia coli*. Eur. J. Biochem. **2:**132–145.

787. **Schwartz, M., and L. LeMinor.** 1975. Occurrence of the bacteriophage lambda receptor in some Enterobacteriaceae. J. Virol. **15:**679–685.

788. **Schwartz, N. B., and D. S. Feingold.** 1972. L-Rhamnulose 1-phosphate aldolase from *Escherichia coli*. III. The role of divalent cations in enzyme activity. Bioinorg. Chem. **2:**75–86.

789. **Schweizer, H., M. Argast, and W. Boos.** 1982. Characteristics of

a binding protein-dependent transport system for *sn*-glycerol-3-phosphate in *Escherichia coli* that is part of the *pho* regulon. J. Bacteriol. **150**:1154–1163.

790. **Schweizer, H., W. Boos, and T. J. Larson.** 1985. Repressor for the *sn*-glycerol-3-phosphate regulon of *Escherichia coli*: cloning of the *glpR* gene and identification of its product. J. Bacteriol. **161**:563–566.

791. **Schwinn, H., and G. V. Schulz.** 1971. Untersuchungen über amylomaltose. III. Kinetische analyse des reaktions mechanismus. Biochim. Biophys. Acta **227**:313–326.

792. **Scripture, J. B., and R. W. Hogg.** 1983. The nucleotide sequences defining the signal peptides of the galactose-binding protein and the arabinose-binding protein. J. Biol. Chem. **258**:10853–10855.

793. **Shamanna, D. K., and K. E. Sanderson.** 1979. Uptake and catabolism of D-xylose in *Salmonella typhimurium* LT2. J. Bacteriol. **139**:64–70.

794. **Shamanna, D. K., and K. E. Sanderson.** 1979. Genetics and regulation of D-xylose utilization in *Salmonella typhimurium* LT2. J. Bacteriol. **139**:71–79.

795. **Shapiro, J. A.** 1966. Chromosomal locations of the gene determining uridine diphosphoglucose formation in *Escherichia coli* K-12. J. Bacteriol. **92**:518–520.

795a.**Shapiro, J. A., and S. L. Adhya.** 1969. The galactose operon of *E. coli* K-12. II. A deletion analysis of operon structure and polarity. Genetics **62**:249–264.

796. **Shattuck-Eidens, D. M., and R. J. Kadner.** 1981. Exogenous induction of the *Escherichia coli* hexose phosphate transport system defined by *uhp-lac* operon fusions. J. Bacteriol. **148**:203–209.

797. **Shattuck-Eidens, D. M., and R. J. Kadner.** 1983. Molecular cloning of the *uhp* region and evidence for a positive activator for expression of the hexose phosphate transport system of *Escherichia coli*. J. Bacteriol. **155**:1062–1070.

798. **Shedlovsky, A., and S. Brenner.** 1963. A chemical basis for the host-induced modification of T-even bacteriophages. Proc. Natl. Acad. Sci. USA **50**:300–305.

799. **Sheinin, R., and B. F. Crocker.** 1961. The induced concurrent formation of alpha-galactosidase and beta-galactosidase in *Escherichia coli* B. Can. J. Biochem. Physiol. **39**:63–72.

800. **Sherman, J. R., and J. Adler.** 1963. Galactokinase from *Escherichia coli*. J. Biol. Chem. **238**:873–878.

801. **Shiota, S., H. Yazyu, and T. Tsuchiya.** 1984. *Escherichia coli* mutants with altered cation recognition by the melibiose carrier. J. Bacteriol. **160**:445–447.

802. **Shiota, S., Y. Yamano, M. Futai, and T. Tsuchiya.** 1985. *Escherichia coli* mutants possessing an Li⁺-resistant melibiose carrier. J. Bacteriol. **162**:106–109.

803. **Shopsis, C. S., R. Engel, and B. E. Tropp.** 1972. Effects of phosphonic acid analogues of glycerol-3-phosphate on the growth of *Escherichia coli*. J. Bacteriol. **112**:408–412.

804. **Shopsis, C. S., W. D. Nunn, R. Engel, and B. E. Tropp.** 1973. Effects of phosphonic acid analogues of glycerol-3-phosphate on the growth of *Escherichia coli*: phospholipid metabolism. Antimicrob. Agents Chemother. **4**:467–473.

805. **Short, S. A., and J. T. Singer.** 1984. Studies on *deo* operon regulation in *Escherichia coli*: cloning and expression of the *deoR* structural gene. Gene **31**:205–211.

806. **Shuman, H. A.** 1982. Active transport of maltose in *Escherichia coli* K12: role of the periplasmic maltose-binding protein and evidence for a substrate recognition site in the cytoplasmic membrane. J. Biol. Chem. **257**:5455–5461.

807. **Shuman, H. A., and J. Beckwith.** 1979. *Escherichia coli* K-12 mutants that allow transport of maltose via the beta-galactoside transport system. J. Bacteriol. **137**:365–373.

808. **Shuman, H. A., and T. J. Silhavy.** 1981. Identification of the *malK* gene product: a peripheral membrane component of the *Escherichia coli* maltose transport system. J. Biol. Chem. **256**:560–562.

809. **Shuman, H. A., T. J. Silhavy, and J. Beckwith.** 1980. Labeling of proteins with beta-galactosidase by gene fusion. Identification of a cytoplasmic membrane component of the *Escherichia coli* maltose transport system. J. Biol. Chem. **255**:168–174.

810. **Shuman, H. A., and N. A. Treptow.** 1985. The maltose-maltodextrin-transport system of *Escherichia coli* K-12, p. 561–575. *In* A. N. Martonosi (ed.), The enzymes of biological membranes, 2nd ed., vol. 3. Plenum Publishing Corp., N.Y.

811. **Shuster, C. W., and K. Rundell.** 1969. Resistance of *Salmonella typhimurium* mutants to galactose death. J. Bacteriol. **100**:103–109.

812. **Silhavy, T. J., I. Hartig-Beecken, and W. Boos.** 1976. Periplasmic protein related to the sn-glycerol-3-phosphate transport

system of *Escherichia coli*. J. Bacteriol. **126**:951–958.

813. **Singer, J. T., C. S. Barbier, and S. Short.** 1985. Identification of the *Escherichia coli deoR* and *cytR* gene products. J. Bacteriol. **63**:1095–1100.

814. **Singh, A. P., and P. D. Bragg.** 1976. Anaerobic transport of amino acids coupled to the glycerol-3-phosphate-fumarate oxidoreductase system in a cytochrome-deficient mutant of *Escherichia coli*. Biochim. Biophys. Acta **423**:450–461.

815. **Skjold, A. C., and D. H. Ezekiel.** 1982. Analysis of lambda insertions in the fucose utilization region of *Escherichia coli* K-12: use of lambda *fuc* and lambda *argA* transducing bacteriophages to partially order the fucose utilization genes. J. Bacteriol. **152**:120–125.

816. **Skjold, A. C., and D. H. Ezekiel.** 1982. Regulation of D-arabinose utilization in *Escherichia coli* K-12. J. Bacteriol. **152**:521–523.

817. **Slater, A. C., M. C. Jones-Mortimer, and H. L. Kornberg.** 1981. L-Sorbose phosphorylation in *Escherichia coli* K-12. Biochim. Biophys. Acta **646**:365–367.

818. **Smiley, J. D., and G. Ashwell.** 1960. Uronic acid metabolism in bacteria. III. Purification and properties of D-altronic acid and D-mannonic acid dehydrases in *Escherichia coli*. J. Biol. Chem. **235**:1517–1519.

819. **Smith, H. W., and Z. Parsell.** 1975. Transmissible substrate-utilizing ability in enterobacteria. J. Gen. Microbiol. **87**:129–140.

820. **Soffer, R. L.** 1961. Enzymatic expression of genetic units of function concerned with galactose metabolism in *Escherichia coli*. J. Bacteriol. **82**:471–478.

821. **Solomon, E., and E. C. C. Lin.** 1972. Mutations affecting the dissimilation of mannitol by *Escherichia coli* K-12. J. Bacteriol. **111**:566–574.

822. **Sprenger, G. A., and J. W. Lengeler.** 1984. L-Sorbose metabolism in *Klebsiella pneumoniae* and Sor⁺ derivatives of *Escherichia coli* K-12 and chemotaxis toward sorbose. J. Bacteriol. **157**:39–45.

823. **Sridhara, S., and T. T. Wu.** 1969. Purification and properties of lactaldehyde dehydrognease in *Escherichia coli*. J. Biol. Chem. **244**:5233–5238.

824. **Sridhara, S., T. T. Wu, T. M. Chused, and E. C. C. Lin.** 1969. Ferrous-activated nicotinamide adenine dinucleotide-linked dehydrogenase from a mutant of *Escherichia coli* capable of growth on 1,2-propanediol. J. Bacteriol. **98**:87–95.

825. **Stephenson, M., and A. R. Trim.** 1938. The metabolism of adenine compounds by *Bact. coli*. Biochem. J. **32**:1740–1751.

826. **Stevens, F. J., and T.-T. Wu.** 1976. Growth on D-lyxose of a mutant strain of *Escherichia coli* K12 using a novel isomerase and enzymes related to D-xylase metabolism. J. Gen. Microbiol. **97**:257–265.

827. **St. Martin, E. J., W. B. Freedberg, and E. C. C. Lin.** 1977. Kinase replacement by a dehydrogenase for *Escherichia coli* glycerol utilization. J. Bacteriol. **131**:1026–1028.

828. **Stock, J., and S. Roseman.** 1971. A sodium-dependent sugar co-transport system in bacteria. Biochem. Biophys. Res. Commun. **44**:132–138.

829. **Stoeber, F.** 1957. Sur la beta-glucuronide-permease d'*Escherichia coli*. C.R. Acad. Sci. Ser. D **244**:1091–1094.

830. **Stoeber, F., A. Lagarde, G. Nemoz, G. Novel, M. Novel, R. Portalier, J. Pouyssegur, and J. Robert-Baudouy.** 1974. Le metabolisme des hexuronides et des hexuronates chez *Escherichia coli* K-12: aspects physiologiques et genetiques de sa regulation. Biochimie **56**:199–213.

831. **Stokes, H. W., P. W. Betts, and B. G. Hall.** 1985. Sequence of the *ebgA* gene of *Escherichia coli*: comparison with the *lacZ* gene. Mol. Biol. Evol. **2**:469–477.

832. **Stokes, H. W., and B. G. Hall.** 1985. Sequence of the *ebgR* gene of *Escherichia coli*: evidence that the EBG and LAC operons are descended from a common ancestor. Mol. Biol. Evol. **2**:478–483.

833. **Strange, P. G., and D. E. Koshland, Jr.** 1976. Receptor interactions in a signalling system: competition between ribose receptor and galactose receptor in the chemotaxis response. Proc. Natl. Acad. Sci. USA **73**:762–766.

834. **Sukhodoletz, V. V., V. P. Galeys, and Y. V. Smirnov.** 1973. The nature of phenotypical reversions of mutants for thymidine phosphorylase in *Escherichia coli*. Genetika **9**:167–169.

835. **Sundaram, T. K.** 1972. *Myo*-inositol catabolism in *Salmonella typhimurium*: enzyme repression dependent on growth history of organism. J. Gen. Microbiol. **73**:209–219.

835a.**Sundararajan, T. A.** 1963. Interference with glycerolkinase induction in mutants of *E. coli* accumulating GAL-1-P. Biochemistry **50**:463–469.

836. **Sundararajan, T. A., A. M. C. Rapin, and H. M. Kalckar.** 1962. Biochemical observations on *E. coli* mutants defective in uridine diphosphoglucose. Proc. Natl. Acad. Sci. USA **48**:2187–2193.

837. **Svenningsen, B. A.** 1975. Regulated *in vitro* synthesis of enzymes

of *deo* operon of *Escherichia coli*: properties of DNA directed system. Mol. Gen. Genet. **137**:289–304.

838. **Svenningsen, B. A.** 1977. *In vitro* regulation of the *deo* operon of *Escherichia coli* at the initiation of transcription. Carlsberg Res. Commun. **42**:517–524.

839. **Synenki, R. M., J. A. Wohlhieter, E. M. Johnson, J. R. Lazere, and L. S. Baron.** 1973. Isolation and characterization of circular deoxyribonucleic acid obtained from lactose-fermenting *Salmonella* strains. J. Bacteriol. **116**:1185–1190.

840. **Szmelcman, S., and M. Hofnung.** 1975. Maltose transport in *Escherichia coli*: involvement of the bacteriophage lambda receptor. J. Bacteriol. **124**:112–118.

841. **Szmelcman, S., M. Schwartz, T. J. Silhavy, and W. Boos.** 1976. Maltose transport in *Escherichia coli* K12. A comparison of transport kinetics in wild type and lambda-resistant mutants with the dissociation constants of the maltose-binding protein as measured by fluorescence quenching. Eur. J. Biochem. **65**:13–19.

842. **Takagi, Y., M. Kanda, and Y. Nakata.** 1959. Studies on a D-galacturonic acid isomerase. Biochim. Biophys. Acta **31**:264–265.

843. **Takagi, Y., and H. Sawada.** 1964. The metabolism of L-rhamnose in *Escherichia coli*. I. L-Rhamnose isomerase. Biochim. Biophys. Acta **92**:10–17.

844. **Takagi, Y., and H. Sawada.** 1964. The metabolism of L-rhamnose in *Escherichia coli*. II. L-Rhamnulose kinase. Biochim. Biophys. Acta **92**:18–25.

845. **Taketo, A., and S. Kuno.** 1972. Internal localization of nucleoside-catabolic enzymes in *Escherichia coli*. J. Biochem. **72**:1557–1563.

846. **Tanaka, K., S. Niiya, and T. Tsuchiya.** 1980. Melibiose transport in *Escherichia coli*. J. Bacteriol. **141**:1031–1036.

847. **Tang, C.-T., R. Engel, and B. E. Tropp.** 1977. L-Glycerol 3-phosphate, a bactericidal agent. Antimicrob. Agents Chemother. **11**:147–153.

848. **Tang, C.-T., F. E. Ruch, Jr., and E. C. C. Lin.** 1979. Purification and properties of a nicotinamide adenine dinucleotide-linked dehydrogenase that serves an *Escherichia coli* mutant for glycerol catabolism. J. Bacteriol. **140**:182–187.

849. **Tang, J. C.-T., R. G. Forage, and E. C. C. Lin.** 1982. Immunochemical properties of NAD+-linked glycerol dehydrogenases from *Escherichia coli* and *Klebsiella pneumoniae*. J. Bacteriol. **152**:1169–1174.

850. **Tang, J. C.-T., E. J. St. Martin, and E. C. C. Lin.** 1982. Derepression of an NAD-linked dehydrogenase that serves an *Escherichia coli* mutant for growth on glycerol. J. Bacteriol. **152**:1001–1007.

851. **Taniguchi, T., M. O'Neill, and B. de Crombrugghe.** 1979. Interaction site of *Escherichia coli* cyclic AMP receptor protein on DNA of galactose operon promoters. Proc. Natl. Acad. Sci. USA **76**:5090–5094.

852. **Taylor, A. L., and M. S. Thoman.** 1964. The genetic map of *Escherichia coli* K-12. Genetics **50**:659–677.

853. **Thanner, F., D. Palm, and S. Shaltiel.** 1975. Hydrophobic and biospecific chromatography in the purification of maltodextrin phosphorylase from *E. coli*. FEBS Lett. **55**:178–182.

854. **Thirion, J. P., and M. Hofnung.** 1972. On some genetic aspects of phage lambda resistance in *E. coli* K12. Genetics **71**:207–216.

855. **Thorner, J. W., and H. Paulus.** 1971. Composition and subunit structure of glycerol kinase from *Escherichia coli*. J. Biol. Chem. **246**:3885–3894.

856. **Thorner, J. W., and H. Paulus.** 1973. Catalytic and allosteric properties of glycerol kinase from *Escherichia coli*. J. Biol. Chem. **248**:3922–3932.

857. **Thorner, J. W., and H. Paulus.** 1973. Glycerol and glycerate kinases, p. 487–508. *In* P. D. Boyer (ed.), The enzymes. Academic Press, Inc., New York.

858. **Tokuda, H., and H. R. Kaback.** 1977. Sodium-dependent methyl 1-thio-beta-D-galactopyranoside transport in membrane vesicles isolated from *Salmonella typhimurium*. Biochemistry **16**:2130–2136.

859. **Torriani, A. M., and J. Monod.** 1949. Sur la reversibilite de la reaction catalysee par l'amylomaltase. C.R. Acad. Sci. Ser. D **228**:718–720.

860. **Trisler, P., and S. Gottesman.** 1984. *lon* transcriptional regulation of genes necessary for capsular polysaccharide synthesis in *Escherichia coli* K-12. J. Bacteriol. **160**:184–191.

861. **Tsuchiya, T., J. Lopilato, and T. H. Wilson.** 1978. Effect of lithium ion on melibiose transport in *Escherichia coli*. J. Membr. Biol. **42**:45–59.

862. **Tsuchiya, T., M. Oho, and S. Shiota-Niiya.** 1983. Lithium ion-sugar cotransport via the melibiose transport system in *Escherichia coli*. J. Biol. Chem. **258**:12765–12767.

863. **Tsuchiya, T., K. Ottina, Y. Moriyama, M. J. Newman, and T. H. Wilson.** 1982. Solubilization and reconstitution of the melibiose carrier from a plasmid-carrying strain of *Escherichia coli*. J. Biol. Chem. **257**:5125–5128.

864. **Tsuchiya, T., J. Raven, and T. H. Wilson.** 1977. Co-transport of Na+ and methyl-beta-D-thiogalactopyranoside mediated by the melibiose transport system of *Escherichia coli*. Biochem. Biophys. Res. Commun. **76**:26–31.

865. **Tsuchiya, T., K. Takeda, and T. H. Wilson.** 1980. H+-substrate cotransport by the melibiose membrane carrier in *Escherichia coli*. Membr. Biochem. **3**:131–146.

866. **Tsuchiya, T., and T. H. Wilson.** 1978. Cation-sugar cotransport in the melibiose transport system of *Escherichia coli*. Membr. Biochem. **2**:63–79.

867. **Valentin-Hansen, P.** 1982. Tandem CRP binding sites in the *deo* operon of *Escherichia coli* K-12. EMBO J. **1**:1049–1054.

868. **Valentin-Hansen, P., H. Aiba, and D. Schümperli.** 1982. The structure of tandem regulatory regions in the *deo* operon of *Escherichia coli* K12. EMBO J. **1**:317–322.

869. **Valentin-Hansen, P., F. Boëtius, K. Hammer-Jespersen, and I. Svendsen.** 1982. The primary structure of *Escherichia coli* K 12 2-deoxyribose 5-phosphate aldolase. Nucleotide sequence of the *deoC* gene and the amino acid sequence of the enzyme. Eur. J. Biochem. **125**:561–566.

870. **Valentin-Hansen, P., K. Hammer-Jespersen, F. Boetius, and I. Svendsen.** 1984. Structure and function of the intercistronic regulatory *deoC-deoA* element of *Escherichia coli* K-12. EMBO J. **3**:179–183.

871. **Valentin-Hansen, P., K. Hammer-Jespersen, and R. S. Buxton.** 1979. Evidence for the existence of three promoters for the *deo* operon of *Escherichia coli* K12 *in vitro*. J. Mol. Biol. **133**:1–17.

872. **Valentin-Hansen, P., B. Svenningsen, A. Munch-Petersen, and K. Hammer-Jespersen.** 1978. Regulation of the *deo* operon in *Escherichia coli*. Mol. Gen. Genet. **159**:191–202.

873. **van Thienen, G. M., P. W. Postma, and K. Van Dam.** 1977. Proton movements coupled to sugar transport via the galactose transport system in *Salmonella typhimurium*. Eur. J. Biochem. **73**:521–527.

874. **Venkateswaran, P. S., and H. C. Wu.** 1972. Isolation and characterization of a phosphonomycin-resistant mutant of *Escherichia coli* K-12. J. Bacteriol. **110**:935–944.

875. **von Dippe, P. J., K.-K. Leung, S. Roy-Burman, and D. W. Visser.** 1975. Deoxycytidine transport in the presence of a cytidine deaminase inhibitor and the transport of uracil in *Escherichia coli* B. J. Biol. Chem. **250**:3666–3671.

876. **von Dippe, P. J., S. Roy-Burman, and D. W. Visser.** 1973. Transport of uridine in *Escherichia coli* B and a showdomycin-resistant mutant. Biochim. Biophys. Acta **318**:105–112.

877. **von Hofsten, B.** 1961. The inhibitory effect of galactosides on the growth of *Escherichia coli*. Biochim. Biophys. Acta **48**:164–171.

878. **von Meyenburg, K., F. G. Hansen, L. D. Nielsen, and P. Jørgensen.** 1977. Origin of replication, *oriC*, of the *Escherichia coli* chromosome: mapping of genes relative to R.*Eco*RI cleavage sites in the *oriC* region. Mol. Gen. Genet. **158**:101–109.

879. **von Meyenburg, K., F. G. Hansen, L. D. Nielsen, and E. Riise.** 1978. Origin of replication, *oriC*, of the *Escherichia coli* chromosome on specialized transducing phages lambda *asn*. Mol. Gen. Genet. **160**:287–295.

880. **von Meyenburg, K., and H. Nikaido.** 1977. Outer membrane of gram-negative bacteria. XVII. Specificity of transport process catalysed by the lambda-receptor protein in *Escherichia coli*. Biochem. Biophys. Res. Commun. **78**:1100–1107.

881. **von Wilcken-Bergmann, B., and B. Müller-Hill.** 1982. Sequence of *galR* gene indicates a common evolutionary origin of *lac* and *gal* repressor in *Escherichia coli*. Proc. Natl. Acad. Sci. USA **79**:2427–2431.

882. **Vorisek, J., and A. Kepes.** 1972. Galactose transport in *Escherichia* and the galactose-binding protein. Eur. J. Biochem. **28**:364–372.

883. **Vyas, N. K., M. N. Vyas, and F. A. Quiocho.** 1983. The 3Å resolution structure of a D-galactose-binding protein for transport and chemotaxis in *Escherichia coli*. Proc. Natl. Acad. Sci. USA **80**:1792–1796.

884. **Wahba, A. J., J. W. Hickman, and G. Ashwell.** 1958. Enzymatic formation of D-tagaturonic acid and D-fructuronic acid. J. Am. Chem. Soc. **80**:2594–2595.

885. **Wallenfels, K., J. Lehman, and O. P. Malhotra.** 1960. Untersuchungen über milchzuckerspaltende Enzyme VII. Die Specifität der B-Galaktosidase von *E. coli* ML 309. Biochem. Z. **333**:209–225.

886. **Wallenfels, K., O. P. Malhotra, and D. Dabich.** 1960. Untersuchungen über milchzuckerspaltende Enzyme VIII. Der Einfluss des Kationen-Milieus auf die Aktivität der β-Galactosidase von *E. coli* ML 309. Biochem. Z. **333**:377–394.

887. **Wallenfels, K., and R. Weil.** 1972. Beta-galactosidase, p. 617–663. *In* P. D. Boyer (ed.), The enzymes, 3rd ed., vol. 7. Academic Press, Inc., New York.

888. **Wallenfels, K., M. L. Zarnitz, G. Laude, H. Bender, and M. Keser.** 1959. Untersuchungen über milchzuckerspaltende Enzyme III. Reinigung, Kristallisation and Eigenschafter der beta-Galaktosidase von *Escherichia coli* ML 309. Biochem. Z. **331:**459–485.

889. **Wandersman, C., M. Schwartz, and T. Ferenci.** 1979. Mutants of *Escherichia coli* impaired in the transport of maltodextrins. J. Bacteriol. **140:**1–13.

890. **Wang, T. P., H. Z. Sable, and J. O. Lampon.** 1950. Enzymatic deamination of cytosine nucleosides. J. Biol. Chem. **184:**17–28.

891. **Waygood, E. B.** 1980. Resolution of the phosphoenolpyruvate:fructose phosphotransferase system of *Escherichia coli* into two components; enzyme II fructose and fructose-induced HPr-like protein (FPr). Can. J. Biochem. **58:**1144–1146.

892. **Waygood, E. B., R. L. Mattoo, and K. G. Peri.** 1984. Phosphoproteins and the phosphoenolpyruvate:sugar phosphotransferase system in *Salmonella typhimurium* and *Escherichia coli*: evidence for III^{Man}, III^{Fru}, III^{Glucitol}, and the phosphorylation of enzyme II^{Mannitol} and enzyme II$^{N-acetylglucosamine}$. J. Cell. Biochem. **25:**139–159.

893. **Weiner, J. H.** 1974. The localization of glycerol-3-phosphate dehydrogenase in *Escherichia coli*. J. Membr. Biol. **15:**1–14.

894. **Weiner, J. H., and L. A. Heppel.** 1972. Purification of the membrane-bound and pyridine nucleotide-independent L-glycerol 3-phosphate dehydrogenase from *Escherichia coli*. Biochem. Biophys. Res. Commun. **47:**1360–1365.

895. **Weiner, J. H., E. Lohmeier, and A. Schryvers.** 1978. Cloning and expression of the glycerol-3-phosphate transport genes of *Escherichia coli*. Can. J. Biochem. **56:**611–617.

896. **West, I. C.** 1970. Lactose transport coupled to proton movements in *Escherichia coli*. Biochem. Biophys. Res. Commun. **41:**655–661.

897. **West, I. C., and P. Mitchell.** 1973. Stoichiometry of lactose-H$^+$ symport across the plasma membrane of *Escherichia coli*. Biochem. J. **132:**587–592.

898. **White, R. J.** 1968. Control of amino sugar metabolism in *Escherichia coli* and isolation of mutants unable to degrade amino sugars. Biochem. J. **106:**847–858.

899. **White, R. J.** 1970. The role of the phosphoenolpyruvate phosphotransferase system in the transport of N-acetyl-D-glucosamine by *Escherichia coli*. Biochem. J. **118:**89–92.

900. **White, R. J., and P. W. Kent.** 1970. An examination of the inhibitory effects of N-iodoacetylglucosamine on *Escherichia coli* and isolation of resistant mutants. Biochem. J. **118:**81–87.

901. **White, R. J., and C. A. Pasternak.** 1967. The purification and properties of N-acetylglucosamine-6-phosphate deacetylase from *Escherichia coli*. Biochem. J. **105:**121–125.

902. **Wiesmeyer, H., and M. Cohn.** 1960. The characterization of the pathway of maltose utilization by *Escherichia coli*. I. Purification and physical chemical properties of the enzyme amylomaltase. Biochim. Biophys. Acta **39:**417–426.

903. **Wiesmeyer, H., and M. Cohn.** 1960. The characterization of the pathway of maltose utilization by *Escherichia coli*. II. General properties and mechanism of action of amylomaltase. Biochim. Biophys. Acta **39:**427–439.

904. **Wiesmeyer, H., and M. Cohn.** 1960. The characterization of the pathway of maltose utilization by *Escherichia coli*. III. A description of the concentrating mechanism. Biochim. Biophys. Acta **39:**440–447.

905. **Willis, D. K., B. E. Uhlin, K. S. Amini, and A. J. Clark.** 1981. Physical mapping of the *srl recA*: analysis of Tn10 generated insertions and deletions. Mol. Gen. Genet. **183:**497–504.

906. **Willis, R. C., and C. E. Furlong.** 1974. Purification and properties of a ribose-binding protein from *Escherichia coli*. J. Biol. Chem. **249:**6926–6929.

907. **Willis, R. C., R. G. Morris, C. Cirakoglu, G. D. Schellenberg, N. H. Gerber, and C. E. Furlong.** 1974. Preparation of the periplasmic binding proteins from *Salmonella typhimurium* and *Escherichia coli*. Arch. Biochem. Biophys. **161:**64–75.

908. **Wilson, D. B.** 1974. The regulation and properties of the galactose transport system in *Escherichia coli* K12. J. Biol. Chem. **249:**553–558.

909. **Wilson, D. B.** 1976. Properties of the entry and exit reactions of the beta-methyl galactoside transport system in *Escherichia coli*. J. Bacteriol. **126:**1156–1165.

910. **Wilson, D. B., and D. S. Hogness.** 1964. The enzymes of the galactose operon in *Escherichia coli*. I. Purification and characterization of uridine diphosphogalactose 4-epimerase. J. Biol. Chem. **239:**2469–2481.

911. **Wilson, D. M., and S. Ajl.** 1957. Metabolism of L-rhamnose by *Escherichia coli*. I. L-Rhamnose isomerase. J. Bacteriol. **73:**410–414.

912. **Wilson, D. M., and S. Ajl.** 1957. Metabolism of L-rhamnose by *Escherichia coli*. II. The phosphorylation of L-rhamnulose. J. Bacteriol. **73:**415–420.

913. **Wilson, D. M., M. Kusch, J. L. Flagg-Newton, and T. H. Wilson.** 1980. Control of lactose transport in *Escherichia coli*. FEBS Lett. **117:**37–44.

914. **Wilson, D. M., R. M. Putzrath, and T. H. Wilson.** 1981. Inhibition of growth of *Escherichia coli* by lactose and other galactosides. Biochim. Biophys. Acta **649:**377–384.

915. **Wilson, T. H., and E. R. Kashket.** 1969. Isolation and properties of thiogalactoside transacetylase-negative mutants of *Escherichia coli*. Biochim. Biophys. Acta **173:**501–508.

916. **Wilson, T. H., and M. Kusch.** 1972. A mutant of *Escherichia coli* K 12 energy-uncoupled for lactose transport. Biochim. Biophys. Acta **255:**786–797.

917. **Wilson, T. H., M. Kusch, and E. R. Kashket.** 1970. A mutant in *Escherichia coli* energy-uncoupled for lactose transport: a defect in the lactose-operon. Biochem. Biophys. Res. Commun. **40:**1409–1414.

918. **Wilson, T. H., and E. C. C. Lin.** 1980. Evolution of membrane bioenergetics. J. Supramol. Struct. **13:**421–446.

919. **Wilson, T. H., and P. C. Mahoney.** 1976. Speculations on the evolution of ion transport mechanisms. Fed. Proc. **35:**2174–2179.

920. **Wilson, T. H., K. Ottina, and D. M. Wilson.** 1982. Melibiose transport in bacteria, p. 33–39. *In* A. Martonosi (ed.), Membranes and transport, vol. 2. Plenum Publishing Corp., New York.

921. **Wilson, T. H., and D. M. Wilson.** 1983. Sugar-cation cotransport systems in bacteria, p. 1–39. *In* E. Elson, W. Frazier, and L. Glaser (ed.), Cell membranes: methods and reviews, vol. 1. Plenum Publishing Corp., New York.

922. **Wiman, M., G. Bortani, B. Kelly, and I. Sasaki.** 1970. Genetic map of *Escherichia coli* C. Mol. Gen. Genet. **107:**1–31.

923. **Winkler, H. H.** 1966. A hexose-phosphate transport system in *Escherichia coli*. Biochim. Biophys. Acta **117:**231–240.

924. **Winkler, H. H.** 1970. Compartmentation in the induction of the hexose-6-phosphate transport system of *Escherichia coli*. J. Bacteriol. **101:**470–475.

925. **Winkler, H. H.** 1971. Kinetics of exogenous induction of the hexose-6-phosphate transport system of *Escherichia coli*. J. Bacteriol. **107:**74–78.

926. **Winkler, H. H.** 1973. Energy coupling of the hexose phosphate transport system in *Escherichia coli*. J. Bacteriol. **116:**203–209.

927. **Wohlhieter, J. A., J. R. Lazere, N. J. Snellings, E. M. Johnson, R. M. Synenki, and L. S. Baron.** 1975. Characterization of transmissible genetic elements from sucrose-fermenting *Salmonella typhimurium*. J. Bacteriol. **122:**401–406.

928. **Wolfe, J. B., B. B. Britton, and H. I. Nakada.** 1957. Glucosamine degradation by *Escherichia coli*. III. Isolation and studies of "phosphoglucosaminisomerase-aminisomerase". Arch. Biochem. Biophys. **66:**333–339.

929. **Wolfe, J. B., R. Y. Morita, and H. I. Nakada.** 1956. Glucosamine degradation by *Escherichia coli*. Observations with lasting cells and dry-cell preparations. Arch. Biochem. Biophys. **63:**480–488.

930. **Wolfe, J. B., and H. I. Nakada.** 1956. Glucosamine degradation by *Escherichia coli*. II. The isomeric conversion of glucosamine 6-PO$_4$ to fructose 6-PO$_4$. Arch. Biochem. Biophys. **64:**489–497.

931. **Wolff, J. B., and N. O. Kaplan.** 1956. Hexitol metabolism in *Escherichia coli*. J. Bacteriol. **71:**557–564.

932. **Wolff, J. B., and N. O. Kaplan.** 1956. D-Mannitol 1-phosphate dehydrogenase from *Escherichia coli*. J. Biol. Chem. **218:**849–869.

933. **Wong, P. T. S., E. R. Kashket, and T. H. Wilson.** 1970. Energy coupling in the lactose transport system of *Escherichia coli*. Proc. Natl. Acad. Sci. USA **65:**63–69.

934. **Woodward, M. J., and H. P. Charles.** 1982. Genes for L-sorbose utilization in *Escherichia coli*. J. Gen. Microbiol. **128:**1969–1980.

935. **Woodward, M. J., and H. P. Charles.** 1983. Polymorphism in *Escherichia coli*: rtl, atl and gat regions behave as chromosomal alternatives. J. Gen. Microbiol. **129:**75–84.

936. **Wovcha, M. G., D. L. Steuerwald, and K. E. Brooks.** 1983. Amplification of D-xylose and D-glucose isomerase activites in *Escherichia coli* by gene cloning. Appl. Environ. Microbiol. **45:**1402–1404.

937. **Wu, H. C., and T. C. Wu.** 1971. Isolation and characterization of a glucosamine-requiring mutant of *Escherichia coli* K-12 defective in glucosamine-6-phosphate synthetase. J. Bacteriol. **105:**455–466.

938. **Wu, H. C. P.** 1966. Endogenous induction of the galactose oper-on and the galactose transport system in *Escherichia coli* K-12. Proc. Natl. Acad. Sci. USA **55**:622–629.

939. **Wu, H. C. P.** 1967. Role of the galactose transport system in the establishment of endogenous induction of the galactose operon in *E. coli*. J. Mol. Biol. **24**:213–223.

940. **Wu, H. C. P., and H. M. Kalckar.** 1966. Endogenous induction of the galactose operon in *Escherichia coli* K12. Proc. Natl. Acad. Sci. USA **55**:622–629.

941. **Wu, H. C. P., and H. M. Kalckar.** 1969. Role of the galactose transport system in the retention of intracellular galactose in *Escherichia coli*. J. Mol. Biol. **41**:109–120.

942. **Wu, T.-T.** 1976. Growth of a mutant of *Escherichia coli* K-12 on xylitol by recruiting enzymes for D-xylose and L-1,2-propanediol metabolism. Biochim. Biophys. Acta **428**:656–663.

943. **Wu, T.-T.** 1976. Growth on D-arabitol of a mutant strain of *Escherichia coli* K12 using a novel dehydrogenase and enzymes related to L-1,2-propanediol and D-xylose metabolism. J. Gen. Microbiol. **94**:246–256.

944. **Wu, T.-T., E. C. C. Lin, and S. Tanaka.** 1968. Mutants of *Aerobacter aerogenes* capable of utilizing xylitol as a novel carbon source. J. Bacteriol. **96**:447–456.

945. **Yarmolinsky, M. B., H. Wiesmeyer, H. M. Kalckar, and E. Jordan.** 1959. Hereditary defects in galactose metabolism in *Escherichia coli* mutants. II. Galactose-induced sensitivity. Proc. Natl. Acad. Sci. USA **45**:1785–1791.

946. **Yashphe, J., and N. O. Kaplan.** 1975. Revertants of *Escherichia coli* mutants defective in the cyclic AMP system. Arch. Biochem. Biophys. **167**:388–392.

947. **Yazyu, H., S. Shiota-Niiya, T. Shimamoto, H. Kanazawa, M. Futai, and T. Tsuchiya.** 1984. Nucleotide sequence of the *melB* gene and characteristics of deduced amino acid sequence of the *melB* gene and characteristics of deduced amino acid sequence of the melibiose carrier in *Escherichia coli*. J. Biol. Chem. **259**:4320–4326.

948. **Zabin, I.** 1963. Crystalline thiogalactoside transacetylase. J. Biol. Chem. **238**:3300–3306.

949. **Zabin, I., and A. V. Fowler.** 1978. Beta-galactosidase, the lactose permease protein, and thiogalactoside transacetylase, p. 89–121. *In* J. H. Miller and W. S. Reznikoff (ed.), The operon. Cold Spring Harbor Laboratory, Cold Spring Harbor, N.Y.

950. **Zabin, I., A. Kepes, and J. Monod.** 1959. On the enzymic acetylation of isopropyl beta-D-thiogalactoside and its association with galactoside-permease. Biochem. Biophys. Res. Commun. **1**:289–292.

951. **Zabin, I., A. Kepes, and J. Monod.** 1962. Thiogalactoside transacetylase. J. Biol. Chem. **237**:253–257.

952. **Zipser, D.** 1963. A study of the urea-produced subunits of beta-galactosidase. J. Mol. Biol. **7**:113–121.

953. **Zlotnikov, K. M., V. V. Sukhodoletz, and G. E. Baumanis.** 1969. The mapping of the genes which control the catabolism of nucleosides in *Escherichia coli* K-12. Genetika **5**:114–119.

954. **Zukin, R. S., P. G. Strange, L. R. Heavey, and D. E. Koshland, Jr.** 1977. Properties of the galactose binding protein of *Salmonella typhimurium* and *Escherichia coli*. Biochemistry **16**:381–386.

955. **Zwaig, N., W. S. Kistler, and E. C. C. Lin.** 1970. Glycerol kinase, the pacemaker for the dissimilation of glycerol in *Escherichia coli*. J. Bacteriol. **102**:753–759.

956. **Zwaig, N., and E. C. C. Lin.** 1966. Feedback inhibition of glyc-erol kinase, a catabolic enzyme in *Escherichia coli*. Science **153**:755–757.

957. **Zwaig, N., and E. C. C. Lin.** 1966. A method for isolating mu-tants resistant to catabolite repression. Biochem. Biophys. Res. Commun. **22**:414–418.

19. Two-Carbon Compounds and Fatty Acids as Carbon Sources[†]

WILLIAM D. NUNN

Department of Molecular Biology and Biochemistry, University of California, Irvine, California 92717

INTRODUCTION

Gram-negative bacteria such as *Escherichia coli* and *Salmonella typhimurium* can utilize fatty acids and acetate as sole carbon and energy sources (25, 44, 68). After entry into the cell, fatty acids are either catabolized or directly incorporated into complex lipids. Fatty acid degradation occurs by the cyclic β-oxidation and thiolytic cleavage of fatty acids, yielding several moles of acetyl coenzyme A (acetyl-CoA). The acetyl-CoA produced is further metabolized, yielding energy and precursors for cellular biosynthesis. A considerable body of information regarding the physiology, genetics, and molecular biology of fatty acid and acetate metabolism has been amassed from studies with *E. coli*. These studies have enabled workers to define the structural and regulatory genes required for the catabolism of fatty acids and acetate. Studies involving the use of recombinant DNA technology, coupled with studies that involve the purification and characterization of enzymes, have enabled workers to correlate the structural genes (*fad*) with their respective gene products. One consequence of these efforts is that our understanding of fatty acid transport processes has been significantly enhanced. These studies have also uncovered a regulatory interaction among

fatty acid degradation, acetate metabolism, and unsaturated fatty acid (UFA) biosynthesis.

The purpose of this chapter is to review the studies which have illuminated the molecular details of fatty acid and acetate catabolism in *E. coli* and *S. typhimurium*. Although no major reviews involving the catabolism of fatty acids, acetate, or both have been written concerning these organisms, this article will not provide a comprehensive review on all aspects of these processes in *E. coli* and *S. typhimurium*. Instead, I shall examine the literature on fatty acid and acetate catabolism with emphasis on studies that use genetic and biochemical manipulations to define the structural and regulatory components involved in these reactions.

GENETIC AND BIOCHEMICAL ANALYSIS OF FATTY ACID OXIDATION

Most of our knowledge of the genetics and biochemistry of fatty acid degradation has been derived from studies with *E. coli*. Growth of wild-type *E. coli* on fatty acids as sole carbon sources occurs only when the fatty acid is 12 or more carbons long (long-chain fatty acids [LCFA]), and then only after a distinct lag period. The synthesis of at least five fatty acid-oxidative (FAO) enzymes (Fig. 1) is coordinately induced when LCFA (C_{12} to C_{18}) are present in the growth medium (44, 45, 69). The genetic studies of Klein et al. (25) and Overath and co-workers (45) were the first evidence that the structural genes encoding the FAO enzymes are located at several sites on the *E. coli* chromosome (Fig. 2) and compose a regulon (referred to as the *fad* regulon). The *fad* regulon is primarily

[†] It is with deep regret that the Editors note the untimely death (1 July 1986) of William D. Nunn. He had written this chapter in its entirety, but some editorial details and all of the proofreading were graciously completed by two of his colleagues at the University of California-Irvine, Rowland H. Davis and Boctor Said. We are grateful for their generous assistance.

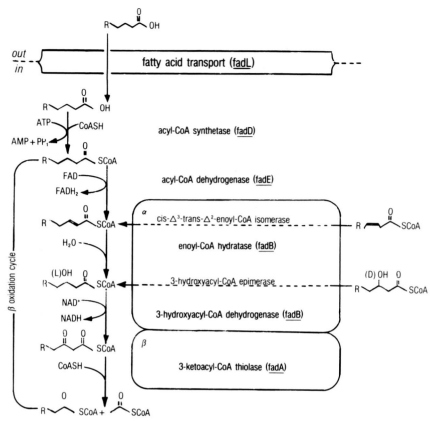

FIG. 1. The cyclic pathway of fatty acid degradation. Principal enzymes of the pathway are listed on the right, along with the respective structural genes of the *fad* regulon. Acetyl-CoA is further metabolized in the tricarboxylic acid cycle.

responsible for the transport, acylation, and β-oxidation of medium-chain fatty acids (MCFA; C_7 to C_{11}) and LCFA (C_{12} to C_{18}). Growth of *E. coli* on short-chain fatty acids (SCFA; C_4 to C_6) requires, in addition to the FAO enzymes, two additional degradative enzymes (Fig. 3) encoded by the *atoD*, *atoA*, and *atoB* genes (49). The latter genes appear to be regulated by the *atoC* gene (49).

In this section, I shall examine the genetic and molecular information regarding the FAO and short-

chain fatty acid oxidative (ATO) enzymes. In addition, we will describe how SCFA, MCFA, and LCFA are taken up by *E. coli*. After the descriptive material on the FAO and ATO enzymes and transport systems, I will relate what is known regarding the mechanism

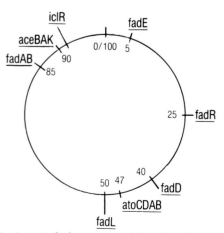

FIG. 2. Genetic linkage map of *E. coli* K-12 showing the location of *fad*, *ato*, and *ace* structural and regulatory genes. Adapted from the revised linkage map of Bachmann (1).

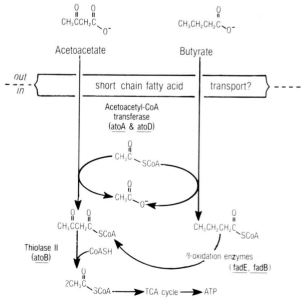

FIG. 3. Pathways of SCFA degradation in *E. coli*.

by which the *fad* and *ato* structural genes are regulated.

Studies on the FAO Enzyme System

The basic pathways by which *E. coli* degrades fatty acids are substantially similar to the β-oxidative pathways present in mammals and other eucaryotic organisms. This pathway is a classical example of the oxidation of a series of homologous substrates through a series of homologous intermediates. Certain features of the pathway are illustrated in Fig. 1. With each turn of the β-oxidation cycle, the fatty acyl-CoA loses a two-carbon fragment as acetyl-CoA and reduces one molecule of flavin adenine dinucleotide (during the acyl-CoA dehydrogenase reaction) and one molecule of NAD (during the 3-hydroxyacyl-CoA dehydrogenase [3-OH acyl-CoA dehydrogenase] reaction). Acetyl-CoA, produced in the CoA-dependent thiolytic cleavage, is further metabolized in the tricarboxylic acid cycle. The other product of the cleavage step, a shortened fatty acyl-CoA molecule, reenters the degradation cycle without further activation (44).

The first step of fatty acid degradation is the activation of the free fatty acid to an acyl-CoA thioester by acyl-CoA synthetase (fatty acid:CoA ligase [AMP forming]; EC 6.2.1.3). This reaction requires two high-energy phosphate equivalents per molecule of free fatty acid activated. Overath et al. (44) have suggested that *E. coli* has one acyl-CoA synthetase with broad specificity for MCFA and LCFA. As supporting evidence, these workers have shown that an *fadD* mutant isolated in their laboratory lacks acyl-CoA synthetase activity for MCFA and LCFA (44). In contrast to the hypothesis of Overath et al. (44), Samuel et al. (56), on the basis of their studies with partially purified acyl-CoA synthetase, have suggested that the acyl-CoA synthetase is a complex of two enzymes, one that activates MCFA and another that activates LCFA. To resolve whether *E. coli* has either a single acyl-CoA synthetase with broad substrate specificity or multiple acyl-CoA synthetases with limited substrate specificity, Kameda and Nunn (24) purified to homogeneity the *E. coli* acyl-CoA synthetase and found that the purified enzyme had broad substrate specificity for both MCFA and LCFA. Although acyl-CoA synthetase was previously believed to be a membrane-associated protein, Kameda and Nunn (24) demonstrated that over 90% of this enzyme was present in cytoplasmic fractions. The molecular weight of the native enzyme was approximately 130,000, and the subunit molecular weight determined by polyacrylamide gel electrophoresis in the presence of sodium dodecyl sulfate was 47,000. These experiments suggested that the enzyme may be a dimer or a trimer composed of apparently identical subunits (24).

In *E. coli*, very little is known about the enzyme responsible for the next step, acyl-CoA dehydrogenase. Although mutants (*fadE*) lacking this enzymatic activity have been isolated and mapped (25), it is not certain whether the loss of the dehydrogenase protein or of an associated flavoprotein is responsible for the observed loss of activity.

The other β-oxidation enzymes are also cytoplasmic in *E. coli*. The remaining β-oxidation enzymes (Fig. 1) are part of a multienzyme complex that has broad substrate specificity. In *E. coli*, two proteins in the β-oxidative pathway are associated with a multienzyme complex which has a molecular mass of 260,000 daltons (3, 43, 50, 51). Five FAO enzyme activities, 3-ketoacyl-CoA thiolase (EC 2.3.1.16), enoyl-CoA hydratase (EC 4.2.1.17), 3-hydroxyacyl-CoA dehydrogenase (EC 1.1.1.35), cis-Δ^3-trans-Δ^2-enoyl-CoA isomerase (EC 5.3.3.3), and 3-hydroxyacyl-CoA epimerase (EC 5.1.2.3) are associated with this multienzyme complex (3, 50, 51). Binstock et al. (3), Pawar and Schulz (49), and Pramanik et al. (51) have purified the complex and found it to have an $\alpha_2 \beta_2$ subunit structure (α, 78,000 daltons; β, 42,000 daltons). Through biochemical characterization of the multienzyme complex from an *E. coli* B strain, these investigators have determined that 3-ketoacyl-CoA thiolase activity is associated with the 42,000-dalton subunit and that the remaining four enzyme activities are associated with the larger, 78,000-dalton subunit (3, 50, 51). For saturated fatty acid oxidation, three final steps of the pathway are needed: enoyl-CoA hydratase, 3-hydroxyacyl-CoA dehydrogenase, and 3-ketoacyl-CoA thiolase. All are catalyzed by the complex (Fig. 1). When UFA are degraded, two additional activities also carried by the complex, cis-Δ^3-trans-Δ^2-enoyl-CoA isomerase and 3-hydroxyacyl-CoA epimerase, are required (Fig. 1). Although no mutants lacking the latter two activities have been isolated, pleiotropic *fadAB* mutants (44, 45), and biochemical studies (50, 51, 59) have demonstrated that the 78,000-dalton protein also contains both cis-Δ^3-trans-Δ^2-enoyl-CoA isomerase and 3-hydroxyacyl-CoA epimerase activities.

Overath et al. (44) have suggested that the genes for the enzymes 3-ketoacyl-CoA thiolase, 3-hydroxyacyl-CoA dehydrogenase, enoyl-CoA hydratase, and possibly 3-hydroxyacyl-CoA epimerase and cis-Δ^3-trans-Δ^2-enoyl-CoA isomerase form an operon. Their evidence for suggesting that the *fadAB* genes constitute an operon was based on the high coordinate induction of 3-ketoacyl-CoA thiolase, 3-hydroxyacyl-CoA dehydrogenase, and enoyl-CoA hydratase, as well as on the genetic mapping results with mutants deficient in (i) all five enzymes (*fadAB*), (ii) 3-keto-CoA thiolase (*fadA*), and (iii) 3-hydroxyacyl-CoA dehydrogenase (44). Since there was no genetic evidence that the *fadAB* genes were organized into an operon, Spratt et al. (59) cloned the *fadAB* genes from *E. coli* to determine the organization of these *fad* structural genes (Fig. 4). The approximate location and orientation of the *fadA* and *fadB* genes were determined by subcloning and by Tn5 mutagenesis (59). These studies supported the hypothesis of Overath et al. (44), that the *fadAB* genes are part of an operon, and showed that the direction of transcription is from *fadA* to *fadB* (59). The *fadA* and *fadB* gene products were identified by maxicell analysis. The *fadA* gene was found to code for the 42,000-dalton β subunit, and the *fadB* gene was found to be associated with the 78,000-dalton α subunit. A hybrid plasmid containing the *fadA* gene expressed 3-ketoacyl-CoA thiolase I activity in *fadA* mutants but not in *fadAB* mutants (Table 1). The latter results were interpreted to mean that the 42,000-dalton β subunit (representing 3-ketoacyl-CoA thiolase I) may not be functional in the absence of the *fadB* gene product, the 78,000-dalton α subunit. Some support for this contention comes from studies which show that *fadA* mutants carrying multicopy *fadA+B+* plasmids ex-

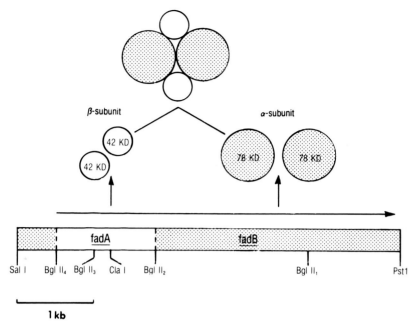

FIG. 4. Structural organization and direction of transcription of the *fadAB* region. The relative location of the *fadA* and *fadB* genes and the direction of transcription are indicated. The endpoints of these genes are not precisely defined and are indicated by a broken line. The protein products (expressed in kilodaltons [KD]) specified by the genes are indicated at the top of the figure. The two protein subunits encoded by the *fadAB* genes form a multifunctional enzyme complex which catalyzes at least five fatty acid-degradative enzyme activities (see text). Restriction sites *Pst*I, *Bgl*II, *Cla*I, and *Sal*I are indicated.

press both the 78,000- and the 42,000-dalton proteins and that they have amplified levels of 3-ketoacyl-CoA thiolase activity, whereas those containing multicopy *fadA*+ plasmids express only the 42,000-dalton protein and have wild-type levels of 3-ketoacyl-CoA thiolase activity (59). It appears that in the latter case the *fadA* mutant, containing the multicopy *fadA*+ plasmid and high levels of the 42,000-dalton β subunit, may be restricted from expressing amplified levels of 3-ketoacyl-CoA thiolase activity because it has only wild-type levels of the 78,000-dalton α subunit. Additional support for this view comes from studies showing that wild-type strains harboring the multicopy *fadA*+ plasmid have levels of 3-ketoacyl-CoA thiolase activity comparable to those of wild-type strains which contain either no plasmid or the plasmid vector pUC9 (W. D. Nunn, unpublished data).

Pawar and Schulz (50) have also shown that the medium-chain enoyl-CoA hydratase but not the long-chain enoyl-CoA hydratase is associated with the *fadAB* multienzyme complex. Although it is not clear where the gene that codes for the soluble long-chain enoyl-CoA hydratase maps, it is most likely linked to the *fadAB* genes. The reason for suggesting this is that during their characterization of clones carrying the *fadAB* genes, Spratt et al. (59) observed an additional approximately 60,000-dalton protein encoded by the plasmid. The synthesis of this protein, like that of the *fadA* and *fadB* gene products, was prevented by polar transposon Tn5 insertions in the *fadA* gene (59). Additional studies that correlate the 60,000-dalton protein with long-chain enoyl-CoA hydratase activity must be performed before definitive proof of the role of this protein can be established.

TABLE 1. FAO enzyme activities in *E. coli* strains containing different *fadAB* plasmids

Plasmid/strain (genotype)	Gene complemented by plasmid	Sp act[a]		
		3-Keto-CoA thiolase	3-Hydroxyacyl-CoA dehydrogenase	Enoyl-CoA hydratase
pBR322				
LE392 (wild type)	None	14	304	1,613
LS6749 (*fadAB*)		0[b]	17	16
LS6595 (*fadA*)		0	280	1,680
pK52/LS6749 (*fadAB*)	*fadAB*	308	1,745	7,004
pK1				
LS6749 (*fadAB*)	*fadA*	0	0	0
LS6595 (*fadA*)		12	ND[c]	ND

[a] Expressed in nanomoles per minute per milligram of protein.
[b] A measurement of zero indicates activity of <0.05.
[c] ND, Not determined.

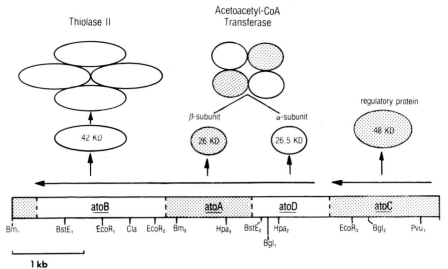

FIG. 5. Physical map of the chromosomal region encoding the *ato* structural and regulatory genes. The relative location of the *atoD*, *atoA*, *atoB*, and *atoC* genes, along with the direction of transcription, are indicated. The endpoints of these genes are not precisely defined. The protein products of the *ato* genes (*atoD*, *atoA*, and *atoB*) enable *E. coli* to degrade SCFA (see text). The *atoC* gene regulates the expression of *atoD*, *atoA*, and *atoB*.

Studies on the ATO Enzyme System

Wild-type *E. coli* can utilize exogenous LCFA as sole carbon and energy source because LCFA induce the synthesis of the FAO enzymes (44, 45, 69). MCFA (C_7 to C_{11}) can serve as substrates for the FAO enzymes but cannot induce the synthesis of these enzymes. Therefore, only fatty acids more than 12 carbons long may be used as the sole carbon sources by the wild type (*fadR$^+$*). Strains that constitutively synthesize the FAO enzymes grow on MCFA and LCFA. The constitutive strains have been given the designation *fadR*.

fadR$^+$ and *fadR* strains cannot utilize SCFA (C_4 to C_6) as sole carbon sources (48, 53, 54, 67). For growth to occur on the SCFA butyrate (C_4) and valerate (C_5), *E. coli* must have constitutive levels of three FAO enzymes (enoyl-CoA hydratase, 3-hydroxyacyl-CoA dehydrogenase, and acyl-CoA dehydrogenase) and at least two enzymes involved with the degradation of the β-keto SCFA acetoacetate (48, 66). The degradation of acetoacetate to acetyl-CoA is a two-step reaction (Fig. 3) that first results in the activation of acetoacetate to acetoacetyl-CoA, which is catalyzed by acetyl-CoA:acetoacetyl-CoA transferase, and then involves subsequent cleavage of acetoacetyl-CoA to acetyl-CoA, catalyzed by 3-ketoacyl-CoA thiolase II (15, 16, 49, 61, 66, 67). These enzymes are highly inducible in the presence of acetoacetate and have substrate specificities for β-keto SCFA.

Earlier studies performed by Pauli and Overath (49) identified the loci responsible for acetoacetate degradation as *atoA*, which encodes acetoacetyl-CoA transferase, and *atoB*, which encodes 3-ketoacyl-CoA thiolase II. These structural genes are closely linked and are located at the 47 min region on the revised *E. coli* chromosomal map (49). The biochemical work of Sramek and Frerman (61, 62) has shown that acetoacetyl-CoA transferase is a tetrameric protein composed of two α and two β subunits and that 3-ketoacyl-

CoA thiolase II is a tetrameric protein composed of four identical subunits.

Although only one gene, *atoA*, had initially been correlated with acetoacetyl-CoA transferase activity (49), the biochemical studies of Sramek and Frerman (61) suggested that two genes, encoding the α and β subunits, may be required for functional acetoacetyl-CoA transferase activity. To resolve this question and to characterize the structural organization and regulation of the *ato* structural genes more thoroughly, L. S. Jenkins and W. D. Nunn (J. Bacteriol., in press) have cloned the *ato* genes. Three *ato* structural genes and one *ato* regulatory gene have been identified on a 6.2-kilobase DNA fragment cloned into pBR322 (Fig. 5). The genes encoded by the 6.2-kilobase *ato* fragment include *atoD* and *atoA*, which encode the two subunits of acetoacetyl-CoA transferase; *atoB*, which encodes 3-ketoacyl-CoA thiolase II; and *atoC*, which encodes a regulatory element for the *ato* system. The proteins encoded by these genes have been identified via in vitro transcription and translation and maxicell protein analysis. *atoD* codes for a 26.5-kilodalton protein, *atoA* codes for a 26-kilodalton protein, and *atoB* codes for a 42-kilodalton protein.

By using subcloning strategies coupled with transposon mutagenesis, Jenkins and Nunn (unpublished results) have established that the *atoD*, *atoA*, and *atoB* structural genes compose an operon, with the direction of transcription from *atoD* to *atoB* (Fig. 5).

TRANSPORT OF FATTY ACIDS

MCFA and LCFA Transport

Before β-oxidation, fatty acids must enter the cell via uptake systems which translocate them across the membrane. Before any rigorous kinetic analyses, the prevailing thought was that fatty acids diffuse through membranes without requiring a protein carrier. However, physiological and kinetic studies performed in

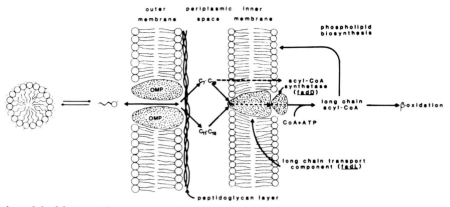

FIG. 6. Proposed model of fatty acid transport in *E. coli* K-12. LCFA (C_{12} to C_{18}) and MCFA (C_7 to C_{11}) traverse the outer membrane via the outer membrane porins (partially encoded by *ompC* and *ompF*). LCFA and MCFA bind the long-chain transport component (*fadL* protein) and traverse the inner membrane concurrently with activation by the acyl-CoA synthetase (*fadD* gene product). MCFA also traverse the inner membrane by a diffusional process.

several laboratories (16, 33, 44) suggest that a carrier mechanism facilitates the entry of LCFA (C_{12} to C_{18}) into *E. coli*. Furthermore, genetic and biochemical studies have implicated at least two proteins, encoded by the *fadD* and *fadL* genes, that are required to deliver exogenous LCFA across the cell membrane to the cytosolic fatty acid-degradative enzymes. The *fadD* gene codes for a peripheral membrane-bound protein, the acyl-CoA synthetase, which has broad-chain-length specificity (24). The *fadL* gene codes for a 43,000-dalton cytoplasmic membrane protein (FLP) which has been implicated to be essential for LCFA transport (see Addendum in Proof). Although the exact role of the *fadL* gene product in the uptake of LCFA is unknown, fatty acid binding studies have demonstrated a correlation between the presence of FLP and LCFA binding activity (39). The *fadL* gene product, FLP, is the first integral membrane protein which has been shown both genetically and biochemically to be involved in the uptake of fatty acids (34, 41, 42). As a working model consistent with results obtained from genetic and biochemical studies (16, 34, 41, 42, 44) on the transport system(s) in *E. coli*, Black et al. (4), Maloy et al. (32), and Nunn et al. (39) have proposed that LCFA are adsorbed to the integral FLP membrane-binding sites, from which they are transferred to the peripheral membrane-bound acyl-CoA synthetase where they are activated and released into the cytosol (Fig. 6). When the *fadL* transport system, which also actively transports MCFA (C_7 to C_{11}), is nonfunctional, MCFA are still able to diffuse across the cell membrane to the acyl-CoA synthetase, where they are activated and released into the cell (Fig. 6). The transport of SCFA (C_4 to C_6) also appears to be carrier mediated (15) but does not require functioning of either the *fadD* or the *fadL* gene products.

Several observations demonstrate that the LCFA enter *E. coli* via an active unidirectional carrier-mediated mechanism(s). (i) The transport of LCFA into wild-type strains has been shown to be a process in which the presumed binding protein can be saturated (16, 34). (ii) Inhibitors which prevent electron transport or uncouple oxidative phosphorylation or both have been shown to block LCFA transport completely (16, 34). (iii) No efflux of transported LCFA

occurred when wild-type *E. coli* strains were washed with unlabeled LCFA (34, 44). (iv) Both the energy of activation and the Q_{10} of LCFA transport are representative of enzyme-mediated processes (34). As indicated above, the *fadD* and *fadL* gene products are required to deliver LCFA across the membrane to the cytosolic fatty acid-degradative enzymes of *E. coli*. Since preliminary studies indicate that the *fadL* gene product does not require energy to adsorb LCFA (W. D. Nunn, unpublished results), the active component required to deliver LCFA into the cytosol appears to be the acyl-CoA synthetase.

The evidence for suggesting that the acyl-CoA synthetase is required for MCFA and LCFA transport into the cell was obtained from the studies of Klein et al. (25) and Frerman and Bennett (16), which showed that *fadD* mutants are unable to accumulate exogenous fatty acids of any length into the cytosol or membrane lipids. These investigators suggested that the acyl-CoA synthetase may be required for a group translocation step in the transport of fatty acids (i.e., vectoral thiol esterification). They based their translocation hypothesis upon studies showing that (i) the chain-length specificity of the acyl-CoA synthetase for fatty acids could be correlated with chain-length specificity for the uptake system, (ii) no efflux of labeled fatty acids occurred when preloaded cells were diluted into excess unlabeled fatty acids, and (iii) no free labeled fatty acids were detected intracellularly. It should be mentioned, however, that neither has the putative translocation product, fatty acyl-CoA, been detected intracellularly. In contrast to Klein et al. (25), Rock and Jackowski (52), on the basis of studies showing that *fadD* strains incorporate approximately 2% of the amount of labeled fatty acids that wild-type cells incorporate, have concluded that the *fadD* gene product is not required for fatty acid transport. The latter studies are difficult to assess because Rock and Jackowski (52) measured only the incorporation of labeled fatty acid into membrane lipids. Since these workers did not show that their *fadD* strain was capable of delivering LCFA into its cytosolic fraction (52), their studies have not definitively disproved the contention of Klein et al. (25) that the *fadD* gene product is essential for delivering fatty acids into the cytosol of *E. coli*. Obviously, more studies, especially in vitro reconstitution experiments, will be required to resolve the role of the acyl-CoA synthetase in fatty acid transport. The fact that mutations in the *fadD* gene disrupt fatty acid transport more drastically than mutations in any of the other *fad* genes, except *fadL*, warrants a more thorough inspection of the involvement of this gene in the transport process.

The *fadL* gene, which is located at 50 min on the *E. coli* linkage map, was identified by Nunn and Simons (41) in 1978 as being essential for LCFA transport. The evidence for the latter suggestion was as follows. (i) *fadL fadR* strains, which are constitutive for the synthesis of FAO enzymes, can oxidize MCFA but not LCFA. (ii) In vitro extracts of *fadR fadL* strains oxidize LCFA at rates comparable to those of extracts from *fadR fadL*⁺ strains. (iii) Toxic LCFA analogs inhibit the growth of *fadL*⁺ strains but not *fadL* strains. (iv) An *fadL* mutation prevents *fabA*(Ts) mutants from satisfying their UFA auxotrophic requirement at nonpermissive temperatures in the presence of UFA. The *fabA*(Ts) allele codes for a thermolabile β-hydroxydecanoyl thioester dehydrase (11).

The *fadL* gene product was first identified by Ginsburgh et al. (19) by comparing the membrane proteins of *fadL*⁺ and *fadL* strains. These workers demonstrated a close correlation between *fadL*⁺-encoded transport activity and the presence of a 33,000-dalton inner membrane protein (19). To show that the *fadL* gene was the structural gene for the 33,000-dalton protein, the *fadL* gene was cloned by Black et al. (4) and shown by in vitro translation to encode the latter protein. Further analysis indicated that the *fadL* gene product (FLP) has an isoelectric point (pI) of 4.6, and that its molecular weight, as determined by sodium dodecyl sulfate gel electrophoresis, was dependent upon sample treatment before electrophoresis. The apparent molecular weight was 33,000 when membrane fractions were heated to 50°C, whereas treatment at 100°C modified the protein such that it migrated more slowly, at an apparent molecular weight of 43,000 (Fig. 7). The heat-modifiable nature of FLP suggests that it may assume different conformations, similar to many other membrane proteins. P. N. Black and W. D. Nunn (unpublished results) have recently purified FLP and have shown that it has a native molecular mass of 130,000 ± 5,000 daltons. If the monomeric molecular mass of FLP is 33,000 daltons, the latter results would suggest that the native form of FLP may be a tetramer of 33,000-dalton subunits. Alternatively, if the monomeric molecular mass of FLP is 43,000 daltons, the native form of FLP may be a trimer of 43,000-dalton subunits.

Early studies to discern the role of the *fadL* gene product in the fatty acid transport process entailed establishing the kinetic parameters of MCFA and LCFA transport in *fadL*⁺ and *fadL* strains. Maloy et al. (34) showed that *fadL*⁺ strains were capable of transporting LCFA by an active saturable process. MCFA uptake by *fadL*⁺ strains characteristically showed Michaelis-Menten curves (i.e., one in which one component was rate limiting) that suggested the presence of both a carrier-mediated and a free diffusion process (34). In contrast, kinetic analysis of MCFA transport by *fadL* strains indicated a nonsaturable process. These results suggest that the *fadL* gene product,

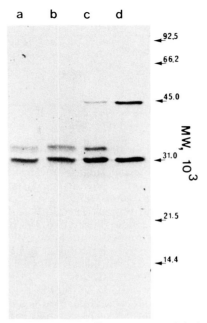

FIG. 7. Fluorographs of [³⁵S]methionine-labeled proteins from maxicells of pACC in LS6922 with different protein solubilization temperatures (4). Lanes: a, 25°C; b, 50°C; c, 70°C; d, 100°C.

although absolutely required for LCFA transport, has some overlapping specificity for MCFA. However, the *fadL* gene product is not required for the transport of MCFA. Since the *fadL* strains were incapable of transporting LCFA (34), these kinetic studies confirmed the contention of Maloy et al. (34), Nunn and Simons (41), and Nunn et al. (42) that the *fadL* gene product was essential for LCFA transport into *E. coli*. Since both the *fadD* and *fadL* gene products are required to deliver LCFA into the cytosol, the kinetic studies did not reveal an unambiguous role for the *fadL* gene product in the LCFA transport process. Therefore, after the identification of the *fadL* gene product as an integral membrane protein (referred to as FLP), the next question that was addressed was whether FLP is a receptor or a translocase or both.

To examine the function of FLP as a receptor, a study which involved comparing LCFA binding activity in *fadD fadL*⁺ and *fadD fadL* strains was performed (39). The *fadD* mutation was included in the latter strains to avoid LCFA transport and LCFA binding contributed by the acyl-CoA synthetase. The binding studies revealed that strains containing a functional *fadL* gene bind significantly more LCFA than do strains containing a defective *fadL* gene (Table 2). In addition, *fadD fadL* strains harboring multicopy *fadL*⁺ plasmids bind 16-fold more LCFA than do *fadD fadL* strains harboring only the plasmid vector (Table 3). Furthermore, whereas an *fadD fadL*⁺ strain has sixfold more energy-independent LCFA binding activity than an *fadD fadL* strain, an *fadD*⁺ *fadL* strain has the same LCFA binding activity as *fadD fadL* strains (Table 2). The latter results imply that, in the LCFA transport process, it is the *fadL* gene product, rather than the *fadD* gene product (acyl-CoA synthetase), that is responsible for sequestering significant quantities of

TABLE 2. [³H]oleate binding by *fadL*, *fadD*, and *fadD fadL* strains[a]

Strain	Genotype	[³H]oleate bound (pmol/mg of protein)
LS6164	*fadL*	65
LS6928	*fadD*[b]	442
LS6929	*fadD fadL*	63

[a] Assayed as described by Nunn et al. (39).

[b] LS6928 contains a wild-type *fadL* gene.

LCFA at sites in the cell membrane for transport. Studies with *fadL* strains harboring plasmids which carry defective *fadL* genes indicate that mutations altering the physical properties of FLP also alter the *fadL* specific LCFA binding activity (39). Collectively, the binding studies suggested that one role of FLP in the transport process is to function as an LCFA receptor. At present, it is not clear whether FLP also functions as an LCFA translocase.

As indicated above, the LCFA binding studies with *fadD fadL*[+], *fadD fadL*, and *fadL* strains suggest that LCFA encounter the *fadL* gene product before being activated by the *fadD* gene product. Further evidence supporting this contention was obtained when an analysis of the lipids bound to the membranes of the *fadD* and *fadL* mutants revealed that over 96% of the LCFA were not esterified (W. D. Nunn, unpublished results). A more extensive analysis of the complex lipid composition of *fadD* and *fadL* strains by Rock and Jackowski (52) has revealed that *fadD* strains, but not *fadL* strains, incorporate about 2% as much exogenous LCFA into their membrane lipids as does their wild-type parent (*fadD*[+] *fadL*[+]). The incorporation of exogenous fatty acids into phospholipids of the wild type originates with the activation of fatty acids by the acyl-CoA synthetase followed by the distribution of the acyl moieties into all phospholipid classes via the *sn*-glycerol-3-phosphate acyltransferase (52). This pathway does not function in *fadD* mutants. Instead, *fadD* mutants, with the aid of a functional *fadL* gene product(s), translocate exogenous fatty acids to an inner membrane phospholipid enzyme system, the acyl-acyl carrier protein synthetase/2-acyl-glycero-

TABLE 3. Binding of [³H]oleate in an *fadD fadL* strain containing various plasmids

Plasmid[a]	[³H]oleate bound (pmol/mg of protein)		
	−BSA[b]	+BSA	+Brij 58[c]
pACYC177	185	64	71
pACC	3,174	995	845
pAEV	3,025	968	817
pACS	1,725	68	82
pACK	1,550	69	78
pACP, pBEB, pABC, pAPP, or pLSS	≤180	≤64	≤68

[a] The plasmids were harbored in the *fadD fadL* strain, LS6929.

[b] The binding assay was performed as described by Nunn et al. (39) in the absence of bovine serum albumin (BSA).

[c] The binding assay was performed as described by Nunn et al. (39), except that bovine serum albumin was replaced with the detergent Brij 58 at a final concentration of 0.5%.

phosphoethanolamine acyltransferase system, which introduces the fatty acid exclusively into position 1 of phosphatidylethanolamine (52). Since *fadL* strains are incapable of incorporating exogenous fatty acids into phospholipids via any route, Rock and Jackowski (52) have suggested that the *fadL* gene product must play a role in translocating these fatty acids to intracellular pools which are accessible to both phospholipid biosynthetic pathways. Since the latter studies do not unequivocally show that exogenous fatty acids can enter the cytosol of *fadD* mutants, it is more likely that the *fadL* gene product delivers exogenous fatty acids to inner membrane pools that are accessible to these pathways.

Mutants which lack acyl-CoA dehydrogenase (*fadE*), 3-ketoacyl-CoA thiolase (*fadA*), 3-hydroxyacyl-CoA dehydrogenase and enoyl-CoA hydratase (*fadB*) activities, or all of these transport significantly less MCFA and LCFA than do their wild-type parents (34, 44). These findings imply that fatty acid transport is coupled to fatty acid oxidation. Although the mechanism(s) is unknown, it is conceivable that transport is reduced in these *fad* mutants as a consequence of feedback inhibition. For example, the accumulation of fatty acyl-CoA intermediates in the *fad* mutants may be inhibitory to the fatty acid transport components (i.e., FLP or acyl-CoA synthetase). β-Oxidation is not an absolute requirement for transport because UFA auxotrophs which are unable to oxidize fatty acids take up and incorporate exogenous UFA into their phospholipids (14, 46). In contrast, conditional UFA auxotrophs [*fabA*(Ts)] which have *fadD* or *fadL* mutations or both become nonviable at nonpermissive temperatures because they cannot incorporate sufficient UFA into phospholipids to satisfy their auxotrophy (W. D. Nunn, unpublished results). The results with the *fabA*(Ts) *fadD* and *fabA*(Ts) *fadL* strains are further testimony that defects in LCFA transport components (encoded by *fadL* and *fadD*) restrict *E. coli* not only from transferring exogenous LCFA across the membrane to the cytosol, where they are degraded, but also from delivering these molecules to sites where bulk membrane phospholipid synthesis occurs. Since *fadE* and *fadAB* mutations do not restrict *fabA*(Ts) strains from incorporating sufficient UFA into their lipids to satisfy their UFA auxotrophy (W. D. Nunn, unpublished data), it is clear that defects in the cytosolic fatty acid-degradative enzymes do not affect the uptake of LCFA as adversely as defects in the integral membrane FLP and peripheral membrane acyl-CoA synthetase.

The process of fatty acid uptake requires that fatty acids traverse both the outer and inner membranes of *E. coli*. In addition to the *fadL* mutation, at least two other mutations (*ompC* and *ompF*) reduce the amount of LCFA entering the cell. The *ompC* and *ompF* genes encode proteins which compose part of the outer membrane (37). Hydrophilic molecules are thought to pass through the outer membrane by a specific carrier or via nonspecific pores formed by the outer membrane porins (37). Kinetic studies by Maloy et al. (34) have shown that strains defective in *ompC* and *ompF* have a reduced ability to transport MCLA and LCFA. These studies suggested that there is a nonspecific *fadL*-independent mechanism which allows fatty acids to traverse the outer membrane. Collectively, these studies have led to a tentative model of LCFA transport, which is illustrated in Fig. 6 (34).

SCFA Transport

E. coli can also transport SCFA (C$_4$ and C$_6$). However, the mechanism(s) by which SCFA are transported differs from that involved in the uptake of MCFA and LCFA, as follows. (i) The acyl-CoA synthetase is not required for the uptake of SCFA (2). (ii) SCFA are not transported by *fadR* strains. (iii) SCFA do not competitively inhibit the uptake of MCFA and LCFA (53, 54). Therefore, it appears that the SCFA uptake system is distinct from the uptake system(s) which concentrates MCFA and LCFA. Evidence suggesting that acetoacetyl-CoA transferase may play a role in the transport of SCFA was obtained from the biochemical studies of Frerman (15). By comparing the uptake of C$_4$ acids into membrane vesicles of *E. coli* strains taken from noninduced and induced (i.e., the presence of acetoacetate) cells, Frerman showed that membrane vesicles from induced cells translocate C$_4$ acids. This uptake was stimulated by ATP and acetyl-CoA and did not occur in membrane vesicles from noninduced cells (15). Frerman also found that significant amounts of acetoacetyl-CoA transferase were associated with the membranes and that uptake was rapidly inhibited by butyryl-CoA and acetate, the products of the acetoacetyl-CoA transferase-catalyzed reaction (Fig. 3). Although these results suggest a role for acetoacetyl-CoA transferase in SCFA, they do not unequivocally establish whether this enzyme is solely responsible for SCFA transport. The fact that genetic studies showing that mutations not only in the *atoA* gene but also in *atoB* and *fadAB* genes reduce SCFA transport suggests that SCFA uptake is also contingent upon subsequent SCFA metabolism. Obviously, more studies will have to be performed to unravel the molecular details of the SCFA transport process.

REGULATION OF FATTY ACID DEGRADATION

The fatty acid-degradative (*fad*) system is primarily responsible for the transport, acylation, and β-oxidation of MCFA (C$_7$ to C$_{11}$) and LCFA (C$_{12}$ to C$_{18}$). The *fad* structural genes, which map at no fewer than four distinct loci on the *E. coli* chromosome (Fig. 2), are regulated by the *fadR* gene (44, 57, 58). SCFA (C$_4$ to C$_6$) metabolism requires, in addition to the FAO enzymes, at least two additional degradative enzymes (Fig. 3), which are encoded for by the *ato* genes. The regulation of the *ato* structural genes appears to be regulated by the *atoC* gene (49).

Control of the *fad* Regulon by the *fadR* Gene

The *fadR*$^+$ gene, a multifunctional regulatory gene at 25.5 min, appears to exert negative control over the *fad* regulon (50, 57, 58) and the *aceBA* operon (33, 35, 36). The *fadR*$^+$ gene is also required for maximal expression of the genes (*fab*) for UFA biosynthesis (40). Considerable evidence has accrued which suggests that the *fadR* gene product is a repressor that exerts control over the *fad* regulon and *ace* operon at the level of transcription (see below). Overath and co-workers (44, 45) and Weeks et al. (68) first established

that LCFA can induce FAO enzymes. MCFA can be degraded by the FAO enzymes but do not induce their synthesis. Mutants able to utilize MCFA as sole carbon sources have been isolated by plating wild-type cells onto minimal medium containing the MCFA decanoate as sole carbon source. Overath and co-workers (44, 45) first showed that mutants obtained in this way were constitutive for the FAO enzymes and that they could rapidly oxidize both MCFA and LCFA. These mutants have been termed *fadR*, and Overath et al. (44) have suggested that the *fadR* gene codes for a regulatory protein, possibly a repressor.

Several lines of evidence now support Overath's original hypothesis, indicating that the *fadR* gene product is a diffusible repressor protein. First, genetic mapping studies showed that the *fadR* gene maps at a locus distinct from all other *fad* loci (57, 65). Second, mutants that harbor transposon-mediated insertion mutations in *fadR* (fadR::transposon) have been shown to be constitutive for FAO enzymes. Genetic studies with strains that were stably merodiploid for the *fadR* gene showed that the *fadR*$^+$ gene is *trans*-dominant to *fadR* (58). Furthermore, although inducible, fatty acid oxidation in *fadR* strains harboring multicopy *fadR*$^+$ plasmids is at least twofold lower than in wild-type strains containing one copy of the *fadR*$^+$ gene (12).

Evidence suggesting that the *fadR* gene product regulates the expression of the *fad* regulon at the level of transcription was obtained from studies with *lac* fusions with the *fad* structural genes (10, 55). The studies of Clark (10), with strains harboring *fadAB-lac* and *fadE-lac* operon fusions, and of Sallus et al. (55), with *fadL-lac* operon fusions, revealed that β-galactosidase activity is constitutive in *fadR fad-lac* strains and inducible by LCFA in *fadR*$^+$ *fad-lac* strains. Furthermore, the expression of β-galactosidase was repressed in these strains under catabolite-repressing growth conditions. Noting that FAO enzyme activity in strains harboring a multicopy *fadR*$^+$ plasmid is fourfold less than wild type under noninducing conditions, DiRusso and Nunn (12) determined whether *fad* gene transcriptional activity was regulated similarly by the multicopy *fadR*$^+$ plasmids. Using strains with *fadE-lac* operon fusions, these workers showed that β-galactosidase activity in strains harboring the multicopy *fadR*$^+$ plasmid was fivefold lower than that found in wild-type (harboring one copy of *fadR*$^+$) strains under noninducing conditions (Table 4). The latter studies suggest that *fad* transcriptional activity can be repressed further by the presence of more copies of the *fadR* gene product. Overall, the *fad-lac* operon fusion studies strongly suggest that the *fadR* gene regulates the *fad* regulon at the level of transcription.

In complementation studies with 24 merodiploids harboring *fadR* alleles on both the main chromosome and the episome, Simons et al. (58) found that all of the merodiploids displayed the *fadR* phenotype. These studies suggested that only one polypeptide was encoded by the *fadR* gene. The latter presumption was confirmed when DiRusso and Nunn (12) cloned the *fadR* gene and identified a 29,000-dalton protein as its gene product.

At present, the inducer of the *fad* regulon has not been identified. Mutants defective in the *fadD* and *fadL* genes are unable to induce the other FAO en-

TABLE 4. Transcriptional control of λφ(*fadE-lacZ*+) by different *fadR* plasmids (15)

Plasmid/strain	*fadR* genotype of λφ(*fadE-lacZ*)	β-Galactosidase sp act[a] with the following growth conditions[b]:		
		TB	TB lac	TB ole
pACYC177				
LS6926	Wild type	173	82	743
LS6927	*fadR*::Tn*10*	932	287	837
pACfadR				
LS6926	Wild type	32	24	877
LS6927	*fadR*::Tn*10*	24	22	150

[a] Nanomoles per minute per milligram of protein.

[b] TB, TB medium supplemented with 0.5% Brij 58; TB lac, TB medium supplemented with 0.5% Brij 58 and 0.4% lactose; TB ole, TB medium supplemented with 0.5% Brij 58 and 5 mM oleate. All were supplemented with ampicillin or kanamycin.

zymes, whereas mutants with lesions in the *fadA*, *fadB*, and *fadE* genes remain inducible for the remaining functional FAO enzymes and transport activities (25, 41). When Klein et al. (25) found that *fadD* mutants were noninducible, these workers postulated that long-chain acyl-CoA thioesters are the in vivo inducers of the *fad* regulon. However, since *fadD* mutants, like *fadL* mutants, are incapable of transporting LCFA into their cytosol, it is more likely that they cannot be induced because the fatty acids never enter the cell. Therefore, the question of whether free fatty acids or their acyl-CoA derivatives are the inducers of the *fad* regulon must await in vitro studies.

Growth of *fadR* mutants in the presence of D-glucose causes a severe repression of *fad* transcriptional activity (48). Pauli et al. (48) reported that this repression is partially relieved by the addition of adenosine 3',5'-phosphate to the growth medium, and also observed that mutants lacking a functional adenosine 3',5'-phosphate receptor protein could not be induced for the FAO enzymes. The expression of the FAO enzymes thus requires both adenosine 3',5'-phosphate and its receptor protein, establishing that the *fad* regulon responds to control by catabolite repression. The ability of other class A catabolites (47) to repress the *fad* genes has not been reported, although glycerol does repress to some extent (45).

E. coli K-12 mutants constitutive for the synthesis of the enzymes of fatty acid degradation synthesize significantly less UFA than do wild-type (*fadR*+) strains (40). The constitutive *fadR* mutants synthesize less UFA than do *fadR*+ strains both in vivo and in vitro. The inability of *fadR* strains to synthesize UFA at rates comparable to those of *fadR*+ strains is phenotypically asymptomatic unless the *fadR* strain also carries a lesion in *fabA*, the structural gene for β-hydroxydecanoyl-thioester dehydrase. Unlike *fadR*+ *fabA*(Ts) mutants, *fadR* *fabA*(Ts) strains synthesize insufficient UFA to support their growth even at low temperatures and, therefore, must be supplemented with UFA at both low and high temperatures (W. D. Nunn, unpublished results). The low UFA levels of *fadR* strains are not due to the constitutive level of fatty acid-degrading enzymes in these strains (40). These results suggest that a functional *fadR* gene is required for the

maximal expression of UFA biosynthesis in *E. coli*. The latter studies have not shed any light on the mechanism(s) by which *fadR* controls UFA synthesis. One possibility is that the *fadR* gene acts as an activator to positively regulate the transcription of the *fabA* gene product. Alternatively, the *fabA* gene product may be allosterically inhibited by an as yet unidentified factor(s) expressed in *fadR* but not *fadR*+ strains.

ato System

The *atoD*, *atoA*, and *atoB* genes encode enzymes responsible for SCFA degradation (Fig. 3). These genes are clustered at the 47 min region on the revised *E. coli* linkage map (1, 49). Acetoacetate serves as metabolic inducer for the *ato* system. When acetoacetate is used as the sole carbon source, a 200- to 300-fold induction of both acetoacetyl-CoA transferase and 3-ketoacyl-CoA thiolase II is observed in wild-type strains (49).

Although acetoacetate can be utilized by *fadR* strains, *fadR* strains cannot grow on the saturated C_4 or C_5 acids because these substrates do not induce the ATO enzymes (59, 53, 66). Only strains which have constitutive levels of the ATO and FAO enzymes can utilize C_4 or C_5 acids as sole carbon source. Constitutive levels of the *ato* gene products result from a mutation(s) in a regulatory gene, *atoC*, which also maps at the 47 min region (49). Mutants (But+) able to utilize C_4 or C_5 acids as sole carbon source are readily selected by plating *fadR* strains onto minimal media containing butyrate (49, 53, 54, 66). Pauli and Overath (49) first showed that most But+ mutants obtained in this way were constitutive for the ATO enzymes. The mutations causing constitutivity of the ATO enzymes are in the *atoC* gene, and the genotype of strains which utilize C_4 and C_5 as sole carbon sources is *fadR atoC*.

All available evidence to date suggests that the *atoC* gene codes for a regulatory protein which exerts positive control of the *ato* structural genes. A single study by Pauli and Overath (49) examined the effect of a single mutation, *atoC49*, on the activity of the *ato* structural gene products in cells merodiploid for *atoC*. These investigators found that all merodiploids which had *atoC*+*A*+*B*+ on the episome and *atoC49A*+*B*− or *atoC49A*−*B*+ on the host chromosome had the But+ phenotype and constitutive levels of acetoacetyl-CoA transferase and 3-ketoacyl-CoA thiolase II activity. Because *atoC49* appeared to be *trans*-dominant to *atoC*+, Pauli and Overath (49) hypothesized that *atoC* codes for a positive regulatory element. Comparisons of the *ato* system with other well-characterized positive regulatory systems (i.e., arabinose [*ara*] operon [7] and D-serine deaminase [*dsd*] system) reveal that many basic regulatory properties are not shared. For example, the frequency of obtaining constitutive mutants in the positive regulatory elements encoded by *araC* and *dsdC* is very low. In contrast, the frequency of obtaining spontaneous *atoC* mutants is relatively high (10^{-5} to 10^{-6}). The majority of *dsdC* and *araC* mutants display low constitutive enzyme levels which are lower than the inducible levels (5, 9, 13, 20). Also, genetic studies suggest that *araC*+ or *dsdC*+ is partially dominant to *araC* or *dsdC*, respectively (5, 9, 13, 20). In contrast, the single *atoC49* mutant investigated displayed high constitutive levels of the ATO enzymes

TABLE 5. ATO enzyme activities in a $\Delta atoCDAB$ host containing $atoC^+D^+A^+B^+$ or $atoC^+D^+A^+$ plasmids or both

Host genotype/plasmid	Growth conditions	Genes complemented by plasmid	Sp act (nmol/min per mg of protein)	
			Acetoacetyl-CoA transferase	3-Ketoacyl-CoA thiolase
Wild type ato^+				
pBR322	Uninduced	None	0	0
pBR322	Induced		48.8	507.8
$\Delta atoCAB$				
pBR322	Induced	None	0	0
pLJ10	Uninduced	atoDAB	0	8.3
pLJ10	Induced		64.5	603.5
pLJ12	Uninduced	atoDA	0	ND[a]
pLJ12	Induced		307.3	0

[a] ND, Not determined.

(greater than inducible levels) (49), and merodiploid studies showed that $atoC49$ was fully *trans*-dominant to $atoC^+$. It also must be noted that Pauli and Overath's merodiploid studies (56) were done with $atoC49$ on the chromosome. Since these mutants were heavily mutagenized to get $atoC\ atoA$ or $atoC\ atoB$ genotypes, it is conceivable that there were other host chromosomal mutations which caused secondary effects on the ato regulatory system.

To determine the nature of the ato regulatory system, Jenkins and Nunn (unpublished data) have characterized the regulation of the ato structural genes in $atoC^u$ (uninducible) and $\Delta atoCDAB$ strains harboring ato plasmids containing in vitro-constructed deletions. The rationale for the latter studies is that the ato plasmids with deletions in a locus encoding a negative controlling element should express the ato structural gene products (acetoacetyl-CoA transferase and 3-ketoacyl-CoA thiolase II) constitutively, whereas ato plasmids with deletions in a locus coding for a positive controlling element should be unable to express the ato structural genes. The preliminary findings from these studies suggest that the ato plasmid pLJ10 has both positive and negative regulatory signals (Jenkins and Nunn, in press). Evidence for suggesting that pLJ10 encodes for a positive regulator was obtained from studies showing that the plasmid pLJ16, deleted for the region to the right of the side designated PVU_1 (Fig. 5), expresses the acetoacetyl-CoA transferase and 3-ketoacyl-CoA thiolase II activity in $atoC^+$ strains but not in $atoC^u$ strains. A comparison of the in vitro translation profiles of pLJ10 and pLJ16 indicates that the pLJ16 plasmid is capable of expressing the $atoD$, $atoA$, and $atoB$ gene products but not the $atoC$ gene product (a 48,000-dalton protein). Although the exact mechanism by which $atoC$ controls the ato structural genes is not evident from these studies, the data suggest that the $atoC$ gene product is required for expression of the ato structural genes. The latter findings, therefore, support Pauli and Overath's contention (49) that the $atoC$ gene product may be an activator. Using ato-$galK$ gene fusions, Jenkins and Nunn (unpublished results) have shown that the $atoC$ gene regulates the ato structural genes at the level of transcription.

Evidence which suggests that the ato plasmid (pLJ10) also encodes for a negative controlling element was obtained from studies which revealed that $atoCDAB$ strains harboring the ato plasmid pLJ12, deleted between the BM_1 to BM_2 sites (Fig. 5) of pLJ10, have 10-fold greater levels of acetoacetyl-CoA transferase activity than $\Delta atoCDAB$ strains harboring pLJ10 (Table 5). Although the exact mechanism of control is not evident from these studies, the latter findings suggest that the BM_1 to BM_2 region of pLJ10 encodes for a component(s) that negatively controls acetoacetyl-CoA transferase activity. Since the BM_1 to BM_2 region is known to encode for 3-ketoacyl-CoA thiolase II (Jenkins and Nunn, in press), it is conceivable that this enzyme exerts an allosteric inhibitory effect on the acetoacetyl-CoA transferase. Alternatively, this enzyme or the product of some as yet unidentified gene within the BM_1 to BM_2 region of pLJ10 may function as a repressor to prevent the expression of the genes for the acetoacetyl-CoA transferase ($atoD$ and $atoA$).

An interesting aspect of SCFA metabolism which requires further investigation is the finding that at least two types of spontaneous But$^+$ mutants can be selected by plating $fadR$ mutants onto minimal media containing butyrate (49, 53, 54). As indicated above, one type constitutively synthesizes the ATO enzymes (49). The second type of But$^+$ mutant, first isolated by Salanitro and Wegener (53, 54), can grow on C_4 or C_5 acids after a 4- to 6-h lag. These workers showed that C_4 acid uptake was inducible in the latter mutants, whereas C_4 uptake was constitutive in $atoC$ mutants (53, 54). To explain the phenotype of their inducible But$^+$ mutant, Salanitro and Wegener (53) proposed that it might have a regulatory mutation that alters the inducer binding site of the regulatory protein. The mutation(s) causing the inducible But$^+$ phenotype was never mapped by Salanitro and Wegener (53, 54). However, Nunn and co-workers (unpublished results) have isolated But$^+$ mutants with the same phenotype as that described by these workers and have determined that the mutation responsible for this defect maps at 27 min on the revised E. coli linkage map (1). Since the $atoC$ gene maps at 47 min (49), the mutation(s) which enables C_4 acids to induce the ato operon does not appear to be directly affecting the product encoded by the $atoC$ gene.

The studies described above suggest that the regulatory system(s) for SCFA metabolism may be considerably more complex than that proposed by Pauli and Overath (49). In addition, studies by Pauli and Overath (49), Salanitro and Wegener (53, 54) and Jenkins and Nunn (unpublished data) indicate that the SCFA *atoC* constitutive mutation somehow affects fatty acid metabolism. Observations for the latter suggestion have come from studies showing that (i) *fadR atoC* strains grow considerably more slowly than their *fadR* parent on MCFA and LCFA (49, 53) and (ii) *fadR atoC* strains have considerably less enoyl-CoA hydratase and 3-hydroxyacyl-CoA dehydrogenase activity than their *fadR* parent (53). The question arising from these phenomena is whether these alterations in fatty acid metabolism are primary or secondary consequences of the *atoC* constitutive mutation.

Alternative routes of SCFA metabolism have been discussed by Wegener and co-workers (69) in a review. Basically, these workers have summarized in vitro studies which show that there are enzymes which appear to condense glyoxylate with various SCFA. Although very interesting, the significance of these pathways in SCFA metabolism is not known at this time.

To degrade odd-chain (C_3 and C_5) SCFA, *E. coli* must be able to metabolize the C_3 acid. Although Wegener and co-workers have proposed pathways by which odd-chain fatty acids are degraded, the precise pathway is unknown. One gene, *prp*, which maps at 97 min, is known to be required for C_3 and C_5 acid metabolism (60). The function of the *prp* gene has not been established.

GENETIC AND BIOCHEMICAL FEATURES OF ACETATE METABOLISM

Glyoxylate Shunt

In addition to the FAO enzymes, the expression of the glyoxylate shunt enzymes (Fig. 8) is also required for the growth of *E. coli* on fatty acids and acetate as sole carbon sources. In wild-type *E. coli*, repression of the *ace* operon is under the control of the two genes, *fadR* and *iclR*. The studies of Maloy and Nunn (35, 36) suggest that the *iclR* and *fadR* genes both regulate the glyoxylate shunt in a *trans*-dominant and synergistic manner at the level of transcription.

Growth on fatty acids as sole carbon source not only requires the transport and degradation of fatty acids

but also the utilization of the two-carbon acetyl-CoA units generated via β-oxidation. Acetyl-CoA is mainly catabolized by the tricarboxylic acid cycle. However, since with each turn of the tricarboxylic acid cycle two carbon atoms are lost as CO_2, no net assimilation of carbon from acetyl-CoA can occur by this means. Therefore, growth on substrates such as fatty acids or acetate which are catabolized solely to acetyl-CoA requires the operation of a separate anaplerotic pathway to replenish necessary intermediates for cellular biosynthesis. This is accomplished by the glyoxylate shunt (Fig. 8). The net effect of the glyoxylate shunt is the formation of 1 mol of dicarboxylic acids from 2 mol of acetyl-CoA. The glyoxylate shunt is the only known anaplerotic pathway allowing growth on acetyl-CoA, and thus it has been observed in all other organisms that utilize fatty acids or acetate as sole carbon source (21).

The two unique enzymes of the glyoxylate shunt, isocitrate lyase and malate synthase A, are normally induced only when *E. coli* is grown on acetate or fatty acids (26, 64). The structural genes for isocitrate lyase (*aceA*) and malate synthase A (*aceB*) map at 90 min on the *E. coli* K-12 chromosome (Fig. 2). Mutations in an adjacent gene, *iclR*, produce constitutive levels of isocitrate lyase and malate synthase. Since the *aceA* and *aceB* genes were found to be closely linked and coordinately expressed, Brice and Kornberg (7) postulated that the *aceA* and *aceB* genes may form an operon controlled by a repressor protein encoded by the *iclR* gene. From genetic and biochemical studies with mutants containing Tn*10* insertions in their *aceA* or *aceB* gene or in both, Maloy and Nunn (36) confirmed that the *aceAB* genes compose an operon which is transcribed from the *aceB* gene to the *aceA* gene.

Other *ace* Enzymes

The glyoxylate shunt diverts isocitrate from the Krebs cycle, bypassing the CO_2-producing steps (Fig. 8). During growth on acetate (or fatty acids), the flow of isocitrate through the glyoxylate bypass is also controlled via the allosteric regulation of isocitrate dehydrogenase (IDH), the Krebs cycle enzyme which competes with isocitrate lyase. The latter control mechanism is accomplished by the phosphorylation of IDH by a bifunctional protein, the IDH kinase-phosphatase (17, 18, 22). Since the phosphorylated form of IDH is completely inactive (6, 28), phosphorylation decreases IDH activity and forces isocitrate through the glyoxylate bypass (30, 38). Recently, LaPorte and co-workers (30) have presented genetic and biochemical evidence which suggests that kinase and phosphatase activities are catalyzed by a single protein which is encoded by the *aceK* gene. The latter gene was mapped in the *ace* operon downstream from both the *aceB* and *aceA* genes (30). Through the cloning and characterization of the *aceK* gene, LaPorte and Chung (27) have presented evidence that this gene is the structural gene for the 66,000-dalton IDH kinase-phosphatase. It is not clear at this time whether this bifunctional enzyme has two distinct active sites per monomer, one catalyzing the kinase reaction and the other catalyzing the phosphatase reaction, or whether one active site per monomer catalyzes both reactions. LaPorte and Chung (27) have proposed that the two

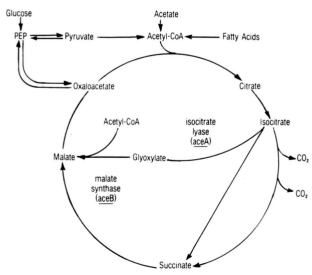

FIG. 8. Glyoxylate shunt in *E. coli* and related reactions.

reactions occur at independent sites and that the phosphatase domain was produced through a partial duplication of a primordial IDH kinase gene, leaving a fused product. Evaluation of this hypothesis must await the determination of the nucleotide sequence of *aceK*. The phosphorylation system can respond to variations in the intracellular levels of IDH. From studies which involved comparing the levels of IDH activity and phosphorylation in strains containing either one or multiple copies of the IDH structural gene, LaPorte et al. (29) have shown that the phosphorylation system can compensate for changes in the cellular level of IDH in excess of 10-fold, maintaining a nearly constant activity for IDH during growth on acetate. Given the role it plays in the glyoxylate bypass reaction, it is not surprising that the *aceK* gene is part of the *ace* operon.

The utilization of acetate, whether for lipid biosynthesis, for oxidation via the tricarboxylic acid cycle, or for replenishing dicarboxylic acids of the tricarboxylic acid cycle via the glyoxylate bypass, requires that it first be activated to acetyl-CoA. Two mechanisms that bring about this conversion have been elucidated. In one mechanism, acetyl-CoA synthetase (acetate:CoA ligase [AMP-forming]; EC 6.2.1.1) catalyzes the acetylation of CoA concomitant with the cleavage of ATP to AMP and inorganic pyrophosphate (32, 63; N. O. Kaplan and F. Lipman, Fed. Proc. **7**:163, 1948). In the second mechanism, two enzymes catalyze (i) the conversion of acetate to acetyl phosphate, with cleavage of ATP to ADP, and (ii) the transfer of the acetyl moiety from acetyl phosphate to CoA, with liberation of P_i.

$$\text{acetate} + \text{ATP} \rightarrow \text{acetyl phosphate} + \text{ADP} \quad (1)$$

$$\text{acetyl phosphate} + \text{CoA} \rightarrow \text{acetyl-CoA} + P_i \quad (2)$$

Acetate kinase (ATP acetate phosphotransferase; EC 2.7.2.1), which is encoded by the *ack* gene, catalyzes the reaction of equation 1, and phosphotransacetylase (acetyl-CoA[CoA]:orthophosphate acetyltransferase; EC 2.3.1.8), encoded by the *pta* gene, catalyzes the reaction in equation 2. The latter two genes have been mapped in both *S. typhimurium* and *E. coli* near *purF*, at 48.5 to 49 min on the respective linkage maps (31). Acetyl phosphate is thought to be an intermediate in the activation of acetate to acetyl-CoA (8). Also, it has been implicated as the energy source for transport systems utilizing periplasmic binding proteins (23).

The levels of acetate kinase and phosphotransacetylase in extracts of wild-type *E. coli* do not vary with different carbon sources (8), suggesting that the expression of the *ack* and *pta* genes is neither induced by acetate nor catabolite repressed by glucose. Mutants defective in either the *ack* or *pta* gene are severely impaired in utilizing acetate as sole carbon source, and when grown on glucose they are incapable of incorporating labeled acetate (8). However, *ack* and *pta* mutants grown on glycerol are capable of incorporating labeled acetate. Brown et al. (8) have performed studies which suggest that an inducible acetyl-CoA synthetase enables *ack* and *pta* mutants to incorporate labeled acetate. Since these mutants are unable to incorporate acetate when grown on glucose (8), the acetyl-CoA synthetase, like the glyoxylate shunt enzymes, appears to be regulated by catabolite repression. Overall, these studies suggest that the acetate kinase and phosphotransacetylase are required for *E. coli* (and presumably *S. typhimurium*) to grow optimally on acetate as a carbon source and to incorporate acetate under catabolite-repressing growth conditions. Furthermore, the studies with the *ack* and *pta* mutants suggest that the acetyl-CoA synthetase provides these organisms with a second route for converting exogenous acetate to acetyl-CoA. It is not clear at this time whether any other enzymes or transport proteins are required for the uptake of acetate, because no mutants deficient in acetyl-CoA synthetase activity have been isolated or characterized.

Regulation of *ace* Operon

Evidence that the *fadR* gene plays a role in acetate metabolism was first noted when Maloy and Nunn (35) observed that acetate was incorporated into *fadR* strains at a considerably greater rate than in isogenic *fadR*[+] strains. Biochemical studies demonstrated that the increased rate of acetate incorporation by *fadR* mutants was not due to an increased rate of macromolecular synthesis or degradation, differences in the activities of enzymes required for acetate transport or oxidation, or differences in the acetyl-CoA pool size. However, the increased rate of acetate incorporation by *fadR* mutants did require a functional glyoxylate shunt. Enzyme studies (33) further showed that the levels of glyoxylate shunt enzymes were elevated in *fadR* mutants under noninducing growth conditions (Table 6). These studies strongly suggested that the

TABLE 6. Specific activities of glyoxylate shunt enzymes in *iclR* and *fadR* mutants[a]

| Strain | Relevant genotype | Sp act (nmol/min per mg of protein) | | | | | |
| | | Isocitrate lyase | | | Malate synthase | | |
		Succinate	Acetate	Succinate + acetate	Succinate	Acetate	Succinate + acetate
K-12	Prototrophic	21	244	70	103	410	186
SM6034	*iclR*	209	320	276	267	594	330
RS3040	*fadR* Tn*10*	172	272	204	371	566	389
SM6042	*iclR fadR* Tn*10*	283	396	386	396	592	402

[a] Experimental assay conditions were as described by Maloy and Nunn (36).

TABLE 7. Mu d(Ap *lac*) operon fusions with the *aceA* gene[a]

Strain (relevant genotype)/growth conditions	β-Galactosidase activity (U)
TL1 (*iclR fadR ΔlacZ*)	
Noninducing	<1
Inducing	<1
SMUD-1 (*aceA*::Mu d(Ap *lac*) *iclR⁺ fadR⁺ ΔlacZ*)	
Noninducing	20
Inducing	310
SMUD-2 (*aceA*::Mu d(Ap *lac*) *iclR fadR⁺ ΔlacZ*)	
Noninducing	300
Inducing	310
SMUD-3 (*aceA*::Mu d(Ap *lac*) *iclR⁺ fadR ΔlacZ*)	
Noninducing	200
Inducing	440
SMUD-4 (*aceA*::Mu d(Ap *lac*) *iclR fadR ΔlacZ*)	
Noninducing	650
Inducing	700

[a] Experimental conditions were as described by Maloy and Nunn (36).

activity of the glyoxylate shunt enzymes is regulated by the *fadR* gene.

To determine whether the *fadR* or *iclR* gene or both regulated the expression of the *ace* operon at the level of transcription, Maloy and Nunn (36) constructed *ace-lac* operon fusions. Studies with these fusions demonstrated that both *iclR* and *fadR* negatively regulate the *ace* operon at the level of transcription. Interestingly, *iclR fadR* mutants have greater levels of the glyoxylate shunt enzymes than do *iclR fadR⁺* and *iclR⁺ fadR* mutants. When these same mutants contain *ace-lac* operon fusions, the *iclR fadR* double mutants also have greater levels of β-galactosidase activity than do the *iclR fadR⁺* and *iclR⁺ fadR* mutants (Table 7). Merodiploid studies demonstrated that the *iclR* and *fadR* genes both regulate the expression of the glyoxylate shunt in a *trans*-dominant manner (Tables

5, 8, and 9). Although several models could be proposed from these studies, the simplest interpretation is that the *iclR* and *fadR* gene products act independently and synergistically to cause repression of the *ace* operon; acting together, they cause the full repression of the *ace* operon.

CONCLUSIONS

Most of our knowledge regarding the utilization of fatty acids, acetate, or both as sole carbon sources stems from genetic and biochemical studies performed with *E. coli*. For this organism to metabolize the two-carbon compound acetate, the glyoxylate shunt enzymes, encoded by the *ace* operon (7, 8, 36), must be expressed for the continual replenishing of dicarboxylic acids drained from the tricarboxylic acid cycle for cellular biosynthesis. In addition, before its metabolism the acetate must be activated to acetyl-CoA by either (i) acetyl-CoA synthetase (8) or (ii) acetate kinase and phosphotransacetylase. As indicated in this chapter, the synthesis of the glyoxylate shunt enzymes and the acetyl-CoA synthetase are induced by the presence of acetate in the growth medium of *E. coli*. The syntheses of these enzymes are also catabolite repressed by glucose. All of the evidence presently available suggests that induction of the *ace* operon when acetate is present is accomplished by the inactivation of the *iclR* gene product(s). The nature of the inducer under these conditions is presumed to be acetyl-CoA, but this has not been definitively established. S. R. Maloy (unpublished data) has been characterizing the *ace* operon in *S. typhimurium* and has found that the genetics of the structural genes in this organism seem to be identical to that in *E. coli*. The regulation of the *ace* operon in *S. typhimurium* also appears to be similar to that in *E. coli*, except for the fact that the expression of glyoxylate shunt enzyme activities is considerably lower in this organism than in *E. coli*.

One of the most interesting findings with respect to the *ace* operon is the fact that the *fadR* gene product negatively regulates its expression (36). Since fatty acids are degraded to acetyl-CoA by the FAO enzymes, it was originally thought that elevated levels of the glyoxylate shunt enzymes in *fadR* mutants might be due to the endogenous buildup of acetate (or acetyl-

TABLE 8. Specific activity of isocitrate lyase in strains merodiploid for the *iclR* gene[a]

Strain	Relevant genotype	Isocitrate lyase sp act (nmol/min per mg of protein)		
		Succinate	Succinate + acetate	Acetate
MM-1	O/*aceA⁺ iclR⁺*	21	98	316
MM-2	O/*aceA⁺ iclR*	206	299	355
MM-4	O/*aceA iclR*	<0.5	<0.5	<0.5
MM-5	F′ *aceA⁺ iclR⁺/aceA⁺ iclR⁺*	31	50	114
MM-6	F′ *aceA⁺ iclR⁺/aceA iclR⁺*	18	45	206
MM-7	F′ *aceA⁺ iclR⁺/aceA iclR*	52	79	305
MM-8	F′ *aceA iclR⁺/aceA⁺ iclR⁺*	3	31	209
MM-9	F′ *aceA iclR⁺/aceA⁺ iclR*	19	69	276
MM-10	F′ *aceA⁺ iclR/aceA iclR⁺*	23	74	298
MM-11	F′ *aceA⁺ iclR/aceA⁺ iclR*	292	326	422
MM-12	F′ *aceA iclR/aceA⁺ iclR⁺*	17	81	310
MM-13	F′ *aceA iclR/aceA⁺ iclR*	269	285	387

[a] Experimental growth and assay conditions were as described by Maloy and Nunn (36).

TABLE 9. Specific activities of glyoxylate shunt enzymes in strains merodiploid for the *fadR* gene

Strain (*fadR* genotype[a])/ growth conditions	Sp act (nmol/min per mg of protein)	
	Isocitrate lyase	Malate synthase
M11-1-S (0/+)		
Succinate	11	42
Acetate	95	198
Oleate	125	253
M2-5-S (0/−)		
Succinate	81	111
Acetate	106	315
Oleate	128	320
M11-1 (+/+)		
Succinate	7	36
Acetate	84	183
Oleate	83	238
M12-1 (+/−)		
Succinate	12	45
Acetate	64	205
Oleate	121	284
M47-1 (−/+)		
Succinate	18	52
Acetate	60	210
Oleate	146	277
M23-1 (−/−)		
Succinate	62	125
Acetate	95	329
Oleate	122	336

[a] *fadR* genotype shown as episomal allele/chromosomal allele. A zero indicates that no episome was present.

CoA) resulting from the intracellular degradation of fatty acids by the constitutive levels of FAO enzymes in these mutants. The latter notion was dispelled when studies with *fadR fad* and *fadR+ fad* mutants revealed that *fadR fad* strains had greater levels of glyoxylate shunt enzymes activities than the parental *fadR+ fad* strains (33, 35, 36). Although the studies described in this chapter indicate that the *ace* operon is under the transcriptional control of the *iclR* and *fadR* genes, the molecular details by which this regulation is accomplished have not been elucidated. For instance, it is not known whether the *fadR* gene product exerts control over the *ace* operon by directly interacting with *cis*-acting *ace* operon regulatory sites or by interacting with the *iclR* gene product(s).

The FAO enzymes, coupled with glyoxylate shunt enzymes, enable *E. coli* to utilize MCFA and LCFA as sole carbon and energy sources. As indicated in this chapter, wild-type *E. coli* can only utilize LCFA as carbon sources because LCFA induce the synthesis of the FAO enzymes. To be capable of utilizing MCFA, *E. coli* must have a nonfunctional *fadR* gene. All of the available data indicate that the latter gene exerts negative transcriptional control over the *fad* regulon. The nature of the inducing substrate and the molecular details by which the *fadR* gene product interacts with the *cis*-acting regulatory sites of the *fad* regulon remain to be determined.

To grow on saturated SCFA as sole carbon source, *E. coli* must constitutively express two ATO enzymes and three FAO enzymes (49). The regulation of the ATO enzymes apparently requires an activator and possibly a repressor (49; Jenkins and Nunn, unpublished

data). Further genetic and recombinant DNA technology, coupled with biochemical characterizations, will undoubtedly enable workers to unravel the molecular details by which these regulators control the *ato* system.

To induce or to be degraded by the FAO enzymes, LCFA must utilize the *fadD* and *fadL* gene products to be delivered to the cytosolic sites where these events occur. The membrane-bound *fadL* gene product, FLP, appears to function as an LCFA receptor (39). It is not clear whether FLP also functions as an LCFA permease. Evidence suggesting that this protein may function as a permease is as follows. (i) Complementation studies suggest that only one gene is required for LCFA binding and transport (W. D. Nunn, unpublished results). (ii) A hybrid plasmid which expresses only FLP enables *fadL* strains to bind and transport LCFA (39). (iii) Mutations in the promoter-distal region of the *fadL* gene result in the synthesis of a physically altered FLP which binds but does not transport LCFA (39). The detailed mechanics by which FLP delivers LCFA across the membrane will most likely be resolved when in vitro transport studies with the FLP purified by Black and Nunn (unpublished results) are conducted in reconstituted membrane vesicles. Such studies may also clarify the role of the *fadD* gene product, acyl-CoA synthetase, in the transport process.

The translocation of MCFA to the cytosolic FAO enzymes can occur actively via *fadL*-mediated assistance or by a diffusional mechanism. In either case, the *fadD* gene product is required for MCFA to be delivered to the other cytosolic FAO enzymes. To date, attempts to isolate mutants which specifically affect the transport of MCFA have failed. A major difficulty in screening for MCFA mutants is the fact that these fatty acids are more toxic to the cell than are LCFA. Therefore, genetic procedures which avoid this technical problem may enable workers to isolate MCFA transport mutants.

A considerable amount of information regarding the FAO enzymes has been gained from studies with *E. coli*. At least seven of the FAO enzymes are part of the *fadAB* multienzyme complex (51, 52). Although biochemical studies suggest that the quaternary structure of this complex is not an artifact (51, 52), it remains to be established whether posttranslational processing occurs for the active complex to form. Furthermore, genetic and recombinant DNA techniques may enable workers to define more precisely the catalytic sites for each of the seven enzyme reactions carried out by the complex. It will also be very interesting to determine the relationship, if any, between the long-chain enoyl-CoA hydratase and the multienzyme complex and to identify the gene which codes for this enzyme. Finally, the nature of the only known mutation (*fadE*) affecting acyl-CoA dehydrogenase activity has never been established. Clearly, more genetic and biochemical studies must be performed before a perspective on the molecular details of the latter enzyme and possible cofactors can be formulated.

The most interesting finding, which has not been further explored, is the study indicating that the *fadR* gene plays a role in UFA biosynthesis (40). Conceptually, it makes sense for endogenous fatty acid syntheses to be restricted when exogenous fatty acids are present in the growth medium of a bacterial cell. Therefore, it is not surprising that a regulator gene such as the *fadR* gene functions not only in the degradation of fatty acids but also in the synthesis of fatty acids. Why this gene controls, directly or indirectly, UFA biosynthesis but not saturated fatty acid biosynthesis is not known. Clearly, the most exciting research in the future will be directed at the molecular details by which *fadR* exerts its global control over UFA biosynthesis, fatty acid degradation, and acetate metabolism.

ACKNOWLEDGMENTS

This work was supported by Public Health Service grant GM22466. I thank Rowland H. Davis for helpful comments on the manuscript and Esther Ervin for the figure illustrations. I also thank Kathy Ruiz for typing the manuscript.

ADDENDUM IN PROOF

Prior to W. D. Nunn's death, it had become clear that the *fadL* protein was localized on the outer membrane, not on the inner membrane as previously thought. This evidence was gained from a purification of the *fadL* protein, the production of an antiserum to it, and the use of this antiserum as a probe for purified inner and outer membranes (P. N. Black, B. Said, C. R. Ghosn, J. V. Beach, and W. D. Nunn, J. Biol. Chem., in press).

LITERATURE CITED

1. **Bachmann, B. J.** 1983. Linkage map of *Escherichia coli* K-12, edition 7. Microbiol. Rev. **47**:180–230.
2. **Bennett, P. M., and W. H. Holms.** 1975. Reversible inactivation of the isocitrate dehydrogenase of *Escherichia coli* ML308 during growth on acetate. J. Gen. Microbiol. **87**:37–51.
3. **Binstock, J. F., A. Pramanik, and H. Schultz.** 1977. Isolation of a multienzyme complex of fatty acid oxidation from *Escherichia coli*. Proc. Natl. Acad. Sci. USA **74**:492–495.
4. **Black, P. N., S. F. Kianian, C. C. DiRusso, and W. D. Nunn.** 1985. Long chain fatty acid transport in *Escherichia coli*: cloning, mapping, and expression of the *fadL* gene. J. Biol. Chem. **260**:1780–1790.
5. **Bloom, F. R., E. McFall, M. C. Young, and A. M. Carothers.** 1975. Positive control in the D-serine deaminase system of *Escherichia coli* K-12. J. Bacteriol. **121**:1092–1101.
6. **Borthwick, A. C., W. H. Holms, and H. G. Nimmo.** 1984. The phosphorylation of *Escherichia coli* isocitrate dehydrogenase in intact cells. Biochem. J. **222**:797–804.
7. **Brice, C. B., and H. L. Kornberg.** 1968. Genetic control of isocitrate lyase activity in *Escherichia coli*. J. Bacteriol. **96**:2185–2186.
8. **Brown, T. D. K., M. C. Jones-Mortimer, and H. L. Kornberg.** 1977. The enzymatic interconversion of acetate and acetyl-coenzyme A in *Escherichia coli*. J. Gen. Microbiol. **102**:327–336.
9. **Carothers, A. M., E. McFall, and S. Palchaudhuri.** 1980. Physical mapping of the *Escherichia coli* D-serine deaminase region: contiguity of the *dsd* structural and regulatory genes. J. Bacteriol. **142**:174–184.
10. **Clark, D.** 1981. Regulation of fatty acid degradation in *Escherichia coli*: analysis by operon fusion. J. Bacteriol. **148**:521–526.
11. **Cronan, J. E., Jr., D. F. Silbert, and D. L. Wulff.** 1972. Mapping of the *fabA* locus for unsaturated fatty acid biosynthesis in *Escherichia coli*. J. Bacteriol. **112**:206–211.
12. **DiRusso, C. C., and W. D. Nunn.** 1985. Construction and characterization of a gene (*fadR*) involved in regulation of fatty acid metabolism in *Escherichia coli*. J. Bacteriol. **161**:583–588.
13. **Englesberg, E., and G. Wilcox.** 1974. Regulation: positive control. Annu. Rev. Genet. **8**:242–292.
14. **Esfahani, M., T. Ioneda, and S. J. Wakil.** 1971. Studies on the control of fatty acid metabolism. J. Biol. Chem. **246**:50–56.
15. **Frerman, F. E.** 1973. The role of acetyl-coenzyme A in the transferase uptake of butyrate by isolated membrane vesicles of *Escherichia coli*. Arch. Biochem. Biophys. **159**:444–452.
16. **Frerman, F. E., and W. Bennett.** 1973. Studies on the uptake of fatty acids by *Escherichia coli*. Arch. Biochem. Biophys. **159**:434–443.
17. **Garnak, M., and H. C. Reeves.** 1979. Phosphorylation of isocitrate dehydrogenase of *Escherichia coli*. Science **203**:1111–1112.
18. **Garnak, M., and H. C. Reeves.** 1979. Purification and properties of phosphorylated isocitrate dehydrogenase of *Escherichia coli*. J. Biol. Chem. **254**:7915–7920.
19. **Ginsburgh, C. L., P. N. Black, and W. D. Nunn.** 1984. Transport of long chain fatty acids in *Escherichia coli*. Identification of a membrane protein associated with the *fadL* gene. J. Biol. Chem. **259**:8437–8443.
20. **Heincz, M. C., and E. McFall.** 1978. Role of the *dsdC* activator in regulation of D-serine deaminase synthesis. J. Bacteriol. **136**:96–103.
21. **Hillier, S., and W. T. Charnetzky.** 1981. Glyoxylate bypass enzymes in *Yersinia* species and multiple forms of isocitrate lyase in *Yersinia pestis*. J. Bacteriol. **145**:452–458.
22. **Holms, W. H., and P. M. Bennett.** 1971. Regulation of isocitrate dehydrogenase activity in *Escherichia coli* on adaptation to acetate. J. Gen. Microbiol. **65**:57–68.
23. **Hong, J., A. C. Hunt, P. S. Masters, and M. A. Lieberman.** 1979. Requirement of acetyl phosphate for the binding protein-dependent transport systems in *Escherichia coli*. Proc. Natl. Acad. Sci. USA **76**:1213–1217.
24. **Kameda, K., and W. D. Nunn.** 1981. Purification and characterization of acyl coenzyme A synthetase from *Escherichia coli*. J. Biol. Chem. **256**:5702–5707.
25. **Klein, K., R. Steinberg, B. Fiethen, and P. Overath.** 1971. Fatty acid degradation in *Escherichia coli*. An inducible system for the uptake of fatty acids and further characterization of *old* mutants. Biochemistry **19**:442–450.
26. **Kornberg, H. L.** 1966. The role and control of the glyoxylate cycle in *Escherichia coli*. Biochem. J. **99**:1–11.
27. **LaPorte, D. C., and T. Chung.** 1985. A single gene codes for the kinase and phosphatase which regulates isocitrate dehydrogenase. J. Biol. Chem. **260**:15291–15297.
28. **LaPorte, D. C., and D. E. Koshland.** 1982. A protein with kinase and phosphatase activities involved in regulation of tricarboxylic acid cycle. Nature (London) **300**:458–460.
29. **LaPorte, D. C., P. E. Thorsness, and R. E. Koshland.** 1985. Compensatory phosphorylation of isocitrate dehydrogenase: a mechanism for adaptation to the intracellular environment. J. Biol. Chem. **260**:10563–10568.
30. **LaPorte, D. C., K. Walsh, and D. E. Koshland.** 1984. The branch point effect: ultrasensitivity and subsensitivity to metabolic control. J. Biol. Chem. **259**:14068–14075.
31. **LeVine, S. M., F. Ardeshir, and G. F.-L. Ames.** 1980. Isolation and characterization of acetate kinase and phosphotransacetylase mutants of *Escherichia coli* and *Salmonella typhimurium*. J. Bacteriol. **143**:1081–1085.
32. **Lipmann, F.** 1944. Enzymatic synthesis of acetyl phosphate. J. Biol. Chem. **155**:55–70.
33. **Maloy, S. R., M. Bohlander, and W. D. Nunn.** 1980. Elevated levels of glyoxylate shunt enzymes in *Escherichia coli* strains constitutive for fatty acid degradation. J. Bacteriol. **143**:720–725.
34. **Maloy, S. R., C. L. Ginsburgh, R. W. Simons, and W. D. Nunn.** 1981. Transport of long and medium chain fatty acids by *Escherichia coli*. J. Biol. Chem. **256**:3735–3742.
35. **Maloy, S. R., and W. D. Nunn.** 1981. Role of gene *fadR* in *Escherichia coli* acetate metabolism. J. Bacteriol. **148**:83–90.
36. **Maloy, S. R., and W. D. Nunn.** 1982. Genetic regulation of the glyoxylate shunt in *Escherichia coli* K-12. J. Bacteriol. **149**:173–180.
37. **Nikaido, H.** 1979. Nonspecific transport through the outer membrane of bacteria, p. 361–408. *In* M. Inouye (ed.), Bacterial outer membranes, biogenesis and functions. John Wiley & Sons, Inc., New York.
38. **Nimmo, G. A., and H. G. Nimmo.** 1984. The regulatory properties of isocitrate dehydrogenase kinase and isocitrate dehydrogenase phosphatase from *Escherichia coli* ML308 and the roles of these activities in the control of isocitrate dehydrogenase. Eur. J. Biochem. **141**:409–414.
39. **Nunn, W. D., R. Colburn, and P. N. Black.** 1985. Transport of long chain fatty acids in *Escherichia coli*: evidence for role of *fadL* gene product as long chain fatty acid receptor. J. Biol. Chem. **261**:167–171.
40. **Nunn, W. D., K. Giffin, D. Clark, and J. E. Cronan, Jr.** 1983. Role for *fadR* in unsaturated fatty acid biosynthesis in *Escherichia coli*. J. Bacteriol. **154**:554–560.
41. **Nunn, W. D., and R. W. Simons.** 1978. Transport of long-chain fatty acids by *Escherichia coli*: mapping and characterization of mutants in the *fadL* gene. Proc. Natl. Acad. Sci. USA **75**:3377–3381.
42. **Nunn, W. D., R. W. Simons, P. A. Egan, and S. R. Maloy.** 1979. Kinetics of the utilization of medium and long chain fatty acids

by a mutant of *Escherichia coli* defective in the *fadL* gene. J. Biol. Chem. **254**:9130–9134.

43. **O'Brien, W. J., and F. E. Frerman.** 1977. Evidence for a complex of three beta-oxidation enzymes in *Escherichia coli*: induction and localization. J. Bacteriol. **132**:532–540.

44. **Overath, P., G. Pauli, and H. U. Schairer.** 1969. Fatty acid degradation in *Escherichia coli*. An inducible acyl-CoA synthetase, the mapping of *old*-mutations, and the isolation of regulatory mutants. Eur. J. Biochem. **7**:559–574.

45. **Overath, P., E. Raufuss, W. Stoffel, and W. Ecker.** 1967. The induction of the enzymes of fatty acid degradation in *Escherichia coli*. Biochem. Biophys. Res. Commun. **29**:28–33.

46. **Overath, P., H. U. Schairer, and W. Stoffel.** 1970. Correlation of *in vivo* phase transitions of membrane lipids in *Escherichia coli*. Proc. Natl. Acad. Sci. USA **67**:606–642.

47. **Pastan, I., and S. Adhya.** 1976. Cyclic adenosine 5'-monophosphate in *Escherichia coli*. Bacteriol. Rev. **40**:527–551.

48. **Pauli, G., R. Ehring, and P. Overath.** 1974. Fatty acid degradation in *Escherichia coli*: requirement of cyclic adenosine monophosphate and cyclic adenosine monophosphate receptor protein for enzyme synthesis. J. Bacteriol. **117**:1178–1183.

49. **Pauli, G., and P. Overath.** 1972. *ato* operon: a highly inducible system for acetoacetate and butyrate degradation in *Escherichia coli*. Eur. J. Biochem. **29**:553–562.

50. **Pawar, S., and H. Schulz.** 1981. The structure of the multienzyme complex of fatty acid oxidation from *Escherichia coli*. J. Biol. Chem. **256**:3894–3899.

51. **Pramanik, A., S. Pawar, E. Antonian, and H. Schultz.** 1979. Five different enzymatic activities are associated with the multienzyme complex of fatty acid oxidation from *Escherichia coli*. J. Bacteriol. **137**:469–473.

52. **Rock, G. O., and S. Jackowski.** 1985. Pathway for incorporation of exogenous fatty acids into phosphatidylethanolamine in *Escherichia coli*. J. Biol. Chem. **260**:12720–12724.

53. **Salanitro, J. P., and W. S. Wegener.** 1971. Growth of *Escherichia coli* on short-chain fatty acids: growth characteristics of mutants. J. Bacteriol. **108**:885–892.

54. **Salanitro, J. P., and W. S. Wegener.** 1971. Growth of *Escherichia coli* on short-chain fatty acids: nature of the transport system. J. Bacteriol. **108**:893–901.

55. **Sallus, L., R. J. Haselbeck, and W. D. Nunn.** 1983. Regulation of fatty acid transport in *Escherichia coli*: analysis by operon fusion. J. Bacteriol. **155**:1450–1454.

56. **Samuel, D., J. Estroumza, and G. Ailhaud.** 1970. Partial purification and properties of acyl CoA synthetase of *Escherichia coli*. Eur. J. Biochem. **12**:576–582.

57. **Simons, R. W., P. A. Egan, H. T. Chute, and W. D. Nunn.** 1980. Regulation of fatty acid degradation in *Escherichia coli*: isolation and characterization of strains bearing insertion and temperature-sensitive mutations in gene *fadR*. J. Bacteriol. **142**:621–632.

58. **Simons, R. W., K. T. Hughes, and W. D. Nunn.** 1980. Regulation of fatty acid degradation in *Escherichia coli*: dominance studies with strains merodiploid in gene *fadR*. J. Bacteriol. **143**:726–730.

59. **Spratt, S. K., P. N. Black, M. M. Ragozzino, and W. D. Nunn.** 1984. Cloning, mapping, and expression of genes involved in the fatty acid-degradative multienzyme complex of *Escherichia coli*. J. Bacteriol. **158**:535–542.

60. **Spratt, S. K., C. L. Ginsburgh, and W. D. Nunn.** 1981. Isolation and genetic characterization of *Escherichia coli* mutants defective in propionate metabolism. J. Bacteriol. **146**:1166–1169.

61. **Sramek, S. J., and F. E. Frerman.** 1975. Purification and properties of *Escherichia coli* coenzyme A transferase. Arch. Biochem. Biophys. **171**:14–26.

62. **Sramek, S. J., and F. E. Frerman.** 1975. *Escherichia coli* coenzyme A transferase: kinetics, catalytic pathway and structure. Arch. Biochem. Biophys. **171**:27–35.

63. **Stadman, E. R., and H. A. Barker.** 1950. Fatty acid synthesis by enzyme preparations of *Clostridium kluyveri*. J. Biol. Chem. **184**:769–793.

64. **Vanderwinkel, E., and M. DeVlieghere.** 1968. Physiologie et genetique de l'isocitritase et des malate synthases chez *Escherichia coli*. Eur. J. Biochem. **5**:81–90.

65. **Vanderwinkel, E., M. De Vlieghere, M. Fontaine, D. Charles, F. Denamur, D. Vandevoorde, and D. DeKegel.** 1976. Septation deficiency and phospholipid perturbation in *Escherichia coli* genetically constitutive for the beta oxidation pathway. J. Bacteriol. **127**:1389–1399.

66. **Vanderwinkel, E., M. De Vlieghere, and J. Vande Meersshe.** 1971. Mutation habilitant *Escherichia coli* à croitre sur acides gras moyens. Eur. J. Biochem. **22**:115–120.

67. **Vanderwinkel, E., P. Furmanski, H. C. Reeves, and S. J. Ajl.** 1968. Growth of *Escherichia coli* on fatty acid: requirement for coenzyme A transferase activity. Biochem. Biophys. Res. Commun. **33**:902–908.

68. **Weeks, G., M. Shapiro, R. O. Burns, and S. J. Wakil.** 1969. Control of fatty acid metabolism. I. Induction of the enzymes of fatty acid oxidation in *Escherichia coli*. J. Bacteriol. **97**:827–836.

69. **Wegener, W. S., H. C. Reeves, R. Rabin, and S. J. Ajl.** 1968. Alternate pathways of metabolism of short-chain fatty acids. Bacteriol. Rev. **32**:1–26.

Section B. Class II Reactions: Conversion of Precursor Metabolites to
Small-Molecule Building Blocks

B1. Biosynthesis of Amino Acids

20. Ammonia Assimilation and the Biosynthesis of Glutamine, Glutamate, Aspartate, Asparagine, L-Alanine, and D-Alanine

LAWRENCE J. REITZER AND BORIS MAGASANIK

Department of Biology, Massachusetts Institute of Technology, Cambridge, Massachusetts 02139

ASSIMILATION OF AMMONIA

Ammonia as Nitrogen Source

Ammonia is the preferred source of nitrogen for the growth of enteric bacteria in a defined minimal medium with glucose as the source of carbon. Although there has been no systematic study of the growth rate of *Escherichia coli* or *Salmonella typhimurium* with various nitrogen sources, we are not aware of any data in the literature or any observation from our laboratory that suggests that any other nitrogen source supports a faster growth rate than ammonia. It is the purpose of this section to describe both how ammonia is assimilated and the relationship between ammonia assimilation and the biosynthesis of glutamate and glutamine in the enteric bacteria, *E. coli* and *S. typhimurium*. Information gained from the study of the related enteric organism, *Klebsiella aerogenes*, has sometimes complemented that from *E. coli* and *S. typhimurium*; therefore, *K. aerogenes* is discussed when appropriate. This subject has been most recently reviewed by Tyler (123) and Magasanik (74).

All cellular nitrogen for the synthesis of macromolecules in the enteric bacteria is derived from the amido group of glutamine, the amino group of glutamate, or directly from incorporation of ammonia. Glutamate provides nitrogen for the synthesis of most of the amino acids, whereas glutamine donates nitrogen for the synthesis of purines, pyrimidines, amino sugars, histidine, tryptophan, asparagine, NAD, and *p*-aminobenzoate. A kilogram of dry weight of *E. coli* contains 11 to 12 g-atoms of total nitrogen; the synthesis of glutamate and its products requires about 10 g-atoms of nitrogen, whereas the synthesis of glutamine and nitrogen-containing compounds which derive nitrogen from the glutamine amide requires 1.3 g-atoms of nitrogen (144). When the ammonium ion concentration of the growth medium is sufficiently high (greater than 1 mM), ammonia is incorporated directly into glutamate, glutamine, and asparagine. However, when the ammonium ion concentration of the growth medium is less than about 0.1 mM, ammonia is incorporated into glutamine only.

The reactions responsible for ammonia assimilation and the synthesis of glutamate and glutamine in ammonia-containing medium are shown below.

$$NH_3 + glutamate + ATP \xrightarrow[Mg^{2+}]{} glutamine + ADP + P_i \quad (1)$$

$$\begin{aligned} Glutamine + 2\text{-ketoglutarate} + NADPH \\ \rightarrow 2\ glutamate + NADP^+ \quad (2) \end{aligned}$$

$$\begin{aligned} NH_3 + 2\text{-ketoglutarate} + NADPH \\ \rightarrow glutamate + NADP^+ \quad (3) \end{aligned}$$

These three reactions are catalyzed by glutamine synthetase, glutamate synthase, and glutamate dehydrogenase, respectively

The reaction catalyzed by glutamine synthetase is the only known biosynthetic route for the synthesis of glutamine. Mutations in *glnA*, the structural gene for glutamine synthetase, result in an absolute requirement for glutamine; thus, the *glnA* gene is the only gene coding for a glutamine synthetase (66, 74, 82). A strain with mutations that result in the loss of both glutamate dehydrogenase and glutamate synthase is a glutamate auxotroph. Thus, the enzymes essential for ammonia assimilation and for the synthesis of glutamine and glutamate are glutamine synthetase, glutamate synthase, and glutamate dehydrogenase.

A strain devoid of glutamate dehydrogenase activity (equation 3) has no detectable phenotype, but a deficiency of glutamate synthase (equation 2) results in the inability to grow when the medium contains a low level of ammonium ion or a nitrogen source which generates ammonia slowly. These observations were made first in *Klebsiella* species and subsequently in *E. coli* and *S. typhimurium* (8, 13, 14, 31, 92, 98). When cells are grown in ammonia-containing medium, both glutamate dehydrogenase and glutamate synthase synthesize glutamate. For cells in medium with a growth rate-limiting source of ammonia, glutamate dehydrogenase is not involved in ammonia assimilation and glutamate formation; instead, glutamine synthetase (equation 1) is the only active ammonia-assimilating enzyme, and glutamate synthase (equation 2) is the only active glutamate-forming enzyme. Therefore, glutamine synthetase has two functions: the synthesis of glutamine and the assimilation of ammonia when the growth of the cell is limited by the availability of ammonia.

About eight times more glutamate than glutamine is required for cellular biosyntheses; consequently, during growth in a medium containing a low level of ammonia or some other source of nitrogen, when ammonia is assimilated exclusively into glutamine by glutamine synthetase, most of the glutamine must be

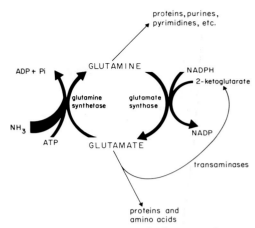

FIG. 1. Ammonia assimilatory cycle.

recycled to glutamate. This means that the glutamine-dependent amidotransferases other than glutamate synthase which convert glutamine to glutamate can only provide about 12% of the cellular glutamate during ammonia-limited growth. Apparently, glutamate synthase produces the remaining 88% of the cellular glutamate.

The reactions catalyzed by glutamine synthetase and glutamate synthase form an ammonia assimilatory cycle (Fig. 1). This cycle allows the net assimilation of ammonia into glutamine via glutamine synthetase and the replenishment and maintenance of an adequate intracellular level of glutamate.

Sources of Nitrogen Other than Ammonia

E. coli and *S. typhimurium* can grow on a variety of organic nitrogen-containing compounds as sole source of nitrogen but cannot grow on any inorganic nitrogen compound except ammonia. However, the related *Klebsiella* species can utilize nitrate, nitrite, or atmospheric diatomic nitrogen. A limited number of organic nitrogen compounds, about 25, can support growth as sole sources of nitrogen. The best surveys are presented by Gutnick et al. (54) for *S. typhimurium* and by Tyler (123) for *E. coli*. Broach et al. (15), Shaibe et al. (114, 115), and Wild et al. (139) sometimes used nitrogen sources that were not included in the surveys. Therefore, the surveys should be considered incomplete and probably strain dependent. Furthermore, variants of *E. coli* have been isolated that can grow on an extended range of nitrogen sources (63).

Growth with these organic nitrogen sources is invariably slower than with ammonia. We term such growth nitrogen limited. Both the level and the specific activity of glutamine synthetase are invariably high, consistent with its role in ammonia assimilation. This observation strongly indicates that the growth-limiting factor is always the rate of ammonia generation from these nitrogen sources and of its subsequent assimilation. The same regulatory system that regulates the level and activity of glutamine synthetase also regulates the level of an ammonia uptake system, thereby increasing the ability of the cell to scavenge a diminishing supply of ammonia (58).

The degradative enzymes and transport systems for a number of these nitrogen sources are frequently induced by growth in a nitrogen-limited medium. These systems, termed Ntr for nitrogen regulated, are involved in the degradation of arginine, ornithine, agmatine, putrescine, and γ-aminobutyrate in *E. coli* (115, 146); the degradation of histidine and urease in *K. aerogenes* (123); the transport of γ-aminobutyrate and glutamine in *E. coli* (59, 142); and the transport of glutamine, arginine, aspartate, lysine, ornithine, glutamate, and histidine in *S. typhimurium* (44, 65). The regulation of the *hisJQMP* and *argT* operons in *S. typhimurium*, which specify components of the histidine and arginine transport systems, was shown to be transcriptional (120). In the most extensive study of the transport systems of amino acids, Kustu et al. (65) showed increased levels of many periplasmic binding proteins that are involved in amino acid transport in *S. typhimurium*. This is the only report that correlates increased transport activity during nitrogen deprivation with the presence of specific proteins.

The same regulatory system controls the activity and level of glutamine synthetase and all of the Ntr systems. The activation during nitrogen-limited growth is mediated by two regulatory proteins, P_{II} and the bifunctional uridylyltransferase/uridylyl-removing enzyme, the activities of which are determined by the intracellular ratio of glutamine/2-ketoglutarate, a sensitive barometer of the ammonia content of the environment. The transcriptional regulation of *glnA* expression and the regulation of glutamine synthetase activity are discussed under Glutamine, below. In summary, nitrogen deprivation elevates the activities of a number of proteins that allow the bacteria to respond to an environmental stress in a highly coordinated fashion.

There are two general classes of nitrogen sources. The first class of compounds, when degraded, generate ammonia; D- and L-serine are members of this class. The second class consists of compounds that are degraded to glutamate, such as proline or aspartate, or that can form glutamate through transamination, such as 2-aminobutyrate. Some amino acids generate both glutamate and ammonia when degraded.

Growth with an ammonia-generating nitrogen source requires a high level of glutamine synthetase for ammonia assimilation and of glutamate synthase for replenishment and synthesis of the glutamate. A strain deficient in glutamate synthase cannot grow on any of the ammonia-generating nitrogen sources except D-serine (98, 123). Apparently, D-serine is degraded so rapidly that enough ammonia is generated for the synthesis of glutamate by glutamate dehydrogenase, which obviates the need for glutamate synthase.

Growth with a glutamate-generating nitrogen source also requires a high level of glutamine synthetase for the utilization of the ammonia derived from glutamate by an unidentified pathway which utilizes neither glutamate dehydrogenase nor aspartase (Reitzer and Magasanik, unpublished data). Paradoxically, a *gltB* mutant, which is deficient in glutamate synthase, cannot grow with most of the glutamate-generating nitrogen sources, such as arginine or proline (98, 123). This is an interesting consequence of the fact that much of the glutamine produced from the assimilation of ammonia must be converted back to glutamate by glutamate synthase. When *gltB* strains are grown with arginine or proline, the rate of ammonia generation limits growth. The ammonia is assimilated by glutamine synthetase to form glutamine and is not converted to glutamate. The accumulation of glutamine represses both glutamine synthetase and other Ntr systems (see Glutamine, below). Arginine is transported and degraded by Ntr systems, but because glutamine accumulates, these systems are repressed and the cell cannot accumulate or degrade the arginine. Genetic suppression of the inability of *gltB* cells to grow with a glutamate-forming source of nitrogen results in the constitutively high synthesis of glutamine synthetase and the Ntr systems (98, 123). Once the glutamate-forming nitrogen source is degraded, the need for glutamate synthesis by glutamate synthase is obviated. Furthermore, growth of a wild-type strain of *S. typhimurium*, with arginine as a nitrogen source, represses glutamate synthase. In other words, glutamate synthase is required for the establishment, but not the maintenance, of steady-state nitrogen-limited growth with a glutamate-forming source of nitrogen. The fact that a *gltB* strain cannot grow on a particular glutamate-forming nitrogen source has been taken to imply that the degradation of that nitrogen source is regulated by nitrogen deprivation (123). A *gltB* strain can grow with glutamate, aspartate, and asparagine as source of nitrogen. This probably means that the degradation of these compounds does not require an Ntr system.

Growth with L-glutamine as the nitrogen source sustains a growth rate almost as rapid as growth with ammonia, but, curiously, the transcription of *glnA* and other Ntr systems is activated, indicating that growth on glutamine is nitrogen limited. It is unexpected that growth with glutamine as the nitrogen source should increase the expression of a gene the product of which synthesizes glutamine. Glutamine per se does not cause any unusual changes in metabolism because cells grown in medium containing both glutamine and ammonia have a low level of glutamine synthetase (74). The glutamate synthase appears to be sufficiently active to deplete the intracellular glutamine pool when cells are grown with glutamine as sole source of nitrogen. This view is supported by the observation that, in a *gltB* strain, growth with L-glutamine as the nitrogen source results in a low level of glutamine synthetase, in consequence of the intracellular accumulation of glutamine (14). The significance of this metabolic oddity is that glutamine-requiring strains and some strains deficient in their ammonia assimilatory capacity can be grown on a nitrogen-limited medium. This has greatly facilitated the study of the response of *E. coli* and *S. typhimurium* to nitrogen starvation.

Pleiotropic Defects in Nitrogen Utilization

Mutations in genes that code for proteins other than glutamine synthetase, glutamate synthase, or their regulatory proteins can cause a pleiotropic inability to grow with a variety of single nitrogen sources other than ammonia. *E. coli* and *K. aerogenes* have two asparagine synthetases, an ammonia-dependent enzyme and a glutamine-dependent enzyme. A deficiency of the glutamine-dependent enzyme in *K. aero-*

genes results in the inability to grow well on any nitrogen source other than ammonia and asparagine; the ammonia-dependent enzyme is insufficient to supply the cell with asparagine in ammonia-restricted medium (103; also see Asparagine).

Analogous phenotypes result from certain mutations in amidotransferases. Some glutamine-dependent amidotransferases have two unequal subunits coded by separate genes; one subunit binds and cleaves glutamine and transfers the amide nitrogen to the other subunit, the synthetase subunit. The synthetase subunit can be shown to synthesize the appropriate product when supplied with ammonia in high concentration. A mutation causing the loss of the glutamine-binding subunit should result in a requirement for ammonia or for the product of the reaction. Because the K_m of synthetase subunits for ammonia tends to be very high (10 to 50 mM), the growth rate should be dependent on the ammonia concentration in the medium. Strains with mutations in the glutamine-binding subunit of carbamylphosphate synthetase have been isolated; optimal growth required a very high level of ammonia (86). Broach et al. (15) isolated a strain of *S. typhimurium* the growth rate of which depended on the concentration of ammonia in the medium. Although they could not identify the enzymatic deficiency, it is undoubtedly in a gene for an amidotransferase. This strain could not grow with a wide variety of single nitrogen sources. The lesson to be learned is that not all pleiotropic nitrogen source nonutilizers are deficient in their ability to assimilate ammonia.

GLUTAMINE

Functions and Properties of Glutamine Synthetase

Glutamine synthetase is essential for the formation of glutamine for protein synthesis; for the synthesis of a number of nitrogen-containing metabolites, such as purines and pyrimidines; and, in addition, for the assimilation of ammonia when the cells are grown in an ammonia-limited medium. Because of the importance of glutamine in cellular metabolism and because of the multiple functions of glutamine synthetase, it is not surprising that both the catalytic activity and the synthesis of glutamine synthetase are highly regulated (reviewed by Tyler [123] and Magasanik [74]).

Glutamine synthetase catalyzes the ATP-dependent synthesis of L-glutamine from ammonia and L-glutamate, as follows: $NH_4 + $ L-glutamate $+ ATP \xrightarrow{Mg^{2+}}$ L-glutamine $+ ADP + P_i$. The purified enzyme is a dodecamer of identical 55,000-dalton subunits. The catalytic activity of glutamine synthetase is regulated by the covalent addition of an AMP group to a tyrosine residue in each of its subunits. The adenylylation of a subunit inactivates that subunit only, and consequently an enzyme with 9 of its 12 subunits adenylylated has 3 active subunits. The activity of glutamine synthetase is measured by what is referred to as the glutamyl transferase reaction. Both adenylylated and unmodified glutamine synthetases catalyze the transfer of a glutamyl residue of glutamine to hydroxylamine in the presence of Mn^{2+}, arsenate, and ADP. At the appropriate pH, which is slightly different in different species of enteric bacteria, the transferase activities of both forms of glutamine synthetase are the same. However, the two forms of glutamine synthetase can be distinguished because 60 mM Mg^{2+} completely inhibits fully adenylylated glutamine synthetase but has no effect on the unadenylated enzyme (5, 53, 66). Because of these properties, the degree of adenylylation can be easily assessed. The discovery of a simple procedure for fixing the degree of adenylylation during the preparation of a sample for assay has greatly aided the study of glutamine formation. The treatment of cells with the detergent cetyltrimethyl ammonium bromide simultaneously kills and increases the permeability of cells and inactivates adenylyltransferase, the enzyme responsible for the addition and removal of adenylyl groups from glutamine synthetase (5).

Glutamine synthetase activity is subject to cumulative feedback inhibition by the products of glutamine metabolism. Each of the following compounds, L-alanine, glycine, histidine, tryptophan, CTP, AMP, carbamyl phosphate, and glucosamine-6-phosphate, inhibits glutamine synthetase partially; together, these compounds can inhibit glutamine synthetase completely. All but two of these compounds require glutamine for their synthesis. Curiously, glutamine itself is not an inhibitor. It is thought that each inhibitor has its own binding site. Stadtman and Ginsburg (118) have made the surprising observation that it is the adenylylated form of glutamine synthetase that is sensitive to this cumulative feedback inhibition. We will discuss the physiological significance of this form of inhibition below.

Stadtman and Levine have recently studied the two-step degradation of glutamine synthetase in *E. coli* and *K. aerogenes* (reviewed by Levine [70]). Glutamine synthetase is initially tagged by the oxidation of a specific histidinyl residue. The oxidation is caused by a mixed function oxidase; both oxygen and reducing equivalents are required. No specific oxidation system has been identified, but a number of enzymatic and nonenzymatic systems can cause the initial oxidation. The oxidation results in a subtle conformational change that makes glutamine synthetase susceptible to what seems to be a specific proteolytic system. However, the time required for the initial oxidation is a significant fraction of one generation (71), a feature that implies that the degradation of glutamine synthetase is not of physiological significance for regulating the rate of glutamine formation.

Regulation of Glutamine Synthetase Activity and Synthesis in Response to Nitrogen Availability

Overview. Enteric bacteria use a system consisting of two proteins, uridylyltransferase/uridylyl-removing enzyme and P_{II}, to transmit the information regarding nitrogen availability to the systems responsible for the appropriate adjustment of glutamine synthetase activity and of the rate of its synthesis.

The state of the cell with regard to excess or deficiency of nitrogen is reflected in the intracellular ratio of glutamine to 2-ketoglutarate. The ready availability of ammonia increases the rate of synthesis of glutamine by glutamine synthetase and the conversion of 2-ketoglutarate to glutamate by glutamate

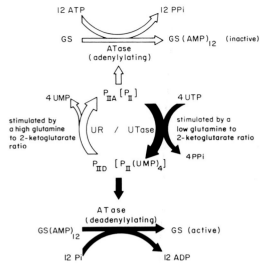

FIG. 2. Covalent modification of glutamine synthetase (GS). UR/UTase, uridylyl-removing enzyme/uridylyltransferase.

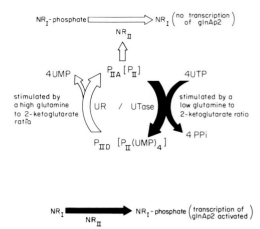

FIG. 3. Covalent modification of NR$_I$, the regulator of glnA transcription. UR/UTase, uridylyl-removing enzyme/uridylyltransferase.

synthase and glutamate dehydrogenase. Conversely, a deficiency of ammonia decreases the rate of glutamine synthesis and the rate of 2-ketoglutarate utilization. Therefore, growth in a minimal medium with an excess of ammonia results in a high glutamine–to–2-ketoglutarate ratio and a low intracellular concentration of a partially adenylylated glutamine synthetase; growth in a medium deficient in ammonia results in a low glutamine–to–2-ketoglutarate ratio and a high intracellular concentration of unmodified glutamine synthetase (113).

The regulation of glutamine synthetase activity is accomplished by the action of three proteins, uridylyltransferase/uridylyl-removing enzyme, P$_{II}$, and adenylyltransferase. The sequence of events that lead to the adenylylation (inactivation) of glutamine synthetase is illustrated by the open arrows in Fig. 2. A high intracellular concentration of glutamine activates the uridylyl-removing enzyme, which causes the deuridylylation of the regulatory protein P$_{II}$. The deuridylylated (unmodified) form of P$_{II}$ is sometimes referred to as P$_{IIA}$ for the form of P$_{II}$ that acts to cause the adenylylation of glutamine synthetase. This unmodified P$_{II}$ interacts with adenylyltransferase, which in turn catalyzes the adenylylation of glutamine synthetase. The sequence of events that lead to the deadenylylation (activation) of glutamine synthetase is illustrated by the solid arrows in Fig. 2. The high intracellular concentration of 2-ketoglutarate activates uridylyltransferase, which transfers a UMP group to each subunit of the regulatory protein P$_{II}$ to form P$_{II}$-UMP or P$_{IID}$ (deadenylylating). P$_{II}$-UMP interacts with adenylyltransferase, which in turn catalyzes the removal of AMP from glutamine synthetase (119).

Studies in the laboratories of B. Magasanik and S. Kustu have established that the regulation of the transcription of glnA, the structural gene for glutamine synthetase, in response to the availability of nitrogen results from the action of uridylyltransferase/uridylyl-removing enzyme, P$_{II}$, NR$_I$ (nitrogen regulator I), and NR$_{II}$ and requires core RNA polymerase

and σ^{60}, rather than the abundant σ^{70} (17, 21, 46, 47a, 55, 57, 63a, 97a; for a review, see reference 74). When the intracellular concentration of 2-ketoglutarate is high, uridylyltransferase converts P$_{II}$ to P$_{II}$-UMP. In the absence of P$_{II}$, NR$_{II}$ catalyzes the conversion of NR$_I$ to NR$_I$-phosphate (94). This phosphorylated protein activates the initiation of transcription at the σ^{60}-dependent promoter glnAp2. On the other hand, when the intracellular concentration of glutamine is high, uridylyl-removing enzyme converts P$_{II}$-UMP to P$_{II}$; P$_{II}$ causes NR$_{II}$ to remove the phosphate group from NR$_I$ phosphate and thus to halt the initiation of transcription at glnAp2 (94) (Fig. 3).

A glossary of genes and proteins involved in the regulation of glutamine formation is presented in Table 1.

The role of uridylyltransferase/uridylyl-removing enzyme and P$_{II}$. The bifunctional uridylyltransferase/uridylyl-removing enzyme catalyzes the covalent uridylylation and deuridylylation of P$_{II}$. The deuridylylation reaction is not a reversal of the uridylylation reaction, as can be seen in the following reactions:

$$\text{P}_{II} + n\text{UTP} \xrightarrow[\text{Mg}^{2+}, \text{ 2-ketoglutarate}]{\text{uridylyltransferase}} \text{P}_{II}\text{-UMP}n + n\text{PP};$$

TABLE 1. Glossary of gln genes and proteins

Gene	Chromosomal location (min)	Product[a]	Target of glutamine synthetase regulation[b]
glnA	87	GS	—
glnB	55	P$_{II}$	CA, T
glnD	4	UTase/UR	CA, T
glnE	?	ATase	CA
rpoN (glnF, ntrA)	70	σ^{60}	T
glnG (ntrC)	87	NR$_I$	T
glnL (ntrB)	87	NR$_{II}$	T

[a] Abbreviations: GS, glutamine synthetase; UTase/UR, uridylyltransferase/uridylyl-removing enzyme; ATase, adenylyltransferase.

[b] CA, Catalytic activity; T, transcription.

$$P_{II} \cdot UMPn \xrightarrow[\text{glutamine}]{\text{uridylyl-removing enzyme}} P_{II} + n\text{UMP}$$

The purified P_{II} protein, molecular weight 44,000, is a tetramer of identical subunits; each subunit can be uridylylated at a specific tyrosyl residue (1). Glutamine stimulates the uridylyl-removing enzyme, and 2-ketoglutarate stimulates uridylyltransferase. Only one polypeptide, molecular weight 95,000 ± 5,000, is present in the purest preparations of uridylyltransferase/uridylyl-removing enzyme. The purified protein is unstable and has a tendency to oligomerize; the most active species is a monomer. By a number of criteria, both uridylyltransferase and uridylyl-removing enzyme activities are on the same polypeptide. First, both activities are copurified, and the ratio of activities is constant. Second, strains with point mutations in glnD, which are deficient in uridylyltransferase, also lack uridylyl-removing enzyme; revertants acquire both activities (3, 38, 48).

The loss of uridylyltransferase/uridylyl-removing enzyme causes only the unmodified form of P_{II} to be present (Fig. 2), which in turn will interact with adenylyltransferase to overadenylylate glutamine synthetase and which will interact with NR_{II} to prevent the transcription of glnA (Fig. 3). Strains of S. typhimurium and K. aerogenes with mutations in glnD, which code for uridylyltransferase/uridylyl-removing enzyme, were isolated by their inability to grow on a variety of organic nitrogen compounds; in E. coli, a mutation in glnD causes glutamine auxotrophy (3, 10, 38). It is unlinked to the genes coding for P_{II} or adenylyltransferase or to any other gene the product of which is known to be involved in the regulation of glutamine synthetase. The phenotype of glnD strains is not solely due to the overadenylylation of glutamine synthetase but also to the lower level of glutamine synthetase, a result of a lower rate of glnA transcription (105). However, the overadenylylation of glutamine synthetase is an important prediction of the regulatory cascade and demonstrates that uridylyltransferase/uridylyl-removing enzyme is involved in the adenylylation of glutamine synthetase in vivo.

A specific type of mutation in the gene for P_{II}, the glnB gene, could theoretically result in an altered P_{II} that would not be uridylylated by uridylyltransferase. Such a mutant would have the same phenotype as a glnD strain lacking uridylyltransferase/uridylyl-removing enzyme. In fact, some glnB mutants of S. typhimurium and K. aerogenes are glutamine auxotrophs (45, 99). The mutation in K. aerogenes was shown to result in an altered form of P_{II} that could interact with adenylyltransferase to adenylylate, but not deadenylylate, glutamine synthetase, presumably because P_{II} could not be uridylylated (39, 40). Again, the phenotype of these mutant strains results from the combined effects of overadenylylation of glutamine synthetase and a lower level of glnA transcription. The overadenylylation of glutamine synthetase in these mutant strains verifies the regulatory scheme depicted in Fig. 2.

The effects of the loss of uridylyltransferase/uridylyl-removing enzyme can be suppressed by the loss of P_{II} (17, 40). Double mutants of E. coli with insertions or deletions in glnD, as well as in glnB, no longer require glutamine for growth and have the same elevated level of glutamine synthetase, irrespective of the presence of ammonia in the growth medium; the lack of P_{II} allows NR_I to exist in the active, phosphorylated form even when the intracellular concentration of glutamine is high (94). The degree of adenylylation of glutamine synthetase in the mutants lacking P_{II} is still determined by the availability of ammonia. The significance of this observation will be discussed below.

It is of some interest that mutants lacking P_{II} but containing uridylyltransferase/uridylyl-removing enzyme can, in response to ammonia deprivation, increase glutamine synthetase to a level higher than that found in the mutant lacking both P_{II} and uridylyltransferase/uridylyl-removing enzyme (17). This elevation of the glutamine synthetase level may reflect the ability of uridylyltransferase to stimulate the conversion of NR_I to NR_I-phosphate by NR_{II}. The molecular mechanism of this putative interaction has not been investigated.

Transcription of glnA: roles of σ^{60}, NR_I, and NR_{II}. The structural gene for glutamine synthetase glnA, is a member of the glnALG operon, where glnL and glnG are the structural genes for NR_{II} and NR_I, respectively (2a, 21, 82a, 97). NR_{II} and NR_I are dimers of identical subunits with respective molecular weights of 68,000 and 110,000 (95, 104). This operon has three promoters: glnAp1, with a transcriptional start site located 187 base pairs (bp) upstream from the translational start site in glnA; glnAp2, with a transcriptional start site 73 bp upstream from the translational start site in glnA; and glnLp, with a transcriptional site located 256 bp downstream from the translational termination site in glnA and 33 bp upstream from the translational start site in glnL (55, 62, 97, 105, 124). The three promoters enable the cell to maintain a low level of glutamine synthetase and NR_I during growth with an excess of nitrogen and to increase the level of glutamine synthetase and NR_I rapidly in response to nitrogen deprivation.

The promoters glnAp1 and glnLp have the nucleotide sequences characteristic of the majority of promoters of enteric bacteria in the regions 10 and 35 bp upstream from their respective transcriptional start sites, and their transcription requires core polymerase with σ^{70}. The initiation of transcription from these promoters is negatively regulated by NR_I and NR_I-phosphate. This repression results from the ability of NR_I to bind to a sequence of nucleotides, GCAN$_6$TG-GTGC, which overlap the transcriptional start site of glnLp, and the −35 RNA polymerase contact site and the transcriptional start site of glnAp1. In addition, the initiation of transcription at glnAp1 is stimulated by catabolite-activating protein and cyclic AMP; a characteristic binding site for catabolite-activating protein–cyclic AMP is located between 65 and 81 bp upstream from the transcriptional start site of glnAp1 (55, 62, 97, 104, 105, 124).

In the case of glnAp2, a nucleotide sequence TTG-GCACAN$_4$TCGCT, located between bp 27 and bp 11 relative to the start site of transcription, has been identified as the promoter. This sequence fits a consensus found in other nitrogen-regulated promoters for which transcription requires σ^{60}, the product of rpoN, rather than the abundant σ^{70}. This dependence is demonstrated by the observation that mutants with lesions in *rpoN* are unable to initiate transcription at *glnAp2* even when deprived of nitrogen. These mutants require glutamine for growth in a glucose-containing medium, as a result of the reduced activation of transcription at *glnAp1* by catabolite-activating protein–cyclic AMP and repression at *glnAp1* by NR_I, formed as a result of transcription initiated at *glnLp*. The glutamine requirement of *rpoN* mutants can be suppressed by substituting a poor carbon source, such as succinate, for glucose, or by a mutation in *glnG*, resulting in the loss of NR_I. In these cases, transcription of *glnA* is initiated at *glnAp1* (55, 57, 105).

In intact cells, the initiation of transcription at *glnAp2* requires NR_I. *glnG* mutants are not glutamine requirers, since they produce glutamine synthetase due to transcription initiated at *glnAp1* (105). However, their level of glutamine synthetase is low, and they are unable to increase it in response to nitrogen deprivation. We have described, in Assimilation of Ammonia, the ability of a variety of amino acids and other nitrogen-containing compounds to serve as sources of nitrogen. The lack of NR_I or of σ^{60} prevents the utilization of most of these compounds in a glucose-containing medium, indicating that NR_I and σ^{60} are essential for the activation of transcription of the corresponding genes in response to nitrogen limitation. The fact that some of these compounds can be used as sole sources of nitrogen in a medium containing a poor source of carbon in place of glucose is an indication of a separate system responsible for the activation of the expression of these genes in response to carbon starvation (74, 105, 123).

The activation of transcription of *glnAp2* has been studied in systems consisting of a DNA template containing *glnAp2* and purified components of *E. coli* or partially purified components of *S. typhimurium*. In either case, it could be shown that the initiation of transcription required core RNA polymerase and σ^{60} (55, 57). In the *E. coli* system, no other component was necessary when the template was supercoiled, but transcription was initiated at a very low rate. The addition of only NR_I had no effect, but the addition of both NR_I and NR_{II} greatly increased initiation of transcription from the supercoiled template and was absolutely required for the initiation of transcription from a linear template (55, 57).

It could be shown that the role of NR_{II} is to catalyze the phosphorylation of NR_I by ATP; NR_I-phosphate is then responsible for the activation of transcription at *glnAp2*. The role of NR_{II} in both kinase and phosphatase activity is confirmed by the properties of NR_{II} isolated from a *glnL* mutant. The mutation results in a high level of glutamine synthetase in cells grown with an excess of nitrogen, even when these cells are *glnD* mutants, unable to produce uridylyltransferase/uridylyl-removing enzyme. The NR_{II} isolated from such a *glnL* mutant catalyzes the phosphorylation of NR_I but fails to be stimulated by P_{II} to remove the phosphate group of NR_I-phosphate (94).

A deletion of *glnL*, which results in the lack of NR_{II}, does not prevent the intact cell from increasing the level of glutamine synthetase in response to nitrogen deprivation through transcription initiated at *glnAp2* (2b). However, this mutation suppresses the effects of mutations in *glnB* and *glnD* on the synthesis of glutamine synthetase (17, 21). It could be shown that, in the intact cell, NR_{II} is required for the rapid initiation of transcription at *glnAp2* when cells replete with nitrogen are deprived of nitrogen and for the rapid cessation of transcription when nitrogen-starved cells are supplied with ammonia (105). In view of the fact that in the in vitro system the phosphorylation of NR_I by NR_{II} is absolutely required for endowing NR_I with the ability to activate transcription at *glnAp2*, it is puzzling that, in intact cells, NR_I is able to activate this transcription in response to nitrogen starvation in the absence of NR_{II}, albeit at a slow rate. It is possible that the cell contains a protein other than NR_{II} which is able to catalyze the phosphorylation of NR_I.

There are additional features in the structure of the *glnALG* operon that explain its efficient operation. Located between the end of *glnA* and the *glnLp* promoter is a canonical Rho-independent terminator, which is presumably responsible for the fact that *glnG* is transcribed at a considerably lower rate than *glnA* when the transcription of both genes is initiated at *glnAp2*. The partial termination of transcription accounts in part for the fact that a cell grown with nitrogen limitation contains approximately 1,000 molecules of the dodecomeric enzyme glutamine synthetase but only approximately 70 molecules of the dimeric regulator NR_I (104, 124). It has also been shown that, when transcription is initiated at *glnLp*, the *glnL*- and *glnG*-specific segments of RNA are translated at approximately the same rate but that, when transcription is initiated at *glnAp2*, the *glnL*-specific RNA segment is translated at a considerably lower rate than the *glnG* segment (T. Hunt and B. Magasanik, unpublished observation). The partial inhibition of the translation of *glnL* RNA, the synthesis of which was initiated at *glnAp2*, may be the result of the presence of a nucleotide sequence in the RNA transcribed from the intercistronic *glnA-glnL* region which is complementary to the ribosome-binding site for NR_{II} (124). The fact that NR_I and NR_{II} are not produced in equimolar amounts is in accord with the role of NR_{II} as a catalyst of NR_I modification.

Considering the operation of the *glnALG* operon, we find that, in cells growing with a great excess of nitrogen, that is, essentially in a condition of carbon and energy limitation, *glnA* is transcribed from the catabolite-activating protein–cyclic AMP-activated promoter *glnAp1*, which is partially repressed by NR_I, resulting in a low level of glutamine synthetase. The cells also maintain a low level of NR_I, approximately five molecules per cell (104), through transcription initiated at *glnLp* (124), which is regulated autogenously through repression exerted by NR_I (62, 97). These cells presumably contain unmodified P_{II}, which prevents the phosphorylation of NR_I by NR_{II} (94). A shift of the cells to a nitrogen-deprived medium causes the activation of uridylyltransferase by the increase in intracellular 2-ketoglutarate; the uridylyltransferase, in turn, converts P_{II} to the innocuous P_{II}-UMP and allows NR_{II} to convert NR_I to the active NR_I-phosphate. At the low level of not more than five molecules per cell, NR_I-phosphate can fully activate the initiation of transcription at *glnAp2* by σ^{60}-RNA polymerase. However, this level of NR_I-phosphate is not adequate for the activation of transcription of other

σ^{60}-dependent, nitrogen-regulated promoters (97). Expression of these Ntr genes is only initiated after the level of intracellular NR_I has been increased, eventually to approximately 70 molecules per cell, through the transcription of *glnG* initiated at *glnAp2*. The increase in the level of NR_I results in complete repression at *glnApl* and *glnLp*. A shift of the cells from nitrogen starvation to nitrogen excess causes the activation of uridylyl-removing enzyme by the increased intracellular glutamine and the removal of UMP from P_{II}-UMP by uridylyl-removing enzyme. The resulting P_{II} causes NR_{II} to remove the phosphate from NR_I-phosphate, bringing to a halt the initiation of transcription at *glnAp2*. Continuing growth in this medium results in the decline of the levels of glutamine synthetase and NR_I by dilution; the level of NR_I eventually approaches approximately five molecules per cell, which lifts the repression at *glnApl* and *glnLp* sufficiently to allow both glutamine synthetase and NR_I to be maintained at their respective low levels (97, 105) (Fig. 3).

These results demand an explanation of the difference between the structure of *glnAp2* and the other Ntr promoters that accounts for the much greater sensitivity of *glnAp2* to activation by NR_I-phosphate. This sensitivity appears to depend on the ability of NR_I, presumably as NR_I-phosphate, to bind to the two sites near *glnApl*, located approximately 110 and 140 bp, respectively, upstream from the start of transcription at *glnAp2*. The deletion of these sites abolishes the activation of transcription by NR_I-phosphate in low intracellular concentration; it decreases, but does not prevent, activation of transcription of *glnAp2* when the level of NR_I is high. It is of particular interest that the binding sites for NR_I can be moved more than 1,000 bp away from their original sites without diminishing the ability of NR_I-phosphate to activate transcription at *glnAp2*. Thus, the NR_I binding sites resemble enhancers in eucaryotic cells (106).

Activation and inactivation of glutamine synthetase: the role of adenylyltransferase. Adenylyltransferase catalyzes the ATP-dependent addition of AMP to a subunit of glutamine synthetase, with the release of PP_i. Each subunit can be adenylylated, so that a molecule of glutamine synthetase can have 12 adenylyl groups covalently attached; glutamine synthetase + nATP $\xrightarrow{Mg^{2+}}$ glutamine synthetase $\cdot AMP_n$ + nPP$_i$. Adenylyltransferase can also catalyze the phosphate-dependent removal of AMP from each subunit of glutamine synthetase; glutamine synthetase $\cdot AMP_n$ + nP$_i$ $\xrightarrow{Mg^{2+}+ATP}$ glutamine synthetase + nADP. Clearly, the deadenylylation reaction is not a reversal of the adenylylation reaction. Adenylyltransferase has been purified and is a monomer with a molecular weight of 115,000. Adenylyltransferase can catalyze both reactions in vitro without the regulatory proteins, P_{II} and uridylyltransferase/uridylyl-removing enzyme (53, 119). This demonstrates that the catalytic potential for adenylylation and deadenylylation resides in adenylyltransferase. Biochemical evidence indicates that the two activities of adenylyltransferase depend on separate sites on the same polypeptide (53). This is consistent with genetic evidence. Most mutant strains lacking the adenylylating activity of adenylyltransferase, which were isolated as *glnD* suppressors, have

also lost deadenylylating activity (3, 108). The physical separation of the two activities of adenylyltransferase was demonstrated by the isolation of an unusual mutant strain of *K. aerogenes* that had lost the adenylylating activity of adenylyltransferase but that still retained normal deadenylylating activity (108).

The adenylylation of glutamine-synthetase by adenylyltransferase without P_{IIA} is stimulated by glutamine; the deadenylylation reaction of adenylyltransferase without P_{IID} is inhibited by glutamine and stimulated by the substrate phosphate (2, 33). Ginsburg and Stadtman (53) showed that the binding sites on adenylyltransferase for glutamine and P_{IIA} are separate, which strongly indicates that glutamine can directly affect adenylyltransferase without the mediation of P_{IIA}. This regulation of purified adenylyltransferase without P_{II} by glutamine is similar to the overall regulation of the cascade system with P_{II} and uridylyltransferase/uridylyl-removing enzyme. Were it not for the fact that strains without uridylyltransferase/uridylyl-removing enzyme do not control the adenylylation of glutamine synthetase normally, it would appear that P_{II} and uridylyltransferase/uridylyl-removing enzyme are redundant. The rate of the deadenylylation reaction of adenylyltransferase without P_{IID} is only 15% of the rate with P_{IID}. The physiological significance of this reaction was initially doubted because the reaction was observed only when adenylyltransferase was adsorbed to a manganese phosphate precipitate; it was thought that highly purified adenylyltransferase behaves anomalously because adenylyltransferase may be part of a multienzyme complex in vivo that is destroyed during purification (2). However, genetic and later biochemical evidence indicates that both reactions of adenylyltransferase without P_{II} are not artifactual. Strains with an insertion in the *glnB* gene, resulting in the loss of P_{II}, have been isolated as suppressors of *glnD* mutations (discussed earlier). In a strain with only the *glnB* mutation (P_{II} deficient), the activity of adenylyltransferase alone determines the degree of adenylylation of glutamine synthetase. When a *glnB* mutant is grown in a nitrogen-rich and nitrogen-limited medium, the adenylylation of glutamine synthetase is high and low, respectively (17, 40). This means that the degree of glutamine synthetase adenylylation in a *glnB* strain is similar to that of a wild-type strain. When a *glnB* strain is shifted from a nitrogen-rich medium to a nitrogen-poor medium, the rate of deadenylylation is very slow compared with that in a wild-type strain, but, when shifted in the opposite direction, the rate of adenylylation is almost as high as that in a wild-type strain (40). These observations suggest that adenylyltransferase alone responds to metabolites that control the entire adenylylation system; a high glutamine/2-ketoglutarate ratio favors adenylylation, and a low ratio favors deadenylylation. Below we discuss the advantages of the control of the adenylylation of glutamine synthetase by the entire adenylylation cascade.

A major function of the adenylylation of glutamine synthetase was revealed by the elegant study of Kustu et al. (64) of the physiological effect of the loss of adenylyltransferase in *S. typhimurium*. A *glnE* mutant lacking adenylyltransferase could grow with a variety of nitrogen and carbon sources and had a normal growth rate with NH_4^+ as the source of nitrogen. When the *glnE* mutation was combined with mutations that result in a constitutively high level of glutamine synthetase, the presence of ammonia in the medium significantly increased the generation time. This implies that the excessive activity of glutamine synthetase was detrimental, possibly by depleting intracellular pools of glutamate or ATP. When cells of a *glnE* strain were shifted from a nitrogen-poor medium to a nitrogen-rich medium by the addition of ammonia, there was a long lag in the growth rate, and the glutamate pools dropped 10-fold and remained low for a long time. The pool of glutamine increased enormously, but that of ATP was constant. For a wild-type strain subjected to the same nutritional shift, there was no lag in the growth rate, the pools of ATP and glutamate remained constant, and the pool of glutamine increased moderately. The difference between the *glnE* mutant and the wild-type strain is that, in the wild-type strain but not in the *glnE* strain, glutamine synthetase becomes adenylylated when ammonia is added. Therefore, an important function of the adenylylation system is to prevent the depletion of glutamate from cells experiencing the transition from a nitrogen-poor environment to an ammonia-containing environment.

Overall regulation of the adenylylation cascade. The most straightforward approach to a quantitative assessment of the factors involved in the overall regulation of the adenylylation cascade would be to purify the individual components and reconstitute the entire system in vitro. There are a number of practical problems with this approach. First, in such a reconstitution it is difficult to determine how much of each component should be used. The ratio of the various components may be an important factor in the response of the enteric system to metabolite levels. Second, highly purified components sometimes exhibit anomalous properties when they are purified away from some potential multiprotein complex; this may be the case for the deadenylylation of glutamine synthetase by purified adenylyltransferase. Sometimes, purified proteins are simply unstable; the purified uridylyltransferase/uridylyl-removing enzyme has a tendency to aggregate and to lose activity.

Stadtman and co-workers have circumvented these problems by studying the quantitative properties of the adenylylation system in situ in detergent-treated cells (90, 91). When cells of *E. coli* are grown in medium with 10 mM glutamine and then treated with the nonionic detergent Lubrol WX, the cells become permeable to small molecules but not to proteins. All of the proteins of the adenylylation system are still active. The entire adenylylation system is called a bicyclic cascade because two reversible modifications, uridylylation-deuridylylation and adenylylation-deadenylylation, control the activity of glutamine synthetase. The biochemical evidence obtained with purified components suggested that the overall degree of adenylylation of glutamine synthetase is determined principally by the ratio of glutamine to 2-ketoglutarate but that about 30 other metabolites can also affect the activities of uridylyltransferase/uridylyl-removing enzyme, P_{II}, and adenylyltransferase (53, 117). In Lubrol WX-treated cells, the ratio of glutamine to 2-ketoglutarate does determine the degree of adeny-

lylation of glutamine synthetase by the bicyclic cascade; a high concentration of glutamine and 2-ketoglutarate causes the adenylylation and deadenylylation, respectively, of glutamine synthetase. Furthermore, these responses are affected by Mn^{2+}, CMP, UTP, and a high pH, all of which are known to affect uridylyltransferase/uridylyl-removing enzyme but not P_{IIA}, P_{IID}, or adenylyltransferase (90). Therefore uridylyltransferase/uridylyl-removing enzyme directs the adenylylation and deadenylylation of glutamine synthetase in situ through P_{II} and adenylyltransferase. However, when Lubrol WX treatment follows growth in medium with 5 mM glutamine, uridylyltransferase/uridylyl-removing enzyme is inactive, and presumably the degree of uridylylation of P_{II} is frozen. In this case, the ability of adenylyltransferase without uridylyltransferase/uridylyl-removing enzyme to adenylylate or deadenylylate glutamine synthetase can be quantitatively assessed. This cascade is termed a monocyclic cascade because only one reversible modification, adenylylation-deadenylylation, controls glutamine synthetase activity (119). It is assumed in these studies that the monocyclic cascade is not affected by the presence of P_{II}. Control of the degree of adenylylation of glutamine synthetase was controlled by the ratio of glutamine to 2-ketoglutarate. The deadenylylation reaction required phosphate as a substrate and 2-ketoglutarate and ATP as allosteric effectors (90, 91). The effect of ATP as an allosteric effector had been previously observed with purified adenylyltransferase (2). The adenylylation of glutamine synthetase by adenylyltransferase without uridylyltransferase/uridylyl-removing enzyme was stimulated by glutamine, a result that is consistent with previous biochemical and genetic evidence.

This study of the mono- and bicyclic cascades in situ and by computer simulation permits the quantitative assessment of the response to changes in the concentrations of the effectors. The major advantage of a bicyclic cascade over a monocyclic cascade is an increase in the sensitivity to alteration in effector (glutamine/2-ketoglutarate) concentration. This has been termed signal amplification. The increased signal sensitivity is largely the result of the sigmoidal response to changes in the levels of effectors. In other words, the response of a bicyclic cascade to a particular change in effectors is faster and larger than the response of a monocyclic cascade. These aspects of regulation have been reviewed (25, 119).

A second advantage of the bicyclic over the monocyclic adenylylation cascade is only evident when the role of uridylyltransferase/uridylyl-removing enzyme and P_{II} in the control of *glnA* (glutamine synthetase) transcription is considered. A longer sequence of regulatory interaction allows the possibility of coupling and coordinating a number of physiological processes. In this case, because uridylyltransferase/uridylyl-removing enzyme and P_{II} regulate both the degree of adenylylation of glutamine synthetase and the transcription of *glnA*, the catalytic activity and level of glutamine synthetase are controlled by the same changes in effector concentrations. Because this cascade control system is bicyclic, small changes in the availability of ammonia in the environment result in a rapid and coordinated cellular response.

Regulation of Glutamine Synthetase by Cumulative Feedback Inhibition

Glutamine synthetase has two functions: the formation of glutamine for the synthesis of protein and other nitrogen compounds and the assimilation of ammonia when the availability of ammonia in the environment is restricted. In an ammonia-rich medium, the level of glutamine synthetase is low, and glutamine synthetase functions primarily for the synthesis of glutamine, whereas in an ammonia-poor, nitrogen-limited medium the level of glutamine synthetase is high, and glutamine synthetase has both functions. The ammonia assimilatory function is quantitatively more significant, as is evident from the fact that almost all of the glutamine synthesized must be reconverted to glutamate by glutamate synthase.

The cumulative feedback inhibition of glutamine synthetase affects only adenylylated glutamine synthetase (118). The physiological significance of this inhibition is evident when the functions of glutamine synthetase are considered. When cells are grown in a nitrogen-limited medium, glutamine synthetase functions primarily to assimilate ammonia. Glutamine synthetase is not adenylylated and not susceptible to feedback inhibition. However, when the cells are grown in an ammonia-containing medium, glutamine synthetase is partially adenylylated and functions primarily in the formation of glutamine for the synthesis of protein and in the formation of some nitrogenous intermediates. Therefore, it is appropriate that glutamine synthetase is susceptible to inhibition by the products of glutamine metabolism only in the ammonia-containing medium when glutamine synthetase is not necessary for ammonia assimilation. As a corollary, part of the function of the adenylylation cascade is to make glutamine synthetase susceptible to cumulative feedback inhibition.

GLUTAMATE

Synthesis of Glutamate

Strains lacking both glutamate synthase and glutamate dehydrogenase have an absolute requirement for glutamate (8, 13, 14, 73, 123). The presence of either enzyme in the cell allows the synthesis of glutamate in ammonia-containing minimal medium. Strains deficient in glutamate dehydrogenase have no growth phenotype, whereas cells deficient in glutamate synthetase, an enzyme required for the assimilation of ammonia when it is present in the medium in low concentration (see Assimilation of Ammonia, above), fail to grow with a variety of nitrogen sources. Strains of E. coli and K. aerogenes lacking glutamate synthase can grow with glutamate, asparagine, aspartate, and D-serine and very slowly with glutamine as sole source of nitrogen (14, 92, 98; for a review, see reference 123). Mutations in strains of S. typhimurium causing the complete loss of glutamate synthase activity result in the inability to utilize arginine as sole source of nitrogen, but these strains were not further characterized with regard to their growth with other nitrogen sources (31). As discussed in the previous sections, the inability of glutamate synthase-deficient strains to grow on some amino acids that give rise to glutamate results from their failure to produce the Ntr systems that degrade these compounds. This failure is due to the inability of these mutants to convert the glutamine, formed by ammonia assimilation, to glutamate. Therefore, glutamate synthase has two functions, the synthesis of glutamate and the removal of glutamine, the primary product of ammonia assimilation during nitrogen-limited growth.

Glutamate Synthase

Properties of the purified enzyme. In 1970, Tempest, Meers, and Brown (84, 121) discovered glutamate synthase, which catalyzes the reductive amination (NADPH dependent) of 2-ketoglutarate with glutamine as the nitrogen donor. They suspected the existence of a route of glutamate synthesis independent of glutamate dehydrogenase, because during the growth of K. aerogenes in a nitrogen-limited medium, glutamate dehydrogenase was repressed, but the intracellular pool of glutamate was normal. The discovery of glutamate synthase established a previously unknown pathway of glutamate formation from ammonia; first, ammonia is incorporated by glutamine synthetase into glutamine, and then the amide of glutamine is transferred to 2-ketoglutarate to form two molecules of glutamate.

The enzyme was subsequently purified from E. coli W and K. aerogenes. The enzyme has two nonidentical subunits in equimolar amounts. In E. coli, the subunits have molecular weights of 53,000 and 135,000; in K. aerogenes, the subunits have molecular weights of 51,500 and 175,000 (88, 122). The native molecular weight of the purified enzyme of E. coli was estimated to be 800,000 by sedimentation equilibrium and gel filtration; the enzyme in K. aerogenes had an $s_{20,w}$ consistent with this molecular weight. However, Miller and Stadtman (88) noticed that the molecular weight (the $s_{20,w}$) was 13S instead of 20S after an early stage of the purification. Furthermore, Sakamoto et al. (112) showed that active glutamate synthase had two molecular weights, 200,000 and 800,000, and two $s_{20,w}$s, 13S and 20S. Therefore, active glutamate synthase is probably a dimer that aggregates during the course of purification. Glutamate synthase has not been purified from S. typhimurium, but antiserum raised against E. coli enzyme precipitates two polypeptides from S. typhimurium with the same molecular weights as those from E. coli (73).

The purified enzyme contains flavin (both flavin adenine dinucleotide and flavin mononucleotide), iron (mostly ferrous), and labile sulfide. A stoichiometry of 1:4:4 was suggested from the data of Miller and Stadtman (88) and from observations in other iron-sulfide–containing flavoproteins. This suggested stoichiometry is at some variance with the data of Trotta et al. (122). The large subunit of the K. aerogenes enzyme was associated with the flavin and the iron-sulfide (122). Flavin adenine dinucleotide, not flavin mononucleotide, has been implicated as the active species of flavin, but the data presented were considered inconclusive (88).

Many glutamine-dependent amidotransferases, including glutamate synthase, hydrolyze glutamine to glutamate and ammonia without the transfer of the amide to the appropriate substrate. For a thorough discussion of amidotransferases, the reader should consult the excellent review of Buchanan (16). The glutaminase activity of purified glutamate synthase can be 10% of the glutamate synthetic activity in *E. coli* and *K. aerogenes* (50, 75, 122). Two lines of evidence indicate that the glutaminase activity is a property of the larger subunit. First, affinity labeling (alkylation) with L-2-amino-4-oxo-5-chloropentanoic acid (chloroketone) binds the large subunit and inactivates both the glutaminase and the glutamate synthase activity (76, 122). Second, the large subunit, separated from the small subunit, has the glutaminase activity (75). The glutaminase activity is not affected by the removal of the flavin, a finding that implies that the flavin is not required for the binding of glutamine (75). Storage at high pH increases the glutaminase activity, a result that has been seen with another amidotransferase, carbamyl phosphate synthetase, and has been taken to imply that the glutaminase activity is evident only after the enzyme has sustained some form of damage (50, 135). Several researchers have commented on the uniqueness of glutamate synthase because the glutamine-binding subunit is the larger of the two subunits. It should be noted that the size of the small subunit of glutamate synthase is not unusual for the large subunit of many amidotransferases (16). Therefore, the large subunit of glutamate synthase is atypically large for the glutamine-binding subunit. This probably results from the fact (described below) that this subunit also participates in electron transfer, which is unusual for an amidotransferase.

Glutamate synthase and other amidotransferases can utilize a high level of ammonia in place of glutamine as nitrogen donor. The rate of the ammonia-dependent reaction depends on the organism that serves as the source of glutamate synthase; the ratio of the ammonia-dependent activity to the glutamine-dependent activity is 10 and 1.4% when the enzyme is from *K. aerogenes* and *E. coli* B, respectively (50, 122). The ammonia-dependent activity is a property of the small subunit by a number of criteria. First, a variety of treatments that inactivate or damage the large subunit have no effect on the ammonia-dependent activity. These treatments include alkylation by the glutamine analog, chloroketone, and removal of the flavin, the iron-sulfide, or both from the large subunit (50, 75–77, 122). Second, the small subunit alone can catalyze the ammonia-dependent reaction (75). It should be noted that the ammonia-dependent reaction is the same reaction as that catalyzed by glutamate dehydrogenase. The available evidence suggests that this activity is not the result of contamination by glutamate dehydrogenase. First, Mantsala and Zalkin (77) were unable to detect glutamate dehydrogenase in their glutamate synthase preparation by a variety of immunological methods. Second, preincubation of purified glutamate synthase at low pH stimulates the ammonia-dependent activity of glutamate synthase but destroys the activity of purified glutamate dehydrogenase (77).

Miller and Stadtman (88) first proposed that the reaction catalyzed by glutamate synthase occurs in two steps: the reduction of the enzyme-bound flavin by NADPH, followed by the reaction of the reduced flavin with 2-ketoglutarate and glutamine to generate oxidized flavin and two glutamates. This view has been supported by all subsequent data. After the chemical (nonenzymatic) reduction of the flavin by dithionate, the flavin is reoxidized by the substrate glutamine and 2-ketoglutarate with the formation of glutamate (50, 77, 88). The iron-sulfur cluster also appears to be involved in the electron transfer (unpublished observation cited in reference 107). As noted before, removal of the flavin and iron does not inhibit the ammonia-dependent reaction but does abolish the glutamine-dependent reaction (50, 76, 77). Furthermore, in the ammonia-dependent reaction, the electrons are transferred directly to 2-ketoglutarate, whereas in the glutamine-dependent reaction they are transferred to water (50). From these results, it has been concluded that the electron transport for the ammonia-dependent reaction occurs by a nonphysiological route.

Kinetic data have been obtained from the enzymes purified from *E. coli* and *K. aerogenes*. The K_ms for glutamine and NADPH are of the order of 230 to 300 μM and 2.2 to 12 μM, respectively. There is a large discrepancy in the K_ms for 2-ketoglutarate; the *E. coli* enzyme has a low K_m, about 5 μM, whereas the *K. aerogenes* enzyme has a K_m of 300 μM. This difference may be the only manifestation of the different-sized large subunits formed by these bacteria. The reader is referred to the references for more kinetic, physiochemical, and enzymological information (11, 76, 88, 107, 122).

From these data, the following scheme for the overall glutamate synthase reaction can be proposed. NADPH binds the small subunit and transfers electrons to the large subunit, which reduces the flavin. 2-Ketoglutarate binds the small subunit, and glutamine binds the large subunit. The glutamine amide is transferred to 2-ketoglutarate, and the reduced flavin reduces a proposed iminoglutarate intermediate to glutamate. Clearly, the loss of either subunit by mutation should result in the inability to catalyze the glutamate synthase reaction.

Characterization of the genes coding for the subunits of glutamate synthase. The mutations that result in the loss of glutamate synthase have been located in the three enteric species and are located at a position corresponding to 69 min on the *E. coli* chromosome (43, 46, 98). Only one locus has been found for mutations that result in the loss of glutamate synthase. This observation suggests that the genes for the two subunits are linked. This locus has been identified as specifying the structural genes in *S. typhimurium* (43). The locus for glutamate synthase has been designated *gltB* in earlier publications; however, recently the designations *gltB* and *gltD* have been proposed for the large and small subunits, respectively (73).

The two genes are part of one transcriptional unit and constitute the *gltBD* operon (49, 72, 73). The glutamate synthase genes were cloned from *E. coli*, and the hybrid ColE1 plasmid overproduced glutamate synthase threefold (29). Using antiserum raised against purified glutamate synthase from *E. coli*, Lozoya et al. (72) showed that this plasmid directed the synthesis of both subunits of glutamate synthase. A deletion analysis of plasmid-borne glutamate synthase and the analysis of polar insertion mutations indicate that both subunits are translated from one transcript and that the large subunit is transcribed first (49, 73). The original mutant strain of *E. coli* analyzed by Berberich (8) failed to synthesize either subunit and probably carries a polar mutation in *gltB* (72). Madonna et al. (73) showed in *S. typhimurium* that loss of only the large subunit is sufficient to cause the characteristic phenotype of *gltB* mutants. Furthermore, a *gltB gdh* strain of *S. typhimurium*, which has the small subunit of glutamate synthase but not the large subunit or glutamate dehydrogenase, requires glutamate for growth (73); therefore, despite the observation that the purified small subunit has glutamate dehydrogenase-like activity, this activity does not provide the intact cell with glutamate.

Regulation of glutamate synthase. The primary form of regulation appears to be repression by glutamate or activation by glutamate deprivation, although very little is known about the mechanism of regulation. Despite the important role of glutamate synthase in the assimilation of ammonia during growth in an ammonia-restricted medium, the level of glutamate synthase does not correlate with the level of glutamine synthetase, the other enzyme required for ammonia assimilation. The level of glutamate synthase is generally highest in ammonia-containing minimal medium. Restriction of ammonia availability results in no change in activity in *E. coli*, slightly higher activity in *S. typhimurium*, and slightly lower activity in *K. aerogenes* (12, 14, 88). In all three bacterial species, glutamate synthase is repressed when the sole source of nitrogen is degraded to or provides glutamate; these nitrogen sources are glutamate, arginine, glutamine, histidine, and proline (6, 12, 14, 88, 109). When ammonia is added to these media, the glutamate synthase is not repressed (6, 12). The presence of ammonia causes repression of the Ntr systems that degrade these nitrogen sources, and the major source of nitrogen for glutamate is then ammonia. Aspartate, which is readily transaminated to glutamate, apparently can rapidly enter *S. typhimurium* in the presence of ammonia because aspartate in the medium can repress glutamate synthase even in the presence of ammonia (12). Glutamate synthase in *K. aerogenes* is also repressed when histidine is the sole source of carbon and nitrogen; histidine is degraded to glutamate (6). It should be noted that this medium is considered nitrogen rich and carbon poor. To summarize these observations, there is no observation inconsistent with the hypothesis that glutamate represses glutamate synthase. The lack of repression by some glutamate-producing nitrogen sources in ammonia-containing medium results from the failure to form the enzymes responsible for their degradation. Glutamate synthase can be repressed in either nitrogen-limited or nitrogen-rich medium, an observation that implies that there is no relation of necessity between the control of synthesis of glutamate synthase and that of glutamine synthetase.

There have been a few observations that suggest that tRNAs may be involved in glutamate synthase regulation. Lapointe et al. (69) showed that, in an *E. coli* strain with a temperature-sensitive glutamyl-tRNA synthetase, the levels of glutamate synthase and glutamine synthetase were 10 times higher than those in the wild-type strain. The strains had been grown in broth medium, which is highly repressive for both enzymes (12). The elevated level of glutamine synthase could be the result of the elevated glutamate synthase level, which could deplete the intracellular glutamine. The elevated level of glutamate synthase may result from an increased level of uncharged glutamyl-tRNA, although no mechanism linking tRNA charging to glutamate synthase regulation has been proposed. Rosenfeld and Brenchley (109) have shown that, in *hisT* strains of *S. typhimurium*, which have an altered pseudouridine synthetase I, resulting in undermodification of two uridines in many anticodon loops of tRNA, the activity of glutamate synthase is decreased twofold in glucose-ammonia (nitrogen-excess) and glucose-arginine (nitrogen-limiting) medium; however, the glutamate synthase activity was not altered in a different nitrogen-limiting medium. The *hisT* strains do not have an observable defect in ammonia assimilation and, in fact, grow faster on some nitrogen sources. Considering the potential pleiotropic nature of such mutations which affect macromolecular synthesis and which could result in the alterations of metabolic pools, it is difficult to make a strong case for the direct involvement of a glutamyl-tRNA in the regulation of glutamate synthase (24).

Glutamate Dehydrogenase

Glutamate dehydrogenase is a completely dispensable enzyme; a strain deficient in glutamate dehydrogenase has no observable growth phenotype (13, 123, 129). Mutants with lesions in *gltB* do not require glutamate if the ammonium ion concentration in the medium is greater than about 1 mM, indicating that glutamate dehydrogenase can synthesize glutamate when provided with sufficient ammonia (123).

Glutamate dehydrogenase has been purified from *E. coli* and *S. typhimurium*. The purified enzyme can use NADPH, but not NADH, for the reduction of 2-ketoglutarate, an observation that suggests a biosynthetic function. The enzyme is a hexamer of identical subunits with a molecular weight of 300,000. The K_ms for ammonia and 2-ketoglutarate are about 1 mM (27, 28, 112, 130).

Mutations resulting in the loss of glutamate dehydrogenase have been mapped to a site located at 27 min on the *E. coli* and *S. typhimurium* chromosomes (6, 98, 110); in *S. typhimurium*, this was shown to be the locus for the structural gene (110). The gene for glutamate dehydrogenase, which has been designated *gdh* (or *gdhA*), *gdhD*, and *gdhA* in *E. coli*, *K. aerogenes*, and *S. typhimurium*, respectively, has been cloned from all three organisms (29, 79, 143). The sequence of the amino terminus of glutamate dehydrogenase from *K. aerogenes* and the complete sequence from the *E. coli* enzyme have been determined (83, 89, 126, 127). There is a high degree of homology between the glutamate dehydrogenases from the enteric bacteria and, surprisingly, enzymes in *Neurospora crassa* and *Saccharomyces cerevisiae* (79, 87, 89, 127).

Knowledge of the start sites for transcription and the sequences for the promoters would be of interest because the regulation of glutamate dehydrogenase differs in enteric bacteria. However, promoters have not been identified beyond sequence homology to other promoters. The glutamate dehydrogenase activity is repressed in *E. coli* by the presence of glutamate in the medium; there is no control by the availability of ammonia in the medium (128). Restricting the extent of the charging of glutamyl-tRNAs does not affect the level of glutamate dehydrogenase, an observation that suggests that glutamyl-tRNAs do not play a role in repression (69). In *K. aerogenes*, growth in a nitrogen-limited medium represses glutamate dehydrogenase, even when exogenous glutamate is not provided (12, 14). The intracellular pool of glutamate is relatively invariant when cells of *K. aerogenes* are grown with an excess or limiting amount of ammonium sulfate (84); therefore, the glutamate pool does not regulate glutamate dehydrogenase in *K. aerogenes*. In *S. typhimurium*, glutamate dehydrogenase is not regulated by the quality of the nitrogen source or the presence of exogenous glutamate. The only regulation seen is when cells are grown in rich broth medium or with a mixture of amino acids (12).

The inverse correlation between the degree of nitrogen starvation and the level of glutamate dehydrogenase activity in *K. aerogenes* has been studied by Bender et al. (6, 7). Certain mutations in *glnL* (*glnL45*) cause constitutively high expression of *glnA* specifying glutamine synthetase and *glnG* specifying NR$_I$ and result in repression of glutamate dehydrogenase. A strain with *glnL45* and *gltB* mutations requires glutamate for growth because glutamate synthase activity is deficient and glutamate dehydrogenase is repressed. One class of revertant strains, selected for the ability to grow on glucose-ammonia, had nonrepressible glutamate dehydrogenase activity but also failed to produce histidase (an Ntr system) during nitrogen limitation (6, 7). A mutation in a previously unidentified gene, *nac* (nitrogen assimilatory control), was identified as being responsible for this phenotype. It is possible that the *nac* gene product plays a role in the response of certain genes to nitrogen limitation and that it specifically acts as a repressor *gdhD*, the structural gene for glutamate dehydrogenase.

ASPARTATE

Aspartate is synthesized as follows from oxaloacetate by transamination with glutamate as the amino donor: oxaloacetate + L-glutamate → L-aspartate + 2-ketoglutarate. This is probably the universal route of its synthesis. Virtually all of the biochemical and genetic studies of aspartate biosynthesis have been made in *E. coli*; it would be surprising if the synthesis of aspartate differed substantially in *S. typhimurium*. Aspartate synthesis has been reviewed most recently by Umbarger (125) and Reitzer (102).

Bacterial transaminases were first studied in *E. coli* by Rudman and Meister (111). They found three general transaminases which could donate the amino group of a variety of amino acids to 2-ketoglutarate to form glutamate. This is the reverse reaction of the transaminase in the growing cell. Some researchers still assay transaminase in this reverse direction, which has caused confusion in the literature. Rudman and Meister (111) designated these activities transaminase A for aspartate, tryptophan, tyrosine, and phenylalanine; transaminase B for leucine, isoleucine, and valine; and transaminase C for the following reaction: L-valine + pyruvate → 2-ketoisovalerate + L-alanine.

Subsequent work has shown that transaminase A is really two enzymes and that they are the only two enzymes that synthesize aspartate. One enzyme is the major aspartate transaminase. It has a low K_m for oxaloacetate, 0.4 mM, and a high K_m for the 2-keto analogs of phenylalanine and tyrosine. The activity of this component of transaminase A is constitutive (26, 52, 80). The second component of transaminase A is tyrosine repressible and has a low K_m for the 2-keto acid analogs of phenylalanine and tyrosine. The K_m of the tyrosine-repressible component for oxaloacetate is high, 3 mM (22, 23, 26, 80, 81, 116, 131). The predominant activity can be considered the high-affinity aspartate transaminase, and the tyrosine-repressible enzyme can be considered the low-affinity aspartate transaminase. There is confusion in the literature as to the nomenclature of these enzymes. We suggest that the high-affinity enzyme be called transaminase A1 and that the low-affinity enzyme be called transaminase A2.

Both aspartate transaminases have been purified to homogeneity. Both are dimers with identical subunits; the molecular weights are 82,000 and 88,000 for the high- and low-affinity aspartate transaminases, respectively (80, 81). The amino acid sequence of the aspartate transaminase A1 from *E. coli* B has been completely determined by Kondo et al. (61). The subunit is composed of 396 amino acids and has a molecular weight of 43,573. The *E. coli* enzyme is 40% homologous to the pig heart isozymes.

There has been only one genetic study of aspartate auxotrophy. A specific requirement for aspartate in ammonia-containing minimal medium results from mutations in both *aspC* and *tyrB*, which cause the loss of transaminases A1 and A2. The double mutant requires aspartate and tyrosine for growth. A look at the transaminases characterized by Rudman and Meister (111) would suggest that a strain deficient in transaminase A would also require phenylalanine for growth. However, transaminase B, the product of the *ilvE* gene, also participates in the synthesis of phenylalanine. Neither an *aspC*$^+$*tyrB* strain nor an *aspC* *tyrB*$^+$ strain requires any amino acid for growth, but the latter strain produces small colonies on minimal medium agar plates. Thus, it may be that the *aspC* gene product, the high-affinity aspartate transaminase, is the predominant aspartate transaminase (52). Two groups have isolated mutants that have lost transaminase A1. The mutations result in the loss of a 44,000-dalton polypeptide, but the mutations were located at different positions on the *E. coli* chromosome (51, 78). Therefore, the structural gene for the high-affinity aspartate transaminase has not yet been identified. It is curious that aspartate cannot be synthesized by a reversal of the following reaction catalyzed by aspartase: aspartate → fumarate + ammonia. The reaction is readily reversible, as is apparent from the fact that the commercial production of aspartate involves the addition of fumarate and ammonia to immobilized cells of *E. coli* (145). However, an aspartate auxotroph is deficient in the two components of transaminase A, not in aspartase. Therefore, aspartate cannot be synthesized by a reversal of aspartase. It is possible that the intracellular concentration of fumarate is too low for aspartate synthesis during growth in ammonia-containing minimal medium, unless fumarate is provided exogenously.

ASPARAGINE

Asparagine auxotrophy in *E. coli* or *K. aerogenes* results from mutations in two unlinked genes, designated *asnA* and *asnB*. An *asnA*[+] *asnB* or an *asnA asnB*[+] mutant does not require asparagine. Cedar and Schwartz (19, 20) partially purified and characterized an ammonia-dependent asparagine synthetase from *E. coli*; in *E. coli* and *K. aerogenes*, *asnA* codes for this enzyme (56, 103). The *asnB* gene of *E. coli* was initially shown to specify an enzyme which could use either glutamine or ammonia as a nitrogen donor (56), but subsequent work in *K. aerogenes* showed that this enzyme is strictly glutamine dependent (103). These recent studies established that bacteria have two asparagine synthetases, a glutamine-dependent enzyme, which is probably universally distributed, and an ammonia-dependent enzyme, which is confined to procaryotes (see references cited in reference 103; for a review, see reference 102).

Reitzer and Magasanik (103) showed that an *asnA*[+] *asnB* mutant strain of *K. aerogenes* could not grow on a variety of slowly metabolized nitrogen sources, such as proline, aspartate, nitrate, or glutamate, but that it grew as well as did the wild-type strain with ammonia or asparagine as nitrogen source. This finding suggested that the *asnB* product, the glutamine-dependent asparagine synthetase, was required for growth in ammonia-free, nitrogen-limited medium and indicated that the *asnA* product could synthesize the asparagine required for growth with ammonia as donor of the amide group. An *asnA asnB*[+] strain has no observable growth phenotype (103). It is likely that the function of these enzymes in *E. coli* and *S. typhimurium* is the same as in *K. aerogenes*. The *asnA* and *asnB* genes are located on the *E. coli* chromosome at 84 and 16 min, respectively (37, 56). Both genes are located in analogous positions on the chromosome of *K. aerogenes* (103).

It should be noted that the synthesis of another nitrogen-containing compound, glutamate, is also catalyzed by separate ammonia- and glutamine-dependent enzymes. In this case, too, loss of the glutamine-dependent reaction results in the inability to grow with a variety of slowly metabolized nitrogen sources, whereas the loss of the ammonia-dependent activity results in no observable phenotype.

The ammonia-dependent asparagine synthetase has been partially purified from *E. coli* and *K. aerogenes* (19, 103). The enzyme has a molecular weight of about 80,000 (19, 103) and a subunit molecular weight of 36,688 (18, 93); therefore, it is a dimer of equal subunits. The nucleotide sequence of the *asnA* gene has been determined (18, 93). The ammonia-dependent asparagine synthetase catalyzes the following reaction: $NH_3 + ATP + aspartate \rightarrow$ asparagine + AMP + PP_i. The formation of AMP and PP_i as products is common to all asparagine synthetases (85, 102). The kinetic properties of this enzyme are described by Cedar and Schwartz (20) and Reitzer and Magasanik (103). The K_m for ammonia is about 0.3 mM at neutral pH, and asparagine is a potent inhibitor.

The glutamine-dependent asparagine synthetase was purified to homogeneity from *K. aerogenes* by Reitzer and Magasanik (103). The purified enzyme is a tetramer of identical subunits; each subunit has a molecular weight of 57,000. The enzyme catalyzed the following reaction: aspartate + ATP + glutamine $\rightarrow$ asparagine + glutamate + AMP + PP_i. The enzyme shared properties with other amidotransferases: it exhibited aspartate-independent glutaminase activity, and a high level of ammonia could replace glutamine as nitrogen donor. Like the ammonia-dependent enzyme, the glutamine-dependent enzyme is inhibited by asparagine. It is clear from these studies that the two asparagine synthetases do not share components.

Asparagine represses both asparagine synthetases in *E. coli* and *K. aerogenes* (19, 56, 103). Recently, the mechanism of asparagine repression of *asnA* of *E. coli* has been elucidated. A 17,000-dalton protein is made from a gene, designated *asnC*, which is adjacent to, but divergently transcribed from, *asnA* (18, 32, 60). The *asnC* gene product activates the transcription of *asnA*; asparagine inhibits the activation of *asnA* transcription by the *asnC* product. The *asnC* gene product autogenously regulates its own synthesis; it represses the transcription of *asnC* with or without asparagine (32, 60). de Wind et al. (32) also showed that an *asnA*[+] *asnB asnC* strain was not an asparagine auxotroph at 37°C but that it required asparagine at 42°C. They suggested that this resulted from the known thermosensitivity of the ammonia-dependent enzyme (56). The *asnC* gene product was also implicated in activation of a high-affinity uptake system for asparagine because an *asnA asnB asnC* strain required a higher concentration of asparagine for growth than did an *asnA asnB asnC*[+] strain. The data of de Wind et al. (32) also suggest that *asnA* is controlled by more than the *asnC* product. First, an *asnA*[+] *asnB asnC* strain does produce a low level of asparagine synthetase. Second, asparagine starvation increases the level of enzyme in this strain. de Wind et al. (32) suggested that the *asnA* product itself, the ammonia-dependent asparagine synthetase, represses *asnA* transcription. Nothing is known about the mechanism of the repression of *asnB* by asparagine.

In *K. aerogenes*, there must be still another way to regulate the transcription of *asnA*. The level of the ammonia-dependent asparagine synthetase is high when cells are grown in ammonia-containing, nitrogen-rich medium, whereas the level of the glutamine-dependent enzyme is low. However, in cells grown in ammonia-free, nitrogen-limited medium, the level of the ammonia-dependent enzyme is low, and that of the glutamine-dependent enzyme is high (103). Nitrogen limitation implies an elevated level of NR_I (nitrogen regulator I), the *glnG* gene product. In strains with mutations that result in a low level of NR_I, the ammonia-dependent enzyme was not repressed. This implies that NR_I represses *asnA* (103), but examination of the nucleotide sequence between the *asnA* and *asnC* genes of *E. coli* showed no likely high-affinity NR_I binding site (32). The effect of NR_I, therefore may be indirect. The level of the glutamine-dependent enzyme of *K. aerogenes* is normally regulated in an NR_I-deficient strain, indicating that NR_I is not involved in the regulation of *asnB* transcription. Therefore, in nitrogen-limited medium, the elevated level of NR_I causes the repression of the ammonia-dependent enzyme, which results in asparagine starvation and derepression of the glutamine-dependent enzyme (103).

L-ALANINE

Several enzymatic pathways lead to the synthesis of L-alanine, as indicated by the fact that so far no L-alanine auxotroph has been isolated. At least two enzymes can synthesize L-alanine, but there may be as many as four.

In many organisms, the formation of L-alanine is catalyzed by a glutamate-pyruvate transaminase. Transfer of the amino group of L-alanine to 2-ketoglutarate could not be detected in the first survey of transaminases in E. coli by Rudman and Meister (111). Subsequent workers have found such an activity in crude extracts, but there has been no attempt to purify the protein (22, 35, 101, 125). While studying the degradation of D-alanine, Raunio and Jenkins (101) observed that a glutamate-pyruvate transaminase was induced by growth with glycerol as a carbon source or by the presence of D- or L-alanine in glucose-containing medium. This suggests a degradative role for an alanine transaminase. It is not clear whether this inducible enzyme is the same one previously measured in crude extracts.

Transaminase C, discovered by Rudman and Meister (111), catalyzes the following reversible trans-amination between L-alanine and valine: valine + pyruvate $\rightleftarrows$ L-alanine + 2-ketoisovalerate. The net reaction catalyzed by transaminase C, together with a reaction catalyzed by transaminase B, glutamate + 2-ketoisovalerate $\rightleftarrows$ 2-ketoglutarate + valine, results in the conversion of pyruvate to L-alanine, with glutamate as the amino donor. A mutant deficient in transaminase C activity was sought by isolating a valine auxotroph, based on the assumption that valine can be synthesized only by transaminases B and C. Transaminase B, the ilvE gene product, catalyzes the synthesis of the three branched-chain amino acids, isoleucine, valine, and leucine, with glutamate as the amino donor. A strain with a polar mutation in ilvE, which results in the failure to produce transaminase B as well as the keto acid precursors for all three branched-chain amino acids, can grow if supplemented with isoleucine and 2-ketoisovalerate, the latter which supplies the keto acid precursors for both valine and leucine. Valine-requiring mutants of such E. coli and S. typhimurium ilvE mutants were isolated after Mu d1 mutagenesis and were found to be deficient in transaminase C activity; this fact demonstrates that transaminase B and transaminase C are the only enzymes that synthesize valine (9, 136). The gene for transaminase C was designated avtA for alanine-valine transaminase. A strain with only the avtA mutation is a prototroph; it does not require L-alanine.

The facts that, in an ilvE mutant, the avtA product alone is responsible for the conversion of 2-ketoisovalerate to valine and that an ilvE avtA strain does not require alanine for growth prove that E. coli can produce alanine by another route. It then is likely that, in an ilvE mutant, a deficiency in the ability to produce alanine by this second pathway should result in a requirement for valine or alanine. In fact, some of the valine requirers isolated were deficient in glutamate-pyruvate transaminase activity (9).

Does transaminase C, the avtA gene product, normally participate in valine or L-alanine biosynthesis?

Berg and co-workers (9) investigated the effect of all 20 amino acids on transaminase C activity in S. typhimurium and showed that L-alanine and leucine, but not valine or D-alanine, repressed transaminase C. A similar conclusion had been reached by Falkinham (36). The repression by leucine in E. coli and S. typhimurium was hypothesized to be caused by its structural similarity to L-alanine (137). It was concluded that L-alanine is the true corepressor of avtA and that transaminase C functions primarily for L-alanine biosynthesis.

In summary, there are at least two pathways of L-alanine biosynthesis. Transaminase C contributes to L-alanine biosynthesis, but an avtA strain has no phenotype. One or possibly two glutamate-pyruvate transaminases are involved in L-alanine synthesis. Finally, there is a hint that a different, pyruvate-independent, pathway for L-alanine biosynthesis does exist. This conclusion was drawn from the observation that the labeling pattern of alanine from double-labeled glucose is inconsistent with the synthesis of L-alanine exclusively from pyruvate. An alternative pathway was not suggested (30).

D-ALANINE

The only pathway for the synthesis of D-alanine, a constituent of the cell wall, is racemization of L-alanine. This was demonstrated by the fact that a D-alanine–requiring strain has mutations in two genes and that both encode alanine racemases. The recent work of Wasserman et al. (134) and Wild et al. (139) has shown that one of the two alanine racemases in S. typhimurium and E. coli is constitutive and serves a biosynthetic role and that the other is inducible by D- or L-alanine and serves a catabolic role. There is a great deal of confusion concerning the designations of the genes for the enzymes of D-alanine metabolism. The gene for the biosynthetic alanine racemase is now designated alr in E. coli and S. typhimurium; the gene for the catabolic racemase is designated dadB in S. typhimurium and dadX in E. coli.

The genes coding for the two alanine racemases are not linked (47, 134, 138). Both dadB and alr have been cloned from S. typhimurium, their nucleotide sequences have been determined, and the alanine racemases have been purified from overproducing strains (34, 47, 133, 134). The genes have 52% sequence homology, and their products have 43% amino acid homology. For both enzymes, the active species is a monomer of 39,000 daltons with 1 mol of pyridoxal 5'-phosphate per monomer. Both enzymes were purified from the soluble fraction of cell extracts, which implies a cytoplasmic localization. These results are consistent with the earlier fractionation studies of Wijsman (138) and Franklin and Venables (41). Previous studies had indicated that the partially purified catabolic alanine racemase from E. coli B or E. coli W was a dimer with a molecular weight of 95,000 to 100,000 and a membrane localization (68, 132).

The level of the biosynthetic alanine racemase, the product of the alr gene, in a dadX or dadB strain is too low to be detected (134, 139). Curiously, the purified biosynthetic enzyme is only about 1.5% as active as the purified catabolic enzyme (34, 133). The biosynthetic alanine racemase is constitutive, and, despite its low specific activity, is capable of supplying the cell with sufficient D-alanine for growth. A $dadB^+$ ($dadX^+$) alr strain has no observable growth phenotype (134, 139).

On the other hand, a $dadB$ alr^+ strain of *S. typhimurium* cannot grow with L-alanine but can grow with D-alanine as sole source of carbon and nitrogen (134). The pathway for the degradation of L-alanine is racemization to D-alanine and oxidation of D-alanine by D-amino acid dehydrogenase to pyruvate and ammonia (41, 100, 101). D-Amino acid dehydrogenase is membrane bound, and the oxidation of D-amino acids is linked to a cytochrome-containing respiratory chain (41, 96, 101, 141). The dehydrogenase has been purified and was shown to have a broad specificity for D-amino acids and to consist of two unequal subunits (96).

The catabolic alanine racemase of *E. coli* and *S. typhimurium* is induced by DL-alanine and repressed by glucose (41, 67, 134, 138, 139). Wild et al. (139) provided evidence that L-alanine is the actual inducer in *E. coli*.

The data of Wild et al. (139) strongly suggest that the genes for the catabolic alanine racemase, *dadX*, and D-amino acid dehydrogenase, *dadA*, form an operon in *E. coli*. Strains with insertions in *dadA* also lack the inducible alanine racemase. Curing of the insertion restored alanine racemase activity, which suggests, but does not prove, that *dadA* and *dadX* are in the same operon. An operon structure for the genes coding for D-amino acid dehydrogenase and the catabolic alanine racemase in *S. typhimurium* is also possible; the genes are linked (134). Further evidence that these genes are part of an operon is the fact that both enzymes are induced by DL-alanine and repressed by glucose in *E. coli* and *S. typhimurium* (140, 141).

There are some studies that show that the genes coding for the catabolic alanine racemase and D-amino acid dehydrogenase are not linked. The *dadA* and *dadB* (or *dadX*) genes discussed so far are all located near *hemA* on their respective chromosome, 25 min on the *E. coli* map (134, 139). However, Beelen et al. (4), Franklin and Venables (41), and Franklin et al. (42) have determined that the gene for D-amino acid dehydrogenase of *E. coli* is located between the *ara* and *leu* operons at 1.5 min. They call the gene for the dehydrogenase *dadB*, which is why Wild et al. (139) designated the *E. coli* gene coding for the inducible alanine racemase *dadX*. These are not genes coding for the separate subunits of the D-amino acid dehydrogenase because Wild et al. (139) showed that a strain with a deletion of the *ara-leu* region of the chromosome still has D-amino acid dehydrogenase activity. The reason for the discrepancy in these data concerning the location of the D-amino acid dehydrogenase gene is not clear. All groups agree that the inducible alanine racemase gene is located at about 25 min on the *E. coli* chromosme (138, 139).

ACKNOWLEDGMENTS

Preparation of this review and studies carried out in this laboratory were supported by Public Health Service research grants GM07446 from the National Institute of General Medical Sciences and AM13894 from the National Institute of Arthritis, Diabetes, and Digestive and Kidney Diseases and by grant PCM84-00291 from the National Science Foundation.

We thank Hilda Harris-Ransom for the preparation of the manuscript.

LITERATURE CITED

1. **Adler, S. P., D. Purich, and E. R. Stadtman.** 1975. Cascade control of *Escherichia coli* glutamine synthetase. Properties of the P_{II} regulatory protein and the uridylyltransferase-uridylyl-removing enzyme. J. Biol. Chem. **250:**6264–6272.
2. **Anderson, W. B., and E. R. Stadtman.** 1971. Purification and functional roles of the P_I and P_{II} components of *Escherichia coli* glutamine synthetase deadenylylation system. Arch. Biochem. Biophys. **143:**428–443.
2a.**Backman, K., Y.-M. Chen, and B. Magasanik.** 1981. Physical and genetic characterization of the *glnA-glnG* region of the *Escherichia coli* chromosome. Proc. Natl. Acad. Sci. USA **78:**3743–3747.
2b.**Backman, K. C., Y.-M. Chen, S. Ueno-Nishio, and B. Magasanik.** 1983. The product of *glnL* is not essential for regulation of bacterial nitrogen assimilation. J. Bacteriol. **154:**516–519.
3. **Bancroft, S., S. G. Rhee, C. Neumann, and S. Kustu.** 1978. Mutations that alter the covalent modification of glutamine synthetase in *Salmonella typhimurium*. J. Bacteriol. **134:**1046–1055.
4. **Beelen, R. H. J., A. M. Feldmann, and H. J. W. Wijsman.** 1973. A regulatory gene and a structural gene for alaninase in *Escherichia coli*. Mol. Gen. Genet. **121:**369–374.
5. **Bender, R. A., K. A. Janssen, A. D. Resnick, M. Blumenberg, F. Foor, and B. Magasanik.** 1977. Biochemical parameters of glutamine synthetase from *Klebsiella aerogenes*. J. Bacteriol. **129:**1001–1009.
6. **Bender, R. A., A. Macaluso, and B. Magasanik.** 1976. Glutamate dehydrogenase: genetic mapping and isolation of regulatory mutants of *Klebsiella aerogenes*. J. Bacteriol. **127:**141–148.
7. **Bender, R. A., P. M. Snyder, R. Bueno, M. Quinto, and B. Magasanik.** 1983. Nitrogen regulation system of *Klebsiella aerogenes*: the *nac* gene. J. Bacteriol. **156:**444–446.
8. **Berberich, M. A.** 1972. A glutamate-dependent phenotype in *E. coli* K12: the result of two mutations. Biochem. Biophys. Res. Commun. **47:**1498–1503.
9. **Berg, C. M., W. A. Whalen, and L. B. Archambault.** 1983. Role of alanine-valine transaminase in *Salmonella typhimurium* and analysis of an *avtA*::Tn5 mutant. J. Bacteriol. **155:**1009–1014.
10. **Bloom, F. R., M. S. Levin, F. Foor, and B. Tyler.** 1978. Regulation of glutamine synthetase formation in *Escherichia coli*: characterization of mutants lacking uridylyltransferase. J. Bacteriol. **134:**569–577.
11. **Bower, S., and H. Zalkin.** 1983. Chemical modification and ligand binding studies with *Escherichia coli* glutamate synthase. Biochemistry **22:**1613–1620.
12. **Brenchley, J. E., C. A. Baker, and L. G. Patil.** 1975. Regulation of the ammonia assimilatory enzymes in *Salmonella typhimurium*. J. Bacteriol. **124:**182–189.
13. **Brenchley, J. E., and B. Magasanik.** 1974. Mutants of *Klebsiella aerogenes* lacking glutamate dehydrogenase. J. Bacteriol. **117:**544–550.
14. **Brenchley, J. E., M. J. Prival, and B. Magasanik.** 1973. Regulation of the synthesis of enzymes responsible for glutamate formation in *Klebsiella aerogenes*. J. Biol. Chem. **248:**6122–6128.
15. **Broach, J., C. Neumann, and S. Kustu.** 1976. Mutant strains (*nit*) of *Salmonella typhimurium* with a pleiotropic defect in nitrogen metabolism. J. Bacteriol. **128:**86–98.
16. **Buchanan, J. M.** 1973. The amidotransferases. Adv. Enzymol. Relat. Areas Mol. Biol. **39:**91–183.
17. **Bueno, R., G. Pahel, and B. Magasanik.** 1985. Role of *glnB* and *glnD* gene products in regulation of the *glnALG* operon of *Escherichia coli*. J. Bacteriol. **164:**816–822.
18. **Buhk, H.-J., and W. Messer.** 1983. The replication origin region of *Escherichia coli*: nucleotide sequence and functional units. Gene **24:**265–279.
19. **Cedar, H., and J. H. Schwartz.** 1969. The asparagine synthetase of *Escherichia coli*. I. Biosynthetic role of the enzyme, purification, and characterization of the reactions products. J. Biol. Chem. **244:**4112–4121.
20. **Cedar, H., and J. H. Schwartz.** 1969. The asparagine synthetase of *Escherichia coli*. II. Studies on mechanism. J. Biol. Chem. **244:**4122–4127.
21. **Chen, Y.-M., K. Backman, and B. Magasanik.** 1982. Characterization of a gene, *glnL*, the product of which is involved in the regulation of nitrogen utilization in *Escherichia coli*. J. Bacteriol. **150:**214–220.
22. **Chesne, S., and J. Pelmont.** 1973. Glutamate-oxaloacétate transaminase d'*Escherichia coli*. I. Purification et spécificité. Biochimie **55:**237–244.
23. **Chesne, S., and J. Pelmont.** 1974. Glutamate-oxaloacétate transaminase d'*Escherichia coli*. II. Propriétés. Biochimie **56:**631–639.

24. **Cheung, A., S. Morgan, K. B. Low, and D. Soll.** 1979. Regulation of the biosynthesis of aminoacyl-transfer ribonucleic acid synthetases and of transfer ribonucleic acid in *Escherichia coli*. VI. Mutants with increased levels of glutaminyl-transfer ribonucleic acid synthetase and of glutamine transfer ribonucleic acid. J. Bacteriol. **139:**176–184.

25. **Chock, P. B., E. Schacter, S. R. Jurgensen, and S. G. Rhee.** 1985. Cyclic cascade systems in metabolic regulation. Curr. Top. Cell. Regul. **27:**3–11.

26. **Collier, R. H., and G. Kohlhaw.** 1972. Nonidentity of the aspartate and the aromatic aminotransferase components of transaminase A in *Escherichia coli*. J. Bacteriol. **112:**365–371.

27. **Coulton, J. W., and M. Kapoor.** 1973. Purification and some properties of the glutamate dehydrogenase of *Salmonella typhimurium*. Can. J. Microbiol. **19:**427–438.

28. **Coulton, J. W., and M. Kapoor.** 1973. Studies on the kinetics and regulation of glutamate dehydrogenase of *Salmonella typhimurium*. Can. J. Microbiol. **19:**439–450.

29. **Covarrubias, A. A., R. Sanchez-Pescador, A. Osaria, F. Bolivar, and F. Bastarrachea.** 1980. ColE1 hybrid plasmids containing *Escherichia coli* genes involved in the biosynthesis of glutamate and glutamine. Plasmid **3:**150–164.

30. **Csonka, L. N.** 1977. Use of ^{3}H and ^{14}C double-labeled glucose to assess *in vivo* pathways of amino acid biosynthesis in *Escherichia coli*. J. Biol. Chem. **252:**3392–3398.

31. **Dendinger, S. M., L. G. Patil, and J. E. Brenchley.** 1980. *Salmonella typhimurium* mutants with altered glutamate dehydrogenase and glutamate synthase activities. J. Bacteriol. **141:**190–198.

32. **de Wind, N., M. de Jong, M. Meijer, and A. R. Stuitje.** 1985. Site-directed mutagenesis of the *Escherichia coli* chromosome near *oriC*: identification and characterization of *asnC*, a regulatory element in *E. coli* asparagine metabolism. Nucleic Acids Res. **13:**8797–8811.

33. **Ebner, E., D. Wolf, C. Gancedo, S. Elsasser, and H. Holzer.** 1970. Glutamine synthetase adenylyltransferase from *Escherichia coli* B. Purification and properties. Eur. J. Biochem. **14:**535–544.

34. **Esaki, N., and C. T. Walsh.** 1986. Biosynthetic alanine racemase of *Salmonella typhimurium*: purification and characterization of the enzyme encoded by the *alr* gene. Biochemistry **25:**3261–3267.

35. **Falkinham, J. O., III.** 1977. *Escherichia coli* K-12 mutant with alternate requirements for vitamin B$_6$ or branched-chain amino acids and lacking transaminase C activity. J. Bacteriol. **130:**566–568.

36. **Falkinham, J. O., III.** 1979. Identification of a mutation affecting alanine-α-ketoisovalerate transaminase activity in *Escherichia coli* K12. Mol. Gen. Genet. **176:**147–149.

37. **Felton, J., S. Michaelis, and A. Wright.** 1980. Mutations in two unlinked genes are required to produce asparagine auxotrophy in *Escherichia coli*. J. Bacteriol. **142:**221–238.

38. **Foor, F., R. J. Cedergren, S. L. Streicher, S. G. Rhee, and B. Magasanik.** 1978. Glutamine synthetase of *Klebsiella aerogenes*: properties of *glnD* mutants lacking uridylyltransferase. J. Bacteriol. **134:**562–568.

39. **Foor, F., K. A. Janssen, and B. Magasanik.** 1975. Regulation of synthesis of glutamine synthetase by adenylylated glutamine synthetase. Proc. Natl. Acad. Sci. USA **72:**4844–4848.

40. **Foor, F., Z. Reuveny, and B. Magasanik.** 1980. Regulation of the synthesis of glutamine synthetase by the P$_{II}$ protein in *Klebsiella aerogenes*. Proc. Natl. Acad. Sci. USA **77:**2636–2640.

41. **Franklin, F. C. H., and W. A. Venables.** 1976. Biochemical, genetic, and regulatory studies of alanine catabolism in *Escherichia coli* K12. Mol. Gen. Genet. **149:**229–237.

42. **Franklin, F. C. H., W. A. Venables, and H. J. W. Wijsman.** 1981. Genetic studies of D-alanine-dehydrogenase-less mutants of *Escherichia coli* K12. Genet. Res. **38:**197–208.

43. **Fuchs, R. L., M. J. Madonna, and J. E. Brenchley.** 1982. Identification of the structural genes for glutamine synthase and genetic characterization of this region of the *Salmonella typhimurium* chromosome. J. Bacteriol. **149:**906–915.

44. **Funanage, V. L., P. D. Ayling, S. M. Dendinger, and J. E. Brenchley.** 1978. *Salmonella typhimurium* LT-2 mutants with altered glutamine synthetase levels and amino acid uptake activities. J. Bacteriol. **136:**588–596.

45. **Funanage, V. L., and J. E. Brenchley.** 1977. Characterization of *Salmonella typhimurium* mutants with altered glutamine synthetase activity. Genetics **86:**513–526.

46. **Gaillardin, C. M., and B. Magasanik.** 1978. Involvement of the product of the *glnF* gene in the autogenous regulation of glutamine synthetase formation in *Klebsiella aerogenes*. J. Bacteriol. **133:**1329–1388.

47. **Galakatos, N. G., E. Daub, D. Botstein, and C. T. Walsh.** 1986. Biosynthetic *alr* alanine racemase from *Salmonella typhimurium*: DNA and protein sequence determination. Biochemistry **25:**3255–3260.

47a.**Garcia, E., S. Bancroft, S. G. Rhee, and S. Kustu.** 1977. The product of a newly identified gene, *glnF*, is required for synthesis of glutamine synthetase in *Salmonella*. Proc. Natl. Acad. Sci. USA **74:**1662–1666.

48. **Garcia, E., and S. G. Rhee.** 1983. Cascade control of *Escherichia coli* glutamine synthetase: purification and properties of P$_{II}$ uridylyltransferase and uridylyl-removing enzyme. J. Biol. Chem. **258:**2246–2253.

49. **Garciarrubio, A., E. Lozoya, A. Covarrubias, and F. Bolivar.** 1983. Structural organization of the genes that encode two glutamate synthase subunits of *Escherichia coli*. Gene **26:**165–170.

50. **Geary, L. E., and A. Meister.** 1977. On the mechanism of glutamine-dependent reductive amination of α-ketoglutarate catalyzed by glutamate synthase. J. Biol. Chem. **252:**3501–3508.

51. **Gelfand, D. H., and N. Rudo.** 1977. Mapping of the aspartate and aromatic amino acid aminotransferase genes *tyrB* and *aspC*. J. Bacteriol. **130:**441–444.

52. **Gelfand, D. H., and R. A. Steinberg.** 1977. *Escherichia coli* mutants deficient in the aspartate and aromatic amino acid aminotransferases. J. Bacteriol. **130:**429–440.

53. **Ginsburg, A., and E. R. Stadtman.** 1973. Regulation of glutamine synthetase in *Escherichia coli*, p. 9–44. *In* S. Prusiner and E. R. Stadtman (ed.), The enzymes of glutamine metabolism. Academic Press, Inc., New York.

54. **Gutnick, D., J. M. Calvo, T. Klopotowski, and B. N. Ames.** 1969. Compounds which serve as the sole source of carbon or nitrogen for *Salmonella typhimurium* LT-2. J. Bacteriol. **100:**215–219.

55. **Hirschman, J., P.-K. Wong, K. Sei, J. Keener, and S. Kustu.** 1985. Products of nitrogen regulatory genes *ntrA* and *ntrC* of enteric bacteria activate *glnA* transcription *in vitro*: evidence that the *ntrA* product is a σ factor. Proc. Natl. Acad. Sci. USA **82:**7525–7529.

56. **Humbert, R., and R. D. Simoni.** 1980. Genetic and biochemical studies demonstrating a second gene coding for asparagine synthetase in *Escherichia coli*. J. Bacteriol. **142:**212–220.

57. **Hunt, T. P., and B. Magasanik.** 1985. Transcription of *glnA* by purified *Escherichia coli* components: core RNA polymerase and the products of *glnF*, *glnG*, and *glnL*. Proc. Natl. Acad. Sci. USA **82:**8453–8457.

58. **Jayakumar, A., I. Schulman, D. MacNeil, and E. M. Barnes, Jr.** 1986. Role of *Escherichia coli* glnALG operon in regulation of ammonium transport. J. Bacteriol. **166:**281–284.

59. **Kahane, S., R. Levitz, and Y. S. Halpern.** 1978. Specificity and regulation of γ-aminobutyrate transport in *Escherichia coli*. J. Bacteriol. **135:**295–299.

60. **Kölling, R., and H. Lother.** 1985. AsnC: an autogenously regulated activator of asparagine synthetase A transcription in *Escherichia coli*. J. Bacteriol. **164:**310–315.

61. **Kondo, K., S. Wakabayashi, T. Yagi, and H. Kagamiyama.** 1984. The complete amino acid sequence of aspartate aminotransferase from *Escherichia coli*: sequence comparison with pig isoenzymes. Biochem. Biophys. Res. Commun. **122:**62–67.

62. **Krajewska-Grynkiewicz, K., and S. Kustu.** 1984. Evidence that nitrogen regulatory gene *ntrC* of *Salmonella typhimurium* is transcribed from the *glnA* promoter as well as from a separate *ntr* promoter. Mol. Gen. Genet. **193:**135–142.

63. **Kuhn, R., and R. L. Somerville.** 1971. Mutant strains of *Escherichia coli* K12 that use D-amino acids. Proc. Natl. Acad. Sci. USA **68:**2484–2487.

63a.**Kustu, S., D. Burton, E. Garcia, L. McCarter, and N. McFarland.** 1979. Nitrogen control in *Salmonella*: regulation by the *glnR* and *glnF* gene products. Proc. Natl. Acad. Sci. USA **76:**4576–4580.

64. **Kustu, S., J. Hirschman, D. Burton, J. Jelesko, and J. C. Meeks.** 1984. Covalent modification of bacterial glutamine synthetase: physiological significance. Mol. Gen. Genet. **197:**309–317.

65. **Kustu, S. G., N. C. McFarland, S. P. Hui, B. Esmond, and G. F.-L. Ames.** 1979. Nitrogen control in *Salmonella typhimurium*: co-regulation of synthesis of glutamine synthetase and amino acid transport systems. J. Bacteriol. **138:**218–234.

66. **Kustu, S. G., and K. McKereghan.** 1975. Mutations affecting glutamine synthetase activity in *Salmonella typhimurium*. J. Bacteriol. **122:**1006–1016.

67. **Lambert, M. P., and F. C. Neuhaus.** 1972. Factors affecting the level of alanine racemase in *Escherichia coli*. J. Bacteriol. **109:**1156–1161.

68. **Lambert, M. P., and F. C. Neuhaus.** 1972. Mechanism of D-cycloserine action: alanine racemase from *Escherichia coli* W. J. Bacteriol. **110:**978–987.

69. **Lapointe, J., G. Delcuve, and L. Duplain.** 1975. Derepressed

levels of glutamate synthase and glutamine synthetase in *Escherichia coli* mutants altered in glutamyl-transfer ribonucleic acid synthetase. J. Bacteriol. 123:843–850.

70. **Levine, R. L.** 1985. Covalent modification of proteins by mixed function oxidation. Curr. Top. Cell. Regul. 27:305–316.

71. **Levine, R. L., C. N. Oliver, R. M. Fulks, and E. R. Stadtman.** 1981. Turnover of bacterial glutamine synthetase: oxidative inactivation precedes proteolysis. Proc. Natl. Acad. Sci. USA 78:2120–2124.

72. **Lozoya, E., R. Sanchez-Pescador, A. Covarrubias, I. Vichido, and F. Bolivar.** 1980. Tight linkage of genes that encode the two glutamate synthase subunits of *Escherichia coli* K-12. J. Bacteriol. 144:616–621.

73. **Madonna, M. J., R. L. Fuchs, and J. E. Brenchley.** 1985. Fine structure analysis of *Salmonella typhimurium* glutamate synthase genes. J. Bacteriol. 161:353–360.

74. **Magasanik, B.** 1982. Genetic control of nitrogen assimilation in bacteria. Annu. Rev. Genet. 16:135–168.

75. **Mantsala, P., and H. Zalkin.** 1976. Active subunits of *Escherichia coli* glutamate synthase. J. Bacteriol. 126:539–541.

76. **Mantsala, P., and H. Zalkin.** 1976. Glutamate synthase. Properties of the glutamine-dependent activity. J. Biol. Chem. 251:3294–3299.

77. **Mantsala, P., and H. Zalkin.** 1976. Properties of apoglutamate synthase and comparison with glutamate dehydrogenase. J. Biol. Chem. 251:3300–3305.

78. **Marcus, M., and Y. S. Halpern.** 1969. The metabolic pathway of glutamate in *Escherichia coli* K-12. Biochim. Biophys. Acta 177:314–320.

79. **Mattaj, I. W., M. J. McPherson, and J. C. Wootton.** 1982. Localisation of a strongly conserved section of coding sequence in glutamate dehydrogenase genes. FEBS Lett. 147:21–25.

80. **Mavrides, C., and W. Orr.** 1974. Multiple forms of plurispecific aromatic:2-oxoglutarate (oxaloacetate) aminotransferase (transaminase A) in *Escherichia coli* and selective repression by L-tyrosine. Biochim. Biophys. Acta 336:70–78.

81. **Mavrides, C., and W. Orr.** 1975. Multispecific aspartate and aromatic amino acid aminotransferases in *Escherichia coli*. J. Biol. Chem. 250:4128–4133.

82. **Mayer, E. P., O. H. Smith, W. W. Fredricks, and M. A. McKinney.** 1975. The isolation and characterization of glutamine-requiring strains of *Escherichia coli* K12. Mol. Gen. Genet. 137:131–142.

82a. **McFarland, N., L. McCarter, S. Artz. amd S. Kustu.** 1981. Nitrogen regulatory locus "*glnR*" of enteric bacteria is composed of cistrons *ntrB* and *ntrC*: identification of their protein products. Proc. Natl. Acad. Sci. USA 78:2135–2139.

83. **McPherson, M. J., and J. C. Wootton.** 1983. Complete nucleotide sequence of the *Escherichia coli gdhA* gene. Nucleic Acids Res. 11:5257–5266.

84. **Meers, J., D. Tempest, and C. Brown.** 1970. Glutamine (amide):2-oxoglutarate aminotransferase (NADP); an enzyme involved in the synthesis of glutamate by some bacteria. J. Gen. Microbiol. 64:187–194.

85. **Meister, A.** 1974. Asparagine synthetase, p. 561–580. *In* P. D. Boyer (ed.), The enzymes, vol. X. Academic Press, Inc., New York.

86. **Mergeay, M., D. Gigot, J. Beckmann, N. Glansdorff, and A. Piérard.** 1974. Physiology and genetics of carbamoylphosphate synthesis in *Escherichia coli* K-12. Mol. Gen. Genet. 133:299–316.

87. **Miller, E. S., and J. E. Brenchley.** 1984. Cloning and characterization of *gdhA*, the structural gene for glutamate dehydrogenase of *Salmonella typhimurium*. J. Bacteriol. 157:171–178.

88. **Miller, R. E., and E. R. Stadtman.** 1972. Glutamate synthase from *Escherichia coli*. An iron-sulfide flavoprotein. J. Biol. Chem. 247:7407–7419.

89. **Mountain, A., M. J. McPherson, A. J. Baron, and J. C. Wootton.** 1985. The *Klebsiella aerogenes* glutamate dehydrogenase (*gdhA*) gene: cloning, high-level expression and hybrid enzyme formation in *Escherichia coli*. Mol. Gen. Genet. 199:141–145.

90. **Mura, U., P. B. Chock, and E. R. Stadtman.** 1981. Allosteric regulation of the state of adenylylation of glutamine synthetase in permeabilized cell preparations of *Escherichia coli*. Studies of monocyclic and bicyclic interconvertible enzyme cascades, *in situ*. J. Biol. Chem. 256:13022–13029.

91. **Mura, U., and E. R. Stadtman.** 1981. Glutamine synthetase adenylylation in permeabilized cells of *Escherichia coli*. J. Biol. Chem. 256:13014–13021.

92. **Nagatani, H., M. Shimizu, and R. C. Valentine.** 1971. The mechanism of ammonia assimilation in nitrogen-fixing bacteria. Arch. Mikrobiol. 79:164–175.

93. **Nakamura, M., M. Yamada, Y. Hirota, K. Sugimoto, A. Oka, and M. Takanami.** 1981. Nucleotide sequence of the *asnA* gene coding for asparagine synthetase of *E. coli* K-12. Nucleic Acids Res. 9:4669–4676.

94. **Ninfa, A. J., and B. Magasanik.** 1986. Covalent modification of the *glnG* product, NR$_I$, by the *glnL* product, NR$_{II}$, regulates the transcription of the *glnALG* operon in *Escherichia coli*. Proc. Natl. Acad. Sci. USA 83:5909–5913.

95. **Ninfa, A. J., S. Ueno-Nishio, T. P. Hunt, B. Robustell, and B. Magasanik.** 1986. Purification of nitrogen regulator II, the product of the *glnL* (*ntrB*) gene of *Escherichia coli*. J. Bacteriol. 168:1002–1004.

96. **Olsiewski, P. J., G. J. Kaczorowski, and C. Walsh.** 1980. Purification and properties of D-amino acid dehydrogenase, an inducible membrane-bound iron-sulfur flavoenzyme from *Escherichia coli* B. J. Biol. Chem. 255:4487–4494.

97. **Pahel, G., D. M. Rothstein, and B. Magasanik.** 1982. Complex *glnA-glnL-glnG* operon of *Escherichia coli*. J. Bacteriol. 150:202–213.

97a. **Pahel, G., and B. Tyler.** 1979. A new *glnA*-linked regulatory gene for glutamine synthetase in *Escherichia coli*. Proc. Natl. Acad. Sci. 76:4544–4548.

98. **Pahel, G., A. D. Zelenetz, and B. M. Tyler.** 1978. *gltB* gene and regulation of nitrogen metabolism by glutamine synthetase in *Escherichia coli*. J. Bacteriol. 133:139–148.

99. **Prival, M. J., J. E. Brenchley, and B. Magasanik.** 1973. Glutamine synthetase and the regulation of histidase formation in *Klebsiella aerogenes*. J. Biol. Chem. 248:4334–4344.

100. **Raunio, R. P., L. D'Ari Straus, and W. T. Jenkins.** 1973. D-Alanine oxidase from *Escherichia coli*: participation in the oxidation of L-alanine. J. Bacteriol. 115:567–573.

101. **Raunio, R. P., and W. T. Jenkins.** 1973. D-Alanine oxidase from *Escherichia coli*: localization and induction by L-alanine. J. Bacteriol. 115:560–566.

102. **Reitzer, L. J.** 1983. Aspartate and asparagine biosynthesis, p. 133–145. *In* K. M. Herrmann and R. L. Somerville (ed.), Amino acids: biosynthesis and genetic regulation. Addison-Wesley, Reading, Mass.

103. **Reitzer, L. J., and B. Magasanik.** 1982. Asparagine synthetases of *Klebsiella aerogenes*: properties and regulation of synthesis. J. Bacteriol. 151:1299–1313.

104. **Reitzer, L. J., and B. Magasanik.** 1983. Isolation of the nitrogen assimilation regulator NR$_I$, the product of the *glnG* gene of *Escherichia coli*. Proc. Natl. Acad. Sci. USA 80:5554–5558.

105. **Reitzer, L. J., and B. Magasanik.** 1985. Expression of *glnA* in *Escherichia coli* is regulated at tandem promoters. Proc. Natl. Acad. Sci. USA 82:1979–1983.

106. **Reitzer, L. J., and B. Magasanik.** 1986. Transcription of *glnA* in *E. coli* is stimulated by activator bound to sites far from the promoter. Cell 45:785–792.

107. **Rendina, A. R., and W. H. Orme-Johnson.** 1978. Glutamate synthase: on the kinetic mechanism of the enzyme from *Escherichia coli* W. Biochemistry 17:5388–5393.

108. **Reuveny, Z., F. Foor, and B. Magasanik.** 1981. Regulation of glutamine synthetase by regulatory protein P$_{II}$ in *Klebsiella aerogenes* mutants lacking adenylyltransferase. J. Bacteriol. 146:740–745.

109. **Rosenfeld, S. A., and J. E. Brenchley.** 1980. Regulation of nitrogen utilization in *hisT* mutants of *Salmonella typhimurium*. J. Bacteriol. 143:801–808.

110. **Rosenfeld, S. A., S. M. Dendinger, C. H. Murphy, and J. E. Brenchley.** 1982. Genetic characterization of the glutamate dehydrogenase gene (*gdhA*) of *Salmonella typhimurium*. J. Bacteriol. 150:795–803.

111. **Rudman, D., and A. Meister.** 1953. Transamination in *Escherichia coli*. J. Biol. Chem. 200:591–604.

112. **Sakamoto, N., A. M. Kotre, and M. A. Savageau.** 1975. Glutamate dehydrogenase from *Escherichia coli*: purification and properties. J. Bacteriol. 124:775–783.

113. **Senior, P. J.** 1975. Regulation of nitrogen metabolism in *Escherichia coli* and *Klebsiella aerogenes*: studies with the continuous-culture technique. J. Bacteriol. 123:407–418.

114. **Shaibe, E., E. Metzer, and Y. S. Halpern.** 1985. Metabolic pathway for the utilization of L-arginine, L-ornithine, agmatine, and putrescine as nitrogen sources in *Escherichia coli* K-12. J. Bacteriol. 163:933–937.

115. **Shaibe, E., E. Metzer, and Y. S. Halpern.** 1985. Control of utilization of L-arginine, L-ornithine, agmatine, and putrescine as nitrogen sources in *Escherichia coli* K-12. J. Bacteriol. 163:938–942.

116. **Silbert, D. F., S. E. Jorgensen, and E. C. C. Lin.** 1963. Repression of transaminase A by tyrosine in *Escherichia coli*. Biochim. Biophys. Acta 73:232–240.

117. **Stadtman, E. R., and P. B. Chock.** 1978. Interconvertible en-

zyme cascades in metabolic regulation. Curr. Top. Cell. Regul. **13**:53–95.

118. **Stadtman, E. R., and A. Ginsburg.** 1974. The glutamine synthetase of *Escherichia coli*: structure and control, p. 755–807. *In* P. D. Boyer (ed.), The enzymes, vol. 10. Academic Press, Inc., New York.

119. **Stadtman, E. R., E. Mura, P. B. Chock, and S. G. Rhee.** 1980. The interconvertible enzyme cascade that regulates glutamine synthetase activity, p. 41–59. *In* J. Mora and R. Palacios (ed.), Glutamine: metabolism, enzymology and regulation. Academic Press, Inc., New York.

120. **Stern, M. J., C. F. Higgins, and G. F.-L. Ames.** 1984. Isolation and characterization of *lac* fusions to two nitrogen-regulated promoters. Mol. Gen. Genet. **195**:219–227.

121. **Tempest, D. W., J. L. Meers, and C. M. Brown.** 1970. Synthesis of glutamate in *Aerobacter aerogenes* by a hitherto unknown route. Biochem. J. **117**:405–407.

122. **Trotta, P. P., K. E. B. Platzer, R. H. Haschemeyer, and A. Meister.** 1974. Glutamine-binding subunit of glutamate synthase and partial reactions catalyzed by this glutamine amidotransferase. Proc. Natl. Acad. Sci. USA **71**:4607–4611.

123. **Tyler, B.** 1978. Regulation of the assimilation of nitrogen compounds. Annu. Rev. Biochem. **47**:1127–1162.

124. **Ueno-Nishio, S., S. Mango, L. J. Reitzer, and B. Magasanik.** 1984. Identification and regulation of the *glnL* operator-promoter of the complex *glnALG* operon of *Escherichia coli*. J. Bacteriol. **160**:379–384.

125. **Umbarger, H. E.** 1978. Amino acid biosynthesis and its regulation. Annu. Rev. Biochem. **47**:533–606.

126. **Valle, F., B. Becerril, E. Chen, P. Seeberg, H. Heyneker, and F. Bolivar.** 1984. Complete nucleotide sequence of the glutamate dehydrogenase gene from *Escherichia coli* K-12. Gene **27**:193–199.

127. **Valle, F., E. Sanvicente, P. Seeberg, A. Covarrubias, R. L. Rodriguez, and F. Bolivar.** 1983. Nucleotide sequence of the promoter and amino-terminal coding region of the glutamate dehydrogenase structural gene of *Escherichia coli*. Gene **23**:199–209.

128. **Varricchio, F.** 1969. Control of glutamate dehydrogenase synthesis in *Escherichia coli*. Biochem. Biophys. Acta **177**:560–564.

129. **Vender, J., and H. V. Rickenberg.** 1964. Ammonia metabolism in a mutant of *Escherichia coli* lacking glutamate dehydrogenase. Biochim. Biophys. Acta **90**:218–220.

130. **Veronese, F. M., E. Boccu, and L. Conventi.** 1975. Glutamate dehydrogenase from *Escherichia coli*: induction, purification and properties of the enzyme. Biochim. Biophys. Acta **377**:217–228.

131. **Wallace, B. J., and J. Pittard.** 1969. Regulator gene controlling enzymes concerned in tyrosine biosynthesis in *Escherichia coli*. J. Bacteriol. **97**:1234–1241.

132. **Wang, E., and C. Walsh.** 1978. Suicide substrates for the alanine racemase of *Escherichia coli* B. Biochemistry **17**:1313–1321.

133. **Wasserman, S. A., E. Daub, P. Grisafi, D. Botstein, and C. T. Walsh.** 1984. Catabolic alanine racemase from *Salmonella typhimurium*: DNA sequence, enzyme purification, and characterization. Biochemistry **23**:5182–5187.

134. **Wasserman, S. A., C. T. Walsh, and D. Botstein.** 1983. Two alanine racemase genes in *Salmonella typhimurium* that differ in structure and function. J. Bacteriol. **153**:1439–1450.

135. **Wellner, V. P., and A. Meister.** 1975. Enhancement of the glutaminase activity of carbamyl phosphate synthetase by alterations in the interaction between the heavy and light subunits. J. Biol. Chem. **250**:3261–3266.

136. **Whalen, W. A., and C. M. Berg.** 1982. Analysis of an *avtA*::Mu *d*1(Ap *lac*) mutant: metabolic role of transaminase C. J. Bacteriol. **150**:739–746.

137. **Whalen, W. A., and C. M. Berg.** 1984. Gratuitous repression of *avtA* in *Escherichia coli* and *Salmonella typhimurium*. J. Bacteriol. **158**:571–574.

138. **Wijsman, H. J. W.** 1972. The characterization of an alanine racemase mutant of *Escherichia coli*. Genet. Res. **20**:269–277.

139. **Wild, J., J. Hennig, M. Lobocka, W. Walczak, and T. Klopotowski.** 1985. Identification of the *dadX* gene coding for the predominant isozyme of alanine racemase in *Escherichia coli* K12. Mol. Gen. Genet. **198**:315–322.

140. **Wild, J., and T. Klopotowski.** 1975. Insensitivity of D-amino acid dehydrogenase synthesis to catabolic repression in *dadR* mutants of *Salmonella typhimurium*. Mol. Gen. Genet. **136**:63–73.

141. **Wild, J., and T. Klopotowski.** 1981. D-Amino acid dehydrogenase of *Escherichia coli* K12: positive selection of mutants defective in enzyme activity and localization of the structural gene. Mol. Gen. Genet. **181**:373–378.

142. **Willis, R. C., K. K. Iwata, and C. E. Furlong.** 1975. Regulation of glutamine transport in *Escherichia coli*. J. Bacteriol. **122**:1032–1037.

143. **Windass, J. D., M. J. Worsey, E. M. Pioli, D. Pioli, P. T. Barth, K. T. Atherton, E. C. Dart, D. Byrom, K. Powell, and P. J. Senior.** 1980. Improved conversion of methanol to single-cell protein by *Methylophilus methylotrophus*. Nature (London) **287**:396–401.

144. **Wohlheuter, R. M., H. Schutt, and H. Holzer.** 1973. Regulation of glutamine synthesis *in vivo* in *Escherichia coli*, p. 45–64. *In* S. Prusiner and E. R. Stadtman (ed.), The enzymes of glutamine metabolism. Academic Press, Inc., New York.

145. **Yoshinaga, F., and S. Nakamori.** 1983. Production of amino acids, p. 405–429. *In* K. H. Herrmann and R. L. Somerville (ed.), Amino acids: biosynthesis and genetic regulation. Addison-Wesley, Reading, Mass.

146. **Zaboura, M., and Y. S. Halpern.** 1978. Regulation of γ-aminobutyric acid degradation in *Escherichia coli* by nitrogen metabolism enzymes. J. Bacteriol. **133**:447–451.

21. Biosynthesis of Arginine and Polyamines

NICOLAS GLANSDORFF

Research Institute of the Centre d'Enseignement et de Recherches des Industries Alimentaires et Chimiques, and Microbiology, Vrije Universiteit Brussel, B-1070 Brussels, Belgium

INTRODUCTION

The biosynthesis of arginine has been a focus of interest for several laboratories during the last 30 years. At the time the operon concept arose (107), it was already clear that the arginine pathway offered a paradigm to study the molecular basis for control of gene expression at a higher level of complexity. Indeed, the arginine genes are not clustered on the chromosome in a single functional unit. Furthermore, the occurrence of a precursor common to the biosynthesis of arginine and the pyrimidines—carbamoylphosphate—posed very early a question of basic physiological interest: what were the metabolic controls coordinating the production of this substance? Another important metabolic interconnection was disclosed with the discovery that arginine and ornithine are precursors in the biosynthesis of polyamines.

Investigations of arginine biosynthesis brought to light basic features of metabolic regulation. Interpreting a series of experiments suggesting that the synthesis of *N*-acetylornithinase was antagonized by arginine, Vogel (248) proposed to use the term "repression" for a "relative decrease resulting from the exposure of cells to a given substance, in the rate of synthesis of a particular apoenzyme," a phenomenon for which a few examples had already been reported at the time (248). Experiments on the kinetics of ornithine carbamoyltransferase formation in batch cultures and in chemostats (89), in bradytrophic mutants, and in cells grown in arginine-free rich medium (173), established that not only added arginine but also the endogenously produced amino acid sets the pace of enzyme synthesis in the pathway.

The first developments regarding the basic type of mechanism involved in this regulation required the isolation of derepressed (*argR*) regulatory mutants (87, 144), and the demonstration by enzyme assays in transient and stable *argR⁺*/*argR⁻* diploids (147, 149) favored the view that the *argR* product was an aporepressor, in keeping with the conclusion reached earlier in the case of *lacI* (179). Based on the results of these dominance tests, and considering the deviation from complete coordinate expression displayed by the scattered arginine genes, Maas (144) and Gorini et al. (87) considered that a unique aporepressor could interact with a family of slightly dissimilar operators. For physiological entities consisting of scattered genes controlled by the same regulatory molecule, Maas and Clark (147) coined the word "regulon."

The research of the next 20 years has essentially borne out this proposition. The transition to the molecular approach required the development of methods that could bypass the problems posed by the scattering of the genes in several functional units and the absence of suitable attachment sites to construct transducing phages. When these difficulties were circumvented, the isolation of specific operator and promoter mutants became possible, and the way was paved towards the isolation of individual genes (see Mutant Isolation and Genetic Organization and cis-Dominant Regulatory Mutations, below). These mutants provided the tools for the demonstration in vivo and in vitro that regulation operates essentially at the level of gene transcription (see Levels of Control, below). The utilization of the newly developed cloning and sequencing methodology led to data that support the current model for modulation of gene activity in the arginine regulon (see Structure of Control Regions of Arginine Genes, below). By a curious twist the results bearing out the "classical" mechanism outlined above came as a surprise to many since, as for the lysine and methionine pathways (70, 213, 223), there was no evidence for regulation by attenuation (see references 60, 192, and 195), in contrast with the conclusions reached in the meantime for many other amino acid pathways (see reference 267).

Genetic and molecular studies of *arg* genes yielded several findings of general interest: (i) the divergent *argECBH* operon, with promoters facing each other over an internal control region (see Structure of Control Regions of Arginine Genes, below); (ii) the tandem promoters of the carbamoylphosphate synthetase operon (see The *carAB* Operon, below), which are, respectively, controlled by arginine and the pyrimidines; (iii) the cryptic, inducible acetylornithine transaminase of *Escherichia coli* which might be related to a degradative pathway operating in other bacteria (see *N*-Acetylornithine-δ-Transaminase, below); (iv) the occurrence in *E. coli* K-12 of a duplicate ornithine carbamoyltransferase gene, *argF*, flanked by insertion sequences and absent in all other *Enterobacteriaceae* investigated so far (see Ornithine Carbamoyltransferase, below); and (v) the homologies displayed by *argF*, its homolog *argI*, and the structural gene (*pyrB*) for the catalytic subunit of aspartate carbamoyltransferase (same section).

Arginine biosynthesis has not been studied as extensively in *Salmonella typhimurium* as in *E. coli* except for the *pyrA* locus (equivalent of *carAB*). The information available points to an overall similarity between this species and *E. coli* in terms of gene organization and regulatory mechanisms. A comprehensive review of arginine biosynthesis and metabolism in all procaryotes will be published elsewhere (R. Cunin, N. Glansdorff, A. Piérard, and V. Stalon, submitted for publication).

ARGININE AND POLYAMINE BIOSYNTHESIS: GENERAL OUTLINE

The purpose of this section is to provide an overall picture of the pathways, their interconnections, the regulatory circuits involved, and the resulting interferences.

Arginine biosynthesis proceeds in eight steps according to the scheme depicted in Fig. 1, starting from glutamate. The first four intermediates are acetylated (see reference 251). Except for *N*-acetylornithine, they penetrate the membrane only weakly (mutants more permeable to *N*-acetylglutamate than the wild type have repeatedly been described). This *N*-acetylation prevents the cyclization which, in the proline pathway, leads from glutamate-γ-semialdehyde to Δ^1-pyrroline carboxylate. In the *Enterobacteriaceae* and in the genus *Bacillus* *N*-acetylornithine is deacylated by an acetylornithinase, whereas in other bacteria and in fungi the acetyl group is recycled on glutamate by an ornithine-glutamate acetyltransferase (see reference 251; Cunin et al., submitted; R. Davis, submitted for publication). To these alternative fates of the acetyl group correspond two strategies of feedback inhibition by which arginine regulates the flow of metabolic intermediates through the pathway. In organisms with an acetyltransferase, *N*-acetylglutamokinase is inhibited (66, 241), and in some organisms *N*-acetylglutamate synthase is inhibited as well (91); in contrast, in species endowed with an acetylornithinase, *N*-acetylglutamate synthase alone is the target enzyme (258; see below). From ornithine onwards, the intermediates of the pathway appear to be common to all organisms investigated; ornithine and citrulline are taken up easily, argininosuccinate poorly.

The synthesis of the eight enzymes is repressed by arginine to various extents in *E. coli* (87, 144, 258; see below) and presumably in *S. typhimurium* as well (see references 3, 77, and 124). A single regulatory gene (*argR*) accounts for this repression in both organisms (3, 77, 87, 124, 144, 258).

Carbamoylphosphate is a precursor common to arginine and the pyrimidines. In both *E. coli* and *S. typhimurium* it is produced by a single synthetase, carbamoylphosphate synthetase, using glutamine as physiological amino group donor (1, 188, 191). This situation contrasts with the existence of separate enzymes specific for arginine and pyrimidine biosynthesis in *Bacillus subtilis* and fungi (see references 58 and 65). The enzyme is inhibited by UMP; this inhibition is antagonized by ornithine and IMP (1, 10. 185). Since arginine controls the formation of ornithine through feedback inhibition of *N*-acetylglutamate synthetase, the antagonistic effects of UMP and ornithine assure a balanced distribution of carbamoylphosphate between the tributary pathways. In both organisms, carbamoylphosphate synthetase is cumulatively repressed by arginine and the pyrimidines (1, 188, 191). The *argR* product is involved in this control in both *E. coli* and *S. typhimurium* (4, 124, 189).

Pyrimidines interfere with the regulation of the arginine pathway at the level of carbamoylphosphate utilization. The growth of *E. coli* *pyr* mutants blocked after the aspartate carbamoyltransferase step in a chemostat limited by uracil results in a marked derepression of the *arg* pathway (81); this effect is not seen in *pyrB* mutants (lacking aspartate carbamoyltransferase). A similar effect has been observed in both *E. coli* and *S. typhimurium*: pyrimidine bradytrophic mutants are derepressed for both the pyrimidine and the arginine pathway (124; A. Piérard and N. Glansdorff, unpublished data). Conversely, adding uracil to

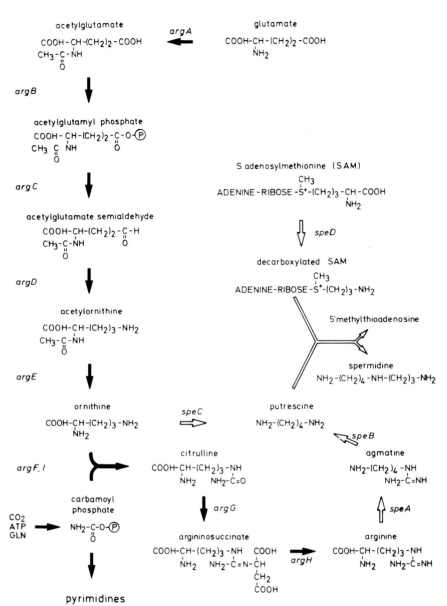

FIG. 1. Biosynthesis of arginine (solid arrows) and polyamines (open arrows).

wild-type cells growing in minimal medium results in partial derepression of arginine biosynthetic enzymes presumably because a slightly excessive inhibition of carbamoylphosphate synthetase by UMP diminishes the supply of carbamoylphosphate for arginine biosynthesis (88). Addition of arginine represses the synthesis of aspartate carbamoyltransferase about twofold, even in an *argR* nonsense mutant. This is probably due to diversion of excess carbamoylphosphate towards the pyrimidine pathway when the pool of ornithine is low (187).

An interesting type of indirect suppression resulting from an interaction between the arginine and proline pathways was brought to light when Itikawa et al. (106; see also reference 130) showed that *proAB* auxotrophs of *E. coli* and *S. typhimurium*, blocked in the conversion of glutamate into glutamate-γ-semialdehyde, could revert to the Pro⁺ phenotype by mutations in the *argD* gene, encoding *N*-acetylornithine-δ-transaminase. *argD* mutants display a leaky arginineless phenotype probably because another transaminase takes over the missing function. When such cells are grown without extraneous arginine, the feedback inhibition of *N*-acetylglutamate synthase is lifted and the whole pathway is derepressed. Under those circumstances, enough *N*-acetylglutamate-γ-semialdehyde is produced and deacylated by the relatively nonspecific *N*-acetylornithinase to feed the proline pathway. In the presence of arginine, this indirect suppression is abolished and the cells return to the Pro⁻ phenotype. Several authors (see *N*-Acetylglutamate Synthase and Mutant Isolation and Genetic Organization, below) have taken advantage of this effect of arginine on *argD proAB* mutants to isolate derepressed mutants or strains in which *N*-acetylglutamate synthase is resistant to arginine.

TABLE 1. Enzymes of arginine and polyamine biosynthesis

Common name	Reaction catalyzed	Gene symbol	Map position E. coli	Map position S. typhimurium
N-Acetylglutamate synthase (EC 2.3.1.1)	Glutamate + acetylCoA → N-acetylglutamate + CoASH[a]	argA	60	61
N-Acetylglutamate kinase (EC 2.7.2.8)	N-Acetylglutamate + ATP → N-acetyl-γ-glutamylphosphate + ADP	argB	88	88
N-Acetylglutamylphosphate reductase (EC 1.2.1.38)	N-Acetyl-γ-glutamylphosphate + NADPH + H$^+$ → N-acetylglutamate-γ-semialdehyde + NADP$^+$	argC	88	88
N-Acetylornithine transaminase (EC 2.6.1.1)	N-Acetylglutamate-γ-semialdehyde + glutamate → N-acetylornithine + α-ketoglutarate	argD	72.5	72
N-Acetylornithinase (EC 3.5.1.16)	N-Acetylornithine + H$_2$O → ornithine + acetate	argE	88	88
Ornithine carbamoyltransferase (EC 2.1.3.3)	Ornithine + carbamoylphosphate → citrulline + P$_i$	argF / argI	6 / 95	98
Argininosuccinate synthetase (EC 6.3.4.5)	Citrulline + aspartate + ATP → argininosuccinate + AMP + PP$_i$	argG	68	69
Argininosuccinase (EC 4.3.2.1)	Argininosuccinate → arginine + fumarate	argH	88	88
Carbamoylphosphate synthetase (EC 6.3.5.5)	Glutamine + 2 ATP + HCO$_3$$^-$ → carbamoylphosphate + glutamate + 2 ADP + P$_i$	carAB[b]	1	2
Arginine decarboxylase (EC 4.1.1.19)	L-Arginine → agmatine + CO$_2$	speA	63	
Agmatine ureohydrolase (EC 3.5.3.11)	Agmatine + H$_2$O → putrescine + urea	speB	63	
Ornithine decarboxylase (EC 4.1.1.17)	L-Ornithine → putrescine + CO$_2$	speC	63.5	
S-Adenosylmethionine decarboxylase (EC 4.1.1.50)	S-Adenosylmethionine → S-adenosyl-5′-δ-methylmercaptopropylamine (decarboxylated S-adenosylmethionine) + CO$_2$	speD	3	
Spermidine synthetase (EC 2.5.1.16)	Decarboxylated S-adenosylmethionine + putrescine → Spermidine + 5′-methylthioadenosine			
Lysine decarboxylase (EC 4.1.1.18)	L-Lysine → cadaverine + CO$_2$	cadA	92	

[a] CoASH, Sulfhydryl coenzyme A.
[b] pyrA in S. typhimurium.

Polyamine biosynthesis has been particularly well studied in *E. coli*. The diamine putrescine can be made either directly by decarboxylation of ornithine or indirectly by decarboxylation of arginine into agmatine followed by hydrolysis of agmatine into putrescine and urea by an agmatine ureohydrolase (165, 166). Urea is not degraded by *E. coli* and is actually an indication of the flux through the agmatine pathway (163). Putrescine and an aminopropyl group from enzymatically decarboxylated *S*-adenosylmethionine give rise to the triamine spermidine. The tetramine spermine is normally not found in *E. coli*. Cadaverine is produced by decarboxylation of lysine. Cadaverine and decarboxylated *S*-adenosylmethionine give rise to *N*-3-aminopropyl-1,5-diaminopentane.

Each of the amino acids arginine and ornithine is a substrate for two decarboxylases (see Polyamine Biosynthesis and Control, below, and reference 227). In each case, the so-called biosynthetic and "constitutive" decarboxylase is the only one to be found in cells grown at physiological pH in minimal medium. At low pH, at high substrate concentrations, and in rich media, distinct, biodegradative decarboxylases are induced. These enzymes probably constitute a defense mechanism against acidity since a mutant lacking arginine decarboxylase is not able to grow at low pH values (E. F. Becker, Jr., Fed. Proc. **26**:812, 1967). Until recently lysine decarboxylase had been considered to be an inducible enzyme exclusively; evidence for another decarboxylase, produced under normal growth conditions, is now available (see reference 259). The enzyme is inhibited by putrescine and spermidine, in keeping with the suggestion (259) that cadaverine could provide a substitute for putrescine under conditions of polyamine deprivation.

From the pattern of regulation operating in the arginine pathway it is clear that in the presence of a concentration of arginine high enough to inhibit ornithine formation, putrescine is made via agmatine exclusively whereas in unsupplemented minimal medium the route from ornithine is preferred (164). In minimal medium, putrescine and spermidine amount to about 30% of the arginine in the cell protein (232).

ARGININE BIOSYNTHETIC ENZYMES

This section summarizes what we know about the enzymes involved in the arginine pathway of *E. coli* and *S. typhimurium*. Arginyl-tRNA synthetase and arginine permease will be considered in the next two sections. In several instances our understanding of enzymology and related observations on metabolite flow have been exploited to select mutants affected in the regulation of enzyme activity or enzyme synthesis. The rationale for these selection procedures is also presented in this section.

Table 1 gives the common and systematic names of the arginine biosynthetic enzymes, the reactions they catalyze, and the symbols and map positions of the corresponding genes.

N-Acetylglutamate Synthase

Using resting cells, Vyas and Maas (258) showed that the unstable enzyme N-acetylglutamate synthase is both repressed and inhibited by arginine. After Haas et al. (91) found that glycerol stabilized the enzyme in extracts of *Pseudomonas aeruginosa*, Leisinger and Haas (139) and Marvil and Leisinger (151) were able to examine the properties of the *E. coli* synthase. This enzyme consists of a single type of subunit of molecular weight (MW) 50,000; arginine and N-acetylglutamate stabilize or induce a hexameric form. Fifty percent feedback inhibition is achieved with 0.02 mM L-arginine. N-Acetylglutamate synthase from *S. typhimurium* is also very sensitive to arginine (3). The analog O-(L-norvalyl-5)isourea is as effective as arginine. Indospicin inhibits the enzyme to some extent.

Ennis and Gorini (73) found that feedback inhibition efficiently controlled the flux of arginine precursors only when the level of arginine biosynthetic enzymes was kept low, a status that is normally achieved by repression in *E. coli* K-12 or W and is an intrinsic property of *E. coli* B (see The Arginine Repressor, below). Derepressed (argR) mutants excrete arginine.

Desensitized argA mutants could be selected for thanks to the growth-inhibitory effect exerted by arginine on argD pro double mutants incubated in minimal medium without proline supplement (see above). This inhibition is due to repression of the first five enzymes of the pathway and to feedback inhibition of N-acetylglutamate synthase. When applied in two steps, the selection first gives slow-growing argR mutants (106, 124) from which feedback-insensitive mutants can be obtained by further selection and screening for colonies able to excrete proline even in the presence of arginine (71). Alternatively, desensitized mutants can be detected directly in a population of mutagenized argD pro argR mutants as proline excreters (71).

N-Acetylglutamokinase

N-Acetylglutamate and ATP are the substrates of N-acetylglutamokinase (253; see references 83 and 253 for assays). Linn et al. (141) obtained an estimation of the MW of the primary polypeptide product of the argB gene (29,000) from gel electrophoresis of the proteins synthesized by UV-irradiated cells infected with λdargECBH transducing phages. This value is compatible with measurements of the lengths of argC-argB deletions (47).

The labile intermediate, N-acetylglutamylphosphate, has not been isolated in the free state, and it is not known whether the kinase and the subsequent N-acetylglutamate semialdehyde dehydrogenase, which are encoded by adjacent genes both in *E. coli* and *S. typhimurium*, form some kind of complex in vivo. In this respect, it is interesting that the two equivalent yeast proteins are produced from one single unit of genetic expression (108, 158).

N-Acetylglutamylphosphate Reductase

N-Acetylglutamylphosphate reductase catalyzes the reduction of N-acetylglutamylphosphate into the corresponding semialdehyde. It has been partially puried (254; see references 83 and 254 for assays). According to Linn et al. (141), the MW of argC-encoded peptide should be around 47,000, which again is compatible with length estimates for argC-argB deletions.

N-Acetylornithine-δ-Transaminase

N-Acetylornithine-δ-transaminase, product of the argD gene, catalyzes the formation of N-acetylornithine and α-ketoglutarate from N-acetylglutamate semialdehyde and glutamate (see reference 252 for assay). The same enzyme also exhibits an ornithine-δ-transaminase activity which probably plays no physiological role (22).

The transaminase has been purified and crystallized (23) with the object of comparing it with the purified product of the argM gene. *E. coli* W (15) and *E. coli* K-12 (T. Eckhardt, Ph.D. thesis, Eidgenössische Technische Hochschule, Zurich, Switzerland, 1975) indeed contain a cryptic function, ascribed to gene argM and unlinked to argD, which can be brought to expression by selecting for suppressors of argD mutants. The active argM enzyme is inducible by arginine; curiously, comparing argR+ and argR− derivatives of argD− argM+ strains suggests that induction is mediated by the argR product itself (15). The respective MWs of the argM and argD transaminases of *E. coli* W are 61,000 and 119,000. These proteins do not cross-react immunologically, though both enzymes are composed of 31,000-dalton (Da) subunits and their tryptic patterns are almost identical (23). The argM protein has ornithine-δ-transaminase activity as well (22). The actual amino acid sequences are not yet known, but the data certainly suggest a common origin for the two genes. It cannot yet be concluded whether wild-type strains contain an inactive inducible transaminase or an inactive argM gene. The inducibility of argM suggests that this gene is a cryptic element of an arginine degradative pathway that is silent and possibly incomplete in the *E. coli* strains analyzed so far. In keeping with this idea, the argM enzyme is able to transaminate succinylornithine, an intermediate in a newly discovered catabolic pathway responsible for the breakdown of arginine in a number of bacteria including *Klebsiella*, another member of the *Enterobacteriaceae* (245). Clearly, our understanding of the mechanisms involved in the evolution of metabolic pathways in procaryotes may gain much from further investigation of both transaminases and of their genetic control.

N-Acetylornithinase

N-Acetylornithinase is characteristic of the so-called "linear pathway" for arginine biosynthesis (see Arginine and Polyamine Biosynthesis: General Outline, above). A superficially similar reaction has been found in organisms recycling the N-acetyl group, such as *Saccharomyces cerevisiae* and *Thermus aquaticus*; it could be ascribed to a carboxypeptidase (67). N-Acetylornithinase hydrolyzes N-acetylornithine into ornithine and acetate (see reference 255). It is dependent on Co^{2+} and a thiol compound (preferably gluta-thione) for maximal activity. The enzyme has been purified to homogeneity by J. Charlier (abstract no. S-05, TM171, FEBS Meeting, Brussels, Belgium, 1983). The molecule appears to be a monomer of MW 62,000 (J. Charlier, personal communication). This estimate, based on molecular filtration experiments, differs from the 52,000-Da value inferred from Linn et al. (141) for the argE product examined by electrophoresis on sodium dodecyl sulfate gels. Both estimates remain compatible with the limits assigned by physical mapping data (47).

N-Acetylornithinase readily deacylates substrates other than N-acetylornithine, including N-acetylglutamate semialdehyde, N-acetylarginine, N-acetyl- and N-formylmethionine, and N-acetylhistidine (see references 16 and 255). The action of the enzyme on N-acetylglutamate semialdehyde explains the suppression of proAB mutants by argD mutations as explained above (see Arginine and Polyamine Biosynthesis: General Outline, above). Two interesting applications derive from the relative lack of specificity of this enzyme. Among mutants resistant to N-acetylnorvaline, Kelker and Maas (121) were able to obtain argR+ revertants from derepressed mutants and argE auxotrophs from the wild type. Baumberg (16) showed that it was possible to obtain regulatory mutants in the arginine pathway by selecting for derivatives of his auxotrophs able to utilize N-acetylhistidine even in the presence of ornithine and arginine.

Ornithine Carbamoyltransferase

Ornithine carbamoyltransferase catalyzes the sixth step of the pathway. With the collateral carbamoylphosphate synthetase (see below) it is the best known enzyme of the regulon. The reaction is usually followed in the forward direction, by far the more favored one, but arsenolytic cleavage of citrulline can be used to measure the reaction in the reverse direction (see reference 136).

A peculiarity of E. coli K-12 is that it contains two ornithine carbamoyltransferase genes (84), argF and argI, the products of which interact to form a family of four trimeric isoenzymes (135). The argI gene or its equivalent is the only one to be found in other E. coli strains (109) or other Enterobacteriaceae (see reference 137), S. typhimurium in particular (224). The kinetic parameters of the E. coli F and I isoenzymes are very similar, but the F protein is much more thermolabile (137).

The kinetic properties of the argI protein have been investigated extensively in E. coli W (136) and S. typhimurium (6). The reaction displays a preferred sequence of substrate binding: carbamoylphosphate is the first to bind, and inorganic phosphate is the last to be released, as in the reaction catalyzed by aspartate carbamoyltransferase (198). In contrast with the E. coli enzymes, S. typhimurium ornithine carbamoyltransferase is moderately inhibited (58%) by arginine at relatively high concentration (5 mM). This situation is unlikely to be of physiological importance per se, in contrast to the situation prevailing in Agrobacterium tumefaciens (247). Knight and Jones (128) have shown that the ornithine carbamoyltransferase of E. coli W displays a kinetically complex activation by orotate

which may be of physiological value in coordinating the flow of metabolites through the arginine and pyrimidine pathways.

An extremely strong and specific inhibition of the reaction is exerted by the bisubstrate analog phosphonoacetylornithine (PALO) (182). E. coli is impervious to PALO but not to the oligopeptide Gly-Gly-PALO (181, 183), which penetrates via the oligopeptide permease, product of the opp gene. PALO is then liberated by an intracellular peptidase and inhibits ornithine carbamoyltransferase. opp and, to a certain extent, argR mutants are resistant to Gly-Gly-PALO. These investigations may have therapeutic implications since they provide an example of "illicit" uptake, by which a substance unable to penetrate the cell membrane does so in covalent association with a pervasive carrier (181).

The equilibrium of ornithine carbamoyltransferase is not favorable to the reverse reaction. Phosphorolysis of citrulline can, however, be demonstrated in vivo: strains defective in carbamoylphosphate synthesis grow very slowly on citrulline as a source of carbamoylphosphate for the pyrimidine pathway (221). This property was used to select for mutants with high ornithine carbamoyltransferase specific activity, such as argG bradytrophs (in the presence of citrulline, they accumulate this amino acid and at the same time exhibit derepression of ornithine carbamoyltransferase) and constitutive argF or argI mutants specifically derepressed by operator mutations or unstable chromosomal rearrangements (114, 115, 138). The same strategy was applied to S. cerevisiae to obtain cis-dominant constitutive ornithine carbamoyltransferase mutants (157).

The occurrence of hybrid F-I isoenzymes suggested that the constituent polypeptides were homologous products of an ancestral duplication (135). Heteroduplex analysis of λ argF and λ argI transducing phages (127) and comparison based on parts of the cognate amino acid (80) and nucleotide sequences (195) supported this idea. The full extent of this homology can be appreciated now that the nucleotide sequences of argF (246) and argI (18) have been determined. The structural parts of the two genes display 86% amino acid and 78.1% nucleotides in common, not unlike the trpA genes of E. coli and S. typhimurium (85.1 and 72.5%, respectively). It is possible that argF has been inherited laterally from a related species, as suggested by the presence of two IS1 elements flanking that gene in E. coli K-12 (105, 268), or that the divergence between argF and argI occurred within a branch leading to present-day E. coli.

The comparison between ornithine carbamoyltransferase and the catalytic subunit of aspartate carbamoyltransferase is of considerable interest from the evolutionary point of view. The overall homology is 35 to 40% in terms of amino acids and shows up in similar structural domains, mainly in the polar moiety and helical regions joining the polar and equatorial domains of aspartate carbamoyltransferase (104, 246). The two proteins therefore appear to have a common origin. An ancestral, possibly ambiguous carbamoyltransferase gene could have been duplicated in near-tandem copies which would have diverged in the course of evolution but have remained associated on the chromosome in certain organisms, as the strong linkage between *argI* and *pyrB* in *E. coli* and *S. typhimurium* suggests.

Argininosuccinate Synthetase

Argininosuccinate synthetase, encoded by the *argC* gene, catalyzes the conversion of citrulline, aspartate, and ATP into argininosuccinate, AMP, and pyrophosphate. The enzyme is not well known among the *Enterobacteriaceae*, but its mammalian and *S. cerevisiae* equivalents have been purified (see references 98 and 204). The yeast enzyme is a tetramer of identical 49,000-MW subunits, while the *E. coli* enzyme appears to consist of a basic polypeptide of similar MW (48,000) as estimated from denaturing gels loaded with extracts of minicells producing a plasmid-encoded *argG* protein (169).

Argininosuccinase

Argininosuccinase, encoded by the *argH* gene, hydrolyzes argininosuccinate into arginine and fumarate. Little is known of the enzyme in the *Enterobacteriaceae* (see references 61 and 256 for assays). The purified mammalian argininosuccinase is a tetramer of 50,000-Da subunits (see reference 204). The *E. coli* enzyme could be similar since extracts of UV-irradiated cells infected with a λ transducing phage carrying the *argH* gene display a 55,000-Da polypeptide when examined by electrophoresis under denaturing conditions on polyacrylamide gels (141).

Carbamoylphosphate Synthetase

In both *E. coli* and *S. typhimurium*, a single carbamoylphosphate synthetase which utilizes glutamine as natural amino group donor provides carbamoylphosphate for arginine and pyrimidine biosynthesis (1, 188, 191). The ammonium ion is a low-affinity nitrogen donor for the enzyme (118); the reaction with ammonium ion would appear to be functional in vivo only when the ammonium ion concentration in the medium is high.

With pure carbamoylphosphate synthetase from *E. coli* B, Anderson and Meister (9) showed that the synthesis of carbamoylphosphate proceeds in four steps: (i) formation of an enzyme-CO_2 complex activated by ATP; (ii) reaction between this complex and glutamine; (iii) transfer of the amido group of glutamine to activated CO_2 and formation of enzyme-bound carbamate; (iv) phosphorylation of carbamate in the presence of a second molecule of ATP and liberation of the carbamoylphosphate formed.

Trotta et al. (236) showed that pure *E. coli* carbamoylphosphate synthetase consists of two subunits, a heavy one (MW 130,000) able to catalyze the synthesis of carbamoylphosphate from HCO_3^-, ATP, and NH_3 (but not from glutamine) and a small one (MW 42,000) which carries the glutamine binding site and displays a glutaminase activity in vitro. Similar conclusions were drawn for the *Salmonella* enzyme (2). The two proteins are encoded in *E. coli*, respectively, by the adjacent *carA* and *carB* genes (156) and in *Salmonella* by equivalent regions of the *pyrA* locus (2). *carA* and *carB* form an operon oriented from *A* to *B*; some nonsense *carB* mutants display curious antipolar effects as yet unexplained (56, 79). The complete sequence of the two genes is known in *E. coli* (174, 196).

From glutamine bound to the small subunit, the amido group is transferred, probably as nascent NH_3, to the site of the large subunit which also accepts the ammonium ion as nitrogen donor. Carbamoylphosphate synthetase is thus basically similar to other amidotransferases involved in the synthesis of amino acids, purines, pyrimidines, or cofactors (see references 95, 186, and 202). It is possible that the evolution of this family of enzymes proceeded by the combination of a primordial glutaminase with various synthetases originally utilizing NH_3 as nitrogen donor (95, 186, 202).

There is a high degree of homology (39% identical residues) between the amino- and carboxy-terminal moieties of the *carB* protein, and it has been suggested that *carB* results from the duplication of a smaller ancestral gene (174). As the enzyme is allosteric (see below), Nyunoya and Lusty have also suggested that the two halves of the *carB* molecule fold as separate domains capable of conformational interactions, each domain possibly carrying one of the two ATP binding sites recognized on carbamoylphosphate synthetase (26, 200).

Carbamoylphosphate synthetase is a highly regulated enzyme. It is subject to feedback inhibition by UMP (1, 188). This inhibition is antagonized by IMP (10), an effect of probable significance for the coordination of purine and pyrimidine pathways, and, most importantly, by ornithine (185). Since arginine curtails ornithine formation by feedback inhibition of *N*-acetylglutamate synthase, the UMP-ornithine antagonism amounts to a double feedback by arginine and UMP. In the presence of ornithine, UMP exerts a partial inhibition; in the absence of ornithine, the inhibition is almost total. The sites of the known effectors of the enzyme are all on the large (*carB*) subunit (236).

Carbamoylphosphate synthetase is thus controlled by several allosteric effectors belonging to different pathways. Anderson and Marvin (7, 8) have shown that the effectors affect the distribution of the enzyme between at least three conformational states, one of them stabilized by the positive effectors and favorable to substrate binding, another one stabilized by the inhibitor UMP. Mutational alterations of carbamoylphosphate synthetase are thus likely to result in a variety of phenotypes. In certain uracil-sensitive *E. coli* mutants such as strain P678M1 (188; G. Leclercq, Ph.D. thesis, Université Libre de Bruxelles, Brussels, 1971), the enzyme displays a much higher apparent K_m for ATP than does the wild-type enzyme so that in the presence of UMP and even in the presence of ornithine, the activity is not high enough to support growth. The apparent K_m for ATP is particularly low in the absence of ornithine. This observation provides an explanation for the partial sensitivity towards arginine also displayed by the mutant.

supJ-hisT double mutants of *S. typhimurium* are also uracil sensitive (27). It is not yet known whether the primary effect of the mutation is structural or regulatory; an explanation in terms of attenuation control (27) now appears unlikely (see The *carAB* Operon, below).

Other mutations at the *car* locus exhibit either a pseudo arginineless phenotype or an arginine-sensitive one (79, 156). The same phenotypes were investigated by Abdelal et al. in corresponding *pyrA* mutants of *S. typhimurium* (4). They proposed that the formation of a complex between carbamoylphosphate synthetase and ornithine carbamoyltransferase is a sine qua non condition for proper assembly and function of carbamoylphosphate synthetase subunits in arginine-sensitive *pyrA* mutants but not in the wild type. Other observations by Abdelal et al. (5) suggest that arginineless *pyrA* mutants display this phenotype because arginine curtails the formation of *N*-acetylornithine which, from the behavior of double *pyrA-argA*, *B*, or *C* mutants, would appear to antagonize the maturation of the mutant carbamoylphosphate synthetase. No such effect was observed in the wild type, however.

ARGINYL-tRNA SYNTHETASE

Arginyl-tRNA synthetase shares an unusual property with glutamyl- and glutaminyl-tRNA synthetases: it does not catalyze ATP-PP$_i$ exchange in the absence of tRNA^{Arg} (155, 178; other references in reference 78). Furthermore, no arginyl-adenylate intermediate accumulates, whether tRNA^{Arg} is present or not (57). It is not clear whether ATP, arginine, and tRNA^{Arg} coreact in one step by a concerted mechanism (143) or whether binding with tRNA^{Arg} is a prerequisite to activate the enzyme and make the formation of arginyl-adenylate possible (155). If the latter is the case, the aminoacyl-AMP would then react immediately with the tRNA^{Arg}. It was shown recently (28) that *E. coli* arginyl-tRNA synthetase catalyzes ATP-PP$_i$ exchange in the presence of the analog tRNA^{Arg}c-c-2'dA (i.e., a tRNA with 2'-deoxy-adenosine at the 3' end), which cannot be acylated. A classic, two-step mechanism is therefore suggested.

The enzyme itself (α protein) is a monomer of MW 60,000 that is copurified with a 40,000-MW "β" protein. The two molecules can best be separated by two successive affinity elutions with a tRNA gradient after absorption of the protein preparation on a column of phosphocellulose (50). This procedure provides a 1,000-fold purified, 95% homogenous enzyme which is more active than previously reported preparations. The biological significance of the β-α interactions, if any, is not known.

The K_m for arginine is 3.5 10^{-6} M both for ATP-PP$_i$ exchange and for esterification (155). The arginine analog canavanine is also esterified by the synthetase. The K_m of canavanine is 4.10^{-4} M, and the V_{max} is twice that found with arginine. Both canavanine and homoarginine inhibit the binding of arginine competitively, but homoarginine is not esterified (155). Whether the enzyme is also inhibited by arginine precursors has been the subject of some debate. The synthetase of *E. coli* W and K-12 was claimed to be inhibited by argininosuccinate, ornithine, and citrulline (see reference 35). As the accumulation of arginine precursors in *arg* mutants blocked at different steps was correlated with derepression of arginine biosynthesis, the results suggested a role for the synthetase in the regulation of this pathway (35). However, subsequent studies on partially purified extracts of *E. coli* W and K-12 (including the strain used by Williams and

colleagues [35]) showed that the enzyme remained insensitive to ornithine and citrulline up to 2 mM (49). At relatively high concentrations of argininosuccinate, the apparent inhibition exerted by this compound could be ascribed to isotopic dilution of the labeled arginine used in the assay by nonradioactive arginine produced from argininosuccinate by argininosuccinase present in the extract (49). At lower concentrations of argininosuccinate (1 μM) the observations made by the two groups remain conflicting, and this discrepancy remains unresolved. The experiments discussed in the section Nature of the Corepressor (below) provide no evidence for involvement of tRNA or the synthetase in the regulation of the arginine pathway.

Mutants affected in arginyl-tRNA synthetase have been described. Besides their intrinsic interest they also have implications regarding a possible involvement of the synthetase or the tRNA^{Arg} in regulation. Canavanine-resistant (Can^r) isolates of *E. coli* with both altered arginyl-tRNA synthetase activities and nonrepressible arginine biosynthetic enzymes have been reported (see reference 35). However, these observations could not be confirmed; double mutants may have been involved. In contrast, other Can^r synthetase mutants, with increased K_m values for arginine or for ATP, were found to be fully repressible by arginine (99, 100); moreover, mutants with a block drastic enough to accumulate endogenous arginine showed enhanced repression. Hence it was assumed, and the assumption still seems valid at present (see Nature of the Corepressor, below) that arginine and not arginyl-tRNA is the real corepressor of the arginine regulon. The mutants isolated by Hirshfield and colleagues provide evidence for a single arginine-activating enzyme in *E. coli*, encoded by the *argS* gene.

Little is known about regulation of the synthesis of arginyl-tRNA synthetase. Like isoleucyl- and phenyl-alanyl-tRNA synthetases, the enzyme becomes permanently derepressed during cognate amino acid starvation (see references 160 and 172). In the case of phenylalanyl-tRNA synthetase, this phenomenon was correlated with the existence of an attenuation mechanism dependent on the concentration of tRNA^{Phe} (237). A positive response to growth rate of the steady-state concentration of several synthetases could be established (171). The approximate number of molecules of arginyl-tRNA synthetase per genome shifts from 192 when the cells use acetate as carbon source, to 510 on glucose-supplemented medium, and to 867 on rich medium. These values may be compared to the 1,500 molecules of ornithine carbamoyltransferase and 2,000 molecules of carbamoylphosphate synthetase that the cell makes on minimal medium supplemented with glucose. These levels of the different synthetases remain in approximate balance with those of tRNA, elongation factors, and ribosomes.

ARGININE PERMEASE

Two kinetically distinct, active uptake systems mediate the transport of arginine and ornithine across the membrane of *E. coli* (see reference 42). One of them, the "low-affinity" system (with a K_m of about 10^{-7} M for arginine), transports arginine and ornithine. Both amino acids repress the formation of this system (46). Another system, repressible by lysine, displays a higher affinity towards arginine (K_m about 10^{-9} M). It also transports lysine and ornithine (46, 211). Mutations affecting the latter system (94) leave only the low-affinity one functional. At arginine concentrations of the order of 10^{-6} M or higher the main route of entry is through the more specific, low-affinity system. An arginine-repressible acetylornithine permease was reported by Vogel (249); its relations with the components involved in arginine transport are not known. Studies of arginine uptake and accumulation have taken considerable advantage from the fact that decarboxylation of this amino acid into agmatine is inhibited by aminooxyacetic acid (262).

Osmotic shock of wild-type *E. coli* cells abolishes the activity of both the low- and the high-affinity systems and liberates periplasmic proteins binding arginine (263). One of these proteins binds arginine, ornithine, and lysine, another one binds arginine (210), and a third one binds arginine and ornithine (42). The physiological importance of the third is attested by a number of arguments. Large quantities of this protein are made by strains obtained by selecting for D-arginine utilization and containing regulatory mutations affecting the low-affinity system (41). Derivatives of such a strain unable to use D-arginine exhibit low transport activity, and some of them are suppressible by nonsense suppressors and display protein fragments consisting of prematurely terminated binding protein (42). Active transport of arginine was shown to require ATP. A current model involves a phosphorylation-dephosphorylation cycle of the arginine-ornithine binding protein, in close association with a membrane-bound carrier (44).

Such an active transport system should consist of at least two structural elements (the periplasmic proteins and carrier[s] more deeply associated with the membrane) and an unpredictable number of regulatory components. To what extent do the mutants isolated up to now identify the genetic determinants of these functions? On the basis of the mutants' kinetic properties and from their pattern of periplasmic binding proteins, several mutations were assigned to a particular element, structural or regulatory. A first mutation (145, 217) conferring resistance to canavanine defined a locus named *argP* near *serA*. The defect curtails transport of all three basic amino acids (46). Another pleiotropic Can^r mutation (211) was genetically linked to the previous one. Since no alterations of the periplasmic binding proteins could be detected in these mutants and since they excreted arginine, the mutations were considered to affect energy coupling (211). Celis (40) studied a third, closely linked Can^r mutation decreasing uptake of arginine and ornithine but not lysine. Both the kinetics of influx and the steady-state accumulation of arginine and ornithine were affected, suggesting a reduction in the number of membrane carriers; the mutation was therefore considered to be regulatory. Obviously, complementation studies between these different Can^r mutations would improve our understanding of the *argP* locus.

Celis (41) also reported the isolation of regulatory mutants, already mentioned above, able to utilize D-arginine. In these strains a four- to fivefold derepressed level of the arginine (and ornithine; 42) binding protein paralleled increased uptake of ornithine

and arginine as well as oversensitivity to canavanine. The arginine transport system in each of these strains was no longer repressible by arginine or ornithine. Derivatives unable to utilize D-arginine (44; see above) were shown to carry mutations in the structural gene for the arginine and ornithine binding protein. Both this structural mutation and the effect leading to derepressed synthesis of the binding protein were located close to each other near the *argA* gene and have been assigned the symbols *abpS* and *abpR*, respectively (43). Again, complementation studies could confirm whether the mutations really affect different genes or, for example, structural and *cis*-dominant controlling elements of the same gene.

Are the regulatory systems that are responsible for repression of transport elements and of arginine biosynthetic enzymes connected in some way? The answer would appear to be "no," since *argR* mutants are repressible for transport and *abpR* mutants are still repressible for the biosynthesis (41).

The picture obtained in *S. typhimurium* differs somewhat from that in *E. coli*. The existence of a high-affinity, arginine-specific system and of a transport system common to arginine, ornithine, and lysine has been documented (203). The same authors showed that the specific system is able to discriminate against ornithine as well as lysine. Homoserine and transhydroxyproline proved to be good inhibitors but not substrates of this system; homoserine was without effect in *E. coli* (40). The effect of these substances in *S. typhimurium* and the absence of effect of ε-*N*-methyl-homoarginine suggest that the secondary nitrogen of arginine may act as hydrogen donor in the recognition process (203).

The lysine-arginine-ornithine binding protein interacts with a membrane-bound receptor involved in the high-affinity histidine-specific transport system (the *hisP* gene product). Transport of arginine for utilization as nitrogen source depends on this interaction. However, *hisP* mutants retain normal high-affinity transport of arginine (131).

THE ARGININE REGULON

Mutant Isolation and Genetic Organization

Most of the structural genes of the arginine pathway have been identified and localized in *E. coli* and *S. typhimurium* by conventional selection and mapping techniques (see references 14 and 214 for results and references). *argM*, coding for the cryptic inducible acetylornithine transaminase, has not been localized with precision; it is carried by episome F'14 (206; Eckhardt, thesis). Forward selection methods have also been devised for mutants blocked in steps A, B, C (53), and E (121). As in fungi (13, 167), *car* mutants could be recovered from a dihydroorotaseless (*pyrC*) strain by selecting for derivatives unable to accumulate the toxic intermediate carbamoylaspartate (D. Charlier, Masters thesis, Vrije Universiteit Brussel, Brussels, 1974). The toxicity of carbamoylaspartate is not understood; it has recently been observed in *S. typhimurium* as well (239).

With the exception of *argF*, a peculiarity of *E. coli* K-12, both *E. coli* and *S. typhimurium* display very similar arrangements of *arg* loci (Table 1). The only genes to be clustered were *argECBH*. In *E. coli* (see *cis*-Dominant Regulatory Mutations and Structure of Control Regions of Arginine Genes, below) *argECBH* was shown to be a divergent operon consisting of two arms, *argE* and *argCBH*, with an internal operator region flanked by two convergent promoters.

Since the development of methods for the construction of Φ80 and λ phages transducing *arg* genes (126, 137, 152, 201; B. Konrad, unpublished data) and of plasmid vectors (summarized in reference 58), most *arg* genes of *E. coli* have now been cloned (18, 58, 169, 206, 246). Physical maps are available for the regions encoding *argD* (206), *argG* (169), *argECBH* (47, 54), *carAB* (56), and *argF* and *argI* (54, 159, 195). Partial sequence data have been published for *argECBH* (192), and complete data are available for *argF* (246), *argI* (18), *carA* (196), and *carB* (174).

Mutants affected in the regulatory gene *argR* were obtained in *E. coli* B and W by selecting for canavanine-resistant derivatives (87, 144). A gene of unknown function, also giving rise to Can^r mutants, has been localized by Maas (146). *argR* mutants of *E. coli* B can be obtained by selecting for resistance to homoarginine (184), which does not affect the K-12 strain. Canavanine does not inhibit growth of *S. typhimurium*. *argR* mutants of both *S. typhimurium* and *E. coli* could also be isolated by indirect suppression of the proline auxotrophy induced by arginine in *argD* mutants (see *N*-Acetylglutamate Synthase, above). *argR* mutants of *E. coli* were also recovered as strains resistant to a mixture of 2-thiouracil and arginine (189), the toxic effect of which has remained unexplained (19). This combination is useful to select for *argR* derivatives of *arg* auxotrophs; the use of auxotrophs also circumvents selection for permease mutants. Another useful method is the selection from histidine or methionine auxotrophs of derivatives able to use the cognate *N*-acetyl amino acid in the presence of ornithine or arginine (see *N*-Acetylornithinase, above). A last method, of general interest for the selection of derepressed mutants, involves selecting for derivatives of leaky mutants (a *carA* mutant was used; 189) that exhibit no further growth lag when transferred from minimal medium supplemented with arginine (and in this case, uracil as well) to minimal medium. The method has been applied with success to isolate *cis*-dominant constitutive mutants of *arg* genes in *S. cerevisiae* (108).

The *argR* gene has been cloned and its sequence has been determined completely (T. Eckhardt, unpublished data; see reference 60 and section below on The Arginine Repressor).

Because several *arg* genes are unlinked to each other, isolating operator mutants in the arginine regulon required ad hoc strategies which will be discussed in the section below on *cis*-Dominant Regulatory Mutations.

Levels of Control

It is now well established that the largest part of the repression response of arginine biosynthetic enzymes is due to transcriptional control. However, whether repression is entirely accounted for by control of initiation of transcription or also involves regulation at a later step is not completely settled.

Control of mRNA translation rather than modulation of mRNA synthesis was advocated early as a mechanism for the regulation of *arg* genes (250; see review by Vogel and Vogel [256]). When transducing phages carrying the *argECBH* genes became available (see Mutant Isolation and Genetic Organization, above), RNA/DNA hybridization allowed a direct test of these hypotheses. Rogers et al. (208) and Cunin and Glansdorff (62) showed that the amount of pulse-labeled *argECBH* hybridizable RNA varied considerably over the range of conditions investigated. The total amplitude of variations, however, appeared lower for mRNA than for enzyme levels. Other studies confirmed the notion of transcriptional control in vivo (see reference 170 for *argA*, *F*, and *I*) and in vitro using *argA*, *ECBH*, *F*, and *I* DNA in coupled transcription-translation systems (207, 218, 219).

More precise investigations with low-background DNA probes for the two arms of the *argECBH* divergent operon established that the amplitude of variation for hybridizable RNA levels is three- to fourfold lower than for enzyme specific activities (59); the phenomenon has recently been reproduced in vitro (269). Such a discrepancy, however, was not observed for the *carAB* operon: there a close correspondence was obtained between enzyme and RNA levels (190).

One interpretation of the mRNA-enzyme discrepancy would have been attenuation control: in repression, the cells would contain a relatively higher proportion of leader mRNA not contributing to enzyme synthesis. Indeed, a similar discrepancy had been reported by Lavallé and De Hauwer (134) for the *trp* operon at a time when the concept of attenuation was not even considered. For a while, the idea was supported by experiments (129) which showed that in repressed cells, hybridizable RNA was shorter (6 to 8S) than in derepressed ones (14 to 23S). However, Bény et al. (21), using several probes, showed that transcription of *argCBH* in a purified in vitro system was not restricted to a proximal leader sequence and that, in vivo, no preferential transcription of operator-proximal sequences occurred under conditions of repression. Sequence data (see references 60 and 196 and Cunin et al., submitted) provide no evidence for attenuation control either, whether in *argECBH*, *carAB*, *argF*, or *argI*.

Another explanation, advocated by McLellan and Vogel (154; see also reference 256), assumed the discrepancy to be due to a greater stability of the mRNA in derepression than in repression; indeed, restricting the arginine supply of an arginine auxotroph increased the chemical half-life of *argECBH* hybridizable RNA by a factor of 3 to 4. Krzyzek and Rogers (129) confirmed that arginine starvation increased the stability of hybridizable *argECBH* RNA but showed that bulk RNA was affected in the same way. They could find no evidence for a specific effect of arginine on the stability of *argECBH* RNA.

Hall and Gallant (93), measuring functional rather than chemical decay of ornithine carbamoyltransferase mRNA, did not detect any difference between repressed and derepressed cells, in contradiction to the results of a previous experiment by McLellan and Vogel (153).

Lavallé (133) proposed a translational regulatory mechanism that could have been compatible with a discrepancy between mRNA and enzyme repression ratios. A kinetic analysis of enzyme repression showed that addition of the corepressor provoked a temporary stagnation in the amount of active enzyme and suggested that mRNA molecules themselves became transiently frozen in an inactive state. The model assumed this effect to be mediated by the repressor and another protein of the regulon. The parameters of this interaction could have been such that the efficiency of translation would have remained comparatively lower under conditions of steady-state repression and thus have given rise to the observed discrepancy. However, closer examination of the phenomenon showed that after addition of the corepressor, the enzyme level actually dropped. This observation suggested an alternative interpretation based on enzyme decay rather than translational arrest (190). This interpretation is in good agreement with the data in the case of carbamoylphosphate synthetase and may be valid also for other examples of apparent translational control reported in the wake of Lavallé's original publication (31).

Posttranscriptional control was invoked by Faane and Rogers (74) to explain that canavanine has little or no effect on the level of hybridizable argECBH RNA but appears to repress the level of translatable messenger by a mechanism involving tRNA synthetase. These experiments are difficult to interpret since incorporation of canavanine into polypeptides produces defective proteins. Furthermore, more recent results obtained by the same group showed that canavanine does not repress mRNA synthesis in vitro (269). Also conjectural is the molecular basis for an increase of repression-derepression range observed in a particular streptomycin-resistant derivative of a partial argR mutant of E. coli W (257). A fine genetic analysis of this strain would be needed to support the proposed interpretation of this effect in terms of specific translational regulation.

In conclusion, no satisfactory explanation has been provided for the argECBH mRNA-enzyme discrepancy which is presently the most tangible evidence for possible second-site control. Though the argECBH cluster is not controlled by a classical attenuation mechanism, the data remain compatible with the notion that in repressed cells, transcripts of a distal portion of the cluster, well beyond the beginning of argC, would be less abundant or stable than those of more proximal segments. The mechanism that would be responsible for the preferential occurrence of these shorter mRNA molecules under conditions of repression remains conjectural. In our opinion, it is possible that the mRNA-enzyme discrepancy results from an entirely different mechanism. In argC there is a weak secondary argE promoter (argEp2) active under conditions of repression (see Structure of Control Regions of Arginine Genes, below); formation of translationally inactive RNA-RNA duplexes by mRNA molecules respectively initiated at argCBHp and at argEp2 could account for the relative excess of unproductive RNA present in repressed cells.

Synthesis of acetylornithinase and argininosuccinase is enhanced by ppGpp, the chemical messenger of the stringent response (129). These investigations correct an earlier report by Yang et al. (266) concluding that synthesis of acetylornithinase was inhibited

TABLE 2. Repression response in the arginine regulon of E. coli K-12

Transcription unit	Repression-derepression ratio[a]
argA	≥250
argCBH	50–70
argE	17
argD	20
argF	150–200
argI	300–400
argG	ND
argR	10
carAB	33[b]

[a] Ratio between specific activities (micromoles per hour per milligram of protein) as assayed in argR cells (no arginine added) and in argR+ cells (100 µg of arginine added per ml). ND, Not determined.
[b] Uracil (100 µg/ml) added to the cultures.

by ppGpp and are in keeping with the positive effect of ppGpp observed on the synthesis of enzymes A and I (76, 120, 238). By in vitro experiments, Zidwick et al. (270) showed that this effect did not involve argR and was not exerted at the level of DNA transcription. More recent experiments from the same laboratory (M. G. Williams, Ph.D. thesis, University of Minnesota, Minneapolis, 1985) indicate that ppGpp acts by reducing the frequency of translational errors, in keeping with earlier observations (see reference 76).

In contrast, ppGpp was shown to inhibit carAB expression partially (29); starvation for an amino acid would therefore not only turn off the synthesis of RNA but would also affect the synthesis of RNA precursors.

Incomplete Coordination of the Repression Response

The scattering of arg genes on the chromosome is accompanied by a certain lack of coordination in their response to repression. Table 2 gives ratios between enzyme levels observed in derepressed (argR) mutants and repressed argR+ cells of E. coli K-12. argA, F, and I display the largest values of all; in terms of RNA levels the corresponding coefficients of variation are at least 10 (170). This pattern is not reproducible from strain to strain, sometimes even between K-12 derivatives. The variations may be due to differences in the repressor or the operator sites, or both, though interstrain variations in the capacity to accumulate arginine (as a function of permeability and rate of decarboxylation into agmatine) have not been excluded. An extreme case is provided by E. coli B, in which the arginine biosynthetic enzymes are actually slightly inducible by arginine (see The Arginine Repressor, below).

The argECBH cluster exhibits coordinated regulation for the enzymes of the last three genes only (83), an observation that could have suggested the existence of two independently regulated transcription units, argE and argCBH. Later findings, however, established that argE and argCBH are divergently transcribed (72, 110, 176, 199) from a control region containing an operator common to both (20, 192, 193). The lack of complete coordination between these two transcriptional units is due to the activity of a weak secondary promoter for argE, argEp2, located outside of the control region (see Structure of Control Regions of Arginine Genes, below; reference 60; and Cunin et al., submitted). Near the border between *argB* and *argH* there is a secondary promoter accounting for a very low level of *argH* expression (one-third of the fully repressed cells; 72).

The physiological significance of this incomplete coordination has not been established. In principle, the scattering of a physiological unit in several subunits of expression offers a greater flexibility than coordinated operon-type control. This flexibility could be of particular value at the point of metabolic branching (as in the case of carbamoylphosphate synthetase, ornithine carbamoyltransferase, and aspartate carbamoyltransferase) or at the level of reactions dependent on substrates available in limiting concentrations. For example, it is known that in *S. cerevisiae* the flux of metabolites through the arginine pathway is limited by the supply of ATP (98). It is not known whether a similar correlation occurs in *E. coli* as well. The quantitative complexity of the system, owing to the numbers of parameters involved, appears to have discouraged a systematic approach to this question until now.

The coefficient of repression of the autogenously controlled *argR* gene is the lowest of all and is probably an overestimate. The significance of this moderate amplitude of variation could be the same as suggested for *trpR* (123). In cells growing on minimal medium the arginine biosynthetic enzymes are still 80 to 90% repressed, yet the concentration of arginine is relatively low, hence the need for an increase in aporepressor concentration to maintain a sufficient level of active repressor.

The Arginine Repressor

The regulation of arginine biosynthesis presents a paradox: while the K-12 strain exhibits high and repressible levels of enzyme activity, the B strain displays lower, slightly inducible enzyme levels. Nevertheless, fully derepressed mutants with comparable specific activities could be obtained from both strains by selecting for resistance to canavanine (87). Further work (112, 119) disclosed that $argR_K$ and $argR_B$ were alleles and specified repressor molecules with different properties. At low intracellular arginine concentrations *E. coli* B is derepressed, although not as extensively as strain K-12; it is repressed at slightly higher concentrations, but becomes derepressed again beyond a certain threshold. The strain B repressor therefore appears to exist in different, interconvertible forms. Excess of arginine would favor a form having reduced affinity for operator sites.

A mutation (111) or a single amino acid substitution brought about by suppression (112) may cause the strain B repressor to behave quantitatively like its strain K-12 counterpart. Furthermore, a particular thermolabile mutant of $argR_B$ is inducible at 25°C but derepressed at 42°C (112), whereas the *E. coli* B wild type becomes repressible in the high temperature range (119).

The genetic aspects of this unifying account were confirmed by the extensive complementation studies of Kadner and Maas (116). In each combination of alleles (wild-type, K12-K12, or K12-B), the allele giv-ing the lowest enzyme level in the presence of arginine was found to be dominant. The *argR* locus appears to contain only one functional unit.

From the physiological and evolutionary points of view, *E. coli* B is an interesting contrast to K-12. The efficiency of feedback inhibition of the *argA* enzyme is assured by the intrinsically low level of enzyme in this strain (see *N*-Acetylglutamate Synthase, above). Yet, *E. coli* B is able to respond to arginine starvation by partial derepression.

Merodiploids have been constructed that combine totally nonfunctional *argR* mutations of *S. typhimurium* with an episome carrying the $argR^+$ allele of *E. coli* K-12 (77). In these organisms the K-12 repressor appears as effective as its *Salmonella* counterpart, implying conservation of regulatory elements in the two species. In diploids prepared with other *argR* alleles of *Salmonella*, only partial repression had been observed, suggesting the formation of partially defective hybrid repressor molecules (125).

Physiological derepression of $argR^+$ cells obtained by limiting arginine supply in various ways (chemostat, bradytrophic mutants, limiting source of arginine; see references 53, 144, and 173) leads to enzyme levels not markedly different from those found in fully derepressed *argR* mutants. This observation contrasts with the situation encountered in tryptophan biosynthesis, in which the starvation for tryptophan results, via the attenuation control mechanism, in much higher levels than in *trpR* mutants.

The strain K-12 repressor has been partially isolated either by a differential labeling technique (242) or by using in vitro protein synthesis to follow the progress of the purification (122, 243). This preparation (up to 70-fold purified) has been used in several studies on in vitro repression of arginine genes in coupled transcription-translation systems (69, 120) or purified transcription systems (63). The gene itself has been cloned and sequenced (T. Eckhardt, unpublished data); the data indicate that the *argR* protein should have an MW of approximately 17,000. In view of Udaka's estimate (242) of 45,000 MW for the holorepressor, it may be that the active *arg* repressor is a dimer.

Unpublished experiments by Udaka (personal communication) on in vitro gene transcription suggest that active arginine repressor is associated with an arginine binding protein that could be involved in arginine uptake.

Estimates of the number of molecules of repressor per cell range from 200 (reference 243) to 67 (51) or 40 (142), values in the same range as those for the *trp* repressor (123). This order of magnitude is in keeping with experiments showing a moderate "escape" synthesis and reduced repressibility of arginine biosynthetic enzymes (including carbamoylphosphate synthetase) in cells carrying multicopy vectors of *arg* genes (54, 55, 82).

Nature of the Corepressor

Hirshfield et al. (100) showed that mutants of arginyl-tRNA synthetase impaired in their charging ability and therefore accumulating arginine displayed reduced levels of ornithine carbamoyltransferase and remained repressible; no correlation was found between repression and level of charging. These findings were in keeping with the notion that arginine and not arginyl-tRNA was the corepressor. This view was strengthened by the work of Leisinger and Vogel (140) and also Celis and Maas (45), who showed that the charging profiles of the five isoacceptor tRNA^{Arg}s appeared to be the same in repressed and derepressed cells. Subsequent reports on derepressed synthetase mutants or on derepression resulting from inhibition of the synthetase by arginine precursors (see reference 35) could not be confirmed or were shown to involve artifacts (see Arginyl-tRNA Synthetase, above).

Confirmation that arginine is able to play the role of corepressor came from in vitro experiments: in a purified system for transcription of the argECBH genes (177), addition of arginine and partially purified repressor free of arginyl-tRNA synthetase provoked repression of hybridizable RNA synthesis (63). In similar experiments, Lissens et al. (142) found that free arginine and repressor repressed transcription of carAB DNA in vitro to the same extent as in vivo. G Bény (Ph.D. thesis, Vrije Universiteit Brussel, Brussels, 1984) was unable to observe any effect of tRNA^{Arg} nor of the synthetase itself on repression of argECBH transcription in vitro. The lack of evidence for involvement of tRNA^{Arg} in the mechanism of repression is paralleled by the absence of attenuation control for the arg genes investigated so far, including the car operon.

Early experiments (86) suggesting that ornithine was able to counteract the repressive effect of arginine in a chemostat operated under conditions of partial derepression could not be reproduced. In chemostats in which the dilution rate was controlled with great accuracy, no effect of ornithine or citrulline could be observed (53; unpublished experiments from this laboratory). Zidwick et al. (269) showed that L-ornithine, L-citrulline, and D-arginine had no effect on in vitro synthesis of argECBH mRNA and of two of the cognate enzymes. Besides, argG bradytrophs accumulating citrulline remain derepressed (138).

cis-Dominant Regulatory Mutations

Isolating operator mutants was an obligatory step in assessing the idea that the arg regulon comprised a set of individual operons responding to the same repressor. Several approaches were followed to isolate cis-dominant constitutive mutations.

Selecting for suppressors of polar argC or argB mutations gave rise to tandem duplications connecting argH or argB and H to other promoters or creating new ones at their novel joints (see references 17 and 47). The same approach provided a deletion (sup-102; see Fig. 2 and next section) which disclosed the existence of an internal operator region (between E and C) and of two flanking promoters facing each other, argEp and argCBHp (72). The deletion indeed appeared to destroy part of argB, the whole of argC, and the promoter for argE while making the expression of argH partly constitutive. This somewhat unexpected arrangement of controlling elements had been considered as an unlikely possibility for another divergent gene cluster, bioABCFD (90); it was nevertheless proved to apply to that case as well (175).

Less extensive alterations of the Oc (operator-constitutive) type were obtained by various means, including relief from repression of streptomycin-in-duced suppression of argI or argC nonsense mutations (110, 112). In the second case the mutants were partially constitutive for both argE and argCBH even though they mapped between argE and argC. The existence of a divergent operon was inferred from this finding, in keeping with the properties of the sup-102 deletion (72). Other argECBH operator-constitutive mutants were obtained by localized mutagenesis (103) and selection for derivatives of his auxotrophs able to achieve high enough levels of N-acetylornithinase to deacetylate N-acetylhistidine even when the arginine regulon was partly repressed (32, 36, 52). This method proved particularly rewarding since it also provided mutations increasing the translational rate of argE (see reference 33 and next section). The same method gave apparently specific argE operator-constitutive mutants; these were later shown to harbor secondary promoters relatively insensitive to repression (see references 60 and 192 and next section).

cis-Dominant mutations enhancing argE expression were also obtained by selecting for derivatives of the sup-102 deletion mutant able to utilize acetylornithine as source of arginine. Some of these strains were of the operator-constitutive type and also affected argH; others were promoter-up mutations or involved the insertion of a mobile promoter such as IS3 (48, 192).

argF and argI operator mutants or unstable rearrangements causing argF constitutivity were isolated from a car deletion mutant as derivatives able to use citrulline as a source of carbamoylphosphate for pyrimidine biosynthesis (see Ornithine Carbamoyltransferase, above; 246).

Last but not least, the construction of operon- and protein-fusion strains where the lacZ gene was expressed from a foreign promoter (39) provided a general approach to isolate regulatory mutants which has been applied to the argECBH gene cluster (20).

DNA sequence data are now available for mutants obtained by every one of these approaches (see references 60 and 246; Fig. 2 and below).

Structure of Control Regions of Arginine Genes

The steps leading to cloning and sequence determination of arg genes from the wild type and control mutants have been described in several papers (18, 159, 192–196, 246), including two recent reviews (58, 60). The results of these investigations are presented in Fig. 2, in which the control regions of genes argECBH, argF, argI, and carAB are compared. The assignment of promoter sites rests on polymerase binding tests, S1 mapping and fingerprint experiments (argECBH), S1 mapping (argF, argI, carAB), and nucleotide sequence determination of promoter mutations (argECBH). The argR promoter (not shown) has been tentatively located according to the effects of DNA insertions on the expression of lacZ in an argR-lacZ fusion strain (T. Eckhardt, unpublished data).

A feature common to all genes investigated so far is the presence of one (argR) or two (all other cases) partly conserved 18-base-pair-long sequences that display hyphenated symmetry and overlap the promoters to various extents. These "ARG boxes" (192, 194) were shown to constitute operator sites by examining the nucleotide sequence changes in operator mutations in argF (246), argECBH (60), and car (D. Charlier

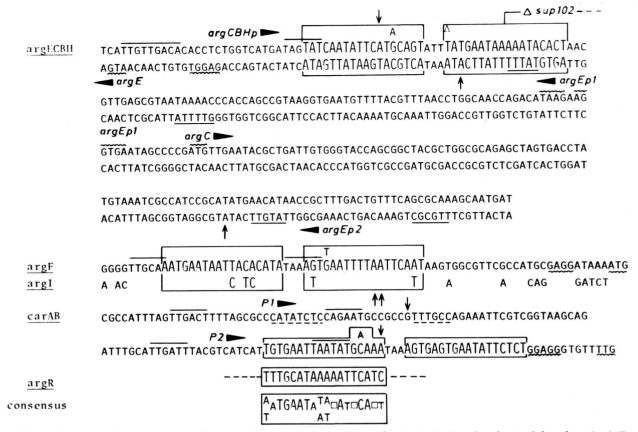

FIG. 2. Structure of control regions for genes involved in the synthesis of arginine (*arg*) and carbamoylphosphate (*car*). For the gene-enzyme relationship, see Table 1. *argR* codes for the repressor of arginine biosynthesis. In the case of *argI*, only the nucleotides differing from the *argF* sequence are mentioned. Promoters: Straight lines above the sequence (or under it for *argEp1* and 2) refer to −35 and −10 regions; promoter polarity is indicated by a horizontal arrowhead, and each transcription start point is indicated by a vertical arrow. Dashed lines in *carAB* indicate homology with the promoter of *pyrBI*, the aspartate carbamoyltransferase operon. Operators: Operator sites (*arg* boxes) are framed. Nucleotide substitutions or deletions (Δ) observed in operator mutants are indicated above the line. Translational signals: Wavy lines indicate sequence complementarities to the 3' OH extremity of 16S RNA and translational start codons.

and M. Rovers, unpublished data from this laboratory). A consensus sequence (Fig. 2) could be derived from their comparison. Recently other pairs of putative ARG boxes were found in the DNA regions 5' to *argD* and *argG* (A. Boyen and F. Van Vliet, unpublished data from this laboratory).

The organization of the nearly identical *argF* and *argI* control regions is by far the simplest. Two ARG boxes separated by three nucleotides overlap a typical promoter site. No sequence upstream of nucleotide −60 appears to be required for expression or regulation of *argF*. The sequences display no signs of attenuation features. It has been known for a long time that repression-derepression ratios for *argF* and *argI* are comparatively high (Table 2). It is therefore worth noting that the extent of dyad symmetry and of overlap with the promoter is greater for *argF* and *argI* than for any other gene of the regulon investigated so far. Mutations altering positions that are both highly conserved among ARG boxes and involved in the internal symmetry of the *argF* and *carAB* operator cause partial constitutivity. What are the respective parts played by the degree of symmetry and the extent of promoter-operator overlap in the tightness of gene control? Since in the *lac* operon control is tight and

the overlap is relatively weak, we assume symmetry to be the most important factor. However, constitutive mutations in *argECBH* (Fig. 2) can result from nucleotide substitution at a position which is neither conserved nor involved in symmetry. Therefore the actual composition of the boxes is also an important factor in repressor-operator interaction.

The nucleotide sequence of the *argECBH* control region fully confirmed the predictions advanced on a formal genetic basis (72): two promoters, *argCBHp* and *argEp1*, face each other and overlap with a pair of ARG boxes separated by three nucleotides. The *sup-102* deletion (Fig. 2) destroys *argEp1*, affects the operator, and leaves *argCBHp* intact. Again, attenuation features are lacking, even though *argCBHp* precedes a 110-nucleotide-long "leader" sequence of unknown function and without distinctive features.

argCBHp presents a paradox: the −35 region contains the consensus sequence TTGACA only 15 nucleotides away from the −10 region, yet it is not a weak promoter. It is possible that the sequence TTGTTG, 18 nucleotides from the Pribnow box, plays a role in polymerase-promoter interactions.

argCBHp and *argEp1* overlap the central operator region to similar extents, yet their respective coefficients of repression are markedly different: 9 (argEp1) and 23 (argCBHp) in terms of mRNA levels, 17 and 60, respectively, in terms of enzyme specific activities. This paradox was explained by the discovery of a weak secondary argE promoter (argEp2) located in argC (Fig. 2) 180 nucleotides upstream from the major transcription initiation site for argE. argE transcription initiated from argEp2 can be detected by S1 mapping in repressed cells (60; J. Piette, Ph.D. thesis, Vrije Universiteit Brussel, Brussels, 1983) and accounts for about half the specific activity of enzyme E under those conditions (A. Boyen, unpublished data from this laboratory). When this background is considered, there is little difference between the individual repression responses of argE and argCBH. Expression from argEp2 appears relatively insensitive to repression, as confirmed by the partially constitutive acetylornithinase synthesis observed in "argEp2-up" mutants (60; Piette, thesis). These observations are in keeping with previous data on trp-lac fusions (205; see also The carAB Operon, below) that suggest that steric hindrance between RNA polymerase and repressor is necessary to achieve efficient repression. Furthermore, experiments by Sens et al. (219) on the order of addition of S30 extracts from $argR^+$ and $argR^-$ cells to in vitro-coupled translation-transcription systems suggest that steric hindrance is also sufficient to explain repression.

Do converging promoters interfere with each other? Initiation at either of the primary promoters overlapping the operator region probably occurs in an alternate fashion. Interferences between convergent promoters appear when the distance between them increases; derepression of argCBHp decreases transcription from argEp2 even when the latter is strengthened by up mutations. Likewise, derepression of argCBHp antagonizes expression from the outward promoter of IS3 inserted 190 nucleotides in front of the argCBH transcription start point (48). It is not clear whether these interferences result from collisions between polymerases, from interactions between complementary mRNA molecules initiated at argCBHp and argEp2, or from the unavailability of argEp2 when being traversed by frequent transcriptional waves initiated at argCBHp.

The sequence analysis of several mutations allowed a fine-structure dissection of the genetic determinants intertwined in the argECBH control region (60, 192; Fig. 2). One of the most interesting results of this study concerns mutations that simultaneously depress the activity of the argCBH promoter and enhance the efficiency of argE translation (32, 33). The mutations are confined to the limits of argCBHp but also modify the sequence of the ribosome binding site for argE translation, between the Shine-Dalgarno box and the 5' end of argE mRNA. The sequence data strongly suggest that maximal translation efficiency depends directly on the composition of proximal RNA sequences not necessarily included in secondary structures. They support the notion of a proximal mRNA "consensus" sequence containing elements interacting with ribosomal components (216).

What is the significance of the presence of two ARG boxes in argECBH, argF, argI, and carAB? One appears sufficient for repressor binding, as in argR (approximate coefficient of autogenous repression, 10) and the sup-102 deletion mutation (residual amplitude of vari-

ation about 8). Two adjacent boxes thus constitute distinct binding sites. It is possible that efficient repression results from cooperative binding at these adjacent sequences. Indeed, a 1-base-pair deletion between the two argECBH boxes causes constitutive expression of both argE and argCBH, suggesting that repressor binding becomes less efficient when the two boxes are put out of register.

The lack of attenuation control in the genes investigated so far could have resulted from the dispersion of a hypothetical ancestral arg operon possibly regulated in this way at the origin. On the other hand, attenuation control may have been lost at each individual arg gene or operon in the course of evolution. Structural reasons could also explain absence of attenuation control: as discussed by Cunin et al. (60), an attenuation mechanism based on arginyl-tRNA may be basically inefficient. In this respect, the sequences of argA, D, and G will be interesting to examine. Preliminary data suggest that expression of argD and G does not involve attenuation control (A. Boyen and F. Van Vliet, unpublished data from this laboratory).

The carAB Operon

Since transcription of the carAB operon is cumulatively repressed by arginine and the pyrimidines, the cognate control region is expected to be more complex than those of other genes of the arg regulon. Indeed, transcription of carAB is initiated at two tandem promoters (29, 196) which differ in their regulatory properties (196).

P1, the upstream promoter, is specifically regulated by the pyrimidines through a mechanism still unknown. Derepression of aspartate carbamoyltransferase in strains carrying a large number of copies of carAB suggests that pyrBI and carAB may respond to a common regulatory factor (82), and indeed there is some homology (indicated in Fig. 2) between P1 and the pyrBI promoter. The two promoters differ, however, in one important respect since the P1 region presents none of the attenuation features found at pyrBI (209, 240). At present the mechanism of pyrimidine control at carABp1 appears complex: in S. typhimurium unlinked mutations (38), some of them affecting RNA polymerase (113) alter the expression of carAB and pyr genes. In addition, both pyrBIp and carABp1 are negatively controlled by ppGpp (29, 238, 240).

P2, the downstream promoter, is regulated specifically by arginine. P2 overlaps a tandem of ARG boxes separated by three nucleotides as in other genes of the arg regulon. Transcription of carAB is controlled by arginine over a 30-fold range, as can be seen by comparing argR cells, grown in the presence of uracil, with $argR^+$ cells provided with arginine and uracil (Table 2). This is not unlike repression of argE (30-fold, contribution of argEp2 excluded) and argCBH (60-fold). There is, however, evidence from S1 nuclease mapping experiments that repression by arginine at P2 is more intense when P1 is also repressed by pyrimidines (Piette, thesis). Furthermore, under conditions of pyrimidine shortage, the synthesis of carbamoylphosphate synthetase is only weakly repressed by addition of arginine (187). Two types of explanations may be considered: an unknown pyrimidine-specific factor cooperates with the arginine repressor to enhance repression at P2, or initiation at P1 facilitates initiation at P2 with an apparent increase of repression at P2 when P1 becomes inhibited. Clearly, the topography of the P1-P2 region would allow binding of RNA polymerase, regulatory factors, or both at one promoter to interfere with initiation at the other one.

Notwithstanding the open question regarding the exact role played by pyrimidines in the regulation of carAB transcription, the cumulative pattern of carAB repression can now be understood at the molecular level. Both promoters would be active under conditions of derepression. Excess pyrimidines would inactivate P1; excess arginine would repress P2. In the latter circumstance P1 would be accessible, but to express the operon the polymerase would have to displace the arginine repressor. Indeed, other examples indicate that transcription initiation at a promoter upstream from a repressor binding site is much less sensitive to repression than is that at a promoter that overlaps with an operator. Examples are the trp-lac fusions studied by Reznikoff et al. (205) and the secondary promoter argEp2 discussed in the previous section. Furthermore, it is possible that RNA polymerase binding at P1 is able to destabilize P2-bound repressor.

The physiological significance of a dual promoter structure is, in the case of carAB, particularly clear. At the level of gene expression it complements the regulatory effects exerted on the enzyme itself. It constitutes an elegant alternative to the existence of independently controlled carbamoylphosphate synthetase isoenzymes as in fungi and B. subtilis (see references 58 and 65).

The two genes of the carAB operon are separated by a short intercistronic space (174, 196), and translation of carA is initiated at the atypical UUG codon (260).

POLYAMINE BIOSYNTHESIS AND CONTROL

The biosynthesis of putrescine, spermidine, and cadaverine is well known in E. coli inasmuch as the cognate enzymes and most of the structural genes have been characterized.

The main problems that remain to be solved concern regulatory mechanisms involved in these pathways and the more general question of the physiological role of polyamines. As discussed in a recent review (227), it is clear that polyamines are required for optimal growth (see also below) and that they must act through their polybasic nature. It is, however, very difficult to evaluate which cellular functions are most directly affected by their presence or their absence. The present section deals mainly with the biosynthetic and regulatory aspects of polyamine biosynthesis. The reader is referred to the above review (see also reference 228) for a thorough discussion of the physiological effects of polyamines and to volume 94 (1983) of Methods in Enzymology for detailed information concerning polyamine biosynthetic and biodegradative enzymes, as well as mutant screening techniques (229). Table 1 gives the gene-enzyme relationships for the steps involved in polyamine biosynthesis.

Both the biodegradative and the biosynthetic arginine decarboxylases have been well characterized. The biodegradative enzyme (75) was purified from E. coli B (see references 25 and 161) and shown to consist of a decamer of MW 820,000. Active biosynthetic arginine decarboxylase is a tetramer of 296,000 Da composed of 70- and 74-kDa monomers (161, 265). It is inhibited by putrescine and spermidine, an effect of probable physiological significance. It is also inhibited by difluoromethylarginine (117) and by aminooxyacetic acid (262), a property that has been particularly useful for studies on arginine transport and accumulation. The "antizymes" (see below) that inhibit ornithine decarboxylase also inhibit arginine decarboxylase, though less efficiently.

Tabor and Tabor (233) have shown that exogenous arginine is preferentially channelled into putrescine before becoming mixed with the endogenous pool. This appears to be due to the localization of the 70-kDa arginine decarboxylase in the inner periplasmic space, the cytoplasmic 74-kDa species being a precursor of this form (37).

Agmatine ureohydrolase, which converts agmatine into putrescine and urea (166), has been purified from E. coli (S. M. Boyle, quoted in reference 227). With arginine decarboxylase this enzyme constitutes the so-called putrescine biosynthetic pathway II (162; see Fig. 1). Mutations in this pathway were isolated by screening for strains defective in the production of urea (162) or by looking for strains requiring putrescine for optimal growth in the presence of arginine, which curtails ornithine synthesis and therefore prevents formation of putrescine by ornithine decarboxylation (101, 148). The first approach gave mutants only partially blocked in speA (the arginine decarboxylase gene) or speB (the ureohydrolase gene) and not requiring putrescine for optimal growth. The second approach delivered tight speA and speB mutants with a reduced growth rate in the absence of putrescine, spermidine, or spermine.

Pathway I leads directly from ornithine to putrescine via the biosynthetic ornithine decarboxylase (165). Like the biodegradative ornithine decarboxylase (75), the biosynthetic enzyme is a dimer of MW about 160,000, absent from most E. coli strains, K-12 included (11). The two proteins may be evolutionarily related (11, 12; see also reference 161). The regulation of biodegradative ornithine decarboxylase is surprisingly complex: the enzyme is inhibited by putrescine and spermidine, activated by GTP, and inhibited by ppGpp (see reference 228), but, in addition, polyamines increase the specific activity of inhibitory proteins called antizymes (96, 132). Three different antizymes have been purified from E. coli (97), one acidic protein of MW 49,000 and two smaller basic proteins of MW 11,000 and 9,000, which account for 95% of the inhibitory activity. Moreover, "antiantizymes" have been described as well, which bind to the antizyme and thus release ornithine decarboxylase from the complex (132). specC mutants define the biosynthetic ornithine decarboxylase gene. They were isolated as derivatives of a speA strain requiring putrescine or spermidine in the absence of arginine (64). Their behavior suggests that putrescine can partially replace spermidine.

S-Adenosylmethionine decarboxylase was purified from E. coli (150, 261). The enzyme has an MW of 108,000 and comprises six identical subunits. It requires covalently bound pyruvate for activity, but no pyridoxal phosphate, and is inhibited by decarboxylated S-adenosylmethionine. The corresponding gene is *speD*. Tight *speD* mutants grow at 75% of the wild-type rate and require spermidine for optimal growth (235).

Pure spermidine synthetase (or putrescine aminopropyl transferase), a dimer of MW 72,000, was obtained from *E. coli* (30, 226). The enzyme also transfers the aminopropyl group to spermidine (to form spermine) or to cadaverine, but much less efficiently than to putrescine; spermine is not normally found in *E. coli*, though small amounts were detected in *speA speB speC* mutants growing in the presence of spermidine (92). A putative transition-state analog, S-adenosyl-1,8-diamino-3-thiooctane, inhibits the enzyme (180). Mutants affecting the synthetases are not yet known, and the cognate genetic locus has not been named.

Combinations of tight *spe* mutations have been used to create strains totally unable to synthesize putrescine and spermidine, to assess the physiological importance of these substances (92). A strain containing deletions affecting *speA*, *speB*, *speC*, and *speD* still grew at one-third the rate observed in the presence of spermidine. The cells were abnormal with respect to phage λ production, the mating ability of Hfr strains, and the adsorption of phage f2 (92). Putrescine was slightly less efficient than spermidine in restoring the growth rate. This partial requirement could be made absolute in a *speA speB speC* mutant by introducing an *rpsL* mutation (234), which affects the S12 ribosomal protein. Changes in ribosomal structure or conformation could therefore explain the stringent polyamine requirement. Indeed, polyamines may be involved in ribosomal structure and in protein synthesis as previous work has already suggested (see references 225, 227, and 228). The absolute requirement for polyamines displayed by the *speA speB speC rpsL* strain is interesting also because this strain should not be impaired in the synthesis of cadaverine, a putative substitute for the other diamines (see below). Estimates of cadaverine pools in the above polyamine mutant as well as investigations on the effect of exogenous cadaverine would therefore be instructive.

Under conditions of putrescine starvation, mutants unable to synthesize this compound produce detectable amounts of cadaverine (a product of lysine decarboxylation) and of its aminopropyl derivatives (92). Furthermore, exogenous cadaverine stimulates the growth of *speB* mutants depleted of polyamines by the addition of arginine (68). It was therefore conceivable (68; see also other references in reference 259) that cadaverine would act as a substitute for other diamines. Data from Leifer (Z. Leifer, Ph.D. thesis, New York University Medical School, New York, 1972) and Goldemberg (85) suggested that under physiological growth conditions cadaverine would be produced by a lysine decarboxylase different from the inducible one studied by Sabo et al. (212; see also reference 24). Evidence for such an enzyme was provided by Wertheimer and Leifer (259), who showed that *E. coli* grown in minimal medium at neutral pH synthesizes a biosynthetic lysine decarboxylase inhibited by putrescine and spermidine. This regulatory pattern is obviously in keeping with the substitute role postulated for cadaverine. The actual evidence for this role is still inconclusive, however. Tabor et al. (231) constructed a *speA speB speC speD* strain also lacking lysine decarboxylase activity (geno-

type *cadA*) and found this organism to be phenotypically identical to the *cadA*+ parent, growing at a rate one-third of that found in the presence of polyamines. However, as noted by Wertheimer and Leifer (259), the identical growth rates of the *cadA* and *cadA*+ versions of the multiple *spe* mutants may reflect the fact that in the *cadA*+ parent the cadaverine pool was already unusually low. It is also not clear which lysine decarboxylase was affected by the *cadA* mutation. The *cadA* mutant was obtained from a previously isolated *cadR* regulatory mutant producing high amounts of a decarboxylase which was assumed to be the inducible one at a time the second enzyme had not yet been disclosed. *cadR* may be identical to *lysP* (197), a gene proposed to code for a regulatory element affecting the synthesis of both a lysine permease and a lysine decarboxylase.

Of the so-called constitutive enzymes involved in the biosynthesis of polyamines, several were in fact shown to be subject to metabolic control. However, observations made with different strains are conflicting and their biological significance therefore remains unclear. The expression of *speA* and *speC* is partially repressed by putrescine (225, 232). According to Boyle's group (215, 264), *speA*, *speB*, and *speC* are negatively and partially controlled by cyclic AMP via the mediation of the cyclic AMP receptor protein so that glucose enhances the expression of all three genes. However, working with strains different from those studied by Boyle and co-workers, Shaibe et al. (220) found that *speB* expression was subject to catabolite repression whereas *speA* and *speC* did not respond to different carbon sources. In addition, *speB* was stimulated by nitrogen limitation which could actually override catabolite repression (220).

Although they are closely linked (146), *speA* and *speB* would appear not to constitute an operon since the insertion of bacteriophage Mu in either gene exerts no polar effect on the other one (92). In keeping with these observations, the expression of *speB* but not of *speA* is enhanced by agmatine (215).

The *speA*, *speB*, *speC*, *speD*, and *cadA* genes of *E. coli* have been cloned (34, 230, 244). The fact that *speA* and *speB* can be expressed independently from each other on separate cloning vectors also argues for their not constituting an operon (unpublished data cited in reference 215).

Little is known about the transport of polyamines across the cell membrane. Putrescine and spermidine appear to be transported into the cell against a concentration gradient (see reference 228), but it is not known how much of the intracellular pool consists of free polyamines. Multiple transport components have been reported for putrescine (168). Curiously, it appears that streptomycin enters the *E. coli* cell via an inducible polyamine transport system (102). Further investigations of polyamine transport might help clarify some apparent contradictions. Whereas Satishchandran and Boyle (215) found none of their *E. coli* K-12 strains able to utilize agmatine as a source of nitrogen, the reverse has been reported by Stalon and Mercenier (222) and Shaibe et al. (220) for other K-12 strains.

SUMMARY AND OUTLOOK

The network of reactions involved in arginine and polyamine biosynthesis is well established, at least in

E. coli. Mechanisms controlling the activity and the synthesis of the relevant enzymes have been identified. At the enzyme level, the most striking type of regulation revealed in the last few years is the antizyme-antiantizyme interaction controlling ornithine decarboxylase. At the gene level, recent research has revealed the outlines of the mechanism responsible for repression of enzyme synthesis by arginine. Interaction between a unique repressor molecule activated by arginine and a family of different but related operators, called ARG boxes, appears to account for the phenomenon including the quantitative variations observed from gene to gene in the repression response. Defining the determinants of these variations will require further studies on interactions between the ARG boxes and the repressor. The latter has to be purified and its possible relationship with elements of the arginine transport system needs to be clarified. The physiological significance of the partially uncoordinated repression response in the arginine regulon remains to be evaluated.

The metabolic control exerted on *arg* genes is largely transcriptional. It is possible that a second-site control accounts for the fact that *argE* and *argCBH* mRNAs display a narrower range of variation than do the cognate enzyme levels, although other explanations are not excluded such as an interaction between the complementary portions of overlapping divergent messengers. At any rate, an attenuation mechanism of the type observed in several amino acid biosynthetic pathways does not appear to play a role in the *arg* regulon nor in the control of carbamoylphosphate synthetase. The role of the *argCBH* leader sequence, if any, remains to be defined. It also remains to be seen whether *argA* and the arginyl-tRNA synthetase gene also lack attenuation control. The same question concerns all *arg* genes of *S. typhimurium*.

Two peculiar genetic arrangements have been disclosed in the course of these studies: the converging promoters involved in the divergent transcription pattern of the *argECBH* cluster and the tandem promoters of the *carAB* operon. The physiological significance of the latter structure is particularly clear. A full understanding of the mechanisms operating in both systems will require investigations on possible interactions between RNA polymerases and regulatory molecules bound at the converging promoters of *argECBH* or the tandem promoters of *carAB*. The pyrimidine-specific element of *carAB* control remains to be defined.

These are some of the most immediate prospects for further studies on the regulatory aspects of the arginine system. Other observations of general interest made with the same biological material also demand further investigations, such as the influence exerted by mRNA sequences 5' to the Shine-Dalgarno box on the efficiency of translation (the case of mutants affecting *argE* translation), the role of cryptic genes in metabolic evolution (the *argD-argM* problem), and the formation of chromosomal rearrangements resulting in tandem or inverted repeats of the *argE* gene. Last but not least, a better understanding of the evolutionary relationship between ornithine and aspartate carbamoyltransferases is required at the structural and functional levels.

ACKNOWLEDGMENTS

Thanks are due to R. Cunin and A. Piérard for reading and discussing the manuscript, to P. Stalon for typing it, and to J. P. Ten Have for preparing the figures.

Work pursued in my laboratory was supported over the years by Belgian research foundations (FNRS-NFWO, FRFC-FKFO, IRSIA-IWONL) and by a Concerted Action between Brussels University and the Belgian State.

LITERATURE CITED

1. **Abdelal, A., and J. L. Ingraham.** 1969. Control of carbamoyl-phosphate synthesis in *Salmonella typhimurium.* J. Biol. Chem. **244:**4033–4038.
2. **Abdelal, A., and J. L. Ingraham.** 1975. Carbamoylphosphate synthetase from *Salmonella typhimurium.* Regulation, subunit composition and function of the subunits. J. Biol. Chem. **250:**4410–4417.
3. **Abdelal, A. T., and O. V. Nainan.** 1979. Regulation of *N*-acetylglutamate synthesis in *Salmonella typhimurium.* J. Bacteriol. **137:**1040–1042.
4. **Abdelal, A. T. H., E. Griego, and J. L. Ingraham.** 1976. Arginine-sensitive phenotype of mutations in *pyrA* of *Salmonella typhimurium*: role of ornithine carbamyltransferase in the assembly of mutant carbamylphosphate synthetase. J. Bacteriol. **128:**105–113.
5. **Abdelal, A. T. H., E. Griego, and J. L. Ingraham.** 1978. Arginine auxotrophic phenotype of mutations in *pyrA* of *Salmonella typhimurium*: role of *N*-acetylornithine in the maturation of mutant carbamylphosphate synthetase. J. Bacteriol. **134:**528–536.
6. **Abdelal, A. T. H., E. H. Kennedy, and O. Nainan.** 1977. Ornithine transcarbamylase from *Salmonella typhimurium*: purification, subunit composition, kinetic analysis, and immunological cross-reactivity. J. Bacteriol. **129:**1387–1396.
7. **Anderson, P. M.** 1977. Binding of allosteric effectors to carbamoylphosphate synthetase from *Escherichia coli.* Biochemistry **16:**587–592.
8. **Anderson, P. M., and S. V. Marvin.** 1970. Effect of allosteric effectors and adenosine triphosphate on the aggregation and rate of inhibition by N-ethyl-maleimide on carbamoylphosphate synthetase of *Escherichia coli.* Biochemistry **9:**171–178.
9. **Anderson, P. M., and A. Meister.** 1966. Bicarbonate-dependent cleavage of adenosine triphosphate and other reactions catalyzed by *Escherichia coli* carbamoylphosphate synthetase. Biochemistry **5:**3157–3163.
10. **Anderson, P. M., and A. Meister.** 1966. Control of *Escherichia coli* carbamoylphosphate synthetase by purine and pyrimidine nucleotides. Biochemistry **5:**3164–3167.
11. **Appelbaum, D. M., J. C. Dunlop, and D. R. Morris.** 1977. Comparison of the biosynthetic and biodegradative ornithine decarboxylases of *Escherichia coli.* Biochemistry **16:**1580–1584.
12. **Appelbaum, D. M., D. L. Sabo, E. H. Fischer, and D. R. Morris.** 1975. Biodegradative ornithine decarboxylase of *Escherichia coli.* Purification, properties and pyridoxal 5'-phosphate binding site. Biochemistry **14:**3675–3681.
13. **Bach, M. L., and F. Lacroute.** 1972. Direct selection techniques for the isolation of pyrimidine auxotrophs in yeast. Mol. Gen. Genet. **115:**126–130.
14. **Bachmann, B.** 1983. Linkage map of *Escherichia coli* K-12, edition 7. Microbiol. Rev. **47:**180–230.
15. **Bacon, D. F., and H. J. Vogel.** 1963. A regulatory gene simultaneously involved in repression and induction. Cold Spring Harbor Symp. Quant. Biol. **28:**437–438.
16. **Baumberg, S.** 1970. Acetylhistidine as substrate for acetylornithinase: a new system for the selection of arginine regulation mutants in *Escherichia coli.* Mol. Gen. Genet. **106:**162–173.
17. **Beeftink, F., R. Cunin, and N. Glansdorff.** 1974. Arginine gene duplication in recombination proficient and deficient strains of *Escherichia coli* K12. Mol. Gen. Genet. **132:**244–253.
18. **Bencini, D. A., J. E. Houghton, T. A. Hoover, K. F. Foltermann, J. R. Wild, and G. A. O'Donovan.** 1983. The DNA sequence of *argI* from *Escherichia coli* K12. Nucleic Acids Res. **11:**8509–8518.
19. **Ben-Ishai, R., M. Lahav, and A. Zamir.** 1964. Control of uracil synthesis in *Escherichia coli.* J. Bacteriol. **87:**1436–1442.
20. **Bény, G., A. Boyen, D. Charlier, W. Lissens, A. Feller, and N. Glansdorff.** 1982. Promoter mapping and selection of operator mutants by using insertion of bacteriophage Mu in the *argECBH* divergent operon of *Escherichia coli* K-12. J. Bacteriol. **151:**62–67.
21. **Bény, G., R. Cunin, N. Glansdorff, A. Boyen, J. Charlier, and N. Kelker.** 1982. Transcription of regions within the divergent

argECBH operon of *Escherichia coli*: evidence for lack of an attenuation mechanism. J. Bacteriol. **151**:58–61.

22. **Billheimer, J. T., H. N. Carnevale, T. Leisinger, T. Eckhardt, and E. E. Jones.** 1976. Ornithine δ-transaminase activity in *Escherichia coli*: its identity with acetylornithine-δ-transaminase. J. Bacteriol. **127**:1315–1323.

23. **Billheimer, J. T., M. Y. Shen, H. N. Carnevale, H. R. Horton, and E. E. Jones.** 1979. Isolation and characterization of acetylornithine δ-transaminase of wild-type *Escherichia coli* W. Comparison with arginine-induced acetylornithine δ-transaminase. Arch. Biochem. Biophys. **105**:401–413.

24. **Boeker, E. A., and E. H. Fischer.** 1983. Lysine decarboxylase (*Escherichia coli* B). Methods Enzymol. **94**:180–184.

25. **Boeker, E. A., and E. E. Snell.** 1972. Aminoacid decarboxylase, p. 217–253. *In* P. D. Boyer (ed.), The enzymes, 3rd ed., vol. 6. Academic Press, Inc., New York.

26. **Boettcher, B. R., and A. Meister.** 1980. Covalent modification of the active site of carbamoylphosphate synthetase by 5′-fluorosulfonylbenzoyladenosine. Direct evidence for two functionally different ATP-binding sites. J. Biol. Chem. **255**:7129–7133.

27. **Bossi, L., T. Kohno, and J. R. Roth.** 1983. Genetic characterization of the *SufJ* frameshift suppressor in *Salmonella typhimurium*. Genetics **103**:31–42.

28. **Bottu, G.** 1983. The role of cations and of the tRNA 3′-terminal end in the activity of *Escherichia coli* arginyl-tRNA synthetase. Arch. Intern. Physiol. Biochim. **91**:B6.

29. **Bouvier, J., J. C. Patte, and P. Stragier.** 1984. Multiple regulatory signals in the control region of the *Escherichia coli carAB* operon. Proc. Natl. Acad. Sci. USA **81**:4139–4143.

30. **Bowan, W. H., C. W. Tabor, and H. Tabor.** 1973. Spermidine biosynthesis. Purification and properties of polyamine transferase from *Escherichia coli*. J. Biol. Chem. **248**:2480–2486.

31. **Boy, E., J. Thèze, and J. C. Patte.** 1973. Transient regulation of enzyme synthesis in *Escherichia coli*. Mol. Gen. Genet. **121**:77–78.

32. **Boyen, A., D. Charlier, M. Crabeel, R. Cunin, S. Palchaudhuri, and N. Glansdorff.** 1978. Studies on the control region of the bipolar *argECBH* operon of *Escherichia coli*. I. Effects of regulatory mutations and IS2 insertions. Mol. Gen. Genet. **161**:185–196.

33. **Boyen, A., J. Piette, R. Cunin, and N. Glansdorff.** 1982. Enhancement of translation efficiency in *Escherichia coli* by mutations in a proximal domain of messenger RNA. J. Mol. Biol. **162**:715–720.

34. **Boyle, S. M., G. D. Markham, E. W. Hafner, J. M. Wright, H. Tabor, and C. W. Tabor.** 1984. Expression of the cloned genes encoding the putrescine biosynthetic enzymes and methionine adenosyltransferase of *Escherichia coli* (*speA*, *speB*, *speC*, and *metK*). Gene **30**:129–136.

35. **Brenchley, J. E., and L. S. Williams.** 1975. Transfer RNA involvement in the regulation of enzyme synthesis. Annu. Rev. Microbiol. **29**:251–274.

36. **Bretscher, A. P., and S. Baumberg.** 1976. Divergent transcription of the *argECBH* cluster of *Escherichia coli* K12. Mutations which alter the control of enzyme synthesis. J. Mol. Biol. **102**:205–220.

37. **Buch, J. K., and S. M. Boyle.** 1985. Biosynthetic arginine decarboxylase in *Escherichia coli* is synthesized as a precursor and located in the cell envelope. J. Bacteriol. **163**:522–527.

38. **Bussey, L. B., and J. L. Ingraham.** 1982. A regulatory gene (*use*) affecting the expression of *pyrA* and certain other pyrimidine genes. J. Bacteriol. **151**:144–152.

39. **Casadaban, M., and S. N. Cohen.** 1979. Lactose genes fused to exogenous promoters in one step using Mu-lac bacteriophage as in vivo probe for transcriptional control sequences. Proc. Natl. Acad. Sci. USA **76**:4520–4533.

40. **Celis, R.** 1977. Properties of an *Escherichia coli* K-12 mutant defective in the transport of arginine and ornithine. J. Bacteriol. **130**:1234–1243.

41. **Celis, R.** 1977. Independent regulation of transport and biosynthesis of arginine in *Escherichia coli* K-12. J. Bacteriol. **130**:1244–1252.

42. **Celis, R.** 1981. Chain-terminating mutants affecting a periplasmic binding protein involved in the active transport of arginine and ornithine in *Escherichia coli*. J. Biol. Chem. **256**:773–779.

43. **Celis, R.** 1982. Mapping of two loci affecting the synthesis and structure of a periplasmic protein involved in arginine and ornithine transport in *Escherichia coli* K-12. J. Bacteriol. **151**:1314–1319.

44. **Celis, T. F. R.** 1984. Phosphorylation in vivo and in vitro of the arginine-ornithine periplasmic transport protein of *Escherichia coli*. Eur. J. Biochem. **145**:403–411.

45. **Celis, T. F. R., and W. K. Maas.** 1971. Studies on the mechanism of repression of arginine biosynthesis in *Escherichia coli*. IV. Further studies on the role of arginine transfer RNA in repression of the enzymes of arginine biosynthesis. J. Mol. Biol. **62**:179–188.

46. **Celis, T. F. R., H. J. Rosenfeld, and W. K. Maas.** 1973. Mutant of *Escherichia coli* K-12 defective in the transport of basic amino acids. J. Bacteriol. **116**:619–626.

47. **Charlier, D., M. Crabeel, R. Cunin, and N. Glansdorff.** 1979. Tandem and inverted repeats of arginine genes in *Escherichia coli* K12. Mol. Gen. Genet. **174**:75–88.

48. **Charlier, D., J. Piette, and N. Glansdorff.** 1982. IS3 can function as a mobile promoter in *E. coli*. Nucleic Acids Res. **10**:5935–5948.

49. **Charlier, J., and E. Gerlo.** 1976. Arginyl-tRNA synthetase from *Escherichia coli*. Influence of arginine biosynthetic precursors on the charging of arginine-acceptor tRNA with [¹⁴C]arginine. Eur. J. Biochem. **70**:137–145.

50. **Charlier, J., and E. Gerlo.** 1979. Arginyl-tRNA synthetase from *Escherichia coli* K12. Purification, properties and sequence of substrate addition. Biochemistry **18**:3171–3178.

51. **Cleary, M. L., R. T. Garvin, and E. James.** 1977. Synthesis of the *Escherichia coli* K12 isoenzymes or ornithine transcarbamylase performed in vitro. Mol. Gen. Genet. **157**:155–165.

52. **Crabeel, M., D. Charlier, A. Boyen, R. Cunin, and N. Glansdorff.** 1974. Mutant selection in the control region of the *argECBH* bipolar operon of *Escherichia coli*. Arch. Intern. Physiol. Biochim. **82**:973–974.

53. **Crabeel, M., D. Charlier, R. Cunin, A. Boyen, N. Glansdorff, and A. Piérard.** 1975. Accumulation of arginine precursors in *Escherichia coli*: effects on growth, enzyme repression, and application to the forward selection of arginine auxotrophs. J. Bacteriol. **123**:898–904.

54. **Crabeel, M., D. Charlier, R. Cunin, and N. Glansdorff.** 1979. Cloning and endonuclease restriction analysis of *argE* and of the control region of the *argECBH* bipolar operon in *Escherichia coli*. Gene **5**:207–231.

55. **Crabeel, M., D. Charlier, N. Glansdorff, S. Palchaudhuri, and W. K. Maas.** 1977. Studies on the bipolar *argECBH* operon of *E. coli*. Characterization of restriction endonuclease fragments obtained from λ *dargECBH* transducing phages and a ColE1 *argECBH* plasmid. Mol. Gen. Genet. **151**:161–168.

56. **Crabeel, M., D. Charlier, G. Weyens, A. Feller, A. Piérard, and N. Glansdorff.** 1980. Use of gene cloning to determine polarity of an operon: genes *carAB* of *Escherichia coli*. J. Bacteriol. **143**:921–925.

57. **Craine, J., and A. Peterkovsky.** 1975. Evidence that arginyladenylate is not an intermediate in the arginyl-tRNA synthetase reaction. Arch. Biochem. Biophys. **168**:343–350.

58. **Cunin, R.** 1983. Regulation of arginine biosynthesis in prokaryotes, p. 53–79. *In* K. Hermann and R. Somerville (ed.), Biotechnology series 3: Amino acid biosynthesis and genetic regulation. Addison-Wesley, New York.

59. **Cunin, R., A. Boyen, P. Pouwels, N. Glansdorff, and M. Crabeel.** 1975. Parameters of gene expression in the bipolar *argECBH* operon of *E. coli* K12. The question of translational control. Mol. Gen. Genet. **140**:51–60.

60. **Cunin, R., T. Eckhardt, J. Piette, A. Boyen, A. Piérard, and N. Glansdorff.** 1983. Molecular basis for modulated regulation of gene expression in the regulon of *Escherichia coli* K12. Nucleic Acids Res. **11**:5007–5019.

61. **Cunin, R., D. Elseviers, G. Sand, G. Freundlich, and N. Glansdorff.** 1969. On the functional organization of the *argECBH* cluster of genes in *Escherichia coli* K12. Mol. Gen. Genet. **106**:32–47.

62. **Cunin, R., and N. Glansdorff.** 1971. Messenger RNA from arginine and phosphoenolpyruvate carboxylase genes in *argR⁺* and *argR⁻* strains of *Escherichia coli* K12. FEBS Lett. **18**:135–137.

63. **Cunin, R., N. Kelker, A. Boyen, H. Lang-Yang, G. Zubay, N. Glansdorff, and W. K. Maas.** 1976. Involvement of arginine in in vitro repression of transcription of arginine genes *C*, *B* and *H* in *Escherichia coli* K12. Biochem. Biophys. Res. Commun. **69**:377–382.

64. **Cunningham-Rundles, S., and W. K. Maas.** 1975. Isolation, characterization, and mapping of *Escherichia coli* mutants blocked in the synthesis of ornithine decarboxylase. J. Bacteriol. **124**:791–799.

65. **Davis, R.** 1983. Arginine synthesis in eukaryotes, p. 81–102. *In* K. Hermann and R. Somerville (ed.), Biotechnology series 3: Amino acid biosynthesis and genetic regulation. Addison-Wesley, New York.

66. **De Deken, R. H.** 1963. Biosynthèse de l'arginine chez la levure.

I. Le sort de la N-α-acétylornithine. Biochim. Biophys. Acta **78**:606–616.

67. **Degryze, E.** 1974. Evidence that yeast acetylornithinase is a carboxypeptidase. FEBS Lett. **43**:285–288.

68. **Dion, A. S., and S. S. Cohen.** 1972. Polyamine stimulation of nucleic acid synthesis in an uninfected and phage-infected polyamine auxotroph of *Escherichia coli* K12. Proc. Natl. Acad. Sci. USA **69**:213–217.

69. **Dohi, M., A. Kikuchi, and L. Gorini.** 1978. Some regulation profiles of ornithine transcarbamylase synthesis in vitro. J. Biochem. **84**:1401–1409.

70. **Duchange, N., M. M. Zakin, P. Ferrara, I. Saint-Girons, I. Park, S. V. Tran, M. C. Py, and G. N. Cohen.** 1983. Sequence of the *metB* structural gene and of the 5′ and 3′ flanking regions of the *metBL* operon. J. Biol. Chem. **258**:14868–14871.

71. **Eckhardt, T., and T. Leisinger.** 1975. Isolation and characterization of mutants with a feedback resistant N-acetylglutamate synthase in *Escherichia coli* K12. Mol. Gen. Genet. **138**:225–232.

72. **Elseviers, D., R. Cunin, N. Glansdorff, S. Baumberg, and E. Ashcroft.** 1972. Control regions within the *argECBH* gene cluster of *Escherichia coli* K12. Mol. Gen. Genet. **117**:349–366.

73. **Ennis, H. L., and L. Gorini.** 1961. Control of arginine biosynthesis in strains of *Escherichia coli* not repressible by arginine. J. Mol. Biol. **3**:439–446.

74. **Faane, R., and P. Rogers.** 1972. Repression of enzymes of arginine biosynthesis by L-canavanine in arginyl-transfer ribonucleic acid synthetase mutants of *Escherichia coli*. J. Bacteriol. **112**:102–113.

75. **Gale, E. F.** 1946. The bacterial amino acid decarboxylases. Adv. Enzymol. **6**:1–32.

76. **Gallant, J. A.** 1979. Stringent control in *E. coli*. Annu. Rev. Genet. **13**:393–415.

77. **Gardner, M. M., D. D. Hennig, and R. A. Kelln.** 1983. Control of *arg* gene expression in *Salmonella typhimurium* by the arginine repressor from *Escherichia coli* K12. Mol. Gen. Genet. **189**:458–462.

78. **Gerlo, E., W. Freist, and J. Charlier.** 1982. Arginyl-tRNA synthetase from *Escherichia coli* K12: specificity with regard to ATP analogs and their magnesium complexes. Hoppe-Seyler's Z. Physiol. Chem. **363**:S365–373.

79. **Gigot, D., M. Crabeel, A. Feller, D. Charlier, W. Lissens, N. Glansdorff, and A. Piérard.** 1980. Patterns of polarity in the *Escherichia coli carAB* gene cluster. J. Bacteriol. **127**:302–308.

80. **Gigot, D., N. Glansdorff, C. Legrain, A. Piérard, V. Stalon, W. Konigsberg, I. Caplier, A. D. Strosberg, and G. Hervé.** 1977. Comparison of the N-terminal sequences of aspartate and ornithine carbamoyltransferases of *Escherichia coli*. FEBS Lett. **81**:28–32.

81. **Glansdorff, N., S. Bourgeois, and J. M. Wiame.** 1962. Interaction de régulation enzymatique entre les biosynthèses de l'arginine et des pyrimidines chez *Escherichia coli*. Arch. Intern. Physiol. Biochim. **70**:149–151.

82. **Glansdorff, N., C. Dambly, S. Palchaudhuri, M. Crabeel, A. Piérard, and P. Halleux.** 1976. Isolation and heteroduplex mapping of a λ transducing phage carrying the structural genes for *Escherichia coli* K-12 carbamoylphosphate synthetase: regulation of enzyme synthesis in lysogens. J. Bacteriol. **127**:302–308.

83. **Glansdorff, N., and G. Sand.** 1965. Coordination of enzyme synthesis in the arginine pathway of *Escherichia coli* K12. Biochim. Biophys. Acta **108**:308–311.

84. **Glansdorff, N., G. Sand, and C. Verhoef.** 1967. The dual genetic control of ornithine transcarbamylase synthesis in *Escherichia coli* K-12. Mutat. Res. **4**:743–751.

85. **Goldemberg, S. H.** 1980. Lysine decarboxylase mutants of *Escherichia coli*: evidence for two enzyme forms. J. Bacteriol. **141**:1428–1431.

86. **Gorini, L.** 1960. Antagonism between substrate and repressor controlling the formation of a biosynthetic enzyme. Proc. Natl. Acad. Sci. USA **46**:682–690.

87. **Gorini, L., W. Gundersen, and M. Burger.** 1961. Genetics of regulation of enzyme synthesis in the arginine biosynthetic pathway of *Escherichia coli*. Cold Spring Harbor Symp. Quant. Biol. **26**:173–182.

88. **Gorini, L., and S. M. Kalman.** 1963. Control by uracil of carbamoylphosphate synthesis in *Escherichia coli*. Biochim. Biophys. Acta **69**:335–360.

89. **Gorini, L., and W. K. Maas.** 1957. The potential for the formation of a biosynthetic enzyme in *Escherichia coli*. Biochim. Biophys. Acta **25**:208–209.

90. **Guha, A., Y. Saturen, and W. Szybalski.** 1971. Divergent orientation of transcription from the biotin locus of *Escherichia coli*. J. Mol. Biol. **56**:53–62.

91. **Haas, D., V. Kurer, and T. Leisinger.** 1972. N-acetylglutamate synthetase of *Pseudomonas aeruginosa*. An assay in vitro and feedback inhibition by arginine. Eur. J. Biochem. **31**:290–295.

92. **Hafner, E. W., C. W. Tabor, and H. Tabor.** 1979. Mutants of *Escherichia coli* that do not contain 1,4-diaminobutane (putrescine) or spermidine. J. Biol. Chem. **254**:12419–12426.

93. **Hall, B. G., and J. A. Gallant.** 1973. On the rate of messenger decay during amino acid starvation. J. Mol. Biol. **73**:121–124.

94. **Hallshall, D. M.** 1975. Overproduction of lysine by mutant strains of *Escherichia coli* with defective transport systems. Biochem. Genet. **13**:109–124.

95. **Hartman, S. C.** 1973. Relationship between glutamine amidotransferases and glutaminases, p. 319–330. *In* S. Prusiner and E. R. Stadtman (ed.), The enzymes of glutamine metabolism. Academic Press, Inc., New York.

96. **Heller, J. S., W. F. Fong, and E. S. Canellakis.** 1976. Induction of a protein inhibitor to ornithine decarboxylase by the end products of its reaction. Proc. Natl. Acad. Sci. USA **73**:1858–1862.

97. **Heller, J. S., R. Rostomily, D. A. Kyriadikis, and E. S. Canellakis.** 1983. Regulation of polyamine biosynthesis in *Escherichia coli* by basic proteins. Proc. Natl. Acad. Sci. USA **80**:5181–5184.

98. **Hilger, F., J. P. Simon, and V. Stalon.** 1979. Yeast argininosuccinate synthetase. Purification, structural and kinetic properties. Eur. J. Biochem. **94**:153–163.

99. **Hirschfield, I. N., and H. P. J. Bloemers.** 1969. The biochemical characterization of two mutant arginyl transfer ribonucleic acid synthetases from *Escherichia coli* K12. J. Biol. Chem. **244**:2911–2916.

100. **Hirschfield, I. N., R. H. De Deken, P. C. Horn, D. A. Hopwood, and W. K. Maas.** 1968. Studies on the mechanism of repression of arginine biosynthesis in *Escherichia coli*. III. Repression of enzymes of arginine biosynthesis in arginyl-tRNA synthetase mutants. J. Mol. Biol. **35**:83–93.

101. **Hirschfield, I. N., H. J. Rosenfeld, Z. Leifer, and W. K. Maas.** 1970. Isolation and characterization of a mutant of *Escherichia coli* blocked in the synthesis of putrescine. J. Bacteriol. **101**:725–730.

102. **Höltje, J. V.** 1978. Streptomycin uptake via an inducible polyamine transport system in *Escherichia coli*. Eur. J. Biochem. **86**:345–351.

103. **Hong, J., and B. N. Ames.** 1979. Localized mutagenesis of any specific small region of the bacterial chromosome. Proc. Natl. Acad. Sci. USA **68**:3158–3162.

104. **Houghton, J. E., D. A. Bencini, G. A. O'Donovan, and J. R. Wild.** 1984. Protein differentiation: a comparison of aspartate transcarbamoylase and ornithine transcarbamoylase from *Escherichia coli* K12. Proc. Natl. Acad. Sci. USA **81**:4864–4868.

105. **Hu, M., and R. C. Deonier.** 1981. Mapping of IS1 elements flanking the *argF* region on the *Escherichia coli* K-12 chromosome. Mol. Gen. Genet. **181**:222–229.

106. **Itikawa, H., S. Baumberg, and H. J. Vogel.** 1968. Enzymic basis for a genetic suppression: accumulation and deacylation of N-acetylglutamate-γ-semialdehyde in enterobacterial mutants. Biochim. Biophys. Acta **159**:547–550.

107. **Jacob, F., and J. Monod.** 1961. Genetic regulatory mechanism in the synthesis of proteins. J. Mol. Biol. **3**:318–356.

108. **Jacobs, P., J. C. Jauniaux, and M. Grenson.** 1980. A cis-dominant regulatory mutation linked to the *argB-argC* gene cluster in *Saccharomyces cerevisiae*. J. Mol. Biol. **139**:691–704.

109. **Jacoby, G. A.** 1971. Mapping the gene determining ornithine transcarbamylase and its operator in *Escherichia coli* B. J. Bacteriol. **108**:645–651.

110. **Jacoby, G. A.** 1972. Control of the *argECBH* cluster in *Escherichia coli*. Mol. Gen. Genet. **177**:337–348.

111. **Jacoby, G. A., and L. Gorini.** 1967. Genetics of control of the arginine pathway in *Escherichia coli*. J. Mol. Biol. **24**:41–50.

112. **Jacoby, G. A., and L. Gorini.** 1969. A unitary account of the repression mechanism of arginine biosynthesis in *Escherichia coli*. I. The genetic evidence. J. Mol. Biol. **39**:73–87.

113. **Jensen, K. F., J. Neuhardt, and L. Schack.** 1982. RNA polymerase involvement in the regulation of expression of *Salmonella typhimurium pyr* genes. Isolation and characterization of a fluorouracil-resistant mutant with high, constitutive expression of the *pyrB* and *pyrE* genes due to a mutation in *rpoBC*. EMBO J. **1**:69–74.

114. **Jessop, A. P., and C. Clugston.** 1985. Amplification of the *argF* region in strain Hfr P4X of *E. coli* K12. Mol. Gen. Genet. **201**:347–350.

115. **Jessop, A. P., and N. Glansdorff.** 1980. Genetic factors affecting recovery of nonpoint mutations in the region of a gene coding for ornithine transcarbamylase: involvement of both the F factor in its chromosomal state and the *recA* gene. Genetics **96**:779–799.

116. **Kadner, R., and W. K. Maas.** 1971. Regulatory gene mutations affecting arginine biosynthesis in *Escherichia coli.* Mol. Gen. Genet. **111:**1–14.

117. **Kallio, A., P. P. McCann, and P. Berg.** 1981. DL-α-(difluoromethyl)-arginine: a potent enzyme-activated irreversible inhibitor of bacterial arginine decarboxylase. Biochemistry **20:**3163–3166.

118. **Kalman, S. M., P. H. Duffield, and T. Brzozowski.** 1965. Identity in *Escherichia coli* of carbamylphosphokinase and an activity which catalyzes amino group transfer from glutamine to ornithine in citrulline synthesis. Biochem. Biophys. Res. Commun. **18:**530–537.

119. **Karlström, O., and L. Gorini.** 1969. A unitary account of the repression mechanisms of arginine biosynthesis in *Escherichia coli.* II. Application to the physiological evidence. J. Mol. Biol. **3:**89–94.

120. **Kelker, N., and T. Eckhardt.** 1977. Regulation of *argA* operon expression in *Escherichia coli* K12: cell-free synthesis of beta-galactosidase under *argA* control. J. Bacteriol. **132:**67–72.

121. **Kelker, N., and W. K. Maas.** 1974. Selection of genetically repressible (*argR*⁺) strains of *E. coli* K-12 from genetically derepressed (*argR*⁻) mutants using acetylnorvaline. Mol. Gen. Genet. **132:**131–135.

122. **Kelker, N., W. K. Maas, H. L. Yang, and G. Zubay.** 1976. In vitro synthesis and repression of argininosuccinase in *Escherichia coli* K-12: partial purification of the arginine repressor. Mol. Gen. Genet. **144:**17–20.

123. **Kelley, R. L., and C. Yanofsky.** 1982. *trp* aporepressor production is controlled by autogenous regulation and inefficient translation. Proc. Natl. Acad. Sci. USA **79:**3120–3124.

124. **Kelln, R. A., and G. A. O'Donovan.** 1976. Isolation and partial characterization of an *argR* mutant of *Salmonella typhimurium.* J. Bacteriol. **128:**528–535.

125. **Kelln, R. A., and V. L. Zak.** 1978. Arginine regulon control in *Salmonella typhimurium-Escherichia coli* hybrid merodiploid. Mol. Gen. Genet. **161:**333–335.

126. **Kikuchi, A., D. Elseviers, and L. Gorini.** 1975. The isolation and characterization of λ transducing phages for *argF, argI,* and adjacent genes. J. Bacteriol. **122:**727–742.

127. **Kikuchi, A., and L. Gorini.** 1975. Similarity of genes *argF* and *argI.* Nature (London) **256:**621–623.

128. **Knight, D. M., and E. E. Jones.** 1977. Regulation of *Escherichia coli* ornithine transcarbamylase by orotate. J. Biol. Chem. **252:**5928–5930.

129. **Krzyzek, R. A., and P. Rogers.** 1976. Effect of arginine on the stability and size of *argECBH* messenger ribonucleic acid in *Escherichia coli.* J. Bacteriol. **126:**365–376.

130. **Kuo, T., and B. A. D. Stocker.** 1969. Suppression of proline requirement of *proA* and *proAB* deletion mutants in *Salmonella typhimurium* by mutation to arginine requirement. J. Bacteriol. **98:**593–598.

131. **Kustu, S. G., and G. F. Ames.** 1973. The *hisP* protein, a known histidine transport component in *Salmonella typhimurium,* is also an arginine transport component. J. Bacteriol. **116:**107–113.

132. **Kyriakidis, D. A., J. S. Heller, and E. S. Canellakis.** 1978. Modulation of ornithine decarboxylase activity in *Escherichia coli* by positive and negative effectors. Proc. Natl. Acad. Sci. USA **75:**4699–4703.

133. **Lavallé, R.** 1970. Regulation at the level of translation in the arginine pathway of *Escherichia coli* K-12. J. Mol. Biol. **51:**449–451.

134. **Lavallé, R., and G. De Hauwer.** 1970. Tryptophan messenger translation in *Escherichia coli.* J. Mol. Biol. **51:**435–437.

135. **Legrain, C., P. Halleux, V. Stalon, and N. Glansdorff.** 1972. The dual genetic control of ornithine carbamoyltransferase in *Escherichia coli:* a case of bacterial hybrid enzymes. Eur. J. Biochem. **27:**93–102.

136. **Legrain, C., and V. Stalon.** 1976. Ornithine carbamoyltransferase from *Escherichia coli* W: purification, structure and steady-state kinetic analysis. Eur. J. Biochem. **63:**289–301.

137. **Legrain, C., V. Stalon, and N. Glansdorff.** 1976. *Escherichia coli* ornithine carbamoyltransferase isoenzymes: evolutionary significance and the isolation of λ*argF* and λ*argI* transducing phages. J. Bacteriol. **128:**35–38.

138. **Legrain, C., V. Stalon, N. Glansdorff, D. Gigot, A. Piérard, and M. Crabeel.** 1976. Structural and regulatory mutations allowing utilization of citrulline or carbamoylaspartate as a source of carbamoylphosphate in *Escherichia coli* K-12. J. Bacteriol. **128:**39–48.

139. **Leisinger, T., and D. Haas.** 1975. N-acetylglutamate synthase of *Escherichia coli:* regulation of synthesis and activity by arginine. J. Biol. Chem. **250:**1690–1693.

140. **Leisinger, T., and H. J. Vogel.** 1969. Repression by arginine in *Escherichia coli:* a comparison of arginyl transfer RNA profiles. Biochim. Biophys. Acta **182:**572–578.

141. **Linn, T., M. Goman, and J. Scaife.** 1979. Lambda transducing bacteriophage carrying deletions of the *argECBH-rpoBC* region of the *Escherichia coli* chromosome. J. Bacteriol. **140:**479–489.

142. **Lissens, W., R. Cunin, N. Kelker, N. Glansdorff, and A. Piérard.** 1980. In vitro synthesis of *Escherichia coli* carbamoylphosphate synthase: evidence for participation of the arginine repressor in cumulative repression. J. Bacteriol. **141:**58–66.

143. **Loftfield, R. B., and E. A. Eigner.** 1969. Mechanism of action of amino acid transfer ribonucleic acid ligases. J. Biol. Chem. **244:**1746–1754.

144. **Maas, W. K.** 1961. Studies on repression of arginine biosynthesis in *Escherichia coli.* Cold Spring Harbor Symp. Quant. Biol. **26:**183–191.

145. **Maas, W. K.** 1965. Genetic defects affecting an arginine permease and repression of arginine synthesis in *Escherichia coli.* Fed. Proc. **24:**1239–1242.

146. **Maas, W. K.** 1972. Mapping of genes involved in the synthesis of spermidine in *Escherichia coli.* Mol. Gen. Genet. **119:**1–9.

147. **Maas, W. K., and A. J. Clark.** 1964. Studies on the mechanism of repression of arginine biosynthesis in *Escherichia coli.* II. Dominance of repressibility in diploids. J. Mol. Biol. **8:**365–370.

148. **Maas, W. K., Z. Leifer, and J. Pointdexter.** 1970. Studies with mutants blocked in the synthesis of polyamines. Ann. N.Y. Acad. Sci. **171:**957–967.

149. **Maas, W. K., R. Maas, J. M. Wiame, and N. Glansdorff.** 1964. Studies on the mechanism of repression of arginine biosynthesis in *Escherichia coli.* I. Dominance of repressibility in zygotes. J. Mol. Biol. **8:**359–364.

150. **Markham, G. D., C. W. Tabor, and H. Tabor.** 1983. S-adenosylmethionine decarboxylase (*Escherichia coli*). Methods Enzymol. **94:**228–230.

151. **Marvil, D. K., and T. Leisinger.** 1977. N-acetylglutamate synthase of *Escherichia coli.* Purification, characterization, and molecular properties. J. Biol. Chem. **252:**3295–3303.

152. **Mazaïtis, A., S. Palchaudhuri, N. Glansdorff, and W. K. Maas.** 1976. Isolation and characterization of *dargECBH* transducing phages and heteroduplex analysis of the *argECBH* cluster. Mol. Gen. Genet. **143:**185–196.

153. **McLellan, W., and H. J. Vogel.** 1970. Translation repression in the arginine system of *Escherichia coli.* Proc. Natl. Acad. Sci. USA **67:**1703–1707.

154. **McLellan, W., and H. J. Vogel.** 1973. Stability of *argECBH* messenger RNA under arginine excess or restriction. Biochem. Biophys. Res. Commun. **55:**1385–1389.

155. **Mehler, A. H., and S. K. Mitra.** 1967. The activation of arginyl transfer ribonucleic acid synthetase by transfer ribonucleic acid. J. Biol. Chem. **242:**5495–5499.

156. **Mergeay, M., D. Gigot, J. Beckmann, N. Glansdorff, and A. Piérard.** 1974. Physiology and genetics of carbamoylphosphate synthesis in *Escherichia coli* K-12. Mol. Gen. Genet. **133:**299–316.

157. **Messenguy, F.** 1976. Regulation of arginine biosynthesis in *Saccharomyces cerevisiae:* isolation of a *cis*-dominant constitutive mutant for ornithine carbamoyltransferase synthesis. J. Bacteriol. **128:**49–55.

158. **Minet, M., J. C. Jauniaux, P. Thuriaux, M. Grenson, and J. M. Wiame.** 1979. Organization and expression of a two-gene cluster in arginine biosynthesis of *Saccharomyces cerevisiae.* Mol. Gen. Genet. **168:**299–308.

159. **Moore, S., R. Garvin, and E. James.** 1981. Nucleotide sequence of the *argF* regulatory region of *Escherichia coli* K12. Gene **16:**119–132.

160. **Morgan, S. D., and D. Söll.** 1978. Regulation of the biosynthesis of amino acids: tRNA ligases and tRNA. Prog. Nucleic Res. Mol. Biol. **21:**181–207.

161. **Morris, D. R., and E. A. Boeker.** 1983. Biosynthetic and biodegradative ornithine and arginine decarboxylases from *Escherichia coli.* Methods Enzymol. **94:**125–134.

162. **Morris, D. R., and C. M. Jorstad.** 1970. Isolation of conditionally putrescine-deficient mutants of *Escherichia coli.* J. Bacteriol. **101:**731–737.

163. **Morris, D. R., and K. L. Koffron.** 1967. Urea production and putrescine biosynthesis in *Escherichia coli.* J. Bacteriol. **94:**1516–1519.

164. **Morris, D. R., and K. L. Koffron.** 1969. Putrescine biosynthesis in *Escherichia coli.* Regulation through pathway selection. J. Biol. Chem. **244:**6094–6099.

165. **Morris, D. R., and A. B. Pardee.** 1965. A biosynthetic ornithine decarboxylase in *Escherichia coli.* Biochem. Biophys. Res. Commun. **20:**697–702.

166. **Morris, D. R., and A. B. Pardee.** 1966. Multiple pathways of putrescine biosynthesis in *Escherichia coli.* J. Biol. Chem. **241:**3129–3135.

167. **Motta, R.** 1967. Méthode de sélection de mutants uracile exigeants au locus ur1 de *Coprinus radiatus.* C. R. Acad. Sci. (Paris) **204:**654–657.

168. **Munro, G. F., C. A. Bell, and M. Lederman.** 1974. Multiple transport components for putrescine in *Escherichia coli.* J. Bacteriol. **118:**952–963.

169. **Nakamura, Y., and H. Uchida.** 1983. Isolation of conditionally lethal amber mutations affecting synthesis of the *nusA* protein of *Escherichia coli.* Mol. Gen. Genet. **190:**196–203.

170. **Natter, W., D. Sens, and E. James.** 1977. Metabolism of arginine-specific messenger ribonucleic acid in *Escherichia coli* K-12. J. Bacteriol. **131:**214–223.

171. **Neidhardt, F. C., P. L. Bloch, S. Pedersen, and S. Reeh.** 1977. Chemical measurement of steady-state levels of ten aminoacyl-transfer ribonucleic acid synthetases in *Escherichia coli.* J. Bacteriol. **129:**378–387.

172. **Neidhardt, F. C., J. Parker, and W. G. McKeever.** 1975. Function and regulation of aminoacyl-tRNA synthetases in prokaryotic and eukaryotic cells. Annu. Rev. Microbiol. **29:**215–250.

173. **Novick, R. P., and W. K. Maas.** 1961. Control by endogenously synthesized arginine of the formation of ornithine transcarbamylase in *Escherichia coli.* J. Bacteriol. **81:**236–240.

174. **Nyunoya, H., and C. J. Lusty.** 1983. The *carB* gene of *Escherichia coli*: a duplicated gene coding for the large subunit of carbamoylphosphate synthetase. Proc. Natl. Acad. Sci. USA **80:**4629–4633.

175. **Otsuka, A., and J. Abelson.** 1978. The regulatory region of the biotin operon in *Escherichia coli.* Nature (London) **276:**689–694.

176. **Panchal, C. J., S. N. Bagchee, and A. Guha.** 1974. Divergent orientation of transcription from the *argECBH* operon of *Escherichia coli.* J. Bacteriol. **117:**675–680.

177. **Pannekoek, H., R. Cunin, A. Boyen, and N. Glansdorff.** 1975. In vitro transcription of the bipolar *argECBH* operon in *Escherichia coli* K12. FEBS Lett. **51:**143–145.

178. **Papas, T. S., and A. Peterkofsky.** 1972. A random sequential mechanism for arginyl transfer ribonucleic acid synthetase of *Escherichia coli.* Biochemistry **11:**4602–4608.

179. **Pardee, A. B., F. Jacob, and J. Monod.** 1959. The genetic control and cytoplasmic expression of "inducibility" in the synthesis of β-galactosidase by *Escherichia coli.* J. Mol. Biol. **1:**165–178.

180. **Pegg, A. E., A. J. Bitoni, P. P. McCann, and J. K. Coward.** 1983. Inhibition of bacterial aminopropyltransferases by S-adenosyl-1,8-diamino-3-thiooctane and by dicylohexamine. FEBS Lett. **155:**192–196.

181. **Penninckx, M.** 1980. "Illicit" uptake of antimetabolites: potential use in antimicrobial chemotherapy. Trends Pharmacol. Sci. **June:**271–272.

182. **Penninckx, M., and D. Gigot.** 1978. Synthesis and interaction with *Escherichia coli* L-ornithine carbamoyltransferase of two potential transition state analogs. FEBS Lett. **88:**94–96.

183. **Penninckx, M., and D. Gigot.** 1979. Synthesis of a peptide from N-δ-(phosphonoacetyl)-L-ornithine. Its antibacterial effect through the specific inhibition of *Escherichia coli* carbamoyltransferase. J. Biol. Chem. **254:**6392–6396.

184. **Peyru, G. M., and W. K. Maas.** 1967. Inhibition of *Escherichia coli* B by homoarginine. J. Bacteriol. **94:**712–718.

185. **Piérard, A.** 1966. Control of the activity of *Escherichia coli* carbamoylphosphate synthetase by antagonistic allosteric effectors. Science **154:**1572–1573.

186. **Piérard, A.** 1983. Evolution des systèmes de synthèse et d'utilisation du carbamoylphosphate, p. 55–61. *In* G. Hervé (ed.), L'évolution des protéines. Masson, Paris.

187. **Piérard, A., N. Glansdorff, D. Gigot, M. Crabeel, P. Halleux, and L. Thiry.** 1976. Repression of *Escherichia coli* carbamoylphosphate synthase: relationship with enzyme synthesis in the arginine and pyrimidine pathways. J. Bacteriol. **127:**291–301.

188. **Piérard, A., N. Glansdorff, M. Mergeay, and J. M. Wiame.** 1965. Control of the biosynthesis of carbamoylphosphate in *Escherichia coli.* J. Mol. Biol. **14:**23–36.

189. **Piérard, A., N. Glansdorff, and J. Yashphe.** 1972. Mutations affecting uridine monophosphate pyrophosphorylase or the *argR* gene in *Escherichia coli.* Effects on carbamylphosphate and pyrimidine biosynthesis and on uracil uptake. Mol. Gen. Genet. **118:**235–245.

190. **Piérard, A., W. Lissens, P. Halleux, R. Cunin, and N. Glansdorff.** 1980. Role of transcriptional regulation and enzyme inactivation in the synthesis of *Escherichia coli* carbamoylphosphate synthase. J. Bacteriol. **141:**382–385.

191. **Piérard, A., and J. M. Wiame.** 1964. Regulation and mutation

192. **Piette, J., R. Cunin, A. Boyen, D. Charlier, M. Crabeel, F. Van Vliet, N. Glansdorff, C. Squires, and C. L. Squires.** 1982. The regulatory region of the divergent *argECBH* operon in *Escherichia coli* K12. Nucleic Acids Res. **10:**8031–8048.

193. **Piette, J., R. Cunin, M. Crabeel, A. Boyen, N. Glansdorff, C. Squires, and C. L. Squires.** 1980. Nucleotide sequence of the control region of the *argECBH* bipolar operon in *Escherichia coli.* Arch. Int. Physiol. Biochim. **88:**B242–B243.

194. **Piette, J., R. Cunin, M. Crabeel, and N. Glansdorff.** 1981. The regulatory region of the *argF* gene of *Escherichia coli* K-12. Arch. Int. Physiol. Biochim. **89:**B127–B128.

195. **Piette, J., R. Cunin, F. Van Vliet, D. Charlier, M. Crabeel, Y. Ota, and N. Glansdorff.** 1982. Homologous control sites and DNA transcription starts in the related *argF* and *argI* sites of *Escherichia coli* K12. EMBO J. **1:**853–857.

196. **Piette, J., H. Nyunoya, C. Lusty, R. Cunin, G. Weyens, M. Crabeel, D. Charlier, N. Glansdorff, and A. Piérard.** 1984. DNA sequence of the *carA* gene and the control region of *carAB*: tandem promoters, respectively controlled by arginine and the pyrimidines, regulate the synthesis of carbamoylphosphate synthetase in *Escherichia coli* K12. Proc. Natl. Acad. Sci. USA **81:**4134–4138.

197. **Popkin, P. S., and W. K. Maas.** 1980. *Escherichia coli* regulatory mutation affecting lysine transport and lysine decarboxylase. J. Bacteriol. **141:**485–492.

198. **Porter, R. W., M. O. Modebe, and G. R. Stark.** 1969. Aspartate transcarbamylase. Kinetic studies of the catalytic subunits. J. Biol. Chem. **244:**1846–1859.

199. **Pouwels, P., R. Cunin, and N. Glansdorff.** 1974. Divergent transcription in the *argECBH* cluster of genes in *Escherichia coli* K-12. J. Mol. Biol. **83:**421–424.

200. **Powers, S. G., and A. Meister.** 1978. Mechanism of the reaction catalyzed by carbamoylphosphate synthetase. Binding of ATP to the two functionally different ATP sites. J. Biol. Chem. **253:**800–803.

201. **Press, R., N. Glansdorff, P. Miner, J. De Vries, R. Kadner, and W. K. Maas.** 1971. Isolation of transducing particles of Φ80 bacteriophage that carry different regions of the *Escherichia coli* genome. Proc. Natl. Acad. Sci. USA **68:**795–798.

202. **Prusiner, S.** 1973. Glutaminases of *Escherichia coli.* Properties, regulation and evolution, p. 293–318. *In* S. Prusiner and E. R. Stadtman (ed.), The enzymes of glutamine metabolism. Academic Press, Inc., New York.

203. **Quay, S., and H. N. Christensen.** 1974. Basis of transport discrimination of arginine from other basic amino acids in *Salmonella typhimurium.* J. Biol. Chem. **249:**7011–7017.

204. **Ratner, S.** 1976. Enzymes of arginine and urea synthesis, p. 181–220. *In* S. Grisolia, R. Bagena, and F. Mayor (ed.), The urea cycle. Academic Press, Inc., New York.

205. **Reznikoff, W. S., J. H. Miller, J. G. Scaife, and J. R. Beckwith.** 1969. A mechanism for repressor action. J. Mol. Biol. **43:**201–213.

206. **Riley, M., and N. Glansdorff.** 1983. Cloning the *Escherichia coli* K-12 *argD* gene specifying acetylornithine δ-transaminase. Gene **24:**335–339.

207. **Rogers, P., T. M. Kaden, and M. Toth.** 1975. Repression of *arg* mRNA synthesis by L-arginine in cell-free extracts of *Escherichia coli.* Biochem. Biophys. Res. Commun. **65:**1284–1291.

208. **Rogers, P., R. Krzyzek, T. M. Kaden, and E. Arfman.** 1971. Effect of arginine and canavanine on arginine messenger RNA synthesis. Biochem. Biophys. Res. Commun. **44:**1220–1226.

209. **Roof, W. D., K. F. Foltermann, and J. R. Wild.** 1982. The organization and regulation of the *pyrBI* operon in *E. coli* includes a rho-dependent attenuator sequence. Mol. Gen. Genet. **187:**391–400.

210. **Rosen, B. P.** 1971. Basic amino acid transport in *Escherichia coli.* J. Biol. Chem. **246:**3653–3662.

211. **Rosen, B. P.** 1973. Basic amino acid transport in *Escherichia coli*: properties of canavanine-resistant mutants. J. Bacteriol. **116:**627–635.

212. **Sabo, D. L., E. A. Boeker, B. Byers, H. Waron, and E. H. Fischer.** 1974. Purification and physical properties of inducible *Escherichia coli* lysine decarboxylase. Biochemistry **13:**662–670.

213. **Saint-Girons, I., N. Duchange, M. Zakin, I. Park, D. Margarita, P. Ferrara, and G. N. Cohen.** 1983. Nucleotide sequence of *metF*, the *E. coli* structural gene for 5-10 methylene tetrahydrofolate reductase and of its control region. Nucleic Acids Res. **11:**6723–6732.

214. **Sanderson, K. E., and J. R. Roth.** 1983. Linkage map of *Salmonella typhimurium*, edition VI. Microbiol. Rev. **47:**410–453.

affecting a glutamine-dependent formation of carbamoylphosphate in *Escherichia coli.* Biochem. Biophys. Res. Commun. **15:**76–81.

215. **Satishchandran, C., and S. M. Boyle.** 1984. Antagonistic transcriptional regulation of the putrescine biosynthetic enzyme agmatine ureohydrolase by cyclic AMP and agmatine in *Escherichia coli.* J. Bacteriol. **157:**552–559.
216. **Scherer, G. F. F., M. D. Walkinshaw, S. Arnott, and D. J. Morré.** 1980. *E. coli* ribosomes have regions with signal character in both the leader and protein coding segments. Nucleic Acids Res. **8:**3895–3907.
217. **Schwartz, J. H., W. K. Maas, and E. J. Simon.** 1959. An impaired concentrating mechanism for amino acids in mutants of *Escherichia coli* resistant to L-canavanine and D-serine. Biochim. Biophys. Acta **32:**582–583.
218. **Sens, D., W. Natter, R. T. Garvin, and E. James.** 1977. Transcription of the *argF* and *argI* genes of the arginine biosynthetic regulon of *E. coli* K12 performed in vitro. Mol. Gen. Genet. **155:**7–18.
219. **Sens, D., W. Natter, and E. James.** 1977. In vitro transcription of the *Escherichia coli* K-12 *argA, argE,* and *argCBH* operons. J. Bacteriol. **130:**642–655.
220. **Shaibe, E., E. Metzer, and Y. S. Halpern.** 1985. Control of utilization of L-arginine, L-ornithine, agmatine, and putrescine as nitrogen sources in *Escherichia coli* K-12. J. Bacteriol. **163:**938–942.
221. **Shepherdson, M., and A. B. Pardee.** 1960. Production and crystallization of aspartate transcarbamylase. J. Biol. Chem. **235:**3233–3237.
222. **Stalon, V., and A. Mercenier.** 1984. L-arginine utilization by *Pseudomonas* species. J. Gen. Microbiol. **130:**69–76.
223. **Stragier, P., D. Olivier, and J. C. Patte.** 1983. Regulation of diaminopimelate decarboxylase synthesis in *Escherichia coli.* II. Nucleotide sequence of the *lysA* gene and its regulatory region. J. Mol. Biol. **168:**321–331.
224. **Syvanen, J. L., and J. R. Roth.** 1972. The structural genes for ornithine transcarbamylase in *Salmonella typhimurium* and *Escherichia coli.* J. Bacteriol. **110:**69–76.
225. **Tabor, C. W., and H. Tabor.** 1976. 1,4-diaminobutane (putrescine), spermidine and spermine. Annu. Rev. Biochem. **45:**285–306.
226. **Tabor, C. W., and H. Tabor.** 1983. Putrescine aminopropyltransferase. Methods Enzymol. **94:**265–269.
227. **Tabor, C. W., and H. Tabor.** 1984. Polyamines. Annu. Rev. Biochem. **53:**749–790.
228. **Tabor, C. W., and H. Tabor.** 1985. Polyamines in microorganisms. Microbiol. Rev. **49:**81–99.
229. **Tabor, C. W., H. Tabor, and E. W. Hafner.** 1983. Mass screening for mutants in the biosynthetic pathway for polyamines in *Escherichia coli:* a general method for mutants in enzymatic reactions producing CO$_2$. Methods Enzymol. **94:**83–90.
230. **Tabor, C. W., H. Tabor, E. W. Hafner, G. D. Markham, and S. M. Boyle.** 1983. Cloning of the *Escherichia coli* genes for the biosynthetic enzymes for polyamines. Methods Enzymol. **94:**117–124.
231. **Tabor, H., E. W. Hafner, and C. Tabor.** 1980. Construction of an *Escherichia coli* strain unable to synthesize putrescine, spermidine, or cadaverine: characterization of two genes controlling lysine decarboxylase. J. Bacteriol. **144:**952–956.
232. **Tabor, H., and C. W. Tabor.** 1969. Formation of 1,4-diaminobutane and of spermidine by an ornithine auxotroph of *Escherichia coli* grown on limiting ornithine or arginine. J. Biol. Chem. **244:**2286–2292.
233. **Tabor, H., and C. W. Tabor.** 1969. Partial separation of two pools of arginine in *Escherichia coli:* preferential use of exogenous rather than endogenous arginine for the biosynthesis of 1,4-diaminobutane. J. Biol. Chem. **244:**6383–6387.
234. **Tabor, H., C. W. Tabor, M. S. Cohn, and E. W. Hafner.** 1981. Streptomycine resistance (*rpsL*) produces an absolute requirement for polyamines for growth of an *Escherichia coli* strain unable to synthesize putrescine and spermidine [Δ(*speA-speB*) Δ*speC*]. J. Bacteriol. **147:**702–704.
235. **Tabor, H., C. W. Tabor, and E. W. Hafner.** 1978. *Escherichia coli* mutants completely deficient in adenosylmethionine decarboxylase and in spermidine biosynthesis. J. Biol. Chem. **253:**3671–3676.
236. **Trotta, P., M. E. Burt, R. H. Haschemeyer, and A. Meister.** 1971. Reversible dissociation of carbamoylphosphate synthetase into a regulated synthesis subunit and a subunit required for glutamine utilization. Proc. Natl. Acad. Sci. USA **68:**2599–2603.
237. **Trudel, M., M. Springer, M. Graffe, G. Fayat, S. Blanquet, and M. Grunberg-Manago.** 1984. Regulation of *E. coli* phenylalanyl-tRNA synthetase operon in vivo. Biochim. Biophys. Acta **782:**10–17.
238. **Turnbough, C. L., Jr.** 1983. Regulation of *Escherichia coli* aspartate transcarbamylase synthesis by guanosine tetraphosphate and pyrimidine ribonucleoside triphosphate. J. Bacteriol. **153:**998–1007.
239. **Turnbough, C. L., Jr., and B. R. Bochner.** 1985. Toxicity of the pyrimidine biosynthetic pathway intermediate carbamylaspartate in *Salmonella typhimurium.* J. Bacteriol. **163:**500–505.
240. **Turnbough, C. L., Jr., K. L. Hicks, and J. P. Donahue.** 1983. Attenuation control of *pyrBI* operon expression in *Escherichia coli* K12. Proc. Natl. Acad. Sci. USA **80:**368–372.
241. **Udaka, S.** 1966. Pathway-specific pattern of control of arginine biosynthesis in bacteria. J. Bacteriol. **91:**617–621.
242. **Udaka, S.** 1970. Isolation of the arginine repressor in *Escherichia coli.* Nature (London) **228:**336–338.
243. **Urm, E., H. Y. Lang, G. Zubay, N. Kelker, W. K. Maas.** 1973. In vitro repression of N-α-acetylornithinase synthesis in *Escherichia coli.* Mol. Gen. Genet. **121:**1–7.
244. **VanBogelen, R. A., V. Vaughn, and F. C. Neidhardt.** 1983. Gene for heat-inducible lysyl-tRNA synthetase (*lysU*) maps near *cadA* in *Escherichia coli.* J. Bacteriol. **153:**1066–1068.
245. **Vander Wauven, C., and V. Stalon.** 1984. Enzymes of arginine degradation in *Pseudomonas cepacia* and *Klebsiella aerogenes.* Arch. Int. Physiol. Biochim. **92:**B67.
246. **Van Vliet, F., R. Cunin, A. Jacobs, J. Piette, D. Gigot, M. Lauwereys, A. Piérard, and N. Glansdorff.** 1984. Evolutionary divergence of genes for ornithine and aspartate carbamoyltransferases—complete sequence and mode of regulation of the *Escherichia coli argF* gene: comparison of *argF* with *argI* and *pyrB.* Nucleic Acids Res. **12:**6277–6289.
247. **Vissers, S., Y. Dessaux, C. Legrain, and J. M. Wiame.** 1981. Feedback inhibition by arginine on ornithine carbamoyltransferase of *Agrobacterium tumefaciens.* Arch. Int. Physiol. Biochim. **89:**B83–B84.
248. **Vogel, H. J.** 1957. Repression and induction as control mechanisms of enzyme biogenesis: the "adaptive" formation of acetylornithinase, p. 276–289. *In* W. D. McElroy and B. Glass (ed.), The chemical basis of heredity. Johns Hopkins Press, Baltimore.
249. **Vogel, H. J.** 1960. Repression of an acetylornithine permeation system. Proc. Natl. Acad. Sci. USA **46:**488–494.
250. **Vogel, H. J.** 1961. Aspects of repression in the regulation of enzyme synthesis: pathway-wide control and enzyme-specific response. Cold Spring Harbor Symp. Quant. Biol. **26:**163–171.
251. **Vogel, H. J.** 1970. Arginine biosynthetic system in *Escherichia coli.* Methods Enzymol. **17A:**249–251.
252. **Vogel, H. J., and E. E. Jones.** 1970. Acetylornithine-δ-aminotransferase (*Escherichia coli*). Methods Enzymol. **17A:**260–264.
253. **Vogel, H. J., and W. L. McLellan.** 1970. *N*-Acetyl-γ-glutaminokinase (*Escherichia coli*). Methods Enzymol. **17A:**251–255.
254. **Vogel, H. J., and W. L. McLellan.** 1970. *N*-Acetylglutamic-γ-semialdehyde dehydrogenase (*Escherichia coli*). Methods Enzymol. **17A:**255–260.
255. **Vogel, H. J., and W. L. McLellan.** 1970. Acetylornithinase (*Escherichia coli*). Methods Enzymol. **17A:**265–269.
256. **Vogel, H. J., and R. H. Vogel.** 1974. Enzymes of arginine biosynthesis and their repressive controls. Adv. Enzymol. **40:**65–90.
257. **Vogel, R. H., E. A. Devine, and H. J. Vogel.** 1978. Evidence for translational repression of arginine biosynthetic enzymes in *Escherichia coli:* altered regulation in a streptomycin-resistant mutant. Mol. Gen. Genet. **162:**157–162.
258. **Vyas, S., and W. K. Maas.** 1963. Feedback inhibition of acetylglutamate synthetase by arginine in *Escherichia coli.* Arch. Biochem. Biophys. **100:**542–546.
259. **Wertheimer, S. J., and Z. Leifer.** 1983. Putrescine and spermidine sensitivity of lysine decarboxylase in *Escherichia coli:* evidence for a constitutive enzyme and its mode of regulation. Biochem. Biophys. Res. Commun. **114:**882–888.
260. **Weyens, G., K. Rose, P. Falmagne, N. Glansdorff, and A. Piérard.** 1985. Synthesis of *Escherichia coli* carbamoylphosphate synthetase initiates at a UUG codon. Eur. J. Biochem. **150:**111–115.
261. **Wickner, R. B., C. W. Tabor, and H. Tabor.** 1970. Purification of adenosylmethionine decarboxylase from *Escherichia coli* W: evidence for covalently bound pyruvate. J. Biol. Chem. **245:**2132–2139.
262. **Wilson, O. H., and J. T. Holden.** 1969. Arginine transport and metabolism in osmotically shocked and unshocked cells of *Escherichia coli* W. J. Biol. Chem. **244:**2737–2742.
263. **Wilson, O. H., and J. T. Holden.** 1969. Stimulation of arginine transport in osmotically shocked *Escherichia coli* W cells by purified arginine-binding protein fractions. J. Biol. Chem. **244:**2743–2749.

264. **Wright, J. M., and S. M. Boyle.** 1982. Negative control of ornithine decarboxylase and arginine decarboxylase by adenosine-3':5'-cyclic monophosphate in *Escherichia coli.* Mol. Gen. Genet. **186:**482–487.
265. **Wu, W. H., and D. R. Morris.** 1973. Biosynthetic arginine decarboxylase from *Escherichia coli.* Purification and properties. J. Biol. Chem. **248:**1687–1695.
266. **Yang, H. L., G. Zubay, E. Urm, G. Reiness, and M. Cashel.** 1974. Effects of guanosine tetraphosphate, guanoside pentaphosphate and B-r-methylenyl-guanosine pentaphosphate on gene expression of *Escherichia coli* in vitro. Proc. Natl. Acad. Sci. USA **71:**63–67.

267. **Yanofsky, C., and R. Kolter.** 1982. Attenuation in amino acid biosynthetic operons. Annu. Rev. Genet. **16:**113–134.
268. **York, M. K., and M. Stodolsky.** 1981. Characterization of P1*argF* derivatives from *Escherichia coli* K-12 transduction. I. IS1 elements flank the *argF* gene segment. Mol. Gen. Genet. **181:**230–240.
269. **Zidwick, M. J., G. Keller, and P. Rogers.** 1984. Regulation and coupling of *argECBH* mRNA and enzyme synthesis in cell extracts of *Escherichia coli.* J. Bacteriol. **159:**640–646.
270. **Zidwick, M. J., J. Korshus, and P. Rogers.** 1984. Positive control of expression of the *argECBH* gene cluster in vitro by guanosine 5'-diphosphate 3'-diphosphate. J. Bacteriol. **159:**647–651.

22. Biosynthesis of Proline

THOMAS LEISINGER

*Mikrobiologisches Institut Eidgenössische Technische Hochschule Zürich, ETH-Zentrum,
CH-8092 Zürich, Switzerland*

INTRODUCTION

Only a relatively small amount of information on proline biosynthesis accumulated during the 1960s and 1970s, when studies on the biochemistry and regulation of amino acid biosyntheses flourished. This lag in progress was due to the difficulties encountered in developing sensitive and specific assays for the first two enzymes of this pathway and to the fact that the enzyme levels in the system seemed not to respond to changes in the concentration of proline in the growth medium. Proline biosynthesis in enterobacteria was therefore not suitable for studies on gene expression, and the main challenges in the system were the elucidation of the enzymology and the formal proof of a pathway in which reactions and intermediates were already generally accepted on the basis of circumstantial evidence.

The pathway from glutamate via glutamate γ-semialdehyde (GSA) and its spontaneous cyclization product, Δ¹-pyrroline-5-carboxylate (P5C) to proline (Fig. 1) was first proposed by Vogel and Davis (49) and was based on the observation that the accumulation of P5C by certain *Escherichia coli* proline auxotrophs permitted growth of another type of proline auxotroph. Isotope dilution experiments showed that unlabeled GSA decreased the incorporation of [^{14}C]glutamate into proline (50) and thus supported the proposed route. A further important stage in the development of the model for proline biosynthesis was the demonstration that feedback regulation of the pathway by proline is at some stage between glutamate and GSA (4). This conversion had been proposed to involve an activated glutamate such as γ-glutamyl phosphate. Baich (3) indeed provided evidence for a

proline-inhibitable enzyme activity catalyzing the ATP-dependent phosphorylation of glutamic acid. An *E. coli* mutant resistant to proline inhibition in vivo contained a γ-glutamyl kinase with markedly lowered sensitivity to proline. This finding supported the existence of a γ-glutamyl kinase specifically involved in the first step of proline biosynthesis and has led to the formulation of the pathway in its present form (Fig. 1). The enzymes of this pathway have also been demonstrated in other microorganisms such as *Pseudomonas aeruginosa* (34, 35) and *Brevibacterium flavum* (52). Proline formation from glutamate is thought to occur in a variety of animal tissues, in which, however, there also exists a second route for proline synthesis that leads from ornithine via P5C to proline (1).

An extensive review by Adams and Frank (1) on the metabolism of proline and the hydroxyprolines in microbes and higher organisms summarizes the state of the art in 1979. Some of the recent progress in proline biosynthesis with emphasis on what is known about the pathway in *E. coli* and *Salmonella typhimurium* has been reviewed by Csonka and Baich (16). During the past few years the application of gene cloning techniques and the renewed interest in proline biosynthesis in connection with the role of proline as an osmoregulator have increased the speed at which the understanding of this pathway progresses.

THE PATHWAY

γ-Glutamyl Kinase

γ-Glutamyl kinase catalyzes the ATP-dependent phosphorylation of the γ-carboxy group of L-glutamic acid. Three types of experimental difficulties inherent

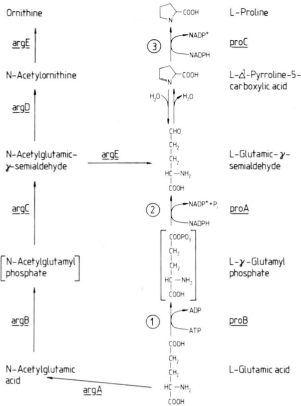

FIG. 1. Proline biosynthetic pathway and its relation to arginine biosynthesis. Reaction 1 is catalyzed by γ-glutamyl kinase (ATP: L-glutamate 5-phosphotransferase; EC 2.7.2.11), reaction 2 by GSA dehydrogenase (L-glutamate-γ-semialdehyde: NADP$^+$ oxidoreductase [phosphorylating]; EC 1.2.1.41), and reaction 3 by P5C reductase (L-proline: NADP$^+$ 5-oxidoreductase; EC 1.5.1.2). *argA* to *argE* represent the structural genes of the enzymes catalyzing steps 1 to 5 of arginine biosynthesis (see chapter 21).

to the system have hampered studies on this reaction in crude extracts and have precluded the purification of the enzyme until recently.

First, the extreme lability of γ-glutamyl phosphate and its tendency to cyclize to 5-oxopyrrolidine-2-carboxylate (46) have prevented the direct identification of the product of the first enzymatic step in any experimental system. However, the enzyme-dependent formation of γ-glutamyl hydroxamate and P$_i$ from glutamate, ATP, Mg^{2+}, and hydroxylamine (3) is compatible with γ-glutamyl phosphate being the first intermediate of the pathway. When the GSA dehydrogenase reaction, the second step in the pathway, is examined with homogeneous enzyme in the reverse of its biosynthetic direction, the end product is 5-oxopyrrolidine-2-carboxylic acid. Since the latter compound arises from γ-glutamyl phosphate, this observation is in accordance with the scheme in Fig. 1, in which γ-glutamyl phosphate is depicted as the activated intermediate between the first and the second biosynthetic steps (26). Considerations on the lability of free γ-glutamyl phosphate have led to the notion that the compound exists in vivo as an enzyme-bound intermediate.

Second, the proline-specific γ-glutamyl kinase catalyzes the same activation reaction of glutamate as

glutamine synthetase. Although the two activities can be differentiated on the basis of their sensitivity to proline (3, 29), the presence of both enzymes in crude extracts of *E. coli* makes the assay of the first enzyme of proline biosynthesis rather imprecise. Using wild-type extracts and the classical assay which is based on the detection of γ-glutamyl hydroxamate, blank values due to glutamine synthetase activity are up to 10-fold higher than the activity due to the proline-specific γ-glutamyl kinase.

The third obstacle to an understanding of γ-glutamyl kinase was its association with the second enzyme of proline biosynthesis. Only recently the work of Smith et al. (45) has conclusively shown that, to be enzymatically active, γ-glutamyl kinase needs to be associated with GSA dehydrogenase (Fig. 2). The existence of a complex catalyzing sequential reactions in proline biosynthesis and ensuring the direct transfer of an unstable intermediate had been suggested before as a result of gel filtration studies with crude extracts (29). It appears that the complex is very labile in vitro. It is now clear that previous attempts to purify γ-glutamyl kinase have led to the dissociation of the two enzymes with a consequent loss of activity of the first enzyme.

The purification of γ-glutamyl kinase from *E. coli* (45) was successful because a strain containing the *proBA* genes in a multicopy expression vector (17) yielded starting material of high specific activity, and the use of a coupled enzyme assay based on the NADPH-dependent reduction of γ-glutamyl phosphate by GSA dehydrogenase throughout the purification procedure satisfied the requirement of γ-glutamyl kinase for the second enzyme in the pathway. In this assay a large excess of highly purified GSA dehydrogenase over γ-glutamyl kinase was used (Fig. 2). Some properties of purified γ-glutamyl kinase are listed in Table 1. It appears to be a hexamer composed

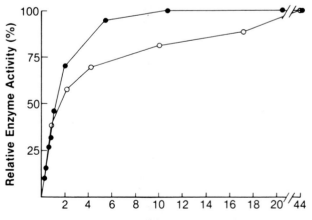

FIG. 2. Effect of highly purified GSA dehydrogenase on the activity of pure γ-glutamyl kinase from *E. coli*. γ-Glutamyl kinase activity, as measured by following the formation of γ-glutamyl hydroxamate (○) or by a coupled assay with GSA dehydrogenase (●), depends on the presence of GSA dehydrogenase. The ratio of GSA dehydrogenase to γ-glutamyl kinase was calculated from the molecular weights of the purified enzymes (see Table 1). The γ-glutamyl kinase assay contained 0.55 μg of γ-glutamyl kinase and from 0 to 27 μg of GSA dehydrogenase. From Smith et al. (45).

TABLE 1. Genes and enzymes of proline biosynthesis in *E. coli*

Gene			Enzyme				Reference
Designation	Map position[a] (min)	Size (base pairs)	Trivial name	Mol wt		Substrate (K_m)	
				Native	Subunit[b]		
proB	5.8	1,101	γ-Glutamyl kinase	236,000	40,000; 38,952	L-Glutamate (—[c]) ATP (0.5 mM)	45
proA	5.8	1,221	GSA dehydrogenase	268,000 189,000	42,000; 43,503 47,000	DL-P5C (2.5 mM) NADP$^+$ (0.05 mM) Phosphate (0.35 mM)	45 27
proC	8.9	807	P5C reductase	280,000 320,000	26,500; 28,112	DL-P5C (0.15 mM) NADPH (0.03 mM)	18 42

[a] Bachmann et al. (2); cf. Hadley et al. (22).

[b] The first number is the subunit molecular weight obtained by protein purification; the second number is the molecular weight deduced from the nucleotide sequence of the corresponding gene.

[c] —, Unknown.

of identical subunits. The activity of the enzyme depends in a nonhyperbolic fashion on the glutamate concentration, and 30 to 40 mM L-glutamate is necessary for half-maximal activity. L-Proline at 0.007 mM or 3,4-dehydroproline at 0.5 mM is needed to inhibit the enzyme activity by 50%. The substrate specificity of purified γ-glutamyl kinase has not been investigated.

GSA Dehydrogenase

GSA dehydrogenase, the second enzyme in the pathway, catalyzes the in vivo NADPH-dependent reduction of γ-glutamyl phosphate to GSA. Since γ-glutamyl phosphate is not available as a substrate in vitro, enzyme activity is measured in the reverse direction by the NADP- and phosphate-dependent oxidation of GSA. The actual substrate used in this assay is P5C, a rather labile compound, which is in equilibrium with the straight-chain GSA that probably serves as the true substrate of the enzyme. DL-P5C is most conveniently prepared by periodate oxidation of δ-hydroxylysine (51). An enzymatic method for the preparation of L-P5C from L-ornithine with a partially purified preparation of ornithine 5-aminotransferase from *P. aeruginosa* has also been described (24).

GSA dehydrogenase from *E. coli* has been purified in two laboratories. When cell material containing a ColE1-*proBA* plasmid was used, 1,200-fold purification was needed to obtain a homogeneous product (27). A more favorable situation was created by using cells carrying an expression vector with the *proBA* region for enzyme purification. A 450-fold amplification of GSA dehydrogenase with a correspondingly lowered purification factor was achieved in this case (17, 45). As evident from the data compiled in Table 1, there exists a discrepancy concerning the molecular weight of the native protein, so that it cannot be decided at this point whether the enzyme is composed of four or six subunits. Studies with the pure enzyme showed that the reaction is highly specific for DL-P5C and NADP, but that a number of divalent anions can substitute for phosphate. The kinetics of the reaction are consistent with a rapid-equilibrium, random-order mechanism (28).

P5C Reductase

The final step in proline biosynthesis, the NADPH-linked reduction of P5C to proline, is catalyzed by P5C reductase. In crude extracts of *E. coli* cells this enzyme interferes with the assay for GSA dehydrogenase activity: both enzymes use P5C as a substrate, and NADPH, a product of the GSA dehydrogenase reaction, serves as a cosubstrate for P5C reductase. To assign the *proA* and *proB* loci to their respective enzymes, a method for the rapid removal of P5C reductase from cell extracts was devised. It permitted the establishment of the gene-enzyme relationships in a number of *proA* and *proB* mutants (25).

P5C reductase from an *E. coli* wild-type strain has been enriched 200-fold to give a partially purified preparation that is sufficiently free of competing enzyme activities to be used in kinetic studies (42). The enzyme exhibits Michaelis-Menten kinetics, permitting determination of its kinetic parameters (Table 1). Proline and NADP, the end products of the reaction, act as competitive inhibitors with approximate K_i values of 15.0 and 0.6 mM, respectively. Since exogenous proline is channeled into the membrane-associated proline degradative pathway (see chapter 94), the physiological significance of P5C reductase inhibition by proline remains obscure. Recently P5C reductase has been purified to homogeneity from an *E. coli* strain harboring the expression plasmid pGW7 *proC*. The 90-fold increase in P5C reductase activity relative to the wild-type level that was obtained in the plasmid-bearing strain greatly facilitated enzyme purification. The homogeneous enzyme preparation was used for the determination of the amino- and carboxy-terminal amino acid sequences and for establishing the amino acid composition, but was not characterized with respect to its catalytic properties (18).

Proline Auxotrophs

Two largely unexplained observations have been made that show that under certain conditions, a deficiency in proline synthesis does increase the survival rate relative to the prototrophic wild type or to mutants blocked in other biosynthetic pathways. They are mentioned here because they may be useful in the isolation of proline auxotrophs. 4-Nitropyridine 1-oxide is bactericidal for *E. coli* wild-type and *proC* cells, but not for *proB* or *proA* mutants. This effect, whose biochemical basis is unknown, can be conveniently used for the selection of *proA* or *proB* mutants (31). Preferential recovery of all three types of proline auxotrophs is also observed upon penicillin enrichment of a mutagenized population. The viable counts of resting cells of most types of auxotrophs are reduced by penicillin treatment, whereas resting cells of proline-requiring mutants for unknown reasons are completely insensitive to the antibiotic (41).

Second Route to GSA

The biosynthesis of arginine from glutamate is initiated by an *N*-acetylation which is thought to prevent the internal cyclization of the *N*-acetylglutamate semialdehyde that is formed in the third biosynthetic step (Fig. 1). Activation of *N*-acetylglutamate and the reduction of *N*-acetylglutamyl phosphate in the arginine pathway are analogous to the first and the second steps of proline biosynthesis. The similarity of the early intermediates in the two pathways forms the basis for the phenotypic suppression of mutations in *proB* or *proA* by mutations in the arginine pathway. This phenomenon has been observed in *E. coli* (5, 32) and in *S. typhimurium* (36).

argD mutants accumulate *N*-acetylglutamate-γ-semialdehyde which is converted into GSA by *N*-acetylornithine deacetylase, the product of *argE* catalyzing the fifth step in arginine biosynthesis (Fig. 1). Such mutants are able to grow in the absence of arginine because some nonspecific aminotransferases substitute for the missing *argD* enzyme. In *proB* (or *proA*) argD mutants, the block in proline biosynthesis is bypassed since the first three enzymes of the arginine pathway and *N*-acetylornithine deacetylase lead to the formation of GSA. Proline biosynthesis in these double mutants is no longer feedback inhibited by proline but is now under repression and feedback control by arginine. The double mutants excrete proline on minimal medium and are resistant to the proline analogs 3,4-dehydroproline and azetidine-2-carboxylate (5). The interrelationship between arginine and proline biosynthesis has been used for the selection of mutants containing feedback-resistant *N*-acetylglutamate synthases, the first enzyme of arginine biosynthesis (20). A nonrepressible *proB* (or *proA*) argD argR mutant grows slowly on minimal medium with arginine. As a result of the partial inhibition of *N*-acetylglutamate synthase by arginine, it does not excrete proline. Cultivation of the triple mutant on minimal medium with arginine enriches for faster growing quadruple mutants containing a feedback-resistant *N*-acetylglutamate synthase. They can be recognized as proline excretors on minimal agar with arginine.

THE *pro* GENES

Map Locations

References to the literature describing the mapping of the genes of proline biosynthesis in *E. coli* are given

by Bachmann et al. (2). A physical map of the chromosomal segment extending from approximately 5.0 to 12.5 min and covering the *pro* genes has been prepared (22) and corresponds well with the locations of *proBA* at 5.8 min and of *proC* at 8.9 min on the genetic map (2). In *S. typhimurium* the three *pro* genes are arranged in a similar fashion at the same position of the linkage map (44). Only recently it was established that in this organism, as in *E. coli*, the *proB* gene codes for γ-glutamyl kinase and *proA* codes for GSA dehydrogenase (38).

Cloning and Sequencing of the *pro* Genes

Several groups have cloned the *E. coli proBA* genes using ColE1 hybrid plasmids from the collection of Clarke and Carbon (12) or an F' *proBA* plasmid as the initial material (17, 23, 38). The *proB* and *proA* genes are on a 3.0-kilobase *Pst*I DNA fragment, for which a detailed restriction map is available (23). Mahan and Csonka (38) found that Tn5 insertions into either *proA* or *proB* of the cloned fragment exerted no polar effect on the other *pro* gene. Because it was possible that transcription had been initiated within the Tn5 insertion, these authors were uncertain whether the *proBA* genes are transcribed from separate promoters or a single promoter in *E. coli*. However, complementation tests with Tn10 insertions into the chromosome of *S. typhimurium* revealed a strong polar effect of *proB* on *proA*, suggesting that the *proBA* genes of this organism are organized in an operon that is transcribed from *proB* to *proA*.

The nucleotide sequence of the entire 3.0-kilobase *Pst*I fragment from *E. coli* has been determined (17). It contains two successive open reading frames that specify amino acid sequences that are correlated with the NH2-terminal and the COOH-terminal amino acid sequences of γ-glutamyl kinase and GSA dehydrogenase, as well as with the amino acid compositions of the two proteins. The following features of the nucleotide sequence are in accordance with the concept of a *proBA* operon. (i) The DNA sequence upstream from the *proB* coding region contains various domains of dyad symmetry and a putative promoter. A putative ribosome binding site is located 5 base pairs upstream of the *proB* translation-initiation codon. (ii) No promoterlike sequences can be found in the sequence preceding the *proA* coding region. (iii) The COOH terminal of *proB* is separated by 14 base pairs from the amino terminal of *proA*. There are a single translation termination codon and two Shine-Dalgarno sites within this intercistronic region. (iv) A possible transcription termination structure at the 3' end of the *proA* gene can be identified, whereas there is no indication for a terminator at the 3' end of the *proB* gene.

The *E. coli proC* gene has been located on a 1.1-kilobase *Hinc*II-*Hae*II fragment of a λ transducing phage, with its *Hinc*II site originating in the λ vector. This hybrid fragment was subcloned into plasmid pACYC177 (6). Determination of its nucleotide sequence (18) revealed an open reading frame for a polypeptide with a molecular weight of 28,112. This value correlates well with the molecular weight of the P5C reductase subunit that was determined by protein purification. Analysis of the amino-terminal se-

TABLE 2. Effects of proline analogs on *E. coli*

Compound	Inhibition of proline transport	Inhibition of γ-glutamyl kinase	Incorporation into protein	Reference
L-Azetidin-2-carboxylate	+	+	+	21
3,4-Dehydro-DL-proline	+	+	+	48
cis-4-Fluoro-L-proline	+	+	+	43
cis-3,4-Methano-L-proline	+	+	−	47
trans-3,4-Methano-L-proline	+	+	−	47
cis-4-Bromo-L-proline	+	−	−	43
cis-4-Chloro-L-proline	+	−	−	43

quence of purified P5C reductase supports the notion that translation starts at the proposed site, which is also preceded by a ribosome binding site. However, the fact that 94 nucleotides on the noncoding strand of the *lacA* structural gene have recently been found to be identical to nucleotides −119 to −25, preceding the translation start of *proC*, casts some doubt on the authenticity of the putative *proC* promoter region and suggests that a segment of *lacA* DNA was mistakenly incorporated into the *proC* region during construction of the *proC* clones used in the sequence studies (30). It may thus be necessary to reestablish the nucleotide sequence of the *proC* promoter region using wild-type *E. coli* as a source of DNA.

REGULATION

Control of Proline Synthesis in the Wild Type

Feedback inhibition of γ-glutamyl kinase was early recognized to be the major control mechanism for proline biosynthesis in *E. coli*. It was first demonstrated in resting cells of a *proC* mutant of *E. coli* by comparing the conversion rate of glutamate into GSA in the presence and absence of proline (4, 46). Later, the sensitivity of the first enzyme to proline was also observed in crude extracts (3, 29). Recent in vitro studies with purified γ-glutamyl kinase by Smith et al. (45) suggest that the activity of this enzyme is modulated not only by proline but also by glutamate and by ADP. The sigmoidal saturation kinetics of γ-glutamyl kinase with respect to glutamate lead to its relative insensitivity to small changes in the concentration of this substrate, thereby ensuring a constant rate of proline biosynthesis at varying intracellular concentrations of glutamate. ADP, a competitive inhibitor with respect to ATP, inhibits the enzyme under conditions of energy depletion and thus may lead to the tuning of proline biosynthesis with the energy charge of the cell. Proline decreases the affinity of the enzyme for glutamate at low or intermediate concentrations of this metabolite. In a mutant with abolished proline sensitivity the modulation of enzyme activity by glutamate and ADP is conserved. Since this mutant excretes proline, the end product of the pathway is thought to be the most important effector for pathway control.

The question of whether the synthesis of the proline enzymes in *E. coli* is subject to repression by the end product of the pathway has been addressed in several laboratories (7, 13, 25, 42, 47). Enzyme levels of cells grown on media without proline or under proline starvation show no significant differences from those obtained with cells grown on media with proline.

These observations indicate the absence of negative control by end product repression. They do not, however, exclude the possibility that other controls, such as autogenous regulation, positive control, or translational regulatory mechanisms, may be involved in the formation of the proline enzymes. For example, the mechanisms leading to a differential increase in the specific activities of γ-glutamyl kinase and GSA dehydrogenase upon cloning the *proBA* operon is unknown. In accordance with the molar excess of GSA dehydrogenase needed for the formation of an enzymatically active complex between the two proteins in vitro (Fig. 2), a three- to sixfold differential increase in specific activity in favor of GSA dehydrogenase is observed in plasmid-bearing strains compared with the wild-type activity (17, 25). Regulation at the translational level would seem likely in this case, since the *proBA* operon is thought to be transcribed into an mRNA encoding both proteins.

Excretion and Overproduction of Proline

Various proline analogs have been shown to interfere at different stages with protein synthesis in *E. coli* (Table 2). In addition to the compounds listed in Table 2, a number of analogs inhibit or act as substrates for prolyl-tRNA synthetases (8, 39). Azetidine-2-carboxylate, 3,4-dehydroproline, and 4-thiazolidine carboxylate, analogs inhibiting γ-glutamyl kinase as well as supporting protein synthesis, have proven useful in the genetic analysis of the proline system (13). Resistance to these agents is based either on a deficiency in proline uptake or on proline overproduction due to defective pathway control. Mutants with resistance of the latter type are easily recognized as proline excretors. In all cases the corresponding mutations are closely linked to *proB* (13). This observation suggests that proline overproduction is the consequence of decreased sensitivity to proline inhibition of γ-glutamyl kinase. In one case this was documented by in vitro studies with purified γ-glutamyl kinase from a 3,4-dehydroproline-resistant, proline-excreting mutant strain (45).

Preliminary data on the use of a genetically engineered *E. coli* strain for the overproduction of L-proline are available. A yield of 27 g of proline per liter with a 40% conversion of glucose to proline is reported (7). This productivity is achieved by a strain that is unable to degrade proline owing to a chromosomal deletion of the gene coding for proline dehydrogenase. In addition, it carries a ColE1-based vector with the cloned *proBA* operon as well as the *proC* gene. Furthermore, the *proB* gene on this plasmid specifies a feedback-resistant γ-glutamyl kinase. Evidently the high yield of proline is only possible in a strain that overproduces glutamate, the precursor of proline. Details on how this is achieved in *E. coli* have not been reported, nor is it known whether the high-productivity strain carries mutations affecting proline excretion.

Excretion indeed seems to be an important step in proline overproduction according to experiments by Rancourt et al. (40). They found that a plasmid-coded mutation in *proB*, causing intracellular proline overproduction, leads to proline excretion only if combined with mutations in genes involved in the active uptake or the catabolism of proline. Some proline excretion was observed in a strain lacking proline dehydrogenase due to a mutation in *putA* or in a strain with a mutation in *putP*, the structural gene for a carrier involved in proline transport. Maximal proline excretion, however (in this case 1 g of proline per liter), is observed exclusively when the plasmid with the feedback-resistant γ-glutamyl kinase is brought into a *putA putP* double mutant (40).

Proline and Osmoregulation

Proline and certain betaines, such as betaine, glycine betaine, and choline, are known to act as osmoprotectants in a wide variety of organisms ranging from bacteria to higher plants (37). Under conditions of osmotic stress these molecules accumulate and protect the cell and its constituents against dehydration. In enteric bacteria this function of proline was discovered by Christian (11), who observed that proline stimulates the growth of *Salmonella oranienburg* in media of elevated osmotic strength. The molecular basis of this effect is being studied in *S. typhimurium* and other enteric bacteria.

Intracellular accumulation of proline for osmoprotection can result either from an increase in net biosynthesis of proline or from an enhanced transport of extracellular proline into the cell. The operation of the former mechanism is illustrated by the fact that *S. typhimurium* harboring an F′ plasmid that encodes a feedback-resistant γ-glutamyl kinase excretes proline and at the same time exhibits increased tolerance to osmotic stress (14, 38). The gene encoding the feedback-resistant enzyme has been cloned on a broadhost-range plasmid vector. When transferred to *E. coli* and *Klebsiella pneumoniae* it conferred both proline overproduction and osmotolerance (33).

Surprisingly, the rate of proline synthesis is not regulated in response to osmotic stress. Media of high osmotic strength, however, do stimulate the uptake of exogenous proline. Two of the three known proline transport systems, namely, PP-II (encoded by the *proP* gene) and PP-III (encoded by the *proU* gene), are stimulated by osmotic stress, whereas PP-I, the third and major proline permease (encoded by the *putP* gene), is not affected by an increase in the osmolarity of the medium (15). More recently it was demonstrated that the primary physiological substrate of the PP-II and PP-III systems is not proline but betaine, which in enteric bacteria acts as a more effective osmoprotectant than proline (9, 10). Expression studies from two laboratories show that both the PP-II and the PP-III system are regulated by the osmolarity of the medium at the levels of synthesis and activity (9,

10, 19). The transcription of these genes in response to osmotic stress was monitored using *S. typhimurium* strains with *lacZ* fusions to genes *proP* and *proU*. Transcription of the *proU* gene increased 10- to 100-fold upon addition of 0.3 M NaCl or equivalent concentrations of some other solutes to the medium. The *proP* gene, however, was regulated only about twofold at the level of transcription. The major control of transport activity in the *proP* system thus must be due to an as yet unknown posttranscriptional regulatory mechanism.

CONCLUSIONS

The use of recombinant DNA techniques in studies on proline biosynthesis in *E. coli* has led to long-overdue answers to a number of questions. At the protein level, the purification of the first enzyme of proline synthesis and the unequivocal demonstration of its aggregation with the second enzyme represent major steps in the understanding of the system. At the level of DNA, the elucidation of the nucleotide sequences of all three structural genes of the pathway opens the way for a functional analysis of the transcription initiation signals. It is yet necessary to determine the effect of deletions in the control regions of *proBA* and *proC* on the efficiency of in vivo and in vitro transcription and to study transcriptional initiation of the *pro* genes. Furthermore, a reexamination of possible pathway-specific control mechanisms for the *proBA* operon acting at the transcriptional or the translational level should be undertaken. The application of refined techniques like the immunochemical assay of gene products or the use of *pro-lacZ* fusion strains may reveal effects that have not been detected with the insensitive enzyme assays used in the past.

ACKNOWLEDGMENTS

I thank F. Bloom and L. Csonka for communicating results and D. Jeenes for discussion and critical reading of the manuscript.

Research in my laboratory was supported by the Swiss National Foundation for Scientific Research, grants no. 3.263.78 and 3.050.81.

LITERATURE CITED

1. **Adams, E., and L. Frank.** 1980. Metabolism of proline and the hydroxyprolines. Annu. Rev. Biochem. **49**:1005–1061.
2. **Bachmann, B. J., K. B. Low, and A. L. Taylor.** 1976. Recalibrated linkage map of *Escherichia coli* K-12. Bacteriol. Rev. **40**:116–167.
3. **Baich, A.** 1969. Proline synthesis in *Escherichia coli*. A proline-inhibitable glutamic acid kinase. Biochim. Biophys. Acta **192**:462–467.
4. **Baich, A., and D. J. Pierson.** 1965. Control of proline synthesis in *Escherichia coli*. Biochim. Biophys. Acta **104**:397–404.
5. **Berg, C. M., and J. J. Rossi.** 1974. Proline excretion and indirect suppression in *Escherichia coli* and *Salmonella typhimurium*. J. Bacteriol. **118**:928–939.
6. **Berg, P. E.** 1981. Cloning and characterization of the *Escherichia coli* gene coding for alkaline phosphatase. J. Bacteriol. **146**:660–667.
7. **Bloom, F., C. J. Smith, J. Jessee, B. Veilleux, and A. H. Deutch.** 1983. The use of genetically engineered strains of *Escherichia coli* for the overproduction of free amino acids: proline as a model system, p. 383–394. *In* K. Downey, R. W. Voellmy, F., Ahmad, and J. Schultz (ed.), Advances in gene technology: molecular genetics of plants and animals. Academic Press, Orlando, Fla.
8. **Busiello, V., M. Di Girolamo, C. Cini, and C. De Marco.** 1980. β-Selenaproline as competitive inhibitor of proline activation. Biochim. Biophys. Acta **606**:347–352.
9. **Cairney, J., I. R. Booth, and C. F. Higgins.** 1985. *Salmonella typhimurium proP* gene encodes a transport system for the osmoprotectant betaine. J. Bacteriol. **164**:1218–1223.
10. **Cairney, J., I. R. Booth, and C. F. Higgins.** 1985. Osmoregulation

of gene expression in *Salmonella typhimurium*: *proU* encodes an osmotically induced betaine transport system. J. Bacteriol. **164**:1224–1232.

11. **Christian, J. H. B.** 1955. The influence of nutrition on the water relations of *Salmonella oranienburg*. Austr. J. Biol. Sci. **8**:75–82.

12. **Clarke, L., and J. Carbon.** 1976. A colony bank containing synthetic ColE1 hybrid plasmids representative of the entire *E. coli* genome. Cell **9**:91–99.

13. **Condamine, H.** 1971. Sur la régulation de la production de proline chez *E. coli* K12. Ann. Inst. Pasteur **120**:126–143.

14. **Csonka, L. N.** 1981. Proline over-production results in enhanced osmotolerance in *Salmonella typhimurium*. Mol. Gen. Genet. **182**:82–86.

15. **Csonka, L. N.** 1982. A third L-proline permease in *Salmonella typhimurium* which functions in media of elevated osmotic strength. J. Bacteriol. **151**:1433–1443.

16. **Csonka, L. N., and A. Baich.** 1983. Proline biosynthesis, p. 35–41. *In* K. M. Herrmann and R. L. Somerville (ed.), Amino acids: biosynthesis and regulation. Addison-Wesley, Reading, Mass.

17. **Deutch, A. H., K. E. Rushlow, and C. J. Smith.** 1984. Analysis of the *Escherichia coli proBA* locus by DNA and protein sequencing. Nucleic Acids Res. **12**:6337–6355.

18. **Deutch, A. H., C. J. Smith, K. E. Rushlow, and P. J. Kretschmer.** 1982. *Escherichia coli* Δ^1-pyrroline-5-carboxylate reductase: gene sequence, protein overproduction and purification. Nucleic Acids Res. **10**:7701–7714.

19. **Dunlap, V. J., and L. N. Csonka.** 1985. Osmotic regulation of L-proline transport in *Salmonella typhimurium*. J. Bacteriol. **163**:296–304.

20. **Eckhardt, T., and T. Leisinger.** 1975. Isolation and characterization of mutants with a feedback resistant *N*-acetylglutamate synthase in *Escherichia coli* K12. Mol. Gen. Genet. **138**:225–232.

21. **Grant, M. M., A. S. Brown, L. M. Corwin, R. F. Troxler, and C. Franzblau.** 1975. Effect of L-azetidine 2-carboxylic acid on growth and proline metabolism in *Escherichia coli*. Biochim. Biophys. Acta **404**:180–187.

22. **Hadley, R. G., H. Ming, M. Timmons, K. Yun, and R. C. Deonier.** 1983. A partial restriction map of the *ProA-purE* region of the *Escherichia coli* K-12 chromosome. Gene **22**:281–287.

23. **Hayzer, D. J.** 1983. Sub-cloning of the wild-type *proAB* region of the *Escherichia coli* genome. J. Gen. Microbiol. **129**:3215–3225.

24. **Hayzer, D. J., R. V. Krishna, and R. Margraff.** 1979. Enzymic synthesis of glutamic acid γ-semialdehyde (Δ^1-pyrroline-5-carboxylate) and *N*-acetyl-L-glutamic acid γ-semialdehyde: isolation and characterization of their 2,4-dinitrophenylhydrazones. Anal. Biochem. **96**:94–103.

25. **Hayzer, D. J., and T. Leisinger.** 1980. The gene-enzyme relationships of proline biosynthesis in *Escherichia coli*. J. Gen. Microbiol. **118**:287–293.

26. **Hayzer, D. J., and T. Leisinger.** 1981. Proline biosynthesis in *Escherichia coli*. Stoichiometry and end-product identification of the reaction catalysed by glutamate semialdehyde dehydrogenase. Biochem. J. **197**:269–274.

27. **Hayzer, D. J., and T. Leisinger.** 1982. Proline biosynthesis in *Escherichia coli*. Purification and characterization of glutamate semialdehyde dehydrogenase. Eur. J. Biochem. **121**:561–565.

28. **Hayzer, D. J., and T. Leisinger.** 1983. Proline biosynthesis in *Escherichia coli*. Kinetic and mechanistic properties of glutamate semialdehyde dehydrogenase. Biochim. Biophys. Acta **742**:391–398.

29. **Hayzer, D. J., and V. Moses.** 1978. The enzymes of proline biosynthesis in *Escherichia coli*. Their molecular weights and the problem of enzyme aggregation. Biochem. J. **173**:219–228.

30. **Hediger, M. A., D. F. Johnson, D. P. Nierlich, and I. Zabin.** 1985. DNA sequence of the lactose operon: the *lacA* gene and the transcriptional termination region. Proc. Natl. Acad. Sci. USA **82**:6414–6418.

31. **Inuzuka, M., H. Toyama, H. Miyano, and M. Tomoeda.** 1976. Specific action of 4-nitropyridine 1-oxide on *Escherichia coli* K-12 Pro+ strains leading to the isolation of proline-requiring mutants:

mechanism of action of 4-nitropyridine 1-oxide. Antimicrob. Agents Chemother. **10**:333–343.

32. **Itikawa, H., S. Baumberg, and H. J. Vogel.** 1968. Enzymic basis for a genetic suppression: accumulation and deacylation of *N*-acetylglutamic γ-semialdehyde in enterobacterial mutants. Biochim. Biophys. Acta **159**:547–550.

33. **Jakowec, M. W., L. T. Smith, and A. M. Dandekar.** 1985. Recombinant plasmid conferring proline overproduction and osmotic tolerance. Appl. Environ. Microbiol. **50**:441–446.

34. **Krishna, R. V., P. Beilstein, and T. Leisinger.** 1979. Biosynthesis of proline in *Pseudomonas aeruginosa*. Properties of γ-glutamyl-phosphate reductase and 1-pyrroline-5-carboxylate reductase. Biochem. J. **181**:223–230.

35. **Krishna, R. V., and T. Leisinger.** 1979. Biosynthesis of proline in *Pseudomonas aeruginosa*. Partial purification and characterization of γ-glutamyl kinase. Biochem. J. **181**:215–222.

36. **Kuo, T.-T., and B. A. D. Stocker.** 1969. Suppression of proline requirement of *proA* and *proAB* deletion mutants in *Salmonella typhimurium* by mutation to arginine requirement. J. Bacteriol. **98**:593–598.

37. **Le Rudulier, D., A. R. Strom, A. M. Dandekar, L. T. Smith, and R. C. Valentine.** 1984. Molecular biology of osmoregulation. Science **224**:1064–1068.

38. **Mahan, M. J., and L. N. Csonka.** 1983. Genetic analysis of the *proBA* genes of *Salmonella typhimurium*: physical and genetic analyses of the cloned *ProB+A+* genes of *Escherichia coli* and of a mutant allele that confers proline overproduction and enhanced osmotolerance. J. Bacteriol. **156**:1249–1262.

39. **Papas, T. S., and A. H. Mehler.** 1970. Analysis of the amino acid binding to the proline transfer ribonucleic acid synthase of *Escherichia coli*. J. Biol. Chem. **245**:1588–1595.

40. **Rancourt, D. E., J. T. Stephenson, G. A. Vickell, and J. M. Wood.** 1984. Proline excretion by *Escherichia coli* K12. Biotechnol. Bioeng. **26**:74–80.

41. **Rossi, J. J., and C. M. Berg.** 1971. Differential recovery of auxotrophs after penicillin enrichment in *Escherichia coli*. J. Bacteriol. **106**:297–300.

42. **Rossi, J. J., J. Vender, C. M. Berg, and W. H. Coleman.** 1977. Partial purification and some properties of Δ^1-pyrroline-5-carboxylate reductase from *Escherichia coli*. J. Bacteriol. **129**:108–114.

43. **Rowland, I., and H. Tristram.** 1975. Specificity of the *Escherichia coli* proline transport system. J. Bacteriol. **123**:871–877.

44. **Sanderson, K. E., and P. E. Hartman.** 1978. Linkage map of *Salmonella typhimurium*, edition V. Microbiol. Rev. **42**:471–519.

45. **Smith, C. J., A. H. Deutch, and K. E. Rushlow.** 1984. Purification and characteristics of a γ-glutamyl kinase involved in *Escherichia coli* proline biosynthesis. J. Bacteriol. **157**:545–551.

46. **Strecker, H. J.** 1957. The interconversion of glutamic acid and proline. I. The formation of Δ^1-pyrroline-5-carboxylic acid from glutamic acid in *Escherichia coli*. J. Biol. Chem. **225**:825–834.

47. **Tristram, H.** 1972. Some aspects of the regulation of amino acid biosynthesis in bacteria. Annu. Proc. Phytochem. Soc. **9**:21–48.

48. **Tristram, H., and C. F. Thurston.** 1966. Biochemistry of proline biosynthesis by proline and proline analogues. Nature (London) **212**:74–75.

49. **Vogel, H. J., and B. D. Davis.** 1952. Glutamic γ-semialdehyde and Δ^1-pyrroline-5-carboxylic acid, intermediates in the biosynthesis of proline. J. Am. Chem. Soc. **74**:109–112.

50. **Vogel, R. H., and M. J. Kopac.** 1959. Glutamic γ-semialdehyde in arginine and proline synthesis of *Neurospora*: a mutant-tracer analysis. Biochim. Biophys. Acta **36**:505–510.

51. **Williams, I., and L. Frank.** 1975. Improved chemical synthesis and enzymatic assay of Δ^1-pyrroline-5-carboxylic acid. Anal. Biochem. **64**:85–97.

52. **Yoshinaga, F., T. Tsuchida, and S. Okumura.** 1975. Purification and properties of glutamate kinases required for L-proline and L-glutamine biosynthesis in *Brevibacterium flavum*. Agr. Biol. Chem. **39**:1269–1273.

23. Biosynthesis of the Branched-Chain Amino Acids

H. E. UMBARGER

Department of Biological Sciences, Purdue University, West Lafayette, Indiana 47907

INTRODUCTION

The biosynthetic pathways leading to the branched-chain amino acids have been reviewed several times over the last 25 years, and these reviews should be consulted by anyone interested in the way our understanding of the biosynthetic steps has been developed (20, 32, 64, 122–125, 127). These reviews have appeared as specific chapters and as parts of broader coverages of amino acid biosynthesis in general. Actually, these pathways illustrate well the way isotope studies, mutant methodology, and enzyme studies were often combined to decipher metabolic pathways in microorganisms. Although most of the reviews cited addressed the biosynthesis of these amino acids in all bacteria, as well as in the lower and higher plants, much of the material covered was actually obtained with either *Salmonella typhimurium* or *Escherichia coli*.

This chapter will review some of the early findings, as well as the recently described modifications in our understanding of the biosynthetic pathways. The chapter will be devoted mainly to a consideration of gene-enzyme relationships and regulation of the pathway at the level of gene expression and at the level of metabolite flow.

GENE-ENZYME RELATIONSHIPS IN THE PATHWAYS TO ISOLEUCINE AND VALINE

Interest in the pathways to isoleucine and valine was stimulated in the early days of nutritional mutant analysis by the fact that *E. coli* and *S. typhimurium* mutants were often found that required both isoleucine and valine for growth. Mutants requiring isoleucine only were less frequent, and strains requiring only valine were very rare indeed. Also of interest was the fact that the growth of one of the commonly used strains of *E. coli*, strain K-12, was inhibited by valine and that the inhibition was antagonized by isoleucine. The common double auxotrophy was readily explained by the finding that the two pathways are parallel and that the parallel steps in the two pathways are catalyzed by common enzymes. The peculiar sensitivity of the K-12 strain to valine is now understood both in terms of enzymology and in terms of the kind of mutation that had occurred in the progenitor of the K-12 strain. The parallel pathways are shown in Fig. 1. Both pathways involve the condensation of an α-keto acid with an active acetaldehyde group derived from pyruvate. In the pathway to valine, the α-keto acid acceptor is pyruvate itself, and acetolactate is formed. For isoleucine, it is α-ketobutyrate. Thus, for

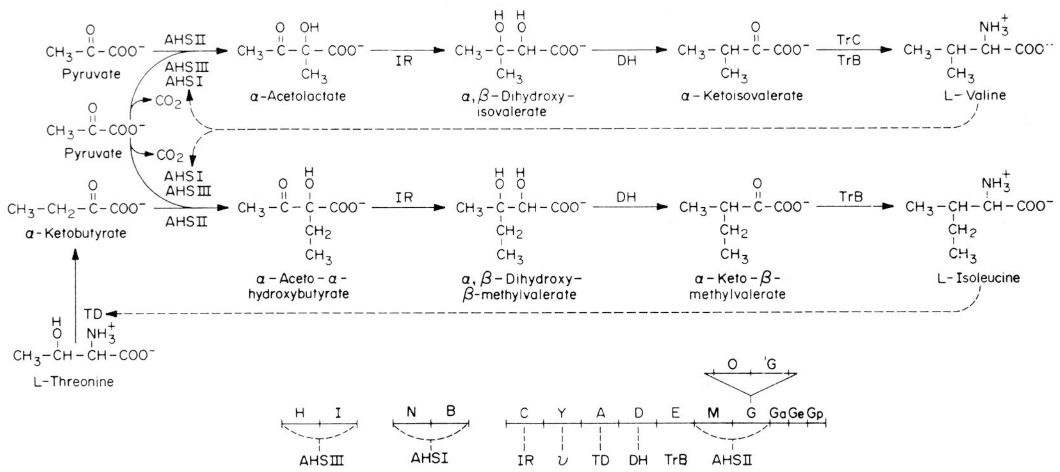

FIG. 1. Biosynthesis of isoleucine and valine. The enzymes catalyzing the biosynthetic steps are abbreviated, and the corresponding structural genes are indicated, as follows: TD (*ilvA*), threonine deaminase; AHSI (*ilvBN*) and AHSIII (*ilvIH*), end product-inhibited acetohydroxy acid synthases; AHSII (*ilvGM*), end product-noninhibited acetohydroxy acid synthase; IR (*ilvC*), acetohydroxy acid isomeroreductase; DH (*ilvD*), dihydroxy acid dehydrase; TrB (*ilvE*), transaminase B; TrC (*avtA*), transaminase C. *ilvG* and *ilvM* are inactive in *E. coli* K-12, owing to a polar lesion in the *ilvO* region of the *ilvG* gene. Mutations in *ilvO* result in *ilvGM* function and increased downstream expression. υ (*ilvY*) is a positive control element required for substrate induction of the *ilvC* gene. Gp, Ge, and Ga are the promoter, leader region, and attenuator, respectively, of the *ilvGMEDA* operon.

valine, the condensation step is the one that serves to draw off a central metabolite specifically into its pathway; i.e., it is the first specific step in the pathway. For isoleucine, another specific step is required since α-ketobutyrate, required for acetohydroxybutyrate formation, is not an intermediate in the central metabolic routes. For this function, both *E. coli* and *S. typhimurium* use threonine deaminase, the product of the *ilvA* gene.

The enzymes forming acetolactate and acetohydroxybutyrate present a complex picture. Three different isozymes of acetohydroxy acid synthase have been demonstrated in *S. typhimurium* and *E. coli*. The activities of two, acetohydroxy acid synthase I and acetohydroxy acid synthase III, are inhibited by valine. Thus, they exhibit the pattern of end product sensitivity found so often among initial enzymes in biosynthetic pathways. The activity of the third isozyme, acetohydroxy acid synthase II, is not sensitive to valine and probably accounts for the fact that many bacteria overproduce valine and excrete it into the medium, particularly late in the growth cycle.

As will be discussed below, an efficient inhibition of this enzymatic step by valine would be expected to prevent the overproduction of valine. However, since the same enzyme is also required for the second step in the isoleucine biosynthetic pathway, it follows that an efficient regulation of this step by valine would be detrimental to isoleucine biosynthesis. Indeed, the inhibition of the growth of the K-12 strain of *E. coli* is due to the fact that it cannot form acetohydroxy acid synthase II, although it can form the two valine-sensitive isozymes. The LT-2 strain of *S. typhimurium* can form isozyme II, and therefore its growth is not inhibited by valine, but it forms only one of the valine-sensitive isozymes, isozyme I (106). Other strains of the two organisms have not been examined in sufficient detail to determine the most common isozyme pattern.

It is possible that still different isozymes will be found among other wild-type organisms, since Robinson and Jackson (96) have reported evidence for additional "cryptic" genes in *E. coli* that specify other acetohydroxy acid synthases after mutations that allow their expression. These isozymes, which seem to be formed only after certain mutations have occurred, will be considered later.

The gene-enzyme relationship among the three isozymes now seems to be clear. Acetohydroxy acid synthase II, purified from *S. typhimurium*, consists of two kinds of subunits, one large (59.3 kilodaltons [kDa]) and one small (9.7 kDa) (97). The gene for the larger subunit had earlier been identified as the *ilvG* gene by analysis of mutants and by nucleotide sequence determination (38, 74, 103). Nucleotide sequence determination had also revealed an open reading frame immediately downstream of the *ilvG* gene which Schloss et al. (97) showed to account for the small subunit and designated *ilvM*. That the *ilvM* gene is essential has recently been shown by Lu and Umbarger (manuscript submitted for publication). In the absence of the *ilvM* protein, no in vitro activity could be measured even with a high-copy (pBR322 derivative)- or moderate-copy (pACYC derivative)-number plasmid carrying the *ilvG* gene alone. However, strains carrying such plasmids could grow slowly in minimal medium supplemented with valine alone.

Early work with acetohydroxy acid synthase mutants, which was based on the assumption of a single gene, had led to the inference that gene *ilvB* lay at one terminus of the *ilv* gene cluster. Later studies by Newman and Levinthal (90) demonstrated that *ilvB* is, in fact, on the opposite side of the chromosomal replication origin from the *ilv* gene cluster. A recent purification of acetohydroxy acid synthase I by Eoyang and Silverman (36) revealed that, along with the *ilvB* product (60 kDa), a smaller protein (9.5 kDa) was purified. Nucleotide sequence determination of the *ilvB* gene region has revealed an open reading frame downstream of the *ilvB* gene which could specify a subunit of 11.1 kDa (40, 133). This gene has been designated *ilvN* and, with *ilvB*, constitutes a single operon.

Recently, Eoyang and Silverman (37) have examined acetohydroxy acid synthase-negative strains of *E. coli* transformed with a high-copy-number plasmid carrying the *ilvBN* operon from which most of the *ilvN* gene had been deleted. They found that the deletion reduced the activity of extracts by about 90%, but the activity was quite resistant to valine inhibition. (Actually, maximal inhibition of acetohydroxy butyrate synthesis by valine was about 40%, whereas acetolactate synthesis was stimulated by valine.)

The analysis of mutants that led DeFelice and colleagues (30) to the discovery of acetohydroxy acid synthase III in *E. coli* first involved mutations that resulted in the loss of sensitivity of this isozyme to valine, which they identified as *ilvH* mutations, and subsequently mutations that resulted in loss of the activity itself, which they designated *ilvI* mutations. Nucleotide sequence determination has demonstrated an operon consisting of the *ilvI* gene accounting for a 61.8-kDa subunit and the *ilvH* gene accounting for a 17.5-kDa subunit, products readily found in minicells containing an *ilvIH* plasmid (107). Furthermore, Squires et al. (105) have demonstrated some homology between the *ilvH* and *ilvM* genes and between the *ilvI* and *ilvG* genes. However, this homology appears to be less than that between the *ilvBN* genes and the *ilvGM* genes (133).

Whether the genes found by nucleotide sequence analysis correspond to the properties described during the analysis of mutants has not been established. However, clones containing an intact *ilvI* gene but only an incomplete *ilvH* gene have been shown to exhibit only very low levels of acetolactate-forming activity (107). Interestingly, the activity was insensitive to valine, an observation in accord with the original hypothesis concerning the role of *ilvH*.

Thus, it appears that for all three isozymes the small subunit is important for maximal catalytic activity of the large subunit and contributes importantly to the regulation of the activity of the two valine-resistant isozymes.

The next step in each pathway is the isomerization and NADPH-dependent reduction of the acetohydroxy acids by the isomeroreductase, specified by the *ilvC* gene, to yield the α,β-dihydroxy acids. These intermediates, in turn, undergo dehydration reactions to yield the α-keto acid precursors of isoleucine and valine. These reactions are catalyzed by dihydroxy acid dehydratase, the product of the *ilvD* gene.

The final step in the two pathways is a transamination between the two α-keto acids and glutamate. Transaminase B, or the branched-chain amino acid transaminase, is absolutely required for isoleucine biosynthesis. It is the product of the *ilvE* gene. The enzyme has been purified from both *S. typhimurium* and *E. coli* and has been shown to be a 182-kDa hexamer of identical subunits (3, 78, 81). Nucleotide sequence of the *ilvE* gene of *E. coli* revealed a calculated subunit molecular weight of 34,055 (72). The enzyme is also important for the formation of leucine and valine. As will be shown below, there is a second

glutamate transaminase that can form leucine. In addition, there is a second transaminase that converts α-ketoisovalerate to valine with alanine as the amino donor. The same enzyme also interconverts α-ketobutyrate and α-aminobutyrate. Because of these additional transaminases, mutants lacking transaminase B exhibit an absolute requirement only for isoleucine but grow much better when all three branched-chain amino acids are present. The alanine-valine transaminase is specified by the *avtA* gene, which is unlinked to the *ilv* gene cluster (135).

Also shown in Fig. 1 is the arrangement of the *ilv* genes on the chromosome of *S. typhimurium* and *E. coli*. The *ilv* gene cluster contains an operon consisting of genes *G*, *M*, *E*, *D*, and *A*. In addition, the cluster also contains the *ilvC* gene and the positive control element that is needed for *ilvC* expression, *ilvY*, which will be described below.

Although the genes for the valine-insensitive acetohydroxy acid synthase II lie within the *ilv* gene cluster, the genes for the two valine-sensitive acetohydroxy acid synthases do not. The *ilvBN* genes specifying acetohydroxy acid synthase I lie about 2.5 min counterclockwise (as the chromosome map is usually drawn) from the *ilv* gene cluster and are thus unlinked to the cluster by either P1 or P22 transduction. The structural genes for acetohydroxy acid synthase III, *ilvI* and *ilvH*, lie on the *ara* side and adjacent to the *leu* operon. This location becomes of interest when the regulatory pattern of the *ilvIH* operon is considered.

The nucleotide sequences of most of the *ilv* genes have been reported. These reports are cited in Table 1.

There is a second route to α-ketobutyrate and isoleucine that has been demonstrated in *E. coli* and that may be important in some strains and under certain growth conditions. This route is via β-methylaspartate, which is presumed to be formed from glutamate by glutamate mutase. Part of the pathway was first demonstrated by Abramsky and Shemin (2), who showed that β-methylaspartate underwent transamination in *E. coli* to yield β-methyloxaloacetate. This unstable β-keto acid was decarboxylated to yield α-ketobutyrate. Later, the formation of β-methylaspar-

TABLE 1. Genes for branched-chain amino acid biosynthetic enzymes for which nucleotide sequences have been reported

Gene	Enzyme or subunit	Reference(s)
ilvB	Acetohydroxy acid synthase I, large subunit	40, 133
ilvN	Acetohydroxy acid synthase I, small subunit	40, 133
ilvG	Acetohydroxy acid synthase II, large subunit	74
ilvM	Acetohydroxy acid synthase II, small subunit	74
ilvI	Acetohydroxy acid synthase III, large subunit	107
ilvH	Acetohydroxy acid synthase III, small subunit	107
ilvY	υ, Positive regulatory element for *ilvC* expression	132
ilvC	Acetohydroxy acid isomeroreductase	132
ilvE	Transaminase B	72
leuD	α-Isopropylmalate isomerase, smaller subunit	45

tate from glutamate a sufficient rate to allow growth of an *ilvA* mutant was demonstrated in the Crookes strain of *E. coli*, provided high concentrations of glutamate were included in the medium (92). More recently, LeMaster and Cronan (79) have studied a multiply blocked strain of *E. coli* in which the tricarboxylic acid cycle and the glycolytic pathways were metabolically isolated from each other so that there was no interconversion between the two pathways. Isotope experiments revealed that about 60% of the isoleucine was made via an intermediate other than threonine and probably involved the β-methyl-aspartate pathway.

REGULATION OF METABOLITE FLOW INTO VALINE AND ISOLEUCINE

The efficiency with which metabolite flow through the pathways to valine and isoleucine is regulated is dependent solely on the sensitivity of the initial enzymes in the two pathways to end product inhibition. For isoleucine, this sensitivity would presumably be the same for any one strain under all growth conditions. In other words, the efficiency of the control appears to be dependent solely on the isoleucine sensitivity specified by the particular *ilvA* gene carried by that strain, since no regulatory subunits or modifying proteins have been found for threonine deaminase (18, 140).

For valine, the regulation can be more complex since there are both valine-sensitive and valine-insensitive acetohydroxy acid synthases. Strains can differ one from another in the particular combination of isozymes that they can form. Furthermore, acetohydroxy acid synthase activity in the same strain can exhibit different sensitivities under different growth conditions since the three predominant isozymes exhibit different repression patterns (see below).

Regulation of Threonine Deaminase Activity

The inhibition of threonine deaminase by isoleucine provides an ideal example of the way carbon flow into a specific arm of a highly branched pathway is controlled by whether there is a need for the product of that branch of the pathway, i.e., when it or a precursor is not available from the medium (121). That the control of carbon flow into the isoleucine pathway was an effective one was shown by the simple but elegant isotope incorporation experiments of the Biophysics Group of the Carnegie Institution of Washington Department of Terrestrial Magnetism (95).

Essentially, *E. coli* B was grown in a minimal medium containing uniformly labeled glucose for one generation. The cell protein was hydrolyzed, and the amino acids were separated by paper chromatography. The radioactivity, determined by a Geiger counter held directly above the amino acid spot, allowed computation of the amount of that amino acid in the cell protein and led to an early idealized composition of *E. coli*. When isoleucine was present in the medium, there was an almost complete suppression of incorporation of glucose carbon into isoleucine. That there was little or no radioactive isoleucine accumulating in the medium indicated that the preferential utilization of exogenous isoleucine was due to

a quenching of isoleucine biosynthesis and not to a rapid exchange and "swamping" of an endogenously formed pool.

Although an inhibition by isoleucine of any step in the pathway would have accounted for these results, only an inhibition of the first (irreversible) step in the pathway could also account for the sparing effect that isoleucine had on the threonine requirement of threonine auxotrophs (120). Although the concept is a simple one, the mechanism of the inhibition is complex and is worthy of further study.

Early studies by Changeux (24) showed that the inhibitor and substrate sites of threonine deaminase of *E. coli* were separate and that the interactions between substrate and inhibitor could be abolished by treating the enzyme with mercuric ions or by heating. He also showed that valine was an activator of the enzyme. Changeux (25) studied the kinetic behavior of an only partially purified enzyme in great detail and interpreted the cooperative binding exhibited by the substrate and inhibitor in terms of the concerted transition model of Monod et al. (88). Indeed, the studies with threonine deaminase were very influential in the synthesis that led to the concept of allosteric interactions and the concerted transition model itself. The model proposed essentially that symmetry between subunits of an enzyme is favored so that all of the subunits of an enzyme will be in either the active or inactive conformation. The two forms of the enzyme are in equilibrium with each other. The cooperative effects observed were accounted for by the assumption that binding of a single inhibitor molecule or a single substrate or activator molecule to the enzyme favored binding of the same ligand to other subunits.

Later kinetic and binding studies by Decidue et al. (29), Hatfield and Burns (57), and Zarlengo et al. (140) with the essentially identical enzyme purified from *S. typhimurium* were interpreted differently. They demonstrated clearly that the enzyme is a tetramer of identical subunits of 48.5 kDa (57, 140). A surprising finding was that only two of the subunits could bind pyridoxal phosphate (16). This half-the-sites reactivity may be an example of a strong negative cooperativity as described by Stallcup and Koshland (108). Another possibility is that the subunits in the disulfide-linked dimer may not be arranged in a symmetrical fashion that allows access of pyridoxal phosphate to both potential coenzyme-binding sites. There are also but two isoleucine (inhibitor) sites. In the tetrameric form of the enzyme a similar event may occur, since there is but a single valine (activator)-binding site.

The interactions between the catalytic, activator, and inhibitor sites on threonine deaminase are complex (29, 62). Binding of isoleucine to the two inhibitor sites in the presence or absence of valine is a noncooperative process. However, to prevent valine binding at the single activator site or to inhibit threonine deamination, both isoleucine sites must be occupied, and the interaction is therefore a function of [isoleucine]^2; i.e., the effect is cooperative. In the absence of isoleucine, the binding of threonine is noncooperative; i.e., binding to one catalytic site is independent of binding to the other. In the presence of isoleucine, binding of threonine is cooperative because threonine itself must bind at the activator site to displace isoleucine before it can bind at the substrate site. If sufficient valine is present, it displaces the isoleucine, and threonine binding is noncooperative. Binding to the activator site by either valine or threonine is necessary only when isoleucine is present. Although it seems clear that the catalytic sites are distinct from the activator and inhibitor sites, it is still unclear whether the activator and inhibitor sites are distinct from each other. It is possible that valine binds at one of the isoleucine sites and that, by a pronounced negative cooperativity, a second molecule of valine (as well as isoleucine) is excluded from the second site.

Regulation of Acetohydroxy Acid Synthase Activity and the Acetohydroxy Acid Synthase Isozymes

Acetohydroxy acid synthase catalyzes the second step in isoleucine biosynthesis, but it also catalyzes the first step in valine biosynthesis and thus might be expected to be inhibited by valine. Indeed, the valine sensitivity of acetohydroxy acid synthase activity in extracts of the K-12 strain of *E. coli* was invoked as evidence that the enzyme was involved in valine biosynthesis (126). At about the same time, the same kind of experiments with *E. coli* B by the Biophysics Group of the Carnegie Institution of Washington, D.C., that showed the quenching of isoleucine biosynthesis by exogenous isoleucine did not support the idea of a significant regulation of valine biosynthesis by valine. Rather, they clearly demonstrated that the utilization of exogenous valine carbons in preference to endogenously synthesized valine was due primarily to swamping of the endogenous pool through rapid exchange between internal and external valine (95). Indeed, their experiments indicated that virtually as much valine was formed in the presence of exogenous valine as was formed in its absence. Although it may not yet be possible to explain this discrepancy completely, we can be fairly certain that the presence of an active acetohydroxy acid synthase II (*ilvGM* product) in *E. coli* B and its absence in the K-12 strain is the most important factor.

Nearly all prototrophic strains of *E. coli* and *S. typhimurium* are able to grow in a minimal medium containing valine and thus presumably contain an active *ilvGM* gene product. In all strains examined, however, there is at least a partial inhibition of total acetohydroxy acid synthase activity by valine, observations implying the presence of at least one valine-sensitive isozyme. In strain LT-2 of *S. typhimurium*, this activity is attributed to acetohydroxy acid synthase I (*ilvBN*), because the *ilvIH* genes of this strain do not yield an active synthase III (106). Although no leucine-repressed *ilvIH* mRNA (see below) has been detected in *S. typhimurium*, a plasmid containing its *ilvIH* region does yield leucine-repressed *ilvIH* mRNA (see below), but not enzyme, when carried in *E. coli* (106). Furthermore, after a plasmid-borne mutation has occurred, the same plasmid DNA yields an active acetohydroxy acid synthase, presumably that due to isozyme III.

At this time, the distribution of the three isozymes in other strains of *S. typhimurium* and *E. coli* is unknown. Nor is it known whether some strains have predominant acetohydroxy acid synthases that arise from still other genes such as those that have been found as cryptic genes in the K-12 strain by Robinson and Jackson (96) (see below).

Until now, most studies on the inhibition of acetohydroxy acid synthase have been made with crude extracts or, at most, only partially purified preparations. Thus, little can be stated concerning the interaction of valine with the enzyme. Since discovery of the isozyme distribution, a few studies have been done with crude systems from mutants containing only a single isozyme. One such study would suggest that isozyme III has a lower affinity for pyruvate than does isozyme I but that isozyme III has a greater affinity for valine (33).

The addition of substrate and cofactors to acetohydroxy acid synthase I during sucrose density gradient centrifugation was shown to cause an apparent increase in its molecular weight (51). This effect was reversed when the inhibitor valine was also present. (Thus, valine may tend to dissociate the aggregated, active form of the enzyme into a less aggregated, inhibited form.) Cross-linking studies on this or other isozymes have not been reported.

Recently, a new "marker" for acetohydroxy acid synthase isozymes was discovered when a newly developed herbicide, sulfometuron methyl, was found to inhibit acetohydroxy acid synthase activity not only of plants but also of bacteria (73). However, the isozymes I of both *S. typhimurium* and *E. coli* were found to be uniquely resistant to the enzyme. The inhibitor appears to be slow binding, since the inhibitory effect increases with continued incubation of the enzyme with inhibitor (97). Ciskanik and Schloss (Abstr. 190th Natl. Meet. Am. Chem. Soc. 1985, 14, p. 3357) have further deduced that sulfometuron methyl binds to the site at which α-ketobutyrate (or the second pyruvate) binds after the first pyruvate has been converted to hydroxyethyl thiamine pyrophosphate.

Now that the three isozymes of acetohydroxy acid synthase have been so well defined genetically and biochemically, it should be possible to correlate mechanistic and kinetic analyses of the individual isozymes and inhibitor interactions with an analysis of mutants in which these parameters have been altered. Such analyses should make more clear the role of the small subunit of each of the isozymes. Finally, it should be possible to define more clearly the nature of the mutations that have been reported to have given rise to the formation of new isozymes specified by cryptic genes and which, if any, have modified isozymes I or III.

An example of a mutation giving rise to loss of valine sensitivity of acetohydroxy acid synthase activity is that at a locus called *ilvF*, at about 54 min on the *E. coli* chromosome (93). This mutation arose in a K-12 derivative that was wild type with respect to the three major isozymes. Unfortunately, the effect of the mutation has not yet been examined in cells lacking the genes for these isozymes, so it is not known whether *ilvG* affects one of these genes or specifies a novel, valine-insensitive enzyme.

A more likely possibility for a cryptic gene specifying a unique enzyme is that described by Robinson and Jackson (96). Although closely linked to *ilvIH*, it appears to have been distinguished from *ilvIH*, and its product appears to be different from the other isozymes that have been described. The cryptic gene in which the mutations arose has been designated *ilvJ*, and its product has been designated acetohydroxy acid synthase IV.

A question closely related to whether these mutations have affected cryptic genes or modified the product or regulation of some previously functioning gene is that concerning the nature of the carboligase reaction itself and the specificity of any of the enzymes catalyzing such reactions for their substrates. Some years ago, Juni and Heym (66) showed that pig heart pyruvate dehydrogenase, as well as that of *E. coli*, formed acetoin (or acetolactate) if the enzymes were incubated with large amounts of pyruvate. α-Ketoglutarate dehydrogenase and pyruvate dehydrogenase of *E. coli* have been shown to catalyze the transfer of active succinic semialdehyde and active acetaldehyde to such acceptors as acetaldehyde, glycolaldehyde, and glyceraldehyde in secondary carboligase reactions giving rise to a variety of derivatives (67, 98, 139). Acetohydroxy acid synthase II of *S. typhimurium* has been shown to use, to a limited extent, α-ketobutyrate as an active aldehyde donor at the site at which the first pyruvate molecule is usually attacked (L. M. Abel, M. H. O'Leary, and J. V. Schloss, Abstr. 190th Natl. Meet. Am. Chem. Soc. 1985, 14, p. 3357). Such observations lead to the idea that perhaps any enzyme containing thiamine pyrophosphate as a prosthetic group can catalyze a carboligase reaction if an α-keto acid can bind at its active site. Thus it is conceivable that an enzyme that normally forms only traces of acetolactate or acetohydroxybutyrate in a side reaction could emerge as a cryptic acetohydroxy acid synthase, after either an "up-promoter" or a mutation affecting its substrate specificity. The techniques to explore these questions are now readily available.

REGULATION OF ENZYME AMOUNT IN ISOLEUCINE AND VALINE BIOSYNTHESIS

In general terms, the enzymes required for isoleucine and valine biosynthesis are subject to multivalent repression by all three branched-chain amino acids (39). In other words, they are repressed when all three are present in excess but derepressed if the supply of one is restricted. Such a restriction can be readily imposed on an *E. coli* or *S. typhimurium* mutant blocked in one step in isoleucine and valine biosynthesis by exploiting the fact that the branched-chain amino acids share the several systems for active transport into the cell (5). Thus, when isoleucine, for example, is supplied at a relatively low concentration (10^{-4} M) and the concentrations of leucine and valine are five to ten times as high, entry of isoleucine is severely restricted, and protein synthesis is limited by isoleucine. Similarly, valine- or leucine-limited growth can be obtained by supplying valine or leucine in relative low amounts. Furthermore, as growth proceeds, the difficulty for the limiting amino acid to enter the cell becomes progressively greater.

Although initially an unexpected observation, in retrospect the involvement of leucine in this control might have been anticipated. The biosynthesis of leucine is initiated by drawing its initial substrates not solely from the pool of common intermediates in the central metabolic routes but also from α-ketoiso-

valerate, the valine precursor, as well. A metabolically analogous pattern is found in the relationship of isoleucine to the pathway to threonine in which isoleucine, as well as threonine, is involved in repression of the threonine biosynthetic enzymes.

The first hint concerning the mechanism of multivalent repression came from the seminal experiments of Eidlic and Neidhardt (35), who showed that at least some of the enzymes of the isoleucine and valine pathway were derepressed when the cells contained a temperature-sensitive valyl tRNA synthetase. These experiments were followed shortly by evidence that mutations affecting the affinity of leucyl tRNA synthetase and isoleucyl tRNA synthetase for the respective amino acids also resulted in derepression of some *ilv* gene products (4, 63, 113). It should be recalled that these experiments, as well as those with other amino acid pathways that were found to be affected by the degree of tRNA charging, did not distinguish between the altered synthetase or the reduced level of product (charged tRNA) as the cause of the derepression. However, they were widely inferred (and, as we now know, correctly so) to indicate that the charged tRNAs were the "signal" metabolites.

Pattern of Multivalent Repression in *S. typhimurium* and *E. coli*

The statement above, that the isoleucine and valine biosynthetic enzymes are subject to a mutivalent control, requires some qualification. As summarized in Fig. 2, it is the pattern for *S. typhimurium* and valine-resistant strains of *E. coli* when total acetohydroxy acid synthase activity is considered. However, when the specific product of the *ilv* genes is examined, it is found that only the *ilvGMEDA* operon is dependent on all three branched-chain amino acids for repression.

In contrast, the acetohydroxy acid synthase I is, in both organisms, derepressed only when either valine or leucine is limiting; it is not derepressed if isoleucine is limiting. Acetohydroxy acid synthase III for-

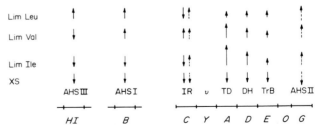

FIG. 2. Pattern of multivalent repression in isoleucine and valine biosynthesis. The arrows pointing upward indicate derepression under the indicated condition (limiting leucine, limiting valine, or limiting isoleucine). The arrows pointing downward indicate repression under the indicated condition. XS, Excess branched-chain amino acids in medium. Other abbreviations are as in the legend to Fig. 1. Broken lines represent behavior in *ilvO* (Val^r, IlvG^+) derivatives of *E. coli* K-12 or in *S. typhimurium*, where it is different from that of the K-12 itself. Size of the upward-pointing arrow is a qualitative indicator of the derepression, relative to that of transaminase B, obtained upon growing cells under various derepressing conditions. υ formation appears to be refractory to *ilv*-specific control.

mation is controlled by the level of leucine alone and is not formed at all by *S. typhimurium*. Thus, when valine is added to a minimal medium culture of the K-12 strain of *E. coli* (which contains only the valine-sensitive isozymes of acetohydroxy acid synthase), isoleucine is no longer formed and thus becomes increasingly more limiting, whereas valine (and leucine) becomes relatively more in excess. This process leads to a progressively tighter repression of both acetohydroxy acid synthases I and III.

Acetohydroxy acid isomeroreductase appears to be trivalently controlled in *S. typhimurium* but regulated by valine alone in the K-12 strain of *E. coli*. This difference can be readily explained by the fact that the isomeroreductase is induced by its substrates, which are formed by their acetohydroxy acid synthases (94). Since *E. coli* K-12 has only the valine-sensitive isozymes, the inducer will be formed only when valine is limiting. In contrast, most other organisms would respond to a limitation of any one of the branched-chain amino acids by derepression of acetohydroxy acid synthase II which, being insensitive to valine, would readily form inducing levels of one of the acetohydroxy acids even when valine is present at high concentrations.

In the sections that follow, the mechanisms underlying the patterns of regulation outlined above will be considered. The trivalently controlled *ilvGMEDA* operon and the bivalently controlled *ilvBN* operon will be considered first. For both, repression is dependent on ample charging of the cognate tRNAs by the repressing amino acids.

Regulation of the *ilvGMEDA* Operon by Attenuation

As was to be expected from the fact that multivalent repression of the *ilvGMEDA* operon was dependent on unrestricted transfer of all three branched-chain amino acids to the cognate tRNAs, control of transcription into the structural genes is by an attenuation of transcription at the end of the leader region (75, 89). The general features and importance of this kind of regulation are reviewed elsewhere in this volume by Landick and Yanofsky (chapter 77), so that it is appropriate to review only the specific features of the *ilvG* leader and its transcription here.

Briefly, the leader is a transcript of about 183 bases that can be terminated at a rho-independent site. The upstream arm of the potential stem and loop of this terminator can also be base paired with a region upstream to form an alternative stem-and-loop structure, the preemptor, or antiterminator. If the preemptor were formed, transcription would not be terminated but would, instead, be continued into the *ilvG* gene and the rest of the operon. Preemptor formation itself can be prevented because the bases of the upstream arm of that stem and loop can be base paired with bases still farther upstream to yield a protector. The formation of the protector stem instead of a preemptor stem would thus allow formation of the intact terminator and presumably assure termination at that site (attenuation of transcription or repression of the operon).

Modulation and selection of these mutually exclusive potential structures by the supply of the branched chain amino acids is possible because the leader specifies a 32-amino acid peptide of which 15 are branched-chain amino acids. Stalling of the lead ribosome at a codon for one of the branched-chain amino acids can be readily pictured as a means to preclude protector formation, assure preemptor formation, and allow transcription into the operon (derepression). If these amino acids are in excess, ribosome stalling would not occur, and translation would proceed to the amber codon of the leader. In so doing, those bases in the leader that provide the upstream arm of the protector would be covered by the ribosome and could not interfere with terminator formation (repression).

Because of the protector stem and loop, transcription will be terminated at the end of the leader when no ribosomes are able to initiate translation, as might occur, for example, in the presence of chloramphenicol.

As emphasized in the general chapter on attenuation control (chapter 77), an important feature in attenuation is thought to be the presence of polymerase pause sites, which presumably allow the relatively more slowly moving, but otherwise unimpeded, lead ribosomes to approach more closely the transcribing polymerases. Evidence for polymerase pausing at one specific site has been demonstrated for the *ilvGMEDA* operon (61). Such pausing would presumably prevent preemptor formation, except for a transcript being translated by a ribosome severely retarded by a limitation for leucine, valine, or isoleucine.

The outline described above is a very simplified one of the attenuation process. A variety of additional stem-and-loop structures, some mutually exclusive of each other, in addition to those described above, can be postulated for the *ilvGMEDA* leader. The importance of these additional structures cannot presently be assessed. Such factors as codon selection, relative rate of transcription and translation, and pause sites could all be involved in whether attenuation or continued transcription into the rest of the operon occurs. What appears to have been selected in the nucleotide sequence of the *ilv* leader regions of *E. coli* and *S. typhimurium* (which are identical in those parts of the sequences that are thought to be responsible for modulation of transcription [54, 114]) is a subtle balance of these factors that results in an efficient, multivalent control of gene expression in the operon.

Attenuation control appears to be the only specific control over expression of the *ilvGMEDA* operon. Loss of the attenuator by deletion results in loss of response to either excess or limiting levels of the branched-chain amino acids (7; J.-W. Chen and H. E. Umbarger, unpublished observations). What apparent control there is may be attributed to minor modulations of gene expression that occur within the operon itself. Indeed, it has been shown that the expression of the *ilvGMEDA* operon is not coordinate (103).

Noncoordinant expression is particularly noted with threonine deaminase, which is derepressed more than are the other products of the operon upon limitation by the branched-chain amino acids, as implied in the diagram in Fig. 2 (103). The differential derepression is greatest with isoleucine limitation and least with leucine limitation. Thus, there is, in effect, a mechanism that allows threonine deaminase, which is needed only for isoleucine biosynthesis, to be formed at the highest rate when isoleucine itself is limiting. The basis for the noncoordinant expression has not been established. A recent nucleotide sequence analysis of the *ilvA* gene has failed to reveal any obvious sequences, suggesting an *ilv*-controlled attenuation in the early region of the *ilvA* gene (E. Garrison, Ph.D. thesis, Purdue University, West Lafayette, Ind., 1986). However, the analysis did reveal that isoleucine residues comprise 3.9% of the total residues in threonine deaminase, whereas leucine and valine comprise 11.5 and 7.9%, respectively. The values for isoleucine, leucine, and valine residues in transaminase B, as deduced from nucleotide sequence data (72), are 6.5, 6.7, and 8%, respectively. Thus, the differential response to limitation of the three amino acids may be due to the relative amounts of these residues in the two proteins. Interestingly, the subunits of the two proteins have an average branched-chain amino acid composition very close to that of *E. coli* bulk protein (chapter 2).

An additional factor in the noncoordinate expression of the *ilvGMEDA* operon is the presence of a promoter within the *ilvM* gene (8, 19, 82, 131). This promoter apparently does not respond to *ilv*-specific regulation and, therefore, dampens the extent of derepression of the *ilvE*, *ilvD*, and *ilvA* genes, relative to the extent of derepression observed when *ilvG* and *ilvM* expression (the valine-resistant acetohydroxy acid synthesis) is determined.

Although *ilvD* and *ilvA* are not coordinately expressed, they are nevertheless translationally coupled, and *ilvD* mutations often cause polar effects on *ilvA* expression. The strongest evidence for such coupling was that of Blazey and Burns (13), who observed that Tn*10* in *ilvD* prevented expression of *ilvA*. Furthermore, when *ilvD* expression was prevented by Tn*10* insertion (in one orientation) in *ilvE*, translation was also prevented in *ilvA*.

Mutations Increasing *ilvGMEDA* Expression

Most mutations causing increased expression of the *ilvGMEDA* operon in *E. coli* and *S. typhimurium* that have been reported until now can be explained on the basis of interference with attenuation control. These include mutations affecting the amount of branched-chain aminoacyl-tRNA synthetase activity (4, 12, 35, 63, 83, 113), tRNA modification enzymes (14, 26, 55, 99), or tRNA isoacceptor profiles (44). The test for whether these mutations have led to derepression by bypassing or modifying attenuation control would be to examine their effect in strains in which the terminator portion of the *ilvGMEDA* leader has been deleted. Unfortunately, this test has not been applied to such mutations.

However, other mutations have been found that lead to enhanced expression of the *ilvGMEDA* operon and that appear to be independent of attenuation (65, 70). Whether these mutations are actually involved in *ilv*-specific regulation (i.e., whether they respond to physiological signals such as end product availability) will be discussed below.

Is There a Special Role for Threonine Deaminase in Regulation of *ilv* Gene Expression?

At one time, an elaborate model was in vogue that portrayed threonine deaminase as the key element in the regulation of *ilv* gene expression (best summarized in reference 17). Much of the case made for such a role for threonine deaminase was based on one *E. coli* mutant (93), now known to have multiple lesions, and the intriguing observation that threonine deaminase, reconstituted from *in vitro*-resolved apoenzyme, could bind leucyl tRNA (58). The binding was abolished after maturation of the reconstituted enzyme by binding with either positive (e.g., valine) or negative (e.g., isoleucine) effectors. There is, however, no evidence that this immature form of the enzyme lies on the in vivo pathway to functional threonine deaminase. Most of the evidence for the model can now be dismissed in light of more recent observations, most of which are adequately accounted for by the attenuation model or by the multiple lesions in the unusual mutant. However, there is one mutation, the *ilvA538* mutation, that remains unexplained and that implies at least an auxiliary or indirect role of *ilvA* in the control of *ilv* gene expression (80). The *ilvA538* mutation is accompanied by greatly reduced expression of the genes for isoleucine and valine biosynthesis and those for isoleucyl and valyl tRNA synthesis (80, 101). These regulatory effects are overcome by Val^r mutations in *ilvH* or in the *ilvO* region of *ilvG* (53, 102). Although there is little experimental evidence, a model has been proposed in which isoleucyl and valyl tRNA synthetases are inhibited by virtue of a complex between synthetase, threonine deaminase, and α-ketobutyrate, making threonine deaminase, in effect, an antiterminator (101). Although such a process could lead to an autocatalytic increase in *ilv* gene expression, an unlikely consequence, the possibility of such an interaction affecting regulation directly or indirectly should be kept in mind.

Regulation of the *ilvBN* Operon

As noted above, total acetohydroxy acid synthase activity is influenced by the branched-chain amino acids differently in *E. coli* and *S. typhimurium*. However, when the *ilvBN* operon specifying acetohydroxy acid synthase I is specifically examined, one finds that its behavior in the two organisms is the same; there is bivalent repression by valine and leucine (derepression upon limiting the supply of either).

The classical results of Eidlic and Neidhardt (35) that showed derepression of acetohydroxy acid synthase activity in *E. coli* upon restricted valyl tRNA formation are now known to be due exclusively to derepression of the *ilvBN* operon. Examination of *E. coli* strains lacking one or the other of the acetohydroxy acid synthase isozymes has indicated that restricted formation of leucyl tRNA also leads to derepression of the *ilvBN* operon (unpublished observations). These physiological experiments led to the expectation that the *ilvB* gene would be controlled by an attenuation mechanism. Nucleotide sequence analysis did, in fact, demonstrate that the expected leader contained an open reading frame for a peptide highly enriched for leucine and valine (41). The nucleotide sequence of the entire operon also accounted for the demonstration by Eoyang and Silverman (36) that acetohydroxy acid synthase I was composed of two kinds of subunit (40, 133). The *ilvB* gene is promoter proximal, and *ilvN* is immediately downstream of the *ilvB* gene.

Although the *ilv*-specific control of the *ilvBN* operon can be readily explained by the attenuation of transcription, the regulation of the operon is more complex. For example, Coukell and Polglase (27) showed that acetohydroxy acid synthase activity of *E. coli* B was catabolite repressed. Later, experiments with *E. coli* K-12 by Whitlow and Polglase (137) showed that acetohydroxy acid synthase became derepressed during growth on acetate or during growth on glucose when cyclic AMP (cAMP) was added. Although there was no change in sensitivity of the enzyme to valine, cell growth on acetate was not inhibited by 1 mM valine. The catabolite repressibility of acetohydroxy acid synthase was also demonstrated by Harvey (56), who noted derepression of the activity in cells grown in a chemostat under conditions of glucose limitation. After the isozymic nature of the acetohydroxy acid synthases became clear, Sutton and Freundlich (111) demonstrated that it was isozyme I that was catabolite repressible.

Although catabolite repressibility of a biosynthetic enzyme may be unexpected, a recent observation reported by Dailey and Cronan (28) allows an appreciation of the importance of catabolite control of *ilvB* expression. These investigators studied a class of mutants of *E. coli* K-12 unable to grow on acetate as the sole carbon source unless isoleucine and valine were present. The block in synthesis was found to be due to a lesion in the *ilvBN* operon. The strain still expressed *ilvIH* during growth on acetate (or oleate), but the enzyme was apparently unable to form acetohydroxy acid at a rate sufficient for growth. The deficiency is probably explained by the fact (see above) that the affinity of acetohydroxy acid synthase III for pyruvate is much lower than that of acetohydroxy acid synthase I (33). The level of pyruvate would, of course, be low during growth on acetate. Under these conditions, acetohydroxy acid synthase I, with its higher affinity for pyruvate, would be essential for valine and isoleucine biosynthesis. This role for the *ilvBN* gene products would be enhanced by the greater expression of the operon that is observed in the presence of elevated cAMP level (111). Indeed, it may be the greater affinity of isozyme I for pyruvate that provided the selection pressure for the cAMP stimulation of *ilvBN* expression. For many carbon sources that give rise to elevated cAMP levels, the pyruvate levels would be relatively low.

In this context, the observations of Whitlow and Polglase (137) that growth of the K-12 strain on acetate is resistant to valine becomes understandable. The isozyme that is less sensitive to valine and has the greater affinity for pyruvate is the one that is derepressed.

Two other effects on expression of the *ilvBN* operon occur in mutants that until now remain unexplained. One is the concerted effect caused by mutations in both the *cpxA* and *cpxB* genes (86). The other is the effect of mutations in either of the genes (*himA* or *himD*) for the host integration factor subunits (43, 46).

Temperature-sensitive lesions in *cpxA* and *cpxB* together lead to loss of acetohydroxy acid synthase I activity upon shift of the cultures to 41°C (85). These mutations appear not to prevent formation of the enzyme but to prevent its function (112). The same lesions prevent DNA donor activity in cells carrying normal F episomes. It was proposed that the lesions also prevent formation of some factor needed for in vivo activity of the enzyme, perhaps a membrane component. This idea is an attractive one because *cpx* mutations do alter the protein composition of the inner and outer bacterial membrane (85). Furthermore, it has been postulated that acetohydroxy acid synthase I activity is influenced by interactions with the inner membrane (11).

The *himA* (α subunit of host integration factor) and *himB* (β subunit) mutations have been shown to affect several functions but were originally identified as mutations preventing λ-mediated, site-specific recombination (integration and excision) (46). In the presence of *himA* or *himD* mutations, transcription of the *ilvBN* operon is strongly inhibited, so that, in the presence of leucine (which represses *ilvIH*) or in the absence of an intact *ilvIH* operon, growth is markedly slowed or completely prevented. The effect on the *ilvGMEDA* operon appeared to be less striking, but its derepression seemed to be prevented.

cAMP-Dependent Transcription and Attenuation in the *ilvBN* Operon

The nucleotide sequence of the *ilvBN* operon and the associated promoter and leader regions account not only for the two subunits specified by the *ilvBN* operon but also for the bivalent control by leucine and valine and the catabolite sensitivity of the operon (41, 133). Transcription of the operon can be attenuated at the end of the leader region to yield a transcript of about 188 bases. This transcript specifies a 32-amino-acid peptide of which 12 are valine or leucine, the regulatory amino acids for the operon. The attenuation model, analogous to those from other attenuation-controlled amino-acid-forming operons, is postulated to depend on formation of a rho-independent terminating stem and loop when the valyl and leucyl tRNA are in ample supply or when there is no protein synthesis but formation of a preemptor when the lead ribosome has stalled at a valine or leucine codon (41, 133).

Of some interest have been the codons in the *ilvBN* leader transcript for amino acids other than valine or leucine. The question arose of whether any of these amino acid codons could also have regulatory effects. Experiments of Hauser and Hatfield (60), based on hybridization with probes binding preattenuation (leader) transcripts and probes binding postattenuation (readthrough) transcripts, indicated that limitation of threonine and alanine led to derepression (deattenuation) of *ilvBN* expression. In contrast, the experiments of Tsui and Freundlich (119), based on enzyme activity measurements, indicated that limitation only for valine or leucine led to derepression. Limitation of any of the other nine amino acids specified by the leader codons did not lead to derepression. The basis for this apparent contradiction is unknown. However, it should be pointed out that the experiments of Hauser and Hatfield (60) were short-term, pulse-labeling experiments, whereas those of Tsui and Freundlich (119) involved longer-term starvation periods.

The sequence in the region of the *ilvBN* promoter reveals a nearly consensus cAMP receptor protein

(CRP)-binding site (40). Interestingly, this site is also an RNA polymerase-binding site but apparently a nonproductive one, since in vitro transcripts have not been observed to be initiated from it (42). The functional RNA polymerase-binding site, downstream of the CRP-binding site, binds RNA polymerase alone less well than does the CRP site, but binding of RNA polymerase by the functional promoter is improved if the CRP site is filled with a CRP-cAMP complex. In vitro transcription experiments are in accord with these binding experiments, as well as with the cAMP control of the *ilvBN* operon (42, 133).

Regulation of the *ilvIH* Operon

The regulation of the *ilvIH* operon, which specifies acetohydroxy acid synthase III, remains unexplained. It is repressed in the presence of leucine (31). For this reason, *E. coli* K-12 mutants that have lost *ilvBN* function are sensitive to leucine. The mechanism for the repression exerted by leucine is unknown.

A hint that *ilvIH*, like the other *ilv* genes, might be subject to an attenuation control came from the report of a regulatory lesion linked to but distinct from the *ilvIH* locus that resulted in a reduced repressibility of the *ilvIH* operon and an altered distribution of leucyl tRNA species (128). However, subsequent nucleotide sequence analysis of the *ilvIH* locus and the upstream region provided no evidence of attenuator-like structures in the leader region (105). The upstream region might be a target for a positive control element, and if that element was one controlled by attenuation, the level of leucyl tRNA might exert regulation indirectly. However, in experiments in this laboratory (data not shown), no derepression of *ilvIH* was noted under conditions of restricted leucyl tRNA synthetase activity that did result in the derepression of the *ilvBN* operon. Thus far, other mutant loci that have been shown to affect expression of one or more of the *ilv* or *leu* genes do not appear to have been examined specifically for their effects on *ilvIH* expression.

An important region for control of the *ilvIH* operon is that between 200 and 350 bases upstream of the transcription start site (59). In vivo, this region is needed for leucine repression, as well as for maximal expression. In vitro, transcription is only poorly transcribed from the major in vivo promoter. Predominant in vitro transcription occurs at three sites upstream of the in vivo promoter. These upstream promoters lie in the region that is needed for enhancement of transcription in vivo. Whether these sites bind RNA polymerase or some leucine-modulated regulatory protein is not yet clear.

Regulation of the *ilvC* Gene by Substrate Induction

The apparent difference between the regulatory behavior of the *ilvC* gene in *S. typhimurium* and that in the K-12 strain of *E. coli* can be explained completely by the fact that its gene product, acetohydroxy acid isomeroreductase, is induced by either of its substrates (94). In *S. typhimurium* and other organisms containing a functional *ilvG* gene, limiting the isoleucine supply of an isoleucine and valine auxotroph results in an accumulation of the isoleucine precursor

preceding the blocked reaction and, hence, increased carbon flow through the pathway. In the K-12 strain, carbon flow is blocked at the acetohydroxy acid synthase step by valine, which is in excess when isoleucine is limiting. With limiting valine, the block is removed, and flow through the valine pathway can occur with a resulting induction of the *ilvC* gene by the increased amount of acetolactate that is formed. In *S. typhimurium* or K-12 derivatives containing a functional *ilvG* gene, there is no absolute block in acetohydroxy acid formation. Since acetohydroxy acid butyrate is also an inducer, limiting isoleucine or limiting leucine, either of which leads to derepression of acetohydroxy acid synthase II activity, results in induction of the *ilvC* gene.

The induction process was first studied in vitro with an *ilvC-lacZ* operon fusion phage which served as template in a coupled transcription-translation system from *E. coli* (138). However, it was found that only strains containing an intact *ilvY* gene (adjacent to *ilvC*) yielded S-30 extracts exhibiting the in vitro induction (130). The isolation of constitutive *ilvY* mutants in which inducer was no longer needed, and the *trans* dominance of the constitutive *ilvY* allele over an *ilvY*+ allele indicated that υ, the *ilvY* product, served as a positive control element (10).

The expression of the *ilvY* gene itself is autogenously controlled (as shown by examining S-30 extracts containing a template bearing both an *ilvY* gene and an *ilvC* gene) (J. D. Falk, Ph.D. thesis, Purdue University, West Lafayette, Ind., 1986). Extracts containing preformed υ produce isomeroreductase if acetohydroxybutyrate, the inducer, is present but if very little υ is present. If the S-30 extract was prepared from a strain containing a deletion covering *ilvY*, υ was formed, but no isomeroreductase was formed (until late in the incubation period, when sufficient υ had accumulated to allow induction).

Nucleotide sequence determination and S1 mapping experiments by Wek and Hatfield (132) revealed that the *ilvC* and *ilvY* genes are divergently transcribed from promoter regions that are partially overlapping. Their model attractively accounts for the apparent reciprocal effect of inducer and υ on *ilvC* and *ilvY* expression. The model may also explain the Tn*10* insertion mutant that prevents expression of both *ilvY* and *ilvC* and that therefore requires both *ilvC* and *ilvY* for complementation (117; Falk, Ph.D. thesis).

Regulation of Expression of the *avtA* Gene

There is an unresolved discrepancy concerning the way expression of the *avtA* gene is regulated. Earlier experiments involving the *ilvE* gene, specifying transaminase B, provided evidence that variants with altered regulation of formation of the alanine-valine transaminase could be selected in an *ilvE* background (87). Whether such mutants appeared depended on how the *ilvE* mutants were grown. An *ilvE* lesion transferred to a wild-type K-12 strain yields organisms with an absolute requirement only for isoleucine, but growth is slow. These isolates grow at the parental growth rate with all three branched-chain amino acids. They initially grow more poorly with either isoleucine-valine or isoleucine-leucine than with isoleucine alone. The slow growth on isoleucine

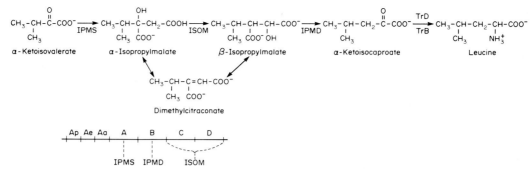

FIG. 3. Biosynthesis of leucine. The enzymes catalyzing the biosynthetic steps are abbreviated, and the corresponding structural genes are indicated, as follows: IPMS (*leuA*), isopropyl malate synthase; ISOM (*leuC* and *leuD*), isopropyl malate isomerase; IPMD (*leuB*), β-isopropyl malate dehydrogenase; TrD (*tyrB*), transaminase D; TrB (*ilvE*), transaminase B. Ap, Ae, and Aa are the promoter, leader region, and attenuator, respectively, of the *leu* operon.

plus valine can be overcome with either leucine or α-ketoisovalerate; the slow growth on isoleucine plus leucine can be overcome by valine. These results led to the hypothesis that the alanine-valine transaminase gene was repressed by either valine or leucine. The *ilv* variants appeared to have an altered control over *avtA* expression in which repression occurred only with valine plus leucine. Such *ilvE* mutants grew well on isoleucine plus valine.

More recent studies by Berg and her colleagues (9) and Whalen and Berg (135, 136) have failed to substantiate this hypothesis. Rather, they find the enzyme to be repressed by leucine or alanine but not by valine. Repression appeared to be caused by any aliphatic amino acid that contained a nonsubstituted β-carbon (e.g., leucine, α-aminobutyrate, norvaline, and alanine, but not L-valine or L-isoleucine [9]). It is not clear, however, whether the *ilvE* background examined for *avtA* expression in these studies had not already undergone the selection for the altered control over *avtA*. Any condition that led to the presence of either valine or leucine but not both might very well have provided the selection pressure for the altered control.

GENE-ENZYME RELATIONSHIPS IN THE PATHWAY TO LEUCINE

A review covering some aspects of the biosynthesis of leucine has recently appeared (20). This pathway was studied initially in *S. typhimurium* and *Neurospora crassa* because of the extensive genetic analyses performed by Margolin (84) and Gross (52). Margolin's studies revealed four genes that constituted an operon designated *leuABCD*. There were polar mutations that indicated a promoter and control region adjacent to the *leuA* gene. The pathway itself had been correctly predicted on the basis of isotope studies to involve a lengthening of the carbon chain of α-ketoisovalerate, the valine precursor, by condensation with acetyl coenzyme A (acetyl-CoA), an isomerization, and an oxidative decarboxylation to yield α-ketoisocaproate (1, 110). α-Ketoisocaproate is the immediate precursor of leucine, to which it is converted by means of either the aromatic transaminase (*tyrB*) or the branched-chain amino acid transaminase (*ilvE*). The pathway is shown in Fig. 3. The nucleotide sequence of only one of the *leu* genes has been reported (Table 1).

The condensation between the acetyl group of acetyl-CoA and α-ketoisocaproate is catalyzed by α-isopropyl malate synthase, the product of the *leuA* gene. The enzyme from *S. typhimurium* has been purified by Kohlhaw and his colleagues (71) and Leary and Kohlhaw (77). The enzyme contains identical 50-kDa subunits. In the presence of both substrates, the enzyme exists as a tetramer. Leucine, the inhibitor (see below), favors dissociation into monomers and dimers (76).

Isopropyl malate isomerase is specified by the two promoter-distal genes in the operon *leuC* and *leuD*. The enzyme is unstable and has not been purified from *E. coli* or *S. typhimurium*. The similarity of the reaction it catalyzes to the aconitase reactions in the tricarboxylic acid cycle suggests that the two proteins might exhibit similarities. An interesting suppressing mutation has been described for the *leuD* mutation of *S. typhimurium* (68). The mutation in the *supQ* locus apparently is one that allows expression of a structural gene, *newD*, the product of which can substitute for the *leuD* product (48). The natural function of the *newD* gene product is unknown.

The final pathway-specific step is catalyzed by the NAD-dependent β-isopropyl malate dehydrogenase, the product of the *leuB* gene. The enzyme was purified from *S. typhimurium* by Parsons and Burns (91) and shown to be a dimer of 35-kDa subunits. The enzyme has a monovalent cation requirement.

This pattern of increasing the chain length by one carbon is sometimes compared with the conversion of oxaloacetate to α-ketoglutarate in the tricarboxylic acid cycle. However, in the usual citrate synthase reaction, the condensation is such (*si* citrate synthase) that the dehydration step in the isomerization of citrate to isocitrate by aconitase occurs between the original α- and β-carbons of the oxaloacetate moiety of citrate rather than between the original α-carbon of oxaloacetate and the original methyl carbon of the acetyl moiety of citrate. Interestingly, in at least some anaerobic organisms in which the conversion of citrate to isocitrate serves only for glutamate biosynthesis, the citrate synthase is of the *re* type (109). If we assume that the stereospecificity of isopropyl malate isomerase is like that of all aconitaselike enzymes, the reaction catalyzed by isopropyl-malate synthase may be like that catalyzed by the *re* citrate synthase. This general type of reaction also occurs in the conversion of α-ketoglutarate to homocitrate, an essential reaction for lysine biosynthesis in fungi (129).

The leucine-forming enzymes exhibit only a limited specificity. For this reason, the pathway has the potential of increasing the chain length of pyruvate, α-ketobutyrate, and even α-ketoisocaproate itself through one or even two cycles of elongation to yield longer keto acids. Although no mutants of either S. typhimurium or E. coli have been described as accumulating these unusual ketoacids or their aminated derivatives (e.g., α-aminobutyrate, norvaline, norleucine, and homoleucine), such mutants of S. marcescens have been described (69).

REGULATION OF CARBON FLOW INTO THE LEUCINE PATHWAY

In the classical experiments on biosynthesis performed by the Carnegie Institution Biophysics Group, leucine was one of the amino acids that caused a nearly complete suppression of endogenous synthesis in E. coli. Because these experiments were performed with cells that had been grown in the absence of leucine and that underwent only one doubling in its presence, the results clearly indicated that leucine prevented carbon flow into the pathway. These results are readily explained by end product inhibition of α-isopropyl malate synthase by leucine, an observation made when the reaction was first demonstrated in S. typhimurium extracts.

Detailed binding studies of inhibitor and the two substrates have been performed by Soper et al. (104) and Teng-Leary and Kohlhaw (115, 116). It is interesting that, in the absence of acetyl-CoA, α-ketoisovalerate can bind to only two sites on the tetramer, whereas in the presence of a second substrate (actually, propionyl-CoA was used in the experiment) all four substrate sites are filled. Half-the-sites availability is also exhibited by leucine binding. No conditions were found that allowed binding of more than two molecules of leucine per tetramer. The inhibition by leucine was nearly competitive with respect to acetyl-CoA but noncompetitive with respect to α-ketoisovalerate. Neither substrate exhibited cooperative binding to the enzyme, but leucine did.

A feedback-resistant mutant form of the enzyme from S. typhimurium has been examined and compared with the wild-type enzyme (104, 115). The altered enzyme lacks several C-terminal amino acids and is about 1,000-fold less sensitive to leucine inhibition (6). Furthermore, the mutant enzyme is not dissociated into monomers or dimers by leucine (115). Thus, the mutation has led to loss of the ability of the dimers to bind leucine preferentially, and all forms of the enzyme seem to have the same reduced affinity for leucine. The high concentration of leucine required to bring about some inhibition of the mutant enzyme evoked positive cooperativity between sites for both substrates. This effect was not observed with the lower leucine concentrations used with the wild-type enzyme.

REGULATION OF ENZYME AMOUNT IN LEUCINE BIOSYNTHESIS

The enzymes of the leucine biosynthetic pathway are repressed when leucine is in excess but are derepressed when leucine is limiting (15). The mechanism by which the operon is regulated has been revealed largely from genetic analysis of mutants with altered regulation of the leucine pathway.

The promoter and operon control regions were demonstrated genetically by the isolation of an S. typhimurium mutant carrying the leu500 mutation (84) and the isolation of derepressed and essentially constitutive mutants that were resistant to trifluoroleucine (22). Both kinds of mutation were closely linked to the operon and were adjacent to the leuA gene. Although it was not possible at that time to interpret the nature of leu operon regulation, the locations of the regulatory region, the promoter, and the structural genes were correctly deduced.

Two other classes of trifluoroleucine-resistant S. typhimurium mutants were found by Calvo et al. (21). These unlinked mutations were of special interest because they caused derepression of both the leucine-forming enzymes and the isoleucine- and valine-forming enzymes. They overproduced and excreted all three branched-chain amino acids. (In contrast, the derepressed mutants with lesions near the promoter end of the leu operon overproduced and excreted only leucine.) One class was later shown to consist of leuS mutants with leucyl tRNA synthetases exhibiting reduced affinities for leucine (4). The second class had lesions in a locus near but distinct from leuS, designated flrB, and appeared to exhibit a slightly altered leucine-isoaccepting-tRNA profile as observed by chromatographic analysis on RPC-5 columns (44). The flrB locus was so named to distinguish it from the flrA locus found by Kline (70) in E. coli B/r. The flrA locus deserves special mention.

The flrA locus lies very close to trpR. The mutants identified by Kline (70) exhibited elevated levels of the leucine, isoleucine, and valine biosynthetic enzymes. The unusual feature of the flrA locus is that, when present in a haploid genome, it leads to a small, two- to threefold derepression of the biosynthetic enzymes. However, if present in a merodiploid with the wild-type allele present in trans, the derepression is about 20-fold. The role of the locus is not known. Nor is it known whether the derepression is one that bypasses the specific, amino-acid-mediated control, thus rendering the several leu and ilv operons constitutive, or one that amplifies expression under both repressing and derepressing conditions. This phenotype has not been found in the K-12 strain, and transduction of the mutant allele to E. coli K-12 does not result in derepression (J. E. Jones and H. E. Umbarger, unpublished observations).

The leu-500 lesion mentioned above has been shown to be the result of an A-to-G transition that destroys the −10 region of the leu promoter (49). Among the mutations suppressing this promoter mutation is the supX (or top) mutation, which affects topoisomerase I and leads to enhanced supercoiling (34, 115). Whether the enhanced supercoiling allows the altered promoter to function or whether it opens a new promoter has not been shown. However, another suppressing mutation is due to a G-to-A transition in the −35 region that converted the −35 region to a sequence more nearly homologous to the consensus −10 sequence than that in the wild type and that thus created a new transcription initiation site (49). However, *in vivo*, this newly generated promoter is catabolite repressible; i.e., cAMP and CRP are required for its function (47).

Regulation of the *leu* Operon by Attenuation

The kinds of mutation that led to enhanced *leu* operon expression clearly indicated that the operon is regulated by attenuation. Nucleotide sequence determination of the DNA in the region of polymerase binding in *S. typhimurium* revealed a leader region of about 160 base pairs. The transcript specified a 28-amino-acid peptide that included four adjacent leucine residues (50). An essentially similar leader sequence was found preceding the *E. coli leu* operon (134). Interestingly, the four leucine codons in the *E. coli leu* leader and three of the four leucine codons in the *S. typhimurium leu* leader are the rarely used leucine codon, CUA. It may be that part of the effectiveness of the *leu* operon is due to the use of these less frequently used codons.

Mutations in the *leu* leader region that led to derepression of the *leu* operon have been shown to be mutations in the GC-rich region that specifies the terminator of the leader transcript (100). These mutations result in a reduced stability of the *leu* terminator and increased readthrough by the polymerase. Strains containing these mutations do not exhibit further derepression when grown in the presence of limiting leucine (100). This observation, along with the fact that the mutations unlinked to the *leu* operon that lead to derepression affect the profiles of either the leucine-accepting tRNA or the leucyl tRNA synthetase, leads to the idea that attenuation in the *leu* operon is the major (if not the only) mechanism affecting leucine-specific regulation of the *leu* operon.

Additional evidence for the major role played by attenuation in expression of the *leu* operon was shown by comparing the wild-type operon with one that contained four threonine codons in the leader transcript in place of the four leucine codons (23). The substitution led to a regulation of the *leu* operon by threonine (actually threonyl tRNA) limitation. There remained a low level of derepression by leucine limitation after the substitution was made (about 5% of that exhibited by wild-type operon). Whether that low level of derepression was due to some other leucine-mediated regulation of the operon (albeit of low efficiency) or a secondary consequence of leucine starvation (such as undermodified tRNAs which might retard translation of the leader) was not determined.

The above statement implying that attenuation is the only mechanism controlling *leu* operon expression must be tempered by the proviso that it is the only mechanism modulated by leucine availability. There is indeed a rich-medium effect on the operon, as there is on all amino acid biosynthetic pathways. It could also be that, within the cell, there are proteins that do decrease (negative control elements) or increase (positive control elements) the frequency of initiation by RNA polymerase at the *leu* promoter. The activities of such proteins, if they exist, do not seem to be influenced by the availability of leucine. Thus, in terms of the concepts underlying this essay, they would be nonspecific. Nevertheless, their activities could clearly be abolished by mutations, to result in either reduced or increased expression of the *leu* operon, respectively. Such mutations, however, would not be expected to alter the regulation that is due to the leucine-specific attenuation control.

ACKNOWLEDGMENT

The work on branched-chain amino acid biosynthesis conducted in this laboratory has been supported for the past 35 years exclusively by grants from the National Institutes of Health and since 1964 by Public Health Service grant GM12522 from the National Institutes of Health.

LITERATURE CITED

1. **Abelson, P. H.** 1954. Amino acid biosynthesis in *Echerichia coli*. Isotopic competition with C¹⁴-glucose. J. Biol. Chem. **206:**335–343.
2. **Abramsky, T., and D. Shemin.** 1965. The formation of isoleucine from β-methylaspartic acid in *Escherichia coli* W. J. Biol. Chem. **240:**2971–2975.
3. **Adams, C. W., R. P. Lawther, and G. W. Hatfield.** 1979. The *ilvEDA* operon of *Escherichia coli* K-12 encodes only one valine-α-ketoglutarate transaminase activity. Biochem. Biophys. Res. Commun. **89:**650–658.
4. **Alexander, R. R., J. M. Calvo, and M. Freundlich.** 1971. Mutants of *Salmonella typhimurium* with an altered leucyl-transfer ribonucleic acid synthetase. J. Bacteriol. **106:**213–219.
5. **Anderson, J. P., and D. L. Oxender.** 1978. Genetic separation of high- and low-affinity transport systems for branched-chain amino acids in *Esherichia coli* K-12. J. Bacteriol. **136:**168–174.
6. **Bartholomew, J. C., and J. M. Calvo.** 1971. α-Isopropylmalate synthase from *Salmonella typhimurium*: carboxypeptidase digestion studies of parent and feedback-insensitive enzymes. Biochim. Biophys. Acta **250:**568–576.
7. **Bennett, D. C., and H. E. Umbarger.** 1984. Isolation and analysis of two *Escherichia coli* K-12 *ilv* attenuator deletion mutants with high-level constitutive expression of an *ilv-lac* fusion operon. J. Bacteriol. **157:**839–845.
8. **Berg, C. M., and K. J. Shaw.** 1981. Organization and regulation of the *ilvGEDA* operon in *Salmonella typhimurium* LT2. J. Bacteriol. **145:**984–989.
9. **Berg, C. M., W. A. Whalen, and L. B. Archambault.** 1983. Role of alanine-valine transaminase in *Salmonella typhimurium* and analysis of an *avtA*::Tn5 mutant. J. Bacteriol. **155:**1009–1014.
10. **Biel, A. J., and H. E. Umbarger.** 1981. Mutations in the *ilvY* gene of *Escherichia coli* K-12 that cause constitutive expression of *ilvC*. J. Bacteriol. **146:**718–724.
11. **Blatt, J. M., and J. H. Jackson.** 1978. Enhanced allosteric regulation of threonine deaminase and acetohydroxy acid synthase from *Escherichia coli* in a permeabilized-cell assay system. Biochim. Biophys. Acta **526:**267–276.
12. **Blatt, J. M., and H. E. Umbarger.** 1972. On the role of isoleucyl-tRNA synthetase in multivalent repression. Biochem. Genet. **6:**99–118.
13. **Blazey, D. L., and R. O. Burns.** 1982. Transcriptional activity of the transposable element Tn*10* in the *Salmonella typhimurium ilvGEDA* operon. Proc. Natl. Acad. Sci. USA **79:**5011–5015.
14. **Bresalier, R. S., A. A. Rizino, and M. Freundlich.** 1975. Reduced maximal levels of derepression of the isoleucine-valine and leucine enzymes in *hisT* mutants of *Salmonella typhimurium*. Nature (London) **253:**279–280.
15. **Burns, R. O., J. Calvo, P. Margolin, and H. E. Umbarger.** 1966. Expression of the leucine operon. J. Bacteriol. **91:**1570–1576.
16. **Burns, R. O., and M. H. Zarlengo.** 1968. Threonine deaminase of *Salmonella typhimurium*. I. Purification and properties. J. Biol. Chem. **243:**178–191.
17. **Calhoun, D. H., and G. W. Hatfield.** 1973. Autoregulation: a role for a biosynthetic enzyme in the control of gene expression. Proc. Natl. Acad. Sci. USA **70:**2757–2761.
18. **Calhoun, D. H., R. A. Rimerman, and G. W. Hatfield.** 1973. Threonine deaminase from *Escherichia coli*. I. Purification and properties. J. Biol. Chem. **248:**3511–3516.
19. **Calhoun, D. C., J. W. Wallen, L. Traub, J. E. Gray, and H.-F. Kung.** 1985. Internal promoter in the *ilvGEDA* transcription unit of *Escherichia coli*. J. Bacteriol. **161:**128–132.
20. **Calvo, J. M.** 1983. Leucine biosynthesis in prokaryotes, p. 267–284. *In* K. M. Herrmann and R. L. Somerville (ed.), Amino acids biosynthesis and genetic regulation. Addison-Wesley Publishing Co., Inc., Reading, Mass.
21. **Calvo, J. M., M. Freundlich, and H. E. Umbarger.** 1969. Regu-

lation of branched-chain amino acid biosynthesis in *Salmonella typhimurium*: isolation of regulatory mutants. J. Bacteriol. **97**:1272–1282.

22. **Calvo, J. M., P. Margolin, and H. E. Umbarger.** 1969. Operator constitutive mutations in the leucine operon of *Salmonella typhimurium*. Genetics **61**:777–787.

23. **Carter, P. W., D. L. Weiss, H. L. Weith, and J. M. Calvo.** 1985. Mutations that convert the four leucine codons of the *Salmonella typhimurium leu* leader to four threonine codons. J. Bacteriol. **162**:943–949.

24. **Changeux, J.-P.** 1961. The feedback control mechanism of biosynthetic L-threonine deaminase by L-isoleucine. Cold Spring Harbor Symp. Quant. Biol. **26**:313–318.

25. **Changeux, J.-P.** 1963. Allosteric interactions on biosynthetic L-threonine deaminase from *E. coli* K-12. Cold Spring Harbor Symp. Quant. Biol. **28**:497–504.

26. **Cortese, R., R. Landsberg, R. A. Vonder Haar, and H. E. Umbarger.** 1974. Pleiotropy of *hisT* mutants blocked in pseudouridine synthesis in tRNA: leucine and isoleucine-valine operons. Proc. Natl. Acad. Sci. USA **71**:1857–1861.

27. **Coukell, M. B., and W. J. Polglase.** 1969. Repression of glucose of acetohydroxy acid synthetase in *Escherichia coli* B. Biochem. J. **111**:273–278.

28. **Dailey, F. E., and J. E. Cronan, Jr.** 1986. Acetohydroxy acid synthase I, a required enzyme for isoleucine and valine biosynthesis in *Escherichia coli* K-12 during growth on acetate as the sole carbon source. J. Bacteriol. **165**:453–460.

29. **Decidue, C. J., J. G. Hoffler, and R. O. Burns.** 1975. Relationship between regulatory sites. J. Biol. Chem. **250**:1563–1570.

30. **DeFelice, M., J. Guardiola, B. Esposito, and M. Iaccarino.** 1974. Structural genes for a newly recognized acetolactate synthase in *Escherichia coli* K-12. J. Bacteriol. **120**:1068–1077.

31. **DeFelice, M., and M. Levinthal.** 1977. The acetohydroxy acid synthase III isoenzyme of *Escherichia coli* K-12: regulation of synthesis by leucine. Biochem. Biophys. Res. Commun. **79**:82–87.

32. **DeFelice, M., M. Levinthal, M. Iaccarino, and J. Guardiola.** 1979. Growth inhibition as a consequence of antagonism between related amino acids: effect of valine on *Escherichia coli* K-12. Microbiol. Rev. **43**:42–58.

33. **DeFelice, M., C. Squires, and M. Levinthal.** 1978. A comparative study of the acetohydroxy acid synthase isoenzymes of *Escherichia coli*. Biochim. Biophys. Acta **541**:9–17.

34. **Dubnau, E., and P. Margolin.** 1972. Suppression of promoter mutations by the pleiotropic *supX* mutations. Mol. Gen. Genet. **117**:91–112.

35. **Eidlic, L., and F. C. Neidhardt.** 1965. Role of valyl-sRNA synthetase in enzyme repression. Proc. Natl. Acad. Sci. USA **53**:539–543.

36. **Eoyang, L., and P. M. Silverman.** 1984. Purification and subunit composition of acetohydroxy acid synthase I from *Escherichia coli* K-12. J. Bacteriol. **157**:184–189.

37. **Eoyang, L., and P. M. Silverman.** 1986. Role of small subunit (IlvN polypeptide) of acetohydroxyacid synthase I from *Escherichia coli* K-12 in sensitivity of the enzyme to valine inhibition. J. Bacteriol. **166**:901–904.

38. **Favre, R., A. Wiater, S. Puppo, M. Iaccarino, R. Noelle, and M. Freundlich.** 1976. Expression of a valine-resistant acetolactate synthase activity mediated by the *ilvO* and *ilvG* genes of *Escherichia coli* K-12. Mol. Gen. Genet. **143**:243–252.

39. **Freundlich, M., R. O. Burns, and H. E. Umbarger.** 1962. Control of isoleucine, valine, and leucine biosynthesis. I. Multivalent repression. Proc. Natl. Acad. Sci. USA **48**:1804–1808.

40. **Friden, P., J. Donegan, J. Mullen, P. Tsui, and M. Freundlich.** 1985. The *ilvB* locus of *Escherichia coli* K-12 is an operon encoding both subunits of acetohydroxyacid synthase I. Nucleic Acids Res. **13**:3979–3993.

41. **Friden, P., T. Newman, and M. Freundlich.** 1982. Nucleotide sequence of the *ilvB* promoter-regulatory region: a biosynthetic operon controlled by attenuation and cyclic AMP. Proc. Natl. Acad. Sci. USA **79**:6156–6160.

42. **Friden, P., P. Tsui, K. Okamoto, and M. Freundlich.** 1984. Interaction of cyclic AMP receptor protein with the *ilvB* biosynthetic operon in *E. coli*. Nucleic Acids Res. **12**:8145–8160.

43. **Friden, P., K. Voelkel, R. Sternglanz, and M. Freundlich.** 1984. Reduced expression of the isoleucine and valine enzymes in integration host factor mutants of *Escherichia coli*. J. Mol. Biol. **172**:573–579.

44. **Friedberg, D., T. W. Mikulka, J. Jones, and J. M. Calvo.** 1974. *flrB*, a regulatory locus controlling branched-chain amino acid biosynthesis in *Salmonella typhimurium*. J. Bacteriol. **118**:942–951.

45. **Friedberg, D., E. R. Rosenthal, J. W. Jones, and J. M. Calvo.** 1985. Characterization of the 3′ end of the leucine operon of *Salmonella typhimurium*. Mol. Gen. Genet. **199**:486–494.

46. **Friedman, D. I., E. J. Olson, D. Carver, and M. Gellert.** 1984. Synergistic effect of *himA* and *gyrB* mutations: evidence that Him functions control expression of *ilv* and *xyl* genes. J. Bacteriol. **157**:484–489.

47. **Friedman, S. B., and P. Margolin.** 1968. Evidence for an altered operator specificity: catabolite repression control of the leucine operon in *Salmonella typhimurium*. J. Bacteriol. **95**:2263–2269.

48. **Fultz, P. N., D. Y. Kwoh, and J. Kemper.** 1979. *Salmonella typhimurium newD* and *Escherichia coli leuC* genes code for a functional isopropylmalate isomerase in *Salmonella typhimurium-Escherichia coli* hybrids. J. Bacteriol. **137**:1253–1262.

49. **Gemmill, R. M., M. Tripp, S. B. Friedman, and J. M. Calvo.** 1984. Promoter mutation causing catabolite repression of the *Salmonella typhimurium* leucine operon. J. Bacteriol. **158**:948–953.

50. **Gemmill, R. M., S. R. Wessler, E. B. Keller, and J. M. Calvo.** 1979. *leu* operon of *Salmonella typhimurium* is controlled by an attenuation mechanism. Proc. Natl. Acad. Sci. USA **76**:4941–4945.

51. **Grimminger, H., and H. E. Umbarger.** 1979. Acetohydroxy acid synthase I of *Escherichia coli*: purification and properties. J. Bacteriol. **137**:846–853.

52. **Gross, S. R.** 1962. On the mechanism of complementation of the *leu-2* locus of Neurospora. Proc. Natl. Acad. Sci. USA **48**:922–930.

53. **Hahn, J. E., and D. H. Calhoun.** 1978. Suppressors of a genetic regulatory mutation affecting isoleucine-valine biosynthesis in *Escherichia coli* K-12. J. Bacteriol. **136**:117–124.

54. **Harms, E., J.-H. Hsu, C. S. Subrahmanyam, and H. E. Umbarger.** 1985. Comparison of the regulatory regions of *ilvGEDA* operons from several enteric organisms. J. Bacteriol. **164**:207–216.

55. **Harris, C. L., L. Lui, S. Sakallah, and R. DeVore.** 1983. Cysteine starvation, isoleucyl-tRNA^{Ile}, and the regulation of the *ilvGEDA* operon of *Escherichia coli*. J. Biol. Chem. **258**:7676–7683.

56. **Harvey, R. J.** 1970. Metabolic regulation in glucose-limited chemostat cultures of *Escherichia coli*. J. Bacteriol. **104**:698–706.

57. **Hatfield, G. W., and R. O. Burns.** 1970. Threonine deaminase from *Salmonella typhimurium*. III. The intermediate substructure. J. Biol. Chem. **245**:787–791.

58. **Hatfield, G. W., and R. O. Burns.** 1970. Specific binding of leucyl transfer RNA to an immature form of L-threonine deaminase: its implications in repression. Proc. Natl. Acad. Sci. USA **66**:1027–1035.

59. **Haughn, G. W., C. H. Squires, M. DeFelice, C. T. Largo, and J. M. Calvo.** 1985. Unusual organization of the *ilvIH* promoter of *Escherichia coli*. J. Bacteriol. **163**:186–198.

60. **Hauser, C. A., and G. W. Hatfield.** 1984. Attenuation of the *ilvB* operon by amino acids reflecting substrates or products of the *ilvB* gene product. Proc. Natl. Acad. Sci. USA **81**:76–79.

61. **Hauser, C. A., J. A. Sharp, L. K. Hatfield, and G. W. Hatfield.** 1985. Pausing of RNA polymerase during in vitro transcription through the *ilvB* and *ilvGEDA* attenuator regions of *Escherichia coli* K12. J. Biol. Chem. **260**:1765–1770.

62. **Hoffler, J. G., and R. O. Burns.** 1978. Threonine deaminase from *Salmonella typhimurium*. Effect of regulatory ligands on the binding of substrates and substrate analogues to the active sites and the differentiation of the activator and inhibitor sites from the active sites. J. Biol. Chem. **253**:1245–1251.

63. **Iaccarino, M., and P. Berg.** 1971. Isoleucine auxotrophy as a consequence of a mutationally altered isoleucyl tRNA synthetase. J. Bacteriol. **105**:527–537.

64. **Iaccarino, M., J. Guardiola, M. DeFelice, and R. Favre.** 1978. Regulation of isoleucine and valine biosynthesis. Curr. Top. Cell. Regul. **14**:29–73.

65. **Johnson, D. I., and R. L. Somerville.** 1983. Evidence that repression mechanisms can exert control over the *thr*, *leu*, and *ilv* operons of *Escherichia coli* K-12. J. Bacteriol. **155**:49–55.

66. **Juni, E., and G. A. Heym.** 1956. Acyloin reactions of pyruvate oxidase. J. Biol. Chem. **218**:365–378.

67. **Kabasik, N. P., D. A. Richert, R. J. Bloom, R. Y. Hsu, and W. W. Westerfeld.** 1972. Pyruvate-glyoxylate carboligase activity of the pyruvate dehydrogenase complex of *Escherichia coli*. Biochemistry **11**:2225–2229.

68. **Kemper, J.** 1974. Evolution of a new gene substituting for the *leuD* gene of *Salmonella typhimurium*: origin and nature of *supQ* and *newD* mutations. J. Bacteriol. **120**:1176–1185.

69. **Kisumi, M., M. Sugiura, and I. Chibata.** 1976. Biosynthesis of norvaline, norleucine, and homoisoleucine in *Serratia marcescens*. J. Biochem. **80**:333–339.

70. **Kline, E. L.** 1972. New amino acid regulatory locus having unusual properties in heterozygous merodiploids. J. Bacteriol. **110**:1127–1134.

71. **Kohlhaw, G., T. R. Leary, and H. E. Umbarger.** 1969. α-Isopropyl-malate synthase from *Salmonella typhimurium*: purification and properties. J. Biol. Chem. **244**:2218–2225.

72. **Kuramitsu, S., T. Ogawa, H. Ogawa, and H. Kagimiyama.** 1985. Branched-chain amino acid aminotransferase of *Escherichia coli*: nucleotide sequence of the *ilvE* gene and the deduced amino acid sequence. J. Biochem. **97**:993–999.

73. **LaRossa, R. A., and J. V. Schloss.** 1984. The sulfonylurea herbicide sulfometuron methyl is an extremely potent and selective inhibitor of acetolactate in *Salmonella typhimurium*. J. Biol. Chem. **259**:8753–8757.

74. **Lawther, R. P., D. H. Calhoun, C. W. Adams, C. A. Hauser, J. Gray, and G. W. Hatfield.** 1981. Molecular basis of valine resistance in *Escherichia coli* K-12. Proc. Natl. Acad. Sci. USA **78**:922–925.

75. **Lawther, R. P., and G. W. Hatfield.** 1980. Multivalent translational control of transcription termination at attenuator of *ilvGEDA* operon of *Escherichia coli* K-12. Proc. Natl. Acad. Sci. USA **77**:1862–1866.

76. **Leary, T. R., and G. Kohlhaw.** 1970. Dissociation of α-isopropyl-malate synthase from *Salmonella typhimurium* by its feedback inhibitor leucine. Biochem. Biophys. Res. Commun. **39**:494–501.

77. **Leary, T. R., and G. B. Kohlhaw.** 1972. α-Isopropylmalate synthase from *Salmonella typhimurium*: analysis of the quaternary structure and its relation to function. J. Biol. Chem. **247**:1089–1095.

78. **Lee-Peng, F.-C., M. A. Hermodson, and G. B. Kohlhaw.** 1979. Transaminase B from *Escherichia coli*: quaternary structure, amino-terminal sequence, substrate specificity, and absence of a separate valine-α-ketoglutarate activity. J. Bacteriol. **139**:339–345.

79. **LeMaster, D. M., and J. E. Cronan, Jr.** 1982. Biosynthetic production of ¹³C-labeled amino acids with site-specific enrichment. J. Biol. Chem. **257**:1224–1230.

80. **Levinthal, M., M. Levinthal, and L. S. Williams.** 1976. The regulation of the *ilvADGE* operon: evidence for positive control by threonine deaminase. J. Mol. Biol. **102**:453–465.

81. **Lipscomb, E. L., H. R. Horton, and F. B. Armstrong.** 1971. Molecular weight, subunit structure, and amino acid composition of the branched-chain amino acid aminotransferase of *Salmonella typhimurium*. Biochemistry **13**:2070–2076.

82. **Lopes, J. M., and R. P. Lawther.** 1986. Analysis and comparison of the internal promoter, pE, of the *ilvGMEDA* operons from *Escherichia coli* K-12 and *Salmonella typhimurium*. Nucleic Acids Res. **14**:2779–2798.

83. **Low, B., F. Gates, T. Goldstein, and D. Söll.** 1971. Isolation and partial characterization of temperature-sensitive *Escherichia coli* mutants with altered leucyl- and seryl-transfer ribonucleic acid synthetases. J. Bacteriol. **108**:742–750.

84. **Margolin, P.** 1963. Genetic fine structure of the leucine operon in *Salmonella*. Genetics **48**:441–457.

85. **McEwen, J., and P. Silverman.** 1980. Genetic analysis of *Escherichia coli* K-12 chromosomal mutants defective in expression of F-plasmid functions: identification of genes *cpxA* and *cpxB*. J. Bacteriol. **144**:60–67.

86. **McEwen, J., and P. Silverman.** 1980. Mutations in genes *cpxA* and *cpxB* of *Escherichia coli* K-12 cause a defect in isoleucine and valine synthases. J. Bacteriol. **144**:68–73.

87. **McGilvray, D., and H. E. Umbarger.** 1974. Regulation of transaminase C synthesis in *Escherichia coli*: conditional leucine auxotrophy. J. Bacteriol. **120**:715–723.

88. **Monod, J., J. Wyman, and J.-P. Changeux.** 1965. On the nature of allosteric transitions: a plausible model. J. Mol. Biol. **12**:88–118.

89. **Nargang, F. E., C. S. Subrahmanyam, and H. E. Umbarger.** 1980. Nucleotide sequence of *ilvGEDA* operon attenuator region of *Escherichia coli*. Proc. Natl. Acad. Sci. USA **77**:1823–1827.

90. **Newman, T. C., and M. Levinthal.** 1979. A new map position for the *ilvB* locus of *Escherichia coli*. Genetics **96**:59–77.

91. **Parsons, S. J., and R. O. Burns.** 1969. Purification and properties of β-isopropylmalate dehydrogenase. J. Biol. Chem. **244**:996–1003.

92. **Phillips, A. T., J. I. Nuss, J. Moosic, and C. Foshay.** 1972. Alternate pathway for isoleucine biosynthesis in *Escherichia coli*. J. Bacteriol. **109**:714–719.

93. **Pledger, W. J., and H. E. Umbarger.** 1973. Isoleucine and valine metabolism in *Escherichia coli*. XXI. Mutations affecting derepression and valine resistance. J. Bacteriol. **114**:183–194.

94. **Ratzkin, B., S. Arfin, and H. E. Umbarger.** 1972. Isoleucine and valine metabolism in *Escherichia coli*. XVIII. Induction of acetohydroxy acid isomeroreductase. J. Bacteriol. **112**:131–141.

95. **Roberts, R. B., D. B. Cowie, P. H. Abelson, E. T. Bolton, and R. J. Britten.** 1955. Studies of biosynthesis in *Escherichia coli*, p. 406–417. Publication 607. Carnegie Institution, Washington, D.C.

96. **Robinson, C. L., and J. H. Jackson.** 1982. New acetohydroxy acid synthase activity from mutational activation of a cryptic gene in *Escherichia coli* K-12. Mol. Gen. Genet. **186**:240–246.

97. **Schloss, J. V., D. E. van Dyk, J. F. Vasta, and R. M. Kutny.** 1985. Purification and properties of *Salmonella typhimurium* acetolactate synthase isozyme II from *Escherichia coli* HB101/pDU9. Biochemistry **24**:4952–4959.

98. **Schlossberg, M. A., R. J. Bloom, D. A. Richert, and W. W. Westerfeld.** 1970. Carboligase activity of 2-ketoglutarate dehydrogenase. Biochemistry **9**:1148–1153.

99. **Searles, L. L., J. W. Jones, M. J. Fournier, N. Grambow, B. Tyler, and J. M. Calvo.** 1986. *Escherichia coli* B/r *leuK* mutant lacking pseudouridine synthase I activity. J. Bacteriol. **166**:341–345.

100. **Searles, L. L., S. R. Wessler, and J. M. Calvo.** 1983. Transcription attenuation is the major mechanism by which the *leu* operon of *Salmonella typhimurium* is controlled. J. Mol. Biol. **163**:377–394.

101. **Singer, P. A., M. Levinthal, and L. S. Williams.** 1984. Synthesis of the isoleucyl- and valyl-tRNA synthetases and the isoleucine-valine biosynthetic enzymes in a threonine deaminase regulatory mutant of *Escherichia coli* K-12. J. Mol. Biol. **175**:39–55.

102. **Singer, P. A., M. Levinthal, and L. S. Williams.** 1984. Reversion of the effects of a threonine deaminase regulatory mutant by a mutation in *ilvH* in *Escherichia coli* K-12. Biochem. Biophys. Res. Commun. **118**:270–277.

103. **Smith, J. M., D. E. Smolin, and H. E. Umbarger.** 1976. Polarity and the regulation of the *ilv* gene cluster in *Escherichia coli* strain K-12. Mol. Gen. Genet. **148**:111–124.

104. **Soper, T. S., G. J. Doellgast, and G. B. Kohlhaw.** 1976. Mechanism of feedback inhibition by leucine: purification and properties of a feedback-resistant α-isopropylmalate synthase. Arch. Biochem. Biophys. **173**:362–374.

105. **Squires, C. H., M. DeFelice, J. Devereux, and J. M. Calvo.** 1983. Molecular structure of *ilvIH* and its evolutionary relationship to *ilvG* in *Escherichia coli* K12. Nucleic Acids Res. **11**:5299–5313.

106. **Squires, C. H., M. DeFelice, C. T. Lago, and J. M. Calvo.** 1983. *ilvHI* locus of *Salmonella typhimurium*. J. Bacteriol. **154**:1054–1061.

107. **Squires, C. H., M. DeFelice, S. R. Wessler, and J. M. Calvo.** 1981. Physical characterization of the *ilvHI* operon of *Escherichia coli* K-12. J. Bacteriol. **147**:797–804.

108. **Stallcup, W. B., and D. E. Koshland, Jr.** 1973. Half-of-the-sites reactivity and negative co-operativity: the case of yeast glyceraldehyde-3-phosphate dehydrogenase. J. Mol. Biol. **80**:41–62.

109. **Stern, J. R., C. S. Hegre, and G. Bambers.** 1966. Glutamate biosynthesis in anaerobic bacteria. II. Stereospecificity of aconitase and citrate synthetase of *Clostridium kluyveri*. Biochemistry **5**:1119–1124.

110. **Strassman, M., L. A. Locke, A. J. Thomas, and S. Weinhouse.** 1956. A study of leucine biosynthesis in *Torulopsis utilis*. J. Am. Chem. Soc. **78**:1599–1602.

111. **Sutton, A., and M. Freundlich.** 1980. Regulation by cyclic AMP of the *ilvB*-encoded biosynthetic acetohydroxy acid synthase in *Escherichia coli* K-12. Mol. Gen. Genet. **178**:179–183.

112. **Sutton, A., T. Newman, J. McEwen, P. M. Silverman, and M. Freundlich.** 1982. Mutations in genes *cpxA* and *cpxB* of *Escherichia coli* K-12 cause a defect in acetohydroxyacid synthase I function in vivo. J. Bacteriol. **151**:976–982.

113. **Szentirmai, A., M. Szentirmai, and H. E. Umbarger.** 1968. Isoleucine and valine metabolism in *Escherichia coli*. XV. Biochemical properties of mutants resistant to thioisoleucine. J. Bacteriol. **95**:1672–1679.

114. **Taillon, M. P., D. A. Gotto, and R. P. Lawther.** 1981. The DNA sequence of the promoter-attenuator of the *ilvGEDA* operon of *Salmonella typhimurium*. Nucleic Acids Res. **9**:3419–3432.

115. **Teng-Leary, E., and G. B. Kohlhaw.** 1973. Mechanism of feedback inhibition by leucine: binding of leucine to wild-type and feedback-resistant α-isopropylmalate synthases and its structural consequences. Biochemistry **12**:2980–2986.

116. **Teng-Leary, E., and G. B. Kohlhaw.** 1975. Binding of α-ketoisovalerate to α-isopropylmalate synthase: half-of-the-sites and all-of-the-sites availability. Biochim. Biophys. Acta **410**:210–219.

117. **Tessman, I., J. S. Fassler, and D. C. Bennett.** 1982. Relative map location of the *rep* and *rho* genes of *Escherichia coli*. J. Bacteriol. **151**:1637–1640.

118. **Trueksis, M., and R. E. Depew.** 1981. Identification and localization of a gene that specifies production of *Escherichia coli*

DNA topoisomerase I. Proc. Natl. Acad. Sci. USA **78**:2164–2168.

119. **Tsui, P., and M. Freundlich.** 1985. Starvation for *ilvB* operon leader amino acids other than leucine or valine does not increase acetohydroxy acid synthase activity in *Escherichia coli*. J. Bacteriol. **162**:1314–1316.

120. **Umbarger, H. E.** 1953. Nutritional requirements of threonineless mutants of *Escherichia coli*. J. Bacteriol. **65**:203–209.

121. **Umbarger, H. E.** 1956. Evidence for a negative feedback mechanism in the biosynthesis of isoleucine. Science **123**:848.

122. **Umbarger, H. E.** 1969. Regulation of the biosynthesis of the branched-chain amino acids. Curr. Top. Cell. Regul. **1**:57–76.

123. **Umbarger, H. E.** 1971. The regulation of enzyme levels in the pathways to the branched-chain amino acids, p. 447–462. *In* Henry J. Vogel (ed.), Metabolic pathways, vol. 5. Metabolic regulation. Academic Press, Inc., New York.

124. **Umbarger, H. E.** 1978. Amino acid biosynthesis and its regulation. Annu. Rev. Biochem. **47**:533–606.

125. **Umbarger, H. E.** 1983. The biosynthesis of isoleucine and valine and its regulation, p. 245–266. *In* K. M. Herrmann and R. L. Somerville (ed.), Amino acids biosynthesis and genetic regulation. Addison-Wesley, Reading, Mass.

126. **Umbarger, H. E., and B. Brown.** 1958. Isoleucine and valine metabolism in *Escherichia coli*. VIII. The formation of acetolactate. J. Biol. Chem. **233**:1156–1160.

127. **Umbarger, H. E., and B. D. Davis.** 1962. Pathways of amino acid biosynthesis, p. 167–251. *In* I. C. Gunsalus and R. Y. Stanier (ed.), The bacteria, vol. 3. Academic Press, Inc., New York.

128. **Ursini, M. V., P. Arcari, and M. DeFelice.** 1981. Acetohydroxy acid synthase isozymes of *Escherichia coli* K-12: a *trans*-acting regulatory locus for *ilvHI* gene expression. Mol. Gen. Genet. **181**:491–496.

129. **Vogel, H. J.** 1964. Distribution of lysine pathways among fungi: evolutionary implications. Am. Nat. **98**:435–446.

130. **Watson, M. D., J. Wild, and H. E. Umbarger.** 1979. Positive control of *ilvC* expression in *Escherichia coli* K-12: identification and mapping of regulatory gene *ilvY*. J. Bacteriol. **139**:1014–1020.

131. **Wek, R. C., and G. W. Hatfield.** 1986. Examination of the internal promoter, P_E, in the *ilvGMEDA* operon of *E. coli* K-12. Nucleic Acids Res. **14**:2763–2777.

132. **Wek, R. C., and G. W. Hatfield.** 1986. Nucleotide sequence and *in vivo* expression of the *ilvY* and *ilvC* genes in *Escherichia coli* K12. Transcription from divergent overlapping promoters. J. Biol. Chem. **261**:2441–2450.

133. **Wek, R. C., C. A. Hauser, and G. W. Hatfield.** 1985. The nucleotide sequence of the *ilvBN* operon of *Escherichia coli*: sequence homologies of the acetohydroxy acid synthase isozymes. Nucleic Acids Res. **13**:3995–4010.

134. **Wessler, S. R., and J. M. Calvo.** 1981. Control of *leu* operon expression in *Escherichia coli* by a transcription attenuation mechanism. J. Mol. Biol. **149**:579–597.

135. **Whalen, W. A., and C. M. Berg.** 1982. Analysis of an *avtA*::Mu d1(Ap *lac*) mutant: metabolic role of transaminase C. J. Bacteriol. **150**:739–746.

136. **Whalen, W. A., and C. M. Berg.** 1984. Gratuitous repression of *avtA* in *Escherichia coli* and *Salmonella typhimurium*. J. Bacteriol. **158**:571–574.

137. **Whitlow, K. J., and W. J. Polglase.** 1974. Relaxation of catabolite repression and loss of valine sensitivity in *Escherichia coli* K-12. FEBS Lett. **43**:64–66.

138. **Wild, J., J. M. Smith, and H. E. Umbarger.** 1977. In vitro synthesis of β-galactosidase with *ilv-lac* fusion deoxyribonucleic acid as template. J. Bacteriol. **132**:876–883.

139. **Yokota, A., and K.-I. Sasajima.** 1985. Formation of 2,3,-dideoxy-hex-4-ulosonic acid by carboligase activity of 2-ketoglutarate dehydrogenase in microorganisms. Inst. Ferment. Osaka Res. Commun. **12**:19–33.

140. **Zarlengo, M. H., G. W. Robinson, and R. O. Burns.** 1968. Threonine deaminase from *Salmonella typhimurium*. II. The subunit structure. J. Biol. Chem. **243**:186–191.

24. Biosynthesis of the Aromatic Amino Acids

A. J. PITTARD

Department of Microbiology, University of Melbourne, Parkville, Victoria 3052, Australia

THE PATHWAY

The pathway of biosynthesis of the aromatic amino acids is shown, represented for convenience in four parts (Fig. 1 through 4). Figure 1 shows the "common pathway" leading to the synthesis of the branch point compound chorismate, and Fig. 2, 3, and 4 show the three terminal pathways in which chorismate is converted to phenylalanine, tyrosine, and tryptophan. The common pathway has in the past sometimes been referred to as the shikimate pathway. There are an additional four terminal pathways leading from chorismate respectively to folate, ubiquinone, menaquinone, and enterochelin, but these will not be considered further.

The identification of the various intermediates in the aromatic pathway was completed by the early 1960s. As in other pathways, studies of auxotrophic mutants were of key importance. The first intermediate to be identified was shikimate; its position in the common pathway was established when Davis showed that this compound alone out of 55 aromatic and hydroaromatic compounds tested could replace the requirement of certain aromatic auxotrophs of *Escherichia coli* for tyrosine, phenylalanine, and tryptophan (57). Other mutants which failed to re-

spond to shikimate accumulated this compound in their culture fluids. In the same study it was established that anthranilate could be substituted for tryptophan, phenylpyruvate could substitute for phenylalanine, and, under appropriate conditions, 4-hydroxyphenylpyruvate could substitute for tyrosine. The accumulation by auxotrophs of intermediates preceding the blocked reaction led to the identification of 3-dehydroquinate (DHQ), dehydroshikimate, and shikimate 3-phosphate, all intermediates in the common pathway (61, 191, 228, 230).

The identification of a phosphorylated compound (shikimate 3-phosphate) in the culture fluids was an exception to the general experience with other phosphorylated intermediates in that the compounds that accumulated in culture fluids were predominantly in the dephosphorylated form. Such was the case with 5-enolpyruvylshikimate (EPS), indoleglycerol, and 1-(*O*-carboxyphenylamino)-1-deoxyribulose (CDR) (60, 73, 89, 183). Neither the phosphorylated intermediates nor their dephosphorylated derivatives serve as growth factors for mutants blocked prior to their formation in a pathway, as the phosphorylated compounds cannot enter the cell and the dephosphorylated derivatives are generally not true intermediates. Shikimate is a notable exception. The branch point

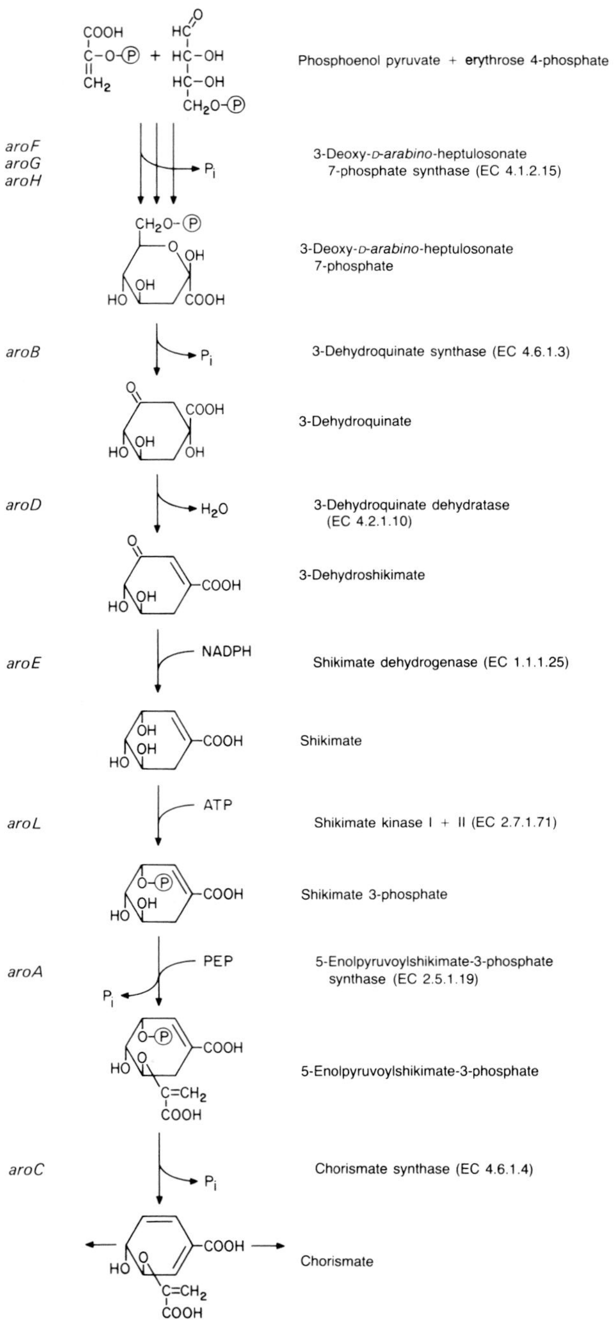

FIG. 1. Common pathway of aromatic amino acid biosynthesis. (Figures 1 through 4 are adapted from K. M. Herrmann and R. L. Somerville, ed., *Amino Acids: Biosynthesis and Genetic Regulation*, Addison-Wesley Publishing Co., with kind permission of the publishers.)

compound, chorismate, also fails to act as a growth factor, presumably because it is not able to enter the cell. The efficiency with which intermediates gain entry varies. Many mutant strains of *E. coli* that respond to shikimate or DHQ nevertheless grow at extremely slow rates. They can, however, give rise to mutants with growth rates approximately the same as observed when the aromatic amino acids are present (27, 61; L. Mason and J. Pittard, unpublished data).

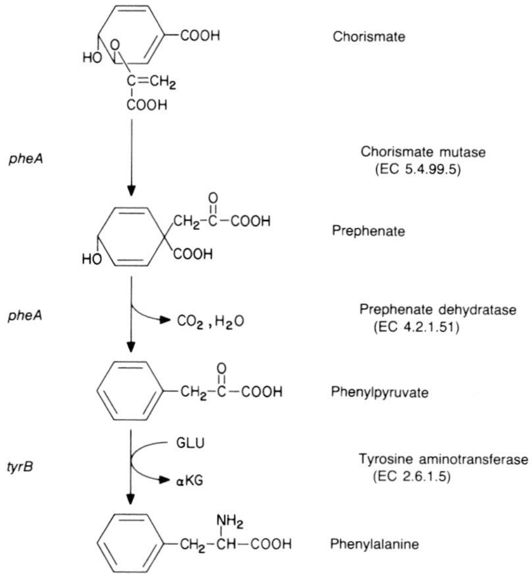

FIG. 2. Terminal pathway of phenylalanine biosynthesis.

Other mutants have been isolated with a particular defect in the transport of shikimate and dehydroshikimate, suggesting that a specific porter system for these compounds may exist (178). Similarly, the ability of tyrosine auxotrophs blocked before 4-hydroxyphenylpyruvate to use this compound for growth is limited and probably affected by the lack of an efficient transport system. Growth on 4-hydroxyphenylpyruvate can be dramatically improved either by the selection of mutants (84) or by growing cells at an acid pH with succinate as carbon source (57).

The early reactions of the common pathway were more difficult to establish. A mutant strain of *E. coli* that accumulated shikimic acid was grown on glucose variously labeled with ^{14}C. The shikimate was isolated and chemically degraded to establish the distribution of specific carbon atoms of glucose in the shikimate

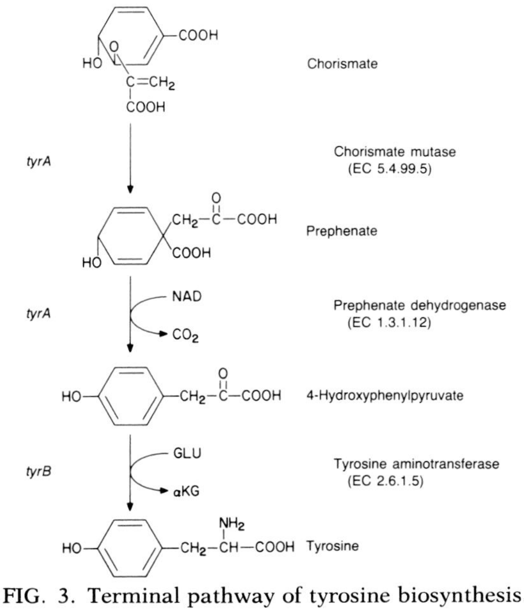

FIG. 3. Terminal pathway of tyrosine biosynthesis.

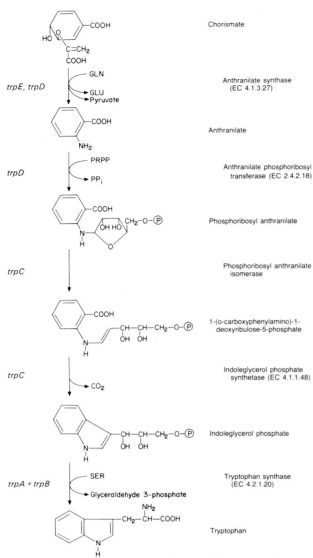

FIG. 4. Terminal pathway of tryptophan biosynthesis.

molecule. The results showed that three of the carbon atoms of shikimate are derived from glycolysis and four come from the pentose phosphate pathway (211).

The nature of these precursor compounds was established when erythrose 4-phosphate became available (16) and it was demonstrated that cell extracts of *E. coli* could convert erythrose 4-phosphate and phosphoenolpyruvate (PEP) to DHQ. Furthermore, on fractionation, an enzyme preparation could be obtained that converted erythrose 4-phosphate and PEP to 2-keto-3-deoxy-D-arabo-heptonate 7-phosphate (209). This compound was later to be termed 3-deoxy-D-*arabino*-heptulosonate 7-phosphate (DAHP).

The identification of chorismate, the branch point compound, required the use of mutant strains blocked in each of the three terminal pathways. A strain of *Aerobacter aerogenes* (62-1) was isolated that was blocked in the second reaction of the tryptophan pathway and in the first reactions each of the phenyl-alanine and tyrosine pathways. Extracts of this strain were able to convert shikimate to anthranilate. When,

however, glutamine was omitted from the reaction mixture, a new compound was formed. This compound could be extracted from the reaction mixture and used as a substrate for anthranilate production by extracts of a mutant blocked after EPS 3-phosphate (EPSP) (91). Subsequent studies showed that the compound could be accumulated by washed cell suspensions of strain 62-1 and that it could be converted enzymically into anthranilate, prephenate, 4-hydroxy-phenylpyruvate, phenylpyruvate, and 4-hydroxyben-zoate. The compound was called chorismic acid (88). After the structure of chorismate had been established, it was shown that it could be formed from EPSP by cell extracts of *E. coli* (161). The position of EPSP in the pathway was further supported by the demonstration that cell extracts produced EPSP from shikimate 3-phosphate and PEP and could convert EPSP to phenylpyruvate (143).

All of the reactions shown in Fig. 1 through 4 have been confirmed in cell extracts, many with purified enzyme preparations. Although much of the early work establishing the pathway used *E. coli* and in some cases *A. aerogenes*, the reactions apply equally to *Salmonella typhimurium*.

THE ENZYMES

The DAHP Synthases

Srinivasan and Sprinson were the first to demonstrate the conversion of erythrose 4-phosphate and PEP to DAHP in cell extracts of *E. coli* (212). It was subsequently shown (206) that this activity comprised at least two enzymes that were distinguishable by their inhibition and repression by phenylalanine and tyrosine, respectively (DAHP synthase [phe] and DAHP synthase [tyr]). A few years later, a third minor enzyme was identified whose synthesis was repressed by tryptophan (DAHP synthase [trp]; 26, 72). This enzyme was subsequently shown to be subject to inhibition by tryptophan (34, 71, 175). Whereas inhibition of DAHP synthase (tyr) and (phe) by tyrosine and phenylalanine was as high as 95%, the inhibition of DAHP synthase (trp) by tryptophan did not exceed 60% (176). Early studies in *E. coli* on the reaction mechanism of DAHP synthase (phe) and DAHP synthase (tyr) involved enzyme preparations with low specific activities. Results with those preparations which suggested a "ping-pong" mechanism of reaction and which gave K_m values in the molar range (159, 215) were later disproved when more highly purified enzyme preparations were used. With a highly purified preparation of DAHP synthase (tyr) from *S. typhimurium*, it was shown that the reaction mechanism was ordered and sequential rather than ping-pong, with PEP being the first substrate to bind (68). Similar results were obtained with a highly purified preparation of DAHP synthase (phe) from *E. coli* K-12 (205) and with DAHP synthase (tyr) purified to homogeneity from *E. coli* K-12 (198). K_m values for PEP of 5.8 μM have been reported for purified DAHP synthase (tyr) (198), and although data were not presented, a similar value has been suggested for purified DAHP synthase (phe) (153). In partial conflict with these results is a value of 80 ± 40 μM obtained with highly purified DAHP synthase (PHE) (205). A value of 14.9

μM has been reported for the partially purified DAHP synthase (tyr) of *S. typhimurium* (68). It has been pointed out (153) that because the intracellular concentration of PEP in *E. coli* never falls below 88 μM (146), the enzyme-PEP complex may be the native form of both DAHP synthase (tyr) and DAHP synthase (phe).

The three enzymes from *E. coli* K-12 and DAHP synthase (tyr) from *S. typhimurium* have been purified to homogeneity. DAHP synthase (tyr) and DAHP synthase (trp) are dimers with subunit molecular weights of about 40,000 (115, 198; M. D. Poling, J. Suzich, J. Shultz, and K. M. Herrmann, Fed. Proc. **40**:1581, 1981). DAHP synthase (phe) on the other hand, is a tetramer with a subunit molecular weight of 35,000 (153).

The genes for these three enzymes have also been cloned in *E. coli*, and the complete nucleotide sequences have been determined (56, 116, 201, 252; G. S. Hudson and B. E. Davidson, manuscript in preparation). The complete amino acid sequence of DAHP synthase (tyr) has also been determined (201). Although the relationship between these genes will be discussed below, the sequence data establish that DAHP synthase (tyr) and DAHP synthase (phe) have polypeptide subunits of 356 and 350 residues, respectively. All three DAHP synthases have been shown to contain 1 mol of iron per mol of native enzyme (152), and DAHP synthase (tyr) contains a short amino acid sequence which is highly homologous with a putative iron-binding sequence in hemerythrin, the oxygen-carrying molecule in the sea worm *Phascolopsis gouldii* (110). This sequence, however, is not conserved in DAHP synthase (trp) and DAHP synthase (phe) (202).

The presence of three isofunctional DAHP synthase enzymes whose activities and rates of synthesis are differentially affected by individual amino acids provides the cell with the ability to modulate the overall rate of synthesis in response to changes in the availability of particular aromatic amino acids. In addition to the feedback inhibition of the three enzymes by the respective amino acids, the synthesis of DAHP synthase (tyr) is repressed by tyrosine or very high levels of phenylalanine, that of DAHP synthase (phe) is repressed by phenylalanine and tryptophan, and that of DAHP synthase (trp) is repressed by tryptophan (28).

When *E. coli* grows in minimal medium, DAHP synthase (phe) comprises about 80% or more of total DAHP synthase activity; however, under conditions in which the aromatic amino acids limit growth, as in minimal medium supplemented with all amino acids except phenylalanine, tyrosine, and tryptophan, derepression of DAHP synthase (tyr) makes it the major enzyme (217). A similar situation occurs when cells are starved for iron (154). DAHP synthase (phe) is also the major isoenzyme present when *S. typhimurium* is grown in minimal medium (68, 115), although there is one report of high levels of DAHP synthase (tyr) under these conditions (127). In both organisms, DAHP synthase (trp) normally accounts for a very small proportion of DAHP synthase activity.

When fully derepressed, the specific activity of DAHP synthase (trp) in cell extracts is only 50 mU/mg of protein, whereas the corresponding values for DAHP synthase (tyr) and DAHP synthase (phe) are 560 and 300 mU/mg (International Units: 1 μmol of substrate used or product formed per min at 37°C) (33, 217). This difference may be due to a relatively inefficient promoter for the gene for DAHP synthase (trp) (252).

In vivo studies with mutant strains containing single isoenzymes have shown that, in the presence of all three aromatic amino acids, the cell retains enough residual activity of both DAHP synthase (phe) and DAHP synthase (trp) to allow for continued synthesis of the aromatic vitamins (227). When cells reach stationary phase, DAHP synthase (tyr) activity decays at a much faster rate than does DAHP synthase (phe) activity, presumably as a result of some specific degradation process (218; A. DeLucia, R. Schoner, and K. M. Herrmann, Int. Cong. Biochem. Abstr. **11**:252, 1979).

DHQ Synthase

DHQ, formerly known as 5-dehydroquinate synthetase, catalyzes the conversion of DAHP to DHQ. DHQ was first identified as an intermediate in aromatic biosynthesis in 1953 (228). It was isolated from the culture fluids of an aromatic auxotroph of *E. coli*. These culture fluids had been previously shown to support the growth of mutants blocked very early in the aromatic pathway. The structure of DHQ was determined, and it was shown that it could be chemically converted to dehydroshikimate by heat at an acid pH (61).

The enzymic conversion of DAHP to DHQ with a partially purified enzyme preparation from *E. coli* was first reported in 1963. It was shown that the enzyme required both NAD$^+$ and Co^{2+} for activity, and a scheme was proposed to account for the molecular conversion of DAHP to DHQ. A key component of this scheme was the oxidation and later reduction of the C-5 of DAHP, involving NAD (210) (Fig. 5).

In 1964, 3,7-dideoxy-D-threo-hepto-2,6-diulosonic acid, the compound immediately preceding DHQ in Fig. 2 (top) (compound III), was synthesized. Although this compound could be converted chemically at pH 11 to DHQ, it was not converted to DHQ enzymically (3). Subsequent studies by Rotenberg and Sprinson (188, 189) and Le Marechal and Azerad (140), using DAHP specifically labeled with tritium, have established that this diketo compound is not formed as an intermediate in the formation of DHQ. A methyl group is never formed at C-7, and the cyclization step involves an interaction between the enol formed in phosphate elimination and the carbonyl at C-2 in an aldolase-type reaction. Kinetic isotope effects were observed at C-5, supporting a mechanism involving oxidation at C-5 by NAD. The finding that all the tritium of labeled DAHP is conserved in DHQ establishes that hydride transfer involved in the subsequent reduction of C-5 uses the same hydrogen atom as was taken from DAHP. NADH is enzyme bound and reduces the keto group at C-5 with the regeneration of NAD (141, 188, 189). The amended scheme is shown in Fig. 2 (bottom).

The DHQ synthase enzyme has been purified to homogeneity from *E. coli* B and shown to be a single polypeptide with a molecular weight of 57,000. The native enzyme is a monomer, and the K_m for DAHP is 33 μM (147). Berlyn and Giles (20) measured the molecular weight of the DHQ synthase enzyme of *E.*

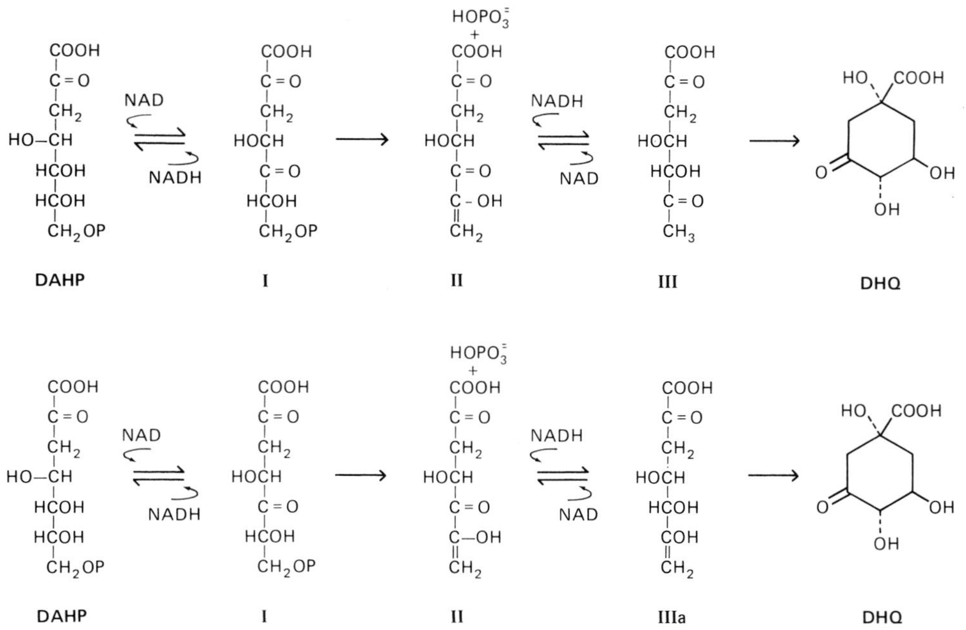

FIG. 5. Hypothetical (top) and current (bottom) schemes for conversion of DAHP to DHQ. Top, after Srinivasan et al. (210); bottom, from Rotenberg and Sprinson (189).

coli K-12 by ultracentrifugation in sucrose density gradients and found it to be 56,000. More recently, Frost et al. (81) have used gene cloning techniques to produce high levels of DHQ synthase for purification. They have purified the enzyme to homogeneity and shown it to be a monomeric protein with an M_r of between 40,000 and 44,000.

As with the other remaining enzymes of the common pathway (with the exception of shikimate kinase), DHQ synthase appears also to be synthesized constitutively. Its synthesis is not repressed by any of the aromatic amino acids or by chorismate, nor is it induced by DAHP. The specific activity of DHQ synthase in extracts of E. coli is 25 mU/mg of protein. It has been calculated that this value is about fivefold greater than would be required to meet the aromatic needs of cells growing with a doubling time of about 1 h (217). Ogino et al (169) have demonstrated that strains of E. coli with feedback-resistant DAHP synthase (tyr) accumulate DAHP, indicating that under these conditions DHQ synthase activity has become rate limiting.

DHQ Dehydratase

DHQ dehydratase, formerly known as dehydroquinase, catalyzes DHQ into 3-dehydroshikimate and introduces the first double bond of the aromatic ring. The reaction exhibits cis stereochemistry in both directions (29, 104, 223). The enzymic reaction was first studied by Mitsuhashi and Davis (158), who partially purified the enzyme from both A. aerogenes and E. coli. They separated DHQ dehydratase from dehydroshikimate reductase (now known as shikimate dehydrogenase) and demonstrated the reversibility of the reaction. The K_m for DHQ was 44 μM (158). Some confusion exists concerning the molecular weight of the enzyme. Berlyn and Giles (20) reported a

molecular weight of 40,000 for the native enzyme from E. coli. However, Kinghorn et al. (130), using the same technique of ultracentrifugation in sucrose density gradients, reported a molecular weight of about 59,000 for native enzyme from wild-type E. coli K-12 and from a derivative that carried the gene for DHQ dehydratase on the plasmid pBR322. The use of minicells to study plasmid-coded proteins revealed a protein band with a molecular weight of 63,000 which appeared to represent native DHQ dehydratase. The coincident appearance of a 31,000-molecular-weight band was hypothesized to represent a subunit of the native enzyme (130). Chaudhuri and Coggins (35) reported that DHQ dehydratase from E. coli is a simple dimeric protein with a subunit M_r of 29,000. On the other hand, in a short article (222) it was reported that the enzyme had been purified to apparent electrophoretic homogeneity from E. coli and that it exists as a tetramer with a minimum molecular weight of 40,000. As previously mentioned for DHQ synthase, DHQ dehydratase appears to be synthesized constitutively. When E. coli and S. typhimurium are grown in minimal medium, DHQ dehydratase specific activities range from 50 to 100 mU/mg (97, 130, 177, 217).

In their original paper, Mitsuhashi and Davis (158) made the observation that mutants of A. aerogenes selected for good growth on quinate and a mutant of E. coli blocked between dehydroshikimate and shikimate produced greatly enhanced levels of DHQ dehydratase. This latter observation was not confirmed in the study of DHQ dehydratase levels in aromatic auxotrophs of E. coli and S. typhimurium (97, 177).

Shikimate Dehydrogenase

The enzyme that converts dehydroshikimate to shikimate was first studied in 1954 with partially purified extracts of *E. coli* W. The reaction was studied in the reverse direction (shikimate to dehydroshikimate) in a coupled reaction with glutathione reductase to oxidize the NADPH that is formed. The reaction was shown to be dependent on HADP and to be specific for shikimate. Shikimate dehydrogenase (formerly dehydroshikimate reductase) activity was not detected in extracts of aromatic auxotrophs that accumulated dehydroshikimate. The K_m of the enzyme for shikimate was 55 μM (236). The reaction has been shown to be stereospecific, involving transfer of hydrogen from the A side of NADPH (54). Estimation of molecular weight by sucrose gradient centrifugation gives a value of 25,000 (20). Recently, Chaudhuri and Coggins (35) have purified shikimate dehydrogenase from *E. coli* to homogeneity and report that it is a monomeric protein with an M_r of 32,000. The constitutive level of this enzyme in cells grown in minimal medium is about 60 mU/mg (217).

Shikimate Kinase

Shikimate kinase catalyzes the formation of shikimate 3-phosphate from shikimate and ATP. Shikimate 3-phosphate was first identified as a possible intermediate in aromatic biosynthesis by Davis and Mingioli (60). Its chemical structure was established by Weiss and Mingioli (230). No mutants blocked in the shikimate kinase reaction were found among aromatic auxotrophs of either *E. coli* (177) or *S. typhimurium* (97). The reasons for this became apparent when it was shown that extracts of *S. typhimurium* produce two separable peaks of shikimate kinase activity when chromatographed on DEAE-cellulose (162). Using ultracentrifugation in sucrose density gradients, Berlyn and Giles (20) confirmed the observation made in *S. typhimurium* and established that *E. coli* also appeared to have two shikimate kinase enzymes. Further study of the *E. coli* enzymes established that the synthesis of one of these (called shikimate kinase II because it was the second peak to be eluted from DEAE-Sephadex) was subject to specific control. In particular, when cells were starved for tyrosine and tryptophan or had an inactive *tyrR* regulator gene, synthesis of shikimate kinase II was derepressed about 10-fold. Shikimate kinase levels vary between 5 and 55 mU/mg. A mutant lacking shikimate kinase II activity was isolated (80), but subsequent attempts to isolate mutants lacking shikimate kinase I activity have failed (B. Ely and J. Pittard, unpublished data). The structural gene for shikimate kinase II (*aroL*) has been cloned (66), and its nucleotide sequence has been determined (65). Shikimate kinase II has been purified to homogeneity, and the amino acid sequence of its amino terminus has been determined (67). Shikimate kinase I has been partially purified (358-fold from cell extracts), and some of its physical parameters have been measured (R. C. De Feyter, Ph.D. thesis, University of Melbourne, Parkville, Victoria, Australia, 1984). Previously, both these enzymes had been reported to have molecular weights of about 20,000 (80). To determine whether the genes for shikimate kinase I and II were closely related, a 0.39-kilobase probe containing most of the *aroL* gene was hybridized to various digests of chromosomal DNA after separation of the fragments by electrophoresis on 0.8% agarose gel and

transfer to nitrocellulose. In each case only a single band was found on the autoradiograph, indicating that *aroL* and the putative gene for shikimate kinase I have less than 60% sequence homology, if any (De Feyter, thesis).

An examination of the kinetics of the reaction using the fully purified shikimate kinase II and the partially purified shikimate kinase I has, however, been more rewarding. The K_m of shikimate kinase II for shikimate is 200 μM. On the other hand, the K_m of shikimate kinase I for shikimate appears to be in excess of 5 mM. This apparent low affinity of shikimate kinase I for shikimate explains why *aroL* mutants which lack shikimate kinase II activity will only grow in minimal medium if the levels of DAHP synthase are high. Neither enzyme is inhibited by any of the end products, and shikimate kinase II has been shown to require Mg^{2+} as a cofactor (De Feyter, thesis).

The existence of isofunctional enzymes has been observed most frequently in the first reaction of a pathway that later branches to multiple end products, e.g., aspartokinases for lysine, methionine, and threonine (39) and DAHP synthases for the aromatic amino acids. It is also true that specific regulation of gene expression most frequently affects genes specifying the first enzyme in a pathway, and it is noteworthy that, of all the genes of the common pathway of aromatic biosynthesis, only the three specifying the three DAHP synthase enzymes and the gene for shikimate kinase II are subject to control.

Does this mean that in the evolutionary past, aromatic biosynthesis used to start with shikimate, or does it indicate an as yet undiscovered pathway diverging from shikimate? Although in a study by Gollub et al. (97) no regulation of shikimate kinase activity was observed in *S. typhimurium*, cells had not been grown under conditions of starvation for tyrosine and tryptophan, and the question of whether or not *S. typhimurium* and *E. coli* differ remains to be established.

The very low affinity of shikimate kinase I for shikimate raises the possibility that this enzyme has another major function concerned with some other essential pathway. Failure to obtain mutants that have lost kinase I activity may reflect a failure to provide the product of this putative kinase I pathway. An alternative explanation for the failure to isolate mutants lacking all kinase activity would be the presence of a third activity in cells which has not been detected in chromatography of cell extracts.

As will be discussed below, the structural gene for shikimate kinase II is organized in an operon that contains one and possibly two additional genes specifying polypeptides of unknown functions. All of these results suggest that there is still much to learn about this particular section of the pathway.

EPSP Synthase

EPSP synthase converts PEP and shikimate 3-phosphate into EPSP. It was the dephosphorylated form of this compound that was first identified in culture fluids of aromatic auxotrophs blocked in the last step of the common pathway (60).

Levin and Sprinson (142, 143) partially purified the enzyme activity from *E. coli* and showed that the product of the reaction was EPSP. The mechanism of this reaction has been studied with a partially purified enzyme from *S. typhimurium* and it has been shown to involve the transfer of the enolpyruvyl grouping unchanged to the acceptor molecule (23), as was first proposed in a scheme by Levin and Sprinson (143). This reaction introduces the three-carbon fragment that is destined to become the side chain of phenylalanine and tyrosine and to be removed again in the synthesis of tryptophan. The enzyme from *E. coli* has been purified to homogeneity (144). The subunit M_r was estimated to be 49,000 by polyacrylamide gel electrophoresis in the presence of sodium dodecyl sulfate, and the native molecular weight was estimated to be 55,000 by gel filtration. Thus, the enzyme is a monomer. A key step in the purification was elution with substrate from a cellulose phosphate column. A mixture of shikimate 3-phosphate and PEP was very effective, whereas PEP alone had no effect and shikimate 3-phosphate alone caused elution over a very broad peak. Lewendon and Coggins concluded that these results agree with an ordered reaction in which shikimate 3-phosphate binds to the enzyme first (144).

The gene for this enzyme, *aroA*, has been cloned, and strains carrying this gene on a multicopy plasmid overproduce EPSP synthase 100-fold (77). The K_m values for PEP and shikimate 3-phosphate are 16 and 2.5 μM. The enzyme isolated from the overproducing strain is indistinguishable from that produced by wild-type *E. coli*. The amino acid sequence of the purified enzyme has been determined (78).

A molecular analysis of the region containing *aroA* has revealed that it is organized in a transcription unit with *serC*. *serC* is the first gene in the transcription unit, and some transcripts terminate between *serC* and *aroA* (76). Genetic analyses of various *S. typhimurium* mutants lacking EPSP synthase has revealed that in that organism also, *serC* and *aroA* constitute a unit of transcription (114). The *aroA* gene from *S. typhimurium* has been cloned, and its nucleotide sequence has been determined (214).

The enzyme, EPSP synthase, is synthesized constitutively in *S. typhimurium* (97) and *E. coli* (217), and cell extracts have a specific activity of about 100 mU/mg of protein. Since the phosphoserine aminotransferase specified by *serC* is also synthesized constitutively (179), it is not clear whether the cell gains any particular advantage from having *serC* and *aroA* in a single transcription unit.

Chorismate Synthase

The chorismate synthase reaction introduces the second double bond into the aromatic ring system and forms the branch point compound chorismate, which serves as the starting substrate for the three terminal pathways of amino acid biosynthesis and for the pathways to ubiquinone, menaquinone, folate, and enterochelin (88, 91). The enzyme has been partially purified from *E. coli*. It is oxygen sensitive and requires a reducing environment such as can be established with a reduced flavin adenine dinucleotide- and NADH-regenerating system in an anaerobic environment. Activity is inhibited by iron chelators and by high levels of Fe^{2+}. The basis for the latter effect is unknown, but it is postulated that Fe^{2+} is a cofactor for chorismate synthase. It is suggested that the reaction is irreversible. A tentative molecular weight for the enzyme of somewhere between 70,000 and 100,000 has been proposed (161).

Synthesis of the enzyme is believed to be constitutive, and in both *S. typhimurium* and *E. coli* the specific activity of chorismate synthase in cell extracts is only 10 to 20% of that of the preceding enzyme, EPSP synthase (97, 217).

It is puzzling that the last reaction of the pathway appears to be rate limiting, unless such a situation facilitates subsequent controls directing chorismate down the various terminal pathways. Clearly more work on this important enzyme and its role in vivo in aromatic biosynthesis is needed.

THE TYROSINE AND PHENYLALANINE PATHWAYS

The Chorismate Mutases

The first reaction of both the phenylalanine and tyrosine pathways involves the conversion of chorismate to prephenate. The structure of prephenate was determined by Weiss et al. (229). This relatively unstable compound can be converted to phenylpyruvate at an acid pH. This reaction provided some puzzling results when phenylalanine auxotrophs were first studied.

Simmonds (204) described a phenylalanine auxotroph that appeared on prolonged culture to provide its own auxotrophic requirements. Davis (59) and Katagiri and Sato (129) investigated this phenomenon and suggested that the substrate of the blocked reaction was being converted chemically to a precursor in phenylalanine biosynthesis. The phenylalanine precursor was shown by Davis to be phenylpyruvate, and the accumulated substrate was prephenate. The enzymes that carry out the synthesis of prephenate have been referred to as chorismate mutases. Cotton and Gibson (44) showed that chromatography of cell extracts of *A. aerogenes* and *E. coli* on DEAE-cellulose gave two well-separated peaks of chorismate mutase activity. Prephenate dehydratase activity was associated with the first peak to be eluted, and prephenate dehydrogenase activity appeared with the second. Prephenate dehydratase carries out the second reaction of the phenylalanine pathway, and prephenate dehydrogenase carries out the second reaction in the tyrosine pathway. Subsequent studies involving the purification of these enzymes have confirmed that in each case both activities are the product of a single bifunctional enzyme (70, 85, 118, 134, 135, 192, 196).

The phenylalanine enzyme is referred to as chorismate mutase-prephenate dehydratase, and the tyrosine enzyme is termed chorismate mutase-prephenate dehydrogenase. Because of many similarities between these enzymes, they will be discussed together. Both enzymes are homodimers with subunit molecular weights of about 40,000 (15, 55, 85, 118, 134, 135, 192, 196). In the case of chorismate mutase-prephenate dehydratase a number of studies indicate that the enzyme has two independent catalytic sites. In both *E. coli* and *S. typhimurium*, mutants have been isolated in which only one of the catalytic activities has been lost (13, 12, 177, 195). The two enzymic activities can be inhibited differentially by chemical modifying reagents (86, 87, 195), by phenylalanine (70), and by substrate analogs (14). Kinetic studies and the use of radiolabeled chorismate suggest that when the reaction is carried out with purified enzyme in vitro, prephenate dissociates from the mutase site and equilibrates with the bulk medium before combining at the dehydratase site (75). Kinetic studies also support a random mechanism involving the formation of two dead-end complexes, E-NADH-prephenate and E-NAD-hydroxyphenylpyruvate (193).

In the case of chorismate mutase-prephenate dehydrogenase, mutants have been isolated that have lost prephenate dehydrogenase activity but retain chorismate mutase activity (62, 184). Although chemical modification of the purified enzyme causes parallel loss of both activities (118, 136), tyrosine differentially inhibits the prephenate dehydrogenase activity (117), NAD activates chorismate mutase activity, and chorismate stimulates prephenate dehydrogenase activity (112). Kinetic and computer simulation studies, on the other hand, have supported a single active site (111). In the light of all these results, it seems likely the two reactions of chorismate mutase-prephenate dehydrogenase are catalyzed at closely situated and interacting active sites (118).

Chorismate mutase-prephenate dehydrogenase and chorismate mutase-prephenate dehydratase activities are inhibited by tyrosine and phenylalanine, respectively, with the most significant inhibition being exerted on the second activity in both cases. Tyrosine can cause up to 95% inhibition of prephenate dehydrogenase activity and, in the presence of NAD, up to 45% inhibition of the associated chorismate mutase activity (117). Inhibition of prephenate dehydratase by phenylalanine approaches 90% in both $E.$ $coli$ and $S.$ $typhimurium$, and the associated mutase activity in both cases is inhibited 55% (70, 196, 197). The synthesis of chorismate mutase-prephenate dehydrogenase is repressed when cells are grown in the presence of tyrosine. Derepressed synthesis of chorismate mutase-prephenate dehydratase occurs when cells are starved for phenylalanine. Both effects will be discussed below.

Because each of these enzymes and anthranilate synthase compete with each other for the same substrate, chorismate, it is of interest to compare the affinities of each for this substrate. The K_m values of highly purified enzyme preparation are as follows. Chorismate mutase-prephenate dehydratase from $E.$ $coli$ has a K_m for chorismate of 45 μM (70), whereas chorismate mutase-prephenate dehydrogenase from the same organism has K_m values of 92 μM for chorismate and 50 μM for prephenate (118). By comparison, the K_m of anthranilate synthase for chorismate is 1.2 μM (10). This very marked difference in affinities would imply that under conditions of chorismate limitation this compound would be preferentially directed down the tryptophan pathway rather than towards phenylalanine or tyrosine. This assumption may well explain why aromatic auxotrophs with incomplete blocks in any of the common pathway reactions seem able to meet their tryptophan requirement but not the requirement for phenylalanine and tyrosine (58).

The Aminotransferases

The last reaction in both the phenylalanine and tyrosine pathways involves the transamination of the respective α-keto acids, phenylpyruvate and 4-hydroxyphenylpyruvate, with glutamate as the amino donor. Early studies on aminotransferases in $E.$ $coli$ had indicated that there were a number of aminotransferase enzymes and that these enzymes had rather broad specificities (36, 42, 150, 151, 160, 190, 203). Phenylalanine and tyrosine could be formed by at least two enzymes, termed the aromatic aminotransferase and the aspartate aminotransferase. The synthesis of the aromatic aminotransferase could be repressed by tyrosine, but the synthesis of the aspartate aminotransferase appeared to be constitutive. There was at least one more enzyme that could function in the conversion of phenylpyruvate to phenylalanine. The analysis of this rather complex situation has been greatly assisted by the isolation of mutants lacking individual aminotransferase enzymes (83, 84, 221), coupled with the development of methods for the separation of aminotransferase activities (150, 151, 181).

There are three enzymes that need to be considered, the branched-chain amino acid aminotransferase specified by $ilvE$, the aromatic aminotransferase specified by $tyrB$, and the aspartate aminotransferase specified by $aspC$. Mutants with lesions in $tyrB$ and $aspC$ require tyrosine but not phenylalanine for growth, whereas mutants lacking all three activities require both phenylalanine and tyrosine (84). The purification of the aspartate and aromatic aminotransferases provides an opportunity to compare their properties, as done by Mavrides and Orr (151) and by Powell and Morrison (181). Both of these studies found that the enzymes were homodimers with subunit molecular weights of approximately 46,000 for the aromatic aminotransferase and 43,000 for the aspartate aminotransferase. Although there are differences in the two reports, both conclude that the amino acid composition of the two enzymes is very similar.

The aspartate aminotransferase has a much higher affinity for aspartate than does the aromatic aminotransferase. On the other hand, the affinity of the aromatic aminotransferase for phenylalanine and tyrosine and their α-keto acids is much greater than that of the aspartate aminotransferase (Table 1; values from the paper of Powell and Morrison [181]). The aminotransferase reaction is freely reversible, and it

TABLE 1. K_m and V_{max} values for aromatic and aspartate aminotransferases

Substrate	Aspartate aminotransferase		Aromatic aminotransferase	
	K_m (μM)	V_{max} (U/mg)	K_m (μM)	V_{max} (U/mg)
2-Oxoglutarate	150		230	
Glutamate	900		280	
Oxaloacetate	580	200	3,800	6
Aspartate	450	53	5,000	13
Phenylpyruvate	650	45	56	18
Phenylalanine	550	9	60	9
4-Hydroxyphenypyruvate	400	30	32	25
Tyrosine	450	9	42	20

seems that under normal physiological conditions phenylalanine and tyrosine syntheses are primarily carried out by the aromatic aminotransferase, the product of the *tyrB* gene. Only when the pool sizes of phenylpyruvate and 4-hydroxyphenylpyruvate become very high will the aspartate aminotransferase start to contribute to the synthesis of these two amino acids. The converse is true when one considers the synthesis of aspartate. It is not known whether the *ilvE*-coded aminotransferase is ever involved in phenylalanine biosynthesis other than in *tyrB aspC* mutant strains. Mavrides and Orr (151) treated purified aromatic aminotransferase with subtilisin and obtained a smaller protein with enhanced aspartate aminotransferase activity. The implication that aspartate aminotransferase is formed by the processing of the aromatic aminotransferase, however, is not supported for a number of reasons (181). Recently, the nucleotide sequences of *aspC*, *ilvE*, and *tyrB* have been determined (138, 139; H. Kagamiyama, S. Kuramitsu, K. Kondo, and K. Inoue, Proc. 13th Int. Cong. Biochem., 1985, p. 516). These studies resolve previous speculations about possible relationships between *aspC* and *tyrB* and their encoded aminotransferases. The coding regions of *aspC* and *tyrB* encode 396 and 397 amino acid residues, respectively. These sequences have 169 residues in common, and it would seem reasonable to assume that gene duplication played a role in their evolution. The deduced amino acid sequence for the aspartate aminotransferase also agrees with the published amino acid sequence for this enzyme purified from *E. coli* B (137). As expected, the gene for the branched-chain aminotransferase *ilvE* encodes a smaller protein of 320 amino acids and does not show homology with either *aspC* or *tyrB*. A comparison of the amino acid sequences of the cytoplasmic and mitochondrial aspartate aminotransferases from pig heart with the aspartate aminotransferase from *E. coli* shows a remarkable 120 invariant residues (137). Ninety-two of these residues are also present in the bacterial aromatic aminotransferase. Kondo et al. (137) argue that the homologies between the bacterial aspartate aminotransferases and each of the pig heart enzymes marginally favor a hypothesis in which the genes for the bacterial enzyme are more closely related to the gene for the mitochondrial enzyme than to the gene for the cytoplasmic enzyme. The surprising feature overall is that it appears that the postulated duplication of an ancestral bacterial gene to form the precursors of *aspC* and *tyrB* occurred very early in evolution, not long after the postulated origin of eucaryotic cells.

Using highly purified preparations of the aromatic aminotransferase, chorismate mutase-prephenate dehydrogenase, and chorismate mutase-prephenate dehydratase, Powell and Morrison (182) have demonstrated protein interactions between the aminotransferase and each of the other two enzymes. This interaction did not occur if the aspartate aminotransferase was substituted for the aromatic aminotransferase. The complex with chorismate mutase-dehydrogenase dissociated in the presence of tyrosine, and the interaction with chorismate mutase-dehydratase required the presence of phenylpyruvate. The interactions, therefore, are quite specific. The extent to which these interactions occur in vivo and the role they play in synthesis have yet to be determined.

THE TRYPTOPHAN PATHWAY

Anthranilate Synthase and Anthranilate Phosphoribosyltransferase (PR Transferase)

As shown in Fig. 4, the first reaction of the terminal pathway of tryptophan biosynthesis involves the conversion of chorismate and glutamine to anthranilate, glutamate, and pyruvate. The enzyme that catalyzes this reaction is called anthranilate synthase (10, 91). Much of the early work has been reviewed (249). The active enzyme has been shown to be an aggregate that is composed of two molecules each of the polypeptides specified by the *trpE* and *trpD* genes. These are referred to as components I and II, respectively. The necessity for both components in the reaction was established in both *E. coli* and *S. typhimurium* by studying mutants with nonsense mutations in these genes (19, 79, 125). The aggregate has been purified from both *E. coli* and *S. typhimurium* (108, 119, 125, 216), and the components have also been purified singly, component I from *E. coli* and *S. typhimurium* and component II from *S. typhimurium* (108, 123, 165, 243). Component I has a molecular weight of about 60,000 and contains the binding site for chorismate. In the absence of component II it cannot catalyze the formation of anthranilate with glutamine as the nitrogen source. It can, however, form anthranilate with a considerably reduced efficiency if provided with ammonia instead of glutamine. Component II also has a molecular weight of about 60,000 and specifies two activities. The first of these, an amidotransferase activity, is required to activate component I in the anthranilate synthase reaction. This activity channels the nitrogen from glutamine to the active site for anthranilate production. Only the aggregate exhibits this activity, which is stimulated by chorismate. Glutamine hydrolysis has been shown to require the prior binding of chorismate to the aggregate (172).

The second activity of component II involves the conversion of anthranilate to anthranilate 5-phosphoribosyl pyrophosphate. It is termed anthranilate phosphoribosyltransferase or PR transferase. This activity can be carried out either by the aggregate or by the purified component II.

When the anthranilate synthase enzyme from *S. typhimurium* is subjected to limited proteolysis with trypsin, PR transferase activity is lost and the size of the aggregate is reduced from about 261,000 to about 141,000. Analysis of this smaller-molecular-weight aggregate shows that component II has been reduced to a polypeptide whose molecular weight is between 15,000 and 19,000. The aggregate with this reduced component II nevertheless retains anthranilate synthase activity including the amidotransferase activity (119). Studies of various *trpD* nonsense mutants reveal that the glutamine binding site of component II is found in the amino-terminal end of the molecule (243).

In *Serratia marcescens* the enzymes amidotransferase and PR transferase are coded for by separate genes, *trpG* and *trpD*. The DNA sequence of the end of *trpG* and the beginning of *trpD*, including the intercistronic region between them, has been compared with the sequences of the corresponding regions of the bifunctional *trpD* genes of *S. typhimurium* and

E. coli. The intercistronic region is 13 base pairs long. A hypothetical deletion of one base from this sequence, along with a limited number of base substitutions which remove the stop codon of *trpG* and the ribosome binding site of *trpD*, allows one to construct a simple scheme in which a gene like the *trpD* gene of *E. coli* or *S. typhimurium* could be formed by a gene fusion involving the separate *trpG* and *trpD* genes of some ancestral strain (166).

Nonsense mutations in *trpE* can be shown to have a much more dramatic effect on the expression of *trpD* than on the more distal genes *trpC*, *trpB*, and *trpA*. The basis for this difference was revealed when it was shown that efficient translation of *trpD* is dependent on the efficient translation of the end of *trpE* (170). A clue to the molecular basis for this translational coupling was discovered when nucleotide sequence analysis showed that the translation stop signal of *trpE* overlaps the translation start signal of *trpD* in the sequence $\overline{TGATG}$ (166). Although the exact mechanism of translational coupling is not understood, the same phenomenon was observed also with the translation of the *trpB* and *trpA* polypeptides, where again the same overlapping codons are found (180).

As has been found for the first enzyme in a number of biosynthetic pathways, the activity of anthranilate synthase is inhibited by the end product of the pathway, in this case, tryptophan (91, 107, 123). Tryptophan inhibits both anthranilate synthase activity and PR transferase activity of the aggregate. Whereas anthranilate synthase can be inhibited 100%, PR transferase inhibition is incomplete, not exceeding 70%. Uncomplexed PR transferase is not sensitive to inhibition by tryptophan. The binding of tryptophan and chorismate is competitive, and both sites are postulated to be present on component I. Mutational studies, however, show that the two binding sites are distinct. Conformational changes involving both components I and II are associated with the binding of substrates or inhibitor and affect anthranilate synthase, amidotransferase, and PR transferase activities (172).

The DNA sequences of the *trpE* gene of both *E. coli* and *S. typhimurium* have been compared. They show a high degree of homology at the amino acid level, with only a 12.5% difference in amino acids. The differences at the DNA level are significantly higher but involve a large number of synonymous codon changes. It is of interest that *trpE* polypeptides from both organisms do not contain any tryptophan residues (248).

PRA Isomerase-InGP Synthase

PRA is converted to 1-(*O*-carboxyphenylamino)1-deoxyribulose 5-phosphate (CDRP), and this compound, in turn is converted to indoleglycerol phosphate (InGP) by a single enzyme specified by the *trpC* gene. PRA and CDRP were first postulated as intermediates in the pathway by Yanofsky (237). CDRP was identified in the dephosphorylated form (CDR) as an accumulation product when certain tryptophan auxotrophs of *A. aerogenes* and *E. coli* were starved for tryptophan (73). By using cell extracts of mutants blocked between anthranilate and InGP, Smith and Yanofsky (208) were able to establish that the phos-

phorylated compound CDRP was an intermediate between anthranilate and InGP. In some mutants unable to form InGP the conversion of CDRP to InGP was blocked, whereas in others the blocked reaction was the conversion of anthranilate to CDRP (208). Partial purification of the enzyme from *E. coli* was achieved in 1960, and it was shown that the reaction CDRP to InGP involved only a single enzyme. Reactions leading to the formation of CDRP were not studied (90).

Doy et al (74) extended the work with cell extracts from auxotrophic strains and established that PRA was an acid-labile intermediate between anthranilate and CDRP. Because of the extreme lability of PRA, its conversion from anthranilate may have been overlooked in previous investigations. The complete purification of the enzyme PRA isomerase-InGP synthetase was achieved in 1966. It was shown to be a single-polypeptide chain with a molecular weight of approximately 45,000, and since the ratio of PRA isomerase activity to InGP synthetase activity was the same in the crude cell extract and in the purified enzyme, it was concluded that both activities resided on the same polypeptide (53).

A genetic and biochemical study of missense mutations in the *trpC* gene established a relationship between the map position of the mutation and the enzymic function lost. Strains with mutations in the proximal half of the *trpC* cistron (closest to *trpD*) accumulated CDR, i.e., had lost InGP synthase activity. On the other hand, when the mutations were in the distal region of *trpC* (closest to *trpB*), anthranilic acid accumulated owing to a loss of PRA isomerase activity. These latter mutants retained InGP synthetase activity, although at a reduced level (207). Limited proteolysis of the purified enzyme has allowed the isolation of large amino-terminal and carboxy-terminal fragments of the enzyme which retain InGP synthetase and PRA isomerase activities, respectively (131). Kinetic studies on the purified enzyme led to the conclusion that the enzyme contained two distinct and nonoverlapping sites for the reaction PRA → CDRP → InGP, and it was proposed that CDRP must dissociate from the enzyme before being converted to InGP. A mixture of mutant enzymes lacking different activities could carry out the overall reaction of converting PRA to InGP. This reconstitution remains one of the few examples of intracistronic complementation not involving an oligomeric protein (48).

It is interesting that in some nonenteric microorganisms these two activities, PRA isomerase and InGP synthetase, are found in two separate polypeptides (45). Further studies involving a fluorescent analog of CDRP and recent X-ray crystallographic data on the purified enzyme are helping to reveal a detailed picture of the reaction mechanisms of this bifunctional enzyme (21, 41, 233). InGP synthetase activity is inhibited by anthranilate (90). The enzyme is unique among the enzymes of the tryptophan pathway in that it is rapidly inactivated in nongrowing cells of *E. coli*. This inactivation may involve more than a single mechanism, as both PRA isomerase and InGP synthetase activities are not always involved to the same extent and the kinetics of restoration of activity suggest partial reversibility, whereas studies with antisera suggest that protein degradation or at least denaturation may be taking place (163). Nucleotide sequence studies in *E. coli* have identified the 1,356 nucleotides of the coding region of *trpC*. The gene is flanked by intercistronic regions of 11 nucleotides (*trpC-trpB*) and 6 nucleotides (*trpD-trpC*) and in this way differs from the *trpE-trpD* and *trpB-trpA* junctions (37). Studies of the *trpC-trpB* intercistronic region in *S. typhimurium* reveal a 9-nucleotide spacer (200).

Tryptophan Synthase

Tryptophan synthase was one of the first enzymes of the tryptophan pathway to be extensively studied and has been the subject of a number of review articles (155, 156, 239, 241). It was initially believed that the major function of the enzyme was to convert indole and serine to tryptophan. The enzyme, however, catalyzes three reactions directly relevant to tryptophan biosynthesis (238, 246):

$$InGP \rightarrow indole + \text{D-glyceraldehyde 3-phosphate} \quad (1)$$

$$Indole + \text{L-serine} \xrightarrow{\text{Pyridoxal phosphate}} \text{L-tryptophan} + H_2O \quad (2)$$

$$InGP + \text{L-serine} \xrightarrow{\text{Pyridoxal phosphate}} \text{L-tryptophan} + \text{D-glyceraldehyde 3-phosphate} + H_2O \quad (3)$$

A consideration of the rates of the three reactions pointed to reaction 3, namely, the conversion of InGP and serine to tryptophan and glyceraldehyde 3'-phosphate, as the major physiological reaction. This idea was supported by the demonstration that free indole was not an intermediate in the reaction (47, 49, 239, 246). Early studies showed that two protein components, A and B, were involved in all three reactions (47, 247). Subsequent work has revealed that tryptophan synthase is a tetramer comprising two α units coded for by the *trpA* gene and a β_2 dimer coded for by the *trpB* gene. The β_2 subunit has two binding sites for pyridoxal phosphate and also two independent sites to which α units can bind (52, 234). The α subunit has the ability to convert InGP to indole and glyceraldehyde 3'-phosphate, and the β_2 subunit can convert indole and serine, in the presence of pyridoxal phosphate, to tryptophan. In both cases, however, the $\alpha_2\beta_2$ complex carries out these reactions at greatly increased rates (18, 47). The α subunit has been purified and crystallized and has a molecular weight of 29,000 (109, 132, 199). The amino acid sequence of the α subunit has been determined (145, 242). As is the case for the *trpE*-coded polypeptide, the α subunit does not contain any tryptophan residues. The analysis of the amino acid substitutions that occurred in certain *trpA* mutants and the map location of the mutations causing these changes provided the first compelling evidence of the colinearity between the gene and the polypeptide encoded by it (242). The β_2 subunit has also been purified and crystallized and has a molecular weight of 89,000 (2, 93, 105, 106, 157). The β_2 subunit and the $\alpha_2\beta_2$ complex catalyze other reactions involving pyridoxal phosphate, but these are not directly relevant to aromatic biosynthesis (156). The binding of pyridoxal phosphate and the

interaction of the subunits has been extensively studied (17, 52, 113, 156, 219, 220). The native $\alpha_2\beta_2$ complex (molecular weight, 147,000) has been purified and crystallized and shown to be identical with the reconstituted tryptophan synthase complex (1, 101).

The tryptophan synthase enzyme has also been used to investigate the evolutionary relationship between *E. coli* and *S. typhimurium*. Because tryptophan synthase is a multienzyme complex, experiments were designed to test whether single components from *S. typhimurium* could replace corresponding components in the *E. coli* complex without destroying activity. A number of experiments involving in vitro and in vivo construction of hybrid complexes confirmed that such substitutions could occur (11, 50, 51, 164).

The nucleotide sequences of both the *trpA* and the *trpB* genes have been determined in *E. coli* and *S. typhimurium* (46, 167). These studies reveal a very close similarity between the polypeptides synthesized by both organisms. In the case of *trpB*-coded polypeptide, more than 96% of the amino acid sequences were identical, and for *trpA*, about 85% were identical. Studies of the nucleotide sequence of the mRNA corresponding to the intercistronic region show that in both organisms the end of the *trpB* gene and the beginning of the *trpA* overlap in a sequence at the mRNA level of $\overline{UGA}UG$. An RNA fragment of 40 nucleotides containing this central AUG codon is protected from nuclease attack by ribosomes and contains an identifiable ribosome binding site (180). As for *trpE* and *trpD*, translational coupling occurs between *trpB* and *trpA* (4).

SYSTEMS FOR TRANSPORTING AROMATIC AMINO ACIDS

Although this chapter is concerned with the biosynthesis of the aromatic amino acids, to understand control of this process it is also necessary to consider the various systems for transporting exogenous aromatic amino acids into the cell. Furthermore, the expression of three transport genes is subject to specific control by the aromatic amino acids. Therefore, these genes and the systems that they specify will be considered here. Ames (7) identified a system in *S. typhimurium* for transporting the three aromatic amino acids and isolated mutants defective in this general transport system (7, 8). A similar general aromatic transport system was identified in *E. coli*, and mutants defective in this system were isolated (25, 173). The only gene identified so far as a component of this system is *aroP*. The *aroP* gene product has been identified in *E. coli* as a polypeptide with a molecular weight of about 37,000 (102). The gene has been mapped by transpositional mutagenesis, and its product has been shown to be normally associated with the cytoplasmic membrane (M. L. Chye, unpublished data). The expression of *aroP* is repressed when cells are grown in the presence of phenylalanine, tyrosine, or tryptophan (232). The work of Ames (7) and Brown (25) also established the existence of transport systems specific for individual amino acids. Mutations in the *tyrP* gene inactivate the tyrosine-specific system (231). This gene has been cloned and shown to specify a polypeptide that is normally associated with the cytoplasmic membrane and has a molecular weight of about 24,500 (235). The expression of this gene is repressed when cells are grown in the presence of tyrosine and enhanced by growth in the presence of phenylalanine or by growth in complex medium such as Casamino Acids (128, 232).

Mutations in the *mtr* gene inactivate a tryptophan-specific transport system (171, 232). The expression of the *mtr* gene has also shown to be induced when cells are grown in the presence of phenylalanine but without tryptophan (232). In each of these cases (*aroP*, *tyrP*, and *mtr*), the modulation of gene expression by aromatic amino acids requires functional TyrR protein. This topic will be discussed further when considering regulation of the pathway.

THE GENES OF THE AROMATIC PATHWAY

The chromosomal locations of the structural genes for the enzymes of the aromatic amino acid biosynthetic pathway, the genes coding for the transport systems, and the pathway's major regulator genes are shown in Table 2. Apart from the fact that the *trp* operon genes lie on an inverted segment of chromosome in *S. typhimurium* and that there appear to be different locations for *pheR*, the overall distribution of genes is the same in both organisms studied. In some cases, a gene may be present on one map but not on the other, e.g., *tyrP*, *aroL* and *aroM*, and *pheP*. It is likely that these genes will be found in the same location in the other organism, but this has not yet been established. The function of the gene *aroM* is not known. However, with *aroL* and an additional open reading frame it comprises a single transcription unit that is under the control of the *tyrR* gene product (66). Previous maps (9, 176) have included an *aroI* gene at min 84 on the *E. coli* chromosome. Some strains carrying mutations at this locus were temperature sensitive but able to grow slowly at the restrictive temperature if the aromatic amino acids were added to the medium. Since attempts to identify the enzymic step that is affected in these mutants have failed, it is possible that the effects of the aromatic amino acids may be indirect (unpublished data). The gene has, accordingly, been left out of the list in Table 2.

The isolation of mutants and the mapping of most of the structural genes has been straightforward. Gene designations were independently allocated to five of the genes for the common pathway in *S. typhimurium* and *E. coli* (97, 168, 177). Unfortunately not all the

TABLE 2. Genes for the biosynthesis and transport of aromatic amino acids

Gene symbol	Function[a]	Map position (min)[b]	
		E. coli	*S. typhimurium*
aroA	5-Enolpyruvylshikimate-3-phosphate synthase (EC 2.5.1.19)	20	19
aroB	3-Dehydroquinate synthase (EC 4.6.1.3)	75	73
aroC	Chorismate synthase (EC 4.6.1.4)	51	47
aroD	Dehydroquinate dehydratase (EC 4.2.1.10)	37	36
aroE	Shikimate dehydrogenase (EC 1.1.1.25)	72	71
aroF	DAHP synthase (tyr) (EC 4.1.2.15)	57	55
aroG	DAHP synthase (phe) (EC 4.1.2.15)	17	17
aroH	DAHP synthase (trp) (EC 4.1.2.15)	37	36
aroL	Shikimate kinase 11 (EC 2.7.1.71)	9	
aroM	Open reading frame in an operon *aroLM*, function unknown	9	
aroP	General aromatic amino acid transport	3	3
aroT	Transport of aromatic amino acids and glycine and alanine	28	35
azaB	Phenylalanine transport	70	
mtr	Tryptophan-specific transport	69	
pheA	Chorismate mutase-prephenate dehydratase (EC 5.4.99.5, EC 4.2.1.51)	57	55
pheP	Phenylalanine-specific transport	13[c]	
pheU	Structural gene for tRNA[Phe]	94.5	
pheV	Structural gene for tRNA[Phe]	12[d]	
pheR	Structural gene for tRNA[Phe][e]	94	64
pheC	Can complement mutations in *pheR*	(66)	
trpA	Tryptophan synthase: A protein (EC 4.2.1.20)	28	34
trpB	Tryptophan synthase: B protein (EC 4.2.1.20)	28	34
trpC	N-(5-Phosphoribosyl) anthranilate isomerase-indole-3-glycerol phosphate synthase (EC 4.1.1.48)	28	34
trpD	Glutamine amidotransferase-phosphoribosyl anthranilate transferase (EC 2.4.2.18)	28	34
trpE	Anthranilate synthase (EC 4.1.3.27)	28	34
trpR	*trpR* aporepressor	0	99
trpS	Tryptophanyl-tRNA synthase (EC 6.1.1.2)	74	
tyrA	Chorismate mutase-prephenate dehydrogenase (EC 5.4.99.5, EC 1.3.1.12)	57	55
tyrB	Aromatic aminotransferase (EC 2.6.1.5)	92	
tyrP	Tyrosine-specific transport	42	
tyrR	TyrR protein	29	32

[a] See the text for a more detailed description of functions.
[b] Map positions are from Bachmann (9) and Sanderson and Roth (194).
[c] P. A. Kasian and J. Pittard, unpublished data.
[d] Caillet, unpublished data.
[e] Narasaiah and Davidson, unpublished data.

gene designations coincided, but this problem has been remedied in the most recent maps (9, 194). The cloning in *E. coli* of *aroF tyrA pheA*, *aroL*, and *aroP* has provided detailed molecular information on the map locations of these genes (66, 102, 116).

The isolation of mutants affected in the DAHP synthase, shikimate kinase, and aminotransferase reactions and the subsequent mapping of the relevant genes proved a little more difficult than that of the other pathway genes. The strategies that were used are briefly summarized.

Since the first reaction of the common pathway is carried out by three isofunctional enzymes, the loss of just one of these will not normally create an aromatic auxotroph. However, because each of these enzymes is subject to specific feedback inhibition and their respective structural genes are subject to specific repression, it is possible to create conditions in which the loss of one enzyme will cause a condition of aromatic auxotrophy. Wallace and Pittard (224, 225) first isolated mutant strains that had lost the DAHP synthase (tyr) activity specified by the *aroF* gene. Such mutants grow well on minimal medium but are unable to grow if the medium is supplemented with both phenylalanine and tryptophan, as these amino acids repress and inhibit the remaining two enzymes. From the *aroF* mutant a second strain was isolated that could not grow if the minimal medium was supplemented with tryptophan. Such mutants had additional mutations in *aroG*, the structural gene for DAHP synthase (phe). From such an *aroF aroG* strain a third mutant was isolated which now lacked all DAHP synthase activity because of an additional mutation in *aroH*, the gene for DAHP synthase (trp). These mutations could then be transferred to other strains by using standard genetic techniques, and their map locations could be determined. A similar strategy was used in *S. typhimurium* by DeLeo and Sprinson (69) and independently in *E. coli* by Brown (24). In the study by Brown and Somerville (28) there were some unexpected results, however. In this study, a mutant with an incomplete block in the tryptophan pathway also exhibited the phenotype of growth on minimal medium but no growth on minimal medium supplemented with phenylalanine and tyrosine. In this case, the two aromatic amino acids so decreased the flow of substrates along the common pathway by feedback inhibition of DAHP synthase (tyr) and (phe) that there was insufficient substrate of the partially blocked reaction in the tryptophan pathway for conversion to occur. In the absence of phenylalanine and tyrosine and the presence of a regulatory mutation which derepresses DAHP synthase (phe) and DAHP synthase (tyr), growth readily occurred in minimal medium.

As the shikimate kinase enzymes are not subject to feedback inhibition, a different strategy had to be used. In wild-type *E. coli*, shikimate kinases I and II appear to be present in the cell at about the same levels. In strains with a mutation in the *tyrR* regulator gene, however, the levels of shikimate kinase II are elevated to the point where this constitutes the major shikimate kinase activity. In such strains, this enzyme is synthesized constitutively (80). When strains of *E. coli* are grown at fast growth rates in rich medium, the presence of multiple replication forks increases the relative abundance of gene copies for genes nearest the origin of replication. If genes are expressed constitutively, it is possible to use a measure of the levels of the enzymes that they specify to indicate their position relative to the fixed origin of the replicating chromosome. Since replication is bidirectional, such studies provide two possible locations equidistant from the origin of replication. In the case of the gene for shikimate kinase II, Tribe et al. (217) used this technique to predict that this gene should be found in a region of the *E. coli* chromosome near either min 13 or 45. By introducing known F-merogenotes into the *tyrR* strain and measuring kinase activity it was clearly shown that the structural gene was somewhere between *proC* and *purE* in the min 13 region. By mutagenizing a *purE* strain and selecting Pur+ revertants, it was hoped to obtain strains with additional mutations in this region. Fortunately, one of these Pur+ revertants was found to have lost the ability to synthesize shikimate kinase II (80). The gene was termed *aroL*, and subsequent work has shown that it is very tightly linked (99%) in transduction with the *proC* gene. The genes *proC* and *aroL* have been cloned on the same fragment, and by insertional mutagenesis they have been shown to be separated by only 0.9 kilobase (66). As previously mentioned, it has also been shown that in *aroF aroG* strains a mutation in *aroL* does produce a phenotype of aromatic auxotrophy.

The precise map location for the genes *pheP*, *tyrP*, and *pheR* also presented difficulties because of the absence of a clean selection for PheP+, TyrP+, or PheR+ phenotypes. This problem, however, was remedied when λ Tn*10* became available (133) and could be used both to mutate a gene and to enable the determination of its map position (99, 128; P. A. Kasian, Ph.D. thesis, University of Melbourne, Parkville, Victoria, Australia, 1985). In the case of the *tyrB* gene it was necessary first to isolate mutants that could utilize 4-hydroxyphenylpyruvate instead of tyrosine before selecting for *tyrB* mutants (84).

The Tryptophan Operon

The genes specifying the enzymes of the tryptophan pathway are arranged in a single operon. Their tight linkage to one another was established in some of the first mapping experiments with the generalized transducing phage P1 to be carried out in *E. coli* (244). Similar tight linkage was established in *S. typhimurium* (22). The subsequent observations that in *E. coli* these genes could be picked up by the specialized transducing phage φ80 (63, 148) and also be incorporated into phage λ (185), coupled with the generation of deletions which extended from the region containing *tonB* into the *trp* operon, resulted in the in vivo cloning of these genes considerably in advance of the development of in vitro methods. The genes of the tryptophan operon and their transcription and translation are described in depth elsewhere in this book (chapter 90).

REGULATION OF THE AROMATIC PATHWAY

Feedback inhibition of various key enzymes of the pathway by the aromatic amino acids has already been discussed. In this section, the various controls on gene expression and the ways in which these have been investigated will be considered. As far as we know currently, all of the controls affect the transcription rather than the translation of the structural genes. Specific regulation of transcription is mediated solely by repressor proteins except for the *trp* operon (126, 240) and the *pheA* gene (99, 116, 251), for which attenuation control has been demonstrated, and *aroF*, for which an as yet undefined control exists in addition to repressor-mediated control (30). Metabolic regulation as described by Rose and Yanofsky (186) exerts a general control on the aromatic pathway (217). There are only three examples of tightly clustered genes on the map (Table 2), namely, the *trp* operon, *aroF tyrA pheA*, and *aroLM*. The *trp* operon is part of a regulon that includes *aroH* and is controlled by the product of the *trpR* gene (24, 33, 40, 124). The *trpR* gene and the interaction of the TrpR protein with its relevant operator targets are discussed in detail elsewhere in this volume. The expression of the tryptophan operon is also controlled by a process of attenuation or premature transcription termination (240). The molecular details of this process have been worked out in a series of very elegant experiments and are also described in detail elsewhere in this volume (chapter 76). Early studies on enzyme repression showed that *pheA* and *tyrA* were independently controlled (44), whereas *aroF* and *tyrA* appeared to be part of a single transcription unit (96, 149, 226). As previously mentioned, the function of *aroM* is currently not known, but insertional mutagenesis and the creation of *lac* fusion strains suggest that it is transcribed from the *aroL* promoter (66). In addition to *trpR*, there are two other regulator genes, *tyrR* and *pheR*. The PheR gene product has only been shown to control the expression of a single gene, *pheA*, whereas the TyrR protein has been shown to control the expression of eight different transcription units. In considering regulation, the *tyrR* gene and the various structural genes under its control are discussed first.

The *tyrR* Regulon

The *tyrR* gene was first identified in *E. coli* when strains were isolated that were resistant to the tyrosine analog 4-aminophenylalanine. The starting strain used in these experiments had only a single functional DAHP synthase enzyme, namely, DAHP synthase (tyr), and since 4-aminophenylalanine acts as a false corepressor of the synthesis of this enzyme but has no activity as a feedback inhibitor, derepressed mutants were readily isolated (226). Similar *tyrR* derivatives were found in *S. typhimurium* among 4-fluorophenylalanine-resistant mutants (94). In both *E. coli* and *S. typhimurium* the mutations were shown to be recessive in diploids (94, 120) and to cause the derepressed synthesis of DAHP synthase (tyr), chorismate mutase-prephenate dehydrogenase, and the aromatic aminotransferase.

The isolation and analysis of putative operator mutants affecting *aroF* and *tyrA* in *S. typhimurium* (96) and in *E. coli* (149) showed that the gene for the aminotransferase was not part of the *aroF-tyrA* operon and added confirmation to the hypothesis that *tyrR* coded for a repressor which acted on a specific operator target controlling the expression of both *aroF* and tyrA.

The discovery that the *tyrR* gene product was also involved in the regulation of the synthesis of DAHP synthase (phe) was partly fortuitous. Im et al. (120) used a selection for resistance to 4-fluorophenylalanine that had been shown to yield mutants derepressed for the single enzyme chorismate mutase-prephenate dehydratase in the hope that they might isolate mutants derepressed for this enzyme and also for DAHP synthase (phe). This expectation was based on the assumption that both genes might be controlled by the same regulator gene. In fact, they isolated mutants derepressed for DAHP synthase (phe) but not for chorismate mutase-prephenate dehydratase. These strains were shown to have mutations in the *tyrR* gene and also to be derepressed for DAHP synthase (tyr), chorismate mutase-prephenate dehydrogenase, and the aromatic aminotransferase. As we now know, the gene for chorismate mutase-prephenate dehydratase is controlled by a separate regulator gene, *pheR*. As previously discussed, Brown and Somerville (28) also isolated a *tyrR* mutant in which both DAHP synthase (phe) and DAHP synthase (tyr) were derepressed, but this mutation in combination with a leaky mutation in the tryptophan pathway occurred in a strain found in a selection intended to isolate a mutant deficient in the enzyme DAHP synthase (trp). An analysis of the early *tyrR* mutants of Wallace and Pittard showed that in these strains repression of the synthesis of DAHP synthase (phe) by phenylalanine was increased rather than decreased (120). Complementation studies between a number of different *tyrR* alleles, however, failed to establish the existence of more than a single cistron (122). The isolation of amber *tyrR* mutants and of a temperature-sensitive *tyrR* mutant of the thermolabile type established unambiguously that the *tyrR* product was a protein that was either the repressor itself or at least a component of the repressor. The isolation of amber mutants and of *tyrR* mutants in which phage Mu had inserted in the *tyrR* gene established that the TyrR protein was not required for cell viability (31). Subsequently it has been shown that the TyrR protein is involved in the repression of the *aroLM* operon (66, 80), the repression of the *aroP* and *tyrP* genes, and the induction or enhancement of the expression of the *tyrP* and *mtr* genes (128, 232). In these latter cases, gene expression was first shown to be subject to control by one or another of the aromatic amino acids, and the role of the TyrR protein was established by analysis of these functions in *tyrR* mutants.

The control of *tyrP* is particularly interesting because in this case the TyrR protein is involved with tyrosine as a repressor and with phenylalanine as an inducer. In the latter case the TyrR protein appears to exert a positive effect on transcription.

The TyrR protein also represses the expression of its own structural gene, *tyrR*. This effect was demonstrated by making a *tyrR-lacZ* operon fusion strain and introducing increasing copies of the wild-type *tyrR* gene on plasmids into the cell (32). The various genes comprising the *tyrR* regulon and the particular amino acid effectors that are involved for each gene are shown in Table 3, in which it is shown that different aromatic amino acids are able to interact with the

TABLE 3. The *tyrR* regulon

Transcription unit	Gene product(s)	Effector molecules[a]
aroF tyrA	DAHP synthase (tyr), chorismate mutase-prephenate dehydrogenase	Tyrosine
tyrB	Aromatic aminotransferase	Tyrosine
tyrP	Tyrosine-specific transport system	Tyrosine, phenylalanine (induction)
aroLM	Shikimate kinase II and unknown	Tyrosine or tryptophan
aroP	General aromatic transport system	Tyrosine or phenylalanine or tryptophan
aroG	DAHP synthase (phe)	Phenylalanine and tryptophan
mtr	Tryptophan-specific transport system	Phenylalanine (induction)
tyrR	TyrR protein	None appears to be required

[a] All effectors involved in repression unless otherwise indicated.

TyrR protein and effect the regulation of different genes in the *tyrR* regulon.

Table 4 provides some data (217, 226, 232) indicating the relative sensitivity of these various genes to repression. For example, the genes for DAHP synthase (tyr), aromatic aminotransferase, the tyrosine-specific permease, and the general aromatic permease are all repressed by the TyrR protein complexed with tyrosine. Cells growing in minimal medium are, however, fully derepressed for aminotransferase and the general aromatic permease, 55% repressed for the tyrosine-specific permease, and 87% repressed for DAHP synthase (tyr). Presumably, therefore, not only do the various operator loci associated with the genes of the *tyrR* regulon differ in their ability to distinguish different forms of the TyrR protein (R-Tyr, R-Phe, etc.), but also, where the same form of repressor is involved, other differences exist in either the affinity between repressor and operator(s) or the effectiveness with which repressor binding can inhibit transcription. The discovery that, under appropriate conditions, *tyrR*[+] strains would grow in the presence of β-thienylalanine whereas *tyrR*[−] strains would not (H. Camakaris, unpublished data) facilitated subsequent work in which the *tyrR* gene was picked up on lambda phage (38) and cloned into pBR322 (43). Its nucleotide sequence has been determined (42a). From the sequence, the molecular weight of the TyrR protein subunit is calculated to be 53,099, in contrast to 63,000 as measured in maxicell preparations. This value makes it considerably larger than the TrpR protein of 12,356 (103). The TyrR protein has not yet been produced in amounts that would allow it to be studied in in vitro systems, although recently the *tyrR* gene has been successfully joined to the *tac* promoter

(6), resulting in significant production of the TyrR protein (E. C. Cornish, unpublished data). Gicquel-Sanzey and Cossart (92) have studied the amino acid sequences of 13 procaryotic regulatory proteins and found two regions of homology. Neither of these is to be found in the *tyrR* protein sequence (42a), although, as discussed below, the operators controlled by *tyrR* share some properties with other procaryotic operators.

Some progress has also been made recently in identifying and characterizing the operator loci with which the TyrR protein interacts. The complete nucleotide sequences including the promoter and operator loci of *aroG*, *tyrR*, *aroF-tyrA*, and *aroLM* have been completed (42a, 56, 65, 116). In addition, the nucleotide sequences of the regions of the *tyrP* and *aroP* genes which include the promoters and operator loci have been determined (P. A. Kasian, B. E. Davidson, and A. J. Pittard, J. Bacteriol., in press; Chye, unpublished data), and specific base-pair changes of operator-constitutive mutants have been determined for *aroF* (82), *tyrP* (Kasian et al., in press), *aroP* (Chye, unpublished data), and *aroG* (N. Baseggio and B. E. Davidson, unpublished data).

The nucleotide sequences of the first two genes to be examined, *aroG* and *tyrR*, had almost identical palindromic sequences overlapping putative promoter sites. Analysis of the relevant sequences in *aroL*, *aroF*, *aroP*, and *tyrP* subsequently revealed the presence of similar sequences. A consensus sequence comprising 17 nucleotides was derived from five of these sequences (65) (see Fig. 6). Sequences bearing a close homology to this have been referred to as TYR boxes (174). A consideration of all 12 TYR boxes in the six different control regions leaves the consensus almost

TABLE 4. Activities of proteins of the *tyrR* regulon

Treatment	Sp act (mU/mg of protein)				Transport activity (nmol of tyrosine per min per mg dry wt)	
	DAHP synthase (tyr)	DAHP synthase (phe)	Shikimate kinase I	Aromatic aminotransferase	General aromatic permease	Tyrosine-specific permease
Repressed[a]	13	40	8	12	1	0
tyrR[+] (minimal medium)	70	230	15	39	6	1.6
Derepressed[b]	560	300	48	37	7	3.6[c]

[a] Cells grown in the presence of the aromatic amino acids.

[b] Values obtained in *tyrR* mutants.

[c] Value of 9.0 nmol/min per mg dry weight was obtained in a *tyrR*[+] strain grown in the presence of phenylalanine.

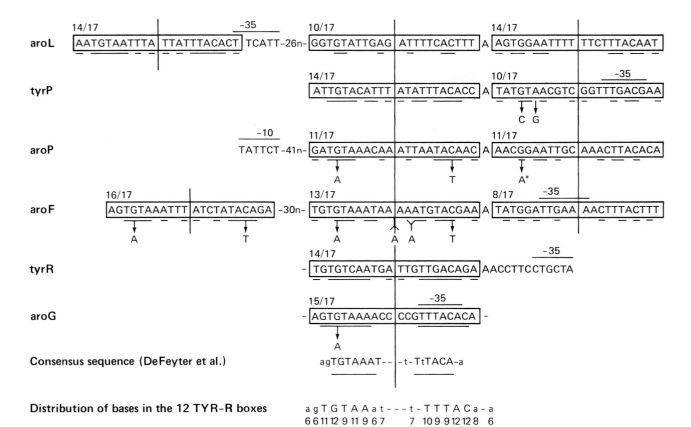

Distribution of bases in the 12 TYR-R boxes

a g T G T A A a t - - - t - T T T A C a - a
6 6 11 12 9 11 9 6 7   7 10 9 9 12 12 8 6

FIG. 6. Comparison of the nucleotide sequences in *aroL*, *tyrP*, *aroP*, *aroF*, *tyrR*, and *aroG* which contain variants of the TYR box that are believed to act as operators for different forms of the TYR repressor. The base changes shown for *tyrP*, *aroF*, *aroP*, and *aroG* are changes which result in a partially constitutive phenotype. The G→A change in the right-hand box of *aroP* (*) has so far only been seen in conjunction with a corresponding G→A change in the left box. Whereas this latter change by itself gives only partial constitutivity, the strain with the double substitution, plus some other base changes outside the TYR box region, expresses the *aroP* gene in a completely constitutive manner (Chye and Pittard, unpublished data). The sequences regarded as variations of the TYR box are enclosed, and the numbers at the left-hand end of each box indicate the homology of that box with the consensus shown below. The consensus sequence was constructed by comparing the relevant sequences in *tyrR* and *aroG*, two of the three sequences in *aroF*, the 14/17 box of *tyrP*, and the sequence overlapping the postulated −35 region in *aroL* (65). Symmetrical sequences are underlined. The distribution of bases in the 12 TYR boxes presented in this figure is also shown.

unchanged (Fig. 6). Whereas *aroG* and *tyrR* each have a single copy of the palindrome that has been referred to as TYR box, *aroF aroL aroP* and *tyrP* appear to contain two or more TYR boxes each.

Two points to note in this respect are, first, that tyrosine is a major corepressor for *aroF*, *aroL*, *aroP*, and *tyrP* but not for *aroG* or *tyrR*, and second, that neither *aroG* nor *tyrR* can be completely repressed in haploid *tyrR+* strains. By increasing the number of copies of *tyrR+* in the cell, more effective repression can be achieved (32).

Figure 6 compares the relevant sequences for these six genes and indicates the position of the boxes, the position of the base changes in mutants, and the relative positions of the −10 or −35 regions of the relevant promoters. In all cases, the position of promoter sequences is supported by data that have established the start point of transcription. It is of interest that *aroP*, which is one of the least sensitive genes to repression, should have TYR boxes with relatively low homology to the consensus and should have these boxes positioned at some distance on the downstream side of the putative promoter.

The four genes whose expression is controlled by tyrosine all have two TYR boxes that are adjacent and are separated in each case by a single adenine residue. *aroF* and *aroL* also have a third TYR box situated some 30 to 31 bases upstream from these two. The mutational studies of Garner and Herrmann (82) clearly implicate this upstream box in the regulation of *aroF*. Because operator mutations were present on multicopy plasmids in haploid *tyrR+* strains in the studies of Garner and Herrmann, it is not possible to evaluate the relative importance of each of the TYR boxes to repression. C. S. Cobbett (unpublished data) has recently demonstrated that deletion of this upstream box results in partial constitutivity of *aroF* expression. Even though the positions of the putative −35 regions of *aroF* and *aroL* differ, in both cases sites that are situated at least 30 bases apart seem to be involved together in regulating initiation of transcription. The most likely explanation for this arrangement would involve cooperative interaction between TyrR molecules binding to the different sites. If this were true, changing the distance between the upstream box and the two downstream boxes should affect repression by changing the relative positions of the TyrR molecules bound at the different sites. Such rearrangements are currently being explored (Cobbett, unpublished data).

Whether the double TYR box arrangement of aroL, aroF, aroP, and tyrP indicates that TyrR-tyrosine can act as a tetramer remains to be seen. Confirmatory evidence that the promoter-proximal box in aroF is involved in repression has been provided by the recent finding that insertion of an adenine residue into the middle of this TYR box results in constitutive expression of aroF (Cobbett, unpublished data). The degree of symmetry in the arms of the various TYR boxes varies considerably and is weak in a number of cases. It is interesting that the single TYR boxes of both tyrR and aroG overlap or are adjacent to the −35 regions and in both cases show a high degree of symmetry.

Gicquel-Sanzey and Cossart (92) have reported that the sequence TGTGT N_{6-10}ACACA, or minor variations of it, occurs in a number of E. coli and phage lambda operator sites. As Fig. 6 shows, the sequence TGT N_{12}ACA is found in most of the TYR boxes. The observation of Garner and Herrmann (82) that, in aroF, deletion or insertion of a base in the middle of the box alters repressibility suggests that this spacing is important. Of the base substitutions so far shown to alter repressibility, all four in the operator loci of aroF, all three in aroP, the only one so far identified in aroG, and one of two mutations in tyrP involve the G or C of this common sequence. A second mutation in tyrP is a substitution of a base adjacent to TGT. The fact that those mutations that interfere with repressibility of tyrP occur in a TYR box that shows a low degree of symmetry and a weak homology with the box consensus sequence demonstrates that we still have a great deal to discover about the specific interactions between the various forms of the TyrR protein and its respective operator targets. For example, we have no information yet on the site to which TyrR-Phe binds to cause enhanced expression of tyrP, although the TYR box other than the one in which mutations have been identified is a tempting possibility. Nevertheless, the tyrR regulon clearly offers an excellent opportunity to examine more closely the specificity of interactions between repressor proteins and their targets. The purification of the TyrR protein, the isolation of more operator-constitutive mutants, the analysis of the phenotypic effects of base changes within and between the TYR boxes, and the isolation of tyrR mutants with different alterations in the binding of TyrR to the various operators should assist in this endeavor.

The finding that the introduction of a mutation in trpS, the gene for the tryptophanyl-tRNA synthetase, caused further elevation of DAHP synthase (tyr) in tyrR mutants indicated that a second system of control affects this operon. The involvement of the trpS gene led to the hypothesis that some form of attenuation in which charged tryptophanyl-tRNA plays a major role may be operating (30). The determination of the nucleotide sequence for the leader region of aroF, however, seems to make this unlikely (116), and the trpS effect remains unresolved.

McCray and Herrmann (154) reported experiments which suggested a possible role for iron in the regulation of aromatic amino acid biosynthesis. When they grew wild-type E. coli under conditions of iron starvation, the levels of a number of enzymes, dihydrobenzoylserine synthase, DAHP synthase (tyr), tryptophan synthase, and prephenate dehydratase, were considerably elevated. In the case of DAHP synthase (tyr), prephenate dehydratase, and tryptophan synthase, this iron deficiency-mediated derepression was completely reversed by the addition of 0.6 μM iron, a concentration at which growth inhibition caused by iron starvation was also fully reversed. DAHP synthase (phe) was not significantly derepressed by iron starvation, although, as shown in Table 4, it is normally derepressed in minimal medium. The derepression of DAHP synthase (tyr), prephenate dehydratase, and tryptophan synthase by iron starvation was prevented if either shikimate or the aromatic amino acids were added to the medium.

McCray and Herrmann suggested that increased levels of DAHP synthase (tyr), prephenate dehydratase, and tryptophan synthase could be the result of altered attenuation control caused by the failure of the cell to fully modify certain tRNAs under conditions of iron starvation. Rosenberg and Gefter (187) had shown the involvement of an iron-containing enzyme in these reactions. The possibility that the observed derepression was an indirect effect caused by the channeling of chorismate down the enterochelin pathway was excluded by introducing a mutation in entC, the gene for isochorismate synthase, and demonstrating that derepression still occurred (154). However, a number of facts remain to be explained. Although derepression still occurs in the entC strain, it is significantly reduced. Furthermore, the derepressed enzyme levels of iron-starved wild-type cells correspond to the fully derepressed levels observed on starvation for the aromatic amino acids or in derepressed mutants. The observation that the addition of either the aromatic amino acids or shikimate to the medium prevents the derepression caused by iron starvation also supports the hypothesis that the primary cause of derepression is depletion of the aromatic amino acid pools. If derepression, on the other hand, were the result of improperly modified tRNA molecules affecting attenuation mechanisms, one would not expect this to be changed by the addition of shikimate or the amino acids to the medium. As mentioned previously, the enzyme chorismate synthase requires iron as a cofactor, and the three DAHP synthases are iron-containing enzymes. It seems likely, therefore, that the derepression observed under conditions of iron starvation is a direct result of amino acid starvation, resulting from both iron deficiency-mediated inactivation of key synthetic enzymes and channeling of chorismate along the enterochelin pathway. Some effects on attenuation may also occur, but these appear to play a minor role.

It has been reported in S. typhimurium that mutants deficient in adenylcyclase activity (cya) have reduced levels of the general aromatic permease, suggesting some form of control by catabolite repression (5). A similar situation probably does not occur in E. coli, as aroP-lacZ operon fusion strains show the same levels of β-galactosidase activity when grown on glycerol or glucose as a carbon source (Chye, unpublished data). Experiments similar to those reported for S. typhimu-

rium using *cya* mutants have not been carried out, however.

The expression of the *tyrP* gene is subject to modulation in *tyrR* strains in which TyrR-mediated repression is absent. The addition of leucine or Casamino Acids (0.2%) to the minimal medium in which the *tyrR* strains are grown results in a twofold and four- to fivefold increase, respectively, in *tyrP* expression (128). An examination of the *tyrP*-specific mRNA transcripts reveals that there are two transcription start points separated by 140 bases. One of these involves a promoter which overlaps one of two adjacent TYR boxes, and transcription from this start point is stimulated by phenylalanine and repressed by tyrosine in *tyrR*⁺ cells. If leucine is present in the medium, transcription from this start point is also enhanced approximately twofold in both *tyrR*⁺ and *tyrR* strains. The second transcription start point, 140 bases upstream from the first, is of major importance when the cells are grown in the presence of Casamino Acids. Stimulation of transcription from this start site does not require functional TyrR protein. We have not identified the promoter locus responsible for this transcription, and as expected, no TYR boxes are found in this region. One cannot detect transcription from this start point when the cells are grown in minimal medium (P. Kasian, J. Pittard, and B. E. Davidson, in press).

Presumably, the existence of a second promoter separated from the TYR boxes and activated in an unknown way by a component(s) of Casamino Acids allows derepression of *tyrP* expression even in the presence of tyrosine. More work is required to understand the mechanism and the physiological significance of this second system of control. No similar effect is observed with *aroP* (Chye, unpublished data).

The *pheA* Gene and Its Regulation

Whether cells of *E. coli* are grown in minimal medium or in minimal medium supplemented with aromatic amino acids, the level of expression of the *pheA* gene is the same. However, starving cells for phenylalanine or introducing specific mutations can result in a 2- to 20-fold increase in the level of *pheA* expression (28, 99, 121). Strains of *E. coli* derepressed for the synthesis of the *pheA* gene product, chorismate mutase-prephenate dehydratase, were found among mutants resistant to 3- or 4-fluorophenylalanine. Genetic studies showed that the mutations in these strains were tightly linked to the *pheA* structural gene and were not recessive in diploids. On this basis such mutations were provisionally identified as mutations in a putative operator locus of the *pheA* gene (121). Attempts to isolate strains with mutations in a putative regulator gene, *pheR*, using the same selection were not successful (S. W. K. Im and J. Pittard, unpublished data). In *S. typhimurium*, on the other hand, the first mutants to be isolated that were derepressed for chorismate mutase-prephenate dehydratase were not operator mutants but mutants with changes in a gene, *pheR*, that was unlinked to the *pheA* structural gene. These mutations were recessive in partial diploids made with the F-merogenote F116 from *E. coli*. This F-merogenote contains the 59- to 65-min region of the *E. coli* chromosome. Crosses with *S. typhimurium* Hfr donors had showed that *pheR*⁺ was located in

this same region of the *S. typhimurium* chromosome (95).

Putative *pheR* mutants were ultimately isolated in *E. coli* after the construction of *pheA-lacZ* operon fusion strains and selection for mutants able to grow on medium containing phenyl-β-D-galactoside as sole carbon source (98, 99). These mutant strains were derepressed for the synthesis of chorismate mutase-prephenate dehydratase and, as was the case in *S. typhimurium*, were converted to a repressible state upon the introduction of the F-merogenote F116. However, when the mutations were mapped by Hfr crosses and P1 transduction using a closely linked copy of Tn*10*, they were found to be in the 93- to 94-min region of the *E. coli* map and to be cotransducible with *ampA* and *purA*. In support of this conclusion, it was shown that the introduction of F-merogenote F117, which covers this 93- to 98-min region of the chromosome, also converted these mutants to a repressible state. The gene identified by these *E. coli* mutants was called *pheR* (99).

Clearly, the mapping presented a dilemma. The F-merogenote F117 presumably carried the wild-type *pheR*⁺ allele, but the nature of the gene on F116, which was tentatively called *pheC* and whose addition to the cell could also convert strains to a repressible phenotype, remained obscure. In search of a functional *pheC* allele, the mutation *pheR372* was introduced into a number of *E. coli* strains including AB312, the Hfr strain from which F116 was derived. All such *pheR372* derivatives were derepressed for *pheA* expression. From this it was concluded that the gene, which was active on F116, was either missing or inactive on the chromosome of these strains, or, alternatively, that it needed to be present in at least two copies per chromosome before it could convert *pheR* strains to a repressible phenotype (J. Gowrishankar, Ph.D. thesis, University of Melbourne, Melbourne, Australia, 1982). Starting with a *pheR* derivative of Hfr AB312, a number of new F-merogenotes were generated. All the F-merogenotes that carried the region from the sex factor to the *metC* gene have the capacity to convert *pheR* mutants to a repressible phenotype (Fig. 7). This then suggests that *pheC* maps somewhere between the origin of Hfr AB312 and *metC* (S. Morrison and J. Pittard, unpublished data).

Earlier attempts to map *pheC* by transduction with P1 propagated on a strain carrying F116 had suggested that *pheC* was strongly linked to the *fuc* locus (see Fig. 7). However, in these experiments all of the transductants with the *pheC*⁺ phenotype were unstable and could be shown to be diploid for the region of chromosome carried by F116 (Gowrishankar, thesis). Subsequent analysis has suggested that in these experiments the locus being mapped was not *pheC* but a "duplication joint" present on F116 (8a; R. Lang, B.Sc. Hons. thesis, University of Melbourne, Melbourne, Australia, 1984). If the previous hypothesis is true and *pheC*⁺ only produces a repressible phenotype in *pheR* cells when present in two or more copies per chromosome, the quite fortuitous presence of a duplication joint on F116, by producing unstable diploids in transductants, makes it possible to demonstrate the presence of *pheC*⁺ by P1 transduction. Stable segregants formed from such unstable diploids are always derepressed for *pheA* expression (Gowrishankar, thesis).

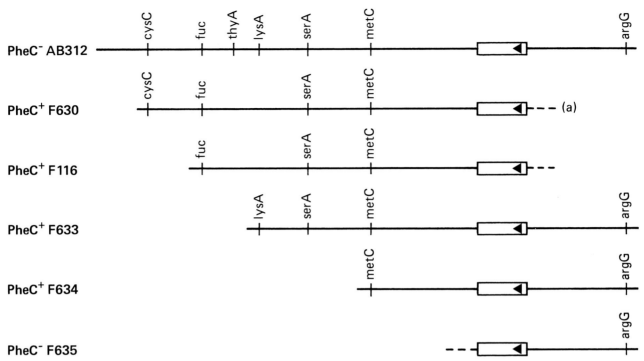

FIG. 7. Map of the chromosomal region in which the putative *pheC* gene is located, showing the PheC⁻ parental strain Hfr AB312 and the PheC⁺ and PheC⁻ F-merogenotes derived from it (Morrison, unpublished data). The phenotypic symbols PheC⁺ and PheC⁻ are used to describe *pheA* expression in each of these *pheR* strains. PheC⁺ is a repressed phenotype and PheC⁻ is derepressed. As discussed in the text, in *pheR* strains a single copy of the *pheC⁺* gene is postulated to result in a PheC⁻ phenotype, whereas two copies of *pheC⁺* produce a PheC⁺ phenotype. It is assumed that one functional copy of *pheC⁺* is present on the chromosome and that the F-merogenotes designated PheC⁺ carry a second copy of *phec⁺*. (a), The presence of regions specified by dotted lines has not yet been established.

A gene which complements the *pheR* mutation has been cloned from the homologous chromosomal region of wild-type *E. coli*. Transposition with Tn*1000* was used to map the gene, and analysis of proteins formed in maxicells suggested that the gene product was a protein of approximate molecular weight of 19,000. This has been referred to as the PheR protein (100).

At this stage, it seemed clear that repression played a major role in the control of the expression of the *pheA* gene and that the mutants that had been genetically characterized had mutations either in a *pheA* operator locus or in the putative repressor gene, *pheR*. There were clearly unresolved questions relating to the map positions of *pheR* in *E. coli* and *S. typhimurium* and the additional gene whose presence was only expressed in diploids, but the general model of repression seemed secure. Further support for repression control came with the identification of three palindromic sequences near the promoter of the *pheA* gene, two of which could potentially act as operator loci (116). It was also reported that in strains that overproduced the TrpR repressor protein, transcription from the *pheA* promoter was repressed approximately 10- to 14-fold (22a). It should, however, be noted that, in wild-type *E. coli*, minimally grown cells are already maximally repressed for *pheA* expression. This reaction with strains that overproduce the TrpR repressor protein is therefore unusual. The result does, however, point to the existence of specific targets on the DNA to

which repressor protein can bind to inhibit *pheA* expression.

Other results indicate that attenuation also has a role to play in the control of *pheA* expression. An analysis of the DNA sequence of the leader region preceding *pheA* revealed the potential for an attenuation control very similar to that reported for the trp operon. The coding region for the leader peptide, in this case, is phenylalanine rich and contains 7 Phe codons in a total of 15. Furthermore, it was shown that readthrough past the termination site of the *phe* attenuator in in vitro transcription was about 40% (251). This model of attenuation was supported by the observation that mutations in the genes *rpoB* or *miaA*, which had been shown to decrease attenuation in the *trp* operon, caused a twofold derepression of *pheA* expression (99). Attenuated transcripts were also demonstrated in vivo (116). It seemed then that, like *trp*, the gene *pheA* was controlled by both repression and attenuation. Repression, as measured by the *pheR* and *pheO* mutants, would account for a 10-fold change in enzyme synthesis and attenuation as measured by the *rpoB* or *mia* mutants, and the in vitro transcription studies would account for a 2- to 3-fold change.

Recent results, however, have called into question some of these conclusions. At the same time, they also offer a possible explanation for some of the puzzling results with *pheR* and *pheC*. The putative *pheR⁺* gene that was cloned by Gowrishankar and Pittard has been sequenced. Analysis of this sequence and comparison with other published sequences establish without a doubt that the cloned gene specifies a tRNA^{Phe} molecule (G. Narasaiah and B. Davidson, unpublished data). Although the origin of the 19,000-molecular-weight protein observed in maxicell experiments is as yet unexplained, this most recent result necessitates a reappraisal of the overall scheme for control of *pheA* expression. In *E. coli* K-12 two genes have already been identified that specify tRNA^{Phe}. Both of these have been cloned, and their nucleotide sequences have been determined. One of these, *pheU* (81a, 199a), lies near min 94.5, and the other, *pheV* (29a, 29b), lies near the *purE* locus at about min 12 (J. Caillet, personal communication). The mapping data available on *pheR* and *pheU* (81a, 99) place them within the same small region of the chromosome but are not accurate enough to indicate whether both studies have involved the same locus.

Although more work is necessary, it is possible at this stage to advance a speculative hypothesis to explain all these results. The hypothesis states that *E. coli* K-12 has either three or four separate genes for the synthesis of tRNA^{Phe}, namely, *pheU*, *pheV*, *pheC*, and *pheR*. There is some uncertainty as to whether *pheU* and pheR are identical, and at the moment there is no evidence to support the proposed function for *pheC* except that it can effectively carry out the same function as *pheR*. The sequence of *pheU*, *pheR*, and *pheV* all specify identical tRNA^{Phe} molecules. Since mutation in one of these genes, e.g., *pheR*, can result in derepression of *pheA* expression, it seems likely that full expression of all the tRNA^{Phe} genes is required to produce sufficient tRNA^{Phe} to ensure that in the presence of excess phenylalanine full attenuation control of *pheA* expression will occur. Perhaps this fine balance has been achieved by some form of evolutionary titration between the numbers of genes for tRNA^{Phe} and the number of Phe codons in the leader transcript. This hypothesis would predict that mutations in *pheV*, *pheC*, or *pheU* (if it is different from *pheR*) would also cause derepression of *pheA* expression and hence would explain the apparent difference between the map positions of phenotypically similar regulatory mutations in strains of *S. typhimurium* and *E. coli*.

The *pheS-pheT* transcription unit which specifies the small and large subunits of phenylalanyl-tRNA synthetase has also been shown to be subject to attenuation control in vitro and in vivo (80a, 208b). In this case, the leader transcript contains five Phe codons, three of which are contiguous. Temperature-sensitive *pheS* mutants are suppressed by multicopy plasmids containing *pheU* (208b), *pheV* (29b), and *pheR* (G. Narasaiah and J. Pittard, unpublished data). If the *pheR* mutations in *E. coli* affect only attenuation, then in wild-type cells approximately 90% of the transcripts initiated from the *pheA* promoter in vivo must terminate at the attenuator. A similar percentage has been reported for the *pheS-pheT* transcription unit (208a).

It is possible to explain previous observations on *pheA* expression within a model in which attenuation is the only form of control. However, much work needs to be done to validate this model, and a molecular analysis of the putative operator mutants of *pheA* is clearly required. Do these strains have deletions or insertions that join the *pheA* structural gene to a new promoter, or do they have specific alterations affecting attenuation control? Finally, could they, in fact, carry operator mutations affecting the activity of an as yet unrecognized repressor? Although the observation that the level of *pheA* expression is the same in *pheR*$^+$ *pheAo351* and *pheR372 pheAo351* strains would seem to argue against this latter possibility, recent experience cautions against too hasty a conclusion.

Evolutionary Relationship between Genes Involved in the Biosynthesis and Transport of the Aromatic Amino Acids

We have discussed a number of unlinked genes that specify either isofunctional proteins or proteins that carry out very similar functions.

Examples of these are the genes for the DAHP synthases, *aroF*, *aroG*, and *aroH*; the genes for chorismate mutase-prephenate dehydratase and chorismate mutase-prephenate dehydrogenase, *pheA* and *tyrA*; the genes for the shikimate kinases, of which only *aroL* has been identified; the two aminotransferase genes, *tyrB* and *aspC*; the transport genes *aroP* and *tyrP*; and the regulator gene *pheR* and the mysterious *pheC*. For those genes for which there is now a complete nucleotide sequence, it is possible with hindsight to see the dangers of leaping to conclusions about gene relationships before complete information is available. On the basis of the sequence of the first 40 amino acids at the amino termini of DAHP synthase (trp), DAHP synthase (tyr), and DAHP synthase (phe) it was concluded that the three enzymes had evolved independently and not from a common ancestor (Poling et al., Fed. Proc. **40**:1581, 1981). However, further analysis of these sequences, involving consideration of possible secondary structures, resulted in the proposition that the three enzymes did arise by divergence from a common ancestral gene for DAHP synthase (202).

When the complete nucleotide sequences of *aroF*, *aroG*, and *aroH* were determined, it became apparent that the three genes were highly homologous (Table 5). There is 50 to 60% identity in the nucleotide sequences of any two of the three genes and a similar degree of identity in the predicted amino acid sequences of the encoded enzymes. When the amino acid sequences are compared using the amino acid-score matrix that makes allowances for structural homologies between amino acids (213), the relatedness of the enzymes is even more pronounced (Fig. 8). Amino acid sequence identity occurs in a number of

TABLE 5. Percentage identity of the *aroF*, *aroG*, and *aroH* genes and their encoded products[a]

Genes	% Identical		
	Nucleotides	Codons	Encoded amino acids
aroF vs. *aroG*	57	25	53
aroF vs. *aroH*	51	26	46
aroG vs. *aroH*	60	30	56
All three	40	10	40

[a] Sources of sequences: *aroG* (56); *aroF* (116, 201); *aroH* (252; Hudson and Davidson, in preparation). The initiator codon has not been included in the comparison.

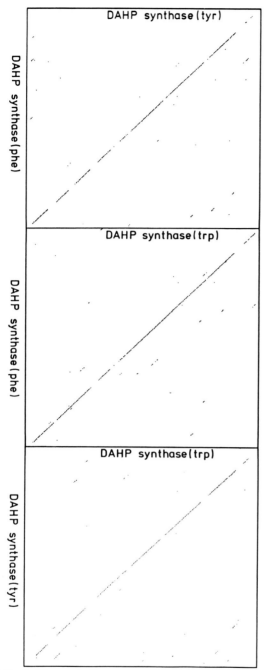

FIG. 8. Diagonal alignments of the amino acid sequences of the three *E. coli* DAHP synthase isoenzymes. The plots were obtained by using the DIAGON method of Staden (213), a window of 11 residues, the Staden amino acid score matrix, and a proportional score of greater than 130. The statistical probability of this proportional score is 0.522×10^{-3} for both the alignments using DAHP synthase (tyr) and 0.537×10^{-3} for the remaining alignment.

ternative proposal by Shultz et al. (201) is that the genes may have evolved through combinations of preexisting parts of different origins. At the moment, there is little evidence to support this latter hypothesis, particularly in the light of the close relationships shown in Fig. 8. The similarity in the relatedness of each pair of genes precludes any conclusion concerning the timing of gene duplication events in the evolution of these three genes.

In a study of tryptic peptides from chorismate mutase-prephenate dehydratase, Baldwin and Davidson (12) observed that some amino acid sequences were repeated in different peptides and postulated that duplication events may have occurred in the evolution of the *pheA* gene. Now that the complete nucleotide sequence of *pheA* has been determined (116), it can be seen that their hypothesis is not supported. On the other hand, a comparison of nucleotide sequences of *pheA* and *tyrA* reveals regions of homology at the amino-terminal ends (116). Of the first 55 amino acid residues of the *tyrA* gene produce, 22 are identical with those of the *pheA* gene product, but practically no homology exists elsewhere in the proteins. There is some evidence to suggest that the mutase activity of these proteins may reside in the amino-terminal portion, and consequently the possibility remains that these genes may have evolved by gene duplication of an ancestral mutase gene followed by fusion to dehydratase and dehydrogenase genes, respectively (116). As mentioned previously, the deduced amino acid sequences for both the aspartate and aromatic aminotransferases show significant homology, and it seems likely that *aspC* and *tyrB* are derived from a common ancestral gene. Insufficient information is available to allow any comments on the other genes that have been mentioned: *aroL* and the second kinase gene, *aroP* and *tyrP*, and *pheR* and *pheC*.

In an examination of the distribution of genes on the *E. coli* chromosome, Zipkas and Riley (250) pointed out that many pairs of genes coding for gene products that were functionally related were found either 90° or 180° apart on the *E. coli* chromosome map. They suggested that this relationship may be the result of two sequential duplications of an ancestral genome, followed by mutation and divergence of the function of the replicate genes. Two of the proposed gene pairs were *tnaA* and *trpB*, the genes for tryptophanase and tryptophan synthase B protein, and *trpE* and either *pheA* or *tyrA*, the genes for anthranilate synthase and chorismate mutase prephenate-dehydratase and chorismate mutase-prephenate dehydrogenase, respectively. Complete nucleotide sequences for all these genes are now available (64, 116, 245), but there are no significant regions of homology between these particular pairs. Thus, if these genes are related as a result of past duplications of the genome, the record of the duplications is no longer present in their nucleotide sequences. It should also be noted that the locations of the genes for the DAHP synthases and aspartokinase-homoserine dehydrogenases, which comprise homologous families, do not conform with the 90° and 180° predictions, although the locations of *aspC* and *tyrB* are approximately 90° apart, and *pheC* and *pheV* are approximately 180° apart with *pheU/pheR* almost midway between them.

large blocks throughout the molecules, with the region of least homology being at the amino terminus, thus explaining the early confusion. The overall homology is strongly suggestive of a divergent evolutionary origin of the three genes from a common ancestral gene (Hudson and Davidson, in preparation). An al-

ACKNOWLEDGMENTS

I wish to thank the following: all my colleagues for permission to quote unpublished results; a number of colleagues for critical reading of the manuscript, with the provision that any errors and omissions that remain are mine; B. E. Davidson for computer analyses of some sequences; and the Australian Research Grants Committee, which has supported research work in my laboratory for a number of years. I also thank J. Feary and M. Gierveld for excellent secretarial assistance.

LITERATURE CITED

1. Adachi, O., L. D. Kohn, and E. W. Mies. 1974. Crystalline $\alpha_2\beta_2$ complexes of tryptophan synthetase of *Escherichia coli*. J. Biol. Chem. 249:7756–7763.

2. Adachi, O., and E. W. Miles. 1974. A rapid method for preparing crystalline β_2 subunit of tryptophan synthetase of *Escherichia coli* in high yield. J. Biol. Chem. 249:5430–5434.

3. Adlersberg, M., and D. B. Sprinson. 1964. Syntheses of 3,7-dideoxy-D-threo-hepto-2,6 diulosonic acid: a study in 5-dehydroquinic acid formation. Biochemistry 3:1855–1860.

4. Aksoy, S., C. L. Squires, and C. Squires. 1984. Translational coupling of the *trpB* and *trpA* genes in the *Escherichia coli* tryptophan operon. J. Bacteriol. 157:363–367.

5. Alper, M. D., and B. N. Ames. 1978. Transport of antibiotics and metabolite analogs by systems under cyclic AMP control: positive selection of *Salmonella typhimurium cya* and *crp* mutants. J. Bacteriol. 133:149–157.

6. Amann, E., J. Brosius, and M. Ptashne. 1983. Vectors bearing a hybrid *trp-lac* promoter useful for regulated expression of cloned genes in *Escherichia coli*. Gene 25:167–178.

7. Ames, G. F. 1964. Uptake of amino acids by *Salmonella typhimurium*. Arch. Biochem. Biophys. 104:1–18.

8. Ames, G. F., and J. R. Roth. 1968. Histidine and aromatic permeases of *Salmonella typhimurium*. J. Bacteriol. 96:1742–1749.

8a. Anderson, R. P., and J. R. Roth. 1977. Tandem genetic duplications in phage and bacteria. Annu. Rev. Microbiol. 31:473–505.

9. Bachmann, B. J. 1983. Linkage map of *Escherichia coli* K-12, edition 7. Microbiol. Rev. 47:180–230.

10. Baker, T. I., and I. P. Crawford. 1966. Anthranilate synthetase. Partial purification and some kinetic studies on the enzyme from *Escherichia coli*. J. Biol. Chem. 241:5577–5584.

11. Balbinder, E. 1964. Intergeneric complementation between A and B components of bacterial tryptophan synthetase. Biochem. Biophys. Res. Commun. 17:770–774.

12. Baldwin, G. S., and B. E. Davidson. 1979. Amino acid sequences of soluble tryptic peptides of chorismate mutase/prephenate dehydratase from *Escherichia coli* K-12. Biochim. Biophys. Acta 579:483–486.

13. Baldwin, G. S., and B. E. Davidson. 1981. A kinetic and structural comparison of chorismate mutase/prephenate dehydratase from mutant strains of *Escherichia coli* K-12 defective in the *pheA* gene. Arch. Biochem. Biophys. 211:66–75.

14. Baldwin, G. S., and B. E. Davidson. 1983. Kinetic studies on the mechanism of chorismate mutase/prephenate dehydratase from *Escherichia coli* K-12. Biochim. Biophys. Acta 742:374–383.

15. Baldwin, G. S., G. H. McKenzie, and B. E. Davidson. 1981. The self-association of chorismate mutase/prephenate dehydratase from *Escherichia coli* K-12. Arch. Biochem. Biophys. 211:76–85.

16. Ballou, C. E., H. O. L. Fischer, and D. L. MacDonald. 1955. The synthesis and properties of D-erythrose-4-phosphate. J. Am. Chem. Soc. 77:5967–5970.

17. Bartholmes, P., H. Balk, and K. Kirschner. 1980. Mechanism of reconstitution of the apoβ2 subunit and the α2apoβ2 complex of tryptophan synthase with pyridoxal 5'-phosphate: kinetic studies. Biochemistry 19:4527–4533.

18. Bartholmes, P., K. Kirschner, and H. P. Gschwind. 1976. Cooperative and non-cooperative binding of pyridoxal 5'-phosphate to tryptophan synthase from *Escherichia coli*. Biochemistry 15:4712–4717.

19. Bauerle, R. H., and P. Margolin. 1966. A multifunctional enzyme complex in the tryptophan pathway of *Salmonella typhimurium*. Cold Spring Harbor Symp. Quant. Biol. 31:203–215.

20. Berlyn, M. B., and N. H. Giles. 1969. Organization of enzymes in the polyaromatic synthetic pathway: separability in bacteria. J. Bacteriol. 99:222–230.

21. Bisswanger, H., K. Kirschner, W. Cohn, V. Hager, and E. Hanssom. 1979. N-(5-Phosphoribosyl) isomerase-indoleglycerolphosphate synthase. 1. A substrate analogue binds to two different binding sites on the bifunctional enzyme from *Escherichia coli*. Biochemistry 18:5946–5953.

22. Blume, A., and E. Balbinder. 1966. The tryptophan operon of *Salmonella typhimurium*. Fine structure analysis by deletion mapping and abortive transduction. Genetics 53:577–592.

22a. Bogosian, G., and R. Somerville. 1983. Trp repressor protein is capable of intruding into other amino acid biosynthetic systems. Mol. Gen. Genet. 191:51–58.

23. Bondinell, W. E., J. Vnek, P. F. Knowles, M. Sprecher, and D. B. Sprinson. 1971. On the mechanism of 5-enolpyruvylshikimate 3-phosphate synthetase. J. Biol. Chem. 246:6191–6196.

24. Brown, K. D. 1968. Regulation of aromatic amino acid biosynthesis in *Escherichia coli* K-12. Genetics 60:31–48.

25. Brown, K. D. 1970. Formation of aromatic amino acid pools in *Escherichia coli* K-12. J. Bacteriol. 104:177–188.

26. Brown, K. D., and C. H. Doy. 1966. Control of three isoenzymic 7-phospho-2-oxo-3-deoxy-D-*arabino*-heptonate-D-erythrose-4-phosphate lyases of *Escherichia coli* W and derived mutants by repressive and "inductive" effects of the aromatic amino acids. Biochim. Biophys. Acta 118:157–172.

27. Brown, K. D., and C. H. Doy. 1976. Transport and utilization of the biosynthetic intermediate shikimic acid in *Escherichia coli*. Biochim. Biophys. Acta 428:550–562.

28. Brown, K. D., and R. L. Somerville. 1971. Repression of aromatic amino acid biosynthesis in *Escherichia coli* K-12. J. Bacteriol. 108:386–399.

29. Butler, J. R., W. L. Alworth, and M. J. Nugent. 1974. Mechanism of dehydroquinase catalyzed dehydration. J. Am. Chem. Soc. 96:1617–1618.

29a. Caillet, J., J. A. Plumbridge, and M. Springer. 1985. Evidence that *pheV*, a gene for tRNA^{Phe} of *E. coli*, is transcribed from tandem promoters. Nucleic Acids Res. 13:3699–3710.

29b. Caillet, J., J. A. Plumbridge, M. Springer, J. Vacher, C. Delamarche, R. N. Buckingham, and M. Grunberg-Manago. 1983. Identification of clones carrying an *E. coli* tRNA^{Phe} gene by suppression of phenylalanyl-tRNA synthetase thermosensitive mutants. Nucleic Acids Res. 11:727–736.

30. Camakaris, H., J. Camakaris, and J. Pittard. 1980. Regulation of aromatic amino acid biosynthesis in *Escherichia coli* K-12: control of the *aroF-tyrA* operon in the absence of repression control. J. Bacteriol. 143:613–620.

31. Camakaris, H., and J. Pittard. 1973. Regulation of tyrosine and phenylalanine biosynthesis in *Escherichia coli* K-12: properties of the *tyrR* gene product. J. Bacteriol. 115:1135–1144.

32. Camakaris, H., and J. Pittard. 1982. Autoregulation of the *tyrR* gene. J. Bacteriol. 150:70–75.

33. Camakaris, J., and J. Pittard. 1971. Repression of 3-deoxy-D-arabinoheptulosonic acid-7-phosphate synthetase (trp) and enzymes of the tryptophan pathway in *Escherichia coli* K-12. J. Bacteriol. 107:406–414.

34. Camakaris, J., and J. Pittard. 1974. Purification and properties of 3-deoxy-D-arabinoheptulosonic acid-7-phosphate synthetase (trp) from *Escherichia coli*. J. Bacteriol. 120:590–597.

35. Chaudhuri, S., and J. R. Coggins. 1985. The purification of shikimate dehydrogenase from *Escherichia coli*. Biochem. J. 226:217–223.

36. Chesne, S., A. Montmitonnet, and J. Pelmont. 1975. Transamination du L-aspartate et de la L-phenylalanine chez *Escherichia coli* K-12. Biochimie 57:1029–1034.

37. Christie, G. E., and T. Platt. 1980. Gene structure in the tryptophan operon of *Escherichia coli*. Nucleotide sequence of *trpC* and the intercistronic regions. J. Mol. Biol. 142:519–530.

38. Cobbett, C. S., and J. Pittard. 1980. Formation of a λ(Tn10) *tyrR$^+$* specialized transducing bacteriophage from *Escherichia coli* K-12. J. Bacteriol. 144:877–883.

39. Cohen, G. 1983. The common pathway to lysine, methionine and threonine, p. 147–172. *In* K. M. Herrmann and R. L. Somerville (ed.), Amino acids: biosynthesis and genetic regulation. Addison-Wesley Publishing, Inc., Reading, Mass.

40. Cohen, G., and F. Jacob. 1959. Sur la repression de la synthèse des enzymes intervenant dans la formation du tryptophane chez *Escherichia coli*. C. R. Acad. Sci. 248:3490–3492.

41. Cohn, W., K. Kirschner, and P. Paul. 1979. N-(5-Phosphoribosyl) anthranilate isomerase-indoleglycerol-phosphate synthase. 2. Fast reaction studies show that a fluorescent substrate analogue binds independently to two different sites. Biochemistry 18:5953–5959.

42. Collier, R. H., and G. Kohlhaw. 1972. Nonidentity of the aspartate and the aromatic aminotransferase components of transaminase A in *Escherichia coli*. J. Bacteriol. 112:365–371.

42a. Cornish, E. C., V. P. Argyropoulos, J. Pittard, and B. E. Davidson. 1986. Structure of the *Escherichia coli* K-12 regulatory gene *tyrR*: nucleotide sequence and sites of initiation and translation. J. Biol. Chem. 261:403–410.

43. **Cornish, E. C., B. E. Davidson, and J. Pittard.** 1982. Cloning and characteristization of *Escherichia coli* K-12 regulator gene *tyrR*. J. Bacteriol. **152:**1276–1279.

44. **Cotton, R. G. H., and F. Gibson.** 1965. The biosynthesis of phenylalanine and tyrosine; enzymes converting chorismic acid into prephenic acid and their relationships to prephenate dehydratase and prephenate dehydrogenase. Biochim. Biophys. Acta **100:**76–88.

45. **Crawford, I. P.** 1975. Gene rearrangements in the evolution of the tryptophan synthetic pathway. Bacteriol. Rev. **39:**87–120.

46. **Crawford, I. P., B. P. Nichols, and C. Yanofsky.** 1980. Nucleotide sequence of the *trpB* gene in *Escherichia coli* and *Salmonella typhimurium*. J. Mol. Biol. **142:**489–502.

47. **Crawford, I. P., and C. Yanofsky.** 1958. On the separation of the tryptophan synthetase of *Escherichia coli* into two protein components. Proc. Natl. Acad. Sci. USA **44:**1161–1170.

48. **Creighton, T. E.** 1970. N-(5'-phosphoribosyl) anthranilate isomerase-indol-3-ylglycerol phosphate synthetase of tryptophan biosynthesis. Biochem. J. **120:**699–707.

49. **Creighton, T. E.** 1970. A steady-state kinetic investigation of the reaction mechanism of the tryptophan synthetase of *Escherichia coli*. Eur. J. Biochem. **13:**1–10.

50. **Creighton, T. E.** 1974. The functional significance of the evolutionary divergence between the tryptophan operons of *Escherichia coli* and *Salmonella typhimurium*. J. Mol. Evol. **4:**121–137.

51. **Creighton, T. E., D. R. Helinski, R. L. Somerville, and C. Yanofsky.** 1966. Comparison of the tryptophan synthetase α subunits of several species of *Enterobacteriaceae*. J. Bacteriol. **91:**1819–1826.

52. **Creighton, T. E., and C. Yanofsky.** 1966. Association of the α and β₂ subunits of the tryptophan synthetase of *Escherichia coli*. J. Biol. Chem. **241:**980–990.

53. **Creighton, T. E., and C. Yanofsky.** 1966. Indole-3-glycerol phosphate synthetase of *Escherichia coli*, an enzyme of the tryptophan operon. J. Biol. Chem. **241:**4616–4624.

54. **Dansette, P., and R. Azerad.** 1974. The shikimate pathway. II. Stereospecificity of hydrogen transfer catalyzed by NADPH-dehydroshikimate reductase of *E. coli*. Biochimie **56:**751–755.

55. **Davidson, B. E., E. H. Blackburn, and T. A. A. Dopheide.** 1972. Chorismate mutase-prephenate dehydratase from *Escherichia coli* K-12. Purification, molecular weight, and amino acid composition. J. Biol. Chem. **247:**4441–4446.

56. **Davies, W. D., and B. E. Davidson.** 1982. The nucleotide sequence of *aroG*, the gene for 3-deoxy-D-arabinoheptulosonate-7-phosphate synthetase (phe) in *Escherichia coli* K-12. Nucleic Acids Res. **10:**4045–4058.

57. **Davis, B. D.** 1950. Aromatic biosynthesis. I. The role of shikimic acid. J. Biol. Chem. **191:**315–325.

58. **Davis, B. D.** 1952. Aromatic biosynthesis. IV. Preferential conversion, in incompletely blocked mutants, of a common precursor of several metabolites. J. Bacteriol. **64:**729–748.

59. **Davis, B. D.** 1953. Autocatalytic growth of a mutant due to accumulation of an unstable phenylalanine precursor. Science **118:**251–252.

60. **Davis, B. D., and E. S. Mingioli.** 1953. Aromatic biosynthesis. VII. Accumulation of two derivatives of shikimic acid by bacterial mutants. J. Bacteriol. **66:**129–136.

61. **Davis, B. D., and U. Weiss.** 1953. Aromatic biosynthesis. VIII. The roles of 5-dehydroquinic acid and quinic acid. Arch. Exp. Pathol. Pharmakol. **220:**S.1–15.

62. **Dayan, J., and D. B. Sprinson.** 1971. Enzyme alterations in tyrosine and phenylalanine auxotrophs of *Salmonella typhimurium*. J. Bacteriol. **108:**1174–1180.

63. **Deeb, G., K. Okamoto, and B. D. Hall.** 1967. Isolation and characterization of non-defective transducing elements of bacteriophage φ80. Virology **31:**288–295.

64. **Deeley, M. C., and C. Yanofsky.** 1981. Nucleotide sequence of the structural gene for tryptophanase of *Escherichia coli* K-12. J. Bacteriol. **147:**787–796.

65. **DeFeyter, R. C., B. E. Davidson, and J. Pittard.** 1986. Nucleotide sequence of the transcription unit containing the *aroL* and *aroM* genes from *Escherichia coli* K-12. J. Bacteriol. **165:**233–239.

66. **DeFeyter, R. C., and J. Pittard.** 196. Genetic and molecular analysis of *aroL*, the gene for shikimate kinase II in *Escherichia coli* K-12. J. Bacteriol. **165:**226–232.

67. **DeFeyter, R. C., and J. Pittard.** 1986. Purification and properties of shikimate kinase II from *Escherichia coli* K-12. J. Bacteriol. **165:**331–333.

68. **De Leo, A. B., J. Dayan, and D. B. Sprinson.** 1973. Purification and kinetics of tyrosine-sensitive 3-deoxy-D-*arabino*-heptulosonic acid 7-phosphate synthetase from *Salmonella*. J. Biol. Chem. **248:**2344–2353.

69. **DeLeo, A. B., and D. B. Sprinson.** 1975. 3-Deoxy-D-arabino-heptulosonic acid 7-phosphate synthase mutants of *Salmonella typhimurium*. J. Bacteriol. **124:**312–1320.

70. **Dopheide, T. A. A., P. Crewther, and B. E. Davidson.** 1972. Chorismate mutase-prephenate dehydratase from *Escherichia coli* K-12. J. Biol. Chem. **247:**4447–4452.

71. **Doy, C. H.** 1967. Tryptophan as an inhibitor of 3-deoxy-*arabino*-heptulosonate 7-phosphate synthetase. Biochem. Biophys. Res. Commun. **26:**187–192.

72. **Doy, C. H., and K. D. Brown.** 1965. Control of aromatic biosynthesis: the multiplicity of 7-phospho-2-oxo-3-deoxy-D-*arabino*-heptonate D-erythrose-4-phosphate-lyase (pyruvate phosphorylating) in *Escherichia coli* W. Biochim. Biophys. Acta **104:**377–389.

73. **Doy, C. H., and F. Gibson.** 1959. 1-(*O*-Carboxyphenylamino)-1-deoxyribulose. A compound formed by mutant strains of *Aerobacter aerogenes* and *Escherichia coli* blocked in the biosynthesis of tryptophan. Biochem. J. **72:**586–597.

74. **Doy, C., A. Rivera, and P. R. Srinivasan.** 1961. Evidence for the enzymatic synthesis of N-(5'-phosphoribosyl) anthranilic acid, a new intermediate in tryptophan biosynthesis. Biochem. Biophys. Res. Commun. **4:**83–88.

75. **Duggleby, R. G., M. K. Snedden, and J. F. Morrison.** 1978. Chorismate mutase-prephenate dehydratase from *Escherichia coli*: active sites of a bifunctional enzyme. Biochemistry **17:**1548–1554.

76. **Duncan, K., and J. R. Coggins.** 1986. The *serC-aroA* operon of *Escherichia coli*. Biochem. J. **234:**49–57.

77. **Duncan, K., A. Lewendon, and J. R. Coggins.** 1984. The purification of 5-enolpyruvylshikimate-3-phosphate synthase from an overproducing strain of *Escherichia coli*. FEBS Lett. **165:**121–127.

78. **Duncan, K., A. Lewendon, and J. R. Coggins.** 1984. The complete amino acid sequence of *Escherichia coli* 5-enolpyruvylshikimate 3-phosphate synthase. FEBS Lett. **170:**59–63.

79. **Egan, A. F., and F. Gibson.** 1966. Partial purification of anthranilate synthase aggregate from *A. aerogenes*. Biochim. Biophys. Acta **130:**276–277.

80. **Ely, B., and J. Pittard.** 1979. Aromatic amino acid biosynthesis: regulation of shikimate kinase in *Escherichia coli* K-12. J. Bacteriol. **138:**933–943.

80a. **Fayat, G., J. Mayaux, C. Saurdot, M. Fromant, M. Springer, M. Grunberg-Manago, and S. Blanquet.** 1983. *Escherichia coli*, phenylalanyl-tRNA synthetase operon region. J. Mol. Biol. **171:**239–261.

81. **Frost, J. W., J. L. Binder, J. T. Kadonaga, and J. R. Knowles.** 1984. Dehydroquinate synthase from *Escherichia coli*: purification, cloning and construction of overproducers of the enzyme. Biochemistry **23:**4470–4475.

81a. **Gallagher, P. J., I. Schwarts, and D. Elseviers.** 1984. Genetic mapping of *pheU*, an *Escherichia coli* gene for phenylalanyl tRNA. J. Bacteriol. **158:**762–763.

82. **Garner, C. C., and K. M. Herrmann.** 1985. Operator mutations of the *Escherichia coli aroF* gene. J. Biol. Chem. **260:**3820–3825.

83. **Gelfand, D. H., and N. Rudo.** 1977. Mapping of the aspartate and aromatic amino acid aminotransferase genes *tyrB* and *aspC*. J. Bacteriol. **130:**441–444.

84. **Gelfand, D. H., and R. A. Steinberg.** 1977. *Escherichia coli* mutants deficient in the aspartate and aromatic amino acid aminotransferases. J. Bacteriol. **130:**429–440.

85. **Gething, M.-J. H., and B. E. Davidson.** 1976. Chorismate mutase/prephenate dehydratase from *Escherichia coli*. 2. Evidence for identical subunits catalysing the two activities. Eur. J. Biochem. **71:**327–336.

86. **Gething, M.-J. H., and B. E. Davidson.** 1977. Chorismate mutase/prephenate dehydratase from *Escherichia coli* K-12. Eur. J. Biochem. **78:**103–110.

87. **Gething, M.-J. H., and B. E. Davidson.** 1977. Chorismate mutase/prephenate dehydratase from *Escherichia coli* K-12. Effects of chemical modification on the enzymic activities and allosteric inhibition. Eur. J. Biochem. **78:**111–117.

88. **Gibson, F.** 1964. Chorismic acid: purification and some chemical and physical studies. Biochem. J. **90:**256–261.

89. **Gibson, F., M. J. Jones, and H. Teltscher.** 1955. Synthesis of indole and anthranilic acid by mutants of *Escherichia coli*. Nature (London) **175:**853–854.

90. **Gibson, F., and C. Yanofsky.** 1960. The partial purification and properties of indole-3-glycerol phosphate synthetase from *Escherichia coli*. Biochim. Biophys. Acta **43:**489–500.

91. **Gibson, M. I., and F. Gibson.** 1964. Preliminary studies on the isolation and metabolism of an intermediate in aromatic biosynthesis: chorismic acid. Biochem. J. **90:**248–256.

92. **Gicquel-Sanzey, B., and P. Cossart.** 1982. Homologies between different procaryotic DNA-binding regulatory proteins and between their sites of action. EMBO J. **1:**591–595.

93. **Goldberg, M. E., T. E. Creighton, R. L. Baldwin, and C. Yanofsky.** 1966. Subunit structure of tryptophan synthetase of *Escherichia coli*. J. Mol. Biol. **21:**71–82.

94. **Gollub, E. G., K. P. Liu, and D. B. Sprinson.** 1973. *tyrR*, a regulatory gene of tyrosine biosynthesis in *Salmonella typhimurium*. J. Bacteriol. **115:**1094–1102.

95. **Gollub, E. G., K. P. Liu, and D. B. Sprinson.** 1973. A regulatory gene of phenylalanine biosynthesis (*pheR*) in *Salmonella typhimurium*. J. Bacteriol. **115:**121–128.

96. **Gollub, E. G., and D. B. Sprinson.** 1969. A regulatory mutation in tyrosine biosynthesis. Biochem. Biophys. Res. Commun. **35:**389–395.

97. **Gollub, E. G., H. Zalkin, and D. B. Sprinson.** 1967. Correlation of genes and enzymes, and studies on regulation of the aromatic pathway in *Salmonella*. J. Biol. Chem. **242:**5323–5328.

98. **Gowrishankar, J., and J. Pittard.** 1982. Construction from Mu d1 (*lac* Ap^r) lysogens of lambda bacteriophage bearing promoter-*lac* fusions: isolation of λ*ppheA-lac*. J. Bacteriol. **150:**1122–1129.

99. **Gowrishankar, J., and J. Pittard.** 1982. Regulation of phenylalanine biosynthesis in *Escherichia coli* K-12: control of transcription of the *pheA* operon. J. Bacteriol. **150:**1130–1137.

100. **Gowrishankar, J., and J. Pittard.** 1982. Molecular cloning of *pheR* in *Escherichia coli* K-12. J. Bacteriol. **152:**1–6.

101. **Gschwind, H. P., V. Gschwind, C. H. Paul, and K. Kirschner.** 1979. Affinity chromotography of tryptophan synthase from *Escherichia coli*: systemic studies with immobilized tryptophanol phosphate. Eur. J. Biochem. **96:**403–416.

102. **Guest, J. R., S. T. Cole, and K. Jeyaseelan.** 1981. Organization and expression of the pyruvate dehydrogenase complex genes of *Escherichia coli* K-12. J. Gen. Microbiol. **127:**65–79.

103. **Gunsalus, R. P., and C. Yanofsky.** 1980. Nucleotide sequence and expression of *Escherichia coli trpR*, the structural gene for the trp aporepressor. Proc. Natl. Acad. Sci. USA **77:**7117–7121.

104. **Hanson, K. R., and I. A. Rose.** 1963. The absolute stereochemical course of citric acid biosynthesis. Proc. Natl. Acad. Sci. USA **50:**981–988.

105. **Hathaway, G. M., and I. P. Crawford.** 1970. Studies on the association of β-chain monomers of *Escherichia coli* tryptophan synthetase. Biochemistry **9:**1801–1808.

106. **Hathaway, G. M., S. Kida, and I. P. Crawford.** 1969. Subunit structure of the B component of *Escherichia coli* tryptophan synthetase. Biochemistry **8:**989–996.

107. **Henderson, E. J., H. Nagano, H. Zalkin, and L. H. Hwang.** 1970. The anthranilate synthetase-anthranilate 5-phosphoribosyl-pyrophosphate phosphoribosyl-transferase aggregate. J. Biol. Chem. **245:**1416–1423.

108. **Henderson, E. J., and H. Zalkin.** 1971. On the composition of anthranilate synthetase-anthranilate 5-phosphoribosyl-pyrophosphate phosphoribosyl transferase from *Salmonella typhimurium*. J. Biol. Chem. **246:**6891–6898.

109. **Henning, J., D. R. Helinski, F. B. Chao, and C. Yanofsky.** 1960. Isolation and crystallization of A protein of tryptophan synthetase of *E. coli*. J. Biol. Chem. **237:**1523–1530.

110. **Herrmann, K. M., J. Shultz, and M. A. Hermodson.** 1980. Sequence homology between tyrosine sensitive 3-deoxy D-*arabino*-heptulosonate 7-phosphate synthase from *Escherichia coli* and hemerythrin from *Sipunculida*. J. Biol. Chem. **255:**7079–7081.

111. **Heyde, E.** 1979. Chorismate mutase-prephenate dehydrogenase from *Aerobacter aerogenes*. Evidence that the two reactions occur at one active site. Biochemistry **18:**2766–2775.

112. **Heyde, E., and J. F. Morrison.** 1978. Kinetic studies on the reactions catalyzed by chorismate mutase-prephenate dehydrogenase from *Aerobacter aerogenes*. Biochemistry **17:**1573–1580.

113. **Högberg-Raibaud, A., and M. E. Goldberg.** 1977. Preparation and characterization of a modified form of β₂ subunit of *Escherichia coli* tryptophan synthetase suitable for investigating protein folding. Proc. Natl. Acad. Sci. USA **74:**442–446.

114. **Hoiseth, S. K., and B. A. D. Stocker.** 1985. Genes *aroA* and *serC* of *Salmonella typhimurium* constitute an operon. J. Bacteriol. **163:**355–361.

115. **Hu, C.-Y., and D. B. Sprinson.** 1977. Properties of tyrosine-inhibitable 3-deoxy-D-arabinoheptulosonic acid-7-phosphate synthase from *Salmonella*. J. Bacteriol. **129:**177–183.

116. **Hudson, G. S., and B. E. Davidson.** 1984. Nucleotide sequence and transcription of the phenylalanine and tyrosine operons of *Escherichia coli* K-12. J. Mol. Biol. **180:**1023–1051.

117. **Hudson, G. S., G. J. Howlett, and B. E. Davidson.** 1983. The binding of tyrosine and NAD$^+$ to chorismate mutase/prephenate dehydrogenase from *Escherichia coli* K-12 and the effects of

118. these ligands on the activity and self-association of the enzyme. J. Biol. Chem. **258:**3114–3120.

118. **Hudson, G. S., V. Wong, and B. E. Davidson.** 1985. Chorismate mutase-prephenate dehydrogenase from *Escherichia coli* K-12. Purification, characterisation and identification of a reactive cysteine. Biochemistry **23:**6240–6249.

119. **Hwang, L. H., and H. Zalkin.** 1971. Multiple forms of anthranilate synthetase-anthranilate 5-phosphoribylpyrophosphate phosphoribosyl-transferase from *Salmonella typhimurium*. J. Biol. Chem. **246:**2338–2345.

120. **Im, S. W. K., H. Davidson, and J. Pittard.** 1971. Phenylalanine and tyrosine biosynthesis in *Escherichia coli* K-12: mutants derepressed for 3-deoxy-D-arabinoheptulosonic acid 7-phosphate dythetase (phe), 3-deoxy-D-arabinoheptulosonic acid 7-phosphate synthetase (tyr), chorismate mutase T-prephenate dehydrogenase, and transminase A. J. Bacteriol. **108:**400–409.

121. **Im, S. W. K., and J. Pittard.** 1971. Phenylalanine biosynthesis in *Escherichia coli* K-12: mutants derepressed for chorismate mutase P-prephenate dehydratase. J. Bacteriol. **106:**784–790.

122. **Im, S. W. K., and J. Pittard.** 1973. Tyrosine and phenylalanine biosynthesis in *Escherichia coli* K-12: complementation between different *tyrR* alleles. J. Bacteriol. **115:**1145–1150.

123. **Ito, J., E. C. Cox, and C. Yanofsky.** 1969. Anthranilate synthetase, an enzyme specified by the tryptophan operon of *Escherichia coli*: purification and characterization of component I. J. Bacteriol. **97:**725–733.

124. **Ito, J., and I. P. Crawford.** 1965. Regulation of the enzymes of the tryptophan pathway in *Escherichia coli*. Genetics **52:**1303–1316.

125. **Ito, J., and C. Yanofsky.** 1966. The nature of the anthranilic acid synthetase complex of *Escherichia coli*. J. Biol. Chem. **241:**4113–4114.

126. **Jackson, E. N., and C. Yanofsky.** 1973. The region between the operator and first structural gene of the tryptophan operon of *Escherichia coli* may have a regulatory function. J. Mol. Biol. **76:**89–101.

127. **Jensen, R. A., D. S. Nasser, and E. W. Nester.** 1967. Comparative control of a branch point enzyme in microorganisms. J. Bacteriol. **94:**1582–1593.

128. **Kasian, P. A., and J. Pittard.** 1984. Construction of a *tyrP-lac* operon fusion strain and its use in the isolation and analyses of mutants derepressed for *tyrP* expression. J. Bacteriol. **160:**175–183.

129. **Katagiri, M., and R. Sato.** 1953. Accumulation of phenylalanine by a phenylalanineless mutant of *Escherichia coli*. Science **118:**250–251.

130. **Kinghorn, J. R., M. Schweizer, N. H. Giles, and S. R. Kushner.** 1981. The cloning and analyses of the *aroD* gene of *E. coli* K-12. Gene **14:**73–80.

131. **Kirschner, K., H. Szadkowski, A. Henschen, and F. Lottspeich.** 1980. Limited proteolysis of N-(5′-phosphoribosyl) anthranilate isomerase: indole glycerol phosphate synthase from *Escherichia coli* yields two different enzymically active, functional domains. J. Mol. Biol. **143:**395–409.

132. **Kirschner, K., R. L. Wiskocil, M. Foehn, and L. Rizean.** 1975. The tryptophan synthase from *Escherichia coli*. Eur. J. Biochem. **60:**513–523.

133. **Kleckner, N., J. Roth, and D. Botstein.** 1977. Genetic engineering *in vivo* using translocatable drug resistance elements. J. Mol. Biol. **116:**125–159.

134. **Koch, G. L. E., D. C. Shaw, and F. Gibson.** 1971. The purification and characterization of chorismate mutase-prephenate dehydrogenase from *Escherichia coli* K-12. Biochim. Biophys. Acta **229:**795–804.

135. **Koch, G. L. E., D. C. Shaw, and F. Gibson.** 1971. Characterization of the subunits of chorismate mutase-prephenate dehydrogenase from *Escherichia coli* K-12. Biochim. Biophys. Acta **229:**805–812.

136. **Koch, G. L. E., D. C. Shaw, and F. Gibson.** 1972. Studies on the relationship between the active sites of chorismate mutase-prephenate dehydrogenase from *Escherichia coli* or *aerobacter aerogenes*. Biochim. Biophys. Acta **258:**719–730.

137. **Kondo, K., S. Wakabayashi, T. Yagi, and H. Kagamiyama.** 1984. The complete amino acid sequence of aspartate aminotransferase from *Escherichia coli*: sequence comparison with pig isoenzymes. Biochem. Biophys. Res. Commun. **122:**62–67.

138. **Kuramitsu, S., T. Ogawa, H. Ogawa, and H. Kagamiyama.** 1985. Branched chain amino acid aminotransferase of *Escherichia coli*: nucleotide sequence of the *ilvE* gene and deduced amino acid sequence. J. Biochem. **97:**993–999.

139. **Kuramitsu, S., S. Okumo, T. Ogawa, H. Ogawa, and H. Kagamiyama.** 1985. Aspartate aminotransferase of *Escherichia*

coli: nucleotide sequence of the *aspC* gene. J. Biochem. **97**:1259–1262.

140. **LeMarechal, P., and R. Azerad.** 1976. The shikimate pathway. III. 3-dehydroquinate synthetase of *E. coli*. Mechanistic studies by kinetic isotope effect. Biochimie **58**:1123–1128.

141. **LeMarechal, P., C. Froussios, M. Level, and R. Azerad.** 1981. Synthesis of phosphono analogues of 3-deoxy-D-*arabino*-hept-2-ulosonic acid 7-phosphate. Carbohydr. Res. **94**:1–10.

142. **Levin, J. G., and D. B. Sprinson.** 1960. The formation of 3-enolpyruvyl shikimate 5-phosphate in extracts of *Escherichia coli*. Biochem. Biophys. Res. Commun. **3**:157–163.

143. **Levin, J. G., and D. B. Sprinson.** 1964. The enzymatic formation and isolation of 3-enolpyruvylshikimate 5-phosphate. J. Biol. Chem. **239**:1142–1150.

144. **Lewendon, A., and J. R. Coggins.** 1983. Purification of 5-enolpyruvyl shikimate 3-phosphate synthase from *Escherichia coli*. Biochem. J. **213**:187–191.

145. **Li, S. L., and C. Yanofsky.** 1972. Amino acid sequences of fity residues from the amino termini of the tryptophan synthetase α chains of several enterobacteriaceae. J. Biol. Chem. **247**:1031–1037.

146. **Lowry, O. H., J. Carter, J. B. Ward, and L. Glaser.** 1971. The effect of carbon and nitrogen sources on the level of metabolic intermediates in *Escherichia coli*. J. Biol. Chem. **246**:6511–6521.

147. **Maitra, U. S., and D. B. Sprinson.** 1978. 5-Dehydro-3-deoxy-D-*arabino*-heptulosonic acid 7-phosphate. An intermediate in the 3-dehydroquinate synthase reaction. J. Biol. Chem. **253**:5426–5430.

148. **Matsushiro, A.** 1963. Specialized transduction of tryptophan markers in *Escherichia coli* K-12 by bacteriophage φ80. Virology **19**:475–482.

149. **Mattern, I. E., and J. Pittard.** 1971. Regulation of tyrosine biosynthesis in *Escherichia coli* K-12: isolation and characterization of operator mutants. J. Bacteriol. **107**:8–15.

150. **Mavrides, C., and W. Orr.** 1974. Multiple forms of plurispecific aromatic: 2-oxoglutarate (oxaloacetate) aminotransferase (transaminase A) in *Escherichia coli* and selective repression by L-tyrosine. Biochim. Biophys. Acta **336**:70–78.

151. **Mavrides, C., and W. Orr.** 1975. Multispecific aspartate and aromatic amino acid aminotransferases in *Escherichia coli*. J. Biol. Chem. **250**:4128–4133.

152. **McCandliss, R. J., and K. M. Herrmann.** 1978. Iron, an essential element for the biosynthesis of aromatic compounds. Proc. Natl. Acad. Sci. USA **75**:4810–4813.

153. **McCandliss, R. J., M. D. Poling, and K. M. Herrmann.** 1978. 3-Deoxy-D-arabino-heptulosonate 7-phosphate synthase. Purification and molecular characterization of the phenylalanine-sensitive isoenzyme from *Escherichia coli*. J. Biol. Chem. **253**:4259–4265.

154. **McCray, J. W., Jr., and K. M. Herrmann.** 1976. Derepression of certain aromatic amino acid biosynthetic enzymes of *Escherichia coli* K-12 by growth in Fe^{3+}-deficient medium. J. Bacteriol. **125**:608–615.

155. **Miles, E. W.** 1979. Tryptophan synthase: structure, function and subunit interaction, p. 127–186. *In* A. Meister (ed.), Advances in enzymology, vol. 49. John Wiley and Sons, New York.

156. **Miles, E. W.** 1980. Tryptophan synthase: structure, function and interaction with D-tryptophan and L-tryptophan, p. 137–147. *In* D. Hayaishi, Y. Ishimara, and R. Kido (ed.), Biochemical and medical aspects of tryptophan metabolism. Elsevier-North Holland, Amsterdam.

157. **Miles, E. W., and M. Moriguchi.** 1977. Tryptophan synthase of *Escherichia coli*. J. Biol. Chem. **252**:6594–6599.

158. **Mitsuhashi, S., and B. D. Davis.** 1954. Aromatic biosynthesis. XII. Conversion of 5-dehydroquinic acid to 5-dehydroshikimic acid by 5-dehydroquinase. Biochim. Biophys. Acta **15**:54–61.

159. **Moldovani, I. S., and G. Denes.** 1968. Mechanism of the action and of the allosteric inhibition of 3-deoxy-D-arabino-heptulosonate 7-phosphate synthase (tyrosine sensitive) of *Escherichia coli* W. Acta Biochim. Biophys. Acad. Sci. Hung. **3**:259–273.

160. **Monnier, N., A. Montmitonnet, S. Chesne, and J. Pelmont.** 1976. Transaminase B d'*Escherichia coli*. I. Purification et premières propriétés. Biochimie **58**:663–675.

161. **Morell, H., M. J. Clark, P. F. Knowles, and D. B. Sprinson.** 1967. The enzymic synthesis of chorismic and prephenic acids from 3-enolpyruvyl shikimic acid 5-phosphate. J. Biol. Chem. **242**:82–90.

162. **Morell, H., and D. B. Sprinson.** 1968. Shikimate kinase isoenzymes in *Salmonella typhimurium*. J. Biol. Chem. **243**:676–677.

163. **Mosteller, R. D., K. R. Nishimoto, and R. V. Goldstein.** 1977. Inactivation and partial degradation of phosphoribosylanthra-nilate isomerase-indoleglycerol phosphate synthetase in non-growing cultures of *Escherichia coli*. J. Bacteriol. **131**:153–162.

164. **Murphy, T. M., and S. E. Mills.** 1969. Immunochemical and enzymatic comparisons of the tryptophan synthase α subunits from five species of Enterobacteriaceae. J. Bacteriol. **97**:1310–1320.

165. **Nagano, H., and H. Zalkin.** 1970. Some physicochemical properties of anthranilate synthetase component I from *Salmonella typhimurium*. J. Biol. Chem. **245**:3097–3103.

166. **Nichols, B. P., G. F. Miozzari, M. Van Cleemput, G. N. Bennett, and C. Yanofsky.** 1980. Nucleotide sequences of the *trpG* regions of *Escherichia coli, Shigella dysenteriae, Salmonella typhimurium* and *Serratia marcescens*. J. Mol. Biol. **142**:503–517.

167. **Nichols, B. P., and C. Yanofsky.** 1979. Nucleotide sequences of *trpA* of *Salmonella typhimurium* and *Escherichia coli*: an evolutionary comparison. Proc. Natl. Acad. Sci. USA **76**:5244–5248.

168. **Nishioka, Y., M. Demerec, and A. Eisenstark.** 1967. Genetic analysis of aromatic mutants of *Salmonella typhimurium*. Genetics **56**:341–351.

169. **Ogino, T., C. Garner, J. L Markley, and K. M. Herrmann.** 1982. Biosynthesis of aromatic compounds: ^{13}C NMR spectroscopy of whole *Escherichia coli* cells. Proc. Natl. Acad. Sci. USA **79**:5828–5832.

170. **Oppenheim, D. S., and C. Yanofsky.** 1980. Translational coupling during expression of the tryptophan operon of *Escherichia coli*. Genetics **95**:785–795.

171. **Oxender, D. L.** 1975. Genetic approaches to the study of transport systems, p. 214–231. *In* H. N. Christensen (ed.), Biological transport, 2nd ed. W. A. Benjamin, Reading, Mass.

172. **Pabst, M. J., J. C. Kuhn, and R. L. Somerville.** 1973. Feedback regulation in the anthranilate aggregate from wild type and mutant strains of *Escherichia coli*. J. Biol. Chem. **248**:901–914.

173. **Piperno, J. R., and D. L. Oxender.** 1968. Amino acid transport systems in *Escherichia coli* K-12. J. Biol. Chem. **243**:5914–5920.

174. **Pittard, J.** 1983. The regulation of the biosynthesis of aromatic amino acids and vitamins in *Escherichia coli* K-12, p. 75–86. *In* Proceedings of the XV International Congress of Genetics. Oxford and I.B.M. Publishing Co., New Delhi.

175. **Pittard, J., J. Camakaris, and B. J. Wallace.** 1969. Inhibition of 3-deoxy-D-arabinoheptulosonic acid-7-phosphate synthetase (trp) in *Escherichia coli*. J. Bacteriol. **97**:1242–1247.

176. **Pittard, J., and F. Gibson.** 1970. The regulation of aromatic amino acids and vitamins. Curr. Top. Cell. Regul. **2**:29–63.

177. **Pittard, J., and B. J. Wallace.** 1966. Distribution and function of genes concerned with aromatic biosynthesis in *Escherichia coli*. J. Bacteriol. **91**:1494–1508.

178. **Pittard, J., and B. J. Wallace.** 1966. Gene controlling the uptake of shikimic acid by *Escherichia coli*. J. Bacteriol. **92**:1070–1075.

179. **Pizer, L. I., and M. L. Potochny.** 1964. Nutritional and regulatory aspects of serine metabolism in *Escherichia coli*. J. Bacteriol. **88**:611–619.

180. **Platt, T., and C. Yanofsky.** 1975. An intercistronic region and ribosome binding site in bacterial messenger RNA. Proc. Natl. Acad. Sci. USA **72**:2399–2403.

181. **Powell, J. T., and J. F. Morrison.** 1978. The purification and properties of the aspartate aminotransferase and aromatic amino acid aminotransferase from *Escherichia coli*. Eur. J. Biochem. **87**:391–400.

182. **Powell, J. T., and J. F. Morrison.** 1979. Enzyme-enzyme interaction and the biosynthesis of the aromatic amino acids in *Escherichia coli*. Biochim. Biophys. Acta **568**:467–474.

183. **Rivera, A., Jr., and P. R. Srinivasan.** 1962. 3-Enolpyruvylshikimate 5-phosphate, an intermediate in the biosynthesis of anthranilate. Proc. Natl. Acad. Sci. USA **48**:864–867.

184. **Rood, J. I., B. Perrot, E. Heyde, and J. F. Morrison.** 1982. Characterisation of monofunctional chorismate mutase/prephenate dehydrogenase enzymes obtained via mutagenesis of recombinant plasmids *in vitro*. Eur. J. Biochem. **124**:513–519.

185. **Rose, J. K., C. L. Squires, C. Yanofsky, H. L. Yang, and G. Zubay.** 1973. Regulation of "in vitro" transcription of the tryptophan operon by purified RNA polymerase in the presence of partially purified repressor and tryptophan. Nature (London) New Biol. **245**:133–137.

186. **Rose, J. K., and C. Yanofsky.** 1972. Metabolic regulation of the tryptophan operon of *Escherichia coli*: repressor-independent regulation of transcription initiation frequency. J. Mol. Biol. **69**:103–118.

187. **Rosenberg, A. H., and M. L. Gefter.** 1969. An iron dependent modification of several transfer RNA species in *Escherichia coli*. J. Mol. Biol. **46**:581–584.

188. **Rotenberg, S. L., and D. B. Sprinson.** 1970. Mechanism and stereochemistry of 5-dehydroquinate synthetase. Proc. Natl. Acad. Sci. USA **67**:1669–1672.

189. **Rotenberg, S. L., and D. B. Sprinson.** 1978. Isotope effects in 3-dehydroquinate synthase and dehydratase. J. Biol. Chem. **253:**2210–2215.

190. **Rudman, D., and A. Meister.** 1953. Transamination in *Escherichia coli.* J. Biol. Chem. **200:**591–604.

191. **Salamon, I. I., and B. D. Davis.** 1953. Aromatic biosynthesis. IX. The isolation of a precursor of shikimic acid. J. Am. Chem. Soc. **75:**5567–5571.

192. **Sampathkumar, P., and J. F. Morrison.** 1982. Chorismate mutase-prephenate dehydrogenase from *Escherichia coli.* Purification and properties of the bifunctional enzyme. Biochim. Biophys. Acta **702:**204–211.

193. **Sampathkumar, P., and J. F. Morrison.** 1982. Chorismate mutase-prephenate dehydrogenase from *Escherichia coli.* Kinetic mechanism of the prephenate dehydrogenase reaction. Biochim. Biophys. Acta **702:**212–219.

194. **Sanderson, K. E., and J. R. Roth.** 1983. Linkage map of *Salmonella typhimurium,* edition VI. Microbiol. Rev. **47:**410–453.

195. **Schmit, J. C., S. W. Artz, and H. Zalkin.** 1970. Chorismate mutase-prephenate dehydrogenase. J. Biol. Chem. **245:**4019–4027.

196. **Schmit, J. C., and H. Zalkin.** 1969. Chorismate mutase-prephenate dehydratase. Partial purification and properties of the enzyme from *Salmonella typhimurium.* Biochemistry **8:**174–181.

197. **Schmit, J. C., and H. Zalkin.** 1971. Chorismate mutase-prephenate dehydratase. Phenylalanine induced dimerization and its relationship to feedback inhibition. J. Biol. Chem. **246:**6002–6010.

198. **Schoner, R., and K. M. Herrmann.** 1976. 3-Deoxy-D-*arabino*-heptulosonate 7-phosphate synthase. J. Biol. Chem. **251:**5440–5447.

199. **Schultz, G. E., and T. E. Creighton.** 1969. Preliminary X-ray diffraction study of the wild-type and mutationally altered tryptophan synthetase α subunit. Eur. J. Biochem. **10:**195–197.

199a. **Schwartz, J., R. Klotsky, P. J. Gallagher, M. Krauskoff, M. A. Q. Siddiqui, J. F. H. Wong, and B. A. Roc.** 1983. Molecular cloning and sequencing of *pheU,* a gene for *Escherichia coli* tRNA[Phe]. Nucleic Acids Res. **11:**4379–4389.

200. **Selker, E., and C. Yanofsky.** 1979. Nucleotide sequence of the *trpC-trpB* intercistronic region from *Salmonella typhimurium.* J. Mol. Biol. **130:**135–143.

201. **Shultz, J., M. A. Hermodson, C. C. Garner, and K. M. Herrmann.** 1984. The nucleotide sequence of the *aroF* gene of *Escherichia coli* and the amino acid sequence of the encoded protein, the tyrosine-sensitive 3-deoxy-D-*arabino*-haptulosinate 7-phosphate synthase. J. Biol. Chem. **259:**9655–9661.

202. **Shultz, J., M. A. Hermodson, and K. M. Herrmann.** 1981. A comparison of the amino-terminal sequences of 3-deoxy-D-*arabino*-heptulosonate-7-phosphate synthase isoenzymes from *Escherichia coli.* FEBS Lett. **131:**108–110.

203. **Silbert, D. F., S. E. Jorgensen, and E. C. C. Lin.** 1963. Repression of transaminase A by tyrosine in *Escherichia coli.* Biochim. Biophys. Acta **73:**232–240.

204. **Simmonds, S.** 1950. The metabolism of phenylalanine and tyrosine in mutant strains of *Escherichia coli.* J. Biol. Chem. **185:**755–762.

205. **Simpson, R. J., and B. E. Davidson.** 1976. Studies on 3-deoxy-D-*arabino*heptulosonate-7-phosphate synthetase (phe) from *Escherichia coli* K-12. Eur. J. Biochem. **70:**501–507.

206. **Smith, L. C., J. M. Ravel, S. R. Lax, and W. Shine.** 1962. The control of 3-deoxy-D-*arabino*heptulosonic acid-7-phosphate synthesis by phenylalanine and tyrosine. J. Biol. Chem. **237:**3566–3570.

207. **Smith, O. H.** 1967. Structure of the *trpC* cistron specifying indoleglycerol phosphate synthetase, and its localization in the tryptophan operon of *Escherichia coli.* Genetics **57:**95–105.

208. **Smith, O. H., and C. Yanofsky.** 1960. 1-(O-carboxyphenylamino)-1-deoxyribulose 5-phosphate. A new intermediate in the biosynthesis of tryptophan. J. Biol. Chem. **235:**2051–2057.

208a. **Springer, M., J. F. Mayaux, G. Fayat, J. A. Plumbridge, M. Graffe, S. Blanquet, and M. Grunberg-Manago.** 1985. Attenuation control of the *Escherichia coli* phenylalanyl tRNA synthetase operon. J. Mol. Biol. **181:**467–478.

208b. **Springer, M., M. Trudd, M. Graffe, J. Plumbridge, G. Fayat, J. Mayaux, C. Saurdot, S. Blanquet, and M. Grunberg-Manago.** 1983. Phenylalanyl-tRNA synthetase operon is controlled by attenuation in vivo. J. Mol. Biol. **171:**279.

209. **Srinivasan, P. R., M. Katagiri, and D. B. Sprinson.** 1959. The conversion of phosphoenolpyruvic acid and d-erythrose 4-phosphate to 5-dehydroquinic acid. J. Biol. Chem. **234:**713–715.

210. **Srinivasan, P. R., J. Rothschild, and D. B. Sprinson.** 1963. The enzymic conversion of 3-deoxy-D-*arabino*-heptulosonic acid 7-phosphate to 5-dehydroquinate. J. Biol. Chem. **238:**3176–3182.

211. **Srinivasan, P. R., H. T. Shigeura, M. Sprecher, D. B. Sprinson, and B. D. Davis.** 1956. The biosynthesis of shikimic acid from D-glucose. J. Biol. Chem. **220:**477–497.

212. **Srinivasan, P. R., and D. B. Sprinson.** 1959. 2-Keto-3-deoxy-D-*arabino*-heptonic acid 7-phosphate synthetase. J. Biol. Chem. **234:**716–722.

213. **Staden, R.** 1982. An interactive graphics program for comparing and aligning of nucleic acid and amino acid sequences. Nucleic Acids. Res. **10:**2951–2961.

214. **Stalker, D., W. Hiatt, and L. Comai.** 1985. A single amino acid substitution in the enzyme 5-enolpyruvylshikimate-3-phosphate synthase confers resistance to the herbicide glyphosate. J. Biol. Chem. **260:**4724–4728.

215. **Staub, M., and S. Denes.** 1969. Purification and properties of the 3-deoxy-D-*arabino*-heptulosonate 7-phosphate synthase (phenylalanine sensitive) of *Escherichia coli* K-12. Biochim. Biophys. Acta **178:**588–598.

216. **Tamir, H., and P. R. Srinivasan.** 1969. Purification and properties of anthranilate synthase from *Salmonella typhimurium.* J. Biol. Chem. **224:**6507–6513.

217. **Tribe, D. E., H. Camakaris, and J. Pittard.** 1976. Constitutive and repressible enzymes of the common pathway of aromatic biosynthesis in *Escherichia coli* K-12: regulation of enzyme synthesis at different growth rates. J. Bacteriol. **127:**1085–1097.

218. **Tribe, D. E., and J. Pittard.** 1979. Hyperproduction of tryptophan by *Escherichia coli*: genetic manipulation of the pathways leading to tryptophan formation. Appl. Environ. Microbiol. **38:**181–190.

219. **Tschopp, J., and K. Kirschner.** 1980. Subunit interactions of tryptophan synthase from *Escherichia coli* as revealed by binding studies with pyridoxal phosphate analogues. Biochemistry **19:**4514–4521.

220. **Tschopp, J., and K. Kirschner.** 1980. Kinetics of cooperative ligand binding to the apoβ$_2$ subunit of tryptophan synthase and its modulation by the α subunit. Biochemistry **19:**4522–4526.

221. **Umbarger, H. E., and J. H. Mueller.** 1951. Isoleucine and valine metabolism of *Escherichia coli.* 1. Growth studies on amino acid-deficient mutants. J. Biol. Chem. **189:**277–285.

222. **Vaz, A. D. N.** 1980. Studies on the enzymatic cis dehydration of 3-dehydroquinic acid catalyzed by *Escherichia coli* 3-dehydroquinate dehydratase. Diss. Abstr. Int. B Sci. Eng. **41:**941–942.

223. **Vaz, A. D. N., J. R. Butler, and M. J. Nugent.** 1975. Dehydroquinase catalyzed dehydration. II. Identification of the reactive conformation of the substrate responsible for syn elimination. J. Am. Chem. Soc. **97:**5914–5915.

224. **Wallace, B. J., and J. Pittard.** 1967. Genetic and biochemical analysis of the isoenzymes concerned in the first reaction of aromatic biosynthesis in *Escherichia coli.* J. Bacteriol. **93:**237–244.

225. **Wallace, B. J., and J. Pittard.** 1967. Chromatography of 3-deoxy-D-arabinoheptulosonic acid-7-phosphate synthetase (trp) on diethylaminoethyl cellulose: a correction. J. Bacteriol. **94:**1279–1280.

226. **Wallace, B. J., and J. Pittard.** 1969. Regulator gene controlling enzymes concerned in tyrosine biosynthesis in *Escherichia coli.* J. Bacteriol. **97:**1234–1241.

227. **Wallace, B. J., and J. Pittard.** 1969. Regulation of 3-deoxy-D-*arabino*-heptulosonic 7-phosphate acid synthetase activity in relation to the synthesis of the aromatic vitamins in *Escherichia coli* K-12. J. Bacteriol. **99:**707–712.

228. **Weiss, U., B. D. Davis, and E. S. Mingioli.** 1953. Aromatic biosynthesis. X. Identification of an early precursor as 5-dehydroquinic acid. J. Am. Chem. Soc. **75:**5572–5576.

229. **Weiss, U., C. Gilvarg, E. S. Mingioli, and B. D. Davis.** 1954. Aromatic biosynthesis. XI. The aromatization step in the synthesis of phenylalanine. Science **119:**774–775.

230. **Weiss, U., and E. S. Mingioli.** 1955. Aromatic biosynthesis. XV. The isolation and identification of shikimic acid 5-phosphate. J. Am. Chem. Soc. **78:**2894–2898.

231. **Whipp, M. J., D. M. Halsall, and A. J. Pittard.** 1980. Isolation and characterization of an *Escherichia coli* K-12 mutant defective in tyrosine- and phenylalanine-specific transport systems. J. Bacteriol. **143:**1–7.

232. **Whipp, M. J., and A. J. Pittard.** 1977. Regulation of aromatic amino acid transport systems in *Escherichia coli* K-12. J. Bacteriol. **132:**453–461.

233. **White, J. L., M. G. Grutter, E. Wilson, C. Thaller, G. C. Ford, J. D. G. Smit, J. N. Jansonius, and K. Kirschner.** 1982. Crystallization and preliminary X-ray crystallographic data of the bifunctional enzyme phosphoribosyl-anthranilate isomerase-indole-3-glycerol phosphate synthase from *Escherichia coli.* FEBS Lett. **148:**87–90.

234. **Wilson, D. A., and I. P. Crawford.** 1965. Purification and properties of the B component of *Escherichia coli* tryptophan synthetase. J. Biol. Chem. **240:**4801–4808.

235. **Wookey, P. J., J. Pittard, S. M. Forrest, and B. E. Davidson.** 1984. Cloning of the *tyrP* gene and further characterization of the tyrosine-specific transport system in *Escherichia coli* K:12. J. Bacteriol. **160:**169–174.

236. **Yaniv, H., and C. Gilvarg.** 1955. Aromatic biosynthesis. XIV. 5-Dehydroshikimic reductase. J. Biol. Chem. **213:**787–795.

237. **Yanofsky, C.** 1956. The enzymic conversion of anthranilic acid to indole. J. Biol. Chem. **223:**171–185.

238. **Yanofsky, C.** 1959. A second reaction catalyzed by the tryptophan synthetase of *Escherichia coli*. Biochim. Biophys. Acta **31:**409–416.

239. **Yanofsky, C.** 1960. The tryptophan synthetase system. Bacteriol. Rev. **24:**221–245.

240. **Yanofsky, C.** 1981. Attenuation in the control of expression of bacterial operons. Nature (London) **289:**751–758.

241. **Yanofsky, C., and I. P. Crawford.** 1972. Tryptophan synthetase, p. 1–31. *In* P. D. Boyer (ed.), The enzymes, 3rd ed., vol. VII. Academic Press, Inc., New York.

242. **Yanofsky, C., G. Drapeau, J. R. Guest, and B. C. Carlton.** 1967. The complete amino acid sequence of the tryptophan synthetase A protein (α subunit) and its colinear relationship with the genetic map of the A gene. Proc. Natl. Acad. Sci. USA **57:**296–298.

243. **Yanofsky, C., V. Horn, M. Bonner, and S. Stasiowski.** 1971. Polarity and enzyme functions in mutants of the first three genes of the tryptophan operon of *Escherichia coli*. Genetics **69:**409–433.

244. **Yanofsky, C., and E. S. Lennox.** 1959. Transduction and recombination study of linkage relationships among the genes controlling tryptophan synthesis in *Escherichia coli*. Virology **8:**425–447.

245. **Yanofsky, C., J. Platt, I. P. Crawford, B. P. Nichols, G. E. Christie, H. Horowitz, M. Van Cleemput, and A. M. Wu.** 1981. The complete nucleotide sequence of the tryptophan operon of *Escherichia coli*. Nucleic Acids Res. **9:**6647–6668.

246. **Yanofsky, C., and M. Rachmeler.** 1958. The exclusion of free indole as an intermediate in the biosynthesis of tryptophan in *Neurospora crassa*. Biochim. Biophys. Acta **28:**640–641.

247. **Yanofsky, C., and J. Stadler.** 1958. The enzymatic activity associated with the protein immunologically related to tryptophan synthetase. Proc. Natl. Acad. Sci. USA **44:**245–253.

248. **Yanofsky, C., and M. van Cleemput.** 1982. Nucleotide sequence of *trpE* of *Salmonella typhimurium* and its homology with the corresponding sequence of *Escherichia coli*. J. Mol. Biol. **155:**235–246.

249. **Zalkin, H.** 1973. Anthranilate synthetase, p. 1–37. *In* A. Meister (ed.), Advances in enzymology, vol. 38. Robert E. Krieger Publishing Co., Melbourne, Fla.

250. **Zipkas, D., and M. Riley.** 1975. Proposal concerning mechanism of evolution of the genome of *Escherichia coli*. Proc. Natl. Acad. Sci. USA **72:**1354–1358.

251. **Zurawski, G., K. Brown, D. Killingly, and C. Yanofsky.** 1978. Nucleotide sequence of the leader region of the phenylalanine operon of *Escherichia coli*. Proc. Natl. Acad. Sci. USA **75:**4271–4275.

252. **Zurawski, G., R. P. Gunsalus, K. D. Brown, and C. Yanofsky.** 1981. Structure and regulation of *aroH*, the structural gene for the tryptophan-repressible 3-deoxy-D-*arabino*-heptulosonic acid-7-phosphate synthetase of *Escherichia coli*. J. Mol. Biol. **145:**47–73.

25. Biosynthesis of Histidine

MALCOLM E. WINKLER

Department of Molecular Biology, Northwestern University Medical School, Chicago, Illinois 60611

INTRODUCTION

The biosynthesis of histidine in *Salmonella typhimurium* and *Escherichia coli* has been an important system in which to study the relationship between the flow of intermediates through a biosynthetic pathway and the control of the genes encoding enzymes that catalyze the steps in the pathway. Mechanisms basic to regulation of biosynthetic pathways, such as feedback inhibition, energy charge, and the setting of basal biosynthetic enzyme levels, have been intensively investigated for histidine biosynthesis. In addition, fundamental concepts in gene regulation, such as polarity, polycistronic organization of mRNA molecules, attenuation, autogenous regulation, and positive control of metabolic regulation, have been formulated to explain aspects of histidine biosynthesis. Finally, knowledge about the histidine biosynthetic pathway and the histidine operon has provided a powerful genetic tool for studying gene duplications, transpositions, the effects of mutagens, and tRNA biosynthesis and function.

The first comprehensive review of histidine biosynthesis was written by Brenner and Ames in 1971 (25). This extraordinary article consolidated information about control of the histidine biosynthetic pathway and posed many of the questions about histidine biosynthesis and *his* operon control that have been the subject of investigation for the past 14 years. A number of these questions have now been answered by a variety of experimental approaches. In this chapter, current data and information are compiled and summarized concerning the histidine biosynthetic pathway and the control of the *his* operon. In addition, new results from the laboratory of C. B. Bruni on the structure of the *his* operon are presented. Recent discoveries from the laboratory of S. W. Artz are described which concern the mechanism of ppGpp-mediated stimulation of transcription from the *his* operon primary promoter. Findings about the structure and function of the *hisR*, *hisS*, *hisT*, *hisU*, and *hisW* genes are related to *his* operon control. In particular, results from K. E. Rudd are included which suggest that *hisW* and a subclass of *hisU* alleles encode the subunits of DNA gyrase. Areas of ongoing research and topics requiring further investigation are noted and discussed. Two excellent reviews concerning aspects of histidine biosynthesis have recently appeared and should be consulted for additional perspectives and information (11, 16).

THE HISTIDINE BIOSYNTHETIC PATHWAY

Intermediates in the Pathway

The pathway and details of histidine biosynthesis appear to be the same in *E. coli* and *S. typhimurium* (16). The 10 steps in this unbranched pathway, which were established by Ames and his associates (see reference 25), include a number of complex and unusual reactions (Fig. 1). For simplicity, the eight enzymes that catalyze the reactions (Table 1) are designated by the genes that encode them in the *his* operon. Two of the proteins, products of the *hisB* and *hisIE* genes, are bifunctional in that each enzyme catalyzes two separate steps (Table 1 and Fig. 1). The initial substrates, phosphoribosylpyrophosphate (PRPP) and ATP, are central to intermediary metabolism and tie histidine biosynthesis into a divergent pathway with the biosynthesis of pyrimidine nucleotides, purine nucleotides, pyridine nucleotides, and tryptophan (104; see the appropriate chapters in this volume). This metabolic interrelationship is indicated by the observation that *S. typhimurium* mutants that have increased expression of the *his* operon and lack feedback control of the histidine biosynthetic pathway require adenine for growth at 42°C (60). The important role played by histidine biosynthesis in cellular metabolism is further underscored by the fact that considerable energy is required for the synthesis of each histidine molecule (25).

The first reaction in the pathway is catalyzed by the *hisG* protein and involves a displacement on C-1 of PRPP by N-1 of the purine ring of ATP (Fig. 1, *a*). This Mg^{2+} ion-dependent condensation releases a pyrophosphate molecule, inverts the ribose moiety derived from PRPP from the α to the β configuration, and is reversible. The role played by inhibition of the *hisG* enzyme in controlling the flow of intermediates through the histidine pathway is discussed in another section. The next steps in the pathway involve irreversible, Mg^{2+} ion-dependent hydrolysis of the *N'*-5'-phosphoribosyl-ATP to *N'*-5'-phosphoribosyl-AMP and pyrophosphate catalyzed by the *hisE* enzyme activity, followed by a ring-opening reaction on the purine ring of the AMP-containing intermediate catalyzed by the *hisI* enzyme activity (Fig. 1, *b*).

The next biosynthetic step is an internal redox reaction known as an Amadori rearrangement, which is catalyzed by the *hisA* gene product (Fig. 1, *c*). The order of the next series of reactions has not been unequivocally established and contains at least one intermediate of unknown structure (97). In the scheme shown in Fig. 1, the product of the Amadori rearrangement is hydrolytically cleaved to yield 5-aminoimidazole-4-carboxamide ribotide. This by-product is an established intermediate of purine biosynthesis and thus forms another link between histidine and purine biosynthesis (Fig. 1, *d*). The other proposed product of this reaction contains five carbon atoms derived from PRPP and single nitrogen and carbon atoms derived from ATP (Fig. 1, brackets). The *hisH* gene product catalyzes addition of a nitrogen atom from glutamine, followed by a ring closure catalyzed by the *hisF* enzyme to form an imidazole group attached to glycerol phosphate (Fig. 1, *e*).

The final, established steps in histidine biosynthesis include a Mn^{2+} ion-dependent dehydration catalyzed by one activity of the bifunctional *hisB* protein followed by a ketonization of the resulting enol, a reversible transamination with a nitrogen atom from glutamate catalyzed by the *hisC* protein, a dephosphorylation of L-histidinol-phosphate catalyzed by the other activity of the *hisB* protein, and an NAD^+-dependent oxidation of the primary hydroxyl group of L-histidinol to give the amino acid end product, L-histidine. In summary, the atoms of histidine are derived from the following precursors through the biosynthetic pathway: three carbon atoms of the amino acid backbone and carbons 4 and 5 of the imidazole ring are from PRPP, the amino group is from glutamate, the nitrogen 3 of the imidazole ring is from glutamine, and carbon 2 and nitrogen 1 of the imidazole ring are from ATP.

Regulation of the Pathway

A bacterium that completely lacks control of histidine biosynthesis will waste about 2.5% of its metabolic energy synthesizing excess histidine when growing with a doubling time of about 50 min (25). Therefore, it is not surprising that *E. coli* and *S. typhimurium* have evolved an elaborate network to control the rate of histidine biosynthesis. The two most important points of control are regulation of the flow of intermediates through the pathway and regulation of the amounts of histidine biosynthetic enzymes produced.

Regulation of the flow of intermediates through the pathway

The flow of intermediates through the histidine biosynthetic pathway can be adjusted by varying the enzymatic activity of the *hisG* protein, which catalyzes the first reaction in the pathway (Fig. 1). Modulation of *hisG* enzyme activity is brought about by three interrelated forms of inhibition: (i) classical, noncompetitive feedback inhibition by histidine; (ii) inhibition by ppGpp in the presence of partially inhibiting concentrations of histidine; and (iii) inhibition by ADP and AMP in response to the overall energy status in the cell. In wild-type bacteria growing in minimal medium, the rate of histidine biosynthesis seems to be controlled primarily by regulation of *hisG* enzymatic activity (see below).

Because of its crucial role, the enzymology of the *hisG* protein has been intensively studied (Table 1). Although *hisG* protein is a hexamer under near-physiological conditions, the aggregation state of the protein is influenced in a complex way by temperature, ionic strength, pH, and the presence of ligands (reviewed in reference 16). The K_m of the enzyme for ATP is much lower than the intracellular ATP concentration, whereas the K_m for PRPP is probably closer to the intracellular PRPP concentration (Table 1 and Table 2); therefore, the rate of histidine biosynthesis most likely is directly affected by variations in the intracellular PRPP pool size. Each subunit of the *hisG* protein hexamer contains one allosteric site for histidine binding, which does not seem to overlap the subunit's active site. High concentrations of histidine totally inhibit the enzyme in a positively cooperative manner. The K_i for histidine is comparable to the intracellular histidine concentration found in bacteria grow-

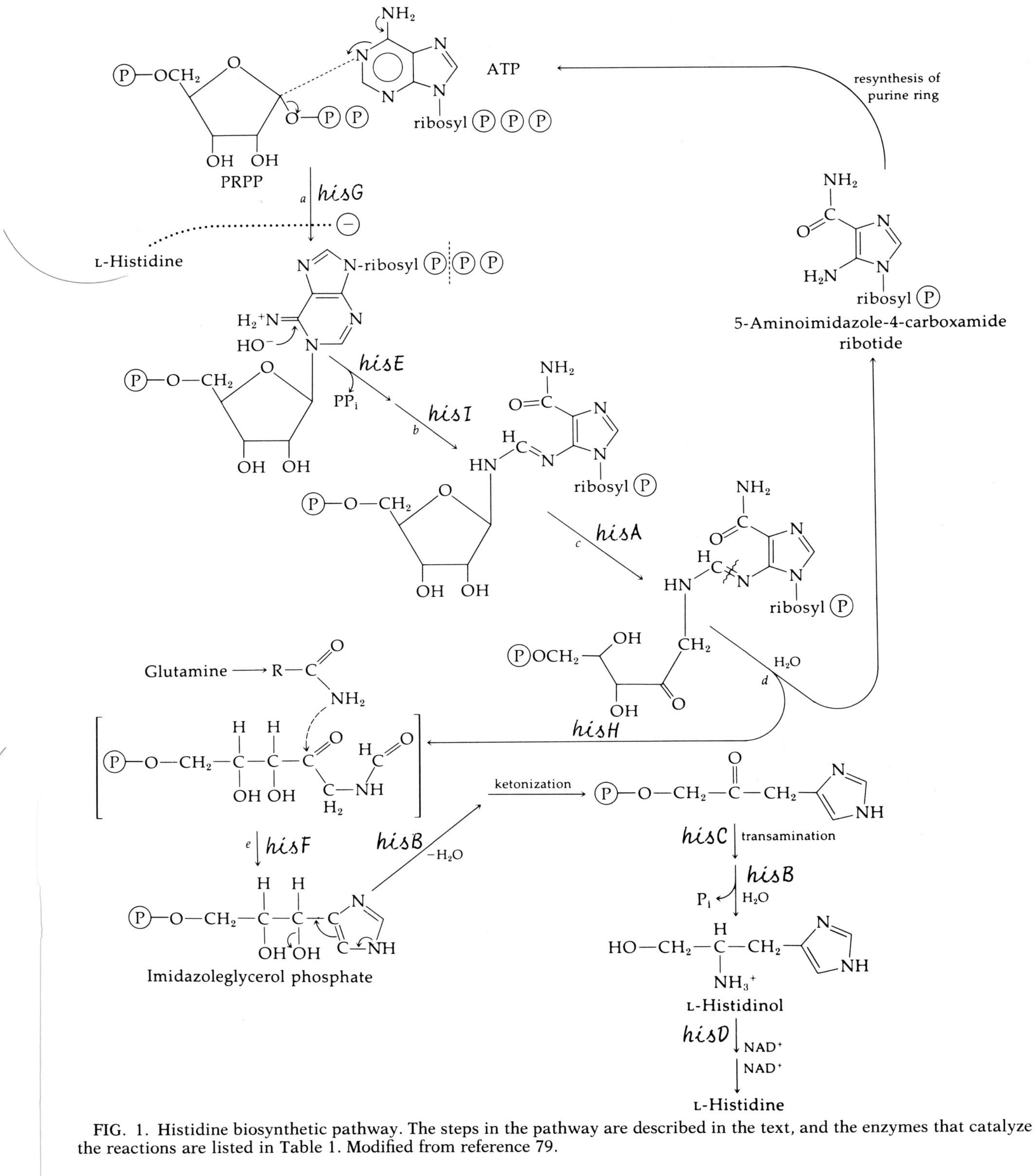

FIG. 1. Histidine biosynthetic pathway. The steps in the pathway are described in the text, and the enzymes that catalyze the reactions are listed in Table 1. Modified from reference 79.

ing in minimal medium containing histidine, an observation that implies substantial inhibition of the rate of biosynthesis through the pathway (Table 1 and Table 2). In contrast, the K_i for histidine is considerably higher than the intracellular histidine concentration found in bacteria that must synthesize histidine.

A number of feedback-resistant and feedback-hypersensitive mutations have been mapped in a region that encodes the carboxy-terminal portion of the hisG protein (100; see references in Hoppe et al. [56]). Feedback-resistant mutants selected for their resistance to the analog 2-thiazolealanine excrete histidine

TABLE 1. Properties of enzymes encoded by the *his* operon and *his* regulatory loci

Enzyme	Gene	Mol wt of gene product[a] (amino acids)	Native mol wt of enzyme[b]	Kinetic parameters for substrates[c]	Inhibitors	References for parameters and inhibitors
ATP phosphoribosyltransferase (EC 2.4.2.17)	*hisG*	33,000 (299)	200,000 (hexamer)	K_m (ATP) = 110 μM K_m (PRPP) = 11 μM	Histidine (K_i = 60–380 μM); 2-thiazoleala-nine; AMP, ADP; AMP + histidine > AMP; ppGpp + histidine > histidine	25, 81–83, 100
Histidinol dehydrogenase (EC 1.1.1.23)	*hisD*	47,000 (434)	83,000 (dimer)	K_m (NAD$^+$) = 1,300 μM K_m (histidinol) = 18 μM		73
Histidinol-phosphate aminotransferase (EC 2.6.1.9)	*hisC*	39,300 (356)	74,200 (dimer)	K_m (imidazole acetol-P) = 120 μM		1
Imidazoleglycerol-phosphate dehydratase (EC 4.2.1.19):histidinol phosphatase (EC 3.1.3.15)	*hisB* (bifunctional)	40,000 (355)	145,000 (tetramer?)	K_m (imidazole glycerol-P) = 700 μM; K_m (histidinol-P) = 300 μM	3-Amino-1,2,4-triazole (inhibits dehydratase only)	24
Glutamine amidotransferase	*hisH*	21,600 (196)	44,000 (dimer)	Unknown		
Phosphoribosylformimino-5-amino-1-phosphoribosyl-4-imidazolecarboxamide isomerase (EC 5.3.1.16)	*hisA*	26,100 (245)	29,000 (monomer)	Unknown		
Cyclase	*hisF*	28,500 (258)	41,000 (dimer?)	Unknown		
Phosphoribosyl-ATP pyrophosphohydrolase:phosphoribosyl-AMP cyclohydrolase (EC 3.5.4.19)	*hisIE* (bifunctional)	22,800 (203)	48,000 (dimer)	Unknown		
Histidyl-tRNA synthetase (EC 6.1.1.21)	*hisS*	40,000 (unknown)	78,000 (dimer)	Aminoacylation: K_m (His) = 6–25 μM K_m (ATP) = 140 μM K_m (tRNA) = 0.04 μM	α-Methylhistidine; histidinol; adenosine, AMP, ADP; AMP + PP$_i$ > AMP; His-tRNA^{His}	25–27, 39, 62
Pseudouridine synthase I	*hisT*	30,399 (270)	33,000 (monomer)	K_m (tRNA) ≅ 1 μM		33

[a] Polypeptide molecular weights are from the following sources: *E. coli his* operon DNA sequence (50; C. B. Bruni, personal communication); purified *S. typhimurium hisD* protein (70); purified *E. coli* and *S. typhimurium* histidyl-tRNA synthetase (36, 62); *E. coli hisT* operon DNA sequence (9).

[b] Enzyme native molecular weights are from the following sources: *hisG*, *hisB*, *hisH*, *hisA*, *hisF*, and *hisIE* (16); *hisD* (29); *hisC* (54); *hisS* (36, 62); *hisT* (H. O. Kammen and C. C. Marvel, personal communication).

[c] Enzyme assay protocols can be found in the references. Additional and updated assay methods can be found in the following references: all of the *his* operon-encoded enzymes (76); *hisG* enzyme (38, 66, 71); *hisD* enzyme (32); *hisB* enzyme (phosphatase) (23, 42); *hisS* enzyme (45); *hisT* enzyme (77). P, Phosphate.

into the growth medium (95). This important observation indicates that feedback inhibition holds histidine biosynthesis far below its full capacity, even when histidine is not supplied as a supplement (25). Some feedback-hypersensitive mutants also have a distinct phenotype; they are growth restricted at 20°C because of severe inhibition of the mutant *hisG* enzyme by histidine at lower temperatures (84, 100).

The rate of the *hisG* enzyme reaction is sensitive to several other molecules whose presence indicates the metabolic state of the cell. The alarmone ppGpp, which is a positive effector of *his* operon transcription (see below), does not inhibit *hisG* enzyme activity by itself; however, in the presence of moderate histidine concentrations (≥25 μM), physiologically significant concentrations of ppGpp (≥200 μM) strongly inhibit

TABLE 2. Parameters of histidine biosynthesis in wild-type *S. typhimurium*

Parameter[a]	Growth condition[b]	Value	References
Intracellular His concentration	Minimal medium	15 μM	6
	Minimal medium + 50 μM His	100 μM	
Percentage of tRNA^{His} charged with His	Minimal medium	77%	72
	Minimal medium + 100 μM His	88%	
Relative *his* operon expression	Minimal medium	≡1.0	25, 106
	Minimal medium + 50 μM His	0.9	
	Rich medium	0.3	
Steady-state rate of His biosynthesis	Minimal medium	≅1 μmol/g (dry wt) per min	25

[a] Additional related parameters: K_m for His is approximately 10^{-8} or 10^{-4} M for the high (*hisP*) or low (*aroP*) affinity transport system, respectively (7). Intracellular concentrations of 3,000 μM for ATP, 31 μM for ppGpp, 2 μM for tRNA^{His}, and 2 μM for histidyl-tRNA synthetase have been estimated for bacteria growing in minimal medium (17, 35).

[b] Glucose is the carbon source in the minimal medium. Rich medium is LB or nutrient broth.

hisG enzyme activity in a positively cooperative manner (82). The synergistic inhibition of *hisG* enzyme by ppGpp and histidine might play a physiological role (11). Starvation of bacteria for amino acids elicits ppGpp accumulation as part of the stringent response (see chapter 87). If the bacteria are starved for an amino acid in the presence of histidine, then the synergistic inhibition of *hisG* enzyme by intracellular ppGpp (≥200 μM) and histidine (≅100 μM) will completely inhibit histidine biosynthesis. In contrast, ppGpp accumulation in bacteria starved for histidine will not inhibit the *hisG* enzyme. It should also be noted that *hisG* enzyme activity will not be strongly inhibited by the intracellular pools of histidine (≅15 μM) and ppGpp (≅30 μM) present in bacteria growing exponentially in minimal medium lacking histidine.

AMP and ADP also bind to the *hisG* enzyme with affinities comparable to ATP. In the absence of histidine, AMP and ADP inhibit *hisG* enzyme activity by competing with ATP for the enzyme's active site (83). Since ATP is an initial substrate and considerable cellular energy is consumed in histidine biosynthesis, inhibition of *hisG* enzyme activity by AMP and ADP has frequently been cited as an example of an energy-utilizing system that responds to the overall energy status in the cell, as expressed by the Atkinson energy charge formula (25, 68). In addition, the presence of histidine causes the *hisG* enzyme to discriminate against its substrate ATP and preferentially bind its coinhibitors AMP and ADP (83). Therefore, inhibition of histidine biosynthesis in response to a decrease in energy charge is greater when histidine is present than when its supply is restricted.

In summary, inhibition of *hisG* enzyme activity by combinations of histidine, ppGpp, AMP, and ADP exerts sensitive control over the rate of histidine biosynthesis in response to a variety of cellular metabolic states. This complex, multivalent control is required because histidine biosynthesis is regulated chiefly by modulating the flow of intermediates through the pathway in wild-type bacteria growing under common culture conditions. This topic is discussed in the next section.

Regulation of the amounts of histidine biosynthetic enzymes

In effect, noncompetitive inhibition by histidine lowers the apparent V_{max} of the *hisG* enzyme reaction and makes it appear as if less total enzyme were present (74). Another way to control the rate of histidine biosynthesis would be to adjust the intracellular concentrations of the histidine biosynthetic enzymes in response to histidine and other metabolites. The structure and regulation of the *his* operon, which encodes all of the histidine biosynthetic enzymes (Fig. 2 and Table 3), has been a subject of investigation for two decades. Results from many studies show that two mechanisms regulate *his* operon expression at the level of transcription: (i) transcription initiations at the *his* operon primary promoter are positively regulated by increasing ppGpp concentrations up to the ppGpp concentration found in cells growing in minimal-glucose medium; and (ii) transcription of the *his* structural genes is regulated by an attenuation mechanism that responds to the intracellular concentration of His-tRNA^{His} (Table 3; see below).

Despite the potential for a wide range of control by attenuation, in vivo *his* operon expression is largely unaffected by the presence of histidine in the growth medium (Table 2). This surprising feature of *his* operon expression in part reflects the fact that histidine addition does not greatly increase the percentage of tRNA^{His} molecules charged with histidine (Table 2). However, even when exogenous histidine is absent, the amount of charged tRNA^{His} is still relatively high; yet, significant readthrough transcription beyond the *his* attenuator occurs. The high basal level of wild-type *his* operon expression is especially apparent from results which show that certain mutants containing defects in the attenuation mechanism fail to express the *his* operon structural genes (59, 61). Clearly a much greater potential exists to limit *his* operon expression by attenuation than is actually used in the bacterium. A mechanism that might contribute to setting the basal level of *his* attenuation is described in the section on regulation of the *his* operon.

The need to maintain relatively large amounts of the histidine biosynthetic enzymes, even in the presence of exogenous histidine, probably evolved in response to the high affinity of the *hisP* protein-mediated transport system (Table 2; 25). Since transport of histidine is so effective at low concentrations, the internal pool of histidine will not decrease significantly until the external histidine supply is almost completely gone. If attenuation greatly lowered the amounts of the histidine biosynthetic enzymes in response to an external histidine supply, then it is

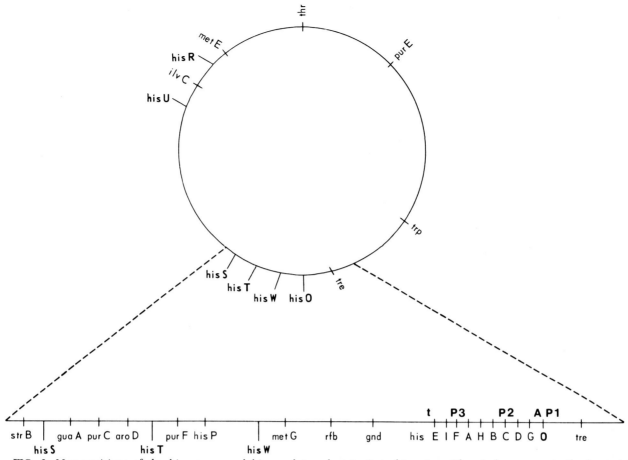

FIG. 2. Map positions of the *his* operon and *his* regulatory loci in *S. typhimurium*. The circle represents the bacterial chromosome and shows the relative positions of the *his* operon and the unlinked *hisR*, *hisS*, *hisT*, *hisU*, and *hisW* loci. The enlargement shows the genetic map around the *his* operon. The order of the genes in the *his* operon is indicated together with the primary promoter (P1), the attenuator (A), the internal promoters (P2, P3), and the terminator at the end of the operon (t). Modified from reference 4.

conceivable that the time between when the internal histidine pool starts to fall and the total histidine supply is exhausted would be too short to allow for adequate transcription and translation of the *his* operon structural genes. Hence, a system has evolved that favors histidine transport at the relatively small expense of maintaining a constant supply of the histidine biosynthetic enzymes (25). The evolutionary chronology for adjusting histidine transport relative to *his* attenuation is a highly speculative, but fascinating subject. Finally, it should be noted that when the bacteria are growing in nutrient-rich medium, there apparently is an advantage in decreasing the amounts of the histidine biosynthetic enzymes by about fourfold (Table 2). This metabolic regulation, which gears histidine biosynthesis to cellular growth rate, appears to be mediated by ppGpp and is independent from His-tRNA^{His}-specific attenuation control (see below; 99, 106).

The Histidine Biosynthetic Enzymes

The properties of the enzymes encoded by the *his* operon are compiled in Table 1. References to assay protocols are contained in a footnote to the table. The enzymology of the three best-characterized histidine biosynthetic enzymes, encoded by the *hisG*, *hisD*, and *hisB* genes, has recently been reviewed (16), and only new developments in this area are presented here. The nucleotide sequence of the entire *E. coli his* operon reveals several interesting details about the structures of the enzymes (C. B. Bruni, personal communication). The bifunctional *hisB* enzyme has been extremely difficult to purify and characterize because of aggregation and proteolysis (98). The nucleotide sequence shows that the actual molecular mass of the *hisB* polypeptide is 40,000 daltons, which closely matches the 46,000-dalton molecular mass determined for the major form of the purified enzyme (98). Minicell experiments containing *E. coli* or *S. typhimurium* DNA templates confirm the molecular weights of the *hisC* and *hisB* polypeptides deduced from the nucleotide sequence (Bruni, personal communication). Therefore, to account for earlier results from purification and antibody-precipitation experiments, it must be assumed that some type of modification or unusual aggregation produces forms of the *hisB* polypeptide that migrate with high apparent molecular weights on sodium dodecyl sulfate-polyacrylamide gels. The seemingly high K_m values of the *hisB* enzyme

TABLE 3. Parameters of *his* operon structure and regulation

Parameter[a]	Comments	Organism[b]	References
Gene order	OGDCBHAF(IE)	Both	53
Size of full-length primary transcript: P1 → t	7,342 nt (DNA sequence)	*E. coli*	C. B. Bruni, personal communication
	≅7,300 nt (Northern blots)	Both	
Size of major terminated leader transcript: P1 → A	177 nt	*S. typhimurium*	47
	180 nt	*E. coli*	48
Size of secondary transcripts:			
P2 → t	≅4,000 nt (Northern blots)	Both	C. B. Bruni, personal communication
P3 → t	≅1,000 nt (Northern blots)		
Established modes of operon regulation	ppGpp-stimulated transcription from P1 and P2	Both	99, 106
	Attenuation control in response to His-tRNA^His concentrations	Both	10, 16, 59
Type of terminator at the end of the operon	Rho independent (bidirectional)	*S. typhimurium*	31; C. B. Bruni, personal communication
	Rho independent (bidirectional?)	*E. coli*	

[a] P1, *his* operon primary promoter; A, *his* attenuator; P2 and P3, *his* operon internal promoters; t, terminator at the end of the *his* operon (see Fig. 2).

[b] Both, Established in both *E. coli* and *S. typhimurium*.

for its two substrates (Table 1) may also reflect a property of the purified protein, since studies with crude extracts reveal greater than sufficient *hisB* enzyme activities to account for the in vivo rate of histidine biosynthesis (Table 2; 25). In addition, the nucleotide sequences indicate that the *E. coli* and *S. typhimurium hisI* and *hisE* enzyme activities are contained in single bifunctional proteins (Bruni, personal communication; Table 1). This finding is consistent with results from genetic analyses of the *S. typhimurium hisIE* region (53) and explains the cosedimentation of the *S. typhimurium hisI* and *hisE* enzyme activities in sucrose gradients (76).

Further analysis of the nucleotide sequence should reveal whether there are evolutionary relationships among the genes in the *E. coli his* operon. Based on limited amino acid sequences, a moderate degree of homology has been observed in the amino-terminal regions of the *hisG*, *hisD*, and *hisC* polypeptides (see reference 16). The extent of homology can now be rigorously determined throughout the coding region of each *his* gene to test the hypothesis that the histidine biosynthetic enzymes all evolved from a common ancestral protein (57). As in the case of the *trp* operon (108), comparisons of *his* operon sequences from different bacterial species should give insights into the evolution of operon structure.

STRUCTURE AND REGULATION OF THE *his* OPERON

The chromosomal locations of the *S. typhimurium his* operon and *his* regulatory loci are depicted in Fig. 2 together with the order of the genes within the *his* operon. Parameters relevant to *his* operon structure and control are collected in Table 3 and give the following general picture of the operon. The eight structural genes are transcribed into a single, polycistronic mRNA molecule (75; Table 3), which extends from the primary promoter (P1) to the strong, rho-independent terminator (t) (Fig. 2). The frequency of

transcription initiations at *hisP1* is positively regulated by a limited range of intracellular ppGpp concentrations. An RNA polymerase molecule that initiates transcription at *hisP1* first transcribes a leader region and must continue transcribing past the attenuator control site (A) if it is to enter the first structural gene of the operon. The *hisP1* promoter, leader region, and attenuator are contained in a genetic locus which traditionally has been designated as the *hisO* region even though the operon is not controlled by a classical repression mechanism (Fig. 2 and 3). Transcription termination or readthrough at the *his* attenuator will occur when the percentage of tRNA^His charged with histidine is high (88%) or low (≅12%), respectively (72). Transcription termination at the *his* attenuator produces a terminated leader transcript of approximately 180 nucleotides, which extends from *hisP1* to a site in the attenuator region. In some instances, transcription initiations can also occur at two internal promoters (*hisP2* and *hisP3*, Fig. 2) located before the start of the *hisB* and *hisIE* genes. These features of *his* operon structure and regulation are described in detail in the next sections.

Structure of the *his* Operon

The primary promoter *hisP1*

The nucleotide sequence of the *S. typhimurium hisO* region is presented in Fig. 3 (11, 13). The start of transcription determined in vitro (47) and the corresponding −35 and −10 regions of the promoter are indicated (see chapter 75). The nucleotide sequence of the *E. coli hisP1* promoter is identical to that of the *S. typhimurium* promoter in the −35 and −10 regions and is similar in the 18-nucleotide (nt) −30 to −14 spacer and in the −6 to +1 spacer (103). The start of transcription at the *E. coli hisP1* promoter in vivo and in vitro is analogous to the position indicated in Fig. 3 (48). Recently, Artz and his associates devised an effective method to isolate mutations in the *S.*

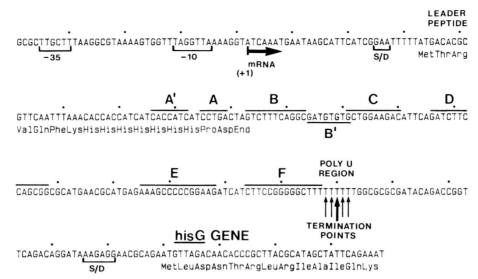

FIG. 3. DNA sequence of the *hisO* region in *S. typhimurium*. The −35 and −10 regions of the *hisP1* promoter and the start point of transcription are indicated. S/D designates the weak and strong Shine-Dalgarno sequences in the translation initiation regions for the *his* leader peptide and *hisG* polypeptide, respectively. A′, A, B, B′, C, D, E, and F refer to segments of the *his* leader transcript that can base pair to form the alternative, mutually exclusive RNA secondary structures shown in Fig. 4. The major (heavy arrow) and minor (light arrows) points of transcription termination in the *his* attenuator region are indicated. Redrawn from reference 11.

typhimurium hisP1 promoter (R. Shand, P. Blum, and S. Artz, Abstr. Annu. Meet. Am. Soc. Microbiol. 1983, H157, p. 132). Using *his::lac* fusions, they identified 14 different mutations that decrease the efficiency of the *hisP1* promoter by 4- to 400-fold compared to the wild type. The mutations alter the sequences of the −35 or −10 regions or the spacing between these two regions (11).

The *hisP1* promoter is unusually strong in vivo and in vitro on supercoiled templates (103). The *E. coli hisP1* promoter has been cloned into a McKenney-Rosenberg *galK* expression vector and found to be about four times stronger than the *gal* promoter in vivo (88). The strength of *hisP1* in vivo is further indicated by the result that the histidine biosynthetic enzymes amount to at least 4% of the total cellular proteins in bacteria deleted for the *his* attenuator (86). Results from in vitro experiments confirm that *hisP1* is strong compared to a pBR322 promoter and show that transcription from *hisP1* is about 20-fold stronger on supercoiled templates than on linear templates (103). This observation could potentially be of physiological significance in view of the suggestion that the *hisW* gene encodes a subunit of DNA gyrase. This topic and several recent discoveries made by S. W. Artz and his associates concerning the effect of ppGpp on transcription from the *hisP1* promoter are discussed elsewhere in this chapter.

The internal promoters *hisP2* and *hisP3*

The positions of two internal promoters have been evolutionarily conserved in the *his* operons of *E. coli* and *S. typhimurium* (51, 92). Using Tn*10* insertions to block transcription from upstream regions in the *S. typhimurium his* operon, it has been possible to map *hisP2* near the end of the *hisC* coding region before the start of *hisB* and to map *hisP3* near the end of the *hisF*

coding region before the start of *hisIE* (92). Analyses of in vivo transcripts using S1 mapping and Northern hybridizations locate the *E. coli hisP2* promoter 108 nt upstream from the end of the *hisC* coding region (51). Experiments in which transcription from *hisP2* was measured in mutants lacking transcription from *hisP1* or in which transcription from *hisP1* or *hisP2* was measured in *galK* expression vectors showed that *hisP2* is about half as strong as *hisP1* (43, 51). An indirect method based on ratios of enzyme activities has suggested that transcription from *hisP2* is insignificant when transcription of the *his* structural genes is equal to or greater than the level found in wild-type bacteria (43). This conclusion needs to be tested directly using cloned regions of the *his* operon as hybridization probes to measure the in vivo rates of *his* mRNA synthesis and degradation.

It has been proposed that the function of the internal promoter in the *trp* operon is to maintain sufficient levels of the *trpC* and *trpB* polypeptides, each of which contains multiple tryptophan residues. The *trpE* and *trpD* polypeptides, which are encoded by genes upstream from the internal promoter, contain few tryptophan residues and would not be seriously depleted by sudden tryptophan starvation (109). An analogous function for the *hisP2* and *hisP3* internal promoters seems unlikely because the *hisG* polypeptide, which is encoded by the first gene in the operon, contains multiple histidine residues (87). In both *E. coli* and *S. typhimurium*, the *hisP2* promoter is metabolically regulated (51, 106); however, the physiological functions of the *his* internal promoters remain unknown.

The *his* operon structural genes

Complete restriction maps and subclones containing different regions of the *E. coli* and *S. typhimurium his* operons have existed for several years (12, 13, 28,

30, 51). From these studies, it was concluded that the *his* operon must be compactly organized in both species. The DNA sequence of the entire *E. coli his* operon has recently been determined (Bruni, personal communication), and sequence determination of the *S. typhimurium his* operon is nearly complete (W. M. Barnes, personal communication). The *E. coli* sequence dramatically confirms the conclusion about compact operon organization. There is only one small 5-base-pair intercistronic region between *hisG* and *hisD* in the entire 7,379-nt *E. coli his* operon; for all the other gene pairs, the translation stop codon of the upstream gene overlaps the translation initiation codon of the next downstream gene (Bruni, personal communication). An intercistronic region between *hisG* and *hisD* (55) and overlapping translational signals between *hisD* and *hisC* (89) have also been reported in *S. typhimurium*. Comparison of the *hisG*-*hisD* intercistronic regions in the two species is of particular interest, because the ≅100-nt *S. typhimurium hisG*-*hisD* border contains a repetitive extragenic palindromic DNA sequence that is found throughout the bacterial chromosome (55, 101). The absence of a repetitive extragenic palindromic sequence in the *E. coli hisG*-*hisD* intercistronic region supports the idea that such sequences arose through transpositions (55).

The presence of overlapping translational signals raises the possibility that extensive translational coupling occurs in the expression of the *his* operon. In the *trp* operon, translational coupling between the *trpE* and *trpD* genes is thought to guarantee equimolar synthesis of the corresponding gene products, which interact to form a multimeric enzyme (109). In *S. typhimurium*, the *hisG*, *hisD*, *hisC*, and *hisA* polypeptides are expressed in molar ratios of 3:1:1:1 (105). Because of the large intercistronic region, direct translational coupling would not be expected between *hisG* and *hisD*. Some level of translational coupling between other genes in the operon could produce the equimolar synthesis of the *hisD*, *hisC*, and *hisA* polypeptides detected in vivo. The mechanism that produces the differential expression between the *hisG* gene and the other genes in the *S. typhimurium his* operon is unknown, but probably does not involve the repetitive extragenic palindromic sequence in the *hisG*-*hisD* intercistronic region (101).

The existence of translational coupling could help to explain other aspects of in vivo *his* operon expression. For example, there is a strong polarity gradient in the *hisG* gene, but essentially no polarity gradients in the internal genes of the operon (44). This pattern of polarity corresponds to the potential for translational coupling between internal genes in the *his* operon. Translational coupling might also influence the kinetics of *his* operon expression after the onset of histidine limitation; perhaps the sequential appearance of the histidine biosynthetic enzymes in cells with low formylating capacity reflects translational coupling, whereas the simultaneous appearance of the histidine biosynthetic enzymes in cells with high formylating capacity indicates a mode of uncoupled, independent translation (15). Before these problems are readdressed, the important question of whether extensive translational coupling occurs in the expression of the *his* operon needs to be examined experimentally.

The terminator (t) at the end of the operon

Transcription termination at the end of the *E. coli* and *S. typhimurium his* operons occurs at a rho-independent terminator (30, 31; Bruni, personal communication; see chapter 76). The *S. typhimurium his* terminator is a symmetrical, mirror-image structure; each strand contains (reading 5' to 3') a G+C-rich inverted repeat followed by several T residues. This mirror-image structure suggested that the terminator might function in both orientations. This prediction has been confirmed both in vivo and in vitro (31). Analysis of in vivo transcription termination points by S1 mapping and Northern hybridizations demonstrated that this structure terminates *his* operon mRNA initiated at the *his* primary and internal promoters and, at the same time, terminates a 1,200-nt-long transcript synthesized from the DNA strand opposite to the one copied into *his* mRNA. Both the *his* mRNA and the *his*-convergent transcripts synthesized in vivo end with polyuridylate residues, as expected for rho-independent transcription termination. Additional experiments established that the *his* terminator functions at greater than 90% efficiency in either orientation in an in vivo expression vector system. Finally, the rho-independent nature and high efficiency of the *his* terminator in both orientations were confirmed in a purified in vitro transcription system. Similar analysis of the *E. coli his* terminator is in progress (Bruni, personal communication).

Regulation of the *his* Operon

Metabolic regulation

As noted in a previous section, *his* operon expression is about fourfold greater in bacteria growing in minimal-glucose medium than in rich medium (Table 2). This inverse relationship between *his* operon expression and cellular growth rate is a form of metabolic regulation that adjusts *his* operon expression in response to the general amino acid supply in the cell (99). Because ppGpp levels were known to vary inversely with growth rate (see chapter 87), ppGpp was examined as a possible positive effector of *his* operon expression (99). In an in vitro coupled transcription-translation system prepared from a *relA* mutant defective in ppGpp synthesis (see chapter 87), addition of 100 µM ppGpp caused a 10-fold increase in expression of the wild-type *his* operon contained on a linear, transducing phage template. Equal levels of ppGpp-mediated stimulation were detected from the wild-type template and from a mutant template deleted for the *his* attenuator, which shows that ppGpp acts independently of attenuation. Furthermore, by uncoupling transcription and translation in the in vitro system, it was possible to show that ppGpp stimulates *his* operon transcription, but not translation. From these results, ppGpp was postulated to be the effector molecule that directly mediates metabolic regulation of *his* operon expression (99).

Results from physiological experiments support the model for the role of ppGpp as a positive effector of *his* operon expression. The increase in expression of the *his* operon in bacteria subjected to sudden histidine starvation in amino acid-rich medium is markedly defective in *relA* mutants compared with *relA*⁺ strains (99). In addition, there is a positive correlation between in vivo *his* operon expression and intracellular ppGpp concentrations, up to the ppGpp level found in bacteria growing in minimal-glucose medium (91, 99, 106, 107); increases in in vivo ppGpp concentrations beyond this level fail to increase *his* operon expression (106). These results support the notion that *his* operon transcription is maximally stimulated at lower than maximum in vivo ppGpp concentrations.

Recently, experiments were performed in which intracellular ppGpp was drastically reduced below the level found in *relA*⁺ bacteria growing in rich medium (P. Blum and S. Artz, Abstr. Annu. Meet. Am. Soc. Microbiol. 1983, H156, p. 131). Results from this study show that attenuator-independent *his* operon expression decreases about 15-fold in a *relA* mutant and increases about 2-fold in a *relA*⁺ strain in response to the decreased or increased ppGpp levels, respectively, induced by addition of the analog serine hydroxamate to the amino acid-rich growth medium. Thus, the full range of the ppGpp-mediated metabolic regulation of in vivo *his* operon expression is at least 30-fold. Finally, a number of physiological experiments have confirmed the conclusion that ppGpp-mediated metabolic regulation and attenuation are independent mechanisms for the control of *his* operon transcription (88, 91, 99, 106, 107). A corollary of this conclusion has also been confirmed, namely, starvation for amino acids other than histidine increases transcription of the *his* operon (106). In these experiments, the amino acid limitation should not have interfered directly with leader peptide synthesis in a way that would increase readthrough transcription of the *his* attenuator (see next section). The effect of ppGpp accumulation on the control of the flow of intermediates through the histidine biosynthetic pathway in response to starvation for histidine or other amino acids was described in an earlier section of this chapter.

Perhaps the most interesting question about *his* operon metabolic regulation concerns the mechanism by which ppGpp stimulates transcription. Strains containing putative mutations in the *hisP1* promoter show altered levels of attenuator-independent *his* operon expression in response to amino acid downshifts (106). These results suggest that ppGpp affects transcription initiation frequencies at the *hisP1* promoter; however, more rigorous interpretations are not possible, because the nucleotide alterations of the mutations are unknown. Artz and his associates have begun a systematic analysis of ppGpp-mediated stimulation of *his* operon transcription, which has led to several important new discoveries. They have used oligonucleotide-directed, site-specific mutagenesis to change the −10 region of the *S. typhimurium hisP1* promoter from TAGGTT (Fig. 3) to the perfect consensus, TATAAT. This mutation has two effects on transcription from supercoiled templates in an in vitro coupled transcription-translation system: (i) transcription from the mutant promoter is increased about 20-fold over that from the wild-type promoter in the absence of added ppGpp; and (ii) concentrations of ppGpp that stimulate transcription from the wild-type *hisP1* promoter by 30-fold fail to stimulate transcription from the mutant promoter (S. W. Artz,

personal communication). This intriguing result strongly suggests that the −10 region of *hisP1* plays some role in the mechanism that responds to ppGpp. Experiments in progress should indicate whether this mutation and the other mutations generated in the *hisP1* promoter affect *his* operon metabolic regulation in vivo (Artz, personal communication). In addition, it will be interesting to learn whether ppGpp alone or in the presence of some other factor increases the transcription initiation frequency from the *hisP1* promoter in a purified in vitro transcription system.

In another series of experiments, Artz and his co-workers have found that 100 μM ppGpp stimulates *his* operon transcription about 30-fold on supercoiled templates, compared to 10-fold on linear templates in their cell-free system (Artz, personal communication). This 30-fold range of ppGpp-stimulated *his* operon transcription detected in vitro matches the full range of *his* operon metabolic regulation found in vivo. However, the concentration of ppGpp that maximally stimulates *his* operon transcription in vitro (≅100 μM) is about threefold greater than recent estimates of the intracellular ppGpp concentration that seems to maximally stimulate *his* operon metabolic regulation (≅30 μM; Table 2; 17). The basis for this discrepancy requires further investigation.

Several other questions remain to be answered about *his* operon metabolic regulation. Comparison of the −10 regions of the *hisP1* promoter and *hisP2* internal promoter shows little similarity. Therefore, it is possible that metabolic regulation at *hisP1* and *hisP2* occurs by different mechanisms in response to different effectors. In addition, the availability of cloned regions of the *his* operon makes it possible to measure the rate of *his* mRNA synthesis and evaluate metabolic regulation of *his* operon transcription during transient nutrient shifts. Finally, it remains to be established whether the mechanism that controls *his* operon metabolic regulation is widespread as originally proposed (99), or limited to a small number of genes and operons.

Attenuation control

Histidine-specific control of *his* operon expression in *E. coli* and *S. typhimurium* is exerted through an attenuation mechanism. Based on the analysis of mutations, attenuation can potentially regulate *his* operon expression over a 200-fold range (25, 42, 61). Therefore, the combination of metabolic regulation (30×) and attenuation (200×) gives a rather extraordinary total potential range of 6,000 for *his* operon regulation. The discovery of *his* operon attenuation paralleled the work of Yanofsky and his associates on *trp* operon attenuation (see chapter 77), and results from the two systems led to a rapid elucidation of the mechanism underlying attenuation. A delightful history of the discovery of *his* operon attenuation was published previously (11). In addition, chapters on the general mechanism of attenuation and attenuation control of the *trp* operon appear in this volume (chapters 77 and 90). Consequently, only a brief description is presented of a slightly updated version of the model for regulation of *his* operon expression by attenuation. Several current developments concerning this model are also discussed.

Model for *his* attenuation. The attenuation model proposes that transcription termination at the *his* attenuator is modulated by synthesis of a peptide encoded by the *his* leader region (13, 16, 40, 59). Except for details, the mechanism that brings about the coupling of transcription termination and translation is formally similar for *his* and *trp* attenuation. This mechanism relies on two critical features of the *his* leader transcript, which precedes the start of the *hisG* coding region (Fig. 3). First, the *his* leader transcript encodes a 16-amino-acid peptide that contains seven tandem histidine residues. Second, a series of mutually exclusive, alternative secondary structures can form in the *his* leader transcript (Fig. 3 and 4). These stem-and-loop structures are formed by pairing between bases in segments A′ and B′, A and B, B and C, C and D, D and E, and E and F in the *his* leader transcript and are designated as A′B′/AB, BC, CD, DE, and EF, respectively (Fig. 3 and 4). Secondary structure EF together with the polyuridylate residues downstream from it constitutes a strong rho-independent terminator. If EF forms, transcription termination occurs at one of the uridylate residues and produces a terminated leader transcript (Fig. 3, Fig. 4, and Table 3). In essence, control by attenuation amounts to varying the frequency at which the EF "terminator" structure forms. Any factor, condition, or mutation that prevents EF terminator formation acts as an "antiterminator" and allows an RNA polymerase molecule to continue transcription into the *his* operon structural genes. In wild-type attenuation, the antiterminator is an alternative RNA secondary structure (DE; Fig. 4) whose frequency of formation is determined by translation of the leader transcript. Transcription and translation of the *his* leader transcript are probably synchronized by transcriptional pausing.

To see how the model works, consider bacteria growing in minimal-glucose medium containing histidine. Under these growth conditions, the intracellular concentration of histidine will approach 100 μM and about 90% of the tRNA^{His} molecules will be charged with histidine (Table 2). An RNA polymerase molecule that initiates transcription at the *hisP1* promoter will start to transcribe the *his* leader region (Fig. 3). A transcribing RNA polymerase molecule will most likely pause near position 101 in the leader transcript as a result of A′B′/AB formation (Fig. 4). Meanwhile, a ribosome will begin to translate the leader transcript and produce the leader peptide. Because the intracellular concentration of His-tRNA^{His} is high, the entire leader peptide will be synthesized, and the translating ribosome will move all the way to the stop codon at position 80 in the transcript. Ribosome movement to this position will disrupt A′B′/AB, thereby releasing the RNA polymerase molecule from its paused state, and will mask segments B and B′ in the transcript. When transcription resumes, structure CD will have an opportunity to form first, preclude formation of structure DE, and allow the EF terminator to form (Fig. 4A). Once transcription termination occurs, the RNA polymerase and terminated leader transcript are probably released spontaneously from the DNA template.

If the translating ribosome rapidly dissociates from the stop codon, A′B′/AB will have an opportunity to

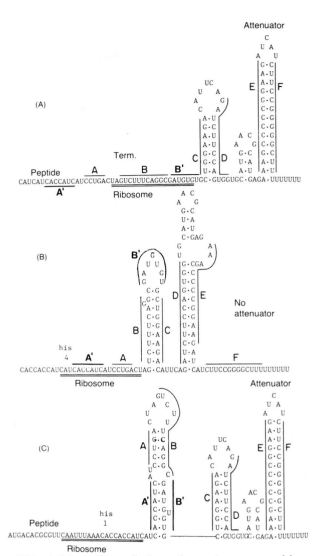

FIG. 4. Mutually exclusive, alternative stem-and-loop structures that form in the *his* leader transcript as part of the attenuation process. Letters designate segments of the leader transcript that base pair to form secondary structures. Double lines indicate the nucleotides masked by a ribosome stopped at a codon. The model for *his* attenuation is described in the text. (A) Transcription termination configuration caused by translation of the leader transcript to the UAG stop codon. (B) Readthrough transcription configuration caused by ribosome stalling at the fourth histidine codon in the leader transcript. (C) Transcription termination configuration caused by ribosome stalling at the Gln codon in the leader transcript. Configuration (C) will also occur in the absence of translation of the leader transcript or after rapid dissociation of the ribosome at the translation stop codon. Modified from reference 61.

form again, prevent structure BC formation, and still allow CD and EF to form (Fig. 4C). On the other hand, rapid dissociation of the ribosome at the instant the RNA polymerase molecule is released from pausing could allow synthesis of segment C before A′B′/AB has a chance to form. This situation would allow BC formation, which will lead to readthrough transcription. BC might also have a chance to form by reequilibration of secondary structures if the transcribing RNA polymerase molecule pauses again in response to CD formation. In either instance, rapid release of ribosomes at the stop codon could account for the relatively high basal level of *his* operon expression detected in vivo in the presence of histidine.

Consider next bacteria moderately starved for histidine. Omission of histidine from minimal-glucose medium does not reduce the concentration of His-tRNA^{His} sufficiently ($\cong$80% charged; Table 2) to cause significant readthrough of the wild-type *his* attenuator (see above). However, mild histidine starvation can be induced genetically (see, for example, reference 11) or by adding the analog aminotriazole, which inhibits one of the *hisB* enzyme activities (Table 1). Under these conditions, a ribosome translating the leader transcript will stall at the tandem histidine codons because the cellular concentration of His-tRNA^{His} is low ($\leq$12% charged; 72). If a ribosome prevents secondary structure formation in a transcript for about 16 nt downstream from a codon in the aminoacyl site (11, 61), then ribosome stalling at histidine codons 3 to 5 will mask segments A and A' and release the RNA polymerase molecule from its paused state. Segment C will then be synthesized, and subsequent BC formation will preclude CD formation, allow the antiterminator DE to form, and allow transcription to continue past the *his* attenuator (Fig. 4B).

Finally, if translation of the leader peptide is prevented by a mutation in the initiation codon, superattenuation results. An RNA polymerase molecule transcribes the leader region, pauses after formation of A'B'/AB, eventually resumes transcription without assistance from a ribosome, and synthesizes segment C and then segment D of the leader transcript. Formation of CD prevents DE formation as the RNA polymerase molecule continues transcription, and the absence of DE allows formation of the EF terminator. Superattenuation should also occur if a translating ribosome stalls upstream of the third histidine codon in the leader transcript, because a ribosome stalled so far upstream in the leader transcript will fail to disrupt A'B'/AB formation (Fig. 4C). It has been reported that in vivo attenuator-dependent *his* operon expression is less during severe histidine starvation than during mild histidine starvation (28). This observation can be explained by superattenuation which results from occasional stalling of ribosomes at the first two histidine codons in response to severe histidine limitation (11).

Evidence and developments. The following features have been well established experimentally for the model of *his* attentuation: (i) response to the intracellular amount of His-tRNA^{His} rather than free histidine (72); (ii) nearly complete evolutionary conservation of the DNA sequence of the *his* leader regions of *E. coli* and *S. typhimurium* (59); (iii) existence of *his* terminated leader and *his* readthrough transcripts in vivo and in vitro (47, 48, 63); (iv) formation of mutually exclusive, alternative RNA secondary structures in the *his* leader transcript (14, 16, 42, 59, 61); (v) translation of the *his* leader to form the encoded peptide (10, 11, 59, 61); (vi) masking of transcript segments by stalled ribosomes (11, 61); and (vii) lack of function of the leader peptide other than to be translated as part of the attenuation mechanism (61). Transcriptional pausing to ensure coupling of trans-

lation and transcription-termination, rapid release of ribosomes setting basal levels of attenuation, and spontaneous release of the termination complex at the attenuator were formulated recently for *trp* attenuation (see chapter 77). These constructs very likely apply to *his* attenuation and require experimental verification. Analysis of transcriptional pausing in the *his* leader region could be especially informative because of the potential for multiple stem-and-loop structures and multiple pause sites.

Several other aspects of the *his* attenuation model require further experimental support. One of the more pressing issues concerns secondary structure formation in the *his* leader transcript. Comparison of nucleotide sequences has revealed that the *his* leader transcript and tRNA^{His} molecule are extremely homologous; furthermore, the *his* leader transcript might assume a "cloverleaf" secondary structure that resembles the folded tRNA^{His} molecule (5). This observation has led to the interesting suggestion that protein molecules like pseudouridine synthase I, histidyl-tRNA synthetase, and the *hisG* protein, which are all known to bind tRNA^{His}, might bind to or even modify the *his* leader transcript and thereby influence *his* attenuation. Experimental verification of this hypothesis requires analysis of secondary structures that form in the purified *his* leader transcript. Direct determination of secondary structures might also help distinguish which of two alternative AB structures forms in the *his* leader transcript (14) and whether a stem-and-loop structure forms in the wild-type or mutant ribosome binding site of the *his* leader transcript (61).

Another issue, which has been partially resolved, concerns the postulated role of the *hisG* enzyme as a regulator of *his* operon expression. Suggestive results from a number of experimental approaches led to the conclusion that the *hisG* enzyme, which binds histidine and His-tRNA^{His} molecules, acts as an autogenous corepressor of *his* operon expression (80). Although this conclusion was instrumental in the formulation of the model of autogenous regulation (49), subsequent genetic and physiological experiments showed conclusively that *hisG* protein is not an essential component of the mechanism for *his* operon regulation (94). Therefore, the only known forms of *his* operon control are metabolic regulation and attenuation. Results from physiological experiments that implicated *hisG* protein as a putative corepressor probably reflect indirect effects on intracellular PRPP, ppGpp, or His-tRNA^{His} concentrations. Nevertheless, the *hisG* enzyme may play a direct, but ancillary role in controlling *his* attenuation by acting as a regulated reservoir for His-tRNA^{His} molecules (67). The question of whether the *hisG* enzyme plays some auxiliary role in *his* operon regulation remains unanswered and needs to be readdressed experimentally.

Role of the *his* Regulatory Loci

One of the early schemes devised for the selection of *S. typhimurium* mutants with high *his* operon expression relied on resistance to a combination of the analogs 3-amino-1,2,4-triazole and 1,2,4-triazole-3-alanine (90). 3-Amino-1,2,4-triazole inhibits the *hisB* dehydratase activity (Table 1) and reduces histidine biosynthesis, while 1,2,4-triazole-3-alanine is mistaken for histidine in the cell and is charged onto tRNA^{His} and incorporated into proteins. Growth of wild-type cells is inhibited by a combination of the analogs, because 3-amino-1,2,4-triazole causes reduction of the histidine supply below a critical level and 1,2,4-triazole-3-alanine incorporation presumably inactivates proteins. High expression of the *his* operon caused by mutations can overcome the effect of the analogs and allow the mutant bacteria to grow.

Because the absolute amount of His-tRNA^{His} controls the level of *his* attenuation, mutants containing defects in tRNA^{His} biosynthesis (*hisR, hisU* [*gyrB*], *hisW* [*gyrA*]), charging (*hisS*), modification (*hisT*), and processing (*hisU* [*rnpA*]) were selected by resistance to the two analogs above. It is beyond the scope of this chapter to describe in detail the molecular genetics of each *his* regulatory locus; however, several new discoveries will be mentioned that are relevant to histidine biosynthesis.

hisR

The *hisR* gene codes for the single cellular species of tRNA^{His} (25). This single-copy gene is part of a tRNA gene cluster whose order is tRNA^{Arg}-tRNA^{His}-tRNA^{Leu}-tRNA^{Pro} in both *E. coli* (58) and *S. typhimurium* (18). The coding region for the tRNA molecules is bounded by a promoter, which has a putative stringency-control discriminator sequence between the −10 region and the start point of transcription, and a rho-independent terminator (18, 58).

Mutations in the *S. typhimurium hisR* promoter reduce the total cellular content of tRNA^{His} molecules by about 50% and thereby cause increased readthrough transcription of the *his* attenuator (22). Two of these mutations are novel. One is a 3-base-pair deletion in the −70 region of the *hisR* promoter which apparently disrupts an enhancerlike sequence. The enhancer sequence may function by assuming an unusual bent DNA conformation which stimulates transcription from the *hisR* promoter. The other *hisR* promoter mutation is a single-base-pair insertion in the −5 region that alters the start and efficiency of transcription initiation. Transcription from the *S. typhimurium hisR* promoter in vitro is also extremely dependent on supercoiling of the DNA template (L. Bossi, personal communication). If this conclusion applies to in vivo transcription as well, it could mean that expression of the *hisR* gene will be reduced in *gyrA* or *gyrB* mutants. This notion is supported by new results from genetic and complementation experiments which suggest that *hisW* corresponds to the *gyrA* gene and a subclass of *hisU* corresponds to the *gyrB* gene (K. E. Rudd, personal communication; see below).

hisS

The *hisS* gene encodes histidyl-tRNA synthetase, which aminoacylates tRNA^{His} molecules with histidine (Table 1). Because attenuation responds to the amount of charged tRNA^{His}, the activity of the histidyl-tRNA synthetase potentially can affect histidine biosynthesis. In wild-type cells growing in minimal-glucose medium without histidine, the K_m of the synthetase for histidine is comparable to the histidine concentration (Tables 1 and 2). In this concentration range, the rate of aminoacylation should be strongly affected by fluctuations in the histidine concentration (25). In addition, aminoacylation of tRNA^{His} is noncompetitively inhibited by AMP (39), competitively inhibited by ADP or adenosine (39), very strongly inhibited by AMP in the presence of pyrophosphate (27), and strongly product inhibited by His-tRNA^{His} (27). This pattern of inhibition has three consequences that could affect the intracellular amount of charged tRNA^{His} and thereby influence the rate of histidine biosynthesis: (i) the activity of histidyl-tRNA synthetase is subject to control by cellular energy charge, like the *hisG* enzyme (26; see above); (ii) product inhibition probably plays a role in setting the percentage of tRNA^{His} molecules that are charged at a given histidine concentration (72); and (iii) only a small fraction of the histidyl-tRNA synthetase molecules actually will be active in bacteria growing under normal conditions because of product inhibition by the high percentage of charged tRNA^{His} molecules (Table 2; 27).

Mutations in the *hisS* gene that affect the level of *his* attenuation act by reducing the percentage of tRNA^{His} molecules charged with histidine (72). These mutations generally lower the activity of the histidyl-tRNA synthetase and decrease the enzyme's affinity for histidine, tRNA^{His}, or ATP (37). Furthermore, histidine biosynthesis seems to be linked directly to the control of the *hisS* gene, since limitation for histidine causes increased expression of *hisS* in vivo (78). Multiple dyad symmetries in the DNA sequence around the in vivo *hisS* promoter of *E. coli* suggest that *hisS* transcription might be autogenously repressed by histidyl-tRNA synthetase in the presence of histidine (41, 46). Additional experiments are needed to test this hypothesis.

hisT

The *hisT* gene encodes pseudouridine synthase I, which catalyzes formation of pseudouridine residues at positions 38, 39, and 40 in the anticodon stem and loop of at least 30 tRNA isoaccepting species, including tRNA^{His} (Table 1; 96, 102). Transcription termination at the *his* attenuator is greatly decreased in *hisT* mutants even though the undermodified tRNA^{His} molecules are charged with histidine to the same extent as in wild-type strains (72). To explain this observation, it was postulated that undermodification of His-tRNA^{His} molecules in *hisT* mutants slows translation of the seven histidine codons contained in the *his* leader transcript and thereby mimics ribosome stalling induced by low concentrations of His-tRNA^{His} molecules in wild-type bacteria moderately starved for histidine (59). Slow translation or ribosome stalling will result in increased transcription readthrough of the *his* attenuator (Fig. 4B). Support for this conjecture has come from kinetic experiments which show that the absence of pseudouridine modifications in *hisT* mutants reduces the general rate of translation elongation by at least 25% and severely reduces translation of mRNA molecules and leader transcripts containing tandem codons (11, 85).

Accumulating evidence suggests that specific undermodifications of tRNA molecules occur in response to cellular stress and may play a regulatory function (see references in Ames et al. [5]). In some genetic backgrounds, starvation for amino acids results in preferential undermodification for pseudouridine residues at positions 38, 39, and 40 in tRNA molecules (64). This state of tRNA undermodification parallels the effect of the *hisT* mutation and may lead to increased *his* operon expression in response to general amino acid deprivation. The relationship between tRNA undermodification induced by environmental stress and expression of operons controlled by attenuation requires further physiological and biochemical study. A related issue concerns the possibility that histidine biosynthesis might be affected by fluctuations in the level of expression of the *hisT* gene itself. Cloning, DNA sequence determination, and gene expression experiments indicate that *hisT* is part of a complex, differentially expressed operon in *E. coli* K-12 (9, 77). Recent genetic experiments show that *pdxB* is located upstream of *hisT* in the operon. This result establishes an unusual link at the transcription and gene organization levels between tRNA pseudouridine modification and vitamin B_6 biosynthesis, which might play an indirect role in the regulation of histidine biosynthesis (P. J. Arps and M. E. Winkler, unpublished data).

hisW

Despite careful physiological characterization, the identity of the *hisW* locus in *S. typhimurium* has remained elusive. Stable RNA accumulation is defective in a cold-sensitive *hisW* mutant at 20°C, which strongly suggests that *hisW* does not encode a tRNA modification enzyme (34). New genetic data show that *gyrA* and *hisW* are linked by greater than 98% in transductional crosses in *S. typhimurium* (K. E. Rudd, personal communication). In addition, a recombinant plasmid that contains the wild-type *gyrA* gene on a 6-kilobase segment of chromosomal DNA complements a cold-sensitive *hisW* mutation (K. E. Rudd and R. Menzel, personal communication). This complementation is abolished when the coding region of the *gyrA* gene is disrupted by insertion of a chloramphenicol resistance "cassette." These results have led to the proposal that *hisW* is equivalent to the *gyrA* gene (K. E. Rudd, personal communication; chapter 87). This proposal is inviting because it is consistent with the pleiotropic phenotypes of *hisW* mutants (25, 34). In particular, the reduction in the total amount of tRNA[His] in *hisW* mutants can be neatly explained if *hisW* is *gyrA*. Because the efficiency of the *hisR* promoter is highly dependent on supercoiling, a defect in DNA gyrase could decrease transcription of the *hisR* gene (K. E. Rudd and L. Bossi, personal communication). Transcription from the *hisP1* promoter also seems to be strongly dependent on supercoiling (see above). However, a defect in supercoiling in the bacterial nucleoid appears to have a greater effect on the *hisR* promoter than on the *hisP1* promoter, since normal *his* operon expression is restored in a *hisW* (*gyrA*) mutant by an episome carrying a single copy of the *hisR* gene (25). Ongoing experiments should definitively show whether *hisW* and *gyrA* are the same gene in *S. typhimurium* (Rudd, personal communication).

hisU

There are at least two subclasses of *hisU* mutations in *S. typhimurium*. One subclass is in the *rnpA* gene, which encodes the protein component of RNase P (18–20). The *hisU* (*rnpA*) mutants are temperature sensitive and contain reduced amounts of mature tRNA[His] at permissive temperatures presumably due to a defect in tRNA processing. The other subclass of *hisU* mutations is probably in the *gyrB* gene (Rudd, personal communication). In *E. coli*, very tight linkage of *dnaA*, *gyrB*, and *rnpA* has been noted (52). Mapping experiments indicate a similar tight linkage of the second *hisU* subclass to *dnaA* and *hisU* (*rnpA*) in *S. typhimurium*. In addition, the pleiotropic phenotypes caused by the second subclass of *hisU* mutations are nearly identical to those of the *hisW* (*gyrA*) mutations (25). Taken together, these preliminary results strongly suggest that this subclass of *hisU* mutations is in the *gyrB* gene. Experiments are now in progress to test this conjecture. Last, it should be noted that some *hisU* mutations do not seem to belong strictly to either subclass and may represent a third *his* regulatory gene in the *hisU* region (Rudd, personal communication).

HISTIDINE BIOSYNTHESIS AS A GENETIC TOOL

The detailed information about histidine biosynthesis described in this chapter forms a powerful system for the analysis of aspects of cellular physiology and genetics. Well-defined deletions in the *his* operon have been used to map the distribution of transposon insertions (65; see chapter 62) and to study formation of chromosomal duplications and inversions (8, 93). Strains containing mutations in the *his* operon were chosen as testers for the detection of mutagens and carcinogens (2, 3) and were used to study frameshift suppression and the influence of codon context on translation (21, 69). The involvement of translation in the attenuation mechanism has led to the identification of the *his* regulatory loci. By basing selections on *his* operon expression, novel classes of mutations have been identified in the *hisR*, *hisW* (*gyrA*), and *hisU* (*gyrB*) genes. In addition, it has been possible to find subclasses of rifampin-resistant mutations and streptomycin-resistant mutations that cause decreased transcription termination at the *his* attenuator (Roth, personal communication). Finally, the involvement of ppGpp in *his* operon expression has led to the identification of new types of mutations that affect *his* operon metabolic regulation (R. Shand and S. W. Artz, Abstr. Annu. Meet. Am. Soc. Microbiol. 1985, H199, p. 141) and the metabolism of ppGpp (91; see chapter 87). Taken together, these examples demonstrate the scope and potential of the histidine biosynthetic system for analyzing fundamental problems in cellular metabolism.

CONCLUSION

Combined genetic, biochemical, and molecular biological approaches have yielded the wealth of information about histidine biosynthesis in *E. coli* and *S. typhimurium* presented in this chapter. In all major respects, histidine biosynthesis seems to be the same in these two closely related enteric species. Analysis of the histidine biosynthetic pathway and the *his* operon has also led to many insights about cellular regulation that go far beyond the topic of histidine biosynthesis. A number of general conclusions, models, and genetic approaches based on knowledge about histidine biosynthesis can be found above. Ongoing experiments aimed at further understanding the mechanisms of control of histidine biosynthesis will undoubtedly continue in this tradition.

ACKNOWLEDGMENTS

I thank S. W. Artz, L. Bossi, C. B. Bruni, J. R. Roth, and K. E. Rudd for unpublished data and conclusions used in this review. I also thank J. R. Roth and K. E. Rudd for helpful discussions, P. J. Arps, S. W. Artz, C. B. Bruni, P. E. Hartman, K. E. Rudd, and P. V. Schoenlein for reading the manuscript, and I. Neal and M. Johnson for help with preparation of the manuscript. I thank my mentors, P. E. Hartman and C. Yanofsky, for the opportunity of working in their laboratories.

Work in my laboratory is supported by grant DMB-8417005 from the National Science Foundation.

LITERATURE CITED

1. **Albritton, W. L., and A. P. Levin.** 1970. Some comparative kinetic data on the enzyme imidazoleacetol phosphate:L-glutamate aminotransferase derived from mutant strains of *Salmonella typhimurium.* J. Biol. Chem. **245:**2525–2528.
2. **Ames, B. N.** 1972. A bacterial system for detecting mutagens and carcinomas, p. 57–66. *In* H. E. Sutton and M. I. Harris (ed.), Mutagenic effects of environmental contaminants. Fogerty International Center Proceedings, no. 10. Academic Press, Inc., New York.
3. **Ames, B. N.** 1979. Identifying environmental chemicals causing mutations and cancer. Science **204:**587–593.
4. **Ames, B. N., and P. E. Hartman.** 1974. The histidine operon of *Salmonella typhimurium,* p. 223–235. *In* R. C. King (ed.), Handbook of genetics, vol. 1. Plenum Publishing Corp., New York.
5. **Ames, B. N., T. H. Tsang, M. Buck, and M. F. Christman.** 1983. The leader mRNA of the histidine attenuator region resembles tRNA^His: possible general regulatory implications. Proc. Natl. Acad. Sci. USA **80:**5240–5242.
6. **Ames, G. F.** 1964. Uptake of amino acids by *Salmonella typhimurium.* Arch. Biochem. Biophys. **104:**1–18.
7. **Ames, G. F.** 1972. Components of histidine transport, p. 409–426. *In* C. F. Fox (ed.), Membrane research. Academic Press, Inc., New York.
8. **Anderson, R. P., C. G. Miller, and J. R. Roth.** 1976. Tandem duplication of the histidine operon observed following generalized transduction in *Salmonella typhimurium.* J. Mol. Biol. **105:**201–218.
9. **Arps, P. J., C. C. Marvel, B. C. Rubin, D. A. Tolan, E. E. Penhoet, and M. E. Winkler.** 1985. Structural features of the *hisT* operon of *Escherichia coli* K-12. Nucleic Acids Res. **13:**5297–5315.
10. **Artz, S. W., and J. R. Broach.** 1975. Histidine regulation in *Salmonella typhimurium:* an activator-attenuator model of gene regulation. Proc. Natl. Acad. Sci. USA **72:**3453–3457.
11. **Artz, S. W., and D. Holzschu.** 1983. Histidine biosynthesis and its regulation, p. 379–404. *In* K. M. Herrmann and R. L. Somerville (ed.), Amino acids: biosynthesis and genetic regulation. Addison-Wesley Publishing Co., Reading, Mass.
12. **Artz, S., D. Holzschu, P. Blum, and R. Shand.** 1983. Use of M13mp phages to study gene regulation, structure, and function: cloning and recombinational analysis of genes of the *Salmonella typhimurium* histidine operon. Gene **26:**147–158.
13. **Barnes, W. M.** 1978. DNA sequence from the histidine operon control region: seven histidine codons in a row. Proc. Natl. Acad. Sci. USA **75:**4281–4285.
14. **Barnes, W. M., and E. Tuley.** 1983. DNA sequence changes of mutations in the histidine operon region that decrease attenuation. J. Mol. Biol. **165:**443–459.
15. **Berberich, M. A., J. S. Kovach, and R. F. Goldberger.** 1967. Chain initiation in a polycistronic message: sequential versus simultaneous derepression of the enzymes for histidine biosynthesis in *Salmonella typhimurium.* Proc. Natl. Acad. Sci. USA **57:**1857–1864.
16. **Blasi, F., and C. B. Bruni.** 1981. Regulation of the histidine operon: translation-controlled transcription termination (a mechanism common to several biosynthetic operons). Curr. Top. Cell. Regul. **19:**1–45.
17. **Bochner, B. R., and B. N. Ames.** 1982. Complete analysis of cellular nucleotides by two-dimensional thin layer chromatography. J. Biol. Chem. **257:**9759–9769.
18. **Bossi, L.** 1983. The *hisR* locus of *Salmonella:* nucleotide sequence and expression. Mol. Gen. Genet. **192:**163–170.
19. **Bossi, L., M. S. Ciampi, and R. Cortese.** 1978. Characterization of a *Salmonella typhimurium hisU* mutant defective in tRNA precursor processing. J. Bacteriol. **134:**612–620.
20. **Bossi, L., and R. Cortese.** 1977. Biosynthesis of tRNA in histidine regulatory mutants of *Salmonella typhimurium.* Nucleic Acids Res. **4:**1945–1956.
21. **Bossi, L., and J. R. Roth.** 1980. The influence of codon context on genetic code translation. Nature (London) **286:**123–127.
22. **Bossi, L., and D. M. Smith.** 1984. Conformational change in the DNA associated with an unusual promoter mutation in a tRNA operon of *Salmonella.* Cell **39:**643–652.
23. **Brady, D. R., and L. L. Houston.** 1972. New assay for histidinol phosphate phosphatase using a coupled reaction. Anal. Biochem. **48:**480–482.
24. **Brady, D. R., and L. L. Houston.** 1973. Some properties of the catalytic sites of imidazoleglycerol phosphate dehydratase-histidinol phosphate phosphatase, a bifunctional enzyme from *Salmonella typhimurium.* J. Biol. Chem. **248:**2588–2592.
25. **Brenner, M., and B. N. Ames.** 1971. The histidine operon and its regulation, p. 349–387. *In* H. J. Vogel (ed.), Metabolic pathways, vol. 5: Metabolic regulation. Academic Press, Inc., New York.
26. **Brenner, M., F. DeLorenzo, and B. N. Ames.** 1970. Energy charge and protein synthesis. Control of aminoacyl transfer ribonucleic acid synthesis. J. Biol. Chem. **245:**450–452.
27. **Brenner, M., J. A. Lewis, D. S. Straus, F. DeLorenzo, and B. N. Ames.** 1972. Histidine regulation in *Salmonella typhimurium.* XIV. Interaction of the histidyl transfer ribonucleic acid synthetase with histidine transfer ribonucleic acid. J. Biol. Chem. **247:**4333–4339.
28. **Bruni, C. B., A. M. Musti, R. Frunzio, and F. Blasi.** 1980. Structural and physiological studies of the *Escherichia coli* histidine operon inserted into plasmid vectors. J. Bacteriol. **142:**32–42.
29. **Burger, E., H. Gorish, and F. Lingens.** 1979. The catalytically active form of histidinol dehydrogenase from *Salmonella typhimurium.* Biochem. J. **181:**771–774.
30. **Carlomagno, M. S., F. Blasi, and C. B. Bruni.** 1983. Gene organization in the distal part of the *Salmonella typhimurium* histidine operon and determination and sequence of the operon transcription terminator. Mol. Gen. Genet. **191:**413–420.
31. **Carlomagno, M. S., A. Riccio, and C. B. Bruni.** 1985. Convergently functional, rho-independent terminator in *Salmonella typhimurium.* J. Bacteriol. **163:**362–368.
32. **Ciesla, Z., F. Salvatore, J. R. Broach, S. W. Artz, and B. N. Ames.** 1975. Histidine regulation in *Salmonella typhimurium.* XVI. A sensitive radiochemical assay for histidinol dehydrogenase. Anal. Biochem. **63:**44–55.
33. **Cortese, R., H. O. Kammen, S. J. Spengler, and B. N. Ames.** 1974. Biosynthesis of pseudouridine in transfer ribonucleic acid. J. Biol. Chem. **249:**1103–1108.
34. **Davis, L., and L. S. Williams.** 1982. Characterization of a cold-sensitive *hisW* mutant of *Salmonella typhimurium.* J. Bacteriol. **151:**867–878.
35. **DeLorenzo, F., and B. N. Ames.** 1970. Histidine regulation in *Salmonella typhimurium.* VII. Purification and general properties of the histidyl transfer ribonucleic acid synthetase. J. Biol. Chem. **245:**1710–1716.
36. **DeLorenzo, F., P. Di Natale, and A. N. Schechter.** 1974. Chemical and physical studies on the structure of the histidyl transfer ribonucleic acid synthetase from *Salmonella typhimurium.* J. Biol. Chem. **249:**908–913.
37. **DeLorenzo, F., D. S. Straus, and B. N. Ames.** 1972. Histidine regulation in *Salmonella typhimurium.* X. Kinetic studies of mutant histidyl transfer ribonucleic acid synthetases. J. Biol. Chem. **247:**2302–2307.
38. **Di Natale, P., and F. DeLorenzo.** 1974. The pyrophosphate exchange reaction in histidyl-tRNA synthetase from *Salmonella typhimurium:* reaction parameters and inhibition by transfer ribonucleic acid. FEBS Lett. **46:**175–179.
39. **Di Natale, P., A. N. Schechter, G. C. Lepore, and F. DeLorenzo.** 1976. Histidyl transfer ribonucleic acid synthetase from *Salmonella typhimurium.* Interaction with substrates and ATP analogues. Eur. J. Biochem. **62:**293–298.
40. **Di Nocera, P. P., F. Blasi, R. Di Lauro, R. Frunzio, and C. B. Bruni.** 1978. Nucleotide sequence of the attenuator region of the histidine operon of *Escherichia coli* K-12. Proc. Natl. Acad. Sci. USA **75:**4276–4280.
41. **Eisenbeis, S. J., and J. Parker.** 1982. The nucleotide sequence of the promoter region of *hisS,* the structural gene for histidyl-tRNA synthetase. Gene **18:**107–114.

42. **Ely, B.** 1974. Physiological studies of *Salmonella* histidine operator-promoter mutants. Genetics **78**:593–606.

43. **Ely, B., and Z. Ciesla.** 1974. Internal promoter P2 of the histidine operon of *Salmonella typhimurium*. J. Bacteriol. **120**:984–986.

44. **Fink, G. R., and R. G. Martin.** 1967. Polarity in the histidine operon. II. J. Mol. Biol. **30**:97–107.

45. **Fishman, S. E., K. R. Kerchief, and J. Parker.** 1979. Specialized lambda transducing bacteriophage which carries *hisS*, the structural gene for histidyl-transfer ribonucleic acid synthetase. J. Bacteriol. **139**:404–410.

46. **Freedman, R., B. Gibson, D. Donovan, K. Biemann, S. Eisenbeis, J. Parker, and P. Schimmel.** 1985. Primary structure of histidine-tRNA synthetase and characterization of *hisS* transcripts. J. Biol. Chem. **260**:10063–10068.

47. **Freedman, R., and P. Schimmel.** 1981. In vitro transcription of the histidine operon. J. Biol. Chem. **256**:10747–10750.

48. **Frunzio, R., C. B. Bruni, and F. Blasi.** 1981. *In vivo* and *in vitro* detection of the leader RNA of the histidine operon of *Escherichia coli* K-12. Proc. Natl. Acad. Sci. USA **78**:2767–2771.

49. **Goldberger, R. F.** 1974. Autogenous regulation of the gene expression. Science **183**:810–816.

50. **Grisolia, V., M. S. Carlomagno, A. G. Nappo, and C. B. Bruni.** 1985. Cloning, structure, and expression of the *Escherichia coli* K-12 *hisC* gene. J. Bacteriol. **164**:1317–1323.

51. **Grisolia, V., A. Riccio, and C. B. Bruni.** 1983. Structure and function of the internal promoter (*hisBp*) of the *Escherichia coli* K-12 histidine operon. J. Bacteriol. **155**:1288–1296.

52. **Hansen, E. B., T. Atlung, F. G. Hansen, O. Skovgaard, and K. von Meyenburg.** 1984. Fine structure genetic map and complementation analysis of mutations in the *dnaA* gene of *Escherichia coli*. Mol. Gen. Genet. **196**:387–396.

53. **Hartman, P. E., Z. Hartman, R. C. Stahl, and B. N. Ames.** 1971. Classification and mapping of spontaneous and induced mutations in the histidine operon of *Salmonella*. Adv. Genet. **16**:1–34.

54. **Henderson, G. B., and E. E. Snell.** 1973. Crystalline L-histidinol phosphate aminotransferase from *Salmonella typhimurium*. Purification and subunit structure. J. Biol. Chem. **248**:1906–1911.

55. **Higgins, C. F., G. F.-L. Ames, W. M. Barnes, J. M. Clement, and M. Hofnung.** 1982. A novel intercistronic regulatory element of prokaryotic operons. Nature (London) **298**:760–762.

56. **Hoppe, I., H. M. Johnston, D. Biek, and J. R. Roth.** 1979. A refined map of the *hisG* gene of *Salmonella typhimurium*. Genetics **92**:17–26.

57. **Horowitz, N. H.** 1965. The evolution of biochemical syntheses—retrospect and prospect, p. 15–23. *In* V. Bryson and H. J. Vogel (ed.), Evolving genes and proteins. Academic Press, Inc., New York.

58. **Hsu, L. M., H. J. Klee, J. Zagorski, and M. J. Fournier.** 1984. Structure of an *Escherichia coli* tRNA operon containing linked genes for arginine, histidine, leucine, and proline tRNA. J. Bacteriol. **158**:934–942.

59. **Johnston, H. M., W. M. Barnes, F. G. Chumley, L. Bossi, and J. R. Roth.** 1980. Model for regulation of the histidine operon of *Salmonella*. Proc. Natl. Acad. Sci. USA **77**:508–512.

60. **Johnston, H. M., and J. R. Roth.** 1979. Histidine mutants requiring adenine: selection of mutants with reduced *hisG* expression in *Salmonella typhimurium*. Genetics **92**:1–15.

61. **Johnston, H. M., and J. R. Roth.** 1981. DNA sequence changes of mutations altering attenuator control of the histidine operon of *Salmonella typhimurium*. J. Mol. Biol. **145**:735–756.

62. **Kalousek, F., and W. H. Konigsberg.** 1974. Purification and characterization of histidyl transfer ribonucleic acid synthetase of *Escherichia coli*. Biochemistry **13**:999–1006.

63. **Kasai, T.** 1974. Regulation of the expression of the histidine operon in *Salmonella typhimurium*. Nature (London) **249**:523–527.

64. **Kitchingman, G. R., and M. J. Fournier.** 1977. Modification-deficient transfer ribonucleic acids from relaxed control *Escherichia coli*: structure of the major undermodified phenylalanine and leucine transfer RNAs produced during leucine starvation. Biochemistry **16**:2213–2220.

65. **Kleckner, N., J. R. Roth, and D. Botstein.** 1977. Genetic engineering *in vivo* using translocatable drug-resistance elements. New methods in bacterial genetics. J. Mol. Biol. **116**:125–159.

66. **Kleeman, J., and S. M. Parsons.** 1975. A sensitive assay for the reverse reaction of the first histidine biosynthetic enzyme. Anal. Biochem. **68**:236–241.

67. **Kleeman, J. E., and S. M. Parsons.** 1977. Inhibition of histidyl-tRNA-adenosine triphosphate phosphoribosyltransferase complex formation by histidine and by guanosine tetraphosphate. Proc. Natl. Acad. Sci. USA **74**:1535–1537.

68. **Klungsoyr, L., J. H. Hagemen, L. Fall, and D. E. Atkinson.** 1968. Interaction between energy charge and product feedback in regulation of biosynthetic enzymes. Biochemistry **7**:4035–4040.

69. **Kohno, T., L. Bossi, and J. R. Roth.** 1983. New suppressors of frameshift mutations in *Salmonella typhimurium*. Genetics **103**:23–29.

70. **Kohno, T., and W. R. Gray.** 1981. Chemical and genetic studies on L-histidinol dehydrogenase of *Salmonella typhimurium*: isolation and structure of tryptic peptides. J. Mol. Biol. **147**:451–464.

71. **Kronenberg, H. M., T. Vogel, and R. F. Goldberger.** 1975. A new and highly sensitive assay for the ATP phosphoribosyltransferase that catalyzes the first step of histidine biosynthesis. Anal. Biochem. **65**:380–388.

72. **Lewis, J. A., and B. N. Ames.** 1972. Histidine regulation in *Salmonella typhimurium*. XI. The percentage of transfer RNA^{His} charged *in vivo* and its relation to the repression of the histidine operon. J. Mol. Biol. **66**:131–142.

73. **Loper, J. C., and E. Adams.** 1965. Purification and properties of histidinol dehydrogenase from *Salmonella typhimurium*. J. Biol. Chem. **240**:788–795.

74. **Martin, R. G.** 1963. The first enzyme in histidine biosynthesis: the nature of feedback inhibition by histidine. J. Biol. Chem. **238**:257–268.

75. **Martin, R. G.** 1963. The one operon-one messenger theory of transcription. Cold Spring Harbor Symp. Quant. Biol. **28**:357–361.

76. **Martin, R. G., M. A. Berberich, B. N. Ames, W. W. Davis, R. F. Goldberger, and J. D. Yourno.** 1971. Enzymes and intermediates of histidine biosynthesis in *Salmonella typhimurium*. Methods Enzymol. **17B**:3–44.

77. **Marvel, C. C., P. J. Arps, B. C. Rubin, H. O. Kammen, E. E. Penhoet, and M. E. Winkler.** 1985. *hisT* is part of a multigene operon in *Escherichia coli* K-12. J. Bacteriol. **161**:60–71.

78. **McGinnis, E., and L. S. Williams.** 1972. Regulation of histidyl-transfer ribonucleic acid synthetase formation in a histidyl-transfer ribonucleic acid synthetase mutant of *Salmonella typhimurium*. J. Bacteriol. **111**:739–744.

79. **Metzler, D. E.** 1977. Biochemistry—the chemical reactions of living cells. Academic Press, Inc., New York.

80. **Meyers, M., M. Levinthal, and R. F. Goldberger.** 1975. *trans*-Recessive mutation in the first structural gene of the histidine operon that results in constitutive expression of the operon. J. Bacteriol. **124**:1227–1235.

81. **Morton, D. P., and S. M. Parsons.** 1976. Biosynthetic direction substrate kinetics and product inhibition studies on the first enzyme of histidine biosynthesis, adenosine triphosphate phosphoribosyltransferase. Arch. Biochem. Biophys. **175**:677–686.

82. **Morton, D. P., and S. M. Parsons.** 1977. Synergistic inhibition of ATP phosphoribosyltransferase by guanosine tetraphosphate and histidine. Biochem. Biophys. Res. Commun. **74**:172–177.

83. **Morton, D. P., and S. M. Parsons.** 1977. Inhibition of ATP phosphoribosyltransferase by AMP and ADP in the absence and presence of histidine. Arch. Biochem. Biophys. **181**:643–648.

84. **O'Donovan, G. A., and J. L. Ingraham.** 1965. Cold-sensitive mutants of *Escherichia coli* resulting from increased feedback inhibition. Proc. Natl. Acad. Sci. USA **54**:451–457.

85. **Palmer, D. T., P. H. Blum, and S. W. Artz.** 1983. Effects of the *hisT* mutation of *Salmonella typhimurium* on translation elongation rate. J. Bacteriol. **153**:357–363.

86. **Parsons, S. M., and D. E. Koshland, Jr.** 1974. A rapid isolation of phosphoribosyladenosine triphosphate synthetase and comparison to native enzyme. J. Biol. Chem. **249**:4104–4109.

87. **Piszkiewicz, D., B. E. Tilley, T. Randmeir, and S. M. Parsons.** 1979. Amino acid sequence of ATP phosphoribosyltransferase of *Salmonella typhimurium*. Proc. Natl. Acad. Sci. USA **76**:1589–1592.

88. **Riccio, A., C. Bruni, M. Rosenberg, M. Gottesman, K. McKenney, and F. Blasi.** 1985. Regulation of single and multicopy *his* operons of *Escherichia coli*. J. Bacteriol. **163**:1172–1179.

89. **Riggs, D., and S. Artz.** 1984. The *hisD-hisC* gene border of the *Salmonella typhimurium* histidine operon. Mol. Gen. Genet. **196**:526–529.

90. **Roth, J. R., D. N. Anton, and P. E. Hartman.** 1966. Histidine regulatory mutants in *Salmonella typhimurium*. I. Isolation and general properties. J. Mol. Biol. **22**:305–323.

91. **Rudd, K. E., B. R. Bochner, M. Cashel, and J. R. Roth.** 1985. Mutations in the *spoT* gene of *Salmonella typhimurium*: effects on *his* operon expression. J. Bacteriol. **163**:534–542.

92. **Schmid, M. B., and J. R. Roth.** 1983. Internal promoters of the *his* operon in *Salmonella typhimurium*. J. Bacteriol. **153**:1114–1119.

93. **Schmid, M. B., and J. R. Roth.** 1983. Genetic methods for analysis and manipulation of inversion mutations in bacteria. Genetics **105**:517–537.

94. **Scott, J. F., J. R. Roth, and S. W. Artz.** 1975. Regulation of

histidine operon does not require *hisG* enzyme. Proc. Natl. Acad. Sci. USA **72:**5021–5025.

95. **Sheppard, D. E.** 1964. Mutants of *Salmonella typhimurium* resistant to feedback inhibition by L-histidine. Genetics **50:**611–623.

96. **Singer, C. E., G. R. Smith, R. Cortese, and B. N. Ames.** 1972. Mutant tRNA^His ineffective in repression and lacking two pseudouridine modifications. Nature (London) New Biol. **238:**72–74.

97. **Smith, D. W. E., and B. N. Ames.** 1964. Intermediates in the early steps of histidine biosynthesis. J. Biol. Chem. **239:**1848–1855.

98. **Staples, M. A., and L. L. Houston.** 1979. Proteolytic degradation of imidazole-glycerolphosphate dehydratase-histidinol phosphatase from *Salmonella typhimurium* and isolation of a resistant bifunctional core enzyme. J. Biol. Chem. **254:**1395–1401.

99. **Stephens, J. C., S. W. Artz, and B. N. Ames.** 1975. Guanosine 5'-diphosphate 3'diphosphate (ppGpp): positive effector for histidine operon transcription and general signal for amino-acid deficiency. Proc. Natl. Acad. Sci. USA **72:**4389–4393.

100. **Sterboul, C. C., J. E. Kleeman, and S. M. Parsons.** 1977. Purification and characterization of a mutant ATP phosphoribosyl-transferase hypersensitive to histidine feedback inhibition. Arch. Biochem. Biophys. **181:**632–642.

101. **Stern, M. J., G. F.-L. Ames, N. H. Smith, E. C. Robinson, and C. F. Higgins.** 1984. Repetitive extragenic palindromic sequences: a major component of the bacterial genome. Cell **37:**1015–1026.

102. **Turnbough, C. L., R. J. Neill, R. Landsberg, and B. N. Ames.** 1979. Pseudouridylation of tRNAs and its role in regulation in *Salmonella typhimurium*. J. Biol. Chem. **254:**5111–5119.

103. **Verde, P., R. Frunzio, P. P. De Nocera, F. Blasi, and C. B. Bruni.** 1981. Identification, nucleotide sequence and expression of the regulatory region of the histidine operon of *Escherichia coli* K-12. Nucleic Acids Res. **9:**2075–2086.

104. **White, M. N., J. Olszowy, and R. L. Switzer.** 1971. Regulation and mechanism of phosphoribosylpyrophosphate synthetase: repression by end products. J. Bacteriol. **108:**122–131.

105. **Whitfield, H. J., Jr., D. L. Gutnick, N. Margolies, R. G. Martin, M. M. Rechler, and M. J. Voll.** 1970. Relative translation frequencies of the cistrons of the histidine operon. J. Mol. Biol. **49:**245–249.

106. **Winkler, M. E., D. J. Roth, and P. E. Hartman.** 1978. Promoter- and attenuator-related metabolic regulation of the *Salmonella typhimurium* histidine operon. J. Bacteriol. **133:**830–843.

107. **Winkler, M. E., R. V. Zawodny, and P. E. Hartman.** 1979. Mutation *spoT* of *Escherichia coli* increases expression of the histidine operon deleted for the attenuator. J. Bacteriol. **139:**993–1000.

108. **Yanofsky, C.** 1984. Comparison of regulatory and structural regions of genes of tryptophan metabolism. Mol. Biol. Evol. **1:**143–161.

109. **Yanofsky, C., T. Platt, I. P. Crawford, B. P. Nichols, G. E. Christie, H. Horowitz, M. Van Cleemput, and A. M. Wu.** 1981. The complete nucleotide sequence of the tryptophan operon of *Escherichia coli*. Nucleic Acids Res. **9:**6647–6668.

26. Biosynthesis of Serine and Glycine

GEORGE V. STAUFFER

Department of Microbiology, University of Iowa, Iowa City, Iowa 52242

INTRODUCTION

Serine, glycine, and one-carbon (C_1) biosynthesis constitutes a major metabolic pathway. Serine is used in cysteine and tryptophan biosynthesis, and 3-phosphoserine, an intermediate in serine biosynthesis, is a possible precursor of pyridoxine (4, 5, 39, 46). Glycine is incorporated into purines and heme-containing compounds (19, 35). C_1 units are required for purine, thymine, histidine, and methionine biosynthesis and for the formyl group of fMet-tRNA$_f$ (2). In addition, methionine is converted to *S*-adenosylmethionine (SAM), a C_1 donor used in numerous other methylation reactions (28). It is estimated that 15% of the carbon assimilated from glucose passes through the serine-glycine pathway (35). Thus, this pathway has a central function in cell metabolism and its regulation is important and complex.

The serine-glycine pathway in *Escherichia coli* and *Salmonella typhimurium* is shown in Fig. 1 (24, 32, 58, 59). The synthesis of serine requires three primary enzymes, 3-phosphoglycerate dehydrogenase, 3-phosphoserine aminotransferase, and 3-phosphoserine phosphatase. The initial reaction is the oxidation of 3-phosphoglycerate to 3-phosphohydroxypyruvate, followed by transamination to 3-phosphoserine and dephosphorylation to serine. The interconversion of serine and glycine occurs by a single enzyme, serine hydroxymethyltransferase (44, 48). This reaction also produces 5,10-methylenetetrahydrofolate (5,10-mTHF), an important contributor of C_1 units in cell metabolism (2, 28). The final reaction is the oxidative cleavage of glycine to NH_3, CO_2, and 5,10-mTHF. This reaction was first described by Sagers and Gunsalus (43) in *Diplococcus glycinophilus*. This enzyme system is not well characterized in either *E. coli* or *S. typhimurium*, but in *D. glycinophilus* the pathway consists of four protein components designated P_1, P_2, P_3, and P_4 (20). Since at least one protein from *E. coli* is capable of interacting with the P_1 protein from *D. glycinophilus*, the two systems may be very similar (D. K. Ransom and R. D. Sagers, Abstr. Annu. Meet.

Am. Soc. Microbiol. 1974, P266, p. 189). The combination of the serine hydroxymethyltransferase reaction and the glycine cleavage (GCV) enzyme system converts both the α and β carbons of serine into 5,10-mTHF and enables their entry into the C_1 pool.

Mutations that produce an enzymatic deficiency in the biosynthetic pathway to serine cause a nutritional requirement satisfied by either serine or glycine (32, 59), whereas mutations that cause a loss of serine hydroxymethyltransferase activity produce a nutritional requirement for glycine that is not satisfied by serine (33). These results provide evidence that serine is the normal precursor of glycine. Single mutations that result in a requirement satisfied by either serine or glycine occur frequently, suggesting that more than one pathway for serine and glycine biosynthesis is unlikely (58, 59). Several observations, however, point to another pathway for serine and glycine biosynthesis in both *E. coli* and *S. typhimurium* that does not utilize the currently recognized enzymes. *E. coli* and *S. typhimurium* serine-glycine auxotrophs grow slowly without a serine or glycine supplement provided that threonine is supplied to the growth medium (25, 33; R. Dalal and J. S. Gots, Bacteriol. Proc., p. 89, 1965) and grow well if leucine, arginine, lysine, threonine, and methionine are supplied to the growth medium (P. Ravnikar and R. Somerville, personal communication). In addition, pseudorevertants have been isolated in serine-glycine auxotrophs that can grow on unsupplemented glucose minimal medium (Ravnikar and Somerville, personal communication). Evidence suggests that threonine is cleaved to glycine and acetyl coenzyme A and that glycine then serves as the precursor of serine via the serine hydroxymethyltransferase reaction (12, 27; Ravnikar and Somerville, personal communication).

FEEDBACK INHIBITION OF SERINE BIOSYNTHESIS

Regulation of serine biosynthesis differs from regulation of other amino acids. Inhibition of the first

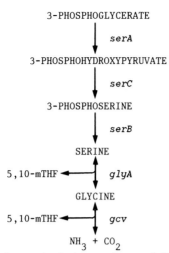

3-PHOSPHOGLYCERATE

$\downarrow$ serA

3-PHOSPHOHYDROXYPYRUVATE

$\downarrow$ serC

3-PHOSPHOSERINE

$\downarrow$ serB

SERINE

5,10-mTHF $\longleftarrow$ $\downarrow$ glyA

GLYCINE

5,10-mTHF $\longleftarrow$ $\downarrow$ gcv

$NH_3 + CO_2$

FIG. 1. Serine and glycine pathway of E. coli and S. typhimurium. The gene designations are as follows: serA, 3-phosphoglycerate dehydrogenase; serC, 3-phosphoserine aminotransferase; serB, 3-phosphoserine phosphatase; glyA, serine hydroxymethyltransferase; gcv, glycine cleavage enzymes.

enzyme in the pathway, 3-phosphoglycerate dehydrogenase, by L-serine is the major form of control (32, 53, 56, 59). L-Serine is a noncompetitive inhibitor of enzyme activity, with 50% inhibition occurring at a concentration of 4×10^{-5} M. The only other amino acids that inhibit enzyme activity are D-serine and glycine, with 50% inhibition occurring at concentrations of 8.5×10^{-4} and 4.8×10^{-3} M, respectively (32).

The enzyme has a molecular weight of about 163,000 and consists of four identical subunits (41, 61). Serine inhibition occurs by an allosteric process involving a conformational change in the enzyme (9, 54). The titration curve for serine is sigmoidal, indicating positive cooperativity in binding (54). Fluorescence quenching titrations indicate that two serine molecules are necessary to inhibit enzyme activity 80 to 90%, while saturating amounts inhibit more than 99% of the activity (10). The actual number of serine binding sites has not been unambiguously determined. Glycine inhibition also occurs by an allosteric process (9). It is not known whether serine and glycine bind to the enzyme at the same or different sites. 3-Phosphoserine phosphatase is inhibited by L-serine, but inhibition is at least 1,000 times less sensitive than inhibition of 3-phosphoglycerate dehydrogenase (32, 59). There is no evidence for inhibition of 3-phosphoserine aminotransferase activity.

Serine hydroxamate is a competitive inhibitor of seryl-tRNA synthetase. A mutant resistant to this analog has a 3-phosphoglycerate dehydrogenase that is resistant to L-serine inhibition and overproduces serine (56). Thus, inhibition of 3-phosphoglycerate dehydrogenase activity by L-serine is a major control point of serine and glycine biosynthesis.

FEEDBACK INHIBITION OF GLYCINE BIOSYNTHESIS

Inhibition of serine hydroxymethyltransferase activity could not be demonstrated in S. typhimurium

when a number of compounds were tested (D-serine, α-methylserine, L-methionine, L-threonine, SAM, purines, and thymine) (G. Stauffer, Ph.D. thesis, The Pennsylvania State University, University Park, 1976). Methionine and SAM also have no effect on serine hydroxymethyltransferase activity in E. coli (22, 45). These results provide evidence against a feedback inhibition effect on the activity of this enzyme.

REGULATION OF FORMATION OF THE SERINE BIOSYNTHETIC ENZYMES

The enzymes of serine biosynthesis appear to be constitutively synthesized in both E. coli and S. typhimurium (23, 32, 59). Their levels are not decreased by the addition of an excess of serine to the growth medium, nor increased by starvation of cells for serine. The substrate, 3-phosphoglycerate, does not induce 3-phosphoglycerate dehydrogenase synthesis (J. C. McKitrick and L. I. Pizer, Abstr. Annu. Meet. Am. Soc. Microbiol. 1973, P141, p. 164). Thus, the enzymes are not regulated by a conventional induction-repression mechanism.

E. coli mutants have been isolated having either high or low seryl-tRNA synthetase activity compared to the level found in the wild-type organism (34, 56). The activities of the enzymes for serine biosynthesis are comparable in the mutants and the wild-type organism. Furthermore, the presence of L-serine hydroxamate in the growth medium, which inhibits charging of seryl-tRNA, does not increase 3-phosphoglycerate dehydrogenase synthesis (23, 56). If seryl-tRNA or seryl-tRNA synthetase were involved in regulation, altered levels of the serine biosynthetic enzymes would be expected under these conditions.

The 3-phosphoglycerate dehydrogenase levels in E. coli are reduced by growth on different carbon sources that give a fourfold range in generation times (23). There appears to be an inverse relationship between generation time and 3-phosphoglycerate dehydrogenase levels, although exceptions are observed. Thus, although the growth rate influences the enzyme levels, it is not the only factor involved. 3-Phosphoglycerate dehydrogenase appears to be the critical enzyme in the serine biosynthetic pathway as the levels of 3-phosphoserine aminotransferase and 3-phosphoserine phosphatase remain constant under similar growth conditions (35).

3-Phosphoglycerate dehydrogenase levels are reduced to 10% of the level found in cells grown in unsupplemented glucose minimal medium by growth in the presence of amino acids not directly related to serine biosynthesis (threonine, methionine, leucine, isoleucine) plus lactate as the carbon source (23, 35). In cells with such marked differences in this enzyme activity, the concentration of serine remains constant (23). Thus, in vivo the regulation of 3-phosphoglycerate dehydrogenase activity by the endogenous serine pool rather than the amount of the biosynthetic enzymes appears to be the major factor controlling intracellular serine levels. It is also possible that the pathway that does not involve 3-phosphoglycerate dehydrogenase produces serine under this growth condition (Ravnikar and Somerville, personal communication).

Exogenous cyclic AMP at 3 mM reduces 3-phosphoglycerate dehydrogenase levels 20% in E. coli K-12 grown on glucose, and the effect is abolished in a strain defective in the cyclic AMP binding protein (23). In addition, an adenylate cyclase mutant grown on glucose has higher levels of the enzyme than the wild-type strain. It is possible that enzyme levels are modulated in part by cyclic AMP acting in a negative manner to reduce enzyme synthesis.

REGULATION OF FORMATION OF THE GLYCINE BIOSYNTHETIC ENZYMES

The enzyme serine hydroxymethyltransferase is responsible for the synthesis of both glycine and 5,10-mTHF from serine (Fig. 1). This reaction is the cell's only source of glycine (33) and its major source of C_1 units (28). Thus, one might expect that regulation of the synthesis of this enzyme would be complex and respond to different environmental conditions. In *S. typhimurium* and *E. coli*, several products of C_1 metabolism (serine, glycine, methionine, purines, thymine) repress enzyme synthesis (11, 15, 22, 26, 47, 55). Purines contribute a considerable proportion of the repressive effect of the total mixture (7). It has also been reported that the intracellular concentration of glycine, alone or together with variations in the concentration of C_1 intermediates, regulates serine hydroxymethyltransferase synthesis in *E. coli* (26). However, an *S. typhimurium glyA* mutant with 40% of the normal serine hydroxymethyltransferase level will not grow in glucose minimal medium supplemented with serine, methionine, adenine, guanine, and thymine because of repression of this enzyme to levels insufficient for glycine synthesis (47, 48). Thus, repression occurs even though a condition of glycine limitation is produced. This observation provides evidence that glycine is not solely responsible for the regulation of the synthesis of serine hydroxymethyltransferase in *S. typhimurium*.

Mutants with an altered glycyl-tRNA synthetase (*glyS* mutants) have been isolated (11). When deprived of glycine, the level of charged tRNA^{Gly} drops to 4% of the level in a control strain, but serine hydroxymethyltransferase levels remain unchanged. Also, when a temperature-sensitive *glyS* mutant is shifted to the nonpermissive temperature, levels of the enzyme remain unchanged. These studies indicate that serine hydroxymethyltransferase is not controlled by the functional state of glycyl-tRNA synthetase or its product glycyl-tRNA, but do not rule out the possibility that a minor species of glycyl-tRNA is involved in regulation.

Significant derepression of serine hydroxymethyltransferase is observed when *metE* and *metF* mutants of *E. coli* (15, 22, 45) and *S. typhimurium* (L. Stauffer and G. Stauffer, unpublished data) are grown on D-methionine sulfoxide as a methionine source. The derepression is not prevented by the addition of glycine, adenosine, guanosine, and thymidine to the growth medium. Thus, the increase is not due to a limitation of these compounds. Similar results have been observed in *E. coli metE* and *metF* mutants grown in methionine-limited chemostat cultures, or in an *E. coli metE* mutant grown in a vitamin B_{12}-limited chemostat culture (7). Only marginal increases in serine hydroxymethyltransferase levels are observed when *metA* and *metB* mutants of *E. coli* (15) and *S. typhimurium* (Stauffer and Stauffer, unpublished data) are grown on D-methionine sulfoxide or in methionine-limited chemostat cultures (7). These results support a role for methionine in the overall production of glycine and C_1 units by regulating serine hydroxymethyltransferase synthesis. Furthermore, it appears to be the methionine requirement for methylation reactions rather than that for protein synthesis that controls synthesis of this enzyme.

When the intracellular concentrations of sulfur-containing amino acids and nucleosides are measured in methionine-limited chemostat cultures, no correlation is found between the absolute intracellular concentration of any single metabolite and the rate of serine hydroxymethyltransferase synthesis (8). A high correlation has been observed between the ratio of homocysteine to SAM and the rate of serine hydroxymethyltransferase synthesis. Thus, homocysteine may act as an inducer and SAM as a corepressor for synthesis of this enzyme, both competing for binding to a repressor molecule. The hypothetical repressor molecule has not been identified.

Purine limitation also results in significant derepression of serine hydroxymethyltransferase in *E. coli* and *S. typhimurium* (7, 26, 47, 51). In purine-limited chemostat cultures levels of this enzyme are not altered by methionine addition, and in methionine-limited chemostat cultures enzyme levels are not altered by purine addition, indicating that, during purine and methionine limitation, a common corepressor or repressor is affected (7).

Purine limitation is of interest since it was proposed that 5-amino 4-imidazole carboxamide riboside 5'-triphosphate (ZTP) is a presumptive alarmone for C_1-folate deficiency (3). ZTP is formed from 5-amino 4-imidazole carboxamide riboside 5'-monophosphate, an intermediate in purine biosynthesis. If ZTP acts as a true alarmone to signal C_1-folate deficiency, its metabolism and its relationship to folate metabolism may be important and related to regulation of the *glyA* gene.

Significant derepression of serine hydroxymethyltransferase is observed when an *E. coli* serine-glycine auxotroph is grown in a serine-limited chemostat culture (7). The addition of purines to a serine-limited chemostat culture reduces levels of the enzyme, but the rate of its synthesis is still twofold higher than the rate observed without serine limitation. These results suggest a role for serine in the regulation of serine hydroxymethyltransferase synthesis, but the control mechanisms that operate during serine- and purine-limited growth may be functioning independently of each other.

A final growth condition that leads to derepression of serine hydroxymethyltransferase in *E. coli* and *S. typhimurium* is the addition of trimethoprim, an inhibitor of dihydrofolate reductase (7, 47, 51). In *E. coli*, the increase in serine hydroxymethyltransferase levels in trimethoprim-supplemented medium can be overcome by the addition of adenosine, serine, methionine, thymidine, and glycine to the growth medium, but not by the addition of these compounds without adenosine (7). This observation suggests that it is the limitation of the supply of purines in trimethoprim medium that results in derepression of enzyme synthesis.

As can be seen from the above discussion, the synthesis of serine hydroxymethyltransferase in *E. coli* and *S. typhimurium* responds to the changing demands for purines, folates, serine, and methionine. These studies did not definitively establish the component that serves as the regulatory signal, nor the regulatory mechanism involved.

Two classes of methionine regulatory mutants (*metK* and *metJ* mutants) have altered regulation of serine hydroxymethyltransferase synthesis. The *metK* gene codes for SAM synthetase (1, 17, 21, 49). The *metJ* gene codes for the aporepressor for the methionine pathway (1, 14, 18, 21). The current model for regulation of the methionine pathway is that SAM interacts with the aporepressor to form an activated repressor complex. The holorepressor then acts on each of the *met* genes, whose operators have different affinities for the holorepressor (42). In *E. coli*, *metK* mutations, but not *metJ* mutations, result in elevated serine hydroxymethyltransferase levels (15, 24). In both *E. coli* and *S. typhimurium*, *metK* and *metJ* mutations prevent complete repression of serine hydroxymethyltransferase synthesis (15, 24, 49). In addition, presumptive *metK* mutants were isolated by a procedure that selects for mutants with elevated levels of serine hydroxymethyltransferase (49). These results indicate that this enzyme is partially regulated as a methionine enzyme and that an altered SAM pool and the *metJ* gene product are part of a regulatory signal for both the methionine pathway and the *glyA* gene. It is worth noting, however, that there are differences between the regulatory system for the *glyA* gene and that for the *met* genes. First, *metJ* mutants are constitutive for the *met* genes in both minimal and supplemented media. In a *metJ* mutant, the *glyA* gene is not constitutive in minimal medium, but is not repressed in supplemented medium (15, 49). The nonrepressibility of *glyA* in *metJ* mutants indicates that the *metJ* gene product is a necessary component of the repression system, but its absence does not signal derepression. Second, prolonged methionine-limited growth of either *metE*, *metF*, or *metA* mutants causes derepression of the *met* genes, but derepression of the *glyA* gene occurs only in the *metE* and *metF* mutants (7, 15). Finally, *metK* mutants are derepressed for serine hydroxymethyltransferase synthesis when cells are grown in minimal medium, but when grown in medium supplemented with glycine, purines, and thymidine, levels of the enzyme are reduced; the methionine biosynthetic enzymes are derepressed under both growth conditions (15). Thus, the methionine component is only one factor in a complicated regulatory mechanism that involves several control elements.

Two additional mutants have been isolated from *S. typhimurium* with elevated levels of serine hydroxymethyltransferase (50). One mutant has sevenfold elevated enzyme levels but shows the normal repression response when grown in supplemented medium. The second mutant has twofold elevated enzyme levels, and enzyme synthesis is not as sensitive to repression when grown in supplemented medium. The new lesions in the mutants are cotransducible with the *glyA* gene and are genetically distinct from the *metK* and the *metJ* mutations that alter the regulation of both the *glyA* gene and the *met* genes. The phenotypes of these two mutants are distinct from each other, suggesting that at least two new control sites exist for the *glyA* gene. The results, however, do not rule out the possibility that both mutations lie in the same gene. Whether these mutations identify a DNA control site or a gene whose product is necessary for the control of enzyme synthesis is unknown.

CLONING AND CHARACTERIZATION OF THE *E. COLI glyA* GENE

Serine hydroxymethyltransferase was partially purified from *E. coli* (22). Its molecular weight is about 170,000, estimated by Sephadex G200 filtration. The *glyA* gene has been cloned from *E. coli* and *S. typhimurium* (51, 60). The nucleotide sequence of the *E. coli glyA* gene has been determined and shown to specify a 417-amino-acid-long polypeptide with a calculated M_r of 45,265 (38). Although the enzyme has not been purified to homogeneity, the results indicate that the enzyme is a tetramer of identical subunits.

The interaction of RNA polymerase with DNA from the *glyA* gene control region protects a stretch of 88 base pairs (bp) from DNase I digestion (Fig. 2) (37). The 5′ end of the *glyA* gene mRNA was mapped within this protected region (37). A Pribnow box sequence is located 5 bp upstream of the transcription initiation

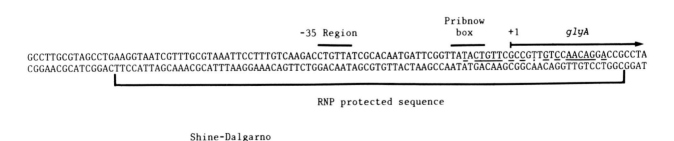

FIG. 2. Nucleotide sequence of the *glyA* gene control region. The bracket indicates the region protected from DNase I digestion by RNA polymerase (RNP). Position +1 indicates the transcription start point. The Pribnow box sequence, −35 sequence, and ribosome binding site are indicated. The 22-bp dyad symmetry segment that could serve as a possible operator is indicated by solid bars between the two strands. The dot indicates the center of the symmetry. The deduced NH₂-terminal amino acid sequence is shown.

site (Fig. 2), with five of six residues identical to the consensus sequence (TATAAT), including the three most conserved residues at positions 1, 2, and 6 (40). In the −35 region of the *glyA* promoter, only three residues are homologous to the consensus −35 RNA polymerase recognition sequence (TTGACA) (40).

There is a 22-bp palindrome centered about the transcription start site and the Pribnow box region (Fig. 2). The presence of sequences with twofold symmetry is a common feature of many procaryotic operators, and some of these have been associated with the binding of regulatory proteins (40). A specific protein that interacts with the *glyA* regulatory region and influences transcription initiation has not been identified, although evidence indicates that such a regulatory molecule exists (8).

A 67-bp leader sequence precedes a possible AUG translation initiation codon for serine hydroxymethyltransferase (Fig. 2) (37). The 67-bp leader sequence does not contain any open reading frames for short leader polypeptides or palindromes that would permit the formation of stable RNA secondary structures. Such sequences have been associated with transcription attenuators (62). In vitro transcription experiments also failed to reveal a terminated *glyA* mRNA transcript (M. Plamann and G. Stauffer, unpublished data). These results make it unlikely that the *glyA* gene is regulated by an attenuation mechanism.

After the coding region there is a 182-nucleotide sequence preceding the proposed transcription termination region for the *glyA* gene (38). The transcription termination region consists of a G-C-rich sequence that could form a stable stem-loop structure once transcribed, followed by an A-T-rich sequence within which transcription appears to terminate (Fig. 3). There is a long region of dyad symmetry and numerous smaller symmetrical regions between the site of translation termination and the proposed transcription termination region (Fig. 3). The longest symmetrical region could form a very stable stem-loop structure once transcribed (ΔG = −62.4 kcal/mol). These symmetrical sequences show remarkable homology with intercistronic repetitive extragenic palindromic

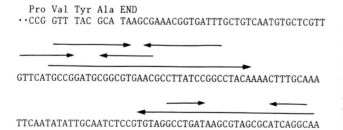

```
Pro Val Tyr Ala END
··CCG GTT TAC GCA TAAGCGAAACGGTGATTTGCTGTCAATGTGCTCGTT


GTTCATGCCGGATGCGGCGTGAACGCCTTATCCGGCCTACAAAACTTTGCAAA


TTCAATATATTGCAATCTCCGTGTAGGCCTGATAAGCGTAGCGCATCAGGCAA


TTTTTCGTTTATGATCATCAAGGCTTCCTTCGGGAAGCCTTTCTACGTTATCG
```
FIG. 3. The 3′ flanking region of the *glyA* gene. The sequence begins 12 bases upstream from the translation termination codon at the end of *glyA* and extends 6 bases beyond the proposed transcription termination region (38). The thin arrows indicate the long and short regions of dyad symmetry, and the thick arrows indicate the proposed transcription termination region. The bar indicates the major 3′ termini of *glyA* mRNA (38).

sequences of other procaryotic operons (13, 16, 38, 52) and may play a role in the regulation of *glyA* gene expression.

A Mu *cts*-generated *glyA* mutant with only 30% of the normal serine hydroxymethyltransferase activity has been isolated (37a). The mutation is *cis* acting, and mapping studies show that Mu *cts* has inserted within the nontranslated distal region of the *glyA* gene, about 35 bases after the translation stop codon. These results show that the nontranslated sequence distal to the *glyA* structural gene is important in maintaining high serine hydroxymethyltransferase levels, but say nothing about the mechanism involved. It is possible that the lower level of this enzyme activity is due to decreased stability of the upstream *glyA* mRNA, but it is unknown whether it is the repetitive extragenic palindromic sequence, the transcription termination signal, or both, that stabilizes *glyA* mRNA.

REGULATION OF THE GCV PATHWAY

It has been estimated that about 60 to 75% of the cell's total need of C_1 units can be accounted for by the conversion of serine to glycine and 5,10-mTHF (6, 57). An additional way of generating C_1 units is by the glycine cleavage (GCV) pathway (43). Such an enzyme system, inducible by exogenous glycine, has been demonstrated in both *E. coli* (24, 36) and *S. typhimurium* (L. Stauffer, M. Plamann, and G. Stauffer, unpublished data). Serine auxotrophs can meet their serine requirement if supplied with exogenous glycine (33). The GCV reaction accounts for the formation of the C_1 units required for the synthesis of serine from glycine under this growth condition. Consistent with this hypothesis is the observation that the loss of the GCV pathway in mutants blocked in serine biosynthesis leads to their inability to use glycine as a serine source (30, 36; F. L. Pizer, J. C. McKitrick, J. L. Stern, D. Share, and L. I. Pizer, Abstr. Annu. Meet. Am. Soc. Microbiol. 1977, K205, p. 220).

Under normal conditions the GCV system acts to balance the cell's glycine and C_1 unit requirements. If the cell's C_1 requirements are greater than its glycine requirements, some of the glycine made via the serine hydroxymethyltransferase reaction is converted to additional C_1 units. Mutants of both *E. coli* and *S. typhimurium* with no detectable GCV activity excrete glycine, consistent with this hypothesis (36; Plamann and Stauffer, unpublished data). Interestingly, the GCV enzymes have only one-tenth the activity of serine hydroxymethyltransferase, so the question arises as to whether glycine is the only other source of C_1 units in *E. coli* and *S. typhimurium*. Mutants grown under conditions where the supply of C_1 units from serine is blocked obtained only 46 to 89% of their C_1 units from glycine (31, 33), and strains defective in both the serine hydroxymethyltransferase reaction and the GCV enzyme system still have a source of C_1 units (30). Other evidence suggests that serine and glycine are the only sources of C_1 units in the folate pathway of *E. coli* (6). Additional studies will be necessary to confirm whether or not other sources of C_1 units exist. The low activity of the GCV enzymes, however, may be necessary to insure that sufficient glycine levels are maintained for other pathways that utilize this amino acid (protein synthesis, purine synthesis, etc.).

There is some disagreement over the question of whether the synthesis of the GCV enzymes is repressed by C_1 compounds. Growth of serine-glycine auxotrophs of *E. coli* in a medium containing serine and the purine base xanthine results in a temporary inability of the cells to grow upon transfer to a medium containing only glycine, presumably owing to the inability of cells to obtain C_1 units from glycine (29). Thus, formation of the GCV enzymes may be controlled by repression exerted by an excess of C_1 units. However, the presence of glycine, serine, and xanthine in the medium allows growth immediately upon transfer to medium containing only glycine, an observation suggesting that glycine induces the GCV enzyme system rather than C_1 units repressing the enzyme system. This idea is supported by the further observation that induction of the GCV enzyme system is not blocked by the addition of adenine, methionine, and SAM to the growth medium (24). However, in wild-type strains of *E. coli* and *S. typhimurium*, the addition of inosine to the growth medium results in a three- and an eightfold reduction of the glycine-induced GCV enzyme levels, respectively (Stauffer and Stauffer, unpublished data). Thus, the differences reported may be due to the type of C_1 compounds added to the growth medium. These results support a role for a purine component in the regulation of the synthesis of the GCV enzymes. Whether induction by glycine and repression by inosine are part of the same regulatory mechanism or function independently is unknown. Additional studies will be necessary to define the inducible character of the GCV system, to determine its significance in the regulation of serine, glycine, and C_1 unit biosynthesis, and to analyze the mechanism by which glycine and inosine exert their regulatory effects.

ACKNOWLEDGMENT

The studies from my laboratory, cited in this review, were supported by Public Health Service grant GM-26878 from the National Institute of General Medical Sciences.

LITERATURE CITED

1. **Ahmed, A.** 1973. Mechanism of repression of methionine biosynthesis in *Escherichia coli*. Mol. Gen. Genet. **123**:299–324.
2. **Blakley, R. L.** 1969. The biochemistry of folic acid and related pteridines. Elsevier/North-Holland Publishing Co., Amsterdam.
3. **Bochner, B. R., and B. N. Ames.** 1982. ZTP (5-amino 4-imidazole carboxamide riboside 5'-triphosphate): a proposed alarmone for 10-formyl-tetrahydrofolate deficiency. Cell **29**:929–937.
4. **Dempsey, W. B.** 1969. Evidence that 3-phosphoserine may be a precursor of vitamin B_6 in *Escherichia coli*. Biochem. Biophys. Res. Commun. **37**:89–93.
5. **Dempsey, W. B.** 1969. 3-Phosphoserine transaminase mutants of *Escherichia coli* B. J. Bacteriol. **100**:1114–1115.
6. **Dev, I. K., and R. J. Harvey.** 1982. Sources of one-carbon units in the folate pathway of *Escherichia coli*. J. Biol. Chem. **257**:1980–1986.
7. **Dev, I. K., and R. J. Harvey.** 1984. Regulation of synthesis of serine hydroxymethyltransferase in chemostat cultures of *Escherichia coli*. J. Biol. Chem. **259**:8394–8401.
8. **Dev, I. K., and R. J. Harvey.** 1984. Role of methionine in the regulation of the synthesis of serine hydroxymethyltransferase in *Escherichia coli*. J. Biol. Chem. **259**:8402–8406.
9. **Dubrow, R., and L. I. Pizer.** 1977. Transient kinetic studies on the allosteric transition of phosphoglycerate dehydrogenase. J. Biol. Chem. **252**:1527–1538.
10. **Dubrow, R., and L. I. Pizer.** 1977. Transient kinetic and deuterium isotope effect studies on the catalytic mechanism of phosphoglycerate dehydrogenase. J. Biol. Chem. **252**:1539–1551.
11. **Folk, W. R., and P. Berg.** 1970. Isolation and partial characterization of *Escherichia coli* mutants with altered glycyl transfer ribonucleic acid synthetases. J. Bacteriol. **102**:193–203.
12. **Fraser, J., and E. B. Newman.** 1975. Derivation of glycine from threonine in *Escherichia coli* K-12 mutants. J. Bacteriol. **122**:810–817.
13. **Gilson, E., J.-M. Clement, D. Brutlag, and M. Hofnung.** 1984. A family of dispersed repetitive extragenic palindromic DNA sequences in *E. coli*. EMBO J. **3**:1417–1422.
14. **Greene, R. C., J. S. V. Hunter, and E. H. Coch.** 1973. Properties of *metK* mutants of *Escherichia coli* K-12. J. Bacteriol. **115**:57–67.
15. **Greene, R. C., and C. Radovich.** 1975. Role of methionine in the regulation of serine hydroxymethyltransferase in *Escherichia coli*. J. Bacteriol. **124**:269–278.
16. **Higgins, C. F., C. F.-L. Ames, W. M. Barnes, J. M. Clement, and M. Hofnung.** 1982. A novel intercistronic regulatory element of prokaryotic operons. Nature (London) **298**:760–762.
17. **Hobson, A. C.** 1974. The regulation of methionine and S-adenosylmethionine biosynthesis and utilization in mutants of *Salmonella typhimurium* with defects in S-adenosylmethionine synthetase. Mol. Gen. Genet. **131**:263–273.
18. **Hobson, A. C., and D. A. Smith.** 1973. S-adenosylmethionine synthetase in methionine regulatory mutants of *Salmonella typhimurium*. Mol. Gen. Genet. **126**:7–18.
19. **Jordan, P. M., and D. Shemin.** 1972. δ-Aminolevulinic acid synthetase, p. 339–356. *In* P. D. Boyer (ed.), The enzymes, vol. VII. Academic Press, Inc., New York.
20. **Klein, S. M., and R. D. Sagers.** 1967. Glycine metabolism. III. A flavin-linked dehydrogenase associated with the glycine cleavage system in *Peptococcus glycinophilus*. J. Biol. Chem. **242**:297–300.
21. **Lawrence, D. A.** 1972. Regulation of the methionine feedback-sensitive enzyme in mutants of *Salmonella typhimurium*. J. Bacteriol. **109**:8–11.
22. **Mansouri, A., J. B. Decter, and R. Silber.** 1972. Studies on the regulation of one-carbon metabolism. II. Repression-derepression of serine hydroxymethyltransferase by methionine in *Escherichia coli* 113-3. J. Biol. Chem. **247**:348–352.
23. **McKitrick, J. C., and L. I. Pizer.** 1980. Regulation of phosphoglycerate dehydrogenase levels and effect on serine synthesis in *Escherichia coli* K-12. J. Bacteriol. **141**:235–245.
24. **Meedel, T. H., and L. I. Pizer.** 1974. Regulation of one-carbon biosynthesis and utilization in *Escherichia coli*. J. Bacteriol. **118**:905–910.
25. **Meinhart, J. O., and S. Simmonds.** 1955. Serine metabolism in a mutant strain of *Escherichia coli* strain K-12. J. Biol. Chem. **213**:329–341.
26. **Miller, B. A., and E. B. Newman.** 1974. Control of serine transhydroxymethylase synthesis in *Escherichia coli* K-12. Can. J. Microbiol. **20**:41–47.
27. **Miller, D. A., and S. Simmonds.** 1957. The metabolism of L-threonine and glycine by *Escherichia coli*. Proc. Natl. Acad. Sci. USA **43**:195–199.
28. **Mudd, S. H., and G. L. Cantoni.** 1964. Biological transmethylation, methyl-group neogenesis and other "one-carbon" metabolic reactions dependent upon tetrahydrofolic acid, p. 1–47. *In* M. Florkin and E. H. Stotz (ed.), Comprehensive biochemistry, vol. 15. Elsevier, Amsterdam.
29. **Newman, E. B., and B. Magasanik.** 1963. The relation of serine-glycine metabolism to the formation of single-carbon units. Biochim. Biophys. Acta **78**:437–448.
30. **Newman, E. B., B. Miller, and V. Kapoor.** 1974. Biosynthesis of single-carbon units in *Escherichia coli* K12. Biochim. Biophys. Acta **338**:529–539.
31. **Pitts, J. D., J. A. Stewart, and G. W. Crosbie.** 1961. Observations on glycine metabolism in *Escherichia coli*. Biochim. Biophys. Acta **50**:361–363.
32. **Pizer, L. I.** 1963. The pathway and control of serine biosynthesis in *Escherichia coli*. J. Biol. Chem. **238**:3934–3944.
33. **Pizer, L. I.** 1965. Glycine synthesis and metabolism in *Escherichia coli*. J. Bacteriol. **89**:1145–1150.
34. **Pizer, L. I., J. McKitrick, and T. Tosa.** 1972. Characterization of a mutant of *E. coli* with elevated levels of seryl-tRNA synthetase. Biochem. Biophys. Res. Commun. **49**:1351–1357.
35. **Pizer, L. I., and M. L. Potochny.** 1964. Nutritional and regulatory aspects of serine metabolism in *Escherichia coli*. J. Bacteriol. **88**:611–619.
36. **Plamann, M. D., W. D. Rapp, and G. V. Stauffer.** 1983. *Escherichia coli* K12 mutants defective in the glycine cleavage enzyme system. Mol. Gen. Genet. **192**:15–20.
37. **Plamann, M. D., and G. V. Stauffer.** 1983. Characterization of the *Escherichia coli* gene for serine hydroxymethyltransferase. Gene **22**:9–18.
37a. **Plamann, M. D., and G. V. Stauffer.** 1985. Characterization of a

cis-acting regulatory mutation that maps at the distal end of the *Escherichia coli glyA* gene. J. Bacteriol. **161**:650–654.
38. **Plamann, M. D., L. T. Stauffer, M. L. Urbanowski, and G. V. Stauffer.** 1983. Complete nucleotide sequence of the *E. coli glyA* gene. Nucleic Acids Res. **11**:2065–2075.
39. **Qureshi, M. A., D. A. Smith, and A. J. Kingsman.** 1975. Mutants of *Salmonella typhimurium* responding to cysteine or methionine: their nature and possible role in the regulation of cysteine biosynthesis. J. Gen. Microbiol. **89**:353–370.
40. **Rosenberg, M., and D. Court.** 1979. Regulatory sequences involved in the promotion and termination of RNA transcription. Annu. Rev. Genet. **13**:319–353.
41. **Rosenbloom, J., E. Sugimoto, and L. I. Pizer.** 1968. The mechanism of end product inhibition of serine biosynthesis. III. Physical and chemical properties of phosphoglycerate dehydrogenase. J. Biol. Chem. **243**:2099–2107.
42. **Rowbury, R. J.** 1983. Methionine biosynthesis and its regulation, p. 191–211. *In* K. M. Herrmann and R. L. Somerville (ed.), Amino acids: biosynthesis and genetic regulation. Addison-Wesley Publishing Co., Reading, Mass.
43. **Sagers, R. D., and I. C. Gunsalus.** 1961. Intermediary metabolism of *Diplococcus glycinophilus*. I. Glycine cleavage and one-carbon interconversions. J. Bacteriol. **81**:541–549.
44. **Scrimgeour, K. G., and F. M. Huennekens.** 1962. Serine hydroxymethylase. Methods Enzymol. **5**:838–843.
45. **Silber, R., and A. Mansouri.** 1971. Regulation of folate-dependent enzymes. Ann. N.Y. Acad. Sci. **186**:55–69.
46. **Smith, O. H., and C. Yanofsky.** 1962. Enzymes involved in the biosynthesis of tryptophan. Methods Enzymol. **5**:794–806.
47. **Stauffer, G. V., C. A. Baker, and J. E. Brenchley.** 1974. Regulation of serine transhydroxymethylase activity in *Salmonella typhimurium*. J. Bacteriol. **120**:1017–1025.
48. **Stauffer, G. V., and J. E. Brenchley.** 1974. Evidence for the involvement of serine transhydroxymethylase in serine and glycine interconversions in *Salmonella typhimurium*. Genetics **77**:185–198.
49. **Stauffer, G. V., and J. E. Brenchley.** 1977. Influence of methionine biosynthesis on serine transhydroxymethylase regulation in *Salmonella typhimurium* LT2. J. Bacteriol. **129**:740–749.
50. **Stauffer, G. V., and J. E. Brenchley.** 1978. Selection of *Salmonella*

typhimurium mutants with altered serine transhydroxymethylase regulation. Genetics **88**:221–233.
51. **Stauffer, G. V., M. D. Plamann, and L. T. Stauffer.** 1981. Construction and expression of hybrid plasmids containing the *Escherichia coli glyA* gene. Gene **14**:63–72.
52. **Stern, M. J., G. F.-L. Ames, N. H. Smith, E. C. Robinson, and C. F. Higgins.** 1984. Repetitive extragenic palindromic sequences: a major component of the bacterial genome. Cell **37**:1015–1026.
53. **Sugimoto, E., and L. I. Pizer.** 1968. The mechanism of end product inhibition of serine biosynthesis. I. Purification and kinetics of phosphoglycerate dehydrogenase. J. Biol. Chem. **243**:2081–2089.
54. **Sugimoto, E., and L. I. Pizer.** 1968. The mechanism of end product inhibition of serine biosynthesis. II. Optical studies of phosphoglycerate dehydrogenase. J. Biol. Chem. **243**:2090–2098.
55. **Taylor, R. T., H. Dickerman, and H. Weissbach.** 1966. Control of one-carbon metabolism in a methionine-B$_{12}$ auxotroph of *Escherichia coli*. Arch. Biochem. Biophys. **117**:405–412.
56. **Tosa, T., and L. I. Pizer.** 1971. Biochemical bases for the antimetabolite action of L-serine hydroxamate. J. Bacteriol. **106**:972–982.
57. **Umbarger, H. E.** 1977. A one-semester project for the immersion of graduate students in metabolic pathways. Biochemical Educ. **5**:67–71.
58. **Umbarger, H. E., and M. A. Umbarger.** 1962. The biosynthetic pathway of serine in *Salmonella typhimurium*. Biochim. Biophys. Acta **62**:193–195.
59. **Umbarger, H. E., M. A. Umbarger, and P. M. L. Siu.** 1963. Biosynthesis of serine in *Escherichia coli* and *Salmonella typhimurium*. J. Bacteriol. **85**:1431–1439.
60. **Urbanowski, M. L., M. D. Plamann, L. T. Stauffer, and G. V. Stauffer.** 1984. Cloning and characterization of the gene for *Salmonella typhimurium* serine hydroxymethyltransferase. Gene **27**:47–54.
61. **Winicov, I., and L. I. Pizer.** 1974. The mechanism of end product inhibition of serine biosynthesis. IV. Subunit structure of phosphoglycerate dehydrogenase and steady state kinetic studies of phosphoglycerate oxidation. J. Biol. Chem. **249**:1348–1355.
62. **Yanofsky, C.** 1981. Attenuation in the control of expression of bacterial operons. Nature (London) **289**:751–758.

27. Biosynthesis of Cysteine

NICHOLAS M. KREDICH

Department of Medicine, Division of Rheumatic and Genetic Diseases, Duke University Medical Center, Durham, North Carolina 27710

INTRODUCTION

Cysteine biosynthesis in *Salmonella typhimurium* and *Escherichia coli* provides an amino acid necessary for the synthesis of proteins, glutathione, methionine, and other sulfur-containing metabolites. It also represents the means by which inorganic sulfur is reduced and "fixed" into organic linkage, a process that is analogous to nitrate reduction and ammonia fixation. Sulfate reduction in these two organisms is "assimilatory," i.e., produces only the amount required for the biosynthesis of sulfur amino acids, and differs in several ways from "dissimilatory" reduction, which is carried out by some anaerobic organisms as part of a respiratory pathway that utilizes sulfate as a terminal electron acceptor (70).

In an environment containing sulfide, cysteine biosynthesis is relatively simple, requiring only conversion of serine to *O*-acetylserine, which then reacts with sulfide. Growth on oxidized forms of inorganic sulfur, e.g., sulfate, calls into play a more complicated set of reactions involved in sulfur uptake and reduction to sulfide. Therefore, much of the machinery of cysteine biosynthesis is dedicated to sulfide synthesis, and the overall pathway is a convergent one, with sulfate uptake and reduction comprising one arm and *O*-acetylserine synthesis the other (Fig. 1). Regulation is achieved through feedback inhibition of *O*-acetylserine synthesis by cysteine and by a system of positive genetic control termed the "cysteine regulon."

In addition to the direct incorporation of sulfide into cysteine, two other pathways of sulfur fixation have been described in *S. typhimurium*. Thiosulfate can react with *O*-acetylserine to give an *S*-sulfonate derivative, which is subsequently reduced to cysteine (Fig. 1) (60). Given an exogenous source of thiosulfate, this pathway is efficient enough to sustain a normal growth rate (61). In addition, cystathionine γ-synthase catalyzes the reaction of *O*-succinylhomoserine and

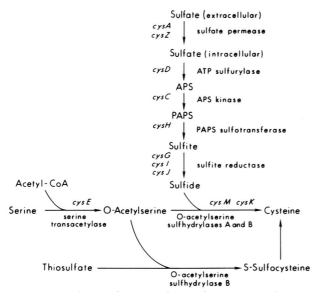

FIG. 1. Pathway of cysteine biosynthesis in *S. typhimurium* and *E. coli*. *O*-Acetylserine sulfhydrylase B and the synthesis of *S*-sulfocysteine from *O*-acetylserine and thiosulfate have only been shown in *S. typhimurium* but are presumed to occur in *E. coli* as well.

sulfide to give homocysteine directly (27), thus bypassing the transsulfuration intermediate cystathionine. Although of importance in plants, this reaction is very slow in *S. typhimurium* and probably of little physiologic significance.

Sulfur Requirements

Wild-type strains grow on a variety of sulfur compounds including sulfate, thiosulfate, sulfite, cysteine sulfinic acid, cysteine, cystine, glutathione, and djenkolic acid. Methionine will satisfy only about one-half the sulfur requirement by sparing the conversion of cysteine to methionine, but cannot serve as sole sulfur source because there is no efficient pathway from methionine to cysteine in *S. typhimurium* and *E. coli* (as there is in fungi, plants, and other higher organisms). Measurements of sulfur uptake and total cellular content indicate that growth on liquid medium requires about 70 μM sulfate or its equivalent to achieve a density of 10^9 cells per ml (49). Sulfur-free medium E (83), prepared by substituting $MgCl_2$ for $MgSO_4$, supports growth to 10% of this density and presumably contains 7 μM sulfate from trace contaminants in the reagent-grade chemicals used in its preparation. The sulfur content of agar or of highly purified agarose precludes obtaining a sulfur-deficient solid medium with these agents.

Cystine Transport

Cysteine transport has not been characterized in detail and is complicated by rapid oxidation of this amino acid to cystine. Cystine enters *S. typhimurium* by three different systems designated CTS-1, CTS-2, and CTS-3 (3). CTS-1 is a saturable transport system with a K_m of 2 μM and a V_{max} in sulfur-limited cultures of 9 nmol/min per mg of cell protein. It is part of the cysteine regulon and is repressed by growth on 0.5 mM L-cystine to a V_{max} of less than 0.8 nmol/min per mg of cell protein. Inhibition of CTS-1 by 2,4-dinitrophenol, by a combination of α-methylglucoside and azide, and by osmotic shock implies the existence of a periplasmic binding protein, as has been described in *E. coli* (see below).

CTS-2 is a higher affinity system with a K_m of 0.1 μM, but has a V_{max} of only 0.3 nmol/min per mg of cell protein. It is not known to be a part of the cysteine regulon. CTS-3 is unsaturable, with a capacity of 0.04 nmol/min per mg of cell protein per μM cystine, and probably represents passive diffusion.

Cultures lacking CTS-1 owing to regulatory or structural gene mutations grow well on 0.5 mM cystine. The capacities of CTS-2 and CTS-3 at 0.04 mM cystine are sufficient for the sulfur requirement of 3.6 nmol of sulfur per min per mg of protein in cells growing with a doubling time of 50 min (49). CTS-1 provides an advantage only at limiting concentrations of cystine and, when derepressed, gives this rate of uptake at 0.5 μM cystine. Regulatory gene mutations causing high-level constitutive expression of CTS-1 lead to cystine sensitivity, which is probably due to excessive accumulation of this amino acid and its reduction to cysteine (19).

E. coli has two cystine transport systems, one termed "general" because of its activity with diaminopimelic acid and cystine analogs, and the other with greater specificity for cystine (7). Both are associated with osmotic shock-releasable, cystine-binding proteins. It is not known how these transport systems are regulated, but loss of the general system and increased activity of the specific system with growth on rich medium suggest that only the former belongs to the cysteine regulon.

BIOSYNTHETIC PATHWAY

Sulfate Transport

Both *E. coli* and *S. typhimurium* have an efficient sulfate permease which also transports thiosulfate. Kinetic studies of the *S. typhimurium* system show an apparent K_m value for sulfate of 36 μM with an intra- to extracellular concentration ratio of 9.1 at 100 μM extracellular sulfate (21). Sulfite, selenate, chromate, and molybdate inhibit sulfate transport, presumably by serving as substrates for sulfate permease (67). Mutations causing loss of sulfate permease activity lie in the *cysA* region in both *S. typhimurium* and *E. coli*. In addition, in *E. coli* a second locus, designated *cysZ* and situated about 10 kilobases (kb) from *cysA*, is required for sulfate uptake (68).

The molecular characteristics of sulfate permease have not been defined, but a periplasmic sulfate-binding protein has been purified and studied in detail (66). This 32,000-dalton protein binds one molecule of sulfate with a K_d of approximately 0.1 μM. Chromate competes for binding, but thiosulfate and molybdate are virtually unreactive with purified protein. Since sulfate permease-deficient mutants are easily isolated by selecting for chromate resistance, the failure of a systematic effort to find any that specifically lack sulfate-binding protein (63) suggests that this protein is not essential for chromate transport. A role for sulfate-binding protein is inferred because its expression, as well as that of sulfate permease, is controlled in parallel with the cysteine regulon (see below). If sulfate-binding protein participates in sulfate transport, its role may be only as a nonessential enhancer of permease efficiency, or perhaps chromate and sulfate transport differ with respect to their requirements for this protein.

Both sulfite and sulfide are excellent sulfur sources in wild type, but little is known about their uptake. Sulfite may be transported by sulfate permease but must enter the cell by other means as well since *cysA* mutants can utilize sulfite. Sulfide probably diffuses into cells by the same mechanism as water.

Sulfate Activation

Sulfate reduction requires initial "activation" to a phosphosulfate, mixed anhydride. In plants and in microbes that carry out dissimilatory sulfate reduction, this activation is a one-step process in which ATP sulfurylase catalyzes the reaction of ATP and sulfate to give pyrophosphate and adenosine 5'-phosphosulfate (APS), which in turn is reduced to either sulfite or a sulfite derivative (75). In *S. typhimurium* and *E. coli* APS itself cannot be reduced and instead is phosphorylated by ATP to give 3'-phosphoadenosine 5'-phosphosulfate (PAPS) in a reaction catalyzed by APS kinase. ATP sulfurylase and APS kinase are specified by *cysD* and *cysC*, which in *S. typhimurium* are closely linked and probably contiguous (20).

PAPS is a high-energy sulfate donor for sulfation reactions in higher organisms but appears not be be used as such in *S. typhimurium* and *E. coli*. Since the K_{eq} of the reversible ATP sulfurylase reaction is extremely low in the forward direction (ca. 10^{-8} M) (72), it follows that pyrophosphatase or significant APS kinase activity, or both, are required to move sulfur past this point in the pathway.

PAPS Reduction

PAPS reduction in *E. coli* (and presumably in *S. typhimurium*) is a two-step process analogous to that occurring in yeasts and plants and in microorganisms that reduce APS directly (75). The enzyme once termed PAPS reductase and specified by *cysH* (22) is now known to be a PAPS-specific sulfotransferase, which transfers the sulfonyl moiety of PAPS to a thiol acceptor to form an acceptor-$S-SO_3^{-1}$ derivative (81). In *E. coli* it appears that thioredoxin serves as the thiol acceptor (82). The immediate fate of thioredoxin-$S-SO_3^{-1}$ is not known, but free sulfite is an eventual product and may result from displacement by a thiol such as glutathione or by the second sulfhydryl group of thioredoxin-$S-SO_3^{-1}$ itself.

It has been suggested (81) that thioredoxin-$S-SO_3^{-1}$ may be reduced by a thiosulfonate reductase to the level of thioredoxin hydrodisulfide (R-S-SH), which could then release free sulfide or serve directly as a sulfide donor for *O*-acetylserine sulfhydrylase. The in vivo significance of such a pathway in *S. typhimurium* and *E. coli* is doubtful, however, since it would eliminate the necessity for sulfite reductase, and it is clear that mutants lacking this enzyme will not grow on sulfate.

Sulfite Reductase

Since mutants lacking sulfite reductase accumulate free sulfite when grown in the presence of sulfate (22) and can utilize neither sulfate nor sulfite for cysteine biosynthesis, sulfite is believed to be an obligatory intermediate of sulfate reduction in *E. coli* and *S. typhimurium*.

The sulfite reductase of *E. coli* is a complex enzyme which carries out the six-electron reduction of sulfite to sulfide without release of intermediates (80). It is composed of two different polypeptides, termed α and β, with molecular weights of about 60,000 and 55,000, and has a subunit structure of $\alpha_8\beta_4$ with a molecular weight of 670,000 (77). Treatment with urea separates holoenzyme into a flavoprotein with subunit structure α_8 and a hemoprotein existing as a monomeric β polypeptide. The α_8 flavoprotein contains four FAD and four FMN and has NADPH-cytochrome *c* reductase activity. The β hemoprotein contains four nonheme iron/sulfide moieties which exist as an Fe_4S_4 cluster and one novel heme group termed siroheme (58, 59), and has reduced methyl viologen-sulfite reductase activity. The separated proteins can be combined in vitro to reconstitute holoenzyme with full NADPH-sulfite reductase activity. Electron flow is believed to proceed as follows: NADPH $\rightarrow$ FAD $\rightarrow$ FMN $\rightarrow$ Fe_4S_4 $\rightarrow$ siroheme $\rightarrow$ sulfite (78). Neither free flavoprotein nor free hemoprotein is found in wild-type extracts, but in *S. typhimurium* mutants lacking the ability to synthesize either one, the other can be demonstrated (79).

The α and β polypeptides are specified by *cysJ* and *cysI*, which are contiguous. The *S. typhimurium cysJ* and *cysI* genes have been sequenced (J. Ostrowski and N. Kredich, unpublished data), and the polypeptides predicted from these sequences have molecular weights and amino acid compositions close to those found by analyses of α and β polypeptides from *E. coli* (77). *cysG* mutants lack sulfite reductase activity and are deficient in nitrite reductase activity (13) and in cobalamin synthesis as well (39). Since both sulfite reductase and nitrite reductase contain siroheme and cobalamin synthesis requires porphyrin ring methylations similar to those that distinguish siroheme, *cysG* is believed to specify an enzyme, probably a methylase, necessary for siroheme and cobalamin synthesis.

Synthesis of *O*-Acetylserine

In *E. coli* and *S. typhimurium*, as well as in plants and other microorganisms, it is clear that *O*-acetylserine, rather than serine, is the immediate precursor of the carbon moiety of cysteine (51). Serine transacetylase catalyzes the synthesis of *O*-acetylserine from serine and acetyl coenzyme A (acetyl-CoA) and is present in crude extracts as a multifunctional enzyme complex called cysteine synthase (48). The other component of the complex is *O*-acetylserine sulfhydrylase A, an enzyme that synthesizes cysteine from *O*-acetylserine and sulfide. The complex has a molecular weight of 309,000, but aggregates to multimers of twice and four times that size. *O*-Acetylserine at concentrations of 10^{-4} M or greater causes cysteine synthase to dissociate to free serine transacetylase and *O*-acetylserine sulfhydrylase-A, which can be separated and analyzed individually. This dissociation is prevented by sulfide. The stoichiometry of complex dissociation indicates that the nonaggregated form is composed of a single 160,000-dalton molecule of serine transacetylase and two 68,000-dalton molecules of O-acetylserine sulfhydrylase A. The subunit structure of serine transacetylase itself has not been well studied but probably consists of four identical polypeptides. O-Acetylserine sulfhydrylase A is composed of two identical 36,000-dalton subunits and contains 2 mol of pyridixal phosphate (5).

The kinetic properties of serine transacetylase are complex, and in studies on the cysteine synthase complex, apparent K_m values for serine and acetyl-CoA were found to be influenced by effectors such as cysteine, ATP, and acetyl-CoA (17). Much of this behavior appears to be due to the effects of these ligands on the state of complex aggregation. Cysteine is a potent inhibitor of serine transacetylase, with a K_i of 10^{-6} M at 0.1 mM acetyl-CoA for both free and complexed enzyme (48). Thus, cysteine regulates its pathway through feedback inhibition of its carbon skeleton precursor. Furthermore, since O-acetylserine is an internal inducer of the cysteine regulon (see below), inhibition of serine transacetylase provides a mechanism by which cysteine can regulate synthesis of its other precursor, sulfide. Serine transacetylase is specified by cysE (40, 51).

O-Acetylserine Sulfhydrylase

Two O-acetylserine sulfhydrylase enzymes, designated -A and -B, have been described in S. typhimurium (5, 6) and are presumed to occur in E. coli as well. Both catalyze the reaction of O-acetylserine with sulfide to give cysteine; O-acetylserine sulfhydrylase-B can also use thiosulfate in place of sulfide to give cysteine thiosulfonate ($R-S-SO_3^{-1}$) as a product (60).

O-Acetylserine sulfhydrylase-A is specified by cysK (26, 33) and is found not only in association with serine transacetylase as cysteine synthase but free as well (48). Levels of O-acetylserine sulfhydrylase-A vary as much as 50-fold depending on the state of derepression of the cysteine regulon, whereas serine transacetylase levels remain almost constant (47). Even the lowest levels of O-acetylserine sulfhydrylase-A provide twice that required to bind all of serine transacetylase. Thus, wild-type extracts always contain both free and complexed O-acetylserine sulfhydrylase-A, whereas serine transacetylase is present only as the cysteine synthase complex. Free serine transacetylase is found, however, in a mutant that overproduces this enzyme (32) and in mutants lacking O-acetylserine sulfhydrylase-A (33).

The kinetic mechanism of O-acetylserine sulfhydrylase-A is "Bi Bi Ping Pong," with O-acetylserine binding followed by release of acetate and with binding of sulfide followed by release of cysteine (14, 15). Both substrates are inhibitory because they can bind to second sites and generate dead-end inhibition complexes. With uncomplexed enzyme, K_m and K_i values are 0.15 and 47 mM for O-acetylserine and 0.066 and 0.013 mM for sulfide. Cysteine also binds to two separate sites and causes S-parabolic, I-linear inhibition with respect to O-acetylserine and inhibition constants of $K_{is1} = 1.0$ mM, $K_{is2} = 12$ mM, and $K_{ii} = 11$ mM. Azaserine (O-diazoacetyl-L-serine) is almost as good a substrate as O-acetylserine; with sulfide as a cosubstrate it gives cysteine and diazoacetic acid as products (31). 1,2,4-Triazole can replace sulfide in the reaction with O-acetylserine (49) and gives 1,2,4-triazole-1-alanine as a product. These reactions with substrate analogs are useful in selection of various cysteine mutants (see below).

Formation of the cysteine synthase complex alters certain kinetic properties of O-acetylserine sulfhydrylase-A, decreasing the V_{max} by 50% and increasing the apparent K_m for O-acetylserine about fourfold (48). Sulfide kinetics appear to be unaltered. Detailed kinetic analysis of the complex is complicated by the fact that O-acetylserine causes it to dissociate to free O-acetylserine sulfhydrylase and this dissociation is, in turn, prevented by the addition of sulfide either before or with O-acetylserine. The $K_{0.5}$ for O-acetylserine in complex dissociation is 0.02 to 0.04 mM; the effect of sulfide on dissociation has not been carefully determined. Thus, any analysis of the cysteine synthase complex would have to take into account not only the kinetic differences between free and complexed O-acetylserine sulfhydrylase-A, but also the effects of substrates on the equilibrium between the two enzyme forms.

Using a complete, two-step reaction system containing cysteine synthase complex, serine, acetyl-CoA, and sulfide, Cook and Wedding (16) have shown that O-acetylserine arising from the serine transacetylase reaction is released prior to its reaction with complexed O-acetylserine sulfhydrylase-A. Their data did show, however, a decrease in the expected lag time between O-acetylserine synthesis and utilization, an observation suggesting that cysteine synthase provides a kinetic advantage by increasing the O-acetylserine concentration in the near vicinity of complexed O-acetylserine sulfhydrylase-A.

S-Sulfocysteine Synthase

S. typhimurium mutants completely lacking O-acetylserine sulfhydrylase-A are cysteine prototrophs owing to the presence of O-acetylserine sulfhydrylase-B, which is specified by cysM (31). This enzyme has recently been found to catalyze the reaction of O-acetylserine and thiosulfate to form S-sulfocysteine ($R-S-SO_3^{-1}$) and is also known as S-sulfocysteine synthase (60, 61). O-Acetylserine sulfhydrylase-A lacks this activity. Purified S-sulfocysteine synthase has a native molecular weight of 55,000 and contains two identical subunits and 2 mol of pyridoxal phosphate. Kinetic data suggest a sequential mechanism with both thiosulfate and sulfide as substrates. K_m values are 2.7 and 3.6 mM for thiosulfate and sulfide and 9 and 7 mM for O-acetylserine in the S-sulfocysteine synthase and sulfhydrylase reactions, respectively.

S-Sulfocysteine has been identified as an intermediate of cysteine biosynthesis in Aspergillus nidulans (62), but is of doubtful significance in aerobic, sulfate-grown cultures of S. typhimurium and E. coli, where thiosulfate has not been identified as an intermediate in sulfate reduction. S-Sulfocysteine synthase is of importance in utilization of exogenous thiosulfate, however, and cysM cysK+ mutants will not grow on this sulfur source (61). The reaction provides a conve-

TABLE 1. Nutritional characteristics of cysteine auxotrophs

Sulfur source	Growth of mutant class:					
	cysA	cysB	cysC cysD cysH	cysG cysI cysJ	cysE	cysM cysK[a]
Sulfate	−	−	−	−	−	−
Thiosulfate	−[b]	−	+	+	−	−[c]
Sulfite or cysteine sulfinic acid	+	−[d]	+	−	−	−
Sulfide	+	+	+	+	−	−
O-Acetylserine						
+ Sulfate	−	−	−	−	+	−
+ Sulfide	+	+	+	+	+	−
Cystine, cysteine, glutathionine, or djenkolate	+	+	+	+	+	+

[a] cysM cysK[+] and cysM[+] cysK strains are Cys[+], although cysM strains are cysteine bradytrophs when grown anaerobically.

[b] cysA strains will grow on high concentrations of thiosulfate.

[c] cysM strains are defective in thiosulfate utilization, while cysK strains are not.

[d] Some cysB strains grow slowly on sulfite owing to low-level expression of sulfite reductase activity.

nient way to incorporate the sulfane moiety of thiosulfate into organic linkage without further reduction of the sulfur source. An additional reductive step is required, however, to generate cysteine from S-sulfocysteine. This could be accomplished by a mechanism described in Penicillium chrysogenum (88), where a thiol such as glutathione reacts with S-sulfocysteine to give sulfite and glutathione-cysteine mixed disulfide as products. The mixed disulfide in turn is reduced by glutathione reductase to give cysteine. In wild-type cells grown on thiosulfate, the sulfite generated from this reaction sequence is reduced by sulfite reductase to sulfide, which is then incorporated into cysteine through the O-acetylserine sulfhydrylase reaction. In sulfite reductase-deficient mutants only the sulfane moiety of thiosulfate can be used for cysteine synthesis (52).

GENETICS

Cysteine auxotrophy is defined operationally as an inability to utilize sulfate as a sulfur source (without the addition of O-acetylserine). Most Cys[−] strains will grow on sulfide and other forms of reduced sulfur and are often characterized by nutritional studies before genetic mapping (Table 1) (12). Owing to the duplication of O-acetylserine sulfhydrylase activities, cysM and cysK mutants are prototrophs, although cysM strains cannot utilize thiosulfate (61) and are cysteine bradytrophs when grown anaerobically (24).

Several methods are available for positive selection of cys mutants including: chromate resistance of cysA mutants (67); 1,2,4-triazole resistance for cysK and constitutive cysB mutants (29, 30) and for a cysE promoter-up mutation (32); azaserine resistance for all cys mutants except cysE (31); and selection for Leu[+] in a leu-500 strain of S. typhimurium for Δ(topA cysB) mutants (57).

Chromosomal Organization

Genes known or suspected to be involved in cysteine biosynthesis (Table 2) are found in five different chromosomal regions, which for convenience will be designated cysG, cysE, cysB, cysAMK, and cysCDHIJ. In addition, it is clear that the cystine transport system CTS-1 is a part of the S. typhimurium cysteine regulon, but the genes for this activity and for the sulfate-binding protein have not been characterized. cysL is a designation given to S. typhimurium mutations that lie near cysA and are selenate resistant (29), but nothing is known of their biochemical defect. cysS is the gene for cysteinyl-tRNA ligase in E. coli and is not linked to other cys genes (8).

cysG region. The cysG region is not specific for cysteine biosynthesis, since it specifies an element involved in both siroheme and cobalamin synthesis. Therefore, it is required for activities of sulfite reductase, nitrite reductase, and cobalamin-dependent enzymes (13, 39). In E. coli cysG is closely linked to nirB, which appears to specify nitrite reductase apoprotein. The gene order is crp-nirB-cysG-aroB at 73 to 74 min in E. coli (54) and crp—cysG-aroB at 72 min in S. typhimurium. The cysG genes of E. coli and S. typhimurium have been cloned (54) (R. Monroe and N. Kredich, unpublished data).

cysE region. The gene order in both S. typhimurium (79 min) and E. coli (80 min) is xyl-mtl-cysE-rfa-pyrE (1, 73). cysE specifies serine transacetylase and is not linked to other known cys genes. Abortive transduction studies indicate there are two cysE cistrons in S.

TABLE 2. Chromosomal location of genes of cysteine biosynthesis

Gene	Location (min)		Cysteine regulon	Activity
	S. typhimurium	E. coli		
cysA	49	52	+	Sulfate permease
cysB	33	28	+	Positive regulatory protein
cysC	60	59	+	APS kinase
cysD	60	59	+	ATP sulfurylase
cysE	79	80	+	Serine transactylase
cysG	72	73	?	Sulfite reductase: siroheme synthesis
cysH	60	59	+	PAPS sulfotransferase
cysI	60	59	+	Sulfite reductase: hemoprotein
cysJ	60	59	+	Sulfite reductase: flavoprotein
cysK	49	52	+	O-Acetylserine sulfhydrylase-A
cysL	50	—[a]	?	Selenate resistance
cysM	49	—[a]	+	O-Acetylserine sulfhydrylase-B
cysS	—[b]	12	?	Cysteinyl-tRNA ligase
cysZ	—[b]	52	?	Sulfate transport
	—[b]	52	+	Sulfate binding protein
	—[b]	—[a]	+	L-Cystine transport: CTS-1

[a] Not identified in E. coli.

[b] Not identified in S. typhimurium.

typhimurium (56), and mapping of a putative promoter mutation suggests that transcription begins at the *pyrE*-proximal side (32). *cysE* has been cloned from *S. typhimurium* (Monroe and Kredich, unpublished data).

cysB region. The *cysB* gene specifies a regulatory protein of positive control required for expression of all other genes of the cysteine regulon except for *cysE*, which appears to be exempt from such control (41, 47). The gene order is *purB—pyrF-cysB-topA-trp—aroD* (33 min) in *S. typhimurium* and *purB—trpA-cysB-pyrF—aroD* in *E. coli* (28 min), where the entire chromosomal region between 25 and 35 min is inverted (34). The *cysB* genes from *S. typhimurium* and *E. coli* have been cloned (38) (D. Hulanicka and N. Kredich, unpublished data), and their nucleotide sequences have been determined (J. Ostrowski, G. Jagura-Burdzy, and N. Kredich, unpublished data). Both have open reading frames corresponding to a 324-residue, 36,000-dalton polypeptide chain with 95% amino acid sequence homology; the nucleotide sequence homology is 80%.

cysAMK region. In *S. typhimurium* the *cysAMK* region consists of the *cysA*, *cysM*, and *cysK* genes (49 min), which specify sulfate permease, *O*-acetylserine sulfhydrylase-B, and *O*-acetylserine sulfhydrylase-A, respectively, together with a cluster of three genes involved in the phosphoenolpyruvate-dependent phosphotransferase system, i.e., *crr*, *ptsI*, and *ptsH*. Abortive transduction studies of *cysA* show three complementation groups (56), and *cysM* is adjacent to if not contiguous with *cysA* (31). Genetic analysis of multisite deletions shows a relative gene order of *crr-ptsI-ptsH-cysK*, in which the genes are probably contiguous, and one-way, three-point crosses in *S. typhimurium* indicate that *cysA* is situated on the *cysK* side of this cluster (18). Conflicting results are obtained, however, from Southern blot analyses of multisite deletions using DNA probes cloned from wild-type (D. Hulanicka, C. Garrett, and N. Kredich, unpublished data). These data place the phosphotransferase genes between *cysA* and *cysK* and, together with functional analyses of subcloned DNA fragments, give a gene order *dsd—cysK-ptsH-ptsI-crr-cysM-cysA—purC*. The distance between *cysA* and *cysK* in DNA cloned from this region is about 9 kb, a value that agrees well with the 8 kb predicted from a P22 cotransduction frequency of 50%.

Analysis of DNA cloned from the analogous *E. coli* (52 min) region shows the same gene order (9, 11) except for *cysM*, which has not been demonstrated in this organism. In addition, a *cysZ* gene has been identified in *E. coli* which is necessary for sulfate transport but distinct from *cysA* (68). The *E. coli* gene order is *dsd-lig—cysZ-cysK-ptsH-ptsI-gsr-cysA—purC*, where *gsr* is identical to the *crr* of *S. typhimurium*.

cysCDHIJ region. The *cysCDHIJ* region is located at 60 min on the *S. typhimurium* chromosome and has the relative gene order *cysC-cysD-cysH-cysI-cysJ* (20). Fine-structure genetic analysis of this region indicates that *cysC* and *cysD*, which specify APS kinase and ATP sulfurylase, respectively, are adjacent to each other and separated from *cysH* by about 25 kb of DNA termed the "silent section" because of its lack of known genetic markers. *cysH* codes for PAPS sulfotransferase, whereas *cysI* and *cysJ* code for the apo-

hemoprotein and apoflavoprotein components of sulfite reductase. Deletions of the entire *cysHIJ* region occur at very high frequency, which suggests an unusual degree of instability of this part of the chromosome (35). A large number of these deletions begin in an area outside of *cysJ* and extend for various distances into the silent section between *cysD* and *cysH*. These have been called "ditto deletions" because of their similarity to one another. In *E. coli* this region as a whole is inverted with respect to other chromosomal loci, with a gene order of *cysCDHIJ-argA-thyA* in contrast to *cysJIHDC-argA-thyA* in *S. typhimurium* (42).

Cym⁻ and *trzB* Strains

Some Cys⁻ strains of *S. typhimurium* will grow on either cysteine or methionine. Many are simply leaky *cys* mutants, for which methionine has a sparing effect on cysteine requirements; others comprise a class termed Cym⁻, with characteristic genetic and biochemical features (71). A Cym⁻ mutation can occur in and affect the activity specified by either *cysA*, *cysC*, *cysD*, *cysG*, *cysH*, *cysI*, or *cysJ*. Methionine somehow restores the deficient activity in such mutants, allowing them to utilize sulfate. Interestingly, fine-structure genetic mapping indicates that most Cym⁻ mutations in the *cysCDHIJ* region are situated between genes or at the ends of genes. Thus, they may lie in regulatory regions. It is not known how methionine causes phenotypic suppression of these mutations.

1,2,4-Triazole-resistant, *cysK* strains of *S. typhimurium* are of two types, *trzA* mutants, which are not unusual, and *trzB* mutants, which are characterized by high-frequency segregation to Trz⁺, poor linkage to nearby genes by Hfr conjugation, and normal linkage by P22-mediated transduction (29). It has been proposed that *trzB* mutations result from a reversible transposition of part or all of *cysK* from the chromosome to an autogenous plasmid.

Selection for resistance to 1,2,4-triazole (Trzʳ) in Cym⁻ strains occasionally gives derivatives designated CTS, which are Cys⁺, i.e., the Cym⁻ phenotype is suppressed (45, 46). CTS strains are unstable and segregate to Cym⁻ at frequencies of 50 to 70% but in some cases can be stabilized by repeated passage. Both segregants and nonsegregants retain the Trzʳ phenotype, which by genetic mapping appears to be due to a *cysK* mutation. It has been suggested that suppression of the Cym⁻ phenotype in CTS strains is caused by the chromosomal insertion of an acridine-curable DNA species originating from the *cysAMK* region, and may represent a secondary consequence of *trzB* mutations. Analyses of genetic linkage in several CTS strains are compatible with an insertion of approximately 9 kb of DNA in the *cysCDHIJ* region (44).

REGULATION

Cysteine biosynthesis in *S. typhimurium* and *E. coli* is regulated in response to end-product availability and virtually ceases in cells grown on cysteine or cystine. This control is attributable to several mechanisms including inhibition of enzyme activity, a system of gene regulation known as the cysteine regulon, and perhaps enzyme degradation as well.

Kinetic Regulation

Feedback inhibition of serine transacetylase by cysteine (48, 51) constitutes a major, and possibly the only physiologically significant, form of kinetic regulation in the cysteine biosynthetic pathway. Cysteine also inhibits O-acetylserine sulfhydrylase A, but with K_i values of 1 to 12 mM (14) compared to the apparent K_i of 1 μM found for serine transacetylase. Inhibition of serine transacetylase serves a dual regulatory purpose since O-acetylserine is both a direct precursor of cysteine and an internal inducer of the cysteine regulon.

Inhibition of sulfate transport by cysteine has been suggested in $E.$ $coli$ (R. J. Ellis, Biochem. J. **93p:**19–20, 1964), but not confirmed in $S.$ $typhimurium$ (21). In vitro studies on the conversion of sulfate to PAPS have shown no significant inhibition by cysteine or sulfide (23), whereas product inhibition of sulfite reductase by sulfide requires concentrations that are unlikely to occur physiologically. In vivo overproduction of sulfide in a mutant constitutive for expression of the cysteine regulon indicates there is little or no kinetic regulation of sulfate uptake and reduction (10).

O-Acetylserine sulfhydrylase A is subject to substrate inhibition by sulfide, but at concentrations lower than the K_m value for activity (14). As a result, a significant decrease in activity occurs only at sulfide concentrations of 0.1 mM or greater. Furthermore, since studies on leaky $cysK$ mutants indicate that less than 2% of wild-type O-acetylserine sulfhydrylase activity is sufficient for normal growth (33), it seems unlikely that kinetic regulation of this enzyme by sulfide is of any physiologic importance.

The Cysteine Regulon

The cysteine regulon consists of those genes required for cysteine biosynthesis and cystine transport that either are regulated in response to the availability of utilizable sulfur or participate in this process. Genes known to be regulated include $cysA$, $cysC$, $cysD$, $cysH$, $cysI$, $cysJ$, $cysK$, $cysM$, and the genes coding for sulfate-binding protein and CTS-1 (Table 2). $cysB$ and $cysE$ participate in regulation, but are themselves insensitive to sulfur availability. $cysB$ may be autoregulated. The $cysJIH$ region of $E.$ $coli$ is regulated at the level of transcription (25), and it is assumed that other genes are similarly controlled. Maximal expression of the cysteine regulon requires three factors: sulfur limitation, O-acetylserine, and the $cysB$ protein.

Sulfur limitation. Growth on minimal medium with djenkolate or glutathione as sulfur source provides a convenient method for sulfur limitation and results in high levels of cysteine biosynthetic enzymes (22). These activities decrease progressively with growth on sulfate, sulfite, sulfide, and cysteine or cystine (47, 69, 87). In $E.$ $coli$ this order of sulfur sources correlates directly with intracellular cysteine and inversely with levels of ATP sulfurylase and APS kinase (86). Presumably, cysteine or a metabolite thereof represses the cysteine regulon. ATP sulfurylase and APS kinase are more sensitive to repression than either sulfite reductase or O-acetylserine sulfhydrylase-A and are very low to unmeasurable in sulfite-grown cells, which do not require sulfate activation. Interestingly, PAPS sulfotransferase, which also is not needed for growth on sulfite, is expressed coordinately with sulfite reductase (47, 69), presumably because $cysH$, $cysI$, and $cysJ$ comprise a single operon (see below).

O-Acetylserine. Sulfur-limited cultures of $cysE$ mutants cannot be derepressed for the biosynthetic pathway unless supplied with an exogenous source of O-acetylserine (43, 47). Therefore, O-acetylserine is considered an internal inducer. Since O-acetylserine synthesis is inhibited by cysteine and its utilization requires sulfide, steady-state levels should be inversely related to sulfur availability and thus provide a signal appropriate for modulating expression of the cysteine regulon.

Given these relationships between O-acetylserine, sulfide, and cysteine, one might wonder whether the need for sulfur limitation is simply to promote O-acetylserine accumulation. Addition of O-acetylserine does increase levels of cysteine biosynthetic enzymes in sulfate- or sulfite-grown cells but has no effect during growth on sulfide or cysteine (47). This suggests that cysteine or sulfide directly blocks the action of O-acetylserine. This interpretation is complicated, however, by the fact that cysteine and sulfide are readily interconverted in the presence of O-acetylserine through the actions of cysteine desulfhydrase (50) and O-acetylserine sulfhydrylase. Therefore, rather than interfering with the action of O-acetylserine, these two sulfur compounds may simply deplete intracellular concentrations of inducer as rapidly as it can enter the cell.

$cysB$ regulatory protein. Most $cysB$ strains, including all known deletions, are Cys⁻ and like $cysE$ mutants are pleiotropic in that they cannot be derepressed for the biosynthetic pathway (22, 41, 47). Serine transacetylase levels are not affected by $cysB$ mutations. Since $cysB^+/cysB$ merodiploids behave as wild type and $cysB$ is distant from other genes of the cysteine regulon, the $cysB$ protein is considered to be a $trans$-acting element of positive control (36, 41). The $cysB$ gene products of $S.$ $typhimurium$ and $E.$ $coli$ have been identified in two-dimensional polyacrylamide gels of crude extracts (2, 55), and more recently the native $cysB$ protein from $S.$ $typhimurium$ has been purified to homogeneity (B. Miller and N. Kredich, unpublished data). The protein is a tetramer of identical 36,000-dalton subunits, but thus far no in vitro activity has been demonstrated.

Several $cysB$ alleles have been described that cause constitutive expression of all or part of the cysteine biosynthetic pathway in the absence of both sulfur limitation and O-acetylserine (namely, a $cysE$ strain grown on cystine) (10, 47). These constitutive alleles of $cysB$ generally give a Cys⁺ phenotype that also includes resistance to 1,2,4-triazole. The most carefully studied allele, $cysB1352$, of $S.$ $typhimurium$ is dominant to $cysB^+$, i.e., a $cysB^+/F'$ $cysB1352$ merodiploid has a constitutive phenotype (37). One mutation, $cysB484$, is of particular interest because the resulting mutant is constitutive only for O-acetylserine sulfhydrylase-A and deficient for other activities (47).

A simple working model for the cysteine regulon consists of a complex of $cysB$ protein and O-acetylserine acting to facilitate initiation of transcription at various *cys* promoters in a manner analogous to that described for other systems of positive control. According to this model, *cysB* constitutive alleles specify mutant proteins that are insensitive to cysteine or sulfide and interact with promoters in the absence of inducer. Presumably *cys* promoters are not all identical, since the *cysB484* protein effects constitute expression of *cysK* but is inactive for other *cys* genes.

Transcriptional units. The *cysJIH* cluster is the only documented multitranscriptional unit in the cysteine regulon. Analyses of polar mutants indicate that transcription begins upstream from *cysJ* and proceeds towards *cysH* (53). This conclusion is supported by studies of a mutant selected for expression of sulfite reductase in a Δ*cysB* background, which is constitutive for expression of both sulfite reductase and PAPS sulfotransferase and carries a mutation mapping at the *cysH*-distal end of *cysJ* (64). This putative promoter mutation indicates the gene order in this region to be promoter-*cysJ*-*cysI*-*cysH*. The behavior of this mutant also demonstrates that *cysG* can be expressed in a Δ*cysB* strain because sulfite reductase activity requires siroheme.

The large distance between *cysCD* and *cysJIH* (~25 kb), together with a lack of polar effects from *cysJIH* ditto deletions, indicates that *cysC* and *cysD* are not part of the *cysJIH* operon. Polarity within the *cysCD* region has not been carefully examined, but the apparent contiguity of *cysC* with *cysD* and the fact that their products catalyze consecutive reactions suggest they might comprise a single operon. *cysA* and *cysM* may also be contiguous, but it is not known whether they share the same promoter. A regulatory mutation in *S. typhimurium* has been described that lies near *cysK* and prevents derepression of both *O*-acetylserine sulfhydrylase A and sulfate permease (84). It is difficult to understand this effect in terms of a single transcriptional unit for *cysA* and *cysK*, since several genes of the phosphotransferase system lie between. It is of interest that, in *E. coli*, sulfate permease activities were between 3 and 50% of wild type in all of 13 *cysK* mutants studied by Waiter and Hulanicka (85), suggesting that *cysK* and *cysZ* belong to the same operon. Perhaps *S. typhimurium* has an as yet undescribed gene for sulfate transport (analogous to *cysZ* in *E. coli*) which is part of an operon that includes *cysK*.

cysB autoregulation. Although *cysE* is independent of control by *cysB* and sulfur availability, the *cysB* gene itself is probably autoregulated. Using *E. coli* strains carrying phage Mu-mediated *cysB-lac* fusions, Jagura-Burdzy and Hulanicka (37) were able to measure β-galactosidase activity that was under control of the *cysB* promoter. Growth on different sulfur sources did not affect β-galactosidase activity, but introduction of either an F' *cysB* plasmid or a pBR322 derivative carrying wild-type *cysB* decreased enzyme levels as much as 10-fold in sulfur-depleted and sulfur-replete cultures. Thus *cysB* appears to be subject to autogenous control by a mechanism independent of the state of sulfur availability.

Dependence on DNA gyrase. In *S. typhimurium*, expression of sulfite reductase and *O*-acetylserine sulfhydrylase A is inhibited by the DNA gyrase inhibitors nalidixic acid and novobiocin, suggesting that *cys* promoters are sensitive to the state of DNA

superhelicity (65). This effect was also noted in a *cysB* constitutive strain, but in a strain carrying a constitutive mutation in the *cysJIH* promoter, sulfite reductase expression was not inhibited by nalidixic acid. It is not clear whether sensitivity to gyrase inhibitors occurs at the level of *cysB* expression or involves other *cys* promoters directly. Genes sensitive to catabolite repression are known to be sensitive to DNA gyrase inhibitors, but expression of the cysteine regulon is insensitive to growth on glucose. Perhaps the interaction of the *cysB* protein with a *cys* promoter has a preference for DNA superhelicity similar to that presumed for the interaction of the cyclic AMP receptor protein with its DNA binding site (74).

Enzyme Degradation

Rapid depletion of *O*-acetylserine by the addition of 1,2,4-triazole to early log-phase *S. typhimurium* results in a loss of ATP sulfurylase and APS kinase activities at rates greater than those predicted by dilution through continued growth (49). Other activities of the pathway are not so affected. Furthermore, levels of ATP sulfurylase and APS kinase are sensitive to growth phase and are low to immeasurable in extracts prepared from late-log- or stationary-phase, sulfur-limited cultures (47). By contrast, *O*-acetylserine sulfhydrylase A and sulfite reductase activities are relatively insensitive to growth phase. Sensitivity to degradation of the first two enzymes of the pathway may be significant in providing a mechanism by which sulfate-grown cells can quickly halt PAPS synthesis when given an exogenous source of reduced sulfur. Accumulation of PAPS is believed to be detrimental to cell survival since stored cultures of *cysH* mutants (lacking PAPS sulfotransferase) rapidly accumulate secondary mutations in *cysA*, *cysC*, and *cysD* (28).

Integration of Regulatory Mechanisms

The combination of kinetic inhibition of serine transacetylase by cysteine and induction of the sulfate-reducing enzymes by *O*-acetylserine allows communication between the two branches of the pathway while maintaining both under the control of cysteine (Fig. 2). During sulfur limitation, *O*-acetylserine levels rise because of decreased utilization and a lack of feedback inhibition on serine transacetylase, thereby stimulating expression of the genes for sulfate reduction and cystine transport. Increased sulfide production and cystine transport then lower steady-state levels of *O*-acetylserine until an appropriate balance is achieved between the two branches of the pathway. One disadvantage of such a system is the lack of kinetic inhibition of sulfate reduction, since cells suddenly shifted to a sulfur source such as sulfide or cysteine theoretically maintain the capacity for sulfate reduction until enzymes are diluted by cell growth. Rapid in vivo turnover of ATP sulfurylase and APS kinase provides a mechanism by which this branch of the pathway can be down-regulated more quickly.

Regulation during Anaerobic Growth

Barrett and Chang (4) have reported that *cysB*, *cysI*, and *cysJ* mutants of *S. typhimurium* can utilize sulfate

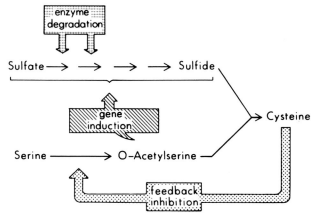

FIG. 2. Regulation of cysteine biosynthesis by a combination of feedback inhibition of serine transacetylase, gene induction through the cysteine regulon, and rapid enzyme turnover (degradation) of ATP sulfurylase and APS kinase.

when grown anaerobically in an atmosphere of H_2 and CO_2. Since *cysG* mutants are Cys⁻ under these conditions, the most likely explanation for the growth of *cysI* and *cysJ* strains is that anaerobiosis allows expression of another siroheme-requiring sulfite reductase, perhaps nitrite reductase, which is known to have sulfite reductase activity. The behavior of *cysB* mutants indicates that factors other than the *cysB* regulatory protein can effect expression of the cysteine regulon during anaerobic growth.

Another difference between aerobic and anaerobic cysteine biosynthesis is illustrated by the behavior of *cysM* mutants, which are Cys⁺ aerobically and cysteine bradytrophs anaerobically (24). Perhaps anaerobic sulfate reduction proceeds by a pathway that generates thiosulfate as a product or that involves carrier-bound intermediates similar to those found in *Chlorella* (76). Carrier-bound sulfide might require *O*-acetylserine sulfhydrylase B to be incorporated into cysteine.

LITERATURE CITED

1. **Bachmann, B. J.** 1983. Linkage map of *Escherichia coli* K-12, edition 7. Microbiol. Rev. **47:**180–230.
2. **Baptist, E. W., S. G. Hallquist, and N. M. Kredich.** 1982. Identification of the *Salmonella typhimurium cysB* gene product by two-dimensional protein electrophoresis. J. Bacteriol. **151:**495–499.
3. **Baptist, E. W., and N. M. Kredich.** 1977. Regulation of L-cystine transport in *Salmonella typhimurium.* J. Bacteriol. **131:**111–118.
4. **Barrett, E. L., and G. W. Chang.** 1979. Cysteine auxotrophs of *Salmonella typhimurium* which grow without cysteine in a hydrogen/carbon dioxide atmosphere. J. Gen. Microbiol. **115:**513–516.
5. **Becker, M. A., N. M. Kredich, and G. M. Tomkins.** 1969. The purification and characterization of O-acetylserine sulfhydrylase A from *Salmonella typhimurium.* J. Biol. Chem. **244:**2418–2427.
6. **Becker, M. A., and G. M. Tomkins.** 1969. Pleiotropy in a cysteine-requiring mutant of *Salmonella typhimurium* resulting from an altered protein-protein interaction. J. Biol. Chem. **244:**6023–6030.
7. **Berger, E. A., and L. A. Heppel.** 1972. A binding protein involved in the transport of cystine and diaminopimelic acid in *Escherichia coli.* J. Biol. Chem. **247:**7684–7694.
8. **Bohman, K., and L. A. Isaksson.** 1979. Temperature-sensitive mutants in cysteinyl-tRNA ligase of *E. coli* K-12. Mol. Gen. Genet. **176:**53–55.
9. **Boronat, A., P. Britton, M. C. Jones-Mortimer, H. L. Kornberg, L. G. Kee, D. Murfitt, and F. Parra.** 1984. Location on the *Escherichia coli* genome of a gene specifying *O*-acetylserine (thiol)-lyase. J. Gen. Microbiol. **130:**673–685.
10. **Borum, P. R., and K. J. Monty.** 1976. Regulatory mutants and control of cysteine biosynthetic enzymes in *Salmonella typhimurium.* J. Bacteriol. **125:**94–101.
11. **Britton, P., A. Boronat, D. A. Hartley, M. C. Jones-Mortimer, H. L. Kornberg, and F. Parra.** 1983. Phosphotransferase-mediated regulation of carbohydrate utilization in *Escherichia coli*: location of the *gsr(tgs)* and *iex(crr)* genes by specialized transduction. J. Gen. Microbiol. **129:**349–356.
12. **Clowes, R. C.** 1958. Nutritional studies of cysteineless mutants of *Salmonella typhimurium.* J. Gen. Microbiol. **118:**140–153.
13. **Cole, J. A., B. M. Newman, and P. White.** 1980. Biochemical and genetic characterization of *nirB* mutants of *Escherichia coli* K12 pleiotropically defective in nitrite and sulfite reduction. J. Gen. Microbiol. **120:**475–483.
14. **Cook, P. F., and R. T. Wedding.** 1976. A reaction mechanism from steady state kinetic studies for O-acetylserine sulfhydrylase from *Salmonella typmimurium.* J. Biol. Chem. **251:**2023–2029.
15. **Cook, P. F., and R. T. Wedding.** 1977. Overall mechanism and rate equation for O-acetylserine sulfhydrylase. J. Biol. Chem. **252:**3459.
16. **Cook, P. F., and R. T. Wedding.** 1977. Initial kinetic characterization of the multifunctional complex, cysteine synthetase. Arch. Biochem. Biophys. **178:**293–302.
17. **Cook, P. F., and R. T. Wedding.** 1978. Cysteine synthetase from *Salmonella typhimurium.* Aggregation, kinetic behavior and effect of modifiers. J. Biol. Chem. **253:**7874–7879.
18. **Cordaro, C. J., and S. Roseman.** 1972. Deletion mapping of the genes coding for HPr and enzyme I of the phosphoenolpyruvate: sugar phosphotransferase system in *Salmonella typhimurium.* J. Bacteriol. **112:**17–29.
19. **Datta, P.** 1967. Regulation of homoserine biosynthesis by L-cysteine, a terminal metabolite of a linked pathway. Proc. Natl. Acad. Sci. USA **58:**635–641.
20. **Demerec, M., D. H. Gillespie, and K. Mizobuchi.** 1963. Genetic structure of the *cysC* region of the *Salmonella* genome. Genetics **48:**997–1009.
21. **Dreyfuss, J.** 1964. Characterization of a sulfate and thiosulfate-transporting system from *Salmonella typhimurium.* J. Biol. Chem. **239:**2292–2297.
22. **Dreyfuss, J., and K. J. Monty.** 1963. The biochemical characterization of cysteine-requiring mutants of *Salmonella typhimurium.* J. Biol. Chem. **238:**1019–1024.
23. **Ellis, R. J., S. K. Humphries, and C. A. Pasternak.** 1964. Repressors of sulfate activation in *Escherichia coli.* Biochem. J. **92:**167–172.
24. **Filutowicz, M., A. Waiter, and D. Hulanicka.** 1982. Delayed inducibility of sulfite reductase in *cysM* mutants of *Salmonella typhimurium* under anaerobic conditions. J. Gen. Microbiol. **128:**1791–1794.
25. **Fimmel, A. L., and R. E. Loughlin.** 1977. Isolation of a λ*dcys* transducing bacteriophage and its use in determining the regulation of cysteine messenger ribonucleic acid synthesis in *Escherichia coli* K-12. J. Bacteriol. **132:**757–763.
26. **Fimmel, A. L., and R. E. Loughlin.** 1977. Isolation and characterization of *cysK* mutants of *Escherichia coli* K12. J. Gen. Microbiol. **103:**37–43.
27. **Flavin, M., and C. Slaughter.** 1967. Enzymatic synthesis of homocysteine or methionine directly from O-succinylhomocysteine. Biochim. Biophys. Acta **132:**400–405.
28. **Gillespie, D., M. Demerec, and H. Itikawa.** 1968. Appearance of double mutants in aged cultures of *Salmonella typhimurium* cysteine-requiring strains. Genetics **59:**433–442.
29. **Hulanicka, D., and T. Klopotowski.** 1972. Mutants of *Salmonella typhimurium* resistant to triazole. Acta Biochim. Pol. **19:**251–260.
30. **Hulanicka, D., T. Klopotowski, and D. A. Smith.** 1972. The effect of triazole on cysteine biosynthesis in *Salmonella typhimurium.* J. Gen. Microbiol. **72:**291–301.
31. **Hulanicka, M. D., S. G. Hallquist, N. M. Kredich, and T. Mojica-A.** 1979. Regulation of *O*-acetylserine sulfhydrylase B by L-cysteine in *Salmonella typhimurium.* J. Bacteriol. **140:**141–146.
32. **Hulanicka, M. D., and N. M. Kredich.** 1976. A mutation affecting expression of the gene coding for serine transacetylase in *Salmonella typhimurium.* Mol. Gen. Genet. **148:**143–148.
33. **Hulanicka, M. D., N. M. Kredich, and D. M. Treiman.** 1974. The structural gene for O-acetylserine sulfhydrylase A in *Salmonella typhimurium.* J. Biol. Chem. **249:**867–872.
34. **Ino, I., and M. Demerec.** 1968. Eneteric hybrids. II. *S. typhimurium-E. coli* hybrids for the *trp-cysB-purF* region. Genetics **59:**167–176.
35. **Itikawa, H., and M. Demerec.** 1967. Ditto deletions in the *cysC* region of the *Salmonella* chromosome. Genetics **55:**63–68.
36. **Jagura, G., D. Hulanicka, and N. M. Kredich.** 1978. Analysis of

merodiploids of the *cysB* region in *Salmonella typhimurium*. Mol. Gen. Genet. **165**:31–38.

37. **Jagura-Burdzy, G., and D. Hulanicka.** 1981. Use of gene fusions to study expression of *cysB*, the regulatory gene of the cysteine regulon. J. Bacteriol. **147**:744–751.

38. **Jagura-Burdzy, G., and N. M. Kredich.** 1983. Cloning and physical mapping of the *cysB* region of *Salmonella typhimurium*. J. Bacteriol. **155**:578–585.

39. **Jeter, R. M., B. M. Olivera, and J. R. Roth.** 1984. *Salmonella typhimurium* synthesizes cobalamin (vitamin B$_{12}$) de novo under anaerobic growth conditions. J. Bacteriol. **159**:206–213.

40. **Jones-Mortimer, M. C.** 1968. Positive control of sulfate reduction in *Escherichia coli*: isolation, characterization and mapping of cysteineless mutants of *E. coli* K12. Biochem. J. **110**:589–595.

41. **Jones-Mortimer, M. C.** 1968. Positive control of sulfate reduction in *Escherichia coli*: the nature of the pleiotropic cysteineless mutants of *E. coli* K12. Biochem. J. **110**:597–602.

42. **Jones-Mortimer, M. C.** 1973. Mapping of structural genes for enzymes of cysteine biosynthesis in *Escherichia coli* K12 and *Salmonella typhimurium* LT2. Heredity **31**:213–221.

43. **Jones-Mortimer, M. C., J. R. Wheldrake, and C. A. Pasternak.** 1968. The control of sulphate reduction in *Escherichia coli* by O-acetyl-L-serine. Biochem. J. **107**:51–53.

44. **Kingsman, A. J.** 1977. The structure of the *cysCDHIJ* region in unstable cysteine or methionine requiring mutants of *Salmonella typhimurium*. Mol. Gen. Genet. **156**:327–332.

45. **Kingsman, A. J., and D. A. Smith.** 1978. The nature of genetic instability in auxotrophs of *Salmonella typhimurium* requiring cysteine or methionine and resistant to inhibition by 1,2,4-triazole. Genetics **89**:439–451.

46. **Kingsman, A. J., D. A. Smith, and M. D. Hulanicka.** 1978. Genetic instability of auxotrophs of *Salmonella typhimurium* requiring cysteine or methionine and resistant to inhibition by 1,2,4-triazole. Genetics **89**:419–437.

47. **Kredich, N. M.** 1971. Regulation of L-cysteine biosynthesis in *Salmonella typhimurium*. I. Effects of growth on varying sulfur sources and O-acetyl-L-serine on gene expression. J. Biol. Chem. **246**:3474–3484.

48. **Kredich, N. M., M. A. Becker, and G. M. Tomkins.** 1969. Purification and characterization of cysteine synthetase, a bifunctional protein complex, from *Salmonella typhimurium*. J. Biol. Chem. **244**:2428–2439.

49. **Kredich, N. M., L. J. Foote, and M. D. Hulanicka.** 1975. Studies on the mechanism of inhibition of *Salmonella typhimurium* by 1,2,4-triazole. J. Biol. Chem. **250**:7324–7331.

50. **Kredich, N. M., B. S. Keenan, and L. J. Foote.** 1972. The purification and subunit structure of cysteine desulfhydrase from *Salmonella typhimurium*. J. Biol. Chem. **247**:7157–7162.

51. **Kredich, N. M., and G. M. Tomkins.** 1966. The enzymatic synthesis of L-cysteine in *Escherichia coli* and *Salmonella typhimurium*. J. Biol. Chem. **241**:4955–4965.

52. **Leinweber, F. J., and K. J. Monty.** 1963. The metabolism of thiosulfate in *Salmonella typhimurium*. J. Biol. Chem. **238**:3775–3780.

53. **Loughlin, R. E.** 1975. Polarity of the *cysJIH* operon of *Salmonella typhimurium*. J. Gen. Microbiol. **86**:275–282.

54. **Macdonald, H., and J. Cole.** 1985. Molecular cloning and functional analysis of the *cysG* and *nirB* genes of *Escherichia coli* K12, two closely-linked genes required for NADH-dependent nitrite reductase activity. Mol. Gen. Genet. **200**:320–334.

55. **Mascarenhas, D. M., and M. D. Yudkin.** 1980. Identification of a positive regulatory protein in *Escherichia coli*: the product of the *cysB* gene. Mol. Gen. Genet. **177**:535–539.

56. **Mizobuchi, K., M. Demerec, and D. H. Gillespie.** 1962. Cysteine mutants of *Salmonella typhimurium*. Genetics **47**:1617–1627.

57. **Mukai, F., and P. Margolin.** 1963. Analysis of unlinked suppressors of an 0° mutation in *Salmonella typhimurium*. Proc. Natl. Acad. Sci. USA **50**:140–148.

58. **Murphy, M. J., and L. M. Siegel.** 1973. Siroheme and sirohydrochlorin. J. Biol. Chem. **248**:6911–6919.

59. **Murphy, M. J., L. M. Siegel, and H. Kamin.** 1973. Reduced nicotinamide adenine dinucleotide phosphate-sulfite reductase of Enterobacteria. J. Biol. Chem. **248**:2801–2814.

60. **Nakamura, T., H. Iwahashi, and Y. Eguchi.** 1984. Enzymatic proof for the identity of the *S*-sulfocysteine synthase and cysteine synthase B of *Salmonella typhimurium*. J. Bacteriol. **158**:1122–1127.

61. **Nakamura, T., Y. Kon, H. Iwahashi, and Y. Eguchi.** 1983. Evidence that thiosulfate assimiliation by *Salmonella typhimurium* is catalyzed by cysteine synthase B. J. Bacteriol. **156**:656–662.

62. **Nakamura, T., and R. Sato.** 1963. Enzymatic synthesis of S-sulfocysteine from thiosulfate and serine. Nature (London) **198**:1198.

63. **Ohta, N., P. R. Galsworthy, and A. B. Pardee.** 1971. Genetics of sulfate transport by *Salmonella typhimurium*. J. Bacteriol. **105**:1053–1062.

64. **Ostrowski, J., and D. Hulanicka.** 1979. Constitutive mutation of *cysJIH* operon in a *cysB* deletion strain of *Salmonella typhimurium*. Mol. Gen. Genet. **175**:145–149.

65. **Ostrowski, J., and D. Hulanicka.** 1981. Effect of DNA gyrase inhibition on gene expression of the cysteine regulon. Mol. Gen. Genet. **181**:363–366.

66. **Pardee, A. B.** 1966. Purification and properties of a sulfate-binding protein from *Salmonella typhimurium*. J. Biol. Chem. **241**:5886–5892.

67. **Pardee, A. B., L. S. Prestidge, M. B. Whipple, and J. Dreyfuss.** 1966. A binding site for sulfate and its relation to sulfate transport into *Salmonella typhimurium*. J. Biol. Chem. **241**:3962–3969.

68. **Parra, F., P. Britton, C. Castle, M. C. Jones-Mortimer, and H. L. Kornberg.** 1983. Two separate genes involved in sulfate transport in *Escherichia coli* K12. J. Gen. Microbiol. **129**:357–358.

69. **Pasternak, C. A., R. J. Ellis, M. C. Jones-Mortimer, and C. E. Crichton.** 1965. The control of sulfate reduction in bacteria. Biochem. J. **96**:270–275.

70. **Peck, H. D., Jr.** 1961. Enzymatic basis for assimilatory and dissimilatory sulfate reduction. J. Bacteriol. **82**:933–939.

71. **Qureshi, M. A., D. A. Smith, and A. J. Kingsman.** 1975. Mutants of *Salmonella typhimurium* responding to cysteine or methionine: their nature and possible role in the regulation of cysteine biosynthesis. J. Gen. Microbiol. **89**:353–370.

72. **Robbins, P. W., and F. Lipmann.** 1958. Enzymatic synthesis of adenosine-5′-phosphosulfate. J. Biol. Chem. **233**:686–690.

73. **Sanderson, K. E., and J. R. Roth.** 1983. Linkage map of *Salmonella typhimurium*, edition VI. Microbiol. Rev. **47**:410–453.

74. **Sanzey, B.** 1979. Modulation of gene expression by drugs affecting deoxyribonucleic acid gyrase. J. Bacteriol. **138**:40–47.

75. **Schiff, J. A.** 1980. Pathways of assimilatory sulphate reduction in plants and microorganisms. CIBA Found. Symp. **72**:49–64.

76. **Schmidt, A., W. R. Abrams, and J. A. Schiff.** 1974. Reduction of adenosine-5′-phosphosulfate to cysteine in extracts from *Chlorella* and mutants blocked for sulfate reduction. Eur. J. Biochem. **147**:423–434.

77. **Siegel, L. M., and P. S. Davis.** 1974. Reduced nicotinamide adenine dinucleotide phosphate-sulfite reductase of Enterobacteria. IV. The *Escherichia coli* hemoflavoprotein: subunit structure and dissociation into hemoprotein and flavoprotein components. J. Biol. Chem. **249**:1587–1598.

78. **Siegel, L. M., P. S. Davis, and H. Kamin.** 1974. Reduced nicotinamide adenine dinucleotide phosphate-sulfite reductase of Enterobacteria. III. The *Escherichia coli* hemoflavoprotein: catalytic parameters and the sequence of electron flow. J. Biol. Chem. **249**:1572–1586.

79. **Siegel, L. M., H. Kamin, D. C. Rueger, R. P. Presswood, and Q. H. Gibson.** 1971. An iron-free sulfite reductase flavoprotein from mutants of *Salmonella typhimurium*, p. 523–553. *In* H. Kamin (ed.), Flavins and flavoproteins. University Park Press, Baltimore.

80. **Siegel, L. M., M. J. Murphy, and H. Kamin.** 1973. Reduced nicotinamide adenine dinucleotide phosphate-sulfite reductase of Enterobacteria. I. The *Escherichia coli* hemoflavoprotein: molecular parameters and prosthetic groups. J. Biol. Chem. **248**:251–261.

81. **Tsang, M. L.-S., and J. A. Schiff.** 1976. Sulfate-reducing pathway in *Escherichia coli* involving bound intermediates. J. Bacteriol. **125**:923–933.

82. **Tsang, M. L.-S., and J. A. Schiff.** 1978. Assimilatory sulfate reduction in an *Escherichia coli* mutant lacking thioredoxin activity. J. Bacteriol. **134**:131–138.

83. **Vogel, H. J., and D. M. Bonner.** 1956. Acetylornithinase of *Escherichia coli*: partial purification and some properties. J. Biol. Chem. **218**:97–106.

84. **Waiter, A., and D. Hulanicka.** 1978. The regulatory *cysK* mutant of *S. typhimurium*. Acta Biochim. Pol. **25**:281–287.

85. **Waiter, A., and D. Hulanicka.** 1979. Properties of *cysK* mutants of *Escherichia coli* K12. Acta Biochim. Pol. **26**:21–28.

86. **Wheldrake, J. F.** 1967. Intracellular concentration of cysteine in *Escherichia coli* and its relation to repression of the sulphate-activating enzymes. Biochem. J. **105**:697–699.

87. **Wheldrake, J. F., and C. A. Pasternak.** 1965. The control of sulfate activation in bacteria. Biochem. J. **96**:276–280.

88. **Woodin, T. S., and I. H. Segel.** 1968. Glutathione reductase-dependent metabolism of cysteine-S-sulfate by *Penicillium chrysegenum*. Biochim. Biophys. Acta **167**:78–88.

28. Biosynthesis of Threonine, Lysine, and Methionine

GEORGES N. COHEN AND ISABELLE SAINT-GIRONS

Unité de Biochimie Cellulaire, Département de Biochimie et Génétique Moléculaire, Institut Pasteur, 75724 Paris Cedex 15, France

THE COMMON PATHWAY AND THREONINE BIOSYNTHESIS

In *Escherichia coli*, diaminopimelate, lysine, methionine, threonine, and isoleucine derive part or all of their carbon atoms from aspartate. This section reviews *E. coli* aspartate kinase I-homoserine dehydrogenase I, aspartate kinase II-homoserine dehydrogenase II, aspartate kinase III, aspartate semialdehyde dehydrogenase, homoserine kinase, and threonine synthase. The first reaction of the common pathway is catalyzed by three distinct aspartate kinases, present in all the *Enterobacteriaceae* examined (41), which differ in the way their synthesis and activity are regulated. Aspartate kinase I is inhibited by threonine, and its synthesis is repressed by threonine plus isoleucine. The synthesis of aspartate kinase II is repressed by methionine. Lysine inhibits the activity and represses the synthesis of aspartate kinase III. The reduction of aspartate semialdehyde to homoserine is catalyzed by two distinct homoserine dehydrogenases. Homoserine dehydrogenase I synthesis is repressed by threonine plus isoleucine, and its activity is inhibited by threonine. Methionine represses the synthesis of homoserine dehydrogenase II. Homo-

serine kinase and threonine synthase expression is repressed by threonine plus isoleucine.

The first three proteins catalyze the synthesis of intermediates common to lysine, threonine, and methionine or to threonine and methionine syntheses (Fig. 1). They are part of what has been called the common pathway (33). The reactions specific to threonine biosynthesis are shown in Fig. 2.

Aspartate Kinase I-Homoserine Dehydrogenase I

In *E. coli* a threonine-regulated aspartate kinase was first recognized by Stadtman et al. (171), and a threonine-regulated homoserine dehydrogenase was recognized 2 years later (133). These two enzymes were considered to be separate entities. Early evidence that the two enzyme activities were part of the same protein (38, 136) was confirmed by Truffa-Bachi et al. (187).

The molecular weight of the native protein as determined by several methods is 355,000 ± 27,000 (59, 172, 187). A variety of chemical and genetic studies indicate that the enzyme has a structure with subunit molecular weights of 86,000 ± 4,000 (32, 59, 172, 187, 192, 197).

429

COOH
|
H₂N–CH
|
CH₂   Aspartate
|
COOH

thrA, metL, lysC ——— ATP   Aspartate kinase I, II, III
(EC 2.7.2.4)

COOH
|
H₂N–CH
|
CH₂   Aspartyl phosphate
|
C–O–Ⓟ
‖
O

asd ——— NADPH   Aspartate semialdehyde
dehydrogenase
(EC 1.2.1.11)

COOH
|
LYS ◄—— H₂N–CH
|
CH₂   Aspartate semialdehyde
|
C–H
‖
O

thrA, metL ——— NADPH   Homoserine
dehydrogenase I, II
(EC 1.1.1.3)

MET   COOH
THR ◄—— H₂N–CH
ILE |
CH₂   Homoserine
|
CH₂OH

FIG. 1. Common biosynthetic pathway to lysine, methionine, and threonine.

Functionally active fragments isolated from a nonsense mutant (aspartate kinase I fragment) or from native enzyme by limited proteolysis (homoserine dehydrogenase I fragment) possess only one or the other activity of the intact polypeptide. The tetrameric aspartate kinase I fragment (M_r about 4 × 53,000) is located in the N-terminal part of the polypeptide, and it is still inhibited by threonine. The dimeric homoserine dehydrogenase fragment (M_r = 2 × 57,000) is in the C-terminal part of the polypeptide chain, and it is insensitive to threonine. There is thus an overlap of about 21,000 daltons between the two fragments. The existence of these two fragments has helped in the early positioning of some tryptic and cyanogen bromide peptides (19, 46, 165; P. Cossart, Ph.D. thesis, University of Paris VII, Paris, France, 1977; L. Sibilli, Ph.D. thesis, University of Paris VII, Paris, France, 1977). About 75% of the protein sequence was known when *thrA* was cloned (45). Nucleotide sequence analysis completed the primary structure of the *thrA* product (102).

The ATG corresponding to Met-249 is preceded by a possible ribosome-binding site, GAGGT. The RNA encoded by this region of *thrA* could potentially form a stem-and-loop structure similar to functional intergenic ribosome binding sites identified in the *trp* operon of *E. coli* (159). The calculated ΔG value for formation of this structure is about −13 kcal (−54 kJ/mol). Thus Met-249 could represent a start codon for the dehydrogenase domain of the polypeptide chain. The ATG encoding Met-249 is immediately preceded by TTG TTG AAG TCG. Each of these four

triplets could be converted into amber termination codons by a single base change. One could speculate that two separate base changes could have created a bifunctional polypeptide chain by abolishing two termination codons frequently found at the end of genes. The subtilisin fragment carrying only the dehydrogenase activity starts at Ser-298 (19). In limited proteolysis, a series of proteases with widely different specificities all cleave within a short segment between Thr-293 and Asp-300 (166). These findings are compatible with the idea that a gene fusion occurred during evolution (192). The presumptive ancestral kinase would have had a molecular weight of about 25,000. Coordinate control of both activities at the transcription and translation level, in addition to common allosteric control of both activities by threonine, could be the evolutionary advantage of a fused protein (34, 39, 131, 191).

The aspartate kinase I-homoserine dehydrogenase I and the isolated fragments refold and regain their enzyme activities after total denaturation by guanidine hydrochloride (51, 73, 74). The reappearance of the kinase activity is a monomolecular process, much faster for the fragment than for the whole protein. The dehydrogenase fragment renatures according to biphasic kinetics, corresponding to the folding of the monomer followed by an association reaction that limits the appearance of the threonine-sensitive activity. The native enzyme refolds in three steps. A monomolecular step leads to a monomeric species with kinase activity. An association step then leads to a dimeric species with kinase and threonine-sensitive dehydrogenase activity. Finally, a second association step leads to a tetrameric species with the two activities, both sensitive to threonine. The folding of this large protein appears as a sequential process during

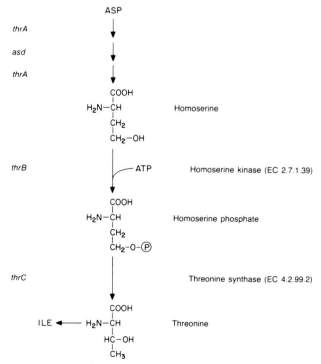

FIG. 2. Threonine biosynthetic pathway.

which the functional properties are regained successively as the protein structure becomes more complex. During this process, the two regions of each polypeptide chain, respectively responsible for the kinase and the dehydrogenase activities, seem to acquire their native conformation rather independently of each other.

Aspartate kinase I-homoserine dehydrogenase I forms tetragonal crystals in the presence of threonine and NADPH. The asymmetric unit is 10^6 Å^3 (10^5 nm^3); the space group is I422 or I4122. If it is assumed that the solvent occupies 53% of the total volume, the asymmetric unit contains one tetrameric molecule (92). The symmetry of the crystal form does not express the presumed symmetrical arrangement of the four identical subunits. Crystals suitable for high-resolution X-ray crystallographic analysis have been obtained recently and are currently under study (J. Janin, personal communication).

The nature and numerology of the ligands of aspartokinase I-homoserine dehydrogenase I, as well as the study of the conformational changes associated with the regulation of the enzyme activity upon threonine binding, have been reviewed in detail elsewhere (33, 34).

The existence of separate domains, although inferred only by indirect and circumstantial evidence, would explain the ability to obtain independent fragments (192) and the intersubunit and intrasubunit interactions in the native enzyme (191, 195). A two-domain model is consistent with both the dimeric nature of the dehydrogenase fragment and the tetrameric nature of the kinase fragment (192). The model is compatible with immunological evidence and with the effects of dissociation on the accessibility of the sulfhydryl groups and the rate of proteolysis (195). The Hill coefficient for threonine inhibition for the kinase fragment provides evidence that the tetrameric fragment retains the allosteric properties of the native tetrameric enzyme. The major conformational changes characteristic of the native enzyme may be largely confined to this region. Using a quite different approach, Mackall and Neet (110) have proposed an essentially identical model. On the basis of refined proteolysis experiments, a hybrid dimeric precursor of the homoserine dehydrogenase fragment has been established as consisting of one native chain and one dehydrogenase fragment (62, 63). Isolation of other active intermediates (193) has led to a triglobular model of the protein, with the third domain responsible for subunit contact (63) (Fig. 3). A careful study of the reversible dissociation of the native protein has led to the conclusion that the active species is the tetramer (194).

Aspartate Kinase II-Homoserine Dehydrogenase II

E. coli K-12 aspartate kinase II-homoserine dehydrogenase II is also a bifunctional protein (131). The two activities remain associated during a 400-fold purification. The molecular weight of the protein, determined by equilibrium sedimentation (52, 60), is 185,000 ± 10,000. The protein is a dimer, composed of two identical subunits of M_r = 88,000. The N terminus is Ser-Val-Ile-Ala-Gln-Ala-Gly-Ala-Lys, and the C terminus is Leu-Leu-COOH (52). The homodimeric nature of this protein implies that the two catalytic activities are carried by the same polypeptide chain. *E. coli* B synthesizes only aspartokinase II; no homoserine dehydrogenase II is detectable, even under conditions of derepression (131).

Limited proteolysis of aspartate kinase II-homoserine dehydrogenase II yields two dimeric fragments. One has dehydrogenase activity (M_r = 2 × 35,000). The other is devoid of catalytic activity (M_r = 2 × 25,000). The smaller fragment, although inactive, remains dimeric. It originates from a different part of the polypeptide chain than the dehydrogenase fragment, and one can regard it as part of the kinase domain. A model for the quaternary structure of aspartate kinase II-homoserine dehydrogenase II correlates the fragmentation pattern to the structure of the native enzyme (49).

The native enzyme and the dehydrogenase fragment refold and renature after denaturation in 6 M guanidine hydrochloride (50). When the intact polypeptide is denatured, the reappearance of both enzyme activities follows biphasic kinetics, with the reassociation of the two chains to form a dimer as rate-limiting step. The two activities reappear in parallel. The homoserine dehydrogenase II fragment also refolds and regains activity after complete denaturation, albeit 20 times faster than the entire protein. Thus the compact region corresponding to the dehydrogenase II fragment is an independent folding unit.

The gene coding for aspartate kinase II-homoserine dehydrogenase II, *metL*, has been cloned, and its

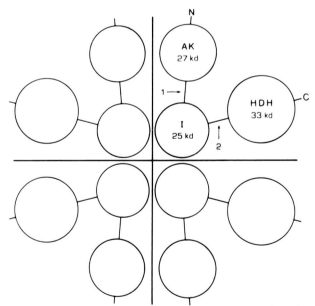

FIG. 3. Domain I is depicted as responsible for the subunit contacts that generate the dimeric structure of the homoserine dehydrogenase fragment and possibly the tetrameric structure of the native protein. The actual polymeric fragments obtained by limited proteolysis or from extraction from an ochre mutant can be visualized as the sum of I and of either of the two catalytically active fragments, yielding the corresponding dimer (HDH$_D$) or tetramer (AK from the nonsense mutant). Arrows indicate sites of cleavages leading to different well-characterized fragments.

sequence has been determined (208). Homologies between *thrA* and *metL* in several sections of the corresponding bifunctional proteins point to a common ancestor for the two genes (208) and corroborate previous immunochemical observations (209).

A recent study (5) has shown that *E. coli* aspartokinase II-homoserine dehydrogenase II is, like aspartokinase I-homoserine dehydrogenase I, composed of three globular domains: the N-terminal domain, endowed with kinase activity, the C-terminal domain, which carries the dehydrogenase activity, and a central inactive domain which separates the two. Thus, the polypeptide chains of the two multifunctional proteins are homologous not only in their sequence but also in their triglobular domain structure.

Neither aspartate kinase II nor homoserine dehydrogenase II is inhibited by methionine, *S*-adenosylmethionine, or threonine or by paired combinations of methionine, threonine, isoleucine, and lysine (60, 131). There is no cooperative binding of the substrates for either of the activities. A homoserine dehydrogenase II activity repressible by methionine has been described in *Salmonella typhimurium* (25).

Aspartate Kinase III

Aspartate kinase III is the third isofunctional protein that furnishes aspartyl phosphate for further metabolism. The enzyme activity is inhibited and its synthesis is repressed by lysine (171). Aspartate kinase III is activated by ammonium and potassium ions (198).

Aspartate kinase III has been purified to homogeneity (106, 185, 196). Contradictory assignments of molecular weights and subunit structures are due to different aggregation states of the enzyme under various conditions (124, 140). The native enzyme in the presence of 0.15 M KCl is a dimer of identical subunits with a molecular weight of 50,000. The dimeric protein is in equilibrium with both a tetrameric species, favored by high ionic strength in the presence of lysine, and a monomeric form, favored by low ionic strength in the absence of lysine. Only the dimeric form of the protein is enzymatically active (139). However, lysine causes a conformational transition, leading to an inactive dimeric species (see below).

The N-terminal and C-terminal amino acids are serine and leucine, respectively (140). Immunological studies reveal no homologies between the lysine-sensitive and the other two aspartate kinases when native enzymes are analyzed (99); however, cross-reactions occur with denatured proteins (122). The nucleotide sequence of the gene testifies to the common origin of the three kinase-encoding genes (27a, 208). A tentative model describes the evolution of the two aspartate kinase-homoserine dehydrogenases, taking into account internal homologies in each aspartate kinase and each homoserine dehydrogenase moiety (64). The knowledge of the sequence of aspartate kinase III leads to a more refined model (27a).

Feedback inhibition by lysine is noncompetitive with respect to aspartate (171). Lysine inhibition is cooperative (130); 0.2 mM lysine, a concentration comparable to that of the intracellular lysine pool, inhibits about 50% of the enzyme activity (185). Certain so-called nonspecific amino acids such as leucine, isoleucine, phenylalanine, and others at much higher concentrations also inhibit aspartate kinase III substantially (134). Synergism of inhibition by lysine and low concentrations of the nonspecific amino acids appears to be physiologically significant. Aspartate kinase III is inhibited allosterically by the lysine analogs β-aminoethylcysteine (185) and selenalysine (56).

The dimeric protein binds 4 mol of lysine (139). The four sites are nonequivalent: two have a K_d of 8 μM, while the remaining two have a K_d of 100 μM. Lysine binding is cooperative, but cooperativity disappears in the presence of leucine or the other nonspecific amino acids. This may explain the synergistic inhibition.

Lysine or leucine binding to aspartate kinase III leads to conformational changes at protein concentrations below those required for tetramerization (139). Experimental kinetic data are in good agreement with theoretical values predicted (115) for a concerted model of a class V allosteric enzyme (121). The kinetic parameters of the state and saturation functions have been determined (115). The results support the contention that aspartate kinase III is one of the best examples of a class V allosteric enzyme.

Several mutant proteins exhibit different degrees of resistance to inhibition by lysine and the nonspecific amino acids (15). All mutations lie within *lysC* without specific clustering (182; J. C. Patte, unpublished data).

Two lines of evidence suggest that feedback inhibition of aspartate kinase III is the main point of regulation for lysine biosynthesis. First, most of the mutants selected on the basis of resistance to growth inhibition by lysine analogs, like aminoethylcysteine or lysine hydroxamate, possess desensitized enzymes coded for by *lysC* alleles (15; J. C. Patte and G. N. Cohen, unpublished results). Such mutants excrete lysine into the growth medium. Second, mutants with elevated expression of *lysC* are as sensitive to growth inhibition by lysine analogs as the nonconstitutive parental strain (17).

Because lysine analog-resistant mutants excrete lysine but not threonine or methionine, it is relevant to inquire whether there is specific channeling of aspartyl phosphate synthesized by aspartate kinase III into diaminopimelate and lysine. Many attempts to demonstrate protein aggregates between aspartate kinases, aspartate semialdehyde dehydrogenase, and dihydrodipicolinate synthase have been unsuccessful) (G. N Cohen, unpublished data; E. W. Westhead, personal communication). It is thus likely that threonine inhibition of homoserine dehydrogenase is sufficient to direct aspartate semialdehyde toward diaminopimelate and lysine.

The adenylylation of aspartate kinase III, tentatively proposed, supposedly varies with the growth phase of the culture (125). Presumably, adenylylation also affects the aggregation state of the enzyme.

Aspartate kinase III is a good substrate for protease II, one of the *E. coli* cytoplasmic proteases (127). Protease II is copurified with aspartate kinase III through several chromatographic steps. In the absence of lysine, protease II cleaves specific peptide bonds in aspartate kinase III, suggesting a possible physiological role for proteolysis in lysine biosynthesis.

Aspartate Semialdehyde Dehydrogenase

Aspartate semialdehyde dehydrogenase catalyzes the reversible substrate-dependent reduction of NADP in the presence of phosphate or arsenate (Fig. 1, reverse reaction). This reaction is formally similar to that catalyzed by glyceraldehyde-3-phosphate dehydrogenase. The aspartate semialdehyde dehydrogenase has been obtained in 90% purity from *E. coli* K-12 grown under conditions of lysine limitation (89). The preparation of the enzyme has been highly improved by starting with a phenotypically derepressed strain (8). The molecular weight of the enzyme is 77,000 made up of a dimer of identical subunits ($M_r = 38,000 \pm 2,000$). The N-terminal sequence is Met-Lys-Asx-Val-Gly-. Each subunit contains three cysteine residues; two are reactive in the native enzyme, and one is partially protected by the substrate. Formation of an acyl-enzyme intermediate has been detected.

The substrate binding site of aspartate semialdehyde dehydrogenase has been labeled with 2-amino-4-oxo-5-chloropentanoate. This analog inactivates the enzyme with pseudo-first-order kinetics and with one-half of the reactivity of the site. Aspartate semialdehyde protects against the inactivation. A single group is labeled at the active site (9). A peptic digest of the labeled enzyme yields the sequence Phe-Val-Gly-Gly-Asp-modified residue-Thr-Val-Ser. Biellmann et al. (9) suggested that the side chain of a histidine residue is modified; however, the nucleotide sequence indicates a cysteine to be the target of modification (88). Aspartate semialdehyde dehydrogenase has also been alkylated with the coenzyme analog chloroacetylpyridine-ADP, which irreversibly inactivates the enzyme with pseudo-first-order kinetics. NADP and NADPH, but not the substrate, protect. The stoichiometry for total inactivation is again 1 mol of analog per dimer (7). The sequence of the entire *asd* gene has been determined (88).

Homoserine Kinase

The *thrB* gene encodes homoserine kinase, which catalyzes the phosphorylation of homoserine. The native protein consists of two identical 29,000-dalton subunits (23). The purified protein requires Mg^{2+} for activity, and K^+ stimulates the enzyme (181). The protein has been purified to homogeneity (23), and the first eight amino acids from the N-terminus have been determined. The entire nucleotide sequence of *thrB* is known (44), and the amino acid sequence of the protein has been deduced.

Homoserine kinase is inhibited by threonine with an apparent K_i of 0.6 mM (23). No product inhibition is observed, nor do the threonine analogs D-threonine, glycyl-threonine, 3- or 4-hydroxy butyrate, 3-aminobutyrate, or O-methyl-threonine, at concentrations of 10 mM, inhibit the enzyme (23). Isoleucine inhibits partially purified preparations of homoserine kinase, but this inhibition is not additive with threonine inhibition (181). Threonine inhibition is competitive with respect to homoserine. A low level of cooperativity ($n = 1.4$) is observed with partially purified homoserine kinase but not with the pure enzyme (181). Since the intracellular homoserine level in *E. coli* is low, threonine may regulate homoserine kinase competitively rather than allosterically (23, 181). The binding of ATP is strictly Michaelian (23).

Threonine Synthase

There are very few biochemical studies on the threonine synthase of *E. coli* (178, 205). The enzyme catalyzes the rather complex conversion of homoserine phosphate to threonine. Like other threonine synthases, the *E. coli* enzyme requires pyridoxal phosphate (35, 66, 126, 167). The apparent K_m for homoserine phosphate is 0.23 mM; the molecular weight as determined by gel filtration is 36,000 (178). A class of azide-dependent mutants requires threonine or pyridoxine for growth. The mutation is in *thrC*, and the threonine synthase exhibits modified kinetics and stability (47). The reaction mechanism has been thoroughly studied in *Neurospora crassa*, unequivocally assigning the positions in threonine into which solvent hydrogen is introduced during the reaction (100). *Bacillus subtilis* threonine synthase catalyzes the conversion of homoserine phosphate to α-ketobutyrate (157). This secondary reaction of enzyme has been used for threonine synthase estimation (B. Burr, personal communication). The first 40 N-terminal residues of the protein and the total nucleotide sequence of the corresponding *thrC* gene have been determined (128). The sequence of the deduced protein shows a molecular weight of 47,000, in agreement with the value estimated by sodium dodecyl sulfate-polyacrylamide gel electrophoresis for the purified protein and obtained for the polypeptide synthesized in minicells.

THE BIOSYNTHETIC BRANCH LEADING TO LYSINE

Bacteria synthesize lysine by a pathway in which the final step is the decarboxylation of diaminopimelate. This compound, as is lysine in some organisms, is an essential building block for peptidoglycan of the bacterial cell wall. This section reviews the biosynthesis of diaminopimelate and lysine in *E. coli*.

The sequence of enzyme-catalyzed steps leading from aspartate semialdehyde to lysine and the structures of the metabolic intermediates are given in Fig. 4. Mutants blocked at almost every recognized enzymatic step have been isolated. They are either diaminopimelate or lysine auxotrophs. Most diaminopimelate auxotrophs undergo lysis in a complex medium unless diaminopimelate is present at low concentrations, presumably because of its requirement for cell wall synthesis. Some diaminopimelate auxotrophs do not lyse in a complex medium (21).

In addition to the specific biosynthetic enzymes of the branch (Fig. 4), three other enzymes play a role in lysine metabolism: aspartate kinase III and aspartate semialdehyde dehydrogenase, which catalyze the first and second reactions of the common pathway (see preceding paragraph), and the catabolic lysine decarboxylase, which converts lysine into cadaverine. Enzymes that mediate diaminopimelate incorporation into peptidoglycan are not discussed here, although their activities could influence diaminopimelate and lysine pools.

The regulatory pattern of lysine biosynthesis described below applies only to *E. coli*. Many differences

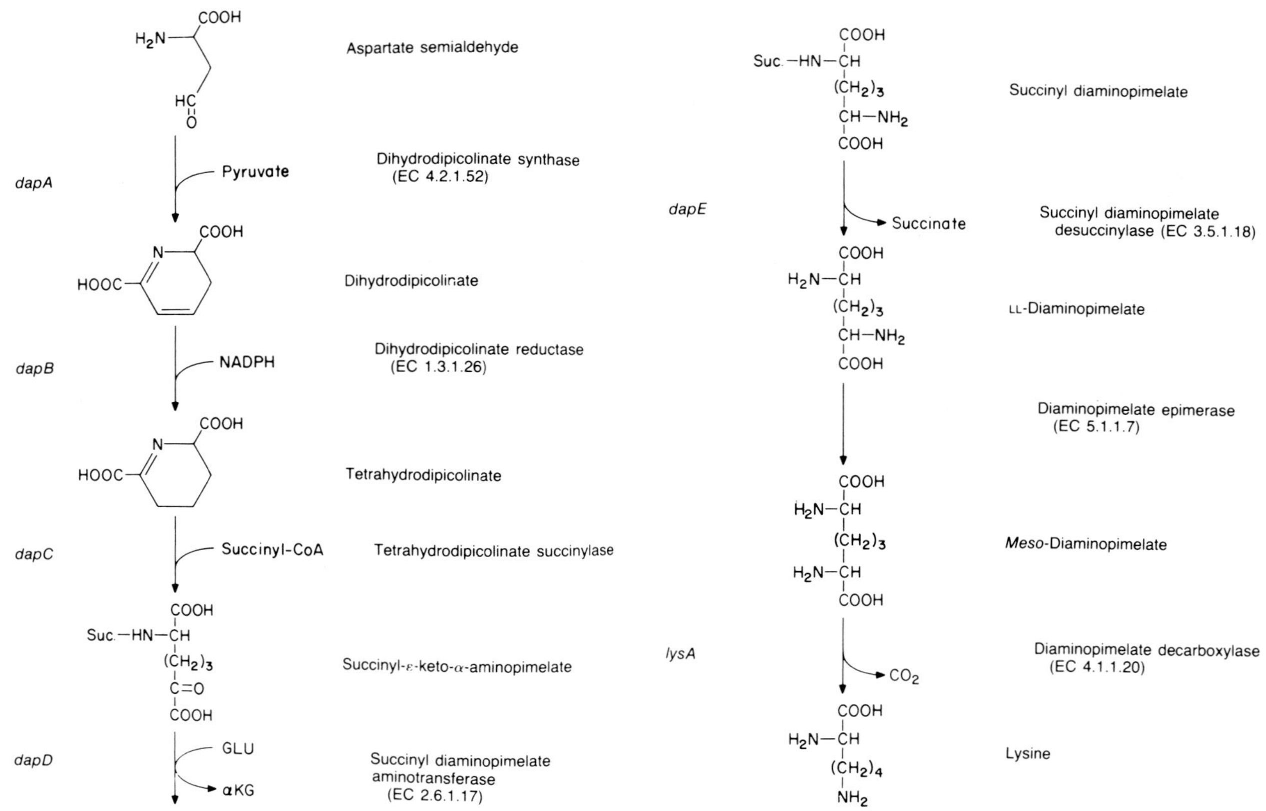

FIG. 4. Biosynthetic pathway for diaminopimelate and lysine.

emerge when other bacterial species are considered (188). One of the most striking features of the diaminopimelate pathway is that the known structural genes are scattered over the *E. coli* chromosome (4). There are no multigenic operons. Although the regulation of the synthesis of some of the lysine enzymes has not been demonstrated and no common regulatory protein has been implicated in the repression of the *lys* and *dap* genes, the term "*lys* regulon" is used here to convey the notion of a physiological unit. The salient properties of the different enzymes of the *lys* regulon are discussed below.

Dihydrodipicolinate Synthase

Dihydrodipicolinate synthase, encoded by the *dapA* gene at min 53 of the *E. coli* chromosome (21, 40), has been purified to homogeneity (160). The native protein has a molecular weight of 134,000. The subunit structure is not known. There is no methionine in the protein.

The enzyme is inhibited by lysine (207). Lysine inhibition is cooperative and noncompetitive toward the substrates (186). The physiological role of lysine inhibition is questionable. Mutants with desensitized aspartate kinase III excrete lysine but not threonine, an unexpected property of a cell with efficient regulation at the branch point. Mutants with desensitized dihydrodipicolinate synthase have not been found, despite the use of strong selective agents (Patte, unpublished data).

However, dihydrodipicolinate synthase must catalyze the rate-limiting step in lysine biosynthesis after aspartate kinase III because increasing the *dapA* gene copy number in a *lysC* mutant with desensitized aspartate kinase III results in a parallel increase in dihydrodipicolinate synthase activity and lysine excretion (48).

Dihydrodipicolinate Reductase

Dihydrodipicolinate reductase, first described by Farkas and Gilvarg (61), is encoded by *dapB* at min 0.5 on the *E. coli* chromosome between the *thr* and *ara* operons (21). The pure protein, with a native molecular weight of 115,000, has a K_m for dihydrodipicolinate of 10^{-5} M; no specific inhibitors of the enzyme activity are known (180). A 32,000-dalton polypeptide has tentatively been identified as the *dapB* gene product, suggesting that the enzyme is a tetramer (111). Its DNA sequence is now known, and *dapB* encodes a polypeptide of 273 residues ($M_r = 28,798$) (12).

Tetrahydrodipicolinate Succinylase and *N*-Succinyl Diaminopimelate Aminotransferase

Tetrahydrodipicolinate succinylase from *E. coli* (79) has not been studied in detail. The aminotransferase from *E. coli* has been partially purified and characterized (137).

Mutants have been isolated that are blocked between aspartate semialdehyde and diaminopimelate but do not fall into the *dapA*, *dapB*, or *dapE* class (21). The mutants affect one or the other of the two remaining metabolic steps. Two loci, designated *dapC* and *dapD*, have been identified at min 3.5 on the *E. coli* chromosome, on either side of *tonA*. The *dapC* class is represented by only one mutant.

The *dapD* gene specifying the aminotransferase is part of a 5.5-kilobase *Eco*RI restriction fragment that has been cloned on a λ-transducing phage and transferred to pBR322 (6, 143). The nucleotide sequence of the *dapD* gene has been determined. The gene encodes a polypeptide of 274 residues ($M_r = 30,040$) (141).

Some auxotrophs simultaneously requiring lysine and methionine can grow on succinate in the absence of the amino acids. These auxotrophs were identified as *suc* mutants (100) with impaired synthesis of succinyl coenzyme A, which participates in lysine and methionine biosynthesis through succinylation reactions.

N-Succinyl Diaminopimelate Desuccinylase

N-Succinyl diaminopimelate desuccinylase (78) has been partially purified (104). The enzyme requires Co^{2+} for activity. The enzyme is encoded by *dapE* at min 52.5 on the *E. coli* chromosome (21, 40).

Diaminopimelate Epimerase

Diaminopimelate epimerase catalyzes the interconversion of LL- and *meso*-diaminopimelate (2, 203). Two mutants have been isolated that could grow on *meso*- but not on LL-diaminopimelate; both mutants were unstable and were lost (E. Work, personal communication; Patte, unpublished data).

Diaminopimelate Decarboxylase

Diaminopimelate decarboxylase catalyzes the conversion of diaminopimelate to lysine (204). The enzyme is specific for *meso*-diaminopimelate; LL-diaminopimelate is not a substrate. The purified protein has a native molecular weight of 200,000 (200) as calculated from its sedimentation coefficient. Mutants devoid of this enzyme activity (53, 55) accumulate diaminopimelate to an extent that is useful for diaminopimelate production.

Diaminopimelate decarboxylase is encoded in *lysA* at min 61 on the *E. coli* chromosome near *thyA* and *galR* (91). A second class of mutants with lesions at the same locus, formerly designated *lysB*, require lysine or pyridoxal phosphate for growth. These mutations are now also classified as *lysA* mutations, since they alter the affinity of the decarboxylase for pyridoxal phosphate in such a way that lysine cannot be synthesized unless a large excess of pyridoxal phosphate is supplied (22).

The *lysA* gene, identified on a λ-transducing phage (161), is part of a 4.4-kilobase *Hin*dIII-*Bam*HI restriction fragment that has been cloned into pBR322 (30). The complete nucleotide sequence of *lysA* has been established (174). It encodes a polypeptide of 420 residues for a predicted molecular weight of 46,099, suggesting the tetrameric nature of the decarboxylase.

METHIONINE BIOSYNTHESIS

The reactions of the specific branch leading to methionine are shown in Fig. 5.

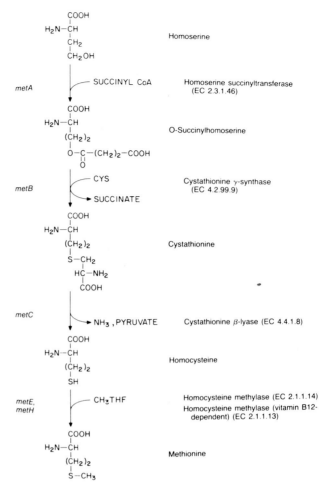

FIG. 5. Pathway of methionine biosynthesis.

Homoserine Succinyltransferase

The first specific step of methionine synthesis is catalyzed by homoserine succinyltransferase, product of the *metA* gene, which transforms homoserine into *O*-succinylhomoserine in the presence of succinyl coenzyme A. The enzyme has been partially purified in *E. coli* (145); its activity is allosterically inhibited in a synergistic way by methionine and *S*-adenosylmethionine (108). It is the only enzyme specific for the methionine branch that is subjected to an allosteric control. The enzyme is inhibited by α-methylmethionine (147). Resistance mutations are localized in the *metA* gene; such mutants overproduce methionine, and their enzyme is insensitive to methionine and *S*-adenosylmethionine (170). The *metA* gene has been cloned (117). Expression in minicells yields a polypeptide of $M_r = 40,000$. Since gel filtration of the partly purified native enzyme yields an M_r of 84,000, homoserine succinyltransferase appears to be a dimer (118). Homoserine succinyltransferase has unusual temperature sensitivity in many *Enterobactericeae*, which appears to limit growth at elevated temperature (144).

Cystathionine-γ-Synthase

O-Succinylhomoserine is transformed to cystathionine in the presence of cysteine. This reaction is catalyzed by cystathionine-γ-synthase, encoded by the *metB* gene. The enzyme has been obtained in the pure state in *S. typhimurium* (101) and in *E. coli* (184). In both cases the enzyme is a tetramer composed of four subunits of $M_r = 40,000$; in *S. typhimurium* it has been shown to contain four pyridoxal phosphate molecules. The *metB* gene of *E. coli* has been cloned, and its sequence has been determined (57). It is composed of 1,158 nucleotides, corresponding to 386 amino acid residues. The deduced N-terminal sequence agrees perfectly with the experimentally determined N-terminal protein sequence (184).

β-Cystathionase (Cystathionine-β-Lyase)

Cystathionine is cleaved to give homocysteine in a reaction catalyzed by β-cystathionase, product of the *metC* gene. This is also a pyridoxalphosphate enzyme which has been purified in *E. coli* and *S. typhimurium*. The *E. coli* enzyme was purified from an *E. coli* strain harboring a multicopy plasmid carrying the *metC* gene. Its pH optimum, substrate specificity, and kinetic parameters have been studied; 3,3,3-trifluoroalanine binds covalently to the enzyme and inhibits it irreversibly. The protein is a hexamer composed of six identical subunits of $M_r = 45,000$, each binding one molecule of pyridoxalphosphate (58). The enzyme from *S. typhimurium* was less well characterized (86).

The *metC* gene was recloned in *E. coli*, and its restriction map was determined; the gene shows a high degree of homology with representative strains of all tribes of the *Enterobacteriaceae* (119). The nucleotide sequence of the gene has been determined; the deduced sequence of the protein (395 residues), supported by the experimentally determined sequence of the first 10 residues, shows strong homology with that of cystathionine-γ-synthase and thus points to a common ancestor for the two proteins (5a).

Vitamin B₁₂-Dependent and Independent Homocysteine Methylases

The two last steps in methionine biosynthesis involve the methylation of homocysteine (75). The methyl donor is ^5N-methyl tetrahydropteroylglutamate ($N \geq 1$), which is derived from $^5N,^{10}N$-methylene tetrahydrofolate by a reductase specified by *metF* (see Part II, Section B1, Chapter 7 and Section B5, Chapter 1). This reaction yields the methyl group donor specific for methionine synthesis. The last step in methionine synthesis is effected by homocysteine methylation. Two enzymes can catalyze this reaction: a transmethylase with an activity dependent on the presence of vitamin B₁₂ (product of the *metH* gene) which can use the mono- or polyglutamate forms of ^5N-methyl tetrahydropteroylglutamate as methyl donor (85); and a vitamin B₁₂-independent transmethylase (product of *metE*) which can utilize as substrate only the polyglutamate forms of ^5N-methyl tetrahydropteroylglutamate ($N \geq 3$) (201). *E. coli*, an exception, possesses both enzymes whereas most organisms possess only one. *E. coli* does not synthesize vitamin B₁₂ and uses one or the other transmethylase according to vitamin B₁₂ availability in the medium. Strains with a mutation in *metE* require either methionine or

vitamin B₁₂ for their growth (54). The two transmethylases have been purified (180a, 201). A word of caution is necessary: vitamin B₁₂ synthesis under anaerobic conditions is proven for *S. typhimurium* and strongly inferred for *E. coli*. The authors of this finding provide evidence that the vitamin B₁₂-independent transmethylase, coded by *metE*, is a rather inefficient enzyme (94).

5,10-Methylene Tetrahydrofolate Reductase

5,10-Methylene tetrahydrofolate reductase has been purified, and some of its properties have been studied (103). The gene from *E. coli* has been cloned, and its sequence has been determined (153). The polypeptide chain has an $M_r = 33,065$. An in vitro synthesis system has identified the N-terminal tripeptide sequence of the enzyme, confirming the data deduced from the nucleotide sequence (163).

S-Adenosylmethionine Synthetase

The ubiquitous *S*-adenosylmethionine synthetase catalyzes the only known route of biosynthesis of the intracellular alkylating agent adenosylmethionine from methionine and ATP, the other products of the reaction being P_i and PP_i (26). Mutations in the *metK* locus at 64 min on the *E. coli* chromosomal map (3) result in reduced synthetase activity (80, 82), and a temperature-sensitive *metK* mutant with a thermolabile adenosylmethionine synthetase activity has been described (87), confirming that *metK* is the structural gene for the enzyme. This gene has been cloned (18), and its sequence has been determined (113). It specifies a polypeptide of 384 residues ($M_r = 41,941$). The results agree with the experimentally determined sequence of the 35 amino-terminal residues of the purified protein, which is actually a homotetramer (114). In addition to the synthetase reaction, the purified enzyme catalyzes a tripolyphosphate reaction stimulated by *S*-adenosylmethionine. The mechanism of the two reactions has been extensively studied (112, 114), and preliminary X-ray diffraction studies have been presented (76).

REGULATION OF GENE EXPRESSION

Threonine Biosynthesis

The three structural genes coding for the threonine biosynthetic enzymes belong to a single operon (183) located at 0 min on the genetic map of *E. coli* (182). These genes, *thrA*, *thrB*, and *thrC*, code respectively for aspartate kinase I-homoserine dehydrogenase I, homoserine kinase, and threonine synthase. The biosynthesis of these enzymes is subject to multivalent repression by threonine and isoleucine (67). Some complex regulatory mutants resistant to an isoleucine analog (179) and to the antibiotic borrelidin (123) suggested, respectively, the involvement of charged isoleucyl- and threonyl-tRNAs in the regulation of the threonine operon. The resistance to the isoleucine analog resulted from three mutations at different loci. However, a single mutation in *ilvS* coding for the isoleucyl-tRNA synthetase leads to the derepression of the threonine operon in *S. typhimurium* (10).

A *thrS* mutant, having a threonyl-tRNA synthetase exhibiting a 200-fold decreased apparent affinity for threonine, has regulatory properties that imply the involvement of charged threonyl-tRNA or the related synthetase in the regulation of the threonine operon. A mutation in *ilvA*, coding for threonine deaminase, leads also to derepression of the *thr* operon, presumably as a result of isoleucine limitation (97).

The genetic fine structure of the threonine operon has been analyzed by deletion mapping (155). Complementation analysis indicates that both *thrB* and *thrC* consist of a single cistron, whereas *thrA* is composed of two cistrons specifying a single polypeptide chain (183). This corroborates the bifunctional character of the *thrA* protein (see Aspartate Kinase I-Homoserine Dehydrogenase I, above).

Regulatory mutants affecting the expression of the threonine operon were obtained by insertion of lambda phage at a secondary site between the threonine promoter and the first structural gene (71, 72); other mutants were isolated on the basis of resistance to α-amino-β-hydroxyvaleric acid and identified by their excretion of threonine into the medium. They are *cis* dominant, lie adjacent to *thrA* (154), and are different from mutants resistant to the same analog affecting the *thrA* structural gene, previously described (37, 40). They have been called *thrO*; the gene sequence is *thrOABC*, as suggested by earlier studies. This order was established by analysis of phage Mu insertion and nonsense mutations in the threonine operon (183) and by analysis of a phage lambda insertion into the *thr* regulatory region (72).

Regulatory mutants have also been isolated from strains bearing *thr-lac* fusions (151a). They are also localized upstream of *thrA* (155).

All these results identify a regulatory region, situated upstream from the first structural gene of the operon, which acts only in *cis* in merodiploids and causes pleiotropic effect on threonine operon expression. Although the mutants were first thought to be classical operator mutants (71, 154), all mutations analyzed to this date reside in the *thr* attenuator.

The sequence of the *thr* operon regulatory region has been determined (68) by using the DNA cloned from lambda transducing phages (70). The sequence was identified by the known amino-terminal sequence of aspartokinase I-homoserine dehydrogenase I (191). The transcription initiation and the promoter region have been determined (68).

Downstream of the transcription initiation site, the threonine regulatory region contains extensive dyad symmetry and a stretch of A-T base pairs characteristic of Rho-independent terminators (146). This region has been called the *thr* attenuator (68) and is analogous to similar regions found in other biosynthetic operons (206). It contains a sequence encoding a leader peptide, containing eight threonine and four isoleucine codons (threonine and isoleucine being the two amino acids which regulate the operon expression). The *thr* attenuator also contains the classical mutually exclusive secondary structures which allow or prevent RNA polymerase transcription of the structural genes, according to the levels of the charged threonyl- and isoleucyl-tRNAs. The attenuation model thus accommodates the results showing that mutations in *ilvS* or *thrS* result in increased expression of the operon (10, 97). The attenuation region contains also an RNA polymerase pausing site of unknown significance (69). An internal promoter, situated within a 61-base-pair (bp) fragment at the 3' end of *thrA*, allows the expression of *thrB* in addition to the major promoter, but with an efficiency lower by at least one order of magnitude (156).

All the constitutive mutations of the "operator" type analyzed to this date are either point mutations (68) or a deletion of the transcription termination signal of the attenuator (129). Among α-amino-β-hydroxyvalerate-resistant mutants (37, 40, 151a, 154), one class does not excrete threonine. In these mutants, homoserine dehydrogenase I activity is increased 10-fold and the operon is not regulated by threonine and isoleucine. The mutations are not located within the *thr* operon or in the *thrS* gene. These mutants (*thrX*) might affect one of the threonyl-tRNA genes (43) or some gene involved in tRNA modification or act in some unknown manner.

In conclusion, regulation of threonine biosynthesis is at the level of enzyme activity and at the level of enzyme synthesis. Whereas aspartate kinase-homoserine dehydrogenase has been thoroughly studied and characterized, the other enzymes of the operon and aspartic semialdehyde dehydrogenase have been less studied, especially threonine synthase. Nucleotide sequence analysis of the entire *thr* operon is complete, including the promoter, attenuator, and termination signals of the operon (68, 128). Nuclease S1 mapping experiments have allowed the precise localization of the termination of the operon mRNA transcript 60 nucleotides downstream of the *thrC* stop codon (C. Parsot, personal communication). The comparison of aspartate kinases and homoserine dehydrogenases has allowed a model of evolution of these enzymes (27a, 64, 208).

The threonine operon has been considered to be regulated exclusively via attenuation mechanisms (68). Recent evidence suggests that repressor-operator interactions also play a role in the transcriptional control of this operon (11, 95). A mutation in the *ileR* gene leads to elevated expression of the *thr* and *ilv* operons (96). The nucleotide sequence of the *ileR* gene, which encodes a *trans*-acting product that negatively affects *thr* and *ilv* expression, has been determined (D. L. Weiss, D. I. Johnson, H. L. Weith, and R. L. Somerville, J. Biol. Chem., in press). Whether the negative effect of the *trans*-acting product is itself regulated by isoleucine, threonine, or some other factor has not yet been established.

Lysine Biosynthesis

The specific activity of aspartate kinase III varies in response to the intracellular lysine concentration. Approximately a 10-fold repression is observed when lysine is added to the growth medium (171), whereas a 15-fold derepression is seen during lysine-limited growth in a chemostat (133). The number of active enzyme molecules, estimated to be 525 per wild-type cell growing in a minimal salts medium (142), may thus vary about 150-fold. In *relA* strains derepression is only twofold (142), implicating guanosine tetraphosphate as an effector of transcription as proposed for general amino acid deficiency (173).

Although lysine is a specific effector, arginine also plays a role in aspartate kinase III regulation. When both amino acids are present simultaneously, arginine prevents lysine from maximally repressing aspartate kinase III synthesis (27). All identified regulatory mutants affecting aspartate kinase III synthesis are cis-dominant constitutive; no repressor-type mutants have yet been found (14, 27). In a lysC-lacZ fusion, β-galactosidase levels reflect the intracellular levels of lysine and arginine (142). This result provides strong support for regulation of aspartate kinase III at the transcriptional level. Although no derepression of the enzyme results when lysyl-tRNA charging is decreased, as in a K_m mutant of lysyl-tRNA synthetase (16), all other results were compatible with attenuation control (206) of lysC expression. However, when lysC-lacZ fusions were obtained and cloned in a multicopy plasmid, the regulatory sequence of lysC was determined, the promoter and the start of transcription were identified, but no signals similar to those described in the case of an attenuation mechanism were found in the long sequence (308 bp) existing between the transcription and translation starts (28).

Aspartate semialdehyde dehydrogenase activity is not inhibited by lysine; however, the synthesis of the enzyme is subject to repression. Excess lysine represses about 50%, and lysine limitation causes a 20-fold to 30-fold derepression. Thus the enzyme level of the cell can vary 40-fold to 60-fold (37). Threonine and methionine may participate with lysine in a multivalent repression of aspartate semialdehyde dehydrogenase (15). Although the effect of lysine limitation is the dominant factor, limitation of the other two metabolites leads to a 1.5-fold derepression over the fully repressed level. An analysis of the regulation of the synthesis of aspartate semialdehyde dehydrogenase in mutants lacking phosphoglucoisomerase activity has led to indirect evidence that glucose 6-phosphate is an effector of this synthesis (14).

Variations of the lysine pool have no effect on the synthesis of the apparently constitutive dihydrodipicolinate synthase (24). The dapA gene is part of a 2.8-kilobase PstI restriction fragment that has been identified in an E. coli library and transferred to pBR322 (143).

Expression of dapB is regulated by lysine (180). Lysine represses threefold to fourfold, whereas growth limitation by diaminopimelate in a chemostat leads to fourfold to sixfold derepression (13). Thus the specific activity of dihydrodipicolinate reductase can vary 15-fold to 20-fold. However, the range of variation is much lower in relA strains (135), indicating guanosine tetraphosphate to be a regulatory signal as discussed previously for aspartate kinase III synthesis. Charging of lysyl-tRNA also affects dapB expression (13, 16). The dapB gene is part of a 7.1-kilobase EcoRI-BamHI restriction fragment that has been identified on λ transducing phages and transferred to pBR322 (111, 143). The regulation of synthesis of the succinylase, the aminotransferase, and aspartate semialdehyde dehydrogenase has been discussed in terms of sequence homologies in their promoter regions, the significance of which is still a matter of conjecture (12).

Expression of dapE (N-succinyl diaminopimelate desuccinylase) is repressed twofold by lysine (40; Patte, unpublished data), suggesting that lysine could be a corepressor for the synthesis of this enzyme. The nucleotide sequence immediately preceding the coding sequence of lysA (diaminopimelate decarboxylase) does not suggest attenuation control for lysA expression (P. Stragier, personal communication).

Lysine represses diaminopimelate decarboxylase (132). In E. coli W the enzyme is induced by diaminopimelate, which may constitute the only effector if lysine acts indirectly by decreasing the diaminopimelate pool (199). However, growth limitation by lysine in a chemostat using dap mutants leads to a large increase in diaminopimelate decarboxylase activity (17). Thus a multiple regulation involving repression by lysine and induction by diaminopimelate agrees well with the double metabolic role of this enzyme: anabolic for lysine biosynthesis and catabolic for diaminopimelate degradation.

Regulation of lysA expression is dependent upon relA. A large diaminopimelate-mediated induction is seen in guanosine tetraphosphate-accumulating relA$^+$ strains but not in relA mutants (135). In addition, derepression of lysA under conditions of lysine limitation is partially abolished in hisT mutants (14), in which several tRNAs are not fully modified (20).

Expression of lysA cloned into pBR322 does not parallel gene dosage; a 50-fold increase in gene copy number yields only a 3-fold increase in diaminopimelate decarboxylase activity (30). This reflects the existence of an activator that is necessary for transcription of lysA and that is present in limiting amounts in the bacterial cell (176). A new class of lys mutations linked to lysA has been identified. These mutations appear to affect the proposed activator that is required for diaminopimelate decarboxylase synthesis. The mutations define a new gene, lysR, located on the thyA-distal side of lysA. The lysR product apparently does not affect the expression of the other genes of the lys regulon. The nucleotide sequence of the gene has been determined (175). Its target is localized on a 73-bp fragment found 48 bp upstream from the lysA coding region (176).

In conclusion, owing to the genetic complexity and the multiplicity of potential effectors, the regulation of the biosynthesis of lysine at the genetic level is far from completely understood. Very few regulatory mutations affecting lysine biosynthesis have been described. The lysX mutations (93), identified by the ability of the mutants to excrete lysine, are located at min 60 of the E. coli chromosome; they could actually be rel mutations, since relA influences the activities of several steps in lysine biosynthesis (135).

Methionine Biosynthesis

The various genes involved in methionine biosynthesis are scattered on the E. coli chromosome (4), and although certain genes are clustered, most constitute independent transcription units. Addition of methionine to the growth medium causes the repression of the synthesis of all the enzymes specific for the methionine branch with the exception of the metH product (42, 149, 150, 202), as well as repression of synthesis of S-adenosylmethionine synthetase, coded by the metK gene (90), and of aspartokinase II-homoserine dehydrogenase II, an enzyme of the common pathway coded by the metL gene (131).

The study of the regulation of methionine synthesis has been clarified by the development of *E. coli* and *S. typhimurium* mutants resistant to methionine analogs such as norleucine (36), α-methylmethionine, and ethionine (107; for reviews, see references 65 and 148). Three classes of mutants have been isolated: (i) α-methylmethionine-resistant mutants with lesions in *metA* affecting the allosteric control of homoserine succinyltransferase (170); (ii) mutants in which the synthesis of the methionine enzymes is constitutive (36, 107, 131) and which bear lesions in an *E. coli* locus called *metJ* at min 88 (177) (for both *S. typhimurium* and *E. coli* [1, 29], *metJ*⁺ is *trans*-dominant to *metJ*, indicating a diffusible product encoded by *metJ*; this interpretation that the *metJ* product is a protein was reinforced by the fact that certain *metJ* mutations were suppressed by nonsense suppressors); and (iii) mutants with lesions in *metK* (64 min on the *E. coli* chromosome). Different types of mutants have been obtained, and notwithstanding the complications introduced by the central role of *S*-adenosylmethionine synthetase, coded by *metK*, certain types of mutants exhibit a phenotype similar to that of the *metJ* mutants (82). An explanation could be that *S*-adenosylmethionine or one of its derivatives could be a corepressor of the regulation system. This possibility cannot be confirmed in vivo since *E. coli* is impermeable to *S*-adenosylmethionine.

It thus appears that regulation of methionine synthesis at the gene level is governed by the *metJ* gene, coding for an aporepressor which controls in a noncoordinated fashion (151) the scattered genes of the methionine regulon. Methionyl-tRNA does not seem to have a role in regulating the methionine genes since *metG* mutants, affected in methionyl-tRNA synthetase, do not produce a defect in their level of expression (1, 84).

The *metJ* gene has been mapped in *E. coli* (177) and found to be close to the *metBLF* cluster (182). Defective phages carrying the entire *metJBLF* cluster have been isolated (98, 138), and the entire cluster has been cloned in plasmid vectors (210). The physical maps have been studied, and the four genes have been localized (210) and their nucleotide sequences have been determined (57, 152, 153, 184, 208). The cluster is organized into three independent transcriptional units as follows.

(i) The *metJ* structural gene, transcribed counterclockwise, occupies 312 bp and is autoregulated (152). A complex 279-bp region is found between the *metJ* and *metB* structural genes. Analysis of the 5′ region of the *metJ* gene shows a regulatory region for two divergent transcriptional units (the *metJ* gene and the *metBL* operon). The promoter of the *metBL* operon has been identified, as well as its transcription start (105). The first results showed the presence in this region of another promoter activity in an orientation opposite to the *metBL* operon (57), but were not sufficient to locate precisely the *metJ* promoter. A refined deletion analysis has allowed the localization of three *metJ* promoters, each corresponding to a different transcript. The first and longest transcript is strongly repressed by the *metJ* product, the second is repressed less strongly, and the third is not repressed at all (105). This region also contains a putative operator region which might be active in the control of both

metBL and *metJ*, which are transcribed on two opposite DNA strands.

(ii) The *metB* and *metL* genes form an operon; their structural genes occupy 1,158 nucleotides for *metB* and 2,427 nucleotides for *metL*. The intracistronic region is 2 bp long. The *metL* gene is followed by a 351-bp segment containing a typical Rho-independent terminator structure (146) preceded by a potential stem-and-loop structure very similar to the consensus structure described for several intercistronic and interoperonic regions (77). This segment also contains the promoter and the transcriptional start of *metF*, which have been identified (153). Independent observations based on insertion mutagenesis by phage Mu or transposon Tn5 show that *metB* and *metL* form an operon (81).

(iii) The *metF* structural gene comprises 888 nucleotides. No typical Rho-independent terminator could be identified downstream of the structural gene; however, about 160 nucleotides beyond the end of the structural gene, some structural features of a Rho-dependent terminator (146) can be identified (153). The expression of *metF* is specifically inhibited in an in vitro system by the *metJ* protein in the presence of *S*-adenosylmethionine. The inhibition is at the level of transcription (164). The same in vitro system has been used to examine the expression of the *metB*, *metL*, and *metJ* genes; again, the inhibition by the *metJ* protein is at the level of transcription (162). Finally, putative operator regions upstream from the structural genes can be surmised by their properties of dyad symmetry. Such structures upstream of *metB* and *metF* show a very significant homology (153). Furthermore, such regions have been observed upstream of the *metA* (116), *metC* (5a), and, less significantly, the *metK* (113) structural genes. By analogy with the *arg* regulon, for which the homologous operator regions were called "ARG boxes," the name "MET boxes" has been proposed for these operator regions (152). Constitutive mutants have been isolated (J. Belfaiza, personal communication) in which discrete point mutations have occurred precisely within the postulated MET boxes, which can thus be defined unambiguously as operators.

Upstream of *metA*, two transcription starting points 74 bp apart have been found. Only one is under negative control by methionine (116). In this case, as in that of the three transcriptional units *metJ*, *metBL*, and *metF*, no structure typical of attenuation control could be detected.

The sequence of the *metC* gene of *E. coli* has been determined (5a). The *metE* gene of the same organism, coding for the B₁₂-independent transmethylase, has been isolated in a λ d*metE* transducing phage at the λ attachment site, cloned into a plasmid able to transform cells devoid of the enzyme (31). The *metE* gene of *S. typhimurium* has been recently cloned. Its expression in minicells yields a polypeptide of $M_r = 92,500$ (158).

The *metJ* gene product (104 amino acid residues; $M_r = 11,996$) was identified in maxicell extracts of a strain carrying the proper plasmid (168). The sequence of the gene has been completely determined (152). Its N-terminal half contains 67% of the basic residues, which are often in tandem. The nucleotide sequences of two *trans*-dominant constitutive alleles

(109) have been determined (152) and correspond respectively to a point mutation (Ala→Thr) and to a nonsense mutation (Trp→amber). A *metJ-lacZ* fusion has been constructed and has provided evidence that *metJ* is autoregulated (152). The first five residues of the pure hybrid protein have been determined and correspond to the deduced amino acid sequence of the *metJ* product (152). The methionine aporepressor is a dimer of the product of the *metJ* gene. The purified dimer protects a short segment of DNA in the common regulatory region of the *metB* and *metJ* genes from hydrolysis by DNase I (169). This segment coincides with the *metB-metJ* box defined above (5a, 152).

The *S. typhimurium metJ* gene has also been cloned, and its sequence has been determined (189, 190). The two deduced protein sequences differ very slightly: a Met residue in position 95 and an Asn residue in position 98 are, respectively, replaced by a Leu and an Asp residue in *S. typhimurium*. Fifteen identical amino acids are specified by different codons, and the termination codons are different (TAA in *E. coli*, TGATAG in *S. typhimurium*). Antisera raised against the purified aporepressor were used to estimate the levels of this protein, which are 0.01% of the proteins of the wild-type cells and 2% of those of an overproducing strain. The parameters of its binding to *S*-adenosylmethionine to the *metF* operator DNA and its effects on in vitro β-galactosidase synthesis under the control of the *metF* promoter have been studied (I. Saint-Girons, J. Belfaiza, Y. Guillou, D. Perrin, N. Guiso, O. Bârzu, and G. N. Cohen, J. Biol. Chem., in press). The *met* aporepressor has been crystallized, and a preliminary diffraction pattern has been obtained (C. Rojas, personal communication).

The relationship between *metJ* (specifying the *met* repressor), *metE*, and *metH* is complex. The expression of *metE* (coding for the B_{12}-independent transmethylase) is repressed by the *metJ* regulatory system when cells are grown in the presence of methionine. The expression of *metE* is also repressed by the *metH* gene product (coding for the B_{12}-independent enzyme) in the presence of vitamin B_{12} (83). Evidence suggests that repression by vitamin B_{12} is due to the increased formation of the transmethylase holoenzyme rather than to increased methionine synthesis (120).

In conclusion, the study of the regulation of methionine biosynthesis is well advanced (163), and rapid progress is to be expected with the use of in vitro systems, in which the *met* repressor will be used to test the MET boxes operator hypothesis, its direct effect on transcription and translation of the *met* structural genes will be verified, and the effect of methionine, *S*-adenosylmethionine, and the vitamin B_{12}-dependent transmethylase will be directly tested.

LITERATURE CITED

1. **Ahmed, A.** 1973. Mechanism of repression of methionine biosynthesis in *Escherichia coli*. I. The role of methionine, S-adenosylmethionine and methionyl transfer ribonucleic acid in repression. Mol. Gen. Genet. **123:**299–324.
2. **Antia, M., D. S. Hoare, and E. Work.** 1957. The steroisomers of αε-diaminopimelic acid. 3. Properties and distribution of diaminopimelic acid decarboxylase, an enzyme causing interconversion of the LL and *meso* isomers. Biochem. J. **65:**448–459.
3. **Bachmann, B. J.** 1983. Linkage map of *Escherichia coli* K-12, edition 7. Microbiol. Rev. **47:**180–230.
4. **Bachmann, B. J., and K. B. Low.** 1980. Linkage map of *Esche-*

5. **Belfaiza, J., A. Fazel, K. Müller, and G. N. Cohen.** 1984. *E. coli* aspartokinase II-homoserine dehydrogenase II polypeptide chain has a triglobular structure. Biochem. Biophys. Res. Commun. **123:**16–20.
5a.**Belfaiza, J., C. Parsot, A. Martel, C. Bouthier de la Tour, D. Margarita, G. N. Cohen, and I. Saint-Girons.** 1986. Evolution in biosynthetic pathways: two enzymes catalyzing consecutive steps in methionine biosynthesis originate from a common ancestor and share a common regulatory region. Proc. Natl. Acad. Sci. USA **83:**867–871.
6. **Bendiak, D. S., and J. D. Friesen.** 1981. Organization of genes in the four minute region of the *Escherichia coli* chromosome: evidence that *rpsB* and *tsf* are co-transcribed. Mol. Gen. Genet. **181:**356–362.
7. **Biellmann, J. F., P. Eid, and C. Hirth.** 1980. Affinity labeling of the aspartate-β-semialdehyde dehydrogenase with an alkylating coenzyme analogue. Eur. J. Biochem. **104:**65–69.
8. **Biellmann, J. F., P. Eid, C. Hirth, and H. Jörnvall.** 1980. Aspartate-β-semialdehydedehydrogenase from *Escherichia coli*. Purification and general properties. Eur. J. Biochem. **104:**53–58.
9. **Biellmann, J. F., P. Eid, C. Hirth, and H. Jörnvall.** 1980. Aspartate-β-semialdehydedehydrogenase from *Escherichia coli*. Affinity labeling with the substrate analogue L-2-amino-4-oxo-5 chloropentanoic acid. An example of half-site reactivity. Eur. J. Biochem. **104:**59–64.
10. **Blatt, J. M., and H. E. Umbarger.** 1972. On the role of isoleucyl-t-RNA synthetase in multivalent repression. Biochem. Genet. **6:**99–118.
11. **Bogosian, G., and R. L. Somerville.** 1983. Trp repressor protein is capable of intruding into other amino acid biosynthetic systems. Mol. Gen. Genet. **191:**51–58.
12. **Bouvier, J., C. Richaud, F. Richaud, J. C. Patte, and P. Stragier.** 1984. Nucleotide sequence and expression of the *Escherichia coli dapB* gene. J. Biol. Chem. **259:**14829–14834.
13. **Boy, E., F. Borne, and J. C. Patte.** 1978. Effect of mutations affecting lysyl-tRNA^{Lys} on the regulation of lysine biosynthesis in *Escherichia coli*. Mol. Gen. Genet. **159:**33–38.
14. **Boy, E., F. Borne, and J. C. Patte.** 1979. Isolation and identification of mutants constitutive for aspartokinase III synthesis in *Escherichia coli* K12. Biochimie **61:**1151–1160.
15. **Boy, E., and J. C. Patte.** 1972. Multivalent repression of aspartic semialdehyde dehydrogenase in *Escherichia coli* K-12. J. Bacteriol. **112:**84–92.
16. **Boy, E., F. Reinisch, C. Richaud, and J. C. Patte.** 1976. Role of lysyl-tRNA in the regulation of lysine biosynthesis in *Escherichia coli* K12. Biochimie **58:**213–218.
17. **Boy, E., C. Richaud, and J. C. Patte.** 1979. Multiple regulation of DAP-decarboxylase synthesis in *Escherichia coli* K12. FEMS Microbiol. Lett. **5:**287–290.
18. **Boyle, S. M., G. D. Markham, E. W. Hafner, J. M. Wright, H. Tabor, and C. W. Tabor.** 1984. Expression of the cloned genes encoding the putrescine biosynthetic enzymes and methionine adenosyltransferase of *Escherichia coli* (*speA, speB, speC* and *metK*). Gene **30:**129–136.
19. **Briley, P. A., L. Sibilli, M. A. Chalvignac, P. Cossart, G. Le Bras, A. de Wolf, and G. N. Cohen.** 1978. The primary structure of *Escherichia coli* K12 aspartokinase I-homoserine dehydrogenase I. Site of proteolytic cleavage by subtilisin. J. Biol. Chem. **253:**8867–8871.
20. **Bruni, C. B., V. Colantuoni, L. Sbordone, R. Cortese, and F. Blasi.** 1977. Biochemical and regulatory properties of *Escherichia coli* K-12 *hisT* mutants. J. Bacteriol. **130:**4–10.
21. **Bukhari, A. I., and A. L. Taylor.** 1971. Genetic analysis of diaminopimelic acid- and lysine-requiring mutants of *Escherichia coli*. J. Bacteriol. **105:**844–854.
22. **Bukhari, A. I., and A. L. Taylor.** 1971. Mutants of *Escherichia coli* with a growth requirement for either lysine or pyridoxine. J. Bacteriol. **105:**988–998.
23. **Burr, B., J. Walker, P. Truffa-Bachi, and G. N. Cohen.** 1976. Homoserine kinase from *Escherichia coli* K12. Eur. J. Biochem. **62:**519–526.
24. **Butour, J. L., B. Felenbok, and J. C. Patte.** 1974. Synthesis of dihydrodipicolinate synthetase in *Escherichia coli* K12. Ann. Microbiol. (Paris) **125b:**459–462.
25. **Cafferata, R. L., and M. Freundlich.** 1969. Evidence for a methionine-controlled homoserine dehydrogenase in *Salmonella typhimurium*. J. Bacteriol. **97:**193–198.
26. **Cantoni, G. L., and J. Durell.** 1957. Activation of methionine for transmethylation. The methionine activating enzyme: studies of the mechanism of the reaction. J. Biol. Chem. **225:**1033–1048.
27. **Cassan, M., E. Boy, F. Borne, and J. C. Patte.** 1975. Regulation

richia coli K12, edition 6. Microbiol. Rev. **44:**1–56.

of the lysine biosynthetic pathway in *Escherichia coli* K12: isolation of a *cis*-dominant constitutive mutant for AK III synthesis. J. Bacteriol. **123**:391–399.

27a. **Cassan, M., C. Parsot, G. N. Cohen, and J. C. Patte.** 1986. Nucleotide sequence of the *lysC* gene encoding the lysine sensitive aspartokinase III of *Escherichia coli* K12: evolutionary pathway leading to three isofunctional enzymes. J. Biol. Chem. **261**:1052–1057.

28. **Cassan, M., J. Ronceray, and J. C. Patte.** 1983. Nucleotide sequence of the promoter region of the *E. coli lysC* gene. Nucleic Acids Res. **11**:6157–6166.

29. **Chater, C.** 1970. Dominance of the wild-type alleles of methionine regulatory genes in *Salmonella typhimurium*. J. Gen. Microbiol. **63**:95–109.

30. **Chenais, J., C. Richaud, J. Ronceray, H. Cherest, Y. Surdin-Kerjan, and J. C. Patte.** 1981. Construction of hybrid plasmids containing the *lysA* gene of *Escherichia coli*: studies of expression in *Escherichia coli* and *Saccharomyces cerevisiae*. Mol. Gen. Genet. **182**:456–461.

31. **Chu, J., R. Shoeman, J. Hart, T. Coleman, A. Mazaitis, N. Kelker, N. Brot, and H. Weissbach.** 1985. Cloning and expression of the *metE* gene in *Escherichia coli*. Arch. Biochem. Biophys. **239**:467–474.

32. **Clark, R. B., and J. W. Ogilvie.** 1972. Aspartokinase I-homoserine dehdyrogenase I of *Escherichia coli* K12. Subunit molecular weight and nicotinamide-adenine dinucleotide phosphate binding. Biochemistry **11**:1278–1282.

33. **Cohen, G. N.** 1983. The common pathway to lysine, methionine and threonine, p. 147–171. *In* K. M. Herrmann and R. L. Somerville (ed.), Amino acids: biosynthesis and genetic regulation. Addison-Wesley Publishing Co., Inc., Reading, Mass.

34. **Cohen, G. N., and A. Dautry-Varsat.** 1980. The aspartokinases-homoserine dehydrogenases of *Escherichia coli*, p. 49–121. *In* H. Bisswanger and E. S. Schmincke-Ott (ed.), Multifunctional proteins. J. Wiley and Sons, New York.

35. **Cohen, G. N., M. L. Hirsch, S. B. Wiesendanger, and B. Nisman.** 1954. Précisions sur la synthèse de la L-thréonine à partir d'acide L-aspartique par des extraits de *E. coli*. C. R. Acad. Sci. Paris **238**:1746–1748.

36. **Cohen, G. N., and F. Jacob.** 1959. Sur la répression de la synthèse des enzymes intervenant dans la formation du tryptophane chez *Escherichia coli*. C. R. Acad. Sci. Paris **248**:3490–3492.

37. **Cohen, G. N., and J. C. Patte.** 1963. Some aspects of the regulation of amino acid biosynthesis in a branched pathway. Cold Spring Harbor Symp. Quant. Biol. **28**:513–516.

38. **Cohen, G. N., J. C. Patte, and P. Truffa-Bachi.** 1965. Parallel modifications caused by mutations in two enzymes concerned with the biosynthesis of threonine in *Escherichia coli*. Biochem. Biophys. Res. Commun. **19**:546–550.

39. **Cohen, G. N., J. C. Patte, P. Truffa-Bachi, and J. Janin.** 1967. Polycephalic proteins. A new pattern in the regulation of branched biosynthetic pathways showing enzyme multiplicity, p. 357–365. *In* Lunteren Symposium on Regulation 1966. Regulation of nucleic acid and protein biosynthesis. Elsevier Publishing Co., Amsterdam.

40. **Cohen, G. N., J. C. Patte, P. Truffa-Bachi, C. Sawas, and M. Doudoroff.** 1965. Repression and end product inhibition in a branched biosynthetic pathway, p. 243–253. Colloque International du C.N.R.S. sur les Mécanismes de Régulation des Activités Cellulaires chez les Micoorganismes, Marseille, July 1963. CNRS, Paris.

41. **Cohen, G. N., R. Y. Stanier, and G. Le Bras.** 1969. Regulation of the biosynthesis of the amino acids of the aspartate family in coliform bacteria and pseudomonads. J. Bacteriol. **99**:791–801.

42. **Cohn, M., G. N. Cohen, and J. Monod.** 1953. L'effet inhibiteur spécifique de la méthionine dans la formation de la méthionine-synthèse chez *E. coli*. C. R. Acad. Sci. Paris **28**:746–748.

43. **Comer, M. M.** 1982. Threonine tRNAs and their genes in *Escherichia coli*. Mol. Gen. Genet. **187**:132–137.

44. **Cossart, P., M. Katinka, and M. Yaniv.** 1981. Nucleotide sequence of the *thrB* gene of *E. coli*, and its two adjacent regions: the *thrAB* and *thrC* junctions. Nucleic Acids Res. **9**:339–347.

45. **Cossart, P., M. Katinka, M. Yaniv, I. Saint-Girons, and G. N. Cohen.** 1979. Construction and expression of a hybrid plasmid containing the *Escherichia coli thrA* and *thrB* genes. Mol. Gen. Genet. **175**:39–44.

46. **Cossart-Gheerbrant, P., L. Sibilli, P. A. Briley, M. A. Chalvignac, G. Le Bras, and G. N. Cohen.** 1978. The primary structure of *Escherichia coli* K12 aspartokinase I-homoserine dehydrogenase I. Isolation and characterization of the peptides produced by cyanogen bromide. Biochim. Biophys. Acta **535**:206–215.

47. **Daniel, J.** 1976. Azide-dependent mutants in *E. coli* K12. Nature (London) **264**:90–92.

48. **Dauce, L. B., M. Boitel, A. M. Deschamps, J. M. Lebeault, K. Sano, K. Takinami, and J. C. Patte.** 1982. Improvement of *E. coli* strains overproducing lysine using recombinant DNA techniques. Eur. J. Appl. Microbiol. Biotechnol. **15**:227–231.

49. **Dautry-Varsat, A., and G. N. Cohen.** 1977. Proteolysis of the bifunctional methionine-repressible aspartokinase II-homoserine dehydrogenase II of *Escherichia coli* K12. Production of an active homoserine dehydrogenase fragment. J. Biol. Chem. **252**:7685–7689.

50. **Dautry-Varsat, A., and J. R. Garel.** 1978. Refolding of a bifunctional enzyme and of its monofunctional fragment. Proc. Natl. Acad. Sci. USA **75**:5979–5982.

51. **Dautry-Varsat, A., and J. R. Garel.** 1981. Independent folding regions in aspartokinase-homoserine dehydrogenase. Biochemistry **20**:1396–1401.

52. **Dautry-Varsat, A., L. Sibilli-Weill, and G. N. Cohen.** 1977. Subunit structure of the methionine-repressible aspartokinase II-homoserine dehydrogenase II from *Escherichia coli* K12. Eur. J. Biochem. **76**:1–6.

53. **Davis, B. D.** 1952. Biosynthetic interrelations of lysine, diaminopimelic acid, and threonine in mutants of *Escherichia coli*. Nature (London) **169**:534–536.

54. **Davis, B. D., and E. S. Mingioli.** 1950. Mutants of *Escherichia coli* requiring methionine or vitamin B_{12}. J. Bacteriol. **60**:17–28.

55. **Dewey, D. C., and E. Work.** 1952. Diaminopimelic acid decarboxylase. Nature (London) **169**:533–534.

56. **Di Girolamo, M., C. De Marco, V. Busiello, and C. Cini.** 1982. Thialysine and selenalysine as allosteric inhibitors of *E. coli* aspartokinase III. Mol. Cell. Biochem. **49**:43–48.

57. **Duchange, N., M. M. Zakin, P. Ferrara, I. Saint-Girons, I. Park, S. V. Tran, M. C. Py, and G. N. Cohen.** 1983. Structure of the *metJBLF* cluster in *E. coli* K12. Sequence of the *metB* structural gene and of the 5' and 3' flanking regions of the *metBL* operon. J. Biol. Chem. **258**:14868–14871.

58. **Dwivedi, C. M., R. C. Ragin, and J. R. Uren.** 1982. Cloning, purification and characterization of β-cystathionase from *Escherichia coli*. Biochemistry **21**:3064–3069.

59. **Falcoz-Kelly, F., J. Janin, J. C. Saari, M. Véron, P. Truffa-Bachi, and G. N. Cohen.** 1972. Revised structure of aspartokinase I-homoserine dehydrogenase I of *Escherichia coli* K12. Evidence for four identical subunits. Eur. J. Biochem. **28**:507–519.

60. **Falcoz-Kelly, F., R. van Rapenbusch, and G. N. Cohen.** 1969. The methionine-sensitive homoserine dehydrogenase and aspartokinase activities of *Escherichia coli* K12. Preparation of the homogeneous protein catalyzing the two activities. Molecular weight of the enzyme and of its subunits. Eur. J. Biochem. **8**:146–152.

61. **Farkas, W., and C. Gilvarg.** 1965. The reduction step in diaminopimelic acid biosynthesis. J. Biol. Chem. **240**:4717–4722.

62. **Fazel, A., Y. Guillou, and G. N. Cohen.** 1983. A hybrid proteolytic fragment of *Escherichia coli* aspartokinase I-homoserine dehydrogenase I: structure, inhibition pattern, dissociation properties and generation of two homodimers. J. Biol. Chem. **258**:13570–13574.

63. **Fazel, A., K. Müller, G. Le Bras, J. R. Garel, M. Véron, and G. N. Cohen.** 1983. A triglobular model for the polypeptide chain of aspartokinase I-homoserine dehydrogenase I of *Escherichia coli* K12. Biochemistry **22**:158–165.

64. **Ferrara, P., N. Duchange, M. M. Zakin, and G. N. Cohen.** 1984. Internal homologies in *Escherichia coli* aspartokinases-homoserine dehydrogenases. Proc. Natl. Acad. Sci. USA **81**:3019–3023.

65. **Flavin, M.** 1975. Methionine biosynthesis, p. 407–503. *In* D. M. Greenberg (ed.), Metabolic pathways, 3rd ed., vol. 7. Metabolism of sulfur compounds. Academic Press, Inc., New York.

66. **Flavin, M., and C. Slaughter.** 1960. Threonine synthetase mechanism: studies with isotopic hydrogen. J. Biol. Chem. **235**:1112–1118.

67. **Freundlich, M.** 1963. Multivalent repression in the biosynthesis of threonine in *Salmonella typhimurium* and *Escherichia coli*. Biochem. Biophys. Res. Commun. **10**:277–282.

68. **Gardner, J. F.** 1979. Regulation of the threonine operon: tandem threonine and isoleucine codons in the control region and translational control of transcription termination. Proc. Natl. Acad. Sci. USA **76**:1706–1710.

69. **Gardner, J. F.** 1982. Initiation, pausing and termination of transcription in the threonine operon regulatory region of *Escherichia coli*. J. Biol. Chem. **257**:3896–3904.

70. **Gardner, J. F., and W. S. Reznikoff.** 1978. Identification and restriction endonuclease mapping of the threonine operon regulatory region. J. Mol. Biol. **126**:241–258.

71. **Gardner, J. F., and O. H. Smith.** 1975. Operator-promoter functions in the threonine operon of *Escherichia coli*. J. Bacteriol. **124**:161–166.

72. **Gardner, J. F., O. H. Smith, W. W. Fredricks, and M. A. McKinney.** 1974. Secondary-site attachment of coliphage lambda near the *thr* operon. J. Mol. Biol. **90**:613–631.

73. **Garel, J. R., and A. Dautry-Varsat.** 1980. The formation of the native structure in the bifunctional enzymes aspartokinases-homoserine dehydrogenases I and II from *E. coli* K12, and in some of their monofunctional fragments, p. 485–499. *In* R. Jaenicke (ed.), Protein folding. Elsevier-North Holland, Amsterdam.

74. **Garel, J. R., and A. Dautry-Varsat.** 1980. Sequential folding of a bifunctional allosteric protein. Proc. Natl. Acad. Sci. USA **77**:3379–3383.

75. **Gibson, F., and D. D. Woods.** 1960. The synthesis of methionine by suspensions of *Escherichia coli*. Biochem. J. **74**:160–162.

76. **Gilliland, G. L., G. D. Markham, and D. R. Davies.** 1983. Adenosylmethionine synthetase from *Escherichia coli*. Crystallization and preliminary X-ray diffraction studies. J. Biol. Chem. **258**:6963–6964.

77. **Gilson, E., J. M. Clément, D. Brutlag, and M. Hofnung.** 1984. A family of dispersed repetitive extragenic palindromic DNA sequences in *E. coli*. EMBO J. **3**:1417–1421.

78. **Gilvarg, C.** 1959. N-succinyl-L-diaminopimelic acid. J. Biol. Chem. **234**:2955–2959.

79. **Gilvarg, C.** 1961. N-succinyl-α-amino-ε-ketopimelic acid. J. Biol. Chem. **236**:1429–1431.

80. **Greene, R. C., J. S. V. Hunter, and E. H. Coch.** 1973. Properties of *metK* mutants of *Escherichia coli* K-12. J. Bacteriol. **115**:57–67.

81. **Greene, R. C., and A. A. Smith.** 1984. Insertion mutagenesis of the *metJBLF* gene cluster of *Escherichia coli*: evidence for an *metBL* operon. J. Bacteriol. **159**:767–769.

82. **Greene, R. C., C. H. Su, and C. J. Holloway.** 1970. S-adenosyl-methionine synthetase deficient mutants of *Escherchia coli* K12 with impaired control of methionine synthesis. Biochem. Biophys. Res. Commun. **38**:1120–1126.

83. **Greene, R. C.. R. D. Williams, H. F. Kung, C. Spears, and H. Weissbach.** 1973. Effect of methionine and vitamin B12 on the activities of methionine biosynthetic enzymes in *metJ* mutants of *Escherichia coli* K12. Arch. Biochem. Biophys. **158**:249–256.

84. **Gross, T. S., and R. J. Rowbury.** 1969. Methionyl-transfer RNA synthetase mutants of *Salmonella typhimurium* which have normal control of the methionine biosynthetic enzymes. Biochim. Biophys. Acta **184**:233–236.

85. **Guest, J. R., S. Friedman, M. A. Foster, G. Tejerina, and D. D. Woods.** 1964. Transfer of the methyl group from N⁵-methyl tetrahydrofolate to homocysteine in *E. coli*. Biochem. J. **92**:497–504.

86. **Guggenheim, S.** 1971. β-Cystathionase (Salmonella). Methods Enzymol. **17B**:439–442.

87. **Hafner, E. W., C. W. Tabor, and H. Tabor.** 1977. Isolation of a *metK* mutant with a temperature-sensitive S-adenosylmethionine synthetase. J. Bacteriol. **132**:832–840.

88. **Haziza, C., P. Stragier, and J. C. Patte.** 1982. Nucleotide sequence of the *asd* gene of *Escherichia coli*: absence of a typical attenuation signal. EMBO J. **1**:379–384.

89. **Hegeman, G. D., G. N. Cohen, and C. Morgan.** 1970. Aspartate semialdehyde dehydrogenase from *E. coli* K12. Methods Enzymol. **17**:708–713.

90. **Holloway, C. T., R. C. Greene, and C. H. Su.** 1970. Regulation of S-adenosylmethionine synthetase in *Escherichia coli*. J. Bacteriol. **104**:734–747.

91. **Jacob, F., and E. L. Wollman.** 1961. Sexuality and the genetics of bacteria. Academic Press, Inc., New York.

92. **Janin, J.** 1974. Crystallization of *Escherichia coli* aspartokinase I-homoserine dehydrogenase I. FEBS Lett. **45**:318–319.

93. **Jenkins, S. J., C. A. Sparkes, and M. L. Jones-Mortimer.** 1974. A gene involved in lysine excretion in the *Escherichia coli* K12. Heredity **32**:409–412.

94. **Jeter, R. M., B. M. Olivera, and J. R. Roth.** 1984. *Salmonella typhimurium* synthesizes cobalamin (vitamin B₁₂) de novo under anaerobic growth conditions. J. Bacteriol. **159**:206–213.

95. **Johnson, D. I., and R. L. Somerville.** 1983. Evidence that repression mechanisms can exert control over the *thr*, *leu*, and *ilv* operons of *Escherichia coli* K-12. J. Bacteriol. **155**:49–55.

96. **Johnson, D. I., and R. L. Somerville.** 1984. New regulatory genes involved in the control of transcription initiation at the *thr* and *ilv* promoters of *Escherichia coli* K12. Mol. Gen. Genet. **195**:70–76.

97. **Johnson, E. J., G. N. Cohen, and I. Saint-Girons.** 1977. Threonyl-transfer ribonucleic acid synthetase and the regulation of the threonine operon in *Escherichia coli*. J. Bacteriol. **129**:66–70.

98. **Johnson, J. R., R. C. Greene, and J. H. Krueger.** 1977. Isolation and characterization of specialized lambda transducing bacteriophage carrying the *metBJF* gene cluster. J. Bacteriol. **131**:795–800.

99. **Kaminski, M., F. Falcoz-Kelly, P. Truffa-Bachi, J. C. Patte, and G. N. Cohen.** 1969. Antigenic independence of the three aspartokinases of *Escherichia coli*. Eur. J. Biochem. **11**:278–282.

100. **Kaplan, M., and M. Flavin.** 1965. Threonine biosynthesis. On the pathway in fungi and bacteria and the mechanism of the isomerization reaction. J. Biol. Chem. **240**:3928–3933.

101. **Kaplan, M., and M. M. Flavin.** 1966. Cystathionine-γ-synthase of Salmonella. Structural properties of a new enzyme in bacterial methionine biosynthesis. J. Biol. Chem. **241**:5781–5789.

102. **Katinka, M., P. Cossart, L. Sibilli, I. Saint-Girons, M. A. Chalvignac, G. Le Bras, G. N. Cohen, and M. Yaniv.** 1980. Nucleotide sequence of the *thrA* gene of *Escherichia coli*. Proc. Natl. Acad. Sci. USA **77**:5730–5733.

103. **Katzen, H. M., and J. M. Buchanan.** 1965. Enzymatic synthesis of the methyl group of methionine. VII. Repression-derepression, purification and properties of 5-10 methylene tetrahydrofolate reductase from *Escherichia coli*. J. Biol. Chem. **240**:825–835.

104. **Kindler, S. H., and C. Gilvarg.** 1960. N-succinyl-L-α-ε-diaminopimelic acid decarboxylase. J. Biol. Chem. **235**:3532–3535.

105. **Kirby, T. W., B. R. Hindenach, and R. C. Greene.** 1986. Regulation of in vivo transcription of the *Escherichia coli* K-12 *metJBLF* cluster. J. Bacteriol. **165**:671–677.

106. **Lafuma, C., C. Gros, and J. C. Patte.** 1970. Regulation of the lysine biosynthetic pathway of *Escherichia coli* K12. Isolation, molecular weight and amino acid analysis of the lysine-sensitive aspartokinase. Eur. J. Biochem. **15**:111–115.

107. **Lawrence, D. A., D. A. Smith, and R. J. Rowbury.** 1968. Regulation of methionine biosynthesis in *Salmonella typhimurium*: mutants resistant to inhibition by analogues of methionine. Genetics **58**:473–492.

108. **Lee, C. W., J. M. Ravel, and W. Shive.** 1966. Multimetabolic control of a biosynthetic pathway by sequential metabolites. J. Biol. Chem. **291**:5479–5480.

109. **Liljestrand-Golden, C. A., and J. R. Johnson.** 1984. Physical organization of the *metJB* component of the *Escherichia coli* K-12 *metJBLF* gene cluster. J. Bacteriol. **157**:413–419.

110. **Mackall, J. C., and K. E. Neet.** 1974. Studies on the quaternary structure of the threonine-sensitive aspartokinase-homoserine dehydrogenase of *Escherichia coli*. A proposed subunit interaction model. Eur. J. Biochem. **42**:275–282.

111. **Mackie, G. A.** 1980. Cloning of fragments of λ*dapB*2 DNA and identification of the *dapB* gene product. J. Biol. Chem. **255**:8928–8935.

112. **Markham, G. D.** 1981. Spatial proximity of two divalent metal ions at the active site of S-adenosylmethionine synthetase. J. Biol. Chem. **256**:1903–1909.

113. **Markham, G. D., J. De Parisis, and J. Gatmaitan.** 1984. The sequence of *metK*, the structural gene for S-adenosylmethionine in *Escherichia coli*. J. Biol. Chem. **259**:14505–14507.

114. **Markham, G. D., E. W. Hafner, C. W. Tabor, and H. Tabor.** 1980. S-adenosylmethionine synthetase from *Escherichia coli*. J. Biol. Chem. **255**:9082–9092.

115. **Mazat, J. P., and J. C. Patte.** 1976. Lysine-sensitive aspartokinase of *Escherichia coli* K12. Synergy and autosynergy in an allosteric V system. Biochemistry **15**:4053–4058.

116. **Michaeli, S., M. Mevarech, and E. Z. Ron.** 1984. Regulatory region of the *metA* gene of *Escherichia coli* K-12. J. Bacteriol. **160**:1158–1162.

117. **Michaeli, S., and E. Z. Ron.** 1981. Construction and physical mapping of plasmids containing the *metA* gene of *Escherichia coli* K12. Mol. Gen. Genet. **182**:349–354.

118. **Michaeli, S., and E. Z. Ron.** 1984. Expression of the *metA* gene of *E. coli* K12 in recombinant plasmids. FEBS Lett. **23**:125–129.

119. **Michaeli, S., and E. Z. Ron.** 1984. The *metC* gene in *Escherichia coli* K12: isolation and studies of relatedness in Enterobacteriaceae. FEMS Microbiol. Lett. **22**:31–33.

120. **Milner, L., C. Whitfield, and H. Weissbach.** 1969. Effects of L-methionine and vitamin B12 on methionine biosynthesis in *Escherichia coli*. Arch. Biochem. Biophys. **133**:413–419.

121. **Monod, J., J. Wyman, and J. P. Changeux.** 1965. On the nature of allosteric transitions: a plausible model. J. Mol. Biol. **12**:88–118.

122. **Mouhli, H., M. M. Zakin, C. Richaud, and G. N. Cohen.** 1980. Detection of the homology among the aspartokinase I-homoserine dehydrogenase I and the aspartokinase III from *E. coli* K12 by immunochemical cross-reactivity between denatured species. Biochem. Int. **1**:403–409.

123. **Nass, G., and J. Thomale.** 1974. Alteration of structure or level of threonyl-tRNA synthetase in borrelidin resistant mutants of *Escherichia coli.* FEBS Lett. **39:**182–186.

124. **Niles, E. G., and E. W. Westhead.** 1973. The variable subunit structure of lysine-sensitive aspartylkinase from *Escherichia coli* TIR-8. Biochemistry **12:**1715–1722.

125. **Niles, E. G., and E. W. Westhead.** 1973. *In vitro* adenylylation of lysine-sensitive aspartylkinase from *Escherichia coli* TIR-8. Biochemistry **12:**1723–1729.

126. **Nisman, B., G. N. Cohen, S. B. Wiesendanger, and M. L. Hirsch.** 1954. Transformation de l'acide aspartique en homosérine et en thréonine par des extraits de *Escherichia coli.* C. R. Acad. Sci. Paris **238:**1342–1344.

127. **Pacaud, M., and C. Richaud.** 1975. Protease II from *Escherichia coli.* Purification and characterization. J. Biol. Chem. **250:**7771–7779.

128. **Parsot, C., P. Cossart, I. Saint-Girons, and G. N. Cohen.** 1983. Nucleotide sequence of *thrC* and of the transcription termination region of the threonine operon in *Escherichia coli* K12. Nucleic Acids Res. **11:**7331–7345.

129. **Parsot, C., I. Saint-Girons, and P. Cossart.** 1982. DNA sequence change of a deletion mutation abolishing attenuation control of the threonine operon of *E. coli* K12. Mol. Gen. Genet. **188:**455–458.

130. **Patte, J. C., and G. N. Cohen.** 1964. Interactions coopératives effecteur-effecteur chez deux enzymes allostériques inhibées de manière noncompétitive. C. R. Acad. Sci. Paris **259:**1255–1258.

131. **Patte, J. C., G. Le Bras, and G. N. Cohen.** 1967. Regulation by methionine of the synthesis of a third aspartokinase and a second homoserine dehydrogenase in *Escherichia coli* K12. Biochim. Biophys. Acta **136:**245–257.

132. **Patte, J. C., T. Loviny, and G. N. Cohen.** 1962. Répression de la décarboxylase de l'acide méso-α-ε-diaminoplimélique par la L-lysine chez *Escherichia coli.* Biochim. Biophys. Acta **58:**359–360.

133. **Patte, J. C., T. Loviny, and G. N. Cohen.** 1963. Rétro-inhibition et répression de l'homosérine déshydrogénase d'*Escherichia coli.* Biochim. Biophys. Acta **67:**16–30.

134. **Patte, J. C., T. Loviny, and G. N. Cohen.** 1965. Effets inhibiteurs coopératifs de la L-lysine avec d'autres amino acides sur une aspartokinase d'*Escherichia coli.* Biochim. Biophys. Acta **99:**523–530.

135. **Patte, J. C., P. Morand, E. Boy, C. Richaud, and F. Borne.** 1980. The *relA* locus and the regulation of lysine biosynthesis in *Escherichia coli.* Mol. Gen. Genet. **179:**319–325.

136. **Patte, J. C., P. Truffa-Bachi, and G. N. Cohen.** 1966. The threonine-sensitive homoserine dehydrogenase and aspartokinase activities of *Escherichia coli.* I. Evidence that the two activities are carried by a single protein. Biochim. Biophys. Acta **128:**426–439.

137. **Peterkofsky, B., and C. Gilvarg.** 1961. N-succinyl-L-diaminopimelic-glutamic transaminase. J. Biol. Chem. **236:**1432–1438.

138. **Press, R., N. Glansdorff, P. Miner, J. de Vries, R. Kadner, and W. K. Maas.** 1971. Isolation of transducing particles of φ80 bacteriophage that carry different regions of the *Escherichia coli* genome. Proc. Natl. Acad. Sci. USA **68:**795–798.

139. **Richaud, C., J. P. Mazat, B. Felenbok, and J. C. Patte.** 1974. The role of lysine and leucine binding on the catalytical and structural properties of aspartokinase III of *Escherichia coli* K12. Eur. J. Biochem. **48:**147–156.

140. **Richaud, C., J. P. Mazat, C. Gros, and J. C. Patte.** 1973. Subunit structure of the lysine-sensitive aspartokinase of *Escherichia coli* K12. Eur. J. Biochem. **40:**619–629.

141. **Richaud, C., F. Richaud, C. Martin, C. Haziza, and J. C. Patte.** 1984. Regulation of expression and nucleotide sequence of the *Escherichia coli dapD* gene. J. Biol. Chem. **259:**14824–14828.

142. **Richaud, F., N. H. Phuc, M. Cassan, and J. C. Patte.** 1980. Regulation of aspartokinase III synthesis in *Escherichia coli:* isolation of mutants containing *lysC-lac* fusions. J. Bacteriol. **143:**513–515.

143. **Richaud, F., C. Richaud, C. Haziza, and J. C. Patte.** 1981. Isolement et purification de gènes d'*Escherichia coli* K12 impliqués dans la biosynthèse de la lysine. C. R. Acad. Sci. Paris **293:**507–512.

144. **Ron, E. Z.** 1975. Growth rate of *Escherichia coli* at elevated temperatures: limitation by methionine. J. Bacteriol. **124:**243–246.

145. **Ron, E. Z., and M. Shani.** 1971. Growth rate of *Escherichia coli* at elevated temperatures: reversible inhibition of homoserine transsuccinylase. J. Bacteriol. **107:**397–400.

146. **Rosenberg, M., and D. Court.** 1979. Regulatory sequences involved in the promotion and termination of RNA transcription. Annu. Rev. Genet. **13:**319–353.

147. **Rowbury, R. J.** 1968. The inhibitory action of α-methylmethionine on *Escherichia coli.* J. Gen. Microbiol. **52:**223–230.

148. **Rowbury, R. J.** 1983. Methionine biosynthesis and its regulation, p. 191–211. *In* K. M. Hermann and R. L. Somerville (ed.), Amino acids: biosynthesis and genetic regulation. Addison-Wesley Publishing Co., Reading, Mass.

149. **Rowbury, R. J., and D. D. Woods.** 1961. Further studies in the repression of methionine synthesis in *Escherichia coli.* J. Gen. Microbiol. **24:**129–144.

150. **Rowbury, R. J., and D. D. Woods.** 1964. Repression by methionine of cystathionine formation in *Escherichia coli.* J. Gen. Microbiol. **35:**145–148.

151. **Rowbury, R. J., and D. D. Woods.** 1966. The regulation of cystathionine formation in *Escherichia coli.* J. Gen. Microbiol. **42:**155–163.

151a. **Saint-Girons, I.** 1978. New regulatory mutations affecting the expression of the threonine operon in *E. coli* K12. Mol. Gen. Genet. **162:**95–100.

152. **Saint-Girons, I., N. Duchange, G. N. Cohen, and M. M. Zakin.** 1984. Structure and autoregulation of the *metJ* regulatory gene in *E. coli.* J. Biol. Chem. **259:**14282–14286.

153. **Saint-Girons, I., N. Duchange, M. M. Zakin, I. Park, D. Margarita, P. Ferrara, and G. N. Cohen.** 1983. Nucleotide sequence of *metF*, the *E. coli* structural gene for 5-10 methylene tetrahydrofolate reductase, and of its control region. Nucleic Acids Res. **11:**6723–6732.

154. **Saint-Girons, I., and D. Margarita.** 1975. Operator-constitutive mutants in the threonine operon of *Escherichia coli* K-12. J. Bacteriol. **124:**1137–1141.

155. **Saint-Girons, I., and D. Margarita.** 1978. Fine structure analysis of the threonine operon in *Escherichia coli* K12. Mol. Gen. Genet. **162:**101–107.

156. **Saint-Girons, I., and D. Margarita.** 1985. Evidence for an internal promoter in the *Escherichia coli* threonine operon. J. Bacteriol. **161:**461–462.

157. **Schildkraut, I., and S. Greer.** 1973. Threonine synthetase-catalyzed conversion of phosphohomoserine to α-ketobutyrate in *Bacillus subtilis.* J. Bacteriol. **115:**777–785.

158. **Schulte, L. L., L. T. Stauffer, and G. V. Stauffer.** 1984. Cloning and characterization of the *Salmonella typhimurium metE* gene. J. Bacteriol. **158:**928–933.

159. **Selker, S., and C. Yanofsky.** 1979. Nucleotide sequence of the *trpC-trpB* intracistronic region from *Salmonella typhimurium.* J. Mol. Biol. **130:**135–143.

160. **Shedlarski, J. G., and C. Gilvarg.** 1970. The pyruvate-aspartic semialdehyde condensing enzyme of *Escherichia coli.* J. Biol. Chem. **245:**1362–1373.

161. **Shimada, K., R. A. Weisberg, and M. E. Gottesman.** 1972. Prophage lambda at unusual chromosomal locations. I. Location of the secondary attachment sites and the properties of the lysogens. J. Mol. Biol. **63:**483–503.

162. **Shoeman, R., T. Coleman, B. Redfield, R. C. Greene, A. A. Smith, I. Saint-Girons, N. Brot, and H. Weissbach.** 1985. Regulation of methionine synthesis in *Escherichia coli:* effect of *metJ* gene product and S-adenosylmethionine on the *in vitro* expression of the *metB, metL* and *metJ* genes. Biochem. Biophys. Res. Commun. **133:**731–739.

163. **Shoeman, R., B. Redfield, T. Coleman, N. Brot, H. Weissbach, R. C. Greene, A. A. Smith, I. Saint-Girons, M. M. Zakin, and G. N. Cohen.** 1985. Regulation of the methionine regulon in *Escherichia coli.* BioEssays **3:**210–213.

164. **Shoeman, R., B. Redfield, T. Coleman, R. C. Greene, A. A. Smith, N. Brot, and H. Weissbach.** 1985. Regulation of methionine synthesis in *Escherichia coli:* effect of *metJ* gene product and S-adenosylmethionine on the expression of the *metF* gene. Proc. Natl. Acad. Sci. USA **82:**3601–3605.

165. **Sibilli, L., P. Cossart, M. A. Chalvignac, P. Briley, J. M. Costrejean, G. Le Bras, and G. N. Cohen.** 1977. The primary structure of *Escherichia coli* K12 aspartokinase I-homoserine dehydrogenase I. Distribution of the methioninyl residues and of the cysteinyl and tryptophanyl tryptic peptides. Biochimie **59:**943–946.

166. **Sibilli, L., G. Le Bras, G. Le Bras, and G. N. Cohen.** 1981. Two regions of the bifunctional protein aspartokinase I-homoserine dehydrogenase I are connected by a short hinge. J. Biol. Chem. **256:**10228–10230.

167. **Skarstedt, M. T., and J. B. Greer.** 1973. Threonine synthase of *Bacillus subtilis.* The nature of an associated dehydratase activity. J. Biol. Chem. **248:**1032–1044.

168. **Smith, A., and R. C. Greene.** 1984. Cloning of the methionine regulatory gene, *metJ*, or *E. coli* K12 and identification of its product. J. Biol. Chem. **259:**14279–14281.

169. **Smith, A. A., R. C. Greene, T. W. Kirby, and B. R. Hindenach.** 1985. Isolation and characterization of the product of the methionine-regulatory gene *metJ* of *Escherichia coli* K12. Proc. Natl. Acad. Sci. USA **82:**6104–6108.

170. **Smith, D. A.** 1961. S-amino acid metabolism and its regulation in *Escherichia coli* and *Salmonella typhimurium*. Adv. Genet. **16:**141–165.

171. **Stadtman, E. R., G. N. Cohen, G. Le Bras, and H. de Robichon-Szulmajster.** 1961. Feedback inhibition and repression of aspartokinase activity in *Escherichia coli* and *Saccharomyces cerevisiae*. J. Biol. Chem. **236:**2033–2038.

172. **Starnes, W. L., P. Munk, S. B. Maul, G. N. Cunningham, D. J. Cox, and W. Shive.** 1972. Threonine-sensitive aspartokinase-homoserine dehydrogenase complex, amino acid composition, molecular weight, and subunit composition of the complex. Biochemistry **11:**677–687.

173. **Stephens, J. C., S. W. Artz, and B. N. Ames.** 1975. Guanosine 5'-diphosphate-3'-diphosphate (ppGpp): positive effector for histidine operon transcription and general signal for amino acid deficiency. Proc. Natl. Acad. Sci. USA **72:**4389–4393.

174. **Stragier, P., O. Danos, and J. C. Patte.** 1983. Regulation of diaminopimelate decarboxylase synthesis in *Escherichia coli*. II. Nucleotide sequence of the *lysA* gene and its regulatory region. J. Mol. Biol. **168:**321–331.

175. **Stragier, P., and J. C. Patte.** 1983. Regulation of diaminopimelate decarboxylase synthesis in *E. coli*. III. Nucleotide sequence and regulation of the *lysR* gene. J. Mol. Biol. **168:**333–350.

176. **Stragier, P., F. Richaud, F. Borne, and J. C. Patte.** 1983. Regulation of diamiopimelate decarboxylase activity in *Escherichia coli*. I. Nucleotide sequence of a *lysR* gene encoding an activator of the *lysA* gene. J. Mol. Biol. **168:**307–320.

177. **Su, C. H., and R. C. Greene.** 1971. Regulation of methionine biosynthesis in *Escherichia coli*: mapping of the *metJ* locus and properties of a *metJ*⁺-*metJ*⁻ diploid. Proc. Natl. Acad. Sci. USA **68:**367–371.

178. **Szczesiul, M., and D. E. Wampler.** 1976. Regulation of a metabolic system *in vitro*: synthesis of threonine from aspartic acid. Biochemistry **15:**2236–2244.

179. **Szentirmai, A., M. Szentirmai, and H. E. Umbarger.** 1968. Isoleucine and valine metabolism of *Escherichia coli*. XV. Biochemical properties of mutants resistant to thiaisoleucine. J. Bacteriol. **95:**1672–1679.

180. **Tamir, H., and C. Gilvarg.** 1974. Dihydrodipicolinic acid reductase. J. Biol. Chem. **249:**3034–3040.

180a.**Taylor, R. T., and H. Weissbach.** 1967. *N*⁵-Methyltetrahydrofolate-homocysteine transmethylase. Partial purification and properties. J. Biol. Chem. **242:**1502–1508.

181. **Thèze, J., L. Kleidman, and I. Saint-Girons.** 1974. Homoserine kinase from *Escherichia coli*: properties, inhibition by L-threonine, and regulation of biosynthesis. J. Bacteriol. **118:**577–581.

182. **Thèze, J., D. Margarita, G. N. Cohen, F. Borne, and J. C. Patte.** 1974. Mapping of the structural genes of the three aspartokinases and of the two homoserine dehydrogenases of *Escherichia coli* K-12. J. Bacteriol. **117:**133–143.

183. **Thèze, J., and I. Saint-Girons.** 1974. Threonine locus of *Escherichia coli* K-12: genetic structure and evidence for an operon. J. Bacteriol. **118:**990–998.

184. **Tran, V. S., E. Schaeffer, O. Bertrand, R. Mariuzza, and P. Ferrara.** 1983. Purification, molecular weight and N-terminal sequence of cystathionine-γ-synthase of *Escherichia coli*. J. Biol. Chem. **258**(Appendix):14872–14873.

185. **Truffa-Bachi, P., and G. N. Cohen.** 1966. La β-aspartokinase sensible à la lysine d'*Escherichia coli*. Purification et propriétés. Biochim. Biophys. Acta **113:**531–541.

186. **Truffa-Bachi, P., J. C. Patte, and G. N. Cohen.** 1967. Sur la dihydrodipicolinate synthétase d'*E. coli* K12. C. R. Acad. Sci. Paris **265:**928–929.

187. **Truffa-Bachi, P., R. van Rapenbusch, J. Janin, C. Gros, and G. N. Cohen.** 1968. The threonine-sensitive homoserine dehydrogenase and aspartokinase activities of *Escherichia coli* K12. 4. Isolation, molecular weight, amino acid analysis and behaviour of the sulfhydryl groups of the protein catalyzing the two activities. Eur. J. Biochem. **5:**73–80.

188. **Umbarger, H. E.** 1978. Amino acid biosynthesis and its regulation. Annu. Rev. Biochem. **47:**533–606.

189. **Urbanowski, M. L., and G. V. Stauffer.** 1985. Cloning and initial characterization of the *metB* and *metJ* genes from *Salmonella typhimurium* LT₂. Gene **35:**187–197.

190. **Urbanowski, M. L., and G. V. Stauffer.** 1985. Nucleotide sequence and biochemical characterization of the *metJ* gene from *Salmonella typhimurium* LT₂. Nucleic Acids Res. **13:**673–685.

191. **Véron, M., and G. N. Cohen.** 1974. Intra- and interprotomeric interactions between the catalytic regions of aspartokinase I-homoserine dehydrogenase I from *Escherichia coli* K12, p. 335–347. *In* E. H. Fisher, E. G. Krebs, H. Neurath, and E. R. Stadtman (ed.), Metabolic interconversion of enzymes. Springer-Verlag, New York.

192. **Véron, M., F. Falcoz-Kelly, and G. N. Cohen.** 1972. The threonine-sensitive homoserine dehydrogenase and aspartokinase activities of *Escherichia coli* K12. The two catalytic activities are carried by two independent regions of the polypeptide chain. Eur. J. Biochem. **28:**520–527.

193. **Véron, M., Y. Guillou, and G. N. Cohen.** 1985. Isolation of the aspartokinase domain of bifunctional aspartokinase I-homoserine dehydrogenase I. FEBS Lett. **181:**381–384.

194. **Véron, M., Y. Guillou, A. Fazel, and G. N. Cohen.** 1985. Reversible dissociation of aspartokinase I-homoserine dehydrogenase I from *Escherichia coli* K12. The active species is the tetramer. Eur. J. Biochem. **151:**521–524.

195. **Véron, M., J. C. Saari, C. Villar-Palasi, and G. N. Cohen.** 1973. The threonine-sensitive homoserine dehydrogenase and aspartokinase activities of *Escherichia coli* K12. Intra and intersubunit interactions between the catalytic regions of the bifunctional enzyme. Eur. J. Biochem. **38:**325–335.

196. **von Dippe, P. J., A. Abraham, C. A. Nelson, and W. G. Smith.** 1972. Kinetic and molecular properties of lysine-sensitive aspartokinase. Quaternary structure, catalytic activity, and feedback control. J. Biol. Chem. **247:**2433–2438.

197. **Wampler, D. E.** 1972. Threonine-sensitive aspartokinase-homoserine dehydrogenase from *Escherichia coli*. Subunit stoichiometry and size of the catalytic unit. Biochemistry **11:**4428–4435.

198. **Wampler, D. E., and E. W. Westhead.** 1968. Two aspartokinases from *Escherichia coli*. Nature of the inhibition and molecular changes accompanying reversible inhibition. Biochemistry **7:**1661–1671.

199. **White, P. J.** 1976. The regulation of diaminopimelate decarboxylase activity in *Escherichia coli* strain W. J. Gen. Microbiol. **96:**51–62.

200. **White, P. J., and B. Kelly.** 1965. Purification and properties of diaminopimelate decarboxylase from *Escherichia coli*. Biochem. J. **96:**75–84.

201. **Whitfield, C. D., E. J. Steers, and H. Weissbach.** 1970. Purification and properties of 5-methyl-tetrahydropteroyltriglutamate-homocysteine transmethylase. J. Biol. Chem. **245:**390–401.

202. **Wijesundera, S., and D. D. Woods.** 1960. Suppression of methionine synthesis in *Escherichia coli* by growth in the presence of this amino acid. J. Gen. Microbiol. **22:**229–241.

203. **Work, E.** 1962. Diaminopimelic racemase. Methods Enzymol. **5:**858–864.

204. **Work, E.** 1962. Diaminopimelic decarboxylase. Methods Enzymol. **5:**864–870.

205. **Wormser, E. H., and A. B. Pardee.** 1958. Regulation of threonine biosynthesis in *Escherichia coli*. Arch. Biochem. Biophys. **78:**416–423.

206. **Yanofsky, C.** 1981. Attenuation in the control of expression of bacterial operons. Nature (London) **289:**751–758.

207. **Yugari, Y., and C. Gilvarg.** 1965. The condensation step in diaminopimelate synthesis. J. Biol. Chem. **240:**4710–4716.

208. **Zakin, M. M., N. Duchange, P. Ferrara, and G. N. Cohen.** 1983. Nucleotide sequence of the *metL* gene of *Escherichia coli*. Its product, the bifunctional aspartokinase II-homoserine dehydrogenase II and the bifunctional product of the *thrA* gene, aspartokinase I-homoserine dehydrogenase I derive from a common ancestor. J. Biol. Chem. **258:**3028–3031.

209. **Zakin, M. M., J. R. Garel, A. Dautry-Varsat, G. N. Cohen, and G. Boulot.** 1978. Detection of the homology among proteins by immunochemical cross-reactivity between denatured antigens. Application to the threonine and methionine regulated aspartokinase-homoserine dehydrogenase from *E. coli* K12. Biochemistry **17:**4318–4323.

210. **Zakin, M. M., R. C. Greene, A. Dautry-Varsat, N. Duchange, P. Ferrara, M. C. Py, D. Margarita, and G. N. Cohen.** 1982. Construction and physical mapping of plasmids containing the *metJBLF* gene cluster of *E. coli* K12. Mol. Gen. Genet. **187:**101–106.

B2. Biosynthesis and Conversions of Nucleotides

29. Purines and Pyrimidines

JAN NEUHARD AND PER NYGAARD

Enzyme Division, University Institute of Biological Chemistry B, DK-1307 Copenhagen, Denmark

OVERVIEW

Cellular Distribution of Purines and Pyrimidines

Purines and pyrimidines are normally not present intracellularly as free bases or nucleosides; they are almost exclusively found as nucleotides, i.e., linked through an N-glycosylic bond to either ribose or 2-deoxyribose phosphate. More than 95% of the nucleotides are found in the acid-insoluble fraction of the cells as nucleic acids. The acid-soluble fraction consists of nucleoside mono-, di-, and triphosphates (NMP, NDP, and NTP, respectively), nucleotide-containing coenzymes, and effectors (Table 1).

NTPs as Precursors

NTPs are the immediate precursors of nucleic acids and certain coenzymes. In addition, ATP and GTP drive many biosynthetic processes. A small fraction of the nucleic acids, i.e., the mRNAs, as well as many of the nucleotide-containing coenzymes, are metabolically unstable. Although they constitute less than 10% of total cellular nucleotides, their rapid turnover results in a higher demand for NTPs than does biosynthesis of stable nucleic acids. However, the synthesis of stable RNA and DNA requires de novo synthesis of purine and pyrimidine nucleotides; the turnover of mRNA and the nucleotide coenzymes does not. The products of mRNA turnover are NMPs and NDPs (Table 2).

Conversion of NMP and NDP to NTP

The eight naturally occurring NTPs can be derived from the corresponding NMPs and NDPs by successive phosphorylations catalyzed by nucleotide kinases.

NMP kinases specific for AMP, GMP, CMP, UMP, and dTMP have been resolved. AMP kinase (95, 107), GMP kinase (208), and dTMP kinase (187) from *Escherichia coli* and UMP kinase (J. Justesen, Ph.D. thesis, University of Copenhagen, Copenhagen, Denmark, 1975) from *Salmonella typhimurium* have been purified to homogeneity. The enzymes for AMP, GMP, and CMP are active with the corresponding 2'-deoxyribonucleotides, whereas UMP kinase is specific for UMP. dTMP kinase can use dUMP but not TMP as substrate. All of these enzymes use MgATP as the preferred phosphate donor. The gene encoding AMP kinase from *E. coli* has been sequenced (36).

Mutations within the structural genes for AMP kinase (*adk* [71]), GMP kinase (*gmk* [B. Jochimsen, Ph.D. thesis, University of Copenhagen, Copenhagen, Denmark, 1979]), UMP kinase (*pyrH* [109, 216]), and dTMP kinase (*tmk* [49]) that result only in temperature-sensitive- or cold-sensitive-growth phenotypes have been described. Thus, these four enzymes are essential for growth. In contrast, studies on *cmk* mutants indicate that CMP kinase is dispensable (20). Certain *adk* mutants are defective in phospholipid synthesis. Genetic and biochemical studies of such mutants suggest that AMP kinase, in addition to its essential role in nucleotide metabolism, may be directly involved in phospholipid synthesis through formation of a complex with the membrane-bound *sn*-glycerol-3-phosphate acyltransferase (72). Based on the observation that *E. coli* F' factors containing the wild-type *pyrH* allele are not tolerated in *S. typhimurium*, it has been proposed that UMP kinase has a similar structural function in *S. typhimurium* and that the *E. coli* enzyme inhibits that function competitively (129). This may also offer an explanation for the osmotic-remedial phenotype of many *pyrH* mutants (276).

Nucleoside diphosphokinase of *S. typhimurium* catalyzes the transfer of the γ-phosphate of any nucleoside 5'-triphosphate to any nucleoside 5'-diphosphate. A phosphorylated form of the enzyme is an intermediate in the reaction (69). The *E. coli* enzyme has been reported to be loosely bound to the outer surface of the cytoplasmic membrane (225), a location that seems incompatible with a function of generating intracellular nucleic acid precursors.

An *S. typhimurium* mutant with an altered nucleoside diphosphokinase has been isolated (70, 224). Studies of this mutant have indicated that the mutation lies in the structural gene (*ndk*). The nucleoside diphosphokinase content of *S. typhimurium* decreases with increasing growth rate (69), but nucleoside diphosphokinase activity probably never limits NTP synthesis, because its activity is considerably higher than the activities of the NMP kinases, even in fast-growing cells (70). The physiological role of the growth rate-dependent variations in nucleoside diphosphokinase activity is obscure.

TABLE 1. Intracellular amounts of nucleotides and PRPP[a]

Compound[b]	μmol/g (dry wt)	Intracellular concn (μM)	Compound[b]	μmol/g (dry wt)	Intracellular concn (μM)
RNA			UTP	2.08	894
AMP	165		UDP	0.22	93
GMP	203		UMP	0.33	142
CMP	126		dTTP	0.18	77
UMP	136				
			ADPGlc	0.01	5
DNA			GDPMan	0.04	18
dAMP	24.7		UDPGlc(Gal)	1.33	570
dGMP	25.4		UDPGlcNAc	0.38	164
dCMP	25.4		UDPAcMurP$_5$	0.55	234
dTMP	24.4		dTDPGlc	0.58	248
ATP	7.00	3,000			
ADP	0.58	250	NAD + NADH	1.88	806
AMP	0.24	105	NADP + NADPH	0.47	200
dATP	0.41	175	FAD	0.12	51
			FMN	0.21	88
GTP	2.15	923			
GDP	0.30	128			
GMP	0.05	20	AcCoA	0.54	231
dGTP	0.28	122	SucCoA	0.03	15
			MalCoA	0.03	15
IMP	0.38	162			
			cAMP	0.01	6
CTP	1.20	515	ppGpp	0.07	31
CDP	0.19	81	pppGpp	0.04	18
dCTP	0.15	65	PRPP	1.24[c]	472

[a] Determined in cells growing exponentially in glucose minimal medium. The values for RNA and DNA are for *E. coli* (108). The other values are for *S. typhimurium* (30). To convert the intracellular concentrations into micromoles per gram (dry weight), we have assumed that 1 g in bacterial dry weight corresponds to 2.33 g of intracellular water.

[b] Abbreviations: Glc, glucose; Man, mannose; Gal, galactose; GlcNAc, *N*-acetylglucosamine; UDPAcMurP$_5$, uridine 5'-diphospho-*N*-acetylmuramyl pentapeptide; FAD, flavin adenine dinucleotide; FMN, flavin mononucleotide; AcCoA, acetyl coenzyme A; SucCoA, succinyl coenzyme A; MalCoA, malonyl coenzyme A; pppGpp, guanosine 5'-triphosphate 3'-diphosphate.

[c] From data of Jensen (116).

Pyruvate kinase also catalyzes the phosphorylation of a wide range of nucleoside 5'-diphosphates (231). Anaerobically grown *E. coli* contains quite low levels of nucleoside diphosphokinase but increased pyruvate kinase activity. It has been suggested that under such conditions pyruvate kinase replaces nucleoside diphosphokinase in synthesizing NTPs.

De Novo Synthesis of NTPs

With the exception of CTP, which is formed by amination of UTP, the ribonucleoside triphosphates

TABLE 2. Metabolic function and turnover of nucleotide-containing compounds

Compound(s)[a]	Product for which compound(s) is required	Product(s) of turnover
mRNA	Proteins	AMP, GMP, CMP, UMP
GTP	Proteins	GDP
NDP sugars[b]	Polysaccharides	NDP
UDPGlcNAc	Peptidoglycans	UDP
UDPAcMurP$_5$	Peptidoglycans	UMP
CDP diglycerides	Phospholipids	CMP
NAD	DNA (ligations)	AMP

[a] GlcNAc, *N*-Acetylglucosamine; UDPAcMurP$_5$, uridine 5'-diphospho-*N*-acetylmuramyl pentapeptide.

[b] N can be A, G, U, and dT.

are derived de novo from the corresponding ribonucleoside monophosphates, as outlined above. Similarly, dTTP is formed by successive phosphorylations of dTMP. The other deoxyribonucleoside triphosphates are derived by reduction of ribonucleoside diphosphates to 2'-deoxyribonucleoside diphosphates, followed by phosphorylation catalyzed by nucleoside diphosphokinase. Thus the sole physiological function of CMP kinase is to rephosphorylate CMP produced by turnover of nucleic acids and of CDP diglycerides.

Ribonucleoside monophosphates may be derived from exogenous purines and pyrimidines via the so-called salvage pathways, or they may be synthesized de novo from simpler precursors. In either case, the ribose 5-phosphate moiety is derived from 5-phosphoribosyl-1-pyrophosphate.

5-PHOSPHORIBOSYL-1-PYROPHOSPHATE

Physiological Function

The synthesis of 5-phosphoribosyl-α-1-pyrophosphate (PRPP) from ATP and ribose 5-phosphate is catalyzed by phosphoribosylpyrophosphate synthetase, as follows: ATP + ribose 5-phosphate → PRPP + AMP. The equilibrium constant of the reaction in the direction of PRPP synthesis is 29 (241). PRPP is required for the biosynthesis of purine and pyrimidine nucleotides, histidine, tryptophan, and pyridine nucleotides. A total of

10 enzymes compete for PRPP as substrate. A culture growing exponentially in glucose-minimal medium with a doubling time of 50 min utilizes about 8 μmol of PRPP per min per mg (dry weight). Of that amount, 30 to 40% is used for the synthesis of purine nucleotides; an equal amount is used for synthesis of pyrimidine nucleotides. Histidine and tryptophan each consume about 10 to 15%; only 1 to 2% is used for the synthesis of nicotinamide coenzymes (116).

PRPP Synthetase

Enzymology and regulation. PRPP synthetase has been purified to homogeneity from both *S. typhimurium* (241) and *E. coli* (105). The molecular mass of the subunit from both sources is 31 kilodaltons (kDa). The native enzyme exists in several stages of aggregation. Under assay conditions, the predominant form has a molecular mass of about 160 kDa (234). There are only minor differences in amino acid composition between the enzymes from the two bacteria (105).

P_i is essential to stability of the enzyme. Magnesium ions and ATP also stabilize the enzyme, but they cannot replace P_i (241). In addition, magnesium ions and P_i are required for activity. Magnesium ions are needed both as chelators of ATP and as free cations. The function of P_i in catalysis is not clear, but it has been shown to bind to the enzyme near the substrate-binding site (77).

Many nucleotides inhibit enzyme activity competitively with ATP. However, for most nucleotides only high, unphysiological concentrations cause significant inhibition (242). The only potent nucleotide inhibitor is ADP; it competes with ATP and is an allosteric inhibitor that binds to a site different from the active site (65). The pattern of ADP inhibition is highly influenced by the concentration of ribose 5-phosphate, which displays substrate inhibition in the presence of ADP (242).

Genetics and regulation of expression. Mutants with altered PRPP synthetase, i.e., *prs* mutants, have been obtained in both *E. coli* and *S. typhimurium* (104, 106, 123). One of the *E. coli* mutants produces a PRPP synthetase with greatly increased K_m values for both ATP and ribose 5-phosphate. The *S. typhimurium prs* mutant has only 16% of the wild-type PRPP synthetase activity, and the mutant enzyme is somewhat more heat labile in crude extract than the parental enzyme.

Pandey and Switzer (211) have isolated an *S. typhimurium* mutant that lacks PRPP synthetase activity in vitro and contains low PRPP pools (20% of wild type). The mutation is linked to *hisW*, and thus does not lie within *prs* (211). The genetic locus affected in the mutant has not been identified, and its function in the synthesis of PRPP synthetase is unknown. The nucleotide sequence of the *prs* gene from *E. coli* has been determined (105).

Expression of *prs* is regulated. By starving for different end products of PRPP metabolism, it was found that only uracil starvation resulted in derepression of PRPP synthetase synthesis. Up to 10-fold increased levels of the enzyme have been observed in partially uracil-starved cultures of *S. typhimurium*. The nature of the repressing metabolite is not known, but studies on mutants specifically blocked in pyrimidine nucleotide interconversion indicated that the effector is a uracil nucleotide different from UMP (209, 269). The level of PRPP synthetase is twofold elevated in an *rpoBC* mutant of *S. typhimurium* that contains fourfold increased UTP pools and derepressed levels of certain pyrimidine de novo enzymes. This observation suggests that syntheses of PRPP synthetase and of the pyrimidine de novo enzymes (see Genetic Regulation of the Pathway, below) may share common regulatory elements (120).

PRPP Pools

Measurements of intracellular PRPP under a variety of different growth conditions have provided some information on how the supply of this essential metabolite is regulated in vivo (13, 106, 116, 118, 230). The intracellular PRPP pool is influenced by the carbon source. In certain *E. coli* strains, proportionality between PRPP pool size and growth rate was observed, whereas such strict correlation was not seen in *S. typhimurium*. Starvation for carbon, nitrogen, or phosphate resulted in rapid depletion of PRPP. Under most of these conditions, a similar variation in the NTP pools was observed. In contrast, conditions that specifically stimulate RNA synthesis caused transient depletion of NTPs and transient elevation of PRPP. Purine starvation caused severe depletion of all of the NTP pools concomitantly with an accumulation of PRPP. Nevertheless, pyrimidine starvation resulted in depletion of PRPP and swelling of the purine nucleotide pools. These results, together with results obtained from starvation experiments with guanine and adenine auxotrophic mutants, demonstrated that conditions that lead to high pools of adenine nucleotide result in reduction of the PRPP pool. The metabolite responsible for these effects could be ADP, which is a negative allosteric effector of PRPP synthetase.

Addition of purine bases to exponential-phase cultures of *E. coli* and *S. typhimurium* results in an immediate shrinkage of the intracellular PRPP pool (Table 3). Adenosine and inosine have the same effect as hypoxanthine, whereas guanosine addition only reduces the pool to 50% (14).

DE NOVO SYNTHESIS OF ATP AND GTP

All intermediates in the pathway by which purine ribonucleotides are synthesized de novo (Fig. 1) are 5-phosphoribosyl derivatives; the first purine nucleotide formed is IMP. AMP and GMP are synthesized from IMP via separate branches of the pathway. From these monophosphates, ATP and GTP are synthesized by phosphorylations, with the intermediate formation of ADP and GDP (see Conversion of NMP to NTP, above).

TABLE 3. Intracellular PRPP pools in cells grown in the presence or absence of purine supplement[a]

Purine base (0.1 mM)	PRPP pool (μmol/g [dry wt]) in:	
	E. coli	*S. typhimurium*
Unsupplemented	1.30	1.24
Adenine	0.04	0.12
Hypoxanthine	0.08	0.32
Guanine	0.17	0.69

[a] Data are taken from Jensen (116).

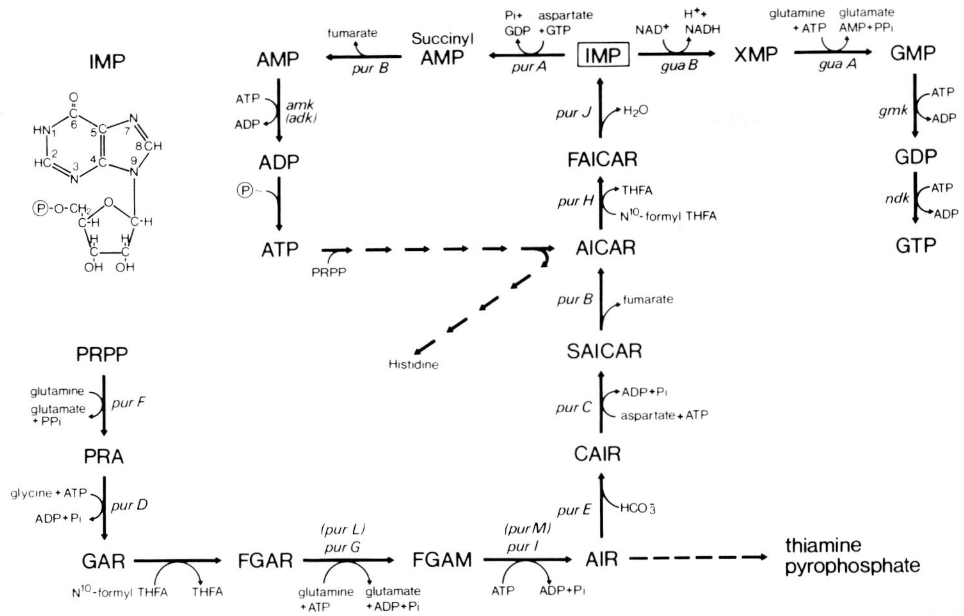

FIG. 1. De novo synthesis of ATP and GTP. The individual enzymes are identified by their gene symbols as used in *S. typhimurium*. The gene symbols in *E. coli* are, in most cases, identical; exceptions are given in parentheses. THFA, tetrahydrofolic acid; PRA, 5-phosphoribosylamine; GAR, 5'-phosphoribosyl-1-glycinamide; FGAM, 5'-phosphoribosyl-*N*-formylglycinamidine; CAIR, 5'-phosphoribosyl-5-aminoimidazole carboxylic acid; SAICAR, 5'-phosphoribosyl-4-(*N*-succino-carboxamide)-5-aminoimidazole; FAICAR, 5'-phosphoribosyl-4-carboxamide-5-formamidoimidazole. For definitions of other abbreviations, see the text.

Biosynthesis of IMP

Characterization of the pathway. The purine ring of IMP is formed in 10 enzyme reactions by stepwise additions of functional groups to an activated form of ribose 5-phosphate, i.e., PRPP.

In the first step, the amide group of glutamine displaces the pyrophosphate group of PRPP to form 5-phosphoribosylamine. This step is catalyzed by PRPP amidotransferase (Table 4). During the reaction, the configuration of carbon atom 1 of ribose undergoes inversion from α to β. In the second step, the amino group of 5-phosphoribosylamine reacts with ATP and the carboxyl group of glycine to form 5'-phosphoribosyl-1-glycinamide and ADP. Glycine supplies atoms 4, 5, and 7 of the purine ring. The free amino group of the glycine residue is formylated by 10-formyltetrahydrofolic acid to give 5'-phosphoribosyl-*N*-formylglycinamide (FGAR) (52). At this point, nitrogen atom 3 of the purine ring is introduced by transfer of the amide group from glutamine catalyzed by FGAR amidotransferase (Table 4). The reaction is energized by ATP, and the product formed is 5'-phosphoribosyl-*N*-formylglycinamidine. An ATP-depend-

TABLE 4. Properties of certain enzymes of purine biosynthesis and interconversion

Enzyme	Gene designation	Origin	Structure	Molecular mass of subunit (kDa)	DNA sequence known?	Activators and inhibitors[a]
Adenylosuccinate synthetase	*purA*	*E. coli*	NK[b]	56[c]	No	GDP(−), AMP(−), ppGpp(−)
AIR synthetase	*purM*	*E. coli*	a_2[d]	38	Yes[d]	
AMP glucosylase	*amn*[e]	*E. coli*	a_6	52	No	ATP(+), P$_i$(−)
FGAR amidotransferase	*purG*	*S. typhimurium*	NK	130[c]	No	
GMP reductase	*guaC*	*S. typhimurium*	a_4	45	No	ATP(−), GTP(+)
GMP synthetase	*guaA*	*E. coli*	a_2	58	Yes	
IMP dehydrogenase	*guaB*	*E. coli*	a_4	54[f]	Yes[f]	GMP(−), ppGpp(−)
PRPP amidotransferase	*purF*	*E. coli*	a_3 or a_4	53	Yes	AMP(−), GMP(−)
SAICAR synthetase[g]	*purC*	*E. coli*	NK	27	No	

[a] −, Inhibitors; +, activators.
[b] NK, Not known.
[c] Molecular mass of the native enzyme.
[d] References 233a and 237a.
[e] Reference 148.
[f] References 247 and 250.
[g] SAICAR, 5'-Phosphoribosyl-4-(*N*-succinocarboxamide)-5-aminoimidazole.

ent ring closure completes the formation of the imidazole ring of 5'-phosphoribosyl-5-aminoimidazole (AIR).

Carbon atom 6 is next introduced from bicarbonate by AIR carboxylase (210) to yield 5'-phosphoribosyl-5-aminoimidazole-4-carboxylic acid, followed by a two-step amination in which the amino group of aspartate is transferred to the carboxylic group of 5'-phosphoribosyl-5-aminoimidazole-4-carboxylic acid. In the first step, the N-acyl aspartate derivative 5'-phosphoribosyl-4-(N-succinocarboxamide)-5-aminoimidazole is formed, catalyzed by 5'-phosphoribosyl-4-(N-succinocarboxamide)-5-aminoimidazole synthetase (Table 4) (212). This reaction requires ATP. Subsequently, fumarate is eliminated, resulting in the formation of 5'-phosphoribosyl-4-carboxamide-5-aminoimidazole (AICAR). This latter reaction is catalyzed by the bifunctional enzyme adenylosuccinate lyase, which also has a function in the synthesis of AMP from IMP (see Conversion of IMP to AMP, below). The final carbon atom of the purine ring (carbon atom 2) is introduced by transfer of the formyl group from 10-formyltetrahydrofolate to the 5-amino group of AICAR to yield 5'-phosphoribosyl-4-carboxamide-5-formamidoimidazole. In the last step, the second ring is closed by elimination of one molecule of water to form IMP. This ring closure does not require ATP.

A more extensive discussion of the purine biosynthetic pathway may be found in reviews by Hartman (85) and Henderson (86).

An important feature of purine biosynthesis is its linkage to histidine biosynthesis (Fig. 1). One carbon atom and one nitrogen atom of the imidazole ring of histidine are derived from the purine ring of ATP, and the purine biosynthetic intermediate AICAR is formed as a by-product.

Inhibitors that interfere with IMP synthesis include the following: the glutamine analogs azaserine and 6-diazo-5-oxo-norleucine, which compete with glutamine for the two amidotransferases of the pathway; folic acid analogs, e.g., aminopterin; and sulfonamides, which inhibit purine biosynthesis indirectly by reducing the supply of folate coenzymes and result in the accumulation of AICAR and its triphosphate derivative (31).

Since most of the intermediates of the pathway are not readily available, few enzymological studies on purine biosynthesis have been conducted. The first enzyme of the pathway, PRPP amidotransferase, has been extensively purified from E. coli (Table 4) (173). The amino acid sequence of the subunit polypeptide has been deduced from the DNA sequence of the purF gene (254). Like a number of other amidotransferases, PRPP amidotransferase can use both ammonia and glutamine as N donors. Binding studies with radiolabeled 6-diazo-5-oxo-norleucine established that the amino-terminal cysteine is the active site residue required for the glutamine amide transfer reaction (253). The enzyme is subject to feedback inhibition by purine nucleotides which act competitively with PRPP. In the presence of GMP, the hyperbolic PRPP saturation curve is converted into a sigmoid curve. AMP and GMP are the strongest inhibitors of the enzyme; in combination, they act synergistically (173).

The second amidotransferase of the pathway, FGAR amidotransferase, has been purified from S. typhimu-

rium (Table 4) (37). Labeling of the active site of this enzyme by the glutamine analog azaserine, followed by degradation of the labeled enzyme, revealed a specific amino acid sequence containing an alkylated cysteine. This amino acid sequence shows significant homology with amino acid sequences found in the glutamine active sites of other amidotransferases (204).

Genetics of the pathway. The structural genes encoding 8 of the 10 enzymes of the pathway have been identified in both organisms. Mutations in purJ have not been identified in E. coli, and mutants defective in the third enzyme, 5'-phosphoribosyl-1-glycinamide formyltransferase, have not been isolated in either organism. Two contiguous and complementing genes, $purE_1$ and $purE_2$, have been implied to specify the sixth enzyme of the pathway, i.e., AIR carboxylase, in S. typhimurium. The genes occur mostly as single units; however, purJ, purH, and purD, as well as $purE_1$ and $purE_2$, appear to constitute operons (74). The purF gene is the distal gene of a polycistronic operon. The first gene of this operon encodes a protein of unknown function (166).

Purine auxotrophic mutants blocked only in the de novo synthesis of IMP can grow with adenine, hypoxanthine, or guanine. In contrast, purB mutants grow only with adenine. Since all intermediates of the pathway are phosphorylated, they cannot be supplied exogenously. The only intermediate that can be supplied exogenously is AICAR, which can be formed from 4-carboxamide-5-aminoimidazole by the salvage enzyme adenine phosphoribosyltransferase.

The reactions leading to the formation of AIR (Fig. 1) are also required for the synthesis of thiamine; hence, mutants blocked in one of the first five steps of purine biosynthesis concurrently have a thiamine requirement. However, other purine-requiring mutants may also display a partial thiamine requirement owing to repression of the pathway by the purines added to satisfy the growth requirement (194). A chromatographic technique which allows identification of the enzyme defect in mutants blocked in any of the first five enzymes of the pathway has been developed (102). Mutants defective in one of the five last enzymes excrete diazotizable amines when starved for purines. By determining the nature of the chromatophore formed, the enzyme defect can be identified. The method does not allow a distinction between purE and purC mutants (76, 267).

Regulation of the pathway. Purine nucleotide biosynthesis is regulated both through control of the activity of PRPP amidotransferase, the first enzyme of the pathway, and by repression and derepression of the pur genes. As described above, PRPP amidotransferase is sensitive to feedback inhibition by AMP and GMP. Direct evidence for the operation of this control in vivo was obtained by measuring the amount of intermediates excreted into the medium by purine auxotrophic mutants. Addition of purines to such cultures caused an immediate decrease in the amount of excreted intermediates (73). The reversal by thiamine of the growth-inhibiting effect of adenine is also in keeping with feedback inhibition by AMP of an early step in purine biosynthesis (87, 177).

The strong reduction in the intracellular concentration of PRPP after addition of purines to cultures of E.

TABLE 5. Effect of exogenous purines on the levels of the four first purine de novo enzymes in *S. typhimurium*[a]

Addition to the medium (0.1 mM)	Enzyme activity (nmol/min per mg of protein) of[b]:			
	PRPP amidotransferase (*purF*)	GAR synthetase (*purD*)	GAR formyltransferase	FGAR amidotransferase (*purG*)
None	66	16	39	10
Adenine	56	19	39	11
Hypoxanthine	15	2	3	3
Guanine	24	8	9	5

[a] From data of Houlberg and Jensen (103); similar results have been obtained with *E. coli* (P. Nygaard and B. Hove-Jensen, unpublished results).
[b] GAR, 5'-Phosphoribosyl-1-glycinamide.

coli and *S. typhimurium* (Table 3) may have a significant effect on the activity of PRPP amidotransferase, since inhibition of the enzyme by purine nucleotides is significantly enhanced at low PRPP concentrations (173).

Expression of the *pur* genes is repressed when preformed purines are added to the growth medium (Table 5) and derepressed when purine auxotrophs are starved for purines. The nature of the purine effector responsible for this control is not known. Koduri and Gots (136) isolated a protein from *E. coli* that binds to *pur* regions of DNA, provided ATP or GTP is simultaneously present. Based on these observations, they suggested that the low-molecular-weight effectors controlling *pur* gene expression are ATP and GTP. In contrast, it has been observed that addition of guanine or hypoxanthine to *S. typhimurium* and *E. coli* mutants defective in the conversion of these purine bases to the corresponding nucleotides caused complete repression of *pur* gene expression and a decrease in the intracellular ATP and GTP pools (103; P. Nygaard, unpublished results). These observations led to the proposal that free hypoxanthine and guanine, rather than purine NTPs, are the true repressing metabolites.

A number of regulatory mutants that show derepressed synthesis of at least eight of the enzymes involved in IMP synthesis have been isolated from *E. coli* (150, 151) and *S. typhimurium* (27, 248). Expression of *purB* is not affected, and it is unknown whether expression of *purC* is altered by these regulatory mutations. Such mutants, designated *purR* mutants, have been obtained in various ways. They were found among mutants resistant to 6-mercaptopurine (see Table 9), they arise rather frequently during growth of *purA* mutants, and they have been selected in *pur::lac* fusion strains on the basis of lactose utilization. None of these regulatory mutations has been characterized genetically.

Conversion of IMP to AMP

AMP is synthesized from IMP in a two-step, aspartate-dependent amination with succinyl-AMP as an intermediate (Fig. 1). The first step is a condensation of aspartate and position 6 of IMP, with GTP acting as a specific dehydrating agent. This reaction is catalyzed by adenylosuccinate synthetase encoded by *purA* (Table 4). The enzyme has been purified from *E. coli* (240). It is inhibited by AMP and GDP, as well as by the IMP analog 6-mercaptopurine ribonucleotide (54, 228). In addition, guanosine-5'-diphosphate 3'-diphosphate (ppGpp) is a potent inhibitor of the enzyme; this indicates that AMP synthesis is under stringent control (239).

The second step is catalyzed by adenylosuccinate lyase (*purB*). Adenylosuccinate lyase is a bifunctional enzyme that catalyzes the removal of fumarate from 5'-phosphoribosyl-4-(*N*-succinocarboxamide)-5-aminoimidazole and from succinyl-AMP to form AICAR and AMP, respectively (29).

Genetics and regulation of expression. The *purA* and *purB* genes are unlinked on the chromosome; *purA* and *purB* mutants require adenine for growth. The regulation of *purA* and *purB* expression has been studied in *E. coli* strains containing *purB::lac* fusions (272). Expression of *purA* is derepressed in response to adenine or guanine starvation in the presence of excess of the other purine base and hypoxanthine. Expression of *purB* responds similarly to adenine starvation in the presence of either hypoxanthine or guanine alone but not if both are present simultaneously. Thus, *purA* and *purB* are regulated separately in *E. coli*. The putative regulatory protein of purine biosynthesis, isolated by Koduri and Gots (136), binds to *purA*-containing DNA in the presence of ATP. Expression of *purB* by *S. typhimurium* is decreased twofold when adenine is added to the medium (29).

Conversion of IMP to GMP

GMP is derived from IMP in two steps. The first step is an NAD-dependent oxidation of IMP in position 2 catalyzed by IMP dehydrogenase (Table 4). The product formed is XMP. The enzyme from *E. coli* consists of four identical subunits with a cysteine residue at the IMP-binding site (67, 68, 141). It is inhibited by GMP and ppGpp (61). GMP inhibition of the *S. typhimurium* enzyme has also been reported (44).

In the second step, XMP is aminated at carbon atom 2 in a glutamine-dependent process requiring ATP; concomitantly, ATP is cleaved to AMP and pyrophosphate (Fig. 1) (213). The reaction is catalyzed by the amidotransferase GMP synthetase (Table 4). The enzyme from *E. coli* is composed of two apparently identical subunits and can use either glutamine or ammonia as the amino donor; both activities are inhibited irreversibly by glutamine analogs (214, 278). A glutamine amide transfer domain has been identified in an amino-terminal segment of the enzyme (277). GMP synthetase is uniquely inhibited by the naturally occurring adenosine analogs psicofuranine and decoyinine (238).

Genetics and regulation of expression. IMP dehydrogenase and GMP synthetase are specified by the two contiguous genes *guaB* and *guaA*, which constitute an operon (142, 197, 233). In *E. coli*, the gene order is promoter-*guaB*-*guaA*, with a control region lying proximal to *guaB* (236). A 68-base-pair intercistronic region separates *guaA* and *guaB* (251). A *guaA* mutant requires guanine, whereas a *guaB* mutant can grow on either guanine or xanthine.

In *E. coli*, the *gua*-operon is under dual control. It is repressed by guanine nucleotides and induced by adenine nucleotides (171, 246). By using *gua::lac* fusion strains, regulatory mutants with increased expression of the *gua* operon have been isolated. The regulatory mutations have been divided into three classes: linked *cis*-acting, linked *trans*-acting, and unlinked *trans*-acting regulatory mutations (249).

Interconversion of Adenine and Guanine Compounds

Adenine nucleotides and guanine nucleotides may be interconverted through the common precursor, IMP. These pathways are particularly important when adenine and guanine compounds are available in the culture medium, and they are essential for purine auxotrophic mutants when only a single purine is supplied.

Conversion of ATP to IMP. When histidine is synthesized de novo, a molar equivalent amount of AICAR, and hence IMP, is formed. Histidine inhibits the first enzyme of the histidine pathway, ATP phosphoribosyltransferase, and it represses the synthesis of all of the histidine biosynthetic enzymes. Thus, exogenous histidine prevents the formation of IMP from ATP (164, 201). The pathway may also be subject to stringent control, because ppGpp is a potent inhibitor of ATP phosphoribosyltransferase, provided a partially inhibitory concentration of histidine is also present (176).

Conversions involving free adenine. AMP can be converted to IMP through the intermediate formation of adenine, as shown in Fig. 2. This conversion is initiated by the hydrolytic cleavage of AMP to adenine

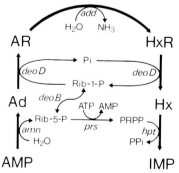

FIG. 2. *The purine nucleoside phosphorylase-dependent pathway for conversion of AMP to IMP. Enzymes are identified by their gene symbols:* add, *adenosine deaminase;* amn, *AMP glycosylase;* deoB, *phosphopentomutase;* deoD, *purine nucleoside phosphorylase;* hpt, *hypoxanthine phosphoribosyltransferase;* prs, *PRPP synthetase. Ad, Adenine; AR, adenosine; Hx, hypoxanthine; HxR, inosine; Rib-1-P, ribose 1-phosphate; Rib-5-P, ribose 5-phosphate.*

and ribose 5-phosphate catalyzed by AMP glycosylase (AMP nucleosidase). The adenine thus formed may be converted to IMP through the intermediate formation of adenosine, inosine, and hypoxanthine catalyzed by the salvage enzymes purine nucleoside phosphorylase, adenosine deaminase, and hypoxanthine phosphoribosyltransferase, as described in more detail under Utilization of Preformed Purine Compounds, below. The ribose 5-phosphate produced in the AMP glycosylase reaction can be converted to ribose 1-phosphate by phosphopentomutase. This conversion promotes the formation of adenosine by pushing the purine nucleoside phosphorylase reaction in the direction of nucleoside formation. In the final step of the pathway, the ribose phosphate residue may serve as a precursor of the PRPP required for IMP formation from hypoxanthine.

AMP glycosylase from *E. coli* is an oligomer of identical subunits (Table 4) (147). It is activated by MgATP and inhibited by P_i. In the presence of either activator or inhibitor it displays positive cooperativity with respect to AMP. The physiological function of this enzyme is not known, but its regulatory properties suggest that it may balance the supply of adenine and guanine nucleotides.

Adenine is also produced in rather large quantities as a result of *S*-adenosylmethionine turnover. The large number of *S*-adenosylmethionine-dependent reactions result in the production of stoichiometric amounts of *S*-adenosylhomocysteine and 5′-methylthioadenosine, which are subsequently hydrolyzed to adenine (201).

Conversion of GMP to IMP. GMP reductase (*guaC*) catalyzes the reductive deamination of GMP to IMP, with NADPH as the electron donor. The enzyme has been purified from both *E. coli* and *S. typhimurium*, and it was found to be specifically inhibited by physiological concentrations of ATP. The inhibition is completely reversed by GTP (Table 4) (165; Jochimsen, Ph.D. thesis). The level of GMP reductase is increased up to 10-fold when guanine is added to the culture medium, and this induction is prevented by adenine (26, 133, 196). In *E. coli*, but not in *S. typhimurium*, this induction requires cyclic AMP and the cyclic AMP receptor protein (25, 26). In *S. typhimurium*, glutamine starvation leads to increased levels of the enzyme (63), and the enzyme is inactivated under conditions of purine starvation (Jochimsen, Ph.D. thesis). Thus, it appears that the conversion of GMP to IMP is regulated by the ratio of guanine nucleotides to adenine nucleotides and that glutamine may play a role in this control.

DE NOVO SYNTHESIS OF UTP AND CTP

Characterization of the Pathway

The pathway responsible for the synthesis of the two RNA precursors UTP and CTP is shown in Fig. 3. It is initiated by the formation of carbamoylphosphate from glutamine, bicarbonate, and ATP catalyzed by carbamoylphosphate synthase (Table 6). This enzyme consists of two nonidentical subunits specified by the *carAB* operon in *E. coli* and the *pyrA* operon in *S. typhimurium*. Carbamoylphosphate is also an intermediate in arginine biosynthesis. Thus, carba-

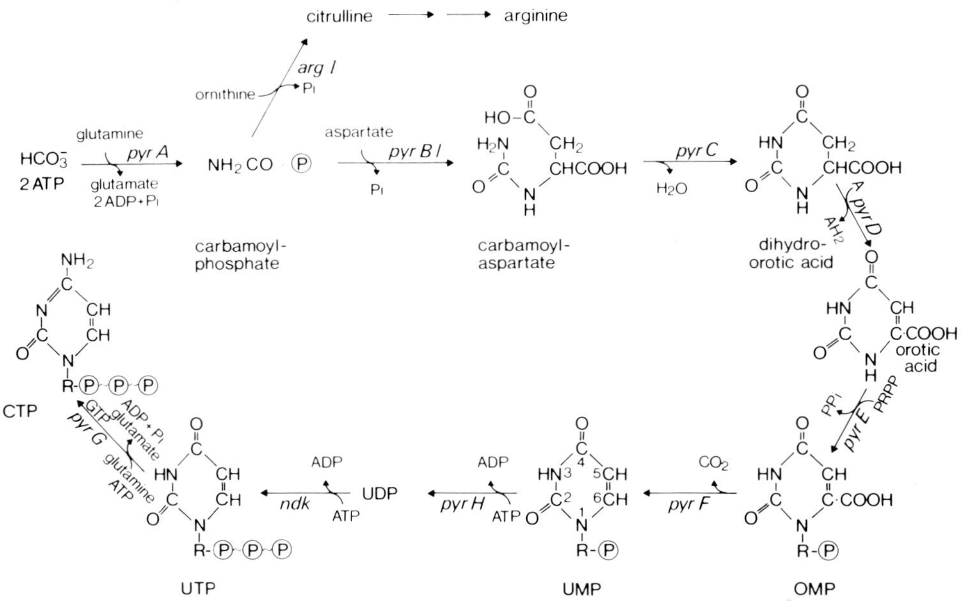

FIG. 3. De novo synthesis of UTP and CTP. The individual enzymes are identified by their gene symbols: *argI*, ornithine carbamoyltransferase; *ndk*, nucleoside diphosphokinase; *pyrA* (*carAB*), carbamoylphosphate synthase; *pyrBI*, aspartate carbamoyltransferase; *pyrC*, dihydroorotase; *pyrD*, dihydroorotate oxidase; *pyrE*, orotate phosphoribosyltransferase; *pyrF*, orotidine 5'-phosphate decarboxylase; *pyrG*, CTP synthetase; *pyrH*, UMP kinase.

moylphosphate synthase has a dual function in metabolism, as is reflected by its regulation.

The first reaction unique to pyrimidine biosynthesis is catalyzed by aspartate carbamoyltransferase (Table 6), a multimeric enzyme containing two kinds of protein subunits. It catalyzes the condensation of carbamoylphosphate with the amino group of aspartate, forming carbamoylaspartate (ureidosuccinate)

TABLE 6. Properties of pyrimidine biosynthetic enzymes

Enzyme/gene description	Origin	Structure/molecular mass of subunit (kDa)	Subunit DNA sequence known?	Activators, inhibitors, and cofactors[a]
Carbomoylphosphate synthase		ab		
carA, carB	E. coli	$a = 40, b = 120$	Yes, yes	UMP($-$), IMP($+$), ornithine($+$), NH_3($+$)
pyrA	S. typhimurium	$a = 45, b = 110$	No, no	UMP($-$), IMP($+$), ornithine($+$), PRPP($+$)
Aspartate carbamoyltransferase		$2c_3\ 3r_2$		
pyrI, pyrB	E. coli	$r^b = 17, c^b = 34$	Yes, yes	CTP($-$), ATP($+$), Zn^{2+}
pyrI, pyrB	S. typhimurium	$r^b = 17, c^b = 33$	Yes,[c] yes[c]	CTP($-$), ATP($+$)
Dihydroorotase		a_2		
pyrC	E. coli	$a = 38.5$	Yes[d]	Zn^{2+}
pyrC	S. typhimurium	$a = 36.5$	Yes	ND
Dihydroorotate oxidase/pyrD	E. coli	$a_2/a = 37$	Yes	Flavin mononucleotide
Orotate phosphoribosyltransferase		a_2		
pyrE	E. coli	$a = 23.4$	Yes	
pyrE	S. typhimurium	$a = 23$	P[e]	
OMP decarboxylase		a_2		
pyrF	E. coli	$a = 27$	Yes[f]	
pyrF	S. typhimurium	$a = 27$	Yes[g]	
CTP synthetase/pyrG	E. coli	$a_4/a = 50$	Yes[h]	GTP($+$), ATP($+$), UTP($+$)

[a] $-$, Inhibitor; $+$, activator; ND, not determined.
[b] r, Regulatory polypeptide chain; c, catalytic polypeptide chain.
[c] R. A. Kelln, personal communication.
[d] L. Lundberg, personal communication.
[e] P, Only part of the DNA sequence has been determined.
[f] C. L. Turnbough, personal communication.
[g] M. Theisen and J. Neuhard, unpublished results.
[h] Reference 264a.

in a reaction driven by the cleavage of the high-energy phosphate anhydride bond of carbamoylphosphate. The structural genes specifying the two different subunits of the enzyme are adjacent on the chromosome and constitute the *pyrBI* operon.

In the two subsequent steps, the pyrimidine orotic acid is formed through cyclization and oxidation. The first reaction shows some resemblance to the formation of a peptide bond, although it is fully reversible. It is catalyzed by dihydroorotase (*pyrC*), a dimeric protein of identical subunits containing one atom of tightly bound zinc per subunit (Table 6) (119, 263). Dihydroorotate oxidase (*pyrD*) is membrane bound and linked to the electron transport system of the cell (128, 244). The enzyme from *E. coli* has been purified to homogeneity, and it was found to consist of two identical, flavin-containing subunits (Table 6) (119, 143). There is evidence that the physiological electron acceptor in the presence of oxygen is ubiquinone (132). In anaerobically grown cells, menaquinone appears to be an obligatory hydrogen carrier between dihydroorotate and fumarate (195).

The nucleotide orotidylic acid (OMP) is formed by the transfer of ribose 5-phosphate from PRPP to orotic acid, and subsequently OMP is decarboxylated to UMP. The reactions are catalyzed by orotate phosphoribosyltransferase (*pyrE*) and OMP decarboxylase (*pyrF*), both of which are dimeric proteins (Table 6) (53, 193, 221). The nucleotide sequences of the entire *E. coli pyrE* gene (221) and the 5' end of the *S. typhimurium pyrE* gene (193) are known. They show large homologies; of the first 50 N-terminal amino acids of orotate phosphoribosyltransferase from the two organisms, only one amino acid is different.

The last steps in the generation of UTP involve phosphorylation of UMP to UDP and of UDP to UTP by the sequential action of UMP kinase (*pyrH*) and nucleoside diphosphokinase (*ndk*) (see Conversion of NMP and NDP to NTP, above). Finally, UTP is converted to CTP by CTP synthetase (*pyrG*), a tetrameric protein of identical subunits (Table 6) (138). The CTP synthetase reaction is an amination of UTP at the expense of ATP. In the presence of the positive effector GTP, glutamine is the preferred nitrogen donor. Thus, the last step in the pyrimidine pathway involves all four RNA precursors.

Genetics of the Pathway

Mutants defective in each step of the pathway have been isolated in *S. typhimurium* and, with the exception of *ndk* mutants, also in *E. coli*. The genetic loci specifying the nine enzymes have been identified and shown to be scattered on the chromosome. The direction of transcription of the *carAB* operon (66), the *pyrBI* operon (174, 227), and *pyrC*, *pyrD*, *pyrE*, and *pyrF* (119, 131) has been established. With the exception of *pyrF*, which is located on a DNA segment inverted in *S. typhimurium* relative to its orientation in *E. coli*, the orientations of the genes are the same in the two organisms.

Mutations in *carAB* (*pyrA*) usually produce a dual requirement for arginine and a pyrimidine, an observation indicating that only a single enzyme is responsible for the production of carbamoylphosphate for both arginine and pyrimidine biosynthesis. Mutations

in *pyrBI*, *pyrC*, *pyrD*, *pyrE*, or *pyrF* cause only pyrimidine auxotrophy. Except for orotate, none of the intermediates of the pathway is permeable to the cells. Orotate will satisfy the growth requirement of *pyrB*, *pyrC*, and *pyrD* mutants, provided glycerol is used as carbon source. Ureidosuccinate (carbamoylaspartate)-permeable mutants (*usp*) have been isolated (146, 243). A *pyrB usp* double mutant will grow on carbamoylaspartate as the sole pyrimidine source.

Mutants blocked in the last two steps of UMP synthesis, i.e., *pyrE* and *pyrF* mutants, accumulate orotic acid in the medium, particularly if limited for pyrimidines. Certain strains of *E. coli* K-12, but not *E. coli* B or *S. typhimurium* LT2, excrete large quantities of orotic acid into the growth medium (163). Genetically, this phenotype is linked to *pyrF* and may indicate that orotate-excreting strains contain a less active (mutated?) OMP decarboxylase (273).

Recently, it has been shown that *pyrE* and *pyrF* both are part of small bicistronic operons (119, 220). The first gene in the *pyrE* operon specifies a 26-kDa protein of unknown function. Since insertion of Mu d1 into this gene does not cause any recognizable phenotype except for Pyr⁻, the protein is probably not essential for growth (193). The *pyrF* gene is the first gene of the *pyrF* operon. The second gene specifies a small 13-kDa protein of unknown function.

Mutants defective in UMP kinase (*pyrH* mutants) and in nucleoside diphosphokinase (*ndk* mutants) are necessarily "leaky" (see Conversion of NMP and NDP to NTP, above). Phenotypically, *pyrH* mutants show increased resistance to a number of toxic pyrimidine analogs, such as 5-fluorouracil, 5-fluorouridine, and 5-fluoroorotate (109, 216, 276), because they excrete large quantities of pyrimidines. Mutants with reduced UMP kinase activity contain low endogenous UTP pools; this in turn causes derepression of the synthesis of the six enzymes involved in UMP synthesis and hence overproduction and excretion of pyrimidines.

Mutants unable to synthesize CTP owing to mutations in the *pyrG* gene require cytidine for growth (190, 216). Thus, CTP synthetase is the only enzyme capable of converting uracil nucleotides to cytosine nucleotides. However, the cytidine requirement can only be satisfied if the strain is defective in cytidine deaminase (*cdd*); i.e., a *pyrG* mutation is incompatible with a *cdd*⁺ allele (19).

Properties of Certain De Novo Enzymes

Carbamoylphosphate synthase (*carAB* [*pyrA*]). Carbamoylphosphate synthase is an allosteric enzyme which catalyzes the synthesis of carbamoylphosphate from bicarbonate and either ammonia or glutamine at the expense of two molecules of ATP. Studies on purified enzyme preparations from both *E. coli* (7–10) and *S. typhimurium* (5) have revealed the following properties. The MgATP saturation curve is sigmoidal, whereas the enzymes display Michaelis-Menten kinetics with respect to glutamine or bicarbonate. Enzyme activity is inhibited by UMP and activated by ornithine, IMP, and, for the *Salmonella* enzyme, also by PRPP. The effectors influence activity by altering the affinity of the enzyme for MgATP.

The enzyme is a monomer that can be reversibly dissociated into two nonidentical subunits (5, 252).

The heavy subunit binds all of the substrates, except glutamine, as well as the effectors. It catalyzes the synthesis of carbamoylphosphate from ammonia but not from glutamine. This ammonia activity is regulated normally and displays sigmoidal MgATP saturation curves. The light subunit binds glutamine and has glutaminase activity.

The *carAB* (*pyrA*) operon contains the structural genes for the light (*carA*) and the heavy (*carB*) subunits oriented from *A* to *B* (5, 48, 66). The nucleotide sequence of the entire operon from *E. coli* has been determined (203, 219, 268). The deduced amino acid sequence of the large subunit exhibits a highly significant homology between the amino- and carboxy-terminal halves; thus, the *carB* gene may have been derived from a smaller ancestral gene by duplication. Comparison of the amino acid sequence of the small subunit with sequences of other amidotransferases led to the identification of a putative catalytic domain centered around a specific cysteine residue in the protein (204).

As one might expect for an enzyme that catalyzes a reaction involving three substrates and several allosteric effectors, a number of different phenotypes are found among mutants affected in carbamoylphosphate synthase (4, 172). The most common phenotype is a dual requirement for arginine and pyrimidine owing to lack of enzyme activity. Auxotrophy for only arginine is another consequence of missense mutations within *carAB* (*pyrA*). A significant class of these mutants is cold sensitive; i.e., they require arginine only at low temperatures (20°C). The cold-sensitive arginine requirement of one such *S. typhimurium* mutant is caused by accumulation of *N*-acetylornithine, an intermediate of the arginine pathway. Apparently, *N*-acetylornithine antagonizes the maturation of the mutant enzyme (2).

Uracil and arginine sensitivity may also result from mutations in *carAB* (*pyrA*). One uracil-sensitive mutation has been shown to cause hypersensitivity of the enzyme to feedback inhibition by UMP (217). The biochemical basis for the arginine-sensitive phenotype was studied in *S. typhimurium* (1). It appeared that ornithine carbamoyltransferase, an enzyme of arginine biosynthesis, is required for proper assembly of the subunits of the mutant enzyme. Thus, arginine inhibits growth by repressing the synthesis of ornithine carbamoyltransferase. Whether ornithine carbamoyltransferase is also involved in the assembly of the wild-type carbamoylphosphate synthase is not known, but the possibility is intriguing.

Aspartate carbamoyltransferase (*pyrBI*). Aspartate carbamoyltransferase from *E. coli* is probably the most extensively studied regulatory enzyme (for reviews, see references 64, 112, 125, and 126). The amino acid sequence (137, 264) and the three-dimensional structure of the enzyme (99) are known. The enzyme is a dodecamer composed of two catalytic trimers (c_3) and three regulatory dimers (r_2). Positive homotropic cooperativity is observed with the substrates aspartate and carbamoylphosphate, and heterotropic effects are induced by the inhibitor CTP and the activator ATP. Treatment with mercurials or heat dissociates the holoenzyme into catalytic (c_3) and regulatory (r_2) subunits. The catalytic subunit is enzymatically active but lacks the homotropic response to the substrates, and it is insensitive to inhibition by CTP. The regulatory subunit is catalytically inactive but contains the common binding site for the allosteric effectors CTP and ATP. Fully active holoenzyme, which shows normal allosteric behavior, may be formed from the isolated subunits under proper conditions (34, 169). Zinc ions are required for proper reconstitution (186).

The cooperative homotropic and heterotropic interactions are caused by changes in the quaternary structure of the holoenzyme, and they involve intricate protein:protein interactions between both homologous chains (c:c and r:r interactions) and heterologous chains (c:r interactions) (38, 126, 139, 274). Modification of aspartate carbamoyltransferase has been used extensively to study the structural basis for catalysis and cooperativity, and both chemically modified (125, 144) and mutationally modified (127, 223, 232) enzymes have been investigated.

Aspartate carbamoyltransferase from *S. typhimurium* appears to be very similar to the *E. coli* enzyme both in structure and in catalytic and regulatory properties (207, 270). Intergeneric hybrid enzymes composed of *E. coli* catalytic subunits with *S. typhimurium* regulatory subunits, and vice versa, show cooperative aspartate saturation curves.

The gene for the catalytic chain, *pyrB*, and the gene for the regulatory chain, *pyrI*, are organized in an operon transcribed from *B* to *I* (215, 227, 243). Fine structure maps of the *pyrB* gene of *E. coli* and *S. typhimurium* have been constructed (114, 243). Two *E. coli pyrBI* mutants have been obtained that produce catalytically active aspartate carbamoyltransferases but that lack substrate cooperative interactions and are insensitive to inhibition by CTP. One of the mutations was in *pyrB* (271), and the other mutation was in *pyrI* (56). A number of feedback-modified *S. typhimurium* mutants in which the enzyme is partially desensitized to inhibition by CTP were isolated by O'Donovan and Gerhart (206). Apart from being cotransducible with the *pyrB* locus, the mutations were not further characterized. Interallelic complementation experiments with *pyrB* mutants of *E. coli* defined four complementation groups (114). The alterations in the amino acid sequence of the catalytic chains in four of the mutants were identified, and certain properties of the altered proteins were established (232). In combination with the known three-dimensional structure of the wild-type enzyme, such studies provide the structural basis for the mechanism of interallelic complementation.

The nucleotide sequence of the entire *pyrBI* operon has been determined, establishing the following order: promoter-leader-*pyrB-pyrI* (100, 232; R. A. Kelln, personal communication). Curiously, the nucleotide sequence downstream from *pyrI* includes a transcriptional termination structure and overlapping transcriptional initiation signals for a gene of unknown function (J. R. Wild, personal communication).

Ornithine carbamoyltransferase, encoded by *argI*, resembles the catalytic subunit of aspartate carbamoyltransferase in several ways. It is a trimer of three identical 38-kDa subunits which catalyzes the carbamoylation of an amino group. Comparison of the DNA sequences of *pyrB* and *argI* (24) indicates that the two proteins are indeed evolutionarily related (101).

CTP synthetase (*pyrG*). CTP synthetase catalyzes the formation of CTP from UTP, ATP, and glutamine. Its activity is greatly stimulated by GTP. Like a number of other amidotransferases, CTP synthetase can utilize ammonia instead of glutamine, although with lower efficiency. Only the enzyme from *E. coli* B has been studied in detail (138, 161).

Under assay conditions, the enzyme exists as a tetramer of identical subunits. In the absence of ATP and UTP, it dissociates to a dimer. The reaction mechanism involves the initial glutamylation of a specific SH group on the enzyme, with the liberation of ammonia. The nascent ammonia immediately reacts with UTP, and in a series of steps that probably involve a phosphorylated amido derivative of UTP, CTP is formed at the expense of one ATP. GTP has no effect on the ammonia reaction, whereas it stimulates the glutamine reaction markedly by increasing the V_{max} of the reaction and decreasing the K_m for glutamine. Furthermore, it was shown that GTP induces conformational changes in a way that results in negative cooperativity.

Conformational changes are also induced by ATP and UTP, both of which show strong cooperative binding to the enzyme. These changes result in conversion of the dimer to the tetramer. In addition, ATP and UTP were shown to be allosteric effectors of the glutamylation reaction, indicating that they induce conformational changes affecting other active sites. Finally, it was found that the inhibitory glutamine analog 6-diazo-5-oxo-norleucine reacted with only half of the available glutamine sites, resulting in complete inactivation of the enzyme for reaction with glutamine, whereas it did not affect the K_m and V_{max} of the sites involved in the ammonia reaction.

The only phenotype described for mutations within *pyrG* is a cytidine requirement. In view of the complex allosteric properties of the CTP synthetase reaction, it should be possible to obtain a variety of regulatory mutants as a consequence of mutations in *pyrG*.

Allosteric Regulation of the Pathway

The pathway is regulated at three strategic points. Carbamoylphosphate synthase is feedback regulated in accordance with its metabolic role. It is inhibited by a pyrimidine, i.e., UMP, and it is stimulated by ornithine, an intermediate of arginine biosynthesis. In the presence of excess pyrimidines, UMP accumulates and inhibits the enzyme. If the supply of carbamoylphosphate becomes limiting for arginine synthesis, ornithine will accumulate and antagonize the inhibition by UMP. In the presence of excess arginine, ornithine is not produced and the enzyme will be controlled solely by UMP. The physiological importance of this control is illustrated by the observation that mutations rendering carbamoylphosphate synthase hypersensitive to inhibition by UMP induce a uracil-sensitivity phenotype (217). Aspartate carbamoyltransferase, the first enzyme specific for pyrimidine synthesis, is feedback inhibited by the end product of the pathway, CTP. Two observations suggest that this allosteric control is operative in vivo. (i) Starvation for cytidine nucleotides leads to overproduction of uracil nucleotides (188, 190). (ii) *pyrB* mutants that contain an enzyme with lowered sensitivity

to CTP inhibition excrete pyrimidines (206). CTP synthetase is regulated by an interplay of NTP effectors. Although it is difficult to rationalize the complex allosteric behavior of the enzyme in vitro with its physiological function, it appears that UTP is a positive effector and that CTP at high substrate concentrations may act as a competitive inhibitor with respect to UTP (161). That regulation of CTP synthetase occurs in vivo is indicated by the observation that the intracellular concentration of UTP is depressed much more than that of CTP during pyrimidine starvation (51, 109).

Addition of purines to a growing culture causes a severe drop in the PRPP pool and transient inhibition of the pyrimidine pathway (13). Low concentrations of PRPP may interfere with pyrimidine biosynthesis by lowering the rate of OMP formation from orotate and PRPP. Alternatively, it may cause a decrease in carbamoylphosphate synthase activity, since PRPP is a positive effector of this enzyme (5).

Genetic Regulation of the Pathway

Expression of the genes and small operons encoding the first six enzymes of the pathway is regulated by pyrimidines. The rate of synthesis of all six enzymes is increased by pyrimidine starvation and reduced by addition of pyrimidines (21, 275). However, the response of the six enzymes is not coordinate (Table 7).

Expression of *carAB* (*pyrA*). The synthesis of carbamoylphosphate synthase is subject to cumulative repression by arginine and a pyrimidine (3, 217). The *argR*-coded arginine repressor, which controls the expression of the arginine regulon, is also involved in cumulative repression; the synthesis of carbamoylphosphate synthase by an *argR* mutant is not repressible by arginine, and its repressibility by pyrimidines is greatly reduced (1, 218). The endogenous corepressor for the arginine control appears to be arginine itself (157).

Transcription of *carAB* is initiated from two adjacent promoters 70 base pairs apart. The Pribnow box of the downstream promoter is part of an 18-base-pair DNA sequence that constitutes the basic arginine operator, and transcription from this promoter is

TABLE 7. Repressibility of the *pyr* genes in *S. typhimurium*[a]

Gene	Derepression/repression ratio[b]	Negative effector(s)
pyrA	23–35	Arginine, cytosine nucleotide
pyrB	150–200	UTP, guanine nucleotide
pyrC	10–15	Cytosine nucleotide (not CMP)
pyrD	15–25	Cytosine nucleotide (not CMP)
pyrE	25–30	UTP, guanine nucleotide
pyrF	5–10	UTP (or UDP), guanine nucleotide

[a] Similar results are found in *E. coli*, except that both uracil and cytosine nucleotides appear to be effective in repression of *carAB*.

[b] Ratios between specific activities in conditions of derepression (obtained by growing pyrimidine auxotrophic strains with UMP as sole pyrimidine source, or by growing *pyrH* mutants in the absence of pyrimidines) and of repression (obtained by growing *pyrH*+ strains in the presence of uracil, cytidine, and, for *pyrA* estimations, also with arginine).

regulated by arginine. Transcription from the upstream promoter is controlled by pyrimidines by an unknown mechanism, and it appears to be subject to stringent control (35, 219). One puzzling observation is that carB, the distal gene of the operon, can be transcribed separately from a promoter located in the terminal part of the coding region of carA (203).

Identification of the nucleotide effectors. By using mutant strains that permit manipulations of individual nucleotide pools, it has been established that expression of pyrB, pyrE, and pyrF is repressed by a uracil nucleotide different from UMP, whereas pyrC and pyrD expression is controlled primarily by a cytosine nucleotide other than CMP (130, 216, 235). The repressing pyrimidine for pyrA in S. typhimurium was claimed to be a cytosine compound (3), whereas derivatives of both cytidine and uridine were implied to participate in carAB repression in E. coli (216). Evidence for the participation of purines in the regulation of pyr gene expression was obtained by Jensen (115), who found that starvation for guanine nucleotides resulted in derepression of pyrB, pyrE, and pyrF expression and repression of pyrC and pyrD expression. Since pyrB expression is subject to stringent control (255), the effect of guanine nucleotide starvation on pyrB, pyrE, and pyrF expression could result from a decrease of the ppGpp pool.

From these data (Table 7), it can be concluded that the genes encoding the first six enzymes of the pathway do not constitute a single regulon.

Regulatory mutants. Mutants showing constitutive synthesis of the pyrimidine biosynthetic enzymes have been isolated by selecting for resistance to toxic pyrimidine analogs. Most of the mutants thus obtained are derepressed for pyrimidine synthesis owing to low, endogenous pools of effector nucleotides. These include bradytrophic pyrH mutants defective in UMP kinase (109, 124, 216) and bradytrophic guaB mutants containing low IMP dehydrogenase activity (115).

Regulatory mutants, in which the altered regulation is not caused by disturbed effector pools, have also been obtained. Some of them are altered in RNA polymerase owing to mutations in the rpoBC gene cluster. Two classes of S. typhimurium rpoBC mutants were characterized. Mutants of the first class show high constitutive expression of pyrB and pyrE, although they contain elevated pools of UTP and CTP (120). RNA polymerase from one such mutant has been purified and found to display a fourfold increased K_m towards UTP in the elongation reaction (117). Mutants of the second class are uracil sensitive owing to hyperrepression of pyrA expression by exogenous uracil (191). These findings establish that RNA polymerase is directly involved in the regulation of pyr gene expression.

Bussey and Ingraham (41) have characterized another uracil-sensitive S. typhimurium mutant, use-1, in which expression of pyrA, pyrC, pyrD, and argI is hyperrepressed by uracil. The mutation was located close to the ilv gene cluster (42). The function of use-1 is not known. However, its properties indicate that it may be a mutated form of a regulatory protein involved in regulating the expression of the three cytosine nucleotide-controlled pyrimidine genes; however, that interpretation would not explain its effect on argI expression.

Nowlan and Kantrowitz (198) have isolated a pyrimidine auxotrophic mutant of E. coli that is defective in both pyrE and pyrF expression. The affected gene, pyrS, was located close to pyrE. They proposed that pyrS codes for an activator protein required for expression of the two genes. If so, this protein must be present in rather large amounts, since there is no titration effect of a high copy number of pyrE on pyrF expression, and vice versa (119, 193).

Regulatory mutants of S. typhimurium showing increased expression of only pyrBI were isolated by Michaels and Kelln (174). They were selected in a stable pyrB::lac fusion strain on the basis of lactose utilization. Partial characterization of a number of these mutants indicated that they had mutations in either pyrH or the promoter-operator of pyrBI. By using a similar approach for other pyr genes, mutants displaying increased expression of pyrC, pyrD, or pyrE have been isolated and characterized. The mutations were linked to the specific pyr genes and were inferred to be cis acting (131).

Regulation of pyrBI and pyrE expression. The primary structures of the DNAs preceding the coding regions of pyrBI (185, 227, 256; R. A. Kelln, personal communication) and pyrE (193, 221) show several common features. Shortly before the start of the structural gene is a rho-independent transcriptional terminator. Another region of dyad symmetry, followed by a thymidylate-rich cluster, precedes the attenuator. The leader RNAs contain open reading frames, which may encode leader peptides. In pyrE, the second gene of an operon, the "leader peptide" is a 26-kDa polypeptide produced constitutively in rather large quantities (220). Based on these structures and on in vitro transcription, it was proposed that tight coupling between the transcribing RNA polymerase and the leading ribosome translating the leader region is required for transcription through the attenuator. To account for pyrimidine control of expression, it was further proposed that translation and transcription are significantly decoupled when the intracellular UTP pool is high. Only when UTP is low is this polarity relieved owing to retardation of the polymerase within uracil-rich sequences. The finding that an rpoBC mutant containing an RNA polymerase with increased K_m for UTP is derepressed for pyrBI and pyrE expression is in accordance with this model (117, 120).

Direct evidence for the involvement of the attenuator structure in the regulation of pyrBI expression has been obtained from studies of mutants which showed that base substitutions in the attenuator stem, which decrease the stability of the putative hairpin (J. R. Wild, personal communication), and deletions, which remove this structure (149), increase expression and decrease repressibility of the operon.

The proposed function of translation in the regulatory process has also been explored experimentally. Roland et al. (226) have shown that changing the ATG initiation codon of the pyrBI leader polypeptide to ACG results in a strong reduction in pyrBI expression and regulation. Bonekamp and co-workers (33, 47) constructed a series of plasmids which contained DNA sequences encoding different short, artificial leader peptides inserted between the lac promoter-operator region and the pyrB and pyrE attenuators. By studying the effect of pyrimidine starvation on the expression of pyrB and pyrE from these plasmids, they were able to show that translation close to or across from the attenuator is required for suppression of attenuation by low UTP. Furthermore, they have shown that infrequently used codons in the region coding for the artificial leader peptides are needed for repression by high UTP (32). This observation suggests that a translational pause site in the leader peptide is required to ensure proper decoupling of translation and transcription under repressing conditions.

At present it is not known whether guanine nucleotide starvation affects *pyrBI* and *pyrE* expression at the level of attenuation or whether other control mechanisms are operative under such conditions.

Regulation of *pyrC* and *pyrD* expression. The nucleotide sequence of the leader regions of *pyrC* from *S. typhimurium* (192) and *pyrD* from *E. coli* (143) have been determined. They do not contain structures resembling attenuators, and they do not encode leader peptides. Thus, attenuation is presumably not involved in the regulation of expression of these two genes. However, the leaders do contain regions of dyad symmetry, which allow the leader transcripts to form rather stable hairpins. Single-base-pair substitutions in the *pyrC* leader that would destabilize the putative hairpin of the transcript cause increased expression of *pyrC* (R. A. Kelln and J. Neuhard, unpublished results). This observation may indicate that these structures have some regulatory function.

Since *pyrC* and *pyrD* expression is regulated coordinately under most conditions (235), it is conceivable that they are controlled by a common regulatory protein specific for a cytosine nucleotide. However, no evidence for such a protein was obtained in experiments that measured the effect of high concentrations of one gene on the expression of the other (119).

SALVAGE PATHWAYS

E. coli and *S. typhimurium* contain a collection of enzymes, the salvage enzymes, which enables the cells to utilize preformed nucleobases and nucleosides for nucleotide synthesis. These enzymes allow circumvention of the respective de novo pathways when precursors are provided. In addition, these enzymes make the pentose moiety of nucleosides and the free amino group of cytosine and adenine nucleosides available as sources of carbon, energy, and nitrogen. Nucleotides present in the growth medium may also be utilized as nucleic acid precursors. However, their utilization requires prior dephosphorylation to nucleosides by periplasmic nucleotidases.

Purine and pyrimidine analogs that become toxic when converted to nucleotides have been used extensively to select for mutants defective in salvage enzymes. Such selection can be done because most analogs are metabolized by the same enzymes that metabolize the natural compounds. Studies of such mutants have contributed significantly to the elucidation of the salvage pathways.

In the following discussion, we summarize the basic characteristics of the purine and pyrimidine salvage pathways. For more details, readers are referred to two recent reviews (189, 201).

Utilization of Preformed Purine Compounds

The purine salvage pathways of *E. coli* and *S. typhimurium* are shown in Fig. 4. Purine auxotrophic mutants blocked in the biosynthesis of IMP can satisfy their entire purine requirement with a variety of purines (Table 8). Our present knowledge of purine salvage is to a large extent based on studies of mutants defective in one or more of the salvage enzymes. A number of these mutants have been obtained in selections involving purine analogs, as shown in Table 9.

Adenine. Adenine is converted to AMP by adenine phosphoribosyltransferase, specified by *apt* (Table 10) (88, 91). The finding that adenine can satisfy the entire purine requirement of a purine auxotroph even in the

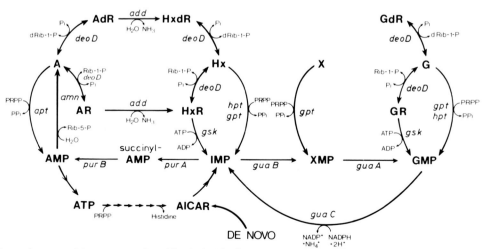

FIG. 4. Purine salvage and interconversion. The individual enzymes are identified by their gene symbols: *add*, adenosine deaminase; *amn*, AMP glycosylase; *apt*, adenine phosphoribosyltransferase; *deoD*, purine nucleoside phosphorylase; *gpt*, guanine phosphoribosyltransferase; *gsk*, guanosine kinase; *guaA*, GMP synthetase; *guaB*, IMP dihydrogenase; *guaC*, GMP reductase; *hpt*, hypoxanthine phosphoribosyltransferase; *purA*, adenylosuccinate synthetase; *purB*, adenylosuccinate lyase. Nucleobases: A, adenine; G, guanine; Hx, hypoxanthine, X, xanthine, dRib-1-P, deoxyribose 1-phosphate; Rib-1-P, ribose 1-phosphate. Ribonucleosides and deoxyribonucleosides are identified by R and dR, respectively.

TABLE 8. Growth rates of *purE* mutants on various purine compounds

Compound(s) added[a]	Doubling time (min) of [b]:	
	E. coli	S. typhimurium
Adenine	43	66
Hypoxanthine	42	50
Guanine	76	60
Xanthine	102	76
Adenosine	42	64
Deoxyadenosine	46	53
Inosine	42	60
Deoxyinosine	42	54
Guanosine	68	68
Deoxyguanosine	68	59
Xanthosine	>400	>400
2'-AMP, 3'-AMP	220	94
5'-AMP	52	184

[a] Purines were added to a concentration of 0.1 mM.
[b] Determined in glucose minimal medium at 37°C.

presence of histidine indicates that adenine can be converted to IMP and then GMP by a route that does not involve histidine biosynthesis. This route (Fig. 2) involves the intermediate formation of adenosine, inosine, and hypoxanthine and the participation of purine nucleoside phosphorylase (*deoD* [Table 10]) (121), adenosine deaminase (*add* [Table 10]) (200), and hypoxanthine phosphoribosyltransferase. The conversion of adenine to IMP via either the histidine pathway or the purine nucleoside phosphorylase-dependent pathway is less efficient in *S. typhimurium* than it is in *E. coli*. Since *purE deoD* double mutants are unable to grow in the presence of adenine plus histidine, it can be concluded that both organisms are devoid of adenine deaminase and AMP deaminase activities. Mutants of *E. coli* have been isolated that have acquired adenine deaminase activity; however, the nature of these mutations has not been established (135).

The synthesis of purine nucleoside phosphorylase is regulated. The *deoD* gene is the distal gene of the *deo* operon, which, in addition, includes the *deoA*, *deoB*, and *deoC* genes specifying thymidine phosphorylase, phosphopentomutase, and deoxyriboaldolase, respectively. Expression of the *deo* operon and of two genes (*nupG* and *tsx*) specifying proteins involved in nucleoside transport (see Transport of Nucleobases and Nucleosides, below) is controlled by two repressor

TABLE 9. Selection of mutants defective in purine salvage enzymes by the use of purine analogs

Analog (concn)	Required genetic background	Mutations selected
8-Azaguanine[a] (1 mM)	Purine prototrophy	gpt, gmk
2,6-Diaminopurine (2 mM)	Purine prototrophy	apt
2-Fluoroadenine (0.1 mM)	Purine prototrophy	apt
2-Fluoroadenosine (0.1 mM)	Purine prototrophy and apt+	deoD
5-Fluorouracil (20 μM) + deoxyadenosine (2 mM)	upp+	deoD
6-Mercaptopurine (2 mM)	Purine prototrophy	hpt, purR

[a] This selection does not apply to E. coli.

proteins specified by the *deoR* and *cytR* genes. In addition, the cyclic AMP receptor protein-cyclic AMP complex is required for that part of the expression that is controlled by the *cytR* repressor. The endogenous inducer of the *deoR*-controlled expression is deoxyribose 5-phosphate, which, in vivo, may be derived from deoxyribonucleosides. Expression of the *deoD* and *deoB* genes is also induced by inosine and guanosine by an unknown mechanism (Table 11) (79, 260).

The synthesis of adenosine deaminase is induced by adenine and hypoxanthine in *E. coli*, whereas it appears to be synthesized constitutively in *S. typhimurium* (Table 11) (202).

Guanine, hypoxanthine, and xanthine. *E. coli* and *S. typhimurium* have two 6-oxopurine phosphoribosyltransferases: hypoxanthine phosphoribosyltransferase (*hpt*) and guanine phosphoribosyltransferase (*gpt* [Table 10]) (75, 90, 122, 158, 210). Both enzymes use hypoxanthine and guanine as substrates, although at different efficiencies, and both are inhibited by ppGpp (92). The nucleotide sequence of the *E. coli gpt* gene has been determined (113). Studies with purine auxotrophic mutants containing additional mutations in *hpt* or *gpt* or both have shown that hypoxanthine phosphoribosyltransferase is primarily involved in hypoxanthine salvage; its activity with guanine is sufficient to salvage guanine for guanine nucleotide synthesis but not for total purine nucleotide synthesis. Guanine phosphoribosyltransferase, on the other hand, can salvage guanine, hypoxanthine, and xanthine for total purine nucleotide synthesis. Studies on purine-requiring *deoD* mutants have indicated that purine nucleoside phosphorylase does not play any significant role in the utilization of hypoxanthine and guanine.

Adenosine and deoxyadenosine. Adenosine and deoxyadenosine are metabolized by two different routes (Fig. 4). The major pathway involves deamination catalyzed by adenosine deaminase, followed by cleavage of inosine (deoxyinosine) to hypoxanthine, catalyzed by purine nucleoside phosphorylase; direct phosphorolysis of adenosine (deoxyadenosine) to adenine is only of minor quantitative importance (Fig. 5). An adenine-requiring mutant, i.e., a *purA* or *purB* mutant, shows a very poor growth yield on adenosine or deoxyadenosine, whereas the growth yield of a *purA add* mutant is the same on adenosine and adenine. Because a *purA add deoD* mutant is unable to grow with adenosine or deoxyadenosine as the sole adenine source, it can be concluded that *E. coli* and *S. typhimurium* are devoid of adenosine kinase activity (94, 159).

Inosine and deoxyinosine. Both inosine and deoxyinosine are rapidly degraded to hypoxanthine by the inducible purine nucleoside phosphorylase. However, a purine-requiring *deoD* mutant in which inosine degradation is blocked by mutation will grow slowly on inosine, indicating that *E. coli* and *S. typhimurium* contain inosine kinase activity. Mutant studies have revealed that inosine kinase is specified by the same gene that specifies guanosine kinase, i.e., the *gsk* gene (122).

Guanosine and deoxyguanosine. Guanosine and deoxyguanosine are rapidly degraded to guanine by purine nucleoside phosphorylase; only small amounts

TABLE 10. Properties of certain salvage enzymes

Enzyme	Gene designation	Origin	Molecular mass (kDa)	Activators and inhibitors[a]
Adenine phosphoribosyltransferase	apt	E. coli	40	AMP(−)
Adenosine deaminase	add	E. coli	29	Coformycin(−)[b]
Cytidine deaminase	cdd	E. coli	54 (2 × 35?)	Tetrahydrouridine(−)[b]
Cytosine deaminase	codA	S. typhimurium	230 (4 × 54)	
Guanine phosphoribosyltransferase	gpt	E. coli	50 (3 × 16.9)	GMP(−), ppGpp(−)
Hypoxanthine phosphoribosyltransferase	hpt	E. coli		ppGpp(−)
Purine nucleoside phosphorylase	deoD	E. coli	140 (6 × 23.7)	
		S. typhimurium	140 (6 × 23.7)	
Thymidine kinase	tdk	E. coli	84 (2 × 42)	dTTP(−), dCTP(+)
Thymidine phosphorylase	deoA	E. coli	90 (2 × 45)	
Uracil phosphoribosyltransferase	upp	E. coli	75 (3 × 23.5)	GTP(+), ppGpp(−), uracil nucleotide(−)
Uridine kinase	udk	E. coli	90	CTP(−), UTP(−)
Uridine phosphorylase	udp	E. coli	176 (8 × 22)	
Xanthosine phosphorylase	xapA	E. coli	205	

[a] −, Inhibitor; +, activator.

[b] Transition state analogs.

of guanosine escape phosphorolysis by being converted directly to GMP by guanosine kinase (K. F. Jensen, Ph.D. thesis, University of Copenhagen, 1977). A purine auxotrophic mutant defective in nucleoside degradation, i.e., a purE deoD double mutant, will grow on guanosine, provided the requirement for adenine nucleotides is satisfied by adenine; however, the mutant will not grow on guanosine alone. The basis of this phenomenon is not understood, but it has been shown that the lack of growth of such a mutant on guanosine can be partially suppressed by gsk mutations that result in the synthesis of an altered guanosine kinase. Complete suppression can be achieved by further mutations in genes specifying enzymes of the de novo purine biosynthetic pathway before purE (Fig. 1), i.e., mutations in purF, purG (purL), purI(purM), or prs (74, 106). In this context, PRPP synthetase (prs) may be regarded as the first enzyme of purine biosynthesis. Finally, it has been found that certain mutations in the rpoB gene, speci-

fying the β-subunit of RNA polymerase, can suppress the lack of growth of an E. coli purD deoD mutant on guanosine (160).

Xanthosine. Under most growth conditions, xanthosine is metabolically inert because E. coli and S. typhimurium lack xanthosine kinase and because xanthosine is not cleaved by purine nucleoside phosphorylase. However, E. coli, but not S. typhimurium,

TABLE 11. Repressibility (inducibility) of genes encoding salvage enzymes

Gene	Derepression/ repression (induction) ratio	Effector(s)[a]
add	30	Adenine(+), hypoxanthine(+)
cdd	80	Cytidine (uridine)[b](+), cAMP(+)
codA	5–10	Pyrimidine nucleotide(−), purine(−), NH₃(−)?
deoA	150–200	Deoxyribose 5-P(+), cAMP(+), cytidine (uridine)(+)[b]
deoD	20	Deoxyribose 5-P(+), cAMP(+), cytidine (uridine)(+)[b], inosine(+), guanosine(+)
udk	5	Pyrimidine nucleotide(−)
udp	30	Cytidine (uridine)[b](+), cAMP(+)
upp	10	Pyrimidine nucleotide(−)
xapA[c]	>200	Xanthosine(+)

[a] +, Positive; −, negative. Deoxyribose 5-P, Deoxyribose 5-phosphate; cAMP, cyclic AMP.

[b] Cytidine and uridine are inducers in S. typhimurium, whereas cytidine is the only inducer in E. coli.

[c] Xanthosine phosphorylase is absent from S. typhimurium.

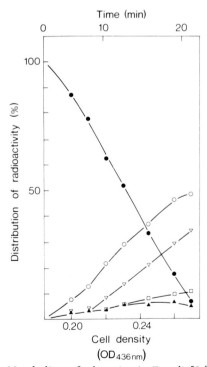

FIG. 5. Metabolism of adenosine in E. coli. [8-¹⁴C]adenosine was added at time zero to exponentially growing cultures of wild-type cells. At different times thereafter, samples of the culture were analyzed for the distribution of label in the medium. ●, Adenosine; ○, inosine; ▽, hypoxanthine; ▲, adenine. □, Label incorporated into whole cells. OD₄₃₆nm, Optical density at 436 nm.

contains a gene, *xapA*, specifying a second purine nucleoside phosphorylase, xanthosine phosphorylase, which can use xanthosine, guanosine, and inosine as substrates (Table 10) (43, 80). Owing to very low intracellular levels of this enzyme, it plays no role in purine salvage, and xanthosine will not serve as a purine (guanine) source in either organism (Table 8). In *E. coli*, high exogenous concentrations of xanthosine induce *xapA* expression (Table 11). This induction is dependent on a functional *xapR* gene product (134), and it is to some extent sensitive to catabolite repression. In glycerol minimal medium, the induction shows a pronounced lag of about three generations. In the absence of other carbon sources, xanthosine will serve as the sole purine, carbon, and energy source.

Purine analogs as purine sources. The purine analogs 2,6-diaminopurine and 6-methylaminopurine are toxic at high concentrations (>1 mM) because they are converted to toxic nucleotides by adenine phosphoribosyltransferase. However, they may also be converted to guanosine and inosine, respectively, by the consecutive action of purine nucleoside phosphorylase and adenosine deaminase. At low, nontoxic concentrations of these analogs (<0.1 mM), they may serve as purine sources for purine auxotrophs, although with low efficiency. By selecting for fast growth on low concentrations of these analogs as the sole purine sources, mutants with increased levels of purine nucleoside phosphorylase (*deoR* mutants) or adenosine deaminase have been obtained (62; P. Nygaard, unpublished results).

Utilization of Preformed Pyrimidine Compounds

The pyrimidine salvage pathways in *E. coli* and *S. typhimurium* are shown in Fig. 6. Each enzyme has been studied in vitro, and each enzyme function has been established through studies of mutant strains with single or multiple blocks in the pathways. The

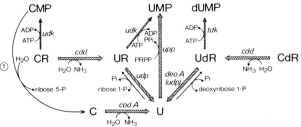

FIG. 6. Pyrimidine salvage pathways. The individual enzymes are identified by the corresponding gene symbols: *cdd*, cytidine deaminase and deoxycytidine deaminase; *codA*, cytosine deaminase; *deoA*, thymidine phosphorylase; *tdk*, thymidine kinase; *udk*, uridine kinase and cytidine kinase; *udp*, uridine phosphorylase; *upp*, uracil phosphoribosyltransferase; ①, ribonucleotide glycosylase. U, Uracil; C, cytosine; UR, uridine; CR, cytidine; UdR, deoxyuridine; CdR, deoxycytidine.

phenotypes of certain of the mutants are summarized in Table 12. Thymine and thymidine cannot serve as total pyrimidine sources but may serve as sources of dTMP (see Utilization of Preformed DNA Precursors, below).

Uracil. Uracil is converted to UMP by uracil phosphoribosyltransferase (*upp*). The enzyme is activated by GTP and inhibited by ppGpp and uracil nucleotides (Table 10) (222). This fact may explain the observation that amino acid starvation of stringent *E. coli* strains causes a severe reduction in the incorporation of uracil into UTP (55). Uracil may also be converted to UMP through the concerted action of uridine phosphorylase and uridine kinase. However, this route operates only if either of the substrates for uridine phosphorylase, i.e., uracil or ribose 1-phosphate, is present at high intracellular concentrations.

Pyrimidine starvation results in 10-fold increased levels of uracil phosphoribosyltransferase, indicating

TABLE 12. Effects of mutations in pyrimidine salvage pathways on the utilization of pyrimidines and on the sensitivity toward 5-fluoropyrimidines

Relevant genotype	Growth[a] on:											
	Pyrimidine source[b]						Analog[c]					
	U	C	UR	CR	UdR	CdR	FU	FC	FUR	FCR	FUdR	FCdR
upp	−	−	+	+	−	−	+	+	−	−	−	−
codA	+	−	+	+	+	+	−	+	−	−	−	−
udk	+	+	+	+	+	+	−	−	−	−	−	−
udp	+	+	+	+	+	+	−	−	−	−	−	−
upp udk[d]	−	−	−	−	−	−	+	+	+	+	−	−
udp udk	+	+	+	+	+	+	−	−	−	ND	−	−
cdd	+	+	+	+	+	−	−	−	−	−	−	+
cdd udk	+	+	+	−	+	−	−	−	−	+	−	+
cdd codA	+	−	+	−	+	−	−	+	−	−	−	+
deoA	+	+	+	+	+	+	−	−	−	−	−	−
deoA udp	+	+	+	+	−	−	−	−	−	−	−	−
deoA udp tdk	+	+	+	+	−	−	−	−	−	−	+	+

[a] +, Growth; −, no growth; ND, not determined.

[b] All strains were pyrimidine auxotrophic. Abbreviations: U, uracil; C, cytosine; UR, uridine; CR, cytidine; UdR, deoxyuridine; CdR, deoxycytidine.

[c] All strains were pyrimidine prototrophic. 5-fluoro-substituted uracil, cytosine, uridine, cytidine, deoxyuridine, and deoxycytidine analogs are denoted FU, FC, FUR, FCR, FUdR, and FCdR, respectively.

[d] This strain contained a *pyrC* mutation which confers a temperature-sensitive pyrimidine requirement. When pyrimidine sources were tested, the strain was grown at 42°C (Pyr⁻). When analog sensitivity was tested, the strain was grown at 30°C (Pyr⁺).

that *upp* expression is controlled by a pyrimidine compound (222). Since uracil phosphoribosyltransferase and orotate phosphoribosyltransferase (*pyrE*) catalyze similar chemical reactions and since the subunit sizes of the two enzymes are practically identical, the question arises of whether the expression of the two genes is controlled in the same way, i.e., by UTP-modulated attenuation (see Genetic Regulation of the Pathway, above).

Cytosine. Cytosine can only be metabolized through deamination to uracil, catalyzed by cytosine deaminase (Table 10) (266). The structural gene specifying the enzyme, *codA*, is located differently in the two organisms. In *E. coli*, it is closely linked to the *lac* operon, i.e., on a section of the chromosome that is absent from *S. typhimurium*, whereas its position on the *S. typhimurium* chromosome is at approximately 70 min. Expression of *codA* is controlled in a complex manner. It is repressed by purines, and it is derepressed by starvation for either pyrimidines or nitrogen (Table 11) (265; L. Andersen, Ph.D. thesis, University of Copenhagen, 1979). The regulatory system responsible for this control has not been identified, but it appears to be involved also in controlling the synthesis of compounds required for cytosine uptake (see Transport of Nucleobases and Nucleosides, below).

Uridine. Uridine is converted to UMP by two different routes: phosphorylation catalyzed by uridine kinase (*udk*) and phosphorolytic cleavage to uracil and ribose 1-phosphate by uridine phosphorylase (*udp*). Uridine kinase phosphorylates uridine and cytidine, with GTP as the preferred phosphate donor. UTP and CTP inhibit the reaction (Table 10) (259). Expression of *udk* is derepressed during pyrimidine starvation, but the nature of the repressing nucleotide has not been established (Table 11).

Uridine phosphorylase shows weak activity with deoxyuridine and thymidine as substrates but no activity towards cytidine (Table 10) (145). The synthesis of uridine phosphorylase is coregulated with the synthesis of a number of other proteins involved in nucleoside transport and catabolism. Thus, expression of the *udp* and *cdd* genes, the *deo* operon (chapter 18), and the genetic loci for three nucleoside transport systems (see Transport of Nucleobases and Nucleosides, below) are negatively controlled by the *cytR*-specified repressor and positively controlled by the cyclic AMP receptor protein-cyclic AMP complex. The inducer of the *cytR*-controlled genes is cytidine in *E. coli*, whereas both cytidine and uridine are effectors in *S. typhimurium* (Table 11) (79, 199).

From the growth yield from uridine of pyrimidine-requiring mutants defective in the ability to utilize uracil (*upp* mutants), it was established (189) that a large fraction (50 to 75%) of the added uridine escapes kinase because it is degraded to uracil. However, if uridine is provided in low concentrations, it is quantitatively converted to UMP via the kinase reaction. Mutants defective in uridine phosphorylase still degrade uridine to some extent. This uridine-degrading activity may eventually be accounted for by thymidine phosphorylase.

Cytidine. Cytidine can be converted directly to CMP by uridine kinase. However, the predominant route for cytidine utilization is deamination by the inducible cytidine deaminase (*cdd* [Table 10]) (11, 12). Thus,

a *pyrG* mutant defective in CTP synthetase cannot grow on cytidine unless it contains an additional mutation in *cdd* (19). Pyrimidine auxotrophic *cdd* mutants are able to grow on cytidine as the sole pyrimidine source, although the growth rate is somewhat reduced. This residual growth is abolished by mutations inactivating either cytosine deaminase (*codA*) or cytidine kinase (*udk* [Table 12]). In accordance with these observations, a ribonucleotide glycosylase which hydrolyzes CMP to cytosine and ribose 5-phosphate has been identified (A. Eisenhardt, Ph.D. thesis, University of Copenhagen, 1971). The synthesis of cytidine deaminase and uridine phosphorylase are coregulated by the *cytR*-coded repressor, as described above.

Deoxyuridine. Deoxyuridine also can be utilized by two different routes. It can be converted directly to dUMP by thymidine kinase (*tdk* [Table 10]), and thereby serve as a precursor for thymidine nucleotides, or it can be cleaved to uracil and deoxyribose 1-phosphate by the inducible enzyme thymidine phosphorylase (*deoA* [Table 10]) (see Utilization of Preformed DNA Precursors, below). Mutants defective in thymidine phosphorylase are still capable of utilizing deoxyuridine as the sole pyrimidine source, whereas *deoA udp* double mutants are not. Thus, uridine phosphorylase can degrade deoxyuridine to some extent (145).

Since the *deoA* gene is part of the *deo* operon, it follows that the synthesis of thymidine phosphorylase is induced by ribonucleosides and deoxyribonucleosides (Table 11) (chapter 18).

Deoxycytidine. *E. coli* and *S. typhimurium* are devoid of deoxycytidine kinase activity, and they are unable to convert deoxycytidine to cytosine. However, they can deaminate deoxycytidine effectively to deoxyuridine. Studies of mutants have shown that this conversion requires a functional cytidine deaminase. Thus, deoxycytidine is metabolically stable in a *cdd* mutant.

Utilization of Nucleotides

E. coli and *S. typhimurium* can use exogenous nucleotides as nucleic acid precursors (152, 258). Thus, 3'-, 5'-, and cyclic-2',3'-ribonucleotides, as well as 5'-deoxyribonucleotides, will satisfy the growth requirement of purine- and pyrimidine-requiring mutants. Since the cytoplasmic membrane is impermeable to nucleotides, they must be dephosphorylated to the corresponding nucleosides by periplasmic enzymes before they can be utilized. A number of periplasmic phosphatases capable of hydrolyzing nucleotides have been identified (189). The importance of these enzymes in nucleotide salvage has been explored by studying mutants defective in one or more of these phosphatases (15, 17, 258). Mutants defective in specific outer membrane proteins involved in pore formation are unable to utilize exogenous nucleotides, although they contain normal levels of the periplasmic phosphatases (16, 261). Thus, specific pores in the outer membrane are required for exogenous nucleotides to reach the hydrolytic enzymes.

The rate-limiting step in the utilization of nucleotides for nucleic acid synthesis appears to be the conversion of nucleotides to nucleosides outside the cytoplasmic membrane. This restriction provides a simple way of feeding cells continuously with low concentrations of nucleosides and thereby of preventing their phosphorolytic cleavage.

Transport of Nucleobases and Nucleosides

Nucleobases. Uptake of exogenous nucleobases is tightly coupled to the metabolic processes that convert them to nucleotides. However, several studies (reviewed by Munch-Petersen and Mygind [181]) suggest that transport of purine bases and of uracil also involves specific transport systems energized by proton motive force (39, 40).

An *E. coli* mutation causing a defect in high-affinity adenine transport (40) lay in *purP*, located away from any known gene for purine metabolism. Adenine uptake, but not the uptake of other purine bases, was severely affected in the mutant. *S. typhimurium* mutants defective in guanine and xanthine uptake have been reported (28). One of the mutations, *guaP*, was located in the region of the chromosome that contains the *guaC* gene; however, whether *guaP* and *guaC* represent two distinct loci was not established.

Existence of a specific transport system for uracil was indicated from studies of *E. coli* mutants defective in uracil uptake (181). The mutations were located in *uraA*, which is closely linked to but distinct from *upp*. A pyrimidine-requiring *uraA* mutant will grow normally on cytosine, whereas the growth rate on uracil at concentrations below 80 μM is strictly dependent on the uracil concentration. Mutants defective in cytosine transport but containing normal levels of cytosine deaminase have been characterized in *E. coli* (155). The mutated gene, *codB*, is closely linked to but clearly distinct from the *codA* gene (50). The transport system is highly specific for cytosine. The synthesis of the transport system is regulated in parallel with the synthesis of cytosine deaminase, and it seems possible that *codA* and *codB* constitute an operon.

Nucleosides. Exogenous nucleosides are rapidly metabolized after their entry into the cells. However, the use of *E. coli* mutants defective in nucleoside metabolism has demonstrated (184) that nucleosides are transported actively and that the concentrating effect of the transport systems is several hundredfold. Two high-affinity transport systems, designated the G system and the C system, have been identified in *E. coli* (for a review, see reference 181). The G system, which transports all nucleosides, is inactivated by mutations in the *nupG* locus located at 64 min. Transport of purine nucleosides by this system is regulated by the intracellular concentration of ATP (183). The C system transports only pyrimidine and adenine nucleosides. Mutations inactivating this system, i.e., *nupC* mutations, are located at 50 min on the chromosome. Thus, *nupC* mutants can utilize all nucleosides as carbon sources, whereas a *nupG* mutant cannot grow on guanosine or deoxyguanosine. Double mutants defective in both systems are unable to grow on nucleosides as sole carbon and energy sources, but they are still able to use these compounds as purine and pyrimidine sources. Transport studies with membrane vesicles derived from mutants containing only one of the two transport systems have indicated that both systems are energized by the proton motive force and that essential proteins of both systems are integrated in the cytoplasmic membrane (182, 183).

The coupling of nucleoside transport and catabolism is further emphasized by the observation that the synthesis of essential components of the transport systems and the synthesis of the nucleoside-catabolizing enzymes are controlled by the same regulatory proteins (180). Thus, expression of the G system is regulated by the *deoR* repressor, the *cytR* repressor, and the cyclic AMP receptor protein-cyclic AMP complex, whereas control of expression of the C system involves only the latter two regulatory proteins.

A 2,000-base-pair DNA fragment that complements all known *nupC* mutations has been cloned. Because this DNA fragment specified only one protein (43 kDa), which was found associated with membranes, it was tentatively concluded that it is the *nupC* gene product (J. Rasmussen, Ph.D. thesis, University of Copenhagen, 1985). The product of the *nupG* locus has not been identified, but it has been found that a *nupG::lacZ* fusion will direct the synthesis of a membrane-associated β-galactosidase (179).

An outer membrane protein, the phage T6 receptor encoded by *tsx*, facilitates the passage of nucleosides through the outer membrane (84, 170). All nucleosides, except cytidine and deoxycytidine, are affected in their transport by a *tsx* mutation. The synthesis of the *tsx* protein is coregulated with the synthesis of the G transport system (140).

Physiological Function of Nucleobase and Nucleoside Salvage

The salvage pathways fulfill three physiological functions. The first is to scavenge exogenous, preformed nucleobases and nucleosides for nucleotide synthesis. The second is to make the pentose moiety of exogenous nucleosides and the amino groups of cytosine and adenine compounds available as sources of carbon, energy, and nitrogen. The third function is to reutilize nucleobases and nucleosides produced endogenously as a result of nucleotide turnover.

Salvage of exogenous nucleobases and nucleosides is beneficial to the cell, provided the de novo synthesis decreases concomitantly. A number of studies have shown that this is indeed the case (14, 46, 178, 235, 279). When nucleobases or nucleosides are present in the growth medium, the contribution from the de novo pathways to nucleotide synthesis is greatly suppressed. This suppression is accomplished by increased feedback inhibition of the de novo pathways and by repression of the synthesis of the de novo enzymes. Accordingly, the endogenous nucleotide pools increase in cells grown in the presence of exogenous nucleobases or nucleosides. However, other factors, such as the pool of PRPP, may also affect this control.

Exogenous nucleosides are predominantly converted to nucleobases before they are utilized for nucleotide synthesis. This catabolism is catalyzed by two nucleoside deaminases and four nucleoside phosphorylases. The ribose 1-phosphate and deoxyribose 1-phosphate liberated in these processes are further converted to intermediates of the pentose phosphate shunt and of glycolysis, respectively, by the action of phosphopentomutase and deoxyriboaldolase (79). In addition, three transport systems are involved in the uptake of exogenous nucleosides. The fact that the synthesis of all of these proteins is inducible by nucleosides emphasizes that catabolism is dominating when nucleosides are abundant in the growth medium. There is some evidence suggesting that when nucleosides are present in low concentrations they escape phosphorolysis and are converted to nucleotides by the corresponding nucleoside kinases (189).

Large quantities of ribonucleoside monophosphates are formed as a result of mRNA turnover, DNA ligations, and the turnover of certain nucleotide-coenzymes. Nucleic acid precursors are not normally excreted during balanced growth; these ribonucleotides are reutilized, presumably by rephosphorylation to the corresponding triphosphates. However, several observations indicate that significant amounts of the ribonucleotides are degraded during normal growth and that the nucleobases and nucleosides thus formed are reconverted to ribonucleotides via the salvage pathways. (i) Pyrimidine prototrophic strains unable to utilize uracil or cytosine owing to mutations in the uracil phosphoribosyltransferase gene or the cytosine deaminase gene excrete uracil and cytosine, respectively (218; J. Neuhard, unpublished observations). (ii) Purine prototrophic strains defective in hypoxanthine phosphoribosyltransferase activity excrete hypoxanthine, and strains unable to salvage adenine excrete adenine during exponential growth (P. Nygaard, unpublished results). (iii) Mutants of *E. coli* defective in cytidine utilization owing to mutations inactivating cytidine deaminase, uridine kinase, and cytidine kinase contain endogenously induced levels of uridine phosphorylase (81, 199). As cytidine is the inducer of uridine phosphorylase synthesis, it follows that the mutants contain elevated cytidine pools. Although the physiological significance of this endogenous turnover of ribonucleotides to nucleobases and nucleosides is not understood yet, the turnover points to a significant role of the salvage pathways even under conditions in which preformed nucleobases and nucleosides are absent from the growth medium.

DEOXYRIBONUCLEOTIDE BIOSYNTHESIS

Ribonucleotide Reduction

The pathways by which the four DNA precursors dATP, dGTP, dCTP, and dTTP are synthesized are shown in Fig. 7. The first step specific for deoxyribo-

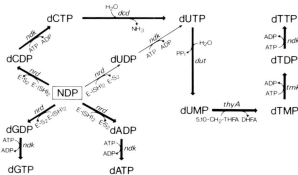

FIG. 7. Deoxyribonucleotide metabolism. The individual enzymes are identified by the corresponding gene symbols: *dcd*, dCTP deaminase; *dut*, dUTPase; *ndk*, nucleoside diphosphokinase; *nrd*, ribonucleoside diphosphate reductase; *thyA*, thymidylate synthase; *tmk*, dTMP kinase. 5,10-CH$_2$-THFA, 5,10-methylene-tetrahydrofolate; DHFA, dihydrofolate. NDP stands for UDP, CDP, ADP, or GDP. E-(SH)$_2$, reduced thioredoxin or glutaredoxin; E-S$_2$, oxidized thioredoxin or glutaredoxin.

nucleotide synthesis is catalyzed by ribonucleoside diphosphate reductase, which converts all four ribonucleoside diphosphates to the corresponding 2'-deoxyribonucleoside diphosphates. Whereas the substrates ADP, GDP, and UDP are intermediates in the synthesis of the corresponding ribonucleoside triphosphates, CDP is not. CDP may be derived either from CTP by the action of nucleoside diphosphokinase or from CMP produced as a result of phospholipid synthesis and mRNA turnover (Table 2).

Ribonucleoside diphosphate reductase. The enzyme from *E. coli* B has been extensively studied by Reichard and co-workers (for a review, see reference 245). It consists of two nonidentical subunits, proteins B1 and B2. The B1 protein is composed of two polypeptides (α and α') of similar size (Table 13). The two monomers have the same carboxy-terminal amino acid but differ in their amino termini. From the primary structure of the *nrdA* gene, which codes for both the α and the α' polypeptide, it appears that both are processing products of a precursor polypeptide (45). The B2 protein is a dimer of identical subunits containing two ferric ions and a free radical located on a tyrosine residue (Table 13). The structural gene for the β-polypeptide is specified by the *nrdB* gene,

TABLE 13. Properties of enzymes involved in deoxyribonucleotide metabolism

Enzyme	Gene designation	Origin	Structure/molecular mass of subunit (kDa)	DNA sequence known?	Activators, inhibitors, and cofactors[a]
Ribonucleoside diphosphate reductase	*nrdA*	*E. coli*	$\alpha\alpha'\beta_2$		
			$\alpha = 87.3$	Yes	ATP(+), dATP(−)
			$\alpha' = 84.5$	Yes	dATP(+), dTTP(+), dGTP(+)
	nrdB		$\beta = 43.4$	Yes	Fe^{3+}
Thioredoxin reductase	*trxB*	*E. coli*	$a_2/a = 33.0$	No	FAD
dCTP deaminase	*dcd*	*S. typhimurium*	82^b	No	dCTP(+), dTTP(−)
dUTPase	*dut*	*E. coli*	$a_4/a = 16.0$	Yes	Zn^{2+}
Thymidylate synthase	*thyA*	*E. coli*	$a_2/a = 30.4$	Yes	
dTMP kinase	*tmk*	*E. coli*	65^b	No	

^{a} +, Activator; −, inhibitor. FAD, Flavin adenine dinucleotide.

^{b} Molecular mass of the native enzyme. The subunit structure is not known.

which is the distal gene of the *nrdAB* operon. In the presence of magnesium ions, proteins B1 and B2 form a 1:1 complex which is catalytically active. The active site of the enzyme is thought to involve ferric ions and the free radical of protein B2 plus thiol groups located on the B1 protein.

Both the activity and the specificity of the enzyme are regulated by NTP effectors through binding to specific allosteric sites on the B1 protein. The overall activity is determined by the positive effector ATP and the negative effector dATP bound to the "activity sites," and the affinity of the enzyme towards the different substrates is determined by the positive effector (ATP, dATP, dTTP, and dGTP) bound in the "substrate affinity site."

Because the only pathway for producing deoxyribonucleotides is via ribonucleotide reduction, only conditional *nrd* mutants can be obtained (175). Studies on a temperature-sensitive *nrdA* mutant showed that the rapid cessation of DNA synthesis that occurred after a shift to the nonpermissive temperature was not accompanied by depletion of the deoxyribonucleoside triphosphate pools (167). This observation may indicate that ribonucleotide reductase is a component of a DNA replication complex that is rendered nonfunctional by the mutation. Expression of the *nrdAB* operon is regulated at the level of transcription (82, 83, 254). Any condition that tends to decrease the cellular DNA/mass ratio results in increased synthesis of *nrd* mRNA. Such conditions may be dTTP limitation, a *rep* mutation, shift-up in growth rate, and inhibition of DNA replication (57, 58).

Hydrogen donors in ribonucleotide reduction. Two low-molecular-weight proteins, i.e., thioredoxin and glutaredoxin, have been implicated as hydrogen donors for ribonucleotide reduction by *E. coli*. Thioredoxin (98, 154, 262) contains an active site disulfide which functions in redox reactions. It is reduced by NADPH in the presence of the flavo-enzyme thioredoxin reductase (Table 13) (245). In vitro, thioredoxin participates in reduction of ribonucleotides, methionine sulfoxide, sulfate, and protein disulfides. In addition, thioredoxin is an essential subunit of bacteriophage T7 DNA polymerase, and it is required for assembly of the filamentous bacteriophages f1 and M13 (154, 229). Since mutants defective in either thioredoxin (*trxA* [168]) or thioredoxin reductase (*trxB* [60, 78]) show no defect in ribonucleotide reduction in vivo, it was concluded that the thioredoxin system is not essential for deoxyribonucleotide synthesis.

Glutaredoxin is reduced by glutathione, which in turn is reduced by the flavo-enzyme glutathione reductase, with NADPH as hydrogen donor (96, 97). So far, mutants defective in glutaredoxin have not been isolated.

Synthesis of dTTP

Three of the DNA precursors, i.e., dATP, dGTP, and dCTP, are derived from the corresponding deoxyribonucleoside diphosphates by phosphorylations catalyzed by nucleoside diphosphokinase. Because thymine deoxyribonucleotides have no ribonucleotide counterpart, additional reactions are required for dTTP synthesis.

The synthesis of thymine deoxyribonucleotides involves the transfer of a methylene group and two reducing equivalents from 5,10-methylenetetrahydrofolate to dUMP, catalyzed by the dimeric enzyme thymidylate synthase (Table 13) (22, 23). The products of the reaction are dTMP and dihydrofolate. 5-Fluoro-dUMP is a potent inhibitor of thymidylate synthase. In the presence of 5,10-methylenetetrahydrofolate, the enzyme binds two equivalents of 5-fluoro-dUMP, forming a very stable, covalent ternary complex. Mutants defective in thymidylate synthase (*thyA*) require thymine or thymidine for growth. Neither the activity of the enzyme nor the expression of the *thyA* gene seems to be regulated.

Biosynthesis of dUMP. *E. coli* and *S. typhimurium* can synthesize dUMP from both dCDP and dUDP (175). The major pathway involves phosphorylation of dCDP, deamination of dCTP, and hydrolysis of dUTP to dUMP. Only 25% of the cellular dUMP is derived from dUDP through phosphorylation followed by hydrolysis of dUTP (Fig. 7). In accordance with its function at a metabolic branch point, dCTP deaminase is regulated (Table 13) (18). It shows positive homotropic cooperativity toward dCTP and is feedback inhibited by dTTP. The next enzyme in the pathway, dUTPase, is a pyrophosphatase that contains zinc ions (Table 13) (237). The primary structure of the structural gene for *E. coli* dUTPase (*dut*) is known (162).

Mutants having the structural gene for dCTP deaminase (*dcd*) deleted have been characterized in both *E. coli* and *S. typhimurium* (175). Phenotypically such mutants are thymine prototrophic, indicating that the dUDP pathway is capable of taking over the entire synthesis of dUMP. *dcd* mutants contain 10-fold increased dCTP pools, whereas their dTTP pools are only 25 to 50% of the normal size. The low dTTP pool may explain how the cells can synthesize sufficient dUMP in the absence of a functional dCDP pathway; the low dTTP pool causes derepression of ribonucleotide reductase synthesis and thereby an increase in the rate of dUDP production.

E. coli mutants defective in dUTPase activity (*dut*) all appear to contain some residual dUTPase activity (89). The most deficient one, which contains about 2% residual dUTPase activity in vitro, is thymine prototrophic. It shows an increased rate of recombination and spontaneous mutation (59, 257). The intracellular concentration of dUTP is 25- to 30-fold increased, whereas the dTTP pool concentration is normal (175). The elevated dUTP/dTTP ratio leads to an increased frequency of misincorporation of uracil for thymine in DNA. Subsequent excision repair of these uracil moieties by uracil-DNA glycosylase (156) causes temporary breaks and gaps in the DNA, events that may explain the increased recombination and mutation frequencies. The finding of normal dTTP pools in the *dut* mutants is surprising, because dUTPase is believed to be essential for the synthesis of dUMP (Fig. 7). One possible explanation is that a low percentage of residual dUTPase activity, combined with a 30-fold increased dUTP pool, may suffice to satisfy the dUMP requirement of the cells; the price that the cell pays is increased incorporation of uracil into its DNA. In fact, a tight *dut* mutation may be lethal owing to fragmentation of the uracil-containing DNA.

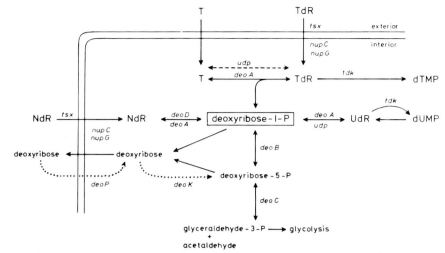

FIG. 8. Metabolism of thymine, thymidine, and deoxyribose 1-phosphate. Enzymes and proteins are identified by the corresponding gene symbols: *deoA*, thymidine phosphorylase; *deoB*, phosphopentomutase; *deoC*, deoxyriboaldolase; *deoD*, purine nucleoside phosphorylase; *deoK*, deoxyribokinase; *deoP*, deoxyribose permease; *nupC*, nucleoside uptake; *nupG*, nucleoside uptake; *tdk*, thymidine kinase; *tsx*, phage T6 receptor and nucleoside uptake; *udp*, uridine phosphorylase. N, Uracil, adenine, hypoxanthine, or guanine. – – –→, Route of minor importance; ·····→, route exclusive for *S. typhimurium*.

Conversion of dTMP to dTTP. Two phosphorylation steps convert dTMP to dTTP. The first step is catalyzed by the highly specific dTMP kinase (Table 13), and the second step is catalyzed by nucleoside diphosphokinase (see Conversion of NMP and NDP to NTP, above).

Utilization of Preformed DNA Precursors

E. coli and *S. typhimurium* lack deoxyadenosine, deoxyguanosine, and deoxycytidine kinase activities and therefore are unable to utilize deoxyadenosine, deoxyguanosine, and deoxycytidine as specific DNA precursors. As described previously, these compounds are catabolized to nucleobases and deoxyribose 1-phosphate. The only deoxyribonucleoside that can serve as a specific DNA precursor is thymidine, which may be phosphorylated to dTMP by the allosterically regulated thymidine kinase (Table 10) (110, 111). Preformed thymine may also be utilized for dTMP synthesis, since it can be converted to thymidine by condensation with deoxyribose 1-phosphate catalyzed by thymidine phosphorylase (Table 10).

Deoxyribose 1-phosphate is produced intracellularly by phosphorolytic cleavage of deoxyribonucleosides catalyzed by purine nucleoside phosphorylase (*deoD*) and thymidine phosphorylase (*deoA*), and it is rapidly catabolized by the sequential action of phosphopentomutase (*deoB*) and deoxyriboaldolase (*deoC*). Expression of the four *deo* genes, which constitute the *deo* operon, is controlled by the *cytR*- and *deoR*-coded repressors (chapter 18). *S. typhimurium* contains a deoxyribokinase and a deoxyribose transport system, both of which are inducible by deoxyribose (93). *E. coli* is devoid of deoxyribose kinase activity. The reactions involved in the metabolism of thymine, thymidine, and deoxyribose 1-phosphate are summarized in Fig. 8.

Utilization of thymine and thymidine by thymine prototrophs. Thymine prototrophic strains do not incorporate exogenous thymine into DNA, because they lack an endogenous deoxyribose 1-phosphate pool. Addition of deoxyribonucleosides will stimulate thymine utilization temporarily by supplying deoxyribose 1-phosphate. Thymidine is readily incorporated into DNA, but the incorporation stops after a short time owing to the rapid breakdown of thymidine to thymine. Since thymidine catabolism is induced by thymidine, high concentrations of thymidine do not promote further incorporation, and *deoR* mutants, which contain high levels of thymidine phosphorylase and the deoxyribose 1-phosphate-catabolizing enzymes, are unable to utilize thymidine as a DNA precursor. In contrast, conditions that decrease the thymidine phosphorylase activity will promote thymidine incorporation. These include *deoA* mutations inactivating thymidine phosphorylase and the addition of uridine, which is a potent inhibitor of the enzyme. Prolonged incorporation of thymidine is also observed by using exogenous dTMP as precursor, because dTMP causes a slow feeding of thymidine, which favors phosphorylation over degradation (175).

Utilization of thymine and thymidine by thymine auxotrophs. Mutants defective in thymidylate synthase acquire the ability to utilize exogenous thymine, although not very effectively. About 150 μM exogenous thymine is required to support normal growth of a *thyA* mutant. The difference between *thyA* mutants and wild-type strains in their abilities to utilize thymine is due to an increased production of deoxyribose 1-phosphate from dUMP in the *thyA* mutants. The dUMP pool in thymine mutants is 50- to 100-fold above normal. Because dUMP production is effectively controlled by the intracellular concentration of dTTP through feedback inhibition of dCTP deaminase and control of ribonucleotide reductase synthesis, it follows that low dTTP pools will promote dUMP accumulation, and thereby deoxyribose 1-phosphate production, through degradation of excess dUMP. An increased deoxyribose 1-phosphate pool, in turn, stimulates the thymidine phosphorylase reaction in the direction of thymidine synthesis (175).

TABLE 14. Thymine requirements of different *thyA* mutants

Genotype	Thymine or thymidine phenotype	Thymine concn required for growth (μM)
thyA	High thymine requirer	150
thyA deoB or *deoC*	Low thymine requirer	15
thyA cytR or *deoR*	Super high thymine requirer	300
thyA deoB cytR or *deoR*	Super low thymine requirer	2–5
thyA deoA	Super high thymine requirer	>500
thyA deoA udp	Thymidine requirer	
thyA deoA cytR	High thymine requirer	150
thyA dut	Thymidine requirer	
thyA dcd	Thymidine requirer	

Secondary mutations that increase the endogenous deoxyribose 1-phosphate pool, combined with an increase in the level of thymidine phosphorylase, enable a *thyA* mutant to utilize thymine more efficiently; i.e., they result in a low thymine requirement. In contrast, mutations that decrease the rate of synthesis of deoxyribose 1-phosphate from dUMP increase the rate of deoxyribose 1-phosphate catabolism, or decrease thymidine phosphorylase activity induces a lowered efficiency of thymine utilization. The effect of different mutations on the thymine concentration required for normal growth of *thyA* mutants is illustrated in Table 14.

Pulse-labeling of DNA. Low concentrations of labeled thymidine of high specific activity have been used extensively as a precursor to measure short-time incorporation into nascent DNA. To interpret such experiments, it is of utmost importance to realize that artifacts may arise owing to known or unknown genetic differences between the strains used. Thus, the presence of unrecognized mutations affecting the uptake of thymidine (*tsx, nupG, nupC*) or mutations that cause alterations in the internal level of thymidine phosphorylase (*cytR, deoR*) may give a false impression of altered DNA synthesis. A more detailed discussion of these experimental problems and how they may be overcome has been presented elsewhere (6, 175, 205).

ACKNOWLEDGMENTS

We express our gratitude to the many colleagues who have been willing to share some of their unpublished data with us.

Our own research has been supported by grants from the Danish Natural Science Research Council.

LITERATURE CITED

1. **Abdelal, A., E. Griego, and J. L. Ingraham.** 1976. Arginine-sensitive phenotype of mutations in *pyrA* of *Salmonella typhimurium*: role of ornithine carbamyltransferase in the assembly of mutant carbamylphosphate synthetase. J. Bacteriol. **128**:105–113.
2. **Abdelal, A., E. Griego, and J. L. Ingraham.** 1978. Arginine auxotrophic phenotype of mutations in *pyrA* of *Salmonella typhimurium*: role of *N*-acetylornithine in the maturation of mutant carbamylphosphate synthetase. J. Bacteriol. **134**:528–536.
3. **Abdelal, A., and J. L. Ingraham.** 1969. Control of carbamyl phosphate synthesis in *Salmonella typhimurium*. J. Biol. Chem. **244**:4033–4038.
4. **Abdelal, A., and J. L. Ingraham.** 1969. Cold sensitivity and other phenotypes resulting from mutation in *pyrA* gene. J. Biol. Chem. **244**:4039–4045.
5. **Abdelal, A., and J. L. Ingraham.** 1975. Carbamylphosphate synthetase from *Salmonella typhimurium*. Regulation, subunit composition, and function of the subunits. J. Biol. Chem. **250**:4410–4417.
6. **Anderson, M. L. M.** 1978. Size distribution of short chain DNA in two strains of *Escherichia coli*, p. 309–319. *In* I. Molineux and M. Kohiyama (ed.), DNA synthesis. Present and future. Plenum Publishing Corp., New York.
7. **Anderson, P. M.** 1977. Binding of allosteric effectors to carbamylphosphate synthetase from *Escherichia coli*. Biochemistry **16**:587–592.
8. **Anderson, P. M., and S. V. Marvin.** 1968. Effect of ornithine, IMP, and UMP on carbamyl phosphate synthetase from *Escherichia coli*. Biochem. Biophys. Res. Commun. **32**:928–934.
9. **Anderson, P. M., and A. Meister.** 1965. Evidence for activated form of carbon dioxide in the reaction catalyzed by *Escherichia coli* carbamyl phosphate synthetase. Biochemistry **4**:2803–2809.
10. **Anderson, P. M., and A. Meister.** 1966. Control of *Escherichia coli* carbamyl phosphate synthetase by purine and pyrimidine nucleotides. Biochemistry **5**:3164–3169.
11. **Ashley, G. W., and P. A. Bartlett.** 1984. Purification and properties of cytidine deaminase from *Escherichia coli*. J. Biol. Chem. **259**:13615–13620.
12. **Ashley, G. W., and P. A. Bartlett.** 1984. Inhibition of *Escherichia coli* cytidine deaminase by a phosphapyrimidine nucleoside. J. Biol. Chem. **259**:13621–13627.
13. **Bagnara, A. S., and L. R. Finch.** 1973. Relationships between intracellular contents of nucleotides and 5-phosphoribosyl 1-pyrophosphate in *Escherichia coli*. Eur. J. Biochem. **36**:422–427.
14. **Bagnara, A. S., and L. R. Finch.** 1974. The effects of bases and nucleosides on the intracellular contents of nucleotides and 5-phosphoribosyl 1-pyrophosphate in *Escherichia coli*. Eur. J. Biochem. **41**:421–430.
15. **Beacham, I. R., and S. Garrett.** 1980. Isolation of *Escherichia coli* mutants (*cpdB*) deficient in periplasmic 2′,3′-cyclic phosphodiesterase and genetic mapping of the *cpdB* locus. J. Gen. Microbiol. **119**:31–34.
16. **Beacham, I. R., D. Haas, and E. Yagil.** 1977. Mutants of *Escherichia coli* "cryptic" for certain periplasmic enzymes: evidence for an alteration of the outer membrane. J. Bacteriol. **129**:1034–1044.
17. **Beacham, I. R., R. Kahana, L. Levy, and E. Yagil.** 1973. Mutants of *Escherichia coli* K-12 "cryptic," or deficient in 5′-nucleotidase (uridine diphosphate-sugar hydrolase) and 3′-nucleotidase (cyclic phosphodiesterase) activity. J. Bacteriol. **116**:957–964.
18. **Beck, C. F., A. R. Eisenhardt, and J. Neuhard.** 1975. Deoxycytidine triphosphate deaminase of *Salmonella typhimurium*. Purification and characterization. J. Biol. Chem. **250**:609–616.
19. **Beck, C. F., and J. L. Ingraham.** 1971. Location on the chromosome of *Salmonella typhimurium* of genes governing pyrimidine metabolism. Mol. Gen. Genet. **111**:303–316.
20. **Beck, C. F., J. Neuhard, E. Thomassen, J. L. Ingraham, and E. Kleker.** 1974. *Salmonella typhimurium* mutants defective in cytidine monophosphate kinase (*cmk*). J. Bacteriol. **120**:1370–1379.
21. **Beckwith, J. R., A. B. Pardee, R. Austrian, and F. Jacob.** 1962. Coordination of the synthesis of the enzymes in the pyrimidine pathway of *E. coli*. J. Mol. Biol. **5**:618–634.
22. **Belfort, M., G. F. Maley, and F. Maley.** 1983. Characterization of the *Escherichia coli thyA* gene and its amplified thymidylate synthetase product. Proc. Natl. Acad. Sci. USA **80**:1858–1861.
23. **Belfort, M., G. F. Maley, J. Pedersen-Lane, and F. Maley.** 1983. Primary structure of the *Escherichia coli thyA* gene and its thymidylate synthase product. Proc. Natl. Acad. Sci. USA **80**:4914–4918.
24. **Bencini, D. A., J. E. Houghton, T. A. Hoover, K. F. Foltermann, J. R. Wild, and G. A. O'Donovan.** 1983. The DNA sequence of *argI* from *Escherichia coli* K12. Nucleic Acids Res. **11**:8509–8518.
25. **Benson, C. E., B. A. Brehmeyer, and J. S. Gots.** 1971. Requirement of cyclic AMP for induction of GMP reductase in *Escherichia coli*. Biochem. Biophys. Res. Commun. **43**:1089–1094.
26. **Benson, C. E., and J. S. Gots.** 1975. Regulation of GMP reductase in *Salmonella typhimurium*. Biochim. Biophys. Acta **403**:47–57.
27. **Benson, C. E., and J. S. Gots.** 1976. Occurrence of a regulatory deficiency in purine biosynthesis among *purA* mutants of *Salmonella typhimurium*. Mol. Gen. Genet. **145**:31–36.
28. **Benson, C. E., D. L. Hornick, and J. S. Gots.** 1980. Genetic separation of purine transport from phosphoribosyltransferase

activity in *Salmonella typhimurium*. J. Gen. Microbiol. **121**:357–364.

29. **Berberich, M. A., and J. S. Gots.** 1965. A structural gene mutation in *Salmonella typhimurium* resulting in repressibility of adenylosuccinase. Proc. Natl. Acad. Sci. USA **54**:1254–1261.
30. **Bochner, B. R., and B. N. Ames.** 1982. Complete analysis of cellular nucleotides by two-dimensional thin layer chromatography. J. Biol. Chem. **257**:9759–9769.
31. **Bochner, B. R., and B. N. Ames.** 1982. ZTP (5-amino 4-imidazole carboxamide riboside 5'-triphosphate): a proposed alarmone for 10-formyl-tetrahydrofolate deficiency. Cell **29**:929–937.
32. **Bonekamp, F., H. D. Andersen, T. Christensen, and K. F. Jensen.** 1985. Codon-defined ribosomal pausing in *Escherichia coli* detected by using the *pyrE* attenuator to probe the coupling between transcription and translation. Nucleic Acids Res. **13**:4113–4123.
33. **Bonekamp, F., K. Clemmesen, O. Karlstrom, and K. F. Jensen.** 1984. Mechanism of UTP-modulated attenuation at the *pyrE* gene of *Escherichia coli*: an example of operon polarity control through the coupling of translation to transcription. EMBO J. **3**:2857–2861.
34. **Bothwell, M., and H. K. Schachman.** 1980. A model for the assembly of aspartate transcarbamoylase from catalytic and regulatory subunits. J. Biol. Chem. **255**:1971–1977.
35. **Bouvier, J., J.-C. Patte, and P. Stragier.** 1984. Multiple regulatory signals in the control region of the *Escherichia coli carAB* operon. Proc. Natl. Acad. Sci. USA **81**:4139–4143.
36. **Brune, M., R. Schumann, and F. Wittinghofer.** 1985. Cloning and sequencing of the adenylate kinase gene (*adk*) of *Escherichia coli*. Nucleic Acids Res. **13**:7139–7151.
37. **Buchanan, J. M., S. Ohnoki, and B. S. Hong.** 1978. 2-Formamido-*N*-ribosylacetamide 5'-phosphate: L-glutamine amido-ligase (adenosine diphosphate). Methods Enzymol. **51**:193–201.
38. **Burns, D. L., and H. K. Schachman.** 1982. Ligand-promoted strengthening of interchain bonding domains in catalytic subunit of aspartate transcarbamoylase. J. Biol. Chem. **257**:12214–12218.
39. **Burton, K.** 1977. Transport of adenine, hypoxanthine, and uracil into *Escherichia coli*. Biochem. J. **168**:195–204.
40. **Burton, K.** 1983. Transport of nucleic acid bases into *Escherichia coli*. J. Gen. Microbiol. **129**:3505–3513.
41. **Bussey, L. B., and J. L. Ingraham.** 1982. A regulatory gene (*use*) affecting the expression of *pyrA* and certain other pyrimidine genes. J. Bacteriol. **151**:144–152.
42. **Bussey, L. B., and J. L. Ingraham.** 1982. Isolation and mapping of a uracil-sensitive mutant of *Salmonella typhimurium*. Mol. Gen. Genet. **185**:513–514.
43. **Buxton, R. S., K. Hammer-Jespersen, and P. Valentin-Hansen.** 1980. A second purine nucleoside phosphorylase in *Escherichia coli* K-12. I. Xanthosine phosphorylase regulatory mutants isolated as secondary-site revertants of a *deoD* mutant. Mol. Gen. Genet. **179**:331–340.
44. **Buzzee, D. H., and A. P. Levin.** 1968. Demonstration of an effector site for the enzyme inosine 5'-phosphate dehydrogenase. Biochem. Biophys. Res. Commun. **30**:673–677.
45. **Carlson, J., J. A. Fuchs, and J. Messing.** 1984. Primary structure of the *Escherichia coli* ribonucleoside diphosphate reductase operon. Proc. Natl. Acad. Sci. USA **81**:4294–4297.
46. **Christopherson, R. I., and L. R. Finch.** 1978. Response of the pyrimidine pathway of *Escherichia coli* K12 to exogenous adenine and uracil. Eur. J. Biochem. **90**:347–358.
47. **Clemmesen, K., F. Bonekamp, O. Karlstrom, and K. F. Jensen.** 1985. Role of translation in the UTP-modulated attenuation at the *pyrBI* operon of *Escherichia coli*. Mol. Gen. Genet. **201**:247–251.
48. **Crabeel, M., D. Charlier, G. Weyens, A. Feller, A. Pierard, and N. Glansdorff.** 1980. Use of gene cloning to determine polarity of an operon: genes *carAB* of *Escherichia coli*. J. Bacteriol. **143**:921–925.
49. **Daws, T. D., and J. A. Fuchs.** 1984. Isolation and characterization of an *Escherichia coli* mutant deficient in dTMP kinase activity. J. Bacteriol. **157**:440–444.
50. **deHaan, P. G., H. S. Felix, and R. Peters.** 1972. Mapping of the gene for cytosine deaminase on the *Escherichia coli* chromosome. Antonie van Leeuwenhoek. J. Microbiol. Serol. **38**:257–263.
51. **Dennis, P. P., and R. K. Herman.** 1970. Pyrimidine pools and macromolecular composition of pyrimidine-limited *Escherichia coli*. J. Bacteriol. **102**:118–123.
52. **Dev, I. K., and R. J. Harvey.** 1978. N^{10}-formyltetrahydrofolate is the formyl donor for glycinamide ribotide transformylase in *Escherichia coli*. J. Biol. Chem. **253**:4242–4244.

53. **Donovan, W. P., and S. R. Kushner.** 1983. Purification and characterization of orotidine-5'-phosphate decarboxylase from *Escherichia coli* K-12. J. Bacteriol. **156**:620–624.
54. **Eyzaguirre, J., and D. E. Atkinson.** 1974. Regulatory effects of purine ribonucleotides on *Escherichia coli* adenylosuccinate synthase. Arch. Biochem. Biophys. **169**:339–343.
55. **Fast, R., and O. Skold.** 1977. Biochemical mechanism of uracil uptake regulation in *Escherichia coli* B. Allosteric effects on uracil phosphoribosyltransferase under stringent conditions. J. Biol. Chem. **252**:7620–7624.
56. **Feller, A., A. Pierard, N. Glansdorff, D. Charlier, and M. Crabeel.** 1981. Mutation of gene encoding regulatory polypeptide of aspartate carbamoyltransferase. Nature (London) **292**:370–373.
57. **Filpula, D., and J. A. Fuchs.** 1977. Regulation of ribonucleoside diphosphate reductase synthesis in *E. coli*. Increased enzyme synthesis as a result of inhibition of deoxyribonucleic acid synthesis. J. Bacteriol. **130**:107–113.
58. **Filpula, D., and J. A. Fuchs.** 1978. Regulation of the synthesis of ribonucleoside diphosphate reductase in *Escherichia coli*: specific activity of the enzyme in relationship to perturbations of DNA replication. J. Bacteriol. **135**:429–435.
59. **Frisch, S. M., J. L. Couch, and D. A. Glaser.** 1978. Mutator activity of a short Okazaki fragment mutant of *Escherichia coli*. J. Bacteriol. **134**:1192–1194.
60. **Fuchs, J.** 1977. Isolation of an *Escherichia coli* mutant deficient in thioredoxin reductase. J. Bacteriol. **129**:967–972.
61. **Gallant, J., J. Irr, and M. Cashel.** 1971. The mechanism of amino acid control of guanylate and adenylate biosynthesis. J. Biol. Chem. **246**:5812–5816.
62. **Garber, B. B., and J. S. Gots.** 1980. Utilization of 2,6-diaminopurine by *Salmonella typhimurium*. J. Bacteriol. **143**:864–871.
63. **Garber, B. B., B. U. Jochimsen, and J. S. Gots.** 1980. Glutamine and related analogs regulate guanosine monophosphate reductase in *Salmonella typhimurium*. J. Bacteriol. **143**:105–111.
64. **Gerhart, J. C.** 1970. A discussion of the regulatory properties of aspartate transcarbamylase from *Escherichia coli*. Curr. Top. Cell. Regul. **2**:275–325.
65. **Gibson, K. J., K. R. Schubert, and R. L. Switzer.** 1982. Binding of the substrates and the allosteric inhibitor adenosine 5'-diphosphate to phosphoribosylpyrophosphate synthetase from *Salmonella typhimurium*. J. Biol. Chem. **257**:2391–2396.
66. **Gigot, D., M. Crabeel, A. Feller, D. Charlier, W. Lissens, N. Glansdorff, and A. Pierard.** 1980. Patterns of polarity in the *Escherichia coli carAB* gene cluster. J. Bacteriol. **143**:914–920.
67. **Gilbert, H. J., and W. T. Drabble.** 1980. Active-site modification of native and mutant forms of inosine 5'-monophosphate dehydrogenase from *Escherichia coli* K12. Biochem. J. **191**:533–541.
68. **Gilbert, H. J., C. R. Lowe, and W. T. Drabble.** 1979. Inosine 5'-monophosphate dehydrogenase of *Escherichia coli*. Purification by affinity chromatography, subunit structure and inhibition by guanosine 5'-monophosphate. Biochem. J. **183**:481–494.
69. **Ginther, C. L., and J. L. Ingraham.** 1974. Nucleoside diphosphokinase of *Salmonella typhimurium*. J. Biol. Chem. **249**:3406–3411.
70. **Ginther, C. L., and J. L. Ingraham.** 1974. Cold-sensitive mutant of *Salmonella typhimurium* defective in nucleoside diphosphokinase. J. Bacteriol. **129**:1020–1026.
71. **Glaser, M., W. Nulty, and P. R. Vagelos.** 1975. Role of adenylate kinase in the regulation of macromolecular biosynthesis in a putative mutant of *Escherichia coli* defective in membrane phospholipid biosynthesis. J. Bacteriol. **123**:128–136.
72. **Goelz, S. E., and J. E. Cronan, Jr.** 1982. Adenylate kinase of *Escherichia coli*: evidence for a functional interaction in phospholipid synthesis. Biochemistry **21**:189–195.
73. **Gots, J. S.** 1971. Regulation of purine and pyrimidine metabolism, p. 225–255. *In* H. J. Vogel (ed.), Metabolic pathways, vol. 5. Academic Press, Inc., New York.
74. **Gots, J. S., C. E. Benson, B. Jochimsen, and K. R. Koduri.** 1977. Microbial models and regulatory elements in the control of purine metabolism. CIBA Found. Symp. **48**:23–41.
75. **Gots, J. S., C. E. Benson, and S. R. Shumas.** 1972. Genetic separation of hypoxanthine and guanine-xanthine phosphoribosyltransferase activity by deletion mutations in *Salmonella typhimurium*. J. Bacteriol. **112**:910–916.
76. **Gots, J. S., F. R. Dalal, and S. R. Shumas.** 1969. Genetic separation of the inosinic acid cyclohydrolase-transformylase complex of *Salmonella typhimurium*. J. Bacteriol. **99**:441–449.
77. **Granot, J., K. J. Gibson, R. L. Switzer, and A. S. Mildvan.** 1980. NMR studies of the nucleotide conformation and the arrangement of substrates and activators on phosphoribosyl pyrophosphate synthetase. J. Biol. Chem. **255**:10931–10937.
78. **Haller, B. L., and J. A. Fuchs.** 1984. Mapping of *trxB*, a mutation

responsible for reduced thioredoxin reductase activity. J. Bacteriol. 159:1060–1062.

79. Hammer-Jespersen, K. 1983. Nucleoside catabolism, p. 203–258. In A. Munch-Petersen (ed.), Metabolism of nucleotides, nucleosides, and nucleobases in microorganisms. Academic Press, Inc. (London), Ltd., London.

80. Hammer-Jespersen, K., R. S. Buxton, and T. D. Hansen. 1980. A second purine nucleoside phosphorylase in Escherichia coli K-12. II. Properties of xanthosine phosphorylase and its induction by xanthosine. Mol. Gen. Genet. 179:341–348.

81. Hammer-Jespersen, K., and A. Munch-Petersen. 1973. Mutants of Escherichia coli unable to metabolize cytidine: isolation and characterization. Mol. Gen. Genet. 126:177–186.

82. Hanke, P. D., and J. A. Fuchs. 1983. Characterization of the mRNA coding for ribonucleoside diphosphate reductase in Escherichia coli. J. Bacteriol. 156:1192–1197.

83. Hanke, P. D., and J. A. Fuchs. 1984. Requirement of protein synthesis for the induction of ribonucleoside diphosphate reductase mRNA in Escherichia coli. Mol. Gen. Genet. 193:327–331.

84. Hantke, K. 1976. Phage T6-Colicin K receptor and nucleoside transport in Escherichia coli. FEBS Lett. 70:109–112.

85. Hartman, S. C. 1970. Purines and pyrimidines, p. 1–68. In D. M. Greenberg (ed.), Metabolic pathways, vol. 4. Academic Press, Inc., New York.

86. Henderson, J. F. 1972. Regulation of purine biosynthesis. American Chemical Society, Washington, D.C.

87. Henderson, J. F. 1980. Inhibition of microbial growth by naturally occurring purine bases and ribonucleosides. Pharmacol. & Ther. 8:605–627.

88. Hershey, H. V., R. Gutstein, and M. W. Taylor. 1982. Cloning and restriction map of the E. coli apt gene. Gene 19:89–92.

89. Hochhauser, S. J., and B. Weiss. 1978. E. coli mutants deficient in deoxyuridine triphosphatase. J. Bacteriol. 134:157–166.

90. Hochstadt, J. 1978. Hypoxanthine phosphoribosyltransferase and guanine phosphoribosyltransferase from enteric bacteria. Methods Enzymol. 51:549–557.

91. Hochstadt, J. 1978. Adenine phosphoribosyltransferase from Escherichia coli. Methods Enzymol. 51:558–567.

92. Hochstadt-Ozer, J., and M. Cashel. 1972. The regulation of purine utilization in bacteria. V. Inhibition of purine phosphoribosyltransferase activities and purine uptake in isolated membrane vesicles by guanosine tetraphosphate. J. Biol. Chem. 247:7067–7072.

93. Hoffee, P. A. 1968. 2-Deoxyribose gene-enzyme complex in Salmonella typhimurium. I. Isolation and enzymatic characterization of 2-deoxyribose-negative mutants. J. Bacteriol. 95:449–457.

94. Hoffmeyer, J., and J. Neuhard. 1971. Metabolism of exogenous purine bases and nucleosides by Salmonella typhimurium. J. Bacteriol. 106:14–24.

95. Holmes, R. K., and M. F. Singer. 1973. Purification and characterization of adenylate kinase as an apparent adenosine triphosphate-dependent inhibitor of ribonuclease II in Escherichia coli. J. Biol. Chem. 248:2014–2021.

96. Holmgren, A. 1979. Glutathione-dependent synthesis of deoxyribonucleotides. Purification and characterization of glutaredoxin from Escherichia coli. J. Biol. Chem. 254:3664–3671.

97. Holmgren, A. 1979. Glutathione-dependent synthesis of deoxyribonucleotides. Characterization of the enzymatic mechanism of Escherichia coli glutaredoxin. J. Biol. Chem. 254:3672–3678.

98. Holmgren, A. 1985. Thioredoxin. Annu. Rev. Biochem. 54:237–271.

99. Honzatko, R. B., J. L. Crawford, H. L. Monaco, J. E. Ladner, B. F. P. Edwards, D. R. Evans, S. G. Warren, D. C. Wiley, R. C. Ladner, and W. N. Lipscomb. 1982. Crystal and molecular structures of native and CTP-ligated aspartate carbamoyltransferase from Escherichia coli. J. Mol. Biol. 160:219–263.

100. Hoover, T. A., W. D. Roof, K. F. Foltermann, G. A. O'Donovan, D. A. Bencini, and J. R. Wild. 1983. Nucleotide sequence of the structural gene (pyrB) that encodes the catalytic polypeptide of aspartate transcarbamoylase of Escherichia coli. Proc. Natl. Acad. Sci. USA 80:2462–2466.

101. Houghton, J. E., D. A. Bencini, G. A. O'Donovan, and J. R. Wild. 1984. Protein differentiation: a comparison of aspartate transcarbamoylase and ornithine transcarbamoylase from Escherichia coli K-12. Proc. Natl. Acad. Sci. USA 81:4864–4868.

102. Houlberg, U., B. Hove-Jensen, B. Jochimsen, and P. Nygaard. 1983. Identification of the enzymatic reactions encoded by the purG and purI genes of Escherichia coli. J. Bacteriol. 154:1485–1488.

103. Houlberg, U., and K. F. Jensen. 1983. Role of hypoxanthine and guanine in regulation of Salmonella typhimurium pur gene expression. J. Bacteriol. 153:837–845.

104. Hove-Jensen, B. 1983. Chromosomal location of the gene encoding phosphoribosylpyrophosphate synthetase in Escherichia coli. J. Bacteriol. 154:177–184.

105. Hove-Jensen, B., K. W. Harlow, C. J. King, and R. L. Switzer. 1986. Phosphoribosylpyrophosphate synthetase of Escherichia coli. Properties of the purified enzyme and primary structure of the prs gene. J. Biol. Chem. 261:6765–6771.

106. Hove-Jensen, B., and P. Nygaard. 1982. Phosphoribosylpyrophosphate synthetase of Escherichia coli. Identification of a mutant enzyme. Eur. J. Biochem. 126:327–332.

107. Huss, R. J., and M. Glaser. 1983. Identification and purification of an adenylate kinase-associated protein that influences the thermolability of adenylate kinase from a temperature-sensitive adk mutant of Escherichia coli. J. Biol. Chem. 258:13370–13376.

108. Ingraham, J. L., O. Maaløe, and F. C. Neidhardt. 1983. Growth of the bacterial cell. Sinauer Associates, Inc., Sunderland, Mass.

109. Ingraham, J. L., and J. Neuhard. 1972. Cold-sensitive mutants of Salmonella typhimurium defective in uridine monophosphate kinase (pyrH). J. Biol. Chem. 247:6259–6265.

110. Iwatsuki, N., and R. Okazaki. 1967. Mechanism of regulation of deoxythymidine kinase of Escherichia coli. I. Effect of regulatory deoxynucleotides on the state of aggregation of the enzyme. J. Mol. Biol. 29:139–154.

111. Iwatsuki, N., and R. Okazaki. 1967. Mechanism of regulation of deoxythymidine kinase of Escherichia coli. II. Effect of temperature on the enzyme activity and kinetics. J. Mol. Biol. 29:155–165.

112. Jacobsen, G. R., and G. R. Stark. 1973. Aspartate transcarbamylases, p. 225–308. In P. D. Boyer (ed.), The enzymes, vol. 9. Academic Press, Inc., New York.

113. Jagadeeswaran, P., C. R. Ashman, S. Roberts, and J. Langenberg. 1984. Nucleotide sequence and analysis of deletion mutants of the Escherichia coli gpt gene in plasmid pSV2gpt. Gene 31:309–313.

114. Jenness, D. D., and H. K. Schachman. 1983. Genetic characterization of the folding domains of the catalytic chains in aspartate transcarbamoylase. J. Biol. Chem. 258:3266–3273.

115. Jensen, K. F. 1979. Apparent involvement of purines in the control of expression of Salmonella typhimurium pyr genes: analysis of a leaky guaB mutant resistant to pyrimidine analogs. J. Bacteriol. 138:731–738.

116. Jensen, K. F. 1983. Metabolism of 5-phosphoribosyl 1-pyrophosphate (PRPP) in Escherichia coli and Salmonella typhimurium, p. 1–25. In A. Munch-Petersen (ed.), Metabolism of nucleotides, nucleosides and nucleobases in microorganisms. Academic Press, Inc. (London), Ltd., London.

117. Jensen, K. F., R. Fast, O. Karlstrom, and J. N. Larsen. 1986. Association of RNA polymerase having increased K_m for ATP and UTP with hyperexpression of the pyrB and pyrE genes of Salmonella typhimurium. J. Bacteriol. 166:857–865.

118. Jensen, K. F., U. Houlberg, and P. Nygaard. 1979. Thin-layer chromatographic methods to isolate ^{32}P-labeled 5-phosphoribosyl-α-1-pyrophosphate (PRPP): determination of cellular PRPP pools and assay of PRPP synthetase activity. Anal. Biochem. 98:254–263.

119. Jensen, K. F., J. N. Larsen, L. Schack, and A. Sivertsen. 1984. Studies on the structure and expression of Escherichia coli pyrC, pyrD, and pyrF using the cloned genes. Eur. J. Biochem. 140:343–352.

120. Jensen, K. F., J. Neuhard, and L. Schack. 1982. RNA polymerase involvement in the regulation of expression of Salmonella typhimurium pyr genes. Isolation and characterization of a fluorouracil-resistant mutant with high, constitutive expression of the pyrB and pyrE genes due to a mutation in rpoBC. EMBO J. 1:69–74.

121. Jensen, K. F., and P. Nygaard. 1975. Purine nucleoside phosphorylase from Escherichia coli and Salmonella typhimurium. Purification and some properties. Eur. J. Biochem. 51:253–265.

122. Jochimsen, B., P. Nygaard, and T. Vestergaard. 1975. Location on the chromosome of Escherichia coli of genes governing purine metabolism. Mol. Gen. Genet. 143:85–91.

123. Jochimsen, B. U., B. Hove-Jensen, B. B. Garber, and J. S. Gots. 1985. Characterization of a Salmonella typhimurium mutant defective in phosphoribosylpyrophosphate synthetase. J. Gen. Microbiol. 131:245–252.

124. Justesen, J., and J. Neuhard. 1975. pyrR identical to pyrH in Salmonella typhimurium: control of expression of the pyr genes. J. Bacteriol. 123:851–854.

125. Kantrowitz, E. R., S. C. Pastra-Landis, and W. N. Lipscomb. 1980. E. coli aspartate transcarbamylase. Part I. Catalytic and regulatory functions. Trends Biochem. Sci. 5:124–128.

126. **Kantrowitz, E. R., S. C. Pastra-Landis, and W. N. Lipscomb.** 1980. *E. coli* aspartate transcarbamylase. Part II. Structure and allosteric interactions. Trends Biochem. Sci. **5:**150–153.

127. **Kantrowitz, E. R., H. W. Reed, R. A. Ferraro, and J. P. Daigneault.** 1981. Analysis of mutant *Escherichia coli* aspartate transcarbamylases isolated from a series of suppressed *pyrB* nonsense strains. J. Mol. Biol. **153:**569–587.

128. **Karibian, D.** 1978. Dihydroorotate dehydrogenase (*Escherichia coli*). Methods Enzymol. **51:**58–63.

129. **Kelln, R. A.** 1984. Evidence for involvement of *pyrH*⁺ of an *Escherichia coli* K-12 F-prime factor in inhibiting construction of hybrid merodiploids with *Salmonella typhimurium*. Can. J. Microbiol. **30:**991–996.

130. **Kelln, R. A., J. J. Kinahan, K. F. Foltermann, and G. A. O'Donovan.** 1975. Pyrimidine biosynthetic enzymes of *Salmonella typhimurium*, repressed specifically by growth in the presence of cytidine. J. Bacteriol. **124:**764–774.

131. **Kelln, R. A., J. Neuhard, and L. Stauning.** 1985. Isolation and characterization of pyrimidine mutants of *Salmonella typhimurium* altered in expression of *pyrC*, *pyrD*, and *pyrE*. Can. J. Microbiol. **31:**981–987.

132. **Kerr, C. T., and R. W. Miller.** 1968. Dihydroorotate-ubiquinone reductase complex of *Escherichia coli* B. J. Biol. Chem. **243:**2963–2968.

133. **Kessler, A. I., and J. S. Gots.** 1985. Regulation of *guaC* expression in *Escherichia coli*. J. Bacteriol. **164:**1288–1293.

134. **Kocharian, A. M., M. A. Melkumian, and S. M. Kocharian.** 1985. Regulatory mutants for synthesis of a second purine nucleoside phosphorylase in *Escherichia coli* K-12. II. Mapping and dominance studies of *pndR* mutations. Mol. Gen. Genet. **21:**220–228.

135. **Kocharian, S. M., A. M. Kocharian, G. O. Meliksetian, and J. I. Akopian.** 1982. Mutants of *Escherichia coli* K-12 utilizing adenine via a new metabolic pathway. Genetika **18:**906–915.

136. **Koduri, R. K., and J. S. Gots.** 1980. A DNA-binding protein with specificity for *pur* genes in *Escherichia coli*. J. Biol. Chem. **255:**9594–9598.

137. **Konigsberg, W. H., and L. Henderson.** 1983. Amino acid sequence of the catalytic subunit of aspartate transcarbamoylase from *Escherichia coli*. Proc. Natl. Acad. Sci. USA **80:**2467–2471.

138. **Koshland, D. E., and A. Levitzki.** 1974. CTP synthetase and related enzymes, p. 539–559. *In* P. D. Boyer (ed.), The enzymes, vol. 10. Academic Press, Inc., New York.

139. **Krause, K. L., K. W. Volz, and W. N. Lipscomb.** 1985. Structure at 2.9Å resolution of aspartate carbamoyltransferase complexed with the bisubstrate analogue, *N*-(phosphonacetyl)-L-aspartate. Proc. Natl. Acad. Sci. USA **82:**1643–1647.

140. **Krieger-Brauer, H. J., and V. Braun.** 1980. Functions related to the receptor protein specified by the *tsx* gene of *Escherichia coli*. Arch. Microbiol. **124:**233–242.

141. **Krishnaiah, K. V.** 1975. Inosine acid 5′-monophosphate dehydrogenase of *Escherichia coli*: purification by affinity chromatography and some properties. Arch. Biochem. Biophys. **170:**567–575.

142. **Lambden, P. R., and W. T. Drabble.** 1973. The *gua* operon of *Escherichia coli* K-12: evidence for polarity from *guaB* to *guaA*. J. Bacteriol. **115:**992–1002.

143. **Larsen, J. N., and K. F. Jensen.** 1985. Nucleotide sequence of the *pyrD* gene of *Escherichia coli* and characterization of the flavoprotein dihydroorotate dehydrogenase. Eur. J. Biochem. **151:**59–65.

144. **Lauritzen, A. M., and W. N. Lipscomb.** 1982. Modification of three active site lysine residues in the catalytic subunit of aspartate transcarbamylase by D- and L-bromosuccinate. J. Biol. Chem. **257:**1312–1319.

145. **Leer, J. C., K. Hammer-Jespersen, and M. Schwartz.** 1977. Uridine phosphorylase from *Escherichia coli*. Physical and chemical characterization. Eur. J. Biochem. **75:**217–224.

146. **Legrain, C., V. Stalon, N. Glansdorff, D. Gigot, A. Pierard, and M. Crabeel.** 1976. Structural and regulatory mutations allowing utilization of citrulline or carbamoylaspartate as a source of carbamoylphosphate in *Escherichia coli* K-12. J. Bacteriol. **128:**39–48.

147. **Leung, H. B., and V. L. Schramm.** 1980. Adenylate degradation in *Escherichia coli*. The role of AMP nucleosidase and properties of the purified enzyme. J. Biol. Chem. **255:**10867–10874.

148. **Leung, H. B., and V. L. Schramm.** 1984. The structural gene for AMP nucleosidase. Mapping, cloning, and overproduction of the enzyme. J. Biol. Chem. **259:**6972–6978.

149. **Levin, H. L., and H. K. Schachman.** 1985. Regulation of aspartate transcarbamoylase synthesis in *Escherichia coli*: analysis of deletion mutations in the promoter region of the *pyrBI* operon. Proc. Natl. Acad. Sci. USA **82:**4643–4647.

150. **Levine, R. A., and M. W. Taylor.** 1981. Selection for purine regulatory mutants in an *E. coli* hypoxanthine phosphoribosyltransferase-guanine phosphoribosyltransferase double mutant. Mol. Gen. Genet. **181:**313–318.

151. **Levine, R. A., and M. W. Taylor.** 1982. Regulation of *purE* transcription in a *purE::lac* fusion strain of *Escherichia coli*. J. Bacteriol. **149:**1041–1049.

152. **Lichtenstein, J., H. D. Barner, and S. S. Cohen.** 1960. The metabolism of exogenously supplied nucleotides by *Escherichia coli*. J. Biol. Chem. **235:**457–465.

153. **Lim, C.-J., D. Geraghty, and J. A. Fuchs.** 1985. Cloning and nucleotide sequence of the *trxA* gene of *Escherichia coli* K-12. J. Bacteriol. **163:**311–316.

154. **Lim, C.-J., B. Haller, and J. A. Fuchs.** 1985. Thioredoxin is the bacterial protein encoded by *fip* that is required for filamentous bacteriophage f1 assembly. J. Bacteriol. **161:**799–802.

155. **Lind, R. M., V. V. Sukhodolets, and Y. V. Smirnov.** 1973. Mutations affecting deamination and transport of cytosine in *Escherichia coli*. Genetika **9:**116–121.

156. **Lindahl, T.** 1979. DNA glycosylases, endonucleases for apurinic/apyrimidinic sites, and base excision-repair. Prog. Nucleic Acid Res. Mol. Biol. **22:**135–192.

157. **Lissens, W., R. Cunin, N. Kelker, N. Glansdorff, and A. Pierard.** 1980. In vitro synthesis of *Escherichia coli* carbamoylphosphate synthase: evidence for participation of the arginine repressor in cumulative repression. J. Bacteriol. **141:**58–66.

158. **Liu, S. W., and G. Milman.** 1983. Purification and characterization of *Escherichia coli* guanine-xanthine phosphoribosyltransferase produced by a high efficiency expression plasmid utilizing a λP_L promoter and CI857 temperature-sensitive repressor. J. Biol. Chem. **258:**7469–7475.

159. **Livshits, V. A.** 1976. Effect of 2,6-diaminopurine resistant mutations on the utilization of adenine and adenosine by adenine-requiring *Escherichia coli* K-12 strains. Genetika **12:**180–182.

160. **Livshits, V. A., and V. V. Sukhodolets.** 1973. On the role of adenine nucleotides in the regulation of purine ribonucleoside utilization by mutants of *Escherichia coli* K-12 defective in purine nucleoside phosphorylase. Genetika **9:**102–111.

161. **Long, C. W., and A. B. Pardee.** 1967. Cytidine triphosphate synthetase of *Escherichia coli* B. I. Purification and kinetics. J. Biol. Chem. **242:**4715–4721.

162. **Lundberg, L. G., H.-O. Thoresson, O. H. Karlstrom, and P.-O. Nyman.** 1983. Nucleotide sequence of the structural gene for dUTPase of *Escherichia coli* K-12. EMBO J. **2:**967–971.

163. **Machida, H., and A. Kuninaka.** 1969. Studies on the accumulation of orotic acid by *Escherichia coli* K12. Agric. Biol. Chem. **33:**868–875.

164. **Magasanik, B., and D. Karibian.** 1960. Purine nucleotide cycles and their metabolic role. J. Biol. Chem. **235:**2672–2681.

165. **Mager, J., and B. Magasanik.** 1960. Guanosine 5′-phosphate reductase and its role in the interconversion of purine nucleotides. J. Biol. Chem. **235:**1474–1478.

166. **Makaroff, C. A., and H. Zalkin.** 1985. Regulation of *Escherichia coli purF*. Analysis of the control region of a *pur* regulon gene. J. Biol. Chem. **260:**10378–10387.

167. **Manwaring, J. D., and J. A. Fuchs.** 1979. Relationship between deoxyribonucleoside triphosphate pools and deoxyribonucleic acid synthesis in an *nrdA* mutant of *Escherichia coli*. J. Bacteriol. **138:**245–248.

168. **Mark, D. F., J. W. Chase, and C. C. Richardson.** 1977. Genetic mapping of *trxA*, a gene affecting thioredoxin in *Escherichia coli* K12. Mol. Gen. Genet. **155:**145–152.

169. **McCarthy, M. P., and N. M. Allewell.** 1983. Thermodynamics of assembly of *E. coli* aspartate transcarbamoylase. Proc. Natl. Acad. Sci. USA **80:**6824–6828.

170. **McKeown, M., M. Kahn, and P. Hanawalt.** 1976. Thymidine uptake and utilization in *Escherichia coli*: a new gene controlling nucleoside transport. J. Bacteriol. **126:**814–822.

171. **Mehra, R. K., and W. T. Drabble.** 1981. Dual control of the *gua* operon of *Escherichia coli* K12 by adenine and guanine nucleotides. J. Gen. Microbiol. **123:**27–37.

172. **Mergeay, M., D. Gigot, J. Beckmann, N. Glansdorff, and A. Pierard.** 1974. Physiology and genetics of carbamoylphosphate synthesis in *Escherichia coli* K12. Mol. Gen. Genet. **133:**299–316.

173. **Messenger, L. J., and H. Zalkin.** 1979. Glutamine phosphoribosylpyrophosphate amidotransferase from *Escherichia coli*. Purification and properties. J. Biol. Chem. **254:**3382–3392.

174. **Michaels, G., and R. A. Kelln.** 1983. Construction and use of *pyr::lac* fusion strains to study regulation of pyrimidine biosynthesis in *Salmonella typhimurium*. Mol. Gen. Genet. **186:**463–470.

175. **Møllgaard, H., and J. Neuhard.** 1983. Biosynthesis of

deoxythymidine triphosphate, p. 149–198. *In* A. Munch-Petersen (ed.), Metabolism of nucleotides, nucleosides, and nucleobases in microorganisms. Academic Press, Inc. (London), Ltd., London.

176. **Morton, D. P., and S. M. Parsons.** 1977. Synergistic inhibition of ATP phosphoribosyltransferase by guanosine tetraphosphate and histidine. Biochem. Biophys. Res. Commun. **74:**325–331.

177. **Moyed, H. S.** 1964. Inhibition of the biosynthesis of the pyrimidine portion of thiamine by adenosine. J. Bacteriol. **88:**1024–1029.

178. **Mueller, K., and H. Bremer.** 1968. Rate of synthesis of messenger ribonucleic acid in *Escherichia coli.* J. Mol. Biol. **38:**329–353.

179. **Munch-Petersen, A., and N. Jensen.** 1985. Labelling of a nucleoside transporting protein with β-galactosidase by gene fusion. *In* M. Schaechter, F. C. Neidhardt, J. L. Ingraham, and O. Kjeldgaard (ed.), The molecular biology of bacterial growth. Jones & Bartlett Publishing, Inc., Boston.

180. **Munch-Petersen, A., and B. Mygind.** 1976. Nucleoside transport in *Escherichia coli* K 12: specificity and regulation. J. Cell. Physiol. **89:**551–559.

181. **Munch-Petersen, A., and B. Mygind.** 1983. Transport of nucleic acid precursors, p. 259–305. *In* A. Munch-Petersen (ed.), Metabolism of nucleotides, nucleosides, and nucleobases in microorganisms. Academic Press, Inc. (London), Ltd., London.

182. **Munch-Petersen, A., B. Mygind, A. Nicolaisen, and N. J. Pihl.** 1979. Nucleoside transport in cells and membrane vesicles from *Escherichia coli* K 12. J. Biol. Chem. **254:**3730–3737.

183. **Munch-Petersen, A., and N. J. Pihl.** 1980. Stimulatory effect of low ATP pools on transport of purine nucleosides in cells of *Escherichia coli.* Proc. Natl. Acad. Sci. USA **77:**2519–2523.

184. **Mygind, B., and A. Munch-Petersen.** 1975. Transport of pyrimidine nucleosides in cells of *Escherichia coli* K 12. Eur. J. Biochem. **59:**365–372.

185. **Navre, M., and H. K. Schachman.** 1983. Synthesis of aspartate transcarbamoylase in *Escherichia coli:* transcriptional regulation of the *pyrB-pyrI* operon. Proc. Natl. Acad. Sci. USA **80:**1207–1211.

186. **Nelbach, M. E., V. P. Pigiet, J. C. Gerhart, and H. K. Schachman.** 1972. A role for zinc in the quaternary structure of aspartate transcarbamylase from *Escherichia coli.* Biochemistry **11:**315–327.

187. **Nelson, D. P., and C. E. Carter.** 1969. Purification and characterization of thymidine 5'-monophosphate kinase from *Escherichia coli* B. J. Biol. Chem. **244:**5254–5262.

188. **Neuhard, J.** 1968. Pyrimidine nucleotide metabolism and pathways of thymidine triphosphate biosynthesis in *Salmonella typhimurium.* J. Bacteriol. **96:**1519–1527.

189. **Neuhard, J.** 1983. Utilization of preformed pyrimidine bases and nucleosides, p. 95–148. *In* A. Munch-Petersen (ed.), Metabolism of nucleotides, nucleosides, and nucleobases in microorganisms. Academic Press, Inc. (London), Ltd., London.

190. **Neuhard, J., and J. Ingraham.** 1968. Mutants of *Salmonella typhimurium* requiring cytidine for growth. J. Bacteriol. **95:**2431–2433.

191. **Neuhard, J., K. F. Jensen, and E. Stauning.** 1982. *Salmonella typhimurium* mutants with altered expression of the *pyrA* gene due to changes in RNA polymerase. EMBO J. **9:**1141–1145.

192. **Neuhard, J., R. A. Kelln, and E. Stauning.** 1986. Cloning and structural characterization of the *Salmonella typhimurium pyrC* gene encoding dihydroorotase. Eur. J. Biochem. **157:**335–342.

193. **Neuhard, J., E. Stauning, and R. A. Kelln.** 1985. Cloning and characterization of the *pyrE* gene and of *pyrE*::Mud1(Ap^R *lac*) fusions from *Salmonella typhimurium.* Eur. J. Biochem. **146:**597–603.

194. **Newell, P. C., and R. G. Tucker.** 1968. Biosynthesis of the pyrimidine moiety of thiamine. A new route of the pyrimidine biosynthesis involving purine intermediates. Biochem. J. **106:**279–287.

195. **Newton, N. A., G. B. Cox, and F. Gibson.** 1971. The function of menaquinone (vitamin K₂) in *Escherichia coli* K-12. Biochim. Biophys. Acta **244:**155–166.

196. **Nijkamp, H. J. J., and P. G. DeHaan.** 1967. Genetic and biochemical studies of the guanosine 5'-monophosphate pathway in *Escherichia coli.* Biochim. Biophys. Acta **145:**31–40.

197. **Nijkamp, H. J. J., and A. A. G. Oskamp.** 1968. Regulation of the biosynthesis of guanosine 5'-monophosphate: evidence for one operon. J. Mol. Biol. **35:**103–109.

198. **Nowlan, S. F., and E. R. Kantrowitz.** 1983. Identification of a *trans*-acting regulatory factor involved in the control of the pyrimidine pathway in *E. coli.* Mol. Gen. Genet. **192:**264–271.

199. **Nygaard, P.** 1973. Nucleoside-catabolizing enzymes in *Salmonella typhimurium.* Induction by ribonucleosides. Eur. J. Biochem. **36:**267–272.

200. **Nygaard, P.** 1978. Adenosine deaminase from *Escherichia coli.* Methods Enzymol. **51:**508–512.

201. **Nygaard, P.** 1983. Utilization of preformed purine bases and nucleosides, p. 27–93. *In* A. Munch-Petersen (ed.), Metabolism of nucleotides, nucleosides, and nucleobases in microorganisms. Academic Press, Inc. (London), Ltd., London.

202. **Nygaard, P.** 1984. Multiple functions of purine nucleoside phosphorylase and adenosine deaminase in *Escherichia coli,* p. 96–125. *In* A. Barth, R. Haschen, F. H. Laibach, H. Possin, and R. L. Schowen (ed.), Molecular and cellular regulation of enzyme activity. Proceedings of Symposium in Halle, German Democratic Republic.

203. **Nyunoya, H. and C. J. Lusty.** 1983. The *carB* gene of *Escherichia coli:* a duplicated gene coding for the large subunit of carbamoyl-phosphate synthetase. Proc. Natl. Acad. Sci. USA **80:**4629–4633.

204. **Nyunoya, H., and C. J. Lusty.** 1984. Sequence of the small subunit of yeast carbamyl phosphate synthetase and identification of its catalytic domain. J. Biol. Chem. **259:**9790–9798.

205. **O'Donovan, G. A.** 1978. Thymidine metabolism in bacteria, p. 219–253. *In* I. Molineux and M. Kohiyama (ed.), DNA synthesis. Present and future. Plenum Publishing Corp., New York.

206. **O'Donovan, G. A., and J. C. Gerhart.** 1972. Isolation and partial characterization of regulatory mutants of the pyrimidine pathway in *Salmonella typhimurium.* J. Bacteriol. **109:**1085–1096.

207. **O'Donovan, G. A., H. Holoubek, and J. C. Gerhart.** 1972. Regulatory properties of intergeneric hybrids of aspartate transcarbamylase. Nature (London) New Biol. **238:**264–266.

208. **Oeschger, M. P., and M. J. Bessman.** 1966. Purification and properties of guanylate kinase from *Escherichia coli.* J. Biol. Chem. **241:**5452–5460.

209. **Olszowy, J., and R. L. Switzer.** 1972. Specific repression of phosphoribosylpyrophosphate synthetase by uridine compounds in *Salmonella typhimurium.* J. Bacteriol. **110:**450–451.

210. **O'Reilly, C., P. D. Turner, P. F. Smith-Keary, and D. J. McConnell.** 1984. Molecular cloning of genes involved in purine biosynthetic and salvage pathways of *Salmonella typhimurium.* Mol. Gen. Genet. **196:**152–157.

211. **Pandey, N. K., and R. L. Switzer.** 1982. Mutant strains of *Salmonella typhimurium* with defective phosphoribosylpyrophosphate synthetase activity. J. Gen. Microbiol. **128:**1863–1871.

212. **Parker, J.** 1984. Identification of the *purC* gene product of *Escherichia coli.* J. Bacteriol. **157:**712–717.

213. **Patel, N., H. S. Moyed, and J. F. Kane.** 1975. Xanthosine-5'-phosphate amidotransferase from *Escherichia coli.* J. Biol. Chem. **250:**2609–2613.

214. **Patel, N., H. S. Moyed, and J. F. Kane.** 1977. Properties of xanthosine 5'-monophosphate-amidotransferase from *Escherichia coli.* Arch. Biochem. Biophys. **178:**652–661.

215. **Pauza, C. D., M. J. Karels, M. Navre, and H. K. Schachman.** 1982. Genes encoding *Escherichia coli* aspartate transcarbamoylase: the *pyrB-pyrI* operon. Proc. Natl. Acad. Sci. USA **79:**4020–4024.

216. **Pierard, A., N. Glansdorff, D. Gigot, M. Crabeel, P. Halleux, and L. Thiry.** 1976. Repression of *Escherichia coli* carbamoylphosphate synthase: relationships with enzyme synthesis in the arginine and pyrimidine pathways. J. Bacteriol. **127:**291–301.

217. **Pierard, A., N. Glansdorff, M. Mergeay, and J. M. Wiame.** 1965. Control of the biosynthesis of carbamoyl phosphate in *Escherichia coli.* J. Mol. Biol. **14:**23–36.

218. **Pierard, A., N. Glansdorff, and J. Yashphe.** 1972. Mutations affecting uridine monophosphate pyrophosphorylase or the *argR* gene in *Escherichia coli.* Effects on carbamoyl phosphate and pyrimidine biosynthesis and on uracil uptake. Mol. Gen. Genet. **118:**235–245.

219. **Piette, J., H. Nyunoya, C. J. Lusty, R. Cunin, G. Weyens, M. Crabeel, D. Charlier, N. Glansdorff, and A. Pierard.** 1984. DNA sequence of the *carA* gene and the control region of *carAB:* tandem promoters, respectively controlled by arginine and the pyrimidines, regulate the synthesis of carbamoyl-phosphate synthetase in *Escherichia coli* K-12. Proc. Natl. Acad. Sci. USA **81:**4134–4138.

220. **Poulsen, P., F. Bonekamp, and K. F. Jensen.** 1984. Structure of the *Escherichia coli pyrE* operon and control of *pyrE* expression by a UTP modulated intercistronic attenuation. EMBO J. **3:**1783–1790.

221. **Poulsen, P., K. F. Jensen, P. Valentin-Hansen, P. Carlsson, and L. G. Lundberg.** 1983. Nucleotide sequence of the *Escherichia coli pyrE* gene and of the DNA in front of the protein-coding region. Eur. J. Biochem. **135:**223–229.

222. **Rasmussen, U. B., B. Mygind, and P. Nygaard.** 1986. Purification and some properties of uracil phosphoribosyltransferase from *Escherichia coli* K-12. Biochim. Biophys. Acta **881:**268–275.

223. **Robey, E. A., and H. K. Schachman.** 1984. Site-specific mutagenesis of aspartate transcarbamoylase. Replacement of tyrosine 165 in the catalytic chain by serine reduces enzymatic activity. J. Biol. Chem. **259:**11180–11183.

224. **Rodriguez, S. B., and J. L. Ingraham.** 1983. Location on the *Salmonella typhimurium* chromosome of the gene encoding nucleoside diphosphokinase (*ndk*). J. Bacteriol. **153:**1101–1103.

225. **Roisin, M. P., and A. Kepes.** 1978. Nucleosidediphosphate kinase of *Escherichia coli*, a periplasmic enzyme. Biochim. Biophys. Acta **526:**418–428.

226. **Roland, K. L., F. E. Powell, and C. L. Turnbough.** 1985. Role of translation and attenuation in the control of *pyrBI* operon expression in *Escherichia coli* K-12. J. Bacteriol. **163:**991–999.

227. **Roof, W. D., K. F. Foltermann, and J. R. Wild.** 1982. The organization and regulation of the *pyrBI* operon in *E. coli* includes a rho-independent attenuator sequence. Mol. Gen. Genet. **187:**391–400.

228. **Rudolph, F. B., and H. J. Fromm.** 1969. Initial rate studies of adenylosuccinate synthetase with product and competitive inhibitors. J. Biol. Chem. **244:**3832–3839.

229. **Russel, M., and P. Model.** 1985. Thioredoxin is required for filamentous phage assembly. Proc. Natl. Acad. Sci. USA **82:**29–33.

230. **Sadler, W. C., and R. L. Switzer.** 1977. Regulation of *Salmonella* phosphoribosylpyrophosphate synthetase activity *in vivo*. Deductions from pool measurements. J. Biol. Chem. **252:**8504–8511.

231. **Saeki, T., M. Hori, and H. Umezawa.** 1974. Pyruvate kinase of *Escherichia coli*. Its role in supplying nucleoside triphosphates in cells under anaerobic conditions. J. Biochem. **76:**631–637.

232. **Schachman, H. K., C. D. Pauza, M. Navre, M. J. Karels, L. Wu, and Y. R. Yang.** 1984. Location of amino acid alterations in mutants of aspartate transcarbamoylase: structural aspects of interallelic complementation. Proc. Natl. Acad. Sci. USA **81:**115–119.

233. **Schafer, M. P., W. H. Hannon, and A. P. Levin.** 1974. In vivo and in vitro complementation between *guaB* and in vivo complementation between *guaA* auxotrophs of *Salmonella typhimurium*. J. Bacteriol. **117:**1270–1279.

233a.**Schrimsher, J. L., F. J. Schendel, J. Stubbe, and J. M. Smith.** 1986. Purification and characterization of aminoimidazole ribonucleotide synthetase from *Escherichia coli*. Biochemistry **25:**4366–4371.

234. **Schubert, K. R., and R. L. Switzer.** 1975. Studies of the quaternary structure and the chemical properties of phosphoribosylpyrophosphate synthetase from *Salmonella typhimurium*. J. Biol. Chem. **250:**7492–7500.

235. **Schwartz, M., and J. Neuhard.** 1975. Control of expression of the *pyr* genes in *Salmonella typhimurium*: effects of variations in uridine and cytidine nucleotide pools. J. Bacteriol. **121:**814–822.

236. **Shimada, K., Y. Fukumaki, and Y. Takagi.** 1976. Expression of the guanine operon of *Escherichia coli* as analyzed by bacteriophage lambda-induced mutations. Mol. Gen. Genet. **147:**203–208.

237. **Shlomai, J., and A. Kornberg.** 1978. Deoxyuridine triphosphatase of *Escherichia coli*. Purification, properties, and use as reagent to reduce uracil incorporation into DNA. J. Biol. Chem. **253:**3305–3312.

237a.**Smith, J. M., and H. A. Daum, III.** 1986. Nucleotide sequence of the *purM* gene encoding 5′-phosphoribosyl-5-aminoimidazole synthetase of *Escherichia coli* K-12. J. Biol. Chem. **261:**10632–10637.

238. **Spector, T., and L. M. Beacham.** 1975. Guanosine monophosphate synthetase from *Escherichia coli* B-96. J. Biol. Chem. **250:**3101–3107.

239. **Stayton, M. M., and H. J. Fromm.** 1977. Guanosine 5′-diphosphate-3′-diphosphate inhibition of adenylosuccinate synthetase. J. Biol. Chem. **254:**2579–2581.

240. **Stayton, M. M., F. B. Rudolph, and H. J. Fromm.** 1983. Regulation, genetics, and properties of adenylosuccinate synthetase: a review. Curr. Top. Cell. Regul. **22:**104–141.

241. **Switzer, R. L.** 1974. Phosphoribosylpyrophosphate synthetase and related pyrophosphokinases, p. 607–628. *In* P. D. Boyer (ed.), The enzymes, vol. 9. Academic Press, Inc., New York.

242. **Switzer, R. L., and D. C. Sogin.** 1973. Regulation and mechanism of phosphoribosylpyrophosphate synthetase. J. Biol. Chem. **248:**1063–1073.

243. **Syvanen, J. M., and J. R. Roth.** 1973. Structural genes for catalytic and regulatory subunits of aspartate transcarbamylase. J. Mol. Biol. **76:**363–378.

244. **Taylor, W. H., and M. L. Taylor.** 1964. Enzymes of the pyrimidine pathway in *Escherichia coli*. II. Intracellular localization and properties of dihydroorotate dehydrogenase. J. Bacteriol. **88:**105–110.

245. **Thelander, L., and P. Reichard.** 1979. Reduction of ribonucleotides. Annu. Rev. Biochem. **48:**133–158.

246. **Thomas, M. S., and W. T. Drabble.** 1984. Molecular cloning and characterization of the *gua* regulatory region of *Escherichia coli* K12. Mol. Gen. Genet. **195:**238–245.

247. **Thomas, M. S., and W. T. Drabble.** 1985. Nucleotide sequence and organization of the *gua* promoter region of *Escherichia coli*. Gene **36:**45–53.

248. **Thomulka, K. W., and J. S. Gots.** 1982. Isolation and characterization of purine regulatory mutants of *Salmonella typhimurium* with an episomal *purE-lac* fusion. J. Bacteriol. **151:**153–161.

249. **Tiedeman, A. A., and J. M. Smith.** 1984. Isolation and characterization of regulatory mutations affecting the expression of the *guaBA* operon of *Escherichia coli* K12. Mol. Gen. Genet. **195:**77–82.

250. **Tiedeman, A. A., and J. M. Smith.** 1985. Nucleotide sequence of the *guaB* locus encoding IMP dehydrogenase of *Escherichia coli* K12. Nucleic Acids Res. **13:**1303–1316.

251. **Tiedeman, A. A., J. M. Smith, and H. Zalkin.** 1985. Nucleotide sequence of the *guaA* gene encoding GMP synthetase of *Escherichia coli* K12. J. Biol. Chem. **260:**8676–8679.

252. **Trotta, P. P., L. M. Pinkus, R. H. Haschemeyer, and A. Meister.** 1974. Reversible dissociation of the monomer of glutamine-dependent carbamyl phosphate synthetase into catalytically active heavy and light subunits. J. Biol. Chem. **249:**492–499.

253. **Tso, J. Y., M. A. Hermodson, and H. Zalkin.** 1982. Glutamine phosphoribosylpyrophosphate amidotransferase from cloned *Escherichia coli purF*. NH$_2$-terminal amino acid sequence, identification of the glutamine site, and trace metal analysis. J. Biol. Chem. **257:**3532–3536.

254. **Tso, J. Y., H. Zalkin, M. van Cleeput, C. Yanofsky, and J. M. Smith.** 1982. Nucleotide sequence of *Escherichia coli purF* and deduced amino acid sequence of glutamine phosphoribosylpyrophosphate amidotransferase. J. Biol. Chem. **257:**3525–3531.

254a.**Tuggle, C. K., and J. A. Fuchs.** 1986. Regulation of the operon encoding ribonucleotide reductase in *Escherichia coli*: evidence for both positive and negative control. EMBO J. **5:**1077–1085.

255. **Turnbough, C. L.** 1983. Regulation of *Escherichia coli* aspartate transcarbamylase synthesis by guanosine tetraphosphate and pyrimidine triphosphates. J. Bacteriol. **153:**998–1007.

256. **Turnbough, C. L., K. L. Hicks, and J. P. Donahue.** 1983. Attenuation control of *pyrBI* operon expression in *Escherichia coli* K-12. Proc. Natl. Acad. Sci. USA **80:**368–372.

257. **Tye, B.-K., P.-O. Nyman, I. R. Lehman, S. Hochhauser, and B. Weiss.** 1977. Transient accumulation of Okazaki fragments as a result of uracil incorporation into nascent DNA. Proc. Natl. Acad. Sci. USA **74:**154–157.

258. **Uerkvitz, W., and C. F. Beck.** 1981. Periplasmic phosphatases in *Salmonella typhimurium* LT2. A biochemical, physiological, and partial genetic analysis of three nucleoside monophosphate dephosphorylating enzymes. J. Biol. Chem. **256:**382–389.

259. **Valentin-Hansen, P.** 1978. Uridine-cytidine kinase from *Escherichia coli*. Methods Enzymol. **51:**308–314.

260. **Valentin-Hansen, P., K. Hammer, J. E. L. Larsen, and I. Svendsen.** 1984. The internal regulated promoter of the *deo* operon of *Escherichia coli* K-12. Nucleic Acids Res. **12:**5211–5224.

261. **van Alphen, W., N. van Selm, and B. Lugtenberg.** 1978. Pores in the outer membrane of *Escherichia coli* K12. Involvement of proteins b and e in the functioning of pores for nucleotides. Mol. Gen. Genet. **159:**75–83.

262. **Wallace, B. J., and S. R. Kushner.** 1984. Genetic and physical analysis of the thioredoxin (*trxA*) gene of *Escherichia coli* K-12. Gene **32:**399–408.

263. **Washabaugh, M. W., and K. D. Collins.** 1984. Dihydroorotase from *Escherichia coli*. Purification and characterization. J. Biol. Chem. **259:**3293–3298.

264. **Weber, K.** 1968. New structural model of *E. coli* aspartate transcarbamylase and the amino-acid sequence of the regulatory polypeptide chain. Nature (London) **218:**1114–1119.

264a.**Weng, M., C. A. Makaroff, and H. Zalkin.** 1986. Nucleotide sequence of *Escherichia coli pyrG* encoding CTP synthetase. J. Biol. Chem. **261:**5568–5574.

265. **West, T. P., and G. A. O'Donovan.** 1982. Repression of cytosine deaminase by pyrimidines in *Salmonella typhimurium*. J. Bacteriol. **149:**1171–1174.

266. **West, T. P., M. S. Shanley, and G. A. O'Donovan.** 1982. Purification and some properties of cytosine deaminase from *Salmo-*

nella typhimurium. Biochim. Biophys. Acta **719**:251–258.

267. **Westby, C. A., and J. S. Gots.** 1969. Genetic blocks and unique features in the biosynthesis of 5'-phosphoribosyl-*N*-formylglycineamide in *Salmonella typhimurium.* J. Biol. Chem. **244**:2095–2102.

268. **Weyens, G., K. Rose, P. Falmagne, N. Glansdorff, and A. Pierard.** 1985. Synthesis of *Escherichia coli* carbamoylphosphate synthetase initiates at a UUG codon. Eur. J. Biochem. **150**:111–115.

269. **White, M. N., J. Olszowy, and R. L. Switzer.** 1971. Regulation and mechanism of phosphoribosylpyrophosphate synthetase: repression by end products. J. Bacteriol. **108**:122–131.

270. **Wild, J. R., K. F. Foltermann, and G. A. O'Donovan.** 1980. Regulatory divergence of aspartate transcarbamoylase within the Enterobacteriaceae. Arch. Biochem. Biophys. **201**:506–517.

271. **Wild, J. R., K. F. Foltermann, W. D. Roof, and G. A. O'Donovan.** 1981. A mutation in the catalytic cistron of aspartate carbamoyltransferase affecting catalysis, regulatory response and holoenzyme assembly. Nature (London) **292**:373–375.

272. **Wolfe, S. A., and J. M. Smith.** 1985. Separate regulation of *purA* and *purB* loci of *Escherichia coli* K-12. J. Bacteriol. **162**:822–825.

273. **Womack, J. E., and G. A. O'Donovan.** 1978. Orotic acid excretion in some wild-type strains of *Escherichia coli* K-12. J. Bacteriol. **136**:825–827.

274. **Yang, Y. R., and H. K. Schachman.** 1980. Communication between catalytic subunits in hybrid aspartate transcarbamoylase molecules: effect of ligand binding to active chains on the conformation of unliganded, inactive chains. Proc. Natl. Acad. Sci. USA **77**:5187–5191.

275. **Yates, R. A., and A. B. Pardee.** 1957. Control by uracil of formation of enzymes required for orotate synthesis. J. Biol. Chem. **227**:667–692.

276. **Zak, V. L., and R. A. Kelln.** 1978. 5-fluoroorotate-resistant mutants of *Salmonella typhimurium.* Can. J. Microbiol. **24**:1339–1345.

277. **Zalkin, H., P. Argos, S. V. L. Narayana, A. A. Tiedeman, and J. M. Smith.** 1985. Identification of a *trpG*-related glutamine amide transfer domain in *Escherichia coli* GMP synthetase. J. Biol. Chem. **260**:3350–3354.

278. **Zalkin, H., and C. D. Truitt.** 1977. Characterization of the glutamine site of *Escherichia coli* guanosine 5'-monophosphate synthetase. J. Biol. Chem. **252**:5431–5436.

279. **Zimmerman, E. F., and B. Magasanik.** 1964. Utilization and interconversion of purine bases and ribonucleosides by *Salmonella typhimurium.* J. Biol. Chem. **239**:293–300.

30. Biosynthesis of Membrane Lipids

JOHN E. CRONAN, JR.,[1] AND CHARLES O. ROCK[2]

Department of Microbiology, University of Illinois, Urbana, Illinois 61801,[1] and Department of Biochemistry,
St. Jude Children's Research Hospital, Memphis, Tennessee 38101[2]

INTRODUCTION

The study of the membrane lipids of *Escherichia coli* and *Salmonella typhimurium* has contributed greatly to our understanding of the synthesis and functions of membrane lipids in general. In addition to the myriad other advantages of studying these organisms, their lipid compositions are among the simplest found in biology. These bacteria contain three major phospholipids: phosphatidylethanolamine (PE), phosphatidylglycerol (PG), and cardiolipin (CL) (also called di-phosphatidylglycerol) (3, 30, 39). Traces of other phospholipids (e.g., monoacylphosphatidylglycerol, phosphatidylserine [PS]) are also found. Phospholipids contain about 90% of the total fatty acyl groups of *E. coli*, the remainder being found in the lipid A component of the lipopolysaccharide, in lipoproteins, and in traces of neutral lipids (diglyceride and free fatty acids) (39).

PE, the major phospholipid of *E. coli* and *S. typhimurium* (and of gram-negative bacteria in general) contains about 75% of the total phospholipid. The relative amounts of the other lipids (PG and CL) depend on the growth phase of the cultures; PG is dominant in log-phase cells, and CL is dominant in stationary-phase cells (3, 30). The phospholipids, neutral lipids, and lipoproteins of *E. coli* have similar fatty acid compositions, consisting of the saturated fatty acids, palmitic (hexadecanoic) acid and myristic (tetradecanoic) acid, and the monounsaturated fatty acids, palmitoleic (*cis*-9-hexadecenoic) acid and *cis*-vaccenic (*cis*-11-octadecenoic) acid (30, 39). Traces of other fatty acids (lauric [dodecanoic] acid, stearic [octadecanoic] acid, and *cis*-7-tetradecenoic acid) are also present in the phospholipids (7, 39). The fatty acid compositions of *E. coli* cultures depend on both growth temperature and growth phase (39). At 37°C, the phospholipids of a typical *E. coli* strain contain about 45% palmitic acid, 2% myristic acid, 35% palmitoleic acid, and 18% *cis*-vaccenic acid. However, in cultures grown at 25°C, the percentages of palmitic acid and *cis*-vaccenic acid are reversed. This temperature-dependent compositional alteration (thermal regulation) is thought to be a mechanism to optimize membrane lipid fluidity. The mechanism of this compositional change is understood in some detail (46) and will be discussed below.

The variation of fatty acid composition with growth phase is due to the conversion of palmitoleic acid and (to a lesser extent) *cis*-vaccenic acid to their cyclopropane derivatives by addition of a methylene group (from *S*-adenosylmethionine) across the double bond of the phospholipid-bound fatty acid (112). This membrane bilayer modification reaction has received some recent attention and will also be discussed below. The acyl moieties of lipid A also change with growth temperature. At 37°C, only 3-hydroxymyristic acid (80% of the total fatty acid) and saturated fatty acids (C_{12}, C_{14}, and C_{16}) are found in lipid A, whereas in cells grown at 15°C, the saturated acids are partially replaced by unsaturated acids (206, 219). It should be noted that most workers in the field have used *E. coli*, and more compositional data are available for this bacterium. However, in cases in which *S. typhimurium* has been examined, the two organisms seem identical (3); thus, information obtained with one organism can be assumed to be directly applicable to the other. The analytical data on the lipid composition of these organisms were essentially complete over 10 years ago and have been reviewed by Cronan and Vagelos (39). This review will not cover the topologic, spatial, and physical aspects of cell membrane phospholipid assembly and function as these aspects are discussed in chapter 5.

INTRACELLULAR LOCALIZATION OF LIPID BIOSYNTHESIS AND OF LIPID-CONTAINING AND LIPID-DERIVED STRUCTURES

All of the phospholipids, neutral lipids, and lipoproteins of *E. coli* and *S. typhimurium* are membrane components; no cytoplasmic, extracellular, or periplasmic lipid-containing structures are known (for a review, see reference 32). Both the phospholipids and the neutral lipids are distributed between the inner (cytoplasmic) membrane and the outer membrane (32). The outer membrane contains roughly one-third of the total phospholipid and has been reported to

consist largely or exclusively of PE (32). In contrast, the phospholipids of the inner membrane have been reported to contain roughly equal amounts of PE and of the two acidic lipids PG and CL. It should be noted, however, that although the outer membrane has generally been reported to be enriched in PE relative to the cytoplasmic membrane, the magnitude of the enrichment varies greatly among reports (32). It seems reasonable to accept the more extreme values because such values are generally found in the more carefully documented reports and because the lack of enrichment can be attributed to transfer of lipids between the two membranes during cellular lysis. Phospholipid fatty acid compositions of the two membranes are similar, although the inner membrane seems somewhat enriched in unsaturated fatty acids. This enrichment can probably be attributed to the increased concentration of PG in the cytoplasmic membrane, which contains a greater fraction of unsaturated fatty acid than do the other phospholipids (32).

Lipid A and the membrane-derived oligosaccharides (MDO) are exclusively localized to the outer membrane (118) and periplasmic space (chapter 43), respectively.

LOCATION OF LIPID BIOSYNTHESIS

The enzymes of lipid biosynthesis are distributed between the cytosol and the inner membrane. The fatty acid synthetic enzymes and the *sn*-glycerol-3-phosphate (G3P) synthase are cytosolic (104, 164), whereas the enzymes of phospholipid synthesis are firmly bound to the inner membrane (12, 166, 218). Two exceptions to these generalizations have been reported. First, the acyl carrier protein (ACP) which carries the growing fatty acid chain and donates the completed acyl chain to G3P was reported to be localized to the inner face of the cytoplasmic membrane on the basis of autoradiography of intact cells (208). However, a more refined colloidal gold-antibody technique has demonstrated that ACP is distributed throughout the cytoplasm of *E. coli* (98). The second exception was the observation that the phospholipid biosynthetic enzyme, PS synthase, was found associated with the ribosomes rather than with the inner membranes of disrupted cells (166, 170). Subsequent work has shown this association to be an artifact of cell disruption. In the presence of the lipid substrate of the enzyme, CDP-diglyceride, the enzyme is released from the ribosomes and binds to the inner membrane lipid bilayer (117).

The purification of the lipid biosynthetic enzymes can be either quite straightforward or very tedious, depending upon the cellular location. The fatty acid biosynthetic enzymes are readily purified by standard chromatographic procedures performed in standard aqueous buffers (164), whereas the purification of the membrane-bound phospholipid biosynthetic enzymes involves detergent solubilization and usually requires at least one highly selective step, such as affinity chromatography or overproduction by molecular cloning, before significant amounts of the pure proteins can be obtained (155, 166, 167). The amino acid sequences of two of these enzymes, G3P acyltransferase (109, 114, 115) and diglyceride kinase (114), have been obtained by nucleotide sequencing of the cloned

TABLE 1. Mutants in *E. coli* lipid biosynthesis[a]

Gene	Enzyme(s) affected	Gene function	Phenotype	Cloned?
fabA	3-Hydroxydecanoyl-ACP dehydrase	Structural	Unsaturated fatty acid auxotroph	Yes
fabB	3-Ketoacyl-ACP synthase I	Structural	Unsaturated fatty acid auxotroph	Yes
fabD	Malonyl-CoA:ACP	Structural	Temperature-sensitive mutant; required both saturated and unsaturated fatty acids	No
fabE	Acetyl-CoA carboxylase	Probably structural	Same as *fabD*	No
fabF	3-Ketoacyl-ACP synthase II	Structural	Altered thermal regulation	Yes
acpS	[ACP]synthase	Probably structural	Requires high intracellular CoA levels	No
cfa	Cyclopropane fatty acid synthase	Structural	Grows normally but lacks cyclopropane fatty acids	Yes
plsB	G3P acyltransferase	Structural	G3P or glycerol auxotroph	Yes
cds	CDP-diacylglycerol	?	pH sensitive	No
cdh	CDP-diacylglycerol	?	None	Yes
pss	PS synthase	Structural	Temperature-sensitive mutant	Yes
psd	PS decarboxylase	Structural	Temperature-sensitive mutant	Yes
pgsA	PGP synthase	Probably structural	None	Yes
pgsB	Disaccharide-1-phosphate synthase	Probably structural	None	Yes
pgsA *pgsB*	Double mutant	Probably structural	Temperature-sensitive mutant	No
cls	CL synthase	Structural	None	Yes
dgk	Diacylglycerol kinase	Structural	Osmotically sensitive	Yes
pgpAB	PGP phosphatase	?	None	Yes
pldA	Detergent-resistant phospholipase A	Structural	None	Yes
pldB	Inner membrane lysophospholipase	Structural	None	Yes
pssR	Overproduction of PS synthase	Regulatory	None	No
dgkR	Overproduction of diacylglycerol kinase	Regulatory	None	No
adk	Adenylase kinase and G3P acyltransferase	Regulatory	Temperature-sensitive mutant; G3P acyltransferase also thermolabile	Yes
cdsS	Stabilizes mutant CDP-diacylglycerol synthase	Regulatory	Suppresses pH sensitivity of *cds* strain	No
fabAUp	Overproduction of 3-hydroxydecanoyl-ACP dehydrase	Regulatory	Overproduction of saturated fatty acids	No

[a] See text for details and references.

structural genes. The latter protein was found to be extremely hydrophobic, whereas the acyltransferase has a considerably lower content of hydrophobic amino acids.

OUTLINE OF LIPID BIOSYNTHESIS IN *E. COLI*

The phospholipid synthetic pathway can be considered to be the convergence of two pathways, a lengthy pathway of fatty acid synthesis (with product divergence between saturated and unsaturated species) and a single-step pathway for synthesis of G3P. The key phospholipid synthetic intermediate, phosphatidic acid, is formed by the acylation of G3P by the acyl-ACP products of fatty acid synthesis and is then coverted into the three major phospholipid classes. A large number of mutants in *E. coli* lipid biosynthesis have been isolated (Table 1).

THE FATTY ACID SYNTHETIC PATHWAY

The modern era of *E. coli* phospholipid enzymology began in the early 1960s, when Vagelos and colleagues discovered that the intermediates of fatty

acid biosynthesis were bound to a heat-stable cofactor termed ACP (for a review, see reference 164). The realization that the reactions in fatty acid biosynthesis could be separated and studied individually precipitated a flurry of activity, primarily by the laboratories of Vagelos, Bloch, and Wakil, and within a few years the structures of all of the intermediates in fatty acid biosynthesis had been elucidated (15, 16, 164, 212, 215). These experiments had a great deal of impact on the lipid metabolism field as a whole because the fatty acid synthases of higher organisms (except plants) are multifunctional protein complexes, and the protein domains catalyzing the individual reactions could not be isolated. This work has been thoroughly reviewed (16, 164, 188, 212, 215). For the present purposes, the reactions of fatty acid synthesis can be conveniently discussed in terms of initiation reactions, elongation reactions, and product diversification.

Initiation

The precursors for fatty acid biosynthesis are derived from the acetyl coenzyme A (CoA) pool. Three

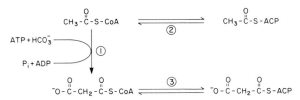

FIG. 1. Initiation phase of fatty acid biosynthesis. The initiation of new acyl chains is accomplished by the action of three enzymes: 1, acetyl-CoA carboxylase; 2, acetyl-CoA:ACP transacylase; and 3, malonyl-CoA:ACP transacylase.

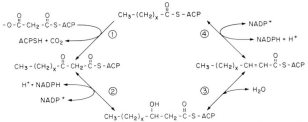

FIG. 2. Elongation cycle of fatty acid biosynthesis. The elongation of a growing acyl chain is accomplished by the action of four enzymes: 1, 3-ketoacyl-ACP synthase; 2, 3-ketoacyl-ACP reductase; 3, 3-hydroxyacyl-ACP dehydrase; and 4, enoyl reductase (*trans*-2-acyl-ACP reductase).

enzyme reactions are involved (Fig. 1). Malonyl-CoA is required for all of the elongation steps and is formed by the carboxylation of acetyl-CoA by acetyl-CoA carboxylase. Acetyl-CoA carboxylase is composed of three individual proteins: biotin carboxylase, biotin carboxyl carrier protein, and carboxyltransferase (for a review, see references 2 and 220). All three proteins have been purified to homogeneity. The carrier protein subunit is the sole biotin-containing protein found in *E. coli* (53, 54) and is one of the least abundant proteins in this bacterium (ca. 10 molecules per cell). Acetyl-ACP, the primer of fatty acid biosynthesis, is formed from acetyl-CoA by the action of acetyl transacylase. This is the only step in which an acetate unit is directly used in fatty acid biosynthesis (164). Similarly, malonyl-ACP is formed from malonyl-CoA by an analogous transacylase (a protein of M_r 35,000) encoded by the *fabD* gene. Conditional lethal (temperature-sensitive) mutants in both acetyl-CoA carboxylase (189) and malonyl transacylase (82) have been isolated. These strains are totally deficient in synthesis of fatty acids at the nonpermissive temperature (81). However, it should be noted that the identity of the defective subunit in the acetyl-CoA carboxylase mutant (called *fabE*) is unclear (189).

Elongation

The initial condensation of malonyl-ACP and acetyl-ACP is catalyzed by the 3-ketoacyl-ACP synthases (condensing enzymes) (Fig. 2). There are two such enzymes in *E. coli* (42), called synthases I and II, which are encoded by the *fabB* and *fabF* genes, respectively. Genetic evidence indicates that either of these two enzymes can catalyze all of the condensation reactions needed to form saturated fatty acids but that they differ in their roles in synthesis of unsaturated fatty acids. Both enzymes are homodimeric species consisting of subunits with M_rs of ca. 43,000 that share little or no protein sequence homology (61). In addition to the synthases, three other enzymes participate in each cycle of chain elongation (164); the general reaction scheme is shown in Fig. 2. First, the condensing enzymes add a C_2 unit from malonyl-ACP. The resulting ketoester is reduced by an NADPH-dependent 3-ketoacyl-ACP reductase, and a water molecule is then removed by the 3-hydroxyacyl-ACP dehydrase. The last step is catalyzed by enoyl-ACP reductase to form a saturated acyl-ACP, which in turn can serve as the substrate for another condensation reaction (Fig. 2). NADPH is probably the preferred cofactor for the

enoyl-ACP reductase, but there may be two enzymes involved, one specific for NADPH and the other specific for NADH (215). No mutants blocked in these last three steps have been isolated. The condensing enzymes catalyze the only irreversible steps in the elongation cycle.

Product Diversification

A specific dehydrase enzyme, 3-hydroxydecanoyl-ACP dehydrase, first described by Bloch and co-workers (for a review, see reference 15), catalyzes a key reaction at the point at which the biosynthesis of unsaturated fatty acids diverges from that of saturated fatty acids (Fig. 3). This dehydrase is an enzyme distinctly different from the 3-hydroxyacyl-ACP dehydrase that participates in the elongation cycle. Only the 3-hydroxydecanoyl-ACP dehydrase is capable of isomerizing *trans*-2-decenoyl-ACP to *cis*-3-decenoyl-ACP (14). Genetic studies have also shown that the two condensing enzymes, 3-ketoacyl-ACP synthases I and II, are responsible for different elongation reactions in this branch. Loss of 3-ketoacyl-ACP synthase I activity results in a complete block in unsaturated fatty acid synthesis, whereas 3-ketoacyl-ACP synthetase II is responsible for the elongation of palmitoleate to *cis*-vaccenate (see below).

The enzymology of the product diversification reactions has been reviewed by us (178) and others (15, 46, 164); thus, we will restrict discussion to four major fatty acid synthetic proteins: ACP, the two 3-ketoacyl-ACP synthases, and 3-hydroxydecanoyl-ACP dehydrase.

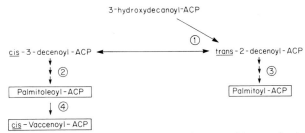

FIG. 3. Product diversification in fatty acid biosynthesis. There are three major fatty acids produced by the *E. coli* fatty acid synthase system. The ratio of these fatty acids is controlled by the activity of three enzymes (see text).

ACP

ACP is a required cofactor in four of the five phases of membrane phospholipid biogenesis; consequently, this protein has received considerable experimental attention. ACP is one of the most abundant proteins in *E. coli* (ca. 5×10^4 molecules per cell); however, this fact is not generally appreciated since ACP is not displayed by the two-dimensional gel electrophoresis system of O'Farrell (145) owing to its small size (M_r 8,847) and acidic pI (pH 4.1). The complete amino acid sequence of ACP has been determined (207), and the protein has a number of characteristic structural features (for a review, see reference 179). The preponderance of acidic residues accounts for its low isoelectric point, high content of α-helical secondary structure, and asymmetric shape (175, 179). The intermediates of fatty acid biosynthesis are attached to the terminal sulfhydryl of the 4'-phosphopantetheine prosthetic group (the sole thiol of ACP), which is in turn attached to the protein via a phosphodiester linkage to Ser-36 located in the fourth position of a predicted β-turn structure (178). The native α-helical secondary structure of ACP is important to its specific interaction with enzymes that utilize ACP or its thioesters (179, 205). High-quality crystals of ACP have recently been obtained from preparations obtained by the improved purification procedure of Rock and Cronan (177); thus, the detailed structure of this important and evolutionarily conserved protein should soon be known (J. Richardson, personal communication).

An enigmatic observation is that, although the protein portion of ACP is metabolically stable (162) (as are most proteins in *E. coli* [71]), the 4'-phosphopantetheine prosthetic group is metabolically active and undergoes considerable turnover in vivo. The prosthetic group turnover cycle is mediated by the action of two enzymes: [ACP]synthase and [ACP]phosphodiesterase (Fig. 4). [ACP]synthase catalyzes the transfer of the 4'-phosphopantetheine moiety of CoA to apo-ACP (51), and [ACP]phosphodiesterase generates apo-ACP by cleaving the prosthetic group from ACP (204). The lack of significant apo-ACP in vivo (100) suggests that [ACP]phosphodiesterase is the rate-determining step in the cycle. Each round of prosthetic group turnover results in the expenditure of the two mole-

cules of ATP required to convert phosphopantetheine to CoA (164). An accurate measurement of the rate of prosthetic group turnover in vivo has been difficult to obtain. Conventional pulse-chase experiments cannot be applied since the CoA pool is both large and metabolically stable and thus cannot be effectively chased. Experiments in which the CoA pool was depleted (by starvation of a pantothenate auxotroph) gave an early estimate of the rate of prosthetic group turnover by measuring the flow of tritium from ACP prelabeled in the prosthetic group into the CoA pool during the recovery from pantothenate starvation (163). However, the metabolic disturbance and complexity of the labeling patterns make it difficult to apply this estimate to normal growth conditions.

Recently, a second approach exploited the effect of heavy isotopes on the solution structure of ACP (the constitutional isotope effect) to differentiate between newly synthesized and preexisting ACP in a pulse-chase experiment (100). The incorporation of deuterium into nonexchangeable positions (e.g., $-CH_2-$) of the ACP amino acid backbone resulted in the destabilization of the protein and increased sensitivity to pH-induced hydrodynamic expansion determined by conformationally sensitive polyacrylamide gel electrophoresis. ACP derivatives more stable than normal ACP migrate more quickly in this partially denaturing system owing to their smaller hydrodynamic radius, whereas ACP derivatives less stable than ACP, such as deuterio-ACP, migrate more slowly owing to their increased molecular volume (175, 177). After prelabeling the ACP prosthetic group with tritiated prosthetic group precursor (β-alanine) during growth on deuterium oxide medium, the cells were chased with ^{14}C-labeled β-alanine, and the rate of prosthetic group turnover was determined by the replacement of ^{3}H with ^{14}C in preexisting deuterio-ACP. These experiments showed that, at low intracellular CoA levels, the rate of ACP prosthetic group turnover was four times faster than the rate of new ACP synthesis. However, at the higher CoA concentrations characteristic of logarithmic growth, turnover was an order of magnitude lower, amounting to 25% of the ACP pool per generation (100).

Although ACP plays an indispensible role in fatty acid biosynthesis, recent work indicates that the size of the ACP pool can be severely depleted without significantly affecting lipid synthesis (99, 161). Normally, a significant pool of inactive apo-ACP does not exist in vivo, but in *E. coli* strains with the *acpS* mutation (abnormal [ACP]synthase function), apo-ACP accumulates and becomes the major form of ACP in vivo. Although the *acpS* mutant contains much less ACP than do wild-type strains, it still has a normal lipid:protein ratio, indicating that ACP concentration per se is not a factor in determining the overall rate of phospholipid biosynthesis (99).

3-Hydroxydecanoyl-ACP dehydrase

The introduction of the double bond into the growing fatty acid chain is catalyzed by 3-hydroxydecanoyl-ACP dehydrase (for a review, see reference 15) (Fig. 3). This enzyme, a homodimer of an 18,000-molecular-weight subunit (protein 28×40 in the classification of Neidhardt et al. [136]), catalyzes the specific

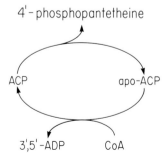

Prosthetic Group Turnover

FIG. 4. Turnover of the prosthetic group of ACP. The prosthetic group is provided by transfer from CoA by [ACP]synthase and is cleaved from the protein by [ACP]-phosphodiesterase. See text for details.

dehydration of 3-hydroxydecanoyl-ACP to a mixture of trans-2-decenoyl-ACP and cis-3-decenoyl-ACP. The double bond of the trans-2 intermediate is reduced to decanoyl-ACP by enoyl-ACP reductase, whereas cis-3-decenoyl-ACP is elongated to form the unsaturated fatty acids of E. coli, palmitoleic acid and cis-vaccenic acid. The dehydration reaction proceeds first by formation of the trans-2-decenoyl-ACP as an enzyme-bound intermediate (for a review, see reference 15). This intermediate sometimes dissociates from the enzyme, and is subsequently converted to saturated fatty acids via enoyl-ACP reductase, whereas the enzyme-bound trans-2-decenoyl-ACP is isomerized to cis-3-decenoyl-ACP by the dehydrase. The phenotype of mutants (fabA) lacking 3-hydroxydecanoyl-ACP dehydrase demonstrates that the isomerase reaction is the activity required for unsaturated fatty acid synthesis (37).

The first mutants blocked in the fatty acid biosynthesis pathway were shown to lack the 3-hydroxydecanoyl-ACP dehydrase (191). These mutants (fabA) require unsaturated fatty acids for growth but normally synthesize saturated fatty acids. In vitro, the mutant enzymes are unable to catalyze the formation of either trans-2-decenoyl-ACP or cis-3-decenoyl-ACP (37, 191). This finding, together with the continued synthesis of saturated fatty acids observed in vivo, indicates that trans-2-decenoyl-ACP can be synthesized by a dehydrase other than 3-hydroxydecanoyl-ACP dehydrase (13). However, this second dehydrase is unable to catalyze the isomerization of trans-2-decenoyl-ACP to cis-3-decenoyl-ACP. This second dehydrase is presumably the enzyme (or enzymes) that catalyzes the dehydration of the 3-hydroxyacyl-ACP molecules of both shorter and longer chain lengths (see above).

Although 3-hydroxydecanoyl-ACP dehydrase catalyzes a required step in unsaturated fatty acid synthesis (15), recent work indicates that the level of enzyme activity normally present does not limit the rate of unsaturated fatty acid synthesis (27). Increasing the intracellular levels of the enzyme either by molecular cloning or by using a regulatory mutant (a putative promotor lesion) does not result in an increased level of unsaturated fatty acids but rather gives a small increase in saturated fatty acid content (27). This result indicates that the supply of cis-3-decenoyl-ACP is not rate limiting for unsaturated fatty acid production and that the dehydrase can also contribute to the trans-2-decenoyl-ACP pool, resulting in an increase in saturated fatty acid synthesis. A possible regulation of fabA gene expression by the regulatory gene of the β-oxidation pathway, fadR, has been reported (143).

3-Ketoacyl-ACP synthase I

Another class of fatty acid mutants (fabB) has the same phenotype as the fabA strains (35). Both fabB and fabA mutants normally synthesize saturated fatty acids but require the addition of unsaturated fatty acids to the medium for growth (26). However, fabB strains contain normal levels of 3-hydroxydecanoyl-ACP dehydrase, and genetic analysis showed that the fabA and fabB mutations define two different genes (35). When first isolated, the fabB mutants seemed to

be an enigma since the same set of fatty acid biosynthetic enzymes (except 3-hydroxydecanoyl-ACP dehydrase) was thought to function in both unsaturated and saturated fatty acid synthesis. Hence, a mutation resulting in the loss of one of these enzymes was expected to block the synthesis of saturated, as well as unsaturated, fatty acids.

The enigma was resolved by enzymological studies showing that two distinct enzymes catalyze the 3-ketoacyl-ACP synthase reaction in E. coli (42). Therefore, the loss of one enzyme could produce a fabB mutant since the second enzyme was available for the elongation steps required in saturated fatty acid synthesis. Indeed, it was soon shown that the fabB gene encodes synthase I (protein 69 × 91 in the scheme of Neidhardt et al. [136]) (60). Therefore, it is clear that synthase I is responsible for a key condensation reaction in unsaturated fatty acid synthesis that synthase II cannot catalyze. The identity of this key step remains uncertain, but the elongation of cis-3-decenoyl-ACP seems to be the most likely candidate. Both synthases I and II are capable of elongating saturated fatty acids. A fabB mutant of S. typhimurium has also been isolated (92).

3-Ketoacyl-ACP synthase II

The discovery of the role of 3-ketoacyl-ACP synthase II in unsaturated fatty acid synthesis was an outcome of the investigation of a phenomenon called thermal control of fatty acid synthesis (see below). In the present context, synthase II is the enzyme responsible for the synthesis of cis-vaccenic acid, the C_{18} unsaturated fatty acid (60). Mutants (fabF) deficient in synthase II activity contain only traces of cis-vaccenic acid. Moreover, in vitro, synthase II is much more active in elongating palmitoleic acid than is synthase I (61). Double mutants carrying an fabF allele and a temperature-sensitive fabB allele are unable to grow at 42°C, even if provided with an appropriate unsaturated fatty acid (61). This is due to a deficiency in the synthesis of saturated fatty acids.

The fatty acid synthetic pathway, therefore, produces three major products to be used in phospholipid synthesis: palmitic acid, palmitoleic acid, and cis-vaccenic acid (Fig. 3). A fourth acid, 3-hydroxymyristic acid, is produced for the synthesis of the lipid A moiety of lipopolysaccharide (118). This C_{14} acid is an intermediate in the synthesis of palmitic acid. It is clear that the acid is produced by the lipid synthetic pathway rather than by fatty acid degradation (which also has a 3-hydroxy acid intermediate) since (i) the hydroxyl group has the D, rather than the L, configuration (118); (ii) mutants defective in fatty acid degradation normally synthesize 3-hydroxymyristate; and (iii) mutations or inhibitors that block fatty acid synthesis also block 3-hydroxymyristate synthesis (81, 216). The mechanism whereby the C_{14} hydroxy acid is rerouted from the fatty acid synthetic pathway for synthesis of lipid A rather than for synthesis of palmitate is unknown but seems likely to involve the competitive balance between the lipid A synthetic enzyme(s) and the enoyl-ACP reductase for 3-hydroxymyristoyl-ACP. This hypothesis is based on the unusual behavior of strains with temperature-sensitive lesions in genes encoding enzymes required for both saturated and unsaturated fatty acid synthesis (*fabD or fabE* strains). These strains grow normally at 30°C, but they fail to grow at 37°C unless supplemented with both a saturated and an unsaturated fatty acid (81, 82; J. E. Cronan, Jr., unpublished data). However, at 42°C these strains fail to grow even in the presence of these supplements. Strains grown at 37°C synthesize fatty acids at about 10% of the normal rate, but at 42°C essentially no synthesis occurs. The residual synthesis at 37°C is almost entirely of 3-hydroxymyristic acid (81, 216); thus, it seems that, under conditions of restricted fatty acid synthetic capacity, the 3-hydroxymyristoyl-ACP intermediate is preferentially utilized by the enzyme(s) that acylates the UDP-glucosamine with 3-hydroxymyristate residues (3a; C. R. H. Raetz, Annu. Rev. Genet., in press).

The fatty acids produced by the synthetic pathway as their ACP thioesters are in an activated form suitable for the direct acylation of G3P and the lipid A precursors. Rock and Jackowski (181) recently demonstrated the accumulation of acyl-ACP molecules in cells blocked in the acylation of G3P (by starvation of a *plsB* strain for G3P), showing a precursor-product relationship. Under these conditions, the acyl-ACP molecules are not stable indefinitely but are eventually degraded to the free acid and ACP (40, 181). This cleavage is presumably mediated by one or both of the two thioesterases present in *E. coli*; these enzymes cleave acyl-CoA esters much more readily than they cleave acyl-ACP molecules (195). Rock and Jackowski (181) detected no accumulation of acyl-CoA esters in the acylation of G3P with endogenously synthesized fatty acids. It should be noted, however, that acyl-CoA molecules function as the acyl donors when fatty acids are taken up from the medium (33, 182).

E. coli and *S. typhimurium* reside in an environment (the intestine) rich in fatty acids and have an inducible β-oxidation system to utilize this carbon source (153). However, long-chain fatty acids are not only degraded but can also be incorporated directly into phospholipid. A major role for CoA thioesters was suggested by the inability of *E. coli* to elongate exogenously supplied fatty acids (190) and by the decreased rate of phospholipid synthesis from exogenous fatty acids found in mutants (*fadD*) defective in acyl-CoA synthetase activity (152). However, Klein et al. (105) and Frerman and Bennet (55) subsequently concluded that acyl-CoA synthetase was absolutely required for fatty acid uptake and proposed a "vectorial acylation" transport mechanism. Therefore, the lack of exogenous fatty acid incorporation into phospholipid in *fadD* mutants could simply be due to a transport defect. This idea was supported by the finding that fatty acids generated intracellularly could be incorporated into phospholipid in the absence of acyl-CoA synthetase (172) and by the subsequent discovery of an acyl-ACP synthetase (173). These observations opened the possibility (31) that incorporation of fatty acids into phospholipid proceeded via acyl-CoA hydrolysis by one or both of the acyl-CoA thioesterases (see below), followed by the ligation of the fatty acids to ACP. Cronan (33) used an [ACP]synthase mutant (*acpS*) to reduce the intracellular ACP concentration to <10% of the normal value and found no effect on the incorporation of exogenous unsaturated fatty acid into phospholipid. This result

suggests that acyl-CoA is the species utilized by the G3P acyltransferase; however, a role for ACP cannot be ruled out since the incorporation pathway may have a high affinity for ACP.

Recently, Rock and Jackowski (182) have provided evidence that acyl-CoA synthetase is not required for fatty acid transport. These workers observed the same marked inhibition of fatty acid incorporation into phospholipid in acyl-CoA (*fadD*) mutants as did previous investigators but also extended this work to show that the *fadD*-independent incorporation was specific for position 1 of PE. These results show that the residual incorporation of fatty acid into phospholipid in *fadD* mutants was not due to the incomplete inhibition of the acyl-CoA-dependent pathway but that rather it reflected fatty acid incorporation via the acyl-ACP synthetase–plus–2-acyl-glycerophosphoethanolamine (GPE) acyltransferase route (see below). Both incorporation pathways are abolished in mutants lacking the fatty acid permease (*fadL*). These data help clarify the physiological role of the acyl-ACP synthetase and define two independent pathways for the incorporation of exogenous fatty acids into phospholipid. This metabolic scheme provides a rationale for all of the known enzymes involved in fatty acid activation and metabolism, except the enigmatic thioesterases (see below).

An experimental oddity can also be explained by the findings described above. Exogenously added 3-hydroxymyristic acid cannot be incorporated into lipid A by either *E. coli* (81) or *S. typhimurium* (216). In *E. coli*, the acid is known to enter the cell and to be converted to acyl-CoA since 3-hydroxymyristate can be utilized as a carbon source (J. E. Cronan, Jr., unpublished data). These data, therefore, indicate that 3-hydroxymyristoyl-CoA is not an intermediate for lipid A synthesis and that a mechanism to ligate 3-hydroxymyristate to ACP for incorporation into lipid A is not present.

Phospholipid Synthesis: Mechanism and Regulation

Synthesis of phosphatidic acid

Biosynthesis of G3P. G3P, which forms the "backbone" of all phospholipid molecules, as well as the polar groups of PG and CL, can be synthesized by two different short pathways. During growth with glycerol as sole carbon source, the enzymes of the glycerol catabolic (*glp*) operon are induced (for a review, see reference 116). A glycerol kinase is one of the enzymes induced; however, during growth on carbon sources other than glycerol, G3P is made by a direct reduction of the glycolytic (and gluconeogenic) intermediate, dihydroxyacetone phosphate, with NADH (104). The enzyme catalyzing this reaction, called the biosynthetic G3P dehydrogenase, or G3P synthase, is the product of the *gpsA* gene (34). This enzyme is very sensitive to inhibition by its product, G3P (50, 104), and this mechanism functions to regulate the intracellular level of G3P, as shown by the isolation of mutants resistant to feedback inhibition (11). These mutants, isolated as extragenic suppressors of *plsB* mutants (see below), contain a G3P synthase enzyme resistant to G3P inhibition and that results in elevated intracellular levels of G3P (11, 104).

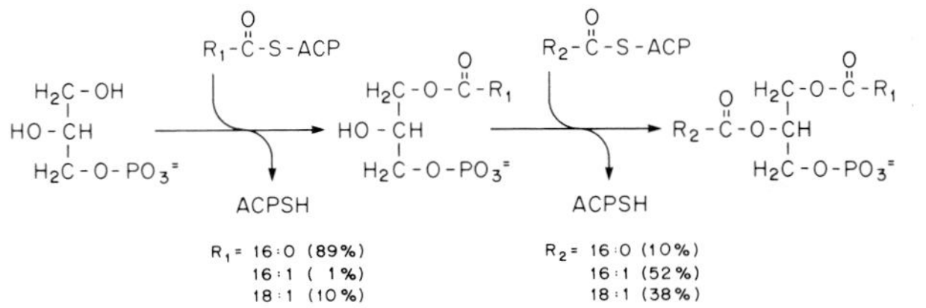

FIG. 5. Transfer to the membrane. The utilization of fatty acids to form the first membrane phospholipid in the pathway is catalyzed by the G3P acyltransferase, and the product of this reaction is acylated by 1-acyl-G3P-acyltransferase. A typical positional distribution of fatty acids (for *E. coli* grown at 37°C) between positions 1 and 2 of the glycerol backbone found in vivo is indicated below each step.

It is clear that G3P synthase is the major (if not the sole) enzyme that produces G3P (in the absence of exogenous glycerol) since *gpsA* mutants require G3P (which enters *E. coli* intact by several transport systems [116, 183]) or glycerol for growth. In the absence of G3P, phospholipid synthesis is >95% inhibited. However, a second minor pathway could account for the residual phospholipid synthesis. It should be noted that, under nonpermissive conditions, *gpsA plsA* double mutants incorporate $^{32}P_i$ into phospholipid at <2% of the normal rate. All of the residual incorporation can be attributed to diglyceride phosphorylation (T. K. Ray and J. E. Cronan, Jr., unpublished data) largely resulting from MDO synthesis (see below). It should also be noted that the rate of G3P synthesis does not limit the rate of phospholipid synthesis in *E. coli*. The intracellular pool can be greatly (>10-fold) increased in vivo without noticeable effect on lipid composition or levels (65).

Acylation of G3P to phosphatidic acid. G3P acyltransferase catalyzes the first acylation of position 1 of G3P. The second fatty acid is added by another enzyme(s), 1-acyl-G3P acyltransferase, to form phosphatidic acid (for a review, see reference 178) (Fig. 5). Phosphatidic acid composes only about 0.1% of the total phospholipid of *E. coli* and turns over rapidly (56), a property consistent with its role as a key intermediate in phospholipid synthesis. The G3P acyltransferase step represents the transition point from soluble to membrane-bound enzymes and intermediates and is the subject of a very large literature (for reviews, see references 31 and 178). Much of this work has focused on the ability of the acyltransfer reaction in vitro to duplicate the positional specificity of fatty acid acylation observed in vivo. The results of these experiments often conflict (31, 178), but in general they indicate that the acyltransferase system does possess the needed specificity. However, the extent of specificity observed depends on the experimental conditions used. This is consistent with in vivo experiments that indicate that acylation specificity is not absolute but can be altered by the supply of acyl donors (see below).

A major advance made in the study of the G3P acyltransferase resulted from the isolation of mutants of *E. coli* with a defective acyltransferase activity (9). These mutants were isolated as G3P auxotrophs and show an increased Michaelis constant for G3P in in vitro acyltransferase assays (9, 10). Thus, these mutants require a greater than normal intracellular G3P concentration, which can be supplied by either exogenous supplementation or release of feedback inhibition of G3P synthesis (see above). Therefore, these mutants are thought to contain a mutant G3P acyltransferase with a decreased affinity for G3P. However, it is also possible that these strains contain no G3P acyltransferase and that the activity seen at high G3P concentration is due to a 1-acyl-G3P acyltransferase (or perhaps a third enzyme) functioning with G3P rather than with 1-acyl-G3P (180). The interpretation of the exact alteration in these mutants is complicated by the report (also evident in the earlier literature) that the *plsB* phenotype depends on two unlinked mutations, one mutation in the *plsB* gene and a second mutation in an unlinked gene termed *plsX*. Both mutations are required for a strain to exhibit a requirement for G3P since strains harboring only the *plsB* or *plsX* lesion do not have a phenotype (110). The apparent K_m defect in G3P acyltransferase activity observed in vitro is associated with the *plsB* mutation, not the *plsX* mutation.

Although the exact alterations produced by the *plsB* and *plsX* mutations are unknown, these strains allowed the molecular cloning and subsequent purification of G3P acyltransferase (109, 115). The acyltransferase could be extracted from its integral inner membrane location with detergent, but only minimal purification was possible from wild-type strains (192). However, hybrid plasmids that suppressed the G3P requirement of *plsB plsX* strains resulted in 10-fold overproduction of G3P acyltransferase. Extraction of the membrane fraction from these strains, followed by three column chromatographic steps, yielded a single protein of M_r 83,000 (109). Each step of the purification was carried out in detergent-containing buffers, but, as is common for membrane enzymes, reconstitution of the protein with phospholipids is required for enzymatic activity. The complete primary sequence (806 residues) of the acyltransferase has been determined by a combination of protein- and DNA-sequencing techniques (114). The single polypeptide catalyzes the formation of 1-acyl-G3P from either acyl-CoA or acyl-ACP acyl donors. From the presence of other markers on the plasmids, it is clear that *plsB*, rather than *plsX*, was cloned; thus, *plsB* encodes the G3P acyltransferase.

In contrast to the first acylation reaction, the identity of the enzyme(s) that catalyzes the second acylation of G3P is unknown. The product of G3P acyltransferase is thought to be 1-acyl-G3P (see above), but the possibility of acyl migration from position 2 to the thermodynamically favored position 1 cannot be entirely excluded. However, Ray and Cronan (unpublished data) have recently isolated a mutant with an abnormally thermolabile 1-acyl-G3P acyltransferase; thus, elucidation of this step should be forthcoming.

Naturally occurring phospholipids are generally characterized as having a saturated fatty acid at position 1 and an unsaturated fatty acid at position 2 of the glycerol backbone (Fig. 5). The substrate specificity of the G3P acyltransferase system is considered the most likely origin of acyl group asymmetry in bacterial phospholipids. Accordingly, higher V_{max} values and lower K_m values are generally found for saturated, rather than unsaturated, acyl donors when either the purified or membrane-bound form of G3P acyltransferase is used as the enzyme source (31, 155, 178). These data demonstrate that the acyltransferase has a substrate specificity consistent with the role of this enzyme in controlling the positional placement of fatty acids on the glycerol backbone. However, two different in vivo experiments demonstrate that acyltransferase specificity is not absolute. First, upon starvation of *E. coli* unsaturated fatty acid auxotrophs (either *fabA* or *fabB*) for the required exogenous unsaturated fatty acid supplement, a significant accumulation of disaturated molecular species of phospholipid is observed (140, 187). Restoration of unsaturated fatty acids to the medium results in the synthesis of molecular species having the typical fatty acid positional asymmetry. Second, strains which overproduce 3-ketoacyl-ACP synthase I have elevated levels of *cis*-vaccenic acid in their phospholipids; this results in a marked increase in diunsaturated species (48). These data demonstrate that the acylation specificity of the G3P acyltransferase is not absolute in vivo but is controlled in part by the supply (either by endogenous synthesis or exogenous supplementation) of fatty acids.

In vitro, the acylation of G3P to phosphatidic acid proceeds readily with either acyl-CoA or acyl-ACP donors. This raises the question of whether both thioesters are acyl donors in vivo. Recent work indicates that endogenously synthesized fatty acids are transferred to G3P as acyl-ACPs since blocking phospholipid synthesis in vivo (via deprivation of a *plsB* or *gpsA* strain for G3P) results in the accumulation of acyl-ACP molecules (no acyl-CoAs are found) (181). However, the major pathway for incorporation of exogenously supplied fatty acids into phospholipids seems to be via acyl-CoA. Long-chain fatty acids are converted to CoA ester concomitant with transport and are incorporated into phospholipid even in cells depleted (>90%) of ACP (by use of an *acpS* mutant) (33). Thus, the acyl donor probably depends on the source of the fatty acids. The ability of the *E. coli* acyltransferases to use acyl-CoA molecules is not shared by other bacteria (the acyltransferases of *Rhodopseudomonas sphaeroides* and *Clostridium butyricum* function only with acyl-ACPs [29, 70]) and probably reflects the abundance of fatty acids present in the gut environment in which *E. coli* lives.

It should be noted that phosphatases that specifically dephosphorylate phosphatidic acid and 1-acyl-G3P are known (208) and that mutants deficient in these enzymes have been isolated (95). The function of these phosphatases remains an enigma, but their involvement in the acylation of G3P is suggested by the observation that certain acyl-ACPs promote the dephosphorylation of G3P, 1-acyl-G3P, and phosphatidic acid (180).

Phospholipid Synthesis—Diversification of Head Groups

The key finding in the pathway of phospholipid synthesis was E. P. Kennedy's demonstration of an activated form of phosphatidic acid, CDP-diglyceride (CDP-diacylglycerol [166]). In *E. coli*, conversion of phosphatidic acid to a mixture of CDP-diglyceride and dCDP-diglyceride is catalyzed by a single enzyme, CDP-diglyceride synthase (56, 57) (Fig. 6). Mutants

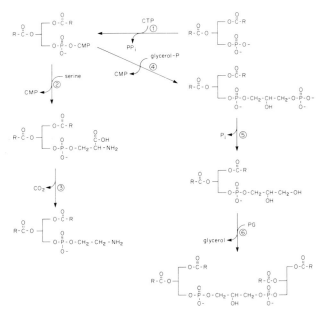

FIG. 6. Synthesis of polar head groups. The three phospholipid species found in *E. coli* are synthesized by a series of reactions catalyzed by six enzymes: 1, CDP-diglyceride synthase (phosphatidate cytidyltransferase); 2, PS synthase; 3, PS decarboxylase; 4, PGP synthase; 5, PGP phosphatase; and 6, CL synthase.

deficient in this activity (which is called *cds*) have lesions at a discrete genetic locus, and despite retaining only 5% of the normal CDP-diglyceride synthase level, the strains grow normally under standard conditions. However, some of these strains accumulate abnormally high levels of phosphatidic acid (up to 5% of the total phospholipid). Phosphatidic acid accumulation seems responsible for the increased sensitivity of these strains to erythromycin and for their poor growth at elevated pH. Shifting the growth medium of a culture of a *cds* mutant from pH 7.0 to 8.5 triggers massive accumulation (up to 25% of the total lipid) of phosphatidic acid and results in drastic inhibition of phospholipid synthesis (57). Mutants with a conditionally altered enzyme have not been isolated, suggesting that CDP-diacylglycerol synthase activity is present in large excess over the minimum amount of enzyme required to sustain phospholipid synthesis. An unlinked mutation that suppresses the phenotype of strains carrying a *cds* mutant allele has also been isolated. Unfortunately, the allele specificity of this unlinked recessive suppressor has not been tested (58).

The CDP-diglyceride generated by CDP-diglyceride synthase can then react with serine to form PS or with G3P to form phosphatidylglycerolphosphate (PGP). The presence of both ribo- and deoxyribo- forms of the liponucleotide could play a role in the determination of the relative rates of PS and PGP synthesis. The ratio of ribo- to deoxyribo- species is about 3:1 (57, 200), a ratio similar to the relative rates of PS and PGP synthesis. However, in vitro, PS synthase can favor either the ribo- substrate or the deoxyribo- substrate, depending on experimental conditions (155, 166), and Ganong and Raetz (57) report that a threefold change in the ratios of the ribo- to the deoxyribo- species has no effect on the ratio of PE synthesis to PG synthesis.

Synthesis of PE

The first step in the synthesis of PE is the formation of PS by PS synthase (155) (Fig. 6). As mentioned above, PS synthase does not fractionate with the inner membrane during standard cellular localization procedures, as do the other phospholipid synthetic enzymes. Surprisingly, the enzyme is found bound to ribosomes (170). This association is an artifact of cell disruption; the ribosome functions essentially as an ion-exchange resin. Upon addition of CDP-diglyceride, PS synthase is dissociated from the ribosome and (if membranes are also present) will associate with micelles containing CDP-diglyceride (117). The enzyme has been purified to homogeneity, and the structural gene encoding the enzyme (*pss*) has been identified (167).

A number of temperature-sensitive *pss* mutants have been isolated (148, 166, 167). Shift to a nonpermissive temperature results in a gradual decline in growth rate until stasis. As expected from the pathway, PE synthesis is inhibited, and the rate of PG and CL synthesis proportionally increases. However, since all of the *pss* strains are leaky at 42°C, the PE content does not decline below about one-third of the normal level (167). Addition of monovalent cations to the medium allows growth to proceed longer at 42°C, which probably reflects stabilization of the aberrant membranes. These strains are also extremely sensitive to antibiotics (especially aminoglycosides) at all temperatures, probably reflecting the altered lipid composition present even at nonpermissive temperatures (168). The phenotypic characters of these strains could be due to the PE deficiency or to the overproduction of the acidic lipids, PG and CL (or to both). Construction and analysis of mutants deficient in both arms of the pathway should allow further analysis of these effects.

The second step in the formation of PE is the decarboxylation of PS, catalyzed by PS decarboxylase. This inner membrane enzyme has a pyruvate prosthetic group that participates in the decarboxylation reaction by forming a Schiff base with PS (155). Mutants (*psd*) with a temperature-sensitive decarboxylase accumulate PS at the nonpermissive temperature (for reviews, see references 155 and 167). Despite the reduced PE levels and the concomitant extreme increase in PS levels (increased 1,000-fold over the levels of wild-type cells), the mutants continue growth for several hours at the nonpermissive temperature. The *psd* gene has been cloned, and strains harboring such clones overproduce the enzyme 30- to 50-fold (167). However, under conditions of extreme overproduction, about half of the enzyme is only loosely bound to the membrane (202).

Synthesis of PG

The enzymes of PG synthesis, PGP synthase and PGP phosphatase (Fig. 6), were readily demonstrated in vitro (166). However, mutants deficient in the synthesis of PGP and PG enzymes have only recently been obtained (95, 138). The dephosphorylation of PGP is catalyzed by either of two enzymes, one of which seems to also act as a phosphatidic acid phosphatase (95).

PGP synthase mutants were readily isolated by Raetz and co-workers by a powerful colony autoradiography method of enzyme assay (for the most recent version, see reference 18). Although these mutants (*pgsA*) contained <5% of the normal activity, there was no detectable growth phenotype or deficiency in PG or CL synthesis. However, starting with *pgsA* strains, mutants (*pgsB*) lacking the residual enzyme activity at 42°C were isolated (138). These double mutants show temperature-dependent growth and PG synthesis (unlike the *pgsA* parent strain). The second mutation is not linked to the *pgsA* locus but lies 36 min away on the *E. coli* genetic map. A surprising phenotype of the *pgsAB* strains is the accumulation at 42°C of two novel glycolipid precursors of lipid A (138) (chapter 31). This finding demonstrates a previously unknown link between the phospholipid and lipopolysaccharide synthetic pathways. This link could be very indirect (e.g., protein modification or processing), but no data are yet available. Both mutations are necessary for all three phenotypes (inhibition of growth and of PG synthesis and lipid A precursor accumulation), and clones carrying both *pgsA* (the probable structural gene) and *pgsB* have been isolated. Further analyses are needed. For example, the allele specificity of the interaction of the *pgsB* allele with various *pgsA* alleles has not yet been reported (137a, 138).

A recent report by Miyazaki et al. (132) indicates that certain *pgsA* alleles are defective in PG synthesis in a *pgsB*+ background. These workers found that *E. coli* mutants defective in the synthesis of both PE and CL (a *pss cls* double mutant) are more defective in growth than are strains defective in the synthesis of only one phospholipid. Upon selection for growth under nonpermissive conditions, revertants were isolated that were defective in PG synthesis. These revertants were shown to be *pgsA* mutants and were *pgsB*+ in that they failed to accumulate lipid A precursors. Under appropriate growth conditions, these strains grew at a normal rate despite having extremely low levels of PG and CL (>95% of the phospholipids were PE). However, these strains did contain high levels of PA and unidentified phospholipids that might compensate for the decrease in the normal acidic lipids, PG and CL. The selection procedure of Miyazaki et al. (132) should allow the isolation of other mutants in the *pgsA* gene. Such mutants, coupled with structural characterization of the unidentified lipids, should definitively test the possibility that *E. coli* can grow without acidic lipids.

It seems clear that PG and CL per se are not required for growth and membrane function in *E. coli*. This conclusion is based on the results described above and on the finding that *E. coli* can substitute phosphatidylmannitol for PG and diphosphatidylmannitol for CL (185). Shibyua et al. (184) found that growth of a *pss* mutant in the presence of high concentrations of mannitol resulted in mannitol-lipid–containing cells that lacked PG and CL and contained <5% of the normal level of PE. Analogous results were obtained with a number of other sugar alcohols. The incorporation of mannitol into phospholipids requires a functional *cls* gene; thus, it is thought to occur by reversal of the CL synthetase pathway, with the sugar alcohol substituting for glycerol. The recent results of

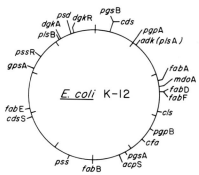

FIG. 7. Location of lipid synthetic genes on the *E. coli* K-12 genetic map.

Miyazaki et al. (132) and Shibuya et al. (184, 185) indicate that *E. coli* can grow normally (given appropriate culture conditions) with extremely abnormal phospholipid compositions. Therefore, it seems difficult to ascribe specific roles for the normal phospholipids in required cellular processes.

CL Synthesis

Kennedy first proposed that CL (also called diphosphatidylglycerol) was synthesized by reaction of CDP-diglyceride with PG. However, various physiological experiments (for a review, see reference 39) indicate that, unlike the other phospholipids, CL synthesis could occur when ATP (and hence CTP and CDP-diglyceride) levels were very low. Thus, another pathway was postulated in which CL was formed by the condensation reaction of two PG molecules to yield CL and free glycerol (Fig. 6). An enzyme catalyzing this reaction was subsequently demonstrated by Hirschberg and Kennedy (86) and Tunaitis and Cronan (200). The physiological relevance of CL synthase was demonstrated by the isolation of a mutant (*cls*) deficient in both the synthesis of CL and the transesterification activity (158). These mutants retain only a low level of CL synthesis and were initially reported to have no growth phenotype. However, recent work suggests that growth of these strains is slow (184) and that it is particularly impaired in alkaline media (B. E. Tropp, personal communication) and in cells deficient in PE synthesis (185). This latter phenotype has been used to clone the *cls* gene (147). Shibuya et al. (185) have recently reported the remarkable finding that *E. coli* accumulates phosphatidylmannitol and diphosphatidylmannitol upon growth in 0.6 M D-mannitol. The formation of the analogs was markedly increased in a *pps* mutant grown at the nonpermissive temperature. This reaction appears to be catalyzed by CL synthase since a *cls* mutant failed to accumulate these lipids and since strains carrying a cloned *cls* gene overproduce the novel lipids. A number of other sugar alcohols also were incorporated into phospholipid. These workers have proposed a pathway involving the reversal of the CL synthase reaction coupled to a lack of specificity of CL synthase. Remarkably, cells containing >93% phosphatidylmannitol and diphosphatidylmannitol are said to grow almost normally (185), although no pertinent data have yet been published.

Genetic Characterization of Lipid Synthesis Mutants

The available mutations in fatty acid, G3P, and phospholipid synthesis define genes that are unlinked, constitutively expressed, and generally distributed around the genetic map (26, 167) (Fig. 7). The only closely linked genes are the *fabD fabF* (203) and *plsB dgk* (114, 115) loci. In each case, the results of genetic mapping were confirmed by the isolation of recombinant DNA clones that carry both genes. The *plsB dgk* clone is the only DNA segment analyzed in detail (114). The nucleotide sequences of both genes have been determined, and transcription of the two genes occurs bidirectionally from a DNA segment located between the two genes and spanning about 170 base pairs. The *fabDF* genes have thus far been localized only to within a ca. 40-kilobase cloned chromosomal segment (M. A. Rawlings, M. L. Narasimhan, and J. E. Cronan, Jr., unpublished data). All of the mutants blocked in fatty acid synthesis seem to be structural gene lesions (although the identity of the acetyl-CoA carboxylase subunit defective in *fabE* mutants is unclear). Likewise, the mutants deficient in phospholipid synthesis are lesions in the genes specifying the enzymes, although this has not been firmly established for the *cds* and *cls* loci. The relationships of the *plsB-plsX* and *pgsA-pgsB* interactions also remain to be defined.

Anomalous Enzymes with Possible Roles in Lipid Synthesis

There are a number of enzymes that interact with either the intermediates or the products of fatty acid and phospholipid synthesis but which currently lack an assigned function in lipid metabolism.

Two soluble enzymes catalyze the hydrolysis of acyl-CoA esters (for a review, see reference 6). These enzymes, called thioesterases I and II, also cleave acyl-ACP thioesters, but the activity on these substrates is 10^3 to 10^4 lower than on the analogous acyl-CoA esters (195). The presence of these enzymes is an enigma since acyl-CoA esters are found to be synthesized by *E. coli* only when the β-oxidation system is operating (105). Hence, it seems probable that these enzymes function as transacylases in vivo and that the hydrolysis assayed in vitro is due to the absence of the physiological acceptor molecule; the enzyme either transfers the acyl chain to water (in lieu of the proper substrate) or forms a readily hydrolyzed acyl-enzyme intermediate. Mutants deficient in thioesterase I or thioesterase II have recently been isolated (134) and should provide answers to the function of these enzymes. The isolation of a mutant missing only thioesterase I strongly suggests that the two enzymes are specified by different genes.

An inner membrane enzyme (acyl-ACP synthetase) which ligates long-chain fatty acids to ACP was demonstrated by Ray and Cronan (173). The function of this enzyme has not been established, but recent results of Rock (174) indicate that the enzyme may be involved in the resynthesis of 2-acyl GPE. The rapid turnover of 2-acyl lysophosphatidyl ethanolamine was demonstrated in *E. coli*, and the cycle is believed to involve transfer of the position-1 fatty acid to the outer membrane lipoprotein (101), followed by reacylation of the phospholipid molecules. Enzymatic evidence strongly suggests that acyl-ACP synthetase plays a major role in the reacylation step and that the acyl-ACP synthesis activity assayed in vitro may be a partial reaction of the intact reacylation enzyme. The isolation of mutants defective in the synthetase activity is needed to confirm these provocative enzymological studies.

A membrane-bound enzyme capable of dephosphorylating lysophosphatidic acid and phosphatidic acid to monoglyceride and diglyceride, respectively, has been reported (208). Icho and Raetz (95) have recently reported a mutant (pgpB) deficient in this activity. These mutants retain a low level of activity (particularly for lysophosphatidic acid cleavage) and have no obvious phenotype.

A CDP-diglyceride hydrolase that cleaves CDP-diglyceride and dCDP-diglyceride to the nucleotide monophosphate and phosphatidic acid is present in E. coli. This enzyme also catalyzes the transfer (cytidylylation) of CMP or dCMP from the liponucleotide to phosphate or any of a variety of phosphomonoesters (18). Bulawa and Raetz (20) recently isolated mutants (cdh) missing this enzyme, and these strains (including some due to Tn10 insertion) lack any phenotype other than a fivefold elevation in the intracellular levels of CDP-diglyceride and dCDP-diglyceride. Surprisingly, a sixfold overproduction of the hydrolase by molecular cloning did not affect the liponucleotide pools or phospholipid synthesis in general (20).

E. coli contains a number of phospholipase activities, the most notable of which is the outer membrane enzyme, phospholipase A₁, which has been purified to homogeneity. The structural gene for this enzyme has been cloned (44, 87) and sequenced (44, 88). This enzyme cleaves either the position 1 fatty acid or the position 2 fatty acid from E. coli phospholipids but does not seem to play an important physiological role since mutants defective in the structural gene, pldA, have no growth phenotype. Moreover, strains harboring a hybrid plasmid bearing the pldA gene greatly overproduce the enzyme but grow normally (44, 87). de Geus et al. (45) have reported the pldA gene product to be the only known outer membrane enzyme synthesized without a posttranslationally cleaved leader peptide. This conclusion was based on the amino acid sequence deduced from the nucleotide sequence and on the fact that synthesis of a precursor form in minicells was not observed. However, the amino-terminal sequence deduced by de Geus et al. (45) differs from that reported by Homma et al. (88) in that the nucleotide sequence of the latter workers contains two additional base pairs. Thus, the deduced sequences of the two genes begin in different reading frames. The sequence reported by Homma et al. (88) seems correct since it agrees with the amino-terminal amino acid sequence of the purified protein. (It should be noted that de Geus et al. [45] reported a blocked amino terminus for the protein.) In the reading frame identified by the N-terminal sequence, Homma et al. (88) found a typical 20-amino-acid signal sequence upstream of the sequence that corresponds to the mature protein. Therefore, it seems likely that a precursor will be demonstrated.

The lysophospholipids produced by the pldA gene product are assumed to be further degraded by either of two lysophospholipases, one (an L2) found in the inner membrane and the other (an L1) a cytosolic enzyme. Mutants deficient in the inner membrane (pldB) enzymes have been isolated (106), and the locus has been cloned (87) along with the closely linked pldA gene. These mutants have no detectable phenotype. (One temperature-sensitive mutant has been isolated, but the lesion has not been shown to be directly related to the lipase defect.) Strains that overproduce the lysophospholipase also have no phenotype. It seems possible that some of these enzymes function in vivo as transacylases rather than as phospholipases (a situation analogous to that proposed above for the thioesterases). Indeed, the inner membrane lysophospholipase activity has been implicated in the synthesis of monoacyl-PG (106). Also, it should be noted that there is no direct evidence for significant degradation of E. coli phospholipids during normal growth.

To complement the phospholipases, numerous lysophospholipid reacylation enzyme activities have been described. 1-Acyl-lysophospholipids are reacylated by a membrane-bound enzyme that utilizes acyl-CoA as the acyl donor (165). The specific activity of this enzyme is low, and the distinct possibility that this reaction is due to the presence of 1-acyl-G3P acyltransferase was not tested. The best-studied reacylation enzyme is 2-acyl-GPE acyltransferase (89, 174, 198). This acyltransferase either utilizes acyl-ACP or activates fatty acids in the presence of ATP, Mg²⁺, and a bound ACP subunit, as evidenced by the complete inhibition of acyltransferase activity by ACP-specific antibodies (174). Acyl-CoAs are not utilized by this enzyme. It has been suggested that these enzymes function to reacylate the lysolipids formed by phospholipase hydrolysis; however, all of the established metabolic activity of membrane phospholipids can be traced to their use in the biosynthesis of other molecules (see below).

A number of transacylation reactions have also been described in envelope fractions of E. coli (90, 91, 139, 214). The acyl donors in these experiments were apparently endogenous membrane phospholipids, and, in most cases, the enzymes responsible for the observed activities were not specified. However, Homma and Nojima (91) concluded that CL synthase is responsible for three of the many products formed in their assays since extracts from a cls mutant did not catalyze these conversions. Therefore, it is possible that most, if not all, of the transacylation reactions described can be attributed to known biosynthetic enzymes acting on poor, nonphysiological substrates in vitro.

In Situ Modification of the Phospholipids: Synthesis of Cyclopropane Fatty Acids

In addition to saturated and unsaturated fatty acids, S. typhimurium, E. coli, and many other bacteria contain cyclopropane fatty acids (67, 69). These acids are formed by a postsynthetic modification of the unsaturated fatty acids in the phospholipids residing in the inner and outer membranes (112). In vitro, neither free fatty acids nor acylthioesters are substrates for cyclopropane fatty acid synthase; only phospholipids dispersed in a micelle are a substrate. The reaction involves the modification of the cis double bond with a CH_2 group (derived from the activated CH_3 group of S-adenosylmethionine) to form a *cis* cyclopropane ring (112).

Cyclopropane fatty acid synthesis occurs primarily as bacterial cultures enter the stationary growth phase (as the cultures cease growth due to oxygen or nutrient limitation); rapidly growing cells synthesize little cyclopropane fatty acid (30, 111). However, the levels of cyclopropane fatty acid synthase vary little during growth (30, 197); thus, the timing is not due to a change in enzyme synthesis. Alterations in substrate levels have also been excluded. Recent work showed that cells carrying a cloned segment of DNA containing the cyclopropane fatty acids synthase gene synthesized the acids throughout exponential growth (75). Despite this modification, the cells grew normally; therefore, the presence of these acids does not antagonize normal growth. The continuous synthesis of cyclopropane fatty acids in strains carrying a cloned *cfa* gene indicates that the regulatory process inhibiting cyclopropane fatty acid synthesis in rapidly growing cells is overcome by increased production of the enzyme. This result suggests a stoichiometric inhibition, such as a protein-protein interaction, rather than interaction with a small molecule. The cloned gene has been used to map the structural gene (*cfa*) for the enzyme (76) and to demonstrate that the enzyme is a tetramer of a 39,000-M_r protein (Grogan and Cronan, Jr., unpublished data).

Many physiological functions have been proposed for cyclopropane fatty acids; however, some *cfa* mutants of *E. coli* completely lack cyclopropane fatty acids (<70 molecules per cell) owing to disruption of the gene encoding the enzyme (76; Grogan and Cronan, Jr., unpublished data). These mutants survive and grow under various environmental stresses (stationary phase, high- and low-O_2 tension, etc.) as well as do strains that accumulate cyclopropane acids in a normal manner (77, 197). Therefore, we are faced with an enigma—cyclopropane fatty acids are widely conserved among bacteria but seem to play no essential role. We can only conclude that they play a vital role in the natural environment that has not yet been duplicated in the laboratory.

The final novel aspect of cyclopropane fatty acid synthesis is the topologic problem. Cyclopropane fatty acid synthase is primarily a soluble enzyme, although traces of activity are found loosely bound to the inner membrane (197). How does the enzyme have access to virtually all of the phospholipids of both the inner and the outer membrane? The enzyme is large (M_r 80,000 to 90,000) and hydrophilic; thus, it seems unlikely that cyclopropane fatty acid synthase can cross lipid bilayers. It seems more probable that the lipid somehow gains access to the enzyme. There is evidence for a flow of phospholipid between the inner and outer membranes (chapter 5), but neither the mechanism of the flow nor that of phospholipid flip-flop from one face of a bilayer to the other is known, although the former process requires an electrochemical gradient (49). The partially purified enzyme has been shown to react with phospholipids on both leaflets of synthetic lipid bilayers in vitro (197). Moreover, the interaction of the enzyme with phospholipid molecules is a spe-

cific reaction; only phospholipids containing unsaturated or cyclopropane chains bind to the enzyme. The recent molecular cloning of the cyclopropane synthase gene should allow complete purification of this enzyme and hence greatly facilitate a detailed analysis of this novel reaction.

Phospholipids as Intermediates in the Synthesis of Other Molecules

Early observations on phospholipid metabolism showed that PG was unstable in pulse-chase experiments in growing cells, whereas PE was stable (39). At first, it was thought that the PG was being degraded. After the discovery that *E. coli* contains CL (3, 30), it was realized that some of the PG turnover was actually the conversion of PG to CL, catalyzed by CL synthase. However, CL synthesis did not account for all of the loss of ^{32}P-labeled PG observed in pulse-chase experiments or explain why the nonacylated (polar group) glycerol was labeled (and chased) more rapidly than the acylated (backbone) glycerol moiety. A nonlipid phosphate-containing compound was sought, and a family of molecules called membrane-derived oligosaccharides (MDO) was discovered (chapter 43). These molecules are glucose polymers derived with *sn*-glycerol-1-phosphate (derived from PG) and, to a lesser extent, with succinate and ethanolamine moieties. These polymers (M_r 4,000 to 5,000) reside in the osmotically active periplasmic space of gram-negative bacteria. The synthesis of the MDO compounds is regulated by the osmotic pressure of the growth medium (decreased osmotic pressure gives an increased rate of MDO synthesis); thus, MDO seem to be involved in osmotic regulation (for a review, see chapter 43).

The discovery of the MDO compounds provided a function for the well-studied but enigmatic enzyme diglyceride kinase (for a review, see reference 167). In the synthesis of MDO, the *sn*-glycerol-1-phosphate polar group of PG is transferred to the oligosaccharide, forming 1,2-diglyceride as the other product. Although direct proof is lacking, it is likely that the ethanolamine moiety is derived from PE, as this is the only known source of ethanolamine in *E. coli* (chapter 43). Diglyceride kinase then phosphorylates the diglyceride to phosphatidic acid, a central intermediate in the phospholipid biosynthetic pathway (see above). Therefore, only the *sn*-glycerol-1-phosphate portion of the PG molecule is consumed in the overall reaction, since the lipid portion of the molecule is recycled into phospholipid (167). It is clear that MDO synthesis is responsible for most of the metabolic instability of the polar group of PG (e.g., references 5 and 121) since this turnover is greatly decreased if MDO synthesis is blocked at the level of oligosaccharide formation (chapter 43). Moreover, the rate of accumulation of diglyceride in strains (*dgk*) lacking the diglyceride kinase depends on the presence of the oligosaccharide acceptor and on the osmotic pressure of the growth medium, and *dgk* mutants grow poorly on media of low osmolarity.

The monoacylated glycerol of PG is also the precursor of a second novel molecule, the bacterial lipoprotein. Certain outer membrane and excreted proteins are proteolytically cleaved only after modification of the protein by addition of a glycerol moiety to a specific cysteine residue close to the amino terminus of the primary translation product (17). A glyceryl thioether bond is formed in which the glycerol is derived from the sn-glycerol-1-phosphate polar group of PG. It should be noted that much less PG-derived sn-glycerol-1-phosphate is consumed in this reaction than is used in MDO synthesis (chapter 43). The glycerolcysteine of the prolipoprotein molecule becomes acylated, as does the cysteine amino group exposed by signal peptidase cleavage. These fatty acids seem to be derived from the membrane phospholipids rather than being derived directly from de novo biosynthesis (for a review, see reference 221). Consistent with this idea, a detailed examination of phospholipid fatty acid turnover has revealed that, whereas position 2 of membrane phospholipids is metabolically stable (38), turnover at position 1 of PE occurs in E. coli (174). The amount of PE turnover is small (3 to 5%) compared with the size of the PE pool but can account for all of the phospholipid fatty acids required to acylate the amino terminus of the lipoprotein (101). The resulting 2-acyl-GPE is then reacylated by 2-acyl-GPE acyltransferase (see above). Thus, an unknown enzyme(s) may transfer fatty acids from a phospholipid to the nonacylated prolipoprotein. A system in which several of these steps occur in vitro has recently been developed and should quickly lead to more detailed knowledge of the synthetic pathway (199).

INHIBITORS OF LIPID SYNTHESIS

Two specific inhibitors of fatty acid synthesis are known: cerulenin (150) and 3-decenoyl-N-acetylcysteamine (15). Cerulenin inevitably inhibits 3-ketoacyl-ACP synthases I and II, with synthase I being considerably more sensitive both in vitro (21) and in vivo (203). This differential sensitivity accounts for the preferential inhibition of unsaturated fatty acid synthesis at low cerulenin concentrations; at high concentrations, the synthesis of all fatty acids (saturated, unsaturated, and 3-hydroxy) is inhibited in both E. coli (20, 150, 203) and S. typhimurium (216). 3-Decenoyl-N-acetylcysteamine is a specific dehydrase-activated ("suicide") inhibitor of 3-hydroxydecanoyl-ACP dehydrase, the fabA gene product (15, 140). Thus, treating cells with this reagent results in a specific and irreversible inhibition of unsaturated fatty acid synthesis (15, 140). Although use of this inhibitor has been limited by its lack of commercial availability, it has been used to isolate E. coli mutants that overproduce 3-hydroxydecanoyl-ACP dehydrase due to a cis-active mutation closely linked to the fabA structural gene (27). Recently, thiolactomyocin was reported to be an inhibitor of E. coli fatty acid synthesis in vivo and in vitro (84, 85). However, the data are of a preliminary nature, and the identities of the enzymes inhibited by thiolactomyocin have not yet been demonstrated.

Two analogs of G3P, dihydroxylbutyryl-1-phosphonate (94, 201) and sn-glycerol-3-phosphorothioate (80, 151), have been used to inhibit phospholipid synthesis in E. coli. Both compounds contain carbon-phosphate bond analogs that are expected to be poorly hydrolyzed by cellular phosphatases. Indeed, Tyhach et al. (201) demonstrated the accumulation of PG

phosphonate in cells grown in the presence of dihydroxybutyryl-1-phosphonate. Curiously, no accumulation of PG-phosphorothioate was observed in cells inhibited by sn-glycerol-phosphorothioate (151). The main problem with utilization of both G3P analogs is their nonspecific inhibition of PE synthesis, as well as of PG synthesis (151, 186, 201). The inhibition of PE synthesis is not understood in either case, but in the case of the bacteriocidal phosphorothionate analog, inhibition may be due to inhibition of CDP-diglyceride synthase (151). Recently, mutants resistant to dihydroxylbutyryl-1-phosphonate were shown to have a lesion in the cls (CL synthase) locus (94).

Ingram and co-workers (96, 97) have studied the effects of ethanol addition on lipid metabolism in growing cultures of E. coli. Addition of ethanol results in an increase in the rate of cis-vaccenic acid synthesis similar to that seen upon decrease in growth temperature. A small inhibition of PE synthesis (with a concomitant increase in PG synthesis) is also seen at high ethanol concentrations. The effect on fatty acid composition was first attributed to an effect on G3P acyltransferase caused by intercalation of ethanol into the membrane phospholipids. However, subsequent data were interpreted as inhibition of saturated fatty acid synthesis resulting in increased incorporation of cis-vaccenic acid into phospholipid. This effect has been primarily localized to an effect on 3-ketoacyl synthase II, but this conclusion is based on the use of nonisogenic strains and indirect in vitro assays. Moreover, subsequent work has shown that ethanol addition triggers both the heat shock response (135) and accumulation of the alarmone, ApppppA (113). These results suggest that more direct experiments are needed to establish the effects of ethanol on phospholipid synthesis.

REGULATION OF THE COMPOSITION OF MEMBRANE PHOSPHOLIPIDS

Physiological Importance of the Regulation of Membrane Phospholipid Composition

An important lesson learned from studies of the various mutants of E. coli blocked in fatty acid synthesis is that the organism tolerates a wide variation in the fatty acid composition of the membrane phospholipids (37). The unsaturated fatty acid auxotrophs (fabA and fabB) can grow if the medium is supplemented with any of several fatty acids. Although saturated fatty acids alone will not support growth, a wide variety of cis unsaturated fatty acids (mono-, di-, or triunsaturated) will support growth of fabA and fabB mutants. Indeed, even unsaturated fatty acids with a centrally located trans double bond (a type of fatty acid not found in E. coli and very rarely found in nature) will suffice. It is clear that the double bond per se plays no chemical role in metabolism. The role of the double bond is to decrease the phase transition temperature of the phospholipid containing the fatty acid. A number of fatty acids lacking double bonds will also support growth. These acids (cis or trans cyclopropane, branched, centrally brominated) do, however, share with double bonds the ability to disrupt the close packing of phospholipid acyl chains, resulting in a lower phase transition temperature (for further discussion, see chapter 5). This property is purely physical in that the presence of a substituent or double bond in the middle of the hydrocarbon chain sterically disrupts strong hydrophobic interactions with other acyl chains (126). It should also be noted that phospholipid phase transition temperatures increase with chain length of the hydrocarbon chain; hence, chain length, as well as the saturated:unsaturated fatty acid ratio, must be controlled (129). However, there are limits to the fatty acid compositions that will permit growth. The finding that *fabA* and *fabB* mutants require an unsaturated or equivalent fatty acid for growth indicates that a membrane composed of phospholipids containing only saturated fatty acids is nonfunctional (for reviews, see references 31 and 37). These mutants undergo cell lysis when deprived of the unsaturated fatty acid supplement. Thus, *E. coli* requires some fluid lipid for a functional membrane, and several laboratories have reported that, if less than half of the membrane phospholipid is in the fluid state, the *E. coli* membrane becomes nonfunctional (1, 102, 154). A similar argument can be made for the importance of some nonfluid lipid for a functional membrane. Mutants (*fabD fabE* or *fabB fabF*) that block fatty acid for growth (81). If only an unsaturated fatty acid is added, the cells leak internal components and eventually lyse (4).

The conclusions from these experiments are straightforward. A functional *E. coli* membrane requires that the composition of the membrane phospholipids be within the limits of the phase transition. If all of the phospholipids are in either the ordered state or the disordered state, the membrane is nonfunctional. However, quite wide variations in fluidity are tolerated; that is, the cells do not have to maintain a precise ratio of fluid to nonfluid lipid to have functional membranes. However, there does seem to be an optimal fluidity, at which cell growth is most rapid. The regulatory systems seem designed for optimizing the fluidity within the tolerated range, rather than for extending the range.

The polar head group compositions also affect the phospholipid phase transitions in model systems. Given identical fatty acid compositions, PE micelles have much higher phase transitions than do PG vesicles, whereas vesicles of the mixed lipids have transitions similar to those of the PG vesicles (129, 159). The molecular structures of PE and CL also impart the ability to form the nonlamellar HII phase to vesicles of those lipids (210). Mixture of micelles of these lipids with a lipid unable to form the HII phase (such as PG) results in a lamellar phase rather than an HII phase. Although the reasons for the different behaviors of these phospholipid molecules are not completely clear, these model system studies do suggest that regulation of polar group composition is needed to optimize membrane fluidity. Only one study on the interrelationship between fatty acid and head group composition has been done in *E. coli*. Pluschke and Overath (159) showed that a *fabB* mutant could grow without an unsaturated supplement if the strain also carried a lesion (*cls*) blocking CL synthesis and other uncharacterized mutation(s). These findings were rationalized by the accumulation of PG (owing to the *cls* mutation), which lowered the phase transition. The

uncharacterized mutation may have been needed to allow the cell to tolerate the larger amounts of myristic acid synthesized. Miyazaki et al. (132) have recently reported that mutants that overproduce PE owing to a defect in PG synthesis grow poorly at low temperatures, a result consistent with the high phase transition of PE.

It should be noted that recent deuterium nuclear magnetic resonance results indicate that the conformation of the lipids within the *E. coli* membrane strongly resembles that seen in synthetic phospholipid vesicles (for a review, see chapter 5).

THERMAL REGULATION OF FATTY ACID SYNTHESIS

E. coli, in common with most (if not all) other organisms, synthesizes phospholipids with a greater proportion of unsaturated fatty acids when grown at low temperatures (e.g., 25°C) rather than at high temperatures (e.g., 42°C) (125). This regulatory system is believed to be designed to ameliorate the effects of temperature change on the physical state of the membrane phospholipids. As discussed in chapter 5 of this volume and in other reviews (37, 126), the proportion of fluid (disordered) lipid to nonfluid (ordered) lipid in cell membranes plays a major role in membrane function. Increased incorporation of unsaturated fatty acids into the phospholipids decreases the temperature at which the transition from ordered to disordered membrane phospholipid occurs, whereas increased incorporation of saturated fatty acids has the opposite effect. The thermal regulatory system can thus adapt the membrane lipids for optimal functioning at new growth temperatures.

A key finding in unraveling the mechanism of thermal regulation in *E. coli* was that, after a temperature shift, neither RNA nor protein synthesis was required for the synthesis of new phospholipid with a fatty acid composition characteristic of the new temperature. This result indicated that thermal regulation was exerted by a protein synthesized at all growth temperatures but active only at low temperatures. A clue to the identity of this protein was the observation that only one of the two *E. coli* unsaturated fatty acids was synthesized in greater quantity at low growth temperature (59). The level of *cis*-vaccenic acid increased, whereas no increase was seen in the level of palmitoleic acid. The subsequent isolation of a mutant (*fabF*) defective in both the conversion of palmitoleic acid to *cis*-vaccenic acid and thermal regulation indicated that a specific enzyme catalyzes this step (62). The *fabF* strain had only traces of *cis*-vaccenic acid in its phospholipids, and its fatty acid composition (chiefly palmitic and palmitoleic acids) did not change with change in growth temperature (23, 62, 120, 203). It was soon demonstrated that *fabF* strains are defective in the gene encoding 3-ketoacyl-ACP synthase II (60). A genetic analysis of strains carrying mutations in both *fabB* (a temperature-sensitive mutant was used) and *fabF* showed that such double mutants were for the synthesis of new phospholipid with a fatty acid composition characteristic of the new temperature. This result indicated that thermal regulation was exerted by a protein synthesized at all growth temperatures but active only at low temperatures. A clue

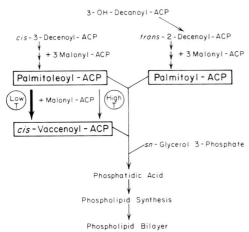

FIG. 8. Thermal regulation of fatty acid biosynthesis. 3-Ketoacyl-ACP synthase II is primarily responsible for the temperature control of *E. coli* fatty acid composition by being more active in the conversion of palmitoleate to *cis*-vaccenate at lower temperatures than at higher temperatures.

to the identity of this protein was the observation that only one of the two *E. coli* unsaturated fatty acids was synthesized in greater quantity at low growth temperature (59). The level of *cis*-vaccenic acid increased, whereas no increase was seen in the level of palmitoleic acid. The subsequent isolation of a mutant (*fabF*) defective in both the conversion of palmitoleic acid to *cis*-vaccenic acid and thermal regulation indicated that a specific enzyme catalyzes this step (62). The *fabF* strain had only traces of *cis*-vaccenic acid in its phospholipids, and its fatty acid composition (chiefly palmitic and palmitoleic acids) did not change with change in growth temperature (23, 62, 120, 203). It was soon demonstrated that *fabF* strains are defective in the gene encoding 3-ketoacyl-ACP synthase II (60). A genetic analysis of strains carrying mutations in both *fabB* (a temperature-sensitive mutant was used) and *fabF* showed that such double mutants were completely defective in fatty acid synthesis, thereby eliminating the possibility of a third 3-ketoacyl-ACP synthase activity capable of synthesizing long-chain acyl-ACP (60).

The genetic analysis of the *fabF* mutants suggested that 3-ketoacyl-ACP synthase II should function much better than synthase I in the elongation of palmitoleoyl-ACP to the 3-keto precursor of *cis*-vaccenoyl-ACP. Moreover, this difference should be accentuated at lower temperatures (60). Since mutants containing either synthase I or synthase II synthesized saturated fatty acids normally, it was expected that either enzyme could catalyze all of the elongation steps required in saturated fatty acid synthesis (61). These predictions were borne out by in vitro studies on the purified enzymes (61); thus, a simple model emerged in which decreased growth temperature alters the activity of synthase II, which in turn regulates the fatty acid composition by producing more *cis*-vaccenoyl-ACP for incorporation into phospholipid (46) (Fig. 8). Though all of the available data were consistent with this model, one dilemma remained. It could be argued that the lack of synthase II resulted in the

loss of thermal regulation or, conversely, that the lack of *cis*-vaccenoyl-ACP, the product of elongation of palmitoleoyl-ACP by synthase II, caused the lack of temperature regulation (thermal regulation would, in this case, be exerted at a later step). Although indirect evidence indicated that the first argument was correct, direct proof was needed. This was provided by an experiment in which normal amounts of *cis*-vaccenic acid were synthesized in the absence of 3-ketoacyl-ACP synthase II with a cloned synthase I gene (48). Although synthase I elongates palmitoleoyl-ACP more poorly than does synthase II, the synthase I reaction proceeds at a measurable rate with this substrate (61). Thus, these investigators reasoned that, if synthase I could be overproduced by molecular cloning onto a multicopy plasmid, appreciable *cis*-vaccenic acid synthesis should occur. The results of this experiment showed that the presence of a plasmid carrying the *fabB* gene resulted in the synthesis of a normal level of *cis*-vaccenic acid in the absence of synthase II. More important, the fatty acid composition of these strains was not altered by growth temperature (48). This experiment demonstrated that 3-ketoacyl-ACP synthase II per se is the only protein responsible for thermal regulation of membrane lipid composition in *E. coli*. A second conclusion is that synthase I catalyzes the rate-limiting reaction in unsaturated fatty acid synthesis (27, 48).

Regulation of Fatty Acid Chain Length

Likely candidates for the site of chain length regulation were the 3-ketoacyl-ACP synthases I and II, which catalyze the elongation reactions. Substrate specificity studies in vitro indicate that one reason why membrane phospholipids are devoid of chains of more than 18 carbons is in part the reduced activity of synthases I and II on C_{18} substrates (61, 74). Moreover, synthase II mutants (*fabF*) are defective in the elongation of palmitoleate to *cis*-vaccenate (62); therefore, synthase II plays a critical role in determining the amount of C_{18} fatty acids in the membrane. Although these data indicated that the condensing enzymes play a significant role in determining chain length, physiological experiments indicate that the level of G3P acyltransferase activity is also important. When phospholipid biosynthesis is slowed or arrested at the acyltransferase step (by using a G3P acyltransferase mutant [*plsB*] or a G3P synthase [*gpsA*] mutant), the fatty acids synthesized had abnormally long chains compared with the normal distribution of fatty acids synthesized in the presence of G3P acyltransferase activity (40). These data indicate that competition between the rate of elongation and the rate of utilization of the acyl-ACPs by the acyltransferase is a significant determinant of fatty acid chain length in *E. coli*.

Regulation of Head Group Composition

Studies with model membranes suggest that regulation of the relative levels of PE, PG, and CL in the membrane should be a parameter important to membrane function (see above). Indeed, the phospholipid head group composition of *E. coli* is unaffected by manipulating the parameters of cell growth (except for the conversion of PE to CL occurring early in the stationary phase) (39, 166). The only means of perturbing the head group composition is by isolating temperature-sensitive mutants deficient in the various phospholipid biosynthetic enzymes (166, 167, 179).

The most striking result of studying these mutants at high temperatures is the rather benign nature of these lesions in comparison to the fatty acid mutants. These strains exhibit an abrupt alteration in phospholipid biosynthesis when shifted to the nonpermissive temperature but continue to grow for several generations. However, the fact that growth continues in the presence of quite abnormal phospholipid composition indicates that most membrane functions are relatively unaffected by perturbing membrane polar head group composition. A possible exception is cell division since several of these mutants form filaments at high temperature (167). However, filament formation is triggered by a number of other physiological perturbations and may be a nonspecific effect.

Two complications of these experiments should be noted. First, the phospholipid compositions of the inner and outer membranes differ (32); thus, one membrane could buffer the effect on the other. Second, a deficiency in the synthesis of one lipid results in increased synthesis of the other species that likewise could buffer the effects of the deficiency. This possibility can be approached by the construction of double or triple mutants, but only two such studies have been reported. Muller and Cronan (133) utilized a *psd cls* double mutant in their studies of the lipid-containing phage PR4, and the double mutant behaved essentially as expected on the basis of properties of the single mutants, indicating that increased CL synthesis does not compensate for the accumulation of PS. Shibuya et al. (184) constructed mutants with the *cls* lesion and a temperature-sensitive *pss* lesion. At nonpermissive temperatures, these strains contained almost 50% PG, and the double mutant grew significantly more poorly than did either single mutant.

As mentioned above, the *plsB*, *pgsA*, *pgsB*, *pss*, and *psd* genes have all been isolated in functional form on plasmid cloning vectors (167, 178). All of these consructions result in a major overproduction of the encoded enzyme but have no readily discernible effect on cellular phospholipid compositions. In all cases, the increase in activity is due to an increased level of the catalytic protein, and (with the possible exception of *psd* [202]) this increased protein seems to reside in its normal cellular location. These results indicate that these biosynthetic enzymes are present in large functional excess in *E. coli* and that they are not rate limiting in the phospholipid biosynthesis. It follows that the phospholipid head group composition of *E. coli* is not determined by the concentration of the various enzymes involved and that control is exerted at another level. However, one result implies that these conclusions may be premature. Sparrow and Raetz (193) have isolated a mutant of *E. coli* that overproduces PS synthase four- to sixfold. This mutation is active in *trans*, does not alter the cellular phospholipid composition, and is unlinked to the *pss* structural gene. Therefore, we are faced with a paradox. *E. coli* seems to have a specific locus that regulates expression of *pss*, whereas fine control over PS synthase levels does not seem to be important for

cellular growth as judged by the cloning studies. A similar regulatory mutant has also been found for diglyceride kinase (*dgk*) (115, 169). There is no evidence for coordinate control over the expression of the lipid synthetic genes of *E. coli*, and these loci are scattered about the genome. The only candidates for a group of genes controlled as an operon are the *fabD* and *fabF* genes (203; Rawlings, Narasimhan, and Cronan, unpublished data). Coordinate control of the *plsB* and *dgk* genes also seems possible despite their divergent transcription (114).

Possible Coupling of Fatty Acid and Phospholipid Synthesis

It seems that the rate of phospholipid synthesis is determined by the rate of fatty acid synthesis, since wild-type *E. coli* cells have very small pools of free fatty acids or other precursors of phospholipid acyl groups (acyl-ACPs or acyl-CoA). Jackowski and Rock (181) have shown that, after inhibition of phospholipid synthesis, acyl-ACP compounds accumulate. These data were obtained by G3P starvation of a *plsB* (or *gpsA*) pantothenate auxotroph that had been prelabeled with a radioactive pantothenate precursor. Although this technique avoids the problem of changing intracellular pools of fatty acid precursors (see below), any fatty acids not bound to a pantetheine compound (ACP or CoA) would not be detected. Previous workers had attempted direct labeling of fatty acids by using [14C]acetate. Mindich (131) first reported that fatty acid synthesis was completely inhibited when phospholipid synthesis was blocked in a *gpsA* strain. However, those experiments failed to account for the influence of fatty acid degradation. Subsequent workers blocked fatty acid degradation but came to different conclusions concerning the level of fatty acids synthesized after cessation of phospholipid synthesis. Cronan et al. (40) reported that fatty acid synthesis continued unchecked and that free fatty acids of abnormal chain length accumulated. However, Nunn et al. (144) reported that fatty acid synthesis was >90% inhibited. The differing results were attributed to the labeled precursors used, [1-14C]acetate or [2,3-14C]succinate (144). Nunn et al. (144) reported that starvation of a *plsB* or *gpsA* strain for glycerol or G3P causes the intracellular acetyl-CoA pool to shrink drastically, yielding erroneously high values for fatty acid synthesis as measured by [14C]acetate incorporation. These workers believed that such pool effects could be avoided by use of [14C]succinate. However, more recent work from the same laboratory showed that the succinate pool was expanded during glycerol starvation (122, 123). These unexpected results are probably due to the interplay among malate synthase, isocitrate lyase, and isocitrate dehydrogenase kinase/phosphatase that function to alter the flux of intermediates through the tricarboxylic acid cycle in cells utilizing the glyoxylate bypass (137). The first two enzymes are known to be coordinately controlled by the negative repressor (*fadR*) of the fatty acid β-oxidation system (123), and it would be interesting to determine whether the novel kinase/phosphatase is also under *fadR* control. Upon glycerol starvation of *gpsA* or *plsB* strains defective in β-oxidation, the free fatty acids produced (21, 40) could neutralize the *fadR* repressor, causing a partial induction of the glyoxylate cycle. Induction of the glyoxylate cycle enzymes results in depletion of the acetyl-CoA pool and lowers the levels of the tricarboxylic acid cycle intermediates (123). In attempting to use the new data to correct the old results, it seems that fatty acid synthesis proceeds in the absence of phospholipid synthesis at 20 to 30% of the normal rate, a value consistent with the results of Jackowski and Rock (181). However, definitive experiments will require a G3P auxotroph also carrying mutational lesions in the β-oxidation system (fadA, fadB, or fadE) and the glyoxylate cycle (aceA or aceB), in addition to careful pool measurements of the relevant metabolites.

There is, as yet, no concrete evidence for regulation of phospholipid or fatty acid synthesis by classical feedback inhibition mechanisms or by repression of enzyme synthesis. Several laboratories (for a review, see references 68 and 160) have reported that addition of exogenous fatty acids to cultures of E. coli results in decreased incorporation of endogenously synthesized fatty acids into phospholipid. These results were interpreted as evidence for feedback inhibition or repression of fatty acid synthesis. However, another possibility was that the exogenous acids competed with the endogenously synthesized acids for incorporation into phospholipid. Indeed, Polacco and Cronan (160) have shown that addition of exogenous oleic acid does not inhibit the synthesis of free fatty acids when phospholipid synthesis is blocked (see above). Moreover, Goldfine (68) has presented strong arguments that inhibition of lipid synthesis by exogenous acids could be disadvantageous for E. coli and C. butyricum. Accumulation of intermediates in the phospholipid synthetic pathway does not inhibit earlier lipid synthetic reactions. Strains accumulating CDP-diglyceride (20), phosphatidic acid (57), PS (83), and PG phosphate (95) have a normal overall rate of lipid synthesis and normal protein:lipid ratios.

Coordination of Phospholipid Production with Energy Metabolism and Macromolecule Biosynthesis

Much of the work on coordination of phospholipid production with energy metabolism and macromolecule biosynthesis was reviewed recently (178), and our present comments will summarize the developments in this area since that review.

REGULATION IN RESPONSE TO ENERGY SUPPLY

The genesis of this work is the observation that temperature-sensitive mutants in the adk (adenylate kinase) gene of E. coli had a specific effect on phospholipid synthesis in vivo and made the G3P acyltransferase of these strains abnormally thermolabile. The adk locus is the structural gene for adenylate kinase (66, 78, 93), and this phenomenon has been attributed to the formation of a functional complex of adenylate kinase and G3P acyltransferase in vivo (66). Some direct data that support this model are as follows. (i) Monospecific antibody raised against adenylate kinase inhibits acyltransferase activity in vitro and precipitates a number of inner membrane polypeptides. (ii)

The thermolability of the acyltransferase of the mutants can be ameliorated by incubation with wild-type adenylate kinase. Several recent papers have indicated that regulation of G3P acyltransferase by adenylate kinase may be quite complex. Huss and Glaser (93) have isolated a soluble protein that specifically associates with adenylate kinase, and Greenaway and Silbert (73) have suggested that the activity of the acyltransferase may depend on an interaction with the acyl-CoA synthetase.

REGULATION IN RESPONSE TO INHIBITION OF MACROMOLECULE SYNTHESIS

In wild-type strains of E. coli and S. typhimurium, inhibition of protein synthesis by starvation for a required amino acid results in a strong inhibition of stable RNA synthesis. Inhibition of RNA synthesis is correlated with the presence of the novel nucleotide, guanosine 5′-diphosphate-3′-diphosphate (ppGpp), that accumulates after starvation of wild-type (rel⁺), but not relA mutant, strains. The inhibition of ribosomal RNA synthesis seems to be due to inhibition of RNA polymerase molecules bound to a certain class of promoters (63). Several laboratories have reported that phospholipid synthesis decreased after the starvation of rel⁺, but not relA, strains, and considerable evidence has accumulated that ppGpp is the effector of lipid synthesis in vivo (179). A regulatory role for the G3P acyltransferase was suggested by Merlie and Pizer (130), who reported that the acyltransferase was inhibited by ppGpp in vitro. However, the need for control at the G3P acyltransferase step was not supported by in vivo experiments that demonstrated a direct effect of the relA gene on fatty acid biosynthesis (142, 194). The view that the early step in fatty acid biosynthesis was the primary site for stringent control was solidified by the finding that, although ppGpp did inhibit acyltransferase when acyl-CoAs were substrates, this nucleotide did not inhibit the G3P acyltransferase when ACP thioesters were the acyl donors (119, 171). A passive role for the G3P acyltransferase in the stringent response, however, was not consistent with the demonstration of relA-mediated control of phospholipid synthesis in strains requiring exogenous fatty acids for growth (140), but at the time of the interpretation this role was unclear because the identity of the acyl donor used by exogenous fatty acids had not been established (178). Recently, exogenous fatty acids have been shown to be incorporated into phospholipid via acyl-CoA, rather than acyl-ACP, intermediates (see above), suggesting that, to diminish phospholipid synthesis from exogenous fatty acids, utilization of acyl-CoAs at the G3P acyltransferase step must be restricted. Thus, there are two sites for the stringent control of phospholipid synthesis in E. coli. (i) Regulation at an early step in fatty acid biosynthesis controls phospholipid production via the de novo acyl-ACP-dependent pathway. (ii) Regulation at the G3P acyltransferase step modulates phospholipid synthesis from exogenous fatty acids.

Coupled Synthesis of Membrane Proteins and Lipids

Most cells maintain a constant ratio of proteins (both whole cell and membrane) to phospholipid;

thus, neither membrane phospholipids nor proteins are overproduced or underproduced. How do most cells maintain themselves so that their membranes become neither too fat nor too thin? The mechanism responsible for this homeostasis is unknown. The simplest mechanism would be a strict coupling of the synthesis of membrane proteins and the synthesis of membrane phospholipids. Indeed, in 1969 and 1970, two groups (for reviews, see references 31, 37, and 154) reported that concomitant phospholipid synthesis was required for assembly of the lactose transporter (an integral cytoplasmic membrane protein) into the *E. coli* membrane. However, other laboratories subsequently found the transporter to be inserted normally in the absence of phospholipid synthesis and demonstrated that the previous results were due to invalid transport assays and extensive nonspecific damage to the cells. Results of McIntyre and Bell (127) showed that the lack of coupling observed with the lactose transporter was also true for bulk protein synthesis. By using mutants blocked in the first step of phospholipid synthesis, these workers found that both the inner and the outer membranes became enriched with protein. The protein:phospholipid ratio of both membranes increased approximately 60%. This work demonstrates that the membrane is not normally saturated with protein and that the synthesis of membrane phospholipid is not required for the synthesis and insertion of bulk membrane protein. The protein-enriched membranes were not lethal to the cell, and the enrichment was quickly normalized upon the restoration of phospholipid synthesis (128).

From these experiments, the question of coupling of the synthesis of membrane proteins and the synthesis of phospholipids seemed to be a dead issue. However, two recent results indicate that synthesis of discrete cytoplasmic membrane proteins can markedly increase the amount of phospholipid synthesized by *E. coli*.

Weiner et al. (217) reported that *E. coli* cells harboring a recombinant plasmid carrying the fumarate reductase genes overproduced the reductase 30-fold and the overproduced enzyme composed over 50% of the total inner membrane protein. At the same time, the amount of phospholipid increased so that the cellular lipid:protein ratio remained constant. The excess membrane was localized in novel tubular structures which branched from the cytoplasmic membrane into the cell cytoplasm. These tubules were oriented in the long axis of the cell and appeared to be composed of an aggregate of fumarate reductase and phospholipid. Despite this membrane overproduction, the phospholipid composition in these cells was quite normal, as were the levels of other membrane-bound enzymes. The synthesis of extra membranes did not begin immediately after the shift to anaerobic conditions (induction of reductase synthesis) but only after a lag. This lag was interpreted as the time needed to saturate the preexisting membrane with fumarate reductase.

In the second study, the overproduced protein was the F_1/F_0 protein-translocating ATPase of the inner membrane (213). Cloning the segment of DNA carrying the eight genes encoding the subunits of this enzyme into pBR322 resulted in a 10- to 12-fold overproduction of enzyme activity in cells harboring the recombinant plasmid. These cells grew more poorly than cells carrying only pBR322. Morphologically, the cells were strikingly similar to those overproducing fumarate reductase; again, the tubular membranes were seen (albeit not as abundantly). Cells that overproduced the enzyme to a lesser extent (fivefold) were found to contain significantly fewer tubular structures. Taken together, these studies suggest a new level of regulation of membrane biogenesis.

Both enzyme complexes are major components of the inner membrane of wild-type cells and are composed of two types of subunits:hydrophobic proteins that anchor the enzyme to the membrane and hydrophilic subunits that constitute the catalytic sites. These similarities do not seem to be accidental, and they suggest mechanisms whereby the massive production of such lipid-binding proteins triggers an increased synthesis of membrane phospholipids. These mechanisms are discussed in more detail in chapter 5, but the finding that the amount of lipid synthesized by *E. coli* can be increased twofold by overproduction of inner membrane proteins indicates that the enzymes of lipid synthesis are all synthesized in functional excess and that regulation of the pathway is not achieved by controlling gene expression of a critical enzyme. Some type of feedback inhibition of the lipid biosynthetic enzymes that would be relieved by the overproduced inner membrane protein seems plausible. For example, some phospholipases and ligases are known to be inhibited when the enzyme and phospholipids become closely packed together, and dilution of the packing relieves the inhibition (chapter 5).

It should be noted that neither the data of Weiner et al. (217) nor those of von Meyenberg et al. (213) preclude the possibility that the increased membrane production is due to the selection of a mutant strain from the transformed population by introduction of the plasmid. It is possible that the plasmid is lethal to wild-type cells but that spontaneous mutants exist in cell populations that can tolerate the overproduced enzyme owing to a propensity for membrane overproduction. Indeed, R. Simoni and co-workers (R. Simoni, personal communication) have failed to observe extra membranes in a number of strains harboring *atp* plasmids. Moreover, Cole and Guest (28) have reported that strains overproducing fumarate reductase due to massive duplication of the chromosomal locus accumulated soluble (rather than membrane-bound) fumarate reductase. In the latter case, however, methods of cellular fractionation different from those of Weiner et al. (217) were used, and no morphological studies were reported. A third case of increased membrane synthesis may be engendered by overproduction of the G3P acyltransferase protein (*plsB*) (218a).

IS PHOSPHOLIPID SYNTHESIS CELL CYCLE DEPENDENT?

It is surprising that the question of dependence of the rate of phospholipid synthesis on the stage of cell division is not answered for *E. coli*. This situation is not for the lack of experiments; nine different groups have reported experiments directed at this question. Three groups (8, 25, 146) found that lipid synthesis increased exponentially (like cell mass) during the division cycle, whereas four other groups (24, 43, 79, 156) reported an abrupt twofold increase in the rate of lipid synthesis coincident with the commencement of cross wall formation. The reasons for these differing results are not at all obvious. A variety of different radioactive labels and synchronization techniques were used. An approach that avoids potential problems of radioactive labels is use of the intrinsic density of the cell. The major components of *E. coli* (protein and RNA) have a density considerably greater than that of lipid and are made continuously during the cell cycle. Thus, an abrupt increase in phospholipid content should transiently decrease cell density. From the work of Koch and Blumberg (107), a decrease of about 1.6 mg/ml can be calculated for a twofold increase in lipid content. However, Kubitschek et al. (108) have reported buoyant density of *E. coli* to vary by less than ±0.3 mg/ml; thus, it seems likely that an abrupt increase in lipid synthesis would have been detected in these experiments. However, it is possible that cell-cycle-dependent synthesis of a more dense molecule would offset a decreased density due to a burst of lipid synthesis. The most recent report in the literature (103) suggests that the conflicting results could be due to differing bacterial strains and growth conditions. These workers examined *E. coli* strains ML30, B/r, and F and three different *E. coli* K-12 strains. They found that all five strains displayed an abrupt twofold increase in the rate of phospholipid synthesis during the cell cycle, but the timing of this rate change in relation to cell division varied greatly from strain to strain and for a given strain depending on the carbon source used. These results indicate that any simultaneity of increased phospholipid synthesis with cell division should be considered to be a coincidence rather than causally or mechanistically related. Indeed, Pierucci and Rickert (157) have reported that both chromosome replication and cell division proceed in the absence of net phospholipid synthesis. It should be noted that several experiments indicate that newly synthesized lipids are inserted into the cell envelope in a disperse manner, although the rapid migration of phospholipid molecules probably would make localization of a discrete site difficult (chapter 5).

Outlook

The framework of lipid biosynthesis in *E. coli* and (mainly by analogy) in *S. typhimurium* is now essentially complete. The genetic investigations have confirmed and extended the pathways, deduced by enzymology, in a satisfying manner. The major future questions concern the regulation of lipid synthesis, both within the pathway and in relation to other cellular components. Mutants altered in such regulatory networks seem likely to be very pleiotropic; thus, the full armament of genetic, physiological, and biochemical advantages of these bacteria will be needed.

LITERATURE CITED

1. **Akutso, H., Y. Akamatsu, T. Shimbo, K. Uehara, K. Takahashi, and T. Kyogoku.** 1980. Evidence for phase separations in the membrane of an osmotically stabilized fatty acid auxotroph of *E. coli* and its biological significance. Biochim. Biophys. Acta **598:**437–446.

2. **Alberts, A., and P. R. Vagelos.** 1972. Acyl-CoA carboxylases, p. 37–82. *In* P. D. Boyer (ed.), The enzymes, 3rd ed., vol. 6. Academic Press, Inc., New York.
3. **Ames, G. F.** 1968. Lipids of *Salmonella typhimurium* and *Escherichia coli*: structure and metabolism. J. Bacteriol. **95:**833–843.
3a.**Anderson, M. S., C. E. Balawa, and C. R. H. Raetz.** 1985. The biosynthesis of gram-negative endotoxin. Formation of lipid A precursors from UDP-GlcNAc in extracts of *Escherichia coli*. J. Biol. Chem. **260:**15536–15541.
4. **Baldassare, J. J., K. B. Rhinehart, and D. F. Silbert.** 1976. Modification of membrane lipid: physical properties in relation to fatty acid structure. Biochemistry **15:**2976–2994.
5. **Ballesta, J. P. G., C. L. de Garcia, and M. Schaechter.** 1973. Turnover of phosphatidylglycerol in *Escherichia coli*. J. Bacteriol. **116:**210–214.
6. **Barnes, E. M., Jr.** 1975. Long-chain fatty acyl thioesterases I and II from *Escherichia coli*. Methods Enzymol. **35:**102–109.
7. **Batchelor, J. G., and J. E. Cronan, Jr.** 1973. Occurrence of *cis*-7-tetradecenoic acid in the envelope phospholipids of *Escherichia coli* K-12. Biochem. Biophys. Res. Commun. **52:**1374–1381.
8. **Bauza, M. T., J. R. DeLoach, J. J. Aguanno, and A. R. Larrabee.** 1976. Acyl carrier protein prosthetic group exchange and phospholipid synthesis in synchronized cultures of a pantothenate auxotroph of *Escherichia coli*. Arch. Biochem. Biophys. **174:**344–349.
9. **Bell, R. M.** 1974. Mutants of *Escherichia coli* defective in membrane phospholipid synthesis: macromolecular synthesis in an *sn*-glycerol 3-phosphate acyltransferase K_m mutant. J. Bacteriol. **117:**1065–1076.
10. **Bell, R. M.** 1975. Mutants of *Escherichia coli* defective in membrane phospholipid synthesis. Properties of wild type and K_m defective *sn*-glycerol-3-phosphate acyltransferase activities. J. Biol. Chem. **250:**7147–7152.
11. **Bell, R. M., and J. E. Cronan, Jr.** 1975. Mutants of *Escherichia coli* defective in membrane phospholipid synthesis. Phenotypic suppression of *sn*-glycerol-3-phosphate acyltransferase K_m mutants by loss of feedback inhibition of the biosynthetic *sn*-glycerol 3-phosphate dehydrogenase. J. Biol. Chem. **250:**7153–7158.
12. **Bell, R. M., R. D. Mavis, M. J. Osborn, and P. R. Vagelos.** 1971. Enzymes of phospholipid metabolism: localization in the cytoplasmic and outer membrane of the cell envelope of *Escherichia coli* and *Salmonella typhimurium*. Biochim. Biophys. Acta **249:**628–635.
13. **Birge, C. H., D. F. Silbert, and P. R. Vagelos.** 1967. A β-hydroxydecanoyl-ACP dehydrase specific for saturated fatty acid synthesis in *E. coli*. Biochem. Biophys. Res. Commun. **29:**808–814.
14. **Birge, C. H., and P. R., Vagelos.** 1972. Acyl carrier protein XVII: purification and properties of β-hydroxyacyl-acyl carrier protein dehydratase. J. Biol. Chem. **247:**4930–4938.
15. **Bloch, K.** 1970. β-Hydroxydecanoyl thioester dehydrase, p. 441–464. *In* P. D. Boyer (ed.), The enzymes, 3rd ed., vol. 5. Academic Press, Inc., New York.
16. **Bloch, K., and D. Vance.** 1977. Control mechanisms in the synthesis of saturated fatty acids. Annu. Rev. Biochem. **46:**236–298.
17. **Braun, V.** 1975. Covalent lipoprotein from the outer membrane of *Escherichia coli*. Biochem. Biophys. Acta **415:**335–377.
18. **Bulawa, C. E., B. R. Ganong, C. P. Sparrow, and C. R. H. Raetz.** 1981. Enzymatic sorting of bacteria colonies on filter paper replicas: detection of labile activities. J. Bacteriol. **148:**391–393.
19. **Bulawa, C. E., J. D. Hermes, and C. R. H. Raetz.** 1983. Chloroform-soluble nucleotides in *Escherichia coli*. Role of CDP-diglyceride in the enzymatic cytidylation of phosphomonoester acceptors. J. Biol. Chem. **258:**14974–14980.
20. **Bulawa, C. E., and C. R. H. Raetz.** 1984. Isolation and characterization of *Escherichia coli* strains defective in CD-diglyceride hydrolyase. J. Biol. Chem. **259:**11257–11264.
21. **Buttke, T. M., and L. O. Ingram.** 1978. Mechanism of ethanol-induced changes in lipid composition of *Escherichia coli*: inhibition of saturated fatty acid synthesis *in vivo*. Biochemistry **17:**637–644.
22. **Buttke, T. M., and L. O. Ingram.** 1978. Inhibition of unsaturated fatty acid synthesis in *Escherichia coli* by the antibiotic cerulenin. Biochemistry **17:**5282–5286.
23. **Buttke, T. M., and L. O. Ingram.** 1980. Ethanol-induced changes in lipid composition of *Escherichia coli*: inhibition of saturated fatty acid synthesis *in vitro*. Arch. Biochem. Biophys. **203:**565–571.
24. **Carty, C. E., and L. O. Ingram.** 1981. Lipid synthesis during the *Escherichia coli* cell cycle. J. Bacteriol. **145:**472–478.

494   CRONAN AND ROCK

25. **Churchward, G. G., and J. B. Holland.** 1976. Envelope synthesis during the cell cycle in *Escherichia coli* B/r. J. Mol. Biol. **105**:245–261.
26. **Clark, D. P., and J. E. Cronan, Jr.** 1981. Bacterial mutants for the study of lipid metabolism. Methods Enzymol. **72**:693–707.
27. **Clark, D. P., D. de Mendoza, M. P. Polacco, and J. E. Cronan, Jr.** 1983. β-Hydroxydecanoyl thioester dehydrase does not catalyse a rate-limiting step in *Escherichia coli* unsaturated fatty acid synthesis. Biochemistry **22**:5897–5902.
28. **Cole, S. T., and J. R. Guest.** 1979. Production of a soluble form of fumarate reductase by multiple gene duplication in *Escherichia coli* K-12. Eur. J. Biochem. **102**:65–71.
29. **Cooper, C. L., and D. R. Lueking.** 1984. Localization and characterization of the *sn*-glycerol 3-phosphate acyltransferase in *Rhodopseudomonas sphaeroides*. J. Lipid Res. **25**:1222–1232.
30. **Cronan, J. E., Jr.** 1968. Phospholipid alterations during growth of *Escherichia coli*. J. Bacteriol. **95**:2054–2061.
31. **Cronan, J. E., Jr.** 1978. Molecular biology of bacterial membrane lipids. Annu. Rev. Biochem. **47**:163–189.
32. **Cronan, J. E., Jr.** 1979. Phospholipid synthesis and assembly, p. 35–65. *In* M. Inouye (ed.), The bacterial outer membrane. John Wiley & Sons, Inc., New York.
33. **Cronan, J. E., Jr.** 1984. Evidence that incorporation of exogenous fatty acids into the phospholipids of *Escherichia coli* does not require acyl carrier protein. J. Bacteriol. **159**:773–775.
34. **Cronan, J. E., Jr., and R. M. Bell.** 1974. Mutants of *Escherichia coli* defective in membrane phospholipid synthesis: mapping of the structural gene for L-glycerol 3-phosphate dehydrogenase. J. Bacteriol. **118**:598–605.
35. **Cronan, J. E., Jr., C. H. Birge, and P. R. Vagelos.** 1969. Evidence for two genes specifically involved in unsaturated fatty acid biosynthesis in *Escherichia coli*. J. Bacteriol. **100**:601–604.
36. **Cronan, J. E., Jr., and E. P. Gelmann.** 1973. An estimate of the minimum amount of unsaturated fatty acid required for growth of *Escherichia coli*. J. Biol. Chem. **248**:1188–1195.
37. **Cronan, J. E., Jr., and E. P. Gelmann.** 1975. Physical properties of membrane lipids: biological relevance and regulation. Bacteriol. Rev. **39**:232–256.
38. **Cronan, J. E., Jr., and J. H. Prestegard.** 1977. Difference decoupling nuclear magnetic resonance: a method to study the exchange of fatty acids between phospholipid molecules. Biochemistry **16**:4738–4742.
39. **Cronan, J. E., Jr., and P. R. Vagelos.** 1972. Metabolism and function of the membrane phospholipids of *Escherichia coli*. Biochim. Biophys. Acta **265**:25–60.
40. **Cronan, J. E., Jr., L. J. Weisberg, and R. G. Allen.** 1975. Regulation of membrane lipid synthesis in *Escherichia coli*: accumulation of free fatty acid of abnormal length during inhibition of phospholipid synthesis. J. Biol. Chem. **250**:5835–5840.
41. **D'Agnolo, G., I. S. Rosenfeld, J. Awaya, S. Omura, and P. R. Vagelos.** 1973. Inhibition of fatty acid synthesis by the antibiotic cerulenin: specific inactivation of β-keto-acyl-acyl carrier protein synthetase. Biochim. Biophys. Acta **326**:155–166.
42. **D'Agnolo, G. D., I. S. Rosenfeld, and P. R. Vagelos.** 1975. Multiple forms of β-ketoacyl-acyl-carrier-protein synthetase in *Escherichia coli*. J. Biol. Chem. **250**:5289–5294.
43. **Daniels, M. J.** 1969. Lipid synthesis in relation to the cell cycle of *Bacillus megaterium* KM and *Escherichia coli*. Biochem. J. **115**:697–701.
44. **de Geus, P., I. van Dee, H. Bergman, J. Tommassen, and G. de Haas.** 1983. Molecular cloning of *pldA*, the structural gene for outer membrane phospholipase of *E. coli* K-12. Mol. Gen. Genet. **190**:150–155.
45. **de Geus, P., H. M. Verkeij, N. H. Riegman, W. P. M. Hoekstra, and G. H. de Haas.** 1984. The pro- and mature forms of *E. coli* K-12 outer membrane phospholipase A are identical. EMBO J. **3**:1799–1802.
46. **de Mendoza, D., and J. E. Cronan, Jr.** 1983. Thermal regulation of membrane lipid fluidity in bacteria. Trends Biochem. Sci. **8**:49–52.
47. **de Mendoza, D., J. L. Garwin, and J. E. Cronan, Jr.** 1982. Overproduction of *cis*-vaccenic acid and altered temperature control of fatty acid synthesis in a mutant of *Escherichia coli*. J. Bacteriol. **151**:1608–1611.
48. **de Mendoza, D., A. K. Ulrich, and J. E. Cronan, Jr.** 1983. Thermal regulation of membrane fluidity in *Escherichia coli*: effects of overproduction of β-keto-acyl-acyl-carrier protein synthase I. J. Biol. Chem. **258**:2098–2101.
49. **Donohue-Rolfe, A. M., and M. Schaechter.** 1980. Translocation of phospholipids from the inner to the outer membrane of *Escherichia coli*. Proc. Natl. Acad. Sci. USA **77**:1867–1871.
50. **Edgar, J. R., and R. M. Bell.** 1978. Biosynthesis in *Escherichia*
coli of *sn*-glycerol-3-phosphate, a precursor of phospholipid. Kinetic characterization of wild type and feedback-resistant forms of the biosynthetic *sn*-glycerol-3-phosphate dehydrogenase. J. Biol. Chem. **253**:6354–6363.
51. **Elovson, J., and P. R. Vagelos.** 1968. Acyl carrier protein X. Acyl carrier protein synthetase. J. Biol. Chem. **243**:3603–3611.
52. **Esmon, B. E., C. R. Kensil, C.-H. C. Cheng, and M. Glaser.** 1980. Genetic analysis of *Escherichia coli* mutants defective in adenylate kinase and *sn*-glycerol 3-phosphate acyltransferase. J. Bacteriol. **141**:405–408.
53. **Fall, R. R.** 1979. Analysis of microbial biotin proteins. Methods Enzymol. **62**:390–398.
54. **Fall, R. R., A. W. Alberts, and P. R. Vagelos.** 1975. Analysis of bacterial biotin-proteins. Biochim. Biophys. Acta **379**:496–503.
55. **Frerman, F. E., and W. Bennett.** 1973. Studies on the uptake of fatty acids by *Escherichia coli*. Arch. Biochem. Biophys. **159**:434–443.
56. **Ganong, B. R., J. M. Leonard, and C. R. H. Raetz.** 1980. Phosphatidic acid accumulation in the membranes of *Escherichia coli* mutants defective in CDP-diglyceride synthetase. J. Biol. Chem. **255**:1623–1629.
57. **Ganong, B. R., and C. R. H. Raetz.** 1983. Massive accumulation of phosphatidic acid in conditionally lethal CDP-diglyceride synthetase mutants and cytidine auxotrophs of *Escherichia coli*. J. Biol. Chem. **257**:389–394.
58. **Ganong, B. R., and C. R. H. Raetz.** 1983. pH-sensitive CDP-diglyceride synthetase mutants of *Escherichia coli*: phenotypic suppression by mutations at a second site. J. Bacteriol. **153**:731–738.
59. **Garwin, J. L., and J. E. Cronan, Jr.** 1980. Thermal modulation of fatty acid synthesis in *Escherichia coli* does not involve de novo synthesis. J. Bacteriol. **141**:1457–1459.
60. **Garwin, J. L., A. L. Klages, and J. E. Cronan, Jr.** 1980. β-Ketoacyl-acyl-carrier protein synthase II of *Escherichia coli*: evidence for function in the thermal regulation of fatty acid synthesis. J. Biol. Chem. **255**:3263–3265.
61. **Garwin, J. L., A. L. Klages, and J. E. Cronan, Jr.** 1980. Structural, enzymatic, and genetic studies of β-ketoacyl-acyl carrier protein synthases I and II of *Escherichia coli*. J. Biol. Chem. **255**:11949–11956.
62. **Gelmann, E. P., and J. E. Cronan, Jr.** 1972. Mutant of *Escherichia coli* deficient in the synthesis of *cis*-vaccenic acid. J. Bacteriol. **112**:381–387.
63. **Glaser, G. P., D. Sarmentor, and M. Cashel.** 1983. Functional interrelationship between two tandem *E. coli* ribosomal RNA promoters. Nature (London) **302**:74–76.
64. **Glaser, M., W. Nulty, and P. R. Vagelos.** 1975. Role of adenylate kinase in the regulation of macromolecular biosynthesis in a putative mutant of *Escherichia coli* defective in membrane phospholipid biosynthesis. J. Bacteriol. **123**:128–136.
65. **Goelz, S. E., and J. E. Cronan, Jr.** 1980. The positional distribution of fatty acids in the phospholipids of *Escherichia coli* is not regulated by *sn*-glycerol 3-phosphate levels. J. Bacteriol. **144**:462–464.
66. **Goelz, S. E., and J. E. Cronan, Jr.** 1981. Adenylate kinase of *Escherichia coli*. Evidence for a functional interaction in phospholipid synthesis. Biochemistry **21**:189–195.
67. **Goldfine, H.** 1972. Comparative aspects of bacterial lipids. Adv. Microb. Physiol. **8**:1–58.
68. **Goldfine, H.** 1979. Why bacteria may not tightly regulate the synthesis of fatty acids in response to exogenous fatty acids, p. 14–16. *In* D. Schlessinger (ed.), Microbiology—1979. American Society for Microbiology, Washington, D.C.
69. **Goldfine, H.** 1982. Lipids of prokaryotes—structure and distribution. Curr. Top. Membr. Transp. **17**:2–44.
70. **Goldfine, H., and G. P. Ailhaud.** 1971. Fatty acyl-acyl carrier protein and fatty acyl CoA as acyl donors in the biosynthesis of phosphatidic acid in *clostridium butyricum*. Biochem. Biophys. Res. Commun. **45**:1127–1133.
71. **Gottesman, S.** 1984. Bacterial regulation: global regulatory networks. Annu. Rev. Biochem. **18**:415–441.
72. **Green, P. R., A. H. Merrill, Jr., and R. M. Bell.** 1981. Membrane phospholipid synthesis in *Escherichia coli*. Purification, reconstitution and characterization of *sn*-glycerol-3-phosphate acyltransferase. J. Biol. Chem. **256**:11151–11159.
73. **Greenaway, D. L. A., and D. F. Silbert.** 1982. Altered acyltransferase activity in *Escherichia coli* associated with mutations in acyl-CoA synthetase. J. Biol. Chem. **258**:13034–13042.
74. **Greenspan, M. D., C. H. Birge, G. Powell, W. S. Hancock, and P. R. Vagelos.** 1970. Enzyme specificity as a factor in regulation of fatty acid chain length in *Escherichia coli*. Science **170**:1203–1204.

75. **Grogan, D. W., and J. E. Cronan, Jr.** 1984. Genetic characterization of the *Escherichia coli* cyclopropane fatty acid (*cfa*) locus and neighboring loci. Mol. Gen. Genet. **196:**367–372.

76. **Grogan, D. W., and J. E. Cronan, Jr.** 1984. Cloning and manipulation of the fatty acid synthase gene of *Escherichia coli*: physiological aspects of enzyme overproduction. J. Bacteriol. **158:**286–295.

77. **Grogan, D. W., and J. E. Cronan, Jr.** 1986. Characterization of *Escherichia coli* mutants completely defective in synthesis of cyclopropane fatty acids. J. Bacteriol. **166:**872–877.

78. **Guise, N., S. Michelson, and O. Barzu.** 1984. Inactivation and proteolysis of heat-sensitive adenylate kinase of *Escherichia coli* CR 341 T28. J. Biol. Chem. **259:**8713–8717.

79. **Hackenbeck, R., and W. Messer.** 1977. Oscillations in the synthesis of cell wall components in synchronized cultures of *Escherichia coli*. J. Bacteriol. **129:**1234–1238.

80. **Hammelburger, J. W., and G. A. Orr.** 1983. Interaction of *sn*-glycerol 3-phosphorothioate with *Escherichia coli*: effect on cell growth and metabolism. J. Bacteriol. **156:**789–799.

81. **Harder, M. E., I. R. Beacham, J. E. Cronan, Jr., K. Beacham, K. L. Honegger, and D. F. Silbert.** 1972. Temperature-sensitive mutants of *Escherichia coli* requiring saturated and unsaturated fatty acids for growth: isolation and properties. Proc. Natl. Acad. Sci. USA **69:**3105–3109.

82. **Harder, M. E., R. C. Ladenson, S. D. Schimmel, and D. F. Silbert.** 1974. Mutants of *Escherichia coli* with temperature-sensitive malonyl coenzyme A-acyl carrier protein transacylase. J. Biol. Chem. **349:**7468–7475.

83. **Hawrot, E., and E. P. Kennedy.** 1978. Phospholipid composition and membrane function in phosphatidylserine decarboxylase mutants of *Escherichia coli*. J. Biol. Chem. **253:**8213–8220.

84. **Hayashi, T., O. Yamamoto, H. Sasaki, A. Kawaguski, and H. Okazaki.** 1983. Mechanism of action of the antibiotic thiolactomycin: inhibition of fatty acid synthesis of *Escherichia coli*. Biochem. Biophys. Res. Commun. **115:**1108–1113.

85. **Hayashi, T., O. Yamamoto, H. Sasaki, and H. Okazaki.** 1984. Inhibition of fatty acid synthesis by the antibiotic thiolactomycin. J. Antibiot. **37:**1456–1461.

86. **Hirschberg, C. B., and E. P. Kennedy.** 1972. Mechanism of the enzymatic synthesis of cardiolipin in *Escherichia coli*. Proc. Natl. Acad. Sci. USA **69:**648–651.

87. **Homma, H., N. Chiba, T. Kobayashi, I. Kudo, K. Inoue, H. Ikeda, M. Sekiguchi, and S. Nojima.** 1984. Characteristics of detergent-resistant phospholipase A overproduced in *E. coli* cells bearing its cloned structural gene. J. Biochem. **96:**1645–1654.

88. **Homma, H., T. Kobayashi, N. Chiba, K. Kansawa, H. Mizushima, I. Kudo, K. Inoue, H. Ikeda, M. Sekiguchi, and S. Nojima.** 1984. The DNA sequence encoding the *pldA* gene, the structural gene for detergent-resistant phospholipase A. J. Biochem. **96:**1655–1664.

89. **Homma, H., M. Nishijima, T. Kobayashi, H. Okuyama, and S. Nojima.** 1981. Incorporation and metabolism of 2-acyl lysophospholipids by *Escherichia coli*. Biochim. Biophys. Acta **663:**1–13.

90. **Homma, H., and S. Nojima.** 1982. Transacylation between diacylphospholipids and 2-acyl lysophospholipids catalyzed by *Escherichia coli* extract. J. Biochem. **91:**1093–1101.

91. **Homma, H., and S. Nojima.** 1982. Synthesis of various phospholipids from 2-acyl lysophospholipids by *Escherichia coli* extract. J. Biochem. **91:**1103–1110.

92. **Hong, J. S., and B. N. Ames.** 1971. Local mutagenesis of any specific small region of the bacterial chromosome. Proc. Natl. Acad. Sci. USA **68:**3158–3162.

93. **Huss, R. J., and M. Glaser.** 1983. Identification and purification of an adenylate kinase-associated protein that influences the thermolability of adenylate kinase from a temperature-sensitive *adk* mutant of *Escherichia coli*. J. Biol. Chem. **258:**13370–13376.

94. **Hwang, Y. W., R. Engel, and B. E. Tropp.** 1984. Correlation of 3,4-dihydroxybutyl 1-phosphonate resistance with a defect in cardiolipin synthesis in *Escherichia coli*. J. Bacteriol. **157:**846–856.

95. **Icho, T., and C. R. H. Raetz.** 1983. Multiple genes for membrane-bound phosphatases in *Escherichia coli* and their action on phospholipid precursors. J. Bacteriol. **153:**722–730.

96. **Ingram, L. O., and T. M. Buttke.** 1984. Effects of alcohols on microorganisms. Adv. Microb. Physiol. **25:**26–47.

97. **Ingram, L. O., B. F. Dickens, and T. M. Buttke.** 1980. Reversible effects of ethanol on *E. coli*. Adv. Exp. Med. Biol. **126:**299–337.

98. **Jackowski, S., H. H. Edwards, D. Davis, and C. O. Rock.** 1985. Localization of acyl carrier protein in *Escherichia coli*. J. Bacteriol. **162:**5–8.

99. **Jackowski, S., and C. O. Rock.** 1983. Ratio of active to inactive forms of acyl carrier protein in *Escherichia coli*. J. Biol. Chem. **258:**16186–16191.

100. **Jackowski, S., and C. O. Rock.** 1984. Turnover of the 4′-phosphopantetheine prosthetic group of acyl carrier protein. J. Biol. Chem. **259:**1891–1895.

101. **Jackowski, S., and C. O. Rock.** 1986. Transfer of fatty acids from the 1-position of phosphatidylethanolamine to the major outer membrane protein of *Escherichia coli*. J. Biol. Chem. **261:**11328–11333.

102. **Jackson, M. B., and J. E. Cronan, Jr.** 1978. An estimate of the minimum amount of fluid lipid required for the growth of *Escherichia coli*. Biochim. Biophys. Acta **512:**472–479.

103. **Joseleau-Petit, D., F. Kepes, and A. Kepes.** 1984. Cyclic changes of the rate of phospholipid synthesis during synchronous growth of *Escherichia coli*. Eur. J. Biochem. **139:**605–610.

104. **Kito, M., and L. I. Pizer.** 1969. Purification and regulatory properties of the biosynthetic L-glycerol-3-phosphate dehydrogenase from *Escherichia coli*. J. Biol. Chem. **244:**3316–3323.

105. **Klein, K., R. Steinberg, B. Fiethen, and P. Overath.** 1971. Fatty acid degradation in *Escherichia coli*. An inducible system for the uptake of fatty acids and further characterization of *old* mutants. Eur. J. Biochem. **19:**442–450.

106. **Kobayashi, T., H. Homma, Y. Natori, I. Kudo, I. Inoue, and S. Nojima.** 1984. Isolation of two kinds of *E. coli* K-12 mutants for lysophospholipase L2: one with an elevated level of the enzyme and the other defective in it. J. Biochem. **96:**137–145.

107. **Koch, A. L., and G. Blumberg.** 1976. Distribution of bacteria in the velocity gradient centrifuge. Biophys. J. **16:**389–405.

108. **Kubitschek, H. E., W. W. Baldwin, and R. Graetzer.** 1983. Buoyant density constancy during the cell cycle of *Escherichia coli*. J. Bacteriol. **155:**1027–1032.

109. **Larson, T. J., V. A. Lightner, P. R. Green, P. Modrich, and R. M. Bell.** 1980. Membrane phospholipid synthesis in *Escherichia coli*. Identification of the *sn*-glycerol-3-phosphate acyltransferase as the *plsB* gene product. J. Biol. Chem. **255:**9421–9426.

110. **Larson, T. J., D. N. Ludtke, and R. M. Bell.** 1984. *sn*-Glycerol-3-phosphate auxotrophy of *plsB* strains of *Escherichia coli*: evidence that a second mutation, *plsX*, is required. J. Bacteriol. **160:**711–717.

111. **Law, J., H. Zalkin, and T. Kaneshiro.** 1963. Transmethylation reactions in bacterial lipids. Biochim. Biophys. Acta **70:**143–151.

112. **Law, J. H.** 1971. Biosynthesis of cyclopropane rings. Acc. Chem. Res. **4:**199–203.

113. **Lee, P. C., B. R. Bochner, and B. N. Ames.** 1983. ApppppA, heat-shock stress, and cell oxidation. Proc. Natl. Acad. Sci. USA **80:**7496–7500.

114. **Lightner, V. A., R. M. Bell, and P. Modrich.** 1983. The DNA sequences encoding *plsB* and *dgk* loci of *Escherichia coli*. J. Biol. Chem. **258:**10856–10861.

115. **Lightner, V. A., T. J. Larson, P. Tailleur, G. D. Kantor, C. R. H. Raetz, R. M. Bell, and P. Modrich.** 1980. Membrane phospholipid synthesis in *Escherichia coli*. Cloning of a structural gene (*plsB*) of the *sn*-glycerol-3-phosphate acyltransferase. J. Biol. Chem. **255:**9413–9420.

116. **Lin, E. C. C.** 1976. Glycerol dissimilation and its regulation of bacteria. Annu. Rev. Microbiol. **30:**535–578.

117. **Louie, K., and W. Dowhan.** 1980. Investigation on the association of phosphatidylserine synthase with the ribosomal component of *Escherichia coli*. J. Biol. Chem. **255:**1124–1127.

118. **Luderitz, O., M. A. Frudenberg, C. Galanos, V. Lehmann, E. T. Rietschel, and D. H. Shaw.** 1982. Lipopolysaccharides of gram-negative bacteria. Curr. Top. Membr. Transp. **17:**79–153.

119. **Lueking, D. R., and H. Goldfine.** 1975. The involvement of guanosine 5′-diphosphate-3′-diphosphate in the regulation of phospholipid biosynthesis in *Escherichia coli*: lack of ppGpp inhibition of acyltransfer from acyl-ACP to *sn*-glycerol 3-phosphate. J. Biol. Chem. **250:**4911–4917.

120. **Lugtenberg, E. J. J., and R. Peters.** 1976. Distribution of lipids in cytoplasmic and outer membranes of *Escherichia coli* K-12. Biochim. Biophys. Acta **441:**38–47.

121. **Luzon, C., and J. P. G. Ballesta.** 1976. Metabolism of phosphatidylglycerol in cell-free extracts of *Escherichia coli*. Eur. J. Biochem. **65:**207–212.

122. **Maloy, S. R., M. Bohlander, and W. D. Nunn.** 1980. Elevated levels of glyoxylate shunt enzymes in *Escherichia coli* strains constitutive for fatty acid degradation. J. Bacteriol. **143:**720–725.

123. **Maloy, S. R., and W. D. Nunn.** 1981. Role of gene *fadR* in *Escherichia coli* acetate metabolism. J. Bacteriol. **148:**83–90.

124. **Marianari, L. A., H. Goldfine, and C. Panos.** 1974. Specificity of cyclopropane fatty acid synthesis in *E. coli*: utilization of isomers of monounsaturated fatty acids. Biochemistry **13:**1978–1983.

125. **Marr, A. G., and J. L. Ingraham.** 1962. Effect of temperature on the composition of fatty acids in *Escherichia coli.* J. Bacteriol. **84:**1260–1267.
126. **McElhaney, R. N.** 1982. Effects of membrane lipids on transport and enzymic activities. Curr. Top. Membr. Transp. **17:**317–380.
127. **McIntyre, T. M., and R. M. Bell.** 1975. Mutants of *Escherichia coli* defective in membrane phospholipid synthesis. Effect of cessation of net phospholipid synthesis on cytoplasmic and outer membranes. J. Biol. Chem. **250:**9053–9059.
128. **McIntyre, T. M., B. K. Chamberlain, R. E. Webster, and R. M. Bell.** 1977. Mutants of *Escherichia coli* defective in membrane phospholipid synthesis. Effects of cessation and reinitiation of phospholipid synthesis on macromolecular synthesis and phospholipid turnover. J. Biol. Chem. **252:**4487–4493.
129. **Melchior, D. L.** 1982. Lipid phase transitions and regulation of membrane fluidity in prokaryotes. Curr. Top. Membr. Transp. **17:**263–316.
130. **Merlie, J. P., and L. I. Pizer.** 1973. Regulation of phospholipid synthesis in *Escherichia coli* by guanosine tetraphosphate. J. Bacteriol. **116:**355–366.
131. **Mindich, L.** 1972. Control of fatty acid synthesis in bacteria. J. Bacteriol. **110:**96–102.
132. **Miyazaki, C., M. Kuroda, A. Ohta, and I. Shibuya.** 1985. Genetic manipulation of membrane phospholipid composition in *Escherichia coli: pgsA* mutants defective in phosphatidylglycerol synthesis. Proc. Natl. Acad. Sci. USA **82:**7530–7534.
133. **Muller, E. D., and J. E. Cronan, Jr.** 1983. The lipid-containing bacteriophage PR4. Effects of altered lipid composition on the virion. J. Mol. Biol. **165:**109–124.
134. **Narasimhan, M. L., J. L. Lampi, and J. E. Cronan, Jr.** 1986. Genetic and biochemical characterization of an *Escherichia coli* K-12 mutant deficient in acyl-coenzyme A thioesterase II. J. Bacteriol. **165:**911–917.
135. **Neidhardt, F. C., R. A. Van Bogelen, and V. Vaughn.** 1984. The genetics and regulation of heat-shock proteins. Annu. Rev. Genet. **18:**295–329.
136. **Neidhardt, F. C., V. Vaughn, T. A. Phillips, and P. L. Block.** 1983. Gene-protein index of *Escherichia coli* K-12. Microbiol. Rev. **47:**231–284.
137. **Nimmo, H. G.** 1984. Control of *Escherichia coli* isocitrate dehydrogenase: an example of protein phosphorylation in a prokaryote. Trends Biochem. Sci. **9:**475–478.
137a. **Nishijima, M., C. H. Bulawa, and C. R. H. Raetz.** 1981. Two interacting mutations causing temperature-sensitive phosphatidylglycerol synthesis in *Escherichia coli* membranes. J. Bacteriol. **145:**113–121.
138. **Nishijima, M., and C. R. H. Raetz.** 1979. Membrane lipid biogenesis in *Escherichia coli:* identification of genetic loci for phosphatidylglycerophosphate synthetase and construction of mutants lacking phosphatidylglycerol. J. Biol. Chem. **254:**7837–7844.
139. **Nishijima, M., T. Sa-eki, Y. Tamori, O. Doi, and S. Nojima.** 1978. Synthesis of acyl phosphatidylglycerol from phosphatidylglycerol in *Escherichia coli* K-12: evidence for the participation of detergent-resistant phospholipase A and heat-labile membrane-bound factors. Biochem. Biophys. Acta **528:**107–108.
140. **Nunn, W. D., and J. E. Cronan, Jr.** 1974. Unsaturated fatty acid synthesis is not required for induction of lactose transport in *Escherichia coli.* J. Biol. Chem. **249:**724–730.
141. **Nunn, W. D., and J. E. Cronan, Jr.** 1974. *rel* gene control of lipid synthesis in *Escherichia coli:* evidence for eliminating fatty acid synthesis as the sole regulatory site. J. Biol. Chem. **249:**3994–3996.
142. **Nunn, W. D., and J. E. Cronan, Jr.** 1976. Evidence for a direct effect on fatty acid synthesis in *relA* gene control of membrane phospholipid synthesis. J. Mol. Biol. **102:**167–172.
143. **Nunn, W. D., K. Griffin, D. Clark, and J. E. Cronan, Jr.** 1983. A role for the *fadR* gene in unsaturated fatty acid biosynthesis in *Escherichia coli.* J. Bacteriol. **154:**554–560.
144. **Nunn, W. D., D. L. Kelley, and N. Y. Stumfall.** 1977. Regulation of fatty acid synthesis during glycerol starvation of glycerol auxotrophs of *Escherichia coli.* J. Bacteriol. **132:**526–531.
145. **O'Farrell, P. H.** 1975. High resolution two-dimensional electrophoresis of proteins. J. Biol. Chem. **250:**4007–4021.
146. **Ohki, M.** 1972. Correlation between metabolism of phosphatidylglycerol and membrane synthesis in *Escherichia coli.* J. Mol. Biol. **68:**249–264.
147. **Ohta, A., T. Obara, Y. Asami, and I. Shibuya.** 1985. Molecular cloning of the *cls* gene responsible for cardiolipin synthesis in *Escherichia coli* and phenotypic consequences of its amplification. J. Bacteriol. **163:**506–514.
148. **Ohta, A., and I. Shibuya.** 1977. Membrane phospholipid synthe-

149. **Ohta, A., K. Waggoner, K. Louie, and W. Dowhan.** 1981. Cloning of genes involved in membrane lipid synthesis: effects of overproduction of phosphatidylserine synthase in *Escherichia coli.* J. Biol. Chem. **256:**2219–2225.
150. **Omura, S.** 1981. Cerulenin. Methods Enzymol. **72:**520–532.
151. **Orr, G. A., J. W. Hammelburger, and G. Henry.** 1983. Interaction of *sn*-glycerol 3-phosphothioate with *Escherichia coli:* in vitro and in vivo incorporation into phospholipids. J. Biol. Chem. **258:**9237–9244.
152. **Overath, P., G. Pauli, and H. U. Schairer.** 1969. Fatty acid degradation in *Escherichia coli:* an inducible acyl-CoA synthetase, the mapping of *old*-mutations, and the isolation of regulatory mutants. Eur. J. Biochem. **7:**559–574.
153. **Overath, P., E.-M. Raufuss, W. Stoffel, and W. Ecker.** 1967. The induction of the enzymes of fatty acid degradation in *Escherichia coli.* Biochem. Biophys. Res. Commun. **29:**28–33.
154. **Overath, P., and L. Thilo.** 1978. Structural and functional aspects of biological membranes revealed by lipid phase transitions. Int. Rev. Biochem. **19:**1–44.
155. **Pieringer, R. A.** 1983. Formation of bacterial glycolipids, p. 255–306. *In* P. D. Boyer (ed.), The enzymes, 3rd ed., vol. 16. Academic Press, Inc., New York.
156. **Pierucci, O.** 1979. Phospholipid synthesis during the cell division cycle of *Escherichia coli.* J. Bacteriol. **138:**453–460.
157. **Pierucci, O., and M. Rickert.** 1985. Duplication of *Escherichia coli* during inhibition of net phospholipid synthesis. J. Bacteriol. **162:**374–382.
158. **Pluschke, G., Y. Hirota, and P. Overath.** 1978. Function of phospholipids in *Escherichia coli:* characterization of a mutant deficient in cardiolipin synthesis. J. Biol. Chem. **253:**5048–5055.
159. **Pluschke, G., and P. Overath.** 1981. Function of phospholipids in *Escherichia coli.* Influence of changes in polar head group composition on the lipid phase transition and characterization of a mutant containing only saturated phospholipid acyl chains. J. Biol. Chem. **256:**3207–3212.
160. **Polacco, M. L., and J. E. Cronan, Jr.** 1977. Mechanisms of the apparent regulation of *Escherichia coli* unsaturated fatty acid synthesis by exogenous oleic acid. J. Biol. Chem. **252:**5488–5490.
161. **Polacco, M. L., and J. E. Cronan, Jr.** 1981. A mutant of *Escherichia coli* conditionally defective in the synthesis of holo-[acyl carrier protein]. J. Biol. Chem. **256:**5750–5754.
162. **Powell, G. L., M. Bauza, and A. R. Larabee.** 1973. The stability of acyl carrier protein in *Escherichia coli.* J. Biol. Chem. **248:**4461–4466.
163. **Powell, G. L., J. Elovson, and P. R. Vagelos.** 1969. Acyl carrier protein. XII. Synthesis and turnover of the prosthetic group of acyl carrier protein *in vivo.* J. Biol. Chem. **244:**5616–5624.
164. **Prescott, D. J., and P. R. Vagelos.** 1972. Acyl carrier protein. Adv. Enzymol. Relat. Areas Mol. Biol. **36:**269–311.
165. **Proulx, P. R., and L. L. M. van Deenen.** 1966. Acylation of lysophosphoglycerides by *Escherichia coli.* Biochim. Biophys. Acta **125:**591–593.
166. **Raetz, C. R. H.** 1978. Enzymology, genetics, and regulation of membrane phospholipid synthesis in *Escherichia coli.* Microbiol. Rev. **42:**614–659.
167. **Raetz, C. R. H.** 1982. Genetic control of phospholipid bilayer assembly, p. 435–477. *In* J. N. Hawthorne and G. B. Ansell (ed.), Phospholipids. Elsevier/North-Holland Publishing Co., Amsterdam.
168. **Raetz, C. R. H., and J. Foulds.** 1977. Envelope composition and antibiotic hypersensitivity of *Escherichia coli* mutants defective in phosphatidylserine synthetase. J. Biol. Chem. **252:**5911–5915.
169. **Raetz, C. R. H., G. D. Kantor, M. Nishijima, and M. L. Jones.** 1981. Isolation of *Escherichia coli* mutants with elevated levels of membrane enzymes. A *trans*-acting mutation controlling diglyceride kinase. J. Biol. Chem. **256:**2109–2122.
170. **Raetz, C. R. H., and E. P. Kennedy.** 1974. Partial purification and properties of phosphatidylserine synthetase from *Escherichia coli.* J. Biol. Chem. **249:**5038–5045.
171. **Ray, T. K., and J. E. Cronan, Jr.** 1975. Acylation of *sn*-glycerol-3-phosphate in *Escherichia coli:* study of reaction with native palmitoyl-acyl carrier protein. J. Biol. Chem. **250:**8422–8427.
172. **Ray, T. K., and J. E. Cronan, Jr.** 1976. Mechanism of phospholipid biosynthesis in *Escherichia coli:* acyl-CoA synthetase is not required for the incorporation of intracellular free fatty acid into phospholipid. Biochem. Biophys. Res. Commun. **69:**506–513.
173. **Ray, T. K., and J. E. Cronan, Jr.** 1976. Activation of long chain fatty acids with acyl carrier protein: demonstration of a new enzyme, acyl-acyl carrier protein synthetase in *Escherichia coli.* Proc. Natl. Acad. Sci. USA **73:**4374–4378.

sis and phenotypic correction of an *Escherichia coli pss* mutant. J. Bacteriol. **132:**434–443.

174. **Rock, C. O.** 1984. Turnover of fatty acids in the 1-position of phosphatidylethanolamine in *Escherichia coli*. J. Biol. Chem. **259:**6188–6194.

175. **Rock, C. O., and J. E. Cronan, Jr.** 1979. Re-evaluation of the solution structure of acyl carrier protein. J. Biol. Chem. **254:**9778–9785.

176. **Rock, C. O., and J. E. Cronan, Jr.** 1979. Solubilization, purification, and salt activation of acyl-acyl carrier protein synthetase from *Escherichia coli*. J. Biol. Chem. **254:**7116–7122.

177. **Rock, C. O., and J. E. Cronan, Jr.** 1981. Acyl carrier protein from *Escherichia coli*. Methods Enzymol. **71:**341–351.

178. **Rock, C. O., and J. E. Cronan, Jr.** 1982. Regulation of bacterial membrane lipid synthesis. Curr. Top. Membr. Transp. **17:**209–233.

179. **Rock, C. O., and J. E. Cronan, Jr.** 1982. Solution structure of acyl carrier protein, p. 333–337. *In* A. Martinosi (ed.), Membranes and transport: a critical review, vol. 1. Plenum Publishing Corp., New York.

180. **Rock, C. O., S. Goelz, and J. E. Cronan, Jr.** 1980. Phospholipid synthesis in *Escherichia coli*: characteristics of fatty acid transfer from acyl-carrier protein to *sn*-glycerol 3-phosphate. J. Biol. Chem. **256:**736–742.

181. **Rock, C. O., and S. Jackowski.** 1982. Regulation of phospholipid synthesis in *Escherichia coli*. Composition of the acyl-acyl carrier protein pool *in vivo*. J. Biol. Chem. **257:**10759–10765.

182. **Rock, C. O., and S. Jackowski.** 1985. Pathways for the incorporation of exogenous fatty acids into phosphatidylethanolamine in *Escherichia coli*. J. Biol. Chem. **260:**12720–12724.

183. **Schweizer, H., M. Argast, and W. Boos.** 1982. Characterization of a binding protein-dependent transport system for *sn*-glycerol 3-phosphate in *Escherichia coli* that is part of the *pho* regulon. J. Bacteriol. **150:**1154–1163.

184. **Shibuya, I., C. Miyazaki, and A. Ohta.** 1985. Alteration of phospholipid composition by combined defects in phosphatidylserine and cardiolipin synthases and physiological consequences in *Escherichia coli*. J. Bacteriol. **161:**1086–1092.

185. **Shibuya, I., S. Yamogoe, C. Miyazaki, H. Matsuzaki, and A. Ohta.** 1985. Biosynthesis of novel acidic phospholipid analogs in *Escherichia coli*. J. Bacteriol. **161:**473–477.

186. **Shopsis, C. S., R. Engel, and B. E. Tropp.** 1974. The inhibition of phosphatidylglycerol synthesis in *Escherichia coli* by 3,4-dihydroxybutyl 1-phosphonate. J. Biol. Chem. **249:**2473–2477.

187. **Silbert, D. F.** 1970. Arrangement of fatty acyl groups in phosphatidylethanolamine from a fatty acid auxotroph of *Escherichia coli*. Biochemistry **9:**3631–3640.

188. **Silbert, D. F.** 1975. Genetic modification of membrane lipid. Annu. Rev. Biochem. **44:**315–339.

189. **Silbert, D. F., T. Pohlman, and A. Chapman.** 1976. Partial characterization of a temperature-sensitive mutation affecting acetyl coenzyme A carboxylase in *Escherichia coli* K-12. J. Bacteriol. **126:**1351–1354.

190. **Silbert, D. F., F. Ruch, and P. R. Vagelos.** 1968. Fatty acid replacement in a fatty acid auxotroph of *Escherichia coli*. J. Bacteriol. **95:**1658–1665.

191. **Silbert, D. F., and P. R. Vagelos.** 1967. Fatty acid mutant of *E. coli* lacking a β-hydroxydecanoyl thioester dehydrase. Proc. Natl. Acad. Sci. USA **58:**1579–1586.

192. **Snider, M. D., and E. P. Kennedy.** 1977. Partial purification of glycerophosphate acyltransferase from *Escherichia coli*. J. Bacteriol. **130:**1072–1083.

193. **Sparrow, C. P., and C. R. H. Raetz.** 1983. A *trans*-acting regulatory mutation that causes overproduction of phosphatidylserine synthase in *Escherichia coli*. J. Biol. Chem. **258:**9963–9967.

194. **Spencer, A., E. Muller, J. E. Cronan, Jr., and T. A. Gross.** 1977. *relA* gene control of the synthesis of lipid A fatty acid moieties. J. Bacteriol. **130:**114–117.

195. **Spencer, A. K., A. D. Greenspan, and J. E. Cronan, Jr.** 1978. Thioesterases I and II of *Escherichia coli*: hydrolysis of native acyl-acyl carrier protein thioesters. J. Biol. Chem. **253:**5922–5926.

196. **Taylor, F., and J. E. Cronan, Jr.** 1976. Selection and properties of *Escherichia coli* mutants defective in the synthesis of cyclopropane fatty acids. J. Bacteriol. **125:**518–523.

197. **Taylor, F. R., and J. E. Cronan, Jr.** 1979. Cyclopropane fatty acid synthase of *Escherichia coli*: stabilization, purification and interaction with phospholipid vesicles. Biochemistry **15:**3292–3300.

198. **Taylor, S. S., and E. C. Heath.** 1969. The incorporation of 3-hydroxy fatty acids into a phospholipid of *Escherichia coli* B. J. Biol. Chem. **244:**6605–6616.

199. **Tokunaga, M., J. M. Loranger, and H. C. Wu.** 1984. Prolipopro-

200. **Tunaitis, E., and J. E. Cronan, Jr.** 1973. Characterization of the cardiolipin synthetase activity of *Escherichia coli* cell envelopes. Arch. Biochem. Biophys. **155:**420–447.

201. **Tyhach, R. J., R. Engel, and B. E. Tropp.** 1976. Metabolic fate of 3,4-dihydroxybutyl 1-phosphonate in *Escherichia coli*. J. Biol. Chem. **251:**6717–6723.

202. **Tyhach, R. J., E. Hawrot, M. Satre, and E. P. Kennedy.** 1979. Increased synthesis of phosphatidylserine decarboxylase in a strain of *Escherichia coli* bearing a hybrid plasmid. Altered association with membrane. J. Biol. Chem. **254:**627–633.

203. **Ulrich, A. K., D. de Mendoza, J. L. Garwin, and J. E. Cronan, Jr.** 1983. Genetic and biochemical analysis of *Escherichia coli* mutants altered in the regulation of membrane lipid composition by temperature. J. Bacteriol. **154:**221–230.

204. **Vagelos, P. R., and A. R. Larrabee.** 1967. Acyl carrier protein IX. Acyl carrier protein hydrolase. J. Biol. Chem. **242:**1776–1781.

205. **Vallari, D. V., and C. O. Rock.** 1982. Role of spermidine in the activity of *sn*-glycerol-3-phosphate acyltransferase from *Escherichia coli*. Arch. Biochem. Biophys. **218:**402–408.

206. **van Alphen, L., B. Lugtenberg, E. T. Rietschel, and C. Mombers.** 1979. Architecture of the outer membrane of *Escherichia coli* K-12. Phase transition of the bacteriophage K3 receptor complex. Eur. J. Biochem. **101:**571–579.

207. **Vanaman, T. C., S. J. Wakil, and R. L. Hill.** 1968. The complete amino acid sequence of the acyl carrier protein of *Escherichia coli*. J. Biol. Chem. **243:**6420–6431.

208. **van den Bosch, H., and P. R. Vagelos.** 1970. Fatty acid CoA and fatty acyl-acyl carrier protein as acyl donors in the synthesis of lysophosphatidate and phosphatidate of *Escherichia coli*. Biochim. Biophys. Acta **218:**233–248.

209. **van den Bosch, H., J. R. Williamson, and P. R. Vagelos.** 1970. Localization of acyl carrier protein in *Escherichia coli*. Nature (London) **288:**338–341.

210. **van Dijck, P. W. M., P. H. J. T. Ververgaert, A. J. Verkleij, L. L. M. Ven Deenen, and J. DeGier.** 1975. Influence of Ca^{+2} and Mg^{+2} on the thermotrophic behavior of permeability properties of liposomes prepared from dimyristoyl phosphatidylglycerol and mixtures of dimyristoyl phosphatidylglycerol and dimyristoyl phosphatidylcholine. Biochim. Biophys. Acta **406:**465–478.

211. **van Golde, L. M. G., H. Schulman, and E. P. Kennedy.** 1973. Metabolism of membrane phospholipids and its relation to a novel class of oligosaccharides in *Escherichia coli*. Proc. Natl. Acad. Sci. USA **70:**1368–1372.

212. **Volpe, J. J., and P. R. Vagelos.** 1976. Mechanisms and regulation of biosynthesis of saturated fatty acids. Physiol. Rev. **56:**339–417.

213. **von Meyenburg, K., B. B. Jorgensen, and B. Reurs.** 1984. Physiological and morphological effect of overproduction of membrane-bound ATP synthase in *Escherichia coli* K-12. EMBO J. **3:**1791–1797.

214. **Vos, M. M., J. A. F. op den Kamp, S. Beckerdite-Quagliata, and P. Elsbach.** 1978. Acylation of monoacylglycerophosphoethanolamine in the inner and outer membranes of the envelope of an *Escherichia coli* K-12 strain and its phospholipase A-deficient mutant. Biochim. Biophys. Acta **508:**165–173.

215. **Wakil, S. J.** 1970. Fatty acid metabolism, p. 1–48. *In* S. J. Wakil (ed.), Lipid metabolism. Academic Press, Inc., New York.

216. **Walenga, R. W., and M. J. Osborn.** 1980. Biosynthesis of lipid A. Formation of acyl-deficient lipopolysaccharides in *Salmonella typhimurium* and *Escherichia coli*. J. Biol. Chem. **258:**4257–4263.

217. **Weiner, J. H., B. D. Lemire, M. L. Elmes, R. D. Bradley, and D. G. Scraba.** 1984. Overproduction of fumarate reductase in *Escherichia coli* induces a novel intracellular lipid-protein organelle. J. Bacteriol. **158:**590–596.

218. **White, D. A., F. A. Albright, W. J. Lennarz, and C. A. Schnaitman.** 1971. Distribution of phospholipid-synthesizing enzymes in the wall and membrane subfractions of the envelope of *Escherichia coli*. Biochim. Biophys. Acta **249:**636–642.

218a. **Wilkinson, W. O., J. P. Walsh, J. M. Corless, and R. M. Bell.** 1986. Crystalline arrays of the *Escherichia coli* sn-glycerol-3-phosphate acyltransferase, an integral membrane protein. J. Biol. Chem. **261:**9951–9958.

219. **Wollenweber, H.-W., S. Schlecht, O. Luderitz, and E. T. Rietschel.** 1983. Fatty acid in lipopolysaccharides of *Salmonella* species grown at low temperature. Identification and position. Eur. J. Biochem. **130:**167–171.

220. **Wood, H. G., and R. E. Barden.** 1977. Biotin enzymes. Annu. Rev. Biochem. **46:**385–413.

221. **Wu, H. C., J.-S. Lai, S. Hayashi, and C.-Z. Giam.** 1982. Biogenesis of membrane lipoproteins in *Escherichia coli*. Biophys. J. **37:**307–315.

tein modification and processing enzymes in *Escherichia coli*. J. Biol. Chem. **259:**3825–3830.

31. Structure and Biosynthesis of Lipid A in *Escherichia coli*

CHRISTIAN R. H. RAETZ

Department of Biochemistry, College of Agricultural and Life Sciences, University of Wisconsin, Madison, Wisconsin 53706

INTRODUCTION

As discussed elsewhere in this volume, *Escherichia coli* is surrounded by two distinct membranes that are separated by the peptidoglycan and the periplasmic space (11). Since there are no internal membranes in *E. coli*, all lipid molecules and related substances are associated with the two membranes of the envelope (35, 39). Several recent reviews have provided detailed accounts of the biochemistry and molecular biology of *E. coli* membrane lipids (29, 36, 39, 45, 54).

The purpose of this chapter is to summarize recent progress that has led to the elucidation of the structure (15, 33, 45, 46, 50, 52) and biosynthesis (1, 7, 37, 38, 42, 52) of lipid A. As discussed in this book by Nikaido and Vaara (chapter 3), lipid A molecules constitute the outer monolayer of the outer membrane of *E. coli*, as well as of most other gram-negative bacteria (11, 13, 30). Lipid A is not only a major molecular constituent of the outer membrane but also the hydrophobic anchor of lipopolysaccharide (Fig. 1) (45). The biochemistry of the polysaccharide domain of lipopolysaccharide (13) is reviewed in chapter 41.

Considering the prominent role of lipid A in the *E. coli* envelope and the pathophysiological importance of lipid A as the toxic and immunostimulatory component of lipopolysaccharide (13, 45), it is surprising that the true covalent biochemistry of the lipid A molecule was unknown before 1983 (45, 46, 52). On the basis of the discovery of monosaccharide precursors of lipid A, Takayama and co-workers (52) first postulated the proper structure for mature lipid A (15, 33, 50) (Fig. 2, bottom), whereas Shiba et al. (16–18) synthesized this and a variety of related molecules by chemical methods.

The three formal domains of lipopolysaccharide (the O antigen, the core region, and the lipid A moiety) are diagrammed in Fig. 1. The lipid A (Fig. 2) domain is especially well conserved among different gram-negative organisms, whereas the O antigens are highly variable (13, 45). Because of its covalent attachment to the core and O-antigen sugars, lipid A is not

recovered in chloroform-methanol extracts of *E. coli* (13, 45).

E. coli K-12 lacks the O-antigen domain entirely (13, 45), but this does not prevent its growth under laboratory conditions. The O-antigen domain is presumed to be important for establishing infections in mammalian organisms (11, 13). Like the O-antigen sugars, most of the core sugars can be deleted by suitable mutations without adverse effects on cell growth (45). However, the two innermost 3-deoxy-D-*manno*-octulosonic acid (KDO) residues (Fig. 1 and 2) are essential, since mutants that are defective in KDO biosynthesis are temperature sensitive (19, 20, 40, 43, 44, 48; J. Sutcliffe, J. Capobianco, and R. Darveau, Fed. Proc. **44:**652, 1985). Given its abundance on the outer surface of the outer membrane (45), mature lipid A (Fig. 2, bottom) is also likely to be essential for cell growth and outer membrane assembly, but firm genetic evidence is lacking.

Lipid A of *E. coli* (Fig. 2) and of most related bacteria is a disaccharide of glucosamine, linked β-1→6 and phosphorylated at positions 1 and 4' (45). It is acylated with four β-hydroxymyristoyl moieties that are attached directly to the glucosamine residues at positions 2, 3, 2', and 3' (15, 33, 45, 46, 50, 52), and it contains two additional short-chain fatty acids, a laurate and a myristate residue, esterified to the nonreducing end (30, 33), as indicated in Fig. 2. About half of the lipid A molecules in the *E. coli* envelope contain an unsubstituted pyrophosphate residue at position 1, as indicated by the dashed line (47, 49). The KDO residues are attached to position 6' (49), and all other sugars (if present) branch off from these KDO moieties (45).

BIOSYNTHESIS OF LIPID A

The elucidation of the biosynthesis of lipid A (1, 7, 37, 38, 42, 52) was greatly facilitated by the discovery, in my laboratory (52), of a novel, fatty acylated monosaccharide compound termed lipid X, which was first isolated as an unknown from certain kinds of phosphatidylglycerol-deficient mutants of *E. coli* (27,

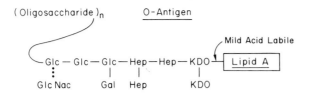

FIG. 1. Three domains of lipopolysaccharide of *E. coli*. Strains of *E. coli* K-12 do not possess O antigens (45). Some core domains may contain a third KDO residue (45). The structures of heptose (Hep) and KDO are reviewed elsewhere (45, 46). Gal, Galactose; Glc, glucose.

28). The relationship between phosphatidylglycerol deficiency and lipid X accumulation remains unknown. The structure of lipid X (Fig. 2), when contrasted with that of mature lipid A (Fig. 2), strongly suggested that it might be a biosynthetic precursor of the reducing end sugar of lipid A (42). Furthermore, the pattern of substitution with β-hydroxymyristoyl moieties on the nonreducing end of lipid A suggested to me the possibility that a nucleotide derivative of lipid X might exist (Fig. 2) that could serve as a precursor for the nonreducing half of lipid A (7, 42). In support of this idea, Bulawa and Raetz (7) were able to show the existence of lipid X (more correctly referred to as 2,3-diacylglucosamine 1-phosphate) and UDP-2,3-diacylglucosamine in wild-type cells of *E. coli*.

Ray et al. (42) provided the first direct evidence for the scheme shown in Fig. 2 by unequivocally demonstrating the existence of an enzyme, in extracts of wild-type cells, that catalyzes the condensation of UDP-2,3-diacylglucosamine with 2,3-diacylglucosamine 1-phosphate to generate a tetraacyldisaccharide 1-phosphate with the typical β-1→6 linkage of mature lipid A. The gene specifying the disaccharide synthase was identified as the *lpxB* (*pgsB*) locus, near min 4 (10, 26, 42). The disaccharide synthase can be overproduced by molecular cloning of *lpxB* (10) and is useful for preparation of lipid A-like molecules (unpublished data).

Since fatty acylated monosaccharides like lipid X and UDP-2,3-diacylglucosamine had not been described previously in *E. coli* or in any other organism, it was important to elucidate their biosynthesis from known metabolic intermediates. Short-term labeling of *E. coli* with $^{32}P_i$, carried out by Anderson et al. (1), revealed that UDP-2,3-diacylglucosamine was actually a precursor of lipid X. Accordingly, it was necessary to postulate a novel system of fatty acyltransferases specific for a UDP-sugar (1). Anderson et al. (1) were able to find a novel set of enyzmes in *E. coli* that utilize β-hydroxymyristoyl acyl carrier protein (Fig. 2) to convert UDP-*N*-acetylglucosamine (UDP-GlcNAc) to UDP-2,3-diacylglucosamine. The first enzyme in this pathway generates a 3-*O*-monoacyl UDP-GlcNAc intermediate (Fig. 2) and is specified by the *lpxA* gene (1, 10). This gene has recently been shown to be in an operon with *lpxB* (the disaccharide synthase) (10). The biological significance of the *lpx* operon is not yet certain, but it may be of importance in the control of *E. coli* envelope assembly.

The finding that UDP-GlcNAc is a key precursor of lipid A reveals that UDP-GlcNAc is situated at a metabolic branchpoint in *E. coli*, leading to either lipid A (1) or peptidoglycan (22). This branching is shown schematically in Fig. 3. The left branch of the scheme shows the initial step in the conversion of UDP-GlcNAc to lipid A (1) (and subsequently to lipopolysaccharide), whereas the right branch illustrates the initial step in the generation of peptidoglycan (22) via UDP-*N*-acetylmuramic acid. In both cases, the biochemistry begins with substitution of the 3-OH of the GlcNAc residue. It seems reasonable to postulate that this branchpoint is a site of regulation, since most of the glucosamine in the *E. coli* envelope is found in either peptidoglycan or lipopolysaccharide.

Unlike the enzymes of glycerophospholipid synthesis, which are largely membrane bound (35, 36, 39), all of the enzymatic reactions shown in Fig. 2 through the disaccharide synthase are recovered in the cytoplas-

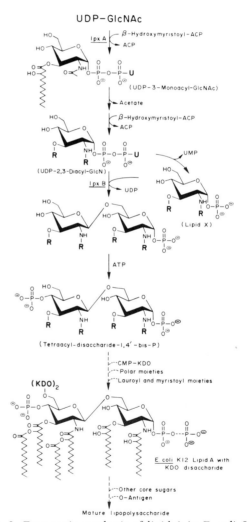

FIG. 2. Enzymatic synthesis of lipid A in *E. coli*. In the structure shown above, R designates a β-hydroxymyristoyl moiety, and U designates uridine. The structure shown at the bottom is the minimal one required for growth and outer membrane assembly (39). A portion of the molecules are substituted with a pyrophosphate residue on the reducing end unit, as indicated by the dashed bond.

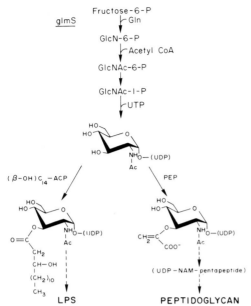

FIG. 3. Biosynthesis and partitioning of UDP-GlcNAc in *E. coli*. UDP-GlcNAc is situated at a branchpoint in *E. coli* leading to either lipid A (see Fig. 2) or peptidoglycan (22). Abbreviations: GlcN, glucosamine; acetyl CoA, acetyl coenzyme A; ACP, acyl carrier protein; PEP, phosphoenolpyruvate; NAM, *N*-acetylmuramic acid; LPS, lipopolysaccharide.

mic fraction (1, 42). Since lipid X is membrane bound (28), however, it is likely that the early enzymes of lipid A biosynthesis function as peripheral membrane proteins on the inner surface of the inner membrane. The 4′ kinase (Fig. 2) is the first integral membrane protein in the pathway (B. L. Ray and C. R. H. Raetz, unpublished data).

All of the reactions indicated by solid arrows (Fig. 2) have been shown to occur in extracts, and the structures of the metabolites have been verified by nuclear magnetic resonance spectroscopy, mass spectrometry, and chemical synthesis (39, 45). Least well characterized are the later stages of the pathway, during which the KDO residues and the "piggy-backed" fatty acids are attached to the nonreducing end. We have recently obtained evidence that the first KDO residue is attached to the tetraacyldisaccharide 1,4′-bisphosphate intermediate (Fig. 2) by means of a predominantly cytoplasmic KDO transferase (K. Hosaka and C. R. H. Raetz, unpublished data). This enzyme preparation adds only one KDO residue and utilizes CMP-KDO as the donor. The second KDO residue is attached by a membrane-associated enzyme (24). The tetraacyldisaccharide 1,4′-bisphosphate intermediate (Fig. 2) is of considerable interest, not only as a precursor in the generation of mature lipid A, but also because it possesses many of the biological activities of lipid A and lipopolysaccharide, especially the immunostimulatory properties (12, 40).

The nature of the fatty acyl donor for the laurate and myristate residues is entirely unknown. However, we have recently developed an in vitro system (Fig. 4) (K. Brozek and C. R. H. Raetz, unpublished data) for converting lipid X (52) to lipid Y (51). The latter monosaccharide bears an additional piggy-backed

palmitoyl moiety that is not present in lipid X (51), and the mechanism of its biosynthesis may be a model for the addition of the laurate and myristate residues to lipid A. In the case of lipid Y, the palmitate is attached by a transacylating enzyme (Fig. 4) that can utilize any glycerophospholipid on the donor, provided that it bears a palmitoyl moiety at position 1 (Brozek and Raetz, unpublished data). The enzyme has a remarkable chain length specificity (Fig. 4), since it does not function with 15- or 17-carbon fatty acyl moieties (Brozek and Raetz, unpublished data). In a formal sense, the lipid X acyltransferase resembles the mammalian lecithin cholesterol acyltransferase (2, 9). The latter enzyme transfers the fatty acyl moiety from position 2 of the phospholipid donor to its acceptor (cholesterol), but it does not display an unusual degree of specificity for the fatty acyl moiety (2, 9).

As discussed below, it will be necessary to verify the scheme described above with a full set of defined mutants. The possibility of additional biosynthetic routes cannot yet be excluded. However, it is likely that the enzymatic synthesis of lipid A in *E. coli* will be a model for that of other gram-negative organisms, since the lipid A's from *Salmonella typhimurium* (33, 50) and *S. minnesota* (32) are virtually identical to that of *E. coli* (15, 52).

An important question concerning envelope biogenesis in *E. coli* is the mechanism by which lipids and proteins are transported from their site of synthesis in the cytoplasm (or on the inner membrane) to the outer membrane. In the case of lipid A, nothing is known about this process. Mulford and Osborn (23) have shown that O-antigen chains (in organisms that have them) are added on the periplasmic surface of the cytoplasmic membrane. This would require that, after completion of the structure shown at the bottom of

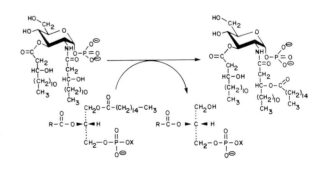

PC ADDED:	C16/C16	C15/C15	C17/C17	C16/C18:1	C18:1/C16	NONE
Rate of X to Y nmol/min/mg	1.8	0	0.05	3.0	0.03	0.01

FIG. 4. Conversion of lipid X to lipid Y by ethanol-extracted membranes of *E. coli*. This system may be a model for the enzymes that add the lauroyl and myristoyl moieties to mature lipid A. The system is reconstituted by incubating 1 mM lipid X, 1 mM phospholipid, ethanol-extracted membranes, and 0.2% Triton X-100 at 37°C (K. Brozek and C. R. H. Raetz, unpublished data). The table at the bottom shows the remarkable specificity of this system for palmitate. As discussed by Takayama et al. (51), the function of lipid Y in the overall scheme of lipid A biosynthesis is still unknown.

Fig. 2 (or perhaps after the further addition of core sugars), a "flip-flop" occurs whereby the nascent lipid A, with its KDO and several core sugars, is translocated to the periplasmic surface of the inner membrane in preparation for addition of O-antigen chains. To understand lipid A translocation, it will be necessary to devise new methods of probing for protein-assisted lipid transport within the inner membrane. A novel approach for studying phospholipid flip-flop has recently been developed by Bishop and Bell (6), who have obtained evidence that glycerolipid "flippases" in microsomes can be assayed as transporters for synthetic short-chain glycerophospholipids, such as dibutyryl glycerophosphocholine, which are entirely hydrophilic and have no affinity for the hydrocarbon interior of lipid bilayers.

MUTANTS DEFICIENT IN THE BIOSYNTHESIS OF LIPID A

We have recently identified two genes that specify enzymes involved in the scheme shown in Fig. 2 (10). These are the UDP-GlcNAc acyltransferase (*lpxA*) and the disaccharide synthase (*lpxB*). The molecular cloning and the nucleotide sequence determination of these genes (10; D. N. Crowell, J. Coleman, and C. R. H. Raetz, unpublished data) indicate that they are situated in an operon near min 4 on the *E. coli* chromosome and transcribed in a clockwise direction. The sequence analysis also indicates the presence of an upstream open reading frame, the biochemical function of which is not yet known.

The *lpxB* gene (originally termed *pgsB*) was initially identified as a mutation that accentuated the phosphatidylglycerol deficiency of certain strains harboring lesions in phosphatidylglycerol phosphate synthase (specified by the *pgsA* gene) (26, 27). Only one mutant allele of *lpxB* is currently available (26, 42). This mutation causes a 100-fold reduction in the specific activity of the lipid A disaccharide synthase. As a consequence of this mutation, cells accumulate several hundred times more 2,3-diacylglucosamine 1-phosphate and its nucleotide derivative than are present in the wild type (7), but the relationship of this perturbation to phosphatidylglycerol deficiency is still unclear. In wild-type cells, intermediates, such as lipid X and UDP-2,3-diacylglucosamine, are present at levels of only 100 to 1,000 molecules per cell (7). This contrasts with lipid A itself, which is present at about 2×10^6 molecules per wild-type cell (39). The determination of the levels of the lipid A pathway intermediates, like determination of those of glycerophospholipid metabolism, is very difficult, because there are very many minor unidentified membrane lipids in *E. coli* from which the relevant intermediates have to be separated (39). It is important to point out that mutants carrying the *lpxB1* allele are still able to generate mature lipid A (K. Takayama, unpublished data), since the accumulation of the precursor molecules probably compensates for the reduction in the level of the disaccharide synthase.

Using localized mutagenesis near min 4, we have recently isolated a mutant with a 30-fold reduction in the level of UDP-GlcNAc acyltransferase (*lpxA*) (C. R. H. Raetz, M. S Anderson, and J. Coleman, unpublished data). This organism is temperature sensi-

tive for growth. The phenotype can be corrected by *lpxA*⁺-bearing plasmids (10), which also overproduce the UDP-GlcNAc acyltransferase. It will be of great interest to determine the effects of lesions in *lpxA* on the total content of *E. coli* lipid A and also to search for secondary mutations capable of suppressing the lipid A deficiency. Because lipid A is so toxic (14, 45), it would be highly desirable to construct special strains of *E. coli* lacking lipid A (or bearing some other lipid in place of lipid A), since such strains would simplify the preparation of proteins destined for injection into animals.

Easy access to all of the enzymes of the pathway of Fig. 2 is also desirable for the semisynthetic preparation of lipid A substructures, some of which have strong immunostimulatory activity (13, 40) or may be useful for antagonizing the toxicity of mature lipopolysaccharide (31). So far, only the genes specifying the UDP-GlcNAc acyltransferase and the disaccharide synthase are known (10).

Mutants defective in the biosynthesis of KDO from arabinose 5-phosphate and phosphoenolpyruvate are available in both *E. coli* and *S. typhimurium* (19, 20, 40, 43, 44, 48; Sutcliffe et al., Fed. Proc. **44**:652, 1985). The latter mutants are better characterized (19, 20, 40, 43, 44, 48). Such mutants are temperature sensitive for growth and are impaired in the formation of KDO at 42°C. Under these conditions, they accumulate large amounts of the tetraacyldisaccharide 1,4'-bisphosphate intermediate shown in Fig. 2 (40, 48). This compound is of interest in its own right because it is strongly immunostimulatory (see below) but only moderately toxic (12). It is unclear whether the temperature sensitivity of mutants altered in KDO biosynthesis is due to the reduced content of lipid A or to the accumulation of precursors that are ordinarily present in very small quantities.

In this system, as in most other lipid systems, the biological function of the molecular diversity is unknown (39). The glucosamine-derived phospholipids shown in Fig. 2 have not been analyzed by X-ray diffraction, but it is likely from model-building studies that they are capable of assuming conformations that would be compatible with insertion into a lipid bilayer (42). The fatty acyl moieties of lipid A and its precursors are a little shorter than those of the classical glycerophospholipids, which are generally 16 to 18 carbons in length (35, 36, 39). On the other hand, the polar moieties of lipid A and its precursors are somewhat larger than the corresponding glycerol 3-phosphate backbone of the classical phospholipids. As has been pointed out previously, lipid X may be viewed as an analog to phosphatidic acid, whereas UDP-2,3-diacylglucosamine is similar, in some respects, to CDP-diglyceride (7).

It will obviously be of great interest to characterize the phenotypes of mutants defective in lipid A biogenesis. It is likely that mutants defective in the formation of the outer monolayer of the outer membrane will show an altered pattern of sensitivity to antibiotics, detergents, and membrane-active molecules, such as colicins. Furthermore, the mechanism by which lipid A interacts with outer membrane proteins will have to be probed by X-ray diffraction. In this regard, the tetraacyldisaccharide 1,4'-bisphosphate precursor (Fig. 2) forms a stoichiometric complex with the *ompF* protein which can be crystallized (R. M. Garavito, personal communication).

PHARMACOLOGICAL IMPORTANCE OF LIPID A AND OF GLUCOSAMINE-DERIVED PHOSPHOLIPIDS

Not only are lipid A and its precursors vital for the assembly of the *E. coli* envelope, but they are also of great interest because of their pathophysiological effects on mammals. The lipid A moiety of lipopolysaccharide is known to have many complicated effects on animal systems, including the induction of shock, the generation of fever, and the activation of lymphocytes and macrophages (13, 45). Recent evidence strongly suggests the involvement of tumor necrosis factor, a protein synthesized by macrophages in response to lipid A, as a key mediator of endotoxin-induced shock (3–5). By mechanisms that are probably independent of tumor necrosis factor, lipid A also stimulates the differentiation of certain pre-B-lymphocyte tumor cells to synthesize kappa chains and surface immunoglobulins (21, 53). Essentially nothing is known about the mechanisms by which lipid A interacts with the membranes of macrophages or lymphocyte tumor cells to trigger the synthesis of specific proteins.

A rational approach to lipid A pharmacology, including the identification of possible receptors, was not feasible before 1983, because the correct covalent structure of lipid A was unknown (46). Since almost all of the pathophysiology associated with lipopolysaccharide is mediated by lipid A (14, 45), it is important to study the effects of lipid A substructure and analogs (39), especially with defined radioactive probes for detecting receptors. A complete knowledge of lipid A pharmacology will be crucial to treatment of many common clinical syndromes, especially endotoxin-induced shock.

The known pharmacological properties of lipid A and its precursors (Fig. 2) can be summarized as follows. Many of the immunostimulatory activities of lipid A are fully manifested in the tetraacyldisaccharide 1,4'-bisphosphate structure shown in Fig. 2 (12, 40), but mature lipid A, by virtue of its piggybacked lauroyl and myristoyl moieties, is an order of magnitude more toxic with regard to its ability to induce shock (14). Interestingly, the monosaccharides, especially lipid X, are entirely nontoxic when injected into sensitive mammals, such as sheep (8, 31). Considerable evidence has now accumulated suggesting that pretreatment of sheep and mice with lipid X renders these animals more resistant to a challenge with a potentially lethal dose of lipopolysaccharide (31; D. Golenbock, R. Proctor, J. Will, and C. R. H. Raetz, unpublished data). Although the mechanism of protection is not known, it may reflect competition for a common target molecule, such as a receptor, on vascular endothelial cells or macrophages. Perhaps the synthesis of tumor necrosis factor is partially blocked. In addition to the amelioration of endotoxic shock, the monosaccharide precursors are endowed with some immunostimulatory activities, as judged by the activation of splenic lymphocytes and the release of prostaglandins from macrophages (25, 41). It would appear that the best therapeutic effects might be obtained with the nontoxic monosaccharides (8, 31) or

perhaps with underacylated or monophosphorylated disaccharides (34).

In summary, the discovery of lipid X (27, 52) and the elucidation of the structure of lipid A (15, 33, 45, 46, 50, 52) have provided important new insights into the assembly of the *E. coli* envelope and into lipid A pharmacology. It will be of interest to extend these findings to other gram-negative bacteria. The functions of lipid A in the outer membrane of *E. coli* are still unknown, but the availability of lipid A and its precursors (Fig. 2) in large quantities will facilitate physical, as well as biological, studies. The coordination of lipid A biosynthesis with glycerophospholipid synthesis, which may be mediated by the partitioning of β-hydroxymyristoyl acyl carrier protein (39), also deserves further study, as does the balanced utilization of UDP-GlcNAc for lipid A (1) and peptidoglycan biosynthesis (22).

ACKNOWLEDGMENTS

The work in this laboratory cited in this review was supported by Public Health Service grants AM 19551 and AM 21722 from the National Institutes of Health.

LITERATURE CITED

1. **Anderson, M. S., C. E. Bulawa, and C. R. H. Raetz.** 1985. The biosynthesis of gram-negative endotoxin: formation of lipid A precursors from UDP-GlcNAc in extracts of *Escherichia coli.* J. Biol. Chem. **260:**15536–15541.
2. **Aron, L., S. Jones, and C. J. Fielding.** 1979. Human plasma lecithin-cholesterol acyltransferase: characterization of cofactor-dependent phospholipase activity. J. Biol. Chem. **253:**7220–7226.
3. **Beutler, B., and A. Cerami.** 1986. Cachectin and tumour necrosis factor as two sides of the same biological coin. Nature (London) **320:**584–588.
4. **Beutler, B., N. Krochin, I. W. Milsark, C. Luedke, and A. Cerami.** 1986. Control of cachectin (tumor necrosis factor) synthesis: mechanisms of endotoxin resistance. Science **232:**977–980.
5. **Beutler, B., I. W. Milsark, and A. C. Cerami.** 1985. Passive immunization against cachectin/tumor necrosis factor protects mice from lethal effect of endotoxin. Science **229:**869–871.
6. **Bishop, W. R., and R. M. Bell.** 1985. Assembly of the endoplasmic reticulum phospholipid bilayer: the phosphatidylcholine transporter. Cell **42:**51–60.
7. **Bulawa, C. E., and C. R. H. Raetz.** 1984. The biosynthesis of gram-negative endotoxin: identification and function of UDP-diacylglucosamine in *Escherichia coli.* J. Biol. Chem. **259:**4846–4851.
8. **Burhop, K. E., R. A. Proctor, R. B. Helgerson, C. R. H. Raetz, J. Starling, and J. A. Will.** 1985. Pulmonary pathophysiological changes in sheep caused by endotoxin precursor, lipid X. J. Appl. Physiol. **59:**1726–1732.
9. **Chung, Y., D. A. Abano, G. M. Floss, and A. M. Scanu.** 1979. Isolation, properties and mechanism of *in vitro* action of lecithin:cholesterol acyltransferase from human plasma. J. Biol. Chem. **254:**7456–7464.
10. **Crowell, D. N., M. S. Anderson, and C. R. H. Raetz.** 1986. Molecular cloning of the genes for lipid A disaccharide synthase and UDP-N-acetylglucosamine in *Escherichia coli.* J. Bacteriol. **168:**152–159.
11. **Davis, B. D., R. Dulbecco, H. N. Eisen, and H. S. Ginsberg.** 1980. Microbiology, 3rd ed. Harper & Row Publishers, Inc., New York.
12. **Galanos, C., V. Lehmann, O. Lüderitz, E. T. Rietschel, O. Westphal, H. Brade, L. Brade, M. A. Freudenberg, T. Hansen-Hugge, T. Lüderitz, G. McKenzie, U. Schade, W. Strittmatter, K.-I. Tanamoto, U. Zähringer, M. Imoto, H. Yoshimura, M. Yamamoto, T. Shimamoto, S. Kusumoto, and T. Shiba.** 1984. Endotoxic properties of chemically synthesized lipid A part structures. Comparison of synthetic lipid A precursor and synthetic analogues with biosynthetic lipid A precursor and free lipid A. Eur. J. Biochem. **140:**221–227.
13. **Galanos, C., O. Lüderitz, E. T. Rietschel, and O. Westphal.** 1977. Newer aspects of the chemistry and biology of bacterial lipopolysaccharides with special reference to their lipid A component, p. 239–335. *In* T. W. Goodwin (ed.), International review of bio-

chemistry: biochemistry of lipids II, vol. 14. University Park Press, Baltimore.

14. **Galanos, C., O. Lüderitz, E. T. Rietschel, O. Westphal, H. Brade, L. Brade, M. Freudenberg, U. Schade, M. Imoto, H. Yoshimura, S. Kusumoto, and T. Shiba.** 1985. Synthetic and natural *Escherichia coli* free lipid A express identical endotoxic activities. Eur. J. Biochem. **148:**1–5.

15. **Imoto, M., S. Kusumoto, T. Shiba, H. Naoki, T. Iwashita, E. T. Rietschel, H.-W. Wollenweber, C. Galanos, and O. Lüderitz.** 1983. Chemical structure of *E. coli* lipid A: linkage site of acyl groups in the disaccharide backbone. Tetrahedron Lett. **24:**4017–4020.

16. **Imoto, M., H. Yoshimura, S. Kusumoto, and T. Shiba.** 1984. Total synthesis of lipid A, active principle of bacterial endotoxin. Proc. Jpn. Acad. **60B:**285–288.

17. **Imoto, M., H. Yoshimura, M. Yamamoto, T. Shimamoto, S. Kusumoto, and T. Shiba.** 1984. Chemical synthesis of phosphorylated tetraacyl disaccharide corresponding to a biosynthetic precursor of lipid A. Tetrahedron Lett. **25:**2667–2670.

18. **Kusumoto, S., M. Yamamoto, and T. Shiba.** 1984. Chemical syntheses of lipid X and lipid Y, acyl glucosamine 1-phosphates isolated from *Escherichia coli* mutants. Tetrahedron Lett. **25:**3727–3730.

19. **Lehmann, V.** 1977. Isolation, purification and properties of an intermediate in 3-deoxy-D-*manno*-octulosonic acid-lipid A. Eur. J. Biochem. **75:**257–266.

20. **Lehmann, V., and E. Rupprecht.** 1977. Microheterogeneity in lipid A demonstrated by a new intermediate in the biosynthesis of 3-deoxy-D-*manno*-octulosonic acid-lipid A. Eur. J. Biochem. **81:**443–452.

21. **Mains, P. E., and C. H. Sibley.** 1983. LPS-non-responsive variants of mouse B cell lymphoma, 70Z/3: isolation and characterization. Somatic Cell Genet. **9:**699–720.

22. **Mirelman, D.** 1979. Biosynthesis and assembly of cell wall peptidoglycan, p. 115–166. *In* M. Inouye (ed.), Bacterial outer membranes. Biogenesis and functions. John Wiley & Sons, Inc., New York.

23. **Mulford, C. A., and M. J. Osborn.** 1983. An intermediate step in translocation of lipopolysaccharide to the outer membrane of *Salmonella typhimurium*. Proc. Natl. Acad. Sci. USA **80:**1159–1163.

24. **Munson, R. S., Jr., N. S. Rasmussen, and M. J. Osborn.** 1978. Biosynthesis of lipid A. Enzymatic incorporation of 3-deoxy-D-mannooctulosonate into a precursor of lipid A in *Salmonella typhimurium*. J. Biol. Chem. **253:**1503–1511.

25. **Nishijima, M., F. Amano, Y. Akamatsu, K. Akagawa, T. Tokunaga, and C. R. H. Raetz.** 1985. Macrophage activation by monosaccharide precursors of lipid A. Proc. Natl. Acad. Sci. USA **82:**282–286.

26. **Nishijima, M., C. E. Bulawa, and C. R. H. Raetz.** 1981. Two interacting mutations causing temperature-sensitive phosphatidylglycerol synthesis in *Escherichia coli* membranes. J. Bacteriol. **145:**113–121.

27. **Nishijima, M., and C. R. H. Raetz.** 1979. Membrane lipid biogenesis in *Escherichia coli*: identification of genes for phosphatidylglycerophosphate synthetase and construction of mutants lacking phosphatidylglycerol. J. Biol. Chem. **254:**7837–7844.

28. **Nishijima, M., and C. R. H. Raetz.** 1981. Characterization of two membrane-associated glycolipids from an *Escherichia coli* mutant deficient in phosphatidylglycerol. J. Biol. Chem. **256:**10690–10696.

29. **Nunn, W. D.** 1986. A molecular view of fatty acid catabolism in *Escherichia coli*. Microbiol. Rev. **50:**179–192.

30. **Osborn, M. J.** 1979. Biosynthesis and assembly of lipopolysaccharide of the outer membrane, p. 15–34. *In* M. Inouye (ed.), Bacterial outer membranes. Biogenesis and functions. John Wiley & Sons, Inc., New York.

31. **Proctor, R. A., J. A. Will, K. E. Burhop, and C. R. H. Raetz.** 1986. Protection of mice against lethal endotoxemia by a lipid A precursor. Infect. Immun. **52:** 905–907.

32. **Qureshi, N., P. Mascagni, E. Ribi, and K. Takayama.** 1985. Monophosphoryl lipid A obtained from lipopolysaccharides of *Salmonella typhimurium* R595. Purification of the dimethyl derivative by high performance liquid chromatography and complete structural determination. J. Biol. Chem. **260:**5271–5278.

33. **Qureshi, N., K. Takayama, D. Heller, and C. Fenselau.** 1983. Position of ester groups in the lipid A backbone of lipopolysaccharides obtained from *Salmonella typhimurium*. J. Biol. Chem. **258:**12947–12951.

34. **Qureshi, N., K. Takayama, and E. Ribi.** 1982. Purification and structural determination of non-toxic lipid A obtained from

lipopolysaccharide of *Salmonella typhimurium*. J. Biol. Chem. **257:**11808–11815.

35. **Raetz, C. R. H.** 1978. Enzymology, genetics, and regulation of membrane phospholipid synthesis in *Escherichia coli*. Microbiol. Rev. **42:**614–659.

36. **Raetz, C. R. H.** 1982. Genetic control of phospholipid bilayer assembly, p. 435–477. *In* J. N. Hawthorne and G. B. Ansell (ed.), Phospholipids, vol. 4. Elsevier's new comprehensive biochemistry. Elsevier Biomedical Press, Amsterdam.

37. **Raetz, C. R. H.** 1984. The enzymatic synthesis of lipid A: molecular structure and biological significance of monosaccharide precursors. Rev. Infect. Dis. **6:**463–472.

38. **Raetz, C. R. H.** 1984. *Escherichia coli* mutants that allow elucidation of the precursors and biosynthesis of lipid A, p. 248–268. *In* E. T. Rietschel (ed.), Chemistry of the endotoxins. Elsevier Biomedical Press, Amsterdam.

39. **Raetz, C. R. H.** 1986. Molecular genetics of membrane phospholipid synthesis. Annu. Rev. Genet. **29:**253–295.

40. **Raetz, C. R. H., S. Purcell, M. V. Meyer, N. Qureshi, and K. Takayama.** 1985. Isolation and characterization of eight lipid A precursors from a KDO-deficient mutant of *Salmonella typhimurium*. J. Biol. Chem. **260:**16080–16088.

41. **Raetz, C. R. H., S. Purcell, and K. Takayama.** 1983. Molecular requirements for B lymphocyte activation by *Escherichia coli* lipopolysaccharide. Proc. Natl. Acad. Sci. USA **80:**4624–4628.

42. **Ray, B. L., G. Painter, and C. R. H. Raetz.** 1984. The biosynthesis of gram-negative endotoxin: formation of lipid A disaccharides from monosaccharide precursors in extracts of *Escherichia coli*. J. Biol. Chem. **259:**4852–4859.

43. **Rick, P. D., L. W.-M. Fung, C. Ho, and M. J. Osborn.** 1977. Lipid A mutants of *Salmonella typhimurium*. Purification and characterization of a lipid A precursor produced by a mutant in 3-deoxy-D-mannooctulosonate-8-phosphate synthetase. J. Biol. Chem. **252:**4904–4912.

44. **Rick, P. D., and M. J. Osborn.** 1977. Lipid A mutants of *Salmonella typhimurium*. Characterization of a conditional lethal mutant in 3-deoxy-D-mannooctulosonate-8-phosphate synthetase. J. Biol. Chem. **252:**4895–4903.

45. **Rietschel, E. T. (ed.).** 1984. Handbook of endotoxin, vol. 1: Chemistry of endotoxin. Elsevier Biomedical Press, Amsterdam.

46. **Rietschel, E. T., H.-W. Wollenweber, Z. Sidorczyk, U. Zähringer, and O. Lüderitz.** 1983. Analysis of the primary structure of lipid A, p. 214. *In* L. Anderson and F. M. Unger (ed.), Bacterial lipopolysaccharides. Structure, synthesis, and biological activities. ACS Symposium Series 231. American Chemical Society, Washington, D.C.

47. **Rosner, M. R., H. G. Khorana, and A. G. Satterthwait.** 1979. The structure of lipopolysaccharide from a heptose-less mutant of *Escherichia coli* K-12. II. The application of ^{31}P-NMR spectroscopy. J. Biol. Chem. **254:**5918–5925.

48. **Strain, S. M., I. M. Armitage, L. Anderson, K. Takayama, N. Qureshi, and C. R. H. Raetz.** 1985. Location of polar substituents and fatty acyl chains on lipid A precursors from a KDO-deficient mutant of *Salmonella typhimurium*: studies by ^{1}H, ^{13}C and ^{31}P nuclear magnetic resonance. J. Biol. Chem. **260:**16089–16098.

49. **Strain, S. M., S. W. Fesik, and I. M. Armitage.** 1983. Characterization of lipopolysaccharide from a heptoseless mutant of *Escherichia coli* by carbon 13 nuclear magnetic resonance. J. Biol. Chem. **258:**2905–2910.

50. **Takayama, K., N. Qureshi, and P. Mascagni.** 1983. Complete structure of lipid A obtained from the lipopolysaccharides of the heptoseless mutant of *Salmonella typhimurium*. J. Biol. Chem. **258:**12801–12803.

51. **Takayama, K., N. Qureshi, P. Mascagni, L. Anderson, and C. R. H. Raetz.** 1983. Glucosamine-derived phospholipids in *Escherichia coli*: structure and chemical modification of a triacyl GlcN-1-P found in a phosphatidylglycerol-deficient mutant. J. Biol. Chem. **258:**14245–14252.

52. **Takayama, K., N. Qureshi, P. Mascagni, M. A. Nashed, L. Anderson, and C. R. H. Raetz.** 1983. Fatty acyl derivatives of glucosamine 1-phosphate in *Escherichia coli* and their relation to lipid A. Complete structure of a diacyl GlcN-1-P found in a phosphatidylglycerol-deficient mutant. J. Biol. Chem. **258:**7379–7385.

53. **Weeks, R. S., P. E. Mains, and C. H. Sibley.** 1984. Comparison of membrane IgM expression in the murine B cell lymphoma 70Z/3 treated with LPS or supernatant containing T cell factors. J. Immunol. **133:**351–358.

54. **Wu, H. C.** 1986. Post-translational modification and processing of membrane proteins in bacteria. *In* M. Inouye (ed.), Bacterial outer membranes as model systems. John Wiley & Sons, Inc., New York.

32. Biosynthesis of Sugar Residues for Glycogen, Peptidoglycan, Lipopolysaccharide, and Related Systems

O. GABRIEL

Department of Biochemistry, Schools of Medicine and Dentistry, Georgetown University, Washington, D.C. 20007

INTRODUCTION

The metabolic reactions that interconvert carbohydrates are numerous and appear complex at first glance. Closer examination of these reactions reveals that the transformation of sugars can be divided into two distinct groups, the interconversion as phosphorylated sugars or when conjugated with nucleotides. Usually, sugar nucleotide biosynthesis is preparatory to incorporation of the carbohydrate residues into more complex structures. Metabolic reactions involved in the conversion of monosaccharides when linked to nucleotides, and their precursor relation to important complex carbohydrate-containing polymers, will be the focus of this chapter. Of primary concern will be the reactions that occur in *Escherichia coli* and *Salmonella typhimurium*, but in some instances reactions of other bacterial species will be mentioned.

The relationship between the pathways of "primary" carbohydrates such as glucose, mannose, and galactose and sugar nucleotides as intermediates in carbohydrate interconversions is shown in Fig. 1. The few primary sugars are transformed into other carbohydrates after conjugation with nucleotides to yield a variety of sugar nucleotide intermediates. In the process of complex carbohydrate polymer biosynthesis, nucleotide sugars have three major functions. First, they serve as intermediates during the biosynthesis of most carbohydrates found in bacterial cells. Second, they serve as precursors or activated forms by acting as glycosyl donors for carbohydrate-containing biopolymers. The third function of sugar nucleotides concerns their role as regulatory mediators of many important biological pathways leading ultimately to the biosynthesis of complex carbohydrate-containing polymers.

The general features of sugar transformations will be discussed first, followed by a brief description of special aspects characteristic for the role of sugar nucleotide precursors in the synthesis of various bacterial biopolymers.

The strategy for the synthesis of carbohydrate polymers requires three individual types of reactions. The first type of reactions concerns the activation of sugar 1-phosphates by reaction with nucleoside triphosphates. These reactions are catalyzed by pyrophosphorylases and result in most instances in formation of nucleoside diphosphosugars. Reasons for the choice of a specific base for a given reaction are at present unclear, but the use of different bases separates different biosynthetic pathways. In addition, this separation provides a way for the independent metabolic control of various pyrophosphorylases. For example, CDP-paratose is a specific inhibitor of CDP-glucose pyrophosphorylase but not of UDP-glucose pyrophosphorylase. Thus, CDP-glucose is only a precursor for dideoxyhexose biosynthesis and not a glucosyl donor for other reactions.

The second type of reaction for biopolymer synthesis concerns various modifications of nucleotide-linked

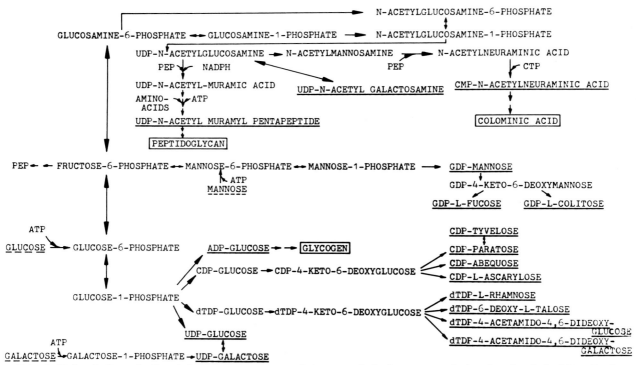

FIG. 1. Unless indicated, all carbohydrates are D-sugars. Primary carbohydrates are underlined with a dashed line. Products of metabolic pathways or precursors for incorporation into carbohydrate-containing biopolymers are underlined. Most of these sugar nucleotides are allosteric effectors for the pyrophosphorylase that commits the carbohydrate to a specific reaction sequence. Carbohydrate polymers are in rectangular boxes. Abbreviations and nomenclature: 4-keto-6-deoxy-glucose, 6-deoxy-D-xylo-4-hexosulose; 4-keto-6-deoxy-mannose, 6-deoxy-D-lyxo-4-hexosulose; L-fucose, 6-deoxy-L-galactose; L-rhamnose, 6-deoxy-L-mannose; L-colitose, 3,6-dideoxy-L-galactose; D-tyvelose, 3,6-dideoxy-D-mannose; D-paratose, 3,6-dideoxy-D-glucose; D-abequose, 3,6-dideoxy-D-galactose; L-ascarylose, 3,6-dideoxy-L-mannose.

sugars, such as epimerizations or deoxysugar synthesis, to be described below.

The transfer of nucleotide-linked sugars formed by the two reactions described above to appropriate acceptors is catalyzed by glycosyltransferases and is the final step in polymer synthesis. During the biosynthesis of peptidoglycan and lipopolysaccharide, a membrane carrier, undecaprenylphosphate, serves as an intermediate prior to transfer to a final acceptor.

ACTIVATION OF CARBOHYDRATES

Pyrophosphorylases catalyze reactions between nucleoside triphosphates and sugar 1-phosphates according to:

$$\text{sugar 1-phosphate} + \text{nucleoside triphosphate} \rightleftharpoons \begin{array}{c}\text{sugar diphosphonucleoside} \\ + \text{ pyrophosphate} \\ PP_i \rightarrow 2\ P_i\end{array}$$

The equilibrium of this reaction in vivo is moved in the direction of sugar nucleotide formation owing to cleavage of pyrophosphate to yield P_i. The enzyme for this cleavage, inorganic pyrophosphatase, occurs in all bacteria. Thus, the overall reaction becomes exergonic and proceeds in favor of sugar nucleotide biosynthesis. The free energy of hydrolysis of the glycosidic linkage of sugar nucleotides is about $-8,000$ cal (ca. $-33,000$ J) per mol and provides the driving force for synthesis of carbohydrate-containing polymers.

SPECIFICITY OF PYROPHOSPHORYLASE REACTIONS

Nucleoside triphosphates of various naturally occurring bases participate in sugar activation. The substrate specificity of a single pyrophosphorylase is limited to a given combination of sugar phosphate and a specific nucleoside triphosphate. Some sugar nucleotides occur only in one unique sugar-base combination (e.g., GDP-L-fucose). Other sugars occur in the same organism in combination with various bases (e.g., UDP-glucose, ADP-glucose, and dTDP-glucose) and are the result of individual specific pyrophosphorylases. These different combinations between sugar and base separate the fate of the sugar moiety and are a means by which the same carbohydrate (e.g., D-glucose) is used in different metabolic pathways (e.g., conversion to D-galactose, glycogen synthesis, or formation of deoxyhexoses).

CONVERSION OF SUGAR NUCLEOTIDES

In many instances, sugars linked to diphosphonucleosides are modified before transfer. A few characteristic examples of sugar transformations that are well understood will be presented below. The initiation of carbohydrate interconversions involves en-

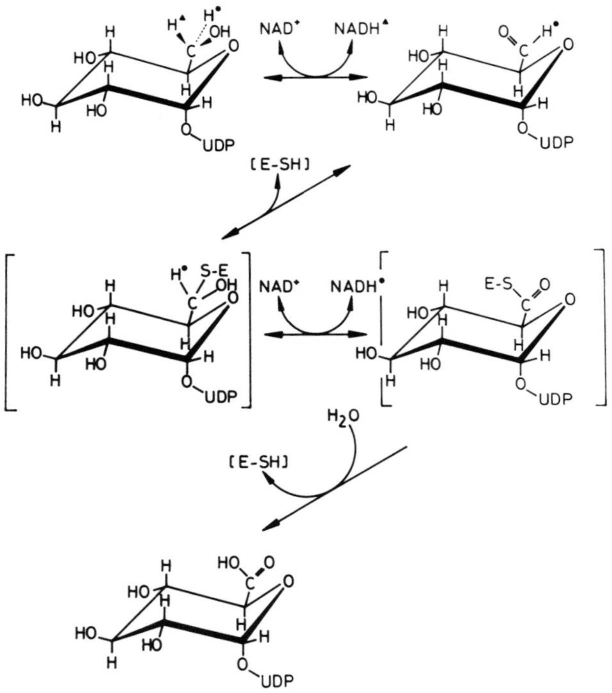

FIG. 2. Biosynthesis of UDP-D-glucuronic acid.

zymes that involve hydrogen transfer with NAD$^+$ as the coenzyme. These types of reactions involve either the oxidation of the parent hexose at carbon 6 to yield the corresponding uronic acid or the oxidation at carbon 4 resulting in formation of a 4-hexosulose intermediate. These latter 4-keto intermediates are common key intermediates of various metabolic pathways and lead to formation of new carbohydrates such as epimers, deoxysugars, or branched carbohydrates. The enzymatic process of hydrogen transfer prior to sugar transformation is ubiquitous for all living systems and was studied extensively in bacterial systems.

BIOSYNTHESIS OF URONIC ACIDS

UDP-glucose dehydrogenase (EC 1.1.1.22) catalyzes a four-electron oxidation at carbon 6 of UDP-glucose to yield UDP-glucuronic acid with concomitant conversion of 2 mol of NAD$^+$ to 2 mol of NADH. The overall irreversible reaction (Fig. 2) proceeds in two steps. The first reversible step leads to UDP-D-glucohexodialdose, producing 1 mol of NADH, followed by a hemithioacetal formation with an essential enzyme

sulfhydryl group. This enzyme intermediate undergoes a second dehydrogenase reaction that yields a second NADH and results in the irreversible release of the product UDP-D-glucuronic acid from the enzyme (34).

The UDP-glucose dehydrogenase from *E. coli* is composed of two subunits that appear to be identical with a molecular weight of 47,000 and that contain two thiol groups per subunit. The *E. coli* enzyme exhibits strictly competitive inhibition with UDP-D-xylose, whereas in eucaryotic cells the same enzyme appears to be under allosteric control by UDP-D-xylose since UDP-D-glucuronic acid is its precursor (36).

EPIMERIZATION OF SUGARS

Inversion of configuration at a single center of chirality appears to be a simple sugar transformation, but considerable effort was expended to establish the reaction mechanism as an oxidation-reduction mechanism mediated by NAD$^+$. Comprehensive reviews concerning sugar nucleotide epimerases have been published (1, 7).

UDP-glucose 4-epimerase (EC 5.1.3.2), as isolated from *E. coli* (45) and *Saccharomyces cerevisiae*, was studied extensively. This enzyme from *E. coli* contains two apparently identical subunits (37) bound to 1 mol of NAD$^+$ and one active site per molecule, which permits classification of this enzyme as a "half-of-the-site" enzyme. The reaction is initiated by removal of carbon-bound hydrogen at carbon 4 and involves intramolecular hydrogen transfer mediated by enzyme-bound NAD$^+$ (Fig. 3). This hydrogen transfer results in formation of a nucleotide 4-hexosulose intermediate with concomitant conversion of enzyme-NAD$^+$ to enzyme-NADH. The conversion to enzyme-NADH causes a conformational change of the enzyme protein whereby the intermediate UDP-D-xylo-4-hexosulose remains tightly bound to the enzyme. The enzyme-NADH donates the hydrogen to the 4-hexosulose as the acceptor, resulting in formation of the 4-epimer, reoxidation to enzyme-NADH$^+$, and release of product. This reaction mechanism is typical for a group of enzymes that catalyze various apparently unrelated sugar transformations (8).

DEOXYHEXOSE BIOSYNTHESIS

The biosynthesis of dTDP-L-rhamnose and dTDP-6-deoxy-L-talose was studied extensively in various strains of *E. coli*. The pathway can be divided into at least three individual steps. First, an irreversible conversion of dTDP-D-glucose to dTDP-6-deoxy-D-xylo-4-

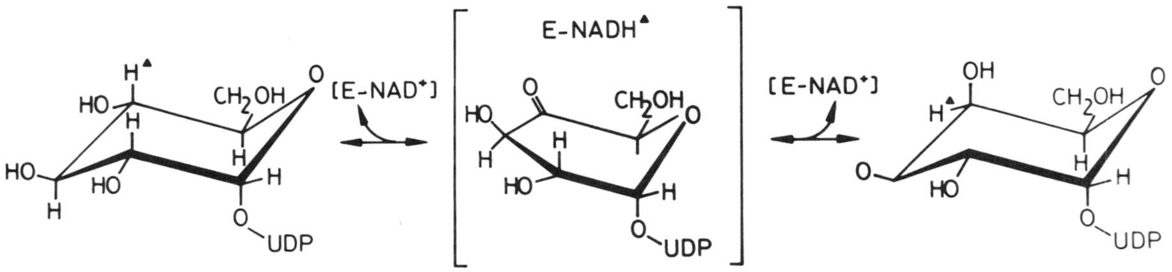

FIG. 3. Reaction mechanism of UDP-galactose-4-epimerase.

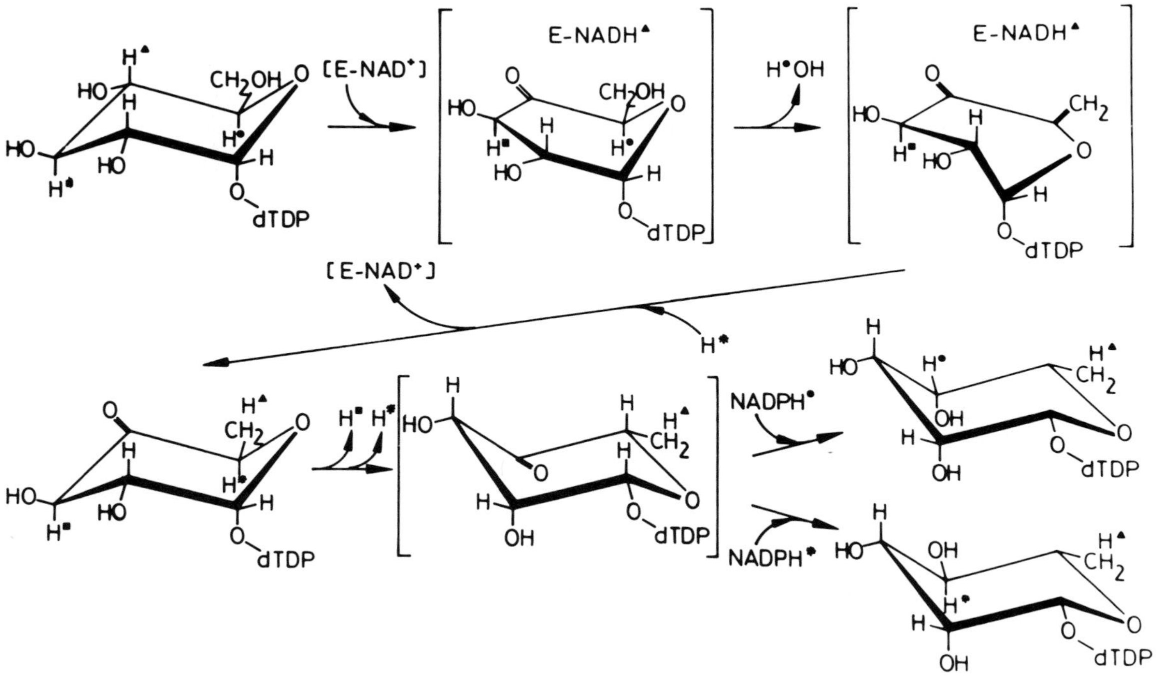

FIG. 4. Biosynthesis of 6-deoxyhexoses.

hexosulose is catalyzed by dTDP-glucose dehydratase (EC 4.2.1.46). This is followed by epimerizations at carbons 3 and 5, most likely via enediol intermediates. The third and final step requires reduction with NADPH and yields either dTDP-L-rhamnose or dTDP-6-deoxy-L-talose, depending on the strains employed. dTDP-glucose dehydratase was isolated from *E. coli* with a molecular weight of 88,000, containing 1 mol of NAD^+ per mol of active enzyme. The subunits of 43,000 molecular weight appear to be identical (43, 44). The reaction mechanism for the enzyme bears similarities to the one described for UDP-galactose-4-epimerase (Fig. 4). Intramolecular hydrogen transfer from carbon 4 to carbon 6 of glucose is mediated by enzyme-bound NAD^+. β-Elimination of water leads to an unsaturated enzyme-bound 4-keto-5,6-glucoseen derivative which acts as hydrogen acceptor for enzyme-NADH. The product dTDP-4-keto-6-deoxy-D-glucose is epimerized at carbons 3 and 5 by enzyme-bound enediol intermediates, as suggested by exchange of carbon-bound hydrogens in these positions. dTDP-4-keto-L-rhamnose-3,5-epimerase can be separated from dTDP-6-deoxy-L-mannose dehydrogenase or dTDP-6-deoxy-L-talose dehydrogenase. These latter dehydrogenases are stereospecific and require NADPH as hydrogen donor. Mixing experiments with 3,5-epimerases and 4-reductases isolated from different *E. coli* strains established that enzymes isolated from different sources can work together but the product of the 6-deoxyhexose pathway is determined by the stereospecificity of the 4-reductase used (9).

BIOSYNTHESIS OF 3,6-DIDEOXYHEXOSES

Details of the pathways for biosynthesis of 3,6-dideoxyhexoses remain to be elucidated, but some information on the conversion of CDP-6-deoxy-D-xylo-4-hexosulose is available. This intermediate is converted to CDP-3,6-dideoxy-D-xylohexosulose by the action of two enzyme proteins and participation of NADPH and pyridoxamine 5-phosphate. In a final step, a specific reductase catalyzes stereospecific reduction using NADPH as hydrogen donor to yield CDP-3,6-dideoxy-D-xylohexose (13).

BIOSYNTHESIS OF 4-ACETAMIDODEOXYHEXOSES

Stereospecific transamination with L-glutamate as the amino group donor and participation of pyridoxal phosphate can yield dTDP-6-deoxy-D-xylohexosulose, dTDP-4-amino-4,6-dideoxy-D-glucose, or dTDP-4-amino-4,6-dideoxy-D-galactose. This step is followed by acetylation, resulting in formation of the corresponding 4-N-acetamido derivatives (22, 27).

SYNTHESIS OF N-ACETYLMURAMIC ACID

The synthesis of N-acetyl-2-amino-2-deoxy-3-O-lactyl-D-glucose starts with a unique reaction between UDP-N-acetyl-D-glucosamine and phosphoenolpyruvate (PEP) (15). Addition of a three-carbon fragment to carbon 3 of the hexosamine yields UDP-N-acetyl-2-amino-2-deoxy-3-enolpyruvyl-D-glucose. Reduction of the enol intermediate with NADPH results in formation of the product, N-acetylmuramic acid (41).

COMMON MECHANISTIC PRINCIPLES

The reaction mechanisms of UDP-galactose-4-epimerase and dTDP-glucose dehydratase described above have great similarities. These enzymes catalyze intramolecular hydrogen transfer mediated by enzyme-bound NAD^+. The various end products formed by these enzymes are the result of different molecular rearrangements that follow the formation of nucleoside diphosphohexosulose intermediates. The role of enzyme-bound NAD^+ as mediator in the intramolecular hydrogen transfer in both the initiation and termination of the reaction is an important functional aspect of these oxidoreductases. The oxidation to the nucleotide hexosulose intermediate is accompanied by reduction of enzyme-NAD^+ to enzyme-NADH, resulting in a conformational change of the enzyme protein. This, in turn, leads to tight binding of nucleotide intermediates to the enzyme protein during the molecular rearrangement. The hydrogen originally removed from the substrate is donated to the last intermediate, thereby regenerating enzyme-NAD^+ and causing release of product from the enzyme. The mechanism requires the presence of enzyme-NAD^+ for catalytic action, whereas enzyme-NADH is inactive because it cannot accept hydrogen from the substrate. Compounds that can donate hydrogen to cause accumulation of enzyme-NADH but fail to permit reoxidation to enzyme-NAD^+ render the enzyme inactive.

The same mechanistic principles were found in additional enzymes that are involved in various sugar transformations, and the mechanistic and structural similarities of these enzymes suggest to me that these enzymes may be evolutionarily related. The working hypothesis is advanced that the common properties reflect the maintenance of "catalytic principles" in terms of enzyme structure and mechanism during the evolutionary process of enzyme catalysis (5).

PATHWAY FOR COLOMINIC ACID BIOSYNTHESIS

E. coli K-235 contains a homopolymer of *N*-acetylneuraminic acid referred to as colominic acid (17). The biosynthesis of colominic acid from simple sugars occurs in several steps and starts with the condensation of *N*-acetyl-D-mannosamine and PEP. This reaction is catalyzed by *N*-acetylneuraminate synthase (EC 4.1.3.19) and was first studied in *Neisseria meningitidis*. By contrast, in mammalian systems the analogous reaction occurs starting with the corresponding hexosamine 6-phosphate, and condensation with PEP leads to *N*-acetylneuraminic acid-9-phosphate (35).

The anomeric hydroxyl group of sialic acid is covalently linked to CMP by the enzyme CMP-sialic acid synthase (EC 2.7.7.43) (18, 19) at the expenditure of CTP, leading to CMP-sialate and PP$_i$. CMP-sialate represents the activated form of the keto sugar sialic acid and is like 2-keto-3-deoxyoctonic (octulosonic) acid (KDO), which also yields a CMP-glycoside. This unique type of activation of the sugar moiety involves a single phosphate group and appears to be confined to 2-keto-3-deoxy sugars (10). By contrast, all other sugar nucleotides have the sugar moiety attached in a pyrophosphate linkage.

The occurrence of CMP-sialic acid was first demonstrated in *E. coli* by Comb and co-workers (2). The CMP-glycoside contains a β-glycosidic linkage between the phosphate group and the anomeric hydroxy group at carbon 2 of sialic acid (3, 16). This is in contrast to other nucleotide-linked D-sugars which contain α-glycosidic linkages and is consistent with the fact that all sialic acid residues in biopolymers are linked in α-glycosidic linkage, which indicates a simple displacement mechanism for sialyltransferase reactions and occurs with inversion of configuration.

E. coli K-235 was also used to characterize the first sialytransferase that produces the homopolymer colominic acid by transfer of the sugar residue from CMP-*N*-acetylneuraminic acid to the nonreducing end of colominic acid (21).

GLYCOGEN SYNTHESIS

The biosynthesis of α-1,4-glycosidic linkages of bacterial glycogen occurs by the intermediary formation of ADP-glucose followed by transfer of the glucose moiety from the sugar nucleotide to an appropriate glucan acceptor. Unlike mammalian glycogen biosynthesis, regulation occurs at the sugar nucleotide level. In general, glycolytic intermediates activate ADP-glucose biosynthesis, and AMP, ADP, and P$_i$ are inhibitors. Glycogen is not required for bacterial growth to occur but is utilized for energy production and as a carbon source when exogenous carbon is not available.

Accumulation of glycogen usually occurs in the presence of excess carbon source, such as glucose, but limitation of other growth requirements such as nitrogen, sulfur, or phosphate.

In contrast to mammalian systems, the precursor for glycogen biosynthesis is ADP-glucose and not UDP-glucose. The enzyme ADP-glucose pyrophosphorylase (glucose-1-phosphate adenylyltransferase; EC 2.7.7.27) catalyzes the reaction:

$$ATP + \text{glucose 1-phosphate} \rightleftharpoons \text{ADP-glucose} + PP_i$$

Glycogen synthase specific for ADP-glucose transfers the glucose moiety of ADP-glucose to an appropriate glucan maltodextrin acceptor in α-1,4-glycosidic linkage. In animals a complex system that involves covalent modification of glycogen synthase is responsible for the regulation of glycogen synthesis. In bacteria the regulation occurs at the ADP-glucose pyrophosphorylase step (30). The action of glycolytic intermediates as activators and AMP, ADP, or P$_i$ as inhibitors indicates that the flux of glucose moieties into glycogen is determined by the energy charge state of the cell. The presence of glycolytic intermediates such as fructose 1,6-bisphosphate signals carbon excess in the cell and results in activation of ADP-glucose and glycogen formation (32).

The fundamentally different modes of mammalian and bacterial glycogen synthesis may at least in part be due to the difference of the glucosyl donor UDP-glucose as opposed to ADP-glucose. In animal systems UDP-glucose serves in addition to glycogen synthesis as a precursor for many other cellular functions, e.g., formation of UDP-galactose, UDP-glucuronic acid, and UDP-xylose and generally the biosynthesis of glycolipids and glycoproteins. By contrast, in bacteria the sole function of ADP-glucose is to serve as precursor for glycogen synthesis. Consequently, this step commits the glucose moiety to become incorporated into glycogen, and ADP-glucose pyrophosphorylase becomes the rate-limiting step.

The evidence for the allosteric regulation of ADP-glucose pyrophosphorylase in bacterial systems was established primarily by studies of mutant strains of

E. coli (14, 31) and *S. typhimurium* (38). It was found that the enzyme has modified allosteric properties. For example, mutant strains have higher affinity for positive effectors like fructose 1,6-bisphosphate and much lower affinity for negative effectors like AMP. Consequently, the mutant strains maintain a higher glycogen steady-state level than does their parent strain.

PRECURSORS FOR PEPTIDOGLYCAN BIOSYNTHESIS

The initial experiments that provided understanding of bacterial peptidoglycan synthesis were carried out using *Staphylococcus aureus*. The important observation that the presence of the antibiotic penicillin in bacterial cell cultures results in accumulation of a family of precursor UDP-*N*-acetylmuramylpeptides facilitated the isolation and identification of these intermediates. Moreover, the striking structural similarities between the peptide sequence of the UDP-intermediates and the peptidoglycan suggested the precursor relationship before the actual biosynthetic pathway for peptidoglycan synthesis was established (12, 28).

The general applicability of this concept was rapidly extended to *E. coli* and *S. typhimurium*. It became apparent that the steady-state levels of sugar nucleotides involved in the biosynthesis of cell wall components are intimately related to their utilization in biosynthetic pathways. For example, strains containing L-rhamnose as a component of lipopolysaccharide (LPS) contained undetectable levels of dTDP-L-rhamnose, whereas rough variants of these strains lacking L-rhamnose in their LPS contain large amounts of dTDP-L-rhamnose (39).

Peptidoglycan biosynthesis occurs in three stages, each of which takes place at a different site of the bacterial cell. Peptidoglycan sugar nucleotide precursors are formed in the cytoplasm (stage 1, the focus of this presentation). Stage 2 is the transfer of the muramylpentapeptide to an undecaprenylphosphate membrane carrier. Stage 3 concerns the transfer of the muramylpentapeptide to the outside of the cell to an acceptor nascent peptidoglycan chain which undergoes cross-linking to render the cell envelope insoluble.

All reactions leading to the completed muramylpentapeptide precursor occur in the cytoplasm. The first step which commits the precursor to undergo the reactions of peptidoglycan synthesis is the transfer of PEP to UDP-*N*-acetylglucosamine, catalyzed by enoylpyruvate transferase (EC 2.5.1.7) and resulting in the formation of the pyruvate-enol ether of UDP-*N*-acetylglucosamine (15). In contrast to other reactions involving PEP, the enol structure is preserved in this step. This is followed by reduction to the D-lactyl ether by the enzyme UDP-*N*-acetylenolpyruvoyl-D-glucosamine reductase (EC 1.1.1.158) to yield UDP-*N*-acetylmuramic acid (40). The addition of amino acids to yield pentapeptide linked to the carboxy group of the lactyl ether residue occurs stepwise to UDP-*N*-acetylmuramic acid, and each step requires a specific enzyme, expenditure of ATP, and participation of Mg^{2+} or Mn^{2+}. The formation of the peptide linkage is not template directed but depends on the specificity of the various synthetases. In this way L-alanine, D-

glutamic acid, a third amino acid residue, and finally a preformed D-alanyl-D-alanine dipeptide unit are added to yield the end product of the cytoplasmic peptidoglycan biosynthesis, UDP-*N*-acetylmuramyl-pentapeptide. The third amino acid residue of the pentapeptide is species specific. *meso*-Diaminopimelic acid is found in *E. coli*, but in *S. aureus*, the subject of most of the initial studies on peptidoglycan structure and biosynthesis, it is replaced by L-lysine. The completed UDP-*N*-acetylmuramylpentapeptide of *E. coli* has the sequence UDP-*N*-acetylmuramoyl-L-alanyl-γ-D-glutamyl-*meso*-2,6-diaminopimeloyl-D-alanyl-D-alanine (39). It should be noted that D-glutamic acid is γ-linked to the third amino acid residue and that during the cross-linking, the third stage of peptidoglycan synthesis, the terminal D-alanine residue is cleaved off. This cleavage results in formation of tetrapeptide units in the final cross-linked peptidoglycan. The synthesis of the dipeptide unit D-alanyl-D-alanine and its addition to the nucleotide tripeptide requires additional comments. Three enzymes are involved in the "alanine branch" of peptidoglycan synthesis (25). Alanine racemase catalyzes the reversible conversion of L-alanine ⇌ D-alanine. The formation of the dipeptide requires the enzyme D-alanyl-D-alanine synthetase (EC 6.3.2.4) and, in the final step, the D-alanyl-D-alanine adding enzyme (EC 6.3.2.15), which yields the product UDP-*N*-acetylmuramoyl-L-alanyl-D-glutamyl-*meso*-2,6-diaminopimeloyl-D-alanyl-D-alanine. The inhibition of this pathway by D-cycloserine is a result of competitive inhibition of the racemase and the D-alanyl-D-alanine synthetase enzyme (11).

CONTROL OF CELL WALL PRECURSOR SYNTHESIS

Osmotically fragile mutants of *E. coli* K-12 with membrane-bound enzyme defects were examined. Elevated levels of UDP-*N*-acetylmuramylpentapeptide and its precursors were expected, but none were found. Inhibition of cell wall biosynthesis of the parental strain with penicillin or vancomycin did not increase the intracellular level of UDP-*N*-acetylmuramylpentapeptide. In contrast, inhibition of cell wall synthesis with D-cycloserine resulted in significant accumulation of UDP-*N*-acetylmuramyltripeptide. This finding strongly suggests that in *E. coli* the level of UDP-*N*-acetylmuramylpentapeptide regulates its own biosynthetic pathway by feedback inhibition. Apparently, the choice of *S. aureus* for the early studies of cell wall biosynthesis and its inhibition by penicillin was fortunate since only few bacterial species lack this feedback control and respond to inhibition by antibiotics like penicillin with increased levels of UDP-*N*-acetylmuramylpentapeptide (11).

LPS SYNTHESIS

The study of LPS biosynthesis is closely related to our understanding of LPS structure, and the continuing updating of structural details indicates that our knowledge in this area is incomplete. Some of the sugar nucleotide precursors of known LPS constituents are postulated but have not yet been isolated. An example is the nucleoside diphosphate derivative of

4-amino-4-deoxy-L-arabinose (42). This sugar is a constituent of the LPS core oligosaccharide region. The early stage of lipid A biosynthesis and the formation of the hexosaminedisaccharide unit in particular are still unclear. A biosynthetic pathway was recently postulated, however (33). These studies, carried out with *E. coli* mutants, indicated the presence of a UDP-2,3-diacylglucosamine derivative that appears to be an early precursor of lipid A biosynthesis. The acyl groups are β-hydroxymiristoyl residues. According to our present knowledge, the remaining core oligosaccharides are added stepwise to the lipid A structure. The first step was demonstrated by the transfer of KDO from CMP-KDO to lipid A by an *E. coli* preparation. The steps to transfer additional KDO residues and L-glycero-D-manno-heptose from ADP-L-glyceromanno-heptose remain to be elucidated. (The synthesis of lipid A is considered in greater detail in chapter 31.)

An important tool in the study of LPS structure and biosynthesis of *S. typhimurium* was mutant strains that lacked the ability to produce sugar nucleotide precursors necessary for LPS synthesis. For example, a mutant unable to produce UDP-galactose owing to the loss of UDP-galactose-4-epimerase produced an incomplete LPS containing only the "core sugars" KDO, glucose, and heptose. These mutants could bypass the metabolic block provided exogenous galactose was added to the medium, resulting in the synthesis of complete LPS. The incomplete LPS of mutant strains is useful for studying in vitro biosynthesis and structure since it serves as acceptor for the addition of individual carbohydrate residues. The concept of the sequential transfer of monosaccharide units while attached to lipid A was demonstrated by various core-defective mutants in *E. coli* and *Salmonella* strains. Thus, the synthesis requirements for the core region of LPS are threefold: an appropriate lipid A acceptor, a sugar nucleotide, and the specific glycosyltransferase (26).

The synthesis of the O-specific antigenic main chains of the LPS consists of repeating units of oligosaccharide and occurs through a mechanism different from that of the core region. In this instance, the individual carbohydrates are preassembled on the same lipid intermediate used in peptidoglycan synthesis, undecaprenyl pyrophosphate. Each repeating oligosaccharide unit of the O antigen is first assembled, while linked covalently to the lipid pyrophosphate intermediate, by stepwise transfer from a sugar nucleotide precursor, catalyzed by an appropriate transferase. After completion of the repeating unit linked to the lipid pyrophosphate complex, the completed oligomer is transferred to lipid A core acceptor. Thus, each cycle results in the preassembly of a repeating unit linked to undecaprenyl pyrophosphate before extension of the O-antigen chain by one oligomeric unit. In Fig. 1 the major metabolic steps involved in the biosynthesis of LPS carbohydrate precursors are shown (e.g., UDP-glucose, UDP-galactose, UDP-*N*-acetylglucosamine, and CDP, GDP, and dTDP derivatives).

ISOLATION, SYNTHESIS, AND ANALYTICAL PROCEDURES FOR SUGAR NUCLEOTIDES

This article was primarily intended to document general principles of the central role of sugar nucleo-

tides as precursors in the biosynthesis of carbohydrate-containing biopolymers. Readers requiring specific information about isolation and synthesis should consult a recently prepared tabulation of naturally occurring sugar nucleotides (6). It should be noted that reliable methods for the organic synthesis of most sugar nucleotides are available (4, 20, 23, 24).

The separation, identification, and quantitative determination of sugar nucleotides is an important requisite for work in this area. Recent advances of modern methods have simplified analytical procedures and require less time. These newer procedures provide considerable advantages over earlier methods, especially in view of the labile nature of the glycosidic linkage between sugar and nucleotide. An elegant example for the rapid analysis of sugar nucleotide pools in *E. coli* and *S. typhimurium* was reported by Payne and Ames (29).

ACKNOWLEDGMENTS

I wish to express my deep appreciation for discussions and help in the preparation of this manuscript to Roland Schauer and Elfriede Schauer, Christian-Albrechts-Universität, D-2300 Kiel, Federal Republic of Germany.

This work was supported in part by Public Health Service grant DE 06744-02 from the National Institutes of Health.

LITERATURE CITED

1. **Adams, E.** 1976. Catalytic aspects of enzymatic racemization. Adv. Enzymol. **44:**69–138.
2. **Comb, D. G., F. Shimizu, and S. Roseman.** 1959. Isolation of cytidine-5'-monophospho-N-acetylneuraminic acid. J. Am. Chem. Soc. **81:**5513–5514.
3. **Comb, D. G., D. R. Watson, and S. Roseman.** 1966. The sialic acids. IX. Isolation of cytidine-5'-monophospho-N-acetylneuraminic acid from *E. coli* K-235. J. Biol. Chem. **241:**5637–5642.
4. **Elbein, A. D.** 1966. Microscale adaptation of the morpholidate procedure for the synthesis of sugar nucleotides. Methods Enzymol. **8:**142–149.
5. **Gabriel, O.** 1978. Common mechanistic and structural properties of enzymes. Trends Biochem. Sci. **3:**193–195.
6. **Gabriel, O.** 1982. Isolation and synthesis of sugar nucleotides. Methods Enzymol. **83:**332–353.
7. **Gabriel, O., H. M. Kalckar, and R. A. Darrow.** 1975. UDP-galactose-4-epimase, p. 85–135. *In* E. Ebner (ed.), Subunit enzymes: biochemistry and function. Marcel Dekker, Inc., New York.
8. **Gabriel, O., and L. Van Lenten.** 1978. The interconversion of monosaccharides. Int. Rev. Biochem. **16:**1–36.
9. **Gaugler, R. W., and O. Gabriel.** 1973. Biological mechanisms involved in the formation of deoxysugars. VII. Biosynthesis of 6-deoxy-L-talose. J. Biol. Chem. **248:**6041–6049.
10. **Ghalambor, M. A., and E. C. Heath.** 1966. The biosynthesis of cell wall lipopolysaccharide in *Escherichia coli*. IV. Purification and properties of cytidine monophosphate 3-deoxy-D-mannooctulosonate synthetase. J. Biol. Chem. **241:**3216–3221.
11. **Ghuysen, J. M., and G. D. Shockman.** 1973. Biosynthesis of peptidoglycan, p. 38–130. *In* L. Leive (ed.), Bacterial membranes and walls. Marcel Dekker, Inc., New York.
12. **Ghuysen, J. M., J. L. Strominger, and D. J. Tipper.** 1968. Bacterial cell walls. Compr. Biochem. **26A:**53–104.
13. **Gonzales-Porque, P., and J. L. Strominger.** 1972. Enzymatic synthesis of cytidine 3,6-dideoxy hexoses. VI. Purification and homogeneity and some properties of cytidine diphosphate-D-glucose oxidoreductase, enzyme E, and enzyme E₃. J. Biol. Chem. **247:**6748–6756.
14. **Govons, S., N. Gentner, E. Greenberg, and J. Preiss.** 1973. Biosynthesis of bacterial glycogen. XI. Kinetic characterization of an altered adenosine diphosphate-glucose synthase from a "glycogen excess" mutant of *Escherichia coli*. J. Biol. Chem. **248:**1731–1740.
15. **Gunetileke, K. G., and R. A. Anwar.** 1968. Biosynthesis of uridine diphospho-N-acetylmuramic acid. II. Purification and properties of pyruvate-uridine diphospho-N-acetyl glucosamine transferase and characterization of uridine diphospho-N-acetylenolpyruvyl glucosamine. J. Biol. Chem. **243:**5770–5778.

16. **Haverkamp, J., T. Spoormaker, L. Dorland, J. F. G. Vliegenthart, and R. Schauer.** 1979. Determination of the β-anomeric configuration of cytidine-5′-monophospho-N-acetylneuraminic acid by ^{13}C NMR spectroscopy. J. Am. Chem. Soc. **101:**4851–4853.

17. **Ito, E., N. Ishimoto, and M. Saito.** 1958. Uridine diphosphate amino-sugar compounds from *Staphylococcus aureus* inhibited by penicillin. Nature (London) **181:**906–907.

18. **Kean, E. L.** 1972. CMP-sialic acid synthetase of nuclei. Methods Enzymol. **28:**413–421.

19. **Kean, E. L., and S. Roseman.** 1966. CMP-sialic acid synthetase. Methods Enzymol. **8:**208–215.

20. **Kochetkov, N. K., and V. N. Shibaev.** 1973. Glycosyl esters of nucleoside pyrophosphates. Adv. Carbohydr. Chem. Biochem. **28:**307–399.

21. **Kundig, F. D., D. Aminoff, and S. Roseman.** 1971. The sialic acids. XII. Synthesis of colominic acid by a sialyltransferase from *Escherichia coli* K-235. J. Biol. Chem. **246:**2543–2550.

22. **Matsuhashi, J., and J. L. Strominger.** 1966. Thymidine diphosphate 4-acetamido 4,6-dideoxyhexose. III. Purification and properties of thymidine diphosphate 4-keto-6-deoxy-D-glucose transaminase from *Escherichia coli* strain B. J. Biol. Chem. **241:**4738–4744.

23. **Michelson, A. M.** 1964. Synthesis of nucleotide anhydrides by anion exchange. Biochim. Biophys. Acta **91:**1–13.

24. **Moffat, J. G.** 1966. Sugar nucleotide synthesis by the phosphomorpholidate procedure. Methods Enzymol. **8:**136–142.

25. **Neuhaus, F. C., C. V. Carpenter, M. P. Lambert, and R. J. Wargel.** 1972. D-cycloserine as a tool in studying the enzymes and the alanine branch of peptidoglycan synthesis, p. 339–362. *In* E. Munoz, F. Ferrandiz, and D. Vazquez (ed.), Proceedings of the Symposium on Molecular Mechanisms of Antibiotic Action on Protein Biosynthesis and Membranes. Elsevier, Amsterdam.

26. **Nikaido, H.** 1973. Biosynthesis and assembly of lipopolysaccharide and the outer membrane layer of gram-negative cell wall, p. 132–202. *In* L. Leive (ed.), Bacterial membranes and walls. Marcel Dekker, Inc., New York.

27. **Ohashi, H., M. Matsuhashi, and S. Matsuhashi.** 1971. Thymidine diphosphate 4-acetamido 4,6-dideoxyhexoses. IV. Purification and properties of thymidine diphosphate 4-keto-6-deoxy-D-glucose transaminase from *Pasteurella pseudotuberculosis*. J. Biol. Chem. **246:**2325–2334.

28. **Park, J. T., and M. J. Johnson.** 1949. Accumulation of labile phosphate in *Staphylococcus aureus* grown in the presence of penicillin. J. Biol. Chem. **179:**585–592.

29. **Payne, S. M., and B. N. Ames.** 1982. A procedure for rapid extraction and high-pressure liquid chromatographic separation of the nucleotides and other small molecules from bacterial cells. Anal. Biochem. **123:**151–161.

30. **Preiss, J.** 1978. Regulation of adenosine diphosphate glucose pyrophosphorylase. Adv. Enzymol. **46:**317–381.

31. **Preiss, J., C. Lammel, and E. Greenberg.** 1976. Biosynthesis of bacterial glycogen. Kinetic studies of glucose-1-P adenyltransferase (EC 2.7.7.27) from a glycogen excess mutant of *Escherichia coli* B. Arch. Biochem. Biophys. **174:**105–119.

32. **Preiss, J., S. G. Yung, and P. A. Baecker.** 1983. Regulation of bacterial glycogen synthesis. Mol. Cell. Biochem. **57:**61–80.

33. **Raetz, C. R. H.** 1984. The enzymatic synthesis of lipid A: molecular structure and biologic function of monosaccharide precursors. Rev. Infect. Dis. **6:**463–471.

34. **Ridley, W. P., J. P. Houchins, and S. Kirkwood.** 1975. Mechanism of action of uridine diphosphoglucose dehydrogenase. Evidence for a second reversible dehydrogenation step involving an essential thiol group. J. Biol. Chem. **250:**8761–8767.

35. **Schauer, R.** 1982. Chemistry, metabolism and biological functions of sialic acids. Adv. Carbohydr. Chem. Biochem. **40:**131–234.

36. **Schiller, J. G., F. Lamy, R. Frazier, and D. S. Feingold.** 1976. UDP-glucose dehydrogenase from *Escherichia coli*. Purification and subunit structure. Biochim. Biophys. Acta **453:**418–425.

37. **Schlesinger, D. H., H. D. Niall, and D. Wilson.** 1974. Localization of translation initiation in the message for *Escherichia coli* UDP-galactose-4-epimerase: the amino terminal sequence. Biochem. Biophys. Res. Commun. **61:**282–289.

38. **Steiner, K. E., and J. Preiss.** 1977. Biosynthesis of bacterial glycogen: genetic and allosteric regulation of glycogen biosynthesis in *Salmonella typhimurium* LT-2. J. Bacteriol. **129:**246–253.

39. **Strominger, J. L.** 1962. Biosynthesis of bacterial cell walls, p. 413–470. *In* I. C. Gunsalus and R. Y. Stainer (ed.), The bacteria: a treatise on structure and function, vol. 3. Academic Press, Inc., New York.

40. **Taku, A., and R. A. Anwar.** 1973. Biosynthesis of uridine diphospho-N-acetylmuramic acid. IV. Activation of uridine diphospho-N-acetylenoylpyruvylglucosamine reductase by monovalent cations. J. Biol. Chem. **248:**4971–4976.

41. **Taku, A., K. G. Gunetileke, and R. A. Anwar.** 1970. Biosynthesis of uridine diphospho-N-acetylmuramic acid. III. Purification and properties of uridine diphospho-N-acetylenoylpyruvylglucosamine reductase. J. Biol. Chem. **245:**5012–5016.

42. **Vaara, M. V., T. Vaara, M. Jensen, I. Helander, M. Nurimen, E. T. Rietschel, and P. H. Mekala.** 1981. Characterization of the lipopolysaccharide from the polymyxin-resistant mutants of *Salmonella typhimurium*. FEBS Lett. **129:**145–149.

43. **Wang, S. F., and O. Gabriel.** 1969. Biological mechanisms involved in the formation of deoxysugars. V. Isolation and crystallization of thymidine diphosphoglucose oxidoreductase from *Escherichia coli* B. J. Biol. Chem. **244:**3430–3437.

44. **Wang, S. F., and O. Gabriel.** 1970. Biological mechanism involved in the formation of deoxysugars. VI. Role and function of enzyme bound nicotinamide adenine dinucleotide in thymidine diphosphate-D-glucose oxidoreductase. J. Biol. Chem. **245:**8–14.

45. **Wilson, D. B., and D. S. Hogness.** 1969. The enzymes of the galactose operon in *Escherichia coli*. II. The subunits of uridine diphosphogalactose-4-epimerase. J. Biol. Chem. **244:**2132–2136.

33. Biosynthesis of the Isoprenoid Quinones Ubiquinone and Menaquinone

RONALD BENTLEY[1] AND R. MEGANATHAN[2]

Department of Biological Sciences, University of Pittsburgh, Pittsburgh, Pennsylvania, 15260,[1] *and Department of Biological Sciences, Northern Illinois University, DeKalb, Illinois 60115*[2]

INTRODUCTION

As is generally the case for gram-negative facultatively anaerobic rods, *Escherichia coli* and *Salmonella typhimurium* contain isoprenoid quinones of both the benzene and the naphthalene series. The chemical structures of these quinones are shown in Fig. 1. In accordance with the IUPAC-IUB recommendations (37), the benzoquinones are termed ubiquinones (Q-*n*) (compound 1 in Fig. 1), and the naphthoquinones are termed either menaquinones (MK-*n*) (compound 2 in Fig. 1) or demethylmenaquinones (DMK-*n*) (compound 3 in Fig. 1). In each case, *n* refers to the number of prenyl units present. The major quinones in *E. coli* are Q-8 and MK-8; minor amounts of Q-1 to Q-7, Q-9, MK-6, MK-7, MK-9, and DMK-8 may also be present (17). The prenyl side chains are known to have the all-*trans* configuration (9). Although the quinone composition of *S. typhimurium* has not been studied in detail, both quinone types must be present because genes for their biosynthesis have been located. These organisms contain neither quinones that have one or more of the prenyl residues of the side chain are reduced, nor MK with more than one methyl group. Methods for the isolation and analysis of the quinones have been reviewed (21, 24, 47, 48, 53, 57–59, 65, 69–71, 76, 77, 81, 84, 85).

Most of the information concerning the biosynthesis of the bacterial Q and MK was obtained with *E. coli* by the use of isotope tracer methodology, the isolation of mutants, the preparation of enzyme extracts, and inductive processes. Owing to space limitations, only a general account can be given here; for more information, several comprehensive reviews should be consulted (10–12, 27, 32, 84). Q and MK have some common structural features that relate to their biosynthetic mechanisms: a quinone nucleus derived from chorismate, a prenyl side chain on the nucleus derived from a prenyl PP_i (and ultimately from mevalonate), and a nuclear methyl group derived from *S*-adenosylmethionine (except in the case of DMK). In addition to these precursors, MK biosynthesis requires 2-ketoglutarate as a precursor, as well as the following cofactors: thiamine PP_i, coenzyme A (CoA), and ATP. The biosynthesis of Q under aerobic conditions has additional requirements for oxygen, flavoprotein, and NADH. It should be noted that the Q biosynthetic pathway in procaryotes differs in several details from that in eucaryotes (28).

Despite the overall similarities in the pathways to Q and MK in *E. coli*, there are important differences.

(i) In the formation of the quinonoid nuclei, the two pathways diverge at the chorismate branchpoint of the shikimate pathway. In each case, a benzenoid aromatic acid is used as the framework on which the rest of the molecule is constructed:

$$\text{Shikimate} \rightarrow \text{Chorismate} \nearrow \frac{\text{4-Hydroxybenzoate} \rightarrow \text{Q}}{\searrow \; o\text{-Succinylbenzoate} \rightarrow \text{MK}}$$

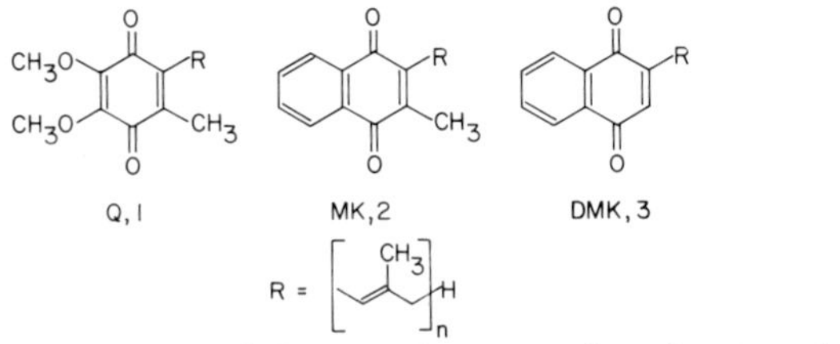

FIG. 1. Structures of the prenylquinones found in *E. coli* and *S. typhimurium*.

(ii) In Q biosynthesis, the prenyl side chain is introduced at an early stage of the pathway, and with retention of the aromatic carboxyl; for MK biosynthesis, prenylation is the next to the last step and is accompanied by a decarboxylation.

(iii) In MK biosynthesis, methylation onto carbon of the nucleus is the last step, whereas in Q biosynthesis the last step is methylation of a hydroxyl group; in Q biosynthesis, a second O methylation and the C methylation take place in the middle portion of the pathway.

(iv) Q biosynthesis, under aerobic conditions, requires the introduction of OH groups by reactions involving oxygen; anaerobic Q biosynthesis and MK biosynthesis utilize oxygen atoms from water.

Q BIOSYNTHESIS

The biosynthetic pathway for Q formation in *E. coli* (Fig. 2) owes much to the work of Gibson and colleagues (27). Sufficient amounts of intermediates were accumulated by using *E. coli* mutants so that structure could be determined. The use of mass spectrometry and nuclear magnetic resonance spectrometry was particularly important (79).

Formation of 4-Hydroxybenzoate (Compound 4 → 5)

Step 1 (Fig. 2), catalyzed by chorismate (pyruvate) lyase, is under the control of the *ubiC* gene (51). It is a straightforward elimination of pyruvate, leading to the formation of 4-hydroxybenzoate. Crude prepara-

tions of this enzyme were obtained from cells of *E. coli* AN246 (which lacks chorismate mutase-prephenate dehydratase and chorismate mutase-prephenate dehydrogenase activities). The *ubiC* gene was transduced into AN246, and a transductant unable to form 4-hydroxybenzoate, designated AN247, was selected. *E. coli* AN247 showed no chorismate lyase activity and consequently required 4-hydroxybenzoate for growth on a succinate minimal medium. A partial purification of the chorismate lyase has been described (J. Lawrence, Ph.D. thesis, Australian National University, 1973).

Prenylation of 4-Hydroxybenzoate (Compound 5 → 6)

The prenylation reaction (step 2), under control of the *ubiA* gene, requires a polyprenyl PP$_i$ as donor and results in the release of PP$_i$. The prenyl PP$_i$ with n isoprene units, is assembled in the usual way from 1 mol of dimethylallyl PP$_i$ and $n - 1$ mol of isopentenyl PP$_i$. However, the required precursor, 3-hydroxy-3-methylglutaryl (HMG)-CoA is apparently not derived from acetoacetyl-CoA and acetyl-CoA by the HMG-CoA synthase reaction (68). Instead, *E. coli* and some other bacteria produce HMG-CoA, by a lengthy process, from acetolactate (Fig. 3). The acetolactate is itself obtained from 1 mol of pyruvate and 1 mol of the acetaldehyde-thiamine PP$_i$ anion; acetolactate is, of course, also used as precursor for the biosynthesis of the branched-chain amino acids, leucine and valine (chapter 23). After formation of 2-ketoisocaproate (the keto acid corresponding to leucine), the usual catabolic reactions of the leucine pathway take place, with formation of HMG-CoA. This proposal accounts for the observation that [2-^{14}C]acetate yields dimethylallyl PP$_i$ and isopentenyl PP$_i$ carrying label at a single position (the HMG-CoA synthase reaction introduces label at two positions). Furthermore, the initially surprising observation that [1-^{14}C]acetate to a lesser extent labels the same position as does the 2-^{14}C-labeled material can be rationalized by assuming the operation of HMG-CoA lyase and, paradoxically, HMG-CoA synthase (68).

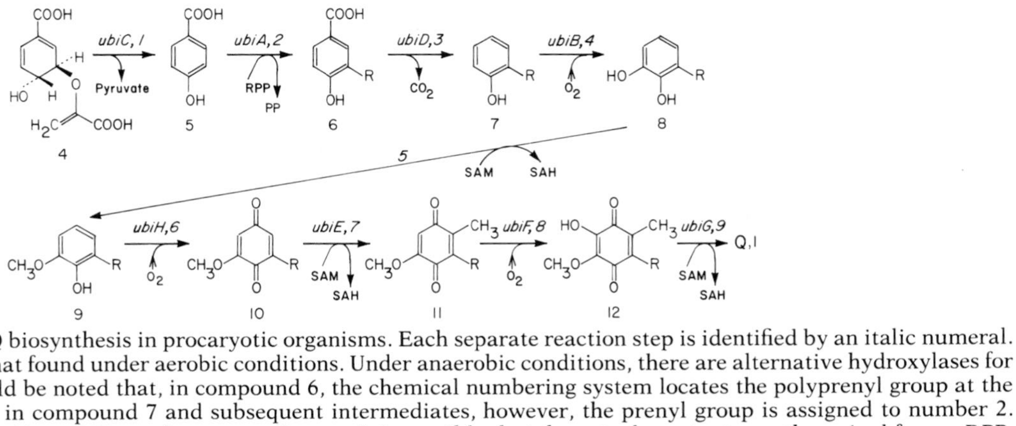

FIG. 2. Pathway for Q biosynthesis in procaryotic organisms. Each separate reaction step is identified by an italic numeral. The pathway drawn is that found under aerobic conditions. Under anaerobic conditions, there are alternative hydroxylases for steps 4, 6, and 8. It should be noted that, in compound 6, the chemical numbering system locates the polyprenyl group at the number 3 carbon atom; in compound 7 and subsequent intermediates, however, the prenyl group is assigned to number 2. Although compounds 10, 11, and 12 are drawn as quinones, it is possible that the actual reactants are the quinol forms. RPP, polyprenyl PP$_i$; SAM, *S*-adenosylmethionine; SAH, *S*-adenosylhomocysteine. Chemical names of the intermediates for Q-8 biosynthesis are as follows: 4, chorismate; 5, 4-hydroxybenzoate; 6, 3-octaprenyl-4-hydroxybenzoate; 7, 2-octaprenylphenol; 8, 2-octaprenyl-6-hydroxyphenol; 9, 2-octaprenyl-6-methoxyphenol; 10, 2-octaprenyl-6-methoxy-1,4-benzoquinone; 11, 2-octaprenyl-3-methyl-6-methoxy-1,4-benzoquinone; 12, 2-octaprenyl-3-methyl-5-hydroxy-6-methoxy-1,4-benzoquinone.

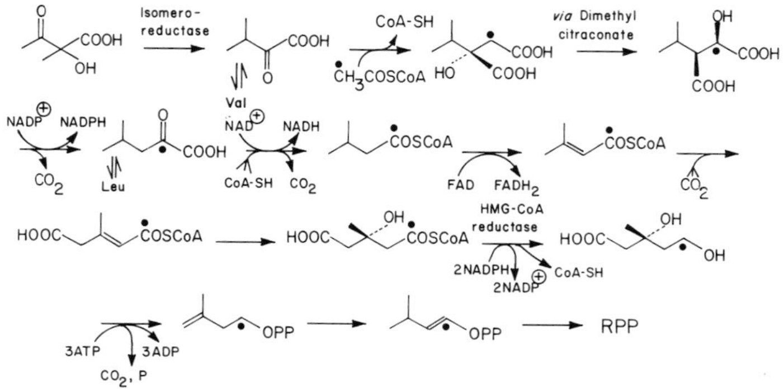

FIG. 3. Acetolactate pathway for the production of HMG-CoA and the intermediates needed for synthesis of the polyprenyl PP$_i$ (RPP). ^{14}C is indicated as ●. PP, PP$_i$; FAD, flavin adenine dinucleotide.

An enzyme preparation containing 4-hydroxybenzoate octaprenyltransferase was studied with an *E. coli* strain, AB2830, blocked in the common pathway of aromatic biosynthesis, *aroC* (87). This strain synthesized the necessary octaprenyl PP$_i$ and also accumulated the all-*trans* C-40 octaprenol (geranylgeraniol), presumably a breakdown product of the actual side chain precursor. Strain AN164, carrying an *aroB* allele, was also used. The enzyme activity and side chain precursor sedimented at 150,000 × *g* (3 h), indicating that both were membrane bound. Mg^{2+} was required for activity.

Similar work on the prenyltransferase used farnesyl, phytyl, or solanesyl PP$_i$ as an exogenous source of the side chain (25). With *E. coli* 83-1, blocked between dehydroquinate and dehydroshikimate, extracts produced 2-octaprenylphenol (compound 7; R = C-40), as well as a 3-polyprenyl-4-hydroxybenzoate (compound 6), when 4-hydroxybenzoate was added. Hence, in addition to the prenyltransferase activity, a decarboxylase was also present (presumably carrying out step 3, compound 6 → 7). The transferase activity was, however, more readily extracted. When *o*-succinylbenzoate was added to the incubation mixtures, the endogenous side chain precursor did not accumulate (it probably was being used for MK biosynthesis; see below). Under these conditions, the further addition of solanesyl PP$_i$ led to formation of a nonaprenyl-4-hydroxybenzoate (compound 6; R = C-45).

As indicated above, the transferase is relatively nonspecific with respect to the prenyl donor; since 4-aminobenzoate can replace 4-hydroxybenzoate as a substrate (25), the same lack of specificity extends to the aromatic component.

Formation of 2-Polyprenylphenol (Compound 6 → 7)

In 1969, cell extracts from *E. coli* AB3311 and AB2154 were shown to catalyze the step 3 reaction, converting 3-octaprenyl-4-hydroxybenzoate to 2-octaprenylphenol (20). Extracts of cells from the *ubiD* strain AN66 (which accumulates 3-octaprenyl-4-hydroxybenzoate) were inactive. As noted above, decarboxylase activity was also observed by El Hachimi et al. (25).

The process used to prepare the cell extracts resulted in a soluble enzyme, since activity was found in the supernatant fluid remaining after centrifugation at 150,000 × *g* (3 h). Protamine sulfate treatment, ammonium sulfate precipitation, and the use of a Bio-Gel A-5M column gave a 24-fold purification (54). The molecular mass of the decarboxylase (carboxy-lyase) was approximately 340,000 daltons. Activity was stimulated by the presence of methanol or ethanol, dithiothreitol, and a phospholipid preparation from cells of strain AN164. A metal requirement (optimally Mg^{2+}) was observed, and supernatant fluids of a cell extract of AN291 (*ubiD*) contained a further, unidentified, stimulatory factor. The enzyme is specific for the decarboxylation of 3-octaprenyl-4-hydroxybenzoate.

The *ubiD* mutants are unusual in that they show a definite level of Q (about 20% of the wild-type levels [20]). It is possible that the mutant is leaky or that there is an alternative pathway for the decarboxylation (55).

Hydroxylation Reactions

The hydroxylation reactions are the conversion of 2-octaprenylphenol to 2-octaprenyl-6-hydroxyphenol (compound 7 → 8), the conversion of 2-octaprenyl-6-methoxyphenol to 2-octaprenyl-6-methoxy-1,4-benzoquinone (compound 9 → 10), and the conversion of 2-octaprenyl-3-methyl-6-methoxy-1,4-benzoquinone to 2-octaprenyl-3-methyl-5-hydroxy-6-methoxy-1,4-benzoquinone (compound 11 → 12). A characteristic property of the Q biosynthetic pathway is the requirement for the introduction of three hydroxyl groups at positions 6, 4, and 5 of the benzene nucleus (steps 4, 6, and 8, respectively). For convenience, these hydroxylations will be considered together, and this description will be followed by a consideration of the methylation reactions.

It has been found that, under aerobic growth conditions, the hydroxyl groups derive from molecular oxygen but that alternative hydroxylases are available under anaerobic conditions, with the oxygen atoms presumably deriving from water. When *E. coli* AB3311 was grown in an atmosphere of $^{18}O_2$, the mass spectrum of the isolated Q was characterized by several prominent peaks with *m/z* values differing from those of normal Q by +6. Thus, three atoms of ^{18}O had been incorporated, and they were shown to be located at positions 4, 5, and 6 (1).

The exact nature of these hydroxylations is not clear. A mutant *hemA* defective in heme synthesis, *E. coli* A1004C, produced Q under aerobic growth conditions (44); no cytochromes were detected in this strain, so the involvement of a cytochrome P-450 monooxygenase system was ruled out. Earlier work had indicated that some, but not all, P-450 inhibitors decreased the conversion of 2-octaprenylphenol to Q. It is possible that a flavin monooxygenase system is involved (43). The three genes *ubiB*, *ubiH*, and *ubiF* specify the aerobic monooxygenases.

Appreciable quantities of Q (50 to 70% of aerobic growth levels) are formed by a wide variety of *E. coli* strains grown anaerobically on glycerol-fumarate medium. In addition, high levels of 2-octaprenylphenol may be present in some cases, although this material does not accumulate significantly under aerobic growth conditions. Although the *ubiB*, *ubiH*, and *ubiF* mutants do not form Q aerobically, growth of these mutants under anaerobic conditions leads to Q biosynthesis (2). Clearly, under anaerobiosis, there must exist three hydroxylation enzymes as alternatives to the aerobic monooxygenases. The nature of these anaerobic hydroxylases remains unknown. The OH group presumably derives ultimately from water; in other systems, such reactions are associated with electron transport systems which are often cytochrome linked.

The compound 2-octaprenyl-6-hydroxyphenol, compound 8, is the putative product of the first hydroxylation, step 4, but it is the one pathway intermediate for which there is only indirect evidence (27). It has not been satisfactorily characterized; furthermore, mutants defective in its methylation (step 5, compound 8 → 9) have not been identified. The products of the two other reactions (steps 6 and 8) are, respectively, 2-octaprenyl-6-methoxy-1,4-benzoquinone, compound 10, and 2-octaprenyl-3-methyl-5-hydroxy-6-methoxy-1,4-benzoquinone, compound 12; these materials have been characterized.

Methylation Reactions

The methylation reactions are the conversion of 2-octaprenyl-6-hydroxyphenol to 2-octaprenyl-6-methoxyphenol (compound 8 → 9), the conversion of 2-octaprenyl-6-methoxy-1,4-benzoquinone to 2-octaprenyl-3-methyl-6-methoxy-1,4-benzoquinone (compound 10 → 11), and the conversion of 2-octaprenyl-3-methyl-5-hydroxy-6-methoxy-1,4-benzoquinone to Q (compound 12 → 1). These methylation reactions (steps 5, 7, and 9) alternate with the three hydroxylations, introducing methyl groups at the 6-OH, at the ring C-3, and at the 5-OH group, respectively. These groups are known to originate from methionine (38), with S-adenosylmethionine being the actual methyl donor. The intermediates, compounds 9 and 11 (2-octaprenyl-6-methoxyphenol and 2-octaprenyl-3-methyl-6-methoxy-1,4-benzoquinone, respectively) have been characterized, and mutants have been isolated that are blocked in the last two methylation reactions (step 7, *ubiE*, and step 9, *ubiG*) (27). The terminal reaction (step 9) leading to Q itself has been studied in cell extracts of *E. coli* AB3311 and AN164, *aroB* (54). The methyltransferase was separated from the membrane fraction by centrifugation and purified 14-fold (protamine sulfate to precipitate 70% of the nucleic acid present, ammonium sulfate precipitation, and the use of a Bio-Gel A-5M column). The molecular mass of the enzyme was about 50,000 daltons. After purification, there was an absolute requirement for dithiothreitol and a divalent metal ion (optimally Zn^{2+}). Although the quinone form of the substrate, compound 12, was added to the incubation mixtures, and although Q itself was isolated, the reaction in this and the other steps shown in Fig. 2 with quinone intermediates may actually involve the quinol forms of the substrates.

Localization of Q Biosynthetic Enzymes in *E. coli*

Although, as previously indicated, a number of the enzymes can be obtained as soluble preparations, all of the enzymes for reaction steps 2 to 9 are believed to be associated with the cytoplasmic membrane (54, 55). A soluble enzyme complex isolated from *E. coli* K-12 cytoplasmic membranes has a molecular mass of 2×10^6 daltons and contains at least 12 proteins (41). It also contains a high level of 2-octaprenylphenol, compound 7, but levels of Q, phospholipid, and other cytoplasmic membrane proteins are low. Since this complex was prepared without detergent treatment, it was thought that it had broken from the membrane as a distinct and native domain. On treatment with a cytoplasmic enzyme (probably a methyltransferase), S-adenosylmethionine, O_2, and other cofactors, all of the 2-octaprenylphenol contained in the complex was converted to Q-8. This complex, therefore, contains the oxygen-dependent Q-8 biosynthesis apparatus. In anaerobically grown cells, this apparatus, which is charged with a high level of 2-octaprenylphenol, may be kept in a standby position; if oxygen becomes available, Q-8 synthesis by the aerobic monooxygenases can be effectively turned on (42).

ubi Genes of *E. coli* and Some Consequences of Defects in Q Biosynthesis

In general, the isolation of *E. coli* mutants defective in Q biosynthesis has depended on the inability of such strains to grow on nonfermentable substrates (e.g., succinate and malate) while retaining the ability to grow on glucose (18). Extensive studies by Gibson and colleagues (27) have shown the following locations for the *ubi* genes: *ubiA* and *ubiC*, 92 min; *ubiB*, *ubiD*, and *ubiE*, 86 min; *ubiH*, 63 min; *ubiG*, 48 min; and *ubiF*, 15 min (5).

The *ubiF* gene is deficient in a strain of *E. coli* (NSW77) which shows a partial resistance to streptomycin (66). This resistance is probably associated with a reduced capability for the intracellular accumulation of antibiotics. Membranes of this strain contain 2-octaprenyl-3-methyl-6-methoxy-1,4-benzoquinone, compound 11, but no Q-8. Furthermore, the previously characterized *ubiF* mutant, AN146, shows a reduced uptake of gentamicin. There is, at present, no direct evidence for a Q function in aminoglycoside antibiotic uptake, and the observations are attributed to the general impairment of respiratory capability.

Mutations in the Q biosynthetic pathway give rise to immotility and lack of flagellum synthesis under aerobic growth conditions; specifically tested were

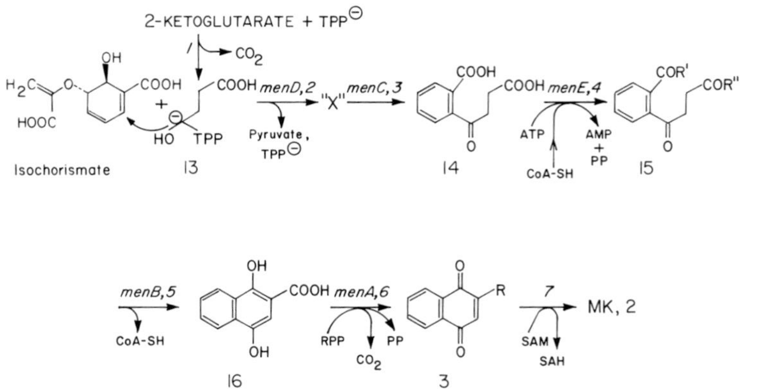

FIG. 4. Menaquinone biosynthetic pathway. Each separate reaction step is identified by a numeral as described in the legend to Fig. 2. Compound 3 (DMK) may be formed initially as the quinol structure. Chemical names of the intermediates are as follows: 14, *o*-succinylbenzoic acid; 15, monoCoA ester of *o*-succinylbenzoic acid; 16, 1,4-dihydroxy-2-naphthoic acid; 3, DMK; PP, PP$_i$.

ubiD, *ubiB*, and *ubiG* mutants (8, 35). The mutant *ubiA men$^+$* was motile under anaerobic growth conditions; however, with defects in both the Q and the MK biosynthetic pathways, mutants were immotile under both anaerobic and aerobic growth conditions. There is obviously a relationship between flagellum synthesis and a functional electron transport chain.

Q Biosynthesis in *S. typhimurium*

There is only fragmentary information concerning the biosynthesis of Q in *S. typhimurium*, usually in relation to other physiological events. As with *E. coli*, flagellum formation in *S. typhimurium* requires a functional Q biosynthetic pathway (7). One mutant isolated in the course of this work, TA1851, phenotypically resembles the *E. coli ubiD* mutants in being deficient in decarboxylation of 2-octaprenyl-4-hydroxybenzoate but still maintaining about 20 to 35% of the wild-type Q level. Possible mechanisms for the observed low level of Q formation in this mutant have been discussed (36).

Studies of pyrimidine metabolism in *S. typhimurium* led to the isolation of a mutant (KR42) which was dependent on carbamylaspartate for resistance to 5-fluorouracil (40, 88). The mutation, arbitrarily designated *cad*, was impaired in energy metabolism and appeared to be located within the corresponding region of the *E. coli* K-12 chromosome containing the *ubiF* gene. Assuming that the genetic and biochemical organization for Q biosynthesis is the same in the two organisms, it is inferred that KR42 is a *ubiF* mutant. Whether the alteration of *pyr* gene expression stems from a direct interrelationship with Q biosynthesis or a more general physiological effect is not known.

Ames and colleagues (3, 4) have isolated many *S. typhimurium* mutants which affect the high-affinity histidine transport system; some of these mutants show growth stimulation in the presence of 4-hydroxybenzoate. They also have 25% of the wild-type level of Q, and this low level is ascribed to the deletion of one or more of the *ubi* genes; the deletion is referred to as *ubiX*. Although the *ubiX* locus is

carried on a cloned 12.4-kilobase fragment of DNA (sa-1), nothing is known about the nature or the number of the genes.

Functions of Q

The role of Q in bacterial motility and flagellum formation has been discussed briefly above. The role of Q in electron transport and nitrate respiration has been extensively discussed (31, 39).

MK BIOSYNTHESIS

As was the case with Q, the MK biosynthetic pathway (Fig. 4) has been elucidated on the basis of tracer experiments, inspired guesses, and, particularly in the case of *E. coli*, the use of mutants and enzyme preparations. Early tracer experiments with various bacteria established roles for methionine and mevalonate, respectively, for the methyl and polyprenyl substituents. This and other work has been comprehensively reviewed by Bentley and Meganathan (12). Similarly, tracer experiments established a role for shikimate and indicated that all seven carbon atoms of this precursor were utilized in construction of the naphthoquinone ring system. The remaining three atoms of the ring system were eventually found to be the three central atoms of either glutamate or its deaminated product, 2-ketoglutarate. Of particular importance was the suggestion of roles for *o*-succinylbenzoate (OSB) (22) and 1,4-dihydroxy-2-naphthoate (DHNA) (16, 72). As will be seen, experimental verification of roles for these intermediates has been obtained.

Formation of OSB, Compound 14, from Isochorismate

Dansette and Azerad (22) showed that OSB supported growth and allowed MK formation in *E. coli* mutants blocked after shikimate. Tracer experiments confirmed that OSB was a good precursor for MK formation (12), and *menB* mutants were found to accumulate OSB (86). In a direct study of OSB biosynthesis from chorismate and glutamate, it was

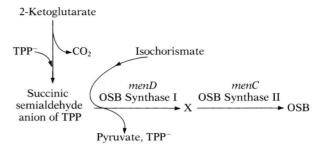

FIG. 5. Proposed pathway for the formation of the succinic semialdehyde-TPP anion, compound 13. Only the thiazole ring component of TPP is shown, and the substituent groups have been omitted.

shown that C-1 of glutamate was lost, whereas C-2 to C-5 were retained (63).

Since *E. coli* mutants blocked in the formation of chorismate do not form MK (19, 22, 52), chorismate was assumed to be the branchpoint for formation of OSB. It was also suggested that isochorismate was a more attractive precursor on chemical grounds (22, 32), and Weische and Leistner (83) have now obtained evidence supporting this hypothesis. Since *E. coli* strains with the *entC* mutation, and thus unable to convert chorismate to isochorismate, are able to form MK (86), it appears that these mutants must be leaky.

The reactions for OSB formation, steps 1 and 2 of Fig. 4, are the least understood portions of the biosynthetic pathway. The overall reaction may be written as follows: isochorismate + 2-ketoglutarate → CO_2 + OSB + pyruvate.

The first evidence for an OSB synthase activity was obtained by Meganathan (60) by using *E. coli* AN154 (a strain blocked in all of the aromatic pathways, except that for MK). Extracts made with a French press were incubated with 2-[U-^{14}C]ketoglutarate and chorismate in the presence of thiamine PP_i (TPP). OSB formed by the extracts was converted to a dimethyl derivative for examination by radio-gas chromatography. The OSB synthase activity has also been obtained in extracts of wild-type organisms (13, 64). It has been postulated that 2-ketoglutarate undergoes a TPP-dependent decarboxylation, with the formation of the succinic semialdehyde anion of TPP (15), and there is evidence that TPP is actually involved in the synthesis of OSB (64). This process, step 1, is the same as that postulated for the first reaction catalyzed by the 2-ketoglutarate dehydrogenase complex (Fig. 5). However, it has been possible to separate the 2-ketoglutarate dehydrogenase complex from enzymes synthesizing OSB by either treatment with protamine sulfate (83) or gel-permeation chromatography on Sepharose CL-6B (56). Other evidence (56) also supports the existence of a separate decarboxylase activity for 2-ketoglutarate during MK formation. The succinic semialdehyde anion of TPP formed by the decarboxylase, compound 13, is believed to react directly with isochorismate (83).

Further information has come from the fact that there are two groups of *E. coli* mutants, *menC* and *menD*, involved in OSB synthesis (29, 30). Moreover, evidence has been obtained for the participation of at least two enzymes (OSB synthase I and II) and for the involvement of an intermediate, termed X (64). Since there are strong indications that X does not contain a pyruvoyl group or a TPP residue a more detailed representation of OSB synthesis is as follows:

Despite an inherent instability, partial purification of X has been possible by high-pressure liquid chromatography. Furthermore, on mild acid treatment, X was found to yield some OSB and a further material, unambiguously identified as succinylbenzene. These properties suggest that X is a cyclohexadiene carboxylic acid containing -OH and -$COCH_2CH_2COOH$ groups. Several possible reaction schemes can be written to account for the formation of various structural isomers having these features and for their conversion to OSB and succinylbenzene; the most probable structure (Fig. 6), on the basis of nuclear magnetic resonance data, is 2-succinyl-6-hydroxy-2,4-cyclohexadiene-1-carboxylic acid (26).

Conversion of OSB to 1,4-Dihydroxy-2-Naphthoate (Compound 14 → 16)

Experimental evidence that DHNA was actually involved in MK biosynthesis was first obtained in 1973 with the normally MK-requiring organism *Bacteroides melaninogenicus* (73). Young (86) showed that *menA* mutants accumulated DHNA, and the conversion of OSB to DHNA was demonstrated in *E. coli* extracts in 1976 (14); it was found to be dependent on the presence of ATP and CoA.

Therefore, it was suggested that an OSB-CoA derivative was formed as an intermediate in the overall reaction (14, 62); the activated intermediate was isolated in 1981 and was shown to be a monoCoA ester (33, 34). With extracts of *Mycobacterium phlei*, two enzyme activities were separated (OSB-CoA synthetase, step 4, and DHNA synthase, step 5) (62). The step 4 reaction was also shown to result in the production of AMP and PP_i. However, in *E. coli*, the techniques used for separation of the two enzyme activities in *M. phlei* (dimethyl sulfoxide-protamine sulfate precipitation) led to loss of DHNA synthase activity.

Although the CoA group was originally suggested to be on the aromatic carboxyl (compound 15; R' = CoA, R" = OH) (14, 62), and although initial evidence favored this suggestion (12, 33, 45), a recent note suggests that the CoA group is, in fact, located on the aliphatic carboxyl (compound 15; R" = CoA, R' = OH) (46). The picture is complicated by the statement that the "enzymically formed OSB monoCoA ester consisted of both the aliphatic (85%) as well as the aromatic (15%) CoA ester" (46) and by the greater instability of

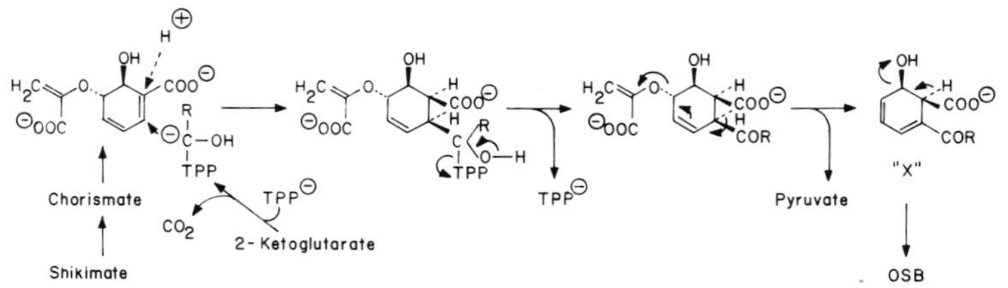

FIG. 6. Possible reaction for the formation of the intermediate, X, and for its conversion of OSB. The formation of succinylbenzene from the probable structure of X (2-succinyl-6-hydroxy-2,4-cyclohexadiene-1-carboxylic acid) can be rationalized readily.

the aliphatic ester. Clarification of this matter will have to await the complete details of the work from Leistner's laboratory.

Conversion of 1,4-Dihydroxy-2-Naphthoate to DMK (Compound 16 → 3)

The conversion of 1,4-dihydroxy-2-naphthoate to DMK (step 6) was demonstrated in intact cells of E. coli in 1975, and cell extracts were also shown to convert DHNA and farnesyl PP_i to MK-3, DMK-3, or both (11). A membrane-bound 1,4-dihydroxy-2-naphthoate octaprenyltransferase was studied in more detail by Shineberg and Young (78). This enzyme has some features in common with the transferase involved in Q biosynthesis (being membrane bound, Mg^{2+} requiring, and relatively nonspecific for the polyprenyl PP_i). Both this enzyme and the 4-hydroxybenzoate octaprenyltransferase probably use a common pool of membrane-bound octaprenyl PP_i (42, 87). In this event, the polyprenyl PP_i is presumably derived by the acetolactate mechanism (Fig. 3); however, this question has not been examined experimentally. Demethylmenaquinol is a likely intermediate, although the exact nature of the conversion of DHNA to DMK is not known. Prenylation and decarboxylation may be concerted since symmetry experiments exclude 1,4-naphthoquinone itself as an intermediate (6). Assuming the demethylmenaquinol to be an intermediate, the formation of DMK presumably occurs by a spontaneous oxidation.

Conversion of DMK to MK (Compound 3 → 2)

The methyltransferase for step 7 requires S-adenosylmethionine; it is known that, in intact organisms, all three hydrogen atoms of methionine are transferred to DMK (38). The S-adenosylmethionine:2-demethylmenaquinone methyltransferase has been studied to some extent in cell extracts of E. coli (14); a transferase isolated from M. phlei showed maximal activity towards DMK-3 and DMK-4 (74). Although not investigated in detail, E. coli 2-45 is believed to be defective in the methylation of DMK to MK; this strain is also a ubiA mutant (25). Methyltransferase and prenyltransferase are apparently membrane-bound enzymes; however, all of the activities leading to DHNA formation have proved to be readily soluble.

men Genes of E. coli and Some Consequences of Defects in MK Biosynthesis

The men mutants of E. coli have been isolated by mutagenesis (12, 18); mutants defective in both MK

and Q biosynthesis have also been obtained (82). The menB, menC, menD, and menE genes form a cluster at 48.5 min on the E. coli linkage map, and the menA gene is located at 88 min. A transducing phage containing functional menC and menB genes and part of the menD gene has been described, and the gene order is believed to be nalA-menC-menB-menD-purF. However, the exact location of menE relative to the other men genes is not known (75).

A major role for bacterial menaquinones in E. coli is in anaerobic electron transport (80). In particular, the use of fumarate as a terminal electron acceptor under anaerobic conditions has been studied. E. coli mutants lacking MK fail to grow anaerobically on glucose unless uracil is added; normally, MK functions as electron acceptor in the oxidation of dihydroorotate and subsequently transfers the acquired electrons to fumarate (67). More recently, Meganathan (61) has studied men mutants of E. coli that are unable to use trimethylamine-N-oxide as an electron acceptor (growth on a complex medium with glucose). Reduction of this electron acceptor was restored by either OSB or DHNA for a menC mutant but was restored only by DHNA for a menB mutant. These results are consistent with the metabolic blocks previously described (for similar work with S. typhimurium, see below).

men Genes in S. typhimurium

In contrast to the extensive work on MK biosynthesis in E. coli, relatively little has been reported with S. typhimurium. Davidson et al. (23) and Kwan and Barrett (49) have isolated and studied mutants which are defective in the reduction of trimethylamine-N-oxide. These mutants showed genotypes which corresponded to menA (TC94), menB (TC99), menC (EB127), and menD (TC88) mutants of E. coli. The menBCD cluster lies between ack/pta and glpT, as is the case in E. coli. In this work, it was found that men mutants, when streaked side by side, did not cross-feed each other. However, cross-feeding with respect to providing intermediates for H_2S production was detected by using S. typhimurium mutants and was detected between S. typhimurium and E. coli men mutants (50).

Functions of MK

Some of the functions of MK were briefly noted above. For details of the role of MK in fumarate reduction and anaerobic respiration, other reviews should be consulted (31, 39).

LITERATURE CITED

1. **Alexander, K., and I. G. Young.** 1978. Three hydroxylations incorporating molecular oxygen in the aerobic biosynthesis of ubiquinone in *Escherichia coli.* Biochemistry **17:**4745–4750.
2. **Alexander, K., and I. G. Young.** 1978. Alternative hydroxylases for the aerobic and anaerobic biosynthesis of ubiquinone in *Escherichia coli.* Biochemistry **17:**4750–4755.
3. **Ames, G. F.-L., K. D. Noel, H. Taber, E. N. Spudich, K. Nikaido, J. Afong, and F. Ardeshir.** 1977. Fine-structure map of the histidine transport genes in *Salmonella typhimurium.* J. Bacteriol. **129:**1289–1297.
4. **Ardeshir, F., and G. F.-L. Ames.** 1980. Cloning of the histidine transport genes from *Salmonella typhimurium* and characterization of an analogous transport system in *Escherichia coli.* J. Supramol. Struct. **13:**117–130.
5. **Bachmann, B. J.** 1983. Linkage map of *Escherichia coli* K-12, edition 7. Microbiol. Rev. **47:**180–230.
6. **Baldwin, R. M., C. D. Snyder, and H. Rapoport.** 1974. Biosynthesis of bacterial menaquinones. Dissymmetry in the naphthalenic intermediates. Biochemistry **13:**1523–1530.
7. **Bar-Tana, J., B. J. Howlett, and R. Hertz.** 1980. Ubiquinone synthetic pathway in flagellation of *Salmonella typhimurium.* J. Bacteriol. **143:**637–643.
8. **Bar-Tana, J., B. J. Howlett, and D. E. Koshland, Jr.** 1977. Flagellar formation in *Escherichia coli* electron transport mutants. J. Bacteriol. **130:**787–792.
9. **Bentley, R.** 1970. Molecular asymmetry in biology, vol. 2, p. 480–486. Academic Press, Inc., New York.
10. **Bentley, R.** 1975. Biosynthesis of quinones. Biosynthesis **3:**181–246.
11. **Bentley, R.** 1975. Biosynthesis of vitamin K and other natural naphthoquinones. Pure Appl. Chem. **41:**47–68.
12. **Bentley, R., and R. Meganathan.** 1982. Biosynthesis of vitamin K (menaquinone) in bacteria. Microbiol. Rev. **46:**241–280.
13. **Bentley, R., and R. Meganathan.** 1983. Vitamin K biosynthesis in bacteria—precursors, intermediates, enzymes, and genes. J. Natl. Prod. (Lloydia) **46:**44–59.
14. **Bryant, R. W., Jr., and R. Bentley.** 1976. Menaquinone biosynthesis: conversion of *o*-succinylbenzoic acid to 1,4-dihydroxy-2-naphthoic acid and menaquinones by *Escherichia coli* extracts. Biochemistry **15:**4792–4796.
15. **Campbell, I. M.** 1969. The roles of alanine, aspartate and glutamate in lawsone biosynthesis in *Impatiens balsamina.* Tetrahedron Lett. **1969:**4777–4780.
16. **Campbell, I. M., D. J. Robins, M. Kelsey, and R. Bentley.** 1971. Biosynthesis of bacterial menaquinones (vitamin K₂). Biochemistry **10:**3069–3078.
17. **Collins, M. D., and D. Jones.** 1981. Distribution of isoprenoid quinone structural types in bacteria and their taxonomic implications. Microbiol. Rev. **45:**316–354.
18. **Cox, G. B., and J. A. Downie.** 1979. Isolation and characterization of mutants of *Escherichia coli* K-12 affected in oxidative phosphorylation or quinone biosynthesis. Methods Enzymol. **56:**106–117.
19. **Cox, G. B., and F. Gibson.** 1964. Biosynthesis of vitamin K and ubiquinone. Relation to the shikimic acid pathway in *Escherichia coli.* Biochim. Biophys. Acta **93:**204–206.
20. **Cox, G. B., I. G. Young, L. M. McCann, and F. Gibson.** 1969. Biosynthesis of ubiquinone in *Escherichia coli* K-12: location of genes affecting the metabolism of 3-octaprenyl-4-hydroxybenzoic acid and 2-octaprenylphenol. J. Bacteriol. **99:**450–458.
21. **Crane, F. L., and R. Barr.** 1971. Determination of ubiquinones. Methods Enzymol. **18:**137–165.
22. **Dansette, P., and R. Azerad.** 1970. A new intermediate in naphthoquinone and menaquinone biosynthesis. Biochem. Biophys. Res. Commun. **40:**1090–1095.
23. **Davidson, A. E., H. E. Fukumoto, C. E. Jackson, E. L. Barrett, and G. W. Chang.** 1979. Mutants of *Salmonella typhimurium* defective in the reduction of trimethylamine oxide. FEMS Microbiol. Lett. **6:**417–420.
24. **Dunphy, P. J., and A. F. Brodie.** 1971. The structure and function of quinones in respiratory metabolism. Methods Enzymol. **18:**407–461.
25. **El Hachimi, Z., O. Samuel, and R. Azerad.** 1974. Biochemical study on ubiquinone biosynthesis in *Escherichia coli.* I. Specificity of *para*-hydroxybenzoate polyprenyltransferase. Biochimie **56:**1239–1247.
26. **Emmons, G. T., I. M. Campbell, and R. Bentley.** 1985. Vitamin K (menaquinone) biosynthesis in bacteria: purification and probable structure of an intermediate prior to *o*-succinylbenzoate. Biochem. Biophys. Res. Commun. **131:**956–960.
27. **Gibson, F.** 1973. Chemical and genetic studies on the biosynthesis of ubiquinone by *Escherichia coli.* Biochem. Soc. Trans. **1:**317–326.
28. **Goewert, R. C., C. J. Sippel, M. F. Grimm, and R. E. Olson.** 1981. Identification of 3-methoxy-4-hydroxy-5-hexaprenylbenzoic acid as a new intermediate in ubiquinone biosynthesis by *Saccharomyces cerevisiae.* Biochemistry **20:**5611–5616.
29. **Guest, J. R.** 1977. Menaquinone biosynthesis: mutants of *Escherichia coli* K-12 requiring 2-succinylbenzoate. J. Bacteriol. **130:**1038–1046.
30. **Guest, J. R.** 1979. Anaerobic growth of *Escherichia coli* K-12 with fumarate as terminal electron acceptor. Genetic studies with menaquinone and fluoroacetate-resistant mutants. J. Gen. Microbiol. **115:**259–271.
31. **Haddock, B. A., and W. A. Hamilton. (ed.).** 1977. Microbial energetics. Cambridge University Press, Cambridge.
32. **Haslam, E.** 1974. The shikimate pathway. John Wiley & Sons, Inc., New York.
33. **Heide, L., S. Arendt, and E. Leistner.** 1982. Enzymatic synthesis, characterization, and metabolism of the coenzyme A ester of *o*-succinylbenzoic acid, an intermediate in menaquinone (vitamin K₂) biosynthesis. J. Biol. Chem. **257:**7396–7400.
34. **Heide, L., and E. Leistner.** 1981. Enzymatic synthesis of the coenzyme-A ester of *o*-succinylbenzoic acid, an intermediate in menaquinone (vitamin K₂) biosynthesis. FEBS Lett. **128:**201–204.
35. **Hertz, R., and J. Bar-Tana.** 1977. Anaerobic electron transport in anaerobic flagellum formation in *Escherichia coli.* J. Bacteriol. **132:**1034–1035.
36. **Howlett, B. J., and J. Bar-Tana.** 1980. Polyprenyl *p*-hydroxybenzoate carboxylyase in flagellation of *Salmonella typhimurium.* J. Bacteriol. **143:**644–651.
37. **IUPAC-IUB Commission on Biochemical Nomenclature.** 1965. Tentative rules. Nomenclature of quinones with isoprenoid side-chains. Biochim. Biophys. Acta **107:**5–10.
38. **Jackman, L. M., I. G. O'Brien, G. B. Cox, and F. Gibson.** 1967. Methionine as the source of methyl groups for ubiquinone and vitamin K: a study using nuclear magnetic resonance and mass spectrometry. Biochim. Biophys. Acta **141:**1–7.
39. **Jones, C. W.** 1982. Bacterial respiration and photosynthesis. American Society for Microbiology, Washington, D.C.
40. **Kelln, R. A., and V. L. Zak.** 1980. A mutation in *Salmonella typhimurium* imparting conditional resistance to 5-fluorouracil and a bioenergetic defect: mapping of *cad.* Mol. Gen. Genet. **179:**677–681.
41. **Knoell, H.-E.** 1979. Isolation of a soluble enzyme complex comprising the ubiquinone-8 synthesis apparatus from the cytoplasmic membrane of *Escherichia coli.* Biochem. Biophys. Res. Commun. **91:**919–925.
42. **Knoell, H.-E.** 1981. Stand-by position of the dioxygen-dependent ubiquinone-8 synthesis apparatus in anaerobically grown *Escherichia coli* K-12. FEMS Microbiol. Lett. **10:**59–62.
43. **Knoell, H.-E.** 1981. On the nature of the monooxygenase system involved in ubiquinone-8 synthesis. FEMS Microbiol. Lett. **10:**63–65.
44. **Knoell, H.-E., R. Kraft, and J. Knappe.** 1978. Dioxygen and temperature dependence of ubiquinone formation in *Escherichia coli*: studies of cells charged with 2-octaprenylphenol. Eur. J. Biochem. **90:**107–112.
45. **Kolkmann, R., G. Knauel, S. Arendt, and E. Leistner.** 1982. Site of activation of *o*-succinylbenzoic acid during its conversion to menaquinone (vitamin K₂). FEBS Lett. **137:**53–56.
46. **Kolkmann, R., and E. Leistner.** 1985. Synthesis and revised structure of the *o*-succinylbenzoic acid coenzyme A ester, an intermediate in menaquinone biosynthesis. Tetrahedron Lett. **26:**1703–1704.
47. **Krivánková, L., and V. Dadák.** 1980. Semimicro extraction of ubiquinone and menaquinone from bacteria. Methods Enzymol. **67:**1111–1114.
48. **Kröger, A.** 1978. Determination of contents and redox states of ubiquinone and menaquinone. Methods Enzymol. **53:**579–591.
49. **Kwan, H. S., and E. L. Barrett.** 1983. Roles for menaquinone and the two trimethylamine oxide (TMAO) reductases in TMAO respiration in *Salmonella typhimurium*: Mu *d*(Ap^r *lac*) insertion mutants in *men* and *tor.* J. Bacteriol. **155:**1147–1155.
50. **Kwan, H. S., and E. L. Barrett.** 1984. Map locations and functions of *Salmonella typhimurium men* genes. J. Bacteriol. **159:**1090–1092.
51. **Lawrence, J., G. B. Cox, and F. Gibson.** 1974. Biosynthesis of ubiquinone in *Escherichia coli* K-12: biochemical and genetic characterization of a mutant unable to convert chorismate into 4-hydroxybenzoate. J. Bacteriol. **118:**41–45.
52. **Leduc, M. M., P. M. Dansette, and R. G. Azerad.** 1970. Incorporation de l'acide shikimique dans le noyau des naphthoquinones

d'origine bactérienne et végétale. Eur. J. Biochem. **15:**428–435.

53. **Leistner, E., and M. H. Zenk.** 1971. Biosynthesis of vitamin K_2. Methods Enzymol. **18:**547–559.

54. **Leppik, R. A., P. Stroobant, B. Shineberg, I. G. Young, and F. Gibson.** 1976. Membrane-associated reactions in ubiquinone biosynthesis. 2-Octaprenyl-3-methyl-5-hydroxy-6-methoxy-1,4-benzoquinone methyltransferase. Biochim. Biophys. Acta **428:**146–156.

55. **Leppik, R. A., I. G. Young, and F. Gibson.** 1976. Membrane-associated reactions in ubiquinone biosynthesis in *Escherichia coli.* 3-Octaprenyl-4-hydroxybenzoate carboxy-lyase. Biochim. Biophys. Acta **436:**800–810.

56. **Marley, M. G., R. Meganathan, and R. Bentley.** 1986. Menaquinone (vitamin K_2) biosynthesis in *Escherichia coli*: synthesis of *o*-succinylbenzoate does not require the decarboxylase activity of the ketoglutarate dehydrogenase complex. Biochemistry **25:**1304–1307.

57. **Mayer, H., and O. Isler.** 1971. Isolation of vitamins K. Methods Enzymol. **18:**469–491.

58. **Mayer, H., and O. Isler.** 1971. Synthesis of ubiquinones. Methods Enzymol. **18:**182–213.

59. **Mayer, H., and O. Isler.** 1971. Synthesis of vitamins K. Methods Enzymol. **18:**491–547.

60. **Meganathan, R.** 1981. Enzymes from *Escherichia coli* synthesize *o*-succinylbenzoic acid, an intermediate in menaquinone (vitamin K_2) biosynthesis. J. Biol. Chem. **256:**9386–9388.

61. **Meganathan, R.** 1984. Inability of *men* mutants of *Escherichia coli* to use trimethylamine-*N*-oxide as an electron acceptor. FEMS Microbiol. Lett. **24:**57–62.

62. **Meganathan, R., and R. Bentley.** 1979. Menaquinone (vitamin K_2) biosynthesis: conversion of *o*-succinylbenzoic acid to 1,4-dihydroxy-2-naphthoic acid by *Mycobacterium phlei* enzymes. J. Bacteriol. **140:**92–98.

63. **Meganathan, R., and R. Bentley.** 1981. Biosynthesis of *o*-succinylbenzoic acid in a *men⁻ Escherichia coli* mutant requires decarboxylation of L-glutamate at the C-1 position. Biochemistry **20:**5336–5340.

64. **Meganathan, R., and R. Bentley.** 1983. Thiamine pyrophosphate requirement for *o*-succinylbenzoic acid synthesis in *Escherichia coli* and evidence for an intermediate. J. Bacteriol. **153:**739–746.

65. **Morimoto, H., and I. Imada.** 1971. Gas chromatography of ubiquinone and related quinones. Methods Enzymol. **18:**169–179.

66. **Muir, M. E., D. R. Hanwell, and B. J. Wallace.** 1981. Characterization of a respiratory mutant of *Escherichia coli* with reduced uptake of aminoglycoside antibiotics. Biochim. Biophys. Acta **638:**234–241.

67. **Newton, N. A., G. B. Cox, and F. Gibson.** 1971. The function of menaquinone (vitamin K_2) in *Escherichia coli* K-12. Biochim. Biophys. Acta **244:**155–166.

68. **Pandian, S., S. Saengchjan, and T. S. Raman.** 1981. An alternative pathway for the biosynthesis of isoprenoid compounds in bacteria. Biochem. J. **196:**675–681.

69. **Ramasarma, T., and J. Jayaraman.** 1971. Metabolism of ubiquinone in the rat. Methods Enzymol. **18:**232–237.

70. **Ramasarma, T., and J. Jayaraman.** 1971. Reverse-phase chromatographic separation of ubiquinone isoprenologs. Methods Enzymol. **18:**165–169.

71. **Redalieu, E., and K. Folkers.** 1971. Assay of coenzyme Q_{10} in blood. Methods Enzymol. **18:**179–181.

72. **Robins, D. J., I. M. Campbell, and R. Bentley.** 1970. Glutamate—a precursor for the naphthalene nucleus of bacterial menaquinones. Biochem. Biophys. Res. Commun. **39:**1081–1086.

73. **Robins, D. J., R. B. Yee, and R. Bentley.** 1973. Biosynthetic precursors of vitamin K as growth promoters for *Bacteroides melaninogenicus.* J. Bacteriol. **116:**965–971.

74. **Samuel, O., and R. Azerad.** 1969. C-methylation of desmethylmenaquinones: specificity of the enzyme system of *Mycobacterium phlei.* FEBS Lett. **2:**336–338.

75. **Shaw, D. J., J. R. Guest, R. Meganathan, and R. Bentley.** 1982. Characterization of *Escherichia coli* mutants defective in conversion of *o*-succinylbenzoate to 1,4-dihydroxy-2-naphthoate. J. Bacteriol. **152:**1132–1137.

76. **Sheppard, A. J.** 1971. Gas chromatography of vitamin K_1. Methods Enzymol. **18:**461–464.

77. **Sheppard, A. J., and W. D. Hubbard.** 1971. Gas chromatography of vitamin K_3. Methods Enzymol. **18:**465–469.

78. **Shineberg, B., and I. G. Young.** 1976. Biosynthesis of bacterial menaquinones: the membrane-associated 1,4-dihydroxy-2-naphthoate octaprenyltransferase of *Escherichia coli.* Biochemistry **15:**2754–2758.

79. **Stroobant, P., I. G. Young, and F. Gibson.** 1972. Mutants of *Escherichia coli* K-12 blocked in the final reaction of ubiquinone biosynthesis: characterization and genetic analysis. J. Bacteriol. **109:**134–139.

80. **Taber, H.** 1980. Functions of vitamin K_2 in microorganisms, p. 177–187. *In* J. W. Suttie (ed.), Vitamin K metabolism and vitamin K-dependent proteins. University Park Press, Baltimore.

81. **Threlfall, D. R., and G. R. Whistance.** 1971. Biosynthesis of phylloquinone. Methods Enzymol. **18:**559–562.

82. **Wallace, B. J., and I. G. Young.** 1977. Role of quinones in electron transport to oxygen and nitrate in *Escherichia coli*. Studies with a *ubiA⁻ menA⁻* double quinone mutant. Biochim. Biophys. Acta **461:**84–100.

83. **Weische, A., and E. Leistner.** 1985. Tetrahedron Lett. **26:**1487–1490.

84. **Weiss, U., and J. M. Edwards.** 1980. The biosynthesis of aromatic compounds. John Wiley & Sons, Inc., New York.

85. **Winrow, M. J., and H. Rudney.** 1971. The biosynthesis of ubiquinone. Methods Enzymol. **18:**214–232.

86. **Young, I. G.** 1975. Biosynthesis of bacterial menaquinones. Menaquinone mutants of *Escherichia coli*. Biochemistry **14:**399–406.

87. **Young, I. G., R. A. Leppik, J. A. Hamilton, and F. Gibson.** 1972. Biochemical and genetic studies on ubiquinone biosynthesis in *Escherichia coli* K-12: 4-hydroxybenzoate octaprenyltransferase. J. Bacteriol. **110:**18–25.

88. **Zak, V. L., and R. A. Kelln.** 1981. A *Salmonella typhimurium* mutant dependent upon carbamyl aspartate for resistance to 5-fluorouracil is specifically affected in ubiquinone biosynthesis. J. Bacteriol. **145:**1095–1098.

34. Biosynthesis of Folic Acid, Riboflavin, Thiamine, and Pantothenic Acid

GENE M. BROWN[1] AND JOANNE M. WILLIAMSON[2]

Massachusetts Institute of Technology, Cambridge, Massachusetts 02139,[1] *and Merck Sharp and Dohme Research Laboratories, Rahway, New Jersey 07065*[2]

FOLIC ACID

The objective in this section is to present information on the biosynthesis of folic acid in the enteric bacteria (*Escherichia coli* and *Salmonella typhimurium*). Consequently, in this chapter we have tried to cover the subject comprehensively from the point of view of what is known in *E. coli*, but the limited scope precluded a complete review of the field. However, much of the work in other organisms was done in parallel with the investigations in *E. coli*, and the conclusions, for the most part, have been the same. For more inclusive treatments of this subject, the reader is directed to other reviews (21, 26, 134).

Biosynthesis of H$_2$-Pteroic Acid and H$_2$-Folic Acid

The enzymatic reactions responsible for the formation, in *E. coli*, of H$_2$-folic acid from *p*-aminobenzoate (*p*-AB), glutamate, and a pteridine precursor are presented in Fig. 1. The two enzymes required for the production of H$_2$-pteroic acid from *p*-AB and the pteridine were separated and purified from extracts of *E. coli*, and the identity of the pteridine precursor was established as 2-amino-4-oxy-6-hydroxymethyl-7,8-dihydropteridine (abbreviated hereafter as H$_2$-pterin-CH$_2$OH) by Brown and co-workers (20, 25, 126, 149). Parallel work with an enzyme system from *Lactobacillus plantarum* by Shiota and collaborators led to the same conclusion (136, 138), and later this compound was shown to be the precursor of H$_2$-folic acid in other

microorganisms and in plants (see the review by Brown and Williamson [26] for the primary references).

The enzyme that catalyzes the synthesis of the pyrophosphate ester of H$_2$-pterin-CH$_2$OH (H$_2$-pterin-CH$_2$OPP) (see reaction A, Fig. 1) has been purified 400-fold from extracts of *E. coli* (126) and has been shown to be relatively heat stable (heating at 100°C for 5 min results in the loss of only 20% of its activity). Its molecular weight is 15,000, and it requires Mg^{2+}. Reaction B, Fig. 1, is catalyzed by a second enzyme, called H$_2$-pteroate synthase (126). This enzyme is not stable to heating, and it also requires Mg^{2+} for activity. H$_2$-pteroate synthase has not been purified to homogeneity in *E. coli*, but it has been purified free from contaminating phosphatases. The reaction can be described as the displacement of PP$_i$ by an attack of the amino group of *p*-AB on H$_2$-pterin-CH$_2$OPP since H$_2$-pteroate and PP$_i$ are known to be produced in equal quantities by the action of the enzyme (126, 137). H$_2$-pteroate synthase activity has been found widely distributed among microorganisms and plants (26).

H$_2$-pteroate synthase is capable of using *p*-aminobenzoylglutamic acid (*p*-ABG) as substrate in place of *p*-AB to yield H$_2$-folate as the product, but *p*-ABG is not used as effectively as *p*-AB. This is true regardless of the source of the enzyme (see review by Brown [21] for a complete discussion). For the enzyme from *E. coli*, the K_m values for *p*-AB and *p*-ABG are 2.5 μM and 1.3 mM, respectively (127). This information, along with the absence of any credible evidence that *p*-ABG

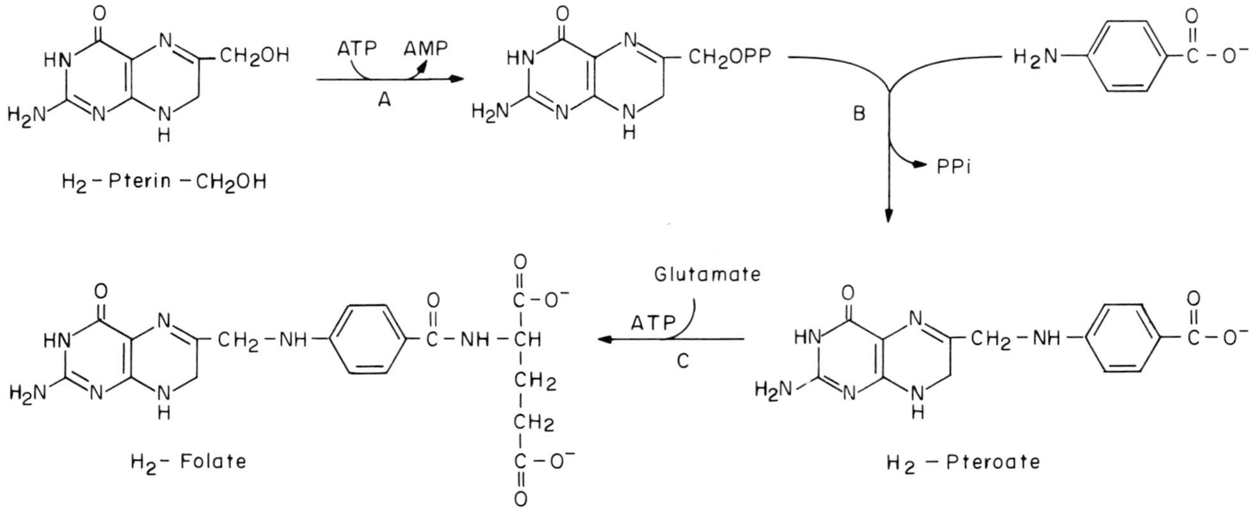

FIG. 1. Enzymatic reactions involved in the conversion of H_2-pterin-CH_2OH, p-AB, and glutamic acid to H_2-folic acid.

is a biosynthetic product (i.e., no p-ABG-requiring mutants of any organism have been obtained, and no enzyme has been found that catalyzes the synthesis of p-ABG from p-AB and glutamate), indicate that H_2-pteroate is a biosynthetic intermediate and that p-ABG is not. Evidence in support of this conclusion was obtained with the discovery in *E. coli* and other microorganisms (52) of an enzyme that causes the formation of H_2-folate from glutamate and H_2-pteroate (reaction C, Fig. 1). This enzyme has been named "H_2-folate synthetase." Neither pteroic acid nor tetrahydropteroic acid can be used as substrate in place of H_2-pteroate. In addition to ATP, this enzyme requires the presence of Mg^{2+} and either NH_4^+ or K^+ for activity (52). Recent evidence has indicated that H_2-folate synthetase from *E. coli* is a multifunctional enzyme in that it also catalyzes the synthesis of polyglutamate forms of folic acid (11, 46, 47). This and some other properties of the enzyme will be discussed below.

The elucidation of the enzymatic process for the formation of H_2-pteroate provided an opportunity to investigate the effects of sulfonamides on the utilization of p-AB. All sulfonamides that inhibit the growth of *E. coli* are effective inhibitors of the enzymatic conversion of p-AB and H_2-pterin-CH_2OPP to H_2-pteroate in the presence of H_2-pteroate synthase (19), although a great variation was observed with respect to the relative effectiveness among several sulfonamides. Sulfathiazole was the most effective in enzyme experiments, and sulfanilamide was least effective. Further investigations revealed that the sulfonamides inhibit the formation of H_2-pteroate by competing with p-AB as substrate for the enzyme (19, 137). A product is formed that is an analog of H_2-pteroate in that it contains a sulfonamide residue in place of the p-AB residue (19). There is no evidence to indicate that such a product is itself an inhibitory compound.

Enzymatic Formation of the Pteridine Portion of Folic Acid

Folic acid is only one of a number of naturally occurring pteridines. Before enzyme work was under-

taken, evidence had accumulated to indicate that either a purine (or purines) is a precursor of pteridines or the two classes of compounds are made by similar biosynthetic pathways. Most of this information was obtained from work with whole organisms in which incorporation of various ^{14}C-labeled compounds was measured. For a more complete presentation of this early work, the reader is directed to the review articles by Brown and Reynolds (22), Brown (21), and Shiota (134).

Evidence derived primarily from *E. coli* indicated initially that either guanosine or a guanine nucleotide is the precursor of the pteridine moiety of folic acid and that during this transformation carbon 8 (C-8) of the guanosine compound is not incorporated into folic acid (124, 125). These observations led to the proposal by Reynolds and Brown (125) of a biosynthetic pathway that was similar to a hypothetical pathway proposed earlier for the biosynthesis of pteridine pigments of insects (12, 151). One of the key intermediates in this speculative pathway is 2-amino-4-oxy-6-(D-*erythro*-1',2',3'-trihydroxypropyl)-7,8-dihydropteridine (known by the trivial name of H_2-neopterin). The oxidized form of this compound, neopterin, had been known to occur in substantial quantities in the larvae of honeybees (123), and the phosphate ester, neopterin phosphate, had been found in *E. coli* (51). In 1964 it was reported that H_2-neopterin could be converted to H_2-pteroate in quite good yields in the presence of ATP, p-AB, and a relatively crude extract of *E. coli* (65), an observation that supported the notion that H_2-neopterin is an intermediate in the biosynthesis of the pteridine portion of folic acid. This work was extended later to indicate that in the presence of an extract of *E. coli*, H_2-neopterin could be converted to H_2-pterin-CH_2OH (64), the pterin known to be the direct precursor of H_2-pteroate and H_2-folate, and also that H_2-neopterin could be produced from a guanine nucleotide.

The formulation of an assay for the utilization of radioactive guanine nucleotide (labeled with ^{14}C exclusively at C-8), based on the measurement of the amount of radioactive one-carbon compound pro-

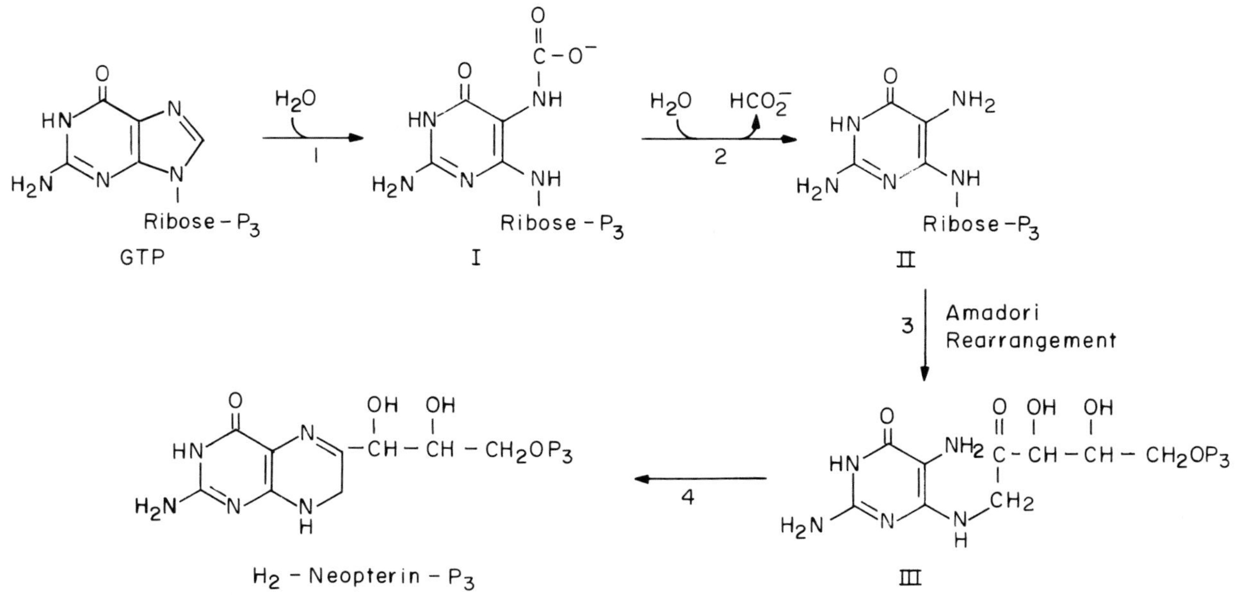

FIG. 2. Reactions and enzyme-bound intermediates involved in the conversion of GTP to H_2-neopterin-P_3 in the presence of the enzyme GTP cyclohydrolase.

duced from C-8 of guanine nucleotide, provided a convenient way to measure the enzymatic utilization of guanine nucleotide without the necessity of coupling the reaction with the enzymatic production of H_2-pteroate. Use of this assay also negated the necessity to include ATP (needed for H_2-pteroate production) in the reaction mixture and thus allowed a decision to be made concerning the identity of the guanine nucleotide used as precursor. With the use of this assay, it was established that the precursor guanine nucleotide is GTP and that formate is the one-carbon compound produced from carbon-8 of GTP (27, 28). The enzyme, which was named GTP cyclohydrolase, was partially purified (ca. 800-fold) from extracts of *E. coli* (28), and the other product of its action was identified as H_2-neopterin triphosphate (abbreviated hereafter as H_2-neopterin-P_3). No cofactors were required, and the enzyme from *E. coli* was shown to be specific for the utilization of GTP (GMP, GDP, and all other purine nucleotides were found not to be substrates). Later, the enzyme was purified to homogeneity from extracts of *E. coli* (164), and many of its properties were described. For example, the molecular weight of the enzyme was determined to be 210,000, and the native enzyme was found to consist of four identical subunits, each of which contains two identical polypeptide chains. Thus, the native enzyme contains eight identical polypeptide chains. The native enzyme can dissociate into its four subunits in the presence of 0.3 M NaCl, but the dissociated form of the enzyme is not catalytically active. The native enzyme is quite heat stable, with a half-life of 7 min at 82°C; a substantial amount of activity (approximately 30%) remains even after the enzyme has been heated at 92°C for 5 min. The dissociated form of the enzyme is much less heat stable, with a half-life of less than 1 min at 82°C. The K_m for GTP was found to be 0.02 μM. Each subunit was shown to have two binding sites for GTP; thus,

each polypeptide chain seems capable of binding one GTP.

From theoretical considerations a reasonable set of reactions (Fig. 2) can be formulated for the conversion of GTP to H_2-neopterin-P_3. These reactions include two successive hydrolytic steps that would result in the production of formate and ribosylated pyrimidine (compound II, Fig. 2). The latter compound would then undergo an Amadori-type rearrangement to convert the ribose unit to a 1-deoxy-2-pentulose moiety (see compound III, Fig. 2), and finally, compound III would undergo ring closure to yield H_2-neopterin-P_3. Since the overall transformation of GTP to H_2-neopterin-P_3 is known to be catalyzed by a single enzyme, proposed intermediates I, II, and III of Fig. 2 might be expected to be enzyme-bound intermediates in this enzymatic process. There is experimental evidence to indicate that the set of reactions shown in Fig. 2 accounts for this enzymatic transformation. Compound I has been prepared synthetically and has been shown to be converted to formate and H_2-neopterin-P_3 in the presence of either *L. plantarum* extracts (135, 139) or the purified GTP cyclohydrolase of *E. coli* (164). That an Amadori-like reaction (reaction 3, Fig. 2) occurs was demonstrated with the observations (161) that 7-methyl-GTP can be used as substrate in place of GTP for the production of formate (although at a rate much slower than that observed with GTP) and that the Amadori rearrangement apparently also can occur since a compound identified as methylated compound III was produced. The presence of the methyl group apparently prevents ring closure so that no dihydropterin can be formed. Thus, when 7-methyl-GTP was provided as substrate, the enzymatic product was the methylated pyrimidine intermediate instead of H_2-neopterin-P_3, the normal product of action with GTP as substrate. These observations provide convincing evidence that GTP cyclohydrolase from *E. coli* catalyzes four chemical reactions, in the

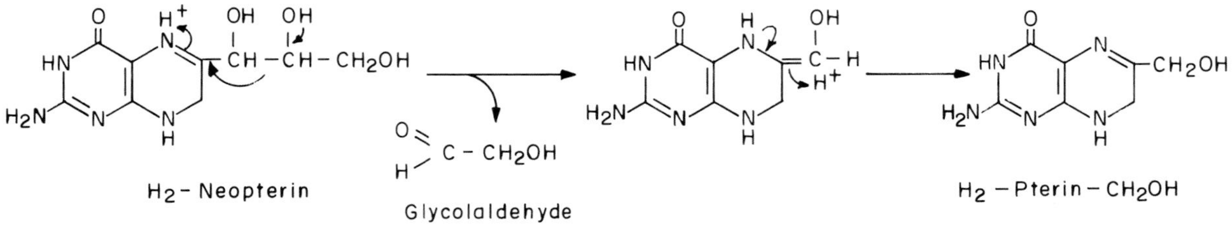

FIG. 3. Conversion of H_2-neopterin to H_2-pterin-CH_2OH, catalyzed by the enzyme dihydroneopterin aldolase.

order shown via reactions 1 through 4 in Fig. 2, and that compounds I, II, and III of Fig. 2 are enzyme-bound intermediates in the conversion of GTP to H_2-neopterin-P_3.

The initial report on the existence of an enzyme in *E. coli* that catalyzes the conversion of H_2-neopterin to H_2-pterin-CH_2OH (64) was followed by purification of the enzyme and the determination of some of its properties (86). The enzyme catalyzes the reaction shown in Fig. 3 to produce H_2-pterin-CH_2OH and glycolaldehyde. This enzyme has been given the name of "dihydroneopterin aldolase" since the reaction it catalyzes resembles those reactions catalyzed by aldolases. The enzyme is heat stable; it can withstand heating at 100°C for 5 min with no loss of activity. The K_m for H_2-neopterin was found to be 9 μM. Neopterin, H_2-neopterin-P_3, and H_2-neopterin monophosphate (H_2-neopterin-P) could not replace H_2-neopterin as substrate. H_2-pterin-CH_2OH, a product of the action of the enzyme, is also a potent inhibitor of its action (K_i = 1.7 μM). The only other pterin that is significantly inhibitory is 6-formyl-H_2-pterin (86).

The action of GTP cyclohydrolase allows the production of H_2-neopterin-P_3, but neither this compound nor H_2-neopterin-P can be converted directly to H_2-pterin-CH_2OH through the action of dihydroneopterin aldolase. Thus, the conclusion can be drawn that a necessary enzymatic process in the biosynthetic pathway is the conversion of H_2-neopterin-P_3 to H_2-neopterin. This process is known to occur in two steps. An enzyme that catalyzes the removal of PP_i from H_2-neopterin-P_3 to yield H_2-neopterin-P was purified 30- to 60-fold from extracts of *E. coli* by Suzuki and Brown (143). This enzyme, which was given the name

of "dihydroneopterin triphosphate pyrophosphohydrolase," is specific for H_2-neopterin-P_3 as substrate (i.e., no nucleoside triphosphate is capable of being used). As expected, Mg^{2+} is needed for activity to be expressed. The K_m for H_2-neopterin-P_3 is 11 μM, and the molecular weight of the enzyme was determined to be approximately 17,000.

The second step in the process for the conversion of H_2-neopterin-P_3 to H_2-neopterin is catalyzed by phosphomonoesterases whose action allows H_2-neopterin-P to be dephosphorylated to H_2-neopterin. Several such phosphomonoesterases active in this process were detected in *E. coli*, but none exhibited specificity for H_2-neopterin-P (143). It seems reasonable to conclude that this reaction can probably be catalyzed in *E. coli* by any one of these several hydrolytic enzymes.

To summarize briefly, the reactions that allow the conversion of GTP to H_2-pterin-CH_2OH, and of the latter compound to H_2-folate, are shown in Fig. 4. The enzymes that catalyze these reactions have all been discovered in *E. coli*, and information about these enzymes is available as described above.

Pteroylpolyglutamates

Pteroylpolyglutamates (also known as "folypolyglutamates") is a term applied to a variety of derivatives of pteroylglutamate (PteGlu) containing a total of from 2 to 12 glutamate residues in γ-peptide linkages. In fact, nearly all of the intracellular "folate" occurs as polyglutamates (in microorganisms and animal cells), with four- to six-glutamate residues the predominant forms, depending on the kind of cells ana-

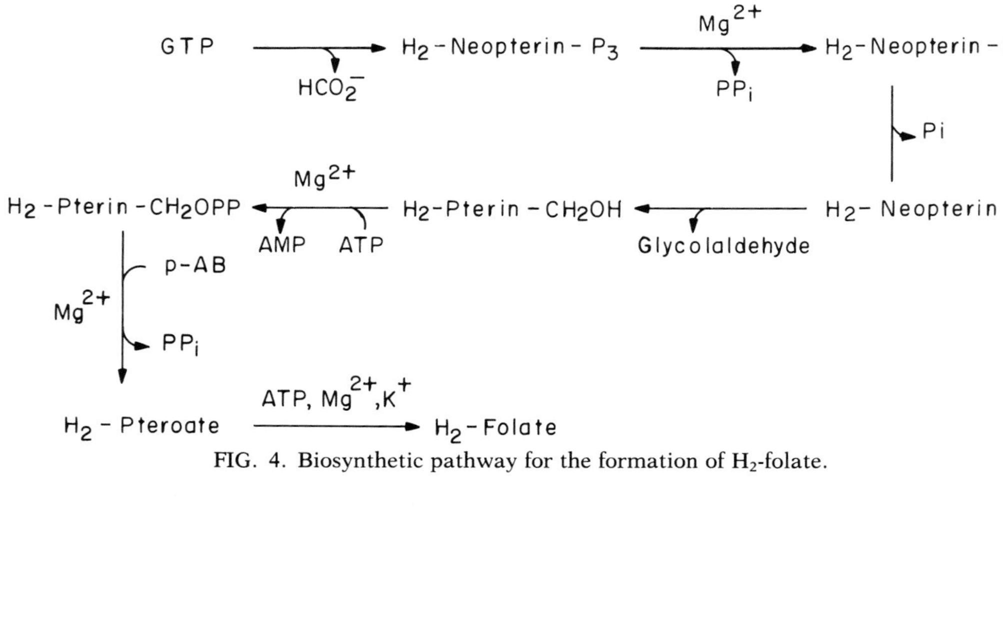

FIG. 4. Biosynthetic pathway for the formation of H_2-folate.

lyzed. Since these pteroylpolyglutamates can occur as the dihydro and tetrahydro forms, and the tetrahydro compounds can exist as various one-carbon derivatives (5-methyl, 4-formyl, 10-formyl, 10-hydroxymethyl, 5,10-methylene, and 5,10-methenyl), analyses for specific compounds containing more than one glutamate residue can be quite complicated, and some of the older literature values may not be very accurate.

The enzymatic production of these pteroylpolyglutamates in animal and microbial systems has been a subject of great interest and relatively intense activity in recent years. Since a comprehensive review of this subject is beyond the scope of this article, only the work on the enzymatic synthesis of pteroylpolyglutamates in *E. coli* will be discussed.

The first information on the enzymatic production of pteroylpolyglutamates appeared in 1964 with the report (52) that in the presence of an extract of *E. coli* H₄-folate and glutamic acid could be converted in small quantities to substances that appeared to be H₄-pteroyldiglutamate (H₄-PteGlu₂) and H₄-pteroyltriglutamate (H₄-PteGlu₃). This report was followed a few years later by a more detailed and extensive investigation of this system in which an enzyme that catalyzes the production of the diglutamate was partially purified and some of its properties were determined (85). Of a number of substrates tested, 10-formyl-H₄-PteGlu was used most effectively, and 5,10-methylene-H₄-PteGlu and H₄-PteGlu were utilized less well. No other folate compound was active as substrate, and no other amino acid (including the dipeptide, γ-glutamylglutamate) could replace glutamate as substrate. ATP, Mg²⁺, and K⁺ (or NH₄⁺) were needed for activity. The product was identified as a diglutamate, and no evidence was found for the formation of a triglutamate (or any other polyglutamate) with the use of this partially purified enzyme. The suggested name of the enzyme was "10-formyl-H₄-pteroyldiglutamate synthetase."

Ferone and co-workers (46, 47) investigated this enzyme further and rather surprisingly found that the enzymatic activities for the synthesis of 10-formyl-H₄-pteroyldiglutamate and for the conversion of H₂-pteroate and glutamate to H₂-folate purified together. From these and later investigations (11), the conclusion was drawn that a bifunctional protein carries both enzymatic activities, although kinetic experiments suggest that the two reactions are catalyzed on independent sites on the protein. The molecular weight of the enzyme is 47,000. The gene (*folC*) for this protein has been cloned (11), and it is now possible to obtain this enzyme from *E. coli* transformants (containing recombinant plasmids) in amounts 100- to 400-fold higher than the amounts present in wild-type strains. It should be mentioned that a similar bifunctional enzyme carrying the same two enzymatic activities has also been obtained from a strain of *Corynebacterium* (130). It is not yet clear whether or not the purified *E. coli* enzyme will catalyze the addition of more than two glutamate residues (R. Ferone, personal communication). Since folate compounds with three or more glutamate residues occur in *E. coli*, the possibility thus exists that more than one enzyme is needed for the formation of these higher polyglutamates in *E. coli*.

RIBOFLAVIN

Since much of the evidence concerning the biosynthesis of riboflavin has been obtained through investigations with microorganisms other than *E. coli* and *S. typhimurium*, this chapter will not be strictly limited to what is known in the enteric bacteria.

From analyses of a number of riboflavin-requiring mutants of *Saccharomyces cerevisiae* by Bacher and co-workers (7, 8, 110), a tentative pathway was proposed for the biosynthesis of riboflavin with guanine (or the nucleoside or a nucleotide of guanine) as the precursor. A great deal of experimental evidence to indicate that a guanine compound is the precursor had accumulated earlier. This evidence has been reviewed thoroughly (116) and will not be presented or further discussed here. Experimental evidence reported since 1974 has led to the modification of the original pathway proposed by Bacher and co-workers; the set of enzymatic reactions now thought to result in the biosynthesis of riboflavin is shown in Fig. 5. Evidence for the individual enzymatic steps is introduced and discussed below.

GTP as the Precursor

The initial evidence to indicate that the precursor is the nucleoside or a nucleotide of guanine was the demonstration that a mutant of *S. typhimurium* that is unable to interconvert guanine and guanosine can incorporate [¹⁴C-ribose]guanosine into riboflavin and GMP without dilution of radioactivity (82). This established that either guanosine or a guanine nucleotide is the precursor and also that the ribose portion of the precursor is retained and converted to the ribityl group of riboflavin. Analyses of the nucleoside and nucleotide pools of *Eromothecium ashbyii* under various conditions of flaviogenesis led to the same conclusions by Mitsuda and co-workers (89, 90).

That GTP is the precursor was suggested by the discovery by Foor and Brown (49) of an enzymatic process in *E. coli* in which C-8 of GTP is eliminated as formate and the compound shown as IV in Fig. 5 is produced. Somewhat later, similar enzymes were shown to exist in other microorganisms (14, 131, 133). The enzyme was named "GTP cyclohydrolase II" by Foor and Brown (49) to distinguish it from the enzyme (called GTP cyclohydrolase I for convenience in this paper) that had been discovered earlier (27, 28) and shown to be involved in the biosynthesis of folic acid. The enzyme has been purified 2,200-fold from extracts of *E. coli*, and some of its properties have been described (49) as follows. GTP cyclohydrolase II requires Mg²⁺ for activity. Formate, pyrophosphate, and 2,5-diamino-6-oxy-4-(5'-phosphoribosylamino)pyrimidine (compound IV, Fig. 5) were identified as products formed from GTP. The K_m of the enzyme for GTP was reported to be 41 μM, its pH optimum is 8.5, and its activity is inhibited by PP$_i$ (approximate K_i of 0.1 mM). All of these properties distinguish the enzyme from GTP cyclohydrolase I, also present in *E. coli*. Other relevant information that confirms the unrelatedness of the two enzymes is that no cross-reactivity to antibodies prepared against GTP cyclohydrolase I could be observed (164).

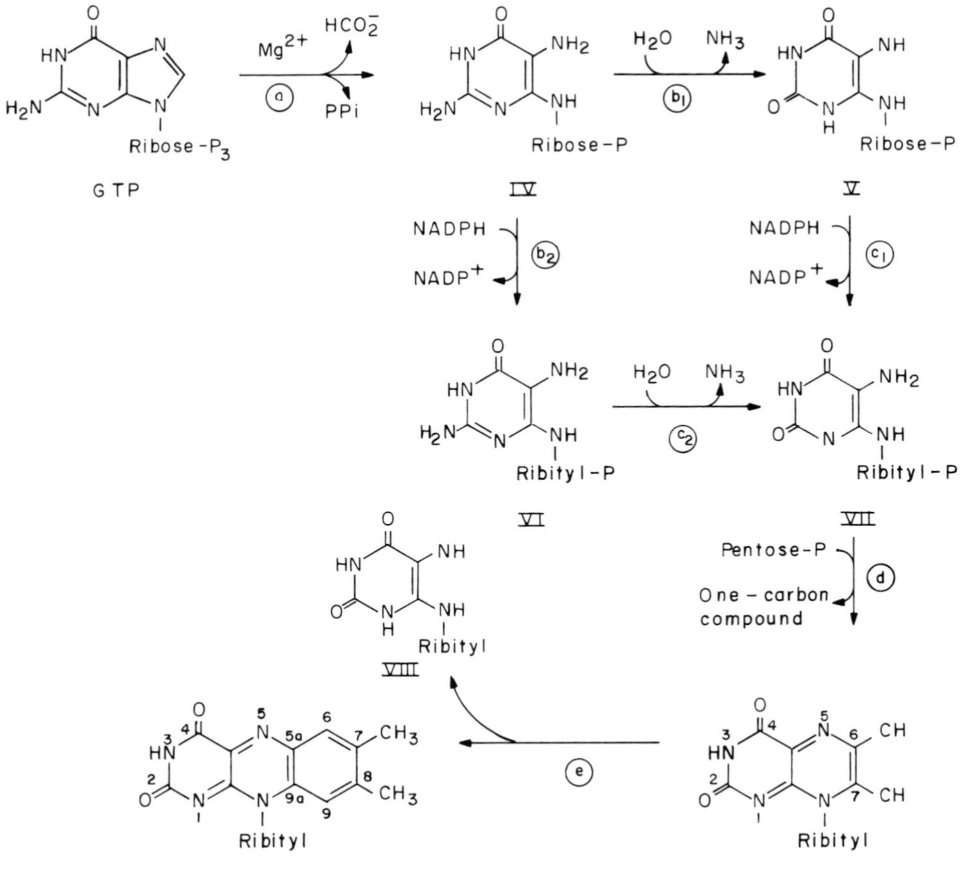

FIG. 5. Reactions and intermediates for the biosynthesis of riboflavin.

Deamination and Reduction

The occurrence of GTP cyclohydrolase II in *E. coli* suggests that it functions as the first step in the biosynthesis of riboflavin, although no direct supporting evidence was presented by Foor and Brown (49). Such evidence became available later with the report by Burrows and Brown (29) of the presence in *E. coli* of two enzymes, a deaminase and a reductase, that catalyze reactions that would be expected to occur in the overall conversion of GTP to riboflavin. These reactions are shown in Fig. 5 as reactions b_1 and c_1. The two enzymes were separated by fractionation of *E. coli* extracts, and some of the properties of the two were reported. The deaminase uses 2,5-diamino-6-oxy-4-(5'-phosphoribosylamino)pyrimidine (shown as compound IV in Fig. 5) as substrate; this compound is the product of GTP cyclohydrolase II. The deaminase was purified 200-fold, and its molecular weight was estimated at 80,000. The enzyme could not act on the dephosphorylated form of compound IV.

The product of action of the deaminase was identified as 5-amino-2,6-dioxy-4-(5'-phosphoribosylamino)pyrimidine, compound V in Fig. 5. This compound was shown to be the substrate for the reductase (reaction c_1, Fig. 5). The reductase was purified approximately 200-fold. Its molecular weight is approximately 37,000, and it cannot utilize as sub-strate the dephosphorylated form of V. NADPH is needed as cosubstrate for the reaction. The product was identified as 5-amino-2,6-dioxy-4-(5'-phospho-ribitylamino)pyrimidine (VII, Fig. 5).

The enzyme work in *E. coli* clearly indicates that the sequence of enzymatic reactions in the pathway is deamination followed by reduction (29). However, evidence initially based on excretion patterns of different classes of riboflavin-requiring mutants suggests that in *S. cerevisiae* the reduction step precedes deamination (7, 88). On the other hand, from genetic and biochemical information in *Bacillus subtilis*, the conclusion is that, as for *E. coli*, deamination precedes reduction (13). Enzymatic work indicates that in eucaryotes such as *S. cerevisiae* (88, 105) and *Ashbya gossypii* (56) reduction precedes deamination (shown by reactions b_2 and c_2 in Fig. 5), whereas in the procaryote *E. coli*, the order is reversed (b_1 and c_1, Fig. 5).

There is no direct evidence in support of the involvement in *E. coli* of GTP cyclohydrolase II, the deaminase, and the reductase in the biosynthesis of riboflavin, although from the nature of the reactions it seems reasonable to believe that these enzymes function in the riboflavin biosynthetic pathway. However, there is evidence in the yeast *Candida guilliermondii* (also called *Pichia guilliermondii*) that a GTP cyclohydrolase, with properties similar to those of the *E. coli* enzyme, functions in the biosynthesis of riboflavin since *rib1* mutants of this yeast are devoid of GTP cyclohydrolase activity (132).

Formation of DMRL

6,7-Dimethyl-8-ribityllumazine (DMRL) has been thought to be an intermediate in the biosynthetic pathway for riboflavin production from the time it was first identified as a substrate for the enzyme (riboflavin synthase) that catalyzes, in a single enzymatic step (reaction e, Fig. 5), the conversion of 2 mol of DMRL to 1 mol of riboflavin and 1 mol of the ribitylpyrimidine compound shown as compound VIII in Fig. 5 (see the review by Plaut [116] for a complete description of this enzymatic process). Thus, the methyl carbons and C-6 and C-7 of DMRL are transferred as a four-carbon unit from DMRL to another mole of DMRL to produce riboflavin in which C-5a, C-9a, C-7, and C-8 are derived from C-6 and C-7 of DMRL and C-6 and C-9 and the methyl carbons are derived from the methyl carbons of DMRL. The step in the riboflavin biosynthetic pathway that has not yet been elucidated is the conversion of 5-amino-2,6-dioxy-4-(5'-phosphoribitylamino)pyrimidine (compound VII in Fig. 5) to DMRL. This reaction clearly involves the addition of a four-carbon unit to compound VII to yield DMRL. However, the identity and the origin of the four-carbon unit have remained elusive until recently. Experimental evidence now indicates that a pentose provides all of the carbons of the four-carbon unit and, surprisingly, that the four-carbon unit is derived from C-1, C-2, C-3, and C-5 of the pentose. These conclusions were reached from work done mostly with microorganisms other than the enteric bacteria, but we will nevertheless briefly summarize the evidence below since it seems to us that the same system is likely to operate in the enteric bacteria.

As a result of a comprehensive study of the incorporation of various ^{13}C-labeled compounds into riboflavin by *A. gossypii*, Bacher and co-workers concluded that the four-carbon unit needed for the formation of DMRL is derived from a pentose (5, 6, 48, 70, 71). This conclusion thus supported a similar conclusion reached from work that showed that C-1 of ribose is efficiently incorporated into riboflavin by *Propionibacterium shermanii* (4). The surprising finding that the carbons of the pentose used to form the four-carbon unit are C-1, C-2, C-3, and C-5 (6, 48, 70) indicates that a rearrangement must occur before or during the formation of DMRL whereby C-4 of the pentose is lost and C-3 and C-5 are joined in covalent linkage.

The work on incorporation of precursors described in the preceding paragraph was done with whole cells. Recent enzymatic evidence suggests that a pentose phosphate is the origin of the four-carbon unit. Thus, it has been reported that in the presence of dialyzed cell extracts of *Pichia* (or *Candida*) *guilliermondii*, DMRL and riboflavin can be produced from 2,5-diamino-6-oxy-4-(5'-phosphoribosylamino)pyrimidine (compound VI in Fig. 5) provided NADPH (presumably needed for reduction of the phosphoribosyl group) and ribose 5-phosphate are present (77). Ribose could not replace ribose 5-phosphate. In a later paper, similar extracts were reported to promote the formation

of DMRL from ribose 5-phosphate and 5-amino-2,6-dioxy-4-(5'-phosphoribitylamino)pyrimidine (compound VII, Fig. 5), or the dephosphorylated form of this compound, in the absence of NADPH (76). Similar results were obtained by Nielson et al. (106), who reported that in the presence of pentose 5-phosphate or pentulose 5-phosphate and extracts of *C. guilliermondii*, either compound VII or its dephosphorylated form could be converted to DMRL. Highest yields were obtained from ribose 5-phosphate and xylulose 5-phosphate; ribulose 5-phosphate and arabinose 5-phosphate were utilized less well. Utilization of various species of ^{14}C-labeled ribulose 5-phosphate indicated that C-1, C-2, C-3, and C-5 were incorporated, but C-4 was not. This is consistent with the labeling patterns observed in the in vivo experiments discussed above.

Most recently, evidence has been obtained for the enzymatic formation of a stable intermediate (from ribose 5-phosphate) that can be used for the enzymatic formation of DMRL (99). In these investigations, two groups of DMRL synthase mutants of *C. guilliermondii* were identified on the basis of in vitro complementation. Extracts of group I mutants promote the Mg^{2+}-dependent conversion of ribose 5-phosphate to a compound which in the presence of extracts of group II mutants and compound VII (or the dephosphorylated form of compound VII) can be converted to DMRL. The latter transformation does not require Mg^{2+}. Since treatment of the intermediate with alkaline phosphatase abolishes its ability to promote the synthesis of DMRL, the conclusion was drawn that the unknown intermediate is a phosphate ester.

The information presented and discussed above indicates that a previous proposal that the four-carbon unit needed for the formation of DMRL is derived from the ribityl group of VII (15, 55) is probably untenable. Although none of the enzyme work reported above to indicate that a pentose phosphate is the source of the four-carbon unit has been done with enteric bacteria, it is not unreasonable to expect that the bacterial system is similar to that of yeasts. In fact, one relevant observation has been made in *S. typhimurium* to indicate that label is not incorporated into this four-carbon unit from guanosine labeled with ^{14}C in the ribose moiety (82). The relevant enzyme work in the enteric bacteria remains to be done.

Riboflavin Synthase

The conversion of DMRL to riboflavin (reaction e, Fig. 5) is catalyzed by the enzyme riboflavin synthase (or riboflavin synthetase). This enzyme is known to occur in a variety of microorganisms (115). Most of the information about riboflavin synthase has been obtained with the use of the enzyme prepared from baker's yeast, from which it has been purified to homogeneity. As shown in Fig. 5, the reaction involves the utilization of 2 mol of DMRL; 1 mol functions as the donor of a four-carbon unit, and the second is the acceptor, with the formation of 1 mol of riboflavin and 1 mol of 5-amino-2,6-dioxy-4-ribitylaminopyrimidine (compound VIII, Fig. 5). The mechanism of this enzymatic transformation has been elucidated by the elegant work of Plaut and co-workers, and since this subject has been thoroughly discussed in the review by Plaut (115), it will not be repeated here. The enzyme is known to occur in *E. coli* (114), although no studies with the *E. coli* enzyme have been undertaken that have been as comprehensive and informative as the work accomplished with the yeast enzyme. The limited available information suggests that the *E. coli* enzyme resembles the enzyme from *S. cerevisiae* in its mechanism of action (114).

THIAMINE

Thiamine, or vitamin B_1, is formed by the condensation of independently synthesized pyrimidine and thiazole moieties. The coenzyme form of the vitamin is thiamine pyrophosphate. Thus, consideration of the biosynthesis of the coenzyme can be divided into three parts: the individual pathways responsible for production of the pyrimidine and thiazole, and the conversion of these compounds to the enzymatically active form.

Biosynthesis of Pyrimidine

Early physiological studies showed that, at high concentrations, adenine or adenosine inhibits the production of the 4-amino-5-hydroxymethyl-2-methylpyrimidine (pyrimidine) moiety of thiamine (16). This suggestion of a link between the biosyntheses of the purines and the pyrimidine was reinforced by the results of later genetic and biochemical experiments. Yura (165) demonstrated that certain single-site mutations in *S. typhimurium* result in a double growth requirement for a purine and the pyrimidine (*ath* mutants). In addition, it was shown in *E. coli* that [^{14}C]formate and [^{14}C]glycine, precursors of the purines, were incorporated into the pyrimidine moiety, whereas other pyrimidines (uracil and orotic acid) and substances known to be precursors of the nucleic acid pyrimidines (aspartate and CO_2) were not (50).

The identity of the molecular link between the biosyntheses of the purines and the pyrimidine moiety was established by the elegant studies of Newell and Tucker in *S. typhimurium* (100–104). These workers showed that 5-aminoimidazole ribonucleotide (AIR) is a common precursor of purines and the pyrimidine of thiamine with the demonstration that this substance can satisfy the requirements for both purine and thiamine in a mutant of *S. typhimurium* selected for permeability to AIR. Furthermore, they found that [^{14}C]AIR was converted by this mutant to the pyrimidine without significant dilution of radioactivity. Methionine was found to be required for the conversion of AIR to pyrimidine, but none of the carbons of methionine was incorporated, an observation consistent with earlier results reported by Goldstein and Brown (50). Finally, Newell and Tucker established that both carbons of glycine are incorporated into the pyrimidine.

Although the details of the reactions involved in the conversion of AIR to the pyrimidine remain unknown, studies on the incorporation of isotopically labeled precursors have provided some information about how this may happen (Fig. 6). Formate is known to be incorporated exclusively into C-2 of the pyrimidine in

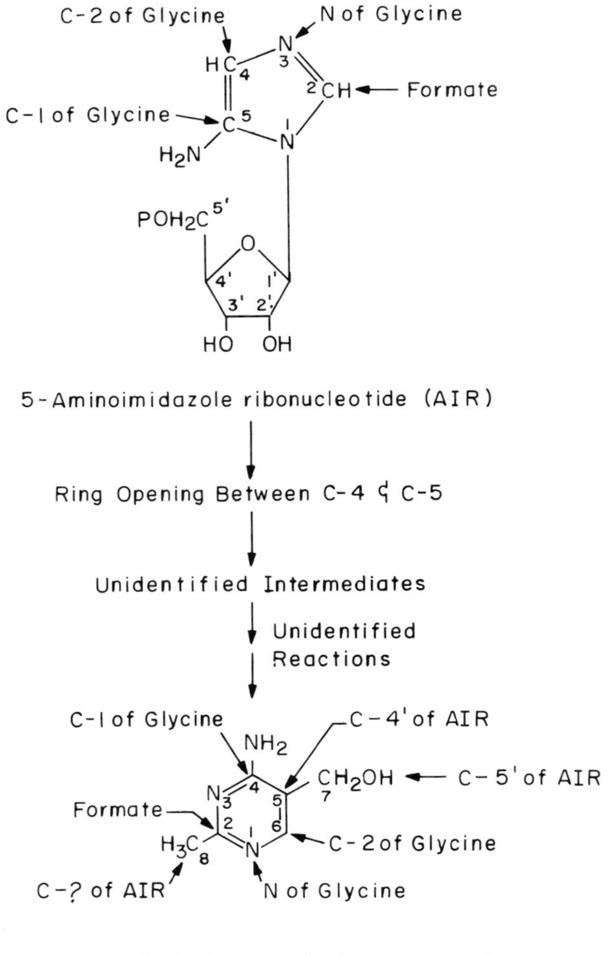

FIG. 6. Biosynthesis of the pyrimidine component of thiamine.

both *E. coli* (73) and *S. typhimurium* (42). Estramareix and Lesieur (40, 42) found in *S. typhimurium* that C-1 and C-2 of glycine become, respectively, C-4 and C-6 of the pyrimidine. These results were also obtained in *E. coli* by White and Rudolph (156), who showed, additionally, that the nitrogen of glycine becomes N-1 of the pyrimidine. It was clear, therefore, that the imidazole ring of AIR had to be opened between C-4 and C-5 and that a two-carbon unit is inserted to form C-5 of the pyrimidine ring and the hydroxymethyl group attached to the ring at that position. What was a mystery for some time, however, was the source of both the two-carbon unit and the methyl group attached to C-2 of the pyrimidine ring.

Recent studies from two laboratories have suggested that in *E. coli* and *S. typhimurium*, AIR is the precursor of all of the carbon atoms of the pyrimidine moiety. Indirect evidence to support this conclusion was obtained by Yamada and Kumaoka (162, 163), who showed that, in *E. coli*, [6-^{14}C]glucose is incorporated specifically into the hydroxymethyl carbon (C-7) of the pyrimidine moiety, whereas [1-^{14}C]glucose is not incorporated. [6-^{14}C]glucose was used to label the ribose pool (and therefore AIR) because of the low permeability to ribose of *E. coli* (39). The dilution of specific radioactivity of [6-^{14}C]glucose in the pyrimidine was similar to that obtained for the ribose moiety of AMP. [U-^{14}C]glucose was found to be incorporated primarily into C-5 and C-7 of the pyrimidine. More direct evidence has been obtained in *S. typhimurium* by Estramareix and Therisod (45), who synthesized [U-^{13}C]AIR and [^{14}C]AIR labeled primarily in the ribose moiety and found that (i) the isotopic composition of the pyrimidine synthesized from [U-^{13}C]AIR was similar to that of the substrate, and (ii) the three carbons of the pyrimidine known not to be derived from the imidazole ring (C-5, C-7, and C-8) were labeled by [^{14}C-ribose]AIR with a specific radioactivity comparable to the substrate, whereas the other carbons were not. Thus, it appears that the ribose moiety of AIR undergoes a cleavage of unknown mechanism to complete the pyrimidine ring. It remains to be established directly that C-4 of AIR becomes C-5 of the pyrimidine and which of the remaining carbons of the ribose becomes the methyl group, C-8. In addition, these results suggest that the product of the biosynthetic pathway may be the pyrimidine monophosphate, but there is, as yet, no experimental evidence in support of this.

As mentioned previously, Newell and Tucker (104) observed that methionine is required for the conversion of AIR to the pyrimidine moiety even though none of the carbon atoms of this amino acid is utilized for pyrimidine biosynthesis. These workers speculated that *S*-adenosylmethionine might be a required cofactor for the transformation. Leder (75) has taken this speculation a bit further and suggested the possibility that both *S*-adenosylmethionine and a B_{12}-containing enzyme are involved in the generation of the methyl group at C-2 of the pyrimidine. Even though this suggestion was originally made on the basis of older incorporation data, it is still an intriguing idea and, in view of the results just discussed on the fragmentation of the ribose moiety of AIR in the conversion to the pyrimidine, one that deserves further study.

Biosynthesis of Thiazole

Progress toward identification of the precursors of the 4-methyl-5-(β-hydroxyethyl)thiazole (thiazole) moiety has been slow. Investigations in this area have thus far been limited to radioactive and stable isotope incorporation experiments with their concomitant difficulties in interpretation. Curiously, the analysis of mutants blocked in the synthesis of the thiazole moiety has yet to play an important role in the elucidation of its biosynthesis. The reason for this is not clear.

Much experimental effort has been focused on the origin of C-2 of the thiazole moiety (see Fig. 7 for the numbering system). Because of an early speculative scheme for thiazole biosynthesis suggested by Harrington and Moggridge (53), methionine was tested by a number of research groups as the precursor of both C-2 and the sulfur atom of the ring. It was not, however, found to be incorporated into thiazole in either *E. coli* or *S. typhimurium* (9, 41; M. Julius and G. M. Brown, unpublished data). Based on the results of feeding experiments with *E. coli*, Nakayama (95) proposed that cysteine might provide both carbons

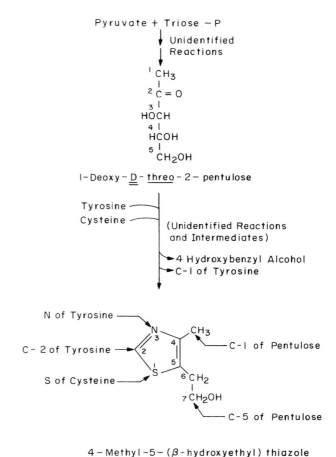

FIG. 7. Biosynthesis of the thiazole component of thiamine.

and sulfur for thiazole biosynthesis. Experimental observations, however, indicate that none of the carbons of cysteine is incorporated into thiazole in either organism (9, 41; Julius and Brown, unpublished data). The question of the origin of the sulfur atom will be considered later in this chapter.

Iwashima and Nose (58) obtained the first evidence suggesting that either phenylalanine or tyrosine is a precursor of the thiazole moiety. They found that addition of phenylalanine to the growth medium of a strain of *E. coli* caused repression of the synthesis of thiamine. This effect was reversed by addition of any one of a number of compounds including tyrosine, shikimic acid, tryptophan, and thiazole, but not by the pyrimidine moiety of thiamine. One interpretation of this observation is that phenylalanine interferes with the synthesis of tyrosine, which could be a precursor of thiazole. Subsequent work from several laboratories has confirmed that this interpretation is correct. Estramareix and Therisod (43) observed that a tyrosine auxotroph of *E. coli* incorporated radioactivity into C-2 of the thiazole moiety from either [U-^{14}C]tyrosine or [2-^{14}C]tyrosine. Similar results were obtained by Bellion et al. (10), who worked with a prototrophic strain of *S. typhimurium*. White and Rudolph (155) have shown that [^{15}N]tyrosine is incorporated into thiazole in *E. coli* B cells. The level of incorporation of ^{15}N observed is consistent with the conclusion that tyrosine is directly utilized for thiazole synthesis. Thus, it seems clear that C-2 and the nitrogen of tyrosine are incorporated as a unit into the thiazole ring. White (153) has provided additional evidence on the mechanism of the cleavage reaction in *E. coli* with the finding that 4-hydroxybenzyl alcohol is a metabolite of tyrosine whose intracellular level is tied to that of thiamine. Incubation of *E. coli* cells with [3,3-^{2}H]tyrosine led to the formation of 4-hydroxybenzyl alcohol containing both deuterium atoms, a result which suggests that the initial product of tyrosine cleavage is an unstable quinone methide that reacts with water to yield the observed product. Finally, the fate of C-1 of tyrosine is not yet clear. Estramareix and Therisod (44) isolated and identified from *E. coli* a new compound, 5(β-hydroxyethyl)-4-methyl-thiazole-2-carboxylic acid (thiazole 2-carboxylate), that supports the growth of a thiazole-requiring strain of *E. coli* and still contains C-1 of tyrosine. Evidence was obtained to suggest that the level of this compound is tied to the intracellular level of thiamine, but neither thiazole 2-carboxylate nor a phosphorylated derivative (or other dihydro derivatives) was found to be decarboxylated by cell extracts of strains prototrophic for thiamine. Thus, a role for this substance as an intermediate in thiazole biosynthesis remains to be established.

Information about the origin of the five-carbon chain (C-4', C-4, C-5, C-6, C-7) of thiazole in *E. coli* has been reported by White (152, 154). From the results of incorporation experiments with ^{2}H-, ^{18}O-, and ^{13}C-carbohydrates and related compounds, he has proposed that the two-carbon unit comprising C-4' and C-4 of thiazole is derived from pyruvate and that the remaining three-carbon unit (C-5, C-6, and C-7) comes from a three-carbon sugar, possibly glyceraldehyde 3-phosphate. White has speculated that pyruvate and glyceraldehyde 3-phosphate are first condensed in an acyloin-type reaction to yield a pentulose which might then react with tyrosine and a sulfur compound in an undefined series of steps to yield thiazole.

Based on White's results and their own finding that *E. coli* excretes a derivative of hydroxyethyl thiazole containing a hydroxyl group at C-6 (146), Therisod et al. (145) synthesized both the *threo* and *erythro* isomers of 1-deoxy-[1,1,1-^{2}H]-D-2-pentulose and tested these compounds as precursors of the thiazole moiety in *E. coli* (see Fig. 7). They found that the *threo* isomer is efficiently incorporated into C-4' of thiazole whereas the *erythro* isomer is not. To rule out the possibility that cleavage of the pentulose had taken place before its incorporation into thiazole, Estramareix and coworkers (32, 33) synthesized [1,1,1,5-^{2}H]-1-deoxy-D-*threo*-2-pentulose and found that 25% of the thiazole synthesized by *E. coli* cells incubated with the labeled pentulose contained deuterium at both C-4' and C-7. These results strongly suggest that the pentulose is utilized directly for thiazole synthesis. However, neither the pentulose itself nor a phosphorylated derivative of the pentulose has yet been detected intracellularly. Estramareix et al. suggested that a Schiff's base adduct between tyrosine and the pentulose is a probable intermediate in thiazole biosynthesis, but further work will be required before definitive statements can be made.

Two temperature-sensitive thiazole-requiring mutants (one in *E. coli* and one in *S. typhimurium*) have been reported that have interesting and similar properties. Iwashima and Nose (59, 60) isolated a temperature-sensitive mutant of *E. coli* which at the nonpermissive temperature requires thiazole for growth on glucose but not on glycerol. Under nonpermissive conditions glycine can partially replace the requirement for thiazole; [2-^{14}C]glycine was incorporated into both the pyrimidine and thiazole portions of thiamine. Parada and Ortega (112) reported the isolation of a temperature-sensitive strain of *S. typhimurium* which requires thiazole or thiamine only when grown on a hexose as the carbon source of 37 to 42°C; above 42°C, thiazole is required irrespective of the carbon source. The growth requirements of both of these mutants are intriguing. Since their isolation preceded the finding of tyrosine as the precursor of C-2 of thiazole, tyrosine metabolism was not examined in these mutants. Such an examination might prove interesting, especially in the *E. coli* mutant where glycine is known to replace the requirement for thiazole. In addition, these mutants might be blocked in the production of the pentulose at the nonpermissive temperature, since provision of a triose satisfies the requirement for the thiazole moiety.

In both *E. coli* and *S. typhimurium*, there is some evidence that cysteine can be used more effectively than methionine as the source of the sulfur atom of thiazole. Estramareix et al. (41) found with a methionine auxotroph of *E. coli* that incorporation of [^{35}S]sulfate into thiamine was lowered by the presence of cysteine or glutathione but was not affected by the presence of methionine or homocysteine. Similar results were obtained in more recent work by DeMoll and Shive (37) who used mass spectrometry to study the effect of addition of methionine and cysteine on incorporation of [^{35}S]sulfate into thiamine. Bellion and Kirkley (9) reported that in *S. typhimurium* the incorporation of [^{35}S]cysteine was not reduced by the addition of nonradioactive methionine, homocysteine, or glutathione. Thus, the experimental observations indicate that either cysteine or H_2S (which could be derived from cysteine) is the most likely precursor of the sulfur atom of thiazole.

Biosynthesis of Thiamine Pyrophosphate

The conclusion that thiamine pyrophosphate is formed from independently synthesized thiazole and pyrimidine moieties was derived largely from early work in organisms other than *E. coli* and *S. typhimurium*. Studies in cell extracts of *S. cerevisiae* by a number of research groups established that four enzymes are required for the synthesis of thiamine monophosphate from the thiazole and pyrimidine (for a review of this work, see Leder [75]): (i) thiazole kinase, which converts thiazole to thiazole monophosphate; (ii) hydroxymethyl pyrimidine kinase, which converts the pyrimidine to its phosphorylated form; (iii) hydroxymethylpyrimidine phosphokinase, which converts the pyrimidine monophosphate to the pyrophosphoryl form; and finally (iv) thiamine phosphate synthase or thiamine phosphate pyrophosphorylase, which condenses the thiazole phosphate and pyrimidine pyrophosphate to yield thiamine monophosphate

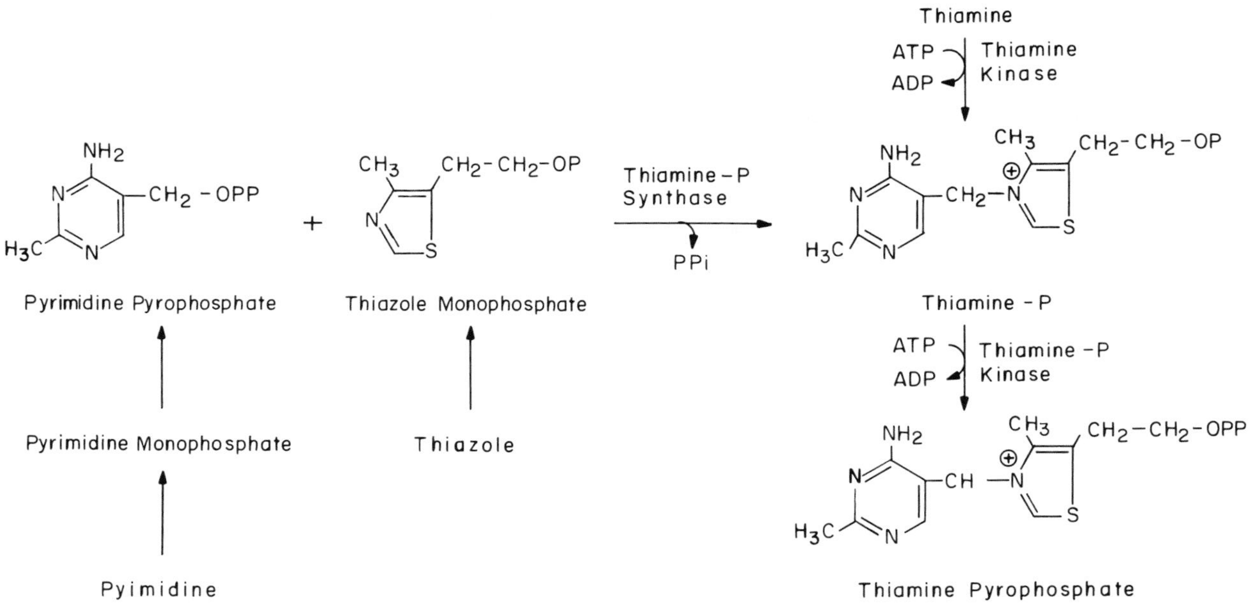

FIG. 8. Enzymatic pathway for the formation of thiamine and thiamine pyrophosphate.

and PP$_i$ (Fig. 8). Nose and co-workers (109) first demonstrated the presence of these four enzymes in *E. coli* and have studied their regulation by intracellular levels of thiamine and other metabolites (61, 66, 68). Thiamine phosphate synthase has been purified about 175-fold from extracts of *E. coli*, and some of its properties have been described (69). A mutant strain of *E. coli* K-12 that lacks this enzyme has been reported (67). None of the other three enzymes has been purified from either *E. coli* or *S. typhimurium*. Since work on the biosynthesis of the pyrimidine and thiazole moieties, already discussed, has suggested that the de novo pathways for their syntheses may yield the phosphorylated derivatives directly, it is possible that the thiazole and pyrimidine kinases are part of a salvage pathway.

In *E. coli*, both biochemical and genetic evidence indicates that thiamine monophosphate is a key intermediate in the biosynthesis of thiamine pyrophosphate. Enzymes catalyzing the conversion of thiamine to thiamine monophosphate (57) and thiamine monophosphate to thiamine pyrophosphate (107, 108) have been detected in cell extracts of *E. coli*. Furthermore, Nakayama and Hayashi (96, 97, 98) have isolated mutants of *E. coli* W that can grow on thiamine phosphates, but not on thiamine. These mutants exhibit two different phenotypes. The first type requires thiamine pyrophosphate for growth and accumulates thiamine monophosphate when provided with thiamine; cell extracts of this mutant contain thiamine kinase but not thiamine phosphate kinase. The second type of mutant requires either thiamine monophosphate or thiamine pyrophosphate for growth and cannot phosphorylate free thiamine; cell extracts of this mutant contain thiamine phosphate kinase but not thiamine kinase. These results establish the importance of thiamine monophosphate as a biosynthetic intermediate in the production of the thiamine pyrophosphate. There are a couple of reports in the

literature of thiamine pyrophosphokinase activity in *E. coli* (91, 92), but their significance is unclear.

PANTOTHENIC ACID AND COENZYME A

Pantothenic acid, one of the B vitamins, is formed by the ATP-dependent condensation of 2,4-dihydroxy-3,3-dimethylbutyric acid (pantoic acid) and 2-aminopropionic acid (β-alanine). Pantothenate is utilized primarily for the biosynthesis of coenzyme A and acyl carrier protein, coenzymes that function in the metabolism of acyl moieties which are bound as thioesters to the sulfhydryl group of the 4'-phosphopantetheine portion of these molecules (3, 129). This review of the biosynthesis of coenzyme A will be divided into three parts: (i) the formation of pantoate and β-alanine; (ii) the synthesis of pantothenate; and (iii) the conversion of pantothenate to coenzyme A. The reactions of this pathway are summarized in Fig. 9.

Biosynthesis of β-Alanine

By the 1950s, indirect evidence had been obtained in support of a number of different pathways as the source of β-alanine for pantothenate biosynthesis. Early studies with bacteria had suggested that β-alanine is derived from aspartic acid by decarboxylation (34, 84, 147). This view was given further support by the findings of Shive and Macow (140) that the growth-inhibitory properties of β-hydroxyaspartic acid for *E. coli* can be reversed by either aspartate, β-alanine, or pantothenate. Based on their observation that the growth of a β-alanine auxotroph was supported by dihydrouracil or β-ureidopropionate (*N*-carbamyl-β-alanine), Slotnick and Weinfeld (141, 142) proposed that β-alanine for pantothenate biosynthesis is derived from the catabolism of uracil. Finally, it seemed possible that β-alanine could be derived by transamination since enzymes from a number of bac-

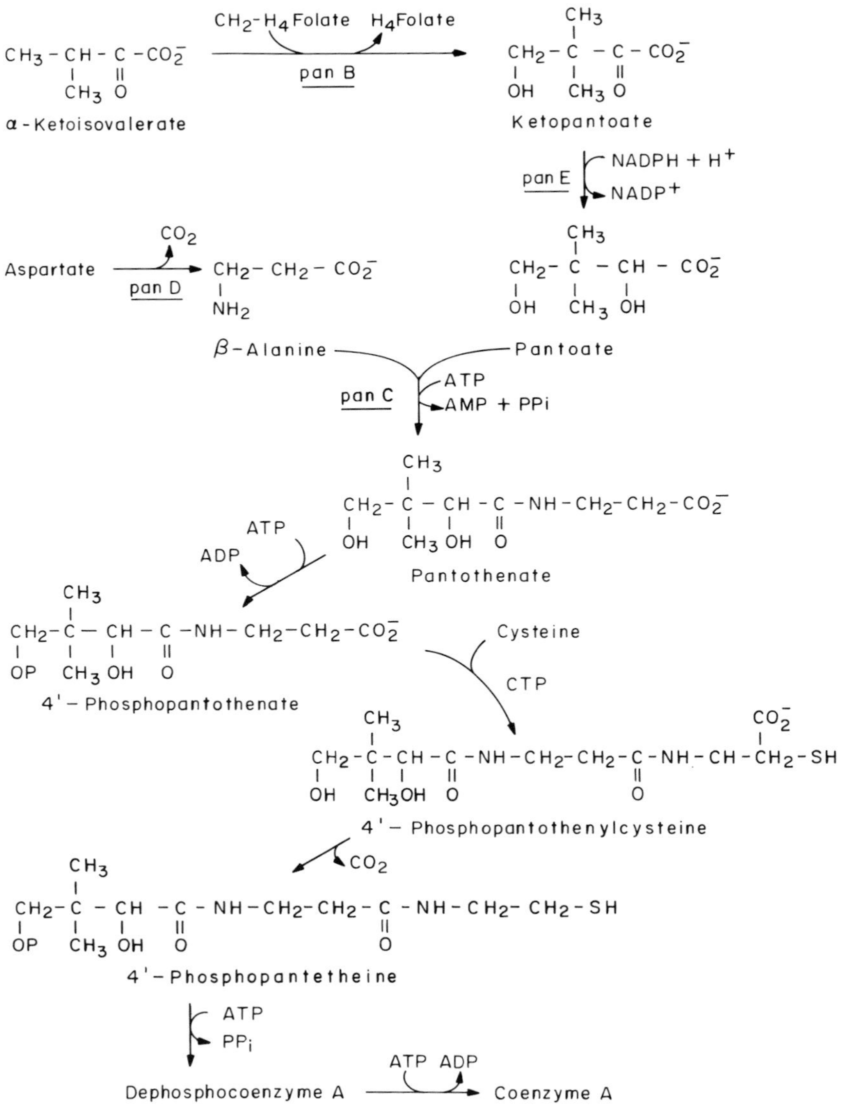

FIG. 9. Biosynthetic pathway for the production of pantothenic acid and coenzyme A.

terial sources, including *E. coli*, had been described that catalyze the conversion of malonic semialdehyde to β-alanine (128). In the 1970s, this problem was reinvestigated by Williamson and Brown (160) and, independently, by Cronan (30). Both of these groups demonstrated that the main pathway for β-alanine biosynthesis in *E. coli* is by α-decarboxylation of aspartic acid, with the findings that (i) *E. coli* contains an enzyme that catalyzes the conversion of aspartic acid to stoichiometric amounts of β-alanine and CO_2, and (ii) this aspartate-1-decarboxylase is missing in mutant strains of *E. coli* (*panD*) that require β-alanine or pantothenate as a nutritional factor, but is present in wild-type strains and revertants of the mutant strains. In addition to this direct evidence in support of asparate as the main precursor of β-alanine, the participation of the uracil catabolic pathway is further ruled out by the fact that Cronan was unable to repeat the observation of Slotnick and Weinfeld that dihydrouracil supports the growth of β-alanine-requiring strains of *E. coli*. Recently, West and co-workers

(150) have shown in *S. typhimurium* that defects in β-alanine synthesis and uracil catabolism are clearly separable by mutation.

Williamson and Brown (159, 160) purified aspartate-1-decarboxylase to apparent homogeneity from extracts of *E. coli* B. One of the most interesting aspects of the structure of the decarboxylase is that it contains a pyruvoyl residue as a prosthetic group rather than pyridoxal phosphate, the coenzyme present in most other bacterial decarboxylases. The presence of a covalently bound pyruvate implies that some posttranslational processing of the protein must occur. Two types of mechanisms are possible for this processing. First, the conversion of proenzyme to active enzyme could be catalyzed by a separate "activating" enzyme. Second, the conversion could be affected by the proenzyme itself in an autocatalytic process. There is evidence that such an autocatalytic process occurs in the activation of the proenzyme form of histidine decarboxylase of *Lactobacillus* sp. strain 30a (122). Further work, involving biochemical and possibly genetic analyses, is necessary to elucidate the mechanism of formation of this prosthetic group in *E. coli*.

Cronan and co-workers (31) have mapped the *panD* gene of *E. coli* K-12 and find that it lies between *panB* (ketopantoate hydroxymethyltransferase) and *panC* (pantothenate synthetase) at min 3 on the chromosome. Ortega and co-workers (111) reported that isolation of the first mutant of *S. typhimurium* requiring β-alanine as a nutritional factor and mapped the mutation to min 89 of the chromosome. More recently, Primerano and Burns (119) have reported the isolation of additional β-alanine-requiring mutants of *S. typhimurium*. In contrast to the results of Ortega et al. (111), these workers found that the *panD* locus is closely linked to *panC* at min 4.5 of the chromosome. A report from Cronan's laboratory (31) shows that decarboxylase activity can be detected in extracts of *S. typhimurium* under conditions used for the assay of the *E. coli* enzyme, but no detailed biochemical characterization of the decarboxylase from *S. typhimurium* has been reported. It is possible that the discrepancy in determination of the map position of *panD* results from the isolation of different types of β-alanine-requiring mutants by these two groups. Further work will be required to resolve this problem.

Biosynthesis of Pantoic Acid

That the precursor of ketopantoic acid is α-ketoisovaleric acid, an intermediate in the biosynthesis of valine, was first suggested by Maas and Vogel (81), who found that whole cells of *E. coli* can synthesize ketopantoate and pantoate from α-ketoisovalerate, whereas a particular pantoate-requiring mutant cannot. McIntosh et al. (87) subsequently reported on the detection of an enzyme activity in cell extracts of *E. coli* that could convert α-ketoisovaleric acid to ketopantoate in the presence of formaldehyde. Previous observations in *Bacterium linens* that *p*-AB could replace pantoate as a growth factor for this strain (120, 121) led these workers to think that tetrahydrofolate would be a cofactor for the formaldehyde-dependent reaction in *E. coli*. However, when the enzyme that catalyzes the condensation of formaldehyde and α-ketoisovalerate was partially purified, no role for tetrahydrofolate was found (87). Moreover, the K_m values for formaldehyde (0.01 M) and α-ketoisovalerate(0.1 M) are so high that the physiological significance of the reaction seemed questionable. Snell and co-workers (117, 118, 144) reinvestigated this problem and found two different enzymes that carry out the synthesis of ketopantoate in *E. coli*. One is similar to that studied by McIntosh et al. (87), and the second, ketopantoate hydroxymethyltransferase, catalyzes the tetrahydrofolate-dependent formation of ketopantoate. The conclusion that the latter enzyme is the one involved in the biosynthesis of pantothenate is based on their findings that (i) the Michaelis constants of the transferase for its substrates are all within the physiological range, and (ii) the transferase is absent from a ketopantoate auxotroph (*E. coli* 99-4; *panB*) whereas the tetrahydrofolate-independent activity is found in the same amounts in the auxotroph as in wild-type *E. coli* (144). Powers and Snell (118) have purified the ketopantoate hydroxymethyltransferase

to homogeneity from extracts of *E. coli*. Their findings that pantoate, pantothenate, and coenzyme A are all allosteric inhibitors of transferase activity suggest that feedback inhibition is an important mode of regulation for this pathway. Regulation by repression of enzyme synthesis appears not to occur (118, 119). The stereochemistry of addition of the hydroxymethyl group to α-ketoisovalerate has also been determined (1).

In early studies, Demerec and co-workers (36) isolated a mutant strain of *S. typhimurium* that responds to α-ketoisovalerate and valine as well as ketopantoate, pantoate, and pantothenate. Since *panB* mutants respond only to ketopantoate, pantoate, and pantothenate, these workers proposed that the mutant, which was given a new designation, *panA*, is defective in the synthesis of α-ketoisovalerate. More recently, however, Cronan has pointed out that this hypothesis would require that the *panA* strain be auxotrophic for valine and leucine as well as pantothenate (30). He suggested that the *panA* mutants may produce an altered hydroxymethyltransferase with a decreased affinity for α-ketoisovalerate. Subsequent studies by Cronan and co-workers (31) in extracts of the *panA* mutant have shown that the strain is deficient in ketopantoate hydroxymethyltransferase activity. Moreover, the residual enzyme activity is unusually labile to purification and storage and shows kinetics for α-ketoisovalerate utilization significantly different from that of the wild-type enzyme. Thus, the *panA* lesion appears to be in the *panB* gene rather than in a separate locus. No *panA* mutants have been reported in *E. coli*.

In *E. coli*, evidence that ketopantoate is the immediate precursor of pantoate was first provided by Lansford and Shive (74), who found that a particular class of pantoate-requiring mutants can utilize ketopantoate in place of pantoate for growth. Enzymes that catalyze the NADPH-dependent reduction of both ketopantoate and ketopantoyllactone have been purified from *E. coli* and characterized by King and Wilken (157, 158). The physiological significance of these activities for pantothenate biosynthesis was unclear, however, because of the lack of mutants in either *E. coli* or *S. typhimurium* that require pantoate or pantothenate, but not ketopantoate, for growth which could be tested for these enzymes. The absence of such mutants prompted Primerano and Burns (119) to investigate the possibility that more than one enzyme are able to catalyze the conversion of ketopantoate to pantoate. These workers found that purified acetohydroxy acid isomeroreductase of *S. typhimurium*, the product of the *ilvC* gene which catalyzes the second common step in isoleucine and valine biosynthesis, binds ketopantoate and catalyzes its reduction at 1/20 the rate of the normal substrate, α-acetolactate. In addition, in *ilvC* strains they were able to isolate mutants blocked in the conversion of ketopantoate to pantoate. These mutants grew in isoleucine-valine medium supplemented with pantoate or pantothenate but not in the same medium supplemented with ketopantoate or β-alanine. This mutation which confers pantoate auxotrophy has been designated *panE*. *panE* strains were found to have significantly reduced amounts of a ketopantoate reductase activity similar to that described by King and

Wilken for *E. coli*, but wild-type amounts of both ketopantoyllactone reductase and ketopantoate hydroxymethyltransferase. None of the *panE* mutations was found to be closely linked with *panB*, *panC*, or *panD*. To demonstrate that the *ilvC* gene functions in vivo to reduce ketopantoate, Primerano and Burns constructed strains that were *panE ilvC*$^+$. These strains were found to require either pantoate or pantothenate for growth only when the isomeroreductase was present at low levels; when the synthesis of this enzyme was induced, *panE ilvC*$^+$ strains required no supplementation. Manch (83) has reported the isolation of a mutant of *E. coli* K-12 which requires pantoate for growth. This mutation is not linked to either *panB*, *panC*, or *panD* and is presumed to be in the reductase. Further enzyme work will be required, however, to verify this hypothesis.

Enzymatic Synthesis of Pantothenic Acid

The enzyme that catalyzes the ATP-dependent synthesis of pantothenic acid from pantoic acid and β-alanine was partially purified from *E. coli* by two independent groups (78, 80, 113). Subsequently, the purification procedure was modified to obtain a homogeneous preparation of the enzyme (93, 94). Pantothenate synthetase requires both a divalent cation (Mg^{2+} or Mn^{2+}) and a monovalent cation (K^+ or NH_4^+) for activity (78). The *E. coli* enzyme has a pH optimum of 10.0 (94); the *S. typhimurium* enzyme has a pH optimum of 8.1 and is inactive at pH 10.0 (31). Maas and Novelli (80) established that AMP and PP_i are formed in stoichiometric amounts with pantothenate as a result of enzyme action. Maas has studied the mechanism of the reaction extensively (W. K. Maas, Fed. Proc. **15**:305, 1956). His results and kinetic analyses of the enzyme reaction done by Miyatake et al. (94) are consistent with the following mechanistic scheme:

$$\text{Pantoate} + \text{ATP} + \text{enzyme}$$
$$\rightleftharpoons \text{enzyme-pantoyl-AMP} + PP_i$$

$$\text{Enzyme-pantoyl-AMP} + \beta\text{-alanine}$$
$$\rightarrow \text{pantothenate} + \text{AMP} + \text{enzyme}$$

It has been concluded that pantothenate synthetase is a product of the *panC* gene in *E. coli* because a *panC* mutant of *E. coli* W that requires pantothenate as a nutritional factor lacks this enzyme (78). A similar conclusion has been drawn for *panC* mutants of *S. typhimurium* and other strains of *E. coli* (31). Constitutive synthesis of pantothenate by *E. coli* results in the excretion of copious amounts of pantothenate when the cells have an adequate supply of β-alanine or its precursor, aspartate (35, 62, 79, 160). It has been estimated that more than 90% of the pantothenate synthesized by *E. coli* is excreted rather than being utilized for coenzyme A biosynthesis (62).

Enzymatic Synthesis of Coenzyme A from Pantothenic Acid

Early work with whole cells of *Lactobacillus arabinosus* (W. S. Pierpoint and D. E. Hughes, Abstr. Congr. Intern. Biochem., 2nd, Paris, abstr. no. 91,

1952) and *Proteus morganii* (23) established that the sulfur-containing portion of coenzyme A, β-mercaptoethylamine, is derived from cysteine. Additional studies with the bacteria *Acetobacter suboxydans* and *Lactobacillus helveticus* (18, 23, 24, 72, 148) and preparations of mammalian liver (2, 54, 148) led to the formulation of the following pathway for coenzyme A biosynthesis:

Pantothenate + cysteine → pantothenylcysteine
Pantothenylcysteine → pantetheine + CO_2
Pantetheine + ATP → phosphopantetheine
Phosphopantetheine + ATP → dephosphocoenzyme A
Dephosphocoenzyme A + ATP → Coenzyme A

Confidence that the initial portion of this pathway is correct was reduced, however, with the finding that bacteria other than *A. suboxydans* and *L. helveticus* cannot decarboxylate pantothenylcysteine (18). Brown (17), therefore, reinvestigated this problem and showed that phosphorylation of all of the proposed intermediates is required before these compounds can serve as substrates for coenzyme A biosynthesis in *E. coli* and *P. morganii*. Evidence was presented to show that the following reactions occur in cell extracts of *E. coli* and other organisms:

Pantothenate + ATP → 4′-phosphopantothenate

4′-Phosphopantothenate + cysteine + CTP
→ phosphopantothenylcysteine

Phosphopantothenylcysteine
→ phosphopantetheine + CO_2

The rest of the pathway is the same as has been established with the other organisms. None of the enzymes of this pathway has been purified from either *E. coli* or *S. typhimurium*.

As described earlier in this chapter, many mutants of both *E. coli* and *S. typhimurium* have been isolated that are unable to synthesize pantothenate. There is only one report, however, of a mutant blocked in the conversion of pantothenate to coenzyme A. Dunn and Snell (38) have isolated a temperature-sensitive mutant of *S. typhimurium* LT2 that cannot phosphorylate pantothenate at the nonpermissive temperature. This genetic locus has been designated *coaA*. Addition of coenzyme A does not permit this strain to grow under the restrictive conditions, a result which suggests that the organism is not permeable to CoA; in *E. coli* K-12, Jackowski and Rock (63) have shown that exogenous 4′-phosphopantetheine is not assimilated into either coenzyme A or acyl carrier protein. Dunn and Snell (38) have demonstrated that extracts of most of the mutant strains contain temperature-sensitive pantothenate kinase activity. One mutant (DD9), however, which shows the same phenotype as the others, appears to contain a wild-type pantothenate kinase. The *coaA* locus has been mapped at min 89 of the chromosome, a location close to that determined for *panD* by Ortega et al. (111). The DD9 mutation also lies close to min 89. Further work will be required to determine whether DD9 contains a more subtly altered pantothenate kinase or is blocked in some other step in coenzyme A biosynthesis.

LITERATURE CITED

1. **Aberhart, D. J.** 1979. Stereochemistry of pantoate biosynthesis from 2-keto-isovalerate. J. Am. Chem. Soc. **101**:1354–1355.

2. **Abiko, Y., T. Suzuki, and M. Shimuzu.** 1967. Pantothenic acid and its related compounds. XI. Biochemical studies. 6. A final stage in the biosynthesis of CoA. J. Biochem. (Tokyo) **61**:309–312.

3. **Alberts, A. W., and P. R. Vagelos.** 1966. Acyl carrier protein. XIII. Studies of acyl carrier protein and coenzyme A in *Escherichia coli* pantothenate and β-alanine auxotrophs. J. Biol. Chem. **241**:5201–5204.

4. **Alworth, W. L., M. F. Dove, and H. N. Baker.** 1977. Biosynthesis of the dimethylbenzene moiety of riboflavin and dimethylbenzimidazole: evidence for the involvement of C-1 of a pentose as a precursor. Biochemistry **16**:526–531.

5. **Bacher, A., Q. LeVan, M. Buhler, P. J. Keller, V. Einicke, and H. G. Floss.** 1982. Biosynthesis of riboflavin. Incorporation of D-[1-^{13}C]ribose. J. Am. Chem. Soc. **104**:3754–3755.

6. **Bacher, A., Q. LeVan, P. J. Keller, and H. G. Floss.** 1983. Biosynthesis of riboflavin. Incorporation of ^{13}C-labelled precursors into the xylene ring. J. Biol. Chem. **258**:13431–13437.

7. **Bacher, A., and F. Lingens.** 1970. Biosynthesis of riboflavin. Formation of 2,5-diamino 6-hydroxy-4-(1'-D-ribitylamino)pyrimidine in a riboflavin auxotroph. J. Biol. Chem. **245**:4647–4652.

8. **Bacher, A., and F. Lingens.** 1971. Biosynthesis of riboflavin. Formation of 6-hydroxy-2,4,5-triaminopyrimidine in *rib7* mutants of *Saccharomyces cerevisiae*. J. Biol. Chem. **246**:7018–7022.

9. **Bellion, E., and D. H. Kirkley.** 1977. The origin of the sulfur atom in thiamine. Biochim. Biophys. Acta **497**:323–328.

10. **Bellion, E., D. H. Kirkley, and J. R. Faust.** 1976. The biosynthesis of the thiazole moiety of thiamine in *Salmonella typhimurium*. Biochim. Biophys. Acta **437**:229–237.

11. **Bognar, A. L., C. Osborne, B. Shane, S. C. Singer, and R. Ferone.** 1985. Folypoly-γ-glutamate synthetase-dihydrofolate synthetase. Cloning and high expression of the *Escherichia coli folC* gene and purification and properties of the gene product. J. Biol. Chem. **260**:5625–5630.

12. **Brenner-Holzach, O., and F. Leuthardt.** 1959. Untersuchungen zur Biosynthese der Pterine bei *Drosophila melanogaster*. Helv. Chim. Acta **42**:2254–2257.

13. **Bresler, S. E., E. A. Glazunov, D. A. Perumov, and T. P. Chernik.** 1977. Riboflavin operon in *Bacillus subtilis*. XIII. Genetic and biochemical study of mutants related to riboflavin intermediates. Genetika **13**:2007–2016.

14. **Bresler, S. E., and D. A. Perumov.** 1979. Riboflavin operon in *Bacillus subtilis*. Regulation of GTP-cyclohydrolase synthesis in strains of different genotypes. Genetika **15**:967–971.

15. **Bresler, S. E., D. A. Perumov, T. P. Chernik, and E. A. Glazunov.** 1976. Investigation of the operon of riboflavin biosynthesis in *Bacillus subtilis*, communication X. Genetic and biochemical study of mutants that accumulate 6-methyl-7-(1',2'-dihydroxyethyl)-8-ribityllumazine. Genetika **12**:83–91.

16. **Brook, M. S., and B. Magasanik.** 1954. The metabolism of purines in *Aerobacter aerogenes*: a study of purineless mutants. J. Bacteriol. **68**:727–733.

17. **Brown, G. M.** 1954. The metabolism of pantothenic acid. J. Biol. Chem. **234**:370–378.

18. **Brown, G. M.** 1957. Pantothenylcysteine, a precursor of pantetheine in *Lactobacillus helveticus*. J. Biol. Chem. **226**:651–661.

19. **Brown, G. M.** 1962. The biosynthesis of folic acid. Inhibition by sulfonamides. J. Biol. Chem. **237**:536–540.

20. **Brown, G. M.** 1970. Enzymic synthesis of pterins and dihydropteroic acid, p. 243–264. *In* K. Iwai, M. Akino, M. Goto, and Y. Iwanami (ed.), Chemistry and biology of pteridines. International Academic Printing Co. Ltd., Tokyo.

21. **Brown, G. M.** 1971. The biosynthesis of pteridines. Adv. Enzymol. Relat. Areas Mol. Biol. **35**:37–77.

22. **Brown, G. M., and J. J. Reynolds.** 1963. Biosynthesis of water-soluble vitamins. Annu. Rev. Biochem. **32**:419–462.

23. **Brown, G. M., and E. E. Snell.** 1953. N-Pantothenylcysteine as a precursor for pantetheine and coenzyme A. J. Am. Chem. Soc. **75**:2782–2783.

24. **Brown, G. M., and E. E. Snell.** 1954. Pantothenic acid conjugates and growth of *Acetobacter suboxydans*. J. Bacteriol. **67**:465–471.

25. **Brown, G. M., R. A. Weisman, and D. A. Molnar.** 1961. Biosynthesis of folic acid. I. Substrate and cofactor requirements for enzymatic synthesis by cell free extracts of *Escherichia coli*. J. Biol. Chem. **236**:2534–2543.

26. **Brown, G. M., and H. Williamson.** 1982. Biosynthesis of riboflavin, folic acid, thiamine, and pantothenic acid. Adv. Enzymol. Relat. Areas Mol. Biol. **53**:345–381.

27. **Burg, A. W., and G. M. Brown.** 1966. The biosynthesis of folic acid. VI. Enzymatic conversion of carbon atom 8 of guanosine triphosphate to formic acid. Biochim. Biophys. Acta **117**:275–278.

28. **Burg, A. W., and G. M. Brown.** 1968. The biosynthesis of folic acid. VIII. Purification and properties of the enzyme that catalyzes the production of formate from carbon atom 8 of guanosine triphosphate. J. Biol. Chem. **243**:2349–2358.

29. **Burrows, R. H., and G. M. Brown.** 1978. Presence in *Escherichia coli* of a deaminase and a reductase involved in biosynthesis of riboflavin. J. Bacteriol. **136**:657–667.

30. **Cronan, J. E.** 1980. β-Alanine synthesis in *Escherichia coli*. J. Bacteriol. **141**:1291–1297.

31. **Cronan, J. E., K. H. Littel, and S. Jackowski.** 1982. Genetic and biochemical analysis of pantothenate biosynthesis in *Escherichia coli* and *Salmonella typhimurium*. J. Bacteriol. **149**:916–922.

32. **David, S., B. Estramareix, J. C. Fischer, and M. Therisod.** 1981. 1-Deoxy-D-*threo*-2-pentulose: the precursor of the five-carbon chain of the thiazole of thiamine. J. Am. Chem. Soc. **102**:7341–7342.

33. **David, S., B. Estramareix, J. C. Fischer, and M. Therisod.** 1982. The biosynthesis of thiamine. Syntheses of [1,1,1,5-^{2}H]-1-deoxy-D-*threo*-2-pentulose and incorporation of this sugar in biosynthesis of thiazole by *Escherichia coli* cells. J. Chem. Soc. Perkin Trans. I, p. 2131–2137.

34. **David, W. E., and H. C. Lichstein.** 1950. Aspartic acid decarboxylase in bacteria. Proc. Soc. Exp. Biol. Med. **73**:216–218.

35. **Davis, B. D.** 1950. Studies on nutritionally deficient bacterial mutants isolated by means of penicillin. Experientia **6**:41–50.

36. **Demerec, M. E., E. L. Lahr, E. Balbinder, T. Miyake, C. Mack, D. Mackay, and J. Ishidu.** 1959. Bacterial genetics. Carnegie Inst. Wash. Year Book **58**:433–440.

37. **DeMoll, E., and W. Shive.** 1985. Determination of the metabolic origin of the sulfur atom in thiamine of *Escherichia coli* by mass spectrometry. Biochem. Biophys. Res. Commun. **132**:217–222.

38. **Dunn, S. D., and E. E. Snell.** 1979. Isolation of temperature-sensitive pantothenate kinase mutants of *Salmonella typhimurium* and mapping of the *coaA* gene. J. Bacteriol. **140**:805–808.

39. **Eggleston, L. V., and H. A. Krebs.** 1959. Permeability of *Escherichia coli* to ribose and ribose nucleotides. Biochem. J. **73**:264–270.

40. **Estramareix, B.** 1970. Biosynthesis of the pyrimidine moiety of thiamine: origin of carbon-6 in *Salmonella typhimurium*. Biochim. Biophys. Acta **208**:170–171.

41. **Estramareix, B., D. Gaudry, and M. Therisod.** 1977. Biosynthesis of the thiamine thiazole in *Escherichia coli*. Biochimie (Paris) **59**:857–859.

42. **Estramareix, B., and M. Lesieur.** 1969. Biosynthesis of the pyrimidine moiety of thiamine: origin of carbons in positions 2 and 4 in *Salmonella typhimurium*. Biochim. Biophys. Acta **192**:375–377.

43. **Estramareix, B., and M. Therisod.** 1972. Tyrosine as a factor in biosynthesis of the thiazole moiety of thiamine in *Escherichia coli*. Biochim. Biophys. Acta **273**:275–282.

44. **Estramareix, B., and M. Therisod.** 1980. Isolation of 5-(β-hydroxyethyl)-4-methylthiazole-2-carboxylic acid, a metabolite related to thiamine biosynthesis in *Escherichia coli*. Biochem. Biophys. Res. Commun. **95**:1017–1022.

45. **Estramareix, B., and M. Therisod.** 1984. Biosynthesis of thiamine: 5-aminoimidazole ribotide as the precursor of all the carbon atoms of the pyrimidine moiety. J. Am. Chem. Soc. **106**:3857–3860.

46. **Ferone, R., S. C. Singer, M. H. Hanlon, and S. Roland.** 1983. Isolation and characterization of an *E. coli* mutant affected in dihydrofolate and folylpolyglutamate-synthetase, p. 585–589. *In* J. A. Blair (ed.), Chemistry and biology of pteridines. Walter de Gruyter, Berlin.

47. **Ferone, R., and A. Warskow.** 1983. Co-purification of dihydrofolate synthesis and N^{10}-formyltetrahydropteroyl-diglutamate synthetase from *E. coli*, p. 167–181. *In* I. D. Goldman, B. Chabner, and J. R. Bertino (ed.), Folyl and antifolyl polyglutamates. Plenum Publishing Corp., New York.

48. **Floss, H. G., Q. LeVan, P. J. Keller, and A. Bacher.** 1983. Biosynthesis of riboflavin. An unusual rearrangement in the formation of 6,7-dimethyl-8-ribityllumazine. J. Am. Chem. Soc. **105**:2494–2495.

49. **Foor, F., and G. M. Brown.** 1975. Purification and properties of guanosine triphosphate cyclohydrolase II from *Escherichia coli*. J. Biol. Chem. **250**:3545–3551.

50. **Goldstein, G. A., and G. M. Brown.** 1963. The biosynthesis of thiamine. V. Studies concerning precursors of the pyrimidine moiety. Arch. Biochem. Biophys. **103**:449–452.

51. **Goto, M., and H. S. Forrest.** 1961. Identification of a new phosphorylated pteridine from *E. coli*. Biochem. Biophys. Res. Commun. **6**:180–183.

52. **Griffin, M. J., and G. M. Brown.** 1964. The biosynthesis of folic acid. III. Enzymatic formation of dihydrofolic acid from dihydropteroic acid and of tetrahydropteroyl-polyglutamic acid compounds from tetrahydrofolic acid. J. Biol. Chem. **239**:310–316.

53. **Harrington, C. R., and R. C. G. Moggridge.** 1940. Experiments on the biogenesis of vitamin B1. Biochem. J. **34**:685–689.

54. **Hoagland, M. B., and G. D. Novelli.** 1954. Biosynthesis of coenzyme A from phosphopantetheine and of pantetheine from pantothenate. J. Biol. Chem. **207**:767–773.

55. **Hollander, I., J. C. Braman, and G. M. Brown.** 1980. Biosynthesis of riboflavin: enzymatic conversion of 5-amino-2,4-dioxy-6-ribitylaminopyrimidine to 6,7-dimethyl-8-ribityl-lumazine. Biochem. Biophys. Res. Commun. **94**:515–521.

56. **Hollander, I., and G. M. Brown.** 1979. Biosynthesis of riboflavin: reductase and deaminase of *Ashbya gossypii*. Biochem. Biophys. Res. Commun. **89**:759–763.

57. **Iwashima, A., and H. Nishino.** 1972. Conversion of thiamine to thiamine monophosphate by cell-free extracts of *Escherichia coli*. Biochim. Biophys. Acta **258**:333–336.

58. **Iwashima, A., and Y. Nose.** 1970. Inhibition by phenylalanine of thiazole biosynthesis in *Escherichia coli*. J. Bacteriol. **104**:1014–1019.

59. **Iwashima, A., and Y. Nose.** 1970. Effect of glycine on thiazole biosynthesis in *Escherichia coli*. J. Bacteriol. **101**:1076–1078.

60. **Iwashima, A., and Y. Nose.** 1971. Incorporation of glycine in pyrimidine and thiazole moieties of thiamine in *Escherichia coli*. Biochim. Biophys. Acta **252**:235–238.

61. **Iwashima, A., K. Takahashi, and Y. Nose.** 1971. Overproduction of hydroxymethylpyrimidine by a thiamine regulatory mutant of *Escherichia coli*. J. Vitaminol. (Kyoto) **17**:43–48.

62. **Jackowski, S., and C. O. Rock.** 1981. Regulation of coenzyme A biosynthesis. J. Bacteriol. **148**:926–932.

63. **Jackowski, S., and C. O. Rock.** 1984. Metabolism of 4′-phosphopantetheine in *Escherichia coli*. J. Bacteriol. **158**:115–120.

64. **Jones, T. H. D., and G. M. Brown.** 1967. The biosynthesis of folic acid. VII. Enzymatic synthesis of pteridines from guanosine triphosphate. J. Biol. Chem. **242**:3989–3997.

65. **Jones, T. H. D., J. J. Reynolds, and G. M. Brown.** 1964. Enzymatic formation of dihydropteroic acid from 2-amino-4-hydroxy-8-trihydroxypropyl-7-8-dihydropteridine. Biochem. Biophys. Res. Commun. **17**:486–489.

66. **Kawasaki, T., A. Iwashima, and Y. Nose.** 1969. Regulation of thiamine biosynthesis in *Escherichia coli*. J. Biochem. (Tokyo) **65**:407–416.

67. **Kawasaki, T., T. Nakata, and Y. Nose.** 1968. Genetic mapping with a thiamine-requiring auxotroph of *Escherichia coli* K-12 defective in thiamine phosphate pyrophosphorylase. J. Bacteriol. **95**:1483–1485.

68. **Kawasaki, T., and Y. Nose.** 1969. Thiamine regulatory mutants in *Escherichia coli*. J. Biochem. (Tokyo) **65**:417–425.

69. **Kayama, Y., and T. Kawasaki.** 1973. Purification and properties of thiaminephosphate pyrophosphorylase of *Escherichia coli*. Arch. Biochem. Biophys. **158**:242–248.

70. **Keller, P. J., Q. LeVan, A. Bacher, and H. G. Floss.** 1983. Biosynthesis of riboflavin. ^{13}C-NMR techniques for the analysis of multiple ^{13}C-labeled riboflavins. Tetrahedron **39**:3471–3481.

71. **Keller, P. J., Q. LeVan, A. Bacher, J. F. Kozlowski, and H. G. Floss.** 1983. Biosynthesis of riboflavin. Analysis of biosynthetically ^{13}C-labelled riboflavin by double-quantum and two-dimensional NMR. J. Am. Chem. Soc. **105**:2505–2507.

72. **King, T. E., and V. H. Cheldelin.** 1953. Pantothenic acid derivatives and growth of *Acetobacter suboxydans*. Proc. Soc. Exp. Biol. Med. **84**:591–593.

73. **Kumaoka, H., and G. M. Brown.** 1967. Biosynthesis of thiamine. VI. Incorporation of formate into carbon atom two of the pyrimidine moiety of thiamine. Arch. Biochem. Biophys. **122**:378–384.

74. **Lansford, E. M., Jr., and W. Shive.** 1952. The microbiological activity of α-keto-β,β-dimethyl-γ-butyrolactone. Arch. Biochem. Biophys. **38**:353–365.

75. **Leder, I.** 1975. Thiamine, biosynthesis and function, p. 57–85. *In* D. M. Greenberg, (ed.), Metabolic pathways, 3rd ed., vol. 7. Academic Press, Inc., New York.

76. **Logvinenko, E. M., G. M. Shavlovsky, and N. Tsarenko.** 1984. The role of iron in regulation of 6,7-dimethyl-8-ribityl-lumazine synthase synthesis in flavinogenic yeasts. Biokhimiya **49**:45–50.

77. **Logvinenko, E. M., G. M. Shavlovsky, A. E. Zakalsky, and I. V. Zakhodylo.** 1982. Biosynthesis of 6,7-dimethyl-8-ribitylluma-zine in yeast extracts of *Pichia guilliermondii*. Biokhimiya **47**:931–936.

78. **Maas, W. K.** 1952. Pantothenate studies. III. Description of the extracted pantothenate-synthesizing enzyme of *Escherichia coli*. J. Biol. Chem. **198**:23–32.

79. **Maas, W. K., and B. D. Davis.** 1950. Pantothenate studies. I. Interference by D-serine and L-aspartic acid with pantothenate synthesis in *Escherichia coli*. J. Bacteriol. **60**:733–745.

80. **Maas, W. K., and G. D. Novelli.** 1953. Synthesis of pantothenic acid by depyrophosphorylation of adenosine triphosphate. Arch. Biochem. Biophys. **43**:236–238.

81. **Maas, W. K., and H. Vogel.** 1953. α-Oxoisovaleric acid, a precursor of pantothenic acid in *Escherichia coli*. J. Bacteriol. **65**:388–393.

82. **Mailänder, B., and A. Bacher.** 1976. Biosynthesis of riboflavin. Structure of the purine precursor and origin of the ribityl side chain. J. Biol. Chem. **251**:3623–3628.

83. **Manch, J. N.** 1981. Mapping of a new *pan* mutation in *Escherichia coli* K-12. Can. J. Microbiol. **27**:1231–1233.

84. **Mardashev, S. R., and R. N. Etinogof.** 1948. The chemical nature of aspartic acid decarboxylase. Biokhimiya **13**:402–408.

85. **Masurekar, M., and G. M. Brown.** 1975. Partial purification and properties of an enzyme from *Escherichia coli* that catalyzes the conversion of glutamic acid and 10-formyltetrahydropteroylglutamic acid to 10-formyltetrahydropteroyl-γ-glutamylglutamic acid. Biochemistry **14**:2424–2430.

86. **Mathis, J. B., and G. M. Brown.** 1970. The biosynthesis of folic acid. XI. Purification and properties of dihydroneopterin aldolase. J. Biol. Chem. **245**:3015–3025.

87. **McIntosh, E. N., M. Purko, and W. A. Wood.** 1957. Ketopantoate formation by a hydroxymethylation enzyme from *Escherichia coli*. J. Biol. Chem. **228**:499–510.

88. **Miersch, J., E. M. Logvinenko, A. E. Zakalsky, G. M. Shavlovsky, and H. Reinbothe.** 1978. Origin of the ribityl side-chain of riboflavin from the ribose moiety of guanosine triphosphate in *Pichia guilliermondii* yeast. Biochim. Biophys. Acta **543**:305–312.

89. **Mitsuda, H., and K. Nakajima.** 1975. Guanosine nucleotide precursor for flavinogenesis of *Eremothecium ashbyii*. J. Nutr. Sci. Vitaminol. **21**:331–345.

90. **Mitsuda, H., K. Nakajima, and T. Nadamoto.** 1977. The immediate nucleotide precursor, guanosine triphosphate, in the riboflavin biosynthetic pathway. J. Nutr. Sci. Vitaminol. **23**:23–34.

91. **Miyata, I.** 1968. Thiamine kinase of *Escherichia coli*. I. The properties of thiamine kinase in the cytoplasmic membrane fraction. Bitamin **38**:55–61.

92. **Miyata, I., T. Kawasaki, and Y. Nose.** 1967. Thiamine kinase in the membrane fraction of *Escherichia coli*. Biochem. Biophys. Res. Commun. **27**:601–606.

93. **Miyatake, K., Y. Nakano, and S. Kitaoka.** 1976. Pantothenate synthetase of *Escherichia coli* B. I. Physicochemical properties. J. Biochem. (Tokyo) **79**:673–678.

94. **Miyatake, K., Y. Nakano, and S. Kitaoka.** 1979. Pantothenate synthetase from *Escherichia coli* [D-pantoate: β-alanine ligase (AMP-forming), EC 6.3.2.1]. Methods Enzymol. **62**:215–219.

95. **Nakayama, H.** 1956. Thiamine auxotrophs of *Escherichia coli*. IV. Thiazole biosynthesis. Vitamins (Kyoto) **11**:169–175.

96. **Nakayama, H., and R. Hayashi.** 1972. Biosynthesis of thiamine pyrophosphate in *Escherichia coli*. J. Bacteriol. **109**:936–938.

97. **Nakayama, H., and R. Hayashi.** 1972. Biosynthetic pathway of thiamine pyrophosphate: a special reference to the thiamine monophosphate-requiring mutant and the thiamine pyrophosphate-requiring mutant of *Escherichia coli*. J. Bacteriol. **112**:1118–1126.

98. **Nakayama, H., and R. Hayashi.** 1979. Isolation and characterization of *Escherichia coli* mutants auxotrophic for thiamine phosphates. Methods Enzymol. **62**:94–101.

99. **Neuberger, G., and A. Bacher.** 1985. Biosynthesis of riboflavin. An aliphatic intermediate in the formation of 6,7-dimethyl-8-ribityllumazine from pentose phosphate. Biochem. Biophys. Res. Commun. **127**:175–181.

100. **Newell, P. C., and R. G. Tucker.** 1966. The derepression of thiamine biosynthesis by adenosine. A tool for investigating this biosynthetic pathway. Biochem. J. **100**:512–516.

101. **Newell, P. C., and R. G. Tucker.** 1966. The control mechanism of thiamine biosynthesis. A model for the study of control of converging pathways. Biochem. J. **100**:517–524.

102. **Newell, P. C., and R. G. Tucker.** 1967. New pyrimidine pathway involved in the biosynthesis of the pyrimidine of thiamine. Nature (London) **215**:1384–1385.

103. **Newell, P. C., and R. G. Tucker.** 1968. Precursors of the pyrimidine moiety of thiamine. Biochem. J. **106**:271–278.

104. **Newell, P. C., and R. G. Tucker.** 1968. Biosynthesis of the py-

rimidine moiety of thiamine. A new route of pyrimidine biosynthesis involving purine intermediates. Biochem. J. 106:279–287.

105. Nielsen, P., and A. Bacher. 1981. Biosynthesis of riboflavin. Characterization of the product of the deaminase. Biochim. Biophys. Acta 662:312–317.

106. Nielson, P., G. Neuberger, H. G. Floss, and A. Bacher. 1984. Biosynthesis of riboflavin. Enzymatic formation of the xylene moiety from [14C]ribulose-5-phosphate. Biochem. Biophys. Res. Commun. 118:814–820.

107. Nishino, H. 1972. Biogenesis of cocarboxylase in Escherichia coli. Partial purification and some properties of thiamine monophosphate kinase. J. Biochem. (Tokyo) 72:1093–1100.

108. Nishino, H., A. Iwashima, and Y. Nose. 1971. Biogenesis of cocarboxylase in Escherichia coli. Novel enzyme catalyzing the formation of thiamine pyrophosphate from thiamine monophosphate. Biochem. Biophys. Res. Commun. 45:363–368.

109. Nose, Y., Y. Tokuda, M. Hirabayashi, and A. Iwashima. 1964. Thiamine biosynthesis from (hydroxymethyl)pyrimidine and thiazole by washed cells and cell extracts of Escherichia coli and its mutants. J. Vitaminol. (Kyoto) 10:105–110.

110. Oltmanns, O., and A. Bacher. 1972. Biosynthesis of riboflavine in Saccharomyces cerevisiae: the role of genes rib₁ and rib₇. J. Bacteriol. 110:818–822.

111. Ortega, M. V., A. Cardenas, and D. Ubiera. 1975. PanD, a new chromosomal locus of Salmonella typhimurium for the biosynthesis of β-alanine. Mol. Gen. Genet. 140:159–164.

112. Parada, J. L., and M. V. Ortega. 1967. Growth inhibition by hexoses of a temperature-sensitive thiazoleless mutant of Salmonella typhimurium. J. Bacteriol. 94:707–711.

113. Pfleiderer, G., A. Kreiling, and T. Wieland. 1960. Über Pantothensäure-synthetase aus E. coli. Biochem. Z. 333:302–307.

114. Plaut, G. W. E. 1963. Studies on the nature of the enzymic conversion of 6,7-dimethyl-8-ribityllumazine to riboflavin. J. Biol. Chem. 238:2225–2243.

115. Plaut, G. W. E. 1971. The biosynthesis of riboflavin. Compr. Biochem. 21:11–45.

116. Plaut, G. W. E., C. M. Smith, and W. L. Alworth. 1974. Biosynthesis of water-soluble vitamins. Annu. Rev. Biochem. 43:899–922.

117. Powers, S. G., and E. E. Snell. 1976. Ketopantoate hydroxymethyltransferase. II. Physical, catalytic, and regulatory properties. J. Biol. Chem. 251:3786–3793.

118. Powers, S. G., and E. E. Snell. 1979. Purification and properties of ketopantoate hydroxymethyltransferase. Methods Enzymol. 62:204–209.

119. Primerano, D. A., and R. O. Burns. 1983. Role of acetohydroxyacid isomeroreductase in biosynthesis of pantothenic acid in Salmonella typhimurium. J. Bacteriol. 153:259–269.

120. Purko, M., W. O. Nelson, and W. A. Wood. 1953. The nutritional equivalence of pantothenate and p-aminobenzoate for the growth of Bacterium linens. J. Bacteriol. 66:561–567.

121. Purko, M., W. O. Nelson, and W. A. Wood. 1954. The role of p-aminobenzoate in pantoate synthesis by Bacterium linens. J. Biol. Chem. 207:51–58.

122. Recsei, P. A., and E. E. Snell. 1984. Pyruvoyl enzymes. Annu. Rev. Biochem. 53:357–387.

123. Rembold, H., and L. Buschmann. 1963. Struktur und Synthese des Neopterins. Chem. Ber. 96:1406–1410.

124. Reynolds, J. J., and G. M. Brown. 1962. Enzymatic formation of the pteridine moiety of folic acid from guanine compounds. J. Biol. Chem. 237:PC2713–PC2715.

125. Reynolds, J. J., and G. M. Brown. 1964. The biosynthesis of folic acid. IV. Enzymatic synthesis of dihydrofolic acid from guanine and ribose compounds. J. Biol. Chem. 239:317–325.

126. Richey, D. P., and G. M. Brown. 1969. The biosynthesis of folic acid. IX. Purification and properties of the enzymes required for the formation of dihydropteroic acid. J. Biol. Chem. 244:1582–1592.

127. Richey, D. P., and G. M. Brown. 1970. A comparison of the effectiveness with which p-aminobenzoic acid and p-aminobenzoylglutamic acid are used as substrate by dihydropteroate synthetase from Escherichia coli. Biochim. Biophys. Acta 222:237–239.

128. Roberts, E., P. Ayengar, and J. Posner. 1953. Transamination of γ-aminobutyric acid and β-alanine in microorganisms. J. Biol. Chem. 203:195–204.

129. Rock, C. O. 1982. Mixed disulfides of acyl carrier protein and coenzyme A with specific soluble proteins in Escherichia coli. J. Bacteriol. 152:1298–1300.

130. Shane, B. 1980. Pteroylpoly(γ-glutamate) synthesis by Corynebacterium species. Purification and properties of folylpoly(γ-glutamate) synthetase. J. Biol. Chem. 255:5655–5662.

131. Shavlovsky, G. M., V. E. Kaschenko, L. V. Koltun, E. M. Logvinenko, and A. E. Zakalsky. 1977. Regulation by iron of the synthesis of GTP-cyclohydrolase, participating in yeast flavinogenesis. Microbiologiya 46:578–580.

132. Shavlovsky, G. M., E. M. Logvinenko, R. Benndorf, L. V. Koltun, V. E. Kaschenko, A. E. Zakalsky, D. Schlee, and H. Reinbothe. 1980. First reaction of riboflavin biosynthesis. Catalysis by a guanine triphosphate cyclohydrolase from yeast. Arch. Microbiol. 124:255–259.

133. Shavlovsky, G. M., E. M. Logvinenko, V. E. Kaschenko, L. V. Koltun, and A. E. Zakalsky. 1976. Detection of the enzyme of the first stage of flavinogenesis, GTP cyclohydrolase, in the yeast Pichia guilliermondii. Dokl. Akad. Nauk SSSR 230:1485–1487.

134. Shiota, T. 1971. The biosynthesis of folic acid and 6-substituted pteridine derivatives. Compr. Biochem. 21:111–152.

135. Shiota, T., C. M. Baugh, and J. Myrick. 1969. The assignment of structure to the formamidopyrimidine nucleoside triphosphate precursor of pteridines. Biochim. Biophys. Acta 192:205–210.

136. Shiota, T., and M. N. Disraely. 1961. Enzymic synthesis of dihydrofolate from 2-amino-4-hydroxy-8-hydroxymethyl-dihydropteridine and p-aminobenzoylglutamate by extracts of Lactobacillus plantarium. Biochim. Biophys. Acta 52:467–473.

137. Shiota, T., M. N. Disraely, and M. P. McCann. 1964. The enzymatic synthesis of folate-like compounds from hydroxymethyldihydropteridine pyrophosphate. J. Biol. Chem. 239:2259–2266.

138. Shiota, T., R. Jackson, and C. M. Baugh. 1970. The biosynthetic pathway of dihydrofolate, p. 265–279. In K. Iwai, M. Akino, M. Goto, and Y. Iwanami (ed.), Chemistry and biology of pteridines. International Academic Printing Co. Ltd., Tokyo.

139. Shiota, T., M. P. Palumbo, and L. Tsai. 1967. A chemically prepared formamidopyrimidine derivative of guanosine triphosphate as a possible intermediate in pteridine biosynthesis. J. Biol. Chem. 242:1961–1969.

140. Shive, W., and J. Macow. 1946. Biochemical transformation as determined by competitive analogue-metabolite growth inhibition. I. Some transformations involving aspartic acid. J. Biol. Chem. 162:451–462.

141. Slotnick, I. J. 1956. Dihydrouracil as a growth factor for a mutant strain of Escherichia coli. J. Bacteriol. 72:276–277.

142. Slotnick, I. J., and H. Weinfeld. 1957. Dihydrouracil as a growth factor for mutant strains of Escherichia coli. J. Bacteriol. 74:122–125.

143. Suzuki, Y., and G. M. Brown. 1974. The biosynthesis of folic acid. XII. Purification and properties of dihydroneopterin triphosphate pyrophosphohydrolase. J. Biol. Chem. 249:2405–2410.

144. Teller, J. H., S. G. Powers, and E. E. Snell. 1976. Ketopantoate hydroxymethyltransferase. I. Purification and role in pantothenate biosynthesis. J. Biol. Chem. 251:3780–3785.

145. Therisod, M., J. C. Fischer, and B. Estramareix. 1981. The origin of the carbon chain in the thiazole moiety of thiamine in Escherichia coli: incorporation of deuterated 1-deoxy-D-threo-2-pentulose. Biochem. Biophys. Res. Commun. 98:374–379.

146. Therisod, M., D. Gaudry, and B. Estramareix. 1978. The biosynthesis of thiamine: isolation of a new thiazolic metabolite of tyrosine. Nouv. J. Chem. 2:119–121.

147. Virtanen, A. I., and T. Laine. 1937. The decarboxylation of D-lysine and L-aspartic acid. Enzymologia 8:266–270.

148. Ward, G. B., G. M. Brown, and E. E. Snell. 1955. Phosphorylation of pantothenic acid and pantetheine by an enzyme from Proteus morganii. J. Biol. Chem. 213:869–876.

149. Weisman, R. A., and G. M. Brown. 1964. The biosynthesis of folic acid. V. Characteristics of the enzyme system that catalyzes the synthesis of dihydropteroic acid. J. Biol. Chem. 239:326–331.

150. West, T. P., T. W. Traut, M. S. Shanley, and G. A. O'Donovan. 1985. A Salmonella typhimurium strain defective in uracil catabolism and β-alanine synthesis. J. Gen. Microbiol. 131:1083–1090.

151. Weygand, F., H. Simon, G. Dahms, M. Waldschmidt, H. J. Shliep, and H. Wacker. 1961. Über die Biogenese des Leucopterins. Angew. Chem. 73:402–407.

152. White, R. H. 1978. Stable isotope studies on the biosynthesis of the thiazole moiety of thiamine in Escherichia coli. Biochemistry 17:3833–3840.

153. White, R. H. 1979. 4-Hydroxybenzylalcohol. A metabolite produced during the biosynthesis of thiamine in Escherichia coli. Biochim. Biophys. Acta 583:55–62.

154. White, R. H. 1980. Synthesis of L-[3-²H,¹⁸O]glycerol and its incorporation into the 4-methyl-5-hydroxyethyl thiazole moiety of thiamine by Escherichia coli. Experientia 36:637–638.

155. White, R. H., and F. B. Rudolph. 1978. The origin of the nitrogen atom in the thiazole ring of thiamine in Escherichia coli. Biochim. Biophys. Acta 542:340–347.

156. White, R. H., and F. B. Rudolph. 1979. Biosynthesis of the

pyrimidine moiety of thiamine in *Escherichia coli*: incorporation of stable isotope-labeled glycines. Biochemistry **18:**2632–2636.

157. **Wilken, D. R., H. L. King, and R. E. Dyar.** 1975. Ketopantoic acid and ketopantoyllactone reductases. J. Biol. Chem. **250:**2311–2314.

158. **Wilken, D. R., H. L. King, and R. E. Dyar.** 1979. Ketopantoyllactone reductases. Methods Enzymol. **62:**209–215.

159. **Williamson, J. M.** 1985. L-Aspartate-α-decarboxylase. Methods Enzymol. **113:**589–595.

160. **Williamson, J. M., and G. M. Brown.** 1979. Purification and properties of L-aspartate-α-decarboxylase, an enzyme that catalyzes the formation of β-alanine in *Escherichia coli*. J. Biol. Chem. **254:**8074–8082.

161. **Wolf, W. A., and G. M. Brown.** 1969. Biosynthesis of folic acid. X. Evidence for an Amadori rearrangement in enzymatic formation of dihydroneopterin triphosphate from GTP. Biochim. Biophys. Acta **192:**468–478.

162. **Yamada, K., and H. Kumaoka.** 1982. Biosynthesis of thiamin. Incorporation of a two-carbon fragment derived from ribose of 5-aminoimidazole ribotide into the pyrimidine moiety of thiamin. Biochem. Int. **5:**771–776.

163. **Yamada, K., and H. Kumaoka.** 1983. Precursor of carbon atom five and hydroxymethyl carbon atom of the pyrimidine moiety of thiamin in *Escherichia coli*. J. Nutr. Sci. Vitaminol. **29:**389–398.

164. **Yim, J. J., and G. M. Brown.** 1976. Characteristics of guanosine triphosphate cyclohydrolase I purified from *Escherichia coli*. J. Biol. Chem. **251:**5087–5094.

165. **Yura, T.** 1956. Evidence of non-identical alleles in purine-requiring mutants of *Salmonella typhimurium*. Carnegie Inst. Wash. Publ. **612:**63–75.

35. Synthesis of Pyridoxal Phosphate

WALTER B. DEMPSEY

Department of Medical and Microbial Genetics, Veterans Administration Medical Center, and Department of Biochemistry, University of Texas Health Science Center, Dallas, Texas 75216

INTRODUCTION

The biosynthesis of the group of compounds known as vitamin B_6, or pyridoxine, has been studied extensively in *Escherichia coli* without yielding any incontrovertible data that would positively identify any compound as a committed obligatory precursor of this family of compounds. Two recent reviews have analyzed all of the pertinent data critically (9, 20). These reviews should be consulted for a discussion of the many problems involved in the attempts to elucidate the pathway.

The terms pyridoxine and vitamin B_6 originally meant exclusively the specific compound designated pyridoxol in Fig. 1. In the absence of any generally agreed-upon name for the entire group of compounds in Fig. 1, both terms have begun to be used to refer to the group without regard to the relative proportions of each member in a sample or to their efficiencies as nutrients. In this review, this usage of these two terms as general names for the group will apply exclusively. Individual compounds will be named as in Fig. 1.

The small amount of pyridoxine in cells makes difficult the elucidation of the biosynthetic pathway. For investigations of the pathway by isotopic methods, the small amount means arduous and tedious isolations of compounds with very low total radioactivity. From what is already known about the pathway, this particular problem is compounded by the fact that many of the suspected precursors are common metabolites which can be very effective in diluting the incorporation of test compounds. Similarly, the small amount of pyridoxine required by cells (360 nmol/g dry weight) (2) and the fact that, as a coenzyme, it cannot be easily depleted by simple starvation have led to confusion in the identification of mutants as true pyridoxineless mutants. Obviously, true pyridoxineless mutants lack a gene required for the biosynthesis of pyridoxine, but mutants with altered transaminases that require greater than normal amounts of pyridoxal phosphate also respond to nutritional supplements with pyridoxine. These and other mutants that hamper the investigation of pyridoxine biosynthesis have been explained in another review (9) and will not be discussed further here.

SALVAGE VERSUS DE NOVO SYNTHESIS

In *E. coli* and in other microorganisms, two pyridoxine pathways can be defined. One is the biosynthetic pathway, and the other is the salvage pathway. The salvage pathway, shown in Fig. 1, serves to convert whatever form of pyridoxine might enter the cell from the environment to the coenzymatically active forms. Although the scheme in Fig. 1 suggests that pyridoxol is the immediate end product of the biosynthetic pathway, this has not been established. It is known, however, that both pyridoxol and pyridoxol-5'-phosphate are accumulated by *pdxH* mutants of *E. coli* (pyridoxol-5'-phosphate oxidase-less) (3). At present, there is no way to tell whether the pyridoxol accumulated by the mutants derives exclusively from hydrolysis of pyridoxol-5'-phosphate.

Apart from pyridoxine oxidase-less mutants, no other mutants are known in the salvage pathway. The existence of the salvage pathway itself is known from the identification of all of the compounds in extracts of *E. coli* (3) and from partial purification of pyridoxine phosphate oxidase (13) and pyridoxal kinase (33) from extracts of *E. coli*.

With regard to the synthesis of the 3-hydroxypyridine ring of pyridoxine, the present status of the biosynthetic pathway for pyridoxine in either *Salmonella typhimurium* or *E. coli* is that no compound has been identified as being on a pathway exclusively committed to pyridoxine synthesis. The emphasis here is on exclusivity and committedness. There is an abundant amount of precise data from genetics and chemistry that severely limit the possible compounds involved. There are also several physiological experiments that seem to give clues about the pathway. All of these data have been recently reviewed, as noted above.

EVIDENCE FROM GENETIC ANALYSIS

Genetic tests have been performed on approximately 250 pyridoxineless mutants of *E. coli* (4). Genetic analysis showed that *pdx* mutants fell into five unlinked groups widely distributed on the chromosome (1). A single gene may constitute each group, but this has not been rigorously established. Complemen-

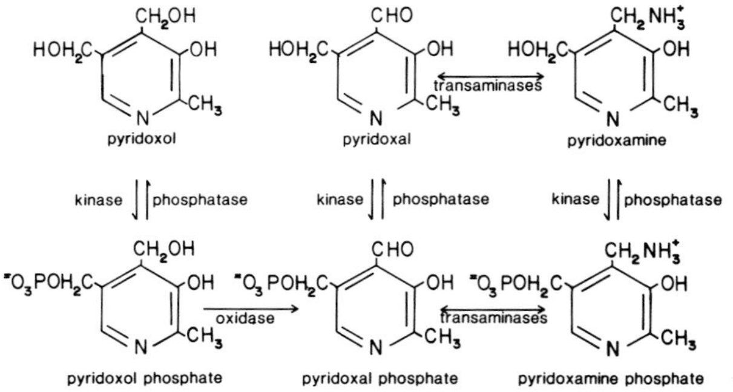

FIG. 1. Salvage pathway for vitamin B_6 in *E. coli*. Pyridoxol cannot be converted directly to pyridoxal because oxidaseless mutants require pyridoxal or pyridoxamine for growth and because they accumulate pyridoxol and pyridoxol phosphate.

tation analyses confirmed the existence of separate groups but did not allow subdivision of the groups (4). Nutritional and enzymatic analyses, on the other hand, did allow identification of genes within some groups. For example, all members of one linkage group (IV) required pyridoxal or pyridoxamine and lacked pyridoxol-5'-phosphate oxidase. These are known as *pdxH* mutants (3, 6). In another group (III), all mutants required both serine and pyridoxine to grow and lacked 3-phosphoserine-oxoglutarate transaminase. These are known as *pdxF* or *serC* mutants (5, 28). (In *S. typhimurium*, the *pdxF* (*serC*) gene is the upstream part of a two-gene operon with *aroA* downstream [21].) A third group (V) contained two phenotypes, PdxJ and PdxK, which are distinguishable because PdxJ mutants are simple pyridoxineless mutants, whereas PdxK mutants grow partially with D- or L-alanine as the sole supplement (8). Although groups III and IV appear to contain single *pdx* genes because no other *pdx* mutations have been found to be linked to the *pdxH* or to the *pdxF* (*serC*) gene, group V may include more than one *pdx* gene if the two phenotypes derive from mutations in different genes.

The remaining two groups (I and II) contain most of the *pdx* mutations isolated in *E. coli*. These two groups were originally subdivided by cross-feeding tests (11). This subdivision of group I into three classes is now known to have been based on two kinds of secondary mutations carried by many of the original mutants rather than on the *pdx* mutation they carried (S. Shimizu and W. B. Dempsey, unpublished observation). Curiously, one set of secondary mutations was 100% linked to *thrA*, and the other set was linked to *ilvA*. (The mutagen in many of these double mutants was UV, not nitrosoguanidine.) When transferred into an isogenic background, all group I mutations behave in the same manner (Shimizu and Dempsey, unpublished observation). Under these conditions, all group I mutants grow normally with either pyridoxine or glycolaldehyde at 30°C and with pyridoxine only at 42°C. Secondary mutations that allow full glycolaldehyde response at 42°C arise easily.

If one assumes that a single enzyme is in each group and that the *pdxK* mutants of group V are variants of *pdxJ* mutants, there is a minimum of four *pdx* genes in the biosynthetic pathway for the 3-hydroxypyridine ring. *E. coli* B differs from the K-12 strain and *S. typhimurium* with regard to one of these genes. In the B strain, mutations in *pdxF* (*serC*) that totally inactivate 3-phosphoserine-oxoglutarate transaminase always have a double requirement for both pyridoxine and serine (5). No suppressor-type mutations of these mutants are ever seen in which only one requirement remains. In the K-12 strain, on the other hand, Pdx$^+$ Ser$^-$ suppressed mutants of *pdxF* (*serC*) mutants arise relatively frequently and gain a serine transaminase activity (28). In addition, in the K-12 strain, the pyridoxine requirement of *pdxF* mutants can always be met by relatively high levels of 3-hydroxypyruvate, and normal amounts of pyridoxine are synthesized by these mutants when fed 3-hydroxypyruvate (28). Glycolaldehyde also replaces pyridoxine for group III mutants if a secondary mutation occurs. The nature of the secondary mutation is unknown (Shimizu and Dempsey, unpublished observation). On the basis of experiments with the B strain, one is forced to conclude that the enzyme specified by the *pdxF* (*serC*) is absolutely required for the biosynthesis of pyridoxine. As noted above, this gene has recently been shown not only to be linked to *aroA* in *S. typhimurium* but also to be the upstream cistron in a two-cistron operon with the *aroA* gene. The *aroA* gene specifies the enzyme of aromatic amino acid biosynthesis that makes the immediate precursor of chorismate, namely, 3-enol-pyruvylshikimate-5-phosphate synthetase. This discovery establishes for the first time a link between serine and pyridoxine biosynthesis and the synthesis of other aromatic rings.

Attention should be called to the observation made above that *pdxF* (*serC*) mutants of *E. coli* B never give rise to the suppressed phenotype, Ser$^-$ Pdx$^+$, whereas mutants of *E. coli* K-12 and *S. typhimurium* do. The simplest interpretation is that the difference resides in one of the already described differences between strains B and K-12. Two known differences that may be relevant are as follows. First, there is only one homoserine dehydrogenase in the B strain, but there are two isozymes of it in the K-12 strain (26). Second, the first enzyme of threonine biosynthesis, homoserine kinase, is strongly inhibited by threonine in the B strain but is insensitive to threonine in the K-12 strain (34).

EVIDENCE FROM NUTRITIONAL EXPERIMENTS

There are some other observations to support the idea that a tenuous linkage exists between the biosynthesis of threonine, branched-chain amino acids, and pyridoxine. One is the discovery that threonine prevents derepression of pyridoxine biosynthesis in *E. coli* B (10). A second observation is that isoleucine and threonine are the amino acids most sensitive to pyridoxine starvation (12). A third observation is that pyridoxine is required along with the branched-chain amino acids to repress the branched-chain pathway (32). A fourth observation is that threonine deaminase mutants cross-feed pyridoxine mutants (Dempsey and Shimizu, unpublished observation). A fifth observation is the probability that the feeding of group I mutants by the mixture of glycolaldehyde and glycine (25) occurs by the formation of the threonine analog 4-hydroxythreonine. All of the observations described above suggest some metabolic linkage between threonine and pyridoxine.

Provocative experiments in the biosynthesis of pyridoxine in *E. coli* were reported almost 30 years ago by Morris et al. (24, 25). Although some of the mutants that Morris worked with may have been multiple mutants (from an inspection of the mutagenic procedure used to isolate them), his work nonetheless led to identification of two compounds that replace pyridoxine requirements in some true pyridoxineless mutants. The two compounds are those mentioned above, namely, hydroxypyruvate and the two-carbon compound glycolaldehyde.

Hydroxypyruvate does feed *pdxF* mutants, as noted above. It behaves as a true precursor in that the response of the mutants to it is instantaneous; all mutants in the group respond to it, and the amount of pyridoxine synthesized by mutants fed hydroxypyruvate is indistinguishable from the amount normally biosynthesized by the wild-type parent. It has not yet been tested as a radiolabeled precursor (28).

Glycolaldehyde, on the other hand, is quite different. None of the mutants that respond to it do so instantaneously; all mutants that respond to it appear to require a secondary mutation to use it, and all mutants that use glycolaldehyde make significantly less pyridoxine with it than is found in the wild-type parent (Shimizu and Dempsey, unpublished observation). It has been tested as a precursor and shown to be incorporated without dilution into pyridoxine (29).

The lack of an instantaneous response to glycolaldehyde gives rise to conflicting reports about the efficacy of glycolaldehyde. The presence or absence of glycine, the concentration, the temperature during the test, and the duration of the test are all critical. In all cases studied, and that includes all of the known group I mutations, the introduction of the *pdx* gene into an isogenic background leads to the following observation. None of the previously responsive mutants respond to glycolaldehyde with or without glycine when tested at 25 mg/liter for 24 h at 42°C by replica plating from an ML master. Longer incubations show some growth and mutant colonies (W. B. Dempsey, unpublished observation).

Group III mutants respond to glycolaldehyde in the same manner that group I mutants do. One presumes that part of the odd response may be transport difficulties and that another part may be that the high reactivity of glycolaldehyde gives various concentrations of the true intermediate under various conditions.

The conclusions from the nutritional and genetic experiment are that there are at least four genes in the biosynthesis of pyridoxine, one of which is the *aroA*-linked *pdxF* (*serC*) gene that encodes 3-phosphoserine-oxoglutarate transaminase; that 3-hydroxypyruvate and glycolaldehyde replace the pyridoxine requirement in some mutants; and that the end product of the pathway is either pyridoxol or its 5'-phosphate ester. The other solid data have come from carefully done isotopic labeling experiments.

EVIDENCE FROM ISOTOPIC LABELING EXPERIMENTS

The only labeling data in this field that have been obtained with scrupulous attention being paid to

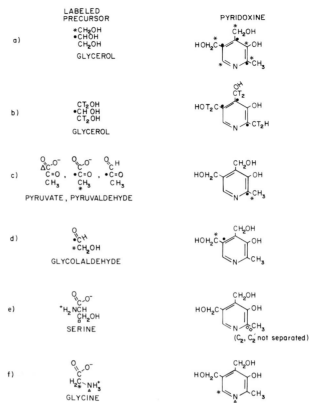

FIG. 2. Summary of the critical isotope incorporation data. Lines a through e show experiments performed with the oxidaseless mutant WG2. The experiments of line d were also performed with the pyridoxineless mutant WG3, and the experiments of line f were performed only with WG3. References: a, 14–18; b, 31; c, 17, 18, 31; d, 15, 19, 29, 30; e, 14, 15; f, 22.

possible artifacts have come from Hill and Spenser and their co-workers (14–20, 22, 30, 31). The labeling experiments that they have reported used two *E. coli* B strains. One, WG3, a group I mutant, was deficient in the synthesis of the basic 3-hydroxypyridine compound but normal for the salvage pathway; the other, WG2, a group IV mutant, was deficient in the salvage pathway but wild type for synthesis of the 3-hydroxypyridine portion. The experiments were of several different designs. With few exceptions, the labeling patterns were determined by isolating and determining the radioactivity in derivatives of each individual carbon atom or of pairs of carbon atoms. In all cases, the labeled pyridoxine was rigorously purified before the controlled degradations were done. This very labor-intensive work, unfortunately, has not led to the unequivocal identification of any precursor but has, of course, showed distribution patterns that limit the possibilities to relatively few. This work has been summarized recently (20).

Labeling experiments show several unequivocal results. When glycerol is labeled and supplied as the sole source of carbon in the culture, *E. coli* makes specifically labeled pyridoxines, according to the patterns shown in Fig. 2. The results are particularly interesting because there is no trace of randomization of the label. This means that the carbon skeleton of glycerol is incorporated into pyridoxine as a derivative that can be formed directly from three-carbon intermediates without entering the citric acid cycle or the pentose cycle.

All of the principal results are shown in Fig. 2. The first five lines show the patterns found in pyridoxine isolated from WG2 fed specifically labeled carbon sources. Labeling by specific compounds tested both by direct incorporation and by isotope-sparing experiments showed the results in lines c and d. It was established that C-1 of pyruvate did not label pyridoxine at all and that C-3 of pyruvate labeled exclusively C-2' of pyridoxine. Similarly exclusive labeling of the C-2' carbon atom was found with C-3–labeled serine and pyruvaldehyde (31).

In addition, C-2 of pyruvate and pyruvaldehyde labeled C-2 of pyridoxine exclusively. [2-^{14}C]glycerol also labels C-2 of pyridoxine (Fig. 2, line a). Vella et al. (30, 31) showed that pyruvate spared this incorporation but that 3-hydroxypyruvate, acetate, and glycolaldehyde did not. The conclusion is that a compound derivable from both serine and pyruvate gives rise to C-2' of pyridoxine.

An argument has been presented by Hill and Spenser (20) that the differential labeling seen in the C-2' atom when [1-^{14}C]glucose or [6-^{14}C]glucose is used suggests that pyruvate derived from dihydroxyacetone phosphate rather than from glyceraldehyde 3-phosphate is the precursor of pyridoxine. The arguments are impossible to evaluate since they do not include any data on the relative activity of shunt versus glycolysis during the incorporation period. It is certainly clear from their data that serine and pyruvate are both closer to the pyridoxine C-2' atom than are the common glycolytic trioses, but it is not established rigorously that either is a committed precursor.

Incorporation of glycolaldehyde specifically into the 5,5' carbons of pyridoxine by the group I mutant WG3 (*pdxB3*) occurs as indicated in Fig. 2, line d (15). This incorporation is not diluted by other metabolites in WG3, undoubtedly because the *pdxB3* mutation blocks synthesis of the compound that glycolaldehyde somehow replaces. As indicated, the strain WG2 also incorporates glycolaldehyde specifically and exclusively into the same positions in pyridoxine, but this incorporation is diluted by other metabolites. The problem with glycolaldehyde as a test compound is that it is highly reactive and is needed in millimolar amounts to make micromolar amounts of pyridoxine. One cannot be certain that it is glycolaldehyde itself that is a precursor of pyridoxine.

Recently, Iwanow et al. (22) have shown by using nuclear magnetic resonance and [2-^{14}C, ^{13}N]glycine that the pyridine N and C-6 of pyridoxine are derived intact from the N and C-2 of glycine in the WG3 strain grown with glycolaldehyde (Fig. 2, line f). This finding may explain an observation of Morris and Woods (25), that glycine stimulates glycolaldehyde feeding. An obvious but unproven route might be through a threonine aldolase-type reaction to form 4-hydroxythreonine, which may thus enter as a unit accompanied by decarboxylation. Genetic data indicate that the glycine needed for this reaction is not an intermediate in normal *E. coli*. Glycine is made exclusively from serine in this organism, and *serA* and *serB* mutants

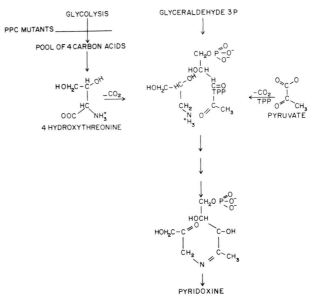

FIG. 3. Hypothetical scheme for pyridoxine biosynthesis. Pyruvate and triose labeling experiments support the structures shown entering from the top and right of the figure. TPP, Thiamine pyrophosphate. PPC mutants lack phosphoenolpyruvate carboxylase and cannot synthesize pyridoxine without glycolaldehyde.

starved for serine (or glycine) make normal or greater than normal (*serB*) amounts of pyridoxine.

A possible scheme for the origins of the carbon and nitrogen atoms of pyridoxine is given in Fig. 3. Pyruvate is shown to be incorporated as a thiamine pyrophosphate derivative to explain both the observation that thiamine is required for pyridoxine biosynthesis (7) and the isotope labeling data that C-1 of pyruvate is not incorporated but that C-2 and C-3 of pyruvate become C-2 and C-2' of pyridoxine, respectively (30, 31). A four-carbon amino acid derivative, 4-hydroxythreonine, is shown as the precursor of the C-5,5', C-6, and N-1 atoms because this compound could be synthesized from glycine and glycolaldehyde to account for the labeling data from those compounds. It is known that mutations in the *ppc* gene prevent pyridoxine biosynthesis and that this failure to synthesize pyridoxine is prevented by glycolaldehyde (7). Therefore, it seems logical that the normal precursor is derived from a four-carbon acid. There is no evidence that the demonstrated normal precursor is 4-hydroxythreonine. It is most likely that the requirement for glutamate or oxoglutarate (7), the utility of 3-hydroxypyruvate, and the necessity of the *pdxF* (*serC*) gene are all involved in the biosynthesis of the normal precursor of this group of atoms.

The discovery that the genes *pdxF* (*serC*) and *aroA* are linked in one operon makes one wonder what controls expression of that operon and ask whether there is a link between the concentrations of amino acids in the cell and pyridoxine biosynthesis. It is known that transcription of the gene encoding the first enzyme of serine biosynthesis (*serA*) is regulated by the concentrations of threonine, isoleucine, leucine, and methionine (23, 27). It is also known that pyridoxine is required to regulate the expression of the genes for branched-chain amino acid biosynthesis (32), that threonine regulates expression of the pyridoxine biosynthetic genes, and that threonine and isoleucine are the amino acids most sensitive to pyridoxine levels in the cell. Now we have an operon that links a gene required for both serine and pyridoxine biosynthesis to one required for aromatic amino acids. Is it possible that the regulation of this serine biosynthetic gene will also require threonine, methionine, isoleucine, and leucine, as well as the aromatic amino acids, serine, and pyridoxine?

Tryptophan synthetase, the last enzyme in tryptophan biosynthesis, catalyzes the incorporation of 1 mol of serine into each mole of tryptophan. This enzyme is a pyridoxal phosphate-requiring enzyme. Perhaps the *aroA* and *serC* (*pdxF*) genes are linked to ensure that there is enough serine and pyridoxine for tryptophan biosynthesis. One would expect that this operon would be controlled by the levels of several amino acids, perhaps as many as those listed above.

LITERATURE CITED

1. **Bachmann, B. J.** 1983. Linkage map of *Escherichia coli* K-12, edition 7. Microbiol. Rev. **47:**180–230.
2. **Dempsey, W. B.** 1965. Control of pyridoxine biosynthesis in *Escherichia coli.* J. Bacteriol. **90:**431–437.
3. **Dempsey, W. B.** 1966. Synthesis of pyridoxine by a pyridoxal auxotroph of *Escherichia coli.* J. Bacteriol. **92:**333–337.
4. **Dempsey, W. B.** 1969. Characterization of pyridoxine auxotrophs of *Escherichia coli*: results of P1 transduction. J. Bacteriol. **97:**1403–1410.
5. **Dempsey, W. B.** 1969. 3-Phosphoserine transaminase mutants of *Escherichia coli* B. J. Bacteriol. **100:**1114–1115.
6. **Dempsey, W. B.** 1971. Control of vitamin B₆ biosynthesis in *Escherichia coli.* J. Bacteriol. **108:**415–421.
7. **Dempsey, W. B.** 1971. Role of vitamin B₆ biosynthetic rate in the study of vitamin B₆ synthesis in *Escherichia coli.* J. Bacteriol. **108:**1001–1007.
8. **Dempsey, W. B.** 1972. *Escherichia coli* mutants that require either pyridoxine or alanine. J. Bacteriol. **111:**838–840.
9. **Dempsey, W. B.** 1980. Biosynthesis and control of vitamin B₆ in *Escherichia coli*, p. 93–111. *In* G. P. Tryfiates (ed.), Vitamin B₆ metabolism and role in growth. Food and Nutrition Press, Westport, Conn.
10. **Dempsey, W. B.** 1982. Threonine prevents derepression of pyridoxine synthesis in *Escherichia coli* B. J. Bacteriol. **150:**1476–1478.
11. **Dempsey, W. B., and P. F. Pachler.** 1966. Isolation and characterization of pyridoxine auxotrophs of *Escherichia coli.* J. Bacteriol. **91:**642–645.
12. **Dempsey, W. B., and K. R. Sims.** 1972. Isoleucine and threonine can prolong protein and ribonucleic acid synthesis in pyridoxine-starved mutants of *Escherichia coli* B. J. Bacteriol. **112:**726–735.
13. **Henderson, H. M.** 1965. The conversion of pyridoxine phosphate into pyridoxal phosphate in *Escherichia coli.* Biochem. J. **95:**775–779.
14. **Hill, R. E., R. N. Gupta, F. J. Rowell, and I. D. Spenser.** 1971. Biosynthesis of pyridoxine. J. Am. Chem. Soc. **93:**518–520.
15. **Hill, R. E., P. Horsewood, I. D. Spenser, and Y. Tani.** 1975. Biosynthesis of vitamin B₆. Incorporation of glycolaldehyde into pyridoxal. J. Chem. Soc. Perkin Trans. **1:**1622–1627.
16. **Hill, R. E., I. Miura, and I. D. Spenser.** 1977. Biosynthesis of vitamin B₆. The incorporation of [1,3-¹³C₂]glycerol. J. Am. Chem. Soc. **99:**4179–4181.
17. **Hill, R. E., F. J. Rowell, R. N. Gupta, and I. D. Spenser.** 1972. Biosynthesis of vitamin B₆. J. Biol. Chem. **247:**1869–1882.
18. **Hill, R. E., and I. D. Spenser.** 1970. Biosynthesis of vitamin B₆. Incorporation of three-carbon units. Science **169:**773–775.
19. **Hill, R. E., and I. D. Spenser.** 1973. Biosynthesis of vitamin B₆. Incorporation of terminally labeled glucose. Can. J. Biochem. **51:**1412–1416.
20. **Hill, R. E., and I. D. Spenser.** 1985. The biosynthesis of vitamin B₆. *In* D. Dolphin, R. Poulso, and O. Avramovic (ed.), Pyridoxal phosphate: chemical, biochemical, and medical aspects. John Wiley & Sons, Inc., New York.
21. **Hoiseth, S. K., and B. A. D. Stocker.** 1985. Genes *aroA* and *serC* of *Salmonella typhimurium* constitute an operon. J. Bacteriol. **163:**355–361.
22. **Iwanow, A., R. E. Hill, B. G. Sayer, and I. D. Spenser.** 1984. Biosynthesis of vitamin B₆. Incorporation of a C-N unit derived from glycine. J. Am. Chem. Soc. **106:**1840–1841.
23. **McKitrick, J. C., and L. I. Pizer.** 1980. Regulation of phosphoglycerate dehydrogenase levels and effect on serine synthesis in *Escherichia coli* K-12. J. Bacteriol. **141:**235–245.
24. **Morris, J. G.** 1959. The synthesis of vitamin B₆ by some mutant strains of *Escherichia coli*. J. Gen. Microbiol. **20:**597–604.
25. **Morris, J. G., and D. D. Woods.** 1959. Interrelationships of serine, glycine, and vitamin B₆ in the growth of mutants of *Escherichia coli*. J. Gen. Microbiol. **20:**576–596.
26. **Patte, J.-C., G. LeBras, and G. N. Cohen.** 1967. Regulation by methionine of the synthesis of a third aspartokinase and a second homoserine dehydrogenase in *Escherichia coli* K12. Biochim. Biophys. Acta **136:**245–257.
27. **Pizer, L. I., and M. L. Potochny.** 1964. Nutritional and regulatory aspects of serine metabolism in *Escherichia coli*. J. Bacteriol. **88:**611–619.
28. **Shimizu, S., and W. B. Dempsey.** 1978. 3-Hydroxypyruvate substitutes for pyridoxine in *serC* mutants of *Escherichia coli* K-12. J. Bacteriol. **134:**944–949.
29. **Tani, Y., and W. B. Dempsey.** 1973. Glycolaldehyde is a precursor of pyridoxal phosphate in *Escherichia coli* B. J. Bacteriol. **116:**341–345.
30. **Vella, G. J., R. E. Hill, B. S. Mootoo, and I. D. Spenser.** 1980. The status of glycolaldehyde in the biosynthesis of vitamin B₆. J. Biol. Chem. **255:**3042–3048.
31. **Vella, G. J., R. E. Hill, and I. D. Spenser.** 1981. Biosynthesis of pyridoxal, the origin of the C₂ unit C'2,-2. J. Biol. Chem. **256:**10469–10474.
32. **Wasmuth, J., H. E. Umbarger, and W. B. Dempsey.** 1973. A role for a pyridoxine derivative in the multivalent repression of the isoleucine and valine biosynthetic enzymes. Biochem. Biophys. Res. Commun. **51:**158–164.
33. **White, R. S., and W. B. Dempsey.** 1970. Purification and properties of vitamin B₆ kinase from *Escherichia coli* B. Biochemistry **9:**4057–4064.
34. **Wormser, E. H., and A. B. Pardee.** 1958. Regulation of threonine biosynthesis in *Escherichia coli*. Arch. Biochem. Biophys. **78:**416–432.

36. Biosynthesis of Biotin and Lipoic Acid

MAX EISENBERG

Department of Biochemistry and Molecular Biophysics, College of Physicians and Surgeons, Columbia University, New York, New York 10032

BIOTIN BIOSYNTHESIS

The earliest clues to possible intermediates in the biotin biosynthetic pathway were derived from growth and feeding experiments with a variety of bacteria, yeast, and fungi. This earlier work has been extensively reviewed (4). The isolation and identification of the intermediate 7-keto-8-aminopelargonic acid (7-KAP) (3, 14) proved to be the key factor in establishing the pathway of biotin biosynthesis. Rolfe and Eisenberg (22) proposed, on the basis of genetic and biochemical evidence, the sequence of reactions indicated in Fig. 1 for the biosynthesis of biotin in *Escherichia coli*. Pimeloyl coenzyme A (CoA), which can be synthesized from a number of precursors, condenses with L-alanine to form 7-KAP. This amino-ketone is transaminated to give 7,8-diaminopelargonic acid (DAPA), which in turn incorporates CO_2 to form dethiobiotin. The introduction of sulfur with the formation of the tetrahydrothiophene ring is the final step in the reaction sequence. Each of these steps is discussed in greater detail. Also included in the scheme are the letters designating the genes coding for the enzymes that catalyze the individual steps of the pathway. The regulation of the biotin operon has been extensively studied in the laboratories of Guha, Campbell, and Eisenberg, and this aspect has recently been reviewed (6).

PIMELIC ACID SYNTHESIS

There has been a great deal of confusion in the literature concerning the role of pimelic acid in the biosynthesis of biotin. This confusion arose primarily from feeding experiments with a variety of organisms which either did or did not respond to added pimelic acid with an increase in biotin excretion. Even after the direct incorporation of the dicarboxylic acid had been demonstrated in two laboratories (4), a report appeared suggesting glutaric acid rather than pimelic acid as the intermediate in biotin biosynthesis. There are two major problems associated with feeding experiments: first, the permeability of the membrane to the substrate; and second, the ability of the organism to activate the substrate. In *E. coli*, pimelic acid does not cross the membrane barrier, nor has it been possible to demonstrate its activation in a crude cell extract by the hydroxamic acid assay. However, Itzumi et al. (13) were able to observe activation indirectly in cell extracts of their strain of *E. coli* by coupling it with the synthesis of 7-KAP. The insensitivity of the hydroxamic acid assay, as well as the presence of a pimeloyl-CoA deacylase, may account for the earlier results. These studies raised the question of whether pimelic acid or its activated form is the end product of de novo synthesis.

Although the reaction sequence for the synthesis of pimelic acid is still unknown, there is substantial indirect evidence that pimeloyl-CoA is the end product and that some of the enzymes of fatty acid synthesis and degradation are involved. In the early feeding experiments with a variety of organisms, lower and higher homologs of pimelic acid, such as malonic, glutaric, and azelaic acids, could enhance the excretion of biotin into the medium. These substrates were probably activated to form the CoA derivative before chain elongation or degradation to pimeloyl-CoA; the enzymes only function with the activated substrates. With other carbon sources, such as carbohydrates and amino acids, a common intermediate in their metabolism is acetyl-CoA, which could serve as the earliest precursor for de novo synthesis of pimeloyl-CoA. That such may indeed be so was strongly suggested by the isotope experiments of Lezius et al. (16). These investigators grew *Achromobacter* IVSW in the presence of $^{14}CO_2$ with isovaleric acid as the main carbon source and determined the distribution of the isotope in the biotin synthesized. The carbonyl carbon of the ureido ring and the carboxyl group of the side chain of biotin had the same high specific activities, suggesting the fixation of CO_2 into these two carbon atoms with little loss of activity. The carbon atoms with low specific activity were thought to arise from acetyl-CoA formed from isovaleric acid degradation, which involves a step for CO_2 fixation. Their results also indicated that the symmetrical pimelic acid molecule could not be an intermediate since the two carbon atoms corresponding to the two carboxyl groups had different

544

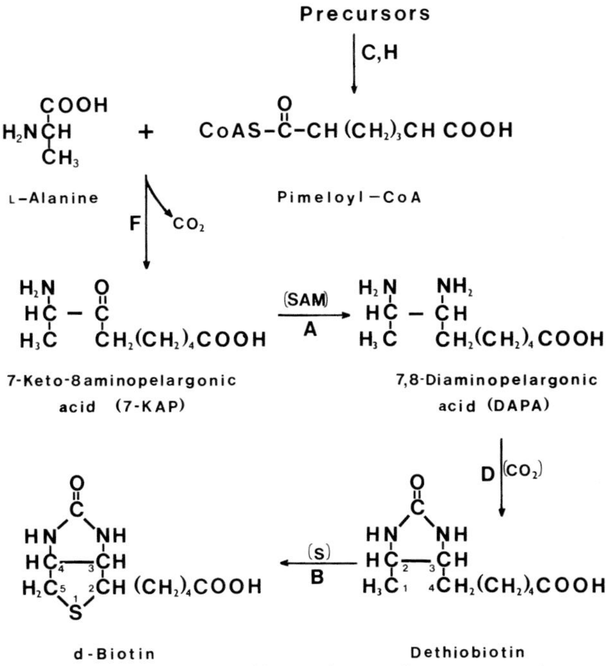

FIG. 1. Pathway of biotin biosynthesis in *E. coli*.

activities. To account for later findings, these investigators proposed a scheme for pimelic acid biosynthesis that involved the condensation of three molecules of malonyl-CoA. The anchor malonyl-CoA would retain the $^{14}CO_2$ fixed as the result of acetyl-CoA carboxylase activity, while the other two would lose the radioactivity during the condensation reaction; the end product would be the activated form of pimelic acid.

Although there is no evidence that malonyl-CoA can serve as an anchor unit in fatty acid biosynthesis, a specific enzyme for this reaction cannot be eliminated. There is still no information on the function of the product of the *bioC* and *bioH* genes, except that a mutation in either gene results in the failure to excrete any of the known intermediates in the biotin pathway. Hence, the products of these genes have been assigned to some early steps before 7-KAP synthesis. If, however, de novo synthesis proceeds through the normal fatty acid pathway that produces even-chained activated intermediates, then a C_1 unit must be introduced at some stage before final chain elongation to form the activated C-7 dicarboxylic acid. This addition may involve the fixation of CO_2, which would account for the asymmetric labeling of the pimelic acid formed. An alternative mechanism would require omega oxidation and loss of a C_1 unit, which are not supported by the isotope data.

7-KAP SYNTHETASE

$$\text{Pimeloyl-CoA} + \text{L-Alanine} \xrightarrow{\text{PLP}} \text{7-KAP} + \text{CoA}$$

The product of this reaction, 7-KAP, was observed in the culture medium of a variety of organisms, including *E. coli*, and was shown by Dhyse and Hertz (2a, 4) to be avidin uncombinable. It was characterized biochemically as an open-chain structure devoid of sulfur before its chemical characterization.

The synthesis of 7-KAP in crude cell extracts was first demonstrated by Eisenberg and Starr (7), who found that the only requirements for this reaction were L-alanine, pyridoxal phosphate (PLP), and pimeloyl-CoA. Whereas serine could replace alanine to a limited extent, cysteine was completely inhibitory. In the previous study by Lezius et al. (16), showing CO_2 fixation into pimelic acid, the investigators determined that [3-^{14}C]cysteine was also incorporated into C-5 of biotin. Since the specific activity of C-5 was identical to the cysteine isolated from the cellular proteins, they assumed that the molecule of cysteine was incorporated as a unit. They postulated a mechanism of biotin biosynthesis in which the first step was the condensation of cysteine with pimeloyl-CoA to form 7-keto-8-amino-9-thiopelargonic acid. Recently, Frappier and Marquet (12) repeated these studies to determine whether this was indeed an alternative pathway for this organism. Using [3-^{14}C, ^{35}S]cysteine, they found no evidence of either label incorporated into biotin or biotin-*d*-sulfoxide. Since radioactive dethiobiotin was readily converted into biotin in this organism, they concluded that the pathway for biotin biosynthesis proposed for *E. coli* was also operative in *Achromobacter* IVSW.

The 7-KAP synthetase is specified by the *bioF* gene, which is coordinately repressed with the other biotin genes in the operon when biotin in excess of 1 ng/ml is added to the growth medium. The enzyme has been purified over 300-fold by column chromatographic procedures (4), and native disc gel electrophoresis of the purified material showed a number of bands with the biological activity associated with the major band. Gel-filtration studies indicated a molecular weight of 45,000. Cleary and Campbell (1), in complementation analysis of the biotin operon, found that *bioF* mutants exhibited intragenic complementation and concluded that the enzyme must be composed of two or more identical subunits. The mechanism of this reaction is similar to that first described for the condensation of succinyl-CoA and glycine to form δ-aminolevulinic acid.

DAPA SYNTHETASE

$$\text{7-KAP} + \text{SAM} \xrightarrow{\text{PLP}} \text{DAPA} + \textit{S}\text{-Adenosyl-2-Oxo-4-} \\ \text{Methylbutyric Acid}$$

The earliest suggestion that the pelargonic acid derivative DAPA may play a role in biotin biosynthesis was made by Du Vigneaud et al. (2c, 4), who found that it had 19% of the biological activity of biotin in supporting the growth of yeast. Pontecorvo later showed that it could completely replace biotin in supporting a biotin auxotroph of *Aspergillus nidulans* (4). These earlier studies were overlooked, and this compound was given serious consideration as an intermediate only when it was observed to accumulate in the growth medium of the *bioA* mutants of *E. coli* (22).

Pai (19) carried out the initial in vitro experiments with crude extracts of *E. coli* but could not directly assay the DAPA formed. However, by coupling the reaction shown above with the subsequent step in the sequence, dethiobiotin synthesis, he was able to monitor the purification of the enzyme and to determine the basic requirements for this reaction. In addition to 7-KAP and PLP, L-methionine was required as the amino group donor and could not be replaced with any other amino acid. ATP and HCO_3^- were also required for the dethiobiotin synthetase reaction. Using a *bioA* mutant selected to grow on DAPA concentrations as low as 25 ng/ml, Eisenberg and Stoner (8) were able to determine directly the DAPA concentration in the culture medium of a *bioD* mutant blocked in the synthesis of dethiobiotin. Glucose enhanced its excretion twofold, and L-methionine was the most effective amino donor. However, in in vitro studies with crude extracts, DAPA could not be detected after the addition of 7-KAP, PLP, and L-methionine unless ATP was also added, suggesting the possible activation of L-methionine. When *S*-adenosyl-L-methionine (SAM) replaced ATP and L-methionine, it proved to be 10 times more effective. Therefore, the activity that Pai observed in the coupled system was due to the addition of ATP for dethiobiotin synthesis, which resulted in L-methionine activation. This was the first report of SAM serving as an amino group donor in addition to its other donor activities.

The enzyme has been purified about 1,000-fold with about 90% purity as estimated by native disc gel electrophoresis (9). The average molecular weight determined from gel-filtration and sucrose gradient centrifugation studies was 94,000. Sodium dodecyl sulfate-polyacrylamide gel electrophoresis showed a single subunit with a molecular weight of 47,000, indicating a dimeric structure. The purified enzyme could not activate L-methionine with ATP to form SAM. The requirement for PLP could only be demonstrated by resolving the enzyme in the presence of SAM, a treatment that resulted in the complete loss of enzymatic activity in about 10 min because of the dissociation of PLP from the enzyme. Activity could be fully restored by addition of PLP to the reaction mixture. Specificity studies indicated that the requirement for SAM was absolute, whereas 7-KAP could be replaced with 8-amino-7-ketopelargonic acid, as well as with 7,8-diketopelargonic acid. Resolution of the enzyme in the presence of SAM also resulted in the dissociation of the enzyme into subunits as determined by gel filtration. Genetic studies showing intragenic complementation among the *bioA* mutants (1), however, suggest that the subunits are identical. The ketone product of the transaminase reaction, *S*-adenosyl-2-oxo-4-methylbutyric acid, could not be detected in the reaction mixture. With the aid of $[2-^{14}C]SAM$, it was determined that the ketone product readily decomposed under the experimental conditions used and gave rise to a ketone fragment, probably 2-oxo-3-butenoic acid, in an amount equal to that of the DAPA formed. Extensive kinetic analysis also provided additional information on the possible conformations of the substrates in the active site of the enzyme (24).

DETHIOBIOTIN SYNTHETASE

$$DAPA + ATP + CO_2 \longrightarrow Dethiobiotin + ADP + P_i$$

The product-precursor relationship of DAPA and dethiobiotin was firmly established by an extensive genetic and biochemical analysis of a large number of biotin mutants of *E. coli* (22). This relationship was first demonstrated experimentally by Krell and Eisenberg (15). Using *bioB*, a biotin auxotroph which is unable to grow on dethiobiotin, they found that the synthesis of dethiobiotin could be enhanced by the addition of DAPA, bicarbonate, and glucose. In a dialyzed crude extract, ATP and Mg^{2+} were required as an energy source, and the addition of $[^{14}C]HCO_3^-$ resulted in the labeling of the ureido carbonyl group of dethiobiotin. The enzyme was purified over 200-fold (15) and was found to be cold labile, losing 80 to 90% of its activity when frozen overnight. The molecular weight was estimated to be 42,000 by gel filtration. Sodium dodecyl sulfate-polyacrylamide gel electrophoresis resulted in one band of 23,000 daltons and thus established its dimeric nature. When the reaction was conducted in the presence of limiting concentrations of either HCO_3^- or CO_2, more dethiobiotin was formed in the presence of CO_2 than in the presence of HCO_3^-. Upon addition of carbonic anhydrase, dethiobiotin formation with CO_2 was reduced, but it was increased with HCO_3^-, indicating that CO_2 was the actual substrate. The stoichiometry of the reaction indicated that only one ATP was required. On the basis of the stoichiometry, a mechanism was proposed in which CO_2 reacts with an amino group nonenzymatically to form a DAPA monocarbamate. The monocarbamate is then activated with ATP to form a substituted carbamyl phosphate which can undergo cyclization as the result of a nucleophilic attack by the neighboring amino group.

BIOTIN SYNTHETASE

$$Dethiobiotin \xrightarrow{(S)} Biotin$$

Since Dittmer et al. (2b) first demonstrated that dethiobiotin was as effective as biotin in supporting the growth of *Saccharomyces cerevisiae*, a large number of studies have appeared in the literature confirming this reaction in resting and growing cells of a variety of organisms (4). In some of this work, the sulfur sources were also explored with the hope of establishing the identity of the immediate sulfur donor. A number of studies were also devoted to elucidating the mechanism of this most interesting reaction with the aid of specifically labeled dethiobiotin. Progress in both areas has been hampered, primarily because a cell-free system for this reaction is still unavailable.

Niimura et al. (4, 18a) were the first to explore possible sulfur donors with sulfur-deficient media and sulfur-starved yeast cells. L-Methionine and L-methionine sulfoxide headed the list of organic sulfur compounds, whereas Na_2S, $NaHSO_3$, and Na_2SO_4 were the most effective inorganic sulfur sources. The sulfur of $L-[^{35}S]methionine$ was found to be incorporated into biotin, but the specific activity was not determined; thus, direct incorporation could not be established. Interestingly, methionine was most effective when preincubated with the cells, suggesting that it was not the immediate sulfur donor. A similar study was also performed with resting cells of *E. coli* ABD 313-136, a *bioR* mutant (5). This strain is fully derepressed and is an overproducer of biotin. The data in Table 1 reveal that, in nonstarved cells, the addition of a number of sulfur sources did not substantially in-

TABLE 1. Effect of sulfur donors on biotin synthetase activity in sulfate-starved and nonstarved resting cells of E. coli ABD 313-136

Sulfur source (nM)	Biotin formed (nmol/ml) in cells that were:	
	Starved	Nonstarved
None	0.03	0.90
$(NH_4)_2S$ (0.5)	0.66	1.03
Na_2SO_4 (0.5)	0.66	1.03
Methionine (5.0)	0.14	0.93
Glutathionine (5.0)	0.74	1.12
Cysteine (5.0)	0.34	0.48
2-Mercaptoethanol (5.0)	0.03	0.46

crease the synthesis of biotin, indicating that the intracellular concentration of the sulfur donor was adequate. In sulfur-starved cells, however, the addition of a fully reduced and fully oxidized form of inorganic sulfur gave a 22-fold increase in biotin synthesis. Glutathione was the most effective organic sulfur source, whereas methionine and cysteine were marginal and 2-mercaptoethanol was completely ineffective. In a comprehensive study by Roberts et al. (21) of biosynthetic pathways in E. coli, the same inorganic sulfur sources were also the most reactive in promoting cellular growth, and cystine was the most effective organic sulfur source. Methionine and cysteine were inactive; the latter was shown not to permeate the cell membrane unless it was first oxidized to cystine. The results given above were corroborated in a recent study by DeMoll and Shive (2) with [35S]-amino acids. In addition, it was found that, in growing cultures of E. coli, the sulfur atoms of [35S]cystine and L-[sulfane-35S]thiocystine were transferred with an efficiency of 75 and 29%, respectively. The latter compound is the trisulfide of cystine, and this degree of incorporation represents 86% of the expected value when all three sulfur atoms are equivalent in their ability to contribute to the synthesis of biotin. Evidence that all sulfur atoms in this compound are probably cycled through cysteine in E. coli was recently provided by White (28). Unfortunately, the high specific activities obtained in the experiments described above do not provide definitive proof that cysteine is the immediate sulfur donor for biotin synthesis.

As indicated previously, the mechanism of incorporation of sulfur is still an open question, not only in biotin but also in other sulfur-containing cofactors. Li et al. (17) attempted to obtain information on this question by feeding a mixture of [14C]dethiobiotin and randomly 3H-labeled dethiobiotin to growing cultures of Aspergillus niger. The biotin sulfone isolated had a 3H/14C ratio 15 to 20% lower than that of the starting mixture. The loss of four tritium atoms could account for the observed percentage loss if uniform tritium labeling is assumed. In addition, they also assumed that the four atoms were lost from the tetrahydrothiophene ring. This loss could be accounted for by the formation of two double bonds in the biosynthesis of biotin. A mechanism for this series of reactions was proposed by Eisenberg (4), as shown in Fig. 2. The figure also includes a second mechanism, involving the loss of three hydrogen atoms as the result of the hydroxylation of the methyl group, and a subsequent

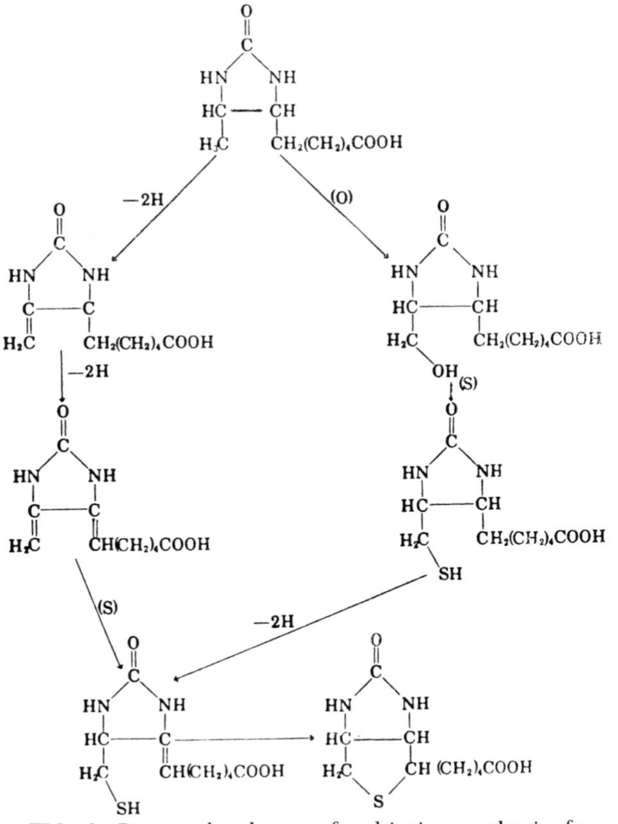

FIG. 2. Proposed schemes for biotin synthesis from dethiobiotin involving loss of either three or four hydrogens.

exchange with reduced sulfur, as in the conversion of serine to cysteine.

Parry (20) synthesized dethiobiotin specifically labeled with tritium to circumvent the uncertainties associated with the use of randomly labeled substrates. The tritiated dethiobiotin was mixed with [9-14C]dethiobiotin and added to growing cultures of A. niger. Biotin was isolated as biotin sulfone methyl ester crystallized to constant specific activity; the 3H/14C ratios are shown in Table 2. The data in experiments 1, 2, and 3 indicate that, in the formation of the tetrahydrothiophene ring, there is no loss of hydrogen at C-2, C-3, and C-5. Thus, it is unlikely that unsaturation is introduced in the bridge carbons (C-2, C-3) or between C-4 and C-5, as indicated in Fig. 2.

TABLE 2. Incorporation of specifically tritiated (± racemic mixture) dethiobiotin into biotin by A. niger

Expt. no.	Labeled precursor	3H/14C ratio for:		Retention (% 3H)
		Precursor	Product	
1	2,3-3H, 10-14C[a]	6.05	5.74	95.0
2	3-3H, 10-14C[b]	2.89	3.03	105.0
3	5-3H, 10-14C	5.72	5.33	93.2
4	1-3H, 10-14C	6.88	4.81	69.9
5	RS-4-3H, 10-14C	5.88	3.10	52.7
6	R-4-3H, 10-14C	3.37	3.07	91.1
7	S-4-3H, 19-14C	6.00	0.42	7.0

[a] Precursor has 58% 3H at C-2, 42% at C-3.
[b] Precursor has 17% 3H at C-2, 83% at C-3.

TABLE 3. Incorporation of dethiobiotin into biotin by *E. coli*

Expt. no.	Labeled precursor	Labeling pattern[a]	
		Precursor	Biotin
1	9-^{14}C, 2,3-^{3}H (^{3}H/^{14}C = 8.2)	85% ^{3}H at C-2, 15% ^{3}H at C-3	101% ^{3}H retention (^{3}H/^{14}C = 8.3)
2	2,3-^{2}H	85% (2-^{2}H$_1$), 15% (3-^{2}H$_1$)	89% (2-^{2}H$_1$), 11% (3-^{2}H$_1$)
3	5-^{2}H$_2$	88% (5-^{2}H$_2$), 12% unlabeled	84% (5-^{2}H$_2$), 14% unlabeled
4	1-^{2}H$_3$	65% (1-^{2}H$_3$), 25% (1-^{2}H$_2$), 10% unlabeled	71% (1-^{2}H$_2$),[b] 16% (1-^{2}H$_1$), 14% unlabeled

[a] Tabulated by Parry (20).
[b] Calculated values are 73.5% (1-^{2}H$_2$), 16.5% (1-^{2}H$_1$), and 10% unlabeled.

Essentially the same conclusions were reached by Frappier et al. (11), who used deuterated dethiobiotin with growing cells of *E. coli*, as shown in Table 3. However, one cannot exclude the possibility that, in the course of desaturation, the tritium or deuterium removed is strongly bound to the enzyme and cannot exchange with the medium. Experiment 4 of Table 2 also shows that 30% of the tritium is lost from [1-^{3}H]dethiobiotin, as would be expected on the basis of the loss of one of the three methyl hydrogens at C-1. The most interesting result is the loss of 47% tritium from (RS)-[4-^{3}H]dethiobiotin (experiment 5), indicating a stereospecific removal of one hydrogen from C-4 during sulfur incorporation. Parry (20) studied the reaction at C-4 in greater detail by synthesizing the chirally labeled forms of dethiobiotin, (R)-4-^{3}H and (S)-4-^{3}H. When added to cultures of *A. niger*, the 4-*pro-S* hydrogen was lost upon the incorporation of sulfur (experiment 7). Since the absolute configuration at C-4 had been previously determined to be S (25), it followed that the introduction of sulfur proceeded with retention of configuration. The alternative mechanism of biotin biosynthesis in Fig. 2, that is, the activation of the methyl group of dethiobiotin by hydroxylation, was also considered by Parry but was dismissed on the basis of stereochemical considerations of a two-step reaction.

Evidence against the intermediary hydroxylated form of dethiobiotin was provided by Frappier et al. (10), who synthesized the racemic mixtures of C-1 and C-4 hydroxy dethiobiotin, as well as the dihydroxy derivative. When added to growing cultures of *E. coli* C124, a biotin auxotroph blocked in the formation of dethiobiotin, none of the three compounds would support the growth of the auxotroph. Permeability could be ruled out as all three compounds entered the cell. This study essentially eliminated the second mechanism of a hydroxylated intermediate in the biosynthetic pathway.

Most recently, Salib et al. (23) reported the isolation of a radioactive intermediate from a reaction mixture containing [2,3-^{3}H]dethiobiotin, [9-^{14}C]dethiobiotin (the numbering has been changed to conform to Parry's system [20]), and resting cells of *E. coli* C124. The compound could only be separated from dethiobiotin by column chromatography after esterification. The isolated radiolabeled compound was readily converted into biotin in growing cultures of *E. coli* C124 and could also support the growth of a number of *bioB* mutants which are blocked in the biotin synthetase reaction. The presence of sulfur in the compound was demonstrated with [^{35}S]sulfate. The intermediate is unstable and is readily transformed into a compound that is also biologically active. The surprising observation is the complete absence of biotin in the reaction mixture, since similar experiments carried out with a *bioR* mutant showed extensive formation of biotin and biotin-*d*-sulfoxide (unpublished observations). Since the unknown intermediate supported the growth of the *bioB* mutants, these investigators suggested that these mutants are blocked in the conversion of dethiobiotin into the unknown compound. At present, there is only one cistron known for the sulfur incorporation step, and that is *bioB*. Unless another gene has escaped previous genetic analysis, these results can best be explained on the basis of a multifunctional protein with independent functional domains. Therefore, all mutants would map at the *bioB* locus regardless of which domain was inactivated. Further genetic and biochemical analysis of this cistron, as well as the development of an in vitro system, is necessary for the further elucidation of this step. Structural determination of the unknown intermediate may provide clues to the mechanism of sulfur incorporation.

LIPOIC ACID BIOSYNTHESIS

Our knowledge of the biosynthesis of alpha (+) lipoic acid (6,8-thiooctic acid) has lagged far behind its early chemical and structural determination. The minute quantities present in the cell and an inability to apply mutant techniques have been in large measure responsible for this delay. Most of our information about its biosynthesis has come from the use of stable and radioactive isotopes, primarily from the laboratories of Parry (20) and White (27).

Lipoic acid and biotin share many common features in terms of function and biosynthesis. Both cofactors function only when covalently linked to their respective holoenzymes by an identical mechanism of activation with ATP to form an acyl-AMP intermediate. Both biosynthetic pathways appear to utilize many of the enzymatic reactions of the fatty acid biosynthetic pathway for the formation of part or all of their carbon skeletons. Finally, the immediate sulfur donor is still to be determined for both pathways, and they may well share the same sulfur donor and have common features in the mechanism of sulfur incorporation.

The earliest information on the biosynthesis of lipoic acid appeared in an abstract by Reed (L. J. Reed, cited in reference 12a) in which octanoic acid was suggested as a precursor. When [1-^{14}C]octanoic acid was fed to *E. coli*, it appeared to be incorporated as a unit. Specificity studies indicated that the C$_6$ mono- and

TABLE 4. Incorporation of labeled octanoic acid into lipoic acid by *E. coli*

Expt. no.	Precursor	$^3H/^{14}C$ ratio for precursor	$^3H/^{14}C$ ratio for lipoic acid	Labeling pattern	% ^3H retention
1	$1\text{-}^{14}C$			>90% label at C-1	
2	$RS\text{-}5\text{-}^3H, 1\text{-}^{14}C$	4.05	4.11		102
3	$RS\text{-}7\text{-}^3H, 1\text{-}^{14}C$	3.95	3.81		96.5
4	$8\text{-}^3H, 1\text{-}^{14}C$	5.02	4.81		95.8
5	$RS\text{-}6\text{-}^3H, 1\text{-}^{14}C$	5.08	2.53		49.8
6	$S\text{-}6\text{-}^3H, 1\text{-}^{14}C$	4.13	3.47		84.0
7	$R\text{-}6\text{-}^3H, 1\text{-}^{14}C$	4.40	0.48		10.9

dicarboxylic acids and 8-hydroxyoctanoic acid were not substrates. The direct precursor role of octanoic acid was later verified by Parry (20), who degraded the lipoic acid and determined that 90% of the label from [1-^{14}C]octanoic acid was retained in C-1. It was subsequently shown by White (27) that octanoic acid (U-^{2}H$_{15}$) is directly incorporated into lipoic acid with the loss of two deuterium atoms which were displaced during the introduction of two sulfhydryl groups.

To obtain further insight into the mechanism of sulfur incorporation, Parry synthesized octanoic acid specifically tritiated at C-5, C-6, C-7, and C-8. Each of the compounds was combined with [1-^{14}C]octanoic acid and administered to growing cultures of *E. coli*. The tritium retention was calculated from the difference in the ^{3}H/^{14}C ratio between the precursor and the lipoic acid isolated. The data from experiments 1 and 2 of Table 4 show that the tritium on C-5 and C-7 is retained, suggesting that the incorporation of sulfur does not occur by the introduction of unsaturation at these two carbon atoms; the same argument was used for the bridge hydrogens of biotin. The tritium in C-8 was also retained (experiment 3), indicating a strong isotope effect associated with the removal of one tritium atom on the introduction of a sulfhydryl group. Sodium (RS)-[6-^{3}H]octanoic acid was incorporated with about 50% loss of tritium (experiment 4), which is the calculated amount for the stereospecific removal of one hydrogen. When the chirally labeled compounds, (6R)- and (6S)-octanoic acid, were added to the growth medium of *E. coli*, only the pro-R hydrogen was lost on the introduction of sulfur (experiment 6). Since the absolute configuration of C-6 had been previously determined to be R (18), the data indicated an inversion of configuration during sulfur incorporation. These results were fully corroborated by White (27). One possible explanation to account for the inversion could involve the hydroxylation of C-6 before the sulfur incorporation step. White (26), therefore, synthesized the 6-hydroxy, 8-hydroxy, 6,8-dihydroxy, and 8-mercapto derivatives of octanoic acid, which were then added to the growing cultures of *E. coli*. None of the hydroxy derivatives were incorporated. However, 8-mercaptooctanoic acid was incorporated to the extent of 19 to 28%, depending on the concentration used. The latter concentration suggested a possible permeability problem that could have accounted for the negative results obtained with the hydroxylated derivatives. This possibility was eliminated when the distribution of the 8-hydroxyoctanoic acid was determined. It was found to be present within the cell in a very small quantity in the nonesterified form. Calculations, however, showed that this amount of substrate would have resulted in a 10%

labeling of the lipoic acid if only 1% was converted. Since less than 0.5% was observed, it essentially eliminated the hydroxy derivatives but not the 8-mercapto derivative as possible intermediates.

The data given above would support the following reaction sequence: acetyl-CoA → octanoic acid → 8-mercaptooctanoic acid → 6,8-thioctic.

LITERATURE CITED

1. **Cleary, P., and A. Campbell.** 1977. Deletion and complementation analysis of the biotin gene cluster of *Escherichia coli*. J. Bacteriol. **112:**830–839.
2. **Demoll, E., and W. Shive.** 1983. The origin of sulfur in biotin. Biochem. Biophys. Res. Commun. **110:**243–249.
2a. **Dhyse, F. G., and R. Hertz.** 1958. Arch. Biochem. Biophys. **74:**7.
2b. **Dittmer, K., D. V. Melville, and V. Du Vigneaud.** 1944. The possible synthesis of biotin from desthiobiotin by yeast and the anti-biotin effect of desthiobiotin for *L. casei*. Science **99:**203–205.
2c. **Du Vigneaud, V., K. Hoffman, and D. V. Melville.** 1942. On the structure of biotin. J. Am. Chem. Soc. **64:**188–189.
3. **Eisenberg, M. A.** 1966. The biosynthesis of biotin in growing yeast cells. The formation of biotin from an early intermediate. Biochem. J. **101:**598–600.
4. **Eisenberg, M. A.** 1973. Biotin: biogenesis, transport and their regulation. Adv. Enzymol. Relat. Areas Mol. Biol. **38:**317–372.
5. **Eisenberg, M. A.** 1982. Mode of action of biotin antimetabolites, actithiazic acid and methyldethiobiotin. Antimicrob. Agents Chemother. **21:**5–10.
6. **Eisenberg, M. A.** 1984. Regulation of the biotin operon. Ann. N.Y. Acad. Sci. **447:**335–349.
7. **Eisenberg, M. A., and C. Starr.** 1968. Synthesis of 7-oxo-8-aminopelargonic acid, a biotin vitamer, in cell-free extracts of *Escherichia coli* biotin auxotrophs. J. Bacteriol. **96:**1291–1297.
8. **Eisenberg, M. A., and G. L. Stoner.** 1971. Biosynthesis of 7,8-diaminopelargonic acid, a biotin intermediate, from 7-keto-8-aminopelargonic acid and S-adenosyl-L-methionine. J. Bacteriol. **108:**1135–1140.
9. **Eisenberg, M. A., and G. L. Stoner.** 1979. 7,8-Diaminopelargonic acid aminotransferase. Methods Enzymol. **62:**342–347.
10. **Frappier, F., G. Guillerm, A. G. Salib, and A. Marquet.** 1979. On the mechanism of conversion of dethiobiotin to biotin in *Escherichia coli*. Discussion of the occurrence of an intermediate hydroxylation. Biochem. Biophys. Res. Commun. **91:**521–527.
11. **Frappier, F., M. Jouany, A. Marquet, A. Oleskar, and J. C. Tabet.** 1982. On the mechanism of the conversion of dethiobiotin to biotin in *E. coli* studies with deuterated precursors, using MS-MS techniques. J. Org. Chem. **47:**2257–2261.
12. **Frappier, F., and A. Marquet.** 1981. On the biosynthesis of biotin in *Achromobacter* IVSW. A reinvestigation. Biochem. Biophys. Res. Commun. **103:**1288–1293.
12a. **Greenberg, D.** 1975. Metabolic pathways VII, p. 91. Academic Press, Inc., New York.
13. **Itzumi, Y., H. Morita, L. Sato, Y. Tani, and K. Ogata.** 1972. Synthesis of biotin-vitamers from pimelic acid and coenzyme A by cell-free extracts of various bacteria. Biochim. Biophys. Acta **264:**210–213.
14. **Iwahara, S., M. Kikuchi, T. Tochikura, and K. Ogata.** 1966. Some properties of avidin-uncombinable unknown biotin-vitamers produced by *Bacillus sp.* and its role in the biosynthesis of desthiobiotin. Agr. Biol. Chem. **30:**304–306.
15. **Krell, K., and M. A. Eisenberg.** 1970. The purification and properties of dethiobiotin synthetase. J. Biol. Chem. **245:**6558–6566.

16. **Lezius, A., E. Ringelman, and F. Lynen.** 1963. Zur Biochemischen function de biotens. IV. Die biosynthese des biotens. Biochem. Z. **336:**510–525.
17. **Li, H., D. B. McCormick, and L. D. Wright.** 1968. Conversion of dethiobiotin to biotin in *Aspergillus niger.* J. Biol. Chem. **243:**6442–6445.
18. **Mislow, K., and W. C. Meluch.** 1956. The stereochemistry of α-lipoic acid. J. Am. Chem. Soc. **78:**5920–5923.
18a. **Nimura, T., T. Suzuki, and Y. Sahashi.** 1964. Gen. Vitamin. **10:**218.
19. **Pai, C. H.** 1971. Biosynthesis of biotin: synthesis of 7,8-diamino-pelargonic acid in cell-free extracts of *Escherichia coli.* J. Bacteriol. **105:**793–800.
20. **Parry, R. J.** 1983. Biosynthesis of some sulfur containing natural products. Investigations of the mechanism of carbon-sulfur bond formation. Tetrahedron **39:**1215–1238.
21. **Roberts, R. B., P. H. Abelson, D. B. Cowie, E. T. Bolton, and R. J. Britten.** 1955. Studies of biosynthesis in *Escherichia coli.* Carnegie Inst. Wash. Publ. **607:**58–94.
22. **Rolfe, B., and M. A. Eisenberg.** 1968. Genetic and biochemical analysis of the biotin loci of *Escherichia coli* K-12. J. Bacteriol. **96:**515–524.
23. **Salib, A. G., F. Frappier, G. Guillerm, and A. Marquet.** 1979. On the mechanism of conversion of dethiobiotin to biotin in *Escherichia coli.* III. Isolation of an intermediate in the biosynthesis of biotin from dethiobiotin. Biochem. Biophys. Res. Commun. **88:**312–319.
24. **Stoner, G. L., and M. A. Eisenberg.** 1975. Biosynthesis of 7,8-diaminopelargonic acid from 7-keto-8-aminopelargonic acid and S-adenosyl-L-methionine. The kinetics of the reaction. J. Biol. Chem. **250:**4037–4043.
25. **Trotter, J., and J. A. Hamilton.** 1966. The absolute configuration of biotin. Biochem. J. **5:**713–714.
26. **White, R. H.** 1980. Biosynthesis of lipoic acid: extent of incorporation of deuterated hydroxy- and thioctanoic acids into lipoic acid. J. Am. Chem. Soc. **102:**6605–6607.
27. **White, R. H.** 1980. Stable isotope studies on the biosynthesis of lipoic acid in *Escherichia coli.* Biochemistry **19:**15–19.
28. **White, R. H.** 1982. Metabolism of L(sulfane-S)thiocystine by *Escherichia coli.* Biochemistry **21:**4271–4275.

37. Synthesis and Use of Vitamin B$_{12}$

R. JETER, J. C. ESCALANTE-SEMERENA, D. ROOF, B. OLIVERA, AND J. ROTH

Department of Biology, University of Utah, Salt Lake City, Utah 84112

INTRODUCTION

Salmonella typhimurium and *Escherichia coli* use vitamin B$_{12}$ as a cofactor for two known enzymes, ethanolamine ammonia-lyase (*eut* [ethanolamine utilization]) and homocysteine methyltransferase (*metH*) (14, 15, 21). Until recently, it was thought that enteric bacteria could not synthesize B$_{12}$ because their ability to perform these two reactions depends on exogenously supplied B$_{12}$. The requirement for exogenous B$_{12}$ is consistent with the existence of a B$_{12}$ transport system that has been characterized in *E. coli* (51). Recently, it has been shown that *S. typhimurium*, and probably *E. coli* as well, is capable of synthesizing B$_{12}$ but does so only under anaerobic conditions (32). This synthesis requires a formidable biosynthetic pathway estimated to involve about 30 enzymes. The existence of this long pathway and its anaerobic expression raises the possibility that B$_{12}$ may play an important role in the physiology of enteric bacteria, particularly under anaerobic conditions. It seems likely that additional B$_{12}$-catalyzed reactions will be found that are important to anaerobic metabolism.

BACKGROUND ON B$_{12}$ STRUCTURE AND SYNTHESIS OF CATALYTIC ACTIVITY

Vitamin B$_{12}$ is one of the largest organic molecules known (excluding polymers) (22). It consists of a highly decorated corrinoid ring attached to D-1-amino-2-propanol esterified to the nucleotide dimethylbenzimidazole (DMB) (Fig. 1). The cobalt atom in the center of the corrinoid ring is coordinated with DMB below the ring (α ligand; *Coα*). Covalently attached to the cobalt is an adenosyl moiety (in the case of adenosyl-B$_{12}$) or a methyl group (in the case of methyl-B$_{12}$). B$_{12}$ is commercially available as cyano-B$_{12}$, which has a CN group attached to cobalt above the ring (β ligand; *Coβ*). The determination of this structure and of its complete organic synthesis has been described in several reviews (25, 27, 53, 58).

Biosynthesis of B$_{12}$ has been roughly worked out in that several landmark intermediates are known and several reaction sequences have been studied in some detail in nonenteric bacteria (3, 6, 7, 11, 20, 22, 23, 30, 34, 35, 39, 42, 44, 48, 49, 50). Progress has been hampered mainly by the scarcity of natural materials. Recent reports in the literature describe synthetic approaches to the production of cobalamin precursors (e.g., C-methylated isobacteriochlorins), the availability of which would greatly facilitate research on the biosynthesis of cobalamin (4, 5, 8, 40). At present, the exact sequence of biosynthetic reactions is unclear. We estimate that about 30 enzymes may be uniquely required for synthesis of B$_{12}$. If many enzymes had multiple functions (especially in ring decoration), this estimate could, of course, be reduced considerably.

A general outline of the synthetic pathway is presented in Fig. 2 (32). We have divided the pathway into three major branches on the basis of the nutritional behavior of mutants (see below). The *cobI* pathway is defined as synthesis of cobinamide, which includes the corrinoid ring, cobalt, and the side chain (D-1-amino-2-propanol) to which DMB becomes attached. This part of the pathway probably involves most of the enzymes; the greatest proportion of *cob* mutants that we have isolated is blocked here. It is only this part of the pathway that fails to function in the presence of O$_2$. This failure could be due to sensitivity of cobinamide precursors to oxidation (which is known to be the case in vitro) (8) or to repression of the pathway in response to oxygen, which has also been seen (see below). Although sensitivity of intermediates to oxygen is a reasonable possibility, many other bacteria are known to synthesize B$_{12}$ aerobically and must be able to protect synthetic intermediates (9, 18, 37).

The second part of the pathway (*cobII*) involves synthesis of DMB. In some organisms (*Propionibacterium* and *Streptomyces* spp.), this pathway uses riboflavin as a starting material (Fig. 2) (29). In other

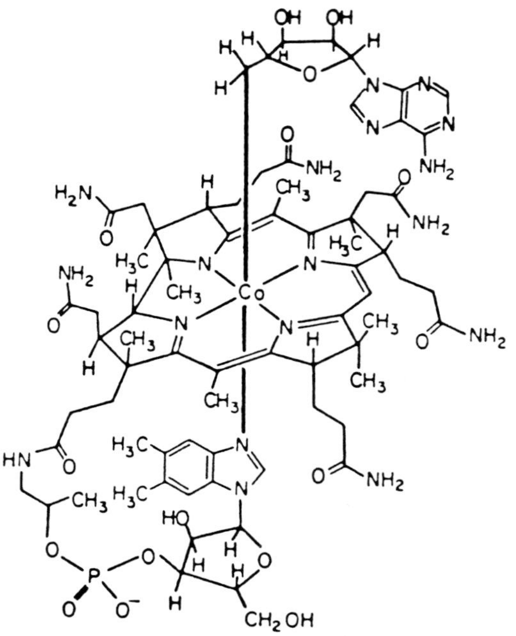

FIG. 1. Structure of adenosyl-B_{12}.

organisms (*Eubacterium* and *Clostridium* spp.), glycine and methionine contribute to DMB synthesis (28, 38). We do not yet know which route is used by *S. typhimurium*.

The third section of the pathway (*cobIII*) involves the use of nicotinic acid mononucleotide as a ribose donor to DMB (24) and the joining of this nucleotide to cobinamide to form B_{12}. One could, in principle, define a fourth part of the pathway, which would involve synthesis of adenosyl-B_{12} from exogenous cyano-B_{12}.

Vitamin B_{12} is involved in catalysis of two general sorts of reactions (26). Adenosyl-B_{12} is used in a set of reactions in which carbon-carbon bonds are rearranged (1). Methyl-B_{12} is involved in methyl transfer reactions (43). One example of each of these two reaction types is known to occur in *Salmonella* spp. and *E. coli*.

ENZYMATIC REACTIONS IN *S. TYPHIMURIUM* AND *E. COLI* THAT USE B_{12} AS A COFACTOR

The two B_{12}-requiring enzymes known in enteric bacteria use different forms of the cofactor. One enzyme, ethanolamine ammonia-lyase (12), is needed for assimilation of ethanolamine as a carbon or nitrogen source or both. The other is one of two enzymes that catalyze the final methyl transfer in methionine biosynthesis.

Ethanolamine Ammonia-Lyase

Ethanolamine ammonia-lyase catalyzes the reaction shown in Fig. 3. The mechanism of this reaction has been reviewed (2). The enzyme requires adenosyl-B_{12} as cofactor (13, 14, 47) and is encoded by the *eut* locus, which lies at 50 min on the *S. typhimurium* chromosome (D. Roof, unpublished results).

Mutants lacking this activity have also been studied in *E. coli* (33). In both *E. coli* and *S. typhimurium*, expression of the structural gene is induced by the simultaneous presence of B_{12} and ethanolamine (10; unpublished data). Apparently, B_{12} is unique among cofactors in enterics in not always being present; this may make it advantageous to produce enzymes requiring B_{12} only if the vitamin is available.

Homocysteine Methyltransferase

Homocysteine methyltransferase is the second B_{12}-dependent enzyme known for enteric bacteria; it is

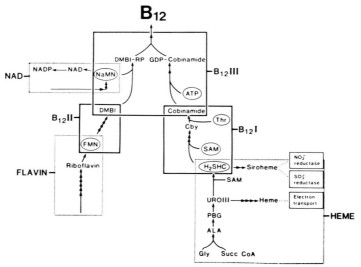

FIG. 2. An outline of cobalamin biosynthesis as it is currently known. Branch I (B_{12}I) of the pathway represents synthesis of the corrinoid ring, branch II (B_{12}II) represents synthesis of DMBI, and branch III (B_{12}III) represents assembly of the several parts to form the mature cobalamin molecule. Encircled compounds are those that make a direct contribution to the final structure. Abbreviations: ALA, 5-aminolevulinic acid; Cby, cobyric acid; DMBI-RP, 1-alpha-D-ribofuranosido-DMBI; FMN, flavin mononucleotide; GDP-cobinamide, guanosine diphosphocobinamide; Gly, glycine; H_2SHC, dihydrosirohydrochlorin; NaMN, nicotinic acid mononucleotide; PBG, porphobilinogen; SAM, *S*-adenosylmethionine; Succ CoA, succinyl coenzyme A; Thr, L-threonine; and UROIII, uroporphyrinogen III.

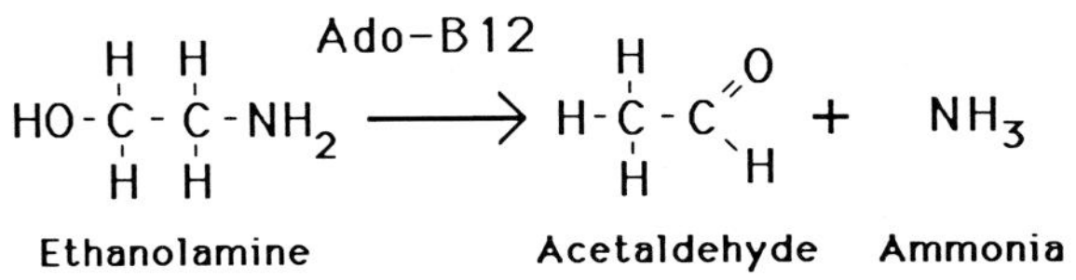

FIG. 3. Reaction catalyzed by the enzyme ethanolamine ammonia-lyase, that is encoded by the *eut* region at min 50 of the *S. typhimurium* chromosome. The enzyme requires adenosyl-B$_{12}$ (ado-B12).

specified by the gene *metH* (16). The *metH* enzyme is one of two enzymes capable of catalyzing the final step in methionine synthesis in *S. typhimurium* (Fig. 4).

The other enzyme (*metE*) catalyzes the same reaction but does not require B$_{12}$ as a cofactor. This second enzyme (*metE*) is much less efficient and has a turnover number 57 times lower than that of the B$_{12}$-dependent enzyme (*metH*) (55, 57). When cells grow with the *metE* enzyme, 5% of their total soluble protein is invested in this inefficient enzyme (52, 56, 57).

The more efficient *metH* enzyme uses methyl-B$_{12}$ as the immediate methyl donor. The enzyme transfers a methyl group from N^5-methyl-tetrahydrofolate to some form of B$_{12}$ and from there to homocysteine (54). The form(s) of B$_{12}$ incorporated into this enzyme in vivo is not clear but, in vitro, cyano-B$_{12}$, OH-B$_{12}$, and methyl-B$_{12}$ can be used.

These enzymes are regulated in response to B$_{12}$ availability. The *metE* gene is turned off and *metH* is turned on when cells are presented with B$_{12}$ (45). Genetic evidence has been reported documenting that the repression of the *metE* gene by B$_{12}$ requires the *metH* holoenzyme and a functional *metF* gene product (5,10-methylenetetrahydrofolate reductase, EC 1.1.1.68). This dual requirement suggests that 5-methyl-tetrahydrofolate needs to bind to the *metH* holoenzyme before repression of the *metE* gene can be achieved (36, 40, 46).

SYNTHESIS OF VITAMIN B$_{12}$ IN *S. TYPHIMURIUM*

The two B$_{12}$-dependent enzymes (*eut* and *metH*) described above act only when exogenous B$_{12}$ is supplied in the medium (for aerobic cultures); their synthesis is induced by B$_{12}$. This led to the belief that B$_{12}$ was not synthesized by *S. typhimurium* or *E. coli* (41).

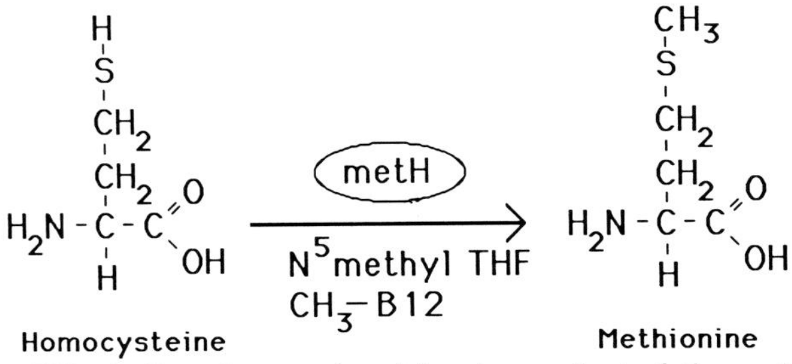

FIG. 4. Reaction catalyzed by the product of the *metH* gene. Methyl-B$_{12}$ is formed on the enzyme, and the methyl group is transferred to homocysteine. S-Adenosylmethionine is required to prime the reaction. N^5-methyl-THF, N^5-methyl-tetrahydrofolate.

TABLE 1. Phenotypes of *metE* and *metH* mutants

| Genotype | Growth under the following conditions[a]: | | | | | | | |
| | With O$_2$ plus: | | | | Without O$_2$ but with: | | | |
	Min	Met	B$_{12}$	Cobin-amide	Min	Met	B$_{12}$	Cobin-amide
metE	−	+	+	+	+	+	+	+
metE metH	−	+	−	−	−	+	−	−
metH	+	+	+	+	+	+	+	+
metA	−	+	−	−	−	+	−	−

[a] +, Growth, or −, no growth on minimal medium plus the indicated nutrient. Min, Minimal medium; Met, methionine.

However, it had been noted that the biosynthetic intermediate cobinamide could serve as a source of B$_{12}$ (19); this made it clear that enteric bacteria must be able to perform many of the B$_{12}$ biosynthetic reactions (including the synthesis of DMB and the joining of cobinamide and DMB, pathways II and III). It should be noted that the conversion of cobinamide to B$_{12}$ occurs aerobically. It had also been reported that minute quantities of B$_{12}$ were synthesized by *S. typhimurium*, but not in sufficient quantities to permit adequate in vivo methionine synthesis (14).

Several years ago we found that complete synthesis of B$_{12}$ does occur in *S. typhimurium*, but only under anaerobic growth conditions (32). The initial evidence for this conclusion involved scoring the growth behavior of *metE* mutants. Such mutants lack the B$_{12}$-independent methyltransferase but still have the B$_{12}$-dependent *metH* enzyme. Therefore, they require either methionine or B$_{12}$ for growth (aerobically). We found that such mutants can grow on minimal medium (without B$_{12}$) under anaerobic conditions. These results are summarized in Table 1 and described below.

The anaerobic methionine synthesis by a *metE* mutant depends on possession of a functional *metH* gene. This suggestive evidence was supported by isolation of mutants unable to synthesize B$_{12}$ anaerobically and presumed to lack B$_{12}$ biosynthetic enzymes (Table 2). These mutants fall into three classes on the basis of their ability to use biosynthetic intermediates supplied in the culture medium as a source of B$_{12}$. Mutants inferred to lack ability to synthesize cobinamide (*cobI*) convert exogenously supplied cobinamide to B$_{12}$. Mutants blocked in synthesis of DMB are classified *cobII*; such mutants can make B$_{12}$ only if supplied with exogenous DMB. A third class of mutants (*cobIII*) are not satisfied by added DMB or cobinamide (or by both together). We infer that they are defective for later reactions involved in joining DMB and cobinamide to form B$_{12}$.

TABLE 2. Phenotypes of *cobI*, *cobII*, and *cobIII* mutants

| Genotype | Anaerobic growth on minimal medium plus[a]: | | | | | |
	Min	Met	B$_{12}$	Cobinamide	DMB	DMB and cobinamide
metE	+	+	+	+	+	+
metE cobI	−	+	+	+	−	+
metE cobII	−	+	+	−	+	+
metE cobIII	−	+	+	−	−	−

[a] +, Growth, or −, no growth on minimal medium plus the indicated nutrient(s). Min, Minimal medium; Met, methionine.

TABLE 3. Production of B_{12} by *S. typhimurium*

Genotype	Cobalamin (B_{12}) production (ng/ml of acid extract) without O_2^a
LT2 (wild type)	1.45
metE	1.25
metE cobI	<0.02
metE cobII	<0.02
metE cobIII	<0.02

^{a} Cobalamin production in the presence of O_2 was <0.02 ng/ml of acid extract for all genotypes.

The indirect evidence described above for B_{12} synthesis was confirmed by direct assays of B_{12} levels in cells grown under aerobic and anaerobic conditions (Table 3). The B_{12} was assayed by a commercial assay involving binding to hog intrinsic factor (a B_{12}-binding protein). Details of this assay have been described (32).

GENES ENCODING THE B_{12} BIOSYNTHETIC ENZYMES

All mutants isolated to date that are defective in B_{12} synthesis bear lesions in a small region represented counterclockwise from the *his* operon at min 42 on the *S. typhimurium* map. These mutants (*cob*) are not cotransducible with *his*, but no essential genes lie between *his* and *cob*. Therefore, *his-cob* deletions can be isolated, and deletions of the intervening region can be made that bring the *his* and *cob* regions close together and permit cotransduction.

Based primarily on deletion mapping crosses with *his-cob* deletion mutations, the map order of functions is as shown in Fig. 5. The *cobIII* and *cobII* mutations appear to lie close together to the right of the *cobI* mutations but not immediately adjacent to *cobI*. Transductional (P22) linkage between *cobI* and *cobIII* mutations or between *cobI* and *cobII* mutations is approximately 20%.

We have determined the direction of transcription of several *cob* genes by determining the orientation of Mu d-*lac* prophages that are cotranscribed with *cob* (31). All *cob* genes are transcribed counterclockwise or in the direction opposite to that of *his* operon transcription. The *cobI* genes appear to constitute a single operon. We conclude this from construction of double mutants of a series of *cobI*::Tn*10* and *cobI*::Mu d insertions. On the basis of polar effects of Tn*10* inserts on Mu d-*lac* expression, one can infer a linear order. Insertions of Tn*10* inferred to lie at the left (promoter-proximal) end of the operon exert polar effects on most *cobI*::Mu d(*lac*) fusions. An example of a polar Tn*10* insertion is *cobII*::Tn*10* in Table 4. Insertions of Tn*10* that are not polar on *cob*::Mu d insertions are inferred to lie further downstream in the operon. Results such as these permit construction of a map that accounts for the behavior of most Tn*10* and Mu d insertion mutations (R. Jeter, unpublished results). Several exceptional Tn*10* insertions appear to be nonpolar in the sense that they express downstream fusions constitutively. By analogy with insertions in the *his* operon, we believe that these Tn*10* insertions express downstream *cob* genes from the outward-directed promoter of Tn*10* (17). An example of such a nonpolar Tn*10* insertion is *cob-80*::Tn*10* in Table 4. The map order of inserts in this operon inferred by the criteria described above is confirmed by preliminary deletion mapping data. We do not know how many transcription units are involved in expression of *cobII* and *cobIII* genes, but at least one operon is involved in each pathway (R. Jeter, unpublished results).

REGULATION OF B_{12} SYNTHESIS

At least three controls seem to affect B_{12} production. One is repression by B_{12}, the second is repression by oxygen, and the third is stimulation by cyclic AMP (cAMP). These effects have been studied primarily with Mu d-*lac* fusions to the *cobI* operon. In such strains, β-galactosidase is produced owing to transcription from the *cobI* operon promoter. Thus, expression of *lac* should reflect expression and regulation of the *cobI* operon. Data on *cobI* regulation in response to oxygen and cAMP are presented in Table 5. All assays reflect levels found in the absence of B_{12}.

Several points regarding regulation of the *cobI* operon should be noted. (i) Transcription is stimulated by both anaerobiosis and exogenous cAMP. (ii) Maximum expression is seen for cells growing under conditions of anaerobic respiration. (iii) Maximal transcription of the operon is obtained in the presence of exogenous cAMP. This effect requires a functional

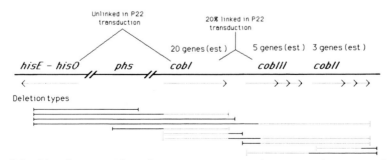

FIG. 5. Genetic map of the *his-cob* region. The *cobI* genes appear to make up a single operon. The *cobII* and *cobIII* genes appear to lie in different transcription units from each other and from *cobI*, but it is not clear how many operons might be involved in each pathway. Deletion types represent classes of deletions; the dotted portion of the horizontal line indicates that a series of deletions of this type have been isolated with endpoints distributed through the region indicated by the dotted portion of the line.

TABLE 4. Regulation of *lac* fusions to *cobI*, *cobII*, and *cobIII* operons by B_{12}

Relevant genotype	β-Galactosidase level $(U/A_{650})^a$ when medium contained:	
	No addition	B_{12}
cobI::*lac*	282	50
cobI::*lac cobI-11*::Tn*10*	1	1
cobI::*lac cobI-80*::Tn*10*	252	242
cobI::*lac cysG*::Tn*10*	177	67
cobII::*lac*	199	74
cobII::*lac cobI-11*::Tn*10*	47	46
cobII::*lac cobI-80*::Tn*10*	162	60
cobII::*lac cysG*::Tn*10*	144	73
cobIII::*lac*	107	54
cobIII::*lac cobI-11*::Tn*10*	42	40
cobIII::*lac cobI-80*::Tn*10*	134	54
cobIII::*lac cysG*::Tn*10*	101	74

^{a} All strains carry mutation *metE205* and were grown anaerobically on medium containing methionine with glycerol as carbon and energy source and fumarate as electron acceptor.

regulator protein (*crp*). (iv) Regulation in response to oxygen is independent of cAMP. This is seen for strain TT11297, which carries a *cya*::Tn*10* insertion and can therefore make no cAMP. This strain also carries a *crp* mutation that allows activation of transcription in the absence of cAMP (Crp*). This strain can express cAMP-dependent operons but makes no cAMP under any growth conditions. In this strain, which should be unable to modulate its cAMP effects, O_2 still shows the full range of regulation of *cobI* transcription. (v) Although exogenous cAMP stimulates substantial transcription of *cobI* under aerobic conditions, no B_{12} production occurs. This suggests the existence of additional mechanisms whereby oxygen prevents B_{12} synthesis.

Expression of the *cobI* operon is repressed approximately fivefold by exogenous B_{12} (Table 4). Expression of the *cobII* and *cobIII* operons appears to be entrained to expression of the *cobI* operon. Strains carrying a strongly polar block early in the *cobI* operon (*cob-11*::Tn*10*) still express the *cobII* and *cobIII* operons but lose regulation in response to B_{12}. Strains carrying a nonpolar Tn*10* insertion early in *cobI* are unaffected for regulation of *cobII* or *cobIII*.

TABLE 5. Regulation of *lac* fusions to the *cobI* operon

Relevant genotype	β-Galactosidase level (U/A_{650}) under the following conditions^a:					
	Glucose		Glucose + cAMP		Glycerol	
	$+O_2$	$-O_2$	$+O_2$	$-O_2$	$+O_2$	+Fumarate
cobI::*lac*	4	17	12	65	8	434
cobI::*lac cya*::Tn*10*	3	6	18	43	NG	NG
cobI::*lac crp*::Tn*10*	3	6	4	13	NG	NG
cobI::*lac crp** *cya*::Tn*10*	4	15	15	45	8	351

^{a} All strains carry mutation *metE205* and have been grown on minimal E medium including methionine in addition to the supplements indicated. NG, No growth.

The data are consistent, with a positive effector of the *cobII* and *cobIII* operons being specified in the *cobI* operon. We think that none of these effects is due to regulation by biosynthetic intermediates in the *cobI* pathway since *cysG* mutants (which fail to make the substrate for the *cobI* pathway) still show normal regulation of *cobII* and *cobIII* operons (Escalante-Semerena, unpublished results) (Table 4).

ACKNOWLEDGMENTS

Our research work is supported by Public Health Service grants GM-23408 and GM-34804 to J.R. and training grant T32 GM-07464-09 to D.R. from the National Institutes of Health and by Damon Runyon-Walter Winchell Cancer Fund postdoctoral fellowship DRG-811 to J.C.E.-S.

LITERATURE CITED

1. **Babior, B. M.** 1975. Cobamides as cofactors. Adenosylcobamide-dependent reactions, p. 141–212. *In* B. M. Babior (ed.), Cobalamin: biochemistry and pathophysiology. John Wiley & Sons, Inc., New York.
2. **Babior, B. M.** 1982. Ethanolamine ammonia-lyase, p. 263–288. *In* D. Dolphin (ed.), B_{12}, vol. 2. John Wiley & Sons, Inc., New York.
3. **Battersby, A. R., M. J. Bushell, C. Jones, N. G. Jones, and A. Pfenniger.** 1981. Biosynthesis of vitamin B_{12}: identity of fragment extruded during ring contraction to the corrin macrocycle. Proc. Natl. Acad. Sci. USA **78**:13–15.
4. **Battersby, A. R., C. J. R. Fookes, and R. J. Snow.** 1984. Synthetic studies relevant to biosynthetic research on vitamin B_{12}. Part 1. Synthesis of C-methylated chlorins based on 1-pyrrolines (3,4-dihydropyrroles). J. Chem. Soc. Perkin Trans. I **1984**:2725–2732.
5. **Battersby, A. R., C. J. R. Fookes, and R. J. Snow.** 1984. Synthetic studies relevant to biosynthetic research on vitamin B_{12}. Part 2. Synthesis of C-methylated chlorins via lactams. J. Chem. Soc. Perkin Trans. I **1984**:2733–2741.
6. **Battersby, A. R., and E. McDonald.** 1982. Biosynthesis of the corrin macrocycle, p. 107–144. *In* D. Dolphin (ed.), B_{12}, vol 1. John Wiley & Sons, Inc., New York.
7. **Battersby, A. R., E. McDonald, M. Thompson, and V. Y. Bykhovsky.** 1978. Biosynthesis of vitamin B_{12}: proof of A-B structure for sirohydrochlorin by its specific incorporation into cobyrinic acid. J. Chem. Soc. **1978**:150–151.
8. **Battersby, A. R., and L. A. Reiter.** 1984. Synthetic studies relevant to biosynthetic research on vitamin B_{12}. Part 3. An approach to isobacteriochlorins via nitrones. J. Chem. Soc. Perkin Trans. I **1984**:2743–2749.
9. **Beck, W. S.** 1982. Biological and medical aspects of vitamin B_{12}, p. 1–30. *In* D. Dolphin (ed.), B_{12}, vol. 1. John Wiley & Sons, Inc., New York.
10. **Blackwell, C. M., and J. M. Turner.** 1978. Microbial metabolism of amino alcohols: formation of coenzyme B_{12}-dependent ethanolamine ammonia-lyase and its concerted induction in *Escherichia coli*. Biochem. J. **176**:751–757.
11. **Boylan, S. A., and E. E. Dekker.** 1981. L-Threonine dehydrogenase. Purification and properties of the homogeneous enzyme from *Escherichia coli* K12. J. Biol. Chem. **256**:1809–1815.
12. **Bradbeer, C.** 1965. The clostridial fermentations of choline and ethanolamine. I. Preparation and properties of cell-free extracts. J. Biol. Chem. **240**:4669–4674.
13. **Bradbeer, C.** 1965. The clostridial fermentations of choline and ethanolamine. II. Requirement for a cobinamide coenzyme by an ethanolamine deaminase. J. Biol. Chem. **240**:4675–4681.
14. **Cauthen, S. E., M. A. Foster, and D. D. Woods.** 1966. Methionine synthesis by extracts of *Salmonella typhimurium*. Biochem. J. **98**:630–635.
15. **Chang, G. W., and J. T. Chang.** 1975. Evidence for the B_{12}-dependent enzyme ethanolamine deaminase in *Salmonella*. Nature (London) **254**:150–151.
16. **Childs, J. D., and D. A. Smith.** 1969. New methionine structural gene in *Salmonella typhimurium*. J. Bacteriol. **100**:377–382.
17. **Ciampi, M. S., M. B. Schmid, and J. R. Roth.** 1982. Transposon Tn*10* provides a promoter for transcription of adjacent sequences. Proc. Natl. Acad. Sci. USA **79**:5016–5020.
18. **Demain, A. L., and R. F. White.** 1971. Porphyrin overproduction by *Pseudomonas denitrificans*: essentiality of betaine and stimulation by ethionine. J. Bacteriol. **107**:456–460.
19. **Ford, J. E., E. S. Holdworth, and S. K. Kon.** 1955. The biosynthesis of vitamin B_{12}-like compounds. Biochem. J. **59**:86–93.

20. **Ford, S. H., and H. C. Friedmann.** 1976. Vitamin B_{12} biosynthesis: *in vitro* formation of cobinamide from cobyric acid and L-threonine. Arch. Biochem. Biophys. **175:**121–130.
21. **Foster, M. A., G. Tejerina, J. R. Guest, and D. D. Woods.** 1964. Two enzymic mechanisms of the methylation of homocysteine by extracts of *Escherichia coli.* Biochem. J. **92:**476–488.
22. **Friedmann, H. C.** 1975. Biosynthesis of corrinoids, p. 75–109. *In* B. M. Babior (ed.), Cobalamin: biochemistry and pathophysiology. Wiley & Sons, Inc., New York.
23. **Friedmann, H. C., and L. M. Cagen.** 1970. Microbial biosynthesis of B_{12}-like compounds. Annu. Rev. Microbiol. **24:**159–208.
24. **Fyfe, J. A., and H. C. Friedmann.** 1969. Vitamin B_{12} biosynthesis. J. Biol. Chem. **244:**1659–1666.
25. **Glusker, J. P.** 1982. X-ray crystallography of B_{12} and cobaloximes, p. 23–106. *In* D. Dolphin (ed.), B_{12}, vol. 1. John Wiley & Sons, Inc., New York.
26. **Halpern, J.** 1985. Mechanisms of coenzyme B_{12}-dependent rearrangements. Science **227:**869–875.
27. **Hodgkin, D. C.** 1979. New and old problems in the structure analysis of vitamin B_{12}, p. 19–36. *In* B. Zagalak and W. Friedrich (ed.), Vitamin B_{12}. de Gruyter, Berlin.
28. **Höllriegl, V., L. Lamm, J. Rowold, J. Hörig, and P. Renz.** 1982. Biosynthesis of vitamin B_{12}. Arch. Microbiol. **132:**155–158.
29. **Hörig, J. A., and P. Renz.** 1980. Biosynthesis of vitamin B_{12}. Eur. J. Biochem. **105:**587–592.
30. **Huennekens, F. M., K. S. Vitols, K. Fujii, and D. W. Jacobsen.** 1982. Biosynthesis of cobalamin coenzymes, p. 145–168. *In* D. Dolphin (ed.), B_{12}, vol. 1. John Wiley & Sons, New York.
31. **Hughes, K. T., and J. R. Roth.** 1984. Conditionally transposition-defective derivative of Mu d1(Amp Lac). J. Bacteriol. **159:**130–137.
32. **Jeter, R. M., B. M. Olivera, and J. R. Roth.** 1984. *Salmonella typhimurium* synthesizes cobalamin (vitamin B_{12}) de novo under anaerobic growth conditions. J. Bacteriol. **159:**206–213.
33. **Jones, P. W., and J. M. Turner.** 1984. A model for the common control of enzymes of ethanolamine catabolism in *Escherichia coli.* J. Gen. Microbiol. **130:**849–860.
34. **Kelley, J. J., and E. E. Dekker.** 1984. D-1-Amino-2-propanol:NAD^+ oxidoreductase. Purification and general properties of the large molecular weight form of the enzyme from *Escherichia coli* K12. J. Biol. Chem. **259:**2124–2129.
35. **Kelley, J. J., and E. E. Dekker.** 1985. Identity of *Escherichia coli* D-1-amino-2-propanol:NAD^+ oxidoreductase with *E. coli* glycerol dehydrogenase but not with *Neisseria gonorrhoeae* 1,2-propanediol:NAD^+ oxidoreductase. J. Bacteriol. **162:**170–175.
36. **Kung, H.-F., C. Spears, R. C. Greene, and H. Weissbach.** 1972. Regulation of the terminal reactions in methionine biosynthesis by vitamin B_{12} and methionine. Arch. Biochem. Biophys. **150:**23–31.
37. **Kusel, J. P., Y. H. Fa, and A. L. Demain.** 1984. Betaine stimulation of vitamin B_{12} biosynthesis of *Pseudomonas denitrificans* may be mediated by an increase in activity of δ-aminolevulinic acid synthase. J. Gen. Microbiol. **130:**835–841.
38. **Lamm, L., G. Heckmann, and P. Renz.** 1982. Biosynthesis of vitamin B_{12} in anaerobic bacteria. Eur. J. Biochem. **122:**569–571.
39. **Mombelli, L., C. Nussbaumer, H. Weber, G. Muller, and D. Arigoni.** 1981. Biosynthesis of vitamin B_{12}. Nature of the volatile fragment generated during formation of the corrin ring system. Proc. Natl. Acad. Sci. USA **78:**11–12.
40. **Montforts, F.-P., S. Ofner, V. Rasetti, A. Eschenmoser, W.-D. Woggon, K. Jones, and A. R. Battersby.** 1979. A synthetic approach to the isobacteriochlorin macrocycle. Angew. Chem. Int. Ed. Engl. **18:**675–677.
41. **Mulligan, J. T., W. Margolin, J. H. Krueger, and G. C. Walker.** 1982. Mutations affecting regulation of methionine biosynthetic genes isolated by use of *met-lac* fusions. J. Bacteriol. **151:**609–619.
42. **Nussbaumer, C., M. Infeld, G. Worner, G. Müller, and D. Arigoni.** 1981. Biosynthesis of vitamin B_{12}: mode of incorporation of factor III into cobyrinic acid. Proc. Natl. Acad. Sci. USA **78:**9–10.
43. **Postow, J. M., and T. C. Stadtman.** 1975. Cobamides as cofactors. Methylcobamides and the synthesis of methionine, methane, and acetate, p. 111–140. *In* B. M. Babior (ed.), Cobalamin: biochemistry and pathophysiology. John Wiley & Sons, Inc., New York.
44. **Rasetti, V., A. Pfaltz, C. Kratky, and A. Eschenmosser.** 1981. Ring contraction of hydroporphinoid to corrinoid complexes. Proc. Natl. Acad. Sci. USA **78:**16–19.
45. **Rowbury, R. J.** 1983. Methionine biosynthesis and its regulation, p. 191–211. *In* K. M. Hermann and R. L. Somerville (ed.), Amino acids: biosyntheses and genetic regulation. Addison-Wesley, Reading, Mass.
46. **Saint-Girons, I., N. Duchange, G. N. Cohen, and M. M. Zakin.** 1984. Structure and autoregulation of the *metJ* regulatory gene in *Escherichia coli.* J. Biol. Chem. **259:**14282–14285.
47. **Scarlett, F. A., and J. M. Turner.** 1976. Microbial metabolism of amino alcohols. Ethanolamine catabolism via coenzyme B_{12}-dependent ethanolamine ammonia-lyase in *Escherichia coli* and *Klebsiella aerogenes.* J. Gen. Microbiol. **95:**173–176.
48. **Scott, A. I.** 1981. Application of CMR spectroscopy to the study of porphyrin and corrin biosynthesis *in vitro* and *in vivo.* Pure Appl. Chem. **53:**1215–1232.
49. **Scott, A. I., A. J. Irwin, L. M. Siegel, and J. N. Schoolery.** 1978. Sirohydrochlorin. Prosthetic group of a sulfite reductase enzyme and its role in the biosynthesis of vitamin B_{12}. J. Am. Chem. Soc. **100:**316–318.
50. **Scott, A. I., N. E. MacKenzie, P. J. Santander, and P. E. Fagerness.** 1984. Biosynthesis of vitamin B_{12}: timing of the methylation steps between Uro'gen III and cobyrinic acid. Bioorg. Chem. **12:**356–362.
51. **Sennett, K. E., L. E. Rosenberg, and I. S. Mellman.** 1981. Transmembrane transport of cobalamin in prokaryotic and eukaryotic cells. Annu. Rev. Biochem. **50:**1053–1086.
52. **Smith, M. W., and F. C. Neidhart.** 1983. Proteins induced by anaerobiosis in *Escherichia coli.* J. Bacteriol. **154:**344–350.
53. **Stevens, R. V.** 1982. The total synthesis of vitamin B_{12}, p. 169–200. *In* D. Dolphin (ed.), B_{12}, vol. 1. John Wiley & Sons, Inc., New York.
54. **Taylor, R. T.** 1982. B_{12}-dependent methionine biosynthesis, p. 307–355. *In* D. Dolphin (ed.), B_{12}, vol. 2. John Wiley & Sons, Inc., New York.
55. **Taylor, R. T., and M. L. Hanna.** 1970. *Escherichia coli* B 5-methyltetrahydrofolate-homocysteine cobalamin methyltransferase: catalysis by a reconstituted methyl-^{14}C-cobalamin holoenzyme and the function of S-adenosyl-L-methionine. Arch. Biochem. Biophys. **137:**453–459.
56. **Taylor, R. T., and H. Weissbach.** 1973. N^5-methyltetrahydrofolate homocysteine methyltransferases, p. 121–165. *In* P. D. Boyer (ed.), The enzymes, vol. 9. Academic Press, Inc., New York.
57. **Whitfield, C. D., E. J. Steers, Jr., and H. Weissbach.** 1970. Purification and properties of 5-methyltetrahydropteroyltriglutamate-homocysteine methyltransferase. J. Biol. Chem. **245:**390–401.
58. **Woodward, R. B.** 1979. Synthetic vitamin B_{12}, p. 37–87. *In* B. Zagalak and W. Friedrich (ed.), Vitamin B_{12}. de Gruyter, Berlin.

38. NAD Biosynthesis and Recycling

GERALD J. TRITZ

Department of Microbiology and Immunology, Kirksville College of Osteopathic Medicine, Kirksville, Missouri 63501

INTRODUCTION

Nicotinamide adenine dinucleotide (NAD) and its phosphorylated derivative (NADP) are ubiquitous in nature, where they participate in cellular oxidation/reduction reactions. In addition, NAD can be degraded at either of two high-energy bonds to produce the energy to drive DNA repair and recombination reactions and to ribosylate proteins. Similarities and differences in NAD metabolism and its regulation exist between procaryotes and eucaryotes; even among the procaryotes, there is a large degree of variation. However, there is a high degree of similarity between NAD metabolism in *Escherichia coli* and that in *Salmonella typhimurium*, and these can and will be discussed as one, with any major differences noted. This is intended as a comprehensive review of pyridine nucleotide metabolism in *E. coli* and *S. typhimurium* only. An excellent review of this subject in both procaryotes and eucaryotes has been published by White (45). Foster and Moat (11) have presented a detailed discussion of pyridine nucleotide metabolism in procaryotes in general.

DE NOVO NAD BIOSYNTHESIS

The majority of research on the de novo biosynthesis of NAD in procaryotes has been accomplished with the *E. coli* system. Beginning in 1960 with the report by Ortega and Brown (32) that glycerol and a dicarboxylic acid are the precursors of the nicotinamide moiety of NAD, the pathway has been defined to generally accepted metabolic sequences (Fig. 1). The basis for the early portion of this biosynthetic sequence is the observation by Steiner et al. (39) that aspartate is modified before its condensation with dihydroxacetone phosphate, along with the reports from Gholson's laboratory (28–30) that this modified product is the result of the action of L-aspartate oxidase on aspartate. L-Aspartate oxidase is linked to flavin adenine dinucleotide, the reduced form of which donates electrons to oxygen. Although this pathway has been termed an anaerobic pathway (11) based upon the observation that a mutation in the gene coding for L-aspartate oxidase is expressed under both aerobic and anaerobic growth conditions, in vitro studies demonstrate that oxygen is the obligate electron acceptor (S. Nasu, F. Wicks, and R. K. Gholson, Fed. Proc. **38**:644, 1979).

The in vitro enzymatic synthesis of quinolinate via iminoaspartate is optimal at pH 8.0 and a temperature of 37°C, and its rate is proportional to L-aspartate oxidase concentration as long as dihydroxyacetone phosphate and quinolinate synthetase concentrations are not limiting. If either or both are limiting, the synthesis of oxaloacetate via iminoaspartate is proportional to the square root of the enzyme concentration, i.e., it is a parabolic function of L-aspartate oxidase activity. Iminoaspartate is an unstable intermediate with a half life of 140 s; it spontaneously decays to oxaloacetate and ammonia in the absence of cellular ability to convert it to the next intermediate in the pathway, quinolinate. At high concentrations (2.5 mM) of oxaloacetate and ammonia (as ammonium sulfate), a level probably never achieved intracellularly, the reaction is reversible (28). However, oxaloacetate can also be recycled to aspartate via aspartate aminotransferase as well as indirectly via the formation of fumarate in the citric acid cycle and the action of aspartate deaminase on fumarate.

L-Aspartate oxidase is specified by the *nadB* gene, which lies at min 55 on the *E. coli* chromosome and min 58 on the *S. typhimurium* chromosome. The enzyme is under the repressive control of the *nadR* gene (43) with NAD as the probable corepressor molecule. NAD and NADH, but not NADP or NADPH, inhibit L-aspartate oxidase activity. Thus the *nadB* gene is the primary site of regulation of the de novo NAD biosynthetic pathway. The L-aspartate oxidase activity also shows a marked substrate activation at substrate concentrations above 1.25 mM; the K_m for aspartate is 0.63 mM at substrate concentrations between 0.25 and 1.25 mM but is 3.33 mM when determined at substrate concentrations of 2.0 to 10.0 mM. A model consistent with this enzyme behavior is that the inactive L-aspartate oxidase is a dimer that must separate into monomers to become active (29):

$$B \cdot B \rightleftharpoons 2B$$
$$\text{inactive} \quad \text{active}$$

557

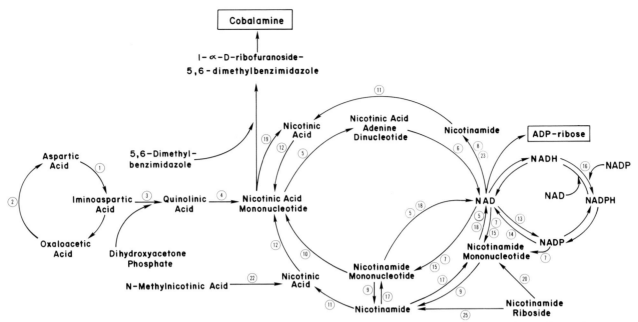

FIG. 1. Pathways for the biosynthesis and cycling of NAD and related compounds. The circled numbers refer to the reaction numbers listed in Table 1.

It has been postulated that the nonlinear behavior of the L-aspartate oxidase activity serves a regulatory function, since it greatly reduces the conversion of L-aspartate to oxaloacetate under conditions in which quinolinate is not being synthesized.

Iminoaspartate is converted to quinolinate by quinolinate synthetase. This enzyme has not been purified owing to its labile nature. Therefore it is not known whether quinolinate synthetase represents a single enzyme or multiple enzymes. The complexity of the conversion, which involves two condensation reactions, the removal of two molecules of water and the removal of P_i, argues for multiple enzymatic reactions (3, 4). However, genetic evidence supports the concept of a single enzymatic activity; only two genes, *nadA* and *nadB*, have been identified as specifying enzymes involved in the conversion of aspartate to quinolinate (42, 44). The *nadA* gene is also under the repressive control of *nadR* (43); its product functions optimally at pH 7.0. Evidence for quinolinate being on the NAD biosynthetic pathway was offered by Andreoli et al. (1).

THE PYRIDINE NUCLEOTIDE CYCLES

Gholson (12) first proposed the existence of a pyridine nucleotide cycle in a diversity of species. Since that time, studies with both *E. coli* and *S. typhimurium* have demonstrated the existence of multiple forms of this cycle (Fig. 1). Quinolinate is introduced into these cycles by conversion to nicotinic acid mononucleotide (NaMN) via quinolinate phosphoribosyltransferase. There is a single route for the conversion of NaMN to NAD, via nicotinic acid adenine dinucleotide (NaADN), part of the Preiss-Handler pathway (33, 34). Once NAD is formed it can be recycled, theoretically, through any of five pyridine nucleotide cycles, three of which are known to be

functional in *E. coli* and *S. typhimurium*. A unique feature of these cycles is the highly conserved nature of the pyridine ring. This is evident in the two-, three-, four-, five-, and six-member cycles commonly termed PNC II, PNC III, PNC IV, PNC V, and PNC VI (see Fig. 1). This pyridine ring cycling is initiated by the enzymatic degradation of NAD by NAD pyrophosphatase, DNA ligase, NAD:protein(ADP ribose) ADPribosyltransferase, NAD glycohydrolase, or a combination of these.

In the two-member cycle, PNC II, there is interconversion of NAD and NMN which is initiated by the enzymatic hydrolysis of the pyrophosphate bond of NAD. This can be accomplished by either NAD pyrophosphatase or DNA ligase. In each case the end result is the formation of two nucleotides, NMN and AMP, and the release of energy. In the DNA ligase reaction this energy is coupled to the sealing of the break in DNA via formation of a phosphodiester bond. There is a direct correlation then between DNA ligase activity and production of NMN from NAD; irradiation of cells to induce DNA damage results in increased ligase activity and NMN production in both *E. coli* and *S. typhimurium*. The presence of NAD pyrophosphatase activity in *S. typhimurium* was indicated in several studies, but the definitive proof of the existence of such an enzyme was shown by Foster (7), who demonstrated the activity in crude extracts and then utilized molecular sieve chromatography to obtain an apparent molecular weight of 120,000 for the enzyme. The partially purified enzyme has a specific activity of 709 nmol of product formed per h per mg of protein, a much higher activity than other pyridine nucleotide cycle enzymes. DNA ligase activity is stimulated by NAD (36) and inhibited by NMN (37). It is pyrophosphate bond cleavage that initiates most NAD turnover in *E. coli* (15) and probably in *S. typhimurium*. However, the relative contribution of each of

TABLE 1. Enzymes involved in NAD biosynthesis and cycling and pyridine nucleotide transport[a]

Reaction no.[b]	Enzyme nomenclature		Reaction	Alternative name	Genetic locus	Map position (min)	
	Name	EC no.[c]				S. typhimurium	E. coli
1	L-Aspartate oxidase	1.4.3.00	L-ASP + FAD → iminoaspartate + FADH$_2$	L-Aspartate → O$_2$ transhydrogenase	nadB	58	55
2	Aspartate aminotransferase	2.6.1.1	GLU + OXAL → ASP + α-ketoglutarate	Glutamate-oxaloacetate transaminase	aspC		20
3	Quinolinate synthetase		Iminoaspartate + DHAP $\xrightarrow{Mg^{2+}}$ QA + H$_3$PO$_4$ + 2H$_2$O		nadA	17	16
4	Quinolinate phosphoribosyl transferase	2.4.2.19	QA + PRPP $\xrightarrow{Mg^{2+}}$ NaMN + PP$_i$ + CO$_2$		nadC	3	2
5	NaMN adenylyltransferase	2.7.7.18	NaMN + ATP $\xrightarrow{Mg^{2+}}$ NaADN + PP$_i$ NMN + ATP $\xrightarrow{Mg^{2+}}$ NAD + PP$_i$	deNAD pyrophosphorylase	nadD	14	
6	Deamido-NAD$^+$: ammonia ligase	6.3.5.1	NaADN + ATP + NH$_3$ $\xrightarrow{Mg^{2+}}$ NAD + PP$_i$ + AMP	NAD synthetase	nadE		
7	NAD (NADP) pyrophosphatase	3.6.1.22	NAD + H$_2$O $\xrightarrow{Mg^{2+}}$ AMP + NMN NADP + H$_2$O $\xrightarrow{Mg^{2+}}$ 2',5'-AMP + NMN				
8	NAD$^+$ glycohydrolase	3.2.2.5	NAD + H$_2$O $\xrightarrow{Mg^{2+}}$ NAm + ADPribose	NAD$^+$ nucleosidase			
9	NMN glycohydrolase	3.2.2.00	NMN + H$_2$O → NAm + ribose 5-phosphate; transport of NMN				
10	NMN amidohydrolase	3.5.1.00	NMN + H$_2$O → NaMN + NH$_3$	NMN deamidase	pncC		
11	NAm amidohydrolase	3.5.1.19	NAm + H$_2$O → NA + NH$_3$; transport of nicotinamide	NAm deamidase	pncA	27	39
12	Nicotinate phosphoribosyltransferase	2.4.2.11	NA + PRPP + ATP $\xrightarrow{Mg^{2+}}$ NaMN + PP$_i$ + ADP + P$_i$; transport of NA	NAMN pyrophosphorylase	pncB	20	
13	NAD$^+$ kinase	2.7.1.23	NAD + ATP $\xrightarrow{Mg^{2+}}$ NADP + ADP				
14	NADP phosphatase	3.1.2.00	NADP + H$_2$O $\xrightarrow{Mg^{2+}}$ NAD + H$_3$PO$_4$				
15	DNA ligase	6.5.1.2	NAD + nicked DNA $\xrightarrow{Mg^{2+}}$ NMN + AMP + DNA		lig		51
16	NADP transhydrogenase	1.6.1.1	NADH + NADP ⇌ NADPH + NAD	Pyridine nucleotide transhydrogenase	pnt		35
17	Nicotinamide phosphoribosyltransferase	2.4.2.12	NAm + PRPP + ATP → NMN + PP$_i$ + ADP + P$_i$	NMN pyrophosphorylase			
18	NMN adenylyltransferase	2.7.7.1	NMN + PRPP + ATP → NAD + PP$_i$ + ADP + P$_i$	NAD pyrophosphorylase			
19	Trans-N-glycosidase		5,6-Dimethylbenzimidazole + NaMN → 1-α-D-ribofuranoside-5,6-dimethylbenzimidazole + Na + H$^+$		cob		41
20	Ribosylnicotinamide kinase	2.7.1.22	NAm-ribose + ATP → NMN + ADP	Nicotinamide ribonucleoside kinase			
21	NMN uptake		Slow NMN transport into the cell		pnuA		
22	N-Methylnicotinic acid demethylase		N-Methylnicotinic acid → NA	Trigonelline demethylase			
23	NAD:protein(ADPribose) ADPribosyltransferase		NAD + H$_2$O + protein → NAm + ribosylated protein	Poly(ADPribose) synthetase			
24	NMN uptake		Rapid NMN transport into the cell		pnuB		
25	Nicotinamide ribonucleoside glycohydrolase		NAm-ribose + H$_2$O → NAm + ribose				

[a] Abbreviations: L-ASP, L-aspartate; FAD, flavin adenine dinucleotide; GLU, glutamate; OXAL, oxaloacetate; DHAP, dihydroxyacetone phosphate; QA, quinolinate; PRPP, phosphoribosyl pyrophosphate; NaMN, nicotinic acid mononucleotide; NaADN, nicotinic acid adenine dinucleotide; NMN, nicotinamide mononucleotide; NAm, nicotinamide; ADPribose, adenosine diphosphate ribose; NA, nicotinic acid; NAm-ribose, nicotinamide ribonucleoside.

[b] Reaction numbers correspond to those in Fig. 1.

[c] Enzyme classification numbers are taken from reference 19.

these enzymes to pyrophosphate bond cleavage in vivo remains to be determined. To complete a PNC II, NMN must be directly converted back to NAD. This specific enzymatic activity (NMN adenyltransferase) occurs in at least one species of bacteria (21) but has not yet been demonstrated in either S. typhimurium or E. coli, except as a function of NaMN adenyltransferase in E. coli. However, NaMN adenyltransferase has a much greater affinity for NaMN than for NMN (6); the rate of NaADN synthesis from NaMN is 17 times faster than the synthesis of NAD from NMN. Available evidence indicates that, if this reaction does occur in vivo, it is quantitatively a very minor pathway. The *nadD* gene, which specifies NaMN adenyltransferase, lies at min 14 on the *S. typhimurium* chromosome. All *nadD* mutants belong to a single complementation group (17).

PNC III is another pathway that requires the direct conversion of NMN to NAD; in this case the pathway is NAD → nicotinamide → NMN → NAD. The release of nicotinamide from NAD occurs through the action of both NAD glycohydrolase and NAD:protein(ADPribose) ADPribosyltransferase. In the former reaction, ADP-ribose is released free, whereas in the latter reaction the ADP-ribose is conjugated to one or more protein species. Hydrolysis of the β-*N*-glycoside linkage of NAD results in the release of energy in the range of −8.2 kcal/mol (47), and this is the driving force for the ADP-ribosylation reactions. Although NAD glycohydrolase activity has been searched for in *E. coli*, it has not yet been found (2) except as an envelope-bound enzyme not having access to the cytoplasm (15). However, Foster (7) was able to detect two distinct NAD glycohydrolase activities in a partially purified extract of *S. typhimurium* soluble fraction. The apparent molecular weights are between 53,000 and 58,000. NAD:protein(ADPribose) ADPribosyltransferase activity has been reported in *E. coli* (38) but not in *S. typhimurium*. This activity is not the result of bacteriophage infection, as has been reported for cells infected with the T4 or N4 viruses, but is an inherent property of the bacterial cell. Thus, even though neither *E. coli* nor *S. typhimurium* has been reported to contain both intracellular NAD glycohydrolase activity and NAD:protein(ADPribose) ADPribosyltransferase activity, they each have the ability to generate nicotinamide from NAD and complete the first step of PNC III.

The second step in this cycle, the conversion of nicotinamide to NMN, occurs via nicotinamide phosphoribosyltransferase action in at least one species of procaryote (21) but has not been reported in either *E. coli* or *S. typhimurium*. Available evidence indicates that if this pathway does exist in these enteric organisms it is a very minor pathway. The observation that double mutants of either organism, with defective L-aspartic oxidase and nicotinamide amidohydrolase, cannot proliferate in a minimal medium containing nicotinamide supports the hypothesis that this pathway does not occur in these species. However, recent studies indicate that nicotinamide amidohydrolase in *E. coli* is located in the periplasmic space and is responsible for the binding, transport, and conversion to nicotinic acid of exogenous nicotinamide (J. J. Rowe, R. D. Lemmon, and G. J. Tritz, manuscript in preparation). These mutants do not accumulate intracellular nicotinamide and thus could not carry out the conversion of nicotinamide to NMN even if the appropriate enzymes were present. Still, in the absence of definitive evidence of the presence of nicotinamide phosphoribosyltransferase and NMN adenyltransferase, the PNC III must be considered lacking in *E. coli* and *S. typhimurium*.

The PNC IV is thought to be the major pathway for the recycling of NAD in both *E. coli* and *S. typhimurium*. This is based on an analysis of ^{32}P labeling patterns in vivo in *E. coli* and the isolation from *E. coli* and *S. typhimurium* of NMN amidohydrolase, which catalyzes the conversion of NMN to NaMN (15, 22).

This enzyme from *E. coli* has a molecular weight of 33,000, a pH optimum of 9.0, linear kinetics, a K_m of 1.35×10^{-5} M, and no deamidation activity toward NAD, NADP, or nicotinamide. The *S. typhimurium* enzyme has a pH optimum of 8.7, exhibits sigmoid kinetics suggestive of allosterism, and is present in the cell at levels 20 to 30 times that of NMN glycohydrolase. In *E. coli* the rate of NAD turnover via PNC IV is approximately eight times more rapid than turnover via the next most rapid cycle, PNC VI. In *S. typhimurium* the PNC IV is twice as effective as PNC VI in recycling NAD.

While this cycling of NAD → NMN → NaMN → NaADN → NAD is initiated by the conversion of NAD into NMN, the relative contribution of DNA ligase and NAD pyrophosphatase to the production of NMN is in doubt. DNA ligase is found in both *E. coli* and *S. typhimurium*. A report by Manlapaz-Fernandez and Olivera (26) indicates that there is a reduced turnover of NAD in a temperature-sensitive DNA ligase mutant of *E. coli*, suggesting a major role for DNA ligase activity in NAD breakdown. Studies by Foster (7) indicate a very high NAD pyrophosphatase activity in *S. typhimurium*; this same enzyme is also found in *E. coli* (6). It may be that these two enzymes each play a prominent role in NAD breakdown at different stages of growth of the cell. In any case, NMN is the major breakdown product of NAD in these two species. Double-label pulse-chase experiments in *E. coli* have established that NMN is deamidated to form NaMN (26). A similar conclusion was reached by Foster and Baskowsky-Foster (8) when they studied the fate of the [^{14}C]nicotinamide moiety of NAD in *S. typhimurium*. Once NaMN is formed it is converted to NAD via NaADN through the Preiss-Handler pathway. The enzyme responsible for this final step in the synthesis of NAD, deamido-NAD$^+$:ammonia ligase, preferentially utilizes ammonia rather than glutamine as the amino donor. This is evidenced by the fact that the K_m for glutamine is 250-fold higher than that for ammonia. The K_m for NaADN is 1.1×10^{-4} M (46).

In PNC V the initiation of cycling is similar to that in PNC III, where the action of either NAD glycohydrolase or NAD:protein(ADPribose) ADPribosyltransferase, or both, is required. This cycle (NAD → nicotinamide → nicotinate → NaMN → NaADN → NAD), in addition to utilizing enzymes already described, also uses nicotinamide amidohydrolase and nicotinic acid phosphoribosyltransferase. Nicotinamide amidohydrolase is specified by the *pncA* gene, which lies at position 16 on the *E. coli* chromosome and at min 17 on the *S. typhimurium* chromosome (9). It catalyzes the deamination of nicotinamide to form nicotinate (40). Nicotinic acid phosphoribosyltransferase is the first enzyme of the Preiss-Handler pathway and is coded for by the *pncB* gene, which has been mapped at position 20 on the *S. typhimurium* chromosome but is unmapped in *E. coli*. This enzyme is repressible in both *E. coli* (18) and *S. typhimurium* (10, 18), but this control is more stringent in *E. coli*. NAD is probably the corepressor molecule. There is no evidence of feedback inhibition of this enzyme in either organism. The other enzymes of PNC V and PNC VI that have been studied exhibit no significant repression. This has led to the hypothesis that nicotinic acid phosphoribosyltransferase represents the major point of control of flux of the pyridine ring in either pyridine nucleotide cycle (PNC V and PNC VI) where nicotinic acid is an intermediate.

PNC VI (NAD → NMN → nicotinamide → nicotinate → NaMN → NaADN → NAD) is similar to PNC II and PNC IV in that initiation of the cycle requires the formation of NMN. The difference between these three pathways lies in the way the NMN is recycled to NAD. PNC VI requires NMN glycohydrolase to release nicotinamide from NMN. This enzyme is found as both a membrane-bound and a cytoplasmic entity (2). The *S. typhimurium* enzyme has a molecular weight of 67,000 and a pH optimum of 8.0. It is inhibited by NAD, NADH, GTP, GMP, AMP, and ADP-ribose, which suggests a competitive inhibition at the catalytic site. Enzyme synthesis is not repressed by growth in nicotinic acid (7). Thus PNC VI is regulated at two sites, nicotinate phosphoribosyltransferase via repression and NMN glycohydrolase via feedback inhibition. The enzymes necessary for PNC VI are all present in both *E. coli* and *S. typhimurium*, and this pathway is, quantitatively, the second most important pyridine nucleotide cycle in these organisms.

Although it is not considered in the context of a pyridine nucleotide cycle, the alternation of NAD and NADP and their reduced forms does represent a kind of cycle in which the entire NAD molecule is circulated. NAD is reduced in any one of hundreds of oxidation-reduction reactions where it functions as a cofactor. NAD can also be phosphorylated via the action of NAD kinase to form NADP. The reversal of this activity is catalyzed by NADP phosphatase. Reducing equivalents are exchanged between NAD and NADP via pyridine nucleotide transhydrogenase, a membrane-bound and energy-linked enzyme composed of two polypeptides of molecular weight 50,000 and 47,000. The *pnt* gene near min 35 on the *E. coli* chromosome codes for this enzyme (5). This enzyme is repressed when the cells are grown in high concentrations of amino acids (16) and is induced by ammonium when the cells are grown in a minimal medium (24). The NAD$^+$ reduction by NADPH is more rapid than the NADP reduction by NADH by a factor of 5 to 10 (23).

FUNCTIONS OF THE PYRIDINE NUCLEOTIDE CYCLE

Foster et al. (10) have suggested that the pyridine nucleotide cycle represents a mechanism for the regulation of levels of NAD. While this is probably true and may represent the major function, at least two other needs are served by the cycle. These cycles represent a mechanism by which certain pyridines not in the cycle may be introduced into that cycle and utilized as a source of NAD. They also represent a mechanism of supplying pyridine nucleotide cycle intermediates for use in other pathways.

The scavenging of pyridine compounds for introduction into the pyridine nucleotide cycle occurs at two points. One of these is the demethylation of *N*-methylnicotinic acid, a major product of nicotinic acid metabolism in humans, into nicotinic acid. This allows enteric organisms to utilize any *N*-methylnicotinic acid excreted into the intestine. *N*-Methylnicotinic acid demethylase activity has recently been reported in *E. coli* (41). In addition, Liu et al. (25) have shown that *S. typhimurium* has the ability to convert

nicotinamide ribonucleoside directly into NMN, without conversion to the free base. This nicotinamide ribonucleoside kinase activity allows the cell to salvage the ribonucleoside via two mechanisms, the second being the hydrolysis of the riboside to yield nicotinamide and ribose.

The pyridine nucleotide cycle also serves as a source of precursors for two known biosynthetic schemes. *S. typhimurium* can synthesize vitamin B$_{12}$ (cobalamin) de novo only under anaerobic conditions (20). In this synthesis, NaMN serves as the specific ribosyl donor in a *trans-N*-glycosidase reaction. Without a cycling of NAD to continually replenish the supply of NaMN, this reaction could not occur. In a similar way, NAD acts as a source of ADP-ribose for the mono (ADP-ribos)ylation of at least two proteins in *E. coli* (38). This latter reaction results in the release of nicotinamide, which can then be recycled.

TRANSPORT OF PYRIDINES

Although early studies indicated that the uptake of nicotinic acid by *E. coli* is via a non-energy-requiring diffusion process (27, 31), more recent work has shown that the uptake of nicotinic acid is dependent on the presence of nicotinate phosphoribosyltransferase and a source of energy. Under optimal conditions, this uptake exhibits saturation kinetics with a K_m of 1.75 μM of nicotinic acid and a V_{max} of 0.116 nmol/min per mg of dry weight. Inhibitors of glycolysis, uncouplers of ATP production, and sodium arsenate reduce vitamin transport (35). The active transport of nicotinamide in *E. coli* was indicated by the isolation of a nicotinamide-binding protein from osmotically shocked cells (14). Further characterization of this uptake system indicates that the binding protein is nicotinamide amidohydrolase and that transport follows kinetics similar to that for nicotinic transport. Furthermore, uptake is reduced by inhibitors of glycolysis and uncouplers of ATP production (Rowe et al., manuscript in preparation).

Gholson et al. (13) demonstrated that *E. coli* is incapable of transporting NAD into the cell but rather produces an extracellular enzyme(s) which degrades NAD to products that cross the cell envelope. It was hypothesized that this compound was nicotinamide. In confirmation of this hypothesis, Hillyard et al. (15) demonstrated an NAD glycohydrolase in *E. coli* which does not have access to the intracellular pool of NAD but does have the ability to degrade extracellular NAD. The only phosphorylated pyridine compound known to cross the cell envelope in *S. typhimurium* is NMN. This transport is mediated by the product of the *pnuA* gene at a relatively slow rate and by the product of the *pnuB* gene at a more rapid rate (25). NMN is also transported by the membrane-associated NMN glycohydrolase which, in the process of transport, converts NMN to nicotinamide. Liu et al. have also shown that nicotinamide riboside crosses the cell envelope, but it is not known whether this is via diffusion or enzyme mediation.

FUTURE DIRECTIONS

Several areas of study need to be pursued to resolve anomalies and completely define the NAD biosynthetic and cycling pathways in both *E. coli* and *S. typhimurium*. In the de novo pathway, the differences between the aerobic and anaerobic biosynthesis of quinolinic acid need elucidation, as do the nature and mechanism of action of quinolinate synthetase. Also, the pH optimum for L-aspartate oxidase of 7.0 is not consistent with a pH optimum of 8.0 for the next enzyme in the pathway, quinolinate synthetase.

The cycling of NAD is complex. Although PNC IV and PNC VI are the predominant pathways, the contribution of PNC V is not known. If a mechanism to convert NMN directly to NAD could be shown, this would validate PNC II and PNC III. There is a need to purify each pyridine nucleotide cycle enzyme from each organism and compare kinetic, regulatory, and physical properties. The compartmentalization of these enzymes has been suggested, but further clarification would more completely elucidate this feature of pyridine metabolism. The genes specifying these enzymes, in many cases, need to be located on the respective chromosomes, and any genetic regulatory mechanisms need to be defined.

Although there is much work remaining, past research has defined the framework of NAD biosynthesis and recycling. The details will yield to future efforts.

LITERATURE CITED

1. **Andreoli, A. J., T. W. Ikeda, T. Nishizuka, and O. Hayaishi.** 1963. Quinolinic acid: a precursor to nicotinamide adenine dinucleotide in *Escherichia coli*. Biochem. Biophys. Res. Commun. **12**:92–97.
2. **Andreoli, A. J., T. W. Okita, R. Bloom, and T. A. Grover.** 1972. The pyridine nucleotide cycle: presence of a nicotinamide mononucleotide specific glycohydrolase in *Escherichia coli*. Biochem. Biophys. Res. Commun. **49**:264–269.
3. **Chen, J.-L., and G. J. Tritz.** 1975. Isolation of a metabolite capable of differentially supporting the growth of nicotinamide adenine dinucleotide auxotrophs of *Escherichia coli*. J. Bacteriol. **121**:212–218.
4. **Chen, J., and G. J. Tritz.** 1976. Detection of precursors of quinolinic acid in *Escherichia coli*. Microbios **16**:207–218.
5. **Clarke, D. M., and P. D. Bragg.** 1985. Cloning and expression of the transhydrogenase gene of *Escherichia coli*. J. Bacteriol. **162**:367–373.
6. **Dahmen, W., B. Webb, and J. Preiss.** 1967. The deamidodiphosphopyridine nucleotide and diphosphopyridine nucleotide pyrophosphorylases of *Escherichia coli* and yeast. Arch. Biochem. Biophys. **120**:440–450.
7. **Foster, J. W.** 1981. Pyridine nucleotide cycle of *Salmonella typhimurium*: in vitro demonstration of nicotinamide adenine dinucleotide glycohydrolase, nicotinamide mononucleotide glycohydrolase, and nicotinamide adenine dinucleotide pyrophosphatase activities. J. Bacteriol. **145**:1002–1009.
8. **Foster, J. W., and A. M. Baskowsky-Foster.** 1980. Pyridine nucleotide cycle of *Salmonella typhimurium*: in vivo recycling of nicotinamide adenine dinucleotide. J. Bacteriol. **142**:1032–1035.
9. **Foster, J. W., D. M. Kinney, and A. G. Moat.** 1979. Pyridine nucleotide cycle of *Salmonella typhimurium*: isolation and characterization of *pncA*, *pncB*, and *pncC* mutants and utilization of exogenous nicotinamide adenine dinucleotide. J. Bacteriol. **137**:1165–1175.
10. **Foster, J. W., D. M. Kinney, and A. G. Moat.** 1979. Pyridine nucleotide cycle of *Salmonella typhimurium*: regulation of nicotinic acid phosphoribosyl transferase and nicotinamide deamidase. J. Bacteriol. **138**:957–961.
11. **Foster, J. W., and A. G. Moat.** 1980. Nicotinamide adenine dinucleotide biosynthesis and pyridine nucleotide cycle metabolism in microbial systems. Microbiol. Rev. **44**:83–105.
12. **Gholson, R. K.** 1966. The pyridine nucleotide cycle. Nature (London) **212**:933–935.
13. **Gholson, R. K., G. J. Tritz, T. S. Matney, and A. J. Andreoli.** 1969. Mode of nicotinamide adenine dinucleotide utilization by *Escherichia coli*. J. Bacteriol. **99**:895–896.
14. **Griffith, T. W., and F. R. Leach.** 1973. The effect of osmotic shock on vitamin transport in *Escherichia coli*. Arch. Biochem. Biophys. **159**:658–663.
15. **Hillyard, D., M. Rechsteiner, P. Manlapaz-Ramos, J. S. Imperial, L. J. Cruz, and B. M. Olivera.** 1981. The pyridine nucleotide cycle. Studies in *Escherichia coli* and the human cell line D98/AH2. J. Biol. Chem. **256**:8491–8497.
16. **Houghton, R. L., R. J. Fisher, and D. R. Sanadi.** 1976. Control of NAD(P)$^+$ transhydrogenase levels in *Escherichia coli*. Arch. Biochim. Biophys. **176**:747–752.
17. **Hughes, K. T., D. Ladika, J. R. Roth, and B. M. Olivera.** 1983. An indispensable gene for NAD biosynthesis in *Salmonella typhimurium*. J. Bacteriol. **155**:213–221.
18. **Imsande, J.** 1964. A comparative study of the regulation of pyridine nucleotide formation. Biochim. Biophys. Acta **82**:445–453.
19. **International Union of Biochemistry.** 1973. Enzyme nomenclature. American Elsevier Publishing Co., Inc., New York.
20. **Jeter, R. M., B. M. Olivera, and J. R. Roth.** 1984. *Salmonella typhimurium* synthesizes cobalamin (vitamin B$_{12}$) de novo under anaerobic growth conditions. J. Bacteriol. **159**:206–213.
21. **Kasarov, L. B., and A. G. Moat.** 1973. Biosynthesis of nicotinamide adenine dinucleotide in *Haemophilus haemoglobinophilus*. Biochim. Biophys. Acta **320**:372–378.
22. **Kinney, D. M., J. W. Foster, and A. G. Moat.** 1979. Pyridine nucleotide cycle of *Salmonella typhimurium*: in vitro demonstration of nicotinamide mononucleotide deamidase and characterization of *pnuA* mutants defective in nicotinamide mononucleotide transport. J. Bacteriol. **140**:607–611.
23. **Liang, A., and R. L. Houghton.** 1980. Structural aspects of the membrane-bound *Escherichia coli* pyridine nucleotide transhydrogenase (EC 1.6.1.1). FEBS Lett. **109**:185–188.
24. **Liang, A., and R. L. Houghton.** 1981. Coregulation of oxidized nicotinamide adenine dinucleotide (phosphate) transhydrogenase and glutamate dehydrogenase activities in enteric bacteria during nitrogen limitation. J. Bacteriol. **146**:997–1002.
25. **Liu, G., J. Foster, P. Manlapaz-Ramos, and B. M. Olivera.** 1982. Nucleoside salvage pathway for NAD biosynthesis in *Salmonella typhimurium*. J. Bacteriol. **152**:1111–1116.
26. **Manlapaz-Fernandez, P., and B. M. Olivera.** 1973. Pyridine nucleotide metabolism in *Escherichia coli*. IV. Turnover. J. Biol. Chem. **248**:5067–5073.
27. **McLaren, J., T. C. Ngo, and B. M. Olivera.** 1973. Pyridine nucleotide metabolism in *Escherichia coli*. III. Biosynthesis from alternative precursors in vivo. J. Biol. Chem. **248**:5144–5149.
28. **Nasu, S., and R. K. Gholson.** 1981. Replacement of the B protein requirement of the *E. coli* quinolinate synthetase system by chemically-generated iminoaspartate. Biochem. Biophys. Res. Commun. **101**:533–539.
29. **Nasu, S., F. D. Wicks, and R. K. Gholson.** 1982. L-aspartate oxidase, a newly discovered enzyme of *Escherichia coli*, is the B protein of quinolinate synthetase. J. Biol. Chem. **257**:626–632.
30. **Nasu, S., F. D. Wicks, and R. K. Gholson.** 1982. The mammalian enzyme which replaces B protein of *E. coli* quinolinate synthetase is D-aspartate oxidase. Biochim. Biophys. Acta **704**:240–252.
31. **Olivera, B. M., and R. Lundquist.** 1971. DNA synthesis in *Escherichia coli* in the presence of cyanide. J. Mol. Biol. **57**:263–277.
32. **Ortega, M. V., and G. M. Brown.** 1960. Precursors of nicotinic acid in *Escherichia coli*. J. Biol. Chem. **235**:2939–2945.
33. **Preiss, J., and P. Handler.** 1958. Biosynthesis of diphosphopyridine nucleotides. I. Identification of the intermediates. J. Biol. Chem. **233**:483–492.
34. **Preiss, J., and P. Handler.** 1958. Biosynthesis of diphosphopyridine nucleotides. II. Enzymatic aspects. J. Biol. Chem. **233**:493–500.
35. **Rowe, J. J., R. D. Lemmon, and G. J. Tritz.** 1985. Nicotinic acid transport in *Escherichia coli*. Microbios **44**:169–184.
36. **Seeberg, E., W. D. Rupp, and P. Strike.** 1980. Impaired incision of ultraviolet-irradiated deoxyribonucleic acid in *uvrC* mutants of *Escherichia coli*. J. Bacteriol. **144**:97–104.
37. **Sharma, S., and R. E. Moses.** 1979. *uvrC* gene function in excision repair in toluene-treated *Escherichia coli*. J. Bacteriol. **137**:397–408.
38. **Skorko, R., and J. Kur.** 1981. ADP-ribosylation of proteins in non-infected *Escherichia coli* cells. Eur. J. Biochem. **116**:317–322.
39. **Steiner, B. M., J. T. Heard, Jr., and G. J. Tritz.** 1980. Modification of aspartate before its condensation with dihydroxyacetone phosphate during quinolinic acid formation in *Escherichia coli*. J. Bacteriol. **141**:989–992.
40. **Sundaram, T. K.** 1967. Biosynthesis of nicotinamide adenine dinucleotide in *Escherichia coli*. Biochem. Biophys. Acta **136**:586–588.
41. **Taguchi, H., and Y. Shimabayashi.** 1983. Findings of trigonelline demethylating enzyme activity in various organisms and some properties of the enzyme from hog liver. Biochem. Biophys. Res. Commun. **113**:569–574.

42. **Taylor, A. L., and C. D. Trotter.** 1967. Revised linkage map of *Escherichia coli*. Bacteriol. Rev. **31:**332–353.

43. **Tritz, G. J., and J. L. R. Chandler.** 1973. Recognition of a gene involved in the regulation of nicotinamide adenine dinucleotide biosynthesis. J. Bacteriol. **144:**128–136.

44. **Tritz, G. J., T. S. Matney, and R. K. Gholson.** 1970. Mapping of the *nadB* locus adjacent to a previously undescribed purine locus in *Escherichia coli* K-12. J. Bacteriol. **102:**377–381.

45. **White, H. B., III.** 1982. Biosynthetic and salvage pathways of pyridine nucleotide coenzymes, p. 1–17. *In* J. Everse, B. M. Anderson, and K.-S. You (ed.), Pyridine nucleotide coenzymes. Academic Press, Inc., New York.

46. **Zalkin, H.** 1985. NAD synthetase. Methods Enzymol. **113:**297–302.

47. **Zatman, L. J., N. O. Kaplan, and S. P. Colowick.** 1953. Inhibition of spleen diphosphopyridine nucleotidase by nicotinamide, an exchange reaction. J. Biol. Chem. **200:**197–212.

Section C. Class III Reactions: Formation and Processing of Polymers

39. DNA Replication

ROGER McMACKEN,[1] LYNN SILVER,[2] AND COSTA GEORGOPOULOS[3]

Department of Biochemistry, The Johns Hopkins University, Baltimore, Maryland 21205[1]; Department of Basic Microbiology, Merck and Co., Rahway, New Jersey 07065[2]; and Department of Cellular, Viral and Molecular Biology, University of Utah Medical Center, Salt Lake City, Utah 84132[3]

INTRODUCTION

Our basic understanding of DNA replication has been derived mostly from studies of *Escherichia coli* and its viruses. This chapter is meant to provide an overview and an attempt at synthesis of genetic, biochemical, and physiological studies which have led to our current view of DNA replication. This review will deal in large part with what has been learned from the *E. coli* DNA replication system, since most of the genetic and biochemical analyses have been done with this organism. In those studies in which direct comparisons have been made between *E. coli* and *Salmonella typhimurium* DNA replication, the two organisms have been shown to behave similarly, including interchangeability among all components examined (261). For a general understanding of the basic principles of the biochemistry of DNA replication, the reader is referred to the unsurpassed texts on the subject written by Kornberg (196, 197). More specialized information and in-depth analysis of some concepts discussed here have been covered in a number of recent reviews (198, 254, 264, 297, 388, 418; chapter 98).

Overview and Historical Background

E. coli is the prototype of a unichromosomal haploid procaryotic organism. Its circular, double-stranded chromosome consists of approximately 4×10^6 base pairs (bp). It has a fixed origin of replication, *oriC*, from which bidirectional replication is initiated. *trans*- and *cis*-acting elements that regulate initiation exert their influence at the origin to modulate the level (or frequency) of initiation. Although circular, the chromosome has a "terminus" where replication forks "collide." These elements are consistent with the replicon hypothesis of Jacob et al. (159). The replicon hypothesis suggests that a genetic element consists of a unit of replication, the replicon, containing a *cis*-acting site at which replication is initiated, the "activator," and a gene(s) specifying a *trans*-acting "initiator" protein(s) which interacts with the activator to initiate replication. Although the enumeration and nature of the elements required for replication of the

E. coli chromosome are finally yielding to study, the mechanism of control of chromosome replication is still unclear. This subject is treated in depth elsewhere in this volume (chapter 98).

A growing cell must increase all its macromolecular components (ribosomes, proteins, cell envelope, and DNA) in a coordinated manner such that a single cell will, in one generation, give rise to identical daughter cells. This notion of balanced growth implies that the rates of synthesis of these components are coupled and that a change in nutritional conditions will lead to the adjustment of the overall rates of synthesis of all macromolecular components to a new rate consistent with the changed environment. The concept of coordination of macromolecular synthesis emerged from the pioneering studies of Maaløe and his co-workers, who studied the effects of nutritional shifts on cellular components. A linchpin of this concept is that the rate of chain elongation (or synthesis of a single macromolecule) for each macromolecule is constant irrespective of growth rate, while the overall rate of synthesis of each macromolecular species varies with growth rate. This seeming paradox is resolved by the recognition that it is the number of chains being elongated (or single molecules being synthesized) that increases (or decreases) with increasing (or decreasing) growth rate. That is, the overall rates of synthesis of macromolecules are controlled by varying the frequency of chain initiation (250).

In this light, DNA replication can be viewed as a process divisible into two distinct stages: elongation, the constitutive progression along the chromosome of replication forks consisting of protein complexes moving in a coordinated manner at a constant rate; and initiation, the tightly regulated process by which elongation complexes are created at a unique origin site on double-stranded DNA (dsDNA). The coordination of the frequency of initiation of chromosome replication with other macromolecular synthesis and with the cell division cycle is a complex function resulting from the integration of various regulatory circuits.

In the 1960s and early 1970s, physiological experiments were carried out which employed starvation of auxotrophic mutants or specific antibiotic inhibitors to block DNA, RNA, and protein synthesis differentially. These experiments showed that to initiate new rounds of replication, both protein synthesis and RNA synthesis are absolutely required prior to the initiation event. Specifically, Maaløe and Hanawalt (249) and Hanawalt et al. (125) showed that cells require RNA or protein synthesis, or both, to initiate replication. Thymine starvation of thymine-requiring auxotrophs of *E. coli* leads to death in cells that are actively synthesizing DNA at the time of starvation (65, 249). Cells prevented from synthesizing RNA and protein by starvation of auxotrophic mutants cease replicating DNA and become immune to thymineless death at times after the onset of starvation consistent with the time needed for cells to complete a round of replication but not to start a new one. Lark et al. (227) showed that DNA radioactively labeled for a short period is not subsequently rereplicated when the culture is starved for RNA and protein precursors. This result indicates that blockage of RNA and protein synthesis prevents initiation of new rounds of replication. Similarly, Lark and Renger (226) used chloramphenicol (an inhibitor of protein synthesis) to show that protein synthesis is required before initiation, presumably to make initiator proteins for a new round of replication. They showed that high concentrations of chloramphenicol (150 µg/ml) were as efficient at inhibiting initiation as amino acid starvation if added >15 min before initiation was to take place. Ward and Glaser (406) obtained the same result with *E. coli* B/r. Lark (222) showed in similar experiments, using rifampin or streptolidigin (both inhibitors of RNA polymerase) to block RNA synthesis, that RNA synthesis is required (separate from the requirement for transcription of translatable mRNA) immediately before or at the time of initiation.

The experiments of Lark (and others) also define a system for synchronization of initiation of DNA replication. Inhibition of protein synthesis, by starvation of amino acid auxotrophs or by antibiotic treatment, leads to completion of ongoing rounds of replication to termination, with little effect on fork velocity. However, new rounds of replication are not initiated. If protein synthesis is resumed, initiation can proceed, but it will not proceed in a synchronous manner in the population; rather, the participation in the initiating population will reflect the state of the population at the onset of inhibition of protein synthesis (372). Amino acid starvation of *E. coli* K-12 and B, but not B/r, will, however, lead to synchronous rounds of cell division (327). To obtain a synchronously initiating population, protein synthesis must be resumed in the absence of DNA synthesis. This treatment allows the accumulation of "initiation mass" (which here can be viewed simply as the synthesis of initiation-specific proteins, but which may also reflect the increase in concentration of small molecule effectors paralleling increase in protein) while preventing the triggering of initiation. After about one generation time, all replication origins should be prepared to "fire." Originally, the reversible inhibition of DNA synthesis was carried out by starving thymine auxotrophs of thymine. More recently, Bouché (37) and others have shown that a very high degree of synchrony can be obtained by using reversible temperature-sensitive mutations in a gene required for initiation, *dnaC* (see below).

A unifying theme from the physiological studies was the conclusion that DNA replication is not the simple process envisioned by Watson and Crick; rather, superimposed on the self-replicating ability of the double helix are the requirements for efficiency, fidelity, regulation, and coupling of chromosome duplication with other macromolecular synthesis and the cell cycle. To approach these problems at the molecular level, efforts were made to define the components of the DNA replication process through isolation of bacterial mutants specifically blocked in this process.

The Genetic Approach

The genetic approach can be used to identify the genes whose products are implicated in a selected process. For a process as important to cell viability as DNA replication, the basic strategy is to define genes and products through isolation of conditional-lethal mutations, generally those yielding thermolabile proteins. In the simplest sense, a conditional-lethal mutation defines a gene that is essential for cell viability under the nonpermissive conditions. However, by limiting the selection to conditional lethals, it is possible that not all genes essential for the process will be selected. Mutations that lead to lowering of efficiency to a lesser (nonlethal) extent will not be easily detected. Backup or secondary mechanisms may exist that are capable of bypassing normal replication processes, making the normally used processes apparently nonessential. The extent to which each mechanism is used could depend on the physiological conditions encountered, e.g., temperature and growth medium. Thus, although the existence of conditional-lethal mutations in a gene is a strong indicator of essentiality of a gene product under the nonpermissive conditions, a nonlethal mutation in a gene, even a nonlethal deletion or null mutation, does not exclude the possibility that such genes may make a contribution to the overall efficiency of the replication process. Hence, exhaustive selection of conditional-lethal mutants may still yield an incomplete catalog of proteins that act in replication. One means of extending the list of gene products implicated in replication is by the isolation of extragenic suppressors, an extremely powerful tool in identifying potentially interacting sets of proteins or functions (36, 132).

Since, as noted above, mutations in genes whose products are essential for DNA replication should be lethal for *E. coli*, conditional-lethal (mostly temperature-sensitive) mutations in these genes were sought. A variety of selection techniques were used. The first involved a search among a collection of thermosensitive mutants unable to form colonies in rich medium for those that specifically blocked the incorporation of radioactive thymine (55, 194). The brute-force method reached its zenith with the development of a computerized automatic procedure enabling the screening of >1.4×10^6 mutagenized bacterial colonies (353). Among the 2,206 temperature-sensitive mutants isolated, 110 were shown to be specifically blocked in DNA synthesis in contrast to protein synthesis.

A second approach took advantage of the lethal effects on cells of the incorporation of 5-bromodeoxyuridine into DNA (35). Bacteria mutagenized at 30°C were shifted en masse to 42°C, allowed to grow for 50 min, and then exposed to 5-bromodeoxyuridine for 3 h at 42°C. Cells were subsequently irradiated with long-wavelength UV light (313 nm) to allow bacteria that incorporated the 5-bromodeoxyuridine into their DNA to "commit suicide." This resulted in an enrichment for mutants unable to replicate their DNA at high temperature. Since mutations that inhibit either RNA or protein synthesis would be expected to result eventually in cessation of DNA synthesis (by blocking initiation), only candidates that were blocked specifically in DNA synthesis, as opposed to RNA and protein synthesis, were further investigated.

A third approach was the isolation of mutant bacteria unable to propagate certain bacteriophages at all temperatures. Among these classes, some were shown to specifically block bacteriophage DNA replication at all temperatures and host replication at high temperature (118, 121, 335, 377). This class was predicted as bacteriophages have a small coding capacity and rely to a great extent on the replication machinery of the host. It should be pointed out, however, that for this class to occur, the alteration of the bacterial function should not appreciably interfere with host replication at the permissive temperature (99).

In agreement with predictions of physiological studies, these mutants were shown to exhibit two diverse phenotypes, "fast stop" and "slow stop" of incorporation of radioactive DNA precursors, as expected for mutations blocking elongation or initiation, respectively (194, 410). However, a mutation in an elongation protein may exhibit a leaky or slow-stop phenotype at the nonpermissive temperature and might be confused with a mutation blocking initiation. Although Kohiyama's group of mutants also contained examples with intermediate behavior, this approach has proved very useful in sorting DNA replication mutants.

These mutations were also genetically characterized by mapping, complementation, and dominance studies (references 145, 147, and 410 and many others). In at least two cases (dnaB and dnaC), it was found that mutations in the same cistron could lead either to a slow- or fast-stop phenotype. This observation may be taken to indicate either that these gene products play a role in both initiation and elongation processes or that proteins of initiation complexes interact with proteins of elongation complexes.

With a battery of mutants in hand, steps in a pathway or process may potentially be temporally ordered by experiments using combinations of mutants (36, 162) or drugs and mutants (126, 337, 445). The study of DNA replication has benefited significantly from the impact of genetics on biochemistry. Since DNA replication is a complex process, purification of proteins by following their known enzymatic activities has been very difficult; much of the recent progress in the biochemical study of DNA replication has come through the use of complementation assays, that is, complementation of extracts of mutant cells by fractionated extracts of wild-type cells.

In the discussion of DNA replication genes that follows, only those genes with products likely to participate directly in the DNA replication process will be discussed, in contrast to those genes acting indirectly, e.g., dnaF (nrdA), which codes for the B1 sub-

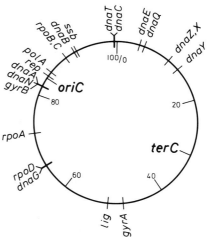

FIG. 1. Chromosomal map locations of *E. coli* genes specifying products essential for DNA replication.

unit of ribonucleotide reductase (100). The chromosomal map location of *E. coli* genes essential for bacterial DNA replication is shown in Fig. 1.

BIOCHEMISTRY OF CHROMOSOMAL DNA REPLICATION

Stages in the Replication of the Bacterial Chromosome

Initial insights into the structure of the replicating *E. coli* chromosome came from the pioneering autoradiographic studies of Cairns (53). Replicating chromosomes were found to resemble the Greek letter theta (Fig. 2); hence, such replication intermediates were termed theta structures. The presence of two forks in the theta structures at first suggested a model in which one fork represented an origin of DNA synthesis on the circular chromosome and the other was the advancing replication fork. Subsequent autoradiographic and physiological studies, however,

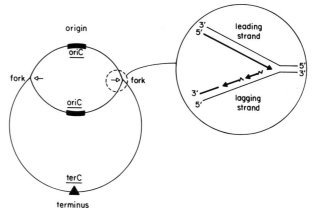

FIG. 2. Simplified view of the replication of the circular *E. coli* chromosome. Replication of the bacterial chromosome, initiated at *oriC*, proceeds bidirectionally towards the terminus, *terC*. The semidiscontinuous nature of DNA synthesis at an advancing replication fork is depicted in the insert.

clearly indicated that the circular *E. coli* chromosome is replicated in a bidirectional fashion from a fixed origin (32, 255, 257, 308, 318, 398, 435), *oriC*, which is located at 83.5 min on the *E. coli* genetic map (see below). Thus, each branch point in the theta structure replication intermediate (Fig. 2) is the result of an advancing replication fork. Because replication forks proceed at approximately the same rate in each direction, replication of the chromosome terminates at a site, *terC* (Fig. 2), which is situated almost diametrically opposite the origin in the region from 30 to 32 min on the *E. coli* genetic map (32, 37, 38, 246).

A combination of genetic and biochemical studies have revealed that replication of the bacterial chromosome is comprised of at least three discrete stages, namely, initiation, elongation, and termination. Because of the complexity of each of these stages, it was not until this decade that biochemists succeeded in establishing in vitro systems that could support all of these enzymatic reactions. Although the biochemical complexity of DNA replication was not appreciated at first, it soon became apparent that even the seemingly simple progression of a replication fork along the *E. coli* chromosome required the resolution of multiple mechanistic problems. For example, detailed studies of cellular and viral DNA polymerases indicated that each synthesized DNA in a $5' \rightarrow 3'$ direction (196). This fact, coupled with the opposing polarities of the two strands of the DNA double helix, focused attention on the problem of how both parental strands could be replicated concurrently at a replication fork moving in a single direction. Okazaki and colleagues resolved this paradox with the demonstration (309) that the lagging strand at a replication fork is synthesized in a retrograde $5' \rightarrow 3'$ direction in short 500- to 2000-bp nascent fragments (termed Okazaki fragments) that are rapidly ligated together.

Not unexpectedly, this finding led to new complications. A semidiscontinuous mode of DNA synthesis at the replication fork (Fig. 2, inset) demands that a new lagging-strand DNA chain be initiated about every second at 37°C. Yet, none of the known cellular DNA polymerases can initiate synthesis of a DNA chain de novo; each is only capable of extending a preexisting polynucleotide chain (196). These findings provided an impetus to the search for cellular replication machinery that can synthesize the necessary primer chains. The inherent stability of the double-helical form of DNA presents still another mechanistic problem that must be solved by the enzymatic machinery responsible for propagation of a replication fork, i.e., the unwinding of the DNA duplex. Other studies suggested the need for additional proteins at the replication fork that (i) act to protect unwound DNA and remove intrastrand secondary structure; (ii) insure the fidelity of DNA synthesis; (iii) guarantee the processivity and speed of fork movement; (iv) coordinate synthesis of the leading and lagging strands; and (v) deal with the intracellular structure of an *E. coli* chromosome that is folded into 50 to 100 discrete topological domains (317, 369, 430, 431) and organized by histonelike proteins into a chromatin structure (45, 123, 331). Thus, by the early 1970s it was evident, both to biochemists and to geneticists, that replication of the bacterial chromosome was a complex process that required the action of multiple proteins.

Isolation of *E. coli* Replication Proteins: Development of In Vitro Replication Systems

As late as 1970, just two of the numerous *E. coli* proteins that directly act in the replication of the bacterial chromosome had been purified to homogeneity and characterized: DNA polymerase I and DNA ligase. Purification of the remaining components was only made possible by the development of relatively physiological in vitro replication systems that utilized natural DNA templates. Two such systems have had the greatest impact on the progress of the biochemistry of *E. coli* DNA replication. Bonhoeffer and colleagues (342) developed an ingenious system in which a highly concentrated lysate of DNA polymerase I-deficient *E. coli* is placed on a permeable cellophane disk layered on the surface of a buffer containing deoxyribonucleoside triphosphates (dNTPs) and small molecule cofactors. DNA synthesis in this particulate system is semiconservative and reflects the continued propagation of replication forks that were established in vivo prior to cell lysis. The cellophane disk system has been useful for characterizing the effects of cofactors, specific antibodies, and inhibitors on propagation of the bacterial replication fork or on the joining of Okazaki fragments. If the cell lysate is prepared from an *E. coli* replication mutant that is thermosensitive for fork movement (i.e., a fast-stop mutant), little or no DNA synthesis is observed in vitro at the restrictive temperature. Under these conditions it is sometimes possible to complement the defective lysate with protein fractions prepared from wild-type cells and to restore the capacity of the system to synthesize DNA. Although in principle this approach can be used as an assay to purify each of the genetically identified components of the elongation machinery, the extreme dilution sensitivity of the cellophane disk system in practice has limited its usefulness in this regard. Furthermore, the utility of the particulate lysate system is severely limited by its inability to respond to the addition of exogenous DNA templates.

The in vitro replication system that has received the most attention from researchers in the field was developed in the Kornberg laboratory. Kornberg and colleagues chose to study the replicative life cycle of various small bacteriophages of *E. coli* that contain circular single-stranded DNA (ssDNA) chromosomes. The limited coding capacity of the single-stranded phage chromosomes, which are each about 5 or 6 kilobases (kb) in length, constrains the phage to rely predominantly on host-coded proteins for their replication machinery. In fact, the first stage in the replication of such viral genomes, the conversion of the single-stranded chromosome to the duplex replicative form (SS → RF conversion) is entirely carried out by bacterial proteins. Thus, mechanistic studies of this conversion provide a window into the process of *E. coli* DNA replication itself. An additional advantage of using these single-stranded genomes as natural templates in vitro arises from the fact that the circular DNA chains contain no free 3'-hydroxyl termini that can be utilized to prime DNA synthesis adventitiously. DNA synthesis in vitro was anticipated, therefore, to be strongly dependent on de novo strand initiation events. For these reasons, the Kornberg laboratory devoted considerable efforts towards developing a soluble extract system that could support the replication of single-stranded phage chromosomes (344, 425).

Development of a useful system depended on the utilization of gently lysed *E. coli* that enabled intact bacterial chromosomes and other particulate matter to be removed by high-speed centrifugation. This soluble *E. coli* extract was found to convert the viral genomes of phages M13, φX174, and G4 efficiently to the duplex replicative form. Two different approaches were employed to purify the proteins that participated in the strand initiation and chain elongation stages of DNA replication. As with the cellophane disk system, it was possible to establish in vitro complementation assays for those required proteins that could be conditionally inactivated with temperature-sensitive mutations. Moreover, with the soluble system it was also possible to set up classical resolution and reconstitution assays, a direct approach whereby a required protein is sorted out by protein resolving techniques and then assayed by its capacity to restore DNA synthesis when it is added back to a specific protein fraction that contains limiting quantities of the essential replication protein.

The power of this soluble in vitro system was soon demonstrated. For example, it was used to verify the suggestion (48) that short RNA transcripts prime the initiation of DNA strands (346, 425). Furthermore, many of the *E. coli* replication proteins that had been identified genetically by that time were found to act in the soluble system to catalyze the enzymatic conversion of added φX174 single strands to the duplex replicative form (345, 346, 415, 420). These proteins included the *E. coli* DnaB, DnaC, and DnaT proteins, the ssDNA binding protein (SSB), the DnaG primase, and the multisubunit DNA polymerase III holoenzyme. The reliance of φX174 DNA replication on this collection of host proteins thus provided the first physiological assay for the purification of many of the proteins required for propagation of a replication fork on the *E. coli* chromosome.

Once the proteins required for replication of the single-stranded phage chromosomes were in hand in a purified form, it was possible to begin the arduous task of determining their molecular roles in the synthesis of the complementary strands of M13, φX174, and G4. It is not within the scope of this review to provide a detailed account of the molecular mechanisms involved in these SS → RF reactions (for a summary see references 196 and 197). However, in the listing of the characteristics of individual *E. coli* replication proteins given below, we of course refer to the properties of individual replication proteins that were first recognized during in vitro studies of phage ssDNA replication.

Studies of the replication of single-stranded phage genomes identified many of the prominent features of the bacterial elongation machinery. Yet, these SS → RF systems could not be used to elucidate mechanisms involved in unwinding of the double helix in advance of the biosynthetic machinery, nor could they be employed to isolate and characterize the class of replication proteins that participates solely in the initiation stage of the bidirectional replication of the *E. coli* chromosome. To remedy this situation, the Kornberg laboratory spent several years attempting to develop a soluble in vitro system that would support the specific initiation of bidirectional DNA replication at the unique chromosomal origin of *E. coli*, *oriC*. This approach was made feasible by the isolation and development of small multicopy plasmids that carry *oriC* as the sole replication origin (398, 435).

Efficient replication of *oriC* plasmids in the soluble enzyme system that was ultimately developed (102) depended on several exacting conditions. A soluble extract of *E. coli* (a freeze-thaw lysate that is clarified by high-speed centrifugation) needed to be precisely fractionated with ammonium sulfate to separate inhibitory factors from essential replication proteins. Furthermore, *oriC*-specific DNA synthesis was only observed if the system was supplemented both with an ATP-regeneration system and with a hydrophilic polymer such as polyvinyl alcohol. The latter reagent is believed to enhance "macromolecular crowding," thus raising the local concentration of the *oriC* template and of essential proteins that may have been present at suboptimal concentrations. The *oriC* system provided the first functional assay for the *E. coli* DnaA protein, the one single protein most closely identified by genetic and physiological experiments with the initiation process at *oriC* (see below). Resolution of the proteins that function in *oriC* DNA replication has not only yielded DnaA protein but has also yielded several proteins that participate in the specificity of, and perhaps even the regulation of, the initiation process in vitro. Recently, as will be described later, the *oriC* system has been reconstituted with a collection of highly purified proteins (104, 301, 394).

PHYSICAL AND FUNCTIONAL PROPERTIES OF PURIFIED *E. COLI* REPLICATION PROTEINS

Genetic and biochemical studies have identified at least 18 different gene products of *E. coli* that participate directly in the replication of the bacterial chromosome. Some of the salient properties of each of these proteins are listed in Table 1. At least 14 other gene products, listed in Table 2, have been implicated as potentially performing some role in *E. coli* DNA replication. A more detailed accounting of the physical and functional properties of many of these proteins is presented below.

Proteins That Directly Participate in the Propagation of a Replication Fork

ssDNA-binding protein (SSB)

The bacteriophage T4 gene *32* protein is the prototype of the class of proteins that bind tightly and cooperatively to ssDNA and destabilize natural duplex DNA (1, 2). The clear dependence of T4 DNA replication in vivo and in vitro on the gene *32* protein (26) prompted a search for a similar protein in *E. coli*. The bacterial ssDNA-binding protein (SSB) was isolated on the basis of its strong affinity for ssDNA (357).

TABLE 1. Essential replication proteins of *E. coli*

Protein	Primary function	Polypeptide mass (kilodaltons)	No. of subunits in native protein	Molecules per cell	Gene	Map position (min)	Phenotype of mutant
DnaA	Origin recognition	52.5	1	200	*dnaA*	83	Slow stop
DnaB	DNA helicase; priming	52.3	6	20	*dnaB*	92	Quick stop
DnaC	Delivery of dnaB protein to *oriC*	29	1	100	*dnaC*	99	Slow or quick stop
SSB	Binding to ssDNA	18.8	4	500	*ssb*	92	Quick stop
primase	RNA primer synthesis	65.6	1	75	*dnaG*	67	Quick stop
DNA polymerase III (holoenzyme)	Replicative DNA polymerase	(740)		20			
α	DNA polymerase	130	(2)		*polC (dnaE)*	4	Quick stop
ε	$3' \rightarrow 5'$ proofreading exonuclease	27.5	(2)		*dnaQ (mutD)*	5	Hypermutation
θ	Unknown	10	1		Unknown		
β	DNA polymerase accessory protein	40.6	(2)		*dnaN*	83	Quick stop
τ	DNA polymerase accessory protein	71.1	(2)		*dnaZX*	11	Quick stop
γ	DNA polymerase accessory protein	52			*dnaZX*	11	Quick stop
δ	DNA polymerase accessory protein	32			*dnaZX* (?)		Quick stop
DNA gyrase	DNA supercoiling; decatenation	374	4				
α	Nicking-closing	97	2	250	*gyrA*	48	Slow stop
β	ATPase	89.8	2	150	*gyrB*	83	Slow stop
DNA polymerase I	Primer removal; gap filling	102	1	300	*polA*	87	DNA repair defective
DNA ligase	Sealing of DNA nicks	74	1	300	*lig*	52	Accumulate nascent fragments
DnaT	Termination of DNA replication	19.3	3	75	*dnaT*	99	Slow stop
RNA polymerase	Transcriptional activation of *oriC*	460	5	3,000	*rpoA, B, C, D*	73, 90, 90, 67	Slow stop

That the *E. coli* SSB also functions in DNA replication was soon demonstrated. Resolution of the proteins required for the in vitro SS → RF conversion of the viral strands of coliphages M13, φX174, and G4 revealed in each instance that SSB participates in the synthesis of the complementary strand (112, 345, 420, 438). The properties and biological roles of *E. coli* SSB and other ssDNA-binding proteins have been recently reviewed (60).

E. coli SSB is a polypeptide of M_r 18,873 (340) that exists in vitro as a stable tetramer, the species that is likely to be the functional form of the protein in vivo. Its affinity for ssDNA is at least 1,000-fold greater than its affinity for double-stranded DNA (dsDNA) (240,

TABLE 2. *E. coli* proteins that may play an auxiliary role in the replication of the bacterial chromosome

Protein	Known function	Polypeptide mass (kilodaltons)	No. of subunits in native protein	Gene
Rep protein	DNA helicase	66	1	*rep*
DNA helicase II	DNA helicase	82.1	1	*uvrD*
RNase H	Excises RNA primer; prevents nonspecific initiation	17.6	1	*rnh*
DNA topoisomerase I	Removes negative supercoils	105	1	*topA*
HU protein	Organizes chromosome structure	9.5	2	*hupA, hupB*
Protein n	Assembly of φX174 primosome	14	2	Unknown
Protein n'	Assembly of φX174 primosome	76	1	Unknown
Protein n"	Assembly of φX174 primosome	17	1	Unknown
DnaJ protein	Heat shock protein; binds DNA	41.1	2	*dnaJ*
DnaK protein	Heat shock protein; 5'-nucleotidyl phosphatase	69.1	1	*dnaK*
GrpE protein	Heat shock protein; modulates DnaK protein activity	24		*grpE*
GroES protein	Phage assembly; suppresses *dnaA*(Ts) mutations	10.5	6–8	*mopB (groES)*
GroEL protein	Phage assembly; suppresses *dnaA*(Ts) mutations	60	14	*mopA (groEL)*
DNA adenine methylase	Mismatch repair; regulates initiation at *oriC*	31	1	*dam*
(Arg-tRNA)	Functions in protein synthesis			*dnaY*

284, 427). The tight binding of SSB to ssDNA enables this protein to lower the melting temperature of duplex DNA significantly (357, 427).

Electron microscopic studies of the interaction of SSB with natural ssDNAs, obtained from filamentous phage such as M13, demonstrated that the contour length of a DNA chain is reduced by about 35% when the strand is completely coated with SSB (63, 357). It has been suggested that this condensation arises from the wrapping of ssDNA around the surface of the SSB tetramer (63, 203). Unfortunately, a determination of the precise nature of this interaction will be a demanding chore, now that it is recognized that SSB utilizes at least two different binding modes in its interaction with ssDNA (124, 240). Both the concentration of salt and the concentration of SSB influence the mode of binding. At NaCl concentrations above 0.2 M, the size of the SSB binding site is about 65 ± 5 nucleotides per tetramer. As observed with the electron microscope, this form of the SSB-DNA complex appears as nucleosomelike beads. Nuclease digestion and other studies suggest a structure in which 145 nucleotides of ssDNA is wrapped around two SSB tetramers and 30 nucleotides of uncoated DNA is present between adjacent octamers (63). A second type of structure, formed in the presence of excess SSB at salt concentrations below 10 mM, appears as an extended but smooth nucleoprotein filament (124). Under these conditions the average SSB binding site consists of 33 ± 3 nucleotides per tetramer (240). Based on the in vitro properties of SSB-ssDNA interactions, it seems probable that in vivo both modes of SSB binding come into play during DNA transactions involving long stretches of ssDNA.

E. coli SSB performs a variety of functions in vitro that may relate to its intracellular roles in chromosomal DNA replication. Although SSB destabilizes duplex DNA, it is not capable of unwinding the double helix at physiological temperatures. However, DNA unwinding catalyzed by several E. coli DNA helicases, including DnaB protein, Rep protein, and helicase III, is greatly stimulated by the binding of SSB to the ssDNA created by helicase action (228, 350, 433). It is believed that the role of the binding protein in these systems is to prevent the separated strands from renaturing. SSB-coated ssDNA is highly resistant to nucleolytic cleavage by various single-strand-specific nucleases of E. coli such as exonuclease I and RecBCD enzyme. Thus, the coating of unwound DNA by SSB at a replication fork may well serve a protective role in vivo.

In model DNA synthesis and replication systems, SSB has been found to have multiple beneficial effects on the activity of the replicative DNA polymerase of E. coli, the DNA polymerase III holoenzyme. On long ssDNA templates, SSB stimulates sevenfold the initial rate of DNA chain polymerization by this replicative DNA polymerase (85). The processivity of the holoenzyme during the replication of primed fd phage viral strands is greatly increased by the presence of saturating concentrations of SSB (215). This boost in polymerase processivity may, in part, reflect the capacity of SSB to markedly reduce the pausing of DNA polymerase III holoenzyme and polymerase subassemblies at regions of potential secondary structure (215). Finally, it has been demonstrated that SSB

enhances the fidelity of DNA polymerase III approximately fivefold (211). Several of the effects of SSB on DNA synthesis are undoubtedly mediated by the capacity of the binding protein to extend the template strand into a rigid configuration, thereby removing inhibitory secondary structure. Of course, it is also possible that direct interactions between SSB and DNA polymerase III holoenzyme are at least partially responsible for the observed enhancement of the synthetic and binding activities of the DNA polymerase.

Interestingly, the physiological roles of E. coli SSB apparently are not restricted simply to stimulating propagation of advancing replication forks. Mechanistic studies of several in vitro replication systems indicate that SSB is a vital participant in the assembly or stabilization of nucleoprotein structures that form at origins of DNA replication prior to the initial priming event. This fact was first demonstrated for the conversion of phage G4 viral strands to the duplex replicative form. Synthesis of a specific RNA primer at the G4 complementary strand origin by the E. coli DnaG primase depends absolutely on the presence of SSB (40). Additional studies showed that primase cannot bind to the G4 viral strand unless it is first coated with saturating levels of SSB (358, 365, 417). SSB has been found to stabilize a nucleoprotein structure formed on negatively supercoiled DNA at the replication origin of bacteriophage λ (C. Alfano and R. McMacken, unpublished data), a structure that contains the λ O and P replication initiators and the E. coli DnaB helicase (76; Alfano and McMacken, unpublished data). SSB may function in a similar manner at oriC to stabilize a nucleoprotein structure formed by DnaA, DnaB, DnaC, and HU proteins in a system reconstituted solely with purified proteins (104, 394).

Two different general mechanisms have been identified for priming DNA synthesis on ssDNA in E. coli. The M13 SS → RF conversion depends on the synthesis of a primer transcript by E. coli RNA polymerase (48, 110, 425). A second mechanism utilizes the E. coli general priming reaction in which DnaB protein and DnaG primase cooperate to synthesize short oligonucleotide transcripts on virtually any naked ssDNA chain (5). In the presence of sufficient SSB to coat all of the DNA of single-stranded chromosomes, initial primer synthesis by each of these priming systems is specifically localized to regions at or near physiological replication origins (5, 6, 112). Thus, coating of unwound DNA, present at replication forks or elsewhere, with SSB is expected to enhance the specificity of DNA replication by precluding cellular priming systems from acting.

DnaB protein

A series of genetic searches for E. coli genes that are essential for replication of the bacterial chromosome led to an early identification of the dnaB gene, owing to the preponderance of conditional-lethal mutations that were found at this locus (see reference 410 for a summary). The product of the dnaB gene was found to be required for the in vitro conversion of φX174 DNA to the duplex replicative form (346). This provided a functional assay that was used to guide the purification of the DnaB protein to homogeneity (324, 392). Although the intracellular concentration of DnaB protein is remarkably low, approximately 10 to 20 molecules per cell, physical studies of the DnaB protein have been made possible by the 100- to 200-fold amplification of its intracellular levels obtained when the *dnaB* gene was cloned into "runaway" plasmid vectors (12; R. McMacken and C. Alfano, unpublished data).

The DnaB protein is a hexamer of 52.3-kilodalton subunits (12, 291, 324). DnaB protein contains a nucleoside triphosphatase activity that is greatly stimulated by ssDNA (7, 324a, 424). In the presence of ribonucleoside triphosphates (rNTPs) or a nonhydrolyzable analog of ATP, DnaB protein binds relatively weakly to ssDNA (7, 324a), covering a region of ~80 nucleotide residues, judging from nuclease protection studies (8). In the presence of rNTPs, the DnaB protein forms a stable complex with the *E. coli* DnaC protein in vitro, a complex in which each subunit of the DnaB hexamer apparently binds to a monomer of DnaC (185, 220, 421). This complex, presumably, is the physiological form of DnaB protein in vivo.

Genetic evidence has also pointed to an in vivo protein-protein interaction between the DnaB and DnaC proteins. R. Maurer et al. (260) have shown that mutations located in the *S. typhimurium dnaB* gene compensate for the Ts phenotype of certain *S. typhimurium dnaC*(Ts) mutations; also, Sclafani and Wechsler (349) have shown that overproduction of DnaC protein suppresses the Ts phenotype of certain *dnaB*(Ts) mutations. The DnaC proteins of *E. coli* and *S. typhimurium* have been shown to be functionally interchangeable both in vivo (260, 261) and in vitro (183). This is not surprising in view of the fact that the *S. typhimurium* DnaB protein is 93% identical to its *E. coli* counterpart at the amino acid level (R. Maurer, personal communication).

When thermosensitive *dnaB* mutants of *E. coli* are placed at a restrictive temperature, cellular DNA replication usually ceases abruptly (410). The obvious conclusion reached from these findings is that the DnaB protein performs a function that is essential for propagation of a replication fork along the bacterial chromosome. The recent discovery that DnaB protein is a DNA helicase (228) suggests that the primary role of DnaB protein in the replication of the *E. coli* chromosome is to unwind duplex DNA in advance of the enzymatic machinery responsible for synthesis of the leading and lagging strands. Unwinding of model, partially duplex substrates by the DnaB protein, as with other known DNA helicases, absolutely depends on the hydrolysis of nucleoside triphosphates. To invade the duplex portion of the substrate, the DnaB protein requires the presence of a 3'-terminal extension of ssDNA in the strand to which it is not bound. It has been determined that the DnaB protein moves in a $5' \rightarrow 3'$ direction along ssDNA, apparently in a processive fashion (228). The presence of SSB in the reaction mixture leads to a significant stimulation of the DnaB helicase activity, which is stimulated further still by the addition of primase and all four rNTPs.

The discovery that the DnaB protein is capable of coupling hydrolysis of rNTPs to movement was first suggested by biochemical studies of its role in the conversion of φX174 viral strands to the duplex replicative form (272). An early step in the synthesis of the complementary strand is a complex multiprotein reaction in which a nucleoprotein structure containing DnaB protein is assembled at the complementary strand origin of φX174 (6, 9, 10, 272, 412, 419, 420). Once assembled, this multiprotein unit, called the φX primosome, moves rapidly and unidirectionally in the $5' \rightarrow 3'$ direction along the viral DNA strand (i.e., in the direction opposite to primer and DNA chain synthesis) (6). The component of the primosome that is responsible for its movement has not been conclusively determined, but is believed to be either the DnaB protein (272) or protein n' (10), each of which possesses nucleoside triphosphatase activity.

Although the DnaB protein presumably functions in vivo to unwind DNA in advance of the DNA synthetic machinery, mechanistic studies of the role of the DnaB helicase in the replication of ssDNA templates indicate that it performs a second crucial role in the replication process. Synthesis of RNA primers by the DnaG primase depends on the presence of bound DnaB protein on the template DNA strand (5, 271, 272). The DnaB protein has been termed a mobile replication promoter (272) because a single molecule of bound DnaB protein, apparently without dissociating, enables primase to synthesize multiple RNA primers at many different sites on the template strand. The precise mechanism by which DnaB protein empowers the DnaG primase to synthesize primer transcripts has not been defined. To date, no genetic or biochemical evidence exists indicating a protein-protein interaction between DnaB protein and DnaG primase. One possibility, suggested by biochemical studies of the interaction of DnaB protein with ssDNA, is that the DnaB protein engineers a domain in the template strand with suitable secondary structure that can be recognized by the DnaG primase (8).

In summary, the available genetic and biochemical evidence suggests that the DnaB protein is the replicative helicase of *E. coli*. As will be discussed in more detail below (see section on the propagation of an *E. coli* replication fork), it is likely that during its action as a helicase the DnaB protein travels processively in a $5' \rightarrow 3'$ direction along the lagging-strand template, joining with primase every second or so to synthesize the primers for the nascent fragments of the lagging strand (228, 272).

DnaG primase

Chromosomal DNA replication is rapidly arrested when mutants of *E. coli* with a temperature-sensitive mutation in the *dnaG* gene are placed at a restrictive temperature (410). The discovery that extracts from such mutants were incapable of supporting the conversion of φX174 and G4 viral strands to the duplex replicative form (344, 415, 438) provided physiological in vitro assays for the purification of the *dnaG* gene product (40, 332, 423). The normal intracellular concentration of DnaG protein is about 75 molecules per cell, a concentration that made its original purification to homogeneity a difficult chore. Multimilligram quantities of this protein can be readily obtained at present, however, with the recent availability of recombinant plasmids carrying the *dnaG* gene that can be used to amplify cellular DnaG protein levels more than 100-fold (429).

The nucleotide sequence of the *dnaG* gene (361) indicates that it specifies a polypeptide of M_r 65,600. Physical studies of the isolated DnaG protein demonstrated that its native form is a monomer (332, 423). The functional role of the DnaG protein in DNA replication was elucidated from studies of the biochemical events that were responsible for the phage G4 SS → RF conversion. Initiation of the complementary strand of G4 DNA turned out to be relatively simple, since this reaction could be performed in an in vitro system that was reconstituted with just three *E. coli* replication proteins, SSB, DnaG primase, and DNA polymerase III holoenzyme (438). A comparison of the G4 reaction to the M13 SS → RF reaction was very illuminating, since a similar collection of proteins was found to be necessary for synthesis of the M13 complementary strand (112). The single difference was a requirement in the case of M13 for RNA polymerase to synthesize the RNA primer transcript. The requirement for RNA polymerase for DNA chain initiation on the M13 chromosome was consistent with the known rifampin sensitivity of M13 viral strand replication. These findings, coupled with the fact that the initial stage of G4 replication is resistant to rifampin, gave a clear indication that the DnaG protein was likely to be a rifampin-resistant RNA polymerase that primes synthesis of the complementary strand.

This notion was confirmed by the demonstration that the DnaG protein, in the presence of SSB and the four rNTPs, transcribes a specific region of the G4 viral strand to produce an oligoribonucleotide that is elongated by the DNA polymerase III holoenzyme into the full-length complementary strand (39, 40). With this discovery the DnaG protein was appropriately renamed primase (332). A clear implication of these results is that the functional role of the DnaG protein in the propagation of a replication fork along the bacterial chromosome is to synthesize the RNA primers needed for initiation of the lagging-strand Okazaki fragments. This hypothesis is consistent with the physiological properties of temperature-sensitive *dnaG* mutants and confirms the suggestion (223) that the *dnaG* gene product acts in the initiation of Okazaki fragments.

Primer synthesis by primase invariably requires the presence of a second protein. In the case of G4 DNA replication, the supporting protein is SSB. In all other replication systems examined to date that require the *E. coli* primase, including *oriC*, phage λ, and φX174, the assisting protein is the bacterial DnaB helicase (23, 229, 271). That the DnaB protein facilitates primer synthesis by primase is most convincingly demonstrated in the *E. coli* general priming reaction (5). In this nonspecific reaction, DnaB protein and primase cooperate in the presence of the four rNTPs to synthesize oligoribonucleotide primers on virtually any naked ssDNA chain. RNA primers synthesized in the general priming reaction are 10 to 60 residues in length and are made in a distributive fashion. It has been reported that primase enhances the stability of a complex of DnaB protein and ssDNA (5), an observation that suggests that the two proteins directly interact. The strict dependence of primase activity on supporting proteins may be a biological mechanism for controlling the specificity and extent of priming.

Several factors influence the length and structure of oligonucleotide primers that are synthesized by primase. One must consider, in this regard, that *E. coli* primase incorporates deoxynucleotides as well as ribonucleotides at all positions except the 5' terminus of the primer (270, 271, 333, 417). Primase appears to bind more tightly to dNTPs than to rNTPs during primer synthesis, but the rate of incorporation of dNTPs is considerably slower (271). As a consequence, primers composed of deoxynucleotides are significantly shorter than oligoribonucleotide primers. The properties of the priming reaction in the φX174 system, in which, as in *E. coli*, primer synthesis depends on the interaction of primase with DnaB protein, suggest that primers synthesized in vivo are composed of a mixture of ribo- and deoxyribonucleotides. Incorporation of deoxynucleotides into primers could conceivably lower the fidelity of DNA replication, since the DnaG primase presumably does not contain a mechanism for editing misincorporation events. A possible explanation for the incorporation of deoxynucleotides by the *E. coli* primase is that this may be a mechanism to insure that the length of each primer is relatively short in vivo. A short primer may be more efficiently removed by cellular nuclease activities and replaced with DNA synthesized by high-fidelity DNA polymerases.

An analysis of primer synthesis in the complete G4 SS → RF enzyme system indicated that the length of the primer covalently linked to the 5' terminus of the cDNA strand was considerably shorter than the primer synthesized in a reaction in which priming was uncoupled from DNA synthesis (333). In the coupled reaction, DNA polymerase III holoenzyme apparently is capable of extending primers as short as a dinucleotide.

DNA polymerase III holoenzyme

The only DNA polymerase activity detectable in crude extracts of wild-type *E. coli* is that of DNA polymerase I. For this reason, it was presumed for many years to be the bacterial replicative DNA polymerase. The discovery by DeLucia and Cairns (70) of the viability of *E. coli polA* mutants, which contain a mutation that eliminates the DNA polymerase activity of DNA polymerase I, forced biochemists to reassess their views on the mechanism of DNA replication in *E. coli* and prompted a search for additional DNA polymerase activities. This search was facilitated by the reduced levels of DNA polymerase activity present in extracts prepared from the Cairns mutant. Shortly thereafter, a number of laboratories reported the identification of two new *E. coli* DNA polymerases, DNA polymerase II (182, 201, 289) and DNA polymerase III (202). Based on the behavior of *polB* mutants, *E. coli* strains that lack DNA polymerase II activity in vitro, it has been concluded that DNA polymerase II does not play a significant role in normal cells in the replication of the bacterial chromosome (54, 144). In contrast, DNA polymerase III was found to be thermosensitive when isolated from temperature-sensitive *dnaE* mutants of *E. coli* (107). *polC*(Ts) [formerly known as *dnaE*(Ts)] mutants rapidly cease DNA synthesis at the nonpermissive temperature (410). These data convincingly demonstrate that DNA polymerase III is the replicative DNA polymerase of *E. coli*.

The assays used for the initial attempts to purify DNA polymerase III utilized gapped DNA templates created by the partial nucleolytic digestion of duplex DNA. It is now recognized that the enzyme purified by this approach represents the catalytic core of DNA polymerase III holoenzyme. Early work demonstrated that DNA polymerase III core enzyme contains a large subunit, α, of M_r 130,000 (237, 316). Later, however, it was discovered during studies of homogeneous core polymerase that two additional subunits, ϵ and θ, of M_r 27,500 and M_r 10,000, respectively, were also present, roughly in stoichiometric amounts (265). Biochemical analysis of the enzymatic properties of the DNA polymerase III core yielded disappointing results. The catalytic enzyme has few of the properties expected of a replicative DNA polymerase. DNA synthesis on long single-stranded templates is feeble and is limited to just a small number of residues per binding event (85, 237).

With the establishment of soluble enzyme systems that could synthesize the full-length complementary strand of phage M13 and ϕX174 chromosomes, it was possible to purify the cellular DNA polymerase that carries out polymerization of these long chains. Using these natural chromosomes as templates instead of nuclease-damaged DNA, the Kornberg laboratory isolated a more physiological and substantially more complex form of DNA polymerase III, termed DNA polymerase III holoenzyme (holoenzyme) (267). The most highly purified preparations of this enzyme have as many as 13 polypeptides (268), of which at least 7 (Table 1) appear to be bona fide subunits as determined by genetic and biochemical studies (268, 269). All of these polypeptides cosediment in glycerol gradients as an 11S complex. A conclusive subunit structure for the DNA polymerase III holoenzyme remains elusive 10 years after its purification to near homogeneity. The pronounced tendency of the holoenzyme to dissociate into partially active subassemblies during purification has resulted in some variation, from preparation to preparation, in polypeptide composition and subunit stoichiometries. The relatively weak stability of the DNA polymerase III holoenzyme has made it possible to purify several of the auxiliary subunits by assaying for their capacity to stimulate DNA synthesis by the catalytic core on primed, long ssDNA (reviewed by Wickner [419]). It seems likely that a better understanding of holoenzyme structure may come from studies aimed at reconstituting the intact holoenzyme from preparations of the purified subunits. This approach has been made feasible by advances in recombinant DNA technology that allow researchers to clone the genes that specify the individual holoenzyme subunits and to amplify greatly the normally meager intracellular levels of these polypeptides.

The genes coding for several of the auxiliary subunits of DNA polymerase III holoenzyme have been identified, with dnaN coding for the β subunit and dnaZ specifying the γ subunit. The genetic history of the E. coli dnaZ gene is somewhat complex, however, and current studies of this locus raise questions about the genetic source and nature of several of the polypeptide subunits of the DNA polymerase III holoenzyme. The E. coli dnaZ gene was originally defined by a mutation causing a fast-stop defect in DNA synthesis

and mapped at 10.5 min (88). Subsequently, Henson et al. (139) reported the discovery of a neighboring gene, dnaX, mutations in which resulted in a similar phenotype. These two genes were originally thought to lie at near but separate loci because plasmids from the Clarke-Carbon collection could be isolated which complemented mutations in only dnaX or dnaZ or both (139).

Hübscher and Kornberg (153) reported that the dnaZ gene product was the γ subunit (M_r 52,000) of the DNA polymerase III holoenzyme because purified γ protein complemented extracts from dnaZ(Ts) strains for ϕX174 DNA replication in vitro. Small DNA fragments, approximately 2 kb in length, were shown to direct the synthesis of two polypeptides, one of M_r 75,000 to 78,000 and the other of M_r 52,000 to 56,000 (187, 290). The M_r 78,000 polypeptide was shown by two-dimensional gel electrophoresis to be identical to the τ subunit of DNA polymerase III holoenzyme and to be related to the M_r 52,000 polypeptide by partial proteolysis (187). Since the smaller M_r 56,000 protein is not made in vitro (290), it is likely that it is derived from an in vivo posttranslational cleavage of the τ subunit. The nucleotide sequence of the dnaXZ-containing segment has been determined and shown to contain only a single substantial open reading frame (93a, 435a). Antibodies raised from a predicted amino-terminal peptide cross-react with both the γ and τ subunits, indicating that the two proteins are related at their amino-terminal ends (Walker, personal communication). It has been reported that the purified δ holoenzyme subunit restores DNA polymerizing activity in a dnaX(Ts) extract (152). However, it is not clear how δ is related, if at all, to the γ and τ subunits. Obviously, both genetic and biochemical approaches will be needed to resolve the many remaining questions about the structure of the DNA polymerase III holoenzyme.

The physical and functional properties of the DNA polymerase III holoenzyme have been recently reviewed (264). In addition to its $5' \rightarrow 3'$ DNA polymerase activity, which is carried on the large M_r 130,000 α subunit (251, 252) specified by the dnaE gene (413), the holoenzyme has been reported to contain two associated nuclease activities, a $3' \rightarrow 5'$ exonuclease and a $5' \rightarrow 3'$ exonuclease. However, in contrast to the situation for DNA polymerase I, neither nuclease activity is carried on the same polypeptide as the DNA polymerase activity. Initial studies demonstrated that the $3' \rightarrow 5'$ exonuclease activity resided in the catalytic core of DNA polymerase III (238, 265). Later, it was proven that this activity is carried by the ϵ subunit of DNA polymerase III (347), which is the product of the dnaQ (mutD) gene (348). The $3' \rightarrow 5'$ exonuclease activity requires ssDNA with 3'-hydroxyl termini and yields 5' mononucleotides as products. There exists strong genetic and biochemical evidence that the role of the $3' \rightarrow 5'$ exonuclease activity in DNA replication in vivo is to improve the fidelity of DNA synthesis by removing noncomplementary nucleotides at the growing point of the DNA chain (69, 73, 83).

It also has been reported that the DNA polymerase III holoenzyme and its catalytic core contain an unusual $5' \rightarrow 3'$ exonuclease activity that requires 5' single-stranded termini and degrades processively into neighboring duplex DNA after digestion of the single-stranded extension (238, 265). A question of the validity of the 5′ → 3′ exonuclease as an intrinsic activity of DNA polymerase III holoenzyme has been raised by two recent studies that failed to find significant 5′ → 3′ exonuclease activity associated with DNA polymerase III preparations (252, 347). It is perhaps notable that no physiological role for the 5′ → 3′ exonuclease activity has been demonstrated and there is no evidence that this activity is involved in the excision of RNA primers. The DNA polymerase III holoenzyme will usually extend a DNA chain one to five nucleotides beyond the 5′ terminus of a RNA primer that is annealed to the template strand in its path (299). The 5′ terminus of the RNA primer is displaced, but not hydrolyzed.

The auxiliary subunits of DNA polymerase III holoenzyme have received increasing attention. In contrast to the limited processivity of the DNA polymerase III catalytic core, the holoenzyme form of this DNA polymerase can rapidly synthesize a 5,000-nucleotide DNA chain on SSB-coated G4 DNA without dissociating (85). This 500-fold-enhanced processivity appears to be one of the major alterations caused by the interaction of the four auxiliary holoenzyme subunits, β, τ, γ, and δ, with the catalytic core. Several of the auxiliary subunits appear to contribute to the increased processivity (85, 86). For example, addition of the τ subunit to the catalytic core to generate DNA polymerase III′ (263) produces a severalfold increase in the processivity of DNA polymerization. In the presence of SSB, DNA polymerase III*, the subassembly of DNA polymerase III holoenzyme that contains τ, γ, and δ and only lacks the β subunit, is more processive than DNA polymerase III′. Finally, the central role of the β subunit as a processivity factor was demonstrated by the finding that the conversion of DNA polymerase III* to the complete holoenzyme induces a greater than 25-fold increase in processivity. Likewise, the processivity of the DNA polymerase III core and that of the free α subunit are each increased by the presence of a large excess of the β subunit (214, 253). These data suggest that the β subunit interacts with α to clamp the DNA polymerase onto the primer-template.

Kuwabara and Uchida (213) have shown that the Ts phenotype of certain dnaE(Ts) mutations (sueA) can be compensated by dominant extragenic suppressor mutations located in the dnaN gene. This result suggests a direct protein-protein interaction between the α and β subunits of the holoenzyme.

After the synthesis of the complementary strand of phage G4 by DNA polymerase III holoenzyme, the β subunit remains bound to the replicative form II product, presumably still as a constituent of the holoenzyme (165). Kinetic studies also indicate that the holoenzyme dissociates very slowly from the completed product. In the presence of excess primed template, a 5,000-nucleotide DNA chain can be synthesized in just 15 s. However, the transfer of the holoenzyme from the replicative form II product to another primed template takes minutes and, furthermore, requires hydrolysis of ATP or dATP (52, 165). The dissociation rate of the polymerase from a completed Okazaki fragment on the lagging strand of a replication fork in vivo must be considerably faster, since there are only 10 to 20 molecules of DNA polymerase III holoenzyme

per cell and two new Okazaki fragments are completed every second. The discovery that holoenzyme recycling to a new primer occurs at a much higher rate when the next available primer is located on the same template partially resolves this apparent problem (52, 298). In this situation the holoenzyme can complete one chain and, within 1 s or less, locate an available primer and initiate synthesis of a new DNA chain.

An additional factor that needs to be considered in regard to the kinetics of recycling of the DNA polymerase III holoenzyme on the lagging-strand side of a replication fork is the strong possibility that this DNA polymerase functions as a dimer at the replication fork, thereby enabling it to synthesize both daughter strands concurrently (197, 264, 266). A significant finding in support of the dimeric holoenzyme model was the observation that the catalytic core of DNA polymerase III dimerizes upon addition of the τ subunit of the holoenzyme (263). More recent studies by the Kornberg laboratory of the subunit composition and stoichiometry of the holoenzyme also support the dimeric holoenzyme hypothesis (253).

McHenry and colleagues (264, 266) have extended this model by suggesting that the two halves of the dimeric holoenzyme are asymmetric with distinct functional properties. This functional asymmetry would be anticipated for a dimeric holoenzyme that replicates both leading and lagging strands simultaneously, since only the polymerase that replicates the lagging strand must repeatedly undergo association-dissociation cycles. The biochemical evidence in support of this model rests on the effect of adenosine 5′-O-(3-thiotriphosphate) (ATPγS) on the formation of initiation complexes by DNA polymerase III holoenzyme with primed DNA. Formation of an active initiation complex by holoenzyme on primed G4 DNA requires the presence of ATP or dATP (51, 417, 426) and the participation of the β subunit of the holoenzyme (164, 416). McHenry and colleagues (166) have demonstrated that ATPγS will substitute for ATP in supporting initiation complex formation. However, the use of ATPγS led to a twofold reduction in the extent of initiation complex formation. Furthermore, the addition of ATPγS to preformed initiation complexes caused 50% of the complexes to rapidly dissociate (166). These data suggest that the holoenzyme preparation could contain two species of DNA polymerase, one which can form an initiation complex in the presence of the ATP analog and a second form that requires ATP for initiation complex formation and that dissociates from the primer in the presence of ATPγS. McHenry has made the intriguing proposal that this functional heterogeneity is not caused by the presence of two independent species of holoenzyme, but instead arises from differences between the two halves of a dimeric holoenzyme (264). If so, it remains to be determined whether the two halves have different subunit compositions or whether the asymmetry could be induced by allosteric interactions.

DNA polymerase I

DNA polymerase I, the prototype of DNA polymerases, was discovered by Arthur Kornberg and colleagues in extracts of E. coli (200) just 3 years after the elucidation of the double-helical structure of duplex

DNA by Watson and Crick. Detailed studies of the structure and mechanism of DNA polymerase I by the Kornberg laboratory provided the conceptual foundation for current research on nucleic acid biosynthesis. The reader is referred to Kornberg's monographs on DNA replication (196, 197) for a detailed summation of the properties of DNA polymerase I.

DNA polymerase I is a single polypeptide of M_r 103,000 (167, 195) that contains three enzymatic activities involved in the synthesis or hydrolysis of phosphodiester bonds (195). It contains the prototypical $5' \rightarrow 3'$ DNA polymerase activity that functions by addition of mononucleotide residues from dNTPs to the 3'-hydroxyl terminus of a primer chain. DNA polymerase I also has two exonuclease activities. Of central importance to the fidelity of DNA synthesis is a $3' \rightarrow 5'$ proofreading exonuclease that cleaves a non-base-paired 3'-hydroxyl terminus to yield 5'-deoxymononucleotides (47). DNA polymerase I also contains a $5' \rightarrow 3'$ exonuclease that only degrades base-paired DNA from a 5' terminus, releasing oligodeoxynucleotides up to 10 residues long (196). After binding at a nick or short gap in duplex DNA, DNA polymerase I can move a nick linearly along the duplex, a reaction that has been termed nick translation. It accomplishes this feat by coordinating DNA synthesis with $5' \rightarrow 3'$ exonucleolytic digestion ahead of the growing DNA chain.

Digestion of DNA polymerase I with certain proteases yields two active fragments: a large fragment of M_r 68,000 from the C-terminal two-thirds of the polypeptide (i.e., the Klenow fragment) that contains the DNA polymerase and $3' \rightarrow 5'$ exonuclease activities, and a small fragment of M_r 35,000 from the N terminus that contains the $5' \rightarrow 3'$ exonuclease activity (46, 180). The crystal structure of the large proteolytic fragment has been recently determined at 3.3 Å (0.33 nm) resolution (312). The structure of this polypeptide has two domains, the smaller of which appears to contain the active site of the $3' \rightarrow 5'$ exonuclease. The larger domain contains the C-terminal portion of the DNA polymerase I polypeptide. The most striking feature of this domain, which presumably contains the active site of the DNA polymerase, is a deep cleft that is of the approximate size and shape for binding B-DNA. In fact, the structure suggests that the enzyme may be able to surround the DNA substrate completely, perhaps providing thereby a mechanism for enhancing the processivity of the enzyme in DNA chain elongation.

The isolation by DeLucia and Cairns (70) of the polA1 mutant of E. coli, a viable mutant that contained less than 1% of the normal intracellular level of DNA polymerase I, at first led to the conclusion that replication of the bacterial chromosome did not depend on the action of DNA polymerase I. Later, however, the isolation of conditional-lethal polA mutants that produce a thermosensitive DNA polymerase I convincingly demonstrated that this DNA polymerase was not dispensable for viability. Physiological studies of several polA mutants, combined with detailed characterization of the enzymatic properties of mutant DNA polymerase I proteins isolated from these strains, suggested that the nick-translation activity of DNA polymerase I is essential for discontinuous DNA replication. Joining of nascent DNA fragments was found to be severely retarded in conditional-lethal polA strains (207, 310, 393), and in each case the primary enzymatic defect in the mutant DNA polymerase was identified as a loss of the enzyme's capacity to carry out its $5' \rightarrow 3'$ exonuclease function (231). Furthermore, the level of RNA-linked DNA fragments was found to be significantly increased in polA(Ts) strains at the restrictive temperature (302). These findings are all consistent with the hypothesis that the $5' \rightarrow 3'$ exonuclease activity of DNA polymerase I functions in vivo to remove RNA primers present on the 5' termini of nascent DNA fragments and that this excision reaction is coordinated with gap-filling DNA synthesis. That DNA polymerase I indeed is capable of coordinating excision of RNA primers with gap filling was demonstrated in an in vitro system by Westergaard et al. (414).

Recently, a second pathway of DNA replication has been reported at nonpermissive temperatures in E. coli strains carrying various dnaE(Ts) alleles (295). This pathway has been shown to be dependent on both PolA+ activity and the presence of a cryptic mutation, called pcbA, which lies near dnaA (49). The exact mechanism by which this dnaE bypass pathway functions has not yet been delineated.

DNA ligase

Prompted by models of DNA recombination and repair that proposed that E. coli contained an enzymatic activity that joined DNA chains, several different laboratories in 1967 reported the nearly simultaneous isolation of E. coli DNA ligase (108, 113, 311). DNA ligase is a single polypeptide of M_r 75,000 that catalyzes the joining of adjacent 3'-hydroxyl and 5'-phosphoryl termini of nucleotides that are hydrogen-bonded to a complementary strand. Synthesis of a new phosphodiester bond in this manner requires the simultaneous splitting of a pyrophosphate bond in NAD, which serves as a cosubstrate in the reaction. The properties and enzymatic mechanism of DNA ligase are reviewed by Lehman (230).

With Okazaki's discovery that at least one of the strands at an E. coli replication fork is synthesized discontinuously via small nascent fragments that are rapidly joined together, it became apparent that DNA ligase, which is the sole E. coli enzyme that can catalyze DNA joining, also functions in the DNA-elongation phase of chromosomal DNA replication. Strong support for this conjecture was yielded by studies of DNA replication in E. coli mutants that produce a thermosensitive DNA ligase. In the conditional-lethal mutant strain with the most severe ligase defect, a pronounced and prolonged accumulation of nascent fragments is observed at the restrictive temperature as compared to the behavior of a wild-type strain (230). This accumulation of nascent fragments accompanies a rapid loss in cell viability. Moreover, the extent of nascent fragment accumulation in vivo in various ligase mutants correlates well with the severity of the ligase defect measured in vitro. At the same time, however, these physiological studies provided clear evidence that DNA ligase is apparently present in E. coli in large excess over the amount needed to sustain replication and repair. Assuming that there are approximately 100 joining events per min per cell during replication of the bacterial chromosome at 30°C and that there is sufficient intracellular DNA ligase to catalyze sealing of approximately 7,500 single-strand breaks per min at 30°C, ligase activity needs to be reduced greater than 50-fold before effects on joining of nascent fragments become apparent (230).

Additional support for a role for DNA ligase in chromosomal DNA replication comes from studies of soluble enzyme replication systems for the replication of M13 and φX174 viral DNAs or *oriC* minichromosomes. In each system, production of covalently closed, circular duplex DNA requires the addition of DNA ligase to the system (104, 196, 414).

Proteins That Act in the Initiation of DNA Replication at *oriC*

DnaA protein

DnaA protein function is essential for the viability of wild-type *E. coli* cells. As discussed in detail later in this chapter, genetic and physiological studies indicate that the DnaA protein acts as the replicon-specific initiator protein first postulated in the replicon model of Jacob et al. (159). The critical importance of the DnaA protein is also underscored by biochemical analysis of the replication of *oriC* minichromosomes in soluble enzyme systems. In these studies the DnaA protein is demonstrated to be the bacterial replication protein that provides the pivotal recognition specificity for DNA sequences at *oriC* (103, 199).

Isolation of the DnaA protein was afforded by the use of specialized transducing phages that carry the *dnaA* gene (58) or by placement of the cloned *dnaA* gene into plasmids under the control of the powerful λ p_L promoter (103). The DnaA protein was isolated in a functional form by assaying for its capacity to specifically bind to *oriC* DNA (58) or for its ability to promote the replication of *oriC* minichromosome plasmids in a soluble enzyme system (103). Purified DnaA protein, which has an M_r of 52,500 (128), is active as a monomer, but has a strong tendency to aggregate into multimeric forms that have a reduced specific activity (103).

DnaA protein binds specifically to multiple sites in the minimal 245-bp *oriC* sequence delimited by Oka et al. (308). The strongest interactions occur at four conserved 9-bp sequences termed DnaA boxes, each of which contains the common sequence 5'-TTAT^C_ACA^C_AA-3' (101) (see Fig. 4 and 5), and at a fifth site, located between the first and second DnaA boxes, which contains a closely related sequence (259). DnaA protein interacts with *oriC* DNA in a highly cooperative fashion, judging from DNase I footprinting and filter-binding analysis (101). The size of the DnaA protein-*oriC* DNA complex, revealed by electron microscopy, suggests that 20 to 30 DnaA protein monomers interact both with *oriC* DNA and with each other to form a large nucleoprotein structure that contains 200 to 250 bp of origin DNA (101, 199). Little is known about how the origin DNA is arranged in this structure, but it has been inferred that the DNA is wrapped on the surface of a DnaA protein core, as judged by the accessibility of *oriC* DNA in the nucleoprotein complex to cleavage by DNase I (101). Clearly, additional biochemical and biophysical studies need to be pursued to provide a more detailed picture of the nature of this key nucleoprotein structure, which is apparently the first intermediate formed in the replication of *oriC* minichromosomes in vitro (23, 101, 199).

As discussed elsewhere in this chapter, DnaA protein also binds to single 9-bp DnaA boxes. Such DnaA boxes are located near the replication origins of plasmids pSC101 and ColE1 and in the regulatory regions of a large number of genes whose products are involved in the metabolism or biosynthesis of DNA (101). Genetic and physiological experiments have demonstrated that the DnaA protein negatively regulates the expression of several of these genes (see below and chapter 98). Both electron microscopic and nuclease protection studies indicate that DnaA protein forms large nucleoprotein structures at a single DnaA box. It is presumed that binding of DnaA protein to its 9-bp recognition site nucleates the polymerization of DnaA protein along the DNA duplex in the regions surrounding the DnaA box (101).

One intriguing feature of the DnaA protein that was only recently detected is its capacity to avidly bind ATP (K_d = 0.03 μM) (199). In the presence of DNA the bound ATP is very slowly hydrolyzed to ADP (turnover number ≤ 1; A. Kornberg, personal communication). ADP also binds with great affinity to DnaA protein (K_d = 0.10 μM) and competes with ATP for the ATP-binding site. In fact, the capacity of the DnaA protein to promote initiation of DNA replication at *oriC* in vitro is completely abolished when this *E. coli* initiator is preincubated with ADP (1 μM) before its addition to the soluble enzyme replication system. The disabled DnaA protein, however, is slowly reactivated ($t_{1/2}$ = 30 min) during a subsequent incubation in the presence of 1 mM ATP (199). The effects of ATP binding and hydrolysis on the replicative and regulatory functions of DnaA protein will undoubtedly receive close scrutiny in the near future.

DnaC protein

The *dnaC* gene product is essential for the viability of *E. coli*. Most temperature-sensitive *dnaC* mutants exhibit a slow-stop DNA replication phenotype when they are placed at a nonpermissive temperature (410), suggesting that the DnaC protein performs an essential role in the initiation of DNA replication at the chromosomal origin. Involvement of the DnaC protein in the DNA elongation process cannot be excluded, however, since a significant number of *dnaC* mutants display a fast-stop phenotype at the restrictive temperature (408). The discovery that the conversion of φX174 viral DNA to the duplex replicative form depends on the *dnaC* gene product elicited the establishment of a physiological assay for the purification of the DnaC protein. Purification of the DnaC protein to homogeneity proved to be a difficult task, however, and was not accomplished for another 10 years, in part owing to its innate instability and to its pronounced tendency to undergo nonspecific hydrophobic interactions with many standard chromatographic resins (184, 220). Fortunately, DnaC protein can be purified in sufficient quantities for biochemical analysis now that plasmid expression vectors containing the *dnaC* gene have been constructed that yield 150-fold overproduction of the DnaC protein (183).

DnaC protein, a monomer of M_r 29,000, has no known enzymatic activities. However, as discussed earlier, the DnaC protein does specifically interact with the DnaB helicase to form a stable $(DnaB)_6$-$(DnaC)_6$ protein complex. DnaC protein has little effect on the ssDNA-dependent ATPase activity of DnaB protein (185). Nevertheless, the physical DnaB-DnaC protein interaction is manifested by a modification of other functional properties of the DnaB protein. For example, the general priming reaction (5) catalyzed by DnaB protein and primase is stimulated severalfold by the presence of DnaC protein (273). This activation may reflect the capacity of DnaC protein to stabilize the binding of the DnaB helicase to ssDNA (8). A better understanding of the roles of the DnaC protein in chromosomal DNA replication may come from an examination of the effect of DnaC protein on the recently discovered helicase activity of DnaB protein.

Although the precise role of the *dnaC* gene product in *E. coli* DNA replication remains to be determined, some clues have been revealed through extensive biochemical studies of φX174 and *oriC* minichromosome replication in purified protein systems. In the case of φX174, the DnaC protein, assisted by proteins n, n', n″, i (DnaT), and SSB, participates in the assembly of DnaB protein into the φX primosome, a multiprotein priming complex that is formed on the φX174 viral strand at the complementary strand origin prior to the initial priming event (9, 185, 272, 412, 419). Likewise, the primary function of the DnaC protein in the replication of *oriC* minichromosomes apparently is to deliver DnaB protein to the *E. coli* replication origin, a process that depends on specific interactions of the DnaB-DnaC protein complex with DnaA protein bound at *oriC* (23, 199, 394). A more detailed description of the molecular events that take place in the initiation of DNA replication at *oriC* is provided later in this chapter.

It is not certain whether DnaC protein plays any role in the several steps identified for φX174 DNA replication that occur subsequent to the assembly of the primosome; these steps include translocation of the primosome along the SSB-coated template, RNA primer synthesis, and DNA chain elongation. Kobori and Kornberg (185), utilizing ^{3}H-DnaC protein, have found that the DnaC protein is not tightly associated with isolated φX174 preprimosome nucleoprotein structures. Yet, such structures have been shown to undergo all later steps in the φX complementary-strand synthesis pathway when supplemented with primase, DNA polymerase III holoenzyme, and dNTPs. Nevertheless, it cannot be concluded that DnaC protein plays no role in the steps that follow φX primosome assembly. This cautionary note is necessary, for if synthesis of the φX174 complementary strand is carried out in a single-stage reaction mixture containing all required proteins, at least three monomers of DnaC protein appear to be specifically associated with the primosome that remains bound to the duplex replicative form product (185, 247).

DNA gyrase

The chromosomes of *E. coli* and *S. typhimurium* are maintained in a torsionally strained state consisting of approximately 50 topologically independent domains of negatively supercoiled DNA (80, 114, 275, 359, 360, 430). Several types of DNA transactions, such as initiation of transcription and initiation of duplex DNA replication and site-specific DNA recombination, have been found to be facilitated in certain cases by activation of the DNA substrate into a negatively supercoiled state (80, 114). DNA gyrase, the enzyme responsible for introducing negative supercoils into bacterial and plasmid chromosomes in *E. coli*, was discovered by Gellert and colleagues (116), being initially detected as an *E. coli* factor that stimulates site-specific recombination by the bacteriophage λ integrase protein in vitro.

Enzymology of DNA gyrase. DNA gyrase catalyzes a number of reactions: (i) ATP-dependent negative supercoiling of covalently closed, circular duplex DNA; (ii) DNA-dependent hydrolysis of ATP; (iii) relaxation of negatively supercoiled DNA in the absence of ATP; and (iv) formation and resolution of catenated and knotted duplex DNA. In each type of reaction, *E. coli* DNA gyrase alters the linking number (the number of times one strand of duplex DNA is wound around the other) in steps of two (44), apparently by making transient double-strand breaks in the DNA and passing another segment of duplex DNA through the enzyme-bridged break, which is then resealed.

Neither the structure of the DNA gyrase-DNA complex nor the mechanics of the strand passage reaction are understood in detail, but prominent features of the protein-DNA interaction have been ascertained. The breaks produced by DNA gyrase are staggered; each break contains a 4-base extension of ssDNA at each 5' terminus. Examination of the DNA sequences present at DNA gyrase cleavage sites has uncovered a limited sequence specificity (92, 239, 288) that appears to be the same both in vivo and in vitro. DNA gyrase protects about 140 bp of DNA from digestion from nucleases (235). Moreover, the nuclease cleavage patterns suggest that most of the protected DNA is wrapped on the surface of the enzyme (235). This wrapping appears to create a positively supercoiled domain in the DNA (236). A transient double-strand break within this right-handed coil, followed by a specifically oriented translocation of another DNA segment from the same DNA molecule through the enzyme-bridged break, would lead to the production of negative supercoils in the DNA substrate. Recent experiments with the technique of transient electric dichroism suggest that a single turn of DNA is wrapped around DNA gyrase in the absence of ATP, with DNA tails entering and exiting the complex in close proximity (M. Gellert, personal communication), a view that is consistent with images of the DNA gyrase-DNA complex obtained by electron microscopy (177). Clear structural changes are observed by transient electric dichroism when the enzyme binds ATP or a nonhydrolyzable analog of ATP. It has been concluded that the binding of ATP probably causes the DNA tails to become wrapped around the protein (Gellert, personal communication). This finding is consistent with the discovery by Cozzarelli and colleagues (374) that the binding of ATP to DNA gyrase elicits conformational changes that yield a single round of supercoiling. They have postulated that nucleotide hydrolysis simply facilitates the return of DNA gyrase to its original conformation, thereby permitting enzyme turnover.

DNA gyrase contains two polypeptides, gyrase A protein (M_r 97,000) and gyrase B protein (M_r 90,000), which are present in equimolar amounts. The enzyme exists in solution as an A_2B_2 tetramer. Shortly after the discovery of DNA gyrase, it was recognized that each of the two kinds of subunit was the target of known antibiotics. The gyrase A protein is selectively inhibited by oxolinic acid, nalidixic acid, and related compounds (115, 375). Coumermycin A_1, novobiocin, chlorobiocin, and related antibiotics block the function of the gyrase B protein (117). These antibiotics have aided in the assignment of DNA gyrase enzymatic activities to the individual subunits, since neither subunit alone demonstrates any significant activity. For example, denaturation of DNA gyrase-duplex DNA complexes formed in the presence of oxolinic acid results in the production of double-strand breaks in which a gyrase A promoter is covalently bonded through a phosphotyrosine linkage to each 5' terminus created by the break (288, 390). Inasmuch as similar phosphotyrosine linkages have been identified as presumptive intermediates in the nicking-closing reactions carried out by other DNA topoisomerases (405), it seems certain that the gyrase A subunit mediates the transient breakage-rejoining reactions catalyzed by DNA gyrase. The DNA-dependent ATPase activity of DNA gyrase resides on the gyrase B protein, judging from the capacity of the gyrase B antagonist coumermycin A_1 to competitively inhibit both ATP hydrolysis and supercoiling (374) and from the recent discovery that heat or urea treatment of gyrase B protein greatly stimulates a latent DNA-independent ATPase activity (262).

Roles of DNA gyrase in DNA replication. DNA gyrase is believed to play a crucial role in each of the three major stages involved in the replication of the *E. coli* chromosome: initiation, propagation, and termination. This conclusion rests in part on the behavior of *E. coli* cells harboring conditional-lethal mutations in the *gyrA* or *gyrB* gene, on the cellular effects of antibiotics that inhibit DNA gyrase function, and on biochemical analysis of the replication of *oriC* minichromosomes in purified protein systems. The in vitro studies will be discussed separately later in this chapter.

Elongation phase of DNA replication. Cairns pointed out, in his description of the circular nature of the bacterial chromosome, that replication of circular duplex DNA would require the presence of a "swivel" to permit the two highly intertwined parental strands to be separated from one another (53). In the absence of a swivel, during each minute of fork movement approximately 10^4 positive supercoils would accumulate in the unreplicated portion of the bacterial chromosome. Positive superhelicity would increase until the rising torsional strain precluded further unwinding of parental DNA at the replication fork. DNA topoisomerases, enzymes which catalyze the reversible breakage and rejoining of DNA strands, are ideally suited to provide the necessary swivel.

Three DNA topoisomerases have been identified thus far in *E. coli*. DNA topoisomerases I (originally designated ω protein) and III are type I DNA topoisomerases, enzymes that make single-strand breaks. However, the available evidence suggests that neither of these type I topoisomerases functions as the major

cellular swivel. Neither DNA topoisomerase I nor DNA topoisomerase III is capable of relaxing positively supercoiled DNA in vitro (405), and furthermore, *E. coli* mutants carrying a deletion of the *topA* gene, which encodes DNA topoisomerase I, grow normally (368). In contrast, there is ample experimental support for the concept that it is DNA gyrase that serves as the swivel that reduces the linking number between the parental strands during replication of the bacterial chromosome.

Numerous studies, summarized by Drlica (80), have investigated the effect on growing cells of antibiotics specific for DNA gyrase. Bacterial DNA synthesis abruptly slows upon the addition to the growth medium of either of the two classes of drugs known to affect DNA gyrase function. Interestingly, the inhibitory effect of the oxolinic acid class of drugs occurs in the absence of any decrease in the number of titratable negative supercoils in the bacterial chromosome. This finding raises the possibility that DNA gyrase functions directly in the propagation of a replication fork. A careful examination of the mode of action of nalidixic and oxolinic acids in vivo, however, demonstrated that these gyrase A protein inhibitors promoted the formation of tight DNA gyrase-DNA complexes that apparently interfere with the movement of replication forks (80, 114). More readily interpretable results come from studies of *E. coli* cells carrying thermosensitive mutations in either the *gyrA* or *gyrB* gene (91, 205). Because DNA synthesis in *gyrA* and *gyrB* mutants ceases more rapidly at the nonpermissive temperature than would be expected for a defect in initiation of chromosome replication, it has been concluded that DNA gyrase plays an obligatory role in the propagation of the replication apparatus along the bacterial chromosome. Precisely how DNA gyrase aids the elongation process cannot be determined from these physiological studies. DNA gyrase could function directly in fork movement by acting as an essential component of the enzymatic machinery. Also, in addition to providing the swivel needed for replication of topologically constrained chromosomes, DNA gyrase may have other indirect roles in the elongation stage of DNA replication. For example, DNA gyrase might facilitate unwinding of the parental DNA chains at the fork or the binding of replication proteins by maintaining the replicating chromosome in an underwound, negatively superhelical state.

Initiation of DNA replication. The concept that DNA gyrase or DNA gyrase-mediated supercoiling also plays an essential role in the initiation of *E. coli* chromosomal DNA replication is supported by both genetic and biochemical studies. Certain temperature-sensitive *gyrB* mutants display normal DNA elongation rates when they are first shifted to a restrictive temperature, but after continued incubation at the nonpermissive temperature they exhibit reduced rates of DNA synthesis, consistent with a defect in the capacity to initiate new rounds of DNA replication (90, 313). The properties of an *E. coli* mutant with a temperature-sensitive *dnaA46* mutation are also consistent with a role for DNA gyrase in the initiation process. The frequencies at which spontaneous mutations conferring resistance to nalidixic acid or coumermycin arise at the *gyrA* and *gyrB* loci, respectively, are 20- to 30-fold lower in a *dnaA46* mutant

(89). This fact may indicate that certain mutations in DNA gyrase cannot coexist with the DnaA46 replication initiator, even at temperatures that permit the mutant initiator to function.

Lother et al. (244) have identified a site near the right-hand border of the 245-bp minimal *oriC* sequence (see Fig. 4) to which DNA gyrase preferentially binds. Since a typical DNA gyrase-DNA complex encompasses 140 bp of DNA, the DNA gyrase bound to this site at *oriC* would also cover at least two of the nearby DnaA protein recognition sequences. It is not known whether DnaA protein and DNA gyrase modulate the binding of each other to the *oriC* region. Also uncertain is the function of the DNA gyrase site at *oriC* in the initiation of chromosomal DNA replication.

Funnell et al. (104) have demonstrated that an early step in the initiation of DNA replication at *oriC* in vitro is the formation of a protein-DNA initiation complex at the origin. Formation of this complex does not require DNA gyrase, but does depend on the presence of negative supercoils in an *oriC* minichromosome. More recent studies from the Kornberg laboratory suggest that localized unwinding of the DNA duplex at or near *oriC*, observed to be an early step in the initiation process (23), also requires a negatively supercoiled template, but does not initially depend on the presence of DNA gyrase (199; T. Baker and A. Kornberg, personal communication). These studies provide the clearest demonstration that DNA gyrase plays an indirect role in the initiation of DNA replication at *oriC*, primarily serving to provide a negatively supercoiled template needed for the binding of certain initiation factors (described in more detail later in this chapter).

Termination of DNA replication. Resolution of two newly replicated *E. coli* chromosomes into individual monomeric chromosomes requires that the linking number of the two parental strands be reduced to exactly zero. There is increasing evidence in both procaryotic and eucaryotic systems (407), however, that at the precise moment when replication of many circular DNA molecules is completed, as many as 10 to 30 intertwinings between the parental strands remain to be eliminated. Thus, highly intertwined catenated dimers may be obligatory intermediates in the replication of many circular duplex DNA molecules, as initially proposed by Skalka and co-workers (363) and by Gefter and Botstein (106). This intertwining may be a consequence of the inaccessibility of unreplicated parental DNA at the chromosome terminus to DNA topoisomerases (376). It is possible that DNA topoisomerases are sterically excluded from binding to the terminus by the presence of leading- or lagging-strand (or both) replication machinery that does not immediately dissociate from terminal sequences following the collision of opposing replication forks.

Both DNA gyrase and DNA topoisomerase I are enzymatically capable of decatenating linked pairs of duplex DNA circles, but the latter protein requires that at least one of the DNA molecules of each pair contain a single-strand interruption or gap. Physiological studies of a temperature-sensitive DNA gyrase mutant of *E. coli* have provided clear evidence that DNA gyrase participates in the segregation of newly replicated daughter chromosomes (366). Incubation of *gyrB41*(Ts) mutants at a restrictive temperature leads to the formation of bacterial nucleoids that contain an increased mass of DNA. When examined by light microscopy, such nucleoids were found to be primarily dumbbell-shaped doublets. Purified doublet nucleoids were converted to singlets upon incubation with DNA gyrase in the absence of DNA synthesis. This evidence suggests that the doublet nucleoids represent two completely replicated daughter chromosomes that are catenated via one or more interlocks and that segregation of the daughter chromosomes is arrested at the nonpermissive temperature due to a lack of intracellular DNA gyrase activity. DNA topoisomerase I, on the other hand, is not obligatorily required for *E. coli* chromosome segregation, a conclusion based on the fact that *topA* deletion mutants grow normally.

Proteins That Act in the Termination of *E. coli* DNA Replication

The termination of DNA replication in *E. coli* is poorly understood. Although it is known that chromosomal DNA replication terminates at *terC*, located in the 30- to 32-min region on the *E. coli* genetic map (32, 37, 38, 206, 246), whether specific DNA sequences participate in the termination process remains to be determined. It is conceivable that replication simply terminates at the point where the two opposing replication forks collide. Little, too, is known about the identity of the proteins that participate in the termination process. As discussed above, DNA gyrase actively segregates interlocked daughter chromosomes that apparently are formed during termination of DNA replication. Physiological and genetic studies of cell division suggest that termination of DNA replication is tightly coupled to cell division in *E. coli* (136), and gene products that may play a role in this coupling have been identified (28, 81). To date, however, the *E. coli* DnaT protein is the only protein identified with a primary cellular function that may involve direct participation in the termination process per se (221).

DnaT protein

Initiation of new rounds of chromosomal DNA replication normally requires both RNA and protein synthesis (see above; see also chapter 98). Some conditions, however, induce a type of bypass replication, termed stable DNA replication (see below; reviewed in reference 224), in which replication continues in the absence of protein and RNA synthesis. Detailed studies of this phenomenon led to a proposal that, during normal replication of the *E. coli* chromosome, a regulatory protein is incorporated into the processive elongation machinery during either the initiation or propagation stages. In this scheme, the presence of the regulatory protein permits the replication complex to be disassembled during the termination of chromosomal DNA replication. Based on this hypothesis, Lark et al. (221) isolated dominant conditional-lethal mutants of *E. coli* that exhibited the expected phenotype, namely, obligatory termination of DNA replication, temperature sensitivity of DNA replication, and aberrant cell division at permissive temperatures.

Transduction mapping of the *dnaT* gene placed it at 99 min on the standard *E. coli* gene map (20), near the *dnaC* gene. More recently, the *dnaT* gene has been cloned, and its sequence has been determined (256). The gene is separated by just two nucleotide pairs from the *dnaC* locus and is probably transcribed in an operon with the *dnaC* gene. This tight linkage is interesting in view of genetic studies that suggest that the DnaC and DnaT proteins may functionally interact (221, 224).

Masai et al. (256) have presented evidence that strongly suggests that *E. coli* protein i (345) is the product of the *dnaT* gene. Protein i, which is also known as factor X (420), acts in the assembly of the φX primosome in the conversion of the φX174 viral strand to the duplex replicative form (11, 272, 412, 419) and participates in the replication of pBR322 DNA in vitro (280). Protein i, a trimer of a M_r 19,300 polypeptide (11, 256), has no intrinsic enzymatic activities. It has been proposed that protein i acts at a stage late in the primosome assembly pathway requiring the participation of the DnaB and DnaC proteins (11, 419). Nevertheless, the precise molecular role of protein i in the primosome assembly reaction remains uncertain. Inasmuch as the complete replication of *oriC* minichromosomes, including termination of DNA replication, can be reconstituted with a set of purified proteins that contains no protein i (104), it may not prove possible to utilize this in vitro system to elucidate the role of the DnaT protein in *E. coli* chromosomal replication.

Auxiliary Replication Proteins

On the basis of genetic, physiological, or biochemical studies, the proteins listed in Table 2 have been implicated as playing possible roles in *E. coli* DNA replication. These proteins are listed as auxiliary replication proteins for the purpose of this review, since convincing evidence has not been presented that any of these proteins acts directly in chromosomal DNA replication as a constituent of the replication machinery.

DNA helicases

The recent discovery that DnaB protein has DNA helicase activity brings the number of ATP-dependent DNA helicases identified in *E. coli* to four. The other helicases are Rep protein, DNA helicase II (the *uvrD* gene product), and DNA helicase III (111). No genetic or biochemical evidence has been reported that links DNA helicase III to *E. coli* DNA metabolism in vivo. The *E. coli* Rep protein, on the other hand, has been shown to be a true replicative DNA helicase in the replication of phage φX174 or phage fd replicative form DNA (109, 350). Once phage initiator proteins have made a sequence-specific cleavage in the viral strand at the replicative form origin, the Rep protein catalytically unwinds the duplex DNA in a $3' \rightarrow 5'$ direction, aided by the binding of *E. coli* SSB to the ssDNA generated by the unwinding reaction (84). The role of the Rep protein in the replication of the *E. coli* chromosome remains to be proven, however, since all *rep* mutants grow normally. It is possible that the Rep protein plays some auxiliary role in fork movement.

This notion is consistent with the finding that the rate of replication fork movement is moderately reduced in a *rep* strain (219).

It has been suggested that the Rep protein cooperates with DNA helicase II to unwind DNA during replication of the bacterial chromosome (382) and that either Rep protein or DNA helicase II alone could drive fork movement. This hypothesis was inspired in part by the finding in the Hoffmann-Berling laboratory that *uvrD-rep* double mutants appear to be inviable and by enzymatic studies that indicated that the Rep protein and DNA helicase II had opposite polarities of movement during DNA unwinding (111). Thus, Rep protein and DNA helicase II were thought to be capable of coordinating DNA unwinding by acting on opposite strands at the replication fork. This scheme apparently is not tenable, however, now that it has been demonstrated that a bona fide replication protein of *E. coli* is a DNA helicase (228) and that the DNA helicase II protein actually translocates in the same direction ($3' \rightarrow 5'$) (258) as does the Rep helicase (434).

DNA helicase II, a protein of M_r 82,000, acts stoichiometrically relative to the DNA template in its unwinding of DNA (209). Each monomer of DNA helicase II binds and melts approximately 5 bp of duplex DNA in a reaction that is coupled to the hydrolysis of ATP. The intracellular level of DNA helicase II is sufficient to unwind 30 kb of DNA (111), certainly enough to drive unwinding at the replication fork, assuming the helicase is recycled as a consequence of DNA replication. Moreover, there is limited biochemical evidence that supports the concept that DNA helicase II plays some role in *E. coli* DNA replication. Antibodies against DNA helicase II partially inhibit *E. coli* DNA synthesis in a crude in vitro system (181), and DNA helicase II stimulates DNA synthesis by DNA polymerase III holoenzyme at an artificial replication fork (208). Nevertheless, a direct role for DNA helicase II in movement of bacterial replication forks remains unlikely in view of the normal replication behavior exhibited by all DNA helicase II mutants (*uvrD*) thus far examined.

Perhaps DNA helicase II performs a role in *E. coli* DNA replication that is similar to the function of the bacteriophage T4 dda protein in T4 DNA replication. The T4 dda protein is an ATP-dependent DNA helicase that, like DNA helicase II, acts in a stoichiometric fashion to unwind duplex DNA (204). Bedinger et al. (27) have found that movement of T4 replication machinery is completely blocked by a single molecule of *E. coli* RNA polymerase that is bound to a promoter on the template. Furthermore, the T4 replication fork slows dramatically to the rate of a transcribing RNA polymerase molecule when it encounters a transcription complex moving in the same direction. Unimpeded fork movement is once again permitted when the T4 dda protein is added to the in vitro replication system, since replication complexes containing this supplementary helicase cause RNA polymerase molecules to dissociate from the DNA (27). It is not known whether movement of the *E. coli* replication apparatus is also sensitive to the presence of tightly bound proteins in its path. In this connection, it is interesting that DNA helicase II has been shown to stimulate the turnover of tightly bound *E. coli* proteins that act in the repair of UV-irradiated DNA (57).

RNase H

RNase H is a protein of M_r 17,600 that catalyzes the hydrolysis of RNA strands that are hybridized to DNA (172). Physiological studies of RNase H (*rnh*) mutants suggest that RNase H may act in vivo to degrade the 5'-terminal portion of primer RNA that is covalently linked to nascent (Okazaki) fragments produced during discontinuous replication of the bacterial chromosome (303). Since the 5' → 3' exonuclease activity of DNA polymerase I is also capable of removing primer RNA both in vivo and in vitro (see above), it is not possible to determine with certainty which enzyme performs this task in wild type cells. It is notable in this context that of the two enzymes, only RNase H is incapable of removing the ribonucleotide residue at a RNA-DNA junction in vitro (29).

A clearly defined role for RNase H in preventing nonspecific initiation of chromosomal DNA replication has emerged from recent genetic and biochemical studies. The presence of functional RNase H in vivo is needed to restrict initiation of *E. coli* DNA replication to the genetically defined origin, *oriC*. In *rnh* mutants of *E. coli*, at least four new sites, collectively termed *oriK*, are utilized as replication origins, although initiation at *oriC* predominates (71). Interestingly, *E. coli rnh* mutants can tolerate deletion of the normal origin of DNA replication at *oriC* and insertional inactivation of the *dnaA* gene (192). Furthermore, in contrast to the situation in wild-type cells, DNA replication in *rnh* mutants continues in the absence of protein synthesis, a phenomenon termed conditional stable DNA replication (see below) (188). This bypass pathway, which initiates DNA replication at *oriK* sequences, requires transcription (188, 190) and active RecA protein (190). It seems likely that "abnormal" initiation of chromosomal DNA replication at *oriK* sites depends on the formation of stable RNA-DNA hybrid structures, structures that are destroyed by RNase H action in wild-type cells. This view is supported by studies of *oriC* minichromosome replication in vitro which have demonstrated that RNase H functions as a specificity factor to suppress DnaA protein-independent and *oriC*-independent replication of supercoiled plasmid DNA templates (170, 304).

RNase H performs a prominent role in the initiation of ColE1 plasmid replication in vitro by producing the RNA primer for leading-strand DNA synthesis (157, 158). In this system RNase H makes a sequence-specific cleavage of a particular mRNA transcript that is hybridized to the origin of ColE1 DNA replication. Could RNase H function in a similar fashion to generate specific primers at the *E. coli* replication origin? Recent genetic and physiological studies suggest that RNase H plays no direct role either in the initiation of DNA replication at *oriC* or in the regulation of initiation frequency (191).

DNA topoisomerase I

The prototype of DNA topoisomerases, DNA topoisomerase I (ω protein), was discovered in 1969, initially as an activity that converted negatively supercoiled DNA to a more relaxed form (404). Classified as a type I DNA topoisomerase, the enzyme transiently breaks the phosphodiester linkage of DNA in one strand and later reseals the break. Because no energy donor is required for this reaction, it was proposed that a covalent protein-DNA intermediate is formed that conserves phosphodiester bond energy for the strand-closing step (404). This proposal has since been confirmed with the demonstration that a tyrosine in DNA topoisomerase I is linked through its phenolic oxygen to the terminal phosphoryl group at the 5' end of the broken DNA strand (405).

DNA topoisomerase I also catalyzes the knotting of circular ssDNA as well as the catenation and decatenation of duplex DNA circles, although in the latter reaction at least one of the circles must contain a single-strand nick or gap (234). These reactions suggest that DNA topoisomerase I can catalyze the passage of ssDNA and dsDNA chains through enzyme-bridged breaks. Thus DNA topoisomerase I could, at least in theory, participate in the resolution of DNA catenanes that may be formed during the termination of *E. coli* DNA replication. It is notable in this connection that DNA topoisomerase I is much more effective than DNA gyrase in segregating multiply intertwined, catenated dimers of pBR322 plasmid DNA produced in an in vitro replication system (281). It has been proposed that DNA topoisomerase I may serve as the "swivelase" that is needed to reduce the linking number between the two parental DNA strands as replication of the circular bacterial chromosome proceeds (404). This notion appears less likely now that it is recognized that DNA topoisomerase I is incapable of relaxing positively supercoiled DNA that is presumably generated during DNA replication (178). Furthermore, the viability of *E. coli* mutants carrying a deletion of the *topA* gene (368), which specifies DNA topoisomerase I, is convincing proof that this enzyme is not obligatorily required for any of the nicking-closing reactions that accompany replication of the bacterial chromosome.

In studies of *oriC* plasmid replication in a system that is partially reconstituted with purified proteins, several factors that improved the specificity of DNA replication were identified. One such factor, identified as DNA topoisomerase I, both suppressed the initiation of DNA synthesis on templates that lacked the *oriC* sequence and conferred DnaA protein dependence on the replication of an *oriC* plasmid (171). The mechanism by which DNA topoisomerase I improves the specificity of DNA replication in this system has not been determined. However, the fact that the procaryotic enzyme cannot be replaced with its eucaryotic counterpart suggests that DNA topoisomerase I does not improve specificity simply by adjusting the superhelical density of the template (171). A role for DNA topoisomerase I in maintaining replication specificity in vivo is suggested by the discovery that genetic inactivation of this nicking-closing enzyme suppresses the conditional lethality of *dnaA*(Ts) mutations, possibly by permitting chromosomal initiations at other origins (245).

HU protein

Reconstitution, with purified proteins, of a system that can specifically replicate *oriC* plasmids led to the isolation and identification of HU protein, a histone-like DNA-binding protein, as a stimulatory factor (75).

HU protein provides two distinct functions in the replication of *oriC* plasmids in vitro. At relatively low concentrations of approximately 40 HU molecules per DNA template circle, HU acts to stimulate DNA synthesis two- or threefold. At higher concentrations, where it is present in a 1:1 mass ratio with the DNA template, HU protein behaves as a specificity factor to suppress DnaA protein-independent initiation of DNA synthesis in a partially reconstituted *oriC* replication system (170). The exact mechanisms by which HU protein exerts these effects have not been determined (see below).

To this point no corroborating genetic evidence has been obtained to confirm that HU protein serves similar roles in vivo, perhaps because it is difficult to isolate conditional-lethal mutations in HU protein that affect DNA synthesis without also interfering with RNA or protein synthesis. Since HU protein is a major protein component of the bacterial nucleoid and may play a predominant role in the organization of the bacterial chromosome in vivo, it seems likely that many conditional-lethal mutations in this DNA-binding protein would have pleiotropic effects on cellular metabolism. Another factor to be considered with any genetic approach to defining the physiology of HU protein function is that the *E. coli* HU protein is composed of two closely related polypeptides, HUα and HUβ, encoded by the *hupA* and *hupB* genes, respectively (173). Inactivation of either subunit alone may not be sufficient to eliminate HU protein function in vivo. This possibility is suggested by the results of enzymological studies of phage λ DNA replication (269). A potent inhibitor of the initiation of λ DNA replication in vitro has been purified from crude extracts of *E. coli* and identified as protein HU. Interestingly, either of the subunits of protein HU is capable of mediating this inhibition (K. Mensa-Wilmot and R. McMacken, unpublished results). Now that the *E. coli hupB* gene has been cloned (174; J. Rouvière-Yaniv, personal communication), it should be possible to employ insertional inactivation or to delete this gene to determine whether both subunits of HU protein are required for cell viability. The availability of a viable *hupB* mutant would be expected to greatly facilitate the isolation of conditional-lethal mutations in the *hupA* gene.

HU protein is a basic, low-molecular-weight DNA-binding protein with many histonelike properties. Initially identified in *E. coli* (330), this protein is highly conserved in procaryotes; nearly identical proteins are found in cyanobacteria, in archaebacteria, and in other eubacteria (111, 217). It has been isolated in many different laboratories and is also referred to as HD proteins, NS-1 and NS-2 protein, BH2 protein, HLPII, and DNA-binding protein II (111). HU protein binds to both ssDNA and dsDNA and to RNA, but at low ionic strength more than 50% of the cellular HU protein is associated with the bacterial nucleoid (329, 396). This finding, coupled with its capacity to condense duplex DNA into nucleosomelike structures (331) and its great abundance (estimated at 2×10^4 to 1×10^5 copies per cell), strongly suggests that the HU protein plays a prominent role in determining the organization of the bacterial chromosome. For these reasons the physical properties of purified HU protein have received close scrutiny.

HU protein is composed of two heat-stable 90-amino acid polypeptides, HUα (equivalent to HU-2 and NS-2) (M_r 9,225) and HUβ (equivalent to HU-1 and NS-1) (M_r 9,535), which are 69% homologous in sequence. The quaternary structure of purified HU protein apparently varies with protein concentration (241). At concentrations of HU below 10 μM the protein is predominantly a dimer or tetramer, whereas at more physiological HU concentrations (above 50 μM), aggregates of HU tetramers become increasingly important. It is notable in this regard that the presence of duplex DNA contributes to the formation of HU aggregates (218). Both homologous aggregates of HUα or HUβ alone and heterologous aggregates can be formed (241). The details of the interaction of *E. coli* HU protein with dsDNA remain to be elucidated, but some clues have come from the determination of the crystal structure of the closely related HU protein of *Bacillus stearothermophilus* (380). The fundamental protein unit of the crystal is a dimer, stabilized by extensive hydrophobic interactions between monomers. The most striking feature of the dimer is the presence of a helical depression approximately 2.5 nm in diameter that has a concave surface that is exactly complementary to the right-handed double helix of B-DNA. On the basis of this and other published data, Tanaka et al. (380) proposed that HU protein interacts nonspecifically with duplex DNA via electrostatic interactions with the sugar-phosphate backbone.

The discovery that HU protein organizes duplex DNA into beaded, nucleosomelike structures (331) further strengthened the notion that HU protein is the procaryotic equivalent of eucaryotic histones. Nevertheless, the manner in which HU protein interacts with dsDNA differs fundamentally from the prototypical histone-DNA interaction. Nuclease digestion of the HU-DNA complex suggested both that the DNA is wrapped on the surface of HU protein and that the helical periodicity of the bound DNA is reduced to an astonishing 8.5 bp per helical turn from the normal periodicity of 10.5 bp per turn (45). This substantial overwinding of the DNA helix effected by HU protein may explain why HU protein raises the T_m of linear duplex DNA by several degrees (278). At the same time it is known that HU protein restrains one negative supercoil per 275 to 290 bp (45, 331). Taken together, the available data suggest a model in which approximately 1.5 left-handed toroidal coils of DNA are wrapped at 60-bp intervals around a tetramer of HU protein (45). However, in contrast to the case for eucaryotic histones, the increased helical pitch elicited by HU binding almost completely compensates for these tight toroidal writhes. The net effect is that at maximal binding of HU protein only about half of the superhelical tension of negatively supercoiled DNA is restrained (45). This mode of DNA organization is compatible with procaryotic DNA metabolism, which is known to be dependent on negative superhelical tension, e.g., for normal DNA replication and gene expression (80, 114).

In contrast to eucaryotic nucleosomes, the procaryotic nucleosomes are apparently very unstable and can only be observed by electron microscopy if special precautions are taken (123). These observations are consistent with the physical properties of the HU-DNA interaction, which is known to be very sensitive to ionic strength (329). Whereas histones remain tightly bound to duplex DNA at salt concentrations approaching 1 M NaCl, HU protein readily dissociates from DNA at low to moderate salt concentrations, with a dissociation half-life of 0.6 min in 50 mM NaCl (45). The picture of procaryotic chromatin that emerges from these studies is one of a very dynamic structure in which individual stretches of DNA are only transiently condensed. Yet, at the same time, the overall structure of the bacterial chromosome is stably maintained, driven by the high intracellular concentration of free HU protein.

Proteins n, n', and n''

A system that can convert the single-stranded chromosome of phage φX174 to the duplex replicative form has been reconstituted with a set of nine bacterial proteins (5, 344, 345, 420). The number of known *E. coli* replication proteins that participate in the synthesis of the φX174 complementary strand has risen to six with the recent identification of protein i as the product of the *dnaT* gene. The three remaining proteins, proteins n, n', and n'', all act in the prepriming stage of φX174 DNA replication by facilitating the assembly of the primosome, a mobile multiprotein priming apparatus (6, 9, 272). Assembly of the primosome is initiated by the binding of protein n' (equivalent to factor Y of the Hurwitz laboratory) to a specific sequence in the φX174 viral strand, called a primosome assembly site (6, 355, 356, 419, 422). It has been suggested that prepriming proteins n and n'' subsequently interact with the bound n' protein to form a larger nucleoprotein complex that serves as the locus for the assembly of DnaB protein into the primosome (9). As discussed earlier, assembly of DnaB protein into the primosome depends on the function of two additional *E. coli* replication proteins, DnaC protein and DnaT protein (protein i). The precise structure of the primosome is uncertain, but it is believed to contain one molecule each of protein n' and DnaB protein as well as other prepriming proteins and primase (9). Once assembled, the primosome translocates processively 5' → 3' along the viral strand and repeatedly synthesizes RNA primer transcripts that are utilized to initiate DNA chains (9, 272). It has been proposed that ATP (or dATP) hydrolysis by protein n' energizes the processive movement of the primosome along the SSB-coated φX174 chromosome (10). However, studies of a primosomelike priming complex that acts in bacteriophage λ DNA replication in vitro suggest instead that DnaB protein may be responsible for translocation of the primosome (228, 229).

Phage φX174 DNA replication, especially in connection with the mode of action of the primosome, has long served as a model for the molecular mechanisms involved in the discontinuous synthesis of the lagging strand during replication of the *E. coli* chromosome (196, 197, 344, 420). Additional support for this model came from studies of the replication of *E. coli* plasmids. Primosome assembly sites have been identified near replication origins present on certain *E. coli* plasmids, including ColE1, pBR322, and the F fertility factor (155, 254, 296, 439). Moreover, Minden and Marians (280) have demonstrated that the *E. coli* primosomal proteins are required for the

initiation of lagging-strand DNA synthesis in a pBR322 replication system that is reconstituted with purified proteins.

Despite all the circumstantial evidence, there is no definitive proof that a primosome containing protein n' functions directly in *E. coli* DNA replication. In fact, several pieces of recently obtained evidence suggest that the n proteins probably are not obligatorily required for replication of the bacterial chromosome. First, these proteins retain functional activity in each of the currently available replication-defective mutants of *E. coli*. Second, no n'-dependent primosome assembly sites are found close to the *E. coli* replication origin (170, 254), although several putative primosome assembly sites are located in the region 2.0 to 3.3 kb clockwise of *oriC* (370, 395). Finally, recent biochemical studies demonstrate that *oriC* plasmids can be completely replicated in a purified protein system that is devoid of proteins n, n', and n'' (see below) (104, 394). Thus, an auxiliary role for the primosome in *E. coli* DNA replication is suggested. In accordance with this view, W. Seufert and W. Messer (352a) have under certain conditions detected strand initiation events in a crude soluble *oriC* replication system near the putative primosome assembly sites located clockwise of *oriC*. Initiation of DNA chains at these sites was blocked by the presence of antibody directed against protein i (DnaT protein). Moreover, this initiation depends on the presence of *oriC* sequences in the template plasmid and on the presence of a chain-terminating dNTP analog in the system. Seufert and Messer suggest that if DNA replication initiated at *oriC* is terminated prematurely for any reason, then replication might be reinitiated by the assembly of a φX174-type primosome at one of the primosome assembly sites (which first must be converted to a single-strand form by DnaB helicase-mediated unwinding that initiates at *oriC*; see below).

DnaJ and DnaK proteins

The *dnaJ* and *dnaK* genes were originally discovered because mutations in them blocked bacteriophage λ DNA replication at all temperatures (118, 121, 335, 377; reviewed in reference 99). It has been shown that: (i) mutations in the λ *P* gene compensate for these blocks, suggesting that the P protein interacts with both the DnaJ and DnaK proteins; (ii) the two genes form an operon at 0.5 min (336, 436); (iii) both DnaJ protein (M_r 41,000) and DnaK protein (M_r 69,000) are heat shock proteins (25, 122), and the DnaK protein is closely related to the Hsp70 heat shock protein family of eucaryotes (24); (iv) mutations in both genes interfere with bacterial growth at high temperature, resulting in an inhibition of both DNA and RNA synthesis (156, 335, 336, 401); (v) the two gene products appear to be dispensable for bacterial growth at low temperature since each can be deleted or insertionally inactivated (K.-H. Paek, G. Walker, E. Craig, and S. Sell, personal communication; C. Georgopoulos, unpublished data); (vi) the DnaK protein exhibits both autophosphorylating and 5'-nucleotidyl phosphatase activities (34, 440, 442); (vii) the DnaK protein has been shown to be a negative modulator of the heat shock response, inasmuch as *dnaK⁻* mutants overproduce heat shock proteins and overproduction of the

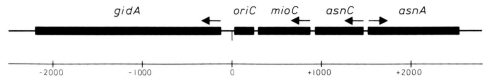

FIG. 3. Genetic and physical map of the *oriC* region of the *E. coli* chromosome. The arrows at the top indicate the directions of transcription. The numbers at the bottom indicate the number of base pairs from position 1, which is defined as the first nucleotide of the *Bam*HI recognition site located immediately counterclockwise (left) of the minimal *oriC* region (274, 308, 373).

DnaK protein dampens the heat shock response (383); and (viii) the DnaJ protein binds to both ssDNA and dsDNA (443). The DnaJ and DnaK proteins both appear to function during a prepriming step in λ DNA replication that involves the transfer of the DnaB helicase onto λ ssDNA near the origin of replication (229, 269). Finally, neither the DnaJ protein nor the DnaK protein is required for the reconstitution of *oriC* plasmid DNA replication in a purified protein system (104).

GrpE protein

The *grpE* gene was also discovered through the analysis of *E. coli* mutants that fail to support λ DNA replication (335). Once again it was found that mutations in the λ *P* gene could suppress the defect in λ DNA replication (335). It has been shown that (i) *grpE* maps at 56 min and specifies an M_r 24,000 protein (334); (ii) *grpE* gene expression is under heat shock regulation (4); (iii) in analogy with the *dnaK* and *dnaJ* genes, mutations in *grpE* block bacterial growth at high temperature, interfering with both RNA and DNA synthesis (4); and (iv) the GrpE and DnaK proteins intimately interact as judged by both genetic and biochemical criteria (C. Johnson, D. Ang, G. N. Chandrasekhar, and C. Georgopoulos, unpublished data).

GroES and GroEL proteins

Mutations in either the *groES* (*mopB*) or *groEL* (*mopA*) gene have been shown to interfere with the proper assembly of the bacteriophage λ prohead (120, 367; reviewed in reference 99). It has been shown that these genes (i) lie at 93 min (119) and (ii) form an operon with the order promoter-*groES(mopB)*-*groEL-(mopA)* (384) under heat shock regulation (292, 385). Furthermore, mutations in either gene result in inability to form colonies at 43°C (119). The bacterial growth defect includes pleiotropic effects on DNA and RNA synthesis (400). The two gene products have been purified to homogeneity (59a, 138, 149). The native form of the MopA (GroEL) protein is a decatetramer of M_r-61,000 subunits (138, 149; C. Woolford, R. Hendrix, K. Tilly, and C. Georgopoulos, unpublished data). The native form of MopB (GroES) protein contains six to eight subunits of M_r 11,000 (59a; Woolford et al., unpublished data). It has been shown that both of these gene products are necessary for *E. coli* growth at physiological temperatures (O. Fayet, T. Ziegelhoffer, and C. Georgopoulos, unpublished results) and that their overproduction leads to the suppression of various *dnaA*(Ts) mutations (87, 163; see below).

The *dnaY* gene

The *E. coli dnaY* gene was originally defined because a mutation in it inhibited DNA replication at 42°C (139). Although it exhibited a slow-stop phenotype at this temperature, it was thought to function in DNA elongation because it immediately decreased the rate of synthesis after a shift to 44°C and it inhibited replication in a toluene-treated cell system. It was subsequently shown, following cloning and mutagenesis studies, that *dnaY* is the structural gene of a minor arginine-accepting tRNA species that recognizes the rare AGA codon (105). The function of this tRNA in DNA replication is not known. Many possibilities have been considered, including a structural role in the replication complex or participation in the expression of a labile replication protein (105).

STRUCTURAL AND PHYSIOLOGICAL STUDIES OF *oriC*

Structure of *oriC*

The original positioning of the *E. coli* origin of DNA replication (*oriC*) at 83.5 min on the genetic map was correctly inferred on the basis of gene dosage by biochemical (32) and genetic (257) experiments. DNA segments derived from this region were subsequently shown (i) to allow an F′ plasmid to be maintained in an Hfr background (142), (ii) to enable a drug resistance element to replicate (435), and (iii) to allow λ specialized transducing phage to replicate in a lambda lysogen (279, 399). Such studies suggested that DNA from this region was capable of autonomous replication. Further cloning, sequence determination, and analysis of neighboring DNA segments revealed the arrangement of the genes in the *oriC* region shown in Fig. 3 (50, 129, 274, 286, 373, 402; summarized in reference 397). Furthermore, advantage was taken of the fact that ColE1-like plasmids are unable to replicate in *polA*(Ts) hosts at the nonpermissive temperature, but do so if they carry the *oriC* region of *E. coli*. This enabled a detailed deletion and point mutation analysis of *oriC* (13, 146, 306–308). The minimal DNA segment capable of autonomous DNA replication was shown to be 245 bp (Fig. 4). Analogous DNA segments derived from five other gram-negative bacteria, including *S. typhimurium* (446, 449), were shown to be capable of replication in *E. coli* and to have a highly conserved sequence (444) (Fig. 5). Approximately half of the 245 nucleotides are conserved among the six organisms tested, and this conservation appears to be regularly spaced (Fig. 5). This type of species comparison, as well as sequencing of variants, produced by site-specific mutagenesis, demonstrated the impor-

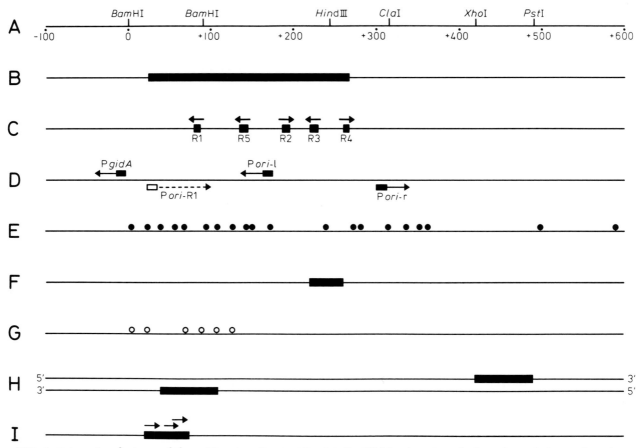

FIG. 4. Interesting features of the *oriC* region. See text for details. (A) Map location of selected restriction endonuclease recognition sites (274, 373). Numbering (+1 position) is from the first nucleotide of the leftmost *Bam*HI recognition sequence (see legend to Fig. 3). (B) Extent of the minimal *E. coli* replication origin, *oriC*, with boundaries at nucleotide positions 23 and 267 (308). (C) Distribution of DnaA box sequences. The arrows indicate the orientation of the 5'-TTAT$^{C}_{A}$CA$^{C}_{A}$A-3' consensus sequence recognized by the DnaA initiator protein. (D) Location of putative promoter elements. The arrows indicate direction of transcription. (E) Distribution of Dam methylase GATC sites. (F) Location of the major DNA gyrase-binding site (244). (G) Distribution of the predominant RNA-DNA transition points for counterclockwise leading-strand synthesis according to Hirose et al. (143) and Kohara et al. (193). (H) Location of the binding sites for an ssDNA-binding protein isolated from *E. coli* membrane fractions (160). (I) Location of the three repeating sequences that may represent binding sites for a DnaB-DnaC protein complex (199).

tance of both the conserved base sequences (since base substitutions in these are more likely to eliminate *oriC* function than base substitution in the unconserved regions) and the distance between the conserved segments (since small deletions or insertions often abolished *oriC* function [13, 146]). These studies suggest that specific proteins may bind to these conserved sites and that the exact nucleotide arrangements may be necessary for the creation of the appropriate spatial arrangement of the DNA-protein complex necessary for initiation. Furthermore, *dnaA*(Ts) mutations in *E. coli* can be suppressed by *dnaA*+-carrying clones of *S. typhimurium* and vice versa (261; O. Skovgaard and F. G. Hansen, personal communication). These results suggest that the DnaA protein of each of the two organisms is not only capable of recognizing the *oriC* region of the other, but is also capable of making all the protein-protein contacts necessary for successful initiation of DNA replication.

The 9-bp sequence 5'-TTAT$^{C}_{A}$CA$^{C}_{A}$A-3' is highly conserved at four positions (designated R1, R2, R3, and R4 [447]) within the minimal *oriC* sequences of *E. coli*

(Fig. 5) and five other gram-negative bacteria (444). The DnaA initiation protein of *E. coli* has been shown to bind these four sites (referred to as "DnaA boxes"), as well as an additional one, R5 (matching the consensus sequence at only seven of nine positions), as judged both by filter-binding and footprinting experiments (101, 103, 259). This binding is highly cooperative and involves approximately 250 bp at *oriC* and 100 bp on a DNA fragment containing a single 9-bp site (101). DnaA boxes occur elsewhere in the genome, including between the two promoters of the *dnaA* gene (see Fig. 8) (with an unusually large number of near matches in the *dnaA* structural gene) and in the *mioC* promoter (which is located immediately clockwise of *oriC*, directing transcription towards *oriC*). Transcription from these two promoters has been shown to be under the negative regulation of the DnaA protein (17, 41, 243). Cloning of the corresponding *Bacillus subtilis oriC* region (287) has also revealed the presence of many DnaA boxes, demonstrating a conservation of this region in both gram-negative and gram-positive organisms.

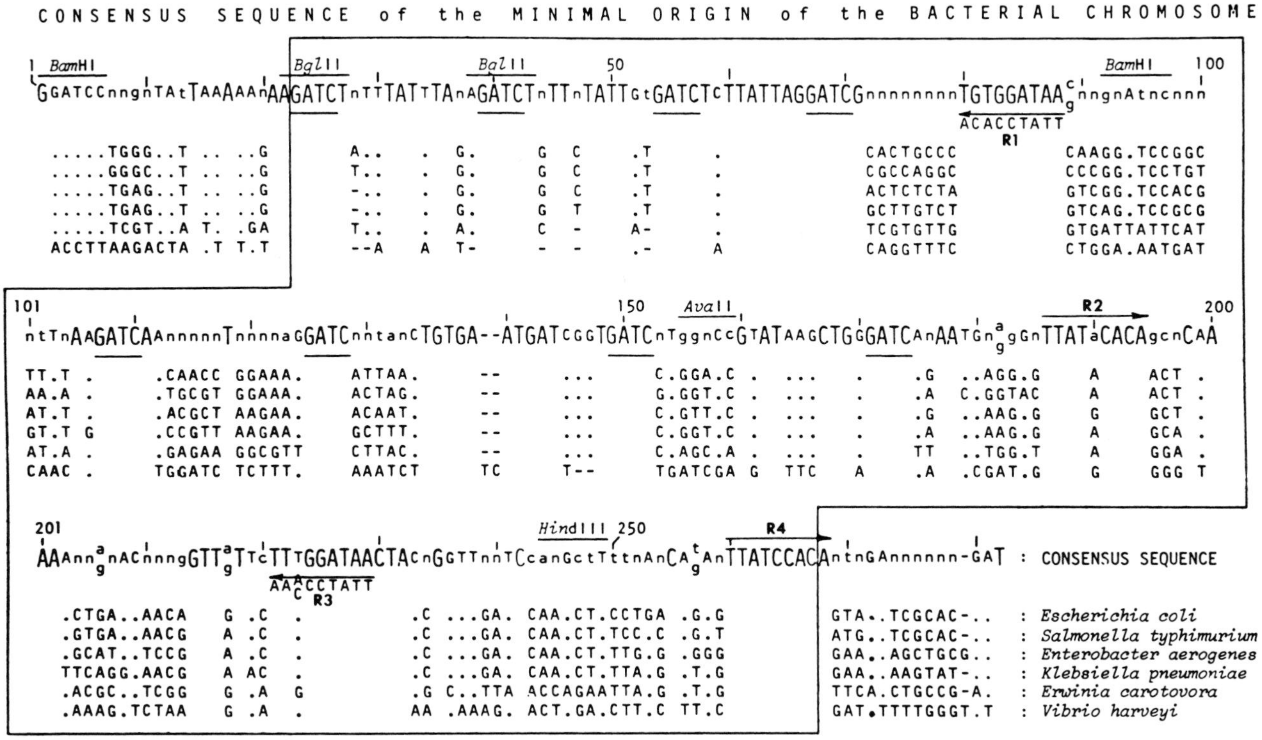

CONSENSUS SEQUENCE of the MINIMAL ORIGIN of the BACTERIAL CHROMOSOME

FIG. 5. Consensus sequence of the minimal origin of the bacterial chromosome. The consensus sequence is derived from six bacterial origin sequences (444). A large capital letter means that the same nucleotide is found in all six origins; a small capital letter means the nucleotide is present in five of the six sequences; a lowercase letter is used when that nucleotide is present in three or four of six bacterial origins but only two different nucleotides are found at that site; and where three or four of the four possible nucleotides, or two different nucleotides plus a deletion, are found at a site, the letter n is used. In the individual origin sequences, − means a deletion relative to the consensus site is present and a dot indicates that the nucleotide in the bacterial sequence is the same as the nucleotide in the consensus sequence. GATC sites are underlined in the consensus sequence, and certain *E. coli* restriction sites are noted. The minimal origin of *E. coli* is enclosed within the box. The numbering of the nucleotide positions is that used for *E. coli*, and the 5′ end is at the upper left. The four related 9-bp DnaA boxes, R1, R2, R3, and R4, are indicated by the arrows, with the 5′→3′ consensus sequence listed below the arrows for those DnaA boxes in the opposite orientation. A related sequence, which may represent a fifth possible DnaA box, R5, is found between *E. coli* nucleotide positions 135 and 143. (Figure courtesy of Judith W. Zyskind and Douglas W. Smith.)

Another notable feature of the *E. coli* minimal *oriC* sequence is the abundance of GATC sites, the substrate of the Dam methylase. At least 8 of these 11 GATC sites have been conserved among the six gram-negative origins examined, with each organism containing from 11 to 14 sites (444). This unusual clustering of GATC sites at *oriC* (Fig. 4) is thought to play an important role in its ability to replicate in vivo and in vitro (154, 276, 362). The in vitro inability of unmethylated *oriC* plasmids to replicate is restored by in vitro methylation by the Dam methylase. Bacteria mutated in the *dam* gene are transformed poorly by plasmids that contain *oriC* and are dependent on *oriC* function for their replication (276, 362). This phenotype is not suppressed by host mutations that suppress mismatch repair and is not exhibited by *E. coli dcm* mutants, suggesting that lack of GATC methylation per se is responsible for this inability to replicate. Surprisingly, when the *oriC*-containing plasmid is grown on *dam* strains, some of the GATC sites in *oriC* appear to be preferentially methylated, as judged by sensitivity to digestion by *Dpn*I restriction endonuclease (362). It is not known whether this is due to residual Dam methylase activity or to another as yet unidentified enzymatic activity.

Kornberg et al. (199) have noted that the region from nucleotides 23 to 66 contains three direct repeats of the consensus 5′-GATCTNTTNTTTT-3′ (Fig. 4) and have preliminary data that suggest that this region serves as a DnaB/DnaC protein complex entry site to the *oriC* region, potentially through interaction with the DnaA protein bound at *oriC* (see below). Note that the Dam methylase sequence GATC is part of this consensus sequence.

A strong DNA gyrase-binding site has been mapped within the minimal *oriC* segment (Fig. 4) near position +235, located between the R3 and R4 DnaA boxes (161, 244). Gyrase binding protects the *Hind*III site from digestion (161), as does DnaA protein itself (103). The binding has been observed with both DNA gyrase holoenzyme and GyrA subunit alone and is observed equally well with supercoiled and linear template DNA. Interestingly, binding to this site, but not to others, is suppressed by both oxolinic acid (an inhibitor of the GyrA subunit) and a nonhydrolyzable ATP analog (244). This result suggests that this site is uniquely recognized by DNA gyrase. Binding to this site may underlie the observed requirement for DNA gyrase in the minimal *oriC* replication system (23, 170).

A membrane protein has been purified from *E. coli* and shown to bind (by filter retention) to two regions of DNA in and near the minimal *oriC* region (160, 161). One site is located near the +70 position (overlapping DnaA box R1) on the 5' → 3' counterclockwise strand, and the other lies near the +460 position on the 5' → 3' clockwise strand (Fig. 4). dsDNA fragments containing the *oriC* DNA region have been shown to be preferentially associated with other membrane proteins (137, 212). Although the significance of this binding has not been established yet, such protein association may be important for attachment of the *oriC* region to the membrane and may play a role in switching from uni- to bidirectional replication in vivo (437).

Potential Transcriptional Signals at *oriC*

The original in vivo studies of Hansen et al. (129) and Morita et al. (286), employing promoter probe vectors, identified three strong promoters in the *oriC* vicinity, P1, P2, and P3. These promoters correspond to those of the *mioC*, *asnA*, and *gidA* genes and direct transcription counterclockwise. In addition, Hansen et al. (129) found a strong terminator of counterclockwise transcription (between +92 and +243) within *oriC*. On the basis of analogous cloning studies and in vitro analysis, other promoters and terminators within or near *oriC* have been identified (242, 285, 326; M.-A. Schauzu et al., Nucleic Acids Res., in press), but the quality and strength of such sites, with a few exceptions, have not been firmly established in vivo. Lother et al. (242) identified the presence of two promoters, Pori-l (with transcription counterclockwise from +170) and Pori-r (with transcription clockwise from +310) in vitro (Fig. 4). Transcripts from each promoter were approximately 110 bases in length (Schauzu et al., in press). However, transcription from these promoters has not been convincingly demonstrated in vivo (168; Schauzu et al., in press). Another promoter, Pori-R1, located between nucleotides +1 and +97, was identified in vivo by Junker et al. (168). Junker et al. (168) confirmed the presence of a terminator within *oriC*, but found that for maximal termination efficiency a contiguous region from +1 to +249 was required. They also identified a terminator of counterclockwise transcription located to the right of *oriC* between nucleotides +243 and +416. In addition to the two terminator signals identified by cloning, 3' ends of transcripts entering the *oriC* region counterclockwise have been localized at a few distinct sites within *oriC* by S1 mapping (168, 326; Schauzu et al., in press).

Hirose et al. (143) and Kohara et al. (193) have studied the RNA-DNA transition points in DNA isolated from *dnaC*(Ts) cells synchronized for initiation of DNA replication. At least nine transition points were discernible, all involving RNA-DNA transitions in the counterclockwise strand, and mostly located in the left half of the *oriC* region (Fig. 4). If replication starts within *oriC*, then this would correspond to leading-strand synthesis in the counterclockwise direction. This is consistent with the finding that the replication fork proceeds counterclockwise during the earliest period of replication (437). However, since Tabata et al. (378) have presented evidence indicating that the 245-bp *oriC* region directs bidirectional replication adjacent (counterclockwise) to *oriC* and no other strong evidence for the start point of DNA polymerization of leading strands in or near *oriC* has been presented, the RNA-DNA transitions found within *oriC* are not necessarily leading-strand start points. Interestingly, the majority of the RNA-DNA transition points occur near or at GATC methylation sites. Furthermore, their position coincides well with the termination sites of the counterclockwise transcript entering *oriC* (probably from the *mioC* promoter) both in vivo and in vitro (325, 326; Schauzu et al., in press).

Recently, it has been shown that transcription from the *mioC* gene appears to be negatively regulated by both DnaA protein (243) and the stringent control mechanism, i.e., low expression under conditions of high ppGpp levels (326). It also appears that the amount of transcription which enters the *oriC* region counterclockwise (from *mioC* and other possible promoters) is decreased in a *dnaA*(Ts) strain and under conditions of high DnaA$^+$ overproduction. This decrease correlates with a decreased rate of chromosome replication (326). Thus the level of transorigin transcription in the counterclockwise direction may be correlated with successful initiation of replication.

The role of counterclockwise transcription through *oriC* has not yet been established. Counterclockwise transcripts terminating at multiple sites could be used to prime leading-strand synthesis. Alternatively, the termination of transcription at these sites might allow priming by DNA primase in close proximity, leading to the observed multiple RNA-DNA junctions. Trans-*oriC* transcription might also serve to activate the *oriC* region for DNA replication in a manner analogous to that observed with transcriptional activation of phage λ DNA replication (79), favoring the formation of a specific protein-DNA structural complex. Alternatively, such transcripts might play a regulatory role by forming RNA-RNA hybrid structures with their complementary transcripts as observed in some plasmid replication systems (387).

It is not clear how the transcriptional activity demonstrated in vivo at *oriC* relates to the ability of the minimal *oriC* system to initiate replication in the absence of RNA polymerase (301, 394).

BIOCHEMICAL ANALYSIS OF *oriC* DNA REPLICATION IN VITRO

oriC DNA Replication in Crude Enzyme Mixtures

The development by Fuller et al. (102) of a soluble enzyme system that supports the specific initiation of DNA replication at the *E. coli* replication origin, *oriC*, ushered in an era of rapid advances in our understanding of the biochemical mechanisms involved in the initiation and propagation of bacterial DNA replication. Futhermore, this discovery greatly aided the development of closely related in vitro systems that could support both the replicative transposition of phage Mu (282) and the initiation of bacteriophage λ DNA replication (3, 391, 428). As discussed below, studies of the latter system complement investigations of *oriC* DNA replication, since the initiation of λ DNA replication is mechanistically similar to the initiation of *E. coli* DNA replication.

As initially developed, the *oriC* replication system utilized bacterial replication proteins contained in an ammonium sulfate fraction prepared from a cleared lysate of gently lysed *E. coli* (102). *oriC* minichromosomes, small superhelical plasmids that contain the intact 245-bp *oriC* region (308) of the bacterial chromosome, were used as templates in this in vitro system. Vigorous *oriC*-dependent DNA synthesis, representing replication of about 40% of the input template DNA, was sustained in this relatively crude system only when reaction mixtures were supplemented with both an ATP-regenerating system and a hydrophilic polymer such as polyethylene glycol or polyvinyl alcohol (102). One or more of the macromolecules required for initiation of DNA synthesis at *oriC* may be present at a limiting concentration in the crude system. The presence of relatively high concentrations of the hydrophilic polymer apparently serves to crowd all macromolecular constituents into an "excluded volume," thereby in effect raising the concentration of the limiting component or components to more optimal levels.

By several criteria, the DNA synthesis observed in the crude *oriC* system represents authentic initiation of *E. coli* DNA replication (102, 169). DNA synthesis in this in vitro system (i) depends on exogenously supplied plasmid DNA that carries a functional *oriC* sequence; (ii) is initiated at or near the *oriC* site present in the plasmid template and proceeds bidirectionally through theta structure intermediates; (iii) depends on the *E. coli* DnaA protein; and (iv) depends on other known *E. coli* replication proteins such as DnaB protein, DnaC protein, SSB, DNA gyrase, and RNA polymerase, judging from the effects on DNA synthesis of antibiotics and antibodies that specifically inhibit each of these individual proteins.

Proteins That Insure Specificity of DNA Replication at *oriC*

The obvious physiological relevance of the crude *oriC* in vitro replication system encouraged investigators to proceed with the fractionation of the system to identify the required components. Their ultimate goal, of course, was to reconstitute the *oriC* initiation reaction solely with purified components so that reliable mechanistic studies of the initiation process could be undertaken. In the meantime, however, studies of partially reconstituted enzyme systems led to the identification of several auxiliary proteins, including RNase H, DNA topoisomerase I, and HU protein, that improved the specificity of *oriC* DNA replication (75, 170, 171, 304). The identification of RNase H as a specificity factor in vitro is consistent with numerous genetic studies (discussed earlier) that demonstrated a role for this enzyme in suppressing *dnaA*-independent and *oriC*-independent initiation of *E. coli* DNA replication in vivo. The molecular mechanisms by which these diverse proteins, consisting of an RNase, a DNA topoisomerase, and a histonelike protein, act to suppress *oriC*-independent initiation of DNA synthesis have not been determined. At present, a likely explanation is that in the partially reconstituted system each of these proteins, especially RNase H, reduces adventitious priming of DNA synthesis by the bacterial RNA polymerase. Perhaps the strongest evidence

for this hypothesis comes from studies of an *oriC* replication system that is reconstituted with a set of purified *E. coli* replication proteins (301). In this system, which does not contain the bacterial RNA polymerase, highly specific initiation of DNA replication at *oriC* can be achieved in an apparently physiological manner in the absence of any auxiliary proteins. Yet, when the purified protein system is supplemented with RNA polymerase, small amounts of RNase H must also be added to maintain strict *oriC* specificity.

It should be noted that both DNA topoisomerase I and HU protein, which act as specificity factors in the partially reconstituted system, act instead as potent inhibitors of the completely reconstituted *oriC* system when they are present at high concentrations (301). This finding underscores a limitation that is inherent in any attempt to reconstitute a protein mixture that specifically initiates DNA replication at the *E. coli* origin in a physiologically relevant manner. Although the need for many replication proteins appears to be absolute under a wide variety of in vitro conditions, the requirement for others appears to be optional. The necessity for certain auxiliary proteins, such as specificity factors and stimulatory factors, depends on the exact composition of the *oriC* protein mixture. For example, in the presence of high levels of HU protein, efficient initiation of DNA replication at *oriC* is observed only if sufficient RNA polymerase is also present to counteract the inhibitory effect of this histonelike protein. RNA polymerase acts as a stimulatory factor in this instance to enhance replication of *oriC* minichromosomes. At the same time, however, RNA polymerase also promotes *oriC*-independent DNA synthesis. This nonspecific priming of DNA replication, in turn, can only be controlled if a specificity factor, such as RNase H, is present. The net effect of the presence of high levels of HU protein in the in vitro replication system, then, is to create a requirement for RNase H so as to observe specific initiation of DNA replication at *oriC*. In view of the pliability of the *oriC* in vitro system in regard to requirements for auxiliary proteins, assignment of physiological roles for such proteins must be supported by corroborating evidence from genetic and physiological studies of *E. coli*.

Enzymatic Studies of Replication of *oriC* Plasmids: Initiation of *E. coli* DNA Replication Reconstituted with Purified Proteins

Rapid progress with the fractionation of the crude system for *oriC* DNA replication has recently culminated in the development of a purified protein mixture that supports the initiation of DNA replication at the *E. coli* origin (23, 104, 199, 301, 394). A minimum of 12 different protein fractions (Table 3) are required to reconstitute a complete cycle of *oriC* plasmid replication, starting and ending with supercoiled DNA molecules (104). The close correspondence of this list of *E. coli* replication proteins, identified through biochemical studies in vitro, with the list of bacterial replication proteins identified via genetic approaches (Table 1) attests to the authenticity of the reconstituted *oriC* enzyme system. Nevertheless, it is likely that additional factors still remain to be identified, since the efficiency of this minimal system is poor. Only

TABLE 3. Proteins required for a complete cycle of *oriC* plasmid replication in vitro

Protein	No. of protein molecules per template molecule
DnaA protein	15
DnaB protein	1
DnaC protein	5
HU protein	7
SSB	36
GyrA protein	10
GyrB protein	6
Primase	1
DNA polymerase III holoenzyme	1
DNA polymerase I	2
RNase H	0.1
DNA ligase	0.3

about one-fifth of the available *oriC* plasmid template molecules are replicated.

Characterization of the replication intermediates formed during the replication of *oriC* plasmids provides further evidence that the reconstituted enzyme system functions in a physiological manner (23, 104, 199). DNA replication is initiated at or near the *oriC* region on the supercoiled template and proceeds bidirectionally through θ structures to generate a pair of multiply catenated daughter molecules in which each daughter plasmid contains one parental strand and one newly synthesized strand. Subsequently, a portion of the catenated pairs are enzymatically segregated in the reconstituted *oriC* system to yield circular plasmid monomers.

Multiple Stages in the Replication of *oriC* Plasmids

Mechanistic studies of the replication of *oriC* plasmids have revealed that the overall reaction can be resolved into several separable stages (199). Figure 6 depicts a hypothetical scheme, based on all of the data that are currently available, for the initiation and propagation stages of *oriC* DNA replication. Briefly, DnaA protein binds to its multiple recognition sites at *oriC* (Fig. 4 and 5) to form a specific nucleoprotein structure that serves as a locus for the transfer of DnaB helicase onto the template DNA. Unwinding of the duplex DNA in the *oriC* region by the helicase

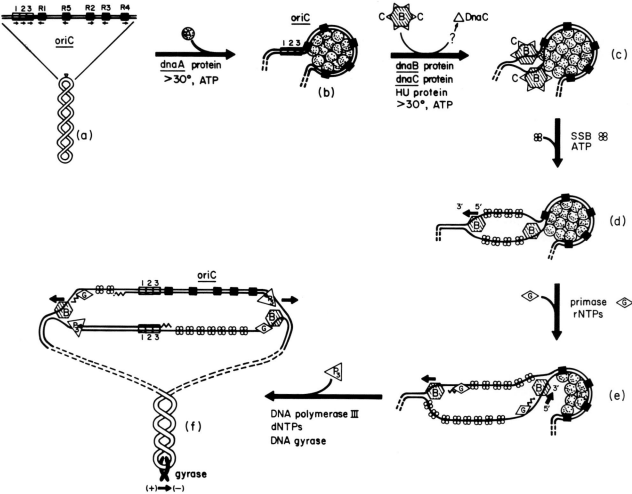

FIG. 6. Hypothetical scheme for the initiation of bidirectional DNA replication at the *E. coli* origin in vitro. See text for details. The closed boxes, R1 through R5, represent recognition sequences for the *E. coli* DnaA protein. The three open boxes (1, 2, 3) represent possible sites for interaction of the DnaB-DnaC protein complex.

action of DnaB protein and the trapping of the unwound DNA by the binding of SSB permits the entry of *E. coli* primase. After priming, DNA chain elongation by DNA polymerase III holoenzyme proceeds rapidly, assisted by the swiveling action of DNA gyrase, which removes positive supercoils created during the unwinding of the two strands of the closed circular template. A more detailed description of each of these stages is presented below and is followed by a discussion of the comprehensive *E. coli* replication scheme presented in Fig. 6.

Stage I: binding of DnaA protein to *oriC*

The *E. coli* replication origin is apparently wrapped or folded into a specific three-dimensional nucleoprotein structure as a consequence of the binding of multiple copies of DnaA protein, perhaps 20 to 30 DnaA monomers, to *oriC* DNA [structure (b) of Fig. 6] (101). In the absence of chromatin structure in the *oriC* region, as occurs in the reconstituted *oriC* replication system, formation of the specialized DnaA-*oriC* complex represents the first stage of the initiation of bacterial DNA replication. Formation of an active *oriC*-DnaA protein structure requires incubation at temperatures of ≥30°C (199, 394). Moreover, once formed, the *oriC*-DnaA complex is cold sensitive and must be maintained at an elevated temperature (≥30°C) to retain replicative activity. Additional evidence for the complexity of the *oriC*-DnaA protein structure comes from studies currently in progress in the Kornberg laboratory. As discussed earlier, the DnaA protein initiator binds avidly to ATP or ADP (199). Thus, DnaA protein can exist in three forms, DnaA·ATP, DnaA·ADP, and DnaA with no bound nucleotide. Although each of the three forms of DnaA protein binds to DnaA recognition sites in *oriC*, only the *oriC*-DnaA·ATP structure retains replicative competence (Kornberg, personal communication).

Much remains to be learned about the assembly and function of a DnaA protein complex at *oriC*. What is the role of ATP in the assembly of a functional DnaA protein structure at the *E. coli* origin? And does the wrapping or folding of *oriC* DNA by this replication initiator potentiate the initiation process? Is each of the copies of DnaA protein bound at *oriC* needed for proper initiation of DNA replication? Even though our understanding of the DnaA protein-origin complex is severely limited, it does seem likely that the functional properties of the *oriC*-DnaA protein complex reflect a requirement for a specific, but tenuous, configuration of protein and DNA.

Stage II: assembly of a prepriming complex containing DnaB helicase

In stage II the DnaB helicase, in the form of a DnaB protein-DnaC protein complex, is delivered to the DnaA protein-encoated *E. coli* origin to form a larger nucleoprotein structure termed a prepriming complex [Fig. 6, structure (c)]. Stable assembly of DnaB helicase at *oriC* (i) requires a negatively supercoiled *oriC* plasmid; (ii) depends on incubation of the required components (Fig. 6) at a temperature of 30°C or greater; and (iii) requires the ATP form of DnaA protein bound at *oriC* (199, 394; Kornberg, personal communication). The prepriming complex is very unstable in the presence of standard Mg²⁺ levels, but is sufficiently stable at low Mg²⁺ concentrations (0.5 mM) to permit its isolation in a form active for replication.

HU protein, an *E. coli* histonelike protein, stimulates the replication of *oriC* plasmids when it is present at relatively low levels (ca. 10 HU dimers per plasmid circle) (75, 301). Recent studies suggest a role for HU protein in the transfer of DnaB helicase into the prepriming complex formed at *oriC*. Production of functional prepriming complexes is stimulated approximately fourfold when HU protein is included along with DnaB protein and DnaC protein during the assembly reaction (T. Baker and A. Kornberg, personal communication). The mechanistic basis for this stimulatory effect of HU protein has not been determined.

The molecular architecture of the prepriming complex is poorly understood. This specialized nucleoprotein structure is known to contain DnaA initiator protein and DnaB helicase (23, 199; T. Baker, B. Funnell, and A. Kornberg, personal communication). Isolated prepriming complex also contains HU protein, but it has not yet been possible to determine whether some or all of this histonelike protein is specifically bound at *oriC* or, in contrast, is randomly bound to other regions of the plasmid template. The requirement for the bacterial DnaC protein in the formation of the prepriming complex at *oriC* is absolute. Nevertheless, DnaC protein may not be an essential component of the prepriming complex formed in the reconstituted *oriC* replication system. It has not been possible to demonstrate unequivocally that DnaC protein is associated with isolated prepriming complexes that retain replicative activity (Baker and Kornberg, personal communication).

As visualized by electron microscopy, the prepriming assembly at the *E. coli* origin is larger than the DnaA-*oriC* nucleoprotein structure and covers a longer stretch of *oriC* (199). With the addition of DnaB helicase to the structure, the prepriming complex extends approximately 50 base pairs further to the left, covering the left boundary of the minimal *oriC* sequence. Kornberg and his colleagues have pointed out that this region of the *E. coli* origin contains three direct repeats of a 13-bp sequence (Fig. 4 and 6) with a consensus structure of 5'-GATCTNTTNTTTT-3' (Fig. 5, base pairs 23 to 66) (199). It seems likely that these sequences represent the sites at which the DnaB protein is positioned prior to the initiation of *E. coli* DNA replication. Additional support for this notion comes from the fact that closely related sequences are found adjacent to initiator binding sites at the replication origins of other chromosomes that depend on the DnaB helicase for their replication, such as plasmid pSC101 and bacteriophage λ.

Stage III: localized unwinding at *oriC* by DnaB helicase

Once the DnaB protein is properly positioned at *oriC* in the prepriming complex as a consequence of its interactions with the DnaA and DnaC initiator proteins, it proceeds to unwind the duplex DNA in the vicinity of the *E. coli* origin (23, 199). SSB binds to the separated sister strands created by the helicase activity of DnaB protein, thereby blocking rapid reassociation of the complementary chains. Action of the DnaB helicase in this stage is obligatorily coupled to the hydrolysis of rNTPs. In contrast to the previous two stages, stage III proceeds at 16°C and does not require incubation of the reaction mixture at an elevated temperature (23). The nucleoprotein structure present on partially unwound *oriC* plasmids has been termed a prepriming complex [Fig. 6, structure (d)] because *E. coli* primase is capable of interacting with this complex to synthesize primers for DNA replication (23).

In the absence of a DNA topoisomerase, unwinding of duplex DNA at *oriC* is limited to a small portion of the covalently closed *oriC* plasmid owing to the topological constraint engendered by the accumulation of positive supercoils in the remainder of the template (199). In the presence of DNA gyrase, which can provide the necessary swiveling activity, and in the absence of both priming and DNA chain elongation, stage III unwinding mediated by the helicase activity of DnaB protein proceeds until approximately 90% of the duplex *oriC* template is unwound (23). Electron microscopic analysis of partially and fully unwound *oriC* plasmids indicates that unwinding is initiated at or near the *E. coli* origin and proceeds in both directions simultaneously at roughly equivalent rates (199). Preliminary studies indicate that the prepriming complex present on highly unwound *oriC* minichromosomes contains approximately two molecules of DnaB protein and lesser amounts of DnaC protein (199). The fate of the DnaA protein, initially bound at *oriC*, is currently under investigation.

The rate of DNA unwinding of *oriC* plasmids is highly temperature dependent, rising sharply with temperature above 35°C (Baker and Kornberg, personal communication). In addition, because topological constraints are induced by DNA helicase action on covalently closed DNA duplexes, the unwinding rate is necessarily influenced by the level of DNA gyrase present in the in vitro system. Initial rates of unwinding of an 8-kb *oriC* plasmid by DnaB helicase approach 350 bp/s at 44°C when about 40 DNA gyrase molecules are present per input plasmid circle. Unwinding rates are more than halved under these conditions by a fourfold reduction in the level of DNA gyrase. Rising topological constraints apparently account for the pronounced slowing of DNA unwinding that is observed as the two bidirectionally moving forks approach collision on the opposite side of the *oriC* plasmid (Baker and Kornberg, personal communication).

Stage IV: RNA priming

oriC-specific synthesis of RNA primers in the reconstituted *oriC* replication system is strongly dependent upon the presence of primase and four other *E. coli* replication proteins, DnaA, DnaB, and DnaC proteins and DNA gyrase (394). Although the bacterial SSB and HU proteins are not required for specific RNA priming in this system, each stimulates RNA synthesis approximately twofold. Not all of these proteins, however, participate directly in the priming reaction. Instead, many function in the assembly of the nucle-

oprotein prepriming complex produced in stage III. It is this complex that is the physiological substrate upon which primase acts. Even the highly underwound form of the prepriming complex (created during stage III by extensive DnaB helicase unwinding of the *oriC* template in the presence of DNA gyrase) is a suitable substrate for primase (23). As will be discussed below, it is probable that the synthesis of RNA primers depends upon productive interactions between primase and a DnaB helicase molecule that is positioned at a fork.

The amount of RNA synthesis observed in the *oriC* system, under conditions in which priming is uncoupled from replication, depends almost exclusively on the level of primase in the reaction mixture. For example, a 15-fold increase in the amount of primase over that needed for optimal *oriC* DNA replication leads to a 10-fold increase in RNA synthesis (394). However, at either rate of RNA synthesis, only a small fraction of the nascent RNA becomes linked to DNA during subsequent DNA replication. This presumably reflects the fact that in this uncoupled system a significant portion of the RNA primers fail to remain annealed to the covalently closed duplex template and hence are unavailable to DNA polymerases.

The lengths of the RNA chains synthesized in reactions coupled to or uncoupled from DNA replication (i.e., in the presence or absence of DNA polymerase III holoenzyme and dNTPs, respectively) have not yet been determined. Neither have the *oriC* template sites used for RNA priming been defined. In this regard, it seems likely that an analysis of the priming sites utilized at the *E. coli* origin and surrounding region should help delimit possible initiation mechanisms. This is because the priming sites mark the paths taken by the DnaB helicase components of the processive priming machinery.

Stage V: DNA chain elongation

DNA polymerase III holoenzyme elongates all DNA chains in the reconstituted *oriC* replication system. The rate of DNA chain elongation appears to be limited by the rate of DNA unwinding by the DnaB helicase. For example, it takes 4 or 5 min to complete the replication of an 8-kb *oriC* plasmid at 30°C when DNA unwinding and replication are coupled. But if instead the *oriC* plasmid is completely unwound (stage III) prior to the addition of DNA polymerase and dNTPs, then full-length (8-kb) chains can be completed in as little as 30 s (Baker and Kornberg, personal communication). In the latter uncoupled reaction it is likely that DNA polymerase III synthesizes long DNA chains without dissociating from the template (85). The numbers, sizes, and distributions of DNA chains synthesized in vitro on an *oriC* template have not been reported for the standard one-stage replication reaction. Thus, it is not known whether multiple nascent (Okazaki) fragments are synthesized on the lagging-strand template when DNA unwinding, priming, and DNA chain elongation are coupled.

Stage VI: termination

Like initiation, termination of replication of *oriC* plasmids is a relatively slow process in the reconstituted *oriC* replication system. Only about a third of the final replication products are monomer circles (104). The remainder of the products have not completed the termination or segregation stages of DNA replication. In part, the sluggishness of the termination stage reflects the fact that replication fork movement slows perceptively as the enzymatic elongation machinery approaches the terminus (104), presumably as a consequence of rising topological constraints. Thus, fully one-third of the final reaction products are θ structures, most of which contain a short stretch (60 to 380 bp) of unreplicated parental DNA. It is likely that these late replicative intermediates are precursors to the other major product of the in vitro reaction, multiply intertwined, catenated dimers (104). Multiple intertwinings between nascent daughter circles would result if the terminal stage of DNA replication proceeds in the absence of DNA topoisomerase action. Under such conditions the multiple topological linkages between the two parental strands that exist in late θ structures are partially or completely transformed into double-stranded intertwines behind the replication fork.

The minimal *oriC* replication system does not contain DNA ligase or DNA polymerase I, two enzymes that function in the covalent joining of nascent DNA fragments in vivo (230). Thus, each daughter molecule synthesized in vitro contains gaps or interruptions and, moreover, presumably contains covalently linked primers at the 5′ termini of all nascent DNA chains as well. As anticipated, addition of DNA polymerase I, RNase H, and DNA ligase as possible termination enzymes to the reaction products of the reconstituted *oriC* replication system results in the production of negatively supercoiled molecules (104). It is possible that additional *E. coli* proteins may function in the termination process, since the efficiency of conversion of *oriC* replication products into covalently closed structures was relatively poor (25%) under optimal conditions.

Hypothetical Scheme for the Initiation and Propagation of *E. coli* DNA Replication

Assembly and activation of DNA replication machinery at *oriC*

Figure 6 depicts a possible scheme for the biochemical mechanisms involved in the initiation and propagation of a round of bidirectional DNA replication at the *E. coli* replication origin. Outlined below is a detailed description of this hypothetical model. The initiating and origin-specific event in the replication of the bacterial chromosome is the binding of multiple copies of DnaA protein to its recognition sites (DnaA boxes R1 through R5) in *oriC*. Additional copies of this bacterial initiator bind to the bound DnaA protein via direct protein-protein interactions, resulting in the formation of a nucleosomelike structure at the origin. In the presence of ATP this specialized nucleoprotein complex acquires a specific and active configuration [Fig. 6, (b)].

Provided that the template is negatively supercoiled, the *oriC*-DnaA protein complex then serves as a locus for the delivery of the DnaB helicase to the bacterial origin to form a preprepriming complex

[Fig. 6, (c)]. It is notable that this step absolutely depends on the prior formation of a tight complex between the DnaB protein and the *E. coli* DnaC replication protein. Aided in some unknown manner by the bacterial histonelike protein HU, two molecules of the DnaB protein hexamer are transiently positioned on the three AT-rich repeats of 13 bp [Fig. 6, (a) and (b)] located near the left boundary of the minimal *oriC* sequence. The preprepriming complex that is formed contains one molecule of DnaB helicase bound to each strand of *oriC* juxtaposed to the DnaA protein aggregate [Fig. 6, (c)].

Once in the proper position, one molecule of DnaB helicase penetrates between the strands of the DNA duplex and, moving away from the DnaA complex in a 5′ → 3′ direction along the strand to which it is bound, initiates localized unwinding of DNA to the left of *oriC* [Fig. 6, (d)]. SSB protein binds to the ssDNA displaced by the DnaB helicase. The other molecule of DnaB helicase, still positioned at *oriC*, begins to move rightward, but its action as a helicase is slowed by the presence of the large DnaA protein complex. This partially unwound nucleoprotein structure represents one form of the prepriming complex described by Kornberg and colleagues (23, 199).

Soon after unwinding of the DNA duplex is initiated, DnaG primase recognizes the DnaB helicases at each fork and is induced to synthesize RNA primers as a result of functional interactions with these DnaB molecules [Fig. 6, (e)]. Simultaneously, the DnaA complex dissociates from *oriC* as the rightward moving helicase translocates across the origin. The destabilization of the DnaA complex at *oriC* during DNA unwinding is suggested by electron microscopic studies of partially unwound prepriming complexes (23, 199). This investigation demonstrated that the entire *oriC* sequence is unwound early after the onset of DnaB helicase action.

The replication scheme presented in Fig. 6 predicts that the first primer for rightward chain elongation is synthesized at a site to the left of *oriC* and that the first primer for leftward chain elongation is produced by primase-helicase interactions at a site or sites located near the R1 DnaA box. These initial primers are extended by the DNA polymerase III holoenzyme to become the respective leading strands [Fig. 6, (f)]. Processive unwinding of the DNA helix by DnaB helicase on the lagging strand at each fork permits both continuous leading-strand synthesis and intermittent synthesis by primase of multiple primers for the lagging-strand nascent (Okazaki) fragments. Thus, in this hypothetical scheme the primers for all leading- and lagging-strand DNA chains are synthesized solely by the DnaG primase. In effect, the first of the multiple primers made by a DnaB helicase-DnaG primase combination moving in one direction is utilized to prime synthesis of the leading strand for DNA replication in the opposite direction [Fig. 6, (e) and (f)].

Many facets of the mechanisms involved in the initiation of bidirectional DNA replication at *oriC* and involved in the propagation of established replication forks remain to be clarified. For example, the replication scheme proposed above ignores the thoroughly documented contribution of RNA polymerase to the initiation process in vivo. Possible roles for RNA

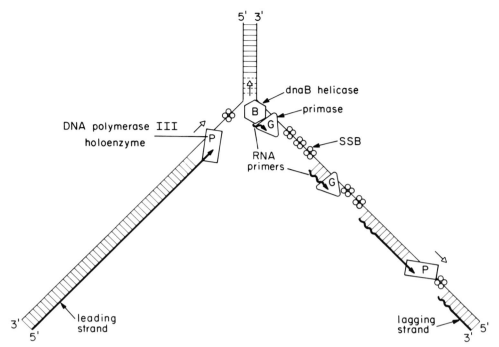

FIG. 7. Hypothetical scheme for the mechanism of action of *E. coli* replication proteins at a replication fork. See text for details.

polymerase in the initiation of *E. coli* DNA replication in vivo are discussed elsewhere in the context of the physiology of the process and in conjunction with studies of the temporal ordering of required steps in vivo.

Another of the many significant questions yet to be answered about the initiation of DNA replication at the bacterial origin concerns the precise role of the *E. coli* DnaC protein in the formation of the prepriming complex [Fig. 6, (c)]. Although there is a dearth of genetic evidence that bears on this point, it is possible that DnaC protein physically interacts with DnaA protein bound at *oriC* and that this interaction is responsible for delivering DnaB helicase to the origin region. This role for DnaC protein is suggested by studies of the early steps of the initiation of bacteriophage λ DNA replication in vitro. The λ P replication protein, a phage analog of the bacterial DnaC protein, also forms a very tight complex with DnaB helicase (179, 273, 419; J. Mallory, Ph.D. thesis, The Johns Hopkins University, Baltimore, Md., 1983). Subsequently, the λ P protein delivers DnaB protein to the phage replication origin as a result of direct physical interactions between the P protein and the λ O initiator protein bound at *oriλ* (76, 269, 441). Like DnaA initiator, λ O protein forms specialized nucleosome-like structures at its cognate replication origin.

Propagation of an *E. coli* replication fork

Our understanding of the mechanism of propagation of DNA replication in *E. coli* was significantly altered by the discovery that the *E. coli* DnaB protein is a DNA helicase (228). This finding has been accompanied by the recent realization (23, 228, 229) that the φX174 primosome model (6, 197) for processive prim-

ing of lagging-strand chains is not directly applicable to *E. coli* DNA replication. A hypothetical scheme (228) for the mechanism of DNA chain growth at an *E. coli* replication fork is presented in Fig. 7. It is proposed that the replication apparatus involved in the initiation of the nascent Okazaki fragments, synthesized on the lagging strands of the *E. coli* chromosome, is relatively simple and consists of four proteins: DnaB helicase, DnaG primase, SSB, and DNA polymerase III holoenzyme. DnaB protein migrates processively in the 5′ → 3′ direction along the lagging-strand template, acting as a DNA helicase to permit continuous chain elongation by the leading-strand DNA polymerase, as well as functioning as a coparticipant with DnaG primase in the synthesis of each of the multiple primers required for production of lagging-strand Okazaki fragments. Thus, in this scheme, the lagging-strand priming machinery migrates in advance of the leading-strand DNA polymerase complex.

The precise nature and structure of the processive lagging- and leading-strand protein complexes at an *E. coli* replication fork remain uncertain. For example, it is not known whether DnaG primase stays tightly coupled to the DnaB helicase during movement of the helicase along the lagging-strand template, or whether, in contrast, the primase repeatedly associates with and dissociates from the moving helicase before and after the priming of each Okazaki fragment (as depicted in Fig. 7). Neither is it known whether a dimeric molecule of DNA polymerase III holoenzyme acts to synthesize both leading- and lagging-strand DNA chains concurrently. Movement of the lagging-strand helicase clearly does not depend on an association with the leading-strand DNA polymerase [for example, as in stage III unwinding, Fig. 6, (d)], but are the movements of the processive leading- and lagging-strand complexes coordinated in some fashion in the context of a physiological replication fork? Finally, recent studies of RNA priming in the λ single-strand replication system (228, 229) indicate that the λ O and P replication proteins function to enhance strongly the processivity of the DnaB helicase in the synthesis of multiple RNA primers on ssDNA templates (269). Given the close parallels between the initiation pathways for *E. coli* and λ DNA replication, it is conceivable that the DnaA and DnaC proteins also act during the initiation process to somehow augment the processivity of the DnaB helicase during DNA unwinding and priming of lagging-strand DNA replication.

GENETIC ANALYSIS OF DnaA PROTEIN FUNCTION

General Considerations on DnaA Protein

Since the DnaA protein plays a key role in the process of initiation of DNA replication, it is important to understand its specific biochemical properties as well as possible interactions with other proteins. Biochemical information about its mode of action is being derived through the use of the minimal DNA replication system developed in Kornberg's laboratory. An approach that might be applicable is the identification of *E. coli* proteins that interact with DnaA protein by use of affinity chromatography, a technique which has been used successfully in the T4 DNA replication system (94). A complementary genetic approach useful for examining functional and in vivo interactions is the identification and characterization of intragenic and extragenic suppressor mutations that correct the phenotypic defect of *dnaA* mutations. This is a powerful tool that has often been used successfully to analyze biochemical and genetic pathways (reviewed in references 36 and 132). In addition, studies of purified mutant DnaA proteins would be very useful for correlating the many genetic and physiological studies of *dnaA* mutant strains with the biochemical defects of the corresponding proteins.

Since DnaA protein is multifunctional, it is useful to summarize here its known biochemical properties to help understand the nature and significance of the genetic suppressor studies summarized below. This is especially important because very little is known about the biochemical defect of the mutant DnaA proteins under study in many laboratories or the level of function remaining in the mutant polypeptides under nonpermissive conditions. The known biochemical properties of the DnaA protein include (i) binding specifically to the DnaA box consensus sequence, found at many places in the chromosome; (ii) a DNA-dependent ATPase activity as well as binding to ADP (199); and (iii) a tendency to self-aggregate (103). This last property may make it possible that DnaA protein bound to DNA at one site interacts with DnaA protein bound at another, leading to torsional stress of the helix if the sites are close or looping out of DNA if the sites are distant (148). Such self-interaction of site-specific DNA-binding proteins while bound to DNA has been implicated recently in several processes including plasmid replication, site-

specific recombination, DNA segment inversion, and transposition.

Some interpretations of experiments done with *dnaA*(Ts) strains proposed an autoregulation of the *dnaA* gene, which is lacking in *dnaA*(Ts) mutants (131, 176). There is, in fact, strong evidence that the *dnaA* gene is autorepressed (41); however, the question of derepression of the *dnaA* gene in *dnaA*(Ts) mutant strains has not been as clearly answered. Evidence supporting autoregulation of *dnaA* at the transcriptional level came from the work of Atlung et al. (15, 17, 19) and Braun et al. (41) using transcriptional and translational fusions of the *dnaA* promoter region to the *lac* or *tet* genes. Their data show that overexpression of DnaA⁺ protein leads to a decrease in expression from the *dnaA* promoter. Their results also point to a derepression of the promoter in various *dnaA*(Ts) mutant backgrounds. While this work strongly implicated the DnaA box located between the two *dnaA* promoters (Fig. 8) as the site of repression, all constructions lacked the many near-matches to the DnaA box sequences that are contained within the *dnaA* structural gene (predicted from its primary sequence) and could play a role in autoregulation. In other words, the mutant DnaA(Ts) proteins might not be able to repress *dnaA* promoter expression because of inadequate binding to the single DnaA box in the fusions, but, if the other potential DnaA protein-binding sequences were present, thus affording some cooperativity between binding sites (perhaps as seen at *oriC*), the mutant proteins might repress adequately. Such an interpretation is in agreement with the results of Sakakibara and Yuasa (338), who measured *dnaA* mRNA and protein in two *dnaA*(Ts) mutants (*dnaA46* and *dnaA167*) and found no significant effect on the synthesis of the mRNA and protein.

Suppressor Studies

Suppression that corrects a defect in a given mutant polypeptide can occur at a variety of levels (for reviews on these concepts see references 36 and 132): (i) restoration of proper folding in a mutant protein,

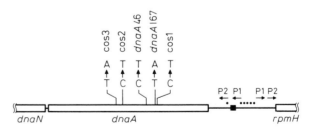

FIG. 8. Genetic and physical map of the *dnaA* gene of *E. coli* and its immediate vicinity. The altered DNA sequences giving rise to the *dnaA167* and *dnaA46* mutations, located respectively at nucleotides 702 and 783 of the *dnaA* structural gene, are depicted (position 1 is defined as the first nucleotide of the initiation codon) (305). The *dnaAcos1*, *-cos2*, and *-cos3* mutations, located at nucleotide positions 699, 986, and 1,086, respectively, were identified by Braun et al. (42). The positions of the DnaA box sequence (filled square) and GATC methylation sites (filled circles) in the *dnaA* promoter region are also indicated. The arrows at the top indicate divergent transcription from the *dnaA* gene and the adjacent *rpmH* gene.

either by an intragenic compensatory mutation at a second site in the same polypeptide chain or by an extragenic compensatory mutation in an interacting protein (such suppressor mutations should be allele specific; however, allele specificity does not guarantee protein-protein interaction, although this is usually taken as evidence for such); (ii) an increase in the level of the mutated polypeptide, achieved by increasing its transcription, diminishing premature termination of its transcript, improving translation of its message, or stabilizing the mutant polypeptide; (iii) elimination or reduction of levels of an inhibitor of its action; (iv) a bypass, either partly or completely, of the action of the mutant by the activation of a new pathway (such as activation of a cryptic pathway, substitution of an altered enzymatic activity, or integrative suppression by another replicon)—such suppressors should be able to survive a complete deletion of the suppressed protein; (v) alteration of the target site or substrate, allowing the mutant protein to function; (vi) overproduction of an interacting component, thus compensating for reduced activity of the mutant protein by mass action. Other possible mechanisms partly related or unrelated to these can be envisioned. The *dnaA* mutants present an ideal genetic system for suppressor analysis since the majority of the Ts$^+$ revertants recovered contain lesions unlinked to the *dnaA* gene (14, 411; see Fig. 9). The most thorough suppressor analysis has been carried out with the *dnaA46* mutation.

Intragenic Suppression

Kellenberger-Gujer et al. (176) identified a Ts$^+$ suppressor of the *dnaA46* mutation, *dnaA46cos*, and showed that it allowed bacterial growth at 42°C but caused cold sensitivity at 30°C. This suppressor mutation was not separable from *dnaA46* by P1 transduction. In the *dnaA46cos* strain, initiation appears normal at high temperature, but when shifted to low temperature, there is a high degree of overinitiation, leading to net increase in DNA. Recently, the nucleotide sequence of the dnaA46cos allele was determined (42) and found to contain three base substitutions, all resulting in amino acid replacements, within the coding sequence of the *dnaA* gene (Fig. 8). Because of the high frequency of occurrence of intragenic suppressors of *dnaA* with the Cos phenotype, it is unlikely that all three point mutations were simultaneously selected. It could be that some of these mutations were selected subsequently during bacterial growth (42). Kellenberger-Gujer et al. (176) proposed that the Cos suppressor compensates for the defect by allowing an overproduction of the weakly active DnaA46 protein at the elevated temperature. Upon subsequent shifting to lower temperature, the overproduced polypeptide would lead to overinitiation and net increase of DNA. However, Braun et al. (42) have presented evidence that the DnaA46cos mutant protein appears to lower rather than to raise the expression of the *dnaA* gene at 37 and 42°C, suggesting an increased affinity of DnaAcos protein for its binding site. Such enhanced binding might result in a more efficient assembly of a replication complex at *oriC*, resulting in the observed higher levels of replication. The presence in a *dnaA46cos* strain of a λ *dv* plasmid (175) or a P1

ban crr prophage (98) suppresses the cold sensitivity of *dnaA46cos* and prevents the overinitiation at low temperature. This suppression is most likely due to the limitation of DnaB protein activity by its demonstrated interactions with the λ P protein (121, 179, 273, 419); and the P1 DnaB-analog protein Ban (68, 300). Kellenberger-Gujer and Podhajska (175) have also shown that strains carrying the *dnaA46* mutation are themselves unable to support the growth of λ *dv*, and this too could be ascribed to the sequestration of DnaB protein by λ P. These results indicate that the overinitiation in bacteria carrying the *dnaA46cos* mutation at low temperature may be due either to a relative excess of DnaB protein at low temperature per se or to a relatively higher affinity of the DnaA46cos protein for DnaB protein. It may be that the *cos* mutation(s) suppresses the *dnaA46* mutation by increasing its affinity for the DnaB protein to a level that allows initiation at the elevated temperature. When shifted to lower temperature, however, this interaction progresses to overinitiation. Since conditions that limit DnaB protein are incompatible with the *dnaA46* mutation, it may be that a consequence of the *dnaA46* mutation is lowered affinity for DnaB protein (see below).

Extragenic Suppression

Bypass mechanisms

Integrative suppression. In a population of *dnaA*(Ts) cells carrying a conjugative plasmid such as F or F'/*lac* (294), ColV2 or R100.1 (293), or R1 (283), selection for growth at 42°C yields stably temperature-resistant clones which have simultaneously become Hfr. This was shown to be due to the integration at various sites into the chromosome of the plasmid in question (294) and the substitution of the plasmid origin for the chromosomal origin as the productive initiation site at high temperature (31, 59). Other plasmids and prophages have been shown to mediate integrative suppression as well, including P1 (62) and mini-P1 (61), P2 (232), and ColE1 (432). It had been assumed that all of these replicons capable of integrative suppression replicate independently of a requirement for the DnaA protein. However, it has recently been shown by Hansen and Yarmolinsky (130) that P1, mini-P1, and mini-F cannot be stably maintained as plasmids in the presence of an interrupted *dnaA* gene (*dnaA*-null) and therefore require DnaA protein for their replication and maintenance. That mini-P1 is capable of integratively suppressing a *dnaA*(Ts) but not a *dnaA*-null strain suggests that the DnaA activity remaining in *dnaA*(Ts) strains is sufficient for mini-P1 replication. However, the situation is more complex, inasmuch as P1 is capable of integratively suppressing *dnaA*-null strains. This last observation suggests the existence of a *dnaA*-independent replication system in P1 not present in mini-P1. One possibility is that this represents a component of the P1 vegetative replication cycle (130).

Inactivation of RNase H. One class of extragenic suppressors (*dasF*, *sdrA*, or *sin*) that fulfills all the criteria defining a bypass mechanism has been identified (14, 188, 233, 389). All *dnaA* alleles, including insertionally inactivated *dnaA* genes, are corrected by these suppressors, and the *oriC* region may be deleted in strains carrying such a suppressor (192). All these suppressors were found to greatly lower the activity of RNase H, the product of the *rnh* gene located at 5 min (56, 150; see Fig. 9). Furthermore, an insertional inactivation of the *rnh* gene (151) leads to suppression of *dnaA46*. In the absence of RNase H activity, it is thought that various mRNA molecules or sites of RNA-DNA hybrids can serve as alternative origins for DNA initiation (71, 304; see above). In such situations, the need for both *oriC* and DnaA protein is thus obviated.

rpoB mutations

A very frequent class of suppressors of *dnaA* alleles was shown to reside in *rpoB* (14, 21, 343), the structural gene for the β subunit of the RNA polymerase (Fig. 9). Among mutants selected as rifampin resistant and bearing lesions in *rpoB*, 8 to 65% can be shown to be capable of suppressing a given *dnaA*(Ts) allele of *E. coli* or *S. typhimurium* (21). This is an interesting class because in vivo RNA transcription is required for initiation of DNA replication (222). Analysis of the pattern of suppression showed that all the Ts alleles tested can be suppressed by at least one *rpoB* mutation, and allele specificity of suppression is observed (14, 15, 19, 21, 343). Although the latter observation suggests the possibility of interaction at the protein level between RNA polymerase and DnaA protein, no direct evidence for such an interaction exists. The fact that a large fraction (60%) of all rifampin-resistant (*rpoB*) mutants exhibit increased supercoiling in the appropriate genetic background (R. Menzel, personal communication) suggests that these RNA polymerase extragenic suppressors may act indirectly. For example, increased supercoiling may aid mutant DnaA proteins in forming the correct initiation complex at *oriC* (see *topA* suppressors, below). The apparent allele specificity of suppression could then be due to the specific supercoiling requirement of each DnaA mutant protein. However, in light of the evidence that *dnaA*(Ts) mutations can lead to altered transcription termination of the *mioC* transcript (325) and in the *trp* attenuator (18) and the fact that some *rpoB-dnaA*(Ts)

mutant combinations are lethal (22, 343), it is tempting to speculate that DnaA protein and RNA polymerase do interact.

Other single-copy suppressors

Early work by Wechsler and Zdzienicka (411) led to the identification of five genetic loci, mutations in which could suppress the temperature-sensitive phenotype of various *dnaA*(Ts) mutations and simultaneously confer a cold sensitivity phenotype for bacterial growth. One of these loci appeared to be in the *dnaA* gene itself, whereas the other four were located elsewhere, including near to but distinct from *dnaA*. The suppressors isolated by Atlung (14) defined additional classes. These include *dasA*, located at 80 min; *dasB* at 83.5 min, near *oriC*; *dasC* at 84 min, near the *rep* gene; *dasE* at 1 min; and *dasG* at 99 min, near *dnaC* and *dnaT* (Fig. 9). None of these genes has been identified as yet. The fact that many of these suppressor mutations simultaneously confer a cold sensitivity phenotype for bacterial growth is interesting inasmuch as a similar cold sensitivity phenotype is seen in *dnaA+/dnaA*(Ts) and *dnaA*(Ts)/*dnaA*(Ts) merodiploids (127, 397) and *dnaA46cos* strains (176), where the cold sensitivity appears to be due to overinitiation at the lower temperatures. Hence it could be that many of these suppressor mutations function by increasing the intracellular levels of the corresponding DnaA(Ts) proteins. Such an increase would restore replication at high temperature but result in overinitiation at low temperature, at which the mutant protein is active.

Some of these suppressors were originally detected as cryptic *das* mutations in a *dnaA46* background (14). Since bacteria carrying the *dnaA46* mutation have been shown to be cold sensitive for growth (314), they may accumulate suppressor mutations enabling them to grow faster. The presence of such suppressors will tend to complicate the genetic analysis. In addition, some of the suppressors are weak and their effects can only be observed in combination with others, e.g., *dasG* in combination with *dasE* (14). The most convincing way to test the individual effects of uncharacterized *das* mutations would be first to introduce them singly into an otherwise wild-type background. Subsequently, *dnaA*(Ts) alleles can be introduced at the permissive temperature by cotransduction with a selectable marker, and the growth patterns can then be analyzed at the restrictive temperature.

Louarn et al. (245) reported that deletions in the *topA* gene (which specifies topoisomerase I) suppress the temperature sensitivity phenotype of *dnaA*(Ts) mutations (Fig. 9). In these *topA* mutants the net DNA supercoiling density is higher than normal despite the presence of compensatory mutations (some of which lie in *gyrA* or *gyrB* [74]) that allow growth of *topA* strains. Thus the effect may be exerted through an indirect mechanism involving (i) differential expression of the *dnaA* gene itself as well as other genes via increased supercoiling or via effects of the compensating *gyr* mutations, or both, or (ii) increased interaction of mutant DnaA proteins with *oriC*. The fact that topoisomerase I plays a discriminatory role in *oriC* replication in vitro by inhibiting nonspecific (*dnaA*-independent) initiation pathways (171) suggests the possibility that such a bypass initiation might occur

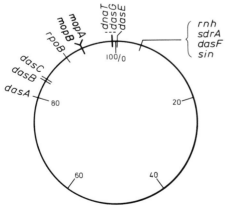

FIG. 9. Chromosomal map location of loci specifying products that directly or indirectly compensate for the effects of mutations in the *dnaA* gene (see chapter 53 and text for details).

in vivo. However, such a bypass mechanism would still require some DnaA protein since *dnaA*(Am) mutations are not suppressed in *topA* deletions (245).

Suppression by overproduction of other gene products

The overproduction of both the GroES and GroEL proteins (99) was shown to be both necessary and sufficient to suppress the Ts phenotype of the *dnaA46* mutation (87, 163; Fig. 9). In the suppressed strains growing at high temperatures, bidirectional replication was shown to start at or near the *oriC* region (87). Furthermore, the need for the DnaA polypeptide was inferred by the allele-specific suppression pattern and inability of *dnaA*(Am) mutants to be suppressed (87, 163). Interestingly, the pattern of suppression of *dnaA*(Ts) mutants was identical to that suppressed by the *rpoB902* allele (127, 163). The mechanism by which the overproduction of GroES and GroEL proteins suppresses *dnaA* mutants is not understood, although it is known that at least a sixfold overproduction is necessary to observe the effect (87). The GroEL protein has been shown to be required for the oligomerization of the phage λ B polypeptide to a preconnector structure that is necessary for the correct assembly of the λ head and is thought to provide a surface for this oligomerization (186). It could be that the suppression of certain *dnaA* mutants is due to an improved ability of the mutant DnaA protein to be oligomerized in the presence of large amounts of GroES and GroEL proteins. Interestingly, most suppressed strains exhibit, simultaneously, a cold sensitivity phenotype the significance of which has been discussed above. That overproduction of the GroEL protein led to suppression of the *dnaA*(Ts) defect was originally shown by Takeda and Hirota (379) in an attempt to clone the *dnaA*+ gene on a multicopy plasmid, although these authors did not identify the gene as such. Projan and Wechsler (322), also attempting to clone the *dnaA*+ gene, identified three additional *E. coli* fragments that when cloned on multicopy plasmids suppressed the defect of the three *dnaA*(Ts) mutations examined. Although the mechanism by which these unidentified cloned segments suppress the *dnaA* defect at high temperature is unknown, it has been shown that an intact DnaA polypeptide is required for suppression (322).

Additional Suppressor Studies

Walker et al. (403) described a class of rare temperature-resistant extragenic suppressors of the *dnaZ-2016*(Ts) mutation, called *dnaA*(SUZ,CS). The *dnaA*-(SUZ,CS) suppressor confers a cold sensitivity (Cs) phenotype and was mapped in the *dnaA* gene (403). It blocks initiation of DNA replication in vivo at low temperature and was shown to be incompatible with additional copies of the *dnaZ*+ gene (33). Furthermore, it was shown to be totally incompatible with *dnaC*(Ts) and *dnaG*(Ts) alleles and partly incompatible with *dnaB*(Ts) and *dnaE*(Ts) alleles. To explain this plethora of phenotypes, it was proposed that the DnaA protein participates not only in the initiation step of DNA replication, but is also part of the replication complex itself (403). Alternatively, the *dnaA*(SUZ,CS)

mutation may exert its suppressing effect by directly or indirectly affecting the intracellular levels of DnaZ protein. For example, lowering of the DnaZ protein levels may lead to restoration of a productive interaction between DnaZ protein and other components of the replication complex that might be limiting. Such a phenomenon was first described by Floor (93) among interacting morphogenetic components. Conversely, an increase in the levels of the mutant DnaZ protein may lead to restoration of function at nonpermissive temperature. Recent evidence indicates that suppressors of *dnaZ*(Ts) mutations can be isolated that lie in the *dnaN* gene (84a). Gene *dnaN* is cotranscribed with *dnaA* (339); hence, mutations in the *dnaA* gene may modulate *dnaN* gene expression to a level compatible with that of the mutant *dnaZ*(Ts) protein.

PHYSIOLOGICAL STUDIES ON DnaA PROTEIN AND INITIATION CONTROL

The replicon model of Jacob et al. (159) was proposed, in part, to explain the coexistence and correct segregation of a low-copy-number plasmid, F, and the host chromosome. The model suggests that initiation occurs when an initiator protein interacts with a fixed activator site on each replicon (chromosome or episome) and further proposes that each replicon is associated with a site on the cell membrane insuring its correct segregation into daughter cells (and explains incompatibility). The control circuitry that times initiation was not specifically defined; possibly it is doubling of particular membrane sites by cell elongation. Studies on the relationship of initiation to the cell cycle led to the finding that initiation takes place, independently of growth rate, at a constant ratio of origins to cell mass (66, 77, 320). Many models were proposed in which initiation was to occur when the concentration of an initiator protein accumulated to a threshold level for interaction with the origin (references 78, 134, 250, and others). Sompayrac and Maaløe (364) specifically proposed that such an initiator protein might be part of an autoregulated operon, such that the concentration (molecules per volume) of the activator is held relatively constant independent of growth rate, with cyclical fluctuations at the time of duplication of the gene. Other models involving a turn-on of initiation by inactivation (328) or dilution of an initiation inhibitor (320) have also been proposed.

The studies discussed in an earlier section showed that de novo synthesis of both protein and RNA is required to initiate a new round of replication. However, it is not clear whether one (or more) of these species is a frequency controller or whether they merely participate in the initiation or replication apparatus or both. As an alternative, Helmstetter (133) proposed that neither RNA nor protein synthesis was required as an immediate trigger for initiation and that the growth of the cell envelope was in some way responsible for regulating initiation frequency.

A priori, there are both positive and negative aspects in initiation regulation to be explained. Origins must fire when cell mass attains some critical level but must not fire again prematurely. That the ability to govern both aspects resides in one controlling element may be too simplistic a view.

Recent studies on the role played by the Dam methylase in initiation of DNA replication via its methylation of adenine in GATC sequences reveal a control circuit which, when viewed in isolation, might appear to explain initiation control. Dam methylation is required for efficient replication of *oriC* (154, 276, 362). If hemimethylated *oriC* will not fire, the refractory period for reinitiation (the negative aspect of initiation control) may be explained. Supporting this, Messer et al. (276) showed that in vivo overproduction of Dam methylase leads to a decrease in interinitiation times. The *dnaA* gene is under both the positive control of methylation and negative autoregulation (41, 43). The *dam* gene itself is under negative control by DnaA protein (D. Smith, personal communication). Hence, an apparently self-consistent control circuit for initiation might be proposed based on these studies. A low level of DnaA protein will permit expression of the *dam* gene and lead to derepression of the *dnaA* gene. Expression of Dam methylase will lead to methylation of the *dnaA* promoter and thereby allow expression of the *dnaA* gene. This amount of DnaA protein will be sufficient for triggering initiation at the origin and repression of both the *dnaA* and *dam* genes. Newly replicated DNA will be hemimethylated and not subject to reinitiation until the DnaA-controlled Dam methylase acts. This model, however, leaves the activity of DnaA protein at the origin unclarified.

Much information on the activity of DnaA protein and its interactions with other elements (DNA, RNA, and protein) can be extracted from the many physiological studies which were carried out in an attempt to ascertain whether the DnaA protein plays a regulatory role in initiation. These studies were designed, in part, to test whether DnaA might be the "initiator" protein of the replicon hypothesis, since the findings that the DnaA protein appears to be a replicon-specific initiation factor (97, 409) that has a positive function in the in vitro initiation system (103) and binds to *oriC* (101, 259) make it an attractive candidate. Its autoregulatory activity might fit the Sompayrac and Maaløe (364) model, as well. In hindsight, however, in light of (i) the fact that the nature of the defects in DnaA mutant proteins is unknown, (ii) recent studies revealing the apparent importance of Dam methylation, and (iii) the many roles of the DnaA protein in regulating the expression of proteins and transcripts including its own, it is probable that the interpretation of many of these studies has been oversimplified. These studies, however, some of which are reviewed below, show the complexity of interactions that must take place for initiation to occur. They indicate that the concentration or activity of a number of different components, including DnaA protein, DnaB protein, RNA polymerase, and a stringently controlled RNA transcript, can be inferred to be rate-limiting, implying that for "normal" initiation to take place, they must be present in some optimal mixture whose formation occurs rarely and regularly enough to be, itself, the frequency controller; or that all components must be "in excess," awaiting an as yet undefined process, such as correct *oriC* methylation or "envelope attachment" or both, which must still be invoked as a frequency controller.

In a *dnaA167*(Ts) strain at permissive temperature, the activity of DnaA protein itself is limiting. According to Fralick (95, 96), who studied a variety of cell parameters in a *dnaA*(Ts) mutant over a range of permissive temperatures, this limitation does not change initiation frequency but rather affects the number of origins which fire at the normal time. Hence, Fralick concluded that it is not DnaA protein that is the frequency controller, but that DnaA protein is required for the formation of productive initiation complexes. The cooperativity of binding exhibited by the DnaA protein at the origin (101, 103) might enable the firing of one origin even in the presence of others. The *dnaA167* allele, however, may be atypical in that it appears to induce some form of stable DNA replication when shifted from nonpermissive to permissive temperature (82).

In a *dnaA5*(Ts) strain grown at permissive temperatures, addition of chloramphenicol to the culture stimulates accumulation of DNA far above that expected for completion of ongoing rounds of replication as is found in a *dnaA*⁺ strain (126, 216, 248, 277, 314, 386). This burst of overreplication has been shown (248) by hybridization studies to be due to overinitiation. Orr et al. (314) showed that, whereas incubation in the presence of chloramphenicol does lead to a burst of initiation, this burst is not seen when cells are amino acid starved except if the cells are *relA*. From this they concluded that overinitiation in a *dnaA*(Ts) strain at permissive temperature by addition of chloramphenicol is not due to inhibition of protein synthesis per se, but is the result of stimulation of stringently controlled transcription which would occur under conditions of chloramphenicol treatment (354) and in amino acid starvation of a *relA* strain, but not a *relA*⁺ one. It was hypothesized (248, 319) that *dnaA*(Ts) mutants are unable to synthesize a stringently controlled RNA that is not normally rate limiting but is rate limiting in a *dnaA*(Ts) mutant, perhaps because of poor interaction between RNA polymerase and the mutant DnaA protein. Another explanation is that the synthesis of this RNA could normally be rate limiting but that in *dnaA* cells there is a limitation of some other protein, possibly another replication component, for origin firing. From studies with the *dnaA46cos* intragenic suppressor of *dnaA46*, discussed in an earlier section, DnaB protein would appear to be a good candidate for such a component. It is likely that the *mioC* promoter (see above) is the stringently controlled promoter implicated in the chloramphenicol effect. However, the overall contribution of raised or lowered DnaA protein levels (or activities) to the ultimate effect of transcription from this promoter on initiation is unclear. The *rpoB* extragenic suppressors of *dnaA*(Ts) mutants (14, 15, 21, 343) and the mutations in *rpoC* (323) and *rpoB* (381), which lead to an increased origin/mass ratio and are temperature sensitive for growth, might reflect the role of RNA polymerase in the transcription and correct termination of the stringent RNA. In light of the possibility that a stringently controlled promoter or transcript might play a role in *E. coli* initiation, it is interesting that in *B. subtilis*, the synthesis of an RNA-DNA copolymer containing *rrnO* RNA sequences is correlated with the transcription requirement for chromosomal initiation (352). Furthermore, the DNA replication in synchronized *B. subtilis* cells appears to be under strict stringent control (351).

Recently, various investigators have addressed the question of the effect of increased intracellular levels of DnaA protein on DNA replication. The results obtained may appear contradictory but might be explained in part by the difference in the ultimate levels of DnaA protein attained in each system.

Churchward et al. (64) found that in a *dnaA46*(Ts) strain at permissive temperature, addition of DnaA$^+$ protein (by induction of cloned *dnaA$^+$* under *lac* promoter control on a plasmid) leads to an overall increase in net synthesis of DNA. This finding was taken to indicate that mutant DnaA protein (or the level of its activity) is limiting, that additional origins will fire with increased DnaA$^+$ protein, and thus there is excess "initiation potential" in such cells. This could be due to relatively higher activity or concentration of another protein or, perhaps, to an increase in transorigin transcription favored by optimizing levels of DnaA$^+$ protein.

Churchward et al. (64) also found that, in a *dnaA$^+$* background, increasing the amount of DnaA$^+$ protein did not lead to a net increase in DNA. This argues that under normal conditions, excess DnaA$^+$ protein does not trigger the firing of origins by its mere concentration and that there is no excess initiation potential in such cells. Results of similar experiments by Atlung et al. (19) (with induction of *dnaA$^+$* under λ p_L control) supported the idea that increasing the level of DnaA$^+$ protein does not lead to a net increase in DNA; however, it does lead to "abortive" initiation, the apparent initiation of forks that do not progress far along the chromosome. This would indicate that, whereas the potential to initiate in such cells is not limiting (there is excess potential), some factor or factors which are necessary for fork progression are limiting. Atlung et al. (19) suggest that it is the high level of DnaA protein, itself, that blocks DNA elongation shortly after initiation. Consistent with such an interpretation, Rokeach et al. (325) reported that even higher DnaA$^+$ protein levels led to a decrease in the overall rate of DNA replication, although they did not specifically address the question of initiation. Alternatively, abortive initiation might be due to a limitation of some part of the replication apparatus or, as suggested by Helmstetter et al. (135), to a topological constraint on fork propagation under these conditions.

Thus, although the DnaA protein is likely to be the origin-specific initiator of the replicon hypothesis, a balance among DnaA protein, DnaB protein, RNA polymerase, topology at the origin, and the level of transorigin transcription, possibly from a stringently controlled promoter, may be critical in the positive control of the initiation event. DnaA protein appears to control many of these factors by regulating its own expression, the levels of both transcription and termination from the stringently controlled *mioC* promoter, and, in all likelihood, the expression of other genes such as Dam methylase which may be directly responsible for timing of the interinitiation period. DnaA protein is certainly pivotal in initiation and in that respect can be considered a controlling element.

Temporal Ordering of Requirements at Initiation

By using thermoreversible temperature-sensitive mutations in the *dnaA*, *dnaC*, and *dnaB* genes, in combination with antibiotics that are known to inhib-

it either protein or RNA synthesis or RNA polymerase, the temporal order of the requirements for each of these proteins in initiation relative to the requirements for protein and RNA synthesis was determined (126, 210, 337, 445, 448). The *dnaB252* mutation is the sole example, so far, of a mutation in *dnaB* that affects the initiation and not the elongation process (448). Thus it is the initiation role of DnaB protein that was explored by these experiments. The theory underlying these experiments is that if a temperature-sensitive mutation leads to a blockage at a specific step in the initiation process, raising the temperature in such a mutant will lead to an arrest at the point in the process where that protein is normally required. With a thermoreversible mutation, return of the culture to the permissive temperature should lead to resumption of initiation. Since it has been shown that initiation is prevented by inhibitors of RNA and protein synthesis, one can determine whether the chloramphenicol- or rifampin-inhibited steps occur before or after the point of arrest of the mutant in question by adding the inhibitor to the culture shortly before the return to permissive temperature. If, for example, protein synthesis is required before the step at which a *dnaA*(Ts) mutant is blocked, then initiation should proceed in the presence of chloramphenicol after the return to permissive temperature; whereas, if protein synthesis is required at the time of or after the arrest, then initiation will not proceed. This type of experiment will indicate only the last step in the process in which the mutated protein is defective. It also does not rule out the participation of a protein at a given step, indicating only the role in which the mutation causes a deficiency.

The results of these experiments (445, 448) indicate that (i) the DnaA, DnaB252, and DnaC proteins act after the chloramphenicol-sensitive step, (ii) the DnaA and DnaB252 proteins act before or during a rifampin-sensitive step, and (iii) DnaC protein acts after this step. This would give the order: protein synthesis → DnaA and DnaB → RNA polymerase transcription initiation → DnaC. However, the result with streptolidigin (an inhibitor of RNA polymerase elongation) indicates that the DnaB252 protein acts before (or during) the streptolidigin-sensitive step but that both DnaA and DnaC proteins act after it. This might be explained if there are two steps inhibited by rifampin, but only the first of these is inhibited by streptolidigin, with the DnaB252 protein being defective in at least the first rifampin-sensitive step and the DnaA(Ts) protein being defective in the second. It could be that DnaB protein acts before or while a specific transcript is initiated and elongated. Furthermore, since rifampin has an antiterminating effect on RNA polymerase (67), it is possible that a transcription-terminating activity of DnaA protein (hypothesized by Atlung and Hansen [18]) is antagonized by rifampin.

This order may be summarized: (i) a chloramphenicol-sensitive step, involving the synthesis of a protein or proteins that are consumed in a round of replication; (ii) initiation and elongation of an RNA transcript, before or during which the activity of DnaB protein that is defective in *dnaB252* strains is required; (iii) the action of DnaA protein before or during a requirement for RNA polymerase that is either inhibited or altered by rifampin, for example, transcription termination; and (iv) DnaC protein acts.

It appears from in vitro evidence (see above) that DnaA protein binding to *oriC* is the earliest step in initiation. However, the temporal ordering experiments imply that the *dnaB252* mutation affects a step preceding the one at which *dnaA*(Ts) mutants are blocked. Thus, a reasonable interpretation of the results of these experiments is that DnaA(Ts) mutant proteins are capable of binding to *oriC* (in apparent contradiction to interpretations of studies on autoregulation of *dnaA* promoter fusions) and that their defect is at a later step. Another important implication of these results is that there are requirements for separate functions of the DnaB and DnaC proteins at initiation. A priori, since the DnaB and DnaC proteins have been shown to form a complex in vitro (421), it might be reasonable to expect that the rare *dnaB* mutant defective in initiation would be defective in the same step as the *dnaC*(Ts) mutations. This does not appear to be the case. Furthermore, Sclafani and Wechsler (349) have shown that the DnaB252 protein's defect in initiation can be suppressed by an overproduction of DnaC protein. All of these results taken together make it likely that DnaB and DnaC proteins do interact during initiation but that the mutations in each define domains with separate functions.

Hypothetical Scheme for Initiation In Vivo

Genetic, biochemical, and physiological data demonstrate not only the complexity of the components necessary for successful initiation at *oriC* but also a degree of uncertainty as to the exact steps of the process in which they participate. Nevertheless, an attempt will be made to synthesize this body of information into a plausible series of events that lead to firing from *oriC* and to correlate, where possible, the scheme derived from in vitro experiments (summarized in Fig. 6) with those inferred from in vivo work. Strikingly, there is no absolute requirement in vitro for RNA polymerase, whereas the in vivo experiments imply a pivotal role for RNA polymerase. In fact, it is this role of RNA polymerase relative to which other protein requirements have been ordered. The in vivo results appear to concern events occurring after DnaA protein binding to *oriC* DNA and before priming by the DnaG primase [see Fig. 6, (b), (c), and (d)]. Since it is the isolation of intermediates from these early steps in the in vitro reaction [Fig. 6, (b) and (c)] that requires elevated temperatures, and RNA polymerase lowers this temperature requirement in vitro (199, 394), it is possible that the in vivo role of RNA polymerase is the localized destabilization of base pairing which is otherwise favored under in vitro conditions.

The steps in a hypothetical scheme for initiation may be summarized as follows. (i) DnaA protein binds to *oriC* [Fig. 6, (b)] perhaps with a strict in vivo requirement for GATC methylation or supercoiling or both. Although the binding of DnaA protein to *oriC* in vitro does not require transcription by RNA polymerase, such transcription might be necessary in vivo if, for example, HU protein or other proteins interfere with DnaA protein binding and transcription acts to

clear the *oriC* region (394). From in vivo results, however, the first demonstrable requirement for transcription is at a step requiring DnaB protein, defective in the *dnaB252* mutant [i.e., stage II formation of the prepriming complex; Fig. 6, (b) → (c)]. (ii) A transcriptional activation of the *oriC* region by leftward transcription through *oriC* from a stringently controlled, DnaA-repressed promoter (such as *mioC*) is required for a productive DnaB-DnaA protein interaction. Such a hypothetical interaction would explain the phenotypes of both the *dnaB252* and *dnaAcos* mutations and is consistent with in vitro electron microscopic studies (199). The *dnaB252* mutation would lower the affinity of DnaB protein for DnaA protein, and the *dnaAcos* mutation would increase such an affinity. (iii) A DnaA protein-RNA polymerase interaction occurs, causing termination or pausing of this transcript [afer stage I and before stage II; Fig. 6, (b) → (c)]. This interaction would be defective in the *dnaA*(Ts) mutants and suppressible by *rpoB* mutations. Although this transcript might be used as a primer for leftward leading-strand synthesis, a more attractive alternative is that pausing or termination of this transcript would prepare the DNA-protein complex for the next crucial step, (iv) the delivery of DnaB protein to ssDNA with the participation of the DnaC protein [stage II, Fig. 6, (c)]. (v) Once positioned, the DnaB protein acts as a helicase [stage III, Fig. 6, (d)], unwinding DNA and allowing entrance of the proteins required for priming and DNA elongation. After forks are formed and progress, the hemimethylation of *oriC* and topological constraints would prevent further rapid initiation. This type of model could explain the need for a balance of activities of DnaA, DnaB, and DnaC proteins, transcription, and methylation. If any one is limiting, the reaction might be favored by an increase in another component.

TERMINATION OF DNA REPLICATION

In organisms such as *E. coli* and *B. subtilis*, DNA replication forks initiated bidirectionally from *oriC* meet diametrically opposite the origin at a region called *terC* (31, 206, 246, 315). That this region is unique, and not merely defined by its position 180° from *oriC*, has been shown by the use of *E. coli* strains in which the functioning origin has been displaced from *oriC* to the site of the integrated replicon P2*sig*5 (206) or R100.1 (72, 246). In such strains, the replication velocity of both forks is slowed drastically near *terC*.

The approximate position of *terC* has been determined by restriction analysis of late-replicating DNA (37, 38) as well as by deletion analysis (140, 141). The *terC* region contains few known genetic loci (20), and *E. coli* can tolerate deletions of approximately 340 kb from this region (141). These deletions relieve the slowing of replication forks through the region and exhibit pleiotropic effects, including slow growth, constitutive SOS induction, and segregation of anucleate cells (141).

Although the function of the *terC* region per se is unknown, it is clear that termination is not a passive process. *cis*- and *trans*-acting elements essential for chromosome partitioning into daughter cells have been described for many plasmid systems. In *B. sub-*

tilis, there is evidence that the chromosome terminus is associated with the cell membrane (341, 371). Such an association may be necessary for proper partitioning of the chromosomes, and it could be that the segregation of anucleate cells in *terC* deletions is due to a defect in proper attachment (141). The process of termination of chromosome replication appears to include the resolution of topological problems incurred in the meeting of forks. As demonstrated in the case of simian virus 40, the replication process results in an intertwining between the two daughter chromosomes (376) which can be resolved by a topoisomerase. Steck and Drlica (366) have shown that *E. coli* gyrase mutations under nonpermissive conditions often result in catenation of chromosomes. The *terC* region might contain specific sequences for efficient DNA gyrase action (92) in resolving these catenated structures. Termination has been proposed as the signal for cell division (136). The *terC* region may thus play a role in coupling DNA replication to the cell cycle. In this respect, it is interesting that the *dicA* and *dicB* loci, both involved in control of cell division, are among the sparse genetic loci in the *terC* region (28).

ADDITIONAL REPLICATION PATHWAYS: INDUCIBLE STABLE DNA REPLICATION

Pritchard and Lark (321) made the observation that when *E. coli* cells were starved for thymine, they developed the capacity to replicate DNA in the absence of further protein synthesis. They ascribed this to the synthesis of a new form of replication complex. It was subsequently shown that many other treatments, including those capable of inducing the SOS response in *E. coli*, led to the induction of this so-called stable DNA replication (SDR) (Kogoma and Lark [189], reviewed by Lark and Lark [224, 225]). Two striking properties of SDR are its dependence on active RecA protein and its mutagenicity. Three classes of mutants were isolated that appeared to affect this system. One type, termed *sdrA*, is constitutive for DNA replication in the absence of protein synthesis (188) and was later shown to be defective in RNase H and not part of the inducible SDR pathway (30). Two types of mutants were isolated that appeared to affect inducible SDR: those unable to assemble it after inducing treatment, termed *dnaT*, and those in which this RecA-dependent replication is the only form of DNA replication, called *sdrT*. Inducible SDR has been shown to require the *dnaB*, *dnaG*, and *dnaE* but not the *dnaA* gene products (224, 225). Some of the *dnaT* mutants were shown to be conditional lethal (exhibiting a delayed arrest of DNA replication) and dominant over *dnaT*⁺ (221). Both the *sdrT* and *dnaT* mutations were shown to lie at 99 min, very close to *dnaC*. It was proposed that these mutations are alleles of a single gene and that the wild-type DnaT protein is an integral part of the replication machinery, its presence being necessary to allow the proper dissociation of the complex at the terminus (224, 225). The SOS-activated form of the RecA protein would modify the wild-type DnaT protein, preventing the proper dissociation of the replication complex at termination, allowing its reutilization without further protein synthesis and endowing it with mutagenic properties, possibly by the incorporation of RecA protein itself. In this model the DnaT mutant protein might be resistant to RecA modification and lead to an easily destabilized replication fork. In an analogous manner, the *sdrT* mutants would be subject to modification (and possible replacement) by nonactivated RecA protein, leading to the formation of RecA-dependent, termination-resistant, error-prone replication (224, 225).

Recently it has been shown that the *dnaT* gene codes for protein i, a protein necessary for primosome assembly in φX174 replication (see above), and is probably the first gene in an operon that includes the adjacent *dnaC* gene (256). In light of this information, it could be that the mutant DnaT protein leads to primosome destabilization or loss owing to its lability and that the altered SdrT protein (assuming that *sdrT* and *dnaT* are the same gene) may lead to increased primosome stability or reassembly at new sites. In this respect, two relevant observations have been made with the *dnaT* and *sdrT* loci. DnaC and DnaT proteins functionally interact since mutations in *dnaT* suppress the severity of the phenotype of certain *dnaC* mutations and vice versa (221, 224). Furthermore, certain mutations in *dnaA* similarly ameliorate the DnaT and SdrT phenotypes (225). One interesting interpretation of these results is that the DnaC, DnaT, and DnaA proteins actively cooperate in the correct assembly of a primosome structure, as discussed above.

CONCLUDING REMARKS

Clearly, it is the integration of the genetic and biochemical approaches to DNA replication that has brought us to our current level of understanding of this complex phenomenon in *E. coli*. It is apparent that many of the unsolved details of bacterial DNA replication will continue to yield to investigations that attempt a synthesis of these two great disciplines. For example, the many classes of mutant DnaA proteins previously characterized by genetic and physiological studies should be prime subjects for biochemical analysis that would be expected to clarify interactions of this key initiator protein with *oriC* DNA and with other replication proteins. In the future, questions concerning the nature of the regulation of initiation of chromosomal DNA replication and investigations into the roles and actions of DnaA protein, RNA polymerase, *oriC* methylation and structure, DNA gyrase, and histonelike proteins and chromatin structure will be of primary interest to the DNA replication field. There is no doubt that studies of *E. coli* DNA replication during the next few years will continue to focus on a wide array of unique and fascinating biochemical problems, problems that will provide some of the most exciting challenges available anywhere in biology.

LITERATURE CITED

1. **Alberts, B. M., F. J. Amodio, M. Jenkins, E. D. Gutmann, and F. L. Ferris.** 1969. Studies with DNA-cellulose chromatography. I. DNA-binding proteins from *Escherichia coli*. Cold Spring Harbor Symp. Quant. Biol. **33:**289–305.
2. **Alberts, B. M., and L. Frey.** 1970. T4 bacteriophage gene *32*: a structural protein in the replication and recombination of DNA. Nature (London) **227:**1313–1318.
3. **Anderl, A., and A. Klein.** 1982. Replication of λdv DNA in vitro. Nucleic Acids Res. **10:**1733–1740.

4. **Ang, D., G. N. Chandrasekhar, M. Zylicz, and C. Georgopoulos.** 1986. The *grpE* gene of *Escherichia coli* codes for heat shock protein B25.3, essential for λ DNA replication at all temperatures and host viability at high temperature. J. Bacteriol. **167:**25–29.

5. **Arai, K., and A. Kornberg.** 1979. A general priming system employing only DnaB protein and primase for DNA replication. Proc. Natl. Acad. Sci. USA **76:**4308–4312.

6. **Arai, K., and A. Kornberg.** 1981. Unique primed start of phage φX174 replication and mobility of the primosome in a direction opposite chain synthesis. Proc. Natl. Acad. Sci. USA **78:** 69–73.

7. **Arai, K., and A. Kornberg.** 1981. Mechanism of DnaB protein. II. ATP hydrolysis by DnaB protein dependent on single- or double-stranded DNA. J. Biol. Chem. **256:**5253–5259.

8. **Arai, K., and A. Kornberg.** 1981. Mechanism of DnaB protein action. III. Allosteric role of ATP in the alteration of structure by DnaB protein in priming replication. J. Biol. Chem. **256:**5260–5266.

9. **Arai, K., R. Low, J. Kobori, J. Shlomai, and A. Kornberg.** 1981. Mechanism of DnaB protein action. V. Association of DnaB protein, n′ and other prepriming proteins in the primosome of DNA replication. J. Biol. Chem. **256:**5273–5280.

10. **Arai, K., R. L. Low, and A. Kornberg.** 1981. Movement and site selection for priming by the primosome in phage φX174 DNA replication. Proc. Natl. Acad. Sci. USA **78:**707–711.

11. **Arai, K., R. McMacken, S. Yasuda, and A. Kornberg.** 1981. Purification and properties of *Escherichia coli* protein i, a prepriming protein in φX174 DNA replication. J. Biol. Chem. **256:** 5281–5286.

12. **Arai, K., S. Yasuda, and A. Kornberg.** 1981. Mechanism of DnaB protein action. I. Crystallization and properties of DnaB protein, an essential replication protein in *Escherichia coli*. J. Biol. Chem. **256:**5247–5252.

13. **Asada, K., K. Sugimoto, A. Oka, M. Takanama, and Y. Hirota.** 1982. Structure of replication origin of the *Escherichia coli* K-12 chromosome: the presence of spacer sequences in the *ori* region carrying information for autonomous replication. Nucleic Acids Res. **10:**3745–3754.

14. **Atlung, T.** 1981. Analysis of seven *dnaA* suppressor loci in *Escherichia coli*. ICN-UCLA Symp. Mol. Cell. Biol. **22:**297–314.

15. **Atlung, T.** 1984. Allele-specific suppression of *dnaA(Ts)* mutations by *rpoB* mutations in *Escherichia coli*. Mol. Gen. Genet. **197:**125–128.

16. **Atlung, T., E. Clausen, and F. G. Hansen.** 1984. Autorepression of the *dnaA* gene of *Escherichia coli*. Adv. Exp. Med. Biol. **179:** 199–207.

17. **Atlung, T., E. S. Clausen, and F. R. Hansen.** 1985. Autoregulation of the *dnaA* gene of *Escherichia coli* K12. Mol. Gen. Genet. **200:**442–450.

18. **Atlung, T., and F. G. Hansen.** 1983. Effect of *dnaA* and *rpoB* mutations on attenuation in the *trp* operon of *Escherichia coli*. J. Bacteriol. **156:**985–992.

19. **Atlung, T., K. V. Rasmussen, E. Clausen, and F. G. Hansen.** 1985. Role of the DnaA protein in control of DNA replication, p. 282–297. *In* M. Schaechter, F. C. Neidhardt, J. L. Ingraham, and N. O. Kjeldgaard (ed.), The molecular biology of bacterial growth. Jones and Bartlett Publishers, Inc., Boston.

20. **Bachmann, B.** 1983. Linkage map of *Escherichia coli* K-12, edition 7. Microbiol. Rev. **47:**180–230.

21. **Bagdasarian, M. M., M. Izakowska, and M. Bagdasarian.** 1977. Suppression of the DnaA phenotype by mutations in the *rpoB* cistron of ribonucleic acid polymerase in *Salmonella typhimurium* and *Escherichia coli*. J. Bacteriol. **130:**577–582.

22. **Bagdasarian, M. M., M. Izakowska, R. Natorff, and M. Bagdasarian.** 1978. The function of RNA polymerase and *dnaA* in the initiation of chromosome replication in *Escherichia coli* and *Salmonella typhimurium*, p. 101–111. *In* I. Molineux and M. Kohiyama (ed.), DNA synthesis present and future. Plenum Publishing Corp., New York.

23. **Baker, T. A., K. Sekimizu, B. E. Funnell, and A. Kornberg.** 1986. Extensive unwinding of the plasmid template during staged enzymatic initiation of DNA replication from the origin of the *Escherichia coli* chromosome. Cell **45:**53–64.

24. **Bardwell, J. C. A., and E. A. Craig.** 1984. Major heat shock genes of *Drosophila* and the *Escherichia coli* heat inducible *dnaK* gene are homologous. Proc. Natl. Acad. Sci. USA **81:**848–852.

25. **Bardwell, J. C. A., K. Tilly, E. Craig, J. King, M. Zylicz, and C. Georgopoulos.** 1986. The nucleotide sequence of the *Escherichia coli* K12 *dnaJ*⁺ gene: a gene that encodes a heat shock protein. J. Biol. Chem. **261:**1782–1785.

26. **Barry, J., and B. M. Alberts.** 1972. In vitro complementation as an assay for new proteins required for bacteriophage T4 DNA replication: purification of the complex specified by T4 genes *44* and *62*. Proc. Natl. Acad. Sci. USA **69:**2717–2721.

27. **Bedinger, P., M. Hochstrasser, C. V. Jongeneel, and B. M. Alberts.** 1983. Properties of the T4 bacteriophage DNA replication apparatus: the T4 dda DNA helicase is required to pass a bound RNA polymerase molecule. Cell **34:**115–123.

28. **Béjar, S., and J.-P. Bouché.** 1985. A new dispensable genetic locus of the terminus region involved in control of cell division in *Escherichia coli*. Mol. Gen. Genet. **201:**146–150.

29. **Berkower, I., J. Leis, and J. Hurwitz.** 1973. Isolation and characterization of an endonuclease from *Escherichia coli* specific for ribonucleic acid in ribonucleic acid-deoxyribonucleic acid hybrid structures. J. Biol. Chem. **248:**5914–5921.

30. **Bialy, H., and T. Kogoma.** 1986. RNase H is not involved in the induction of stable DNA replication in *Escherichia coli*. J. Bacteriol. **165:**321–323.

31. **Bird, R. E., M. Chandler, and L. Caro.** 1976. Suppression of an *Escherichia coli dnaA* mutation by the integrated R factor R100.1: change of chromosome replication origin in synchronized cultures. J. Bacteriol. **126:**1215–1223.

32. **Bird, R. E., J. M. Louarn, J. Martuscelli, and L. Caro.** 1972. Origin and sequence of chromosome replication in *Escherichia coli*. J. Mol. Biol. **70:**549–566.

33. **Blinkowa, A., W. G. Haldenwang, J. A. Ramsey, J. M. Henson, D. A. Mullen, and J. R. Walker.** 1983. Physiological properties of cold-sensitive suppressor mutations of a temperature-sensitive *dnaZ* mutant of *Escherichia coli*. J. Bacteriol. **153:**66–75.

34. **Bochner, B. R., M. Zylicz, and C. Georgopoulos.** 1986. *Escherichia coli* DnaK protein possesses a 5′-nucleotidase activity that is inhibited by ApppppA. J. Bacteriol. **168:**931–935.

35. **Bonhoeffer, F., and H. Schaller.** 1965. A method for selective enrichment of mutants based on the high UV sensitivity of DNA containing 5-bromouracil. Biochem. Biophys. Res. Commun. **20:** 93–97.

36. **Botstein, D., and R. Maurer.** 1982. Genetic approaches to the analysis of microbial development. Annu. Rev. Genet. **16:**61–83.

37. **Bouché, J.-P.** 1982. Physical map of a 470×10³ base-pair region flanking the terminus of DNA replication in the *Escherichia coli* K12 genome. J. Mol. Biol. **154:**1–20.

38. **Bouché, J.-P., J. P. Gélugne, J. Louarn, J.-M. Louarn, and K. Kaiser.** 1982. Relationship between the physical and genetic maps of a 470×10³ base-pair region around the terminus of *Escherichia coli* K12 DNA replication. J. Mol. Biol. **154:**21–32.

39. **Bouché, J.-P., L. Rowen, and A. Kornberg.** 1978. The RNA primer synthesized by primase initiates phage G4 DNA replication. J. Biol. Chem. **253:**765–769.

40. **Bouché, J.-P., K. Zechel, and A. Kornberg.** 1975. *dnaG* gene product, a rifampicin-resistant RNA polymerase, initiates the conversion of a single stranded coliphage DNA to its duplex replicative form. J. Biol. Chem. **250:**5995–6001.

41. **Braun, R. E., K. O'Day, and A. Wright.** 1985. Autoregulation of the DNA replication gene *dnaA* in *E. coli* K12. Cell **40:**159–169.

42. **Braun, R. E., K. O'Day, and A. Wright.** 1987. A genetic analysis of the structure and function of the DNA replication gene *dnaA* in *E. coli* K-12, p. 511–520. *In* T. J. Kelly and R. McMacken (ed.), DNA replication and recombination. Alan R. Liss, Inc., New York.

43. **Braun, R. E., and A. Wright.** 1986. DNA methylation differentially enhances the expression of one of the two *E. coli dnaA* promoters *in vivo* and *in vitro*. Mol. Gen. Genet. **202:**246–250.

44. **Brown, P. O., and N. R. Cozzarelli.** 1979. A sign inversion mechanism for enzymatic supercoiling of DNA. Science **206:** 1081–1083.

45. **Broyles, S. S., and D. E. Pettijohn.** 1986. Interaction of the *Escherichia coli* HU protein with DNA. Evidence for formation of nucleosome-like structures with altered DNA helical pitch. J. Mol. Biol. **187:**47–60.

46. **Brutlag, D., M. R. Atkinson, P. Setlow, and A. Kornberg.** 1969. An active fragment of DNA polymerase produced by proteolytic cleavage. Biochem. Biophys. Res. Commun. **37:**982–989.

47. **Brutlag, D., and A. Kornberg.** 1972. Enzymatic synthesis of deoxyribonucleic acid. XXXVI. A proofreading function for the 3′ → 5′ exonuclease activity in deoxyribonucleic acid polymerases. J. Biol. Chem. **247:**241–248.

48. **Brutlag, D., R. Schekman, and A. Kornberg.** 1971. A possible role for RNA polymerase in the initiation of M13 DNA synthesis. Proc. Natl. Acad. Sci. USA **68:**2826–2829.

49. **Bryan, S. K., and R. E. Moses.** 1984. Map location of the *pcbA* mutation and physiology of the mutant. J. Bacteriol. **158:**216–221.

50. **Buhk, H.-J., and W. Messer.** 1983. The replication origin of *E. coli*: nucleotide sequence and functional units. Gene **24:**265–279.

51. **Burgers, P. M. J., and A. Kornberg.** 1982. ATP activation of DNA polymerase III holoenzyme of *Escherichia coli*. I. ATP-dependent formation of an initiation complex with primed template. J. Biol. Chem. **257:**11468–11473.

52. **Burgers, P. M. J., and A. Kornberg.** 1983. The cycling of *Escherichia coli* DNA polymerase III holoenzyme in replication. J. Biol. Chem. **258:**7669–7675.

53. **Cairns, J.** 1963. The bacterial chromosome and its manner of replication as seen by autoradiography. J. Mol. Biol. **6:**208–213.

54. **Campbell, J. L., L. Soll, and C. C. Richardson.** 1972. Isolation and partial characterization of a mutant of *Escherichia coli* deficient in DNA polymerase II. Proc. Natl. Acad. Sci. USA **69:** 2090–2094.

55. **Carl, P. L.** 1970. *Escherichia coli* mutants with temperature sensitive synthesis of DNA. Mol. Gen. Genet. **109:**107–122.

56. **Carl, P. L., L. Bloom, and R. J. Crouch.** 1980. Isolation and mapping of a mutation in *Escherichia coli* with altered levels of ribonuclease H. J. Bacteriol. **144:**28–35.

57. **Caron, P. R., S. R. Kushner, and L. Grossman.** 1985. Involvement of helicase II (*uvrD* gene product) and DNA polymerase I in excision mediated by the uvrABC protein complex. Proc. Natl. Acad. Sci. USA **82:**4925–4929.

58. **Chakraborty, T., K. Yoshinaga, H. Lother, and W. Messer.** 1982. Purification of the *E. coli dnaA* gene product. EMBO J. **1:**1545–1549.

59. **Chandler, M., L. Silver, and L. Caro.** 1977. Suppression of an *Escherichia coli dnaA* mutation by the integrated R factor R100.1: origin of chromosome replication during exponential growth. J. Bacteriol. **131:**421–430.

59a. **Chandrasekhar, G. N., K. Tilly, C. Woolford, R. Hendrix, and C. Georgopoulos.** 1986. Purification and properties of the GroES morphogenetic protein of *Escherichia coli*. J. Biol. Chem. **261:**12414–12419.

60. **Chase, J. W., and K. R. Williams.** 1986. Single-stranded DNA binding proteins required for DNA replication. Annu. Rev. Biochem. **55:**103–136.

61. **Chattoraj, D. K., K. Cordes, and A. Abeles.** 1984. Plasmid P1 replication: negative control by repeated DNA sequences. Proc. Natl. Acad. Sci. USA **81:**6456–6460.

62. **Chesney, R. H., J. R. Scott, and D. Vapnek.** 1979. Integration of the plasmid prophages P1 and P7 into the chromosome of *Escherichia coli*. J. Mol. Biol. **130:**161–173.

63. **Chrysogelos, S., and J. Griffith.** 1982. *Escherichia coli* single-strand binding protein organizes single-stranded DNA in nucleosome-like units. Proc. Natl. Acad. Sci. USA **79:**5803–5807.

64. **Churchward, G., P. Holmans, and H. Bremer.** 1983. Increased expression of the *dnaA* gene has no effect on DNA replication in a *dnaA*⁺ strain of *Escherichia coli*. Mol. Gen. Genet. **192:**506–508.

65. **Cohen, S. S., and H. D. Barner.** 1954. Studies on unbalanced growth in *E. coli*. Proc. Natl. Acad. Sci. USA **40:**885–892.

66. **Cooper, S., and C. E. Helmstetter.** 1968. Chromosome replication and the division cycle of *Escherichia coli*. J. Mol. Biol. **31:** 519–540.

67. **Cromie, K. D., and R. S. Hayward.** 1984. Evidence for rifampicin-promoted readthrough of a fully rho-dependent transcriptional terminator. Mol. Gen. Genet. **193:**532–534.

68. **D'Ari, R. A., A. Jaffé-Brochet, D. Touati-Schwartz, and M. Yarmolinsky.** 1975. A *dnaB* analogue specified by bacteriophage P1. J. Mol. Biol. **94:**341–366.

69. **Degnen, G. E., and E. C. Cox.** 1974. Conditional mutator gene in *Escherichia coli*: isolation, mapping, and effector studies. J. Bacteriol. **117:**477–487.

70. **DeLucia, P., and J. Cairns.** 1969. Isolation of an *E. coli* strain with a mutation affecting DNA polymerase. Nature (London) **224:**1164–1166.

71. **de Massy B., O. Fayet, and T. Kogoma.** 1984. Multiple origins usage for DNA replication in *sdrA*(*rnh*) mutants of *Escherichia coli* K-12. Initiation in the absence of *oriC*. J. Mol. Biol. **178:**227–236.

72. **de Massy, B., J. Patte, J. M. Louarn, and J.-P. Bouché.** 1984. *oriX*: a new replication origin in *E. coli*. Cell **36:**221–227.

73. **DiFrancesco, R., S. K. Bhatnagar, A. Brown, and M. J. Bessman.** 1984. The interaction of DNA polymerase III and the product of the *Escherichia coli* mutator gene, *mutD*. J. Biol. Chem. **259:**5567–5573.

74. **DiNardo, S., K. A. Voelkel, R. Sternglanz, A. E. Reynolds, and A. Wright.** 1982. *Escherichia coli* DNA topoisomerase I mutants have compensatory mutations in DNA gyrase genes. Cell **31:**43–51.

75. **Dixon, N. E., and A. Kornberg.** 1984. Protein HU in the enzymatic replication of the chromosomal origin of *Escherichia coli*. Proc. Natl. Acad. Sci. USA **81:**424–428.

76. **Dodson, M., J. Roberts, R. McMacken, and H. Echols.** 1985. Specialized nucleoprotein structures at the origin of replication of bacteriophage λ: complexes with λ O, λ P and *Escherichia coli* DnaB proteins. Proc. Natl. Acad. Sci. USA **82:**4678–4682.

77. **Donachie, W. D.** 1968. Relationship between cell size and time of initiation of DNA replication. Nature (London) **219:**1077–1079.

78. **Donachie, W. E., and M. Masters.** 1969. Temporal control of gene expression in bacteria, p. 37–76. *In* G. M. Padilla, G. L. Whitson, and I. L. Cameron (ed), The cell cycle: gene-enzyme interactions. Academic Press, Inc., New York.

79. **Dove, W. F., H. Inokuchi, and W. Stevens.** 1971. Replication control in phage lambda, p. 747–771. *In* A. D. Hershey (ed.), The bacteriophage lambda. Cold Spring Harbor Laboratory, Cold Spring Harbor, N.Y.

80. **Drlica, K.** 1984. Biology of bacterial deoxyribonucleic acid topoisomerases. Microbiol. Rev. **48:**273–289.

81. **Dwek, R., S. Or-Gad, S. Rosenhak, and E. Z. Ron.** 1984. Two new cell division mutants in *Escherichia coli* map near the terminus of chromosome replication. Mol. Gen. Genet. **193:**379–381.

82. **Eberle, H., and N. Forrest.** 1982. Regulation of DNA synthesis and capacity for initiation in DNA temperature sensitive mutants of *Escherichia coli*. II. Requirements for acquisition and expression of initiation capacity. Mol. Gen. Genet. **186:**66–70.

83. **Echols, H., C. Lu, and P. M. J. Burgers.** 1983. Mutator strains of *Escherichia coli*, *mutD* and *dnaQ*, with defective exonucleolytic editing by DNA polymerase III holoenzyme. Proc. Natl. Acad. Sci. USA **80:**2189–2192.

84. **Eisenberg, S., J. Griffith, and A. Kornberg.** 1977. φX174 *cistron A* protein is a multifunctional enzyme in DNA replication. Proc. Natl. Acad. Sci. USA **74:**3198–3202.

84a. **Engstrom, J., A. Wong, and R. Maurer.** 1986. Interaction of DNA polymerase III α and β subunits *in vivo* in *Salmonella typhimurium*. Genetics **113:**499–515.

85. **Fay, P. J., K. O. Johanson, C. S. McHenry, and R. Bambara.** 1981. Size classes of products synthesized processively by DNA polymerase III and DNA polymerase III holoenzyme of *Escherichia coli*. J. Biol. Chem. **256:**976–983.

86. **Fay, P. J., K. O. Johanson, C. S. McHenry, and R. A. Bambara.** 1982. Size classes of products synthesized processively by two subassemblies of *Escherichia coli* DNA polymerase II holoenzyme. J. Biol. Chem. **257:**5692–5699.

87. **Fayet, O., J.-M. Louarn, and C. Georgopoulos.** 1986. Suppression of the *Escherichia coli dnaA46* mutation by amplification of the *groES* and *groEL* genes. Mol. Gen. Genet. **202:**435–445.

88. **Filip, C. C., J. S. Allen, R. A. Gustafson, R. G. Allen, and J. R. Walker.** 1974. Bacterial cell division regulation: characterization of the *dnaH* locus of *Escherichia coli*. J. Bacteriol. **119:**443–449.

89. **Filutowicz, M.** 1980. Requirement for DNA gyrase for the initiation of chromosome replication in *Escherichia coli* K12. Mol. Gen. Genet. **177:**301–309.

90. **Filutowicz, M., and P. Jonczyk.** 1981. Essential role of the *gyrB* gene product in the transcriptional event coupled to the *dnaA*-dependent initiation of *Escherichia coli* chromosome replication. Mol. Gen. Genet. **183:**134–138.

91. **Filutowicz, M., and P. Jonczyk.** 1983. The *gyrB* gene product functions in both initiation and chain polymerization of *Escherichia coli* chromosome replication: suppression of the initiation deficiency in *gyrB*-ts mutants by a class of *rpoB* mutations. Mol. Gen. Genet. **191:**282–287.

92. **Fisher, L. M., H. A. Barot, and M. E. Cullen.** 1986. DNA gyrase complex with DNA: determinants for site-specific DNA breakage. EMBO J. **5:**1411–1418.

93. **Floor, E.** 1970. Interaction of morphogenetic genes of bacteriophage T4. J. Mol. Biol. **47:**293–306.

93a. **Flower, A. M., and C. S. McHenry.** 1986. The adjacent *dnaZ* and *dnaX* genes of *Escherichia coli* are contained within one continuous open reading frame. Nucleic Acids Res. **14:**8091–8101.

94. **Formosa, T., R. L. Burke, and B. M. Alberts.** 1983. Affinity purification of bacteriophage T4 proteins essential for DNA replication and genetic recombination. Proc. Natl. Acad. Sci. USA **80:**2442–2446.

95. **Fralick, J. A.** 1977. Characterization of chromosomal replication in a *dnaA* mutant, p. 71–83. *In* I. Molineux and M. Kohiyama (ed.), DNA synthesis present and future. Plenum Publishing Corp., New York.

96. **Fralick, J. A.** 1978. Studies on the regulation of initiation of chromosome replication in *Escherichia coli*. J. Mol. Biol. **122:**271–286.

97. **Frey, J., M. Chandler, and L. Caro.** 1979. The effect of an *Escherichia coli dnaA*ts mutation on the replication of the plasmids colE1, pSC101, R100.1 and RTF-TC. Mol. Gen. Genet. **174:**1177–1126.

98. **Frey, J., M. Chandler, and L. Caro.** 1984. Overinitiation in a *dnaAcos* mutant of *Escherichia coli* K12. Evidence for dnaA-dnaB interactions. J. Mol. Biol. **179:**271–286.

99. **Friedman, D. I., E. R. Olson, K. Tilly, C. Georgopoulos, I. Herskowitz, and F. Banuett.** 1984. Interactions of bacteriophage and host macromolecules in the growth of bacteriophage λ. Microbiol. Rev. **48:**299–325.

100. **Fuchs, J. A., H. O. Karlstrom, H. R. Warner, and P. Reichard.** 1972. Defective gene product in *dnaF* mutant of *Escherichia coli*. Nature (London) New Biol. **238:**69–71.

101. **Fuller, R. S., B. E. Funnell, and A. Kornberg.** 1984. The DnaA protein complex with the *E. coli* chromosomal replication origin (*oriC*) and other DNA sites. Cell **38:**889–900.

102. **Fuller, R. S., J. M. Kaguni, and A. Kornberg.** 1981. Enzymatic replication of the origin of the *Escherichia coli* chromosome. Proc. Natl. Acad. Sci. USA **78:**7370–7374.

103. **Fuller, R. S., and A. Kornberg.** 1983. Purified DnaA protein in initiation of replication at the *Escherichia coli* chromosomal origin of replication. Proc. Natl. Acad. Sci. USA **80:**5817–5821.

104. **Funnell, B. E., T. A. Baker, and A. Kornberg.** 1986. Complete enzymatic replication of plasmids containing the origin of the *Escherichia coli* chromosome. J. Biol. Chem. **261:**5616–5624.

105. **Garcia, G. M., P. K. Mar, D. A. Mullin, J. R. Walker, and N. E. Prather.** 1986. The *E. coli dnaY* gene encodes an arginine transfer RNA. Cell **45:**453–459.

106. **Gefter, M.** 1975. DNA replication. Annu. Rev. Biochem. **44:**45–78.

107. **Gefter, M., Y. Hirota, T. Kornberg, J. A. Wechsler, and C. Barnaux.** 1971. Analysis of DNA polymerase II and III in mutants of *Escherichia coli* thermosensitive for DNA synthesis. Proc. Natl. Acad. Sci. USA **68:**3150–3153.

108. **Gefter, M. L., A. Becker, and J. Hurwitz.** 1967. The enzymatic repair of DNA. I. Formation of circular λ DNA. Proc. Natl. Acad. Sci. USA **58:**240–247.

109. **Geider, K., I. Bäumel, and T. F. Meyer.** 1982. Intermediate stages in enzymatic replication of bacteriophage fd duplex DNA. J. Biol. Chem. **257:**6488–6493.

110. **Geider, K., E. Beck, and H. Schaller.** 1978. An RNA transcribed from DNA at the origin of phage fd single strand to replicative form conversion. Proc. Natl. Acad. Sci. USA **75:**645–649.

111. **Geider, K., and H. Hoffmann-Berling.** 1981. Proteins controlling the helical structure of DNA. Annu. Rev. Biochem. **50:**233–260.

112. **Geider, K., and A. Kornberg.** 1974. Conversion of the M13 viral single strand to the double-stranded replicative forms by purified proteins. J. Biol. Chem. **249:**3999–4005.

113. **Gellert, M.** 1967. Formation of covalent circles of lambda DNA by *E. coli* extracts. Proc. Natl. Acad. Sci. USA **57:**148–155.

114. **Gellert, M.** 1981. DNA topoisomerases. Annu. Rev. Biochem. **50:**879–910.

115. **Gellert, M., K. Mizuuchi, M. H. O'Dea, T. Itoh, and J.-I. Tomizawa.** 1977. Nalidixic acid resistance: a second genetic character involved in DNA gyrase activity. Proc. Natl. Acad. Sci. USA **74:**4772–4776.

116. **Gellert, M., K. Mizuuchi, M. H. O'Dea, and H. Nash.** 1976. DNA gyrase: an enzyme that introduces superhelical turns into DNA. Proc. Natl. Acad. Sci. USA **73:**3872–3876.

117. **Gellert, M., M. H. O'Dea, T. Itoh, and J.-I. Tomizawa.** 1976b. Novobiocin and coumermycin inhibit DNA supercoiling catalyzed by DNA gyrase. Proc. Natl. Acad. Sci. USA **73:**4474–4478.

118. **Georgopoulos, C. P.** 1977. A new bacterial gene (*groPC*) which affects λ DNA replication. Mol. Gen. Genet. **151:**35–39.

119. **Georgopoulos, C. P., and H. Eisen.** 1974. Bacterial mutants which block phage assembly. J. Supramol. Struct. **2:**349–359.

120. **Georgopoulos, C. P., R. W. Hendrix, S. Casjens, and A. D. Kaiser.** 1973. Host participation in bacteriophage lambda head assembly. J. Mol. Biol. **76:**45–60.

121. **Georgopoulos, C. P., and I. Herskowitz.** 1971. *Escherichia coli* mutants in lambda DNA synthesis, p. 553–564. *In* A. D. Hershey (ed.), The bacteriophage lambda. Cold Spring Harbor Laboratory, Cold Spring Harbor, N.Y.

122. **Georgopoulos, C. P., K. Tilly, D. Drahos, and R. Hendrix.** 1982. The B66.0 protein of *Escherichia coli* is the product of the *dnaK* gene. J. Bacteriol. **149:**1175–1177.

123. **Griffith, J. D.** 1976. Visualization of prokaryotic DNA in a regularly condensed chromatin-like fiber. Proc. Natl. Acad. Sci. USA **73:**563–567.

124. **Griffith, J. D., L. D. Harris, and J. Register III.** 1985. Visualization of SSB-ssDNA complexes active in the assembly of stable recA-DNA filaments. Cold Spring Harbor Symp. Quant. Biol. **49:**553–558.

125. **Hanawalt, P. C., O. Maaloe, D. J. Cummings, and M. Schaechter.** 1961. The normal DNA replication cycle. J. Mol. Biol. **3:**156–165.

126. **Hanna, M. H., and P. L. Carl.** 1975. Reinitiation of deoxyribonucleic acid synthesis by deoxyribonucleic acid initiation mutants of *Escherichia coli*: role of ribonucleic acid synthesis, protein synthesis, and cell division. J. Bacteriol. **121:**219–226.

127. **Hansen, E. B., T. Atlung, F. G. Hansen, O. Skovgaard, and K. von Meyenburg.** 1984. Fine structure genetic map and complementation analysis of mutations in the *dnaA* gene of *Escherichia coli*. Mol. Gen. Genet. **196:**387–396.

128. **Hansen, E. B., F. G. Hansen, and K. von Meyenburg.** 1982. The nucleotide sequence of the *dnaA* gene and the first part of the *dnaN* gene of *Escherichia coli* K-12. Nucleic Acids Res. **10:**7373–7375.

129. **Hansen, E. B., S. Koefoed, K. von Meyenburg, and T. Atlung.** 1981. Transcription and translation events in the *oriC* region of the *E. coli* chromosome. ICN-UCLA Symp. Mol. Cell. Biol. **22:**37–55.

130. **Hansen, E. B., and M. Yarmolinsky.** 1986. Host participation in plasmid maintenance: dependence upon *dnaA* of replicons derived from P1 and F. Proc. Natl. Acad. Sci. USA **83:**4423–4427.

131. **Hansen, F. G., and K. V. Rasmussen.** 1977. Regulation of the *dnaA* product in *Escherichia coli*. Mol. Gen. Genet. **155:**219–225.

132. **Hartman, P. E., and J. R. Roth.** 1973. Mechanisms of suppression. Adv. Genet. **17:**1–105.

133. **Helmstetter, C. E.** 1974. Initiation of chromosome replication in *Escherichia coli*. II. Analysis of the control mechanism. J. Mol. Biol. **84:**21–36.

134. **Helmstetter, C. E., S. Cooper, O. Pierucci, and E. Revelas.** 1969. On the bacterial life sequence. Cold Spring Harbor Symp. Quant. Biol. **33:**809–822.

135. **Helmstetter, C. E., C. A. Krajewski, A. C. Leonard, and M. Weinberger.** 1986. Discontinuity in DNA replication during expression of accumulated initiation potential in *dnaA* mutants of *Escherichia coli*. J. Bacteriol. **165:**631–637.

136. **Helmstetter, C. E., and O. Pierucci.** 1968. Cell division during inhibition of deoxyribonucleic acid synthesis. J. Bacteriol. **95:**1627–1633.

137. **Hendrickson, W. G., T. Kusano, H. Yamaki, R. Balakrishnan, M. King, J. Murchie, and M. Schaechter.** 1982. Binding of the origin of replication of *Escherichia coli* to the outer membrane. Cell **30:**915–923.

138. **Hendrix, R. W.** 1979. Purification and properties of GroE, a host protein involved in bacteriophage assembly. J. Mol. Biol. **129:**375–392.

139. **Henson, J. M., H. Chu, C. A. Irwin, and J. R. Walker.** 1979. Isolation and characterization of *dnaX* and *dnaY* temperature-sensitive mutants of *Escherichia coli*. Genetics **92:**1041–1059.

140. **Henson, J. M., B. Kopp, and P. L. Kuempel.** 1984. Deletion of 60 kilobase pairs of DNA from the *terC* region of the chromosome of *Escherichia coli*. Mol. Gen. Genet. **192:**263–268.

141. **Henson, J. M., and P. L. Kuempel.** 1985. Deletion of the terminus region (340 kilobase pairs of DNA) from the chromosome of *Escherichia coli*. Proc. Natl. Acad. Sci. USA **82:**3766–3770.

142. **Hiraga, S.** 1976. Novel F prime factors able to replicate in *Escherichia coli* Hfr strains. Proc. Natl. Acad. Sci. USA **73:**198–202.

143. **Hirose, S., S. Hiraga, and T. Okazaki.** 1983. Initiation site of deoxyribonucleotide polymerization at the replication origin of the *Escherichia coli* chromosome. Mol. Gen. Genet. **189:**422–431.

144. **Hirota, Y., M. Gefter, and L. Mindich.** 1972. A mutant of *Escherichia coli* defective in DNA polymerase III activity. Proc. Natl. Acad. Sci. USA **69:**3238–3242.

145. **Hirota, Y., J. Mordoh, and F. Jacob.** 1970. On the process of cell division in *Escherichia coli*. III. Thermosensitive mutants of *Escherichia coli* altered in the process of DNA initiation. J. Mol. Biol. **53:**422–431.

146. **Hirota, Y., A. Oka, K. Sugimoto, K. Asada, H. Sasaki, and M. Takanami.** 1981. *Escherichia coli* origin of replication: structural organization of the region essential for autonomous replication and the recognition frame model. ICN-UCLA Symp. Mol. Cell. Biol. **22:**1–12.

147. **Hirota, Y., A. Ryter, and F. Jacob.** 1968. Thermosensitive mutants of *E. coli* affected in the process of DNA synthesis and cellular division. Cold Spring Harbor Symp. Quant. Cell Biol. **33:**677–693.

148. **Hochschild, A., and M. Ptashne.** 1986. Cooperative binding of λ repressors to sites separated by integral turns of the DNA helix. Cell **44:**681–687.

149. **Hohn, T., B. Hohn, A. Engel, M. Wurtz, and P. R. Smith.** 1979.

Isolation and characterization of the host protein GroE involved in bacteriophage lambda assembly. J. Mol. Biol. **129**:359–373.

150. **Horiuchi, T., H. Maki, M. Maruyama, and M. Sekiguchi.** 1981. Identification of the *dnaQ* gene product and localization of the structural gene for RNase H of *Escherichia coli* by cloning of the genes. Proc. Natl. Acad. Sci. USA **78**:3770–3774.

151. **Horiuchi, T., H. Maki, and M. Sekiguchi.** 1984. RNase H-defective mutants of *Escherichia coli*: a possible discriminatory role of RNase H in initiation of DNA replication. Mol. Gen. Genet. **195**:17–22.

152. **Hübscher, U., and A. Kornberg.** 1979. The δ subunit of *Escherichia coli* DNA polymerase III holoenzyme is the *dnaX* gene product. Proc. Natl. Acad. Sci. USA **76**:6284–6288.

153. **Hübscher, U., and A. Kornberg.** 1980. The DnaZ protein, the gamma subunit of DNA polymerase III holoenzyme of *Escherichia coli*. J. Biol. Chem. **255**:11698–11703.

154. **Hughes, P., F.-Z. Squali-Houssaini, P. Forterre, and M. Kohiyama.** 1984. *In vitro* replication of a *dam* methylated and non-methylated *oriC* plasmid. J. Mol. Biol. **176**:155–159.

155. **Imber, R., R. L. Low, and D. S. Ray.** 1983. Identification of a primosome assembly site in the region of the *ori2* replication origin of the *Escherichia coli* mini-F plasmid. Proc. Natl. Acad. Sci. USA **80**:7132–7136.

156. **Itikawa, H., and J. Ryu.** 1979. Isolation and characterization of a temperature-sensitive *dnaK* mutant of *Escherichia coli* B. J. Bacteriol. **138**:339–344.

157. **Itoh, T., and J.-I. Tomizawa.** 1980. Formation of an RNA primer for initiation of replication of ColE1 DNA by ribonuclease H. Proc. Natl. Acad. Sci. USA **77**:2450–2454.

158. **Itoh, T., and J.-I. Tomizawa.** 1982. Purification of ribonuclease H as a factor for initiation of *in vitro* ColE1 DNA replication. Nucleic Acids Res. **10**:5949–5965.

159. **Jacob, F., S. Brenner, and F. Cuzin.** 1963. On the regulation of DNA replication in bacteria. Cold Spring Harbor Symp. Quant. Biol. **28**:329–348.

160. **Jacq, A., M. Kohiyama, H. Lother, and W. Messer.** 1983. Recognition sites for a membrane derived DNA binding protein preparation in the *E. coli* replication origin. Mol. Gen. Genet. **191**:460–465.

161. **Jacq, A., H. Lother, W. Messer, and M. Kohiyama.** 1980. Isolation of a membrane protein having an affinity to the replication origin of *E. coli*. ICN-UCLA Symp. Mol. Cell. Biol. **19**:189–197.

162. **Jarvik, J., and D. Botstein.** 1975. Conditional lethal mutations that suppress genetic defects in morphogenesis by altering structural proteins. Proc. Natl. Acad. Sci. USA **72**:2738–2742.

163. **Jenkins, A. J., J. B. Marsh, I. R. Oliver, and M. Masters.** 1986. A DNA fragment containing the *groE* genes can suppress mutations in the *Escherichia coli dnaA* gene. Mol. Gen. Genet. **202**:446–454.

164. **Johanson, K. O., and C. S. McHenry.** 1980. Purification and characterization of the β subunit of the DNA polymerase II holoenzyme of *Escherichia coli*. J. Biol. Chem. **255**:10984–10990.

165. **Johanson, K. O., and C. S. McHenry.** 1982. The β subunit of the DNA polymerase III holoenzyme becomes inaccessible to antibody after formation of an initiation complex with primed DNA. J. Biol. Chem. **257**:12310–12315.

166. **Johanson, K. O., and C. S. McHenry.** 1984. Adenosine 5'-O-(3-thiotriphosphate) can support the formation of an initiation complex between the DNA polymerase III holoenzyme and primed DNA. J. Biol. Chem. **259**:4589–4595.

167. **Joyce, C. M., W. S. Kelley, and N. D. F. Grindley.** 1982. Nucleotide sequence of the *Escherichia coli polA* gene and primary structure of DNA polymerase I. J. Biol. Chem. **257**:1958–1964.

168. **Junker, D. E., L. A. Rokeach, D. Ganea, A. Chiaramello, and J. W. Zyskind.** 1986. Transcription termination with the *Escherichia coli* origin of DNA replication, *oriC*. Mol. Gen. Genet. **203**:101–109.

169. **Kaguni, J. M., R. S. Fuller, and A. Kornberg.** 1982. Enzymatic replication of *E. coli* chromosomal origin is bidirectional. Nature (London) **296**:623–627.

170. **Kaguni, J. M., and A. Kornberg.** 1984. Replication initiated at the origin (*oriC*) of the *E. coli* chromosome reconstituted with purified enzymes. Cell **38**:183–190.

171. **Kaguni, J. M., and A. Kornberg.** 1984. Topoisomerase I confers specificity in enzymatic replication of the *Escherichia coli* chromosomal origin. J. Biol. Chem. **259**:8578–8583.

172. **Kanaya, S., and R. J. Crouch.** 1983. DNA sequence of the gene coding for *Escherichia coli* ribonuclease H. J. Biol. Chem. **258**:1276–1281.

173. **Kano, Y., M. Wada, T. Nagase, and F. Imamoto.** 1986. Genetic characterization of the gene *hupB* encoding the HU-1 protein of *Escherichia coli*. Gene **45**:37–44.

174. **Kano, Y., S. Yoshino, M. Wada, K. Yokoyama, M. Nobuhara, and F. Imamoto.** 1985. Molecular cloning and nucleotide sequence of the *HU-1* gene of *Escherichia coli*. Mol. Gen. Genet. **201**:360–362.

175. **Kellenberger-Gujer, G., and A. J. Podhajska.** 1978. Interactions between the plasmid lambda-dv and *Escherichia coli dnaA* mutants. Mol. Gen. Genet. **162**:17–22.

176. **Kellenberger-Gujer, G., A. J. Podhajska, and L. Caro.** 1978. A cold sensitive mutant of *E. coli* which overinitiates chromosome replication at low temperature. Mol. Gen. Genet. **162**:9–16.

177. **Kirchhausen, T., J. C. Wang, and S. C. Harrison.** 1985. DNA gyrase and its complexes with DNA: direct observation by electron microscopy. Cell **41**:933–943.

178. **Kirkegaard, K., and J. C. Wang.** 1985. Bacterial DNA topoisomerase I can relax positively supercoiled DNA containing a single-stranded DNA loop. J. Mol. Biol. **185**:625–637.

179. **Klein, A., E. Lanka, and E. Schuster.** 1980. Isolation of a complex between the *P* protein of phage λ and *dnaB* protein of *Escherichia coli*. Eur. J. Biochem. **105**:1–6.

180. **Klenow, H., and I. Henningsen.** 1970. Selective elimination of the exonuclease activity of the deoxyribonucleic acid polymerase from *Escherichia coli* B by limited proteolysis. Proc. Natl. Acad. Sci. USA **65**:168–175.

181. **Klinkert, M.-Q., A. Klein, and M. Abdel-Monem.** 1980. Studies on the functions of DNA helicase I and DNA helicase II of *Escherichia coli*. J. Biol. Chem. **255**:9746–9752.

182. **Knippers, R.** 1970. DNA polymerase II. Nature (London) **228**:1050–1053.

183. **Kobori, J. A., and A. Kornberg.** 1982. The *Escherichia coli dnaC* gene product. I. Overproduction of the DnaC proteins of *Escherichia coli* and *Salmonella typhimurium* by cloning into a high copy number plasmid. J. Biol. Chem. **257**:13757–13762.

184. **Kobori, J. A., and A. Kornberg.** 1982. The *Escherichia coli dnaC* gene product. II. Purification, physical properties and role in replication. J. Biol. Chem. **257**:13763–13769.

185. **Kobori, J. A., and A. Kornberg.** 1982. The *Escherichia coli dnaC* gene product. III. Properties of the DnaB-DnaC protein complex. J. Biol. Chem. **257**:13770–13775.

186. **Kochan, J., and H. Murialdo.** 1983. Early intermediates in bacteriophage lambda prohead assembly. II. Identification of biologically active intermediates. Virology **131**:100–115.

187. **Kodaira, M., S. B. Biswas, and A. Kornberg.** 1983. The *dnaX* gene encodes the DNA polymerase III holoenzyme tau subunit, precursor of the gamma subunit, the *dnaZ* gene product. Mol. Gen. Genet. **192**:80–86.

188. **Kogoma, T.** 1978. A novel *Escherichia coli* mutant capable of DNA replication in the absence of protein synthesis. J. Mol. Biol. **121**:55–69.

189. **Kogoma, T., and K. G. Lark.** 1975. Characterization of the replication of *Escherichia coli* DNA in the absence of protein synthesis: stable DNA replication. J. Mol. Biol. **94**:243–256.

190. **Kogoma, T., K. Skarsted, E. Boye, K. von Meyenburg, and H. B. Steen.** 1985. RecA protein acts at the initiation of stable DNA replication in *rnh* mutants of *Escherichia coli* K-12. J. Bacteriol. **163**:439–444.

191. **Kogoma, T., N. L. Subia, and K. von Meyenburg.** 1985. Function of ribonuclease H in initiation of DNA replication in *Escherichia coli* K-12. Mol. Gen. Genet. **200**:103–109.

192. **Kogoma, T., and K. von Meyenburg.** 1983. The origin of replication, *oriC*, and the DnaA protein are dispensable in stable DNA replication (SDR) mutants of *Escherichia coli* K-12. EMBO J. **2**:463–468.

193. **Kohara, Y., N. Todoh, X. Jiang, and T. Okazaki.** 1985. The distribution and properties of RNA primed initiation sites of DNA synthesis at the replication origin of *Escherichia coli* chromosome. Nucleic Acids Res. **13**:6847–6866.

194. **Kohiyama, M., D. Cousin, A. Ryter, and F. Jacob.** 1966. Mutants thermosensibles d'*Escherichia coli* K-12. I. Isolement et caractérisation rapide. Ann. Inst. Pasteur **110**:465–486.

195. **Kornberg, A.** 1969. Active center of DNA polymerase. Science **163**:1410–1418.

196. **Kornberg, A.** 1980. DNA replication. W. H. Freeman and Co., San Francisco.

197. **Kornberg, A.** 1982. 1982 supplement to DNA replication. W. H. Freeman and Co., San Francisco.

198. **Kornberg, A.** 1984. Enzyme studies of replication of the *Escherichia coli* chromosome. Adv. Exp. Med. Biol. **179**:3–16.

199. **Kornberg, A., T. A. Baker, L. L. Bertsch, D. Bramhill, B. E. Funnell, R. S. Lasken, H. Maki, S. Maki, K. Sekinizu, and E. Wahle.** 1987. Enzymatic studies of replication of *oriC* plasmids, p. 137–149. *In* T. J. Kelly and R. McMacken (ed.), DNA replication and recombination. Alan R. Liss, Inc., New York.

200. **Kornberg, A., I. R. Lehman, M. J. Bessman, and E. S. Simms.** 1956. Enzymic synthesis of deoxyribonucleic acid. Biochim. Biophys. Acta **21**:197–198.
201. **Kornberg, T., and M. Gefter.** 1970. DNA synthesis in cell-free extracts of a DNA polymerase-defective mutant. Biochem. Biophys. Res. Commun. **40**:1348–1355.
202. **Kornberg, T., and M. Gefter.** 1971. Purification and DNA synthesis in cell free extracts: properties of DNA polymerase II. Proc. Natl. Acad. Sci. USA **68**:761–764.
203. **Krauss, G., H. Sindermann, U. Schomberg, and G. Maass.** 1981. *Escherichia coli* single-stranded deoxyribonucleic acid binding protein: stability, specificity and kinetics of complexes with oligonucleotides and deoxyribonucleic acid. Biochemistry **20**:5346–5352.
204. **Krell, H., H. Durwald, and H. Hoffmann-Berling.** 1979. A DNA-unwinding enzyme induced in bacteriophage T4-infected cells. Eur. J. Biochem. **93**:387–395.
205. **Kreuzer, K. N., and N. R. Cozzarelli.** 1979. *Escherichia coli* mutants thermosensitive for deoxyribonucleic acid gyrase subunit A: effects on deoxyribonucleic acid replication, transcription, and bacteriophage growth. J. Bacteriol. **140**:424–435.
206. **Kuempel, P., S. Duerr, and N. Seeley.** 1977. Terminus region of the chromosome of *Escherichia coli* inhibits replication forks. Proc. Natl. Acad. Sci. USA **74**:3927–3931.
207. **Kuempel, P. L., and G. E. Veomett.** 1970. A possible function of DNA polymerase in chromosome replication. Biochem. Biophys. Res. Commun. **41**:973–980.
208. **Kuhn, B., and M. Abdel-Monem.** 1982. DNA synthesis at a fork in the presence of DNA helicases. Eur. J. Biochem. **125**:63–68.
209. **Kuhn, B., M. Abdel-Monem, H. Krell, and H. Hoffmann-Berling.** 1979. Evidence for two mechanisms for DNA unwinding catalyzed by DNA helicases. J. Biol. Chem. **254**:11343–11350.
210. **Kung, F.-C., and D. A. Glaser.** 1978. *dnaA* acts before *dnaC* in the initiation of DNA replication. J. Bacteriol. **133**:755–762.
211. **Kunkel, T., R. R. Meyer, and L. A. Loeb.** 1979. Single-strand binding protein enhances fidelity of DNA synthesis *in vitro*. Proc. Natl. Acad. Sci. USA **76**:6331–6335.
212. **Kusano, T., D. Steinmetz, W. G. Hendrickson, J. Murchie, M. King, A. Benson, and M. Schaechter.** 1984. Direct evidence for specific binding of the replication origin of the *Escherichia coli* chromosome to the membrane. J. Bacteriol. **158**:313–316.
213. **Kuwabara, N., and H. Uchida.** 1981. Functional cooperation of the *dnaE* and *dnaN* gene products in *Escherichia coli*. Proc. Natl. Acad. Sci. USA **78**:5764–5767.
214. **LaDuca, R. J., J. J. Crute, C. S. McHenry, and R. A. Bambara.** 1986. The β subunit of the *Escherichia coli* DNA polymerase III holoenzyme interacts functionally with the catalytic core in the absence of other subunits. J. Biol. Chem. **261**:7550–7557.
215. **LaDuca, R. J., P. J. Fay, C. Chuang, C. S. McHenry, and R. A. Bambara.** 1983. Site specific pausing of deoxyribonucleic acid synthesis catalyzed by four forms of *Escherichia coli* DNA polymerase III. Biochemistry **22**:5177–5188.
216. **LaDuca, R. J., and C. E. Helmstetter.** 1983. Expression of accumulated capacity for initiation of chromosome and minichromosome replication in *dnaA* mutants of *Escherichia coli*. J. Bacteriol. **154**:1371–1380.
217. **Laine, B., D. Belaiche, H. Khanaka, and P. Sautiere.** 1983. Primary structure of the DNA-binding protein HRm from *Rhizobium meliloti*. Eur. J. Biochem. **131**:325–331.
218. **Lammi, M., M. Paci, C. L. Pon, M. A. Losso, A. Miano, R. T. Pawlik, G. L. Gianfranceschi, and C. O. Gualerzi.** 1984. Proteins from the prokaryotic nucleoid: biochemical and ¹H NMR studies on three bacterial histone-like proteins, p. 467–477. *In* U. Hübscher and S. Spadari (ed.), Proteins involved in DNA replication. Plenum Publishing Corp., New York.
219. **Lane, H. E. D., and D. T. Denhardt.** 1975. The *rep* mutation. IV. Slower movement of replication forks in *E. coli* strains. J. Mol. Biol. **97**:99–112.
220. **Lanka, E., and H. Schuster.** 1983. The DnaC protein of *Escherichia coli*. Purification, physical properties and interaction with DnaB protein. Nucleic Acids Res. **11**:987–997.
221. **Lark, C. A., J. Riazzi, and K. G. Lark.** 1978. *dnaT*, a dominant conditional lethal mutation affecting DNA replication in *Escherichia coli*. J. Bacteriol. **136**:1008–1017.
222. **Lark, K. G.** 1972. Evidence for direct involvement of RNA in the initiation of DNA replication in *E. coli* 15T⁻. J. Mol. Biol. **64**:47–60.
223. **Lark, K. G.** 1972. Genetic control over the initiation of the synthesis of the short deoxynucleotide chains in *E. coli*. Nature (London) New Biol. **240**:237–240.
224. **Lark, K. G., and C. A. Lark.** 1979. *recA*⁺-dependent DNA replication in the absence of protein synthesis: characteristics of a dominant lethal replication mutant, *dnaT* and requirement for *recA*⁺ function. Cold Spring Harbor Symp. Quant. Biol. **43**:537–549.
225. **Lark, K. G., C. A. Lark, and E. A. Meenan.** 1981. *recA*-dependent DNA replication in *E. coli*: interaction between *recA*-dependent and normal DNA replication genes. ICN-UCLA Symp. Mol. Cell. Biol. **22**:337–360.
226. **Lark, K. G., and H. Renger.** 1969. Initiation of DNA replication in *E. coli* 15T⁻: chronological dissection of three physiological processes required for initiation. J. Mol. Biol. **42**:221–236.
227. **Lark, K. G., T. Repko, and E. J. Hoffman.** 1963. The effect of amino acid deprivation on subsequent deoxyribonucleic acid replication. Biochim. Biophys. Acta **76**:9–24.
228. **LeBowitz, J. H., and R. McMacken.** 1986. The *Escherichia coli* DnaB replication protein is a DNA helicase. J. Biol. Chem. **261**:4738–4748.
229. **LeBowitz, J. H., M. Zylicz, C. Georgopoulos, and R. McMacken.** 1985. Initiation of DNA replication on single-stranded DNA templates catalyzed by purified replication proteins of bacteriophage λ and *Escherichia coli*. Proc. Natl. Acad. Sci. USA **82**:3988–3992.
230. **Lehman, I. R.** 1976. DNA ligase: structure, mechanism, and function. Science **186**:790–797.
231. **Lehman, I. R., and D. G. Uyemura.** 1976. DNA polymerase I: essential replication enzyme. Science **193**:963–969.
232. **Lindahl, G., Y. Hirota, and F. Jacob.** 1971. On the process of cellular division in *Escherichia coli*: replication of the bacterial chromosome under control of prophage P2. Proc. Natl. Acad. Sci. USA **68**:2407–2411.
233. **Lindahl, G., and T. Lindahl.** 1984. Initiation of DNA replication in *Escherichia coli*: RNase H-deficient mutants do not require the *dnaA* function. Mol. Gen. Genet. **196**:283–289.
234. **Liu, L. F.** 1983. DNA topoisomerases—enzymes that catalyze the breaking and rejoining of DNA. Crit. Rev. Biochem. **15**:1–24.
235. **Liu, L. F., and J. C. Wang.** 1978. DNA-DNA gyrase complex: the wrapping of the DNA duplex outside the enzyme. Cell **15**:979–984.
236. **Liu, L. F., and J. C. Wang.** 1978. *Micrococcus luteus* DNA gyrase: active components and a model for its supercoiling of DNA. Proc. Natl. Acad. Sci. USA **75**:2098–2102.
237. **Livingston, D. M., D. C. Hinkle, and C. C. Richardson.** 1975. Deoxyribonucleic acid polymerase III of *Escherichia coli*. Purification and properties. J. Biol. Chem. **250**:461–469.
238. **Livingston, D. M., and C. C. Richardson.** 1975. DNA polymerase III of *E. coli*: characterization of associated exonuclease activities. J. Biol. Chem. **250**:470–478.
239. **Lockshon, D., and D. R. Morris.** 1985. Sites of reaction of *Escherichia coli* DNA gyrase on pBR322 *in vivo* as revealed by oxolinic acid-induced plasmid linearization. J. Mol. Biol. **181**:63–74.
240. **Lohman, T. M., and T. B. Overman.** 1985. Two binding modes in *Escherichia coli* single strand binding protein-single stranded DNA complexes. Modulation by NaCl concentration. J. Biol. Chem. **260**:3594–3603.
241. **Losso, M. A., R. T. Pawlik, M. A. Canonaco, and C. O. Gualerzi.** 1986. Proteins from the prokaryotic nucleoid. A protein-protein cross-linking study on the quaternary structure of *Escherichia coli* DNA-binding protein NS (HU). Eur. J. Biochem. **155**:27–32.
242. **Lother, H., H.-J. Buhk, G. Morelli, B. Heimann, T. Chakraborty, and W. Messer.** 1981. Genes, transcriptional units and functional sites in and around the *E. coli* replication origin. ICN-UCLA Symp. Mol. Cell. Biol. **22**:57–77.
243. **Lother, H., R. Kolling, C. Kücherer, and M. Schauzu.** 1985. DnaA protein regulated transcription: effects on the *in vitro* replication of *Escherichia coli* minichromosomes. EMBO J. **4**:555–560.
244. **Lother, H., R. Lurz, and E. Orr.** 1984. DNA binding and antigenic specifications of DNA gyrase. Nucleic Acids Res. **12**:901–914.
245. **Louarn, J., J.-P. Bouché, J. Patte, and J. M. Louarn.** 1984. Genetic inactivation of topoisomerase I suppresses a defect in initiation of chromosome replication in *Escherichia coli*. Mol. Gen. Genet. **195**:170–174.
246. **Louarn, J., J. Patte, and J. M. Louarn.** 1977. Evidence for a fixed termination site of chromosome replication in *Escherichia coli* K12. J. Mol. Biol. **115**:295–314.
247. **Low, R. L., K. Arai, and A. Kornberg.** 1981. Conservation of the primosome in successive stages of φX174 DNA replication. Proc. Natl. Acad. Sci. USA **78**:1436–1440.
248. **Lycett, G. W., E. Orr, and R. H. Pritchard.** 1980. Chloramphenicol releases a block in initiation of chromosome replication in a

dnaA strain of *Escherichia coli* K12. Mol. Gen. Genet. **178:**329–336.

249. **Maaløe, O., and P. Hanawalt.** 1961. Thymine deficiency and the normal DNA replication cycle. J. Mol. Biol. **2:**144–155.

250. **Maaløe, O., and N. O. Kjeldgaard.** 1966. Control of macromolecular synthesis. W. A. Benjamin, New York.

251. **Maki, H., T. Horiuchi, and A. Kornberg.** 1985. The polymerase subunit of DNA polymerase III of *Escherichia coli*. I. Amplification of the *dnaE* gene product and polymerase activity of the α subunit. J. Biol. Chem. **260:**12982–12986.

252. **Maki, H., and A. Kornberg.** 1985. The polymerase subunit of DNA polymerase III of *Escherichia coli*. II. Purification of the α subunit, devoid of nuclease activities. J. Biol. Chem. **260:**12987–12992.

253. **Maki, H., S. Maki, R. S. Lasken, and A. Kornberg.** 1987. DNA polymerase III holoenzyme: subunits and functions, p. 63–73. *In* T. J. Kelly and R. McMacken (ed.), DNA replication and recombination. Alan R. Liss, Inc., New York.

254. **Marians, K.** 1984. Enzymology of DNA replication in prokaryotes. Crit. Rev. Biochem. **17:**153–215.

255. **Marsh, R. C., and A. Worcel.** 1977. A DNA fragment containing the origin of replication of the *Escherichia coli* chromosome. Proc. Natl. Acad. Sci. USA **74:**2720–2724.

256. **Masai, H., M. W. Bond, and K.-I. Arai.** 1986. Cloning of the *Escherichia coli* gene for primosomal protein i: the relationship to *dnaT*, essential for chromosomal DNA replication. Proc. Natl. Acad. Sci. USA **83:**1256–1260.

257. **Masters, M., and P. Broda.** 1971. Evidence for the bidirectional replication of the *Escherichia coli* chromosome. Nature (London) New Biol. **232:**137–140.

258. **Matson, S. W.** 1986. *Escherichia coli* helicase II (*uvrD* gene product) translocates unidirectionally in a 3′ to 5′ direction. J. Biol. Chem. **261:**10169–10175.

259. **Matsui, M., A. Oka, M. Takanami, S. Yasuda, and Y. Hirota.** 1985. Sites of DnaA protein-binding in the replication origin of the *Escherichia coli* K-12 chromosome. J. Mol. Biol. **184:**529–533.

260. **Maurer, R., B. C. Osmond, and D. Botstein.** 1984. Genetic analysis of DNA replication in bacteria: *dnaB* mutations that suppress *dnaC* mutations and *dnaQ* mutations that suppress *dnaE* mutations in *Salmonella typhimurium*. Genetics **108:**25–38.

261. **Maurer, R., B. C. Osmond, E. Shektman, A. Wong, and D. Botstein.** 1984. Functional interchangeability of DNA replication genes in *Salmonella typhimurium* and *Escherichia coli* demonstrated by a general complementation procedure. Genetics **108:**1–23.

262. **Maxwell, A., and M. Gellert.** 1984. The DNA dependence of the ATPase activity of DNA gyrase. J. Biol. Chem. **259:**14472–14480.

263. **McHenry, C. S.** 1982. Purification and characterization of DNA polymerase III′. Identification of τ as a subunit of the DNA polymerase III holoenzyme. J. Biol. Chem. **257:**2657–2663.

264. **McHenry, C. S.** 1985. DNA polymerase III holoenzyme of *Escherichia coli*: components and function of a true replicative complex. Mol. Cell. Biochem. **66:**71–85.

265. **McHenry, C. S., and W. Crow.** 1979. DNA polymerase III of *Escherichia coli*: purification and identification of subunits. J. Biol. Chem. **254:**1748–1753.

266. **McHenry, C. S., and K. O. Johanson.** 1984. DNA polymerase III holoenzyme of *Escherichia coli*: an asymmetric dimeric replicative complex containing distinguishable leading and lagging strand polymerases, p. 315–319. *In* U. Hübscher and S. Spadari (ed.), Proteins involved in DNA replication. Plenum Publishing Corp., New York.

267. **McHenry, C. S., and A. Kornberg.** 1977. DNA polymerase III holoenzyme of *Escherichia coli*: purification and resolution into subunits. J. Biol. Chem. **252:**6478–6484.

268. **McHenry, C. S., R. Oberfelder, K. Johanson, H. Tomasiewicz, and M. A. Franden.** 1987. Structure and mechanism of the DNA polymerase III holoenzyme, p. 47–61. *In* T. J. Kelly and R. McMacken (ed.), DNA replication and recombination. Alan R. Liss, Inc., New York.

269. **McMacken, R., C. Alfano, B. Gomes, J. H. LeBowitz, K. Mensa-Wilmot, J. D. Roberts, and M. Wold.** 1987. Biochemical mechanisms in the initiation of bacteriophage λ DNA replication, p. 227–245. *In* T. J. Kelly and R. McMacken (ed.), DNA replication and recombination. Alan R. Liss, Inc., New York.

270. **McMacken, R., J.-P. Bouché, S. L. Rowen, J. H. Weiner, K. Ueda, L. Thelander, C. McHenry, and A. Kornberg.** 1977. RNA priming of DNA replication, p. 15–29. *In* H. J. Vogel (ed.), Nucleic acid-protein recognition. Academic Press, Inc., New York.

271. **McMacken, R., and A. Kornberg.** 1978. A multienzyme system for priming the replication of φX174 viral DNA. J. Biol. Chem. **253:**3313–3319.

272. **McMacken, R., K. Ueda, and A. Kornberg.** 1977. Migration of *Escherichia coli* DnaB protein on the template DNA strand as a mechanism in initiating DNA replication. Proc. Natl. Acad. Sci. USA **74:**4190–4194.

273. **McMacken, R., M. S. Wold, J. H. LeBowitz, J. D. Roberts, J. B. Mallory, J. A. K. Wilkinson, and C. Loehrlein.** 1983. Initiation of DNA replication *in vitro* promoted by the bacteriophage λ O and P replication proteins, p. 819–848. *In* N. Cozzarelli (ed.), Mechanisms of DNA replication and recombination. Alan R. Liss, Inc., New York.

274. **Meijer, M., E. Beck, F. G. Hansen, H. E. N. Bergmans, W. Messer, K. von Meyenburg, and H. Schaller.** 1979. Nucleotide sequence of the origin of replication of the *E. coli* chromosome. Proc. Natl. Acad. Sci. USA **76:**580–584.

275. **Menzel, R., and M. Gellert.** 1983. Regulation of the genes for *E. coli* DNA gyrase: homeostatic control of DNA supercoiling. Cell **34:**105–113.

276. **Messer, W., U. Bellekes, and H. Lother.** 1985. Effect of dam methylation on the activity of the *E. coli* replication origin, *oriC*. EMBO J. **4:**1327–1332.

277. **Messer, W., L. Dankwarth, R. Tippe-Schindler, J. R. Womack, and G. Zahn.** 1975. Regulation of the initiation of DNA replication in *E. coli*. Isolation of I-RNA and the control of I-RNA synthesis. ICN-UCLA Symp. Mol. Cell. Biol. **3:**602–617.

278. **Miano, A., M. A. Losso, G. L. Gianfranceschi, and C. O. Gualerzi.** 1982. Proteins from the prokaryotic nucleoid. I. Effect of NS1 and NS2 (HU) proteins on the thermal stability of DNA. Biochemistry Int. **5:**415–422.

279. **Miki, T., S. Hiraga, T. Nagata, and T. Yura.** 1978. Bacteriophage λ carrying the *Escherichia coli* chromosomal region of the replication origin. Proc. Natl. Acad. Sci. USA **75:**5099–5013.

280. **Minden, J. S., and K. J. Marians.** 1985. Replication of pBR322 DNA *in vitro* with purified proteins. Requirement for topoisomerase I in the maintenance of template specificity. J. Biol. Chem. **260:**9316–9325.

281. **Minden, J. S., M. Mok, and K. J. Marians.** 1987. Studies on the replication of pBR322 DNA *in vitro* with purified proteins, p. 247–263. *In* T. J. Kelly and R. McMacken (ed.), DNA replication and recombination. Alan R. Liss, Inc., New York.

282. **Mizuuchi, K.** 1983. In vitro transposition of bacteriophage Mu: a biochemical approach to a novel replication reaction. Cell **35:**785–794.

283. **Molin, S., and K. Nordström.** 1980. Control of plasmid R1 replication: functions involved in replication, copy number control, incompatibility, and switch-off of replication. J. Bacteriol. **141:**111–120.

284. **Molineux, I., J. A. Pauli, and M. L. Gefter.** 1975. Physical studies of the interaction between the *Escherichia coli* DNA binding protein and nucleic acids. Nucleic Acids Res. **2:**1821–1837.

285. **Morelli, G., H.-J. Buhk, C. Fisseau, H. Lother, K. Yoshinaga, and W. Messer.** 1981. Promoters in the region of the *E. coli* replication origin. Mol. Gen. Genet. **184:**255–259.

286. **Morita, M., K. Sugimoto, A. Oka, M. Takanami, and Y. Hirota.** 1981. Mapping of promoters in the replication origin of the *E. coli* chromosome. ICN-UCLA Symp. Mol. Cell. Biol. **22:**29–35.

287. **Moriya, S., N. Ogasawara, and H. Yoshikawa.** 1985. Structure and function of the region of the replication origin of the *Bacillus subtilis* chromosome. III. Nucleotide sequence of some 10,000 base pairs in the origin region. Nucleic Acids Res. **13:**2251–2265.

288. **Morrison, A., and N. R. Cozzarelli.** 1979. Site-specific cleavage of DNA by *E. coli* DNA gyrase. Cell **17:**175–184.

289. **Moses, R. E., and C. C. Richardson.** 1970. A new DNA polymerase activity of *Escherichia coli*. I. Purification and properties of the activity present in *E. coli* polA1. Biochem. Biophys. Res. Commun. **41:**1557–1564.

290. **Mullin, D. A., C. L. Woldringh, J. M. Henson, and J. R. Walker.** 1985. Cloning of the *Escherichia coli* dnaZX region and identification of its products. Mol. Gen. Genet. **192:**73–79.

291. **Nakayama, N., N. Arai, M. W. Bond, Y. Kaziro, and K. Arai.** 1984. Nucleotide sequence of *dnaB* and the primary structure of the DnaB protein from *Escherichia coli*. J. Biol. Chem. **259:**97–101.

292. **Neidhardt, F. C., T. A. Philips, R. A. VanBogelen, M. W. Smith, Y. Georgalis, and A. R. Subramanian.** 1981. Identity of the B56.5 protein, the A protein, and the *groE* gene product of *Escherichia coli*. J. Bacteriol. **145:**513–520.

293. **Nishimura, A., Y. Nishimura, and L. Caro.** 1973. Isolation of Hfr strains from R⁺ and ColV2⁺ strains of *Escherichia coli* and derivation of an R′*lac* factor by transduction. J. Bacteriol. **116:**1107–1112.

294. **Nishimura, Y., L. Caro, C. M. Berg, and Y. Hirota.** 1971. Chro-

mosome replication in *Escherichia coli* K-12. IV. Control of chromosome replication and cell division by an integrated episome. J. Mol. Biol. **55**:441–456.

295. **Niwa, O., S. Bryan, and R. E. Moses.** 1981. Alternate pathways of DNA replication: DNA polymerase I-dependent replication. Proc. Natl. Acad. Sci. USA **78**:7024–7027.

296. **Nomura, N., and D. Ray.** 1980. Expression of a DNA strand initiation sequence of ColE1 plasmid in a single-stranded DNA phage. Proc. Natl. Acad. Sci. USA **77**:6566–6570.

297. **Nossal, N. G.** 1983. Prokaryotic DNA replication systems. Annu. Rev. Biochem. **53**:581–615.

298. **O'Donnell, M. E., and A. Kornberg.** 1985. Dynamics of DNA polymerase III holoenzyme of *Escherichia coli* in replication of a multiprimed template. J. Biol. Chem. **260**:12875–12883.

299. **O'Donnell, M. E., and A. Kornberg.** 1985. Complete replication of templates by *Escherichia coli* DNA polymerase III holoenzyme. J. Biol. Chem. **260**:12884–12889.

300. **Ogawa, T.** 1975. Analysis of the *dnaB* function of *Escherichia coli* K12 and the *dnaB*-like function of P1 prophage. J. Mol. Biol. **94**:327–340.

301. **Ogawa, T., T. A. Baker, A. van der Ende, and A. Kornberg.** 1985. Initiation of enzymatic replication of the origin of the *Escherichia coli* chromosome: contribution of RNA polymerase and primase. Proc. Natl. Acad. Sci. USA **82**:3562–3566.

302. **Ogawa, T., and T. Okazaki.** 1980. Discontinuous DNA replication. Annu. Rev. Biochem. **49**:421–457.

303. **Ogawa, T., and T. Okazaki.** 1984. Function of RNase H in DNA replication revealed by RNase H defective mutants of *Escherichia coli*. Mol. Gen. Genet. **193**:231–237.

304. **Ogawa, T., G. G. Pickett, T. Kogoma, and A. Kornberg.** 1984. RNase H confers specificity in the *dnaA*-dependent initiation of replication at the unique origin of the *Escherichia coli* chromosome *in vivo* and *in vitro*. Proc. Natl. Acad. Sci. USA **81**:1040–1044.

305. **Ohmori, H., M. Kimura, and Y. Sakakibara.** 1984. Structural analysis of the *dnaA* and *dnaN* genes of *Escherichia coli*. Gene **28**:159–170.

306. **Oka, A., H. Sasaki, K. Sugimoto, and M. Takanami.** 1984. Sequence organization of the replication origin of the *Escherichia coli* K-12 chromosome. J. Mol. Biol. **176**:443–458.

307. **Oka, A., K. Sugimoto, H. Sasaki, and M. Takanami.** 1982. An *in vitro* method of generating base substitutions in preselected regions of plasmid DNA: application to structural analysis of the replication origin of the *Escherichia coli* K-12 chromosome. Gene **19**:59–69.

308. **Oka, A., K. Sugimoto, M. Takanami, and Y. Hirota.** 1980. Replication origin of the *E. coli* K12 chromosome: the size and structure of the minimum DNA segment carrying information for autonomous replication. Mol. Gen. Genet. **178**:9–20.

309. **Okazaki, R., T. Okazaki, K. Sakabe, K. Sugimoto, and A. Sugino.** 1968. Mechanism of DNA chain growth. I. Possible discontinuity and unusual secondary structure of newly synthesized chains. Proc. Natl. Acad. Sci. USA **59**:598–605.

310. **Olivera, B. M., and F. Bonhoeffer.** 1974. Replication of *Escherichia coli* requires DNA polymerase I. Nature (London) **250**:513–514.

311. **Olivera, B. M., and I. R. Lehman.** 1967. Linkage of polynucleotides through phosphodiester bonds by an enzyme from *Escherichia coli*. Proc. Natl. Acad. Sci. USA **57**:1426–1433.

312. **Ollis, D. L., P. Brick, R. Hamlin, N. G. Xuong, and T. A. Steitz.** 1985. Structure of large fragment of *Escherichia coli* DNA polymerase I complexed with dTMP. Nature (London) **313**:762–766.

313. **Orr, E., N. F. Fairweather, I. B. Holland, and R. H. Pritchard.** 1979. Isolation and characterization of a strain carrying a conditional lethal mutation in the *cou* gene of *Escherichia coli* K-12. Mol. Gen. Genet. **177**:103–112.

314. **Orr, E., P. A. Meacock, and R. H. Pritchard.** 1978. Genetic and physiological properties of an *Escherichia coli* strain carrying the *dnaA* mutation T46, p. 85–99. *In* I. Molineux and M. Kohiyama (ed.), DNA synthesis present and future. Plenum Publishing Corp., New York.

315. **O'Sullivan, M. A., and G. Anagnostopoulos.** 1982. Replication terminus of the *Bacillus subtilis* chromosome. J. Bacteriol. **135**:135–143.

316. **Otto, B., F. Bonhoeffer, and H. Schaller.** 1973. Purification and properties of DNA polymerase III. Eur. J. Biochem. **34**:440–447.

317. **Pettijohn, D., and R. Hecht.** 1973. RNA molecules bound to the folded bacterial genome stabilize DNA folds and segregate domains of supercoiling. Cold Spring Harbor Symp. Quant. Biol. **38**:31–40.

318. **Prescott, D. M., and P. L. Kuempel.** 1972. Bidirectional replication of the chromosome in *Escherichia coli*. Proc. Natl. Acad. Sci. USA **69**:2842–2846.

319. **Pritchard, R. H.** 1985. Control of chromosome replication in bacteria, p. 277–282. *In* D. R. Helinski, S. N. Cohen, D. B. Clewell, D. A. Jackson, and A. Hollaender (ed.), Plasmids in bacteria. Plenum Publishing Corp., New York.

320. **Pritchard, R. H., P. T. Barth, and J. Collins.** 1969. Control of DNA synthesis in bacteria. Symp. Soc. Gen. Microbiol. **19**:263–297.

321. **Pritchard, R. H., and K. G. Lark.** 1964. Induction of replication by thymine starvation at the chromosome origin in *Escherichia coli*. J. Mol. Biol. **9**:288–307.

322. **Projan, S. L., and J. A. Wechsler.** 1981. Isolation and analysis of multicopy extragenic suppressors of *dnaA* mutations. J. Bacteriol. **145**:861–866.

323. **Rasmussen, K. V., T. Atlung, G. Kerszman, G. E. Hansen, and F. G. Hansen.** 1983. Conditional change of DNA replication control in an RNA polymerase mutant of *Escherichia coli*. J. Bacteriol. **154**:443–451.

324. **Reha-Krantz, L. J., and J. Hurwitz.** 1978. The *dnaB* gene product of *Escherichia coli*. I. Purification, homogeneity, and physical properties. J. Biol. Chem. **253**:4043–4050.

324a.**Reha-Krantz, L. J., and J. Hurwitz.** 1978. The *dnaB* gene product of *Escherichia coli*. II. Single-stranded DNA-dependent ribonucleoside triphosphatase activity. J. Biol. Chem. **253**:4051–4057.

325. **Rokeach, L. A., A. Chiaramello, D. E. Junker, K. Crain, A. Nourani, M. Jannatipour, and J. W. Zyskind.** 1987. Effects of DnaA protein on replication and transcription events at the *Escherichia coli* origin of replication, *oriC*, p. 415–427. *In* T. J. Kelly and R. McMacken (ed.), DNA replication and recombination. Alan R. Liss, Inc., New York.

326. **Rokeach, L. A., and J. W. Zyskind.** 1986. RNA terminating within the *Escherichia coli* origin of replication: stringent regulation and control by DnaA protein. Cell **46**:763–771.

327. **Ron, E. Z., S. Rozenhak, and N. Grossman.** 1975. Synchronization of cell division in *Escherichia coli* by amino acid starvation: strain specificity. J. Bacteriol. **123**:374–376.

328. **Rosenberg, B. H., L. F. Cavalieri, and G. Ungers.** 1969. The negative control mechanism for *E. coli* DNA replication. Proc. Natl. Acad. Sci. USA **63**:1410–1417.

329. **Rouvière-Yaniv, J.** 1978. Localization of the HU protein on the *Escherichia coli* nucleoid. Cold Spring Harbor Symp. Quant. Biol. **42**:439–447.

330. **Rouvière-Yaniv, J., and F. Gros.** 1975. Characterization of a novel, low-molecular-weight DNA-binding protein from *Escherichia coli*. Proc. Natl. Acad. Sci. USA **72**:3428–3432.

331. **Rouvière-Yaniv, J., and M. Yaniv.** 1979. *E. coli* DNA binding protein HU forms nucleosome-like structure with circular, double-stranded DNA. Cell **17**:265–274.

332. **Rowen, L., and A. Kornberg.** 1978. Primase, the DnaG protein of *Escherichia coli*. An enzyme which starts DNA chains. J. Biol. Chem. **253**:758–764.

333. **Rowen, L., and A. Kornberg.** 1978. A ribo-deoxyribonucleotide primer synthesized by primase. J. Biol. Chem. **253**:770–774.

334. **Saito, H., Y. Nakamura, and H. Uchida.** 1978. A transducing lambda phage carrying *grpE*, a bacterial gene necessary for lambda DNA replication and two ribosomal protein genes *rpsP*(S16) and *rplI8*(L19). Mol. Gen. Genet. **165**:247–256.

335. **Saito, H., and H. Uchida.** 1977. Initiation of the DNA replication of bacteriophage lambda in *Escherichia coli* K12. J. Mol. Biol. **113**:1–25.

336. **Saito, H., and H. Uchida.** 1978. Organization and expression of the *dnaJ* and *dnaK* genes of *Escherichia coli* K12. Mol. Gen. Genet. **164**:1–8.

337. **Saitoh, T., and S. Hiraga.** 1975. Initiation of DNA replication in *Escherichia coli*. III. Genetic analysis of the *dna* mutant exhibiting rifampicin-sensitive resumption of replication. Mol. Gen. Genet. **137**:249–261.

338. **Sakakibara, Y., and S. Yuasa.** 1982. Continuous synthesis of the *dnaA* gene product of *Escherichia coli* in the cell cycle. Mol. Gen. Genet. **186**:87–94.

339. **Sako, T., and Y. Sakakibara.** 1980. Coordinate expression of the *Escherichia coli* *dnaA* and *dnaN* genes. Mol. Gen. Genet. **179**:521–526.

340. **Sancar, A., K. R. Williams, J. W. Chase, and W. D. Rupp.** 1981. Sequences of the *ssb* gene and protein. Proc. Natl. Acad. Sci. USA **78**:4274–4278.

341. **Sargent, M. G., and M. J. Monteiro.** 1982. Characterization of the chromosomal terminus of *B. subtilis* and its attachment to the cell membrane, p. 181–195. *In* A. T. Ganesan, S. Chang, and J. A. Hoch (ed.), Molecular cloning and gene regulation in

bacilli. Academic Press, Inc., New York.

342. **Schaller, H., B. Otto, V. Nusslein, J. Huf, R. Herrmann, and F. Bonhoeffer.** 1972. Deoxyribonucleic acid replication *in vitro*. J. Mol. Biol. **63:**183–200.

343. **Schaus, N., K. O'Day, and A. Wright.** 1981. Suppression of amber mutations in the *dnaA* gene of *Escherichia coli* K-12 by secondary mutations in *rpoB*. ICN-UCLA Symp. Mol. Cell. Biol. **22:**315–323.

344. **Schekman, R., A. Weiner, and A. Kornberg.** 1974. Multienzyme systems of DNA replication. Science **186:**987–993.

345. **Schekman, R., J. H. Weiner, A. Weiner, and A. Kornberg.** 1975. Ten proteins required for conversion of φX174 single-stranded DNA to duplex form *in vitro*. J. Biol. Chem. **250:**5859–5865.

346. **Schekman, R., W. Wickner, O. Westergaard, D. Brutlag, K. Geider, L. L. Bertsch, and A. Kornberg.** 1972. Initiation of DNA synthesis: synthesis of φX174 replicative form requires RNA synthesis resistant to rifampicin. Proc. Natl. Acad. Sci. USA **69:**2691–2695.

347. **Scheuermann, R. H., and H. Echols.** 1984. A separate editing exonuclease for DNA replication: the ε subunit of *Escherichia coli* DNA polymerase III holoenzyme. Proc. Natl. Acad. Sci. USA **81:**7747–7751.

348. **Scheuermann, R. H., S. Tam, P. M. J. Burgers, C. Lu, and H. Echols.** 1983. Identification of the ε-subunit of *Escherichia coli* DNA polymerase III holoenzyme as the *dnaQ* gene product: a fidelity subunit for DNA replication. Proc. Natl. Acad. Sci. USA **80:**7085–7089.

349. **Sclafani, R. A., and J. A. Wechsler.** 1981. Deoxyribonucleic acid initiation mutation *dnaB252* is suppressed by elevated *dnaC*+ gene dosage. J. Bacteriol. **146:**418–421.

350. **Scott, J. F., S. Eisenberg, L. L. Bertsch, and A. Kornberg.** 1977. A mechanism of duplex DNA replication revealed by enzymatic studies of φX174: catalytic strand separation in advance of replication. Proc. Natl. Acad. Sci. USA **74:**193–197.

351. **Seror, S. J., F. Vannier, A. Levine, and G. Henckes.** 1986. Stringent control of initiation of chromosome replication in *Bacillus subtilis*. Nature (London) **321:**709–710.

352. **Seror-Laurent, S. J., and G. Henckes.** 1985. An RNA-DNA copolymer whose synthesis is correlated with the transcriptional requirement for chromosomal initiation in *Bacillus subtilis* contains ribosomal RNA sequences. Proc. Natl. Acad. Sci. USA **82:**3586–3590.

352a.**Seufert, W., and W. Messer.** 1986. Initiation of *Escherichia coli* minichromosome replication at *oriC* and at protein n' recognition sites. Two modes for initiating DNA synthesis *in vitro*. EMBO J. **5:**3401–3406.

353. **Sevastopoulos, C. G., C. T. Wehr, and D. A. Glaser.** 1977. Large-scale automated isolation of *Escherichia coli* mutants with thermosensitive DNA replication. Proc. Natl. Acad. Sci. USA **74:**3485–3489.

354. **Shen, V., and H. Bremer.** 1977. Chloramphenicol-induced changes in the synthesis of ribosomal, transfer, and messenger ribonucleic acids in *Escherichia coli* B/r. J. Bacteriol. **130:**1098–1108.

355. **Shlomai, J., and A. Kornberg.** 1980. An *Escherichia coli* replication protein that recognizes a unique sequence within a hairpin region in φX174 DNA. Proc. Natl. Acad. Sci. USA **77:**799–803.

356. **Shlomai, J., and A. Kornberg.** 1980. A prepriming DNA replication enzyme of *Escherichia coli*. I. Purification of protein n': a sequence-specific, DNA-dependent ATPase. J. Biol. Chem. **255:**6789–6793.

357. **Sigal, N., H. Delius, T. Kornberg, M. L. Gefter, and B. M. Alberts.** 1972. A DNA-unwinding protein isolated from *Escherichia coli*. Its interaction with DNA and DNA polymerases. Proc. Natl. Acad. Sci. USA **69:**3537–3541.

358. **Sims, J., and E. W. Benz.** 1980. Initiation of DNA replication by the *Escherichia coli* DnaG protein: evidence that tertiary structure is involved. Proc. Natl. Acad. Sci. USA **77:**900–904.

359. **Sinden, R. R., J. O. Carlson, and D. E. Pettijohn.** 1980. Torsional tension in the DNA double helix measured with trimethylpsoralen in living *E. coli* cells: analogous measurements in insect and human cells. Cell **21:**773–783.

360. **Sinden, R. R., and D. E. Pettijohn.** 1981. Chromosomes in living *Escherichia coli* cells are segregated into domains of supercoiling. Proc. Natl. Acad. Sci. USA **78:**224–228.

361. **Smiley, B. L., J. R. Lupski, P. S. Svec, R. McMacken, and G. N. Godson.** 1982. Sequences of the *Escherichia coli dnaG* primase gene and regulation of its expression. Proc. Natl. Acad. Sci. USA **79:**4550–4554.

362. **Smith, D. W., A. M. Garland, G. Herman, R. E. Enns, T. A. Baker, and J. W. Zyskind.** 1985. Importance of state of methylation of *oriC* GATC sites in initiation of DNA replication in *Escherichia coli*. EMBO J. **4:**1319–1326.

363. **Sogo, J. M., M. Greenstein, and A. Skalka.** 1976. The circle mode of replication of bacteriophage lambda: the role of covalently closed templates and the formation of mixed catenated dimers. J. Mol. Biol. **103:**537–562.

364. **Sompayrac, L., and O. Maaløe.** 1973. Autorepressor model for control of DNA replication. Nature (London) New Biol. **241:**133–135.

365. **Stayton, M. M., and A. Kornberg.** 1983. Complexes of *Escherichia coli* primase with the replication origin of G4 phage DNA. J. Biol. Chem. **258:**13205–13212.

366. **Steck, T. R., and K. Drlica.** 1984. Bacterial chromosome segregation: evidence for DNA gyrase involvement in decatenation. Cell **36:**1081–1088.

367. **Sternberg, N.** 1973. Properties of a mutant of *Escherichia coli* defective in bacteriophage λ head formation (*groE*). II. The propagation of phage. J. Mol. Biol. **76:**25–44.

368. **Sternglanz, R., S. DiNardo, K. A. Voelkel, Y. Nishimura, Y. Hirota, K. Becherer, L. Zumstein, and J. C. Wang.** 1981. Mutations in the gene coding for *Escherichia coli* DNA topoisomerase I affecting transcription and transposition. Proc. Natl. Acad. Sci. USA **78:**2747–2751.

369. **Stonington, O., and D. Pettijohn.** 1971. The folded genome of *Escherichia coli* isolated in a protein-DNA-RNA complex. Proc. Natl. Acad. Sci. USA **68:**6–9.

370. **Stuitje, A. R., P. J. Weisbeek, and M. Meijer.** 1984. Initiation signals for complementary strand DNA synthesis in the region of the replication origin of the *Escherichia coli* chromosome. Nucleic Acids Res. **12:**3321–3332.

371. **Sueoka, N., and W. G. Quinn.** 1968. Membrane attachment of the chromosome replication origin in *Bacillus subtilis*. Cold Spring Harbor Symp. Quant. Biol. **33:**695–705.

372. **Sueoka, N., and H. Yoshikawa.** 1965. The chromosome of *Bacillus subtilis*. I. Theory of marker frequency analysis. Genetics **52:**747–757.

373. **Sugimoto, K., A. Oka, H. Sugisaki, M. Takanami, A. Nishimura, Y. Yasuda, and Y. Hirota.** 1979. Nucleotide sequence of *Escherichia coli* K-12 replication origin. Proc. Natl. Acad. Sci. USA **76:**575–579.

374. **Sugino, A., N. P. Higgins, P. O. Brown, C. L. Peebles, and N. R. Cozzarelli.** 1978. Energy coupling in DNA gyrase and the mechanism of action of novobiocin. Proc. Natl. Acad. Sci. USA **75:**4838–4842.

375. **Sugino, A., C. L. Peebles, K. N. Kreuzer, and N. R. Cozzarelli.** 1977. Mechanism of action of nalidixic acid: purification of *Escherichia coli nalA* gene product and its relationship to DNA gyrase and a novel nicking-closing enzyme. Proc. Natl. Acad. Sci. USA **74:**4767–4771.

376. **Sundin, O., and A. Varshavsky.** 1981. Arrest of segregation leads to accumulation of highly intertwined catenated dimers: dissection of the final stages of SV40 DNA replication. Cell **25:**659–669.

377. **Sunshine, M., M. Feiss, J. Stuart, and J. Jochem.** 1977. A new host gene (*groPC*) necessary for lambda DNA replication. Mol. Gen. Genet. **151:**27–34.

378. **Tabata, S., A. Oka, K. Sugimoto, M. Takanami, S. Yasuda, and Y. Hirota.** 1983. The 245 base-pair *oriC* sequence of the *E. coli* chromosome directs bidirectional DNA replication at an adjacent region. Nucleic Acids Res. **11:**2617–2626.

379. **Takeda, Y., and Y. Hirota.** 1982. Suppressor genes of a *dnaA* temperature sensitive mutation in *Escherichia coli*. Mol. Gen. Genet. **187:**67–71.

380. **Tanaka, I., K. Appelt, J. Dijk, S. W. White, and K. S. Wilson.** 1984. 3-Å resolution structure of a protein with histone-like properties in prokaryotes. Nature (London) **310:**376–381.

381. **Tanaka, M., H. Ohmori, and S. Hiraga.** 1983. A novel type of *E. coli* mutant with increased chromosome copy number. Mol. Gen. Genet. **192:**51–60.

382. **Taucher-Scholz, G., M. Abdel-Monem, and H. Hoffmann-Berling.** 1983. Functions of DNA helicases in *Escherichia coli*, p. 65–76. *In* N. R. Cozzarelli (ed.), Mechanisms of DNA replication and recombination. Alan R. Liss, Inc., New York.

383. **Tilly, K., N. McKittrick, M. Zylicz, and C. Georgopoulos.** 1983. The DnaK protein modulates the heat shock response of *Escherichia coli*. Cell **34:**641–646.

384. **Tilly, K., H. Murialdo, and C. Georgopoulos.** 1981. Identification of a second *Escherichia coli groE* gene whose product is necessary for bacteriophage morphogenesis. Proc. Natl. Acad. Sci. USA **78:**1629–1633.

385. **Tilly, K., R. A. VanBogelen, C. Georgopoulos, and F. C. Neidhardt.** 1983. Identification of the heat-inducible protein C15.4 as the *groES* gene product in *Escherichia coli*. J. Bacteriol. **154:**1505–1507.

386. **Tippe-Schindler, R., G. Zahn, and W. Messer.** 1979. Control of

the initiation of DNA replication in *Escherichia coli*. I. Negative control of initiation. Mol. Gen. Genet. **168**:185–195.

387. **Tomizawa, J.-I., and T. Itoh.** 1981. Plasmid colE1 incompatibility determined by interaction of RNA I with primer transcript. Proc. Natl. Acad. Sci. USA **78**:6096–6100.

388. **Tomizawa, J.-I., and G. Selzer.** 1979. Initiation of DNA synthesis in *Escherichia coli*. Annu. Rev. Biochem. **48**:554–599.

389. **Torrey, T. A., T. Atlung, and T. Kogoma.** 1984. *dnaA* suppressor (*dasF*) mutants of *Escherichia coli* are stable DNA replication (*sdrA/rnh*) mutants. Mol. Gen. Genet. **196**:350–355.

390. **Tse, Y.-C., K. Kirkegaard, and J. C. Wang.** 1980. Covalent bonds between protein and DNA. Formation of phosphotyrosine linkage between certain DNA topoisomerases and DNA. J. Biol. Chem. **255**:5560–5565.

391. **Tsurimoto, T., and K. Matsubara.** 1982. Replication of λdv plasmid *in vitro* by purified λ O and P proteins. Proc. Natl. Acad. Sci. USA **79**:7639–7643.

392. **Ueda, K., R. McMacken, and A. Kornberg.** 1978. DnaB protein of *Escherichia coli*: purification and role in the replication of φX174 DNA. J. Biol. Chem. **253**:261–269.

393. **Uyemura, D., and I. R. Lehman.** 1976. Biochemical characterization of mutant forms of DNA polymerase I from *Escherichia coli*. J. Biol. Chem. **251**:4078–4084.

394. **van der Ende, A., T. A. Baker, T. Ogawa, and A. Kornberg.** 1985. Initiation of enzymatic replication at the origin of the *Escherichia coli* chromosome: primase as the sole priming enzyme. Proc. Natl. Acad. Sci. USA **82**:3954–3958.

395. **van der Ende, A., R. Teertstra, H. G. A. M. van der Avoort, and P. J. Weisbeek.** 1983. Initiation signals for complementary strand DNA synthesis on single-stranded plasmid DNA. Nucleic Acids Res. **11**:4957–4975.

396. **Varshavsky, A. J., S. A. Nedospasov, V. V. Bakayev, T. G. Bakayeva, and G. P. Georgiev.** 1977. Histone-like proteins in the purified *Escherichia coli* deoxyribonucleoprotein. Nucleic Acids Res. **4**:2725–2745.

397. **von Meyenburg, K., F. G. Hansen, T. Atlung, L. Boe, I. G. Clausen, B. van Deurs, E. B. Hansen, B. B. Jorgensen, F. Jorgensen, O. Michelson, J. Nielsen, P. E. Pedersen, K. V. Rasmussen, E. Riise, and O. Skovgaard.** 1985. Facets of the chromosomal origin of replication *oriC*, p. 260–281. *In* M. Schaechter, F. C. Neidhardt, J. L. Ingraham, and N. O. Kjeldgaard (ed.), The molecular biology of bacterial growth. Jones and Bartlett Publishers, Inc., Boston.

398. **von Meyenburg, K., F. G. Hansen, L. D. Nielsen, and P. Jorgensen.** 1977. Origin of replication, *oriC*, of the *Escherichia coli* chromosome: mapping of genes relative to *Eco* RI cleavage sites in the *oriC* region. Mol. Gen. Genet. **158**:101–109.

399. **von Meyenburg, K., F. G. Hansen, L. D. Nielsen, and E. Riise.** 1978. Origin of replication, *oriC*, of the *Escherichia coli* chromosome on specialized transducing phages λ *tna*. Mol. Gen. Genet. **160**:287–295.

400. **Wada, M., and H. Itikawa.** 1984. Participation of *Escherichia coli* K-12 *groE* gene products in the synthesis of cellular DNA and RNA. J. Bacteriol. **157**:694–696.

401. **Wada, M., Y. Kadokami, and H. Itikawa.** 1982. Thermosensitive synthesis of DNA and RNA in *dnaJ* mutants of *Escherichia coli* K12. Jpn. J. Genet. **57**:407–413.

402. **Walker, J. E., N. J. Gay, M. Saraste, and A. N. Eberle.** 1984. DNA sequence around the *Escherichia coli* unc operon. Biochem. J. **224**:799–815.

403. **Walker, J. R., J. A. Ramsey, and W. G. Haldenwang.** 1982. Interaction of the *Escherichia coli* dnaA initiation protein with the DnaZ polymerization protein *in vivo*. Proc. Natl. Acad. Sci. USA **79**:3340–3344.

404. **Wang, J. C.** 1971. Interaction between DNA and an *Escherichia coli* protein ω. J. Mol. Biol. **55**:523–533.

405. **Wang, J. C.** 1985. DNA topoisomerases. Annu. Rev. Biochem. **54**:665–697.

406. **Ward, C. B., and D. A. Glaser.** 1969. Analysis of the chloramphenicol-sensitive and chloramphenicol-resistant steps in the initiation of DNA synthesis in *E. coli* B/r. Proc. Natl. Acad. Sci. USA **64**:905–912.

407. **Wasserman, S. A., and N. R. Cozzarelli.** 1986. Biochemical topology: applications to DNA recombination and replication. Science **232**:951–960.

408. **Wechsler, J. A.** 1975. Genetic and phenotypic characterization of dnaC mutations. J. Bacteriol. **121**:594–599.

409. **Wechsler, J. A.** 1978. The genetics of *Escherichia coli* DNA replication, p. 49–70. *In* I. Molineux and M. Kohiyama (ed.), DNA synthesis present and future. Plenum Publishing Corp., New York.

410. **Wechsler, J. A., and J. D. Gross.** 1971. *Escherichia coli* mutants

411. **Wechsler, J. A., and M. Zdzienicka.** 1975. Cryolethal suppressors of thermosensitive *dnaA* mutations, p. 624–639. *In* M. Goulian and P. Hanawalt (ed.), DNA synthesis and its regulation. W. A. Benjamin, Menlo Park, Calif.

412. **Weiner, J. H., R. McMacken, and A. Kornberg.** 1976. Isolation of an intermediate which precedes *dnaG* RNA polymerase participation in enzymatic replication of bacteriophage φX174 DNA. Proc. Natl. Acad. Sci. USA **73**:752–756.

413. **Welch, M. M., and C. S. McHenry.** 1982. Cloning and identification of the product of the *dnaE* gene of *Escherichia coli*. J. Bacteriol. **152**:351–356.

414. **Westergaard, O., D. Brutlag, and A. Kornberg.** 1973. Initiation of deoxyribonucleic acid synthesis. IV. Incorporation of the ribonucleic acid primer into the phage replicative form. J. Biol. Chem. **248**:1361–1364.

415. **Wickner, R. B., M. Wright, S. Wickner, and J. Hurwitz.** 1972. Conversion of φX174 and fd single-stranded DNA to replicative forms in extracts of *Escherichia coli*. Proc. Natl. Acad. Sci. USA **69**:3233–3237.

416. **Wickner, S.** 1976. Mechanism of DNA elongation catalyzed by *Escherichia coli* DNA polymerase III, DnaZ protein, and DNA elongation factors I and III. Proc. Natl. Acad. Sci. USA **73**:3511–3515.

417. **Wickner, S.** 1977. DNA or RNA priming of bacteriophage G4 DNA synthesis by *Escherichia coli* DnaG protein. Proc. Natl. Acad. Sci. USA **74**:2815–2819.

418. **Wickner, S.** 1978. DNA replication proteins of *E. coli*. Annu. Rev. Biochem. **47**:1163–1191.

419. **Wickner, S.** 1979. DNA replication proteins of *Escherichia coli* and phage λ. Cold Spring Harbor Symp. Quant. Biol. **43**:303–310.

420. **Wickner, S., and J. Hurwitz.** 1974. Conversion of φX174 viral DNA to double-stranded forms by *E. coli* proteins. Proc. Natl. Acad. Sci. USA **71**:4120–4124.

421. **Wickner, S., and J. Hurwitz.** 1975. Interaction of *Escherichia coli* dnaB and dnaC(D) gene products *in vitro*. Proc. Natl. Acad. Sci. USA **72**:921–925.

422. **Wickner, S., and J. Hurwitz.** 1975. Association of φX174 DNA-dependent ATPase activity with an *E. coli* protein, replication factor Y, required for *in vitro* synthesis of φX174 DNA. Proc. Natl. Acad. Sci. USA **72**:3342–3346.

423. **Wickner, S., M. Wright, and J. Hurwitz.** 1973. Studies on *in vitro* DNA synthesis. Purification of the *dnaG* gene product from *Escherichia coli*. Proc. Natl. Acad. Sci. USA **70**:1613–1618.

424. **Wickner, S., M. Wright, and J. Hurwitz.** 1974. Association of DNA-dependent and independent ribonucleoside triphosphatase activities with the *dnaB* gene product of *Escherichia coli*. Proc. Natl. Acad. Sci. USA **71**:783–787.

425. **Wickner, W., D. Brutlag, R. Schekman, and A. Kornberg.** 1972. RNA synthesis initiates *in vitro* conversion of M13 DNA to its replicative form. Proc. Natl. Acad. Sci. USA **69**:965–969.

426. **Wickner, W., and A. Kornberg.** 1973. DNA polymerase III star requires ATP to start synthesis on a primed DNA. Proc. Natl. Acad. Sci. USA **70**:3679–3683.

427. **Williams, K. R., E. K. Spicer, M. B. LoPresti, R. A. Guggenheimer, and J. W. Chase.** 1983. Limited proteolysis studies of the *Escherichia coli* single-strand DNA binding protein. Evidence for a functionally homologous domain in both the *Escherichia coli* and T4 DNA binding proteins. J. Biol. Chem. **258**:3346–3355.

428. **Wold, M. S., J. B. Mallory, J. D. Roberts, J. H. LeBowitz, and R. McMacken.** 1982. Initiation of bacteriophage λ DNA replication *in vitro* with purified λ replication proteins. Proc. Natl. Acad. Sci. USA **79**:6176–6180.

429. **Wold, M. S., and R. McMacken.** 1982. Regulation of expression of the *Escherichia coli* dnaG gene and amplification of the DnaG primase. Proc. Natl. Acad. Sci. USA **79**:4907–4911.

430. **Worcel, A., and E. Burgi.** 1972. On the structure of the folded chromosome of *Escherichia coli*. J. Mol. Biol. **71**:127–147.

431. **Worcel, A., and E. Burgi.** 1974. Properties of a membrane-attached form of the folded chromosome of *Escherichia coli*. J. Mol. Biol. **82**:91–105.

432. **Yamaguchi, T., and J.-I. Tomizawa.** 1980. Establishment of *Escherichia coli* cells with an integrated high copy number plasmid. Mol. Gen. Genet. **178**:525–533.

433. **Yarronton, G. T., R. H. Dos, and M. L. Gefter.** 1979. Enzyme catalyzed DNA unwinding. Mechanism of action of helicase III. J. Biol. Chem. **254**:12002–12006.

434. **Yarronton, G. T., and M. L. Gefter.** 1979. Enzyme-catalyzed DNA unwinding studies on *Escherichia coli* Rep protein. Proc. Natl. Acad. Sci. USA **76**:1658–1662.

temperature-sensitive for DNA synthesis. Mol. Gen. Genet. **113**:273–284.

435. **Yasuda, S., and Y. Hirota.** 1977. Cloning and mapping of the replication origin of *Escherichia coli*. Proc. Natl. Acad. Sci. USA **74:**5458–5462.

435a. **Yin, K.-C., A. Blinkowa, and J. R. Walker.** 1986. Nucleotide sequence of the *Escherichia coli* replication gene *dnaZX*. Nucleic Acids Res. **14:**6541–6549.

436. **Yochem, J., H. Uchida, M. Sunshine, H. Saito, C. Georgopoulos, and M. Feiss.** 1978. Genetic analysis of two genes, *dnaJ* and *dnaK*, necessary for *Escherichia coli* and bacteriophage lambda DNA replication. Mol. Gen. Genet. **164:**9–14.

437. **Yoshimoto, M., H. Kambe-Honjoh, K. Nagai, and G. Tomora.** 1986. Early replicative intermediates of *Escherichia coli* chromosome isolated from a membrane complex. EMBO J. **5:**787–791.

438. **Zechel, K., J.-P. Bouché, and A. Kornberg.** 1975. Replication of phage G4: a novel and simple system for the initiation of deoxyribonucleic acid synthesis. J. Biol. Chem. **250:**4684–4689.

439. **Zipursky, S. L., and K. J. Marians.** 1980. Identification of two *E. coli* factor Y effector sites near the origins of replication of the plasmids ColE1 and pBR322. Proc. Natl. Acad. Sci. USA **77:**6521–6525.

440. **Zylicz, M., and C. Georgopoulos.** 1984. Purification and properties of the *Escherichia coli* DnaK replication protein. J. Biol. Chem. **259:**8820–8825.

441. **Zylicz, M., I. Gorska, K. Taylor, and C. Georgopoulos.** 1984. Bacteriophage λ replication proteins: formation of a mixed oligomer and binding to the origin of λ DNA. Mol. Gen. Genet. **196:**401–406.

442. **Zylicz, M., J. H. LeBowitz, R. McMacken, and C. Georgopoulos.** 1983. The DnaK protein of *Escherichia coli* possesses an ATPase and autophosphorylating activity and is essential in an *in vitro* DNA replication system. Proc. Natl. Acad. Sci. USA **80:**6431–6435.

443. **Zylicz, M., T. Yamomoto, N. McKittrick, S. Sell, and C. Georgopoulos.** 1985. Purification and properties of the DnaJ replication protein of *Escherichia coli*. J. Biol. Chem. **260:**7591–7598.

444. **Zyskind, J. W., J. M. Cleary, W. S. Brusilow, N. E. Harding, and D. W. Smith.** 1983. Chromosomal replication origin from the marine bacterium *Vibrio harveyi* functions in *Escherichia coli: oriC* consensus sequence. Proc. Natl. Acad. Sci. USA **80:**1164–1168.

445. **Zyskind, J. W., L. T. Deen, and D. W. Smith.** 1977. Temporal sequence of events during the initiation process in *Escherichia coli* deoxyribonucleic acid replication: roles of the *dnaA* and *dnaC* gene products and ribonucleic acid polymerase. J. Bacteriol. **129:**1466–1475.

446. **Zyskind, J. W., L. T. Deen, and D. W. Smith.** 1979. Isolation and mapping of plasmids containing the *S. typhimurium* origin of DNA replication. Proc. Natl. Acad. Sci. USA **76:**3079–3101.

447. **Zyskind, J. W., N. E. Harding, Y. Takeda, J. M. Cleary, and D. W. Smith.** 1981. The DNA replication origin of the enterobacteriaceae. ICN-UCLA Symp. Mol. Cell. Biol. **22:**13–25.

448. **Zyskind, J. W., and D. W. Smith.** 1977. Novel *Escherichia coli dnaB* mutant: direct involvement of the *dnaB252* gene product in the synthesis of an origin-ribonucleic acid species during initiation of a round of deoxyribonucleic acid replication. J. Bacteriol. **129:**1476–1486.

449. **Zyskind, J. W., and D. W. Smith.** 1980. Nucleotide sequence of the *Salmonella typhimurium* origin of DNA replication. Proc. Natl. Acad. Sci. USA **77:**2460–2464.

40. Protein Synthesis

JOHN W. B. HERSHEY

Department of Biological Chemistry, School of Medicine, University of California, Davis, California 95616

INTRODUCTION

Overview

Protein synthesis is the metabolic process whereby amino acids are linked together by peptide bonds to form long linear polypeptides. The formation of a peptide bond itself is relatively simple in chemical terms: an activated carboxyl group of one amino acid condenses with an amino group of another amino acid. However, proteins comprise up to 20 different amino acids arranged in precise sequences as determined by the genetic material. Protein synthesis therefore involves a complex apparatus which translates information encoded in nucleic acid sequences into amino acid sequences in proteins, and is a key component in the overall pathway of gene expression.

A broad overview of protein synthesis follows. An amino acid is activated by formation of a high-energy ester bond to a specific tRNA in an ATP-consuming reaction catalyzed by an aminoacyl-tRNA synthetase. Aminoacyl-tRNAs are linearly ordered by interacting with a template, the mRNA; in effect, the tRNA molecule acts as an adapter between the amino acid and the mRNA. Such interactions occur on the surface of a cellular organelle called the ribosome. The polymerization process is divided conceptually into three phases: initiation, elongation, and termination. During initiation, a unique initiator aminoacyl-tRNA binds at a precise region of the mRNA, thereby specifying the phase in which the mRNA is translated. Elongation is a cyclic process whereby subsequent aminoacyl-tRNA molecules bind to the ribosome as dictated by the mRNA. In each case, a peptide bond is formed by transfer of the aminoacyl (or peptidyl) moiety to the amino group of the incoming aminoacyl-tRNA, and the ribosome progresses down the mRNA. During termination, the mRNA signals the binding of a factor that results in the hydrolysis of the completed peptidyl-tRNA. Polymerization on the ribosome is promoted by soluble protein factors specific for the three phases of protein synthesis. Thus the protein grows from its N terminus towards its C terminus, and the sequence of amino acids is colinear with the nucleic acid sequence in the mRNA. Energy is consumed during the charging of the amino acids to their tRNAs and during the polymerization process. More than 150 macromolecular elements are involved in the translational machinery.

The Genetic Code

Protein synthesis is the translation into protein of information encoded in mRNA. An amino acid is specified by three nucleotides in the mRNA, called a codon, which interact with a specific aminoacyl-tRNA. Thus the genetic code is a triplet code, with the codons arranged contiguously but not overlapping along the mRNA. Since there are four different bases in mRNA, there are 64 possible codons but there are only 20 different amino acids. Most amino acids are specified by more than one codon (Table 1). Although the code is redundant, it is unambiguous in that no codon specifies more than one amino acid. Three codons, UAA, UAG, and UGA, do not code for any amino acids but rather signal termination of protein synthesis.

Analysis of how amino acid codons are grouped in Table 1 indicates that the first two letters of the code word are most important in specifying the amino acid. For some amino acids, the third position can be any of the four letters, whereas for other amino acids, it is either the two pyrimidines or the two purines. Only methionine and tryptophan have a single code word. Redundancy in the code allows some flexibility in the nucleic acid sequence of a mRNA. However, in a particular organism not all codons for a given amino acid are used with equal frequency, some being preferred over others.

Physiological Context

Protein synthesis is a dominant process in rapidly growing bacterial cells. The translational machinery comprises about half the dry weight of the cell, and the process consumes up to 80% of the cell's energy (164). Since the rate of protein synthesis appears to be tightly coupled to growth rate, the regulation of the overall rate of translation is of vital importance to the cell. To understand translational control, we need to have a precise knowledge of the components involved in protein synthesis and how these components interact. In this chapter I will evaluate progress over the past 25 years in an effort to define the machinery and elucidate its mechanism in terms of basic chemical principles. Although the broad outlines of protein synthesis are well established, mechanistic details are still lacking and much work remains to be done. Mechanisms of translational control are described by Gold and Stormo in chapter 78.

SYNTHESIS OF AMINOACYL-tRNAs

Transfer RNAs (tRNAs) are small molecules which play the adaptor role between amino acids and mRNA. Their existence was predicted by Crick in 1955 (48), and soon afterwards they were discovered by Hoagland et al. (107). After the generation of chromatographic methods for their fractionation, various purified species were prepared, and one of these, *Saccharomyces cerevisiae* tRNA[Ala], was sequenced by Holley et al. (108). Since then, more than 600 molecules of tRNA from a great variety of species have been purified and sequenced, and from X-ray crystallographic evidences their three-dimensional structure has been elucidated. The enzymes responsible for charging the amino acids to their tRNAs, called aminoacyl-tRNA synthetases, were discovered in 1955 (106), and the aminoacylation reaction has been studied in considerable detail.

The formation and function of aminoacyl-tRNAs are central to the process of protein synthesis. The charg-

TABLE 1. The genetic code

First position	Second position				Third position
	U	C	A	G	
U	Phe UUU	Ser UCU	Tyr UAU	Cys UGU	U
	Phe UUC	Ser UCC	Tyr UAC	Cys UGC	C
	Leu UUA	Ser UCA	stop UAA	stop UGA	A
	Leu UUG	Ser UCG	stop UAG	Trp UGG	G
C	Leu CUU	Pro CCU	His CAU	Arg CGU	U
	Leu CUC	Pro CCC	His CAC	Arg CGC	C
	Leu CUA	Pro CCA	Gln CAA	Arg CGA	A
	Leu CUG	Pro CCG	Gln CAG	Arg CGG	G
A	Ile AUU	Thr ACU	Asn AAU	Ser AGU	U
	Ile AUC	Thr ACC	Asn AAC	Ser AGC	C
	Ile AUA	Thr ACA	Lys AAA	Arg AGA	A
	Met AUG	Thr ACG	Lys AAG	Arg AGG	G
G	Val GUU	Ala GCU	Asp GAU	Gly GGU	U
	Val GUC	Ala GCC	Asp GAC	Gly GGC	C
	Val GUA	Ala GCA	Glu GAA	Gly GGA	A
	Val GUG	Ala GCG	Glu GAG	Gly GGG	G

ing reaction is in principle the step during which the RNA code is "translated" into the amino acid sequence, since amino acids are converted into RNA derivatives which then are ordered on the mRNA template by RNA-RNA interactions. tRNAs are involved in the two major steps of protein synthesis where accuracy is established, namely, the charging reaction and the binding of the aminoacyl-tRNAs to mRNA-programmed ribosomes. Because tRNAs are small, and because each interacts with a number of different proteins, principles of RNA structure and nucleic acid-protein interactions can be advantageously studied with these molecules. In this section I will review the structure of tRNAs and how they are charged with amino acids, and I will discuss current experimental approaches to major unsolved questions. The function of aminoacyl-tRNAs and how accurate charging is achieved will be addressed in later sections of this chapter. The expression and control of tRNA and synthetase genes are described in chapters 10 and 86 of this volume.

The Aminoacylation Reaction

The attachment of an amino acid to its cognate tRNA occurs in two steps, represented in the following simplified equations:

$$aa_i + ATP + RS_i \rightleftharpoons RS_i \cdot aa_i{-}AMP + PP_i$$
$$RS \cdot aa_i{-}AMP + tRNA_i \rightleftharpoons aa_i{-}tRNA_i + AMP + RS_i$$
$$aa_i + ATP + tRNA_i \rightleftharpoons aa_i{-}tRNA_i + AMP + PP_i$$

In the first step, the aminoacyl-tRNA synthetase (RS_i) activates the amino acid (aa_i) by forming a mixed anhydride, aminoacyl-adenylate, with the liberation of PP_i. The enzyme-bound aminoacyl-adenylate intermediate then reacts with tRNA, forming an aminoacyl ester with one of the vicinal hydroxyl groups of the 3'-terminal adenosine of the tRNA. The overall reaction is therefore the formation of an ester linkage accompanied by ATP hydrolysis to AMP and PP_i (reviewed by Ofengand [188]). The first reaction is usually assayed either by measuring aminoacyl-AMP

formation by gel filtration, or by measuring an amino acid-dependent $^{32}PP_i$ exchange into ATP. The second reaction is readily monitored by measuring aminoacyl-tRNA formation by cold trichloroacetic acid precipitation of radiolabeled amino acids. Each aminoacyl-tRNA synthetase binds its cognate amino acid and tRNA with a high degree of specificity. How the enzyme achieves this discrimination is addressed later in this chapter; how it recognizes tRNA molecules is treated below.

The reactions described above are reversible, with an overall equilibrium constant near 1. This indicates that the ester bond formed is of high energy, comparable to the PP_i bond of ATP. The ester bond is readily hydrolyzed, due to the presence of the positively charged amino group of the amino acid and the vicinal hydroxyl group of the adenosine. The PP_i produced in the reaction is subsequently cleaved to P_i, which may help to shift the reaction towards the right, resulting in more fully charged tRNAs. In effect, the aminoacylation reaction utilizes two high-energy phosphate bonds per amino acid attached. The initial ester product formed is either at the 2' or the 3' hydroxyl group, depending on the enzyme involved (see Table 2). Examination of model compounds suggests that the aminoacyl group migrates rapidly ($t_{1/2}$ in milliseconds) between the two hydroxyls (90), quickly establishing an equilibrium mixture of about 70% 3'-hydroxyl and 30% 2'-hydroxyl esters. However, a recent report claims that the rate of aminoacyl migration is 1 to 8 s^{-1}, which is substantially slower than the rate of protein synthesis (262). The site of initial aminoacylation appears to be the same for a given synthetase from a number of different species; such conservation of mechanism suggests special significance for the site of initial aminoacylation (251). Further work is required to determine whether the site of amino acid attachment influences the process of protein synthesis or plays some other role in the cell.

The affinities of the various synthetases and their cognate amino acids vary considerably, most falling in the range of 5 to 150 μM (Table 2). K_m values are even smaller in the overall reaction and are highly dependent on ionic strength. The high affinities for most amino acids help assure that when amino acids are limiting, they are used primarily for protein synthesis rather than for other metabolic processes. The affinities for tRNAs in the presence of Mg^{2+} generally are even higher, in the 1 to 0.01 μM range. For many synthetases, the aminoacyl-adenylate reaction occurs much faster than the overall rate of aminoacylation; therefore, subsequent steps must be rate limiting. Studies with isoleucine and valine tRNA synthetases indicate that in these cases release of newly formed aminoacyl-tRNA product is rate limiting (65, 305). However, for tyrosine tRNA synthetase the rate constants for aminoacyl-adenylate formation and transfer are quite comparable, and both may contribute to the rate-limiting step (72). It seems prudent to consider the detailed mechanism of each aminoacyl-tRNA synthetase individually, even though the overall reactions are so similar. More detailed descriptions of the mechanism of aminoacyl-tRNA formation are found in papers by Ofengand (188) and Schimmel and Söll (244).

TABLE 2. Aminoacyl-tRNA synthetases from *E. coli*[a]

Enzyme[b]	M_r ($\times 10^3$)	Subunit Structure	Subunit M_r	Level (mol per cell)	K_m (µM) Amino acids	K_m (µM) ATP	K_m (µM) tRNA	Amino acid attachment site
AlaRS	380	α_4	95,000		1,700	8	0.65	3'
ArgRS	60	α	60,000	691	10		2	2', 3'
AsnRS	100		100,000					3'
AspRS	124	α_2	62,000					3'
CysRS	44	α	44,000					2', 3'
GlnRS	63	α	63,023	1,013	150		0.2	2'
GluRS	52	α	52,000	873	100		0.08	2'
GlyRS	226	$\alpha_2\beta_2$	33,000 80,000	682	160	42	0.2	3'
HisRS	84	α_2	42,000		6	320		3'
IleRS	114	α	114,000	1,226	6	30	0.01	2', 3'
LeuRS	104	α	104,000	875	120	80	0.03	2'
LysRS	108	α_2	54,000 60,500	425	3,600	300		3'
MetRS	152	α_2	76,000		20	430	3	2', 3'
PheRS	248	$\alpha_2\beta_2$	36,757 87,024	992	10	300	0.2	2'
ProRS	94	α_2	47,000					2', 3'
SerRS	100	α_2	50,000		75	100	0.1	2', 3'
ThrRS	144	α_2	73,906	527				3'
TrpRS	74	α_2	37,000		20	300	0.4	2', 3'
TyrRS	95	α_2	47,500		25	4	0.03	2'
ValRS	110	α	110,000	536	128	30	0.05	2'

[a] Data are collected from a variety of sources, including Ofengand (188) and Neihardt et al. (177). The kinetic parameters reported here were determined under a variety of conditions; therefore, the values for the various enzymes cannot be compared directly and should only be used as a general indicator of the range of affinities observed.

[b] RS, tRNA synthetase.

tRNA Species and Primary Structure

About 15% of a bacterial cell's total RNA is tRNA, an amount which corresponds to about 10 to 15 tRNA molecules per ribosome, or about 400,000 molecules per bacterium. tRNAs range in size from 74 to 94 nucleotides and have a mass of about 25,000 daltons. They are composed of the usual nucleotides found in RNA, except that about 10% of them are modified post-transcriptionally. Over 50 different modified nucleotides have been identified in tRNA molecules. Descriptions of the modified nucleotides and the enzymes involved fall outside the scope of this chapter; instead, the reader should consult extensive reviews by Nishimura (181) and Bjork (17). Bulk tRNA is fractionated into pure species by a variety of techniques, e.g., by reverse-phase or ion-exchange chromatography (31, 206) and by two-dimensional polyacrylamide gel electrophoresis (76). Fractionation indicates that generally there are a number of different species specific for each amino acid. tRNAs charged with the same amino acid are called isoacceptor tRNAs; isoacceptor families range in size from two to six members. Since mature tRNAs can differ both in the sequence of the bases in the primary transcript and in the types and extents of base modification, convention dictates that different species are defined as those differing in primary sequence only. tRNA species are named by indicating the cognate amino acid as a superscript following the tRNA, and the specific number of the isoacceptor family is indicated by a subscript. For example, species 2 of valine tRNA is designated $tRNA_2^{Val}$. Estimates of the number of different tRNA molecules present in a bacterial cell based on fractionation patterns range from 60 to 75

species. Up to 54 tRNA genes have been mapped in *Escherichia coli*; a recent description of these genes is available (74). However, it is not yet known precisely how many tRNA species are found in *E. coli* cells.

The sequence determination of $tRNA^{Ala}$ reported by Holley et al. (108) utilized spectrophotometric methods to detect oligonucleotides and required milligram quantities of pure tRNA. Modern methods of sequencing utilize radiolabeled tRNAs or fragments thereof and require only microgram amounts of starting material. As a result, it is possible to sequence a tRNA molecule in only a few weeks, starting from a pure species extracted from a two-dimensional gel. However, sequencing is more complicated than that for most other RNA molecules, due to the large number of modified bases which must be identified. The following general strategy (215) is often followed. First, modified nucleotides are identified by two-dimensional thin-layer chromatographic analysis. Then oligonucleotides from complete pancreatic and T_1 RNase digestions are separated, labeled, and sequenced. Finally, a total sequence is determined with end-labeled tRNA by partial cleavage reactions followed by polyacrylamide gel electrophoresis to generate ladder patterns. Such procedures are usually sufficient to establish a precise sequence, although occasionally a more complex analysis is needed for certain species. The sequences of over 600 species of tRNA are collected (252); of these, 33 are tRNAs isolated from *E. coli*.

Secondary and Tertiary Structure of tRNAs

After the first sequence determination of a tRNA, it became apparent that the tRNA might fold into a

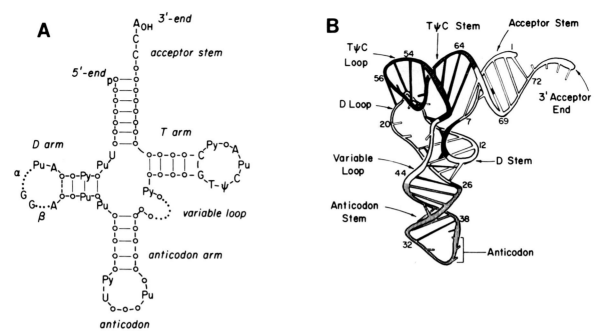

FIG. 1. Structure of tRNA. (A) Secondary structure of tRNA, showing constant and variable regions. Nucleotide residues are shown by circles except for invariant bases, which are identified by letters. Regions with variable numbers of nucleotides are shown by dotted lines. Symbols: Py, pyrimidine; Pu, purine; ψ, pseudouridine; T, ribothymidine. Adapted from Rich (226). (B) Three-dimensional structure of *S. cerevisiae* tRNA[Phe], based on X-ray crystallographic analyses. The ribose phosphate backbone is depicted as a coiled tube, and base-pair interactions are shown as cross ranges. Taken from Rich (225).

number of helical hairpin and loop conformations. Such a two-dimensional secondary structure resembles a cloverleaf (Fig. 1). All subsequently determined tRNA sequences possess the potential to form a similar structure, suggesting that the cloverleaf conformation may be common to all tRNAs. Constant features of tRNA structures are: (i) the 3′ terminus is always CpCpA, to which the amino acid attaches; (ii) the 5′ terminus is pG and is near the 3′ terminus due to the formation of a 7-base-pair stem; (iii) the anticodon is centrally located in a seven-member loop with a stem structure of 5 base pairs; (iv) another arm, called the T arm, contains a 5-base-pair stem and a seven-member loop with the constant sequence TψC; and (v), the third arm, called the D arm, has a variable-length stem and loop containing dihydrouridine residues. Other conserved features are seen as well and are discussed below. Some variability is observed in the D arm, and more is found in the region between the anticodon and T arms, called the variable loop, which contains from 3 to 21 bases.

Evidence for how the tRNA molecule is folded in three-dimensional space is obtained from X-ray diffraction analysis of tRNA crystals of *S. cerevisiae* tRNA[Phe] (140, 230). The molecule is shaped like an L, with the anticodon at one end and the amino acid attachment site at the other (Fig. 1). Each arm of the L-shaped molecule is about 60 Å (6 nm) long and 2 nm in diameter; the aminoacyl end and the anticodon are therefore separated by about 8 nm. All of the base-pair interactions predicted by the cloverleaf structure are found in the X-ray structure. The aminoacyl and T stems are stacked on each other to form one continuous double-helical arm, as are those of the anticodon and D arms. Many of the constant nucleotides interact

in this structure, thereby providing an explanation for their conservation (reviewed by Rich and RajBhandary [227] and Goddard [80]). One of the most striking features of the structure is the interaction between nucleotides which involved nonclassical hydrogen bonding. For example, the interaction of G_{18} (in the D loop) with ψ_{55} (in the T loop) involves hydrogen bonds between the hydrogens of the exocyclic amino group and N-3 of G_{18} and an exocyclic oxygen of ψ_{55}. Other examples involve the interaction of a base with two other nucleotides. The form of the three-dimensional structure indicates that RNAs can generate a great variety of stabilizing interactions. Although we know the rules which govern classical Watson-Crick hydrogen bonding as occurs in double-stranded DNA and RNA, we cannot predict so easily the nonclassical interactions and therefore cannot readily predict from a sequence alone what tertiary structure an RNA will assume.

X-ray diffraction studies also have been carried out on a number of other tRNA species: *S. cerevisiae* tRNA[Gly] (302); *S. cerevisiae* tRNA[Asp] (174); and *S. cerevisiae* and *E. coli* initiator tRNA[Met] (243, 300). The striking result is that the tertiary structures resemble *S. cerevisiae* tRNA[Phe] and one another quite closely. Some interesting differences are discernible, however. *E. coli* tRNA[Met] differs in the conformation of the anticodon loop and the acceptor stem, perhaps thereby conferring its uniqueness as the initiator tRNA. In *S. cerevisiae* tRNA[Asp], the anticodon is self-complementary and interacts in the crystal structure. The tRNA[Asp] conformation differs significantly in the elbow region of the L-shaped molecule; base pairing between the T and D loops is lacking, and these sequences are more open or exposed. Furthermore, tRNA^{Gly} crystallizes in two distinct conformations, depending on the solvent used. It is tempting to propose that the various tRNA structures stabilized in the crystals reflect different conformational states actually assumed during protein synthesis. One of the current challenges for X-ray crystallographers is to define the possible differences in tRNA structures which are caused by specific interactions with other macromolecular components of translation.

Solution studies of tRNA also provide information concerning the structure and dynamics of these molecules. The thermal unfolding of tRNAs has been studied by T-jump relaxation kinetics, in which an absorbance increase is measured as a function of time (49). Increasingly, however, high-resolution nuclear magnetic resonance (NMR) techniques are being applied to the study of tRNA structure (reviewed by Reid [220]). The extreme low field region of the proton NMR spectrum (-15 to -11 ppm) contains the ring N-H or imino protons. Since these are involved in base pairing, changes in their resonances reflect changes in H-bonding. Redfield and colleagues (218) developed pulsed Fourier transform NMR methods that suppress the resonance of H_2O and allow the use of double-resonance and time-resolved techniques. It is now possible to assign all of the imino proton signals of the spectrum to specific nucleotides of a tRNA, as has been done recently for tRNA^{Val} (98). Since each H-bonding interaction can in principle be monitored independently, NMR is a powerful tool for defining conformational changes occurring during thermal unfolding or interactions with other molecules. For example, it has provided direct evidence for the existence of a G-U base pair in S. cerevisiae tRNA^{Phe}. Unfolding of this tRNA was shown to initiate by melting the A-U base pairs of the acceptor stem and the extra loop interactions with the D loop and stem, followed by melting of the rest of the acceptor stem; subsequently, the D stem, anticodon stem, and the T stem melt sequentially (130).

Aminoacyl-tRNA Synthetases

The aminoacyl-tRNA synthetases are the class of enzymes that catalyze the attachment of an amino acid to its cognate tRNAs. Aminoacyl-tRNA synthetases were first identified in mammalian tissues (106). In E. coli, there is a single aminoacyl-tRNA synthetase for each amino acid, except for Lys, where two genes (lysS and lysU) appear to code for four forms of the enzyme (105). The enzymes as a class are of moderate abundance in bacteria; each is present in about one copy per 20 ribosomes, or as 800 molecules per cell grown in rich medium (177). They play an obviously important role in the pathway and accuracy of protein synthesis and are of great interest for studying protein-nucleic acid interactions. In addition, some aminoacyl-tRNA synthetases possess other functions, for example: alanine and threonine tRNA synthetases affect the expression of their encoding genes (see chapter 86); lysine tRNA synthetase and a number of other synthetases catalyze the synthesis of 5′,5′′′-diadenosine tetraphosphate, possibly an important regulatory metabolite in cells (309); still others are involved in amino acid transport (173) or the posttranslational aminoacylation of proteins (55).

To study the structure and function of aminoacyl-tRNA synthetases, pure proteins are required. Baldwin and Berg (11) first purified isoleucine tRNA synthetase from E. coli, using as an assay the attachment of radioactive isoleucine to tRNA. Subsequently, essentially all of the synthetases have been identified and characterized (Table 2). Because all aminoacyl-tRNA synthetases catalyze very similar reactions and interact with tRNAs possessing comparable tertiary structures, it was anticipated that the size and overall structural design of each enzyme might be similar, possibly reflecting a common progenitor. However, the E. coli enzymes vary enormously in size, from tryptophan tRNA synthetase (2×37 kilodaltons [kDa]) to alanine tRNA synthetase (4×95 kDa). Furthermore, these proteins fall into four classes of quaternary structure, α_1, α_2, α_4, and $\alpha_2\beta_2$, and their subunit polypeptides vary from 33 to 110 kDa. A number of synthetases interact with Zn^{2+}: the isoleucine and methionine enzymes contain one molecule of bound Zn^{2+} per active site. Zn^{2+} inhibits the aminoacylation activity of phenylalanine tRNA synthetases, whereas it stimulates its synthesis of 5′,5′′′-diadenosine tetraphosphate (168). The large size of some of these proteins and the difficulty in obtaining large amounts of them due to their low cellular abundance and the ease of their denaturation hampered and complicated their sequencing by conventional protein chemistry. Recently, about 10 of the genes encoding aminoacyl-tRNA synthetases have been cloned and sequenced (see chapter 86). Recombinant DNA techniques not only provide a route to determining the primary sequences of the enzymes, but also allow the construction of overproducing strains for the preparation of large amounts of protein.

Early studies on the synthetases suggested that a unifying principle of structure could be discerned, namely, that each enzyme is constructed from partially mutated repeats of a 35- to 50-kDa polypeptide (reviewed by Ofengand [188]). Peptide analyses indicated that the larger polypeptides contained internal repeats, but this view can no longer be supported. For example, the primary structure of methionine tRNA synthetase (677 residues or 76 kDa per monomer) contains no significant repeats, suggesting that this protein did not evolve by gene duplication (52). Nor do the primary sequences of the aminoacyl-tRNA synthetases determined to date show extensive homologies with one another, ruling out the notion that these enzymes are highly conserved proteins originating from a common progenitor. The strongest homologies yet detected between E. coli enzymes are an 11-amino acid match between the isoleucine and methionine tRNA synthetases and, to a lesser extent (8 of 11), with glutamine tRNA synthetase (284). A 44% homology over about 400 amino acids has been reported for the E. coli and S. cerevisiae methionine tRNA synthetases, with higher homology within the nucleotide binding domain (282). These homologies occur in the N-terminal region of the enzymes and in the mononucleotide-binding domain, which suggests a common precursor for this function. The C-terminal domain of methionine tRNA synthetase thought to be involved in the binding of the 3′ end of the tRNA shows some homologies with comparable regions in the isoleucine, tryptophan, and tyrosine enzymes (111).

An alternative hypothesis is that each synthetase contains a core responsible for the aminoacylation reaction, comparable in size to the smallest enzyme, tryptophan tRNA synthetase (35 kDa; 334 amino acids). Larger synthetases appear to have added onto the core protein domains which may modulate core activity or serve some other function (122). When E. coli methionine tRNA synthetase (a dimer of 76-kDa proteins) is treated with trypsin, a monomeric N-terminal 60-kDa fragment can be isolated which retains activity (40, 143). The cleaved C-terminal domain appears to be essential for dimer formation but not for aminoacylation activity. Because proteolysis generates only a limited set of products, Schimmel and co-workers have constructed in vitro specific gene deletions to study the minimal size required for alanine tRNA synthetase activities (123). An N-terminal, truncated protein of only 385 residues (out of 875) catalyzes the formation of aminoacyl-adenylates; aminoacylation of tRNA requires 461 residues, and oligomerization requires 808 residues. Again, the N-terminal region of alanine tRNA synthetase possesses the enzymic activity, and the C-terminal region possesses that for oligomerization. The DNA-binding region of the protein utilized in regulating the alanine tRNA synthetase gene (see chapter 86) is not yet known.

A number of aminoacyl-tRNA synthetases from a variety of species have been crystallized and analyzed by X-ray diffraction techniques (reviewed by Blow and Brick [18]). The first reported crystallization was with S. cerevisiae lysine tRNA synthetase, but the crystals were very sensitive to radiation (a rather common problem) and therefore were not suitable for analysis. A truncated form of E. coli methionine tRNA synthetase was crystallized and analyzed at 0.25-nm resolution; an elongated, biglobular structure with three domains was deduced (310). Another approach to obtaining stable crystals is to employ synthetases from the thermophile Bacillus stearothermophilus. A high-resolution structure at 0.21 nm has been obtained for tyrosine tRNA synthetase (18), and crystals of the tryptophan enzyme are being studied (35).

Analysis of tyrosine tRNA synthetase indicates three domains: the N-terminal domain (residues 1 through 220), which is highly ordered; an α-helical domain (residues 250 through 320); and a C-terminal domain (99 residues), which is unresolved presumably because it is unordered. The N-terminal domain contains a classical mononucleotide binding fold (23) where tyrosyl-adenylate binds as shown by cocrystallization of the substrate or its analogs. There is a Cys residue involved in the active site of tyrosine tRNA synthetase whose blocking by sulfhydryl reagents inhibits the enzyme. X-ray diffraction data show the Cys-35 residue in the binding site for the tyrosyl-adenylate, where it forms a weak H-bond with adenine. Site-directed mutagenesis was employed to change the Cys-35 to Ser, Ala, or Gly; the enzyme remained active, although the K_m for ATP was increased (293). With the methionine tRNA synthetase fragment, a distinctive mononucleotide-binding fold also is seen in the N-terminal domain, the putative aminoadenylate site (228). Comparison of the E. coli methionine tRNA synthetase structure with that of B. stearothermophilus tyrosine tRNA synthetase (16) shows considerable homology at the level of tertiary structure, especially in the putative mononucleotide-binding domains. It is possible that many synthetases possess similar domains of secondary and tertiary structure even though extensive homologies in primary structure are not discernible.

Recognition of tRNAs by Synthetases

Each aminoacyl-tRNA synthetase recognizes and aminoacylates the members of its cognate isoacceptor tRNA family, yet highly discriminates against all other tRNAs. This recognition event is one of the key events in translation, since information in a nucleic acid is first linked to an amino acid. Kinetic analyses of tRNA binding to synthetases (reviewed by Schimmel and Söll [244]) showed a rapid, nearly diffusion-limited binding; association constants for cognate tRNAs generally fall in the vicinity of 10^7 M^{-1}, whereas noncognate species bind more poorly. The binding reaction is entropy driven, suggesting large solvation changes typical of protein-nucleic acid interactions. However, some noncognate tRNAs bind with an association constant only one or two orders of magnitude less than the cognate species. In these cases, the reaction velocity of aminoacylation is 10^3- to 10^4-fold faster for the cognate tRNAs, thereby achieving a high overall discrimination. Whether or not a covalent Michael adduct between an aminoacyl-tRNA synthetase and its tRNA forms, as has been suggested for alanine tRNA synthetase (254, 255), is unclear.

What is the molecular basis of such tRNA-synthetase interactions, and what features of tRNA structure are recognized by the synthetase that allow discrimination? These questions are vital topics of on-going research, not only because of their importance in understanding the mechanism of translation, but also because the system is a superb model of protein-nucleic acid interactions in general and is amenable to experimentation.

Two general approaches have been used to study these questions: (i) specific aspects of the covalent structure of tRNAs are correlated with the ability to be recognized by a synthetase; and (ii) a direct characterization of tRNA-synthetase complexes is made by physical and chemical methods. In the first, a wide variety of strategies were employed. Researchers sought to identify a short region of RNA sequence shared by isoacceptor species, but common patterns failed to emerge. Instead, changes in tRNA structure that cause an alteration in binding by the synthetase were sought by genetic mutation of tRNA genes. However, the power of the genetic approach to studying tRNA structure/function was found to be severely limited by the fact that even small alterations in tRNA structure often lead to failure of the cell to process the primary transcript to mature tRNA. An alternative approach is to employ chemical and enzymological methods to cut away or replace precise regions of the tRNA structure. Such studies, described in detail by Schimmel and Söll (244) and Ofengand (188), indicate that the aminoacyl stem and terminus, the D stem, and the anticodon all are important for synthetase recognition. A change in a single nucleotide can shift tRNA recognition from one synthetase to another. Some enzymes (e.g., methionine and valine tRNA synthetases) interact with the anticodon region, whereas others (e.g., phenylalanine tRNA synthetase) do not. More recently, site-directed mutagenesis of tRNA genes, coupled with their expression in in vitro transcription systems (thereby obviating the necessity of processing), is enabling researchers to methodically alter tRNA structure and deduce those features important for synthetase binding (161). A severe limitation in these approaches derives from the possibility that a change in one region of the tRNA causes secondary changes elsewhere in the molecule that cause the altered reactivity.

The second approach, physical studies, has provided important information about these interactions, with respect to both the tRNA and the synthetase structures. In early work, tRNA-synthetase complexes were probed to determine tRNA sites accessible to oligonucleotide binding or nuclease digestion (footprinting); alternatively, the two macromolecules were cross-linked with bifunctional reagents or photochemically, and the sites of cross-links were identified. The results (reviewed by Ebel et al. [63]) indicate that many synthetases span the tRNA molecule from the anticodon to the 3' terminus, interact with the D stem, and therefore appear to cover the inner concave surface of the three-dimensional L conformation. Another relevant biophysical approach is neutron scattering, which provides information on the radius of gyration of either protein or RNA. Studies with valine tRNA synthetase in the absence of Mg^{2+} and at low ion strength show that the $tRNA^{Val}$ binds close to the center of mass of the complex and induces a conformational change in the enzyme, which then appears more contracted (308). However, binding of Phe-tRNA ($Phe-tRNA^{Phe}$) to phenylalanine tRNA synthetase in the presence of Mg^{2+} does not alter the radius of gyration of the protein (54).

A more direct approach to determining the structural elements involved in tRNA-synthetase interactions is to crystallize the complex and analyze the crystals by X-ray diffraction techniques. A complex of aspartate tRNA synthetase and $tRNA^{Asp}$ from *S. cerevisiae* has been crystallized (159); unfortunately, the crystals have a large unit cell which is unfavorable for X-ray diffraction work but in spite of this may provide a structure of 0.7-nm resolution. The X-ray crystallographic approach nevertheless remains the most powerful method available; its success awaits suitable crystals for analysis.

INITIATION

Initiation of protein synthesis is a series of reactions which occur before the first peptide bond is formed. Ribosomes dissociate into 30S and 50S subunits; a 30S preinitiation complex is formed with mRNA and a unique initiator tRNA, N-formyl-methionyl (fMet)-$tRNA_f^{Met}$ (fMet-tRNA); and the 50S subunit joins to form the 70S initiation complex, which is capable of entering the elongation phase of protein synthesis. These reactions are promoted by three proteins, called initiation factors IF1, IF2 and IF3, and by GTP, which is hydrolyzed to GDP and P_i. During initiation, two important requirements of protein synthesis are served: (i) the initiation region of a particular cistron on a mRNA is recognized and selected, thereby dedicating a ribosome to the translation of the cistron; and (ii) the process of protein synthesis is "phased" so that the mRNA is translated by the ribosome in the proper reading frame. Thus the formation of initiation complexes must be precise, and the process must distinguish correct initiation regions from all other sequences in the mRNA.

Initiation is presumed to be the phase of protein synthesis that is rate limiting in the bacterial cell. Evidence for this assertion is weak, however. In cases where synthesis rates do not correspond to mRNA levels, such differences can be rationalized in terms of the apparent "strengths" of the initiation signals in the mRNA. Furthermore, mutations in initiation regions often result in large changes in the rate of synthesis of the protein, some leading to more rapid synthesis while others lead to reduced synthesis. However, precise measurements of in vivo initiation rates are lacking. Nevertheless, because of known examples of translational control operating at the level of initiation in bacteria (see chapter 78), there is high interest in elucidating the pathway and mechanism of initiation and in identifying the rate-limiting step. In this section I will review the components of initiation and their pathway and mechanism of interaction. For greater details, recent reviews should be consulted (82, 91, 92, 145, 166).

Components

An essential step in defining the pathway of initiation of protein synthesis is identification of the components of the reactions. A genetic approach, involving the isolation of mutants defective in protein synthesis and the identification of the genes or their products, has provided useful information concerning the aminoacyl-tRNA synthetases, ribosomal proteins, and elongation factors, but not about initiation factors because very few mutations affecting initiation have been found. The only viable approach has been to reconstruct active in vitro systems for initiation, using purified components. Cell-free bacterial cell lysates are capable of incorporating amino acids into protein, measured using radioactive amino acid precursors. Nirenberg and Matthaei (180) showed that the synthetic mRNA poly(U) stimulated ribosomes to synthesize polyphenylalanine. Fractionation of the lysate by ultracentrifugation results in a pellet of ribosomes and a supernatant (called S-100) containing proteins and tRNA; both fractions are required for protein synthesis. When the ribosomes are suspended in high-salt buffer and centrifuged again, the pellet fraction combined with the S-100 is active for polyphenylalanine synthesis, but it cannot translate viral RNAs unless the high-salt ribosomal wash fraction is added. From the wash, three initiation factors were isolated which are required for the translation of viral mRNA. Simplified in vitro assays for the initiation phase were developed, namely, the binding of radiolabeled fMet-tRNA to 70S ribosomes or the synthesis of fMet-puromycin, a model reaction of the first peptide bond-forming reaction. The components required for these reactions are listed in Table 3. De-

TABLE 3. Components for in vitro assays

Component	Protein synthesis	fMet-tRNA–70S binding
Ribosomes (washed)	+	+
Initiation factors	+	+
Elongation factors (in S-100)	+	
Release factors (in S-100)	+	
Aminoacyl-tRNAs		
20 tRNA synthetases (in S-100)	+	
All tRNAs (in S-100)	+	
20 amino acids (in S-100)	+	
fMet-tRNA	(+)	+
mRNA	+	+
Energy: ATP + GTP	+	+
Salts: Mg^{2+}, K^+	+	+
Reducing agent	+	+
Buffer (ca. pH 7.4)	+	+

scriptions of the structure and functional characteristics of the ribosomes, initiator tRNA, mRNA, and initiation factors follow.

Ribosomes. Bacterial ribosomes comprise three rRNAs and 53 proteins. The composition and structure of these particles have been reviewed in detail (297, 298) and are described elsewhere in this volume (chapter 10). Functionally active ribosomes are purified from cell lysates by ultracentrifugation. The initial ribosomal pellet is heavily contaminated with numerous nonribosomal proteins, which may be removed by recentrifugation in a high-salt buffer. Frequently, high-salt-washed ribosomes are deficient in ribosomal proteins and are only partially active. Upon dissociation of the subunits in low-Mg^{2+} buffer, they may undergo conformational changes leading to further inactivation. Noll and Noll (183) developed a method for obtaining highly purified ribosomal subunits that avoids low-Mg^{2+} and high-salt conditions. Such preparations, called A-type ribosomes or "tight couples," associate to 70S ribosomes at 6 mM Mg^{2+}, but readily dissociate at 1 to 2 mM Mg^{2+}. They are "vacant couples," lacking bound peptidyl-tRNA and mRNA fragments, and are nearly 100% active in a variety of in vitro assays for protein synthesis. Nevertheless, even such preparations exhibit heterogeneity in their ribosomal protein composition.

fMet-tRNA. All bacterial proteins are synthesized initially with an fMet N terminus, followed by the removal of the formyl group and sometimes the methionine posttranslationally. There are at least two species of tRNA^{Met}: one, tRNA$_f^{Met}$, is involved uniquely in initiation (167); the other, tRNA$_m^{Met}$ is involved in supplying methionine to internal positions in the protein (44). Both tRNAs recognize the unique AUG code word for methionine and both are aminoacylated by the same synthetase, methionine tRNA synthetase, whereas only Met-tRNA$_f^{Met}$ is formylated with N^{10}-formyl-tetrahydrofolic acid and a transformylase activity (57). The sequences of tRNA$_f^{Met}$ and tRNA$_m^{Met}$ are strikingly different except in their anticodons, which are C-A-U and N^4-acetyl-C-A-U, respectively (12). The initiator species possess an unmodified A$_{37}$ residue adjacent to the 3' base of the anticodon, and its 5'-terminal C residue cannot form hydrogen bonds to the aminoacyl stem.

In contrast, tRNA$_m^{Met}$ resembles most other tRNAs in possessing a hypermodified A$_{37}$ residue adjacent to the anticodon and a 5'-terminal G residue hydrogen bonded to its stem. The nonmodified A$_{37}$ may allow tRNA$_f^{Met}$ to interact with initiator codons such as GUG which differ in the first code letter. However, whether the non-hydrogen-bonded 5'-terminal C residue plays a role is doubtful since no change in function is seen on its conversion to a U which can bind to the stem. In further attempts to identify unique aspects of the tRNA$_f^{Met}$ structure which could account for its specific interactions, initiator tRNA sequences from a variety of bacterial and eucaryotic species were compared; four G-C base pairs were found in identical positions (30). One of these (in the aminoacyl stem) appears to be required for recognition by the transformylase enzyme. *E. coli* tRNA$_f^{Met}$ has been crystallized, and its three-dimensional structure has been determined (300). The structure resembles the L conformation of other tRNAs, but the conformation of the anticodon loop is unusual, a structure consistent with S1 nuclease sensitivity studies (301). As with most tRNAs, the precise structural features that confer specific properties to the molecule and allow discrimination by interacting proteins are not yet known.

The formylation of Met-tRNA$_f^{Met}$ is thought to be obligatory for efficient initiation in *E. coli*. Although a mutant *E. coli* strain was found to initiate protein synthesis in vivo in the absence of formylation (100), it was later shown to have a structural change in tRNA$_f^{Met}$ that may be essential for its activity (13). A requirement for formylation also has been demonstrated in vitro with a DNA-dependent transcription/translation assay (148).

mRNA. Procaryotic mRNAs frequently are polycistronic, possessing a number of translational start and stop signals which give rise to several discrete protein products from the same mRNA. mRNAs begin with a 5'-triphosphate, and up to half of the molecules whose transcription is completed are polyadenylated at the 3' terminus (84). Because translation is "coupled" to transcription, a ribosome initiates soon after the first initiation site of the proximal cistron emerges from the RNA polymerase, before transcription terminates and the nascent mRNA leaves the DNA template. We are concerned here with the structure of the initiation sites of these cistrons.

In early studies, regions of mRNAs protected from nuclease digestion by initiation complexes were sequenced to elucidate the ribosome binding sites (256). As mRNA sequences become available, they are being analyzed statistically to distinguish features common to initiation regions (reviewed by Gold et al. [82]). The results of these efforts show that proximal cistrons possess a 5' nontranslated region and most genes begin with the initiator codon AUG. However, AUG is not used invariantly as the initiator codon; GUG is employed in about 10% of the mRNAs, UUG is used less frequently, and AUU is found in one case. The triplets AUG, GUG, and UUG promote fMet-tRNA binding to 70S ribosomes in vitro (44), and all three function as reinitiation sites following nonsense codons. In vitro analysis of the efficiency of initiation with synthetic mRNAs indicates that GUG, UUG, and ACG may function as initiator codons, but their efficiencies are substantially less than that of AUG (266). It is not entirely apparent why so many different initiator codons are used or what purpose they serve. Perhaps the in vivo efficiency of initiation is modulated by their use, but this has not yet been demonstrated directly.

Further analysis of ribosome binding site sequences led to the discovery of a purine-rich region with 10 nucleotides upstream from the initiator codon. The canonical sequence is GGAGG and is named the Shine-Dalgarno (S-D) region after its discoverers (248). The S-D region is thought to base pair with the 3' terminus of 16S rRNA in the 30S ribosomal subunit. Its function in initiation is discussed in greater detail below. Further nonrandom sequences are a preferred A at position −3, the preferred sequences GCUA or AAAA directly following the initiator codon (positions +4 to +7), and very few G or A residues between the initiator codon and the S-D region (reviewed by Kozak [145] and Gold et al. [82]). No other features of mRNA sequence or secondary structure affecting initiation are discernible at this time. Presumably the "strength" of an mRNA's cistron is determined by the choice of elements grouped in the ribosome binding site, resulting in a wide range of possible translational efficiencies.

Initiation factors. Fractionation of the proteins in the ribosomal high-salt wash fraction of *E. coli* cells resulted in the identification of three initiation factor activities essential for maximal stimulation of fMet-tRNA binding to 70S ribosomes in the presence of mRNA (120, 221). Classical ion-exchange and molecular-sieve chromatography have been used to purify the three factors to near homogeneity (104). About 5 mg of IF1 (8.1 kDa), 30 mg of IF2α (97.3 kDa), 10 mg of IF2β (79.7 kDa), and 10 mg of IF3 (20.7 kDa) are obtained from 1 kg (wet weight) of *E. coli* cells. Immunoblotting of cell lysate proteins fractionated by two-dimensional polyacrylamide gel electrophoresis (113) indicated that only a single form of IF1 and IF3 exists in *E. coli* MRE600 and that these proteins precisely correspond to the purified factors. Two forms of IF2 corresponding to IF2α and IF2β are detected in a 2:1 ratio, but smaller forms are readily generated if precautions against proteolysis are not taken. Amino acid sequences have been determined for IF1 (212) and IF3 (24) by classical protein chemical procedures. Subsequently, the genes for all three factors have been cloned and sequenced (see chapter 86), confirming the primary structures for IF1 and IF3 and providing the sequences for IF2α and IF2β (241). Both IF2 forms are expressed from the same gene by two independent translational initiation sites (209). The levels of initiation factors in exponentially growing cells are coordinately regulated with those of ribosomes; about 0.15 to 0.20 molecule for each factor is present per ribosome (112). All three factors can be overexpressed in cells carrying multiple gene copies. This should allow the easy preparation of large amounts of the factors for physical and chemical studies (see, for example, reference 59).

Knowledge about the functions of initiation factors comes from in vitro assays for initiation (reviewed by Maitra et al. [166] and Gualerzi and Pon [93]). IF2 is implicated in fMet-tRNA binding to 30S ribosomes

TABLE 4. Initiation factors

Factor	M_r	No. of amino acids	Gene	Cell level: factors per ribosome	Function
IF1	8,119	71	*infA*	0.18	Promotes IF2 and IF3 functions
IF2α	97,300	889	*infB*	0.10	Binds fMet-tRNA and GTPase
IF2β	79,700	732	*infB*	0.05	
IF3	20,668	181	*infC*	0.14	Ribosome dissociation; mRNA binding

and possesses a ribosome-dependent GTPase activity. IF3 prevents the association of ribosomal subunits and is required for the binding of natural mRNAs. The role of IF1 is less clear, but it appears to enhance the rates of ribosome dissociation and association and the activities of the other initiation factors. Both IF2α and IF2β are active in simple assays for initiation, and both stimulate coat protein synthesis with bacteriophage R17 RNA. However, only IF2α promotes the synthesis of β-galactosidase in a DNA-linked assay system (67), and only IF2α binds to RNA-Sepharose (58). The presence of both forms of IF2 in numerous species of the *Enterobacteriaceae* (114) suggests a special purpose for the two forms, but different functional roles are not yet demonstrated.

To elucidate the mechanism of action of initiation factors, their protein structures have been studied. Some physical and chemical characteristics of the factors are summarized in Table 4. IF1 and IF3 are small basic proteins, whereas IF2α and IF2β are larger, acidic factors. Pawlik et al. (205) predicted from the IF1 primary sequence that this factor has a low content of α-helix (5 to 11%) but possibly a higher content of β-sheets. Proton NMR spectroscopy also suggests a rather simple structure with high flexibility and internal mobility. Although the crystallization of a protein called IF1 has been reported (154), it appears from the amino acid composition that the protein is not authentic IF1. The primary structures of IF2α and IF2β show regions of homology with the GTP-binding domains of elongation factor Tu (EF-Tu) and EF-G. IF2α possesses a small N-terminal domain attached to a much larger C-terminal domain by a highly charged linker sequence with curious repeating sequences (241), whereas IF2β lacks the N-terminal domain and half of the linker region. IF3 appears to be elongated, with an axial ratio of 3.5:1 (93). Proton NMR spectroscopy indicates that IF3 has a highly folded structure with large hydrophobic regions. Gualerzi and coworkers have identified a variety of amino acid residues implicated in initiation factor functions by using chemical modification and spectrophotometric techniques (94, 210). These studies, along with knowledge of the primary sequences, form a basis for more detailed studies on the structure and function of these proteins. However, none of the initiation factors has yet been crystallized. Therefore we lack the precise three-dimensional structures which are necessary for understanding how these proteins act at the molecular level.

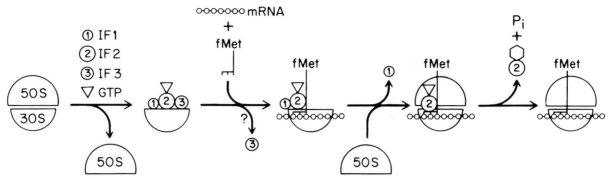

FIG. 2. Pathway of initiation. The steps of the pathway are described in the text. Symbols used are identified in the figure except for GDP (○).

Pathway of Initiation Complex Formation

The pathway for the formation of the 70S initiation complex involves numerous individual reactions, the large majority of which are noncovalent in character: dissociation of ribosomes into their subunits; binding of fMet-tRNA, mRNA, and 30S subunits to one another; and the ribosomal binding and release of the three initiation factors. The only covalent step is the hydrolysis of GTP to GDP and P_i. I propose here a simplified model for the temporal course of events (Fig. 2), describe the experimental basis for the model, and discuss possible alternate pathways.

Ribosome dissociation. Ribosome dissociation into subunits occurs between individual rounds of protein synthesis. This was demonstrated in vivo by measuring subunit exchange between "light" and "heavy" ribosomes in cells shifted from a medium containing heavy isotopes (^{3}H, ^{13}C, ^{15}N) to one with light isotopes (136). The idea that initiation may involve ribosomal subunits rather than the 70S particle was reinforced by showing that fMet-tRNA binding in vitro proceeds only with subunits (185).

A 70S ribosome is in dynamic equilibrium with its subunits, as follows:

$$70S \underset{k_{-1}}{\overset{k_1}{\rightleftharpoons}} 30S + 50S$$

The extent of dissociation is measured by light-scattering techniques (115) or by sucrose gradient centrifugation (183). Both k_1 and k_{-1} are greatly influenced by the Mg^{2+} concentration. At about 5 mM Mg^{2+} (thought to approximate the physiological concentration), ribosomal subunits are primarily in the associated state, whereas at Mg^{2+} concentrations lower than 1 mM, dissociation occurs. The equilibrium is also temperature dependent; the association reaction is exothermic ($\Delta H = -19.9$ kcal [ca. 83.3 kJ] mol^{-1}), and the entropy change is negative. These results suggest that the subunits interact primarily between their RNAs. Initiation factors also influence the equilibrium position of the ribosomes (81). IF3 readily binds to 30S subunits and thereby greatly decreases the rate of k_{-1}. In effect, IF3 acts mechanistically as an antiassociation factor rather than as a dissociation factor. IF1 increases the rates of both k_1 and k_{-1}, thereby enhancing the rate of attainment, but not the position, of equilibrium. IF2 appears to promote the association of subunits. In the presence of all three initiation factors, 70S ribosomes are largely dissociated at 5 mM Mg^{2+}.

Initiation factor binding. The binding of initiation factors to ribosomes or their subunits was first shown by preparing radioactive initiation factors and analyzing binding mixtures by sucrose density gradient centrifugation (103). When tested alone, IF3 binds well to 30S subunits, whereas IF2 binding is weaker and no binding is detected with IF1. However, one molecule of each of the three factors binds tightly when all are added to 30S subunits (reviewed by Maitra et al. [166]). These findings were confirmed and quantitated by using fluorescence polarization measurements to detect ribosome complexes with fluorescence-labeled factors (285, 312). The equilibrium association constants for IF1 and IF2 with 30S subunits are strongly enhanced by the presence of the other factors; all three factors exhibit values between 3×10^7 and 3×10^8 M^{-1}. Little or no binding of factors to 50S subunits occurs. From the association constants and the in vivo concentrations of initiation factors (ca. 1 μM) and free ribosomes (those not in polysomes; ca. 1 μM), it is calculated that the native 30S subunits are essentially saturated with all three initiation factors in intact cells. The order of binding of the initiation factors to 30S subunits is not known. This will require determination of the rates of binding of each factor in the various combinations possible. The rate of IF3 binding to 30S subunits has been measured by stopped-flow fluorescence spectrophotometry (88); rate-constant values determined at 6 mM Mg^{2+} for binding (25 μM^{-1} s^{-1}) and release (0.88 s^{-1}) indicate rapid binding and are consistent with the equilibrium constant measured by static titrations.

Initiation factors are capable of interacting with other components of the initiation pathway. The formation of a complex of IF2 and fMet-tRNA in the absence of ribosomes can be detected by glycerol gradient centrifugation, protection against deacylation, and a variety of other techniques (reviewed by Maitra et al. [166].) This raises the possibility that fMet-tRNA is presented to the 30S ribosome as a ternary complex with IF2 and GTP, as opposed to the alternative view that fMet-tRNA binds to the trifactor-30S ribosomal complex. This problem can best be resolved by a kinetic analysis of the two possible pathways. Gualerzi and Wintermeyer have used stopped-flow fluorescence techniques to measure the rate of acetylphenylalanine (AcPhe)-tRNA binding to poly(U)-programmed 30S subunits. The results indicate that IF2 interacts first with the 30S subunit, not with the tRNA (C. Gualerzi and W. Wintermeyer, personal communication). IF3 is capable of binding to RNAs and therefore may interact with mRNA before binding to the ribosome. IF3 is reported to recognize the initiator codon, AUG, at physiological salt concentration (290), but generally appears to bind RNAs nonspecifically (291). No compelling evidence exists for a role of IF3 in mRNA recognition in the absence of ribosomes. The binding of IF2 and IF3 to fMet-tRNA and mRNA, respectively, in the absence of ribosomes may reflect interactions that actually occur on the surface of the 30S subunit, as discussed below.

Formation of the 30S preinitiation complex. The 30S preinitiation complex consists of a 30S subunit, fMet-tRNA, mRNA, GTP, and the initiation factors. Its formation is readily measured with radioactive fMet-tRNA or mRNA; the complex is separated from the labeled reagents by filtration through nitrocellulose or by sucrose density gradient centrifugation. It is presumed that the anticodon of the fMet-tRNA hydrogen bonds directly to the initiator codon in the mRNA. Therefore, both the selection of mRNAs and their precise phasing for translation are accomplished at this step. The major conceptual issues are as follows: (i) How is the initiator region of the mRNA recognized, and what interactions are involved in mRNA binding? (ii) How do the initiation factors promote preinitiation complex formation? (iii) What is the order of binding of fMet-tRNA and mRNA, and is the order obligatory or random? The first two issues are addressed in detail later in this section. We are concerned here primarily with the third.

Evidence has been obtained for the formation of functional binary complexes of either 30S-mRNA or 30S–fMet-tRNA. A complex forms between bacteriophage MS2 RNA and 30S subunits which subsequently can be "chased" into 70S ribosomal complexes actively synthesizing MS2 coat protein, even in the presence of inhibitors of mRNA binding (276). Formation of the MS2 RNA binary complex (in the absence of initiation factors) is inhibited by an oligodeoxynucleotide (8-mer: AAGGAGGT) which binds to the 3' terminus of 16S rRNA if the oligomer is added before, but not after, the MS2 RNA (10). These experiments suggest that the binary mRNA-30S subunit complexes are intermediates in the initiation pathway. In contrast, Kaempfer and colleagues argue that an fMet-tRNA–30S subunit complex forms before mRNA binding and that the binary complex is an obligate intermediate (124, 125). They showed that such binary complexes can be chased into productive 70S initiation complexes, whereas bacteriophage R17 RNA-30S binary complexes are nonproductive.

Kinetic analysis of 30S preinitiation complex formation with either AcPhe-tRNA and poly(U) or fMet-tRNA and poly(A,U,G) led Gualerzi and co-workers to propose that the tRNA and mRNA bind independently through a random-order mechanism (95). In contrast, Ellis and Conway (66) recently measured the initial rates of 30S preinitiation complex formation at limiting fMet-tRNA and R17 RNA concentrations and concluded that a nonrandom order occurs, with fMet-tRNA binding preceding mRNA binding. In kinetic studies with both synthetic and natural (viral) mRNAs, the rate-limiting step is not the binding of either tRNA or mRNA, but rather is a rearrangement step following the formation of the ternary complex. The stability of the 30S preinitiation complex is significantly greater after the rearrangement or conformational change, and the tRNA or mRNA do not rapidly exchange with their free, unbound forms. Wintermeyer and Gualerzi (294) studied the kinetics of AcPhe-tRNA binding to poly(U)-programmed 30S subunits by using a fluorescence stopped-flow technique. Formation of the 30S preinitiation complex at 20°C in the presence of the three initiation factors and GTP was resolved into a fast, apparently second-order step ($k_{binding} = 5 \times 10^6$ M^{-1} s^{-1}; $k_{dissociation} = 1.4$ s^{-1}) followed by a slower rearrangement step ($k < 0.1$ s^{-1}). In summary, more work is required to resolve whether the binding of fMet-tRNA and mRNA is ordered or random and to elucidate the nature of the slow rearrangement or conformational change in the 30S preinitiation complex. It will be necessary to define these parameters for a number of mRNA species, since mRNAs differing in the primary and secondary structures of their initiator regions may behave quite differently.

Another issue is the role played by the initiation factors in the binding of fMet-tRNA and mRNA to 30S subunits. In the fast kinetic measurement for AcPhe-tRNA binding (294), the rate constant for binding is enhanced several hundred-fold by the three initiation factors and GTP. Either IF2-GTP or IF3, but not IF1 alone, stimulates significantly. IF3 also affects the rate of synthetic mRNA binding and dissociation, as well as the rate-limiting rearrangement step (95). Gualerzi and colleagues interpret the effects of both IF2 and IF3 as allosteric consequences of their binding to the 30S subunit (211, 294). There is controversy whether or not IF3 remains bound to the 30S preinitiation complex after fMet-tRNA binding. Voorma and co-workers (275) argue that IF3 binding on the 30S subunit is mutually exclusive of IF2 + fMet-tRNA, because IF3 causes the release of IF2–fMet-tRNA from 30S complexes as measured by sucrose density gradient centrifugation of glutaraldehyde-fixed particles. On the other hand, Gualerzi et al. (94) measured the rate of N-ethylmaleimide modification of IF3, which is slower in the ribosome-bound state than when free, and observed no change upon binding of fMet-tRNA. Continued IF3 binding or its release upon fMet-tRNA binding has not yet been demonstrated by fluorescence techniques, and the issue remains ambiguous.

Formation of the 70S initiation complex. A 50S ribosomal subunit joins the 30S preinitiation complex to form a 70S initiation complex. GTP is hydrolyzed and initiation factors are released, leaving fMet-tRNA and mRNA bound to the 70S particle. The fMet-tRNA is located in the P site, as indicated by its reactivity with puromycin, and the complex is competent to enter the elongation phase of protein synthesis. The 50S junction step is rapid ($k = 2.2 \times 10^7$ M^{-1} s^{-1}) compared to the rate of formation of 30S preinitiation complexes (at least fourfold slower). Complete 30S preinitiation complexes cannot be detected at measurable concentrations in cells. The junction reaction is essentially irreversible (89).

The formation of 70S initiation complexes involves a number of individual reactions which have been studied separately. The hydrolysis of high-energy bonds is prevented when nonhydrolyzable analogs such as guanylyl-5'-methylenediphosphonate (GMP-PCP) or guanylyl-5'-iminodiphosphonate (GMP-PNP) are used in place of GTP. These analogs support the formation of the 30S preinitiation complex and its junction with the 50S subunit, but the resulting fMet-tRNA–70S complex is not reactive with puromycin because IF2 remains bound. Upon 50S junction, but before GTP hydrolysis, IF1 and likely IF3 are ejected from the ribosome (94, 103). IF2 release involves GTP hydrolysis and may be the rate-limiting step in the formation of 70S initiation complexes. IF1 appears to increase the rate of IF2 release after GTP hydrolysis (14, 260), although the mechanism for this effect is not known. The three initiation factors ejected at the junction step are then free to recycle through another round of initiation.

The role of GTP in the initiation pathway is not entirely clear. It stimulates the formation of 30S preinitiation complexes, but is not required for their stability or subsequent junction reaction with 50S subunits. 30S preinitiation complexes formed with either GTP or GMP-PCP can be depleted of the nucleotide. Upon addition of 50S subunits, 70S initiation complexes form, initiation factors are ejected, and the fMet-tRNA is reactive with puromycin (60). Thus GTP hydrolysis is not required to place fMet-tRNA in the ribosomal P site. Consistent with this view, it has been shown that the bound mRNA is not translocated on the ribosome after IF2-mediated GTP hydrolysis (147, 267). Rather, GTP hydrolysis appears to cause a more rapid release of IF2 from the 70S initiation complex. The precise manner in which GTP hydrolysis and IF1 promote the release of IF2 is not yet clear. A kinetic analysis of the individual steps of the pathway could provide such needed information.

Mechanistic Aspects of Initiation

mRNA recognition. As described earlier, an mRNA forms a ternary complex with fMet-tRNA and 30S subunits, interacting in the complex at a specific initiator site in the mRNA sequence. Examination of rates of bacteriophage protein synthesis as a function of mRNA level shows clearly that mRNAs differ by several orders of magnitude in their efficiency to initiate (217). I will examine here the process of mRNA binding to ribosomes, to clarify the molecular basis for how one mRNA initiator site is "stronger" than another.

Most work on this question has centered on aspects of mRNA sequence and structure which influence initiation efficiency. Examination of the sequences of ribosome binding sites confirmed the presence of the initiator codon (usually AUG) and led to the discovery of the S-D sequence (GGAGG) which is complementary to the 3' terminus of 16S rRNA (CCUCC) in the 30S subunit (Fig. 3). Variations in the number of complementary bases in the S-D region (3 to 9 bases) or in the distance separating it from the initiator codon (5 to 9 bases) might account for differences in mRNA strengths. Much evidence has accumulated in support of the idea that the S-D region hydrogen bonds to the 16S rRNA (for detailed reviews, see references 82 and 145). Three kinds of biochemical evidence support the theory. (i) The 3'-terminal sequence of 16S rRNA is highly conserved and appears to be accessible to mRNA. (ii) Oligonucleotides complementary to the 3' terminus of 16S rRNA inhibit translation (264). (iii) a complex between the ribosomal binding site of mRNA and a 3'-terminal frag-

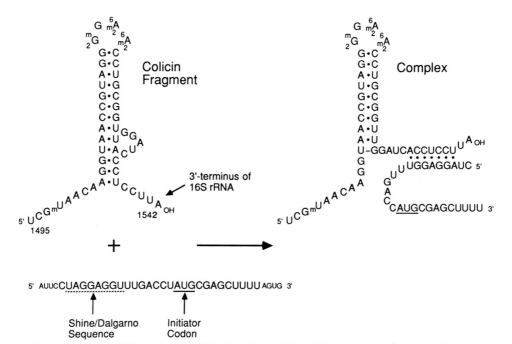

FIG. 3. Interaction between mRNA and 16S rRNA. Binding of the S-D sequence of the mRNA (underlined) with the 3'-terminal sequence of 16S rRNA is shown. Base pairs are shown by dots. Adapted from Steitz et al. (258).

ment of 16S rRNA was isolated specifically from initiation complexes (257). Further evidence comes from analyses of mRNA structurally altered by genetic mutations or by recombinant DNA techniques. Changes in the S-D region which reduce complementarity with rRNA severely restrict translation (61). Variations in the distance separating the S-D region and the initiator codon also affect translational efficiency (229). Finally, it was shown that mRNA secondary structure in the S-D region reduces translation (45, 119).

Although the evidence for involvement of the S-D sequence in initiation is overwhelming, there remained some uncertainty that the region truly interacted directly with the 3' terminus of 16S rRNA rather than with some other region of rRNA or mRNA. More definitive proof has been obtained recently by preparing ribosomes with altered sequences in their 3'-terminal 16S rRNA. de Boer et al. (53) generated translationally inactive mRNAs, carrying altered S-D sequences which fail to complement wild-type rRNA, and constructed ribosomes with 3' termini designed to complement the altered S-D sequences. The specialized ribosomes were shown to actively translate the mutant mRNAs, thereby demonstrating that the 3'-terminal region of 16S rRNA interacts with the S-D sequences. Jacob, Santer, and Dahlberg (reported by Dahlberg [51]) altered the 16S rRNA gene at 1 base pair near the 3' terminus and then expressed the gene in vivo from a multicopy plasmid. They found that mRNAs better matching the altered ribosomes were translated more efficiently, whereas those matching less well expressed more poorly. In summary, the length of complementarity and accessibility of the S-D sequence, its distance from the initiator codon, and the initiator codon itself all contribute to the strength of an mRNA. An S-D interaction with rRNA is not absolutely essential for mRNA binding to ribosomes, however, since a few mRNAs entirely lack the S-D region (e.g., the phage lambda cI repressor) yet are translated, albeit rather inefficiently.

A statistical analysis of known initiation regions, as well as analyses of mutations which affect translational efficiency, suggest that other features of the mRNA structure may influence mRNA binding to ribosomes (reviewed in detail by Gold et al. [82] and Kozak [145]). A computer-assisted survey of initiation regions shows that A or U is preferred at every position between the S-D sequence and the initiator codon and, specifically, that A is preferred at position −3. The sequence GCUA or AAAA is favored following the initiator codon (i.e., positions +4 to +7) (259). Evidence exists for the nonrandom appearance of other nucleotides between −21 and +13. Further evidence for the possible involvement of still more regions of mRNA comes from genetic studies involving mutations which affect translational efficiency but lie outside the S-D region and initiator codon (reviewed by Gold et al. [82]). Some of these effects can be rationalized in terms of secondary structure which masks the S-D sequence or initiator codon, whereas others cannot. It is clear from work on the RNA bacteriophages that mRNA secondary structure affects the translational efficiency of a cistron (reviewed by Iserentant and Fiers [119]). More work is required to identify those portions of an mRNA that contribute to its

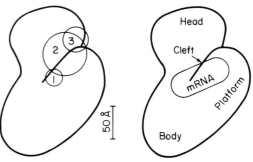

FIG. 4. Active sites on 30S ribosomal subunits. Model depicts the interface side of the subunit, based on a consensus model proposed by Printz et al. (213). Initiation factors are drawn to scale, assuming spherical structures, and are placed on the model according to Boileau et al. (20) and Liljas (155). The region for mRNA binding is based on data reviewed by Liljas (155) and on the text.

binding to ribosomes and to identify with what structures in the 30S ribosomal subunit or associated proteins each interacts. Since it is conceivable that individual mRNAs utilize different modes of binding, this may be a particularly difficult problem to solve.

Structure of ribosomal initiation complexes. The structure of the ribosome is discussed in detail by Noller and Nomura (chapter 10). We are concerned here with identifying where on the ribosomal surface the initiation components bind and what sorts of interactions occur. Such information is essential for a complete description of the initiation process and for an understanding of the molecular mechanisms involved.

From the discussion above on mRNA binding, it is already established that the 3' terminus of the 16S rRNA is important in the initiation pathway. The 3' terminus maps to the cleft region of the 30S subunit model, and the pyrimidine-rich region which interacts with the S-D region of mRNA is thought to lie between the tip of the "thumb" and the cleft (Fig. 4). The decoding site is deep in the cleft (see The Elongation Cycle, below) and must be close to where the anticodon of fMet-tRNA binds to the initiator codon. The sites of initiation factor binding have been elucidated by cross-linking the factors to ribosomal proteins with cleavable bifunctional reagents and identifying ribosomal proteins cross-linked to each factor (20). The results place all three factors near the cleft region. When bound to the 30S particle, IF1 and IF3 cross-link to IF2 but not to each other, suggesting that each lies contiguous to IF2 but possibly not to the other. S1 and S21 map in this region also and are essential for mRNA binding. Since the factors bind to the same surface of the 30S subunit as does the 50S subunit, a competition between factors and 50S subunits could occur based on steric hindrance alone. This is not true for IF2, however, since the factor does not inhibit formation of 70S initiation complexes, yet remains bound until GTP hydrolysis occurs. It is noteworthy that all three factors cross-link to the oxidized 3' terminus of 16S rRNA. Finally, IF3 can be cross-linked by *trans*-diaminedichloroplatinum(II) to another region of the 16S rRNA, residues 819 through 859 in the central domain (B. Ehresmann, personal communication). It is possible that in the absence of

IF3, the 3' terminus of 16S rRNA interacts with the purine-rich sequence in this region, making the 3' terminus unavailable for binding to mRNAs. IF3 binding to the 819–859 region may result in the release of the 3'-terminal domain, thereby allowing it to bind to the S-D region of mRNAs.

The localization of fMet-tRNA on the 30S subunit also has been studied by cross-linking techniques. Using electrophilic analogs of AcPhe-tRNA as models of the initiator tRNA, Girshovich et al. (78) detected cross-linking to a number of ribosomal proteins that reside in the head and cleft region of the 30S subunit model. Experiments with fMet-tRNA derivatives and 70S ribosomes show that the acceptor end of the tRNA binds in the peptidyltransferase center (see section below), as expected (101). The channel in which the mRNA resides and the region thought to accommodate the codon-anticodon interaction are described in detail in a subsequent section. A precise knowledge of the rRNAs and ribosomal proteins involved and the arrangement of these components is not yet available. As models of the 30S subunit are refined, especially with respect to the three-dimensional arrangement of the RNA in relation to the proteins, we can expect to be able to localize more precisely the initiation components on the ribosomal surface.

After junction of the 50S subunit to form a 70S initiation complex, the fMet-tRNA appears to reside in the ribosomal P site. A detailed description of this site is given in the next section. I shall restrict this discussion to the ribosomal elements involved in the GTPase activity of IF2. This activity requires the presence of 50S subunits. 50S subunits depleted of protein L7/L12 fail to stimulate IF2-catalyzed hydrolysis of GTP (70, 157). The L7/L12 molecules have been localized to a stalklike protuberance of the 50S subunit which in the 70S ribosome is rather far removed from the apparent IF2 binding site on the 30S subunit (see Fig. 6). However, IF2 can be cross-linked to L7/L12 (102), suggesting close proximity of these proteins in the 70S initiation complex.

THE ELONGATION CYCLE

Overview

A nascent protein chain is extended by one amino acid during each turn of the elongation cycle. The process is conveniently divided into three steps, the binding of aminoacyl-tRNA, peptide bond formation, and translocation. These reactions take place on the surface of the 70S ribosome. Ribosomes contain two tRNA-binding sites: the P site, where peptidyl-tRNA binds before peptide bond formation, and the A site, where aminoacyl-tRNA binds. The aminoacyl-tRNA first forms a ternary complex with elongation factor EF-Tu and GTP. The ternary complex then binds to a ribosome already complexed with peptidyl-tRNA and mRNA. During the binding step, a specific ternary complex among the numerous species available is selected on the basis of an interaction between the mRNA's codon in the A site and the tRNA anticodon. After hydrolysis of GTP and release of EF-Tu, peptide bond formation occurs by transfer of the peptidyl moiety to the amino group of the aminoacyl-tRNA, which is located in the A site. The translocation step,

promoted by EF-G and GTP hydrolysis, involves the movement of the peptidyl-tRNA–mRNA complex from the A site to the P site and results in a ribosomal complex ready for another cycle of elongation. These steps, together with the recycling of EF-Tu catalyzed by another factor, EF-Ts, are shown in Fig. 5. From an examination of the pathway, we conclude that peptides grow from the N terminus towards the C terminus. The mRNA is read sequentially as a nonoverlapping triplet code from the 5' terminus towards the 3' terminus, and at least two molecules of GTP are hydrolyzed per peptide bond formed.

Although the overall pathway of elongation of protein synthesis has been known for over 2 decades (for recent reviews, see references 15, 42, and 289), a number of issues continue to be actively investigated. For example, what is the molecular mechanism of aminoacyl-tRNA binding to ribosomes, and how is the accuracy established? Much recent kinetic and structural work sheds light on this problem, and progress is described below in this section. The problem of how the ribosome discriminates between different aminoacyl-tRNAs is discussed later in the section on Translational Accuracy. Another major issue is elucidation of the mechanism of translocation. Both problems require a detailed knowledge of the structure of the ribosome and its ligands, as well as an understanding of the basic rules for nucleic acid-protein interactions.

Components

The major macromolecular components of the elongation cycle are: the 70S ribosome, whose structure is described in detail by Noller and Nomura (chapter 10); mRNA, aminoacyl-tRNA, and peptidyl-tRNA, described in previous sections; and the elongation factors EF-Tu, EF-Ts, and EF-G (Table 5). I shall examine the structure and properties of these factors below to better clarify their function. A brief description of the tRNA-binding properties of ribosomes also is given.

Elongation factor EF-Tu. Two elongation factors, EF-T and EF-G, were isolated from the postribosomal supernatant (S-100) of bacterial extracts and detected by their stimulation of polyphenylalanine synthesis by ribosomes, Phe-tRNA, and poly(U) (1). Further fractionation of EF-T yielded EF-Tu, an unstable protein, and EF-Ts, a more stable factor (163). The role of EF-Tu is to promote the binding of aminoacyl-tRNA to the A site of the ribosome and to hydrolyze GTP. Its function involves the interaction of the factor with a large number of other translational components: aminoacyl-tRNAs, ribosomes, EF-Ts, and GTP. To understand the function of EF-Tu, I shall first discuss its physical properties, which are summarized in Table 5 (see also reference 288). The protein is readily purified to homogeneity (5, 169) and has been sequenced (132). There are two genes for EF-Tu, *tufA* and *tufB*; both have been sequenced (3, 307), confirming the amino acid sequence at the DNA level. The gene products are 393 amino acids long and differ only at the C-terminal amino acid: *tufA* codes for a Gly; *tufB* codes for a Ser. *tufA* is expressed about 3.5 times as much as *tufB* (219).

EF-Tu is the most abundant protein in *E. coli*, comprising about 5% of the soluble protein and exhibiting a factor/ribosome ratio of up to 7:1 in rapidly

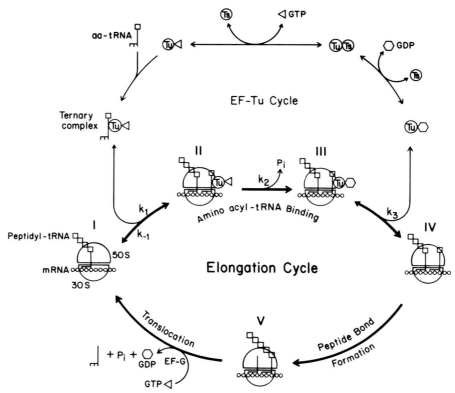

FIG. 5. Pathway of elongation. The steps of the cyclic process of elongation are shown with heavy arrows, and certain rate constants are shown that are discussed in the text. The steps involved in the recycling of EF-Tu through the action of EF-Ts are shown with light arrows. All symbols are defined in the figure.

growing cells. The EF-Tu/tRNA ratio appears to be about 1:1, suggesting that EF-Tu may be able to complex all of the aminoacyl-tRNAs in the cell. The protein is acetylated at the N-terminal Ser, and about 60% of the molecules are monomethylated at Lys-56. The factor normally is isolated as a complex with GDP. EF-Tu alone is very unstable, but is readily stabilized by GTP, GDP, GMP-PCP, GMP-PNP, or EF-Ts. The binary EF-Tu–GDP complex crystallizes readily in a variety of forms. The best crystals for X-ray diffraction work are those obtained after nicking the complex with trypsin at Arg-44, Arg-58, or both (134, 135, 175). High-resolution electron density maps of nicked EF-Tu suggest three structural domains (43). Domain I possesses a tight structure and contains the guanine nucleotide-binding site (133,

150). The arrangement of α-helices and β-sheets and the sequences of amino acids between them are typical of mononucleotide-binding sites in other proteins (234), among them EF-G, IF-2, and the family of *ras* oncogenes. The other domains of EF-Tu are looser in structure and therefore more difficult to define precisely. These may be involved in binding to aminoacyl-tRNA or ribosomes, at which time a tighter, more defined structure may result. The diversity of interactions with nucleic acids and proteins makes a detailed understanding of EF-Tu structure highly desirable. Much effort is being directed to preparing suitable crystals of intact (rather than nicked) EF-Tu and its complexes with aminoacyl-tRNAs and EF-Ts.

Elongation factor EF-Ts. EF-Ts is a protein of about 30 kDa which promotes guanosine nucleotide exchange on EF-Tu (see below). It is readily purified as an EF-Tu–EF-Ts complex which yields free EF-Ts upon treatment with GDP. EF-Ts is coded by the gene *tsf* and has been sequenced (2).

Elongation factor EF-G. EF-G binds transiently to the ribosome, thereby catalyzing the hydrolysis of GTP and the translocation step of protein synthesis. It is a single polypeptide chain of 80 kDa, coded by the *E. coli* gene *fusA*. EF-G has been purified to homogeneity and crystallized (201), but no X-ray diffraction studies have been reported. The factor is an acidic protein and is extremely sensitive to thiol reagents (182). The cellular level of EF-G is one molecule per ribosome, or about 0.5% of the soluble protein (85). The gene for EF-G has been sequenced, from which an amino acid sequence has been derived (311). The

TABLE 5. Elongation factors

Factor	M_r	No. of amino acids	Gene	Cell level: factors per ribosome	Function
EF-Tu	43,195	393	*tufA*	2–7	Bind aminoacyl-tRNA to A site; GTPase
	43,225	393	*tufB*	2–7	
EF-Ts	30,257	282	*tsf*	1	GDP/GTP exchange on EF-Tu
EF-G	77,444	703	*fus*	1	Translocation; GTPase

protein binds GTP or GDP; a mononucleotide-binding domain containing homologies with EF-Tu and IF2 has been identified. The EF-G affinity for GTP is quite low ($K_a = 2.7 \times 10^4$ M^{-1}) (9), and a facile exchange of mononucleotides occurs without the assistance of other factors. This contrasts with the guanine nucleotide-binding properties of EF-Tu, as described below. However, as with EF-Tu, EF-G contains the active site for the ribosome-dependent GTPase activity, since the factor exhibits a low level of GTPase activity in the absence of ribosomes when 20% 2-propanol is added (56).

tRNA-binding sites on ribosomes. The interactions of tRNA derivatives on the 70S ribosome have been explained by a two-site model proposed over 20 years ago by Watson (283). The model proposes that at least two molecules of tRNA can bind to the ribosome simultaneously. In one site, called the P site, peptidyl-tRNA is found just before peptide bond formation. The other, named the A site, is where aminoacyl-tRNA first binds. A definition of these sites in precise structural terms is just now becoming possible (see below, Location of Ribosomal Active Sites). Historically, the two sites have been distinguished by functional tests. Aminoacyl-tRNA derivatives bound in the P site are capable of reacting with puromycin, whereas those in the A site are not. Puromycin is an analog of the 3' terminus of aminoacyl-tRNA, binds to a part of the A site, and interacts with the peptidyltransferase active site. Using puromycin reactivity, researchers have characterized the sites of binding of a variety of tRNA derivatives. The P site has a relatively high affinity for stripped (nonaminoacylated) tRNA and peptidyl-tRNA, whereas the A site prefers aminoacyl-tRNA. These binding interactions occur at Mg^{2+} concentrations from 10 to 20 mM and do not require auxiliary factors. Three-site models for tRNA binding have been proposed in which the third site is either an entry site, where aminoacyl-tRNA binds prior to A-site binding (151), or an exit site where stripped tRNA remains bound after translocation but before ejection (223). These three-site models are discussed in the sections below. Controversies about the number of binding sites are due in large part to the apparent heterogeneity of ribosome populations and to the difficulty in evaluating the number of active particles.

Pathway

Ternary complex formation. EF-Tu binds GTP and aminoacyl-tRNA to form a ternary complex which is the intermediate that binds to the ribosome. Ternary complex formation proceeds in two steps: EF-Tu forms a binary complex with GTP, and the binary complex then interacts with aminoacyl-tRNA. Research in this area has focused on two major problems, the competition of GDP and GTP for binding to EF-Tu and its sensitivity to energy charge, and the molecular basis for EF-Tu recognition of aminoacyl-tRNAs.

EF-Tu is normally isolated as a binary complex with GDP; it binds GDP ($K_d = 5 \times 10^{-9}$ M) more tightly than GTP ($K_d = 4 \times 10^{-7}$ M) (6). The binding (or release) of guanine nucleotides to EF-Tu is readily detected by trapping radioactive nucleotides in binary complexes on nitrocellulose filters. GTP binding causes a change in the conformation of EF-Tu, resulting in altered rates of ^{3}H exchange (214) or fluorescent dye emission (47) compared to the EF-Tu–GDP complex. The rate of exchange of GDP in the binary complex, which is equal to the rate of dissociation of the complex, is of the order 10^{-2} s^{-1}, a rate much too slow to accommodate the overall rate of protein synthesis in vivo. EF-Ts greatly stimulates the rate of exchange by displacing the nucleotide and in turn being displaced itself by another nucleotide:

$$\text{EF-Tu–GDP} + \text{EF-Ts} \underset{\text{GTP}}{\overset{\text{GDP}}{\rightleftharpoons}} \text{EF-Tu–EF-Ts} \rightleftharpoons \text{EF-TU–GTP} + \text{EF-Ts}$$

Kinetic studies indicate that intermediates comprising EF-Tu–EF-Ts and the guanine nucleotide exist as well (116, 232). The effect of EF-Ts on the exchange rate guarantees that this step is not rate limiting for protein synthesis. However, EF-Ts does not affect the equilibrium position of the reaction above, which lies far to the left when GTP and GDP are present at about the same concentrations. In actively metabolizing cells, GDP is converted to GTP, thereby pushing the equilibrium toward the right, and the GTP binary complex predominates. The position of equilibrium is clearly sensitive to the energy charge of the cell, which can in principle modulate the overall rate of elongation at this step.

EF-Tu–GTP (but not EF-Tu–GDP) binds to aminoacyl-tRNAs to form a ternary complex. The formation of ternary complexes can be measured by nitrocellulose filtration with labeled GTP, where the ternary complex fails to adsorb to the filter. Alternatively, an aminoacyl-tRNA in a ternary complex is more resistant to alkaline (208) or RNase (126) hydrolysis than is the free form, allowing measurement of the amount of complex formed (142). The formation for ternary complexes requires a nonacylated amino group in the aminoacyl-tRNA; acylaminoacyl-tRNAs or nonaminoacylated tRNAs bind poorly (216), although hydroxyacyl-tRNAs form complexes (69). EF-Tu does not form a stable ternary complex with fMet-tRNA, whether or not the formyl group is present (245), although a rather unstable complex with the nonformylated species has been detected recently (263). Louie et al. (160) measured the relative affinities of a large number of species of aminoacyl-tRNA for EF-Tu–GTP and found that these vary 12-fold (Gln-tRNA is the strongest; Val-tRNA is the weakest). The dissociation constants fall in the range of 10^{-8} to 10^{-9} M (15). Given the cellular concentrations of EF-Tu and tRNAs, we conclude that essentially all aminoacyl-tRNAs are present as ternary complexes in intact cells.

The formation of ternary complexes offers a superb system for studying nucleic acid-protein interactions. The EF-Tu molecule is of intermediate size, X-ray crystallographic information on its three-dimensional structure is being generated, and the protein binds tightly to a large number of different tRNA derivatives. A great variety of experiments have been designed to elucidate what structural features of the tRNA and protein molecules are involved in the binding. Besides the α-amino group of the aminoacyl-tRNA, EF-Tu appears to recognize the CCA 3'-terminal sequence (190) and the sequence of the base pairs in the acceptor stem (126). Footprinting experiments with a variety of RNases indicate that EF-Tu–GTP protects both the acceptor and T stems and the extra arm, but causes enhanced cutting in the anticodon stem (292). All initial cuts lie on the side of the L-shape where the extra arm is exposed, whereas enhanced cuts lie on the opposite side. On the other hand, chemical modification of tRNA phosphates with ethylnitrosourea showed no change in pattern upon ternary complex formation, suggesting that tRNA binding to EF-Tu may be rather loose (279). Cross-linking (137) and NMR studies support the footprinting results.

The regions of EF-Tu which interact with the aminoacyl-tRNA are more difficult to identify. Residues essential for tRNA binding can be identified by correlating loss of activity with derivatization or by observing protection upon tRNA binding. Such studies implicated Cys-81, His-66, and His-118 (131, 153). Cross-linking of ε-bromoacetyl-Lys-tRNA to His-66 confirms the proximity of this residue to the binding site (129, 236). These residues are found near the GTP-binding region in domain I (43). A more definitive solution to the problem of defining the contact surfaces of EF-Tu and aminoacyl-tRNA may come from an X-ray diffraction study of crystals of the ternary complex. Unfortunately, no crystals have yet been reported which are suitable for high-resolution analysis.

Aminoacyl-tRNA binding to ribosomes. From the point of view of informational decoding and translational accuracy, the binding phase of the elongation cycle is most important and interesting. There are at least three reaction steps in the pathway of binding of an aminoacyl-tRNA to an mRNA-programmed ribosome which precede the peptide bond-forming reaction. First, the aminoacyl-tRNA in the form of its ternary complex with EF-Tu and GTP binds to the A site in a reversible reaction. Next, GTP is hydrolyzed to GDP and P_i; third, EF-Tu–GDP dissociates from the particle (Fig. 5). I shall begin by examining these steps from a kinetic viewpoint. I shall also discuss structural aspects of EF-Tu and the aminoacyl-tRNAs involved in these interactions on the ribosomal surface. The reader should keep in mind that the mass of the 70S ribosome is two orders greater than that of a tRNA molecule, although the diameters of the two differ by only about three- to sixfold. Even though our knowledge of ribosome structure still lags behind that of tRNA, much progress has been made recently. A detailed analysis of the A site in terms of ribosome structure will be described later in this section.

The selection and binding of an aminoacyl-tRNA are determined primarily by the codon-anticodon interaction. The interaction is an antiparallel hydrogen bonding between the three bases in the mRNA codon and the three bases in the anticodon of the tRNA. Our knowledge of the structure of the decoding site on the ribosome is summarized below. Since the tRNA-mRNA interaction serves exclusively in the selection of the aminoacyl-tRNA bound, it is essential that the interaction discriminate accurately between the different species. The problem of how the aminoacyl-tRNAs are distinguished is discussed in detail later (Translational Accuracy).

It is surprising that the rates of ribosomal binding of all species of ternary complex differ by no more than an order of magnitude. Thompson and Dix (268) have measured many of the rate constants for the reactions shown in Fig. 5. The rate of ternary complex binding, k_1, is 5×10^6 M^{-1} s^{-1} for a tRNA matching the codon of the mRNA and 0.4×10^6 M^{-1} s^{-1} for a near-cognate tRNA. Their rates of dissociation from the ribosome differ substantially, however; k_{-1} is 2.7 $\times 10^{-3}$ s^{-1} and 6 s^{-1}, respectively. Wintermeyer and Robertson (296) determined the rates of aminoacyl-tRNA binding to mRNA-programmed ribosomes by using fluorescence stopped-flow techniques and found kinetic evidence for two binding steps, a rapid initial binding ($k = 17 \times 10^6$ M^{-1} s^{-1}) followed by a slower step, likely a conformation change ($k = 1.8$ s^{-1}). Since these experiments were performed in the absence of EF-Tu, their relevance to ternary complex binding is difficult to assess. Using rapid mixing and chemical quench methods, Thompson et al. (269, 270) also determined the rates of GTP hydrolysis, k_2. The rate constant for both cognate and noncognate tRNAs is 20 s^{-1} (at 5°C), indicating that the precision of codon-anticodon fit does not influence GTP hydrolysis. The GTPase rate is much faster than the dissociation rate for cognate ternary complexes. As a result, equilibrium conditions are not attained during the reversible ternary complex-binding reaction, but rather the aminoacyl-tRNA is carried forward in the reaction pathway. It is thought that the back reaction for GTP hydrolysis is so slow that in effect it is an irreversible step. The rate of ejection of EF-Tu–GDP has not been determined directly, but this rate, together with that of peptide bond formation, was measured in the same system for determining GTP hydrolysis rates, giving a composite rate constant of 0.36 s^{-1}. Thompson et al. argue that of the two steps, EF-Tu–GDP release is probably the slower and is therefore the rate-limiting step of the elongation cycle. It should be noted that measurement of the composite rate constant was at 5°C, which perhaps explains why its value is significantly lower than the in vivo rate of amino acid polymerization, which is at least 15 to 20 s^{-1}. It is difficult to evaluate the relevance of kinetic measurements made at 5°C or the validity of extrapolating the values to higher temperatures.

The stable binding of ternary complex to the A site occurs with nonhydrolyzable GTP analogs such as GMP-PCP and GMP-PNP. In such cases, GTP hydrolysis is prevented, EF-Tu fails to be released, and the transfer of the peptidyl moiety from P site-bound peptidyl-tRNA to the aminoacyl-tRNA is prevented. Apparently, EF-Tu prevents the 3'-terminal aminoacyl group from reaching the peptidyltransferase center, for the presence of EF-Tu does not inhibit puromycin from reacting at this site. However, the actual function of EF-Tu is not entirely clear. The binding of aminoacyl-tRNA and the elongation of protein synthesis can proceed in the absence of EF-Tu, but higher Mg^{2+} concentrations are required and the rate of elongation is much slower. The stimulation of aminoacyl-tRNA binding at low Mg^{2+} concentrations and the specificity for charged versus stripped tRNAs are important roles for the factor. EF-Tu may also be involved in assuring a high fidelity of productive aminoacyl-tRNA binding by preventing peptide bond formation through its slow release, thereby providing time for noncognate aminoacyl-tRNAs to dissociate from the ribosome. (See section below on Translational Accuracy for a detailed discussion on this point.)

Considerable effort has been directed to elucidating how EF-Tu and the aminoacyl-tRNA interact with the ribosome. EF-Tu–GTP interacts directly with ribosomes, rather than exclusively through tRNA, since it binds as a binary complex to 70S ribosomes with nearly the same kinetic constants as the ternary complex (270a). The regions of EF-Tu in direct contact with the ribosome have not yet been identified. EF-Tu possesses two RNA-binding sites (277), which suggests that the factor may bind simultaneously to both aminoacyl-tRNA and peptidyl-tRNA. Binding to the ribosome activates the GTPase function of EF-Tu. That the active site for GTP hydrolysis resides in EF-Tu has been shown by the stimulation of GTPase activity when the antibiotic kirromycin is added to pure EF-Tu (reviewed by Parmeggiani and Swart [202]). Kirromycin inhibits protein synthesis by binding to EF-Tu and preventing its release upon GTP hydrolysis. It has been used extensively to characterize EF-Tu functions and to obtain mutant forms of the factor.

Peptide bond formation. Peptide bond formation involves an acyl group O-to-N migration, resulting in the conversion of an ester into an amide. The reaction has been reviewed in detail by Harris and Pestka (99). Peptidyl-tRNA in the P site is the ester that donates its peptidyl group to an acceptor, the amino group of aminoacyl-tRNA bound in the A site (Fig. 5). The reaction is reversible, but product formation is favored without the input of additional energy. Besides the hydrolysis of GTP, catalyzed by EF-Tu and EF-G, peptide bond formation is the only other covalent reaction in the elongation cycle. The reaction is catalyzed by peptidyltransferase, an enzymic activity which is an integral part of the 50S subunit (272). Peptidyltransferase is conveniently measured by the puromycin reaction, in which puromycin acts as the acceptor substrate, forming peptidyl-puromycin. However, the synthesis of di- or tripeptides can be measured also, although these assays are more complex, involving aminoacyl-tRNA binding and translocation steps as well.

Peptidyltransferase activity can be detected with fragments of peptidyl-tRNA and 50S subunits, in the absence of 30S subunits and templates, if the reactions are run in the presence of ethanol (171). This simplifed reaction has been used to define the structural requirements for the donor and acceptor substrates. The 3′-terminal CCA oligomer carrying an acylaminoacyl group is the smallest fragment which can act was a donor substrate, although its reduced activity indicates that other regions of the tRNA are involved also. The integrity of the ribose in the terminal adenosine and acylation of the aminoacyl group are essential features. The acceptor substrate, on the other hand, can be as small as an aminoacyl-nucleoside, as indicated by the reactivity of puromycin. Although the charged α-amino group is important, α-hydroxyacyl-tRNAs participate as both acceptors and donors in the peptidyltransferase reaction (68). The 3′-O-aminoacyl isomer is preferred over the 2′ isomer as acceptor substrate, although the opposite preference is expressed in ternary complex binding to the A site (253). This had led Sprinzl and co-workers to propose a transacylation step following the release of EF-Tu–GDP and preceding peptide bond formation. The preferred donor substrate appears to be the 3′ isomer.

The enzymic mechanism for peptide bond formation remains poorly elucidated. Besides the binding of the substrates, catalysis likely involves the kinds of general acid and base catalysis seen with proteases (which catalyze in effect the reverse reaction). Since proteases are frequently small proteins, it is reasonable to expect that the peptidyltransferase activity might reside in a single 50S ribosomal protein. A number of experimental approaches have been used to define the peptidyltransferase center on the ribosome. Electrophilic or photoactivatable derivatives of aminoacyl-tRNAs bound either in the P or A sites have been cross-linked to a number of ribosomal proteins or to rRNA (reviewed by Ofengand et al. [146, 193]). Similarly, cross-linking derivatives of antibiotics such as puromycin or chloramphenicol have implicated specific proteins. Partial reconstitution of ribosomal protein-deficient 50S cores suggested that L16 may be the peptidyltransferase protein. However, a mutant strain lacking L16 is competent for protein synthesis, as are strains lacking numerous other ribosomal proteins. As a result, no single ribosomal protein has been implicated as the peptidyltransferase enzyme. Instead, the activity appears to be due to a number of components in the ribosome and may even involve rRNA directly.

Translocation. Translocation is the movement of peptidyl-tRNA and its associated mRNA from the A site to the P site on the ribosomal surface (Fig. 5) (reviewed by Brot [26] and Spirin [250]). The reaction is catalyzed by elongation factor EF-G and involves GTP hydrolysis. EF-G forms a binary complex with GTP ($K_a = 2.7 \times 10^4 \ M^{-1}$), which then binds to 70S ribosomes containing stripped tRNA in the P site and peptidyl-tRNA in the A site. EF-G–GTP binding promotes the release of stripped tRNA from the A site and the translocation of peptidyl-tRNA to the P site (162). After GTP hydrolysis, EF-G–GDP itself is ejected from the ribosome. The result is peptidyl-tRNA located in the P site and a new mRNA codon exposed in the A site.

The translocation reaction is studied by constructing mRNA-ribosome complexes with peptidyl-tRNA in the A site and stripped tRNA in the P site. For example, 70S ribosomes are incubated with fMet-tRNA, $AUG(U)_n$, Phe-tRNA, initiation factors, EF-Tu, and GTP, but not EF-G. fMet is transferred to Phe-tRNA to form fMet-Phe-tRNA^{Phe} in the A site and stripped tRNA$_f^{Met}$ in the P site. EF-G–GTP-promoted translation is usually measured by the reactivity of fMet-Phe-tRNA with puromycin, indicative of P site occupancy, or sometimes by the ejection of tRNA$_f^{Met}$. Another strategy for forming the ribosomal complex is to use nonenzymatic (high-Mg^{2+}) conditions for tRNA binding. 70S ribosomes are incubated with poly(U) and one equivalent of stripped tRNA^{Phe}, which binds to the P site; added AcPhe-tRNA binds to the A site. Its translocation is then measured as above. Nearly all studies of translocation have used puromycin reactivity; a more structural determination of movement from A to P sites is not yet practical.

EF-G–GTP binds to 70S (or 50S) ribosomes lacking tRNA derivatives and subsequently hydrolyzes GTP. This GTPase reaction is said to be "uncoupled" from protein synthesis. In the uncoupled reaction, the interactions of EF-G with GTP and ribosomes are strictly ordered: EF-G binds GTP first and, after hydrolysis, dissociates from the 70S ribosome as a binary EF-G–GDP complex (231). The bindings of EF-G and EF-Tu are mutually exclusive; EF-G–GDP must dissociate before the ternary complex can bind.

To elucidate the mechanism of translocation, a number of researchers have studied the ribosome-binding affinities of tRNA derivatives to the A and P sites. Holschuh and Gassen used analytical ultracentrifugation and nitrocellulose filtration assays to demonstrate that the binding of AcPhe-tRNA to the P site is 20-fold stronger than to the A site (109). When pretranslocation complexes were constructed by incubating 70S ribosomes with $AUGU_7$, fMet-[^{3}H]tRNA, Phe-tRNA^{Phe}, GTP, and EF-Tu, subsequent addition of EF-G–GTP and [^{14}C]Phe-tRNA^{Phe} resulted in release of [^{3}H]tRNA$_f^{Met}$ and binding of [^{14}C]Phe-tRNA at the same rates. Holschuh and Gassen concluded that EF-G–GTP binding causes the release of stripped tRNA$_f^{Met}$ from the P site, followed by a rapid translocation of fMet-Phe-tRNA and the subsequent binding of [^{14}C]Phe-tRNA to the A site. They propose that the movement of peptidyl-tRNA is a directed diffusion from the low-affinity A site to the high-affinity P site while the codon-anticodon interaction is maintained. In this view, no energy is required for translocation, consistent with the fact that peptidyl-tRNA in the A site is converted to a puromycin-reactive state by EF-G–GMP-PCP (118).

Rheinberger and Nierhaus (222) proposed an alternative mechanism for translocation, based on a three-site tRNA binding model of ribosome function. The third nonoverlapping site, called an exit site (E site) is thought to contain stripped tRNA after the translocation reaction. Evidence for three tRNA binding sites comes from nonenzymatic, high-Mg^{2+} binding of labeled tRNA^{Phe} derivatives monitored by nitrocellulose filtration in the presence of poly(U). Consistent with other workers, Rheinberger and Nierhaus found that one equivalent of Phe-tRNA bound exclusively at the A site. Results with Ac-Phe-tRNA binding indicates that only one molecule binds, preferably to the P site, but to the A site if the P site is already occupied by tRNA. Stripped tRNA^{Phe} binds to both the E and P sites with equal affinity, even more tightly than AcPhe-tRNA binds to the P site, but also to the A site at saturation (three tRNAs per ribosome). When [^{14}C]tRNA^{Phe} in the P site and Ac[^{3}H]Phe-tRNA in the A site were treated with EF-G–GTP, translocation of Ac[^{3}H]Phe-tRNA was observed, but no release of [^{14}C]tRNA occurred (178). Nierhaus argued that on translocation, stripped tRNA moves from the P to the E site; its release requires still another round of the elongation cycle and thus a second translocation step.

Spirin (249) tested the two- and three-site models by using ribosomes bound to matrix-coupled poly(U), thereby working with a ribosome population that was presumably 100% active. Ribosomes synthesizing poly(Phe) were brought to the posttranslocation state

by treatment with EF-G–GTP. After the EF-G was washed away, the column material was incubated with [^{14}C]Phe-[^{3}H]tRNA, EF-Tu, and GTP, washed, and subsequently treated with EF-G–GTP. A second cycle of elongation reactions was then carried out stepwise. After the second EF-G–GTP treatment, 81% of the [^{3}H]tRNA was released, consistent with the two-site model and inconsistent with the three-site model.

Wintermeyer and co-workers (295) employed fluorescence techniques to measure tRNA binding during the translocation process. Using a fluorescent derivative of tRNA^{Phe} bound in the P site and AcPhe-tRNA in the A site, they measured the time course of fluorescence polarization after addition of EF-G–GTP. A rapid drop in polarization indicated that much of the tRNA^{Phe} derivative was released from the ribosome. Complete release was obtained by the addition of either cognate or noncognate tRNA. The latter release by exchange was thought to occur from the E site, where tRNA binding occurs with low affinity and is not significantly influenced by the mRNA. Stopped-flow fluorescence measurements were used to determine the rate constants for translocation of AcPhe-tRNA from the A to P sites and for the release of stripped tRNA. Fluorescence changes were usually biphasic, yielding a fast and a slow rate constant. The fast rate constant was about $3\ s^{-1}$ for translocation of AcPhe-tRNA and about $4\ s^{-1}$ for the release of tRNA^{Phe}. The similarity of rate constants suggests that the release and translocation reactions are coupled. The slower rate constants observed in these single-turnover experiments are too slow to belong to the main pathway of elongation, since poly(Phe) synthesis under the experimental conditions employed proceeds at about three amino acids per ribosome per s. Interestingly, Wintermeyer et al. observed that the rate constant for translocation of AcPhe-Phe-tRNA is $10\ s^{-1}$, significantly faster than that of the peptidyl-tRNA analog AcPhe-tRNA.

In summary, most experimental evidence is consistent with the two-tRNA site model of elongation. Although a third E site can be demonstrated, its characteristics suggest that it plays no important role in protein synthesis. The question of whether tRNA release precedes peptidyl-tRNA movement to the P site or whether these reactions are concerted remains unresolved. Nor is it clear just how the peptidyl-tRNA–mRNA complex "diffuses" from the A to the P site. A portion of the binary complex may move first, e.g., the 3'-terminal peptidyl moiety, followed by movement of the anticodon-mRNA portion. A deeper understanding of this movement and the role played by EF-G–GTP requires detailed knowledge of the structure of the ribosome and its tRNA complexes.

Location of Ribosomal Active Sites

A complete elucidation of the elongation steps of protein synthesis must include a description of the structure of the ribosome and where the various macromolecular ligands bind on its surface. Ribosome structure is described in detail in this volume (chapter 10) and has been review extensively (41, 155, 184, 298). Furthermore, a current publication provides a detailed description of the structure and func-

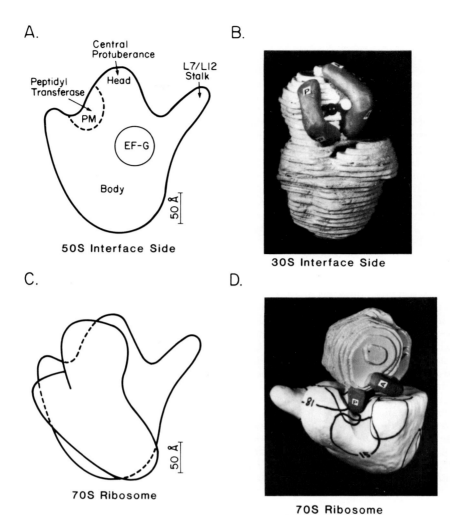

FIG. 6. Active sites on ribosomes. (A and C) Consensus models (213) of the 50S and 70S ribosomes based primarily on the electron microscopy work of Lake (151). Identification of the sites for puromycin (PM) binding, the peptidyltransferase center, and EF-G binding is based on data described by Liljas (155) and Traut et al. (273). (B and D) From Ofengand et al. (192), depicting where the two aminoacyl-tRNAs bind. In these models, the anticodons bind in the cleft of the 30S subunit and the aminoacyl ends bind to the heads of the 30S and 50S subunits.

tion of bacterial ribosomes (97). I will restrict this section to a brief overview of experimental approaches and results related to defining where the various translational components bind to ribosomes. The reader should keep in mind the large size of the 70S ribosomal particle (2.7×10^6 Da) and the facts that about one-third of its dry weight is protein, two-thirds is RNA, and 40% of its volume is water. The particle is not likely to be rigid or static, but rather may be quite dynamic and capable of numerous conformational states.

Elongation factors. As noted earlier, EF-Tu and EF-G are mutually exclusive in their binding to ribosomes. That the factors may possess overlapping binding sites is reinforced by a number of observations. The binding of both is inhibited by thiostrepton, an antibiotic which binds to L11 and rRNA (50). Futhermore, L11 is labeled by a photoaffinity derivative of GDP or GTP bound to EF-G (165). Both elongation factors exhibit reduced binding and GTPase activity when 50S subunits lacking L7/L12 are employed (27). Protein-protein cross-linking experiments

identified ribosomal proteins L1, L5, L7/L12, L15, L20, L30, and L33 at or near the binding site for EF-Tu (242) and S12, L6, L7/L12, and L14 near that for EF-G (273). The preponderance of evidence localizes both factors to the region of the 70S ribosome lying in the cavity between the two ribosomal subunits, as depicted in Fig. 6. This view is reinforced by electron microscopy of 50S–EF-G cross-linked complexes, which shows EF-G close to the "stalk" protuberance (79). However, a different view of the site of EF-Tu binding based on electron microscopy has been proposed recently (152); in this model, EF-Tu is placed on the exterior surface of the 30S subunit rather than near the 30S-50S interface.

Peptidyltransferase center. From functional studies, we may conclude that the 3′ termini of aminoacyl-tRNA and peptidyl-tRNA bind to the 50S subunit at the peptidyltransferase center. A variety of studies have been carried out to identify ribosomal proteins near the 3′ terminus of bound tRNA. Affinity labeling with N-substituted aminoacyl-tRNAs and antibiotic analogs implicated a number of 50S ribosomal proteins and two 30S proteins. L16 is most readily labeled from the A site, whereas L2 and L27 are more frequently labeled from the P site (207). Reconstitution of 50S cores lacking a number of ribosomal proteins suggests that L16 may be the protein most responsible for peptidyltransferase activity (265), although single-omission experiments implicated seven additional proteins (96). These proteins map primarily to the central protuberance and third projection (the one not containing L7/L12) of the 50S subunit and in the head region of the 30S subunit (Fig. 6). Confirmation of these regions has been obtained by direct visualization of cross-linked tRNAs or antibiotic derivatives by immunoelectron microscopy (87, 195).

mRNA-binding and -decoding sites. A number of chemically or photochemically reactive oligo- and polynucleotides have been used to affinity label ribosomal components in the mRNA-binding site (reviewed by Ofengand [189]). 30S ribosomal proteins S1, S3, S4, S5, S12, S18, and S21 were identified, as well as the 16S rRNA. These proteins map to the platform or "thumb" of the 30S particle (Fig. 4 and 6). Cross-linking the anticodon region of bound tRNA also identifies where mRNA and the decoding sites are located. Ofengand and co-workers found that UV irradiation of some tRNAs led to cross-linking to rRNA (246). Irradiation of P-site-bound acetylvaline (AcVal)-tRNA, which contains a carboxymethoxy-U_{34} (at the 5' base of the anticodon), results in a covalent linkage between the modified uridine and C_{1400} of the 16S rRNA. Interestingly, C_{1400} is in the center of a highly conserved sequence of 16 nucleotides (1392–1408) which are believed to be single stranded (see chapter 10). C_{1400} is cross-linked when Val-tRNA, derivatized with a 2-nitrophenyl azide group attached to the carboxymethoxy-U_{34} residue, is irradiated in the A site (191). Residue C_{1400} is localized by immunoelectron microscopy deep in the cleft of the 30S subunit (Fig. 4 and 6), overlapping a part of the mRNA-binding domain.

The two-tRNA ribosomal complex. The problem of defining exactly how peptidyl-tRNA and aminoacyl-tRNA are bound to the 70S ribosome before peptide bond formation has elicited considerable interest and activity. Models are restricted by the fact that both tRNAs must be adjacent at their anticodon ends, since they are binding to adjacent codons, and likely are adjacent at their aminoacyl ends if peptide bond formation is a direct transfer of the peptidyl moiety to the aminoacyl-tRNA. Further structural information has come from studying the cross-linking of Phe-tRNA carrying a reactive azidophenacyl-4-thiouridine$_8$ located near the elbow of the tRNA structure; cross-linking from the A site to S19 has been observed (156). Thus three contacts with the 30S subunit are defined: the 3' terminus (S24), the elbow (S19), and the anticodon loop (C_{1400}), allowing a rather precise placement of A-site-bound tRNA on the 30S subunit surface (Fig. 6). In the Ofengand model (Fig. 6), P-site-bound peptidyl-tRNA lies near the center of the 30S subunit, forming an angle of 45° between the planes of the two tRNA molecules (191). The relative orientation of the two bound tRNAs is primarily a function of the distance separating their respective elbows, given the need to hold the two anticodon loops and 3' termini in close proximity to each other. The distance separating

the elbows in the ribosome-bound state has been measured by fluorescence energy transfer methods with fluorescent derivatives either at 4-thiouridine$_8$ (128) or at residue 37 (204) of tRNA^{Phe}. Results indicate a distance of about 3.5 nm, consistent with a planar angle of 60 ± 30°. If the two tRNAs are correctly situated in the Ofengand model, it follows that the mRNA must make an abrupt U-turn between the A and P sites.

The discussion above has centered primarily on tRNA-binding sites involving ribosomal proteins. However, it is likely that rRNA also play a major role in tRNA binding. Considerable effort has gone into determining whether or not the TψC region of the tRNA T loop binds to a complementary region in 5S rRNA (reviewed by Ofengand [189] and Noller [184]). This putative interaction does not appear to be required, since deletion of the complementary sequence in the 5S rRNA nevertheless results in active ribosomes (197). That other regions of rRNA may be important is indicated by the recent findings of Burma et al. (29), who showed that a naked 16S-23S rRNA complex, together with the 5S rRNA-L5-L18-L25 complex, is capable of binding two tRNAs. Furthermore, Burma et al. claim that this ribosomal protein-deficient complex catalyzes polyphenylalanine synthesis, albeit slowly. Since methods are now available for identifying specific sites of cross-linking or derivatization in RNAs, and since a more detailed view of the placement of rRNA in the ribosome structure is emerging, it should soon be possible to define with precision the involvement of rRNA in tRNA binding.

TERMINATION

Overview

Termination of protein synthesis involves the hydrolysis of peptidyl-tRNA and thereby the release of the completed protein from the ribosome (reviewed by Caskey et al. [36, 38]). Termination occurs when the ribosome reaches one of three termination codons, UAA, UAG, or UGA. These signals in the mRNA, called nonsense codons, were first identified by genetic approaches (25). Termination requires the action of soluble proteins, called release factors. There are two release factors, RF-1 and RF-2, which recognize termination codons: RF-1 is specific for codons UAA and UAG, and RF-2 is specific for UAA and UGA. A third factor, RF-3, stimulates the activities of RF-1 and RF-2 but does not recognize termination codons. RF-1 or RF-2 binds to a ribosome carrying peptidyl-tRNA in the P site and its cognate termination codon in the A site. Release factor binding promotes the hydrolysis of peptidyl-tRNA by peptidyltransferase. The subsequent release of stripped tRNA and mRNA from the ribosome requires the action of another protein, called ribosome release factor (RRF).

Release factor activity was first detected by Ganosa (77). Capecchi (32) developed an assay for termination which involves a stepwise synthesis of hexapeptidyl-tRNA from a mutant bacteriophage R17 RNA template with a termination codon (UAG) at the seventh position. He then proceeded to isolate a substance that stimulates peptide release and identified it as a

TABLE 6. Release factors

Factor	M_r	No. of amino acids	Cell level: factors per ribosome	Function
RF-1	35,911	323	0.02	Recognizes UAA, UAG
RF-2	38,404	339	0.02	Recognizes UAA, UGA
RF-3	46,000			GTPase; stimulates RF-1 and RF-2
RRF	23,500			Ejection of tRNA

protein. Caskey et al. (39) constructed a more convenient assay involving the hydrolysis of fMet-tRNA bound to a ribosome with AUG in the P site; the reaction is dependent on a termination codon triplet and release factors. The two research groups purified RF-1 and RF-2 to homogeneity and identified a third factor, RF-3 (Table 6). None of these protein preparations contains RNA, indicating that termination codon recognition is a protein-nucleic acid interaction, not an RNA-RNA interaction. The genes for RF-1 and RF-2 have been cloned and sequenced (46; see also chapter 86). From the DNA sequences, open reading frames were identified which code for proteins of 35,911 Da (RF-1) and 38,404 Da (RF-2). Both are acidic proteins, contain common antigenic determinants, and share some sequence homologies. The levels of the release factors have been estimated at 500 to 700 molecules per cell (141), about one and two orders of magnitude less than the initiation and elongation factors, respectively. Beginning with the cloned genes, it should be possible to construct overproducing strains of bacteria and thereby prepare larger amounts of these rather low-abundance proteins.

Pathway and Mechanism

A schematic representation of the termination pathway is shown in Fig. 7. The first step begins with peptidyl-tRNA in the P site and an exposed termination codon in the A site. The release factor which recognizes the specific termination codon binds to the ribosome. The specificity of release factor recognition has been studied by measuring the binding of radioactive termination codon triplets to ribosomes containing bound fMet-tRNA, AUG, and RF-1 or RF-2 (247). Attempts to detect specific oligonucleotide interactions with release factors by equilibrium dialysis in the absence of ribosomes have been made, but only modest discrimination was seen (33). Together, the results are consistent with the view that a release factor protein recognizes its cognate codons in the A site, but a knowledge of the molecular basis for the interaction is lacking. In the presence of RF-3, the binding of RF-1 or RF-2 with the triplet termination codon is enhanced. However, RF-3 has no effect on the V_{max} of peptidyl-tRNA hydrolysis (170).

Although the binding of a release factor to the ribosome promotes the hydrolysis of peptidyl-tRNA, the bulk of the available evidence points to the peptidyltransferase center as the active site for the reaction. Conditions that alter peptide bond formation, such as reduced monovalent cations or the presence of a variety of antibiotics (e.g., chloramphenicol, sparsomycin, amicetin), in parallel affect peptidyl-tRNA hydrolysis. Peptidyltransferase is capable of catalyzing the transfer of the peptidyl moiety to ethanol, thereby forming an ester (37). The antibiotic linocin inhibits peptide bond and ester formation but stimulates peptidyl-tRNA hydrolysis (in the absence of release factors). Caskey et al. (38) proposed that during termination peptidyltransferase is modified by the release factor to enhance the hydrolytic capability of the enzyme. Recently, Caskey and co-workers have been able to inhibit the activity of RF-2 (i.e., its stimulation of peptidyl-tRNA hydrolysis) without affecting its binding to ribosomes (C. T. Caskey, personal communication).

After hydrolysis and release of the completed protein, the 70S ribosome contains bound release factor, mRNA, and stripped tRNA in the P site. The rate of release of release factors is enhanced by RF-3. Since GDP causes dissociation of RF-1– or RF-2–[³H]UAA–ribosome complexes formed with RF-3, GTP may play a role in factor dissociation upon termination. In eucaryotic cells, the single release factor possesses GTPase activity and promotes termination catalytically in the presence of GTP but not GMP-PCP. It is tempting to propose that GTP also plays a role in bacterial termination and that the codon recognition and GTPase functions have been separated into discrete proteins. More experiments are needed to establish whether RF-3 and GTP act as proposed.

The release of mRNA and stripped tRNA requires the action of RRF. Kaji and co-workers purified RRF and showed that it stimulates the release of labeled mRNAs from ribosomes (194). RRF has a mass of

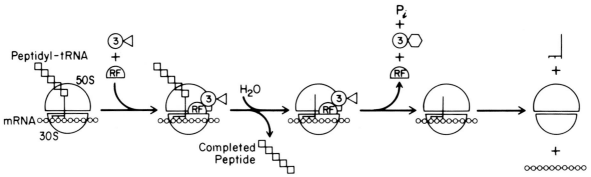

FIG. 7. Pathway of termination. Steps of the pathway are described in the text. Symbols are defined in the figure and in Fig. 5.

23,500 Da as determined by gel electrophoresis, but its sequence has not been determined, nor has its gene been cloned. It stimulates the in vitro synthesis of proteins in crude cell-free systems and is required for the synthesis of β-galactosidase in the highly fractionated DNA-dependent assay of Weissbach and coworkers (148). RRF possesses the interesting function of preventing reinitiation directly following termination codons. Using mutant bacteriophage amb2 R17 RNA, which contains an amber mutation in the seventh codon of the coat cistron, Ryoji et al. (239, 240) detected significant amounts of coat protein synthesis in an in vitro translation system depleted of RRF, but observed inhibition when RRF was added to the system. The presence of coat protein was due to reinitiation at the codon following the amber codon; the initiation event did not involve fMet-tRNA and presumably occurred because RRF-dependent dissociation of the mRNA-ribosome complex did not occur upon termination at the amber codon.

Translational Readthrough

Occasionally, the translational machinery fails to terminate protein synthesis at a stop (nonsense) codon, but inserts an amino acid instead, and protein synthesis continues until another in-frame termination codon is reached. This phenomenon, called translational readthrough or nonsense suppression, occurs at a significant frequency when the cell contains a suppressor tRNA, a tRNA altered by mutation in its anticodon so that it recognizes specific nonsense codons. However, readthrough is observed in special instances in cells which do not possess altered suppressor tRNAs. For example, cells infected with the RNA bacteriophage Qβ make a minor capsid protein, IIb, which is larger than, but related to, the major capsid coat protein. Protein IIb is produced at low frequency when translation of the coat cistron fails to terminate at the UGA termination codon. The normal cellular Trp-tRNA^{Trp} reads the UGA codon and inserts tryptophan, and translation continues for another 270 codons to complete the IIb protein (172). Other examples of readthrough in E. coli are expression of the gene for ribosomal protein S7 (196) and regulation of the tryptophan operon (144). Since readthrough also has been observed in eucaryotic cells, it is possible that this little-studied phenomenon may be more widespread than previously believed and may play a role in the overall regulation of gene expression.

TRANSLATIONAL ACCURACY

The synthesis of functional proteins depends on accurate translation as well as accurate transcription of genetic information. If on the average one error occurs per thousand amino acids incorporated, the probability of correctly synthesizing an average-size protein (40 kDa) would be about 0.7, whereas that for β-galactosidase (116 kDa) is only 0.3 (28). Cells may minimize the effect of errors by rapidly degrading faulty proteins (83), but such a mechanism may not be extensively utilized in E. coli (198). To achieve the synthesis of active enzymes and to conserve energy by minimizing the costly synthesis of defective proteins, cells must have evolved mechanisms of transcription and translation capable of achieving accuracies greater than one error in 10^3 to 10^4. The error rate is about 1.4×10^{-4} for transcription of DNA into RNA in E. coli (233) and is thought to be lower than that for translation. We are therefore concerned with how translational fidelity is achieved, with special emphasis on missense-type errors.

As indicated in earlier sections of this chapter, two steps of protein synthesis are important for the accurate translation of the genetic code: the selection of the amino acid for aminoacylation of tRNA by the synthetase, and the selection of an aminoacyl-tRNA for binding to the mRNA's codon in the ribosomal A site. Errors at either step will result in the incorporation of a different amino acid at a specific site in the protein. I shall review here work designed to measure the in vivo rates of missense errors and discuss how the translational apparatus functions to minimize such mistakes. I will also examine frameshift errors and how they may be utilized in special instances of gene expression.

Missense Error Rates

The first estimate of the in vivo missense error rate of protein synthesis was made by Loftfield and Vanderjagt (158), who measured valine incorporation into specific peptides of α-globin and chicken oviduct ovalbumin that normally contain no valine. Valine replaced threonine at a frequency of 2×10^{-4}, but misincorporation in place of isoleucine was not detectable. Edelmann and Gallant (64) measured the incorporation of cysteine into the E. coli protein flagellin (a protein lacking cysteine) and found 6×10^{-4} mol of cysteine per mol of flagellin. Similar results were obtained recently by measuring cysteine incorporation into the bacteriophage T7 0.3 protein (224). The rates of misincorporation of tryptophan or histidine into bacteriophage Qβ coat protein (which lacks the two amino acids) were measured at about 10^{-3} per codon translated (139). The above experiments require a rigorous purification of the proteins studied, a precise determination of the specific activities of the aminoacyl-tRNA pools, and a high degree of radiochemical purity of the amino acids utilized, conditions not readily achieved. In the examples with bacterial systems, the specific sites of erroneous incorporation were not identified. Bouadloun et al. (22) measured the rate of misincorporation of cysteine at the unique arginine codon of E. coli ribosomal protein L7/L12 and at the unique tryptophan codon of S6, obtaining error rates of about 1×10^{-3} and 3×10^{-3} to 4×10^{-3}, respectively. The experimental approach cleverly exploited the fact that correctly synthesized L7/L12 or S6 can be cleaved at Arg or Trp, whereas molecules substituted with labeled Cys at those sites are not cleaved. Another approach to estimating error rates is to measure isoelectric variants by two-dimensional isoelectric focusing and sodium dodecyl sulfate-polyacrylamide gel electrophoresis. Parker et al. (199) examined bacteriophage MS2 coat protein for misincorporation of lysine for asparagine. Error rates of 1.3×10^{-3} and 2×10^{-4} were detected for codons 3 and 12 in the protein. In summary, in vivo error rates appear to be in the range of 10^{-3} to 10^{-5}, although even higher rates may occur at specific sites.

The error frequencies measured above were made in wild-type bacteria growing in rich media. Misincorporation rates are substantially higher in amino acid-starved cells, where a rate of 0.1 has been seen at the codons for the starved amino acid (200). Certain aminoglycoside antibiotics including streptomycin, kanamycin, and neomycin also cause an increase in misreading of the genetic code. Finally, mutations in the genes for components of the translational machinery alter the fidelity of protein synthesis, in some cases causing an increase and in other cases a decrease in error rate.

The rate of missense incorporation of amino acids also has been studied in vitro systems. Analyses with synthetic polynucleotides as mRNAs often show quite high rates of misincorporation of amino acids (reviewed by Yarus [304]). Jelenc and Kurland have developed a cell-free translation system that closely mimics in vivo fidelity and efficiency (127). They employ purified ribosomes, elongation factors, poly(U), AcPhe-tRNA for more efficient initiation, tRNA^{Phe}, phenylalanine tRNA synthetase, ATP, GTP, an energy-generating system, and a complex mixture of polyvalent ions designed to resemble those found in intact cells. The system at 37°C polymerizes phenylalanine at the rate of about 10 peptide bonds per s per active ribosome (comparable to the in vivo rate of about 16), with a leucine missense frequency of 4×10^{-4} (281). From poly(UG) initiated with AcVal-tRNA, poly(Cys-Val) synthesis proceeds with a missense substitution rate of 10^{-4} for Trp at the Cys codons and 10^{-3} for Met at the Val codons (4). Systems utilizing synthetic homopolymers may more readily deplete the pool of charged cognate aminoacyl-tRNAs due to the high frequency of their utilization and thereby cause abnormally high error rates (34). This problem was solved by using a properly constructed purified system (4). These highly active and accurate in vitro systems are especially useful in probing the molecular mechanisms of translational fidelity as described below.

Misacylation of tRNAs

The problem. In the attachment of an amino acid to tRNA catalyzed by the aminoacyl-tRNA synthetases, two selections are made: the correct amino acid and a member of the cognate isoacceptor tRNA family. tRNA molecules are relatively large and complex, providing ample opportunity for the synthetase proteins to distinguish one from another, even though the precise mechanism for such discrimination is not yet known. In contrast, amino acids are small molecules, many differing from another by a single methylene group (e.g., Gly and Ala, Ser and Thr, Val and Ile, Asp and Glu). One can readily understand how alanine tRNA synthetase can exclude amino acids larger than Ala from a tight-fitting binding cavity by steric hindrance, but how can it discriminate against Gly, which is one methylene group smaller? Pauling (203) calculated that the binding energy of a methylene group can contribute only a factor of 100 to 200 to the specificity of binding. How then does a synthetase distinguish one amino acid from another with a discrimination factor of about 10^4?

Proofreading or editing pathways. Analysis of amino acid discrimination in the first step of the aminoacylation reaction, the formation of an aminoacyl-adenylate, generally proceeds with a discrimination ratio between cognate and near-cognate amino acids of only about 10^2. Norris and Berg (186) studied isoleucine tRNA synthetase complexes with either Ile-AMP or Val-AMP. When tRNA^{Ile} was added, the Ile-AMP complex generated Ile-tRNA and AMP, whereas the Val-AMP complex resulted in the hydrolysis of the Val-AMP. Apparently, isoleucine tRNA synthetase is able to distinguish between cognate and near-cognate aminoacyl-adenylates further down the pathway of aminoacylation. Hopfield (110) proposed a kinetic proofreading mechanism in which discrimination occurs at two binding reactions involving, first, the amino acid before aminoacyl-adenylate formation, and second, the aminoacyl-adenylate before its reaction with tRNA. With two binding equilibria discriminating at 100-fold, separated by an energy-consuming, kinetically irreversible step, an overall discriminatory factor of 10^4 would be possible. Another type of proofreading process was suggested by the observation that the synthetases are weak esterases capable of hydrolyzing aminoacyl-tRNAs (65, 303). In this mechanism, tRNA charged with a noncognate amino acid would be hydrolyzed at a rate greater than that with the cognate amino acid.

The esterase editing mechanism has been rigorously established for a number of aminoacyl-tRNA synthetases. Fersht and Kaethner have shown that valine tRNA synthetase will attach threonine or α-aminobutyric acid to tRNA^{Val} with a discriminatory factor of only 10^2 (73). However, the misacylated tRNAs are rapidly and specifically hydrolyzed by valine tRNA synthetase before they dissociate from the enzyme. Fersht and Dingwall (71) suggested that a double-sieve mechanism is used. The enzyme first sorts the amino acids by size and possibly by chemical characteristics (e.g., charged or polar groups, etc.), excluding those larger than the cognate amino acid, and attaches the cognate and smaller amino acids to the tRNA, using some discrimination in the binding reaction. Finally, it sorts the aminoacyl-tRNAs by size, excluding the cognate aminoacyl-tRNA but hydrolyzing those carrying the smaller amino acids. The esterase site appears to be different from the aminoacylation site, since substrates of the aminoacylation reaction do not inhibit the esterase function (72, 303). A similar chemical proofreading mechanism was proposed for the *S. cerevisiae* isoleucine tRNA synthetase (280), as well as for a number of other *S. cerevisiae* aminoacyl-tRNA synthetases (117).

An editing function before transfer of the amino acid to tRNA also can occur with some enzymes. In this mechanism, near-cognate aminoacyl-adenylates are hydrolyzed more rapidly than cognate aminoacyl-adenylates. Jakubowski and Fersht (121) showed that isoleucine and methionine tRNA synthetases establish accuracy by their pretransfer editing mechanism. They argue that pretransfer proofreading by dissociation of the aminoacyl-adenylate from the enzyme (Hopfield's kinetic proofreading hypothesis) is not likely to be an important mechanism, however.

Errors of Aminoacyl-tRNA–mRNA Recognition

The problem. I have described the binding of
aminoacyl-tRNA (in ternary complex with EF-Tu and
GTP) to the ribosomal A site. The selection of a
particular species of aminoacyl-tRNA is dictated by
the mRNA and involves an interaction between the
mRNA codon and the anticodon of the tRNA. This
interaction follows Watson-Crick hydrogen bonding
rules, except for the interaction at the third code
letter, where some looseness or "wobble" allows
nonclassical interactions. The problem is to explain
discriminations of tRNA binding of 10^3 to 10^4 on the
basis of the RNA triplet-triplet interaction.

The strengths of RNA base-pair interactions in
aqueous solution have been calculated from the melting temperatures of synthetic double-helical RNAs
(274). These values are vastly insufficient to account
for the discrimination observed. Either the ribosome
magnifies the energy differences between correct and
near-correct interactions, possibly by creating a non-
aqueous environment in the decoding site, or some
kind of proofreading mechanism must apply. The
former possibility seems plausible, given the location
of the decoding site deep in the cleft of the 30S
subunit. To test discrimination in the binding step,
Thompson and Karim (271) compared the dissociation equilibrium constant for poly(U)-programmed
ribosomes with those for Phe-tRNA^{Phe} (codon-
anticodon pairing UUU:GAA) and Leu-tRNA$_2^{Leu}$
(UUU:GAG), which differ only by their interaction at
the first code letter. The constant for Phe-tRNA^{Phe} is 8
$\times 10^{-10}$ M, whereas that for Leu-tRNA$_2^{Leu}$ is 1.6×10^{-5}
M, a 2×10^4-fold discrimination. Thus, at the thermodynamic level, it appears that the ribosome is capable
of discriminating near-cognate aminoacyl-tRNAs
solely on the basis of binding affinities. However, this
capability to discriminate so highly requires establishing near-equilibrium conditions, a situation not
realized in vivo. We shall therefore examine the possibility that proofreading mechanisms also apply to
protein synthesis.

Proofreading mechanisms. Two kinds of proofreading mechanisms were proposed independently by
Hopfield (110) and Ninio (179). Hopfield proposed that
discrimination of aminoacyl-tRNAs could occur at
two places in the elongation pathway: at ternary
complex binding and after the EF-Tu–catalyzed
GTPase reaction. Noncognate aminoacyl-tRNAs would
dissociate from either complex more rapidly than
cognate tRNAs, gaining at each intermediate a discriminatory factor of about 10^2. In this mechanism,
energy is employed to separate the two binding states
by a kinetically irreversible reaction and to generate a
high-energy aminoacyl-tRNA–ribosome complex which
could not be formed by the back reaction of the
proposed second dissociation. Ninio used a probablis-
tic approach, focusing on the time aminoacyl-tRNA
remains bound to the ribosome and the effects of
delaying the time of subsequent peptide bond formation. The Hopfield and Ninio schemes are analogous
mathematically and both provide two exit sites for the
substrate, but they differ in the position of the irreversible step relative to the peptide-forming step.

Thompson and co-workers have generated evidence
in support of these proofreading hypotheses. Using

both cognate and near-cognate aminoacyl-tRNAs,
they measured the rate constants for reactions in the
elongation pathway (Fig. 5). Poly(U)-programmed
ribosomes were mixed with either Phe-tRNA^{Phe} or
Leu-tRNA$_2^{Leu}$, and single or multiple turnover kinetics
were measured by rapid mixing and chemical quench
methods (reviewed by Yarus and Thompson [306]).
Values for both tRNA type lead to the following
conclusions. The rate of Leu-tRNA$_2^{Leu}$ binding (k_1) is at
least 1/12 that of Phe-tRNA^{Phe}, indicating rather little
discrimination at that step. On the other hand, the
rate of dissociation (k_{-1}) is at least 3,000 times faster
for the near-cognate species, thus providing the capability of high discrimination. However, the rate of the
EF-Tu–catalyzed GTPase reaction (k_2) is rather similar for both tRNAs and is essentially comparable in
magnitude to the dissociation rate (k_{-1}) for Leu-
tRNA$_2^{Leu}$. This means that the ribosome provides
insufficient time to exploit the discriminatory potential of the ternary complex binding reaction, but
rather elects to progress down the pathway rapidly.
The rate of peptide bond formation is comparable for
both tRNAs and is equal to the rate of release of
EF-Tu–GDP in the absence of tRNA (270a). It is
therefore possible that release of EF-Tu–GDP is the
overall rate-limiting step. Thompson postulates that
the slow release of EF-Tu allows ample opportunity
for noncognate aminoacyl-tRNAs to dissociate, thus
providing the second discriminatory binding interaction postulated by Hopfield. A direct measure of the
rates of release of cognate and near-cognate aminoacyl-tRNAs after GTP hydrolysis, or of the actual rate
of peptide bond formation, has not yet been accomplished.

A different in vitro approach to determining
whether proofreading is used during protein synthesis
has been taken by Kurland and co-workers (238).
Using poly(U)-programmed ribosomes initiated with
AcPhe-tRNA^{Phe}, a system exhibiting high efficiency
and fidelity when physiological salt concentrations
are used, these workers limited the steady-state rate of
protein synthesis by omitting EF-Ts. In the absence of
EF-Ts, most of the EF-Tu is present in the inactive
EF-Tu–GDP complex, and the rate of protein synthesis
is governed by the rate of uncatalyzed GTP-GDP
exchange on EF-Tu. They then altered the ratio of
cognate to near-cognate aminoacyl-tRNAs and measured the number of amino acids incorporated. The
results indicated that leucine incorporation requires
50 times as many EF-Tu cycles as phenylalanine
incorporation. This is consistent with the proofreading mechanism, where Leu-tRNA^{Leu} is rejected after
GTP hydrolysis 50 times more frequently than is
Phe-tRNA^{Phe}.

Factors influencing fidelity. Mutations in ribosomal
proteins and the presence of some antibiotics influence the accuracy of translation (reviewed in reference 306). The *str* and *ram* mutants studied by Gorini
(86) indicate that ribosomal accuracy can be altered
in either direction, which means that protein synthesis is not maximized for accuracy. Therefore the ribosome is intrinsically capable of much greater accuracy, were it to slow the rate of protein synthesis and
thus allow noncognate aminoacyl-tRNAs to dissociate
more completely. However, slowing the rate of EF-
Tu–GDP release and peptide bond formation also

would result in increased dissociation of both cognate and noncognate aminoacyl-tRNAs, thereby increasing the energy cost per peptide bond formed. A group of bacterial strains that are mutated in *rpsL* (the gene for ribosomal protein S12) show the streptomycin-resistant phenotype and exhibit a slower rate of protein synthesis and a lower frequency of nonsense suppression (19). Ribosomes from the mutant bacteria show enhanced proofreading (rejection) of Leu-tRNA$_2^{Leu}$ in the poly(U)-programmed system of Kurland and co-workers (19), as well as increased dissipative loss of cognate ternary complexes. Slower growth of such mutant strains may be due to the kinetically less efficient ribosomes and to the higher energy cost per peptide bond synthesized. Streptomycin-dependent strains also exhibit slow growth and slow elongation rates in the absence of streptomycin. The stimulation of growth by streptomycin may result from enhanced translational efficiency due to a decrease in proofreading (237). There is general agreement that accuracy and speed are mutually antagonistic needs of bacterial cells. Apparently, a certain level of error rate (i.e., in the vicinity of 10^{-3} to 10^{-4}) is tolerated so that protein synthesis can proceed more rapidly and cheaply. Since protein synthesis is rate limiting for cell growth in rich media, this parameter is clearly an important factor in species viability.

Recent studies indicate that EF-Tu is involved in nonsense suppression (278). Various *lacI-lacZ* fusions carried on F′ episomes were constructed with one of the three terminator codons placed in the reading frame of *lacI*. The synthesis of β-galactosidase (dependent on readthrough of the terminator codon) was measured in strains with altered *tuf* genes. Readthrough is reduced when *tufB* is inactivated, but it is restored by introducing *tuf* genes on plasmids. This suggests that gene dosage, or the level of EF-Tu, influences readthrough. When strains carrying mutant forms of *tufA* and *tufB* were tested, readthrough was enhanced only if both EF-Tu genes were mutated. This surprising result implies that nonsense suppression requires the combined action of two different mutant EF-Tu species.

The concentrations of polycations also influence fidelity. As noted by Jelenc and Kurland (127), Mg^{2+}, spermidine, and putrescine are particularly important, and increases in Mg^{2+} concentration increase the error frequency. Since many in vitro translation systems employ high Mg^{2+} concentrations and lack spermidine and putrescine, they can be expected to operate with lower fidelity than in vivo systems. Mutant cell lines with low concentrations of polyamines are affected in protein synthesis (261).

Another influence on fidelity is codon context (reviewed by Buckingham and Grosjean [28] and Yarus and Thompson [306]). This has been studied in detail by measuring the efficiency of nonsense suppression or by analyzing misincorporation of amino acids in vitro (34) or in vivo (176). Nucleotides adjacent to the misread codon influence the error frequency. However, it is not clear whether the tRNA interacts with more than three nucleotides, or whether the effect is due to the presence of a specific species of tRNA in the P site where the interaction of the two tRNAs influences the rate of peptide bond formation.

Finally, changing the ratio of cognate to noncognate aminoacyl-tRNAs could be expected to greatly stimulate error frequencies by amino acid starvation. Such an increase in error frequency has been observed in cells unable to synthesize ppGpp, but not in wild-type cells (187, 200). How ppGpp prevents an increase in error frequency is not known. The ratio of cognate to noncognate aminoacyl-tRNAs depends directly on the levels of tRNA isoacceptor species in the cell. It is possible that error frequencies therefore are influenced by codon usage in the mRNAs.

Frameshift Errors

Frameshift errors strongly affect the protein product, since translation in a different reading frame leads to the incorporation of entirely different amino acids and to the likely encounter of a termination codon. Evidence for frameshift errors is readily detected by measuring the activity levels of a protein whose gene carries a frameshift mutation. The synthesis of significant amounts of β-galactosidase in strains carrying a frameshift mutation in *lacZ* demonstrates that frameshift errors occur in vivo (7). Another system, the *rIIB* gene of bacteriophage T4, also has been studied in detail (287). It was found that factors which influence missense errors (e.g., ribosomal mutations, amino acid starvation) also affect the frequency of frameshift errors, suggesting a correlation between mistranslation and frame shifting (149). The mechanism for frameshift errors is not known, but a number of possibilities have been proposed. Weiss (286) analyzed those cases in which the "shifty" tRNA and codon are known and proposed that the universal U$_{33}$ adjacent to the anticodon may act as the third member of a shifty tRNA anticodon. Both two-base and four-base translocations can then be rationalized on the basis of the ratchet mechanism of translocation proposed by Woese (299).

An alternative approach is to examine frameshift suppressor tRNAs (reviewed by Roth [235]). An unusual example is the *sufJ* mutation in *Salmonella typhimurium* that results in an extra base in the anticodon loop of a tRNA^{Thr} (21). In this case, four-base codon recognition in the A site does not occur, but on translocation the mRNA effectively moves by four bases.

There are several cases of frameshift hot spots in vivo which shed light on the mechanism. Examples include the bacteriophage T7 major capsid protein (62), *trpE* in *S. typhimurium* (8), and an *S. cerevisiae* mitochondrial cytochrome *c* oxidase gene (75) in which frameshift events occur at a frequency of 1×10^{-2} to 5×10^{-2}. Another example comes from the *E. coli* gene for release factor RF-2. Analysis of the DNA sequence of the gene indicates that the factor is coded by two different open reading frames (46). The first reading frame contains the termination signal UGA at codon 26. Amino acid sequence analysis of RF-2 indicates that a frameshift seems to occur at the CUU codon recognized by a shifty tRNA (Leu-tRNA$_{CUN}$) just before the UGA, resulting in avoidance of the termination signal and continued translation in the second reading frame.

Evidence also exists for a frameshift event during translation of the coat cistron of bacteriophage MS2 that results in premature termination of coat synthesis followed by initiation of translation at the beginning of the overlapping lysis cistron (138). This intriguing possibility was ruled out when site-directed mutagenesis of the putative out-of-phase termination codon resulted in no change in lysis gene expression (J. van Duin, personal communcation). It remains possible that a frameshift occurs downstream from the termination codon in the region where the coat and lysis coding sequences overlap, thereby generating a fusion protein containing both coat and lysis sequences.

PERSPECTIVES

A striking feature of the mechanism of protein synthesis is the dearth of covalent bond-making or -breaking reactions. The esterification of amino acids with their tRNAs, the $O \rightarrow N$ acyl shift that is the peptide bond-forming reaction, hydrolysis of pyrophosphate linkages in ATP and GTP, and the hydrolysis of peptidyl-tRNA are the only covalent reactions in the pathway. A second striking feature is the complexity of the translational apparatus and the overwhelming abundance of noncovalent interactions. The major challenge in understanding the molecular mechanism of protein synthesis is to explain the RNA-RNA, RNA-protein, and protein-protein interactions in terms of detailed molecular structures and the basic chemical principles governing them.

Our knowledge of the structure of the ribosome has been increasing steadily and has advanced to the point where questions concerning structure and function can be asked in sharp detail. The primary structures of nearly all of the macromolecular components of translation are known. The important functional role played by rRNA in the process is already recognized, and a description of the spatial arrangement of rRNA in ribosomal particles relative to the overall shape and the protein components is anticipated. X-ray diffraction studies of ribosomal crystals may help refine the structure, although a detailed analysis at the atomic level appears beyond immediate reach. On the other hand, recombinant DNA techniques now enable researchers to prepare large quantities of factors and other translational components. Certainly these proteins in principle are amenable to X-ray analysis, and some are already being studied. The ability to alter precisely the structure of the RNA and protein components by in vitro mutagenesis and then to reconstitute an active translation system in vitro will be important in future structure-function studies. Although the anticipated advances in our knowledge of the structures of the ribosome and its complexes are exciting, it is essential to recognize that these complexes are not static, but likely are exceedingly dynamic. A fluctuating ribosome may be difficult to define, yet this characteristic is surely important for its function.

Another area of research that is required for a more complete understanding of protein synthesis is kinetics. Some important studies have begun to define the kinetic parameters of the various reactions in the pathway, and these have shed light on the process of proofreading. A future goal is to provide rate constants for all of the partial reactions in the pathway of protein synthesis. The approach is particularly impor-

tant for understanding translational control, which needs to be explained in terms of alterations of rate-limiting reactions.

Most of our knowledge concerning the mechanism of protein synthesis is based on in vitro studies with purified components. Classical genetics has been of rather limited use in characterizing this complex pathway. Through the use of recombinant DNA techniques, we now have the opportunity to try to construct mutant components in vitro and to construct bacterial strains carrying such altered genes. An exciting prospect is to begin to study protein synthesis in vivo. For example, studies involving inactivating a component (e.g., a protein factor) and observing the effects of protein synthesis and other aspects of cellular metabolism in the intact cell are now possible. Such studies should provide confirmation or revisions of our understanding of translation and should help to elucidate how protein synthesis relates to and is integrated with other metabolic processes in the bacterial cell.

ACKNOWLEDGMENTS

I thank M. Grunberg-Manago, S. Blanquet, J.-P. Ebel, R. Traut, and J. Ingraham for helpful suggestions during the preparation of the text.

This work was supported by grants NP-70 from the American Cancer Society and INT8312982 from the National Science Foundation.

LITERATURE CITED

1. **Allende, J. E., R. Monro, and F. Lipmann.** 1964. Resolution of the E. coli amino acyl sRNA transfer factor into two complementary fractions. Proc. Natl. Acad. Sci. USA **51**:1211–1216.
2. **An, G., D. S. Bendisck, L. A. Mamelack, and J. D. Friesen.** 1981. Organization and nucleotide sequence of a new ribosomal protein operon in E. coli containing the genes for ribosomal protein S2 and elongation factor Ts. Nucleic Acids Res. **9**:4163–4171.
3. **An, G., and J. D. Friesen.** 1980. The nucleotide sequence of tufB and four nearby tRNA structural genes of E. coli. Gene **12**:33–39.
4. **Andersson, S. G. E., R. H. Buckingham, and C. G. Kurland.** 1984. Does codon-anticodon composition influence ribosome functions? EMBO J. **3**:91–94.
5. **Arai, K., M. Kawakita, and Y. Kaziro.** 1972. Studies on polypeptide elongation factors from E. coli. Purification of factors Tu-guanosine diphosphate, Ts and Tu-Ts, and crystallization of Tu-guanosine diphosphate and Tu-Ts. J. Biol. Chem. **247**:7029–7037.
6. **Arai, K., M. Kawakita, and Y. Kaziro.** 1974. Studies on the polypeptide elongation factors from E. coli. Properties of various complexes containing EF-Tu and EF-Ts. J. Biochem. (Tokyo) **76**:293–306.
7. **Atkins, J. F., D. Elseviers, and L. Gorini.** 1972. Low activity of β-galactosidase in frameshift mutants of E. coli. Proc. Natl. Acad. Sci. USA **69**:1192–1195.
8. **Atkins, J. F., B. P. Nichols, and S. Thompson.** 1983. The nucleotide sequence of the first externally suppressible −1 frameshift mutant and of some nearby leaky frameshift mutants. EMBO J. **2**:1345–1350.
9. **Baca, O. G., M. S. Rohrbach, and J. W. Bodley.** 1976. Equilibrium measurements of the interactions of guanine nucleotides with E. coli elongation factor G and the ribosome. Biochemistry **15**:4570–4574.
10. **Backendorf, C., G. P. Overbeck, J. H. van Boom, G. van der Marel, G. Veeneman, and J. van Duin.** 1980. Role of 16S RNA in ribosome messenger recognition. Eur. J. Biochem. **110**:599–604.
11. **Baldwin, A. W., and P. Berg.** 1966. Purification and properties of isoleucyl ribonucleic acid synthetase from E. coli. J. Biol. Chem. **241**:831–838.
12. **Barrell, B. G., and B. F. C. Clark.** 1974. Handbook of nucleic acid sequences, p. 32–35. Joynson-Bruvvers, Oxford.
13. **Baumstark, B. R., L. L. Spremulli, U. L. RajBhandary, and G. M. Brown.** 1977. Initiation of protein synthesis without formylation in a mutant of Escherichia coli that grows in the absence of tetrahydrofolate. J. Bacteriol. **129**:457–471.
14. **Benne, R., N. Naaktgeboren, J. Gubbens, and H. O. Voorma.**

1973. Recycling of initiation factors IF1, IF2, and IF3. Eur. J. Biochem. **32:**372–380.

15. **Bermek, E.** 1978. Mechanisms in polypeptide chain elongation on ribosomes. Prog. Nucleic Acid Res. Mol. Biol. **21:**63–100.

16. **Bhat, T. N., D. M. Blow, P. Brick, and J. Nyborg.** 1982. Tyrosyl-tRNA synthetase forms a mononucleotide-binding fold. J. Mol. Biol. **158:**699–709.

17. **Bjork, G. R.** 1984. Modified nucleosides in RNA—their formation and function, p. 291–300. *In* D. Apirion (ed.), Processing of RNA. CRC Press, Inc., Boca Raton, Fla.

18. **Blow, D. M., and P. Brick.** 1985. Aminoacyl-tRNA synthetase, p. 441–469. *In* F. A. Jurnak and A. McPherson (ed.), Biological macromolecules and assemblies, vol. 2: Nucleic acids and interactive proteins. John Wiley and Sons, Inc., New York.

19. **Bohman, K., T. Ruusala, P. C. Jelenc, and G. C. Kurland.** 1984. Kinetic impairment of restrictive streptomycin-resistant ribosomes. Molec. Gen. Genet. **198:**90–99.

20. **Boileau, G., P. Butler, J. W. B. Hershey, and R. R. Traut.** 1983. Direct cross-links between initiation factors 1, 2, and 3 and ribosomal proteins promoted by 2-iminothiolane. Biochemistry **22:**3162–3170.

21. **Bossi, L., and D. M. Smith.** 1984. Suppressor *sufJ*: a novel type of tRNA mutant that induces translational frameshifting. Proc. Natl. Acad. Sci. USA **81:**6105–6109.

22. **Bouadloun, F., D. Donner, and C. G. Kurland.** 1983. Codon-specific missense errors *in vivo*. EMBO J. **2:**1351–1356.

23. **Branden, C.-I.** 1980. Relation between structure and function of α/β-proteins. Q. Rev. Biophys. **13:**317–339.

24. **Brauer, D., and B. Wittmann-Liebold.** 1977. The primary structure of the initiation factor IF-3 from *E. coli*. FEBS Lett. **79:**269–275.

25. **Brenner, S., A. O. W. Stretton, and S. Kaplan.** 1965. Genetic code: the "nonsense" triplets for chain termination and their suppression. Nature (London) **226:**994–998.

26. **Brot, N.** 1977. Translocation, p. 375–411. *In* H. Weissbach and S. Pestka (ed.), Molecular mechanisms of protein synthesis. Academic Press, Inc., New York.

27. **Brot, N., R. Marcel, E. Yamasaki, and H. Weissbach.** 1973. Further studies on the role of 50 S ribosomal protein synthesis. J. Biol. Chem. **248:**6952–6956.

28. **Buckingham, R. H., and H. Grosjean.** 1985. The accuracy of messenger RNA: transfer RNA recognition, p. 83–126. *In* D. J. Galas, T. B. L. Kirkwood, and R. Rosenberger (ed.), Accuracy of molecular processes. Chapman and Hall, London.

29. **Burma, D. P., D. S. Tewari, and A. K. Srivastava.** 1985. Ribosomal activity of the 16S-23S RNA complex. Arch. Biochem. Biophys. **239:**427–435.

30. **Calagan, J. L., R. Pirtle, I. Pirtle, M. Keshdan, H. Vreman, and B. Dudock.** 1980. Homology between chloroplast and prokaryotic initiator tRNA. J. Biol. Chem. **255:**9981–9984.

31. **Cantoni, G. L., and D. R. Davies.** 1971. Procedures in nucleic acid research, Harper, New York.

32. **Capecchi, M. R.** 1967. Polypeptide chain termination *in vitro*: Isolation of a release factor. Proc. Natl. Acad. Sci. USA **58:**1144–1151.

33. **Capecchi, M. R., and H. A. Klein.** 1969. Characterization of three proteins involved in polypeptide chain termination, Cold Spring Harbor Symp. Quant. Biol. **34:**469–477.

34. **Carrier, M. J., and R. H. Buckingham.** 1984. An effect of codon context in the mistranslation of UGU codons *in vitro*. J. Mol. Biol. **175:**29–38.

35. **Carter, C. W., Jr., and C. W. Carter.** 1979. Protein crystallization using incomplete factorial experiments. J. Biol. Chem. **254:**12219–12223.

36. **Caskey, C. T.** 1977. Peptide chain termination, p. 443–465. *In* H. Weissback and S. Pestka (ed.), Molecular mechanisms of protein biosynthesis. Academic Press, Inc., New York.

37. **Caskey, C. T., and A. L. Beaudet.** 1971. Antibiotic inhibitors of peptide termination, p. 326–336. *In* E. Munoz, F. Garcias-Fernandez, and D. Vasquez (ed.), Molecular mechanisms of antibiotic action on protein biosynthesis and membranes. Elsevier, Amsterdam.

38. **Caskey, C. T., W. C. Forrester, and W. Tate.** 1984. Peptide chain termination, p. 149–160. *In* B. F. C. Clark and H. U. Petersen (ed.), Gene expression: the translational step and its control. Munksgaard, Copenhagen.

39. **Caskey, C. T., E. Scolnick, T. Caryk, and M. Nirenberg.** 1968. Sequential translation of trinucleotide codons for the initiation and termination of protein synthesis. Science **162:**135–138.

40. **Cassio, D., and J. P. Waller.** 1971. Modification of methionyl-tRNA synthetase by proteolytic cleavage and properties of the trypsin-modified enzyme. Eur. J. Biochem. **20:**283–300.

41. **Chambliss, G., G. R. Craven, J. Davies, K. Davis, L. Kahan, and M. Nomura (ed.).** 1980. Ribosomes: structure, function, and genetics. University Park Press, Baltimore.

42. **Clark, B. F. C.** 1980. The elongation step of protein biosynthesis. Trends Biochem. Sci. **5:**207–210.

43. **Clark, B. F. C., T. F. M. la Cour, K. M. Nielsen, J. Nyborg, H. U. Petersen, G. E. Siboska, and F. P. Wikman.** 1984. Structure of bacterial elongation factor EF-Tu and its interaction with aminoacyl-tRNA, p. 127–145. *In* B. F. C. Clark and H. U. Petersen (ed.), Gene expression: the translational step and its control. Munksgaard, Copenhagen.

44. **Clark, B. F. C., and K. Marcker.** 1966. The role of N-formyl-methionyl-sRNA in protein biosynthesis. J. Mol. Biol. **17:**394–406.

45. **Coleman, J., M. Inouye, and K. Nakamura.** 1985. Mutations upstream of the ribosome-binding site affect translational efficiency. J. Mol. Biol. **181:**139–143.

46. **Craigen, W. J., R. G. Cook, W. P. Tate, and C. T. Caskey.** 1985. Bacterial peptide chain release factors: conserved primary structure and possible frameshift regulation of release factor 2. Proc. Natl. Acad. Sci. USA **82:**3616–3620.

47. **Crane, L. J., and D. L. Miller.** 1974. Guanosine triphosphate and guanosine diphosphate as conformation-determining molecules. Differential interaction of a fluorescent probe with the guanosine nucleotide complexes of bacterial elongation factor Tu. Biochemistry **13:**933–939.

48. **Crick, F. H. C.** 1957. Discussion. Biochem. Soc. Symp. **14:**25–26.

49. **Crothers, D. M., and P. E. Cole.** 1978. Conformation changes of tRNA, p. 196–247. *In* S. Altman (ed.), Transfer RNA. MIT Press, Cambridge, Mass.

50. **Cundliffe, E.** 1980. Antibiotics and prokaryotic ribosomes: action, interaction and resistance, p. 555–581. *In* G. Chambliss, G. R. Craven, J. Davies, K. Davis, L. Kahan, and M. Nomura (ed.), Ribosomes: structure, function and genetics. University Park Press, Baltimore.

51. **Dahlberg, A.** 1986. Site directed mutagenesis of *E. coli* rRNA, p. 686–698. *In* B. Hardesty and G. Kramer (ed.), Structure, function and genetics of ribosomes. Springer-Verlag, New York.

52. **Dardel, F., G. Fayat, and S. Blanquet.** 1984. Molecular cloning and primary sequencing of the *Escherichia coli* methionyl-tRNA synthetase gene. J. Bacteriol. **160:**1115–1122.

53. **de Boer, H., P. Ng, and A. Hui.** 1985. Synthesis of specialized ribosomes in *E. coli*. UCLA Symp. Mol. Cell. Biol. New Ser. **30:**419–437.

54. **Dessen, P. A. Ducruix, C. Hountondji, R. P. May, and S. Blanquet.** 1983. Neutron scattering study of the binding of tRNA^Phe to *E. coli* phenylalanyl-tRNA synthetase. Biochemistry **22:**281–284.

55. **Deutch, C. E., R. C. Scarpulla, and R. L. Soffer.** 1978. Posttranslational NH$_2$-terminal aminoacylation. Curr. Top. Cell. Regul. **13:**1–28.

56. **De Venditis, E., M. Mariorosario, and V. Bocchini.** 1986. The elongation factor G carries a catalytic site for GTP hydrolysis, which is revealed by using 2-propanol in the absence of ribosomes. J. Biol. Chem. **261:**4445–4450.

57. **Dickerman, H. W., E. Steers, Jr., B. G. Redfield, and H. Weissbach.** 1967. Methionyl soluble RNA transformylase. I. Purification and partial characterization. J. Biol. Chem. **242:**1522–1525.

58. **Domogatskii, C. P., T. N. Vlasik, and T. A. Bezlepkina.** 1979. RNA binding activity of prokaryotic initiation factors. Dokl. Akad. Nauk SSSR. **248:**240–243.

59. **Dondon, J., J. A. Plumbridge, J. W. B. Hershey, and M. Grunberg-Manago.** 1985. Overproduction and purification of initiation factor IF2 and pNUSA proteins from a recombinant plasmid-bearing strain. Biochimie **67:**643–649.

60. **Dubnoff, J. S., A. H. Lockwood, and U. Maitra.** 1972. Studies on the role of guanosine triphosphate in polypeptide chain initiation in *E. coli*. J. Biol. Chem. **247:**2884–2894.

61. **Dunn, J. J., E. Buzash-Pollert, and F. W. Studier.** 1978. Mutations of bacteriophage T7 that affect initiation of synthesis of gene 0.3 protein. Proc. Natl. Acad. Sci. USA **75:**2741–2745.

62. **Dunn, J. J., and F. W. Studier.** 1983. Complete nucleotide sequence of bacteriophage T7 DNA and the location of T7 genetic elements, J. Mol. Biol. **166:**477–535.

63. **Ebel, J. P., M. Renaud, A. Dietrich, G. Fesiolo, G. Keitle, O. O. Favorova, S. Vassilenko, M. Baltginger, R. Ehrlich, P. Remy, J. Bonnet, and R. Giege.** 1979. Interaction between tRNA and aminoacyl tRNA synthetase in the valine and phenylalanine systems from yeast, p. 325–343. *In* P. R. Schimmel, D. Söll, and J. N. Abelson (ed.), Transfer RNA: structure, properties and recognition. Cold Spring Harbor Laboratory, Cold Spring Harbor, N.Y.

64. **Edelmann, P., and J. Gallant.** 1977. Mistranslation in *E. coli.* Cell **10**:131–137.
65. **Eldred, E. W., and P. R. Schimmel.** 1972. Investigation of the transfer of amino acid from a tRNA synthetase-aminoacyl adenylate complex to tRNA. Biochemistry **11**:17–23.
66. **Ellis, S., and T. W. Conway.** 1984. Initial velocity kinetic analysis of 30S initiation complex formation in an *in vitro* translation system derived from *E. coli.* J. Biol. Chem. **259**:7607–7614.
67. **Eskin, B., B. Treadwell, B. Redfield, C. Spears, H. Kung, and H. Weissbach.** 1978. Activity of different forms of initiation factor 2 in the *in vitro* synthesis of β-galactosidase. Arch. Biochem. Biophys. **189**:531–534.
68. **Fahnestock, S., and A. Rich.** 1971. Ribosome-catalyzed polyester formation. Science **173**:340–343.
69. **Fahnestock, S., H. Weissbach, and A. Rich.** 1972. Formation of a ternary complex of phenyllactyl-tRNA with transfer factor Tu and GTP. Biochim. Biophys. Acta **269**:62–66.
70. **Fakunding, J. L., R. R. Traut, and J. W. B. Hershey.** 1973. Dependence of initiation factor IF2 activity on proteins L7 and L12 from *E. coli* 50 S ribosomes. J. Biol. Chem. **248**:8555–8559.
71. **Fersht, A. R., and C. Dingwall.** 1979. Evidence for the double-sieve editing mechanism for selection of amino acids in protein synthesis. Steric exclusion of isoleucine by valyl-tRNA synthetases. Biochemistry **18**:2627–2631.
72. **Fersht, A. R., and M. M. Kaethner.** 1976. Mechanism of aminoacylation of tRNA. Proof of the aminoacyl adenylate pathway for the isoleucyl- and tyrosyl-tRNA synthetases from *E. coli* K12. Biochemistry **15**:818–823.
73. **Fersht, A. R., and M. Kaethner.** 1976. Enzyme hyperspecificity. Rejection of threonine by the valyl-tRNA synthetase by misacylation and hydrolytic editing. Biochemistry **15**:3342–3346.
74. **Fournier, M. J., and H. Ozeki.** 1985. Structure and organization of the transfer ribonucleic acid genes of *Escherichia coli* K-12. Microbiol. Rev. **49**:379–397.
75. **Fox, T. D., and B. Weiss-Brummer.** 1980. Leaky +1 and −1 frameshift mutations at the same site in a yeast mitochondrial gene. Nature (London) **288**:60–63.
76. **Fradin, A., H. Gruhl, and H. Feldmann.** 1975. Mapping of yeast tRNAs on two-dimensional electrophoresis on polyacrylamide gels. FEBS Lett. **50**:185–189.
77. **Ganosa, M. C.** 1966. Polypeptide chain termination in cell-free extracts of *E. coli.* Cold Spring Harbor Symp. Quant. Biol. **31**:273–278.
78. **Girshovich, A. S., E. S. Bochkareva, and V. A. Pozdnyakov.** 1974. Affinity labeling of functional centers of *E. coli* ribosomes. Acta Biol. Med. German. **33**:639–648.
79. **Girshovich, A. S., T. V. Kurtskhalia, Y. A. Ovchinnikov, and V. D. Vasiliev.** 1981. Localization of the elongation factor G on *E. coli* ribosomes. FEBS Lett. **130**:54–59.
80. **Goddard, J. P.** 1977. The structures and functions of tRNA. Prog. Biophys. Mol. Biol. **32**:233–308.
81. **Godefroy-Colburn, T., A. D. Wolfe, J. Dondon, M. Grunberg-Manago, P. Dessen, and D. Pantaloni.** 1975. Light-scattering studies showing the effect of initiation factors on the reversible dissociation of *E. coli* ribosomes. J. Mol. Biol. **94**:461–478.
82. **Gold, L., D. Pribnow, T. Schneider, S. Shinedling, B. S. Singer, and G. Stormo.** 1981. Translational initiation in prokaryotes. Annu. Rev. Microbiol. **35**:365–403.
83. **Goldberg, A. L., and A. C. St. John.** 1976. Intracellular protein degradation in mammalian and bacterial cells: part 2. Annu. Rev. Biochem. **45**:747–803.
84. **Gopalakrishna, Y., D. Langley, and N. Sarkar.** 1981. Detection of high levels of polyadenylate-containing RNA in bacteria by use of a single-step RNA isolation procedure. Nucleic Acids Res. **9**:3545–3554.
85. **Gordon, J.** 1970. Regulation of the *in vivo* synthesis of the polypeptide chain elongation factors in *E. coli.* Biochemistry **9**:912–917.
86. **Gorini, L.** 1974. Streptomycin and misreading of the genetic code, p. 791–803. *In* M. Nomura, A. Tissieres, and P. Lengyel (ed.), Ribosomes. Cold Spring Harbor Laboratory, Cold Spring Harbor, N.Y.
87. **Gornicki, P., K. Nurse, W. Hellmann, M. Boublik, and J. Ofengand.** 1984. High resolution localization of the tRNA anticodon interaction site on the *E. coli* 30 S ribosomal subunit. J. Biol. Chem. **259**:10493–10498.
88. **Goss, D. L., L. P. Parkhurst, and A. J. Wahba.** 1982. Kinetic studies on the interaction of chain initiation factor 3 with 70S *E. coli* ribosomes and subunits. J. Biol. Chem. **257**:10119–10127.
89. **Gottlieb, M., and B. D. Davis.** 1975. Irreversible step in the formation of initiation complexes of *E. coli.* Biochemistry **14**:1047–1051.
90. **Griffin, B. E., M. Jarman, C. B. Reese, J. E. Sulston, and D. R. Trentham.** 1966. Some observations relating to acyl mobility in aminoacyl soluble ribonucleic acids. Biochemistry **5**:3638–3649.
91. **Grunberg-Manago, M.** 1980. Initiation of protein synthesis as seen in 1979, p. 445–477. *In* G. Chambliss, G. R. Craven, J. Davies, K. Davis, L. Kahan, and M. Nomura (ed.), Ribosomes: structure, function and genetics. University Park Press, Baltimore.
92. **Grunberg-Manago, M., R. H. Buckingham, B. S. Cooperman, and J. W. B. Hershey.** 1978. Structure and function of the translational machinery. Symp. Soc. Gen. Microbiol. **28**:27–110.
93. **Gualerzi, C., and C. Pon.** 1981. Protein biosynthesis in prokaryotic cells: mechanism of 30S initiation complex formation in *E. coli*, p. 805–826. *In* M. Balaben (ed.), Structural aspects of recognition and assembly in biological macromolecules, vol. II. ISS, Rehovot, Israel.
94. **Gualerzi, C., C. L. Pon, R. T. Pinolik, M. A. Camonaco, M. Paci, and W. Wintermeyer.** 1986. Role of the initiation factors in *E. coli* translational initiation, p. 621–641. *In* B. Hardesty and G. Kramer (ed.), Structure, function, and genetics of ribosomes. Springer-Verlag, New York.
95. **Gualerzi, C., G. Risuleo, and C. L. Pon.** 1977. Initial rate kinetic analysis of the mechanism of initiation complex formation and the role of initiation factor IF-3. Biochemistry **16**:1684–1689.
96. **Hampl, H., H. Schulze, and K. H. Nierhaus.** 1981. Ribosomal components from *E. coli* 50S subunits involved in the reconstruction of peptidyltransferase activity. J. Biol. Chem. **257**:2284–2288.
97. **Hardesty, B., and G. Kramer (ed.).** 1986. Structure, function, and genetics of ribosomes. Springer-Verlag, New York.
98. **Hare, D. R., N. S. Ribeiro, D. E. Wemmer, and B. R. Reid.** 1985. Complete assignment of the imino protons of *E. coli* valine tRNA: two-dimensional NMR studies in water. Biochemistry **24**:4300–4306.
99. **Harris, R. J., and S. Pestka.** 1977. Peptide bond formation, p. 413–442. *In* H. Weissbach and S. Pestka (ed.), Molecular mechanisms of protein biosynthesis. Academic Press, Inc., New York.
100. **Harvey, R. J.** 1973. Growth and initiation of protein synthesis in *Escherichia coli* in the presence of trimethoprim. J. Bacteriol. **114**:309–322.
101. **Hauptmann, R., A. P. Czernilofsky, H. O. Voorma, G. Stoeffler, and E. Kuechler.** 1974. Identification of a protein at the ribosomal donor site by affinity labeling. Biochem. Biophys. Res. Commun. **56**:331–337.
102. **Heimark, R. L., J. W. B. Hershey, and R. R. Traut.** 1976. Cross-linking of initiation factor IF2 to proteins L7/L12 in 70S ribosomes of *E. coli.* Arch. Biochem. Biophys. **182**:626–638.
103. **Hershey, J. W. B., K. F. Dewey, and R. E. Thach.** 1969. Purification and properties of initiation factor f-1. Nature (London) **222**:944–947.
104. **Hershey, J. W. B., J. Yanov, K. Johnston, and J. L. Fakunding.** 1977. Purification and characterization of protein synthesis initiation factors IF1, IF2, IF3 from *E. coli.* Arch. Biochem. Biophys. **182**:626–638.
105. **Hirshfield, I. N., P. L. Bloch, R. A. Van Bogelen, and F. C. Neidhardt.** 1981. Multiple forms of lysyl-transfer ribonucleic acid synthetase in *Escherichia coli.* J. Bacteriol. **146**:345–451.
106. **Hoagland, M. B.** 1955. An enzymic mechanism for amino acid activation in animal tissues. Biochim. Biophys. Acta **16**:288–289.
107. **Hoagland, M. B., P. C. Zamecnik, and M. L. Stephenson.** 1957. Intermediate reactions in protein biosynthesis. Biochim. Biophys. Acta **24**:215–216.
108. **Holley, R. W., J. Apgar, G. A. Everett, J. T. Madison, M. Marquisee, S. H. Merrill, J. R. Penswick, and A. Zamir.** 1965. Structure of a ribonucleic acid. Science **147**:1462–1465.
109. **Holschuh, K., and H. G. Gassen.** 1982. Mechanism of translocation. Binding equilibria between the ribosome, mRNA analogues, and cognate tRNAs. J. Biol. Chem. **257**:1987–1992.
110. **Hopfield, J. J.** 1974. Kinetic proofreading: a new mechanism of reducing errors in biosynthetic processes requiring high specificity. Proc. Natl. Acad. Sci. USA **71**:4135–4139.
111. **Hountondji, C., F. Lederer, P. Dessen, and S. Blanquet.** 1986. *E. coli* tyrosyl and methionyl-tRNA synthetases display sequence similarity at the binding site for the 3′-end of tRNA. Biochemistry **25**:16–21.
112. **Howe, J. G., and J. W. B. Hershey.** 1981. A sensitive immunoblotting method for measuring protein synthesis initiation factor levels in lysates of *E. coli.* J. Biol. Chem. **256**:12836–12839.
113. **Howe, J. G., and J. W. B. Hershey.** 1982. Immunochemical analysis of molecular forms of protein synthesis initiation factors in crude cell lysates of *E. coli.* Arch. Biochem. Biophys. **214**:446–451.

114. **Howe, J. G., and J. W. B. Hershey.** 1984. The rate of evolutionary divergence of initiation factors IF2 and IF3 in various bacterial species determined quantitatively by immunoblotting. Arch. Microbiol. **140:**187–192.

115. **Hui Bon Hoa, G., M. Graffe, and M. Grunberg-Manago.** 1977. Thermodynamic studies of the reversible association of *E. coli* ribosomal subunits. Biochemistry **16:**2800–2805.

116. **Hwang, Y.-N., and D. L. Miller.** 1985. A study of the kinetic mechanism of elongation factor Ts. J. Biol. Chem. **260:** 11498–11502.

117. **Igloi, G. L., F. von der Haar, and F. Cramer.** 1978. Aminoacyl-tRNA synthetases from yeast: generality of chemical proofreading in the prevention of misaminoacylation of tRNA. Biochemistry **17:**3459–3468.

118. **Inoue-Yokosawa, N., C. Ishikawa, and Y. Kaziro.** 1974. The role of guanosine triphosphate in translocation reaction catalyzed by elongation factor G. J. Biol. Chem. **249:**4321–4323.

119. **Iserentant, D., and W. Fiers.** 1980. Secondary structure of mRNA and efficiency of translation initiation. Gene **9:**1–12.

120. **Iwasaki, K., S. L. Sabol, A. J. Wahba, and S. Ochoa.** 1968. Translation of the genetic message. Role of initiation factors in formation of the chain initiation complex with *E. coli* ribosomes. Arch. Biochem. Biophys. **125:**542–547.

121. **Jakubowski, H., and A. R. Fersht.** 1981. Alternative pathways for editing non-cognate amino acids by aminoacyl-tRNA synthetases. Nucleic Acids Res. **9:**3105–3117.

122. **Jasin, M., L. Regan, and P. Schimmel.** 1983. Modular arrangement of functional domains along the sequence of an aminoacyl tRNA synthetase. Nature (London) **306:**441–447.

123. **Jasin, M., L. Regan, and P. Schimmel.** 1985. Two mutations in the dispensable part of alanine tRNA synthesis which affect the catalytic activity. J. Biol. Chem. **260:**2226–2230.

124. **Jay, G., and R. Kaempfer.** 1974. Sequence of events in initiation of translation: a role for initiator tRNA in the recognition of mRNA. Proc. Natl. Acad. Sci. USA **71:**3199–3203.

125. **Jay, G., and R. Kaempfer.** 1975. Initiation of protein synthesis. Binding of mRNA. J. Biol. Chem. **250:**5742–5748.

126. **Jekowsky, E., P. Schimmel, and D. L. Miller.** 1977. Isolation, characterization and structural implications of a nuclease-digested complex of aminoacyl tRNA and *E. coli* elongation factor Tu. J. Mol. Biol. **114:**451–458.

127. **Jelenc, P. C., and C. G. Kurland.** 1979. Nucleoside triphosphate regeneration decreases the frequency of translation errors. Proc. Natl. Acad. Sci. USA **76:**3174–3178.

128. **Johnson, A. E., H. J. Adkins, E. A. Matthews, and C. R. Cantor.** 1982. Distance moved by tRNA during translocation from the A site to the P site on the ribosome. J. Mol. Biol. **156:**113–140.

129. **Johnson, A. E., D. L. Miller, and C. R. Cantor.** 1978. Functional covalent complex between elongation factor Tu and an analog of lysyl-tRNA. Proc. Natl. Acad. Sci. USA **75:**3075–3079.

130. **Johnston, P. D., and A. G. Redfield.** 1979. Proton FTNMR studies of tRNA structure and dynamics, p. 191–206. *In* P. R. Schimmel, D. Söll, and J. N. Abelson (ed.), Transfer RNA: structure, properties, and recognition. Cold Spring Harbor Laboratory, Cold Spring Harbor, N.Y.

131. **Jonak, J., T. E. Petersen, B. Meloun, and I. Rychlik.** 1984. Histidine residues in elongation factor EF-Tu from *E. coli* protected by aminoacyl-tRNA against photo-oxidation. Eur. J. Biochem. **144:**295–303.

132. **Jones, M. D., T. E. Petersen, K. M. Nielsen, S. Magnusson, L. Sottrup-Jensen, K. Gausing, and B. F. C. Clark.** 1980. The complete amino-acid sequence of elongation factor Tu from *E. coli.* Eur. J. Biochem. **108:**507–526.

133. **Jurnak, F.** 1985. Structure of the GDP domain of EF-Tu and location of the amino acids homologous to *ras* oncogene proteins. Science **230:**32–36.

134. **Jurnak, F., A. McPherson, A. H. J. Wang, and A. Rich.** 1980. Biochemical and structural studies of the tetragonal crystalline modification of the *E. coli* elongation factor Tu. J. Biol. Chem. **255:**6751–6757.

135. **Kabsch, W., W. N. Gast, G. E. Schultz, and R. Leberman.** 1977. Low resolution structure of partially trypsin-degraded polypeptide elongation factor, EF-Tu, from *E. coli.* J. Mol. Biol. **117:**999–1012.

136. **Kaempfer, R.** 1974. The ribosome cycle, p. 679–704. *In* M. Nomura, A. Tissieres, and P. Lengyel (ed.), Ribosomes. Cold Spring Harbor Laboratory, Cold Spring Harbor, N.Y.

137. **Kao, T. H., D. L. Miller, M. Abo, and J. Ofengand.** 1983. Formation and properties of a covalent complex between elongation factor Tu and Phe-tRNA bearing a photoaffinity probe of its 3-(3-amino-3-carboxypropyl)-uridine residue. J. Mol. Biol. **166:**383–405.

138. **Kastelein, R. A., E. Remaut, W. Fiers, and J. van Duin.** 1982. Lysis gene expression of RNA phage MS2 depends on frameshift during translation of the overlapping coat protein gene. Nature (London) **295:**35–41.

139. **Khazaie, K., V. H. Buchanan, and R. F. Rosenberger.** 1984. The accuracy of Qβ RNA translation. 1. Errors during the synthesis of Qβ proteins by intact *E. coli* cells. Eur. J. Biochem. **144:**485–489.

140. **Kim, S. H., F. L. Suddath, G. J. Quigley, A. McPherson, J. L. Sussman, A. H. J. Wang, N. C. Seeman, and A. Rich.** 1974. Three-dimensional tertiary structure of yeast phenylalanine tRNA. Science **185:**435–440.

141. **Klein, H. A., and M. R. Capecchi.** 1971. Polypeptide chain termination. Purification of the release factors, R1 and R2 from *E. coli.* J. Biol. Chem. **246:**1055–1061.

142. **Knowlton, R. G., and M. Yarus.** 1980. Discrimination between aminoacyl groups on *su*⁺7 tRNA by elongation factor Tu. J. Mol. Biol. **139:**721–732.

143. **Koch, G. L. E., and C. J. Bruton.** 1974. The subunit structure of methionyl-tRNA synthetase from *E. coli.* FEBS Lett. **40:**180–182.

144. **Kopelowitz, J., R. Schoulaker-Schwarz, A. Lebanon, and H. Engelberg-Kulka.** 1984. Modulation of *E. coli* tryptophan (*trp*) attenuation by the UGA read-through process. Mol. Gen. Genet. **196:**541–545.

145. **Kozak, M.** 1983. Comparison of initiation of protein synthesis in procaryotes, eucaryotes, and organelles. Microbiol. Rev. **47:**1–45.

146. **Kuechler, E., and J. Ofengand.** 1979. Affinity labeling of tRNA binding sites on ribosomes, p. 413–444. *In* P. R. Schimmel, D. Söll, and J. N. Abelson (ed.), Transfer RNA: structure, properties, and recognition. Cold Spring Harbor Laboratory, Cold Spring Harbor, N.Y.

147. **Kuechler, E., and A. Rich.** 1970. Position of the initiator and peptidyl sites in the *E. coli* ribosome. Nature (London) **225:**920–924.

148. **Kung, H. F., B. V. Treadwell, C. Spears, P. C. Tai, and H. Weissbach.** 1977. DNA-directed synthesis *in vitro* of β-galactosidase: requirement for a ribosome release factor. Proc. Natl. Acad. Sci. USA **74:**3217–3221.

149. **Kurland, C. G.** 1979. Reading frame errors on ribosomes, p. 98–108. *In* J. E. Celis and J. D. Smith (ed.), Nonsense mutations and tRNA suppressors. Academic Press, Inc., New York.

150. **la Cour, F. M., J. Nyborg, S. Thirup, and B. F. C. Clark.** 1985. Structural details of the binding of guanosine diphosphate to elongation factor Tu from *E. coli* as studied by X-ray crystallography. EMBO J. **4:**2385–2388.

151. **Lake, J. A.** 1977. Aminoacyl-tRNA binding at the recognition site is the first step of the elongation cycle of protein synthesis. Proc. Natl. Acad. Sci. USA **74:**1903–1907.

152. **Langen, J. A., and J. A. Lake.** 1986. Elongation factor Tu localized on the exterior surface of the small ribosomal subunit. J. Mol. Biol. **187:**617–621.

153. **Laursen, R. A., S. Nagarkatti, and D. L. Miller.** 1977. Amino acid sequence of elongation factor Tu. Characterization and alignment of the cyanogen bromide fragments and location of the cysteine residues. FEBS Lett. **80:**103–106.

154. **Lee-Huang, S., M. A. G. Sillero, and S. Ochoa.** 1971. Isolation and properties of crystalline initiation factor F1 from *E. coli* ribosomes. Eur. J. Biochem. **18:**536–543.

155. **Liljas, A.** 1982. Structural studies of ribosomes. Prog. Biophys. Mol. Biol. **40:**161–228.

156. **Lin, F. L., L. Kahan, and J. Ofengand.** 1984. Crosslinking of phenylalanyl-tRNA to the ribosomal A site via a photoaffinity probe attached to the 4-thiouridine residue is exclusively to ribosomal protein S19. J. Mol. Biol. **172:**77–86.

157. **Lockwood, A. H., U. Maitra, N. Brot, and H. Weissbach.** 1974. The role of ribosomal proteins L7 and L12 in polypeptide chain initiation in *E. coli.* J. Biol. Chem. **249:**1213–1218.

158. **Loftfield, R. B., and D. Vanderjagt.** 1972. The frequency of errors in protein biosynthesis. Biochem. J. **128:**1353–1356.

159. **Lorber, B., R. Giege, J. P. Ebel, C. Berthet, J. C. Thierry, and D. Moras.** 1983. Crystallization of a tRNA-aminoacyl-tRNA synthetase complex. Characterization and first crystallographic data. J. Biol. Chem. **258:**8429–8435.

160. **Louie, A., N. S. Ribeiro, B. R. Reid, and F. Jurnak.** 1984. Relative affinities of all *E. coli* aminoacyl-tRNAs for elongation factor Tu-GTP. J. Biol. Chem. **259:**5010–5016.

161. **Lowary, P., J. Sampson, J. Milligan, D. Groebe, and O. C. Uhlenbeck.** 1986. A better way to make RNA for physical studies, p.69–76. *In* P. van Knippenberg and K. Hilbers (ed.), Structure

and dynamics of RNA. NATO Advanced Research Workshop. Plenum Publishing Corp., New York.

162. **Lucas-Lenard, J., and A. L. Haenni.** 1969. Release of tRNA during peptide chain elongation. Proc. Natl. Acad. Sci. USA **63:**93–97.

163. **Lucas-Lenard, J., and F. Lipmann.** 1966. Separation of three microbial amino acid polymerization factors. Proc. Natl. Acad. Sci. USA **55:**1562–1566.

164. **Maaloe, O.** 1979. Regulation of the protein-synthesizing machinery in ribosomes, tRNA, factors and so on, p. 487–542. *In* R. F. Goldberger (ed.), Biological regulation and development, vol. 1. Plenum Publishing Corp., New York.

165. **Maasen, J. A., and W. Moller.** 1978. Elongation factor G-dependent binding of a photoreactive GTP analogue to *E. coli* ribosomes results in labeling of protein L11. J. Biol. Chem. **253:**2777–2783.

166. **Maitra, U., E. A. Stringer, and A. Chandhuri.** 1982. Initiation factors in protein biosynthesis. Annu. Rev. Biochem. **51:**869–900.

167. **Marcker, K., and F. Sanger.** 1964. N-formyl-methionyl-s-RNA. J. Mol. Biol. **8:**835–840.

168. **Mayaux, J.-F., and S. Blanquet.** 1981. Binding of zinc to *E. coli* phenylalanine transfer ribonucleic acid synthetase. Comparison with other aminoacyl transfer ribonucleic acid synthetases. Biochemistry **20:**4647–4654.

169. **Miller, D. L., and H. Weissbach.** 1970. Studies on the purification and properties of factor Tu from *E. coli.* Arch. Biochem. Biophys. **141:**26–37.

170. **Milman, G., J. Goldstein, E. Scolnick, and C. T. Caskey.** 1969. Peptide chain termination. III. Stimulation of *in vitro* termination. Proc. Natl. Acad. Sci. USA **63:**183–190.

171. **Monro, R. E.** 1971. Ribosomal peptidyltransferase: the fragment reactions. Methods Enzymol. **20:**472–481.

172. **Moore, C. H., F. Farron, D. Bohnert, and C. Weissmann.** 1971. Possible origin of a minor virus specific protein (A₁) in Qβ particles. Nature (London) New Biol. **234:**204–206.

173. **Moore, P. A., D. W. Jayme, and D. L. Oxender.** 1977. A role for aminoacyl-tRNA synthetase in the regulation of amino acid transport in mammalian cell lines. J. Biol. Chem. **252:**7427–7430.

174. **Moras, D., M. B. Comarmond, J. Fischer, R. Weiss, J. C. Thierry, J. P. Ebel, and R. Giege.** 1980. Crystal structure of yeast tRNA^Asp. Nature (London) **288:**669–674.

175. **Morikawa, K., T. F. M. la Cour, J. Nyborg, K. M. Rasmussen, D. L. Miller, and B. F. C. Clark.** 1978. High resolution X-ray crystallographic analysis of a modified form of the elongation factor Tu: guanosine diphosphate complex. J. Mol. Biol. **125:**325–338.

176. **Murgola, E. J., F. T. Pagel, and K. A. Hijazi.** 1984. Codon context effect in missense suppression. J. Mol. Biol. **175,** 19–27.

177. **Neidhardt, F. C., P. L. Bloch, S. Pedersen, and S. Reeh.** 1977. Chemical measurement of steady-state levels of ten aminoacyl-transfer ribonucleic acid synthetases in *Escherichia coli.* J. Bacteriol. **129:**378–387.

178. **Nierhaus, K. H.** 1984. New aspects of the ribosomal elongation cycle. Mol. Cell. Biochem. **61:**63–81.

179. **Ninio, J.** 1975. Kinetic amplification of enzyme discrimination. Biochimie **57:**587–595.

180. **Nirenberg, M. W., and J. H. Matthaei.** 1961. The dependence of cell-free protein synthesis in *E. coli* upon naturally occurring or synthetic polyribonucleotides. Proc. Natl. Acad. Sci. USA **47:**1588–1602.

181. **Nishimura, S.** 1979. Modified nucleosides in tRNA, p. 59–79. *In* P. R. Schimmel, D. Söll, and J. N. Abelson (ed.), Cold Spring Harbor Laboratory, Cold Spring Harbor, N.Y.

182. **Nishizuka, Y., and F. Lipmann.** 1966. The interrelationship between guanosine triphosphatase and amino acid polymerization. Arch. Biochem. Biophys. **116:**344–351.

183. **Noll, M., and H. Noll.** 1976. Structural dynamics of bacterial ribosomes. V. Magnesium-dependent dissociation of tight couples into subunits: measurements of dissociation constants and exchange rates. J. Mol. Biol. **105:**111–127.

184. **Noller, H. F.** 1984. Structure of ribosomal RNA. Annu. Rev. Biochem. **53:**119–162.

185. **Nomura, M., and C. Lowry.** 1967. Phage f2 RNA-directed binding of formylmethionyl-tRNA to ribosomes and the role of 30S ribosomal subunits in the initiation of protein synthesis. Proc. Natl. Acad. Sci. USA **58:**946–953.

186. **Norris, A. T., and P. Berg.** 1964. Mechanism of aminoacyl RNA synthesis: studies with isolated aminoacyl adenylate complexes of isoleucyl RNA synthetase. Proc. Natl. Acad. Sci. USA **52:**330–337.

187. **O'Farrell, P. H.** 1978. The suppression of defective translation

188. **Ofengand, J.** 1977. tRNA and aminoacyl-tRNA synthetases, p. 7–79. *In* H. Weissbach and S. Pestka (ed.), Molecular mechanisms of protein biosynthesis. Academic Press, Inc., New York.

189. **Ofengand, J.** 1980. The topography of tRNA binding sites on the ribosome, p. 497–529. *In* G. Chambliss, G. R. Craven, J. Davies, K. Davis, L. Kahan, and M. Nomura (ed.), Ribosomes: structure, function, and genetics. University Park Press, Baltimore.

190. **Ofengand, J., and C. M. Chen.** 1972. Inactivation of Tu factor-guanosine triphosphate recognition and ribosome-binding ability by terminal oxidation-reduction of yeast phenylalanine tRNA. J. Biol. Chem. **247:**2049–2058.

191. **Ofengand, J., J. Ciesiolka, P. Gornicki, and K. Nurse.** 1985. Structural relationships in the tRNA-ribosomes complex, p. 61–73. *In* S. Ebashi (ed.), Cellular regulation and malignant growth. Springer-Verlag, Berlin.

192. **Ofengand, J., J. Ciesiolka, and K. Nurse.** 1986. Ribosomal RNA at the decoding site of the tRNA-ribosome complex, p. 473–494. *In* B. Hardesty and G. Kramer (ed.), Structure, function and genetics of ribosomes. Springer-Verlag, New York.

193. **Ofengand, J., P. Gornicki, K. Nurse, and M. Boublik.** 1984. On the structural organization of the tRNA-ribosome complex, p. 293–315. *In* B. F. C. Clark and H. U. Petersen (ed.), Gene expression: the translational step and its control. Munksgaard, Copenhagen.

194. **Ogawa, K., and A. Kaji.** 1975. Requirement for ribosome-releasing factor for the release of ribosomes at the termination codon. Eur. J. Biochem. **58:**411–419.

195. **Olsen, H. M., A. W. Nicholson, B. S. Cooperman, and D. G. Glitz.** 1985. Localization of sites of photoaffinity labeling of the large subunit of *E. coli* ribosomes by an arylazide derivative of puromycin. J. Biol. Chem. **260:**10326–10331.

196. **Olsson, M. D., and L. A. Isaksson.** 1980. Analysis of rpsD mutations of *E. coli.* IV. Accumulation of minor forms of protein S7(K) in ribosomes of rpsD mutant strains due to translational read-through. Mol. Gen. Genet. **177:**485–491.

197. **Pace, B., E. A. Matthews, K. D. Johnson, C. R. Cantor, and N. R. Pace.** 1982. Conserved 5S rRNA complement to tRNA is not required for protein synthesis. Proc. Natl. Acad. Sci. USA **79:**36–40.

198. **Parker, J.** 1981. Mistranslated proteins in *E. coli.* J. Biol. Chem. **256:**9770–9773.

199. **Parker, J., T. C. Johnston, and P. Borgia.** 1980. Mistranslation in cells infected with the bacteriophage MS2: direct evidence of Lys for Asn substitution. Mol. Gen. Genet. **180:**275–281.

200. **Parker, J., J. W. Pollard, J. D. Friesen, and C. P. Stanners.** 1978. Stuttering: high level mistranslation in animal and bacterial cells. Proc. Natl. Acad. Sci. USA **75:**1091–1095.

201. **Parmeggiani, A.** 1968. Crystalline transfer factor from *E. coli.* Biochem. Biophys. Res. Commun. **30:**613–619.

202. **Parmeggiani, A., and G. W. M. Swart.** 1985. Mechanism of action of kirromycin-like antibiotics. Annu. Rev. Microbiol. **39:**557–577.

203. **Pauling, L.** 1958. The probability of errors in the process of synthesis of protein molecules, p. 597. *In* Festschrift Arthur Stoll. Birkhauser Verlag, Basel.

204. **Paulsen, H., J. M. Robertson, and W. Wintermeyer.** 1983. Topological arrangement of two tRNAs on the ribosome. Fluorescence energy transfer measurements between A and P site-bound tRNA^Phe. J. Mol. Biol. **167:**411–426.

205. **Pawlik, R. T., J. Littlechild, C. Pon, and C. Gualerzi.** 1981. Purification and properties of *E. coli* translational initiation factors. Biochem. Int. **2:**421–428.

206. **Pearson, R. L., J. F. Weiss, and A. D. Kelmers.** 1971. Improved separation of tRNAs on polychlorotrifluoroethylene-supported reversed-phase chromatography columns. Biochim. Biophys. Acta **228:**770–774.

207. **Pellegrini, M., and C. R. Cantor.** 1977. Affinity labeling of ribosomes, p. 203–244. *In* H. Weissbach and S. Pestka (ed.), Molecular mechanisms of protein biosynthesis. Academic Press, Inc., New York.

208. **Pingoud, A., C. Urbanke, G. Krauss, F. Peters, and G. Maass.** 1977. Ternary complex formation between elongation factor Tu, GTP and aminoacyl-tRNA: an equilibrium study. Eur. J. Biochem. **78:**403–409.

209. **Plumbridge, J. A., F. Deville, C. Sacerdot, H. H. Petersen, Y. Cenatiempo, A. Cozzone, M. Grunberg-Manago, and J. W. B. Hershey.** 1985. Two translational initiation sites in the infB gene are used to express initiation factor IF2α and IF2β in *E. coli.* EMBO J. **4:**223–229.

210. **Pon, C., S. Cannistraro, A. Giovane, and C. Gualerzi.** 1982. Structure-function relationship in *E. coli* initiation factors.

by ppGpp and its role in the stringent response. Cell **15:**545–547.

Environment of the Cys residues and evidence for a hydrophobic region in IF3 by fluorescence and ESR spectroscopy. Arch. Biochem. Biophys. **217**:47–57.

211. **Pon, C. L., R. T. Pawlik, and C. Gualerzi.** 1982. The topographical localization of IF3 on *E. coli* 30S ribosomal subunits as a clue to its way of functioning. FEBS Lett. **137**:163–167.

212. **Pon, C. L., B. Wittmann-Liebold, and C. Gualerzi.** 1979. Elucidation of the primary structure of initiation factor IF-1. Structure-function relationships in *E. coli* initiation factors. FEBS Lett. **101**:157–160.

213. **Printz, J. B., R. R. Gutell, and R. A. Garrett.** 1983. A consensus model of the *E. coli* ribosome. Trends Biochem. Sci. **8**:359–363.

214. **Printz, M. P., and D. L. Miller.** 1973. Evidence for conformational changes in elongation factor Tu induced by GTP and GDP. Biochem. Biophys. Res. Commun. **53**:149–156.

215. **RajBhandary, U. L., J. E. Heckman, S. Yin, B. Alzner-DeWeerd, and E. Ackerman.** 1979. Recent developments in tRNA sequencing methods as applied to analyses of mitochondrial tRNAs, p. 3–17. *In* P. R. Schimmel, D. Söll, and J. N. Abelson (ed.), Transfer RNA: structure, properties, and recognition. Cold Spring Harbor Laboratory, Cold Spring Harbor, N.Y.

216. **Ravel, J. M., R. L. Shore, and V. Shive.** 1967. Evidence for a guanine nucleotide–aminoacyl-RNA complex as an intermediate in the enzymatic transfer of an aminoacyl-RNA to ribosomes. Biochem. Biophys. Res. Commun. **29**:68–73.

217. **Ray, P. N., and M. L. Pearson.** 1975. Functional inactivation of bacteriophage λ morphogenetic gene mRNA. Nature (London) **253**:647–650.

218. **Redfield, A. G.** 1978. Proton nuclear magnetic resonance in aqueous solutions. Methods Enzymol. **49**:253–270.

219. **Reeh, S., and S. Pedersen.** 1978. Regulation of *E. coli* elongation factor synthesis *in vivo*, p. 89–98. *In* B. F. C. Clark, H. Klenow, and J. Zeuthen (ed.), FEBS Meeting, 11th, Copenhagen, 1977, vol. 43. Pergamon Press, Oxford.

220. **Reid, B. R.** 1981. NMR studies on RNA structure and dynamics. Annu. Rev. Biochem. **50**:969–996.

221. **Revel, M., G. Brawerman, J. C. Lelong, and F. Gros.** 1968. Function of three protein factors and ribosomal subunits in the initiation of protein synthesis in *E. coli*. Nature (London) **219**:1016–1021.

222. **Rheinberger, H. J., and K. H. Nierhaus.** 1983. Testing an alternative model for the ribosomal peptide elongation cycle. Proc. Natl. Acad. Sci. USA **80**:4213–4217.

223. **Rheinberger, H. J., H. Sternbach, and K. H. Nierhaus.** 1981. Three tRNA binding sites on *E. coli* ribosomes. Proc. Natl. Acad. Sci. USA **78**:5310–5314.

224. **Rice, J. B., R. T. Libby, and J. N. Reeve.** 1984. Mistranslation of the mRNA encoding bacteriophage T7-0.3 protein. J. Biol. Chem. **259**:6505–6510.

225. **Rich, A.** 1977. The molecular structure of tRNA and its interaction with synthetases, p. 281–291. *In* H. J. Vogel (ed.), Nucleic acid recognition. Academic Press, Inc., New York.

226. **Rich, A.** 1978. Transfer RNA: Three-dimensional structure and biological function. Trends Biochem. Sci. **3**:34–37.

227. **Rich, A., and U. L. RajBhandary.** 1976. Transfer RNA: molecular structure, sequence, and properties. Annu. Rev. Biochem. **45**:805–860.

228. **Risler, J. L., C. Zelwer, and S. Brunie.** 1981. Methionyl-tRNA synthetase shows the nucleotide binding fold observed in dehydrogenases. Nature (London) **292**:384–386.

229. **Roberts, T. M., I. Bikel, R. R. Yokum, D. M. Livingston, and M. Ptashne.** 1979. Synthesis of simian virus 40 t antigen in *E. coli*. Proc. Natl. Acad. Sci. USA **76**:5596–5600.

230. **Robertus, J. D., J. E. Ladner, J. T. Finch, D. Rhodes, R. S. Brown, B. F. C. Clark, and A. Klug.** 1974. Structure of yeast phenylalanine tRNA at 3Å resolution. Nature (London) **250**:546–551.

231. **Rohrbach, M. S., and J. W. Bodley.** 1976. Steady state kinetic analysis of the mechanism of GTP hydrolysis catalyzed by *E. coli* elongation factor G and the ribosome. Biochemistry **15**:4565–4569.

232. **Romero, G., V. Chan, and R. L. Biltonen.** 1985. Kinetics and thermodynamics of the interaction of elongation factor Tu with elongation factor Ts, guanine nucleotides and aminoacyl-tRNA. J. Biol. Chem. **260**:6167–6174.

233. **Rosenberger, R. F., and G. Foskett.** 1981. An estimate of the frequency of *in vivo* transcriptional errors at a nonsense codon in *E. coli*. Mol. Gen. Genet. **183**:561–563.

234. **Rossmann, M. G., R. M. Garavito, and W. Eventoff.** 1977. Conformational adaptation among dehydrogenases. p. 3–30. *In* H. Sund (ed.), Pyridine nucleotide-dependent dehydrogenases. DeGruyter, Berlin.

235. **Roth, J. R.** 1981. Frameshift suppression. Cell **24**:610–602.

236. **Rubin, J. R., K. Morikawa, J. Nyborg, T. F. M. la Cour, B. F. C. Clark, and D. L. Miller.** 1981. Structural features of the GDP binding site of elongation factor Tu from *E. coli* as determined by x-ray diffraction. FEBS Lett. **129**:177–179.

237. **Ruusala, T., D. Andersson, M. Ehrenberg, and C. G. Kurland.** 1984. Hyper-accurate ribosomes inhibit growth. EMBO J. **3**:2575–2580.

238. **Ruusala, T., M. Ehrenberg, and C. G. Kurland.** 1982. Is there proofreading during polypeptide synthesis? EMBO J. **1**:741–745.

239. **Ryoji, M., R. Berland, and A. Kaji.** 1981. Reinitiation of translation from the triplet next to the amber termination codon in the absence of ribosome-releasing factor. Proc. Natl. Acad. Sci. USA **78**:5973–5977.

240. **Ryoji, M., J. W. Karpen, and A. Kaji.** 1981. Further characterization of ribosome releasing factor and evidence that it prevents ribosomes from reading through a termination codon. J. Biol. Chem. **256**:5798–5801.

241. **Sacerdot, C., P. Dessen, J. W. B. Hershey, J. A. Plumbridge, and M. Grunberg-Manago.** 1984. Sequence of the initiation factor IF2 gene: unusual protein features and homologies with elongation factors. Proc. Natl. Acad. Sci. USA **81**:7787–7791.

242. **San Jose, C., C. G. Kurland, and G. Stoffler.** 1976. The protein neighborhood of ribosome-bound elongation factor Tu. FEBS Lett. **71**:133–137.

243. **Schevitz, R. W., A. D. Podjarny, N. Kirshnamachari, J. J. Hughes, P. B. Sigler, and J. L. Sussman.** 1979. Crystal structure of a eukaryotic initiator tRNA. Nature (London) **278**:188–190.

244. **Schimmel, P. R., and D. Söll.** 1979. Aminocyl-tRNA synthetase: general features and recognition of transfer RNAs. Annu. Rev. Biochem. **48**:601–648.

245. **Schulman, L. H., and M. O. Her.** 1973. Recognition of altered *E. coli* formylmethionine tRNA by bacterial T factor. Biochem. Biophys. Res. Commun. **51**:275–282.

246. **Schwartz, I., and J. Ofengand.** 1978. Photochemical cross-linking of unmodified acetylvalyl-tRNA to 16S RNA at the ribosomal P site. Biochemistry **17**:2524–2530.

247. **Scolnick, E. M., and C. T. Caskey.** 1969. Peptide chain termination. V. The role of release factors in mRNA terminator codon recognition. Proc. Natl. Acad. Sci. USA **64**:1235–1241.

248. **Shine, J., and L. Dalgarno.** 1974. The 3' terminal sequence of *E. coli* 16S ribosomal RNA: complementarity to nonsense triplets and ribosome binding sites. Proc. Natl. Acad. Sci. USA **71**:1342–1346.

249. **Spirin, A. S.** 1984. Testing the classical two-tRNA-site model for the ribosomal elongation cycle. FEBS Lett. **165**:280–284.

250. **Spirin, A. S.** 1985. Ribosomal translocation: facts and models. Prog. Nucleic Acids Res. Mol. Biol. **32**:75–112.

251. **Sprinzl, M., and F. Cramer.** 1975. Site of aminoacylation of tRNAs from *E. coli* with respect to the 2' or 3'-hydroxyl group of the terminal adenosine. Proc. Natl. Acad. Sci. USA **72**:3049–3053.

252. **Sprinzl, M., and D. H. Gauss.** 1984. Compilation of tRNA sequences. Nucleic Acids Res. **12**:r1–r57.

253. **Sprinzl, M., and T. Wagner.** 1979. Role of the 2',3'-isomerization of aminoacyl-tRNA during ribosomal protein synthesis, p. 473–485. *In* P. R. Schimmel, D. Söll, and J. N. Abelson (ed.), Transfer RNA: structure, properties, and recognition. Cold Spring Harbor Laboratory, Cold Spring Harbor, N.Y.

254. **Starzyk, R. M., S. W. Koontz, and P. Schimmel.** 1982. A covalent adduct between the uracil ring and the active site of an aminoacyl tRNA synthetase. Nature (London) **298**:136–140.

255. **Starzyk, R., H. Schoemaker, and P. Schimmel.** 1985. Covalent enzyme RNA complex. A transfer RNA modification that prevents a covalent enzyme interaction also prevents aminoacylation. Proc. Natl. Acad. Sci. USA **81**:339–342.

256. **Steitz, J. A.** 1969. Polypeptide chain initiation: nucleotide sequences of the three ribosomal binding sites in bacteriophage R17 RNA. Nature (London) **244**:957–964.

257. **Steitz, J. A., and K. Jakes.** 1975. How ribosomes select initiator regions in mRNA: base pair formation between the 3' terminus of 16S rRNA and the mRNA during initiation of protein synthesis in *E. coli*. Proc. Natl. Acad. Sci. USA **72**:4734–4738.

258. **Steitz, J. A., K. U. Sprague, R. C. Yuan, M. Laughrea, P. B. Moore, and A. J. Wahba.** 1977. RNA-RNA and protein-RNA interactions during the initiation of protein synthesis, p. 491–508. *In* H. J. Vogel (ed.), Nucleic acid-protein recognition. Academic Press, Inc., New York.

259. **Stormo, G. D., T. Schneider, and L. M. Gold.** 1982. Characterization of translational initiation sites in *E. coli*. Nucleic Acids Res. **10**:2971–2996.

260. **Stringer, E. A., P. Sarkar, and U. Maitra.** 1977. Function of IF1

in the binding and release of IF2 from ribosomal initiation complexes in *E. coli*. J. Biol. Chem. **252**:1739–1744.

261. **Tabor, H., C. W. Tabor, M. S. Cohn, and E. W. Hafner.** 1981. Streptomycin resistance (*rpsL*) produces an absolute requirement for polyamines for growth of an *Escherichia coli* strain unable to synthesize putrescine and spermidine [Δ(*speA-speB*) Δ*speC*]. J. Bacteriol. **147**:702–704.

262. **Taiji, M., S. Yokozama, and T. Miyazawa.** 1983. Transacylation rates of (aminoacyl) adenosine moiety at the 3′ terminus of aminoacyl tRNA. Biochemistry **22**:3220–3225.

263. **Tanada, S., M. Kawakami, K. Nishio, and S. Takemura.** 1982. Interaction of aminoacyl-tRNA with bacterial elongation factor Tu:GTP complex: effects of the amino group of amino acid esterified to tRNA, the amino acid side chain, and tRNA structure. J. Biochem. (Tokyo) **91**:291–299.

264. **Taniguchi, T., and C. Weissmann.** 1978. Inhibition of Qβ RNA 70S ribosome initiation complex formation by an oligonucleotide complementary to the 3′ terminal region of *E. coli* 16S ribosomal RNA. Nature (London) **275**:770–772.

265. **Teraoka, H., and K. H. Nierhaus.** 1978. Proteins from *E. coli* ribosomes involved in the binding of erythromycin. J. Mol. Biol. **126**:185–193.

266. **Thach, R. E., T. A. Sundararajan, K. F. Dewey, J. C. Brown, and P. Doty.** 1966. Translation of synthetic messenger RNA. Cold Spring Harbor Symp. Quant. Biol. **31**:85–97.

267. **Thach, S. S., and R. E. Thach.** 1971. Translocation of messenger RNA and "accommodation" of fMet-tRNA. Proc. Natl. Acad. Sci. USA **68**:1791–1795.

268. **Thompson, R. C., and D. B. Dix.** 1982. Accuracy of protein biosynthesis. A kinetic study of the reaction of poly (U)-programmed ribosomes with a leucyl-tRNA$_2$ elongation factor Tu GTP complex. J. Biol. Chem. **257**:6677–6682.

269. **Thompson, R. C., D. B. Dix, and J. F. Eccleston.** 1980. Single turnover kinetic studies of GTP hydrolysis and peptide bond formation in the elongation factor Tu-dependent binding of aminoacyl-tRNA to *E. coli* ribosomes. J. Biol. Chem. **255**:11088–11090.

270. **Thompson, R. C., D. B. Dix, R. D. Gerson, and A. M. Karim.** 1981. A GTPase reaction accompanying the rejection of Leu-tRNA$_2$ by UUU-programmed ribosomes. J. Biol. Chem. **256**:81–86.

270a. **Thompson, R. C., D. B. Dix, and A. M. Karim.** 1986. The reaction of ribosomes with elongation factor Tu·GTP complexes. Aminoacyl-tRNA-independent reactions in the elongation cycle determine the accuracy of protein synthesis. J. Biol. Chem. **261**:4868–4874.

271. **Thompson, R. C., and A. M. Karim.** 1982. The accuracy of protein biosynthesis is limited by its speed: high fidelity selection by ribosomes of aminoacyl-tRNA ternary complexes containing GTP [γS]. Proc. Natl. Acad. Sci. USA **79**:4922–4926.

272. **Traut, R. R., and R. E. Monro.** 1964. The puromycin reaction and its relation to protein synthesis. J. Mol. Biol. **10**:63–72.

273. **Traut, R. R., D. S. Tewari, A. Sommer, G. R. Gavino, H. M. Olson, and D. G. Glitz.** 1986. Protein topography of ribosomal functional domains: effects of monoclonal antibodies to different epitopes in *E. coli* protein L7/L12 on ribosome function and structure, p. 206–308. *In* B. Hardesty and G. Kramer (ed.), Structure, function and genetics of ribosomes. Springer-Verlag, New York.

274. **Uhlenbeck, O. C., F. H. Martin, and P. Doty.** 1971. Self-complementary oligoribonucleotides: effects of helix defects and guanylic acid-cytidylic acid base pairs. J. Mol. Biol. **57**:217–229.

275. **van der Hofstad, G. A. J. M., A. Buitenhek, L. Bosch, and H. O. Voorma.** 1978. Initiation factor IF-3 and the binary complex between initiation factor IF-2 and fMet-tRNA are mutually exclusive on the 30S ribosomal subunit. Eur. J. Biochem. **89**:213–220.

276. **van Duin, J., G. P. Overbeek, and C. Backendorf.** 1980. Functional recognition of phage RNA by 30-S ribosomal subunits in the absence of initiator tRNA. Eur. J. Biochem. **110**:593–597.

277. **van Noort, J. M., B. Kraal, and L. Bosch.** 1985. A second tRNA binding site on elongation factor Tu is induced while the factor is bound to the ribosome. Proc. Natl. Acad. Sci. USA **82**:3212–3216.

278. **Vijgenboom, E., T. Vink, B. Kraal, and L. Bosch.** 1985. Mutants of the elongation factor EF-Tu, a new class of nonsense suppressors. EMBO J. **4**:1049–1052.

279. **Vlassov, V. V., R. Giege, and J. P. Ebel.** 1981. Tertiary structure of tRNAs in solution monitored by phosphodiester modification with ethylnitrosourea. Eur. J. Biochem. **119**:51–59.

280. **von der Haar, F., and F. Cramer.** 1976. Hydrolytic action of aminoacyl-tRNA synthetases from baker's yeast: "chemical proofreading" preventing acylation of tRNA^{Ile} with misactivated valine. Biochemistry **15**:4131–4138.

281. **Wagner, E. G. H., P. C. Jelenc, M. Ehrenberg, and C. G. Kurland.** 1982. Rate of elongation of polyphenylalanine *in vitro*. Eur. J. Biol. **122**:193–197.

282. **Walter, P., J. Gangloff, J. Bonnet, Y. Boulanger, J. P. Ebel, and F. Fasiolo.** 1983. Primary structure of the *Saccharomyces cerevisiae* gene for methionyl tRNA synthetase. Proc. Natl. Acad. Sci. USA **80**:2437–2441.

283. **Watson, J. D.** 1964. The synthesis of proteins upon ribosomes. Bull. Soc. Chem. Biol. **46**:1399–1425.

284. **Webster, T., H. Tsai, M. Kula, G. Mackie, and P. Schimmel.** 1984. Specific sequence homology and three-dimensional structure of an aminoacyl transfer RNA synthetase. Science **226**:1315–1317.

285. **Weiel, J., and J. W. B. Hershey.** 1982. The binding of fluorescein-labeled protein synthesis initiation factor 2 to *E. coli* 30S ribosomal subunits determined by fluorescence polarization. J. Biol. Chem. **257**:1215–1220.

286. **Weiss, R. B.** 1984. Molecular model of ribosome frameshifting. Proc. Natl. Acad. Sci. USA **81**:5797–5801.

287. **Weiss, R., and J. Gallant.** 1983. Mechanism of ribosome frameshifting during translation of the genetic code. Nature (London) **302**:389–393.

288. **Weissbach, H.** 1980. Soluble factors in protein synthesis, p. 377–411. *In* G. Chambliss, G. R. Craven, J. Davies, K. Davis, L. Kahan, and M. Nomura (ed.), Ribosomes: structure, function, and genetics. University Park Press, Baltimore.

289. **Weissbach, H., and S. Pestka (ed.).** 1977. Molecular mechanisms of protein biosynthesis. Academic Press, Inc., New York.

290. **Wickstrom E.** 1974. *E. coli* initiation factor IF3 binding to AUG and AUG-containing single strands and hairpin loops, and nonspecific binding to polymers. Biochim. Biophys. Acta **349**:125–130.

291. **Wickstrom, E.** 1981. Physical parameters of *E. coli* translational initiation factor 3 binding to poly(A). FEBS Lett. **128**:154–156.

292. **Wikman, F. P., G. E. Siboska, H. U. Petersen, and B. F. C. Clark.** 1982. The site of interaction of aminoacyl-tRNA with elongation factor Tu. EMBO J. **1**:1095–1100.

293. **Wilkinson, A. J., A. R. Fersht, D. M. Blow, and G. Winter.** 1983. Site-directed mutagenesis as a probe of enzyme structure and catalysis: tyrosyl-tRNA synthetase cysteine-35 to glycine-35 mutation. Biochemistry **22**:3581–3586.

294. **Wintermeyer, W., and C. Gualerzi.** 1983. Effect of *E. coli* initiation factors on the kinetics of N-Acphe-tRNA binding to 30S ribosomal subunits. A fluorescence stopped flow study. Biochemistry **22**:690–694.

295. **Wintermeyer, W., R. Lill, H. Paulsen, and J. M. Robertson.** 1986. Mechanism of ribosomal translocation, p. 523–530. *In* B. Hardesty and G. Kramer (ed.), Structure, function, and genetics of ribosomes. Springer-Verlag, New York.

296. **Wintermeyer, W., and J. M. Robertson.** 1982. Transient kinetics of tRNA binding to the ribosomal A and P sites: observation of a common intermediate complex. Biochemistry **21**:2246–2252.

297. **Wittmann, H. G.** 1982. Components of bacterial ribosomes. Annu. Rev. Biochem. **51**:155–183.

298. **Wittmann, H. G.** 1983. Architecture of prokaryotic ribosomes. Annu. Rev. Biochem. **52**:35–65.

299. **Woese, C.** 1970. Molecular mechanisms of translation: a reciprocating ratchet mechanism. Nature (London) **226**:817–820.

300. **Woo, N. H., B. Roe, and A. Rich.** 1980. Three-dimensional structure of *E. coli* initiator tRNA^{Met}. Nature (London) **286**:346–351.

301. **Wrede, P., N. H. Woo, and A. Rich.** 1979. Initiator tRNAs have a unique anticodon loop conformation. Proc. Natl. Acad. Sci. USA **76**:3289–3293.

302. **Wright, H. T., P. C. Manor, K. Buerhing, R. L. Karpel, and J. R. Fresco.** 1979. The structure of baker's yeast tRNA^{Gly}: a second tRNA conformation, p. 145–160. *In* P. R. Schimmel, D. Söll, and J. N. Abelson (ed.), Transfer RNA: structure, properties, and recognition. Cold Spring Harbor Laboratory, Cold Spring Harbor, N.Y.

303. **Yarus, M.** 1972. Phenylalanyl-tRNA synthetase and isoleucyl-tRNA^{Phe}: a possible verification mechanism for aminoacyl-tRNA. Proc. Natl. Acad. Sci. USA **69**:1915–1919.

304. **Yarus, M.** 1979. The accuracy of translation. Prog. Nucleic Acids Res. Mol. Biol. **23**:195–225.

305. **Yarus, M., and P. Berg.** 1969. Recognition of tRNA by isoleucyl-tRNA synthetase. Effect of substrates on the dynamics of tRNA-enzyme interaction. J. Mol. Biol. **42**:171–189.

306. **Yarus, M., and R. C. Thompson.** 1984. Precision of protein biosynthesis, p. 23–63. *In* J. Beckwith, J. Davies, and J. Gallant (ed.), Gene function in prokaryotes. Cold Spring

Harbor Laboratory, Cold Spring Harbor, N.Y.

307. **Yokota, T., H. Sugisaki, M. Takanami, and Y. Kaziro.** 1980. The nucleotide sequence of the cloned *tufA* gene of *E. coli*. Gene **12**:25–31.

308. **Zaccai, G., P. Morin, B. Jacrot, D. Moras, J. C. Thierry, and R. Giege.** 1979. Interaction of valyl-tRNA synthetase with RNAs and conformational changes of the enzyme. J. Mol. Biol. **129**:483–500.

309. **Zamecnik, P. C.** 1983. Diadenosine-5′,5′′′-P¹,P⁴-tetraphosphate (Ap₄A): its role in cellular metabolism. Anal. Biochem. **134**:1–10.

310. **Zelwer, C., J. L. Risler, and S. Brunie.** 1982. Crystal structure of *E. coli* methionyl-tRNA synthetase at 2.5 Å resolution. J. Mol. Biol. **155**:63–81.

311. **Zengel, J. M., R. H. Archer, and L. Lindahl.** 1984. The nucleotide sequence of the *E. coli fus* gene, coding for elongation factor G. Nucleic Acids Res. **12**:2181–2192.

312. **Zucker, F. W., and J. W. B. Hershey.** 1986. Binding of *Escherichia coli* protein synthesis initiation factor IF1 to 30S ribosomal subunits measured by fluorescence polarization. Biochemistry **25**:3682–3690.

41. Lipopolysaccharide Biosynthesis

PAUL D. RICK

Department of Microbiology, Uniformed Services University of the Health Sciences, Bethesda, Maryland 20814-4799

INTRODUCTION

The lipopolysaccharides (endotoxins) of *Salmonella* species, *Escherichia coli*, and other gram-negative bacteria are major components of the outer membrane of these organisms. These unique molecules have been isolated from a wide variety of gram-negative bacteria, and their structural features have been studied extensively. However, our current understanding of the genetics and biosynthesis of lipopolysaccharides has been derived principally from studies of the salmonellae and, to a lesser extent, *E. coli*. Accordingly, the following discussion will focus primarily on lipopolysaccharide biosynthesis in *Salmonella typhimurium*. More specifically, only the biosynthesis of the lipopolysaccharide core and O-antigen regions will be considered here; the biosynthesis of the lipid A region is the topic of chapter 31.

Numerous reviews have been published that deal with various aspects of this topic in considerable detail, and they are cited throughout this discussion. Thus, it is the purpose of this chapter to summarize the major features of lipopolysaccharide core and O-antigen synthesis as they are currently understood. An attempt will also be made to point out areas in which our knowledge is still incomplete.

STRUCTURE OF LIPOPOLYSACCHARIDE

General Structural Features

The general structural organization of lipopolysaccharide molecules is exemplified by the lipopolysaccharide of *S. typhimurium* (Fig. 1). The lipopolysaccharide molecule consists of a complex phosphorylated heteropolysaccharide that is covalently linked to a unique glucosamine-containing lipid, lipid A. The polysaccharide portion has been divided into two major regions, an internal core region and the peripheral O-antigen or O side chain region. The core region has been further divided into an inner core, or backbone, region and an outer core region. The O antigen constitutes the immunodominant portion of the molecule, and the structural determinants of O side chains provide the basis for the serological classification of the *Enterobacteriaceae* according to the Kaufmann-White scheme (36).

Wild-type strains of *Salmonella* species, *E. coli*, and other gram-negative bacteria capable of synthesizing a complete lipopolysaccharide molecule are called S (smooth). These organisms react with specific antibodies directed against their O side chains, and in general they form smooth colonies on solid media. In contrast, strains that are unable to synthesize lipopolysaccharide molecules possessing O side chains are called R (rough) since in many cases they exhibit rough colony morphology. Rough or R mutants result from mutations affecting any one of several specific steps in lipopolysaccharide biosynthesis. These mutants are commonly classified on the basis of the structure or chemotype of the incomplete lipopolysaccharide that they synthesize (Fig. 1). Thus, mutants defective in either the synthesis of O side chains or the subsequent transfer of O side chains to a completed core (see Biosynthesis of O Antigen) synthesize a chemotype Ra lipopolysaccharide. Such mutants are

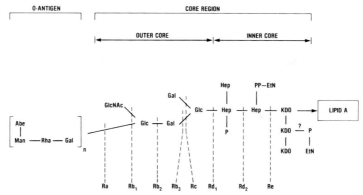

FIG. 1. Structure of *S. typhimurium* lipopolysaccharide. Abbreviations: Abe, abequose; Man, D-mannose; Rha, L-rhamnose; Gal, D-galactose; GlcNAc, N-acetyl-D-glucosamine; Glc, D-glucose; Hep, L-glycero-D-mannoheptose; KDO, 3-deoxy-D-mannooctulosonic acid; EtN, ethanolamine; P, phosphate. All sugars are in the pyranose configuration. Additional structural details are provided in Fig. 2 to 5. The lipopolysaccharide structures and corresponding chemotypes of mutants defective at various stages of core biosynthesis are indicated by the broken lines and associated letter designations, i.e., Ra, Rb₁, etc., respectively.

also referred to collectively as Ra mutants, even though the Ra chemotype can result from a lesion in any one of several distinct genetic loci.

Mutations that affect core biosynthesis also result in R mutants since such mutations preclude synthesis of all portions of the lipopolysaccharide distal to the site of the defect (synthesis proceeds from lipid A toward the O antigen). For example, mutants defective in UDP-galactose-4-epimerase are unable to convert UDP-glucose to UDP-galactose, the donor of galactosyl residues in lipopolysaccharide biosynthesis. As a consequence, these mutants synthesize an incomplete, or chemotype Rc, lipopolysaccharide lacking all sugars distal to the first glucosyl residue.

Although wild-type *Salmonella* species devote a considerable expenditure of energy to the synthesis of the polysaccharide chains of lipopolysaccharide, all portions of the molecule distal to the 3-deoxy-D-mannooctulosonic acid (KDO) residues are nonessential to cell growth and function (see Biosynthesis of the Inner Core). Mutants that synthesize lipopolysaccharides of chemotype Rd₂, Rd₁, or Re (deep-rough mutants) are viable, and they appear to incorporate the incomplete lipopolysaccharides into the outer membrane in normal amounts (92). Indeed, recent studies indicate that the number of lipopolysaccharide molecules per unit cell surface is increased in such mutants (23). However, deep-rough mutants are characteristically more sensitive to various antibiotics, detergents, and dyes (44, 98). It is now clear that the sensitivity of deep-rough mutants to many of these agents is due in large measure to compositional and organizational changes in the outer membrane that result from the defect in lipopolysaccharide synthesis (59).

Structure of the O Antigen

Detailed information on the structure of O antigens has been obtained primarily from studies of the lipopolysaccharides of the salmonellae, shigellae, klebsiellae, and *E. coli* (for reviews, see references 33, 45, and 105). These studies have revealed a number of characteristics that serve to distinguish the O-antigen region from other parts of the lipopolysaccharide molecule. O side chains are typically composed of oligosaccharide repeat units. Individual repeat units may consist of linear trisaccharides or pentasaccharides, or they may be polymers of branched oligosaccharides containing from four to six sugars. The component monosaccharides include commonly occurring neutral sugars, amino sugars, and in some cases rather unusual sugars, such as 6-deoxyhexoses or 3,6-dideoxyhexoses.

The great degree of immunological diversity attributable to O antigens in many cases reflects pronounced structural differences. However, relatively minor differences in O-antigen structure can also be detected immunologically. Thus, although the O-antigen repeat unit of *S. typhimurium* (group B) is structurally similar to the O-antigen repeat unit of *Salmonella typhi* (group D₁) (Fig. 2), they are immunologically distinct. Even more striking in this regard are the O-antigen repeat units of *Salmonella anatum* (group E₁) and *Salmonella newington* (group E₂), which differ in structure only by the occurrence of an acetyl group in the former and by the difference in the anomeric configuration of the linkage between repeat units. The structures of numerous O-antigen repeat units from a variety of organisms are included in the reviews cited above.

In addition to the differences in O-antigen structure that are responsible for distinguishing between O serogroups, variations in the structure of O side chains may also occur within a given O serogroup. Accordingly, the basic O side chain repeat unit of a given strain may be structurally modified so that it is both chemically and serologically distinct from that of the parental strain. Many of these modifications occur by the process of phage conversion. Phage conversion involves the lysogenization of a host strain by a temperate bacteriophage capable of directing specific structural alterations in O-antigen repeat units. For example, lysogenization of *S. typhimurium* with bacteriophage P22 or iota results in the incorporation into individual O side chain units of a glucosyl substituent that is α1,6-linked to the main chain galactosyl residue (93). As a consequence, the O side chain units of the parental strain (O factors 4 and 12)

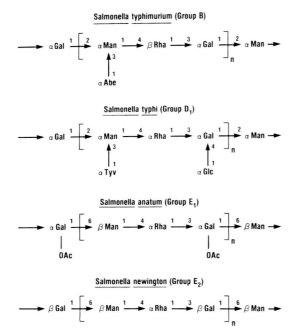

FIG. 2. Structures of O-antigen repeat units synthesized by various *Salmonella* species. The abbreviations for sugars are the same as those described in the legend to Fig. 1. Other abbreviations: Tyv, tyvelose; OAc, O-acetyl.

are converted to O side chain units that are serologically distinct (O factors, 1, 4, and 12). Similarly, lysogenization of *S. typhimurium* with phage φ27 results in conversion of the -α-Gal-1,2-Man- linkage to -α-Gal-1,6-Man- with the attendant introduction of O-factor 27 specificity (2, 93).

Other modifications of O side chain structure occur that are dictated by host genetic determinants. In certain strains of *S. typhimurium*, the abequosyl residues are O acetylated (29), thus giving rise to O-factor 5 specificity. The gene responsible for O-factor 5 specificity (*oafA*) is located at min 46 on the revised map of the *Salmonella* chromosome (37; see also chapter 54). Another type of host-determined modification of O side chain structure occurs by a process known as form variation (35). Form-variable strains are capable of effecting reversible modifications of O side chain structure. One example of form variation in *S. typhimurium* is the reversible incorporation of glucosyl residues into the O side chain repeat unit (99). The glucosyl branches, α1,4-linked to the main chain galactose, confer O-factor 12_2 specificity. Organisms that have the glucosyl branches are regarded as being in the positive state, and those that express factor 12_2 poorly are regarded as being in the negative state. The variation between factors 12_2^+ and 12_2^- is controlled by the *oafR* gene (46). However, the genetics of form variations are still not completely understood. The genetics and biochemistry of O side chain modifications have been reviewed previously (58, 79, 109); the reader is referred to these sources for further discussion.

Early estimates of the average degree of polymerization of O side chains in *S. typhimurium* ranged from 4 repeat units to 8 to 10 repeat units (43, 60). However, recent studies in several laboratories (24, 54, 70) have

revealed that the degree of polymerization varies more widely than was previously suspected. Lipopolysaccharide molecules with from 0 to approximately 40 repeat units have been detected after the analysis of both cell envelopes and isolated lipopolysaccharides by sodium dodecyl sulfate-polyacrylamide gel electrophoresis.

Structure of the Outer Core

Structural variation of the outer core, or hexose region, within the *Enterobacteriaceae* is considerably less pronounced than is the case for O side chains. Thus, the structure of the outer core is essentially invariant within the salmonellae (34, 45), and similar structures have been determined for various strains of *E. coli*. The outer core characteristically consists of a branched pentasaccharide (Fig. 3); the component sugars frequently include 2-acetamido-2-deoxy-D-glucose, as well as glucose and galactose. It is of interest that 2-acetamido-2-deoxy-D-glucose appears to be a variable component of the outer core of some organisms, e.g., *E. coli* K-12 (34) and *E. coli* strains possessing the R2 core (26). However, the basis of this variation is not understood.

Structure of the Inner Core

The inner core, or backbone region, is characterized by the sugars L-glycero-D-mannoheptose and KDO; KDO serves as the structural bridge between lipid A and the polysaccharide portion. This region also contains phosphate and ethanolamine. The basic structure of the heptose region (Fig. 4) is believed to be rather uniform among the *Enterobacteriaceae*. Indeed, essentially the same structure has been found for *Salmonella* species (25, 34) and *E. coli* (3, 34). Howev-

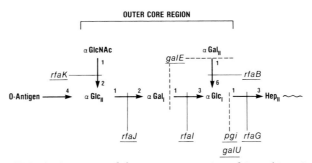

FIG. 3. Structure of the outer core region of *S. typhimurium* lipopolysaccharide. The abbreviations for sugars are the same as those described in the legend to Fig. 1. The points of termination of the incomplete core synthesized by mutants with a defective UDP-galactose-4-epimerase (*galE*), phosphoglucoisomerase (*pgi*), or UDP-glucose pyrophosphorylase (*galU*) are indicated by the broken lines (see also Fig. 1). The structural genes presumed to be responsible for the indicated glycosyltransferase activities are as follows. *rfaG*, UDP-glucose:lipopolysaccharide glucosyltransferase I; *rfaB*, UDP-galactose:lipopolysaccharide α1,6-galactosyltransferase; *rfaI*, UDP-galactose:lipopolysaccharide α1,3-galactosyltransferase; *rfaJ*, UDP-glucose:lipopolysaccharide glucosyltransferase II; *rfaK*, UDP-N-acetylglucosamine:lipopolysaccharide glucosaminyltransferase. See Creeger and Rothfield (10) for additional nomenclature.

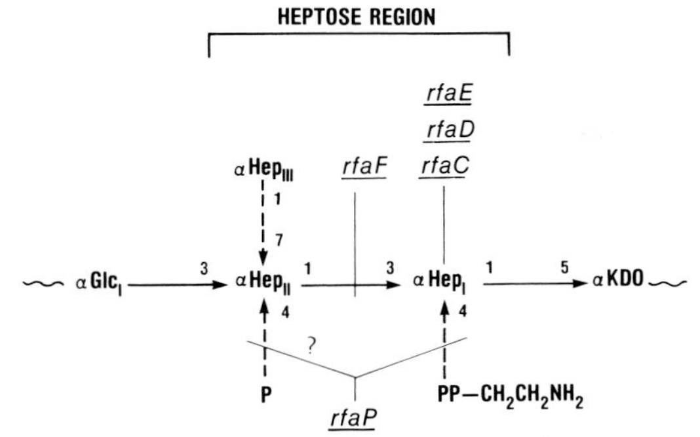

HEPTOSE REGION

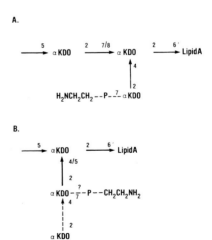

FIG. 4. Structure of the heptose region of *S. typhimurium* lipopolysaccharide. The abbreviations for sugars are the same as those given in the legend to Fig. 1. Genes involved in the biosynthesis of the heptose region are indicated. The *rfaD* and *rfaF* loci are presumed to be the structural genes for ADP-L-glycero-D-mannoheptose-6-epimerase and ADP-heptose:lipopolysaccharide heptosyltransferase II, respectively. Mutations in the *rfaC* gene affect the formation of the proximal heptose residue, whereas mutations in the *rfaE* gene prevent the incorporation of any heptose residue into the inner core. However, the specific functions of these genes are not known. The *rfaP* gene is responsible for the phosphorylation of the inner core; mutants with a defective *rfaP* gene have only been described in *S. minnesota*. The broken lines indicate substituents that are variable and that occur in nonstoichiometric amounts.

er, heterogeneity does exist in this region as indicated by the nonstoichiometric occurrence of pyrophosphorylethanolamine (39). In addition, the degree of substitution of the heptose main chain with the branch chain heptose has been found to vary from 20 to 90% in *Salmonella* lipopolysaccharides (25).

Early studies on the structure of lipopolysaccharides synthesized by *Salmonella* R mutants (13) concluded that the individual lipopolysaccharide chains are cross-linked to one another by either phosphodiester or pyrophosphate bridges located in the inner core region. However, subsequent experiments have indicated otherwise (25, 50). Furthermore, similar proposals that the lipid A units are cross-linked by pyrophosphate bridges also appear to be incorrect (50, 83, 84).

Recent scrutiny of the KDO region has necessitated a reevaluation of the generally accepted structure of this region. It was originally proposed (12, 72) that the KDO residues of *Salmonella* and *E. coli* lipopolysaccharides occur as a branched trisaccharide consisting of a single side chain KDO, substituted nonstoichiometrically at C-7 with phosphorylethanolamine, and two KDO residues in the main chain (Fig. 5A). However, recent investigations of the KDO-lipid A region of an Re mutant of *E. coli*, by use of ^{13}C nuclear magnetic resonance spectroscopy, revealed the occurrence of only two KDO residues (95, 96). Both residues of the KDO disaccharide were found to be in the α-anomeric configuration, and the sequence αKDO-(2 → 5)αKDO-(2 - -→ lipid A was tentatively assigned. Separate studies using chemical methods also revealed the occurrence of only two KDO residues in the lipopolysaccharides obtained from Re mutants of *Salmonella* species and *E. coli* (5, 6). However, unlike in the spectroscopy studies, the KDO disaccharide was

found to have the sequence αKDO-(2 → 4)αKDO-(2 - -→ lipid A.

Although the occurrence of a KDO disaccharide appears to be a structural characteristic of lipopolysaccharides obtained from heptoseless mutants, a different structure has been found for the KDO region of lipopolysaccharides containing a more complete core. Thus, evidence has been obtained which suggests the occurrence of a linear KDO trisaccharide in the lipopolysaccharide of a *Salmonella* Rb₂ mutant (5, 6). The trisaccharide was assigned the sequence αKDO(2 → 4)αKDO(2 → 4)αKDO(2 - -→ lipid A, in which the reducing terminal KDO residue is substituted at position 5 with heptose, and it occurs as the sole main chain KDO residue. In addition, the nonreducing terminal KDO residue of the trisaccharide appears to occur in nonstoichiometric amounts (4).

A revised structure of the KDO region based on the recent data described above is shown in Fig. 5B. The location of phosphorylethanolamine has not been established. However, nonstoichiometric substitution of the second KDO with phosphorylethanolamine as indicated seems likely since KDO-7-phosphorylethanolamine occurs in Re mutants, and such mutants appear to possess only a KDO disaccharide. In addition, substitution of the main chain KDO with phosphorylethanolamine is not consistent with earlier structural data obtained by chemical degradation of the KDO region (12). Structural variations in this region, beyond those already cited, are further indicated by the occurrence of KDO residues substituted with L-rhamnose and D-galactose in the lipopolysaccharides of *E. coli* K-12 (6, 73) and *E. coli* O100*(26), respectively. Recent studies also indicate that the main chain KDO is linked to position 6' of the glucosamine disaccharide of lipid A and not to position 3' as was originally believed (96).

A.

$$\longrightarrow \overset{5}{} \alpha\text{KDO} \overset{2}{\underset{}{\longrightarrow}} \overset{7/8}{} \alpha\text{ KDO} \overset{2}{\underset{}{\longrightarrow}} \overset{6'}{} \text{Lipid A}$$

$$\text{H}_2\text{NCH}_2\text{CH}_2\text{-}\text{-}\overset{7}{\text{P}}\text{-}\alpha\text{KDO}$$

B.

$$\longrightarrow \overset{5}{} \alpha\text{KDO} \overset{2}{\underset{}{\longrightarrow}} \overset{6'}{} \text{Lipid A}$$

$$\alpha\text{KDO}\text{-}\overset{?}{\underset{}{\text{P}}}\text{-}\text{-}\text{CH}_2\text{CH}_2\text{NH}_2$$

$$\alpha\text{KDO}$$

FIG. 5. Structure of the KDO region. (A) Previously accepted structure of the KDO trisaccharide of *Salmonella* species and *E. coli*. (B) Revised structure of the KDO region (see text for discussion). The nonreducing terminal KDO of the revised structure does not appear to be a component of the lipopolysaccharides synthesized by heptoseless mutants. Phosphorylethanolamine is a variable component and occurs in nonstoichiometric amounts. In addition, the location of phosphorylethanolamine in the revised structure has not been firmly established.

BIOSYNTHESIS OF THE CORE AND O-ANTIGEN REGIONS

Biosynthesis of the Inner Core

Biosynthesis of the KDO region. Studies on the assembly of lipopolysaccharide into the outer membrane of *S. typhimurium* have established that lipopolysaccharide is synthesized in the inner or cytoplasmic membrane and subsequently translocated to the outer membrane (64–66). However, detailed information regarding the topology of specific biosynthetic events in the inner membrane is lacking . In addition, the molecular mechanisms of lipopolysaccharide translocation and integration into the outer membrane remain to be established. In contrast, many of the individual enzymatic steps involved in lipopolysaccharide synthesis have been characterized. Indeed, the steps involved in the biosynthesis of the outer core and O-antigen regions of lipopolysaccharide are now reasonably well established. Progress in this area was greatly facilitated by the availability of mutants blocked at various steps in the biosynthesis of these regions. In contrast, our knowledge of the steps involved in the synthesis of the inner core has lagged behind, and it is still incomplete. In particular, elucidation of the mechanism of biosynthesis of the KDO region was frustrated by the longstanding inability to isolate mutants defective in the synthesis of this region. Thus, it was assumed that the KDO region was in some way essential for normal cell growth and function.

A study of the biosynthesis and functional significance of the KDO region was eventually made possible by the isolation of mutants of *S. typhimurium* conditionally defective in KDO synthesis (*kdsA*) (40, 76). The synthesis of KDO involves three sequential reactions (22, 41, 42), as shown below.

$$\text{D-Ribulose-5-P} \rightleftarrows \text{D-arabinose-5-P}$$

$$\text{D-Arabinose-5-P} + \text{P-enolpyruvate} \rightarrow \text{KDO-8-P} + \text{P}_i$$

$$\text{KDO-8-P} \rightarrow \text{KDO} + \text{P}_i$$

These three reactions are catalyzed by the enzymes D-ribulose-5-phosphate isomerase, KDO-8-phosphate synthetase, and KDO-8-phosphate phosphatase, respectively. Finally, free KDO is converted to CMP-KDO, the donor of KDO residues for inner core biosynthesis (21). The *kdsA* mutants were found to have a temperature-sensitive KDO-8-phosphate synthetase owing to a lesion in the structural gene for the enzyme. Under nonpermissive conditions, these mutants accumulate a KDO-deficient and incomplete precursor of lipid A (38, 75, 77). The precursor molecule consists of a phosphorylated glucosamine disaccharide that contains the full complement of *O*- and *N*-β-hydroxymyristoyl residues found in the complete lipid A molecule. However, unlike complete lipid A, the precursor lacks ester-linked 12:0 and 14:0 fatty acids.

Expression of the temperature-sensitive lesion in KDO synthesis results in growth stasis (76). The same phenotype has also been observed for temperature-sensitive CTP:CMP-KDO cytidyltransferase mutants (*kdsB*) of *S. typhimurium* (M. J. Osborn, personal communication). However, the mechanism of growth stasis is unknown. It is possible that a complete KDO-lipid A region is required for some essential function of the outer membrane or for the assembly of a functional outer membrane. Alternatively, the extensive accumulation of lipid A precursor in the cytoplasmic membrane under nonpermissive conditions (66) may interfere with essential cytoplasmic membrane function(s). Thus, growth stasis may not be directly related to an alteration of outer membrane structure or function.

The accumulation of an *O*-acyl–deficient lipid A precursor under nonpermissive conditions provides support for the earlier suggestion (28) that KDO is incorporated at a stage in lipid A synthesis before the complete incorporation of *O*-fatty acyl residues. Indeed, the results of in vitro experiments have clearly established that the underacylated lipid A precursor acts as an efficient acceptor for the transfer of KDO residues from CMP-KDO. Munson and co-workers (55) demonstrated that an enzyme system localized in the cell envelope fraction of *S. typhimurium* catalyzes the transfer of two residues of KDO from CMP-KDO into the purified precursor. Structural studies tentatively identified the product as containing a KDO disaccharide with the sequence KDO(2 → 4)KDO. The CMP-KDO:lipid A KDO transferase system was partially purified after extraction of the enzyme(s) from cell envelopes with nonionic detergent at alkaline pH. Although no evidence was obtained for the existence of two separate transferase activities, it seems unlikely that the synthesis of this region involves a mechanism whereby the KDO disaccharide is first synthesized as a unit and subsequently transferred into nascent lipid A.

Walenga and Osborn (100) later obtained direct evidence that the transfer of KDO to lipid A in vivo precedes the addition of saturated fatty acid residues. Thus, when mutants with a temperature-sensitive KDO-8-phosphate synthetase were shifted from nonpermissive to permissive temperature, the accumulated lipid A precursor was converted to a transient KDO-containing and fatty acyl-deficient intermediate that was indistinguishable from the in vitro product obtained by enzymatic addition of KDO to lipid A precursor. These studies also demonstrated that the transient intermediate was rapidly converted to lipopolysaccharide. Indeed, subsequent investigations (101) revealed that the further addition of core sugars to the acyl-deficient intermediate does not require prior incorporation of the missing saturated fatty acids.

It is clear that additional work is required to establish more firmly the mechanism of synthesis of the KDO region. In this regard, it is of interest that the structure of the KDO disaccharide synthesized in vitro is in agreement with the structure determined for the KDO region of lipopolysaccharides isolated from heptoseless mutants (4, 6, 72). However, the point in the biosynthetic sequence at which the third or nonreducing KDO residue of the KDO trisaccharide is incorporated is not known. The occurrence of a linear KDO trisaccharide in the lipopolysaccharide of Rb$_2$ mutants (5, 6) makes it tempting to speculate that further completion of the backbone region, and possibly incorporation of sugars into the outer core region, are prerequisites for the incorporation of the third KDO residue. Accordingly, it is of interest to determine whether the KDO trisaccharide occurs in deep-rough mutants that synthesize Rd_2 and Rd_1 chemotype lipopolysaccharides. Alternatively, it has been suggested that the incorporation of a third KDO residue may require the prior incorporation of phosphorylethanolamine into the KDO disaccharide (55). However, this seems unlikely since the occurrence of KDO-7-phosphorylethanolamine has been demonstrated in the lipopolysaccharides of heptoseless Salmonella mutants (12), and these mutants appear to synthesize only a KDO disaccharide (5, 6).

Biosynthesis of the heptose region. Unlike the biosynthesis of KDO, our knowledge of the pathway of L-glycero-D-mannoheptose synthesis is still incomplete. Eidels and Osborn (14), using transketolase mutants of S. typhimurium, established that sedoheptulose-7-phosphate is a precursor of the heptose moieties of lipopolysaccharide. These investigators subsequently demonstrated the enzymatic conversion of sedoheptulose-7-phosphate to D-glycero-D-mannoheptose-7-phosphate in cell extracts (15), and they postulated the following biosynthetic pathway for heptose synthesis, in which NDP and NTP represent nucleoside diphosphate and nucleoside triphosphate, respectively:

$$\text{Sedoheptulose-7-P} \xrightleftharpoons{\text{(isomerase)}} \text{D-glycero-D-manno-heptose-7-P}$$

$$\text{D-Glycero-D-manno-heptose-7-P} \xrightleftharpoons{\text{(mutase)}} \text{D-glycero-D-mannoheptose-1-P}$$

$$\text{D-Glycero-D-manno-heptose-1-P} + \text{NTP} \xrightleftharpoons{\text{(NDP-heptose synthase)}} \text{NDP-D-glycero-D-mannoheptose} + \text{P-P}_i$$

$$\text{NDP-D-glycero-D-mannoheptose} \xrightleftharpoons{\text{(epimerase)}} \text{NDP-L-glycero-D-mannoheptose}$$

Evidence in support of this pathway was subsequently provided by the isolation and characterization of mutants of E. coli that synthesize lipopolysaccharides containing D-glycero-D-mannoheptose (8). The mutant phenotype was due to a single mutation at a locus (designated rfaD) located at 80 min on the E. coli chromosome. Mutants lacking a functional rfaD gene product also accumulate ADP-D-glycero-D-mannoheptose (7), and the rfaD locus was subsequently identified as the structural gene for the epimerase that catalyzes the conversion of ADP-D-glycero-D-mannoheptose to ADP-L-glycero-D-mannoheptose. It is anticipated that the identification of ADP-L-glycero-D-mannoheptose as the donor of heptose residues will facilitate further characterization of the specific enzymatic steps involved in the synthesis of this region.

Incorporation of phosphate and ethanolamine. An in vitro system has been used to study the incorporation of phosphate residues into the heptose region of Salmonella minnesota (48, 49, 51). A soluble enzyme fraction from wild-type S. minnesota catalyzed the incorporation of radioactive phosphate from $[\gamma\text{-}^{32}P]$-ATP into the lipopolysaccharide of mutants defective in the incorporation of phosphate into the heptose region. The phosphate was incorporated as a phosphomonoester into position 4 of the first heptosyl

residue (39, 48). Although Rd_1P^- lipopolysaccharide (Rd_1 lipopolysaccharide lacking phosphate in the heptose region) served as an acceptor of phosphate, the incorporation of radioactive phosphate into RcP^- lipopolysaccharide (Rc lipopolysaccharide lacking phosphate in the heptose region) occurred to a significantly greater extent. Thus, it was postulated that the in vivo incorporation of phosphate into the heptose region follows the addition of the first glucose residue of the outer core. The results of another study (25) further suggest that addition of this glucosyl residue also precedes the incorporation of the third, or branch, heptose, as well as the addition of phosphate to the second heptose. The inability of the mutants described above to phosphorylate the heptosyl region of the inner core was due to a lesion in the rfaP locus. Although such mutants have only been described in S. minnesota, it is assumed that a similar locus exists in S. typhimurium and E. coli, as well as in other gram-negative bacteria.

The results of a recent experiment (27) suggest that the linking of the pyrophosphorylethanolamine to position 4 of the first heptose residue results from the transfer of the phosphorylethanolamine head group of phosphatidylethanolamine (PE) to the phosphate located in that position. However, there exists no direct evidence for this reaction, and the stage of lipopolysaccharide synthesis at which this putative reaction occurs is not known. Nothing is known concerning the mechanism of incorporation of ethanolamine into the KDO region.

Biosynthesis of the Outer Core

The biosynthesis of the outer core region of S. typhimurium lipopolysaccharide has been studied extensively (for reviews, see references 58, 68, and 88). Synthesis of this region involves a series of membrane-bound glycosyltransferases that catalyze the sequential transfer of sugars from nucleotide-sugar donors to the nonreducing terminus of the growing polysaccharide chain. Initial studies of the pathway were facilitated by the isolation of mutants defective in the biosynthesis of specific nucleotide sugars. The first such mutants had a lesion in the structural gene for UDP-galactose-4-epimerase (designated galE) (19, 20). In the absence of exogenously supplied galactose, these mutants are unable to synthesize UDP-galactose, the donor of galactose residues for outer core and O-antigen synthesis. Expression of the galE lesion results in the synthesis of an Rc lipopolysaccharide lacking all sugars distal to the first galactose residue (Fig. 1 and 2). Mutants unable to synthesize UDP-glucose were also used to study outer core synthesis. These mutants were deficient in either phosphoglucoisomerase (18) or UDP-glucose pyrophosphorylase (97) owing to lesions in the pgi or galU structural genes, respectively. Mutants of this general class synthesize an Rd_1 lipopolysaccharide lacking all sugars beyond the heptose region of the inner core. Phenotypic repair of the defects in lipopolysaccharide synthesis in the mutants described above can be affected by appropriate supplementation of the growth medium. For example, addition of galactose or glucose to UDP-galactose-4-epimerase-negative mutants or phosphoglucoisomerase-negative mutants, respectively, allows a bypass of the enzymatic defect and results in the synthesis of wild-type lipopolysaccharide.

The incomplete lipopolysaccharides synthesized by the nucleotide-sugar-deficient mutants have been used extensively as acceptors in studies of the individual glycosyltransferases involved in outer core biosynthesis (Fig. 3). The presence of glucosyltransferase I was demonstrated by the incorporation of glucose into endogenous Rd$_1$ lipopolysaccharide when membranes of a phosphoglucoisomerase-negative mutant of *S. typhimurium* were incubated with UDP-glucose (86). Similarly, endogenous Rc lipopolysaccharide present in the membranes of UDP-galactose-4-epimerase-negative mutants served as an acceptor of galactose when the membranes were incubated with UDP-galactose (57, 67, 86). In the latter case, only α1,3-galactosyltransferase activity was observed (82); repeated attempts to demonstrate the existence of α1,6-galactosyltransferase activity in *S. typhimurium*, as predicted by the structure of the outer core, were uniformly unsuccessful. However, the recent isolation of mutants defective in this activity has enabled the demonstration of this glycosyltransferase in extracts of wild-type strains (106).

The presence of glucosyltransferase II and *N*-acetyl-glucosaminyltransferase activities have also been demonstrated in cell-free preparations of *S. typhimurium* (63). Glucosyltransferase II was demonstrated in membranes of *galE* mutants, as determined by the incorporation of glucose from UDP-glucose into endogenous Rc lipopolysaccharide which was dependent on the prior incorporation of galactose from UDP-galactose. Similarly, the incorporation of *N*-acetylglucosamine, as catalyzed by *N*-acetyl-glucosaminyltransferase, required the prior incorporation of galactose and glucose into the acceptor Rc lipopolysaccharide. The products of both reactions were characterized, and in each case the sugars were incorporated exclusively into the outer core in accordance with the structure of this region (63, 68).

The glycosyltransferases involved in outer core synthesis are membrane associated. Therefore, since the cell envelope fraction contains both glycosyltransferase activity and acceptor lipopolysaccharide, they cannot be used in studies to examine substrate and acceptor specificity. However, significant amounts of enzyme activity are released into the soluble fraction after the sonic disruption of cells (86). Accordingly, the acceptor specificities of soluble glucosyltransferase I and α1,3-galactosyltransferase have been demonstrated with either heat-inactivated membranes, containing endogenous lipopolysaccharide acceptors, or mixtures of purified lipopolysaccharide and PE. In this manner, glucosyltransferase I and α1,3-galactosyltransferase were shown to be active only when Rd$_1$ and Rc lipopolysaccharides were used as acceptors, respectively. In addition, the availability of the solubilized enzymes facilitated their purification; the glucosyltransferase I has been purified approximately 1,000-fold (53), and the α1,3-galactosyltransferase has been purified about 6,000-fold (16).

It is of interest that mutants of *S. typhimurium* lacking the α1,3-galactosyltransferase synthesize a lipopolysaccharide with an abbreviated outer core containing a single galactose linked to position 6 of the proximal glucose (61). Thus, it appears that the prior incorporation of galactose linked to position 3 of the proximal glucose is not a prerequisite for α1,6-galactosyltransferase activity. In contrast, mutants lacking the α1,6-galactosyltransferase synthesize two populations of lipopolysaccharide; one contains a core region with only the proximal glucose (Rc chemotype), and the other contains a complete outer core but lacks the α1,6-linked branch galactose (106). These observations have led to the suggestion that the substrate specificity of α1,3-galactosyltransferase is broad enough to catalyze some transfer of D-galactose from UDP-galactose to unsubstituted proximal glucose residues in vivo. However, prior incorporation of the α1,6-linked branch galactose appears to render the lipopolysaccharide molecule a more favorable acceptor. These data suggest a sequence of reactions whereby the α1,6-linked branch galactose is incorporated before the incorporation of the α1,3-linked galactose.

On the basis of results described in this section, a tentative sequence of steps for the biosynthesis of the core region is outlined in Fig. 6. It is interesting that a general characteristic of the pathway appears to be the initiation of synthesis of one region of the core before the complete assembly of the immediately preceding region. For example, the initial incorporation of KDO residues occurs at a stage of lipid A synthesis before the complete incorporation of *O*-fatty acyl residues. Similarly, recent studies on the structure of the KDO region indicate that synthesis of the heptose region is initiated before the incorporation of the nonreducing terminal KDO residue. Finally, it appears that the proximal glucose residue of the outer core is incorporated into the growing polymer before the incorporation of phosphate, phosphorylethanolamine, and the branched heptose into the heptose region. Although the significance of this apparent biosynthetic pattern remains to be established, it may be related in some unknown manner to the regulation of lipopolysaccharide synthesis. It must be emphasized that this pathway only represents a working model that is based, at least in part, on the results of in vitro experiments that may not reflect events as they actually occur in vivo.

Genetics of Core Biosynthesis

Mutants of *S. typhimurium* defective in several of the core glycosyltransferases, as well as other enzymes involved in core synthesis, are members of a group of rough mutants designated rough A or *rfa*. These mutants have been classified as such on the basis of a variety of phenotypic criteria which in each case involve an alteration of core structure (for reviews, see references 47 and 94). Exceptions to this classification are *rfaL* mutants that are defective in O-antigen ligase (see Biosynthesis of O Antigen) and that synthesize an Ra lipopolysaccharide containing a complete core. Most of the *rfa* genes have been mapped in a small region of the chromosome of *S. typhimurium* between *cysE* and *pyrE* at 79 min; the remainder are located outside of this region and are also distant from one another (see chapter 54).

Although many of the *rfa* genes are believed to be structural genes for glycosyltransferases involved in core synthesis, only mutations in the *rfaG* (61) and

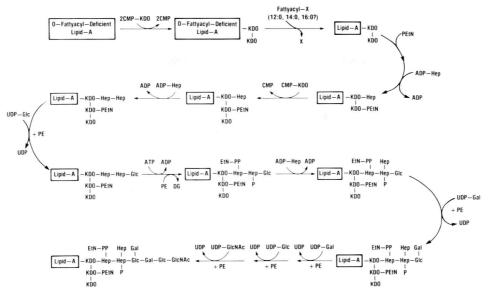

FIG. 6. Postulated sequence of reactions for the synthesis of the lipopolysaccharide core region. Abbreviations are as described in the legend to Fig. 1. Additional abbreviations: PE, phosphatidylethanolamine; DG, diglyceride.

rfaB (106) loci (Fig. 3) have been correlated with a loss of a specific glycosyltransferase activity in *S. typhimurium* as measured in vitro. In addition, hybrid ColE1 2plasmids containing the *rfaG* gene of *E. coli* have been identified (10). The *rfaH* gene was previously included in this group and was believed to be the structural gene for the α1,3-galactosyltransferase. However, a recent study (11) indicates otherwise since mutations in the *rfaH* gene decrease the amount of galactosyltransferase activity without altering the apparent structure of the enzyme(s). (No attempt was made in these studies to distinguish between α1,6-galactosyltransferase and α1,3-galactosyltransferase activities.) Mutations in the *Salmonella rfaH* gene also result in defects in F-factor function similar to those described for *sfrB* mutants of *E. coli*, and it has been suggested that *sfrB* and *rfaH* represent homologous genes in the two organisms (90). Indeed, a hybrid ColE1 plasmid carrying the *sfrB* gene from *E. coli* was capable of restoring galactosyltransferase activity in an *rfaH* mutant of *S. typhimurium* (9). Furthermore, since *sfrB* mutants of *E. coli* display a pronounced reduction in the activities of at least two glycosyltransferases involved in outer core synthesis (11), it has been concluded that the *rfaH* (*sfrB*) gene acts as a positive regulatory element in expression of multiple glycosyltransferases in *E. coli* and *S. typhimurium*. However, the mechanism of this regulation is not known.

The functions of the *rfaD* and *rfaP* genes (Fig. 4) have been described earlier (Biosynthesis of the Inner Core). The functions of the *rfaC* and *rfaE* genes are unknown. Mutants defective in these loci are unable to either synthesize heptose or incorporate heptosyl residues into the core. Mutants defective in the *rfaF* gene are unable to incorporate the distal heptosyl residue into the core. Thus, *rfaF* mutants are presumed to be defective in the transferase that catalyzes this reaction. However, this presumption has not

been verified experimentally owing to a previous lack of knowledge concerning the identity of the donor of heptosyl residues in core synthesis. The recent identification of ADP-L-glycero-D-mannoheptose as the likely donor (7) should now make this possible.

The Role of Phospholipids in Core Biosynthesis

Early attempts to characterize the reactions catalyzed by crude preparations of soluble α1,3-galactosyltransferase and glucosyltransferase I revealed that purified lipopolysaccharide preparations were unable to serve as acceptors (85). In addition, the extraction of heat-inactivated cell envelopes with lipid solvents (under conditions which did not remove endogenous lipopolysaccharides) also destroyed acceptor activity. These results suggested that a lipid factor(s) distinct from lipopolysaccharide was required for enzyme activity. Indeed, when the extracted lipid fraction was heated with the inactive acceptors and then allowed to cool slowly, the acceptor activity of these fractions was restored. Fractionation of the lipid extract identified the active component as PE. Other naturally occurring and synthetic phospholipids were also shown to be active in restoring acceptor activity to purified lipopolysaccharides (87). The structural requirements for activity were found to reside in the nature of the acyl chain and the component linked to the phosphate. Thus, phosphatidic acid, phosphatidylglycerol, cardiolipin, and phosphatidylserine were all found to be active to various degrees. However, phosphatidylcholine, which is not a usual component of gram-negative bacterial membranes, was completely devoid of activity. The occurrence of unsaturation or of cyclopropane groups in the fatty acyl chains linked to the glycerophosphate backbone was also critical for activity. For example, catalytic reduction of active PE preparations rendered them inactive.

Initial investigations (89, 103) into the role of phospholipid in the reaction catalyzed by crude preparations of α1,3-galactosyltransferase provided evidence that phospholipid (represented by PL) first interacts with acceptor lipopolysaccharide in the following sequence:

$$\text{LPS (galactose deficient)} + \text{PL} \longrightarrow \text{LPS} \cdot \text{PL}$$

$$\text{Enzyme} + \text{LPS} \cdot \text{PL} \xrightarrow{\text{Mg}^{2+}} \text{enzyme} \cdot \text{LPS} \cdot \text{PL}$$

$$\begin{array}{l} \text{Enzyme} \cdot \text{LPS} \cdot \text{PL} \\ + \text{UDP-galactose} \end{array} \longrightarrow \begin{array}{l} \text{galactosyl-LPS} \cdot \text{PL} \\ (+ \text{UDP} + \text{enzyme}) \end{array}$$

Indeed, the formation of binary complexes of PE and acceptor lipopolysaccharide, as well as ternary complexes that included enzyme, were demonstrated by equilibrium density gradient centrifugation. A more detailed examination of these interactions was carried out with purified α1,3-galactosyltransferase (16, 17). Incubation of acceptor lipopolysaccharide with PE before addition of the purified enzyme was required to obtain significant activity. There was a pronounced increase in both the rate and yield of the reaction over that obtained in the absence of PE. Furthermore, the $V_{\max}$ of the reaction was increased 13-fold in the presence of PE. Equilibrium density gradient centrifugation studies revealed that the purified enzyme was also able to form binary complexes with various PE preparations (17). In each case, binary complex formation was strictly dependent on the nature of the phospholipid fatty acyl chains. Certain species of PE (e.g., didecanoylphosphatidylethanolamine) only formed binary complexes with lipopolysaccharide, whereas others (e.g., dipalmitoylphosphatidylethanolamine) only formed a complex with enzyme. Only those species able to form complexes with both enzyme and lipopolysaccharide (i.e., those containing unsaturated fatty acids or fatty acyl chains with cyclopropane groups) were active in the transferase reaction. The extent of activation of the α1,3-galactosyltransferase reaction was found to be a function of the PE/lipopolysaccharide ratio rather than of the absolute concentration of phospholipid (17). These observations suggest that PE functions primarily through its interaction with lipopolysaccharide and not through a direct effect on enzyme.

A model has been proposed (17) that stipulates that full catalytic activity requires specific binding of the enzyme to both PE and the polysaccharide chains of the acceptor. According to the model, binding of PE to lipopolysaccharide increases the accessibility of the polysaccharide chains to the enzyme and also provides a second specific binding site for the enzyme. Indeed, the effect of PE on the $V_{\max}$ of the reaction suggests that binding of the enzyme to the PE:lipopolysaccharide binary complex results in direct activation of the enzyme.

A similar, if not identical, role of phospholipid has been demonstrated for the reaction catalyzed by purified glucosyltransferase I (53). A requirement for PE was demonstrated, and the structural features of the PE required for activity were the same as described for the α1,3-galactosyltransferase. Compelling evidence for a primary interaction of PE with lipopoly-

saccharide was provided by the isolation of a binary complex of these two components that was active when enzyme and UDP-glucose were added. Furthermore, the product of the reaction was isolated and shown to be an efficient acceptor of galactosyl residues from UDP-galactose catalyzed by purified α1,3-galactosyltransferase. Indeed, subsequent studies (31) succeeded in the isolation of a functional quaternary complex composed of PE, Rd_1 lipopolysaccharide, glucosyltransferase I, and α1,3-galactosyltransferase. Sequential transfer of glucose and galactose to the lipopolysaccharide was demonstrated after the addition of UDP-glucose and UDP-galactose, respectively. Incorporation of galactose was strictly dependent on the prior incorporation of glucose. Formation of a functional quaternary complex was not dependent on the order of addition of the enzymes to the binary complex of PE and lipopolysaccharide. Thus, a concerted mechanism of enzyme insertion into the complex seems unlikely. However, the occurrence of both enzymes in the same particle was not proven.

It is of interest that the α1,3-galactosyltransferase can participate in the formation of a functional complex even though the complex does not contain the specific acceptor lipopolysaccharide (Rc chemotype) required for activity of the enzyme. This suggests that binding of the enzyme to the binary complex may also involve interactions with a region of the lipopolysaccharide other than the distal acceptor site, e.g., the inner core region. This may explain why the enzyme is inhibited by the polysaccharide portion of nonsubstrate lipopolysaccharides (16).

The use of monolayer techniques has provided additional information regarding the organization of phospholipid, lipopolysaccharide, and glycosyltransferases to form a functional complex (80, 81; for a review, see reference 88). In this technique, a monolayer film of phospholipid is formed over an aqueous subphase resulting in molecules oriented at the air-water interface with their polar head groups solvated in the aqueous phase and their acyl chains directed outward. Penetration of molecules into the monolayer is detected by changes in surface pressure.

Unlike aqueous suspensions, penetration of PE monolayers by Rc lipopolysaccharide did not require heating (80). The resulting binary complex was functional, as demonstrated by the addition of galactosyl residues to the lipopolysaccharide when UDP-galactose and purified α1,3-galactosyltransferase were subsequently injected into the aqueous subphase. These observations suggest that the interaction of lipopolysaccharide with PE monolayers closely reflects native structures. Several lines of evidence favor a side-by-side organization of lipopolysaccharide and PE in the monolayer with polar groups facing the aqueous phase and nonpolar groups facing outward. For example, the addition of lipopolysaccharide to the monolayer resulted in an increase in surface pressure, which reflects an increase in the number of molecules on the surface. In addition, penetration of lipopolysaccharide into the monolayer occurred at a significantly increased rate and to a greater extent when the fatty acyl chains were unsaturated or contained cyclopropane groups. The specificity for the nonpolar region suggests an interaction of the fatty acyl residues with the lipopolysaccharide (presumably the fatty acyl chains of lipid A). Finally, the interaction of lipopolysaccharide with the monolayer did not result in a significant change in surface potential, as would be expected if adsorption of lipopolysaccharide to the polar head groups in the undersurface of the film had occurred.

Purified α1,3-galactosyltransferase was also observed to penetrate into the monolayer in the absence of lipopolysaccharide (81). However, unlike the interactions between lipopolysaccharide and PE monolayers (80), the ability of the enzyme to do so did not reflect a specificity for species of PE previously shown to be active in aqueous suspensions, i.e., those with fatty acyl chains that were unsaturated or contained cyclopropane groups. This difference has led to the consideration that the enzyme primarily interacts with the phospholipid polar head group and may not penetrate into the nonpolar region. It seems reasonable to assume that an interaction of the enzyme with the polar region in the aqueous subphase must take place for the transferase reaction to occur. However, the available data do not rule out the possibility that a specific interaction also occurs that involves the nonpolar region of the complex formed between lipopolysaccharide and active PE. Indeed, the results of an earlier study (17) are consistent with this possibility.

Although a functional ternary complex of α1,3-galactosyltransferase, PE, and lipopolysaccharide has been demonstrated, only about two galactose residues were transferred to acceptor per mole of enzyme (81). In contrast, in aqueous suspension, the yield is not related to the amount of enzyme present, and a considerably greater yield per mole of enzyme is realized. Thus, it has been concluded that in monolayers the lipopolysaccharide and enzyme remain immobile or fixed in position with respect to one another and that the transfer of galactose only occurs to an adjacent acceptor. This view also predicts that the enzyme remains as a structural component of the system after catalysis. In contrast, aqueous suspensions may contain a large population of particles containing phospholipid and lipopolysaccharide, thus allowing multiple encounters of enzyme and substrate.

A tentative model for the biosynthesis of the core has been proposed by Rothfield and Romeo (88). The basic features of the model are depicted schematically in Fig. 7. The model proposes that addition of sugars to the nascent core region occurs sequentially and is catalyzed by a series of membrane-associated glycosyltransferases. The lipopolysaccharide molecules are integrated into the inner leaflet of the cytoplasmic membrane such that the fatty acyl chains of lipid A are dissolved in the hydrocarbon interior. The polar groups of the lipid A disaccharide are aligned with the polar head groups of phospholipids, and the nascent polysaccharide chains (depicted in Fig. 7 as being of Rd₁ chemotype) are oriented toward the cytoplasm or interior. The fluid nature of the hydrocarbon interior allows mobility or lateral movement of lipopolysaccharide in the plane of the membrane. In contrast, the glycosyltransferases are envisioned as being considerably less mobile, and they remain in a relatively fixed position. According to the model, lateral movement of lipopolysaccharide occurs until a specific complex is formed among the lipopolysaccharide, the phospho-

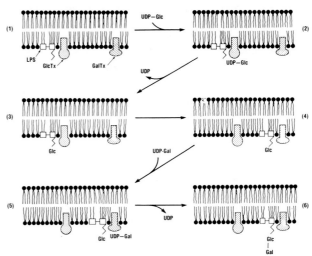

FIG. 7. Tentative model for the mechanism of biosynthesis of the core region of lipopolysaccharide. The sequential transfer of glucose and galactose into an Rd₁-lipopolysaccharide acceptor, depicted in steps 1 through 6, is based on the model proposed by Rothfield and Romeo (88). The lipid A region of the incomplete lipopolysaccharide is integrated into the inner leaflet of the cytoplasmic membrane. The individual glycosyltransferases may also be inserted into the hydrophobic interior of the membrane as shown here for glycosyltransferase I (GlcTx) and galactosyltransferase (GalTx) enzymes. Alternatively, the glycosyltransferase enzymes may be peripheral membrane proteins that interact primarily with the polar head groups of phospholipids, as proposed in the original model. Details concerning the model are provided in the text.

lipid, and the glycosyltransferase for which the acceptor is a substrate. After transfer of the sugar, the affinity of enzyme for the lipopolysaccharide is decreased, and the lipopolysaccharide becomes a substrate for the next glycosyltransferase in the sequence. Thus, in a sense, the lipopolysaccharide is "pulled" to the next enzyme in the sequence by virtue of the higher affinity of the enzyme for the newly generated substrate. As the lipopolysaccharide molecule moves from one glycosyltransferase to the next, the core region is elongated, and the previous glycosyltransferase in the sequence is made available for the next molecule of acceptor. It has also been suggested that the enzymes might be arranged in the proper sequence and in close proximity so as to minimize movement of the lipopolysaccharide and to promote efficient elongation of the polysaccharide chains. Indeed, the comparatively poor efficiency of galactose incorporation into acceptor lipopolysaccharide in the monolayer studies may reflect, at least in part, the absence of such an arrangement in the monolayers.

Biosynthesis of O Antigen

The biosynthesis of O antigen has been studied in considerable detail (for reviews, see references 58, 79, and 109). Much of our knowledge concerning the sequence of reaction involved in this process stems from studies of the pathway as it occurs in *S. typhimurium* (group B) and *S. newington* (group E). The O antigens of these organisms are structurally similar, and all available evidence indicates that their mechanisms of assembly are identical.

The syntheses of O antigen and of the lipopolysaccharide core region occur by fundamentally different mechanisms. As described above, elongation of the core region occurs by the sequential transfer of individual monosaccharides from nucleotide sugar donors to the nonreducing terminus of the growing polysaccharide chain. In contrast, the O antigen is assembled independently as a lipid-linked polymer and subsequently transferred to the nonreducing terminal glucose of the completed core. The biosynthetic sequence, as it occurs in *S. typhimurium*, is shown in Fig. 8. This process can be divided into three phases. The first phase involves synthesis of a single tetrasaccharide repeat unit covalently linked to a unique membrane-bound lipid commonly referred to as either antigen-carrier lipid or glycosyl-carrier lipid (GCL). The latter designation will be used here. During the second phase, the individual repeat units are polymerized to form the completed O antigen still linked to GCL. Finally, the last step results in the transfer of the newly synthesized polymer from GCL to the completed core. All of these steps are carried out by a series of membrane-bound enzymes.

The discovery of lipid-linked intermediates in O-antigen synthesis (102, 108) followed closely the discovery that lipid-linked intermediates are also involved in peptidoglycan synthesis (1). Structural characterization of the lipids involved in these processes (30, 107) revealed that they are both phosphomonoesters of a C_{55} polyisoprenoid alcohol with the following structure:

$$CH_3-C=CH_2-CH_2-\left[CH_2-C=CH-CH_2\right]_9 -CH_2-C=CH-CH_2O-P-O-$$

with CH_3 methyl substituents and the terminal group as $\overset{O}{\underset{O-}{\overset{\|}{P}}}$

Similar, if not identical, lipids are involved in the synthesis of capsular polysaccharides, teichuronic acids, and other polymers in bacteria. Related polyisoprenoid compounds, the dolichols, function in the synthesis of the oligosaccharide moieties of glycoproteins in eucaryotic cells.

The first reaction of the pathway (Fig. 8) is the transfer of galactose-1-phosphate from UDP-galactose to GCL-phosphate to form galactose-pyrophosphoryl-GCL (69). This reaction is somewhat unusual since it involves the cotransfer of sugar and phosphate as a unit; consequently, UMP is released instead of UDP. The reaction has an apparent equilibrium constant of approximately 0.5, and it is freely reversible. Accordingly, the high energy of the pyrophosphoryl-galactose linkage of the donor is conserved in the formation of galactose-pyrophosphoryl-GCL. The remainder of the repeat unit is formed by the sequential incorporation of L-rhamnose, D-mannose, and abequose catalyzed by a series of membrane-bound transferases that use TDP-rhamnose, GDP-mannose, and CDP-abequose, respectively, as donors.

The synthesis of individual repeat units is followed by their polymerization to yield the complete O antigen still linked to the GCL by pyrophosphate linkage. The mechanism of this process, catalyzed by O-antigen polymerase, was elucidated by an elegant series of

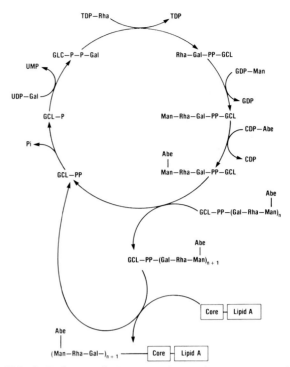

FIG. 8. Pathway of O-antigen biosynthesis in *S. typhimurium*.

in vivo and in vitro experiments conducted by Robbins and co-workers (78). The results of these experiments clearly demonstrated that elongation of the lipid-linked O antigen in *S. newington* occurs by growth at the reducing terminus. Thus, a glycosidic linkage is formed between the reducing terminal galactose of the nascent O antigen and the nonreducing terminal mannosyl residue of a newly synthesized trisaccharide repeat unit (Fig. 9). Generally speaking, the polymerization process resembles the mechanisms

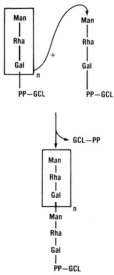

FIG. 9. Mechanism of polymerization of O-antigen repeat units as it occurs in *Salmonella* group E strains. The same mechanism is believed to occur in *S. typhimurium*.

involved in the synthesis of fatty acids and proteins, and in many respects GCL is functionally analogous to both acyl-carrier protein and tRNA. As mentioned previously, all of the intermediates in O-antigen synthesis are membrane bound, as are the enzymes involved in their synthesis. Thus, a polymerization process involving growth of the nonreducing terminus would presumably require a mechanism that would maintain the nonreducing terminus of the growing chain in close association with the membrane. The observed mechanism of polymerization, i.e., growth at the reducing terminus, obviates this requirement and serves to greatly increase the efficiency of chain elongation by ensuring that the reactive components are in close proximity.

The final step in the biosynthetic sequence is catalyzed by the enzyme O antigen:lipopolysaccharide ligase. This enzyme catalyzes the transfer of completed O antigen from GCL to the completed lipopolysaccharide core with the release of GCL-pyrophosphate. During this reaction, the terminal reducing galactose of the O side chain becomes linked to the nonreducing terminal glucose of the core by 1,4-linkage. Thus, the high energy of the GCL-pyrophosphate sugar linkage is used for polymerization of individual repeat units, as well as for joining two macromolecules.

As discussed in previous studies (24, 54, 70), the degree of polymerization of O side chains synthesized by cells of a given species varies considerably. Furthermore, the relative amounts of each size class that are transferred to the lipopolysaccharide core are not equal. These observations suggest that although the ligase enzyme(s) is able to catalyze the transfer of O side chains of various degrees of polymerization, there exists a preference for certain chain lengths. However, the mechanism responsible for this discrimination is not known.

Certain strains of bacteria synthesize O antigens by a mechanism that differs from the one described above. For example, the O antigens of *E. coli* O8 and O9 are α-mannans (71, 74). The repeating unit of the O8 antigen is → 3)αMan-(1→2)αMan-(1→2)αMan-(1→, whereas the repeat unit of the O9 antigen is → 3)αMan-(1→2)αMan-(1→2)αMan-(1→2)αMan-(1→3)αMan-(1→. The synthesis of these polymers occurs by the sequence of reactions summarized below (32):

$$GCL\text{-}P + UDP\text{-}glucose \overset{Mg^{2+}}{\rightleftharpoons} GCL\text{-}PP\text{-}Glc + UMP$$

$$\begin{matrix} GCL\text{-}PP\text{-}Glc \\ + nGDP\text{-}mannose \end{matrix} \longrightarrow \begin{matrix} GCL\text{-}PP\text{-}Glc\text{-}(Man)_n \\ + nGDP \end{matrix}$$

The first reaction is analogous to the first step of O-antigen synthesis in *Salmonella* groups B and E. However, subsequent reactions do not appear to involve the synthesis and polymerization of individual GCL-linked repeat units. In contrast, synthesis of the entire mannan chain occurs by the sequential transfer of mannose residues from GDP-mannose to the nonreducing terminus of the elongating polymers. Thus, although the chain-elongation process occurs in the manner of conventional polysaccharide synthesis, in some unknown manner the mechanism results in the synthesis of a polymer containing repetitive oligosaccharide sequences.

The O-antigen:lipopolysaccharide ligase reaction as it occurs in *E. coli* O9 is also unique. This reaction results in the transfer of the completed α-mannan, as well as the reducing terminal glucose, to a lipopolysaccharide acceptor with an incomplete R1 core (pre-core) lacking one glucose residue (104). The product of the reaction is a lipopolysaccharide with a complete R1 core covalently linked to O antigen. Thus, the nonreducing terminal glucose of the R1 core is cotransferred together with the completed O9 antigen.

The synthesis of the O8 and O9 antigens, as well as of several other O antigens, is dependent on the *rfe* gene(s). The *rfe* gene is also required for the synthesis of enterobacterial common antigen. However, the function(s) of the *rfe* gene in either of these processes is not understood.

The synthesis of O antigens in vitro is inhibited by the antibiotic bacitracin (32, 62). Therefore, it is assumed that the regeneration of GCL-P (Fig. 8) requires the action of a specific bacitracin-sensitive pyrophosphatase analogous to the enzyme involved in peptidoglycan synthesis (91).

A discussion of the topology of lipopolysaccharide synthesis is included in chapter 3. However, it is of interest that the results of recent studies using immunoelectron microscopy techniques (52) suggest the possibility that the ligase reaction, as well as perhaps the polymerase reaction, may occur on the periplasmic face of the cytoplasmic membrane. This mechanism is attractive since it avoids the apparent problems involved in moving either completed O antigen or lipopolysaccharide molecules across the cytoplasmic membrane before translocation of lipopolysaccharide to the outer membrane. However, the mechanisms involved in the translocation of lipopolysaccharide and its subsequent assembly into the outer membrane remain to be established.

Genetics of O-Antigen Synthesis

Most of the genes required for the synthesis of O side chain specific components are located in the *his*-linked *rfb* cluster located at 44 to 45 min on the *Salmonella* chromosomal map (for reviews, see references 47, 58, and 94). This region includes information for nine of the enzymes involved in the synthesis of GDP-mannose, TDP-rhamnose, and CDP-abequose, as well as for the four glycosyltransferases required for the synthesis of a single O repeat unit. Those genes required for the synthesis of sugars that are utilized for purposes other than for O-antigen synthesis, e.g., the structural gene for UDP-galactose-4-epimerase (*galE*) and phosphomannose isomerase (*pmi*), are located outside of this region.

The *rfc* gene, located near *trp* at 32 min, is believed to be the structural gene for the O-antigen polymerase. *Salmonella* mutants with a defective *rfc* gene synthesize lipopolysaccharides with a single O repeat unit (56). As a consequence, the phenotype of these mutants is intermediate between wild-type smooth (S) strains and mutant rough (R) strains, thus they have been designated semirough (SR). Mutations in either the *rfaL* gene, located in the *rfa* gene cluster at 79 min, or the *rfbT* gene result in the synthesis of an Ra lipopolysaccharide, even though such mutants are able to synthesize GCL-linked O antigen. Thus, it has been concluded that the products of both of these genes are required for O-antigen:

lipopolysaccharide ligase activity, although the nature of this interaction is not known.

CONCLUDING REMARKS

The work of a large number of laboratories over the past 20 years has led to the description of many of the individual steps and genetic loci involved in lipopolysaccharide synthesis in *Salmonella* species and *E. coli*. However, a detailed understanding of certain steps, e.g., the mechanism of assembly of the KDO region, has not yet been achieved. Furthermore, the specific functions of several genes involved in lipopolysaccharide synthesis (*rfaH*, *rfaC*, *rfaE*, *rfe*, etc.) remain to be established, and it is clear that we are still far from a complete understanding of the genetics of lipopolysaccharide synthesis. Nevertheless, we now have a reasonably coherent view of the major features of this process, and it is anticipated that additional details concerning the mechanism of less understood steps will be forthcoming.

In contrast to the biosynthesis of lipopolysaccharide, several other aspects of lipopolysaccharide biochemistry and molecular biology either remain unexplored or have received attention from only a limited number of laboratories. Thus, essentially nothing is known concerning the regulation of lipopolysaccharide synthesis at either the genetic level or the level of specific enzymes. Furthermore, our knowledge concerning the organization of enzymes involved in lipopolysaccharide biosynthesis is very limited. The pioneering work of Rothfield and co-workers suggests the possibility that the glycosyltransferase enzymes involved in core biosynthesis are organized or closely arranged in the cytoplasmic membrane so as to facilitate a specific sequence of reaction. However, the functional organization of these enzymes in this manner has not been directly demonstrated. In this regard, it appears that certain steps of lipopolysaccharide synthesis, i.e., O-antigen:lipopolysaccharide ligase, occur on the periplasmic face of the cytoplasmic membrane. Moreover, the possibility has not been excluded that the polymerization of O antigen, and perhaps even earlier steps of lipopolysaccharide synthesis, also take place at this subcellular location. In each case, the transport of substrates or intermediates across the cytoplasmic membrane to the site of their utilization is required. However, the mechanism(s) involved in this process, as well as the overall topology of lipopolysaccharide synthesis, remains to be established.

One of the most exciting areas for future investigation concerns the mechanism of translocation of lipopolysaccharide to the outer membrane. The elegant studies of Osborn and co-workers (64, 65) have clearly established that this molecule is synthesized in the cytoplasmic membrane and then irreversibly translocated to the outer membrane. However, the mechanisms of translocation and integration into the outer membrane have not been elucidated.

ACKNOWLEDGMENTS

This work was supported by grant RO7302 from the Uniformed Services University of the Health Sciences and by Public Health Service grant AI21309 from the National Institutes of Health.

I am grateful to Henry Wu for his helpful suggestions and critical reading of the manuscript. The assistance of Susan Wolski in the preparation of this manuscript is also appreciated.

LITERATURE CITED

1. **Anderson, J. S., M. Matsuhashi, M. A. Haskin, and J. L. Strominger.** 1965. Lipid phosphoacetylmuramyl-pentapeptide and lipid-phosphodisaccharide-pentapeptide: presumed membrane transport intermediates in cell wall synthesis. Proc. Natl. Acad. Sci. USA **53:**881–889.
2. **Bagdian, G., O. Lüderitz, and A. M. Staub.** 1966. Immunochemical studies on *Salmonella*. XI. Chemical modification correlated with conversion of group B *Salmonella* by bacteriophage 27. Ann. N.Y. Acad. Sci. **133:**405–424.
3. **Blache, D., M. Bruneteau, and G. Michel.** 1981. Structure de la région heptose du lipopolysaccharide de *Escherichia coli* K12 CR34. Eur. J. Biochem. **113:**563–568.
4. **Brade, H., and C. Galanos.** 1983. Common lipopolysaccharide specificity: new type of antigen residing in the inner core region of S- and R-form lipopolysaccharides from different families of gram-negative bacteria. Infect. Immun. **42:**250–256.
5. **Brade, H., C. Galanos, and O. Lüderitz.** 1983. Isolation of a 3-deoxy-D-mannooctulosonic acid disaccharide from *Salmonella minnesota* rough-form lipopolysaccharides. Eur. J. Biochem. **131:**201–203.
6. **Brade, H., and E.-T. Rietschel.** 1984. α-2→4-Interlinked 3-deoxy-D-manno-octulosonic acid disaccharide. A common constituent of enterobacterial lipopolysaccharides. Eur. J. Biochem. **145:**231–236.
7. **Coleman, W. G.** 1983. The *rfaD* gene codes for ADP-L-glycero-D-mannoheptose-6-epimerase. J. Biol. Chem. **258:**1985–1990.
8. **Coleman, W. G., Jr., and L. Leive.** 1979. Two mutations which affect the barrier function of the *Escherichia coli* K-12 outer membrane. J. Bacteriol. **139:**899–910.
9. **Creeger, E. S., J. F. Chen, and L. I. Rothfield.** 1979. Cloning of genes for bacterial glycosyltransferases. II. Selection of a hybrid plasmid carrying the *rfaH* gene. J. Biol. Chem. **254:**811–815.
10. **Creeger, E. S., and L. I. Rothfield.** 1979. Cloning of genes for bacterial glycosyltransferases. I. Selection of hybrid plasmids carrying genes for two glycosyltransferases. J. Biol. Chem. **254:**804–810.
11. **Creeger, E. S., T. Schulte, and L. I. Rothfield.** 1984. Regulation of membrane glycosyltransferases by the *sfrB* and *rfaH* genes of *Escherichia coli* and *Salmonella typhimurium*. J. Biol. Chem. **259:**3064–3069.
12. **Dröge, W., V. Lehmann, O. Lüderitz, and O. Westphal.** 1970. Structural investigations on the 2-keto-3-deoxyoctonate region of lipopolysaccharides. Eur. J. Biochem. **14:**175–184.
13. **Dröge, W., E. Ruschmann, O. Lüderitz, and O. Westphal.** 1968. Biochemical studies on lipopolysaccharides of *Salmonella* R mutants. 4. Phosphate groups linked to heptose units and their absence in some R lipopolysaccharides. Eur. J. Biochem. **4:**134–138.
14. **Eidels, L., and M. J. Osborn.** 1971. Lipopolysaccharide and aldoheptose biosynthesis in transketolase mutants of *Salmonella typhimurium*. Proc. Natl. Acad. Sci. USA **68:**1673–1677.
15. **Eidels, L., and M. J. Osborn.** 1974. Phosphoheptose isomerase, first enzyme in the biosynthesis of aldohaptose in *Salmonella typhimurium*. J. Biol. Chem. **249:**5642–5648.
16. **Endo, A., and L. Rothfield.** 1969. Studies of a phospholipid-requiring bacterial enzyme. I. Purification and properties of uridine diphosphate galactose: lipopolysaccharide α-3-galactosyltransferase. Biochemistry **8:**3500–3507.
17. **Endo, A., and L. Rothfield.** 1969. Studies of a phospholipid-requiring bacterial enzyme. II. The role of phospholipid in the uridine diphosphate galactose: lipopolysaccharide α-3-galactosyltransferase reaction. Biochemistry **8:**3508–3515.
18. **Fraenkel, D., M. J. Osborn, and B. L. Horecker.** 1963. Metabolism and cell wall structure of a mutant of *Salmonella typhimurium* deficient in phosphoglucose isomerase. Biochem. Biophys. Res. Commun. **11:**423–428.
19. **Fukasawa, T., and H. Nikaido.** 1961. Galactose-sensitive mutants of *Salmonella*. II. Bacteriolysis induced by galactose. Biochim. Biophys. Acta **48:**470–483.
20. **Fukasawa, T., and H. N. Nikaido.** 1960. Formation of phage receptors induced by galactose in a galactose-sensitive strain of *Salmonella*. Virology **11:**508–510.
21. **Ghalambor, M. A., and E. C. Heath.** 1966. The biosynthesis of cell wall lipopolysaccharide in *Escherichia coli*. J. Biol. Chem. **241:**3222–3227.
22. **Ghalambor, M. A., E. M. Levine, and E. C. Heath.** 1966. The biosynthesis of cell wall lipopolysaccharide in *Escherichia coli*. III. Isolation and characterization of 3-deoxyoctulosonic acid. J. Biol. Chem. **241:**3207–3215.
23. **Gmeiner, J., and S. Schlecht.** 1979. Molecular organization of

the outer membrane of *Salmonella typhimurium*. Eur. J. Biochem. **93**:609–620.

24. **Goldman, R. C., and L. Leive.** 1980. Heterogeneity of antigenic-side-chain length in lipopolysaccharide from *Escherichia coli* O111 and *Salmonella typhimurium* LT2. Eur. J. Biochem. **107**:145–153.

25. **Hämmerling, G., V. Lehmann, and O. Lüderitz.** 1973. Structural studies on the heptose region of *Salmonella* lipopolysaccharides. Eur. J. Biochem. **38**:453–458.

26. **Hämmerling, G., O. Lüderitz, O. Westphal, and P. H. Mäkelä.** 1971. Structural investigation on the core polysaccharide of *Escherichia coli* O100. Eur. J. Biochem. **22**:331–344.

27. **Hasin, M., and E. P. Kennedy.** 1982. Role of phosphatidylethanolamine in the biosynthesis of pyrophosphoethanolamine residues in the lipopolysaccharide of *Escherichia coli*. J. Biol. Chem. **257**:12475–12477.

28. **Heath, E. C., R. M. Mayer, R. D. Edstrom, and C. A. Beaudreau.** 1966. Structure and biosynthesis of the cell wall lipopolysaccharide of *Escherichia coli*. Ann. N.Y. Acad. Sci. **133**:315–333.

29. **Hellerqvist, C. G., B. Lindberg, S. Svensson, T. Holme, and A. A. Lindberg.** 1968. Structural studies on the O-specific side-chains of the cell-wall lipopolysaccharide from *Salmonella typhimurium* 395 MS. Carbohydr. Res. **8**:43–55.

30. **Higashi, Y., J. L. Strominger, and C. C. Sweeley.** 1967. Structure of a lipid intermediate in cell wall peptidoglycan synthesis: a derivative of a C_{55} isoprenoid alcohol. Proc. Natl. Acad. Sci. USA **57**:1878–1884.

31. **Hinckley, A., E. Müller, and L. Rothfield.** 1972. Reassembly of a membrane-bound multienzyme system. I. Formation of a particle containing phosphatidylethanolamine, lipopolysaccharide, and two glycosyltransferases. J. Biol. Chem. **247**:2623–2628.

32. **Jann, K., G. Goldemann, C. Weisgerber, C. Wolf-Ullisch, and S. Kanegasaki.** 1982. Biosynthesis of the O9 antigen of *Escherichia coli*. Eur. J. Biochem. **127**:157–164.

33. **Jann, K., and B. Jann.** 1977. Bacterial polysaccharide antigens, p. 247–287. *In* I. W. Sutherland (ed.), Surface carbohydrates of the prokaryotic cell. Academic Press, Inc., New York.

34. **Jansson, P. E., A. A. Lindberg, B. Lindberg, and R. Wollin.** 1981. Structural studies on the hexose region of the core lipopolysaccharides from *Enterobacteriaceae*. Eur. J. Biochem. **115**:571–577.

35. **Kauffmann, F.** 1941. A typhoid variant and a new serological variation in the *Salmonella* group. Acta Pathol. Microbiol. Scand. **17**:134.

36. **Kauffmann, F.** 1966. The bacteriology of *Enterobacteriaceae*. E. Munksgaard, Copenhagen.

37. **Kishi, K., and S. Iseki.** 1973. Genetic analysis of the O antigens in *Salmonella*. II. Heredity of the O antigens 4, 5, and 9 of *Salmonella*. Jpn. J. Genet. **48**:133–136.

38. **Lehmann, V.** 1977. Isolation, purification, and properties of an intermediate in 3-deoxy-D-manno-octulosonic acid–lipid A biosynthesis. Eur. J. Biochem. **75**:257–266.

39. **Lehmann, V., O. Lüderitz, and O. Westphal.** 1971. The linkage of pyrophosphorylethanolamine to heptose in the core of *Salmonella minnesota* lipopolysaccharides. Eur. J. Biochem. **21**:339–347.

40. **Lehmann, V., E. Rupprecht, and M. J. Osborn.** 1977. Isolation of mutants conditionally blocked in the biosynthesis of the 3-deoxy-D-mannooctulosonic-acid–lipid A part of lipopolysaccharides derived from *Salmonella typhimurium*. Eur. J. Biochem. **76**:41–49.

41. **Levine, D. H., and E. Racker.** 1959. Condensation of arabinose-5-phosphate and phosphorylenolpyruvate by 2-keto-3-deoxy-8-phosphooctonic acid synthetase. J. Biol. Chem. **234**:2532–2539.

42. **Lim, R., and S. S. Cohen.** 1966. D-Phosphoarabinoisomerase and D-ribulokinase in *Escherichia coli*. J. Biol. Chem. **241**:4304–4315.

43. **Lindberg, A. A., and T. Holme.** 1972. Evaluation of some extraction methods for the preparation of bacterial lipopolysaccharides by structural analysis. Acta Pathol. Microbiol. Scand. Sect. B **80**:751–759.

44. **Lindsay, S. A., B. Wheeler, K. E. Sanderson, J. W. Costerton, and K.-J. Cheng.** 1973. The release of alkaline phosphatase and of lipopolysaccharide during the growth of rough and smooth strains of *Salmonella typhimurium*. Can. J. Microbiol. **19**:335–343.

45. **Lüderitz, O., O. Westphal, A. M. Staub, and H. Nikaido.** 1971. Isolation and chemical and immunological characterization of bacterial lipopolysaccharides, p. 145–233. *In* G. Weinbaum, S. Kadis, and S. Ajl (ed.), Microbial toxins, vol. 4. Bacterial endotoxins. Academic Press, Inc., New York.

46. **Mäkelä, P. H.** 1973. Glucosylation of lipopolysaccharide in *Salmonella*: mutants negative for O antigen factor 12_2. J. Bacteriol. **116**:847–856.

47. **Mäkelä, P. H., and B. A. D. Stocker.** 1981. Genetics of the bacterial cell surface, p. 219–264. *In* S. W. Glover and D. A. Hopwood (ed.), Genetics as a tool in microbiology. Cambridge University Press, New York.

48. **Mühlradt, P.** 1969. Biosynthesis of *Salmonella* lipopolysaccharide. The *in vitro* transfer of phosphate to the heptose moiety of the core. Eur. J. Biochem. **11**:241–248.

49. **Mühlradt, P., H. J. Risse, O. Lüderitz, and O. Westphal.** 1968. Biochemical studies on lipopolysaccharides of *Salmonella* R mutants. 5. Evidence for a phosphorylating enzyme in lipopolysaccharide biosynthesis. Eur. J. Biochem. **4**:139–145.

50. **Mühlradt, P., V. Wray, and V. Lehmann.** 1977. A ^{31}P-nuclear magnetic-resonance study of the phosphate groups in lipopolysaccharide and lipid A from *Salmonella*. Eur. J. Biochem. **81**:193–203.

51. **Mühlradt, P. F.** 1970. Biosynthesis of *Salmonella* lipopolysaccharide. Studies on the transfer of glucose, galactose, and phosphate to the core in a cell-free system. Eur. J. Biochem. **18**:20–27.

52. **Mulford, C. A., and M. J. Osborn.** 1983. An intermediate step in translocation of lipopolysaccharide to the outer membrane of *Salmonella typhimurium*. Proc. Natl. Acad. Sci. USA **80**:1159–1163.

53. **Müller, E., A. Hinckley, and L. Rothfield.** 1972. Studies of phospholipid-requiring bacterial enzymes. III. Purification and properties of uridine diphosphate glucose: lipopolysaccharide glucosyltransferase I. J. Biol. Chem. **247**:2614–2622.

54. **Munford, R. S., C. L. Hall, and P. D. Rick.** 1980. Size heterogeneity of *Salmonella typhimurium* lipopolysaccharides in outer membranes and culture supernatant membrane fragments. J. Bacteriol. **144**:630–640.

55. **Munson, R. S., N. S. Rasmussen, and M. J. Osborn.** 1978. Biosynthesis of lipid A. Enzymatic incorporation of 3-deoxy-D-mannooctulosonate into a precursor of lipid A in *Salmonella typhimurium*. J. Biol. Chem. **253**:1503–1511.

56. **Naide, Y., H. Nikaido, P. H. Mäkelä, R. G. Wilkenson, and B. A. D. Stocker.** 1965. Semirough strains of *Salmonella*. Proc. Natl. Acad. Sci. USA **52**:2614–2622.

57. **Nikaido, H.** 1962. Studies on the biosynthesis of cell wall polysaccharide in mutant strains of *Salmonella*. II. Proc. Natl. Acad. Sci. USA **48**:1542–1548.

58. **Nikaido, H.** 1973. Biosynthesis and assembly of lipopolysaccharide and the outer membrane layer of gram-negative cell wall, p. 131–208. *In* L. Leive (ed.), Bacterial membranes and walls. Marcel Dekker, Inc., New York.

59. **Nikaido, H.** 1979. Nonspecific transport through the outer membrane, p. 361–407. *In* M. Inouye (ed.), Bacterial outer membranes. John Wiley & Sons, Inc., New York.

60. **Nurminen, M., C. E. Hellerqvist, V. V. Valtonen, and P. H. Mäkelä.** 1971. The smooth lipopolysaccharide character of 1,4,(5),12 and 1,9,12 transductants formed as hybrids between groups B and D of *Salmonella*. Eur. J. Biochem. **22**:500–505.

61. **Osborn, M. J.** 1968. Biochemical characterization of mutants of *Salmonella typhimurium* lacking glucosyl or galactosyl lipopolysaccharide transferases. Nature (London) **217**:957–960.

62. **Osborn, M. J.** 1969. Structure and biosynthesis of the bacterial cell wall. Annu. Rev. Biochem. **38**:501–538.

63. **Osborn, M. J., and L. D'Ari.** 1964. Enzymatic incorporation of *N*-acetylglucosamine into cell wall lipopolysaccharide in a mutant strain of *Salmonella typhimurium*. Biochem. Biophys. Res. Commun. **16**:568–575.

64. **Osborn, M. J., J. E. Gander, and E. Parisi.** 1972. Mechanism of assembly of the outer membrane of *Salmonella typhimurium*. Site of synthesis of lipopolysaccharide. J. Biol. Chem. **247**:3973–3986.

65. **Osborn, M. J., J. E. Gander, E. Parisi, and J. Carson.** 1972. Mechanism of assembly of the outer membrane of *Salmonella typhimurium*. Isolation and characterization of cytoplasmic and outer membrane. J. Biol. Chem. **247**:3962–3972.

66. **Osborn, M. J., P. D. Rick, and N. S. Rasmussen.** 1980. Mechanism of assembly of the outer membrane of *Salmonella typhimurium*. Translocation and integration of an incomplete mutant lipid A into the outer membrane. J. Biol. Chem. **255**:4246–4251.

67. **Osborn, M. J., S. M. Rosen, L. Rothfield, and B. L. Horecker.** 1962. Biosynthesis of bacterial lipopolysaccharide. I. Enzymatic incorporation of galactose in a mutant strain of *Salmonella*. Proc. Natl. Acad. Sci. USA **48**:1831–1838.

68. **Osborn, M. J., and L. I. Rothfield.** 1971. Biosynthesis of the core region of lipopolysaccharide, p. 331–350. *In* G. Weinbaum, S. Kadis, and S. Ajl (ed.), Microbial toxins, vol. 4. Bacterial endotoxins. Academic Press, Inc., New York.

69. **Osborn, M. J., and R. Y. Tze-Yen.** 1968. Biosynthesis of bacterial

lipopolysaccharide. VII. Enzymatic formation of the first intermediate in biosynthesis of the O-antigen of *Salmonella typhimurium*. J. Biol. Chem. **243:**5145–5152.

70. **Palva, E. T., and P. H. Mäkelä.** 1980. Lipopolysaccharide heterogeneity in *Salmonella typhimurium* analyzed by sodium dodecyl sulfate/polyacrylamide gel electrophoresis. Eur. J. Biochem. **107:**137–143.

71. **Prehm, P., B. Jann, and K. Jann.** 1976. The O9 antigen of *Escherichia coli*. Structure of the polysaccharide chain. Eur. J. Biochem. **67:**53–56.

72. **Prehm, P., S. Stirm, B. Jann, and K. Jann.** 1975. Cell-wall lipopolysaccharide from *Escherichia coli* B. Eur. J. Biochem. **56:**41–55.

73. **Prehm, P., S. Stirm, B. Jann, K. Jann, and H. G. Bowman.** 1976. Cell wall lipopolysaccharides of ampicillin-resistant mutants of *Escherichia coli* K-12. Eur. J. Biochem. **66:**369–377.

74. **Reske, K., and K. Jann.** 1972. The O8 antigen of *Escherichia coli*. Structure of the polysaccharide chain. Eur. J. Biochem. **31:**320–328.

75. **Rick, P. D., L. W.-M. Fung, C. Ho, and M. J. Osborn.** 1977. Lipid A mutants of *Salmonella typhimurium*. Purification and characterization of a lipid A precursor produced by a mutant in 3-deoxy-D-mannooctulosonate-8-phosphate synthetase. J. Biol. Chem. **252:**4904–4912.

76. **Rick, P. D., and M. J. Osborn.** 1977. Lipid A mutants of *Salmonella typhimurium*. Characterization of a conditional lethal mutant in 3-deoxy-D-mannooctulosonate-8-phosphate synthetase. J. Biol. Chem. **252:**4895–4903.

77. **Rick, P. D., and D. A. Young.** 1982. Isolation and characterization of a temperature-sensitive lethal mutant of *Salmonella typhimurium* that is conditionally defective in 3-deoxy-D-mannooctulosonate-8-phosphate synthesis. J. Bacteriol. **150:**447–455.

78. **Robbins, P. W., D. Bray, M. Dankert, and A. Wright.** 1967. Direction of chain growth in polysaccharide synthesis. Science **158:**1536–1540.

79. **Robbins, P. W., and A. Wright.** 1971. Biosynthesis of O-antigens, p. 351–368. *In* G. Weinbaum, S. Kadis, and S. Ajl (ed.), Microbial toxins, vol. 4. Bacterial endotoxins. Academic Press, Inc., New York.

80. **Romeo, D., A. Girard, and L. Rothfield.** 1970. Reconstitution of a functional membrane enzyme system in a monomolecular film. I. Formation of a mixed monolayer of lipopolysaccharide and phospholipid. J. Mol. Biol. **53:**475–490.

81. **Romeo, D., A. Hinckley, and L. Rothfield.** 1970. Reconstitution of a functional membrane enzyme system in a monomolecular film. II. Formation of a functional ternary film of lipopolysaccharide, phospholipid, and transferase enzyme. J. Mol. Biol. **53:**491–501.

82. **Rosen, S. M., M. J. Osborn, and B. L. Horecker.** 1964. Biosynthesis of bacterial lipopolysaccharide. III. Characterization of the galactose incorporation product. J. Biol. Chem. **239:**3196–3200.

83. **Rosner, M. R., H. G. Khorana, and A. C. Satterthwait.** 1979. The structure of lipopolysaccharide from a heptose-less mutant of *Escherichia coli* K-12. II. The application of ³¹P NMR spectroscopy. J. Biol. Chem. **254:**5918–5925.

84. **Rosner, M. R., J. Tang, I. Barzilay, and H. G. Khorana.** 1979. Structure of the lipopolysaccharide from an *Escherichia coli* heptose-less mutant. I. Chemical degradation and identification of products. J. Biol. Chem. **254:**5906–5917.

85. **Rothfield, L., and B. L. Horecker.** 1964. The role of cell wall lipid in the biosynthesis of bacterial lipopolysaccharide. Proc. Natl. Acad. Sci. USA **52:**939–946.

86. **Rothfield, L., M. J. Osborn, and B. L. Horecker.** 1964. Biosynthesis of bacterial lipopolysaccharide. II. Incorporation of glucose and galactose catalyzed by particulate and soluble enzymes in *Salmonella*. J. Biol. Chem. **239:**2788–2795.

87. **Rothfield, L., and M. Pearlman.** 1966. The role of cell envelope phospholipid in the enzymatic synthesis of bacterial lipopolysaccharide. Structural requirements of the phospholipid molecule. J. Biol. Chem. **241:**1386–1392.

88. **Rothfield, L., and D. Romeo.** 1971. Role of lipids in the biosynthesis of the bacterial cell envelope. Bacteriol. Rev. **35:**14–38.

89. **Rothfield, L., and M. Takeshita.** 1965. The role of cell envelope phospholipid in the enzymatic synthesis of bacterial lipopolysaccharide: binding of transferase enzymes to a lipopolysaccha-

ride-lipid complex. Biochem. Biophys. Res. Commun. **20:**521–527.

90. **Sanderson, K. E., and B. A. D. Stocker.** 1981. Gene *rfaH*, which affects lipopolysaccharide core structure in *Salmonella typhimurium*, is required also for expression of F-factor functions. J. Bacteriol. **146:**535–541.

91. **Siewert, G., and J. L. Strominger.** 1967. Bacitracin: an inhibitor of the dephosphorylation of lipid pyrophosphate, an intermediate in biosynthesis of the peptidoglycan of cell walls. Proc. Natl. Acad. Sci. USA **57:**767–773.

92. **Smit, J., Y. Kamio, and H. Nikaido.** 1975. Outer membrane of *Salmonella typhimurium*: chemical analysis and freeze-fracture studies with lipopolysaccharide mutants. J. Bacteriol. **124:**942–958.

93. **Staub, A. M., and G. Bagdian.** 1966. Études immunochimiques sur les *Salmonella*. XII. Analyse immunologique des facteurs 27ₐ, 27_B et 27_D. Ann. Inst. Pasteur (Paris) **110:**849–860.

94. **Stocker, B. A. D., and P. H. Mäkelä.** 1971. Genetic aspects of biosynthesis and structure of *Salmonella* lipopolysaccharide, p. 369–438. *In* G. Weinbaum, S. Kadis, and S. Ajl (ed.), Microbial toxins, vol. 4. Bacterial endotoxins. Academic Press, Inc., New York.

95. **Strain, S. M., S. W. Fesik, and I. M. Armitage.** 1983. Characterization of lipopolysaccharide from a heptoseless mutant of *Escherichia coli* by carbon 13 nuclear magnetic resonance. J. Biol. Chem. **258:**2906–2910.

96. **Strain, S. M., S. W. Fesik, and I. M. Armitage.** 1983. Structure and metal-binding properties of lipopolysaccharides from heptoseless mutants of *E. coli* studied by ¹³C and ³¹P nuclear magnetic resonance. J. Biol. Chem. **258:**13466–13477.

97. **Sundararajan, T. A., A. M. C. Rapin, and H. M. Kalckar.** 1962. Biochemical observations on *Escherichia coli* mutants defective in uridine diphosphoglucose. Proc. Natl. Acad. Sci. USA **48:**2187–2192.

98. **Tamaki, S., T. Sato, and M. Matsuhashi.** 1971. Role of lipopolysaccharides in antibiotic resistance and bacteriophage adsorption of *Escherichia coli* K-12. J. Bacteriol. **105:**968–975.

99. **Tinelli, R., and A. M. Staub.** 1960. Analyse de l'antigene O₁₂ du tableau de Kauffmann-White. Bull. Soc. Chim. Biol. **42:**583–599.

100. **Walenga, R. W., and M. J. Osborn.** 1980. Biosynthesis of lipid A. *In vivo* formation of an intermediate containing 3-deoxy-D-mannooctulosonate in a mutant of *Salmonella typhimurium*. J. Biol. Chem. **255:**4252–4256.

101. **Walenga, R. W., and M. J. Osborn.** 1980. Biosynthesis of lipid A. Formation of acyl-deficient lipopolysaccharides in *Salmonella typhimurium* and *Escherichia coli*. J. Biol. Chem. **255:**4257–4263.

102. **Weiner, I. M., T. Higuchi, L. Rothfield, M. Saltmarsh-Andrew, M. J. Osborn, and B. L. Horecker.** 1965. Biosynthesis of bacterial lipopolysaccharide. V. Lipid-linked intermediates in the biosynthesis of the O-antigen groups of *Salmonella typhimurium*. Proc. Natl. Acad. Sci. USA **54:**228–235.

103. **Weiser, M. M., and L. Rothfield.** 1968. The reassociation of lipopolysaccharide, phospholipid, and transferase enzymes of the bacterial cell envelope. Isolation of binary and ternary complexes. J. Biol. Chem. **243:**1320–1328.

104. **Weisgerber, C., B. Jann, and K. Jann.** 1984. Biosynthesis of the O9 antigen in *Escherichia coli*. Core structure of *rfe* mutant as indication of assembly mechanism. Eur. J. Biochem. **140:**553–556.

105. **Wilkenson, S. G.** 1977. Composition and structure of bacterial lipopolysaccharides, p. 97–175. *In* I. W. Sutherland (ed.), Surface carbohydrates of the prokaryotic cell. Academic Press, Inc., New York.

106. **Wollin, R., E. S. Creeger, L. I. Rothfield, B. A. D. Stocker, and A. A. Lindberg.** 1983. *Salmonella typhimurium* mutants defective in UDP-D-galactose: lipopolysaccharide α1,6-D-galactosyltransferase. J. Biol. Chem. **258:**3769–3774.

107. **Wright, A., M. Dankert, P. Fennessey, and P. W. Robbins.** 1967. Characterization of a polyisoprenoid compound functional in O-antigen biosynthesis. Proc. Natl. Acad. Sci. USA **57:**1798–1803.

108. **Wright, A., M. Dankert, and P. W. Robbins.** 1965. Evidence for an intermediate stage in the biosynthesis of the *Salmonella* O-antigen. Proc. Natl. Acad. Sci. USA **54:**235–241.

109. **Wright, A., and S. Kanegasaki.** 1971. Molecular aspects of lipopolysaccharides. Physiol. Rev. **51:**748–784.

42. Murein Synthesis

JAMES T. PARK

Department of Molecular Biology and Microbiology, Tufts University School of Medicine, Boston, Massachusetts 02111

INTRODUCTION

Murein, the peptidoglycan of the bacterial cell wall, is a unique polymer in terms of composition, structure, formation, and function. The murein sacculus in most procaryotic organisms serves the following two essential functions. (i) It preserves the integrity of the protoplast. Without an intact murein sacculus, the cytoplasmic membrane is subjected to stress and ruptures as water enters the cell to equalize the osmotic pressure. (ii) It is intimately involved in the cell division cycle. The formation of septa represents the specialized synthesis of entirely new murein in a precise location to form the poles of the daughter cells while retaining the integrity of the sacculus. Without a murein sacculus, the cell, or spheroplast, even when protected from osmotic shock, has great difficulty in dividing.

An incidental role of the murein sacculus is to determine the shape of the cell. Thus, whenever the shape of an *Escherichia coli* cell changes, one can be certain that the metabolism of the murein sacculus has been altered.

The composition and structure of murein is presented in greater detail in chapter 4, but to review briefly the principal features of the polymer, it consists of linear polysaccharide strands composed of alternating units of *N*-acetylglucosamine (NAcGlc) and *N*-acetylmuramic acid (NAcMur) (Fig. 1). The linkages between sugars are all β(1→4). The strands contain an average of 30 disaccharides, which is equivalent to a length of about 30 nm. Each strand terminates in a 1,6-anhydromuramic acid residue which is nonreducing. The individual linear strands may have a fourfold helical structure (4, 8). Each muramic acid residue initially carries on its D-lactyl group a short peptide, L-alanyl-D-isoglutamyl-*meso*-diaminopimelyl-D-alanyl-D-alanine, though one or both

D-alanine residues are lost during synthesis of the sacculus, and the predominant form is a tetrapeptide. Neighboring glycan strands are linked to each other usually via a peptide bond between the D-alanine at position 4 of one peptide and the amino group on the D-isomeric carbon of *meso*-diaminopimelic acid of a peptide of an adjacent strand, although up to 20% of the cross-links directly join one diaminopimelic acid to another. About one-half of the disaccharide-peptide repeating units of the polymer (termed muropeptides) are involved in cross-links between strands. The strands are arranged parallel to each other and run around the circumference of the cell perpendicular to the axis. Since the strands average only 30 nm in length, up to 80 strands stacked end to end would be required to traverse the circumference of a typical cell. Fewer strands may be required if significant gaps exist between succeeding strand ends.

Thus, the murein sacculus of *E. coli* may be considered one giant molecule consisting of many short polysaccharide strands arranged in parallel, with adjoining strands cross-linked to each other via short peptides.

BIOSYNTHESIS OF MUREIN

Recent general reviews of the biosynthesis of murein (21, 30, 37) are available. More-recent work on the properties of membrane-bound enzymes involved in the terminal steps of murein biosynthesis has also been reviewed (18, 32, 40). The formation and maintenance of the murein sacculus in the shape of the cell, during the doubling of its length, formation of septa, and division into two daughter cells without loss of its integrity, is obviously a complex and carefully controlled process.

In this chapter, what is known of the biosynthetic pathway and the enzymes involved in the synthesis of

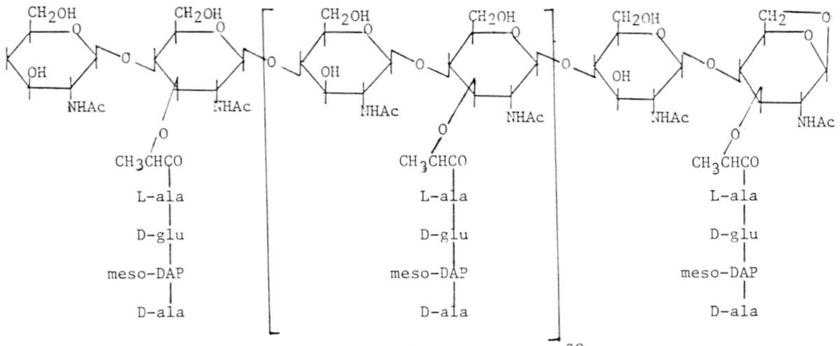

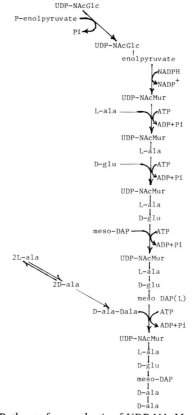

FIG. 1. Linear strand of murein with an average length of 30 muropeptides terminating in 1,6-anhydromuramic acid. Ac, Acetyl; DAP, diaminopimelic acid.

murein will be discussed. The final stages of sacculus assembly and the steps involved in elongation and septation of the sacculus are considered in chapter 4.

The overall pathway for the biosynthesis of murein was defined over 15 years ago largely on the basis of studies in gram-positive bacteria, yet, surprisingly, relatively few of the enzymes involved in this process in *E. coli* have been purified and studied in detail. Consequently, the pathway for synthesis is based on the demonstration that extracts of or envelope preparations from *E. coli* will process each precursor to the next reaction product according to this scheme. Although there is little reason to doubt the overall scheme, one can hardly take much satisfaction from the state of our ignorance concerning the enzymes involved.

The synthesis of murein occurs in three stages, as is the general case for synthesis of heteropolysaccharides. Additionally, during the third stage, transpeptidation reactions take place simultaneously to cross-link the newly synthesized strands to the sacculus. The three stages are as follows. (i) Nucleotide-activated precursors, in this case UDP–*N*-acetylmuramyl-L-alanyl-D-isoglutamyl-(L)-*meso*-diaminopimelyl-D-alanyl-D-alanine (UDP-NAcMur pentapeptide) and UDP-NAcGlc, are synthesized. These precursors are found in the cytoplasm and are synthesized by cytoplasmic enzymes. (ii) The repeating unit of murein on bactoprenyl phosphate is assembled. The bactoprenyl derivatives and all of the enzymes which synthesize and utilize these substrates are bound to the cytoplasmic membrane. (iii) The final stage is polymerization of the repeating units to form linear strands of murein and cross-linking of the strands to preexisting strands by transpeptidation. These enzymes are also associated with the cytoplasmic membrane, but in contrast to those utilizing cytoplasmic precursors their active sites are located on the outer surface of the cytoplasmic membrane. That is, they function in the periplasm where the murein sacculus is located and where the growing sacculus serves as substrate for these enzymes.

SYNTHESIS OF UDP-NAcMur PENTAPEPTIDE

UDP-NAcGlc is the branch point from which precursors unique to murein are derived (Fig. 2). UDP-NAcMur is formed by the condensation of phosphoenolpyruvate with NAcGlc to form a 3-enolpyruvate ether of UDP-NAcGlc, which is subsequently reduced

to the 3-*O*-D-lactyl ether. These reactions have been studied in detail by Anwar and co-workers (12, 34, 35) with enzymes purified from *Enterobacter cloacae*. The reductase from *E. coli* has been purified to homogeneity and shown to lack flavin adenine dinucleotide, which is believed to be a prosthetic group in the *E. cloacae* enzyme. The *E. coli* reductase, like that of *E. cloacae*, has a molecular weight of about 35,000, a pH optimum near 8, a K_m for the substrates of about 5×10^{-6} M, and a requirement for a monovalent cation (K^+ or NH_4^+) and is inhibited by sulfhydryl reagents (1).

UDP-NAcMur pentapeptide is formed from UDP-NAcMur by sequential addition of the amino acids by

FIG. 2. Pathway for synthesis of UDP-NAcMur pentapeptide from UDP-NAcGlc. P-enolpyruvate, Phosphoenolpyruvate; Pi, P_i; DAP, diaminopimelic acid.

specific enzymes. The enzymes are quite specific for the substrate and the amino acid to be added. In each case, the reaction is driven by ATP and requires a divalent cation, either Mg^{2+} or Mn^{2+}. D-Alanyl-D-alanine is synthesized separately and added to UDP-NAcMur tripeptide to form UDP-NAcMur pentapeptide. D-Alanine is formed from L-alanine by a racemase (Fig. 2). A specific D-alanine:D-alanine ligase catalyzes the formation of D-alanyl-D-alanine, and a specific adding enzyme, or ligase, ligates the dipeptide to the precursor to form UDP-NAcMur pentapeptide. Interestingly, though the D-alanine:D-alanine ligase and the UPD-NAcMur tripeptide:D-alanine-D-alanyl ligase are not absolutely specific, the former is quite strict in its requirement for D-alanine as the N-terminal amino acid, and the latter is quite specific in utilizing only dipeptides which contain a C-terminal D-alanine. Hence, if the enzyme is used sequentially, only the correct UDP-NAcMur pentapeptide can be formed (23). It should be noted that the studies of D-alanine cited were concerned with enzymes from *Streptococcus faecalis*. No similar studies have been undertaken in *E. coli*. In fact, none of the enzymes of *E. coli* involved in synthesis of the nucleotide-linked and lipid-linked precursors have been studied in purified form.

No highly purified amino acid adding enzyme involved in synthesis of murein precursors has been studied in any bacterium. In investigated cases, it is known that ATP is hydrolyzed to ADP and P_i. It would be interesting to determine the type of activated intermediate involved in these reactions. Five of the eight reactions required for synthesis of UDP-NAcMur pentapeptide from UDP-NAcGlc must form activated intermediates with the concomitant hydrolysis of ATP to ADP plus P_i.

CELLULAR POOL LEVELS OF PRECURSORS OF MUREIN AND THEIR REGULATION

Mengin-Lecreulx et al. (19, 20) have determined the concentration of the various substrates for murein synthesis present in the cytoplasm of *E. coli*. This is the only comprehensive study published on this subject, and their results for mid-log-phase cells grown in minimal glucose medium are given in Table 1. Concentrations found in cells grown in glucose-rich medium were, for the most part, similar. It is noteworthy that the immediate precursors of murein, UDP-NAcGlc and UDP-NAcMur pentapeptide, are present in relatively high concentrations compared with the other nucleotide-bound intermediates. These investigators also explored the various enzymatic activities involved in UDP-NAcMur pentapeptide synthesis as expressed in crude extracts and showed that their K_m values were such that the substrate available was saturating in all cases. They found no evidence that pool levels were controlled by feedback inhibition by the precursors at the concentrations normally present. On the other hand, the enzyme activities present in the crude extracts were just sufficient to provide the UDP-NAcMur required for murein synthesis during exponential growth. As pointed out by Mengin-Lecreulx et al. (19), a 10^{-5} M pool concentration was equivalent to about 10,000 molecules per cell, whereas several million muropeptides are incorporated into

TABLE 1. Pool levels of cytoplasmic precursors of murein in *E. coli* K-12 cells grown in minimal glucose medium[a]

Precursor	Pool level, concn (M)[b]
UDP-NAcGlc	12.5×10^{-5}
UDP-NAcGlc-enolpyruvate	2.5×10^{-6}
UDP-NAcMur	4.6×10^{-5}
UDP-NAcMur-L-Ala	1.4×10^{-5}
UDP-NAcMur dipeptide	1.1×10^{-5}
UDP-NAcMur tripeptide	0.6×10^{-5}
UDP-NAcMur pentapeptide	17.5×10^{-5}
D-Alanine	5×10^{-4}
L-Alanine	5×10^{-3}
D-Glutamic acid	1×10^{-3}
L-Glutamic acid	8.8×10^{-3}
Diaminopimelic acid	7.5×10^{-4}
D-Alanyl-D-alanine	2.5×10^{-4}

[a] Modified data of Mengin-Lecreulx et al. (19, 20).
[b] Calculated by assuming the water content of the cytoplasm to be 1.2×10^{-12} ml.

murein each generation. This indicates that the pool turnover rate must be very rapid for some of the intermediates.

In contrast to the apparent lack of regulation of the nucleotide-bound precursors, D-alanine:D-alanine ligase is effectively regulated by product inhibition. This regulation is necessary, since without it the entire pool of L-alanine would be converted to D-alanyl-D-alanine.

ASSEMBLY OF THE MUROPEPTIDE REPEATING UNIT OF *E. COLI* MUREIN: THE BACTOPRENYL CYCLE

The two reactions required for formation of the repeating unit and also the reactions required for regeneration of the bactoprenyl phosphate carrier and polymerization of the glycan are shown in Fig. 3.

The repeating unit of *E. coli* murein is relatively simple. That is, it consists of NAcGlc β(1→4)-linked to NAcMur pentapeptide. No modifications, such as amidation or ligation of additional amino acids, common in various gram-positive bacteria, are required. These modifications, if needed, occur during assembly on the bactoprenyl carrier.

Assembly takes place on a lipid carrier called bactoprenyl phosphate. By mass spectrometry, it was shown to be a C_{55}-isoprenoid alcohol phosphate, undecaprenyl phosphate (13, 42). Bactoprenyl phosphate appears to be involved in the synthesis of all heteropolysaccharides of bacteria, including mureins and the O antigens of gram-negative bacteria. There is a limited amount of bactoprenyl phosphate in the cell. Hence, interruption of synthesis of one polymer traps its precursor on bactoprenyl phosphate and eventually leads to limited availability of the carrier and hence to reduced synthesis of other heteropolysaccharides. The bactoprenyl cycle in which reutilization of the carrier occurs is an important feature of the synthesis of murein and the other heteropolysaccharides.

The first reaction of the bactoprenyl cycle is the transfer of phospho-NAcMur pentapeptide from UDP-NAcMur pentapeptide to bactoprenyl phosphate, with the release of UMP (Fig. 3). Since both the substrate

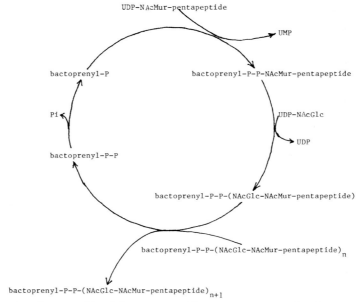

FIG. 3. The bactoprenyl cycle: assembly of the muropeptide repeating unit and polymerization of the glycan. Pi, P_i.

and the product of this reaction are pyrophosphate derivatives, the precursor retains its activated state, and the reaction is freely reversible. The second reaction, which completes the formation of the muropeptide, is the transfer of NAcGlc from UDP-NAcGlc to bactoprenyl-P-P-NAcMur pentapeptide to form the disaccharide pentapeptide repeating unit (Fig. 3). The other product of this reaction is UDP, and the reaction is essentially irreversible. The activated disaccharide pentapeptides are then utilized to form murein during the third stage of synthesis, and bactoprenyl pyrophosphate is released. Bactoprenyl phosphate is regenerated by a membrane-bound pyrophosphatase to complete the cycle (Fig. 3).

POLYMERIZATION AND CROSS-LINKING OF THE REPEATING UNITS

It is probably an historical accident that so little is known about the cytoplasmic enzymes of *E. coli* that synthesize UDP-NAcMur pentapeptide. The nucleotide-activated precursors of murein were originally discovered in staphylococci (24–27). Consequently, the biosynthetic pathway for its synthesis was worked out primarily by studying staphylococcal enzymes (for a review, see reference 33). A major impetus for the study of murein biosynthesis was the fact that penicillin somehow interfered with cell wall synthesis (28). Since murein was an essential polymer, unique to procaryotes, its biosynthesis proved to be a prime target for a variety of antibiotics in addition to penicillin and all other β-lactam antibiotics. It was known at that time that penicillin does not interfere with synthesis of the cytoplasmic precursors of murein or with polymerization of the glycan. The primary target of penicillin proved to be the transpeptidation reaction that cross-links the glycan strands together (36, 41). However, this reaction, first demonstrated with intact staphylococci, could not be demonstrated in vitro with cell-free preparations. The first successful

in vitro transpeptidation and demonstration of its sensitivity to penicillin used membranes from *E. coli* (2, 16). Consequently, many of the studies of the terminal stages of murein synthesis have used *E. coli*. In fact, to date all of the studies of purified transpeptidase and glycan polymerase activities have been done with *E. coli* enzymes. The so-called DD-carboxypeptidases, the principal activity of which is the hydrolysis of the D-alanyl-D-alanine peptide bond, have been investigated in a variety of bacteria as well. They are present in relatively larger amounts (~1,500 molecules per cell) and retain good activity when purified. In contrast, the transpeptidases are usually present at a level of about 100 copies per cell and, upon purification from *E. coli*, exhibit low activity relative to that found in vivo. To purify these enzymes and to obtain sufficient quantities for study, cloning of the relevant genes proved rewarding. This technology is, of course, most advanced and readily available in *E. coli*.

Study of the polymerization and cross-linking of murein by *E. coli*, and by inference study of it in most other rod-shaped bacteria, is complicated by the fact that two or more enzymes are present in the cell which are capable of carrying out these processes. In addition, *E. coli* has multiple enzymes, each capable of both transglycosylation and transpeptidation (18). These are bifunctional, or "two-headed," enzymes with two active sites.

The transglycosylation reaction for lengthening the linear polysaccharide strand is believed to proceed as illustrated in Fig. 3 in all cases. A glycan chain, still attached to bactoprenyl pyrophosphate, is transferred to C-4 of NAcGlc of a newly positioned bactoprenyl-P-P-(NAcGlc-NAcMur pentapeptide), leading to growth of the glycan at the reducing end (39). This is formally similar to growth of a peptide chain during protein synthesis on ribosomes. As in protein synthesis, the advantage of this mechanism is that the active substrates are always in close proximity to the enzymes.

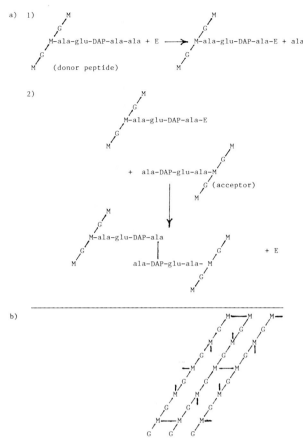

FIG. 4. (a) The transpeptidation reaction. (b) Murein strands with fourfold helix allowing cross-links only between every fourth muropeptide. G, NAcGlc; M, NAcMur; E, transpeptidase; —, cross-links; |, tetrapeptides or pentapeptides; DAP, diaminopimelic acid.

In the case of murein synthesis, the enzyme is membrane bound, and the substrates of the transglycosylation reaction are attached to bactoprenyl pyrophosphate, which is embedded in the cytoplasmic membrane.

Transpeptidation occurs as shown in Fig. 4a. By virtue of being incorporated into a new, growing strand, muropeptide that still carries a pentapeptide is positioned opposite a muropeptide of an adjacent strand in such a way that transpeptidation can take place from the donor pentapeptide to the acceptor muropeptide. This usually carries a tetrapeptide but, theoretically at least, could also be a suitable acceptor when carrying a tri- or pentapeptide, depending on the specificity of the transpeptidase involved. The acceptor specificity of the transpeptidases of *E. coli* has not been investigated; investigation may be impractical because of the potential difficulty of doing so under natural conditions. Transpeptidation involves cleavage of the D-alanyl-D-alanine bond of the pentapeptide with the concomitant formation of an acyl-enzyme intermediate and the release of the terminal D-alanine. The activated intermediate reacts with the amino group (D) of diaminopimelic acid of the available tetrapeptide on the neighboring glycan strand to form a new peptide bond and to cross-link the two

strands. It may be noted that this reaction must take place in the periplasm, where presumably there is no readily available source of energy to drive the reaction. Instead, the energy of the D-Ala-D-Ala bond of the donor pentapeptide is used to drive the transpeptidation reaction. It may also be noted that if the helical model for a strand of murein is correct, once the first cross-link from a new strand to the older, existing strand has taken place only every fourth muropeptide along the strand will be in the proper position to cross-link again to this same acceptor strand (Fig. 4b). This is consistent with the fact that only about 50% of the muropeptides are involved in cross-links.

THE MULTIPLICITY OF MUREIN TRANSPEPTIDASES IN *E. COLI*

The transpeptidation reaction was originally deduced from the observation that murein formed by staphylococci in the presence of penicillin contained excess alanine; i.e., D-alanyl-D-alanine remained intact, and an equivalent amount of acceptor amino groups accumulated (41). On the basis of a different measure of cross-linking, Tipper and Strominger (36) soon thereafter came to the same conclusion, that penicillin must inhibit a transpeptidation reaction. They went on to point out that penicillin may act as a structural analog of D-alanyl-D-alanine so that the β-lactam acylates the active site of the hypothetical transpeptidase and thus inactivates the enzyme. As a result of the work of the past 15 years from a number of laboratories, it is now apparent that any enzyme that metabolizes a D-D peptide bond is likely to be sensitive to β-lactam antibiotics and may be inactivated by them. One evidence of inactivation is that penicillin acylates the enzyme stoichiometrically. That is, penicillin binds to the protein. It has been recognized for many years that 1,000 or 2,000 molecules of penicillin specifically bind to bacteria (8a); the significance of this was somewhat clouded by the fact that penicillin-resistant bacteria also bound similar amounts of penicillin. Edwards and Park (10) demonstrated that the rate of binding of a given β-lactam antibiotic correlates with its biological activity against staphylococci; this served to emphasize that binding was related to the mode of action of β-lactam antibiotics. This finding was somewhat fortuitous since it was subsequently found that staphylococci and other bacteria contain multiple penicillin-binding proteins (PBPs) (5). Spratt (31) showed that *E. coli* contains six PBPs, the largest one of which proved to be a mixture of two separate PBPs (32a). One or more enzymatic activities have now been demonstrated for these proteins (for reviews, see references 18 and 40). In every case, the substrate involved was either a D-alanyl-D-alanine bond or a D-alanyl-*meso*-(D)-diaminopimelyl bond. A summary of some known properties of these penicillin-sensitive enzymes is given in Table 2. Three additional PBPs, PBP-1C, PBP-7 and PBP-8, are included because their high sensitivity to many β-lactams indicates a potential role in murein metabolism, though little is known concerning them at the present time (G. Botta and J. T. Park, unpublished data).

Three of the high-molecular-weight PBPs, PBP-1A, PBP-1Bs, and PBP-3, are bifunctional, transglycosyl-

TABLE 2. *Properties of the penicillin-sensitive enzymes of* E. coli *involved in the terminal stages of murein synthesis*

PBP	Mol wt[a]	Molecules/cell[a]	Known enzymatic activities	Possible function
1A	92,000	100	Transglycosylase-transpeptidase	Murein synthesis during cell elongation
1Bs	90,000	120	Transglycosylase-transpeptidase	Murein synthesis during cell elongation
1C[b]				
2	66,000	20	Transpeptidase	Growth in rod shape, cell elongation
3	60,000	50	Transglycosylase-transpeptidase	Murein synthesis during cell septation
4	49,000	110	DD-Endopeptidase, DD-carboxypeptidase	Cross-link hydrolysis during cell elongation
5	42,000	1,800	DD-Carboxypeptidase	Destruction of unutilized pentapeptide to assure that mature murein serves only as acceptor during transpeptidation to form cross-links
6	40,000	600	DD-Carboxypeptidase	Destruction of unutilized pentapeptide
7[b]				
8[b]				

[a] Estimates modified from data of Spratt (31a, 32).
[b] PBPs that have not yet been studied. PBP-8 appears to be an outer membrane protein (29).

ase-transpeptidase enzymes (Table 2). Since each of these enzymes is present in cells at a concentration of about 100 molecules or less, their isolation was greatly facilitated by cloning the relevant genes and overproducing the gene product. All three bifunctional enzymes form cross-linked murein in vitro from the lipid-linked precursor: bactoprenyl-P-P-NAcGlc-NAcMur (pentapeptide) (18). With PBP-1B, about 23% cross-linkage occurred, which is the normal percentage in intact cells. With PBP-3, cross-linkage was only 6 to 10%, and with PBP-1A, cross-linkage increased gradually over time to a level of 35%, indicating that polymerization and cross-linking were not coordinated under in vitro conditions (18). PBP-1A also caused hyper-cross-linking, with the production of muropeptide trimers. In *E. coli*, about 5% of the muropeptides are present as trimers (11).

In addition to the three bifunctional enzymes, an additional transglycosylase that synthesizes un-cross-linked murein and that does not bind penicillin has recently been purified (12a).

All of the in vitro reactions require the presence of filter paper, glycerol, or methanol. These are thought to serve as initial acceptors for the transglycosylase reaction. The rate of reaction observed in vitro is very slow. Even the most active of these enzymes, the transglycosylase mentioned above, functions at less than 1% of the rate required by the cell. PBP-1B in vitro is actually 20-fold less active than this. Judging from the observation that mutants defective in PBP-1B synthesize murein slowly, PBP-1B is considered to be one of the major transglycosylases of the cell. For this reason, as well as the very low activities of the enzymes in vitro, it is not possible to assess their relative importance to the cell from these results. What seems clear is that the in vitro situation, which does not provide the murein sacculus as an acceptor in its proper position relative to the membrane-bound enzyme, cannot reflect accurately the in vivo potential of an enzyme.

PBP-2 has also been shown indirectly to have transpeptidase activity. Ishino et al. (15) were able to detect the transpeptidase activity of PBP-2 by using membrane preparations from a strain of *E. coli* that overproduces PBP-2. The activity was demonstrated after the inactivation of all previously known transpeptidases with cefmetazole, which binds to all PBPs except PBP-2. Under these conditions, the residual transpeptidase activity is presumably due to PBP-2. This cross-linking reaction was completely inhibited by 1 μg of mecillinam per ml, which at this low concentration binds only to PBP-2.

DD-CARBOXYPEPTIDASES

The high-molecular-weight transglycosylase-transpeptidase enzymes are believed to be responsible for net synthesis of murein in vivo, although their activity in vitro is extremely low and difficult to demonstrate. This contrasts with the three low-molecular-weight DD-carboxypeptidases, PBP-4, PBP-5, and PBP-6 (Table 2), which have no essential role in the synthesis of the murein sacculus but the enzymatic activity of which is high and readily demonstrated in vitro. The DD-carboxypeptidases have been purified, and PBP-5, the most abundant PBP of *E. coli*, has been studied extensively (40). Strains lacking the DD-carboxypeptidase activity of both PBP-4 and PBP-5 have been constructed (33a). These strains grow normally although they lack about 95% of the total DD-carboxypeptidase of the cell. The remaining activity is contributed by PBP-6. Strains with deletions in either PBP-5 or PBP-6 were found to grow normally (32). A strain lacking all three DD-carboxypeptidases has yet to be constructed.

It is known that a very small fraction of the muropeptides of the murein sacculus retain their terminal D-alanine. The half-life of a newly incorporated pentapeptide is estimated to be less than 1 min (8b). Since no more than 25% of the pentapeptides are used for cross-linking, it is evident that DD-carboxypeptidases are active in vivo. Presumably, these enzymes contribute to the orderly incorporation of murein into the sacculus by limiting the time during which a new strand retains its pentapeptide and hence its potential to cross-link. Nevertheless, it is puzzling that *E. coli* retains three enzymes with this activity, considering the fact that the individual DD-carboxypeptidases are not essential.

ACTIVE-SITE STUDIES

The murein transpeptidases and the DD-carboxypeptidases all utilize the D-alanyl-D-alanine moiety of the

TABLE 3. Genes involved in murein metabolism in *E. coli*[a]

Locus[b]	Gene symbol[b]	Function or enzymatic activities or both
2	*ftsI* (*pbpB, sep*)	PBP-3 = transglycosylase, transpeptidase bifunctional enzyme required for septal murein synthesis
2	*murE*	*meso*-Diaminopimelic acid adding enzyme
2	*murF*	D-Alanyl-D-alanine adding enzyme
2	*murG*	Murein biosynthesis function unknown
2	*murC*	L-Alanine adding enzyme
2	*ddl*	D-Alanine:D-alanine ligase
2	*envA*	Required for cell separation
4	*mrcB* (*ponB*)	PBP-1B = transglycosylase, transpeptidase bifunctional enzyme involved in murein synthesis during cell elongation
15	*dacA*	PBP-5 = D-alanine carboxypeptidase
15	*rodA*	Required for growth with a rod-shaped sacculus
15	*pbpA*	PBP-2 = transpeptidase required for growth with a rod-shaped sacculus
36	*lpp*	Murein lipoprotein structural gene
51[c]	*amiA*	N-Acetylmuramyl-L-alanine amidase
51[d]	*mepA*	Penicillin-insensitive murein DD-endopeptidase structural gene
69	*dacB*	PBP-4 = DD-endopeptidase involved in elongation of the murein sacculus; DD-carboxypeptidase
75	*mrcA* (*ponA*)	PBP-1A = transglycosylase, transpeptidase bifunctional enzyme involved in elongation of the murein sacculus
81	*envC*	Required for cell separation
90	*mrbA*	UDP-N-acetylglucosaminyl-3-enolpyruvate reductase
(90)[e]	*murA*	UDP-GlcNAc:phosphoenolpyruvate transferase
92[d]	*mepB*	Penicillin-insensitive murein DD-endopeptidase
(93)	*alr*	Alanine racemase

[a] This table includes only mapped genes for which a clear activity or function has been defined. For a more extensive list of genes involved in cell division or morphogenesis, see reference 9.

[b] Reference 3.

[c] Reference 38.

[d] Reference 14.

[e] Reference 43.

pentapeptide as substrate and are believed to form an acyl-enzyme intermediate, with release of the terminal D-alanine during the course of the reaction. Penicillin presumably binds to the active sites since it is an analog of acyl D-alanyl-D-alanine and acylates these penicillin-sensitive enzymes, thereby inactivating them (32, 40). Thus, radioactive penicillin can be used to tag the active-site peptides. By using this technique, the active-site peptides of PBP-1A, PBP-1B, and PBP-3 have recently been sequenced (17), and the sequences have been compared with the complete amino acid sequences of PBP-1A and PBP-1B (7), of PBP-3 (22), and of PBP-5 (6). PBP-1A, PBP-1B, and PBP-3 have quite similar sequences around the active-site serine to which penicillin is bound. This serine is included in the sequence Gly-Ser-X-X-Lys-Pro in all three proteins. In the D-alanine carboxypeptidases, PBP-5 and PBP-6 of *E. coli* and PBP-5 of *Bacillus subtilis*, the active-site serine is in the sequence Ala-Ser-X-Thr-Lys-X-Met-Thr, in which X is usually a hydrophobic amino acid (7). Thus, the sequence Ser-X-X-Lys is conserved in all of the PBPs of *E. coli* thus far examined. Interestingly, this sequence is also found in the class A and class C β-lactamases, suggesting a common ancestor (40).

The active-site serine, Ser-465, Ser-510, and Ser-307 of PBP-1A, PBP-1B, and PBP-3, respectively, occurs near the middle of these bifunctional enzymes. It is assumed that the active site of the transglycosylase enzyme is located near the amino terminus, though there is no evidence for this at present. The active-site serine of the D-alanine carboxypeptidases and the

β-lactamases is located near the amino terminus of these smaller proteins.

GENES INVOLVED IN METABOLISM OF MUREIN

Genes and map positions of proteins for which there is a known role or activity related to murein metabolism are listed in Table 3. Although over 20 genes are listed, those for the following known or suspected enzymatic activities have yet to be identified: (i) D-glutamic acid synthesis, (ii) D-glutamic acid adding enzyme, (iii) UDP-NAcMur pentapeptide:bactoprenyl phosphate phospho-NAcMur pentapeptide transferase, (iv) UDP-NAcGlc:bactoprenol pyrophosphoryl NAcMur pentapeptide NAcGlc transferase, (v) PBP-6 (D-alanine carboxypeptidase), (vi) PBP-1C, (vii) PBP-7, (viii) PBP-8, (ix) murein transglycosylase (1,6-anhydromuramidase), (x) D-alanine carboxypeptidase II (LD-carboxypeptidase), (xi) diaminopimelic acid-lipoprotein transpeptidase, and (xii) diaminopimelic acid-diaminopimelic acid transpeptidase.

In addition to the genes listed above and in Table 3, there are additional genes needed for processing or transport of these gene products and for cell septation, all of which bear on murein metabolism indirectly. Donachie et al. (9) have arranged over 45 genes required for the cell division cycle of *E. coli* into nine classes on the basis of the stage of the cycle in which they may function. Their discussion of these "morphogenes" by class function is both stimulating and revealing. It is clear that our understanding of the events of the cell cycle, especially in terms of growth of the sacculus, septation, and coordination of these processes, is still quite primitive, but there is good reason to believe that the molecular events that constitute the cell cycle in *E. coli* will be understood in the foreseeable future.

UNANSWERED QUESTIONS

The principal purpose of this chapter is to summarize what is known of the enzymes required for net synthesis of murein. What should be clear from this review is that relatively little is known about most of the 11 or more enzymes involved in synthesizing the lipid-linked precursor, bactoprenyl pyrophosphoryl NAcGlc NAcMur pentapeptide, from UDP-NAcGlc. This situation arose because the pathway was worked out first in gram-positive bacteria and could be confirmed in *E. coli* without studying the enzymes in pure form. Obviously, this leaves most questions about the properties of the individual enzymes unanswered.

Regulation of the activity of these enzymes is also open to conjecture. The enzymes are not very sensitive to inhibition by their product or related effectors, although the most sensitive, phosphoenolpyruvate: UDP-NAcGlc pyruvyl transferase, is inhibited about 50% by a 10-fold increase over the normal intracellular concentration of product or of UDP-NAcMur pentapeptide (20). It is interesting that the absolute amount of each of the enzymes involved in synthesis of UDP-NAcMur pentapeptide is just sufficient to allow growth of the sacculus in step with other cell components (19, 20); however, nothing is known of their regulation at the level of transcription.

The remaining reactions required for synthesis of murein, namely polymerization and cross-linking, have only been studied with purified enzymes from *E. coli*. This stems from the fact that the cross-linking reaction was first demonstrated in broken cell preparations of *E. coli* and that the amount of transpeptidase per cell was so small that cloning of the genes of the PBPs that were potential transpeptidases became necessary.

Although the activity of the isolated transpeptidases proved to be only 0.05 to 1% of that expressed in the cell, study of the isolated PBPs brought out the very important and unexpected fact that PBP-1A, PBP-1B, and PBP-3 are bifunctional enzymes able to polymerize the glycan strands, as well as to cross-link strands to each other by transpeptidation (18). Since the penicillin-binding activity of these proteins was high, and since this is considered an indication that the proteins are not denatured, the low transpeptidase activity may result from the lack of suitable acceptor or from a different orientation of the enzyme compared with that of the membrane-bound enzyme associated with the murein sacculus. The specificity of the transpeptidases for an acceptor has not been investigated; it is also possible that one or another requires either a tetrapeptide or tripeptide as acceptor.

Questions relating to the normal function of the various transpeptidases and DD-carboxypeptidases, though largely unexplored, are discussed in chapter 4, which is concerned with assembly of the murein sacculus.

ACKNOWLEDGMENT

This work was supported by Public Health Service grant AI05090 from the National Institutes of Health.

LITERATURE CITED

1. **Anwar, R. A., and M. Vlaovic.** 1979. Purification of UDP-*N*-acetylenol-pyruvylglucosamine reductase from *Escherichia coli* by affinity chromatography, its subunit structure and the absence of flavin as the prosthetic group. Can. J. Biochem. **57:**188–196.
2. **Araki, Y., A. Shimada, and E. Ito.** 1966. Effect of penicillin on cell wall mucopeptide synthesis in *Escherichia coli* particulate system. Biochem. Biophys. Res. Commun. **23:**518–525.
3. **Bachmann, B. J.** 1983. Linkage map of *Escherichia coli* K-12, edition 7. Microbiol. Rev. **47:**180–230.
4. **Barnickel, G., D. Naumann, H. Bradaczek, H. Labischinski, and P. Giesbrecht.** 1983. Computer aided molecular modeling of the three-dimensional structure of bacterial peptidoglycan, p. 61–66. *In* R. Hakenbeck, J.-V. Holtje, and H. Labischinski (ed.), The target of penicillin. Walter de Gruyter & Co., Berlin.
5. **Blumberg, P. M., and J. L. Strominger.** 1972. Five penicillin-binding components occur in *Bacillus subtilis* membranes. J. Biol. Chem. **247:**8107–8113.
6. **Broome-Smith, J., A. Edelman, and B. G. Spratt.** 1983. Sequence of penicillin-binding protein 5 of *Escherichia coli*, p. 403–408. *In* R. Hakenbeck, J.-V. Holtje, and H. Labischinski (ed.), The target of penicillin. Walter de Gruyter & Co., Berlin.
7. **Broome-Smith, J. K., A. Edelman, S. Yousif, and B. G. Spratt.** 1985. The nucleotide sequences of the *ponA* and *ponB* genes encoding penicillin-binding proteins 1A and 1B of *Escherichia coli* K-12. Eur. J. Biochem. **147:**437–446.
8. **Burge, R. E., A. G. Fowler, and D. A. Reaveley.** 1977. Structure of the peptidoglycan of bacterial cell walls. J. Mol. Biol. **117:**927–953.
8a. **Cooper, P. D.** 1956. Site of action of radiopenicillin. Bacteriol. Rev. **20:**28–48.
8b. **de Petro, M. A., and U. Schwarz.** 1981. Heterogeneity of newly inserted and preexisting murein in the sacculus of *Escherichia coli*. Proc. Natl. Acad. Sci. USA **78:**5856–5860.
9. **Donachie, W. D., K. J. Begg, and N. F. Sullivan.** 1984. Morphogenes of *Escherichia coli*, p. 27–62. *In* R. Losick and L. Shapiro (ed.), Microbial development. Cold Spring Harbor Laboratory, Cold Spring Harbor, N.Y.
10. **Edwards, J. R., and J. T. Park.** 1969. Correlation between growth inhibition and the binding of various penicillins and cephalosporins to *Staphylococcus aureus*. J. Bacteriol. **99:**459–462.
11. **Gmeiner, J.** 1980. Identification of peptide-cross-linked trisaccharide peptide trimers in murein of *Escherichia coli*. J. Bacteriol. **143:**510–512.
12. **Gunetileke, K. G., and R. A. Anwar.** 1968. Biosynthesis of uridine diphospho-*N*-acetyl muramic acid. II. Purification and properties of pyruvate-uridine diphospho-*N*-acetylglucosamine transferase and characterization of uridine diphospho-*N*-acetylenolpyruvylglucosamine. J. Biol. Chem. **243:**5770–5778.
12a. **Hara, H., T. Ueda, and H. Suzuki.** 1983. Glycan polymerase with no penicillin-binding activity in *Escherichia coli*, p. 583–588. *In* R. Hakenbeck, J.-V. Holtje, and H. Labischinski (ed.,), The target of penicillin. Walter de Gruyter & Co., Berlin.
13. **Higashi, Y., J. L. Strominger, and C. C. Sweeley.** 1967. Biosynthesis of the peptidoglycan of bacterial cell walls. XXI. Isolation of free C_{55}-isoprenoid alcohol and of lipid intermediates in peptidoglycan synthesis from *Staphylococcus aureus*. J. Biol. Chem. **245:**3697–3702.
14. **Iida, K., Y. Hirota, and U. Schwarz.** 1983. Mutants of *Escherichia coli* defective in penicillin-insensitive murein DD endopeptidase. Mol. Gen. Genet. **189:**215–221.
15. **Ishino, F., S. Tamaki, B. G. Spratt, and M. Matsuhashi.** 1982. A mecillinam-sensitive peptidoglycan cross-linking reaction in *Escherichia coli*. Biochem. Biophys. Res. Commun. **109:**689–696.
16. **Izaki, K., M. Matsuhashi, and J. L. Strominger.** 1966. Glycopeptide transpeptidase and D-alanine carboxypeptidase:penicillin-sensitive enzymatic reactions. Proc. Natl. Acad. Sci. USA **55:**656–663.
17. **Keck, W., B. Glauner, U. Schwarz, J. K. Broome-Smith, and B. G. Spratt.** 1985. Sequences of the active-site peptides of three of the high-M_r penicillin-binding proteins of *Escherichia coli* K-12. Proc. Natl. Acad. Sci. **82:**1999–2003.
18. **Matsuhashi, M., J. Nakagawa, S. Tomioka, F. Ishino, and S. Tamaki.** 1982. Mechanism of peptidoglycan synthesis by penicillin-binding proteins in bacteria and effect of antibiotics, p. 297–301. *In* S. Mitsuhashi (ed.), Drug resistance in bacteria—genetics, biochemistry and molecular biology. Japan Scientific Societies Press, Tokyo.

19. **Mengin-Lecreulx, D., B. Flouret, and J. van Heijenoort.** 1982. Cytoplasmic steps of peptidoglycan synthesis in *Escherichia coli*. J. Bacteriol. **151:**1109–1117.

20. **Mengin-Lecreulx, D., B. Flouret, and J. van Heijenoort.** 1983. Pool levels of UDP N-acetylglucosamine and UDP-N-acetylglucosamine-enolpyruvate in *Escherichia coli* and correlation with peptidoglycan synthesis. J. Bacteriol. **154:**1284–1290.

21. **Mirelman, D.** 1979. Biosynthesis and assembly of cell wall peptidoglycan, p. 115–166. *In* M. Inouye (ed.), Bacterial outer membranes. John Wiley & Sons, Inc., New York.

22. **Nakamura, M., I. N. Muruyama, M. Soma, J.-I. Kato, H. Suzuki, and Y. Hirota.** 1983. On the process of cellular division in *Escherichia coli*: nucleotide sequence of the gene for penicillin-binding protein 3. Mol. Gen. Genet. **191:**1–9.

23. **Neuhaus, F. C., C. V. Carpenter, M. P. Lambert, and R. J. Wargel.** 1972. D-Cycloserine as a tool in studying the enzymes in the alanine branch of peptidoglycan synthesis, p. 339–362. *In* E. Munoz, F. Garcia-Ferrandiz, and D. Vazquez (ed.), Molecular mechanisms of antibiotic action in protein biosynthesis and membranes. Elsevier Applied Science Publishers Ltd., London.

24. **Park, J. T.** 1951. The uridine-5′-pyrophosphate compounds found in penicillin-treated *Staphylococcus aureus* cells, p. 93–98. *In* W. D. McElroy and B. Glass (ed.), Phosphorus metabolism, I. The Johns Hopkins University Press, Baltimore.

25. **Park, J. T.** 1952. Uridine-5′-pyrophosphate derivatives. I. Isolation from *Staphylococcus aureus*. J. Biol. Chem. **194:**877–884.

26. **Park, J. T.** 1952. Uridine-5′-pyrophosphate derivatives. II. A structure common to three derivatives. J. Biol. Chem. **194:**885–895.

27. **Park, J. T.** 1952. Uridine-5′-pyrophosphate derivatives. III. Amino acid-containing derivatives. J. Biol. Chem. **194:**897–904.

28. **Park, J. T., and J. L. Strominger.** 1957. Mode of action of penicillin. Biochemical basis for the mechanism of action of penicillin and for its selective toxicity. Science **125:**99–101.

29. **Rodríguez-Tebar, A., J. A. Barbas, and D. Vázquez.** 1985. Location of some proteins involved in peptidoglycan synthesis and cell division in the inner and outer membranes of *Escherichia coli*. J. Bacteriol. **161:**243–248.

30. **Rogers, H. J., H. R. Perkins, and J. B. Ward.** 1980. Microbial cell walls and membranes. Chapman & Hall, Ltd., London.

31. **Spratt, B. G.** 1975. Distinct penicillin binding proteins involved in the division, elongation, and shape of *Escherichia coli* K-12.

31a.**Spratt, B. G.** 1977. Properties of the penicillin-binding proteins of *Escherichia coli* K-12. Eur. J. Biochem. **72:**341–352.

32. **Spratt, B. G.** 1983. Penicillin-binding proteins and the future of β-lactam antibiotics. J. Gen. Microbiol. **129:**1247–1260.

32a.**Spratt, B. G., V. Jobanputra, and U. Schwartz.** 1977. Mutants of *Escherichia coli* which lack a component of penicillin-binding protein 1 are viable. FEBS Lett. **79:**374–378.

33. **Strominger, J. L.** 1970. Penicillin-sensitive enzymatic reactions in bacterial cell wall synthesis. Harvey Lect. **64:**179–213.

33a.**Suzuki, H., Y. Nishimura, and Y. Hirota.** 1978. On the process of cellular division in *Escherichia coli*: a series of mutants altered in the penicillin-binding proteins. Proc. Natl. Acad. Sci. USA **75:**664–668.

34. **Taku, A., and R. A. Anwar.** 1973. Biosynthesis of uridine diphospho-N-acetylmuramic acid. IV. Activation of uridine diphospho-N-acetylenolpyruvylglucosamine reductase by monovalent cations. J. Biol. Chem. **248:**4971–4976.

35. **Taku, A., K. G. Gunetileke, and R. A. Anwar.** 1970. Biosynthesis of uridine diphospho-N-acetylmuramic acid. III. Purification and properties of uridine diphospho-N-acetylenolpyruvylglucosaminereductase. J. Biol. Chem. **245:**5012–5016.

36. **Tipper, D., and J. L. Strominger.** 1965. Mechanism of action of penicillins: a proposal based on their structural similarity to acyl-D-alanyl-D-alanine. Proc. Natl. Acad. Sci. USA **54:**1133–1141.

37. **Tipper, D. J., and A. Wright.** 1979. The structure and biosynthesis of bacterial cell walls, p. 291–426. *In* J. R. Sokatch and L. N. Ornstein (ed.), The bacteria, vol VII. Academic Press, Inc., New York.

38. **Tomioka, S., T. Nikaido, T. Miyakawa, and M. Matsuhashi.** 1983. Mutation of the N-acetylmuramyl-L-alanine amidase gene of *Escherichia coli* K-12. J. Bacteriol. **156:**463–465.

39. **Ward, J. B., and H. R. Perkins.** 1973. The direction of glycan synthesis in a bacterial peptidoglycan. Biochem. J. **135:**721–728.

40. **Waxman, D. J., and J. L. Strominger.** 1983. Penicillin-binding proteins and the mechanism of action of β-lactam antibiotics. Annu. Rev. Biochem. **52:**825–869.

41. **Wise, E., Jr., and J. T. Park.** 1965. Penicillin: its basic site of action as an inhibitor of a peptide cross-linking reaction in cell wall mucopeptide synthesis. Proc. Natl. Acad. Sci. USA **54:**75–81.

42. **Wright, A., M. Dankert, P. Fennessey, and P. W. Robbins.** 1967. Characterization of a polyisoprenoid compound functional in O-antigen biosynthesis. Proc. Natl. Acad. Sci. USA **57:**1798–1803.

43. **Wu, H. C., and Venkateswaran.** 1974. Fosfomycin-resistant mutant of *Escherichia coli*. Ann. N.Y. Acad. Sci. **235:**587–592.

43. Membrane-Derived Oligosaccharides

EUGENE P. KENNEDY

Department of Biological Chemistry, Harvard Medical School, Boston, Massachusetts 02115

INTRODUCTION

In the first detailed investigation of the metabolism of membrane phospholipids in *Escherichia coli*, Kanfer and Kennedy (13) measured the turnover of the principal membrane phospholipids under conditions of steady-state, logarithmic growth. Pulse-chase experiments with $^{32}P_i$ revealed that radioactivity was steadily lost from the hydrophilic headgroup of phosphatidylglycerol, whereas ^{32}P incorporated into the headgroup of phosphatidylethanolamine was metabolically stable under these conditions. The results suggested some essential function of phosphatidylglycerol that requires the continuous renewal of its hydrophilic headgroup. In a further investigation of this phenomenon, van Golde et al. (30) labeled cells of *E. coli* K-12 with [2-^{3}H]glycerol. They discovered that the phosphoglycerol headgroup of phosphatidylglycerol, labeled with either ^{32}P or [2-^{3}H]glycerol, was continuously transferred as a unit to a novel type of water-soluble oligosaccharides which, from their relation to membrane phospholipids, were called membrane-derived oligosaccharides (MDO).

PROPERTIES OF MDO

Structure

The MDO of *E. coli* are a heterogeneous family of closely related oligosaccharides containing glucose as the sole sugar and substituted with *sn*-1-phosphoglycerol, phosphoethanolamine, and *O*-succinyl ester residues. The discovery of phosphoethanolamine residues in MDO (15) was surprising in view of the metabolic stability of phosphatidylethanolamine in

the experiments of Kanfer and Kennedy (13). The phosphoethanolamine of MDO is almost certainly derived from phosphatidylethanolamine, and the factors that regulate the ratio of phosphoglycerol to phosphoethanolamine in MDO are presently unknown.

The multiple substitution of species of MDO with *sn*-1-phosphoglycerol and *O*-succinyl ester residues gives these molecules variable net negative charge. This variation is the basis for their separation into fractions designated MDO A, B, and C by chromatography on DEAE-cellulose at pH 7.4 (30). Each of these DEAE fractions can be further subfractionated by chromatography on Dowex-1 acetate at pH 3.7, giving rise to species of MDO designated A-1, A-2, etc. (30). Subfractionation during this second step presumably reflects varying substitution with *O*-succinyl ester residues, the carboxyl groups of which are completely ionized at pH 7.4 during chromatography on DEAE but not at pH 3.7 during chromatography on Dowex.

In addition to heterogeneity of substitution, species of MDO also vary in the size of the glucose-containing backbones. After removal of the succinyl residues and dephosphorylation with HF under conditions of minimum cleavage of sugar residues, MDO species were converted to fluorescent derivatives (6). High-pressure liquid chromatography led to the separation of distinct species (Fig. 1). The size of MDO molecules appears to range from 6 to 12 glucose units per mol, with the principal species containing 8 or 9 glucose residues.

The glucose units of MDO are joined by β-1→2 and β-1→6 linkages (25). The structure is highly branched. The backbone probably consists of β-1→2-linked glucose units (32) to which the branches are attached by

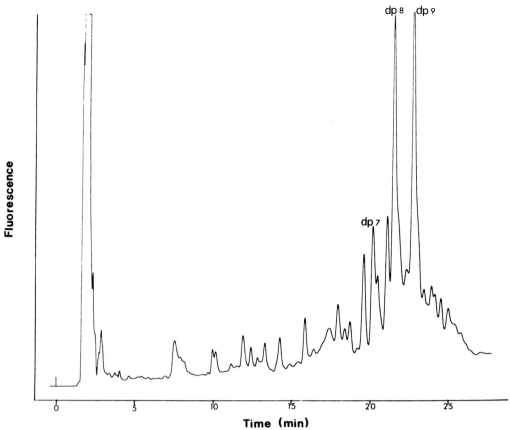

FIG. 1. Heterogeneity of size of MDO fraction B. MDO fraction B (30) from frozen cells of *E. coli* K-12 (Grain Processing Corp.) was treated with mild alkali at room temperature to remove succinyl esters, followed by treatment with 50% HF at 0°C to remove phosphodiester substituents. The dephosphorylated product was purified by chromatography on Sephadex G-25. Fluorescent derivative of 2-aminopyridine were prepared as described by Coles et al. (6) and separated on a column of amino Spherisorb (250 by 4.6 mm), eluted with a gradient of increasing polarity (20 to 60%, vol/vol, of water in acetonitrile). The degree of polymerization (dp) of separated peaks was determined by mass spectrometry (6).

β-1→6 linkages as in the tentative formulation of Fig. 2. The glucose unit at the reducing terminus is known to be linked to the remainder of the oligosaccharide through its 2-position (25).

The phosphoglycerol and phosphoethanolamine residues are attached via phosphodiester bonds to the 6-position of glucose units in MDO. The phosphoglycerol residues are the *sn*-1 stereoisomers (15), consistent with their origin from the headgroups of phosphatidylglycerol (12, 13). Mutants blocked in the transfer of phosphoglycerol to MDO have a higher level of phosphoethanolamine in their MDO (8), suggesting that the phosphoglycerol and phos-

phoethanolamine transferases compete for the same sites on MDO.

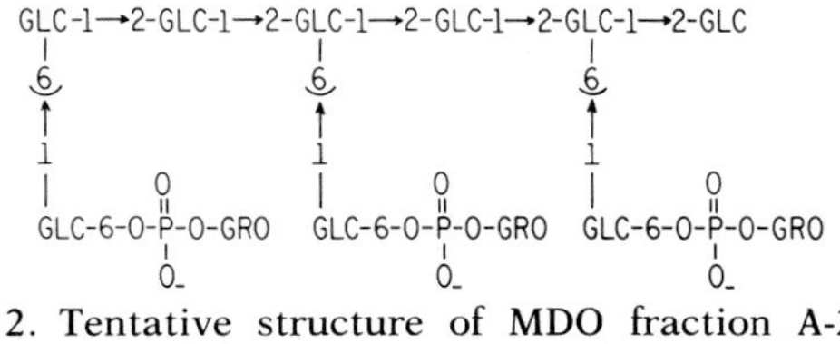

Some species are multiply substituted with succinyl *O*-ester residues, contributing substantially to the net anionic charge at neutral pH. The point of attachment of the succinyl residues has not yet been determined.

MDO fraction A has been subfractionated and analyzed by Kennedy et al. (15) and Schneider and Kennedy (24). Fraction A-1, after separation of Dowex-1 resin at pH 3.7, was found to contain an average of one phosphoglycerol residue and two succinyl residues per mole with a total estimated net negative charge of 3 at neutral pH. About half of the molecules in this subfraction also contain phosphoethanolamine residues, which do not contribute to the net negative charge at neutral pH. Fraction A-2 contains three *sn*-1-phosphoglycerol residues per mole, with no detectable succinate or phosphoethanolamine.

A tentative structure, consistent with what is known about MDO A-2, is shown in Fig. 2. It should be emphasized that some details, such as the exact pattern of branching, are arbitrarily presented.

The molecular weight of the tripotassium salt of the MDO A-2 with the structure shown in Fig. 2 is calcu-

GLC-1→2-GLC-1→2-GLC-1→2-GLC-1→2-GLC-1→2-GLC
        |6 |6 |6
        ↑   ↑   ↑
        | O | O | O
        | ‖ | ‖ | ‖
  GLC-6-O-P-O-GRO   GLC-6-O-P-O-GRO   GLC-6-O-P-O-GRO
            | | |
            O_   O_   O_

FIG. 2. Tentative structure of MDO fraction A-2. The precise linkage of the glucose residues has not been determined and is arbitrarily presented.

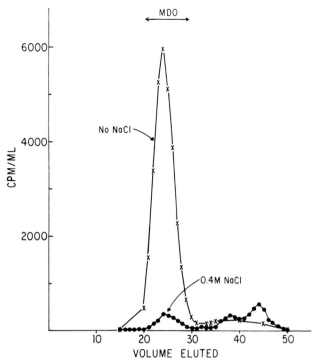

FIG. 3. Osmotic regulation of the biosynthesis of MDO. Cells of strain BB26-36 *plsB* were grown in the presence of [2-³H]glycerol, in medium of low osmolarity, with or without the addition of 0.4 M NaCl. The labeled MDO were extracted and chromatographed on a column of Sephadex G-25. From Kennedy (14) with permission of the publishers.

lated to be 2,305. MDO of fractions B and C (30) have a higher number of anionic substituents and presumably somewhat higher molecular weights.

Intracellular Localization

The MDO of *E. coli* are localized in the periplasm, from which they can be readily extracted by treatment of the cells with EDTA under conditions avoiding osmotic shock (28). This finding suggests that the function of MDO may be related to some biological process taking place in the cell envelope, such as the transport of nutrients, chemotaxis, motility, or the preservation of cellular integrity.

Osmotic Regulation of the Biosynthesis of MDO

Munro and Bell (18) reported that the rate of turnover of membrane phospholipids in *E. coli* is related to the osmolarity of the medium in which the cells are growing. Schulman and Kennedy (26) found that the turnover of membrane phospholipids in *E. coli* is principally, although not entirely, the result of the transfer of their polar headgroups to MDO. It was then discovered (14) that the biosynthesis of MDO in *E. coli* is strikingly regulated by the osmolarity of the medium in which cells are growing. In experiments represented in Fig. 3, the synthesis of MDO in cells grown in low osmolarity was 16 times higher than that in cells grown in the same medium with added 0.4 M sodium chloride. Other solutes at comparable osmolarities

yielded similar results, making it clear that the observed effect is an osmotic one and not specific for sodium chloride.

Distribution of MDO-Like Oligosaccharides in Gram-Negative Bacteria

Schulman and Kennedy (28) reported the presence of oligosaccharides of intermediate molecular weight, with properties similar to those of MDO of *E. coli*, in a number of other gram-negative bacteria, not only of the enteric group, such as *Salmonella anatum*, but also in *Pseudomonas aeruginosa*, suggesting a rather wide distribution of such oligosaccharides in gram-negative bacteria. No MDO-like materials could be detected in the few gram-positive species that were tested.

Because of the economic and biological significance of plant infection by the two closely related genera *Agrobacterium* and *Rhizobium*, much attention has been directed to the surface carbohydrates of these bacteria. The cyclic β-1→2 glucans produced by these organisms resemble the MDO of *E. coli* in their periplasmic localization (1), their intermediate size, and the β-1→2 glucan structure. Because the synthesis of MDO in *E. coli* is osmotically regulated, Miller et al. (17a) tested the production of cyclic glucan in *Agrobacterium tumefaciens* C-58 as a function of the osmolarity of the growth medium. After growth in a medium of low osmolarity (115 mosmol/kg of water), the periplasmic cell-associated glucan was 18 μg/mg of cell protein. The addition of 0.4 M sodium chloride to the medium reduced the glucan to a level of 1.2 μg/mg of cell protein. Comparable amounts of other osmotically active solutes produced similar effects. This response closely parallels the regulation of MDO synthesis in *E. coli* and suggests a general role of periplasmic oligosaccharides in the osmotic adaptation of gram-negative genera as ecologically diverse as *Agrobacterium* spp. and the enteric bacteria.

BIOSYNTHESIS OF MDO

A working model outlining what is presently known of the biosynthesis of MDO in *E. coli* is shown in Fig. 4.

Synthesis of Polyglucose Chains

Schulman and Kennedy (27) reported genetic and biochemical evidence that UDP-glucose is an essential intermediate in the biosynthesis of MDO. Weissborn and Kennedy (32) discovered a novel UDP-glucose-

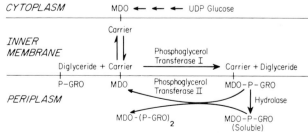

FIG. 4. Working model for MDO biosynthesis. From Jackson et al. (11) with permission of the publishers.

requiring glucosyltransferase system that catalyzes the elongation of β-1→2-linked polyglucose chains. The system also requires a primer such as the disaccharide sophorose or octyl-β-D-glucoside, a trypsin-sensitive membrane fraction, and a heat-stable protein derived from the soluble cytoplasmic fraction, as well as magnesium ions.

The heat-stable protein component of the system has recently been purified to homogeneity and has been found to be identical with the acyl carrier protein of *E. coli* (29a). This finding was very surprising because the function of acyl carrier protein in the transglucosylation system bears no resemblance to its previously known roles in the biosynthesis of fatty acids and phospholipids. It has been suggested (29a) that the role of acyl carrier proteins in the synthesis of cell surface carbohydrates may be important not only in *E. coli* but also in *Rhizobium* species, where such carbohydrates may be involved in cell signalling essential for bacterium-plant symbiosis.

Weissborn and Kennedy (32) suggested that the physiological substrate that is elongated by the glucosyltransferase system in vivo is nascent MDO linked to a lipid carrier, possibly a phosphorylated form of undecaprenol. Such lipid-linked sugars are known to be intermediates in the assembly of other complex carbohydrates targeted for sites external to the cell membrane in *E. coli* and in *Salmonella typhimurium* (31, 33). Although no direct evidence for such a carrier in MDO biosynthesis has as yet been obtained, free glucose is not elongated by the enzyme system of Weissborn and Kennedy (32), indicating the need for some type of activation of glucose prior to the elongation reactions.

Only β-D-glucosides, and not α-D-glucosides, function effectively as primers in the elongation system. This specificity appears to reflect the fact that the linkages between glucose residues in MDO are β. Of the potential primers tested, octyl-β-D-glucoside was the most effective; this is consistent with the notion that the physiological carrier is also a derivative of a hydrocarbon, such as undecaprenol.

Although work on the characterization of the glucosyltransferase system to date has centered on the use of model substrates rather than the physiologically occurring nascent MDO chains, genetic evidence derived from the study of the *mdoA* locus, described below, offers strong support for the conclusion that the glucosyltransferase is an integral part of the machinery for the synthesis of MDO in vivo.

The products of the glucosyltransferase system in vitro are linear oligomers containing only β-1→2-linked glucose units, suggesting that the β-1→2 backbone of an MDO molecule is synthesized first, and the branches are introduced subsequently. Branching may possibly involve UDP-glucose as a donor in a reaction catalyzed by an enzyme specific for the formation of β-1→6 links, or may alternatively involve a rearrangement of glucose units originally linked to the polymer through β-1→2 linkages. The formation of branch points in glycogen is well known to involve the latter type of process.

The glucose polymers produced in vitro by the glucosyltransferase system are of two discrete classes of size. The system thus does not synthesize polymers by independent addition of one glucose unit at a time

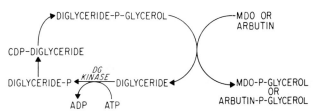

FIG. 5. Biosynthesis of MDO and the diglyceride cycle. From Jackson et al. (11) with the permission of the publishers.

to a pool of intermediates, since this would lead to products with a continuous range of size, but rather by some processive mechanism. The principal products have approximately the same size as MDO molecules, but oligosaccharides that appear to be a multiple of that size are also found. It appears that some chain-terminating event that limits the size of MDO chains in vivo functions only imperfectly in the in vitro enzyme system.

The synthesis of β-1→2 linear glucan as an essential intermediate further underscores the resemblance between the MDO of *E. coli* and the β-1→2 glucans of *Rhizobium* and *Agrobacterium* spp.

Transfer of Phosphoglycerol Residues to MDO

The enzymic transfer of phosphoglycerol residues from phosphatidylglycerol to MDO or to certain synthetic β-glucoside acceptors has been studied by Jackson and Kennedy (12). The products were shown to be *sn*-1-2-diglyceride and β-glucoside-6-phosphoglycerol. The enzyme, designated phosphoglycerol transferase I, is localized in the inner or cytoplasmic membrane of *E. coli* (12) with its catalytic face on the external or periplasmic face of the inner membrane (4).

MDO Biosynthesis and the Diglyceride Cycle

The continuous transfer of phosphoglycerol residues from phosphatidylglycerol to MDO leads to the generations of *sn*-1-2-diglyceride (Fig. 5). The rephosphorylation of *sn*-1,2-diglyceride at the expense of ATP is catalyzed by the enzyme diglyceride kinase, found in *E. coli* by Pieringer and Kunnes (19). Diglyceride kinase does not play a primary role in the de novo synthesis of glycerophosphatides in *E. coli*, as shown by the pulse-chase experiments of Chang and Kennedy (5). Phosphatidic acid for the de novo synthesis of phospholipids is primarily generated by the acylation of glycerophosphate (2).

Mutants defective in diglyceride kinase have been isolated and characterized by Raetz and Newman (20, 21). The biosynthesis of MDO appears to be the principal (although not the sole) source of diglyceride in *E. coli*. Mutants with lesions in the *dgk* locus, with no detectable activity of diglyceride kinase, accumulate large amounts of diglyceride and therefore grow very slowly in medium of low osmolarity (20). Because the biosynthesis of MDO occurs at maximal rates under these conditions, this result suggests that the transfer of phosphoglycerol residues to MDO is the major source of diglyceride in *E. coli*. Raetz and Newman (21) also studied diglyceride accumulation in a strain doubly defective in the *dgk* and *pgi* genes. *pgi* mutants cannot synthesize glucose and produce MDO only when glucose is added to the medium (27). Diglyceride accumulation in the double mutant (*dgk pgi*) was similarly found to require added glucose (21), further supporting the conclusion that MDO synthesis is the principal source of diglyceride.

Phosphoglycerol transferase I has its active site on the outer aspect of the inner membrane (4) and is accessible to substrates such as arbutin when added to the medium. This observation led to the prediction that the growth of *dgk* mutants should be sensitive to arbutin even in medium of high osmolarity because the transfer of phosphoglycerol residues to this substrate should also lead to the accumulation of high levels of diglyceride (Fig. 5). This prediction was confirmed and led to the isolation of *mdoB* mutants (11).

Although the biosynthesis of MDO is probably the major route for the generation of diglyceride, smaller amounts may also be generated by the transfer of phosphoethanolamine from phosphatidylethanolamine to lipopolysaccharide (10).

Periplasmic Phosphoglycerol Transferase II

A periplasmic enzyme, now designated phosphoglycerol transferase II, catalyzes the interchange of phosphoglycerol residues among soluble forms of MDO or certain synthetic β-glucoside model substrates (9). The periplasmic enzyme does not recognize phosphatidylglycerol as a donor of *sn*-1-phosphoglycerol residues. It is thought to function in the formation of multiply substituted MDO species in the periplasm (Fig. 4).

When MDO chains still linked to their lipid carrier appear on the external face of the inner membrane, the membrane-localized phosphoglycerol transferase I catalyzes the transfer of phosphoglycerol from phosphatidylglycerol to these nascent chains. Substitution of the MDO chains by phosphoglycerol, or by succinate, appears to be a signal for cleavage of MDO from the carrier, liberating soluble MDO into the periplasm, since unsubstituted MDO is not observed in significant amounts in pulse-chase experiments. Periplasmic phosphoglycerol transferase II, however, may transfer the phosphoglycerol residue from the nascent MDO, before its cleavage from the carrier, to soluble MDO molecules already in the periplasm. The reaction regenerates unsubstituted, carrier-linked MDO, which may act as acceptor for another phosphoglycerol residue derived from phosphatidylglycerol in the reaction catalyzed by transferase I.

This scheme finds some support from study of the *mdoB* mutants described below. Such mutants, defective in transferase I, contain the same levels of phosphoglycerol transferase II as wild type. The MDO produced by the *mdoB* mutants contains no phosphoglycerol (8, 11). Phosphoglycerol transferase II therefore does not function in the primary transfer of phosphoglycerol from phosphatidylglycerol; more probably it functions in the secondary transfer reactions leading to multiple substituents of MDO with phosphoglycerol.

Although the physiological function of transferase II appears to be the transfer of phosphoglycerol residues, rather than their hydrolysis, at low concentrations of acceptor the enzyme acts as a cyclase/hydro-

lase with the liberation of cyclic *sn*-1-(3)-2-phosphoglycerol (9), suggesting the following mechanism:

$$MDO\text{-}phosphoglycerol \overset{1}{\rightleftharpoons} enzyme\text{-}phosphoglycerol + MDO \\ + enzyme$$

$$2 \downarrow \text{ Slow}$$

$$enzyme + cyclic\ phosphoglycerol$$

REGULATION OF MDO BIOSYNTHESIS

The *mdoA* Locus

Strain T10GP, a derivative of *E. coli* K-12 originally constructed by G. Pluschke, was fortuitously observed by C. R. H. Raetz to be defective in the synthesis of MDO (Raetz, personal communication). Bohin and Kennedy (4) confirmed this observation and obtained evidence that the mutant was blocked at an early stage in the assembly of the polyglucose chains of MDO. These workers found the mutant gene, designated *mdoA*, to be located near 23 min on the *E. coli* map, closely linked to *pyrC* (3).

Weissborn and Kennedy (32) showed that *mdoA* mutants are defective in the membrane component of the glucosyltransferase system described above, but contain normal amounts of the heat-stable protein factor.

No distinct phenotype is associated with *mdoA* mutations that completely block the synthesis of MDO. Such mutants grow about as well in liquid medium of low osmolarity as do wild-type isogenic strains. In certain backgrounds, the *mdoA* lesion may lead to reduced efficiency of plating on agar medium of low osmolarity (M. K. Rumley and E. P. Kennedy, unpublished data).

The *mdoB* Locus

Jackson et al. (11) mutagenized strain RZ60 *dgk6* with transposon Tn*10*. Growth of this *dgk* strain is inhibited by arbutin added to the medium, because that substrate acts as acceptor for phosphoglycerol residues (Fig. 5) as discussed above. A further mutation caused by the insertion of Tn*10* that leads to the inactivation of phosphoglycerol transferase I should render *dgk* cells resistant to arbutin. Such arbutin-resistant mutants were isolated by Jackson et al. (11), who found them to lack detectable activity of phosphoglycerol transferase I. The mutant gene was designated the *mdoB* locus. Mapping experiments indicate that *mdoB* is closely linked to *serB* near min 99, with the probable gene order *mdoB serB thr* in the clockwise direction. These mutants, selected for their inability to transfer phosphoglycerol residues to arbutin in vivo, were also found to lack detectable phosphoglycerol transferase I activity by the in vitro assay of Jackson and Kennedy (11). These results strongly support the conclusions that phosphoglycerol transferase I is an essential enzyme for MDO biosynthesis and that the transfer of phosphoglycerol residues to arbutin in vivo is a valid assay of phosphoglycerol transferase I.

Rotering et al. (23) used an autoradiographic method to screen colonies of *E. coli* immobilized on filter paper for their content of cell products derived from [2-^{3}H]glycerol, after extracting phospholipids from the cells with organic solvents. MDO is the principal glycerol-labeled component of the cell after removal of phospholipids. This procedure led to the isolation of mutants lacking sn-1-phosphoglycerol substituents on MDO.

One such mutation was further characterized and mapped by Fiedler and Rotering (8). It was designated as an mdoB mutation on the basis of its map position near min 99, closely similar to the locus defined by the mdoB mutation of Jackson et al. (11). MDO produced by the mutant had little or no sn-1-phosphoglycerol substituents. Interestingly, it was shown by Fiedler and Rotering (8) that there appeared to be a compensatory increase in the level of phosphoethanolamine substituents in MDO.

As in the studies of Jackson et al. (11), Fiedler and Rotering (8) found that the mdoB mutants also lack the ability to transfer sn-1-phosphoglycerol residues to arbutin in vivo.

Mechanism of Osmoregulation

The biosynthesis of MDO is almost completely suppressed in cells growing in medium of high osmolarity (14). The mechanism of this regulation theoretically could be principally at the genetic level. The enzymes required for MDO biosynthesis might be synthesized only during growth in medium of low osmolarity, and not in medium of high osmolarity. Alternatively, the enzymes for MDO biosynthesis might be present constitutively, even during growth in medium of high osmolarity, but be regulated by effectors that in turn are governed by the osmolarity of the medium. Combinations of the two modes of regulation are, of course, also possible.

Kennedy (14) suggested that regulation at the genetic level might play an important role in setting the level of MDO synthesis because the increase in MDO synthesis upon transition of high to low osmolarity appears to require protein synthesis, as indicated by studies with the inhibitor chloramphenicol. More recent studies, however, indicate that this suggestion is probably incorrect. It now seems more probable that the observed effect of chloramphenicol is an indirect one.

Bohin and Kennedy (4) examined the regulation of phosphoglycerol transferase I in living cells growing in medium of high or low osmolarity. These workers took advantage of the fact that the enzyme can be assayed by measuring the transfer of phosphoglycerol to arbutin, a reaction that has been established as a valid measure of its activity (4, 8). The transfer of phosphoglycerol to added arbutin was remarkably constant over a wide range of osmolarities. This finding indicates that the enzyme is present constitutively and, furthermore, in an active form, even in cells grown in medium of high osmolarity. Because much more phosphoglycerol can be transferred to arbutin than is transferred to MDO during growth at low osmolarity, it is clear that activity of the enzyme is limited by the availability of MDO polyglucose chains to function as acceptor. This result indicates that the rate-making steps for the synthesis of MDO are at an early stage in the biosynthetic process, in the formation of the polyglucose chains. This conclusion also appears reasonable in the light of the general princi-ple that metabolic regulation is usually exerted at an early stage of a given biosynthetic process.

Bohin and Kennedy (4) also measured the production of MDO chains in cells making a transition from low to high osmolarity. Although the experiments did not allow for a sensitive kinetic analysis of early time points in the process, it appeared that as the osmolarity of the medium was increased the rate of synthesis of MDO was down-regulated without detectable time lag. A mechanism for the shutoff of MDO synthesis that requires simple dilution of the level of enzymes present in cells grown at low osmolarity could be excluded. More probably, the enzymes of MDO biosynthesis are present constitutively in cells grown either at high or at low osmolarity, and their activity is regulated by effectors in the cell which reflect the osmolarity of the growth medium.

PROBLEMS AND PROSPECTS

To live and grow, cells require soluble substances in the cytoplasm with a minimum total concentration of about 300 mosmol (7, 29). Because plasma membranes are freely permeable to water, all cells face a fundamental problem of osmoregulation. In medium containing solutes at concentrations higher than about 300 mosmol, cells will undergo shrinkage (plasmolysis) because of the movement of water down its gradient of activity, whereas in medium less than about 300 mosmol, cells will tend to swell.

Osmoregulation is clearly vitally important for mammals, as shown by the intricate and elaborate mechanisms, requiring the effective function of kidney and heart, that are employed for the close regulation of the osmolarity of extracellular fluids. In the absence of the regulation of extracellular fluid that has been developed by mammals, plants and bacteria face two distinct types of osmotic challenge.

Bacterial and plant cells growing in media of high osmolarity respond to this type of osmotic stress by increasing their intracellular content of potassium (7, 17) or glucose and its derivatives (22), and also accumulate high levels of a variety of other substances that have been termed "compatible" solutes. The higher level of osmolarity of the cytoplasm caused by the accumulation of these solutes balances that of the medium, preventing decrease of cell volume. Osmotic adaptation of this type is particularly important for plants growing in surroundings of high salinity, or during periods of drought. The enormous economic importance of an understanding of such osmoregulation in plants has been stressed in the recent review by Le Redulier and Valentine (16).

A second, and less widely studied, type of osmotic challenge occurs when cells with their minimum cytoplasmic osmolarity of about 300 mosmol find themselves in a very dilute medium. Under these conditions, water must flow into the cell, until at equilibrium its flow is resisted by a hydrostatic pressure of about 6.4 atm. It is well established that the peptidoglycan layer of the E. coli cell envelope plays an essential role in maintaining the structural integrity of the cell against such pressures. Lesions in the peptidoglycan caused by treatment of cells with lysozyme in the presence of EDTA, or by treatment of growing cells with penicillin, cause the swelling of cells and their lysis. This function of peptidoglycan is even more striking and obvious in gram-positive bacteria.

In an important study, Stock et al. (29) reported that the periplasmic space of *E. coli* appears to have an osmolarity approximately equal to that of the cytoplasm. For cells growing in medium of low osmolarity, this means that the periplasmic space must also contain solutes that are impermeable to the outer membrane, totaling approximately 300 mosmol. It appears that the principal osmotically active solutes in the periplasmic space of cells grown in medium of low osmolarity are MDO molecules (14), which on the average have a net negative charge of about 5 U/mol. Because MDO molecules are localized in the periplasmic space and cannot diffuse through the outer membrane, they constitute a kind of osmotic buffer, reducing the turgor pressure to which the inner cytoplasmic membrane must be subjected. Furthermore, these MDO molecules with their large net negative charge must contribute substantially to the Donnan potential shown by Stock et al. (29) to be present across the outer membrane of *E. coli*, with the periplasmic space negative with respect to the medium.

Several facts suggest a significant role of periplasmic oligosaccharides in osmotic adaptation in gram-negative bacteria.

(i) MDO biosynthesis requires an estimated 10 to 15 enzymes, probably synthesized constitutively. Under conditions of low osmolarity in the medium, MDO constitute about 5% of the cell mass and thus utilize a corresponding fraction of the total biosynthetic capacity of the cell.

(ii) Periplasmic oligosaccharides with the general properties of MDO are widely distributed in gram-negative bacteria, not only in the enteric bacteria such as *E. coli* and *S. typhimurium*, but also in *Pseudomonas* spp. (28) and the ecologically very different *Agrobacterium* and *Rhizobium* genera. In the latter genera, the synthesis of periplasmic oligosaccharides is also osmotically regulated (17a). These considerations, as well as the periplasmic localization of these oligosaccharides, suggest some important function associated with the cell envelope. At present this function remains mysterious.

The possibility was considered that MDO might be needed as a kind of osmotic buffer for the structural integrity of cells of *E. coli* in medium of low osmolarity. This has proved not to be the case, as indicated by the continuing growth of *mdoA* mutants in medium of low osmolarity (4). Indeed, as mentioned above, no distinct phenotype of MDO-less cells has been yet detected.

The difficulty in finding the phenotypic expression of mutations that block the synthesis of MDO may not be altogether unexpected. There appears to be a considerable redundancy in processes localized in the cell envelope, as, for example, in the multiple transport systems for the uptake of iron, potassium, and magnesium. The function of any one of these multiple systems is seen only after a stepwise elimination of one or more of the alternative systems. Similarly, it is possible that the synthesis of MDO is only one of several mechanisms for coping with medium of low osmolarity. The function of MDO may be seen more clearly only in mutants in which the alternative mechanisms of adaptation have been eliminated.

Although *E. coli* and other enteric bacteria may usually be considered to live in a highly protected environment in the gut, it should not be forgotten that in a natural ecological setting, passage from host to host may involve long periods of survival in medium of very low osmolarity, such as drinking water. In the course of evolution, processes leading to improved adjustment to this type of osmotic stress may well have had great adaptive value.

The fundamental mechanisms by which cells recognize the osmolarity of the medium in which they are growing, and adapt to it by appropriate signaling systems, are completely unknown. From this point of view, the study of the enzymology of the biosynthesis of MDO, particularly the early stages of the formation of the polyglucose backbone, seems of particular interest. The discovery of the effectors that regulate the activity of enzymes carrying out the initial stages of MDO assembly may shed light on the more general problem of how cells recognize the osmolarity of the medium in which they are growing, and adapt to it.

ACKNOWLEDGMENT

Work in my laboratory on the membrane-derived oligosaccharides of *E. coli* has been supported by Public Health Service grants GM19822 and GM22057 from the National Institute of General Medical Sciences.

LITERATURE CITED

1. **Abe, N., A. Amemura, and S. Higashi.** 1982. Studies on cyclic β-1,2-glucan obtained form periplasmic space of *Rhizobium trifolii* cells. Plant Soil **64:**315–324.
2. **Bell, R. M.** 1974. Mutants of *Escherichia coli* defective in membrane phospholipid synthesis: macromolecular synthesis in an sn-glycerol 3-phosphate acyltransferase K_m mutant. J. Bacteriol. **117:**1065–1076.
3. **Bohin, J.-P., and E. P. Kennedy.** 1984. Mapping of a locus (*mdoA*) that affects biosynthesis of membrane-derived oligosaccharides in *Escherichia coli.* J. Bacteriol. **157:**956–967.
4. **Bohin, J.-P., and E. P. Kennedy.** 1984. Regulation of the synthesis of membrane-derived oligosaccharides in *Escherichia coli.* J. Biol. Chem. **259:**9390–9393.
5. **Chang, Y.-Y., and E. P. Kennedy.** 1967. Pathways for the synthesis of glycerophosphatides in *Escherichia coli.* J. Biol. Chem. **242:** 516–519.
6. **Coles, E., V. N. Reinhold, and S. A. Carr.** 1985. Fluorescent labeling of carbohydrates and analysis of HPLC: derivative comparison using mannosidosis oligosaccharides. Carbohydr. Res. **139:**1–11.
7. **Epstein, W., and L. A. Laimins.** 1980. Potassium transport in *Escherichia coli:* diverse systems with common control by osmotic forces. Trends Biochem. Sci. **5:**21–23.
8. **Fiedler, W., and H. Rotering.** 1985. Characterization of an *Escherichia coli mdoB* mutant strain unable to transfer sn-1-phosphoglycerol to membrane-derived oligosaccharides. J. Biol. Chem. **160:**4799–4806.
9. **Goldberg, D. E., M. K. Rumley, and E. P. Kennedy.** 1981. Biosynthesis of membrane-derived oligosaccharides: a periplasmic phosphoglycerol transferase. Proc. Natl. Acad. Sci. USA **78:** 5513–5517.
10. **Hasin, M., and E. P. Kennedy.** 1982. Role of phosphatidylethanolamine in the biosynthesis of pyrophosphoethanolamine residues in the lipopolysaccharide of *Escherichia coli.* J. Biol. Chem. **257:**12475–12477.
11. **Jackson, B. J., J.-P. Bohin, and E. P. Kennedy.** 1984. Biosynthesis of membrane-derived oligosaccharides: characterization of *mdoB* mutants defective in phosphoglycerol transferase I activity. J. Bacteriol. **160:**976–981.
12. **Jackson, B. J., and E. P. Kennedy.** 1983. The biosynthesis of membrane-derived oligosaccharides: a membrane-bound phosphoglycerol transferase. J. Biol. Chem. **258:**2394–2398.
13. **Kanfer, J., and E. P. Kennedy.** 1963. Metabolism and function of

bacterial lipids. I. Metabolism of phospholipids in *Escherichia coli* B. J. Biol. Chem. **238**:2293–2298.

14. **Kennedy, E. P.** 1982. Osmotic regulation and the biosynthesis of membrane-derived oligosaccharides in *Escherichia coli*. Proc. Natl. Acad. Sci. USA **79**:1092–1095.

15. **Kennedy, E. P., M. K. Rumley, H. Schulman, and L. M. G. van Golde.** 1976. Identification of *sn*-glycerol-1-phosphate and phosphoethanolamine residues linked to the membrane-derived oligosaccharides in *Escherichia coli*. J. Biol. Chem. **251**:4208–4213.

16. **Le Redulier, D., and R. C. Valentine.** 1982. Genetic engineering in agriculture: osmoregulation. Trends Biochem. Sci. **7**:431–433.

17. **Measures, J. C.** 1975. Role of amino acids in osmoregulation of non-halophilic bacteria. Nature (London) **257**:398–400.

17a.**Miller, K. J., E. P. Kennedy, and V. N. Reinhold.** 1986. Osmotic adaptation by gram-negative bacteria: possible role for periplasmic oligosaccharides. Science **231**:48–51.

18. **Munro, G. F., and C. A. Bell.** 1973. Effects of external osmolarity on phospholipid metabolism in *Escherichia coli* B. J. Bacteriol. **116**:257–262.

19. **Pieringer, R. A., and R. S. Kunnes.** 1965. The biosynthesis of phosphatidic acid and lysophosphatidic acid by glyceride phosphokinase pathways in *Escherichia coli*. J. Biol. Chem. **240**:2833–2838.

20. **Raetz, C. R. H., and K. F. Newman.** 1978. Neutral lipid accumulation in the membranes of *Escherichia coli* mutants lacking diglyceride kinase. J. Biol. Chem. **253**:3822–3887.

21. **Raetz, C. R. H., and K. F. Newman.** 1979. Diglyceride kinase mutants of *Escherichia coli*: inner membrane association of 1,2-diglyceride and its relation to synthesis of membrane-derived oligosaccharides. J. Bacteriol. **137**:860–868.

22. **Roller, S. D., and G. D. Anagnostopoulos.** 1982. Accumulation of carbohydrate by *Escherichia coli* B/r/l during growth at low water activity. J. Appl. Bacteriol. **52**:425–434.

23. **Rotering, H., W. Fiedler, W. Rollinger, and V. Barun.** 1984. Procedure of the identification of *Escherichia coli* mutants affected in components containing glycerol derived from phospholipid turnover: isolation of mutants lacking glycerol in membrane-derived oligosaccharides (MDO). FEMS Microbiol. Lett. **22**:61–68.

24. **Schneider, J. E., and E. P. Kennedy.** 1978. A novel phosphodiesterase from *Aspergillus niger* and its application to the study of membrane-derived oligosaccharides and other glycerol-containing biopolymers. J. Biol. Chem. **252**:7738–7743.

25. **Schneider, J. E., V. Reinhold, M. K. Rumley, and E. P. Kennedy.** 1979. Structural studies of the membrane-derived oligosaccharides of *Escherichia coli*. J. Biol. Chem. **254**:10135–10138.

26. **Schulman, H., and E. P. Kennedy.** 1977. Relation of turnover of membrane phospholipids to synthesis of membrane-derived oligosaccharides of *Escherichia coli*. J. Biol. Chem. **252**:4250–4255.

27. **Schulman, H., and E. P. Kennedy.** 1977. Identification of UDP-glucose as an intermediate in the biosynthesis of membrane-derived oligosaccharides of *Escherichia coli*. J. Biol. Chem. **252**:6299–6303.

28. **Schulman, H., and E. P. Kennedy.** 1979. Localization of membrane-derived oligosaccharides in the outer envelope of *Escherichia coli* and their occurrence in other gram-negative bacteria. J. Bacteriol. **137**:686–688.

29. **Stock, J. B., B. Rauch, and S. Roseman.** 1977. Periplasmic space in *Salmonella typhimurium* and *Escherichia coli*. J. Biol. Chem. **252**:7850–7861.

29a.**Therisod, H., A. C. Weissborn, and E. P. Kennedy.** 1986. An essential function for acyl carrier protein in the biosynthesis of membrane-derived oligosaccharides of *Escherichia coli*. Proc. Natl. Acad. Sci. USA **83**:7236–7240.

30. **van Golde, L. M. G., H. Schulman, and E. P. Kennedy.** 1973. Metabolism of membrane phospholipids and its relation to a novel class of oligosaccharides in *Escherichia coli*. Proc. Natl. Acad. Sci. USA **70**:1368–1372.

31. **Weisgerber, C., and K. Jann.** 1982. Glucosyldiphospho-undecaprenol, the mannose acceptor in the synthesis of the O9 antigen of *Escherichia coli*. Eur. J. Biochem. **127**:165–168.

32. **Weissborn, A. C., and E. P. Kennedy.** 1984. Biosynthesis of membrane-derived oligosaccharides. J. Biol. Chem. **259**:12644–12651.

33. **Wright, A.** 1971. Mechanism of conversion of *Salmonella* O antigen by bacteriophage ε^{34}. J. Bacteriol. **105**:927–936.

44. Protein Degradation and Proteolytic Modification

CHARLES G. MILLER

Department of Molecular Biology and Microbiology, Case Western Reserve University, Cleveland, Ohio 44106

INTRODUCTION

Intracellular hydrolysis of peptide bonds is important in both growing and starving bacterial cells. Defective or "abnormal" polypeptides of various types are selectively degraded in growing cells, and extensive breakdown of "normal" protein occurs in nongrowing cells undergoing a variety of physiological stresses. Many proteins are proteolytically modified to produce either a mature form or a form with altered activity. The cell has evolved a complex apparatus to ensure the efficiency and specificity of these processes. This chapter will summarize our current understanding of intracellular proteolysis and proteolytic modification. For a more complete picture of intracellular proteolysis, this chapter should be read with chapter 79 by S. Gottesman, which considers proteolytic regulatory mechanisms. Earlier reviews on intracellular proteolysis are by Mandelstam (60, 61), Pine (98), Goldberg and Dice (33), Miller (71), Goldberg and St. John (34), and Mount (81).

PROTEIN DEGRADATION

Early Observations on Protein Degradation in *E. coli*

The first inferences concerning protein degradation in *Escherichia coli* were drawn from experiments designed to find out whether β-galactosidase induction involves de novo synthesis or the activation of a precursor protein (41, 104). In these experiments, cells labeled by several generations of growth in radioactive medium were transferred to nonradioactive medium before induction of β-galactosidase. When the specific activity of purified β-galactosidase was measured, no significant incorporation of label into the newly synthesized enzyme was found. Apparently, in growing *E. coli* β-galactosidase induction does not involve activation of a precursor, and in addition, protein degradation must be very slow relative to the rate of protein synthesis. Hogness et al. (41) suggested that this result casts doubt on the notion that proteins even in mammalian cells are in a "dynamic state" of continual breakdown and resynthesis. Rotman and Spiegelman (104), however, explicitly noted that if the absolute rate of protein turnover in *E. coli* were comparable to that in animal cells this experiment would not have been sensitive enough to detect protein turnover.

Conditions that allowed detection of intracellular protein degradation in *E. coli* were discovered a few years later. Mandelstam (59) found that, although he could detect little protein breakdown in growing cells (<1%/h), starvation for nitrogen led to an easily measured degradation rate (~4 to 5%/h). These results led to the hypothesis that the major function of protein degradation in *E. coli* is to provide amino acids for the synthesis of proteins required for adaptation under starvation or shift-down conditions (60, 61).

That growing cells have the capacity to carry out extensive protein degradation was demonstrated by Willetts (136) and by Pine (96, 97). In a detailed series of experiments, Pine measured protein degradation in growing cells during short time intervals after a pulse of radioactive amino acid. He found that ~5% of the incorporated label was released as free amino acid in about 1 min. The rate of degradation of the pulse-labeled protein fell steadily with continued growth, with an average rate of breakdown of ~2.5%/h. Pine suggested that the initial rapid turnover represents biosynthetic maturation of proteins and that the declining rate of degradation on further growth reflects the heterogeneity of the protein population undergoing degradation. These results showed that, in contrast to previous interpretations, protein degradation is "a normal and integral reaction of the growing cell" (97).

Protein Degradation in Growing Cells

Measurement of protein degradation. Estimates of the rate of protein degradation in growing cells depend on how experimental measurements are made. Frequently cells are labeled by growing several generations in a radioactive amino acid or by procedures in which a pulse of radioactive amino acid is followed by several generations of growth before degradation measurements are made. Either of these labeling protocols leads to incorporation of most of the label into relatively stable proteins. The degradation of these proteins is usually followed by measuring production of acid-soluble radioactivity after resuspension of the cells in medium containing sufficient unlabeled amino acid to block reincorporation of the labeled amino acid produced by degradation. Exchange of the labeled amino acid produced is not rate limiting (82), and the products of protein degradation in wild-type strains are free amino acids (59, 97). These experiments estimate rates of stable protein degradation of ~1%/h under standard growth conditions (minimal glucose medium with aeration at 37°C). In most experiments the extent of protein turnover has not been accurately measured. In glucose-limited chemostat-grown cells (doubling time, 4.5 h), at least 58% of the originally incorporated label was released as acid-soluble radioactivity during 72 h of growth (114). It appears that most of the cell's proteins can be degraded and that earlier estimates that only about 30% of the protein is

subject to degradation under any conditions (96, 98) may not be correct.

Experiments that follow the degradation of proteins labeled by a short pulse of radioactive amino acid give a picture rather different from that obtained from long-term labeling or pulse-chase experiments. Pine (97) found that at least 5% of the label incorporated in a 7-s pulse was released during a 45-s chase period. Uncertainties concerning the efficiency of exchange of radioactive amino acid for cold amino acid on this time scale make this a minimum estimate. This observation suggests that a substantial fraction of the peptide bonds synthesized are rapidly degraded in growing cells. Even higher estimates of the size of this rapidly degraded fraction come from studies of peptidase-deficient mutants of *Salmonella typhimurium* (142). When a mutant lacking peptidases, N, A, B, and D is given a short pulse of [^{14}C]leucine, approximately 20% of the radioactivity is rapidly converted to a heterogeneous mixture of small, nonreutilizable peptides. Only peptides small enough to be soluble in trichloroacetic acid would have been detected in this experiment, so this is a minimum estimate of the fraction of the peptide bonds synthesized that are rapidly turned over. These observations suggest that peptide bond hydrolysis is a major metabolic process in growing bacterial cells.

Degradation of individual proteins. All of the experiments described so far measure only bulk protein degradation. It is important to remember that only a few individual protein species comprise most of the "protein" of *E. coli* (77). Fewer than 50 "proteins" make up more than 50% of the "protein." A large fraction of the proteins of *E. coli* could be unstable without much affecting the overall protein degradation rate. Several studies of the degradation of individual *E. coli* proteins have been carried out. Mosteller and co-workers (80) followed 184 proteins observable on two-dimensional gels made from exponentially growing batch cultures of *E. coli*. Of these proteins, 47 were classified as "unstable" with half-lives estimated between 2 and 23 h. In another study using two-dimensional gels but a different labeling procedure, only three proteins were found to be degraded relatively rapidly (53). These studies cannot distinguish between proteins that are degraded and those that are modified in any way that affects their position on the two-dimensional gel. In addition, they would not have detected proteins with half-lives less than ~0.5 to 1 h. One normal *E. coli* protein, the product of the *sulA* gene (see chapter 79), has a half-life of 1.2 min (78).

It seems reasonable to conclude that, in growing cells, proteins that comprise most of the total protein of *E. coli* have half-lives significantly longer than a generation. There are no data to rule out the possibility that a great many other proteins are degraded with half-lives significantly shorter than a generation time.

Selective degradation of abnormal proteins. The ability of growing *E. coli* cells to carry out selective protein degradation is most clearly seen in the degradation of structurally altered proteins. Such "abnormal" proteins can be generated in several ways. Incomplete proteins are produced as a result of chain termination mutations (37), internal initiation of translation (69), or puromycin treatment (32, 96). Full-length proteins with structural defects can result from missense mutations (5, 147), incorporation of amino acid analogs (32, 96), or mistranslation caused by *ram* (ribosomal ambiguity) mutations or by missense suppressors (32). Full-length, normal proteins may be recognized as abnormal if they are unable to assemble into a structure of which they are normally a part (21). Proteins that may be structurally normal in one organism may be viewed as foreign and abnormal when produced from cloned DNA in *E. coli* (see below). The increased rate of protein degradation observed at higher temperatures (96) may reflect an increase in the concentration of abnormal conformations assumed by normal proteins, although other factors may also be involved (see below).

Pine (96) first recognized that *E. coli* can selectively degrade abnormal proteins during growth. Experiments of Goldberg (32) clearly characterized the features of the process. A typical experimental procedure involves labeling with a radioactive amino acid during exposure to a sublethal concentration of puromycin. When the puromycin is removed and the cells are suspended in medium containing cold amino acid to prevent reincorporation of label, breakdown of the protein made in the presence of the drug can be followed by measuring production of acid-soluble radioactivity. A significant fraction of the labeled protein is rapidly ($t_{1/2} < 30$ min) broken down in growing cells (32). The stability of proteins synthesized before puromycin treatment is unaffected by exposure to the drug, demonstrating the selectivity of the process. Similar procedures can be used to study the degradation of proteins containing amino acid analogs such as canavanine. Most of the protein made in the presence of the analog is rapidly and selectively broken down to free amino acids.

In contrast to the situation with normal proteins, it has been possible to study the degradation of several individual abnormal proteins. Chain termination mutations allow the production in the cell of specific abnormal proteins whose degradation can be directly observed. A large (approximately 12 amino acids shorter than wild type) fragment of β-galactosidase is produced in strains carrying the ochre mutation *lacZX90*. The half-life of the X90 polypeptide measured directly from sodium dodecyl sulfate-polyacrylamide gel electrophoresis is approximately 7.5 min at 37°C (37). Decay rates of other N-terminal fragments of β-galactosidase have been estimated from the rate of loss of their ability to serve as α donors in in vitro complementation assays (56). These latter experiments cannot give the rates of the initial proteolytic cut for most of these peptides since only a small N-terminal fragment is required for complementation activity. They do show, however, that nearly all β-galactosidase nonsense polypeptides are much less stable than the complete protein and that many short fragments are very rapidly degraded ($t_{1/2} < 10$ min). It has also been possible to detect the production of a polypeptide that appears to be an intermediate in the degradation of the X90 fragment (49, 66, 67).

Protein Breakdown in Nongrowing Cells

It is generally thought that the rate of protein degradation is increased by starvation for a required nutrient. It is worth emphasizing again that this is true only for stable proteins. It is not clear that the absolute rate of peptide bond hydrolysis is increased by starvation. As Pine (98, 99) has pointed out, it is more reasonable to think that in starvation a protein fraction that is relatively resistant to degradation in growing cells becomes available for breakdown.

Stresses that lead to increased degradation of normal proteins. Growth arrest initiated in a variety of ways can elicit an increase in the rate of stable protein degradation. Starvation for carbon (59, 60), for nitrogen (59, 60), or for certain required amino acids (99) are all effective. Removal of required inorganic nutrients (phosphate, magnesium, potassium) also leads to increased rates of stable protein degradation (113). A transient increase in stable protein degradation is observed during diauxic lag (135). None of these treatments increases the rate of stable protein degradation to more than approximately 5%/h (99). This rate is not sustained and declines during further starvation. From 25 to 30% of the labeled protein can be degraded during carbon starvation without loss of viability (99, 142). Nitrogen starvation leads to even more extensive (up to 50%) degradation (unpublished data).

Selectivity in starvation-induced protein degradation. One of the most interesting problems related to starvation-induced protein degradation is the question of selectivity. Does the cell selectively degrade proteins in response to particular types of nutritional stress? Surprisingly, there is little evidence for such selectivity. In experiments similar to those used to follow the degradation of individual proteins during growth (two-dimensional gels), R. D. Mosteller (personal communication) has found few proteins that are degraded more rapidly than the bulk protein even under conditions where 40 to 50% of the total protein is degraded. Degradation during starvation is selective in the sense that proteins not degraded during growth are degraded in starvation. There may also be selective degradation in response to starvation for divalent cations as a result of ribosome instability (113). The question of physiologically significant selectivity in starvation degradation must be considered open, however.

How does starvation increase the degradation rate of normal proteins? The increase in stable protein degradation with starvation could result either from alterations in the proteolytic apparatus or from changes in the susceptibility of the proteins to degradation. There is no clear-cut evidence for the synthesis of new proteolytic activities in response to starvation. There is good evidence that the late steps of the protein degradation pathway in which peptides are broken down to free amino acids are common to all types of protein degradation and that the enzymes involved are active in both growing and starving cells (72, 73). It is possible that altered ATP levels or changes in the levels of ppGpp may alter the activity of the proteolytic apparatus, but no satisfying theory is at hand (see below). It may be that certain proteins become more susceptible to degradation during starvation because their substrates are present at diminished levels. Enzymes are frequently found to be protected by their substrates from proteolysis. This idea is somewhat hard to test, but it would seem to predict highly selective degradation during starvation, and such selectivity has not been observed. Another possibility that has not received much attention is that the proteolytic apparatus is made available for normal proteins by starvation because the production of its normal substrates (perhaps signal sequences, prematurely terminated polypeptides, and other products of translational mistakes, see below) has been shut down. If the extent of rapid peptide bond degradation is as great as experiments with peptidase-deficient mutants indicate (see above), a very substantial flow through the degradation apparatus is occurring in growing cells. Perhaps interruption of this flow allows the proteolytic apparatus to cleave substrates for which it has low affinity, such as normal proteins.

GENES AND ENZYMES INVOLVED IN PROTEOLYSIS

The *lon* Gene and Its Product

The first mutation found to affect a proteolytic process was discovered by Bukhari and Zipser (9). Mutations obtained by a selection for stabilization of *lacZ* nonsense fragments were originally assigned to a locus called *deg*. In *deg* strains, β-galactosidase fragments showed substantially increased stability (up to a 10-fold increase in half-life). The map position of the *deg* mutations and the fact that the strains produced mucoid colonies suggested that they were alleles of the *lon* locus. Mutations at *lon* had been isolated years earlier as UV sensitives that grew as long filaments after UV exposure. The *lon* mutations also caused a puzzling group of additional phenotypic consequences.

It now seems clear that the *lon* locus codes for a protease that is stimulated by ATP and that the primary defect in *lon* strains is the absence of this activity. The role of this protease in a variety of regulatory circuits is described in chapter 79, and only its involvement in general degradation processes will be described here. Strains with *lon* mutations degrade all types of abnormal proteins more slowly than their lon$^+$ parents, but they are not defective in the degradation of normal, stable proteins elicited by starvation (34, 63). The *lon* gene has been cloned by Zehnbauer and Markovitz (146). An independently isolated *lon* clone has been used to generate a group of transposon insertions into the *lon* gene (63). These mutations lead to the complete loss of the *lon* protease but are not lethal when introduced into the cell (see below).

The *lon* enzyme has been purified by Markovitz and co-workers (13) and characterized by these workers and by Chung and Goldberg (18). The enzyme has several distinct activities, as follows. (i) It is a protease that attacks casein, globin, denatured albumin, and glucagon but not native hemoglobin, albumin, or insulin. Acid-soluble products are produced from protein substrates. The degradation of protein substrates requires the presence of Mg^{2+} and ATP, which is hydrolyzed. The enzyme will cleave certain small peptide substrates, however, in the presence of nonhydrolyzable ATP analogs including inorganic triphosphate and PP$_i$ (36). Protein substrates for the enzyme can activate hydrolysis of small peptide substrates in the absence of ATP or ATP analogs (134). The

protease activity is sensitive to serine protease inhibitors and to metal chelators (19) and is strongly inhibited by ADP (133). The action of the *lon* protease on β-galactosidase fragments or other substrates that it presumably attacks in vivo has not been studied. (ii) The *lon* enzyme is an ATPase. This ATP hydrolysis is stimulated by protein substrates but not by nonsubstrate proteins (133). (iii) The *lon* product is a DNA-binding protein (145). Addition of DNA (but not RNA) stimulates hydrolysis of protein substrates and ATP hydrolysis (17). Exactly how this unusual assortment of properties bears on the role of the *lon* enzyme in the cell is not completely clear.

Other Soluble Endoproteases

Goldberg and co-workers (35) have identified other endoproteases from extracts of *E. coli*. Some of the properties of these enzymes are shown in Table 1. Protease Pi seems to be the activity also called protease III (15). Mutations leading to the loss of protease III were isolated by screening extracts of microcultures for failure to degrade "auto-α," a small N-terminal fragment of β-galactosidase (16). These mutations do not have any readily identifiable phenotypic consequences. No mutations affecting the other activities have yet been reported. These enzymes overlap in specificity, and it may be necessary to isolate strains missing several activities before their physiological functions can be understood.

Other *E. coli* activities have been detected by using chromogenic N-blocked amino acid esters that are substrates for the classical mammalian pancreatic endoproteases (Table 1). Protease I and protease II have been purified and characterized by Pacaud and co-workers (89–91). The purified preparations show only weak proteolytic activity with most protein substrates, and some workers have questioned whether they are really proteases (48). *S. typhimurium* and *E. coli* mutants lacking both protease I and protease II have been isolated (38, 48, 75). These mutants grow normally and are not defective in the degradation of abnormal proteins or in starvation-induced degradation of normal proteins. *S. typhimurium* and *E. coli* contain several other esterases, some of which hydrolyze N-blocked amino acid esters with specificity for the amino acid residue (39; P. Osdoby and C. Miller, manuscript in preparation). Little is known about either the biochemical properties or the physiological functions of these activities.

Membrane-Associated Endoproteases

Several proteolytic phenomena suggest that the cell membrane contains endoproteolytic activity. Nitrate reductase can be released from *E. coli* cytoplasmic membranes by what appears to be a proteolytic process (57) because it is inhibited by a protease inhibitor (*p*-aminobenzamidine) and because the solubilized product contains a subunit with altered electrophoretic mobility (sodium dodecyl sulfate-polyacrylamide gel electrophoresis). The activity responsible for release of nitrate reductase from the cytoplasmic membrane is localized in the outer membrane (58). Protease activity associated with the *E. coli* outer membrane has also been found to cleave alkaline

TABLE 1. Proteases and peptidases[a]

Enzyme	Substrates[b]	M_r ($\times 10^3$) (subunits)[c]	Inhibitors	Gene locus (map units)	References
Soluble endoproteases					
Do	Globin and casein	540 (10)	D		18, 35
Re	Globin and casein	82 (1)	D, E, O, TPCK		35
Mi	Globin and casein	110	D, E, O		35
Fa	Globin and casein	110	D, E, O, TPCK		35, 119
So	Globin and casein	140 (2)	D, TPCK		35
La (requires ATP)	Globin and casein	450 (4)	D, E, NEM	lon (10)	13, 19, 35
Ci	Insulin	125 (2)	E, O		35
Pi (periplasmic)	Insulin, auto-α	110 (1)	E, O, PHMB	ptr (61)	15, 16, 35
Protease I	NAPNE		D		48, 75, 90, 91
Protease II	BAEE		D, TLCK		38, 89
ISP-L-Eco	Z-Ala-Ala-Leu PNA	55	D, E		116
Membrane-associated endoproteases					
Protease IV[d]	Z-Val ONP	34	D		87, 88
Protease V	Z-Phe ONP		D		87, 88
Signal peptidase I	Precursors of secreted proteins	36		lep (55)	138
Signal peptidase II	Prolipoproteins	18		lsp (1)	23, 121
Peptidases					
Dipeptidases					
Peptidase D	Broad specificity			pepD (7)	70, 76
Peptidase Q	X-Pro		E	pepQ (84)	65
Peptidase E	Asp-X	35		pepE (90)	10
Peptidase G	Gly-Gly			pepG	—[e]
Aminotripeptidase					
Peptidase T	Tripeptides			pepT (25)	115
Aminopeptidases					
Peptidase N	Broad specificity, naphthylamides	87 (1)	Peptides, amino acids	pepN (20)	64, 70, 76
Peptidase A	Broad specificity	323 (6)	E, Zn²⁺	pepA (97)	70, 76, 130
Peptidase B	Broad specificity			pepB (53)	72, 76
Peptidase P	X-Pro-Y	230	E, Pro-X peptides	pepP (63)	65, 141
Peptidase M	Met-X-Y[f]	34	E	pepM (3)	—[g]
C-terminal exopeptidases					
Dipeptidylcarboxypeptidase	Minimally tetrapeptide	97	Captopril, dipeptides	dcp (28)	22, 128, 140
Oligopeptidase A	Minimally pentapeptide			optA (77)	127

[a] Abbreviations: D, diisopropylfluorophosphate; E, EDTA; O, o-phenanthroline; TPCK, N-tosyl-phenylalanine chloromethyl ketone; NEM, N-ethyl maleimide; PHMB, p-hydroxymercuribenzoate; NAPNE, N-acetyl-phenylalanine β-naphthyl ester; Z, benzyloxycarbonyl; ONP, p-nitrophenyl ester; TLCK, N-tosyllysine chloromethyl ketone; BAEE, N-benzoylarginine ethyl ester; PNA, p-nitroanilide.

[b] Typical substrates or substrates used to assay the activity are given. Other substrates may also be attacked.

[c] Number of subunits given in parentheses if known.

[d] Another membrane-associated activity apparently different from this one has also been called protease IV (103).

[e] K. L. Strauch and C. G. Miller, unpublished data.

[f] The identity of the second amino acid determines susceptibility to hydrolysis. Y may be an amino acid or a polypeptide.

[g] Miller, et al., submitted for publication.

phosphatase to yield an active enzyme of decreased size (12), to degrade extensively colicin Ia (6), and to produce a slightly smaller, inactive form of the ferric enterobactin receptor (27). An E. coli outer membrane preparation will also function as a plasminogen activator (54). Cleavage of colicin Ia apparently does not play a role in the action of this toxin (6). The physiological significance of the ferric enterobactin receptor cleavage is doubtful since it is observed only in E. coli K-12 and not in other closely related strains (27).

Two activities, detected with reactive ester substrates (Table 1), have been purified from E. coli membranes by Pacaud (87, 88). One of these, protease IV, is localized in the cytoplasmic membrane. The other, protease V, is present in both the cytoplasmic and the outer membrane. The presence of multiple diisopropylfluorophosphate-binding proteins in the E.

coli outer membrane suggests that there may be several serine hydrolases in this compartment of the cell (79).

Peptidases

E. coli and S. typhimurium contain a surprising number of enzymes that hydrolyze small peptides. These enzymes have been defined by biochemical methods (94, 108, 118) and by the isolation of mutations that lead to their loss (72). The specificities of the activities for which mutations have been isolated are shown in Table 1. All of these enzymes are able to attack small peptides, and all except dipeptidylcarboxypeptidase and oligopeptidase A produce amino acids as products. The products of these two C-terminal exopeptidases are smaller peptides, not free amino acids.

There is clear evidence that some of these enzymes are part of the pathway of protein degradation (142). In the absence of peptidases N, A, B, and D, both the rate and extent of carbon starvation-induced protein turnover are reduced. Even more significantly, the products of protein breakdown in peptidase-deficient mutants are mainly peptides rather than free amino acids. These enzymes participate in the degradation pathways for all types of proteins (73). The participation of restricted-specificity enzymes in the protein breakdown pathway is also required. In the absence of peptidases P and Q (X-Pro–specific peptidases) extensive protein breakdown takes place, but most of the proline originally present in protein ends up in small proline peptides (74). The properties of the peptidase-deficient mutants and the specificities of the peptidases have led to a proposed pathway for the terminal steps of protein degradation (72). Peptides generated by endoproteolytic cuts (responsible enzymes unknown) are attacked at their N termini by aminopeptidases and at their C termini by oligopeptidase A and dipeptidylcarboxypeptidase. For continued N-terminal degradation, peptidase P must act whenever an X-Pro peptide bond is encountered. The aminopeptidases (which also act on dipeptides), specific dipeptidases (e.g., peptidase D), or both are necessary to complete the degradation of the di- and tripeptides produced by the two C-terminal exopeptidases.

In addition to their roles in degrading intracellular protein, this same family of peptidases also hydrolyzes peptides supplied from outside the cell and allows these peptides to be used as nutritional sources (72). They also provide a mechanism for detoxifying inhibitory peptides (72). The action of these peptidases can also lead to the release of toxic amino acids or other toxic compounds from peptides carried into the cell by peptide transport systems (2, 25, 29).

FACTORS INVOLVED IN REGULATING INTRACELLULAR PROTEOLYSIS

Energy Availability

A large number of studies have shown that conditions that limit the cell's capacity for generating metabolic energy block protein degradation. This seems to be true for all types of degradation and all types of cells (34). In one of the most detailed of these studies, Olden and Goldberg (86) obtained evidence that, to block protein degradation, the restriction on energy production must be sufficiently severe to lower the intracellular ATP concentration to about 5 to 10% that of growing cells. In practical terms this means that inhibitors of respiration (e.g., azide or cyanide) will not inhibit protein degradation in media containing glucose since nonrespiratory energy generation pathways produce sufficient energy to support protein degradation. Protein degradation appears to be stimulated by less severe reductions in ATP levels. ATP concentrations 30 to 50% of those in growing cells correlate with an approximately twofold increase in the rate of protein breakdown (112).

The molecular basis for this energy requirement is not understood. There is evidence that ATP itself or some other "high-energy" compound, rather than an "energized membrane state," is involved (86). There are several ways such a compound could act. It might be required directly by the rate-limiting protease. The characterization of the *lon* protease as an ATP-dependent activity has provided support for this possibility. It has been shown, however, that the complete absence of the *lon* enzyme does not abolish either the selectivity or the energy requirement for abnormal protein degradation (although reducing its rate) and does not affect at all the degradation of normal proteins. This does not eliminate the possibility that proteases require ATP directly, but it does require the discovery of energy-dependent proteases other than the *lon* enzyme. Another possibility is that energy is required for some kind of marking reaction. This theory proposes an energy-dependent reaction that converts a protein to a form more susceptible to proteolysis. According to this hypothesis, neither the selectivity nor the energy dependence of proteolysis necessarily resides in a proteolytic step. Hershko and co-workers (see reference 40 for a review) have characterized a rabbit reticulocyte system in which a small polypeptide, ubiquitin ($M_r = 8,500$), is conjugated in an ATP-dependent reaction to amino groups of potential substrates for proteolysis. These ubiquitinylated proteins are assumed to be marked for recognition by specific proteases. The system required for these reactions is complex, and not all the components have been isolated. Impressive support for the relevance of the ubiquitin pathway to events in the cell has come from the discovery of a variant (ts85) of a mouse cell line that is temperature sensitive for conjugation of ubiquitin to substrate proteins (26). At the nonpermissive temperature these cells are deficient in the degradation of abnormal proteins (20). Although *E. coli* extracts will not conjugate ubiquitin to substrate proteins, it is possible that another type of marking reaction could serve a similar function (40).

The Stringent Response

Sussman and Gilvarg (117) showed that a strain carrying a *relA* mutation cannot increase the rate of normal protein degradation in response to amino acid starvation. Mutations in *relA* do not affect the increase that results from starvation for a carbon source or for required inorganic nutrients (34). Since accumulation of ppGpp is one of the hallmarks of the stringent response, Goldberg and co-workers have studied the possible roles of the accumulation of this nucleotide in increasing the rate of protein degradation in response to starvation. For carbon and amino acid starvation, increases in ppGpp levels are correlated with increased protein degradation. The increase in ppGpp levels in carbon starved cells occurs by a *relA*-independent pathway. It is possible that ppGpp plays a direct role in regulating protein degradation in response to amino acid or energy limitation. It has been suggested that ppGpp could act by stimulating the proteolytic apparatus or by increasing the susceptibility of protein substrates to degradation. There is no in vitro proteolytic system that ppGpp has been shown to affect, however. Starvation for inorganic nutrients (potassium, P_i, magnesium) also increases the degradation of normal proteins, but this increased rate is not correlated with an increase in ppGpp levels (113).

It has been proposed (113) that the increased degradation in response to ion starvation may represent enhanced degradation of a different group of proteins from those degraded during carbon or amino acid starvation. Deprivation of these ions leads to an increased breakdown of ribosomal proteins presumably because the ribosomes fall apart in the absence of stabilizing ions and the free ribosomal proteins are susceptible to proteolysis.

Rafaeli-Eshkol and Hershko (101) have proposed a different explanation of the effect of relA mutations in protein degradation. The stringent response plays a major role in maintaining the fidelity of translation in amino acid-starved cells (28). A great deal of evidence indicates that in an amino acid-starved relA mutant several types of translational errors occur at relatively high frequency (28). Rafaeli-Eshkol and Hershko (101) suggested that protein synthesis is directly coupled to protein degradation in some way that is affected by the protein synthesis defect in relA mutants. This idea has been criticized (111, 129) on the assumption that such a coupling necessarily implies that some important component of the proteolytic apparatus or a specific regulator must be continuously synthesized and rapidly turned over. One possible way such a coupling could occur seems not to have been considered. Although not all of the translational errors occurring in amino acid-starved relA strains lead to the production of hyperdegradable proteins (92), it seems possible that a sufficient level of mistranslated abnormal protein accumulates in amino acid-starved relA cells to compete with the degradation of normal proteins. Competition between abnormal proteins accumulated during growth and the degradation of normal proteins elicited by starvation has been observed (34). The failure of relA strains to increase protein degradation when starved for amino acids may be a secondary consequence of defective protein synthesis, not an indication of a fundamental role for this regulatory network in protein degradation.

The Heat Shock Response

Cells respond to a variety of stresses by inducing a group of proteins called heat shock proteins. This response is regulated by the htpR locus, which is required for induction of heat shock proteins (83). This locus also affects the degradation of abnormal proteins (4, 30). Two proteolytically unstable mutant proteins (a mutant RNA polymerase sigma subunit and the lacZ X90 termination fragment) are at least 10-fold more stable in a strain that produces less htpR product than the wild type. Both of these proteins are substrates for lon-dependent degradation, and lon is known to be a heat shock gene controlled by htpR (30, 95). There is considerable evidence to indicate that the reduced degradation rate in the htpR mutant cannot be explained by lower levels of lon protease, however (4). In addition, the effect of mutations in both lon and htpR on degradation rates was greater than that of lon mutations alone. One interesting possibility is that htpR regulates the production of other components of the proteolytic apparatus distinct from the lon protease. The possibility that the appearance of aberrant proteins may be the common feature of the stresses that induce the heat shock response is suggested by the observation that the production of several types of abnormal proteins (including certain mammalian proteins) leads to the htpR-dependent induction of the heat shock genes (31). These new findings suggest that the increase in protein degradation at high temperature observed many years ago (96) may reflect the induction of increased proteolytic capacity in addition to an increased probability of denaturation.

Phage Genes That Affect Protein Degradation

The observation that fragments of bacteriophage proteins appear to be stable in infected cells (11) led Simon and co-workers (110) to study the effect of phage infection on protein degradation. They found that infection with T4, T5, and T7 inhibited the degradation of puromycin fragments and analog-containing proteins, but not the degradation of normal proteins. A region of the T4 genome encoding the pin (proteolysis inhibition) function has been cloned in a plasmid vector in E. coli. The presence of this plasmid inhibits the degradation of puromycin fragments, allows increased plating efficiency of a λ Ots mutant (presumably by stabilizing the missense gene product), and seems to increase the yield of a cloned mammalian protein (109). Neither the nature of the pin gene product (if a single gene is sufficient to confer pin function) nor its target of action is known.

Avoiding Unwanted Proteolysis In Vivo and In Vitro

One of the goals of work on proteolysis is to be able to control the process. The practical utility of this control has increased considerably with the widespread use of E. coli as host for expression of cloned genes from other organisms. These proteins may be degraded or proteolytically modified (or fail to be modified) to such an extent that yields of the desired material are reduced. Human preproinsulin, for example, has a half-life in E. coli of approximately 2 min (120).

A straightforward approach to avoiding unwanted proteolysis is to use strains carrying one or more of the mutations known to affect proteolysis to express the desired protein. Strains with lon mutations have been most often used for this purpose. The potential general usefulness of such strains is suggested by the observation that a lon mutation stabilized (ca. threefold) large hybrid proteins produced from lacZ fusion genes in λgt11 (144). The presence of a lon mutation does not always solve the problem of protein instability, however. The absence of a lon effect on preproinsulin degradation has been documented (24), and considerable anecdotal information communicated to me from a variety of sources suggests that lon mutations frequently do not have the desired stabilizing effect. The heat shock regulatory locus, htpR, affects protein degradation via a pathway that appears to be partially lon independent (see above). In at least one case (human somatomedin-C), lon and htpR mutations led to approximately additive stabilizing effects (8). Although the T4 pin region cloned into E. coli has been reported to increase yields of human fibroblast interferon (109), this construct does not seem to have been widely used to stabilize cloned proteins. None of these mutations seems likely to stabilize small polypeptides many of which probably do not require energy-dependent pathways for their degradation (46). Strains deficient in peptidases have not been useful for blocking degradation since even in the most deficient strains available the remaining enzymes catalyze considerable breakdown to acid-soluble fragments (142).

Another strategy for avoiding proteolysis involves making potential substrates inaccessible to proteolytic attack rather than altering the proteolytic apparatus. One way to do this is to export the protein to the periplasmic space. Talmadge and Gilbert (120) reported that fusion proteins derived from preproinsulin are at least 10 times more stable when exported into the periplasm. The stability of some mutant secreted proteins seems to be increased by inhibiting their secretion, allowing them to accumulate in the cytoplasmic membrane (52). Another way to remove potential substrates from proteases is to sequester them in insoluble intracellular inclusion bodies. A variety of proteins overexpressed in E. coli appear to form intracellular aggregates and this aggregation usually protects from proteolysis (47, 93). Overexpression of even the classical substrate for the E. coli abnormal protein-degrading system, the lacZ X90 fragment, protects it from proteolytic degradation (14). It has been suggested that the stabilization to proteolysis obtained by fusing the foreign gene to an E. coli sequence (see, e.g., reference 45) may result from the increased tendency of the fusion protein to aggregate (107). The formation of insoluble aggregates of abnormal proteins was first observed in canavanine-treated cells (106).

Undesired proteolytic cleavage can also occur in vitro, plaguing the purification of proteins that are susceptible to proteolysis. Approaches to avoiding such artifacts have been thoughtfully reviewed by Pringle (100). There is apparently no guaranteed method, but as our knowledge of the properties of the enzymes that are involved increases (see Table 1) it should be possible to make more intelligent use of inhibitors and to separate the proteases more quickly from their potential substrates.

PROTEOLYTIC MODIFICATION

Some proteolytic processes have as their function not the degradation of unwanted proteins to free amino acids but rather the production of proteins with altered properties. The types of proteolytic modification known to occur in E. coli include removal of N-terminal methionine from certain newly synthesized proteins and removal of signal sequences from proteins that are transported. The inactivation (and possibly activation) of regulatory molecules by proteolysis could also be considered as an example of proteolytic modification (see chapter 79). I know of no examples of a zymogen → enzyme conversion in E. coli or S. typhimurium.

Removal of N-Terminal Methionine

Although all translation is thought to initiate with N-formylmethionine (fMet), not all proteins contain either fMet or Met at their N termini. Early studies of the N-terminal amino acids of E. coli proteins showed that the distribution of N termini in the bulk protein of E. coli is highly nonrandom (131), with Ala, Ser, and

Thr comprising most of the non-Met N-terminal amino acids. A survey of a number of microorganisms showed that there is considerable variability in the amount of N-terminal Met, but again Ala, Ser, and Thr were the most common N-terminal amino acids aside from methionine (105). Since these experiments involved bulk protein, the observed N-terminal distribution could have been heavily influenced by a few protein species at high concentration. It is now possible, however, to compare amino acid sequences with DNA sequences for several genes. A survey of the available data involving genes and proteins from E. coli and other organisms has supported earlier proposals that the nature of the second amino acid determines whether or not N-terminal methionine is removed (126): Ala, Ser, Gly, Pro, Thr, or Val as second amino acids permit N-terminal Met to be removed, whereas Arg, Asn, Asp, Gln, Glu, Ile, Leu, Lys, and Met do not. These rules seem to apply in all organisms for which data are available (126).

The stepwise nature of N-terminal modification was demonstrated by Adams (1), who showed that E. coli extracts contain a formylase that deformylates fMet peptides. No activity capable of cleaving fMet from the N termini of peptides was found. Adams concluded that N-terminal modification requires the action of a Met-specific aminopeptidase on the deformylated product of the formylase. Attempts to identify this peptidase have usually involved the use of small N-terminal Met peptides to search for hydrolase activity in cell extracts. The presence in these extracts of broad-specificity peptidases that hydrolyze these peptides makes this approach difficult. Vogt (130) purified and characterized an E. coli aminopeptidase that rapidly hydrolyzes Met-Ala-Ser. This enzyme attacks many peptides with N-terminal amino acids other than methionine, however, and it was able to proceed beyond the N-terminal amino acid, cleaving its substrates completely to free amino acids. Vogt concluded, therefore, that it was not likely to be involved in N-terminal Met removal. In addition, mutants (pepA) lacking this enzyme (76, 77) are completely viable. Earlier reports of an aminopeptidase associated with ribosomes (124) probably result from contamination of ribosomal preparations with aggregates of this peptidase which form under low-salt conditions (130). Another report of a Met-specific enzyme (7) describes an activity that could not be shown to hydrolyze substrates larger than dipeptides.

The observation that a mutant S. typhimurium strain lacking four broad-specificity aminopeptidases (peptidases N, A, B, and T) uses some N-terminal Met tripeptides but cannot grow on a variety of other peptides suggested that these mutants may contain a Met-specific peptidase (115). The presence of such a peptidase, peptidase M, has been observed in extracts of these mutants (C. Miller, et al., submitted for publication). This activity removes Met from tripeptides with Ala, Ser, Thr, or Gly as the second amino acid but not with Leu or Met in this position. Mutants that overproduce (approximately 30-fold) this activity have been isolated. This enzyme is a strong candidate for the long-postulated activity involved in N-terminal maturation in the cell.

The function of N-terminal Met removal is not understood. I know of only one E. coli enzyme that requires N-terminal maturation for activity. Zalkin and co-workers (125) have shown that glutamine phosphoribosylpyrophosphate amidotransferase, the first enzyme of de novo purine synthesis, requires an amino-terminal Cys residue for its glutamine amide transfer function. This residue is generated by removal of a single N-terminal methionine from a precursor protein. Other N-terminal modifications of unknown function occur after removal of Met from a precursor. Elongation factor Tu, for example, undergoes acetylation of N-terminal Ser after removal of the original N-terminal Met (3). Obviously not all N-terminal modifications require prior Met removal since some mature N termini are generated by signal peptidases.

Removal of Signal Sequences

Most secreted proteins are synthesized with an N-terminal signal peptide that is removed by proteolytic cleavage before the protein attains its mature form. The significance of this leader peptide and its removal are treated elsewhere in this volume. In this section I will discuss only the properties of the enzymes known to be involved in signal sequence removal and what is known about the degradation of the signal peptide that is produced in this reaction.

Two different signal peptidases from *E. coli* have been characterized. One of these, signal peptidase I (Table 1), was studied by Wickner and co-workers (137, 148). This activity can remove signal sequences from a variety of secreted proteins, but does not hydrolyze standard protease substrates. Peptide analogs of the cleavage site are not effective inhibitors of signal peptidase I (138). This observation and the genetic analysis of cleavage requirements (50) suggest that the signal peptidase recognizes a complex structure formed by the signal peptide. Another signal peptidase specifically removes signal peptides from prolipoproteins (23, 121, 122). This enzyme, signal peptidase II (Table 1), requires that its substrates be glyceride modified (123). Genes specifying both of these enzymes have been cloned, and their nucleotide sequences have been determined (44, 137).

Signal peptides, once removed, do not accumulate but are rapidly degraded. This degradation can be observed in an isolated envelope fraction from cells which have accumulated high levels of unprocessed lipoprotein precursor (42). The addition of protease inhibitors (antipain, leupeptin, chymostatin, and elastatinal) blocks degradation and allows the intact signal peptide to accumulate. Degradation of signal peptides occurs only after they are cleaved from the precursor. Ichihara et al. (43) have presented evidence that a single activity (signal peptide peptidase) is responsible for this degradation. Based on copurification and inhibitor specificity, it appears that signal peptide peptidase is the membrane-associated protease IV (Table 1; 87, 88). There are clearly other cytoplasmic enzymes that degrade signal peptides in vitro (132). Two activities that degrade the lipoprotein signal peptide have been characterized (84). About 90% of the signal peptide-degrading activity in the extract is oligopeptidase A (Table 1; 127). The remainder of the activity is protease Do (Table 1; 18). It is possible that protease IV initiates degradation of the peptide by making an endoproteolytic cut that releas-

es fragments of the signal peptide into the cytoplasm, where soluble enzymes complete the process. Alternatively, each of these enzymes may alone be able to degrade the signal peptide.

Signal peptide degradation may offer the possibility of studying steps in the degradation pathway with structurally defined substrates. It seems likely that at least the soluble peptidases involved in signal peptide degradation are also used in other degradation pathways and that an understanding of how they act on signal peptides and their fragments may illuminate these other functions.

PHYSIOLOGICAL SIGNIFICANCE OF INTRACELLULAR PROTEOLYSIS

Extensive proteolysis occurs in normally growing bacterial cells, not just in nongrowing, starved cells. Estimates based on peptide accumulation in peptidase-deficient strains suggest that approximately 20% of the peptide bonds synthesized are rapidly hydrolyzed (143). The inputs into this rapidly degraded pool have not been identified, but it is likely that no one process is responsible. Clearly, signal peptides cleaved from exported proteins and attenuator peptides produced by translation of regulatory sequences (139) are rapidly degraded and must contribute to this pool. An especially interesting possibility is that there is a significant contribution from polypeptides that are produced by various kinds of translational errors. The premature release of peptidyl tRNA during translation is well documented (68). This release may be part of a translational proofreading mechanism (68) or simply accidental. Other kinds of translational errors might be generated from out-of-phase initiation of translation, translational frameshifting, readthrough of termination signals, or premature termination of translation. This last possibility has been given particular support by Manley's observation (62) that approximately 30% of the translational initiation events of *lacZ* lead to incomplete products. A strong case can be made that selective forces have produced a translational apparatus with less than maximum accuracy (51). It is possible that a major role of the elaborate protein degradation system is to make sure that the products of various aberrant translational events are quickly and efficiently disposed of.

The importance of proteolysis in starving or downshifted cells has been reinforced by observations on mutants unable to complete the protein degradation process. When downshifted, multiply peptidase-deficient mutants show greatly increased lag times relative to those of their peptidase-containing parents (142). Studies of the long-term viability during carbon starvation of cells lacking peptidases show that these mutants lose viability approximately three times faster than their wild-type parent (102). These observations are consistent with Mandelstam's ideas (60, 61) concerning the role of protein turnover in nongrowing cells.

CONCLUSION AND OUTLOOK

The bacterial cell contains a complex apparatus that allows the complete breakdown of many different types of polypeptides. This breakdown is important in both growing and nongrowing cells. The components of the system are beginning to be recognized, but many questions remain unanswered. We still do not know the roles of individual enzymes in the early steps of the breakdown pathway. The *lon* protease is clearly important in the degradation of abnormal proteins, but the enzymes involved in other breakdown processes are only beginning to be defined. The role of metabolic energy in proteolysis is not clear. The key question of selectivity—what determines whether or not a particular polypeptide will be degraded—cannot be answered. Also, we do not know where the substrates for the rapid degradation occurring in growing cells are coming from. Protein degradation is clearly an area of procaryotic molecular biology of general biological significance in which much remains to be learned.

ACKNOWLEDGMENT

My work was supported by a grant (AI 10333) from the National Institute for Allergy and Infectious Diseases.

LITERATURE CITED

1. Adams, J. M. 1968. On the release of the formyl group from nascent protein. J. Mol. Biol. 33:571–589.
2. Ames, B. N., G. F.-L. Ames, J. D. Young, D. Tsuchiya, and J. Lecocq. 1973. Illicit transport: the oligopeptide permease. Proc. Natl. Acad. Sci. USA 70:456–458.
3. Arai, K., B. F. C. Clark, L. Duffy, M. D. Jones, Y. Kaziro, R. A. Laursen, J. L'Italien, D. L. Miller, S. Nagarkatti, S. Nakamura, K. M. Nielsen, T. E. Petersen, K. Takahashi, and M. Wade. 1980. Primary structure of elongation factor Tu from *Escherichia coli*. Proc. Natl. Acad. Sci. USA 77:1326–1330.
4. Baker, T., A. D. Grossman, and C. A. Gross. 1984. A gene regulating the heat shock response in *Escherichia coli* also affects proteolysis. Proc. Natl. Acad. Sci. USA 81:6779–6783.
5. Berquist, P. L., and P. Truman. 1978. Degradation of missense mutant β-galactosidase proteins in *Escherichia coli* K-12. Mol. Gen. Genet. 164:105–108.
6. Bowles, L. K., and J. Konisky. 1981. Cleavage of colicin Ia by the *Escherichia coli* K-12 outer membrane is not mediated by the colicin Ia receptor. J. Bacteriol. 145:668–671.
7. Brown, J. L. 1973. Purification and properties of dipeptidase M from *Escherichia coli* B. J. Biol. Chem. 248:409–416.
8. Buell, G., M.-F. Schulz, G. Selzer, A. Chollet, N. R. Movva, D. Semon, S. Escanez, and E. Kawashima. 1985. Optimizing the expression in *E. coli* of a synthetic gene encoding somatomedin-C (IGF-I). Nucleic Acids Res. 13:1923–1938.
9. Bukhari, A. I., and D. Zipser. 1973. Mutants of *Escherichia coli* with a defect in the degradation of nonsense fragments. Nature (London) New Biol. 243:238–241.
10. Carter, T. H., and C. G. Miller. 1984. Aspartate-specific peptidases in *Salmonella typhimurium*: mutants deficient in peptidase E. J. Bacteriol. 159:453–459.
11. Celis, J. E., J. D. Smith, and S. Brenner. 1973. Correlation between genetic and translational maps of gene 23 in bacteriophage T4. Nature (London) New Biol. 241:130–132.
12. Chang, C. N., H. Inouye, P. Model, and J. Beckwith. 1980. Processing of alkaline phosphatase precursor to the mature enzyme by an *Escherichia coli* inner membrane preparation. J. Bacteriol. 142:726–728.
13. Charette, M. F., G. W. Henderson, and A. Markovitz. 1981. ATP hydrolysis-dependent protease activity of the *lon* (*capR*) protein of *Escherichia coli* K12. Proc. Natl. Acad. Sci. USA 78:4728–4732.
14. Cheng, Y.-S. E., D. Y. Kwoh, T. J. Kwoh, B. C. Saltvedt, and D. Zipser. 1981. Stabilization of a degradable protein by its overexpression in *Escherichia coli*. Gene 14:121–130.
15. Cheng, Y.-S. E., and D. Zipser. 1979. Purification and characterization of protease III from *Escherichia coli*. J. Biol. Chem. 254:4698–4706.
16. Cheng, Y.-S. E., D. Zipser, C.-Y. Cheng, and S. J. Rolseth. 1979. Isolation and characterization of mutations in the structural gene for protease III (*ptr*). J. Bacteriol. 140:125–130.
17. Chung, C. H., and A. L. Goldberg. 1982. DNA stimulates ATP-dependent proteolysis and protein-dependent ATPase activity of protease La from *Escherichia coli*. Proc. Natl. Acad. Sci. USA 79:795–799.
18. Chung, C. H., and A. L. Goldberg. 1983. Purification and characterization of protease So, a cytoplasmic serine protease in *Escherichia coli*. J. Bacteriol. 154:231–238.
19. Chung, H. A., and A. L. Goldberg. 1981. The product of the *lon* (*capR*) gene in *Escherichia coli* is the ATP-dependent protease, protease La. Proc. Natl. Acad. Sci. USA 78:4931–4935.
20. Ciechanover, A., D. Finley, and A. Varshavsky. 1984. Ubiquitin dependence of selective protein degradation demonstrated in the mammalian cell cycle mutant ts85. Cell 37:57–66.
21. Dennis, P. P. 1974. Synthesis and stability of individual ribosomal proteins in the presence of rifampicin. Mol. Gen. Genet. 134:39–47.
22. Deutch, C. E., and R. L. Soffer. 1978. *Escherichia coli* mutants defective in dipeptidyl carboxypeptidase. Proc. Natl. Acad. Sci. USA 75:5998–6001.
23. Dev, I. K., and P. H. Ray. 1984. Rapid assay and purification of a unique signal peptidase that processes the prolipoprotein from *Escherichia coli* B. J. Biol. Chem. 259:11114–11120.
24. Emerick, A. W., B. L. Bertolani, A. Ben-Bassat, T. J. White, and M. W. Konrad. 1984. Expression of a β-lactamase preproinsulin fusion protein in *Escherichia coli*. Biotechnology 2:165–168.
25. Fickel, T. E., and C. Gilvarg. 1973. Transport of impermeant substances in *E. coli* by way of oligopeptide permease. Nature (London) New Biol. 241:161–163.
26. Finley, D., A. Ciechanover, and A. Varshavsky. 1984. Thermolability of ubiquitin activating enzyme from the mammalian cell cycle mutant ts85. Cell 37:43–55.
27. Fiss, E. H., P. Stanley-Samuelson, and J. B. Neilands. 1982. Properties and proteolysis of ferric enterobactin outer membrane receptor in *Escherichia coli* K12. Biochemistry 21:4517–4522.
28. Gallant, J. A. 1979. Stringent control in *E. coli*. Annu. Rev. Genet. 13:393–415.
29. Gibson, M. M., M. Price, and C. F. Higgins. 1984. Genetic characterization and molecular cloning of the tripeptide permease (*tpp*) genes of *Salmonella typhimurium*. J. Bacteriol. 160:122–130.
30. Goff, S. A., L. P. Casson, and A. L. Goldberg. 1984. Heat shock regulatory gene *htpR* influences rates of protein degradation and expression of the *lon* gene in *Escherichia coli*. Proc. Natl. Acad. Sci. USA 81:6647–6651.
31. Goff, S. A., and A. L. Goldberg. 1985. Production of abnormal proteins in *E. coli* stimulates transcription of *lon* and other heat shock genes. Cell 41:587–595.
32. Goldberg, A. L. 1972. Degradation of abnormal proteins in *Escherichia coli*. Proc. Natl. Acad. Sci. USA 69:422–426.
33. Goldberg, A. L., and J. F. Dice. 1974. Intracellular protein degradation in mammalian and bacterial cells. Annu. Rev. Biochem. 43:835–869.
34. Goldberg, A. L., and A. C. St. John. 1976. Intracellular protein degradation in mammalian and bacterial cells: part 2. Annu. Rev. Biochem. 45:747–803.
35. Goldberg, A. L., K. H. S. Swamy, C. H. Chung, and F. Larrimore. 1982. Proteases in *Escherichia coli*. Methods Enzymol. 80:680–702.
36. Goldberg, A. L., and L. Waxman. 1985. The role of ATP hydrolysis in the breakdown of proteins and peptides by protease La from *Escherichia coli*. J. Biol. Chem. 260:12029–12034.
37. Goldschmidt, R. 1970. *In vivo* degradation of nonsense fragments in *E. coli*. Nature (London) 228:1151–1154.
38. Heiman, C., and C. G. Miller. 1978. *Salmonella typhimurium* mutants lacking protease II. J. Bacteriol. 135:588–594.
39. Heiman, C., and C. G. Miller. 1978. Acylaminoacid esterase mutants of *Salmonella typhimurium*. Mol. Gen. Genet. 164:57–62.
40. Hershko, A., and A. Ciechanover. 1982. Mechanisms of intracellular protein breakdown. Annu. Rev. Biochem. 51:335–364.
41. Hogness, D. S., M. Cohn, and J. Monod. 1955. Studies on the induced synthesis of β-galactosidase in *Escherichia coli*: the kinetics and mechanism of sulfur incorporation. Biochim. Biophys. Acta 16:99–116.
42. Hussain, M., Y. Ozawa, S. Ichihara, and S. Mizushima. 1982. Signal peptide digestion in *Escherichia coli*. Eur. J. Biochem. 129:233–239.
43. Ichihara, S., N. Beppu, and S. Mizushima. 1984. Protease IV, a cytoplasmic membrane protein of *Escherichia coli*, has signal peptide peptidase activity. J. Biol. Chem. 259:9853–9857.
44. Innis, M. A., M. Tokunaga, M. E. Williams, J. M. Loranger, S.-Y. Chang, S. Chang, and H. C. Wu. 1984. Nucleotide sequence of the *Escherichia coli* prolipoprotein signal peptidase (*lsp*) gene. Proc. Natl. Acad. Sci. USA 81:3708–3712.

45. **Itakura, K., T. Hirose, R. Crea, A. D. Riggs, H. L. Heyneker, F. Bolivar, and H. W. Boyer.** 1977. Expression in *Escherichia coli* of a chemically synthesized gene for the hormone somatostatin. Science **198**:1056–1063.

46. **Kemshead, J. T., and A. R. Hipkiss.** 1976. Degradation of abnormal proteins in *Escherichia coli*: differential proteolysis *in vitro* of *E. coli* alkaline phosphatase cyanogen bromide cleavage products. Eur. J. Biochem. **71**:185–192.

47. **Kleid, D. G., D. Yansura, B. Small, D. Dowbenko, D. M. Moore, M. J. Grubman, P. D. McKercher, D. O. Morgan, B. H. Robertson, and H. L. Bachrach.** 1981. Cloned viral protein vaccine for foot and mouth disease: responses in cattle and swine. Science **214**:1125–1129.

48. **Kowit, J. D., W.-N. Choy, S. P. Champe, and A. L. Goldberg.** 1976. Role and location of protease I from *Escherichia coli*. J. Bacteriol. **128**:776–784.

49. **Kowit, J. D., and A. L. Goldberg.** 1977. Intermediate steps in the degradation of a specific abnormal protein in *Escherichia coli*. J. Biol. Chem. **252**:8350–8357.

50. **Kuhn, A., and W. Wickner.** 1985. Conserved residues of the leader peptide are essential for cleavage by leader peptidase. J. Biol. Chem. **260**:15914–15918.

51. **Kurland, C. G., and M. Ehrenberg.** 1984. Optimization of translation accuracy. Prog. Nucleic Acid Res. Mol. Biol. **31**:191–219.

52. **Landick, R. C., C. J. Daniels, and D. L. Oxender.** 1983. Influence of membrane potential on the insertion and transport of proteins in bacterial membranes. Methods Enzymol. **97**:146–153.

53. **Larrabee, K. L., J. O. Phillips, G. J. Williams, and A. R. Larrabee.** 1980. The relative rates of protein synthesis and degradation in a growing culture of *Escherichia coli*. J. Biol. Chem. **225**:4125–4130.

54. **Leytus, S. P., L. K. Bowles, J. Konisky, and W. R. Mangel.** 1981. Activation of plasminogen to plasmin by a protease associated with the outer membrane of *Escherichia coli*. Proc. Natl. Acad. Sci. USA **78**:1485–1489.

55. **Lifson, E. R., L. Lindahl, and J. M. Zengel.** 1986. Precursor for elongation factor Tu from *Escherichia coli*. J. Bacteriol. **165**:474–482.

56. **Lin, S., and I. Zabin.** 1972. β-Galactosidase: rates of synthesis and degradation of incomplete chains. J. Biol. Chem. **247**:2205–2211.

57. **MacGregor, C. H.** 1975. Solubilization of *Escherichia coli* nitrate reductase by a membrane-bound protease. J. Bacteriol. **137**:574–583.

58. **MacGregor, C. H., C. W. Bishop, and J. E. Blech.** 1979. Localization of proteolytic activity in the outer membrane of *Escherichia coli*. J. Bacteriol. **137**:574–583.

59. **Mandelstam, J.** 1958. Turnover of protein in growing and nongrowing populations of *Escherichia coli*. Biochem. J. **169**:110–119.

60. **Mandelstam, J.** 1960. The intracellular turnover of protein and nucleic acids and its role in biochemical differentiation. Bacteriol. Rev. **24**:289–308.

61. **Mandelstam, J.** 1963. Protein turnover and its function in the economy of the cell. Ann. N.Y. Acad. Sci. **102**:621–636.

62. **Manley, J. L.** 1978. Synthesis and degradation of termination and premature-termination fragments of β-galactosidase *in vitro* and *in vivo*. J. Mol. Biol. **125**:407–432.

63. **Maurizi, M. R., P. Trisler, and S. Gottesman.** 1985. Insertional mutagenesis of the *lon* gene in *Escherichia coli*: *lon* is dispensable. J. Bacteriol. **164**:1124–1135.

64. **McCamen, M. T., and M. R. Villarejo.** 1982. Structural and catalytic properties of peptidase N from *Escherichia coli* K-12. Arch. Biochem. Biophys. **213**:384–394.

65. **McHugh, G. L., and C. G. Miller.** 1974. Isolation and characterization of proline peptidase mutants of *Salmonella typhimurium*. J. Bacteriol. **120**:364–371.

66. **McKnight, J. L., and V. A. Fried.** 1981. Limited proteolysis: early steps in the processing of large premature termination fragments of β-galactosidase in *Escherichia coli*. J. Biol. Chem. **256**:9652–9661.

67. **McKnight, J. L., and V. A. Fried.** 1983. A novel proteolytic activity apparently initiating degradation of β-galactosidase nonsense fragments in *in vitro* extracts of *Escherichia coli*. J. Biol. Chem. **258**:7550–7555.

68. **Menninger, J. R.** 1977. Ribosome editing and the error catastrophe hypothesis of cellular ageing. Mech. Ageing Devel. **6**:131–142.

69. **Michels, C. A., and D. Zipser.** 1969. Mapping of polypeptide reinitiation sites within β-galactosidase structural gene. J. Mol. Biol. **41**:341–347.

70. **Miller, C. G.** 1975. Genetic mapping of *Salmonella typhimurium* peptidase mutations. J. Bacteriol. **122**:171–176.

71. **Miller, C. G.** 1975. Peptidases and proteases of *Escherichia coli* and *Salmonella typhimurium*. Annu. Rev. Microbiol. **219**:485–504.

72. **Miller, C. G.** 1985. Genetics and physiological roles of *Salmonella typhimurium* peptidases, p. 346–349. *In* L. Leive (ed.), Microbiology—1985. American Society for Microbiology, Washington, D.C.

73. **Miller, C. G., and L. Green.** 1981. Degradation of abnormal proteins in peptidase-deficient mutants of *Salmonella typhimurium*. J. Bacteriol. **147**:925–930.

74. **Miller, C. G., and L. Green.** 1983. Degradation of proline peptides in peptidase-deficient strains of *Salmonella typhimurium*. J. Bacteriol. **153**:350–356.

75. **Miller, C. G., C. Heiman, and C. Yen.** 1976. Mutants of *Salmonella typhimurium* deficient in an endoprotease. J. Bacteriol. **127**:490–497.

76. **Miller, C. G., and K. MacKinnon.** 1974. Peptidase mutants of *Salmonella typhimurium*. J. Bacteriol. **120**:355–363.

77. **Miller, C. G., and G. Schwartz.** 1978. Peptidase-deficient mutants of *Escherichia coli*. J. Bacteriol. **135**:603–611.

78. **Mizusawa, S., and S. Gottesman.** 1983. Protein degradation in *Escherichia coli*: the *lon* gene controls the stability of *sulA* protein. Proc. Natl. Acad. Sci. USA **80**:358–362.

79. **Morona, R., and P. Reeves.** 1984. Detection of several diisopropylfluorophosphate-binding proteins in the outer membrane of *Escherichia coli* K12. FEMS Microbiol. Lett. **23**:179–182.

80. **Mosteller, R. D., R. V. Goldstein, and K. R. Nishimoto.** 1980. Metabolism of individual proteins in exponentially growing *Escherichia coli*. J. Biol. Chem. **255**:2524–2532.

81. **Mount, D. W.** 1980. The genetics of protein degradation in bacteria. Annu. Rev. Genet. **14**:279–319.

82. **Nath, K., and A. I. Koch.** 1970. Protein degradation in *Escherichia coli*. I. Measurement of rapidly and slowly degrading components. J. Biol. Chem. **245**:2889–2900.

83. **Neidhardt, F. C., R. A. VanBogelen, and V. Vaughn.** 1984. Genetics and regulation of heat-shock proteins. Annu. Rev. Genet. **18**:295–329.

84. **Novak, P., P. H. Ray, and I. K. Dev.** 1986. Localization and purification of two enzymes from *Escherichia coli* capable of hydrolyzing a signal peptide. J. Biol. Chem. **261**:420–427.

85. **O'Farrell, P. H.** 1975. High resolution two-dimensional electrophoresis of proteins. J. Biol. Chem. **250**:4007–4021.

86. **Olden, K., and A. L. Goldberg.** 1978. Studies of the energy requirement for intracellular protein degradation in *Escherichia coli*. Biochim. Biophys. Acta **542**:385–398.

87. **Pacaud, M.** 1982. Identification and localization of two membrane-bound esterases from *Escherichia coli*. J. Bacteriol. **149**:6–14.

88. **Pacaud, M.** 1982. Purification and characterization of two novel proteolytic enzymes in membranes of *Escherichia coli*: protease IV and protease V. J. Biol. Chem. **257**:4333–4339.

89. **Pacaud, M., and C. Richaud.** 1975. Protease II from *Escherichia coli*. Purification and characterization. J. Biol. Chem. **250**:7771–7779.

90. **Pacaud, M., L. Sibilli, and G. LeBras.** 1976. Protease I from *Escherichia coli*. Some physicochemical properties and substrate specificity. Eur. J. Biochem. **69**:141–151.

91. **Pacaud, M., and J. Uriel.** 1971. Isolation and some properties of a proteolytic enzyme from *Escherichia coli* (protease I). Eur. J. Biochem. **23**:435–442.

92. **Parker, J.** 1981. Mistranslated protein in *Escherichia coli*. J. Biol. Chem. **256**:9770–9773.

93. **Paul, D. C., R. M. VanFrank, W. L. Muth, J. W. Ross, and D. C. Williams.** 1983. Immunocytochemical demonstration of human proinsulin chimeric polypeptide within cytoplasmic inclusion bodies of *Escherichia coli*. Eur. J. Cell Biol. **31**:171–174.

94. **Payne, J. W.** 1975. Microbial peptidohydrolases, p. 282–364. *In* D. M. Matthews and J. W. Payne (ed.), Peptide transport in protein nutrition. North-Holland Publishing Co., Amsterdam.

95. **Phillips, T. A., R. A. VanBogelen, and F. C. Neidhardt.** 1984. *lon* gene product of *Escherichia coli* is a heat-shock protein. J. Bacteriol. **159**:283–287.

96. **Pine, M. J.** 1967. Response of intracellular proteolysis to alteration of bacterial protein and the implications in metabolic regulation. J. Bacteriol. **93**:1527–1533.

97. **Pine, M. J.** 1970. Steady-state measurement of the turnover of amino acid in the cellular protein of growing *Escherichia coli*: existence of two kinetically distinct reactions. J. Bacteriol. **103**:207–215.

98. **Pine, M. J.** 1972. Turnover of intracellular proteins. Annu. Rev. Microbiol. **26**:103–126.

99. **Pine, M. J.** 1973. Regulation of intracellular proteolysis in *Escherichia coli*. J. Bacteriol. **115**:107–116.

100. **Pringle, J. R.** 1975. Methods for avoiding proteolytic artifacts in studies of enzymes and other proteins from yeasts. Methods Cell Biol. **12:**149–184.
101. **Rafaeli-Eshkol, D., and A. Hershko.** 1974. Regulation of intracellular protein breakdown in stringent and relaxed strains of *E. coli*. Cell **2:**31–35.
102. **Reeve, C. A., A. T. Backman, and A. Matin.** 1984. Role of protein degradation in the survival of carbon-starved *Escherichia coli* and *Salmonella typhimurium*. J. Bacteriol. **157:**758–763.
103. **Regnier, P.** 1981. The purification of protease IV of *E. coli* and the demonstration that it is a proteolytic enzyme. Biochem. Biophys. Res. Commun. **99:**1369–1376.
104. **Rotman, B., and S. Spiegelman.** 1954. On the origin of the carbon in the induced synthesis [of] β-galactosidase in *Escherichia coli*. J. Bacteriol. **68:**419–429.
105. **Sarimo, S. S., and M. J. Pine.** 1969. Taxonomic comparison of the amino termini of microbial cell proteins. J. Bacteriol. **98:**368–374.
106. **Schachtele, C. F., D. L. Anderson, and P. Rogers.** 1968. Mechanism of canavanine death in *Escherichia coli*. II. Membrane-bound canavanyl-protein and nuclear disruption. J. Mol. Biol. **33:**861–872.
107. **Shen, S.-H.** 1984. Multiple joined genes prevent product degradation in *Escherichia coli*. Proc. Natl. Acad. Sci. USA **81:**4627–4631.
108. **Simmonds, S., K. S. Szeto, and C. G. Fletterick.** 1976. Soluble tri- and dipeptidases in *Escherichia coli* K12. Biochemistry **15:**261–271.
109. **Simon, L. D., B. Randolph, N. Irwin, and G. Binkowski.** 1983. Stabilization of proteins by a bacteriophage T4 gene cloned in *Escherichia coli*. Proc. Natl. Acad. Sci. USA **80:**2059–2062.
110. **Simon, L. D., K. Tomczak, and A. C. St. John.** 1978. Bacteriophages inhibit degradation of abnormal proteins in *E. coli*. Nature (London) **275:**424–428.
111. **St. John, A. C., K. Conklin, E. Rosenthal, and A. L. Goldberg.** 1978. Further evidence for the involvement of charged t-RNA and guanosine tetraphosphate in the control of protein degradation in *Escherichia coli*. J. Biol. Chem. **253:**3945–3951.
112. **St. John, A. C., and A. L. Goldberg.** 1978. Effects of reduced energy production on protein degradation, guanosine tetraphosphate, and RNA synthesis in *Escherichia coli*. J. Biol. Chem. **253:**2705–2711.
113. **St. John, A. C., and A. L. Goldberg.** 1980. Effects of starvation for potassium and other inorganic ions on protein degradation and ribonucleic acid synthesis in *Escherichia coli*. J. Bacteriol. **143:**1223–1233.
114. **St. John, A. C., K. Jakubas, and D. Beim.** 1979. Degradation of proteins in steady-state cultures of *Escherichia coli*. Biochim. Biophys. Acta **586:**537–544.
115. **Strauch, K. L., and C. G. Miller.** 1983. Isolation and characterization of *Salmonella typhimurium* mutants lacking a tripeptidase (peptidase T). J. Bacteriol. **159:**763–771.
116. **Strongin, A. Y., D. I. Gorodetsky, and V. M. Stepanov.** 1979. The study of *Escherichia coli* proteases. Intracellular serine protease of *E. coli*—an analog of *Bacillus* proteases. J. Gen. Microbiol. **110:**443–451.
117. **Sussman, A. J., and C. Gilvarg.** 1969. Protein turnover in amino acid-starved strains of *Escherichia coli* K12 differing in their ribonucleic acid control. J. Biol. Chem. **244:**6304–6306.
118. **Sussman, A. J., and C. Gilvarg.** 1971. Peptide transport and metabolism in bacteria. Annu. Rev. Biochem. **40:**397–408.
119. **Swamy, K. H. S., C. H. Chung, and A. L. Goldberg.** 1983. Isolation and characterization of protease Do from *Escherichia coli*: a large serine protease containing multiple subunits. Arch. Biochim. Biophys. **224:**543–554.
120. **Talmadge, K., and W. Gilbert.** 1982. Cellular location affects protein stability in *Escherichia coli*. Proc. Natl. Acad. Sci. USA **79:**1830–1833.
121. **Tokunaga, M., J. M. Loranger, S.-Y. Chang, M. Regue, S. Chang, and H. C. Wu.** 1985. Identification of prolipoprotein signal peptidase and genomic organization of the *lsp* gene in *Escherichia coli*. J. Biol. Chem. **260:**5610–5616.
122. **Tokunaga, M., J. M. Loranger, P. B. Wolfe, and H. C. Wu.** 1982. Prolipoprotein signal peptidase in *Escherichia coli* is distinct from the M13 precoat protein signal peptidase. J. Biol. Chem. **257:**9922–9925.
123. **Tokunaga, M., H. Tokunaga, and H. C. Wu.** 1982. Post-translational modification and processing of *Escherichia coli* prolipoprotein *in vitro*. Proc. Natl. Acad. Sci. USA **79:**2255–2259.
124. **Tsai, C. S., and A. T. Matheson.** 1965. Purification and properties of a ribosomal peptidase from *Escherichia coli* B. Can. J. Biochem. **43:**1643–1652.
125. **Tso, J. Y., M. A. Hermodson, and H. Zalkin.** 1982. Glutamine phosphoribosylpyrophosphate amidotransferase from cloned *Escherichia coli purF*. J. Biol. Chem. **257:**3532–3536.
126. **Tsunasawa, S., J. W. Stewart, and F. S. Sherman.** 1985. Amino-terminal processing of mutant forms of yeast iso-1-cytochrome c: the specificities of methionine aminopeptidase and acetyl-transferase. J. Biol. Chem. **260:**5382–5391.
127. **Vimr, E. R., L. Green, and C. G. Miller.** 1983. Oligopeptidase-deficient mutants of *Salmonella typhimurium*. J. Bacteriol. **153:**1259–1265.
128. **Vimr, E. R., and C. G. Miller.** 1983. Dipeptidyl carboxypeptidase-deficient mutants of *Salmonella typhimurium*. J. Bacteriol. **153:**1252–1258.
129. **Voellmy, R., and A. L. Goldberg.** 1980. Adenosine-5'-diphosphate-3'-diphosphate (ppGpp) and the regulation of protein breakdown in *Escherichia coli*. J. Biol. Chem. **255:**1008–1014.
130. **Vogt, V. M.** 1970. Purification and properties of an aminopeptidase from *Escherichia coli*. J. Biol. Chem. **245:**4760–4769.
131. **Waller, J.-P.** 1963. The NH$_2$-terminal residues of the proteins from cell-free extracts of *E. coli*. J. Mol. Biol. **7:**483–496.
132. **Watts, C., P. Silver, and W. Wickner.** 1981. Membrane assembly from purified components. II. Assembly of M13 procoat into liposomes reconstituted with purified leader peptidase. Cell **25:**347–353.
133. **Waxman, L., and A. L. Goldberg.** 1982. Protease La from *Escherichia coli* hydrolyzes ATP and proteins in a linked fashion. Proc. Natl. Acad. Sci. USA **79:**4883–4887.
134. **Waxman, L., and A. L. Goldberg.** 1986. Selectivity of intracellular proteolysis: protein substrates activate the ATP-dependent protease (La). Science **232:**500–503.
135. **Willetts, N. S.** 1965. Protein degradation during diauxic growth of *Escherichia coli*. Biochem. Biophys. Res. Commun. **20:**692–696.
136. **Willetts, N. S.** 1967. Intracellular protein breakdown in growing cells of *Escherichia coli*. Biochem. J. **103:**462–466.
137. **Wolfe, P. B., P. Silver, and W. Wickner.** 1982. The isolation of homogeneous leader peptidase from a strain of *Escherichia coli* which overproduces the enzyme. J. Biol. Chem. **257:**7898–7902.
138. **Wolfe, P. B., C. Zwizinski, and W. Wickner.** 1983. Purification and characterization of leader peptidase form *Escherichia coli*. Methods Enzymol. **97:**40–46.
139. **Yanofsky, C., and R. Kolter.** 1982. Attenuation in amino acid biosynthetic operons. Annu. Rev. Genet. **16:**113–134.
140. **Yaron, A.** 1976. Dipeptidyl carboxypeptidase from *Escherichia coli*. Methods Enzymol. **45:**599–610.
141. **Yaron, A., and A. Berger.** 1970. Aminopeptidase-P. Methods Enzymol. **19:**521–534.
142. **Yen, C., L. Green, and C. G. Miller.** 1980. Degradation of intracellular protein in *Salmonella typhimurium* peptidase mutants. J. Mol. Biol. **143:**21–33.
143. **Yen, C., L. Green, and C. G. Miller.** 1980. Peptide accumulation during growth of peptidase deficient mutants. J. Mol. Biol. **143:**35–47.
144. **Young, R. A., and R. W. Davis.** 1983. Efficient isolation of genes using antibody probes. Proc. Natl. Acad. Sci. USA **80:**1194–1198.
145. **Zehnbauer, B. A., E. C. Foley, G. W. Henderson, and A. Markovitz.** 1981. Identification and purification of the *lon*$^+$ (*capR*$^+$) gene product, a DNA-binding protein. Proc. Natl. Acad. Sci. USA **78:**2043–2047.
146. **Zehnbauer, B. A., and A. Markovitz.** 1980. Cloning of gene *lon* (*capR*) of *Escherichia coli* K-12 and identification of polypeptides specified by the cloned deoxyribonucleic acid fragment. J. Bacteriol. **143:**852–863.
147. **Zipser, D., and P. Bhavsar.** 1976. Missense mutations in the *lacZ* gene that result in degradation of β-galactosidase structural protein. J. Bacteriol. **127:**1538–1542.
148. **Zwizinski, C., and W. Wickner.** 1980. Purification and characterization of a leader (signal) peptidase from *Escherichia coli*. J. Biol. Chem. **255:**7973–7977.

45. DNA Restriction and Modification Systems

THOMAS A. BICKLE

Department of Microbiology, Biozentrum, Basel University, CH-4056 Basel, Switzerland

INTRODUCTION

The phenomenon of DNA restriction and modification was discovered in *Escherichia coli* more than three decades ago by Bertani and Weigle (6) in the course of experiments with the phages P2 and λ. They found that phages grown on *E. coli* K-12 plated with equal efficiency on strains K-12 and C, but the converse was not true: phage grown on *E. coli* C plated with high efficiency on *E. coli* C but poorly on *E. coli* K-12. Moreover, a single cycle of growth in *E. coli* C was sufficient to remove the ability to grow well in *E. coli* K-12. The molecular explanation for this Lamarckian effect came from a series of experiments conducted in Arber's laboratory in the 1960s (2, 16; reviewed in references 1 and 3). The DNA of phage grown in *E. coli* C is degraded by a DNA sequence-specific endonuclease present in *E. coli* K-12. When the phage are grown in *E. coli* K-12, the DNA is protected from the endonuclease because a DNA methylase present in *E. coli* K-12 and absent from *E. coli* C methylates the sequences recognized by the endonuclease, rendering them resistant to digestion. Arber coined the term "host-controlled restriction and modification of DNA" to describe this phenomenon. The endonuclease involved is a restriction endonuclease, and the methylase is a modification methylase. The primary function of the modification methylase is to protect the cell's own genome from restriction. Restriction and modification systems (R-M systems) are common throughout the procaryotic world, and this chapter will focus on those systems that have been found in *E. coli*, *Salmonella typhimurium*, and their close relatives. Several comprehensive reviews on restriction and modification have been published recently (7, 17, 36–38, 51). These reviews should be consulted for details of the enzymology of R-M systems and for references to the early literature.

DIFFERENT CLASSES OF R-M SYSTEMS

Once the first restriction enzymes had been purified, it became apparent that they fell into at least two groups which Boyer (11) called types I and II. The type I enzymes had complicated cofactor requirements and needed Mg^{2+}, ATP, and S-adenosylmethionine (AdoMet) to cut DNA. The type II enzymes had much simpler requirements and would digest DNA with no cofactor other than Mg^{2+}. Later it became apparent that the type I ATP-requiring enzymes included two fundamentally different kinds of enzyme, and the old type I class was split into types I and III (29). Members of all three classes are found in different strains of *E. coli* and *S. typhimurium*, although strains can be found that have no detectable R-M system.

Several R-M systems have been found that either have not been classified or do not fit the classification given above. These include the LT and SA systems of *S. typhimurium* (14, 15). Bullas et al. (12) have suggested that the distribution of the different R-M systems in the *Salmonella* genospecies allows the different species to be grouped in a manner very similar to that proposed by Borman et al. (9).

The major features of the three classes of R-M systems are shown in Table 1. They will be described in increasing order of complexity.

Type II Systems

The type II R-M systems are enzymatically the simplest. The restriction enzymes in all cases studied are oligomers (generally dimers) of a single subunit and catalyze a single enzymatic reaction: the endonucleolytic cleavage of DNA at a fixed position within (or, for a few enzymes, close to) the DNA sequence that they recognize. The modification methylases are normally monomeric enzymes and catalyze the transfer of methyl groups from AdoMet to specific positions within the recognition sequence. The methylated residue is either N_6-methyladenosine or 5-methylcytosine, depending on the R-M system. It is of interest that many bacteria produce sequence-specific DNA methylases but do not have restriction enzymes recognizing the same sequence (chapter 46). Some of

TABLE 1. Characteristics of R-M systems

Characteristic	System type		
	Type I	Type II	Type III
Subunits for restriction	3	1	2
Subunits for modification	3 or 2	1	1 or 2
Cofactors for restriction[a]	Mg^{2+}, ATP, AdoMet	Mg^{2+}	Mg^{2+}, ATP (AdoMet)
Cofactors for modification[a]	AdoMet (ATP, Mg^{2+})	AdoMet	AdoMet, Mg^{2+}
Other enzymatic activities	ATPase, topoisomerase	None	None
Distance between recognition and cleavage sites	Variable and great	Sites coincide	25 to 30 base pairs

[a] Cofactors listed in parentheses stimulate the reaction but are not absolutely required.

these enzymes might be the relics of R-M systems that have lost their restriction endonucleases by mutation. Type II R-M systems are quite widely distributed in *E. coli*, but none have so far been discovered in *S. typhimurium*. The restriction specificities that have been found to date are shown in Table 2. Most of these were found as the result of a systematic survey by A. Janulaitis.

Type III Systems

The type III enzymes are numerically the smallest class, containing for the moment three members, two of them in *E. coli*. However, it is possible that many more type III R-M systems exist; they are difficult to detect in crude lysates, and no systematic search for them has ever been made. The restriction enzymes contain two different subunits and have an absolute requirement for ATP to cleave DNA. The rate of cleavage is increased by the addition of AdoMet, but under these conditions the enzymes also act as modification methylases, and methylation and cleavage are competing reactions. Once a recognition site has been methylated, it can no longer serve for cleavage. Most unusually for ATP-requiring enzymes, ATP is not detectably hydrolyzed during the cleavage reaction. Although the restriction enzymes are bifunctional and also catalyze DNA modification, type III modification methylases with no restriction activity have also been purified. These are tetramers (D. Hornby, personal communication) of the smaller of the two subunits found in the restriction enzymes (23). Unlike most DNA methylases, the type III methylases require Mg^{2+} for activity.

The two type III R-M systems that have been found in *E. coli* are *Eco*P1, coded by the P1 prophage, and *Eco*P15, coded by a plasmid found in *E. coli* 15T⁻ which has homology with the P1 prophage. These enzymes recognize the asymmetric sequences AGACC (*Eco*P1; 4) and CAGCAG (*Eco*P15; 24) and cut the DNA some 25 base pairs to the right of the sequences to leave short, single-stranded 5' protrusions. A peculiarity of these systems is that modified DNA is methylated in one strand of the DNA only (on adenosyl residues). When the DNA replicates, one daughter DNA molecule inherits the parental methyl group and is fully modified, while the other daughter molecule is unmethylated and ought to be a substrate for restriction. This would of course be lethal, and we do not know how this lethality is avoided. Most other R-M systems have both strands of the DNA methylated in the recognition site; newly replicated DNA thus al-

ways has one strand methylated, and this is sufficient to protect against restriction.

Type I Systems

With one possible exception, *Haemophilus influenzae* (22), type I R-M systems have only been found in *E. coli* and the genus *Salmonella*. These are the most complex of all the R-M systems. Indeed, they are among the most complex of all known enzymes in the number of enzymatic reactions that they catalyze; functioning as endonuclease, methylase, ATPase, and topoisomerase. The enzymes contain three different subunits, and the restriction endonuclease activity is completely dependent on the presence of AdoMet and ATP. AdoMet is only required early in the reaction pathway and acts as an allosteric effector which transforms the enzyme to a form that binds the recognition sites. Restriction requires ATP hydrolysis (shown by experiments with nonhydrolyzable ATP analogs), but the hydrolysis continues long after cleavage has ceased, some 10,000 ATP molecules being hydrolyzed per double-strand cleavage event. The enyzme does not turn over in the cleavage reaction, so that each enzyme molecule only cuts the DNA once. A peculiarity of the type I enzymes is that they cleave the DNA at positions that are random with respect to their recognition sites and can be very far from the sites. Cleavage 7,000 base pairs from the closest site has been detected. They do this by forming looped intermediates in which the enzyme is simultaneously bound to

TABLE 2. Type II restriction enzymes in *E. coli*

Sequence recognized[a]	Enzymes	References
GᴬGCᵀC (G/C, A/C)	*Eco* 24, *Eco* 25, *Eco* 26, *Eco* 35, *Eco* 40, *Eco* 41	Janulaitis[b]
CCᴬGG	*Eco*RII, *Eco* 27, *Eco* 38	Janulaitis; 8
GATATC	*Eco*RV, *Eco* 32	Janulaitis; 31, 47
CTGCAG	*Eco* 36, *Eco* 48, *Eco* 49	Janulaitis
GGNCC	*Eco* 39, *Eco*47II	Janulaitis; 27
GGᴬCC	*Eco*47I	Janulaitis; 27
AGCGCT	*Eco*47III	Janulaitis; 27
GGᵀᴬₐₜCC	*Eco* 50	Janulaitis
GAGCTC	*Eco*ICRI	Janulaitis
GAATTC	*Eco*RI	25
AAGCTT	*Eco*VIII	35

[a] Sequences are written 5' to 3', and only one strand is shown. N signifies that any nucleotide can occupy this position in the sequence.
[b] Cited in reference 30.

TABLE 3. Type I restriction recognition sites

Enzyme	Sequence recognized[a]	References
EcoB	TGA*NNNNNNNNNTGCT ACTNNNNNNNNNACGA*	33, 44, 48
EcoK	AA*CNNNNNNGTGC TTGNNNNNNCACG*	28
EcoD	TTA*NNNNNNNGTCY AATNNNNNNNCAGR*	42
SB	GAG*NNNNNNRTAYG CTCNNNNNNYATRC*	41
SP	AA*CNNNNNNNGTRC TTGNNNNNNNCAYG*	41
SQ	AA*CNNNNNNNRTAYG TTGNNNNNNNYATRC*	40
EcoA	GAG*NNNNNNNGTCA CTCNNNNNNNCAGT*	50
EcoDXXI	TCANNNNNNNATTC AGTNNNNNNNTAAG	43a
EcoR124	GAANNNNNNRTCG CTTNNNNNNYAGC	—[b]
EcoR124/3	GAANNNNNNNRTCG CTTNNNNNNNYAGC	—[b]

[a] N, Any nucleotide; R, either purine, Y, either pyrimidine. The adenosyl residues methylated in the modification reaction are marked (where known) with asterisks.
[b] Price and Bickle, unpublished data.

both the recognition and the cleavage site. The loop is formed by the enzyme "pumping" the DNA past it in a manner analogous to DNA topoisomerases forming supercoiled intermediates using the energy of ATP hydrolysis (17a, 52).

As with the type III systems, a modification methylase devoid of restriction activity can also be isolated. These enzymes contain two of the three subunits of the restriction enzyme and are active in the presence of only AdoMet; ATP does not affect the reaction (34, 49).

The DNA sequences recognized by the type I enzymes are shown in Table 3. These sequences all have the characteristic feature that they are split in two by a spacer of nonspecific sequence 6, 7, or 8 base pairs long, depending on the system. Many of the sequences are degenerate, some of them in two positions, and these enzymes therefore recognize several sequences. This can have some interesting consequences. For example, DNA modified by the SP system is resistant to EcoK because one of the two SP sequences is the EcoK sequence.

The type I R-M enzymes all methylate adenosyl residues. Insofar as they have been determined, these methylatable adenosines are indicated by asterisks in Table 3. They are all situated either 10 or 11 base pairs apart, and this offers a clue as to why the sequences

have nonspecific spacers and how the enzymes "see" the sequences. The methylatable N_6 position of adenosine is in the major groove of B-form DNA. When the type I recognition sequences are cast on a double helix such that the methylatable positions are visible, all of the specific base pairs are in two successive major grooves with the nonspecific spacer sequence tucked away in the intervening minor groove.

GENETICS OF R-M SYSTEMS

Type II Systems

None of the genes for type II R-M systems have been shown to be chromosomal; in the few cases to be investigated, they have been plasmid borne (10, 35, 45). It is likely that all of the genes are carried on plasmids, and the restriction specificity of a strain can therefore be changed by the loss or acquisition of plasmids.

No homology has yet been found between the genes coding for different type II systems. Perhaps surprisingly, no strong homologies are found when the endonuclease and methylase genes of the same system are compared (10, 21, 43). It might be expected that as both the methylase and the endonuclease bind to the same DNA sequence, this would be reflected in some features of their primary structure, but this is not the case. The transcriptional organization of the genes is also different for different R-M systems. The EcoRI genes are arranged in tandem and transcribed in the same direction, whereas the EcoRV genes are transcribed divergently from two promoters that lie between them (10, 21, 43).

Type III Systems

The EcoP1 and EcoP15 R-M systems are coded by two adjacent genes called mod and res. Both genes are transcribed from a promoter located before mod, and there is evidence for a second promoter between the genes which would lead to transcription of the res gene (26). The EcoP1 and EcoP15 genes show a great deal of homology. The two res genes are almost totally homologous; a small region of partial homology is detectable by heteroduplex analysis towards the beginning of the gene, and a few restriction site polymorphisms have been detected. The mod genes have a highly conserved carboxyl-terminal half, while the rest of the genes show no detectable homology by heteroduplex analysis (26). As might be expected from the high degree of homology between the structural genes, mutations in either system can be complemented by healthy alleles from the other.

Type I Systems

The type I R-M systems so far studied fall into three families. Within each family there is homology between the structural genes, and mutations in one member of the family can be complemented by wild-type alleles from another member. Between families there is no homology between the structural genes, and complementation does not occur. I propose that these families be named types Ia, Ib, and Ic. The type Ia family, whose structural genes map at 98.5 min on the *E. coli* chromosome (5), contains the majority of the type I systems so far investigated, including the well-studied *Eco*B and *Eco*K systems. The type Ib family has two known members, *Eco*A and *Eco*E (18a, 39, 48a, 50). The type Ic family comprises the plasmid-coded systems R124, R124/3, and *Eco*DXX1 (19, 43a; C. Price, Ph.D. thesis, University of Newcastle, Newcastle-upon-Tyne, U.K., 1984). All three families of enzymes share the general characteristics described earlier. They are all three subunit enzymes, although the size of the subunits differs between the families and most features of the reaction mechanism are similar.

For the type Ia and Ic families the genetic organization is also the same (18a, 46, 48a; C. Price and T. A. Bickle, unpublished data). The three subunits are encoded by three contiguous genes called *hsdR*, *hsdM*, and *hsdS*. The *hsdR* gene product is only required for restriction and is absent from the two-subunit modification methylase. The *hsdM* gene product is required for modification, and the *hsdS* gene provides system-specific DNA recognition in both the restriction and modification reactions. The gene order is *hsdR*-*hsdM*-*hsdS*, and the genes are transcribed from left to right. The *hsdR* gene has its own promoter, and the *hsdM* and *hsdS* genes are transcribed from a single promoter situated between *hsdR* and *hsdM*.

The *hsdR* and *hsdM* genes of *Eco*K and *Eco*B are almost identical as judged by DNA hybridization and immunochemical methods. However, hybridization probes covering the *hsdS* gene showed much weaker homology (39). A DNA sequence analysis of the *hsdS* genes of *Eco*B, *Eco*K, and another type Ia enzyme, *Eco*D, permitted a precise analysis of the relationship between these genes (20). The genes range in length from 1,335 to 1,425 base pairs and contain only two short homologous regions, one of about 100 base pairs towards the middle of the gene and the other 250 base pairs at the very end of the gene. The rest of the genes are totally nonhomologous, a remarkable result considering that the genes are in a region of the chromosome that is otherwise highly conserved.

Despite the limited homology between the *hsdS* genes of the different type Ia systems, one recombinant between two systems has been found that generated a new sequence specificity. This is the *Salmonella* SQ system, which arose as the result of a recombination between the SB and SP *hsd* genes (13) within the central 100-base-pair conserved region (18). Because the SQ recognition sequence is a hybrid between those of SP and SB (Table 3; 40, 41) the recombination event has reassorted two regions of the protein that bind to the recognition sequence. This kind of recombination may be very important for the evolution of new restriction specificities in bacterial populations. One response of bacteriophages to R-M systems in their hosts is to mutate the recognition sites in their own genomes, making them resistant to restriction (32). It is thus an advantage for the bacteria to be able to change the specificity of their R-M systems from time to time.

ACKNOWLEDGMENTS

I am very grateful to Noreen Murray and Andrjev Piekarowicz for sharing results and to Bruno Suri for his comments on the manuscript.

Work from my laboratory has been supported by grants from the Swiss National Science Foundation.

LITERATURE CITED

1. **Arber, W.** 1968. Host controlled restriction and modification of bacteriophage. Symp. Soc. Gen. Microbiol. **18:**296–314.
2. **Arber, W., and D. Dussoix.** 1962. Host specificity of DNA produced by *Escherichia coli*. I. Host controlled modification of bacteriophage λ. J. Mol. Biol. **5:**18–36.
3. **Arber, W., and S. M. Linn.** 1969. DNA modification and restriction. Annu. Rev. Biochem. **38:**467–500.
4. **Bächi, B., J. Reiser, and V. Pirrotta.** 1979. Methylation and cleavage sequences of the *Eco*P1 restriction-modification enzyme. J. Mol. Biol. **128:**143–163.
5. **Bachmann, B. J.** 1983. Linkage map of *Escherichia coli* K-12, edition 7. Microbiol. Rev. **47:**180–230
6. **Bertani, G., and J. J. Weigle.** 1953. Host-controlled variation in bacterial viruses. J. Bacteriol. **65:**113–121.
7. **Bickle, T. A.** 1982. The ATP-dependent restriction endonucleases, p. 85–108. *In* S. M. Linn and R. J. Roberts (ed.), Nucleases. Cold Spring Harbor Press, Cold Spring Harbor, N.Y.
8. **Bigger, C. H., K. Murray, and N. E. Murray.** 1973. Recognition sequence of a restriction enzyme. Nature (London) New Biol. **244:**7–10.
9. **Borman, E. K., C. A. Stuart, and K. M. Wheeler.** 1945. Taxonomy of the family *Enterobacteriaceae*. J. Bacteriol. **48:**351–367.
10. **Bougueleret, L., M. Schwarzstein, A. Tsugita, and M. Zabeau.** 1984. Characterization of the genes coding for the *Eco*RV restriction and modification system of *Escherichia coli*. Nucleic Acids Res. **12:**3659–3676.
11. **Boyer, H. W.** 1971. DNA restriction and modification in bacteria. Annu. Rev. Microbiol. **25:**153–176.
12. **Bullas, L. R., C. Colson, and B. Neufeld.** 1980. Deoxyribonucleic acid restriction and modification systems in *Salmonella*: chromosomally located systems of different serotypes. J. Bacteriol. **141:**275–292.
13. **Bullas, L. R., C. Colson, and A. Van Pel.** 1976. DNA restriction and modification systems in *Salmonella*. SQ, a new system derived by recombination between the SB system of *Salmonella typhimurium* and the SP system of *Salmonella potsdam*. J. Gen. Microbiol. **95:**166–172.
14. **Colson, C., A. M. Colson, and A. Van Pel.** 1970. Chromosomal location of host specificity in *Salmonella typhimurium*. J. Gen. Microbiol. **60:**265–271.
15. **Colson, C., and A. Van Pel.** 1974. DNA restriction and modification systems in *Salmonella*. I. SA and SB, two *Salmonella typhimurium* systems determined by genes with a chromosomal location comparable to that of the *Escherichia coli* genes. Mol. Gen. Genet. **129:**325–337.
16. **Dussoix, D., and W. Arber.** 1962. Host specificity of DNA produced by *Escherichia coli*. II. Control over acceptance of DNA from infecting phage λ. J. Mol. Biol. **5:**37–49.
17. **Endlich, B., and S. M. Linn.** 1981. Type I restriction enzymes, p. 137–156. *In* P. D. Boyer (ed.), The enzymes, vol. 14A. Academic Press, Inc., New York.
17a.**Endlich, B., and S. Linn.** 1985. The DNA restriction endonuclease of *Escherichia coli* B. I. Studies of the DNA translocation and the ATPase activities. J. Biol. Chem. **260:**5720–5728.
18. **Fuller-Pace, F. V., L. R. Bullas, H. Delius, and N. E. Murray.** 1984. Genetic recombination can generate altered restriction specificity. Proc. Natl. Acad. Sci. USA **81:**6095–6099.
18a.**Fuller-Pace, F. V., G. M. Cowan, and N. E. Murray.** 1985. *Eco*A and *Eco*E: alternatives to the *Eco*K family of type I restriction and modification systems of *Escherichia coli*. J. Mol. Biol. **186:**65–75.
19. **Glover, S. W., K. Firman, G. Watson, C. Price, and S. Donaldson.** 1983. The alternate expression of two restriction and modification systems. Mol. Gen. Genet. **190:**65–69.
20. **Gough, J. A., and N. E. Murray.** 1983. Sequence diversity among related genes for recognition of specific targets in DNA molecules. J. Mol. Biol. **166:**1–19.
21. **Greene, P. J., M. Gupta, H. W. Boyer, W. E. Brown, and J. M. Rosenberg.** 1981. Sequence analysis of the DNA encoding *Eco*RI endonuclease and methylase. J. Biol. Chem. **256:**2143–2153.
22. **Gromkova, R., and S. Goodgal.** 1976. Biological properties of a *Haemophilus influenzae* restriction enzyme, *Hind*I. J. Bacteriol. **127:**848–854.
23. **Hadi, S. M., B. Bächi, S. Iida, and T. A. Bickle.** 1983. DNA restriction-modification enzymes of phage P1 and plasmid p15B: subunit functions and structural homologies. J. Mol. Biol. **165:**19–34.
24. **Hadi, S. M., B. Bächi, J. C. W. Shepherd, R. Yuan, K. Ineichen, and T. A. Bickle.** 1979. DNA recognition and cleavage by the *Eco*P15 restriction endonuclease. J. Mol. Biol. **134:**655–666.

25. **Hedgpeth, J., H. M. Goodman, and H. W. Boyer.** 1972. DNA nucleotide sequence restricted by the RI endonuclease. Proc. Natl. Acad. Sci. USA **69**:3448–3452.

26. **Iida, S., J. Meyer, B. Bächi, M. Carlemalm, S. Schrickel, T. A. Bickle, and W. Arber.** 1983. The DNA restriction-modification genes of phage P1 and plasmid p15B: structure and in vitro transcription. J. Mol. Biol. **165**:1–18.

27. **Janulaitis, A., M. Petrusyte, and V. Butkus.** 1983. Three sequence specific endonucleases from *Escherichia coli* RFL 47. FEBS Lett. **161**:213–216.

28. **Kan, N. C., J. A. Lautenberger, M. H. Edgell, and C. A. Hutchison III.** 1979. The nucleotide sequence recognized by the *Escherichia coli* K12 restriction and modification enzymes. J. Mol. Biol. **130**:191–209.

29. **Kauc, L., and A. Piekarowicz.** 1978. Purification and properties of a new restriction endonuclease from *Haemophilus influenzae* Rf. Eur. J. Biochem. **92**:417–426.

30. **Kessler, C., P. S. Neumaier, and W. Wolff.** 1985. Recognition sequences of restriction endonucleases and methylases—a review. Gene **33**:1–102.

31. **Kholmina, G. V., B. A. Rebentish, Y. S. Skoblov, A. A. Mironov, N. K. Yankovskii, Y. I. Kozlov, L. I. Glatman, A. F. Moroz, and V. G. Debabov.** 1980. Isolation and characterization of a new site specific endonuclease, *Eco*RV. Dokl. Akad. Nauk USSR **253**:495–497.

32. **Krüger, D. H., and T. A. Bickle.** 1983. Bacteriophage survival: multiple mechanisms for avoiding the deoxyribonucleic acid restriction systems of their hosts. Microbiol. Rev. **47**:345–360.

33. **Lautenberger, J. A., N. C. Kan, D. Lackey, S. M. Linn, M. H. Edgell, and C. A. Hutchison III.** 1978. Recognition site of *Escherichia coli* B restriction enzyme on φXsB1 and simian virus 40 DNAs: an interrupted sequence. Proc. Natl. Acad. Sci. USA **75**:2271–2275.

34. **Lautenberger, J. A., and S. M. Linn.** 1972. The deoxyribonucleic acid modification and restriction enzymes of *Escherichia coli* B. I. Purification, subunit structure and catalytic properties of the modification methylase. J. Biol. Chem. **247**:6176–6182.

35. **Mise, K., and K. Nakajima.** 1984. Isolation of restriction enzyme *Eco*VIII, an isoschizomer of *Hin*dIII produced by *Escherichia coli* E158–68. Gene **30**:79–85.

36. **Modrich, P.** 1979. Structures and mechanisms of DNA restriction and modification enzymes. Q. Rev. Biophys. **12**:315–369.

37. **Modrich, P.** 1982. Studies on sequence recognition by type II restriction and modification enzymes. Crit. Rev. Biochem. **13**:287–323.

38. **Modrich, P., and R. J. Roberts.** 1982. Type II restriction and modification enzymes, p. 109–154. *In* S. M. Linn and R. J. Roberts (ed.), Nucleases. Cold Spring Harbor Laboratory, Cold Spring Harbor, N.Y.

39. **Murray, N. E., J. A. Gough, B. Suri, and T. A. Bickle.** 1982. Structural homologies among type I restriction-modification systems. EMBO J. **1**:535–539.

40. **Nagaraja, V., J. C. W. Shepherd, and T. A. Bickle.** 1985. The evolution of DNA sequence specificity: a recombinant restriction enzyme has a hybrid recognition sequence. Nature (London) **316**:371–372.

41. **Nagaraja, V., J. C. W. Shepherd, T. Pripfl, and T. A. Bickle.** 1985. Two type I restriction enzymes from *Salmonella* species: purification and DNA recognition sequences. J. Mol. Biol. **182**:579–585.

42. **Nagaraja, V., M. Stieger, C. Nager, S. M. Hadi, and T. A. Bickle.** 1985. The nucleotide sequence recognized by the *Escherichia coli* D type I restriction and modification enzyme. Nucleic Acids Res. **13**:389–399.

43. **Newman, A. K., R. A. Rubin, S. H. Kim, and P. Modrich.** 1981. DNA sequence of structural genes for *Eco*RI DNA restriction and modification enzymes. J. Biol. Chem. **256**:2131–2139.

43a. **Piekarowicz, A., and J. D. Goguen.** 1986. The DNA sequence recognized by the *Eco*DXXI restriction endonuclease. Eur. J. Biochem. **154**:295–298.

44. **Ravetch, J. V., K. Horiuchi, and N. D. Zinder.** 1978. Nucleotide sequence of the recognition site for the restriction/modification enzyme of *Escherichia coli* B. Proc. Natl. Acad. Sci. USA **75**:2266–2270.

45. **Roulland-Doussoix, D., R. Yoshimori, P. Greene, M. Betlach, H. M. Goodman, and H. W. Boyer.** 1975. R factor-controlled restriction and modification of deoxyribonucleic acid, p. 187–198. *In* D. Schlessinger (ed.), Microbiology—1974. American Society for Microbiology, Washington, D. C.

46. **Sain, B., and N. E. Murray.** 1980. The *hsd* (host specificity) genes of *Escherichia coli* K12. Mol. Gen. Genet. **180**:35–46.

47. **Schildkraut, I., C. D. B. Banner, C. S. Rhodes, and S. Parekh.** 1984. The cleavage site for the restriction endonuclease *Eco*RV is 5'-GAT/ATC-3'. Gene **27**:327–329.

48. **Sommer, R., and H. Schaller.** 1979. Nucleotide sequence of the recognition site of the B-specific restriction modification system in *Escherichia coli*. Mol. Gen. Genet. **168**:331–335.

48a. **Suri, B., and T. A. Bickle.** 1985. *Eco*A, the first member of a new family of type I restriction-modification systems: gene organization and enzymatic activities. J. Mol. Biol. **186**:77–85.

49. **Suri, B., V. Nagaraja, and T. A. Bickle.** 1984. Bacterial DNA modification. Curr. Top. Microbiol. Immunol. **108**:1–9.

50. **Suri, B., J. C. W. Shepherd, and T. A. Bickle.** 1984. The *Eco*A restriction and modification system of *Escherichia coli* 15T⁻: enzyme structure and DNA recognition sequence. EMBO J. **3**:575–579.

51. **Yuan, R.** 1981. Structure and function of multifunctional restriction endonucleases. Annu. Rev. Biochem. **50**:285–315.

52. **Yuan, R., D. L. Hamilton, and J. Burckhardt.** 1980. DNA translocation by the restriction enzyme from *Escherichia coli* K. Cell **20**:237–244.

46. Methylation of DNA

M. G. MARINUS

Department of Pharmacology, University of Massachusetts Medical School, Worcester, Massachusetts 01605

METHYLATED BASES IN DNA

The DNA of *Escherichia coli* contains minor amounts of 6-methyladenine (6-meAde) and 5-methylcytosine (5-meCyt) (11, 29). These are present at about 1.5 mol of 6-meAde per 100 mol of adenine and 0.75 mol of 5-meCyt per 100 mol of cytosine. These modified bases are produced by the action of three enzymes (methylases) which will be designated in this article by the genes that produce them. The *hsd* (host specificity) methylase of *Escherichia coli* K-12 produces less than 1% of the total 6-meAde, and it is part of the K-12 restriction-modification system. This enzyme has been described in detail in chapter 45.

The *dam* (DNA adenine methylation) methylase forms most of the 6-meAde in *E. coli* DNA. This conclusion is based on the loss of more than 99% of the 6-meAde residues in *dam* mutants (32). The nucleotide substrate sequence methylated by the enzyme is 5′-GATC-3′, which is palindromic, and methylation occurs on both strands (10, 13, 19). In *E. coli*, all -GATC-sites are methylated, with the possible exception of a small region of newly synthesized daughter DNA (4, 18, 32, 41). The distribution of -GATC- sites appears to be random (once every 256 base pairs) in bulk *E. coli* DNA. The theoretical frequency of 6-meAde on this basis would be 1.56 mol%, a value close to what is normally found experimentally.

Restriction enzyme *Dpn*I specifically cleaves only methylated -GATC- sequences, whereas *Mbo*I (*Dpn*II) cleaves only the unmethylated sequence (10, 18, 19). These endonucleases are useful, therefore, to determine the extent of -GATC- methylation of a particular DNA together with, for example, *Sau*3A, which cleaves at this sequence regardless of its adenine methylation status.

The *dcm* (DNA cytosine methylation) methylase and the *Eco*RII methylase/endonuclease recognize the same substrate sequence (5′-CC^A_TGG-3′) (5, 34). This sequence should be found once every 512 base pairs, a value in good agreement with that obtained experimentally for *E. coli* K-12. The restriction endonuclease *Eco*RII cleaves only the unmethylated sequence, making it useful for determining the extent of DNA cytosine methylation. With the possible exception of newly synthesized daughter DNA, all *dcm* sites are methylated in *E. coli* (4, 32, 41). No 5-meCyt residues can be detected in a mutant strain deleted for the *dcm* gene (2), showing that the *dcm* gene is responsible for all detectable 5-meCyt.

Methylated bases in DNA, and therefore the enzymes responsible for their formation, are not required for viability of *E. coli*. This conclusion is based upon the existence of a mutant strain which has no detectable 6-meAde or 5-meCyt in DNA due to mutations in the *dam*, *dcm*, and *hsd* genes (31).

Salmonella typhimurium contains a *dam* gene which cross-hybridizes with the *E. coli* coding sequence under high-stringency conditions (6), suggesting that the genes, and hence their products, are quite similar. As expected, -GATC- sites in DNA are modified to contain 6-meAde. *S. typhimurium* DNA is also resistant to *Eco*RII, showing that *dcm* sites are methylated (12).

DNA METHYLATION GENES

The *hsd* genes which specify the modification methylase of *E. coli* map at 99 min on the genetic map (1). The *dcm* gene maps at 43 min and the *dam* gene is located at 74 min. The known DNA methylation genes are thus unlinked to one another. All *dcm* mutations map to a single gene and are recessive. Similarly, all *dam* mutations, except one, are in a single complementation group and are recessive to the wild-type allele. The exceptional mutation is very "leaky," which makes it difficult to assign conclusively within the same complementation group (2, 27).

The *dam* gene has been identified as an 834-base-pair sequence (6) which should produce a product of molecular weight about 31,000. This agrees with the observed value of the purified protein (15). The total amino acid composition of the purified enzyme also agrees with that predicted from the DNA sequence (6).

Further studies using directed mutagenesis should allow identification of the active site and other parameters of substrate specificity and enzymatic activity.

Little is known about how the *dam* gene is regulated. The enzyme activity from a *dam::lacZ* fusion is similar in exponential and stationary-phase cultures and in various mutant strains [*dam*⁺, *dam*, *recA*, *recB*, *recC*, *lexA*, *uvrD*, and *dnaA*(Ts)]. Underproduction or overproduction of the enzyme results in hypermutability which may be disadvantageous to the cell (33). This phenotypic trait shows that wild-type *E. coli* must regulate *dam* gene activity precisely. There are no sites which bear a close resemblance to the −35 and −10 consensus sequences to common *E. coli* promoters (14), which may imply that gene regulation is controlled by a *trans*-acting activator(s). A *dnaA* binding site is present immediately before the beginning of the structural gene, and the *dnaA* protein can prevent transcription of the *dam* gene in vitro (D. Smith, personal communication). In vivo, however, there is no change in β-galactosidase activity in a *dnaA*(Ts) strain carrying a *dam::lacZ* fusion on a plasmid at permissive and nonpermissive temperatures (J. Arraj and M. G. Marinus, unpublished data). There is also no difference in β-galactosidase activity from the above plasmid between a *dnaA*(Ts) strain and a *dnaA*⁺ revertant derived from it. Other *dnaA* alleles should be tested, however, since they might yield a positive result.

The specific activity of the methylase produced from *dam* plasmids also varies with the amount of chromosomal DNA 5′ to the gene. Increasing the upstream sequence from 200 to 300 base pairs increases enzyme activity 5- to 10-fold (Arraj and Marinus, unpublished data). Direct measurements of transcription in vitro and in vivo are planned.

DNA METHYLASES

DNA methylases transfer the methyl group from *S*-adenosyl-L-methionine to specific residues in double-stranded DNA (4). In *E. coli* the substrate is hemimethylated DNA. That is, the parental strand is methylated, and methyl transfer occurs only onto the newly synthesized unmethylated daughter DNA strand (3, 20). The methylation of DNA is rapid; undermethylated DNA could not be detected in the minimum time allowed by the sensitivity of the method. In this minimum time period less than 3% (equivalent to 120 kilobases) of the DNA is replicated (3, 20). Urieli-Shoval et al. (48) found that pulse-labeled DNA was completely methylated as determined by its sensitivity to *Dpn*I. They concluded that methylation does not lag behind the replication fork by more than 3,000 base pairs. The methylation level of Okazaki fragments in DNA ligase-deficient strains was found to be 0.96 mol% compared to 1.4 mol% for bulk DNA, suggesting that methylation occurs rapidly after synthesis (28). On the other hand, Lyons and Schendel (26) found that DNA was half methylated after about 7 min, equivalent to approximately 9% of the generation time. Their assay involved high-pressure liquid chromatography analysis of ³²P-labeled mononucleotides derived from digestion of DNA. The prolonged time for methylation after synthesis of DNA as observed by Lyons and Schendel (26) is surprising. It

would predict that a detectable amount of unmethylated DNA should be observed in bulk DNA isolated from *E. coli* under conditions in which the cells are killed as rapidly as possible. All other data from studies using methylation assays, endonuclease sensitivity, and shifts in density of bromouracil-containing DNA do not support this prediction. At present the bulk of the experimental evidence is in favor of methylation occurring rapidly after synthesis.

The *dam* and *dcm* methylases probably exist as a complex in vivo because attempts at purification of one led to considerable contamination with the other (4). Another problem in purifying these enzymes is their low levels in *E. coli*. The successful purification of the *dam* methylase in quantity relied on the use of a *dcm* mutant as a source of enzyme and recombinant DNA technology to clone the *dam* gene (15). The use of an expression vector (33) allows an easy 200- to 500-fold amplification of enzyme activity in vivo. The *dam* methylase has evolved from an enzyme of academic interest to a commercially available reagent (New England Biolabs).

The enzyme has been purified 3,000-fold to 95% purity and has the following properties (15). It is a single polypeptide chain of molecular weight 31,000, has an $s_{20,w}$ of 2.8S and a Stokes radius of 2.4 nm, and exists in solution as a monomer: *S*-Adenosyl-L-methionine has no effect on enzyme aggregation. The enzyme has a turnover number of 19 methyl transfers per min, and the apparent K_m for -GATC- sites is 3.6 nmol. Double-stranded DNA is a better methyl acceptor than denatured DNA, and there is little difference in the rate of methylation between unmethylated or hemimethylated DNA. The enzyme transfers one methyl group per DNA binding event. The properties of the *dam* methylase clearly identify it as a type II modification enzyme, compared to the type I *hsd* methylase (see chapter 45).

In contrast to the *dam* methylase, little is known about the *dcm* enzyme or its regulation due to its low level in cells. The recent success in cloning the *dcm* gene, however, should open the door to overproduction and purification of the methylase (A. Bhagwat, personal communication). The gene was cloned into pBR322 as an 11-kilobase *Sau*3A partial fragment after selection for plasmids resistant to *Eco*RII digestion. A *dcm* mutant strain harboring the plasmid is resistant to the cytotoxic effect of 5-azacytidine, whereas the same strain containing pBR322 is not. There is no homology of the *dcm* gene with the gene for the *Eco*RII methylase, based on DNA hybridization.

BIOLOGICAL FUNCTIONS FOR 5-meCyt

In discussing possible functions for 5-meCyt, it is worthwhile remembering that in contrast to *E. coli* K-12, *E. coli* B is genetically *dcm*⁻ and that a K-12 strain deleted for the *dcm* gene shows no obvious phenotype. Unmethylated *dcm* sites are substrates for the *Eco*RII restriction endonuclease (11), suggesting that one function may be protection of DNA from group N plasmids which produce these restriction endonucleases. Such plasmids, however, appear to turn on the modification enzyme before the restriction protein after transfer into a naive cell.

Coulondre et al. (7) showed that hotspots for spontaneous amber mutations occurred at *dcm* sites, converting them from -CCAGG- to -CTAGG-. This was explained by proposing that deamination of 5-meCyt yields thymine which will yield an amber mutation if the -TAG- triplet is read in frame. Deamination of cytosine yields uracil, which is removed by a specific repair system (22). Mutagenic hotspots for amber mutations are not found in *dcm⁻* strains, and creation of a *dcm* site results in formation of a mutagenic hotspot. There may be an additional sequence determinant, however, because not all *dcm* sites capable of yielding ambers in the *lacI* gene are hotspots.

Deamination of 5-meCyt in DNA should yield a heteroduplex with a T/G mismatch. Such a mismatch within a *dcm* site can be repaired in genetic crosses by a very short patch (VSP) DNA repair pathway (21). The VSP system is quite distinct from the *dam* methylation-dependent mismatch repair pathway (M. Lieb, personal communication). The VSP system, however, could be a mechanism to prevent the formation of mutations via deamination of 5-meCyt. If so, then inactivation of the VSP system should lead to an increase in the spontaneous mutation frequency in *dcm⁺* but not *dcm⁻* strains. The existence of the VSP system also raises the question as to the repairability of heteroduplexes containing hemimethylated sites as well as methylated C/G base pairs.

Additional functions for 5-meCyt involving regulation of gene activity, recombination, and repair are not well understood and have been reviewed in detail by Marinus (29).

BIOLOGICAL FUNCTIONS FOR 6-meAde

In contrast to *dcm* mutant strains, *dam* mutants have a variety of phenotypic traits. These include: increased spontaneous mutability; increased sensitivity to certain chemicals and UV light; a hyper-recombination phenotype; increased spontaneous induction of lysogenic bacteriophages; inviability of *dam recA, dam recB, dam recC, dam recJ, dam lexA, dam polA,* and *dam ruv* double mutants; increased single-strand breaks in *dam lig* (DNA ligase) bacteria at the nonpermissive temperature; increased transposition and other rearrangements of Tn10; increased precise excision (Tex phenotype) of Tn10; modest derepression of certain genes of the SOS regulon; increased expression of *trpR*; and suppression of some *dam* phenotypes by second-site mutation in *mutS* and *mutL*.

This bewildering array of phenotypic traits suggests that DNA adenine methylation has multiple functions. Those functions for which evidence exists are described below, but there probably are others not yet discovered. Other speculative models have been discussed elsewhere (29).

Regulation of Gene Activity

The expression of the *mom* (modification of Mu) gene of bacteriophage Mu requires *dam* methylation (for review, see reference 16). The level of *mom* RNA is decreased at least 20-fold in *dam⁻* versus *dam⁺* strains (12). The effect of *dam* methylation resides in three -GATC- sites located upstream of the *mom* promoter and structural gene. Deletion of one or more of these sequences abolishes *dam* dependence. This dependence is also lost if the *mom* gene is under control of another promoter. The basis for this unique mechanism of gene control is still not understood.

A *dam* site overlaps with the −10 region of the inward promoter of Tn10 (14), the product of which is a transposase. In *dam* mutants, transposase expression is increased about 10-fold (42). This correlates with the isolation of one class of *E. coli* mutants which exhibit an increased transposition frequency. These are *dam⁻* strains. Transposons Tn5 (J. Yin and W. Reznikoff, personal communication) and Tn903 (T. Weinert and N. Grindley, personal communication) also have transposase expression affected by DNA methylation. In addition to *dam* methylation of the promoter region of transposase, a *dam* site is also present at the inner termini of both IS10 and IS50. Methylation of both sites contributes strongly to reduction of transposition in wild-type *E. coli*.

There are several *E. coli* genes whose expression is increased in *dam* mutants (45). One of these is *trpR*, which codes for the tryptophan operon apo-repressor; its activity is increased two- to threefold (30). The increased activity of the *trpR* promoter in a *dam* strain is likely due to lack of methylation at a -GATC- site in the −35 region. It appears that promoter activity is enhanced on a hemimethylated template which may exist at the replication fork. In this context, the *trpR* gene is between the origin of replication and the *trp* operon, perhaps to ensure that a burst of repressor synthesis occurs before replication of the *trp* operator.

The expression of the *recA, lexA, sulA, uvrA, uvrB,* and *dinF* genes, but not *uvrD*, is increased in *dam⁻* relative to *dam⁺* strains (8, 38). The increase varies from two- to sixfold, depending on the gene, and only some of the genes are fully induced. The genes listed are all part of the SOS regulon (23, 49), indicating that in the absence of DNA methylation *E. coli* adjusts the basal level of SOS expression relative to wild type. This increased expression appears to be required for viability of *dam⁻* cells and is mediated by the inactivation of the SOS repressor (LexA protein). The inducing signal in *dam⁻* cells causing low-level inactivation of LexA protein is not known, but it may be lesions arising from overlapping mismatch repair tracts. No *dam* methylation sites are part of or overlap the promoter regions of the *recA, lexA, uvrA,* or *uvrB* genes (14).

The above data indicate that gene expression can be affected by *dam* methylation either directly by methylation of regulatory sites or indirectly as in the case of the SOS regulon. At present the physiological relevance of these observations is unclear since fully unmethylated DNA does not occur in normal *E. coli*. Hemimethylated DNA may occur at the replication fork and during DNA repair and recombination. Transposition and transcription of IS10 from hemimethylated DNA, in the correct orientation, is as efficient as from unmethylated DNA, whereas in hemimethylated DNA of the opposite orientation, transposition and transcription are inhibited to the same extent as fully methylated DNA (42). If hemimethylated DNA at the replication fork is transcriptionally more active than fully methylated DNA, then this would be a mechanism for the cell to control the production of a protein which is needed in small amounts only once in the cell cycle. The transposases of Tn*10*, Tn*5*, and Tn*903* and the *trpR* gene product would seem to fit this category. It should also be pointed out that no essential genes can be regulated solely via DNA methylation since *dam* mutants are viable.

Mismatch Repair

The most direct and convincing evidence for the involvement of methylation in mismatch repair comes from in vitro-constructed heteroduplexes of phage DNA (24, 39). Heteroduplexes containing a mismatch were constructed with one strand methylated, both strands methylated, or neither strand methylated. If neither strand was methylated, repair occurred on either DNA strand. The unmethylated strand was preferentially repaired in heteroduplexes containing one methylated and one unmethylated strand. Surprisingly, no repair was observed when both strands of the heteroduplex were methylated (24, 39). DNA adenine methylation, therefore, determines strand selectivity for this mismatch repair system. The methylation-dependent mismatch repair may be involved in ensuring genetic fidelity in *E. coli* by repair of replication errors in hemimethylated DNA at the replication fork. The time at which methylation occurs on newly synthesized DNA, however, is still in dispute (see above).

In contrast to the above, Fishel and Kolodner (9) have presented evidence in *E. coli* for the repair of heteroduplexes which are fully methylated, indicating the existence of at least two mechanisms for repair of heteroduplexes. The relationship of the VSP system (see above) to that described by Fishel and Kolodner (9) is not clear; they may be the same. Further information on mismatch repair can be found in a review by Weinstock (chapter 60). The contribution of *dam* methylation to mismatch repair has also been reviewed recently by Radman and Wagner (40) and Marinus (29).

Mismatch repair mutations (*uvrD*, *mutH*, *mutL*, *mutS*, and *dam*) enhance excision of transposon Tn*10* (25). There is, however, no readily apparent model to account for this behavior, although the idea that specific lesions generated in the absence of repair or misrepair somehow enhance excision of transposons seems plausible.

Recent data (43) have shown that the *dam* gene of bacteriophage T4 methylates the same nucleotide sequence as the *E. coli* enzyme. The T4 *dam* gene also complements all *E. coli dam* mutant phenotypes except spontaneous mutagenesis. This indicates that it is methylation of DNA per se which is important. For spontaneous mutagenesis, however, the data may be indicative of a direct role for the *E. coli dam* methylase in the process.

Initiation of DNA Replication

The origin of replication (*oriC*) of the *E. coli* K-12 chromosome contains 10 times more -GATC- sequences than expected on a random basis (35, 46). Methylation of these sites does not appear to be required for initiation of chromosomal replication, since *dam* mutants are viable and the *oriC* region of a *dam-3* strain contains no methylated -GATC- sites (47).

On the other hand, plasmids (minichromosomes) containing the *oriC* region cannot be recovered (36) or are recovered at low frequency (44) after transformation of *dam* mutants. Further efficient replication of such minichromosomes in vitro requires *dam* methylation (36, 44). These data suggest that methylation of *oriC* is required for efficient initiation of chromosomal replication. If so, then why are *dam* mutants viable? Smith et al. (44) suggest that there is residual methylation in *dam* mutants. Such methylation must be confined to the *ori* region. Alternatively, initiation in *dam* mutants may occur at a secondary origin(s) rather than *oriC* or may initiate at *oriC* by an unusual mechanism (36). Initiation is *dnaA* dependent in *dam* strains because a *dam dnaA*(Ts) strain is temperature sensitive for growth (as is its *dam*⁺ parent; Arraj and Marinus, unpublished data). To our knowledge the site of chromosomal initiation in *dam* strains has not been determined, but the above observation suggests a *dnaA*-dependent mode. Two promoters containing -GATC- sites within their −10 regions have been identified within *oriC*. Perhaps excessive transcription from these promoters in the absence of methylation interferes with initiation of chromosome replication more severely in minichromosomes than its normal counterpart. Clearly, further experimentation is required to clarify the role of methylation in initiation of replication.

PRACTICAL APPLICATIONS

Strains of *E. coli* which are *dam*⁻, *dcm*⁻, or both are used widely for the preparation of unmethylated plasmid and phage DNA. Absence of *dam* and *dcm* methylation in DNA results in more efficient digestion by certain restriction endonucleases (29). In addition, use of a *dcm* mutant for DNA isolation avoids loss of bands corresponding to 5-meCyt in the Maxam and Gilbert DNA sequencing procedure (37).

Complications can occur in site-directed mutagenesis due to mismatch repair leading to low recovery of the desired mutant class. This can be avoided by transforming DNA preparations directly into mismatch repair-deficient (*mutH*⁻, *mutL*⁻, *mutS*⁻) strains. Alternatively, commercially available DNA adenine methylase can be used to fully methylate DNA before transfection or to transform DNA into a recipient containing pTP166 (33). This plasmid can be regulated to produce large amounts of *dam* methylase in the cell, which methylates DNA before mismatch repair. A simple way to prevent loss of desired markers is to prepare template DNA in *dam* mutants. Site-directed mutagenesis, using mutant primers, is carried out in the usual way, and the product is transformed into a *dam* (or *dam*⁺) strain. Using template DNA prepared from plasmids containing the M13 origin in *dam* cells, efficiencies of around 40% of the desired mutational change can be obtained (D. Twitchell and A. Poteete, personal communication). Use of M13 templates in avoiding loss of mutations has been described by Kramer et al. (17).

OUTLOOK

The recent cloning of the *dcm* gene should aid in determination of its function and regulation. The role of DNA methylation in modulating gene expression in procaryotes is greater than previously suspected. Several promoters of bacteriophage P1 appear to be influenced by *dam* methylation, and this phage appears to code for a protein analogous to the *E. coli dam* protein. In addition, packaging of bacteriophage P1 DNA into virions may be sensitive to *dam* methylation since the DNA controlling packaging contains *dam* sequences (N. Sternberg, personal communication). The identification of *E. coli* genes whose expression can be affected by methylation is just beginning, as is the study of the role of modified bases in gene expression. Regulation of transcription by DNA methylation may be a mechanism which the cell uses to produce proteins that are required only once each generation. If so, then it should be possible to identify such genes by genetic means. Some properties and phenotypes of *dam* mutants cannot be explained satisfactorily by current models, which implies that other functions for methylation may be awaiting discovery. In this context, the role of the *dam* gene of bacteriophage T4 is not understood (S. Hattman, personal communication). The role of DNA methylation in organisms other than *E. coli* or *S. typhimurium* has also not been defined. DNA and protein sequence analysis and site-specific mutagenesis of DNA methylation genes from various organisms should lead to the identification of domains for binding of *S*-adenosyl-L-methionine and for binding to DNA. Such knowledge should allow the isolation of variant methylases with altered nucleotide specificity.

LITERATURE CITED

1. **Bachmann, B. J.** 1983. Linkage map of *Escherichia coli* K-12, edition 7. Microbiol. Rev. **47:**180–230.
2. **Bale, A., M. d'Alarcao, and M. G. Marinus.** 1979. Characterization of DNA adenine methylation mutants of *Escherichia coli* K-12. Mutat. Res. **59:**157–165.
3. **Billen, D.** 1968. Methylation of the bacterial chromosome: an event at the "replication point"? J. Mol. Biol. **31:**477–486.
4. **Borek, E., and P. R. Srinivasan.** 1966. The methylation of nucleic acids. Annu. Rev. Biochem. **35:**275–297.
5. **Boyer, H. W., L. T. Chow, A. Dugaiczyk, J. Hedgpeth, and H. M. Goodman.** 1973. DNA substrate for the EcoRII restriction endonuclease and modification methylase. Nature (London) New Biol. **244:**40–43.
6. **Brooks, J. E., R. M. Blumenthal, and T. R. Gingeras.** 1983. The isolation and characterization of the *Escherichia coli* DNA adenine methylase (dam) gene. Nucleic Acids Res. **11:**837–851.
7. **Coulondre, C., J. M. Miller, P. J. Farrabaugh, and W. Gilbert.** 1978. Molecular basis of base substitution hotspots in *Escherichia coli.* Nature (London) **274:**775–780.
8. **Craig, R. J., J. A. Arraj, and M. G. Marinus.** 1984. Induction of damage inducible (SOS) repair in *dam* mutants of *Escherichia coli* exposed to 2-aminopurine. Mol. Gen. Genet. **194:**539–540.
9. **Fishel, R. A., and R. Kolodner.** 1984. An *Escherichia coli* cell-free system that catalyzes the repair of symmetrically methylated heteroduplex DNA. Cold Spring Harbor Symp. Quant. Biol. **49:**603–609.
10. **Geier, G. E., and P. Modrich.** 1979. Recognition sequence of the *dam* methylase of *Escherichia coli* K-12 and mode of cleavage of DpnI endonuclease. J. Biol. Chem. **254:**1408–1413.
11. **Hattman, S.** 1981. DNA methylation, p. 15A, 517–548. *In* P. D. Boyer (ed.), The enzymes. Academic Press, Inc., New York.
12. **Hattman, S.** 1982. DNA methyltransferase-dependent transcription of the phage Mu *mom* gene. Proc. Natl. Acad. Sci. USA **79:**5518–5521.
13. **Hattman, S., J. E. Brooks, and M. Masurekar.** 1978. Sequence specificity of the PI-modification methylase (M. EcoP) and the DNA methylase (M. Eco dam) controlled by the *E. coli dam*-gene. J. Mol. Biol. **126:**367–380.
14. **Hawley, D. K., and W. McClure.** 1983. Compilation and analysis of *Escherichia coli* promoter DNA sequences. Nucleic Acids Res. **11:**2237–2255.
15. **Herman, G. E., and P. Modrich.** 1982. *Escherichia coli dam* methylase. Physical and catalytic properties of the homogeneous enzymes. J. Biol. Chem. **257:**2605–2612.
16. **Kahmann, R.** 1984. The *mom* gene of bacteriophage Mu, p. 29–47. *In* T. Trautner (ed.), Methylation of DNA. Springer-Verlag, Berlin.
17. **Kramer, B., W. Kramer, and H.-J. Fritz.** 1984. Different base/base mismatches are corrected with different efficiencies by the methyl-directed DNA mismatch-repair system of *E. coli.* Cell **38:**879–887.
18. **Lacks, S., and B. Greenberg.** 1975. A deoxyribonuclease of *Diplococcus pneumoniae* specific for methylated DNA. J. Biol. Chem. **250:**4060–4066.
19. **Lacks, S., and B. Greenberg.** 1977. Complementary specificity of restriction endonuclease of *Diplococcus pneumoniae* with respect to DNA methylation. J. Mol. Biol. **114:**153–168.
20. **Lark, C.** 1968. Studies on in vivo methylation of DNA in *Escherichia coli* 15T. J. Mol. Biol. **31:**389–399.
21. **Lieb, M.** 1983. Specific mismatch correction in bacteriophage lambda crosses by very short patch repair. Mol. Gen. Genet. **191:**118–125.
22. **Lindahl, T.** 1982. DNA repair enzymes. Annu. Rev. Biochem. **51:**61–87.
23. **Little, J. W., and D. W. Mount.** 1982. The SOS regulatory system of *Escherichia coli.* Cell **29:**11–22.
24. **Lu, A.-L., S. Clark, and P. Modrich.** 1983. Methyl-directed repair of DNA base pair mismatches in vitro. Proc. Natl. Acad. Sci. USA **80:**4639–4643.
25. **Lundblad, V., and N. Kleckner.** 1984. Mismatch repair mutations of *Escherichia coli* K-12 enhance transposition excision. Genetics **109:**3–19.
26. **Lyons, S. M., and P. F. Schendel.** 1984. Kinetics of methylation in *Escherichia coli* K-12. J. Bacteriol. **159:**421–423.
27. **Marinus, M. G.** 1973. Location of DNA methylation genes on the *Escherichia coli* K-12 genetic map. Mol. Gen. Genet. **127:**47–55.
28. **Marinus, M. G.** 1976. Adenine methylation of Okazaki fragments in *Escherichia coli.* J. Bacteriol. **128:**853–854.
29. **Marinus, M. G.** 1984. Methylation of procaryotic DNA, p. 81–109. *In* A. Razin, M. Cedar, and A. D. Riggs (ed.), DNA methylation. Springer Verlag, New York.
30. **Marinus, M. G.** 1985. DNA methylation influences *trpR* promoter activity in *Escherichia coli* K-12. Mol. Gen. Genet. **200:**185–186.
31. **Marinus, M. G., M. Carraway, A. Z. Frey, L. Brown, and J. A. Arraj.** 1983. Insertion mutations in the *dam* gene of *Escherichia coli* K-12. Mol. Gen. Genet. **192:**288–289.
32. **Marinus, M. G., and N. R. Morris.** 1973. Isolation of deoxyribonucleic acid methylase mutants of *Escherichia coli* K-12. J. Bacteriol. **114:**1143–1150.
33. **Marinus, M. G., A. Poteete, and J. A. Arraj.** 1984. Correlation of DNA adenine methylase activity with spontaneous mutability in *Escherichia coli* K-12. Gene **28:**123–125.
34. **May, M. S., and S. Hattman.** 1975. Analysis of bacteriophage deoxyribonucleic acid sequences methylated by host- and R-factor-controlled enzymes. J. Bacteriol. **123:**768–770.
35. **Meijer, M., E. Beck, G. Hansen, H. E. N. Bergman, W. Messer, K. von Meyenburg, and H. Schaller.** 1979. Nucleotide sequence of the origin of replication of the *Escherichia coli* K-12 chromosome. Proc. Natl. Acad. Sci. USA **76:**580–584.
36. **Messer, W., U. Bellekes, and H. Lother.** 1985. Effect of *dam* methylation on the activity of the *E. coli* replication origin, *oriC.* EMBO J. **4:**1327–1332.
37. **Ohmori, H., J. I. Tomizawa, and A. Maxam.** 1978. Detection of 5-methylcytosine in DNA sequences. Nucleic Acids Res. **5:**1479–1486.
38. **Peterson, K. R., K. F. Wertman, D. W. Mount, and M. G. Marinus.** 1985. Viability of *Escherichia coli* K-12 DNA adenine methylase *(dam)* mutants requires increased expression of specific genes in the SOS regulon. Mol. Gen. Genet. **201:**14–19.
39. **Pukkila, P., J. Peterson, G. Herman, P. Modrich, and M. Meselson.** 1983. Effects of high levels of DNA adenine methylation on methyl directed mismatch repair in *E. coli.* Genetics **104:**571–582.
40. **Radman, M., and R. Wagner.** 1984. Effects of DNA methylation on mismatch repair, mutagenesis and recombination, p. 23–38. *In* T. Trautner (ed.), Methylation of DNA. Springer-Verlag, Berlin.
41. **Razin, A., S. Urieli, Y. Pollack, Y. Greenbaum, and G. Glazer.** 1980. Studies on the biological role of DNA methylation. IV. Mode of methylation in *E. coli.* Nucleic Acids Res. **81:**1783–1792.
42. **Roberts, D., B. C. Hoopes, W. McClure, and N. Kleckner.** 1985. IS10 transposition is regulated by DNA adenine methylation. Cell **43:**117–130.
43. **Schlagman, S. L., S. Hattman, and M. G. Marinus.** 1986. Direct role of the *Escherichia coli* Dam DNA methyltransferase in

methylation-directed mismatch repair. J. Bacteriol. **165:**896–900.

44. **Smith, D. W., A. M. Garland, G. Herman, R. E. Enns, T. A. Baker, and J. W. Zyskind.** 1985. Importance of state of methylation of *oriC* GATC sites in initiation of DNA replication in *Escherichia coli.* EMBO J. **4:**1319–1327.

45. **Sternberg, N.** 1985. Evidence that adenine methylation influences DNA-protein interactions in *Escherichia coli.* J. Bacteriol. **164:**490–493.

46. **Sugimoto, K., A. Oka, H. Sugisaki, M. Takanami, A. Nishimura, S. Yasuda, and Y. Hirota.** 1979. Nucleotide sequence of *Escherichia coli* K-12 replication origin. Proc. Natl. Acad. Sci. USA **76:**575–579.

47. **Szyf, M., Y. Greenbaum, S. Urieli-Shoval, and A. Razin.** 1982. Studies on the biological role of DNA methylation. V. The pattern of *E. coli* DNA methylation. Nucleic Acids Res. **10:**7247–7259.

48. **Urieli-Shoval, S., Y. Greenbaum, and A. Razin.** 1983. Sequence and substrate specificity of isolated DNA methylases from *Escherichia coli* C. J. Bacteriol. **153:**274–280.

49. **Walker, G. C.** 1984. Mutagenesis and inducible responses to deoxyribonucleic acid damage in *Escherichia coli.* Microbiol. Rev. **48:**60–93.

47. Processing of RNA Transcripts

THOMAS C. KING AND DAVID SCHLESSINGER

Washington University School of Medicine, St. Louis, Missouri 63110

INTRODUCTION

The *Escherichia coli* cell contains at least 70 species of stable RNA which are included within at least 27 transcription units. About 20% of the average transcript is composed of extra spacer sequences that are processed away and degraded as mature RNA species are formed. During this processing, cotranscribed species are also separated from one another. Here, we compile the known RNAs and their map locations, as well as known processing reactions and enzymes. We then turn to a consideration of the factors that influence the stability of RNA segments. Finally, we discuss the possible functions of transcribed spacers and RNA metabolism.

RNA SPECIES

The RNA species in *E. coli* have been categorized in numerous ways. A distinction between stable and unstable RNA is often made on the basis of both theoretical and empirical considerations. Operationally, an RNA species that disappears rapidly in a pulse-chase experiment is termed unstable, whereas a species that persists is termed stable. Theoretically, the distinction is between structural RNA (stable) and mRNA (unstable). In general, theoretical and empiric

distinctions agree for a given RNA species (i.e., mRNA is empirically unstable), but this type of categorization is not easily applied to some RNA species. In particular, it is now clear that some RNA species can serve in both structural and regulatory roles (e.g., the *divE* gene specifies a serine tRNA that controls gene expression and cell division as well as functioning as a tRNA; see below).

Another common distinction is made between genes under stringent and relaxed control. Stringent control is associated with classical structural RNAs (rRNA and tRNA) and with mRNAs coding for ribosomal proteins and RNA polymerase components. Stringent control is characteristic of stable, or structural, RNA (except for a specific subclass of mRNA), whereas relaxed control is characteristic of unstable, or nonstructural, RNA. However, both 6S RNA, a metabolically stable small RNA, and spot 42 RNA, a metabolically unstable small RNA, are under relaxed control. Clearly, then, metabolic stability of an RNA species does not always imply that the species has a structural role or that its expression is controlled by the stringent response. In this discussion, we consider four categories of RNA species: tRNA, rRNA, mRNA, and small RNAs. As will be seen, a few RNA species defy this simple categorization and appear under multiple headings. Of these groups, the small RNAs are by far the most structurally and functionally

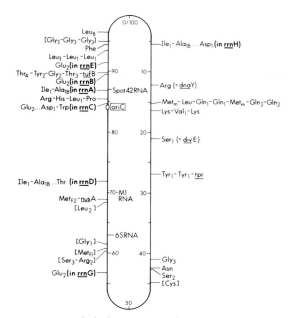

FIG. 1. *E. coli* linkage map. The seven rRNA operons (*rrnA*, *rrnB*, *rrnC*, *rrnD*, *rrnE*, *rrnG*, and *rrnH*) are marked with boldface letters. tRNA genes are indicated; those for which DNA sequence data have not been obtained are bracketed (see reference 43 for details). *dnaY* and *divE*, corresponding to an arginine and a serine tRNA, respectively, are indicated, as are 6S RNA, spot 42 RNA, and M1 RNA.

heterogeneous, as they are classified as such by their length only—arbitrarily less than 200 nucleotides.

tRNA

tRNA genes are distributed throughout the *E. coli* chromosome and are found in operons containing rRNA genes, protein genes, and other tRNA genes. The linkage map locations of 53 tRNA genes in *E. coli* have been determined (43, 61) and are diagrammed in Fig. 1. These genes reside in 27 separate operons with diverse organization. Of the tRNA genes, 26% occur in rRNA operons, 42% occur in multicistronic operons containing only tRNA genes, 19% occur in monocistronic operons, and at least 13% occur in three multifunction operons containing protein-coding genes as well as tRNA genes (2, 60, 63, 112). This latter type of organization is important in assessing the type and number of processing reactions required to generate mature tRNA species. If a protein-coding sequence follows a tRNA gene, some type of endonucleolytic processing reaction is required to separate the two cistrons before 3′ maturation of the tRNA by an exonuclease (see below). In no case is a tRNA gene transcribed to produce a mature species as a primary transcript; all tRNAs require 5′ and 3′ processing to generate their mature termini. The amount of each isoacceptor species has been measured in *E. coli* (65).

rRNA

Seven rRNA operons exist in *E. coli*; their distribution in the chromosome is indicated in Fig. 1. All seven operons have very similar (17) organization, and their nucleotide sequences are known. In each sequence,

the small subunit RNA gene (16S rRNA) is located nearest the 5′ end of the operon. The gene for the large subunit RNA (23S rRNA) is located distal to the 16S rRNA gene (65), and the 5S rRNA gene is the nearest gene to the 3′ end (Fig. 2). The three mature rRNA sequences are separated by spacer sequences which are removed during processing. All rRNA operons contain tRNA genes in the spacer between 16S and 23S rRNA, and some contain tRNA genes distal to 5S rRNA. The spacer regions flanking the rRNA cistrons are highly conserved between operons. Long inverted repeats, flanking both 16S and 23S rRNA, have the potential to form double-stranded stems at the bases of 16S and 23S rRNA (16). These double-stranded regions are predicted to be stable in vivo (135) and have been observed directly by electron microscopy (76). Each operon contains two tandem promoters, one of which is responsive to control by PPGPP, and the other which is subject to growth rate control. Initial cleavages separate 16S and 23S RNA, usually before transcription of the operon is complete (3, 73). As with tRNA, no mature rRNA species are formed as primary transcripts.

mRNA

mRNAs are heterogeneous in size and organization. Approximately 950 mRNA genes have been assigned on the *E. coli* linkage map at present; of these, 25% occur in multicistronic operons composed of genes with related functions (5). As noted above, a small number of genes occur in operons with tRNA and other small RNA genes (2, 60, 63). Organizationally, ribosomal protein (r-protein) genes are somewhat distinct. They are under stringent control, unlike most

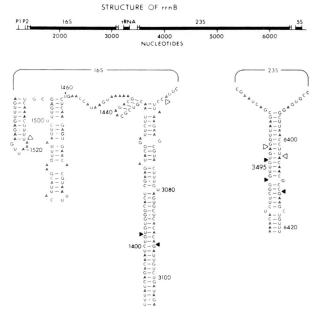

FIG. 2. A typical rRNA operon, shown schematically. The secondary structure of the spacer sequences bordering 16S and 23S rRNA are shown, corresponding to the lowest energy state predicted by computer analyses (see reference 76). Open triangles show the position of mature termini; closed triangles show the positions of RNase III cleavage sites.

other mRNA genes, and 75% of them occur in multi-cistronic operons composed mainly of r-protein genes. This arrangement has regulatory significance, since the translation of many, if not all, r-protein mRNAs is controlled by feedback inhibition by one of the r-protein products of a given operon (41, 98).

Unlike primary rRNA and tRNA transcripts, mRNA processing has only been shown to occur in a few transcriptional units and has not been shown to be of functional significance in any case (see below). Highly conserved sequence elements 35 nucleotides in length that may control mRNA expression or stability or both have been identified at the 3' ends of mRNA operons and in intercistronic regions (57). These repetitive extragenic palindromic (REP) sequences tend to occur as inverted repeats with the potential to form a single, energetically stable, double-stranded stem. More than 500 copies reside in the *E. coli* genome, but none have been identified in its phages (126). These sequences are clearly distinct from transcription termination sites, although both contain double-stranded structures. The significance of these sequences is considered below.

Small RNAs

Small RNAs compose the most heterogeneous group of RNA species considered here. They include regulatory RNAs, enzymatic RNAs, RNA primers, RNA fragments derived from RNA-processing reactions, and small RNAs of unknown function, as well as two tRNAs that apparently subserve functions in addition to their isoacceptor function during translation.

RI RNA has a regulatory function in the control of plasmid copy number for ColE1-derived plasmids. It is a 108-nucleotide RNA, synthesized from the L strand of ColE1 DNA, and hybridizes to and prevents RNase H processing of primer RNA by displacing the primer from the H strand. Failure to process the primer RNA to the proper length results in failure of initiation of DNA replication (129). No comparable regulation system has been identified in *E. coli*, although RNase H is required for *E. coli* viability (68).

mRNA-interfering complementary (MIC) RNA is a small (107-nucleotide) regulatory RNA that is located at 48 min on the *E. coli* chromosome, adjacent to the *ompC* gene (which codes for an outer membrane protein). *ompC* gene expression is coordinately regulated with *ompF* expression (producing another outer membrane protein) in such a way that the sum of *ompF* and *ompC* gene products is a constant. MIC RNA is transcribed from its own operon in a direction opposite to that of *ompC*. The MIC RNA sequence is complementary to the leader region of the *ompF* message and apparently hybridizes to it, inhibiting its translation. It is hypothesized that MIC RNA transcription is coordinate with *ompC* transcription, thus down-regulating *ompF* expression when more *ompC* is produced (91).

M1 RNA is the only RNA in *E. coli* known to have an enzymatic function. It is the catalytically active component of RNase P, a ribonucleoprotein complex composed of M1 RNA (375 nucleotides; located at 70 min on the *E. coli* map [108]) and a protein of 18.5 kilodaltons (kDa), the gene for which is located at 83 min (1). Both RNA and protein moieties are required for full RNase P activity, but the primary transcript of

M1 RNA is catalytically active on tRNA precursors under certain experimental conditions (51). An RNA species with a sequence nearly identical to that of *E. coli* M1 RNA has been isolated from *Salmonella typhimurium*. This RNA species catalyzes the 5' maturation of tRNA precursors in vitro in the absence of protein (6). M1 RNA is transcribed from a single operon under stringent control. Termination occurs approximately 40 nucleotides distal to the mature M1 RNA 3' terminus, but the 5' ends of the primary transcript and mature M1 RNA are identical (108). The 3' flanking region of this operon contains a 113-base-pair sequence reiterated 3.5 times and also contains multiple short open reading frames. The significance of this unusual organization is not clear, but it resembles the arrangement at the 3' end of the *tyrT* operon (107, 112).

RNA primers are required for the initiation of DNA replication in *E. coli* and for discontinuous replication of the lagging strand. Primers for discontinuous DNA replication average 2 to 3 nucleotides in length in *E. coli* but may be up to 10 nucleotides long in some of its phages (78). These primers have random sequence. These very small RNAs probably never exist as free RNA molecules but remain covalently linked to the newly synthesized DNA until they are degraded. Primers for the initiation of *E. coli* DNA replication at *oriC* are expected to have specific nucleotide sequences. These have not been defined as yet, and the lengths of these primers are unknown (99).

Small RNA fragments derived from the processing of larger RNA transcripts are probably very rapidly degraded to mononucleotides. They have not been extensively studied and are not known to serve any function after they are excised from precursor RNAs. Most of these RNAs probably exist as single-stranded species before degradation but, in at least one case, double-stranded RNA fragments are predicted. RNase III cleavage of the double-stranded stems at the base of 16S and 23S rRNA precursors should generate these double-stranded fragments.

6S RNA is an RNA of unknown function 180 nucleotides in length. It is transcribed from a single operon located at 63 min on the *E. coli* chromosome. Its transcription is under relaxed control. This operon contains an open reading frame 75 nucleotides distal to the mature 3' end of 6S RNA, which could specify a single protein of 180 amino acids. 6S RNA is cotranscribed with this open reading frame and so requires 3' processing. In addition, its transcription initiation site is located six to eight nucleotides proximal to the mature 5' terminus, implying 5' processing as well. 6S RNA is rather stable (although its half-life is not known precisely) and exists as a ribonucleoprotein complex sedimenting at 10S. Its function is unknown, but it seems to be dispensable (58, 79).

Spot 42 RNA is a small unstable RNA composed of 109 unmodified nucleotides. It is transcribed from a single copy gene located at 86 min on the *E. coli* chromosome. The gene includes a rho-independent termination site, and the mature species is apparently identical to the primary transcript. The RNA contains an open reading frame that could specify 15 hydrophobic amino acids, including three leucines (109, 113). Its sequence shows marked homology to that of tRNA$_1^{Leu}$, and it may act as a tRNA analog during leucine starvation (R. M. Williamson and J. H. Jackson, Fed. Proc. **44:**1814). It is present at 150 to 200 copies per cell, and its concentration is down-regulated by cyclic AMP. Tenfold overproduction of this RNA by a multicopy plasmid results in cells with an increased generation time and retarded adaptation to changes in culture media (109, 114).

4.5S RNA is an RNA 114 nucleotides in length present as a single copy gene in the *E. coli* chromosome. Its map location is not known. It is transcribed from an operon with an open reading frame specifying 116 amino acids, located 111 nucleotides distal to the mature 3' terminus of 4.5S RNA. Processing is required to generate the 3' terminus; similarly, the mature 5' terminus is located 24 nucleotides distal to the transcription start site. Its transcription is under stringent control. It is stable and present at approximately the level of an individual tRNA species (1,000 copies per cell). A secondary structure can be constructed for this RNA involving 75% of the nucleotides in a single double-stranded stem with a calculated ΔG value of -60 kcal/mol (1 cal = 4.184 J) (59). Its function is unknown, but it is essential, since cells cured of a prophage containing their only functional 4.5S RNA gene are nonviable (18).

divE is the genetic locus in which maps a temperature-sensitive mutation that results in the inability of cells to synthesize several cell-cycle-specific proteins or to divide at nonpermissive temperatures. The mutation responsible for this phenotype is a change from A to G at position 10 in the D loop of tRNA$_1^{Ser}$. The mutant is suppressible by a normal tRNA$_1^{Ser}$ (128). This tRNA is located at approximately 22 min on the *E. coli* chromosome and is transcribed from an operon with an open reading frame specifying a 23.5-kDa protein located 3' to the tRNA gene. It is not clear how this mutation produces the pleiotropic phenotype observed. Its effects may be mediated by changes in the translational function of the tRNA or by some direct regulatory mechanism unrelated to protein synthesis (128).

dnaY is the locus of another temperature-sensitive mutation in which DNA synthesis is blocked at nonpermissive temperatures. This mutation is in an arginine tRNA gene located at 12 min on the *E. coli* chromosome (43, 95).

RNA PROCESSING AND TURNOVER REACTIONS

Processing reactions are understood in detail in a few instances. In most cases, the precursors and products are characterized, but the enzymatic activities involved, the abundance of various precursor species, and the rates of processing reactions are not known. Some reactions clearly represent processing events (e.g., maturation of the 3' end of tRNA) in that they must occur before a molecule is functional. In other cases, processing is not clearly distinct from turnover (e.g., processing of the mRNA *rplJL-rpoBC* has no defined effect on translation, but it may affect turnover). The specificity of a given reaction does not distinguish turnover from processing (e.g., functional inactivation of mRNA occurs at very specific locations in *lac* mRNA). Furthermore, processing and turnover cannot be distinguished as one being endonucleolytic and the other being exonucleolytic, since maturation

of the 3' end of tRNA precursors and final chemical degradation of mRNA are both mediated by exonucleases. Therefore, the only real distinction between processing and turnover is operational (in terms of the functional significance of the reaction). Since the purpose of many processing reactions is uncertain, and since it is conceivable that one reaction may serve both to mature and to destabilize an RNA species, we do not attempt to distinguish processing reactions rigorously from turnover reactions. Processing and turnover reactions are discussed together in this section, and a distinction between the two is only made when the function of a particular reaction seems clear.

tRNA

tRNA-processing enzymes are among the best characterized processing activities in *E. coli*. Some have been assigned different names by different investigators; all of the enzymes thought to be involved in tRNA processing, and suggestions concerning which of them may represent equivalent activities, are included in Table 1.

As mentioned before, some tRNA precursors are multimeric and so might require endonucleolytic cleavage to separate individual tRNA molecules before final maturation. Since all tRNA molecules are eventually cleaved at their 5' end by the endonuclease RNase P (34), it is formally possible that RNase P could also separate multicistronic tRNA precursors (as well as generate mature 5' ends). In some cases, it is clear that RNase P makes the initial cleavage that separates dimeric tRNA precursors (119). There is also evidence that tRNAs in some multicistronic precursors are separated before RNase P cleavage (116).

As mentioned above, in transcripts in which a protein-coding sequence occurs distal to a tRNA gene, some type of endonucleolytic cleavage is required 3' to the tRNA, since final maturation of the 3' terminus is carried out by an exonuclease (see below). Another endonuclease, RNase PC, is absolutely required to separate T4 phage tRNA precursors before final processing (48). One phage T4 tRNA requires a cleavage by RNase III on its 5' side before it becomes a substrate for RNase P (118).

RNase P is an endonuclease that generates the mature 5' terminus of all tRNA species in *E. coli*. It recognizes mature tRNA sequences rather than precursor sequences, correctly processing tRNA precursors from different species (1). Modification of the mature tRNA structure has been shown to result in delayed or inaccurate processing of the affected precursor (46, 123). RNase P is not obligately the first or last step in tRNA maturation, since species with either mature or immature 3' ends are substrates for it. However, it has been shown that different substrates are cleaved more efficiently than are others by RNase P. For example, in a dimeric T4 tRNA precursor with two RNase P cleavage sites, the internal site is cleaved twice as rapidly in vitro and is greatly favored in vivo (119). Even tRNA precursors lacking 3' terminal CCA are poorer substrates for isolated M1 RNA than are species containing CCA (52). The fact that different substrates are cleaved more or less rapidly is consistent with the notion that the initial separation of multicistronic tRNA precursors may be

TABLE 1. *E. coli* nucleases

Nuclease	Function(s)	Reference(s)
Exoribonucleases[a]		
RNase II	Single-strand-specific nuclease involved in mRNA chemical decay	37
Polynucleotide phosphorylase	mRNA chemical decay	37
RNase R	Possible role in mRNA metabolism	70
RNase D	3′ Processing of tRNA precursors	31
RNase BN	3′ Processing of tRNA precursors in phage T4	4
RNase T	"End turnover" at 3′ end of mature tRNA species	36
Endoribonucleases		
RNase III	Initial processing of rRNA precursor, minor role in tRNA processing in T4, possible role in mRNA metabolism, etc.	10, 111
RNase H	Degrades RNA of RNA:DNA hybrids	30
RNase P	5′ Processing of tRNA precursors	51
RNase E	Initial 3′ and 5′ processing of 5S rRNA precursor	90
RNase PC	Separation of multimeric T4 tRNA precursors	48
RNase I	Turnover of rRNA	96, 106
Uncharacterized activities		
RNase "M16S"	Final 5′ and 3′ processing of 16S rRNA; may be two distinct activities	33, 55
RNase "M23S"	Final 5′ and 3′ processing of 23S rRNA; likely two distinct activities	122
RNase "M5S"	Final 5′ and 3′ processing of 5S rRNA in *E. coli*	40
RNase P2	Endonucleolytic separation of multimeric tRNA precursors; the three	34, 44
RNase O	activities may or may not be distinct	
RNase F		
RNase N	Endonuclease	89

[a] All of the exoribonucleases listed are 3′ → 5′.

mediated by an activity other than RNase P. For instance, in multicistronic transcripts that are a poor substrate for RNase P, another endonuclease might cleave more rapidly than RNase P, whether or not there is an absolute requirement for it.

No quantitative information is available about the rate of cleavage by RNase P in vivo; however, precursor tRNA species with immature 5′ ends are not isolated from wild-type *E. coli* strains. Precursors do accumulate when cells are overloaded with tRNA precursors (e.g., when extra tRNA precursors are transcribed from a multicopy plasmid) or when abnormal tRNA precursors are present (e.g., when there is a mutation in the mature tRNA sequence leading to a slower rate of processing by RNase P (see above) (46, 123). This suggests that, normally, RNase P cleavage must proceed with a short half-reaction time, due to either an abundance of enzyme or a high enzyme-turnover number. The number of molecules of RNase P per cell has not been measured.

The mature 3′ terminus of tRNA species is generated exonucleolytically, but the activity or activities involved have not been unambiguously defined. Some specificity for this processing reaction seems likely, since the exonuclease involved does not simply remove 3′ nucleotides processively until a double-stranded region in the tRNA is encountered. On the other hand, the tertiary structure of the tRNA and of the exonuclease could interact sterically to block degradation near the appropriate 3′ terminus. Cudney and Deutscher (31) compared RNase II and RNase D in their ability to mature tRNA 3′ ends and found that RNase D, a nonprocessive nuclease, generated the correct mature terminus, whereas RNase II, a processive nuclease, removed one to two additional nucleotides. The extra nucleotides removed by RNase II could be replaced by tRNA nucleotidyl transferase,

but a mutant deficient in that enzyme shows no detectable alteration in tRNA processing (35). Since mature and precursor tRNAs are exposed to RNase II in the cell, at least some tRNA precursors may be processed by RNase II. Although RNase D would seem to be the major 3′ processing activity, a deletion mutant (for the RNase D gene) is viable and has no noticeable alterations in tRNA processing (12). Clearly, another enzyme can substitute.

Primary tRNA transcripts in *E. coli* contain the entire mature sequence of each tRNA molecule, but in some *E. coli* phages a 3′ CCA must be added posttranscriptionally to some phage-encoded tRNAs, as is found with all tRNAs in mammalian cells (45). Another *E. coli* enzyme, RNase BN, is required for the 3′ trimming of these phage tRNA precursors before CCA addition (since a mutation in this enzyme greatly reduces the efficiency of phage infection [4]). The nature of this reaction is different from that for the 3′ maturation of *E. coli* precursors in that three additional nucleotides must be removed from the 3′ end of the phage precursors. It has been suggested that RNase BN may substitute for RNase D in the RNase D deletion mutant described above. However, a mutant deficient in RNase D, RNase BN, RNase I, and RNase II apparently processes tRNA normally (137). (Since this mutant does not have a deletion of the RNase BN gene, it is possible that the allele was sufficiently leaky to allow complementation by residual RNase BN activity. This explanation seems unlikely, since the mutation results in failure of phage growth. Nevertheless, since a different reaction is required for phage tRNA maturation [i.e., the removal of three additional 3′ nucleotides], it is possible that residual RNase BN activity suffices to mature *E. coli* tRNA precursors.) Which of these activities is dominant in wild-type *E. coli* and whether additional uncharacterized activities are involved in 3' maturation of tRNA precursors remain unsettled.

RNase T is a nonprocessive endonuclease that removes the 3' terminal adenosine residue from mature tRNA species in vitro (36). This reaction has been called end turnover, and its significance is uncertain; however, it does not lead to tRNA degradation. As mentioned above, the adenosine removed can be replaced by tRNA nucleotidyl transferase. This RNase T reaction is at least somewhat specific, in that RNase T is much less active with tRNA species having sequences other than CCA at their 3' end. RNase T is not efficient at maturing normal tRNA precursors; thus, it is doubtful that it could substitute for RNase D or BN. Other nonspecific nucleases may be capable of similar actions, for instance RNase II (see above).

In no instance has bulk tRNA degradation been observed. Even after UV irradiation or during productive lambda infection, which leads to cell death, tRNA molecules remain intact (11, 101). This suggests that tRNA structure is completely resistant to nucleases present in *E. coli*; the argument that structure is protective is strengthened by the findings that tRNA species with particular mutations that affect their secondary structure are degraded (123).

rRNA

The processing of rRNA is complex, involving multiple cleavage reactions at the 5' and 3' ends of each species. Some reactions do occur or at least can occur in vitro with naked pre-rRNA, whereas others require that critical r-proteins be associated with the pre-rRNA. Some of the final processing reactions may even require that the preribosome engage in protein synthesis (20, 54, 84, 136).

In wild-type *E. coli*, the first cleavage to occur is the RNase III-mediated separation of 16S and 23S rRNA from each other and from 5S rRNA and any tRNAs present in the operon. (RNase III is a dimer of 50 kDa [10]. A similar activity is found in *S. typhimurium* [127].) The RNase III reactions occur quite rapidly, even before transcription of the entire operon is complete (3). For this reason, the unprocessed primary transcript encompassing the entire operon, 30S pre-rRNA, is not found in wild-type *E. coli*.

The RNase III cleavage reactions occur in regions of double-stranded RNA structure formed by complementary sequences flanking both 16S and 23S rRNA. Specific cleavage sites have been defined in these double-stranded regions in which most (see below) RNase III cleavages occur. These sites are adjacent to small single-stranded bubbles in the double-stranded stem. These sites are similar to other known RNase III cleavage sites, although no primary or secondary structural features clearly define an RNase III cleavage site in any system. It has been suggested that the tertiary structure of the cleavage site is the critical factor in RNase III recognition (110, 111). The major cleavages occur at staggered positions in each double-stranded stem (Fig. 2) (16); thus, cleavage is thought to involve two separate reactions in each stem. This view is strengthened by the observation that, at one RNase III cleavage site in T7 mRNA (39) and at the λ *sib* cleavage site (26), RNase III cleaves only one strand of a double-stranded stem. RNase III cleavage

occurs in vitro at the same sites as it does in vivo with naked precursor rRNA, so that there is no dependence on r-protein binding for this reaction (10). The specificity of RNase III cleavage is critically dependent on the concentration of monovalent cations; many additional cleavage sites appear at low ionic strength (10).

RNase III cleavage results in qualitatively different precursor species for 16S and 23S RNA. The pre-16S species produced retains long precursor RNA sequences at both its 5' (115 nucleotides) and 3' (33 nucleotides) ends, whereas the pre-23S species produced is much more similar to mature 23S rRNA, with only 7 nucleotides at its 5' end and about 8 nucleotides at its 3' end.

Much of the information about this cleavage reaction is derived from the study of a mutant deficient in RNase III activity, AB301/105 (38). To date, only one mutant allele of RNase III has been isolated. This allele is a point mutation, and its leakiness in vivo is not well defined. Experiments to determine the effect of a deletion of the RNase III gene (or of a temperature-sensitive lesion in RNase III activity, etc.) have not been reported.

It seems likely that some RNase III activity is absolutely required, since the one mutant allele has markedly pleiotropic effects on plasmid replication, on phage growth, and on the quantity of various *E. coli* proteins synthesized, as well as on rRNA processing (24, 47, 56). Detailed study of the processing of rRNA in this mutant by S1 nuclease analysis (73) has shown that RNase III processing in the stem surrounding 16S RNA does not occur but that the 16S RNA is matured normally and at the same rate as would occur if the RNase III cleavage had occurred. The situation is quite different for 23S RNA. Again, the major RNase III cleavages found in wild-type *E. coli* do not occur, but the normal mature 5' and 3' termini are not generated. Instead, most of the functional 23S rRNA in the mutant is a longer pre-rRNA species, though residual or altered RNase III activity in the mutant cleaves the 23S precursor at one alternate 5' site (74, 122). (It is inferred that these unusual 5' termini result from RNase III cleavage because the same species is found in wild-type *E. coli* at low levels and is formed by the action of purified RNase III in vitro.) Presumably, the mutation in the RNase III allele in the mutant prevents it from cleaving at its major cleavage sites but not at some usually minor cleavage sites. More importantly, these results indicate that the normal maturation of *E. coli* 23S rRNA is absolutely dependent on prior processing by RNase III.

Detailed measurements of the rate of RNase III cleavage in vivo in wild-type *E. coli* show a reaction half-time of 0.2 to 0.5 min in rich media at 37°C, consistent with cleavage occurring before the completion of transcription of the operon (73). Furthermore, experiments in which rifampin was used to block transcription showed that RNase III cleavage normally occurs at the 5' end of a 23S rRNA precursor molecule before the 3' end of that molecule is transcribed. This result implies that an intramolecular double-stranded stem does not serve as the recognition site for RNase III. The simplest explanation for this result is that intermolecular hybridization between the 5' end of a 23S precursor undergoing transcription and the 3' end of a completed 23S transcript forms a double-stranded recognition site for RNase III. The possibility that some other base-pairing scheme forms the RNase III site has not been excluded.

The cleavages that generate the mature termini of 16S and 23S rRNA from RNase III-cleaved intermediates are less well characterized. Enzymatic activities that mature the 5' and 3' ends of 16S rRNA have been partially purified (32, 33, 55), and a mutant with altered maturation at the 5' end of 16S RNA has been isolated (32). No enzymatic activities have been identified that are responsible for the maturation of 23S rRNA.

The absolute rates at which 16S rRNA maturation cleavages occur have been measured (73). The reaction half-times for maturation at the 5' end range from 2 to 4 min in wild-type cells under various growth conditions. This rate implies that from 4 to 8% of the total 16S rRNA 5' ends in the cell are immature at any time.

As mentioned previously, the rate of maturation of the 5' end of 16S rRNA is the same in wild-type cells as it is in RNase III-deficient cells in which RNase III cleavage of 16S rRNA precursors does not occur. (The reaction half times for 16S RNA maturation in *E. coli* ABL1 and AB301/105, both harboring the same RNase III, *rnc-105* allele, were not significantly different from those in wild-type cells.) A mutant that accumulates a novel 5' precursor of 16S rRNA has been partially characterized (33). This mutant, the "BUMMER" strain, accumulates a novel species accounting for up to 50% of 16S RNA 5' ends under some conditions. This precursor, 16.3S RNA, has 66 extra nucleotides at its 5' end (32). Extracts from the parental strain convert this species to mature 16S rRNA. This reaction requires that the 16.3S RNA be in the form of 30S subunits or 70S ribosomes and demonstrates that the reaction is dependent on ribosome assembly. Analysis of the reaction products demonstrates that the reaction is endonucleolytic. The activity was partially purified and had a molecular mass of approximately 70 kDa. The origin of 16.3S rRNA was not determined, although rifampin chase experiments suggested that it was a true precursor of mature 16S rRNA. This novel species could result from a nonspecific cleavage (of a primary transcript already cleaved by RNase III) at an accessible (single-stranded) point, exposed in the secondary structure of the precursor (Fig. 2).

Detailed measurements of the kinetics of rRNA processing in the BUMMER strain show normal 23S rRNA processing and unaltered kinetics of RNase III cleavage of 16S RNA precursors (73). The novel species accounted for 23% of the 16S 5' ends under the conditions used. Kinetic experiments were consistent with the 16.3S RNA being formed from the RNase III-cleaved precursor. The half-reaction time for the conversion of 16.3S RNA to mature 16S RNA was nearly 10 min, almost three times slower than the rate of formation of mature 16S RNA in wild-type cells under comparable conditions. All of these results support the ideas that the BUMMER strain is partially deficient in the activity responsible for the maturation of the 5' end of 16S rRNA, that this reaction is endonucleolytic, and that it depends on ribosome assembly.

The structural features that this activity recognizes and the extent to which they exist in the mature or precursor RNA sequences are unknown. Certain mutant rRNAs with alterations in the mature rRNA sequence are not processed, but this could result from a failure of ribosome assembly (since this is a prerequisite to cleavage), as well as from a direct effect on RNA structure at the cleavage site (125).

The reaction that generates the mature 3' terminus of 16S rRNA is less well characterized than that generating the 5' terminus. An activity of approximately 45 kDa that converts a ribonucleoprotein particle with an immature 3' end to mature 16S RNA has been partially purified (55). It is thought to be an endonuclease, since intermediates with multiple intermediate 3' ends have not been detected. The reaction half times for this cleavage range from 6 to 9 min in wild-type cells under various conditions (73), generally about twofold slower than maturation of the 5' end of 16S rRNA. The rate of 3' end maturation is not slowed in mutants deficient in RNase III (as was found for 5' maturation). However, the rate of this reaction is markedly slowed (threefold) in the BUMMER mutant, which is apparently deficient in 5' maturation activity (73). There are at least two explanations for this observation. Some have suggested that the 5' and 3' activities are identical (103). The molecular masses of crude preparations are not consistent (70 versus 45 kDa); however, neither activity has been sufficiently purified to test this hypothesis rigorously. Second, it is possible that there is an obligate (or at least markedly preferred) order of processing, the 5' step occurring first. No test has been made of this hypothesis.

Maturation of 23S rRNA is complex. Although there are predominant mature 5' and 3' termini, other termini exist in wild-type cells (including at least one species with four fewer nucleotides at its 5' end than the predominant mature species [74, 122]). The rates of conversion of RNase III-cleaved 23S pre-RNA to mature 23S RNA have not been determined precisely, but these precursors are rare in wild-type cells, constituting much less than 5% of the total number of 5' and 3' ends (73).

As noted above, normal maturation of 23S rRNA does not occur in cells harboring a mutant RNase III gene. Instead, discrete species with attached precursor sequences from 20 to over 97 nucleotides in length are found at the 5' end (in addition to the species 4 nucleotides shorter than the conventional mature 23S rRNA). At the 3' end, a discrete species with 53 extra nucleotides is found in addition to larger species. All of these species except the one lacking four nucleotides of mature 5' sequence could be produced by nonspecific nuclease action at sites relatively exposed in the secondary structure of the precursor RNA (74). Addition of purified RNase III to these species results in the production of the major RNase III cleavage products expected (species with three and seven extra nucleotides at the 5' end and eight extra nucleotides at the 3' end [122]).

All four of these RNase III cleavage reactions occur in wild-type cells, yielding small amounts of the species with +7, +3, and −4 nucleotides at the 5' end and +8 nucleotides at the 3' end. If a wild-type cell extract is added to a cell extract from the mutant under conditions that support protein synthesis, the major mature 5' terminus is produced in addition to small amounts of the other species. If the same is done in Tris-Mg^{2+} buffer, only RNase III cleavages occur. This result suggests that protein synthesis or at least conditions favorable for protein synthesis are needed for maturation of the 5' end of 23S rRNA. Since the RNase III-deficient mutants are viable and contain immature 23S 5' ends, ribosomes containing at least some, and probably most, of these species are capable of protein synthesis. This observation is also consistent with the finding that the amount of each of these species is the same in free ribosomes as it is in polysomes.

Examination of the 23S rRNA 3' ends yields a different result. Addition of wild-type extract to mutant ribosomes results in the production of ragged termini ranging in length from that of the RNase III-cleaved species to a length two nucleotides shorter than the usual 3' terminus. This result strongly suggests that an exonuclease acts on the RNase-III cleaved precursor to yield the mature 3' terminus. Again, all of the intermediate species formed in vitro are observed at low levels in wild-type cells. This reaction is accelerated by, but not completely dependent upon, protein synthetic conditions. One explanation for this dependence is that maturation at the 5' end (leaving the 3' precursor segment single stranded) may potentiate 3' exonucleolytic activity.

In summary, recent work suggests that the 5' and 3' termini of 23S RNA are somewhat heterogeneous, with several discrete 5' species and multiple 3' termini (probably due to exonucleolytic trimming). It is also suggested that an endonucleolytic activity that is dependent on prior RNase III cleavage and on protein synthetic conditions is responsible for generating the major 5' terminus in wild-type cells. Finally, 3' maturation of 23S RNA appears to be mediated by an exonucleolytic activity that is dependent on prior RNase III cleavage.

Processing of 5S rRNA also requires multiple steps in *E. coli*. In an RNase E mutant, a 5S rRNA precursor (9S RNA) accumulates. This precursor is composed of 5S rRNA with approximately 85 extra 5' nucleotides extending to the RNase III cleavage site near the 3' end of 23S rRNA and extra 3' nucleotides extending to the terminator of the operon (90). This 3' segment would contain trailer tRNAs if they were present in the operon (and if they had not been removed by RNase P). This 9S precursor is not observed in wild-type cells, since it is rapidly cleaved by RNase E to yield 5S rRNA with three extra nucleotides at both its 5' and 3' ends. RNase E has been partially purified and has a molecular mass of about 70 kDa; it acts on 9S RNA in vitro to produce the same 5S RNA precursor found in vivo (90). A mutation in RNase E has effects that are rather pleiotropic (e.g., it inhibits processing of M1 RNA in addition to its effect on 5S RNA processing; see below).

Other activities evidently complete the processing of 5S RNA, but these have not been characterized in *E. coli*. Multiple 5' species with one, two, or three extra nucleotides (125) accumulate in the absence of protein synthesis. These species are also observed shortly after the pulse-labeling of cells and disappear thereafter (66), implying that they are true precursors of 5S RNA. These precursor species are found in polysomes, indicating that final processing occurs in polysomes

(and that these precursors are probably capable of functioning in protein synthesis (40).

5S RNA processing is much better understood in *Bacillus subtilis*, in which processing is mediated by a single enzyme, RNase M5, and occurs in a one-step process. RNase M5 is composed of two subunits, only one of which is catalytically active. The other apparently binds to pre-5S RNA (102) and promotes the proper conformation for cleavage (124). Cleavage occurs in a double-stranded region and separates adjacent nucleotides in the two strands. It is not known whether the two cleavages occur simultaneously, but species with one mature and one immature end are not observed, implying that the enzyme is not released from the substrate before the second cleavage. RNase M5 has been shown to recognize mature 5S RNA sequences within the pre-5S rRNA, as the addition of different synthetic nucleotides to the 3' or 5' end of mature 5S RNA does not affect the precision of RNase M5 cleavage (103). No similar activity has been identified in *E. coli*.

Turnover of rRNA occurs to a very limited extent if at all in growing *E. coli* (9). Intact ribosomes are resistant to all nucleases present in the cytoplasm. The only nuclease known to be capable of attacking intact ribosomes is RNase I, a periplasmic enzyme (96). This endonuclease is released into the cytoplasm only under adverse circumstances (9, 106). Its release is apparently controlled by several other genes (64). Presumably, after initial attack by RNase I, other nucleases, such as RNase II, participate in bulk rRNA degradation. RNase I has been shown to cleave 50S subunits first at a particular site in the three-dimensional structure of the ribosome (at the central protuberance) (106). Free rRNA is susceptible to nucleolytic attack by nucleases other than RNase I. Furthermore, rRNA molecules with deletions or point mutations are degraded, probably because they cannot form ribosomes (125).

mRNA

The only processing reactions known to exert a regulatory effect on *E. coli* mRNAs are those that initiate mRNA turnover. Clearly, mRNA turnover has a major effect on gene expression in *E. coli*. Different mRNA species have very different half-lives and so produce very different yields of protein molecules per mRNA molecule (8, 131).

Even though processing reactions aside from turnover are not currently known to be important in *E. coli*, they are of importance in some of its phages, and since the machinery for these reactions is specified by the *E. coli* genome, it seems likely that some of these reactions have significance in *E. coli* (13). For this reason, a number of novel processing reactions in phage systems are discussed before treatment of mRNA metabolism in *E. coli* proper.

Recently, an intervening sequence has been identified in the thymidylate synthetase gene of bacteriophage T4. This 1,017-nucleotide sequence separates the protein-coding sequence into a 5' portion coding for 183 amino acids, followed by a stop codon at the beginning of the intron and 102 3' amino acids preceded by a methionine codon. The methionine codon at the beginning of the 3' exon is not found in the protein product. No evidence has been found of post-translational joining of two separate protein fragments, and preliminary results suggest that splicing occurs at the mRNA level, removing the intron before translation. Bacteriophage T4 specifies an RNA ligase activity, so it would seem capable of carrying out an mRNA-splicing reaction. However, the message is expressed appropriately when cloned in *E. coli*, suggesting that *E. coli* also contains the necessary machinery for splicing (23). Autocatalytic splicing, as characterized in *Tetrahymena* spp., has not been excluded (21), however. Indeed, when *Tetrahymena* spp. splice sites are cloned into *E. coli*, accurate and efficient splicing occurs (105). No other examples of introns interrupting coding sequences are known in *E. coli* or its phages.

During T7 infection, some phage mRNAs are synthesized as long single transcripts from a multicistronic operon. The individual mRNA species are usually separated by RNase III cleavage posttranscriptionally. The sites of cleavage are well characterized and consist of double-stranded regions with structural features usually associated with RNase III recognition sites. Although these cleavages are not required for the translation of most individual mRNAs (133), cleavage probably enhances the translational efficiency of at least some (38, 56). The kinetics of these cleavage reactions have not been characterized (118).

In phage lambda, expression of the *int* gene is under the control of an RNA structural element, the *sib* site, located near the 3′ end of its gene (25). This element consists of a double-stranded stem which contains a cleavage site for RNase III. When the *int* gene is transcribed from the p_L promoter, the association of N protein with RNA polymerase allows transcription to proceed through this double-stranded structure, which would otherwise serve as a terminator. Transcription of the complete stem creates the RNase III site which is cleaved. The 3′ end generated by RNase III is susceptible to attack by a 3′ → 5′ exonuclease which rapidly destroys the *int* message, decreasing *int* expression. The *int* gene can also be transcribed from the p_I promoter, in which case N protein is not associated with the RNA polymerase and transcription terminates in the double-stranded *sib* site. Termination here results in an incompletely formed stem that is not a substrate for RNase III. Consequently, the mRNA is not cleaved by RNase III and is much less susceptible to exonucleolytic attack, allowing for increased *int* expression. This mechanism for controlling mRNA stability, and therefore expression, is thus far unique (26, 50), but suggestive.

One clear example of mRNA processing has been defined in *E. coli*. This involves the cleavage of the *rplJL-rpoBC* transcript in the intercistronic region separating cistrons coding for r-proteins from those coding for RNA polymerase subunits. This cleavage does not occur in RNase III-deficient cells, again implicating this endonuclease. A double-stranded stem consistent with an RNase III site can be drawn at the appropriate site in the primary sequence. This cleavage or the lack of it has no clear effect on expression of any of these messages (see below) (7). There is a suggestion that intercistronic cleavages occur in the *trp* operon and that they are involved in turnover (82; see reference 117).

Finally, mRNA processing of a kind must occur in bifunctional transcripts containing structural RNA cistrons as well as mRNA cistrons (2, 60, 63). Processing of these bifunctional transcripts has not been characterized in detail, but the processing activity or activities that act on the structural RNA should serve to separate the mRNA cistron from the structural RNA. Whether additional mRNA processing occurs in these situations is unknown.

The factors that control mRNA turnover in *E. coli* are not well understood. A useful distinction that has been made is that between functional inactivation and chemical decay. Functional inactivation is an alteration of an mRNA species that results in it being unsuitable for further translation in the absence of bulk mRNA degradation. Chemical decay, on the other hand, refers to the degradation of an mRNA species to mononucleotides. These two processes are theoretically and experimentally dissociable in at least some cases. For example, a temperature-sensitive mutant exists in which the chemical half-life of bulk mRNA is increased at the nonpermissive temperature although the functional half-life of specific messages (in regard to translation) is not affected (22, 100).

Functional inactivation is generally thought to involve a cleavage or other reaction near the 5′ end of an mRNA molecule, which results in destruction of its ribosome-binding site or of a 5′ coding sequence. Chemical decay is thought to result largely from the action of 3′ → 5′ exonucleases, starting from the 3′ end of the message, from internal, endonuclease-induced breaks, or from both (81). Since translation of an mRNA by ribosomes is thought to protect the mRNA from nucleolytic attack, functional inactivation would be expected to promote chemical decay by blocking the protective effect of ribosomes. Specific endonucleolytic cleavages have been implicated in the functional decay of *lac* mRNA. Discrete cleavage products are produced when *lac* mRNA is incubated with RNase III in vitro (121), and functional inactivation of *lac* mRNA in vivo is slowed in mutants deficient in RNase III (120). One of these cleavages occurs near the 5′ terminus of the mRNA, at a site consistent with a *lac* mRNA 5′ terminus demonstrated in cells (19). The other observed cleavages occurred in regions of potential secondary structure 200 to 250 nucleotides from the 5′ end. It is not certain which, if any, of these in vitro cleavages is important for *lac* mRNA functional inactivation in vivo. Discrete endonucleolytic cleavages near the 5′ end of *trp* mRNA have been observed as well (69). These results taken together suggest a role for RNase III in message metabolism.

It is extremely important to note that other mechanisms of functional inactivation are still possible, including ones that are nonnucleolytic. For example, the binding of a protein or of RNA to the ribosome-binding site of an mRNA could block further ribosome addition and then expose the message to decay. This possibility must be taken more seriously with the recent work of Nomura and collaborators (98), who have demonstrated that certain r-proteins bind to initiation sites to shut down the further translation of mRNAs for those proteins. This form of autoregulation controls the translation of 16% of total cellular protein and so is not an insignificant mechanism. In another case, a small RNA (MIC RNA; see above) has been shown to inhibit mRNA translation. Similar regulation by translational inhibitors may apply to other groups of mRNAs.

Strong support for the notion that the binding of proteins at or near the initiation site for protein synthesis can determine the fate of an mRNA comes from the T4 phage gene 32 mRNA (49). In this case, the gene product shows autoregulation, providing another example of nonnucleolytic functional inactivation. In addition, a phage-encoded factor markedly stabilizes this mRNA. Stabilization is dependent on a specific 5' leader sequence in the mRNA, and transfer of this 5' leader sequence to an otherwise unmodified *lac* operon mRNA confers 10-fold stabilization on the hybrid RNA! Although the factor that mediates stabilization is phage encoded, the result is no less compelling and is highly suggestive as a possible means of control in *E. coli*.

These results are most consistent with models in which the rate-determining event in mRNA decay is functional inactivation, with a critical competition and interplay between proteins that determine the loading rate of ribosomes at initiation codons. The direct action of a nuclease (or the preliminary [intrinsically reversible] binding of an inactivating [or stabilizing] factor followed by nuclease action) would terminate ribosome addition and could be followed by chemical destruction of the mRNA. Thus, for example, RNase III cleavage might follow a functional inactivation event mediated by the binding of another protein.

A crucial observation is that chemical decay of many mRNA cistrons occurs in a net 5' → 3' direction, even though no 5' → 3' exonuclease activity has ever been purified from *E. coli* (81, 92). This net directionality implies the cooperation of an endonuclease with a 3' → 5' exonuclease (probably RNase II and polynucleotide phosphorylase; see below) in message metabolism. Multiple internal endonucleolytic cleavages beginning near the 5' end and progressing 3', coupled with 3' → 5' exonucleolytic degradation from exposed 3' ends, could account for the observed pattern. If the presence of ribosomes on mRNA serves as protection from endonucleolytic attack, as has been suggested, then a 5' cleavage (resulting in functional inactivation) would block ribosome initiation and progressively expose the mRNA to nucleases in a 5' → 3' direction. In *trp* mRNA, as might be expected, nonsense mutations leading to premature ribosome release destabilize the mRNA distal to the mutation (93).

These results must be reconciled with the observation that mRNA decay is paradoxically slowed when protein synthesis is blocked by chloramphenicol (27, 62). The slowing or blockage of ribosomes on mRNA by chloramphenicol could result in the same form of protection as active translation of the message if, for instance, only the density or spacing of ribosomes on an mRNA, or the fraction of time that an initiation site was occupied by ribosomes, was important for inactivation. Again, detailed interpretation of results depends on understanding the nature of the initial inactivation events. The endonucleolytic cleavages occurring in a cistron might be specific or nonspecific. As noted above for *lac*, there is a suggestion of specific cleavage sites, but the cleavages may not be the initial events in decay.

The model described above depends heavily on the protection of mRNA by ribosomes. Consistent with that notion, mRNA is labilized when ribosomes are released by puromycin action (reference 28 and references therein). This suggests that messages with more efficient ribosome-binding sites might be more stable. This strong inference has not been tested yet. If the efficiency of ribosome initiation does affect mRNA half-life, then the peptide yield from a given mRNA should be extremely sensitive to changes in the affinity of the ribosome-binding site, as an increase in affinity would increase both the rate of initiation of protein synthesis and the half-life of the mRNA species. Model calculations suggest a roughly exponential dependence of mRNA functional half-life on the rate of ribosome initiation (data not shown).

In one case, mRNA decay has been shown to occur in a net 3' → 5', rather than a net 5' → 3', direction. *ompA* mRNA has a very long half-life (15 min) compared with most mRNAs (0.5 to 2 min) and decays in a 3' → 5' direction. (In this mRNA, coding sequences are as stable as noncoding sequences! Thus, this mRNA may be a model case, or it may, given its very long half-life, represent an unusual case [131]). In another case (T4 gene 32 mRNA, discussed above [49]), a 5' leader sequence clearly controls the stability of the mRNA.

Metabolism of the individual cistrons in multicistronic messages appears to be largely independent. In some operons, the most 5' cistron has the longer half-life, whereas in other operons the more 3' cistrons are more stable (8, 14, 42, 67, 72, 104). Therefore, there is no net 5' → 3' or 3' → 5' direction to the decay of multicistronic mRNAs. This result strengthens the idea that endonucleolytic cleavages or other individual functional inactivation events within coding sequences or at ribosome-binding sites are important for decay. If a simple 3' → 5' exonuclease activity were responsible for all decay, then the more 3' messages would always have a shorter half-life than 5' messages in a multicistronic operon. In the one case in which processing of a multicistronic operon is known (*rplJL-rpoBC*), RNase III-mediated separation of the cistrons has no effect on the half-life of the 5' messages (7). If a 3' → 5' exonuclease were the primary activity involved, RNase III cleavage should decrease the half-life of the 5' cistrons.

Regardless of the initial events in decay, if endonucleolytic cleavage within coding sequences often occurs during mRNA degradation, a significant number of ribosomes translating a message should be blocked on mRNA fragments with no termination codon. This would result in the release of peptidyl tRNA species with partially completed proteins attached. These incomplete protein species could be released from tRNA by peptidyl tRNA hydrolase (86) and subsequently degraded. Estimation of the maximum number of peptidyl tRNAs formed as a fraction of the total number of peptides made can be inferred from the rates of accumulation of peptidyl tRNAs in a mutant deficient in peptidyl tRNA hydrolase (87, 88), from the fraction of incomplete β-galactosidase chains formed in vivo or in vitro (85), or from the fraction of newly formed protein, which turns over very rapidly after its formation (134). All three methods suggest that up to 10 to 20% of translational products end prematurely.

The fraction of these incomplete proteins that result from translation of partially degraded mRNA species is unknown. However, these estimates place an upper limit on the percentage of ribosomes that become blocked on previously cleaved mRNA molecules.

These results do not allow a very clear distinction between models of decay in which a 5' (or near-5') cleavage occurs first and those in which 3' → 5' exonucleolytic degradation is the earlier event. The translational yield of mRNAs in *E. coli* is 20 to 30 peptides per mRNA, on average (71, 83). This implies that up to six ribosomes would be caught on an average truncated mRNA if the 20% estimate (see above) were correct. If all mRNAs were initially inactivated at their 3' terminus (by a 3' → 5' exonuclease), then all of the ribosomes loaded on each mRNA at the time of inactivation would release incomplete peptides. Most mRNAs fully loaded with ribosomes, and especially longer mRNAs, are expected to have more than six ribosomes bound at a given time (71). Therefore, it seems unlikely that initial inactivation occurs at the 3' terminus. The argument becomes stronger if a large number of abortive translations are due to misincorporation of amino acids or attenuation rather than to mRNA fragmentation. If this were the case, initial mRNA inactivation would necessarily occur near the 5' terminus.

A tacit assumption of the overall 5' → 3' decay mechanisms is that 3' → 5' exonucleolytic decay beginning at the 3' terminus is slow relative to inactivating endonuclease cleavages. Secondary structural features at 3' termini could serve as a barrier to an exonuclease. In particular, double-stranded features are a part of most termination signals and could be important. The major 3' terminus of *trp* mRNA has been shown to be generated in this manner. Most termination occurs at *trp* t', but most transcripts are found to have 3' termini located at a more proximal secondary structural feature, *trp* t. This species can be produced in vitro by RNase II-mediated exonucleolytic degradation from the *trp* t' terminator to the *trp* t hairpin (94). An interesting possibility is that REP sequences could serve as barriers to 3' decay. These highly conserved sequence elements are found at the ends of operons and in intercistronic regions in many operons. They often are present as inverted repeats capable of forming long, double-stranded stems. They have been shown not to be sites for RNase III cleavage; thus, it is unlikely that they have a function similar to the *sib* site in λ retroregulation. They also do not appear to act as terminators, and the expression of distal genes is little affected by the insertion of a REP sequence proximal to them. Recent work suggests that the deletion of REP sequences decreases the half-life of mRNA cistrons located 5' to the sequence. This result is consistent with REP sequences serving as barriers to 3' → 5' exonuclease degradation (as appears to occur in the *trp* operon). These highly conserved sequences may serve other roles as well.

Experimental evidence implicates RNase II and polynucleotide phosphorylase as the major exonucleolytic activities involved in mRNA chemical decay. Temperature-sensitive mutants in both of these activities show chemical decay of mRNA species at the nonpermissive temperature. In one study, functional decay was slowed as well (75), but in another study the cells showed little effect until long after temperature shift, suggesting that functional decay was not greatly affected (although it was not specifically assessed [37]). Other exonucleases may participate in mRNA decay in *E. coli*, but their roles appear to be minor compared with those of RNase II and polynucleotide phosphorylase. Mutants deficient in one of these enzymes do not show any obvious changes in mRNA metabolism, suggesting that the two are largely complementary.

The identity of endonucleases involved in mRNA turnover is less certain. RNase III is implicated in several systems, and its role in mRNA metabolism may be more general than is currently appreciated. RNase III mutants (which retain some RNase III activity) show altered expression of many different proteins. This could result from altered mRNA metabolism or from the abnormal 50S ribosomal subunits present in these strains (see above). No other endonucleases are clearly involved in message metabolism, although a number of partially purified activities could have a role in functional or chemical decay or both (Table 1). In particular, RNase N has been speculated to play a role in mRNA metabolism (89).

Small RNAs

In general, little is known about the processing or turnover of small RNAs in *E. coli*. A number of them (RI RNA, MIC RNA, and spot 42 RNA) require no processing, as the primary transcript is identical to the mature species. 6S RNA requires posttranscriptional processing; however, the activities involved are not known, and the rates of reaction have not been measured. M1 RNA processing is blocked in an RNase E temperature-sensitive mutant, but purified RNase E does not cleave the M1 RNA precursor (53), leaving its mechanism of processing in doubt as well. A 4.5S RNA precursor accumulates in an RNase P-deficient mutant, implying that it is processed by RNase P at the base of its long double-stranded stem (15). *divE* and *dnaY* are presumably processed as are other tRNA species.

As far as turnover is concerned, M1 RNA, 6S RNA, and 4.5S RNA are stable (though M1 RNA with a particular point mutation has a much decreased half-life [115]). *divE* and *dnaY* are probably stable (as they are tRNAs), but no experimental information is available. RI RNA and MIC RNA are presumably unstable, as their role is clearly regulatory, but no information is available concerning their fate. Spot 42 RNA has a half-life of about 20 min at 37°C. The pathway and enzymes responsible for its degradation are unknown.

The fate of RNA primers for DNA replication is not certain. RNase H can remove all but the last nucleotide of each primer but cannot cleave the bond between RNA and DNA. DNA polymerase I is capable of degrading RNA, as well as DNA, so that it is a likely candidate. The issue is currently unsettled, however (30).

RNA fragments derived from the processing of rRNA and tRNA are rapidly degraded, most likely by RNase II and polynucleotide phosphorylase. The fragments with extensive secondary structure (e.g., the RNase III stems bracketing 16S and 23S rRNA) may be longer lived, requiring other enzymes for their degradation.

RNase III is capable of hydrolyzing double-stranded RNA to form oligonucleotides under some conditions (29, 80) and may contribute to the degradation of the double-stranded stems removed from precursor RNAs.

FUNCTION OF RNA PROCESSING

The function of processing is clear for some RNA species since precursor species are not functional. In other cases, the function of processing is less obvious. Similarly, the function of turnover is relatively clear for mRNA but could subserve multiple functions or be an undesirable side effect for other RNA species.

tRNA

The function of tRNA processing is clear, as unprocessed tRNA precursors cannot function in protein synthesis. It seems likely that the existence of extra sequences on nascent tRNAs, in part, reflects a limitation of the transcriptional apparatus, as in no case does a tRNA primary transcript begin or end at the mature termini of the tRNA. Since mature tRNAs have only a single known secondary structure, which is easily recovered after thermal denaturation, it is unlikely that precursor sequences have any role in determining this structure. The functional significance of cotranscribed mRNA genes found in three tRNA operons is uncertain. This coupling may be a means of controlling the relative abundance of the cotranscribed species. Whether tRNA processing serves other functions is less clear (see below).

rRNA

As with tRNAs, transcription of a mature rRNA species as a primary transcript may not be possible with the *E. coli* transcription apparatus. However, not all, and possibly very few, of the rRNA-processing reactions are required for the resultant rRNA species to function. Other functions of processing for rRNA and possibly for other stable RNAs must be postulated.

Usually, the existence of signals for cleavages, methylation, etc., in extra sequences has been invoked as a rationalization for processing; it has even been suggested, for example, that ribose methylation of pre-rRNA may function by providing signals for processing cleavages! Rather than arguing in a circle that extra sequences are formed so that they can catalyze their own removal, we have argued instead that extra transcribed sequences are in themselves important and that processing removes the extra sequences after a mature conformation is adopted (73).

At least six kinds of functions for transcribed spacers and processing can thus be envisaged.

Equimolar synthesis of rRNA species. Extra sequences permit the cotranscription of rRNA species. The organization of *rrn* gene transcripts ensures that equal quantities of the 5S, 16S, and 23S rRNA species found in 70S ribosomes are formed. This argument is suggestive but not compelling, since many proteins are formed in equimolar amounts from different promoters; many tRNAs are formed in monocistronic or polycistronic transcripts without any obvious relationship to cell physiology (43), and polarity effects (125) might still act to produce unequal amounts of 16S and 23S rRNAs. Nevertheless, the argument has a force similar to that for the existence of polycistronic mRNAs for many biosynthetic operons, and the cotranscription of certain structural RNAs with specific mRNA species suggests the possibility of a functional relationship between them (see below).

Signals for regulatory events. The extra sequences in RNAs may have special metabolic roles, although there is currently no experimental support for this notion.

Promotion of secondary structure formation. Since transcribed spacer sequences are eliminated after ribosomes are formed, any biological role they have must be fulfilled during ribosome formation. In fact, pre-rRNA seems to form ribosomal particles in vitro more easily than does mature rRNA (97, 132). Spacer sequences could facilitate ribosome formation by promoting alternate conformational features in rRNA. In particular, the structure of 16S rRNA in the 30S pre-rRNA contains several very large loops that would interrupt certain shorter-range base-paired stems known to exist in the secondary structure of the mature rRNA (77). Formation of 30S ribosomes from mature 16S rRNA in vitro has been found to require a greater activation energy than is available in intact cells (130), and the alternate conformational features in pre-rRNA may allow a more favorable kinetic route.

Verification of functionality: quality control. The signals for processing steps can be very complex; especially for the final steps, the demands on secondary and tertiary structure of rRNA and tRNA can be very exacting (see above). Thus, processing both verifies that a newly forming ribosome is developing normally and ensures that the final structure will be rigorously standardized. For example, single point mutations in 16S rRNA can prevent the correct processing of the RNA to its fully functional form (125). If either conformation or sequence is aberrant, processing tends to fail and prevents a distorted precursor from becoming an aberrant mature RNA chain.

Time of production: quantity control. Theoretically, the slowing of processing steps could lead to the accumulation of precursors of an RNA species, reducing the mature fraction of RNA for that species. In this way, processing could regulate the rate of production of individual types of RNA. Although the rates of different processing reactions are not precisely coordinated in cells growing at different rates (73), the fraction of any RNA species in the form of precursors is usually relatively small (<10%). This observation suggests that the rate of processing could subserve a regulatory function.

Integration of cellular processes. Ordinarily, every kinetic process in a cell is arranged so that particular steps are required before others can occur; however, in a number of cases, the blockage of later steps leads to a feedback regulation of earlier steps. An example is the inhibition of rRNA synthesis and tRNA synthesis when protein synthesis is blocked. The final processing of rRNA may be analogous in that it occurs in polysomes (54, 84, 136) and may require active protein synthesis (20, 122). At present, it is not clear how extensive the coupling of protein synthesis and rRNA processing may be, but it is conceivable that the rate of one can directly influence the rate of the other.

Instability and turnover. Because *E. coli* contains enzymes capable of catalyzing the turnover of any cellular RNA in vitro (44), it is clear that special measures are required to render some species of RNA metabolically stable. Mutations in tRNA structure that labilize the molecule in vivo (123) demonstrate that the intact normal secondary structure is required for stability. It seems likely that, during transcription and further biosynthetic reactions, extra transcribed sequences might help to prevent degradation of unfinished RNA species before their final resistant form is attained.

mRNA

Although processing of mRNA, including splicing, clearly has a regulatory function in mammalian cells (in which it determines the rate of production of a species and can even produce more than one mRNA from a single transcript by alternative pathways), similar events in bacteria are rare. Again, cotranscription of related cistrons as polycistronic mRNAs requires the existence of spacer sequences. Signals for attenuation and leader peptide production are other possible functions of spacer sequences (see above). However, the metabolism of mRNA is dominated not by processing reactions that lead to its maturation but by processing reactions that lead to its degradation.

The chief reasons for mRNA instability are, first, the rapid control of the protein synthetic repertoire of cells, so that shifts in response to changing metabolic needs can be rapid and decisive; and, second, the differential translation yield from different mRNA species, so that the balance of the rates of ribosome initiation and translation with the rates of mRNA inactivation and degradation determine just how much of a given protein is made from a certain amount of the transcribed mRNA.

In the metabolism of unstable mRNA, extra transcribed sequences can be involved in several ways which have already been mentioned: the provision of structural blocks to nuclease action or sites of nuclease attack. How important such sites are in translated sequences compared with untranslated sequences has not been determined, but it seems likely that, as in the lambda *sib* and phage gene 32 cases, sites in untranslated sequences will be quite important.

Small RNAs

As the number of small RNAs with defined function increases, their stability and any functions of processing should become clearer. For example, it is obvious that the MIC RNA must be metabolically unstable, but, for M1 RNA, only speculative possibilities similar to some of those discussed above for tRNA are available to guess at possible functions for extra transcribed sequences or processing steps. The role of cotranscribed open reading frames found in a number of these RNA species is also speculative.

CONCLUSION

It now seems increasingly likely that evolution initially developed extra transcribed sequences as important functional features. Such sequences can promote particular conformations or can help to protect an incomplete RNA chain until it is in its final form. According to this view, processing reactions were developed later, to remove extra sequences and generate an RNA molecule in its final mature form. The processing cleavages and trimming might then be analogous to modifications like methylation, fine tuning reactions to improve the performance of an already partially functional RNA (as in the case of the 23S pre-rRNA). In turn, the processing steps become a series of checkpoints to verify that the assembly of an RNA or of an RNA:protein complex has proceeded correctly before the final production of the maximally active cellular species. It is important to distinguish function from stability; some structural RNAs can be unstable, and some mRNAs can be long lived. But in each case, the structure of an RNA transcript serves as a blueprint for its own maturation and, where appropriate, for its own turnover.

ADDENDUM IN PROOF

An extension of this treatment to other species can be found elsewhere (T. C. King, R. Sirdeshmukh, and D. Schlessinger, Microbiol. Rev. 50:428–451, 1986).

LITERATURE CITED

1. **Altman, S., C. Guerrier-Takada, H. M. Frankfort, and H. D. Robertson.** 1982. RNA-processing nucleases, p. 243–274. *In* S. M. Linn and R. J. Roberts (ed.), Nucleases. Cold Spring Harbor Laboratory, Cold Spring Harbor, N.Y.
2. **Altman, S., P. Model, G. H. Dixon, and M. A. Wosnick.** 1981. An *E. coli* gene coding for a protamine-like protein. Cell **26:**299–304.
3. **Apirion, D., and P. Gegenheimer.** 1984. Molecular biology of RNA processing in prokaryotic cells, p. 36–52. *In* D. Apirion (ed.), Processing of RNA. CRC Press, Inc., Boca Raton, Fla.
4. **Asha, P. K., R. T. Blouin, R. Zaniewski, and M. P. Deutscher.** 1983. Ribonuclease BN: identification and partial characterization of a new tRNA processing enzyme. Proc. Natl. Acad. Sci. USA **80:**3301–3304.
5. **Bachmann, B. J.** 1983. Linkage map of *Escherichia coli* K-12, edition 7. Microbiol. Rev. **47:**180–230.
6. **Baer, M., and S. Altman.** 1985. A catalytic RNA and its gene from *Salmonella typhimurium*. Science **228:**999–1002.
7. **Barry, G., C. Squires, and L. Squires.** 1980. Attenuation and processing of RNA from the *rplJL-rpoBC* transcription unit of *Escherichia coli*. Proc. Natl. Acad. Sci. USA **77:**3331–3335.
8. **Belasco, J. G., J. T. Beatty, C. W. Adams, A. Von Gabain, and S. N. Cohen.** 1985. Differential expression of photosynthesis genes in *R. capsulata* results from segmental differences in stability within the polycistronic *rxcA* transcript. Cell **40:**171–181.
9. **Ben-Hamida, F., and D. Schlessinger.** 1966. Synthesis and turnover of ribonucleic acid in *Escherichia coli* starving for nitrogen. Biochim. Biophys. Acta **119:**183–191.
10. **Birenbaum, M., D. Schlessinger, and S. Hashimoto.** 1978. RNase III cleavage of *Escherichia coli* rRNA precursors: fragment release and dependence on salt concentration. Biochemistry **17:**298–307.
11. **Blanchetot, A., E. Hajnsdorf, and A. Favre.** 1984. Metabolism of tRNA in near-ultraviolet illuminated *Escherichia coli*. The tRNA repair hypothesis. Eur. J. Biochem. **139:**547–552.
12. **Blouin, R. T., R. Zaniewski, and M. P. Deutscher.** 1983. Ribonuclease D is not essential for the normal growth of *Escherichia coli* or bacteriophage T4 or for the biosynthesis of a T4 suppressor tRNA. J. Biol. Chem. **258:**1423–1426.
13. **Blumer, K. J., and D. A. Steege.** 1984. mRNA processing in *Escherichia coli*: an activity encoded by the host processes bacteriophage f1 mRNA. Nucleic Acids Res. **12:**1847–1861.
14. **Blundell, M., E. Craig, and D. Kennell.** 1972. Decay rates of different mRNA in *E. coli* and models of decay. Nature (London) New Biol. **238:**46–49.
15. **Bothwell, A. L., R. L. Garber, and S. Altman.** 1976. Nucleotide sequence and *in vitro* processing of a precursor molecule to *Escherichia coli* 4.5S RNA. J. Biol. Chem. **251:**7709–7716.
16. **Bram, R. J., R. A. Young, and J. A. Steitz.** 1980. The ribonucle-

ase III site flanking 23S sequences in the 30S ribosomal precursor RNA of *E. coli*. Cell **19**:393–401.

17. **Brosius, J., T. J. Dull, D. D. Sleeter, and H. F. Noller.** 1981. Gene organization and primary structure of a ribosomal RNA operon from *Escherichia coli*. J. Mol. Biol. **148**:107–127.

18. **Brown, S., and M. J. Fournier.** 1984. The 4.5S RNA gene of *Escherichia coli* is essential for cell growth. J. Mol. Biol. **178**:533–550.

19. **Cannistraro, V. J., and D. Kennell.** 1985. The 5′ ends of *E. coli lac* mRNA. J. Mol. Biol. **182**:241–248.

20. **Ceccarelli, A., G. P. Potto, F. Altruda, C. Perlo, L. Silengo, E. Turco, and G. Mangiarotti.** 1978. Immature 50S subunits in *Escherichia coli* polysomes. FEBS Lett. **93**:348–350.

21. **Cech, T. R., A. S. Zaug, and P. J. Grabowski.** 1981. *In vitro* splicing of the ribosomal RNA precursor of Tetrahymena: involvement of a guanosine nucleotide in the excision of the intervening sequence. Cell **27**:487–496.

22. **Chanda, P. K., M. Ono, M. Kuwano, and H.-F. Kung.** 1985. Cloning, sequence analysis, and expression of alteration of mRNA stability gene (*ams*) of *E. coli*. J. Bacteriol. **161**:446–449.

23. **Chu, F. K., G.-F. Maley, F. Maley, and M. Belfort.** 1984. Intervening sequences in the thymidylate synthase gene of bacteriophage T4. Proc. Natl. Acad. Sci. USA **81**:3049–3053.

24. **Conrad, B. E., and J. L. Campbell.** 1979. Role of plasmid-coded RNA and ribonuclease III in plasmid DNA replication. Cell **18**:61–71.

25. **Court, D., T. F. Huang, and A. B. Oppenheim.** 1983. Deletion analysis of the retroregulatory site for the lambda int gene. J. Mol. Biol. **166**:233–240.

26. **Court, D., U. Schneissner, M. Rosenberg, A. Oppenheim, G. Guarneros, and C. Montanez.** 1983. Processing of λ *int* RNA: mechanism for gene control, p. 78–81. *In* D. Schlessinger (ed.), Microbiology—1983. American Society for Microbiology, Washington, D.C.

27. **Cremer, K., and D. Schlessinger.** 1974. Ca²⁺ ions inhibit messenger ribonucleic acid degradation, but permit messenger ribonucleic acid transcription and translation in deoxyribonucleic acid-coupled systems from *Escherichia coli*. J. Biol. Chem. **249**:4730–4736.

28. **Cremer, K., L. Silengo, and D. Schlessinger.** 1974. Polypeptide and polyribosome formation in *E. coli* treated with chloramphenicol. J. Bacteriol. **118**:582–591.

29. **Crouch, R. F.** 1974. Ribonuclease III does not degrade deoxyribonucleic acid-ribonucleic acid hybrids. J. Biol. Chem. **249**:1314–1316.

30. **Crouch, R. J., and M. L. Dirksen.** 1982. Ribonucleases H, p. 211–241. *In* S. M. Linn and R. J. Roberts (ed.), Nucleases. Cold Spring Harbor Laboratory, Cold Spring Harbor, N.Y.

31. **Cudney, H., and M. P. Deutscher.** 1980. Apparent involvement of ribonuclease D in the 3′ processing of tRNA precursors. Proc. Natl. Acad. Sci. USA **77**:837–841.

32. **Dahlberg, A. E., J. E. Dahlberg, E. Lund, H. Tokimatsu, A. B. Rabson, P. C. Calvert, F. Reynolds, and M. Zahalak.** 1978. Processing of the 5′ end of *Escherichia coli* 16S ribosomal RNA. Proc. Natl. Acad. Sci. USA **75**:3598–3602.

33. **Dahlberg, A. E., H. Tokimatsu, M. Zahalak, F. Reynolds, P. Calvert, and A. B. Rabson.** 1977. Processing of the 17S precursor ribosomal RNA, p. 509–517. *In* J. H. Vogel (ed.), Nucleic acid protein recognition. Academic Press, Inc., New York.

34. **Deutscher, M. P.** 1984. Processing of tRNA in prokaryotes and eukaryotes. Crit. Rev. Biochem. **17**:45–71.

35. **Deutscher, M. P., J. J. C. Lin, and J. A. Evans.** 1977. Transfer RNA metabolism in *Escherichia coli* cells deficient in tRNA nucleotidyl transferase. J. Mol. Biol. **117**:1081–1099.

36. **Deutscher, M. P., C. W. Marlor, and R. Zaniewski.** 1984. Ribonuclease T: new exoribonuclease possibly involved in end-turnover of tRNA. Proc. Natl. Acad. Sci. USA **81**:4290–4293.

37. **Donovan, W. P., and S. R. Kushner.** 1986. Polynucleotide phosphorylase and ribonuclease II are required for cell viability and mRNA turnover in *Escherichia coli* K12. Proc. Natl. Acad. Sci. USA **83**:120–124.

38. **Dunn, J. J., and F. W. Studier.** 1973. T7 early RNAs and *Escherichia coli* ribosomal RNAs are cut from large precursor RNAs *in vivo* by ribonuclease III. Proc. Natl. Acad. Sci. USA **70**:3296–3300.

39. **Dunn, J. J., and F. W. Studier.** 1981. Nucleotide sequence from the genetic left end of bacteriophage T7 DNA to the beginning of gene 4. J. Mol. Biol. **148**:303–330.

40. **Feunteun, J., B. K. Jordan, and R. Monier.** 1972. Study of the maturation of 5S RNA precursors in *Escherichia coli*. J. Mol. Biol. **70**:465–474.

41. **Fiil, N. P., J. D. Friesen, W. L. Downing, and P. P. Dennis.** 1980.

Posttranscriptional regulatory mutants in a ribosomal protein-RNA polymerase operon of *E. coli*. Cell **19**:837–844.

42. **Forchhammer, J., E. N. Jackson, and C. Yanofsky.** 1972. Different half-lives of messenger RNA corresponding to different segments of the tryptophan operon of *Escherichia coli*. J. Mol. Biol. **71**:687–699.

43. **Fournier, M. J., and H. Ozeki.** 1985. Structure and organization of the transfer ribonucleic acid genes of *Escherichia coli* K-12. Microbiol. Rev. **49**:379–397.

44. **Frankfort, H. M., and H. D. Robertson.** 1982. Appendix C, p. 359–366. *In* S. M. Linn and R. J. Roberts (ed.), Nucleases. Cold Spring Harbor Laboratory, Cold Spring Harbor, N.Y.

45. **Fukada, K., and J. Abelson.** 1980. DNA sequence of a T4 transfer RNA gene cluster. J. Mol. Biol. **139**:377–391.

46. **Furdon, P. J., C. Guerrier-Takada, and S. Altman.** 1983. A G43 to U43 mutation in *E. coli* tRNA tyrsu3⁺ which affects processing by RNase P. Nucleic Acids Res. **11**:1491–1505.

47. **Gitelman, D. R., and D. Apirion.** 1980. The synthesis of some proteins is affected in RNA processing mutants of *Escherichia coli*. Biochem. Biophys. Res. Commun. **96**:1063–1070.

48. **Goldfarb, A., and V. Daniel.** 1980. An *Escherichia coli* endonuclease responsible for primary cleavage of *in vitro* transcripts of bacteriophage T4 tRNA gene cluster. Nucleic Acids Res. **8**:4501–4516.

49. **Gorski, K., J. Roch, P. Prentki, and H. M. Kirsch.** 1985. The stability of bacteriophage T4 gene 32 mRNA: a 5′ leader sequence that can stabilize mRNA transcripts. Cell **43**:461–469.

50. **Guarneros, G., C. Montanez, T. Fernandez, and D. Court.** 1982. Posttranscriptional control of bacteriophage λ int gene expression from a site distal to the gene. Proc. Natl. Acad. Sci. USA **79**:238–242.

51. **Guerrier-Takada, C., and S. Altman.** 1984. Catalytic activity of an RNA molecule prepared by transcription *in vitro*. Science **223**:285–286.

52. **Guerrier-Takada, C., W. H. McClain, and S. Altman.** 1984. Cleavage of tRNA precursors by the RNA subunit of *E. coli* ribonuclease P (M1 RNA) is influenced by 3′ proximal CCA in the substrates. Cell **38**:219–224.

53. **Gurevitz, M., S. K. Jain, and D. Apirion.** 1983. Identification of a precursor molecule for the RNA moiety of processing enzyme RNase P. Proc. Natl. Acad. Sci. USA **80**:4450–4454.

54. **Harvey, R. J.** 1975. Association of nascent ribosomal ribonucleic acid with polyribosomes in *Escherichia coli*. J. Bacteriol. **124**:1330–1343.

55. **Hayes, F., and M. Vasseur.** 1976. Processing of the 17-S *Escherichia coli* precursor RNA in the 27-S pre-ribosomal particle. Eur. J. Biochem. **61**:433–442.

56. **Hercules, K., M. Schweiger, and W. Sauerbier.** 1974. Cleavage by RNase III converts T3 and T7 early precursor RNA into translatable message. Proc. Natl. Acad. Sci. USA **71**:840–844.

57. **Higgens, C. F., G. F. Ames, W. M. Barnes, J. M. Clement, and M. Hofnung.** 1982. A novel intercistronic regulatory element of prokaryotic operons. Nature (London) **298**:760–762.

58. **Hsu, L., J. Zagorski, W. Huang, and M. J. Fournier.** 1985. The *Escherichia coli* 6S RNA gene is part of a dual function transcription unit. J. Bacteriol. **161**:1162–1170.

59. **Hsu, L. M., J. Zagorski, and M. J. Fournier.** 1984. Cloning and sequence analysis of the *Escherichia coli* 4.5S RNA gene. J. Mol. Biol. **178**:509–531.

60. **Hudson, L., J. Rossi, and A. Landy.** 1981. Dual function transcript specifying tRNA and mRNA. Nature (London) **294**:422–427.

61. **Ikemura, T., and H. Ozeki.** 1977. Gross map location of *Escherichia coli* transfer RNA genes. J. Mol. Biol. **117**:419–446.

62. **Imamoto, F.** 1973. Diversity of regulation of genetic transcription. I. Effect of antibiotics which inhibit the process of translation on RNA metabolism in *Escherichia coli*. J. Mol. Biol. **74**:113–136.

63. **Ishii, S., K. Kuroki, and F. Imamoto.** 1984. tRNA Met f2 genes in the leader region of the *nusA* operon in *Escherichia coli*. Proc. Natl. Acad. Sci. USA **81**:409–413.

64. **Ito, R., and Y. Ohnishi.** 1983. The roles of RNA polymerase and RNase I in stable RNA degradation in *Escherichia coli* carrying the srnB⁺ gene. Biochim. Biophys. Acta **739**:27–34.

65. **Jakubowski, H., and E. Goldman.** 1984. Quantities of individual aminoacyl-tRNA families and their turnover in *Escherichia coli*. J. Bacteriol. **158**:769–776.

66. **Jordan, B. R., J. Feuteun, and R. Monier.** 1970. Identification of a 5S RNA precursor in exponentially growing *Escherichia coli* cells. J. Mol. Biol. **50**:605–615.

67. **Joseph, D., A. Danchin, and A. Ullmann.** 1978. Modulation of the lactose operon mRNA turnover by inhibitors of dihydrofolate

reductase. Biochem. Biophys. Res. Commun. **84**:769–776.

68. **Kanaya, S., and R. J. Crouch.** 1984. The *rnh* gene is essential for growth of *Escherichia coli.* Proc. Natl. Acad. Sci. USA **81**:3447–3451.

69. **Kano, Y., and F. Imamoto.** 1979. Evidence for endonucleolytic cleavage at the 5′ proximal segment of the *trp* messenger RNA in *Escherichia coli.* Mol. Gen. Genet. **172**:25–30.

70. **Kasai, T., R. S. Gupta, and D. Schlessinger.** 1977. Exoribonucleases in wild type *Escherichia coli* and RNase II-deficient mutants. J. Biol. Chem. **252**:8950–8956.

71. **Kennell, D., and H. Riezman.** 1977. Transcription and translation frequencies of the *Escherichia coli lac* operon. J. Mol. Biol. **114**:1–21.

72. **Kepes, A.** 1967. Sequential transcription and translation in the lactose operon of *Escherichia coli.* Biochim. Biophys. Acta **138**:107–123.

73. **King, T. C., and D. Schlessinger.** 1983. S1 nuclease mapping analysis of ribosomal RNA processing in wild type and processing deficient *Escherichia coli.* J. Biol. Chem. **258**:12034–12042.

74. **King, T. C., R. Sirdeshmukh, and D. Schlessinger.** 1984. RNase III cleavage is obligate for maturation but not for function of *Escherichia coli* pre-23S rRNA. Proc. Natl. Acad. Sci. USA **81**:185–188.

75. **Kinscherf, T. G., and D. Apirion.** 1975. Polynucleotide phosphorylase can participate in decay of mRNA in *Escherichia coli* in the absence of RNase II. Mol. Gen. Genet. **139**:357–362.

76. **Klein, B. K., A. Staden, and D. Schlessinger.** 1985. Electron microscopy of secondary structures, partially denatured precursor and mature *Escherichia coli* 16S and 23S rRNA. J. Biol. Chem. **260**:8114–8120.

77. **Klein, B. K., A. Staden, and D. Schlessinger.** 1985. Alternate conformations in *Escherichia coli* 16S ribosomal RNA. Proc. Natl. Acad. Sci. USA **82**:3539–3542.

78. **Kornberg, A.** 1982. Supplement to *DNA Replication*, p. S104. W. H. Freeman & Co., San Francisco.

79. **Lee, C. A., M. J. Fournier, and J. Beckwith.** 1985. The 6S RNA of *Escherichia coli* is not essential for growth or protein secretion. J. Bacteriol. **161**:1156–1161.

80. **Libonati, M., A. Carsana, and A. Furia.** 1980. Double-stranded RNA. Mol. Cell. Biochem. **31**:147–164.

81. **Lim, L. W., and D. Kennell.** 1979. Models for decay of *Escherichia coli lac* messenger RNA and evidence for inactivating cleavages between its messages. J. Mol. Biol. **135**:369–390.

82. **Lim, L. W., and D. Kennell.** 1980. Evidence for random endonucleolytic cleavages between messages in decay of *Escherichia coli trp* mRNA. J. Mol. Biol. **141**:227–233.

83. **Mangiarotti, G., and D. Schlessinger.** 1967. Polyribosome metabolism in *Escherichia coli.* II. Formation and lifetime of messenger RNA molecules, ribosomal subunit couples and polyribosomes. J. Mol. Biol. **29**:395–418.

84. **Mangiarotti, G., E. Turco, A. Ponzetto, and F. Altruda.** 1974. Precursor 16S RNA in active 30S ribosomes. Nature (London) **247**:147–148.

85. **Manley, J. L.** 1978. Synthesis and degradation of termination and premature-termination fragments of β-galactosidase *in vitro* and *in vivo.* J. Mol. Biol. **125**:407–432.

86. **Menninger, J. R.** 1976. Peptidyl transfer RNA dissociates during protein synthesis from ribosomes of *Escherichia coli.* J. Biol. Chem. **251**:3392–3398.

87. **Menninger, J. R.** 1978. The accumulation as peptidyl-transfer RNA of isoaccepting transfer RNA families in *Escherichia coli* with temperature-sensitive peptidyl transfer RNA hydrolase. J. Biol. Chem. **253**:68608–68613.

88. **Menninger, J. R., and D. P. Orto.** 1982. Erythromycin, carbomycin, and spiramycin inhibit protein synthesis by stimulating the dissociation of peptidyl tRNA from ribosomes. Antimicrob. Agents Chemother. **21**:811–818.

89. **Misra, T. K., and D. Apirion.** 1978. Characterization of an endoribonuclease, RNase N, from *Escherichia coli.* J. Biol. Chem. **253**:5594–5599.

90. **Misra, T. K., and D. Apirion.** 1979. RNase E, an RNA processing enzyme in *Escherichia coli.* J. Biol. Chem. **254**:11154–11159.

91. **Mizuno, T., M. Chou, and M. Inouye.** 1984. A unique mechanism regulating gene expression: translational inhibition by a complementary RNA transcript (mic RNA). Proc. Natl. Acad. Sci. USA **81**:1966–1970.

92. **Morikawa, N., and F. Imamoto.** 1969. Degradation of tryptophan messenger. Nature (London) **223**:37–40.

93. **Morse, D. E., and C. Yanofsky.** 1969. Polarity and the degradation of mRNA. Nature (London) **224**:329–331.

94. **Mott, J. E., J. L. Galloway, and T. Platt.** 1985. The mature 3′ end of *Escherichia coli* tryptophan operon mRNA is generated by processing after rho dependent termination. EMBO J. **4**:1887–1891.

95. **Mullin, D. A., G. M. Garcia, and J. R. Walker.** 1984. An *E. coli* DNA fragment 118 base pairs in length provides dnaY⁺ complementary activity. Cell **37**:669–674.

96. **Neu, H. C., and L. A. Heppel.** 1954. Some observations on the "latent" ribonuclease of *Escherichia coli.* Proc. Natl. Acad. Sci. USA **51**:1267–1274.

97. **Nikolaev, N., K. Glazier, and D. Schlessinger.** 1985. Cleavage by ribonuclease III of the complex of 30S pre-ribosomal RNA and ribosomal proteins of *Escherichia coli.* J. Mol. Biol. **94**:301–304.

98. **Nomura, M., R. Gourse, and G. Baughman.** 1984. Regulation of the synthesis of ribosomes and ribosomal components. Annu. Rev. Biochem. **53**:75–117.

99. **Okazaki, T., S. Hirose, A. Fujiyama, and Y. Kohara.** 1980. Mapping of initiation sites of DNA replication on prokaryotic genomes, p. 429–447. *In* D. Alberts (ed.), Mechanistic studies of DNA replication and genetic recombination. Academic Press, Inc., New York.

100. **Ono, M., and M. Kuwano.** 1979. A conditional lethal mutation in an *Escherichia coli* strain with a longer chemical lifetime of messenger RNA. J. Mol. Biol. **129**:343–357.

101. **Ono, T., and Y. Ohnishi.** 1981. Degradation of ribosomal RNA in bacteriophage lambda lysogens after thermal induction. Microbiol. Immunol. **25**:433–444.

102. **Pace, B., D. A. Stahl, and N. R. Pace.** 1984. The catalytic element of a ribosomal RNA-processing complex. J. Biol. Chem. **259**:11454–11458.

103. **Pace, N. R.** 1984. Protein-polynucleotide recognition and the RNA processing nucleases in prokaryotes, p. 1–24. *In* D. Apirion (ed.), Processing of RNA. CRC Press, Inc., Boca Raton, Fla.

104. **Pastushok, C., and D. Kennell.** 1974. Residual polarity and transcription translation coupling during recovery from chloramphenicol or fusidic acid. J. Bacteriol. **117**:631–640.

105. **Price, J. V., and T. R. Cech.** 1985. Coupling of *Tetrahymena* ribosomal RNA splicing to β-galactosidase expression in *Escherichia coli.* Science **228**:719–722.

106. **Raziuddin, D. Chatterji, B. Ghosh, and D. P. Burma.** 1979. Site of action of RNase I on the 50 S ribosome of *Escherichia coli* and the association of the enzyme with the partially degraded subunit. J. Biol. Chem. **254**:10575–10578.

107. **Reed, R. E., and S. Altman.** 1983. Repeated sequences and open reading frames in the 3′ flanking region of the gene for the RNA subunit of *Escherichia coli* ribonuclease P. Proc. Natl. Acad. Sci. USA **80**:5359–5363.

108. **Reed, R. E., M. F. Baer, C. Guerrier-Takada, H. Donis-Keller, and S. Altman.** 1982. Nucleotide sequence of the gene encoding the RNA subunit (M1 RNA) of ribonuclease P from *Escherichia coli.* Cell **30**:627–636.

109. **Rice, P. W., and J. E. Dahlberg.** 1982. A gene between *polA* and *gluA* retards growth of *Escherichia coli* when present in multiple copies: physiological effects of the gene for spot 42 RNA. J. Bacteriol. **152**:1195–1210.

110. **Robertson, H. D.** 1976. Structure and function of RNA processing signals, p. 549–568. *In* H. J. Vogel (ed.), Nucleic acid-protein recognition. Academic Press, Inc., New York.

111. **Robertson, H. D.** 1982. *Escherichia coli* ribonuclease III cleavage sites. Cell **30**:669–672.

112. **Rossi, J., J. Egan, L. Hudson, and A. Landy.** 1981. The tyrT locus: termination and processing of a complex transcript. Cell **26**:305–314.

113. **Sahagan, B. G., and J. E. Dahlberg.** 1979. A small unstable RNA molecule of *Escherichia coli*: spot 42 RNA. I. Nucleotide sequence analysis. J. Mol. Biol. **131**:573–592.

114. **Sahagan, B. G., and J. E. Dahlberg.** 1979. A small unstable RNA molecule of *Escherichia coli*: spot 42 RNA. II. Accumulation and distribution. J. Mol. Biol. **131**:593–605.

115. **Sakamoto, H., M. Kimura, F. Nagawa, and Y. Shimura.** 1983. Nucleotide sequence and stability of the RNA component of RNase P from a temperature-sensitive mutant of *E. coli.* Nucleic Acids Res. **11**:8237–8251.

116. **Sakano, H., and M. Shimura.** 1975. Sequential processing of precursor tRNA molecules in *Escherichia coli.* Proc. Natl. Acad. Sci. USA **72**:3369–3373.

117. **Schlessinger, D., K. A. Jacobs, R. S. Gupta, Y. Kano, and F. Imamoto.** 1977. Decay of individual *Escherichia coli trp* messenger RNA molecules is sequentially ordered. J. Mol. Biol. **110**:421–439.

118. **Schmidt, F. J.** 1984. Processing of bacteriophage-coded RNA species, p. 64–89. *In* D. Apirion (ed.), Processing of RNA. CRC Press, Inc., Boca Raton, Fla.

119. **Schmidt, F. J., and W. H. McClain.** 1978. Alternate orders of

processing by RNase P occur *in vitro* but not *in vivo*. J. Biol. Chem. **253**:4730–4734.

120. **Shen, V., M. Cynamon, B. Daugherty, H. F. Kung, and D. Schlessinger.** 1981. Functional inactivation of *lac* alpha-peptide mRNA by a factor that purifies with *Escherichia coli* RNase III. J. Biol. Chem. **256**:1896–1902.

121. **Shen, V., F. Imamoto, and D. Schlessinger.** 1982. RNase III cleavage of *Escherichia coli* beta-galactosidase and tryptophan operon mRNA. J. Bacteriol. **150**:1489–1494.

122. **Sirdeshmukh, R., and D. Schlessinger.** 1985. Ordered processing of *Escherichia coli* 23S rRNA *in vitro*. Nucleic Acids Res. **13**:5041–5054.

123. **Smith, J. D.** 1974. Mutants which allow accumulation of tRNA[tyr] precursor molecules, p. 1–11. *In* J. J. Dunn (ed.), Processing of RNA. Brookhaven National Laboratory, Upton, N.Y.

124. **Stahl, D. A., B. Pace, T. March, and N. R. Pace.** 1984. The ribonucleoprotein substrate for a ribosomal RNA processing nuclease. J. Biol. Chem. **259**:11448–11453.

125. **Stark, M. J. R., R. J. Gregory, R. L. Gourse, D. L. Thurlow, C. Zwieb, R. A. Zimmerman, and A. E. Dahlberg.** 1984. Effects of site-directed mutations in the central domain of 16S rRNA upon r-protein binding, RNA processing and 30S subunit assembly. J. Mol. Biol. **178**:1015–1026.

126. **Stern, M. J., G. F. Ames, N. H. Smith, E. C. Robinson, and C. F. Higgins.** 1984. Repetitive extragenic palindromic sequences: a major component of the bacterial genome. Cell **37**:1015–1026.

127. **Suryanarayana, T., and D. P. Burma.** 1975. Substrate specificity of *Salmonella typhimurium* RNase III and the nature of the products formed. Biochim. Biophys. Acta **407**:459–468.

128. **Tamura, F., S. Nishimura, and M. Ohki.** 1984. The *E. coli divE*

129. **Tomizawa, J., T. Itoh, G. Selzer, and T. Som.** 1981. Inhibition of ColE1 RNA primer formation by a plasmid-specified small RNA. Proc. Natl. Acad. Sci. USA **78**:1421–1425.

130. **Traub, P., and M. Nomura.** 1969. Structure and function of *Escherichia coli* ribosomes. VI. Mechanism of assembly of 30S ribosomes studied *in vitro*. J. Mol. Biol. **40**:391–413.

131. **Von Gabain, A., J. G. Belasco, J. L. Schoffel, A. C. Y. Chang, and S. N. Cohen.** 1982. Decay of mRNA in *Escherichia coli*: investigation of the fate of specific segments of transcripts. Proc. Natl. Acad. Sci. USA **80**:653–657.

132. **Wireman, J. W., and P. S. Sheperd.** 1974. *In vitro* assembly of 30S ribosomal particles from precursor 16S RNA of *Escherichia coli*. Nature (London) **247**:552–554.

133. **Yamada, Y., and D. Nakada.** 1976. Translation of T7 RNA *in vitro* without cleavage by RNase III. J. Virol. **18**:1155–1159.

134. **Yen, C., L. Green, and C. G. Miller.** 1980. Peptide accumulation during growth of peptidase-deficient mutants. J. Mol. Biol. **143**:35–48.

135. **Young, R. A., and J. A. Steitz.** 1978. Complementary sequences 1700 nucleotides apart form a ribonuclease III cleavage site in *Escherichia coli* ribosomal precursor RNA. Proc. Natl. Acad. Sci. USA **75**:3593–3597.

136. **Yuki, A.** 1971. Detection of precursor 16S ribosomal RNA in the polysome fraction of *Escherichia coli*. J. Mol. Biol. **61**:739–744.

137. **Zaniewski, R., E. Petkaitis, and M. P. Deutscher.** 1984. A multiple mutant of *Escherichia coli* lacking the exoribonucleases RNase II, RNase D and RNase BN. J. Biol. Chem. **259**:11651–11653.

mutation which differentially inhibits synthesis of certain proteins is tRNA[Ser]. EMBO J. **3**:1103–1107.

48. Modification of Stable RNA

GLENN R. BJÖRK

Department of Microbiology, University of Umeå, S-901 87 Umeå, Sweden

INTRODUCTION

Synthesis of stable RNA (rRNA, tRNA) constitutes a major biosynthetic pathway in the cell. Both rRNA and tRNA, which are fundamental to protein synthesis, contain modified nucleosides. These are derivatives of the four ordinary nucleosides adenosine (A), guanosine (G), cytidine (C), and uridine (U) (Fig. 1). Since bacterial mRNA does not contain modified nucleosides (75), it was earlier believed that a unique feature of all stable RNAs was the presence of such nucleosides. However, it has now been established that 5S rRNA does not contain any modified nucleosides (22). During maturation of both rRNA and tRNA, several nucleosides are enzymatically modified in a specific manner. All modification reactions, except the formation of queuosine and inosine, occur at the polynucleotide level, i.e., after the polymerization catalyzed by the DNA-dependent RNA polymerase. Figure 1 shows structures of a few modified nucleosides (e.g., 1-methylguanosine [m¹G] contains a methyl group in position 1 of guanosine, etc.). A more complete catalog of structures of modified nucleosides has been published (78). Earlier results and some other aspects of RNA modification, which are not discussed here, are dealt with in some recent reviews (12, 37, 47, 62, 79, 80, 134).

Abbreviations

Abbreviations used in this chapter are as follows. ψ, Pseudouridine; D, dihydrouridine; I, inosine; Q, queuos-ine, or 7-(4,5-*cis*-dihydroxy-2-cyclopenten-1-yl aminomethyl)-7-deazaguanosine; Am, Gm, Cm, Um, 2'-*O*-ribose-methylated derivatives of the corresponding nucleosides. m¹A, 1-methyladenosine; m²A, 2-methyladenosine; m⁶A, 6-methyladenosine; m⁶₂A, 6-dimethyladenosine; m⁵C, 5-methylcytidine; m¹G, 1-methylguanosine; m²G, 2-methylguanosine; m⁷G, 7-methylguanosine; m⁵U, 5-methyluridine; mo⁵U, 5-methoxyuridine; mt⁶A, *N*-[9-(β-D-ribofuranosyl)purin-6-yl-methylcarbamoyl]threonine; t⁶A, *N*-[9-(β-D-ribofuranosyl)purin-6-yl]carbamoylthreonine; s²C, 2-thiocytidine; s²U, 2-thiouridine; s⁴U, 4-thiouridine; ac⁴C, *N⁴*-acetylcytidine; acp³U, 3-(3-amino-3-carboxypropyl)uri-dine; cmo⁵U, uridine-5-oxyacetic acid; mcmo⁵U, methylester of cmo⁵U; mnm⁵s²U, 5-methylaminomethyl-2-thiouridine; cmnm⁵s²U, 5-carboxymethylamino-methyl-2-thiouridine. i⁶A, *N*-6-(Δ²-isopentenyl)adenosine; ms²i⁶A, *N*-6-(Δ²-isopentenyl)-2-methylthioadenosine; ms²io⁶A, *N*-6-(Δ²-4-hydroxy-isopetenyl)-2-methylthioadenosine. tRNA-modifying enzymes catalyzing the formation of m⁵U, ψ, etc., are designated tRNA(m⁵U) methyltransferase, tRNA(ψ)synthetase, etc.

rRNA MODIFICATION

Presence and Synthesis of Modified Nucleosides in rRNA

The 16S rRNA contains 10 modified nucleosides, and 23S rRNA contains 14. Only m⁷G and Gm are present in both rRNA species (Table 1). The modified nucleosides are present in specific domains of the 16S

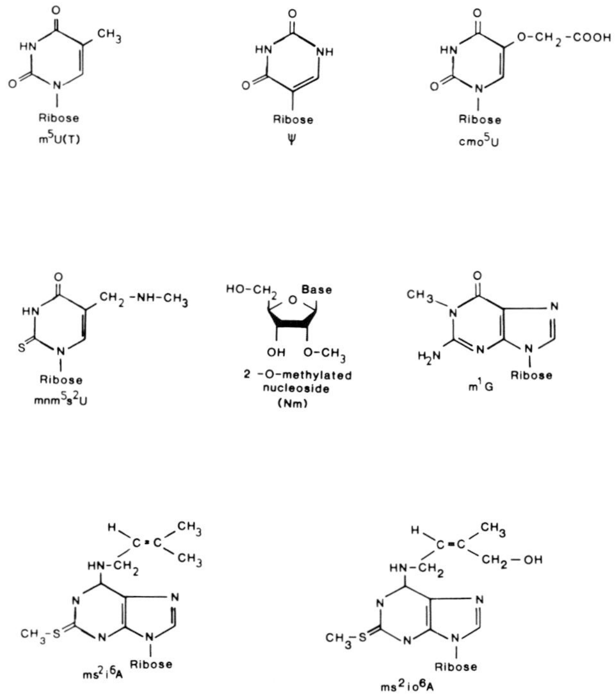

FIG. 1. Structure of some modified nucleosides.

and 23S rRNAs. (Table 1; 20, 131). Modifications in both 16S and 23S rRNA are found mainly in single-stranded regions in the proposed secondary structure. The sequences around the modified nucleosides are evolutionarily conserved, indicating they are likely to be of functional significance. This is especially true for the two m_2^6As, which are present in the 3' end of the rRNA of the small ribosomal subunit in all organisms so far analyzed (125).

The modification of rRNA occurs during the assembly process. While the methylation of 16S rRNA mainly is a late event, the methylation of 23S rRNA is completed at an early stage during maturation (34, 52,

TABLE 1. Modified nucleosides in E. coli rRNA[a]

rRNA	Location (nucleotide)	Domain	rRNA	Location (nucleotide)	Domain
16S			ψ	746	b
m^7G	526	a	m^5U	747	b
m^2G	965	c	m^6A	1618	c
m^5C	966	c	ψ	1911	d
m^2G	1206	c	U^+	1915	d
m^4Cm	1401	c	ψ	1917	d
m^5C	1406	c	m^5U	1939	d
Gm	1496	c	m^6A	2030	d or e
m^2G	1515	c	m^7G	2069	e
m_2^6A	1517	c	Gm	2251	e
m_2^6A	1518	c	U^{++}	2449	e
Total	1541		Cm	2498	e
23S			Um	2552	e
m^1G	745	b	Total	2904	

[a] Data taken from Brosius et al. (21) and van Charldorp et al. (124) for 16S rRNA and from Branlandt et al. (19) for 23S rRNA. The locations of the modified nucleosides are calculated from the 5' end of the 16S and 23S rRNAs. The assignment of domains is according to Brimacombe et al. (20). U^+ and U^{++} are unidentified uridine derivatives.

110). At least some enzymatic modifications of 16S RNA require the presence of specific ribosomal proteins (115). Methylation, however, is not required for the correct processing of 16S and 23S rRNAs (31).

Assuming that one enzyme is required for each modification, 24 rRNA-modifying enzymes would exist in Escherichia coli/Salmonella typhimurium. Of these, only three enzymes, those catalyzing the formation of m^1G, m^2G, and m^6A, have been partly characterized. One enzyme, the rRNA(m_2^6A)methyltransferase, has been purified to homogeneity (Table 2). This enzyme can use only 30S subunits lacking m_2^6A as substrate, whereas the rRNA(m^2G)methyltransferase modifies only unassembled 23S rRNA. In contrast, the rRNA(m^1G)methyltransferase is able to use 23S rRNA as well as 50S and 70S ribosomal particles as substrates.

Genetics of rRNA Modification

Mutants defective in the synthesis of m^2G, m^5C, and m^1G in 23S rRNA have been characterized but not located on the E. coli chromosomal map (13, 14). The ksgA mutant, which is resistant to the antibiotic kasugamycin, lacks the two m_2^6As present in the 3' end of the 16s rRNA (53). The ksgA gene is the structural gene for the rRNA(m_2^6A)methyltransferase (C. P. J. J. van Buul, thesis, University of Leiden, Leiden, The Netherlands, 1985). This gene is located at 0.9 min and is transcribed counterclockwise on the E. coli chromosomal map (3, 9).

Function of Modified Nucleosides in rRNA

Mutants deficient in m_2^6A (16S rRNA), m^1G, m^2G, and m^5C (23S rRNA) are all viable. The 3' end of 16S rRNA harbors a sequence complementary to a sequence (Shine-Dalgarno) upstream of the initiation codon on the mRNA. This 16S rRNA sequence plays a pivotal role in the initiation of translation. Close to it there are two m_2^6As in the loop of the hairpin structure, 14 bases from the 3' end of the 16S rRNA. A mutant (ksgA) lacking these m_2^6As grows slower than wild-type cells in certain growth media (57). The 30S subunits lacking m_2^6A need more initiation factor 3 for optimal binding of formylmethionyl-tRNA$_f^{Met}$. The equilibrium constant for the reaction 30S + 50S $\rightleftarrows$ 70S is dependent on the presence of m_2^6A (92). Furthermore, the presence of m_2^6A destabilizes the hairpin structure, perhaps to facilitate intramolecular interactions (123). A mutant lacking m_2^6A displays an increased leakiness of certain nonsense and frameshift mutations, suggesting that the two m_2^6As also influence translational fidelity (122). Since the presence of m_2^6A has been shown to influence the binding of initiation factor 3, ribosomal protein S21, and the methyltransferase itself, the functional effects observed may be brought about by changes in the binding of these proteins.

Using specific antibodies, it has been shown that the m_2^6A is located in the interface between the 30S and 50S subunits and that the m^7G is located near the junction of the upper one-third and lower two-thirds of the 30S subunit (93, 118). The 50S subunit from mutants lacking m^1G in 23S rRNA might function as a substrate for the rRNA(m^1G)methyltransferase (58).

TABLE 2. rRNA methyltransferases from *E. coli*

rRNA methyl-transferase	Substrates	$M_r{}^a$ ($\times 10^3$)	Effect[b] of treatment with:				Mutants available	References
			Divalent ions	Monovalent ions	Polyamines	Protein influencing activity		
m_2^6A	30S	30 (D)	Mg (+)	K (±) NH$_4$ (±)	ND[c]	S21 (−) IF3 (−)	*ksgA*	91
m^1G	23S rRNA 50S, 70S	ND	Mg (−)	K (±) NH$_4$ (±)	Spermidine (−) Putrescine (±)	Factor A1 (−) Factor B (+)	*rrmA1*	58
m^2G	23S rRNA	ND	Mg (+)	ND	ND		Yes	58
m^6A	23S rRNA, DAPR[d]	31.6 (N)	Mg (+)	K (+)	Spermidine (+)		No	2, 109

[a] Molecular weight determinations in the presence (D) or absence (N) of denaturing agent.
[b] Stimulation (+), inhibition (−), or no effect (±) in the presence of the respective agent.
[c] ND, Not determined.
[d] DAPR, β-9-Ribosyl-2,6-diaminopurine, an adenosine analog.

These results show that m^7G, m^1G, and m_2^6A are exposed on the surface of the ribosome. Since most modified nucleosides in both rRNA species are present in single-stranded regions of the proposed secondary structure, all may be exposed on the surface of the ribosome, allowing them to come in contact with different components associated and interacting with the ribosome.

tRNA MODIFICATION

Presence of Modified Nucleosides in tRNA

tRNA is the most heavily modified nucleic acid in the bacterial cell, and several modified nucleosides have been identified among the 42 tRNA species from *E. coli/S. typhimurium* so far sequenced (111). Some of these (e.g., m^2A, m^6A, ms^2i^6A, s^2C, s^4U, mnm^5s^2U, cmo^5U, $mcmo^5U$, and $cnmm^5s^2U$) are present only in eubacterial tRNAs, whereas others (e.g., m^5U, ψ) are found in both eubacteria and eucaryotes (12). Figure 2

shows that only some specific positions, mainly single-stranded regions, are modified. All tRNA species contain m^5U in position 54 and ψ in position 55. Positions 34 and 37 contain a modified nucleoside in 51 and 67% of all tRNA chains, respectively (Table 3). No unmodified G has been found in position 37.

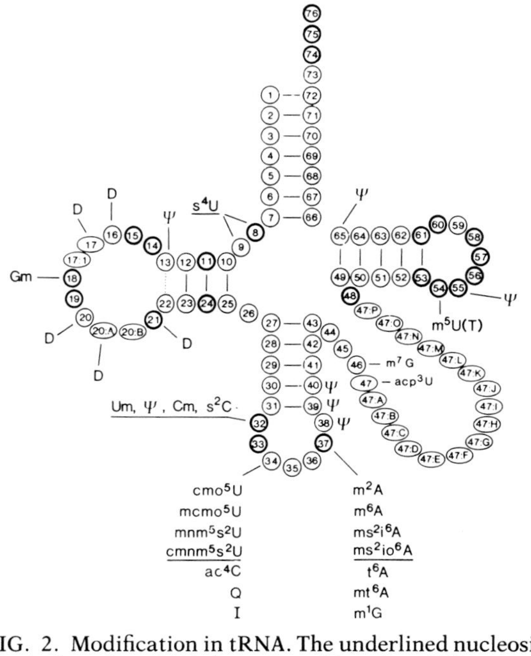

Some modified nucleosides (ψ, s^4U, D) are found in more than one position. It is known that the formation of ψ in positions 38, 39, and 40 is catalyzed by the *hisT* gene product [tRNA(ψ)synthetase I], while another enzyme is involved in the synthesis of ψ55 (108). Not only are positions 34 and 37 often modified, but there are also many different modified nucleosides found in these positions (Fig. 2). Certain patterns can be seen in the modification of nucleosides in position 34 (the "wobble position"; (Fig. 3). No unmodified U or A units have been identified in this position. All codons with a C as the second base and the codons for valine (GUN) are read by tRNAs having an unmodified G or (m)cmo^{5}U as the wobble nucleoside. The Q and s^2U derivatives are present in tRNAs reading codons of the type NA$_C^U$ and NA$_A^G$, respectively.

The modification at position 37 has also some clear regularities. Codons starting with U and A are usually read by tRNAs having an i^6A or t^6A derivative in position 37, respectively. Codons starting with C or G are read by tRNAs having m^1G, m^2A, m^6A, or an unmodified A in this position (Table 4).

tRNA from *E. coli* also harbors selenium-containing modified nucleosides of which 5-methylaminomethyl-2-selenouridine is the most prominent and likely to be present in tRNAs specific for lysine and glutamate in the wobble position (32, 129, 130).

tRNA-Modifying Enzymes

Six tRNA-modifying enzymes have been purified to more than 50% purity (Table 5). These small acidic proteins are present in small amounts in the cell, and they require SH groups for activity. They are usually stimulated by Mg^{2+}, except for the tRNA(m^1G)- and the tRNA(mnm^5s^2U)methyltransferases. The latter enzyme is very sensitive to Mg^{2+} and is 50% inhibited at about 2 mM Mg^{2+}. Of the enzymes so far purified to homogeneity, this is the largest tRNA-modifying enzyme, and it also possesses two enzymatic activities (T. G. Hagervall and G. R. Björk, unpublished results). Although most modifying enzymes are composed of

FIG. 2. Modification in tRNA. The underlined nucleosides are only found in tRNA from *S. typhimurium* (27).

TABLE 3. Modified nucleosides in *S. typhimurium* tRNA[a]

Modified nucleoside	Proposed position(s)	mol/mol of tRNA
s^4U	8,9	0.75
ψ	13,32,38,39,40,55,65	1.7
D	16,17,20,20:A;21	1.6
Gm	18	0.20
m^1A	$22,58^b$	0.02
Cm	32	0.12
s^2C	32	0.08
Um	32	0.05
$mcmo^5U$	34	0.12
mnm^5s^2U	34	0.12
cmo^5U	34	0.11
I	34	0.07
Q	34	0.04
$cmnm^5s^2U$	34^c	0.03
ac^4C	34	0.02
m^2A	37	0.21
m^1G	37	0.19
ms^2io^6A	37	0.09
m^6A	37	0.06
t^6A	37	0.04
mt^6A	37	0.03
ms^2i^6A	37	0.03
i^6A	37	0.02
m^7G	46	0.71
acp^3U	47	0.08
m^5U	54	1.0

[a] Levels of modified nucleosides were calculated from Buck et al. (24). m^5U is present once in all eubacterial tRNA species so far sequenced (111); the levels of the different modified nucleosides are therefore normalized to the level of m^5U. The nucleoside m^2G was also observed by Buck et al. (24), but it is not included here, since it has so far not been identified in any sequenced eubacterial tRNA. Furthermore, three unidentified modified nucleosides were also found. Since D does not absorb well at A_{254} it was not included in the analysis by Buck et al. (24). Therefore, the indicated level of D was calculated from the frequency of its occurrence among the sequenced tRNAs (111). In *E. coli* tRNA, ms^2io^6A is replaced by ms^2i^6A. The level of ms^2i^6A in *E. coli* tRNA is 0.05 mol/mol of tRNA (27).

[b] Suggested position based on the finding that m^1A is present in these positions in tRNA from some G[+] bacteria (111).

[c] Assignment to position 34 is based on its presence in this position in $tRNA^{Lys}$ from *Bacillus subtilis* (132).

one polypeptide in the active state, the partially purified tRNA(s^4U)synthetase is composed of two subunits; this enzyme also requires pyridoxal 5-phosphate for activity (69).

Mutations in the *trmA* and *trmD* genes, which affect the synthesis of m^5U and m^1G in tRNA, respectively, do not influence the synthesis of the same nucleosides in rRNA. Furthermore, a mutation in the *rrmA* gene, which affects the synthesis of m^1G in rRNA, does not affect the synthesis of m^1G in tRNA (13, 15). Thus, the enzymes are specific for tRNA. They are also position specific, since mutations in the *hisT* gene affect the synthesis of ψ only in positions 38, 39, or 40, and not in 55 (108).

Do the tRNA-modifying enzymes also recognize certain sequences surrounding the nucleoside to be modified? tRNAs reading codons starting with U or A have ms^2io^6A/ms^2i^6A and mt^6A/t^6A in position 37, respectively (Table 4). The sequence of A36-A37-A38 and a 5-base-pair anticodon stem may therefore be

involved in the recognition of A37 by the isopentenylating enzyme, while the recognition sequence for the t^6A-producing enzyme may be U36-A37-A38 (119). In fact, two serine tRNAs reading codons starting with U have an unmodified A37. They both possess the A36-A37-A38 sequence, but do not have the 5-base-pair anticodon stem postulated to be involved in the recognition mechanism by the isopentenylating enzyme (48). Carbon et al. (30) have shown by site-directed mutagenesis that the main determining se-

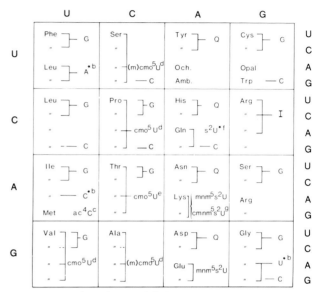

FIG. 3. Nucleosides modified in position 34 (the wobble position). Superscript lowercase letters indicate notes as follows. (a) The coding properties for the tRNA containing the indicated wobble nucleoside are shown; e.g., $tRNA^{Phe}$ contains G34 and reads codons UU_C^U. When the tRNA is only reading one codon, the wobble nucleoside is shown at the place where the codon is located according to the codon table; e.g., $tRNA_I^{Leu}$ contains C34 and reads only CUG. Dashed line indicates uncertain or low codon recognition. (b) A^*, C^*, and U^* are unidentified derivatives of the indicated major nucleosides. (c) Elongator $tRNA_I^{Met}$ has ac^4C, while initiator $tRNA_m^{Met}$ has an unmodified C. (d) Using neutral pH during hydrolysis of tRNA, the methyl ester ($mcmo^5U$) of cmo^5U has been identified (11, 24). The sequencing methods commonly used involve extreme pH values, so that the $mcmo^5U$, if present, is likely to have been hydrolyzed to cmo^5U. Therefore, the presence of cmo^5U in a sequenced tRNA can also have originated from $mcmo^5U$ in the native tRNA. The tRNA($mcmo^5U$)methyltransferase apparently uses $tRNA^{Ser}$ and $tRNA^{Ala}$ as substrate (94); thus these tRNAs have been assigned $(m)cmo^5U$ instead of cmo^5U as reported by Sprinzl and Gauss (111). When no wobble base is indicated, no tRNA suggested to read the indicated codons has been sequenced. (e) Only one $tRNA^{Thr}$ has been sequenced. However, at least two $tRNA^{Thr}$ species exist in *S. typhimurium* (120). The $tRNA^{Thr}$ from *B. subtilis*, as well as the $tRNA^{Ala}$ and $tRNA^{Val}$, has mo^5U in its wobble position (111). The coding capacity for mo^5U is similar to that of cmo^5U (76). Therefore the predicted wobble base in the $tRNA^{Thr}$ is either cmo^5U or $mcmo^5U$. (f) $tRNA_{NUG}^{Gln}$ has an unidentified s^2U derivative in position 34 which has coding properties similar to those of mnm^5s^2U. (g) $cmnm^5s^2U$ is present in tRNA from *S. typhimurium* (24). It has not so far been identified in any sequenced tRNA from *E. coli* or *S. typhimurium*. The assignment to $tRNA^{Lys}$ is based on its presence in $tRNA_I^{Lys}$ from *B. subtilis* (132).

TABLE 4. Modification at position 37 (3' side of the anticodon) in different tRNA species and their coding capacity

Codon nucleotide			Amino acid inserted	Nucleoside in position 37
1st	2nd	3rd		
U	N	N	Phe, Leu, Ser, Tyr, Cys, Trp	ms^{2}i^{6}A/ms^{2}io^{6}A, A[a]
C	Py	N	Leu, Pro	m^{1}G[b]
	Pu	N	His, Gln, Arg	m^{2}A
A	N	N	Ile, Met, Thr, Asn, Lys, Ser	t^{6}A, mt^{6}A,[c] A[d]
G	C/G	N	Ala, Gly	A
	U	N	Val	m^{6}A, A[e]
	A	N	Asp, Glu	m^{2}A

[a] Two serine tRNAs reading UCC and UCU codons contain an unmodified A (48).

[b] Leucine tRNAs have an unidentified guanosine in position 37. However, mutation in the *trmD* gene, which is the structural gene for the tRNA (m^{1}G)methyltransferase, affects the elution properties of tRNA$_1^{Leu}$, tRNA$_2^{Leu}$, and tRNA$_3^{Leu}$ (K. J. Hjalmarsson, unpublished results). Therefore, the modified G must at least be a derivative of m^{1}G.

[c] mt^{6}A has so far only been identified in one tRNA^{Thr}. No tRNA^{Arg} coding for AGA/G has been sequenced so far; thus, it is not known whether the tRNA^{Arg} will contain mt^{6}A or t^{6}A.

[d] tRNA$_f^{Met}$ has an unmodified A in position 37.

[e] While tRNA$_1^{Val}$ contains m^{6}A, tRNA$_2^{Val}$ contains an unmodified A.

quence for the Q-inserting enzyme is U33-G34-U35. By comparing sequenced tRNAs, Tsang et al. (119) have suggested recognition sequences for other modifying enzymes as well.

Wild-type tRNA^{Trp} has the anticodon sequence U33-C34-C35-A36-ms^{2}i^{6}A37-A38. The *glyT*(SP-9)UGG-specific suppressor has the identical sequence, but the hypermodification is absent (96). This suppressor has an additional A in the anticodon loop, which therefore consists of eight bases instead of the normal seven. Thus, the insertion of an extra A changes the conformation such that the i^{6}A-forming enzyme is unable to use the tRNA with an expanded anticodon loop as substrate, although the required primary sequence is present. An efficient ochre suppressor derivative of a glycine tRNA has no ms^{2}i^{6}A37, although it has a 7-base-pair anticodon loop and the proposed recognition sequence for the isopentenylating enzyme (77). Therefore, other structural requirements for this enzyme than those so far recognized must be operating. These results, as well as others (66, 73, 106), suggest that both the primary sequence and the three-dimensional structure of the tRNA are parts of the recognition signal for tRNA-modifying enzymes. The importance of the two parameters for the recognition signal may be different among the various tRNA-modifying enzymes.

The formation of Q occurs on the monomeric level. Queuine, the base of Q, is inserted into tRNA by a tRNA transglycosylase which replaces a guanine for queuine without destroying the phosphate-sugar backbone (84). Since inosine biosynthesis in rat liver occurs by inserting the preformed modified base hypoxanthine into tRNA, a similar mechanism for the formation of inosine in bacteria may be operating (38).

Genetics of tRNA-Modifying Enzymes

Besides the 26 modified nucleosides in Table 3, several unidentified modified nucleosides have also been found in tRNA, leading to the estimate that at least 30 different modified nucleosides may exist in bacterial tRNA. Some, like ψ, are present in more than one position in the tRNA, and several different enzymes are likely to be involved in their synthesis. Other modified nucleosides, like ms^{2}io^{6}A and mnm^{5}s^{2}U, have a complicated structure, and therefore more than one enzyme is probably involved in their respective syntheses. From such considerations one can infer that at least 45 different tRNA-modifying enzymes may be present in *E. coli/S. typhimurium*. Assuming an average gene size of 1 kilobase of DNA for a modifying enzyme, at least 45 kilobases of DNA would be required to be involved in tRNA modification. Thus, a substantial amount of genetic information (about 1% of the total DNA content) is devoted to tRNA modification. Making the same estimation for the 24 rRNA-modifying enzymes possibly present in these bacteria, about 2% of the genetic information is devoted to stable RNA modification. However, since each gene seems to be expressed at a low level, the energetic load for the cell is not extensive.

Mutants defective in tRNA modification have been isolated using different screening and selection procedures. The map locations of the 11 genes likely to be structural genes for tRNA-modifying enzymes are summarized in Fig. 4. In most cases one gene governs the synthesis of one modified nucleoside. However, the formation of s^{4}U and mnm^{5}s^{2}U requires more than one gene. The synthesis of mnm^{5}s^{2}U is governed by the *asuE* (25.3 min), *trmE* (83 min), *trmF* (83 min), and *trmC* (50 min) genes. The *asuE* gene is suggested to be involved in the synthesis of s^{2}U (113). The product in the tRNA from *trmE* and *trmF* mutants is s^{2}U, and it was suggested that the *trmF* gene may be a regulatory gene (39). Furthermore, the polypeptide synthesized from the *trmC* gene has two enzymatic activities, since two different derivatives, S2 and S1, of mnm^{5}s^{2}U34 are accumulated in tRNA from *trmC1* and *trmC2* mutants, respectively (50; Hagervall and Björk, unpublished results). Thus, the sequential formation of mnm^{5}s^{2}U34 may be:

$$\overset{asuE}{U34 \longrightarrow s^2U34} \overset{trmE}{\longrightarrow} \overset{trmC1}{S2 \longrightarrow} \overset{trmC2}{S1 \longrightarrow mnm^5s^2U34}$$

The formation of s^{4}U requires at least two genes, *nuvA* and *nuvC*, which may code for two polypeptides constituting the tRNA(s^{4}U)synthetase. Since only 11 of the proposed 45 potential structural genes for tRNA-modifying enzymes have been identified, it is obvious that more mutants defective in tRNA modification must be isolated before a complete picture of the complex biosynthesis as well as the function of the modified nucleosides in tRNA will emerge.

Formation of Modified Nucleosides in tRNA

Regulation and gene organization. During the maturation of tRNA, modification of some nucleotides takes place at distinct stages (4, 102). The multimeric precursor for tRNA$_1^{Leu}$ contains m^{5}U, ψ, and D but not

TABLE 5. Characteristics of tRNA-modifying enzymes

Enzyme producing:	M_r^a (×10³)	K_m (AdoMet)^b (μM)	K_m (tRNA) (μM)	K_i (AdoHcy)^c (μM)	pH optimum	Mg²⁺	K, Na, or NH₄	Polyamine	Purity (%)	Effect of SH reagent^d:	pI	Molecules per genome equivalent^e	References
m¹G	46 (N), 32 (D)	5	20	6	8–8.5	−	−	+	>95	pCMB (−)	5.2	80	55
m⁵U	42 (D), 56 (N)	17	0.08	Inhibited	8.0	+	+	+	>95	pCMB (−)	4.7	100	—^f
	42 (D), 38 (N)	12.5	1.1	Inhibited	8.4	ND^g	+	+	>95	ND	4.8	ND	45
mnm⁵s²U	80 (D)	ND	ND	ND	ND	−	ND	ND	>95	pCMB (−)		78	—^h
	80 (N)	ND	ND	Inhibited	7.8	−	ND	ND	ND	ND	ND	ND	114
mcmo⁵U	52 (D), 50 (N)	2.0	0.3	Inhibited	7.5–9.0	±	+	ND	50	pCMB (−)	ND	90	95
Q	46 (D), 58 (D)	0.014^i	ND	NA^i	7.0	+	+	ND	>95	ND	4.6–4.8	ND	84
ψ	50 (D), 58 (N)	NA	ND	NA	ND	ND		ND	90	IAcA (−)	4.1	ND	6^k

[a] Molecular weight determination in the presence (D) or absence (N) of denaturing agent.
[b] AdoMet, S-Adenosylmethionine.
[c] AdoHcy, S-Adenosyl-L-homocysteine.
[d] Inhibition (−) of enzyme activity upon addition of p-chlorobenzoate (pCMB) or iodoacetamide (IAcA).
[e] Calculated at a growth rate of $k = 1.0$.
[f] T. Ny, P. Lindström, T. G. Hagervall, and G. R. Björk, unpublished data.
[g] ND, Not determined.
[h] Hagervall and Björk, unpublished data.
[i] K_m for preQ.
[j] NA, Not applicable.
[k] The DNA sequence of the *hisT* gene from *E. coli* has recently been determined (7). According to the DNA sequence, the molecular weight of the tRNA (ψ) synthetase I from *E. coli* is 30,399. The basis for the discrepancy between this molecular weight and that given in the present table is not known.

m¹G37. The m¹G is, however, present in the monomeric precursor which is still lacking Gm18 (107). In fact, ribose methylation seems to be a late event, since ribose-methylated nucleosides have never been found in multimeric precursors. These results, however, are based on analyses of mutants defective in the maturation process, and it is not clear what is the preferred and normal substrate. In fact, it has been suggested that the preferred substrate for the tRNA methyltransferases is RNA precursors of a size similar to mature tRNAs (35, 103). Thus, some tRNA modification reactions can occur at the precursor level, but the rate may be low compared to that of the precursor trimming reaction.

The level of tRNA(m⁵U)methyltransferase increases with increasing growth rate, whereas the levels of tRNA(m¹G)- and tRNA(mnm⁵s²U)methyltransferases as well as that of the tRNA(ψ)synthetase I (*hisT* gene product) are invariant with growth rate (6, 81). Furthermore, the tRNA(m⁵U)methyltransferase is stringently regulated, while the other two tRNA methyltransferases are not (82). In fact, the tRNA(m⁵U)methyltransferase is regulated like stable RNA under all physiological conditions so far analyzed. The tRNA(m¹G)- and tRNA(mnm⁵s²U)methyltransferases are regulated more like other proteins made in low amounts in the cell (88). Thus, although the enzymes are all involved in the modification of tRNA they are regulated differently and only the tRNA(m⁵U)methyltransferase is regulated in a manner similar to its substrate, the tRNA.

The *trmA* gene is monocistronic and is transcribed counterclockwise on the *E. coli* chromosomal map (Fig. 4; 68). The sequence surrounding the Pribnow box, as well as the box itself, is homologous to the corresponding sequence of the P1 promoter of the rRNA operons (P. H. R. Lindstrom, unpublished results). This sequence similarity may explain the similar regulatory behavior of *trmA* and the rRNA genes.

The *trmD* gene is part of a four-polypeptide operon (Fig. 4). Surprisingly, the first and last genes in the *trmD* operon code for ribosomal protein S16 (*rpsP*) and L19 (*rplS*), respectively (29). These two ribosomal proteins are made in 100-fold larger amounts than the tRNA(m¹G)methyltransferase. Furthermore, the synthesis of this enzyme is not regulated like that of the surrounding ribosomal proteins. Measurements of transcription under several physiological conditions revealed only one large transcript constituting the whole *trmD* operon. (A. S. Byström, unpublished results). Introduction of a multicopy plasmid containing the *trmD* operon results in overproduction of both the mRNA and the protein constituents, suggesting that the *trmD* operon, unlike other ribosomal protein operons, is not regulated through translational repression (P. M. Wikström, A. S. Byström, and G. R. Björk, unpublished results). The mechanism(s) behind the differential expression as well as the different regulatory behavior of the genes within the operon must operate entirely at the posttranscriptional level.

The *hisT* gene is downstream of a gene expressed 10- to 14-fold higher than the *hisT* gene, and therefore, like the *trmD* gene, it is part of a differentially expressed multigene operon. The upstream gene (*pdx*) is involved in the biosynthesis of vitamin B₆. Thus, a genetic link between pyridoxal phosphate biosynthe-

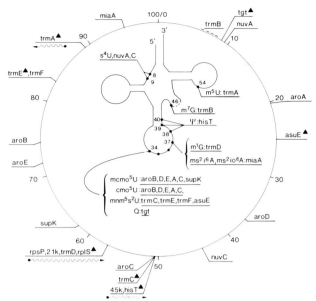

FIG. 4. Known genes involved in tRNA modifications. The positions of the different modified nucleosides in the tRNA are indicated along with the corresponding genes (9) which are involved in their synthesis. The 45K gene of the *hisT* operon has been sequenced and should code for a 36,364-dalton polypeptide (7). This gene has recently been shown to be involved in vitamin B_6 biosynthesis (Winkler and Arps, Abstr. Annu. Meet. Am. Soc. Microbiol. 1986, H53, p. 136). The wavy lines under the genes (operons) indicate the direction of transcription. When the arrow is directed to the right or left, the transcription is with or towards the DNA replication, respectively. The DNA replication is bidirectional and starts at 85 min and terminates at about 32 min on the chromosomal map (9). The sign ▲ indicates that the genes have been cloned. References for gene locations: *miaA* (40, 43); *trmE* and *trmF* (39); *trmC* (50); *asuE* (113); *trmD* (28).

sis and tRNA modification exists. The *hisT* operon is likely to contain more than two cistrons; transcriptional studies have suggested that the operon might be regulated at different levels (7, 71; M. E. Winkler and P. J. Arps, Abstr. Annu. Meet. Am. Soc. Microbiol. 1986, H53, p. 136).

Metabolic aspects. The synthesis of the methylated and thiolated nucleosides as well as t^6A/mt^6A is directly dependent on the availability of methionine, cysteine, and threonine, since these amino acids are direct precursors to these kinds of modified nucleosides. However, some physiological stress conditions also influence the modification of tRNA. Some genetic alterations, other than mutations in the structural genes for the tRNA modifying enzymes, also influence the modification of tRNA (Table 6).

Starvation for leucine or arginine results in severe unbalanced growth which induces D16 and ms^2i^6A37 deficiency in $tRNA^{Phe}$ and $\psi38$ and D17 deficiency in $tRNA^{Leu}$ (65). However, other modified nucleosides of these tRNAs are present in normal amounts. Imposing unbalanced growth by means other than removing a required amino acid, e.g., by the addition of antibiotics to growing cells, also results in the appearance of undermodified tRNA (56, 61). However, unbalanced growth per se does not necessarily have to be responsible for the appearance of undermodified tRNA, since

amino acid limitation during balanced growth can also have that effect (116). A mutation in *ilvU* changes the concentration of isoaccepting $tRNA^{Ile}$ and $tRNA^{Val}$ and also influences the regulation of the isoleucyl-tRNA ligase. The $ilvU^+$ gene product may not only induce derepression of isoleucyl-tRNA ligase but may also retard the conversion of isoaccepting species, most likely by controlling the modification of these species of tRNAs (41).

Three different kinds of auxotrophic mutants (Aro⁻, Thi⁻, and Thy⁻) are, surprisingly, also deficient in tRNA modification (Table 6). A thymine-requiring mutant (Thy⁻) suppresses nonsense as well as frameshift mutations, and it is likely that some tetrahydrofolate-dependent methylation of tRNA may be involved (54). However, an unidentified gene upstream of the *thyA* gene overlaps the latter gene, and the suppressor phenotype of some of the Thy⁻ mutations may be due to mutations in the overlapping sequence of these two genes (10). If so, the coupling of tRNA modification and thymidine biosynthesis may not be metabolic but genetic (cf. organization of the *hisT* gene). A thiamine-requiring mutant (Thi⁻) also lacks s^4U. The mutant lacks factor C of the tRNA(s^4U)synthetase, and the mutation is located in the *nuvC* region of the chromosome. This factor C may also be a subunit of some unknown enzyme in the biosynthesis of thiamine (101). All mutants defective in the common pathway of aromatic amino acids (Fig. 5; *aroA*, *D*, *E*, *B*, and *C*) are deficient in $cmo^5U/mcmo^5U$ in their tRNA. Addition of shikimic acid (SA in Fig. 5) to the growth medium leads to the resumption of $cmo^5U/mcmo^5U$ synthesis by an *aroD* mutant, but not by an *aroC* mutant. Furthermore, mutations in the different pathways branching out from chorismic acid do not influence the level of $cmo^5U/mcmo^5U$ in the tRNA (11; Björk, unpublished results). Thus, chorismic acid itself or some unknown metabolite thereof must play a key role in the formation of $cmo^5U34/mcmo^5U34$ in tRNA.

Enterochelin is an iron chelator compound which is secreted upon iron limitation. Such iron-starved cells also contain i^6A37 instead of ms^2i^6A in the tRNA (100, 128). The same change in tRNA modification occurs

TABLE 6. Physiological conditions and genetic alterations, other than changes in structural genes for tRNA-modifying enzymes, which influence both intermediary metabolism and tRNA modification

Influence	Modified nucleoside affected	References
Genetic alterations		
Aro⁻	$cmo^5U/mcmo^5U$	11
Thy⁻	Suggested to be an unknown methylated nucleoside	54
Thi⁻	s^4U	101
ilvU	Controls modification of $tRNA^{Ile}/tRNA^{Val}$	41
Physiological conditions		
Amino acid limitation	$D16, D17, ms^2i^6A37$ $\psi38$	65,116
Iron limitation	ms^2i^6A37, ms^2io^6A37	23, 100, 128
Oxygen supply	ms^2io^6A37	23

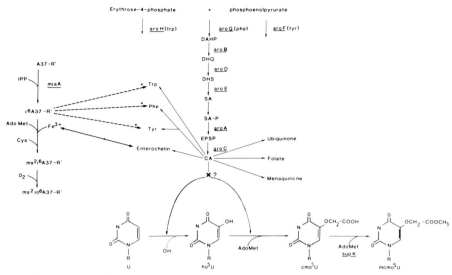

FIG. 5. Interrelationship between biosynthesis of aromatic amino acids and tRNA modification. Arrows denote the stimulation of transport of the corresponding amino acids by the cognate tRNA containing an unmodified A37 (25). The formation of ms^2io^6A occurs only in *S. typhimurium* under aerobic conditions (23).

when cells of *E. coli* grow in body fluids where iron-binding proteins are present. This alteration in tRNA modification may be related to adaptation of *E. coli* growth in body fluids and to bacterial pathogenicity (46). Deficiency of the thiomethyl group of ms^2i^6A in tRNA stimulates the transport of the three aromatic amino acids (25). The *miaA* mutant, which lacks ms^2i^6A, also overproduces enterochelin (23). Thus, the synthesis of enterochelin and the aromatic amino acids is intimately coupled to the formation of ms^2i^6A (Fig. 5).

In summary, both genetic alterations and physiological stress conditions have established links between the formation of modified nucleosides and different parts of intermediary metabolism. Such links may be regulatory (23, 49, 62, 120). The formation of ms^2i^6A may sense the iron concentration in the cell; the hydroxylation step in the formation of ms^2io^6A may be a regulator of aerobiosis in *S. typhimurium*. The mechanism for such a regulation may be through attenuation in which the reading of the leader mRNA is strongly influenced by the presence of the modified nucleoside and the codon context (see Functional Aspects, below).

Functional Aspects

It is apparent that certain positions in tRNA are more prone to be modified than others (Fig. 2). In the three-dimensional structure of *Saccharomyces cerevisiae* tRNA^{Phe}, none of the modified nucleosides present appears to be essential for maintaining the structure (64). The presence of modified groups increases the acceptable surface area of the tRNA by 20%, suggesting that the modified nucleosides are recognized by various proteins, nucleic acids, or both. Most mutants defective in tRNA modification appear to be viable, and some species of *Mycoplasma* lack many of the modified nucleosides. Thus, most modified nucleosides in tRNA are likely to be nonessential for cell growth. The function of modified nucleosides may be to increase or decrease various interactions of the tRNA and therefore to be an important fine tuning of tRNA function. Furthermore, the modifying enzymes themselves may also have a regulatory function beside catalyzing the formation of modified nucleoside (1).

Function of modified nucleosides in position 37. Table 4 shows that all tRNAs reading codons starting with U or A have ms^2i^6A/ms^2io^6A and t^6A/mt^6A in position 37, respectively. The weak U-A/A-U interaction between position 36 and the first position of the codon may be stabilized by the hypermodification present in position 37. In accordance with this, unmodified A or methylated A or G is present in position 37 of tRNAs reading codons starting with C or G (Table 4).

E. coli tRNA^{Ile}, deficient in t^6A, has a reduced ability to bind to poly(A,U,C)- or poly(A,U,U)-programmed ribosomes (74). From measurements of codon-anticodon binding as well as binding to poly(A,G)-programmed ribosomes or binding of free AGA triplets, it was concluded that t^6A stabilizes U-A and U-G base pairs adjacent to the 5' side of the modified nucleoside, most probably by an increased stacking interaction (127).

tRNA^{Phe} from iron-starved cells (thus containing i^6A instead of ms^2i^6A) is much less efficient in poly(U)-directed polyphenylalanine synthesis in vitro. However, no effect of the modification deficiency was observed when natural mRNA was used to direct protein synthesis in vitro (26). Cells growing in iron-limiting medium (and thus containing ms^2-deficient tRNA) derepress the tryptophan operon, most likely due to a lower ability to read the two contiguous tryptophan codons in the *trp* leader mRNA. The ms^2-deficient tRNA may therefore read contiguous codons, such as those present in poly(U) and in leader mRNAs, less well than codons not reiterated, such as in genes coding for proteins.

miaA mutants of *E. coli* and *S. typhimurium* were isolated as having a derepressed tryptophan operon and as harboring an antisuppressor mutation to *supF*, respectively (40, 133). These mutants lack ms^2i^6A37/ms^2io^6A37 and have an unmodified A at position 37. The *miaA1* mutant of *S. typhimurium* has a large (up to 50%) reduction in growth rate, reduced polypeptide chain elongation rate in vivo, and altered regulation of several biosynthetic operons (40). Lack of ms^2i^6A/ms^2io^6A reduces the efficiency of suppression in vivo as well as tRNA binding to the ribosome in vitro (18, 44, 90). This modification, however, is not essential for activity of the tRNA, since the *miaA* mutant is viable and since an ochre suppressor derivative of tRNA^{Gly} lacks ms^2i^6A but is still an efficient suppressor (77). Thus, the effect of ms^2i^6A might be quantitatively different depending on the tRNA (18). Suppressor tRNA lacking ms^2i^6A/ms^2io^6A is more sensitive to the sequence surrounding the codon than is wild-type tRNA, suggesting that ms^2i^6A/ms^2io^6A is directly involved in determining the intrinsic codon context sensitivity of the tRNA (18). The error level is reduced 10-fold in a *miaA* mutant when a mismatch occurs in the third codon position but is unaffected when this mismatch occurs in the first codon position (18, 121). Since an improved accuracy requires energy (67), the decreased cellular yield (40) and peptidyl release (89) of the *miaA1* mutant may be reasonable. The ms^2i^6A modification stabilizes the anticodon, probably by improving the stacking of nucleoside 37 (121). Thus, the effect of ms^2i^6A/ms^2io^6A plays an important part in efficiency and fidelity of translation, as well as in the sensitivity of the tRNA to the sequence surrounding the codon. This last aspect renders the element of specificity to tRNA modification which is a prerequisite for it to be a regulatory device (See Metabolic aspects, above).

Little information exists about the function of the modified nucleosides present in position 37 of tRNAs reading codons starting with C or G. Some of the modified nucleosides like m^1G and m^6A should prevent hydrogen binding to nucleosides on the 5′ side of the codon since these methyl groups prevent Watson-Crick base pairing. A mutant (*trmD3*) of *S. typhimurium*, which at 41°C is lacking m^1G37 but not at 30°C, has a reduction in growth rate only at 41°C. However, the *trmD3* mutation also induces frameshifting at runs of C only at 41°C, suggesting that m^1G37 influences the reading frame maintenance (Björk, unpublished results).

Function of ψ in the anticodon region. A mutation in the *hisT* gene results in a deficiency of ψ in the anticodon stem (positions 38, 39, and 40) in many tRNA species, among them tRNA^{His} (33, 108, 120). The histidine leader mRNA contains seven histidine codons in a row which are read inefficiently by tRNA^{His} lacking ψ38 and ψ39, and this leads to derepression of the histidine operon (60). The growth rate of a *hisT* mutant is reduced (up to 30%) as well as the polypeptide chain elongation rate and the error level (86, 87). Furthermore, a *hisT* mutation also influences the regulation of several amino acid biosynthetic operons, most probably through an attenuation mechanism (120). Lack of ψ in the anticodon stem reduces the efficiency of suppression, but has no or only a minor effect in sensing the sequence surrounding the codon (17; Björk, unpublished results).

Function of modified nucleosides in position 34 (wobble position). Many different modified nucleosides are present in the wobble position (Fig. 2 and 3). The presence of $mcmo^5U/cmo^5U$ enables the tRNA to read codons ending with U in addition to G and A, i.e., the cmo^5- modification increases the normal wobble capacity of U. This modification is present only in tRNAs reading families of four codons specifying the same amino acid (Fig. 3). Such codon families are read in addition by at least one other tRNA species, which reads codons ending with U or C, according to the wobble hypothesis. It is, therefore, not obvious why these modifications are present. However, an Aro$^-$ mutant which lacks $cmo^5U/mcmo^5U$ in tRNA grows 20% slower than Aro$^+$ cells in rich medium, indicating that the presence of cmo^5U is important under some physiological conditions (Table 7; Björk, unpublished results). This wobble base is within 0.4 nm of a pyrimidine in 16S rRNA when the tRNA is in the P site on the ribosome (83). This may extend the anticodon stack into the 16S rRNA and stabilize the tRNA-ribosome interaction. The importance of the modification as such in this interaction has not been elucidated.

A recessive UGA suppressor (*supK*) was isolated which also suppresses some frameshift mutations (8, 98). The *supK* mutants are deficient in a tRNA methyltransferase which most likely catalyzes the formation of $mcmo^5U$ in some tRNA species (99) (see Fig. 5). These results suggest that a tRNA having cmo^5U instead of $mcmo^5U$ is able to cause suppression of UGA or frameshifts mutations. However, the suppressing agent has so far not been identified as a tRNA species. The mechanism behind this suppression is still unknown.

Elongator tRNA$_m^{Met}$ of *E. coli* contains ac^4C in position 34 while initiator tRNA$_f^{Met}$ contains C. By chemically removing the ac^4 modification it was shown that the tRNA$_m^{Met}$ lacking ac^4C binds to AUG-programmed ribosomes almost twice as well as the ac^4C-containing tRNA$_m^{Met}$ (112). However, the presence of ac^4C decreases the misreading in vitro of AUA (Ile). Thus, the primary function of ac^4 modification appears to be to reduce the misreading of the AUA (Ile) codon. This is achieved by a somewhat reduced efficiency in reading the AUG (Met) codon.

The hypermodified nucleoside Q is found in the wobble position of tRNAs specific for tyrosine, histidine, asparagine, and aspartic acid (Fig. 3). This wobble nucleoside has a slightly higher affinity for U than for C (51). A mutant *tgt* which lacks Q dies more rapidly in stationary phase than wild-type cells and is also unable to make nitrate reductase anaerobically. Q modification was suggested as an important factor in the expression of nitrate reductase (59). However, it was not ruled out that this mutant harbors additional mutations, which could explain the observed pleiotropic effects.

tRNAs containing mnm^5s^2U read codons of the type NA_G^A, where N can be C, A, or G (Fig. 3). The mnm^5s^2U may restrict the wobble capacity of U and thus explain why these tRNAs preferentially read codons ending with A (104). The presence of the s^2 modification is not involved in this reaction but has been implicated in increasing the stacking interaction within the anticodon (72, 105, 126). Two mutants, carrying *trmC1* and *trmC2*, respectively, are both de-

TABLE 7. Function of modified nucleosides in tRNA[a]

Modified nucleoside	Results obtained in vitro	Mutant analyzed	Phenotypic effects of mutants analyzed
m^5U54	Thermal stability, fidelity, A-site binding	trmA	Small (4%) reduction in growth rate
$\psi38/\psi39$		hisT	Reduced (up to 30%) growth rate, derepression, antisuppression, reduced CGR_p,[a] lower error level
ms^2i^6A37, ms^2io^6A37	Ribosomal binding	miaA	Reduced (up to 50%) growth rate, derepression, repression, antisuppression, increased codon context sensitivity, lower error level, reduce CGR_p, reduced cellular yield
t^6A37	Stabilizes AC-AC interaction		
m^1G37		trmD	Reduced (up to 40%) growth rate, frameshifting at runs of C
mnm^5s^2U34	Restricted binding to G	trmC, trmE,F, asuE	Antisuppression, increased codon context sensitivity
Q34		tgt	Dies in stationary phase; unable to synthesize nitrate reductase
ac^4C34	Reduced misreading of A		
cmo^5U34	Increased wobble capacity	Aro⁻	Reduced (20%) growth rate
s^4U		nuvA,C	UV resistant

[a] CGR_p, Polypeptide chain growth rate. See text for references.

fective in the synthesis of mnm^5s^2U, which is normally present in $tRNA^{Lys}_{UAA}$ (supG). The efficiency of $tRNA^{Lys}_{UAA}$ is reduced in the trmC mutants, and the undermodified $tRNA^{Lys}_{UAA}$ is more sensitive to the codon context than normal $tRNA^{Lys}_{UAA}$ (49). Mutants (trmE, trmF) having s^2U instead of mnm^5s^2U in their tRNA are less able to read UAG compared to UAA. Such undermodified tRNA also binds less well in vitro to AAG than to AAA (39). Recently, an asuE mutant, probably deficient in the thiolation of mnm^5s^2U, was isolated as an antisuppressor to supL ($tRNA^{Lys}_{UAA}$) (113). Thus, the lack of fully modified mnm^5s^2U affects the efficiency of the tRNA and also makes the tRNA more sensitive to the sequence surrounding the codon.

Function of modified nucleosides in places other than the anticodon region. $tRNA^{Phe}$ contains acp^3U in position 47. Chemical derivatization of acp^3U does not influence the tRNA in the aminoacylation reaction or in polyphenylalanine synthesis in vitro (42). Specific chemical reduction of m^7G46 disrupts the C13-G22-m^7G46 base triple, which leads to a slightly less ordered structure (5). The finding that a mutant (trmB) defective in the formation of m^7G also grows slower than trmB⁺ cells further supports the idea that m^7G stabilizes the structure of tRNA (70). There is no difference in growth rate between a mutant completely lacking s^4U in its tRNA and the wild type. However, s^4U is involved in the photoprotection phenomenon (97, 117). Thus, modification in other parts of tRNA seems to be involved in the stabilization of the molecule, but perhaps most of them may participate in specific interactions with different macromolecules. Therefore much more specific assays might be necessary to reveal the functions of these modified nucleosides.

Ribothymidine (m^5U, rT) is present in all bacterial tRNAs. However, an *E. coli* mutant (trmA5) completely lacking m^5U in its tRNA is viable but is outgrown by wild-type cells in a mixed-population experiment (16). The difference in growth rate between the wild type and the mutant is 4% (Björk,

unpublished results). tRNA lacking m^5U melts at lower temperature than wild-type tRNA (36). Lack of m^5U increases leucine misincorporation during poly (U)-directed polyphenylalanine synthesis (63). It has also been suggested that modification of m^5U and the formylation of initiator $tRNA_f^{Met}$ may be a regulatory device (62).

OUTLOOK

Although much has been learned about the formation and function of modified nucleosides in stable RNA, our knowledge is still scanty. Only a minority of the predicted genes involved in modification of stable RNA have been identified. Future studies should try to widen our knowledge in this area. New and improved techniques, including utilization of transposons and Mu d phages, DNA sequencing methods, and improved in vitro translation systems, will make it possible to ask more specific questions about the interplay between RNA modification and intermediary metabolism as well as functional aspects of stable RNA modification. Well-characterized mutants defective in stable RNA modification will be increasingly valuable when these methods are more widely used, and one can expect such studies to contribute to our understanding of the function and metabolism of stable RNA modification.

ACKNOWLEDGMENTS

Work in my laboratory was supported by the Swedish Cancer Society (project no. 680), the National Science Foundation (project no. BBU-2930), and the Swedish Board of Technical Development.

Critical reading of the manuscript by J. Ericson, T. Hagervall, K. Kjellin-Stråby, and M. Wikström, Umeå, and L. A. Isaksson, Uppsala, is gratefully acknowledged.

LITERATURE CITED

1. **Ames, B. N., T. H. Tsang, M. Buck, and M. F. Christman.** 1983. The leader mRNA of the histidine attenuator region resembles $tRNA^{His}$: possible general regulatory implications. Proc. Natl. Acad. Sci. USA **80:**5240–5242.
2. **Anderson, W. M., Jr., C. N. Remy, and J. E. Sipe.** 1973. Ribo-

somal ribonucleic acid-adenine (N⁶-) methylase of *Escherichia coli* strain B: ionic and substrate site requirements. J. Bacteriol. **114**:988–998.

3. **Andrésson, O. S., and J. E. Davies.** 1980. Some properties of the ribosomal RNA methyltransferase encoded by *ksgA* and the polarity of *ksgA* transcription. Mol. Gen. Genet. **179**:217–222.

4. **Apirion, D.** 1983. RNA processing in a unicellular microorganism: implications for eukaryotic cells. Prog. Nucleic Acids Res. Mol. Biol. **30**:1–40.

5. **Arcari, P., and S. M. Hecht.** 1978. Isoenergetic hydride transfer. A reversible tRNA modification with concomitant alteration of biochemical properties. J. Biol. Chem. **253**:8278–8284.

6. **Arena, F., G. Ciliberto, S. Ciampi, and R. Cortese.** 1978. Purification of pseudouridylate synthetase I from *Salmonella typhimurium.* Nucleic Acids Res. **5**:4523–4536.

7. **Arps, P. J., C. C. Marrel, B. C. Rubin, D. A. Tolan, E. E. Penhoet, and M. E. Winkler.** 1985. Structural features of the *hisT* operon of *Escherichia coli* K-12. Nucleic Acids Res. **13**:5297–5315.

8. **Atkins, J. F., and S. Ryce.** 1974. UGA and non-triplet suppressor reading of the genetic code. Nature (London) **249**:527–530.

9. **Bachmann, B. J.** 1983. Linkage map of *Escherichia coli* K-12, edition 7. Microbiol. Rev. **47**:180–230.

10. **Belfort, M., and J. Pedersen-Lane.** 1984. Genetic system for analyzing *Escherichia coli* thymidylate synthetase. J. Bacteriol. **160**:371–378.

11. **Björk, G. R.** 1980. A novel link between the biosynthesis of aromatic amino acids and transfer RNA modification in *Escherichia coli.* J. Mol. Biol. **140**:391–410.

12. **Björk, G. R.** 1984. Modified nucleosides in RNA—their formation and function, p. 291–330. *In* D. Apiron (ed.), Processing of RNA. CRC Press, Inc. Boca Raton, Fla.

13. **Björk, G. R., and L. A. Isaksson.** 1970. Isolation of mutants of *Escherichia coli* lacking 5-methyluracil in transfer ribonucleic acid or 1-methylguanine in ribosomal RNA. J. Mol. Biol. **51**:83–100.

14. **Björk, G. R., and K. Kjellin-Stråby.** 1978. General screening procedure for RNA modificationless mutants: isolation of *Escherichia coli* strains with specific defects in RNA methylation. J. Bacteriol. **133**:499–507.

15. **Björk, G. R., and K. Kjellin-Stråby.** 1978. *Escherichia coli* mutants with defects in the biosynthesis of 5-methylaminomethyl-2-thio-uridine or 1-methylguanosine in their tRNA. J. Bacteriol. **133**:508–517.

16. **Björk, G. R., and F. C. Neidhardt.** 1975. Physiological and biochemical studies on the function of 5-methyluridine in the transfer ribonucleic acid of *Escherichia coli.* J. Bacteriol. **124**:99–111.

17. **Bossi, L., and J. R. Roth.** 1980. The influence of codon context in genetic code translation. Nature (London) **286**:123–127.

18. **Bouadloun, F., T. Srichaiyo, L. A. Isaksson, and G. R. Björk.** 1986. Influence of modification next to the anticodon in tRNA on codon context sensitivity of translational suppression and accuracy. J. Bacteriol. **166**:1022–1027.

19. **Branlant, C., A. Krol, M. A. Machatt, J. Ponyet, and J.-P. Ebel.** 1980. Primary and secondary structures of *Escherichia coli* MRE600 23S ribosomal RNA. Comparison with models of secondary structure for maize chloroplast 23S SrRNA and large portions of mouse and human 16S mitochondrial rRNAs. Nucleic Acids Res. **9**:4303–4324.

20. **Brimacombe, R., P. Maly, and C. Zwieb.** 1983. The structure of ribosomal RNA and its organization relative to ribosomal protein. Prog. Nucleic Acids. Res. Mol. Biol. **28**:1–48.

21. **Brosius, J., M. L. Palmer, P. J. Kennedy, and H. F. Noller.** 1978. Complete nucleoside sequence of a 16S ribosomal RNA gene from *Escherichia coli.* Proc. Natl. Acad. Sci. USA **75**:4801–4805.

22. **Brownlee, G. G., F. Sanger, and B. G. Barrel.** 1968. The sequence of 5S ribosomal ribonucleic acid. J. Mol. Biol. **34**:379–412.

23. **Buck, M., and B. N. Ames.** 1984. A modified nucleotide in tRNA as a possible regulator of aerobiosis: synthesis of cis-2-methyl-thioribosylzeatin in the tRNA of *Salmonella.* Cell **36**:523–531.

24. **Buck, M., M. Connick, and B. N. Ames.** 1983. Complete analysis of tRNA-modified nucleosides by high-performance liquid *11a* chromatography: the 29 modified nucleosides of *Salmonella typhimurium* and *Escherichia coli* tRNA. Anal. Biochem. **129**:1–13.

25. **Buck, M., and E. Griffiths.** 1981. Regulation of aromatic amino acid transport by tRNA: role of 2-methyl-N⁶-(Δ²-isopentenyl)adenosine. Nucleic Acids Res. **9**:401–414.

26. **Buck, M., and E. Griffiths.** 1982. Iron mediated methylthiolation of tRNA as a regulator of operon expression in *Escherichia coli.* Nucleic Acids Res. **10**:2609–2624.

27. **Buck, M., J. A. McCloskey, B. Basile, and B. N. Ames.** 1982. cis-2-Methylthio-ribosylzeatin (ms²io⁶A) is present in the transfer RNA of *Salmonella typhimurium,* but not *Escherichia coli.* Nucleic Acids Res. **10**:5649–5662.

28. **Byström, A. S., and G. R. Björk.** (1982). Chromosomal location and cloning of the gene (trmD) responsible for the synthesis of tRNA(m¹G)methyltransferase in *Escherichia coli* K-12. Mol. Gen. Genet. **188**:440–446.

29. **Byström, A. S., K. J. Hjalmarsson, P. M. Wikström, and G. R. Björk.** 1983. The nucleotide sequence of an *Escherichia coli* operon containing genes for the tRNA(m¹G)methyltransferase, the ribosomal proteins S16 and L19 and a 21-K polypeptide. EMBO J. **2**:899–905.

30. **Carbon, P., E. Haumont, S. de Henau, G. Keith, and H. Grosjean.** 1983. Site-directed *in vitro* replacement of nucleosides in the anticodon loop of tRNA: application requirements for queuine insertase activity. EMBO J. **2**:1093–1097.

31. **Chelbi-Alix, M. K., A. Expert-Bezancou, F. Hayes, J.-H. Alix, and C. Brandlant.** 1981. Properties of ribosomes and ribosomal RNAs synthesized by *Escherichia coli* grown in the presence of ethionine. Normal maturation of ribosomal RNA in the absence of methylation. Eur. J. Biochem. **115**:627–634.

32. **Ching, W.-W., B. Alzner-deWeerd, and J. Stadtman.** 1985. A selenium-containing nucleoside at the first position of the anticodon in seleno-tRNA^Glu from *Clostridium stichlandii.* Proc. Natl. Acad. Sci. USA **82**:347–350.

33. **Cortese, R., R. Landsberg, R. A. von der Haar, H. E. Umbarger, and B. N. Ames.** 1974. Pleiotropy of *hisT* mutants blocked in pseudouridine synthesis in tRNA: leucine and isoleucine-valine operons. Proc. Natl. Acad. Sci. USA **71**:1857–1861.

34. **Dahlberg, J. E., N. Nikolaev, and D. Schlessinger.** 1974. Post-transcriptional modification of nucleotides in *E. coli* ribosomal RNAs. Brookhaven Symp. **26**:194–200.

35. **Davis, A. R., and D. P. Nierlich.** 1974. The methylation of transfer RNA in *Escherichia coli.* Biochim. Biophys. Acta **374**:23–37.

36. **Davenloo, P., M. Sprinzl, K. Watanabe, M. Albani, and H. Kersten.** 1979. Role of ribothymidine in the thermal stability of transfer RNA as monitored by proton magnetic resonance. Nucleic Acids Res. **6**:1571–1581.

37. **Dirheimer, G.** 1983. Chemical nature, properties, location, and physiological and pathological variations of modified nucleosides in tRNAs. Recent Results Cancer Res. **84**:15–46.

38. **Elliott, M. S., and R. W. Trewyn.** 1984. Inosine biosynthesis in transfer RNA by an enzymatic insertion of hypoxanthine. J. Biol. Chem. **259**:2407–2410.

39. **Elseviers, D., L. A. Petrullo, and P. J. Gallagher.** 1984. Novel *E. coli* mutants deficient in biosynthesis of 5-methylaminomethyl-2-thiouridine. Nucleic Acids Res. **12**:3521–3534.

40. **Ericson, J. U., and G. R. Björk.** 1986. Pleiotropic effects induced by modification deficiency next to the anticodon of tRNA from *Salmonella typhimurium* LT2. J. Bacteriol. **166**:1013–1021.

41. **Fayerman, J. T., M. Coker Vann, L. S. Williams, and H. E. Umbarger.** 1979. *ilvU,* a locus in *Escherichia coli* affecting the derepression of isoleucyl-tRNA synthetase and the RPC-5 chromatographic profiles of tRNA^Ile and tRNA^Val. J. Biol. Chem. **254**:9429–9440.

42. **Friedman, S.** 1979. The effect of chemical modification of 3-(3-amino-3-carboxypropyl)uridine on tRNA function. J. Biol. Chem. **254**:7111–7115.

43. **Gallagher, P. J., I. Schwartz, and D. Elseviers.** 1984. Genetic mapping of *pheU,* an *Escherichia coli* gene for phenylalanine tRNA. J. Bacteriol. **158**:762–763.

44. **Gefter, M. L., and R. L. Russel.** 1969. Role of modification in tyrosine transfer RNA: a modified base affecting ribosome binding. J. Mol. Biol. **39**:145–157.

45. **Greenberg, R., and B. Dudock.** 1980. Isolation and characterization of m⁵U-methyltransferase from *Escherichia coli.* J. Biol. Chem. **255**:8296–8302.

46. **Griffiths, E., and J. Humphreys.** 1978. Alterations in tRNAs containing 2-methyl-N⁶-(Δ²-isopentenyl)adenosine during growth of enteropathogenic *Escherichia coli* in the presence of iron-binding proteins. Eur. J. Biochem. **82**:503–513.

47. **Grosjean, H., and H. Chantrenne.** 1980. On codon-anticodon interaction, p. 347–367. *In* F. Chapeville and A. O. Haenni (ed.), Molecular biology, biochemistry and biophysics, vol. 32: Chemical recognition in biology. Springer-Verlag, New York.

48. **Grosjean, H., K. Nicoghosian, E. Haumont, D. Söll, and R. Cedergren.** 1985. Nucleotide sequence of two serine tRNAs with a GGA anticodon: structure-function relationships in the serine family of *E. coli* tRNAs. Nucleic Acids Res. **13**:5697–5705.

49. **Hagervall, T. G., and G. R. Björk.** 1984. Undermodification in the first position of the anticodon of *supG*-tRNA reduces trans-

lational efficiency. Mol. Gen. Genet. **196**:194–200.

50. **Hagervall, T. G., and G. R. Björk.** 1984b. Genetic mapping and cloning of the gene (*trmC*) responsible for the synthesis of tRNA(mnm⁵s²U)methyltransferase in *Escherichia coli* K12. Mol. Gen. Genet. **196**:201–207.

51. **Harada, F., and S. Nishimura.** 1972. Possible anticodon sequences of tRNA^His, tRNA^Asn, tRNA^Asp from *Escherichia coli* B. Universal presence of nucleoside Q in the first position of the anticodons of these transfer ribonucleic acids. Biochemistry **11**:301–308.

52. **Hayes, F., D. Hayes, P. Fellner, and C. Ehresmann.** 1971. Additional nucleotide sequences in precursor 16S ribosomal RNA from *Escherichia coli*. Nature (London) New Biology **232**:54–55.

53. **Helser, T. L., J. E. Davis, and J. E. Dahlberg.** 1971. Change in methylation of 16S ribosomal RNA associated with mutation to kasugamycin resistance in *Escherichia coli*. Nature (London) New Biol. **233**:12–14.

54. **Herrington, M. B., A. Kohli, and P. H. Lapchak.** 1984. Suppression by thymidine-requiring mutants of *Escherichia coli* K-12. J. Bacteriol. **157**:126–129.

55. **Hjalmarsson, K. J., A. S. Byström, and G. R. Björk.** 1983. Purification and characterization of tRNA(m¹G)methyltransferase from *Escherichia coli*, strain K12. J. Biol. Chem. **258**:1343–1351.

56. **Huang, P. C., and M. B. Mann.** 1974. Comparative fingerprint and composition analysis of the three forms of ³²P-labeled phenylalanine tRNA from chloramphenicol-treated *Escherichia coli*. Biochemistry **13**:4704–4710.

57. **Igarashi, K., K. Kishida, K. Kashiwagi, I. Tatokoro, T. Kakegawa, and S. Hirose.** 1981. Relationship between methylation of adenine near the 3' end of 16-S ribosomal RNA and the activity of 30-S ribosomal subunits. Eur. J. Biochem. **113**:587–593.

58. **Isaksson, L. A.** 1973. Partial purification of ribosomal RNA(m¹G)- and rRNA(m²G)methylase from *Escherichia coli* and demonstration of some proteins affecting their apparent activity. Biochim. Biophys. Acta **312**:122–133.

59. **Jänel, G., U. Michelsen, S. Nishimura, and H. Kersten.** 1984. Queusine modification in tRNA and expression of the nitrate reductase in *Escherichia coli*. EMBO J. **3**:1603–1608.

60. **Johnston, M. H., W. M. Barnes, F. G. Chumley, L. Bossi, and J. R. Roth.** 1980. Model for regulation of the histidine operon of *Salmonella*. Proc. Natl. Acad. Sci. USA **77**:508–512.

61. **Juares, H., A. C. Skjold, and C. Hedgcoth.** 1975. Precursor relationship of phenylalanine transfer ribonucleic acid from *Escherichia coli* treated with chloramphenicol or starved for iron, methionine, or cysteine. J. Bacteriol. **121**:44–54.

62. **Kersten, H.** 1984. On the biological significance of modified nucleosides in tRNA. Prog. Nucleic Acids Res. Mol. Biol. **31**:58–114.

63. **Kersten, H., M. Albani, E. Männlein, R. Praisler, P. Wurmbach, and K. H. Nierhaus.** 1981. On the role of ribosylthymine in prokaryotic tRNA function. Eur. J. Biochem. **114**:451–456.

64. **Kim, S.-H.** 1979. Crystal structure of yeast tRNA^Phe and general structural features of other tRNAs, p. 83–100. *In* P. R. Schimmel, D. Söll, and J. N. Abelson (ed.), Transfer RNA: structure, properties and recognition. Cold Spring Harbor Laboratory, Cold Spring Harbor, N.Y.

65. **Kitchingman, G. R., and M. J. Fournier.** 1977. Modification-deficient transfer ribonucleic acids from relaxed control *Escherichia coli*: structure of the major undermodified phenylalanine and leucine transfer RNAs produced during leucine starvation. Biochemistry **16**:2213–2220.

66. **Kuchino, Y., T. Seno, and S. Nishimura.** 1971. Fragmented *E. coli* methionine tRNA_f as methyl acceptor for rat liver tRNA methylase: alteration of the site of methylation by conformational change of tRNA structure resulting from fragmentation. Biochem. Biophys. Res. Commun. **43**:476–483.

67. **Kurland, C. G., and M. Ehrenberg.** 1984. Optimization of translation accuracy. Prog. Nucleic Acids Res. Mol. Biol. **31**:191–219.

68. **Lindström, P. H. R., D. Stüber, and G. R. Björk.** 1985. Genetic organization and transcription from the gene (*trmA*) responsible for synthesis of tRNA (uracil-5)-methyltransferase by *Escherichia coli*. J. Bacteriol. **164**:1117–1123.

69. **Lipset, M.** 1972. Biosynthesis of 4-thiourylate. Participation of sulfurtransferase containing pyridscal-5-phosplate. J. Biol. Chem. **247**:1458–1461.

70. **Marinus, M. G., N. R. Morris, D. Söll, and T. C. Kwong.** 1975. Isolation and partial characterization of three *Escherichia coli* mutants with altered transfer ribonucleic acid methylases. J. Bacteriol. **122**:257–265.

71. **Marvel, C. C., P. J. Arps, B. C. Rubin, H. O. Kammen, E. E.**

72. **Penhoet, and M. E. Winkler.** 1985. *hisT* is part of a multigene operon in *Escherichia coli* K-12. J. Bacteriol. **161**:60–71.

72. **Mazumdar, S. K., W. Saenger, and K. H. Scheit.** 1974. Molecular structure of poly-2-thiouridylic acid, a double helix with non-equivalent polynucleotide chains. J. Mol. Biol. **85**:213–229.

73. **McClain, W. H., and J. G. Seidman.** 1975. Genetic perturbations that reveal tertiary conformation of tRNA precursor molecules. Nature (London) **257**:106–110.

74. **Miller, J. P., Z. Hussain, and M. P. Schweizer.** 1976. The involvement of the anticodon adjacent modified nucleoside N-[9-(β-D-ribofuranosyl)-purin-6-ylcarbamoyl]-threonine in the biological function of *E. coli* tRNA^ile. Nucleic Acids Res. **3**:1185–1201.

75. **Moore, P. B.** 1966. Methylation of messenger RNA in *Escherichia coli*. J. Mol. Biol. **18**:38–47.

76. **Murao, K., T. Hasegawa, and H. Ishikura.** 1976. 5-Methoxyuridine: a new minor constituent located in the first position of the anticodon of tRNA^Ala, tRNA^Thr and tRNA^Val from *Bacillus subtilis*. Nucleic Acids Res. **3**:2851–2860.

77. **Murgola, E. J., N. E. Prather, F. T. Pagel, B. H. Mims, and K. A. Hijazi.** 1984. Missense and nonsense suppressor derived from a glycine tRNA by nucleotide insertion and deletion *in vivo*. Mol. Gen. Genet. **193**:76–81.

78. **Nishimura, S.** 1979. Structures of modified nucleosides found in tRNA, p. 547–549. *In* P. R. Schimmel, D. Söll, and J. N. Abelson (ed.), Transfer RNA: structure, properties, and recognition. Cold Spring Harbor Laboratory, Cold Spring Harbor, N.Y.

79. **Nishimura, S.** 1979. Modified nucleosides in tRNA, p. 59–79. *In* P. R. Schimmel, D. Söll, and J. N. Abelson (ed.), Transfer RNA: structure, properties, and recognition. Cold Spring Harbor Laboratory, Cold Spring Harbor, N.Y.

80. **Nishimura, S.** 1983. Structure, biosynthesis and function of queuosine in transfer RNA. Prog. Nucleic Acids Res. Mol. Biol. **28**:50–73.

81. **Ny, T., and G. R. Björk.** 1980. Growth rate-dependent regulation of transfer ribonucleic acid (5-methyluridine)methyltransferase in *Escherichia coli* B/r. J. Bacteriol. **141**:67–73.

82. **Ny, T., J. Thomale, K. Hjalmarsson, G. Nas, and G. R. Björk.** 1980. Non-coordinate regulation of enzymes involved in transfer RNA metabolism in *Escherichia coli*. Biochim. Biophys. Acta **607**:277–284.

83. **Ofengand, J., R. Liou, J. Kohut III, I. Schwartz, and R. A. Zimmerman.** 1979. Covalent cross-linking of transfer ribonucleic acid to the ribosomal P site. Mechanism and site of reaction in transfer ribonucleic acid. Biochemistry **18**:4322–4332.

84. **Okada, N., and S. Nishimura.** 1979. Isolation and characterization of a guanine insertion enzyme, a specific tRNA transglycosylase, from *Escherichia coli*. J. Biol. Chem. **254**:3061–3066.

85. **Okada, N., S. Noguchi, H. Kasai, N. Shindo-Okada, T. Ohgi, T. Goto, and S. Nishimura.** 1979. Novel mechanism of post-transcriptional modification of tRNA. Insertion of bases of Q precoursors into tRNA by a specific tRNA transglycosylase reaction. J. Biol. Chem. **254**:3067–3073.

86. **Palmer, D. T., P. H. Blum, and S. W. Artz.** 1983. Effects of the *hisT* mutation of *Salmonella typhimurium* on translation elongation rate. J. Bacteriol. **153**:357–363.

87. **Parker, J.** 1982. Specific mistranslation in *hisT* mutants of *Escherichia coli*. Mol. Gen. Genet. **187**:405–409.

88. **Pedersen, S., P. L. Bloch, S. Reeh, and F. C. Neidhardt.** 1978. Patterns of protein synthesis in *E. coli*: a catalog of the amount of 140 individual proteins at different growth rates. Cell **14**:179–190.

89. **Petrullo, L. A., and D. Elseviers.** 1986. The effect of a 2-methyltio-N⁶-isopentenyladenosine deficiency on peptidyl-tRNA release in *Escherichia coli*. J. Bacteriol. **165**:608–611.

90. **Petrullo, L. A., P. J. Gallagher, and D. Elseviers.** 1983. The role of 2-methylthio-N⁶-isopentenyl-adenosine in readthrough and suppression of nonsense codons in *Escherichia coli*. Mol. Gen. Genet. **190**:289–294.

91. **Polderman, B., L. Roza, and P. H. Van Knippenberg.** 1979. Studies on the function of two adjacent N⁶,N⁶-dimethyladenosines near the 3' end of 16S ribosomal RNA of *Escherichia coli*. III. Purification and properties of the methylating enzyme and methylase-30S interactions. J. Biol. Chem. **254**:9094–9100.

92. **Polderman, B., C. P. J. J. van Buul, and P. H. van Knippenberg.** 1979. Studies on the function of two adjacent N⁶N⁶-dimethyladenosines near the 3' end of 16S ribosomal RNA of *Escherichia coli*. II. The effect of the absence of the methyl groups on initiation of protein biosynthesis. J. Biol. Chem. **254**:9090–9094.

93. **Politz, S. M., and D. G. Glitz.** 1977. Ribosome structure: localization of N⁶,N⁶-dimethyladenosine by electron microscopy of a

94. **Pope, W. T., A. Brown, and R. H. Reeves.** 1978. The identification of the tRNA substrates for the *supK* methylase. Nucleic Acids Res. **5**:1041–1057.

95. **Pope, W. T., R. H. Reeves.** 1978. Purification and characterization of a tRNA methylase from *Salmonella typhimurium*. J. Bacteriol. **136**:191–200.

96. **Prather, N. E., E. J. Murgola, and B. H. Mims.** 1981. Nucleotide insertion in the anticodon loop of a glycine transfer RNA causes missense suppression. Proc. Natl. Acad. Sci. USA **78**:7408–7411.

97. **Ramabhadran, T. V., and J. Jagger.** 1976. Mechanism of growth delay included in *Escherichia coli* by near ultraviolet radiation. Proc. Natl. Acad. Sci. USA **73**:59–63.

98. **Reeves, R. H., and J. R. Roth.** 1971. A recessive UGA suppressor. J. Mol. Biol. **56**:523–533.

99. **Reeves, R. H., and J. R. Roth.** 1975. Transfer ribonucleic acid methylase deficiency found in UGA suppressor strains. J. Bacteriol. **124**:332–340.

100. **Rosenberg, A. H., and M. L. Gefter.** 1969. An iron-dependent modification of several transfer RNA species in *Escherichia coli*. J. Mol. Biol. **46**:581–584.

101. **Ryals, J., R.-Y. Hsu, M. N. Lipsett, and H. Bremer.** 1982. Isolation of single-site *Escherichia coli* mutants deficient in thiamine and 4-thiouridine syntheses: identification of a *nuvC* mutant. J. Bacteriol. **151**:899–904.

102. **Sakano, H., Y. Shimura, and H. Ozeki.** 1974. Selective modification of nucleosides of tRNA precursors accumulated in a temperature-sensitive mutant of *Escherichia coli*. FEBS Lett. **48**:117–121.

103. **Schaefer, K. P., S. Altman, and D. Söll.** 1973. Nucleotide modification *in vitro* of the precursor of tRNA^{Tyr} of *Escherichia coli*. Proc. Natl. Acad. Sci. USA **70**:3626–3630.

104. **Sekiya, T., K. Takeishi, and T. Ukita.** 1969. Specificity of yeast glutamic acid transfer RNA for codon recognition. Biochim. Biophys. Acta **182**:411–426.

105. **Sen, G. C., and H. P. Ghosh.** 1976. Role of modified nucleosides in tRNA: effect of modification of the 2-thiouridine derivative located at the 5'-end of the anticodon of yeast transfer RNA$_2$^{Lys}. Nucleic Acids Res. **3**:523–535.

106. **Shersneva, L. P., T. V. Venkstern, and A. Bayev.** 1973. A study of transfer of methylation. Biochim. Biophys. Acta **294**:250–262.

107. **Shimura, Y., H. Sakano, S. Kubokawa, F. Nagawa, and H. Ozeki.** 1980. tRNA precursors in RNase P mutants, p. 43–58. *In* D. Söll, J. N. Abelson, and P. R. Schimmel (ed.), Transfer RNA: biological aspects. Cold Spring Harbor Laboratory, Cold Spring Harbor, N.Y.

108. **Singer, C. E., G. R. Smith, R. Cortese, and B. N. Ames.** 1972. Mutant tRNA^{His} ineffective in repression and lacking two pseudouridine modifications. Nature (London) New Biol. **238**:72–74.

109. **Sipe, J. E., W. M. Anderson, Jr., C. N. Remy, and S. H. Love.** 1972. Characterization of S-adenosylmethionine: ribosomal ribonucleic acid-adenine (N^6-) methyltransferase of *Escherichia coli* strain B. J. Bacteriol. **110**:81–91.

110. **Sogin, M. L., K. J. Pechman, L. Zablen, B. J. Lewis, and C. R. Woese.** 1972. Observations on the post-transcriptionally modified nucleotides in the 16S ribosomal ribonucleic acid. J. Bacteriol. **112**:13–16.

111. **Sprinzl, M., and D. H. Gauss.** 1984. Complication of tRNA sequences. Nucleic Acids Res. **12**:r1–r57.

112. **Stern, L., and L. H. Schulman.** 1978. The role of the minor base N^4-acetylcytidine in the function of the *Escherichia coli* noninitiator methionine transfer RNA. J. Biol. Chem. **253**:6132–6139.

113. **Sullivan, M. A., J. F. Cannon, F. H. Webb, and R. M. Bock.** 1985. Antisuppressor mutation in *Escherichia coli* defective in biosynthesis of 5-methylaminomethyl-2-thiouridine. J. Bacteriol. **161**:368–376.

114. **Taya, Y., and S. Nishimura.** 1977. Purification and properties of tRNA methylase specific for synthesis of 5-methylaminomethyl-2-thiouridine, p. 251–257. *In* F. Salvatore, E. Borek, V. Zappia, H. G. Williams-Ashman, and F. Schlenk (ed.), The biochemistry of adenosylmethionine. Columbia University Press, New York.

115. **Thammana, P., and W. A. Held.** 1974. Methylation of 16S RNA during ribosome assembly *in vitro*. Nature (London) **251**:682–686.

116. **Thomale, J., and G. Nass.** 1978. Alteration of the intracellular concentration of aminoacyl-tRNA synthetases and isoaccepting tRNAs during amino-acid-limited growth in *Escherichia coli*. Eur. J. Biochem. **85**:407–418.

117. **Thomas, G., and A. Favre.** 1980. 4-Thiouridine triggers both growth delay induced by near-ultraviolet light and photoprotection. Eur. J. Biochem. **113**:67–74.

118. **Trempe, M. R., K. Ohgi, and D. G. Glitz.** 1982. Ribosome structure. Localization of 7-methylguanosine in the small subunits of *Escherichia coli* and chloroplast ribosomes by immunoelectron microscopy. J. Biol. Chem. **257**:9822–9824.

119. **Tsang, T. H., M. Buck, and B. N. Ames.** 1983. Sequence specificity of tRNA-modifying enzymes. An analysis of 258 tRNA sequences. Biochim. Biophys. Acta **741**:180–196.

120. **Turnbough, C. L., Jr., R. J. Neill, R. Landsberg, and B. N. Ames.** 1979. Pseudouridylation of tRNAs and its role in regulation in *Salmonella typhimurium*. J. Biol. Chem. **254**:5111–5119.

121. **Vacher, H., H. Grosjean, C. Houssier, and P. H. Buckingham.** 1984. The effect of point mutations affecting *Escherichia coli* tryptophan tRNA on anticodon-anticodon interactions and on UGA suppression. J. Mol. Biol. **177**:329–342.

122. **van Buul, C. P. J. J., W. Visser, and P. H. van Knippenberg.** 1984. Increased translational fidelity caused by the antibiotic kasugamycin and ribosomal ambiguity in mutants harboring the *ksgA* gene. FEBS Lett. **177**:119–124.

123. **van Charldorp, R., H. A. Heus, and P. H. Knippenberg.** 1981. Adenosine dimethylation of 16S ribosomal RNA: effect of the methylgroups on local conformational stability as deduced from electrophoretic mobility of RNA fragments in denaturing polyacrylamide gels. Nucleic Acids Res. **9**:267–275.

124. **van Charldorp, R., H. A. Haus, and P. H. van Knippenberg.** 1981. 16S ribosomal RNA of *Escherichia coli* contains a N^2-methylguanosine at 27 nucleotides from the 3'-end. Nucleic Acids Res. **9**:2717–2725.

125. **van Charldorp, R., and P. H. van Knippenberg.** 1982. Sequence, modified nucleotides and secondary structure at the 3'-end of small ribosomal subunit RNA. Nucleic Acids Res. **10**:1149–1158.

126. **Weissenbach, J., and G. Dirheimer.** 1978. Pairing properties of the methylester of 5-carboxymethyl uridine in the wobble position of yeast tRNA$_3$^{Arg}. Biochem. Biophys. Acta **518**:530–534.

127. **Weissenbach, J., and H. J. Grosjean.** 1981. Effect of threonylcarbamoyl modification (t^{6}A) in yeast tRNA$_{III}$^{Arg} on codon-anticodon and anticodon-anticodon interactions. A thermodynamic and kinetic evaluation. Eur. J. Biochem. **116**: 207–213.

128. **Wettstein, F. O., and G. S. Stent.** 1968. Physiologically induced changes in the property of phenylalanine tRNA in *Escherichia coli*. J. Mol. Biol. **38**:25–40.

129. **Wittwer, A. J.** 1983. Specific incorporation of selenium into lysine and glutamate-accepting tRNAs from *Escherichia coli*. J. Biol. Chem. **258**:8637–8641.

130. **Wittwer, A. J., L. Tsai, W. M. Ching, and T. C. Stadtman.** 1984. Identification and synthesis of a naturally occurring selenonucleoside in bacterial tRNAs: 5-[(methylamino)methyl]-2-selenouridine. Biochemistry **23**:4650–4655.

131. **Woese, C. R., R. Gutell, R. Gupta, and H. F. Noller.** 1983. Detailed analysis of the higher-order structure of 16S-like ribosomal ribonucleic acids. Microbiol. Rev. **47**:621–669.

132. **Yamada, Y., K. Murao, and H. Ishikura.** 1981. 5-(Carboxymethylaminomethyl)-2-thiouridine, a new modified nucleoside found at the first letter position of the anticodon. Nucleic Acids Res. **9**:1933–1939.

133. **Yanofsky, C., and L. Soll.** 1977. Mutations affecting tRNA^{Trp} and its charging and their effect on regulation of transcription termination at the attenuator of the tryptophan operon. J. Mol. Biol. **113**:663–677.

134. **Yarus, M.** 1982. Translational efficiency of transfer RNAs: uses of an extended anticodon. Science **218**:646–652.

Section D. Utilization of Energy for Cell Activities

49. Motility and Chemotaxis

ROBERT M. MACNAB

Department of Molecular Biophysics and Biochemistry, Yale University, New Haven, Connecticut 06511

INTRODUCTION

As will be evident from other chapters in this book, bacteria like *Escherichia coli* and *Salmonella typhimurium* have considerable metabolic and physiological versatility that enables them to survive and grow in a wide range of environmental conditions. Nonetheless, there is a potential further advantage to being able to migrate to the most favorable environment available. This capability, the tactic response to a variety of environmental information, is highly developed in both species. It requires a means of movement, a means of sensing the environment, and a means of modulating movement in response to this information.

The motility and taxis systems of *E. coli* and *S. typhimurium* resemble each other closely. Unless otherwise stated, information given in this chapter applies to both species; where a gene symbol or other designation is followed by a subscript E or S, it refers specifically to *E. coli* or *S. typhimurium*, respectively.

The motility of *E. coli* and *S. typhimurium* derives from the use of flagella, organelles that consist of a helical filament driven by a rotary motor at the cell surface. The translational motion (swimming) of the cell that results from filament rotation (24, 26, 180) is interrupted at frequent intervals by an episode of chaotic angular motion (tumbling), and the cell proceeds in this way in a random walk through its environment (Fig. 1). Where certain types of gradient information are present, the relative probabilities of swimming and tumbling are altered, and the cell biases its random walk by (statistically) extending swimming intervals in what may operationally be termed a "favorable" direction (27).

Rotation of the flagellar motor is reversible, and swimming and tumbling are accomplished by use of the two rotational senses (109). Thus the tactic response fundamentally devolves from control of the relative probabilities of the counterclockwise (CCW) and clockwise (CW) states of the motor.

This chapter will consider the operation of the motor apparatus, the various means by which motility and taxis are measured, the types of stimuli to which *E. coli* and *S. typhimurium* respond, the genetics of motility and chemotaxis, the nature of the primary reception process, and the subsequent events in communicating information from receptor to the motor output.

Description of the motor apparatus will emphasize its mechanism of energy transduction and switching;

for strictly structural aspects of the motor apparatus, the reader is referred to chapter 7. In terms of the process of sensory reception, there is, for some chemoeffectors, overlap with the process of binding prior to transport; for some others (the phosphotransferase system [PTS] sugars), there is overlap with the uptake process itself. In this volume, relevant chapters in this regard include 11 and 50.

For further information and discussion concerning bacterial motility and chemotaxis, the reader is referred to the recent reviews in references 71, 102, 120, 122, 147, 153, and 206.

MOTOR FUNCTION

A number of the quantitative characteristics of the motor that will be alluded to throughout this section are summarized in Table 1.

Motor Structure

The flagellum is an organelle consisting of (i) the external filament that performs hydrodynamic work on the aqueous medium, (ii) a connecting hook, and (iii) the basal body plus other morphologically unidentified structures, which together constitute the motor. Components that are actively involved in energy transduction and switching will be discussed later in this chapter. See chapter 7 (especially Fig. 1 to 3, 5, and 6 and Table 1) for details concerning flagellar structure and assembly.

Swimming

The flagellar motor operates as a reversible rotary device to cause rotation of the external flagellar filament, which is normally a left-handed helix.

When rotated in the CCW sense, the helical wave travels from proximal to distal and exerts a pushing motion on the cell. Hydrodynamic and mechanical forces cause the flagellar filaments from all around the cell body (see Fig. 1 and 2 of chapter 7) to sweep around the cell onto a common axis, which is usually parallel to the cell body axis. The filaments then operate in a concerted bundle (Fig. 2a and b) to propel the cell and produce swimming (12, 116). Rotation speed of the filaments within the bundle is above the flicker frequency of the eye and is probably around 100 Hz (H. C. Berg, personal communication). Typical swimming speed at room temperature in buffer is around 25 μm s^{-1} (27, 124).

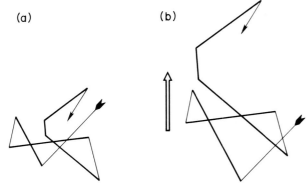

(a)   (b)

FIG. 1. Idealized trajectory of a swimming cell of *E. coli* or *S. typhimurium* (a) in an isotropic medium and (b) in a unidirectional gradient of attractant (open arrow). In either case, the cell swims in a straight line, randomizes its direction by tumbling, swims again, tumbles, etc., yielding a three-dimensional random walk. The effect of an attractant gradient is to extend (by tumble suppression) the mean duration of swimming segments in an up-gradient direction (heavy lines); segments in the down-gradient direction (fine lines) are not appreciably shortened. To illustrate this, the cells (a) and (b) start at the same point (arrow tail) and execute the same number of tumbles with the same resulting directional changes, but the cell (b) in the gradient has up-gradient segments extended by a factor $\cos\theta$ where θ is the angle with respect to the gradient direction. It therefore displays migration in the gradient direction by executing a time-biased random walk. From reference 118, with permission.

Tumbling

The other motility mode of *E. coli* and *S. typhimurium*, tumbling, occurs when the filaments are rotated in the opposite sense, CW. Under these conditions, the helical wave would be expected to travel from distal to proximal. The situation, however, is complicated by structural changes that take place in the filament (126). In chapter 7 it is described how filaments have a polymorphic capability, including right-handed as well as left-handed examples. CW rotation of the filaments places them under right-handed torsional load and initiates, at the proximal end, a polymorphic transition to a right-handed waveform (called "curly" because it has about half the normal wavelength); as long as CW rotation continues, the transition proceeds outward toward the distal end. A heteromorphous filament is characterized by a curly proximal segment, a normal distal segment, and an abrupt angle between the helical axes of the two (Fig. 2c). Tumbling then occurs because the flagellar filaments are partly pulling and partly pushing and have a poorly defined orientation, as is shown schematically in Fig. 2d. The cell is more or less randomly reoriented in the process, ready for the next interval of swimming.

Parenthetically, it is interesting that in the absence of polymorphic transitions, the probable consequence of motor reversal would be jamming of the flagellar bundle (116), and also that abnormally prolonged periods of CW rotation are not only wasteful of time once effective randomization of orientation has occurred, but have a counterproductive effect by con-

verting the flagella into a propulsive right-handed bundle that responds to tactic signals in an inverse manner compared to the wild-type response (91; see section below on mutant phenotypes).

Motor Energetics and Dynamics

Several of the present statements regarding flagellar motor energetics and dynamics refer to observations made (for technical reasons) on species other than *E. coli* and *S. typhimurium*. One major source of knowledge has been a motile strain of *Streptococcus* that, conveniently, lacks endogenous energy sources (cited in reference 132); another has been *Bacillus subtilis*. There is no reason to believe that the characteristics of flagellar motors will be fundamentally different in different species. A useful summary of the properties of flagellar motors is given in reference 28.

The flagellar motor operates, not by precession or a conformational wave, but rather by a true rotary mechanism (26, 109, 180). In other words, there is an indefinite cycling of the azimuthal position of the filament with respect to the cell, and thus the filament may be regarded as a "wheel." Both senses of rotation are possible and are equivalent in terms of torque and other characteristics (24).

TABLE 1. Geometry, energetics, and dynamics of flagellation and motility[a]

No. of flagella per cell	From 0 to around 15; typically around 8
Length of flagellar filaments	From zero to around 20 μm; typically around 5 to 10 μm
Origin of filaments on cell surface	Random (peritrichous flagellation)
Energy source	Proton motive force across cell membrane
Threshold for rotation	Ca. 25 mV
Linear range	Ca. 25 to 125 mV
Saturation range	Above ca. 125 mV
Flagellar rotation speed	Load dependent until motor saturates; on free-swimming cells, saturated speed is ca. 100 Hz[b]
No. of protons per revolution	Not known; minimum of ca. 300
Efficiency	Not known, but possibly quite high, especially at moderate to high load
Torque output per motor	Ca. 10^{-11} dyne cm
Power output per motor	Ca. 10^{-16} W at 20 Hz
Power per cell	Ca. 10^{-15} W under normal swimming conditions
Cost to cell of flagellar operation	Ca. 0.1% of total energy expenditure under growth conditions
Cost to cell of flagellar synthesis	Ca. 2% of biosynthetic energy expenditure

[a] Data are based on information in chapter 7 and on references 24, 46, 90, 117, 131, 132, and 176.
[b] Berg, personal communication.

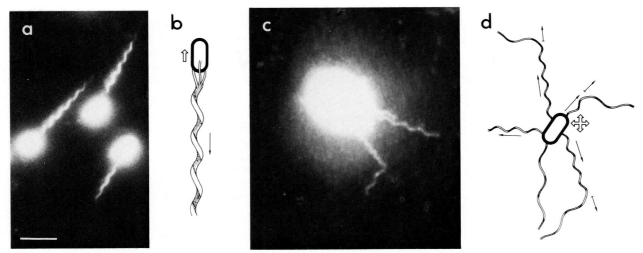

FIG. 2. Behavior of flagella of *E. coli* or *S. typhimurium* during swimming and tumbling. (a and b) Swimming: With the motors in CCW rotation, the flagellar filaments form a propulsive bundle, with wave propagation proceeding from proximal to distal. (c and d) Tumbling: CW rotation of the motors causes the normal left-handed form of the filament to undergo a polymorphic transition to the curly right-handed form. (c) Filament in the heteromorphous state, midway through such a transition. While the filaments are undergoing such transitions, the bundle is dispersed and the cell body moves chaotically, end over end; i.e., the cell tumbles and reorients randomly, ready for the next swimming interval. Bar, 5 μm. From references 91 and 126, with permission.

The energy source for rotation is the transmembrane proton potential, or proton motive force (61, 90, 108, 131, 132, 134, 162, 176); motor rotation has been demonstrated in the laboratory with an artificially imposed proton motive force, as well as with naturally energized cells. Both chemical potential (ΔpH) and electrical potential ($\Delta\psi$) can be used, apparently with equal efficiency.

The source of the proton motive force varies. Under respiratory conditions, it is the electron transport chain, in which case the proton motive force is above the saturating value for motility, so that swimming speed is insensitive to the actual value (90). Under conditions of anaerobic glycolysis, ATP hydrolysis via the proton-linked membrane ATPase is responsible; at least on a short time scale after oxygen removal, the ATPase apparently generates a subsaturating proton motive force and therefore motility, though present, is appreciably diminished (unpublished data).

From the amount of hydrodynamic work performed, it can be estimated that at least several hundred protons must pass through the motor per revolution (24). Although there are likely to be quantized elementary steps in the motion, they have thus far escaped detection (25).

There are several reasons for suspecting that the motor may be a highly efficient device, operating close to equilibrium: it has constant torque characteristics over a wide range of loads (31); it lacks either a temperature dependence or a deuterium isotope dependence (88); and it can operate equally with an artificially imposed proton motive force of either polarity (29). However, most of the above statements refer to relatively high-load conditions (tethered cells). Measurements on free cells—of swimming speed at different proton motive force values (90, 176) and of filament rotation speeds (Berg, personal communication) at different viscosities—indicate saturation and load independence, respectively. Under these conditions, the motors may be operating far from equilibrium.

A related question is whether the stoichiometric coupling between proton flux and motor rotation is tight or loose; i.e., is the number of protons per revolution fixed or does it depend on speed or load? Close-to-equilibrium operation of course implies tight coupling. Far-from-equilibrium operation, however, is consistent with either tight coupling (as with an enzymatic reaction) or loose coupling (as has been postulated in certain transport processes; 54). Advocates of both positions exist (28, 119, 144), but until measurements of proton flux through the motor can be made, the question will remain unanswered.

Since the motor can be driven by a diffusion potential, one might expect it to show a rotational diffusional drift even in the absence of a mean proton motive force. This is not observed, and only when a mean potential is present is there any detectable rotation (90, 176). Yet the threshold value for this potential is not large compared to the thermal energy (kT) (29). From this, one can conclude that one-proton events are not sufficient to cause an elementary stepping event; i.e., motor rotation involves the concerted action of at least several protons.

Mechanism of Energy Transduction

Flagellar rotation in certain alkalophilic species can be driven by sodium potential (74, 75). Although this is not true of *E. coli* and *S. typhimurium*, whose motors are proton-driven, it seems prudent to view with caution any proposed mechanism that obligatorily involves hydrogen bonding, if we assume a broadly similar mechanism regardless of the ion used.

There is no evidence that clearly suggests or favors any specific mechanism for the flagellar motor, but there are various experimental observations (such as constant torque, saturation, temperature independence) with which a successful model must be compatible.

In one detailed model that has been published (28, 88), the proton potential is used to gate and therefore bias a diffusional random walk of force-generating units within the motor. The energy is proposed to be stored via the elastic tethering of these units, with the stored elastic potential almost equal to the driving proton potential, so that the device is operating close to equilibrium. Other classes of models (61, 138, 144) involve the conversion of the proton potential via specific binding sites into an electrical potential between components of the motor so that the motor operates as an electrostatically driven device. In at least one such proposed mechanism (144), the transfer of protons is not obligatorily coupled to rotation; i.e., the mechanism is a loosely coupled one. Yet another proposal (220) is for a proton-generated conformational change (axial twist of the central rod) with cyclic attachment at the proximal and distal ends.

Motor Switching

The flagellar motors of well-energized cells of *E. coli* and *S. typhimurium* rotate incessantly, and they switch continuously, whether or not there is gradient information in the cell's environment. Also, provided they are not forced to interact with each other mechanically or hydrodynamically, the switching of one motor is not synchronized with that of any other (80, 123); in other words, each motor is an autonomous unit.

In a constant environment, the switching probability per unit time is constant. For a tethered wild-type cell, mean intervals are typically around 1 to 2 s for both the CCW and CW directions (30). However, with free cells, there is considerable asymmetry in the mean intervals for swimming and tumbling (ca. 1 s and 0.1 s, respectively; 27). A quantitative correlation between motor rotation probabilities and free swimming behavior has not been achieved as yet. Part of the discrepancy between the measurements on free and tethered cells probably derives from the presence of mechanical interactions among flagellar filaments in the former case, so that switching probabilities are interdependent (cf. discussions in references 80 and 89). It is also possible that switching probabilities may be load dependent.

CHARACTERIZATION OF MOTILITY AND TACTIC RESPONSES

Assays

Although this book is not intended to be a reference source for experimental techniques, it will be helpful to indicate briefly how motility and taxis are studied in the laboratory, as this will enable a better appreciation of the results that have been obtained. For more details concerning methodology, see reference 121.

Swarm plates

When a chemotactic strain of *E. coli* or *S. typhimurium* is inoculated onto the center of a semisolid agar plate containing nutrients which it can metabolize and to which it is tactically responsive, the growing population of cells swarms outward from the center, following the gradient which it has created. This provides a convenient and powerful means of selecting for good motility and taxis and also for screening mutants.

Capillary assay

Cells will preferentially enter one compartment from another if there is a suitable gradient between the two compartments. A widely used application of this is the capillary assay (2); in this procedure a glass capillary tube containing medium plus attractant is immersed in a cell suspension in medium without attractant, and after a suitable interval the contents of the capillary are plated out and counted. The results provide a population-average response to the imposed spatial gradient of attractant.

A related approach (48, 216) involves imposing a quantitatively defined gradient (e.g., exponential or step) in a cuvette and then monitoring the redistribution of a population of cells as a function of time.

Microscopic observation of free-swimming cells

Microscopic observation of motility can provide a lot of information to the experienced eye, permitting rough estimates of speed, tumbling frequency, etc. For more quantitative information, it is necessary to track individual cells and record their position as a function of time. A simple example of this involves video recording and playback, with either manual or automatic analysis. Another approach involves actual tracking in real time, using feedback circuitry to lock onto the cell's image (27). The information is very detailed but applies only to a single cell; accumulation of large amounts of data is needed to provide a population average.

A specialized example of microscopic observation is the visualization of the individual flagellar filaments on a cell, using high-intensity dark-field light microscopy (115, 125). This approach permits statements to be made regarding several filaments simultaneously and also permits study of the consequences of motor rotation on filament operation (for example, polymorphic transitions; 126).

Microscopic observation of tethered cells

Provided a bacterium can be attached ("tethered") to glass by only a single flagellar filament, the operation of the corresponding flagellar motor can be monitored via the rotation of the cell body (180). Precise statements then become possible regarding the CCW versus CW state of that motor, a situation which is generally impossible on free-swimming cells. Cell behavior can be monitored visually or by automatic optical detection devices and in real time or on video recordings. One caveat to this approach is that it assumes that motor switching properties are the same as they would be for the rotating filament on a free cell; i.e., it assumes that switching events are load independent. The validity of this assumption has not been verified.

Temporal assay

The bacterial sensory system is responsive to temporal gradient information, which is normally generated by the cell's movement through spatial gradients, but which may also be imposed directly.

Direct imposition of a temporal gradient, commonly as an abrupt step in concentration either by simple addition or by more sophisticated means, provides a powerful assay (124), permitting decomposition of the response into its major phases of excitation and adaptation (see below).

Empirical Description of the Response

Pure temporal gradient stimulation

When cells are subjected to a temporal stimulus (concentration jump of an attractant or repellent; 124, 216), there follows a characteristic three-part response (Fig. 3) consisting of a latency phase of ca. 0.2 s before there is any detectable change in behavior (173), a rapid excitation phase during which the frequency of motor reversals deviates from its steady-state value, and an adaptation phase (lasting from seconds to minutes, depending on stimulus size) during which the motors progressively relax or adapt to more or less the same steady state that existed before the stimulus. If the stimulus is "favorable" (increase of attractant or decrease of repellent), the excitation consists of deviation toward more CCW rotation; if it is "unfavorable" (decrease of attractant or increase of repellent), the excitation consists of deviation toward more CW rotation. The signals that are responsible for excitation and adaptation will be considered later in this chapter.

Spatial gradient stimulation

The natural stimulus that a cell experiences consists of a temporal gradient that is generated by virtue of the cell's motion through a spatial gradient. In this context, if it has been accumulating excitatory information faster than it can adapt, it will either suppress or enhance tumbling, depending on whether the in-

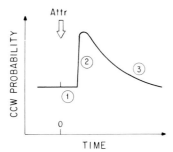

FIG. 3. Time course of the response of a bacterium to temporal stimulation by the sudden addition of an attractant at time zero. After an initial latency phase (1) of about 0.2 s, the excitation phase (2) manifests itself as a rapid increase in the probability that the flagellar motors will be in the CCW versus the CW rotational state. The degree of excitation progressively decreases during the adaptation phase (3), which can last from seconds to minutes, depending on the magnitude of the stimulus. From reference 120, with permission.

formation produced by its current direction of travel is favorable or unfavorable (27). Over a period of time, this pattern results in a net movement in the favorable direction (Fig. 1).

STIMULI DETECTED

The stimuli to which *E. coli* and *S. typhimurium* respond may be categorized in a variety of ways: attractant versus repellent, chemical versus nonchemical, organic versus inorganic, and so on. A later section will deal with the receptors that are responsible for initiating the responses, but a brief summary of the types of stimuli that generate tactic responses is given here and in Table 2. Figure 4 represents these inputs schematically.

Energy-Linked Stimuli

One of the earliest recognized and most striking responses is that towards oxygen: the aerotactic response (for review, see references 206 and 207). Oxygen and alternative electron acceptors such as fumarate and nitrate (111) are mechanistically distinct from organic attractants such as serine or galactose. It is true of both classes that there is a specific binding protein for the attractant, but in the case of oxygen it is the metabolic use of the attractant as an electron acceptor within the electron transport chain that generates the signal for the tactic response (177). The range of oxygen concentrations over which a positive aerotactic response can be elicited is centered around 0.7 μM (110), the K_m of the terminal cytochrome (cytochrome *o*). The primary consequence of electron transport is proton translocation across the cell membrane, and therefore it is reasonable to hypothesize that perturbation of proton motive force is the actual stimulus; measurements made with *S. typhimurium* (177) and also with the aerobic photosynthetic bacterium *Rhodobacter sphaeroides* (14), formerly *Rhodopseudomonas sphaeroides*, support this hypothesis.

A related phenomenon is attraction to proline, a response which occurs as a consequence of the oxidation of proline by a membrane-associated electron transport system with oxygen as the terminal electron acceptor (42).

The phototactic response of photosynthetic bacteria operates in a like manner, by perturbation of light-driven electron transport and hence proton motive force (14).

On the basis of the above description of aerotaxis, one might expect that any metabolic perturbation that affects proton motive force should also constitute a tactic stimulus. A clear demonstration of this, in a context outside that of electron transport, remains to be made.

Oxygen, at sufficiently high concentrations (greater than that in air), acts as a repellent rather than an attractant. It has been suggested (207) that the effect may stem from radical anion production.

Light

E. coli and *S. typhimurium* do not show any physiologically significant responses to light, which, considering that they are nonphotosynthetic organisms,

TABLE 2. *Stimuli to which* E. coli *and* S. typhimurium *are tactically responsive*

Stimulus	Comments	References
Energy-linked attractant stimuli		
Electron acceptors (oxygen, fumarate, nitrate)	Consequence of utilization of oxygen or other electron acceptor, not merely of binding.	110, 111, 177, 206, 207
Electron donors (proline)	Ditto for electron donors. Possibly the flagellar motor itself is responsible for detection.	42
Any increase in proton motive force?		60, 206
Energy-linked repellent stimuli		
High oxygen concn	Free-radical inhibition of cellular processes?	207
High-intensity blue light	Not physiologically significant; occurs only at very high light levels. May be disruption of flavin-mediated electron transport.	125, 208, 209
pH attractant stimuli		
Alkaline internal pH (membrane-permeant weak bases such as trimethylamine)	Weak-base attractants act by taking up protons in the cytoplasm.	136, 164
pH repellent stimuli		
Acidic external pH	There appear to be pH-sensitive sites on the external and cytoplasmic faces of Tsr	93, 164, 188, 217
Alkaline external pH		
Acidic internal pH (membrane-permeant weak acids such as acetate)	Weak-acid repellents act by releasing protons in the cytoplasm	93, 164, 188, 217
Temperature attractant/repellent stimuli		
Temperature increase/decrease	Temperature increase is attractant, decrease is repellent. Competitively inhibited by Tsr chemoattractants.	3, 77, 78, 128
Organic chemical attractants		
Amino acids (serine, aspartate)	Strongest of the organic attractants.	43, 73, 135, 136
Dipeptides (E. coli)	Tripeptides are weakly detected.	128a
Carboxylic acids (malate, citrate [S. typhimurium], magnesium-citrate [S. typhimurium])	E. coli can neither transport nor respond to citrate. In S. typhimurium, the responses to citrate and to magnesium-citrate appear to be distinct.	43, 73, 79, 92, 135, 136
Sugars (glucose, galactose, ribose, maltose [E. coli], PTS sugars)	PTS sugars must be transported and phosphorylated to be detected; others do not.	5, 6, 114, 135, 159
Membrane-permeant weak bases (see under pH)		
Organic chemical repellents		
Hydrophobic amino acids (leucine, isoleucine, valine, tryptophan)	Mechanism unknown.	216, 217
Aliphatic alcohols (ethanol, isopropanol)	Mechanism unknown.	217
Membrane-permeant weak acids (see under pH)		
Phenol (S. typhimurium)	Phenol is repellent for S. typhimurium, weak attractant for E. coli.	112, 216, 217; Y. Imae and R. M. Macnab, unpublished data
Miscellaneous (indole, glycerol, ethylene glycol)	Indole and related compounds through Tsr and other MCPs. Glycerol and ethylene glycol through all major MCPs.	145, 217; Parkinson, personal communication
Inorganic chemical attractants	None clearly established, excepting O_2, and NH_3 as membrane-permeant weak base	
Inorganic chemical repellents		
Co^{2+} and Ni^{2+} ions (E. coli)	S. typhimurium is not repelled by these cations and is weakly attracted to them when present as a complex with citrate.	79, 217
S^{2-} and mercaptans (E. coli)	Only established for E. coli.	217

is not surprising. They do, however, respond to intense blue light, perhaps because of disruption of flavin-linked aspects of electron transport (125, 208, 209). This response, though not a normal physiological one, has proved useful as an experimental tool for inducing tumbling.

pH

The pH of the external environment acts as a tactic stimulus. When tested by the capillary assay, *E. coli* migrates away from either acidic or alkaline pH values, toward a pH value around neutrality (217). In

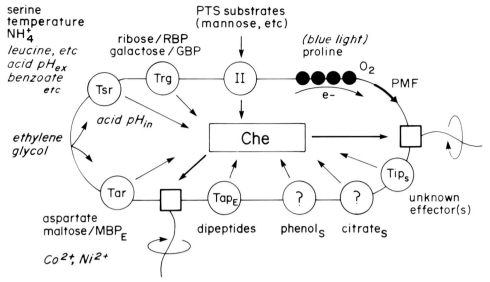

FIG. 4. Sensory stimuli for a cell of *E. coli* or *S. typhimurium*. Attractants (roman letters) and repellents (italics) operate through the receptors indicated. The information is processed by cytoplasmic components of the chemotaxis apparatus (Che) and transmitted to the flagella, where the sense of rotation is modulated to cause migration of the cell to more favorable environments. In the case of attractants, specific binding occurs, directly or as a complex with a periplasmic binding protein (e.g., RBP). Repellents may act indirectly, affecting the state of the membrane environment of the receptors (as with ethylene glycol) or transferring protons across the membrane and thus altering cytoplasmic pH (as with benzoate); the attractant ammonium ion acts by reverse proton transfer. Several of the receptors, such as Tsr and Tar, are methyl-accepting proteins and function in taxis exclusively. Others, notably the enzymes II of the sugar PTS, are transport components. Perturbation of the proton motive force (PMF) generated by the electron transport chain is also a tactic stimulus; this is the mode of action of oxygen (the electron acceptor) and proline (an electron donor). Blue light is only effective as a repellent at high intensities and is probably not a significant stimulus in the natural environment. Subscripts E and S indicate a feature peculiar to *E. coli* or *S. typhimurium*, respectively.

temporal gradient assays, increases and decreases in pH act as attractant and repellent stimuli, respectively, regardless of whether the step is in the acid-to-neutral or neutral-to-alkaline range, although the responses can be mixed (93, 164). This discrepancy between results of spatial and temporal assays may reflect different time scales of the gradients sensed by the cell.

Changes in the internal pH of the cell also act as a tactic stimulus. The most notable examples are where an organic acid or base can permeate the membrane only in its uncharged form (i.e., a protonated acid such as CH_3COOH or a deprotonated base such as CH_3NH_2) and then either release or sequester a proton upon entry into the cytoplasm (93, 164, 188). Thus weak-acid repellents and weak-base attractants act as agents for lowering or raising the cytoplasmic pH. There is no evidence that they act by binding to any specific receptor.

Temperature

Within a range of temperature from at least 20 to 37°C, *E. coli* and *S. typhimurium* respond to temperature increase as an attractant stimulus and to temperature decrease as a repellent stimulus (3, 78, 128). The maximum sensitivity seems to be for gradients centered around 30°C, although this may be influenced by growth temperature, since it has been found that the unstimulated frequency of tumbling is maximal at that value (137). Interestingly, chemotaxis was

found to have a temperature optimum at 30°C (2). For a review of thermosensing in bacteria, see reference 77.

Organic Chemical Attractants

With a few exceptions such as fumarate (111), proline (42), and membrane-permeant bases (164), organic chemical attractants act by binding to specific receptors and initiating a signal process that is peculiar to the chemotaxis system and is not a general feature of metabolism. However, as might be expected, the most powerful attractants are compounds that are central to the cell's metabolism: amino acids, oligopeptides, sugars, and carboxylic acids. It is interesting that only a small subset of the 20 amino acids that are incorporated into protein elicit an appreciable tactic response.

One major difference between *E. coli* and *S. typhimurium* is the presence of a response to citrate in the latter species only (79, 92). A citrate response on the part of *E. coli* would confer no survival advantage, since this species (unlike *S. typhimurium*) cannot transport citrate. The identity of the *S. typhimurium* citrate chemoreceptor remains unknown.

Another difference, less easy to rationalize, is that maltose is an attractant for *E. coli* only (6, 101), even though both species can transport and utilize the compound. As will be discussed below, the relevant transducer (Tar, to be discussed more extensively later) of *E. coli*, but not that of *S. typhimurium*, interacts successfully with the maltose/maltose-binding protein complex (47, 139).

Organic Chemical Repellents

The identity of organic chemicals that are repellents and their mechanism of action are much less well defined than for attractants (217). Most repellents act at much higher concentrations (typically millimolar) than do attractants (typically micromolar), and there is still no documented example of a receptor in the sense of a protein to which a given organic repellent binds specifically to elicit the response. Mention has already been made of the action of weak-acid repellents in perturbing cytoplasmic pH. This is the only case for which a mechanism has been elucidated (93, 164).

It is noteworthy that all of the organic repellents are molecules that are significantly hydrophobic, yet have at least a small degree of polarity (e.g., phenol, leucine). They may well act by perturbing the state of the membrane or, more specifically, the membrane environment of specific sensory transducers; there is clear evidence (to do with methylation levels, see below) that they do affect these molecules. In some cases, notably glycerol and ethylene glycol, any class of transducer will serve (145); indole may also fall in a similar category, since more than one class of transducer can serve (J. S. Parkinson, personal communication); in others, notably the leucine-isoleucine-valine family of amino acids, only one transducer class (Tsr) will serve (191).

An interesting difference between the two species is that phenol, which is a strong repellent for *S. typhimurium* (112, 216), is an attractant for *E. coli* (217), although the latter response may not be significant under natural conditions. This difference derives from the existence of a transducer in *S. typhimurium* that, in *E. coli*, appears either to be lacking or to have different properties. The weak attractant response of *E. coli* is primarily mediated by the Tar_E transducer; a similar response exists in *S. typhimurium*, but is normally masked by the repellent response of the phenol-specific transducer (Y. Imae and R. M. Macnab, unpublished data).

Inorganic Chemical Attractants

Divalent cations act as attractants for *S. typhimurium*, but not for *E. coli*. This response, originally believed to be associated with the membrane ATPase, has now been linked to the sensing of citrate as a chelated complex (79). With the exception of the aerotactic response and the attractant response to NH_4^+ ion, which (as NH_3) acts as a membrane-permeant weak base (164; see pH taxis, above), and the possible exception of weak attractant responses to Mg^{2+} and Ca^{2+} in certain strains of *E. coli* (79, 230), there is no clear evidence for attraction to inorganic compounds per se in either species.

Inorganic Chemical Repellents

The divalent cations Co^{2+} and Ni^{2+}, which at quite low concentrations inhibit motility unless chelated, act as repellents for *E. coli* (217); their action is mediated by a specific transducer (Tar_E). Sulfide ion also is a repellent, at least for *E. coli* (217).

GENETICS

The use of mutants has played and continues to play a crucial role in elucidating the tactic behavior of *E. coli* and *S. typhimurium*, especially so because of the advanced state of molecular genetic techniques for these species.

Several methods have been used for the isolation of chemotaxis mutants, as follows: (i) picking cells from the center of swarm plates made with complex media or (for isolation of mutants defective with respect to a specific compound) media containing that compound as the carbon source (18) (this method has the disadvantage that it may yield strains with multiple mutations; Parkinson, personal communication); (ii) mixing cells with warm soft agar, allowing solidification to occur, and inspecting swarm morphology of individual colonies (the mini-swarm method; Parkinson, personal communication); (iii) spreading cells on soft nutrient gelatin agar, which allows sufficient motility to be able to distinguish wild-type and mutant colonies (227); (iv) layering an attractant into a liquid gradient and selecting cells that fail to move into that region (21); (v) maintaining cells in stationary phase for several days, whereupon the culture, for unknown reasons, becomes enriched for nonflagellate, nonmotile, and nonchemotactic mutants (59, 225).

The various phenotypic classes that have been recognized will be described next, followed by the current state of knowledge regarding motility- and chemotaxis-linked genes and their organization.

Types of Mutants

Nonflagellate

Mutants lacking flagella (Fla⁻ phenotype) have defects in *fla* (*flb*) genes and are described in chapter 7.

Nonmotile

Nonmotile (paralyzed) mutants (Mot⁻ phenotype) have morphologically normal flagella, but these fail to rotate even though the driving force (proton motive force) is normal. In all cases examined so far, the flagella do not seem to be physically jammed, since they can be made to rotate by an externally applied torque (81, 225). They are therefore presumed to be defective in the energy-transducing mechanism.

Generally nonchemotactic

Chemotaxis-defective mutants (Che⁻ phenotype) have actively rotating flagella, but fail to give a chemotactic response. Those that are generally nonchemotactic show a failure of response either to all stimuli or at least to a broad range of stimuli. The defect may be partial or total; the degree to which it manifests itself depends, among other things, on the assay used. Usually the defect is in a component of the sensory transduction apparatus that is shared by many receptor classes, but in a few cases there is a dominant effect exerted by a receptor-specific component. A striking feature of most generally nonchemotactic mutants is that even in the absence of stimulation they can be distinguished from wild type on the basis of their tumbling frequency (or CW/CCW rotation probability if tethered). Two major subclasses of generally nonchemotactic mutants can be recognized, as follows.

CCW biased. CCW-biased mutants swim for extended periods and seldom or never tumble; they are also referred to as "smooth" mutants because of the impression their motility creates in the light microscope. Under high-intensity dark-field illumination (115), they are seen to maintain their flagella in a coordinated bundle (see Fig. 2a) for extended periods, or, in some cases, indefinitely. In a swarm plate assay, they show only a slight degree of spread in spite of the fact that they are fully motile; this emphasizes the critical importance of the tactic response not only for migration in the "right" direction, but also for macroscopically significant movement in any direction. In liquid media, smooth mutants show an enhanced sedimentation compared to wild-type cells, which is believed to be caused by a passive orientation under gravity, accompanied by active translational motion at the swimming velocity. This has proved to be useful for selection of mutants (21; see above).

CW biased. CW-biased mutants have a high tumbling frequency compared to wild type. By high-intensity dark-field light microscopy (115), the flagella are seen to form bundles only briefly and to spend a high proportion of time dispersed and undergoing polymorphic transitions (see Fig. 2c and d; 126). In certain extreme cases, tethered-cell assays of a mutant indicate close to 100% CW rotation, and free cells are observed to enter into an "inverse" swimming mode in which the flagellar filaments, rotating CW, are fully converted by mechanical torsion into a right-handed polymorphic form and develop into a propulsive bundle; such mutants show inverted tactic responses (attraction to repellents and repulsion from attractants) because, for them, enhancement of CCW rotation causes enhanced tumbling by partially converting the filaments back toward the normal waveform (91).

Semispecifically nonchemotactic (transducer-defective)

Some mutants, selected on the basis of their failure to respond to a given chemical (say, serine), are found to be chemotactically defective to a limited range of other stimuli, including ones that might not be thought analogous. They are therefore defective in a component that is shared by a subset of possible stimuli. For example, a given mutant may be defective in serine taxis and weak acid taxis, but normal for aspartate taxis (191).

In some cases, there may also be some perturbation of the unstimulated tumbling frequency, leading to a partially defective response to all stimuli (e.g., *tsr* [CheD] mutants; 152).

Specifically nonchemotactic

Specifically nonchemotactic mutants show normal responses to all but a quite restricted set of stimuli that are closely analogous. As an illustration, a given mutant may be defective in taxis toward D-glucose and D-galactose, but not to other stimuli (68). Such mutants have been found to be defective in the primary receptor for the stimuli to which they cannot respond.

Genes and Genomic Organization

Of the 60 or so genes that are known to be involved in motility and chemotaxis of *E. coli* and *S. typhimurium*, a large number are involved in flagellar structure and assembly only. These genes and their products are described in chapter 7. Table 3 of the present chapter contains a complete compilation of known flagellar, motility, and chemotaxis genes in genome order.

The present description is confined to those genes involved in the communication of sensory information to the motor or in the energy transduction or switching processes of the flagellar motor, i.e., genes that have the possibility of conferring nonchemotactic (Che⁻) or paralyzed (Mot⁻) mutant phenotype. These genes include *motA* and *motB* and three *fla* genes whose products are believed to constitute the flagellar switch; these genes are also considered in chapter 7 in the context of flagellar structure and assembly.

Motility and chemotaxis genes, except for some that code for receptors and transducers, are found in two clusters called motility regions II and III at 42 to 43 min (*E. coli*) and 40 min (*S. typhimurium*). (Region I contains flagellar genes, but no motility or chemotaxis genes.)

The order of motility and chemotaxis genes and their organization into operons is to some extent indicative of functional relationships. The following are examples. One operon (*flbB*$_E$ *flaI*$_E$/*flaK*$_S$ *flaE*$_S$) contains only two genes, both of which are exclusively involved in positive regulation of the entire motility/chemotaxis system (179). The genes for the methyltransferase and methylesterase (*cheR* and *cheB*, respectively; 194, 199) are adjacent. The only two "pure" *mot* genes, *motA* and *motB* (181), are also adjacent.

Genes for sensory receptors

Many of the genes that function in the primary reception process (Table 4) also function in transport. In such cases, the latter function may be considered to be the primary one—there is no value to detecting an attractant if it cannot be assimilated. Not surprisingly, such genes are located in transport-linked regions of the chromosome. In this category are the genes coding for periplasmic binding proteins for galactose, ribose, maltose, and dipeptides, as well as the enzyme II genes for a variety of phosphotransferase system (PTS) sugars. The principal exceptions to the linkage between chemotaxis and transport are those genes that code for the methyl-accepting chemotaxis proteins (MCPs), which are primary receptors for amino acids and secondary receptors for other attractants; these receptors play no role in transport, and the genes are located either within the major chemotaxis/motility clusters (the *tar* and *tap*$_E$ genes in region II) or in isolated locations (*tsr* and *trg*).

TABLE 3. Genomic organization of flagellar, motility, and chemotaxis genes of *E. coli* and *S. typhimurium*[a]

E. coli Map position	*E. coli* Gene	*S. typhimurium* Gene	*S. typhimurium* Map position	*E. coli* Map position	*E. coli* Gene	*S. typhimurium* Gene	*S. typhimurium* Map position	*E. coli* Map position	*E. coli* Gene	*S. typhimurium* Gene	*S. typhimurium* Map position	*E. coli* Map position	*E. coli* Gene	*S. typhimurium* Gene	*S. typhimurium* Map position
16	*nagE*	*nag*	15	31	*trg*	*trg*	?	42	*flaD*	*flaL*	40	47	*gatA*	?	
24	*flaU*	*flaFI*	23	40	*ptsM*	*ptsM*	?		?	*nml*		52	*ptsH*	*ptsH*	49
	flbA	*flaFII*		41	*flaH*	*flaC*	40		*hag*	*H1*			*ptsI*	*ptsI*	
	flaW	*flaFIII*			*flaG*	*flaM*			*flbC*	*flaV*			*crr*	*crr*	
	flaV	*flaFIV*			*cheZ*	*cheZ*		43	*flaN*	*flaAI*			—	*rh1*	56
	flaK	*flaFV*			*cheY*	*cheY*			*flaBI*	*flaAII.1*			—	*H2*	
	flaX	*flaFVI*			*cheB*	*cheB*			*flaBII*	*flaAII.2*			—	*hin*	
	flaL	*flaFVII*			*cheR*	*cheR*			*flaBIII*	*flaAII.3*		58	*gutA*	*gutA*	59?
	flaY	*flaFVIII*			*tap*	—			*flaC*	*flaAIII*		79?	*dpp*	?	
	flaM	*flaFIX*			*tar*	*tar*			*flaO*	*flaS*		81	*mtlA*	*mtlA*	78
	flaZ	*flaFX*			*cheW*	*cheW*			*flaE*	*flaR*		83	*bglC*	?	
	flaS	*flaW*			*cheA*	*cheA*			*flaAI*	?		84	*rbsB*	*rbsP*	82
	flaT	*flaU*			*motB*	*motB*			*flaAII*	*flaQ*		92	*malE*	*malE*	91
24	*ptsG*	*ptsG*	26		*motA*	*motA*			*motD*	*flaN*		99	*tsr*	*tsr*	ca. 99
					flaI	*flaE*			*flbD* —?—	*flaP*					
					flbB	*flaK*			*flaR*	*flaB*					
					?	[*flaY*]			*flaQ*	*flaD*					
				42	[*uvrC*]	[*uvrC*]	40		*flaP*	*flaX*					
								45	*mglB*	*mglB*	?				
								46	*fruA*	*fruA*	?				

(Region I: *flaU* through *flaT* in *E. coli*, *flaFI* through *flaU* in *S. typhimurium*. Region II: *flaH* through *flbB*. Region III: *flaD* through *flaP*. H2: *rh1*, *H2*, *hin*.)

[a] Note the following. (i) Information collated from a number of sources, including references 16, 17, 22, 23, 53, 97, 98, 106, 107, 151, 154, 159, 171, 178, 185, 223, and 225–227. (ii) Homologous genes have been placed in horizontal register; homology between *flbD*$_E$ and *flaP*$_S$ has not been established. (iii) Genes connected by a heavy vertical line are known or believed to be adjacent on the chromosome. Known operons are indicated by arrows; the *hin* operon can, as a consequence of site-specific inversion, exist with either polarity. (iv) A number adjacent to a gene indicates the approximate map position in minutes of that gene and those below it. (v) Where for a given species a gene is indicated by a ?, its existence has not been established; where indicated by —, it has been established that it does not exist. (vi) Where a map position is indicated by a number followed by a ?, the existence of the function is established and there are related genes at this position; where the position is indicated solely by a ?, the existence of the function is established, but the relative map position is a postulate based on the position of the homologous gene in the other species. (vii) No *tap* gene has been found in *S. typhimurium*. The *tip*$_S$ gene has not been mapped, but is not adjacent to *tar* (as *tap*$_E$ is); no gene corresponding to *tip*$_S$ is known in *E. coli*. (viii) *flaY* is known to be near *flaK*, but its exact location is not known (S. Yamaguchi, personal communication).

Genes for sensory transduction components

Genes involved in communication between receptors/transducers and the motor are listed in Table 5. Of these, the ones that communicate between the MCP receptors/transducers and the motor (*cheA*, *cheB*, *cheR*, *cheW*, *cheY*, and *cheZ*) all fall within region II. Also listed in Table 5 are genes (*ptsH*, *ptsI*, and *crr*) that are involved in phosphoryl transfer in the PTS (159) and that may also be responsible for relaying signals to the flagellar motors regarding the concentrations of PTS substrates such as fructose (but see discussion of PTS taxis, below). The latter three genes are found in a distinct region of the chromosome, not in region II.

Genes for energy-transducing and switching components of the flagellar motor

Finally, those genes involved in the function of the motor and its interface with the sensory system are listed in Table 6. They are the two motility genes (*motA* and *motB*, in region II) and three flagellar genes involved in energy transduction and switching (*flaBII*$_E$/*flaAII.2*$_S$, *flaAII*$_E$/*flaQ*$_S$, and *motD*$_E$/*flaN*$_S$, all in region III). Other flagellar genes not involved in energy transduction or switching are described in chapter 7.

Regulation of Gene Expression

Regulation of flagellar gene expression is discussed in chapter 7, where it is noted that the expression of all operons dedicated to motility and chemotaxis is under a hierarchy of control, with an operon that is itself cyclic AMP dependent acting as the master element of the control. Sensory components are at the final stage of this control, so that they are not synthesized unless flagellar assembly has been achieved (96). In the promoter regions of several operons that have been sequenced, examples of symmetries and common sequences have been noted (105; P. Matsumura, personal communication); these may prove to be important in the regulation of expression of the flagellar/motility system.

Receptors that are also components of transport systems are under the regulatory control of these systems, not that of motility.

COMPONENTS OF THE SENSORY TRANSDUCTION SYSTEM

Receptors

Early inferences regarding the existence of receptors derived in part from competition experiments (1):

TABLE 4. Genes and gene products involved in reception/transduction in *E. coli* and *S. typhimurium*

Gene				Gene product			References
Symbol	Map location (min)	Operon	Induc./ constit.[a]	Mol wt	Location	Function and comments	
Periplasmic primary receptors							
$mglB_E$ $mglB_S$	45 ?	mgl_E ?	Induc.	33,387 33×10^3	Periplasm	Glucose- and galactose-binding protein; functions both in transport and in chemotaxis; binds to Trg (MCPIII)	6, 57, 64, 68, 70, 99, 135, 148, 149, 202, 229, 231
dpp_E	79?	dpp_E	?	49×10^3	Periplasm	Dipeptide binding protein; believed to function both in transport and chemotaxis; believed to bind to Tap (MCPIV)	128a
$rbsB_E$ $rbsP_S$	84 82	rbs_E rbs_S	Induc.	30×10^3 29×10^3	Periplasm	Ribose-binding protein; functions both in transport and in chemotaxis; binds to Trg (MCPIII)	6, 10, 70, 71, 76, 99, 202, 224, 229
$malE_E$ $[malE_S]$	92 91	$malE_E$ $malE_S$	Induc.	40,661 ?	Periplasm	Maltose-binding protein; functions both in transport and in chemotaxis; binds to Tar_E ($MCPII_E$) but not to Tar_S; hence no natural maltose response in *S. typhimurium*	6, 47, 57, 67, 71, 129, 130, 139, 191
Integral membrane methyl-accepting receptors/transducers							
tsr_E tsr_S	99 ca. 99	tsr_E ?	Constit.	57,483 60×10^3	Cell membrane	Methyl-accepting chemotaxis protein (Tsr, MCPI); primary receptor and transducer for serine, external pH, weak acid repellents (cytoplasmic pH), temperature, hydrophobic amino acids, indole; certain alleles are dominant, causing CW suppression	39, 43, 72, 84, 104, 105, 141, 152, 163, 164, 182, 188, 191, 221
tar_E tar_S	42 40	$meche_E$ $meche_S$	Constit.	59,965 59,416	Cell membrane	Methyl-accepting chemotaxis protein (Tar, MCPII); primary receptor and transducer for aspartate, Co^{2+}, and Ni^{2+}; secondary receptor, and transducer, for maltose (*E. coli* only)	35, 43, 58, 71, 104, 105, 139, 140, 163, 169, 182, 186, 187, 191, 211, 213, 221
trg_E trg_S	31 ?	trg_E ?	Constit.	57,965 58×10^3	Cell membrane	Methyl-accepting chemotaxis protein (Trg, MCPIII); secondary receptor, and transducer, for ribose and galactose	36, 65, 70, 87, 99, 202; Macnab, unpublished data
tap_E	42	$meche_E$	Constit.	57,645	Cell membrane	Methyl-accepting chemotaxis protein (Tap_E, $MCPIV_E$); secondary receptor, and transducer, for dipeptides; absent in equivalent chromosomal location of *S. typhimurium*	38, 105, 128a, 186, 187, 222
tip_S	?	?	Constit.	60×10^3	Cell membrane	Methyl-accepting chemotaxis protein (Tip_S); similar methylation characteristics to other MCPs; mediates response to unidentified component of complex media	170
Integral membrane PTS receptors/transducers							
$nagE_E$ nag_S	16 15?	nag	Induc.	? ?	Cell membrane	$EnzII^{Nag}$ of PTS; chemoreceptor for *N*-acetylglucosamine	6, 113, 114, 159
$ptsG_E$ $ptsG_S$	24 26	$ptsG_E$ $ptsG_S$	Constit.	? 40×10^3	Cell membrane	$EnzII^{Glc}$ of PTS; chemoreceptor for glucose; also known as *glcA*	5, 6, 113, 114, 159
$ptsM_E$ $ptsM_S$	40 ?	$ptsM_E$?	Constit.	? ?	Cell membrane	$EnzII^{Man}$ of PTS; chemoreceptor for mannose; also known as *manA*	5, 6, 113, 114, 159
$fruA_E$ $?_S$	46	fru_E?	Induc.	?	Cell membrane	$EnzII^{Fru}$ of PTS; chemoreceptor for fructose	6, 113, 114, 159
$gatA_E$ $?_S$	47	$gatC_E$	Induc.	?	Cell membrane	$EnzII^{Gat}$ of PTS; chemoreceptor for galactitol	6, 113, 114, 159
$gutA_E$ $gutA_S$	58 59?	gut_E ?	Induc.	? ?	Cell membrane	$EnzII^{Gut}$ of PTS; chemoreceptor for glucitol (sorbitol); also known as *srlA*	6, 113, 114, 159

Continued on next page

TABLE 4—*Continued*

Gene				Gene product			References
Symbol	Map location (min)	Operon	Induc./ constit.[a]	Mol wt	Location	Function and comments	
Integral membrane PTS receptors/transducers							
mtlA$_E$	81	*mtlC*$_E$	Induc.	?	Cell membrane	EnzII^{Mtl} of PTS; chemoreceptor for mannitol	6, 113, 114, 159
mtlA$_S$	78	*mtl*$_S$		67,893			
bglC$_E$	83	*bgl*$_E$	Induc.	?	Cell membrane	EnzII^{Bgl} of PTS; chemoreceptor for β-glucosides	6, 113, 114, 159
?$_S$							
[*scrA*$_S$]					Cell membrane	EnzII^{Scr} of PTS; chemoreceptor for sucrose; plasmid-encoded from *S. typhimurium*, not a natural gene of *E. coli* K-12	6, 113, 159
scrA$_S$	?	*scr*	Induc.	?			
sorA		*sor*	Induc.	?	Cell membrane	EnzII^{Sor} of PTS; chemoreceptor for sorbose; plasmid encoded from *Klebsiella pneumoniae*, not a natural gene of *E. coli* K-12	6, 113, 159, 190

[a] Induc., Inducible; Constit., constitutive.

if compound A at high and uniform concentration inhibited the response toward a gradient of compound B, while leaving the response to a gradient of compound C intact, it was inferred that A and B shared a common receptor distinct from that responsible for detection of compound C. The existence of mutants that were specifically nonchemotactic was also used as strong circumstantial evidence for receptors. These early studies provided much useful information, although in some cases (especially with repellents) the receptor has not been identified and may not exist, at least in the usual sense of a relatively specific binding protein. Ultimately, the only valid demonstration of a receptor is its isolation and characterization. This has been achieved now in a number of instances, including the receptors for the strongest of the organic attractants, namely, L-serine and L-aspartate, and for a number of sugars.

The known receptors are the products of the genes listed in Table 4, which also provides information on molecular weight and cellular location, where these are known. Table 7, which describes the major chemical attractants for *E. coli* and *S. typhimurium*, also provides information on the receptors responsible for their detection and on quantitative aspects such as the concentrations for threshold and half-maximal response.

Periplasmic binding proteins for sugars

There are three periplasmic proteins, products of the *mglB*, *rbsB*$_E$/*rbsP*$_S$, and *malE*$_E$ genes, that bind sugars (glucose/galactose, ribose, and maltose, respectively) with high affinity and specificity and transfer them to membrane-associated transport systems. When associated with their ligands, they can also bind to integral membrane proteins that are dedicated members of the chemotaxis system, namely, the methyl-accepting chemotaxis proteins Trg (for MglB and RbsB) and Tar$_E$ (for MalE$_E$). The proteins are of medium size, are readily soluble in aqueous media, and have been purified to homogeneity.

Spectroscopic studies (229) have indicated that extensively delocalized conformational changes take place upon ligand binding by these receptors, changes that are presumably responsible for the differential affinity for the corresponding MCP.

Periplasmic binding protein for dipeptides

Dipeptides are recognized by the Dpp protein, a periplasmic protein which recognizes a variety of dipeptides and conveys them to a poorly characterized uptake system (158). The specificity of Dpp is moderately broad with respect to the residues within the dipeptide; the strongest attractants (glycyl-L-leucine and L-leucylglycine) are listed in Table 7, but there are many others that are appreciably detected. Tripeptides are poor attractants.

As with the periplasmic sugar-binding proteins described above, Dpp interacts with a secondary receptor, in this case Tap$_E$, which acts as the transducer for dipeptide stimuli (128a).

Integral membrane methyl-accepting receptors/transducers

There exist integral membrane proteins which are not involved in transport but do act as specific receptors for chemotactic effectors. Unlike the periplasmic binding proteins described above, their role extends beyond that of recognition, because subsequent to binding they initiate events (their own methylation and other as yet unidentified events) in the cytoplasm. Thus they are called transducers as well as receptors. Only their role as receptors will be discussed in this section.

There are four well-characterized proteins of this type: Tsr (taxis to serine and repellents; also known as MCPI), Tar (taxis to aspartate and repellents; MCPII), Trg (taxis to ribose and galactose; MCPIII), and Tap$_E$ (taxis-associated protein; MCPIV$_E$); a fifth, Tip$_S$ (taxis-involved protein), is less understood.

They are all of quite similar molecular weight (ca. 60,000) and show extensive sequence homology (based on gene sequencing; 36, 37, 105, 169), especially towards the C terminus. It seems likely that they are organized (Fig. 5) with a single membrane-spanning

TABLE 5. Genes and gene products involved in intermediate signal processing in *E. coli* and *S. typhimurium*

Gene			Gene product			References
Symbol	Map location (min)	Operon	Mol wt	Location	Function and comments	
$cheA_E$	42	$mocha_E$	59,976/ 70,892	Cytoplasm	Enables CW rotation; coded for by two overlapping genes with same reading frame and same termination point; both proteins believed necessary for proper chemotactic function; inhibits demethylation activity of CheB; forms complex with CheW	150, 166, 183, 189, 192; E. Kofoid and J. S. Parkinson, personal communication
$cheA_S$	40	$mocha_S$?	?			
$cheB_E$	42	$meche_E$	37,465	Cytoplasm	Methylesterase, deamidase; demethylates MCP methylglutamate residues and deamidates MCP glutamine residues; enhances CCW rotation, probably by decreasing CW excitation signal; interacts with CheZ; may function as an oligomer; $cheB_S$ formerly called $cheX_S$	63, 69, 82, 83, 85, 142, 150, 166, 167, 174, 183, 184, 199, 228
$cheB_S$	40	$meche_S$	37,500			
$cheR_E$	42	$meche_E$	32,748	Cytoplasm	Methyltransferase; methylates MCP glutamate residues; enhances CW rotation, probably by increasing CW excitation signal; interacts with CheY; $cheR_E$ formerly called $cheX_E$	40, 62, 94, 142, 157, 166, 183, 194, 219
$cheR_S$	40	$meche_S$	38,000			
$cheW_E$	42	$mocha_E$	18,083	Cytoplasm	Enables CW rotation; forms complex with CheA	142, 166, 183
$cheW_S$	40	$mocha_S$?	?			
$cheY_E$	42	$meche_E$	14,096	Cytoplasm	Component of fast excitation signal? Interacts with CheR and flagellar switch complex proteins (FlaAII_E, FlaBII_E/FlaAII.2_S); binds S-adenosylmethionine; produced in high copy number; inhibits methylation; enhances CW rotation, even in absence of transducers or other Che proteins; enables CW rotation of flagellar envelopes; homologous to N-terminal region of CheB; $cheY_S$ formerly called $cheQ_S$	44, 133, 142, 156, 166, 183, 196a; Ravid et al., in press; Yamaguchi et al., submitted
$cheY_S$	40	$meche_S$	13,980			
$cheZ_E$	42	$meche_E$	23,974	Cytoplasm	Interacts with CheB and flagellar switch complex proteins (FlaAII_E/FlaQ_S, FlaBII_E/FlaAII.2_S, FlaN_S); enhances CCW rotation; post-translationally methylated; controls excitation latency period and signal range; may be responsible for inactivation of excitation signal; $cheZ_S$ formerly called $cheT_S$	34, 142, 155, 156, 166, 172, 173, 182, 183; Yamaguchi et al., submitted
$cheZ_S$	40	$meche_S$	?			

Genes that may be involved in intermediate signal processing for PTS sugars

$ptsH_E$	52	pts_E	?	Cytoplasm	HPr of PTS; accepts phosphoryl group from enzyme I and transfers it to either enzyme II or enzyme III	159, 160
$ptsH_S$	49	pts_S	9,017			
$ptsI_E$	52	pts_E	68×10^3	Cytoplasm	Enzyme I of PTS; accepts phosphoryl group from phosphoenolpyruvate and transfers it to HPr	159, 160
$ptsI_S$	49	pts_S	?			
crr_E	52	crr_E	15,556	Cytoplasm	Enzyme III^{Glc} of PTS; accepts phosphoryl group from HPr and transfers it to enzyme II^{Glc}	159, 160
crr_S	49	crr_S	?			

segment at the N terminus, an external aqueous domain, a second membrane-spanning segment near the middle of the sequence, and a cytoplasmic domain composed of the C-terminal region (105, 140). Thus the homology at the C terminus would be reasonable in terms of common interactions with cytoplasmic components such as the methylation enzymes, while the divergence at the N terminus would be reasonable in terms of the need for chemoeffector-binding sites with different specificity. That the N terminus does confer chemoeffector specificity has been verified directly by constructing chimeric *tsr-tar* genes with the

5' portion of the gene coding for the N-terminal half of, say, Tsr and the 3' portion coding for the C-terminal half of Tar; the product of such a chimeric gene should, and does, have Tsr-like receptor specificity (104).

As receptors, these proteins have several different modes of operation. In the first, they act as conventional receptors, binding specific chemoeffectors with high affinity. In the second, they act as secondary receptors for sugars, which they recognize indirectly via the periplasmic binding proteins described above. In the third, they act as detectors of environmental

TABLE 6. *Genes and gene products involved in flagellar energization and switching in* E. coli *and* S. typhimurium

Gene			Gene product			References
Symbol	Map location (min)	Operon	Mol wt	Location	Function and comments	
$motA_E$ $motA_S$	42 40	$mocha_E$ $mocha_S$	31,974 32,000	Cell membrane	Necessary for motor rotation; extremely hydrophobic; not site limited, even when overproduced	51, 56, 166, 181, 227; M. L. Wilson and R. M. Macnab, unpublished data
$motB_E$ $motB_S$	42 40	$mocha_E$ $mocha_S$	34,189 ?	Cell membrane	Necessary for motor rotation; can activate preexisting flagella; site limited when overproduced	33, 56, 166, 196, 227
$flaBII_E$ $flaAII.2_S$	43 40	$flaBI_E$?	36,788 ?	Cell membrane?	Necessary for any detectable flagellar structure; switch and energy-transducing component of motor; interacts with CheY, CheZ, $FlaN_S$, and $FlaQ_S$; gene also known as $motC_S$, $cheV$	23, 50, 56, 156, 162, 218, 225, 227; Yamaguchi et al., submitted
$flaAII_E$ $flaQ_S$	43 40	$flaAI_E$?	37,806 ?	Cell membrane?	Necessary for any detectable flagellar structure; switch and energy-transducing component of motor; interacts with $CheY_E$, CheZ, $FlaAII.2_S$, and $FlaN_S$; gene also known as $motE$, $cheC$; $cheC_S$ formerly called $cheU_S$	45, 91, 105a, 150, 155, 156, 162, 168a, 225, 227; Yamaguchi et al., submitted
$motD_E$ $flaN_S$	43 40	$flaAI_E$?	14,388 ?	Cell membrane?	Necessary for any detectable flagellar structure; switch and energy-transducing component of motor; interacts with $CheZ_S$, $FlaAII.2_S$, and $FlaQ_S$; $flaN_S$ gene also known as $motD_S$	225, 227; Yamaguchi et al., submitted

parameters such as pH and temperature. In the fourth, they are "receptors" for a variety of hydrophobic compounds, but it seems doubtful that they do so in a conventional manner by binding the compounds specifically (one indication of this is that L- and D-amino acids are equally effective; 217); little is known of the mechanism by which they sense such compounds, although it may be that their interaction with the membrane is perturbed.

Tsr or MCPI. Tsr (MCPI) is responsible for recognizing L-serine, which, along with L-aspartate, is one of the most powerful attractants for both *E. coli* and *S. typhimurium*. Tsr detects L-serine in two different concentration ranges (ca. 5 and 300 μM; 72), but it has not been established whether these correspond to affinities for distinct binding sites. It is also responsible for temperature sensing (127) and for the sensing of acidic pH conditions, both external and cytoplasmic (93, 164, 188). (The latter aspect is the means by which membrane-permeant weak acid repellents such as acetate are sensed.) Tsr is also the "receptor" responsible for detection of other hydrophobic repellents (191). It is seen therefore to be a receptor/transducer of great versatility and of central importance to taxis in *E. coli* and *S. typhimurium*. A further indication of its importance is that certain alleles (sometimes called $cheD_E$ and $cheS_S$) are dominant, giving an unstimulated phenotype that is so strongly CCW biased that the cells are unable to respond even to non-Tsr stimuli (152; A. F. Russo and D. E. Koshland, Jr., personal communication). The lack of a functional Tsr can have other unusual effects, includ-

ing the inversion of certain responses (for example, to weak acids; 141).

It will be of great interest to know how this one protein can detect so many stimuli. One indication that they may all initiate the same consequent events is that L-serine acts as a competitive inhibitor of thermotaxis (127). Two such utterly different stimuli cannot be competitive in the usual sense of competitive binding, and so the phenomenon must be understood in terms of an effect of serine binding that prevents the same effect being accomplished by temperature.

Tar or MCPII. Tar is another important receptor/transducer, since it mediates one of the strongest chemotactic responses, the attractant response to L-aspartate. Like Tsr, it is also responsible for detection of other classes of stimuli, namely, the attractant maltose and the inorganic repellents Co^{2+} and Ni^{2+} (191).

Maltose is detected as its complex to a periplasmic binding protein, MalE (67); i.e., Tar is a secondary receptor for maltose. Affinity chromatography (95, 165) and allele-specific suppression of *malE* mutations by mutations in *tar* (130) provide evidence for the direct interaction of the two proteins. The response is absent in *S. typhimurium*, because Tar_S either lacks a binding site for $MalE_S$ (or $MalE_E$ in artificial constructions) or fails to generate the necessary signals upon binding. $MalE_S$ is normal, since it can initiate a response via Tar_E, as has been demonstrated by physically introducing $MalE_S$ into the periplasm of a *malE* mutant of *E. coli* (47) and also by expressing tar_E on a

TABLE 7. Principal chemical attractants for *E. coli* and *S. typhimurium*[a]

Compound	Threshold (M)	Peak[b] ($\times 10^3$)	K_d or K_m[c]	Receptor	Transducer	References
Amino acids						
L-Alanine$_E$	5×10^{-5}	180	?	Tsr	Tsr	43, 73
L-Alanine$_S$	(Poor attractant)					
L-Aspartate$_E$	3×10^{-8}	310	5×10^{-6}	Tar	Tar	10, 43, 71, 73, 135
L-Aspartate$_S$	2×10^{-6}	1,600	6×10^{-6}			
L-Cysteine$_E$	1×10^{-6}	130	?	Tsr	Tsr	73, 135
L-Cysteine$_S$	6×10^{-6}	160				
L-Glutamate$_E$	3×10^{-5}	220	?	Tar	Tar	43, 73
L-Glutamate$_S$	(Poor attractant)					
Glycine$_E$	6×10^{-5}	160	?	Tsr	Tsr	43, 73, 135
Glycine$_S$	(Poor attractant)					
L-Serine$_E$	1×10^{-7}	310	$3 \times 10^{-6}/3 \times 10^{-4}$	Tsr	Tsr	43, 48, 72, 73, 135
L-Serine$_S$	2×10^{-6}	500	5×10^{-6}			
Dipeptides						
Glycyl-L-leucine$_E$	1×10^{-5}	33	?	Tap$_E$	Tap$_E$	128a
L-Leucylglycine$_E$	$<1 \times 10^{-5}$	29	?			
Other dipeptides$_E$						
Amino acid analogs						
α-Aminoisobutyrate$_E$	2×10^{-5}	300	?	Tsr	Tsr	43, 72, 136
α-Aminoisobutyrate$_S$	?	?	5×10^{-3}			
α-Methylaspartate$_E$	5×10^{-7}	550	?	Tar	Tar	8, 10, 136
α-Methylaspartate$_S$	3×10^{-6}	120	5×10^{-4}			
Carboxylic acids						
Citrate$_E$	(Nonattractant)					79, 92
Citrate$_S$	3×10^{-5}	70	1×10^{-4}	?	?	79, 92
Magnesium citrate$_S$	1×10^{-5}	?	1×10^{-5}	?	?	79
L-Malate$_E$	6×10^{-4}	100	?	Tar	Tar	43, 92, 135, 136
L-Malate$_S$	?	220	5×10^{-3}	?	?	
Sugars						
N-Acetylglucosamine$_E$	1×10^{-5}	60	?	NagE	?	6, 57, 113, 159
N-Acetylglucosamine$_S$						
D-Fructose$_E$	1×10^{-5}	50	?	FruA	?	6, 57, 135, 159
D-Fructose$_S$	1×10^{-5}	60	?		?	
Galactitol$_E$	2×10^{-5}	30	?	GatA$_E$	?	6, 113, 159
D-Galactose$_E$	1×10^{-6}	120	5×10^{-7}	MglB	Trg	6, 57, 70, 99, 135,
D-Galactose$_S$	1×10^{-6}	85	4×10^{-7}			148, 149, 202, 231
D-Glucitol$_E$	1×10^{-5}	50	?	GutA	?	6, 113, 159
D-Glucose$_E$	1×10^{-6}	140	2×10^{-7}	MglB	Trg	5, 6, 135, 231
	3×10^{-6}	80	2×10^{-5}	PtsG	?	5, 159
	3×10^{-6}	60	7×10^{-6}	PtsM	?	5, 159
D-Glucose$_S$	1×10^{-6}	65	2×10^{-7}	MglB	Trg	
Maltose$_E$	3×10^{-6}	120	2×10^{-6}	MalE$_E$	Tar$_E$	6, 47, 57, 67, 71, 129,
Maltose$_S$	(Nonattractant)					130, 139, 191
D-Mannitol$_E$	7×10^{-6}	40	?	MtlA	?	6, 57, 113, 159
D-Mannitol$_S$						
D-Mannose$_E$	1×10^{-5}	80	3×10^{-5}	PtsM	?	5, 6, 135, 159
D-Mannose$_S$	1×10^{-5}	70	?		?	
Methyl-β-D-glucoside$_E$	3×10^{-6}	150	?	MglB$_E$	Trg$_E$	6, 113
	$>3 \times 10^{-4}$	?	?	BglC$_E$	?	
D-Ribose$_E$	7×10^{-6}	70	1×10^{-7}	RbsB$_E$	Trg	6, 9, 10, 70,
D-Ribose$_S$	1×10^{-6}	40	3×10^{-7}	RbsP$_S$		71, 99, 224
Trehalose$_E$	6×10^{-6}	120		Tre$_E$	?	6, 57, 159
Trehalose$_S$	(Nonattractant)					

[a] The most potent attractants are given in boldface type.

[b] Number of cells accumulating in a capillary containing a high concentration of the attractant.

[c] K_d is the dissociation constant of the attractant for its receptor, and K_m is the "behavioral Michaelis constant," i.e., the attractant concentration for half-maximal response. Values of the two constants are similar.

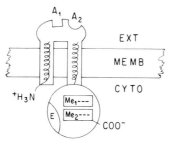

FIG. 5. Organization of an MCP receptor/transducer in the membrane, as predicted from the hydrophobicity of the primary sequence; there is also a substantial body of supporting evidence for this prediction. A very short cytoplasmic N terminus is followed by a membrane-spanning region, an external domain that contains the binding sites (A_1, A_2, etc.) for attractants, a second membrane-spanning region, and a cytoplasmic domain which contains both the region (E) responsible for generating the excitation signal to the flagellar motors and two methylated regions (Me_1 and Me_2) that function in adaptation to sensory stimuli. The two membrane-spanning regions are likely to be alpha helical, with the residues that can be methylated presenting a linear array on one side of the helices. From reference 120, with permission.

plasmid in a *tar* mutant of *S. typhimurium* (139). The *tar* gene has been sequenced in both species (105, 169). The translated sequences are quite similar, with 70% identity in the N-terminal portion that includes the external domain within which effector binding sites should be found; for comparison, Tsr_E and Tar_E show 33% identity in the same region. Many of the differences are quite conservative (such as Ile for Val, Asp for Glu), but somewhere among them must be the basis for this surprising difference in responsiveness. Perhaps chimeric constructions or site-directed mutagenesis will help to clarify this interesting point.

It is not known whether there are specific binding sites for the inorganic ions Co^{2+} and Ni^{2+}.

The Tar protein of *S. typhimurium* has been purified and studied extensively (35, 58, 140). It contains a large amount of alpha helix, especially in the N-terminal part of the molecule. When solubilized in a mixture of detergent, lipid, and glycerol, it retains activity, including ligand-stimulated methylation. Judged by molecular exclusion chromatography, it functions as a tetramer of apparent molecular weight 248,000.

Trg or MCPIII. Trg has a more restricted repertoire than either Tsr or Tar. Its only known effectors are glucose, galactose, and ribose (or close analogs thereof) (65, 70). It acts as the secondary receptor for these sugars by recognizing the corresponding binding-protein/sugar complex: MglB/glucose, MglB/galactose, or RbsB/ribose. Mutations that cause galactose-taxis defects while leaving ribose taxis normal are located in the N-terminal half of the Trg protein (36), i.e., the portion that is presumed to be external to the cell; it is reasonable therefore to suppose that the defect is in the binding site for the MglB protein. If this is correct, it would appear that this binding site is at least partially distinct from that for the RbsB protein.

Of the family of MCP transducers, Trg is the most divergent, especially in its N-terminal portion, where the homology to Tar, for example, is not statistically

significant. The suggestion has been made (36) that the *trg* gene may be a fusion of the 5' end of the gene for a component of a sugar transport system and the 3' end of the gene for a sensory transducer such as Tar.

Tap_E or $MCPIV_E$. Until recently, Tap_E was something of a mystery (38, 186, 187, 222), but it has now been shown to be a bona fide transducer, mediating responses to dipeptides (128a).

The tap_E gene is immediately adjacent to the tar_E gene, and the genes have extensive homology, although somewhat less than tsr_E and tar_E; there is no analogous gene in *S. typhimurium* adjacent to tar_S. Tap_E is present in much lower copy number than Tsr or Tar and therefore makes a small contribution to overall methylation levels; nonetheless, it has similar methylation characteristics to the other MCPs. Tap_E is a secondary receptor and transducer for dipeptides; the primary receptor is thought to be the periplasmic Dpp protein (see above).

Tip_S. The Tip_S protein was discovered (45) by probing a library of *S. typhimurium* genomic DNA for any sequences homologous to the 3' half of the tar_S gene, since that had been shown to be the most highly conserved part among the various known transducer genes. In this way, a gene was cloned that coded for a 63-kilodalton protein with methyl-accepting capability and the ability to confer approximately wild-type motility on *E. coli* Tsr⁻ Tar⁻ Tap⁻ strains, which are normally CCW biased. Such transformants also exhibited limited but reproducible swarming ability on tryptone plates, suggesting that there might be an unidentified chemoeffector in this complex medium. So far, this effector has not been identified. The tip_S gene has not been located on the chromosome, but it is not at the equivalent position to tap_E. In spite of this, it has been suggested (45) that Tip_S and Tap_E have some properties in common and may be related.

Receptors for PTS sugars

Many sugars are taken up by a system known as the phosphoenolpyruvate:carbohydrate PTS; see reference 159 and chapter 11 of this volume for review. This system consists of an initial sugar-specific component, called an enzyme II, and shared components that operate in concert to translocate and phosphorylate the sugar. Thus they differ from simple transport systems that transfer their substrates unchanged across a membrane. It is the enzymes II that act as the initial recognition components for chemotaxis towards PTS sugars (5). Thus the *nagE* gene product is the enzyme II for *N*-acetylglucosamine and recognizes that sugar for both transport and chemotaxis (114, 159). No mutants have been isolated to date that retain the transport capability but lack the chemotactic response.

The mechanism by which PTS chemotactic signals are conveyed to the motor will be discussed later.

"Receptors" for proton motive force sensing

As discussed earlier in the section on stimuli, responses to oxygen and other electron acceptors require the appropriate components of the electron transport chain (cytochrome *o* for oxygen, fumarate reductase for fumarate, nitrate reductase for nitrate), but these components cannot really be regarded as receptors. Changes in their extent of occupancy are relevant only if they result in a change of proton motive force.

All of the available evidence points to the existence of a device that monitors the value of the proton motive force and transmits information in that regard to the flagella. This hypothetical device has been called a proton motive force sensor (or "protometer"). None of the known receptors or transducers plays this role; specifically, the chemotaxis methylation system, including the MCPs, does not participate in proton motive force sensing (49, 143). There is currently no information regarding the identity of the proton motive force sensor; a simple and attractive hypothesis is that it is the motor itself. Perhaps pertinent to this suggestion is the fact that the steady-state switching probabilities of the motor are dependent on the magnitude of the proton motive force (89). It has also been suggested (60) that the proton motive force sensor may be the glucose PTS; the finding (B. L. Taylor, personal communication) that *ptsG* mutants have normal aerotaxis, however, argues against this suggestion.

One obstacle to identifying the proton motive force sensor is the absence of any mutants specifically defective in proton motive sensing, in spite of efforts to select for such mutants. This failure may reflect the nature of the proton motive force sensor itself—for example, the device might be important for regulation of proton motive force, in which case defects could be lethal. Alternatively, if the motor is its own proton motive force sensor it may not be possible to interfere with this function without perturbing the unstimulated switching probability, in which case mutants would be scored as generally nonchemotactic; it is possible that some switch mutants (defective in the *flaBII*$_E$/*flaAII.2*$_S$, *flaAII*$_E$/*flaQ*$_S$, or *motD*$_E$/*flaN*$_S$ genes) might be in this category.

Transducers

We now turn to the question of how tactic stimuli generate sensory information in the cell. In this section we shall deal with those transmembrane components such as the MCPs that have come to be known as the sensory transducers (although one could really apply that term to later components as well). Thus we are interested here in the consequences for the transducer of its interaction with the stimulus, whether that stimulus is direct binding of a chemoeffector, binding of an effector-receptor complex, or some other parameter such as temperature. A later section will treat subsequent events that eventually impinge on the motor.

There are only three known categories of transducers, methylation-dependent transducers or MCPs, PTS-related transducers, and the unidentified transducers associated with proton motive force sensing. These will be discussed in turn.

Methylation-dependent transducers (MCPs)

MCPs were first identified (and received their name) on the basis of their ability to accept methyl radiolabel posttranslationally (100). The identification was an extension of previous attempts to find the substrate for a presumed methylation reaction involving methionine (4) or, more directly, *S*-adenosylmethionine (15, 20).

The predicted organization of MCPs in the cell membrane is described above. There is thought to be only one transmembrane-helical segment directly linking the two domains, although the N terminus is also presumed to span the membrane and might by noncovalent interactions affect the cytoplasmic domain.

An obvious question is how such a meager connection could enable external binding events to influence the cytoplasmic domain where the sensory signals to the motor are presumed to initiate. There have been two main hypotheses in this regard, the chain-pulling (or chain-twisting) hypothesis and the multimer hypothesis.

Early experiments (41) indicated that the MCPs might exist as a tetramer, and recent findings with Tar$_S$ support the idea (58). Binding of effector could then affect the degree of association or cause a quaternary structural shift. However, the tetramer is stable in both the presence and absence of aspartate, and binding to aspartate is noncooperative (S. L. Mowbray and D. E. Koshland, Jr., personal communication). Thus, in spite of the evidence for the existence of a tetramer, the role that it plays in signal transduction is unclear. Recent data on the suppression of mutational defects in the N-terminal membrane-spanning region by second-site mutations mapping to the cytoplasmic domain suggest that tertiary structural interactions between the N terminus and the cytoplasm may be important (M. Simon, personal communication).

The C-terminal domain shows a high degree of homology among the transducers, including several substantial stretches of identity (105). The regions containing the highest homology do not include the sites for methylation/demethylation, suggesting that there are currently unknown functions associated with the transducers where sequence conservation is critical. Many mutants defective in excitation signalling have lesions in this region. About 35 residues may be removed from the C terminus while leaving excitation function—but not adaptation function (see below)—intact; a longer truncation (80 residues) destroys both functions (103, 169).

The methylation sites fall into two clusters, centered around residues 300 and 490 of the approximately 540-residue sequence (82, 84, 211, 213), and each containing several methylatable residues. Within these clusters is the strong suggestion of a repeating motif of (Glx)-Glx-(Glx)-xxx-Ala-Pol-Hph-xxx, where Glx represents Glu or Gln, the methylatable residue is underlined, Pol represents a polar residue such as Ser or Thr, Hph represents a hydrophobic residue such as Met or Val, and xxx is any residue. It is suggested that this might be organized into an alpha helix with the methylatable residues presenting themselves on the same face every second turn (213). Thus an array would be available to the enzymes.

As will be discussed further below, the MCP transducers have a substantial capacity, independent of their modification by the chemotaxis methylation/demethylation reactions, to transduce stimuli into information for the motor.

PTS-related transducers

Sensory transduction of PTS sugar stimuli does not require the participation of the chemotaxis methylation system (49, 143). Because multiple mutants in the latter system are commonly heavily CCW biased, the chemotactic response towards PTS sugars may not be normal, but there does not seem to be any direct involvement of methylation. Strains defective in all known classes of MCP still respond to PTS stimuli, as do *cheR* mutants defective in the MCP methyltransferase. Furthermore, no PTS-sugar stimulation of methylation has been observed.

What then are the transducers for PTS stimuli? As was mentioned above, the enzymes II are the receptors (5, 114, 159). Since they are the only components of the PTS that are believed to be transmembrane, then we may legitimately regard them as both receptors and transducers. In that regard they are analogous to MCPs I and II with respect to amino acid taxis. However, mutants defective in those components responsible for supplying the high-energy phosphoryl group to the sugar, namely, enzyme I and HPr (and in some cases, sugar-specific enzymes III), are generally defective in PTS taxis (160). Thus, whereas neither translocation nor modification of amino acids occurs as part of sensory transduction by MCPs, the translocation and phosphorylation of the PTS sugar appear to be intrinsic to the sensory transduction event. It is possible, however, that the integrity of the PTS transport system is needed for chemotaxis but that transport itself is not needed. One could imagine that interaction of the enzyme II with the remainder of the system is required to place it in the appropriate conformation for transmission of sensory information.

It is not known whether the cytoplasmic signal generated by sugar translocation/phosphorylation is the delivery of the modified sugar to the cytoplasm, or whether it is the perturbation of the state of the macromolecular components of the PTS. Since no PTS taxis mutants have been found to contain lesions outside the PTS, the latter alternative seems more likely.

Other transducers

As discussed above under the categories of stimuli and their reception, proton motive force appears to be monitored by some device, possibly the motor itself. How this information is transduced is completely unknown.

Other Chemotaxis-Related Proteins

One of the advantages of studying behavior in *E. coli* or *S. typhimurium* is that, given a selection procedure (such as failure to migrate on a swarm plate), it is relatively easy to identify the macromolecular components that are responsible for the proper functioning of the system. I shall describe in this section what is known regarding these proteins and later try to draw them together into an overall scheme.

There are six such proteins that can be classified as "pure" chemotaxis components (Table 5), in the sense that they do not (with one exception to be described below) play any known role outside chemotaxis, nor do they affect motor function in any other way than modulating its switching probabilities. These are the CheA, CheB, CheR, CheW, CheY, and CheZ proteins. All are soluble cytoplasmic proteins, although they probably all at some time interact with membrane-bound components. There are a number of reasons for considering that CheR and CheB, CheA and CheW, and CheY and CheZ have pairwise-related roles, and they will therefore be considered as pairs in the following sections.

CheR (methyltransferase) and CheB (methylesterase)

The CheR and CheB proteins are the simplest to categorize; they are the methyltransferase and methylesterase enzymes that are responsible for modulating the level of methylation of the MCPs (194, 199). The CheB protein has deamidation activity also (69, 82, 168, 174), converting certain glutamine residues to glutamate in a reaction that is irreversible and subjects those residues from then on to the major aspect of CheR/CheB function, namely, cycles of methylation with *S*-adenosylmethionine as the donor (15, 20, 94, 219) and demethylation with methanol as the product (199, 214) (Fig. 6). The proteins are quite substrate and residue specific, acting only on certain glutamate and glutamine (CheB) residues on the MCPs. Although among these residues there is a hierarchy of preference such that one residue may statistically be more highly methylated than another, there does not seem to be a preferred order (84; D. E. Koshland, Jr., T. C. Terwilliger, and E. A. Wang, J. Biol. Chem., in press), although this point is not universally accepted (193). The activity of the methylesterase is decreased and increased by attractant and repellent addition, respectively (62, 85, 86, 192, 200, 215). It is suggested

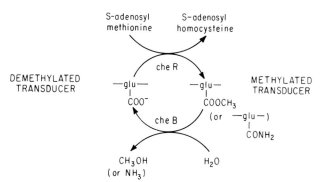

FIG. 6. The methylation/demethylation cycle undergone by bacterial sensory transducers (MCPs). The side-chain carboxyl of certain glutamyl residues is methylated by the methyltransferase enzyme (CheR protein), using *S*-adenosylmethionine as the methyl donor, and demethylated by the methylesterase enzyme (CheB protein). The CheB protein also acts as a deamidase, converting certain glutamine residues to glutamate, whereupon they can enter the methylation/demethylation cycle. From reference 120, with permission.

(86) that there are global effects of this kind, which affect the activity of all MCP classes and not just those that have been stimulated by chemoeffector. However, it would seem that the regulation must to a considerable degree be substrate specific as otherwise transducers would change their methylation levels, and signalling properties, in response to unlinked stimuli.

CheB appears in two forms in the cell, one corresponding to the full gene size and the other to the 3' three-fifths (184). Both forms are seen immediately upon cell lysis, suggesting they are not artifactually produced. Not only does the smaller polypeptide retain esterase activity, it is much more active than the larger form. The activation can be stimulated by cleavage of the larger form with proteases such as trypsin. It is not fully established that the normal pathway involves only proteolytic cleavage, but an alternative explanation—alternate translational initiation sites, as with *cheA* (see below)—appears not to apply (J. Stock, personal communication). The control of the relative amounts of the two forms and its role in regulation of the methylation kinetics is unknown.

The CheR$_E$ protein has recently been shown to possess an additional activity, unlinked to chemotaxis: it mediates the conversion of glutathione to *S*-methylglutathione (212), a reaction whose significance to *E. coli* is unknown.

CheA and CheW

The roles of CheA and CheW are not well understood. Since mutants defective in either protein are CCW biased, one assumes that they are in some way important in enabling the CW state of the motor. The two proteins appear to form a complex, since, in the presence of CheA, CheW was retained on an affinity column prepared with anti-CheA antibody (Matsumura, personal communication). The possible existence of a nucleotide-binding site on CheA is suggested by the fact that it binds strongly to a Cibacron blue column and that *S*-adenosylmethionine can displace it from such columns (Matsumura, personal communication). Both CheA and CheW cause an activation of methylesterase activity (F. W. Dahlquist, personal communication) and in this way confer a global character to the regulation of the MCP transducers.

The CheA protein has the unusual characteristic that it exists in two forms, which derive from translation from two in-frame translational start sites 291 bases apart, so that the longer form contains an additional 97 amino acids (189; E. Kofoid and J. S. Parkinson, personal communication). It is thought that both forms function in sensory transduction, but it is not known why two forms are necessary.

CheY and CheZ

CheY and CheZ have opposite effects on motor rotation: *cheY* mutants are CCW biased, whereas *cheZ* mutants are CW biased, suggesting that CheY is important for CW rotation and CheZ for CCW rotation (156). In measurements with whole cells (44; Berg, personal communication) and with envelopes lacking cytoplasm (S. Ravid, P. Matsumura, and M. Eisenbach, Proc. Natl. Acad. Sci. USA, in press), it has been demonstrated that CheY alone is sufficient to confer CW bias on the flagellar motor and that increasing amounts of CheY increase that bias (44; Ravid et al., in press). Intergenic suppression studies (156; S. Yamaguchi, S.-I. Aizawa, M. Kihara, M. Isomura, C. J. Jones, and R. M. Macnab, submitted for publication) suggest that both CheY and CheZ interact with the motor directly, acting on components that are believed to form a flagellar switch (see below).

CheY is produced at a high stoichiometry relative to other proteins of the sensory transduction apparatus (five to six times greater than Tar or CheZ, for example, both of which are coded for by genes in the same operon as *cheY*; 52). One possible interpretation of this would be that it is present as a dynamic pool interacting with more limited amounts of the transducers and motors.

CheY, like CheA (see above), binds to Cibacron blue columns and can be eluted with *S*-adenosylmethionine (133), suggesting a possible role of nucleotide-CheY interactions in chemotaxis. In biochemical purification protocols, CheY appears to exist in two forms (196a), only one of which is active in generating changes in the rotational bias of the switch (Matsumura, personal communication); whether the two forms are an artifact or whether they are important in the sensory transduction process is not yet known.

The sequence of CheY, obtained by translation of the corresponding gene sequence, has considerable homology to the N-terminal sequence of several other, larger proteins (196a). These include the methylesterase (CheB) and other proteins (OmpR, Dye, and *Bacillus subtilis* Spo0A) which are involved in the regulation of aspects of cell function unrelated to chemotaxis. It is unclear whether the CheY-CheB homologies reflect related roles in the sensory transduction process or related regulatory functions at the transcriptional level (as applies to the other homologous proteins) or are evolutionary relics.

CheZ appears to play important roles in the kinetics of all three phases of the response to a temporal stimulus: *cheZ* mutants show a longer latency phase (173) and slower excitation and adaptation phases (34) than wild type. Also, they show a greater spatial range, so that in experiments with single aseptate filamentous cells, a locally applied stimulus caused a perturbation of motor behavior over a greater distance than it did with wild type (172). It is suggested that the role of CheZ may be to cancel or inactivate the excitation signal to prevent its exerting its effect over time scales that would be too long to be useful to the cell. This model seems plausible, although it does not readily explain the extended latency period and slower excitation of *cheZ* mutants.

It has been reported (182) that CheZ, like the MCP transducers, is methylated. However, the nature of the methyl modification, and its role in the sensory transduction process, has not been further clarified.

MOTOR COMPONENTS

Here we describe only those components of the flagellar apparatus that are directly involved in rotation and switching. For a description of the entire organelle, see chapter 7.

MotA and MotB Proteins

There are two proteins, MotA and MotB, that are needed for flagellar rotation (mutant phenotype is paralyzed) but are not part of the flagellar basal body and are not necessary for its assembly (chapter 7). These proteins, unlike the ones to be described in the next section, have no effect on switching probabilities. Since the flagella on Mot mutants can be made to rotate when external torque is applied (81), the organelles are not simply stuck, and so it is likely that the role of the MotA and MotB proteins in enabling motor rotation is an active one.

Both proteins are integral to the cell membrane (166), but overproduction experiments demonstrate that this is an intrinsic property only in the case of MotA (M. L. Wilson and R. M. Macnab, unpublished data), since overproduced MotB remains in the cytoplasm (196). Presumably at the natural stoichiometry, other motor components stabilize MotB in the membrane. This interpretation is supported by the translated gene sequences (51, 196): MotA is much more hydrophobic than MotB.

Resurrection experiments have been carried out that demonstrate that preexisting flagella start to rotate when Mot proteins are synthesized from plasmid genes (33); thus, if the Mot proteins are part of the motor, their association with it must be rather easily attained.

Although the manner in which the MotA and MotB proteins function in motor rotation is not known, it seems likely on the basis of the evidence given above that they are in the membrane in the vicinity of the flagellar basal body (see Fig. 6 of chapter 7), perhaps as a circlet of force-generating units interacting with the switch proteins (see below). Such circlets, containing around 10 to 12 subunits, have been observed in freeze-fracture preparations of E. coli (S. Khan, personal communication), although their relationship to the Mot proteins remains to be established.

Switch Proteins

There are three flagellar proteins that are important not only structurally, but also for the mechanisms of rotation and switching: the FlaBII$_E$/FlaAII.2$_S$, FlaAII$_E$/FlaQ$_S$, and MotD$_E$/FlaN$_S$ proteins. They share with the MotA and MotB proteins the property of enabling rotation in some active way, since the flagella of paralyzed mutants are able to rotate when external torque is applied (81, 225). However, they differ from the Mot proteins in several major regards.

These are Fla proteins, meaning that they are essential for flagellar assembly—indeed they are among the earliest assembled components (203, 204)—but they are not part of the basal body (7, 45). They may not be integral to the cell membrane, but they do associate with it, as evidenced by retention in cytoplasm-free cell envelopes (161). Missense mutants may be nonflagellate, paralyzed, or nonchemotactic, depending on the particular mutation; nonchemotactic mutants can be either CCW biased or CW biased (225). Because of their importance for motor switching, these proteins have been referred to as "switch proteins" (122, 153).

The various aspects of their function are to a considerable degree segregated into different regions of the primary sequence. Detailed mapping of mutations shows that regions that are important for assembly, rotation, and switching are in general distinct (225).

Genetic evidence suggests that the three proteins may form a "switch complex," since partial or complete phenotypic compensation for a mutation in one of the genes can be achieved by a second mutation in one of the other two genes (Yamaguchi et al., submitted). Such pseudoreversion, or intergenic complementation, also applies between the switch genes and the cheY and cheZ genes (156; Yamaguchi et al., submitted). The possible significance of this will be discussed later.

The combined evidence concerning these proteins suggests that they may be mounted to the cytoplasmic face of the basal body to form the flagellar switch and may interact both with cytoplasmic signals and the remainder of the energy-transducing components of the motor, possibly including MotA and MotB (see Fig. 6 of chapter 7). Support for the latter interaction has been obtained from intergenic suppression data (Yamaguchi et al., submitted).

THE SENSORY TRANSDUCTION PROCESS

The sensory transduction process consists of the communication of information from a variety of receptors to the flagellar motors, the consequent alteration of the state of the motors, and perhaps the reverse flow of information from the motors back to the receptors. The known macromolecular components of this system are described above. Here I consider the processes that these components undergo as they function to control the flagellar motors. I will also discuss whether any nonmacromolecular components play a role in the system.

Recall that the response to a temporal stimulus consists of three phases, latency, excitation, and adaptation (Fig. 3). The latency phase is not fully understood, but may simply consist of the delay between generation of a signal by the transducers and its receipt at the motors. A variety of mutants have been demonstrated to have defects with respect to excitation (e.g., CheD mutants; 39) but not adaptation; others are defective in adaptation but not excitation (103). Thus these two aspects of the response are at least partially separable, even though both are mediated by the same transducer.

Excitation

The molecular identity of the excitation signal is still unknown, perhaps the most important gap in current knowledge of bacterial chemotaxis. Several possibilities, such as membrane potential, Ca^{2+}, and cyclic nucleotides, have been proposed (32, 146, 205) but have not withstood further scrutiny. (A summary of the evidence in this regard is given in reference 55 and R. C. Tribhuwan, M. S. Johnson, and B. L. Taylor, submitted for publication). ATP is an absolute requirement for CW rotation and hence for chemotaxis, but does not appear itself to be the signal (13, 175, 210). S-Adenosylmethionine is required as the methyl donor for transducer methylation (15, 20) and hence adaptation, but the evidence is against its being involved in signaling (175).

If the interpretation of the latency phase as a kinetic delay in transmission of the signal is correct, its length (0.2 s) suggests that diffusion of a macromolecular species may be involved (172). There are four Che proteins (CheA, CheW, CheY, and CheZ) whose roles are not well established, but for which there is evidence of either static or dynamic formation of complexes with each other or the flagellar switch (156; Matsumura, personal communication; Yamaguchi et al., submitted). It is a reasonable hypothesis that some or all of these proteins may shuttle between the transducers and the motors to convey excitation information.

Adaptation

The process of adaptation occurs on a time scale of seconds to minutes, depending on the strength of the excitation (30, 195). There are at least two different systems for adaptation; one depends on methylation processes, and the other is independent of methylation processes.

Methylation-dependent adaptation

Stimuli that act on MCP transducers, such as conventional chemical attractants, repellents, pH, and temperature, use these same transducers in the adaptation phase of the response in a process that involves modulation of methylation levels of the transducers (62, 63). Immediately after excitation, the activity of the methylesterase is altered (activated by repellents, inhibited by attractants), and presumably the relevant transducer becomes more available for the methylating and demethylating enzymes (200). The process is under some circumstances independent of the remainder of the chemotaxis system, since MCPs undergo changes in methylation to an extent and at a rate that is the same in the absence of CheA, CheW, CheY, and CheZ as in their presence (Russo and Koshland, personal communication). The activation of CheB upon repellent stimulation, however, requires CheA and CheW function (192; Dahlquist, personal communication). A comparable analysis of possible effects on CheR remains to be carried out.

It is not clearly established what the role of multiple methylation is, but it may enable the cell to adapt to a wider range of environmental stimuli than would be possible with single-site methylation. A theoretical model that explicitly addresses this idea has been presented (19).

Methylation-independent adaptation

Stimuli—notably, the PTS sugars and oxygen—that do not employ MCP transducers as the means of generating the excitation signal also do not affect their levels of methylation. Thus the mechanism of adaptation must be quite different (143). In the case of the PTS sugars, it is tempting to suggest that phosphorylation levels of one or other of the proteins in the transport system may play a role analogous to methylation of the MCPs (159).

It has recently become evident that, even with respect to MCP-mediated stimuli, there must be a parallel system for adaptation that is methylation independent (197, 198, 201). As one striking illustration of this, one can take a mutant that is totally defective in CheR (methyltransferase) activity and select for a swarming response on tryptone plates. The resulting double mutant, with a second-site mutation in CheB (methylesterase) activity, gives a chemotactic response which, though not as strong as that of wild type, is substantial. Phenotypically, it is found that CheR mutants are CCW biased, CheB mutants are CW biased, and the partially adapting double mutant has a switch bias that is more or less close to wild type. Thus the failure of a CheR mutant to adapt is not entirely due to the failure of methylation directly; at least part of the failure derives from the steady-state switching bias that the lesion confers. However, it seems reasonable to assume that the chemotaxis methylation system is the principal means of adaptation to MCP-linked stimuli.

The Overall Sensory Transduction Process

A summary of the overall process of sensory transduction is given in Fig. 7 for both methylation-dependent and methylation-independent systems. With respect to the latter, which are responsible for PTS sugar stimuli and for oxygen and other stimuli mediated by changes in proton motive force, one can say little at present about the mechanism of sensory transduction except that it does not involve the process of methylation.

Signals emanating from the methyl-accepting receptors/transducers (MCPs such as Tar) occur as a result of external binding of an attractant, either as a small molecule directly or as a small molecule attached to its periplasmic binding protein. The binding event causes a transmembrane propagation of information within the receptor, and an excitation signal is generated that travels to the switch complex of the flagellar motor.

This excitation signal, which has not been positively identified, is likely to involve a dynamic relay of protein interactions, specifically CheY and CheZ. It may be that they can bind to the receptor themselves (with affinities that are attractant dependent) or that other species such as CheA and CheW bind and in doing so affect the state of CheY and CheZ. Since both CheA and CheY may have S-adenosylmethionine-binding sites, modulation of interactions could involve the state of the proteins in this regard. Likewise, if the CheA-CheW complex seen in vitro is dissociable, the process of association/dissociation may contribute. One thing that seems fairly certain is that, after a delay of ca. 0.2 s that presumably reflects the time for diffusion of the relevant proteins between receptor and motor, a change occurs in the relative (and possibly absolute) amounts of CheY and CheZ that are bound to the flagellar switch. An attractant stimulus enhances the extent of CheZ binding and diminishes the extent of CheY binding, and the motor becomes more CCW biased; i.e., excitation behavior occurs.

Meanwhile, the receptor is progressively undergoing further changes. Methylation activity has increased, demethylation activity has decreased, and hence the net methylation state of the receptor is asymptotically approaching a new and higher steady-state level. This change is rather receptor specific, implying that the state of occupancy by attractant not only generates an excitation signal to the motor but also makes the receptor susceptible to methylation

METHYLATION-DEPENDENT
SYSTEM

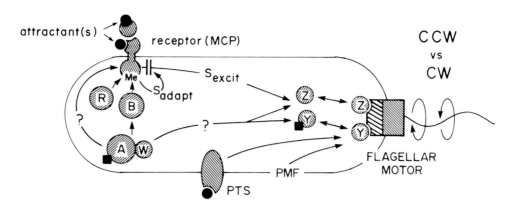

OTHER SYSTEMS

FIG. 7. Process of sensory transduction in *E. coli* and *S. typhimurium*. Sensory information is generated at various types of receptors: the methyl-accepting receptors/transducers (MCPs), the enzymes II of the sugar PTS, and the proton motive force (PMF) generating electron transport chain. The MCP-type receptors bind specific chemicals (solid circles), such as aspartate, or maltose complexed to its periplasmic binding protein. The excitation signal (S_{excit}) causes enhanced CCW rotation of the flagellar motor if the stimulus is an increase in attractant concentration. The signal probably comprises a change in the association state and small-molecule-binding state (solid square) of various chemotaxis proteins (CheY, CheZ, CheA, CheW), as discussed in the text. The adaptation signal (S_{adapt}) progressively cancels S_{excit}, and the motor returns to its unstimulated CCW/CW ratio; the process involves a change in the methylation state (Me) of the MCPs, mediated by a methyltransferase (CheR) and a methylesterase (CheB). Binding of CheZ and CheY to the flagellar switch (diagonal lines) places it in the CCW and CW state, respectively. The relative sizes of the various proteins are approximately indicated by the diameters of the corresponding circles.

changes. There are also effects on enzyme activity, notably that of the methylesterase, mediated by CheA and CheW. These may occur because the state of these proteins, singly or as a complex, is affected directly by transducer state or indirectly via the excitation signal. The effect of the changing methylation state of the receptor is that the effective excitation signal is progressively reduced, eventually reaching the prestimulus level, so that the rotational bias of the motors is restored to the value it had before stimulation and the cell is behaviorally adapted. In principle, there might be chemically distinct signals to the motor for excitation and adaptation, with the algebraic sum determining motor state. However, it seems more likely that adaptation represents a feedback inhibition of the receptor's ability to generate the excitation signal; either the excited state of the receptor (attractant bound, new methylation level not reached), or the excitation signal itself, could be the "adaptation signal" to the receptor to start inhibiting the generation of the excitation signal.

A description similar to the above, but with signs reversed, applies to a negative stimulus such as attractant withdrawal or repellent addition.

MOTILITY AND TAXIS IN THE NATURAL ENVIRONMENT

There can be no doubt that motility and taxis play an important role in the survival of *E. coli* and *S. typhimurium* and of the numerous other bacterial species that are motile. Any system that requires upwards of 60 genes and a fairly heavy protein biosynthetic load (notably the flagellar filament and the receptors) has to

justify its retention by conferring substantial benefit to the cell. The repertoire of stimuli to which *E. coli* and *S. typhimurium* respond is approximately what one might expect given the metabolic capabilities of these species. Being capable of respiratory metabolism, with its more efficient use of reduced organic substrates, these bacteria would derive benefit from a positive aerotactic response. Being capable of metabolizing most common sugars, amino acids, and carboxylic acids, and of using many of them for anabolic purposes also, the positive tactic responses to such compounds makes sense. (It is interesting that only two of the common amino acids, L-aspartate and L-serine, are really strong attractants, perhaps because amino acids in the natural environment usually derive from hydrolyzed protein and these two amino acids act as an indicator of the availability of the rest.) Likewise, the limited ability of *E. coli* and *S. typhimurium* to utilize many aromatic compounds, and the potential for hydrophobic compounds to interfere with membrane structure and in some cases membrane potential or cytoplasmic pH, makes the repellent responses to compounds like benzoate, phenol, and indole reasonable. (By contrast, *Pseudomonas* spp., with a very different metabolic pattern, are attracted to a number of these compounds [66].)

However, although the above scenario is plausible, there has been relatively little work done to establish experimentally the role of motility and taxis in the natural environment. With enteric bacteria, one should really speak of environments in the plural, to include at least the large intestine of the mammalian host and the open-water system that the bacterium finds itself in after excretion.

The possession of motility by itself may convey little

advantage to the cell unless there is tactic responsiveness. One indication of this is the finding that fully motile, but nonchemotactic, mutants of *S. typhimurium* were much less capable of invading intestinal mucosal tissue than were wild-type cells (11).

CONCLUSIONS

Wild-type strains of *E. coli* and *S. typhimurium* are motile and have a quite extensive range of stimuli to which they respond by moving in the direction of increasing stimulus intensity (in the case of attractants) or decreasing stimulus intensity (repellents). The mechanism involves sensing temporal gradients of the stimulus; constant stimulus intensity does not affect steady-state behavior. The flagellar motor itself operates on a rotary principle and is driven by proton motive force across the cell membrane. The motor is capable of both senses of rotation. CCW rotation causes swimming; CW rotation causes tumbling and reorientation. Spontaneous switching between CCW and CW rotation occurs and is modulated by the environmental stimuli so that the cells swim for longer intervals when they are pointing in a favorable direction than when they are pointing in an unfavorable direction.

The genetics of this system is complex but relatively well understood. Many of the genes involved in motor function and sensory transduction have been cloned and sequenced. Especially in the case of the receptors/transducers, sequence analysis of mutants has begun to produce functional information at a molecular level. The manner by which sensory information is conveyed between receptors and motors is still not understood, but activity in this area of research is high and rapid progress is being made.

Ecological information concerning motility and taxis is scarce. We do not know precisely how this system contributes to survival, although the broad picture makes sense. It is also evident that such a complex system would only be retained if it conferred a major advantage to the cell.

ACKNOWLEDGMENTS

I acknowledge the many colleagues who have communicated manuscripts and results before publication. M. Eisenbach, P. Matsumura, J. S. Parkinson, and B. L. Taylor provided helpful comments and criticism. May Kihara assisted in assembling the data tabulated in this chapter.

Work in my laboratory has been supported by Public Health Service grant AI12202 from the National Institutes of Health.

LITERATURE CITED

1. **Adler, J.** 1969. Chemoreceptors in bacteria. Science **166:** 1588–1597.
2. **Adler, J.** 1973. A method for measuring chemotaxis and use of the method to determine optimum conditions for chemotaxis by *Escherichia coli*. J. Gen. Microbiol. **74:**77–91.
3. **Adler, J.** 1976. The sensing of chemicals by bacteria. Sci. Am. **234:**40–47.
4. **Adler, J., and M. M. Dahl.** 1967. A method for measuring the motility of bacteria and for comparing random and non-random motility. J. Gen. Microbiol. **46:**161–173.
5. **Adler, J., and W. Epstein.** 1974. Phosphotransferase-system enzymes as chemoreceptors for certain sugars in *Escherichia coli* chemotaxis. Proc. Natl. Acad. Sci. USA **71:**2895–2899.
6. **Adler, J., G. L. Hazelbauer, and M. M. Dahl.** 1973. Chemotaxis toward sugars in *Escherichia coli*. J. Bacteriol. **115:**824–847.
7. **Aizawa, S.-I., G. E. Dean, C. J. Jones, R. M. Macnab, and S. Yamaguchi.** 1985. Purification and characterization of the flag-ellar hook-basal body complex of *Salmonella typhimurium*. J. Bacteriol. **161:**836–849.
8. **Aksamit, R. R., B. J. Howlett, and D. E. Koshland, Jr.** 1975. Soluble and membrane-bound aspartate-binding activities in *Salmonella typhimurium*. J. Bacteriol. **123:**1000–1005.
9. **Aksamit, R., and D. E. Koshland, Jr.** 1972. A ribose binding protein of *Salmonella typhimurium*. Biochem. Biophys. Res. Commun. **48:**1348–1353.
10. **Aksamit, R. R., and D. E. Koshland, Jr.** 1974. Identification of the ribose binding protein as the receptor for ribose chemotaxis in *Salmonella typhimurium*. Biochemistry **13:**4473–4478.
11. **Allweiss, B., J. Dostal, K. E. Carey, T. F. Edwards, and R. Freter.** 1977. The role of chemotaxis in the ecology of bacterial pathogens of mucosal surfaces. Nature (London) **266:**448–450.
12. **Anderson, R. A.** 1975. Formation of the bacterial flagellar bundle, p. 45–56. *In* T. Y.-T. Wu, C. J. Brokaw, and C. J. Brennen (ed.), Swimming and flying in nature, vol. 1. Plenum Publishing Corp., New York.
13. **Arai, T.** 1981. Effect of arsenate on chemotactic behavior of *Escherichia coli*. J. Bacteriol. **145:**803–807.
14. **Armitage, J. P., C. Ingham, and M. C. W. Evans.** 1985. Role of proton motive force in phototactic and aerotactic responses of *Rhodopseudomonas sphaeroides*. J. Bacteriol. **161:**967–972.
15. **Armstrong, J. B.** 1972. An S-adenosylmethionine requirement for chemotaxis in *Escherichia coli*. Can. J. Microbiol. **18:** 1695–1701.
16. **Armstrong, J. B., and J. Adler.** 1969. Location of genes for motility and chemotaxis on the *Escherichia coli* genetic map. J. Bacteriol. **97:**156–161.
17. **Armstrong, J. B., and J. Adler.** 1969. Complementation of nonchemotactic mutants of *Escherichia coli*. Genetics **61:**61–66.
18. **Armstrong, J. B., J. Adler, and M. M. Dahl.** 1967. Nonchemotactic mutants of *Escherichia coli*. J. Bacteriol. **93:**390–398.
19. **Asakura, S., and H. Honda.** 1984. Two-state model for bacterial chemoreceptor proteins: the role of multiple methylation. J. Mol. Biol. **176:**349–367.
20. **Aswad, D. W., and D. E. Koshland, Jr.** 1975. Evidence for an S-adenosylmethionine requirement in the chemotactic behavior of *Salmonella typhimurium*. J. Mol. Biol. **97:**207–223.
21. **Aswad, D., and D. E. Koshland, Jr.** 1975. Isolation, characterization and complementation of *Salmonella typhimurium* chemotaxis mutants. J. Mol. Biol. **97:**225–235.
22. **Bachmann, B. J.** 1983. Linkage map of *Escherichia coli* K-12, edition 7. Microbiol. Rev. **47:**180–230.
23. **Bartlett, D. H., and P. Matsumura.** 1984. Identification of *Escherichia coli* region III flagellar gene products and description of two new flagellar genes. J. Bacteriol. **160:**577–585.
24. **Berg, H. C.** 1974. Dynamic properties of bacterial flagellar motors. Nature (London) **249:**77–79.
25. **Berg, H. C.** 1976. Does the flagellar rotary motor step?, p. 47–56. *In* R. Goldman, T. Pollard, and J. Rosenbaum (ed.), Cell motility. Cold Spring Harbor Laboratory, Cold Spring Harbor, N.Y.
26. **Berg, H. C., and R. A. Anderson.** 1973. Bacteria swim by rotating their flagellar filaments. Nature (London) **245:**380–382.
27. **Berg, H. C., and D. A. Brown.** 1972. Chemotaxis in *Escherichia coli* analysed by three-dimensional tracking. Nature (London) **239:**500–504.
28. **Berg, H. C., and S. Khan.** 1983. A model for the flagellar rotary motor, p. 486–497. *In* H. Sund and C. Veeger (ed.), Mobility and recognition in cell biology. Walter de Gruyter, Berlin.
29. **Berg, H. C., M. D. Manson, and M. P. Conley.** 1982. Dynamics and energetics of flagellar rotation in bacteria, p. 1–31. *In* W. B. Amos and J. G. Duckett (ed.), Prokaryotic and eukaryotic flagella. Cambridge University Press, Cambridge.
30. **Berg, H. C., and P. M. Tedesco.** 1975. Transient response to chemotactic stimuli in *Escherichia coli*. Proc. Natl. Acad. Sci. USA **72:**3235–3239.
31. **Berg, H. C., and L. Turner.** 1979. Movement of microorganisms in viscous environments. Nature (London) **278:**349–351.
32. **Black, R. A., A. C. Hobson, and J. Adler.** 1980. Involvement of cyclic GMP in intracellular signaling in the chemotactic response of *Escherichia coli*. Proc. Natl. Acad. Sci. USA **77:**3879–3883.
33. **Block, S. M., and H. C. Berg.** 1984. Successive incorporation of force-generating units in the bacterial rotary motor. Nature (London) **309:**470–472.
34. **Block, S. M., J. E. Segall, and H. C. Berg.** 1982. Impulse responses in bacterial chemotaxis. Cell **31:**215–226.
35. **Bogonez, E., and D. E. Koshland, Jr.** 1985. Solubilization of a vectorial transmembrane receptor in functional form: aspartate receptor of chemotaxis. Proc. Natl. Acad. Sci. USA **82:**4891–4895.
36. **Bollinger, J., C. Park, S. Harayama, and G. L. Hazelbauer.** 1984.

Structure of the Trg protein: homologies with and differences from other sensory transducers of *Escherichia coli*. Proc. Natl. Acad. Sci. USA **81**:3287–3291.

37. **Boyd, A., K. Kendall, and M. I. Simon.** 1983. Structure of the serine chemoreceptor in *Escherichia coli*. Nature (London) **301**:623–626.
38. **Boyd, A., A. Krikos, and M. Simon.** 1981. Sensory transducers of *E. coli* are encoded by homologous genes. Cell **26**:333–343.
39. **Callahan, A. M., and J. S. Parkinson.** 1985. Genetics of methyl-accepting chemotaxis proteins in *Escherichia coli*: *cheD* mutations affect the structure and function of the Tsr transducer. J. Bacteriol. **161**:96–104.
40. **Chelsky, D., and F. W. Dahlquist.** 1980. Structural studies of methyl-accepting chemotaxis proteins of *Escherichia coli*: evidence for multiple methylation sites. Proc. Natl. Acad. Sci. USA **77**:2434–2438.
41. **Chelsky, D., and F. W. Dahlquist.** 1980. Chemotaxis in *Escherichia coli*: associations of protein components. Biochemistry **19**:4633–4639.
42. **Clancy, M., K. A. Madill, and J. M. Wood.** 1981. Genetic and biochemical requirements for chemotaxis to L-proline in *Escherichia coli*. J. Bacteriol. **146**:902–906.
43. **Clarke, S., and D. E. Koshland, Jr.** 1979. Membrane receptors for aspartate and serine in bacterial chemotaxis. J. Biol. Chem. **254**:9695–9702.
44. **Clegg, D. O., and D. E. Koshland, Jr.** 1984. The role of a signaling protein in bacterial sensing: behavioral effects of increased gene expression. Proc. Natl. Acad. Sci. USA **81**:5056–5060.
45. **Clegg, D. O., and D. E. Koshland, Jr.** 1985. Identification of a bacterial sensing protein and effects of its elevated expression. J. Bacteriol. **162**:398–405.
46. **Coakley, C. J., and M. E. J. Holwill.** 1972. Propulsion of microorganisms by three-dimensional flagellar waves. J. Theor. Biol. **35**:525–542.
47. **Dahl, M. K., and M. D. Manson.** 1985. Interspecific reconstitution of maltose transport and chemotaxis in *Escherichia coli* with maltose-binding protein from various enteric bacteria. J. Bacteriol. **164**:1057–1063.
48. **Dahlquist, F. W., P. Lovely, and D. E. Koshland, Jr.** 1972. Quantitative analysis of bacterial migration in chemotaxis. Nature (London) New Biol. **236**:120–123.
49. **Dang, C. V., M. Niwano, J.-I. Ryu, and B. L. Taylor.** 1986. Interaction between methylation-dependent and methylation-independent pathways for chemotaxis and aerotaxis. J. Bacteriol. **166**:275–280.
50. **Dean, G. E., S.-I. Aizawa, and R. M. Macnab.** 1983. *flaAII* (*motC, cheV*) of *Salmonella typhimurium* is a structural gene involved in energization and switching of the flagellar motor. J. Bacteriol. **154**:84–91.
51. **Dean, G. E., R. M. Macnab, J. Stader, P. Matsumura, and C. Burks.** 1984. Gene sequence and predicted amino acid sequence of the *motA* protein, a membrane-associated protein required for flagellar rotation in *Escherichia coli*. J. Bacteriol. **159**:991–999.
52. **DeFranco, A. L., and D. E. Koshland, Jr.** 1981. Molecular cloning of chemotaxis genes and overproduction of gene products in the bacterial sensing system. J. Bacteriol. **147**:390–400.
53. **DeFranco, A. L., J. S. Parkinson, and D. E. Koshland, Jr.** 1979. Functional homology of chemotaxis genes in *Escherichia coli* and *Salmonella typhimurium*. J. Bacteriol. **139**:107–114.
54. **Eddy, A. A.** 1980. Slip and leak models of gradient-coupled solute transport. Biochem. Soc. Trans. **8**:271–273.
55. **Eisenbach, M., Y. Margolin, and S. Ravid.** 1985. Excitatory signaling in bacterial chemotaxis, p. 43–61. *In* M. Eisenbach and M. Balaban (ed.), Sensing and response in microorganisms. Elsevier, Amsterdam.
56. **Enomoto, M.** 1966. Genetic studies of paralyzed mutants in *Salmonella*. I. Genetic fine structure of the *mot* loci in *Salmonella typhimurium*. Genetics **54**:715–726.
57. **Fahnestock, M., and D. E. Koshland, Jr.** 1979. Control of the receptor for galactose taxis in *Salmonella typhimurium*. J. Bacteriol. **137**:758–763.
58. **Foster, D. L., S. L. Mowbray, B. K. Jap, and D. E. Koshland, Jr.** 1985. Purification and characterization of the aspartate chemoreceptor. J. Biol. Chem. **260**:11706–11710.
59. **Fujita, H., S. Yamaguchi, T. Taira, and T. Iino.** 1981. A simple method for the isolation of flagellar shape mutants in *Salmonella*. J. Gen. Microbiol. **125**:213–216.
60. **Glagolev, A. N.** 1984. Bacterial $\Delta\bar{\mu}H^+$-sensing. Trends Biochem. Sci. **9**:397–400.
61. **Glagolev, A. N., and V. P. Skulachev.** 1978. The proton pump is a molecular engine of motile bacteria. Nature (London) **272**:280–282.

62. **Goy, M. F., M. S. Springer, and J. Adler.** 1977. Sensory transduction in *Escherichia coli*: role of a protein methylation reaction in sensory adaptation. Proc. Natl. Acad. Sci. USA **74**:4964–4968.
63. **Goy, M. F., M. S. Springer, and J. Adler.** 1978. Failure of sensory adaptation in bacterial mutants that are defective in a protein methylation reaction. Cell **15**:1231–1240.
64. **Harayama, S., J. Bollinger, T. Iino, and G. L. Hazelbauer.** 1983. Characterization of the *mgl* operon of *Escherichia coli* by transposon mutagenesis and molecular cloning. J. Bacteriol. **153**:408–415.
65. **Harayama, S., P. Engstrom, H. Wolf-Watz, T. Iino, and G. L. Hazelbauer.** 1982. Cloning of *trg*, a gene for a sensory transducer in *Escherichia coli*. J. Bacteriol. **152**:372–383.
66. **Harwood, C. S., M. Rivelli, and L. N. Ornston.** 1984. Aromatic acids are chemoattractants for *Pseudomonas putida*. J. Bacteriol. **160**:622–628.
67. **Hazelbauer, G. L.** 1975. Maltose chemoreceptor of *Escherichia coli*. J. Bacteriol. **122**:206–214.
68. **Hazelbauer, G. L., and J. Adler.** 1971. Role of the galactose binding protein in chemotaxis of *Escherichia coli* toward galactose. Nature (London) New Biol. **230**:101–104.
69. **Hazelbauer, G. L., and P. Engstrom.** 1981. Multiple forms of methyl-accepting chemotaxis proteins distinguished by a factor in addition to multiple methylation. J. Bacteriol. **145**:35–42.
70. **Hazelbauer, G. L., and S. Harayama.** 1979. Mutants in transmission of chemotactic signals from two independent receptors of *E. coli*. Cell **16**:617–625.
71. **Hazelbauer, G. L., and S. Harayama.** 1983. Sensory transduction in bacterial chemotaxis. Int. Rev. Cytol. **81**:33–70.
72. **Hedblom, M. L., and J. Adler.** 1980. Genetic and biochemical properties of *Escherichia coli* mutants with defects in serine chemotaxis. J. Bacteriol. **144**:1048–1060.
73. **Hedblom, M. L., and J. Adler.** 1983. Chemotactic response of *Escherichia coli* to chemically synthesized amino acids. J. Bacteriol. **155**:1463–1466.
74. **Hirota, N., and Y. Imae.** 1983. Na$^+$-driven flagellar motors of an alkalophilic *Bacillus* strain YN-1. J. Biol. Chem. **258**:10577–10581.
75. **Hirota, N., M. Kitada, and Y. Imae.** 1981. Flagellar motors of alkalophilic *Bacillus* are powered by an electrochemical potential gradient of Na$^+$. FEBS Lett. **132**:278–280.
76. **Iida, A., S. Harayama, T. Iino, and G. L. Hazelbauer.** 1984. Molecular cloning and characterization of genes required for ribose transport and utilization in *Escherichia coli* K-12. J. Bacteriol. **158**:674–682.
77. **Imae, Y.** 1985. Molecular mechanism of thermosensing in bacteria, p. 73–81. *In* M. Eisenbach and M. Balaban (ed.), Sensing and response in microorganisms. Elsevier, Amsterdam.
78. **Imae, Y., T. Mizuno, and K. Maeda.** 1984. Chemosensory and thermosensory excitation in adaptation-deficient mutants of *Escherichia coli*. J. Bacteriol. **159**:368–374.
79. **Ingolia, T. D., and D. E. Koshland, Jr.** 1979. Response to a metal ion-citrate complex in bacterial sensing. J. Bacteriol. **140**:798–804.
80. **Ishihara, A., J. E. Segall, S. M. Block, and H. C. Berg.** 1983. Coordination of flagella on filamentous cells of *Escherichia coli*. J. Bacteriol. **155**:228–237.
81. **Ishihara, A., S. Yamaguchi, and H. Hotani.** 1981. Passive rotation of flagella on paralyzed *Salmonella typhimurium* (*mot*) mutants by external rotatory driving force. J. Bacteriol. **145**:1082–1084.
82. **Kehry, M. R., M. W. Bond, M. W. Hunkapiller, and F. W. Dahlquist.** 1983. Enzymatic deamidation of methyl-accepting chemotaxis proteins in *Escherichia coli* catalyzed by the *cheB* gene product. Proc. Natl. Acad. Sci. USA **80**:3599–3603.
83. **Kehry, M. R., and F. W. Dahlquist.** 1982. Adaptation in bacterial chemotaxis: *cheB*-dependent modification permits additional methylations of sensory transducer proteins. Cell **29**:761–772.
84. **Kehry, M. R., and F. W. Dahlquist.** 1982. The methyl-accepting chemotaxis proteins of *Escherichia coli*. Identification of the multiple methylation sites on methyl-accepting chemotaxis protein I. J. Biol. Chem. **257**:10378–10386.
85. **Kehry, M. R., T. G. Doak, and F. W. Dahlquist.** 1984. Stimulus-induced changes in methylesterase activity during chemotaxis in *Escherichia coli*. J. Biol. Chem. **259**:11828–11835.
86. **Kehry, M. R., T. G. Doak, and F. W. Dahlquist.** 1985. Sensory adaptation in bacterial chemotaxis: Regulation of demethylation. J. Bacteriol. **163**:983–990.
87. **Kehry, M. R., P. Engstrom, F. W. Dahlquist, and G. L. Hazelbauer.** 1983. Multiple covalent modifications of *trg*, a sensory transducer of *Escherichia coli*. J. Biol. Chem. **258**:5050–5055.

88. **Khan, S., and H. C. Berg.** 1983. Isotope and thermal effects in chemiosmotic coupling to the flagellar motor of *Streptococcus.* Cell **32:**913–919.

89. **Khan, S., and R. M. Macnab.** 1980. The steady-state counterclockwise/clockwise ratio of bacterial flagellar motors is regulated by protonmotive force. J. Mol. Biol. **138:**563–597.

90. **Khan, S., and R. M. Macnab.** 1980. Proton chemical potential, proton electrical potential and bacterial motility. J. Mol. Biol. **138:**599–614.

91. **Khan, S., R. M. Macnab, A. L. DeFranco, and D. E. Koshland, Jr.** 1978. Inversion of a behavioral response in bacterial chemotaxis: explanation at the molecular level. Proc. Natl. Acad. Sci. USA **75:**4150–4154.

92. **Kihara, M., and R. M. Macnab.** 1979. Chemotaxis of *Salmonella typhimurium* toward citrate. J. Bacteriol. **140:**297–300.

93. **Kihara, M., and R. M. Macnab.** 1981. Cytoplasmic pH mediates pH taxis and weak-acid repellent taxis of bacteria. J. Bacteriol. **145:**1209–1221.

94. **Kleene, S. J., M. L. Toews, and J. Adler.** 1977. Isolation of glutamic acid methyl ester from an *Escherichia coli* membrane protein involved in chemotaxis. J. Biol. Chem. **252:**3214–3218.

95. **Koiwai, O., and H. Hayashi.** 1979. Studies on bacterial chemotaxis. IV. Interaction of maltose receptor with a membrane-bound chemosensing component. J. Biochem. **86:**27–34.

96. **Komeda, Y.** 1982. Fusions of flagellar operons to lactose genes on a Mu *lac* bacteriophage. J. Bacteriol. **150:**16–26.

97. **Komeda, Y., K. Kutsukake, and T. Iino.** 1980. Definition of additional flagellar genes in *Escherichia coli* K-12. Genetics **94:**277–290.

98. **Komeda, Y., M. Silverman, and M. Simon.** 1977. Genetic analysis of *Escherichia coli* K-12 region I flagellar mutants. J. Bacteriol. **131:**801–808.

99. **Kondoh, H., C. B. Ball, and J. Adler.** 1979. Identification of a methyl-accepting chemotaxis protein for the ribose and galactose chemoreceptors of *Escherichia coli.* Proc. Natl. Acad. Sci. USA **76:**260–264.

100. **Kort, E. N., M. F. Goy, S. H. Larsen, and J. Adler.** 1975. Methylation of a membrane protein involved in bacterial chemotaxis. Proc. Natl. Acad. Sci. USA **72:**3939–3943.

101. **Koshland, D. E., Jr.** 1980. Bacterial chemotaxis as a model behavioral system. Distinguished Lecture Series of the Society of General Physiologists, vol. 2. Raven Press, New York.

102. **Koshland, D. E., Jr.** 1981. Biochemistry of sensing and adaptation in a simple bacterial system. Annu. Rev. Biochem. **50:**765–782.

103. **Koshland, D. E., Jr., A. F. Russo, and N. I. Gutterson.** 1983. Information processing in a sensory system. Cold Spring Harbor Symp. Quant. Biol. **48:**805–810.

104. **Krikos, A., M. P. Conley, A. Boyd, H. C. Berg, and M. I. Simon.** 1985. Chimeric chemosensory transducers of *Escherichia coli.* Proc. Natl. Acad. Sci. USA **82:**1326–1330.

105. **Krikos, A., N. Mutoh, A. Boyd, and M. I. Simon.** 1983. Sensory transducers of *E. coli* are composed of discrete structural and functional domains. Cell **33:**615–622.

105a.**Kuo, S. C., and D. E. Koshland, Jr.** 1986. Sequence of the *flaA* (*cheC*) locus of *Escherichia coli* and discovery of a new gene. J. Bacteriol. **166:**1007–1012.

106. **Kutsukake, K., and T. Iino.** 1985. Refined genetic analysis of the region II *che* mutants in *Salmonella typhimurium.* Mol. Gen. Genet. **199:**406–409.

107. **Kutsukake, K., T. Iino, Y. Komeda, and S. Yamaguchi.** 1980. Functional homology of *fla* genes between *Salmonella typhimurium* and *Escherichia coli.* Mol. Gen. Genet. **178:**59–67.

108. **Larsen, S. H., J. Adler, J. J. Gargus, and R. W. Hogg.** 1974. Chemomechanical coupling without ATP: the source of energy for motility and chemotaxis in bacteria. Proc. Natl. Acad. Sci. USA **71:**1239–1243.

109. **Larsen, S. H., R. W. Reader, E. N. Kort, W.-W. Tso, and J. Adler.** 1974. Change in direction of flagellar rotation is the basis of the chemotactic response in *Escherichia coli.* Nature (London) **249:**74–77.

110. **Laszlo, D. J., B. L. Fandrich, A. Sivaram, B. Chance, and B. L. Taylor.** 1984. Cytochrome *o* as a terminal oxidase and receptor for aerotaxis in *Salmonella typhimurium.* J. Bacteriol. **159:**663–667.

111. **Laszlo, D. J., and B. L. Taylor.** 1981. Aerotaxis in *Salmonella typhimurium*: role of electron transport. J. Bacteriol. **145:**990–1001.

112. **Lederberg, J.** 1956. Linear inheritance in transductional clones. Genetics **41:**845–871.

113. **Lengeler, J.** 1975. Mutations affecting transport of the hexitols D-mannitol, D-glucitol, and galactitol in *Escherichia coli* K-12:

isolation and mapping. J. Bacteriol. **124:**26–38.

114. **Lengeler, J., A.-M. Auburger, R. Mayer, and A. Pecher.** 1981. The phosphoenolpyruvate-dependent carbohydrate: phosphotransferase system enzymes II as chemoreceptors in chemotaxis of *Escherichia coli* K12. Mol. Gen. Genet. **183:**163–170.

115. **Macnab, R. M.** 1976. Examination of bacterial flagellation by dark-field microscopy. J. Clin. Microbiol. **4:**258–265.

116. **Macnab, R. M.** 1977. Bacterial flagella rotating in bundles: a study in helical geometry. Proc. Natl. Acad. Sci. USA **74:**221–225.

117. **Macnab, R. M.** 1978. Bacterial motility and chemotaxis: the molecular biology of a behavioral system. Crit. Rev. Biochem. **5:**291–341.

118. **Macnab, R. M.** 1979. Chemotaxis in bacteria, p. 310–334. *In* W. Haupt and M. E. Feinleib (ed.), Encyclopedia of plant physiology, new series, vol. 7. Springer-Verlag, Berlin.

119. **Macnab, R. M.** 1983. An entropy-driven engine—the bacterial flagellar motor, p. 147–160. *In* A. Oplatka and M. Balaban (ed.), Biological structures and coupled flows. Academic Press, Inc., New York.

120. **Macnab, R. M.** 1985. Transmembrane signalling in bacterial chemotaxis, p. 455–487. *In* P. Cohen and M. Houslay (ed.), Molecular mechanisms of transmembrane signalling. Elsevier, Amsterdam.

121. **Macnab, R. M.** 1985. The proton-driven bacterial flagellar motor. Methods Enzymol. **125:**563–581.

122. **Macnab, R. M., and S.-I. Aizawa.** 1984. Bacterial motility and the bacterial flagellar motor. Annu. Rev. Biophys. Bioeng. **13:**51–83.

123. **Macnab, R. M., and D. P. Han.** 1983. Asynchronous switching of flagellar motors on a single bacterial cell. Cell **32:**109–117.

124. **Macnab, R. M., and D. E. Koshland, Jr.** 1972. The gradient-sensing mechanism in bacterial chemotaxis. Proc. Natl. Acad. Sci. USA **69:**2509–2512.

125. **Macnab, R. M., and D. E. Koshland, Jr.** 1974. Bacterial motility and chemotaxis: light-induced tumbling response and visualization of individual flagella. J. Mol. Biol. **84:**399–406.

126. **Macnab, R. M., and M. K. Ornston.** 1977. Normal-to-curly flagellar transitions and their role in bacterial tumbling. Stabilization of an alternative quaternary structure by mechanical force. J. Mol. Biol. **112:**1–30.

127. **Maeda, K., and Y. Imae.** 1979. Thermosensory transduction in *Escherichia coli*: inhibition of the thermoresponse by L-serine. Proc. Natl. Acad. Sci. USA **76:**91–95.

128. **Maeda, K., Y. Imae, J.-I. Shioi, and F. Oosawa.** 1976. Effect of temperature on motility and chemotaxis of *Escherichia coli.* J. Bacteriol. **127:**1039–1046.

128a.**Manson, M. D., V. Blank, G. Brade, and C. F. Higgins.** 1986. Peptide chemotaxis in *E. coli* involves the Tap signal transducer and the dipeptide permease. Nature (London) **321:**253–256.

129. **Manson, M. D., W. Boos, P. J. Bassford, Jr., and B. A. Rasmussen.** 1985. Dependence of maltose transport and chemotaxis on the amount of maltose-binding protein. J. Biol. Chem. **260:**9727–9733.

130. **Manson, M. D., and M. Kossmann.** 1986. Mutations in *tar* suppress defects in maltose chemotaxis caused by specific *malE* mutations. J. Bacteriol. **165:**34–40.

131. **Manson, M. D., P. M. Tedesco, and H. C. Berg.** 1980. Energetics of flagellar rotation in bacteria. J. Mol. Biol. **138:**541–561.

132. **Manson, M. D., P. Tedesco, H. C. Berg, F. M. Harold, and C. van der Drift.** 1977. A protonmotive force drives bacterial flagella. Proc. Natl. Acad. Sci. USA **74:**3060–3064.

133. **Matsumura, P., J. J. Rydel, R. Linzmeier, and D. Vacante.** 1984. Overexpression and sequence of the *Escherichia coli cheY* gene and biochemical activities of the CheY protein. J. Bacteriol. **160:**36–41.

134. **Matsuura, S., J.-I. Shioi, and Y. Imae.** 1977. Motility in *Bacillus subtilis* driven by an artificial protonmotive force. FEBS Lett. **82:**187–190.

135. **Melton, T., P. E. Hartman, J. P. Stratis, T. L. Lee, and A. T. Davis.** 1978. Chemotaxis of *Salmonella typhimurium* to amino acids and some sugars. J. Bacteriol. **133:**708–716.

136. **Mesibov, R., and J. Adler.** 1972. Chemotaxis toward amino acids in *Escherichia coli.* J. Bacteriol. **112:**315–326.

137. **Miller, J. B., and D. E. Koshland, Jr.** 1977. Membrane fluidity and chemotaxis: effects of temperature and membrane lipid composition on the swimming behavior of *Salmonella typhimurium* and *Escherichia coli.* J. Mol. Biol. **111:**183–201.

138. **Mitchell, P.** 1984. Bacterial flagellar motors and osmoelectric molecular rotation by an axially transmembrane well and turnstile mechanism. FEBS Lett. **176:**287–294.

139. **Mizuno, T., N. Mutoh, S. M. Panasenko, and Y. Imae.** 1986.

Acquisition of maltose chemotaxis in *Salmonella typhimurium* by the introduction of the *Escherichia coli* chemosensory transducer gene. J. Bacteriol. **165**:890–895.

140. **Mowbray, S. L., D. L. Foster, and D. E. Koshland, Jr.** 1985. Proteolytic fragments identified with domains of the aspartate chemoreceptor. J. Biol. Chem. **260**:11711–11718.

141. **Muskavitch, M. A., E. N. Kort, M. S. Springer, M. F. Goy, and J. Adler.** 1978. Attraction by repellents: an error in sensory information processing by bacterial mutants. Science **201**:63–65.

142. **Mutoh, N., and M. I. Simon.** 1986. Nucleotide sequence corresponding to five chemotaxis genes in *Escherichia coli*. J. Bacteriol. **165**:161–166.

143. **Niwano, M., and B. L. Taylor.** 1982. Novel sensory adaptation mechanism in bacterial chemotaxis to oxygen and phosphotransferase substrates. Proc. Natl. Acad. Sci. USA **79**:11–15.

144. **Oosawa, F., and S. Hayashi.** 1983. Coupling between flagellar motor rotation and proton flux in bacteria. J. Phys. Soc. Japan **52**:4019–4028.

145. **Oosawa, K., and Y. Imae.** 1983. Glycerol and ethylene glycol: members of a new class of repellents of *Escherichia coli* chemotaxis. J. Bacteriol. **154**:104–112.

146. **Ordal, G. W.** 1977. Calcium ion regulates chemotactic behaviour in bacteria. Nature (London) **270**:66–67.

147. **Ordal, G. W.** 1985. Bacterial chemotaxis: biochemistry of behavior in a single cell. Crit. Rev. Microbiol. **12**:95–130.

148. **Ordal, G. W., and J. Adler.** 1974. Isolation and complementation of mutants in galactose taxis and transport. J. Bacteriol. **117**:509–516.

149. **Ordal, G. W., and J. Adler.** 1974. Properties of mutants in galactose taxis and transport. J. Bacteriol. **117**:517–526.

150. **Parkinson, J. S.** 1976. *cheA*, *cheB*, and *cheC* genes of *Escherichia coli* and their role in chemotaxis. J. Bacteriol. **126**:758–770.

151. **Parkinson, J. S.** 1978. Complementation analysis and deletion mapping of *Escherichia coli* mutants defective in chemotaxis. J. Bacteriol. **135**:45–53.

152. **Parkinson, J. S.** 1980. Novel mutations affecting a signaling component for chemotaxis of *Escherichia coli*. J. Bacteriol. **142**:953–961.

153. **Parkinson, J. S.** 1981. Genetics of bacterial chemotaxis, p. 265–290. *In* S. W. Glover and D. A. Hopwood (ed.), Genetics as a tool in microbiology, vol. 31. Cambridge University Press, Cambridge.

154. **Parkinson, J. S., and G. L. Hazelbauer.** 1983. Bacterial chemotaxis: molecular genetics of sensory transduction and chemotactic gene expression, p. 293–318. *In* J. Beckwith, J. E. Davies, and J. A. Gallant (ed.), Gene function in prokaryotes. Cold Spring Harbor Laboratory, Cold Spring Harbor, N.Y.

155. **Parkinson, J. S., and S. R. Parker.** 1979. Interaction of the *cheC* and *cheZ* gene products is required for chemotactic behavior in *Escherichia coli*. Proc. Natl. Acad. Sci. USA **76**:2390–2394.

156. **Parkinson, J. S., S. R. Parker, P. B. Talbert, and S. E. Houts.** 1983. Interactions between chemotaxis genes and flagellar genes in *Escherichia coli*. J. Bacteriol. **155**:265–274.

157. **Parkinson, J. S., and P. T. Revello.** 1978. Sensory adaptation mutants of *E. coli*. Cell **15**:1221–1230.

158. **Perry, D., and C. Gilvarg.** 1984. Spectrophotometric determination of affinities of peptides for their transport systems in *Escherichia coli*. J. Bacteriol. **160**:943–948.

159. **Postma, P. W., and J. Lengeler.** 1985. Phosphoenolpyruvate:carbohydrate phosphotransferase system of bacteria. Microbiol. Rev. **49**:232–269.

160. **Postma, P. W., and S. Roseman.** 1976. The bacterial phosphoenolpyruvate:sugar phosphotransferase system. Biochim. Biophys. Acta **457**:213–257.

161. **Ravid, S., and M. Eisenbach.** 1984. Direction of flagellar rotation in bacterial cell envelopes. J. Bacteriol. **158**:222–230.

162. **Ravid, S., and M. Eisenbach.** 1984. Minimal requirements for rotation of bacterial flagella. J. Bacteriol. **158**:1208–1210.

163. **Reader, R. W., W.-W. Tso, M. S. Springer, M. F. Goy, and J. Adler.** 1979. Pleiotropic aspartate taxis and serine taxis mutants of *Escherichia coli*. J. Gen. Microbiol. **111**:363–374.

164. **Repaske, D. R., and J. Adler.** 1981. Change in intracellular pH of *Escherichia coli* mediates the chemotactic response to certain attractants and repellents. J. Bacteriol. **145**:1196–1208.

165. **Richarme, G.** 1982. Interaction of the maltose-binding protein with membrane vesicles of *Escherichia coli*. J. Bacteriol. **149**:662–667.

166. **Ridgway, H. F., M. Silverman, and M. I. Simon.** 1977. Localization of proteins controlling motility and chemotaxis in *Escherichia coli*. J. Bacteriol. **132**:657–665.

167. **Rollins, C. M., and F. W. Dahlquist.** 1980. Methylation of chemotaxis-specific proteins in *Escherichia coli* cells permeable to S-adenosylmethionine. Biochemistry **19**:4627–4632.

168. **Rollins, C., and F. W. Dahlquist.** 1981. The methyl-accepting chemotaxis proteins of *E. coli*: a repellent-stimulated, covalent modification, distinct from methylation. Cell **25**:333–340.

168a.**Rubik, B. A., and D. E. Koshland, Jr.** 1978. Potentiation, desensitization, and inversion of response in bacterial sensing of chemical stimuli. Proc. Natl. Acad. Sci. USA **75**:2820–2824.

169. **Russo, A. F., and D. E. Koshland, Jr.** 1983. Separation of signal transduction and adaptation functions of the aspartate receptor in bacterial sensing. Science **220**:1016–1020.

170. **Russo, A. F., and D. E. Koshland, Jr.** 1986. Identification of the *tip*-encoded receptor in bacterial sensing. J. Bacteriol. **165**:276–282.

171. **Sanderson, K. E., and J. R. Roth.** 1983. Linkage map of *Salmonella typhimurium*, edition VI. Microbiol. Rev. **47**:410–453.

172. **Segall, J. E., A. Ishihara, and H. C. Berg.** 1985. Chemotactic signaling in filamentous cells of *Escherichia coli*. J. Bacteriol. **161**:51–59.

173. **Segall, J. E., M. D. Manson, and H. C. Berg.** 1982. Signal processing times in bacterial chemotaxis. Nature (London) **296**:855–857.

174. **Sherris, D., and J. S. Parkinson.** 1981. Posttranslational processing of methyl-accepting chemotaxis proteins in *Escherichia coli*. Proc. Natl. Acad. Sci. USA **78**:6051–6055.

175. **Shioi, J.-I., R. J. Galloway, M. Niwano, R. E. Chinnock, and B. L. Taylor.** 1982. Requirement of ATP in bacterial chemotaxis. J. Biol. Chem. **257**:7969–7975.

176. **Shioi, J.-I., S. Matsuura, and Y. Imae.** 1980. Quantitative measurements of proton motive force and motility in *Bacillus subtilis*. J. Bacteriol. **144**:891–897.

177. **Shioi, J., and B. L. Taylor.** 1984. Oxygen taxis and proton motive force in *Salmonella typhimurium*. J. Biol. Chem. **259**:10983–10988.

178. **Silverman, M., and M. Simon.** 1974. Positioning flagellar genes in *Escherichia coli* by deletion analysis. J. Bacteriol. **117**:73–79.

179. **Silverman, M., and M. Simon.** 1974. Characterization of *Escherichia coli* flagellar mutants that are insensitive to catabolite repression. J. Bacteriol. **120**:1196–1203.

180. **Silverman, M., and M. Simon.** 1974. Flagellar rotation and the mechanism of bacterial motility. Nature (London) **249**:73–74.

181. **Silverman, M., and M. Simon.** 1976. Operon controlling motility and chemotaxis in *E. coli*. Nature (London) **264**:577–580.

182. **Silverman, M., and M. Simon.** 1977. Chemotaxis in *Escherichia coli*: methylation of *che* gene products. Proc. Natl. Acad. Sci. USA **74**:3317–3321.

183. **Silverman, M., and M. Simon.** 1977. Identification of polypeptides necessary for chemotaxis in *Escherichia coli*. J. Bacteriol. **130**:1317–1325.

184. **Simms, S. A., M. G. Keane, and J. Stock.** 1985. Multiple forms of the CheB methylesterase in bacterial chemosensing. J. Biol. Chem. **260**:10161–10168.

185. **Simon, M., J. Zieg, M. Silverman, G. Mandel, and R. Doolittle.** 1980. Phase variation: evolution of a controlling element. Science **209**:1370–1374.

186. **Slocum, M. K., and J. S. Parkinson.** 1983. Genetics of methyl-accepting chemotaxis proteins in *Escherichia coli*: organization of the *tar* region. J. Bacteriol. **155**:565–577.

187. **Slocum, M. K., and J. S. Parkinson.** 1985. Genetics of methyl-accepting chemotaxis proteins in *Escherichia coli*: null phenotypes of the *tar* and *tap* genes. J. Bacteriol. **163**:586–594.

188. **Slonczewski, J. L., R. M. Macnab, J. R. Alger, and A. M. Castle.** 1982. Effects of pH and repellent tactic stimuli on protein methylation levels in *Escherichia coli*. J. Bacteriol. **152**:384–399.

189. **Smith, R. A., and J. S. Parkinson.** 1980. Overlapping genes at the *cheA* locus of *Escherichia coli*. Proc. Natl. Acad. Sci. USA **77**:5370–5374.

190. **Sprenger, G. A., and J. W. Lengeler.** 1984. L-Sorbose metabolism in *Klebsiella pneumoniae* and Sor⁺ derivatives of *Escherichia coli* K-12 and chemotaxis toward sorbose. J. Bacteriol. **157**:39–45.

191. **Springer, M. S., M. F. Goy, and J. Adler.** 1977. Sensory transduction in *Escherichia coli*: two complementary pathways of information processing that involve methylated proteins. Proc. Natl. Acad. Sci. USA **74**:3312–3316.

192. **Springer, M. S., and B. Zanolari.** 1984. Sensory transduction in *Escherichia coli*: regulation of the demethylation rate by the CheA protein. Proc. Natl. Acad. Sci. USA **81**:5061–5065.

193. **Springer, M. S., B. Zanolari, and P. A. Pierzchala.** 1982. Ordered methylation of the methyl-accepting chemotaxis proteins of *Escherichia coli*. J. Biol. Chem. **257**:6861–6866.

194. **Springer, W. R., and D. E. Koshland, Jr.** 1977. Identification of a protein methyltransferase as the *cheR* gene product in the bacterial sensing system. Proc. Natl. Acad. Sci. USA **74**:533–537.

195. **Spudich, J. L., and D. E. Koshland, Jr.** 1975. Quantitation of the sensory response in bacterial chemotaxis. Proc. Natl. Acad. Sci. USA **72:**710–713.

196. **Stader, J., P. Matsumura, D. Vacante, G. E. Dean, and R. M. Macnab.** 1986. Nucleotide sequence of the *Escherichia coli motB* gene and site-limited incorporation of its product into the cytoplasmic membrane. J. Bacteriol. **166:**244–252.

196a. **Stock, A., D. E. Koshland, Jr., and J. Stock.** 1985. Homologies between the *Salmonella typhimurium* CheY protein and proteins involved in the regulation of chemotaxis, membrane protein synthesis, and sporulation. Proc. Natl. Acad. Sci. USA **82:**7989–7993.

197. **Stock, J., A. Borczuk, F. Chiou, and J. Burchenal.** 1985. Compensatory mutations in receptor function: a re-evaluation of the role of methylation in bacterial chemotaxis. Proc. Natl. Acad. Sci. USA **82:**8364–8368.

198. **Stock, J., G. Kersulis, and D. E. Koshland, Jr.** 1985. Neither methylating nor demethylating enzymes are required for bacterial chemotaxis. Cell **42:**683–690.

199. **Stock, J. B., and D. E. Koshland, Jr.** 1978. A protein methylesterase involved in bacterial sensing. Proc. Natl. Acad. Sci. USA **75:**3659–3663.

200. **Stock, J. B., and D. E. Koshland, Jr.** 1981. Changing reactivity of receptor carboxyl groups during bacterial sensing. J. Biol. Chem. **256:**10826–10833.

201. **Stock, J. B., A. M. Maderis, and D. E. Koshland, Jr.** 1981. Bacterial chemotaxis in the absence of receptor carboxyl methylation. Cell **27:**37–44.

202. **Strange, P. G., and D. E. Koshland, Jr.** 1976. Receptor interactions in a signalling system: competition between ribose receptor and galactose receptor in the chemotaxis response. Proc. Natl. Acad. Sci. USA **73:**762–766.

203. **Suzuki, T., T. Iino, T. Horiguchi, and S. Yamaguchi.** 1978. Incomplete flagellar structures in nonflagellate mutants of *Salmonella typhimurium*. J. Bacteriol. **133:**904–915.

204. **Suzuki, T., and Y. Komeda.** 1981. Incomplete flagellar structures in *Escherichia coli* mutants. J. Bacteriol. **145:**1036–1041.

205. **Szmelcman, S., and J. Adler.** 1976. Change in membrane potential during bacterial chemotaxis. Proc. Natl. Acad. Sci. USA **73:**4387–4391.

206. **Taylor, B. L.** 1983. Role of proton motive force in sensory transduction in bacteria. Annu. Rev. Microbiol. **37:**551–573.

207. **Taylor, B. L.** 1983. How do bacteria find the optimal concentration of oxygen? Trends Biochem. Sci. **8:**438–441.

208. **Taylor, B. L., and D. E. Koshland, Jr.** 1975. Intrinsic and extrinsic light responses of *Salmonella typhimurium* and *Escherichia coli*. J. Bacteriol. **123:**557–569.

209. **Taylor, B. L., J. B. Miller, H. M. Warrick, and D. E. Koshland, Jr.** 1979. Electron acceptor taxis and blue light effect on bacterial chemotaxis. J. Bacteriol. **140:**567–573.

210. **Taylor, B. L., R. C. Tribhuwan, E. H. Rowsell, J. M. Smith, and J. Shioi.** 1985. Role of ATP and cyclic nucleotides in bacterial chemotaxis, p. 63–71. *In* M. Eisenbach and M. Balaban (ed.), Sensing and response in microorganisms. Elsevier, Amsterdam.

211. **Terwilliger, T. C., E. Bogonez, E. A. Wang, and D. E. Koshland, Jr.** 1983. Sites of methyl esterification involved in bacterial chemotaxis. J. Biol. Chem. **258:**9608–9611.

212. **Terwilliger, T. C., G. D. Bollag, D. W. Sternberg, Jr., and D. E. Koshland, Jr.** 1986. *S*-Methyl glutathione synthesis is catalyzed by the *cheR* methyltransferase in *Escherichia coli*. J. Bacteriol. **165:**958–963.

213. **Terwilliger, T. C., and D. E. Koshland, Jr.** 1984. Sites of methyl esterification and deamination on the aspartate receptor involved in chemotaxis. J. Biol. Chem. **259:**7719–7725.

214. **Toews, M. L., and J. Adler.** 1979. Methanol formation *in vivo* from methylated chemotaxis proteins in *Escherichia coli*. J. Biol. Chem. **254:**1761–1764.

215. **Toews, M. L., M. F. Goy, M. S. Springer, and J. Adler.** 1979. Attractants and repellents control demethylation of methylated chemotaxis proteins in *Escherichia coli*. Proc. Natl. Acad. Sci. USA **76:**5544–5548.

216. **Tsang, N., R. Macnab, and D. E. Koshland, Jr.** 1973. Common mechanism for repellents and attractants in bacterial chemotaxis. Science **181:**60–63.

217. **Tso, W.-W., and J. Adler.** 1974. Negative chemotaxis in *Escherichia coli*. J. Bacteriol. **118:**560–576.

218. **Tsui-Collins, A. L., and B. A. D. Stocker.** 1976. *Salmonella typhimurium* mutants generally defective in chemotaxis. J. Bacteriol. **128:**754–765.

219. **Van der Werf, P., and D. E. Koshland, Jr.** 1977. Identification of a γ-glutamyl methyl ester in bacterial membrane protein involved in chemotaxis. J. Biol. Chem. **252:**2793–2795.

220. **Wagenknecht, T.** 1986. A plausible mechanism for flagellar rotation in bacteria. FEBS Lett. **196:**193–197.

221. **Wang, E. A., and D. E. Koshland, Jr.** 1980. Receptor structure in the bacterial sensing system. Proc. Natl. Acad. Sci. USA **77:**7157–7161.

222. **Wang, E. A., K. L. Mowry, D. O. Clegg, and D. E. Koshland, Jr.** 1982. Tandem duplication and multiple functions of a receptor gene in bacterial chemotaxis. J. Biol. Chem. **257:**4673–4676.

223. **Warrick, H. M., B. L. Taylor, and D. E. Koshland, Jr.** 1977. Chemotactic mechanism of *Salmonella typhimurium*: preliminary mapping and characterization of mutants. J. Bacteriol. **130:**223–231.

224. **Willis, R. C., and C. E. Furlong.** 1974. Purification and properties of a ribose-binding protein from *Escherichia coli*. J. Biol. Chem. **249:**6926–6929.

225. **Yamaguchi, S., H. Fujita, A. Ishihara, S.-I. Aizawa, and R. M. Macnab.** 1986. Subdivision of flagellar genes of *Salmonella typhimurium* into regions responsible for assembly, rotation, and switching. J. Bacteriol. **166:**187–193.

226. **Yamaguchi, S., H. Fujita, T. Taira, K. Kutsukake, H. Homma, and T. Iino.** 1984. Genetic analysis of three additional *fla* genes in *Salmonella typhimurium*. J. Gen. Microbiol. **130:**3339–3342.

227. **Yamaguchi, S., T. Iino, T. Horiguchi, and K. Ohta.** 1972. Genetic analysis of *fla* and *mot* cistrons closely linked to *H1* in *Salmonella abortusequi* and its derivatives. J. Gen. Microbiol. **70:**59–75.

228. **Yonekawa, H., H. Hayashi, and J. S. Parkinson.** 1983. Requirement of the *cheB* function for sensory adaptation in *Escherichia coli*. J. Bacteriol. **156:**1228–1235.

229. **Zukin, R. S., P. R. Hartig, and D. E. Koshland, Jr.** 1979. Effect of an induced conformational change on the physical properties of two chemotactic receptor molecules. Biochemistry **18:**5599–5605.

230. **Zukin, R. S., and D. E. Koshland, Jr.** 1976. Mg^{2+}, Ca^{2+}-dependent adenosine triphosphatase as receptor for divalent cations in bacterial sensing. Science **193:**405–408.

231. **Zukin, R. S., P. G. Strange, L. R. Heavey, and D. E. Koshland, Jr.** 1977. Properties of the galactose binding protein of *Salmonella typhimurium* and *Escherichia coli*. Biochemistry **16:**381–386.

50. ATP-Coupled Solute Transport Systems

BARRY P. ROSEN

Department of Biological Chemistry, University of Maryland School of Medicine, Baltimore, Maryland 21201

INTRODUCTION

Both *Escherichia coli* and *Salmonella typhimurium* have multiple systems for the transport of sugars, amino acids, and ions (41). Many are secondary porters, that is, coupled to previously formed ion gradients of protons or sodium. These have been discussed previously in detail (chapter 17). The topic of this chapter is primary transport, specifically systems that directly utilize the chemical energy of the β-γ phosphoryl bond of ATP. The only bacterial porter in which the coupling of ATP hydrolysis to solute pumping has been demonstrated is the H$^+$-translocating ATPase, or F$_0$F$_1$ (19). Although ATP-coupled K$^+$ pumping has not been directly observed, the linkage between the Kdp K$^+$ transport system and the K$^+$-stimulated ATPase is clear (14).

In addition to these two ion-translocating ATPases, there are two other classes of transport systems in which an in vivo dependency on bond energy has been demonstrated: the periplasmic binding protein systems and the plasmid-mediated arsenical resistance systems. The evidence that they are ATP-coupled pumps is circumstantial but does suggest that they are primary, rather than secondary, transport systems.

IN VIVO DETERMINATION OF ATP DEPENDENCY

How is the distinction made between primary and secondary transport? When transport can be demonstrated in vitro, the distinction can be made most clearly. Transport of many solutes has been shown to be secondarily coupled to proton or sodium gradients in isolated membrane vesicles (chapter 17). Conversely, proton pumping by the F$_0$F$_1$ has been shown to be directly coupled to ATP hydrolysis in everted membrane vesicles (19). When transport has not been or cannot be measured in vitro, the coupling to metabolic energy sources must be implied from in vivo studies.

The potential sources of energy for transport must be evaluated individually to make any judgments about in vivo energy coupling. Berger (2) devised a method for segregating the chemical intermediates of metabolism from electrochemical gradients. The procedure involved (i) depletion of endogenous energy reserves and (ii) isolation of metabolic pathways through the use of mutants or inhibitors or both. Long-term energy stores in bacteria (for example, glycogen) are difficult to deplete just by incubation of cells in carbon-free medium. When a protonophoric uncoupler is added, the cells rapidly use their energy reserves to fuel a futile cycle of respiratory-driven proton extrusion and uncoupler-mediated proton uptake. For these purposes the uncoupler 2,4-dinitrophenol proved best because it could be washed out of the cells easily after starvation. By monitoring both the rate of endogenous respiration and the intracellular ATP levels, Berger and Heppel (3) found that, for most *E. coli* strains, approximately 10 h of starvation was required for adequate depletion of reserves. Only 1 to 2 h was necessary to starve *unc* mutants, which are defective in the F$_0$F$_1$.

The two main sources of cellular energy are chemical energy and electrochemical energy. The currency of chemical energy is ATP, and electrochemical energy in bacteria is mostly in the form of proton gradients, in which the driving force is the proton motive force (PMF). During glucose metabolism, ATP is formed through glycolytic reactions. Respiration of the end products of glycolysis by electron transport chain reactions results in proton extrusion and generation of a PMF. ATP and PMF are in equilibrium with each other by the reaction catalyzed by the F$_0$F$_1$, this process being known as oxidative phosphorylation. Energy-starved cells fed glucose produce ATP by glycolysis and oxidative phosphorylation and a PMF by respiration. When fed only respiratory substrates, they produce a PMF that equilibrates with the ATP pools. To differentiate between ATP- and PMF-coupled systems, this equilibrium must be broken. This uncoupling is accomplished by addition of an inhibitor of the F_0F_1, such as N,N'-dicyclohexylcarbodiimide or, more conveniently, by using *unc* mutants. An energy-depleted *unc* mutant fed a respiratory substrate such as succinate generates a PMF without ATP production. Such cells fed glucose produce ATP only glycolytically. They can still create a PMF through respiration, but this can be prevented with inhibitors of the electron transport chain, such as cyanide.

Thus, an energy-depleted *unc* mutant fed succinate has only a PMF available for transport. Cells fed glucose in the presence of cyanide have only chemical bond energy available. Note that ATP and chemical bond energy are used here interchangeably. From in vivo studies it is difficult to distinguish between various high-energy compounds, since they are usually in rapid equilibrium with each other. Under most conditions the ATP pool is the largest of the high-energy-phosphate bond compounds, and in most energy-transfer reactions ATP is used as the activated form of high-energy phosphate. It is reasonable, therefore, to consider transport reactions using high-energy-phosphate bond chemical energy as ATP linked, keeping in mind that the actual direct donor of chemical energy may be another compound in equilibrium with the ATP pool. Similarly, dependence on the PMF does not necessarily imply direct coupling with protons. The electrochemical proton gradient can be converted into an electrochemical sodium gradient through a reaction catalyzed by the Na^+/H^+ antiporter, and the sodium motive force can couple directly to some transport systems. On the basis of these criteria, a number of transport systems in *E. coli* and *S. typhimurium* can be considered ATP linked, as discussed below.

ION-TRANSLOCATING ATPases

H$^+$-Translocating ATPase

Much is known about the F_0F_1, including the subunit structure, the nucleotide sequence of the genes for each subunit, the structure of the operon encoding the polypeptides of the complex, and, to some degree, the arrangement and function of the polypeptides in the membrane (19). It is the only bacterial pump which has been characterized in vitro as both an ATPase enzyme and an ATP-driven ion transport system. This complex has been described in chapter 17 and will not be considered in further detail here.

To review its role in cellular physiology, it is the enzyme that equilibrates the PMF and ATP (44). Under anaerobic conditions it functions in its hydrolytic mode, splitting ATP, pumping protons, and generating a PMF. This role supplies energy for vectorial systems such as secondary porters and the motility apparatus. When oxidants such as oxygen or fumarate are available, respiratory-driven proton pumping fulfills this role. Since the free energy available from respiration exceeds that of ATP hydrolysis, the net direction of proton pumping is reversed, and the reaction of F_0F_1 favors synthesis of ATP, that is, oxidative phosphorylation.

K$^+$-Translocating ATPase

E. coli has two major transport systems for the accumulation of potassium: the Kdp and Trk systems (14). The Kdp system is the one better characterized. It is a high-affinity ($K_m = 2$ μM) potassium transport system that can be derepressed by growth in low potassium. It is specified by an operon consisting of three structural genes: *kdpA*, *kdpB*, and *kdpC*, and a positive regulator gene, *kdpD*. The enzyme is composed of three types of subunits: the *kdpA*, *kdpB*, and *kdpC* proteins, although the total number of each subunit in the active complex is not known. The molecular weights of the subunits were estimated from electrophoretic separations to be 47,000, 90,000, and 22,000 for the *kdpA*, *kdpB*, and *kdpC* proteins, respectively (29), but the sizes from the nucleotide sequences are 58,189, 72,112, and 20,267, respectively (21). All three subunits are found in the inner membrane. From analysis of the nucleotide sequences of the genes, only the *kdpB* protein has the hydropathic profile of an intrinsic membrane polypeptide. The *kdpA* protein is exposed to the periplasm, and mutations in it produce alterations in the K_m but not in the V_{max}, suggesting that it contains the substrate recognition site. The *kdpB* protein probably contains the catalytic site for ATPase activity and is predicted to have six or seven membrane-spanning loops. Nothing is known about the function of the *kdpC* protein.

Using an experimental approach derived from that of Berger (2), Rhoads and Epstein (37) demonstrated that this system functions as an ATP-driven potassium pump in vivo. In vitro potassium pumping activity has not been measured, but derepressed cells exhibit K^+-stimulated ATPase activity. Since the K_m for potassium is in the micromolar region, the amount of potassium contaminating most solutions is sufficient to saturate the enzyme. Thus, in membrane vesicles from wild-type strains, no potassium stimulation of ATPase activity could be observed. Using mutants with increased K_ms for K^+, Epstein et al. (15) demonstrated K^+-stimulated ATPase activity clearly related to the Kdp system.

The Kdp ATPase shares properties characteristic of eucaryotic cation-translocating ATPases of the E_1E_2 type such as the Ca^{2+}-ATPase of sarcoplasmic reticulum (6). For example, the *kdpB* proteins is phosphorylated by ATP, forming an acyl-phosphate intermediate during the catalytic cycle. Recently, Hesse et al. (21) found homology between the Kdp ATPase and the sarcoplasmic reticulum Ca^{2+}-ATPase. They compared the amino acid sequence of *kdpB*, predicted from the nucleotide sequence, with the partial protein sequence reported for the Ca^{2+}-ATPase of rabbit muscle. Regions of considerable homology were observed (Fig. 1). Sequence similarities were especially striking in two regions: in and around the location of the aspartyl residue, which is phosphorylated during the catalytic cycle, and around the region known to contain the ATP-binding site of the Ca^{2+}-ATPase. Since no sequences for the membrane-spanning portions of the Ca^{2+}-ATPase are available, it is not known whether the homology is confined to the regions of the proteins involved in catalysis. Unquestionably, the two proteins share a common ancestor.

The other *E. coli* potassium transport system, the Trk system, also has an ATP dependency (37). However, a PMF is also required for potassium accumulation by this system, so that the roles of the two energy sources are not clear. Rhoads and Epstein (37) sug-

A

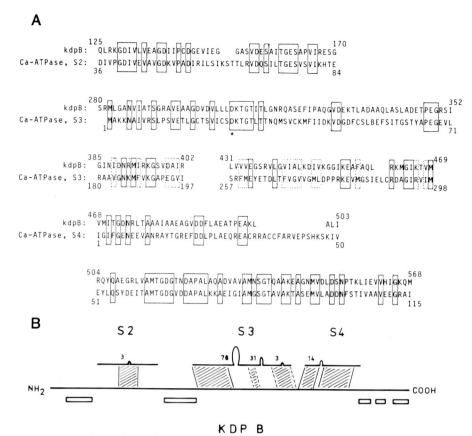

K D P B

FIG. 1. Protein sequence homology of the *kdpB* protein and the Ca²⁺-ATPase of sarcoplasmic reticulum. (A) The amino acid sequence of the *kdpB* protein (682 residues), predicted from the nucleotide sequence, was compared with the protein sequences of three fragments (S2 [116 residues], S3 [298 residues], and S4 [122 residues]) of the Ca²⁺-ATPase. Identical residues are enclosed by boxes. Conservative replacements, enclosed by dotted lines, are shown for the less certain homologies to fragment S3. The standard one-letter code is used. (B) Regions of homology of the *kdpB* protein to the S2, S3, and S4 fragments of the Ca²⁺-ATPase are shown with each peptide drawn to scale. Regions of homology are joined by shaded bands. Regions where homology is less certain are joined by bands enclosed with broken lines. Loops represent places where a deletion of the number of residues shown to the left of the loop would bring the fragment of the Ca²⁺-ATPase into alignment with the *kdpB* protein. Boxes beneath the line represent hydrophobic regions of the *kdpB* protein that are candidates for membrane-spanning sequences (from data of Hesse et al. [21]).

gested that the Trk system is coupled to the PMF but that it is regulated by ATP. Recent experiments by Stewart et al. (50) show a lack of temporal correlation between depletion of ATP pools and loss of Trk activity, suggesting that ATP is not the immediate energy donor. These investigators suggest that ATP acts as an activator, so that when cellular energy supplies are low, lack of ATP results in inactivation of Trk, and K⁺ does not leak out of the cells.

OTHER ATP-LINKED TRANSPORT SYSTEMS

Transport Systems Utilizing Periplasmic Binding Proteins

Several decades ago, it became clear that *E. coli* and *S. typhimurium* utilize a variety of different types of transport systems (41). One class was detectable in membrane vesicles and used a PMF for energy. A second class could be measured only in intact cells, not in membrane vesicles. Even in cells, the catalytic activity of these systems was not always present; treatments that increased the permeability of the

outer membrane produced loss of transport activity. The classic example was the effect of the osmotic shock treatment of Neu and Heppel (34) on transport of some sugars, amino acids, and ions. In this procedure, the cells are first treated with a hypertonic solution of Tris and EDTA in sucrose. This treatment both removes outer membrane lipopolysaccharide and plasmolyzes the cells. Next, the cells are subjected to rapid hypotonic shock by suspension or dilution into water or dilute MgCl₂, causing release into the medium of the contents of the periplasm, about 10 to 15% of the total cell protein, without affecting the viability of the cells. These periplasmic proteins include a variety of catabolic enzymes, including alkaline phosphatase, ribonuclease, and, in cells bearing ampicillin resistance plasmids, β-lactamase. In many respects, the periplasm resembles the lysosomal compartment of eucaryotic cells.

Osmotically shocked cells retain transport activity for those systems observable in membrane vesicles. Activities not found in membrane vesicles are similarly lost in osmotically shocked cells. Piperno and Oxender (36) noted that, concomitant with loss of leucine

TABLE 1. ATP-linked transport systems

Transport system	Reference(s)
ATPases	
Proton	19
Potassium	37
Periplasmic[a]	
Arginine	3
α,ε-Diaminopimelic acid	3
Glutamine	3
Glycylglycine	11
Histidine	3
Leucine	54
Methionine	28
Galactose	53
Maltose	16
Ribose	12
Plasmid-encoded	
Arsenate	33, 49
Arsenite	42

[a] Includes only those periplasmic systems demonstrated to be ATP linked.

transport activity during osmotic shock, cells release a leucine-binding protein into the medium. The protein has the same substrate specificities and affinities as the missing transport system, implicating it as a component of the transport system. Subsequent genetic evidence supports the original hypothesis. Since then, a considerable number of these osmotic-shock-sensitive systems, or periplasmic systems, have been shown to have periplasmic components (43). Berger (2) studied the energetics of several periplasmic systems and demonstrated that intracellular ATP is required but that transport activity is independent of the PMF. Other periplasmic systems have been added to the list of ATP-dependent systems (Table 1), and probably all transport systems with periplasmic binding proteins have a similar mechanism of energy coupling.

In addition to their participation in transport, several periplasmic binding proteins also have been shown to be components of chemotaxis systems (chapter 49). Biochemical reconstitution experiments with purified binding proteins have confirmed the participation of these proteins in transport and chemotaxis. Treatments that remove the binding proteins cause loss of both transport and chemotactic response. When binding protein mutants are treated with calcium in a procedure not unlike that used for uptake of DNA in genetic transformation (4), the cells take binding proteins into the periplasmic space, and both transport activity (4) and chemotaxis (5) are restored. Thus, it seems likely that binding proteins provide the signal recognition elements of branching pathways for transmembrane communication of cytosol with the external environment.

The nature of the energetics of periplasmic systems is a major unanswered question. It is not known whether the porters are ATPases or even if ATP is the direct energy donor. Attempts have been made to identify the true physiological energy donor. Hong (24) isolated pleiotropic mutants defective in, among other things, the activities of periplasmic transport systems. These mutants had normal ATP levels and

were capable of generating a PMF. Subsequently, the effects of the mutations were shown to be indirect (25). In that study, though, a relationship between acetyl phosphate levels and the activity of periplasmic systems was demonstrated, suggesting that acetyl phosphate might be an energy donor. More recent studies have shown that when NAD^+ was incorporated into right-side-out membrane vesicles, addition of glutamine-binding protein allowed for glutamine transport (26, 27). The energetics of this transport were complicated. A PMF was sufficient for glutamine transport. However, NAD^+ was required for a succinate-generated PMF to be effective, and contaminating amounts of phosphorylated compounds, including ATP and acetyl phosphate, were produced, presumably by oxidative phosphorylation and subsequent metabolic pathways. But neither ATP nor acetyl phosphate alone was sufficient. Although only limited conclusions can be drawn from this study, it may prove to be a worthwhile experimental system for analysis of the energetics of periplasmic systems.

Periplasmic binding proteins are apparently soluble proteins. They have never been found in a membrane-bound form or associated with any membrane-bound protein. Yet, by its nature, a transport system must have one or more membrane components. Additional genes have been identified for several transport systems, including the β-methylgalactoside (38), arabinose (10), maltose (45), leucine (35), arginine (8, 40), and histidine (1) systems. Aside from the binding proteins, the other components appear to be membrane associated (Table 2). The organization of these components has been reviewed recently (1, 45). The complexity of these systems can be compared with the simplicity of the PMF-linked systems. In the best-characterized example of a PMF-linked transport system, the lactose carrier consists of a single 46,500-dalton polypeptide which contains all of the information necessary for H^+-lactose cotransport (55). On the other hand, the periplasmic systems show a structural complexity comparable to the K^+-translocating ATPase. Both types of systems are composed of three membrane-bound polypeptides. In each, at least one polypeptide is exposed on the periplasmic side of the membrane and is involved in substrate recognition. Also in each, another subunit is exposed on the cytoplasmic side of the membrane and has a nucleotide-binding site and, perhaps, ATPase catalytic activity.

What is the role of the periplasmic binding protein in energy transduction? Celis (7) has recently proposed a model for the operation of the arginine transport system of E. coli in which the periplasmic arginine-binding protein undergoes a phosphorylation-dephosphorylation cycle. After incubation of cells with ^{32}P, the arginine-binding protein could be isolated in two forms, one phosphorylated and the other not. When a crude periplasmic solution was incubated with [γ-^{32}P]ATP, phosphorylation of the arginine-binding protein was observed. The phosphorylation was catalyzed by a periplasmic protein kinase specific for the arginine-binding protein, producing an acyl phosphate derivative. Of interest was the observation that the phosphorylated form of the binding protein had 50-fold less affinity for arginine than did the unmodified form (K_d = 5 μM for the phosphoprotein, compared with 0.1 μM for the nonphosphorylated

TABLE 2. Components of ATP-linked transport systems

Transport gene system (reference[s])	Gene product (daltons)	Location	Comments
Potassium (Kdp [21])			
kdpA	59,189	Inner membrane	Periplasm exposure, soluble sequence
kdpB	72,112	Inner membrane	Six to seven membrane-spanning loops, active site
kdpC	20,267	Inner membrane	
Histidine (1, 22)			
hisJ	26,127	Periplasm	Histidine-binding protein
hisM	26,404	Inner membrane	
hisP	28,738	Inner membrane	
hisQ	24,573	Inner membrane	Soluble sequence, homology with malK and ndh, nucleotide-binding site
Maltose (13, 18, 45, 46)			
malE	40,661	Periplasm	Maltose-binding protein
malF	56,947	Inner membrane	Six to eight membrane-spanning loops
malG	Unknown	Inner membrane	
malK	40,700	Inner membrane	Soluble sequence, homology with hisP and ndh, predicted nucleotide-binding site

protein). This observation suggested a catalytic mechanism in which ATP coupling alters the affinity of the transport system specifically on the periplasmic side of the membrane. How ATP could reach the periplasmic space is not known. Since the periplasm contains phosphatases and nucleotidases, free ATP would not be expected to exist there. Likewise, there are no data that suggest that the binding proteins are ever in direct contact with the cytosol. In addition, no other binding was found in a phosphorylated form, a condition for a generalized model. The relevance of the phosphorylation of the arginine-binding protein to the catalytic event of transport has yet to be elucidated.

Since ATP is available in the cytosol, another possibility is that another component of periplasmic systems interacts with ATP on the cytoplasmic side of the membrane. The histidine transport system of *S. typhimurium* requires three inner membrane polypeptides: the *hisM*, *hisP*, and *hisQ* proteins (1). Genetic evidence suggests that the binding protein interacts with the *hisP* protein. Although no *in vitro* enzymatic activity has been found for any of the membrane components of a periplasmic system, Hobson et al. (23) have recently reported that the *hisM* and *hisP* proteins become labeled when inner membrane vesicles are reacted with 8-azido ATP, a photoreactive derivative of ATP. It may be that these polypeptides have nucleotide-binding sites that are involved in the catalytic mechanism of the transport system. Another possibility is that these polypeptides evolved from a precursor with a nucleotide-binding site but with an entirely different function, in which case homologies need not imply similar catalytic mechanisms. The *hisP* protein shares homology with the two other membrane-bound proteins: the *malK* protein, which has been proposed to be the energy-coupling unit of the periplasmic maltose transport system, and the *ndh* protein of NADH dehydrogenase (20). The *malK* and *ndh* proteins also share significant homology, a finding that has led to the suggestion that the energy coupling in the periplasmic systems is by redox coupling rather than by ATPase coupling. However, the regions of homology between *ndh* and *malK* are different from the regions of homology between *ndh* and

hisP. One interpretation of these results is that all three proteins have a common ancestor. Since the *ndh* and *hisP* proteins both have nucleotide-binding sites, the ancestral protein may have had such a site. However, NADH did not compete with binding of 8-azido ATP to the *hisP* protein; thus the sites on the *hisP* and *ndh* proteins may be different. In the absence of an assay for measurement of enzymatic activity in vitro, the significance of the binding of 8-azido ATP to the *hisP* protein is unclear. The only conclusion that can be drawn at this point is that these membrane proteins have common ancestry.

If the membrane components form the energy-transducing portion of these transport systems, is the binding protein actually necessary? A quite plausible model different from that of Celis has been suggested by Shuman (45, 46) in which the periplasmic components are not absolutely required for transport (Fig. 2). In this model the binding protein is involved in substrate recognition only, increasing the affinity of the transport system but not participating in the energy-coupling portion of the transport reaction. Robbins and Rotman (39) and Robbins et al. (38) described the transport properties of strains with point mutations in the β-methylgalactoside transport system. These mutants lacked functional galactose-binding protein but were still capable of low-affinity transport through the β-methylgalactoside system, suggesting that the galactose-binding protein was not essential for transport. However, since the mutations were not deletions, low-affinity galactose-binding proteins may have been present in those cells.

In a similar study, Shuman (46) used a *malE* deletion to eliminate the maltose-binding protein. Cells carrying this deletion were unable to utilize maltose for growth and showed no low-affinity maltose transport. In these cells, the maltose-binding protein was required for transport. Mal⁺ pseudorevertants (still lacking the binding protein) were isolated. Some of the secondary mutations were outside of the *mal* region and probably altered other transport systems to allow for recognition of maltose. Other mutations were mapped in *mal* by P1 transduction, and mutant cells containing them could transport maltose. One mutant was shown to have the same substrate speci-

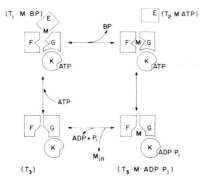

FIG. 2. Model for the accumulation of maltose across the cytoplasmic membrane. In this model, the *malF*, *malG*, and *malK* gene products form a complex in the cytoplasmic membrane. The *malE* gene product, the maltose-binding protein, interacts with this complex from the external periplasmic side of the membrane. M indicates a molecule of substrate, maltose or longer maltodextrin. The cycle proceeds clockwise as drawn through a series of conformational changes in the *malF* and *malG* proteins. The substrate recognition site formed by the *malF* and *malG* proteins can exist in three states: exposed to the periplasmic surface of the membrane (T1 and T2) or exposed to the cytosolic surface of the membrane (T3). The binding protein-substrate complex interacts with the T1 conformation, with transfer of the substrate to the substrate-binding site of the membrane complex and release of the binding protein. Energy derived from ATP binding and hydrolysis to the *malK* protein drives a conformational change from the T2 state to the T3 state, where the substrate-binding site, now exposed to the cytosol, has diminished affinity for substrate, and substrate is released into the cytosol. Exchange of ADP for ATP returns the complex to the T1 conformation (adapted from data of Shuman [46]).

ficity and inducibility as the wild type. One important difference was that the affinity of the transport system for maltose was three orders of magnitude less than that of the wild type. In addition, maltose transport could be assayed in spheroplasts. Spheroplasts, like osmotically shocked cells, have lost their periplasmic proteins and do not exhibit transport via periplasmic systems. Thus, maltose transport in the revertant was independent of a periplasmic protein, presumably by alteration of one of the membrane proteins to allow it to recognize maltose rather than the maltose-maltose-binding-protein complex. Furthermore, transport in spheroplasts of some of the mutants was actually inhibited by addition of maltose-binding protein, suggesting that maltose was binding to the same site as the binding protein (H. A. Shuman, personal communication).

More recent studies demonstrate that the mutations lie in two of the three membrane proteins of the maltose transport (H. A. Shuman, personal communication). Some mutations affect the *malF* gene product, an integral membrane protein of molecular weight 56,947 that spans the membrane six to eight times (18). Others affect the *malG* protein, which is also a membrane protein. A different sort of *malG* mutation resulted in the *malK* protein becoming cytosolic (47). It is possible that the mutation affects assembly of the complex. On the other hand, from the nucleotide sequence, the *malK* protein is predicted to have a composition similar to those of soluble proteins (20).

Thus, its membrane location may be through binding to the *malG* protein. Interestingly, none of the *malE* Mal⁺ pseudorevertants lie in *malK*. If the *malK* protein is localized only on the inner face of the cytosolic membrane, then mutations in the *malK* gene should not produce a Mal⁺ phenotype, assuming that such a phenotype results from a change in portions of those proteins which are exposed to the periplasm.

Thus, Shuman's model (Fig. 2) envisions that translocation of maltose is accomplished by two integral membrane proteins, the *malF* and *malG* gene products. The role of the periplasmic binding protein, the *malE* gene product, is to bind maltose in the periplasm and to transfer it to a substrate-binding site on a complex formed from the *malF and malG* proteins. In the pseudorevertants, this site would be altered to recognize free maltose (with low affinity) rather than maltose bound to the *malE* protein. The role of the *malK* protein might be energy coupling, perhaps as an ATPase. Several observations support this idea. First, as discussed above, *malK* has homology with *ndh* and *hisP*, which specify proteins that bind nucleotides. In fact, from the nucleotide sequence of *malK*, the *malK* protein can be predicted to have a nucleotide-binding site (H. A. Shuman, personal communication). Second, some Mal⁺ pseudorevertants of *malK* mutants have regained facilitated diffusion of maltose but are unable to concentrate the sugar (H. A. Shuman, personal communication). These may be *malG* or *malF* mutants that can catalyze an energy-independent movement of maltose across the membrane.

It may be noted that there is similarity between Shuman's model for the maltose transport system and Epstein's model for the K⁺-ATPase. Both have two membrane proteins that form the substrate-binding site (*kdpA–kdpC* and *malG–malF* proteins). Both have membrane proteins with predicted nucleotide-binding sites (*malK* protein) or ATPase activity (*kdpB* protein), and both systems show the same ATP dependency in vivo. The *mal* system has a periplasmic binding protein not present in the K⁺ transport system, but the *mal* system in Shuman's pseudorevertants transports maltose with only the three membrane proteins and without a binding protein. On the other hand, from the hydropathic profiles generated from the DNA sequences of the membrane proteins, there are distinct differences between the proteins of the two systems. The *malF* protein is predicted to have six to eight membrane-spanning loops, whereas its functional analog in the Kdp system, the *kdpA* protein, has the profile of a soluble protein. The *malK* protein has a soluble character and is peripherally associated with the membrane, whereas its presumed equivalent in Kdp, the *kdpB* protein, is an integral membrane protein with six or seven membrane-spanning loops. Thus, the physical similarities of the periplasmic systems with the Kdp system most likely reflect evolutionary convergent mechanisms.

Plasmid-Encoded Transport Systems

One mechanism of genetic exchange among bacterial species is conjugal transfer of plasmid DNA. The types of information encoded by these plasmids are frequently resistances to antibiotics and heavy metals (17). One mechanism of plasmid-mediated resistance is synthesis of a transport system for excretion of the toxic substance. For example, tetracycline (30, 31) and cadmium (51, 52) resistances are both the result of synthesis of substrate/proton antiport systems for the extrusion of the toxic agents from the cytosol. Everted membrane vesicles prepared from resistant cells have exhibited tetracycline/proton or cadmium/proton exchange activities. These are secondary transport systems coupled to the PMF.

Recently, plasmid-mediated resistances to arsenate and arsenite have been shown to be due to an ATP-linked anion pump which catalyzes removal of arsenicals from resistant cells (33, 42, 49). The genes for the arsenical pump are contained in a single *ars* operon, inducible by AsO_4^{3-}, AsO_2^- or SbO^+ (48). Resistances to arsenite and antimonate have not been separated and are probably due to the same mechanism (9). On the other hand, arsenate resistance requires a gene product that is not required for arsenite resistance, possibly a modifier subunit. The energy coupling of the pump was investigated by a modification of the method of Berger (33, 42). In these studies, energy-depleted cells of an *unc* strain were loaded with $^{74}AsO_4^{3-}$ or $^{74}AsO_2^-$ and efflux initiated by dilution of cells into arsenical-free medium containing various energy sources. Extrusion of the arsenicals was independent of the presence of a PMF but required phosphate bond energy. There was a temporal relationship between the rate of ATP synthesis and arsenate extrusion, suggesting that ATP or something in rapid equilibrium with ATP was the direct energy donor.

The *ars* operon has been cloned into small recombinant plasmids (9, 32). The structural genes of the operon are contained in a 4.3-kilobase region. The promoter-proximal region specifies arsenite resistance and transport. The sequence of the entire operon has been determined (C.-M. Chen, T. K. Misra, S. Silver, and B. P. Rosen, J. Biol. Chem., in press). Three open reading frames, *arsA*, *arsB*, and *arsC*, were found. From genetic evidence with Tn*5* and Mu insertions, the ArsA and ArsB proteins comprise the arsenite pump, whereas the ArsC protein appears to modify the substrate specificity to allow pumping of arsenate.

The *arsA* open reading frame specifies a soluble protein of 63 kilodaltons. The nucleotide sequence of the *arsA* gene shows internal homology suggestive of a gene duplication. The amino acid sequence also exhibits two regions of internal homology. Within each is a sequence homologous with the adenylate-binding site of known nucleotide-binding proteins, such as the β subunit of bacterial, mitochondrial, and chloroplast F_1 and the iron-containing subunit of bacterial nitrogenase. The ArsA protein was purified from cytosol and shown to bind $[\alpha\text{-}^{32}P]ATP$ (U. Weigel and B. P. Rosen, manuscript in preparation). In addition, it binds to dye-agarose affinity chromatography columns specific for nucleotide-binding proteins. Finally, purified ArsA protein has arsenite-stimulated ATPase activity. These results suggest that the ArsA protein may be the catalytic subunit of the anion pump.

The *arsB* open reading frame specifies a membrane protein of 45.5 kilodaltons. Mini-Mu phage transposition was used to create gene fusions between the *arsB* and *lacZ* genes (M. J. D. San Francisco and B. P. Rosen, manuscript in preparation). By immunoprecipitation and immunoblotting with anti-β-galactosidase

serum, this strain was shown to produce a high-molecular-weight hybrid protein localized in the inner membrane and not found in the cytosol. It is proposed that the ArsB protein comprises the anion channel of the arsenical pump. Finally, the *arsC* open reading frame specifies a 15.8-kilodalton soluble protein. The ArsC protein was purified from cytosol. Insertion into or deletion of all or part of the *arsC* gene results in loss of arsenate transport and resistance without affecting arsenite transport or resistance. A working hypothesis is that the ArsA and ArsB proteins comprise an arsenite pump, with the ArsC subunit able to modify the substrate specificity, allowing recognition of arsenate. Thus, the *ars* operon specifies an anion pump that provides resistance to arsenicals.

SUMMARY

Three types of transport activities have been demonstrated to reply on phosphate bond energy: ion-translocating ATPases, transport systems utilizing periplasmic binding proteins, and certain plasmid-encoded transport systems. The ion-translocating ATPases are of two distinct types: the F_0F_1 and the E_1E_2. Both use ATP as the direct energy donor for transport, but the catalytic mechanisms are very different. The F_0F_1 transports only protons and is found in all bacteria, mitochondria, and chloroplasts. The E_1E_2 type, of which the K^+-ATPase is the only known example in *E. coli*, all have a catalytic subunit that undergoes a cycle of phosphorylation and dephosphorylation of an aspartyl residue. The specificity for the cationic substrate differs. E_1E_2 enzymes have been identified with substrates K^+, Ca^{2+}, H^+, Na^+ exchanging for K^+, and H^+ exchanging for K^+. The F_0F_1 and K^+-ATPase of *E. coli* are proving to be good model systems for the study of transport and bioenergetics. The periplasmic transport systems and the plasmid-encoded efflux systems have proven to be more difficult to study, and less is known about their mechanisms.

ACKNOWLEDGMENTS

I thank the following individuals for manuscripts and preprints: G. F. L. Ames, I. R. Booth, J. M. Brass, W. Epstein, R. T. F. Celis, and H. A. Shuman. Discussions with R. T. F. Celis, F. M. Harold, D. L. Oxender, and H. A. Shuman were especially valuable.

LITERATURE CITED

1. **Ames, G. F.-L., and C. F. Higgins.** 1983. The organization, mechanism of action, and evolution of periplasmic transport systems. Trends Biochem. Sci. **8:**97–100.
2. **Berger, E. A.** 1973. Different mechanisms for energy coupling for the active transport of proline and glutamine in *Escherichia coli*. Proc. Natl. Acad. Sci. USA **70:**1514–1518.
3. **Berger, E. A., and L. A. Heppel.** 1974. Different mechanisms of energy coupling for the shock-sensitive and shock-resistant amino acid permeases of *Escherichia coli*. J. Biol. Chem. **249:**7747–7755.
4. **Brass, J. M., W. Boos, and R. Hengge.** 1981. Reconstitution of maltose transport in *malB* mutants of *Escherichia coli* through calcium-induced disruptions of the outer membrane. J. Bacteriol. **146:**10–17.
5. **Brass, J. R., and M. D. Manson.** 1984. Reconstitution of maltose chemotaxis in *Escherichia coli* by addition of maltose-binding protein to calcium-treated cells of maltose regulon mutants. J. Bacteriol. **15:**881–890.
6. **Carafoli, E., G. Inesi, and B. P. Rosen.** 1984. Calcium transport across biological membranes, p. 129–185. *In* H. Sigel (ed.), Metal ions in biological systems. Marcel Dekker, Inc., New York.

7. **Celis, R. T. F.** 1984. Phosphorylation *in vivo* and *in vitro* of the arginine-ornithine periplasmic transport protein of *Escherichia coli*. Biochim. Biophys. Acta **145**:403–411.

8. **Celis, R. T. F., H. J. Rosenfeld, and W. K. Maas.** 1973. Mutant of *Escherichia coli* K-12 defective in the transport of basic amino acids. J. Bacteriol. **116**:619–626.

9. **Chen, C.-M., H. L. T. Mobley, and B. P. Rosen.** 1985. Separate resistances to arsenate and arsenite (antimonate) encoded by the arsenate resistance operon of R factor R773. J. Bacteriol. **161**:758–763.

10. **Clark, A. F., and R. W. Hogg.** 1981. High-affinity arabinose transport mutants of *Escherichia coli*: isolation and gene location. J. Bacteriol. **14**:920–924.

11. **Cowell, J. L.** 1974. Energetics of glycylglycine transport in *Escherichia coli*. J. Bacteriol. **120**:139–146.

12. **Curtis, S. J.** 1974. Mechanism of energy coupling for transport of D-ribose in *Escherichia coli*. J. Bacteriol. **120**:295–303.

13. **Duplay, P., H. Bedouelle, A. Fowler, I. Zabin, W. Saurin, and M. Hofnung.** 1984. Sequences of the malE gene and of its product, the maltose-binding protein of *Escherichia coli* K12. J. Biol. Chem. **259**:10606–10613.

14. **Epstein, W., and L. Laimins.** 1980. Potassium transport in *Escherichia coli*: diverse systems with common control by osmotic forces. Trends Biochem. Sci. **5**:21–23.

15. **Epstein, W., V. Whitelaw, and J. Hesse.** 1978. A K^+ transport ATPase in *Escherichia coli*. J. Biol. Chem. **253**:6666–6668.

16. **Ferenci, T., W. Boos, M. Schwartz, and S. Szmelcman.** 1977. Energy-coupling of the transport system of *Escherichia coli* dependent on maltose-binding protein. Eur. J. Biochem. **75**:187–193.

17. **Foster, T. J.** 1983. Plasmid-determined resistance to antimicrobial drugs and toxic metal ions in bacteria. Microbiol. Rev. **47**:361–409.

18. **Froshauer, S., and J. Beckwith.** 1984. The nucleotide sequence of the gene for *malF* protein, an inner membrane component of the maltose transport system of *Escherichia coli*: repeated DNA sequences are found in the malE–malF intercistronic region. J. Biol. Chem. **259**:10896–10903.

19. **Futai, M., and H. Kanazawa.** 1983. Structure and function of protontranslocating ATPase (F_0F_1): biochemical and molecular biological approaches. Microbiol. Rev. **47**:285–313.

20. **Gilson, E., H. Nikaido, and M. Hofnung.** 1982. Sequence of the malK gene in *E. coli* K12. Nucleic Acids Res. **10**:7449–7458.

21. **Hesse, J. E., L. Weiczorek, K., Altendorf, A. S. Reicin, E. Dorus, and W. Epstein.** 1984. Sequence homology between two membrane transport ATPases, the kdp-ATPase of *Escherichia coli* and the Ca^{2+}-ATPase of sarcoplasmic reticulum. Proc. Natl. Acad. Sci. USA **81**:4746–4750.

22. **Higgins, C. F., P. D. Haag, K. Nikaido, F., Ardeshir, G. Garcia, and G. F.-L. Ames.** 1982. Complete nucleotide sequence and identification of membrane components of histidine transport system of *Salmonella typhimurium*. Nature (London) **298**:723–727.

23. **Hobson, A. C., R. Weatherwax, and G. F.-L. Ames.** 1984. ATP-binding sites in the membrane components of the histidine permease, a periplasmic transport system. Proc. Natl. Acad. Sci. USA **81**:7333–7337.

24. **Hong, J.-S.** 1977. An *ecf* mutation in *Escherichia coli* pleiotropically affecting energy coupling in active transport but not generation or maintenance of membrane potential. J. Biol. Chem. **252**:8582–8588.

25. **Hong, J.-S., A. G. Hunt, P. S. Masters, and M. A. Lieberman.** 1979. Requirement of acetyl phosphate for the binding protein-dependent transport systems in *Escherichia coli*. Proc. Natl. Acad. Sci. USA **76**:1213–1217.

26. **Hunt, A. G., and J.-S. Hong.** 1981. The reconstitution of binding protein-dependent active transport of glutamine in isolated membrane vesicles from *Escherichia coli*. J. Biol. Chem. **256**:11988–11991.

27. **Hunt, A. G., and J.-S. Hong.** 1983. Properties and characterization of binding protein dependent active transport of glutamine in isolated membrane vesicles of *Escherichia coli*. Biochemistry **22**:844–850.

28. **Kadner, R. J., and H. H. Winkler.** 1975. Energy coupling for methionine transport in *Escherichia coli*. J. Bacteriol. **123**:985–991.

29. **Laimins, L. A., D. B. Rhoads, K.-H. Altendorf, and W. Epstein.** 1978. Identification of the structural proteins of an ATP-driven potassium transport system in *Escherichia coli*. Proc. Natl. Acad. Sci. USA **75**:3216–3219.

30. **Levy, S. B., and L. McMurry.** 1978. Plasmid-determined tetracycline resistance involves new transport systems for tetracycline. Nature (London) **276**:90–92.

31. **McMurry, L., R. E. Petrucci, and S. B. Levy.** 1980. Active efflux of tetracycline encoded by four genetically different tetracycline resistance determinants in *Escherichia coli*. Proc. Natl. Acad. Sci. USA **77**:3974–3977.

32. **Mobley, H. L. T., C.-M. Chen, S. Silver, and B. P. Rosen.** 1983. Cloning and expression of R-factor mediated arsenate resistance in *Escherichia coli*. Mol. Gen. Genet. **191**:421–426.

33. **Mobley, H. L. T., and B. P. Rosen.** 1982. Energetics of plasmid-mediated arsenate efflux in *Escherichia coli*. Proc. Natl. Acad. Sci. USA **79**:6119–6122.

34. **Neu, H. C., and L. A. Heppel.** 1965. The release of enzymes from *Escherichia coli* by osmotic shock and during the formation of spheroplasts. J. Biol. Chem. **240**:3685–3692.

35. **Oxender, D. L., S. C. Quay, and J. J. Anderson.** 1980. Regulation of amino acid transport, p. 153–169. *In* J. W. Payne (ed.), Microorganisms and nitrogen sources. John Wiley & Sons Ltd., Chichester, United Kingdom.

36. **Piperno, J. R., and D. L. Oxender.** 1966. Amino acid-binding protein released from *Escherichia coli* by osmotic shock. J. Biol. Chem. **241**:5732–5734.

37. **Rhoads, D. B., and W. Epstein.** 1977. Energy coupling to net K^+ transport in *Escherichia coli* K12. J. Biol. Chem. **252**:1394–1401.

38. **Robbins, A. R., R. Guzman, and B. Rotman.** 1976. Roles of individual mgl gene products in the β-methylgalactoside transport system of *Escherichia coli* K12. J. Biol. Chem. **251**:3112–3116.

39. **Robbins, A. R., and B. Rotman.** 1975. Evidence for binding protein-independent substrate translocation by the methylgalactoside transport system of *Escherichia coli*. Proc. Natl. Acad. Sci. USA **72**:423–427.

40. **Rosen, B. P.** 1973. Basic amino acid transport in *Escherichia coli*: properties of canavanine-resistant mutants. J. Bacteriol. **116**:627–635.

41. **Rosen, B. P. (ed.).** 1978. Bacterial transport. Marcel Dekker, Inc., New York.

42. **Rosen, B. P., and M. G. Borbolla.** 1984. A plasmid-encoded arsenite pump produces arsenite resistance in *Escherichia coli*. Biochem. Biophys. Res. Commun. **124**:760–765.

43. **Rosen, B. P., and L. A. Heppel.** 1973. Present status of binding proteins that are released from gram-negative bacteria by osmotic shock, p. 209–240. *In* L. Leive (ed.), Bacterial membranes and walls. Marcel Dekker, Inc., New York.

44. **Rosen, B. P., and E. R. Kashket.** 1978. Energetics of active transport, p. 559–620. *In* B. Rosen (ed.), Bacterial transport. Marcel Dekker, Inc., New York.

45. **Shuman, H. A.** 1982. The maltose-maltodextrin transport system of *Escherichia coli*. Ann. Microbiol. (Paris) **133A**:153–159.

46. **Shuman, H. A.** 1982. Active transport of maltose in *Escherichia coli*: role of the periplasmic maltose-binding protein and evidence for a substrate recognition site in the cytoplasmic membrane. J. Biol. Chem. **257**:5455–5461.

47. **Shuman, H. A., and T. H. Silhavy.** 1981. Identification of the malK gene product: a peripheral membrane component of the *Escherichia coli* maltose transport system. J. Biol. Chem. **256**:560–562.

48. **Silver, S., K. Budd, K. M. Leahy, W. V. Shaw, D. Hammond, R. P. Novick, G. R. Willsky, M. H. Malamy, and H. Rosenberg.** 1981. Inducible plasmid-determined resistance to arsenate, arsenite, and antimony(III) in *Escherichia coli* and *Staphylococcus aureus*. J. Bacteriol. **146**:983–996.

49. **Silver, S., and D. Keach.** 1982. Energy-dependent arsenate efflux: the mechanism of plasmid mediated resistance. Proc. Natl. Acad. Sci. USA **79**:6114–6118.

50. **Stewart, L. M. D., E. P. Bakker, and I. R. Booth.** 1985. Energy coupling to K^+ uptake via the Trk system in *Escherichia coli*: the role of ATP. J. Gen. Microbiol. **131**:77–85.

51. **Tynecka, Z., Z. Gos, and J. Zajac.** 1981. Reduced cadmium transport determined by a resistance plasmid in *Staphylococcus aureus*. J. Bacteriol. **147**:305–312.

52. **Tynecka, Z., Z. Gos, and J. Zajac.** 1981. Energy-dependent efflux of cadmium coded by a plasmid resistance determinant in *Staphylococcus aureus*. J. Bacteriol. **147**:313–319.

53. **Wilson, D. B.** 1974. Source of energy for the *Escherichia coli* galactose transport systems induced by galactose. J. Bacteriol. **120**:866–871.

54. **Wood, J. M.** 1975. Leucine transport in *Escherichia coli*: the resolution of multiple transport systems and their coupling to metabolic energy. J. Biol. Chem. **250**:4477–4485.

55. **Wright, J. K., and P. Overath.** 1984. Purification of the lactose:H^+ carrier of *Escherichia coli* and characterization of galactoside binding and transport. Eur. J. Biochem. **138**:497–508.

51. Osmotic-Shock-Sensitive Transport Systems

CLEMENT E. FURLONG

Departments of Genetics and Medicine, Division of Medical Genetics, Center for Inherited Diseases, University of Washington, Seattle, Washington 98195

NUTRIENT TRANSPORT SYSTEMS OF *E. COLI* AND *S. TYPHIMURIUM*

Nutrient utilization by *Escherichia coli* and *Salmonella typhimurium* involves transport across the outer membrane, movement across the periplasmic space, and transport through the plasma membrane. Some substrates may be hydrolyzed by lysosomal-like enzymes in the periplasmic space before transport across the plasma membrane. Nutrient accumulation in *E. coli* and *S. typhimurium* has been extensively studied over the past three decades. The early studies indicated that cells were able to accumulate nutrients against a concentration gradient and led to the concept of permeases (59). Subsequent genetic and biochemical studies in the past two decades have led to the identification of the protein components of a number of specific nutrient transport systems, as well as the genes that specify these proteins. The biochemical identification and subcellular localization of the protein components, coupled with genetic studies, has led to the realization that several different mechanisms are utilized for nutrient accumulation. A convenient way to categorize the different systems is on the basis of their mechanisms, reflected in the organization of the protein components of the specific system.

The simplest system is exemplified by the facilitated diffusion system for glycerol transport. Presumably, a single protein facilitates the movement of glycerol across the lipid bilayer (Fig. 1, system 1). A high-affinity kinase then phosphorylates (traps) the internal glycerol (182).

The next most complicated system is typified by the lactose and proline transport systems, in which, again, a single protein catalyzes the movement of substrate across the lipid bilayer. However, by coupling the transport with energy, the cell is able to transport nutrients against concentration gradients (Fig. 1, system 3). This type of transport system is often referred to as membrane bound or osmotic shock insensitive, since the one required protein is an integral membrane protein that is not released by osmotic-shock treatment (147, 148).

The systems referred to as osmotic shock sensitive are composed of a periplasmic substrate-binding protein (referred to as simply a binding protein) and usually three additional membrane-associated components. Further, an outer membrane protein (a porin) may be considered as a component of specific systems (Fig. 1, system 4). The periplasmic substrate-binding proteins are released by an osmotic-shock procedure. Since they are required for transport through this type of system, such systems are often referred to as osmotic-shock-sensitive transport systems or periplasmic transport systems. The osmotic-shock procedure involves initiating the plasmolysis of cells by sucrose in the presence of EDTA, harvesting the cells, and rapidly resuspending them in cold deionized water or in a dilute solution of $MgCl_2$ (210, 312). This procedure selectively releases the periplasmic protein fraction, which consists of the substrate binding proteins of the osmotic-shock-sensitive transport systems along with the lysosomal-like bacterial enzymes (210). The substrate-binding proteins are usually single-subunit proteins with molecular weights ranging between 23,000 and 52,000, with most around 30,000 (Table 1). From the crystallographic studies completed so far, all of the binding proteins seem to have similar two domain structures, with the substrate-binding site located between the two halves (e.g., reference 238). Other common properties include resistance to heat, broad pH optima, and high substrate affinities with binding constants usually in the micromolar range. The observed in vitro substrate affinities and specificities usually correspond well with the kinetic properties of the specific transport systems (222). Most binding proteins are very resistant to heat and proteases.

A fourth type of system (type 2 in Fig. 1) transports mainly hexoses and hexitols. This type of system requires membrane-bound and cytoplasmic protein components. The high-energy metabolite phosphoenolpyruvate serves as the energy source for this type of transport system. Through a series of transfers of a high-energy phosphate, the substrate is phosphorylated as it crosses the membrane. This type of system is usually referred to as a phosphotransferase system (PTS); the mechanism is generally termed a group translocation-type mechanism (67, 271).

This chapter will focus on the osmotic-shock-sensitive or binding protein-dependent transport systems of *E. coli* and *S. typhimurium*. The chapter is organized into sections on periplasmic transport systems for amino acids, peptides, sugars, anions, and vitamins. Since many nutrients are accumulated through more than one type of system, a short discussion of the multiplicity of transport systems for the nutrients of interest will be presented. Short sections on the genetics of the systems, the properties of the isolated binding proteins, the regulation of the systems, the energetics of specific systems, and reconstitution studies are included. The properties of the binding proteins and associated transport systems are summarized in Table 1, along with references for the amino acid and nucleotide sequences where known.

A couple of general points that are applicable to most, if not all, of the osmotic-shock-sensitive transport systems should be made. The evidence for the involvement of the binding proteins in the uptake of their respective ligands has been provided by biochemical, genetic, and molecular genetic experiments. The role of the membrane-associated proteins has been much more difficult to establish because of the association of these proteins with the plasma membrane and because the membrane-associated proteins

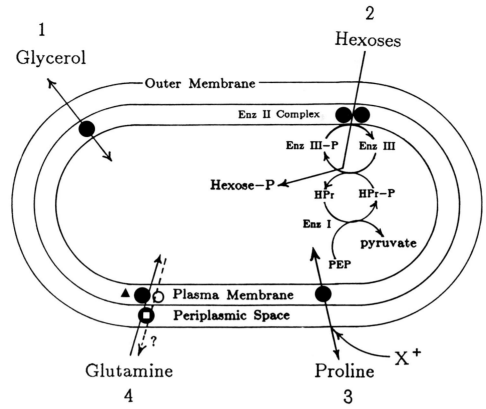

FIG. 1. Schematic representation of the different types of transport systems present in gram-negative bacteria. System 1, typified by the glycerol facilitator, is a facilitated diffusion system (182). System 2 is a phosphotransferase uptake system and is typified by the transport systems for hexoses and hexitols. The salient point about this type of transport system is that it requires both cytoplasmic and membrane-associated components. The nutrients accumulated through this system are said to be taken up by a mechanism of group translocation, in which the incoming substrate is phosphorylated through the interaction of a small phosphorylated protein (HPr) with the membrane-associated proteins. Phosphoenolpyruvate serves as the source of high-energy phosphate. The reader is referred to recent reviews for a detailed description of these interesting systems (67, 99, 271). System 3 is typified by the proline and lactose transport systems. This type of system appears to depend on a single membrane-associated protein for function and derives its energy from chemiosmotic gradients (255, 288; chapter 50). Proton cotransport is most often observed with these systems, although Na^+ cotransport is observed with some systems. In contrast to the other types of energy-coupled systems, the membrane-associated transport systems are active in osmotically shocked cells and membrane vesicles. Several excellent reviews include descriptions of the membrane-bound-type systems (81, 147, 148, 255, 287, 316). The fourth type of system is represented by the binding protein-dependent transport systems (81, 222, 255, 287, 316). This type of system requires a soluble periplasmic substrate recognition component and membrane-associated protein components. The periplasmic protein is initially synthesized with an amino-terminal signal sequence that "guides" the protein to the periplasmic space, where it is processed to a very stable mature substrate recognition component (usually referred to as a binding protein). Recent reviews and papers describe our present state of knowledge of binding protein secretion (140, 205, 240, 241). In addition to the periplasmic binding protein, two or more membrane-associated proteins are required for the function of this fourth type of transport system. The binding proteins may be released from cells by a gentle osmotic shock procedure (13) (or by a recently developed simple chloroform extraction procedure [13]). Thus, these systems are often referred to as osmotic-shock-sensitive systems. The binding protein-dependent transport systems appear to utilize some form of high-energy phosphate that may be ATP or a closely associated high-energy metabolite (35, 37, 132). The binding protein-dependent transport systems of *E. coli* and *S. typhimurium* will be the major focus of this discussion. As noted throughout this discussion, a given nutrient may be, and most often is, transported through one or more of these different types of systems. For each nutrient, the multiplicity of systems available for uptake is discussed briefly to set the role of the binding protein-dependent system(s) in proper perspective.

are present in much lower quantities than their periplasmic counterparts. Most of the evidence for the involvement of the membrane-associated proteins in the function of specific binding protein-dependent transport systems has come from genetic and molecular genetic experiments.

A number of early studies on reconstituting transport by adding binding protein to osmotically shocked cells were reported (18, 20, 28, 199, 317). However, several laboratories reported difficulties in reproducing these observations (81, 88, 109, 222). At least three problems contributed to the execution and interpretation of these studies (81). One of these was that the nitrocellulose filters used to retain the "reconstituted" cells during uptake measurements also retained the added binding protein that had not been reincorporated into the cells; this caused high "background" values. Indeed, the protein-ligand complex binds to

TABLE 1. Binding proteins from *E. coli* and *S. typhimurium*[a]

Binding protein for:	Organism (purification)	Mol wt (×10³)	pI	K_d (μM)
Amino acids				
Arginine-ornithine	*E. coli* (51, 52, 257, 317)	28–33	5.1 (51, 257)	Arg 0.03–0.1 (51, 257)
Lysine-arginine-ornithine (LAO)	*E. coli* (254)	26–30	5.1	Arg 0.15, Lys 3, Orn 5
	S. typhimurium (169)	26 (112)		
Cystine	*E. coli* (36, 219)	27–28 (36)		Cys 0.01 (36)
Glutamine	*E. coli* (305, 306, 313)	23–29 (305, 306)	8.6 (305, 306)	0.15–0.3 (305, 306)
	S. typhimurium (169)	23 (169)		
Glutamate-aspartate	*E. coli* (29, 275, 309)	29–32 (29, 309)	9.1–9.7 (29, 309)	0.8–6 (29, 309)
	S. typhimurium (4, 169)	30 (4, 169)		Asp 1 (169)
Histidine	*E. coli*[c]	25–31[c] (21)		0.8
	S. typhimurium (181, 258)	25 (112, 181, 258)	5.5 (181)	0.15–1.5 (181, 258)
Leucine-isoleucine-valine-threonine (LIVT)	*E. coli* (19, 79, 234)	36.7 (220)	4.8 (79)	0.2–2 (19, 79, 234)
	S. typhimurium (214)	35–39 (214)	4.94 (214)	Leu 0.43, Ile 0.15, Val 0.89 (214)
Leucine-specific	*E. coli* (79, 82)	37 (82, 209)		0.7 (82)
	S. typhimurium (34)	34–38 (34)	4.74 (34)	0.54
Peptides				
Oligopeptide	*E. coli*	52 (116)		
	S. typhimurium	52 (116)		
Sugars				
Arabinose	*E. coli* (126, 128, 277)	33 (238)	0.2–2 (128, 277)	
Galactose-glucose	*E. coli* (19, 126)	32 (41)		1 (19)
	S. typhimurium (324)	33 (324)	0.38 (324)	
Maltose	*E. coli* (158, 159)	40.7 (72)	1 (299)	
Ribose	*E. coli* (77, 308)	29.5 (308)	0.13 (308)	
	S. typhimurium (5)	29	0.33	
Xylose	*E. coli* (3, 63)	37 (3)	7.4 (3)	0.6 (3)
Anions				
Citrate	*S. typhimurium* (298)	28	6.1	1–2.6
Phosphate	*E. coli* (199)	34 (190, 296)		0.8 (199)
sn-Glycerol-3-phosphate	*E. coli* (22)	45	7	0.2
Sulfate	*S. typhimurium* (226)	34.7 (142)		0.02 (226)
Vitamins				
B₁₂	*E. coli* (partial) (300)	22 (300)		0.005 (44)
Thiamine	*E. coli* (partial) (144, 212)			0.03–0.1 (144, 212)

[a] Parentheses indicate references.
[b] Genetic or biochemical identification or both. K, ×10³ molecular weight.
[c] L. A. Beck and C. E. Furlong, unpublished data.
[d] C. E. Furlong, unpublished data.

the nitrocellulose filters to such an extent as to be useful for a binding assay (305). Second, the osmotic shock procedure removes the entire small-molecule pool of the cell (47), and some of the initial reconstitution studies were carried out without an added energy source to drive the substrate uptake. Thus, anything that would provide energy to the shocked cells would stimulate uptake (81). Finally, the nature of the multiplicity of the different systems was not appreciated, so that the presence of shock-insensitive transport systems complicated the interpretation of the results (81). Ideally, to study reconstitution, a strain with only a single osmotic-shock-sensitive transport system for the nutrient in question should be used. Recently, significant advances in reconstitut-

ing binding protein-dependent transport have been made. These are discussed in the sections that follow.

Most, if not all, of the binding protein-dependent transport systems (as well as other transport systems) are subject to one or more forms of regulation: specific induction; repression; general carbon, nitrogen, sulfur, or phosphate control; or combinations of any of these. The term "regulon" was originally proposed to describe systems that were under the control of a single repressor (187), but has gained acceptance as a term to describe collections of operons that share even positive control elements (e.g., reference 303). As described in the specific sections that follow, the expression of many of the transport systems appears to be under the control of one or more regulons.

TABLE 1—*Continued*

K_m for transport (μM)	Map position of structural gene(s) (min)	Amino acid sequence	Gene sequence	Signal sequence	Membrane-associated proteins[b]
Arg 0.15, Orn 4.0 (51)	60 (52)				
Arg 0.5					
	46 (112, 273)	Inferred (7)	(112)	Inferred (112)	Same as histidine
0.1–0.3 (36)					
0.5–0.8 (305, 306)	17.7 (197)	Partial (195)			
0.5 (275)					
	50 (26)				
0.04–0.4 (166)	46 (112,273)	(125)	(112)	Inferred (112)	HisQ 24.5K, HisM 26K, HisP 28.7K (115)
0.1–0.5 (234)	74.5 (209)	(220)	(172)	(172)	LivH 30K, LivM 27K, LivG 23K (209)
0.3–1 (215)					
0.3[d]	74.5 (209)		(172)	(172, 223)	Same as LIV-binding protein system (209)
	27 (116)				(OppB) (116); (OppC) (118); OppD 37K (in *S. typhimurium*)
	34 (116)				
3–8 (48, 277)	45 (26, 56, 124)	(129)	(127)	Inferred (127)	
0.5 (266)	45 (26, 85)	(191)		Inferred (280)	MG1A 52K, Mg1C 38K (96, 264)
1 (299)	91 (26, 122)	(72)	(72)	(72, 143)	MalG 24K?, MalF 57K, MalK 41K, LamB 47K, MolA 13K? (71)
0.3 (308)	84 (26)	(68)	(68)	Inferred (92)	RbsA 50K, RbsC 27K (135)
	82 (273)				
5 (3)					
3	59 (289)				40K? (156)
0.2 (259)	74 (26, 259)		(190, 296)	(296)	PstC 40K, PstA 28K, PstB 29K, PhoU 26K (297)
2 (278)	75.3 (279)				
36 (69)	(See text)				
0.01 (300)					BtuB (105, 106)
0.83 (154)					

AMINO ACID PERMEASES

Branched-Chain Amino Acid Transport

Branched-chain amino acid permeases

In *E. coli*, the branched-chain amino acids are transported by at least three kinetically distinct transport systems, an osmotic-shock-resistant, membrane-bound system (LIV-II) and two binding protein-dependent systems, LIV-I and LS (or leucine specific) (81, 209). The membrane-bound LIV-II system has a low affinity for all of the branched-chain amino acids (15). The osmotic-shock-sensitive, branched-chain amino acid transport systems are organized such that one set of membrane protein components is shared by two different genetically linked periplasmic binding proteins (209). One periplasmic binding protein binds D- or L-leucine (leucine specific, or LS-binding protein) (81, 209), whereas another binds L-leucine, L-isoleu-cine, L-valine, L-threonine, L-alanine, or L-serine (LIV- or LIVT-binding protein) (209). An isoleucine-preferring protein has also been described (7, 78, 239). The present working model for binding protein-dependent branched-chain amino acid transport suggests that the two binding proteins (leucine- and LIV-binding proteins) deliver their bound amino acids to a membrane complex composed of at least the H, M, and G proteins for transport into the cytosol (82). It is not entirely clear whether to refer to leucine- and LIV-binding protein-dependent transport as a single system or systems, since the binding proteins catalyze kinetically distinct transport through shared membrane protein components. Here, we will consider the two well-characterized binding protein-dependent transport activities as kinetically separate but genetically and biochemically interdependent systems. The histidine, lysine-arginine-ornithine (LAO) transport systems are organized in a similar manner (see below).

Genetics of the branched-chain amino acid permeases

The genes that specify the branched-chain amino acid-binding protein-dependent transport system(s) lie at min 74.5 on the *E. coli* chromosome (82). The order of the genes encoding the shock-sensitive branched-chain amino acid transport system(s) is: *livJ, livL, livK, livH, livM,* and *livG*. The *livJ* gene specifies the LIV-binding protein (M_r, 39,000); *livK* codes for the leucine-binding protein (M_r, 39,000); the function of the *livL* is not yet known; and *livH, livM,* and *livG* each specify separate membrane-bound proteins (M_r, 30,000, 28,000, and 22,000, respectively) (82). The gene for the LIV-II transport system, *livP*, is located at 76 min on the *E. coli* chromosome (15).

E. coli branched-chain amino acid-binding proteins

E. coli. Both the LIV- (19, 234) and leucine- (82) binding proteins have been purified to homogeneity and crystallized. The amino acid sequence of LIV-binding protein has been determined (220). It contains one S-S bridge (203). The nucleotide sequence of the genes that specify both proteins has been determined (172). Both proteins are initially synthesized as precursor proteins with NH_2-terminal 23-amino acid signal peptides (223, 224) (Table 1). The crystal structure for the LIV-binding protein has been determined at 0.3-nm resolution. It reveals two globular domains connected by three strands, with a single substrate-binding site located in the cleft. The crystal structure is similar to those found with other binding proteins (274).

The membrane-associated components, as with the other binding protein-dependent systems, are present in much lower levels than the binding proteins. They have not yet been purified or characterized extensively (82).

Several lines of evidence suggest that the LIV-binding protein undergoes a conformational change upon ligand binding. The temperature dependence of substrate binding indicated important substrate-induced conformational changes that were corroborated by scanning calorimetry (87). These studies are consistent with substrate binding inducing a conformational change that increases intramolecular hydrophobic interactions and results in a conformation with higher temperature stability. Fluorescence quenching studies also indicated that the protein-ligand complex was more stable than the apoprotein (38). Difference UV spectra, fluorescence spectra, circular dichroism spectra, and Raman spectra were used to follow substrate binding to the LIV-binding protein (301). The authors concluded that this protein has both low- and high-affinity binding sites and that one tryptophan and perhaps one tyrosine are present in the high-affinity binding site. More recent studies indicate that Trp-18 and Tyr-18 may be involved in ligand binding by leucine- and LIV-binding proteins, respectively. Surface-enhanced Raman spectra suggest the participation of Tyr, Trp, and Phe residues in complex formation in the leucine-binding protein (207).

S. typhimurium. Two branched-chain amino acid-binding proteins have been isolated recently from *S. typhimurium* (214). These proteins (LIVT- and leucine-binding proteins) appear to be very similar to the two branched-chain amino acid-binding proteins isolated from *E. coli* (Table 1).

Regulation of the branched-chain amino acid permeases

E. coli. The regulation of expression of the binding protein-dependent branched-chain amino acid transport system(s) is complex. The level of the LIV-II system does not appear to be totally repressed by growth in rich medium (171, 319). On the other hand, the presence of leucine in the medium significantly decreases the expression of the LIV-binding protein. One mechanism of control appears to be transcriptional repression. The genes *livK* through *livG* appear to be regulated as one operon, and the gene for *livJ* is regulated as a separate operon (16). One mutant has been isolated (*livR*) in which the LIV-binding protein is derepressed in cells grown in the presence of high leucine. Another mutation (*lstR*) derepresses the leucine-specific protein. These mutations map at min 20 on the *E. coli* chromosome (16). Analysis of the sequence upstream from the LIV-binding protein gene suggests the possibility of rho-dependent transcription attenuation (172). In vitro transcription of *livJ* required the nucleotide ppGpp, consistent with the absence of derepression of LIV-binding protein synthesis in a *relA* mutant (16, 225). Mutations in the *hisT* gene which prevent the conversion of uridine to pseudouridine during the maturation of tRNAs for leucine, histidine, and tyrosine result in a lack of repression and derepression of LIV-I transport (225). A temperature-sensitive leucyl-tRNA synthetase mutant synthesized four- to sixfold higher levels of LIV-binding protein than controls, suggesting a possible role of the uncharged mature leucyl-tRNA in the positive control of the transport operon (225). The primary regulatory mechanism is thought to be transcriptional repression by the *livR* gene product (172).

S. typhimurium. Physiological studies with *S. typhimurium* suggest, as expected, that the regulation of branched-chain amino acid transport in this organism is similar to that observed in *E. coli* (161).

Energetics of binding protein-dependent branched-chain amino acid transport

Differences in substrate specificity and susceptibility to osmotic shock allowed Wood (319) to investigate the energetics of the binding protein-dependent branched-chain amino acid transport systems. She concluded that, like other binding protein-dependent transport systems, the binding protein-dependent amino acid transport systems require ATP (319).

Cystine Transport

Cystine permeases

The transport of cystine has been best characterized in *E. coli* W (36, 178). This strain has at least two cystine transport systems, a cystine-specific system

$(K_m = 3 \times 10^{-7}$ M) and a cystine general system $(K_m = 2 \times 10^{-8}$ M) which also transports diaminopimelic acid and is inhibited by several diamino-dicarboxylic acid analogs (36). The two systems have different substrate specificities and probably different mechanisms of transport. The general system is osmotic shock sensitive, whereas the specific system appears to be osmotic shock resistant. However, when cells are grown in rich medium, transport through the specific system increases and appears to become osmotic shock sensitive (54). It may well be that another system is induced under these conditions. The mutant strain E. coli D₂W has only the general osmotic-shock-sensitive cystine transport system, probably at a derepressed level (36).

Cystine-binding protein

Cystine-binding protein with properties consistent with the general cystine transport system has been purified to homogeneity from E. coli strains W (36) and W3092 (219) (Table 1). Binding activity has been partially purified from E. coli MR83 (312). The pure binding protein had the same specificity observed for the general cystine transport system (36, 219, 312). In general, α,α'-diaminodicarboxylic acid analogs with carbon bridges of three or four units between the α,α' carbons were effective inhibitors of binding and transport. Selenocystine-resistant mutants derived from strain D₂W lost both cystine transport and the binding protein (312). Interestingly, the cystine-binding protein was resolved into two components when purified by DEAE-cellulose chromatography. However, no physical or kinetic differences were observed between these two activities (36). As noted below, the citrate-binding protein of S. typhimurium also behaved in the same manner owing to a cyclization of an amino-terminal glutamine residue (156). The reason for the two components observed with the cystine-binding protein has not been elucidated, although it may be speculated that this protein may be similar to the citrate-binding protein or that the protein-ligand complex may fractionate differently than apoprotein.

Novel use of the cystine-binding protein

The high specificity and affinity of the cystine-binding protein has been exploited for developing a "radioimmuno-like assay" for measuring cystine levels in fibroblasts cultured from patients with the cystine storage disease cystinosis (219). The ability to run many assays in a single day facilitated the in vitro development of drugs for treating this genetic disease (146). As noted below, a similar assay using the glutamine-binding protein has been developed for glutamine.

Glutamine Transport

Glutamine permease

Glutamine is one of the relatively few nutrients so far investigated that is transported primarily through a single transport system (305, 306, 311). The transport system appears to be specific for glutamine or the analogs γ-glutamylhydrazide, γ-glutamylhydroxamate (305, 306), and methionine sulfoxide (25). Os-

motic shock reduces E. coli glutamine transport by 90% and releases a glutamine-binding protein with the same specificity and affinity as the transport system (305, 306).

Glutamine-binding protein

The glutamine-binding protein has been purified from both S. typhimurium (169) and E. coli (305) (Table 1). The sequence of the 40 amino-terminal residues of the glutamine-binding protein has been reported (195). The E. coli glutamine-binding protein is specific for glutamine. None of the other naturally occurring amino acids inhibits glutamine binding. However, the same analogs that inhibit glutamine transport (γ-glutamylhydrazide and γ-glutamylhydroxamate) inhibit glutamine binding (305, 306).

Fluorescence (306) and proton magnetic resonance (167) spectroscopy, as well as microcalorimetry (194), indicate that the glutamine-binding protein undergoes a substrate-induced conformational change.

Genetics of glutamine transport

Glutamine transport-negative, binding protein-negative mutants of both E. coli (197) and S. typhimurium (169) have been isolated on the basis of analog resistance to γ-glutamylhydrazide. Curiously, E. coli glnP mutants fail to grow on glutamate as sole carbon source (197). The genes encoding the glutamine permease lie at 17.7 min on the E. coli chromosome (197).

Reconstitution of glutamine transport

The binding protein-dependent glutamine transport system is ideal for studying the reconstitution of binding protein-dependent transport since there are apparently no other high-affinity glutamine transport systems (some glutamine is probably transported through the glutamate systems after hydrolysis by the periplasmic asparaginase). Hong and co-workers have successfully reconstituted binding protein-dependent glutamine transport in both spheroplasts (196) and membrane vesicles (132, 133). To eliminate the problem of removing all the glutamine-binding protein from the vesicle preparations, they prepared spheroplasts and vesicles from an overproducer strain that had a point mutation in the gene for the glutamine-binding protein. Interestingly, it was necessary to incorporate NAD in the vesicles for successful reconstitution when the energy source was a metabolite that could be metabolized to pyruvate, but not when succinate served as the energy source (133). Modifications of either the sole histidine or one of the two tryptophan residues did not affect glutamine binding, but did destroy the interaction of the binding protein with its membrane receptors, rendering it ineffective in reconstituting glutamine transport in vesicles (134). This approach to reconstitution was also used successfully with galactose-binding protein-dependent transport (see below).

Energetics of glutamine transport

In addition to being an excellent system for studying reconstitution, binding protein-dependent glutamine transport is also an excellent system for investigating the energetics of binding protein-dependent transport, since virtually all glutamine transport is through this system (305, 306). In early studies with *E. coli*, Berger and Heppel concluded that the binding protein-dependent systems are coupled directly to phosphate bond energy, whereas the membrane-bound transport systems are driven by an energy-rich membrane state (35, 36). Experiments by Hong and co-workers suggested that acetyl phosphate (131) and succinate or pyruvate or both (or metabolites derived from succinate or pyruvate) might be the energy source for binding protein-dependent glutamine transport (134). The electrochemical proton gradient also appeared to be important; however, Hong et al. felt that it served a function other than direct energization. Recent experiments by Rosenberg and co-workers (260) suggest that the energization of binding protein-dependent transport is complex and involves the proton motive force and as yet unidentified compound(s) (see also Phosphate Transport, below).

Regulation of glutamine transport

In *S. typhimurium*, the levels of the binding protein-dependent transport systems for glutamine, histidine, glutamate-aspartate, and arginine, as well as the levels of glutamine synthetase, are regulated by the availability of nitrogen in the growth medium (39, 169). Low available nitrogen leads to an increase in all of these activities. In *E. coli*, glutamine synthetase and the glutamine transport system appear to be subject to the same regulation observed in *S. typhimurium*. The details of the general mechanism of nitrogen regulation are discussed in more detail in Histidine Transport, below.

Novel uses of glutamine-binding protein

The high specificity and affinity of the glutamine-binding protein allowed Willis and Seegmiller (313) to use this protein to develop a radioimmuno-like assay for measuring the levels of glutamine in physiological samples (see also Cystine-binding protein, above).

Glutamate-Aspartate Transport

Glutamate-aspartate permeases

Glutamate-aspartate transport in *E. coli* and *S. typhimurium* is complicated by the multiplicity of systems that transport these two amino acids. The affinities of the respective systems are so similar that a kinetic analysis of wild-type cells does not suggest the degree of multiplicity that exists (275). The multiplicity of transport systems has been most completely characterized in *E. coli*, which has at least five separate systems involved in the uptake of these two amino acids (275). The kinetics of each of the glutamate transport systems have been analyzed in appropriate mutant strains. One membrane-bound, sodium-independent system is specific for aspartate (K_m = 3.5 μM) (155), and another is specific for glutamate (K_m = 5 μM) or aspartate (K_m = 5 μM) (275). Both of these systems transport the toxic analog beta-hydroxyaspartate (155, 275). A sodium-dependent system

transports only glutamate (K_m = 1.5 μM, in the presence of sodium ion) or its analog, methylglutamate (149, 275). The dicarboxylic acid transport system (dct) transports four-carbon dicarboxylic acids including aspartate (K_m = 30 μM) (183). Only one of the five systems is a typical osmotic-shock-sensitive, binding protein-dependent transport system (K_m for glutamate = 0.5 μM; K_m for aspartate = 0.5 μM) (275). Glutamate-aspartate-binding proteins have been purified from both *E. coli* and *S. typhimurium*.

Glutamate-aspartate-binding proteins

E. coli. Glutamate-aspartate-binding proteins have been purified from strains W (275, 309) and K-12 (29). The amino acid compositions are very similar, with a number of differences reported (perhaps within the usual error range). This protein has the highest isoelectric point of any of the known bacterial binding proteins (pH 9.7) (Table 1), and it binds 1 mol of glutamate or aspartate with equal affinity (309). The binding protein has a buried disulfide and may be reversibly denatured as long as the disulfide bond is not reduced while the protein is in the unfolded state (309). Like other binding proteins, the glutamate-aspartate-binding protein has a very broad pH optimum of binding (pH 5 to 9) (29, 309). Also like other binding proteins, it appears to undergo a conformational change upon binding ligand. Both its excitation and emission spectra for tryptophan demonstrated a blue shift with ligand binding, suggesting the movement of a tryptophan residue(s) to a more hydrophobic environment (309).

Although the detailed genetics of binding protein-dependent glutamate-aspartate transport have yet to be worked out, *E. coli* mutants with 1.6-fold (29) and 8-fold (275) increases in the level of glutamate-aspartate-binding protein and corresponding increases in transport have been described.

S. typhimurium. Glutamate-aspartate-binding protein has been purified from *S. typhimurium* LT-2 (4, 169) (Table 1). It is very similar to its *E. coli* counterpart, and antiserum prepared against the *E. coli* protein cross-reacts with the *S. typhimurium* protein (169). Aksamit et al. were interested in the possibility that the glutamate-aspartate-binding protein might serve as a chemoreceptor. Comparison of the properties of the binding protein with an extracted membrane-bound aspartate-binding activity and the aspartate chemotactic response led them to conclude that neither the glutamate-aspartate-binding protein nor the membrane aspartate-binding activity was the receptor for the aspartate chemotactic response (4).

Regulation of binding protein-dependent glutamate-aspartate transport

Ideally, both the mechanism and regulation of binding protein-dependent glutamate-aspartate should be investigated in strains with all of the other glutamate-aspartate systems genetically eliminated. However, despite the fact that strains bearing only the binding protein-dependent system have yet to be constructed, some information exists on the regulation of the glutamate-aspartate-binding protein and presumably the entire transport system. It is known that both carbon and nitrogen availability affect the level of the protein and probably its corresponding transport system as well. Protein levels in glucose-grown *E. coli* cells were three to five times lower than in succinate-grown cells (85). As noted above, genetic and biochemical studies by Kustu et al. (169) demonstrated that the glutamate-aspartate-binding protein, as well as glutamine-, histidine-, and LAO-binding proteins, were regulated in *S. typhimurium* by the level of nitrogen availability (probably as a regulon). (See Histidine Transport, below, for a more detailed discussion on general nitrogen regulation.)

Histidine Transport

Histidine permeases

As noted above, Ames and co-workers have extensively investigated histidine transport in *S. typhimurium*. At least five separate transport systems take up histidine in this organism (8). A binding protein-dependent histidine transport system has been characterized genetically, biochemically, and at the DNA sequence level (9).

Genetics of histidine transport

An operon at 47 min on the recalibrated *S. typhimurium* map (273) encodes the periplasmic binding protein (J protein) and three membrane-bound proteins, the Q, M, and P proteins, which comprise the osmotic-shock-sensitive histidine transport system (9, 12). In addition, the membrane proteins also interact with the lysine-arginine-ornithine (LAO)-binding protein, the gene for which is located immediately upstream from the histidine transport gene complex (168). The complete nucleotide sequence of the entire histidine transport operon has been determined (90, 113, 115). Furthermore, the sequence of the regulatory region for the histidine transport operon (*dhuA*) has been determined, as well as that of the upstream gene that specifies the LAO-binding protein gene and its regulatory sequence (*argTr*). The LAO and His J protein genes apparently arose by gene duplication and genetic drift (112). The nucleotide sequences of the respective genes show 71% homology, and the proteins themselves show 70% homology. Of the 72 amino acid differences, 31 are conservative replacements. One very interesting mutation [Δ (*his5643*)] results in a chimeric protein formed by the fusion of the amino end of the LAO protein with the carboxyl end of the J protein. Interestingly, the chimeric protein has the binding properties of the LAO protein (112).

Another interesting observation regarding the evolution of the binding protein-dependent membrane transport systems is the extensive homology between the inferred sequence for the MalK membrane protein (see Maltose Transport, below) and that for the histidine P protein (90). Alignment of the inferred P amino acid sequence with the NH$_2$-terminal two-thirds of that of the maltose K protein showed a 32% homology, with another 35% of the residues being functionally similar. Also interesting has been the finding that the HisP and Q proteins as well as the MalK protein can be affinity labeled by an ATP analog. A possible homology of the HisP protein and the ATPase was also noted (121).

Histidine-binding proteins

S. typhimurium. The histidine-binding protein from *S. typhimurium* has been purified to homogeneity (181, 258), and its amino acid sequence has been determined (125). The sequence can also be inferred from the nucleotide sequence (112). The properties of this binding protein are summarized in Table 1. The sequence for the *S. typhimurium* LAO-binding protein has been inferred from its gene sequence. Both proteins appear to have 22-amino acid signal sequences that are removed during protein processing.

Nuclear magnetic resonance (120, 253) and fluorescence (253, 323) studies indicate that the native histidine-binding protein undergoes a significant conformational change upon binding histidine, whereas a mutant J protein appears to have an altered response to ligand binding (120, 253, 323). These results are in agreement with the concept that interaction of the binding proteins with the membrane-bound protein components of the binding protein-dependent nutrient transport systems requires a changed binding protein conformation brought about by ligand binding (322). It is also interesting that a mutation in the His J protein can be complemented by a mutation in the His P protein, suggesting a direct interaction between these two proteins (14).

E. coli. The *E. coli* histidine-binding protein has been purified to homogeneity from strain 7, a derivative of K10. Its properties are summarized in Table 1 (L. A. Beck and C. E. Furlong, unpublished data). Ardeshir and Ames (21) characterized several *E. coli* histidine transport mutants and found them to be similar to the *S. typhimurium* histidine transport mutants extensively characterized by Ames and co-workers. The *E. coli* histidine-binding protein appears to be somewhat more acidic than its *S. typhimurium* counterpart, whereas the *E. coli* M protein appeared to have the same pI as the *S. typhimurium* M protein (21).

Regulation of the binding protein-dependent histidine transport operon

Space permits only a brief discussion of the elegant studies on the regulation of the histidine and LAO transport operons and the adjacent, related LAO-binding protein operon, as well as an even briefer discussion of the studies on the entire system of general nitrogen regulation, which has been reviewed recently (189).

Both the binding protein-dependent histidine and LAO transport systems are regulated by nitrogen availability. The transcriptional regulation of these two operons has been studied through the use of Mu d1-mediated *lac* fusions, in which the *lac* structural genes are brought under the transcriptional control of the promoters of interest (294). Transcriptional regulation of the operon in question is evaluated by the rates of β-galactosidase. These studies indicated that nitrogen regulation occurred at the level of transcription. The LAO-binding protein operon was found to respond to both catabolite repression and nitrogen limitation, whereas the histidine transport operon responded only to nitrogen regulation. Investigation of the interplay of nitrogen and carbon regulation indicated that the cyclic AMP (cAMP) receptor protein (CRP)-cAMP complex interfered with the gene products effecting nitrogen regulation (189). The regulatory region for *argT* (*argTr*) contains a sequence that is highly homologous to other CRP-binding sites. Comparison of sequences of other known nitrogen-regulated promoters with *argTr* and *dhuA* sequences suggested proposed sites of nitrogen regulation (9, 294). Sequencing of the *dhua* promoter-up mutation confirmed the probable upstream location of the *hisJ* promoter (113, 176). At least three genes (*ntrA*, *ntrB*, and *ntrC* in *S. typhimurium* or *glnF*, *glnL*, and *glnG* in *E. coli*) are involved in nitrogen regulation of transport systems (189). Recent DNA binding and footprinting studies of the histidine transport operon promoter region (*dhuA*) indicate that the *ntrC* dimeric gene product binds to a 20-base-pair region of dyad symmetry about 180 base pairs 5′ to the translational start site of the *hisJ* gene. A conserved seven-base sequence found in the promoter regions of several nitrogen-regulated operons (221) was included in the *ntrC* binding site. S1 mapping indicates that the transcription start site within the promoter region (*dhuA*) is at position −48. The *ntrC* gene product did not bind to the promoter region (*argT*) of the LAO-binding protein gene (11). The transcription start site for the LAO operon was shown by S1 mapping to be at position −60 (11).

The level of the histidine-binding protein is approximately 30-fold higher than the levels of the membrane components of the transport system. The possibility that the location of two repetitive extragenic palindromic sequences (111, 114) between *hisJ* and *hisQ* might be responsible for the differences in periplasmic and membrane protein levels through transcriptional or translational regulation has been thoroughly investigated (292, 293). Mu d1-mediated *lac* fusions and quantitation of the M protein in appropriate constructions with deleted and rearranged repetitive extragenic palindromic sequences indicated that the presence of these sequences was not responsible for the observed differences in protein levels. Experiments that varied the levels of the respective protein components of the binding protein-dependent histidine transport system indicated that the membrane-bound components are, in general, rate limiting for histidine transport (292).

Reconstitution of histidine transport

Histidine transport has been reconstituted in a *hisJ* deletion mutant lacking the histidine-binding protein by a procedure developed by Brass and co-workers (45). (See also Reconstitution of Maltose Transport and Chemotaxis, below.)

Basic Amino Acid Transport

Basic amino acid permeases

The kinetics, genetics, and biochemistry of the basic amino acid transport systems of *E. coli* and *S. typhimurium* have been examined. A complete picture of all of the systems transporting lysine, arginine, and ornithine has not yet emerged. However, the studies with *E. coli* and *S. typhimurium* have been quite complementary. As described above, the studies of the shock-sensitive histidine transport system of *S. typhimurium* led to an extensive characterization of the general LAO shock-sensitive transport system. In *E. coli*, on the other hand, examination of basic amino acid transport led to the characterization of the LAO and arginine-ornithine transport systems. The systems are probably quite comparable in the two organisms; however, different approaches have led to the generation of complementing information about the systems present in the two organisms. In designing experiments to investigate basic amino acid transport in either organism, it will be well worth reviewing the information available for both organisms.

In *E. coli*, the basic amino acids appear to be transported through (i) a membrane-bound, lysine-specific system, (ii) the binding protein-dependent LAO system (254), and (iii) a binding protein-dependent arginine-ornithine system (52, 254, 257). The relationship of the LAO systems to the histidine transport system has not yet been determined for *E. coli*, but presumably it will be similar if not identical to that found in *S. typhimurium* (see Histidine Transport, above).

LAO-binding proteins from *E. coli* and *S. typhimurium*

The LAO-binding protein has been purified to homogeneity from *E. coli* (254) and partially purified from *S. typhimurium* (169) (as noted above). The properties of the LAO-binding proteins are summarized in Table 1. As described in Histidine Transport, above, the nucleotide sequences for both the regulatory region and structural gene for the *S. typhimurium* LAO-binding protein have been determined (112, 113, 115). Also, as noted above, a deletion resulted in the formation of a fused chimeric protein consisting of the NH$_2$-terminal region of the LAO-binding protein and the carboxyl terminus of the histidine-binding protein. This protein had the binding properties of the LAO-binding protein (112).

Genetics and regulation of the LAO system

As noted above, the *S. typhimurium* LAO-binding protein is regulated by both carbon and nitrogen availability. The details of this regulation and the genetic organization of the *S. typhimurium* LAO system are discussed in above.

E. coli arginine-ornithine-binding protein

An arginine-specific binding protein has been purified to homogeneity from *E. coli* (257), and its properties are summarized in Table 1. Initial studies with this protein suggested that it was specific for arginine. None of the other naturally occurring amino acids inhibited the binding of labeled arginine to the protein (257). Antiserum prepared against the arginine-specific binding protein did not cross-react against serum prepared against the LAO-binding protein, or vice versa. The protein contains two half-cystine residues, apparently in disulfide linkage (257).

Celis isolated an arginine-ornithine-binding protein and made the interesting observation that this protein bound ornithine at 22°C but not at 4°C (52). Thus, it is probable, judging from examination of the data in both reports, that the arginine-ornithine-binding protein is the same protein as the arginine-specific binding protein. It is interesting that Rosen (254, 257), Wilson and Holden (317), and Celis (52) all resolved at least three arginine-binding activities by DEAE-cellulose chromatography. However, only two of these activities have been characterized.

Genetics of arginine-ornithine transport

Celis isolated a series of chain-terminating mutants in the structural gene for the arginine-ornithine-binding protein (52), as well as an overproducer mutant (51). The properties of the mutants and mutant proteins provided evidence for the involvement of the binding protein in the uptake of these two amino acids. The regulatory and structural gene loci for the arginine-ornithine-binding protein lie between *argA* and *dsdA* on the *E. coli* map (26, 53). Other genetic studies suggest a possible common component shared by the two shock-sensitive basic amino acid transport systems (50, 55, 256).

Energetics of arginine transport

Recently, Celis found that the arginine-ornithine-binding protein could be phosphorylated in vivo and in vitro (54). The phosphorylated protein had a lower affinity for substrate (K_d = 5 μM) than the unphosphorylated protein (K_d = 0.1 μM). A periplasmic protein phosphokinase was purified that appeared to be specific for the arginine-ornithine and LAO proteins. The phosphorylation reaction required ATP and appeared to involve the phosphorylation of an acyl group. This system provides a testable model for the role of direct binding protein phosphorylation in the energetics of binding protein-dependent transport.

Reconstitution of arginine transport

Wilson and Holden reported a partial stimulation of arginine transport by DEAE-cellulose-purified periplasmic arginine-binding fractions in osmotically shocked *E. coli* W cells (317). However, Rosen was unable to confirm their observations (257).

PEPTIDE TRANSPORT

Peptide Permeases

Peptides are transported by three separate systems in *S. typhimurium*, a binding protein-dependent oligopeptide permease, a tripeptide permease, and a dipeptide permease (89, 116, 117, 123, 145). The exact number and nature of the *E. coli* peptide transport systems have yet to be determined (229, 230, 232, 233). However, it is clear that in *E. coli* the outer membrane porins OmpF and OmpC are required for the movement of peptides across the outer membrane and exhibit some specificity for substrates (17). The best-characterized peptide transport system is the oligopeptide transport system of *E. coli* and *S. typhimurium* (89, 116, 117, 123, 145). This system transports di- and oligopeptides with up to five amino acid residues (233). The tripeptide transport system of *S. typhimurium* has recently been characterized (89, 145). The dipeptide transport system is the least characterized of the three *S. typhimurium* peptide transport systems. It has a preference for dipeptides, but also transports some tripeptides (89).

Genetics of Peptide Transport

The *S. typhimurium* tripeptide transport system is specified by two genes, a positive regulatory gene (*tppA*, identical to *ompB*) and a single structural gene (*tppB*), which lie at 74 min and 27 min, respectively, on the *S. typhimurium* chromosome (89, 273). The oligopeptide transport system is typical of a binding protein-dependent system. It is specified by four genes, *oppA*, *oppB*, *oppC*, and *oppD*, which constitute an operon that lies at 34 min on the *S. typhimurium* chromosome and 27 min on the *E. coli* chromosome (117, 180). The *oppA* gene encodes a 52-kilodalton (kDa) periplasmic protein. The other genes (*oppB*, *oppC*, and *oppD*) appear to code for membrane-associated gene products. The four genes specifying the *S. typhimurium* oligopeptide permease have been cloned, and the *oppD* gene has been sequenced. Extensive homology was noted between the inferred protein products of the *oppD* gene and those of the *hisP*, *malK*, and *pstB* genes (proposed peripheral membrane energy-coupling proteins of the histidine, maltose, and phosphate shock-sensitive systems, respectively). The *oppD* gene product was characterized as a fusion protein with β-galactosidase. From the inferred protein sequence, two conserved regions of homology appeared to be involved in nucleotide (ATP) binding. Binding of the fusion protein to Cibacron Blue and its labeling with the nucleotide affinity analog 5'-p-fluorosulfonylbenzoyl-adenosine supported the concept that the *oppD* gene product has an ATP-binding site (118).

The gene for dipeptide permease lies at min 80 on the *S. typhimurium* chromosome (cited in reference 89) and at min 13 on the *E. coli* genome (26). The *E. coli ompC* and *ompF* genes map at min 48 and 21, respectively (26).

Oligopeptide-Binding Protein

A 52,000 M_r periplasmic protein has been identified as the probable substrate recognition component of the oligopeptide permease (110, 116). Mutants deleted for part or all of the *oppA* gene are deficient in this protein. Point mutations in the *oppA* gene led to an altered electrophoretic mobility of the 52,000 M_r protein. An *oppA* mutant of *E. coli* was also missing the 52,000 M_r protein (116). A 56,000 M_r oligopeptide-binding protein has been purified to homogeneity from osmotic-shock fluid of *E. coli* (94). It has a pI of 5.95 and binds 1 mol of labeled Ala-Phe-Gly with a K_d of 0.1 μM. Transport mutants devoid of the binding protein have been isolated (94).

Regulation of Peptide Transport

The transcriptional regulation of the tripeptide and oligopeptide permeases was examined using Mu d1-

mediated *lac* operon fusions. The oligopeptide permease is constitutively expressed. The tripeptide permease is induced by anaerobiosis or by exogenous leucine and is under the positive control of the *ompB* locus (145).

Energetics of Peptide Transport

The energetics of peptide transport have not yet been studied in isogenic strains bearing only one of the three peptide transport systems so far described. However, the sensitivity of peptide transport to osmotic shock and to arsenate inhibition suggests that at least a major component of peptide transport is carried out by a typical binding protein-dependent system (60, 231, 232).

SUGAR TRANSPORT SYSTEMS

The transport of sugars in *E. coli* and *S. typhimurium* may be catalyzed by systems using any one of the three basic mechanisms of energy-dependent transport (Fig. 1). The PTS and membrane-bound-type transport systems will not be discussed here since the main focus of this overview is binding protein-dependent transport. In *E. coli*, arabinose and galactose are each taken up by a membrane-bound-type system and a binding protein-dependent system. The membrane-bound systems are quite similar in properties and lie close together on the genetic map. The binding protein-dependent systems for these two sugars are also very similar in properties and map close together as well. *S. typhimurium* appears to be missing the binding protein-dependent arabinose system. Both maltose and ribose appear to be taken up primarily by single specific, high-affinity, binding protein-dependent systems. However, an additional, yet uncharacterized, low-affinity system also appears to be present for transporting ribose. A number of strain differences have been observed; however, the systems described below should provide a reasonable picture of the kinds of osmotic-shock-sensitive systems likely to be encountered in many if not most strains.

Arabinose Transport

Arabinose permeases

Arabinose transport in *E. coli* is mediated by at least three transport systems, a binding protein-dependent arabinose system, a binding protein-dependent methylgalactoside (or galactose-binding protein-dependent) system (Mgl system), and a membrane-bound arabinose (AraE) transport system (73, 127, 277). The Mgl transport system is described in the next section. *S. typhimurium* appears to have only the low-affinity membrane-bound (AraE) system (269).

Genetics of binding protein-dependent arabinose transport in *E. coli*

The high-affinity arabinose-binding protein transport system is specified by at least two genes that lie at min 45 on the *E. coli* chromosome (56, 124). The *araF* gene encodes the arabinose-binding protein and

is located 5′ to the *araG* gene, which presumably specifies a 37-kDa membrane protein component of this transport system (162). The amino terminus of this component has been inferred by nucleotide sequence analysis (127, 277). The binding-protein gene specifies a 23-amino acid signal sequence which is preceded by a ribosome-binding site 5 nucleotides upstream from the initiation codon. Like the histidine transport operon, the binding-protein gene is followed by a 22-base stem-loop structure and then another ribosome-binding site which precedes the initiation codon for the *araG* gene. The low-affinity arabinose transport system of *E. coli* is specified by the *araE* gene, which lies at min 61 on the *E. coli* chromosome (269). It is interesting that for each arabinose system, there is a related galactose transport system that lies close by (27, 64). The protein product of the *araE* gene has been identified as a membrane protein in both *E. coli* (M_r, 36,000 to 38,000) (188) and *S. typhimurium* (M_r, 41,000) (269).

Arabinose-binding protein

The arabinose-binding protein has been purified from both *E. coli* B/r (128, 277) and *E. coli* K-12 (128) (Table 1). This binding protein has been crystallized, and its structure has been determined (238). The amino acid sequence has been determined (129) and confirmed by crystallographic analysis (238) and by nucleotide sequence determination of the *araF* gene (127). The precursor form of the protein has also been purified (318). As noted above, its amino acid sequence has been inferred from the nucleotide sequence (127, 277).

Recently, the arabinose-binding protein crystal structure has been refined to 0.17 nm by Quiocho and Vyas (238). The high-resolution refinement of the protein structure resulted in an elegant definition of the sugar-binding site with the interesting observation that the positioning of an aspartate residue at position 90 allows the protein to bind either sugar anomer with equal affinity. The determination of the binding site also provides much new information on the general nature of ligand binding and the stabilization of allosteric protein transitions. As noted above, the binding proteins so far studied in detail all appear to have two domain structures with the binding site located between the two domains. Binding of the ligand appears to close the cleft, preparing the protein for interaction with the membrane components of the transport system (193, 238).

In addition to the overall structural similarities of the binding proteins, as noted above, the arabinose-binding protein has sequence homologies with all of the other gram-negative bacterial sugar-binding proteins (24, 72) as well as the Lac and Gal repressor proteins (206). However, unlike the other sugar-binding proteins, the arabinose-binding protein does not appear to function as a sugar receptor for chemotactic responses (104).

Evidence for structural changes in the arabinose-binding protein occurring upon ligand binding has been provided by spectroscopy (200, 228), calorimetry (76), small-angle X-ray scattering (211), and nuclear magnetic resonance (57).

Regulation of arabinose transport

The expression of both high- and low-affinity transport systems is dependent on the presence of the *araC* gene product (73, 127, 177). The *araC* gene is represented at 1 min on the *E. coli* map (27). The *araC* gene and its promoter are organized as a divergent operon head to head with the *araBAD* operon which specifies ribulokinase, ribulose epimerase, and ribulose isomerase (73, 127, 177). In the absence of arabinose, the AraC protein, in its capacity as a repressor, binds to the region between the two divergent operons, blocking the binding of the positive effector (cAMP-CRP complex) of the carbon regulon. In the presence of arabinose, the repressor form of the *araC* protein is converted into a positive activator that acts in concert with the cAMP-CRP complex and RNA polymerase to initiate transcription of the arabinose regulon. In addition to regulating the expression of the two divergent operons at min 1 that specify the isomerase (*araA*), kinase (*araB*), and epimerase (*araD*) and the *araC* protein, the *araC* protein also regulates the expression of the *araE* gene at 61 min and the genes for the arabinose-binding protein-dependent system at 45 min. These three operons under the control of the *araC* gene product constitute the arabinose regulon (73, 127). Interestingly, the AraF system, like the MglP (see below) system, appears to be more sensitive to catabolite repression than its membrane-bound counterpart (64).

Energetics of arabinose transport

The presence of transport systems with different mechanisms simultaneously taking up arabinose in wild-type cells made it critical to generate mutants possessing only a single system to investigate the energetics of the individual systems. Daruwalla et al. generated mutants possessing either the AraE (membrane-bound) system or the AraF (binding protein-dependent) system and examined the energetics of each system in the strain having only that system (64). The AraE system appeared to be energized as a typical membrane-bound-type system. Proton symport appeared to drive arabinose uptake in whole cells or in membrane vesicles. On the other hand, the AraF system was, like all of the other shock-sensitive systems, sensitive to inhibition by arsenate and may have a more direct coupling to a high-energy phosphate compound.

Methyl-β-Galactoside Transport

Galactose permeases

There are a half-dozen or more transport systems in *E. coli* that accumulate galactose. These include the lactose permease, the melibiose transport system, the two arabinose transport systems, the membrane-bound galactose permease (GalP), and the MglP or galactose-binding protein transport system (41, 164, 262, 263, 265, 287, 315). *S. typhimurium* (236) and *E. coli* (165) mutants devoid of the known galactose-transporting systems are still able to accumulate galactose by facilitated diffusion through the PTS. As noted above, the membrane-bound galactose system is analogous to, and is specified by a gene near that of, the membrane-bound arabinose transport system, whereas the MglP transport system is analogous to and is specified by genes lying near the arabinose-binding protein transport system (64). Thus, of the many galactose transport systems, only two appear to be shock sensitive, the arabinose-binding protein and Mgl systems. The binding protein-dependent galactose transport system was designated the methylgalactoside permease (MglP) since it also transports the analog methyl-β-D-galactoside, whereas the membrane-bound system does not transport this analog (263). The arabinose-binding protein system is described above. As are the arabinose transport systems, the binding protein system is a high-affinity system, whereas the membrane-bound system is a low-affinity system (64, 263, 287, 315).

Genetics of galactose transport

Kinetic and genetic analyses demonstrated that the MglP and GalP systems were the two major galactose transport systems in *E. coli* (164, 262, 263, 287, 315). The *galP* gene, specifying the membrane-bound system, lies at 64 min (26, 249), and the genes specifying the Mgl system lie at 45 min (85). Three genes encode the proteins that comprise the osmotic-shock-sensitive MglP system (96, 217, 218, 251, 252, 264, 265). The *mglA* gene specifies a 52-kDa membrane-bound protein (96, 264), *mglB* specifies the 36-kDa galactose-binding protein, and *mglC* encodes a 38-kDa membrane-bound protein (96). These three proteins appear to be necessary and sufficient to specify a functional Mgl transport system (96). However, some transport may be catalyzed by the MglA-MglC complex (252, 265).

The nucleotide sequence of the 5' end of the gene specifying the galactose-binding protein (*mglB*) has been determined, and the probable 23-amino acid signal sequence has been deduced from this sequence (280). Genetic (251) and molecular (96) analysis of the cloned operon has established the gene order as *mglB*, *mglA*, *mglC*.

Recently, Müller et al. characterized the *S. typhimurium mgl* operon. They found that this operon is similar to those of all the other shock-sensitive transport systems. The *mgl* operon contains four genes: *mglB* specifies a 33-kDa galactose-binding protein, *mglA* specifies a 51-kDa membrane-associated protein, *mglE* codes for a 21-kDa membrane protein, and *mglC* specifies a 29-kDa protein. A 38-kDa protein appears to be a degradation product of the 51-kDa protein.

Galactose-binding proteins

E. coli. The galactose-binding protein of *E. coli* has been purified to homogeneity (19, 126), sequenced (191), crystallized (302), and extensively characterized. It binds 1 mol of galactose ($K_d = 0.48 \mu M$) or 1 mol of glucose ($K_d = 0.21 \mu M$) per mol of protein (324). Like the maltose-binding protein, the galactose-binding protein can be purified as dimers that are dissociated by ligand (247). Its properties are summarized in Table 1. Defective proteins have also been described (40, 100). As noted above, the galactose-binding protein has been shown to share sequence homologies not only with all of the other characterized *E. coli* sugar-binding proteins, but also with the Gal and Lac repressor proteins. The homology with the ribose-binding protein is particularly interesting since these two proteins share the same receptor site on the *trg* receptor protein in mediating chemotactic responses to their respective sugars in both *E. coli* and *S. typhimurium* (97, 103, 104, 163, 217, 218, 295) (see below). The high-resolution crystal structures of these two proteins will be particularly interesting to compare when they become available since, on the one hand, they each interact with the membrane protein complex of their own specific transport systems, and on the other hand, they both interact with the *trg* transducer protein of the chemotaxis response system (104).

S. typhimurium. Galactose-binding protein has been purified to homogeneity from *S. typhimurium* (324). It is very similar to the *E. coli* protein in properties and immunoreactivity (Table 1). It binds one molecule of galactose (K_d = 0.38 μM) or glucose (K_d = 0.17 μM).

Conformational changes of galactose-binding protein

A change in the conformation of the *E. coli* galactose-binding protein upon ligand binding has been suggested by several different approaches. Binding of glucose, galactose, or glycerolgalactoside enhanced the fluorescence of the protein (42). Also, a blue shift was observed with galactose as the ligand but not with glucose as the ligand. Subsequent fluorescence studies indicated that a tryptophan residue is probably involved in the binding site (198). Gel electrophoretic studies indicated that the mobility of galactose-binding protein was changed by substrate binding (42).

Conformational changes accompanying ligand binding were also observed with the *S. typhimurium* galactose-binding protein by the use of a distant reporter group attached to a methionine residue of the protein away from the active site (322).

Prediction of secondary structure based on protein sequence suggested that the galactose-binding protein is folded similarly to the proteins for arabinose and ribose (158).

Galactose chemotaxis

In addition to serving as the substrate recognition component of the binding protein-dependent galactose transport system, the galactose-binding protein also serves as the attractant receptor for the chemotactic response to galactose (and as one of two glucose receptors) in both *E. coli* (100) and *S. typhimurium* (295). Of the three gene products required for galactose transport, only the binding protein is required for chemotaxis toward galactose (104, 218). Mutations in the structural gene for the galactose-binding protein can affect transport and chemotaxis, transport only, or chemotaxis only (100, 104, 218). One report of reconstituting galactose chemotaxis has appeared (100). As noted above, both the galactose- and ribose-binding proteins interact with the same signal transducer protein, the *trg* gene product. This protein

appears to undergo two types of modification in regulating the chemotactic response, methylation-demethylation and probable deamidation (of glutamine residues) (157).

Reconstitution of galactose-binding protein-dependent transport

As noted in the introduction, an early report described the reconstitution of galactose uptake in osmotically shocked *E. coli* cells (20). However, such experiments proved difficult to reproduce in many laboratories (88, 109, 222). An understanding of the multiplicity of systems present as well as the protein composition and localization of the components of the different systems has provided a better basis for attempting to reconstitute binding protein-dependent transport. As noted in the introduction, some progress has been made recently in developing systems that can be convincingly reconstituted. Galactose transport has been reconstituted in membrane vesicles prepared from *E. coli* S185–27, a binding protein-negative mutant (*mglB⁻*). The reconstituted transport was sensitive to inhibition by uncouplers and metabolic inhibitors (265).

Regulation of galactose-binding protein-dependent transport

As can be imagined, the regulation of the Mgl transport system has been difficult to examine, with so many different systems capable of taking up galactose and an internal pool of galactose inducing several of the systems in galactokinase-negative cells (320). Oxender pointed out the importance of using mutants with the same genetic background for characterizing the galactose transport systems (222). (This point is valid for properly characterizing any given system when the labeled nutrient used to characterize that system may be simultaneously transported by one or more additional systems). Galactose transport has been studied in such strains (64, 107, 315). Daruwalla et al. (64), using such isogenic strains, confirmed Wilson's earlier observations (315) that the MglP system is more sensitive to catabolite repression than is the GalP system. Wilson had also determined that in most *E. coli* cells, galactose uptake is mediated primarily by the MglP and GalP systems (315). The GalP system in *E. coli* (315) and *S. typhimurium* (179) appears to be controlled as part of a regulon that also includes the galactose metabolic enzymes. Mutations in the gene (*galR*) that encodes the repressor protein result in constitutive expression of the galactose enzymes and the GalP system (179, 315). The expression of the GalP system appears not to be strongly repressed by growth on glucose (64, 315). On the other hand, the MglP system is strongly repressed by growth on glucose and is under separate control from the "Gal regulon" (85, 251, 315). The MglP system is induced by fucose or galactose (263). A gene (*galR*) that appears to regulate MglP has been tentatively located between 56 and 74 min on the *E. coli* linkage map (179). Isolation of what appear to be operative constitutive mutants suggests that the MglP system is subject to negative control (251). Wilson's studies (315) suggest that there is an interaction between the

GalP and MglP systems with respect to substrate transport.

Energetics of galactose-binding protein-dependent transport

The isogenic strains developed by Daruwalla et al., one containing the MglP system and the other the GalP system, were used to demonstrate that the GalP system is typical of a membrane-bound type of transport system (64). The GalP system appeared to be energized by a chemiosmotic potential as indicated by sensitivity to uncouplers of oxidative phosphorylation and alkalinization of the medium during galactose transport. On the other hand, in agreement with other workers (41, 287, 315), Daruwalla et al. found that MglP was, like other binding protein systems, sensitive to inhibition by arsenate and relatively insensitive to uncouplers of oxidative phosphorylation.

Maltose Transport

Maltose permease

As noted above, the binding protein-dependent maltose transport system of *E. coli* is one of the best characterized of the *E. coli* binding protein-dependent transport systems (71, 108). Most, if not all, of the protein components have been identified, and the genes that specify these proteins have been mapped. In addition, a considerable amount of information is available regarding the regulation of the respective genes and the interaction of the protein products that make up the maltose transport system (71, 108, 284). The binding protein-dependent maltose transport system is ideal for investigating the mechanism of binding protein-dependent transport since it is the primary system through which maltose is transported in normal *E. coli* strains (108).

Genetics of the maltose transport system

The genes that specify the proteins of the maltose transport system are found on two divergent operons in the *malB* region at min 91 on the *E. coli* chromosome (26, 71). As noted in Table 1, six proteins appear to be associated with the binding protein-dependent maltose transport system (71). These proteins are specified by two operons in the *malB* region. One operon contains *E*, *F*, and *G* genes. The sequence of the *malE* gene has been determined (72). It encodes the 41-kDa maltose-binding protein, which has been purified to homogeneity (158, 159), sequenced (72), and crystallized (237). In addition to serving as the high-affinity substrate recognition component of the transport system, the binding protein serves as the attractant receptor for the chemotactic response to maltose (104). The nucleotide sequence of the *malF* gene has been determined (75). The gene codes for a 57-kDa integral cytoplasmic membrane protein, the F protein, which has been isolated by immunological techniques (286). The *malG* gene specifies a 24-kDa protein that appears to be associated with the inner membrane (108). A possible coding sequence for the amino terminus of this protein has been identified (286). A second operon, separated from the first by 272 base pairs, contains the *malK*, *lamB*, and *molA* genes. The nucleotide sequence of the *malK* gene has been determined (91). *malK* codes for a 41-kDa protein that appears to require the MalG protein for association with the inner membrane (31, 285). This protein contains a possible internal homology, as well as homologies with the histidine M protein and the *E. coli* NADH dehydrogenase (91). The nucleotide sequence of the *lamB* gene has been determined (58). This gene specifies a 49-kDa integral outer membrane protein (λ receptor) that interacts with the maltose-binding protein and facilitates the transport of maltose and maltodextrins through the outer membrane (108). The λ receptor has been purified, sequenced, and crystallized (86, 242). It has a trimeric (208) structure and appears to be typical of outer membrane porins (210). In addition to serving as an outer membrane pore, the LamB protein is the receptor for a number of bacterial phages including λ. The *molA* gene specifies an exported 13-kDa protein whose function is not yet known.

Relatively little work has been done on the analogous maltose transport system in *S. typhimurium*. Several of the *mal* genes have, however, been mapped (273).

Maltose-binding protein

As noted above, the *E. coli* maltose-binding protein has been well characterized (Table 1). It is synthesized with a 26-amino acid signal sequence that is removed during processing (34, 71, 241). The precursor form of the protein has also been purified, and the sequence of its NH_2 terminus has been determined (143). As noted above, comparison of the amino acid sequence of this protein with other known binding protein primary sequences indicates a significant degree of homology between the maltose-binding protein and the *E. coli* binding proteins for arabinose, galactose, and ribose (72). The binding protein appears to be specific for the α, 1→4 glycosidic bond of maltose (and higher maltodextrins) (170, 299). Interestingly, the protein appears to dimerize in the absence of maltose. The dimers can be dissociated by maltose (245).

The direct interactions of maltose-binding protein with the λ receptor (32) and with the cytoplasmic membrane (246) have been demonstrated. The interaction with cytoplasmic membrane vesicles was enhanced by a membrane potential (248) and abolished in mutants defective in the *tar* gene product (the methyl-accepting chemotaxis protein II, or MCPII) (248). The interaction with the λ receptor was decreased by mutations in either the gene for the λ receptor (185) or the gene for the protein (33).

Conformational changes of maltose-binding protein

Fluorescence spectroscopy has been used to investigate possible conformational changes accompanying ligand binding to the maltose-binding protein. In one study, it was found that ligand binding produced fluorescence quenching and either red or blue shifts depending on the ligand bound (299). The results of yet another study were interpreted to suggest that maltose binding to the protein causes the movement of one tryptophan residue to a more hydrophobic environment and another to a more hydrophilic environment. The attachment of a "reporter group" to maltose-binding protein also indicated that a conformational change accompanied ligand binding (321).

Maltose chemotaxis

The maltose-binding protein has been identified as the attractant receptor for the chemotactic response toward maltose (101, 104). Binding protein-maltose complex interacts with a membrane transducer protein coded by the *tar* gene that also serves as the attractant receptor for the aspartate chemotactic response (104). As noted above, the interaction of labeled binding protein with membrane vesicles containing the *tar* protein can be detected (246, 248). Mutants devoid of the binding protein do not demonstrate chemotaxis toward maltose, whereas other mutants have lost maltose transport but not maltose-binding activity or the maltose chemotactic response (101). Effects of *lamB* mutations on maltose chemotaxis appear to be indirect and due simply to decreased access of maltose to the periplasmic space in such mutants (102).

Reconstitution of maltose transport and chemotaxis

Brass and co-workers have developed an excellent general approach for reconstituting binding protein-dependent transport (45) and chemotaxis (46). The first important consideration was the use of deletion mutants that lacked a specific binding protein but synthesized normal levels of the other protein components of the system in question. Both binding protein-dependent maltose transport and chemotaxis can be reconstituted in such mutants through the use of Ca^{2+} treatment of exponentially growing cells. The use of appropriate mutant strains also allowed for the estimation of the affinity of the binding protein for the outer membrane λ receptor and the cytoplasmic membrane components of the transport system. The same general approach has been used to reconstitute *E. coli* galactose-binding protein-dependent transport and *S. typhimurium* histidine transport (45).

Regulation of the maltose transport system

The two divergent maltose operons are transcriptionally regulated by two positive regulators, the cAMP–cAMP-binding protein complex and the complex of the *malT* gene product with maltose. The expression of the *malT* gene is, in turn, regulated by cAMP (71, 108). An as yet not understood down-regulation between *malE* and *malF* is observed, whereas an up-regulation is observed between *malK* and *lamB*. Both the *ompB* region and the cell cycle appear to be involved in the regulation of expression of the maltose transport system (108). The maltose transport system, once synthesized, appears to be inhibited by interaction with glucose enzyme III of the PTS. The PTS also affects cAMP levels and thus functions at yet another level of regulation (270).

Energetics of maltose transport

The energetics of binding protein-dependent maltose transport have been examined in a maltose me-tabolism-deficient mutant and in an ATPase mutant (74). The maltose transport system is an excellent candidate for such studies since it is the primary system for the transport of maltose (108). Ferenci and co-workers considered both the entrance and exit of maltose. Maltose transport is considerably less sensitive to arsenate inhibition than other binding protein-dependent systems. Like other binding protein-dependent systems, maltose uptake is relatively insensitive to inhibition by uncouplers of oxidative phosphorylation. Curiously, maltose exit is more sensitive to uncouplers than is maltose entry, an observation suggesting that the two processes are mediated by different systems.

Ribose Transport

Ribose permeases of *S. typhimurium* and *E. coli*

Both *S. typhimurium* (5, 6) and *E. coli* K-12 (80, 83, 308) have high-affinity, ribose-specific, binding protein-dependent transport systems. *E. coli* B/r appears to have two cryptic ribose operons, one at 83 min, similar to strain K-12, and one at 2 min that appears to have arisen through a transposition event from the 83-min operon (1, 2). Both *E. coli* K-12 and *S. typhimurium* mutants devoid of the high-affinity, shock-sensitive ribose transport system lose all high-affinity ribose transport. However, several such mutants can grow on ribose (6, 83, 135, 184), in agreement with an early report that *E. coli* X289 had two ribose transport systems (65). The second ribose transport system has not yet been characterized.

Genetics of the *E. coli* binding protein-dependent ribose transport system

The entire *E. coli* ribose operon (mapped at 83 min on the *E. coli* K-12 chromosome) has been cloned and partially characterized (34a, 48a, 131a, 135, 184). The operon specifies the high-affinity, ribose-binding protein-dependent transport system and includes the genes (in order of transcription) *rbsD*, *rbsA*, *rbsC*, *rbsB*, and *rbsK* (34a, 135). The proteins specified by these genes appear to be: a 15,294 M_r membrane-associated protein of 139 amino acids (*rbsD* gene product), a 54,978 M_r membrane-associated protein of 501 amino acids (*rbsA* gene product), a very hydrophobic 33,398 M_r integral membrane protein of 321 amino acids (34a), a 29,000 M_r ribose-binding protein of 271 amino acids (92), and ribokinase (M_r 32,295) (131a, 135). Analysis of the predicted amino acid sequence specified by *rbsA* indicated an internal homology as well as homology with proteins of the histidine (*hisP* gene product), maltose (*malK* gene product), and phosphate (*pstB* gene product) transport systems (48a). The operon is followed by a gene, *rbsR*, which appears to produce a repressor protein for the ribose operon (34a, 131a, 184).

The *S. typhimurium* ribose-binding protein-dependent transport system is represented at min 82 on the linkage map of *S. typhimurium* (273).

Ribose-binding proteins

S. typhimurium. The ribose-binding protein from *S. typhimurium* has been purified to homogeneity (5, 6);

its properties are summarized in Table 1. The protein has also been crystallized and subjected to a preliminary X-ray analysis (204). The crystals are rectangular parallelopipeds with the symmetry of space group $P2_1$; unit cell dimensions of a = 6.44 nm, b = 6.06 nm, c = 6.28 nm, and β = 91.25°; and two molecules of M_r 29,000 per asymmetric unit. The amino acid sequence of the protein has been partially determined. From the sequence completed, only six differences were noted between the *S. typhimurium* and the *E. coli* ribose-binding proteins (49, 92).

E. coli. The ribose-binding protein from *E. coli* has been purified to homogeneity (77, 308). Part of its amino acid sequence has been determined, and the nucleotide sequence of the structural gene for the protein has been determined in its entirety (92). The upstream ribosome-binding site has also been identified (80). Comparison of the nucleotide sequence with the amino terminus of the mature protein suggests that the ribose-binding protein is synthesized with a 25-amino acid signal sequence that is removed during processing (92).

Comparison of the *E. coli* ribose-binding protein sequence with those of other binding proteins indicated a high degree of homology with the binding proteins for arabinose, galactose (24), and maltose (72). One region of high homology between ribose- and galactose-binding proteins was suggested as a possible site of interaction with the chemotactic receptor MCPIII (24). Secondary structural predictions based on sequence data suggested that the overall folding of the sugar-binding proteins is probably quite similar (24).

Reconstitution of ribose transport

More than 90% of the ribose transport observed in ribose-induced *E. coli* K-12 cells is sensitive to osmotic shock, which suggests, in agreement with the mutant studies mentioned above, that virtually all of the high-affinity ribose transport is catalyzed by the ribose-binding protein-dependent system (62, 80). It was possible to reconstitute binding protein-dependent ribose transport in spheroplasts derived from ribose-induced cells but not in spheroplasts from noninduced cells (84, 250), using the procedure developed by Rosenberg and co-workers (88). The reconstituted transport was energy dependent (84). Interestingly, the *E. coli* ribose-binding protein could be used to reconstitute binding protein-dependent ribose transport in spheroplasts derived from induced cells of *S. typhimurium*, despite the fact that the *S. typhimurium* protein has a significantly higher pI than that from *E. coli* (Table 1) (250).

Ribose chemotaxis

Ribose-binding protein has been shown to be the ribose receptor for the chemotactic response in both *S. typhimurium* (5, 6) and *E. coli* (83). Competition studies indicated that the ribose receptor competed with the galactose receptor (galactose-binding protein) in mediating chemotactic responses (295). Mutants deficient in both ribose and galactose chemotaxis but not transport (218) led to the identification of a methyl-accepting chemotaxis protein (MCPIII, the *trg* gene

product) that interacts with ligand complexes of either ribose- or galactose-binding protein in mediating chemotactic responses to the respective sugar (163). The binding protein, but not other protein components of the transport system, is required for methylation (163). *trg* mutants retained the two binding proteins as well as binding protein-dependent galactose and ribose transport; however, they lost the chemotactic response to both attractants (97, 103).

Regulation of ribose transport

The ribose operon appears to be negatively regulated by a repressor protein specified by a gene that is downstream from the ribokinase gene *rbsK*, the last structural gene in the ribose operon (184).

Energetics of ribose transport

Typical of shock-sensitive transport systems, ribose-binding protein-dependent ribose uptake was relatively insensitive to uncouplers of oxidative phosphorylation when uptake was driven by energy sources that could generate ATP. However, ribose transport was severely inhibited by arsenate (62). Interestingly, ribose uptake was more resistant to inhibition by FCCP (cyanide-*p*-trifluoromethoxyphenylhydrazone) than was glutamine uptake. Thus, there may be different efficiencies of ATP (or energy) coupling by the different shock-sensitive systems.

Xylose Transport

Xylose permeases

The xylose transport system(s) of *E. coli* has not been sufficiently characterized to know with certainty whether there is more than one. The available evidence suggests that there are at least two, if not three, transport systems that take up xylose. Biphasic kinetics of xylose transport have been reported. Shamanna and Sanderson observed K_m values for xylose transport of 24 and 110 μM (281). One system partially characterized by Lam et al. appears to be a xylose-specific, membrane-bound-type system with a K_m of 23.9 μM (170). The energetics of the system, together with its activity in membrane vesicles, provide convincing evidence for its not being a binding protein-dependent system. Ahlem et al. also reported two K_m values for xylose uptake in *E. coli*, 5 and 25 μM (3). Kinetic characterization of xylose uptake in shocked and unshocked cells indicated that the high-affinity system is probably associated with a binding protein-dependent xylose transport system (281).

Kinetic studies suggest the presence of only one xylose transport system with a K_m value of 410 μM in *S. typhimurium* (281).

Genetics of xylose transport

The genetics of xylose transport in *E. coli* and *S. typhimurium* have not been analyzed in as much detail as the genetics of the other sugar transport systems. In *S. typhimurium*, the genes for the enzymes of xylose metabolism (kinase and isomerase) appear to lie together as a cluster at 78 min (281, 282). The order of the genes appears to be *xylT* (transport gene[s]), *xylR* (positive regulatory gene), *xylB* (xylose isomerase), and *xylA* (xylulokinase) (282). In *E. coli*, the xylose utilization genes lie at 80 min on the *E. coli* chromosome; however, the order appears to be *xylR/T*, *xylA*, and *xylB* (192, 261). The nucleotide sequences of the xylose isomerase (175, 276) and xylulokinase (175) genes have been determined. The *E. coli* xylose transport genes have not yet been characterized.

Xylose-binding protein

A xylose-binding protein has been purified from *E. coli* (Table 1) (3, 63) which specifically binds xylose with a K_d of 0.63 µM. The analog methyl-α-D-xylofuranoside is also bound with a K_d of 33 µM, suggesting that the protein binds the isomer with the furanose configuration. The intrinsic fluorescence increased 32% in the presence of xylose. This observation was interpreted as providing evidence for the presence of a tryptophan in the area of the sugar-binding site. Oxidation of one of two tryptophan residues completely abolished xylose-binding activity. Like many other binding proteins, the xylose-binding protein has a broad pH optimum (pH 5 to 8), with 50% binding activity at pH values of 3.8 and 9.5.

Regulation of xylose transport

Xylose transport and utilization in both *S. typhimurium* (282) and *E. coli* (66, 261) appear to be positively regulated. The mechanism of regulation, however, remains to be solved, as do the number and nature of the xylose transport systems present in *E. coli*.

Energetics of xylose transport

Studies on the energetics of xylose transport in whole cells and membrane vesicles clearly demonstrate that *E. coli* has a membrane-bound type of xylose transport system with a K_m of 23.9 µM (170). The lack of inhibition of xylose transport by arsenate in whole cells suggests that the membrane-bound system is a major transport system in the strain studied (VL17). On the other hand, the biphasic kinetics observed in *E. coli* K-12, along with the susceptibility of the high-affinity (K_m = 5 µM) system to osmotic shock and the isolation from this strain of a xylose-binding protein (3, 63), suggest that *E. coli* probably has at least two mechanisms of accumulating xylose. The high K_m value (110 µM) (281) suggests that xylose may be entering through another sugar transport system as well.

ANION TRANSPORT

Citrate Transport

Citrate permeases

E. coli. Ordinarily, *E. coli* only utilizes citrate anaerobically provided that a second carbon source is available for generating reducing power (186). The lack of a citrate transport system has been proposed as the reason for the lack of citrate utilization, since *E. coli* has the enzymes necessary for citrate metabolism

(174). However, there are two routes by which *E. coli* can acquire the ability to utilize citrate aerobically as a sole source of carbon and energy. Mutations in *citA* and *citB* (linked to the *gal* operon) result in the semiconstitutive production of a chromosomally encoded citrate transport system (95). This system appears to recognize citrate, isocitrate, and *cis*- and *trans*-aconitate, with a K_m value of 36 µM for citrate (95). The second means by which *E. coli* acquires the ability to transport citrate is via a *cit*⁺ plasmid (119, 141). The plasmid encodes a 35,000 M_r membrane-associated protein (119). The K_m of the plasmid-encoded citrate transport system is 126 µM (243). Both the chromosomally specified and the plasmid-specified systems are active in membrane vesicles and appear to be typical of membrane-bound types of systems. They differ somewhat in substrate specificity (243).

S. typhimurium. In contrast to *E. coli*, *S. typhimurium* readily utilizes citrate as a carbon and energy source. *S. typhimurium* appears to have four distinct inducible citrate transport systems, TCT I, TCT II, TCT III, and TCT IV (289). TCT I is induced by citrate, isocitrate, or *cis*-aconitate (139) and appears to be a typical binding protein-dependent transport system (156, 298). The TCT II system is induced by the same inducers, is stable only at high pH values, and transports citrate or *cis*-aconitate (139, 156). The TCT III system is induced by tricarballylate and transports *cis*-aconitate, citrate, and tricarballylate. TCT IV probably transports only citrate (139).

Genetics of the binding protein-dependent *S. typhimurium* citrate transport system

The genes for TCT I lie at 59 min (156, 289) on the *S. typhimurium* chromosome (between *nalB* and *pheA*) (289). Four distinct regions have been identified, including the probable promoter region. Region 3 specifies the citrate-binding protein. Resistance to 3-nitroisocitrate specifically selects binding protein-deficient (*tctC*) mutants (156). A 21,000 M_r outer membrane protein appears to be involved in citrate uptake; however, specific mutants deficient for this protein have not yet been isolated (156). Cloning of the *tct* genes into *E. coli* has resulted in the identification of a 40,000 M_r membrane-associated protein. The gene encoding this protein has not yet been identified (156).

S. typhimurium citrate-binding protein

A citrate-binding protein from *S. typhimurium* has been purified to homogeneity and crystallized (298), and its amino acid sequence has been determined (W. Kay and M. Hermodson, personal communication); its properties are summarized in Table 1. Two components with citrate-binding activity were resolved during fractionation. It was subsequently shown that a cyclized amino-terminal glutamine residue accounted for the second form (156). It is interesting that Na⁺ ions were required for optimal substrate binding. The binding protein was specific for citrate and isocitrate. Binding could be inhibited by the analogs fluorocitrate, nitrocitrate, or nitroisocitrate (156, 298).

Regulation of citrate transport in *S. typhimurium*

Since isogenic strains each containing a specific citrate transport system have not yet been constructed, a detailed picture of the regulation of the citrate transport systems has not yet emerged. It is clear, however, that the systems are inducible (139, 156, 298) and that catabolite repression plays a role in regulating the expression of at least the TCT I system (156, 298).

Phosphate Transport

Phosphate permeases

At least four transport systems are involved in the uptake of phosphate in *E. coli* (290). Of the four systems, two are primarily involved in the transport of P_i (259) and two are involved in the transport of organo-phosphate metabolites. The two major routes of P_i uptake include a low-affinity membrane-bound system (Pit) (314) and a high-affinity binding protein-dependent system (Pst or phosphate-specific transport system) (199). The transport systems for L-α-glycerol-phosphate (GlpT) and glucose 6-phosphate (Uhp) also appear to transport P_i, but not at rates sufficient to support growth in strains missing both of the inorganic phosphate transport systems (*pst⁻ pit⁻*) (290).

Genetics of phosphate transport

The gene(s) that specifies the membrane-bound transport system (Pit) lies at min 77 on the *E. coli* chromosome (26). The three genes (*phoS, phoT [pstA]*, and *pstB*) that specify the phosphate-binding protein-dependent system (Pst) are represented at min 84 on the *E. coli* genetic map (26). The phosphate-binding protein is encoded by the *phoS* gene and is synthesized with a 25-amino acid NH_2-terminal signal peptide that is removed during maturation. The mature protein is composed of 321 amino acid residues with a calculated molecular weight of 34,000 (190, 296). Four other downstream open reading frames have also been cloned and sequenced, although the role of each in the Pst transport system has not yet been established (297). The order of genes is *phoS* (which specifies the phosphate-binding protein); *pstC* (which specifies a 40,000 M_r peripheral membrane protein [pI = 7.0]), *pstA* (which specifies a 28,000 M_r integral membrane protein); *pstB* (which specifies a 29,000 M_r peripheral membrane protein [pI = 6.5]) and *phoU* (which specifies a 26,000 M_r peripheral membrane protein [pI = 5.4]).

Phosphate-binding protein

The *E. coli* phosphate-binding protein has been purified to homogeneity (Table 1) (199). The NH_2 terminus of the mature protein has been determined (190, 296), as has the NH_2 terminus of the pre-phosphate-binding protein (190). The amino acid sequence data are in complete agreement with the nucleotide sequence.

Regulation of phosphate transport

Two of the most interesting aspects of the binding protein-dependent phosphate transport system (Pst system) are its role in regulating other genes and the regulation to which it is subject. In *E. coli* K-12, more than 20 genes (unlinked and linked) are induced by phosphate starvation. Included among these are: *phoA* (alkaline phosphatase gene), *phoS, phoT (pstA)*, and *pstB* (the genes of the Pst system), *ugpA* and *ugpB* (two genes of the binding protein-dependent L-α-glycerol-phosphate transport system), and *phoE* (the gene for outer membrane porin E) (303). At least three genes (aside from those specifying the Pst system) are involved in controlling the Pho regulon: *phoB* appears to be a positive regulator; *phoR* appears to exert both positive and negative control; and, in some cases, the positive control of *phoR* may be replaced by the *phoM* gene (283).

The mechanism of regulation of the Pho regulon by the genes of the Pst system has not yet been elucidated. It is not yet clear whether the derepression of levels of the other gene products of the Pho regulon results from a decreased level of intracellular phosphate in Pst mutants or whether the gene products of the Pst system themselves play a direct role in controlling the Pho regulon. It is interesting that mutations in the other genes of the Pst system lead to constitutive synthesis of the phosphate-binding protein (202). It is also noteworthy that the Pit system allows free exchange of accumulated phosphate, whereas the binding protein-dependent Pst system does not (259).

Recent nucleotide sequence analysis of the regulatory regions of the *phoS, phoE*, and *phoA* operons provides some interesting insights into the possible molecular mechanisms of the coordinate control of the phosphate regulon (190, 296). Common sequences were noted between *phoE* (nucleotides −99 to −83), *phoS* (nucleotides −106 to −90), and *phoA* (nucleotides −80 to −64). Each common sequence is followed by sequences homologous to the Pribnow box. Since several possibilities for transcriptional initiation of the *phoS* gene have been identified, it will be necessary to carry out S1 mapping of the promoter region to determine the actual point of initiation (190). A possible transcription terminator has been identified following the *phoS* gene (190, 296). The exploration of the transcriptional and translational regulation of the membrane-associated Pst system genes will indeed be interesting.

Reconstitution of binding protein-dependent phosphate transport

Gerdes et al. were able to reconstitute binding protein-dependent phosphate transport in spheroplasts derived from *E. coli* K10 cells (88). These cells were devoid of the membrane-bound Pit system, so that a major contribution to the signal-to-noise problem encountered in many early reconstitution attempts was avoided. Furthermore, the preparation of spheroplasts apparently allowed the binding protein easy access to the plasma membrane. Addition of purified phosphate-binding protein to the spheroplast preparations resulted in a 14-fold stimulation of the near-base-line phosphate transport observed in the spheroplast preparations (88). This reconstitution of phosphate transport through the *pst*-dependent system by Gerdes and co-workers is particularly noteworthy in that their approach turned out to be easily reproducible in other laboratories (84, 196).

Energetics of phosphate transport

The energetics of phosphate transport have been examined in *E. coli* strains bearing single phosphate transport systems. Strain AN710 carries only the Pit system, and strain AN1088 has only the Pst system (260). Phosphate transport requires both H^+ and K^+ gradients (267, 268). Studies with energy-starved cells demonstrated that the Pit system could use either glucose or succinate as an energy source but, in both cases, less effectively at higher pH values (greater than pH 6). The Pit system was very sensitive to inhibition by the uncoupler carbonyl cyanide *M*-chlorophenyl-hydrazone as expected for a membrane-bound type of system. The Pst system was able to utilize glucose effectively as an energy source at pH values up to 8; however, when succinate served as the energy source, it was less effective at higher pH values. Carbonyl cyanide *M*-chlorophenylhydrazone also inhibited energy-driven Pst transport. Succinate-driven phosphate uptake was inhibited at all studied pH values, whereas inhibition of glucose-driven phosphate uptake was especially effective below pH 6.5.

sn-Glycerol 3-Phosphate Transport

Glycerolphosphate permeases

E. coli has two systems for the transport of glycerolphosphate. One system (*glpT*) (98) is active in membrane vesicles and appears in many ways to be a typical membrane-bound type of transport system (43). However, it does have associated with it a periplasmic protein component (GLPT protein) whose function is as yet unknown. This system is induced by glycerol or glycerolphosphate in the medium (61). The second *sn*-glycerolphosphate transport system appears to be a typical binding protein-dependent transport system (22, 278). The hexose phosphate transport system has also been reported to take up glycerol 3-phosphate and some of its analogs (93).

Genetics of the glycerolphosphate transport systems

A gene at 75.3 min on the *E. coli* chromosome (*ugpB*, or a proximal gene) specifies the *sn*-glycerol 3-phosphate-binding protein (26, 279). Mutants devoid of the binding protein are defective in *uhp*-dependent glycerolphosphate transport. As for other osmotic-shock-sensitive transport systems, additional protein components appear to be required for a functional transport system (279). A linked gene (*ugpA*) probably specifies another protein component of the *ugp*-dependent system. If the *ugp*-dependent transport system is like the other binding protein-dependent systems, it would not be surprising if further analysis of the system were to uncover one or more additional genes related to *ugp*-dependent glycerolphosphate transport. The genes for the GlpT system lie at 49 min on the *E. coli* chromosome (26).

Glycerolphosphate-binding protein

A glycerolphosphate-binding protein has been purified to homogeneity from *E. coli* (Table 1) (22). The protein has a K_d for glycerolphosphate of 0.2 μM and

appears not to bind phosphate or fosfomycin, substrates of the *glpT*-dependent transport system. As with many of the other binding proteins, a change in fluorescence is observed upon substrate binding (22).

Regulation of the ugp-dependent sn-glycerol 3-phosphate transport system

The *ugp*-dependent glycerolphosphate transport system, like the phosphate transport system, is a member of the phosphate regulon. It appears to be coregulated with outer membrane protein E, alkaline phosphatase, and phosphate-binding protein (23, 278, 303, 304). Recent studies of phosphate starvation-inducible (*psi*) promoters, using the Mu d1-directed *lacZ* fusion method, identified over 18 unlinked *psi* promoters in *E. coli*. Two of the *psi* promoters, *psiB* and *psiC*, were linked to the *ugpA,B* region. Furthermore, these studies suggested that catabolite repression also probably plays a role in the regulation of *ugp*-dependent transport (303, 304).

Sulfate Transport

Sulfate permease(s)

S. typhimurium. It is thought that sulfate and thiosulfate are transported into *S. typhimurium* via a single system with a K_m value of 36 μM for sulfate uptake (69). Considering that the sulfate-binding protein was the first bacterial binding protein isolated, it is indeed curious that the structural gene for this protein has not yet been identified or mapped. Some possible reasons for this are discussed below.

E. coli. Sulfate transport in *E. coli*, as in *S. typhimurium*, has been reported to be catalyzed by a single system (150, 291). The *E. coli* system has not been as extensively studied as the *S. typhimurium* system. K_m values of 2.5 μM (150) and 50 μM (291) have been reported. The variation in reported K_m values may be due to the possibility that different systems are being measured or to the fact that the K_d of the sulfate-binding protein varies with ionic strength (226). The activity of the *E. coli* system was reported to be 10-fold lower than that of the *S. typhimurium* sulfate transport system (150). However, the activity of the *E. coli* system was determined at a substrate level 2.5 times below the K_m value, so it is possible that the activity of the *E. coli* system is perhaps only twofold lower than that of the *S. typhimurium* sulfate permease.

Genetics of sulfate transport

S. typhimurium. It is generally thought that the mutants defective in the sulfate transport system are divided into three complementation groups (201) with lesions at 49 min on the *S. typhimurium* chromosome (216). However, it is interesting that despite a considerable effort involving the generation of mutants (on the basis of chromate resistance) in all three groups (CysAa, CysAb, and CysAc), binding protein-negative mutants have not yet been found (216). One reasonable possibility for the lack of success in generating specific binding protein-negative mutants is that the *cysA* gene(s) (in addition to the *cysB* gene; see below) may be involved in the positive regulation of two separate sulfate transport systems. The differences in binding protein and transport specificities lend support to this possibility. Thiosulfate and vanadate each have different patterns of inhibition of transport versus binding (227). Since so many other nutrients, including phosphate, are transported through a membrane-bound system and a binding protein-dependent system(s), it is possible that such is the case for sulfate transport. Interestingly, growth of cells on sulfate significantly reduced the level of binding, whereas the level of sulfate transport was not significantly reduced (216). If there are indeed two sulfate transport systems, they would have nearly identical K_m values. If both systems were subject to the same positive regulation, which in addition to the $cysB$ gene product required one or more gene products from the $cysA$ region, then it would be expected that the genetic analysis of such a system would yield the results observed. It is also possible that the membrane-bound components of a single system may allow transport in the absence of binding protein. Future genetic and molecular genetic analyses of the sulfate transport system(s) should elucidate the mechanism and multiplicity (if any) of sulfate transport.

E. coli. Chromate-resistant $cysA$ mutants appear to be analogous to the *S. typhimurium* mutants and have lesions at 52 min, next to the *pstI* locus (150).

Sulfate-binding protein

The sulfate-binding protein of *S. typhimurium* has been purified to homogeneity (226), its amino acid sequence has been determined (136–138, 142), and its crystal structure has been analyzed (173) (Table 1). Interestingly, the K_d for sulfate was dependent on the ionic strength (226). As mentioned above, this could, and probably should, lead to different apparent K_m values for whole-cell transport, depending on the medium used for the transport. Chromate quenched the native tryptophan fluorescence when bound to sulfate-binding protein. Competition of the quenching with sulfate suggested that the protein had about the same affinity for these two anions. A curious property of the sulfate-binding protein is its ability to emulsify silicon oil (226).

Binding of sulfate to the protein caused changes in fluorescence, circular dichroism, and optical rotary dispersion spectra. No changes were observed by nuclear magnetic resonance when sulfate was added (173). (The protein was probably not stripped of endogenous sulfate before measurements were made.)

Regulation of sulfate transport in *S. typhimurium*

Sulfate transport is regulated at the levels of expression and function. Expression of the sulfate transport system(s) in *S. typhimurium* is regulated as part of a sulfate regulon (216). Growth of cells on djenkolate simulates conditions of sulfur starvation. Cells grown on djenkolate as a sulfur source may produce sulfate-binding protein up to 1% of total cell protein (226). Cells grown on sulfate produce very little sulfate-binding protein; however, sulfate transport is not as depressed as the level of the binding protein (216). Expression of the binding protein as well as several of the enzymes involved in the synthesis of cysteine is positively regulated by the gene product of the $cysB$ gene (216).

The transport system(s) also appears to be subject to a type of feedback regulation. The time course of sulfate uptake shows the intracellular level of sulfate rising and falling as a function of time (69, 70). Transport was inhibited in mutants that accumulated the high-energy intermediate 3'-phosphoadenosine-5'-phosphosulfate (216).

VITAMIN TRANSPORT

Vitamin B$_{12}$ Transport

Vitamin B$_{12}$ permease

Vitamin B$_{12}$ is transported into *E. coli* by an osmotic-shock-sensitive transport system that requires an outer membrane receptor (130).

Genetics of B$_{12}$ transport

The outer membrane binding protein is specified by the *btuB* gene represented at 88 min of the *E. coli* linkage map (30). As noted below, the gene specifying this protein has been cloned and sequenced (105, 106). The *tonB* gene product also is involved in the uptake of cobalamin (244). The sequence of the *tonB* gene has been determined (235), and the gene is located at 28 min on the *E. coli* chromosome (26). The *btuC* gene, which is located at 37.7 min on the *E. coli* chromosome (26, 30), is thought to encode a cytoplasmic membrane component of the cobalamin transport system (130). The evidence linking the periplasmic protein to the transport system is circumstantial (see below).

Vitamin B$_{12}$-binding proteins

Two vitamin B$_{12}$-binding proteins have been identified. A periplasmic protein (M_r 22,000) binds cobalamin with a K_d of 5 nM and has a binding specificity that closely matches that of the energy-dependent cobalamin transport system (300, 307). A cyanocobalamin-binding protein has been partially purified from *E. coli* (Table 1) (300). The properties of this protein closely resemble those of the energy-dependent cyanocobalamin transport system (300, 307). The evidence for the role of this protein in vitamin B$_{12}$ transport is the sensitivity of the transport system to osmotic shock (300), the arsenate sensitivity of uptake (244), and the specificity of the binding protein versus that of the transport system (44). There appear to be only three copies of the periplasmic protein per cell (130, 307).

An outer membrane protein (M_r 60,000) binds cobalamin with a K_d of 0.3 nM (130, 160). In addition to serving as the outer membrane B$_{12}$ receptor, the 66,000 M_r protein also functions in the uptake of ferric iron and serves as a receptor for the E colicins and the phage BF23 (68, 130). There are approximately 200 to 300 copies of the outer membrane proteins per cell (307). The gene that specifies the outer membrane binding protein was cloned (106), and its nucleotide sequence was determined (105). It encodes a 66,000 M_r

outer membrane protein. The open reading frame suggests a 614-amino acid polypeptide with a 20-residue leader sequence (105).

Thiamine Transport

Thiamine permease(s)

The multiplicity of thiamine transport systems is not clear from the published data. *E. coli* K-12 cells transport thiamine with a K_m of 8.3×10^{-7} M (153, 154). Reduction of thiamine transport by osmotic shock has been reported (212).

Thiamine-binding protein

A thiamine-binding protein has been partially purified from osmotic shock fluids of *E. coli* strains K-12 (144) ($K_d = 1 \times 10^{-7}$ M) and W (152) ($K_d = 2.9 \times 10^{-8}$ M).

Genetics of thiamine transport

One thiamine transport-deficient mutant has been reported (152). This mutant will, however, grow on elevated levels of thiamine. Thus, it is not clear whether the remaining transport is through part of the binding protein-dependent system, or through yet another thiamine transport system.

Reconstitution of thiamine transport

A partial reconstitution of thiamine transport in osmotically shocked cells by added binding protein has been reported (212).

Regulation of thiamine transport

The thiamine transport system appears to be regulated by the intracellular level of level of thiamine pyrophosphate (151, 153). Growth of cells in the presence of adenine, which inhibits thiamine biosynthesis, results in a derepression of the transport system (151).

CONCLUDING COMMENTS

In looking through Table 1, it is clear that there are a number of interesting "loose ends" that would make excellent dissertation or postdoctoral projects. In addition to the loose ends, there are certainly new binding protein-dependent transport systems that have yet to be characterized. It is hoped that new investigators in this field will find the approaches used by the investigators whose work is reviewed herein useful in designing their own experiments in the field of binding protein-dependent nutrient transport.

LITERATURE CITED

1. **Abou-Sabé, M., J. Pilla, D. Hazuda, and A. Ninfa.** 1982. Evolution of the D-ribose operon of *Escherichia coli* B/r. J. Bacteriol. **150**:762–769.
2. **Abou-Sabé, M., and J. Richman.** 1973. On the regulation of D-ribose metabolism in *E. coli* B/r. II. Chromosomal location and fine structural analysis of the D-ribose permease by P1 transduction. Mol. Gen. Genet. **122**:303–312.
3. **Ahlem, C., W. Huisman, G. Nesland, and A. S. Dahms.** 1982. Purification and properties of a periplasmic D-xylose-binding protein from *Escherichia coli*. J. Biol. Chem. **257**:2926–2931.
4. **Aksamit, R. R., B. J. Howlett, and D. E. Koshland, Jr.** 1975. Soluble and membrane-bound aspartate-binding activities in *Salmonella typhimurium*. J. Bacteriol. **123**:1000–1005.
5. **Aksamit, R. R., and D. E. Koshland.** 1972. A ribose binding protein of *Salmonella typhimurium*. Biochem. Biophys. Res. Commun. **48**:1348–1353.
6. **Aksamit, R. R., and D. E. Koshland.** 1974. Identification of the ribose binding protein as the receptor for ribose chemotaxis in *Salmonella typhimurium*. Biochemistry **13**:4473–4478.
7. **Amanuma, H., J. Itoh, and Y. Anraku.** 1976. Transport of sugars and amino acids in bacteria. XVII. On the existence and nature of substrate amino acids bound to purified branched chain amino acid-binding proteins of *Escherichia coli*. J. Biochem. **79**:1167–1182.
8. **Ames, G. F.-L.** 1972. Components of histidine transport. ICN-UCLA Symp. Mol. Cell. Biol. **1**:409–426.
9. **Ames, G. F.-L.** 1984. Histidine transport system of *Salmonella typhimurium*, p. 13–16. *In* L. Leive and D. Schlessinger (ed.), Microbiology—1984. American Society for Microbiology, Washington, D.C.
10. **Ames, G. F.-L., and J. E. Lever.** 1972. The histidine-binding protein J is a component of histidine transport. J. Biol. Chem. **247**:4309–4316.
11. **Ames, G. F.-L., and K. Nikaido.** 1985. Nitrogen regulation in *Salmonella typhimurium*: identification of an *ntrC* protein-binding site and definition of a consensus binding sequence. EMBO J. **4**:539–547.
12. **Ames, G. F.-L., K. D. Noel, H. Taber, E. N. Spudich, K. Nikaido, J. Afong, and F. Ardeshir.** 1977. Fine-structure map of the histidine transport genes in *Salmonella typhimurium*. J. Bacteriol. **129**:1289–1297.
13. **Ames, G. F.-L., C. Prody, and S. Kustu.** 1984. Simple, rapid, and quantitative release of periplasmic proteins by chloroform. J. Bacteriol. **160**:1181–1183.
14. **Ames, G. F.-L., and E. N. Spudich.** 1976. Protein-protein interaction in transport: periplasmic histidine-binding protein J interacts with P protein. Proc. Natl. Acad. Sci. USA **73**:1877–1881.
15. **Anderson, J. J., and D. L. Oxender.** 1978. Genetic separation of high- and low-affinity transport systems for branched-chain amino acids in *Escherichia coli*. J. Bacteriol. **136**:168–174.
16. **Anderson, J. J., S. C. Quay, and D. L. Oxender.** 1976. Mapping of two loci affecting the regulation of branched-chain amino acid transport in *Escherichia coli* K-12. J. Bacteriol. **126**:80–90.
17. **Andrews, J. C., and S. A. Short.** 1985. Genetic analysis of *Escherichia coli* oligopeptide transport mutants. J. Bacteriol. **161**:484–492.
18. **Anraku, Y.** 1967. The reduction and restoration of galactose transport in osmotically shocked cells of *Escherichia coli*. J. Biol. Chem. **242**:793–800.
19. **Anraku, Y.** 1968. Transport of sugars and amino acids in bacteria. I. Purification and specificity of the galactose- and leucine-binding proteins. J. Biol. Chem. **243**:3116–3122.
20. **Anraku, Y.** 1968. Transport of sugars and amino acids in bacteria. III. Studies on the restoration of active transport. J. Biol. Chem. **243**:3128–3135.
21. **Ardeshir, F. A., and G. F. Ames.** 1980. Cloning of the histidine transport genes from *Salmonella typhimurium* and characterization of an analogous transport system in *Escherichia coli*. J. Supramol. Struct. **13**:117–130.
22. **Argast, M., and W. Boos.** 1979. Purification and properties of the *sn*-glycerol 3-phosphate-binding protein of *Escherichia coli*. J. Biol. Chem. **254**:10931–10935.
23. **Argast, M., and W. Boos.** 1980. Co-regulation in *Escherichia coli* of a novel transport system for *sn*-glycerol-3-phosphate and outer membrane protein Ic (e, E) with alkaline phosphatase and phosphate-binding protein. J. Bacteriol. **143**:142–150.
24. **Argos, P., W. C. Mahoney, M. A. Hermodson, and M. Hanei.** 1981. Structural prediction of sugar-binding proteins functional in chemotaxis and transport. J. Biol. Chem. **256**:4357–4362.
25. **Ayling, P. D.** 1981. Methionine sulfoxide is transported by high-affinity methionine and glutamine transport systems in *Salmonella typhimurium*. J. Bacteriol. **148**:514–520.
26. **Bachmann, B.** 1984. Linkage map of *Escherichia coli*, p. 145–161. *In* S. J. O'Brien (ed.), Genetic maps—1984, vol. 3. Cold Spring Harbor Laboratory, Cold Spring Harbor, N.Y.
27. **Bachmann, B. J., K. B. Low, and A. L. Taylor.** 1976. Recalibrated linkage map of *Escherichia coli* K-12. Bacteriol. Rev. **40**:116–167.
28. **Barash, H., and Y. Halpern.** 1971. Glutamate-binding protein and its relation to glutamate transport in *Escherichia coli*. Biochem. Biophys. Res. Commun. **45**:681–688.

29. **Barash, H., and Y. S. Halpern.** 1975. Purification and properties of glutamate binding protein from the periplasmic space of *Escherichia coli* K-12. Biochim. Biophys. Acta **386**:168–180.

30. **Bassford, P. J., Jr., and R. J. Kadner.** 1977. Genetic analysis of components involved in vitamin B_{12} uptake in *Escherichia coli*. J. Bacteriol. **132**:796–805.

31. **Bavoil, P., M. Hofnung, and H. Nikaido.** 1980. Identification of a cytoplasmic membrane-associated component of the maltose transport system of *Escherichia coli*. J. Biol. Chem. **255**:8366–8369.

32. **Bavoil, P., and H. Nikaido.** 1981. Physical interaction between the phage lambda receptor protein and the carrier immobilized maltose-binding protein of *Escherichia coli*. J. Biol. Chem. **256**:11385–11388.

33. **Bavoil, P., C. Wandersman, M. Schwartz, and H. Nikaido.** 1983. A mutant form of maltose-binding protein of *Escherichia coli* deficient in its interaction with the bacteriophage lambda receptor protein. J. Bacteriol. **155**:919–921.

34. **Bedouelle, P., J. Bassford, Jr., A. V. Fowler, I. Zabin, J. Beckwith, and M. Hofnung.** 1980. Mutations which alter the function of the signal sequence of the maltose binding protein of *Escherichia coli*. Nature (London) **285**:78–81.

34a. **Bell, A. W., S. D. Buckel, J. M. Groarke, J. N. Hope, D. H. Kingsley, and M. A. Hermodson.** 1986. The nucleotide sequences of the *rbsD*, *rbsA*, and *rbsC* genes of *Escherichia coli* K12. J. Biol. Chem. **261**:7652–7658.

35. **Berger, E. A.** 1973. Different mechanisms for energy coupling for the active transport of proline and glutamine in *Escherichia coli*. Proc. Natl. Acad. Sci. USA **70**:1514–1518.

36. **Berger, E. A., and L. A. Heppel.** 1972. A binding protein involved in the transport of cystine and diaminopimelic acid in *Escherichia coli*. J. Biol. Chem. **247**:7684–7755.

37. **Berger, E. A., and L. A. Heppel.** 1974. Different mechanisms of energy coupling for the shock-sensitive and shock-resistant amino acid permeases of *Escherichia coli*. J. Biol. Chem. **249**:7747–7755.

38. **Berman, K., and P. D. Boyer.** 1972. Characteristics of the reversible heat, solvent, and detergent denaturation of leucine binding protein. Biochemistry **25**:4650–4657.

39. **Betteridge, P. R., and P. D. Ayling.** 1976. The regulation of glutamine transport and glutamine synthetase in *Salmonella typhimurium*. J. Gen. Microbiol. **95**:324–334.

40. **Boos, W.** 1972. Structurally defective galactose-binding protein isolated from a mutant negative in the β-methylgalactoside permease system of *Escherichia coli*. J. Biol. Chem. **247**:5414–5424.

41. **Boos, W.** 1974. Pro and contra transport carriers; the role of the galactose-binding protein in the β-methylgalactoside transport system of *Escherichia coli*. Curr. Top. Membr. Transp. **5**:51–136.

42. **Boos, W., A. S. Gordon, R. E. Hall, and H. D. Price.** 1972. Transport properties of the galactose-binding protein of *Escherichia coli*. Substrate-induced conformational changes. J. Biol. Chem. **247**:917–924.

43. **Boos, W., I. Hartig-Beecken, and K. Altendorf.** 1977. Purification and properties of a periplasmic protein related to *sn*-glycerol-3-phosphate transport in *Escherichia coli*. Eur. J. Biochem. **72**:571–581.

44. **Bradbeer, C., J. S. Kenley, D. R. Di Masi, and J. Leighton.** 1978. Transport of vitamin B12 in *Escherichia coli*. Corrinoid specificities of the periplasmic B12-binding protein and of energy dependent B12 transport. J. Biol. Chem. **253**:1347–1352.

45. **Brass, J. M., U. Ehmann, and B. Bukau.** 1983. Reconstitution of maltose transport in *Escherichia coli*: conditions affecting import of maltose-binding protein into the periplasm of calcium-treated cells. J. Bacteriol. **155**:97–106.

46. **Brass, J. M., and M. D. Manson.** 1984. Reconstitution of maltose chemotaxis in *Escherichia coli* by addition of maltose-binding protein to calcium-treated cells of maltose regulon mutants. J. Bacteriol. **157**:881–890.

47. **Britten, R. J., and F. T. McClure.** 1962. The amino acid pool in *Escherichia coli*. Bacteriol. Rev. **26**:292–335.

48. **Brown, C. E., and R. W. Hogg.** 1972. A second transport system for L-arabinose in *Escherichia coli* B/r. J. Bacteriol. **111**:606–613.

48a. **Buckel, S. D., A. W. Bell, J. K. M. Rao, and M. A. Hermodson.** 1986. An analysis of the structure of the product of the *rbsA* gene of *Escherichia coli* K12. J. Biol. Chem. **261**:7659–7662.

49. **Buckenmeyer, G. K., and M. A. Hermodson.** 1983. Appendix: the amino acid sequence of D-ribose-binding protein from *Salmonella typhimurium*. J. Biol. Chem. **258**:12957.

50. **Celis, R. T. F.** 1977. Properties of an *Escherichia coli* K-12 mutant defective in the transport of arginine and ornithine. J. Bacteriol. **130**:1234–1243.

51. **Celis, R. T. F.** 1977. Independent regulation of transport and biosynthesis of arginine in *Escherichia coli* K-12. J. Bacteriol. **130**:1244–1252.

52. **Celis, R. T. F.** 1981. Chain-terminating mutants affecting a periplasmic binding protein involved in the active transport of arginine and ornithine in *Escherichia coli*. J. Biol. Chem. **256**:773–779.

53. **Celis, R. T. F.** 1981. Mapping of two loci affecting the synthesis and structure of a periplasmic protein involved in arginine and ornithine transport in *Escherichia coli* K-12. J. Bacteriol. **151**:1314–1319.

54. **Celis, R. T. F.** 1984. Phosphorylation *in vivo* and *in vitro* of the arginine-ornithine periplasmic transport protein of *Escherichia coli*. Eur. J. Biochem. **145**:403–411.

55. **Celis, R. T. F., H. J. Rosenfeld, and W. K. Maas.** 1973. Mutant of *Escherichia coli* K-12 defective in the transport of basic amino acids. J. Bacteriol. **116**:619–626.

56. **Clark, A., and R. W. Hogg.** 1981. High-affinity arabinose transport mutants of *Escherichia coli*: isolation and gene location. J. Bacteriol. **147**:920–924.

57. **Clark, A. F., T. A. Gerken, and R. W. Hogg.** 1982. Proton nuclear magnetic resonance spectroscopy and ligand binding dynamics of the *Escherichia coli* L-arabinose binding. Biochemistry **21**:2227–2233.

58. **Clement, J. M., and M. Hofnung.** 1981. Gene sequence of the lambda receptor, an outer membrane protein of *E. coli* K12. Cell **27**:507–514.

59. **Cohen, G. N., and J. Monod.** 1957. Bacterial permeases. Bacteriol. Rev. **21**:169–194.

60. **Cowell, J. L.** 1974. Energetics of glycylglycine transport in *Escherichia coli*. J. Bacteriol. **120**:139–146.

61. **Cozzarelli, N. R., W. B. Freedberg, and E. C. C. Lin.** 1968. Genetic control of the L-α-glycerolphosphate system in *E. coli*. J. Mol. Biol. **31**:371–387.

62. **Curtis, S. J.** 1974. Mechanism of energy coupling for transport of D-ribose in *Escherichia coli*. J. Bacteriol. **120**:295–303.

63. **Dahms, S. A., W. Huisman, G. Neslund, and C. Ahlem.** 1982. D-xylose-binding protein from *Escherichia coli*. Methods Enzymol. **90**:473–476.

64. **Daruwalla, K. R., A. T. Paxton, and J. F. Henderson.** 1981. Energization of the transport systems for arabinose and comparison with galactose transport in *Escherichia coli*. Biochem. J. **200**:611–627.

65. **David, J., and H. Wiesmeyer.** 1970. Regulation of ribose metabolism in *Escherichia coli*. I. The ribose catabolic pathway. Biochim. Biophys. Acta **208**:45–55.

66. **David, J. D., and H. Wiesmeyer.** 1970. Control of xylose metabolism in *Escherichia coli*. Biochim. Biophys. Acta **201**:497–499.

67. **Dills, S. S., A. Apperson, M. R. Schmidt, and M. H. Saier, Jr.** 1980. Carbohydrate transport in bacteria. Microbiol. Rev. **44**:385–418.

68. **Di Masi, D. R., J. C. White, C. A. Schnaitman, and C. Bradbeer.** 1973. Transport of vitamin B_{12} in *Escherichia coli*: common receptor sites for vitamin B_{12} and the E colicins on the outer membrane of the cell envelope. J. Bacteriol. **115**:506–513.

69. **Dreyfuss, J.** 1965. Characterization of a sulfate-thiosulfate-transport system in *Salmonella typhimurium*. J. Biol. Chem. **239**:2292–2297.

70. **Dreyfuss, J., and A. B. Pardee.** 1966. Regulation of sulfate transport in *Salmonella typhimurium*. J. Bacteriol. **91**:2275–2280.

71. **Duplay, P., H. Bedouelle, A. Charbit, J. M. Clement, D. E. Gilson, W. Saurin, and M. Hofnung.** 1984. The *malB* region in *Escherichia coli* K-12: gene structure and expression, p. 29–33. *In* L. Leive and D. Schlessinger (ed.), Microbiology—1984. American Society for Microbiology, Washington, D.C.

72. **Duplay, P., H. Bedouelle, A. Fowler, I. Zabin, W. Saurin, and M. Hofnung.** 1984. Sequences of the *malE* gene and of its product, the maltose-binding protein of *Escherichia coli* K12. J. Biol. Chem. **259**:10606–10613.

73. **Englesberg, E., and G. Wilcox.** 1974. Regulation: positive control. Annu. Rev. Genet. **8**:219–241.

74. **Ferenci, T., W. Boos, M. Schwartz, and S. Szmelcman.** 1977. Energy coupling of the transport system of *Escherichia coli* dependent on maltose-binding protein. Eur. J. Biochem. **75**:187–193.

75. **Froshauer, S., and J. Beckwith.** 1984. The nucleotide sequence of the gene for *malF* protein, an inner membrane component of the maltose transport system of *Escherichia coli*: repeated DNA sequences are found in the *malE-malF* intercistronic region. J. Biol. Chem. **259**:10896–10903.

76. **Fukuda, H., J. M. Sturtevant, and F. A. Quiocho.** 1983. Thermo-

dynamics of the binding of L-arabinose and of D-galactose to the L-arabinose-binding protein of *Escherichia coli*. J. Biol. Chem. **258**:13193–13198.

77. **Furlong, C. E.** 1982. Ribose binding protein from *Escherichia coli*. Methods Enzymol. **90**:467–472.

78. **Furlong, C. E., C. Cirakoglu, R. C. Willis, and P. A. Santy.** 1973. A simple preparative polyacrylamide disc gel electrophoresis apparatus: purification of three branched-chain amino acid binding proteins from *Escherichia coli*. Anal. Biochem. **51**:297–311.

79. **Furlong, C. E., and L. A. Heppel.** 1971. Leucine binding proteins from *Escherichia coli*. Methods Enzymol. **17B**:639–643.

80. **Furlong, C. E., and A. Iida.** 1984. Ribose transport and reconstitution in *Escherichia coli*, p. 61–65. *In* L. Leive and D. Schlessinger (ed.), Microbiology—1984. American Society for Microbiology, Washington, D.C.

81. **Furlong, C. E., and G. D. Schellenberg.** 1980. Characterization of membrane proteins, p. 89–123. *In* J. W. Payne (ed.), Microorganisms and nitrogen sources. John Wiley & Sons, New York.

82. **Furlong, C. E., and J. H. Weiner.** 1970. Purification of a leucine-specific binding protein from *Escherichia coli*. Biochem. Biophys. Res. Commun. **38**:1076–1083.

83. **Galloway, D. R., and C. E. Furlong.** 1977. The role of the ribose binding protein in transport and chemotaxis in *Escherichia coli* K-12. Arch. Biochem. Biophys. **184**:496–504.

84. **Galloway, D. R., and C. E. Furlong.** 1979. Reconstitution of ribose binding protein-dependent ribose transport in spheroplasts of *Escherichia coli*. Arch. Biochem. Biophys. **197**:158–162.

85. **Ganesan, A. K., and B. Rotman.** 1965. Transport systems for galactose and galactosides in *Escherichia coli*. I. Genetic determination and regulation of the methyl-β-galactoside permease. J. Mol. Biol. **16**:42–50.

86. **Garavito, R. M., J. A. Jenkins, J. M. Neuhaus, A. P. Pugsley, and J. P. Rosenbusch.** 1982. Structural investigations of outer membrane proteins from *Escherichia coli*. Ann. Microbiol. (Paris) **133A**:37–41.

87. **Gaudin, C., B. Marty, M. Ragot, J. C. Sari, and J. P. Belaich.** 1980. Thermodynamic studies of binding proteins: effects of temperature variations on substrate binding and conformation of the leucine-isoleucine-valine binding protein of *Escherichia coli*. Biochemie **62**:741–746.

88. **Gerdes, R. G., K. P. Strickland, and H. Rosenberg.** 1977. Restoration of phosphate transport by the phosphate-binding protein in spheroplasts of *Escherichia coli*. J. Bacteriol. **131**:512–518.

89. **Gibson, M. M., M. Price, and C. F. Higgins.** 1984. Genetic characterization and molecular cloning of the tripeptide permease (*tpp*) genes of *Salmonella typhimurium*. J. Bacteriol. **160**:122–130.

90. **Gilson, E., C. F. Higgins, M. Hofnung, G. F.-L. Ames, and H. Nikaido.** 1982. Extensive homology between membrane-associated components of histidine and maltose transport systems of *Salmonella typhimurium* and *Escherichia coli*. J. Biol. Chem. **257**:9915–9918.

91. **Gilson, E., H. Nikaido, and M. Hofnung.** 1982. Sequence of the *malK* gene in *E. coli*. Nucleic Acids Res. **10**:7449–7458.

92. **Groarke, J. M., W. C. Mahoney, J. N. Hope, C. E. Furlong, F. T. Robb, H. Zalkin, and M. A. Hermodson.** 1983. The amino acid sequence of D-ribose binding protein from *Escherichia coli* K12. J. Biol. Chem. **258**:12952–12956.

93. **Guth, A., R. Engel, and B. E. Tropp.** 1980. Uptake of glycerol 3-phosphate and some of its analogs by the hexose phosphate transport system of *Escherichia coli*. J. Bacteriol. **143**:538–539.

94. **Guyer, C. A., D. G. Morgan, N. Osheroff, and J. V. Staros.** 1985. Purification and characterization of a periplasmic oligopeptide binding protein from *Escherichia coli*. J. Biol. Chem. **260**:10812–10818.

95. **Hall, B. G.** 1982. Chromosomal mutation for citrate utilization by *Escherichia coli* K-12. J. Bacteriol. **151**:269–273.

96. **Harayama, S., J. Bollinger, T. Iino, and G. L. Hazelbauer.** 1983. Characterization of the *mgl* operon of *Escherichia coli* by transposon mutagenesis and molecular cloning. J. Bacteriol. **153**:408–415.

97. **Harayama, S., E. T. Palve, and G. L. Hazelbauer.** 1979. Transposon-insertion mutants of *Escherichia coli* K12 defective in a component common to galactose and ribose chemotaxis. Mol. Gen. Genet. **171**:193–203.

98. **Hayashi, S.-I., J. P. Koch, and E. C. C. Lin.** 1964. Active transport of L-α-glycerolphosphate in *Escherichia coli*. J. Biol. Chem. **239**:3098–3105.

99. **Hayes, J. B.** 1978. Group translocation transport systems, p. 43–102. *In* B. P. Rosen (ed.), Bacterial transport. Marcel Dekker, New York.

100. **Hazelbauer, G., and J. Adler.** 1971. Role of galactose-binding protein in chemotaxis of *Escherichia coli*. Nature (London) New Biol. **230**:101–104.

101. **Hazelbauer, G. L.** 1975. Maltose chemoreceptor of *Escherichia coli*. J. Bacteriol. **122**:206–214.

102. **Hazelbauer, G. L.** 1975. Role of the receptor for bacteriophage lambda in the functioning of the maltose chemoreceptor of *Escherichia coli*. J. Bacteriol. **124**:119–126.

103. **Hazelbauer, G. L., and S. Harayama.** 1979. Mutants in transmission of chemotactic signals from two independent receptors of *E. coli*. Cell **16**:617–625.

104. **Hazelbauer, G. L., and S. Harayama.** 1983. Sensory transduction in bacterial chemotaxis. Int. Rev. Cytol. **81**:33–70.

105. **Heller, K., and R. J. Kadner.** 1985. Nucleotide sequence of the gene for the vitamin B$_{12}$ receptor protein in the outer membrane of *Escherichia coli*. J. Bacteriol. **161**:904–908.

106. **Heller, K., B. J. Mann, and R. J. Kadner.** 1985. Cloning and expression of the gene for the vitamin B$_{12}$ receptor protein in the outer membrane of *Escherichia coli*. J. Bacteriol. **161**:896–903.

107. **Henderson, J. P. F., R. A. Giddens, and M. C. Jones-Mortimer.** 1977. Transport of galactose, glucose and their analogues by *Escherichia coli*. Biochem. J. **162**:309–320.

108. **Hengge, R., and W. Boos.** 1983. Maltose and lactose transport in *Escherichia coli*. Examples of two different types of concentrative transport systems. Biochim. Biophys. Acta **737**:443–478.

109. **Heppel, L. A., B. P. Rosen, I. Friedberg, E. A. Berger, and J. H. Weiner.** 1972. Studies on binding proteins, periplasmic enzymes and active transport in *Escherichia coli*, p. 133–156. *In* J. F. Woessner, Jr., and F. Huijing (ed.), The molecular basis of biological transport. Academic Press, Inc., New York.

110. **Higgins, C. F.** 1984. Peptide transport systems of *Salmonella typhimurium* and *Escherichia coli*, p. 17–20. *In* L. Leive and D. Schlessinger (ed.), Microbiology—1984. American Society for Microbiology, Washington, D.C.

111. **Higgins, C. F.** 1984. Repetitive extragenic palindromic sequences: a major component of the bacterial genome. Cell **37**:1015–1026.

112. **Higgins, C. F., and G. F.-L. Ames.** 1981. Two periplasmic transport proteins which interact with a common membrane receptor show extensive homology: complete nucleotide sequences. Proc. Natl. Acad. Sci. USA **78**:6038–6042.

113. **Higgins, C. F., and G. F.-L. Ames.** 1982. Regulatory regions of two transport operons under nitrogen control: nucleotide sequences. Proc. Natl. Acad. Sci. USA **79**:1083–1087.

114. **Higgins, C. F., G. F.-L. Ames, W. M. Barnes, J. M. Clement, and M. Hofnung.** 1982. A novel intercistronic regulatory element of prokaryotic operons. Nature (London) **298**:760–762.

115. **Higgins, C. F., P. D. Haag, K. Nikaido, F. Ardeshir, G. Garcia, and G. F.-L. Ames.** 1982. Complete nucleotide sequence and identification of membrane components of the histidine transport operon of *S. typhimurium*. Nature (London) **298**:723–727.

116. **Higgins, C. F., and M. M. Hardie.** 1983. Periplasmic protein associated with the oligopeptide permeases of *Salmonella typhimurium* and *Escherichia coli*. J. Bacteriol. **155**:1434–1438.

117. **Higgins, C. F., M. M. Hardie, D. Jamieson, and L. M. Powell.** 1983. Genetic map of the *opp* (oligopeptide permease) locus of *Salmonella typhimurium*. J. Bacteriol. **153**:830–836.

118. **Higgins, C. F., I. D. Hiles, K. Whalley, and D. J. Jamieson.** 1985. Nucleotide binding by membrane components of bacterial periplasmic binding protein-dependent transport systems. EMBO J. **4**:1033–1040.

119. **Hirato, T., M. Shinagawa, N. Ishiguro, and G. Sato.** 1984. Polypeptide involved in the *Escherichia coli* plasmid-mediated citrate transport system. J. Bacteriol. **160**:421–426.

120. **Ho, C., Y.-H. Giza, S. Takahashi, K. E. Ugen, P. F. Cottam, and S. R. Dowd.** 1980. A proton nuclear magnetic resonance investigation of histidine-binding protein J of *Salmonella typhimurium*: a model for transport of L-histidine across cytoplasmic membrane. J. Supramol. Struct. **13**:131–145.

121. **Hobson, A. C., R. Weatherwax, and G. F.-L. Ames.** 1984. ATP-binding sites in the membrane components of histidine permease, a periplasmic transport system. Proc. Natl. Acad. Sci. USA **81**:7333–7337.

122. **Hofnung, M., D. Hatfield, and M. Schwartz.** 1974. *malB* region in *Escherichia coli* K-12: characterization of new mutations. J. Bacteriol. **117**:40–47.

123. **Hogarth, B. G., and C. F. Higgins.** 1983. Genetic organization of the oligopeptide permease (*opp*) locus of *Salmonella typhimurium* and *Escherichia coli*. J. Bacteriol. **153**:1548–1551.

124. **Hogg, R. W.** 1977. L-arabinose transport and the L-arabinose binding protein of *Escherichia coli*. J. Supramol. Struct. **6**:411–417.

125. **Hogg, R. W.** 1981. The amino acid sequence of the histidine

binding protein of *Salmonella typhimurium*. J. Biol. Chem. **256:**1935–1939.

126. **Hogg, R. W.** 1982. L-arabinose- and D-galactose-binding proteins from *Escherichia coli*. Methods Enzymol. **90:**463–467.

127. **Hogg, R. W.** 1984. High-affinity L-arabinose transport: the *araF,G* operon in *Escherichia coli*, p. 38–41. *In* L. Leive and D. Schlessinger (ed.), Microbiology—1984. American Society for Microbiology, Washington, D.C.

128. **Hogg, R. W., and E. Englesberg.** 1969. L-Arabinose binding protein from *Escherichia coli* B/r. J. Bacteriol. **100:**423–432.

129. **Hogg, R. W., and M. A. Hermodson,** 1977. The amino acid sequence of the L-arabinose binding protein from *Escherichia coli* B/R. J. Biol. Chem. **252:**4135–4141.

130. **Holroyd, C. D., and C. Bradbeer.** 1984. Cobalamin transport in *Escherichia coli*, p. 21–23. *In* L. Leive and D. Schlessinger (ed.), Microbiology—1984. American Society for Microbiology, Washington, D.C.

131. **Hong, J.-S., A. G. Hunt, P. S. Masters, and M. A. Lieberman.** 1979. Requirement of acetyl phosphate for the binding protein-dependent transport systems in *Escherichia coli*. Proc. Natl. Acad. Sci. USA **76:**1213–1217.

131a. **Hope, J. N., A. W. Bell, M. A. Hermodson, and J. M. Groarke.** 1986. Ribokinase from *Escherichia coli* K12. J. Biol. Chem. **261:**7663–7668.

132. **Hunt, A. G., and J.-S. Hong.** 1981. The reconstitution of binding protein-dependent active transport of glutamine in isolated membrane vesicles from *Escherichia coli*. J. Biol. Chem. **256:**11988–11991.

133. **Hunt, A. G., and J.-S. Hong.** 1983. Properties and characterization of binding protein dependent active transport of glutamine in isolated membrane vesicles of *Escherichia coli*. Biochemistry **22:**844–850.

134. **Hunt, A. G., and J.-S. Hong.** 1983. Involvement of histidine and tryptophan residues of glutamine binding protein in the interaction with membrane-bound components of the glutamine transport system of *Escherichia coli*. Biochemistry **22:**851–854.

135. **Iida, A., S. Harayama, T. Iino, and G. L. Hazelbauer.** 1984. Molecular cloning and characterization of genes required for ribose transport and utilization in *Escherichia coli* K-12. J. Bacteriol. **158:**674–682.

136. **Imagawa, T.** 1972. Studies on the primary structure of the sulfate binding protein from *Salmonella typhimurium*. II. Thermolysin digestion. J. Biochem. (Tokyo) **72:**911–925.

137. **Imagawa, T., S. Suzuki, and A. Tsugita.** 1972. Studies on the primary structure of the sulfate binding protein from *Salmonella typhimurium*. III. Digestions with pepsin and dilute hydrochloric acid. J. Biochem. **72:**927–949.

138. **Imagawa, T., and A. Tsugita.** 1972. Studies on the primary structure of sulfate binding protein from *Salmonella typhimurium*. I. Tryptic digestion. J. Biochem. (Tokyo) **72:**889–910.

139. **Imai, K.** 1975. Isolation of tricarboxylate transport-negative mutants of *Salmonella typhimurium*. J. Gen. Appl. Microbiol. **21:**127–134.

140. **Inouye, M., and S. Halegoua.** 1980. Secretion and membrane localization of proteins in *Escherichia coli*. Crit. Rev. Biochem. **7:**339–371.

141. **Ishiguro, N., C. Oka, Y. Hanazawa, and G. Sato.** 1979. Plasmids in *Escherichia coli* controlling citrate-utilizing ability. Appl. Environ. Microbiol. **38:**956–964.

142. **Isihara, H., and R. W. Hogg.** 1980. Amino acid sequence of the sulfate-binding protein from *Salmonella typhimurium*. J. Biol. Chem. **255:**4614–4618.

143. **Ito, K.** 1982. Purification of the precursor form of maltose-binding protein, a periplasmic protein of *Escherichia coli*. J. Biol. Chem. **257:**9895–9897.

144. **Iwashima, A., A. Matsuura, and Y. Nose.** 1971. Thiamine-binding protein of *Escherichia coli*. J. Bacteriol. **108:**1419–1421.

145. **Jamison, D., and C. F. Higgins.** 1984. Anaerobic and leucine-dependent expression of a peptide transport gene in *Salmonella typhimurium*. J. Bacteriol. **160:**131–136.

146. **Jonas, A., and J. A. Schneider.** 1982. Plasma cysteamine concentrations in children treated for cystinosis. J. Pediatr. **100:**321–323.

147. **Kaback, H. R.** 1974. Transport studies in bacterial membrane vesicles. Science **186:**882–892.

148. **Kaback, H. R.** 1986. Active transport in *Escherichia coli*, p. 387–407. *In* T. E. Andreoli, J. F. Hoffman, D. D. Fanestil, and S. G. Schultz (ed.), Physiology of membrane disorders. Plenum Publishing Corp., New York.

149. **Kahane, S., M. Marcus, H. Barash, Y. S. Halpern, and H. R. Kaback.** 1975. Sodium-dependent glutamate transport in membrane vesicles of *Escherichia coli* K12. FEBS Lett. **56:**235–239.

150. **Karbonowska, H., A. Waiter, and D. Hulanicka.** 1977. Sulfate permease of *Escherichia coli* K12. Acta Biochim. Pol. **24:**329–334.

151. **Kawasaki, T., and K. Esaki.** 1971. Thiamine uptake in *Escherichia coli*. III. Regulation of thiamine uptake in *Escherichia coli*. Arch. Biochem. Biophys. **142:**163–169.

152. **Kawasaki, T., A. Iwashima, and Y. Nose.** 1969. Regulation of thiamine metabolism in *Escherichia coli*. J. Biochem. (Tokyo) **65:**407–416.

153. **Kawasaki, T., I. Miyata, K. Esaki, and Y. Nose.** 1969. Thiamine uptake in *Escherichia coli*. I. General properties of thiamine uptake system in *Escherichia coli*. Arch. Biochem. Biophys. **131:**223–230.

154. **Kawasaki, T., I. Miyata, and Y. Nose.** 1969. Thiamine uptake in *Escherichia coli*. II. The isolation and properties of a mutant of *Escherichia coli* defective in thiamine uptake. Arch Biochem. Biophys. **131:**231–237.

155. **Kay, W. W.** 1971. Two aspartate transport systems in *Escherichia coli*. J. Biol. Chem. **246:**7373–7382.

156. **Kay, W. W., J. M. Somers, G. D. Sweet, and K. A. Widenhorn.** 1984. Tricarboxylate transport systems: the *tct* operon in *Salmonella typhimurium*, p. 34–37. *In* L. Leive and D. Schlessinger (ed.), Microbiology—1984. American Society for Microbiology, Washington, D.C.

157. **Kehry, M. R., P. Engström, F. W. Dahlquist, and G. L. Hazelbauer.** 1983. Multiple covalent modifications of Trg, a sensory transducer of *Escherichia coli*. J. Biol. Chem. **258:**5050–5055.

158. **Kellermann, O., and S. Szmelcman.** 1974. Active transport of maltose in *Escherichia coli* K12. Eur. J. Biochem. **47:**139–149.

159. **Kellermann, O. K., and T. Ferenci.** 1982. Maltose-binding protein from *Escherichia coli*. Methods Enzymol. **90:**459–463.

160. **Kenley, J. S., M. Leighton, and C. Bradbeer.** 1978. Transport of vitamin B12 in *Escherichia coli*. Corrinoid specificity of the outer membrane receptor. J. Biol. Chem. **253:**1341–1346.

161. **Kiratani, K., and K. Ohnishi.** 1977. Repression and inhibition of transport systems for branched-chain amino acids in *Salmonella typhimurium*. J. Bacteriol. **129:**589–598.

162. **Kolodrubetz, D., and R. Schleif.** 1981. L-Arabinose transport systems in *Escherichia coli* K-12. J. Bacteriol. **148:**472–479.

163. **Kondoh, H., C. B. Ball, and J. Adler.** 1979. Identification of a methyl-accepting chemotaxis protein for the ribose and galactose chemoreceptors of *Escherichia coli*. Proc. Natl. Acad. Sci. USA **76:**260–264.

164. **Kornberg, H. L.** 1976. Genetics in the study of carbohydrate transport by bacteria. Sixth Griffith Memorial Lecture. J. Gen. Microbiol. **96:**1–16.

165. **Kornberg, H. L., and C. Riordan.** 1976. Uptake of galactose into *Escherichia coli* by facilitated diffusion. J. Gen. Microbiol. **94:**75–89.

166. **Krajewska-Grynkiewicz, K., W. Walczak, and T. Kłopotowski.** 1971. Mutants of *Salmonella typhimurium* able to utilize D-histidine as a source of L-histidine. J. Bacteriol. **105:**28–37.

167. **Kreishman, G. P., D. E. Robertson, and C. Ho.** 1973. PMR studies of the substrate induced conformational change of glutamine binding protein from *E. coli*. Biochem. Biophys. Res. Commun. **53:**18–23.

168. **Kustu, S. G., and G. F.-L. Ames.** 1973. The *hisP* protein, a known histidine transport component in *Salmonella typhimurium*, is also an arginine transport component. J. Bacteriol. **116:**107–113.

169. **Kustu, S. G., N. C. McFarland, S. P. Hui, B. Esmon, and G. F. Ames.** 1979. Nitrogen control in *Salmonella typhimurium*: coregulation of synthesis of glutamine synthetase and amino acid transport systems. J. Bacteriol. **138:**218–234.

170. **Lam, V. M. S., K. Daruwalla, P. J. F. Henderson, and M. C. Jones-Mortimer.** 1980. Proton-linked D-xylose transport in *Escherichia coli*. J. Bacteriol. **143:**396–402.

171. **Landick, R.** 1984. Regulation of LIV-I transport system gene expression, p. 71–74. *In* L. Leive and D. Schlessinger (ed.), Microbiology—1984. American Society for Microbiology, Washington, D.C.

172. **Landick, R., and D. L. Oxender.** 1985. The complete nucleotide sequences of the *Escherichia coli* LIV-BP and LS-BP genes. Implications for the mechanism of high-affinity branched-chain amino acid transport. J. Biol. Chem. **260:**8257–8261.

173. **Langridge, L. R., H. Shinagawa, and A. B. Pardee.** 1970. Sulfate-binding protein from *Salmonella typhimurium*: physical properties. Science **169:**59–61.

174. **Lara, F. J. S., and J. L. Stokes.** 1952. Oxidation of citrate by *Escherichia coli*. J. Bacteriol. **63:**415–420.

175. **Lawlis, V. B., M. S. Dennis, E. Y. Chen, D. H. Smith, and D. J. Henner.** 1984. Cloning and sequencing of the xylose isomerase

and xylulose kinase genes of *Escherichia coli*. Appl. Environ. Microbiol. **47**:15–21.

176. **Lee, G. S., and G. F.-L. Ames.** 1984. Analysis of promoter mutations in the histidine transport operon of *Salmonella typhimurium*: use of hybrid M13 bacteriophages for cloning, transformation, and sequencing. J. Bacteriol. **159**:1000–1005.

177. **Lee, N.** 1980. Molecular aspects of *ara* regulation, p. 389–409. *In* J. H. Miller and W. S. Reznikoff (ed.), The operon, 2nd ed. Cold Spring Harbor Laboratory, Cold Spring Harbor, N.Y.

178. **Leive, L., and B. D. Davis,** 1965. The transport of diaminopimelate and cystine in *Escherichia coli*. J. Biol. Chem. **240**:4362–4369.

179. **Lengler, J., K. O. Hermann, H. J. Unsöld, and W. Boos.** 1971. The regulation of the β-methylgalactoside transport system and of the galactose binding protein of *Escherichia coli*. Eur. J. Biochem. **19**:457–470.

180. **Lenny, A. B., and P. Margolin.** 1980. Locations of the *opp* and *supX* genes of *Salmonella typhimurium* and *Escherichia coli*. J. Bacteriol. **143**:747–752.

181. **Lever, J. A.** 1972. Purification and properties of a component of histidine transport in *Salmonella typhimurium*. J. Biol. Chem. **247**:4317–4326.

182. **Lin, E. C. C.** 1978. Glycerol dissimilation and its regulation in bacteria. Annu. Rev. Microbiol. **30**:535–578.

183. **Lo, T. C. Y.** 1977. The molecular mechanism of dicarboxylic acid transport in *Escherichia coli* K12. J. Supramol. Struct. **7**:463–480.

184. **Lopilato, J. E., J. L. Garwin, S. D. Emr, T. J. Silhavy, and J. R. Beckwith.** 1984. D-Ribose metabolism in *Escherichia coli* K-12: genetics, regulation, and transport. J. Bacteriol. **158**:665–673.

185. **Luckey, M., and H. Nikaido.** 1983. Specificity of diffusion of solutes through channels produced by phage lambda receptor protein of *Escherichia coli*. J. Bacteriol. **153**:1056–1059.

186. **Lutgens, M., and G. Gottshalk.** 1980. Why a co-substrate is required for anaerobic growth of *Escherichia coli* on citrate. J. Gen. Microbiol. **119**:63–70.

187. **Maas, W. K., and A. J. Clark.** 1964. Studies on the mechanism of repression of arginine biosynthesis in *Escherichia coli*. II. Dominance of repressibility in diploids. J. Mol. Biol. **8**:365–370.

188. **MacPhearson, A. J. S., M. C. Jones-Mortimer, and P. J. F. Henderson.** 1981. Identification of the AraE transport protein of *Escherichia coli*. Biochem. J. **196**:269–283.

189. **Magasanik, B.** 1982. Genetic control of nitrogen assimilation in bacteria. Annu. Rev. Genet. **16**:135–168.

190. **Magota, K., N. Otsuji, T. Miki, T. Horiuchi, S. Tsunasawa, J. Kondo, F. Sakiyama, M. Amemura, T. Morita, H. Shinagawa, and A. Nakata.** 1984. Nucleotide sequence of the *phoS* gene, the structural gene for the phosphate-binding protein of *Escherichia coli*. J. Bacteriol. **157**:909–917.

191. **Mahoney, W. C., R. W. Hogg, and M. A. Hermodson.** 1981. The amino acid sequence of the D-galactose-binding protein from *Escherichia coli*. J. Biol. Chem. **256**:4350–4356.

192. **Maleszka, R., P. Y. Wang, and H. Schneider.** 1982. A col E1 hybrid plasmid containing *Escherichia coli* genes complementing D-xylose negative mutants of *Escherichia coli* and *Salmonella typhimurium*. Can. J. Biochem. **60**:144–151.

193. **Mao, B., M. R. Pear, A. McCammon, and F. Quiocho.** 1982. Hinge-bending in L-arabinose-binding protein. The "venus's-flytrap" model. J. Biol. Chem. **257**:1131–1133.

194. **Marty, B., C. Gaudin, M. Ragot, A. Belaich, and J. Belaich.** 1979. Microcalorimetric study of glutamine fixation on the glutamine-binding protein of *Escherichia coli*. Biochem. Biophys. Res. Commun. **86**:1118–1125.

195. **Marty, B., M. Bruschi, M. Ragot, and C. Gaudin.** 1981. Séquence de l'extrémité N terminale de la protéine périplasmique affine pour la glutamine d'*Escherichia coli* K 12. Comparaison avec des autres protéines périplasmiques. C.R. Acad. Sci. Ser. III **292**:987–989.

196. **Masters, P. S., and J.-S. Hong.** 1981. Reconstitution of binding protein dependent active transport of glutamine in spheroplasts of *Escherichia coli*. Biochemistry **20**:4900–4904.

197. **Masters, P. S., and J.-S. Hong.** 1981. Genetics of the glutamine transport system in *Escherichia coli*. J. Bacteriol. **147**:805–819.

198. **McGowan, E. B., T. J. Silhavy, and W. Boos.** 1974. Involvement of a tryptophan residue in the binding site of *Escherichia coli* galactose-binding protein. Biochemistry **13**:993–999.

199. **Medveczky, N., and H. Rosenberg.** 1970. The phosphate-binding protein of *Escherichia coli*. Biochim. Biophys. Acta **211**:158–168.

200. **Miller, D. M., J. S. Olson, J. W. Pflugrath, and F. A. Quiocho.** 1983. Rates of ligand binding to periplasmic proteins involved in bacterial transport and chemotaxis. J. Biol. Chem. **258**:13665–13672.

201. **Mizobuchi, K., M. Demerec, and D. H. Gillespie.** 1962. Cysteine mutants of *Salmonella typhimurium*. Genetics **47**:1617–1627.

202. **Morita, T., M. Amemura, K. Makino, H. Shinagawa, M. Magota, N. Otsuji, and A. Nakata.** 1983. Hyperproduction of phosphate-binding protein, *phoS*, and pre-*phoS* proteins in *Escherichia coli* carrying a cloned *phoS* gene. Eur. J. Biochem. **130**:427–435.

203. **Moroz, I. N., V. A. Grinkevich, N. M. Arzamazova, N. A. Potapenko, Z. A. Akimenko, I. V. Nazimov, and N. A. Aldanova.** 1980. Primary structure of the LIV-binding protein from *E. coli*. III. Peptides from selective tryptic and limited acid hydrolyses. Determination of the disulfide bridge in the protein molecule. Bioorg. Khim. **6**:9–20.

204. **Mowbray, S. L., and G. A. Petsko.** 1982. Preliminary X-ray data for the ribose binding protein from *Salmonella typhimurium*. J. Mol. Biol. **160**:545–547.

205. **Müller, M., and G. Blobel.** 1984. *In vitro* translocation of bacterial proteins across the plasma membrane of *Escherichia coli*. Proc. Natl. Acad. Sci. USA **81**:7421–7425.

205a.**Müller, N., H.-G. Heine, and W. Boos.** 1985. Characterization of the *Salmonella typhimurium mgl* operon and its gene products. J. Bacteriol. **163**:37–45.

206. **Müller-Hill, B.** 1983. Sequence homology between Lac and Gal Repressors and three sugar-binding periplasmic proteins. Nature (London) **302**:163–264.

207. **Nabiev, I., S. Trakhanov, A. Surin, T. Vorotyntseva, E. Efremov, and V. Pletnev.** 1982. Optical spectroscopy study of substrate binding by leucine specific and leucine-isoleucine-valine binding proteins from *E. coli*, p. 467–471. *In* W. Voelter and E. Wuensch (ed.), Chemistry of peptides and proteins, vol. 1. Walter de Gruyter & Co., Berlin.

208. **Nakae, T., and J. N. Ishii.** 1982. Molecular weights and subunit structure of LamB proteins. Ann. Microbiol. (Paris) **133A**:21–25.

209. **Nazos, P. M., T. Z. Su, R. Landick, and D. L. Oxender.** Branched-chain amino acid transport in *Escherichia coli*, p. 24–28. *In* L. Leive and D. Schlessinger (ed.), Microbiology—1984. American Society for Microbiology, Washington, D.C.

210. **Neu, H. C., and L. A. Heppel.** 1965. The release of enzymes from *Escherichia coli* by osmotic shock and during the formation of spheroplasts. J. Biol. Chem. **240**:3685–3692.

211. **Newcomer, M. E., B. A. Lewis, and F. A. Quiocho.** 1981. The radius of gyration of L-arabinose-binding-protein decreases upon binding of ligand. J. Biol. Chem. **256**:13218–13222.

212. **Nishimune, T., and R. Hayashi.** 1971. Thiamine-binding protein and thiamine uptake by *Escherichia coli*. Biochim. Biophys. Acta **224**:573–583.

213. **Novotny, C., and E. Englesberg.** 1969. The arabinose transport system of *Escherichia coli* B/r. Biochim. Biophys. Acta **117**:217–230.

214. **Ohnishi, K., and K. Kiritani.** 1983. Purification and properties of two binding proteins for branched-chain amino acids in *Salmonella typhimurium*. J. Biochem. (Tokyo) **94**:433–441.

215. **Ohnishi, K., K. Murata, and K. Kiratani.** 1980. A regulatory transport mutant for branched-chain amino acids in *Salmonella typhimurium*. Jpn. J. Genet. **55**:349–359.

216. **Ohta, N., P. R. Galsworthy, and A. B. Pardee.** 1971. Genetics of sulfate transport by *Salmonella typhimurium*. J. Bacteriol. **105**:1053–1062.

217. **Ordal, G. W., and J. Adler.** 1974. Isolation and complementation of mutants in galactose taxis and transport. J. Bacteriol. **117**:509–516.

218. **Ordal, G. W., and J. Adler.** 1974. Properties of mutants in galactose taxis and transport. J. Bacteriol. **117**:517–526.

219. **Oshima, R. G., R. C. Willis, C. E. Furlong, and J. A. Schneider.** 1974. Binding assays for amino acids: the utilization of a cystine binding protein from *Escherichia coli* for the determination of acid-soluble cystine in small physiological samples. J. Biol. Chem. **249**:6033–6039.

220. **Ovchinnikov, Y. A., N. A. Aldanova, V. A. Grinkevich, N. M. Arzamazova, and I. N. Moroz.** 1977. The primary structure of a leu, ile, and val (LIV)-binding protein form *Escherichia coli*. FEBS Lett. **78**:313–316.

221. **Ow, D. W., V. Sundaresan, D. M. Rothstein, S. E. Brow, and F. M. Ausubel.** 1983. Promoters regulated by the *glnG* (*ntrC*) and *nifA* gene products share a heptameric consensus sequence in the -15 region. Proc. Natl. Acad. Sci. USA **80**:2524–2528.

222. **Oxender, D. L.** 1972. Membrane transport. Annu. Rev. Biochem. **41**:777–814.

223. **Oxender, D. L., J. J. Anderson, C. J. Daniels, R. Landick, R. P. Gunsalus, G. Zurawski, and C. Yanofsky.** 1980. Amino-terminal sequence and processing of the precursor of the leucine-specific binding protein, and evidence for conformational differences between the precursor and the mature form. Proc. Natl. Acad. Sci. USA **77**:2005–2009.

224. **Oxender, D. L., R. Landick, P. Nazos, and B. R. Copeland.** 1984. Secretion of periplasmic transport components, p. 4–7. *In* L. Leive and D. Schlessinger (ed.), Microbiology—1984. American Society for Microbiology, Washington, D.C.

225. **Oxender, D. L., S. C. Quay, and J. J. Anderson.** 1980. Regulation of amino acid transport, p. 153–169. *In* J. W. Payne (ed.), Microorganisms and nitrogen sources. John Wiley & Sons Ltd., New York.

226. **Pardee, A. B.** 1966. Purification and properties of a sulfate-binding protein from *Salmonella typhimurium*. J. Biol. Chem. **241:**5886–5892.

227. **Pardee, A. B., L. S. Prestidge, M. B. Whipple, and J. Dreyfuss.** 1966. A binding site for sulfate and its relation to sulfate transport into *Salmonella typhimurium*. J. Biol. Chem. **241:** 3962–3969.

228. **Parsons, R. G., and R. W. Hogg.** 1974. Crystallization and characterization of the L-arabinose protein of *Escherichia coli* B/r. J. Biol. Chem. **249:**3602–3607.

229. **Payne, J. W.** 1968. Oligopeptide transport in *Escherichia coli*: specificity with respect to side chain and distinction from dipeptide transport. J. Biol. Chem. **243:**3395–3403.

230. **Payne, J. W.** 1980. Transport and utilization of peptides by bacteria, p. 211–256. *In* J. W. Payne (ed.), Microorganisms and nitrogen sources. John Wiley & Sons Ltd., New York.

231. **Payne, J. W.** 1980. Energetics of peptide transport in bacteria, p. 359–377. *In* J. W. Payne (ed.), Microorganisms and nitrogen sources. John Wiley & Sons Ltd., New York.

232. **Payne, J. W., and G. Bell.** 1979. Direct determination of the properties of peptide transport systems is *Escherichia coli*. J. Bacteriol. **137:**447–455.

233. **Payne, J. W., and C. Gilvarg.** 1968. Size restriction on peptide utilization in *Escherichia coli*. J. Biol. Chem. **243:**6291–6299.

234. **Penrose, W. R., G. E. Nichoalds, J. R. Piperno, and D. L. Oxender.** 1968. Purification and properties of a leucine-binding protein from *Escherichia coli*. J. Biol. Chem. **243:**5921–5928.

235. **Postle, K., and R. F. Good.** 1983. DNA sequence of the *Escherichia coli tonB* gene. Proc. Natl. Acad. Sci. USA **80:**5235–5239.

236. **Postma, P. W.** 1976. Involvement of the phosphotransferase system in galactose transport in *Salmonella typhimurium*. FEBS Lett. **61:**49–53.

237. **Quiocho, F. A., W. E. Meador, and J. W. Pflugrath.** 1979. Preliminary crystallographic data of receptors for transport and chemotaxis in *Escherichia coli*: D-galactose and maltose-binding proteins. J. Mol. Biol. **133:**181–184.

238. **Quiocho, F. A., and N. K. Vyas.** 1984. Novel stereospecificity of the L-arabinose-binding protein. Nature (London) **310:**381–386.

239. **Rahmanian, M., D. Claus, and D. L. Oxender.** 1973. Multiplicity of leucine transport systems in *Escherichia coli* K-12. J. Bacteriol. **116:**1258–1266.

240. **Randall, L. L., and J. J. S. Hardy.** 1984. Export of protein in bacteria: dogma and data. Mod. Cell Biol. **3:**1–20.

241. **Randall, L. L., and S. J. S. Hardy.** 1984. Export of protein in bacteria. Microbiol. Rev. **48:**290–298.

242. **Randall-Hazelbauer, L., and M. Schwartz.** 1973. Isolation of the bacteriophage lambda receptor from *Escherichia coli*. J. Bacteriol. **116:**1436–1446.

243. **Reynolds, C. H., and S. Silver.** 1983. Citrate utilization by *Escherichia coli*. J. Bacteriol. **156:**1019–1924.

244. **Reynolds, P. R., G. P. Mottur, and C. Bradbeer.** 1980. Transport of vitamin B12 in *Escherichia coli*. Some observations on the roles of the gene products of *btuC* and *tonB*. J. Biol. Chem. **255:**4313–4319.

245. **Richarme, G.** 1982. Associative properties of the *Escherichia coli* galactose binding protein and maltose binding protein. Biochem. Biophys. Res. Commun. **105:**476–481.

246. **Richarme, G.** 1982. Interaction of the maltose-binding protein with membrane vesicles of *Escherichia coli*. J. Bacteriol. **149:**662–667.

247. **Richarme, G.** 1983. Associative properties of the *Escherichia coli* galactose binding protein and maltose-binding protein. Biochim. Biophys. Acta **748:**99–108.

248. **Richarme, G., J. M. Meury, and J. Bouvier.** 1982. Membrane potential stimulated binding of the maltose-binding protein to membrane vesicles of *Escherichia coli*. Ann. Microbiol. (Paris) **133A:**199–204.

249. **Riordan, C., and H. L. Kornberg.** 1977. Location of *galP*, a gene which specifies galactose permease activity, on the *Escherichia coli* linkage map. Proc. R. Soc. Lond. Ser. B **198:**401–410.

250. **Robb, F. T., and C. E. Furlong.** 1980. Reconstitution of binding protein dependent ribose transport in spheroplasts derived from a binding protein negative *Escherichia coli* K12 mutant and from *Salmonella typhimurium*. J. Supramol. Struct. **13:**183–190.

251. **Robbins, A. R.** 1975. Regulation of the *Escherichia coli* methylgalactoside system by gene *mglD*. J. Bacteriol. **123:**69–74.

252. **Robbins, A. R., R. Guzman, and B. Rotman.** 1976. Roles of individual *mgl* gene products in the β-methylgalactoside transport system of *Escherichia coli* K12. J. Biol. Chem. **251:**3112–3116.

253. **Robertson, D. E., P. A. Kroon, and C. Ho.** 1977. Nuclear magnetic resonance and fluorescence studies of substrate-induced conformational changes of histidine-binding protein J of *Salmonella typhimurium*. Biochemistry **16:**1443–1451.

254. **Rosen, B. P.** 1971. Basic amino acid transport in *Escherichia coli*. J. Biol. Chem. **246:**3653–3662.

255. **Rosen, B. P. (ed.).** 1978. Bacterial transport. Marcel Dekker, New York.

256. **Rosen, B. P.** 1973. Basic amino acid transport in *Escherichia coli*: properties of canavanine-resistant mutants. J. Bacteriol. **116:**627–635.

257. **Rosen, B. P.** 1973. Basic amino acid transport in *Escherichia coli*. Purification and properties of an arginine specific binding protein. J. Biol. Chem. **248:**1211–1218.

258. **Rosen, B. P., and F. Vasington.** 1971. Purification and characterization of a histidine-binding protein from *Salmonella typhimurium* and its relationship to the histidine permease system. J. Biol. Chem. **246:**5351–5360.

259. **Rosenberg, H., R. G. Gerdes, and K. Chegwidden.** 1977. Two systems for the uptake of phosphate in *Escherichia coli*. J. Bacteriol. **131:**505–511.

260. **Rosenberg, H., C. M. Hardy, and B. P. Surin.** 1984. Energy coupling to phosphate transport in *Escherichia coli*, p. 50–52. *In* L. Leive and D. Schlessinger (ed.), Microbiology—1984. American Society for Microbiology, Washington, D.C.

261. **Rosenfeld, S. A., P. E. Stevis, and N. W. Y. Ho.** 1984. Cloning and characterization of the *xyl* genes from *Escherichia coli*. Mol. Gen. Genet. **194:**410–415.

262. **Rotman, B.** 1959. Separate permeases for the accumulation of methyl-β-D-galactoside and methyl-β-D-thiogalactosides in *Escherichia coli*. Biochim. Biophys. Acta **32:**599–601.

263. **Rotman, B., A. K. Ganesan, and R. Guzman.** 1968. Transport systems for galactose and galactosides in *Escherichia coli*. II. Substrate and inducer specificities. J. Mol. Biol. **36:**247–260.

264. **Rotman, B., and R. Guzman.** 1982. Identification of the *mglA* gene product in the β-methylgalactoside transport system of *Escherichia coli* using plasmid DNA deletions generated *in vitro*. J. Biol. Chem. **257:**9030–9034.

265. **Rotman, B., and R. Guzman.** 1984. Galactose-binding protein-dependent transport in reconstituted *Escherichia coli* membrane vesicles, p. 57–60. *In* L. Leive and D. Schlessinger (ed.), Microbiology—1984. American Society for Microbiology, Washington, D.C.

266. **Rotman, B., and J. Radojkovic.** 1964. Galactose transport in *Escherichia coli*. J. Biol. Chem. **239:**3153–3156.

267. **Russell, R. M., and H. Rosenberg.** 1979. Linked transport of phosphate, potassium ions and protons in *Escherichia coli*. Biochem. J. **184:**13–21.

268. **Russell, R. M., and H. Rosenberg.** 1980. The nature of the link between potassium transport and phosphate transport in *Escherichia coli*. Biochem. J. **188:**715–723.

269. **Russo, R. J., J.-H. Lee, P. Clarke, and G. Wilcox.** 1984. Identification of the *araE* gene product of *Salmonella typhimurium* LT2, p. 42–46. *In* L. Leive and D. Schlessinger (ed.), Microbiology—1984. American Society for Microbiology, Washington, D.C.

270. **Saier, M.** 1984. Maltose transport, p. 75–78. *In* L. Leive and D. Schlessinger (ed.), Microbiology—1984. American Society for Microbiology, Washington, D.C.

271. **Saier, M. H., Jr.** 1985. Mechanisms and regulation of carbohydrate transport in bacteria. Academic Press, Inc., New York.

272. **Saier, M. H., Jr., F. G. Bromberg, and S. Roseman.** 1973. Characterization of constitutive galactose permease mutants in *Salmonella typhimurium*. J. Bacteriol. **113:**512–514.

273. **Sanderson, K., and J. R. Roth.** 1984. Linkage map of *Salmonella typhimurium*, p. 131–144. *In* S. J. O'Brien (ed.), Genetic maps—1984, vol. 3. Cold Spring Harbor Laboratory, Cold Spring Harbor, N.Y.

274. **Saper, M. A., and F. A. Quiocho.** 1983. Leucine, isoleucine, valine-binding protein from *Escherichia coli*. J. Biol. Chem. **258:**11057–11062.

275. **Schellenberg, G. D., and C. E. Furlong.** 1977. Resolution of the multiplicity of the glutamate and aspartate transport systems of *Escherichia coli*. J. Biol. Chem. **252:**9055–9064.

276. **Schellenberg, G. D., A. Sarthy, A. E. Larson, M. P. Backer, J. W. Crabb, M. Lidstrom, B. D. Hall, and C. E. Furlong.** 1984. Xylose isomerase from *Escherichia coli*. J. Biol. Chem. **259:**6826–6832.

277. **Schleif, R.** 1969. An L-arabinose binding protein and arabinose permeation in *Escherichia coli.* J. Mol. Biol. **46:**185–196.

278. **Schweizer, H., M. Argast, and W. Boos.** 1982. Characteristics of a binding protein-dependent transport system for *sn*-glycerol-3-phosphate in *Escherichia coli* that is part of the *pho* regulon. J. Bacteriol. **150:**1154–1163.

279. **Schweizer, H., T. Grussenmeyer, and W. Boos.** 1982. Mapping of two *ugp* genes coding for the *pho* regulon-dependent *sn*-glycerol-3-phosphate transport system of *Escherichia coli.* J. Bacteriol. **150:**1164–1171.

280. **Scripture, J. B., and R. W. Hogg.** 1983. The nucleotide sequences defining the signal peptides of the galactose-binding protein and the arabinose-binding protein. J. Biol. Chem. **258:**10853–10855.

281. **Shamanna, D. K., and K. E. Sanderson.** 1979. Uptake and catabolism of D-xylose in *Salmonella typhimurium* LT2. J. Bacteriol. **139:**64–70.

282. **Shamanna, D. K., and K. E. Sanderson.** 1979. Genetics and regulation of D-xylose utilization in *Salmonella typhimurium* LT2. J. Bacteriol. **139:**71–79.

283. **Shinagawa, H., K. Makino, and A. Nakata.** 1983. Regulation of the *pho* regulon in *Escherichia coli* K-12. Genetic and physiological regulation of the positive regulatory gene *phoB.* J. Mol. Biol. **168:**477–488.

284. **Shuman, H. A.** 1982. The maltose-maltodextrin transport system of *Escherichia coli.* Ann. Microbiol. (Paris) **133A:**153–159.

285. **Shuman, H. A., and T. J. Silhavy.** 1981. Identification of the *malk* gene product: a peripheral membrane component of the *Escherichia coli* maltose transport system. J. Biol. Chem. **256:**560–562.

286. **Shuman, H. A., T. J. Silhavy, and J. R. Beckwith.** 1980. Labelling of proteins with β-galactosidase by gene fusion. Identification of a cytoplasmic membrane component of the *Escherichia coli* maltose transport system. J. Biol. Chem. **255:**168–174.

287. **Silhavy, T. J., T. Ferenci, and W. Boos.** 1978. Sugar transport systems in *Escherichia coli,* p. 127–169. *In* B. P. Rosen (ed.), Bacterial transport. Marcel Dekker, New York.

288. **Simoni, R. D., and P. W. Postma.** 1975. The energetics of bacterial transport. Annu. Rev. Biochem. **44:**523–554.

289. **Sommers, J. M., G. D. Sweet, and W. W. Kay.** 1981. Fluorocitrate resistant tricarboxylate transport mutants of *Salmonella typhimurium.* Mol. Gen. Genet. **181:**338–345.

290. **Sprague, G. F., Jr., R. M. Bell, and J. E. Cronan, Jr.** 1975. A mutant of *Escherichia coli* auxotrophic for organic phosphates: evidence for two defects in inorganic phosphate transport. Mol. Gen. Genet. **143:**71–77.

291. **Springer, S. E., and R. E. Huber.** 1972. Evidence for a sulfate transport system in *Escherichia coli* K-12. FEBS Lett. **27:**13–15.

292. **Stern, M. J., and G. F.-L. Ames.** 1984. Regulation of histidine transport in *Salmonella typhimurium,* p. 79–83. *In* L. Leive and D. Schlessinger (ed.), Microbiology—1984. American Society for Microbiology, Washington, D.C.

293. **Stern, M. J., G. F.-L. Ames, N. H. Smith, E. C. Robinson, and C. F. Higgins.** 1984. Repetitive extragenic palindromic sequences: a major component of the bacterial genome. Cell **37:**1015–1026.

294. **Stern, M. J., C. F. Higgins, and G. F.-L. Ames.** 1984. Isolation and characterization of *lac* fusions to two nitrogen-regulated promoters. Mol. Gen. Genet. **195:**219–227.

295. **Strange, P. G., and D. E. Koshland, Jr.** 1976. Receptor interactions in a signalling system: competition between ribose receptor and galactose receptor in the chemotaxis response. Proc. Natl. Acad. Sci. USA **73:**762–766.

296. **Surin, B. P., D. A. Jans, A. L. Fimmel, D. C. Shaw, G. B. Cox, and H. Rosenberg.** 1984. Structural gene for the phosphate-repressible phosphate-binding protein of *Escherichia coli* has its own promoter: complete nucleotide sequence of the *phoS* gene. J. Bacteriol. **157:**772–778.

297. **Surin, B. P., H. Rosenberg, and G. B. Cox.** 1985. Phosphate-specific transport system of *Escherichia coli:* nucleotide sequence and gene-polypeptide relationships. J. Bacteriol. **161:**189–198.

298. **Sweet, G. D., J. M. Somers, and W. W. Kay.** 1979. Purification and properties of a citrate-binding transport component, the C protein of *Salmonella typhimurium.* Can. J. Biochem. **57:**710–715.

299. **Szmelcman, S., M. Schwartz, T. J. Silhavy, and W. Boos.** 1965. Maltose transport in *Escherichia coli* K12. A comparison of transport kinetics in wild-type and λ-resistant mutants with dissociation constants of the maltose binding protein as measured by fluorescence quenching. Eur. J. Biochem. **65:**13–19.

300. **Taylor, R. T., S. A. Norrell, and M. L. Hanna.** 1972. Uptake of cyanocobalamin by *Escherichia coli* B: some characteristics and evidence for a binding protein. Arch. Biochem. Biophys. **148:**366–381.

301. **Vorotyntseva, T. I., A. M. Surin, S. D. Trakhanov, I. R. Nabiev, and V. K. Antonov.** 1981. Spectral properties of the leucine-isoleucine-valine-binding protein and its complexes with substrates. Bioorg. Khim. **7:**35–46.

302. **Vyas, N. K., N. M. Vyas, and F. A. Quiocho.** 1983. The 3Å resolution structure of a D-galactose-binding protein for transport and chemotaxis in *Escherichia coli.* Proc. Natl. Acad. Sci. USA **80:**1792–1976.

303. **Wanner, B. L.** 1983. Overlapping and separate controls on the phosphate regulon in *Escherichia coli.* J. Mol. Biol. **166:**283–308.

304. **Wanner, B. L., and R. McSharry.** 1982. Phosphate-controlled gene expression in *Escherichia coli* K12 using Mu*d1*-directed *lacZ* fusions. J. Mol. Biol. **158:**347–363.

305. **Weiner, J. H., C. E. Furlong, and L. A. Heppel.** 1971. A binding protein for L-glutamine and its relation to active transport in *E. coli.* Arch. Biochem. Biophys. **142:**715–717.

306. **Weiner, J. H., and L. A. Heppel.** 1971. A binding protein for glutamine and its relation to active transport in *E. coli.* J. Biol. Chem. **246:**6933–6941.

307. **White, J. C., P. M. Di Girolamo, M. L. Fu, Y. A. Preston, and C. Bradbeer.** 1973. Transport of vitamin B12 in *Escherichia coli.* Location and properties of the initial B12-binding site. J. Biol. Chem. **248:**3978–3986.

308. **Willis, R. C., and C. E. Furlong.** 1974. Purification and properties of a ribose-binding protein from *Escherichia coli.* J. Biol. Chem. **249:**6926–6929.

309. **Willis, R. C., and C. E. Furlong.** 1975. Purification and properties of a periplasmic glutamate-aspartate binding protein from *Escherichia coli* strain W3092. J. Biol. Chem. **250:**2574–2580.

310. **Willis, R. C., and C. E. Furlong.** 1975. Interactions of a glutamate-aspartate binding protein with the glutamate transport system of *Escherichia coli.* J. Biol. Chem. **250:**2581–2586.

311. **Willis, R. C., K. K. Iwata, and C. E. Furlong.** 1975. Regulation of glutamine transport in *Escherichia coli.* J. Bacteriol. **122:**1032–1037.

312. **Willis, R. C., R. G. Morris, C. Cirakoglu, G. D. Schellenberg, N. H. Gerber, and C. E. Furlong.** 1974. Preparation of the periplasmic binding proteins from *Salmonella typhimurium* and *Escherichia coli.* Arch. Biochem. Biophys. **161:**64–75.

313. **Willis, R. C., and J. E. Seegmiller.** 1976. A filtration assay specific for the determination of small quantities of L-glutamine. Anal. Biochem. **72:**66–77.

314. **Willsky, G. R., and M. H. Malamy.** 1974. The loss of the *phoS* periplasmic protein leads to a change in the specificity of a constitutive inorganic phosphate transport system in *E. coli.* Biochem. Biophys. Res. Commun. **60:**225–233.

315. **Wilson, D. B.** 1974. The regulation and properties of the galactose transport system in *Escherichia coli* K12. J. Biol. Chem. **249:**553–558.

316. **Wilson, D. B.** 1978. Cellular transport mechanisms. Annu. Rev. Biochem. **47:**933–965.

317. **Wilson, O. H., and J. T. Holden.** 1969. Stimulation of arginine transport in osmotically shocked *Escherichia coli* W cells by purified arabinine-binding protein fractions. J. Biol. Chem. **244:**2743–2749.

318. **Wilson, V. G., and R. W. Hogg.** 1980. The NH2-terminal sequence of a precursor form of the arabinose binding protein. J. Biol. Chem. **255:**6745–6750.

319. **Wood, J. M.** 1975. Leucine transport in *Escherichia coli:* the resolution of multiple transport systems and their coupling to metabolic energy. J. Biol. Chem. **250:**4477–4485.

320. **Wu, H. C. P., and H. Kalckar.** 1966. Endogenous induction of the galactose operon in *Escherichia coli* K12. Proc. Natl. Acad. Sci. USA **55:**622–629.

321. **Zukin, R. S.** 1979. Evidence for a conformational change in *Escherichia coli* maltose receptor by excited-state fluorescence lifetime data. Biochemistry **18:**2139–2145.

322. **Zukin, R. S., P. R. Hartig, and D. E. Koshland.** 1977. Use of a distant reporter group as evidence for a conformational change on a sensory receptor. Proc. Natl. Acad. Sci. USA **74:**1932–1936.

323. **Zukin, R. S., H. M. Steinman, and R. E. Hirsch.** 1984. Use of the distant reporter group method to study bacterial sensory receptors and transport proteins, p. 53–56. *In* L. Leive and D. Schlessinger (ed.), Microbiology—1984. American Society for Microbiology, Washington, D.C.

324. **Zukin, R. S., P. G. Strange, L. R. Heavey, and D. E. Koshland, Jr.** 1977. Properties of the galactose binding protein of *Salmonella typhimurium* and *Escherichia coli.* Biochemistry **16:**381–386.

52. Growth Yield and Energy Distribution

D. W. TEMPEST AND O. M. NEIJSSEL

Laboratorium voor Microbiologie, Universiteit Amsterdam, 1018 WS Amsterdam, The Netherlands

INTRODUCTION

Although quantitative studies of microbial growth, and of growth energetics, were in progress some 50 years ago or more (47, 55), it is widely acknowledged that present-day theories regarding the relationship between substrate concentration, growth rate, and yield value stem from the classical studies of Monod (30). In these studies, quantitative measurements of the growth of *Escherichia coli, Salmonella typhimurium,* and *Bacillus subtilis* in batch culture revealed that the equivalent dry weight of organisms formed per gram of carbon substrate metabolized (the yield value) was remarkably constant. Thus, when growing in a simple salts medium on a variety of related carbon substrates (hexoses, pentoses, polyalcohols, and disaccharides), *E. coli* expressed yield values that ranged between 0.21 and 0.28 g (equivalent dry weight) of cells formed per g of substrate consumed. Corresponding cultures of *S. typhimurium* expressed slightly (though consistently) lower values. Taken together, the constancy of these data indicated the presence in organisms of mechanisms that precisely partition the flow of carbon substrate between catabolic (energy-generating) and anabolic (energy-consuming) processes such as to allow growth to proceed with a fixed overall efficiency. In this connection, however, Monod (30) realized that not all of the energy generated by catabolism necessarily would be consumed by anabolic processes and postulated that a small amount might be needed for cell maintenance (ration d'entretien). But, from his observations, he was forced to conclude that, with actively growing cultures, the latter requirement was too small to be detected by the methods then available.

The relationship between substrate concentration, growth rate, and respiration rate was subsequently studied by Schultz and Lipe (43), who used continuous-culture techniques that allowed growth rate to be varied over a wide range by varying the dilution rate. From these studies, in which an unnamed strain of *E. coli* was used, they were able to conclude that the maintenance rate of glucose consumption was indeed small but that it was sufficient to affect markedly the yield value of a glucose-limited culture at low dilution (=growth) rates. When corrected for maintenance, the (maximum) growth yield was found to be 0.53 g of cells per g of glucose. This value was more than double that reported by Monod (30), but is in accord with many subsequent measurements made with different strains of *E. coli* growing in glucose-limited chemostat culture (Table 1).

The results obtained by Schultz and Lipe (43) confirmed and extended those of Herbert (16) and Marr et al. (27), who had shown that the yield value with respect to carbon substrate consumed was not a constant but that it varied with the growth rate (Table 1). A similar variation also was found, not surprisingly, in the oxygen yield value (i.e., grams [dry weight] of cells formed per mole of oxygen consumed) and was ascribed to the same cause, namely a requirement of metabolic energy for cell maintenance (37). Thus, it became clear that meaningful comparisons of yield values, and their interpretation in physiological or bioenergetic terms or both, could only be made taking into account the metabolic requirements of maintenance functions. Unfortunately, the precise nature of these maintenance processes and their minimum energetic requirements were then (as now) largely unresolved. Hence, it still is not clear whether these maintenance energy requirements are quantitatively independent of growth rate, as originally assumed by Pirt (37) and others (15, 16, 27), or whether they vary with growth rate, as is now thought probable (34, 38, 58).

TABLE 1. Glucose consumption rates ($q_{glucose}$) and yield values ($Y_{glucose}$) obtained with glucose-limited cultures of several strains of *E. coli*[a]

| Parameter | Glucose consumption rate (mmol of glucose/h per g [dry wt] of cells) or yield value (g [dry wt] of cells formed/mol of glucose consumed) for the following *E. coli* strain[b]: | | | | | | | |
| | C (PC-1000) | | B/r (Phabagen) | | ATCC 9001 | | Unnamed | |
	$q_{glucose}$	$Y_{glucose}$	$q_{glucose}$	$Y_{glucose}$	$q_{glucose}$	$Y_{glucose}$	$q_{glucose}$	$Y_{glucose}$
Dilution rate (h^{-1})								
0.1	1.56	64.1	1.50	66.7	1.47	70.9	1.25	80.0
0.3	4.08	73.5	3.93	76.4	3.95	76.0	3.56	84.3
0.5	6.60	75.8	6.36	78.7	6.45	77.5	5.92	84.5
0.7	9.12	76.8	8.86	79.0	8.93	78.4	8.27	84.6
Apparent $Y_{glucose}^{max}$[c]		79.4		81.5		79.8		85.4
Apparent $m_{glucose}$[d]		0.30		0.26		0.17		0.07

[a] To facilitate direct comparison, the reported values at each dilution rate were derived from linear plots of $q_{glucose}$ versus dilution rate.

[b] Data for *E. coli* C were previously unpublished, data for strain B/r are from M. P. M. Leegwater (personal communication), data for the type strain (ATCC 9001) are from Neijssel et al. (31), and data for the unnamed strain were derived from the data of Schultz and Lipe (43).

[c] Calculated maximum growth yield (i.e., corrected for maintenance).

[d] Maintenance rate of glucose consumption; the extrapolated rate of glucose consumption at zero growth rate.

Thus, it is not easy to assess the significance of small differences in the yield value expressed by related organisms growing under comparable conditions, but large consistent differences ought to be interpretable in physiological terms.

It is the purpose of this chapter to concentrate on those conditions that markedly affect the growth yield with respect to carbon substrate and oxygen consumption (Y_{sub} and Y_{O_2}, respectively) or ATP generation (Y_{ATP}) and to assess their physiological implications. To do this, however, it is useful to consider first the overall relationship between catabolism and anabolism which the yield value embodies (Fig. 1). Here, the uptake of carbon substrate into the cell and its subsequent metabolism are depicted as generating initially a flow of intermediary metabolites and reductant. Aerobically, part of the reductant so formed is oxidized by way of the respiratory chain to generate a flow of useable energy (ATP); this energy, along with the remaining reductant, is used to convert intermediary metabolites into the monomers and polymers that culminate in growth. Any potential shortfall in reductant or energy can, of course, be readily avoided by further oxidation of some of the intermediary metabolites to CO_2. However, because the main carriers of reductant (the pyridine nucleotides) and of energy (the adenine nucleotides) are present within the cell only in relatively low concentration (13), any potential surplus of reductant or of energy cannot be accommodated by a change in carrier concentration and must be rapidly disposed of in some way. Thus, whereas with organisms growing aerobically a potential redox imbalance can be circumvented by the organisms expressing a higher respiration rate, if respiration is coupled to ATP generation, then solving a problem of redox imbalance in this way must generate an energy overplus that cannot be stored at the level of the adenine nucleotides. This is not a hypothetical problem, for there are at least two circumstances in which catabolism seemingly is extensively dissociated from anabolism: first, with washed cell suspensions, which often oxidize carbon substrate at a high rate while clearly the organisms cannot grow; second, with carbon substrate sufficient, cultures which frequently are found to catabolize substrate at rates far higher than those of corresponding carbon-limited cultures (33, 34). In both of these cases, an excess of energy is generated that must be dissipated at a high rate (presumably as heat) by growth-unassociated processes. It follows, therefore, that despite indications to the contrary (i.e., the relative constancy of yield values of batch cultures), microbial cells must have an inherent capacity either to uncouple respiration from ADP phosphorylation or else to turn over the ATP pool at a high rate in the absence of biosynthesis. This assumption, then, raises general questions regarding the regulation of energy fluxes in microorganisms and the nature of energy coupling

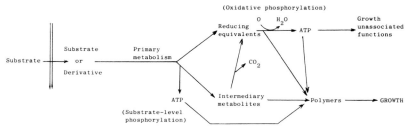

FIG. 1. Schematic representation of the pathways of carbon and energy flow in aerobic bacteria.

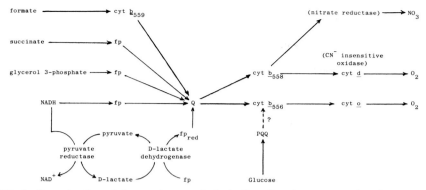

FIG. 2. Organization of the respiratory chain in *E. coli* and some associated processes.

between catabolic and anabolic processes in actively growing cells. To answer these questions, it is appropriate to consider the following factors briefly: (i) the mechanisms of energy generation and how they might be caused to vary, (ii) the multifarious nature of energy-consuming processes, and (iii) the possible nature of energy-spilling reactions.

ENERGY-GENERATING SYSTEMS

Oxidative Phosphorylation

The efficiency of respiratory energy conservation (in terms of moles of ATP equivalents generated per gram-atom of oxygen reduced) depends upon the composition of the respiratory chain and the number of proton-translocating segments. Extensive studies of the respiratory chain composition of several *E. coli* strains (20, 22) suggest that they lack a *c*-type component and that they hence contain only two proton-translocating segments. This conclusion, moreover, is in accord with → H$^+$/O ratios measured with suspensions of three strains of *E. coli* oxidizing endogenous substrate, which ranged from 3.53 to 4.00 (2, 22, 23, 25, 60).

All known respiratory systems show some evidence of branching. However, the extent to which this occurs is extremely variable, and its physiological significance is not always clear (62). Extensive branching generally occurs at the level of the primary oxidases and is manifest in the ability of cells to oxidize directly a variety of substrates in addition to NADH. Furthermore, branching invariably is present at the terminal end of the respiratory chain in those organisms, such as *E. coli*, that can utilize anaerobically terminal electron acceptors such as fumarate and nitrate. In these cases, synthesis of the appropriate reductases is accompanied by formation of significant amounts of the appropriate cytochrome oxidases (11).

Under highly aerobic conditions, the terminal end of the *E. coli* respiratory chain seemingly is a linear sequence; however, under conditions of oxygen limitation, or when low concentrations of cyanide are present in the growth medium, an additional terminal oxidase (a *d*-type cytochrome) is formed, along with a second *b*-type cytochrome (b_{558}). It has been suggested that cytochromes b_{558} and *d* compose a second pathway of electron flow from the reduced quinone to oxygen, one that has a higher affinity for oxygen and that is cyanide insensitive (Fig. 2). What is not clear, however, is whether this second pathway contains a proton-translocating segment similar to that of the pathway from reduced quinone to oxygen via cytochromes b_{556} and *o* (for opposing views, see references 28 and 41).

Apart from the NADH-linked dehydrogenases and those linked to flavins, *E. coli* has been found to synthesize a glucose dehydrogenase apoenzyme that is active in the presence of 2,7,9-tricarboxy-1H-pyrrolo (2,3-f) quinoline-4,5-dione (PQQ) (18). Thus, in the presence of PQQ, though not in its absence, whole cells or cell homogenates oxidize glucose directly to gluconic acid. Strains of *Klebsiella aerogenes*, in contrast to all strains of *E. coli* so far tested, are able to synthesize PQQ and have a functionally competent glucose dehydrogenase. Aerobically, this enzyme appears to be synthesized constitutively, but it is induced to high levels in cells that are stressed energetically (35). A similar pattern of regulation has been found in *E. coli* with respect to synthesis of the apoenzyme, and it seems likely that, in competent cells (i.e., those that can synthesize PQQ or cases in which PQQ is added to the growth medium), this enzyme functions as a low-impedance high-flux energy-generating system. Consistent with this conclusion is the finding that oxidation of glucose by membrane vesicles of *E. coli*, supplemented with PQQ, can drive the uptake of amino acids (58). Whether other PQQ-requiring dehydrogenases are synthesized by *K. aerogenes* and *E. coli* remains to be determined; however, it is relevant to mention in passing that, like *E. coli* strains, those strains of *S. typhimurium* so far examined also synthesize a glucose dehydrogenase apoenzyme (R. W. J. Hommes, personal communication).

Whereas studies of cytochrome content and composition of *E. coli*, along with → H$^+$/O ratio measurements, provide strong evidence for the presence of only two sites of energy conservation on the respiratory chain, it is not clear whether the proton motive force generated by respiration invariably is dissipated in doing useful work. In other words, it is not yet known whether natural mechanisms exist that can uncouple (partially or totally) the process of oxidative phosphorylation. We return to this question later.

Substrate-Level Phosphorylation

When growing anaerobically on glucose, in the absence of an added electron acceptor, *E. coli* effects a mixed-acid (branched-pathway) fermentation in which

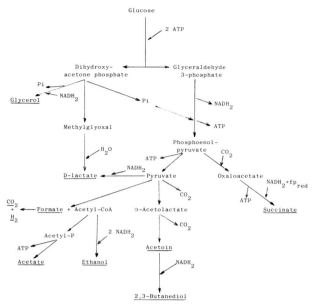

FIG. 3. Pathways of product formation in *E. coli* grown anaerobically on glucose. Data are those of Gottschalk (9). Acetyl-CoA, Acetyl coenzyme A.

the principal products are lactate, ethanol, acetate, and formate (or H_2 plus CO_2, at acid pH values). Thus, energy is conserved (as ATP) principally at the levels of 3-phosphoglycerate kinase, pyruvate kinase, and acetate kinase (Fig. 3). Significant amounts of succinate also often are found (Table 2) that are derived from endogenously generated fumarate, and, since fumarate reduction is coupled to respiratory chain-linked oxidation of NADH or formate, the formation of succinate is accompanied by the generation of a proton motive force that can drive ADP phosphorylation or (more probably) solute uptake processes or both (J. Boonstra, Ph.D. thesis, University Groningen, The Netherlands, 1978). It follows, therefore, that whereas the formation of acetate is accompanied by a net production of 2 mol of ATP per mol, the formation of lactate, ethanol, and succinate each effects production of just 1 mol of ATP per mol of product.

TABLE 2. Mixed acid fermentation of glucose as effected by *E. coli*[a]

Product or recovery	mmol of product/100 mmol of glucose fermented at pH:	
	6.2	7.8
2,3-Butanediol	0.3	0.26
Acetoin	0.06	0.19
Glycerol	1.42	0.32
Ethanol	49.8	50.5
Acetic acid	36.5	38.7
Formic acid	2.43	86.0
Lactic acid	79.5	70.0
Succinic acid	10.7	14.8
Carbon dioxide	88.0	1.75
Hydrogen	75.0	0.26
Carbon recovery (%)	92.0	90.0

[a] Data of Blackwood et al. (1).

Clearly, it would be advantageous (in terms of net ATP gain) to ferment glucose solely to acetate and formate. Redox considerations, however, require a concomitant formation of a product more reduced than lactate because acetate production is accompanied by a net formation of reductant (NADH). Production of ethanol fulfills this requirement in *E. coli* and *K. aerogenes*. Hence, in fermenting glucose, a maximum ATP gain would be achieved in producing equimolar amounts of ethanol and acetate and would amount to 3 mol of ATP per mol of glucose fermented. It follows, therefore, that because anaerobic cultures generally are considered to be limited in their growth by the availability of energy (ATP), one would expect glucose to be fermented entirely to ethanol and acetate (plus formate, of course). Surprisingly, this was not found with batch cultures of *E. coli* (Table 2) or *Aerobacter cloacae* (17) or with glucose-sufficient chemostat cultures of *K. aerogenes* (49). This condition was apparent only with glucose-limited chemostat cultures of some organisms (M. J. Teixeira de Mattos, Ph.D. thesis, University Amsterdam, The Netherlands, 1984). Taken at face value, these observations suggest that organisms growing on glucose anaerobically in batch culture (as in glucose-sufficient chemostat culture) are not energy limited. On the other hand, a high rate of cell synthesis, such as is manifest in batch cultures, demands a high rate of ATP generation, and the presence of a branched fermentation capacity in an organism may actually serve to facilitate this high rate of ATP generation, albeit with a concomitant decrease in the efficiency of catabolic energy conservation (in terms of net moles of ATP formed per mole of glucose fermented).

Product Efflux

On the basis of theoretical considerations, it was postulated by Michels et al. (29) that strictly fermentative bacteria might be able to generate an electrochemical proton gradient ($\Delta\bar{\mu}_{H^+}$) across their cytoplasmic membrane by the efflux of L-lactate on a protonated carrier (with an H^+/lactate ratio of (n) > 1). Subsequently, this energy-recycling model was tested with intact *Streptococcus cremoris* cells and with membrane vesicles, and compelling evidence indicated that lactate efflux could, under some conditions, indeed generate a $\Delta\bar{\mu}_{H^+}$ that could drive uptake and accumulation of leucine (36). However, the H^+:lactate stoichiometry (n) was not constant; it varied with both the magnitude of the transmembrane lactate gradient and the medium pH value (B. ten Brink, Ph.D. thesis, University Groningen, The Netherlands, 1984). Thus, the effective energy gain also varied with these parameters, and lactate excretion could even require an expenditure of energy when the external pH is low and when the extracellular lactate concentration is high.

Studies with *E. coli* membrane vesicles provided experimental evidence that L-lactate translocation (both uptake and efflux) is carrier mediated (54). Here, the H^+:lactate stoichiometry was found to vary between 1 and 2, depending on the external pH value, again consistent with a postulated protonation of the lactate carrier protein. Significantly, L-lactate efflux from membrane vesicles that were preloaded with 50

TABLE 3. Theoretical ATP requirements of *E. coli* growing in a simple salts medium with different carbon and energy sources[a]

Product or parameter	ATP requirement (mmol/g of cells) of cells growing on[b]:			
	Glucose	Malate	Lactate	Acetate
Protein	20.5	28.5	33.9	42.7
Polysaccharide	2.05	5.1	7.1	9.2
RNA	5.86	7.0	8.5	10.1
Lipid	0.15	2.5	2.7	5.0
DNA	1.05	1.3	1.6	1.9
Transport	5.21	20.0	20.0	30.6
Theoretical Y_{ATP}^{max}	28.8	15.4	13.4	10.0
Experimentally assessed Y_{ATP}^{max}[c]	13.9		9.5	7.1

[a] Data of Stouthamer (46).

[b] The macromolecular composition of the cells was taken to be 52.4% protein, 16.6% polysaccharide, 15.7% RNA, 9.4% lipid, and 3.2% DNA.

[c] Reference 46.

mM potassium L-lactate in 50 mM potassium phosphate buffer (pH 6.6) and suspended in the same buffer containing 8.8 μM L-[^{14}C]proline effected an 11-fold accumulation of the amino acid inside the vesicles. This accumulation was completely inhibited by the uncoupler *p*-trifluoromethoxyphenylhydrozone and could be shown not to be the result of either a potassium or an osmotic gradient. Moreover, efflux of lactate from the vesicles could be shown to lead to the generation of an electrical potential ($\Delta\psi$) equal to −55 mV, as assessed by the accumulation of the lipophilic cation tetraphenylphosphonium (54).

Though it has yet to be shown that carrier-mediated efflux of a metabolic end product, such as lactate, from intact growing *E. coli* cells can be energetically favorable, evidence obtained with membrane vesicles is sufficient to undermine confidence in assessments of the net energy gain (in terms of ATP equivalents) associated with fermentation processes.

ENERGY-CONSUMING PROCESSES

Having briefly considered the nature of the known and potential energy-generating systems present in *E. coli* and how their activities might be modulated, it is now appropriate to review the major energy-consuming processes.

Cell Synthesis

The principal known energy-consuming reactions occurring in the growing microbial cell are those associated with solute transport, monomer formation, and, most particularly, polymer synthesis. A detailed analysis of the theoretical ATP requirements for these processes, for organisms growing in a simple salts medium on a variety of carbon substrates, was provided by Stouthamer (45, 46) and is reproduced in Table 3.

On the basis of a rather atypical macromolecular composition of *E. coli* (i.e., a high polysaccharide content and a relatively low protein content), it was concluded that the minimum amount of ATP needed

for cell synthesis varied from 34.8 mmol/g [dry weight] of cells for growth on glucose to 99.5 mmol/g [dry weight] of cells for growth on acetate. Translated into yield coefficients, these values gave theoretical Y_{ATP}^{max} values ranging from 28.8 to 10 g [dry weight] of cells per mol of ATP. These values are, respectively, 207 and 43% higher than the experimentally derived values for Y_{ATP}^{max} for *E. coli* growing aerobically on glucose and acetate, assuming two sites of respiratory chain energy conservation (8, 46). Hence, either the energy requirements for transport and for monomer and polymer synthesis had been grossly underestimated or, as seems more probable, a substantial amount of energy is needed for growth-associated processes other than those specified above.

A similar discrepancy between theoretical and probable Y_{ATP} values previously had been noted by Gunsalus and Shuster (10), who were unable to pinpoint the cause but who postulated a possible dissipation of energy by ATPase mechanisms. Harder et al. (12) also addressed this question and offered the provocative suggestion that biological energy converters well might have evolved to work under conditions of maximal energy output (biomass formation) and that they consequently need to function at a reduced efficiency in much the same way as do steam engines and an electrical power plant. They went on to point out that recent developments in the application of nonequilibrium thermodynamics to energy conversion in biological systems have indicated that this conceivably might be the case (14, 61).

So far as we are aware, no definitive experiments have yet been performed that identify unequivocally how this apparent excess energy is consumed. But notwithstanding the aforementioned imponderable, other processes that consume energy and lead to an apparent lowering of the yield values are known to occur in microbial cells.

Extracellular Product Formation

Whereas most of the polymers synthesized by microbial cells are contained within the plasma membrane and envelope structure, some may be secreted into the medium. This is particularly the case with polysaccharides, which, under some conditions and with particular species, may account for a substantial part of the carbon substrate consumed. For example, with an ammonia-limited chemostat culture of *K. aerogenes* growing on glucose at a rate of 0.17 h^{-1}, more than one-third of the substrate consumed was excreted into the medium as polysaccharide (33). With similarly limited cultures of *E. coli* (PC-1000), also growing at a low rate (0.18 h^{-1}), less extracellular polysaccharide accumulated, but polysaccharide still accounted for 14% of the total glucose consumed (Table 4). Though polyglucose is energetically less expensive to synthesize than protein, RNA, and DNA (i.e., 12.4 mmol of ATP per g as compared with, respectively, 39.1, 37.3, and 33.0 mmol of ATP per g [46]), a substantial production of this compound clearly would consume a very significant portion of the ATP generated by catabolism.

Proteins are not commonly secreted by *E. coli* and *K. aerogenes*. Nevertheless, it is conceivable that enzymes and binding proteins normally located in the

TABLE 4. Comparison of glucose consumed and products formed by variously limited aerobic chemostat cultures of *E. coli* PC-1000[a]

Limiting substrate/dilution rate (h^{-1})	Amt of glucose used (mg-atoms of carbon)	Amt of the following product formed (mg-atoms of carbon)[b]:						Carbon recovery (%)	$Y_{glucose}$ (g/mol)
		CO_2	Acetic acid	Pyruvic acid	D-Lactic acid	2-Oxoglutaric acid	Extracellular polysaccharide		
Glucose									
0.18	40.5	16.1						89	74.1
0.30	40.1	15.7						89	74.8
Ammonia									
0.18	84.6	29.2	9.8	4.6		6.3	11.9	97	35.5
0.34	54.9	17.1	5.3			5.2	9.1	103	54.6
Potassium									
0.15	161.7	62.9	20.2	23.1	4.9		3.8	83	18.6
0.30	106.9	35.1	16.5	2.9	0.8		20.0	89	28.1

[a] To facilitate comparison, all values for glucose used and products formed have been adjusted to a cell production rate of 20 mg-atoms of carbon per h (33).

[b] The amount of cells formed was 20 mg-atoms of carbon under all of the conditions shown.

periplasmic space may, under some conditions, diffuse into the extracellular fluids. Thus, protein was found to accumulate in the extracellular fluids of chemostat cultures of *K. aerogenes* that were sulfate, magnesium, or potassium limited when growing on glucose at a low rate (33, 52). The extracellular protein from the sulfate-limited culture was isolated by ammonium sulfate precipitation, dialyzed, and analyzed by sodium dodecyl sulfate-polyacrylamide gel electrophoresis. Only two major bands and one minor band could be detected, indicating that this protein was not the product of cell lysis. Hydrolysis of the isolated protein with 6 N HCl and analysis of its amino acid composition revealed an excess of acidic amino acids over basic amino acids and a virtual absence of methionine and cysteine. Because extracellular proteins often are low in cysteine (39), one might reasonably conclude that this protein probably was either periplasmic in origin or else derived from the outer membrane. However, irrespective of its origin, synthesis of this extracellular protein must have been energetically expensive.

Metabolite Excretion

Batch cultures of *E. coli*, or carbon substrate-limited chemostat cultures growing aerobically in a defined simple salts medium, generally metabolize substrate solely to cells and CO_2. However, carbon substrate-sufficient chemostat cultures often are found to metabolize substrate inefficiently in that substantial amounts of partially oxidized products of substrate catabolism accumulate in the medium (Table 4). Acetate almost invariably is found (33), along with various amounts of pyruvate, 2-oxoglutarate, and D-lactate. Hence, the yield value with respect to carbon substrate consumed is markedly lowered. In general, formation of these so-called overflow metabolites does not consume energy. On the contrary, these metabolites often (perhaps invariably) are associated with an increased energy flux, as manifest in an increased rate of respiration.

When cells are growing anaerobically, accumulation of metabolites in the extracellular fluid is a sine qua non. However, when growing at a fixed rate, glucose-sufficient chemostat cultures of *K. aerogenes*

were found to consume glucose at enhanced rates as compared with a glucose-limited culture (49). Under conditions of excess glucose, a branched fermentation pattern was manifest with a concomitant lowering of the overall efficiency of ATP generation (i.e., net moles of ATP formed per mole of glucose fermented). Nevertheless, this decrease in the efficiency factor was smaller than the increase in glucose consumption rate, and, again, just as with aerobic cultures, the apparent Y_{ATP} values were low. To the best of our knowledge, no comparable data for *E. coli* cultures are available in the literature.

It remains a matter for speculation whether, in these anaerobic cultures, the formation of some products is energetically favorable, neutral, or unfavorable. For example, the formation of D-lactate might be accompanied by a net ATP gain (as would be the case if it arose from pyruvate and, particularly, if lactate excretion generated a proton motive force), or it might implicate a severe energy drain if it arose via dihydroxyacetone phosphate and methylglyoxal (5). Moreover, product excretion against a large concentration gradient well might consume energy.

Maintenance Functions

As mentioned previously, Monod (30) conceived that a small amount of energy released by substrate catabolism must be dissipated in the maintenance of cell integrity and viability. But although this concept now has been repeatedly validated experimentally, it still is not clear what precise physiological functions constitute maintenance. This issue is clouded and confused by the widely disparate maintenance values cited in the literature for different strains of the same species growing in putatively identical media (Table 1) as well as for a single strain growing on a range of different carbon substrates (15). In this latter study, exceptionally high values of the apparent oxygen consumption rate for maintenance (m_O) were reported for the growth of *E. coli* B on glycerol, succinate, glutamate, and acetate (respectively, 10.2, 13.1, 17.1, and 25.9 mg-atoms of O per h per g [dry weight] of cells) as compared with values of 0.9 and 1.8 mg-atoms of O per h per g (dry weight) of cells for growth on glucose and galactose, respectively. Moreover, at the other extreme, microcalorimetric studies of glucose-limited chemostat cultures of *E. coli* K-12 (21) indicated a maintenance rate of heat evolution of only 0.02 kcal/h per g (dry weight) of cells (1 cal = 4.184 J), which would equate with an oxygen consumption rate of no more than 0.36 mg-atom of O per h per g (dry weight) of cells.

As yet, there would appear to be no rational explanation for these widely differing maintenance rates. Clearly, if the quantitatively major components of maintenance were, say, turnover of macromolecules, maintenance of ion gradients, and motility, then there would not seem to be any obvious reason why the maintenance requirement of an acetate-limited culture should be almost 30-fold that of a glucose-limited one.

High apparent maintenance rates of oxygen or carbon substrate consumption or both are commonplace with chemostat cultures in which growth is limited by the availability of an anabolic substrate (e.g., source of nitrogen, sulfur, or phosphorus) or an essential cation (K^+ or Mg^{2+}). Here, the question becomes one of the extent to which the uptake of excess carbon substrate can be modulated such as to meet the minimum biosynthetic and bioenergetic demands of cell synthesis. From the fact that washed suspensions of bacteria, which clearly cannot grow, often will oxidize substrates such as glucose at a high rate, one may safely conclude that such regulation, if present, is by no means stringent. Consequently, it seems reasonable to suppose that the high rates of oxygen or carbon substrate consumption or both expressed by slowly growing carbon substrate-sufficient chemostat cultures of bacteria do not necessarily and invariably result from some enhanced specific maintenance energy requirement, but more likely result from a partial uncoupling of catabolism from anabolism.

It is abundantly obvious, then, that to proceed further in analyzing yield values in physiological terms one needs to know more about the possible mechanisms that might be invoked in dissociating catabolism from anabolism and how they might be modulated. We address these questions in the next section.

ENERGY-DISSIPATING REACTIONS

Gunsalus and Shuster (10) drew attention to the fact that, when bacterial growth in a batch culture ceases owing to the depletion of an essential nutrient other than the energy source, substrate degradation may continue at a full or reduced rate. They went on to state that "The condition of energy release without coupled growth must be understood if variables in growth yield and substrate carbon use for non-energy functions are to be understood" and then offered three possible mechanisms. These mechanisms were as follows: (i) accumulation of polymeric products, either in storage form or as unusable waste; (ii) dissipation as heat by "ATPase mechanisms"; and (iii) activation of shunt mechanisms that either bypass energy-yielding reactions or require a greater expenditure of energy for priming. We have already dealt with the energetic consequences of extracellular polymeric product formation, but more recent evidence pertaining to the latter two postulated mechanisms is reviewed briefly below.

Possible ATPase Mechanisms

As already mentioned, it is still not clear whether respiration can proceed at a high rate without generating a membrane potential or, if not, whether the membrane potential can be dissipated without doing useful work. However, with a low-K^+ glucose-limited chemostat culture of *K. aerogenes* (19), as well as with a similar culture of *E. coli* (unpublished data), the rate of respiration both varied and appeared to correlate closely with the magnitude of the transmembrane K^+ gradient. Hence, in this specific case, it is reasonable to postulate that much respiratory energy was indeed being dissipated at the level of the membrane as a consequence of an induced K^+ leakage current (53). But such a mechanism is unlikely to account for the high rates of carbon substrate metabolism or respiration expressed by chemostat cultures that were limited in their growth by the availability of anabolic substrates such as sulfate, ammonia, or phosphate (33, 34). Here, energy must be dissipated in some other way, and one possible mechanism is so-called futile cycling (10, 24). Thus, when two compounds are interconverted by reactions that are irreversible, and when the enzymes catalyzing these reactions are both active, there is futile cycling in which dissipation of energy occurs without a corresponding change in metabolites.

One possible futile cycle that has been studied in *E. coli* is the interconversion of fructose 6-phosphate and fructose 1,6-bisphosphate effected by phosphofructokinase and fructose 1,6-bisphosphatase, respectively (3, 7). However, with batch cultures of this organism growing on either glucose or gluconeogenic substrates, little if any futile cycling could be observed. On the other hand, the growth conditions used were not those under which one would expect extensive futile cycling to occur. In this connection, it was found by Scrutton and Utter (44) that phosphofructokinase is inhibited by ATP and activated by AMP, whereas fructose 1,6-bisphosphatase is inhibited by AMP and activated by ATP. Hence, a high rate of ATP generation by respiration under conditions in which ATP utilization by biosynthetic reactions is constrained well might be expected to activate futile cycling at this level.

E. coli is also known to synthesize both glutamine synthetase and glutaminase, which again, at least in theory, could effect a futile cycling with concomitant hydrolysis of ATP (40). Moreover, because glutamine synthetase is highly active in ammonia-limited cells (26) and because significant amounts of glutaminase also are present (51), some futile cycling would be expected to occur in such cells unless the activity of glutaminase was constrained in some way. Of relevance here is the observation that the activities of both enzymes are markedly influenced by the energy charge (40); however, the patterns of regulation are such as to suggest that when the energy charge is high the glutaminase activity would be relatively low (i.e., about 20% of the maximal activity), though nevertheless significant.

A possible explanation for the presence of glutaminase in ammonia-limited cells is that, by releasing

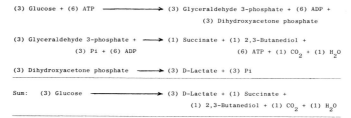

FIG. 4. *Pattern of fermentation of glucose by* K. aerogenes *implicating no net synthesis of ATP.*

ammonia, it allows the key enzyme of ammonia assimilation (glutamine synthetase) to continue working at a significant rate in the virtual absence of exogenous ammonia. This, one might argue, permits the uptake system to remain in a highly reactive state with respect to transient changes in the supply of growth-limiting nutrient. This hypothesis could, of course, be tested by determining whether the intracellular pool level of ammonia was maintained at a significant value after an interruption in the supply of nutrients to an ammonia-limited culture. But to the best of our knowledge, this experiment has not yet been performed.

If, as suggested above, some futile cycles serve as sensors for the uptake of growth-limiting nutrients, one might expect similar reactions to be associated with common nutrients such as sulfate and phosphate; indeed, it has been shown that, when growing on sulfate, *E. coli* has two enzymes (adenosine 5'-phosphosulfate kinase and 3'(2'),5'-diphosphonucleoside 3'-phosphohydrolase) that are potentially capable of acting in concert as a futile cycle (57). Again, it is a well-established fact that phosphatases are synthesized to high levels by phosphate-limited cells (56); however, if these enzymes function intracellularly, their activities must be highly regulated. In this connection, Meyerhof (see reference 10) postulated the presence of a labile phosphatase in yeast cells that functioned in allowing washed suspensions of these organisms to carry out a normal alcoholic fermentation in the absence of growth. Probably of relevance here is the methylglyoxal bypass, the key enzyme of which (methylglyoxal synthase) is a phosphate-releasing enzyme though not, strictly speaking, a phosphatase. Significantly, the activity of methylglyoxal synthase is strongly inhibited by moderate concentrations of phosphate (<1 mM), and it has been suggested that this enzyme actually serves to recycle phosphate under conditions in which the intracellular phosphate concentration is low (5). A somewhat broader interpretation of the functioning of the methylglyoxal bypass is presented in the next section.

Metabolic Uncoupling

It was mentioned previously that a branched fermentation capability allows organisms to vary the rate of ATP generation relative to the rate of glucose fermentation, but only to a limited extent (i.e., between 2 and 3 mol of ATP per mol of glucose fermented). Therefore, a question arises as to whether organisms can more extensively uncouple fermentation from ADP phosphorylation or even totally uncouple the two processes. In this latter connection, evidence

has been presented that this uncoupling of both processes can be achieved by invoking the reactions of the methylglyoxal bypass (48) (Fig. 3). Thus, in a slowly growing glucose-limited anaerobic culture of this organism, the sole products of glucose fermentation were acetate and ethanol (plus formate, CO_2, and H_2). But when pulsed with a cell-saturating concentration of glucose, the pattern of glucose fermentation changed, and 50% of the extra glucose consumed was converted to D-lactate. Because in the short term (20 min) there was a large increase in the glucose consumption rate but no increase in the growth rate and because the key enzymes of the methylglyoxal bypass were present in substantial amounts it was concluded that the formation of D-lactate occurred exclusively via methylglyoxal. The other products formed, after the glucose pulse, were succinate and 2,3-butanediol, and, by drawing out the fermentation scheme (Fig. 4), it became clear that this extra glucose was indeed being fermented with no concomitant net formation of ATP. In other words, by invoking the reactions of the methylglyoxal bypass, fermentation of this extra glucose was totally uncoupled from ATP generation.

Because the enzymes that effect a conversion of dihydroxyacetone phosphate to D-lactate, via methylglyoxal, are formed constitutively in *E. coli* (5, 6), there seems to be little doubt that this organism also can totally uncouple glycolysis from ADP phosphorylation. Indeed, one might speculate that the primary function of the methylglyoxal bypass is to effect metabolic uncoupling under conditions in which the energy charge and glycolytic flux rate within the cell are high. But be that as it may, the fact that *E. coli*, as well as *K. aerogenes*, has the potential to synthesize D-lactate from either pyruvate (via an NADH-linked pyruvate reductase) or dihydroxyacetone phosphate (via methylglyoxal synthase and glyoxylase) renders difficult an accurate assessment of the energy balance in those anaerobic cultures in which this metabolite is a major product. For example, a homolactic fermentation of glucose (to D-lactate) could yield, maximally, 2 mol of ATP per mol of glucose if only the pyruvate reductase was active or, minimally, zero ATP if 50% of the lactate arose via methylglyoxal.

It should be pointed out that, anaerobically, no more than 50% of the glucose carbon can be converted to D-lactate via the methylglyoxal bypass since the conversion of glucose to the two triose phosphates consumes 2 mol of ATP and these must be recovered by glycolysis. Aerobically, however, it is conceivable that all of the glucose carbon could be converted to D-lactate via methylglyoxal and that the D-lactate could then be oxidized to pyruvate to regenerate the ATP invested in phosphorylating glucose and fructose

6-phosphate. Indeed, since the oxidation of D-lactate to pyruvate by the respiratory chain-linked dehydrogenase implicates just one site of energy conservation, at least in theory, glucose can be aerobically catabolized solely to pyruvate without any net gain of biologically useable energy.

Relevant in the context mentioned above is the observation that glucose-limited aerobic chemostat cultures of *K. aerogenes* and *E. coli* can immediately accelerate the rate of glucose consumption when suddenly relieved of the growth limitation (32). There was a concomitant marked increase in the respiration rate but no corresponding increase in the growth rate, indicating that excess energy was being dissipated in some way. Significantly, the extra glucose consumed was not oxidized completely, and substantial amounts of pyruvate and acetate were excreted into the medium (M. P. M. Leegwater, Ph.D. thesis, University Amsterdam, The Netherlands, 1983). Clearly, if the pyruvate arose from D-lactate, as suggested above, one could partially account for the energetic uncoupling; however, the formation of acetate from pyruvate would yield NADH, and therefore a question arises as to whether this could be oxidized without generating a proton motive force. We know of no such mechanism or postulated mechanism, but both *K. aerogenes* and *E. coli* have constitutively two enzymes (an NADH-linked pyruvate reductase that generates D-lactate and a D-lactate dehydrogenase that is a flavoprotein) that together can effect the oxidation of NADH without implicating site 1 of the respiratory chain (Fig. 2). Of significance here is the fact that pyruvate reductase is homotropic with respect to pyruvate; hence, an accumulation of pyruvate within the cell would promote this bypass reaction. However, whether this mode of NADH oxidation actually functions in the cell remains to be established.

GENERAL CONCLUSIONS

Microbial growth is the product of a very large number of interconnected enzyme-catalyzed reactions; the fact that, in any particular environment, cell synthesis proceeds with a more or less constant efficiency indicates that a substantial measure of control must be exercised over the fluxes of intermediary metabolites and precursor substances involved in polymer synthesis. Moreover, in chemoheterotrophic organisms such as *E. coli* and *S. typhimurium*, processes of regulation are further complicated by the fact that the energy needed for biosynthesis necessarily must be derived from the breakdown of carbon substrate that simultaneously is being assimilated into cell substance. Hence, one might expect mechanisms operating in these cells to precisely partition the flow of intermediary metabolites between catabolic (energy-generating) and anabolic (energy-consuming) reactions. Thus, one could envisage control systems to exist within the cell that would act at specific branch points between, respectively, catabolic and anabolic pathways of metabolism and which would be "tuned" to the overall energy status of the cell. That some such controls indeed are present within the microbial cell is abundantly obvious from the widely reported involvement of adenine nucleotides as control elements in intermediary metabolism (4). Indeed,

the mode of action of these regulatory processes (e.g., allosteric effectors) leads not unreasonably to the concept of stringent coupling between ATP synthesis and growth which manifests itself in a precise yield value. However, that such a concept is, in general, untenable is obvious from observations extending back over many years (10, 42, 50) in that energy (ATP) generation can occur at a high rate under conditions in which cell synthesis is severely constrained. Clearly, there is no obligatory coupling between catabolism and anabolism, and herein lies the source of much difficulty in attempting to interpret yield data in energetic terms. It is obvious, therefore, that further progress in evaluating the energetics of microbial growth (particularly aerobic growth) hinges critically both on the acquisition of a better understanding of those energy-spilling processes extant within the cell and on their associated regulatory mechanisms. In this chapter we have attempted to make a start along these lines.

LITERATURE CITED

1. **Blackwood, A. C., A. C. Neish, and G. A. Ledingham.** 1956. Dissimilation of glucose at controlled pH values by pigmented and non-pigmented strains of *Escherichia coli*. J. Bacteriol. **72**:497–499.
2. **Brice, J. M., J. F. Law, D. J. Meyer, and C. W. Jones.** 1974. Energy conservation in *Escherichia coli* and *Klebsiella pneumoniae*. Biochem. Soc. Trans. **2**:523–526.
3. **Chambost, J. P., and D. G. Fraenkel.** 1980. The use of 6-¹⁴C-labelled glucose to assess futile cycling in *Escherichia coli*. J. Biol. Chem. **255**:2867–2869.
4. **Chapman, A. G., and D. E. Atkinson.** 1977. Adenine nucleotide concentrations and turnover rates. Their correlation with biological activity in bacteria and yeast. Adv. Microb. Physiol. **15**:253–306.
5. **Cooper, R. A.** 1984. Metabolism of methylglyoxal in microorganisms. Annu. Rev. Microbiol. **38**:49–68.
6. **Cooper, R. A., and A. Anderson.** 1970. The formation and catabolism of methylglyoxal during glycolysis in *Escherichia coli*. FEBS Lett. **11**:273–276.
7. **Daldal, F., and D. G. Fraenkel.** 1983. Assessment of a futile cycle involving reconversion of fructose 6-phosphate to fructose 1,6-bisphosphate during gluconeogenic growth of *Escherichia coli*. J. Bacteriol. **153**:390–394.
8. **Farmer, I. S., and C. W. Jones.** 1976. The energetics of *Escherichia coli* during aerobic growth in continuous culture. Eur. J. Biochem. **67**:115–122.
9. **Gottschalk, G.** 1979. Bacterial metabolism. Springer-Verlag, New York.
10. **Gunsalus, I. C., and C. W. Shuster.** 1961. Energy yielding metabolism in bacteria, p. 1–58. *In* I. C. Gunsalus and R. Y. Stanier (ed.), The bacteria, vol. 2. Academic Press, Inc., New York.
11. **Haddock, B. A., J. A. Downie, and P. B. Garland.** 1976. Kinetic characterization of the membrane-bound cytochromes of *Escherichia coli* grown under a variety of conditions by using a stopped-flow dual wavelength spectrophotometer. Biochem. J. **154**:285–294.
12. **Harder, W., J. P. van Dijken, and J. A. Roels.** 1981. Utilization of energy in methylotrophs, p. 258–269. *In* H. Dalton (ed.), Microbial growth on C₁ compounds. Heyden & Son Ltd., London.
13. **Harrison, D. E. F.** 1976. The regulation of respiration rate in growing bacteria. Adv. Microb. Physiol. **14**:243–313.
14. **Hellingwerf, K. J., J. S. Lolkema, R. Otto, O. M. Neijssel, A. Stouthamer, W. Harder, K. van Dam, and H. V. Westerhoff.** 1982. Energetics of microbial growth: an analysis of the relationship between growth and its mechanistic basis by mosaic non-equilibrium thermodynamics. FEMS Microbiol. Lett. **15**:7–17.
15. **Hempfling, W. P., and S. E. Mainzer.** 1975. Effects of varying the carbon source limiting growth on yield and maintenance characteristics of *Escherichia coli* in continuous culture. J. Bacteriol. **123**:1076–1087.
16. **Herbert, D.** 1958. Some principles of continuous culture, p. 381–396. *In* Recent progress in microbiology. Proceedings of the International Congress of Microbiology, Stockholm.
17. **Hernandez, E., and M. J. Johnson.** 1967. Anaerobic growth yields

of *Aerobacter cloacae* and *Escherichia coli*. J. Bacteriol. **94**:991–995.

18. **Hommes, R. W. J., P. W. Postma, O. M. Neijssel, D. W. Tempest, P. Dokter, and J. A. Duine.** 1984. Evidence of a quinoprotein glucose dehydrogenase apoenzyme in several strains of *Escherichia coli*. FEMS Microbiol. Lett. **24**:329–333.

19. **Hueting, S., T. de Lange, and D. W. Tempest.** 1979. Energy requirement for maintenance of the transmembrane potassium gradient in *Klebsiella aerogenes* NCTC 418: a continuous culture study. Arch. Microbiol. **123**:183–188.

20. **Ingledew, W. J., and R. K. Poole.** 1984. The respiratory chains of *Escherichia coli*. Microbiol. Rev. **48**:222–271.

21. **Ishikawa, Y., and M. Shoda.** 1983. Calorimetric analysis of *Escherichia coli* in continuous culture. Biotechnol. Bioeng. **25**:1817–1827.

22. **Jones, C. W.** 1977. Aerobic respiratory systems in bacteria, p. 23–59. *In* B. A. Haddock and W. A. Hamilton (ed.), Microbial energetics: 27th symposium of the Society for General Microbiology. Cambridge University Press, Cambridge.

23. **Jones, C. W., J. M. Brice, A. J. Downs, and J. W. Drozd.** 1975. Bacterial respiration-linked proton translocation and its relationship to respiratory chain composition. Eur. J. Biochem. **52**:265–271.

24. **Katz, J., and R. Rognstad.** 1978. Futile cycling in glucose metabolism. Trends Biochem. Sci. **3**:171–174.

25. **Lawford, H. G., and B. A. Haddock.** 1973. Respiration-driven proton translocation in *Escherichia coli*. Biochem. J. **136**:217–220.

26. **Magasanik, B., M. J. Prival, and J. E. Brenchley.** 1973. Glutamine synthetase, regulator of the synthesis of glutamate-forming enzymes, p. 65–70. *In* S. Prusiner and E. R. Stadtman (ed.), The enzymes of glutamine metabolism. Academic Press, Inc., New York.

27. **Marr, A. G., E. H. Nilson, and D. J. Clark.** 1963. The maintenance requirement of *Escherichia coli*. Ann. N.Y. Acad. Sci. **102**:536–548.

28. **Meyer, D. J., and C. W. Jones.** 1973. Oxidative phosphorylation in bacteria which contain different cytochrome oxidases. Eur. J. Biochem. **36**:144–151.

29. **Michels, P. A. M., J. P. J. Michels, J. Boonstra, and W. N. Konings.** 1979. Generation of electrochemical proton gradient in bacteria by the excretion of metabolic end-products. FEMS Microbiol. Lett. **5**:357–364.

30. **Monod, J.** 1942. Recherches sur la croissance des cultures bactériennes. Hermann, Editeurs des Sciences et des Arts, Paris.

31. **Neijssel, O. M., G. P. M. A. Hardy, J. C. Lansbergen, D. W. Tempest, and R. W. O'Brien.** 1980. Influence of growth environment on the phosphoenolpyruvate: glucose phosphotransferase activities of *Escherichia coli* and *Klebsiella aerogenes*: a comparative study. Arch. Microbiol. **125**:175–179.

32. **Neijssel, O. M., S. Hueting, and D. W. Tempest.** 1977. Glucose transport capacity is not the rate-limiting step in the growth of some wild-type strains of *Escherichia coli* and *Klebsiella aerogenes* in chemostat culture. FEMS Microbiol. Lett. **2**:1–3.

33. **Neijssel, O. M., and D. W. Tempest.** 1975. The regulation of carbohydrate metabolism in *Klebsiella aerogenes* NCTC 418 organisms growing in chemostat culture. Arch. Microbiol. **106**:251–258.

34. **Neijssel, O. M., and D. W. Tempest.** 1976. Bioenergetic aspects of aerobic growth of *Klebsiella aerogenes* NCTC 418 in carbon-limited and carbon-sufficient chemostat culture. Arch. Microbiol. **107**:215–221.

35. **Neijssel, O. M., D. W. Tempest, P. W. Postma, J. A. Duine, and J. Frank.** 1983. Glucose metabolism by K⁺-limited *Klebsiella aerogenes*: evidence for the involvement of a quinoprotein glucose dehydrogenase. FEMS Microbiol. Lett. **20**:35–39.

36. **Otto, R., A. Sonnenberg, H. Veldkamp, and W. N. Konings.** 1980. Lactate efflux induced electrical potential in membrane vesicles of *Streptococcus cremoris*. Proc. Natl. Acad. Sci. USA **77**:5502–5506.

37. **Pirt, S. J.** 1965. The maintenance energy of bacteria in growing cultures. Proc. R. Soc. Lond. B Biol. Sci. **163**:224–231.

38. **Pirt, S. J.** 1982. Maintenance energy: a general model for energy-limited and energy-sufficient growth. Arch. Microbiol. **113**:300–302.

39. **Pollock, M. R.** 1962. Exoenzymes, p. 121–178. *In* I. C. Gunsalus and R. Y. Stanier (ed.), The bacteria, vol. 4. Academic Press, Inc., New York.

40. **Prusiner, S.** 1973. Glutaminases of *Escherichia coli*: properties, regulation and evolution, p. 293–316. *In* S. Prusiner and E. R. Stadtman (ed.), The enzymes of glutamine metabolism. Academic Press, Inc., New York.

41. **Rice, C. W., and W. P. Hempfling.** 1978. Oxygen-limited continuous culture and respiratory energy conservation in *Escherichia coli*. J. Bacteriol. **134**:115–124.

42. **Rosenberger, R. F., and S. R. Elsden.** 1960. The yields of *Streptococcus faecalis* grown in continuous culture. J. Gen. Microbiol. **22**:726–739.

43. **Schultz, K. L., and R. S. Lipe.** 1964. Relationship between substrate concentration, growth rate and respiration rate of *Escherichia coli* in continuous culture. Arch. Mikrobiol. **48**:1–20.

44. **Scrutton, M. C., and M. F. Utter.** 1968. The regulation of glycolysis and glyconeogenesis in animal tissues. Annu. Rev. Biochem. **37**:249–302.

45. **Stouthamer, A. H.** 1977. Energetic aspects of the growth of microorganisms, p. 285–315. *In* B. A. Haddock and W. A. Hamilton (ed.), Microbial energetics: 27th Symposium of the Society for General Microbiology. Cambridge University Press, Cambridge.

46. **Stouthamer, A. H.** 1979. The search for correlation between theoretical and experimental growth yields. Int. Rev. Biochem. **21**:1–47.

47. **Teissier, G.** 1936. Les lois quantitatives de la croissance. Ann. Physiol. Veg. (Paris) **12**:527–586.

48. **Teixeira de Mattos, M. J., H. Streekstra, and D. W. Tempest.** 1984. Metabolic uncoupling of substrate-level phosphorylation in anaerobic glucose-limited chemostat cultures of *Klebsiella aerogenes* NCTC 418. Arch. Microbiol. **139**:260–264.

49. **Teixeira de Mattos, M. J., and D. W. Tempest.** 1983. Metabolic and energetic aspects of the growth of *Klebsiella aerogenes* NCTC 418 on glucose in anaerobic chemostat culture. Arch. Microbiol. **134**:80–85.

50. **Tempest, D. W.** 1978. The biochemical significance of microbial growth yields: a reassessment. Trends Biochem. Sci. **3**:180–184.

51. **Tempest, D. W., J. L. Meers, and C. M. Brown.** 1970. Synthesis of glutamate in *Aerobacter aerogenes* by a hitherto unknown route. Biochem. J. **117**:405–407.

52. **Tempest, D. W., and O. M. Neijssel.** 1978. Eco-physiological aspects of microbial growth in aerobic nutrient-limited environments. Adv. Microb. Ecol. **3**:105–153.

53. **Tempest, D. W., and O. M. Neijssel.** 1984. The status of Y_{ATP} and maintenance energy as biologically interpretable phenomena. Annu. Rev. Microbiol. **38**:459–486.

54. **ten Brink, B., and W. N. Konings.** 1980. Generation of an electrochemical proton gradient by lactate efflux in *Escherichia coli* membrane vesicles. Eur. J. Biochem. **111**:59–66.

55. **Terroine, E., and R. Wurmser.** 1922. L'énergie de croissance. I. Le développement de l'*Aspergillus niger*. Bull. Soc. Chim. Biol. **4**:519.

56. **Torriani, A.** 1960. Influence of inorganic phosphate in the formation of phosphatases by *Escherichia coli*. Biochim. Biophys. Acta **38**:460–469.

57. **Tsang, M. L.-S., and J. Schiff.** 1976. Sulfate reducing pathways in *Escherichia coli* involving bound intermediates. J. Bacteriol. **125**:923–933.

58. **van Schie, B. J., K. J. Hellingwerf, J. P. van Dijken, M. G. L. Elferink, J. M. van Dijl, J. G. Kuenen, and W. N. Konings.** 1985. Energy transduction by electron transfer via a pyrrolo-quinoline quinone-dependent glucose dehydrogenase in *Escherichia coli*, *Pseudomonas aeruginosa*, and *Acinetobacter calcoaceticus* (var. *lwoffi*). J. Bacteriol. **163**:493–499.

59. **van Verseveld, H. W., W. R. Chesbro, M. Braster, and A. H. Stouthamer.** 1984. Eubacteria have 3 growth modes keyed to nutrient flow. Consequences for the concept of maintenance and maximal growth yield. Arch. Microbiol. **137**:176–184.

60. **West, I. C., and P. Mitchell.** 1972. Proton-coupled β-galactosidase translocation in non-metabolising *Escherichia coli*. J. Bioenerg. **3**:445–462.

61. **Westerhoff, H. V., J. S. Lolkema, O. Otto, and K. J. Hellingwerf.** 1982. Thermodynamics of growth. Non-equilibrium thermodynamics of bacterial growth; the phenomenological and mosaic approach. Biochim. Biophys. Acta **683**:181–220.

62. **White, D. C., and P. R. Sinclair.** 1971. Branched electron transport in bacteria. Adv. Microb. Physiol. **5**:173–211.

SUBJECT INDEX

i

Cytosine
  deamination, 1045
  salvage, 462
  transport, 463
Cytosine-cytosine dimers, 1044
Cytosine deaminase, 460–461, 464
Cytosine methylase, 1047
Cytosine-thymine dimers, 1044
cytR genes, 256–258, 459, 462–467, 813, 879, 992, 1171

dac genes, 669, 813
dad genes
  dadA, 180, 317, 813, 879, 992, 1077
  dadB, 316–317, 813, 879, 1086
  dadQ, 813
  dadR, 879
  dadX, 316–317
DAHP, 370
DAHP synthase, 370–373, 379–384, 387–388, 1457
dam gene, 174, 570, 587, 599, 697–701, 813, 880,
    1037–1039, 1047, 1065, 1081, 1225–1226, 1234–1235,
    1569
Dam methylase, 587, 599–600
Dam methylase GATC site, 586–587
dAMP, cellular content, 5, 447
dap genes
  dapA, 434, 438, 813, 880, 1080
  dapB, 434, 438, 813, 880, 1414
  dapC, 434–435, 813, 880
  dapD, 435, 813, 880
  dapE, 438, 813, 1080, 1171
  dapF, 880
das genes, 597
dATP, 1350
  cellular content, 447
Daughter-strand gap repair, 1353
dcd gene, 464–465, 467, 813, 880, 993
dCDP-diglyceride, 485
dcm gene, 700, 813, 880, 993, 1047, 1225–1226
dcm methylase, 697–699
dAMP, cellular content, 5, 447
dcp gene, 684, 813, 880
dct genes, 42, 261, 813, 880, 993
dCTP, cellular content, 447
dCTP deaminase, 464–466
ddl gene, 669, 813
Deamido-NAD adenylyltransferase, 559
Deamido-NAD ammonia ligase, 560
Deamination reactions, in riboflavin biosynthesis, 526–527
3-Decenoyl-*N*-acetylcysteamine, 487, 1003
Decoyinine, 451
Deep rough mutants, 8–9, 13–14, 18, 649, 653
deg gene, 1310
Dehydrase, 145
α-Dehydrobiotin, 991
Dehydrogenases, cytoplasmic membrane, 34–37, 39
Dehydroproline, 349, 998, 1514
Dehydroquinate, 368–370
Dehydroquinate dehydratase, 372, 379
Dehydroquinate synthase, 371–372, 379
Dehydroshikimate, 369
Dehydroshikimate reductase, *see* Shikimate dehydrogenase
Dehydroxybutylphosphonate, 1000
Deletion mapping, 1164, 1186
Deletions, 972, 1019–1023
  and cotransduction, 1158–1159
  detection, 1025–1026
  selection, 1001
  transposable element-associated, 1065–1066, 1093–1096
del gene, 813
Demand theory, 1627
Demethylmenaquinone, 512

conversion to menaquinone, 518
formation, 518
denB gene, 1162
deo gene, 1171
  deoA, 256–258, 459–461, 466–467, 813, 880, 993, 1414
  deoB, 257–258, 452, 459, 466–467, 814, 880, 993
  deoC, 257–258, 459, 466–467, 814, 880, 993
  deoD, 257–258, 452, 458–460, 466, 814, 880, 993
  deoK, 258, 466, 880
  deoP, 258, 466, 880
  deoR, 256, 258, 459, 461, 463, 466–467, 814, 880, 993
deo operon, 258, 459, 462, 466, 1234, 1320
Deoxyadenosine, 257, 993
  salvage, 459
3-Deoxy-D-arabino-heptulosonate-7-phosphate, *see* DAHP
Deoxycholate, 999
Deoxycytidine, 257
  salvage, 462
Deoxycytidine deaminase, 461–462
Deoxydihydroxyphosphonylmethyl fructose, 1000
3-Deoxy-3-fluoroglucose, 998
2-Deoxygalactitol, 994
2-Deoxygalactose, 994, 1504–1505
Deoxyglucose, 998
  transport, 130–131
2-Deoxyglucose 6-phosphate, 1000
Deoxyguanosine, salvage, 459–460
6-Deoxyhexoses, 649
  biosynthesis, 506–507
Deoxyinosine, salvage, 459
2-Deoxy-2-iodoacetamidoglucose, 997
3-Deoxy-D-manno-octulosonic acid, 498–501, 649–652
3-Deoxy-D-manno-octulosonic acid transferase, 500
Deoxynucleosides, 256–258
Deoxyriboaldolase, 58, 257–258, 459, 463, 466
Deoxyribokinase, 466
Deoxyribomutase, 257–258
Deoxyribonucleotides, biosynthesis, 464–465
Deoxyribose, 256
  transport, 466
Deoxyribose permease, 466
Deoxyribose 1-phosphate, 257, 466–467
Deoxyribose 5-phosphate, 257
Deoxyuridine, salvage, 462
desD gene, 970
Detergents, 13–14, 16
Dethiobiotin synthetase, 546
Developmental regulation, 1237
dgd gene, 814, 993
dgk gene, 476, 484, 486, 490, 675–676
  dgkA, 814
  dgkR, 476, 814
dGMP, cellular content, 5, 447
dgo genes, 260, 814, 993
dgsA gene, 814, 993
dGTP, cellular content, 447
dhb gene, 880, 993
dhuA gene, 777–778, 880, 993, 1324
2,3-Diacylglucosamine 1-phosphate, 499
Diacylglycerol kinase, 476
7,8-Diaminopelargonic acid, 544
7,8-Diaminopelargonic acid synthetase, 545–546
Diaminopimelate decarboxylase, 435, 438
Diaminopimelic acid, 23–27, 509
  biosynthesis, 433–435
  cellular content, 665
  transport, 763
Diaminopimelic acid adding enzyme, 669
Diaminopimelic acid-binding protein, 58
2,6-Diaminopurine, 459, 461, 991, 993
Diarrhea, 1627, 1639–1640
6-Diazo-5-oxo-norleucine, 450, 456